PLACE
STAMP
HERE

RSMeans
Quarterly Updates
700 Longwater Drive
Norwell, MA 02061
USA

Tips for Customizing Your RSMeans Cost Data By Location

The cost data found throughout the book reflects national averages.

Customize your cost data for over 930 locations throughout the U.S. and Canada.

- ☑ To adjust costs to your region, use the ***Location Factors*** found in the back of the book.
- ☑ To compare costs from region-to-region, use the ***City Cost Indexes*** found in the back of the book.
- ☑ For additional guidance, refer to the ***How RSMeans Data Works*** section towards the front of the book.

RSMeans
Reed Construction Data

Metric Conversion Formulas

Length

cm = 0.3937 in.	in. = 2.5400 cm
meter = 3.2808 ft.	ft. = 0.3048 m
meter = 1.0936 yd.	yd. = 0.9144 m
km = 0.6214 mile	mile = 1.6093 km

Area

sq. cm = 0.1550 sq. in.	sq.in. = 6.4516 sq. cm
sq. m = 10.7639 sq. ft.	sq. ft. = 0.0929 sq. m
sq. m = 1.1960 sq. yd.	sq. yd. = 0.8361 sq. m
hectare = 2.4710 acres	acre = 0.4047 hectare
sq. km = 0.3861 sq. mile	sq. mile = 2.5900 sq. km

Weight

gram = 15.4324 grains	grain = 0.0648 g
gram = 0.0353 oz.	oz. = 28.3495 g
kg = 2.2046 lbs.	lb. = 0.4536 kg

Weight, cont.

kg = 0.0011 ton (short)	ton (short) = 907.1848 kg
ton (met.) = 1.1023 ton (short)	ton (short) = 0.9072 ton (met.)
ton (met.) = 0.9842 ton (large)	ton (long) = 1.0160 ton (met.)

Common Units of Measure

Abbreviation	Measurement	Formula
SF	Square feet	Length (in feet) x Width (in feet)
SY	Square yards	Square feet / 9
CF	Cubic feet	Length (in feet) x Width (in feet) x Depth (in feet)
CY	Cubic yards	Cubic feet / 27
BF	Board foot	Length (in inches) x width (in inches) x thickness (in inches)/144
MBF	Thousand board foot	Board foot / 1,000
LF	Linear feet	12 inches of the item
SFCA	Square foot of contact area	A square foot of concrete formwork
SQ	Square	100 square feet
Ton	Ton	2,000 pounds

$144.95 per copy (in United States)
Price is subject to change without prior notice.

0042

The *2012 Commercial Renovation Cost Data*, 33rd Annual Edition, and all of the RSMeans annual cost data books are dedicated to the memory of our respected friend and colleague, Hsiao-Cheng (John) Chiang, PE.

2012

RSMeans
A division of Reed Construction Data
Construction Publishers & Consultants
700 Longwater Drive
Norwell, MA 02061
USA
1-800-334-3509
www.rsmeans.com

Printed in the United States of America
ISSN 2161-4628
ISBN 978-1-936335-44-2

RSMeans Commercial Renovation Cost Data

33RD ANNUAL EDITION

This book is printed using soy-based printing ink.
This book is recyclable.

PRINTED WITH SOY INK

Related RSMeans Products and Services

This book is aimed primarily at commercial renovation projects. The information contained within has provisions for estimating small jobs with small quantities of work and large jobs that encompass major renovations. As with all RSMeans cost data, with reasonable exercise of judgment, this book can be used for estimating jobs with a wide range of costs.

The engineers at RSMeans suggest the following products and services as companion information resources to *RSMeans Commercial Renovation Cost Data:*

Construction Cost Data Books

Assemblies Cost Data 2012
Square Foot Costs 2012

Reference Books

Estimating Building Costs
RSMeans Estimating Handbook
Green Building: Project Planning & Estimating
How to Estimate with RSMeans Data
Plan Reading & Material Takeoff
Project Scheduling & Management for Construction
Repair & Remodeling Estimating Methods
Universal Design

Seminars and In-House Training

Unit Price Estimating
Means CostWorks® Training
RSMeans Data for Job Order Contracting (JOC)
Practical Project Management for Construction Professionals
Scheduling with MSProject for Construction Professionals
Mechanical & Electrical Estimating

RSMeans on the Internet

Visit RSMeans at **www.rsmeans.com** for a one-stop portal for the most reliable and current resources available on the market. Here you can learn about our more than 20 annual cost data books, also available on CD and online versions, as well as about RSMeans seminars.

RSMeans Electronic Data

Get the information found in RSMeans cost books electronically on CD or on the Web at **MeansCostWorks.com**.

RSMeans Business Solutions

Cost engineers and analysts conduct facility life cycle and benchmark research studies and predictive cost modeling, as well as offer consultation services for conceptual estimating, real property management, and job order contracting. Research studies are designed with the application of proprietary cost and project data from Reed Construction Data and RSMeans' extensive North American databases. Analysts offer building product manufacturers qualitative and quantitative market research, as well as time/motion studies for new products, and Web-based dashboards for market opportunity analysis. Clients are from both the public and private sectors, including federal, state, and municipal agencies; corporations; institutions, construction management firms; hospitals; and associations.

RSMeans for Job Order Contracting (JOC)

Best practice job order contracting (JOC) services support the owner in the development of their program. RSMeans develops all required contracting documents, works with stakeholders, and develops qualified and experienced contractor lists. RSMeans JOCWorks™ software is the best of class in the functional tools that meet JOC specific requirements. RSMeans data is organized to meet JOC cost estimating requirements.

Construction Costs for Software Applications

Over 25 unit price and assemblies cost databases are available through a number of leading estimating and facilities management software providers (listed below). For more information see the "Other RSMeans Products" pages at the back of this publication.

RSMeansData™ is also available to federal, state, and local government agencies as multi-year, multi-seat licenses.

- 4Clicks-Solutions, LLC
- ArenaSoft Estimating
- Assetworks
- Beck Technology
- BSD – Building Systems Design, Inc.
- CMS – Construction Management Software
- Corecon Technologies, Inc.
- CorVet Systems
- Hard Dollar Corporation
- Maxwell Systems
- MC^2
- Parsons Corporation
- ProEst
- Sage Timberline Office
- UDA Technologies, Inc.
- US Cost, Inc.
- VFA – Vanderweil Facility Advisers
- WinEstimator, Inc.
- R & K Solutions
- Tririga

Table of Contents

Foreword

Our Mission

Since 1942, RSMeans has been actively engaged in construction cost publishing and consulting throughout North America.

Today, more than 65 years after RSMeans began, our primary objective remains the same: to provide you, the construction and facilities professional, with the most current and comprehensive construction cost data possible.

Whether you are a contractor, owner, architect, engineer, facilities manager, or anyone else who needs a reliable construction cost estimate, you'll find this publication to be a highly useful and necessary tool.

With the constant flow of new construction methods and materials today, it's difficult to find the time to look at and evaluate all the different construction cost possibilities. In addition, because labor and material costs keep changing, last year's cost information is not a reliable basis for today's estimate or budget.

That's why so many construction professionals turn to RSMeans. We keep track of the costs for you, along with a wide range of other key information, from city cost indexes . . . to productivity rates . . . to crew composition . . . to contractor's overhead and profit rates.

RSMeans performs these functions by collecting data from all facets of the industry and organizing it in a format that is instantly accessible to you. From the preliminary budget to the detailed unit price estimate, you'll find the data in this book useful for all phases of construction cost determination.

The Staff, the Organization, and Our Services

When you purchase one of RSMeans' publications, you are, in effect, hiring the services of a full-time staff of construction and engineering professionals.

Our thoroughly experienced and highly qualified staff works daily at collecting, analyzing, and disseminating comprehensive cost information for your needs. These staff members have years of practical construction experience and engineering training prior to joining the firm. As a result, you can count on them not only for accurate cost figures, but also for additional background reference information that will help you create a realistic estimate.

The RSMeans organization is always prepared to help you solve construction problems through its variety of data solutions, including online, CD, and print book formats, as well as cost estimating expertise available via our business solutions, training, and seminars.

Besides a full array of construction cost estimating books, RSMeans also publishes a number of other reference works for the construction industry. Subjects include construction estimating and project and business management, special topics such as green building and job order contracting, and a library of facility management references.

In addition, you can access all of our construction cost data electronically in convenient CD format or on the Web. Visit **MeansCostWorks.com** for more information on our 24/7 online cost data.

What's more, you can increase your knowledge and improve your construction estimating and management performance with an RSMeans construction seminar or in-house training program. These two-day seminar programs offer unparalleled opportunities for everyone in your organization to become updated on a wide variety of construction-related issues.

RSMeans is also a worldwide provider of construction cost management and analysis services for commercial and government owners.

In short, RSMeans can provide you with the tools and expertise for constructing accurate and dependable construction estimates and budgets in a variety of ways.

Robert Snow Means Established a Tradition of Quality That Continues Today

Robert Snow Means spent years building RSMeans, making certain he always delivered a quality product.

Today, at RSMeans, we do more than talk about the quality of our data and the usefulness of our books. We stand behind all of our data, from historical cost indexes to construction materials and techniques to current costs.

If you have any questions about our products or services, please call us toll-free at 1-800-334-3509. Our customer service representatives will be happy to assist you. You can also visit our Web site at **www.rsmeans.com**

How the Book Is Built: An Overview

The Construction Specifications Institute (CSI) and Construction Specifications Canada (CSC) have produced the 2010 edition of MasterFormat™, a system of titles and numbers used extensively to organize construction information.

All unit price data in the RSMeans cost data books is now arranged in the 50-division MasterFormat™ 2010 system.

A Powerful Construction Tool

You have in your hands one of the most powerful construction tools available today. A successful project is built on the foundation of an accurate and dependable estimate. This book will enable you to construct just such an estimate.

For the casual user the book is designed to be:

- quickly and easily understood so you can get right to your estimate.
- filled with valuable information so you can understand the necessary factors that go into the cost estimate.

For the regular user, the book is designed to be:

- a handy desk reference that can be quickly referred to for key costs.
- a comprehensive, fully reliable source of current construction costs and productivity rates so you'll be prepared to estimate any project.
- a source book for preliminary project cost, product selections, and alternate materials and methods.

To meet all of these requirements, we have organized the book into the following clearly defined sections.

New: Quick Start

See our new "Quick Start" instructions on the following page to get started right away.

Estimating with RSMeans Unit Price Cost Data

Please refer to these steps for guidance on completing an estimate using RSMeans unit price cost data.

How to Use the Book: The Details

This section contains an in-depth explanation of how the book is arranged . . . and how you can use it to determine a reliable construction cost estimate. It includes information about how we develop our cost figures and how to completely prepare your estimate.

Unit Price Section

All cost data has been divided into the 50 divisions according to the MasterFormat system of classification and numbering. For a listing of these divisions and an outline of their subdivisions, see the Unit Price Section Table of Contents.

Estimating tips are included at the beginning of each division.

Assemblies Section

The cost data in this section has been organized in an "Assemblies" format. These assemblies are the functional elements of a building and are arranged according to the 7 elements of the UNIFORMAT II classification system. For a complete explanation of a typical "Assemblies" page, see "How to Use the Assemblies Cost Tables."

Reference Section

This section includes information on Equipment Rental Costs, Crew Listings, Historical Cost Indexes, City Cost Indexes, Location Factors, Reference Tables, Change Orders, Square Foot Costs, and a listing of Abbreviations.

Equipment Rental Costs: This section contains the average costs to rent and operate hundreds of pieces of construction equipment.

Crew Listings: This section lists all of the crews referenced in the book. For the purposes of this book, a crew is composed of more than one trade classification and/or the addition of power equipment to any trade classification. Power equipment is included in the cost of the crew. Costs are shown both with bare labor rates and with the installing contractor's overhead and profit added. For each, the total crew cost per eight-hour day and the composite cost per labor-hour are listed.

Historical Cost Indexes: These indexes provide you with data to adjust construction costs over time.

City Cost Indexes: All costs in this book are U.S. national averages. Costs vary because of the regional economy. You can adjust costs by CSI Division to over 700 locations throughout the U.S. and Canada by using the data in this section.

Location Factors: You can adjust total project costs to over 900 locations throughout the U.S. and Canada by using the data in this section.

Reference Tables: At the beginning of selected major classifications in the Unit Price section are reference numbers shown in a shaded box. These numbers refer you to related information in the Reference Section. In this section, you'll find reference tables, explanations, estimating information that support how we develop the unit price data, technical data, and estimating procedures.

Change Orders: This section includes information on the factors that influence the pricing of change orders.

Square Foot Costs: This section contains costs for 59 different building types that allow you to make a rough estimate for the overall cost of a project or its major components.

Abbreviations: A listing of abbreviations used throughout this book, along with the terms they represent, is included in this section.

Index

A comprehensive listing of all terms and subjects in this book will help you quickly find what you need when you are not sure where it occurs in MasterFormat.

The Scope of This Book

This book is designed to be as comprehensive and as easy to use as possible. To that end we have made certain assumptions and limited its scope in two key ways:

1. We have established material prices based on a national average.
2. We have computed labor costs based on a 30-city national average of union wage rates.

For a more detailed explanation of how the cost data is developed, see "How to Use the Book: The Details."

Project Size/Type

The material prices in Means data cost books are "contractor's prices." They are the prices that contractors can expect to pay at the lumberyards, suppliers/distributers warehouses, etc. Small orders of speciality items would be higher than the costs shown, while very large orders, such as truckload lots, would be less. The variation would depend on the size, timing, and negotiating power of the contractor. The labor costs are primarily for new construction or major renovation rather than repairs or minor alterations.

With reasonable exercise of judgment, the figures can be used for any building work.

Absolute Essentials for a Quick Start

If you feel you are ready to use this book and don't think you will need the detailed instructions that begin on the following page, this Absolute Essentials for a Quick Start page is for you. These steps will allow you to get started estimating in a matter of minutes.

1 Scope

Think through the project that you will be estimating, and identify the many individual work tasks that will need to be covered in your estimate.

2 Quantify

Determine the number of units that will be required for each work task that you identified.

3 Pricing

Locate individual Unit Price line items that match the work tasks you identified. The Unit Price Section Table of Contents that begins on page 1 and the Index in the back of the book will help you find these line items.

4 Multiply

Multiply the Total Incl O&P cost for a Unit Price line item in the book by your quantity for that item. The price you calculate will be an estimate for a completed item of work performed by a subcontractor. Keep adding line items in this manner to build your estimate.

5 Project Overhead

Include project overhead items in your estimate. These items are needed to make the job run and are typically, but not always, provided by the General Contractor. They can be found in Division 1. An alternate method of estimating project overhead costs is to apply a percentage of the total project cost.

Include rented tools not included in crews, waste, rubbish handling, and cleanup.

6 Estimate Summary

Include General Contractor's markup on subcontractors, General Contractor's office overhead and profit, and sales tax on materials and equipment.

Adjust your estimate to the project's location by using the City Cost Indexes or Location Factors found in the Reference Section.

Editors' Note: We urge you to spend time reading and understanding the supporting material in the front of this book. An accurate estimate requires experience, knowledge, and careful calculation. The more you know about how we at RSMeans developed the data, the more accurate your estimate will be. In addition, it is important to take into consideration the reference material in the back of the book such as Equipment Listings, Crew Listings, City Cost Indexes, Location Factors, and Reference Numbers.

Estimating with RSMeans Unit Price Cost Data

Following these steps will allow you to complete an accurate estimate using RSMeans Unit Price cost data.

1 Scope Out the Project

- Identify the individual work tasks that will need to be covered in your estimate.
- The Unit Price data in this book has been divided into 50 Divisions according to the CSI MasterFormat 2010—their titles are listed on the back cover of your book.
- Think through the project and identify those CSI Divisions needed in your estimate.
- The Unit Price Section Table of Contents on page 1 may also be helpful when scoping out your project.
- Experienced estimators find it helpful to begin with Division 2 and continue through completion. Division 1 can be estimated after the full project scope is known.

2 Quantify

- Determine the number of units required for each work task that you identified.
- Experienced estimators include an allowance for waste in their quantities. (Waste is not included in RSMeans Unit Price line items unless so stated.)

3 Price the Quantities

- Use the Unit Price Table of Contents, and the Index, to locate individual Unit Price line items for your estimate.
- Reference Numbers indicated within a Unit Price section refer to additional information that you may find useful.
- The crew indicates who is performing the work for that task. Crew codes are expanded in the Crew Listings in the Reference Section to include all trades and equipment that comprise the crew.
- The Daily Output is the amount of work the crew is expected to do in one day.
- The Labor-Hours value is the amount of time it will take for the crew to install one unit of work.
- The abbreviated Unit designation indicates the unit of measure upon which the crew, productivity, and prices are based.
- Bare Costs are shown for materials, labor, and equipment needed to complete the Unit Price line item. Bare costs do not include waste, project overhead, payroll insurance, payroll taxes, main office overhead, or profit.
- The Total Incl O&P cost is the billing rate or invoice amount of the installing contractor or subcontractor who performs the work for the Unit Price line item.

4 Multiply

- Multiply the total number of units needed for your project by the Total Incl O&P cost for each Unit Price line item.
- Be careful that your take off unit of measure matches the unit of measure in the Unit column.
- The price you calculate is an estimate for a completed item of work.
- Keep scoping individual tasks, determining the number of units required for those tasks, matching each task with individual Unit Price line items in the book, and multiply quantities by Total Incl O&P costs.
- An estimate completed in this manner is priced as if a subcontractor, or set of subcontractors, is performing the work. The estimate does not yet include Project Overhead or Estimate Summary components such as General Contractor markups on subcontracted work, General Contractor office overhead and profit, contingency, and location factor.

5 Project Overhead

- Include project overhead items from Division 1 – General Requirements.
- These items are needed to make the job run. They are typically, but not always, provided by the General Contractor. Items include, but are not limited to, field personnel, insurance, performance bond, permits, testing, temporary utilities, field office and storage facilities, temporary scaffolding and platforms, equipment mobilization and demobilization, temporary roads and sidewalks, winter protection, temporary barricades and fencing, temporary security, temporary signs, field engineering and layout, final cleaning and commissioning.
- Each item should be quantified, and matched to individual Unit Price line items in Division 1, and then priced and added to your estimate.
- An alternate method of estimating project overhead costs is to apply a percentage of the total project cost, usually 5% to 15% with an average of 10% (see General Conditions, page ix).
- Include other project related expenses in your estimate such as:
  - Rented equipment not itemized in the Crew Listings
  - Rubbish handling throughout the project (see 02 41 19.19)

6 Estimate Summary

- Include sales tax as required by laws of your state or country.
- Include the General Contractor's markup on self-performed work, usually 5% to 15% with an average of 10%.
- Include the General Contractor's markup on subcontracted work, usually 5% to 15% with an average of 10%.
- Include General Contractor's main office overhead and profit.
- RSMeans gives general guidelines on the General Contractor's main office overhead (see section 01 31 13.50 and Reference Number R013113-50).
- RSMeans gives no guidance on the General Contractor's profit.
- Markups will depend on the size of the General Contractor's operations, his projected annual revenue, the level of risk he is taking on, and on the level of competition in the local area and for this project in particular.
- Include a contingency, usually 3% to 5%, if appropriate.
- Adjust your estimate to the project's location by using the City Cost Indexes or the Location Factors in the Reference Section.
- Look at the rules on the pages for How to Use the City Cost Indexes to see how to apply the Indexes for your location.
- When the proper Index or Factor has been identified for the project's location, convert it to a multiplier by dividing it by 100, and then multiply that multiplier by your estimated total cost. The original estimated total cost will now be adjusted up or down from the national average to a total that is appropriate for your location.

Editors' Notes:

1) *We urge you to spend time reading and understanding the supporting material in the front of this book. An accurate estimate requires experience, knowledge, and careful calculation. The more you know about how we at RSMeans developed the data, the more accurate your estimate will be. In addition, it is important to take into consideration the reference material in the back of the book such as Equipment Listings, Crew Listings, City Cost Indexes, Location Factors, and Reference Numbers.*
2) *Contractors who are bidding or are involved in JOC, DOC, SABER, or IDIQ type contracts are cautioned that workers' compensation insurance, federal and state payroll taxes, waste, project supervision, project overhead, main office overhead, and profit are not included in bare costs. The coefficient or multiplier must cover these costs.*

How to Use the Book: The Details

What's Behind the Numbers? The Development of Cost Data

The staff at RSMeans continually monitors developments in the construction industry in order to ensure reliable, thorough, and up-to-date cost information. While overall construction costs may vary relative to general economic conditions, price fluctuations within the industry are dependent upon many factors. Individual price variations may, in fact, be opposite to overall economic trends. Therefore, costs are constantly tracked and complete updates are published yearly. Also, new items are frequently added in response to changes in materials and methods.

Costs—$ (U.S.)

All costs represent U.S. national averages and are given in U.S. dollars. The RSMeans City Cost Indexes can be used to adjust costs to a particular location. The City Cost Indexes for Canada can be used to adjust U.S. national averages to local costs in Canadian dollars. No exchange rate conversion is necessary.

G The processes or products identified by the green symbol in this publication have been determined to be environmentally responsible and/or resource-efficient solely by the RSMeans engineering staff. The inclusion of the green symbol does not represent compliance with any specific industry association or standard.

Material Costs

The RSMeans staff contacts manufacturers, dealers, distributors, and contractors all across the U.S. and Canada to determine national average material costs. If you have access to current material costs for your specific location, you may wish to make adjustments to reflect differences from the national average. Included within material costs are fasteners for a normal installation. RSMeans engineers use manufacturers' recommendations, written specifications, and/or standard construction practice for size and spacing of fasteners. Adjustments to material costs may be required for your specific application or location. Material costs do not include sales tax.

Labor Costs

Labor costs are based on the average of wage rates from 30 major U.S. cities. Rates are determined from labor union agreements or prevailing wages for construction trades for the current year. Rates, along with overhead and profit markups, are listed on the inside back cover of this book.

- If wage rates in your area vary from those used in this book, or if rate increases are expected within a given year, labor costs should be adjusted accordingly.

Labor costs reflect productivity based on actual working conditions. In addition to actual installation, these figures include time spent during a normal weekday on tasks such as, material receiving and handling, mobilization at site, site movement, breaks, and cleanup.

Productivity data is developed over an extended period so as not to be influenced by abnormal variations and reflects a typical average.

Equipment Costs

Equipment costs include not only rental, but also operating costs for equipment under normal use. The operating costs include parts and labor for routine servicing such as repair and replacement of pumps, filters, and worn lines. Normal operating expendables, such as fuel, lubricants, tires, and electricity (where applicable), are also included. Extraordinary operating expendables with highly variable wear patterns, such as diamond bits and blades, are excluded. These costs are included under materials. Equipment rental rates are obtained from industry sources throughout North America—contractors, suppliers, dealers, manufacturers, and distributors.

Rental rates can also be treated as reimbursement costs for contractor-owned equipment. Owned equipment costs include depreciation, loan payments, interest, taxes, insurance, storage and major repairs.

Equipment costs do not include operators' wages; nor do they include the cost to move equipment to a job site (mobilization) or from a job site (demobilization).

Equipment Cost/Day—The cost of power equipment required for each crew is included in the Crew Listings in the Reference Section (small tools that are considered as essential everyday tools are not listed out separately). The Crew Listings itemize specialized tools and heavy equipment along with labor trades. The daily cost of itemized equipment included in a crew is based on dividing the weekly bare rental rate by 5 (number of working days per week) and then adding the hourly operating cost times 8 (the number of hours per day). This Equipment Cost/Day is shown in the last column of the Equipment Rental Cost pages in the Reference Section.

Mobilization/Demobilization—The cost to move construction equipment from an equipment yard or rental company to the job site and back again is not included in equipment costs. Mobilization (to the site) and demobilization (from the site) costs can be found in the Unit Price Section. If a piece of equipment is already at the job site, it is not appropriate to utilize mob./demob. costs again in an estimate.

Overhead and Profit

Total Cost including O&P for the *Installing Contractor* is shown in the last column on both the Unit Price and the Assemblies pages of this book. This figure is the sum of the bare material cost plus 10% for profit, the bare labor cost plus total overhead and profit, and the bare equipment cost plus 10% for profit. Details for the calculation of Overhead and Profit on labor are shown on the inside back cover and in the Reference Section of this book. (See the "How to Use the Unit Price Pages" for an example of this calculation.)

General Conditions

Cost data in this book is presented in two ways: Bare Costs and Total Cost including O&P (Overhead and Profit). General Conditions, or General Requirements, of the contract, should also be added to the Total Cost including O&P when applicable. Costs for General Conditions are listed in Division 1 of the Unit Price Section and the Reference Section of this book. General Conditions for the *Installing Contractor* may range from 0%

to 10% of the Total Cost including O&P. For the *General* or *Prime Contractor*, costs for General Conditions may range from 5% to 15% of the Total Cost including O&P, with a figure of 10% as the most typical allowance.

Factors Affecting Costs

Costs can vary depending upon a number of variables. Here's how we have handled the main factors affecting costs.

Quality—The prices for materials and the workmanship upon which productivity is based represent sound construction work. They are also in line with U.S. government specifications.

Overtime—We have made no allowance for overtime. If you anticipate premium time or work beyond normal working hours, be sure to make an appropriate adjustment to your labor costs.

Productivity—The productivity, daily output, and labor-hour figures for each line item are based on working an eight-hour day in daylight hours in moderate temperatures. For work that extends beyond normal work hours or is performed under adverse conditions, productivity may decrease. (See the section in "How To Use the Unit Price Pages" for more on productivity.)

Size of Project—The size, scope of work, and type of construction project will have a significant impact on cost. Economies of scale can reduce costs for large projects. Unit costs can often run higher for small projects.

Location—Material prices in this book are for metropolitan areas. However, in dense urban areas, traffic and site storage limitations may increase costs. Beyond a 20-mile radius of large cities, extra trucking or transportation charges may also increase the material costs slightly. On the other hand, lower wage rates may be in effect. Be sure to consider both of these factors when preparing an estimate, particularly if the job site is located in a central city or remote rural location.

In addition, highly specialized subcontract items may require travel and per-diem expenses for mechanics.

Other Factors—

- season of year
- contractor management
- weather conditions
- local union restrictions
- building code requirements
- availability of:
    - adequate energy
    - skilled labor
    - building materials
- owner's special requirements/restrictions
- safety requirements
- environmental considerations

Unpredictable Factors—General business conditions influence "in-place" costs of all items. Substitute materials and construction methods may have to be employed. These may affect the installed cost and/or life cycle costs. Such factors may be difficult to evaluate and cannot necessarily be predicted on the basis of the job's location in a particular section of the country. Thus, where these factors apply, you may find significant but unavoidable cost variations for which you will have to apply a measure of judgment to your estimate.

Rounding of Costs

In general, all unit prices in excess of $5.00 have been rounded to make them easier to use and still maintain adequate precision of the results. The rounding rules we have chosen are in the following table.

Prices from . . .	Rounded to the nearest . . .
$.01 to $5.00	$.01
$5.01 to $20.00	$.05
$20.01 to $100.00	$.50
$100.01 to $300.00	$1.00
$300.01 to $1,000.00	$5.00
$1,000.01 to $10,000.00	$25.00
$10,000.01 to $50,000.00	$100.00
$50,000.01 and above	$500.00

How Subcontracted Items Affect Costs

A considerable portion of all large construction jobs is usually subcontracted. In fact, the percentage done by subcontractors is constantly increasing and may run over 90%. Since the workers employed by these companies do nothing else but install their particular product, they soon become expert in that line. The result is, installation by these firms is accomplished so efficiently that the total in-place cost, even adding the general contractor's overhead and profit, is no more, and often less, than if the principal contractor had handled the installation himself/herself. Companies that deal with construction specialties are anxious to have their product perform well and, consequently, the installation will be the best possible.

Contingencies

The allowance for contingencies generally provides for unforeseen construction difficulties. On alterations or repair jobs, 20% is not too much. If drawings are final and only field contingencies are being considered, 2% or 3% is probably sufficient, and often nothing need be added. Contractually, changes in plans will be covered by extras. The contractor should consider inflationary price trends and possible material shortages during the course of the job. These escalation factors are dependent upon both economic conditions and the anticipated time between the estimate and actual construction. If drawings are not complete or approved, or a budget cost is wanted, it is wise to add 5% to 10%. Contingencies, then, are a matter of judgment. Additional allowances for contingencies are shown in Division 1.

Important Estimating Considerations

The "productivity," or daily output, of each craftsman includes mobilization and cleanup time, break time and plan layout time, as well as an allowance to carry stock from the storage trailer or location on the job site up to 200' into the building and to the first or second floor. If material has to be transported over greater distances or to higher floors, an additional allowance should be considered by the estimator. An allowance has also been included in the piping and fittings installation time for leak checking and minor tightening. Equipment installation time includes the following applicable items: positioning, leveling and securing the unit in place, connecting all associated piping, ducts, vents, etc., which shall have been estimated separately, connecting to an adjacent power source, filling/bleeding, startup, adjusting the controls up and down to ensure proper response, setting the integral controls/valves/regulators/thermostats for proper operation (does not include external building type control systems, DDC systems, etc.), explaining/training owner's operator, and warranty. A reasonable breakdown of the labor costs is as follows:

1. Movement into building, installation/setting of equipment	35%
2. Connection to piping/duct/power, etc.	25%
3. Filling/flushing/cleaning/touchup, etc.	15%
4. Startup/running adjustments	5%
5. Training owner's representative	5%
6. Warranty/call back/service	15%

Note that cranes or other lifting equipment are not included on any lines in the Mechanical divisions. For example, if a crane is required to lift a heavy piece of pipe into place high above a gym floor, or to put a rooftop unit on the roof of a four-story building, etc., it must be added. Due to the potential for extreme variation—from nothing additional required to a major crane or helicopter—we feel that including a nominal amount for "lifting contingency" would be useless and detract from the accuracy of the estimate. When using equipment rental from RSMeans, remember to include the cost of the operator(s).

Estimating Labor-Hours

The labor-hours expressed in this publication are based on Average Installation time, using an efficiency level of approximately 60%-65% (see item 7), which has been found reasonable and acceptable by many contractors. The book uses this national efficiency average to establish a consistent benchmark. For bid situations, adjustments to this efficiency level should be the responsibility of the contractor bidding the project. The unit labor-hour is divided in the following manner. A typical day for a journeyman might be:

1. Study Plans	3%	14.4 min.
2. Material Procurement	3%	14.4 min.
3. Receiving and Storing	3%	14.4 min
4. Mobilization	5%	24.0 min.
5. Site Movement	5%	24.0 min.
6. Layout and Marking	8%	38.4 min.
7. Actual Installation	64%	307.2 min.
8. Cleanup	3%	14.4 min.
9. Breaks, Nonproductive	6%	28.8 min.
	100%	480.0 min.

If any of the percentages expressed in this breakdown do not apply to the particular work or project situation, then that percentage or a portion of it may be deducted from or added to labor-hours.

Final Checklist

Estimating can be a straightforward process provided you remember the basics. Here's a checklist of some of the steps you should remember to complete before finalizing your estimate.

Did you remember to . . .

- factor in the City Cost Index for your locale?
- take into consideration which items have been marked up and by how much?
- mark up the entire estimate sufficiently for your purposes?
- read the background information on techniques and technical matters that could impact your project time span and cost?
- include all components of your project in the final estimate?
- make use of Minimum Labor/Equipment Charges for Small Quantities (see the following page for more details)?
- double check your figures for accuracy?
- call RSMeans if you have any questions about your estimate or the data you've found in our publications?

Remember, RSMeans stands behind its publications. If you have any questions about your estimate . . . about the costs you've used from our books . . . or even about the technical aspects of the job that may affect your estimate, feel free to call the RSMeans editors at 1-800-334-3509.

Using Minimum Labor/Equipment Charges for Small Quantities

Estimating small construction or repair tasks often creates situations in which the quantity of work to be performed is very small. When this occurs, the labor and/or equipment costs to perform the work may be too low to allow for the crew to get to the job, receive instructions, find materials, get set up, perform the work, clean up, and get to the next job. In these situations, the estimator should compare the developed labor and/or equipment costs for performing the work (e.g., quantity x labor and/or equipment costs) with the *"minimum labor/equipment charge"* within that Unit Price section of the book.(These "minimum labor/equipment charge" line items appear only in *Facilities Construction Cost Data* and *Commercial Renovation Cost Data.*)

If the labor and/or equipment costs developed by the estimator are LOWER THAN the book. *"minimum labor/equipment charge"* listed at the bottom of specific sections of Unit Price costs, the estimator should adjust the developed costs upward to the *"minimum labor/equipment charge."* The proper use of a *"minimum labor/equipment charge"* results in having enough money in the estimate to cover the contractor's higher cost of performing a very small amount of work during a partial workday.

A *"minimum labor/equipment charge"* should be used only when the task being estimated is the only task the crew will perform at the job site that day. If, however, the crew will be able to perform other tasks at the job site that day, the use of a *"minimum labor/equipment charge"* is not appropriate.

08 52 Wood Windows

08 52 10 – Plain Wood Windows

08 52 10.40 Casement Window			Crew	Daily Output	Labor-Hours	Unit	2012 Bare Costs Material	Labor	Equipment	Total	Total Incl O&P
0010	**CASEMENT WINDOW**, including frame, screen and grilles										
0100	Avg. quality, bldrs. model, 2'-0" x 3'-0" H, dbl. insulated glass	G	1 Carp	10	.800	Ea.	284	35.50		319.50	375
0150	Low E glass	G		10	.800		259	35.50		294.50	345
0200	2'-0" x 4'-6" high, double insulated glass	G		9	.889		355	39		394	455
0250	Low E glass	G		9	.889		370	39		409	470
9000	Minimum labor/equipment charge		1 Carp	3	2.667	Job		118		118	193

Example:

Establish the bid price to install two casement windows. Assume installation of 2′ x 4′ metal clad windows with insulating glass [Unit Price line number 08 52 10.40 0250], and that this is the only task this crew will perform at the job site that day.

Solution:

Step One —Develop the Bare Labor Cost for this task:

Bare Labor Cost = 2 windows @ $39.00/each = $78.00

Step Two —Evaluate the *"minimum labor/equipment charge"* for this Unit Price section against the developed Bare Labor Cost for this task:

"Minimum labor/equipment charge" = $118.00 (compare with $78.00)

Step Three —Choose to adjust the developed labor cost upward to the *"minimum labor/equipment charge."*

Step Four —Develop the bid price for this task (including O&P):

Add together the marked-up Bare Material Cost for this task and the marked-up *"minimum labor/equipment charge"* for this Unit Price section.

2 x ($370.00 + 10%) + ($118.00 + 63.9%)
= 2 x ($370.00 + $37.00) + ($118.00 + $75.40)
= 2 x ($407.00) + $193.40
= $814.00 + $193.40
= $1007.40

ANSWER: $1007.40 is the correct bid price to use. This sum takes into consideration the Material Cost (with 10% for profit) for these two windows, plus the *"minimum labor/equipment charge"* (with O&P included) for this section of the Unit Price

Unit Price Section

Table of Contents

Table of Contents (cont.)

Table of Contents (cont.)

How RSMeans Data Works

All RSMeans unit price data is organized in the same way.

It is important to understand the structure, so that you can find information easily and use it correctly.

RSMeans **Line Numbers** consist of 12 characters, which identify a unique location in the database for each task. The first 6 or 8 digits conform to the Construction Specifications Institute MasterFormat 2010. The remainder of the digits are a further breakdown by RSMeans in order to arrange items in understandable groups of similar tasks. Line numbers are consistent across all RSMeans publications, so a line number in any RSMeans product will always refer to the same unit of work.

RSMeans engineers have created **reference** information to assist you in your estimate. If there is information that applies to a section, it will be indicated at the start of the section. In this case, R033105-10 provides information on the proportionate quantities of formwork, reinforcing, and concrete used in cast-in-place concrete items such as footings, slabs, beams, and columns. The reference section is located in the back of the book on the pages with a gray edge.

RSMeans **Descriptions** are shown in a hierarchical structure to make them readable. In order to read a complete description, read up through the indents to the top of the section. Include everything that is above and to the left that is not contradicted by information below. For instance, the complete description for line 03 30 53.40 3550 is "Concrete in place, including forms (4 uses), reinforcing steel, concrete, placement, and finishing unless otherwise indicated; Equipment pad (3000 psi), 4' x 4' x 6" thick."

When using **RSMeans data**, it is important to read through an entire section to ensure that you use the data that most closely matches your work. Note that sometimes there is additional information shown in the section that may improve your price. There are frequently lines that further describe, add to, or adjust data for specific situations.

03 30 Cast-In-Place Concrete

03 30 53 – Miscellaneous Cast-In-Place Concrete

03 30 53.40 Concrete In Place

0010	**CONCRETE IN PLACE**	R033105-10
0020	Including forms (4 uses), reinforcing steel, concrete, placement,	R033105-20
0050	and finishing unless otherwise indicated	R033105-50
0300	Beams (3500 psi), 5 kip per L.F., 10′ span	R033105-65
0350	25′ span	R033105-70
0500	Chimney foundations (5000 psi), industrial, minimum	R033105-85
0510	Maximum	
0700	Columns, square (4000 psi), 12″ x 12″, minimum reinforcing	
3500	Add per floor for 3 to 6 stories high	
3520	For 7 to 20 stories high	
3540	Equipment pad (3000 psi), 3′ x 3′ x 6″ thick	
3550	4′ x 4′ x 6″ thick	
3560	5′ x 5′ x 8″ thick	
3570	6′ x 6′ x 8″ thick	

The data published in RSMeans print books represents a "national average" cost. This data should be modified to the project location using the **City Cost Indexes** or **Location Factors** tables found in the reference section (see pages 702–750). Use the location factors to adjust estimate totals if the project covers multiple trades. Use the city cost indexes (CCI) for single trade projects or projects where a more detailed analysis is required. All figures in the two tables are derived from the same research. The last row of data in the CCI, the weighted average, is the same as the numbers reported for each location in the location factor table.

Crews include labor or labor and equipment necessary to accomplish each task. In this case, Crew C-14H is used. RSMeans selects a crew to represent the workers and equipment that are typically used for that task. In this case, Crew C-14H consists of one carpenter foreman (outside), two carpenters, one rodman, one laborer, one cement finisher, and one gas engine vibrator. Details of all crews can be found in the reference section.

Crews

Crew No.	Bare Costs		Incl. Subs O&P		Cost Per Labor-Hour	
Crew C-14H	**Hr.**	**Daily**	**Hr.**	**Daily**	**Bare Costs**	**Incl. O&P**
1 Carpenter Foreman (outside)	\$46.10	\$368.80	\$70.95	\$567.60	\$43.48	\$66.89
2 Carpenters	44.10	705.60	67.85	1085.60		
1 Rodman (reinf.)	49.10	392.80	78.40	627.20		
1 Laborer	35.10	280.80	54.00	432.00		
1 Cement Finisher	42.35	338.80	62.30	498.40		
1 Gas Engine Vibrator		27.40		30.14	.57	.63
48 L.H., Daily Totals		\$2114.20		\$3240.94	\$44.05	\$67.52

The **Daily Output** is the amount of work that the crew can do in a normal 8-hour workday, including mobilization, layout, movement of materials, and cleanup. In this case, crew C-14H can install thirty 4' x 4' x 6" thick concrete pads in a day. Daily output is variable, based on many factors, including the size of the job, location, and environmental conditions. RSMeans data represents work done in daylight (or adequate lighting) and temperate conditions.

Bare Costs are the costs of materials, labor, and equipment that the installing contractor pays. They represent the cost, in U.S. dollars, for one unit of work. They do not include any markups for profit or labor burden.

Crew	Daily Output	Labor-Hours	Unit	2012 Bare Costs Material	Labor	Equipment	Total	Total Incl O&P
C-14A	15.62	12.804	C.Y.	305	565	47.50	917.50	1,275
"	18.55	10.782		320	480	40	840	1,125
C-14C	32.22	3.476		144	147	.83	291.83	385
"	23.71	4.724		168	200	1.13	369.13	495
C-14A	11.96	16.722		350	740	62	1,152	1,600
	31800	.002			.07	.02	.09	.12
	21200	.003			.10	.03	.13	.20
C-14H	45	1.067	Ea.	44	46.50	.61	91.11	121
	30	1.600		66	69.50	.91	136.41	181
	18	2.667		118	116	1.52	235.52	310
	14	3.429		159	149	1.95	309.95	405

The **Total Incl O&P column** is the total cost, including overhead and profit, that the installing contractor will charge the customer. This represents the cost of materials plus 10% profit, the cost of labor plus labor burden and 10% profit, and the cost of equipment plus 10% profit. It does not include the general contractor's overhead and profit. Note: See the inside back cover for details of how RSMeans calculates labor burden.

The **Total column** represents the total bare cost for the installing contractor, in U.S. dollars. In this case, the sum of \$66 for material + \$69.50 for labor + \$.91 for equipment is \$136.41.

The figure in the **Labor Hours** column is the amount of labor required to perform one unit of work—in this case the amount of labor required to construct one 4' x 4' equipment pad. This figure is calculated by dividing the number of hours of labor in the crew by the daily output (48 labor hours divided by 30 pads = 1.6 hours of labor per pad). Multiply 1.600 times 60 to see the value in minutes: 60 x 1.6 = 96 minutes. Note: the labor hour figure is not dependent on the crew size. A change in crew size will result in a corresponding change in daily output, but the labor hours per unit of work will not change.

All RSMeans unit cost data includes the typical **Unit of Measure** used for estimating that item. For concrete-in-place the typical unit is cubic yards (C.Y.) or each (Ea.). For installing broadloom carpet it is square yard, and for gypsum board it is square foot. The estimator needs to take special care that the unit in the data matches the unit in the take-off. Unit conversions may be found in the Reference Section.

How RSMeans Data Works (continued)

Sample Estimate

This sample demonstrates the elements of an estimate, including a tally of the RSMeans data lines, and a summary of the markups on a contractor's work to arrive at a total cost to the owner. The RSMeans Location Factor is added at the bottom of the estimate to adjust the cost of the work to a specific location.

Work Performed: The body of the estimate shows the RSMeans data selected, including line number, a brief description of each item, its take-off unit and quantity, and the bare costs of materials, labor, and equipment. This estimate also includes a column titled "SubContract." This data is taken from the RSMeans column "Total Incl O&P," and represents the total that a subcontractor would charge a general contractor for the work, including the sub's markup for overhead and profit.

General Requirements: This item covers project-wide needs provided by the general contractor. These items vary by project, but may include temporary facilities and utilities, security, testing, project cleanup, etc. For small projects a percentage can be used, typically between 5% and 15% of project cost. For large projects the costs may be itemized and priced individually.

Bonds: Bond costs should be added to the estimate. The figures here represent a typical performance bond, ensuring the owner that if the general contractor does not complete the obligations in the construction contract the bonding company will pay the cost for completion of the work.

Location Adjustment: RSMeans published data is based on national average costs. If necessary, adjust the total cost of the project using a location factor from the "Location Factor" table or the "City Cost Index" table. Use location factors if the work is general, covering multiple trades. If the work is by a single trade (e.g., masonry) use the more specific data found in the "City Cost Indexes."

Project Name:	Pre-Engineered Steel Building	Architect: As Shown		
Location:	Anywhere, USA			
Line Number	**Description**	**Qty**	**Unit**	**Material**
03 30 53.40 3950	Strip footing, 12" x 36"	33.33	C.Y.	$4,699.53
03 30 53.40 3940	Strip footing, 12" x24"	14.8	C.Y.	$2,175.60
03 31 05.35 0300	Concrete ready mix, 4000 psi, for slab	185.2	C.Y.	$19,631.20
03 31 05.70 4300	Placing concrete, slab on grade	185.2	C.Y.	$0.00
03 11 13.65 3000	Concrete forms	500	L.F.	$190.00
03 22 05.50 0200	WWF	150	C.S.F.	$3,975.00
03 39 23.13 0300	Curing	150	C.S.F.	$787.50
03 35 29.30 0250	Concrete finishing	15000	S.F.	$0.00
03 35 29.35 0160	Control joints	650	L.F.	$84.50
Division 03	**Subtotal**			**$31,543.33**
08 33 23.10 0100	10' x 10' OH Door	8	Ea.	$14,000.00
08 33 23.10 3300	Add for enamel finish	800	S.F.	$1,216.00
Division 8	**Subtotal**			**$15,216.00**
13 34 19.50 1000	Pre-Engineered Steel Building	15,000	SF Flr.	
13 34 19.50 6050	Framing for doors, 3' x 7'	4	Opng.	$0.00
13 34 19.50 6100	Framing for doors, 10' x 10'	8	Opng.	$0.00
13 34 19.50 6200	Framing for windows, 3' x 4'	6	Opng.	$0.00
13 34 19.50 5750	PESB doors	4	Opng.	$1,740.00
13 34 19.50 7750	PESB windows	6	Opng.	$2,100.00
13 34 19.50 6550	Gutters	380	L.F.	$2,679.00
13 34 19.50 8650	Roof vent	15	Ea.	$382.50
13 34 19.50 6900	Insulation	25,000	S.F.	$12,750.00
Division 13	**Subtotal**			**$19,651.50**
Estimate Subtotal				**$66,410.83**

Gen. Requirements	**4,648.76**
Sales Tax	**3,320.54**
Subtotal	**74,380.13**
GC O & P	**7,438.01**
Subtotal	**81,818.14**
Contingency	**@5%**
Subtotal	
Bond $12/1000 +10% O&P	
Subtotal	
Location Adjustment	**Factor**
Grand Total	

This estimate is based on an interactive spreadsheet. A copy of this spreadsheet is located on the RSMeans website at **http://www.reedconstructiondata.com/rsmeans/extras/546011**. You are free to download it and adjust it to your methodology.

This example shows the cost to construct a pre-engineered steel building. The foundation, doors, windows, and insulation will be installed by the general contractor. A subcontractor will install the structural steel, roofing, and siding.

	01/01/12			
Labor	**Equipment**	**SubContract**	**EstimateTotal**	
$2,383.10	$14.33	$0.00		
$1,324.60	$7.99	$0.00		
$0.00	$0.00	$0.00		
$2,666.88	$87.04	$0.00		
$1,010.00	$0.00	$0.00		
$3,450.00	$0.00	$0.00		
$795.00	$0.00	$0.00		
$7,500.00	$300.00	$0.00		
$247.00	$58.50	$0.00		
$19,376.58	**$467.86**	**$0.00**	**$51,387.77**	
$4,080.00	$0.00	$0.00		
$0.00	$0.00	$0.00		
$4,080.00	**$0.00**	**$0.00**	**$19,296.00**	
		$287,250.00		
$0.00	$0.00	$1,900.00		
$0.00	$0.00	$8,400.00		
$0.00	$0.00	$3,450.00		
$572.00	$0.00	$0.00		
$483.00	$61.80	$0.00		
$851.20	$0.00	$0.00		
$2,685.00	$0.00	$0.00		
$7,000.00	$0.00	$0.00		
$11,591.20	**$61.80**	**$301,000.00**	**$33[illegible],304.50**	
$35,047.78	**$529.66**	**$301,000.00**	**$402,988.27**	
2,453.34	**37.08**	**21,070.00**		**Gen. Requirements**
	26.48	**7,525.00**		**Sales Tax**
37,501.12	**593.22**	**329,595.00**		**Sub Total**
20,475.61	**59.32**	**32,959.50**		**GC O & P**
57,976.74	**652.54**	**362,554.50**	**$503,001.92**	**Subtotal**
			25,150.10	**Contingency**
			$528,152.02	**Subtotal**
			6,971.61	**Bond**
			$535,123.63	**Subtotal**
102.30			**12,307.84**	**Location Adjustment**
			$547,431.47	**Grand Total**

Sales Tax: If the work is subject to state or local sales taxes, the amount must be added to the estimate. Sales tax may be added to material costs, equipment costs, and subcontracted work. In this case, sales tax was added in all three categories. It was assumed that approximately half the subcontracted work would be material cost, so the tax was applied to 50% of the subcontract total.

GC O&P: This entry represents the general contractor's markup on material, labor, equipment, and subcontractor costs. RSMeans' standard markup on materials, equipment, and subcontracted work is 10%. In this estimate, the markup on the labor performed by the GC's workers uses "Skilled Workers Average" shown in Column F on the table "Installing Contractor's Overhead & Profit," which can be found on the inside-back cover of the book.

Contingency: A factor for contingency may be added to any estimate to represent the cost of unknowns that may occur between the time that the estimate is performed and the time the project is constructed. The amount of the allowance will depend on the stage of design at which the estimate is done, and the contractor's assessment of the risk involved. Refer to section 01 21 16.50 for contigency allowances.

Estimating Tips

01 20 00 Price and Payment Procedures

- Allowances that should be added to estimates to cover contingencies and job conditions that are not included in the national average material and labor costs are shown in section 01 21.
- When estimating historic preservation projects (depending on the condition of the existing structure and the owner's requirements), a 15%-20% contingency or allowance is recommended, regardless of the stage of the drawings.

01 30 00 Administrative Requirements

- Before determining a final cost estimate, it is a good practice to review all the items listed in Subdivisions 01 31 and 01 32 to make final adjustments for items that may need customizing to specific job conditions.
- Requirements for initial and periodic submittals can represent a significant cost to the General Requirements of a job. Thoroughly check the submittal specifications when estimating a project to determine any costs that should be included.

01 40 00 Quality Requirements

- All projects will require some degree of Quality Control. This cost is not included in the unit cost of construction listed in each division. Depending upon the terms of the contract, the various costs of inspection and testing can be the responsibility of either the owner or the contractor. Be sure to include the required costs in your estimate.

01 50 00 Temporary Facilities and Controls

- Barricades, access roads, safety nets, scaffolding, security, and many more requirements for the execution of a safe project are elements of direct cost. These costs can easily be overlooked when preparing an estimate. When looking through the major classifications of this subdivision, determine which items apply to each division in your estimate.
- Construction Equipment Rental Costs can be found in the Reference Section in section 01 54 33. Operators' wages are not included in equipment rental costs.
- Equipment mobilization and demobilization costs are not included in equipment rental costs and must be considered separately in section 01 54 36.50.
- The cost of small tools provided by the installing contractor for his workers is covered in the "Overhead" column on the "Installing Contractor's Overhead and Profit" table that lists labor trades, base rates and markups and, therefore, is included in the "Total Incl. O&P" cost of any Unit Price line item. For those users who are constrained by contract terms to use only bare costs, there are two line items in section 01 54 39.70 for small tools as a percentage of bare labor cost. If some of those users are further constrained by contract terms to refrain from using line items from Division 1, you are advised to cover the cost of small tools within your coefficient or multiplier.

01 70 00 Execution and Closeout Requirements

- When preparing an estimate, thoroughly read the specifications to determine the requirements for Contract Closeout. Final cleaning, record documentation, operation and maintenance data, warranties and bonds, and spare parts and maintenance materials can all be elements of cost for the completion of a contract. Do not overlook these in your estimate.

Reference Numbers

Reference numbers are shown in shaded boxes at the beginning of some major classifications. These numbers refer to related items in the Reference Section. The reference information may be an estimating procedure, an alternate pricing method, or technical information.

Note: Not all subdivisions listed here necessarily appear in this publication.

01 11 Summary of Work

01 11 31 – Professional Consultants

01 11 31.10 Architectural Fees

	01 11 31.10 Architectural Fees	Crew	Daily Output	Labor-Hours	Unit	2012 Bare Costs Material	Labor	Equipment	Total	Total Incl O&P
0010	**ARCHITECTURAL FEES** R011110-10									
0020	For new construction									
0060	Minimum				Project					4.90%
0090	Maximum									16%
0100	For alteration work, to $500,000, add to new construction fee									50%
0150	Over $500,000, add to new construction fee									25%

01 11 31.20 Construction Management Fees

	01 11 31.20 Construction Management Fees	Crew	Daily Output	Labor-Hours	Unit	Material	Labor	Equipment	Total	Total Incl O&P
0010	**CONSTRUCTION MANAGEMENT FEES**									
0060	For work to $100,000				Project					10%
0070	To $250,000									9%
0090	To $1,000,000									6%
0100	To $5,000,000									5%
0110	To $10,000,000									4%

01 11 31.30 Engineering Fees

	01 11 31.30 Engineering Fees	Crew	Daily Output	Labor-Hours	Unit	Material	Labor	Equipment	Total	Total Incl O&P
0010	**ENGINEERING FEES** R011110-30									
0020	Educational planning consultant, minimum				Project					.50%
0100	Maximum				"					2.50%
0400	Elevator & conveying systems, minimum				Contrct					2.50%
0500	Maximum									5%
1000	Mechanical (plumbing & HVAC), minimum									4.10%
1100	Maximum									10.10%
1200	Structural, minimum				Project					1%
1300	Maximum				"					2.50%

01 11 31.75 Renderings

	01 11 31.75 Renderings	Crew	Daily Output	Labor-Hours	Unit	Material	Labor	Equipment	Total	Total Incl O&P
0010	**RENDERINGS** Color, matted, 20" x 30", eye level,									
0020	1 building, minimum				Ea.	2,000			2,000	2,200
0050	Average					2,800			2,800	3,075
0100	Maximum					4,500			4,500	4,950
1000	5 buildings, minimum					4,000			4,000	4,400
1100	Maximum					8,000			8,000	8,800

01 21 Allowances

01 21 16 – Contingency Allowances

01 21 16.50 Contingencies

	01 21 16.50 Contingencies	Crew	Daily Output	Labor-Hours	Unit	Material	Labor	Equipment	Total	Total Incl O&P
0010	**CONTINGENCIES**, Add to estimate									
0020	Conceptual stage				Project					25%
0050	Schematic stage									20%
0100	Preliminary working drawing stage (Design Dev.)									15%
0150	Final working drawing stage									8%

01 21 53 – Factors Allowance

01 21 53.50 Factors

	01 21 53.50 Factors	Crew	Daily Output	Labor-Hours	Unit	Material	Labor	Equipment	Total	Total Incl O&P
0010	**FACTORS** Cost adjustments R012153-10									
0100	Add to construction costs for particular job requirements									
0500	Cut & patch to match existing construction, add, minimum				Costs	2%	3%			
0550	Maximum					5%	9%			
0800	Dust protection, add, minimum					1%	2%			
0850	Maximum					4%	11%			
1100	Equipment usage curtailment, add, minimum					1%	1%			
1150	Maximum					3%	10%			
1400	Material handling & storage limitation, add, minimum					1%	1%			

01 21 Allowances

01 21 53 – Factors Allowance

01 21 53.50 Factors		Crew	Daily Output	Labor-Hours	Unit	Material (2012 Bare Costs)	Labor (2012 Bare Costs)	Equipment (2012 Bare Costs)	Total (2012 Bare Costs)	Total Incl O&P
1450	Maximum				Costs	6%	7%			
1700	Protection of existing work, add, minimum					2%	2%			
1750	Maximum					5%	7%			
2000	Shift work requirements, add, minimum						5%			
2050	Maximum						30%			
2300	Temporary shoring and bracing, add, minimum					2%	5%			
2350	Maximum					5%	12%			
2400	Work inside prisons and high security areas, add, minimum						30%			
2450	Maximum						50%			

01 21 55 – Job Conditions Allowance

01 21 55.50 Job Conditions		Crew	Daily Output	Labor-Hours	Unit	Material	Labor	Equipment	Total	Total Incl O&P
0010	**JOB CONDITIONS** Modifications to applicable									
0020	cost summaries									
0100	Economic conditions, favorable, deduct				Project				2%	2%
0200	Unfavorable, add								5%	5%
0300	Hoisting conditions, favorable, deduct								2%	2%
0400	Unfavorable, add								5%	5%
0500	General Contractor management, experienced, deduct								2%	2%
0600	Inexperienced, add								10%	10%
0700	Labor availability, surplus, deduct								1%	1%
0800	Shortage, add								10%	10%
0900	Material storage area, available, deduct								1%	1%
1000	Not available, add								2%	2%
1100	Subcontractor availability, surplus, deduct								5%	5%
1200	Shortage, add								12%	12%
1300	Work space, available, deduct								2%	2%
1400	Not available, add								5%	5%

01 21 63 – Taxes

01 21 63.10 Taxes		Crew	Daily Output	Labor-Hours	Unit	Material	Labor	Equipment	Total	Total Incl O&P
0010	**TAXES** R012909-80									
0020	Sales tax, State, average				%	5.07%				
0050	Maximum R012909-85					9.25%				
0200	Social Security, on first $110,100 of wages						7.65%			
0300	Unemployment, combined Federal and State, minimum						.80%			
0350	Average						7.80%			
0400	Maximum						14.36%			

01 31 Project Management and Coordination

01 31 13 – Project Coordination

01 31 13.20 Field Personnel		Crew	Daily Output	Labor-Hours	Unit	Material	Labor	Equipment	Total	Total Incl O&P
0010	**FIELD PERSONNEL**									
0020	Clerk, average				Week		420		420	650
0180	Project manager, minimum						1,850		1,850	2,850
0200	Average						2,125		2,125	3,275
0220	Maximum						2,425		2,425	3,750
0240	Superintendent, minimum						1,800		1,800	2,775
0260	Average						1,975		1,975	3,050
0280	Maximum						2,250		2,250	3,475

01 31 Project Management and Coordination

01 31 13 – Project Coordination

01 31 13.30 Insurance

	01 31 13.30 Insurance	Crew	Daily Output	Labor-Hours	Unit	Material	2012 Bare Costs Labor	Equipment	Total	Total Incl O&P
0010	**INSURANCE** R013113-40									
0020	Builders risk, standard, minimum				Job					.24%
0050	Maximum R013113-60									.64%
0200	All-risk type, minimum									.25%
0250	Maximum									.62%
0400	Contractor's equipment floater, minimum				Value					.50%
0450	Maximum				"					1.50%
0600	Public liability, average				Job					2.02%
0810	Workers' compensation & employer's liability									
2000	Range of 35 trades in 50 states, excl. wrecking, min.				Payroll		1.80%			
2100	Average						13.70%			
2200	Maximum						124.10%			

01 31 13.40 Main Office Expense

	01 31 13.40 Main Office Expense	Crew	Daily Output	Labor-Hours	Unit	Material	Labor	Equipment	Total	Total Incl O&P
0010	**MAIN OFFICE EXPENSE** Average for General Contractors									
0020	As a percentage of their annual volume									
0030	Annual volume to $300,000, minimum				% Vol.				20%	
0040	Maximum								30%	
0060	To $500,000, minimum								17%	
0070	Maximum								22%	
0080	To $1,000,000, minimum								16%	
0090	Maximum								19%	
0110	To $3,000,000, minimum								14%	
0120	Maximum								16%	
0130	To $5,000,000, minimum								8%	
0140	Maximum								10%	

01 31 13.50 General Contractor's Mark-Up

	01 31 13.50 General Contractor's Mark-Up	Crew	Daily Output	Labor-Hours	Unit	Material	Labor	Equipment	Total	Total Incl O&P
0010	**GENERAL CONTRACTOR'S MARK-UP** on Change Orders									
0200	Extra work, by subcontractors, add				%					10%
0250	By General Contractor, add									15%
0400	Omitted work, by subcontractors, deduct all but									5%
0450	By General Contractor, deduct all but									7.50%
0600	Overtime work, by subcontractors, add									15%
0650	By General Contractor, add									10%

01 31 13.60 Installing Contractor's Main Office Overhead

	01 31 13.60 Installing Contractor's Main Office Overhead	Crew	Daily Output	Labor-Hours	Unit	Material	Labor	Equipment	Total	Total Incl O&P
0010	**INSTALLING CONTRACTOR'S MAIN OFFICE OVERHEAD** R013113-50									
0020	As percent of direct costs, minimum				%				5%	
0050	Average								16%	
0100	Maximum								30%	

01 31 13.80 Overhead and Profit

	01 31 13.80 Overhead and Profit	Crew	Daily Output	Labor-Hours	Unit	Material	Labor	Equipment	Total	Total Incl O&P
0010	**OVERHEAD & PROFIT** Allowance to add to items in this R013113-50									
0020	book that do not include Subs O&P, average				%				30%	
0100	Allowance to add to items in this book that									
0110	do include Subs O&P, minimum				%					5%
0150	Average									10%
0200	Maximum									15%
0290	Typical, by size of project, under $50,000								40%	
0310	$50,000 to $100,000								35%	
0320	$100,000 to $500,000								25%	
0330	$500,000 to $1,000,000								20%	

01 41 Regulatory Requirements

01 41 26 – Permit Requirements

01 41 26.50 Permits		Crew	Daily Output	Labor-Hours	Unit	Material	2012 Bare Costs Labor	Equipment	Total	Total Incl O&P
0010	**PERMITS**									
0020	Rule of thumb, most cities, minimum				Job					.50%
0100	Maximum				"					2%

01 45 Quality Control

01 45 23 – Testing and Inspecting Services

01 45 23.50 Testing

		Crew	Daily Output	Labor-Hours	Unit	Material	Labor	Equipment	Total	Total Incl O&P
0010	**TESTING** and Inspecting Services									
1800	Compressive test, cylinder, delivered to lab, ASTM C 39				Ea.				12	13
1900	Picked up by lab, minimum								14	15
1950	Average								18	20
2000	Maximum								27	30
2200	Compressive strength, cores (not incl. drilling), ASTM C 42								36	40
2300	Patching core holes				↓				22	24
4730	Soil testing									
4735	Soil density, nuclear method, ASTM D2922				Ea.				35	38.50
4740	Sand cone method ASTM D1556								27	30
4750	Moisture content, ASTM D 2216								9	10
4780	Permeability test, double ring infiltrometer								500	550
4800	Permeability, var. or constant head, undist., ASTM D 2434								227	250
4850	Recompacted								250	275
4900	Proctor compaction, 4" standard mold, ASTM D 698								123	135
4950	6" modified mold								68	75
5100	Shear tests, triaxial, minimum								410	450
5150	Maximum								545	600
5550	Technician for inspection, per day, earthwork								320	350
5650	Bolting								400	440
5750	Roofing								480	530
5790	Welding				↓				480	530
5820	Non-destructive metal testing, dye penetrant				Day				310	340
5840	Magnetic particle								310	340
5860	Radiography								450	495
5880	Ultrasonic				↓				310	340
6000	Welding certification, minimum				Ea.				91	100
6100	Maximum				"				250	275
7000	Underground storage tank									
7500	Volumetric tightness test ,<=12,000 gal.				Ea.				435	480
7510	<=30,000 gal.				"				615	675
7600	Vadose zone (soil gas) sampling, 10-40 samples, min.				Day				1,375	1,500
7610	Maximum				"				2,275	2,500
7700	Ground water monitoring incl. drilling 3 wells, min.				Total				4,550	5,000
7710	Maximum				"				6,375	7,000
8000	X-ray concrete slabs				Ea.				182	200

01 51 Temporary Utilities

01 51 13 – Temporary Electricity

01 51 13.80 Temporary Utilities		Crew	Daily Output	Labor-Hours	Unit	2012 Bare Costs Material	Labor	Equipment	Total	Total Incl O&P
0010	**TEMPORARY UTILITIES**									
0100	Heat, incl. fuel and operation, per week, 12 hrs. per day	1 Skwk	100	.080	CSF Flr	36.50	3.63		40.13	46
0200	24 hrs. per day	″	60	.133		69.50	6.05		75.55	86.50
0350	Lighting, incl. service lamps, wiring & outlets, minimum	1 Elec	34	.235		2.80	12.15		14.95	22
0360	Maximum	″	17	.471		5.85	24.50		30.35	44
0400	Power for temp lighting only, 6.6 KWH, per month								.92	1.01
0430	11.8 KWH, per month								1.65	1.82
0450	23.6 KWH, per month								3.30	3.63
0600	Power for job duration incl. elevator, etc., minimum								47	51.50
0650	Maximum								110	121

01 52 Construction Facilities

01 52 13 – Field Offices and Sheds

01 52 13.20 Office and Storage Space		Crew	Daily Output	Labor-Hours	Unit	Material	Labor	Equipment	Total	Total Incl O&P
0010	**OFFICE AND STORAGE SPACE**									
0020	Office trailer, furnished, no hookups, 20′ x 8′, buy	2 Skwk	1	16	Ea.	8,550	725		9,275	10,600
0250	Rent per month					144			144	158
0300	32′ x 8′, buy	2 Skwk	.70	22.857		13,600	1,050		14,650	16,700
0350	Rent per month					185			185	203
0400	50′ x 10′, buy	2 Skwk	.60	26.667		22,300	1,200		23,500	26,600
0450	Rent per month					279			279	305
0500	50′ x 12′, buy	2 Skwk	.50	32		27,000	1,450		28,450	32,100
0550	Rent per month					365			365	400
0700	For air conditioning, rent per month, add					41.50			41.50	45.50
0800	For delivery, add per mile				Mile	4.60			4.60	5.05
0890	Delivery each way				Ea.	200			200	220
1200	Storage boxes, 20′ x 8′, buy	2 Skwk	1.80	8.889		2,625	405		3,030	3,525
1250	Rent per month					72			72	79
1300	40′ x 8′, buy	2 Skwk	1.40	11.429		3,325	520		3,845	4,500
1350	Rent per month					94			94	103

01 54 Construction Aids

01 54 16 – Temporary Hoists

01 54 16.50 Weekly Forklift Crew		Crew	Daily Output	Labor-Hours	Unit	Material	Labor	Equipment	Total	Total Incl O&P
0010	**WEEKLY FORKLIFT CREW**									
0100	All-terrain forklift, 45′ lift, 35′ reach, 9000 lb. capacity	A-3P	.20	40	Week		1,800	2,400	4,200	5,475

01 54 19 – Temporary Cranes

01 54 19.50 Daily Crane Crews		Crew	Daily Output	Labor-Hours	Unit	Material	Labor	Equipment	Total	Total Incl O&P
0010	**DAILY CRANE CREWS** for small jobs, portal to portal									
0100	12-ton truck-mounted hydraulic crane	A-3H	1	8	Day		380	825	1,205	1,500
0200	25-ton	A-3I	1	8			380	955	1,335	1,650
0300	40-ton	A-3J	1	8			380	1,175	1,555	1,900
0400	55-ton	A-3K	1	16			710	1,625	2,335	2,900
0500	80-ton	A-3L	1	16			710	2,275	2,985	3,625
0600	100-ton	A-3M	1	16			710	2,275	2,985	3,625

01 54 19.60 Monthly Tower Crane Crew		Crew	Daily Output	Labor-Hours	Unit	Material	Labor	Equipment	Total	Total Incl O&P
0010	**MONTHLY TOWER CRANE CREW**, excludes concrete footing									
0100	Static tower crane, 130′ high, 106′ jib, 6200 lb. capacity	A-3N	.05	176	Month		8,425	22,100	30,525	37,600

01 54 23 – Temporary Scaffolding and Platforms

01 54 23.60 Pump Staging		Crew	Daily Output	Labor-Hours	Unit	Material	2012 Bare Costs Labor	Equipment	Total	Total Incl O&P
0010	**PUMP STAGING**, Aluminum R015423-20									
0200	24′ long pole section, buy				Ea.	330			330	365
0300	18′ long pole section, buy					257			257	282
0400	12′ long pole section, buy					173			173	190
0500	6′ long pole section, buy					91.50			91.50	100
0600	6′ long splice joint section, buy					67.50			67.50	74.50
0700	Pump jack, buy					136			136	149
0900	Foldable brace, buy					55			55	60.50
1000	Workbench/back safety rail support, buy					71.50			71.50	78.50
1100	Scaffolding planks/workbench, 14″ wide x 24′ long, buy					665			665	735
1200	Plank end safety rail, buy					288			288	315
1250	Safety net, 22′ long, buy					320			320	355
1300	System in place, 50′ working height, per use based on 50 uses	2 Carp	84.80	.189	C.S.F.	5.70	8.30		14	19.95
1400	100 uses		84.80	.189		2.86	8.30		11.16	16.80
1500	150 uses		84.80	.189		1.92	8.30		10.22	15.75

01 54 23.70 Scaffolding		Crew	Daily Output	Labor-Hours	Unit	Material	Labor	Equipment	Total	Total Incl O&P
0010	**SCAFFOLDING** R015423-10									
0015	Steel tube, regular, no plank, labor only to erect & dismantle									
0090	Building exterior, wall face, 1 to 5 stories, 6′-4″ x 5′ frames	3 Carp	8	3	C.S.F.		132		132	217
0200	6 to 12 stories	4 Carp	8	4			176		176	289
0301	13 to 20 stories	5 Clab	8	5			176		176	288
0460	Building interior, wall face area, up to 16′ high	3 Carp	12	2			88		88	145
0560	16′ to 40′ high		10	2.400			106		106	174
0800	Building interior floor area, up to 30′ high		150	.160	C.C.F.		7.05		7.05	11.55
0900	Over 30′ high	4 Carp	160	.200	″		8.80		8.80	14.45
0906	Complete system for face of walls, no plank, material only rent/mo				C.S.F.	36			36	39.50
0908	Interior spaces, no plank, material only rent/mo				C.C.F.	3.42			3.42	3.76
0910	Steel tubular, heavy duty shoring, buy									
0920	Frames 5′ high 2′ wide				Ea.	81.50			81.50	90
0925	5′ high 4′ wide					93			93	102
0930	6′ high 2′ wide					93.50			93.50	103
0935	6′ high 4′ wide					109			109	120
0940	Accessories									
0945	Cross braces				Ea.	15.85			15.85	17.40
0950	U head, 8″ x 8″					19.10			19.10	21
0955	J-head, 4″ x 8″					13.90			13.90	15.30
0960	Base plate, 8″ x 8″					15.50			15.50	17.05
0965	Leveling jack					33.50			33.50	36.50
1000	Steel tubular, regular, buy									
1100	Frames 3′ high 5′ wide				Ea.	75			75	82.50
1150	5′ high 5′ wide					88			88	96.50
1200	6′-4″ high 5′ wide					118			118	129
1350	7′-6″ high 6′ wide					158			158	173
1500	Accessories cross braces					15.50			15.50	17.05
1550	Guardrail post					16.40			16.40	18.05
1600	Guardrail 7′ section					6.30			6.30	6.95
1650	Screw jacks & plates					21.50			21.50	24
1700	Sidearm brackets					25.50			25.50	28.50
1750	8″ casters					31			31	34
1800	Plank 2″ x 10″ x 16′-0″					24.50			24.50	27
1900	Stairway section					275			275	305
1910	Stairway starter bar					32			32	35

01 54 Construction Aids

01 54 23 – Temporary Scaffolding and Platforms

01 54 23.70 Scaffolding		Crew	Daily Output	Labor-Hours	Unit	2012 Bare Costs Material	Labor	Equipment	Total	Total Incl O&P
1920	Stairway inside handrail				Ea.	53			53	58
1930	Stairway outside handrail					81.50			81.50	90
1940	Walk-thru frame guardrail					41.50			41.50	46
2000	Steel tubular, regular, rent/mo.									
2100	Frames 3′ high 5′ wide				Ea.	5			5	5.50
2150	5′ high 5′ wide					5			5	5.50
2200	6′-4″ high 5′ wide					5.15			5.15	5.65
2250	7′-6″ high 6′ wide					7			7	7.70
2500	Accessories, cross braces					1			1	1.10
2550	Guardrail post					1			1	1.10
2600	Guardrail 7′ section					1			1	1.10
2650	Screw jacks & plates					2			2	2.20
2700	Sidearm brackets					2			2	2.20
2750	8″ casters					8			8	8.80
2800	Outrigger for rolling tower					3			3	3.30
2850	Plank 2″ x 10″ x 16′-0″					6			6	6.60
2900	Stairway section					40			40	44
2940	Walk-thru frame guardrail					2.50			2.50	2.75
3000	Steel tubular, heavy duty shoring, rent/mo.									
3250	5′ high 2′ & 4′ wide				Ea.	5			5	5.50
3300	6′ high 2′ & 4′ wide					5			5	5.50
3500	Accessories, cross braces					1			1	1.10
3600	U - head, 8″ x 8″					1			1	1.10
3650	J - head, 4″ x 8″					1			1	1.10
3700	Base plate, 8″ x 8″					1			1	1.10
3750	Leveling jack					2			2	2.20
5700	Planks, 2″ x 10″ x 16′-0″, labor only to erect & remove to 50′ H	3 Carp	72	.333			14.70		14.70	24
5800	Over 50′ high	4 Carp	80	.400			17.65		17.65	29

01 54 23.75 Scaffolding Specialties		Crew	Daily Output	Labor-Hours	Unit	Material	Labor	Equipment	Total	Total Incl O&P
0010	**SCAFFOLDING SPECIALTIES**									
1200	Sidewalk bridge, heavy duty steel posts & beams, including									
1210	parapet protection & waterproofing									
1220	8′ to 10′ wide, 2 posts	3 Carp	15	1.600	L.F.	32	70.50		102.50	151
1230	3 posts	"	10	2.400	"	48.50	106		154.50	228
1500	Sidewalk bridge using tubular steel									
1510	scaffold frames, including planking	3 Carp	45	.533	L.F.	5.60	23.50		29.10	44.50
1600	For 2 uses per month, deduct from all above					50%				
1700	For 1 use every 2 months, add to all above					100%				
1900	Catwalks, 20″ wide, no guardrails, 7′ span, buy				Ea.	133			133	146
2000	10′ span, buy					186			186	205
3720	Putlog, standard, 8′ span, with hangers, buy					66.50			66.50	73
3730	Rent per month					10			10	11
3750	12′ span, buy					99			99	109
3755	Rent per month					15			15	16.50
3760	Trussed type, 16′ span, buy					228			228	251
3770	Rent per month					20			20	22
3790	22′ span, buy					274			274	300
3795	Rent per month					30			30	33
4000	7 step					710			710	780
4100	Rolling towers, buy, 5′ wide, 7′ long, 10′ high					1,200			1,200	1,325
4200	For 5′ high added sections, to buy, add					207			207	228
4300	Complete incl. wheels, railings, outriggers,									

01 54 Construction Aids

01 54 23 – Temporary Scaffolding and Platforms

01 54 23.75 Scaffolding Specialties		Crew	Daily Output	Labor-Hours	Unit	Material	2012 Bare Costs Labor	Equipment	Total	Total Incl O&P
4350	21′ high, to buy				Ea.	2,025			2,025	2,225
4400	Rent/month = 5% of purchase cost				"	113			113	124

01 54 23.80 Staging Aids		Crew	Daily Output	Labor-Hours	Unit	Material	Labor	Equipment	Total	Total Incl O&P
0010	**STAGING AIDS** and fall protection equipment									
0100	Sidewall staging bracket, tubular, buy				Ea.	45			45	49.50
0110	Cost each per day, based on 250 days use				Day	.18			.18	.20
0200	Guard post, buy				Ea.	19.85			19.85	22
0210	Cost each per day, based on 250 days use				Day	.08			.08	.09
0300	End guard chains, buy per pair				Pair	27			27	30
0310	Cost per set per day, based on 250 days use				Day	.18			.18	.20
1010	Cost each per day, based on 250 days use				"	.04			.04	.04
1100	Wood bracket, buy				Ea.	16.80			16.80	18.50
1110	Cost each per day, based on 250 days use				Day	.07			.07	.07
2010	Cost per pair per day, based on 250 days use				"	.39			.39	.43
2100	Steel siderail jack, buy per pair				Pair	97.50			97.50	107
2110	Cost per pair per day, based on 250 days use				Day	.39			.39	.43
3010	Cost each per day, based on 250 days use				"	.20			.20	.22
3100	Aluminum scaffolding plank, 20″ wide x 24′ long, buy				Ea.	785			785	865
3110	Cost each per day, based on 250 days use				Day	3.15			3.15	3.46
4000	Nylon full body harness, lanyard and rope grab				Ea.	172			172	189
4010	Cost each per day, based on 250 days use				Day	.69			.69	.76
4100	Rope for safety line, 5/8″ x 100′ nylon, buy				Ea.	56			56	61.50
4110	Cost each per day, based on 250 days use				Day	.22			.22	.25
4200	Permanent U-Bolt roof anchor, buy				Ea.	31			31	34
4300	Temporary (one use) roof ridge anchor, buy				"	32			32	35.50
5000	Installation (setup and removal) of staging aids									
5010	Sidewall staging bracket	2 Carp	64	.250	Ea.		11.05		11.05	18.10
5020	Guard post with 2 wood rails	"	64	.250			11.05		11.05	18.10
5030	End guard chains, set	1 Carp	64	.125			5.50		5.50	9.05
5100	Roof shingling bracket		96	.083			3.67		3.67	6
5200	Ladder jack	↓	64	.125			5.50		5.50	9.05
5300	Wood plank, 2″ x 10″ x 16′	2 Carp	80	.200			8.80		8.80	14.45
5310	Aluminum scaffold plank, 20″ x 24′	"	40	.400			17.65		17.65	29
5410	Safety rope	1 Carp	40	.200			8.80		8.80	14.45
5420	Permanent U-Bolt roof anchor (install only)	2 Carp	40	.400			17.65		17.65	29
5430	Temporary roof ridge anchor (install only)	1 Carp	64	.125	↓		5.50		5.50	9.05

01 54 26 – Temporary Swing Staging

01 54 26.50 Swing Staging		Crew	Daily Output	Labor-Hours	Unit	Material	Labor	Equipment	Total	Total Incl O&P
0010	**SWING STAGING**, 500 lb. cap., 2′ wide to 24′ long, hand operated									
0020	steel cable type, with 60′ cables, buy				Ea.	4,900			4,900	5,375
0030	Rent per month				"	490			490	540
0600	Lightweight (not for masons) 24′ long for 150′ height,									
0610	manual type, buy				Ea.	10,200			10,200	11,200
0620	Rent per month					1,025			1,025	1,125
0700	Powered, electric or air, to 150′ high, buy					25,600			25,600	28,200
0710	Rent per month					1,800			1,800	1,975
0780	To 300′ high, buy					26,000			26,000	28,600
0800	Rent per month					1,825			1,825	2,000
1000	Bosun's chair or work basket 3′ x 3.5′, to 300′ high, electric, buy					10,400			10,400	11,400
1010	Rent per month				↓	730			730	800
2200	Move swing staging (setup and remove)	E-4	2	16	Move		800	61	861	1,550

01 54 Construction Aids

01 54 36 – Equipment Mobilization

01 54 36.50 Mobilization

01 54 36.50 Mobilization		Crew	Daily Output	Labor-Hours	Unit	Material	2012 Bare Costs Labor	Equipment	Total	Total Incl O&P
0010	**MOBILIZATION** (Use line item again for demobilization) R015433-10									
0015	Up to 25 mi haul dist (50 mi RT for mob/demob crew)									
0020	Dozer, loader, backhoe, excav., grader, paver, roller, 70 to 150 H.P.	B-34N	4	2	Ea.		70.50	126	196.50	253
0100	Above 150 HP	B-34K	3	2.667			94	283	377	460
0300	Scraper, towed type (incl. tractor), 6 C.Y. capacity		3	2.667			94	283	377	460
0400	10 C.Y.		2.50	3.200			113	340	453	560
0600	Self-propelled scraper, 15 C.Y.		2.50	3.200			113	340	453	560
0700	24 C.Y.		2	4			141	425	566	695
0900	Shovel or dragline, 3/4 C.Y.		3.60	2.222			78.50	236	314.50	385
1000	1-1/2 C.Y.		3	2.667			94	283	377	460
1100	Small equipment, placed in rear of, or towed by pickup truck	A-3A	8	1			34.50	17.40	51.90	74.50
1150	Equip up to 70 HP, on flatbed trailer behind pickup truck	A-3D	4	2			68.50	62	130.50	179
2000	Crane, truck-mounted, up to 75 ton, (driver only, one-way)	1 Eqhv	7.20	1.111			53		53	84
2100	Crane, truck-mounted, over 75 ton	A-3E	2.50	6.400			266	55.50	321.50	485
2200	Crawler-mounted, up to 75 ton	A-3F	2	8			330	435	765	1,000
2300	Over 75 ton	A-3G	1.50	10.667			445	660	1,105	1,425
2500	For each additional 5 miles haul distance, add						10%	10%		
3000	For large pieces of equipment, allow for assembly/knockdown									
3100	For mob/demob of micro-tunneling equip, see Section 33 05 23.19									
3200	For mob/demob of pile driving equip, see Section 31 62 19.10									

01 54 39 – Construction Equipment

01 54 39.70 Small Tools

01 54 39.70 Small Tools		Crew	Daily Output	Labor-Hours	Unit	Material	Labor	Equipment	Total	Total Incl O&P
0010	**SMALL TOOLS** R013113-50									
0020	As % of contractor's bare labor cost for project, minimum				Total		.50%			
0100	Maximum				"		2%			

01 55 Vehicular Access and Parking

01 55 23 – Temporary Roads

01 55 23.50 Roads and Sidewalks

01 55 23.50 Roads and Sidewalks		Crew	Daily Output	Labor-Hours	Unit	Material	Labor	Equipment	Total	Total Incl O&P
0010	**ROADS AND SIDEWALKS** Temporary									
0050	Roads, gravel fill, no surfacing, 4" gravel depth	B-14	715	.067	S.Y.	4.15	2.49	.47	7.11	9.10
0100	8" gravel depth	"	615	.078	"	8.30	2.89	.54	11.73	14.45

01 56 Temporary Barriers and Enclosures

01 56 13 – Temporary Air Barriers

01 56 13.60 Tarpaulins

01 56 13.60 Tarpaulins		Crew	Daily Output	Labor-Hours	Unit	Material	Labor	Equipment	Total	Total Incl O&P
0010	**TARPAULINS**									
0020	Cotton duck, 10 oz. to 13.13 oz. per S.Y., minimum				S.F.	.78			.78	.86
0050	Maximum					.80			.80	.88
0100	Polyvinyl coated nylon, 14 oz. to 18 oz., minimum					.48			.48	.53
0150	Maximum					.68			.68	.75
0200	Reinforced polyethylene 3 mils thick, white					.15			.15	.17
0300	4 mils thick, white, clear or black					.20			.20	.22
0400	5.5 mils thick, clear					.23			.23	.25
0500	White, fire retardant					.35			.35	.39
0600	12 mils, oil resistant, fire retardant					.28			.28	.31
0700	8.5 mils, black					.57			.57	.63
0720	Steel reinforced polyethylene, 4 mils thick					.53			.53	.58

01 56 Temporary Barriers and Enclosures

01 56 13 – Temporary Air Barriers

01 56 13.60 Tarpaulins		Crew	Daily Output	Labor-Hours	Unit	Material	2012 Bare Costs Labor	Equipment	Total	Total Incl O&P
0730	Polyester reinforced w/integral fastening system 11 mils thick				S.F.	1.07			1.07	1.18
0740	Mylar polyester, non-reinforced, 7 mils thick					1.17			1.17	1.29

01 56 13.90 Winter Protection

		Crew	Daily Output	Labor-Hours	Unit	Material	Labor	Equipment	Total	Total Incl O&P
0010	**WINTER PROTECTION**									
0100	Framing to close openings	2 Clab	750	.021	S.F.	.39	.75		1.14	1.66
0200	Tarpaulins hung over scaffolding, 8 uses, not incl. scaffolding		1500	.011		.25	.37		.62	.89
0250	Tarpaulin polyester reinf. w/integral fastening system 11 mils thick		1600	.010		.80	.35		1.15	1.46
0300	Prefab fiberglass panels, steel frame, 8 uses		1200	.013		.85	.47		1.32	1.71

01 56 16 – Temporary Dust Barriers

01 56 16.10 Dust Barriers, Temporary

		Crew	Daily Output	Labor-Hours	Unit	Material	Labor	Equipment	Total	Total Incl O&P
0010	**DUST BARRIERS, TEMPORARY**									
0020	Spring loaded telescoping pole & head, to 12′, erect and dismantle	1 Clab	240	.033	Ea.		1.17		1.17	1.92
0025	Cost per day (based upon 250 days)				Day	.24			.24	.26
0030	To 21′, erect and dismantle	1 Clab	240	.033	Ea.		1.17		1.17	1.92
0035	Cost per day (based upon 250 days)				Day	.39			.39	.42
0040	Accessories, caution tape reel, erect and dismantle	1 Clab	480	.017	Ea.		.59		.59	.96
0045	Cost per day (based upon 250 days)				Day	.05			.05	.05
0060	Foam rail and connector, erect and dismantle	1 Clab	240	.033	Ea.		1.17		1.17	1.92
0065	Cost per day (based upon 250 days)				Day	.10			.10	.11
0070	Caution tape	1 Clab	384	.021	C.L.F.	2.65	.73		3.38	4.11
0080	Zipper, standard duty		60	.133	Ea.	8	4.68		12.68	16.45
0090	Heavy duty		48	.167	″	9.50	5.85		15.35	20
0100	Polyethylene sheet, 4 mil		37	.216	Sq.	3.23	7.60		10.83	16
0110	6 mil		37	.216	″	4.54	7.60		12.14	17.45
1000	Dust partition, 6 mil polyethylene, 1″ x 3″ frame	2 Carp	2000	.008	S.F.	.27	.35		.62	.88
1080	2″ x 4″ frame	″	2000	.008	″	.29	.35		.64	.90

01 56 23 – Temporary Barricades

01 56 23.10 Barricades

		Crew	Daily Output	Labor-Hours	Unit	Material	Labor	Equipment	Total	Total Incl O&P
0010	**BARRICADES**									
0020	5′ high, 3 rail @ 2″ x 8″, fixed	2 Carp	20	.800	L.F.	5.30	35.50		40.80	64
0150	Movable	″	30	.533	″	4.35	23.50		27.85	43.50
0300	Stock units, 6′ high, 8′ wide, plain, buy				Ea.	475			475	520
0350	With reflective tape, buy				″	540			540	595
0400	Break-a-way 3″ PVC pipe barricade									
0410	with 3 ea. 1′ x 4′ reflectorized panels, buy				Ea.	160			160	176
0500	Plywood with steel legs, 32″ wide					99.50			99.50	109
0600	Warning signal flag tree, 11′ high, 2 flags, buy					330			330	365
0800	Traffic cones, PVC, 18″ high					10			10	11
0850	28″ high					11.60			11.60	12.80
0900	Barrels, 55 gal., with flasher	1 Clab	96	.083		83	2.92		85.92	96.50
1000	Guardrail, wooden, 3′ high, 1″ x 6″, on 2″ x 4″ posts	2 Carp	200	.080	L.F.	1.15	3.53		4.68	7.05
1100	2″ x 6″, on 4″ x 4″ posts	″	165	.097		2.26	4.28		6.54	9.50
1200	Portable metal with base pads, buy					15.25			15.25	16.75
1250	Typical installation, assume 10 reuses	2 Carp	600	.027		2.50	1.18		3.68	4.68
1300	Barricade tape, polyethylene, 7 mil, 3″ wide x 500′ long roll				Ea.	25			25	27.50

01 56 26 – Temporary Fencing

01 56 26.50 Temporary Fencing

		Crew	Daily Output	Labor-Hours	Unit	Material	Labor	Equipment	Total	Total Incl O&P
0010	**TEMPORARY FENCING**									
0020	Chain link, 11 ga, 5′ high	2 Clab	400	.040	L.F.	4.50	1.40		5.90	7.25
0100	6′ high		300	.053		5.30	1.87		7.17	8.90
0200	Rented chain link, 6′ high, to 1000′ (up to 12 mo.)		400	.040		6.30	1.40		7.70	9.25

01 56 Temporary Barriers and Enclosures

01 56 26 – Temporary Fencing

01 56 26.50 Temporary Fencing		Crew	Daily Output	Labor-Hours	Unit	2012 Bare Costs Material	Labor	Equipment	Total	Total Incl O&P
0250	Over 1000' (up to 12 mo.)	2 Clab	300	.053	L.F.	3.29	1.87		5.16	6.70
0350	Plywood, painted, 2" x 4" frame, 4' high	A-4	135	.178		5.45	7.50		12.95	18.20
0400	4" x 4" frame, 8' high	"	110	.218		10.20	9.20		19.40	26
0500	Wire mesh on 4" x 4" posts, 4' high	2 Carp	100	.160		9.25	7.05		16.30	22
0550	8' high	"	80	.200	↓	13.90	8.80		22.70	30

01 56 29 – Temporary Protective Walkways

01 56 29.50 Protection

		Crew	Daily Output	Labor-Hours	Unit	Material	Labor	Equipment	Total	Total Incl O&P
0010	**PROTECTION**									
0020	Stair tread, 2" x 12" planks, 1 use	1 Carp	75	.107	Tread	3.83	4.70		8.53	11.90
0100	Exterior plywood, 1/2" thick, 1 use		65	.123		1.41	5.45		6.86	10.45
0200	3/4" thick, 1 use		60	.133	↓	2.10	5.90		8	11.95
2200	Sidewalks, 2" x 12" planks, 2 uses		350	.023	S.F.	.64	1.01		1.65	2.35
2300	Exterior plywood, 2 uses, 1/2" thick		750	.011		.24	.47		.71	1.03
2400	5/8" thick		650	.012		.31	.54		.85	1.23
2500	3/4" thick	↓	600	.013	↓	.35	.59		.94	1.35

01 56 32 – Temporary Security

01 56 32.50 Watchman

		Crew	Daily Output	Labor-Hours	Unit	Material	Labor	Equipment	Total	Total Incl O&P
0010	**WATCHMAN**									
0020	Service, monthly basis, uniformed person, minimum				Hr.				25	27.50
0100	Maximum								45.50	50
0200	Person and command dog, minimum								31	34
0300	Maximum				↓				54.50	60
0500	Sentry dog, leased, with job patrol (yard dog), 1 dog				Week				290	320
0600	2 dogs				"				390	430
0800	Purchase, trained sentry dog, minimum				Ea.				1,375	1,500
0900	Maximum				"				2,725	3,000

01 58 Project Identification

01 58 13 – Temporary Project Signage

01 58 13.50 Signs

		Crew	Daily Output	Labor-Hours	Unit	Material	Labor	Equipment	Total	Total Incl O&P
0010	**SIGNS**									
0020	High intensity reflectorized, no posts, buy				S.F.	34			34	37.50

01 71 Examination and Preparation

01 71 23 – Field Engineering

01 71 23.13 Construction Layout

		Crew	Daily Output	Labor-Hours	Unit	Material	Labor	Equipment	Total	Total Incl O&P
0010	**CONSTRUCTION LAYOUT**									
1100	Crew for layout of building, trenching or pipe laying, 2 person crew	A-6	1	16	Day		705	78.50	783.50	1,225
1200	3 person crew	A-7	1	24			1,150	78.50	1,228.50	1,925
1400	Crew for roadway layout, 4 person crew	A-8	1	32	↓		1,500	78.50	1,578.50	2,500

01 74 Cleaning and Waste Management

01 74 13 – Progress Cleaning

01 74 13.20 Cleaning Up		Crew	Daily Output	Labor-Hours	Unit	Material	2012 Bare Costs Labor	Equipment	Total	Total Incl O&P
0010	**CLEANING UP**									
0020	After job completion, allow, minimum				Job					.30%
0040	Maximum				"					1%
0052	Cleanup of floor area, continuous, per day, during const.	A-5	16	1.125	M.S.F.	1.75	39.50	3.30	44.55	70
0100	Final by GC at end of job	"	11.50	1.565	"	2.78	55	4.59	62.37	97.50

01 76 Protecting Installed Construction

01 76 13 – Temporary Protection of Installed Construction

01 76 13.20 Temporary Protection

		Crew	Daily Output	Labor-Hours	Unit	Material	Labor	Equipment	Total	Total Incl O&P
0010	**TEMPORARY PROTECTION**									
0020	Flooring, 1/8" tempered hardboard, taped seams	2 Carp	1500	.011	S.F.	.38	.47		.85	1.19
0030	Peel away carpet protection	1 Clab	3200	.003	"	.11	.09		.20	.26

01 91 Commissioning

01 91 13 – General Commissioning Requirements

01 91 13.50 Building Commissioning

		Crew	Daily Output	Labor-Hours	Unit	Material	Labor	Equipment	Total	Total Incl O&P
0010	**BUILDING COMMISSIONING**									
0100	Basic building commissioning, minimum				%					.25%
0150	Maximum									.50%
0200	Enhanced building commissioning, minimum									.50%
0250	Maximum									1%

01 93 Facility Maintenance

01 93 09 – Facility Equipment

01 93 09.50 Moving Equipment

		Crew	Daily Output	Labor-Hours	Unit	Material	Labor	Equipment	Total	Total Incl O&P
0010	**MOVING EQUIPMENT**, Remove and reset, 100' distance,									
0020	No obstructions, no assembly or leveling unless noted									
0100	Annealing furnace, 24' overall	B-67	4	4	Ea.		182	72.50	254.50	365
0200	Annealing oven, small		14	1.143			52	20.50	72.50	105
0240	Very large		1	16			725	289	1,014	1,475
0400	Band saw, small		12	1.333			60.50	24	84.50	122
0440	Large		8	2			91	36	127	183
0500	Blue print copy machine		7	2.286			104	41.50	145.50	210
0600	Bonding mill, 6"		7	2.286			104	41.50	145.50	210
0620	12"		6	2.667			121	48	169	244
0640	18"		4	4			182	72.50	254.50	365
0660	24"		2	8			365	145	510	730
0700	Boring machine (jig)	B-68	7	3.429			156	41.50	197.50	292
0800	Bridgeport mill, standard	B-67	14	1.143			52	20.50	72.50	105
1000	Calibrator, 6 unit	"	14	1.143			52	20.50	72.50	105
1100	Comparitor, bench top	2 Clab	14	1.143			40		40	66
1140	Floor mounted	B-67	7	2.286			104	41.50	145.50	210
1200	Computer, desk top	2 Clab	25	.640			22.50		22.50	37
1300	Copy machine	"	25	.640			22.50		22.50	37
1500	Deflasher	B-67	14	1.143			52	20.50	72.50	105
1600	Degreaser, small		14	1.143			52	20.50	72.50	105
1640	Large 24' overall		1	16			725	289	1,014	1,475

01 93 09.50 Moving Equipment		Crew	Daily Output	Labor-Hours	Unit	Material	2012 Bare Costs Labor	Equipment	Total	Total Incl O&P
1700	Desk with chair	2 Clab	25	.640	Ea.		22.50		22.50	37
1800	Dial press	B-67	7	2.286			104	41.50	145.50	210
1900	Drafting table	2 Clab	14	1.143			40		40	66
2000	Drill press, bench top	"	14	1.143			40		40	66
2040	Floor mounted	B-67	14	1.143			52	20.50	72.50	105
2080	Industrial radial	"	7	2.286			104	41.50	145.50	210
2100	Dust collector, portable	2 Clab	25	.640			22.50		22.50	37
2140	Stationary, small	B-67	7	2.286			104	41.50	145.50	210
2180	Stationary, large	"	2	8			365	145	510	730
2300	Electric discharge machine	B-68	7	3.429			156	41.50	197.50	292
2400	Environmental chamber walls, including assembly	4 Clab	18	1.778	L.F.		62.50		62.50	102
2600	File cabinet	2 Clab	25	.640	Ea.		22.50		22.50	37
2800	Grinder/sander, pedestal mount	B-67	14	1.143			52	20.50	72.50	105
3000	Hack saw, power	2 Clab	24	.667			23.50		23.50	38.50
3100	Hydraulic press	B-67	14	1.143			52	20.50	72.50	105
3500	Laminar flow tables	"	14	1.143			52	20.50	72.50	105
3600	Lathe, bench	2 Clab	14	1.143			40		40	66
3640	6"	B-67	14	1.143			52	20.50	72.50	105
3680	10"		13	1.231			56	22.50	78.50	113
3720	12"		12	1.333			60.50	24	84.50	122
4000	Milling machine		8	2			91	36	127	183
4100	Molding press, 25 ton		5	3.200			145	58	203	293
4140	60 ton		4	4			182	72.50	254.50	365
4180	100 ton		2	8			365	145	510	730
4220	150 ton		1.50	10.667			485	193	678	975
4260	200 ton		1	16			725	289	1,014	1,475
4300	300 ton		.75	21.333			970	385	1,355	1,950
4700	Oil pot stand		14	1.143			52	20.50	72.50	105
5000	Press, 10 ton		14	1.143			52	20.50	72.50	105
5040	15 ton		12	1.333			60.50	24	84.50	122
5080	20 ton		10	1.600			72.50	29	101.50	146
5120	30 ton		8	2			91	36	127	183
5160	45 ton		6	2.667			121	48	169	244
5200	60 ton		4	4			182	72.50	254.50	365
5240	75 ton		2.50	6.400			291	116	407	585
5280	100 ton		2	8			365	145	510	730
5500	Raised floor, including assembly	2 Carp	250	.064	S.F.		2.82		2.82	4.63
5600	Rolling mill, 6"	B-67	7	2.286	Ea.		104	41.50	145.50	210
5640	9"		6	2.667			121	48	169	244
5680	12"		4	4			182	72.50	254.50	365
5720	13"		3.50	4.571			208	82.50	290.50	415
5760	18"		2	8			365	145	510	730
5800	25"		1	16			725	289	1,014	1,475
6000	Sander, floor stand		14	1.143			52	20.50	72.50	105
6100	Screw machine		7	2.286			104	41.50	145.50	210
6200	Shaper, 16"		14	1.143			52	20.50	72.50	105
6300	Shear, power assist		4	4			182	72.50	254.50	365
6400	Slitter, 6"		14	1.143			52	20.50	72.50	105
6440	8"		13	1.231			56	22.50	78.50	113
6480	10"		12	1.333			60.50	24	84.50	122
6520	12"		11	1.455			66	26.50	92.50	133
6560	16"		10	1.600			72.50	29	101.50	146
6600	20"		8	2			91	36	127	183

01 93 09 – Facility Equipment

01 93 09.50 Moving Equipment		Crew	Daily Output	Labor-Hours	Unit	Material	2012 Bare Costs Labor	Equipment	Total	Total Incl O&P
6640	24"	B-67	6	2.667	Ea.		121	48	169	244
6800	Snag and tap machine		7	2.286			104	41.50	145.50	210
6900	Solder machine (auto)	↓	7	2.286			104	41.50	145.50	210
7000	Storage cabinet metal, small	2 Clab	36	.444			15.60		15.60	25.50
7040	Large		25	.640			22.50		22.50	37
7100	Storage rack open, small		14	1.143			40		40	66
7140	Large	↓	7	2.286			80		80	132
7200	Surface bench, small	B-67	14	1.143			52	20.50	72.50	105
7240	Large	"	5	3.200			145	58	203	293
7300	Surface grinder, large wet	B-68	5	4.800			219	58	277	410
7500	Time check machine	2 Clab	14	1.143			40		40	66
8000	Welder, 30 kVA (bench)		14	1.143			40		40	66
8100	Work bench with chair	↓	25	.640	↓		22.50		22.50	37

Division Notes

		CREW	DAILY OUTPUT	LABOR-HOURS	UNIT	2012 BARE COSTS MAT.	LABOR	EQUIP.	TOTAL	TOTAL INCL O&P

Estimating Tips

02 30 00 Subsurface Investigation

In preparing estimates on structures involving earthwork or foundations, all information concerning soil characteristics should be obtained. Look particularly for hazardous waste, evidence of prior dumping of debris, and previous stream beds.

02 41 00 Demolition and Structure Moving

The costs shown for selective demolition do not include rubbish handling or disposal. These items should be estimated separately using RSMeans data or other sources.

- Historic preservation often requires that the contractor remove materials from the existing structure, rehab them, and replace them. The estimator must be aware of any related measures and precautions that must be taken when doing selective demolition and cutting and patching. Requirements may include special handling and storage, as well as security.
- In addition to Subdivision 02 41 00, you can find selective demolition items in each division. Example: Roofing demolition is in Division 7.

02 42 10 Building Deconstruction

This section provides costs for the careful dismantling and recycling of most of low-rise building materials.

02 56 13 Containment of Hazardous Waste

This section addresses on-site hazardous waste disposal costs.

02 80 00 Hazardous Material Disposal/ Remediation

This subdivision includes information on hazardous waste handling, asbestos remediation, lead remediation, and mold remediation. See reference R028213-20 and R028319-60 for further guidance in using these unit price lines.

02 91 10 Monitoring Chemical Sampling, Testing Analysis

This section provides costs for on-site sampling and testing hazardous waste.

Reference Numbers

Reference numbers are shown in shaded boxes at the beginning of some major classifications. These numbers refer to related items in the Reference Section. The reference information may be an estimating procedure, an alternate pricing method, or technical information.

Note: Not all subdivisions listed here necessarily appear in this publication.

02 21 Surveys

02 21 13 – Site Surveys

02 21 13.09 Topographical Surveys		Crew	Daily Output	Labor-Hours	Unit	Material	2012 Bare Costs Labor	Equipment	Total	Total Incl O&P
0010	**TOPOGRAPHICAL SURVEYS**									
0020	Topographical surveying, conventional, minimum	A-7	3.30	7.273	Acre	18.90	350	24	392.90	615
0100	Maximum	A-8	.60	53.333	″	58	2,500	131	2,689	4,225

02 21 13.13 Boundary and Survey Markers		Crew	Daily Output	Labor-Hours	Unit	Material	Labor	Equipment	Total	Total Incl O&P
0010	**BOUNDARY AND SURVEY MARKERS**									
0300	Lot location and lines, large quantities, minimum	A-7	2	12	Acre	33.50	580	39.50	653	1,000
0320	Average	″	1.25	19.200		53.50	925	63	1,041.50	1,625
0400	Small quantities, maximum	A-8	1	32		71.50	1,500	78.50	1,650	2,600

02 32 Geotechnical Investigations

02 32 13 – Subsurface Drilling and Sampling

02 32 13.10 Boring and Exploratory Drilling		Crew	Daily Output	Labor-Hours	Unit	Material	Labor	Equipment	Total	Total Incl O&P
0010	**BORING AND EXPLORATORY DRILLING**									
0020	Borings, initial field stake out & determination of elevations	A-6	1	16	Day		705	78.50	783.50	1,225
0100	Drawings showing boring details				Total		310		310	390
0200	Report and recommendations from P.E.						720		720	900
0300	Mobilization and demobilization	B-55	4	6			209	246	455	610
0350	For over 100 miles, per added mile		450	.053	Mile		1.86	2.18	4.04	5.45
0600	Auger holes in earth, no samples, 2-1/2″ diameter		78.60	.305	L.F.		10.65	12.50	23.15	31
0650	4″ diameter		67.50	.356			12.40	14.55	26.95	36
0800	Cased borings in earth, with samples, 2-1/2″ diameter		55.50	.432		22	15.05	17.70	54.75	68
0850	4″ diameter		32.60	.736		34.50	25.50	30	90	113
1000	Drilling in rock, "BX" core, no sampling	B-56	34.90	.458			18.30	40.50	58.80	74
1050	With casing & sampling		31.70	.505		22	20	45	87	106
1200	"NX" core, no sampling		25.92	.617			24.50	55	79.50	100
1250	With casing and sampling		25	.640		26.50	25.50	57	109	133
1400	Borings, earth, drill rig and crew with truck mounted auger	B-55	1	24	Day		835	985	1,820	2,450
1450	Rock using crawler type drill	B-56	1	16	″		640	1,425	2,065	2,575
1500	For inner city borings add, minimum									10%
1510	Maximum									20%

02 41 Demolition

02 41 13 – Selective Site Demolition

02 41 13.15 Hydrodemolition		Crew	Daily Output	Labor-Hours	Unit	Material	Labor	Equipment	Total	Total Incl O&P
0010	**HYDRODEMOLITION**									
0015	Hydrodemolition, concrete pavement, 4000 PSI, 2″ depth	B-5	500	.112	S.F.		4.33	2.63	6.96	9.90
0120	4″ depth		450	.124			4.81	2.92	7.73	11
0130	6″ depth		400	.140			5.40	3.29	8.69	12.35
0410	6000 PSI, 2″ depth		410	.137			5.30	3.21	8.51	12.10
0420	4″ depth		350	.160			6.20	3.76	9.96	14.15
0430	6″ depth		300	.187			7.20	4.38	11.58	16.50
0510	8000 PSI, 2″ depth		330	.170			6.55	3.99	10.54	15
0520	4″ depth		280	.200			7.75	4.70	12.45	17.65
0530	6″ depth		240	.233			9	5.50	14.50	20.50

02 41 13.17 Demolish, Remove Pavement and Curb		Crew	Daily Output	Labor-Hours	Unit	Material	Labor	Equipment	Total	Total Incl O&P
0010	**DEMOLISH, REMOVE PAVEMENT AND CURB** R024119-10									
5010	Pavement removal, bituminous roads, 3″ thick	B-38	690	.058	S.Y.		2.30	1.71	4.01	5.60
5050	4″ to 6″ thick		420	.095			3.78	2.82	6.60	9.20
5100	Bituminous driveways		640	.063			2.48	1.85	4.33	6.05

02 41 Demolition

02 41 13 – Selective Site Demolition

02 41 13.17 Demolish, Remove Pavement and Curb		Crew	Daily Output	Labor-Hours	Unit	Material	2012 Bare Costs Labor	Equipment	Total	Total Incl O&P
5200	Concrete to 6″ thick, hydraulic hammer, mesh reinforced	B-38	255	.157	S.Y.		6.25	4.64	10.89	15.15
5300	Rod reinforced		200	.200			7.95	5.90	13.85	19.30
5400	Concrete, 7″ to 24″ thick, plain		33	1.212	C.Y.		48	36	84	117
5500	Reinforced		24	1.667	″		66	49.50	115.50	161
5590	Minimum labor/equipment charge		6	6.667	Job		265	197	462	640
5600	With hand held air equipment, bituminous, to 6″ thick	B-39	1900	.025	S.F.		.94	.11	1.05	1.64
5700	Concrete to 6″ thick, no reinforcing		1600	.030			1.11	.13	1.24	1.95
5800	Mesh reinforced		1400	.034			1.27	.15	1.42	2.23
5900	Rod reinforced		765	.063			2.32	.27	2.59	4.08
5990	Minimum labor/equipment charge	B-38	6	6.667	Job		265	197	462	640
6000	Curbs, concrete, plain	B-6	360	.067	L.F.		2.55	.93	3.48	5.15
6100	Reinforced		275	.087			3.34	1.21	4.55	6.75
6200	Granite		360	.067			2.55	.93	3.48	5.15
6300	Bituminous		528	.045			1.74	.63	2.37	3.51
6390	Minimum labor/equipment charge		6	4	Job		153	55.50	208.50	310

02 41 13.33 Minor Site Demolition

		Crew	Daily Output	Labor-Hours	Unit	Material	Labor	Equipment	Total	Total Incl O&P
0010	**MINOR SITE DEMOLITION** R024119-10									
0015	No hauling, abandon catch basin or manhole	B-6	7	3.429	Ea.		131	47.50	178.50	265
0020	Remove existing catch basin or manhole, masonry		4	6			230	83.50	313.50	460
0030	Catch basin or manhole frames and covers, stored		13	1.846			70.50	25.50	96	143
0040	Remove and reset		7	3.429			131	47.50	178.50	265
0100	Roadside delineators, remove only	B-80	175	.183			6.90	3.82	10.72	15.40
0110	Remove and reset	″	100	.320			12.10	6.70	18.80	27
0400	Minimum labor/equipment charge	B-6	4	6	Job		230	83.50	313.50	460
0800	Guiderail, corrugated steel, remove only	B-80A	100	.240	L.F.		8.40	2.64	11.04	16.70
0850	Remove and reset	″	40	.600	″		21	6.60	27.60	42
0860	Guide posts, remove only	B-80B	120	.267	Ea.		10	1.94	11.94	18.35
0870	Remove and reset	B-55	50	.480	″		16.75	19.65	36.40	49
0890	Minimum labor/equipment charge	2 Clab	4	4	Job		140		140	230
0900	Hydrants, fire, remove only	B-21A	5	8	Ea.		350	89.50	439.50	655
0950	Remove and reset	″	2	20	″		875	224	1,099	1,650
0990	Minimum labor/equipment charge	2 Plum	2	8	Job		435		435	680
1000	Masonry walls, block, solid	B-5	1800	.031	C.F.		1.20	.73	1.93	2.75
1200	Brick, solid		900	.062			2.41	1.46	3.87	5.50
1400	Stone, with mortar		900	.062			2.41	1.46	3.87	5.50
1500	Dry set		1500	.037			1.44	.88	2.32	3.30
1600	Median barrier, precast concrete, remove and store	B-3	430	.112	L.F.		4.18	5.30	9.48	12.55
1610	Remove and reset	″	390	.123	″		4.60	5.85	10.45	13.85
1650	Minimum labor/equipment charge	A-1	4	2	Job		70	16.50	86.50	133
2900	Pipe removal, sewer/water, no excavation, 12″ diameter	B-6	175	.137	L.F.		5.25	1.91	7.16	10.60
2930	15″-18″ diameter	B-12Z	150	.160			6.30	11	17.30	22.50
2960	21″-24″ diameter		120	.200			7.85	13.75	21.60	28
3000	27″-36″ diameter		90	.267			10.50	18.35	28.85	37
3200	Steel, welded connections, 4″ diameter	B-6	160	.150			5.75	2.09	7.84	11.60
3300	10″ diameter		80	.300			11.50	4.18	15.68	23
3390	Minimum labor/equipment charge		3	8	Job		305	111	416	615
3500	Railroad track removal, ties and track	B-13	330	.170	L.F.		6.45	2.11	8.56	12.75
3600	Ballast	B-14	500	.096	C.Y.		3.56	.67	4.23	6.55
3700	Remove and re-install, ties & track using new bolts & spikes		50	.960	L.F.		35.50	6.70	42.20	65.50
3800	Turnouts using new bolts and spikes		1	48	Ea.		1,775	335	2,110	3,275
3890	Minimum labor/equipment charge		5	9.600	Job		355	67	422	655
4000	Sidewalk removal, bituminous, 2-1/2″ thick	B-6	325	.074	S.Y.		2.83	1.03	3.86	5.70

02 41 Demolition

02 41 13 – Selective Site Demolition

02 41 13.33 Minor Site Demolition		Crew	Daily Output	Labor-Hours	Unit	Material	2012 Bare Costs Labor	Equipment	Total	Total Incl O&P
4050	Brick, set in mortar	B-6	185	.130	S.Y.		4.97	1.81	6.78	10.05
4100	Concrete, plain, 4"		160	.150			5.75	2.09	7.84	11.60
4200	Mesh reinforced		150	.160			6.15	2.23	8.38	12.35
4290	Minimum labor/equipment charge	B-39	12	4	Job		148	17.45	165.45	260

02 41 13.60 Selective Demolition Fencing		Crew	Daily Output	Labor-Hours	Unit	Material	Labor	Equipment	Total	Total Incl O&P
0010	**SELECTIVE DEMOLITION FENCING** R024119-10									
1600	Fencing, barbed wire, 3 strand	2 Clab	430	.037	L.F.		1.31		1.31	2.14
1650	5 strand	"	280	.057			2.01		2.01	3.29
1700	Chain link, posts & fabric, 8' to 10' high, remove only	B-6	445	.054			2.07	.75	2.82	4.17

02 41 16 – Structure Demolition

02 41 16.13 Building Demolition		Crew	Daily Output	Labor-Hours	Unit	Material	Labor	Equipment	Total	Total Incl O&P
0010	**BUILDING DEMOLITION** Large urban projects, incl. 20 mi. haul R024119-10									
0011	No foundation or dump fees, C.F. is vol. of building standing									
0020	Steel	B-8	21500	.003	C.F.		.12	.14	.26	.34
0050	Concrete		15300	.004			.16	.19	.35	.47
0080	Masonry		20100	.003			.12	.15	.27	.36
0100	Mixture of types		20100	.003			.12	.15	.27	.36
0500	Small bldgs, or single bldgs, no salvage included, steel	B-3	14800	.003			.12	.15	.27	.37
0600	Concrete		11300	.004			.16	.20	.36	.48
0650	Masonry		14800	.003			.12	.15	.27	.37
0700	Wood		14800	.003			.12	.15	.27	.37
0750	For buildings with no interior walls, deduct								50%	
1000	Demoliton single family house, one story, wood 1600 S.F.	B-3	1	48	Ea.		1,800	2,275	4,075	5,400
1020	3200 S.F.		.50	96			3,600	4,550	8,150	10,800
1200	Demoliton two family house, two story, wood 2400 S.F.		.67	71.964			2,700	3,400	6,100	8,100
1220	4200 S.F.		.38	128			4,800	6,075	10,875	14,400
1300	Demoliton three family house, three story, wood 3200 S.F.		.50	96			3,600	4,550	8,150	10,800
1320	5400 S.F.		.30	160			5,975	7,575	13,550	18,100
5000	For buildings with no interior walls, deduct								50%	

02 41 16.17 Building Demolition Footings and Foundations		Crew	Daily Output	Labor-Hours	Unit	Material	Labor	Equipment	Total	Total Incl O&P
0010	**BUILDING DEMOLITION FOOTINGS AND FOUNDATIONS** R024119-10									
0200	Floors, concrete slab on grade,									
0240	4" thick, plain concrete	B-9	500	.080	S.F.		2.84	.42	3.26	5.10
0280	Reinforced, wire mesh		470	.085			3.02	.45	3.47	5.45
0300	Rods		400	.100			3.55	.52	4.07	6.40
0400	6" thick, plain concrete		375	.107			3.79	.56	4.35	6.80
0420	Reinforced, wire mesh		340	.118			4.18	.62	4.80	7.55
0440	Rods		300	.133			4.73	.70	5.43	8.50
1000	Footings, concrete, 1' thick, 2' wide	B-5	300	.187	L.F.		7.20	4.38	11.58	16.50
1080	1'-6" thick, 2' wide		250	.224			8.65	5.25	13.90	19.80
1120	3' wide		200	.280			10.80	6.60	17.40	25
1140	2' thick, 3' wide		175	.320			12.35	7.50	19.85	28.50
1200	Average reinforcing, add								10%	10%
1220	Heavy reinforcing, add								20%	20%
2000	Walls, block, 4" thick	1 Clab	180	.044	S.F.		1.56		1.56	2.56
2040	6" thick		170	.047			1.65		1.65	2.71
2080	8" thick		150	.053			1.87		1.87	3.07
2100	12" thick		150	.053			1.87		1.87	3.07
2200	For horizontal reinforcing, add								10%	10%
2220	For vertical reinforcing, add								20%	20%
2400	Concrete, plain concrete, 6" thick	B-9	160	.250			8.90	1.31	10.21	16
2420	8" thick		140	.286			10.15	1.49	11.64	18.30

02 41 Demolition

02 41 16 – Structure Demolition

	02 41 16.17 Building Demolition Footings and Foundations	Crew	Daily Output	Labor-Hours	Unit	Material	2012 Bare Costs Labor	Equipment	Total	Total Incl O&P
2440	10″ thick	B-9	120	.333	S.F.		11.85	1.74	13.59	21.50
2500	12″ thick	↓	100	.400			14.20	2.09	16.29	26
2600	For average reinforcing, add								10%	10%
2620	For heavy reinforcing, add				↓				20%	20%
9000	Minimum labor/equipment charge	A-1	2	4	Job		140	33	173	267

02 41 16.19 Minor Building Deconstruction

	02 41 16.19 Minor Building Deconstruction	Crew	Daily Output	Labor-Hours	Unit	Material	Labor	Equipment	Total	Total Incl O&P
0010	**MINOR BUILDING DECONSTRUCTION** R024119-10									
0011	For salvage, avg house									
0225	2 story, pre 1970 house, 1400 S.F., labor for all salv mat, min G	6 Clab	25	1.920	SF Flr.		67.50		67.50	111
0230	Maximum G	″	15	3.200	″		112		112	184
0235	Salvage of carpet, tackless G	2 Clab	2400	.007	S.F.		.23		.23	.38
0240	Wood floors, incl denailing and packaging G	3 Clab	270	.089	″		3.12		3.12	5.10
0260	Wood doors and trim, standard G	1 Clab	16	.500	Ea.		17.55		17.55	29
0280	Base or cove mouldings, incl denailing and packaging G		500	.016	L.F.		.56		.56	.92
0320	Closet shelving and trim, incl denailing and packaging G	↓	18	.444	Set		15.60		15.60	25.50
0340	Kitchen cabinets, uppers and lowers, prefab type G	2 Clab	24	.667	L.F.		23.50		23.50	38.50
0360	Kitchen cabinets, uppers and lowers, built-in G		12	1.333	″		47		47	76.50
0370	Bath fixt, incl toilet, tub, vanity, and med cabinet G	↓	3	5.333	Set		187		187	305

02 41 19 – Selective Demolition

02 41 19.13 Selective Building Demolition

	02 41 19.13 Selective Building Demolition	Crew	Daily Output	Labor-Hours	Unit	Material	Labor	Equipment	Total	Total Incl O&P
0010	**SELECTIVE BUILDING DEMOLITION**									
0020	Costs related to selective demolition of specific building components									
0025	are included under Common Work Results (XX 05)									
0030	in the component's appropriate division.									

02 41 19.16 Selective Demolition, Cutout

	02 41 19.16 Selective Demolition, Cutout	Crew	Daily Output	Labor-Hours	Unit	Material	Labor	Equipment	Total	Total Incl O&P
0010	**SELECTIVE DEMOLITION, CUTOUT** R024119-10									
0020	Concrete, elev. slab, light reinforcement, under 6 C.F.	B-9	65	.615	C.F.		22	3.22	25.22	39.50
0050	Light reinforcing, over 6 C.F.		75	.533	″		18.95	2.79	21.74	34
0200	Slab on grade to 6″ thick, not reinforced, under 8 S.F.		85	.471	S.F.		16.70	2.46	19.16	30
0250	8 – 16 S.F.	↓	175	.229	″		8.10	1.20	9.30	14.60
0255	For over 16 S.F. see Line 02 41 16.17 0400									
0600	Walls, not reinforced, under 6 C.F.	B-9	60	.667	C.F.		23.50	3.49	26.99	43
0650	6 – 12 C.F.	″	80	.500	″		17.75	2.62	20.37	32
0655	For over 12 C.F. see Line 02 41 16.17 2500									
1000	Concrete, elevated slab, bar reinforced, under 6 C.F.	B-9	45	.889	C.F.		31.50	4.65	36.15	56.50
1050	Bar reinforced, over 6 C.F.		50	.800	″		28.50	4.18	32.68	51
1200	Slab on grade to 6″ thick, bar reinforced, under 8 S.F.		75	.533	S.F.		18.95	2.79	21.74	34
1250	8 – 16 S.F.	↓	150	.267	″		9.45	1.39	10.84	17.05
1255	For over 16 S.F. see Line 02 41 16.17 0440									
1400	Walls, bar reinforced, under 6 C.F.	B-9	50	.800	C.F.		28.50	4.18	32.68	51
1450	6 – 12 C.F.	″	70	.571	″		20.50	2.99	23.49	37
1455	For over 12 C.F. see Lines 02 41 16.17 2500 and 2600									
2000	Brick, to 4 S.F. opening, not including toothing									
2040	4″ thick	B-9	30	1.333	Ea.		47.50	6.95	54.45	85
2060	8″ thick		18	2.222			79	11.60	90.60	142
2080	12″ thick		10	4			142	21	163	256
2400	Concrete block, to 4 S.F. opening, 2″ thick		35	1.143			40.50	6	46.50	73
2420	4″ thick		30	1.333			47.50	6.95	54.45	85
2440	8″ thick		27	1.481			52.50	7.75	60.25	94.50
2460	12″ thick		24	1.667			59	8.70	67.70	107
2600	Gypsum block, to 4 S.F. opening, 2″ thick		80	.500			17.75	2.62	20.37	32
2620	4″ thick	↓	70	.571	↓		20.50	2.99	23.49	37

02 41 Demolition

02 41 19 – Selective Demolition

02 41 19.16 Selective Demolition, Cutout		Crew	Daily Output	Labor-Hours	Unit	Material	2012 Bare Costs Labor	Equipment	Total	Total Incl O&P
2640	8" thick	B-9	55	.727	Ea.		26	3.80	29.80	46.50
2800	Terra cotta, to 4 S.F. opening, 4" thick		70	.571			20.50	2.99	23.49	37
2840	8" thick		65	.615			22	3.22	25.22	39.50
2880	12" thick	↓	50	.800	↓		28.50	4.18	32.68	51
4000	For toothing masonry, see Section 04 01 20.50									
6000	Walls, interior, not including re-framing,									
6010	openings to 5 S.F.									
6100	Drywall to 5/8" thick	1 Clab	24	.333	Ea.		11.70		11.70	19.20
6200	Paneling to 3/4" thick		20	.400			14.05		14.05	23
6300	Plaster, on gypsum lath		20	.400			14.05		14.05	23
6340	On wire lath	↓	14	.571	↓		20		20	33
7000	Wood frame, not including re-framing, openings to 5 S.F.									
7200	Floors, sheathing and flooring to 2" thick	1 Clab	5	1.600	Ea.		56		56	92
7310	Roofs, sheathing to 1" thick, not including roofing		6	1.333			47		47	76.50
7410	Walls, sheathing to 1" thick, not including siding		7	1.143	↓		40		40	66
8500	Minimum labor/equipment charge	↓	4	2	Job		70		70	115

02 41 19.18 Selective Demolition, Disposal Only

		Crew	Daily Output	Labor-Hours	Unit	Material	Labor	Equipment	Total	Total Incl O&P
0010	**SELECTIVE DEMOLITION, DISPOSAL ONLY** R024119-10									
0015	Urban bldg w/salvage value allowed									
0020	Including loading and 5 mile haul to dump									
0200	Steel frame	B-3	430	.112	C.Y.		4.18	5.30	9.48	12.55
0300	Concrete frame		365	.132			4.92	6.25	11.17	14.80
0400	Masonry construction		445	.108			4.04	5.10	9.14	12.15
0500	Wood frame	↓	247	.194	↓		7.25	9.20	16.45	22

02 41 19.19 Selective Facility Services Demolition

		Crew	Daily Output	Labor-Hours	Unit	Material	Labor	Equipment	Total	Total Incl O&P
0010	**SELECTIVE FACILITY SERVICES DEMOLITION**, Rubbish Handling R024119-10									
0020	The following are to be added to the demolition prices									
0400	Chute, circular, prefabricated steel, 18" diameter	B-1	40	.600	L.F.	52	21.50		73.50	92.50
0440	30" diameter	"	30	.800	"	46.50	28.50		75	98
0600	Dumpster, weekly rental, 1 dump/week, 6 C.Y. capacity (2 Tons)				Week	460			460	505
0700	10 C.Y. capacity (3 Tons)					535			535	590
0725	20 C.Y. capacity (8 Tons) R024119-20					630			630	695
0800	30 C.Y. capacity (7 Tons)					810			810	890
0840	40 C.Y. capacity (10 Tons)				↓	860			860	945
0900	Alternate pricing for dumpsters									
0910	Delivery, average for all sizes				Ea.	80			80	88
0920	Haul, average for all sizes					200			200	220
0930	Rent per day, average for all sizes					20			20	22
0940	Rent per month, average for all sizes				↓	80			80	88
0950	Disposal fee per ton, average for all sizes				Ton	82			82	90
2000	Load, haul, dump and return, 50' haul, hand carried	2 Clab	24	.667	C.Y.		23.50		23.50	38.50
2005	Wheeled		37	.432			15.20		15.20	25
2040	51' to 100' haul, hand carried		16.50	.970			34		34	56
2045	Wheeled		25	.640			22.50		22.50	37
2080	Over 100' haul, add per 100 L.F., hand carried		35.50	.451			15.80		15.80	26
2085	Wheeled		54	.296			10.40		10.40	17.05
2120	In elevators, per 10 floors, add		140	.114			4.01		4.01	6.60
2130	Load, haul, dump and return, up to 50' haul, incl. up to 5 rsr stair, hand		23	.696			24.50		24.50	40
2135	Wheeled		35	.457			16.05		16.05	26.50
2140	6 – 10 riser stairs, hand carried		22	.727			25.50		25.50	42
2145	Wheeled		34	.471			16.50		16.50	27
2150	11 – 20 riser stairs, hand carried	↓	20	.800	↓		28		28	46

02 41 19.19 Selective Facility Services Demolition

	02 41 19.19 Selective Facility Services Demolition	Crew	Daily Output	Labor-Hours	Unit	Material	2012 Bare Costs Labor	Equipment	Total	Total Incl O&P
2155	Wheeled	2 Clab	31	.516	C.Y.		18.10		18.10	29.50
2160	21 – 40 riser stairs, hand carried		16	1			35		35	57.50
2165	Wheeled		24	.667			23.50		23.50	38.50
2170	100′ haul, incl. 5 riser stair, hand carried		15	1.067			37.50		37.50	61.50
2175	Wheeled		23	.696			24.50		24.50	40
2180	6 – 10 riser stair, hand carried		14	1.143			40		40	66
2185	Wheeled		21	.762			26.50		26.50	44
2190	11 – 20 riser stair, hand carried		12	1.333			47		47	76.50
2195	Wheeled		18	.889			31		31	51
2200	21 – 40 riser stair, hand carried		8	2			70		70	115
2205	Wheeled		12	1.333			47		47	76.50
2210	Over 100′ haul, add per 100 L.F., hand carried		35.50	.451			15.80		15.80	26
2215	Wheeled		54	.296			10.40		10.40	17.05
2220	For each additional flight of stairs, up to 5 risers, add		550	.029	Flight		1.02		1.02	1.67
2225	6 – 10 risers, add		275	.058			2.04		2.04	3.35
2230	11 – 20 risers, add		138	.116			4.07		4.07	6.65
2235	21 – 40 risers, add		69	.232			8.15		8.15	13.35
3000	Loading & trucking, including 2 mile haul, chute loaded	B-16	45	.711	C.Y.		25.50	13	38.50	56
3040	Hand loading truck, 50′ haul	"	48	.667			24	12.20	36.20	52.50
3080	Machine loading truck	B-17	120	.267			10	5.75	15.75	22.50
3120	Wheeled 50′ and ramp dump loaded	2 Clab	24	.667			23.50		23.50	38.50
5000	Haul, per mile, up to 8 C.Y. truck	B-34B	1165	.007			.24	.50	.74	.94
5100	Over 8 C.Y. truck	"	1550	.005			.18	.38	.56	.70

02 41 19.20 Selective Demolition, Dump Charges

	02 41 19.20 Selective Demolition, Dump Charges	Crew	Daily Output	Labor-Hours	Unit	Material	Labor	Equipment	Total	Total Incl O&P
0010	**SELECTIVE DEMOLITION, DUMP CHARGES** R024119-10									
0020	Dump charges, typical urban city, tipping fees only									
0100	Building construction materials				Ton	82			82	90
0200	Trees, brush, lumber					70			70	77
0300	Rubbish only					70			70	77
0500	Reclamation station, usual charge					82			82	90

02 41 19.21 Selective Demolition, Gutting

	02 41 19.21 Selective Demolition, Gutting	Crew	Daily Output	Labor-Hours	Unit	Material	Labor	Equipment	Total	Total Incl O&P
0010	**SELECTIVE DEMOLITION, GUTTING** R024119-10									
0020	Building interior, including disposal, dumpster fees not included									
0500	Residential building									
0560	Minimum	B-16	400	.080	SF Flr.		2.85	1.46	4.31	6.25
0580	Maximum	"	360	.089	"		3.17	1.62	4.79	7
0900	Commercial building									
1000	Minimum	B-16	350	.091	SF Flr.		3.26	1.67	4.93	7.20
1020	Maximum		250	.128	"		4.56	2.34	6.90	10
3000	Minimum labor/equipment charge		4	8	Job		285	146	431	625

02 41 19.25 Selective Demolition, Saw Cutting

	02 41 19.25 Selective Demolition, Saw Cutting	Crew	Daily Output	Labor-Hours	Unit	Material	Labor	Equipment	Total	Total Incl O&P
0010	**SELECTIVE DEMOLITION, SAW CUTTING** R024119-10									
0015	Asphalt, up to 3″ deep	B-89	1050	.015	L.F.	.34	.60	.41	1.35	1.78
0020	Each additional inch of depth	"	1800	.009		.11	.35	.24	.70	.94
1200	Masonry walls, hydraulic saw, brick, per inch of depth	B-89B	300	.053		.42	2.11	2.46	4.99	6.55
1220	Block walls, solid, per inch of depth	"	250	.064		.42	2.53	2.95	5.90	7.75
2000	Brick or masonry w/hand held saw, per inch of depth	A-1	125	.064		.34	2.25	.53	3.12	4.63
5000	Wood sheathing to 1″ thick, on walls	1 Carp	200	.040			1.76		1.76	2.89
5020	On roof	"	250	.032			1.41		1.41	2.31
9000	Minimum labor/equipment charge	A-1	2	4	Job		140	33	173	267

02 41 Demolition

02 41 19 – Selective Demolition

02 41 19.27 Selective Demolition, Torch Cutting		Crew	Daily Output	Labor-Hours	Unit	Material	2012 Bare Costs Labor	Equipment	Total	Total Incl O&P
0010	SELECTIVE DEMOLITION, TORCH CUTTING R024119-10									
0020	Steel, 1" thick plate	1 Clab	360	.022	L.F.	.37	.78		1.15	1.69
0040	1" diameter bar	"	210	.038	Ea.		1.34		1.34	2.19
1000	Oxygen lance cutting, reinforced concrete walls									
1040	12" to 16" thick walls	1 Clab	10	.800	L.F.		28		28	46
1080	24" thick walls	"	6	1.333	"		47		47	76.50
1090	Minimum labor/equipment charge	E-25	2	4	Job		205	48.50	253.50	435
1100	See Section 05 05 21.10									

02 42 Removal and Salvage of Construction Materials

02 42 10 – Building Deconstruction

02 42 10.10 Estimated Salvage Value or Savings		Crew	Daily Output	Labor-Hours	Unit	Material	Labor	Equipment	Total	Total Incl O&P
0010	ESTIMATED SALVAGE VALUE OR SAVINGS									
0015	Excludes material handling, packaging, container costs and									
0020	transportation for salvage or disposal									
0050	All Items in Section 02 42 10.10 are credit deducts and not costs									
0100	Copper Wire Salvage Value G				Lb.				1.30	1.30
0110	Disposal Savings G								.03	.03
0200	Copper Pipe Salvage Value G								1.34	1.34
0210	Disposal Savings G								.03	.03
0300	Steel Pipe Salvage Value G								.06	.06
0310	Disposal Savings G								.03	.03
0400	Cast Iron Pipe Salvage Value G								.06	.06
0410	Disposal Savings G								.03	.03
0500	Steel Doors or Windows Salvage Value G								.06	.06
0510	Aluminum G								.66	.66
0520	Disposal Savings G								.03	.03
0600	Aluminum Siding Salvage Value G								.49	.49
0630	Disposal Savings G								.03	.03
0640	Wood Siding (no lead or asbestos) G				C.Y.				10.90	10.90
0800	Clean Concrete Disposal Savings G				Ton				60	60
0850	Asphalt Shingles Disposal Savings G				"				60	60
1000	Wood wall framing clean salvage value G				M.B.F.				50	50
1010	Painted G								40	40
1020	Floor framing G								50	50
1030	Painted G								40	40
1050	Roof framing G								50	50
1060	Painted G								40	40
1100	Wood beams salvage value G								50	50
1200	Wood framing and beams disposal savings G				Ton				60	60
1220	Wood sheating and sub-base flooring G								60	60
1230	Wood wall paneling (1/4 inch thick) G								60	60
1300	Wood panel 3/4-1 inch thick low salvage value G				S.F.				.50	.50
1350	high salvage value G				"				2	2
1400	Disposal savings G				Ton				60	60
1500	Flooring tongue and groove 25/32 inch thick low salvage value G				S.F.				.50	.50
1530	High salvage value G				"				1	1
1560	Disposal savings G				Ton				60	60
1600	Drywall or sheet rock salvage value G								20	20
1650	Disposal savings G								60	60

02 42 10 – Building Deconstruction

02 42 10.20 Deconstruction of Building Components			Crew	Daily Output	Labor-Hours	Unit	Material	2012 Bare Costs Labor	Equipment	Total	Total Incl O&P
0010	**DECONSTRUCTION OF BUILDING COMPONENTS**										
0012	Buildings one or two stories only										
0015	Excludes material handling, packaging, container costs and										
0020	transportation for salvage or disposal										
0050	Deconstruction of Plumbing Fixtures										
0100	Wall hung or countertop lavatory	G	2 Clab	16	1	Ea.		35		35	57.50
0110	Single or double compartment kitchen sink	G		14	1.143			40		40	66
0120	Wall hung urinal	G		14	1.143			40		40	66
0130	Floor mounted	G		8	2			70		70	115
0140	Floor mounted water closet	G		16	1			35		35	57.50
0150	Wall hung	G		14	1.143			40		40	66
0160	Water fountain, free standing	G		16	1			35		35	57.50
0170	Wall hung or deck mounted	G		12	1.333			47		47	76.50
0180	Bathtub, steel or fiberglass	G		10	1.600			56		56	92
0190	Cast iron	G		8	2			70		70	115
0200	Shower, single	G		6	2.667			93.50		93.50	153
0210	Group	G		7	2.286			80		80	132
0300	Deconstruction of Electrical Fixtures										
0310	Surface mount incandescent fixtures	G	2 Clab	48	.333	Ea.		11.70		11.70	19.20
0320	Fluorescent, 2 lamp	G		32	.500			17.55		17.55	29
0330	4 lamp	G		24	.667			23.50		23.50	38.50
0340	Strip Fluorescent, 1 lamp	G		40	.400			14.05		14.05	23
0350	2 lamp	G		32	.500			17.55		17.55	29
0400	Recessed drop-in fluorescent fixture, 2 lamp	G		27	.593			21		21	34
0410	4 lamp	G		18	.889			31		31	51
0500	Deconstruction of appliances										
0510	Cooking stoves	G	2 Clab	26	.615	Ea.		21.50		21.50	35.50
0520	Dishwashers	G	"	26	.615	"		21.50		21.50	35.50
0600	Deconstruction of millwork and trim										
0610	Cabinets, wood	G	2 Carp	40	.400	L.F.		17.65		17.65	29
0620	Countertops	G		100	.160	"		7.05		7.05	11.55
0630	Wall paneling, 1 inch thick	G		500	.032	S.F.		1.41		1.41	2.31
0640	Ceiling trim	G		500	.032	L.F.		1.41		1.41	2.31
0650	Wainscoting	G		500	.032	S.F.		1.41		1.41	2.31
0660	Base, 3/4" to 1" thick	G		600	.027	L.F.		1.18		1.18	1.93
0700	Deconstruction of doors and windows										
0710	Doors, wrap, interior, wood, single, no closers	G	2 Carp	21	.762	Ea.	5.75	33.50		39.25	61.50
0720	Double	G		13	1.231		11.50	54.50		66	102
0730	Solid core, single, exterior or interior	G		10	1.600		5.75	70.50		76.25	122
0740	Double	G		8	2		11.50	88		99.50	158
0810	Windows, wrap, wood, single	G		21	.762		5.75	33.50		39.25	61.50
0812	with no casement or cladding										
0820	with casement and/or cladding	G	2 Carp	18	.889	Ea.	5.75	39		44.75	71
0900	Deconstruction of interior finishes										
0910	Drywall for recycling	G	2 Clab	1775	.009	S.F.		.32		.32	.52
0920	Plaster wall, first floor	G		1775	.009			.32		.32	.52
0930	Second floor	G		1330	.012			.42		.42	.69
1000	Deconstruction of roofing and accessories										
1010	Built-up roofs	G	2 Clab	570	.028	S.F.		.99		.99	1.62
1020	Gutters, facia and rakes	G	"	1140	.014	L.F.		.49		.49	.81
2000	Deconstruction of wood components										
2010	Roof sheeting	G	2 Clab	570	.028	S.F.		.99		.99	1.62

02 42 Removal and Salvage of Construction Materials

02 42 10 – Building Deconstruction

02 42 10.20 Deconstruction of Building Components

Line	Description		Crew	Daily Output	Labor-Hours	Unit	Material	2012 Bare Costs Labor	2012 Bare Costs Equipment	Total	Total Incl O&P
2020	Main roof framing	G	2 Clab	760	.021	L.F.		.74		.74	1.21
2030	Porch roof framing	G		445	.036			1.26		1.26	2.07
2040	Beams 4" x 8"	G	B-1	375	.064			2.29		2.29	3.75
2050	4" x 10"	G		300	.080			2.86		2.86	4.69
2055	4" x 12"	G		250	.096			3.43		3.43	5.65
2060	6" x 8"	G		250	.096			3.43		3.43	5.65
2065	6" x 10"	G		200	.120			4.29		4.29	7.05
2070	6" x 12"	G		170	.141			5.05		5.05	8.30
2075	8" x 12"	G		126	.190			6.80		6.80	11.15
2080	10" x 12"	G		100	.240			8.60		8.60	14.05
2100	Ceiling joists	G	2 Clab	800	.020			.70		.70	1.15
2150	Wall framing, interior	G		1230	.013			.46		.46	.75
2160	Sub-floor	G		2000	.008	S.F.		.28		.28	.46
2170	Floor joists	G		2000	.008	L.F.		.28		.28	.46
2200	Wood siding (no lead or asbestos)	G		1300	.012	S.F.		.43		.43	.71
2300	Wall framing, exterior	G		1600	.010	L.F.		.35		.35	.58
2400	Stair risers	G		53	.302	Ea.		10.60		10.60	17.35
2500	Posts	G		800	.020	L.F.		.70		.70	1.15
3000	Deconstruction of exterior brick walls										
3010	Exterior brick walls, first floor	G	2 Clab	200	.080	S.F.		2.81		2.81	4.60
3020	Second floor	G		64	.250	"		8.80		8.80	14.40
3030	Brick chiminey	G		100	.160	C.F.		5.60		5.60	9.20
4000	Deconstruction of concrete										
4010	Slab on grade, 4" thick, plain concrete	G	B-9	500	.080	S.F.		2.84	.42	3.26	5.10
4020	Wire mesh reinforced	G		470	.085			3.02	.45	3.47	5.45
4030	Rod reinforced	G		400	.100			3.55	.52	4.07	6.40
4110	Foundation wall, 6" thick, plain concrete	G		160	.250			8.90	1.31	10.21	16
4120	8" thick	G		140	.286			10.15	1.49	11.64	18.30
4130	10" thick	G		120	.333			11.85	1.74	13.59	21.50
9000	Deconstruction process, support equipment as needed										
9010	Daily use, portal to portal, 12-ton truck-mounted hydraulic crane crew	G	A-3H	1	8	Day		380	825	1,205	1,500
9020	Daily use, skid steer and operator	G	A-3C	1	8			360	281	641	875
9030	Daily use, backhoe 48 H. P., operator and labor	G	"	1	8			360	281	641	875

02 42 10.30 Deconstruction Material Handling

Line	Description		Crew	Daily Output	Labor-Hours	Unit	Material	2012 Bare Costs Labor	2012 Bare Costs Equipment	Total	Total Incl O&P
0010	**DECONSTRUCTION MATERIAL HANDLING**										
0012	Buildings one or two stories only										
0100	Clean and stack brick on pallet	G	2 Clab	1200	.013	Ea.		.47		.47	.77
0200	Haul 50' and load rough lumber up to 2" x 8" size	G		2000	.008	"		.28		.28	.46
0210	Lumber larger than 2" x 8"	G		3200	.005	B.F.		.18		.18	.29
0300	Finish wood for recycling stack and wrap per pallet	G		8	2	Ea.	46	70		116	166
0350	Light fixtures			6	2.667		83	93.50		176.50	244
0375	Windows			6	2.667		78	93.50		171.50	239
0400	Miscellaneous materials			8	2		23	70		93	141
1000	See Section 02 41 19.19 for bulk material handling										

02 43 Structure Moving

02 43 13 – Structure Relocation

02 43 13.13 Building Relocation		Crew	Daily Output	Labor-Hours	Unit	Material	2012 Bare Costs Labor	Equipment	Total	Total Incl O&P
0010	**BUILDING RELOCATION**									
0011	One day move, up to 24′ wide									
0020	Reset on new foundation, patch & hook-up, average move				Total					11,200
0040	Wood or steel frame bldg., based on ground floor area G	B-4	185	.259	S.F.		9.20	2.50	11.70	17.80
0060	Masonry bldg., based on ground floor area G	"	137	.350			12.45	3.37	15.82	24
0200	For 24′ to 42′ wide, add									15%
0220	For each additional day on road, add G	B-4	1	48	Day		1,700	460	2,160	3,275
0240	Construct new basement, move building, 1 day									
0300	move, patch & hook-up, based on ground floor area G	B-3	155	.310	S.F.	9.20	11.60	14.70	35.50	45

02 58 Snow Control

02 58 13 – Snow Fencing

02 58 13.10 Snow Fencing System		Crew	Daily Output	Labor-Hours	Unit	Material	Labor	Equipment	Total	Total Incl O&P
0010	**SNOW FENCING SYSTEM**									
7001	Snow fence on steel posts 10′ O.C., 4′ high	B-1	500	.048	L.F.	1.09	1.72		2.81	4.01

02 65 Underground Storage Tank Removal

02 65 10 – Underground Tank and Contaminated Soil Removal

02 65 10.30 Removal of Underground Storage Tanks		Crew	Daily Output	Labor-Hours	Unit	Material	Labor	Equipment	Total	Total Incl O&P
0010	**REMOVAL OF UNDERGROUND STORAGE TANKS** R026510-20									
0011	Petroleum storage tanks, non-leaking									
0100	Excavate & load onto trailer									
0110	3000 gal. to 5000 gal. tank G	B-14	4	12	Ea.		445	83.50	528.50	815
0120	6000 gal. to 8000 gal. tank G	B-3A	3	13.333			500	320	820	1,150
0130	9000 gal. to 12000 gal. tank G	"	2	20			750	480	1,230	1,750
0190	Known leaking tank, add				%				100%	100%
0200	Remove sludge, water and remaining product from tank bottom									
0201	of tank with vacuum truck									
0300	3000 gal. to 5000 gal. tank G	A-13	5	1.600	Ea.		71.50	147	218.50	275
0310	6000 gal. to 8000 gal. tank G		4	2			89.50	184	273.50	345
0320	9000 gal. to 12000 gal. tank G		3	2.667			119	245	364	455
0390	Dispose of sludge off-site, average				Gal.				6	6.60
0400	Insert inert solid CO_2 "dry ice" into tank									
0401	For cleaning/transporting tanks (1.5 lb./100 gal. cap) G	1 Clab	500	.016	Lb.	.94	.56		1.50	1.95
0403	Insert solid carbon dioxide, 1.5 lb./100 gal. G	"	400	.020	"	.94	.70		1.64	2.18
0503	Disconnect and remove piping G	1 Plum	160	.050	L.F.		2.73		2.73	4.25
0603	Transfer liquids, 10% of volume G	"	1600	.005	Gal.		.27		.27	.43
0703	Cut accessway into underground storage tank G	1 Clab	5.33	1.501	Ea.		52.50		52.50	86.50
0813	Remove sludge, wash and wipe tank, 500 gal. G	1 Plum	8	1			54.50		54.50	85
0823	3,000 gal. G		6.67	1.199			65.50		65.50	102
0833	5,000 gal. G		6.15	1.301			71		71	111
0843	8,000 gal. G		5.33	1.501			82		82	128
0853	10,000 gal. G		4.57	1.751			95.50		95.50	149
0863	12,000 gal. G		4.21	1.900			104		104	162
1020	Haul tank to certified salvage dump, 100 miles round trip									
1023	3000 gal. to 5000 gal. tank				Ea.				745	820
1026	6000 gal. to 8000 gal. tank								850	935
1029	9,000 gal. to 12,000 gal. tank								1,000	1,100
1100	Disposal of contaminated soil to landfill									

02 65 Underground Storage Tank Removal

02 65 10 – Underground Tank and Contaminated Soil Removal

02 65 10.30 Removal of Underground Storage Tanks			Crew	Daily Output	Labor-Hours	Unit	Material	2012 Bare Costs Labor	Equipment	Total	Total Incl O&P
1110	Minimum					C.Y.				125	140
1111	Maximum					"				400	440
1120	Disposal of contaminated soil to										
1121	bituminous concrete batch plant										
1130	Minimum					C.Y.				70	77
1131	Maximum					"				110	120
1203	Excavate, pull, & load tank, backfill hole, 8,000 gal. +	G	B-12C	.50	32	Ea.		1,325	2,500	3,825	4,875
1213	Haul tank to certified dump, 100 miles rt, 8,000 gal. +	G	B-34K	1	8			282	850	1,132	1,400
1223	Excavate, pull, & load tank, backfill hole, 500 gal.	G	B-11C	1	16			655	335	990	1,425
1233	Excavate, pull, & load tank, backfill hole, 3,000 – 5,000 gal.	G	B-11M	.50	32			1,300	730	2,030	2,900
1243	Haul tank to certified dump, 100 miles rt, 500 gal.	G	B-34L	1	8	↓		360	211	571	795
2010	Decontamination of soil on site incl poly tarp on top/bottom										
2011	Soil containment berm, and chemical treatment										
2020	Minimum	G	B-11C	100	.160	C.Y.	7.10	6.55	3.34	16.99	22
2021	Maximum	G	"	100	.160		9.20	6.55	3.34	19.09	24.50
2050	Disposal of decontaminated soil, minimum									125	140
2055	Maximum					↓				400	440

02 81 Transportation and Disposal of Hazardous Materials

02 81 20 – Hazardous Waste Handling

02 81 20.10 Hazardous Waste Cleanup/Pickup/Disposal		Crew	Daily Output	Labor-Hours	Unit	Material	Labor	Equipment	Total	Total Incl O&P
0010	**HAZARDOUS WASTE CLEANUP/PICKUP/DISPOSAL**									
0100	For contractor rental equipment, i.e., Dozer,									
0110	Front end loader, Dump truck, etc., see Sec. 01 54 33									
1000	Solid pickup									
1100	55 gal. drums				Ea.				225	250
1120	Bulk material, minimum				Ton				170	190
1130	Maximum				"				575	635
1200	Transportation to disposal site									
1220	Truckload = 80 drums or 25 C.Y. or 18 tons									
1260	Minimum				Mile				3.85	4.25
1270	Maximum				"				7	7.10
3000	Liquid pickup, vacuum truck, stainless steel tank									
3100	Minimum charge, 4 hours									
3110	1 compartment, 2200 gallon				Hr.				125	140
3120	2 compartment, 5000 gallon				"				180	200
3400	Transportation in 6900 gallon bulk truck				Mile				7.70	8.50
3410	In teflon lined truck				"				9.90	10.90
5000	Heavy sludge or dry vacuumable material				Hr.				125	140
6000	Dumpsite disposal charge, minimum				Ton				125	140
6020	Maximum				"				400	440

02 82 Asbestos Remediation

02 82 13 – Asbestos Abatement

02 82 13.39 Asbestos Remediation Plans and Methods		Crew	Daily Output	Labor-Hours	Unit	Material	2012 Bare Costs Labor	Equipment	Total	Total Incl O&P
0010	**ASBESTOS REMEDIATION PLANS AND METHODS**									
0100	Building Survey-Commercial Building				Ea.				2,000	2,200
0200	Asbestos Abatement Remediation Plan				″				1,200	1,325

02 82 13.41 Asbestos Abatement Equip.

02 82 13.41		Crew	Daily Output	Labor-Hours	Unit	Material	Labor	Equipment	Total	Total Incl O&P
0010	**ASBESTOS ABATEMENT EQUIP.** R028213-20									
0011	Equipment and supplies, buy									
0200	Air filtration device, 2000 CFM				Ea.	885			885	975
0250	Large volume air sampling pump, minimum					365			365	400
0260	Maximum					530			530	585
0300	Airless sprayer unit, 2 gun					4,900			4,900	5,400
0350	Light stand, 500 watt					46.50			46.50	51.50
0400	Personal respirators									
0410	Negative pressure, 1/2 face, dual operation, min.				Ea.	26			26	28.50
0420	Maximum					32.50			32.50	35.50
0450	P.A.P.R., full face, minimum					109			109	120
0460	Maximum					199			199	219
0470	Supplied air, full face, incl. air line, minimum					191			191	210
0480	Maximum					395			395	435
0500	Personnel sampling pump					226			226	248
1500	Power panel, 20 unit, incl. G.F.I.					565			565	620
1600	Shower unit, including pump and filters					1,200			1,200	1,300
1700	Supplied air system (type C)					5,425			5,425	5,975
1750	Vacuum cleaner, HEPA, 16 gal., stainless steel, wet/dry					1,200			1,200	1,325
1760	55 gallon					1,950			1,950	2,150
1800	Vacuum loader, 9 – 18 ton/hr.					92,000			92,000	101,000
1900	Water atomizer unit, including 55 gal. drum					235			235	258
2000	Worker protection, whole body, foot, head cover & gloves, plastic					8.90			8.90	9.80
2500	Respirator, single use					29.50			29.50	32.50
2550	Cartridge for respirator					7			7	7.70
2570	Glove bag, 7 mil, 50″ x 64″					17.60			17.60	19.35
2580	10 mil, 44″ x 60″					5.50			5.50	6.05
3000	HEPA vacuum for work area, minimum					1,400			1,400	1,525
3050	Maximum					1,475			1,475	1,600
6000	Disposable polyethylene bags, 6 mil, 3 C.F.					2.35			2.35	2.59
6300	Disposable fiber drums, 3 C.F.					17.25			17.25	19
6400	Pressure sensitive caution labels, 3″ x 5″					3.22			3.22	3.54
6450	11″ x 17″					7.15			7.15	7.90
6500	Negative air machine, 1800 CFM					820			820	905

02 82 13.42 Preparation of Asbestos Containment Area

02 82 13.42		Crew	Daily Output	Labor-Hours	Unit	Material	Labor	Equipment	Total	Total Incl O&P
0010	**PREPARATION OF ASBESTOS CONTAINMENT AREA**									
0100	Pre-cleaning, HEPA vacuum and wet wipe, flat surfaces	A-9	12000	.005	S.F.	.02	.26		.28	.44
0200	Protect carpeted area, 2 layers 6 mil poly on 3/4″ plywood	″	1000	.064		2	3.13		5.13	7.25
0300	Separation barrier, 2″ x 4″ @ 16″, 1/2″ plywood ea. side, 8′ high	2 Carp	400	.040		3	1.76		4.76	6.20
0310	12′ high		320	.050		3	2.21		5.21	6.90
0320	16′ high		200	.080		2	3.53		5.53	8
0400	Personnel decontam. chamber, 2″ x 4″ @ 16″, 3/4″ ply ea. side		280	.057		4	2.52		6.52	8.55
0450	Waste decontam. chamber, 2″ x 4″ studs @ 16″, 3/4″ ply ea. side		360	.044		4	1.96		5.96	7.60
0500	Cover surfaces with polyethylene sheeting									
0501	Including glue and tape									
0550	Floors, each layer, 6 mil	A-9	8000	.008	S.F.	.05	.39		.44	.68
0551	4 mil		9000	.007		.03	.35		.38	.60
0560	Walls, each layer, 6 mil		6000	.011		.05	.52		.57	.89

02 82 Asbestos Remediation

02 82 13 – Asbestos Abatement

02 82 13.42 Preparation of Asbestos Containment Area		Crew	Daily Output	Labor-Hours	Unit	2012 Bare Costs Material	Labor	Equipment	Total	Total Incl O&P
0561	4 mil	A-9	7000	.009	S.F.	.03	.45		.48	.76
0570	For heights above 12′, add						20%			
0575	For heights above 20′, add						30%			
0580	For fire retardant poly, add					100%				
0590	For large open areas, deduct					10%	20%			
0600	Seal floor penetrations with foam firestop to 36 sq. in.	2 Carp	200	.080	Ea.	7.60	3.53		11.13	14.20
0610	36 sq. in. to 72 sq. in.		125	.128		15.25	5.65		20.90	26
0615	72 sq. in. to 144 sq. in.		80	.200		30.50	8.80		39.30	48
0620	Wall penetrations, to 36 square inches		180	.089		7.60	3.92		11.52	14.85
0630	36 sq. in. to 72 sq. in.		100	.160		15.25	7.05		22.30	28.50
0640	72 sq. in. to 144 sq. in.		60	.267		30.50	11.75		42.25	53
0800	Caulk seams with latex	1 Carp	230	.035	L.F.	.15	1.53		1.68	2.68
0900	Set up neg. air machine, 1-2k CFM/25 M.C.F. volume	1 Asbe	4.30	1.860	Ea.		91		91	146

02 82 13.43 Bulk Asbestos Removal		Crew	Daily Output	Labor-Hours	Unit	Material	Labor	Equipment	Total	Total Incl O&P
0010	**BULK ASBESTOS REMOVAL**									
0020	Includes disposable tools and 2 suits and 1respirator filter/day/worker									
0100	Beams, W 10 x 19	A-9	235	.272	L.F.	.84	13.30		14.14	22.50
0110	W 12 x 22		210	.305		.94	14.90		15.84	25
0120	W 14 x 26		180	.356		1.10	17.40		18.50	29
0130	W 16 x 31		160	.400		1.24	19.55		20.79	33
0140	W 18 x 40		140	.457		1.42	22.50		23.92	37.50
0150	W 24 x 55		110	.582		1.80	28.50		30.30	48
0160	W 30 x 108		85	.753		2.33	37		39.33	61.50
0170	W 36 x 150		72	.889		2.75	43.50		46.25	73
0200	Boiler insulation		480	.133	S.F.	.48	6.50		6.98	11.05
0210	With metal lath, add				%				50%	
0300	Boiler breeching or flue insulation	A-9	520	.123	S.F.	.38	6		6.38	10.10
0310	For active boiler, add				%				100%	
0400	Duct or AHU insulation	A-10B	440	.073	S.F.	.23	3.56		3.79	6
0500	Duct vibration isolation joints, up to 24 sq. in. duct	A-9	56	1.143	Ea.	3.54	56		59.54	94
0520	25 sq. in. to 48 sq. in. duct		48	1.333		4.13	65		69.13	110
0530	49 sq. in. to 76 sq. in. duct		40	1.600		4.96	78.50		83.46	131
0600	Pipe insulation, air cell type, up to 4″ diameter pipe		900	.071	L.F.	.22	3.48		3.70	5.85
0610	4″ to 8″ diameter pipe		800	.080		.25	3.91		4.16	6.55
0620	10″ to 12″ diameter pipe		700	.091		.28	4.47		4.75	7.50
0630	14″ to 16″ diameter pipe		550	.116		.36	5.70		6.06	9.55
0650	Over 16″ diameter pipe		650	.098	S.F.	.31	4.82		5.13	8.10
0700	With glove bag up to 3″ diameter pipe		200	.320	L.F.	6.05	15.65		21.70	31.50
1000	Pipe fitting insulation up to 4″ diameter pipe		320	.200	Ea.	.62	9.80		10.42	16.45
1100	6″ to 8″ diameter pipe		304	.211		.65	10.30		10.95	17.25
1110	10″ to 12″ diameter pipe		192	.333		1.03	16.30		17.33	27
1120	14″ to 16″ diameter pipe		128	.500		1.55	24.50		26.05	41
1130	Over 16″ diameter pipe		176	.364	S.F.	1.13	17.80		18.93	29.50
1200	With glove bag, up to 8″ diameter pipe		75	.853	L.F.	6.55	41.50		48.05	74
2000	Scrape foam fireproofing from flat surface		2400	.027	S.F.	.08	1.30		1.38	2.19
2100	Irregular surfaces		1200	.053		.17	2.61		2.78	4.37
3000	Remove cementitious material from flat surface		1800	.036		.11	1.74		1.85	2.92
3100	Irregular surface		1000	.064		.14	3.13		3.27	5.20
4000	Scrape acoustical coating/fireproofing, from ceiling		3200	.020		.06	.98		1.04	1.64
5000	Remove VAT and mastic from floor by hand		2400	.027		.08	1.30		1.38	2.19
5100	By machine	A-11	4800	.013		.04	.65	.01	.70	1.11
5150	For 2 layers, add				%				50%	

02 82 Asbestos Remediation

02 82 13 – Asbestos Abatement

02 82 13.43 Bulk Asbestos Removal		Crew	Daily Output	Labor-Hours	Unit	Material	Labor (2012 Bare Costs)	Equipment	Total	Total Incl O&P
6000	Remove contaminated soil from crawl space by hand	A-9	400	.160	C.F.	.50	7.85		8.35	13.15
6100	With large production vacuum loader	A-12	700	.091	″	.28	4.47	1.05	5.80	8.65
7000	Radiator backing, not including radiator removal	A-9	1200	.053	S.F.	.17	2.61		2.78	4.37
8000	Cement-asbestos transite board	2 Asbe	1000	.016		.14	.78		.92	1.41
8100	Transite shingle siding	A-10B	750	.043		.21	2.09		2.30	3.59
8200	Shingle roofing	″	2000	.016		.08	.78		.86	1.35
8250	Built-up, no gravel, non-friable	B-2	1400	.029		.08	1.01		1.09	1.75
8260	Bituminous flashing	1 Rofc	300	.027		.08	1		1.08	1.86
8300	Asbestos millboard	2 Asbe	1000	.016		.08	.78		.86	1.35
9000	For type B (supplied air) respirator equipment, add				%					10%

02 82 13.44 Demolition In Asbestos Contaminated Area		Crew	Daily Output	Labor-Hours	Unit	Material	Labor	Equipment	Total	Total Incl O&P
0010	**DEMOLITION IN ASBESTOS CONTAMINATED AREA**									
0200	Ceiling, including suspension system, plaster and lath	A-9	2100	.030	S.F.	.09	1.49		1.58	2.50
0210	Finished plaster, leaving wire lath		585	.109		.34	5.35		5.69	8.95
0220	Suspended acoustical tile		3500	.018		.06	.89		.95	1.50
0230	Concealed tile grid system		3000	.021		.07	1.04		1.11	1.75
0240	Metal pan grid system		1500	.043		.13	2.09		2.22	3.51
0250	Gypsum board		2500	.026		.08	1.25		1.33	2.10
0260	Lighting fixtures up to 2′ x 4′		72	.889	Ea.	2.75	43.50		46.25	73
0400	Partitions, non load bearing									
0410	Plaster, lath, and studs	A-9	690	.093	S.F.	.86	4.54		5.40	8.25
0450	Gypsum board and studs	″	1390	.046	″	.14	2.25		2.39	3.78
9000	For type B (supplied air) respirator equipment, add				%					10%

02 82 13.45 OSHA Testing		Crew	Daily Output	Labor-Hours	Unit	Material	Labor	Equipment	Total	Total Incl O&P
0010	**OSHA TESTING**									
0100	Certified technician, minimum				Day				200	220
0110	Maximum								300	330
0120	Industrial hygienist, minimum								300	330
0130	Maximum								400	440
0200	Asbestos sampling and PCM analysis, NIOSH 7400, minimum	1 Asbe	8	1	Ea.	2.75	49		51.75	81.50
0210	Maximum		4	2		3	97.50		100.50	160
1000	Cleaned area samples		8	1		2.63	49		51.63	81.50
1100	PCM air sample analysis, NIOSH 7400, minimum		8	1		30.50	49		79.50	113
1110	Maximum		4	2		3.20	97.50		100.70	161
1200	TEM air sample analysis, NIOSH 7402, minimum								90	120
1210	Maximum								360	450

02 82 13.46 Decontamination of Asbestos Containment Area		Crew	Daily Output	Labor-Hours	Unit	Material	Labor	Equipment	Total	Total Incl O&P
0010	**DECONTAMINATION OF ASBESTOS CONTAINMENT AREA**									
0100	Spray exposed substrate with surfactant (bridging)									
0200	Flat surfaces	A-9	6000	.011	S.F.	.36	.52		.88	1.24
0250	Irregular surfaces		4000	.016	″	.31	.78		1.09	1.60
0300	Pipes, beams, and columns		2000	.032	L.F.	.56	1.57		2.13	3.14
1000	Spray encapsulate polyethylene sheeting		8000	.008	S.F.	.31	.39		.70	.97
1100	Roll down polyethylene sheeting		8000	.008	″		.39		.39	.63
1500	Bag polyethylene sheeting		400	.160	Ea.	.77	7.85		8.62	13.45
2000	Fine clean exposed substrate, with nylon brush		2400	.027	S.F.		1.30		1.30	2.10
2500	Wet wipe substrate		4800	.013			.65		.65	1.05
2600	Vacuum surfaces, fine brush		6400	.010			.49		.49	.79
3000	Structural demolition									
3100	Wood stud walls	A-9	2800	.023	S.F.		1.12		1.12	1.80
3500	Window manifolds, not incl. window replacement		4200	.015			.75		.75	1.20
3600	Plywood carpet protection		2000	.032			1.57		1.57	2.52

02 82 Asbestos Remediation

02 82 13 – Asbestos Abatement

02 82 13.46 Decontamination of Asbestos Containment Area		Crew	Daily Output	Labor-Hours	Unit	2012 Bare Costs Material	Labor	Equipment	Total	Total Incl O&P
4000	Remove custom decontamination facility	A-10A	8	3	Ea.	15	147		162	253
4100	Remove portable decontamination facility	3 Asbe	12	2	"	12.75	97.50		110.25	171
5000	HEPA vacuum, shampoo carpeting	A-9	4800	.013	S.F.	.05	.65		.70	1.11
9000	Final cleaning of protected surfaces	A-10A	8000	.003	"		.15		.15	.24

02 82 13.47 Asbestos Waste Pkg., Handling, and Disp.

		Crew	Daily Output	Labor-Hours	Unit	Material	Labor	Equipment	Total	Total Incl O&P
0010	**ASBESTOS WASTE PACKAGING, HANDLING, AND DISPOSAL**									
0100	Collect and bag bulk material, 3 C.F. bags, by hand	A-9	400	.160	Ea.	2.35	7.85		10.20	15.20
0200	Large production vacuum loader	A-12	880	.073		.80	3.56	.83	5.19	7.50
1000	Double bag and decontaminate	A-9	960	.067		2.35	3.26		5.61	7.85
2000	Containerize bagged material in drums, per 3 C.F. drum	"	800	.080		17.25	3.91		21.16	25.50
3000	Cart bags 50′ to dumpster	2 Asbe	400	.040	↓		1.95		1.95	3.14
5000	Disposal charges, not including haul, minimum				C.Y.				60	66
5020	Maximum				"				350	385
5100	Remove refrigerant from system	1 Stpi	40	.200	Lb.		11.15		11.15	17.35
9000	For type B (supplied air) respirator equipment, add				%					10%

02 82 13.48 Asbestos Encapsulation With Sealants

		Crew	Daily Output	Labor-Hours	Unit	Material	Labor	Equipment	Total	Total Incl O&P
0010	**ASBESTOS ENCAPSULATION WITH SEALANTS**									
0100	Ceilings and walls, minimum	A-9	21000	.003	S.F.	.27	.15		.42	.54
0110	Maximum		10600	.006		.42	.30		.72	.94
0200	Columns and beams, minimum		13300	.005		.27	.24		.51	.68
0210	Maximum		5325	.012	↓	.47	.59		1.06	1.47
0300	Pipes to 12″ diameter including minor repairs, minimum		800	.080	L.F.	.37	3.91		4.28	6.70
0310	Maximum	↓	400	.160	"	1.04	7.85		8.89	13.75

02 83 Lead Remediation

02 83 19 – Lead-Based Paint Remediation

02 83 19.21 Lead Paint Remediation Plans and Methods

		Crew	Daily Output	Labor-Hours	Unit	Material	Labor	Equipment	Total	Total Incl O&P
0010	**LEAD PAINT REMEDIATION PLANS AND METHODS**									
0100	Building Survey-Commercial Building				Ea.				2,000	2,200
0200	Lead Abatement Remediation Plan								1,200	1,325
0300	Lead Paint Testing, AAS Analysis								50	55
0400	Lead Paint Testing, X-Ray Fluorescence				↓				50	55

02 83 19.23 Encapsulation of Lead-Based Paint

		Crew	Daily Output	Labor-Hours	Unit	Material	Labor	Equipment	Total	Total Incl O&P
0010	**ENCAPSULATION OF LEAD-BASED PAINT**									
0020	Interior, brushwork, trim, under 6″	1 Pord	240	.033	L.F.	2.46	1.27		3.73	4.74
0030	6″ to 12″ wide		180	.044		3.28	1.70		4.98	6.30
0040	Balustrades		300	.027		1.98	1.02		3	3.81
0050	Pipe to 4″ diameter		500	.016		1.19	.61		1.80	2.29
0060	To 8″ diameter		375	.021		1.57	.82		2.39	3.03
0070	To 12″ diameter		250	.032		2.36	1.22		3.58	4.55
0080	To 16″ diameter		170	.047	↓	3.47	1.80		5.27	6.70
0090	Cabinets, ornate design		200	.040	S.F.	2.97	1.53		4.50	5.70
0100	Simple design	↓	250	.032	"	2.36	1.22		3.58	4.55
0110	Doors, 3′ x 7′, both sides, incl. frame & trim									
0120	Flush	1 Pord	6	1.333	Ea.	30.50	51		81.50	115
0130	French, 10 – 15 lite		3	2.667		6.05	102		108.05	170
0140	Panel		4	2		36.50	76.50		113	162
0150	Louvered	↓	2.75	2.909	↓	33.50	111		144.50	215
0160	Windows, per interior side, per 15 S.F.									
0170	1 to 6 lite	1 Pord	14	.571	Ea.	21	22		43	58

02 83 19.23 Encapsulation of Lead-Based Paint

	02 83 19.23 Encapsulation of Lead-Based Paint	Crew	Daily Output	Labor-Hours	Unit	2012 Bare Costs Material	Labor	Equipment	Total	Total Incl O&P
0180	7 to 10 lite	1 Pord	7.50	1.067	Ea.	23	41		64	90.50
0190	12 lite		5.75	1.391		31	53		84	119
0200	Radiators		8	1		73.50	38.50		112	142
0210	Grilles, vents		275	.029	S.F.	2.15	1.11		3.26	4.15
0220	Walls, roller, drywall or plaster		1000	.008		.59	.31		.90	1.14
0230	With spunbonded reinforcing fabric		720	.011		.68	.43		1.11	1.43
0240	Wood		800	.010		.74	.38		1.12	1.42
0250	Ceilings, roller, drywall or plaster		900	.009		.68	.34		1.02	1.29
0260	Wood		700	.011		.84	.44		1.28	1.62
0270	Exterior, brushwork, gutters and downspouts		300	.027	L.F.	1.98	1.02		3	3.81
0280	Columns		400	.020	S.F.	1.47	.77		2.24	2.84
0290	Spray, siding		600	.013	"	.99	.51		1.50	1.90
0300	Miscellaneous									
0310	Electrical conduit, brushwork, to 2" diameter	1 Pord	500	.016	L.F.	1.19	.61		1.80	2.29
0320	Brick, block or concrete, spray		500	.016	S.F.	1.19	.61		1.80	2.29
0330	Steel, flat surfaces and tanks to 12"		500	.016		1.19	.61		1.80	2.29
0340	Beams, brushwork		400	.020		1.47	.77		2.24	2.84
0350	Trusses		400	.020		1.47	.77		2.24	2.84

02 83 19.26 Removal of Lead-Based Paint

	02 83 19.26 Removal of Lead-Based Paint	Crew	Daily Output	Labor-Hours	Unit	Material	Labor	Equipment	Total	Total Incl O&P
0010	**REMOVAL OF LEAD-BASED PAINT** R028319-60									
0011	By chemicals, per application									
0050	Baseboard, to 6" wide	1 Pord	64	.125	L.F.	.66	4.78		5.44	8.40
0070	To 12" wide		32	.250	"	1.32	9.55		10.87	16.70
0200	Balustrades, one side		28	.286	S.F.	1.32	10.95		12.27	18.90
1400	Cabinets, simple design		32	.250		1.32	9.55		10.87	16.70
1420	Ornate design		25	.320		1.32	12.25		13.57	21
1600	Cornice, simple design		60	.133		1.32	5.10		6.42	9.60
1620	Ornate design		20	.400		5.05	15.30		20.35	30
2800	Doors, one side, flush		84	.095		1	3.64		4.64	6.90
2820	Two panel		80	.100		1.32	3.83		5.15	7.55
2840	Four panel		45	.178		1.32	6.80		8.12	12.30
2880	For trim, one side, add		64	.125	L.F.	.66	4.78		5.44	8.40
3000	Fence, picket, one side		30	.267	S.F.	1.32	10.20		11.52	17.75
3200	Grilles, one side, simple design		30	.267		1.32	10.20		11.52	17.75
3220	Ornate design		25	.320		1.32	12.25		13.57	21
4400	Pipes, to 4" diameter		90	.089	L.F.	1	3.40		4.40	6.55
4420	To 8" diameter		50	.160		2	6.10		8.10	11.95
4440	To 12" diameter		36	.222		3	8.50		11.50	16.85
4460	To 16" diameter		20	.400		4	15.30		19.30	29
4500	For hangers, add		40	.200	Ea.	2.54	7.65		10.19	15
4800	Siding		90	.089	S.F.	1.17	3.40		4.57	6.75
5000	Trusses, open		55	.145	SF Face	1.85	5.55		7.40	10.95
6200	Windows, one side only, double hung, 1/1 light, 24" x 48" high		4	2	Ea.	25.50	76.50		102	150
6220	30" x 60" high		3	2.667		34	102		136	201
6240	36" x 72" high		2.50	3.200		41	122		163	240
6280	40" x 80" high		2	4		51	153		204	300
6400	Colonial window, 6/6 light, 24" x 48" high		2	4		51	153		204	300
6420	30" x 60" high		1.50	5.333		68	204		272	400
6440	36" x 72" high		1	8		102	305		407	600
6480	40" x 80" high		1	8		102	305		407	600
6600	8/8 light, 24" x 48" high		2	4		51	153		204	300
6620	40" x 80" high		1	8		102	305		407	600

02 83 Lead Remediation

02 83 19 – Lead-Based Paint Remediation

02 83 19.26 Removal of Lead-Based Paint		Crew	Daily Output	Labor-Hours	Unit	Material	Labor	Equipment	Total	Total Incl O&P
						2012 Bare Costs				
6800	12/12 light, 24" x 48" high	1 Pord	1	8	Ea.	102	305		407	600
6820	40" x 80" high		.75	10.667		136	410		546	800
6840	Window frame & trim items, included in pricing above									
9000	Minimum labor/equipment charge	1 Pord	3	2.667	Job		102		102	163

02 85 Mold Remediation

02 85 16 – Mold Remediation Preparation and Containment

02 85 16.40 Mold Remediation Plans and Methods

		Crew	Daily Output	Labor-Hours	Unit	Material	Labor	Equipment	Total	Total Incl O&P
0010	MOLD REMEDIATION PLANS AND METHODS									
0020	Initial inspection, average 3 bedroom home				Total				250	275
0030	Average 5 bedroom home								400	440
0040	Testing, air sample each								125	138
0050	Swab sample								125	138
0060	Tape sample								125	140
0070	Post remediation air test								125	140
0080	Mold abatement plan, average 3 bedroom home								1,200	1,300
0090	Average 5 bedroom home								1,600	1,750
0100	Packup & removal of contents, average 3 bedroom home, excl storage								8,000	8,800
0110	Average 5 bedroom home, excl storage								15,000	16,500
0600	For demolition in mold contaminated areas, see Section 02 85 33.50									
0610	For personal protection equipment, see Section 02 82 13.41									

02 85 16.50 Preparation of Mold Containment Area

		Crew	Daily Output	Labor-Hours	Unit	Material	Labor	Equipment	Total	Total Incl O&P
0010	PREPARATION OF MOLD CONTAINMENT AREA									
0100	Pre-cleaning, HEPA vacuum and wet wipe, flat surfaces	A-9	12000	.005	S.F.	.02	.26		.28	.44
0300	Separation barrier, 2" x 4" @ 16", 1/2" plywood ea. side, 8' high	2 Carp	400	.040		3	1.76		4.76	6.20
0310	12' high		320	.050		3	2.21		5.21	6.90
0320	16' high		200	.080		2	3.53		5.53	8
0400	Personnel decontam. chamber, 2" x 4" @ 16", 3/4" ply ea. side		280	.057		4	2.52		6.52	8.55
0450	Waste decontam. chamber, 2" x 4" studs @ 16", 3/4" ply each side		360	.044		4	1.96		5.96	7.60
0500	Cover surfaces with polyethylene sheeting									
0501	Including glue and tape									
0550	Floors, each layer, 6 mil	A-9	8000	.008	S.F.	.06	.39		.45	.70
0551	4 mil		9000	.007		.04	.35		.39	.61
0560	Walls, each layer, 6 mil		6000	.011		.09	.52		.61	.94
0561	4 mil		7000	.009		.07	.45		.52	.80
0570	For heights above 12', add						20%			
0575	For heights above 20', add						30%			
0580	For fire retardant poly, add					100%				
0590	For large open areas, deduct					10%	20%			
0600	Seal floor penetrations with foam firestop to 36 sq. in.	2 Carp	200	.080	Ea.	7.60	3.53		11.13	14.20
0610	36 sq. in. to 72 sq. in.		125	.128		15.25	5.65		20.90	26
0615	72 sq. in. to 144 sq. in.		80	.200		30.50	8.80		39.30	48
0620	Wall penetrations, to 36 square inches		180	.089		7.60	3.92		11.52	14.85
0630	36 sq. in. to 72 sq. in.		100	.160		15.25	7.05		22.30	28.50
0640	72 sq. in. to 144 sq. in.		60	.267		30.50	11.75		42.25	53
0800	Caulk seams with latex caulk	1 Carp	230	.035	L.F.	.15	1.53		1.68	2.68
0900	Set up neg. air machine, 1-2k CFM/25 M.C.F. volume	1 Asbe	4.30	1.860	Ea.		91		91	146

02 85 Mold Remediation

02 85 33 – Removal and Disposal of Materials with Mold

02 85 33.50 Demolition in Mold Contaminated Area		Crew	Daily Output	Labor-Hours	Unit	2012 Bare Costs Material	Labor	Equipment	Total	Total Incl O&P
0010	**DEMOLITION IN MOLD CONTAMINATED AREA**									
0200	Ceiling, including suspension system, plaster and lath	A-9	2100	.030	S.F.	.09	1.49		1.58	2.50
0210	Finished plaster, leaving wire lath		585	.109		.34	5.35		5.69	8.95
0220	Suspended acoustical tile		3500	.018		.06	.89		.95	1.50
0230	Concealed tile grid system		3000	.021		.07	1.04		1.11	1.75
0240	Metal pan grid system		1500	.043		.13	2.09		2.22	3.51
0250	Gypsum board		2500	.026		.08	1.25		1.33	2.10
0255	Plywood		2500	.026	↓	.08	1.25		1.33	2.10
0260	Lighting fixtures up to 2′ x 4′	↓	72	.889	Ea.	2.75	43.50		46.25	73
0400	Partitions, non load bearing									
0410	Plaster, lath, and studs	A-9	690	.093	S.F.	.86	4.54		5.40	8.25
0450	Gypsum board and studs		1390	.046		.14	2.25		2.39	3.78
0465	Carpet & pad		1390	.046	↓	.14	2.25		2.39	3.78
0600	Pipe insulation, air cell type, up to 4″ diameter pipe		900	.071	L.F.	.22	3.48		3.70	5.85
0610	4″ to 8″ diameter pipe		800	.080		.25	3.91		4.16	6.55
0620	10″ to 12″ diameter pipe		700	.091		.28	4.47		4.75	7.50
0630	14″ to 16″ diameter pipe		550	.116	↓	.36	5.70		6.06	9.55
0650	Over 16″ diameter pipe	↓	650	.098	S.F.	.31	4.82		5.13	8.10
9000	For type B (supplied air) respirator equipment, add				%					10%

Division Notes

	CREW	DAILY OUTPUT	LABOR-HOURS	UNIT	2012 BARE COSTS				TOTAL INCL O&P
					MAT.	LABOR	EQUIP.	TOTAL	

Estimating Tips

General

- Carefully check all the plans and specifications. Concrete often appears on drawings other than structural drawings, including mechanical and electrical drawings for equipment pads. The cost of cutting and patching is often difficult to estimate. See Subdivision 03 81 for Concrete Cutting, Subdivision 02 41 19.16 for Cutout Demolition, Subdivision 03 05 05.10 for Concrete Demolition, and Subdivision 02 41 19.23 for Rubbish Handling (handling, loading and hauling of debris).
- Always obtain concrete prices from suppliers near the job site. A volume discount can often be negotiated, depending upon competition in the area. Remember to add for waste, particularly for slabs and footings on grade.

03 10 00 Concrete Forming and Accessories

- A primary cost for concrete construction is forming. Most jobs today are constructed with prefabricated forms. The selection of the forms best suited for the job and the total square feet of forms required for efficient concrete forming and placing are key elements in estimating concrete construction. Enough forms must be available for erection to make efficient use of the concrete placing equipment and crew.
- Concrete accessories for forming and placing depend upon the systems used. Study the plans and specifications to ensure that all special accessory requirements have been included in the cost estimate, such as anchor bolts, inserts, and hangers.
- Included within costs for forms-in-place are all necessary bracing and shoring.

03 20 00 Concrete Reinforcing

- Ascertain that the reinforcing steel supplier has included all accessories, cutting, bending, and an allowance for lapping, splicing, and waste. A good rule of thumb is 10% for lapping, splicing, and waste. Also, 10% waste should be allowed for welded wire fabric.
- The unit price items in the subdivisions for Reinforcing In Place, Glass Fiber Reinforcing, and Welded Wire Fabric include the labor to install accessories such as beam and slab bolsters, high chairs, and bar ties and tie wire. The material cost for these accessories is not included; they may be obtained from the Accessories Division.

03 30 00 Cast-In-Place Concrete

- When estimating structural concrete, pay particular attention to requirements for concrete additives, curing methods, and surface treatments. Special consideration for climate, hot or cold, must be included in your estimate. Be sure to include requirements for concrete placing equipment, and concrete finishing.
- For accurate concrete estimating, the estimator must consider each of the following major components individually: forms, reinforcing steel, ready-mix concrete, placement of the concrete, and finishing of the top surface. For faster estimating, Subdivision 03 30 53.40 for Concrete-In-Place can be used; here, various items of concrete work are presented that include the costs of all five major components (unless specifically stated otherwise).

03 40 00 Precast Concrete
03 50 00 Cast Decks and Underlayment

- The cost of hauling precast concrete structural members is often an important factor. For this reason, it is important to get a quote from the nearest supplier. It may become economically feasible to set up precasting beds on the site if the hauling costs are prohibitive.

Reference Numbers

Reference numbers are shown in shaded boxes at the beginning of some major classifications. These numbers refer to related items in the Reference Section. The reference information may be an estimating procedure, an alternate pricing method, or technical information.

Note: Not all subdivisions listed here necessarily appear in this publication.

03 01 Maintenance of Concrete

03 01 30 – Maintenance of Cast-In-Place Concrete

03 01 30.62 Concrete Patching		Crew	Daily Output	Labor-Hours	Unit	Material	Labor	Equipment	Total	Total Incl O&P
						2012 Bare Costs				
0010	**CONCRETE PATCHING**									
0100	Floors, 1/4" thick, small areas, regular grout	1 Cefi	170	.047	S.F.	1.23	1.99		3.22	4.48
0150	Epoxy grout	"	100	.080	"	7.30	3.39		10.69	13.30
2000	Walls, including chipping, cleaning and epoxy grout									
2100	1/4" deep	1 Cefi	65	.123	S.F.	7.30	5.20		12.50	16.20
2150	1/2" deep		50	.160		14.55	6.80		21.35	26.50
2200	3/4" deep		40	.200		22	8.45		30.45	37.50
9000	Minimum labor/equipment charge		4.50	1.778	Job		75.50		75.50	118

03 05 Common Work Results for Concrete

03 05 05 – Selective Concrete Demolition

03 05 05.10 Selective Demolition, Concrete		Crew	Daily Output	Labor-Hours	Unit	Material	Labor	Equipment	Total	Total Incl O&P
0010	**SELECTIVE DEMOLITION, CONCRETE** R024119-10									
0012	Excludes saw cutting, torch cutting, loading or hauling									
0050	Break up into small pieces, minimum reinforcing	B-9	24	1.667	C.Y.		59	8.70	67.70	107
0060	Average reinforcing		16	2.500			89	13.10	102.10	160
0070	Maximum reinforcing		8	5			178	26	204	320
0150	Remove whole pieces, up to 2 tons per piece	E-18	36	1.111	Ea.		54.50	26.50	81	128
0160	2 – 5 tons per piece		30	1.333			65.50	31.50	97	153
0170	5 – 10 tons per piece		24	1.667			82	39.50	121.50	192
0180	10 – 15 tons per piece		18	2.222			109	53	162	255
0250	Precast unit embedded in masonry, up to 1 C.F.	D-1	16	1			40		40	64
0260	1 – 2 C.F.		12	1.333			53		53	85.50
0270	2 – 5 C.F.		10	1.600			64		64	103
0280	5 – 10 C.F.		8	2			79.50		79.50	128
0990	For hydrodemolition see Section 02 41 13.15									
1910	Minimum labor/equipment charge	B-9	2	20	Job		710	105	815	1,300

03 05 13 – Basic Concrete Materials

03 05 13.20 Concrete Admixtures and Surface Treatments		Crew	Daily Output	Labor-Hours	Unit	Material	Labor	Equipment	Total	Total Incl O&P
0010	**CONCRETE ADMIXTURES AND SURFACE TREATMENTS**									
0040	Abrasives, aluminum oxide, over 20 tons				Lb.	1.57			1.57	1.73
0050	1 to 20 tons					1.73			1.73	1.90
0070	Under 1 ton					1.80			1.80	1.98
0100	Silicon carbide, black, over 20 tons					2.34			2.34	2.57
0110	1 to 20 tons					2.69			2.69	2.96
0120	Under 1 ton					2.80			2.80	3.08
0200	Air entraining agent, .7 to 1.5 oz. per bag, 55 gallon drum				Gal.	11.60			11.60	12.75
0220	5 gallon pail					15.95			15.95	17.55
0300	Bonding agent, acrylic latex, 250 S.F. per gallon, 5 gallon pail					21			21	23
0320	Epoxy resin, 80 S.F. per gallon, 4 gallon case					59.50			59.50	65.50
0400	Calcium chloride, 50 lb. bags, TL lots				Ton	670			670	735
0420	Less than truckload lots				Bag	20			20	22
0500	Carbon black, liquid, 2 to 8 lb. per bag of cement				Lb.	7.55			7.55	8.30
0600	Colors, integral, 2 to 10 lb. per bag of cement, minimum					2.43			2.43	2.67
0610	Average					3.13			3.13	3.44
0620	Maximum					4.92			4.92	5.40
0920	Dustproofing compound, 250 S.F./gal., 5 gallon pail				Gal.	7.75			7.75	8.55
1010	Epoxy based, 125 S.F./gal., 5 gallon pail				"	58.50			58.50	64
1100	Hardeners, metallic, 55 lb. bags, natural (grey)				Lb.	.90			.90	.99
1200	Colors					1.48			1.48	1.63

03 05 Common Work Results for Concrete

03 05 13 – Basic Concrete Materials

	03 05 13.20 Concrete Admixtures and Surface Treatments	Crew	Daily Output	Labor-Hours	Unit	2012 Bare Costs Material	Labor	Equipment	Total	Total Incl O&P
1300	Non-metallic, 55 lb. bags, natural grey				Lb.	.52			.52	.57
1320	Colors					.79			.79	.87
1600	Sealer, hardener and dustproofer, epoxy-based, 125 S.F./gal., 5 gallon, min				Gal.	58.50			58.50	64
1620	5 gallon pail, maximum					64			64	70.50
1630	Sealer, solvent-based, 250 S.F./gal., 55 gallon drum					20			20	22
1640	5 gallon pail					27			27	29.50
1650	Sealer, water based, 350 S.F., 55 gallon drum					19.80			19.80	22
1660	5 gallon pail					23			23	25.50
1900	Set retarder, 100 S.F./gal., 1 gallon pail					19.10			19.10	21
2000	Waterproofing, integral 1 lb. per bag of cement				Lb.	1.76			1.76	1.94
2100	Powdered metallic, 40 lb. per 100 S.F., minimum					2.09			2.09	2.30
2120	Maximum					2.92			2.92	3.21
3000	For integral colors, 2500 psi (5 bag mix)									
3100	Red, yellow or brown, 1.8 lb. per bag, add				C.Y.	22			22	24
3200	9.4 lb. per bag, add					114			114	126
3400	Black, 1.8 lb. per bag, add					28			28	31
3500	7.5 lb. per bag, add					117			117	129
3700	Green, 1.8 lb. per bag, add					44.50			44.50	48.50
3800	7.5 lb. per bag, add					185			185	203

	03 05 13.85 Winter Protection	Crew	Daily Output	Labor-Hours	Unit	Material	Labor	Equipment	Total	Total Incl O&P
0010	**WINTER PROTECTION**									
0012	For heated ready mix, add, minimum				C.Y.	4.25			4.25	4.68
0050	Maximum				"	5.30			5.30	5.85
0100	Temporary heat to protect concrete, 24 hours, minimum	2 Clab	50	.320	M.S.F.	695	11.25		706.25	785
0150	Maximum	"	25	.640	"	905	22.50		927.50	1,025
0200	Temporary shelter for slab on grade, wood frame/polyethylene sheeting									
0201	Build or remove, minimum	2 Carp	10	1.600	M.S.F.	259	70.50		329.50	400
0210	Maximum	"	3	5.333	"	325	235		560	745
0710	Electrically, heated pads, 15 watts/S.F., 20 uses, minimum				S.F.	.23			.23	.26
0800	Maximum				"	.39			.39	.43

03 11 Concrete Forming

03 11 13 – Structural Cast-In-Place Concrete Forming

	03 11 13.20 Forms In Place, Beams and Girders	Crew	Daily Output	Labor-Hours	Unit	Material	Labor	Equipment	Total	Total Incl O&P
0010	**FORMS IN PLACE, BEAMS AND GIRDERS** R031113-40									
0500	Exterior spandrel, job-built plywood, 12" wide, 1 use R031113-60	C-2	225	.213	SFCA	2.51	9.15		11.66	17.75
0650	4 use		310	.155		.82	6.65		7.47	11.80
1000	18" wide, 1 use		250	.192		2.20	8.25		10.45	15.90
1150	4 use		315	.152		.72	6.55		7.27	11.50
1500	24" wide, 1 use		265	.181		2	7.80		9.80	14.95
1650	4 use		325	.148		.65	6.35		7	11.10
2000	Interior beam, job-built plywood, 12" wide, 1 use		300	.160		3.03	6.85		9.88	14.60
2150	4 use		377	.127		.99	5.45		6.44	10.05
2500	24" wide, 1 use		320	.150		2.05	6.45		8.50	12.80
2650	4 use		395	.122		.66	5.20		5.86	9.30
3000	Encasing steel beam, hung, job-built plywood, 1 use		325	.148		2.57	6.35		8.92	13.20
3150	4 use		430	.112		.83	4.79		5.62	8.75
9000	Minimum labor/equipment charge	2 Carp	2	8	Job		355		355	580

03 11 Concrete Forming

03 11 13 – Structural Cast-In-Place Concrete Forming

03 11 13.25 Forms In Place, Columns		Crew	Daily Output	Labor-Hours	Unit	2012 Bare Costs Material	Labor	Equipment	Total	Total Incl O&P
0010	**FORMS IN PLACE, COLUMNS** R031113-40									
0500	Round fiberglass, 4 use per mo., rent, 12" diameter	C-1	160	.200	L.F.	8.15	8.35		16.50	22.50
0550	16" diameter R031113-60		150	.213		9.70	8.95		18.65	25.50
0600	18" diameter		140	.229		10.85	9.55		20.40	27.50
0650	24" diameter		135	.237		13.50	9.90		23.40	31
0700	28" diameter		130	.246		15.10	10.30		25.40	33.50
0800	30" diameter		125	.256		15.75	10.70		26.45	35
0850	36" diameter		120	.267		21	11.15		32.15	41.50
1500	Round fiber tube, recycled paper, 1 use, 8" diameter G		155	.206		1.51	8.65		10.16	15.80
1550	10" diameter G		155	.206		2.10	8.65		10.75	16.45
1600	12" diameter G		150	.213		2.44	8.95		11.39	17.35
1650	14" diameter G		145	.221		3.54	9.25		12.79	19.05
1700	16" diameter G		140	.229		4.29	9.55		13.84	20.50
1720	18" diameter G		140	.229		5	9.55		14.55	21
1750	20" diameter G		135	.237		6.85	9.90		16.75	24
1800	24" diameter G		130	.246		8.75	10.30		19.05	26.50
1850	30" diameter G		125	.256		12.90	10.70		23.60	31.50
1900	36" diameter G		115	.278		16.15	11.65		27.80	37
1950	42" diameter G		100	.320		40	13.40		53.40	66
2000	48" diameter G		85	.376		46	15.75		61.75	77
2200	For seamless type, add					15%				
3000	Round, steel, 4 use per mo., rent, regular duty, 14" diameter G	C-1	145	.221	L.F.	15.45	9.25		24.70	32
3050	16" diameter G		125	.256		15.75	10.70		26.45	35
3100	Heavy duty, 20" diameter G		105	.305		17.40	12.75		30.15	40
3150	24" diameter G		85	.376		18.90	15.75		34.65	47
3200	30" diameter G		70	.457		22	19.15		41.15	55.50
3250	36" diameter G		60	.533		23.50	22.50		46	62
3300	48" diameter G		50	.640		69.50	27		96.50	121
3350	60" diameter G		45	.711		42.50	30		72.50	96
4500	For second and succeeding months, deduct					50%				
5000	Job-built plywood, 8" x 8" columns, 1 use	C-1	165	.194	SFCA	2.62	8.10		10.72	16.20
5050	2 use		195	.164		1.49	6.85		8.34	12.90
5100	3 use		210	.152		1.05	6.40		7.45	11.60
5150	4 use		215	.149		.87	6.25		7.12	11.15
5500	12" x 12" columns, 1 use		180	.178		2.38	7.45		9.83	14.80
5550	2 use		210	.152		1.31	6.40		7.71	11.90
5600	3 use		220	.145		.95	6.10		7.05	11.05
5650	4 use		225	.142		.77	5.95		6.72	10.60
6000	16" x 16" columns, 1 use		185	.173		2.31	7.25		9.56	14.40
6050	2 use		215	.149		1.26	6.25		7.51	11.60
6100	3 use		230	.139		.93	5.80		6.73	10.60
6150	4 use		235	.136		.76	5.70		6.46	10.20
6500	24" x 24" columns, 1 use		190	.168		2.57	7.05		9.62	14.35
6550	2 use		216	.148		1.41	6.20		7.61	11.70
6600	3 use		230	.139		1.03	5.80		6.83	10.70
6650	4 use		238	.134		.83	5.65		6.48	10.10
7000	36" x 36" columns, 1 use		200	.160		1.84	6.70		8.54	13.05
7050	2 use		230	.139		1.06	5.80		6.86	10.70
7100	3 use		245	.131		.74	5.45		6.19	9.75
7150	4 use		250	.128		.60	5.35		5.95	9.45
7400	Steel framed plywood, based on 50 uses of purchased									
7420	forms, and 4 uses of bracing lumber									

03 11 Concrete Forming

03 11 13 – Structural Cast-In-Place Concrete Forming

03 11 13.25 Forms In Place, Columns		Crew	Daily Output	Labor-Hours	Unit	2012 Bare Costs Material	Labor	Equipment	Total	Total Incl O&P
7500	8" x 8" column	C-1	340	.094	SFCA	1.82	3.94		5.76	8.45
7550	10" x 10"		350	.091		1.59	3.83		5.42	8
7600	12" x 12"		370	.086		1.35	3.62		4.97	7.45
7650	16" x 16"		400	.080		1.05	3.35		4.40	6.65
7700	20" x 20"		420	.076		.93	3.19		4.12	6.25
7750	24" x 24"		440	.073		.67	3.04		3.71	5.75
7755	30" x 30"		440	.073		.86	3.04		3.90	5.95
7760	36" x 36"		460	.070		.75	2.91		3.66	5.60
9000	Minimum labor/equipment charge	2 Carp	2	8	Job		355		355	580

03 11 13.35 Forms In Place, Elevated Slabs		Crew	Daily Output	Labor-Hours	Unit	Material	Labor	Equipment	Total	Total Incl O&P
0010	**FORMS IN PLACE, ELEVATED SLABS** R031113-40									
1000	Flat plate, job-built plywood, to 15' high, 1 use	C-2	470	.102	S.F.	3.18	4.38		7.56	10.70
1150	4 use R031113-60		560	.086		1.03	3.68		4.71	7.20
1500	15' to 20' high ceilings, 4 use		495	.097		1.03	4.16		5.19	7.95
2000	Flat slab, drop panels, job-built plywood, to 15' high, 1 use		449	.107		3.57	4.59		8.16	11.45
2150	4 use		544	.088		1.16	3.79		4.95	7.50
2250	15' to 20' high ceilings, 4 use		480	.100		2.08	4.29		6.37	9.35
3500	Floor slab, with 1-way joist pans, 1 use		415	.116		5.35	4.97		10.32	14
3650	4 use		500	.096		2.92	4.12		7.04	9.95
4500	With 2-way waffle domes, 1 use		405	.119		5.45	5.10		10.55	14.35
4550	4 use		470	.102		3.04	4.38		7.42	10.55
5000	Box out for slab openings, over 16" deep, 1 use		190	.253	SFCA	3.93	10.85		14.78	22
5050	2 use		240	.200	"	2.16	8.60		10.76	16.45
5500	Shallow slab box outs, to 10 S.F.		42	1.143	Ea.	9.75	49		58.75	91.50
5550	Over 10 S.F. (use perimeter)		600	.080	L.F.	1.31	3.43		4.74	7.10
6000	Bulkhead forms for slab, with keyway, 1 use, 2 piece		500	.096		1.85	4.12		5.97	8.80
6100	3 piece (see also edge forms)		460	.104		1.97	4.48		6.45	9.50
6200	Slab bulkhead form, 4-1/2" high, exp metal, w/keyway & stakes G	C-1	1200	.027		1.13	1.12		2.25	3.07
6210	5-1/2" high G		1100	.029		1.25	1.22		2.47	3.38
6215	7-1/2" high G		960	.033		1.47	1.39		2.86	3.91
6220	9-1/2" high G		840	.038		1.67	1.59		3.26	4.45
6500	Curb forms, wood, 6" to 12" high, on elevated slabs, 1 use		180	.178	SFCA	1.29	7.45		8.74	13.60
6550	2 use		205	.156		.71	6.55		7.26	11.50
6600	3 use		220	.145		.52	6.10		6.62	10.55
6650	4 use		225	.142		.42	5.95		6.37	10.20
7000	Edge forms to 6" high, on elevated slab, 4 use		500	.064	L.F.	.17	2.68		2.85	4.58
7500	Depressed area forms to 12" high, 4 use		300	.107		.88	4.46		5.34	8.25
7550	12" to 24" high, 4 use		175	.183		1.19	7.65		8.84	13.85
8000	Perimeter deck and rail for elevated slabs, straight		90	.356		9.75	14.90		24.65	35
8050	Curved		65	.492		13.35	20.50		33.85	48.50
8500	Void forms, round plastic, 8" high x 3" dia G		450	.071	Ea.	1.89	2.98		4.87	6.95
8650	8" diameter G		375	.085	"	6	3.57		9.57	12.45
9000	Minimum labor/equipment charge	2 Carp	2	8	Job		355		355	580

03 11 13.40 Forms In Place, Equipment Foundations		Crew	Daily Output	Labor-Hours	Unit	Material	Labor	Equipment	Total	Total Incl O&P
0010	**FORMS IN PLACE, EQUIPMENT FOUNDATIONS** R031113-40									
0020	1 use	C-2	160	.300	SFCA	2.93	12.90		15.83	24
0050	2 use R031113-60		190	.253		1.61	10.85		12.46	19.55
0100	3 use		200	.240		1.17	10.30		11.47	18.20
0150	4 use		205	.234		.96	10.05		11.01	17.55
9000	Minimum labor/equipment charge	1 Carp	3	2.667	Job		118		118	193

03 11 Concrete Forming

03 11 13 – Structural Cast-In-Place Concrete Forming

03 11 13.45 Forms In Place, Footings

	03 11 13.45 Forms In Place, Footings	Crew	Daily Output	Labor-Hours	Unit	2012 Bare Costs Material	Labor	Equipment	Total	Total Incl O&P
0010	**FORMS IN PLACE, FOOTINGS** R031113-40									
0020	Continuous wall, plywood, 1 use	C-1	375	.085	SFCA	5.95	3.57		9.52	12.35
0150	4 use R031113-60	"	485	.066	"	1.93	2.76		4.69	6.65
1500	Keyway, 4 use, tapered wood, 2" x 4"	1 Carp	530	.015	L.F.	.18	.67		.85	1.28
1550	2" x 6"	"	500	.016	"	.27	.71		.98	1.45
3000	Pile cap, square or rectangular, job-built plywood, 1 use	C-1	290	.110	SFCA	2.36	4.62		6.98	10.15
3150	4 use		383	.084		.77	3.50		4.27	6.60
5000	Spread footings, job-built lumber, 1 use		305	.105		1.90	4.39		6.29	9.30
5150	4 use		414	.077		.62	3.23		3.85	6
9000	Minimum labor/equipment charge	1 Carp	3	2.667	Job		118		118	193

03 11 13.47 Forms In Place, Gas Station Forms

	03 11 13.47 Forms In Place, Gas Station Forms	Crew	Daily Output	Labor-Hours	Unit	Material	Labor	Equipment	Total	Total Incl O&P
0010	**FORMS IN PLACE, GAS STATION FORMS**									
0050	Curb fascia, with template, 12 ga. steel, left in place, 9" high G	1 Carp	50	.160	L.F.	13.10	7.05		20.15	26
1000	Sign or light bases, 18" diameter, 9" high G		9	.889	Ea.	82.50	39		121.50	156
1050	30" diameter, 13" high G		8	1	"	131	44		175	217
1990	Minimum labor/equipment charge		2	4	Job		176		176	289
2000	Island forms, 10' long, 9" high, 3'- 6" wide G	C-1	10	3.200	Ea.	365	134		499	625
2050	4' wide G		9	3.556		380	149		529	660
2500	20' long, 9" high, 4' wide G		6	5.333		610	223		833	1,025
2550	5' wide G		5	6.400		635	268		903	1,125
9000	Minimum labor/equipment charge	1 Carp	3	2.667	Job		118		118	193

03 11 13.50 Forms In Place, Grade Beam

	03 11 13.50 Forms In Place, Grade Beam	Crew	Daily Output	Labor-Hours	Unit	Material	Labor	Equipment	Total	Total Incl O&P
0010	**FORMS IN PLACE, GRADE BEAM** R031113-40									
0020	Job-built plywood, 1 use	C-2	530	.091	SFCA	2.68	3.89		6.57	9.30
0150	4 use R031113-60	"	605	.079	"	.87	3.41		4.28	6.55
9000	Minimum labor/equipment charge	2 Carp	2	8	Job		355		355	580

03 11 13.65 Forms In Place, Slab On Grade

	03 11 13.65 Forms In Place, Slab On Grade	Crew	Daily Output	Labor-Hours	Unit	Material	Labor	Equipment	Total	Total Incl O&P
0010	**FORMS IN PLACE, SLAB ON GRADE** R031113-40									
1000	Bulkhead forms w/keyway, wood, 6" high, 1 use R031113-60	C-1	510	.063	L.F.	.87	2.63		3.50	5.25
1400	Bulkhead form for slab, 4-1/2" high, exp metal, incl keyway & stakes G		1200	.027		1.13	1.12		2.25	3.07
1410	5-1/2" high G		1100	.029		1.25	1.22		2.47	3.38
1420	7-1/2" high G		960	.033		1.47	1.39		2.86	3.91
1430	9-1/2" high G		840	.038		1.67	1.59		3.26	4.45
2000	Curb forms, wood, 6" to 12" high, on grade, 1 use		215	.149	SFCA	1.97	6.25		8.22	12.35
2150	4 use		275	.116	"	.64	4.87		5.51	8.70
3000	Edge forms, wood, 4 use, on grade, to 6" high		600	.053	L.F.	.29	2.23		2.52	3.98
3050	7" to 12" high		435	.074	SFCA	.62	3.08		3.70	5.75
3500	For depressed slabs, 4 use, to 12" high		300	.107	L.F.	.58	4.46		5.04	7.95
3550	To 24" high		175	.183		.75	7.65		8.40	13.40
4000	For slab blockouts, to 12" high, 1 use		200	.160		.60	6.70		7.30	11.65
4050	To 24" high, 1 use		120	.267		.76	11.15		11.91	19.15
4100	Plastic (extruded), to 6" high, multiple use, on grade		800	.040		7.90	1.67		9.57	11.45
5000	Screed, 24 ga. metal key joint, see Section 03 15 05.25									
5020	Wood, incl. wood stakes, 1" x 3"	C-1	900	.036	L.F.	.68	1.49		2.17	3.19
5050	2" x 4"		900	.036	"	.67	1.49		2.16	3.18
6000	Trench forms in floor, wood, 1 use		160	.200	SFCA	1.56	8.35		9.91	15.40
6150	4 use		185	.173	"	.51	7.25		7.76	12.40
8760	Void form, corrugated fiberboard, 4" x 12", 4' long G		3000	.011	S.F.	2.64	.45		3.09	3.63
8770	6" x 12", 4' long		3000	.011		3.10	.45		3.55	4.14
8780	1/4" thick hardboard protective cover for void form	2 Carp	1500	.011		.57	.47		1.04	1.40
9000	Minimum labor/equipment charge	1 Carp	2	4	Job		176		176	289

03 11 13 – Structural Cast-In-Place Concrete Forming

03 11 13.85 Forms In Place, Walls			Crew	Daily Output	Labor-Hours	Unit	Material	2012 Bare Costs Labor	Equipment	Total	Total Incl O&P
0010	**FORMS IN PLACE, WALLS**	R031113-10									
0100	Box out for wall openings, to 16" thick, to 10 S.F.		C-2	24	2	Ea.	21	86		107	165
0150	Over 10 S.F. (use perimeter)	R031113-40	"	280	.171	L.F.	1.83	7.35		9.18	14.05
0250	Brick shelf, 4" w, add to wall forms, use wall area abv shelf										
0260	1 use	R031113-60	C-2	240	.200	SFCA	1.97	8.60		10.57	16.25
0350	4 use			300	.160	"	.79	6.85		7.64	12.10
0500	Bulkhead, wood with keyway, 1 use, 2 piece			265	.181	L.F.	1.60	7.80		9.40	14.50
0600	Bulkhead forms with keyway, 1 piece expanded metal, 8" wall	G	C-1	1000	.032		1.47	1.34		2.81	3.82
0610	10" wall	G		800	.040		1.67	1.67		3.34	4.58
0620	12" wall	G		525	.061		2	2.55		4.55	6.40
0700	Buttress, to 8' high, 1 use		C-2	350	.137	SFCA	3.08	5.90		8.98	13.05
0850	4 use			480	.100	"	1.02	4.29		5.31	8.15
1000	Corbel or haunch, to 12" wide, add to wall forms, 1 use			150	.320	L.F.	1.91	13.75		15.66	24.50
1150	4 use			180	.267	"	.62	11.45		12.07	19.45
2000	Wall, job-built plywood, to 8' high, 1 use			370	.130	SFCA	2.18	5.55		7.73	11.55
2050	2 use			435	.110		1.39	4.74		6.13	9.30
2100	3 use			495	.097		1.01	4.16		5.17	7.90
2150	4 use			505	.095		.82	4.08		4.90	7.60
2400	Over 8' to 16' high, 1 use			280	.171		2.40	7.35		9.75	14.70
2450	2 use			345	.139		1.02	5.95		6.97	10.90
2500	3 use			375	.128		.73	5.50		6.23	9.80
2550	4 use			395	.122		.60	5.20		5.80	9.20
2700	Over 16' high, 1 use			235	.204		2.07	8.75		10.82	16.70
2750	2 use			290	.166		1.14	7.10		8.24	12.90
2800	3 use			315	.152		.83	6.55		7.38	11.60
2850	4 use			330	.145		.67	6.25		6.92	11
4000	Radial, smooth curved, job-built plywood, 1 use			245	.196		1.92	8.40		10.32	15.90
4150	4 use			335	.143		.63	6.15		6.78	10.80
4200	Below grade, job-built plywood, 1 use			225	.213		2.17	9.15		11.32	17.40
4210	2 use			225	.213		1.20	9.15		10.35	16.30
4220	3 use			225	.213		1	9.15		10.15	16.10
4230	4 use			225	.213		.71	9.15		9.86	15.80
4600	Retaining wall, battered, job-built plyw'd, to 8' high, 1 use			300	.160		1.55	6.85		8.40	12.95
4750	4 use			390	.123		.50	5.30		5.80	9.20
4900	Over 8' to 16' high, 1 use			240	.200		1.69	8.60		10.29	15.95
5050	4 use			320	.150		.55	6.45		7	11.15
5750	Liners for forms (add to wall forms), A.B.S. plastic										
5800	Aged wood, 4" wide, 1 use		1 Carp	256	.031	SFCA	3.12	1.38		4.50	5.70
5820	2 use			256	.031		1.72	1.38		3.10	4.15
5830	3 use			256	.031		1.25	1.38		2.63	3.63
5840	4 use			256	.031		1.01	1.38		2.39	3.38
5900	Fractured rope rib, 1 use			192	.042		4.52	1.84		6.36	8
5925	2 use			192	.042		2.49	1.84		4.33	5.75
5950	3 use			192	.042		1.81	1.84		3.65	5
6000	4 use			192	.042		1.47	1.84		3.31	4.63
6100	Ribbed, 3/4" deep x 1-1/2" O.C., 1 use			224	.036		4.52	1.57		6.09	7.55
6125	2 use			224	.036		2.49	1.57		4.06	5.30
6150	3 use			224	.036		1.81	1.57		3.38	4.57
6200	4 use			224	.036		1.47	1.57		3.04	4.20
6300	Rustic brick pattern, 1 use			224	.036		3.12	1.57		4.69	6
6325	2 use			224	.036		1.72	1.57		3.29	4.47
6350	3 use			224	.036		1.25	1.57		2.82	3.95

03 11 Concrete Forming

03 11 13 – Structural Cast-In-Place Concrete Forming

03 11 13.85 Forms In Place, Walls		Crew	Daily Output	Labor-Hours	Unit	Material	2012 Bare Costs Labor	Equipment	Total	Total Incl O&P
6400	4 use	1 Carp	224	.036	SFCA	1.01	1.57		2.58	3.70
6500	3/8" striated, random, 1 use		224	.036		3.12	1.57		4.69	6
6525	2 use		224	.036		1.72	1.57		3.29	4.47
6550	3 use		224	.036		1.25	1.57		2.82	3.95
6600	4 use		224	.036		1.01	1.57		2.58	3.70
7500	Lintel or sill forms, 1 use		30	.267		2.62	11.75		14.37	22
7560	4 use		37	.216		.85	9.55		10.40	16.60
7800	Modular prefabricated plywood, based on 20 uses of purchased									
7820	forms, and 4 uses of bracing lumber									
7860	To 8′ high	C-2	800	.060	SFCA	.90	2.58		3.48	5.20
8060	Over 8′ to 16′ high		600	.080		.94	3.43		4.37	6.70
8600	Pilasters, 1 use		270	.178		2.52	7.65		10.17	15.30
8660	4 use		385	.125		.82	5.35		6.17	9.70
9475	For elevated walls, add						10%			
9480	For battered walls, 1 side battered, add					10%	10%			
9485	For battered walls, 2 sides battered, add					15%	15%			
9900	Minimum labor/equipment charge	2 Carp	2	8	Job		355		355	580

03 15 Concrete Accessories

03 15 05 – Concrete Forming Accessories

03 15 05.02 Anchor Bolt Accessories

Line	Description	Crew	Daily Output	Labor-Hours	Unit	Material	Labor	Equipment	Total	Total Incl O&P
0010	**ANCHOR BOLT ACCESSORIES**									
0015	For anchor bolts set in fresh concrete, see Section 05 05 23.05									
8150	Anchor bolt sleeve, plastic, 1" diam. bolts	1 Carp	60	.133	Ea.	7.90	5.90		13.80	18.35
8500	1-1/2" diameter		28	.286		14	12.60		26.60	36
8600	2" diameter		24	.333		17.55	14.70		32.25	43.50
8650	3" diameter		20	.400		32	17.65		49.65	64.50

03 15 05.25 Expansion Joints

Line	Description	Crew	Daily Output	Labor-Hours	Unit	Material	Labor	Equipment	Total	Total Incl O&P
0010	**EXPANSION JOINTS**									
0020	Keyed, cold, 24 ga., incl. stakes, 3-1/2" high G	1 Carp	200	.040	L.F.	.96	1.76		2.72	3.95
0050	4-1/2" high G		200	.040		1.13	1.76		2.89	4.13
0100	5-1/2" high G		195	.041		1.25	1.81		3.06	4.35
0150	7-1/2" high G		190	.042		1.47	1.86		3.33	4.66
0160	9-1/2" high G		185	.043		1.67	1.91		3.58	4.97
0300	Poured asphalt, plain, 1/2" x 1"	1 Clab	450	.018		.65	.62		1.27	1.74
0350	1" x 2"		400	.020		2.60	.70		3.30	4.01
0500	Neoprene, liquid, cold applied, 1/2" x 1"		450	.018		3	.62		3.62	4.32
0550	1" x 2"		400	.020		12	.70		12.70	14.35
0700	Polyurethane, poured, 2 part, 1/2" x 1"		400	.020		1.33	.70		2.03	2.61
0750	1" x 2"		350	.023		5.30	.80		6.10	7.15
0900	Rubberized asphalt, hot or cold applied, 1/2" x 1"		450	.018		.36	.62		.98	1.42
0950	1" x 2"		400	.020		1.45	.70		2.15	2.75
1100	Hot applied, fuel resistant, 1/2" x 1"		450	.018		.54	.62		1.16	1.61
1150	1" x 2"		400	.020		2.18	.70		2.88	3.54
2000	Premolded, bituminous fiber, 1/2" x 6"	1 Carp	375	.021		.44	.94		1.38	2.02
2050	1" x 12"		300	.027		2.01	1.18		3.19	4.14
2140	Concrete expansion joint, recycled paper and fiber, 1/2" x 6" G		390	.021		.39	.90		1.29	1.91
2150	1/2" x 12" G		360	.022		.78	.98		1.76	2.47
2250	Cork with resin binder, 1/2" x 6"		375	.021		.83	.94		1.77	2.45
2300	1" x 12"		300	.027		2.32	1.18		3.50	4.48
2500	Neoprene sponge, closed cell, 1/2" x 6"		375	.021		1.93	.94		2.87	3.66

03 15 Concrete Accessories

03 15 05 – Concrete Forming Accessories

03 15 05.25 Expansion Joints

	03 15 05.25 Expansion Joints		Crew	Daily Output	Labor-Hours	Unit	Material	2012 Bare Costs Labor	Equipment	Total	Total Incl O&P
2550	1" x 12"		1 Carp	300	.027	L.F.	7.25	1.18		8.43	9.95
2750	Polyethylene foam, 1/2" x 6"			375	.021		.62	.94		1.56	2.22
2800	1" x 12"			300	.027		1.99	1.18		3.17	4.12
3000	Polyethylene backer rod, 3/8" diameter			460	.017		.05	.77		.82	1.31
3050	3/4" diameter			460	.017		.07	.77		.84	1.34
3100	1" diameter			460	.017		.14	.77		.91	1.41
3500	Polyurethane foam, with polybutylene, 1/2" x 1/2"			475	.017		.95	.74		1.69	2.27
3550	1" x 1"			450	.018		2.40	.78		3.18	3.93
3750	Polyurethane foam, regular, closed cell, 1/2" x 6"			375	.021		.70	.94		1.64	2.31
3800	1" x 12"			300	.027		2.50	1.18		3.68	4.68
4000	Polyvinyl chloride foam, closed cell, 1/2" x 6"			375	.021		1.90	.94		2.84	3.63
4050	1" x 12"			300	.027		6.55	1.18		7.73	9.15
4250	Rubber, gray sponge, 1/2" x 6"			375	.021		1.60	.94		2.54	3.30
4300	1" x 12"			300	.027		5.75	1.18		6.93	8.30
4400	Redwood heartwood, 1" x 4"			400	.020		1.16	.88		2.04	2.73
4450	1" x 6"			375	.021		1.74	.94		2.68	3.45
5000	For installation in walls, add							75%			
5250	For installation in boxouts, add							25%			

03 15 05.70 Shores

	03 15 05.70 Shores		Crew	Daily Output	Labor-Hours	Unit	Material	Labor	Equipment	Total	Total Incl O&P
0010	**SHORES**										
0020	Erect and strip, by hand, horizontal members										
0500	Aluminum joists and stringers	G	2 Carp	60	.267	Ea.		11.75		11.75	19.30
0600	Steel, adjustable beams	G		45	.356			15.70		15.70	25.50
0700	Wood joists			50	.320			14.10		14.10	23
0800	Wood stringers			30	.533			23.50		23.50	38.50
1000	Vertical members to 10' high	G		55	.291			12.85		12.85	21
1050	To 13' high	G		50	.320			14.10		14.10	23
1100	To 16' high	G		45	.356			15.70		15.70	25.50
1500	Reshoring	G		1400	.011	S.F.	.50	.50		1	1.38
1600	Flying truss system	G	C-17D	9600	.009	SFCA		.40	.08	.48	.74
1760	Horizontal, aluminum joists, 6-1/4" high x 5' to 21' span, buy	G				L.F.	13.40			13.40	14.70
1770	Beams, 7-1/4" high x 4' to 30' span	G				"	15.65			15.65	17.25
1810	Horizontal, steel beam, W8x10, 7' span, buy	G				Ea.	47.50			47.50	52
1830	10' span	G					56			56	61.50
1920	15' span	G					96			96	106
1940	20' span	G					135			135	149
1970	Steel stringer, W8x10, 4' to 16' span, buy	G				L.F.	5.60			5.60	6.15
3000	Rent for job duration, aluminum joist @ 2' O.C., per mo	G				SF Flr.	.33			.33	.37
3050	Steel W8x10	G					.14			.14	.15
3060	Steel adjustable	G					.14			.14	.15
3500	#1 post shore, steel, 5'-7" to 9'-6" high, 10000# cap., buy	G				Ea.	128			128	141
3550	#2 post shore, 7'-3" to 12'-10" high, 7800# capacity	G					148			148	163
3600	#3 post shore, 8'-10" to 16'-1" high, 3800# capacity	G					162			162	178
5010	Frame shoring systems, steel, 12000#/leg, buy										
5040	Frame, 2' wide x 6' high	G				Ea.	93.50			93.50	103
5250	X-brace	G					15.85			15.85	17.40
5550	Base plate	G					15.50			15.50	17.05
5600	Screw jack	G					33.50			33.50	36.50
5650	U-head, 8" x 8"	G					19.10			19.10	21

03 15 05.85 Stair Tread Inserts

	03 15 05.85 Stair Tread Inserts		Crew	Daily Output	Labor-Hours	Unit	Material	Labor	Equipment	Total	Total Incl O&P
0010	**STAIR TREAD INSERTS**										
0015	Cast iron, abrasive, 3" wide	G	1 Carp	90	.089	L.F.	10	3.92		13.92	17.45

03 15 Concrete Accessories

03 15 05 – Concrete Forming Accessories

03 15 05.85 Stair Tread Inserts			Crew	Daily Output	Labor-Hours	Unit	2012 Bare Costs Material	Labor	Equipment	Total	Total Incl O&P
0020	4″ wide	G	1 Carp	80	.100	L.F.	13.30	4.41		17.71	22
0040	6″ wide	G		75	.107		20	4.70		24.70	29.50
0050	9″ wide	G		70	.114		30	5.05		35.05	41.50
0100	12″ wide	G		65	.123		40	5.45		45.45	53
0300	Cast aluminum, compared to cast iron, deduct						10%				
0500	Extruded aluminum safety tread, 3″ wide	G	1 Carp	75	.107		11.45	4.70		16.15	20.50
0550	4″ wide	G		75	.107		15.30	4.70		20	24.50
0600	6″ wide	G		75	.107		23	4.70		27.70	32.50
0650	9″ wide to resurface stairs	G		70	.114		34.50	5.05		39.55	46.50
1700	Cement fill for pan-type metal treads, plain		1 Cefi	115	.070	S.F.	1.89	2.95		4.84	6.70
1750	Non-slip		″	100	.080	″	2.08	3.39		5.47	7.60

03 15 13 – Waterstops

03 15 13.50 Waterstops		Crew	Daily Output	Labor-Hours	Unit	Material	Labor	Equipment	Total	Total Incl O&P
0010	**WATERSTOPS**, PVC and Rubber									
0020	PVC, ribbed 3/16″ thick, 4″ wide	1 Carp	155	.052	L.F.	1.11	2.28		3.39	4.95
0050	6″ wide		145	.055		2.05	2.43		4.48	6.25
0500	With center bulb, 6″ wide, 3/16″ thick		135	.059		1.75	2.61		4.36	6.20
0550	3/8″ thick		130	.062		3.43	2.71		6.14	8.20
0600	9″ wide x 3/8″ thick		125	.064		5.05	2.82		7.87	10.20

03 21 Reinforcement Bars

03 21 10 – Uncoated Reinforcing Steel

03 21 10.60 Reinforcing In Place			Crew	Daily Output	Labor-Hours	Unit	Material	Labor	Equipment	Total	Total Incl O&P
0015	**REINFORCING IN PLACE**, 50-60 ton lots, A615 Grade 60	R032110-10									
0020	Includes labor, but not material cost, to install accessories										
0030	Made from recycled materials	G									
0102	Beams & Girders, #3 to #7	G	4 Rodm	3200	.010	Lb.	.51	.49		1	1.38
0152	#8 to #18	G		5400	.006		.51	.29		.80	1.05
0202	Columns, #3 to #7	G		3000	.011		.51	.52		1.03	1.43
0252	#8 to #18	G		4600	.007		.51	.34		.85	1.13
0402	Elevated slabs, #4 to #7	G		5800	.006		.53	.27		.80	1.04
0502	Footings, #4 to #7	G		4200	.008		.51	.37		.88	1.18
0552	#8 to #18	G		7200	.004		.51	.22		.73	.92
0602	Slab on grade, #3 to #7	G		4200	.008		.49	.37		.86	1.16
0702	Walls, #3 to #7	G		6000	.005		.51	.26		.77	1
0752	#8 to #18	G		8000	.004		.51	.20		.71	.89
0900	For other than 50 – 60 ton lots										
1000	Under 10 ton job, #3 to #7, add						25%	10%			
1010	#8 to #18, add						20%	10%			
1050	10 – 50 ton job, #3 to #7, add						10%				
1060	#8 to #18, add						5%				
1100	60 – 100 ton job, #3 to #7, deduct						5%				
1110	#8 to #18, deduct						10%				
1150	Over 100 ton job, #3 to #7, deduct						10%				
1160	#8 to #18, deduct						15%				
2000	Unloading & sorting, add to above		C-5	100	.560	Ton		27	6.95	33.95	52
2200	Crane cost for handling, add to above, minimum			135	.415			19.95	5.15	25.10	38.50
2210	Average			92	.609			29.50	7.60	37.10	56.50
2220	Maximum			35	1.600			77	19.90	96.90	148
2400	Dowels, 2 feet long, deformed, #3	G	2 Rodm	520	.031	Ea.	.41	1.51		1.92	2.97

03 21 Reinforcement Bars

03 21 10 – Uncoated Reinforcing Steel

03 21 10.60 Reinforcing In Place

	03 21 10.60 Reinforcing In Place		Crew	Daily Output	Labor-Hours	Unit	2012 Bare Costs Material	Labor	Equipment	Total	Total Incl O&P
2410	#4	G	2 Rodm	480	.033	Ea.	.72	1.64		2.36	3.52
2420	#5	G		435	.037		1.12	1.81		2.93	4.25
2430	#6	G		360	.044		1.62	2.18		3.80	5.40
2450	Longer and heavier dowels, add	G		725	.022	Lb.	.54	1.08		1.62	2.40
2500	Smooth dowels, 12" long, 1/4" or 3/8" diameter	G		140	.114	Ea.	.70	5.60		6.30	10.10
2520	5/8" diameter	G		125	.128		1.23	6.30		7.53	11.85
2530	3/4" diameter	G		110	.145		1.52	7.15		8.67	13.55
2600	Dowel sleeves for CIP concrete, 2-part system										
2610	Sleeve base, plastic, for 5/8" smooth dowel sleeve, fasten to edge form		1 Rodm	200	.040	Ea.	.67	1.96		2.63	4.01
2615	Sleeve, plastic, 12" long, for 5/8" smooth dowel, snap onto base			400	.020		1.56	.98		2.54	3.36
2620	Sleeve base, for 3/4" smooth dowel sleeve			175	.046		.56	2.24		2.80	4.36
2625	Sleeve, 12" long, for 3/4" smooth dowel			350	.023		1.23	1.12		2.35	3.22
2630	Sleeve base, for 1" smooth dowel sleeve			150	.053		.86	2.62		3.48	5.30
2635	Sleeve, 12" long, for 1" smooth dowel			300	.027		1.28	1.31		2.59	3.59
2700	Dowel caps, visual warning only, plastic, #3 to #8		2 Rodm	800	.020		.29	.98		1.27	1.96
2720	#8 to #18			750	.021		.64	1.05		1.69	2.46
2750	Impalement protective, plastic, #4 to #9			800	.020		1.79	.98		2.77	3.61
9000	Minimum labor/equipment charge		1 Rodm	4	2	Job		98		98	164

03 21 10.70 Glass Fiber Reinforced Polymer Bars

	03 21 10.70 Glass Fiber Reinforced Polymer Bars	Crew	Daily Output	Labor-Hours	Unit	Material	Labor	Equipment	Total	Total Incl O&P
0010	**GLASS FIBER REINFORCED POLYMER BARS**									
0020	Includes labor, but not material cost, to install accessories									
0050	#2 bar, .043 lb./L.F.	4 Rodm	9500	.003	L.F.	.35	.17		.52	.67
0100	#3 bar, .092 lb./L.F.		9300	.003		.48	.17		.65	.81
0150	#4 bar, .160 lb./L.F.		9100	.004		.70	.17		.87	1.06
0200	#5 bar, .258 lb./L.F.		8700	.004		1.05	.18		1.23	1.46
0250	#6 bar, .372 lb./L.F.		8300	.004		1.38	.19		1.57	1.84
0300	#7 bar, .497 lb./L.F.		7900	.004		1.73	.20		1.93	2.23
0350	#8 bar, .620 lb./L.F.		7400	.004		2.32	.21		2.53	2.90
0400	#9 bar, .800 lb./L.F.		6800	.005		2.92	.23		3.15	3.60
0450	#10 bar, 1.08 lb./L.F.		5800	.006		3.50	.27		3.77	4.30
0500	For Bends, add per bend				Ea.	1.25			1.25	1.38

03 22 Fabric and Grid Reinforcing

03 22 05 – Uncoated Welded Wire Fabric

03 22 05.50 Welded Wire Fabric

	03 22 05.50 Welded Wire Fabric		Crew	Daily Output	Labor-Hours	Unit	Material	Labor	Equipment	Total	Total Incl O&P
0010	**WELDED WIRE FABRIC** ASTM A185	R032205-30									
0020	Includes labor, but not material cost, to install accessories										
0030	Made from recycled materials	G									
0050	Sheets										
0100	6 x 6 - W1.4 x W1.4 (10 x 10) 21 lb. per C.S.F.	G	2 Rodm	35	.457	C.S.F.	13.80	22.50		36.30	52.50
0200	6 x 6 - W2.1 x W2.1 (8 x 8) 30 lb. per C.S.F.	G		31	.516		21.50	25.50		47	66
0300	6 x 6 - W2.9 x W2.9 (6 x 6) 42 lb. per C.S.F.	G		29	.552		24	27		51	71.50
0400	6 x 6 - W4 x W4 (4 x 4) 58 lb. per C.S.F.	G		27	.593		33	29		62	85
0500	4 x 4 - W1.4 x W1.4 (10 x 10) 31 lb. per C.S.F.	G		31	.516		20.50	25.50		46	65
0600	4 x 4 - W2.1 x W2.1 (8 x 8) 44 lb. per C.S.F.	G		29	.552		26	27		53	73.50
0650	4 x 4 - W2.9 x W2.9 (6 x 6) 61 lb. per C.S.F.	G		27	.593		41.50	29		70.50	94
0700	4 x 4 - W4 x W4 (4 x 4) 85 lb. per C.S.F.	G		25	.640		51.50	31.50		83	110
0750	Rolls										
0800	2 x 2 - #14 galv., 21 lb./C.S.F., beam & column wrap	G	2 Rodm	6.50	2.462	C.S.F.	44	121		165	250
0900	2 x 2 - #12 galv. for gunite reinforcing	G	"	6.50	2.462	"	65	121		186	273
9000	Minimum labor/equipment charge		1 Rodm	4	2	Job		98		98	164

03 24 Fibrous Reinforcing

03 24 05 – Reinforcing Fibers

03 24 05.30 Synthetic Fibers		Crew	Daily Output	Labor-Hours	Unit	Material	2012 Bare Costs Labor	Equipment	Total	Total Incl O&P
0010	**SYNTHETIC FIBERS**									
0100	Synthetic fibers, add to concrete				Lb.	5.20			5.20	5.75
0110	1-1/2 lb. per C.Y.				C.Y.	8.05			8.05	8.85

03 24 05.70 Steel Fibers

Line	Description	Crew	Daily Output	Labor-Hours	Unit	Material	Labor	Equipment	Total	Total Incl O&P
0010	**STEEL FIBERS**									
0150	Steel fibers, add to concrete G				Lb.	.90			.90	.99
0155	25 lb. per C.Y. G				C.Y.	22.50			22.50	25
0160	50 lb. per C.Y. G					45			45	49.50
0170	75 lb. per C.Y. G					69.50			69.50	76
0180	100 lb. per C.Y. G					90			90	99

03 30 Cast-In-Place Concrete

03 30 53 – Miscellaneous Cast-In-Place Concrete

03 30 53.40 Concrete In Place

Line	Description	Crew	Daily Output	Labor-Hours	Unit	Material	Labor	Equipment	Total	Total Incl O&P
0010	**CONCRETE IN PLACE** R033105-10									
0020	Including forms (4 uses), reinforcing steel, concrete, placement, R033105-20									
0050	and finishing unless otherwise indicated R033105-70									
0300	Beams (3500 psi), 5 kip per L.F., 10′ span R033105-85	C-14A	15.62	12.804	C.Y.	305	565	47.50	917.50	1,325
0350	25′ span	″	18.55	10.782		320	480	40	840	1,175
0500	Chimney foundations (5000 psi), industrial, minimum	C-14C	32.22	3.476		144	147	.83	291.83	400
0510	Maximum	″	23.71	4.724		168	200	1.13	369.13	510
0700	Columns, square (4000 psi), 12″ x 12″, minimum reinforcing	C-14A	11.96	16.722		350	740	62	1,152	1,675
0720	Average reinforcing		10.13	19.743		550	875	73.50	1,498.50	2,100
0740	Maximum reinforcing		9.03	22.148		830	980	82.50	1,892.50	2,600
0800	16″ x 16″, minimum reinforcing		16.22	12.330		274	545	46	865	1,250
0820	Average reinforcing		12.57	15.911		470	705	59	1,234	1,725
0840	Maximum reinforcing		10.25	19.512		730	865	72.50	1,667.50	2,300
0900	24″ x 24″, minimum reinforcing		23.66	8.453		231	375	31.50	637.50	905
0920	Average reinforcing		17.71	11.293		415	500	42	957	1,325
0940	Maximum reinforcing		14.15	14.134		665	625	52.50	1,342.50	1,825
1000	36″ x 36″, minimum reinforcing		33.69	5.936		205	263	22	490	680
1020	Average reinforcing		23.32	8.576		365	380	32	777	1,075
1040	Maximum reinforcing		17.82	11.223		620	495	42	1,157	1,550
1200	16″ diameter, minimum reinforcing		31.49	6.351		272	281	23.50	576.50	785
1220	Average reinforcing		19.12	10.460		475	465	39	979	1,325
1240	Maximum reinforcing		13.77	14.524		720	645	54	1,419	1,900
1300	20″ diameter, minimum reinforcing		41.04	4.873		274	216	18.15	508.15	675
1320	Average reinforcing		24.05	8.316		460	370	31	861	1,150
1340	Maximum reinforcing		17.01	11.758		725	520	43.50	1,288.50	1,700
1400	24″ diameter, minimum reinforcing		51.85	3.857		257	171	14.35	442.35	580
1420	Average reinforcing		27.06	7.391		460	325	27.50	812.50	1,075
1440	Maximum reinforcing		18.29	10.935		710	485	40.50	1,235.50	1,625
1500	36″ diameter, minimum reinforcing		75.04	2.665		258	118	9.90	385.90	490
1520	Average reinforcing		37.49	5.335		440	236	19.85	695.85	890
1540	Maximum reinforcing		22.84	8.757		690	390	32.50	1,112.50	1,425
1900	Elevated slab (4000 psi), flat slab with drops, 125 psf Sup. Load, 20′ span	C-14B	38.45	5.410		239	239	19.30	497.30	675
1950	30′ span		50.99	4.079		261	180	14.55	455.55	600
2100	Flat plate, 125 psf Sup. Load, 15′ span		30.24	6.878		221	305	24.50	550.50	770
2150	25′ span		49.60	4.194		234	185	14.95	433.95	580
2300	Waffle const., 30″ domes, 125 psf Sup. Load, 20′ span		37.07	5.611		239	248	20	507	690
2350	30′ span		44.07	4.720		224	209	16.85	449.85	605

03 30 53 – Miscellaneous Cast-In-Place Concrete

03 30 53.40 Concrete In Place		Crew	Daily Output	Labor-Hours	Unit	Material	Labor	Equipment	Total	Total Incl O&P
						2012 Bare Costs				
2500	One way joists, 30" pans, 125 psf Sup. Load, 15' span	C-14B	27.38	7.597	C.Y.	281	335	27	643	890
2550	25' span		31.15	6.677		267	295	24	586	805
2700	One way beam & slab, 125 psf Sup. Load, 15' span		20.59	10.102		237	445	36	718	1,025
2750	25' span		28.36	7.334		229	325	26	580	810
2900	Two way beam & slab, 125 psf Sup. Load, 15' span		24.04	8.652		231	385	31	647	915
2950	25' span		35.87	5.799		202	256	20.50	478.50	665
3100	Elevated slabs, flat plate, including finish, not									
3110	including forms or reinforcing									
3150	Regular concrete (4000 psi), 4" slab	C-8	2613	.021	S.F.	1.34	.84	.27	2.45	3.13
3200	6" slab		2585	.022		1.97	.85	.28	3.10	3.83
3250	2-1/2" thick floor fill		2685	.021		.87	.82	.27	1.96	2.55
3300	Lightweight, 110# per C.F., 2-1/2" thick floor fill		2585	.022		1.16	.85	.28	2.29	2.94
3400	Cellular concrete, 1-5/8" fill, under 5000 S.F.		2000	.028		.78	1.09	.36	2.23	3.01
3450	Over 10,000 S.F.		2200	.025		.75	.99	.33	2.07	2.78
3500	Add per floor for 3 to 6 stories high		31800	.002			.07	.02	.09	.13
3520	For 7 to 20 stories high		21200	.003			.10	.03	.13	.21
3540	Equipment pad (3000 psi), 3' x 3' x 6" thick	C-14H	45	1.067	Ea.	44	46.50	.61	91.11	125
3550	4' x 4' x 6" thick		30	1.600		66	69.50	.91	136.41	188
3560	5' x 5' x 8" thick		18	2.667		118	116	1.52	235.52	320
3570	6' x 6' x 8" thick		14	3.429		159	149	1.95	309.95	420
3580	8' x 8' x 10" thick		8	6		335	261	3.42	599.42	800
3590	10' x 10' x 12" thick		5	9.600		575	415	5.45	995.45	1,325
3800	Footings (3000 psi), spread under 1 C.Y.	C-14C	28	4	C.Y.	166	169	.96	335.96	460
3825	1 C.Y. to 5 C.Y.		43	2.605		192	110	.63	302.63	390
3850	Over 5 C.Y.		75	1.493		177	63	.36	240.36	298
3900	Footings, strip (3000 psi), 18" x 9", unreinforced	C-14L	40	2.400		121	98.50	.70	220.20	295
3920	18" x 9", reinforced	C-14C	35	3.200		145	135	.77	280.77	380
3925	20" x 10", unreinforced	C-14L	45	2.133		118	87.50	.62	206.12	274
3930	20" x 10", reinforced	C-14C	40	2.800		137	118	.67	255.67	345
3935	24" x 12", unreinforced	C-14L	55	1.745		117	72	.51	189.51	247
3940	24" x 12", reinforced	C-14C	48	2.333		136	98.50	.56	235.06	315
3945	36" x 12", unreinforced	C-14L	70	1.371		114	56.50	.40	170.90	217
3950	36" x 12", reinforced	C-14C	60	1.867		131	79	.45	210.45	274
4000	Foundation mat (3000 psi), under 10 C.Y.		38.67	2.896		202	122	.70	324.70	425
4050	Over 20 C.Y.		56.40	1.986		178	84	.48	262.48	335
4200	Wall, free-standing (3000 psi), 8" thick, 8' high	C-14D	45.83	4.364		165	192	16.25	373.25	515
4250	14' high		27.26	7.337		200	320	27.50	547.50	775
4260	12" thick, 8' high		64.32	3.109		151	136	11.55	298.55	400
4270	14' high		40.01	4.999		161	219	18.60	398.60	560
4300	15" thick, 8' high		80.02	2.499		144	110	9.30	263.30	350
4350	12' high		51.26	3.902		144	171	14.50	329.50	455
4500	18' high		48.85	4.094		162	180	15.25	357.25	490
4520	Handicap access ramp (4000 psi), railing both sides, 3' wide	C-14H	14.58	3.292	L.F.	305	143	1.88	449.88	570
4525	5' wide		12.22	3.928		315	171	2.24	488.24	625
4530	With 6" curb and rails both sides, 3' wide		8.55	5.614		315	244	3.20	562.20	750
4535	5' wide		7.31	6.566		320	286	3.74	609.74	820
4650	Slab on grade (3500 psi), not including finish, 4" thick	C-14E	60.75	1.449	C.Y.	118	63	.45	181.45	234
4700	6" thick	"	92	.957	"	114	41.50	.30	155.80	194
4701	Thickened slab edge (3500 psi), for slab on grade poured									
4702	monolithically with slab; depth is in addition to slab thickness;									
4703	formed vertical outside edge, earthen bottom and inside slope									
4705	8" deep x 8" wide bottom, unreinforced	C-14L	2190	.044	L.F.	3.18	1.80	.01	4.99	6.45
4710	8" x 8", reinforced	C-14C	1670	.067		5.50	2.83	.02	8.35	10.70

03 30 Cast-In-Place Concrete

03 30 53 – Miscellaneous Cast-In-Place Concrete

03 30 53.40 Concrete In Place		Crew	Daily Output	Labor-Hours	Unit	2012 Bare Costs Material	Labor	Equipment	Total	Total Incl O&P
4715	12″ deep x 12″ wide bottom, unreinforced	C-14L	1800	.053	L.F.	6.55	2.19	.02	8.76	10.80
4720	12″ x 12″, reinforced	C-14C	1310	.086		10.75	3.61	.02	14.38	17.70
4725	16″ deep x 16″ wide bottom, unreinforced	C-14L	1440	.067		11.05	2.74	.02	13.81	16.70
4730	16″ x 16″, reinforced	C-14C	1120	.100		16.10	4.23	.02	20.35	24.50
4735	20″ deep x 20″ wide bottom, unreinforced	C-14L	1150	.083		16.85	3.43	.02	20.30	24
4740	20″ x 20″, reinforced	C-14C	920	.122		23	5.15	.03	28.18	34
4745	24″ deep x 24″ wide bottom, unreinforced	C-14L	930	.103		24	4.24	.03	28.27	33
4750	24″ x 24″, reinforced	C-14C	740	.151		32	6.40	.04	38.44	46
4751	Slab on grade (3500 psi), incl. troweled finish, not incl. forms									
4760	or reinforcing, over 10,000 S.F., 4″ thick	C-14F	3425	.021	S.F.	1.29	.84	.01	2.14	2.77
4820	6″ thick		3350	.021		1.89	.86	.01	2.76	3.46
4840	8″ thick		3184	.023		2.59	.91	.01	3.51	4.31
4900	12″ thick		2734	.026		3.88	1.06	.01	4.95	5.95
4950	15″ thick		2505	.029		4.88	1.15	.01	6.04	7.20
5000	Slab on grade (3000 psi), incl. textured finish, not incl. forms									
5001	or reinforcing, 4″ thick	C-14G	2873	.019	S.F.	1.32	.77	.01	2.10	2.70
5010	6″ thick		2590	.022		2.07	.85	.01	2.93	3.66
5020	8″ thick		2320	.024		2.70	.95	.01	3.66	4.50
5200	Lift slab in place above the foundation, incl. forms,									
5210	reinforcing, concrete (4000 psi) and columns, minimum	C-14B	2113	.098	S.F.	6.55	4.35	.35	11.25	14.70
5250	Average		1650	.126		7.15	5.55	.45	13.15	17.50
5300	Maximum		1500	.139		7.80	6.15	.50	14.45	19.15
5500	Lightweight, ready mix, including screed finish only,									
5510	not including forms or reinforcing									
5550	1:4 (2500 psi) for structural roof decks	C-14B	260	.800	C.Y.	136	35.50	2.86	174.36	210
5600	1:6 (3000 psi) for ground slab with radiant heat	C-14F	92	.783		133	31.50	.30	164.80	197
5650	1:3:2 (2000 psi) with sand aggregate, roof deck	C-14B	260	.800		133	35.50	2.86	171.36	207
5700	Ground slab (2000 psi)	C-14F	107	.673		133	27	.26	160.26	189
5900	Pile caps (3000 psi), incl. forms and reinf., sq. or rect., under 10 C.Y.	C-14C	54.14	2.069		159	87.50	.50	247	320
5950	Over 10 C.Y.		75	1.493		152	63	.36	215.36	270
6000	Triangular or hexagonal, under 10 C.Y.		53	2.113		122	89.50	.51	212.01	281
6050	Over 10 C.Y.		85	1.318		138	55.50	.32	193.82	243
6200	Retaining walls (3000 psi), gravity, 4′ high see Section 32 32	C-14D	66.20	3.021		140	133	11.25	284.25	385
6250	10′ high		125	1.600		134	70	5.95	209.95	269
6300	Cantilever, level backfill loading, 8′ high		70	2.857		154	125	10.65	289.65	385
6350	16′ high		91	2.198		147	96.50	8.20	251.70	330
6800	Stairs (3500 psi), not including safety treads, free standing, 3′-6″ wide	C-14H	83	.578	LF Nose	5.10	25	.33	30.43	47
6850	Cast on ground		125	.384	″	4.35	16.70	.22	21.27	32.50
7000	Stair landings, free standing		200	.240	S.F.	4.12	10.45	.14	14.71	21.50
7050	Cast on ground		475	.101	″	3.37	4.39	.06	7.82	10.95
9000	Minimum labor/equipment charge	2 Carp	1	16	Job		705		705	1,150

03 31 Structural Concrete

03 31 05 – Normal Weight Structural Concrete

03 31 05.25 Concrete, Hand Mix

	03 31 05.25 Concrete, Hand Mix	Crew	Daily Output	Labor-Hours	Unit	2012 Bare Costs Material	Labor	Equipment	Total	Total Incl O&P
0010	**CONCRETE, HAND MIX** for small quantities or remote areas									
0050	Includes bulk local aggregate, bulk sand, bagged Portland cement,									
0060	and water, using gas powered cement mixer									
0125	2500 psi	C-30	135	.059	C.F.	3.60	2.08	1.19	6.87	8.70
0130	3000 psi		135	.059		3.88	2.08	1.19	7.15	9
0135	3500 psi		135	.059		4.03	2.08	1.19	7.30	9.15
0140	4000 psi		135	.059		4.21	2.08	1.19	7.48	9.35
0145	4500 psi		135	.059		4.41	2.08	1.19	7.68	9.55
0150	5000 psi	↓	135	.059	↓	4.74	2.08	1.19	8.01	9.90
0300	Using pre-bagged dry mix and wheelbarrow (80-lb. bag = 0.6 C.F.)									
0340	4000 psi	1 Clab	48	.167	C.F.	5.80	5.85		11.65	15.95

03 31 05.35 Normal Weight Concrete, Ready Mix

	03 31 05.35 Normal Weight Concrete, Ready Mix	Crew	Daily Output	Labor-Hours	Unit	Material	Labor	Equipment	Total	Total Incl O&P
0010	**NORMAL WEIGHT CONCRETE, READY MIX**, delivered									
0012	Includes local aggregate, sand, Portland cement, and water									
0015	Excludes all additives and treatments R033105-20									
0020	2000 psi				C.Y.	91.50			91.50	101
0100	2500 psi					94			94	103
0150	3000 psi					102			102	112
0200	3500 psi					99.50			99.50	110
0300	4000 psi					103			103	113
0350	4500 psi					106			106	116
0400	5000 psi					109			109	120
0411	6000 psi					124			124	136
0412	8000 psi					202			202	223
0413	10,000 psi					287			287	315
0414	12,000 psi					345			345	380
1000	For high early strength cement, add					10%				
1300	For winter concrete (hot water), add					4.25			4.25	4.68
1400	For hot weather concrete (ice), add					9.35			9.35	10.25
1410	For mid-range water reducer, add					4.13			4.13	4.54
1420	For high-range water reducer/superplasticizer, add					6.35			6.35	6.95
1430	For retarder, add					2.71			2.71	2.98
1440	For non-Chloride accelerator, add					4.83			4.83	5.30
1450	For Chloride accelerator, per 1%, add					3.28			3.28	3.61
1460	For fiber reinforcing, synthetic (1 lb./C.Y.), add					6.65			6.65	7.30
1500	For Saturday delivery, add				↓	8.85			8.85	9.70
1510	For truck holding/waiting time past 1st hour per load, add				Hr.	87.50			87.50	96.50
1520	For short load (less than 4 C.Y.), add per load				Ea.	112			112	124
2000	For all lightweight aggregate, add				C.Y.	45%				

03 31 05.70 Placing Concrete

	03 31 05.70 Placing Concrete	Crew	Daily Output	Labor-Hours	Unit	Material	Labor	Equipment	Total	Total Incl O&P
0010	**PLACING CONCRETE** R033105-70									
0020	Includes labor and equipment to place, strike off and consolidate									
0050	Beams, elevated, small beams, pumped	C-20	60	1.067	C.Y.		40	12.85	52.85	79
0100	With crane and bucket	C-7	45	1.600			61	26	87	128
0200	Large beams, pumped	C-20	90	.711			27	8.55	35.55	53
0250	With crane and bucket	C-7	65	1.108			42	18.10	60.10	88
0400	Columns, square or round, 12" thick, pumped	C-20	60	1.067			40	12.85	52.85	79
0450	With crane and bucket	C-7	40	1.800			68.50	29.50	98	144
0600	18" thick, pumped	C-20	90	.711			27	8.55	35.55	53
0650	With crane and bucket	C-7	55	1.309			50	21.50	71.50	104
0800	24" thick, pumped	C-20	92	.696			26	8.40	34.40	51.50
0850	With crane and bucket	C-7	70	1.029	↓		39	16.80	55.80	82

03 31 Structural Concrete

03 31 05 – Normal Weight Structural Concrete

03 31 05.70 Placing Concrete		Crew	Daily Output	Labor-Hours	Unit	Material	2012 Bare Costs Labor	Equipment	Total	Total Incl O&P
1000	36" thick, pumped	C-20	140	.457	C.Y.		17.25	5.50	22.75	34
1050	With crane and bucket	C-7	100	.720			27.50	11.75	39.25	57.50
1400	Elevated slabs, less than 6" thick, pumped	C-20	140	.457			17.25	5.50	22.75	34
1450	With crane and bucket	C-7	95	.758			29	12.40	41.40	60
1500	6" to 10" thick, pumped	C-20	160	.400			15.10	4.82	19.92	30
1550	With crane and bucket	C-7	110	.655			25	10.70	35.70	52.50
1600	Slabs over 10" thick, pumped	C-20	180	.356			13.40	4.28	17.68	26
1650	With crane and bucket	C-7	130	.554			21	9.05	30.05	44
1900	Footings, continuous, shallow, direct chute	C-6	120	.400			14.65	.46	15.11	24.50
1950	Pumped	C-20	150	.427			16.10	5.15	21.25	31.50
2000	With crane and bucket	C-7	90	.800			30.50	13.10	43.60	63.50
2100	Footings, continuous, deep, direct chute	C-6	140	.343			12.55	.39	12.94	21
2150	Pumped	C-20	160	.400			15.10	4.82	19.92	30
2200	With crane and bucket	C-7	110	.655			25	10.70	35.70	52.50
2400	Footings, spread, under 1 C.Y., direct chute	C-6	55	.873			32	.99	32.99	53
2450	Pumped	C-20	65	.985			37	11.85	48.85	73
2500	With crane and bucket	C-7	45	1.600			61	26	87	128
2600	Over 5 C.Y., direct chute	C-6	120	.400			14.65	.46	15.11	24.50
2650	Pumped	C-20	150	.427			16.10	5.15	21.25	31.50
2700	With crane and bucket	C-7	100	.720			27.50	11.75	39.25	57.50
2900	Foundation mats, over 20 C.Y., direct chute	C-6	350	.137			5	.16	5.16	8.30
2950	Pumped	C-20	400	.160			6.05	1.93	7.98	11.85
3000	With crane and bucket	C-7	300	.240			9.15	3.92	13.07	19.05
3200	Grade beams, direct chute	C-6	150	.320			11.70	.36	12.06	19.45
3250	Pumped	C-20	180	.356			13.40	4.28	17.68	26
3300	With crane and bucket	C-7	120	.600			23	9.80	32.80	48
3500	High rise, for more than 5 stories, pumped, add per story	C-20	2100	.030			1.15	.37	1.52	2.26
3510	With crane and bucket, add per story	C-7	2100	.034			1.31	.56	1.87	2.73
3700	Pile caps, under 5 C.Y., direct chute	C-6	90	.533			19.55	.61	20.16	32.50
3750	Pumped	C-20	110	.582			22	7	29	43
3800	With crane and bucket	C-7	80	.900			34.50	14.70	49.20	71.50
3850	Pile cap, 5 C.Y. to 10 C.Y., direct chute	C-6	175	.274			10.05	.31	10.36	16.70
3900	Pumped	C-20	200	.320			12.05	3.85	15.90	24
3950	With crane and bucket	C-7	150	.480			18.30	7.85	26.15	38
4000	Over 10 C.Y., direct chute	C-6	215	.223			8.20	.25	8.45	13.60
4050	Pumped	C-20	240	.267			10.05	3.21	13.26	19.85
4100	With crane and bucket	C-7	185	.389			14.80	6.35	21.15	31
4300	Slab on grade, up to 6" thick, direct chute	C-6	110	.436			16	.50	16.50	26.50
4350	Pumped	C-20	130	.492			18.55	5.95	24.50	36.50
4400	With crane and bucket	C-7	110	.655			25	10.70	35.70	52.50
4600	Over 6" thick, direct chute	C-6	165	.291			10.65	.33	10.98	17.70
4650	Pumped	C-20	185	.346			13.05	4.17	17.22	25.50
4700	With crane and bucket	C-7	145	.497			18.90	8.10	27	39.50
4900	Walls, 8" thick, direct chute	C-6	90	.533			19.55	.61	20.16	32.50
4950	Pumped	C-20	100	.640			24	7.70	31.70	47.50
5000	With crane and bucket	C-7	80	.900			34.50	14.70	49.20	71.50
5050	12" thick, direct chute	C-6	100	.480			17.60	.55	18.15	29
5100	Pumped	C-20	110	.582			22	7	29	43
5200	With crane and bucket	C-7	90	.800			30.50	13.10	43.60	63.50
5300	15" thick, direct chute	C-6	105	.457			16.75	.52	17.27	27.50
5350	Pumped	C-20	120	.533			20	6.40	26.40	39.50
5400	With crane and bucket	C-7	95	.758			29	12.40	41.40	60
5600	Wheeled concrete dumping, add to placing costs above									

03 31 Structural Concrete

03 31 05 – Normal Weight Structural Concrete

03 31 05.70 Placing Concrete		Crew	Daily Output	Labor-Hours	Unit	Material	2012 Bare Costs Labor	Equipment	Total	Total Incl O&P
5610	Walking cart, 50′ haul, add	C-18	32	.281	C.Y.		9.95	1.77	11.72	18.25
5620	150′ haul, add		24	.375			13.25	2.37	15.62	24
5700	250′ haul, add		18	.500			17.65	3.16	20.81	32.50
5800	Riding cart, 50′ haul, add	C-19	80	.113			3.97	1.18	5.15	7.80
5810	150′ haul, add		60	.150			5.30	1.57	6.87	10.45
5900	250′ haul, add		45	.200			7.05	2.10	9.15	13.90
9000	Minimum labor/equipment charge	C-6	2	24	Job		880	27.50	907.50	1,450

03 35 Concrete Finishing

03 35 29 – Tooled Concrete Finishing

03 35 29.30 Finishing Floors

		Crew	Daily Output	Labor-Hours	Unit	Material	Labor	Equipment	Total	Total Incl O&P
0010	**FINISHING FLOORS**									
0012	Finishing requires that concrete first be placed, struck off & consolidated									
0015	Basic finishing for various unspecified flatwork									
0100	Bull float only	C-10	4000	.006	S.F.		.24		.24	.38
0125	Bull float & manual float		2000	.012			.48		.48	.76
0150	Bull float, manual float, & broom finish, w/edging & joints		1850	.013			.52		.52	.82
0200	Bull float, manual float & manual steel trowel		1265	.019			.76		.76	1.21
0210	For specified Random Access Floors in ACI Classes 1, 2, 3 and 4 to achieve									
0215	Composite Overall Floor Flatness and Levelness values up to F35/F25									
0250	Bull float, machine float & machine trowel (walk-behind)	C-10C	1715	.014	S.F.		.56	.03	.59	.92
0300	Power screed, bull float, machine float & trowel (walk-behind)	C-10D	2400	.010			.40	.05	.45	.69
0350	Power screed, bull float, machine float & trowel (ride-on)	C-10E	4000	.006			.24	.06	.30	.45
0352	For specified Random Access Floors in ACI Classes 5, 6, 7 and 8 to achieve									
0354	Composite Overall Floor Flatness and Levelness values up to F50/F50									
0356	Add for two-dimensional restraightening after power float	C-10	6000	.004	S.F.		.16		.16	.25
0358	For specified Random or Defined Access Floors in ACI Class 9 to achieve									
0360	Composite Overall Floor Flatness and Levelness values up to F100/F100									
0362	Add for two-dimensional restraightening after bull float & power float	C-10	3000	.008	S.F.		.32		.32	.51
0364	For specified Superflat Defined Access Floors in ACI Class 9 to achieve									
0366	Minimum Floor Flatness and Levelness values of F100/F100									
0368	Add for 2-dim'l restraightening after bull float, power float, power trowel	C-10	2000	.012	S.F.		.48		.48	.76
0400	Integral topping and finish, using 1:1:2 mix, 3/16″ thick	C-10B	1000	.040		.11	1.52	.25	1.88	2.84
0450	1/2″ thick		950	.042		.29	1.60	.26	2.15	3.19
0500	3/4″ thick		850	.047		.44	1.79	.29	2.52	3.68
0600	1″ thick		750	.053		.59	2.03	.33	2.95	4.27
0800	Granolithic topping, laid after, 1:1:1-1/2 mix, 1/2″ thick		590	.068		.33	2.58	.42	3.33	4.98
0820	3/4″ thick		580	.069		.49	2.62	.43	3.54	5.25
0850	1″ thick		575	.070		.65	2.64	.43	3.72	5.45
0950	2″ thick		500	.080		1.30	3.04	.50	4.84	6.85
9100	Minimum labor/equipment charge	C-10	2	12	Job		480		480	765

03 35 29.35 Control Joints, Saw Cut

		Crew	Daily Output	Labor-Hours	Unit	Material	Labor	Equipment	Total	Total Incl O&P
0010	**CONTROL JOINTS, SAW CUT**									
0100	Sawcut control joints in green concrete									
0120	1″ depth	C-27	2000	.008	L.F.	.03	.34	.07	.44	.65
0140	1-1/2″ depth		1800	.009		.05	.38	.08	.51	.74
0160	2″ depth		1600	.010		.07	.42	.09	.58	.84
0180	Sawcut joint reservoir in cured concrete									
0182	3/8″ wide x 3/4″ deep, with single saw blade	C-27	1000	.016	L.F.	.05	.68	.15	.88	1.28
0184	1/2″ wide x 1″ deep, with double saw blades		900	.018		.10	.75	.16	1.01	1.47
0186	3/4″ wide x 1-1/2″ deep, with double saw blades		800	.020		.20	.85	.19	1.24	1.75

03 35 Concrete Finishing

03 35 29 – Tooled Concrete Finishing

03 35 29.35 Control Joints, Saw Cut		Crew	Daily Output	Labor-Hours	Unit	2012 Bare Costs Material	Labor	Equipment	Total	Total Incl O&P
0190	Water blast joint to wash away laitance, 2 passes	C-29	2500	.003	L.F.		.11	.03	.14	.21
0200	Air blast joint to blow out debris and air dry, 2 passes	C-28	2000	.004			.17	.01	.18	.28
0300	For backer rod, see Section 07 91 23.10									
0340	For joint sealant, see Sections 03 15 05.25 or 07 92 13.20									
0900	For replacement of joint sealant, see Section 07 01 90.81									

03 35 29.60 Finishing Walls		Crew	Daily Output	Labor-Hours	Unit	Material	Labor	Equipment	Total	Total Incl O&P
0010	**FINISHING WALLS**									
0020	Break ties and patch voids	1 Cefi	540	.015	S.F.	.03	.63		.66	1.03
0050	Burlap rub with grout	"	450	.018		.03	.75		.78	1.22
0300	Bush hammer, green concrete	B-39	1000	.048			1.78	.21	1.99	3.12
0350	Cured concrete	"	650	.074			2.74	.32	3.06	4.80
0700	Sandblast, light penetration	E-11	1100	.029		.74	1.15	.19	2.08	3.05
0750	Heavy penetration	"	375	.085		1.49	3.38	.56	5.43	8.20
0850	Grind form fins flush	1 Clab	700	.011	L.F.		.40		.40	.66
9000	Minimum labor/equipment charge	C-10	2	12	Job		480		480	765

03 35 33 – Stamped Concrete Finishing

03 35 33.50 Slab Texture Stamping		Crew	Daily Output	Labor-Hours	Unit	Material	Labor	Equipment	Total	Total Incl O&P
0010	**SLAB TEXTURE STAMPING**									
0050	Stamping requires that concrete first be placed, struck off, consolidated,									
0060	bull floated and free of bleed water. Decorative stamping tasks include:									
0100	Step 1 - first application of dry shake colored hardener	1 Cefi	6400	.001	S.F.	.39	.05		.44	.51
0110	Step 2 - bull float		6400	.001			.05		.05	.08
0130	Step 3 - second application of dry shake colored hardener		6400	.001		.19	.05		.24	.29
0140	Step 4 - bull float, manual float & steel trowel	3 Cefi	1280	.019			.79		.79	1.25
0150	Step 5 - application of dry shake colored release agent	1 Cefi	6400	.001		.09	.05		.14	.18
0160	Step 6 - place, tamp & remove mats	3 Cefi	2400	.010		1.41	.42		1.83	2.22
0170	Step 7 - touch up edges, mat joints & simulated grout lines	1 Cefi	1280	.006			.26		.26	.42
0300	Alternate stamping estimating method includes all tasks above	4 Cefi	800	.040		2.09	1.69		3.78	4.95
0400	Step 8 - pressure wash @ 3000 psi after 24 hours	1 Cefi	1600	.005			.21		.21	.33
0500	Step 9 - roll 2 coats cure/seal compound when dry	"	800	.010		.53	.42		.95	1.25

03 35 43 – Polished Concrete Finishing

03 35 43.10 Polished Concrete Floors		Crew	Daily Output	Labor-Hours	Unit	Material	Labor	Equipment	Total	Total Incl O&P
0010	**POLISHED CONCRETE FLOORS**									
0015	Processing of cured concrete to include grinding, honing,									
0020	and polishing of interior floors with 22" segmented diamond									
0025	planetary floor grinder (2 passes in different directions per grit)									
0100	Removal of pre-existing coatings, dry, with carbide discs using									
0105	dry vacuum pick-up system, final hand sweeping									
0110	Glue, adhesive or tar	J-4	1.60	15	M.S.F.	19.20	600	114	733.20	1,100
0120	Paint, epoxy, 1 coat		3.60	6.667		19.20	266	50.50	335.70	500
0130	2 coats		1.80	13.333		19.20	530	101	650.20	980
0200	Grinding and edging, wet, including wet vac pick-up and auto									
0205	scrubbing between grit changes									
0210	40-grit diamond/metal matrix	J-4A	1.60	20	M.S.F.	31	775	226	1,032	1,525
0220	80-grit diamond/metal matrix		2	16		31	620	180	831	1,225
0230	120-grit diamond/metal matrix		2.40	13.333		31	515	150	696	1,025
0240	200-grit diamond/metal matrix		2.80	11.429		31	445	129	605	885
0300	Spray on dye or stain (1 coat)	1 Cefi	16	.500		271	21		292	330
0400	Spray on densifier/hardener (2 coats)	"	8	1		296	42.50		338.50	390
0410	Auto scrubbing after 2nd coat, when dry	J-4B	16	.500			17.55	11.15	28.70	41.50
0500	Honing and edging, wet, including wet vac pick-up and auto									

03 35 Concrete Finishing

03 35 43 – Polished Concrete Finishing

03 35 43.10 Polished Concrete Floors		Crew	Daily Output	Labor-Hours	Unit	Material	2012 Bare Costs Labor	Equipment	Total	Total Incl O&P
0505	scrubbing between grit changes									
0510	100-grit diamond/resin matrix	J-4A	2.80	11.429	M.S.F.	31	445	129	605	885
0520	200-grit diamond/resin matrix	"	2.80	11.429	"	31	445	129	605	885
0530	Dry, including dry vacuum pick-up system, final hand sweeping									
0540	400-grit diamond/resin matrix	J-4A	2.80	11.429	M.S.F.	31	445	129	605	885
0600	Polishing and edging, dry, including dry vac pick-up and hand									
0605	sweeping between grit changes									
0610	800-grit diamond/resin matrix	J-4A	2.80	11.429	M.S.F.	31	445	129	605	885
0620	1500-grit diamond/resin matrix		2.80	11.429		31	445	129	605	885
0630	3000-grit diamond/resin matrix	↓	2.80	11.429		31	445	129	605	885
0700	Auto scrubbing after final polishing step	J-4B	16	.500	↓		17.55	11.15	28.70	41.50

03 37 Specialty Placed Concrete

03 37 13 – Shotcrete

03 37 13.30 Gunite (Dry-Mix)		Crew	Daily Output	Labor-Hours	Unit	Material	Labor	Equipment	Total	Total Incl O&P
0010	**GUNITE (DRY-MIX)**									
0020	Typical in place, 1" layers, no mesh included	C-16	2000	.028	S.F.	.33	1.09	.15	1.57	2.29
0300	Typical in place, including mesh, 2" thick, minimum		1000	.056		1.31	2.19	.30	3.80	5.30
0350	Maximum		500	.112		1.31	4.38	.60	6.29	9.15
0500	4" thick, minimum		750	.075		1.96	2.92	.40	5.28	7.30
0550	Maximum	↓	350	.160		1.96	6.25	.86	9.07	13.15
0900	Prepare old walls, no scaffolding, minimum	C-10	1000	.024			.96		.96	1.53
0950	Maximum	"	275	.087			3.48		3.48	5.55
1100	For high finish requirement or close tolerance, add, minimum						50%			
1150	Maximum				↓		110%			
9000	Minimum labor/equipment charge	C-10	1	24	Job		960		960	1,525

03 37 13.60 Shotcrete (Wet-Mix)		Crew	Daily Output	Labor-Hours	Unit	Material	Labor	Equipment	Total	Total Incl O&P
0010	**SHOTCRETE (WET-MIX)**									
0020	Wet mix, placed @ up to 12 C.Y. per hour, 3000 psi	C-8C	80	.600	C.Y.	106	23	5.35	134.35	160
0100	Up to 35 C.Y. per hour	C-8E	240	.200	"	95.50	7.65	2.06	105.21	120
1010	Fiber reinforced, 1" thick	C-8C	1740	.028	S.F.	.79	1.06	.25	2.10	2.86
1020	2" thick		900	.053		1.58	2.06	.47	4.11	5.60
1030	3" thick		825	.058		2.38	2.24	.52	5.14	6.80
1040	4" thick	↓	750	.064	↓	3.17	2.47	.57	6.21	8.10

03 39 Concrete Curing

03 39 13 – Water Concrete Curing

03 39 13.50 Water Curing		Crew	Daily Output	Labor-Hours	Unit	Material	Labor	Equipment	Total	Total Incl O&P
0010	**WATER CURING**									
0015	With burlap, 4 uses assumed, 7.5 oz.	2 Clab	55	.291	C.S.F.	11.80	10.20		22	29.50
0100	10 oz.	"	55	.291	"	21	10.20		31.20	40.50
9000	Minimum labor/equipment charge	1 Clab	5	1.600	Job		56		56	92

03 39 23 – Membrane Concrete Curing

03 39 23.13 Chemical Compound Membrane Concrete Curing		Crew	Daily Output	Labor-Hours	Unit	Material	Labor	Equipment	Total	Total Incl O&P
0010	**CHEMICAL COMPOUND MEMBRANE CONCRETE CURING**									
0300	Sprayed membrane curing compound	2 Clab	95	.168	C.S.F.	7.45	5.90		13.35	17.90
0700	Curing compound, solvent based, 400 S.F./gal., 55 gallon lots				Gal.	20			20	22
0720	5 gallon lots				↓	27			27	29.50

03 39 Concrete Curing

03 39 23 – Membrane Concrete Curing

03 39 23.13 Chemical Compound Membrane Concrete Curing		Crew	Daily Output	Labor-Hours	Unit	Material	2012 Bare Costs Labor	Equipment	Total	Total Incl O&P
0800	Curing compound, water based, 250 S.F./gal., 55 gallon lots				Gal.	19.80			19.80	22
0820	5 gallon lots					23			23	25.50

03 39 23.23 Sheet Membrane Concrete Curing

		Crew	Daily Output	Labor-Hours	Unit	Material	Labor	Equipment	Total	Total Incl O&P
0010	**SHEET MEMBRANE CONCRETE CURING**									
0200	Curing blanket, burlap/poly, 2-ply	2 Clab	70	.229	C.S.F.	17.10	8		25.10	32

03 41 Precast Structural Concrete

03 41 05 – Precast Concrete Members

03 41 05.10 Precast Beams

		Crew	Daily Output	Labor-Hours	Unit	Material	Labor	Equipment	Total	Total Incl O&P
0010	**PRECAST BEAMS** R034105-30									
0011	L-shaped, 20′ span, 12″ x 20″	C-11	32	2.250	Ea.	1,900	109	57	2,066	2,350
1000	Inverted tee beams, add to above, small beams					15%				
1050	Large beams					20%				
1200	Rectangular, 20′ span, 12″ x 20″	C-11	32	2.250		1,425	109	57	1,591	1,825
1250	18″ x 36″		24	3		2,625	146	76	2,847	3,225
1300	24″ x 44″		22	3.273		3,625	159	83	3,867	4,375
1400	30′ span, 12″ x 36″		24	3		3,300	146	76	3,522	3,975
1450	18″ x 44″		20	3.600		4,725	175	91.50	4,991.50	5,625
1500	24″ x 52″		16	4.500		6,300	218	114	6,632	7,475
1600	40′ span, 12″ x 52″		20	3.600		6,100	175	91.50	6,366.50	7,125
1650	18″ x 52″		16	4.500		7,275	218	114	7,607	8,525
1700	24″ x 52″		12	6		8,425	291	152	8,868	9,950
2000	"T" shaped, 20′ span, 12″ x 20″		32	2.250		2,475	109	57	2,641	2,975
2050	18″ x 36″		24	3		4,475	146	76	4,697	5,275
2100	24″ x 44″		22	3.273		6,200	159	83	6,442	7,200
2200	30′ span, 12″ x 36″		24	3		5,575	146	76	5,797	6,475
2250	18″ x 44″		20	3.600		8,000	175	91.50	8,266.50	9,225
2300	24″ x 52″		16	4.500		10,700	218	114	11,032	12,300
2500	40′ span, 12″ x 52″		20	3.600		10,300	175	91.50	10,566.50	11,700
2550	18″ x 52″		16	4.500		12,400	218	114	12,732	14,100
2600	24″ x 52″		12	6		14,300	291	152	14,743	16,400

03 41 05.15 Precast Columns

		Crew	Daily Output	Labor-Hours	Unit	Material	Labor	Equipment	Total	Total Incl O&P
0010	**PRECAST COLUMNS** R034105-30									
0020	Rectangular to 12′ high, small columns	C-11	120	.600	L.F.	100	29	15.25	144.25	179
0050	Large columns	"	96	.750	"	173	36.50	19.05	228.55	278

03 41 13 – Precast Concrete Hollow Core Planks

03 41 13.50 Precast Slab Planks

		Crew	Daily Output	Labor-Hours	Unit	Material	Labor	Equipment	Total	Total Incl O&P
0010	**PRECAST SLAB PLANKS** R034105-30									
0020	Prestressed roof/floor members, grouted, solid, 4″ thick	C-11	2400	.030	S.F.	6.05	1.46	.76	8.27	10.10
0050	6″ thick		2800	.026		6.50	1.25	.65	8.40	10.10
0100	Hollow, 8″ thick		3200	.023		7.15	1.09	.57	8.81	10.45
0150	10″ thick		3600	.020		7.40	.97	.51	8.88	10.45
0200	12″ thick		4000	.018		7.90	.87	.46	9.23	10.75

03 41 23 – Precast Concrete Stairs

03 41 23.50 Precast Stairs

		Crew	Daily Output	Labor-Hours	Unit	Material	Labor	Equipment	Total	Total Incl O&P
0010	**PRECAST STAIRS**									
0020	Precast concrete treads on steel stringers, 3′ wide	C-12	75	.640	Riser	129	28	8.15	165.15	197
0300	Front entrance, 5′ wide with 48″ platform, 2 risers		16	3	Flight	485	131	38.50	654.50	785
0350	5 risers		12	4		765	174	51	990	1,175

03 41 Precast Structural Concrete

03 41 23 – Precast Concrete Stairs

03 41 23.50 Precast Stairs		Crew	Daily Output	Labor-Hours	Unit	Material	Labor	Equipment	Total	Total Incl O&P
						2012 Bare Costs				
0500	6′ wide, 2 risers	C-12	15	3.200	Flight	535	139	41	715	860
0550	5 risers		11	4.364		845	190	55.50	1,090.50	1,300
1200	Basement entrance stairs, steel bulkhead doors, minimum	B-51	22	2.182		1,400	77	9.60	1,486.60	1,675
1250	Maximum	″	11	4.364		2,350	154	19.20	2,523.20	2,850

03 45 Precast Architectural Concrete

03 45 13 – Faced Architectural Precast Concrete

03 45 13.50 Precast Wall Panels

		Crew	Daily Output	Labor-Hours	Unit	Material	Labor	Equipment	Total	Total Incl O&P
0010	**PRECAST WALL PANELS**									
0750	20′ x 10′, 6″ thick, smooth gray	C-11	1400	.051	S.F.	35.50	2.49	1.31	39.30	45
2200	Fiberglass reinforced cement with urethane core									
2210	R20, 8′ x 8′, minimum	E-2	750	.075	S.F.	22	3.60	2	27.60	32.50
2220	Maximum	″	600	.093	″	33	4.50	2.49	39.99	47

03 48 Precast Concrete Specialties

03 48 43 – Precast Concrete Trim

03 48 43.40 Precast Lintels

		Crew	Daily Output	Labor-Hours	Unit	Material	Labor	Equipment	Total	Total Incl O&P
0010	**PRECAST LINTELS**, smooth gray, prestressed, stock units only									
0800	4″ wide, 8″ high, x 4′ long	D-10	28	1.143	Ea.	22.50	49.50	16	88	122
0850	8′ long		24	1.333		57.50	58	18.70	134.20	177
1000	6″ wide, 8″ high, x 4′ long		26	1.231		33	53.50	17.25	103.75	140
1050	10′ long		22	1.455		88	63	20.50	171.50	221
1200	8″ wide, 8″ high, x 4′ long		24	1.333		41	58	18.70	117.70	159
1250	12′ long		20	1.600		134	69.50	22.50	226	284
1275	For custom sizes, types, colors, or finishes of precast lintels, add					150%				

03 48 43.90 Precast Window Sills

		Crew	Daily Output	Labor-Hours	Unit	Material	Labor	Equipment	Total	Total Incl O&P
0010	**PRECAST WINDOW SILLS**									
0600	Precast concrete, 4″ tapers to 3″, 9″ wide	D-1	70	.229	L.F.	12.25	9.10		21.35	28
0650	11″ wide		60	.267		16	10.65		26.65	34.50
0700	13″ wide, 3 1/2″ tapers to 2 1/2″, 12″ wall		50	.320		16	12.75		28.75	38

03 51 Cast Roof Decks

03 51 13 – Cementitious Wood Fiber Decks

03 51 13.50 Cementitious/Wood Fiber Planks

		Crew	Daily Output	Labor-Hours	Unit	Material	Labor	Equipment	Total	Total Incl O&P
0010	**CEMENTITIOUS/WOOD FIBER PLANKS**									
0050	Plank, beveled edge, 1″ thick	2 Carp	1000	.016	S.F.	2.39	.71		3.10	3.79
0100	1-1/2″ thick		975	.016		3.15	.72		3.87	4.66
0150	T & G, 2″ thick		950	.017		3.60	.74		4.34	5.20
0200	2-1/2″ thick		925	.017		4.22	.76		4.98	5.90
1000	Bulb tee, sub-purlin and grout, 6′ span, add	E-1	5000	.005		2.22	.23	.02	2.47	2.88
1100	8′ span	″	4200	.006		2.50	.28	.03	2.81	3.27

03 52 Lightweight Concrete Roof Insulation

03 52 16 – Lightweight Insulating Concrete

03 52 16.13 Lightweight Cellular Insulating Concrete		Crew	Daily Output	Labor-Hours	Unit	2012 Bare Costs Material	Labor	Equipment	Total	Total Incl O&P
0010	LIGHTWEIGHT CELLULAR INSULATING CONCRETE R035216-10									
0020	Portland cement and foaming agent G	C-8	50	1.120	C.Y.	144	44	14.30	202.30	244

03 52 16.16 Lightweight Aggregate Insulating Concrete		Crew	Daily Output	Labor-Hours	Unit	Material	Labor	Equipment	Total	Total Incl O&P
0010	LIGHTWEIGHT AGGREGATE INSULATING CONCRETE R035216-10									
0100	Poured vermiculite or perlite, field mix,									
0110	1:6 field mix G	C-8	50	1.120	C.Y.	246	44	14.30	304.30	355
0200	Ready mix, 1:6 mix, roof fill, 2" thick G		10000	.006	S.F.	1.36	.22	.07	1.65	1.93
0250	3" thick G	↓	7700	.007	"	2.05	.28	.09	2.42	2.81

03 54 Cast Underlayment

03 54 13 – Gypsum Cement Underlayment

03 54 13.50 Poured Gypsum Underlayment		Crew	Daily Output	Labor-Hours	Unit	Material	Labor	Equipment	Total	Total Incl O&P
0010	POURED GYPSUM UNDERLAYMENT									
0400	Underlayment, gypsum based, self-leveling 2500 psi, pumped, 1/2"	C-8	24000	.002	S.F.	.33	.09	.03	.45	.55
0500	3/4"		20000	.003		.50	.11	.04	.65	.77
0600	1"	↓	16000	.004		.67	.14	.04	.85	1.01
1400	Hand placed, 1/2"	C-18	450	.020		.33	.71	.13	1.17	1.67
1500	3/4"	"	300	.030	↓	.50	1.06	.19	1.75	2.50

03 54 16 – Hydraulic Cement Underlayment

03 54 16.50 Cement Underlayment		Crew	Daily Output	Labor-Hours	Unit	Material	Labor	Equipment	Total	Total Incl O&P
0010	CEMENT UNDERLAYMENT									
2510	Underlayment, P.C based self-leveling, 4100 psi, pumped, 1/4"	C-8	20000	.003	S.F.	1.43	.11	.04	1.58	1.79
2520	1/2"		19000	.003		2.85	.12	.04	3.01	3.37
2530	3/4"		18000	.003		4.28	.12	.04	4.44	4.94
2540	1"		17000	.003		5.70	.13	.04	5.87	6.50
2550	1-1/2"	↓	15000	.004		8.55	.15	.05	8.75	9.70
2560	Hand placed, 1/2"	C-18	450	.020		2.85	.71	.13	3.69	4.44
2610	Topping, P.C. based self-level/dry 6100 psi, pumped, 1/4"	C-8	20000	.003		2.16	.11	.04	2.31	2.59
2620	1/2"		19000	.003		4.32	.12	.04	4.48	4.98
2630	3/4"		18000	.003		6.45	.12	.04	6.61	7.35
2660	1"		17000	.003		8.65	.13	.04	8.82	9.75
2670	1-1/2"	↓	15000	.004		12.95	.15	.05	13.15	14.55
2680	Hand placed, 1/2"	C-18	450	.020	↓	4.32	.71	.13	5.16	6.05

03 63 Epoxy Grouting

03 63 05 – Grouting of Dowels and Fasteners

03 63 05.10 Epoxy Only		Crew	Daily Output	Labor-Hours	Unit	Material	Labor	Equipment	Total	Total Incl O&P
0010	EPOXY ONLY									
1500	Chemical anchoring, epoxy cartridge, excludes layout, drilling, fastener									
1530	For fastener 3/4" diam. x 6" embedment	2 Skwk	72	.222	Ea.	5.10	10.10		15.20	22
1535	1" diam. x 8" embedment		66	.242		7.65	11		18.65	26.50
1540	1-1/4" diam. x 10" embedment		60	.267		15.30	12.10		27.40	36.50
1545	1-3/4" diam. x 12" embedment		54	.296		25.50	13.45		38.95	50
1550	14" embedment		48	.333		30.50	15.15		45.65	58
1555	2" diam. x 12" embedment		42	.381		41	17.30		58.30	73
1560	18" embedment	↓	32	.500	↓	51	22.50		73.50	93

03 64 Injection Grouting

03 64 23 – Epoxy Injection Grouting

03 64 23.10 Crack Repair		Crew	Daily Output	Labor-Hours	Unit	2012 Bare Costs Material	Labor	Equipment	Total	Total Incl O&P
0010	**CRACK REPAIR**									
0100	Epoxy injection, 1/8" wide, 12" deep	B-9	80	.500	L.F.	5.90	17.75	2.62	26.27	38.50
0110	1/4" wide, 12" deep		60	.667		11.75	23.50	3.49	38.74	56
0200	Latex injection, 1/8" wide, 12" deep		100	.400		2.68	14.20	2.09	18.97	29
0210	1/4" wide, 12" deep		75	.533		5.35	18.95	2.79	27.09	40

03 81 Concrete Cutting

03 81 13 – Flat Concrete Sawing

03 81 13.50 Concrete Floor/Slab Cutting

		Crew	Daily Output	Labor-Hours	Unit	Material	Labor	Equipment	Total	Total Incl O&P
0010	**CONCRETE FLOOR/SLAB CUTTING**									
0300	Saw cut concrete slabs, plain, up to 3" deep	B-89	1060	.015	L.F.	.12	.60	.41	1.13	1.53
0320	Each additional inch of depth		3180	.005		.04	.20	.14	.38	.51
0400	Mesh reinforced, up to 3" deep		980	.016		.14	.65	.44	1.23	1.66
0420	Each additional inch of depth		2940	.005		.05	.22	.15	.42	.55
0500	Rod reinforced, up to 3" deep		800	.020		.17	.79	.54	1.50	2.04
0520	Each additional inch of depth		2400	.007		.06	.26	.18	.50	.68
0590	Minimum labor/equipment charge		2	8	Job		315	215	530	740

03 81 16 – Track Mounted Concrete Wall Sawing

03 81 16.50 Concrete Wall Cutting

		Crew	Daily Output	Labor-Hours	Unit	Material	Labor	Equipment	Total	Total Incl O&P
0010	**CONCRETE WALL CUTTING**									
0800	Concrete walls, hydraulic saw, plain, per inch of depth	B-89B	250	.064	L.F.	.44	2.53	2.95	5.92	7.75
0820	Rod reinforcing, per inch of depth		150	.107	"	.63	4.22	4.92	9.77	12.85
0890	Minimum labor/equipment charge		2	8	Job		315	370	685	910

03 82 Concrete Boring

03 82 13 – Concrete Core Drilling

03 82 13.10 Core Drilling

		Crew	Daily Output	Labor-Hours	Unit	Material	Labor	Equipment	Total	Total Incl O&P
0010	**CORE DRILLING**									
0020	Reinf. conc slab, up to 6" thick, incl. bit, layout & set up									
0100	1" diameter core	B-89A	17	.941	Ea.	.18	38	6.50	44.68	69.50
0150	Each added inch thick in same hole, add		1440	.011		.03	.45	.08	.56	.84
0200	2" diameter core		16.50	.970		.29	39	6.70	45.99	71
0250	Each added inch thick in same hole, add		1080	.015		.05	.60	.10	.75	1.13
0300	3" diameter core		16	1		.41	40.50	6.90	47.81	73.50
0350	Each added inch thick in same hole, add		720	.022		.07	.89	.15	1.11	1.70
0500	4" diameter core		15	1.067		.49	43	7.35	50.84	78.50
0550	Each added inch thick in same hole, add		480	.033		.49	1.34	.23	2.06	2.98
0700	6" diameter core		14	1.143		.81	46	7.85	54.66	84.50
0750	Each added inch thick in same hole, add		360	.044		.13	1.79	.31	2.23	3.41
0900	8" diameter core		13	1.231		1.03	49.50	8.50	59.03	91.50
0950	Each added inch thick in same hole, add		288	.056		.17	2.24	.38	2.79	4.26
1100	10" diameter core		12	1.333		1.50	53.50	9.20	64.20	99.50
1150	Each added inch thick in same hole, add		240	.067		.25	2.68	.46	3.39	5.15
1300	12" diameter core		11	1.455		1.65	58.50	10	70.15	108
1350	Each added inch thick in same hole, add		206	.078		.27	3.13	.54	3.94	6
1500	14" diameter core		10	1.600		2.07	64.50	11	77.57	119
1550	Each added inch thick in same hole, add		180	.089		.34	3.58	.61	4.53	6.90
1700	18" diameter core		9	1.778		2.75	71.50	12.25	86.50	134
1750	Each added inch thick in same hole, add		144	.111		.46	4.47	.77	5.70	8.65

03 82 Concrete Boring

03 82 13 – Concrete Core Drilling

03 82 13.10 Core Drilling		Crew	Daily Output	Labor-Hours	Unit	2012 Bare Costs Material	Labor	Equipment	Total	Total Incl O&P
1754	24" diameter core	B-89A	8	2	Ea.	3.89	80.50	13.80	98.19	150
1756	Each added inch thick in same hole, add		120	.133		.65	5.35	.92	6.92	10.45
1760	For horizontal holes, add to above						20%	20%		
1770	Prestressed hollow core plank, 8" thick									
1780	1" diameter core	B-89A	17.50	.914	Ea.	.25	37	6.30	43.55	67
1790	Each added inch thick in same hole, add		3840	.004		.03	.17	.03	.23	.33
1794	2" diameter core		17.25	.928		.39	37.50	6.40	44.29	68.50
1796	Each added inch thick in same hole, add		2880	.006		.05	.22	.04	.31	.46
1800	3" diameter core		17	.941		.54	38	6.50	45.04	70
1810	Each added inch thick in same hole, add		1920	.008		.07	.34	.06	.47	.68
1820	4" diameter core		16.50	.970		.65	39	6.70	46.35	71.50
1830	Each added inch thick in same hole, add		1280	.013		.08	.50	.09	.67	1
1840	6" diameter core		15.50	1.032		1.08	41.50	7.10	49.68	77
1850	Each added inch thick in same hole, add		960	.017		.13	.67	.11	.91	1.37
1860	8" diameter core		15	1.067		1.37	43	7.35	51.72	79.50
1870	Each added inch thick in same hole, add		768	.021		.17	.84	.14	1.15	1.72
1880	10" diameter core		14	1.143		2	46	7.85	55.85	86
1890	Each added inch thick in same hole, add		640	.025		.25	1.01	.17	1.43	2.10
1900	12" diameter core		13.50	1.185		2.20	47.50	8.15	57.85	89.50
1910	Each added inch thick in same hole, add		548	.029		.27	1.18	.20	1.65	2.44
1999	Drilling, core, minimum labor/equipment charge		2	8	Job		320	55	375	585
3010	Bits for core drill, diamond, premium, 1" diameter				Ea.	73.50			73.50	81
3020	2" diameter					116			116	128
3030	3" diameter					163			163	179
3040	4" diameter					195			195	215
3060	6" diameter					325			325	355
3080	8" diameter					410			410	455
3110	10" diameter					600			600	660
3120	12" diameter					660			660	725
3140	14" diameter					825			825	910
3180	18" diameter					1,100			1,100	1,200
3240	24" diameter					1,550			1,550	1,700

03 82 16 – Concrete Drilling

03 82 16.10 Concrete Impact Drilling		Crew	Daily Output	Labor-Hours	Unit	Material	Labor	Equipment	Total	Total Incl O&P
0010	**CONCRETE IMPACT DRILLING**									
0050	Up to 4" deep in conc/brick floor/wall, incl. bit & layout, no anchor									
0100	Holes, 1/4" diameter	1 Carp	75	.107	Ea.	.05	4.70		4.75	7.75
0150	For each additional inch of depth, add		430	.019		.01	.82		.83	1.35
0200	3/8" diameter		63	.127		.05	5.60		5.65	9.25
0250	For each additional inch of depth, add		340	.024		.01	1.04		1.05	1.71
0300	1/2" diameter		50	.160		.05	7.05		7.10	11.60
0350	For each additional inch of depth, add		250	.032		.01	1.41		1.42	2.32
0400	5/8" diameter		48	.167		.08	7.35		7.43	12.15
0450	For each additional inch of depth, add		240	.033		.02	1.47		1.49	2.43
0500	3/4" diameter		45	.178		.10	7.85		7.95	12.95
0550	For each additional inch of depth, add		220	.036		.03	1.60		1.63	2.66
0600	7/8" diameter		43	.186		.13	8.20		8.33	13.60
0650	For each additional inch of depth, add		210	.038		.03	1.68		1.71	2.79
0700	1" diameter		40	.200		.15	8.80		8.95	14.60
0750	For each additional inch of depth, add		190	.042		.04	1.86		1.90	3.08
0800	1-1/4" diameter		38	.211		.23	9.30		9.53	15.45
0850	For each additional inch of depth, add		180	.044		.06	1.96		2.02	3.27

03 82 16 – Concrete Drilling

03 82 16.10 Concrete Impact Drilling		Crew	Daily Output	Labor-Hours	Unit	2012 Bare Costs Material	Labor	Equipment	Total	Total Incl O&P
0900	1-1/2" diameter	1 Carp	35	.229	Ea.	.35	10.10		10.45	16.95
0950	For each additional inch of depth, add	↓	165	.048	↓	.09	2.14		2.23	3.61
1000	For ceiling installations, add						40%			

Division Notes

		CREW	DAILY OUTPUT	LABOR-HOURS	UNIT	2012 BARE COSTS MAT.	LABOR	EQUIP.	TOTAL	TOTAL INCL O&P

Estimating Tips

04 05 00 Common Work Results for Masonry

- The terms *mortar* and *grout* are often used interchangeably, and incorrectly. Mortar is used to bed masonry units, seal the entry of air and moisture, provide architectural appearance, and allow for size variations in the units. Grout is used primarily in reinforced masonry construction and is used to bond the masonry to the reinforcing steel. Common mortar types are M(2500 psi), S(1800 psi), N(750 psi), and O(350 psi), and conform to ASTM C270. Grout is either fine or coarse and conforms to ASTM C476, and in-place strengths generally exceed 2500 psi. Mortar and grout are different components of masonry construction and are placed by entirely different methods. An estimator should be aware of their unique uses and costs.
- Mortar is included in all assembled masonry line items. The mortar cost, part of the assembled masonry material cost, includes all ingredients, all labor, and all equipment required. Please see reference number R040513-10.
- Waste, specifically the loss/droppings of mortar and the breakage of brick and block, is included in all masonry assemblies in this division. A factor of 25% is added for mortar and 3% for brick and concrete masonry units.
- Scaffolding or staging is not included in any of the Division 4 costs. Refer to Subdivision 01 54 23 for scaffolding and staging costs.

04 20 00 Unit Masonry

- The most common types of unit masonry are brick and concrete masonry. The major classifications of brick are building brick (ASTM C62), facing brick (ASTM C216), glazed brick, fire brick, and pavers. Many varieties of texture and appearance can exist within these classifications, and the estimator would be wise to check local custom and availability within the project area. For repair and remodeling jobs, matching the existing brick may be the most important criteria.
- Brick and concrete block are priced by the piece and then converted into a price per square foot of wall. Openings less than two square feet are generally ignored by the estimator because any savings in units used is offset by the cutting and trimming required.
- It is often difficult and expensive to find and purchase small lots of historic brick. Costs can vary widely. Many design issues affect costs, selection of mortar mix, and repairs or replacement of masonry materials. Cleaning techniques must be reflected in the estimate.
- All masonry walls, whether interior or exterior, require bracing. The cost of bracing walls during construction should be included by the estimator, and this bracing must remain in place until permanent bracing is complete. Permanent bracing of masonry walls is accomplished by masonry itself, in the form of pilasters or abutting wall corners, or by anchoring the walls to the structural frame. Accessories in the form of anchors, anchor slots, and ties are used, but their supply and installation can be by different trades. For instance, anchor slots on spandrel beams and columns are supplied and welded in place by the steel fabricator, but the ties from the slots into the masonry are installed by the bricklayer. Regardless of the installation method, the estimator must be certain that these accessories are accounted for in pricing.

Reference Numbers

Reference numbers are shown in shaded boxes at the beginning of some major classifications. These numbers refer to related items in the Reference Section. The reference information may be an estimating procedure, an alternate pricing method, or technical information.

Note: Not all subdivisions listed here necessarily appear in this publication.

04 01 Maintenance of Masonry

04 01 20 – Maintenance of Unit Masonry

04 01 20.20 Pointing Masonry		Crew	Daily Output	Labor-Hours	Unit	2012 Bare Costs Material	Labor	Equipment	Total	Total Incl O&P
0010	**POINTING MASONRY**									
0300	Cut and repoint brick, hard mortar, running bond	1 Bric	80	.100	S.F.	.53	4.41		4.94	7.70
0320	Common bond		77	.104		.53	4.58		5.11	8
0360	Flemish bond		70	.114		.56	5.05		5.61	8.70
0400	English bond		65	.123		.56	5.45		6.01	9.35
0600	Soft old mortar, running bond		100	.080		.53	3.53		4.06	6.30
0620	Common bond		96	.083		.53	3.67		4.20	6.50
0640	Flemish bond		90	.089		.56	3.92		4.48	6.90
0680	English bond		82	.098		.56	4.30		4.86	7.55
0700	Stonework, hard mortar		140	.057	L.F.	.70	2.52		3.22	4.83
0720	Soft old mortar		160	.050	"	.70	2.21		2.91	4.32
1000	Repoint, mask and grout method, running bond		95	.084	S.F.	.70	3.71		4.41	6.75
1020	Common bond		90	.089		.70	3.92		4.62	7.05
1040	Flemish bond		86	.093		.74	4.10		4.84	7.40
1060	English bond		77	.104		.74	4.58		5.32	8.20
2000	Scrub coat, sand grout on walls, minimum		120	.067		2.85	2.94		5.79	7.85
2020	Maximum		98	.082		3.96	3.60		7.56	10.15
9000	Minimum labor/equipment charge		3	2.667	Job		118		118	189

04 01 20.30 Pointing CMU

		Crew	Daily Output	Labor-Hours	Unit	Material	Labor	Equipment	Total	Total Incl O&P
0010	**POINTING CMU**									
0300	Cut and repoint block, hard mortar, running bond	1 Bric	190	.042	S.F.	.22	1.86		2.08	3.23
0310	Stacked bond		200	.040		.22	1.76		1.98	3.08
0600	Soft old mortar, running bond		230	.035		.22	1.53		1.75	2.71
0610	Stacked bond		245	.033		.22	1.44		1.66	2.56

04 01 20.50 Toothing Masonry

		Crew	Daily Output	Labor-Hours	Unit	Material	Labor	Equipment	Total	Total Incl O&P
0010	**TOOTHING MASONRY**									
0500	Brickwork, soft old mortar	1 Clab	40	.200	V.L.F.		7		7	11.50
0520	Hard mortar		30	.267			9.35		9.35	15.35
0700	Blockwork, soft old mortar		70	.114			4.01		4.01	6.60
0720	Hard mortar		50	.160			5.60		5.60	9.20
9000	Minimum labor/equipment charge		4	2	Job		70		70	115

04 01 30 – Unit Masonry Cleaning

04 01 30.20 Cleaning Masonry

		Crew	Daily Output	Labor-Hours	Unit	Material	Labor	Equipment	Total	Total Incl O&P
0010	**CLEANING MASONRY**									
0200	Chemical cleaning, new construction, brush and wash, minimum	D-1	1000	.016	S.F.	.05	.64		.69	1.09
0220	Average		800	.020		.08	.80		.88	1.37
0240	Maximum		600	.027		.10	1.06		1.16	1.83
0260	Light restoration, minimum		800	.020		.11	.80		.91	1.40
0270	Average		400	.040		.17	1.59		1.76	2.75
0280	Maximum		330	.048		.22	1.93		2.15	3.35
0300	Heavy restoration, minimum		600	.027		.10	1.06		1.16	1.82
0310	Average		400	.040		.15	1.59		1.74	2.74
0320	Maximum		250	.064		.20	2.55		2.75	4.33
0400	High pressure water only, minimum	C-29	500	.016			.56	.13	.69	1.06
0420	Average		375	.021			.75	.17	.92	1.41
0440	Maximum		250	.032			1.12	.25	1.37	2.12
0800	High pressure water and chemical, minimum		450	.018		.13	.62	.14	.89	1.32
0820	Average		300	.027		.20	.94	.21	1.35	1.98
0840	Maximum		200	.040		.27	1.40	.32	1.99	2.94
1200	Sandblast, wet system, minimum	J-6	1750	.018		.50	.72	.12	1.34	1.82
1220	Average		1100	.029		.74	1.14	.19	2.07	2.85
1240	Maximum		700	.046		.99	1.79	.30	3.08	4.28

04 01 Maintenance of Masonry

04 01 30 – Unit Masonry Cleaning

04 01 30.20 Cleaning Masonry		Crew	Daily Output	Labor-Hours	Unit	Material	2012 Bare Costs Labor	Equipment	Total	Total Incl O&P
1400	Dry system, minimum	J-6	2500	.013	S.F.	.50	.50	.08	1.08	1.44
1420	Average		1750	.018		.74	.72	.12	1.58	2.09
1440	Maximum		1000	.032		.99	1.25	.21	2.45	3.32
1800	For walnut shells, add					.50			.50	.55
1820	For corn chips, add					.50			.50	.55
2000	Steam cleaning, minimum	A-1H	750	.011			.37	.09	.46	.71
2020	Average		625	.013			.45	.11	.56	.86
2040	Maximum		375	.021			.75	.18	.93	1.43
4000	Add for masking doors and windows	1 Clab	800	.010		.05	.35		.40	.64
4200	Add for pedestrian protection				Job					10%
9000	Minimum labor/equipment charge	D-4	2	16	"		640	63.50	703.50	1,100

04 05 Common Work Results for Masonry

04 05 05 – Selective Masonry Demolition

04 05 05.10 Selective Demolition

Line	Description	Crew	Daily Output	Labor-Hours	Unit	Material	Labor	Equipment	Total	Total Incl O&P
0010	**SELECTIVE DEMOLITION** R024119-10									
0200	Bond beams, 8" block with #4 bar	2 Clab	32	.500	L.F.		17.55		17.55	29
0300	Concrete block walls, unreinforced, 2" thick		1200	.013	S.F.		.47		.47	.77
0310	4" thick		1150	.014			.49		.49	.80
0320	6" thick		1100	.015			.51		.51	.84
0330	8" thick		1050	.015			.53		.53	.88
0340	10" thick		1000	.016			.56		.56	.92
0360	12" thick		950	.017			.59		.59	.97
0380	Reinforced alternate courses, 2" thick		1130	.014			.50		.50	.81
0390	4" thick		1080	.015			.52		.52	.85
0400	6" thick		1035	.015			.54		.54	.89
0410	8" thick		990	.016			.57		.57	.93
0420	10" thick		940	.017			.60		.60	.98
0430	12" thick		890	.018			.63		.63	1.03
0440	Reinforced alternate courses & vertically 48" OC, 4" thick		900	.018			.62		.62	1.02
0450	6" thick		850	.019			.66		.66	1.08
0460	8" thick		800	.020			.70		.70	1.15
0480	10" thick		750	.021			.75		.75	1.23
0490	12" thick		700	.023			.80		.80	1.32
1000	Chimney, 16" x 16", soft old mortar	1 Clab	55	.145	C.F.		5.10		5.10	8.35
1020	Hard mortar		40	.200			7		7	11.50
1030	16" x 20", soft old mortar		55	.145			5.10		5.10	8.35
1040	Hard mortar		40	.200			7		7	11.50
1050	16" x 24", soft old mortar		55	.145			5.10		5.10	8.35
1060	Hard mortar		40	.200			7		7	11.50
1080	20" x 20", soft old mortar		55	.145			5.10		5.10	8.35
1100	Hard mortar		40	.200			7		7	11.50
1110	20" x 24", soft old mortar		55	.145			5.10		5.10	8.35
1120	Hard mortar		40	.200			7		7	11.50
1140	20" x 32", soft old mortar		55	.145			5.10		5.10	8.35
1160	Hard mortar		40	.200			7		7	11.50
1200	48" x 48", soft old mortar		55	.145			5.10		5.10	8.35
1220	Hard mortar		40	.200			7		7	11.50
1250	Metal, high temp steel jacket, 24" diameter	E-2	130	.431	V.L.F.		21	11.50	32.50	49.50
1260	60" diameter	"	60	.933			45	25	70	108
1280	Flue lining, up to 12" x 12"	1 Clab	200	.040			1.40		1.40	2.30

04 05 Common Work Results for Masonry

04 05 05 – Selective Masonry Demolition

04 05 05.10 Selective Demolition		Crew	Daily Output	Labor-Hours	Unit	2012 Bare Costs Material	Labor	Equipment	Total	Total Incl O&P
1282	Up to 24" x 24"	1 Clab	150	.053	V.L.F.		1.87		1.87	3.07
2000	Columns, 8" x 8", soft old mortar		48	.167			5.85		5.85	9.60
2020	Hard mortar		40	.200			7		7	11.50
2060	16" x 16", soft old mortar		16	.500			17.55		17.55	29
2100	Hard mortar		14	.571			20		20	33
2140	24" x 24", soft old mortar		8	1			35		35	57.50
2160	Hard mortar		6	1.333			47		47	76.50
2200	36" x 36", soft old mortar		4	2			70		70	115
2220	Hard mortar		3	2.667			93.50		93.50	153
2230	Alternate pricing method, soft old mortar		30	.267	C.F.		9.35		9.35	15.35
2240	Hard mortar		23	.348	"		12.20		12.20	20
3000	Copings, precast or masonry, to 8" wide									
3020	Soft old mortar	1 Clab	180	.044	L.F.		1.56		1.56	2.56
3040	Hard mortar	"	160	.050	"		1.76		1.76	2.88
3100	To 12" wide									
3120	Soft old mortar	1 Clab	160	.050	L.F.		1.76		1.76	2.88
3140	Hard mortar	"	140	.057	"		2.01		2.01	3.29
4000	Fireplace, brick, 30" x 24" opening									
4020	Soft old mortar	1 Clab	2	4	Ea.		140		140	230
4040	Hard mortar		1.25	6.400			225		225	370
4100	Stone, soft old mortar		1.50	5.333			187		187	305
4120	Hard mortar		1	8			281		281	460
5000	Veneers, brick, soft old mortar		140	.057	S.F.		2.01		2.01	3.29
5020	Hard mortar		125	.064			2.25		2.25	3.68
5100	Granite and marble, 2" thick		180	.044			1.56		1.56	2.56
5120	4" thick		170	.047			1.65		1.65	2.71
5140	Stone, 4" thick		180	.044			1.56		1.56	2.56
5160	8" thick		175	.046			1.60		1.60	2.63
5400	Alternate pricing method, stone, 4" thick		60	.133	C.F.		4.68		4.68	7.65
5420	8" thick		85	.094	"		3.30		3.30	5.40
9000	Minimum labor/equipment charge		2	4	Job		140		140	230

04 05 13 – Masonry Mortaring

04 05 13.10 Cement		Crew	Daily Output	Labor-Hours	Unit	Material	Labor	Equipment	Total	Total Incl O&P
0010	**CEMENT**									
0100	Masonry, 70 lb. bag, T.L. lots				Bag	9.85			9.85	10.85
0150	L.T.L. lots					10.45			10.45	11.50
0200	White, 70 lb. bag, T.L. lots					15.55			15.55	17.15
0250	L.T.L. lots					17.75			17.75	19.55

04 05 13.20 Lime		Crew	Daily Output	Labor-Hours	Unit	Material	Labor	Equipment	Total	Total Incl O&P
0010	**LIME**									
0020	Masons, hydrated, 50 lb. bag, T.L. lots				Bag	9.10			9.10	10
0050	L.T.L. lots					10			10	11
0200	Finish, double hydrated, 50 lb. bag, T.L. lots					9.90			9.90	10.90
0250	L.T.L. lots					10.90			10.90	12

04 05 13.30 Mortar		Crew	Daily Output	Labor-Hours	Unit	Material	Labor	Equipment	Total	Total Incl O&P
0010	**MORTAR**									
0100	Type M, 1:1:6 mix	1 Brhe	143	.056	C.F.	5.15	1.99		7.14	8.85
0200	Type N, 1:3 mix		143	.056		4.50	1.99		6.49	8.15
0300	Type O, 1:3 mix		143	.056		6.45	1.99		8.44	10.30
0400	Type PM, 1:1:6 mix, 2500 psi		143	.056		6.85	1.99		8.84	10.75
0500	Type S, 1/2:1:4 mix		143	.056		7.25	1.99		9.24	11.20
2000	With portland cement and lime									

04 05 Common Work Results for Masonry

04 05 13 – Masonry Mortaring

04 05 13.30 Mortar		Crew	Daily Output	Labor-Hours	Unit	Material	2012 Bare Costs Labor	Equipment	Total	Total Incl O&P
2100	Type M, 1:1/4:3 mix	1 Brhe	143	.056	C.F.	8.80	1.99		10.79	12.90
2200	Type N, 1:1:6 mix, 750 psi		143	.056		6.80	1.99		8.79	10.65
2300	Type O, 1:2:9 mix (Pointing Mortar)		143	.056		7.50	1.99		9.49	11.45
2400	Type PL, 1:1/2:4 mix, 2500 psi		143	.056		7.45	1.99		9.44	11.40
2500	Type K, 1:3:12 mix, 75 psi		143	.056		6.45	1.99		8.44	10.30
2600	Type S, 1:1/2:4 mix, 1800 psi		143	.056		9.15	1.99		11.14	13.25
2650	Pre-mixed, type S or N					4.94			4.94	5.45
2700	Mortar for glass block	1 Brhe	143	.056		10.35	1.99		12.34	14.60
2900	Mortar for Fire Brick, 80 lb. bag, T.L. Lots				Bag	25.50			25.50	28

04 05 16 – Masonry Grouting

04 05 16.30 Grouting		Crew	Daily Output	Labor-Hours	Unit	Material	Labor	Equipment	Total	Total Incl O&P
0010	**GROUTING**									
0011	Bond beams & lintels, 8″ deep, 6″ thick, 0.15 C.F. per L.F.	D-4	1480	.022	L.F.	.64	.87	.09	1.60	2.18
0020	8″ thick, 0.2 C.F. per L.F.		1400	.023		.99	.91	.09	1.99	2.64
0050	10″ thick, 0.25 C.F. per L.F.		1200	.027		1.06	1.07	.11	2.24	2.99
0060	12″ thick, 0.3 C.F. per L.F.		1040	.031		1.27	1.23	.12	2.62	3.50
0200	Concrete block cores, solid, 4″ thk., by hand, 0.067 C.F./S.F. of wall	D-8	1100	.036	S.F.	.28	1.48		1.76	2.69
0210	6″ thick, pumped, 0.175 C.F. per S.F.	D-4	720	.044		.74	1.78	.18	2.70	3.85
0250	8″ thick, pumped, 0.258 C.F. per S.F.		680	.047		1.09	1.88	.19	3.16	4.42
0300	10″ thick, pumped, 0.340 C.F. per S.F.		660	.048		1.44	1.94	.19	3.57	4.90
0350	12″ thick, pumped, 0.422 C.F. per S.F.		640	.050		1.79	2	.20	3.99	5.40
0500	Cavity walls, 2″ space, pumped, 0.167 C.F./S.F. of wall		1700	.019		.71	.75	.07	1.53	2.07
0550	3″ space, 0.250 C.F./S.F.		1200	.027		1.06	1.07	.11	2.24	2.99
0600	4″ space, 0.333 C.F. per S.F.		1150	.028		1.41	1.11	.11	2.63	3.45
0700	6″ space, 0.500 C.F. per S.F.		800	.040		2.12	1.60	.16	3.88	5.05
0800	Door frames, 3′ x 7′ opening, 2.5 C.F. per opening		60	.533	Opng.	10.60	21.50	2.12	34.22	48
0850	6′ x 7′ opening, 3.5 C.F. per opening		45	.711	″	14.80	28.50	2.83	46.13	65
2000	Grout, C476, for bond beams, lintels and CMU cores		350	.091	C.F.	4.23	3.66	.36	8.25	10.90
9000	Minimum labor/equipment charge	1 Bric	2	4	Job		176		176	284

04 05 19 – Masonry Anchorage and Reinforcing

04 05 19.05 Anchor Bolts		Crew	Daily Output	Labor-Hours	Unit	Material	Labor	Equipment	Total	Total Incl O&P
0010	**ANCHOR BOLTS**									
0015	Installed in fresh grout in CMU bond beams or filled cores, no templates									
0020	Hooked, with nut and washer, 1/2″ diam., 8″ long	1 Bric	132	.061	Ea.	1.39	2.67		4.06	5.80
0030	12″ long		131	.061		1.54	2.69		4.23	6.05
0040	5/8″ diameter, 8″ long		129	.062		3	2.74		5.74	7.70
0050	12″ long		127	.063		3.70	2.78		6.48	8.55
0060	3/4″ diameter, 8″ long		127	.063		3.70	2.78		6.48	8.55
0070	12″ long		125	.064		4.62	2.82		7.44	9.65

04 05 19.16 Masonry Anchors		Crew	Daily Output	Labor-Hours	Unit	Material	Labor	Equipment	Total	Total Incl O&P
0010	**MASONRY ANCHORS**									
0020	For brick veneer, galv., corrugated, 7/8″ x 7″, 22 Ga.	1 Bric	10.50	.762	C	13	33.50		46.50	68.50
0100	24 Ga.		10.50	.762		9.10	33.50		42.60	64
0150	16 Ga.		10.50	.762		25.50	33.50		59	82
0200	Buck anchors, galv., corrugated, 16 gauge, 2″ bend, 8″ x 2″		10.50	.762		49	33.50		82.50	108
0250	8″ x 3″		10.50	.762		51	33.50		84.50	110
0300	Adjustable, rectangular, 4-1/8″ wide									
0350	Anchor and tie, 3/16″ wire, mill galv.									
0400	2-3/4″ eye, 3-1/4″ tie	1 Bric	1.05	7.619	M	385	335		720	965
0500	4-3/4″ tie		1.05	7.619		415	335		750	995
0520	5-1/2″ tie		1.05	7.619		450	335		785	1,025

04 05 Common Work Results for Masonry

04 05 19 – Masonry Anchorage and Reinforcing

04 05 19.16 Masonry Anchors		Crew	Daily Output	Labor-Hours	Unit	2012 Bare Costs Material	Labor	Equipment	Total	Total Incl O&P
0550	4-3/4" eye, 3-1/4" tie	1 Bric	1.05	7.619	M	425	335		760	1,000
0570	4-3/4" tie		1.05	7.619		460	335		795	1,050
0580	5-1/2" tie		1.05	7.619		500	335		835	1,100
0660	Cavity wall, Z-type, galvanized, 6" long, 1/8" diam.		10.50	.762	C	23	33.50		56.50	79
0670	3/16" diameter		10.50	.762		21	33.50		54.50	77.50
0680	1/4" diameter		10.50	.762		38	33.50		71.50	95.50
0850	8" long, 3/16" diameter		10.50	.762		23.50	33.50		57	79.50
0855	1/4" diameter		10.50	.762		46	33.50		79.50	105
1000	Rectangular type, galvanized, 1/4" diameter, 2" x 6"		10.50	.762		67.50	33.50		101	128
1050	4" x 6"		10.50	.762		82.50	33.50		116	145
1100	3/16" diameter, 2" x 6"		10.50	.762		38.50	33.50		72	96.50
1150	4" x 6"		10.50	.762		44	33.50		77.50	103
1200	Mesh wall tie, 1/2" mesh, hot dip galvanized									
1400	16 gauge, 12" long, 3" wide	1 Bric	9	.889	C	83	39		122	154
1420	6" wide		9	.889		121	39		160	196
1440	12" wide		8.50	.941		196	41.50		237.50	283
1500	Rigid partition anchors, plain, 8" long, 1" x 1/8"		10.50	.762		200	33.50		233.50	274
1550	1" x 1/4"		10.50	.762		235	33.50		268.50	315
1580	1-1/2" x 1/8"		10.50	.762		220	33.50		253.50	296
1600	1-1/2" x 1/4"		10.50	.762		275	33.50		308.50	360
1650	2" x 1/8"		10.50	.762		260	33.50		293.50	340
1700	2" x 1/4"		10.50	.762		241	33.50		274.50	320
2000	Column flange ties, wire, galvanized									
2300	3/16" diameter, up to 3" wide	1 Bric	10.50	.762	C	78.50	33.50		112	141
2350	To 5" wide		10.50	.762		85.50	33.50		119	148
2400	To 7" wide		10.50	.762		92	33.50		125.50	155
2600	To 9" wide		10.50	.762		98	33.50		131.50	162
2650	1/4" diameter, up to 3" wide		10.50	.762		99.50	33.50		133	163
2700	To 5" wide		10.50	.762		107	33.50		140.50	172
2800	To 7" wide		10.50	.762		120	33.50		153.50	186
2850	To 9" wide		10.50	.762		130	33.50		163.50	197
2900	For hot dip galvanized, add					35%				
4000	Channel slots, 1-3/8" x 1/2" x 8"									
4100	12 gauge, plain	1 Bric	10.50	.762	C	196	33.50		229.50	270
4150	16 gauge, galvanized	"	10.50	.762	"	134	33.50		167.50	202
4200	Channel slot anchors									
4300	16 gauge, galvanized, 1-1/4" x 3-1/2"				C	51			51	56
4350	1-1/4" x 5-1/2"					60			60	66
4400	1-1/4" x 7-1/2"					68			68	75
4500	1/8" plain, 1-1/4" x 3-1/2"					121			121	133
4550	1-1/4" x 5-1/2"					129			129	142
4600	1-1/4" x 7-1/2"					141			141	155
4700	For corrugation, add					65			65	71.50
4750	For hot dip galvanized, add					35%				
5000	Dowels									
5100	Plain, 1/4" diameter, 3" long				C	47			47	51.50
5150	4" long					52			52	57.50
5200	6" long					64			64	70
5300	3/8" diameter, 3" long					62			62	68.50
5350	4" long					77			77	85
5400	6" long					89			89	98
5500	1/2" diameter, 3" long					90.50			90.50	99.50
5550	4" long					107			107	118

04 05 19 – Masonry Anchorage and Reinforcing

04 05 19.16 Masonry Anchors

	04 05 19.16 Masonry Anchors	Crew	Daily Output	Labor-Hours	Unit	2012 Bare Costs Material	Labor	Equipment	Total	Total Incl O&P
5600	6" long				C	139			139	153
5700	5/8" diameter, 3" long					124			124	137
5750	4" long					153			153	168
5800	6" long					210			210	231
6000	3/4" diameter, 3" long					154			154	170
6100	4" long					196			196	215
6150	6" long					280			280	310
6300	For hot dip galvanized, add					35%				

04 05 19.26 Masonry Reinforcing Bars

	04 05 19.26 Masonry Reinforcing Bars	Crew	Daily Output	Labor-Hours	Unit	2012 Bare Costs Material	Labor	Equipment	Total	Total Incl O&P
0010	**MASONRY REINFORCING BARS**									
0015	Steel bars A615, placed horiz., #3 & #4 bars	1 Bric	450	.018	Lb.	.49	.78		1.27	1.80
0020	#5 & #6 bars		800	.010		.49	.44		.93	1.25
0050	Placed vertical, #3 & #4 bars		350	.023		.49	1.01		1.50	2.16
0060	#5 & #6 bars		650	.012		.49	.54		1.03	1.41
0500	Joint reinforcing, ladder type, mill std galvanized									
0600	9 ga. sides, 9 ga. ties, 4" wall	1 Bric	30	.267	C.L.F.	14.65	11.75		26.40	35
0650	6" wall		30	.267		14.05	11.75		25.80	34.50
0700	8" wall		25	.320		14.20	14.10		28.30	38
0750	10" wall		20	.400		15.30	17.65		32.95	45.50
0800	12" wall		20	.400		15.45	17.65		33.10	45.50
1000	Truss type									
1100	9 ga. sides, 9 ga. ties, 4" wall	1 Bric	30	.267	C.L.F.	19.45	11.75		31.20	40.50
1150	6" wall		30	.267		19.75	11.75		31.50	41
1200	8" wall		25	.320		20.50	14.10		34.60	45.50
1250	10" wall		20	.400		19.10	17.65		36.75	49.50
1300	12" wall		20	.400		19.90	17.65		37.55	50.50
1500	3/16" sides, 9 ga. ties, 4" wall		30	.267		22	11.75		33.75	43.50
1550	6" wall		30	.267		27	11.75		38.75	49
1600	8" wall		25	.320		28.50	14.10		42.60	53.50
1650	10" wall		20	.400		29	17.65		46.65	60
1700	12" wall		20	.400		33	17.65		50.65	65
2000	3/16" sides, 3/16" ties, 4" wall		30	.267		30.50	11.75		42.25	52.50
2050	6" wall		30	.267		31	11.75		42.75	53.50
2100	8" wall		25	.320		32.50	14.10		46.60	58
2150	10" wall		20	.400		34	17.65		51.65	66
2200	12" wall		20	.400		30.50	17.65		48.15	62
2500	Cavity truss type, galvanized									
2600	9 ga. sides, 9 ga. ties, 4" wall	1 Bric	25	.320	C.L.F.	22.50	14.10		36.60	47.50
2650	6" wall		25	.320		22.50	14.10		36.60	47
2700	8" wall		20	.400		33.50	17.65		51.15	65.50
2750	10" wall		15	.533		27	23.50		50.50	68
2800	12" wall		15	.533		26.50	23.50		50	67
3000	3/16" sides, 9 ga. ties, 4" wall		25	.320		33.50	14.10		47.60	59
3050	6" wall		25	.320		30.50	14.10		44.60	56
3100	8" wall		20	.400		28	17.65		45.65	59.50
3150	10" wall		15	.533		29.50	23.50		53	70
3200	12" wall		15	.533		35	23.50		58.50	76.50
3500	For hot dip galvanizing, add				Ton	450			450	495

04 05 23 – Masonry Accessories

04 05 23.13 Masonry Control and Expansion Joints

	04 05 23.13 Masonry Control and Expansion Joints	Crew	Daily Output	Labor-Hours	Unit	2012 Bare Costs Material	Labor	Equipment	Total	Total Incl O&P
0010	**MASONRY CONTROL AND EXPANSION JOINTS**									
0020	Rubber, for double wythe 8" minimum wall (Brick/CMU)	1 Bric	400	.020	L.F.	1.88	.88		2.76	3.49

04 05 Common Work Results for Masonry

04 05 23 – Masonry Accessories

04 05 23.13 Masonry Control and Expansion Joints		Crew	Daily Output	Labor-Hours	Unit	Material	2012 Bare Costs Labor	Equipment	Total	Total Incl O&P
0025	"T" shaped	1 Bric	320	.025	L.F.	1.15	1.10		2.25	3.05
0030	Cross-shaped for CMU units		280	.029		1.48	1.26		2.74	3.66
0050	PVC, for double wythe 8" minimum wall (Brick/CMU)		400	.020		1.38	.88		2.26	2.94
0120	"T" shaped		320	.025		.71	1.10		1.81	2.56
0160	Cross-shaped for CMU units		280	.029		.95	1.26		2.21	3.08

04 05 23.19 Masonry Cavity Drainage, Weepholes, and Vents

Line	Description	Crew	Daily Output	Labor-Hours	Unit	Material	Labor	Equipment	Total	Total Incl O&P
0010	**MASONRY CAVITY DRAINAGE, WEEPHOLES, AND VENTS**									
0020	Extruded aluminum, 4" deep, 2-3/8" x 8-1/8"	1 Bric	30	.267	Ea.	31.50	11.75		43.25	53.50
0050	5" x 8-1/8"		25	.320		41	14.10		55.10	68
0100	2-1/4" x 25"		25	.320		71.50	14.10		85.60	102
0150	5" x 16-1/2"		22	.364		62.50	16.05		78.55	94.50
0200	6" x 16-1/2"		22	.364		83.50	16.05		99.55	118
0250	7-3/4" x 16-1/2"		20	.400		71	17.65		88.65	107
0400	For baked enamel finish, add					35%				
0500	For cast aluminum, painted, add					60%				
1000	Stainless steel ventilators, 6" x 6"	1 Bric	25	.320		126	14.10		140.10	161
1050	8" x 8"		24	.333		132	14.70		146.70	170
1100	12" x 12"		23	.348		153	15.35		168.35	193
1150	12" x 6"		24	.333		134	14.70		148.70	171
1200	Foundation block vent, galv., 1-1/4" thk, 8" high, 16" long, no damper		30	.267		24.50	11.75		36.25	46
1250	For damper, add					8.20			8.20	9

04 21 Clay Unit Masonry

04 21 13 – Brick Masonry

04 21 13.13 Brick Veneer Masonry

Line	Description	Crew	Daily Output	Labor-Hours	Unit	Material	Labor	Equipment	Total	Total Incl O&P
0010	**BRICK VENEER MASONRY**, T.L. lots, excl. scaff., grout & reinforcing R042110-20									
0015	Material costs incl. 3% brick and 25% mortar waste									
2000	Standard, sel. common, 4" x 2-2/3" x 8", (6.75/S.F.) R042110-50	D-8	230	.174	S.F.	3.61	7.10		10.71	15.35
2020	Standard, red, 4" x 2-2/3" x 8", running bond (6.75/S.F.)		220	.182		4.19	7.40		11.59	16.50
2050	Full header every 6th course (7.88/S.F.)		185	.216		4.88	8.80		13.68	19.50
2100	English, full header every 2nd course (10.13/S.F.)		140	.286		6.25	11.65		17.90	25.50
2150	Flemish, alternate header every course (9.00/S.F.)		150	.267		5.55	10.85		16.40	23.50
2200	Flemish, alt. header every 6th course (7.13/S.F.)		205	.195		4.42	7.95		12.37	17.65
2250	Full headers throughout (13.50/S.F.)		105	.381		8.30	15.50		23.80	34
2300	Rowlock course (13.50/S.F.)		100	.400		8.30	16.30		24.60	35
2350	Rowlock stretcher (4.50/S.F.)		310	.129		2.81	5.25		8.06	11.55
2400	Soldier course (6.75/S.F.)		200	.200		4.19	8.15		12.34	17.70
2450	Sailor course (4.50/S.F.)		290	.138		2.81	5.60		8.41	12.15
2600	Buff or gray face, running bond, (6.75/S.F.)		220	.182		4.43	7.40		11.83	16.75
2700	Glazed face brick, running bond		210	.190		11.45	7.75		19.20	25
2750	Full header every 6th course (7.88/S.F.)		170	.235		13.35	9.60		22.95	30
3000	Jumbo, 6" x 4" x 12" running bond (3.00/S.F.)		435	.092		4.90	3.74		8.64	11.40
3050	Norman, 4" x 2-2/3" x 12" running bond, (4.5/S.F.)		320	.125		5.75	5.10		10.85	14.50
3100	Norwegian, 4" x 3-1/5" x 12" (3.75/S.F.)		375	.107		4.15	4.34		8.49	11.55
3150	Economy, 4" x 4" x 8" (4.50/S.F.)		310	.129		4.02	5.25		9.27	12.85
3200	Engineer, 4" x 3-1/5" x 8" (5.63/S.F.)		260	.154		3.54	6.25		9.79	14
3250	Roman, 4" x 2" x 12" (6.00/S.F.)		250	.160		5.75	6.50		12.25	16.85
3300	SCR, 6" x 2-2/3" x 12" (4.50/S.F.)		310	.129		5.25	5.25		10.50	14.20
3350	Utility, 4" x 4" x 12" (3.00/S.F.)		450	.089		4.51	3.62		8.13	10.75
3400	For cavity wall construction, add						15%			
3450	For stacked bond, add						10%			

04 21 Clay Unit Masonry

04 21 13 – Brick Masonry

04 21 13.13 Brick Veneer Masonry		Crew	Daily Output	Labor-Hours	Unit	Material	Labor (2012 Bare Costs)	Equipment (2012 Bare Costs)	Total (2012 Bare Costs)	Total Incl O&P
3500	For interior veneer construction, add						15%			
3550	For curved walls, add						30%			
9000	Minimum labor/equipment charge	D-1	2	8	Job		320		320	515

04 21 13.14 Thin Brick Veneer

Line	Description	Crew	Daily Output	Labor-Hours	Unit	Material	Labor	Equipment	Total	Total Incl O&P
0010	**THIN BRICK VENEER**									
0015	Material costs incl. 3% brick and 25% mortar waste									
0020	On & incl. metal panel support sys, modular, 2-2/3" x 5/8" x 8", red	D-7	92	.174	S.F.	8.75	6.35		15.10	19.60
0100	Closure, 4" x 5/8" x 8"		110	.145		8.75	5.30		14.05	17.95
0110	Norman, 2-2/3" x 5/8" x 12"		110	.145		8.95	5.30		14.25	18.15
0120	Utility, 4" x 5/8" x 12"		125	.128		8.95	4.66		13.61	17.15
0130	Emperor, 4" x 3/4" x 16"		175	.091		10.15	3.33		13.48	16.45
0140	Super emperor, 8" x 3/4" x 16"		195	.082		10.15	2.99		13.14	15.90
0150	For L shaped corners with 4" return, add				L.F.	9.25			9.25	10.20
0200	On masonry/plaster back-up, modular, 2-2/3" x 5/8" x 8", red	D-7	137	.117	S.F.	3.95	4.25		8.20	11
0210	Closure, 4" x 5/8" x 8"		165	.097		3.95	3.53		7.48	9.90
0220	Norman, 2-2/3" x 5/8" x 12"		165	.097		4.15	3.53		7.68	10.10
0230	Utility, 4" x 5/8" x 12"		185	.086		4.15	3.15		7.30	9.50
0240	Emperor, 4" x 3/4" x 16"		260	.062		5.35	2.24		7.59	9.40
0250	Super emperor, 8" x 3/4" x 16"		285	.056		5.35	2.05		7.40	9.10
0260	For L shaped corners with 4" return, add				L.F.	9.25			9.25	10.20
0270	For embedment into pre-cast concrete panels, add				S.F.	14.40			14.40	15.85

04 21 13.15 Chimney

Line	Description	Crew	Daily Output	Labor-Hours	Unit	Material	Labor	Equipment	Total	Total Incl O&P
0010	**CHIMNEY**, excludes foundation, scaffolding, grout and reinforcing									
0100	Brick, 16" x 16", 8" flue	D-1	18.20	.879	V.L.F.	21.50	35		56.50	80
0150	16" x 20" with one 8" x 12" flue		16	1		34	40		74	102
0200	16" x 24" with two 8" x 8" flues		14	1.143		50.50	45.50		96	129
0250	20" x 20" with one 12" x 12" flue		13.70	1.168		38.50	46.50		85	118
0300	20" x 24" with two 8" x 12" flues		12	1.333		57	53		110	149
0350	20" x 32" with two 12" x 12" flues		10	1.600		68	64		132	178

04 21 13.18 Columns

Line	Description	Crew	Daily Output	Labor-Hours	Unit	Material	Labor	Equipment	Total	Total Incl O&P
0010	**COLUMNS**, solid, excludes scaffolding, grout and reinforcing									
0050	Brick, 8" x 8", 9 brick per V.L.F.	D-1	56	.286	V.L.F.	5.45	11.40		16.85	24.50
0100	12" x 8", 13.5 brick per V.L.F.		37	.432		8.15	17.25		25.40	36.50
0200	12" x 12", 20 brick per V.L.F.		25	.640		12.10	25.50		37.60	54.50
0300	16" x 12", 27 brick per V.L.F.		19	.842		16.35	33.50		49.85	72
0400	16" x 16", 36 brick per V.L.F.		14	1.143		22	45.50		67.50	97.50
0500	20" x 16", 45 brick per V.L.F.		11	1.455		27	58		85	124
0600	20" x 20", 56 brick per V.L.F.		9	1.778		34	71		105	152
0700	24" x 20", 68 brick per V.L.F.		7	2.286		41	91		132	193
0800	24" x 24", 81 brick per V.L.F.		6	2.667		49	106		155	225
1000	36" x 36", 182 brick per V.L.F.		3	5.333		110	213		323	460
9000	Minimum labor/equipment charge		2	8	Job		320		320	515

04 21 13.30 Oversized Brick

Line	Description	Crew	Daily Output	Labor-Hours	Unit	Material	Labor	Equipment	Total	Total Incl O&P
0010	**OVERSIZED BRICK**, excludes scaffolding, grout and reinforcing									
0100	Veneer, 4" x 2.25" x 16"	D-8	387	.103	S.F.	6.45	4.21		10.66	13.80
0105	4" x 2.75" x 16"		412	.097		5.80	3.95		9.75	12.70
0110	4" x 4" x 16"		460	.087		5.30	3.54		8.84	11.55
0120	4" x 8" x 16"		533	.075		6.10	3.05		9.15	11.65
0125	Loadbearing, 6" x 4" x 16", grouted and reinforced		387	.103		9.05	4.21		13.26	16.70
0130	8" x 4" x 16", grouted and reinforced		327	.122		9.45	4.98		14.43	18.35
0135	6" x 8" x 16", grouted and reinforced		440	.091		8.70	3.70		12.40	15.50
0140	8" x 8" x 16", grouted and reinforced		400	.100		9.55	4.07		13.62	17.05

04 21 Clay Unit Masonry

04 21 13 – Brick Masonry

04 21 13.30 Oversized Brick		Crew	Daily Output	Labor-Hours	Unit	Material	2012 Bare Costs Labor	Equipment	Total	Total Incl O&P
0145	Curtainwall/reinforced veneer, 6" x 4" x 16"	D-8	387	.103	S.F.	13.20	4.21		17.41	21.50
0150	8" x 4" x 16"		327	.122		15.70	4.98		20.68	25.50
0155	6" x 8" x 16"		440	.091		13.30	3.70		17	20.50
0160	8" x 8" x 16"	↓	400	.100		15.80	4.07		19.87	24
0200	For 1 to 3 slots in face, add					15%				
0210	For 4 to 7 slots in face, add					25%				
0220	For bond beams, add					20%				
0230	For bullnose shapes, add					20%				
0240	For open end knockout, add					10%				
0250	For white or gray color group, add					10%				
0260	For 135 degree corner, add				↓	250%				

04 21 13.35 Common Building Brick

		Crew	Daily Output	Labor-Hours	Unit	Material	Labor	Equipment	Total	Total Incl O&P
0010	**COMMON BUILDING BRICK**, C62, TL lots, material only R042110-20									
0020	Standard, minimum				M	350			350	385
0050	Average (select)				"	430			430	475

04 21 13.40 Structural Brick

		Crew	Daily Output	Labor-Hours	Unit	Material	Labor	Equipment	Total	Total Incl O&P
0010	**STRUCTURAL BRICK** C652, Grade SW, incl. mortar, scaffolding not incl.									
0100	Standard unit, 4-5/8" x 2-3/4" x 9-5/8"	D-8	245	.163	S.F.	5.15	6.65		11.80	16.40
0120	Bond beam		225	.178		9.65	7.25		16.90	22.50
0140	V cut bond beam		225	.178		10.45	7.25		17.70	23
0160	Stretcher quoin, 5-5/8" x 2-3/4" x 9-5/8"		245	.163		10.10	6.65		16.75	22
0180	Corner quoin		245	.163		11.30	6.65		17.95	23
0200	Corner, 45 deg, 4-5/8" x 2-3/4" x 10-7/16"	↓	235	.170	↓	11.70	6.95		18.65	24

04 21 13.45 Face Brick

		Crew	Daily Output	Labor-Hours	Unit	Material	Labor	Equipment	Total	Total Incl O&P
0010	**FACE BRICK** Material Only, C216, TL lots R042110-20									
0300	Standard modular, 4" x 2-2/3" x 8", minimum				M	515			515	565
0350	Maximum R042110-50					675			675	745
0450	Economy, 4" x 4" x 8", minimum					775			775	855
0500	Maximum					1,000			1,000	1,100
0510	Economy, 4" x 4" x 12", minimum					1,225			1,225	1,350
0520	Maximum					1,600			1,600	1,750
0550	Jumbo, 6" x 4" x 12", minimum					1,475			1,475	1,625
0600	Maximum					1,925			1,925	2,125
0610	Jumbo, 8" x 4" x 12", minimum					1,475			1,475	1,625
0620	Maximum					1,925			1,925	2,125
0650	Norwegian, 4" x 3-1/5" x 12", minimum					980			980	1,075
0700	Maximum					1,275			1,275	1,400
0710	Norwegian, 6" x 3-1/5" x 12", minimum					1,425			1,425	1,575
0720	Maximum					1,875			1,875	2,050
0850	Standard glazed, plain colors, 4" x 2-2/3" x 8", minimum					1,550			1,550	1,725
0900	Maximum					2,025			2,025	2,225
1000	Deep trim shades, 4" x 2-2/3" x 8", minimum					1,925			1,925	2,125
1050	Maximum					2,150			2,150	2,375
1080	Jumbo utility, 4" x 4" x 12"					1,350			1,350	1,500
1120	4" x 8" x 8"					1,675			1,675	1,850
1140	4" x 8" x 16"					4,950			4,950	5,450
1260	Engineer, 4" x 3-1/5" x 8", minimum					520			520	570
1270	Maximum					700			700	770
1350	King, 4" x 2-3/4" x 10", minimum					460			460	505
1360	Maximum					560			560	620
1400	Norman, 4" x 2-3/4" x 12"					460			460	505
1450	Roman, 4" x 2" x 12"				↓	460			460	505

04 21 Clay Unit Masonry

04 21 13 – Brick Masonry

04 21 13.45 Face Brick		Crew	Daily Output	Labor-Hours	Unit	2012 Bare Costs Material	Labor	Equipment	Total	Total Incl O&P
1500	SCR, 6" x 2-2/3" x 12"				M	460			460	505
1550	Double, 4" x 5-1/3" x 8"					460			460	505
1600	Triple, 4" x 5-1/3" x 12"					460			460	505
1770	Standard modular, double glazed, 4" x 2-2/3" x 8"					2,275			2,275	2,500
1850	Jumbo, colored glazed ceramic, 6" x 4" x 12"					2,400			2,400	2,650
2050	Jumbo utility, glazed, 4" x 4" x 12"					4,500			4,500	4,950
2100	4" x 8" x 8"					5,300			5,300	5,825
2150	4" x 16" x 8"					6,200			6,200	6,825
2170	For less than truck load lots, add					15			15	16.50
2180	For buff or gray brick, add					16			16	17.60
3050	Used brick, minimum					380			380	420
3100	Maximum					350			350	385
3150	Add for brick to match existing work, minimum					5%				
3200	Maximum				↓	50%				

04 21 26 – Glazed Structural Clay Tile Masonry

04 21 26.10 Structural Facing Tile		Crew	Daily Output	Labor-Hours	Unit	Material	Labor	Equipment	Total	Total Incl O&P
0010	**STRUCTURAL FACING TILE**, std. colors, excl. scaffolding, grout, reinforcing									
0020	6T series, 5-1/3" x 12", 2.3 pieces per S.F., glazed 1 side, 2" thick	D-8	225	.178	S.F.	8.40	7.25		15.65	21
0100	4" thick		220	.182		11.50	7.40		18.90	24.50
0150	Glazed 2 sides		195	.205		14.95	8.35		23.30	30
0250	6" thick		210	.190		17.10	7.75		24.85	31.50
0300	Glazed 2 sides		185	.216		20.50	8.80		29.30	36.50
0400	8" thick	↓	180	.222	↓	22.50	9.05		31.55	39.50
9000	Minimum labor/equipment charge	D-1	2	8	Job		320		320	515

04 21 29 – Terra Cotta Masonry

04 21 29.10 Terra Cotta Masonry Components		Crew	Daily Output	Labor-Hours	Unit	Material	Labor	Equipment	Total	Total Incl O&P
0010	**TERRA COTTA MASONRY COMPONENTS**									
0020	Coping, split type, not glazed, 9" wide	D-1	90	.178	L.F.	10.40	7.10		17.50	23
0100	13" wide		80	.200		15.75	7.95		23.70	30
0200	Coping, split type, glazed, 9" wide		90	.178		17.40	7.10		24.50	30.50
0250	13" wide	↓	80	.200	↓	22.50	7.95		30.45	38
0500	Partition or back up blocks, scored, in C.L. lots									
0700	Non-load bearing 12" x 12", 3" thick, special order	D-8	550	.073	S.F.	26	2.96		28.96	33.50
0850	8" thick		400	.100		11.05	4.07		15.12	18.75
1000	Load bearing, 12" x 12", 4" thick, in walls		500	.080		7.60	3.26		10.86	13.60
1400	8" thick, in walls	↓	400	.100	↓	11.05	4.07		15.12	18.75
9000	Minimum labor/equipment charge	D-1	2	8	Job		320		320	515

04 22 Concrete Unit Masonry

04 22 10 – Concrete Masonry Units

04 22 10.11 Autoclave Aerated Concrete Block		Crew	Daily Output	Labor-Hours	Unit	Material	Labor	Equipment	Total	Total Incl O&P
0010	**AUTOCLAVE AERATED CONCRETE BLOCK**, excl. scaffolding, grout & reinforcing									
0050	Solid, 4" x 12" x 24", incl mortar G	D-8	600	.067	S.F.	1.39	2.71		4.10	5.90
0060	6" x 12" x 24" G		600	.067		2.08	2.71		4.79	6.65
0070	8" x 8" x 24" G		575	.070		2.78	2.83		5.61	7.60
0080	10" x 12" x 24" G		575	.070		3.40	2.83		6.23	8.30
0090	12" x 12" x 24" G	↓	550	.073	↓	4.16	2.96		7.12	9.35

04 22 10.14 Concrete Block, Back-Up		Crew	Daily Output	Labor-Hours	Unit	Material	Labor	Equipment	Total	Total Incl O&P
0010	**CONCRETE BLOCK, BACK-UP**, C90, 2000 psi									
0020	Normal weight, 8" x 16" units, tooled joint 1 side									

04 22 Concrete Unit Masonry

04 22 10 – Concrete Masonry Units

04 22 10.14 Concrete Block, Back-Up

04 22 10.14 Concrete Block, Back-Up		Crew	Daily Output	Labor-Hours	Unit	2012 Bare Costs Material	Labor	Equipment	Total	Total Incl O&P
0050	Not-reinforced, 2000 psi, 2″ thick	D-8	475	.084	S.F.	1.39	3.43		4.82	7.05
0200	4″ thick		460	.087		1.58	3.54		5.12	7.45
0300	6″ thick		440	.091		2.19	3.70		5.89	8.35
0350	8″ thick		400	.100		2.49	4.07		6.56	9.30
0400	10″ thick		330	.121		3.16	4.93		8.09	11.45
0450	12″ thick	D-9	310	.155		3.56	6.15		9.71	13.85
1000	Reinforced, alternate courses, 4″ thick	D-8	450	.089		1.72	3.62		5.34	7.70
1100	6″ thick		430	.093		2.34	3.79		6.13	8.65
1150	8″ thick		395	.101		2.65	4.12		6.77	9.55
1200	10″ thick		320	.125		3.31	5.10		8.41	11.85
1250	12″ thick	D-9	300	.160		3.71	6.40		10.11	14.35
9000	Minimum labor/equipment charge	D-1	2	8	Job		320		320	515

04 22 10.16 Concrete Block, Bond Beam

No.	Description	Crew	Daily Output	Labor-Hours	Unit	Material	Labor	Equipment	Total	Total Incl O&P
0010	**CONCRETE BLOCK, BOND BEAM**, C90, 2000 psi									
0020	Not including grout or reinforcing									
0125	Regular block, 6″ thick	D-8	584	.068	L.F.	2.34	2.79		5.13	7.05
0130	8″ high, 8″ thick	″	565	.071		2.52	2.88		5.40	7.40
0150	12″ thick	D-9	510	.094		3.53	3.75		7.28	9.95
0525	Lightweight, 6″ thick	D-8	592	.068		2.55	2.75		5.30	7.25
2000	Including grout and 2 #5 bars									
2100	Regular block, 8″ high, 8″ thick	D-8	300	.133	L.F.	4.55	5.45		10	13.75
2150	12″ thick	D-9	250	.192	″	6.10	7.65		13.75	19.05
9000	Minimum labor/equipment charge	D-1	2	8	Job		320		320	515

04 22 10.19 Concrete Block, Insulation Inserts

No.	Description	Crew	Daily Output	Labor-Hours	Unit	Material	Labor	Equipment	Total	Total Incl O&P
0010	**CONCRETE BLOCK, INSULATION INSERTS**									
0100	Styrofoam, plant installed, add to block prices									
0200	8″ x 16″ units, 6″ thick				S.F.	1.14			1.14	1.25
0250	8″ thick					1.14			1.14	1.25
0300	10″ thick					1.93			1.93	2.12
0350	12″ thick					2			2	2.20
0500	8″ x 8″ units, 8″ thick					.94			.94	1.03
0550	12″ thick					1.14			1.14	1.25

04 22 10.23 Concrete Block, Decorative

No.	Description	Crew	Daily Output	Labor-Hours	Unit	Material	Labor	Equipment	Total	Total Incl O&P
0010	**CONCRETE BLOCK, DECORATIVE**, C90, 2000 psi									
1000	Fluted high strength									
1100	8″ x 16″ x 4″ thick, flutes 1 side,	D-8	345	.116	S.F.	4.20	4.72		8.92	12.20
1150	Flutes 2 sides		335	.119		5.10	4.86		9.96	13.40
1200	8″ thick		300	.133		6.60	5.45		12.05	16
1250	For special colors, add					.41			.41	.45
1400	Deep grooved, smooth face									
1450	8″ x 16″ x 4″ thick	D-8	345	.116	S.F.	2.75	4.72		7.47	10.60
1500	8″ thick	″	300	.133	″	4.70	5.45		10.15	13.90
1600										
4000	Slump block									
4100	4″ face height x 16″ x 4″ thick	D-1	165	.097	S.F.	3.53	3.86		7.39	10.10
4150	6″ thick		160	.100		4.85	3.99		8.84	11.75
4200	8″ thick		155	.103		5.05	4.11		9.16	12.15
4250	10″ thick		140	.114		10	4.55		14.55	18.35
4300	12″ thick		130	.123		10.25	4.90		15.15	19.15
5000	Split rib profile units, 1″ deep ribs, 8 ribs									
5100	8″ x 16″ x 4″ thick	D-8	345	.116	S.F.	3.49	4.72		8.21	11.45
5150	6″ thick		325	.123		4.01	5		9.01	12.45

04 22 Concrete Unit Masonry

04 22 10 – Concrete Masonry Units

04 22 10.23 Concrete Block, Decorative

	04 22 10.23 Concrete Block, Decorative	Crew	Daily Output	Labor-Hours	Unit	Material	Labor	Equipment	Total	Total Incl O&P
						2012 Bare Costs				
5200	8″ thick	D-8	300	.133	S.F.	4.66	5.45		10.11	13.90
5250	12″ thick	D-9	275	.175	↓	5.50	6.95		12.45	17.20
9000	Minimum labor/equipment charge	D-1	2	8	Job		320		320	515

04 22 10.24 Concrete Block, Exterior

	04 22 10.24 Concrete Block, Exterior	Crew	Daily Output	Labor-Hours	Unit	Material	Labor	Equipment	Total	Total Incl O&P
0010	**CONCRETE BLOCK, EXTERIOR**, C90, 2000 psi									
0020	Reinforced alt courses, tooled joints 2 sides									
0100	Normal weight, 8″ x 16″ x 6″ thick	D-8	395	.101	S.F.	2.57	4.12		6.69	9.45
0200	8″ thick		360	.111		3.34	4.52		7.86	10.95
0250	10″ thick	↓	290	.138		4.30	5.60		9.90	13.80
0300	12″ thick	D-9	250	.192	↓	4.27	7.65		11.92	17
9000	Minimum labor/equipment charge	D-1	2	8	Job		320		320	515

04 22 10.26 Concrete Block Foundation Wall

	04 22 10.26 Concrete Block Foundation Wall	Crew	Daily Output	Labor-Hours	Unit	Material	Labor	Equipment	Total	Total Incl O&P
0010	**CONCRETE BLOCK FOUNDATION WALL**, C90/C145									
0050	Normal-weight, cut joints, horiz joint reinf, no vert reinf									
0200	Hollow, 8″ x 16″ x 6″ thick	D-8	455	.088	S.F.	2.65	3.58		6.23	8.65
0250	8″ thick		425	.094		2.96	3.83		6.79	9.40
0300	10″ thick	↓	350	.114		3.63	4.65		8.28	11.50
0350	12″ thick	D-9	300	.160	↓	4.04	6.40		10.44	14.70
1000	Reinforced, #4 vert @ 48″									
1100	Hollow, 8″ x 16″ block, 4″ thick	D-8	455	.088	S.F.	2.55	3.58		6.13	8.55
1125	6″ thick		445	.090		3.49	3.66		7.15	9.75
1150	8″ thick		415	.096		4.15	3.92		8.07	10.85
1200	10″ thick	↓	340	.118		5.15	4.79		9.94	13.40
1250	12″ thick	D-9	290	.166	↓	5.90	6.60		12.50	17.10
9000	Minimum labor/equipment charge	D-1	2	8	Job		320		320	515

04 22 10.30 Concrete Block, Interlocking

	04 22 10.30 Concrete Block, Interlocking	Crew	Daily Output	Labor-Hours	Unit	Material	Labor	Equipment	Total	Total Incl O&P
0010	**CONCRETE BLOCK, INTERLOCKING**									
0100	Not including grout or reinforcing									
0200	8″ x 16″ units, 2,000 psi, 8″ thick	D-1	245	.065	S.F.	2.72	2.60		5.32	7.20
0300	12″ thick	″	220	.073		4.03	2.90		6.93	9.10
0400	Including grout & reinforcing, 8″ thick	D-4	245	.131		7.35	5.25	.52	13.12	17
0450	12″ thick	″	220	.145	↓	8.90	5.80	.58	15.28	19.70
9000	Minimum labor/equipment charge	D-1	2	8	Job		320		320	515

04 22 10.32 Concrete Block, Lintels

	04 22 10.32 Concrete Block, Lintels	Crew	Daily Output	Labor-Hours	Unit	Material	Labor	Equipment	Total	Total Incl O&P
0010	**CONCRETE BLOCK, LINTELS**, C90, normal weight									
0100	Including grout and horizontal reinforcing									
0200	8″ x 8″ x 8″, 1 #4 bar	D-4	300	.107	L.F.	4.87	4.27	.42	9.56	12.65
0250	2 #4 bars		295	.108		5.10	4.34	.43	9.87	13.05
1000	12″ x 8″ x 8″, 1 #4 bar		275	.116		6.80	4.66	.46	11.92	15.45
1150	2 #5 bars	↓	270	.119	↓	7.30	4.74	.47	12.51	16.10
9000	Minimum labor/equipment charge	D-1	2	8	Job		320		320	515

04 22 10.34 Concrete Block, Partitions

	04 22 10.34 Concrete Block, Partitions	Crew	Daily Output	Labor-Hours	Unit	Material	Labor	Equipment	Total	Total Incl O&P
0010	**CONCRETE BLOCK, PARTITIONS**, excludes scaffolding									
1000	Lightweight block, tooled joints, 2 sides, hollow									
1100	Not reinforced, 8″ x 16″ x 4″ thick	D-8	440	.091	S.F.	1.64	3.70		5.34	7.75
1150	6″ thick		410	.098		2.37	3.97		6.34	9
1200	8″ thick		385	.104		2.88	4.23		7.11	9.95
1250	10″ thick	↓	370	.108		3.48	4.40		7.88	10.95
1300	12″ thick	D-9	350	.137		4.33	5.45		9.78	13.55
1500	Reinforced alternate courses, 4″ thick	D-8	435	.092		1.75	3.74		5.49	7.95
1600	6″ thick	↓	405	.099	↓	2.48	4.02		6.50	9.15

04 22 Concrete Unit Masonry

04 22 10 – Concrete Masonry Units

04 22 10.34 Concrete Block, Partitions		Crew	Daily Output	Labor-Hours	Unit	Material	2012 Bare Costs Labor	Equipment	Total	Total Incl O&P
1650	8″ thick	D-8	380	.105	S.F.	2.99	4.28		7.27	10.20
1700	10″ thick		365	.110		3.60	4.46		8.06	11.15
1750	12″ thick	D-9	345	.139		4.45	5.55		10	13.85
4000	Regular block, tooled joints, 2 sides, hollow									
4100	Not reinforced, 8″ x 16″ x 4″ thick	D-8	430	.093	S.F.	1.49	3.79		5.28	7.75
4150	6″ thick		400	.100		2.10	4.07		6.17	8.85
4200	8″ thick		375	.107		2.40	4.34		6.74	9.65
4250	10″ thick		360	.111		3.07	4.52		7.59	10.70
4300	12″ thick	D-9	340	.141		3.47	5.65		9.12	12.85
4500	Reinforced alternate courses, 8″ x 16″ x 4″ thick	D-8	425	.094		1.63	3.83		5.46	7.95
4550	6″ thick		395	.101		2.25	4.12		6.37	9.10
4600	8″ thick		370	.108		2.56	4.40		6.96	9.90
4650	10″ thick		355	.113		3.49	4.59		8.08	11.25
4700	12″ thick	D-9	335	.143		3.62	5.70		9.32	13.20
9000	Minimum labor/equipment charge	D-1	2	8	Job		320		320	515

04 22 10.38 Concrete Brick		Crew	Daily Output	Labor-Hours	Unit	Material	Labor	Equipment	Total	Total Incl O&P
0010	**CONCRETE BRICK**, C55, grade N, type 1									
0100	Regular, 4 x 2-1/4 x 8	D-8	660	.061	Ea.	.50	2.47		2.97	4.52
0125	Rusticated, 4 x 2-1/4 x 8		660	.061		.56	2.47		3.03	4.59
0150	Frog, 4 x 2-1/4 x 8		660	.061		.54	2.47		3.01	4.56
0200	Double, 4 x 4-7/8 x 8		535	.075		.88	3.04		3.92	5.85

04 22 10.44 Glazed Concrete Block		Crew	Daily Output	Labor-Hours	Unit	Material	Labor	Equipment	Total	Total Incl O&P
0010	**GLAZED CONCRETE BLOCK** C744									
0100	Single face, 8″ x 16″ units, 2″ thick	D-8	360	.111	S.F.	9.20	4.52		13.72	17.45
0200	4″ thick		345	.116		9.45	4.72		14.17	18
0250	6″ thick		330	.121		11.55	4.93		16.48	20.50
0300	8″ thick		310	.129		9.95	5.25		15.20	19.40
0350	10″ thick		295	.136		12.50	5.50		18	22.50
0400	12″ thick	D-9	280	.171		13.30	6.85		20.15	25.50
0700	Double face, 8″ x 16″ units, 4″ thick	D-8	340	.118		13.85	4.79		18.64	23
0750	6″ thick		320	.125		16.25	5.10		21.35	26
0800	8″ thick		300	.133		17.05	5.45		22.50	27.50
1500	Cove base, 8″ x 16″, 2″ thick		315	.127	L.F.	13.60	5.15		18.75	23.50
1550	4″ thick		285	.140		8.65	5.70		14.35	18.75
1600	6″ thick		265	.151		9.30	6.15		15.45	20
1650	8″ thick		245	.163		9.75	6.65		16.40	21.50
9000	Minimum labor/equipment charge	D-1	2	8	Job		320		320	515

04 23 Glass Unit Masonry

04 23 13 – Vertical Glass Unit Masonry

04 23 13.10 Glass Block		Crew	Daily Output	Labor-Hours	Unit	Material	Labor	Equipment	Total	Total Incl O&P
0010	**GLASS BLOCK**									
0100	Plain, 4″ thick, under 1,000 S.F., 6″ x 6″	D-8	115	.348	S.F.	28.50	14.15		42.65	54.50
0150	8″ x 8″		160	.250		15.80	10.20		26	34
0160	end block		160	.250		52.50	10.20		62.70	74
0170	90 deg corner		160	.250		56.50	10.20		66.70	78.50
0180	45 deg corner		160	.250		49	10.20		59.20	70
0200	12″ x 12″		175	.229		24	9.30		33.30	41.50
0210	4″ x 8″		160	.250		16.20	10.20		26.40	34
0220	6″ x 8″		160	.250		15.65	10.20		25.85	33.50

04 23 Glass Unit Masonry

04 23 13 – Vertical Glass Unit Masonry

04 23 13.10 Glass Block		Crew	Daily Output	Labor-Hours	Unit	Material	2012 Bare Costs Labor	Equipment	Total	Total Incl O&P
0300	1,000 to 5,000 S.F., 6" x 6"	D-8	135	.296	S.F.	28	12.05		40.05	50
0350	8" x 8"		190	.211		15.50	8.55		24.05	31
0400	12" x 12"		215	.186		23.50	7.55		31.05	38
0410	4" x 8"		215	.186		6.40	7.55		13.95	19.25
0420	6" x 8"		215	.186		6.10	7.55		13.65	18.90
0700	For solar reflective blocks, add					100%				
1000	Thinline, plain, 3-1/8" thick, under 1,000 S.F., 6" x 6"	D-8	115	.348	S.F.	19.90	14.15		34.05	45
1050	8" x 8"		160	.250		11	10.20		21.20	28.50
1400	For cleaning block after installation (both sides), add		1000	.040		.11	1.63		1.74	2.74
9000	Minimum labor/equipment charge	D-1	2	8	Job		320		320	515

04 27 Multiple-Wythe Unit Masonry

04 27 10 – Multiple-Wythe Masonry

04 27 10.10 Cornices		Crew	Daily Output	Labor-Hours	Unit	Material	Labor	Equipment	Total	Total Incl O&P
0010	**CORNICES**									
0110	Face bricks, 12 brick/S.F.	D-1	30	.533	SF Face	5.95	21.50		27.45	40.50
0150	15 brick/S.F.		23	.696	"	7.15	27.50		34.65	52.50
9000	Minimum labor/equipment charge		1.50	10.667	Job		425		425	685
04 27 10.30 Brick Walls										
0010	**BRICK WALLS**, including mortar, excludes scaffolding									
0800	Face brick, 4" thick wall, 6.75 brick/S.F.	D-8	215	.186	S.F.	4.12	7.55		11.67	16.75
0850	Common brick, 4" thick wall, 6.75 brick/S.F. R042110-50		240	.167		2.98	6.80		9.78	14.20
0900	8" thick, 13.50 bricks per S.F.		135	.296		6.20	12.05		18.25	26
1000	12" thick, 20.25 bricks per S.F.		95	.421		9.35	17.15		26.50	38
1050	16" thick, 27.00 bricks per S.F.		75	.533		12.70	21.50		34.20	49
1200	Reinforced, face brick, 4" thick wall, 6.75 brick/S.F.		210	.190		4.29	7.75		12.04	17.20
1220	Common brick, 4" thick wall, 6.75 brick/S.F.		235	.170		3.15	6.95		10.10	14.60
1250	8" thick, 13.50 bricks per S.F.		130	.308		6.55	12.50		19.05	27
1300	12" thick, 20.25 bricks per S.F.		90	.444		9.85	18.10		27.95	40
1350	16" thick, 27.00 bricks per S.F.		70	.571		13.35	23.50		36.85	52
9000	Minimum labor/equipment charge	D-1	2	8	Job		320		320	515

04 41 Dry-Placed Stone

04 41 10 – Dry Placed Stone

04 41 10.10 Rough Stone Wall		Crew	Daily Output	Labor-Hours	Unit	Material	Labor	Equipment	Total	Total Incl O&P
0011	**ROUGH STONE WALL**, Dry									
0012	Dry laid (no mortar), under 18" thick G	D-1	60	.267	C.F.	11.65	10.65		22.30	30
0100	Random fieldstone, under 18" thick G	D-12	60	.533		11.65	21.50		33.15	47.50
0150	Over 18" thick G	"	63	.508		14	20.50		34.50	48.50
0500	Field stone veneer G	D-8	120	.333	S.F.	10.25	13.55		23.80	33.50
0510	Valley stone veneer G		120	.333		10.25	13.55		23.80	33.50
0520	River stone veneer G		120	.333		10.25	13.55		23.80	33.50
0600	Rubble stone walls, in mortar bed, up to 18" thick G	D-11	75	.320	C.F.	14.10	13.40		27.50	37
9000	Minimum labor/equipment charge	D-1	2	8	Job		320		320	515

04 43 Stone Masonry

04 43 10 – Masonry with Natural and Processed Stone

04 43 10.10 Bluestone

	04 43 10.10 Bluestone	Crew	Daily Output	Labor-Hours	Unit	Material	Labor (2012 Bare Costs)	Equipment	Total	Total Incl O&P
0010	**BLUESTONE**, cut to size									
0500	Sills, natural cleft, 10″ wide to 6′ long, 1-1/2″ thick	D-11	70	.343	L.F.	13.35	14.40		27.75	37.50
0550	2″ thick		63	.381		14.60	15.95		30.55	41.50
0600	Smooth finish, 1-1/2″ thick		70	.343		25	14.40		39.40	50.50
0650	2″ thick		63	.381		30	15.95		45.95	58.50
0800	Thermal finish, 1-1/2″ thick		70	.343		30	14.40		44.40	56
0850	2″ thick		63	.381		35	15.95		50.95	64
1000	Stair treads, natural cleft, 12″ wide, 6′ long, 1-1/2″ thick	D-10	115	.278		30	12.10	3.90	46	56.50
1050	2″ thick		105	.305		30	13.25	4.27	47.52	58.50
1100	Smooth finish, 1-1/2″ thick		115	.278		15	12.10	3.90	31	40
1150	2″ thick		105	.305		15	13.25	4.27	32.52	42
1300	Thermal finish, 1-1/2″ thick		115	.278		15	12.10	3.90	31	40
1350	2″ thick		105	.305		15	13.25	4.27	32.52	42
9000	Minimum labor/equipment charge	D-1	2.50	6.400	Job		255		255	410

04 43 10.45 Granite

	04 43 10.45 Granite	Crew	Daily Output	Labor-Hours	Unit	Material	Labor	Equipment	Total	Total Incl O&P
0010	**GRANITE**, cut to size									
0050	Veneer, polished face, 3/4″ to 1-1/2″ thick									
0150	Low price, gray, light gray, etc.	D-10	130	.246	S.F.	26.50	10.70	3.45	40.65	50.50
0180	Medium price, pink, brown, etc.		130	.246		29.50	10.70	3.45	43.65	53.50
0220	High price, red, black, etc.		130	.246		42	10.70	3.45	56.15	67.50
2500	Steps, copings, etc., finished on more than one surface									
2550	Minimum	D-10	50	.640	C.F.	91	28	8.95	127.95	154
2600	Maximum	″	50	.640	″	146	28	8.95	182.95	214
9000	Minimum labor/equipment charge	D-1	2	8	Job		320		320	515

04 43 10.55 Limestone

	04 43 10.55 Limestone	Crew	Daily Output	Labor-Hours	Unit	Material	Labor	Equipment	Total	Total Incl O&P
0010	**LIMESTONE**, cut to size									
0020	Veneer facing panels									
0750	5″ thick, 5′ x 14′ panels	D-10	275	.116	S.F.	52.50	5.05	1.63	59.18	67.50
1000	Sugarcube finish, 2″ Thick, 3′ x 5′ panels		275	.116		35	5.05	1.63	41.68	48.50
1050	3″ Thick, 4′ x 9′ panels		275	.116		38	5.05	1.63	44.68	52
1200	4″ Thick, 5′ x 11′ panels		275	.116		41	5.05	1.63	47.68	55
1400	Sugarcube, textured finish, 4-1/2″ thick, 5′ x 12′		275	.116		45	5.05	1.63	51.68	59.50
1450	5″ thick, 5′ x 14′ panels		275	.116		53.50	5.05	1.63	60.18	68.50
2000	Coping, sugarcube finish, top & 2 sides		30	1.067	C.F.	68	46.50	14.95	129.45	165
2100	Sills, lintels, jambs, trim, stops, sugarcube finish, average		20	1.600		68	69.50	22.50	160	211
2150	Detailed		20	1.600		68	69.50	22.50	160	211
2300	Steps, extra hard, 14″ wide, 6″ rise		50	.640	L.F.	41.50	28	8.95	78.45	100
3000	Quoins, plain finish, 6″ x 12″ x 12″	D-12	25	1.280	Ea.	38	51.50		89.50	125
3050	6″ x 16″ x 24″	″	25	1.280	″	50.50	51.50		102	139
9000	Minimum labor/equipment charge	D-1	2	8	Job		320		320	515

04 43 10.60 Marble

	04 43 10.60 Marble	Crew	Daily Output	Labor-Hours	Unit	Material	Labor	Equipment	Total	Total Incl O&P
0011	**MARBLE**, ashlar, split face, 4″ + or - thick, random									
0040	Lengths 1′ to 4′ & heights 2″ to 7-1/2″, average	D-8	175	.229	S.F.	17.75	9.30		27.05	34.50
0100	Base, polished, 3/4″ or 7/8″ thick, polished, 6″ high	D-10	65	.492	L.F.	11.80	21.50	6.90	40.20	54.50
0300	Carvings or bas relief, from templates, average		80	.400	S.F.	144	17.35	5.60	166.95	192
0350	Maximum		80	.400	″	335	17.35	5.60	357.95	405
1000	Facing, polished finish, cut to size, 3/4″ to 7/8″ thick									
1050	Average	D-10	130	.246	S.F.	21	10.70	3.45	35.15	44
1100	Maximum	″	130	.246	″	36.50	10.70	3.45	50.65	61
2500	Flooring, polished tiles, 12″ x 12″ x 3/8″ thick									
2510	Thin set, average	D-11	90	.267	S.F.	11.20	11.20		22.40	30.50
2600	Maximum		90	.267		99.50	11.20		110.70	127

04 43 Stone Masonry

04 43 10 – Masonry with Natural and Processed Stone

04 43 10.60 Marble		Crew	Daily Output	Labor-Hours	Unit	2012 Bare Costs Material	Labor	Equipment	Total	Total Incl O&P
2700	Mortar bed, average	D-11	65	.369	S.F.	10.60	15.50		26.10	36.50
2740	Maximum		65	.369		99.50	15.50		115	134
3500	Thresholds, 3′ long, 7/8″ thick, 4″ to 5″ wide, plain	D-12	24	1.333	Ea.	19.85	54		73.85	109
3550	Beveled		24	1.333	″	22.50	54		76.50	111
3700	Window stools, polished, 7/8″ thick, 5″ wide		85	.376	L.F.	20.50	15.20		35.70	47
9000	Minimum labor/equipment charge	D-1	2	8	Job		320		320	515

04 43 10.75 Sandstone or Brownstone		Crew	Daily Output	Labor-Hours	Unit	Material	Labor	Equipment	Total	Total Incl O&P
0011	**SANDSTONE OR BROWNSTONE**									
0100	Sawed face veneer, 2-1/2″ thick, to 2′ x 4′ panels	D-10	130	.246	S.F.	21	10.70	3.45	35.15	44
0150	4″ thick, to 3′-6″ x 8′panels		100	.320		21	13.90	4.48	39.38	50
0300	Split face, random sizes		100	.320		14	13.90	4.48	32.38	42.50
9000	Minimum labor/equipment charge	D-1	2.50	6.400	Job		255		255	410

04 43 10.80 Slate		Crew	Daily Output	Labor-Hours	Unit	Material	Labor	Equipment	Total	Total Incl O&P
0010	**SLATE**									
0040	Pennsylvania - blue gray to black									
0050	Vermont - unfading green, mottled green & purple, gray & purple									
0100	Virginia - blue black									
3100	Stair landings, 1″ thick, black, clear	D-1	65	.246	S.F.	20	9.80		29.80	38
3200	Ribbon	″	65	.246	″	22	9.80		31.80	40
3500	Stair treads, sand finish, 1″ thick x 12″ wide									
3550	Under 3 L.F.	D-10	85	.376	L.F.	22	16.35	5.25	43.60	56
3600	3 L.F. to 6 L.F.	″	120	.267	″	24	11.55	3.74	39.29	48.50
3700	Ribbon, sand finish, 1″ thick x 12″ wide									
3750	To 6 L.F.	D-10	120	.267	L.F.	20	11.55	3.74	35.29	44.50
4000	Stools or sills, sand finish, 1″ thick, 6″ wide	D-12	160	.200		11.50	8.05		19.55	25.50
4200	10″ wide		90	.356		17.75	14.35		32.10	42.50
4400	2″ thick, 6″ wide		140	.229		18.50	9.20		27.70	35.50
4600	10″ wide		90	.356		29	14.35		43.35	55
4800	For lengths over 3′, add					25%				
9000	Minimum labor/equipment charge	D-1	2.50	6.400	Job		255		255	410

04 43 10.85 Window Sill		Crew	Daily Output	Labor-Hours	Unit	Material	Labor	Equipment	Total	Total Incl O&P
0010	**WINDOW SILL**									
0020	Bluestone, thermal top, 10″ wide, 1-1/2″ thick	D-1	85	.188	S.F.	16.05	7.50		23.55	30
0050	2″ thick		75	.213	″	18.75	8.50		27.25	34
0100	Cut stone, 5″ x 8″ plain		48	.333	L.F.	11.90	13.30		25.20	34.50
0200	Face brick on edge, brick, 8″ wide		80	.200		5.45	7.95		13.40	18.80
0400	Marble, 9″ wide, 1″ thick		85	.188		8.50	7.50		16	21.50
0900	Slate, colored, unfading, honed, 12″ wide, 1″ thick		85	.188		17.10	7.50		24.60	31
0950	2″ thick		70	.229		24	9.10		33.10	40.50
9000	Minimum labor/equipment charge	1 Bric	2	4	Job		176		176	284

04 54 Refractory Brick Masonry

04 54 10 – Refractory Brick Work

04 54 10.20 Fire Clay		Crew	Daily Output	Labor-Hours	Unit	Material	2012 Bare Costs Labor	Equipment	Total	Total Incl O&P
0010	**FIRE CLAY**									
0020	Gray, high duty, 100 lb. bag				Bag	59			59	65
0050	100 lb. drum, premixed (400 brick per drum)				Drum	73			73	80.50

04 57 Masonry Fireplaces

04 57 10 – Brick or Stone Fireplaces

04 57 10.10 Fireplace

		Crew	Daily Output	Labor-Hours	Unit	Material	Labor	Equipment	Total	Total Incl O&P
0010	**FIREPLACE**									
0100	Brick fireplace, not incl. foundations or chimneys									
0110	30" x 29" opening, incl. chamber, plain brickwork	D-1	.40	40	Ea.	535	1,600		2,135	3,175
0200	Fireplace box only (110 brick)	"	2	8	"	176	320		496	710
0300	For elaborate brickwork and details, add					35%	35%			
0400	For hearth, brick & stone, add	D-1	2	8	Ea.	197	320		517	730
0410	For steel angle, damper, cleanouts, add		4	4		138	159		297	410
0600	Plain brickwork, incl. metal circulator		.50	32		1,025	1,275		2,300	3,175
0800	Face brick only, standard size, 8" x 2-2/3" x 4"		.30	53.333	M	535	2,125		2,660	4,025
0900	Stone fireplace, fieldstone, add				SF Face	14.05			14.05	15.50
1000	Cut stone, add				"	15.50			15.50	17.05
9000	Minimum labor/equipment charge	D-1	2	8	Job		320		320	515

04 71 Manufactured Brick Masonry

04 71 10 – Simulated or Manufactured Brick

04 71 10.10 Simulated Brick

		Crew	Daily Output	Labor-Hours	Unit	Material	Labor	Equipment	Total	Total Incl O&P
0010	**SIMULATED BRICK**									
0020	Aluminum, baked on colors	1 Carp	200	.040	S.F.	3.53	1.76		5.29	6.75
0050	Fiberglass panels		200	.040		3.87	1.76		5.63	7.15
0100	Urethane pieces cemented in mastic		150	.053		6.70	2.35		9.05	11.20
0150	Vinyl siding panels		200	.040		2.67	1.76		4.43	5.85

04 72 Cast Stone Masonry

04 72 10 – Cast Stone Masonry Features

04 72 10.10 Coping

		Crew	Daily Output	Labor-Hours	Unit	Material	Labor	Equipment	Total	Total Incl O&P
0010	**COPING**, stock units									
0050	Precast concrete, 10" wide, 4" tapers to 3-1/2", 8" wall	D-1	75	.213	L.F.	18.15	8.50		26.65	33.50
0100	12" wide, 3-1/2" tapers to 3", 10" wall		70	.229		19.60	9.10		28.70	36
0150	16" wide, 4" tapers to 3-1/2", 14" wall		60	.267		23	10.65		33.65	42
0300	Limestone for 12" wall, 4" thick		90	.178		15.35	7.10		22.45	28.50
0350	6" thick		80	.200		23	7.95		30.95	38.50
0500	Marble, to 4" thick, no wash, 9" wide		90	.178		23	7.10		30.10	37
0550	12" wide		80	.200		35	7.95		42.95	51.50
0700	Terra cotta, 9" wide		90	.178		5.60	7.10		12.70	17.55
0750	12" wide		80	.200		9.20	7.95		17.15	23
0800	Aluminum, for 12" wall		80	.200		12.50	7.95		20.45	26.50
9000	Minimum labor/equipment charge		2	8	Job		320		320	515

04 72 20 – Cultured Stone Veneer

04 72 20.10 Cultured Stone Veneer Components		Crew	Daily Output	Labor-Hours	Unit	Material	2012 Bare Costs Labor	Equipment	Total	Total Incl O&P
0010	**CULTURED STONE VENEER COMPONENTS**									
0110	On wood frame and sheathing substrate, random sized cobbles, corner stones	D-8	70	.571	V.L.F.	12.20	23.50		35.70	51
0120	Field stones		140	.286	S.F.	8.85	11.65		20.50	28.50
0130	Random sized flats, corner stones		70	.571	V.L.F.	11.95	23.50		35.45	50.50
0140	Field stones		140	.286	S.F.	10.10	11.65		21.75	30
0150	Horizontal lined ledgestones, corner stones		75	.533	V.L.F.	12.20	21.50		33.70	48.50
0160	Field stones		150	.267	S.F.	8.85	10.85		19.70	27
0170	Random shaped flats, corner stones		65	.615	V.L.F.	12.20	25		37.20	54
0180	Field stones		150	.267	S.F.	8.85	10.85		19.70	27
0190	Random shaped/textured face, corner stones		65	.615	V.L.F.	12.20	25		37.20	54
0200	Field stones		130	.308	S.F.	8.85	12.50		21.35	29.50
0210	Random shaped river rock, corner stones		65	.615	V.L.F.	12.20	25		37.20	54
0220	Field stones		130	.308	S.F.	8.85	12.50		21.35	29.50
0240	On concrete or CMU substrate, random sized cobbles, corner stones		70	.571	V.L.F.	11.50	23.50		35	50
0250	Field stones		140	.286	S.F.	8.50	11.65		20.15	28
0260	Random sized flats, corner stones		70	.571	V.L.F.	11.30	23.50		34.80	50
0270	Field stones		140	.286	S.F.	9.75	11.65		21.40	29.50
0280	Horizontal lined ledgestones, corner stones		75	.533	V.L.F.	11.50	21.50		33	47.50
0290	Field stones		150	.267	S.F.	8.50	10.85		19.35	27
0300	Random shaped flats, corner stones		70	.571	V.L.F.	11.50	23.50		35	50
0310	Field stones		140	.286	S.F.	8.50	11.65		20.15	28
0320	Random shaped/textured face, corner stones		65	.615	V.L.F.	11.50	25		36.50	53
0330	Field stones		130	.308	S.F.	8.50	12.50		21	29.50
0340	Random shaped river rock, corner stones		65	.615	V.L.F.	11.50	25		36.50	53
0350	Field stones	↓	130	.308	S.F.	8.50	12.50		21	29.50
0360	Cultured stone veneer, #15 felt weather resistant barrier	1 Clab	3700	.002	Sq.	5.65	.08		5.73	6.35
0370	Expanded metal lath, diamond, 2.5 lb./S.Y., galvanized	1 Lath	85	.094	S.Y.	3.02	3.69		6.71	9.10
0390	Water table or window sill, 18" long	1 Bric	80	.100	Ea.	10.60	4.41		15.01	18.75

Division Notes

	CREW	DAILY OUTPUT	LABOR-HOURS	UNIT	2012 BARE COSTS MAT.	LABOR	EQUIP.	TOTAL	TOTAL INCL O&P

Estimating Tips

05 05 00 Common Work Results for Metals

- Nuts, bolts, washers, connection angles, and plates can add a significant amount to both the tonnage of a structural steel job and the estimated cost. As a rule of thumb, add 10% to the total weight to account for these accessories.
- Type 2 steel construction, commonly referred to as "simple construction," consists generally of field-bolted connections with lateral bracing supplied by other elements of the building, such as masonry walls or x-bracing. The estimator should be aware, however, that shop connections may be accomplished by welding or bolting. The method may be particular to the fabrication shop and may have an impact on the estimated cost.

05 12 23 Structural Steel

- Steel items can be obtained from two sources: a fabrication shop or a metals service center. Fabrication shops can fabricate items under more controlled conditions than can crews in the field. They are also more efficient and can produce items more economically. Metal service centers serve as a source of long mill shapes to both fabrication shops and contractors.
- Most line items in this structural steel subdivision, and most items in 05 50 00 Metal Fabrications, are indicated as being shop fabricated. The bare material cost for these shop fabricated items is the "Invoice Cost" from the shop and includes the mill base price of steel plus mill extras, transportation to the shop, shop drawings and detailing where warranted, shop fabrication and handling, sandblasting and a shop coat of primer paint, all necessary structural bolts, and delivery to the job site. The bare labor cost and bare equipment cost for these shop fabricated items is for field installation or erection.
- Line items in Subdivision 05 12 23.40 Lightweight Framing, and other items scattered in Division 5, are indicated as being field fabricated. The bare material cost for these field fabricated items is the "Invoice Cost" from the metals service center and includes the mill base price of steel plus mill extras, transportation to the metals service center, material handling, and delivery of long lengths of mill shapes to the job site. Material costs for structural bolts and welding rods should be added to the estimate. The bare labor cost and bare equipment cost for these items is for both field fabrication and field installation or erection, and include time for cutting, welding and drilling in the fabricated metal items. Drilling into concrete and fasteners to fasten field fabricated items to other work are not included and should be added to the estimate.

05 20 00 Steel Joist Framing

- In any given project the total weight of open web steel joists is determined by the loads to be supported and the design. However, economies can be realized in minimizing the amount of labor used to place the joists. This is done by maximizing the joist spacing, and therefore minimizing the number of joists required to be installed on the job. Certain spacings and locations may be required by the design, but in other cases maximizing the spacing and keeping it as uniform as possible will keep the costs down.

05 30 00 Steel Decking

- The takeoff and estimating of metal deck involves more than simply the area of the floor or roof and the type of deck specified or shown on the drawings. Many different sizes and types of openings may exist. Small openings for individual pipes or conduits may be drilled after the floor/roof is installed, but larger openings may require special deck lengths as well as reinforcing or structural support. The estimator should determine who will be supplying this reinforcing. Additionally, some deck terminations are part of the deck package, such as screed angles and pour stops, and others will be part of the steel contract, such as angles attached to structural members and cast-in-place angles and plates. The estimator must ensure that all pieces are accounted for in the complete estimate.

05 50 00 Metal Fabrications

- The most economical steel stairs are those that use common materials, standard details, and most importantly, a uniform and relatively simple method of field assembly. Commonly available A36 channels and plates are very good choices for the main stringers of the stairs, as are angles and tees for the carrier members. Risers and treads are usually made by specialty shops, and it is most economical to use a typical detail in as many places as possible. The stairs should be pre-assembled and shipped directly to the site. The field connections should be simple and straightforward to be accomplished efficiently, and with minimum equipment and labor.

Reference Numbers

Reference numbers are shown in shaded boxes at the beginning of some major classifications. These numbers refer to related items in the Reference Section. The reference information may be an estimating procedure, an alternate pricing method, or technical information. *Note:* Not all subdivisions listed here necessarily appear in this publication.

05 05 Common Work Results for Metals

05 05 05 – Selective Metals Demolition

05 05 05.10 Selective Demolition, Metals		Crew	Daily Output	Labor-Hours	Unit	Material	2012 Bare Costs Labor	Equipment	Total	Total Incl O&P
0010	**SELECTIVE DEMOLITION, METALS** R024119-10									
0015	Excludes shores, bracing, cutting, loading, hauling, dumping									
0020	Remove nuts only up to 3/4" diameter	1 Sswk	480	.017	Ea.		.82		.82	1.53
0030	7/8" to 1-1/4" diameter		240	.033			1.64		1.64	3.06
0040	1-3/8" to 2" diameter		160	.050			2.47		2.47	4.59
0060	Unbolt and remove structural bolts up to 3/4" diameter		240	.033			1.64		1.64	3.06
0070	7/8" to 2" diameter		160	.050			2.47		2.47	4.59
0140	Light weight framing members, remove whole or cut up, up to 20 lb.		240	.033			1.64		1.64	3.06
0150	21 – 40 lb.	2 Sswk	210	.076			3.76		3.76	7
0160	41 – 80 lb.	3 Sswk	180	.133			6.60		6.60	12.25
0170	81 – 120 lb.	4 Sswk	150	.213			10.55		10.55	19.55
0230	Structural members, remove whole or cut up, up to 500 lb.	E-19	48	.500			24	19.80	43.80	65
0240	1/4 – 2 tons	E-18	36	1.111			54.50	26.50	81	128
0250	2 – 5 tons	E-24	30	1.067			52	23	75	119
0260	5 – 10 tons	E-20	24	2.667			129	48.50	177.50	286
0270	10 – 15 tons	E-2	18	3.111			150	83	233	360
0340	Fabricated item, remove whole or cut up, up to 20 lb.	1 Sswk	96	.083			4.11		4.11	7.65
0350	21 – 40 lb.	2 Sswk	84	.190			9.40		9.40	17.45
0360	41 – 80 lb.	3 Sswk	72	.333			16.45		16.45	30.50
0370	81 – 120 lb.	4 Sswk	60	.533			26.50		26.50	49
0380	121 – 500 lb.	E-19	48	.500			24	19.80	43.80	65
0390	501 – 1000 lb.	"	36	.667			32.50	26.50	59	86.50
0500	Steel roof decking, uncovered, bare	B-2	5000	.008	S.F.		.28		.28	.47
2950	Minimum labor/equipment charge	E-19	2	12	Job		580	475	1,055	1,550

05 05 21 – Fastening Methods for Metal

05 05 21.10 Cutting Steel

		Crew	Daily Output	Labor-Hours	Unit	Material	Labor	Equipment	Total	Total Incl O&P
0010	**CUTTING STEEL**									
0020	Hand burning, incl. preparation, torch cutting & grinding, no staging									
0100	Steel to 1/2" thick	E-25	320	.025	L.F.		1.28	.30	1.58	2.72
0150	3/4" thick		260	.031			1.58	.37	1.95	3.35
0200	1" thick		200	.040			2.05	.49	2.54	4.35
9000	Minimum labor/equipment charge		2	4	Job		205	48.50	253.50	435

05 05 21.15 Drilling Steel

		Crew	Daily Output	Labor-Hours	Unit	Material	Labor	Equipment	Total	Total Incl O&P
0010	**DRILLING STEEL**									
1910	Drilling & layout for steel, up to 1/4" deep, no anchor									
1920	Holes, 1/4" diameter	1 Sswk	112	.071	Ea.	.07	3.53		3.60	6.65
1925	For each additional 1/4" depth, add		336	.024		.07	1.18		1.25	2.26
1930	3/8" diameter		104	.077		.08	3.80		3.88	7.15
1935	For each additional 1/4" depth, add		312	.026		.08	1.27		1.35	2.44
1940	1/2" diameter		96	.083		.09	4.11		4.20	7.75
1945	For each additional 1/4" depth, add		288	.028		.09	1.37		1.46	2.64
1950	5/8" diameter		88	.091		.12	4.49		4.61	8.50
1955	For each additional 1/4" depth, add		264	.030		.12	1.50		1.62	2.91
1960	3/4" diameter		80	.100		.15	4.94		5.09	9.30
1965	For each additional 1/4" depth, add		240	.033		.15	1.64		1.79	3.23
1970	7/8" diameter		72	.111		.20	5.50		5.70	10.40
1975	For each additional 1/4" depth, add		216	.037		.20	1.83		2.03	3.62
1980	1" diameter		64	.125		.21	6.15		6.36	11.70
1985	For each additional 1/4" depth, add		192	.042		.21	2.06		2.27	4.05
1990	For drilling up, add						40%			
2000	Minimum labor/equipment charge	1 Carp	4	2	Job		88		88	145

05 05 21 – Fastening Methods for Metal

05 05 21.90 Welding Steel		Crew	Daily Output	Labor-Hours	Unit	Material	Labor	Equipment	Total	Total Incl O&P
						2012 Bare Costs				
0010	**WELDING STEEL**, Structural R050521-20									
0020	Field welding, 1/8" E6011, cost per welder, no oper. engr	E-14	8	1	Hr.	4.22	51.50	15.20	70.92	117
0200	With 1/2 operating engineer	E-13	8	1.500		4.22	73.50	15.20	92.92	152
0300	With 1 operating engineer	E-12	8	2		4.22	96	15.20	115.42	187
0500	With no operating engineer, 2# weld rod per ton	E-14	8	1	Ton	4.22	51.50	15.20	70.92	117
0600	8# E6011 per ton	"	2	4		16.85	205	61	282.85	465
0800	With one operating engineer per welder, 2# E6011 per ton	E-12	8	2		4.22	96	15.20	115.42	187
0900	8# E6011 per ton	"	2	8		16.85	385	61	462.85	750
1200	Continuous fillet, stick welding, incl. equipment									
1300	Single pass, 1/8" thick, 0.1#/L.F.	E-14	150	.053	L.F.	.21	2.74	.81	3.76	6.20
1400	3/16" thick, 0.2#/L.F.		75	.107		.42	5.50	1.62	7.54	12.45
1500	1/4" thick, 0.3#/L.F.		50	.160		.63	8.20	2.43	11.26	18.65
1610	5/16" thick, 0.4#/L.F.		38	.211		.84	10.80	3.20	14.84	24.50
1800	3 passes, 3/8" thick, 0.5#/L.F.		30	.267		1.05	13.70	4.05	18.80	31
2010	4 passes, 1/2" thick, 0.7#/L.F.		22	.364		1.48	18.65	5.55	25.68	42
2200	5 to 6 passes, 3/4" thick, 1.3#/L.F.		12	.667		2.74	34	10.15	46.89	77.50
2400	8 to 11 passes, 1" thick, 2.4#/L.F.		6	1.333		5.05	68.50	20.50	94.05	155
2600	For all position welding, add, minimum						20%			
2700	Maximum						300%			
2900	For semi-automatic welding, deduct, minimum						5%			
3000	Maximum						15%			
4000	Cleaning and welding plates, bars, or rods									
4010	to existing beams, columns, or trusses	E-14	12	.667	L.F.	1.05	34	10.15	45.20	76
9000	Minimum labor/equipment charge	"	4	2	Job		103	30.50	133.50	225

05 05 23 – Metal Fastenings

05 05 23.05 Anchor Bolts		Crew	Daily Output	Labor-Hours	Unit	Material	Labor	Equipment	Total	Total Incl O&P
0010	**ANCHOR BOLTS**									
0015	Made from recycled materials									
0025	Single bolts installed in fresh concrete, no templates									
0030	Hooked w/nut and washer, 1/2" diameter, 8" long G	1 Carp	132	.061	Ea.	1.39	2.67		4.06	5.90
0040	12" long G		131	.061		1.54	2.69		4.23	6.10
0050	5/8" diameter, 8" long G		129	.062		3	2.74		5.74	7.80
0060	12" long G		127	.063		3.70	2.78		6.48	8.60
0070	3/4" diameter, 8" long G		127	.063		3.70	2.78		6.48	8.60
0080	12" long G		125	.064		4.62	2.82		7.44	9.75
0090	2-bolt pattern, including job-built 2-hole template, per set									
0100	J-type, incl. hex nut & washer, 1/2" diameter x 6" long	1 Carp	21	.381	Set	4.55	16.80		21.35	32.50
0110	12" long		21	.381		5.15	16.80		21.95	33
0120	18" long		21	.381		6.10	16.80		22.90	34
0130	3/4" diameter x 8" long		20	.400		9.50	17.65		27.15	39.50
0140	12" long		20	.400		11.35	17.65		29	41.50
0150	18" long		20	.400		14.10	17.65		31.75	44.50
0160	1" diameter x 12" long		19	.421		19.90	18.55		38.45	52.50
0170	18" long		19	.421		23.50	18.55		42.05	56.50
0180	24" long		19	.421		28.50	18.55		47.05	61.50
0190	36" long		18	.444		38	19.60		57.60	74
0200	1-1/2" diameter x 18" long		17	.471		60	21		81	100
0210	24" long		16	.500		71.50	22		93.50	115
0300	L-type, incl. hex nut & washer, 3/4" diameter x 12" long		20	.400		10.75	17.65		28.40	41
0310	18" long		20	.400		13.15	17.65		30.80	43.50
0320	24" long		20	.400		15.60	17.65		33.25	46
0330	30" long		20	.400		19.25	17.65		36.90	50

05 05 23 – Metal Fastenings

05 05 23.05 Anchor Bolts		Crew	Daily Output	Labor-Hours	Unit	Material	Labor	Equipment	Total	Total Incl O&P
						2012 Bare Costs				
0340	36" long	1 Carp	20	.400	Set	21.50	17.65		39.15	53
0350	1" diameter x 12" long		19	.421		16.95	18.55		35.50	49
0360	18" long		19	.421		20.50	18.55		39.05	53
0370	24" long		19	.421		25	18.55		43.55	58
0380	30" long		19	.421		29	18.55		47.55	62.50
0390	36" long		18	.444		33	19.60		52.60	68.50
0400	42" long		18	.444		39.50	19.60		59.10	75.50
0410	48" long		18	.444		44	19.60		63.60	80.50
0420	1-1/4" diameter x 18" long		18	.444		30	19.60		49.60	65
0430	24" long		18	.444		35	19.60		54.60	71
0440	30" long		17	.471		40.50	21		61.50	78.50
0450	36" long	↓	17	.471	↓	45.50	21		66.50	84.50
1000	4-bolt pattern, including job-built 4-hole template, per set									
1100	J-type, incl. hex nut & washer, 1/2" diameter x 6" long G	1 Carp	19	.421	Set	7	18.55		25.55	38
1110	12" long G		19	.421		8.25	18.55		26.80	39.50
1120	18" long G		18	.444		10.10	19.60		29.70	43
1130	3/4" diameter x 8" long G		17	.471		16.85	21		37.85	52.50
1140	12" long G		17	.471		20.50	21		41.50	56.50
1150	18" long G		17	.471		26	21		47	62.50
1160	1" diameter x 12" long G		16	.500		38	22		60	77.50
1170	18" long G		15	.533		45	23.50		68.50	88
1180	24" long G		15	.533		54.50	23.50		78	98.50
1190	36" long G		15	.533		74	23.50		97.50	120
1200	1-1/2" diameter x 18" long G		13	.615		118	27		145	175
1210	24" long G		12	.667		140	29.50		169.50	203
1300	L-type, incl. hex nut & washer, 3/4" diameter x 12" long G		17	.471		19.35	21		40.35	55.50
1310	18" long G		17	.471		24	21		45	60.50
1320	24" long G		17	.471		29	21		50	66
1330	30" long G		16	.500		36.50	22		58.50	76
1340	36" long G		16	.500		41	22		63	81.50
1350	1" diameter x 12" long G		16	.500		32	22		54	71
1360	18" long G		15	.533		39	23.50		62.50	81.50
1370	24" long G		15	.533		48	23.50		71.50	91
1380	30" long G		15	.533		56	23.50		79.50	100
1390	36" long G		15	.533		64	23.50		87.50	109
1400	42" long G		14	.571		77	25		102	127
1410	48" long G		14	.571		86.50	25		111.50	137
1420	1-1/4" diameter x 18" long G		14	.571		58	25		83	105
1430	24" long G		14	.571		68.50	25		93.50	117
1440	30" long G		13	.615		79	27		106	131
1450	36" long G	↓	13	.615	↓	89.50	27		116.50	143

05 05 23.15 Chemical Anchors

05 05 23.15 Chemical Anchors		Crew	Daily Output	Labor-Hours	Unit	Material	Labor	Equipment	Total	Total Incl O&P
0010	**CHEMICAL ANCHORS**									
0020	Includes layout & drilling									
1430	Chemical anchor, w/rod & epoxy cartridge, 3/4" diam. x 9-1/2" long	B-89A	27	.593	Ea.	8.55	24	4.08	36.63	53
1435	1" diameter x 11-3/4" long		24	.667		15.50	27	4.59	47.09	66
1440	1-1/4" diameter x 14" long		21	.762		32	30.50	5.25	67.75	91
1445	1-3/4" diameter x 15" long		20	.800		60	32	5.50	97.50	125
1450	18" long		17	.941		72	38	6.50	116.50	148
1455	2" diameter x 18" long		16	1		87	40.50	6.90	134.40	169
1460	24" long	↓	15	1.067	↓	113	43	7.35	163.35	202

05 05 23 – Metal Fastenings

05 05 23.20 Expansion Anchors			Crew	Daily Output	Labor-Hours	Unit	2012 Bare Costs Material	Labor	Equipment	Total	Total Incl O&P
0010	**EXPANSION ANCHORS**										
0100	Anchors for concrete, brick or stone, no layout and drilling										
0200	Expansion shields, zinc, 1/4" diameter, 1-5/16" long, single	G	1 Carp	90	.089	Ea.	.42	3.92		4.34	6.90
0300	1-3/8" long, double	G		85	.094		.54	4.15		4.69	7.40
0400	3/8" diameter, 1-1/2" long, single	G		85	.094		.73	4.15		4.88	7.60
0500	2" long, double	G		80	.100		1.17	4.41		5.58	8.55
0600	1/2" diameter, 2-1/16" long, single	G		80	.100		1.40	4.41		5.81	8.80
0700	2-1/2" long, double	G		75	.107		1.82	4.70		6.52	9.70
0800	5/8" diameter, 2-5/8" long, single	G		75	.107		2.54	4.70		7.24	10.50
0900	2-3/4" long, double	G		70	.114		2.59	5.05		7.64	11.10
1000	3/4" diameter, 2-3/4" long, single	G		70	.114		2.79	5.05		7.84	11.30
1100	3-15/16" long, double	G		65	.123		5	5.45		10.45	14.40
2100	Hollow wall anchors for gypsum wall board, plaster or tile										
2300	1/8" diameter, short	G	1 Carp	160	.050	Ea.	.20	2.21		2.41	3.84
2400	Long	G		150	.053		.23	2.35		2.58	4.11
2500	3/16" diameter, short	G		150	.053		.36	2.35		2.71	4.26
2600	Long	G		140	.057		.49	2.52		3.01	4.67
2700	1/4" diameter, short	G		140	.057		.53	2.52		3.05	4.71
2800	Long	G		130	.062		.53	2.71		3.24	5.05
3000	Toggle bolts, bright steel, 1/8" diameter, 2" long	G		85	.094		.18	4.15		4.33	7
3100	4" long	G		80	.100		.22	4.41		4.63	7.50
3200	3/16" diameter, 3" long	G		80	.100		.22	4.41		4.63	7.50
3300	6" long	G		75	.107		.32	4.70		5.02	8.05
3400	1/4" diameter, 3" long	G		75	.107		.28	4.70		4.98	8
3500	6" long	G		70	.114		.45	5.05		5.50	8.75
3600	3/8" diameter, 3" long	G		70	.114		.67	5.05		5.72	9
3700	6" long	G		60	.133		1.18	5.90		7.08	10.95
3800	1/2" diameter, 4" long	G		60	.133		1.69	5.90		7.59	11.50
3900	6" long	G		50	.160		2.08	7.05		9.13	13.85
4000	Nailing anchors										
4100	Nylon nailing anchor, 1/4" diameter, 1" long		1 Carp	3.20	2.500	C	13.45	110		123.45	196
4200	1-1/2" long			2.80	2.857		16.60	126		142.60	225
4300	2" long			2.40	3.333		19.50	147		166.50	263
4400	Metal nailing anchor, 1/4" diameter, 1" long	G		3.20	2.500		15.75	110		125.75	198
4500	1-1/2" long	G		2.80	2.857		22	126		148	231
4600	2" long	G		2.40	3.333		26.50	147		173.50	270
8000	Wedge anchors, not including layout or drilling										
8050	Carbon steel, 1/4" diameter, 1-3/4" long	G	1 Carp	150	.053	Ea.	.36	2.35		2.71	4.26
8100	3-1/4" long	G		140	.057		.48	2.52		3	4.66
8150	3/8" diameter, 2-1/4" long	G		145	.055		.44	2.43		2.87	4.47
8200	5" long	G		140	.057		.77	2.52		3.29	4.98
8250	1/2" diameter, 2-3/4" long	G		140	.057		.89	2.52		3.41	5.10
8300	7" long	G		125	.064		1.53	2.82		4.35	6.30
8350	5/8" diameter, 3-1/2" long	G		130	.062		1.60	2.71		4.31	6.20
8400	8-1/2" long	G		115	.070		3.40	3.07		6.47	8.80
8450	3/4" diameter, 4-1/4" long	G		115	.070		2.38	3.07		5.45	7.65
8500	10" long	G		95	.084		5.40	3.71		9.11	12.05
8550	1" diameter, 6" long	G		100	.080		8.60	3.53		12.13	15.25
8575	9" long	G		85	.094		11.20	4.15		15.35	19.10
8600	12" long	G		75	.107		12.05	4.70		16.75	21
8650	1-1/4" diameter, 9" long	G		70	.114		24	5.05		29.05	34.50
8700	12" long	G		60	.133		30.50	5.90		36.40	43

05 05 Common Work Results for Metals

05 05 23 – Metal Fastenings

05 05 23.20 Expansion Anchors			Crew	Daily Output	Labor-Hours	Unit	2012 Bare Costs Material	Labor	Equipment	Total	Total Incl O&P
8750	For type 303 stainless steel, add						350%				
8800	For type 316 stainless steel, add						450%				
8950	Self-drilling concrete screw, hex washer head, 3/16" diam. x 1-3/4" long	G	1 Carp	300	.027	Ea.	.19	1.18		1.37	2.14
8960	2-1/4" long	G		250	.032		.20	1.41		1.61	2.53
8970	Phillips flat head, 3/16" diam. x 1-3/4" long	G		300	.027		.18	1.18		1.36	2.13
8980	2-1/4" long	G		250	.032		.20	1.41		1.61	2.53
9000	Minimum labor/equipment charge			4	2	Job		88		88	145

05 05 23.30 Lag Screws

Line	Description		Crew	Daily Output	Labor-Hours	Unit	Material	Labor	Equipment	Total	Total Incl O&P
0010	**LAG SCREWS**										
0020	Steel, 1/4" diameter, 2" long	G	1 Carp	200	.040	Ea.	.07	1.76		1.83	2.97
0200	1/2" diameter, 3" long	G		130	.062		.49	2.71		3.20	4.99
0300	5/8" diameter, 3" long	G		120	.067		1.01	2.94		3.95	5.95

05 05 23.40 Machinery Anchors

Line	Description		Crew	Daily Output	Labor-Hours	Unit	Material	Labor	Equipment	Total	Total Incl O&P
0010	**MACHINERY ANCHORS**, heavy duty, incl. sleeve, floating base nut,										
0020	lower stud & coupling nut, fiber plug, connecting stud, washer & nut.										
0030	For flush mounted embedment in poured concrete heavy equip. pads.										
0200	Stud & bolt, 1/2" diameter	G	E-16	40	.400	Ea.	68.50	20	3.04	91.54	116
0300	5/8" diameter	G		35	.457		76	23	3.47	102.47	130
0500	3/4" diameter	G		30	.533		87.50	27	4.05	118.55	151
0600	7/8" diameter	G		25	.640		95.50	32	4.86	132.36	170
0800	1" diameter	G		20	.800		100	40.50	6.10	146.60	192
0900	1-1/4" diameter	G		15	1.067		133	53.50	8.10	194.60	256

05 05 23.50 Powder Actuated Tools and Fasteners

Line	Description		Crew	Daily Output	Labor-Hours	Unit	Material	Labor	Equipment	Total	Total Incl O&P
0010	**POWDER ACTUATED TOOLS & FASTENERS**										
0020	Stud driver, .22 caliber, buy, minimum					Ea.	95			95	105
0100	Maximum					"	370			370	405
0300	Powder charges for above, low velocity					C	5.10			5.10	5.65
0400	Standard velocity						7			7	7.70
0600	Drive pins & studs, 1/4" & 3/8" diam., to 3" long, minimum	G	1 Carp	4.80	1.667		4.75	73.50		78.25	126
0700	Maximum	G	"	4	2		11.40	88		99.40	158

05 05 23.70 Structural Blind Bolts

Line	Description		Crew	Daily Output	Labor-Hours	Unit	Material	Labor	Equipment	Total	Total Incl O&P
0010	**STRUCTURAL BLIND BOLTS**										
0100	1/4" diameter x 1/4" grip	G	1 Sswk	240	.033	Ea.	1.24	1.64		2.88	4.42
0150	1/2" grip	G		216	.037		1.33	1.83		3.16	4.86
0200	3/8" diameter x 1/2" grip	G		232	.034		1.75	1.70		3.45	5.10
0250	3/4" grip	G		208	.038		1.83	1.90		3.73	5.55
0300	1/2" diameter x 1/2" grip	G		224	.036		5.15	1.76		6.91	8.90
0350	3/4" grip	G		200	.040		5.60	1.97		7.57	9.80
0400	5/8" diameter x 3/4" grip	G		216	.037		8.25	1.83		10.08	12.50
0450	1" grip	G		192	.042		9.50	2.06		11.56	14.25

05 05 23.87 Weld Studs

Line	Description		Crew	Daily Output	Labor-Hours	Unit	Material	Labor	Equipment	Total	Total Incl O&P
0010	**WELD STUDS**										
0020	1/4" diameter, 2-11/16" long	G	E-10	1120	.014	Ea.	.33	.72	.34	1.39	2.08
0100	4-1/8" long	G		1080	.015		.31	.75	.36	1.42	2.12
0200	3/8" diameter, 4-1/8" long	G		1080	.015		.36	.75	.36	1.47	2.18
0300	6-1/8" long	G		1040	.015		.47	.77	.37	1.61	2.36
9000	Minimum labor/equipment charge		1 Sswk	2	4	Job		197		197	365

05 12 Structural Steel Framing

05 12 23 – Structural Steel for Buildings

05 12 23.05 Canopy Framing

	05 12 23.05 Canopy Framing		Crew	Daily Output	Labor-Hours	Unit	Material	2012 Bare Costs Labor	Equipment	Total	Total Incl O&P
0010	**CANOPY FRAMING**										
0020	6" and 8" members, shop fabricated	G	E-4	3000	.011	Lb.	1.50	.53	.04	2.07	2.68
9000	Minimum labor/equipment charge		1 Sswk	1	8	Job		395		395	735

05 12 23.10 Ceiling Supports

	05 12 23.10 Ceiling Supports		Crew	Daily Output	Labor-Hours	Unit	Material	Labor	Equipment	Total	Total Incl O&P
0010	**CEILING SUPPORTS**										
1000	Entrance door/folding partition supports, shop fabricated	G	E-4	60	.533	L.F.	25	26.50	2.03	53.53	79
1100	Linear accelerator door supports	G		14	2.286		114	114	8.70	236.70	345
1200	Lintels or shelf angles, hung, exterior hot dipped galv.	G		267	.120		17.05	5.95	.46	23.46	30.50
1250	Two coats primer paint instead of galv.	G		267	.120		14.80	5.95	.46	21.21	28
1400	Monitor support, ceiling hung, expansion bolted	G		4	8	Ea.	395	400	30.50	825.50	1,200
1450	Hung from pre-set inserts	G		6	5.333		425	266	20.50	711.50	990
1600	Motor supports for overhead doors	G		4	8		202	400	30.50	632.50	995
1700	Partition support for heavy folding partitions, without pocket	G		24	1.333	L.F.	57	66.50	5.05	128.55	192
1750	Supports at pocket only	G		12	2.667		114	133	10.15	257.15	385
2000	Rolling grilles & fire door supports	G		34	.941		49	47	3.58	99.58	144
2100	Spider-leg light supports, expansion bolted to ceiling slab	G		8	4	Ea.	163	199	15.20	377.20	565
2150	Hung from pre-set inserts	G		12	2.667	"	175	133	10.15	318.15	450
2400	Toilet partition support	G		36	.889	L.F.	57	44.50	3.38	104.88	149
2500	X-ray travel gantry support	G		12	2.667	"	195	133	10.15	338.15	475

05 12 23.17 Columns, Structural

	05 12 23.17 Columns, Structural		Crew	Daily Output	Labor-Hours	Unit	Material	Labor	Equipment	Total	Total Incl O&P
0010	**COLUMNS, STRUCTURAL**	R051223-10									
0015	Made from recycled materials	G									
0020	Shop fab'd for 100-ton, 1-2 story project, bolted connections										
0800	Steel, concrete filled, extra strong pipe, 3-1/2" diameter		E-2	660	.085	L.F.	41.50	4.09	2.27	47.86	55.50
0830	4" diameter			780	.072		46	3.46	1.92	51.38	59
0890	5" diameter			1020	.055		55	2.65	1.47	59.12	66.50
0930	6" diameter			1200	.047		72.50	2.25	1.25	76	85.50
0940	8" diameter			1100	.051		72.50	2.46	1.36	76.32	86
1500	Steel pipe, extra strong, no concrete, 3" to 5" diameter	G		16000	.004	Lb.	1.25	.17	.09	1.51	1.78
3300	Structural tubing, square, A500GrB, 4" to 6" square, light section	G		11270	.005	"	1.25	.24	.13	1.62	1.96
8090	For projects 75 to 99 tons, add					All	10%				
8092	50 to 74 tons, add						20%				
8094	25 to 49 tons, add						30%	10%			
8096	10 to 24 tons, add						50%	25%			
8098	2 to 9 tons, add						75%	50%			
8099	Less than 2 tons, add						100%	100%			
9000	Minimum labor/equipment charge		1 Sswk	1	8	Job		395		395	735

05 12 23.20 Curb Edging

	05 12 23.20 Curb Edging		Crew	Daily Output	Labor-Hours	Unit	Material	Labor	Equipment	Total	Total Incl O&P
0010	**CURB EDGING**										
0020	Steel angle w/anchors, shop fabricated, on forms, 1" x 1", 0.8#/L.F.	G	E-4	350	.091	L.F.	1.58	4.56	.35	6.49	10.55
0300	4" x 4" angles, 8.2#/L.F.	G		275	.116	"	13	5.80	.44	19.24	25.50
9000	Minimum labor/equipment charge			4	8	Job		400	30.50	430.50	775

05 12 23.40 Lightweight Framing

	05 12 23.40 Lightweight Framing		Crew	Daily Output	Labor-Hours	Unit	Material	Labor	Equipment	Total	Total Incl O&P
0010	**LIGHTWEIGHT FRAMING**										
0015	Made from recycled materials	G									
0200	For load-bearing steel studs see Section 05 41 13.30										
0400	Angle framing, field fabricated, 4" and larger	G	E-3	440	.055	Lb.	.73	2.73	.28	3.74	6.15
0450	Less than 4" angles	G		265	.091		.75	4.53	.46	5.74	9.75
0600	Channel framing, field fabricated, 8" and larger	G		500	.048		.75	2.40	.24	3.39	5.55
0650	Less than 8" channels	G		335	.072		.75	3.58	.36	4.69	7.90
1000	Continuous slotted channel framing system, shop fab, minimum	G	2 Sswk	2400	.007		3.88	.33		4.21	4.87
1200	Maximum	G	"	1600	.010		4.38	.49		4.87	5.75

05 12 Structural Steel Framing

05 12 23 – Structural Steel for Buildings

05 12 23.40 Lightweight Framing			Crew	Daily Output	Labor-Hours	Unit	2012 Bare Costs Material	Labor	Equipment	Total	Total Incl O&P
1250	Plate & bar stock for reinforcing beams and trusses	G				Lb.	1.38			1.38	1.51
1300	Cross bracing, rods, shop fabricated, 3/4" diameter	G	E-3	700	.034		1.50	1.72	.17	3.39	5.05
1310	7/8" diameter	G		850	.028		1.50	1.41	.14	3.05	4.43
1320	1" diameter	G		1000	.024		1.50	1.20	.12	2.82	4.01
1330	Angle, 5" x 5" x 3/8"	G		2800	.009		1.50	.43	.04	1.97	2.50
1350	Hanging lintels, shop fabricated, average	G		850	.028		1.50	1.41	.14	3.05	4.43
1380	Roof frames, shop fabricated, 3'-0" square, 5' span	G	E-2	4200	.013		1.50	.64	.36	2.50	3.19
1400	Tie rod, not upset, 1-1/2" to 4" diameter, with turnbuckle	G	2 Sswk	800	.020		1.63	.99		2.62	3.62
1420	No turnbuckle	G		700	.023		1.56	1.13		2.69	3.82
1500	Upset, 1-3/4" to 4" diameter, with turnbuckle	G		800	.020		1.63	.99		2.62	3.62
1520	No turnbuckle	G		700	.023		1.56	1.13		2.69	3.82
9000	Minimum labor/equipment charge			2	8	Job		395		395	735

05 12 23.45 Lintels			Crew	Daily Output	Labor-Hours	Unit	Material	Labor	Equipment	Total	Total Incl O&P
0010	**LINTELS**										
0015	Made from recycled materials	G									
0020	Plain steel angles, shop fabricated, under 500 lb.	G	1 Bric	550	.015	Lb.	.96	.64		1.60	2.09
0100	500 to 1000 lb.	G		640	.013		.94	.55		1.49	1.92
0200	1,000 to 2,000 lb.	G		640	.013		.91	.55		1.46	1.89
0300	2,000 to 4,000 lb.	G		640	.013		.89	.55		1.44	1.87
0500	For built-up angles and plates, add to above	G					.31			.31	.34
0700	For engineering, add to above						.13			.13	.14
0900	For galvanizing, add to above, under 500 lb.						.30			.30	.33
0950	500 to 2,000 lb.						.27			.27	.30
1000	Over 2,000 lb.						.25			.25	.27
2000	Steel angles, 3-1/2" x 3", 1/4" thick, 2'-6" long	G	1 Bric	47	.170	Ea.	13.50	7.50		21	27
2100	4'-6" long	G		26	.308		24.50	13.55		38.05	48.50
2500	3-1/2" x 3-1/2" x 5/16", 5'-0" long	G		18	.444		36	19.60		55.60	71
2600	4" x 3-1/2", 1/4" thick, 5'-0" long	G		21	.381		31	16.80		47.80	61
2700	9'-0" long	G		12	.667		56	29.50		85.50	109
2800	4" x 3-1/2" x 5/16", 7'-0" long	G		12	.667		54	29.50		83.50	107
2900	5" x 3-1/2" x 5/16", 10'-0" long	G		8	1		87	44		131	167
9000	Minimum labor/equipment charge			4	2	Job		88		88	142

05 12 23.60 Pipe Support Framing			Crew	Daily Output	Labor-Hours	Unit	Material	Labor	Equipment	Total	Total Incl O&P
0010	**PIPE SUPPORT FRAMING**										
0020	Under 10#/L.F., shop fabricated	G	E-4	3900	.008	Lb.	1.68	.41	.03	2.12	2.63
0200	10.1 to 15#/L.F.	G		4300	.007		1.65	.37	.03	2.05	2.54
0400	15.1 to 20#/L.F.	G		4800	.007		1.63	.33	.03	1.99	2.44
0600	Over 20#/L.F.	G		5400	.006		1.60	.30	.02	1.92	2.33

05 12 23.65 Plates			Crew	Daily Output	Labor-Hours	Unit	Material	Labor	Equipment	Total	Total Incl O&P
0010	**PLATES**										
0015	Made from recycled materials	G									
0020	For connections & stiffener plates, shop fabricated										
0050	1/8" thick (5.1 lb./S.F.)	G				S.F.	6.40			6.40	7
0100	1/4" thick (10.2 lb./S.F.)	G					12.75			12.75	14.05
0300	3/8" thick (15.3 lb./S.F.)	G					19.15			19.15	21
0400	1/2" thick (20.4 lb./S.F.)	G					25.50			25.50	28
0450	3/4" thick (30.6 lb./S.F.)	G					38.50			38.50	42
0500	1" thick (40.8 lb./S.F.)	G					51			51	56
2000	Steel plate, warehouse prices, no shop fabrication										
2100	1/4" thick (10.2 lb./S.F.)	G				S.F.	8.65			8.65	9.50

05 12 Structural Steel Framing

05 12 23 – Structural Steel for Buildings

05 12 23.75 Structural Steel Members

	05 12 23.75 Structural Steel Members	Crew	Daily Output	Labor-Hours	Unit	2012 Bare Costs Material	Labor	Equipment	Total	Total Incl O&P
0010	**STRUCTURAL STEEL MEMBERS** R051223-10									
0015	Made from recycled materials G									
0020	Shop fab'd for 100-ton, 1-2 story project, bolted connections									
0102	W 6 x 9 R051223-15 G	E-2	600	.093	L.F.	12.40	4.50	2.49	19.39	24.50
0302	W 8 x 10 G		600	.093		13.75	4.50	2.49	20.74	26
0502	x 31 G		550	.102		42.50	4.91	2.72	50.13	58.50
0702	W 10 x 22 G		600	.093		30.50	4.50	2.49	37.49	44.50
0902	x 49 G		550	.102		67.50	4.91	2.72	75.13	85.50
1102	W 12 x 16 G		880	.064		22	3.07	1.70	26.77	31.50
1302	x 22 G		880	.064		30.50	3.07	1.70	35.27	41
1502	x 26 G		880	.064		36	3.07	1.70	40.77	47
1902	W 14 x 26 G		990	.057		36	2.73	1.51	40.24	46
2102	x 30 G		900	.062		41.50	3	1.66	46.16	52.50
2702	W 16 x 26 G		1000	.056		36	2.70	1.50	40.20	46
2902	x 31 G	↓	900	.062		42.50	3	1.66	47.16	54
8490	For projects 75 to 99 tons, add					10%				
8492	50 to 74 tons, add					20%				
8494	25 to 49 tons, add					30%	10%			
8496	10 to 24 tons, add					50%	25%			
8498	2 to 9 tons, add					75%	50%			
8499	Less than 2 tons, add				↓	100%	100%			
9000	Minimum labor/equipment charge	E-2	2	28	Job		1,350	750	2,100	3,225

05 12 23.77 Structural Steel Projects

	05 12 23.77 Structural Steel Projects	Crew	Daily Output	Labor-Hours	Unit	Material	Labor	Equipment	Total	Total Incl O&P
0010	**STRUCTURAL STEEL PROJECTS**									
0015	Made from recycled materials G									
0020	Shop fab'd for 100-ton, 1-2 story project, bolted connections									
0700	Offices, hospitals, etc., steel bearing, 1 to 2 stories R050521-20 G	E-5	10.30	7.767	Ton	2,500	380	157	3,037	3,600
1300	Industrial bldgs., 1 story, beams & girders, steel bearing G		12.90	6.202		2,500	305	125	2,930	3,425
1400	Masonry bearing R051223-10 G		10	8		2,500	390	162	3,052	3,625
1600	1 story with roof trusses, steel bearing G		10.60	7.547		2,950	370	153	3,473	4,075
1700	Masonry bearing R051223-15 G	↓	8.30	9.639		2,950	470	195	3,615	4,325
5390	For projects 75 to 99 tons, add					10%				
5392	50 to 74 tons, add					20%				
5394	25 to 49 tons, add					30%	10%			
5396	10 to 24 tons, add					50%	25%			
5398	2 to 9 tons, add					75%	50%			
5399	Less than 2 tons, add				↓	100%	100%			

05 15 Wire Rope Assemblies

05 15 16 – Steel Wire Rope Assemblies

05 15 16.70 Temporary Cable Safety Railing

	05 15 16.70 Temporary Cable Safety Railing	Crew	Daily Output	Labor-Hours	Unit	Material	Labor	Equipment	Total	Total Incl O&P
0010	**TEMPORARY CABLE SAFETY RAILING**, Each 100′ strand incl.									
0020	2 eyebolts, 1 turnbuckle, 100′ cable, 2 thimbles, 6 clips									
0025	Made from recycled materials G									
0100	One strand using 1/4″ cable & accessories G	2 Sswk	4	4	C.L.F.	196	197		393	580
0200	1/2″ cable & accessories G	″	2	8	″	410	395		805	1,175

05 21 Steel Joist Framing

05 21 13 – Deep Longspan Steel Joist Framing

05 21 13.50 Deep Longspan Joists			Crew	Daily Output	Labor-Hours	Unit	Material	Labor	Equipment	Total	Total Incl O&P
							2012 Bare Costs				
0010	**DEEP LONGSPAN JOISTS**										
3010	DLH series, 40-ton job lots, bolted cross bridging, shop primer										
3015	Made from recycled materials	G									
3020	Spans to 144′ (shipped in 2 pieces), minimum	G	E-7	16	5	Ton	1,650	244	109	2,003	2,375
3040	Average	G		13	6.154		1,775	300	134	2,209	2,650
3100	Maximum	G		11	7.273		2,150	355	158	2,663	3,200
3500	For less than 40-ton job lots										
3502	For 30 to 39 tons, add						10%				
3504	20 to 29 tons, add						20%				
3506	10 to 19 tons, add						30%				
3507	5 to 9 tons, add						50%	25%			
3508	1 to 4 tons, add						75%	50%			
3509	Less than 1 ton, add						100%	100%			
4010	SLH series, 40-ton job lots, bolted cross bridging, shop primer										
4020	Spans to 200′ (shipped in 3 pieces), minimum	G	E-7	16	5	Ton	1,625	244	109	1,978	2,350
4040	Average	G		13	6.154		1,825	300	134	2,259	2,725
4060	Maximum	G		11	7.273		2,175	355	158	2,688	3,225
6100	For less than 40-ton job lots										
6102	For 30 to 39 tons, add						10%				
6104	20 to 29 tons, add						20%				
6106	10 to 19 tons, add						30%				
6107	5 to 9 tons, add						50%	25%			
6108	1 to 4 tons, add						75%	50%			
6109	Less than 1 ton, add						100%	100%			

05 21 16 – Longspan Steel Joist Framing

05 21 16.50 Longspan Joists			Crew	Daily Output	Labor-Hours	Unit	Material	Labor	Equipment	Total	Total Incl O&P
0010	**LONGSPAN JOISTS**										
2000	LH series, 40-ton job lots, bolted cross bridging, shop primer										
2015	Made from recycled materials	G									
2020	Spans to 96′, minimum	G	E-7	16	5	Ton	1,550	244	109	1,903	2,250
2040	Average	G		13	6.154		1,700	300	134	2,134	2,575
2080	Maximum	G		11	7.273		2,000	355	158	2,513	3,025
2600	For less than 40-ton job lots										
2602	For 30 to 39 tons, add						10%				
2604	20 to 29 tons, add						20%				
2606	10 to 19 tons, add						30%				
2607	5 to 9 tons, add						50%	25%			
2608	1 to 4 tons, add						75%	50%			
2609	Less than 1 ton, add						100%	100%			

05 21 19 – Open Web Steel Joist Framing

05 21 19.10 Open Web Joists			Crew	Daily Output	Labor-Hours	Unit	Material	Labor	Equipment	Total	Total Incl O&P
0010	**OPEN WEB JOISTS**										
0015	Made from recycled materials	G									
0020	K series, 40-ton lots, horiz. bridging, spans to 30′, shop primer, minimum	G	E-7	15	5.333	Ton	1,350	260	116	1,726	2,100
0050	Average	G		12	6.667		1,525	325	145	1,995	2,425
0080	Maximum	G		9	8.889		1,825	435	193	2,453	3,025
0410	Span 30′ to 50′, minimum	G		17	4.706		1,325	230	102	1,657	2,000
0440	Average	G		17	4.706		1,500	230	102	1,832	2,175
0460	Maximum	G		10	8		1,600	390	174	2,164	2,650
0800	For less than 40-ton job lots										
0802	For 30 to 39 tons, add						10%				
0804	20 to 29 tons, add						20%				

05 21 Steel Joist Framing

05 21 19 – Open Web Steel Joist Framing

05 21 19.10 Open Web Joists			Crew	Daily Output	Labor-Hours	Unit	2012 Bare Costs Material	Labor	Equipment	Total	Total Incl O&P
0806	10 to 19 tons, add						30%				
0807	5 to 9 tons, add						50%	25%			
0808	1 to 4 tons, add						75%	50%			
0809	Less than 1 ton, add						100%	100%			
1010	CS series, 40-ton job lots, horizontal bridging, shop primer										
1020	Spans to 30′, minimum	G	E-7	15	5.333	Ton	1,425	260	116	1,801	2,150
1040	Average	G		12	6.667		1,575	325	145	2,045	2,475
1060	Maximum	G	↓	9	8.889	↓	1,850	435	193	2,478	3,050
1500	For less than 40-ton job lots										
1502	For 30 to 39 tons, add						10%				
1504	20 to 29 tons, add						20%				
1506	10 to 19 tons, add						30%				
1507	5 to 9 tons, add						50%	25%			
1508	1 to 4 tons, add						75%	50%			
1509	Less than 1 ton, add						100%	100%			
6400	Individual steel bearing plate, 6″ x 6″ x 1/4″ with J-hook	G	1 Bric	160	.050	Ea.	7.50	2.21		9.71	11.80
9000	Minimum labor and equipment, bar joists		F-6	1	40	Job		1,650	610	2,260	3,350

05 31 Steel Decking

05 31 13 – Steel Floor Decking

05 31 13.50 Floor Decking

Code	Description		Crew	Daily Output	Labor-Hours	Unit	Material	Labor	Equipment	Total	Total Incl O&P
0010	**FLOOR DECKING** R053100-10										
0015	Made from recycled materials	G									
5100	Non-cellular composite decking, galvanized, 1-1/2″ deep, 16 gauge	G	E-4	3500	.009	S.F.	2.95	.46	.03	3.44	4.13
5120	18 gauge	G		3650	.009		2.39	.44	.03	2.86	3.48
5140	20 gauge	G		3800	.008		1.90	.42	.03	2.35	2.91
5200	2″ deep, 22 gauge	G		3860	.008		1.65	.41	.03	2.09	2.62
5300	20 gauge	G		3600	.009		1.83	.44	.03	2.30	2.87
5400	18 gauge	G		3380	.009		2.34	.47	.04	2.85	3.49
5500	16 gauge	G		3200	.010		2.91	.50	.04	3.45	4.17
5700	3″ deep, 22 gauge	G		3200	.010		1.80	.50	.04	2.34	2.95
5800	20 gauge	G		3000	.011		2.01	.53	.04	2.58	3.24
5900	18 gauge	G		2850	.011		2.49	.56	.04	3.09	3.83
6000	16 gauge	G	↓	2700	.012	↓	3.31	.59	.05	3.95	4.79
9000	Minimum labor/equipment charge		1 Sswk	1	8	Job		395		395	735

05 31 23 – Steel Roof Decking

05 31 23.50 Roof Decking

Code	Description		Crew	Daily Output	Labor-Hours	Unit	Material	Labor	Equipment	Total	Total Incl O&P
0010	**ROOF DECKING**										
0015	Made from recycled materials	G									
2100	Open type, 1-1/2″ deep, Type B, wide rib, galv., 22 gauge, under 50 sq.	G	E-4	4500	.007	S.F.	1.74	.35	.03	2.12	2.60
2400	Over 500 squares	G	″	5100	.006	″	1.25	.31	.02	1.58	1.99

05 31 33 – Steel Form Decking

05 31 33.50 Form Decking

Code	Description		Crew	Daily Output	Labor-Hours	Unit	Material	Labor	Equipment	Total	Total Incl O&P
0010	**FORM DECKING**										
0015	Made from recycled materials	G									
6100	Slab form, steel, 28 gauge, 9/16″ deep, Type UFS, uncoated	G	E-4	4000	.008	S.F.	1.32	.40	.03	1.75	2.22
6200	Galvanized	G	″	4000	.008	″	1.17	.40	.03	1.60	2.06

05 35 Raceway Decking Assemblies

05 35 13 – Steel Cellular Decking

05 35 13.50 Cellular Decking			Crew	Daily Output	Labor-Hours	Unit	Material	2012 Bare Costs Labor	Equipment	Total	Total Incl O&P
0010	**CELLULAR DECKING**										
0015	Made from recycled materials	G									
0200	Cellular units, galv, 1-1/2" deep, Type BC, 20-20 gauge, over 15 squares	G	E-4	1460	.022	S.F.	6.95	1.09	.08	8.12	9.70
0400	3" deep, Type NC, galvanized, 20-20 gauge	G		1375	.023		7.60	1.16	.09	8.85	10.65
1000	4-1/2" deep, Type JC, galvanized, 18-20 gauge	G		1100	.029		10.60	1.45	.11	12.16	14.50
1500	For acoustical deck, add						15%				
1900	For multi-story or congested site, add							50%			

05 41 Structural Metal Stud Framing

05 41 13 – Load-Bearing Metal Stud Framing

05 41 13.05 Bracing

Line	Description		Crew	Daily Output	Labor-Hours	Unit	Material	Labor	Equipment	Total	Total Incl O&P
0010	**BRACING**, shear wall X-bracing, per 10' x 10' bay, one face										
0015	Made of recycled materials	G									
0120	Metal strap, 20 ga x 4" wide	G	2 Carp	18	.889	Ea.	19.80	39		58.80	86.50
0130	6" wide	G		18	.889		33	39		72	101
0160	18 ga x 4" wide	G		16	1		34	44		78	110
0170	6" wide	G		16	1		50	44		94	128
0410	Continuous strap bracing, per horizontal row on both faces										
0420	Metal strap, 20 ga x 2" wide, studs 12" O.C.	G	1 Carp	7	1.143	C.L.F.	59.50	50.50		110	148
0430	16" O.C.	G		8	1		59.50	44		103.50	138
0440	24" O.C.	G		10	.800		59.50	35.50		95	124
0450	18 ga x 2" wide, studs 12" O.C.	G		6	1.333		81.50	59		140.50	186
0460	16" O.C.	G		7	1.143		81.50	50.50		132	172
0470	24" O.C.	G		8	1		81.50	44		125.50	162

05 41 13.10 Bridging

Line	Description		Crew	Daily Output	Labor-Hours	Unit	Material	Labor	Equipment	Total	Total Incl O&P
0010	**BRIDGING**, solid between studs w/1-1/4" leg track, per stud bay										
0015	Made from recycled materials	G									
0200	Studs 12" O.C., 18 ga x 2-1/2" wide	G	1 Carp	125	.064	Ea.	.99	2.82		3.81	5.70
0210	3-5/8" wide	G		120	.067		1.20	2.94		4.14	6.15
0220	4" wide	G		120	.067		1.26	2.94		4.20	6.20
0230	6" wide	G		115	.070		1.65	3.07		4.72	6.85
0240	8" wide	G		110	.073		2.05	3.21		5.26	7.50
0300	16 ga x 2-1/2" wide	G		115	.070		1.26	3.07		4.33	6.45
0310	3-5/8" wide	G		110	.073		1.53	3.21		4.74	6.95
0320	4" wide	G		110	.073		1.62	3.21		4.83	7.05
0330	6" wide	G		105	.076		2.08	3.36		5.44	7.80
0340	8" wide	G		100	.080		2.61	3.53		6.14	8.65
1200	Studs 16" O.C., 18 ga x 2-1/2" wide	G		125	.064		1.27	2.82		4.09	6.05
1210	3-5/8" wide	G		120	.067		1.53	2.94		4.47	6.50
1220	4" wide	G		120	.067		1.62	2.94		4.56	6.60
1230	6" wide	G		115	.070		2.11	3.07		5.18	7.35
1240	8" wide	G		110	.073		2.62	3.21		5.83	8.15
1300	16 ga x 2-1/2" wide	G		115	.070		1.62	3.07		4.69	6.85
1310	3-5/8" wide	G		110	.073		1.96	3.21		5.17	7.40
1320	4" wide	G		110	.073		2.08	3.21		5.29	7.55
1330	6" wide	G		105	.076		2.67	3.36		6.03	8.45
1340	8" wide	G		100	.080		3.35	3.53		6.88	9.50
2200	Studs 24" O.C., 18 ga x 2-1/2" wide	G		125	.064		1.84	2.82		4.66	6.65
2210	3-5/8" wide	G		120	.067		2.22	2.94		5.16	7.25
2220	4" wide	G		120	.067		2.34	2.94		5.28	7.40
2230	6" wide	G		115	.070		3.06	3.07		6.13	8.40

05 41 Structural Metal Stud Framing

05 41 13 – Load-Bearing Metal Stud Framing

05 41 13.10 Bridging			Crew	Daily Output	Labor-Hours	Unit	Material	2012 Bare Costs Labor	Equipment	Total	Total Incl O&P
2240	8" wide	G	1 Carp	110	.073	Ea.	3.80	3.21		7.01	9.40
2300	16 ga x 2-1/2" wide	G		115	.070		2.34	3.07		5.41	7.60
2310	3-5/8" wide	G		110	.073		2.84	3.21		6.05	8.35
2320	4" wide	G		110	.073		3.01	3.21		6.22	8.55
2330	6" wide	G		105	.076		3.87	3.36		7.23	9.75
2340	8" wide	G		100	.080		4.85	3.53		8.38	11.15
3000	Continuous bridging, per row										
3100	16 ga x 1-1/2" channel thru studs 12" O.C.	G	1 Carp	6	1.333	C.L.F.	54	59		113	156
3110	16" O.C.	G		7	1.143		54	50.50		104.50	142
3120	24" O.C.	G		8.80	.909		54	40		94	125
4100	2" x 2" angle x 18 ga, studs 12" O.C.	G		7	1.143		82.50	50.50		133	174
4110	16" O.C.	G		9	.889		82.50	39		121.50	156
4120	24" O.C.	G		12	.667		82.50	29.50		112	139
4200	16 ga, studs 12" O.C.	G		5	1.600		106	70.50		176.50	232
4210	16" O.C.	G		7	1.143		106	50.50		156.50	199
4220	24" O.C.	G		10	.800		106	35.50		141.50	174

05 41 13.25 Framing, Boxed Headers/Beams

			Crew	Daily Output	Labor-Hours	Unit	Material	Labor	Equipment	Total	Total Incl O&P
0010	**FRAMING, BOXED HEADERS/BEAMS**										
0015	Made from recycled materials	G									
0200	Double, 18 ga x 6" deep	G	2 Carp	220	.073	L.F.	5.70	3.21		8.91	11.50
0210	8" deep	G		210	.076		6.25	3.36		9.61	12.40
0220	10" deep	G		200	.080		7.65	3.53		11.18	14.20
0230	12" deep	G		190	.084		8.35	3.71		12.06	15.30
0300	16 ga x 8" deep	G		180	.089		7.25	3.92		11.17	14.40
0310	10" deep	G		170	.094		8.75	4.15		12.90	16.40
0320	12" deep	G		160	.100		9.50	4.41		13.91	17.70
0400	14 ga x 10" deep	G		140	.114		10.10	5.05		15.15	19.35
0410	12" deep	G		130	.123		11.05	5.45		16.50	21
1210	Triple, 18 ga x 8" deep	G		170	.094		9.10	4.15		13.25	16.80
1220	10" deep	G		165	.097		11	4.28		15.28	19.10
1230	12" deep	G		160	.100		12.05	4.41		16.46	20.50
1300	16 ga x 8" deep	G		145	.110		10.50	4.87		15.37	19.55
1310	10" deep	G		140	.114		12.65	5.05		17.70	22
1320	12" deep	G		135	.119		13.75	5.25		19	23.50
1400	14 ga x 10" deep	G		115	.139		13.75	6.15		19.90	25
1410	12" deep	G		110	.145		15.20	6.40		21.60	27.50

05 41 13.30 Framing, Stud Walls

			Crew	Daily Output	Labor-Hours	Unit	Material	Labor	Equipment	Total	Total Incl O&P
0010	**FRAMING, STUD WALLS** w/top & bottom track, no openings,										
0020	Headers, beams, bridging or bracing										
0025	Made from recycled materials	G									
4100	8' high walls, 18 ga x 2-1/2" wide, studs 12" O.C.	G	2 Carp	54	.296	L.F.	9.40	13.05		22.45	32
4110	16" O.C.	G		77	.208		7.55	9.15		16.70	23.50
4120	24" O.C.	G		107	.150		5.65	6.60		12.25	17
4130	3-5/8" wide, studs 12" O.C.	G		53	.302		11.25	13.30		24.55	34.50
4140	16" O.C.	G		76	.211		9	9.30		18.30	25
4150	24" O.C.	G		105	.152		6.75	6.70		13.45	18.45
4160	4" wide, studs 12" O.C.	G		52	.308		11.75	13.55		25.30	35.50
4170	16" O.C.	G		74	.216		9.40	9.55		18.95	26
4180	24" O.C.	G		103	.155		7.10	6.85		13.95	19.05
4190	6" wide, studs 12" O.C.	G		51	.314		14.90	13.85		28.75	39
4200	16" O.C.	G		73	.219		11.95	9.65		21.60	29
4210	24" O.C.	G		101	.158		9	7		16	21.50

05 41 13 – Load-Bearing Metal Stud Framing

05 41 13.30 Framing, Stud Walls		Crew	Daily Output	Labor-Hours	Unit	Material	Labor	Equipment	Total	Total Incl O&P
						2012 Bare Costs				
4220	8" wide, studs 12" O.C. G	2 Carp	50	.320	L.F.	18.15	14.10		32.25	43
4230	16" O.C. G		72	.222		14.60	9.80		24.40	32
4240	24" O.C. G		100	.160		11	7.05		18.05	23.50
4300	16 ga x 2-1/2" wide, studs 12" O.C. G		47	.340		11.15	15		26.15	37
4310	16" O.C. G		68	.235		8.85	10.40		19.25	27
4320	24" O.C. G		94	.170		6.55	7.50		14.05	19.50
4330	3-5/8" wide, studs 12" O.C. G		46	.348		13.30	15.35		28.65	39.50
4340	16" O.C. G		66	.242		10.55	10.70		21.25	29
4350	24" O.C. G		92	.174		7.80	7.65		15.45	21
4360	4" wide, studs 12" O.C. G		45	.356		14.10	15.70		29.80	41
4370	16" O.C. G		65	.246		11.15	10.85		22	30
4380	24" O.C. G		90	.178		8.25	7.85		16.10	22
4390	6" wide, studs 12" O.C. G		44	.364		17.55	16.05		33.60	46
4400	16" O.C. G		64	.250		13.95	11.05		25	33.50
4410	24" O.C. G		88	.182		10.35	8		18.35	24.50
4420	8" wide, studs 12" O.C. G		43	.372		21.50	16.40		37.90	51
4430	16" O.C. G		63	.254		17.20	11.20		28.40	37.50
4440	24" O.C. G		86	.186		12.80	8.20		21	27.50
5100	10' high walls, 18 ga x 2-1/2" wide, studs 12" O.C. G		54	.296		11.30	13.05		24.35	34
5110	16" O.C. G		77	.208		8.95	9.15		18.10	25
5120	24" O.C. G		107	.150		6.60	6.60		13.20	18.05
5130	3-5/8" wide, studs 12" O.C. G		53	.302		13.50	13.30		26.80	37
5140	16" O.C. G		76	.211		10.70	9.30		20	27
5150	24" O.C. G		105	.152		7.90	6.70		14.60	19.65
5160	4" wide, studs 12" O.C. G		52	.308		14.10	13.55		27.65	38
5170	16" O.C. G		74	.216		11.20	9.55		20.75	28
5180	24" O.C. G		103	.155		8.25	6.85		15.10	20.50
5190	6" wide, studs 12" O.C. G		51	.314		17.85	13.85		31.70	42
5200	16" O.C. G		73	.219		14.15	9.65		23.80	31.50
5210	24" O.C. G		101	.158		10.50	7		17.50	23
5220	8" wide, studs 12" O.C. G		50	.320		21.50	14.10		35.60	47
5230	16" O.C. G		72	.222		17.25	9.80		27.05	35
5240	24" O.C. G		100	.160		12.80	7.05		19.85	25.50
5300	16 ga x 2-1/2" wide, studs 12" O.C. G		47	.340		13.50	15		28.50	39.50
5310	16" O.C. G		68	.235		10.60	10.40		21	28.50
5320	24" O.C. G		94	.170		7.70	7.50		15.20	21
5330	3-5/8" wide, studs 12" O.C. G		46	.348		16.10	15.35		31.45	42.50
5340	16" O.C. G		66	.242		12.65	10.70		23.35	31.50
5350	24" O.C. G		92	.174		9.20	7.65		16.85	22.50
5360	4" wide, studs 12" O.C. G		45	.356		17	15.70		32.70	44
5370	16" O.C. G		65	.246		13.35	10.85		24.20	32.50
5380	24" O.C. G		90	.178		9.70	7.85		17.55	23.50
5390	6" wide, studs 12" O.C. G		44	.364		21	16.05		37.05	50
5400	16" O.C. G		64	.250		16.65	11.05		27.70	36.50
5410	24" O.C. G		88	.182		12.15	8		20.15	26.50
5420	8" wide, studs 12" O.C. G		43	.372		26	16.40		42.40	55.50
5430	16" O.C. G		63	.254		20.50	11.20		31.70	41
5440	24" O.C. G		86	.186		15	8.20		23.20	30
6190	12' high walls, 18 ga x 6" wide, studs 12" O.C. G		41	.390		21	17.20		38.20	51
6200	16" O.C. G		58	.276		16.35	12.15		28.50	38
6210	24" O.C. G		81	.198		11.95	8.70		20.65	27.50
6220	8" wide, studs 12" O.C. G		40	.400		25.50	17.65		43.15	57
6230	16" O.C. G	↓	57	.281	↓	19.90	12.40		32.30	42.50

05 41 Structural Metal Stud Framing

05 41 13 – Load-Bearing Metal Stud Framing

05 41 13.30 Framing, Stud Walls			Crew	Daily Output	Labor-Hours	Unit	Material	2012 Bare Costs Labor	Equipment	Total	Total Incl O&P
6240	24" O.C.	G	2 Carp	80	.200	L.F.	14.60	8.80		23.40	30.50
6390	16 ga x 6" wide, studs 12" O.C.	G		35	.457		24.50	20		44.50	60
6400	16" O.C.	G		51	.314		19.35	13.85		33.20	44
6410	24" O.C.	G		70	.229		13.95	10.10		24.05	32
6420	8" wide, studs 12" O.C.	G		34	.471		30.50	21		51.50	67.50
6430	16" O.C.	G		50	.320		24	14.10		38.10	49.50
6440	24" O.C.	G		69	.232		17.20	10.25		27.45	35.50
6530	14 ga x 3-5/8" wide, studs 12" O.C.	G		34	.471		23.50	21		44.50	60
6540	16" O.C.	G		48	.333		18.50	14.70		33.20	44.50
6550	24" O.C.	G		65	.246		13.30	10.85		24.15	32.50
6560	4" wide, studs 12" O.C.	G		33	.485		25	21.50		46.50	62.50
6570	16" O.C.	G		47	.340		19.45	15		34.45	46
6580	24" O.C.	G		64	.250		14	11.05		25.05	33.50
6730	12 ga x 3-5/8" wide, studs 12" O.C.	G		31	.516		33	23		56	73.50
6740	16" O.C.	G		43	.372		25.50	16.40		41.90	55
6750	24" O.C.	G		59	.271		17.85	11.95		29.80	39.50
6760	4" wide, studs 12" O.C.	G		30	.533		35	23.50		58.50	77
6770	16" O.C.	G		42	.381		27	16.80		43.80	57
6780	24" O.C.	G		58	.276		19	12.15		31.15	41
7390	16' high walls, 16 ga x 6" wide, studs 12" O.C.	G		33	.485		32	21.50		53.50	70
7400	16" O.C.	G		48	.333		24.50	14.70		39.20	51
7410	24" O.C.	G		67	.239		17.55	10.55		28.10	36.50
7420	8" wide, studs 12" O.C.	G		32	.500		39.50	22		61.50	79.50
7430	16" O.C.	G		47	.340		30.50	15		45.50	58
7440	24" O.C.	G		66	.242		21.50	10.70		32.20	41.50
7560	14 ga x 4" wide, studs 12" O.C.	G		31	.516		32	23		55	73
7570	16" O.C.	G		45	.356		25	15.70		40.70	53
7580	24" O.C.	G		61	.262		17.65	11.55		29.20	38.50
7590	6" wide, studs 12" O.C.	G		30	.533		40.50	23.50		64	83
7600	16" O.C.	G		44	.364		31.50	16.05		47.55	61
7610	24" O.C.	G		60	.267		22.50	11.75		34.25	44
7760	12 ga x 4" wide, studs 12" O.C.	G		29	.552		45.50	24.50		70	90
7770	16" O.C.	G		40	.400		35	17.65		52.65	67.50
7780	24" O.C.	G		55	.291		24.50	12.85		37.35	47.50
7790	6" wide, studs 12" O.C.	G		28	.571		57.50	25		82.50	105
7800	16" O.C.	G		39	.410		44	18.10		62.10	78
7810	24" O.C.	G		54	.296		31	13.05		44.05	55.50
8590	20' high walls, 14 ga x 6" wide, studs 12" O.C.	G		29	.552		50	24.50		74.50	95
8600	16" O.C.	G		42	.381		38.50	16.80		55.30	69.50
8610	24" O.C.	G		57	.281		27	12.40		39.40	50
8620	8" wide, studs 12" O.C.	G		28	.571		54	25		79	101
8630	16" O.C.	G		41	.390		41.50	17.20		58.70	74
8640	24" O.C.	G		56	.286		29.50	12.60		42.10	53
8790	12 ga x 6" wide, studs 12" O.C.	G		27	.593		71	26		97	121
8800	16" O.C.	G		37	.432		54	19.05		73.05	91
8810	24" O.C.	G		51	.314		37.50	13.85		51.35	63.50
8820	8" wide, studs 12" O.C.	G		26	.615		86	27		113	139
8830	16" O.C.	G		36	.444		65.50	19.60		85.10	104
8840	24" O.C.	G		50	.320		45.50	14.10		59.60	73
9000	Minimum labor/equipment charge			4	4	Job		176		176	289

05 42 Cold-Formed Metal Joist Framing

05 42 13 – Cold-Formed Metal Floor Joist Framing

05 42 13.05 Bracing		Crew	Daily Output	Labor-Hours	Unit	Material	Labor	Equipment	Total	Total Incl O&P
						2012 Bare Costs				
0010	**BRACING**, continuous, per row, top & bottom									
0015	Made from recycled materials G									
0120	Flat strap, 20 ga x 2" wide, joists at 12" O.C. G	1 Carp	4.67	1.713	C.L.F.	62	75.50		137.50	193
0130	16" O.C. G		5.33	1.501		60	66		126	175
0140	24" O.C. G		6.66	1.201		58	53		111	151
0150	18 ga x 2" wide, joists at 12" O.C. G		4	2		80.50	88		168.50	234
0160	16" O.C. G		4.67	1.713		79	75.50		154.50	211
0170	24" O.C. G	↓	5.33	1.501	↓	77.50	66		143.50	195

05 42 13.10 Bridging		Crew	Daily Output	Labor-Hours	Unit	Material	Labor	Equipment	Total	Total Incl O&P
0010	**BRIDGING**, solid between joists w/1-1/4" leg track, per joist bay									
0015	Made from recycled materials G									
0230	Joists 12" O.C., 18 ga track x 6" wide G	1 Carp	80	.100	Ea.	1.65	4.41		6.06	9.05
0240	8" wide G		75	.107		2.05	4.70		6.75	9.95
0250	10" wide G		70	.114		2.55	5.05		7.60	11.05
0260	12" wide G		65	.123		2.90	5.45		8.35	12.10
0330	16 ga track x 6" wide G		70	.114		2.08	5.05		7.13	10.55
0340	8" wide G		65	.123		2.61	5.45		8.06	11.75
0350	10" wide G		60	.133		3.23	5.90		9.13	13.20
0360	12" wide G		55	.145		3.71	6.40		10.11	14.60
0440	14 ga track x 8" wide G		60	.133		3.26	5.90		9.16	13.25
0450	10" wide G		55	.145		4.04	6.40		10.44	14.95
0460	12" wide G		50	.160		4.67	7.05		11.72	16.70
0550	12 ga track x 10" wide G		45	.178		5.90	7.85		13.75	19.35
0560	12" wide G		40	.200		6.05	8.80		14.85	21
1230	16" O.C., 18 ga track x 6" wide G		80	.100		2.11	4.41		6.52	9.55
1240	8" wide G		75	.107		2.62	4.70		7.32	10.60
1250	10" wide G		70	.114		3.27	5.05		8.32	11.85
1260	12" wide G		65	.123		3.71	5.45		9.16	13
1330	16 ga track x 6" wide G		70	.114		2.67	5.05		7.72	11.20
1340	8" wide G		65	.123		3.35	5.45		8.80	12.60
1350	10" wide G		60	.133		4.14	5.90		10.04	14.20
1360	12" wide G		55	.145		4.75	6.40		11.15	15.75
1440	14 ga track x 8" wide G		60	.133		4.17	5.90		10.07	14.25
1450	10" wide G		55	.145		5.20	6.40		11.60	16.20
1460	12" wide G		50	.160		6	7.05		13.05	18.15
1550	12 ga track x 10" wide G		45	.178		7.55	7.85		15.40	21
1560	12" wide G		40	.200		7.75	8.80		16.55	23
2230	24" O.C., 18 ga track x 6" wide G		80	.100		3.06	4.41		7.47	10.60
2240	8" wide G		75	.107		3.80	4.70		8.50	11.85
2250	10" wide G		70	.114		4.73	5.05		9.78	13.45
2260	12" wide G		65	.123		5.35	5.45		10.80	14.80
2330	16 ga track x 6" wide G		70	.114		3.87	5.05		8.92	12.50
2340	8" wide G		65	.123		4.85	5.45		10.30	14.25
2350	10" wide G		60	.133		6	5.90		11.90	16.25
2360	12" wide G		55	.145		6.85	6.40		13.25	18.05
2440	14 ga track x 8" wide G		60	.133		6.05	5.90		11.95	16.30
2450	10" wide G		55	.145		7.50	6.40		13.90	18.75
2460	12" wide G		50	.160		8.65	7.05		15.70	21
2550	12 ga track x 10" wide G		45	.178		10.95	7.85		18.80	25
2560	12" wide G	↓	40	.200	↓	11.25	8.80		20.05	27

05 42 13 – Cold-Formed Metal Floor Joist Framing

05 42 13.25 Framing, Band Joist			Crew	Daily Output	Labor-Hours	Unit	Material	Labor (2012 Bare Costs)	Equipment (2012 Bare Costs)	Total (2012 Bare Costs)	Total Incl O&P
0010	**FRAMING, BAND JOIST** (track) fastened to bearing wall										
0015	Made from recycled materials	G									
0220	18 ga track x 6″ deep	G	2 Carp	1000	.016	L.F.	1.34	.71		2.05	2.64
0230	8″ deep	G		920	.017		1.67	.77		2.44	3.10
0240	10″ deep	G		860	.019		2.08	.82		2.90	3.63
0320	16 ga track x 6″ deep	G		900	.018		1.70	.78		2.48	3.16
0330	8″ deep	G		840	.019		2.13	.84		2.97	3.72
0340	10″ deep	G		780	.021		2.64	.90		3.54	4.38
0350	12″ deep	G		740	.022		3.02	.95		3.97	4.89
0430	14 ga track x 8″ deep	G		750	.021		2.66	.94		3.60	4.46
0440	10″ deep	G		720	.022		3.30	.98		4.28	5.25
0450	12″ deep	G		700	.023		3.81	1.01		4.82	5.85
0540	12 ga track x 10″ deep	G		670	.024		4.82	1.05		5.87	7.05
0550	12″ deep	G		650	.025		4.95	1.09		6.04	7.25

05 42 13.30 Framing, Boxed Headers/Beams			Crew	Daily Output	Labor-Hours	Unit	Material	Labor	Equipment	Total	Total Incl O&P
0010	**FRAMING, BOXED HEADERS/BEAMS**										
0015	Made from recycled materials	G									
0200	Double, 18 ga x 6″ deep	G	2 Carp	220	.073	L.F.	5.70	3.21		8.91	11.50
0210	8″ deep	G		210	.076		6.25	3.36		9.61	12.40
0220	10″ deep	G		200	.080		7.65	3.53		11.18	14.20
0230	12″ deep	G		190	.084		8.35	3.71		12.06	15.30
0300	16 ga x 8″ deep	G		180	.089		7.25	3.92		11.17	14.40
0310	10″ deep	G		170	.094		8.75	4.15		12.90	16.40
0320	12″ deep	G		160	.100		9.50	4.41		13.91	17.70
0400	14 ga x 10″ deep	G		140	.114		10.10	5.05		15.15	19.35
0410	12″ deep	G		130	.123		11.05	5.45		16.50	21
0500	12 ga x 10″ deep	G		110	.145		13.30	6.40		19.70	25
0510	12″ deep	G		100	.160		14.70	7.05		21.75	28
1210	Triple, 18 ga x 8″ deep	G		170	.094		9.10	4.15		13.25	16.80
1220	10″ deep	G		165	.097		11	4.28		15.28	19.10
1230	12″ deep	G		160	.100		12.05	4.41		16.46	20.50
1300	16 ga x 8″ deep	G		145	.110		10.50	4.87		15.37	19.55
1310	10″ deep	G		140	.114		12.65	5.05		17.70	22
1320	12″ deep	G		135	.119		13.75	5.25		19	23.50
1400	14 ga x 10″ deep	G		115	.139		14.65	6.15		20.80	26
1410	12″ deep	G		110	.145		16.10	6.40		22.50	28
1500	12 ga x 10″ deep	G		90	.178		19.45	7.85		27.30	34.50
1510	12″ deep	G		85	.188		21.50	8.30		29.80	37.50

05 42 13.40 Framing, Joists			Crew	Daily Output	Labor-Hours	Unit	Material	Labor	Equipment	Total	Total Incl O&P
0010	**FRAMING, JOISTS**, no band joists (track), web stiffeners, headers,										
0020	Beams, bridging or bracing										
0025	Made from recycled materials	G									
0030	Joists (2″ flange) and fasteners, materials only										
0220	18 ga x 6″ deep	G				L.F.	1.75			1.75	1.93
0230	8″ deep	G					2.06			2.06	2.26
0240	10″ deep	G					2.43			2.43	2.67
0320	16 ga x 6″ deep	G					2.14			2.14	2.36
0330	8″ deep	G					2.56			2.56	2.82
0340	10″ deep	G					3			3	3.30
0350	12″ deep	G					3.40			3.40	3.74
0430	14 ga x 8″ deep	G					3.21			3.21	3.53
0440	10″ deep	G					3.72			3.72	4.09

05 42 Cold-Formed Metal Joist Framing

05 42 13 – Cold-Formed Metal Floor Joist Framing

05 42 13.40 Framing, Joists			Crew	Daily Output	Labor-Hours	Unit	Material	Labor	Equipment	Total	Total Incl O&P
0450	12″ deep	G				L.F.	4.22			4.22	4.64
0540	12 ga x 10″ deep	G					5.40			5.40	5.95
0550	12″ deep	G					6.15			6.15	6.75
1010	Installation of joists to band joists, beams & headers, labor only										
1220	18 ga x 6″ deep		2 Carp	110	.145	Ea.		6.40		6.40	10.50
1230	8″ deep			90	.178			7.85		7.85	12.85
1240	10″ deep			80	.200			8.80		8.80	14.45
1320	16 ga x 6″ deep			95	.168			7.45		7.45	12.20
1330	8″ deep			70	.229			10.10		10.10	16.55
1340	10″ deep			60	.267			11.75		11.75	19.30
1350	12″ deep			55	.291			12.85		12.85	21
1430	14 ga x 8″ deep			65	.246			10.85		10.85	17.80
1440	10″ deep			45	.356			15.70		15.70	25.50
1450	12″ deep			35	.457			20		20	33
1540	12 ga x 10″ deep			40	.400			17.65		17.65	29
1550	12″ deep			30	.533			23.50		23.50	38.50
9000	Minimum labor/equipment charge			4	4	Job		176		176	289

05 42 13.45 Framing, Web Stiffeners			Crew	Daily Output	Labor-Hours	Unit	Material	Labor	Equipment	Total	Total Incl O&P
0010	**FRAMING, WEB STIFFENERS** at joist bearing, fabricated from										
0020	Stud piece (1-5/8″ flange) to stiffen joist (2″ flange)										
0025	Made from recycled materials	G									
2120	For 6″ deep joist, with 18 ga x 2-1/2″ stud	G	1 Carp	120	.067	Ea.	2.07	2.94		5.01	7.10
2130	3-5/8″ stud	G		110	.073		2.30	3.21		5.51	7.80
2140	4″ stud	G		105	.076		2.22	3.36		5.58	7.95
2150	6″ stud	G		100	.080		2.43	3.53		5.96	8.45
2160	8″ stud	G		95	.084		2.49	3.71		6.20	8.85
2220	8″ deep joist, with 2-1/2″ stud	G		120	.067		2.27	2.94		5.21	7.30
2230	3-5/8″ stud	G		110	.073		2.48	3.21		5.69	7.95
2240	4″ stud	G		105	.076		2.43	3.36		5.79	8.15
2250	6″ stud	G		100	.080		2.66	3.53		6.19	8.75
2260	8″ stud	G		95	.084		2.86	3.71		6.57	9.25
2320	10″ deep joist, with 2-1/2″ stud	G		110	.073		3.20	3.21		6.41	8.75
2330	3-5/8″ stud	G		100	.080		3.53	3.53		7.06	9.70
2340	4″ stud	G		95	.084		3.50	3.71		7.21	9.95
2350	6″ stud	G		90	.089		3.78	3.92		7.70	10.60
2360	8″ stud	G		85	.094		3.84	4.15		7.99	11.05
2420	12″ deep joist, with 2-1/2″ stud	G		110	.073		3.38	3.21		6.59	8.95
2430	3-5/8″ stud	G		100	.080		3.70	3.53		7.23	9.85
2440	4″ stud	G		95	.084		3.63	3.71		7.34	10.10
2450	6″ stud	G		90	.089		3.97	3.92		7.89	10.80
2460	8″ stud	G		85	.094		4.27	4.15		8.42	11.50
3130	For 6″ deep joist, with 16 ga x 3-5/8″ stud	G		100	.080		2.42	3.53		5.95	8.45
3140	4″ stud	G		95	.084		2.41	3.71		6.12	8.75
3150	6″ stud	G		90	.089		2.61	3.92		6.53	9.30
3160	8″ stud	G		85	.094		2.78	4.15		6.93	9.85
3230	8″ deep joist, with 3-5/8″ stud	G		100	.080		2.68	3.53		6.21	8.75
3240	4″ stud	G		95	.084		2.64	3.71		6.35	9
3250	6″ stud	G		90	.089		2.89	3.92		6.81	9.65
3260	8″ stud	G		85	.094		3.12	4.15		7.27	10.25
3330	10″ deep joist, with 3-5/8″ stud	G		85	.094		3.67	4.15		7.82	10.85
3340	4″ stud	G		80	.100		3.76	4.41		8.17	11.40
3350	6″ stud	G		75	.107		4.03	4.70		8.73	12.15

05 42 Cold-Formed Metal Joist Framing

05 42 13 – Cold-Formed Metal Floor Joist Framing

05 42 13.45 Framing, Web Stiffeners			Crew	Daily Output	Labor-Hours	Unit	Material (2012 Bare Costs)	Labor (2012 Bare Costs)	Equipment (2012 Bare Costs)	Total (2012 Bare Costs)	Total Incl O&P
3360	8" stud	G	1 Carp	70	.114	Ea.	4.24	5.05		9.29	12.90
3430	12" deep joist, with 3-5/8" stud	G		85	.094		4	4.15		8.15	11.20
3440	4" stud	G		80	.100		3.94	4.41		8.35	11.60
3450	6" stud	G		75	.107		4.32	4.70		9.02	12.45
3460	8" stud	G		70	.114		4.66	5.05		9.71	13.40
4230	For 8" deep joist, with 14 ga x 3-5/8" stud	G		90	.089		3.48	3.92		7.40	10.30
4240	4" stud	G		85	.094		3.54	4.15		7.69	10.70
4250	6" stud	G		80	.100		3.84	4.41		8.25	11.45
4260	8" stud	G		75	.107		3.61	4.70		8.31	11.65
4330	10" deep joist, with 3-5/8" stud	G		75	.107		4.88	4.70		9.58	13.05
4340	4" stud	G		70	.114		4.83	5.05		9.88	13.55
4350	6" stud	G		65	.123		5.30	5.45		10.75	14.75
4360	8" stud	G		60	.133		4.88	5.90		10.78	15
4430	12" deep joist, with 3-5/8" stud	G		75	.107		5.20	4.70		9.90	13.40
4440	4" stud	G		70	.114		5.30	5.05		10.35	14.05
4450	6" stud	G		65	.123		5.75	5.45		11.20	15.20
4460	8" stud	G		60	.133		5.40	5.90		11.30	15.60
5330	For 10" deep joist, with 12 ga x 3-5/8" stud	G		65	.123		5.15	5.45		10.60	14.60
5340	4" stud	G		60	.133		5.30	5.90		11.20	15.45
5350	6" stud	G		55	.145		5.85	6.40		12.25	16.90
5360	8" stud	G		50	.160		6.35	7.05		13.40	18.55
5430	12" deep joist, with 3-5/8" stud	G		65	.123		5.75	5.45		11.20	15.20
5440	4" stud	G		60	.133		5.55	5.90		11.45	15.75
5450	6" stud	G		55	.145		6.35	6.40		12.75	17.50
5460	8" stud	G	↓	50	.160	↓	7.25	7.05		14.30	19.55

05 42 23 – Cold-Formed Metal Roof Joist Framing

05 42 23.05 Framing, Bracing			Crew	Daily Output	Labor-Hours	Unit	Material	Labor	Equipment	Total	Total Incl O&P
0010	**FRAMING, BRACING**										
0015	Made from recycled materials	G									
0020	Continuous bracing, per row										
0100	16 ga x 1-1/2" channel thru rafters/trusses @ 16" O.C.	G	1 Carp	4.50	1.778	C.L.F.	54	78.50		132.50	189
0120	24" O.C.	G		6	1.333		54	59		113	156
0300	2" x 2" angle x 18 ga, rafters/trusses @ 16" O.C.	G		6	1.333		82.50	59		141.50	188
0320	24" O.C.	G		8	1		82.50	44		126.50	164
0400	16 ga, rafters/trusses @ 16" O.C.	G		4.50	1.778		106	78.50		184.50	245
0420	24" O.C.	G	↓	6.50	1.231	↓	106	54.50		160.50	205

05 42 23.10 Framing, Bridging			Crew	Daily Output	Labor-Hours	Unit	Material	Labor	Equipment	Total	Total Incl O&P
0010	**FRAMING, BRIDGING**										
0015	Made from recycled materials	G									
0020	Solid, between rafters w/1-1/4" leg track, per rafter bay										
1200	Rafters 16" O.C., 18 ga x 4" deep	G	1 Carp	60	.133	Ea.	1.62	5.90		7.52	11.45
1210	6" deep	G		57	.140		2.11	6.20		8.31	12.45
1220	8" deep	G		55	.145		2.62	6.40		9.02	13.40
1230	10" deep	G		52	.154		3.27	6.80		10.07	14.70
1240	12" deep	G		50	.160		3.71	7.05		10.76	15.65
2200	24" O.C., 18 ga x 4" deep	G		60	.133		2.34	5.90		8.24	12.20
2210	6" deep	G		57	.140		3.06	6.20		9.26	13.50
2220	8" deep	G		55	.145		3.80	6.40		10.20	14.65
2230	10" deep	G		52	.154		4.73	6.80		11.53	16.30
2240	12" deep	G	↓	50	.160	↓	5.35	7.05		12.40	17.45

05 42 23 – Cold-Formed Metal Roof Joist Framing

05 42 23.50 Framing, Parapets

	05 42 23.50 Framing, Parapets		Crew	Daily Output	Labor-Hours	Unit	Material	Labor (2012 Bare Costs)	Equipment (2012 Bare Costs)	Total (2012 Bare Costs)	Total Incl O&P
0010	**FRAMING, PARAPETS**										
0015	Made from recycled materials	G									
0100	3′ high installed on 1st story, 18 ga x 4″ wide studs, 12″ O.C.	G	2 Carp	100	.160	L.F.	5.90	7.05		12.95	18.05
0110	16″ O.C.	G		150	.107		5.05	4.70		9.75	13.25
0120	24″ O.C.	G		200	.080		4.16	3.53		7.69	10.35
0200	6″ wide studs, 12″ O.C.	G		100	.160		7.55	7.05		14.60	19.85
0210	16″ O.C.	G		150	.107		6.45	4.70		11.15	14.80
0220	24″ O.C.	G		200	.080		5.35	3.53		8.88	11.70
1100	Installed on 2nd story, 18 ga x 4″ wide studs, 12″ O.C.	G		95	.168		5.90	7.45		13.35	18.70
1110	16″ O.C.	G		145	.110		5.05	4.87		9.92	13.55
1120	24″ O.C.	G		190	.084		4.16	3.71		7.87	10.65
1200	6″ wide studs, 12″ O.C.	G		95	.168		7.55	7.45		15	20.50
1210	16″ O.C.	G		145	.110		6.45	4.87		11.32	15.10
1220	24″ O.C.	G		190	.084		5.35	3.71		9.06	12
2100	Installed on gable, 18 ga x 4″ wide studs, 12″ O.C.	G		85	.188		5.90	8.30		14.20	20
2110	16″ O.C.	G		130	.123		5.05	5.45		10.50	14.45
2120	24″ O.C.	G		170	.094		4.16	4.15		8.31	11.35
2200	6″ wide studs, 12″ O.C.	G		85	.188		7.55	8.30		15.85	22
2210	16″ O.C.	G		130	.123		6.45	5.45		11.90	16
2220	24″ O.C.	G		170	.094		5.35	4.15		9.50	12.70

05 42 23.60 Framing, Roof Rafters

	05 42 23.60 Framing, Roof Rafters		Crew	Daily Output	Labor-Hours	Unit	Material	Labor (2012 Bare Costs)	Equipment (2012 Bare Costs)	Total (2012 Bare Costs)	Total Incl O&P
0010	**FRAMING, ROOF RAFTERS**										
0015	Made from recycled materials	G									
0100	Boxed ridge beam, double, 18 ga x 6″ deep	G	2 Carp	160	.100	L.F.	5.70	4.41		10.11	13.50
0110	8″ deep	G		150	.107		6.25	4.70		10.95	14.60
0120	10″ deep	G		140	.114		7.65	5.05		12.70	16.65
0130	12″ deep	G		130	.123		8.35	5.45		13.80	18.10
0200	16 ga x 6″ deep	G		150	.107		6.45	4.70		11.15	14.80
0210	8″ deep	G		140	.114		7.25	5.05		12.30	16.20
0220	10″ deep	G		130	.123		8.75	5.45		14.20	18.50
0230	12″ deep	G		120	.133		9.50	5.90		15.40	20
1100	Rafters, 2″ flange, material only, 18 ga x 6″ deep	G					1.75			1.75	1.93
1110	8″ deep	G					2.06			2.06	2.26
1120	10″ deep	G					2.43			2.43	2.67
1130	12″ deep	G					2.80			2.80	3.08
1200	16 ga x 6″ deep	G					2.14			2.14	2.36
1210	8″ deep	G					2.56			2.56	2.82
1220	10″ deep	G					3			3	3.30
1230	12″ deep	G					3.40			3.40	3.74
2100	Installation only, ordinary rafter to 4:12 pitch, 18 ga x 6″ deep		2 Carp	35	.457	Ea.		20		20	33
2110	8″ deep			30	.533			23.50		23.50	38.50
2120	10″ deep			25	.640			28		28	46.50
2130	12″ deep			20	.800			35.50		35.50	58
2200	16 ga x 6″ deep			30	.533			23.50		23.50	38.50
2210	8″ deep			25	.640			28		28	46.50
2220	10″ deep			20	.800			35.50		35.50	58
2230	12″ deep			15	1.067			47		47	77
8100	Add to labor, ordinary rafters on steep roofs							25%			
8110	Dormers & complex roofs							50%			
8200	Hip & valley rafters to 4:12 pitch							25%			
8210	Steep roofs							50%			
8220	Dormers & complex roofs							75%			

05 42 Cold-Formed Metal Joist Framing

05 42 23 – Cold-Formed Metal Roof Joist Framing

05 42 23.60 Framing, Roof Rafters			Crew	Daily Output	Labor-Hours	Unit	2012 Bare Costs Material	Labor	Equipment	Total	Total Incl O&P
8300	Hip & valley jack rafters to 4:12 pitch							50%			
8310	Steep roofs							75%			
8320	Dormers & complex roofs							100%			
9000	Minimum labor/equipment charge		2 Carp	4	4	Job		176		176	289

05 42 23.70 Framing, Soffits and Canopies

			Crew	Daily Output	Labor-Hours	Unit	Material	Labor	Equipment	Total	Total Incl O&P
0010	**FRAMING, SOFFITS & CANOPIES**										
0015	Made from recycled materials	G									
0130	Continuous ledger track @ wall, studs @ 16" O.C., 18 ga x 4" wide	G	2 Carp	535	.030	L.F.	1.08	1.32		2.40	3.35
0140	6" wide	G		500	.032		1.41	1.41		2.82	3.86
0150	8" wide	G		465	.034		1.75	1.52		3.27	4.41
0160	10" wide	G		430	.037		2.18	1.64		3.82	5.10
0230	Studs @ 24" O.C., 18 ga x 4" wide	G		800	.020		1.03	.88		1.91	2.58
0240	6" wide	G		750	.021		1.34	.94		2.28	3.02
0250	8" wide	G		700	.023		1.67	1.01		2.68	3.49
0260	10" wide	G		650	.025		2.08	1.09		3.17	4.07
1000	Horizontal soffit and canopy members, material only										
1030	1-5/8" flange studs, 18 ga x 4" deep	G				L.F.	1.40			1.40	1.54
1040	6" deep	G					1.76			1.76	1.94
1050	8" deep	G					2.14			2.14	2.35
1140	2" flange joists, 18 ga x 6" deep	G					2			2	2.20
1150	8" deep	G					2.35			2.35	2.59
1160	10" deep	G					2.77			2.77	3.05
4030	Installation only, 18 ga, 1-5/8" flange x 4" deep		2 Carp	130	.123	Ea.		5.45		5.45	8.90
4040	6" deep			110	.145			6.40		6.40	10.50
4050	8" deep			90	.178			7.85		7.85	12.85
4140	2" flange, 18 ga x 6" deep			110	.145			6.40		6.40	10.50
4150	8" deep			90	.178			7.85		7.85	12.85
4160	10" deep			80	.200			8.80		8.80	14.45
6010	Clips to attach facia to rafter tails, 2" x 2" x 18 ga angle	G	1 Carp	120	.067		.98	2.94		3.92	5.90
6020	16 ga angle	G	"	100	.080		1.25	3.53		4.78	7.15
9000	Minimum labor/equipment charge		2 Carp	4	4	Job		176		176	289

05 44 Cold-Formed Metal Trusses

05 44 13 – Cold-Formed Metal Roof Trusses

05 44 13.60 Framing, Roof Trusses

			Crew	Daily Output	Labor-Hours	Unit	Material	Labor	Equipment	Total	Total Incl O&P
0010	**FRAMING, ROOF TRUSSES**										
0015	Made from recycled materials	G									
0020	Fabrication of trusses on ground, Fink (W) or King Post, to 4:12 pitch										
0120	18 ga x 4" chords, 16' span	G	2 Carp	12	1.333	Ea.	65.50	59		124.50	169
0130	20' span	G		11	1.455		82	64		146	195
0140	24' span	G		11	1.455		98.50	64		162.50	213
0150	28' span	G		10	1.600		115	70.50		185.50	242
0160	32' span	G		10	1.600		131	70.50		201.50	260
0250	6" chords, 28' span	G		9	1.778		144	78.50		222.50	287
0260	32' span	G		9	1.778		165	78.50		243.50	310
0270	36' span	G		8	2		185	88		273	350
0280	40' span	G		8	2		206	88		294	370
1120	5:12 to 8:12 pitch, 18 ga x 4" chords, 16' span	G		10	1.600		75	70.50		145.50	199
1130	20' span	G		9	1.778		93.50	78.50		172	232
1140	24' span	G		9	1.778		112	78.50		190.50	253
1150	28' span	G		8	2		131	88		219	289

05 44 Cold-Formed Metal Trusses

05 44 13 – Cold-Formed Metal Roof Trusses

05 44 13.60 Framing, Roof Trusses			Crew	Daily Output	Labor-Hours	Unit	Material	2012 Bare Costs Labor	Equipment	Total	Total Incl O&P
1160	32′ span	G	2 Carp	8	2	Ea.	150	88		238	310
1250	6″ chords, 28′ span	G		7	2.286		165	101		266	345
1260	32′ span	G		7	2.286		188	101		289	370
1270	36′ span	G		6	2.667		212	118		330	425
1280	40′ span	G		6	2.667		235	118		353	450
2120	9:12 to 12:12 pitch, 18 ga x 4″ chords, 16′ span	G		8	2		93.50	88		181.50	248
2130	20′ span	G		7	2.286		117	101		218	294
2140	24′ span	G		7	2.286		140	101		241	320
2150	28′ span	G		6	2.667		164	118		282	375
2160	32′ span	G		6	2.667		187	118		305	400
2250	6″ chords, 28′ span	G		5	3.200		206	141		347	455
2260	32′ span	G		5	3.200		235	141		376	490
2270	36′ span	G		4	4		265	176		441	580
2280	40′ span	G		4	4		294	176		470	615
4900	Minimum labor/equipment charge			4	4	Job		176		176	289
5120	Erection only of roof trusses, to 4:12 pitch, 16′ span		F-6	48	.833	Ea.		34.50	12.75	47.25	70
5130	20′ span			46	.870			36	13.30	49.30	73
5140	24′ span			44	.909			37.50	13.90	51.40	76.50
5150	28′ span			42	.952			39.50	14.55	54.05	80
5160	32′ span			40	1			41	15.30	56.30	84
5170	36′ span			38	1.053			43.50	16.10	59.60	88
5180	40′ span			36	1.111			46	17	63	93
5220	5:12 to 8:12 pitch, 16′ span			42	.952			39.50	14.55	54.05	80
5230	20′ span			40	1			41	15.30	56.30	84
5240	24′ span			38	1.053			43.50	16.10	59.60	88
5250	28′ span			36	1.111			46	17	63	93
5260	32′ span			34	1.176			48.50	18	66.50	99
5270	36′ span			32	1.250			51.50	19.15	70.65	105
5280	40′ span			30	1.333			55	20.50	75.50	112
5320	9:12 to 12:12 pitch, 16′ span			36	1.111			46	17	63	93
5330	20′ span			34	1.176			48.50	18	66.50	99
5340	24′ span			32	1.250			51.50	19.15	70.65	105
5350	28′ span			30	1.333			55	20.50	75.50	112
5360	32′ span			28	1.429			59	22	81	120
5370	36′ span			26	1.538			63.50	23.50	87	129
5380	40′ span			24	1.667			68.50	25.50	94	140
9000	Minimum labor/equipment charge			2	20	Job		825	305	1,130	1,675

05 51 Metal Stairs

05 51 13 – Metal Pan Stairs

05 51 13.50 Pan Stairs			Crew	Daily Output	Labor-Hours	Unit	Material	Labor	Equipment	Total	Total Incl O&P
0010	**PAN STAIRS**, shop fabricated, steel stringers										
0015	Made from recycled materials	G									
0200	Cement fill metal pan, picket rail, 3′-6″ wide	G	E-4	35	.914	Riser	465	45.50	3.47	513.97	600
0300	4′-0″ wide	G		30	1.067		525	53	4.05	582.05	685
0350	Wall rail, both sides, 3′-6″ wide	G		53	.604		355	30	2.29	387.29	450
1500	Landing, steel pan, conventional	G		160	.200	S.F.	62	9.95	.76	72.71	87.50

05 51 Metal Stairs

05 51 16 – Metal Floor Plate Stairs

05 51 16.50 Floor Plate Stairs			Crew	Daily Output	Labor-Hours	Unit	2012 Bare Costs Material	Labor	Equipment	Total	Total Incl O&P
0010	**FLOOR PLATE STAIRS**, shop fabricated, steel stringers										
0015	Made from recycled materials	G									
0400	Cast iron tread and pipe rail, 3′-6″ wide	G	E-4	35	.914	Riser	500	45.50	3.47	548.97	640
0450	4′-0″ wide	G		30	1.067		525	53	4.05	582.05	685
0475	5′-0″ wide	G		28	1.143		575	57	4.34	636.34	745
0500	Checkered plate tread, industrial, 3′-6″ wide	G		28	1.143		310	57	4.34	371.34	450
0550	Circular, for tanks, 3′-0″ wide	G		33	.970		345	48.50	3.68	397.18	475
0600	For isolated stairs, add							100%			
0800	Custom steel stairs, 3′-6″ wide, minimum	G	E-4	35	.914		465	45.50	3.47	513.97	600
0810	Average	G		30	1.067		620	53	4.05	677.05	785
0900	Maximum	G		20	1.600		775	80	6.10	861.10	1,000
1100	For 4′ wide stairs, add						5%	5%			
1300	For 5′ wide stairs, add						10%	10%			

05 51 19 – Metal Grating Stairs

05 51 19.50 Grating Stairs			Crew	Daily Output	Labor-Hours	Unit	Material	Labor	Equipment	Total	Total Incl O&P
0010	**GRATING STAIRS**, shop fabricated, steel stringers, safety nosing on treads										
0015	Made from recycled materials	G									
0020	Grating tread and pipe railing, 3′-6″ wide	G	E-4	35	.914	Riser	310	45.50	3.47	358.97	430
0100	4′-0″ wide	G		30	1.067	″	405	53	4.05	462.05	550
9000	Minimum labor/equipment charge			2	16	Job		800	61	861	1,550

05 51 23 – Metal Fire Escapes

05 51 23.25 Fire Escapes			Crew	Daily Output	Labor-Hours	Unit	Material	Labor	Equipment	Total	Total Incl O&P
0010	**FIRE ESCAPES**, shop fabricated										
0200	2′ wide balcony, 1″ x 1/4″ bars 1-1/2″ O.C., with railing	G	2 Sswk	10	1.600	L.F.	55	79		134	208
0400	1st story cantilevered stair, standard, with railing	G		.50	32	Ea.	2,275	1,575		3,850	5,425
0500	Cable counterweighted, with railing	G		.40	40	″	2,100	1,975		4,075	6,000
0700	36″ x 40″ platform & fixed stair, with railing	G		.40	40	Flight	1,000	1,975		2,975	4,775
0900	For 3′-6″ wide escapes, add to above						100%	150%			

05 51 23.50 Fire Escape Stairs			Crew	Daily Output	Labor-Hours	Unit	Material	Labor	Equipment	Total	Total Incl O&P
0010	**FIRE ESCAPE STAIRS**, portable										
0100	Portable ladder					Ea.	127			127	139

05 51 33 – Metal Ladders

05 51 33.13 Vertical Metal Ladders			Crew	Daily Output	Labor-Hours	Unit	Material	Labor	Equipment	Total	Total Incl O&P
0010	**VERTICAL METAL LADDERS**, shop fabricated										
0015	Made from recycled materials	G									
0020	Steel, 20″ wide, bolted to concrete, with cage	G	E-4	50	.640	V.L.F.	65	32	2.43	99.43	134
0100	Without cage	G		85	.376		36	18.75	1.43	56.18	76
0300	Aluminum, bolted to concrete, with cage	G		50	.640		111	32	2.43	145.43	185
0400	Without cage	G		85	.376		48	18.75	1.43	68.18	89.50
9000	Minimum labor/equipment charge			2	16	Job		800	61	861	1,550

05 51 33.16 Inclined Metal Ladders			Crew	Daily Output	Labor-Hours	Unit	Material	Labor	Equipment	Total	Total Incl O&P
0010	**INCLINED METAL LADDERS**, shop fabricated										
0015	Made from recycled materials	G									
3900	Industrial ships ladder, steel, 24″ W, grating treads, 2 line pipe rail	G	E-4	30	1.067	Riser	187	53	4.05	244.05	310
4000	Aluminum	G	″	30	1.067	″	206	53	4.05	263.05	330

05 51 33.23 Alternating Tread Ladders			Crew	Daily Output	Labor-Hours	Unit	Material	Labor	Equipment	Total	Total Incl O&P
0010	**ALTERNATING TREAD LADDERS**, shop fabricated										
0015	Made from recycled materials	G									
1350	Alternating tread stair, 56/68°, steel, standard paint color	G	2 Sswk	50	.320	V.L.F.	258	15.80		273.80	315
1360	Non-standard paint color	G		50	.320		292	15.80		307.80	350

05 51 Metal Stairs

05 51 33 – Metal Ladders

05 51 33.23 Alternating Tread Ladders			Crew	Daily Output	Labor-Hours	Unit	Material	Labor	Equipment	Total	Total Incl O&P
							2012 Bare Costs				
1370	Galvanized steel	G	2 Sswk	50	.320	V.L.F.	300	15.80		315.80	360
1380	Stainless steel	G		50	.320		435	15.80		450.80	505
1390	68°, aluminum	G		50	.320		315	15.80		330.80	380

05 52 Metal Railings

05 52 13 – Pipe and Tube Railings

05 52 13.50 Railings, Pipe

			Crew	Daily Output	Labor-Hours	Unit	Material	Labor	Equipment	Total	Total Incl O&P
0010	**RAILINGS, PIPE**, shop fab'd, 3'-6" high, posts @ 5' O.C.										
0015	Made from recycled materials	G									
0020	Aluminum, 2 rail, satin finish, 1-1/4" diameter	G	E-4	160	.200	L.F.	35	9.95	.76	45.71	58
0030	Clear anodized	G		160	.200		43	9.95	.76	53.71	67
0040	Dark anodized	G		160	.200		48	9.95	.76	58.71	72.50
0080	1-1/2" diameter, satin finish	G		160	.200		41.50	9.95	.76	52.21	65
0090	Clear anodized	G		160	.200		46.50	9.95	.76	57.21	70.50
0100	Dark anodized	G		160	.200		51	9.95	.76	61.71	75.50
0140	Aluminum, 3 rail, 1-1/4" diam., satin finish	G		137	.234		53	11.65	.89	65.54	81
0150	Clear anodized	G		137	.234		66	11.65	.89	78.54	95
0160	Dark anodized	G		137	.234		73	11.65	.89	85.54	103
0200	1-1/2" diameter, satin finish	G		137	.234		63	11.65	.89	75.54	92
0210	Clear anodized	G		137	.234		72	11.65	.89	84.54	101
0220	Dark anodized	G		137	.234		78.50	11.65	.89	91.04	109
0500	Steel, 2 rail, on stairs, primed, 1-1/4" diameter	G		160	.200		23.50	9.95	.76	34.21	45.50
0520	1-1/2" diameter	G		160	.200		25.50	9.95	.76	36.21	47.50
0540	Galvanized, 1-1/4" diameter	G		160	.200		32	9.95	.76	42.71	54.50
0560	1-1/2" diameter	G		160	.200		36	9.95	.76	46.71	59
0580	Steel, 3 rail, primed, 1-1/4" diameter	G		137	.234		35	11.65	.89	47.54	61
0600	1-1/2" diameter	G		137	.234		36.50	11.65	.89	49.04	62.50
0620	Galvanized, 1-1/4" diameter	G		137	.234		49	11.65	.89	61.54	76.50
0640	1-1/2" diameter	G		137	.234		57	11.65	.89	69.54	85
0700	Stainless steel, 2 rail, 1-1/4" diam. #4 finish	G		137	.234		103	11.65	.89	115.54	135
0720	High polish	G		137	.234		167	11.65	.89	179.54	206
0740	Mirror polish	G		137	.234		210	11.65	.89	222.54	253
0760	Stainless steel, 3 rail, 1-1/2" diam., #4 finish	G		120	.267		156	13.30	1.01	170.31	198
0770	High polish	G		120	.267		258	13.30	1.01	272.31	310
0780	Mirror finish	G		120	.267		315	13.30	1.01	329.31	370
0900	Wall rail, alum. pipe, 1-1/4" diam., satin finish	G		213	.150		19.75	7.50	.57	27.82	36
0905	Clear anodized	G		213	.150		24.50	7.50	.57	32.57	41
0910	Dark anodized	G		213	.150		29	7.50	.57	37.07	46.50
0915	1-1/2" diameter, satin finish	G		213	.150		22	7.50	.57	30.07	38.50
0920	Clear anodized	G		213	.150		27.50	7.50	.57	35.57	45
0925	Dark anodized	G		213	.150		34	7.50	.57	42.07	52
0930	Steel pipe, 1-1/4" diameter, primed	G		213	.150		14	7.50	.57	22.07	30
0935	Galvanized	G		213	.150		20.50	7.50	.57	28.57	37
0940	1-1/2" diameter	G		176	.182		14.50	9.05	.69	24.24	33.50
0945	Galvanized	G		213	.150		20.50	7.50	.57	28.57	37
0955	Stainless steel pipe, 1-1/2" diam., #4 finish	G		107	.299		82.50	14.90	1.14	98.54	120
0960	High polish	G		107	.299		168	14.90	1.14	184.04	214
0965	Mirror polish	G		107	.299		199	14.90	1.14	215.04	248
2000	2-line pipe rail (1-1/2" T&B) with 1/2" pickets @ 4-1/2" O.C.,										
2005	attached handrail on brackets										
2010	42" high aluminum, satin finish, straight & level	G	E-4	120	.267	L.F.	193	13.30	1.01	207.31	238

05 52 Metal Railings

05 52 13 – Pipe and Tube Railings

05 52 13.50 Railings, Pipe			Crew	Daily Output	Labor-Hours	Unit	2012 Bare Costs Material	Labor	Equipment	Total	Total Incl O&P
2050	42" high steel, primed, straight & level	G	E-4	120	.267	L.F.	119	13.30	1.01	133.31	157
4000	For curved and level rails, add						10%	10%			
4100	For sloped rails for stairs, add					↓	30%	30%			
9000	Minimum labor/equipment charge		1 Sswk	2	4	Job		197		197	365

05 52 16 – Industrial Railings

05 52 16.50 Railings, Industrial

			Crew	Daily Output	Labor-Hours	Unit	Material	Labor	Equipment	Total	Total Incl O&P
0010	**RAILINGS, INDUSTRIAL**, shop fab'd, 3'-6" high, posts @ 5' O.C.										
0020	2 rail, 3'-6" high, 1-1/2" pipe	G	E-4	255	.125	L.F.	25.50	6.25	.48	32.23	40
0200	For 4" high kick plate, 10 gauge, add	G					5.30			5.30	5.85
0500	For curved level rails, add						10%	10%			
0550	For sloped rails for stairs, add					↓	30%	30%			
9000	Minimum labor/equipment charge		1 Sswk	2	4	Job		197		197	365

05 56 Metal Castings

05 56 13 – Metal Construction Castings

05 56 13.50 Construction Castings

			Crew	Daily Output	Labor-Hours	Unit	Material	Labor	Equipment	Total	Total Incl O&P
0010	**CONSTRUCTION CASTINGS**										
0020	Manhole covers and frames, see Section 33 44 13.13										
0100	Column bases, cast iron, 16" x 16", approx. 65 lb.	G	E-4	46	.696	Ea.	145	34.50	2.64	182.14	227
0200	32" x 32", approx. 256 lb.	G		23	1.391	"	540	69.50	5.30	614.80	730
0600	Miscellaneous C.I. castings, light sections, less than 150 lb.	G		3200	.010	Lb.	2.24	.50	.04	2.78	3.43
1300	Special low volume items	G	↓	3200	.010	"	3.89	.50	.04	4.43	5.25

05 58 Formed Metal Fabrications

05 58 23 – Formed Metal Guards

05 58 23.90 Window Guards

			Crew	Daily Output	Labor-Hours	Unit	Material	Labor	Equipment	Total	Total Incl O&P
0010	**WINDOW GUARDS**, shop fabricated										
0015	Expanded metal, steel angle frame, permanent	G	E-4	350	.091	S.F.	21.50	4.56	.35	26.41	32.50
0025	Steel bars, 1/2" x 1/2", spaced 5" O.C.	G	"	290	.110	"	15	5.50	.42	20.92	27
0030	Hinge mounted, add	G				Opng.	43.50			43.50	48
0040	Removable type, add	G				"	27.50			27.50	30.50
0050	For galvanized guards, add					S.F.	35%				
0070	For pivoted or projected type, add						105%	40%			
0100	Mild steel, stock units, economy	G	E-4	405	.079		5.90	3.94	.30	10.14	14.15
0200	Deluxe	G		405	.079	↓	12	3.94	.30	16.24	21
0400	Woven wire, stock units, 3/8" channel frame, 3' x 5' opening	G		40	.800	Opng.	160	40	3.04	203.04	253
0500	4' x 6' opening	G	↓	38	.842		255	42	3.20	300.20	365
0800	Basket guards for above, add	G					220			220	242
1000	Swinging guards for above, add	G				↓	75			75	82.50
9000	Minimum labor/equipment charge		1 Sswk	2	4	Job		197		197	365

05 58 25 – Formed Lamp Posts

05 58 25.40 Lamp Posts

			Crew	Daily Output	Labor-Hours	Unit	Material	Labor	Equipment	Total	Total Incl O&P
0010	**LAMP POSTS**										
0020	Aluminum, 7' high, stock units, post only	G	1 Carp	16	.500	Ea.	67.50	22		89.50	111
0100	Mild steel, plain	G		16	.500	"	65	22		87	108
9000	Minimum labor/equipment charge		↓	4	2	Job		88		88	145

05 71 Decorative Metal Stairs

05 71 13 – Fabricated Metal Spiral Stairs

05 71 13.50 Spiral Stairs			Crew	Daily Output	Labor-Hours	Unit	Material	2012 Bare Costs Labor	Equipment	Total	Total Incl O&P
0010	**SPIRAL STAIRS**										
1805	Shop fabricated, custom ordered										
1810	Aluminum, 5′-0″ diameter, plain units	G	E-4	45	.711	Riser	560	35.50	2.70	598.20	685
1820	Fancy units	G		45	.711		1,050	35.50	2.70	1,088.20	1,250
1900	Cast iron, 4′-0″ diameter, plain units	G		45	.711		505	35.50	2.70	543.20	625
1920	Fancy Units	G		25	1.280		690	64	4.86	758.86	885
2000	Steel, industrial checkered plate, 4′ diameter	G		45	.711		505	35.50	2.70	543.20	625
2200	Stock units, 6′-0″ diameter	G		40	.800		610	40	3.04	653.04	745
3100	Spiral stair kits, 12 stacking risers to fit exact floor height										
3110	Steel, flat metal treads, primed, 3′-6″ dia	G	2 Carp	1.60	10	Flight	1,250	440		1,690	2,075
3120	4′-0″ dia	G		1.45	11.034		1,425	485		1,910	2,375
3130	4′-6″ dia	G		1.35	11.852		1,575	525		2,100	2,575
3140	5′-0″ dia	G		1.25	12.800		1,700	565		2,265	2,800
3210	Galvanized, 3′-6″ dia	G		1.60	10		1,475	440		1,915	2,350
3220	4′-0″ dia	G		1.45	11.034		1,700	485		2,185	2,675
3230	4′-6″ dia	G		1.35	11.852		1,900	525		2,425	2,925
3240	5′-0″ dia	G		1.25	12.800		2,050	565		2,615	3,175
3310	Checkered plate tread, primed, 3′-6″ dia	G		1.45	11.034		1,475	485		1,960	2,425
3320	4′-0″ dia	G		1.35	11.852		1,675	525		2,200	2,675
3330	4′-6″ dia	G		1.25	12.800		1,825	565		2,390	2,925
3340	5′-0″ dia	G		1.15	13.913		2,000	615		2,615	3,175
3410	Galvanized, 3′-6″ dia	G		1.45	11.034		1,775	485		2,260	2,750
3420	4′-0″ dia	G		1.35	11.852		2,000	525		2,525	3,050
3430	4′-6″ dia	G		1.25	12.800		2,175	565		2,740	3,325
3440	5′-0″ dia	G		1.15	13.913		2,375	615		2,990	3,625
3510	Red oak covers on flat metal treads, 3′-6″ dia			1.35	11.852		2,550	525		3,075	3,675
3520	4′-0″ dia			1.25	12.800		2,825	565		3,390	4,025
3530	4′-6″ dia			1.15	13.913		3,025	615		3,640	4,350
3540	5′-0″ dia			1.05	15.238		3,325	670		3,995	4,750

05 73 Decorative Metal Railings

05 73 16 – Wire Rope Decorative Metal Railings

05 73 16.10 Cable Railings

			Crew	Daily Output	Labor-Hours	Unit	Material	Labor	Equipment	Total	Total Incl O&P
0010	**CABLE RAILINGS**, with 316 stainless steel 1 x 19 cable, 3/16″ diameter										
0015	Made from recycled materials	G									
0100	1-3/4″ diameter stainless steel posts x 42″ high, cables 4″ OC	G	2 Sswk	25	.640	L.F.	37	31.50		68.50	99

05 73 23 – Ornamental Railings

05 73 23.50 Railings, Ornamental

			Crew	Daily Output	Labor-Hours	Unit	Material	Labor	Equipment	Total	Total Incl O&P
0010	**RAILINGS, ORNAMENTAL**, 3′-6″ high, posts @ 6′ O.C.										
0020	Bronze or stainless, hand forged, plain	G	2 Sswk	24	.667	L.F.	101	33		134	173
0100	Fancy	G		18	.889		174	44		218	274
0200	Aluminum, panelized, plain	G		24	.667		16.50	33		49.50	79
0300	Fancy	G		18	.889		33	44		77	118
0400	Wrought iron, hand forged, plain	G		24	.667		67.50	33		100.50	136
0500	Fancy	G		18	.889		125	44		169	219
0550	Steel, panelized, plain	G		24	.667		16.90	33		49.90	79.50
0560	Fancy			18	.889		25	44		69	109
0600	Composite metal/wood/glass, plain			18	.889		112	44		156	205
0700	Fancy			12	1.333		224	66		290	370
9000	Minimum labor/equipment charge		1 Sswk	2	4	Job		197		197	365

05 75 Decorative Formed Metal

05 75 13 – Columns

05 75 13.10 Aluminum Columns			Crew	Daily Output	Labor-Hours	Unit	2012 Bare Costs Material	Labor	Equipment	Total	Total Incl O&P
0010	**ALUMINUM COLUMNS**										
0015	Made from recycled materials	G									
0020	Aluminum, extruded, stock units, no cap or base, 6" diameter	G	E-4	240	.133	L.F.	10.65	6.65	.51	17.81	24.50
0100	8" diameter	G	"	170	.188		14.10	9.40	.72	24.22	33.50
0500	For square columns, add to column prices above					↓	50%				

Division Notes

	CREW	DAILY OUTPUT	LABOR-HOURS	UNIT	2012 BARE COSTS MAT.	LABOR	EQUIP.	TOTAL	TOTAL INCL O&P

Estimating Tips

06 05 00 Common Work Results for Wood, Plastics, and Composites

- Common to any wood-framed structure are the accessory connector items such as screws, nails, adhesives, hangers, connector plates, straps, angles, and hold-downs. For typical wood-framed buildings, such as residential projects, the aggregate total for these items can be significant, especially in areas where seismic loading is a concern. For floor and wall framing, the material cost is based on 10 to 25 lbs. per MBF. Hold-downs, hangers, and other connectors should be taken off by the piece.

  Included with material costs are fasteners for a normal installation. RSMeans engineers use manufacturer's recommendations, written specifications, and/or standard construction practice for size and spacing of fasteners. Prices for various fasteners are shown for informational purposes only. Adjustments should be made if unusual fastening conditions exist.

06 10 00 Carpentry

- Lumber is a traded commodity and therefore sensitive to supply and demand in the marketplace. Even in "budgetary" estimating of wood-framed projects, it is advisable to call local suppliers for the latest market pricing.
- Common quantity units for wood-framed projects are "thousand board feet" (MBF). A board foot is a volume of wood, 1" x 1' x 1', or 144 cubic inches. Board-foot quantities are generally calculated using nominal material dimensions—dressed sizes are ignored. Board foot per lineal foot of any stick of lumber can be calculated by dividing the nominal cross-sectional area by 12. As an example, 2,000 lineal feet of 2 x 12 equates to 4 MBF by dividing the nominal area, 2 x 12, by 12, which equals 2, and multiplying by 2,000 to give 4,000 board feet. This simple rule applies to all nominal dimensioned lumber.
- Waste is an issue of concern at the quantity takeoff for any area of construction. Framing lumber is sold in even foot lengths, i.e., 10', 12', 14', 16', and depending on spans, wall heights, and the grade of lumber, waste is inevitable. A rule of thumb for lumber waste is 5%–10% depending on material quality and the complexity of the framing.
- Wood in various forms and shapes is used in many projects, even where the main structural framing is steel, concrete, or masonry. Plywood as a back-up partition material and 2x boards used as blocking and cant strips around roof edges are two common examples. The estimator should ensure that the costs of all wood materials are included in the final estimate.

06 20 00 Finish Carpentry

- It is necessary to consider the grade of workmanship when estimating labor costs for erecting millwork and interior finish. In practice, there are three grades: premium, custom, and economy. The RSMeans daily output for base and case moldings is in the range of 200 to 250 L.F. per carpenter per day. This is appropriate for most average custom-grade projects. For premium projects, an adjustment to productivity of 25%–50% should be made, depending on the complexity of the job.

Reference Numbers

Reference numbers are shown in shaded boxes at the beginning of some major classifications. These numbers refer to related items in the Reference Section. The reference information may be an estimating procedure, an alternate pricing method, or technical information.

Note: Not all subdivisions listed here necessarily appear in this publication.

06 05 Common Work Results for Wood, Plastics, and Composites

06 05 05 – Selective Wood and Plastics Demolition

06 05 05.10 Selective Demolition Wood Framing		Crew	Daily Output	Labor-Hours	Unit	Material	2012 Bare Costs Labor	Equipment	Total	Total Incl O&P
0010	SELECTIVE DEMOLITION WOOD FRAMING R024119-10									
0100	Timber connector, nailed, small	1 Clab	96	.083	Ea.		2.92		2.92	4.80
0110	Medium		60	.133			4.68		4.68	7.65
0120	Large		48	.167			5.85		5.85	9.60
0130	Bolted, small		48	.167			5.85		5.85	9.60
0140	Medium		32	.250			8.80		8.80	14.40
0150	Large		24	.333			11.70		11.70	19.20
2958	Beams, 2" x 6"	2 Clab	1100	.015	L.F.		.51		.51	.84
2960	2" x 8"		825	.019			.68		.68	1.12
2965	2" x 10"		665	.024			.84		.84	1.38
2970	2" x 12"		550	.029			1.02		1.02	1.67
2972	2" x 14"		470	.034			1.19		1.19	1.96
2975	4" x 8"	B-1	413	.058			2.08		2.08	3.41
2980	4" x 10"		330	.073			2.60		2.60	4.26
2985	4" x 12"		275	.087			3.12		3.12	5.10
3000	6" x 8"		275	.087			3.12		3.12	5.10
3040	6" x 10"		220	.109			3.90		3.90	6.40
3080	6" x 12"		185	.130			4.64		4.64	7.60
3120	8" x 12"		140	.171			6.15		6.15	10.05
3160	10" x 12"		110	.218			7.80		7.80	12.80
3162	Alternate pricing method		1.10	21.818	M.B.F.		780		780	1,275
3170	Blocking, in 16" OC wall framing, 2" x 4"	1 Clab	600	.013	L.F.		.47		.47	.77
3172	2" x 6"		400	.020			.70		.70	1.15
3174	In 24" OC wall framing, 2" x 4"		600	.013			.47		.47	.77
3176	2" x 6"		400	.020			.70		.70	1.15
3178	Alt method, wood blocking removal from wood framing		.40	20	M.B.F.		700		700	1,150
3179	Wood blocking removal from steel framing		.36	22.222	"		780		780	1,275
3180	Bracing, let in, 1" x 3", studs 16" OC		1050	.008	L.F.		.27		.27	.44
3181	Studs 24" OC		1080	.007			.26		.26	.43
3182	1" x 4", studs 16" OC		1050	.008			.27		.27	.44
3183	Studs 24" OC		1080	.007			.26		.26	.43
3184	1" x 6", studs 16" OC		1050	.008			.27		.27	.44
3185	Studs 24" OC		1080	.007			.26		.26	.43
3186	2" x 3", studs 16" OC		800	.010			.35		.35	.58
3187	Studs 24" OC		830	.010			.34		.34	.55
3188	2" x 4", studs 16" OC		800	.010			.35		.35	.58
3189	Studs 24" OC		830	.010			.34		.34	.55
3190	2" x 6", studs 16" OC		800	.010			.35		.35	.58
3191	Studs 24" OC		830	.010			.34		.34	.55
3192	2" x 8", studs 16" OC		800	.010			.35		.35	.58
3193	Studs 24" OC		830	.010			.34		.34	.55
3194	"T" shaped metal bracing, studs at 16" OC		1060	.008			.27		.27	.43
3195	Studs at 24" OC		1200	.007			.23		.23	.38
3196	Metal straps, studs at 16" OC		1200	.007			.23		.23	.38
3197	Studs at 24" OC		1240	.006			.23		.23	.37
3200	Columns, round, 8' to 14' tall		40	.200	Ea.		7		7	11.50
3202	Dimensional lumber sizes	2 Clab	1.10	14.545	M.B.F.		510		510	835
3250	Blocking, between joists	1 Clab	320	.025	Ea.		.88		.88	1.44
3252	Bridging, metal strap, between joists		320	.025	Pr.		.88		.88	1.44
3254	Wood, between joists		320	.025	"		.88		.88	1.44
3260	Door buck, studs, header & access., 8' high 2" x 4" wall, 3' wide		32	.250	Ea.		8.80		8.80	14.40
3261	4' wide		32	.250			8.80		8.80	14.40
3262	5' wide		32	.250			8.80		8.80	14.40

06 05 05 – Selective Wood and Plastics Demolition

06 05 05.10 Selective Demolition Wood Framing		Crew	Daily Output	Labor-Hours	Unit	2012 Bare Costs Material	Labor	Equipment	Total	Total Incl O&P
3263	6′ wide	1 Clab	32	.250	Ea.		8.80		8.80	14.40
3264	8′ wide		30	.267			9.35		9.35	15.35
3265	10′ wide		30	.267			9.35		9.35	15.35
3266	12′ wide		30	.267			9.35		9.35	15.35
3267	2″ x 6″ wall, 3′ wide		32	.250			8.80		8.80	14.40
3268	4′ wide		32	.250			8.80		8.80	14.40
3269	5′ wide		32	.250			8.80		8.80	14.40
3270	6′ wide		32	.250			8.80		8.80	14.40
3271	8′ wide		30	.267			9.35		9.35	15.35
3272	10′ wide		30	.267			9.35		9.35	15.35
3273	12′ wide		30	.267			9.35		9.35	15.35
3274	Window buck, studs, header & access, 8′ high 2″ x 4″ wall, 2′ wide		24	.333			11.70		11.70	19.20
3275	3′ wide		24	.333			11.70		11.70	19.20
3276	4′ wide		24	.333			11.70		11.70	19.20
3277	5′ wide		24	.333			11.70		11.70	19.20
3278	6′ wide		24	.333			11.70		11.70	19.20
3279	7′ wide		24	.333			11.70		11.70	19.20
3280	8′ wide		22	.364			12.75		12.75	21
3281	10′ wide		22	.364			12.75		12.75	21
3282	12′ wide		22	.364			12.75		12.75	21
3283	2″ x 6″ wall, 2′ wide		24	.333			11.70		11.70	19.20
3284	3′ wide		24	.333			11.70		11.70	19.20
3285	4′ wide		24	.333			11.70		11.70	19.20
3286	5′ wide		24	.333			11.70		11.70	19.20
3287	6′ wide		24	.333			11.70		11.70	19.20
3288	7′ wide		24	.333			11.70		11.70	19.20
3289	8′ wide		22	.364			12.75		12.75	21
3290	10′ wide		22	.364			12.75		12.75	21
3291	12′ wide		22	.364			12.75		12.75	21
3400	Fascia boards, 1″ x 6″		500	.016	L.F.		.56		.56	.92
3440	1″ x 8″		450	.018			.62		.62	1.02
3480	1″ x 10″		400	.020			.70		.70	1.15
3490	2″ x 6″		450	.018			.62		.62	1.02
3500	2″ x 8″		400	.020			.70		.70	1.15
3510	2″ x 10″		350	.023			.80		.80	1.32
3610	Furring, on wood walls or ceiling		4000	.002	S.F.		.07		.07	.12
3620	On masonry or concrete walls or ceiling		1200	.007	"		.23		.23	.38
3800	Headers over openings, 2 @ 2″ x 6″		110	.073	L.F.		2.55		2.55	4.19
3840	2 @ 2″ x 8″		100	.080			2.81		2.81	4.60
3880	2 @ 2″ x 10″		90	.089			3.12		3.12	5.10
3885	Alternate pricing method		.26	30.651	M.B.F.		1,075		1,075	1,775
3920	Joists, 1″ x 4″		1250	.006	L.F.		.22		.22	.37
3930	1″ x 6″		1135	.007			.25		.25	.41
3940	1″ x 8″		1000	.008			.28		.28	.46
3950	1″ x 10″		895	.009			.31		.31	.51
3960	1″ x 12″		765	.010			.37		.37	.60
4200	2″ x 4″	2 Clab	1000	.016			.56		.56	.92
4230	2″ x 6″		970	.016			.58		.58	.95
4240	2″ x 8″		940	.017			.60		.60	.98
4250	2″ x 10″		910	.018			.62		.62	1.01
4280	2″ x 12″		880	.018			.64		.64	1.05
4281	2″ x 14″		850	.019			.66		.66	1.08
4282	Composite joists, 9-1/2″		960	.017			.59		.59	.96

06 05 05 – Selective Wood and Plastics Demolition

06 05 05.10 Selective Demolition Wood Framing		Crew	Daily Output	Labor-Hours	Unit	Material	2012 Bare Costs Labor	Equipment	Total	Total Incl O&P
4283	11-7/8"	2 Clab	930	.017	L.F.		.60		.60	.99
4284	14"		897	.018			.63		.63	1.03
4285	16"		865	.019			.65		.65	1.06
4290	Wood joists, alternate pricing method		1.50	10.667	M.B.F.		375		375	615
4500	Open web joist, 12" deep		500	.032	L.F.		1.12		1.12	1.84
4505	14" deep		475	.034			1.18		1.18	1.94
4510	16" deep		450	.036			1.25		1.25	2.05
4520	18" deep		425	.038			1.32		1.32	2.17
4530	24" deep		400	.040			1.40		1.40	2.30
4550	Ledger strips, 1" x 2"	1 Clab	1200	.007			.23		.23	.38
4560	1" x 3"		1200	.007			.23		.23	.38
4570	1" x 4"		1200	.007			.23		.23	.38
4580	2" x 2"		1100	.007			.26		.26	.42
4590	2" x 4"		1000	.008			.28		.28	.46
4600	2" x 6"		1000	.008			.28		.28	.46
4601	2" x 8 or 2" x 10"		800	.010			.35		.35	.58
4602	4" x 6"		600	.013			.47		.47	.77
4604	4" x 8"		450	.018			.62		.62	1.02
5400	Posts, 4" x 4"	2 Clab	800	.020			.70		.70	1.15
5405	4" x 6"		550	.029			1.02		1.02	1.67
5410	4" x 8"		440	.036			1.28		1.28	2.09
5425	4" x 10"		390	.041			1.44		1.44	2.36
5430	4" x 12"		350	.046			1.60		1.60	2.63
5440	6" x 6"		400	.040			1.40		1.40	2.30
5445	6" x 8"		350	.046			1.60		1.60	2.63
5450	6" x 10"		320	.050			1.76		1.76	2.88
5455	6" x 12"		290	.055			1.94		1.94	3.18
5480	8" x 8"		300	.053			1.87		1.87	3.07
5500	10" x 10"		240	.067			2.34		2.34	3.84
5660	Tongue and groove floor planks		2	8	M.B.F.		281		281	460
5682	Rafters, ordinary, 16" OC, 2" x 4"		880	.018	S.F.		.64		.64	1.05
5683	2" x 6"		840	.019			.67		.67	1.10
5684	2" x 8"		820	.020			.68		.68	1.12
5685	2" x 10"		820	.020			.68		.68	1.12
5686	2" x 12"		810	.020			.69		.69	1.14
5687	24" OC, 2" x 4"		1170	.014			.48		.48	.79
5688	2" x 6"		1117	.014			.50		.50	.82
5689	2" x 8"		1091	.015			.51		.51	.84
5690	2" x 10"		1091	.015			.51		.51	.84
5691	2" x 12"		1077	.015			.52		.52	.86
5795	Rafters, ordinary, 2" x 4" (alternate method)		862	.019	L.F.		.65		.65	1.07
5800	2" x 6" (alternate method)		850	.019			.66		.66	1.08
5840	2" x 8" (alternate method)		837	.019			.67		.67	1.10
5855	2" x 10" (alternate method)		825	.019			.68		.68	1.12
5865	2" x 12" (alternate method)		812	.020			.69		.69	1.13
5870	Sill plate, 2" x 4"	1 Clab	1170	.007			.24		.24	.39
5871	2" x 6"		780	.010			.36		.36	.59
5872	2" x 8"		586	.014			.48		.48	.79
5873	Alternate pricing method		.78	10.256	M.B.F.		360		360	590
5885	Ridge board, 1" x 4"	2 Clab	900	.018	L.F.		.62		.62	1.02
5886	1" x 6"		875	.018			.64		.64	1.05
5887	1" x 8"		850	.019			.66		.66	1.08
5888	1" x 10"		825	.019			.68		.68	1.12

06 05 05.10 Selective Demolition Wood Framing		Crew	Daily Output	Labor-Hours	Unit	Material	2012 Bare Costs Labor	Equipment	Total	Total Incl O&P
5889	1″ x 12″	2 Clab	800	.020	L.F.		.70		.70	1.15
5890	2″ x 4″		900	.018			.62		.62	1.02
5892	2″ x 6″		875	.018			.64		.64	1.05
5894	2″ x 8″		850	.019			.66		.66	1.08
5896	2″ x 10″		825	.019			.68		.68	1.12
5898	2″ x 12″		800	.020			.70		.70	1.15
5900	Hip & valley rafters, 2″ x 6″		500	.032			1.12		1.12	1.84
5940	2″ x 8″		420	.038			1.34		1.34	2.19
6050	Rafter tie, 1″ x 4″		1250	.013			.45		.45	.74
6052	1″ x 6″		1135	.014			.49		.49	.81
6054	2″ x 4″		1000	.016			.56		.56	.92
6056	2″ x 6″		970	.016			.58		.58	.95
6070	Sleepers, on concrete, 1″ x 2″	1 Clab	4700	.002			.06		.06	.10
6075	1″ x 3″		4000	.002			.07		.07	.12
6080	2″ x 4″		3000	.003			.09		.09	.15
6085	2″ x 6″		2600	.003			.11		.11	.18
6086	Sheathing from roof, 5/16″	2 Clab	1600	.010	S.F.		.35		.35	.58
6088	3/8″		1525	.010			.37		.37	.60
6090	1/2″		1400	.011			.40		.40	.66
6092	5/8″		1300	.012			.43		.43	.71
6094	3/4″		1200	.013			.47		.47	.77
6096	Board sheathing from roof		1400	.011			.40		.40	.66
6100	Sheathing, from walls, 1/4″		1200	.013			.47		.47	.77
6110	5/16″		1175	.014			.48		.48	.78
6120	3/8″		1150	.014			.49		.49	.80
6130	1/2″		1125	.014			.50		.50	.82
6140	5/8″		1100	.015			.51		.51	.84
6150	3/4″		1075	.015			.52		.52	.86
6152	Board sheathing from walls		1500	.011			.37		.37	.61
6158	Subfloor, with boards		1050	.015			.53		.53	.88
6160	Plywood, 1/2″ thick		768	.021			.73		.73	1.20
6162	5/8″ thick		760	.021			.74		.74	1.21
6164	3/4″ thick		750	.021			.75		.75	1.23
6165	1-1/8″ thick		720	.022			.78		.78	1.28
6166	Underlayment, particle board, 3/8″ thick	1 Clab	780	.010			.36		.36	.59
6168	1/2″ thick		768	.010			.37		.37	.60
6170	5/8″ thick		760	.011			.37		.37	.61
6172	3/4″ thick		750	.011			.37		.37	.61
6200	Stairs and stringers, minimum	2 Clab	40	.400	Riser		14.05		14.05	23
6240	Maximum	″	26	.615	″		21.50		21.50	35.50
6300	Components, tread	1 Clab	110	.073	Ea.		2.55		2.55	4.19
6320	Riser		80	.100	″		3.51		3.51	5.75
6390	Stringer, 2″ x 10″		260	.031	L.F.		1.08		1.08	1.77
6400	2″ x 12″		260	.031			1.08		1.08	1.77
6410	3″ x 10″		250	.032			1.12		1.12	1.84
6420	3″ x 12″		250	.032			1.12		1.12	1.84
6590	Wood studs, 2″ x 3″	2 Clab	3076	.005			.18		.18	.30
6600	2″ x 4″		2000	.008			.28		.28	.46
6640	2″ x 6″		1600	.010			.35		.35	.58
6720	Wall framing, including studs, plates and blocking, 2″ x 4″	1 Clab	600	.013	S.F.		.47		.47	.77
6740	2″ x 6″		480	.017	″		.59		.59	.96
6750	Headers, 2″ x 4″		1125	.007	L.F.		.25		.25	.41
6755	2″ x 6″		1125	.007			.25		.25	.41

06 05 Common Work Results for Wood, Plastics, and Composites

06 05 05 – Selective Wood and Plastics Demolition

06 05 05.10 Selective Demolition Wood Framing		Crew	Daily Output	Labor-Hours	Unit	Material	2012 Bare Costs Labor	Equipment	Total	Total Incl O&P
6760	2" x 8"	1 Clab	1050	.008	L.F.		.27		.27	.44
6765	2" x 10"		1050	.008			.27		.27	.44
6770	2" x 12"		1000	.008			.28		.28	.46
6780	4" x 10"		525	.015			.53		.53	.88
6785	4" x 12"		500	.016			.56		.56	.92
6790	6" x 8"		560	.014			.50		.50	.82
6795	6" x 10"		525	.015			.53		.53	.88
6797	6" x 12"		500	.016			.56		.56	.92
8000	Soffit, T & G wood		520	.015	S.F.		.54		.54	.89
8010	Hardboard, vinyl or aluminum		640	.013			.44		.44	.72
8030	Plywood	2 Carp	315	.051			2.24		2.24	3.67
9000	Minimum labor/equipment charge	1 Clab	4	2	Job		70		70	115
9500	See Section 02 41 19.23 for rubbish handling									

06 05 05.20 Selective Demolition Millwork and Trim		Crew	Daily Output	Labor-Hours	Unit	Material	Labor	Equipment	Total	Total Incl O&P
0010	**SELECTIVE DEMOLITION MILLWORK AND TRIM** R024119-10									
1000	Cabinets, wood, base cabinets, per L.F.	2 Clab	80	.200	L.F.		7		7	11.50
1020	Wall cabinets, per L.F.	"	80	.200	"		7		7	11.50
1060	Remove and reset, base cabinets	2 Carp	18	.889	Ea.		39		39	64.50
1070	Wall cabinets	"	20	.800	"		35.50		35.50	58
1100	Steel, painted, base cabinets	2 Clab	60	.267	L.F.		9.35		9.35	15.35
1120	Wall cabinets		60	.267	"		9.35		9.35	15.35
1200	Casework, large area		320	.050	S.F.		1.76		1.76	2.88
1220	Selective		200	.080	"		2.81		2.81	4.60
1500	Counter top, minimum		200	.080	L.F.		2.81		2.81	4.60
1510	Maximum		120	.133			4.68		4.68	7.65
1550	Remove and reset, minimum	2 Carp	50	.320			14.10		14.10	23
1560	Maximum	"	40	.400			17.65		17.65	29
2000	Paneling, 4' x 8' sheets	2 Clab	2000	.008	S.F.		.28		.28	.46
2100	Boards, 1" x 4"		700	.023			.80		.80	1.32
2120	1" x 6"		750	.021			.75		.75	1.23
2140	1" x 8"		800	.020			.70		.70	1.15
3000	Trim, baseboard, to 6" wide		1200	.013	L.F.		.47		.47	.77
3040	Greater than 6" and up to 12" wide		1000	.016			.56		.56	.92
3080	Remove and reset, minimum	2 Carp	400	.040			1.76		1.76	2.89
3090	Maximum	"	300	.053			2.35		2.35	3.86
3100	Ceiling trim	2 Clab	1000	.016			.56		.56	.92
3120	Chair rail		1200	.013			.47		.47	.77
3140	Railings with balusters		240	.067			2.34		2.34	3.84
3160	Wainscoting		700	.023	S.F.		.80		.80	1.32
9000	Minimum labor/equipment charge	1 Clab	4	2	Job		70		70	115

06 05 23 – Wood, Plastic, and Composite Fastenings

06 05 23.10 Nails		Crew	Daily Output	Labor-Hours	Unit	Material	Labor	Equipment	Total	Total Incl O&P
0010	**NAILS**, material only, based upon 50# box purchase									
0020	Copper nails, plain				Lb.	10			10	11
0400	Stainless steel, plain					7.80			7.80	8.60
0500	Box, 3d to 20d, bright					1.25			1.25	1.38
0520	Galvanized					1.77			1.77	1.95
0600	Common, 3d to 60d, plain					1.30			1.30	1.43
0700	Galvanized					1.56			1.56	1.72
0800	Aluminum					5.80			5.80	6.40
1000	Annular or spiral thread, 4d to 60d, plain					1.80			1.80	1.98
1200	Galvanized					3			3	3.30

06 05 Common Work Results for Wood, Plastics, and Composites

06 05 23 – Wood, Plastic, and Composite Fastenings

06 05 23.10 Nails		Crew	Daily Output	Labor-Hours	Unit	Material	Labor	Equipment	Total	Total Incl O&P
						2012 Bare Costs				
1400	Drywall nails, plain				Lb.	3.59			3.59	3.95
1600	Galvanized					2.19			2.19	2.41
1800	Finish nails, 4d to 10d, plain					1.36			1.36	1.50
2000	Galvanized					1.79			1.79	1.97
2100	Aluminum					5.80			5.80	6.40
2300	Flooring nails, hardened steel, 2d to 10d, plain					3.10			3.10	3.41
2400	Galvanized					4			4	4.40
2500	Gypsum lath nails, 1-1/8", 13 ga. flathead, blued					3.59			3.59	3.95
2600	Masonry nails, hardened steel, 3/4" to 3" long, plain					2.20			2.20	2.42
2700	Galvanized					3.75			3.75	4.13
2900	Roofing nails, threaded, galvanized					1.63			1.63	1.79
3100	Aluminum					4.80			4.80	5.30
3300	Compressed lead head, threaded, galvanized					2.47			2.47	2.72
3600	Siding nails, plain shank, galvanized					2			2	2.20
3800	Aluminum					5.80			5.80	6.40
5000	Add to prices above for cement coating					.11			.11	.12
5200	Zinc or tin plating					.14			.14	.15
5500	Vinyl coated sinkers, 8d to 16d					.65			.65	.72

06 05 23.50 Wood Screws		Crew	Daily Output	Labor-Hours	Unit	Material	Labor	Equipment	Total	Total Incl O&P
0010	**WOOD SCREWS**									
0020	#8 x 1" long, steel				C	3.41			3.41	3.75
0100	Brass					10.50			10.50	11.55
0600	#10, 2" long, steel					5.85			5.85	6.45
0700	Brass					24.50			24.50	27
1500	#12, 3" long, steel					11.65			11.65	12.80

06 05 23.60 Timber Connectors		Crew	Daily Output	Labor-Hours	Unit	Material	Labor	Equipment	Total	Total Incl O&P
0010	**TIMBER CONNECTORS**									
0100	Connector plates, steel, with bolts, straight	2 Carp	75	.213	Ea.	28	9.40		37.40	46
0110	Tee, 7 gauge		50	.320		32	14.10		46.10	58
0120	T-Strap, 14 gauge, 12" x 8" x 2"		50	.320		32	14.10		46.10	58
0150	Anchor plates, 7 ga, 9" x 7"		75	.213		28	9.40		37.40	46
0200	Bolts, machine, sq. hd. with nut & washer, 1/2" diameter, 4" long	1 Carp	140	.057		.74	2.52		3.26	4.94
0300	7-1/2" long		130	.062		1.34	2.71		4.05	5.90
0500	3/4" diameter, 7-1/2" long		130	.062		3.12	2.71		5.83	7.90
0610	Machine bolts, w/nut, washer, 3/4" diam., 15" L, HD's & beam hangers		95	.084		5.80	3.71		9.51	12.50
0800	Drilling bolt holes in timber, 1/2" diameter		450	.018	Inch		.78		.78	1.29
0900	1" diameter		350	.023	"		1.01		1.01	1.65
1100	Framing anchor, angle, 3" x 3" x 1-1/2", 12 ga		175	.046	Ea.	2.40	2.02		4.42	5.95
1150	Framing anchors, 18 gauge, 4-1/2" x 2-3/4"		175	.046		2.40	2.02		4.42	5.95
1160	Framing anchors, 18 gauge, 4-1/2" x 3"		175	.046		2.40	2.02		4.42	5.95
1170	Clip anchors plates, 18 gauge, 12" x 1-1/8"		175	.046		2.40	2.02		4.42	5.95
1250	Holdowns, 3 gauge base, 10 gauge body		8	1		23	44		67	97.50
1260	Holdowns, 7 gauge 11-1/16" x 3-1/4"		8	1		23	44		67	97.50
1270	Holdowns, 7 gauge 14-3/8" x 3-1/8"		8	1		23	44		67	97.50
1275	Holdowns, 12 gauge 8" x 2-1/2"		8	1		23	44		67	97.50
1300	Joist and beam hangers, 18 ga. galv., for 2" x 4" joist		175	.046		.67	2.02		2.69	4.04
1400	2" x 6" to 2" x 10" joist		165	.048		1.28	2.14		3.42	4.92
1600	16 ga. galv., 3" x 6" to 3" x 10" joist		160	.050		2.72	2.21		4.93	6.60
1700	3" x 10" to 3" x 14" joist		160	.050		4.52	2.21		6.73	8.60
1800	4" x 6" to 4" x 10" joist		155	.052		2.82	2.28		5.10	6.85
1900	4" x 10" to 4" x 14" joist		155	.052		4.62	2.28		6.90	8.85
2000	Two-2" x 6" to two-2" x 10" joists		150	.053		3.85	2.35		6.20	8.10

06 05 Common Work Results for Wood, Plastics, and Composites

06 05 23 – Wood, Plastic, and Composite Fastenings

06 05 23.60 Timber Connectors

06 05 23.60 Timber Connectors		Crew	Daily Output	Labor-Hours	Unit	Material	2012 Bare Costs Labor	Equipment	Total	Total Incl O&P
2100	Two-2" x 10" to two-2" x 14" joists	1 Carp	150	.053	Ea.	4.30	2.35		6.65	8.60
2300	3/16" thick, 6" x 8" joist		145	.055		59.50	2.43		61.93	69.50
2400	6" x 10" joist		140	.057		62	2.52		64.52	72.50
2500	6" x 12" joist		135	.059		64.50	2.61		67.11	75.50
2700	1/4" thick, 6" x 14" joist		130	.062		67.50	2.71		70.21	78.50
2900	Plywood clips, extruded aluminum H clip, for 3/4" panels					.22			.22	.24
3000	Galvanized 18 ga. back-up clip					.17			.17	.19
3200	Post framing, 16 ga. galv. for 4" x 4" base, 2 piece	1 Carp	130	.062		15.05	2.71		17.76	21
3300	Cap		130	.062		21	2.71		23.71	27.50
3500	Rafter anchors, 18 ga. galv., 1-1/2" wide, 5-1/4" long		145	.055		.45	2.43		2.88	4.49
3600	10-3/4" long		145	.055		1.34	2.43		3.77	5.45
3800	Shear plates, 2-5/8" diameter		120	.067		2.19	2.94		5.13	7.25
3900	4" diameter		115	.070		5.20	3.07		8.27	10.75
4000	Sill anchors, embedded in concrete or block, 25-1/2" long		115	.070		11.65	3.07		14.72	17.90
4100	Spike grids, 3" x 6"		120	.067		.88	2.94		3.82	5.80
4400	Split rings, 2-1/2" diameter		120	.067		1.80	2.94		4.74	6.80
4500	4" diameter		110	.073		2.73	3.21		5.94	8.25
4550	Tie plate, 20 gauge, 7" x 3 1/8"		110	.073		2.73	3.21		5.94	8.25
4560	Tie plate, 20 gauge, 5" x 4 1/8"		110	.073		2.73	3.21		5.94	8.25
4575	Twist straps, 18 gauge, 12" x 1 1/4"		110	.073		2.73	3.21		5.94	8.25
4580	Twist straps, 18 gauge, 16" x 1 1/4"		110	.073		2.73	3.21		5.94	8.25
4600	Strap ties, 20 ga., 2 -1/16" wide, 12 13/16" long		180	.044		.87	1.96		2.83	4.17
4700	Strap ties, 16 ga., 1-3/8" wide, 12" long		180	.044		.87	1.96		2.83	4.17
4800	21-5/8" x 1-1/4"		160	.050		2.72	2.21		4.93	6.60
5000	Toothed rings, 2-5/8" or 4" diameter		90	.089		1.63	3.92		5.55	8.25
5200	Truss plates, nailed, 20 gauge, up to 32' span		17	.471	Truss	11.85	21		32.85	47
5400	Washers, 2" x 2" x 1/8"				Ea.	.37			.37	.41
5500	3" x 3" x 3/16"				"	.98			.98	1.08
9000	Minimum labor/equipment charge	1 Carp	4	2	Job		88		88	145

06 05 23.70 Rough Hardware

06 05 23.70 Rough Hardware		Crew	Daily Output	Labor-Hours	Unit	Material	Labor	Equipment	Total	Total Incl O&P
0010	**ROUGH HARDWARE**, average percent of carpentry material									
0020	Minimum				Job	.50%				
0200	Maximum				"	1.50%				

06 05 23.80 Metal Bracing

06 05 23.80 Metal Bracing		Crew	Daily Output	Labor-Hours	Unit	Material	Labor	Equipment	Total	Total Incl O&P
0010	**METAL BRACING**									
0302	Let-in, "T" shaped, 22 ga. galv. steel, studs at 16" O.C.	1 Carp	580	.014	L.F.	.75	.61		1.36	1.83
0402	Studs at 24" O.C.		600	.013		.75	.59		1.34	1.79
0502	16 ga. galv. steel straps, studs at 16" O.C.		600	.013		.97	.59		1.56	2.03
0602	Studs at 24" O.C.		620	.013		.97	.57		1.54	2

06 11 Wood Framing

06 11 10 – Framing with Dimensional, Engineered or Composite Lumber

06 11 10.01 Forest Stewardship Council Certification

06 11 10.01 Forest Stewardship Council Certification		Crew	Daily Output	Labor-Hours	Unit	Material	Labor	Equipment	Total	Total Incl O&P
0010	**FOREST STEWARDSHIP COUNCIL CERTIFICATION**									
0020	For Forest Stewardship Council (FSC) cert dimension lumber, add [G]					65%				

06 11 10.02 Blocking

06 11 10.02 Blocking		Crew	Daily Output	Labor-Hours	Unit	Material	Labor	Equipment	Total	Total Incl O&P
0010	**BLOCKING**									
2600	Miscellaneous, to wood construction									
2620	2" x 4"	1 Carp	.17	47.059	M.B.F.	490	2,075		2,565	3,950
2625	Pneumatic nailed		.21	38.095		490	1,675		2,165	3,300

06 11 Wood Framing

06 11 10 – Framing with Dimensional, Engineered or Composite Lumber

06 11 10.02 Blocking

	06 11 10.02 Blocking	Crew	Daily Output	Labor-Hours	Unit	2012 Bare Costs Material	Labor	Equipment	Total	Total Incl O&P
2660	2" x 8"	1 Carp	.27	29.630	M.B.F.	565	1,300		1,865	2,775
2665	Pneumatic nailed		.33	24.242		565	1,075		1,640	2,375
2720	To steel construction									
2740	2" x 4"	1 Carp	.14	57.143	M.B.F.	490	2,525		3,015	4,675
2780	2" x 8"		.21	38.095	"	565	1,675		2,240	3,375
9000	Minimum labor/equipment charge		4	2	Job		88		88	145

06 11 10.04 Wood Bracing

	06 11 10.04 Wood Bracing	Crew	Daily Output	Labor-Hours	Unit	Material	Labor	Equipment	Total	Total Incl O&P
0010	**WOOD BRACING**									
0012	Let-in, with 1" x 6" boards, studs @ 16" O.C.	1 Carp	150	.053	L.F.	.72	2.35		3.07	4.65
0202	Studs @ 24" O.C.	"	230	.035	"	.72	1.53		2.25	3.30

06 11 10.06 Bridging

	06 11 10.06 Bridging	Crew	Daily Output	Labor-Hours	Unit	Material	Labor	Equipment	Total	Total Incl O&P
0010	**BRIDGING**									
0012	Wood, for joists 16" O.C., 1" x 3"	1 Carp	130	.062	Pr.	.51	2.71		3.22	5
0017	Pneumatic nailed		170	.047		.51	2.08		2.59	3.96
0102	2" x 3" bridging		130	.062		.50	2.71		3.21	5
0107	Pneumatic nailed		170	.047		.50	2.08		2.58	3.95
0302	Steel, galvanized, 18 ga., for 2" x 10" joists at 12" O.C.		130	.062		1.72	2.71		4.43	6.35
0352	16" O.C.		135	.059		1.85	2.61		4.46	6.30
0402	24" O.C.		140	.057		1.85	2.52		4.37	6.15
0602	For 2" x 14" joists at 16" O.C.		130	.062		1.85	2.71		4.56	6.50
0902	Compression type, 16" O.C., 2" x 8" joists		200	.040		1.78	1.76		3.54	4.85
1002	2" x 12" joists		200	.040		1.78	1.76		3.54	4.85

06 11 10.10 Beam and Girder Framing

	06 11 10.10 Beam and Girder Framing	Crew	Daily Output	Labor-Hours	Unit	Material	Labor	Equipment	Total	Total Incl O&P
0010	**BEAM AND GIRDER FRAMING** R061110-30									
1000	Single, 2" x 6"	2 Carp	700	.023	L.F.	.53	1.01		1.54	2.24
1005	Pneumatic nailed		812	.020		.53	.87		1.40	2.01
1020	2" x 8"		650	.025		.75	1.09		1.84	2.61
1025	Pneumatic nailed		754	.021		.75	.94		1.69	2.36
1040	2" x 10"		600	.027		1.01	1.18		2.19	3.04
1045	Pneumatic nailed		696	.023		1.01	1.01		2.02	2.77
1060	2" x 12"		550	.029		1.28	1.28		2.56	3.50
1065	Pneumatic nailed		638	.025		1.28	1.11		2.39	3.21
1080	2" x 14"		500	.032		1.70	1.41		3.11	4.18
1085	Pneumatic nailed		580	.028		1.70	1.22		2.92	3.86
1100	3" x 8"		550	.029		2.12	1.28		3.40	4.43
1120	3" x 10"		500	.032		2.68	1.41		4.09	5.25
1140	3" x 12"		450	.036		3.27	1.57		4.84	6.15
1160	3" x 14"		400	.040		3.85	1.76		5.61	7.15
1170	4" x 6"	F-3	1100	.036		2.50	1.63	.56	4.69	6
1180	4" x 8"		1000	.040		3.40	1.79	.61	5.80	7.35
1200	4" x 10"		950	.042		4.25	1.89	.64	6.78	8.45
1220	4" x 12"		900	.044		5.50	1.99	.68	8.17	10.05
1240	4" x 14"		850	.047		7.15	2.11	.72	9.98	12.05
2000	Double, 2" x 6"	2 Carp	625	.026		1.07	1.13		2.20	3.02
2005	Pneumatic nailed		725	.022		1.07	.97		2.04	2.77
2020	2" x 8"		575	.028		1.50	1.23		2.73	3.66
2025	Pneumatic nailed		667	.024		1.50	1.06		2.56	3.38
2040	2" x 10"		550	.029		2.01	1.28		3.29	4.31
2045	Pneumatic nailed		638	.025		2.01	1.11		3.12	4.02
2060	2" x 12"		525	.030		2.55	1.34		3.89	5
2065	Pneumatic nailed		610	.026		2.55	1.16		3.71	4.71
2080	2" x 14"		475	.034		3.39	1.49		4.88	6.15

06 11 Wood Framing

06 11 10 – Framing with Dimensional, Engineered or Composite Lumber

06 11 10.10 Beam and Girder Framing		Crew	Daily Output	Labor-Hours	Unit	2012 Bare Costs Material	Labor	Equipment	Total	Total Incl O&P
2085	Pneumatic nailed	2 Carp	551	.029	L.F.	3.39	1.28		4.67	5.85
3000	Triple, 2" x 6"		550	.029		1.60	1.28		2.88	3.86
3005	Pneumatic nailed		638	.025		1.60	1.11		2.71	3.57
3020	2" x 8"		525	.030		2.25	1.34		3.59	4.68
3025	Pneumatic nailed		609	.026		2.25	1.16		3.41	4.38
3040	2" x 10"		500	.032		3.02	1.41		4.43	5.65
3045	Pneumatic nailed		580	.028		3.02	1.22		4.24	5.30
3060	2" x 12"		475	.034		3.83	1.49		5.32	6.65
3065	Pneumatic nailed		551	.029		3.83	1.28		5.11	6.30
3080	2" x 14"		450	.036		5.10	1.57		6.67	8.15
3085	Pneumatic nailed		522	.031		5.10	1.35		6.45	7.80
9000	Minimum labor/equipment charge	1 Carp	2	4	Job		176		176	289

06 11 10.12 Ceiling Framing

Line	Description	Crew	Daily Output	Labor-Hours	Unit	Material	Labor	Equipment	Total	Total Incl O&P
0010	**CEILING FRAMING**									
6000	Suspended, 2" x 3"	2 Carp	1000	.016	L.F.	.33	.71		1.04	1.53
6050	2" x 4"		900	.018		.33	.78		1.11	1.65
6100	2" x 6"		800	.020		.53	.88		1.41	2.04
6150	2" x 8"		650	.025		.75	1.09		1.84	2.61
9000	Minimum labor/equipment charge	1 Carp	4	2	Job		88		88	145

06 11 10.14 Posts and Columns

Line	Description	Crew	Daily Output	Labor-Hours	Unit	Material	Labor	Equipment	Total	Total Incl O&P
0010	**POSTS AND COLUMNS**									
0100	4" x 4"	2 Carp	390	.041	L.F.	1.64	1.81		3.45	4.77
0150	4" x 6"		275	.058		2.50	2.57		5.07	6.95
0200	4" x 8"		220	.073		3.40	3.21		6.61	9
0250	6" x 6"		215	.074		3.62	3.28		6.90	9.40
0300	6" x 8"		175	.091		4.82	4.03		8.85	11.90
0350	6" x 10"		150	.107		6.50	4.70		11.20	14.85
9000	Minimum labor/equipment charge	1 Carp	2	4	Job		176		176	289

06 11 10.18 Joist Framing

Line	Description	Crew	Daily Output	Labor-Hours	Unit	Material	Labor	Equipment	Total	Total Incl O&P
0010	**JOIST FRAMING** R061110-30									
2002	Joists, 2" x 4"	2 Carp	1250	.013	L.F.	.33	.56		.89	1.29
2007	Pneumatic nailed		1438	.011		.33	.49		.82	1.16
2100	2" x 6"		1250	.013		.53	.56		1.09	1.52
2105	Pneumatic nailed		1438	.011		.53	.49		1.02	1.39
2152	2" x 8"		1100	.015		.75	.64		1.39	1.88
2157	Pneumatic nailed		1265	.013		.75	.56		1.31	1.74
2202	2" x 10"		900	.018		1.01	.78		1.79	2.40
2207	Pneumatic nailed		1035	.015		1.01	.68		1.69	2.23
2252	2" x 12"		875	.018		1.28	.81		2.09	2.72
2257	Pneumatic nailed		1006	.016		1.28	.70		1.98	2.55
2302	2" x 14"		770	.021		1.70	.92		2.62	3.37
2307	Pneumatic nailed		886	.018		1.70	.80		2.50	3.18
2352	3" x 6"		925	.017		1.35	.76		2.11	2.74
2402	3" x 10"		780	.021		2.68	.90		3.58	4.43
2452	3" x 12"		600	.027		3.27	1.18		4.45	5.55
2502	4" x 6"		800	.020		2.50	.88		3.38	4.20
2552	4" x 10"		600	.027		4.25	1.18		5.43	6.60
2602	4" x 12"		450	.036		5.50	1.57		7.07	8.60
2607	Sister joist, 2" x 6"		800	.020		.53	.88		1.41	2.04
2608	Pneumatic nailed		960	.017		.53	.74		1.27	1.80
2612	2" x 8"		640	.025		.75	1.10		1.85	2.64
2613	Pneumatic nailed		768	.021		.75	.92		1.67	2.34

06 11 10 – Framing with Dimensional, Engineered or Composite Lumber

06 11 10.18 Joist Framing		Crew	Daily Output	Labor-Hours	Unit	Material	2012 Bare Costs Labor	Equipment	Total	Total Incl O&P
2617	2" x 10"	2 Carp	535	.030	L.F.	1.01	1.32		2.33	3.27
2618	Pneumatic nailed		642	.025		1.01	1.10		2.11	2.91
2622	2" x 12"		455	.035		1.28	1.55		2.83	3.94
2627	Pneumatic nailed		546	.029		1.28	1.29		2.57	3.52
3000	Composite wood joist 9-1/2" deep		.90	17.778	M.L.F.	1,625	785		2,410	3,050
3010	11-1/2" deep		.88	18.182		1,825	800		2,625	3,325
3020	14" deep		.82	19.512		2,525	860		3,385	4,175
3030	16" deep		.78	20.513		3,125	905		4,030	4,900
4000	Open web joist 12" deep		.88	18.182		2,900	800		3,700	4,525
4010	14" deep		.82	19.512		3,150	860		4,010	4,875
4020	16" deep		.78	20.513		3,200	905		4,105	5,000
4030	18" deep		.74	21.622		3,350	955		4,305	5,250
6000	Composite rim joist, 1-1/4" x 9-1/2"		.90	17.778		1,750	785		2,535	3,200
6010	1-1/4" x 11-1/2"		.88	18.182		2,250	800		3,050	3,800
6020	1-1/4" x 14-1/2"		.82	19.512		2,525	860		3,385	4,175
6030	1-1/4" x 16-1/2"		.78	20.513		3,100	905		4,005	4,875
9000	Minimum labor/equipment charge	1 Carp	4	2	Job		88		88	145

06 11 10.24 Miscellaneous Framing		Crew	Daily Output	Labor-Hours	Unit	Material	Labor	Equipment	Total	Total Incl O&P
0010	**MISCELLANEOUS FRAMING**									
2002	Firestops, 2" x 4"	2 Carp	780	.021	L.F.	.33	.90		1.23	1.84
2007	Pneumatic nailed		952	.017		.33	.74		1.07	1.58
2102	2" x 6"		600	.027		.53	1.18		1.71	2.52
2107	Pneumatic nailed		732	.022		.53	.96		1.49	2.17
5002	Nailers, treated, wood construction, 2" x 4"		800	.020		.33	.88		1.21	1.82
5007	Pneumatic nailed		960	.017		.33	.74		1.07	1.58
5102	2" x 6"		750	.021		.63	.94		1.57	2.23
5107	Pneumatic nailed		900	.018		.63	.78		1.41	1.98
5122	2" x 8"		700	.023		.88	1.01		1.89	2.62
5127	Pneumatic nailed		840	.019		.88	.84		1.72	2.35
5202	Steel construction, 2" x 4"		750	.021		.33	.94		1.27	1.91
5222	2" x 6"		700	.023		.63	1.01		1.64	2.34
5242	2" x 8"		650	.025		.88	1.09		1.97	2.75
7002	Rough bucks, treated, for doors or windows, 2" x 6"		400	.040		.63	1.76		2.39	3.58
7007	Pneumatic nailed		480	.033		.63	1.47		2.10	3.10
7102	2" x 8"		380	.042		.88	1.86		2.74	4.01
7107	Pneumatic nailed		456	.035		.88	1.55		2.43	3.51
8000	Stair stringers, 2" x 10"		130	.123		1.01	5.45		6.46	10
8100	2" x 12"		130	.123		1.28	5.45		6.73	10.30
8150	3" x 10"		125	.128		2.68	5.65		8.33	12.20
8200	3" x 12"		125	.128		3.27	5.65		8.92	12.85
8870	Laminated structural lumber, 1-1/4" x 11-1/2"		130	.123		2.25	5.45		7.70	11.35
8880	1-1/4" x 14-1/2"		130	.123		2.53	5.45		7.98	11.70
9000	Minimum labor/equipment charge	1 Carp	4	2	Job		88		88	145

06 11 10.26 Partitions		Crew	Daily Output	Labor-Hours	Unit	Material	Labor	Equipment	Total	Total Incl O&P
0010	**PARTITIONS**									
0020	Single bottom and double top plate, no waste, std. & better lumber									
0182	2" x 4" studs, 8' high, studs 12" O.C.	2 Carp	80	.200	L.F.	3.96	8.80		12.76	18.80
0187	12" O.C., pneumatic nailed		96	.167		3.96	7.35		11.31	16.40
0202	16" O.C.		100	.160		3.24	7.05		10.29	15.10
0207	16" O.C., pneumatic nailed		120	.133		3.24	5.90		9.14	13.20
0302	24" O.C.		125	.128		2.52	5.65		8.17	12
0307	24" O.C., pneumatic nailed		150	.107		2.52	4.70		7.22	10.45

06 11 Wood Framing

06 11 10 – Framing with Dimensional, Engineered or Composite Lumber

06 11 10.26 Partitions		Crew	Daily Output	Labor-Hours	Unit	2012 Bare Costs Material	Labor	Equipment	Total	Total Incl O&P
0382	10′ high, studs 12″ O.C.	2 Carp	80	.200	L.F.	4.68	8.80		13.48	19.60
0387	12″ O.C., pneumatic nailed		96	.167		4.68	7.35		12.03	17.20
0402	16″ O.C.		100	.160		3.78	7.05		10.83	15.70
0407	16″ O.C., pneumatic nailed		120	.133		3.78	5.90		9.68	13.80
0502	24″ O.C.		125	.128		2.88	5.65		8.53	12.40
0507	24″ O.C., pneumatic nailed		150	.107		2.88	4.70		7.58	10.85
0582	12′ high, studs 12″ O.C.		65	.246		5.40	10.85		16.25	24
0587	12″ O.C., pneumatic nailed		78	.205		5.40	9.05		14.45	21
0602	16″ O.C.		80	.200		4.32	8.80		13.12	19.20
0607	16″ O.C., pneumatic nailed		96	.167		4.32	7.35		11.67	16.80
0701	24″ O.C.		100	.160		3.24	7.05		10.29	15.10
0706	24″ O.C., pneumatic nailed		120	.133		3.24	5.90		9.14	13.20
0782	2″ x 6″ studs, 8′ high, studs 12″ O.C.		70	.229		6.45	10.10		16.55	23.50
0787	12″ O.C., pneumatic nailed		84	.190		6.45	8.40		14.85	21
0802	16″ O.C.		90	.178		5.30	7.85		13.15	18.65
0807	16″ O.C., pneumatic nailed		108	.148		5.30	6.55		11.85	16.50
0902	24″ O.C.		115	.139		4.10	6.15		10.25	14.55
0907	24″ O.C., pneumatic nailed		138	.116		4.10	5.10		9.20	12.90
0982	10′ high, studs 12″ O.C.		70	.229		7.60	10.10		17.70	25
0987	12″ O.C., pneumatic nailed		84	.190		7.60	8.40		16	22
1002	16″ O.C.		90	.178		6.15	7.85		14	19.60
1007	16″ O.C., pneumatic nailed		108	.148		6.15	6.55		12.70	17.45
1102	24″ O.C.		115	.139		4.69	6.15		10.84	15.20
1107	24″ O.C., pneumatic nailed		138	.116		4.69	5.10		9.79	13.55
1182	12′ high, studs 12″ O.C.		55	.291		8.80	12.85		21.65	30.50
1187	12″ O.C., pneumatic nailed		66	.242		8.80	10.70		19.50	27
1202	16″ O.C.		70	.229		7.05	10.10		17.15	24.50
1207	16″ O.C., pneumatic nailed		84	.190		7.05	8.40		15.45	21.50
1302	24″ O.C.		90	.178		5.30	7.85		13.15	18.65
1307	24″ O.C., pneumatic nailed		108	.148		5.30	6.55		11.85	16.50
1402	For horizontal blocking, 2″ x 4″, add		600	.027		.36	1.18		1.54	2.33
1502	2″ x 6″, add		600	.027		.59	1.18		1.77	2.57
1600	For openings, add		250	.064			2.82		2.82	4.63
1702	Headers for above openings, material only, add				B.F.	.62			.62	.68
9000	Minimum labor/equipment charge	1 Carp	4	2	Job		88		88	145

06 11 10.28 Porch or Deck Framing		Crew	Daily Output	Labor-Hours	Unit	Material	Labor	Equipment	Total	Total Incl O&P
0010	**PORCH OR DECK FRAMING**									
0100	Treated lumber, posts or columns, 4″ x 4″	2 Carp	390	.041	L.F.	.82	1.81		2.63	3.87
0110	4″ x 6″		275	.058		2.11	2.57		4.68	6.55
0120	4″ x 8″		220	.073		4.32	3.21		7.53	10
0130	Girder, single, 4″ x 4″		675	.024		.82	1.05		1.87	2.61
0140	4″ x 6″		600	.027		2.11	1.18		3.29	4.25
0150	4″ x 8″		525	.030		4.32	1.34		5.66	6.95
0160	Double, 2″ x 4″		625	.026		.70	1.13		1.83	2.62
0170	2″ x 6″		600	.027		1.31	1.18		2.49	3.37
0180	2″ x 8″		575	.028		1.82	1.23		3.05	4.01
0190	2″ x 10″		550	.029		2.41	1.28		3.69	4.75
0200	2″ x 12″		525	.030		3.22	1.34		4.56	5.75
0210	Triple, 2″ x 4″		575	.028		1.05	1.23		2.28	3.17
0220	2″ x 6″		550	.029		1.96	1.28		3.24	4.26
0230	2″ x 8″		525	.030		2.73	1.34		4.07	5.20
0240	2″ x 10″		500	.032		3.61	1.41		5.02	6.30

06 11 10 – Framing with Dimensional, Engineered or Composite Lumber

06 11 10.28 Porch or Deck Framing		Crew	Daily Output	Labor-Hours	Unit	2012 Bare Costs Material	Labor	Equipment	Total	Total Incl O&P
0250	2" x 12"	2 Carp	475	.034	L.F.	4.83	1.49		6.32	7.75
0260	Ledger, bolted 4' O.C., 2" x 4"		400	.040		.46	1.76		2.22	3.39
0270	2" x 6"		395	.041		.75	1.79		2.54	3.76
0280	2" x 8"		390	.041		1	1.81		2.81	4.07
0300	2" x 12"		380	.042		1.69	1.86		3.55	4.90
0310	Joists, 2" x 4"		1250	.013		.35	.56		.91	1.32
0320	2" x 6"		1250	.013		.66	.56		1.22	1.65
0330	2" x 8"		1100	.015		.91	.64		1.55	2.06
0340	2" x 10"		900	.018		1.21	.78		1.99	2.62
0350	2" x 12"		875	.018		1.45	.81		2.26	2.92
0440	Balusters, square, 2" x 2"		660	.024		.36	1.07		1.43	2.14
0450	Turned, 2" x 2"		420	.038		.48	1.68		2.16	3.27
0460	Stair stringer, 2" x 10"		130	.123		1.21	5.45		6.66	10.25
0470	2" x 12"		130	.123		1.45	5.45		6.90	10.50
0480	Stair treads, 1" x 4"		140	.114		1.65	5.05		6.70	10.05
0490	2" x 4"		140	.114		.35	5.05		5.40	8.65
0500	2" x 6"		160	.100		.53	4.41		4.94	7.85
0510	5/4" x 6"		160	.100		.90	4.41		5.31	8.25
0520	Turned handrail post, 4" x 4"		64	.250	Ea.	32	11.05		43.05	53
0530	Lattice panel, 4' x 8', 1/2"		1600	.010	S.F.	.80	.44		1.24	1.60
0535	3/4"		1600	.010	"	1.20	.44		1.64	2.04
0540	Cedar, posts or columns, 4" x 4"		390	.041	L.F.	2.83	1.81		4.64	6.10
0550	4" x 6"		275	.058		5.25	2.57		7.82	9.95
0560	4" x 8"		220	.073		8.85	3.21		12.06	15
0800	Decking, 1" x 4"		550	.029		1.81	1.28		3.09	4.09
0810	2" x 4"		600	.027		3.62	1.18		4.80	5.90
0820	2" x 6"		640	.025		6.30	1.10		7.40	8.75
0830	5/4" x 6"		640	.025		4.08	1.10		5.18	6.30
0840	Railings and trim, 1" x 4"		600	.027		1.81	1.18		2.99	3.92
0860	2" x 4"		600	.027		3.62	1.18		4.80	5.90
0870	2" x 6"		600	.027		6.30	1.18		7.48	8.90
0920	Stair treads, 1" x 4"		140	.114		1.81	5.05		6.86	10.25
0930	2" x 4"		140	.114		3.62	5.05		8.67	12.25
0940	2" x 6"		160	.100		6.30	4.41		10.71	14.20
0950	5/4" x 6"		160	.100		4.08	4.41		8.49	11.75
0980	Redwood, posts or columns, 4" x 4"		390	.041		6.40	1.81		8.21	10
0990	4" x 6"		275	.058		11.20	2.57		13.77	16.50
1000	4" x 8"		220	.073		20.50	3.21		23.71	28
1280	Railings and trim, 1" x 4"		600	.027		1.17	1.18		2.35	3.21
1310	2" x 6"		600	.027		7.50	1.18		8.68	10.20
1420	Alternative decking, wood/plastic composite, 5/4" x 6" G		640	.025		3	1.10		4.10	5.10
1440	1" x 4" square edge fir		550	.029		1.66	1.28		2.94	3.92
1450	1" x 4" tongue and groove fir		450	.036		1.41	1.57		2.98	4.12
1460	1" x 4" mahogany		550	.029		2.51	1.28		3.79	4.86
1462	5/4" x 6" PVC		550	.029		3.01	1.28		4.29	5.40
1465	Framing, porch or deck, alt deck fastening, screws, add	1 Carp	240	.033	S.F.		1.47		1.47	2.41
1470	Accessories, joist hangers, 2" x 4"		160	.050	Ea.	.67	2.21		2.88	4.36
1480	2" x 6" through 2" x 12"		150	.053		1.28	2.35		3.63	5.25
1530	Post footing, incl excav, backfill, tube form & concrete, 4' deep, 8" dia	F-7	12	2.667		11.15	106		117.15	185
1540	10" diameter		11	2.909		16.55	115		131.55	207
1550	12" diameter		10	3.200		22	127		149	232

06 11 Wood Framing

06 11 10 – Framing with Dimensional, Engineered or Composite Lumber

06 11 10.30 Roof Framing

Code	06 11 10.30 Roof Framing	Crew	Daily Output	Labor-Hours	Unit	Material (2012 Bare Costs)	Labor (2012 Bare Costs)	Equipment (2012 Bare Costs)	Total (2012 Bare Costs)	Total Incl O&P
0010	**ROOF FRAMING** R061110-30									
2000	Fascia boards, 2" x 8"	2 Carp	225	.071	L.F.	.75	3.14		3.89	6
2100	2" x 10"		180	.089		1.01	3.92		4.93	7.55
5000	Rafters, to 4 in 12 pitch, 2" x 6", ordinary		1000	.016		.53	.71		1.24	1.75
5060	2" x 8", ordinary		950	.017		.75	.74		1.49	2.05
5250	Composite rafter, 9-1/2" deep		575	.028		1.62	1.23		2.85	3.79
5260	11-1/2" deep		575	.028		1.82	1.23		3.05	4.02
5300	Hip and valley rafters, 2" x 6", ordinary		760	.021		.53	.93		1.46	2.11
5360	2" x 8", ordinary		720	.022		.75	.98		1.73	2.44
5540	Hip and valley jacks, 2" x 6", ordinary		600	.027		.53	1.18		1.71	2.52
5600	2" x 8", ordinary		490	.033		.75	1.44		2.19	3.19
5761	For slopes steeper than 4 in 12, add						30%			
5780	Rafter tie, 1" x 4", #3	2 Carp	800	.020	L.F.	.41	.88		1.29	1.90
5800	Ridge board, #2 or better, 1" x 6"		600	.027		.72	1.18		1.90	2.72
5820	1" x 8"		550	.029		1	1.28		2.28	3.20
5840	1" x 10"		500	.032		1.30	1.41		2.71	3.74
5860	2" x 6"		500	.032		.53	1.41		1.94	2.90
5880	2" x 8"		450	.036		.75	1.57		2.32	3.40
5900	2" x 10"		400	.040		1.01	1.76		2.77	4
5920	Roof cants, split, 4" x 4"		650	.025		1.64	1.09		2.73	3.58
5940	6" x 6"		600	.027		3.62	1.18		4.80	5.90
5960	Roof curbs, untreated, 2" x 6"		520	.031		.53	1.36		1.89	2.81
5980	2" x 12"		400	.040		1.28	1.76		3.04	4.29
6000	Sister rafters, 2" x 6"		800	.020		.53	.88		1.41	2.04
6020	2" x 8"		640	.025		.75	1.10		1.85	2.64
6040	2" x 10"		535	.030		1.01	1.32		2.33	3.27
6060	2" x 12"		455	.035		1.28	1.55		2.83	3.94
9000	Minimum labor/equipment charge	1 Carp	4	2	Job		88		88	145

06 11 10.32 Sill and Ledger Framing

Code	06 11 10.32 Sill and Ledger Framing	Crew	Daily Output	Labor-Hours	Unit	Material	Labor	Equipment	Total	Total Incl O&P
0010	**SILL AND LEDGER FRAMING**									
2002	Ledgers, nailed, 2" x 4"	2 Carp	755	.021	L.F.	.33	.93		1.26	1.89
2052	2" x 6"		600	.027		.53	1.18		1.71	2.52
2102	Bolted, not including bolts, 3" x 6"		325	.049		1.35	2.17		3.52	5.05
2152	3" x 12"		233	.069		3.27	3.03		6.30	8.55
2602	Mud sills, redwood, construction grade, 2" x 4"		895	.018		2.26	.79		3.05	3.78
2622	2" x 6"		780	.021		3.40	.90		4.30	5.20
4002	Sills, 2" x 4"		600	.027		.33	1.18		1.51	2.29
4052	2" x 6"		550	.029		.53	1.28		1.81	2.69
4082	2" x 8"		500	.032		.75	1.41		2.16	3.14
4202	Treated, 2" x 4"		550	.029		.33	1.28		1.61	2.47
4222	2" x 6"		500	.032		.63	1.41		2.04	3
4242	2" x 8"		450	.036		.88	1.57		2.45	3.54
4402	4" x 4"		450	.036		.78	1.57		2.35	3.43
4422	4" x 6"		350	.046		2.06	2.02		4.08	5.55
4462	4" x 8"		300	.053		4.26	2.35		6.61	8.55
4481	4" x 10"		260	.062		5.85	2.71		8.56	10.90
9000	Minimum labor/equipment charge	1 Carp	4	2	Job		88		88	145

06 11 10.34 Sleepers

Code	06 11 10.34 Sleepers	Crew	Daily Output	Labor-Hours	Unit	Material	Labor	Equipment	Total	Total Incl O&P
0010	**SLEEPERS**									
0100	On concrete, treated, 1" x 2"	2 Carp	2350	.007	L.F.	.29	.30		.59	.81
0150	1" x 3"		2000	.008		.43	.35		.78	1.05
0200	2" x 4"		1500	.011		.33	.47		.80	1.14

06 11 Wood Framing

06 11 10 – Framing with Dimensional, Engineered or Composite Lumber

06 11 10.34 Sleepers		Crew	Daily Output	Labor-Hours	Unit	Material	2012 Bare Costs Labor	Equipment	Total	Total Incl O&P
0250	2" x 6"	2 Carp	1300	.012	L.F.	.63	.54		1.17	1.58
9000	Minimum labor/equipment charge	1 Carp	4	2	Job		88		88	145

06 11 10.36 Soffit and Canopy Framing		Crew	Daily Output	Labor-Hours	Unit	Material	Labor	Equipment	Total	Total Incl O&P
0010	**SOFFIT AND CANOPY FRAMING**									
1002	Canopy or soffit framing , 1" x 4"	2 Carp	900	.018	L.F.	.41	.78		1.19	1.74
1042	1" x 8"		750	.021		1	.94		1.94	2.64
1102	2" x 4"		620	.026		.33	1.14		1.47	2.23
1142	2" x 8"		500	.032		.75	1.41		2.16	3.14
1202	3" x 4"		500	.032		.85	1.41		2.26	3.25
1242	3" x 10"		300	.053		2.68	2.35		5.03	6.80
9000	Minimum labor/equipment charge	1 Carp	4	2	Job		88		88	145

06 11 10.38 Treated Lumber Framing Material		Crew	Daily Output	Labor-Hours	Unit	Material	Labor	Equipment	Total	Total Incl O&P
0010	**TREATED LUMBER FRAMING MATERIAL**									
0100	2" x 4"				M.B.F.	500			500	550
0110	2" x 6"					630			630	695
0120	2" x 8"					660			660	725
0130	2" x 10"					700			700	770
0140	2" x 12"					780			780	860
0200	4" x 4"					590			590	645
0210	4" x 6"					1,025			1,025	1,125
0220	4" x 8"					1,600			1,600	1,750

06 11 10.40 Wall Framing		Crew	Daily Output	Labor-Hours	Unit	Material	Labor	Equipment	Total	Total Incl O&P
0010	**WALL FRAMING** R061110-30									
0100	Door buck, studs, header, access, 8' H, 2" x 4" wall, 3' W	1 Carp	32	.250	Ea.	13.95	11.05		25	33.50
0110	4' wide		32	.250		15	11.05		26.05	34.50
0120	5' wide		32	.250		18.35	11.05		29.40	38
0130	6' wide		32	.250		19.85	11.05		30.90	40
0140	8' wide		30	.267		27	11.75		38.75	49.50
0150	10' wide		30	.267		36.50	11.75		48.25	60
0160	12' wide		30	.267		52	11.75		63.75	76.50
0170	2" x 6" wall, 3' wide		32	.250		20.50	11.05		31.55	40.50
0180	4' wide		32	.250		21.50	11.05		32.55	42
0190	5' wide		32	.250		25	11.05		36.05	45.50
0200	6' wide		32	.250		26.50	11.05		37.55	47
0210	8' wide		30	.267		33.50	11.75		45.25	56.50
0220	10' wide		30	.267		43	11.75		54.75	67
0230	12' wide		30	.267		58.50	11.75		70.25	84
0240	Window buck, studs, header & access, 8' high 2" x 4" wall, 2' wide		24	.333		14.55	14.70		29.25	40
0250	3' wide		24	.333		17.20	14.70		31.90	43
0260	4' wide		24	.333		18.95	14.70		33.65	45
0270	5' wide		24	.333		22.50	14.70		37.20	48.50
0280	6' wide		24	.333		25	14.70		39.70	51.50
0300	8' wide		22	.364		34	16.05		50.05	64
0310	10' wide		22	.364		44.50	16.05		60.55	75.50
0320	12' wide		22	.364		62	16.05		78.05	94.50
0330	2" x 6" wall, 2' wide		24	.333		23	14.70		37.70	49
0340	3' wide		24	.333		26	14.70		40.70	52.50
0350	4' wide		24	.333		27.50	14.70		42.20	54.50
0360	5' wide		24	.333		31.50	14.70		46.20	58.50
0370	6' wide		24	.333		34.50	14.70		49.20	62
0380	7' wide		24	.333		41	14.70		55.70	69.50
0390	8' wide		22	.364		45	16.05		61.05	76

06 11 Wood Framing

06 11 10 – Framing with Dimensional, Engineered or Composite Lumber

06 11 10.40 Wall Framing		Crew	Daily Output	Labor-Hours	Unit	Material	2012 Bare Costs Labor	Equipment	Total	Total Incl O&P
0400	10′ wide	1 Carp	22	.364	Ea.	56	16.05		72.05	88
0410	12′ wide		22	.364		74.50	16.05		90.55	109
2002	Headers over openings, 2″ x 6″	2 Carp	360	.044	L.F.	.53	1.96		2.49	3.80
2007	2″ x 6″, pneumatic nailed		432	.037		.53	1.63		2.16	3.27
2052	2″ x 8″		340	.047		.75	2.08		2.83	4.23
2057	2″ x 8″, pneumatic nailed		408	.039		.75	1.73		2.48	3.67
2100	2″ x 10″		320	.050		1.01	2.21		3.22	4.73
2105	2″ x 10″, pneumatic nailed		384	.042		1.01	1.84		2.85	4.12
2152	2″ x 12″		300	.053		1.28	2.35		3.63	5.25
2157	2″ x 12″, pneumatic nailed		360	.044		1.28	1.96		3.24	4.61
2202	4″ x 12″		190	.084		5.50	3.71		9.21	12.15
2207	4″ x 12″, pneumatic nailed		228	.070		5.50	3.09		8.59	11.10
2252	6″ x 12″		140	.114		10.15	5.05		15.20	19.45
5002	Plates, untreated, 2″ x 3″		850	.019		.33	.83		1.16	1.73
5007	2″ x 3″, pneumatic nailed		1020	.016		.33	.69		1.02	1.50
5022	2″ x 4″		800	.020		.33	.88		1.21	1.81
5027	2″ x 4″, pneumatic nailed		960	.017		.33	.74		1.07	1.57
5041	2″ x 6″		750	.021		.53	.94		1.47	2.13
5046	2″ x 6″, pneumatic nailed		900	.018		.53	.78		1.31	1.88
5122	Studs, 8′ high wall, 2″ x 3″		1200	.013		.33	.59		.92	1.33
5127	2″ x 3″, pneumatic nailed		1440	.011		.33	.49		.82	1.17
5142	2″ x 4″		1100	.015		.33	.64		.97	1.41
5147	2″ x 4″, pneumatic nailed		1320	.012		.33	.53		.86	1.24
5162	2″ x 6″		1000	.016		.53	.71		1.24	1.75
5167	2″ x 6″, pneumatic nailed		1200	.013		.53	.59		1.12	1.55
5182	3″ x 4″		800	.020		.85	.88		1.73	2.39
5187	3″ x 4″, pneumatic nailed		960	.017		.85	.74		1.59	2.15
8200	For 12′ high walls, deduct						5%			
8220	For stub wall, 6′ high, add						20%			
8240	3′ high, add						40%			
8250	For second story & above, add						5%			
8300	For dormer & gable, add						15%			
9000	Minimum labor/equipment charge	1 Carp	4	2	Job		88		88	145

06 11 10.42 Furring		Crew	Daily Output	Labor-Hours	Unit	Material	Labor	Equipment	Total	Total Incl O&P
0010	**FURRING**									
0012	Wood strips, 1″ x 2″, on walls, on wood	1 Carp	550	.015	L.F.	.25	.64		.89	1.33
0017	Pneumatic nailed		710	.011		.25	.50		.75	1.09
0300	On masonry		495	.016		.25	.71		.96	1.45
0400	On concrete		260	.031		.25	1.36		1.61	2.50
0600	1″ x 3″, on walls, on wood		550	.015		.34	.64		.98	1.43
0607	Pneumatic nailed		710	.011		.34	.50		.84	1.19
0700	On masonry		495	.016		.34	.71		1.05	1.55
0800	On concrete		260	.031		.34	1.36		1.70	2.60
0850	On ceilings, on wood		350	.023		.34	1.01		1.35	2.03
0857	Pneumatic nailed		450	.018		.34	.78		1.12	1.67
0900	On masonry		320	.025		.34	1.10		1.44	2.19
0950	On concrete		210	.038		.34	1.68		2.02	3.13
9000	Minimum labor/equipment charge		4	2	Job		88		88	145

06 11 10.44 Grounds		Crew	Daily Output	Labor-Hours	Unit	Material	Labor	Equipment	Total	Total Incl O&P
0010	**GROUNDS**									
0020	For casework, 1″ x 2″ wood strips, on wood	1 Carp	330	.024	L.F.	.25	1.07		1.32	2.03
0100	On masonry		285	.028		.25	1.24		1.49	2.31

06 11 Wood Framing

06 11 10 – Framing with Dimensional, Engineered or Composite Lumber

06 11 10.44 Grounds

	06 11 10.44 Grounds	Crew	Daily Output	Labor-Hours	Unit	2012 Bare Costs Material	Labor	Equipment	Total	Total Incl O&P
0200	On concrete	1 Carp	250	.032	L.F.	.25	1.41		1.66	2.59
0400	For plaster, 3/4" deep, on wood		450	.018		.25	.78		1.03	1.57
0500	On masonry		225	.036		.25	1.57		1.82	2.85
0600	On concrete		175	.046		.25	2.02		2.27	3.58
0700	On metal lath		200	.040		.25	1.76		2.01	3.17
9000	Minimum labor/equipment charge		4	2	Job		88		88	145

06 12 Structural Panels

06 12 10 – Structural Insulated Panels

06 12 10.10 OSB Faced Panels

	06 12 10.10 OSB Faced Panels		Crew	Daily Output	Labor-Hours	Unit	Material	Labor	Equipment	Total	Total Incl O&P
0010	**OSB FACED PANELS**										
0100	Structural insul. panels, 7/16" OSB both faces, EPS insul, 3-5/8" T	G	F-3	2075	.019	S.F.	2.79	.86	.30	3.95	4.80
0110	5-5/8" thick	G		1725	.023		3.13	1.04	.35	4.52	5.50
0120	7-3/8" thick	G		1425	.028		3.41	1.26	.43	5.10	6.25
0130	9-3/8" thick	G		1125	.036		3.73	1.59	.54	5.86	7.30
0140	7/16" OSB one face, EPS insul, 3-5/8" thick	G		2175	.018		2.97	.82	.28	4.07	4.92
0150	5-5/8" thick	G		1825	.022		3.40	.98	.34	4.72	5.70
0160	7-3/8" thick	G		1525	.026		3.79	1.18	.40	5.37	6.50
0170	9-3/8" thick	G		1225	.033		4.22	1.46	.50	6.18	7.55
0190	7/16" OSB - 1/2" GWB faces , EPS insul, 3-5/8" T	G		2075	.019		2.94	.86	.30	4.10	4.96
0200	5-5/8" thick	G		1725	.023		3.37	1.04	.35	4.76	5.80
0210	7-3/8" thick	G		1425	.028		3.75	1.26	.43	5.44	6.65
0220	9-3/8" thick	G		1125	.036		4.40	1.59	.54	6.53	8.05
0240	7/16" OSB - 1/2" MRGWB faces , EPS insul, 3-5/8" T	G		2075	.019		2.94	.86	.30	4.10	4.96
0250	5-5/8" thick	G		1725	.023		3.37	1.04	.35	4.76	5.80
0260	7-3/8" thick	G		1425	.028		3.75	1.26	.43	5.44	6.65
0270	9-3/8" thick	G		1125	.036		4.40	1.59	.54	6.53	8.05
0300	For 1/2" GWB added to OSB skin, add	G					.62			.62	.68
0310	For 1/2" MRGWB added to OSB skin, add	G					.62			.62	.68
0320	For one T1-11 skin, add to OSB-OSB	G					1.42			1.42	1.56
0330	For one 19/32" CDX skin, add to OSB-OSB	G					.72			.72	.79
0500	Structural insulated panel, 7/16" OSB both sides, straw core										
0510	4-3/8" T, walls (w/sill, splines, plates)	G	F-6	2400	.017	S.F.	6.75	.69	.26	7.70	8.80
0520	Floors (w/splines)	G		2400	.017		6.75	.69	.26	7.70	8.80
0530	Roof (w/splines)	G		2400	.017		6.75	.69	.26	7.70	8.80
0550	7-7/8" T, walls (w/sill, splines, plates)	G		2400	.017		10.20	.69	.26	11.15	12.65
0560	Floors (w/splines)	G		2400	.017		10.20	.69	.26	11.15	12.65
0570	Roof (w/splines)	G		2400	.017		10.20	.69	.26	11.15	12.65

06 12 19 – Composite Shearwall Panels

06 12 19.10 Steel and Wood Composite Shearwall Panels

	06 12 19.10 Steel and Wood Composite Shearwall Panels	Crew	Daily Output	Labor-Hours	Unit	Material	Labor	Equipment	Total	Total Incl O&P
0010	**STEEL & WOOD COMPOSITE SHEARWALL PANELS**									
0020	Anchor bolts, 36" long (must be placed in wet concrete)	1 Carp	150	.053	Ea.	33	2.35		35.35	40
0030	On concrete, 2 x 4 & 2 x 6 walls, 7' - 10' high, 360 lb. shear, 12" wide	2 Carp	8	2		380	88		468	565
0040	715 lb. shear, 15" wide		8	2		440	88		528	625
0050	1860 lb. shear, 18" wide		8	2		470	88		558	665
0060	2780 lb. shear, 21" wide		8	2		560	88		648	765
0070	3790 lb. shear, 24" wide		8	2		640	88		728	845
0080	2 x 6 walls, 11' to 13' high, 1180 lb. shear, 18" wide		6	2.667		580	118		698	835
0090	1555 lb. shear, 21" wide		6	2.667		725	118		843	995
0100	2280 lb. shear, 24" wide		6	2.667		805	118		923	1,075

06 12 Structural Panels

06 12 19 – Composite Shearwall Panels

06 12 19.10 Steel and Wood Composite Shearwall Panels		Crew	Daily Output	Labor-Hours	Unit	2012 Bare Costs Material	Labor	Equipment	Total	Total Incl O&P
0110	For installing above on wood floor frame, add									
0120	Coupler nuts, threaded rods, bolts, shear transfer plate kit	1 Carp	16	.500	Ea.	50	22		72	91
0130	Framing anchors, angle (2 required)	"	96	.083	"	2.40	3.67		6.07	8.65
0140	For blocking see Section 06 11 10.02									
0150	For installing above, first floor to second floor, wood floor frame, add									
0160	Add stack option to first floor wall panel				Ea.	55			55	60.50
0170	Threaded rods, bolts, shear transfer plate kit	1 Carp	16	.500		60	22		82	102
0180	Framing anchors, angle (2 required)	"	96	.083		2.40	3.67		6.07	8.65
0200	For installing stacked panels, balloon framing									
0210	Add stack option to first floor wall panel				Ea.	55			55	60.50
0220	Threaded rods, bolts kit	1 Carp	16	.500	"	35	22		57	74.50

06 13 Heavy Timber Construction

06 13 23 – Heavy Timber Framing

06 13 23.10 Heavy Framing

		Crew	Daily Output	Labor-Hours	Unit	Material	Labor	Equipment	Total	Total Incl O&P
0010	**HEAVY FRAMING**									
0020	Beams, single 6" x 10"	2 Carp	1.10	14.545	M.B.F.	1,400	640		2,040	2,600
0100	Single 8" x 16"		1.20	13.333	"	1,975	590		2,565	3,150
0202	Built from 2" lumber, multiple 2" x 14"		900	.018	B.F.	.73	.78		1.51	2.09
0212	Built from 3" lumber, multiple 3" x 6"		700	.023		.90	1.01		1.91	2.64
0222	Multiple 3" x 8"		800	.020		1.06	.88		1.94	2.62
0232	Multiple 3" x 10"		900	.018		1.07	.78		1.85	2.47
0242	Multiple 3" x 12"		1000	.016		1.09	.71		1.80	2.36
0252	Built from 4" lumber, multiple 4" x 6"		800	.020		1.25	.88		2.13	2.83
0262	Multiple 4" x 8"		900	.018		1.28	.78		2.06	2.69
0272	Multiple 4" x 10"		1000	.016		1.28	.71		1.99	2.56
0282	Multiple 4" x 12"		1100	.015		1.38	.64		2.02	2.56
0292	Columns, structural grade, 1500f, 4" x 4"		450	.036	L.F.	1.69	1.57		3.26	4.43
0302	6" x 6"		225	.071		3.88	3.14		7.02	9.40
0402	8" x 8"		240	.067		7.20	2.94		10.14	12.70
0502	10" x 10"		90	.178		12.10	7.85		19.95	26
0602	12" x 12"		70	.229		18.35	10.10		28.45	36.50
0802	Floor planks, 2" thick, T & G, 2" x 6"		1050	.015	B.F.	1.05	.67		1.72	2.26
0902	2" x 10"		1100	.015		1.05	.64		1.69	2.21
1102	3" thick, 3" x 6"		1050	.015		1.25	.67		1.92	2.48
1202	3" x 10"		1100	.015		1.25	.64		1.89	2.43
1402	Girders, structural grade, 12" x 12"		800	.020		1.53	.88		2.41	3.13
1502	10" x 16"		1000	.016		2.40	.71		3.11	3.80
2302	Roof purlins, 4" thick, structural grade		1050	.015		1.28	.67		1.95	2.50
9000	Minimum labor/equipment charge	1 Carp	2	4	Job		176		176	289

06 15 Wood Decking

06 15 16 – Wood Roof Decking

06 15 16.10 Solid Wood Roof Decking		Crew	Daily Output	Labor-Hours	Unit	Material	Labor	Equipment	Total	Total Incl O&P
						2012 Bare Costs				
0010	**SOLID WOOD ROOF DECKING**									
0400	Cedar planks, 3" thick	2 Carp	320	.050	S.F.	7	2.21		9.21	11.25
0500	4" thick		250	.064		9.30	2.82		12.12	14.90
0702	Douglas fir, 3" thick		320	.050		2.94	2.21		5.15	6.85
0802	4" thick		250	.064		3.92	2.82		6.74	8.95
1002	Hemlock, 3" thick		320	.050		3.28	2.21		5.49	7.25
1102	4" thick		250	.064		4.36	2.82		7.18	9.45
1302	Western white spruce, 3" thick		320	.050		1.92	2.21		4.13	5.75
1402	4" thick		250	.064		2.55	2.82		5.37	7.45
9000	Minimum labor/equipment charge	1 Carp	2	4	Job		176		176	289

06 15 23 – Laminated Wood Decking

06 15 23.10 Laminated Roof Deck		Crew	Daily Output	Labor-Hours	Unit	Material	Labor	Equipment	Total	Total Incl O&P
0010	**LAMINATED ROOF DECK**									
0020	Pine or hemlock, 3" thick	2 Carp	425	.038	S.F.	3.83	1.66		5.49	6.95
0100	4" thick		325	.049		5.10	2.17		7.27	9.15
0300	Cedar, 3" thick		425	.038		4.20	1.66		5.86	7.35
0400	4" thick		325	.049		5.35	2.17		7.52	9.40
0600	Fir, 3" thick		425	.038		3.88	1.66		5.54	7
0700	4" thick		325	.049		5.15	2.17		7.32	9.25
9000	Minimum labor/equipment charge	1 Carp	3	2.667	Job		118		118	193

06 16 Sheathing

06 16 23 – Subflooring

06 16 23.10 Subfloor		Crew	Daily Output	Labor-Hours	Unit	Material	Labor	Equipment	Total	Total Incl O&P
0010	**SUBFLOOR** R061636-20									
0011	Plywood, CDX, 1/2" thick	2 Carp	1500	.011	SF Flr.	.47	.47		.94	1.29
0015	Pneumatic nailed		1860	.009		.47	.38		.85	1.14
0102	5/8" thick		1350	.012		.62	.52		1.14	1.54
0107	Pneumatic nailed		1674	.010		.62	.42		1.04	1.37
0202	3/4" thick		1250	.013		.70	.56		1.26	1.70
0207	Pneumatic nailed		1550	.010		.70	.46		1.16	1.52
0302	1-1/8" thick, 2-4-1 including underlayment		1050	.015		1.53	.67		2.20	2.78
0440	With boards, 1" x 6", S4S, laid regular		900	.018		1.57	.70		2.35	3.02
0452	1" x 8", laid regular		1000	.016		1.60	.71		2.31	2.92
0462	Laid diagonal		850	.019		1.60	.83		2.43	3.12
0502	1" x 10", laid regular		1100	.015		1.64	.64		2.28	2.85
0602	Laid diagonal		900	.018		1.64	.78		2.42	3.09
8990	Subfloor adhesive, 3/8" bead	1 Carp	2300	.003	L.F.	.12	.15		.27	.38
9000	Minimum labor/equipment charge	"	4	2	Job		88		88	145

06 16 26 – Underlayment

06 16 26.10 Wood Product Underlayment		Crew	Daily Output	Labor-Hours	Unit	Material	Labor	Equipment	Total	Total Incl O&P
0010	**WOOD PRODUCT UNDERLAYMENT** R061636-20									
0030	Plywood, underlayment grade, 3/8" thick	2 Carp	1500	.011	SF Flr.	.73	.47		1.20	1.57
0080	Pneumatic nailed		1860	.009		.73	.38		1.11	1.42
0102	1/2" thick		1450	.011		.77	.49		1.26	1.65
0107	Pneumatic nailed		1798	.009		.77	.39		1.16	1.49
0202	5/8" thick		1400	.011		.96	.50		1.46	1.89
0207	Pneumatic nailed		1736	.009		.96	.41		1.37	1.73
0302	3/4" thick		1300	.012		1.17	.54		1.71	2.18
0306	Pneumatic nailed		1612	.010		1.17	.44		1.61	2.01

06 16 Sheathing

06 16 26 – Underlayment

06 16 26.10 Wood Product Underlayment			Crew	Daily Output	Labor-Hours	Unit	Material	Labor	Equipment	Total	Total Incl O&P
							2012 Bare Costs				
0502	Particle board, 3/8" thick		2 Carp	1500	.011	SF Flr.	.35	.47		.82	1.16
0507	Pneumatic nailed			1860	.009		.35	.38		.73	1.01
0602	1/2" thick			1450	.011		.36	.49		.85	1.20
0607	Pneumatic nailed			1798	.009		.36	.39		.75	1.04
0802	5/8" thick			1400	.011		.46	.50		.96	1.34
0807	Pneumatic nailed			1736	.009		.46	.41		.87	1.18
0902	3/4" thick			1300	.012		.65	.54		1.19	1.61
0907	Pneumatic nailed			1612	.010		.65	.44		1.09	1.44
0955	Particleboard, 100% recycled straw/wheat, 4' x 8' x 1/4"	G		1450	.011	S.F.	.26	.49		.75	1.09
0960	4' x 8' x 3/8"	G		1450	.011		.39	.49		.88	1.23
0965	4' x 8' x 1/2"	G		1350	.012		.54	.52		1.06	1.45
0970	4' x 8' x 5/8"	G		1300	.012		.64	.54		1.18	1.59
0975	4' x 8' x 3/4"	G		1250	.013		.74	.56		1.30	1.74
0980	4' x 8' x 1"	G		1150	.014		.99	.61		1.60	2.10
0985	4' x 8' x 1-1/4"	G		1100	.015		1.16	.64		1.80	2.33
1102	Hardboard, underlayment grade, 4' x 4', .215" thick			1500	.011	SF Flr.	.57	.47		1.04	1.40
9000	Minimum labor/equipment charge		1 Carp	4	2	Job		88		88	145

06 16 36 – Wood Panel Product Sheathing

06 16 36.10 Sheathing			Crew	Daily Output	Labor-Hours	Unit	Material	Labor	Equipment	Total	Total Incl O&P
0010	**SHEATHING**	R061110-30									
0012	Plywood on roofs, CDX										
0032	5/16" thick		2 Carp	1600	.010	S.F.	.55	.44		.99	1.32
0037	Pneumatic nailed	R061636-20		1952	.008		.55	.36		.91	1.19
0052	3/8" thick			1525	.010		.47	.46		.93	1.28
0057	Pneumatic nailed			1860	.009		.47	.38		.85	1.14
0102	1/2" thick			1400	.011		.47	.50		.97	1.35
0103	Pneumatic nailed			1708	.009		.47	.41		.88	1.20
0202	5/8" thick			1300	.012		.62	.54		1.16	1.57
0207	Pneumatic nailed			1586	.010		.62	.45		1.07	1.41
0302	3/4" thick			1200	.013		.70	.59		1.29	1.73
0307	Pneumatic nailed			1464	.011		.70	.48		1.18	1.56
0502	Plywood on walls with exterior CDX, 3/8" thick			1200	.013		.47	.59		1.06	1.48
0507	Pneumatic nailed			1488	.011		.47	.47		.94	1.30
0602	1/2" thick			1125	.014		.47	.63		1.10	1.55
0607	Pneumatic nailed			1395	.011		.47	.51		.98	1.35
0702	5/8" thick			1050	.015		.62	.67		1.29	1.78
0707	Pneumatic nailed			1302	.012		.62	.54		1.16	1.57
0802	3/4" thick			975	.016		.70	.72		1.42	1.96
0807	Pneumatic nailed			1209	.013		.70	.58		1.28	1.73
0840	Oriented strand board, 7/16" thick	G		1400	.011		.39	.50		.89	1.26
0845	Pneumatic nailed	G		1736	.009		.39	.41		.80	1.10
0846	1/2" thick	G		1325	.012		.39	.53		.92	1.30
0847	Pneumatic nailed	G		1643	.010		.39	.43		.82	1.13
0852	5/8" thick	G		1250	.013		.79	.56		1.35	1.80
0857	Pneumatic nailed	G		1550	.010		.79	.46		1.25	1.62
1000	For shear wall construction, add							20%			
1200	For structural 1 exterior plywood, add					S.F.	10%				
1402	With boards, on roof 1" x 6" boards, laid horizontal		2 Carp	725	.022		1.57	.97		2.54	3.33
1502	Laid diagonal			650	.025		1.57	1.09		2.66	3.51
1702	1" x 8" boards, laid horizontal			875	.018		1.60	.81		2.41	3.08
1802	Laid diagonal			725	.022		1.60	.97		2.57	3.36
2000	For steep roofs, add							40%			

06 16 Sheathing

06 16 36 – Wood Panel Product Sheathing

06 16 36.10 Sheathing		Crew	Daily Output	Labor-Hours	Unit	2012 Bare Costs Material	Labor	Equipment	Total	Total Incl O&P
2200	For dormers, hips and valleys, add				S.F.	5%	50%			
2402	Boards on walls, 1″ x 6″ boards, laid regular	2 Carp	650	.025		1.57	1.09		2.66	3.51
2502	Laid diagonal		585	.027		1.57	1.21		2.78	3.71
2702	1″ x 8″ boards, laid regular		765	.021		1.60	.92		2.52	3.27
2802	Laid diagonal		650	.025		1.60	1.09		2.69	3.54
2852	Gypsum, weatherproof, 1/2″ thick		1050	.015		.38	.67		1.05	1.52
2900	With embedded glass mats		1100	.015		.75	.64		1.39	1.88
3000	Wood fiber, regular, no vapor barrier, 1/2″ thick		1200	.013		.56	.59		1.15	1.58
3100	5/8″ thick		1200	.013		.71	.59		1.30	1.74
3300	No vapor barrier, in colors, 1/2″ thick		1200	.013		.68	.59		1.27	1.71
3400	5/8″ thick		1200	.013		.72	.59		1.31	1.75
3600	With vapor barrier one side, white, 1/2″ thick		1200	.013		.55	.59		1.14	1.57
3700	Vapor barrier 2 sides, 1/2″ thick		1200	.013		.77	.59		1.36	1.81
3800	Asphalt impregnated, 25/32″ thick		1200	.013		.28	.59		.87	1.27
3850	Intermediate, 1/2″ thick		1200	.013		.20	.59		.79	1.18
9000	Minimum labor/equipment charge	1 Carp	2	4	Job		176		176	289

06 17 Shop-Fabricated Structural Wood

06 17 33 – Wood I-Joists

06 17 33.10 Wood and Composite I-Joists		Crew	Daily Output	Labor-Hours	Unit	Material	Labor	Equipment	Total	Total Incl O&P
0010	**WOOD AND COMPOSITE I-JOISTS**									
0100	Plywood webs, incl. bridging & blocking, panels 24″ O.C.									
1200	15′ to 24′ span, 50 psf live load	F-5	2400	.013	SF Flr.	1.84	.59		2.43	3
1300	55 psf live load		2250	.014		2.08	.63		2.71	3.33
1400	24′ to 30′ span, 45 psf live load		2600	.012		2.87	.55		3.42	4.05
1500	55 psf live load		2400	.013		3.55	.59		4.14	4.87

06 17 53 – Shop-Fabricated Wood Trusses

06 17 53.10 Roof Trusses		Crew	Daily Output	Labor-Hours	Unit	Material	Labor	Equipment	Total	Total Incl O&P
0010	**ROOF TRUSSES**									
0100	Fink (W) or King post type, 2′-0″ O.C.									
0200	Metal plate connected, 4 in 12 slope									
0210	24′ to 29′ span	F-3	3000	.013	SF Flr.	1.50	.60	.20	2.30	2.84
0300	30′ to 43′ span		3000	.013		2.18	.60	.20	2.98	3.59
0400	44′ to 60′ span		3000	.013		2.36	.60	.20	3.16	3.78
0700	Glued and nailed, add					50%				

06 18 Glued-Laminated Construction

06 18 13 – Glued-Laminated Beams

06 18 13.20 Laminated Framing		Crew	Daily Output	Labor-Hours	Unit	Material	Labor	Equipment	Total	Total Incl O&P
0010	**LAMINATED FRAMING**									
0020	30 lb., short term live load, 15 lb. dead load									
0200	Straight roof beams, 20′ clear span, beams 8′ O.C.	F-3	2560	.016	SF Flr.	1.99	.70	.24	2.93	3.59
0300	Beams 16′ O.C.		3200	.013		1.44	.56	.19	2.19	2.70
0500	40′ clear span, beams 8′ O.C.		3200	.013		3.80	.56	.19	4.55	5.30
0600	Beams 16′ O.C.		3840	.010		3.11	.47	.16	3.74	4.36
0800	60′ clear span, beams 8′ O.C.	F-4	2880	.017		6.55	.74	.38	7.67	8.80
0900	Beams 16′ O.C.	″	3840	.013		4.87	.55	.29	5.71	6.55
1100	Tudor arches, 30′ to 40′ clear span, frames 8′ O.C.	F-3	1680	.024		8.50	1.07	.36	9.93	11.50
1200	Frames 16′ O.C.	″	2240	.018		6.70	.80	.27	7.77	8.95

06 18 Glued-Laminated Construction

06 18 13 – Glued-Laminated Beams

06 18 13.20 Laminated Framing		Crew	Daily Output	Labor-Hours	Unit	Material	2012 Bare Costs Labor	Equipment	Total	Total Incl O&P
1400	50′ to 60′ clear span, frames 8′ O.C.	F-4	2200	.022	SF Flr.	9.20	.97	.50	10.67	12.20
1500	Frames 16′ O.C.		2640	.018		7.80	.80	.42	9.02	10.35
1700	Radial arches, 60′ clear span, frames 8′ O.C.		1920	.025		8.60	1.11	.57	10.28	11.85
1800	Frames 16′ O.C.		2880	.017		6.60	.74	.38	7.72	8.85
2000	100′ clear span, frames 8′ O.C.		1600	.030		8.90	1.33	.69	10.92	12.70
2100	Frames 16′ O.C.		2400	.020		7.80	.88	.46	9.14	10.55
2300	120′ clear span, frames 8′ O.C.		1440	.033		11.85	1.47	.76	14.08	16.25
2400	Frames 16′ O.C.		1920	.025		10.80	1.11	.57	12.48	14.30
2600	Bowstring trusses, 20′ O.C., 40′ clear span	F-3	2400	.017		5.35	.75	.26	6.36	7.35
2700	60′ clear span	F-4	3600	.013		4.79	.59	.31	5.69	6.55
2800	100′ clear span		4000	.012		6.80	.53	.28	7.61	8.60
2900	120′ clear span		3600	.013		7.25	.59	.31	8.15	9.30
3100	For premium appearance, add to S.F. prices					5%				
3300	For industrial type, deduct					15%				
3500	For stain and varnish, add					5%				
3900	For 3/4″ laminations, add to straight					25%				
4100	Add to curved					15%				
4300	Alternate pricing method: (use nominal footage of									
4310	components). Straight beams, camber less than 6″	F-3	3.50	11.429	M.B.F.	2,950	510	175	3,635	4,275
4400	Columns, including hardware		2	20		3,175	895	305	4,375	5,275
4600	Curved members, radius over 32′		2.50	16		3,250	715	245	4,210	5,025
4700	Radius 10′ to 32′		3	13.333		3,225	600	204	4,029	4,725
4900	For complicated shapes, add maximum					100%				
5100	For pressure treating, add to straight					35%				
5200	Add to curved					45%				
6000	Laminated veneer members, southern pine or western species									
6050	1-3/4″ wide x 5-1/2″ deep	2 Carp	480	.033	L.F.	3.19	1.47		4.66	5.90
6100	9-1/2″ deep		480	.033		4.56	1.47		6.03	7.40
6150	14″ deep		450	.036		7.30	1.57		8.87	10.60
6200	18″ deep		450	.036		9.90	1.57		11.47	13.40
6300	Parallel strand members, southern pine or western species									
6350	1-3/4″ wide x 9-1/4″ deep	2 Carp	480	.033	L.F.	4.56	1.47		6.03	7.40
6400	11-1/4″ deep		450	.036		5.70	1.57		7.27	8.80
6450	14″ deep		400	.040		7.30	1.76		9.06	10.95
6500	3-1/2″ wide x 9-1/4″ deep		480	.033		15.85	1.47		17.32	19.85
6550	11-1/4″ deep		450	.036		19.15	1.57		20.72	23.50
6600	14″ deep		400	.040		23.50	1.76		25.26	28.50
6650	7″ wide x 9-1/4″ deep		450	.036		31.50	1.57		33.07	37.50
6700	11-1/4″ deep		420	.038		39.50	1.68		41.18	46.50
6750	14″ deep		400	.040		46.50	1.76		48.26	54.50
9000	Minimum labor/equipment charge	F-3	2.50	16	Job		715	245	960	1,450

06 22 13 – Standard Pattern Wood Trim

06 22 13.15 Moldings, Base		Crew	Daily Output	Labor-Hours	Unit	Material	Labor	Equipment	Total	Total Incl O&P
						2012 Bare Costs				
0010	**MOLDINGS, BASE**									
5100	Classic profile, 5/8" x 5-1/2", finger jointed and primed	1 Carp	250	.032	L.F.	1.26	1.41		2.67	3.69
5105	Poplar		240	.033		1.39	1.47		2.86	3.94
5110	Red oak		220	.036		1.99	1.60		3.59	4.82
5115	Maple		220	.036		3.14	1.60		4.74	6.10
5120	Cherry		220	.036		3.73	1.60		5.33	6.75
5125	3/4 x 7-1/2", finger jointed and primed		250	.032		1.63	1.41		3.04	4.10
5130	Poplar		240	.033		2.16	1.47		3.63	4.78
5135	Red oak		220	.036		2.89	1.60		4.49	5.80
5140	Maple		220	.036		4.18	1.60		5.78	7.25
5145	Cherry		220	.036		4.98	1.60		6.58	8.15
5150	Modern profile, 5/8" x 3-1/2", finger jointed and primed		250	.032		.86	1.41		2.27	3.25
5155	Poplar		240	.033		.86	1.47		2.33	3.35
5160	Red oak		220	.036		1.36	1.60		2.96	4.12
5165	Maple		220	.036		2.20	1.60		3.80	5.05
5170	Cherry		220	.036		2.38	1.60		3.98	5.25
5175	Ogee profile, 7/16" x 3", finger jointed and primed		250	.032		.51	1.41		1.92	2.87
5180	Poplar		240	.033		.56	1.47		2.03	3.02
5185	Red oak		220	.036		.76	1.60		2.36	3.46
5200	9/16" x 3-1/2", finger jointed and primed		250	.032		.51	1.41		1.92	2.87
5205	Pine		240	.033		1.23	1.47		2.70	3.76
5210	Red oak		220	.036		2.32	1.60		3.92	5.20
5215	9/16" x 4-1/2", red oak		220	.036		3.47	1.60		5.07	6.45
5220	5/8" x 3-1/2", finger jointed and primed		250	.032		.79	1.41		2.20	3.18
5225	Poplar		240	.033		.86	1.47		2.33	3.35
5230	Red oak		220	.036		1.36	1.60		2.96	4.12
5235	Maple		220	.036		2.20	1.60		3.80	5.05
5240	Cherry		220	.036		2.38	1.60		3.98	5.25
5245	5/8" x 4", finger jointed and primed		250	.032		1.02	1.41		2.43	3.43
5250	Poplar		240	.033		1.11	1.47		2.58	3.63
5255	Red oak		220	.036		1.52	1.60		3.12	4.30
5260	Maple		220	.036		2.46	1.60		4.06	5.35
5265	Cherry		220	.036		2.58	1.60		4.18	5.45
5270	Rectangular profile, oak, 3/8" x 1-1/4"		260	.031		1.36	1.36		2.72	3.72
5275	1/2" x 2-1/2"		255	.031		1.91	1.38		3.29	4.37
5280	1/2" x 3-1/2"		250	.032		2.31	1.41		3.72	4.85
5285	1" x 6"		240	.033		2.77	1.47		4.24	5.45
5290	1" x 8"		240	.033		3.68	1.47		5.15	6.45
5295	Pine, 3/8" x 1-3/4"		260	.031		.52	1.36		1.88	2.79
5300	7/16" x 2-1/2"		255	.031		.78	1.38		2.16	3.13
5305	1" x 6"		240	.033		1.06	1.47		2.53	3.57
5310	1" x 8"		240	.033		1.44	1.47		2.91	3.99
5315	Shoe, 1/2" x 3/4", primed		260	.031		.43	1.36		1.79	2.69
5320	Pine		240	.033		.30	1.47		1.77	2.74
5325	Poplar		240	.033		.33	1.47		1.80	2.77
5330	Red oak		220	.036		.44	1.60		2.04	3.11
5335	Maple		220	.036		.61	1.60		2.21	3.30
5340	Cherry		220	.036		.69	1.60		2.29	3.39
5345	11/16" x 1-1/2", pine		240	.033		.68	1.47		2.15	3.16
5350	Caps, 11/16" x 1-3/8", pine		240	.033		.74	1.47		2.21	3.22
5355	3/4" x 1-3/4", finger jointed and primed		260	.031		.62	1.36		1.98	2.90
5360	Poplar		240	.033		.87	1.47		2.34	3.37
5365	Red oak	↓	220	.036	↓	1	1.60		2.60	3.73

06 22 13 – Standard Pattern Wood Trim

06 22 13.15 Moldings, Base		Crew	Daily Output	Labor-Hours	Unit	2012 Bare Costs Material	Labor	Equipment	Total	Total Incl O&P
5370	Maple	1 Carp	220	.036	L.F.	1.37	1.60		2.97	4.14
5375	Cherry		220	.036		3.09	1.60		4.69	6.05
5380	Combination base & shoe, 9/16" x 3-1/2" & 1/2" x 3/4", pine		125	.064		1.53	2.82		4.35	6.30
5385	Three piece oak, 6" high		80	.100		4.21	4.41		8.62	11.90
5390	Including 3/4" x 1" base shoe		70	.114		4.57	5.05		9.62	13.25
5395	Flooring cant strip, 3/4" x 3/4", pre-finished pine		260	.031		.47	1.36		1.83	2.74
5400	For pre-finished, stain and clear coat, add					.40			.40	.44
5405	Clear coat only, add					.35			.35	.39
9000	Minimum labor/equipment charge	1 Carp	4	2	Job		88		88	145

06 22 13.30 Moldings, Casings		Crew	Daily Output	Labor-Hours	Unit	Material	Labor	Equipment	Total	Total Incl O&P
0010	**MOLDINGS, CASINGS**									
0085	Apron, 9/16" x 2-1/2", pine	1 Carp	250	.032	L.F.	1.63	1.41		3.04	4.10
0090	5/8" x 2-1/2", pine		250	.032		2.17	1.41		3.58	4.70
0110	5/8" x 3-1/2", pine		220	.036		2.17	1.60		3.77	5
0300	Band, 11/16" x 1-1/8", pine		270	.030		1	1.31		2.31	3.24
0310	11/16" x 1-1/2", finger jointed and primed		270	.030		.83	1.31		2.14	3.05
0320	Pine		270	.030		1.06	1.31		2.37	3.30
0330	11/16" x 1-3/4", finger jointed and primed		270	.030		.56	1.31		1.87	2.75
0350	Pine		270	.030		1.25	1.31		2.56	3.51
0355	Beaded, 3/4" x 3-1/2", finger jointed and primed		220	.036		1.02	1.60		2.62	3.75
0360	Poplar		220	.036		.96	1.60		2.56	3.68
0365	Red oak		220	.036		1.36	1.60		2.96	4.12
0370	Maple		220	.036		2.20	1.60		3.80	5.05
0375	Cherry		220	.036		2.67	1.60		4.27	5.55
0380	3/4" x 4", finger jointed and primed		220	.036		1.02	1.60		2.62	3.75
0385	Poplar		220	.036		1.24	1.60		2.84	3.99
0390	Red oak		220	.036		1.67	1.60		3.27	4.47
0395	Maple		220	.036		2.30	1.60		3.90	5.15
0400	Cherry		220	.036		2.90	1.60		4.50	5.80
0405	3/4" x 5-1/2", finger jointed and primed		200	.040		1.02	1.76		2.78	4.01
0410	Poplar		200	.040		1.67	1.76		3.43	4.73
0415	Red oak		200	.040		2.34	1.76		4.10	5.45
0420	Maple		200	.040		3.23	1.76		4.99	6.45
0425	Cherry		200	.040		3.71	1.76		5.47	6.95
0430	Classic profile, 3/4" x 2-3/4", finger jointed and primed		250	.032		.68	1.41		2.09	3.06
0435	Poplar		250	.032		.86	1.41		2.27	3.25
0440	Red oak		250	.032		1.20	1.41		2.61	3.63
0445	Maple		250	.032		1.80	1.41		3.21	4.29
0450	Cherry		250	.032		2.05	1.41		3.46	4.56
0455	Fluted, 3/4" x 3-1/2", poplar		220	.036		.96	1.60		2.56	3.68
0460	Red oak		220	.036		1.36	1.60		2.96	4.12
0465	Maple		220	.036		2.20	1.60		3.80	5.05
0470	Cherry		220	.036		2.67	1.60		4.27	5.55
0475	3/4" x 4", poplar		220	.036		1.24	1.60		2.84	3.99
0480	Red oak		220	.036		3.02	1.60		4.62	5.95
0485	Maple		220	.036		2.30	1.60		3.90	5.15
0490	Cherry		220	.036		2.90	1.60		4.50	5.80
0495	3/4" x 5-1/2", poplar		200	.040		1.67	1.76		3.43	4.73
0500	Red oak		200	.040		2.34	1.76		4.10	5.45
0505	Maple		200	.040		4.58	1.76		6.34	7.95
0510	Cherry		200	.040		3.71	1.76		5.47	6.95
0515	3/4" x 7-1/2", poplar		190	.042		1.38	1.86		3.24	4.55

06 22 13 – Standard Pattern Wood Trim

06 22 13.30 Moldings, Casings

	06 22 13.30 Moldings, Casings	Crew	Daily Output	Labor-Hours	Unit	Material	Labor	Equipment	Total	Total Incl O&P
						2012 Bare Costs				
0520	Red oak	1 Carp	190	.042	L.F.	3.24	1.86		5.10	6.60
0525	Maple		190	.042		4.89	1.86		6.75	8.45
0530	Cherry		190	.042		5.85	1.86		7.71	9.45
0535	3/4" x 9-1/2", poplar		180	.044		3.56	1.96		5.52	7.10
0540	Red oak		180	.044		5.40	1.96		7.36	9.15
0545	Maple		180	.044		7.75	1.96		9.71	11.75
0550	Cherry		180	.044		8.45	1.96		10.41	12.50
0555	Modern profile, 9/16" x 2-1/4", poplar		250	.032		.49	1.41		1.90	2.85
0560	Red oak		250	.032		.64	1.41		2.05	3.01
0565	11/16" x 2-1/2", finger jointed & primed		250	.032		.58	1.41		1.99	2.95
0570	Pine		250	.032		1.33	1.41		2.74	3.77
0575	3/4" x 2-1/2", poplar		250	.032		.76	1.41		2.17	3.14
0580	Red oak		250	.032		.98	1.41		2.39	3.39
0585	Maple		250	.032		1.55	1.41		2.96	4.01
0590	Cherry		250	.032		1.96	1.41		3.37	4.46
0595	Mullion, 5/16" x 2", pine		270	.030		.99	1.31		2.30	3.23
0600	9/16" x 2-1/2", fingerjointed and primed		250	.032		.60	1.41		2.01	2.97
0605	Pine		250	.032		1.41	1.41		2.82	3.86
0610	Red oak		250	.032		2.80	1.41		4.21	5.40
0615	1-1/16" x 3-3/4", red oak		220	.036		6.85	1.60		8.45	10.15
0620	Ogee, 7/16" x 2-1/2", poplar		250	.032		.44	1.41		1.85	2.79
0625	Red oak		250	.032		.62	1.41		2.03	2.99
0630	9/16" x 2-1/4", finger jointed and primed		250	.032		.44	1.41		1.85	2.79
0635	Poplar		250	.032		.49	1.41		1.90	2.85
0640	Red oak		250	.032		.64	1.41		2.05	3.01
0645	11/16" x 2-1/2", finger jointed and primed		250	.032		.54	1.41		1.95	2.90
0700	Pine		250	.032		1.32	1.41		2.73	3.76
0730	11/16" x 3-1/2", finger jointed and primed		220	.036		1.82	1.60		3.42	4.63
0750	Pine		220	.036		2.18	1.60		3.78	5.05
0755	3/4" x 2-1/2", finger jointed and primed		250	.032		.62	1.41		2.03	2.99
0760	Poplar		250	.032		.76	1.41		2.17	3.14
0765	Red oak		250	.032		.98	1.41		2.39	3.39
0770	Maple		250	.032		1.55	1.41		2.96	4.01
0775	Cherry		250	.032		1.96	1.41		3.37	4.46
0780	3/4" x 3-1/2", finger jointed and primed		220	.036		.87	1.60		2.47	3.59
0785	Poplar		220	.036		.96	1.60		2.56	3.68
0790	Red oak		220	.036		1.36	1.60		2.96	4.12
0795	Maple		220	.036		2.20	1.60		3.80	5.05
0800	Cherry		220	.036		2.67	1.60		4.27	5.55
4700	Square profile, 1" x 1", teak		215	.037		2.01	1.64		3.65	4.90
4800	Rectangular profile, 1" x 3", teak		200	.040		5.70	1.76		7.46	9.20
9000	Minimum labor/equipment charge		4	2	Job		88		88	145

06 22 13.35 Moldings, Ceilings

	06 22 13.35 Moldings, Ceilings	Crew	Daily Output	Labor-Hours	Unit	Material	Labor	Equipment	Total	Total Incl O&P
0010	**MOLDINGS, CEILINGS**									
0600	Bed, stock pine, 9/16" x 1-3/4"	1 Carp	270	.030	L.F.	1.19	1.31		2.50	3.45
0650	9/16" x 2"		240	.033		1.42	1.47		2.89	3.97
1200	Cornice molding, stock pine, 9/16" x 1-3/4"		330	.024		1.19	1.07		2.26	3.06
1300	9/16" x 2-1/4"		300	.027		1.45	1.18		2.63	3.53
2400	Cove scotia, stock pine, 9/16" x 1-3/4"		270	.030		1.11	1.31		2.42	3.36
2500	11/16" x 2-3/4"		255	.031		2.18	1.38		3.56	4.67
2600	Crown, stock pine, 9/16" x 3-5/8"		250	.032		2.68	1.41		4.09	5.25
2700	11/16" x 4-5/8"		220	.036		4.52	1.60		6.12	7.60

06 22 Millwork

06 22 13 – Standard Pattern Wood Trim

06 22 13.35 Moldings, Ceilings		Crew	Daily Output	Labor-Hours	Unit	Material	2012 Bare Costs Labor	Equipment	Total	Total Incl O&P
9000	Minimum labor/equipment charge	1 Carp	4	2	Job		88		88	145

06 22 13.40 Moldings, Exterior

Code	Description	Crew	Daily Output	Labor-Hours	Unit	Material	Labor	Equipment	Total	Total Incl O&P
0010	**MOLDINGS, EXTERIOR**									
1500	Cornice, boards, pine, 1" x 2"	1 Carp	330	.024	L.F.	.25	1.07		1.32	2.03
1700	1" x 6"		250	.032		.72	1.41		2.13	3.10
2000	1" x 12"		180	.044		1.63	1.96		3.59	5
2200	Three piece, built-up, pine, minimum		80	.100		1.38	4.41		5.79	8.75
2300	Maximum		65	.123		3.93	5.45		9.38	13.20
3000	Corner board, sterling pine, 1" x 4"		200	.040		.80	1.76		2.56	3.77
3100	1" x 6"		200	.040		1.06	1.76		2.82	4.06
3350	Fascia, sterling pine, 1" x 6"		250	.032		1.06	1.41		2.47	3.48
3370	1" x 8"		225	.036		1.44	1.57		3.01	4.15
3400	Trim, back band, 11/16" x 1-1/16"		250	.032		.98	1.41		2.39	3.39
3500	Casing, 11/16" x 4-1/4"		250	.032		3.20	1.41		4.61	5.85
3600	Crown, 11/16" x 4-1/4"		250	.032		4.50	1.41		5.91	7.25
3610	Casing, 1-1/4" x 2"		200	.040		2.50	1.76		4.26	5.65
3620	Finger jointed, 1-1/4" x 2"		200	.040		1.05	1.76		2.81	4.05
3700	1" x 4" and 1" x 6" railing w/balusters 4" O.C.		22	.364		25.50	16.05		41.55	54.50
3800	Insect screen framing stock, 1-1/16" x 1-3/4"		395	.020		2.38	.89		3.27	4.08
4100	Verge board, sterling pine, 1" x 4"		200	.040		.80	1.76		2.56	3.77
4200	1" x 6"		200	.040		1.06	1.76		2.82	4.06
4300	2" x 6"		165	.048		2.12	2.14		4.26	5.85
4400	2" x 8"	↓	165	.048	↓	2.88	2.14		5.02	6.70
4700	For redwood trim, add					200%				
9000	Minimum labor/equipment charge	1 Carp	4	2	Job		88		88	145

06 22 13.45 Moldings, Trim

Code	Description	Crew	Daily Output	Labor-Hours	Unit	Material	Labor	Equipment	Total	Total Incl O&P
0010	**MOLDINGS, TRIM**									
0200	Astragal, stock pine, 11/16" x 1-3/4"	1 Carp	255	.031	L.F.	1.41	1.38		2.79	3.82
0250	1-5/16" x 2-3/16"		240	.033		2.57	1.47		4.04	5.25
0800	Chair rail, stock pine, 5/8" x 2-1/2"		270	.030		1.68	1.31		2.99	3.99
0900	5/8" x 3-1/2"		240	.033		2.65	1.47		4.12	5.35
1000	Closet pole, stock pine, 1-1/8" diameter		200	.040		1.22	1.76		2.98	4.23
1100	Fir, 1-5/8" diameter		200	.040		2.36	1.76		4.12	5.50
3300	Half round, stock pine, 1/4" x 1/2"		270	.030		.31	1.31		1.62	2.48
3350	1/2" x 1"	↓	255	.031	↓	.73	1.38		2.11	3.07
3400	Handrail, fir, single piece, stock, hardware not included									
3450	1-1/2" x 1-3/4"	1 Carp	80	.100	L.F.	2.50	4.41		6.91	10
3470	Pine, 1-1/2" x 1-3/4"		80	.100		2.38	4.41		6.79	9.85
3500	1-1/2" x 2-1/2"		76	.105		2.67	4.64		7.31	10.55
3600	Lattice, stock pine, 1/4" x 1-1/8"		270	.030		.51	1.31		1.82	2.70
3700	1/4" x 1-3/4"		250	.032		.92	1.41		2.33	3.32
3800	Miscellaneous, custom, pine, 1" x 1"		270	.030		.38	1.31		1.69	2.55
3900	1" x 3"		240	.033		1.13	1.47		2.60	3.65
4100	Birch or oak, nominal 1" x 1"		240	.033		.47	1.47		1.94	2.92
4200	Nominal 1" x 3"		215	.037		1.40	1.64		3.04	4.23
4400	Walnut, nominal 1" x 1"		215	.037		.56	1.64		2.20	3.31
4500	Nominal 1" x 3"		200	.040		1.69	1.76		3.45	4.75
4700	Teak, nominal 1" x 1"		215	.037		2.47	1.64		4.11	5.40
4800	Nominal 1" x 3"		200	.040		7.40	1.76		9.16	11.05
4900	Quarter round, stock pine, 1/4" x 1/4"		275	.029		.31	1.28		1.59	2.44
4950	3/4" x 3/4"		255	.031	↓	.70	1.38		2.08	3.04
5600	Wainscot moldings, 1-1/8" x 9/16", 2' high, minimum	↓	76	.105	S.F.	11.55	4.64		16.19	20.50

06 22 Millwork

06 22 13 – Standard Pattern Wood Trim

06 22 13.45 Moldings, Trim		Crew	Daily Output	Labor-Hours	Unit	Material	Labor	Equipment	Total	Total Incl O&P
						2012 Bare Costs				
5700	Maximum	1 Carp	65	.123	S.F.	20.50	5.45		25.95	31.50
9000	Minimum labor/equipment charge	↓	4	2	Job		88		88	145
06 22 13.50 Moldings, Window and Door										
0010	**MOLDINGS, WINDOW AND DOOR**									
2800	Door moldings, stock, decorative, 1-1/8″ wide, plain	1 Carp	17	.471	Set	44.50	21		65.50	83
2900	Detailed		17	.471	″	87.50	21		108.50	131
2960	Clear pine door jamb, no stops, 11/16″ x 4-9/16″		240	.033	L.F.	4.73	1.47		6.20	7.60
3150	Door trim set, 1 head and 2 sides, pine, 2-1/2 wide		12	.667	Opng.	22.50	29.50		52	72.50
3170	3-1/2″ wide		11	.727	″	37	32		69	93
3250	Glass beads, stock pine, 3/8″ x 1/2″		275	.029	L.F.	.38	1.28		1.66	2.52
3270	3/8″ x 7/8″		270	.030		.50	1.31		1.81	2.69
4850	Parting bead, stock pine, 3/8″ x 3/4″		275	.029		.45	1.28		1.73	2.60
4870	1/2″ x 3/4″		255	.031		.56	1.38		1.94	2.89
5000	Stool caps, stock pine, 11/16″ x 3-1/2″		200	.040		2.65	1.76		4.41	5.80
5100	1-1/16″ x 3-1/4″		150	.053	↓	4.15	2.35		6.50	8.45
5300	Threshold, oak, 3′ long, inside, 5/8″ x 3-5/8″		32	.250	Ea.	9	11.05		20.05	28
5400	Outside, 1-1/2″ x 7-5/8″	↓	16	.500	″	38	22		60	78
5900	Window trim sets, including casings, header, stops,									
5910	stool and apron, 2-1/2″ wide, minimum	1 Carp	13	.615	Opng.	34.50	27		61.50	82
5950	Average		10	.800		40.50	35.50		76	103
6000	Maximum		6	1.333	↓	64.50	59		123.50	168
9000	Minimum labor/equipment charge	↓	4	2	Job		88		88	145
06 22 13.60 Moldings, Soffits										
0010	**MOLDINGS, SOFFITS** R061110-30									
0200	Soffits, pine, 1″ x 4″	2 Carp	420	.038	L.F.	.41	1.68		2.09	3.20
0210	1″ x 6″		420	.038		.72	1.68		2.40	3.54
0220	1″ x 8″		420	.038		1	1.68		2.68	3.85
0230	1″ x 10″		400	.040		1.30	1.76		3.06	4.32
0240	1″ x 12″		400	.040		1.63	1.76		3.39	4.68
0250	STK cedar, 1″ x 4″		420	.038		.48	1.68		2.16	3.28
0260	1″ x 6″		420	.038		.83	1.68		2.51	3.66
0270	1″ x 8″		420	.038		1.29	1.68		2.97	4.17
0280	1″ x 10″		400	.040		1.99	1.76		3.75	5.10
0290	1″ x 12″		400	.040	↓	2.69	1.76		4.45	5.85
1000	Exterior AC plywood, 1/4″ thick		420	.038	S.F.	.62	1.68		2.30	3.43
1050	3/8″ thick		420	.038		.73	1.68		2.41	3.55
1100	1/2″ thick	↓	420	.038		.77	1.68		2.45	3.60
1150	Polyvinyl chloride, white, solid	1 Carp	230	.035		2.04	1.53		3.57	4.75
1160	Perforated	″	230	.035	↓	2.04	1.53		3.57	4.75
9000	Minimum labor/equipment charge	2 Carp	5	3.200	Job		141		141	231

06 25 Prefinished Paneling

06 25 13 – Prefinished Hardboard Paneling

06 25 13.10 Paneling, Hardboard		Crew	Daily Output	Labor-Hours	Unit	Material	2012 Bare Costs Labor	Equipment	Total	Total Incl O&P
0010	**PANELING, HARDBOARD**									
0050	Not incl. furring or trim, hardboard, tempered, 1/8" thick G	2 Carp	500	.032	S.F.	.38	1.41		1.79	2.73
0100	1/4" thick G		500	.032		.55	1.41		1.96	2.92
0300	Tempered pegboard, 1/8" thick G		500	.032		.45	1.41		1.86	2.81
0400	1/4" thick G		500	.032		.64	1.41		2.05	3.01
0600	Untempered hardboard, natural finish, 1/8" thick G		500	.032		.38	1.41		1.79	2.73
0700	1/4" thick G		500	.032		.46	1.41		1.87	2.82
0900	Untempered pegboard, 1/8" thick G		500	.032		.39	1.41		1.80	2.74
1000	1/4" thick G		500	.032		.42	1.41		1.83	2.77
1200	Plastic faced hardboard, 1/8" thick G		500	.032		.68	1.41		2.09	3.06
1300	1/4" thick G		500	.032		.84	1.41		2.25	3.23
1500	Plastic faced pegboard, 1/8" thick G		500	.032		.62	1.41		2.03	2.99
1600	1/4" thick G		500	.032		.74	1.41		2.15	3.12
1800	Wood grained, plain or grooved, 1/4" thick, minimum G		500	.032		.58	1.41		1.99	2.95
1900	Maximum G		425	.038		1.12	1.66		2.78	3.95
2100	Moldings for hardboard, wood or aluminum, minimum		500	.032	L.F.	.39	1.41		1.80	2.74
2200	Maximum		425	.038	"	1.09	1.66		2.75	3.92
9000	Minimum labor/equipment charge	1 Carp	2	4	Job		176		176	289

06 25 16 – Prefinished Plywood Paneling

06 25 16.10 Paneling, Plywood		Crew	Daily Output	Labor-Hours	Unit	Material	Labor	Equipment	Total	Total Incl O&P
0010	**PANELING, PLYWOOD** R061636-20									
2400	Plywood, prefinished, 1/4" thick, 4' x 8' sheets									
2410	with vertical grooves. Birch faced, minimum	2 Carp	500	.032	S.F.	.90	1.41		2.31	3.30
2420	Average		420	.038		1.36	1.68		3.04	4.25
2430	Maximum		350	.046		1.99	2.02		4.01	5.50
2600	Mahogany, African		400	.040		2.54	1.76		4.30	5.70
2700	Philippine (Lauan)		500	.032		1.09	1.41		2.50	3.51
2900	Oak or Cherry, minimum		500	.032		2.12	1.41		3.53	4.64
3000	Maximum		400	.040		3.25	1.76		5.01	6.45
3200	Rosewood		320	.050		4.63	2.21		6.84	8.70
3400	Teak		400	.040		3.25	1.76		5.01	6.45
3600	Chestnut		375	.043		4.82	1.88		6.70	8.40
3800	Pecan		400	.040		2.08	1.76		3.84	5.20
3900	Walnut, minimum		500	.032		2.78	1.41		4.19	5.35
3950	Maximum		400	.040		5.25	1.76		7.01	8.70
4000	Plywood, prefinished, 3/4" thick, stock grades, minimum		320	.050		1.25	2.21		3.46	5
4100	Maximum		224	.071		5.45	3.15		8.60	11.10
4300	Architectural grade, minimum		224	.071		4.02	3.15		7.17	9.55
4400	Maximum		160	.100		6.15	4.41		10.56	14
4600	Plywood, "A" face, birch, V.C., 1/2" thick, natural		450	.036		1.91	1.57		3.48	4.67
4700	Select		450	.036		2.08	1.57		3.65	4.86
4900	Veneer core, 3/4" thick, natural		320	.050		2.01	2.21		4.22	5.85
5000	Select		320	.050		2.25	2.21		4.46	6.10
5200	Lumber core, 3/4" thick, natural		320	.050		3.02	2.21		5.23	6.95
5500	Plywood, knotty pine, 1/4" thick, A2 grade		450	.036		1.65	1.57		3.22	4.39
5600	A3 grade		450	.036		2.08	1.57		3.65	4.86
5800	3/4" thick, veneer core, A2 grade		320	.050		2.13	2.21		4.34	5.95
5900	A3 grade		320	.050		2.40	2.21		4.61	6.25
6100	Aromatic cedar, 1/4" thick, plywood		400	.040		2.10	1.76		3.86	5.20
6200	1/4" thick, particle board		400	.040		1.02	1.76		2.78	4.01
9000	Minimum labor/equipment charge	1 Carp	2	4	Job		176		176	289

06 25 Prefinished Paneling

06 25 26 – Panel System

06 25 26.10 Panel Systems		Crew	Daily Output	Labor-Hours	Unit	Material	2012 Bare Costs Labor	Equipment	Total	Total Incl O&P
0010	PANEL SYSTEMS									
0100	Raised panel, eng. wood core w/wood veneer, std., paint grade	2 Carp	300	.053	S.F.	12.20	2.35		14.55	17.25
0110	Oak veneer		300	.053		20	2.35		22.35	26
0120	Maple veneer		300	.053		24.50	2.35		26.85	30.50
0130	Cherry veneer		300	.053		29	2.35		31.35	36
0300	Class I fire rated, paint grade		300	.053		14.25	2.35		16.60	19.50
0310	Oak veneer		300	.053		23.50	2.35		25.85	30
0320	Maple veneer		300	.053		30	2.35		32.35	37
0330	Cherry veneer		300	.053		38.50	2.35		40.85	46
0510	Beadboard, 5/8" MDF, standard, primed		300	.053		6.60	2.35		8.95	11.10
0520	Oak veneer, unfinished		300	.053		10.05	2.35		12.40	14.90
0530	Maple veneer, unfinished		300	.053		11.20	2.35		13.55	16.15
0610	Rustic paneling, 5/8" MDF, standard, maple veneer, unfinished	↓	300	.053	↓	12.70	2.35		15.05	17.80

06 26 Board Paneling

06 26 13 – Profile Board Paneling

06 26 13.10 Paneling, Boards		Crew	Daily Output	Labor-Hours	Unit	Material	Labor	Equipment	Total	Total Incl O&P
0010	PANELING, BOARDS									
6400	Wood board paneling, 3/4" thick, knotty pine	2 Carp	300	.053	S.F.	1.89	2.35		4.24	5.95
6500	Rough sawn cedar		300	.053		2.86	2.35		5.21	7
6700	Redwood, clear, 1" x 4" boards		300	.053		4.96	2.35		7.31	9.30
6900	Aromatic cedar, closet lining, boards	↓	275	.058	↓	2	2.57		4.57	6.40
9000	Minimum labor/equipment charge	1 Carp	2	4	Job		176		176	289

06 43 Wood Stairs and Railings

06 43 13 – Wood Stairs

06 43 13.20 Prefabricated Wood Stairs		Crew	Daily Output	Labor-Hours	Unit	Material	Labor	Equipment	Total	Total Incl O&P
0010	PREFABRICATED WOOD STAIRS									
0100	Box stairs, prefabricated, 3'-0" wide									
0110	Oak treads, up to 14 risers	2 Carp	39	.410	Riser	92.50	18.10		110.60	132
0600	With pine treads for carpet, up to 14 risers	"	39	.410	"	59.50	18.10		77.60	95
1100	For 4' wide stairs, add				Flight	25%				
1550	Stairs, prefabricated stair handrail with balusters	1 Carp	30	.267	L.F.	75	11.75		86.75	101
1700	Basement stairs, prefabricated, pine treads									
1710	Pine risers, 3' wide, up to 14 risers	2 Carp	52	.308	Riser	59.50	13.55		73.05	88
4000	Residential, wood, oak treads, prefabricated		1.50	10.667	Flight	1,200	470		1,670	2,100
4200	Built in place	↓	.44	36.364	"	1,800	1,600		3,400	4,600
4400	Spiral, oak, 4'-6" diameter, unfinished, prefabricated,									
4500	incl. railing, 9' high	2 Carp	1.50	10.667	Flight	3,625	470		4,095	4,750
9000	Minimum labor/equipment charge	"	3	5.333	Job		235		235	385

06 43 13.40 Wood Stair Parts		Crew	Daily Output	Labor-Hours	Unit	Material	Labor	Equipment	Total	Total Incl O&P
0010	WOOD STAIR PARTS									
0020	Pin top balusters, 1-1/4", oak, 34"	1 Carp	96	.083	Ea.	7.80	3.67		11.47	14.60
0030	38"		96	.083		9.85	3.67		13.52	16.85
0040	42"		96	.083		12.10	3.67		15.77	19.30
0050	Poplar, 34"		96	.083		3.05	3.67		6.72	9.35
0060	38"		96	.083		3.62	3.67		7.29	10
0070	42"	↓	96	.083	↓	8.55	3.67		12.22	15.45

06 43 Wood Stairs and Railings

06 43 13 – Wood Stairs

06 43 13.40 Wood Stair Parts		Crew	Daily Output	Labor-Hours	Unit	Material	Labor	Equipment	Total	Total Incl O&P
						2012 Bare Costs				
0080	Maple, 34"	1 Carp	96	.083	Ea.	4.90	3.67		8.57	11.40
0090	38"		96	.083		6.20	3.67		9.87	12.80
0100	42"		96	.083		7.60	3.67		11.27	14.35
0130	Primed, 34"		96	.083		2.85	3.67		6.52	9.15
0140	38"		96	.083		3.62	3.67		7.29	10
0150	42"		96	.083		4.42	3.67		8.09	10.85
0180	Box top balusters, 1-1/4", oak, 34"		60	.133		10.85	5.90		16.75	21.50
0190	38"		60	.133		12.50	5.90		18.40	23.50
0200	42"		60	.133		12.95	5.90		18.85	24
0210	Poplar, 34"		60	.133		8.65	5.90		14.55	19.15
0220	38"		60	.133		9.50	5.90		15.40	20
0230	42"		60	.133		10	5.90		15.90	20.50
0240	Maple, 34"		60	.133		11.90	5.90		17.80	23
0250	38"		60	.133		13.80	5.90		19.70	25
0260	42"		60	.133		14.20	5.90		20.10	25.50
0290	Primed, 34"		60	.133		6.95	5.90		12.85	17.30
0300	38"		60	.133		8	5.90		13.90	18.45
0310	42"		60	.133		8.35	5.90		14.25	18.85
0340	Square balusters, cut from lineal stock, pine, 1-1/16" x 1-1/16"		180	.044	L.F.	1.32	1.96		3.28	4.66
0350	1-5/16" x 1-5/16"		180	.044		1.70	1.96		3.66	5.10
0360	1-5/8" x 1-5/8"		180	.044		2.82	1.96		4.78	6.30
0370	Turned newel, oak, 3-1/2" square, 48" high		8	1	Ea.	94.50	44		138.50	177
0380	62" high		8	1		138	44		182	225
0390	Poplar, 3-1/2" square, 48" high		8	1		52	44		96	130
0400	62" high		8	1		75	44		119	155
0410	Maple, 3-1/2" square, 48" high		8	1		66	44		110	145
0420	62" high		8	1		93	44		137	175
0430	Square newel, oak, 3-1/2" square, 48" high		8	1		50	44		94	128
0440	58" high		8	1		70	44		114	150
0450	Poplar, 3-1/2" square, 48" high		8	1		49	44		93	127
0460	58" high		8	1		54	44		98	132
0470	Maple, 3" square, 48" high		8	1		50	44		94	128
0480	58" high		8	1		62	44		106	141
0490	Railings, oak, minimum		96	.083	L.F.	10.40	3.67		14.07	17.45
0500	Average		96	.083		12.50	3.67		16.17	19.75
0510	Maximum		96	.083		15.50	3.67		19.17	23
0520	Maple, minimum		96	.083		15	3.67		18.67	22.50
0530	Average		96	.083		15.25	3.67		18.92	23
0540	Maximum		96	.083		15.50	3.67		19.17	23
0550	Oak, for bending rail, minimum		48	.167		21	7.35		28.35	35
0560	Average		48	.167		22	7.35		29.35	36
0570	Maximum		48	.167		23	7.35		30.35	37.50
0580	Maple, for bending rail, minimum		48	.167		23	7.35		30.35	37
0590	Average		48	.167		25	7.35		32.35	39
0600	Maximum		48	.167		27	7.35		34.35	41.50
0610	Risers, oak, 3/4" x 8", 36" long		80	.100	Ea.	18	4.41		22.41	27
0620	42" long		80	.100		21	4.41		25.41	30.50
0630	48" long		80	.100		24	4.41		28.41	34
0640	54" long		80	.100		27	4.41		31.41	37
0650	60" long		80	.100		30	4.41		34.41	40.50
0660	72" long		80	.100		36	4.41		40.41	47
0670	Poplar, 3/4" x 8", 36" long		80	.100		11.10	4.41		15.51	19.45
0680	42" long	1 Carp	80	.100		12.95	4.41		17.36	21.50

06 43 Wood Stairs and Railings

06 43 13 – Wood Stairs

06 43 13.40 Wood Stair Parts		Crew	Daily Output	Labor-Hours	Unit	Material	Labor	Equipment	Total	Total Incl O&P
						2012 Bare Costs				
0690	48″ long	1 Carp	80	.100	Ea.	14.80	4.41		19.21	23.50
0700	54″ long		80	.100		16.65	4.41		21.06	25.50
0710	60″ long		80	.100		18.50	4.41		22.91	28
0720	72″ long		80	.100		22	4.41		26.41	32
0730	Pine, 1″ x 8″, 36″ long		80	.100		3	4.41		7.41	10.55
0740	42″ long		80	.100		3.50	4.41		7.91	11.10
0750	48″ long		80	.100		4	4.41		8.41	11.65
0760	54″ long		80	.100		4.50	4.41		8.91	12.20
0770	60″ long		80	.100		5	4.41		9.41	12.75
0780	72″ long		80	.100		6	4.41		10.41	13.85
0790	Treads, oak, no returns, 1-1/32″ x 11-1/2″ x 36″ long		32	.250		32	11.05		43.05	53
0800	42″ long		32	.250		37.50	11.05		48.55	59
0810	48″ long		32	.250		42.50	11.05		53.55	65
0820	54″ long		32	.250		48	11.05		59.05	71
0830	60″ long		32	.250		53.50	11.05		64.55	76.50
0840	72″ long		32	.250		64	11.05		75.05	88.50
0850	Mitred return one end, 1-1/32″ x 11-1/2″ x 36″ long		24	.333		46	14.70		60.70	74.50
0860	42″ long		24	.333		53.50	14.70		68.20	83
0870	48″ long		24	.333		61.50	14.70		76.20	91.50
0880	54″ long		24	.333		69	14.70		83.70	100
0890	60″ long		24	.333		76.50	14.70		91.20	109
0900	72″ long		24	.333		92	14.70		106.70	125
0910	Mitred return two ends, 1-1/32″ x 11-1/2″ x 36″ long		12	.667		49	29.50		78.50	102
0920	42″ long		12	.667		57	29.50		86.50	111
0930	48″ long		12	.667		65.50	29.50		95	120
0940	54″ long		12	.667		73.50	29.50		103	129
0950	60″ long		12	.667		81.50	29.50		111	138
0960	72″ long		12	.667		98	29.50		127.50	156
0970	Starting step, oak, 48″, bullnose		8	1		192	44		236	284
0980	Double end bullnose		8	1		345	44		389	455
1030	Skirt board, pine, 1″ x 10″		55	.145	L.F.	1.30	6.40		7.70	11.95
1040	1″ x 12″		52	.154	″	1.63	6.80		8.43	12.90
1050	Oak landing tread, 1-1/16″ thick		54	.148	S.F.	8.25	6.55		14.80	19.80
1060	Oak cove molding		96	.083	L.F.	1.45	3.67		5.12	7.60
1070	Oak stringer molding		96	.083	″	2.75	3.67		6.42	9.05
1090	Rail bolt, 5/16″ x 3-1/2″		48	.167	Ea.	2.75	7.35		10.10	15.10
1100	5/16″ x 4-1/2″		48	.167		2.75	7.35		10.10	15.10
1120	Newel post anchor		16	.500		13.50	22		35.50	51
1130	Tapered plug, 1/2″		240	.033		.30	1.47		1.77	2.74
1140	1″		240	.033		.40	1.47		1.87	2.85
9000	Minimum labor/equipment charge		3	2.667	Job		118		118	193

06 43 16 – Wood Railings

06 43 16.10 Wood Handrails and Railings		Crew	Daily Output	Labor-Hours	Unit	Material	Labor	Equipment	Total	Total Incl O&P
0010	**WOOD HANDRAILS AND RAILINGS**									
0020	Custom design, architectural grade, hardwood, minimum	1 Carp	38	.211	L.F.	9.50	9.30		18.80	25.50
0100	Maximum		30	.267		50	11.75		61.75	74.50
0300	Stock interior railing with spindles 4″ O.C., 4′ long		40	.200		32.50	8.80		41.30	50
0400	8′ long		48	.167		32.50	7.35		39.85	47.50
9000	Minimum labor/equipment charge		3	2.667	Job		118		118	193

06 44 Ornamental Woodwork

06 44 19 – Wood Grilles

06 44 19.10 Grilles		Crew	Daily Output	Labor-Hours	Unit	2012 Bare Costs Material	Labor	Equipment	Total	Total Incl O&P
0010	**GRILLES** and panels, hardwood, sanded									
0020	2′ x 4′ to 4′ x 8′, custom designs, unfinished, minimum	1 Carp	38	.211	S.F.	56	9.30		65.30	77
0050	Average		30	.267		65	11.75		76.75	91
0100	Maximum		19	.421		74	18.55		92.55	112
9000	Minimum labor/equipment charge		2	4	Job		176		176	289

06 44 33 – Wood Mantels

06 44 33.10 Fireplace Mantels

Line	Description	Crew	Daily Output	Labor-Hours	Unit	Material	Labor	Equipment	Total	Total Incl O&P
0010	**FIREPLACE MANTELS**									
0015	6″ molding, 6′ x 3′-6″ opening, minimum	1 Carp	5	1.600	Opng.	228	70.50		298.50	365
0100	Maximum		5	1.600		410	70.50		480.50	570
0300	Prefabricated pine, colonial type, stock, deluxe		2	4		1,250	176		1,426	1,700
0400	Economy		3	2.667		530	118		648	780
9000	Minimum labor/equipment charge		3	2.667	Job		118		118	193

06 44 33.20 Fireplace Mantel Beam

Line	Description	Crew	Daily Output	Labor-Hours	Unit	Material	Labor	Equipment	Total	Total Incl O&P
0010	**FIREPLACE MANTEL BEAM**									
0020	Rough texture wood, 4″ x 8″	1 Carp	36	.222	L.F.	8.30	9.80		18.10	25
0100	4″ x 10″		35	.229	″	10.90	10.10		21	28.50
0300	Laminated hardwood, 2-1/4″ x 10-1/2″ wide, 6′ long		5	1.600	Ea.	110	70.50		180.50	237
0400	8′ long		5	1.600	″	150	70.50		220.50	281
0600	Brackets for above, rough sawn		12	.667	Pr.	10	29.50		39.50	59
0700	Laminated		12	.667	″	15	29.50		44.50	64.50
9000	Minimum labor/equipment charge		4	2	Job		88		88	145

06 44 39 – Wood Posts and Columns

06 44 39.10 Decorative Beams

Line	Description	Crew	Daily Output	Labor-Hours	Unit	Material	Labor	Equipment	Total	Total Incl O&P
0010	**DECORATIVE BEAMS**									
0020	Rough sawn cedar, non-load bearing, 4″ x 4″	2 Carp	180	.089	L.F.	1.45	3.92		5.37	8.05
0100	4″ x 6″		170	.094		2.18	4.15		6.33	9.20
0200	4″ x 8″		160	.100		2.90	4.41		7.31	10.45
0300	4″ x 10″		150	.107		4.27	4.70		8.97	12.40
0400	4″ x 12″		140	.114		5.65	5.05		10.70	14.45
0500	8″ x 8″		130	.123		5.80	5.45		11.25	15.30
9000	Minimum labor/equipment charge	1 Carp	3	2.667	Job		118		118	193

06 44 39.20 Columns

Line	Description	Crew	Daily Output	Labor-Hours	Unit	Material	Labor	Equipment	Total	Total Incl O&P
0010	**COLUMNS**									
0050	Aluminum, round colonial, 6″ diameter	2 Carp	80	.200	V.L.F.	18.60	8.80		27.40	35
0100	8″ diameter		62.25	.257		21.50	11.35		32.85	42
0200	10″ diameter		55	.291		25.50	12.85		38.35	49
0250	Fir, stock units, hollow round, 6″ diameter		80	.200		20.50	8.80		29.30	37
0300	8″ diameter		80	.200		24	8.80		32.80	41
0350	10″ diameter		70	.229		33.50	10.10		43.60	53
0400	Solid turned, to 8′ high, 3-1/2″ diameter		80	.200		9	8.80		17.80	24.50
0500	4-1/2″ diameter		75	.213		14	9.40		23.40	31
0600	5-1/2″ diameter		70	.229		18.40	10.10		28.50	36.50
0800	Square columns, built-up, 5″ x 5″		65	.246		15.50	10.85		26.35	35
0900	Solid, 3-1/2″ x 3-1/2″		130	.123		9	5.45		14.45	18.80
1600	Hemlock, tapered, T & G, 12″ diam., 10′ high		100	.160		34.50	7.05		41.55	49.50
1700	16′ high		65	.246		65.50	10.85		76.35	90
1900	10′ high, 14″ diameter		100	.160		97	7.05		104.05	118
2000	18′ high		65	.246		88	10.85		98.85	115
2200	18″ diameter, 12′ high		65	.246		141	10.85		151.85	173
2300	20′ high		50	.320		109	14.10		123.10	143

06 44 Ornamental Woodwork

06 44 39 – Wood Posts and Columns

06 44 39.20 Columns		Crew	Daily Output	Labor-Hours	Unit	2012 Bare Costs Material	Labor	Equipment	Total	Total Incl O&P
2500	20" diameter, 14' high	2 Carp	40	.400	V.L.F.	159	17.65		176.65	204
2600	20' high	↓	35	.457	↓	173	20		193	223
2800	For flat pilasters, deduct				↓	33%				
3000	For splitting into halves, add				Ea.	105			105	116
4000	Rough sawn cedar posts, 4" x 4"	2 Carp	250	.064	V.L.F.	3.32	2.82		6.14	8.30
4100	4" x 6"		235	.068		6	3		9	11.50
4200	6" x 6"		220	.073		10.30	3.21		13.51	16.60
4300	8" x 8"	↓	200	.080	↓	20	3.53		23.53	28
9000	Minimum labor/equipment charge	1 Carp	3	2.667	Job		118		118	193

06 48 Wood Frames

06 48 13 – Exterior Wood Door Frames

06 48 13.10 Exterior Wood Door Frames and Accessories

		Crew	Daily Output	Labor-Hours	Unit	Material	Labor	Equipment	Total	Total Incl O&P
0010	**EXTERIOR WOOD DOOR FRAMES AND ACCESSORIES**									
0400	Exterior frame, incl. ext. trim, pine, 5/4 x 4-9/16" deep	2 Carp	375	.043	L.F.	5.55	1.88		7.43	9.20
0420	5-3/16" deep		375	.043		8.85	1.88		10.73	12.85
0440	6-9/16" deep		375	.043		7.70	1.88		9.58	11.55
0600	Oak, 5/4 x 4-9/16" deep		350	.046		11.70	2.02		13.72	16.20
0620	5-3/16" deep		350	.046		12.90	2.02		14.92	17.45
0640	6-9/16" deep		350	.046		15.30	2.02		17.32	20
1000	Sills, 8/4 x 8" deep, oak, no horns		100	.160		4.75	7.05		11.80	16.80
1020	2" horns		100	.160		17.10	7.05		24.15	30.50
1040	3" horns		100	.160		17.10	7.05		24.15	30.50
1100	8/4 x 10" deep, oak, no horns		90	.178		5.30	7.85		13.15	18.65
1120	2" horns		90	.178		21	7.85		28.85	36.50
1140	3" horns		90	.178	↓	21	7.85		28.85	36.50
2000	Exterior, colonial, frame & trim, 3' opng., in-swing, minimum		22	.727	Ea.	330	32		362	420
2010	Average		21	.762		525	33.50		558.50	630
2020	Maximum		20	.800		1,125	35.50		1,160.50	1,300
2100	5'-4" opening, in-swing, minimum		17	.941		395	41.50		436.50	505
2120	Maximum		15	1.067		1,125	47		1,172	1,325
2140	Out-swing, minimum		17	.941		395	41.50		436.50	505
2160	Maximum		15	1.067		1,125	47		1,172	1,325
2400	6'-0" opening, in-swing, minimum		16	1		395	44		439	510
2420	Maximum		10	1.600		1,125	70.50		1,195.50	1,375
2460	Out-swing, minimum		16	1		395	44		439	510
2480	Maximum		10	1.600	↓	1,125	70.50		1,195.50	1,375
2600	For two sidelights, add, minimum		30	.533	Opng.	64	23.50		87.50	109
2620	Maximum		20	.800	"	795	35.50		830.50	935
2700	Custom birch frame, 3'-0" opening		16	1	Ea.	220	44		264	315
2750	6'-0" opening		16	1		330	44		374	440
2900	Exterior, modern, plain trim, 3' opng., in-swing, minimum		26	.615		38.50	27		65.50	87
2920	Average		24	.667		46	29.50		75.50	98.50
2940	Maximum	↓	22	.727	↓	55	32		87	113

06 48 16 – Interior Wood Door Frames

06 48 16.10 Interior Wood Door Jamb and Frames

		Crew	Daily Output	Labor-Hours	Unit	Material	Labor	Equipment	Total	Total Incl O&P
0010	**INTERIOR WOOD DOOR JAMB AND FRAMES**									
3000	Interior frame, pine, 11/16" x 3-5/8" deep	2 Carp	375	.043	L.F.	5.10	1.88		6.98	8.70
3020	4-9/16" deep		375	.043		5.95	1.88		7.83	9.65
3200	Oak, 11/16" x 3-5/8" deep	↓	350	.046	↓	7.90	2.02		9.92	12

06 48 Wood Frames

06 48 16 – Interior Wood Door Frames

06 48 16.10 Interior Wood Door Jamb and Frames		Crew	Daily Output	Labor-Hours	Unit	2012 Bare Costs Material	Labor	Equipment	Total	Total Incl O&P
3220	4-9/16" deep	2 Carp	350	.046	L.F.	7.90	2.02		9.92	12
3240	5-3/16" deep		350	.046		11.85	2.02		13.87	16.35
3400	Walnut, 11/16" x 3-5/8" deep		350	.046		8.25	2.02		10.27	12.40
3420	4-9/16" deep		350	.046		9	2.02		11.02	13.20
3440	5-3/16" deep		350	.046		9	2.02		11.02	13.20
3600	Pocket door frame		16	1	Ea.	74.50	44		118.50	154
3800	Threshold, oak, 5/8" x 3-5/8" deep		200	.080	L.F.	3.56	3.53		7.09	9.70
3820	4-5/8" deep		190	.084		4.13	3.71		7.84	10.65
3840	5-5/8" deep		180	.089		6.50	3.92		10.42	13.60
9000	Minimum labor/equipment charge	1 Carp	4	2	Job		88		88	145

06 49 Wood Screens and Exterior Wood Shutters

06 49 19 – Exterior Wood Shutters

06 49 19.10 Shutters, Exterior		Crew	Daily Output	Labor-Hours	Unit	Material	Labor	Equipment	Total	Total Incl O&P
0010	**SHUTTERS, EXTERIOR**									
0012	Aluminum, louvered, 1'-4" wide, 3'-0" long	1 Carp	10	.800	Pr.	62	35.50		97.50	126
0400	6'-8" long		9	.889		124	39		163	202
1000	Pine, louvered, primed, each 1'-2" wide, 3'-3" long		10	.800		107	35.50		142.50	176
1100	4'-7" long		10	.800		145	35.50		180.50	217
1250	Each 1'-4" wide, 3'-0" long		10	.800		110	35.50		145.50	178
1350	5'-3" long		10	.800		164	35.50		199.50	239
1500	Each 1'-6" wide, 3'-3" long		10	.800		114	35.50		149.50	183
1600	4'-7" long		10	.800		160	35.50		195.50	234
1610	Door blinds, 6'-9" long 1'-3" wide		9	.889		200	39		239	285
1615	1' - 6" wide		9	.889		220	39		259	305
1620	Hemlock, louvered, 1'-2" wide, 5'-7" long		10	.800		175	35.50		210.50	251
1630	Each 1'-4" wide, 2'-2" long		10	.800		110	35.50		145.50	178
1640	3'-0" long		10	.800		110	35.50		145.50	178
1650	3'-3" long		10	.800		118	35.50		153.50	188
1660	3'-11" long		10	.800		134	35.50		169.50	206
1670	4'-3" long		10	.800		131	35.50		166.50	202
1680	5'-3" long		10	.800		163	35.50		198.50	238
1690	5'-11" long		10	.800		185	35.50		220.50	261
1700	Door blinds, 6'-9" long, each 1'-3" wide		9	.889		186	39		225	269
1710	1'-6" wide		9	.889		200	39		239	285
1720	Hemlock, solid raised panel, each 1'-4" wide, 3'-3" long		10	.800		177	35.50		212.50	253
1730	3'-11" long		10	.800		207	35.50		242.50	286
1740	4'-3" long		10	.800		224	35.50		259.50	305
1750	4'-7" long		10	.800		242	35.50		277.50	325
1760	4'-11" long		10	.800		263	35.50		298.50	345
1770	5'-11" long		10	.800		300	35.50		335.50	390
1800	Door blinds, 6'-9" long, each 1'-3" wide		9	.889		335	39		374	435
1900	1'-6" wide		9	.889		370	39		409	470
2500	Polystyrene, solid raised panel, each 1'-4" wide, 3'-3" long		10	.800		41.50	35.50		77	104
2700	4'-7" long		10	.800		52	35.50		87.50	115
4500	Polystyrene, louvered, each 1'-2" wide, 3'-3" long		10	.800		28	35.50		63.50	89
4600	4'-7" long		10	.800		35	35.50		70.50	96.50
4750	5'-3" long		10	.800		40	35.50		75.50	102
4850	6'-8" long		9	.889		47.50	39		86.50	117
6000	Vinyl, louvered, each 1'-2" x 4'-7" long		10	.800		34.50	35.50		70	96
6200	Each 1'-4" x 6'-8" long		9	.889		48.50	39		87.50	118

06 49 Wood Screens and Exterior Wood Shutters

06 49 19 – Exterior Wood Shutters

06 49 19.10 Shutters, Exterior		Crew	Daily Output	Labor-Hours	Unit	Material (2012 Bare Costs)	Labor (2012 Bare Costs)	Equipment (2012 Bare Costs)	Total (2012 Bare Costs)	Total Incl O&P
9000	Minimum labor/equipment charge	1 Carp	4	2	Job		88		88	145

06 51 Structural Plastic Shapes and Plates

06 51 13 – Plastic Lumber

06 51 13.10 Recycled Plastic Lumber

Line	Description		Crew	Daily Output	Labor-Hours	Unit	Material	Labor	Equipment	Total	Total Incl O&P
0010	**RECYCLED PLASTIC LUMBER**										
4000	Sheeting, recycled plastic, black or white, 4′ x 8′ x 1/8″	G	2 Carp	1100	.015	S.F.	1.56	.64		2.20	2.77
4010	4′ x 8′ x 3/16″	G		1100	.015		2.25	.64		2.89	3.53
4020	4′ x 8′ x 1/4″	G		950	.017		2.79	.74		3.53	4.29
4030	4′ x 8′ x 3/8″	G		950	.017		3.89	.74		4.63	5.50
4040	4′ x 8′ x 1/2″	G		900	.018		4.99	.78		5.77	6.80
4050	4′ x 8′ x 5/8″	G		900	.018		7	.78		7.78	9
4060	4′ x 8′ x 3/4″	G		850	.019		8.50	.83		9.33	10.70
4070	Add for colors	G				Ea.	5%			5%	6%
8500	100% recycled plastic, var colors, NLB, 2″ x 2″	G				L.F.	1.62			1.62	1.78
8510	2″ x 4″	G					3.38			3.38	3.72
8520	2″ x 6″	G					5.30			5.30	5.80
8530	2″ x 8″	G					7.25			7.25	8
8540	2″ x 10″	G					10.60			10.60	11.65
8550	5/4″ x 4″	G					3.38			3.38	3.72
8560	5/4″ x 6″	G					2.99			2.99	3.29
8570	1″ x 6″	G					2.58			2.58	2.84
8580	1/2″ x 8″	G					2.82			2.82	3.10
8590	2″ x 10″ T & G	G					7.60			7.60	8.35
8600	3″ x 10″ T & G	G					12.25			12.25	13.45
8610	Add for premium colors	G					20%			20%	22%

06 63 Plastic Railings

06 63 10 – Plastic (PVC) Railings

06 63 10.10 Plastic Railings

Line	Description	Crew	Daily Output	Labor-Hours	Unit	Material	Labor	Equipment	Total	Total Incl O&P
0010	**PLASTIC RAILINGS**									
0100	Horizontal PVC handrail with balusters, 3-1/2″ wide, 36″ high	1 Carp	96	.083	L.F.	23.50	3.67		27.17	31.50
0150	42″ high		96	.083		26	3.67		29.67	35
0200	Angled PVC handrail with balusters, 3-1/2″ wide, 36″ high		72	.111		26.50	4.90		31.40	37
0250	42″ high		72	.111		29.50	4.90		34.40	40.50
0300	Post sleeve for 4 x 4 post		96	.083		12.05	3.67		15.72	19.25
0400	Post cap for 4 x 4 post, flat profile		48	.167	Ea.	10.40	7.35		17.75	23.50
0450	Newel post style profile		48	.167		21	7.35		28.35	35
0500	Raised corbeled profile		48	.167		23	7.35		30.35	37
0550	Post base trim for 4 x 4 post		96	.083		11.50	3.67		15.17	18.65

06 65 10.10 PVC Trim, Exterior		Crew	Daily Output	Labor-Hours	Unit	2012 Bare Costs Material	Labor	Equipment	Total	Total Incl O&P
0010	**PVC TRIM, EXTERIOR**									
0100	Cornerboards, 5/4" x 6" x 6"	1 Carp	240	.033	L.F.	11.10	1.47		12.57	14.60
0110	Door/window casing, 1" x 4"		200	.040		1.31	1.76		3.07	4.33
0120	1" x 6"		200	.040		2.06	1.76		3.82	5.15
0130	1" x 8"		195	.041		2.72	1.81		4.53	5.95
0140	1" x 10"		195	.041		3.45	1.81		5.26	6.75
0150	1" x 12"		190	.042		4.21	1.86		6.07	7.65
0160	5/4" x 4"		195	.041		1.68	1.81		3.49	4.82
0170	5/4" x 6"		195	.041		2.65	1.81		4.46	5.90
0180	5/4" x 8"		190	.042		3.47	1.86		5.33	6.85
0190	5/4" x 10"		190	.042		4.43	1.86		6.29	7.90
0200	5/4" x 12"		185	.043		4.85	1.91		6.76	8.50
0210	Fascia, 1" x 4"		250	.032		1.31	1.41		2.72	3.75
0220	1" x 6"		250	.032		2.06	1.41		3.47	4.58
0230	1" x 8"		225	.036		2.72	1.57		4.29	5.55
0240	1" x 10"		225	.036		3.45	1.57		5.02	6.35
0250	1" x 12"		200	.040		4.21	1.76		5.97	7.50
0260	5/4" x 4"		240	.033		1.68	1.47		3.15	4.26
0270	5/4" x 6"		240	.033		2.65	1.47		4.12	5.35
0280	5/4" x 8"		215	.037		3.47	1.64		5.11	6.50
0290	5/4" x 10"		215	.037		4.43	1.64		6.07	7.55
0300	5/4" x 12"		190	.042		4.85	1.86		6.71	8.40
0310	Frieze, 1" x 4"		250	.032		1.31	1.41		2.72	3.75
0320	1" x 6"		250	.032		2.06	1.41		3.47	4.58
0330	1" x 8"		225	.036		2.72	1.57		4.29	5.55
0340	1" x 10"		225	.036		3.45	1.57		5.02	6.35
0350	1" x 12"		200	.040		4.21	1.76		5.97	7.50
0360	5/4" x 4"		240	.033		1.68	1.47		3.15	4.26
0370	5/4" x 6"		240	.033		2.65	1.47		4.12	5.35
0380	5/4" x 8"		215	.037		3.47	1.64		5.11	6.50
0390	5/4" x 10"		215	.037		4.43	1.64		6.07	7.55
0400	5/4" x 12"		190	.042		4.85	1.86		6.71	8.40
0410	Rake, 1" x 4"		200	.040		1.31	1.76		3.07	4.33
0420	1" x 6"		200	.040		2.06	1.76		3.82	5.15
0430	1" x 8"		190	.042		2.72	1.86		4.58	6.05
0440	1" x 10"		190	.042		3.45	1.86		5.31	6.85
0450	1" x 12"		180	.044		4.21	1.96		6.17	7.85
0460	5/4" x 4"		195	.041		1.68	1.81		3.49	4.82
0470	5/4" x 6"		195	.041		2.65	1.81		4.46	5.90
0480	5/4" x 8"		185	.043		3.47	1.91		5.38	6.95
0490	5/4" x 10"		185	.043		4.43	1.91		6.34	8
0500	5/4" x 12"		175	.046		4.85	2.02		6.87	8.65
0510	Rake trim, 1" x 4"		225	.036		1.31	1.57		2.88	4.01
0520	1" x 6"		225	.036		2.06	1.57		3.63	4.84
0560	5/4" x 4"		220	.036		1.68	1.60		3.28	4.48
0570	5/4" x 6"		220	.036		2.65	1.60		4.25	5.55
0610	Soffit, 1" x 4"	2 Carp	420	.038		1.31	1.68		2.99	4.19
0620	1" x 6"		420	.038		2.06	1.68		3.74	5
0630	1" x 8"		420	.038		2.72	1.68		4.40	5.75
0640	1" x 10"		400	.040		3.45	1.76		5.21	6.70
0650	1" x 12"		400	.040		4.21	1.76		5.97	7.50
0660	5/4" x 4"		410	.039		1.68	1.72		3.40	4.67
0670	5/4" x 6"		410	.039		2.65	1.72		4.37	5.75

06 65 Plastic Simulated Wood Trim

06 65 10 – PVC Trim

06 65 10.10 PVC Trim, Exterior		Crew	Daily Output	Labor-Hours	Unit	2012 Bare Costs Material	Labor	Equipment	Total	Total Incl O&P
0680	5/4" x 8"	2 Carp	410	.039	L.F.	3.47	1.72		5.19	6.65
0690	5/4" x 10"		390	.041		4.43	1.81		6.24	7.85
0700	5/4" x 12"	↓	390	.041	↓	4.85	1.81		6.66	8.30

Division Notes

		CREW	DAILY OUTPUT	LABOR-HOURS	UNIT	2012 BARE COSTS MAT.	LABOR	EQUIP.	TOTAL	TOTAL INCL O&P

Estimating Tips

07 10 00 Dampproofing and Waterproofing

- Be sure of the job specifications before pricing this subdivision. The difference in cost between waterproofing and dampproofing can be great. Waterproofing will hold back standing water. Dampproofing prevents the transmission of water vapor. Also included in this section are vapor retarding membranes.

07 20 00 Thermal Protection

- Insulation and fireproofing products are measured by area, thickness, volume or R-value. Specifications may give only what the specific R-value should be in a certain situation. The estimator may need to choose the type of insulation to meet that R-value.

07 30 00 Steep Slope Roofing
07 40 00 Roofing and Siding Panels

- Many roofing and siding products are bought and sold by the square. One square is equal to an area that measures 100 square feet.

This simple change in unit of measure could create a large error if the estimator is not observant. Accessories necessary for a complete installation must be figured into any calculations for both material and labor.

07 50 00 Membrane Roofing
07 60 00 Flashing and Sheet Metal
07 70 00 Roofing and Wall Specialties and Accessories

- The items in these subdivisions compose a roofing system. No one component completes the installation, and all must be estimated. Built-up or single-ply membrane roofing systems are made up of many products and installation trades. Wood blocking at roof perimeters or penetrations, parapet coverings, reglets, roof drains, gutters, downspouts, sheet metal flashing, skylights, smoke vents, and roof hatches all need to be considered along with the roofing material. Several different installation trades will need to work together on the roofing system. Inherent difficulties in the scheduling and coordination of various trades must be accounted for when estimating labor costs.

07 90 00 Joint Protection

- To complete the weather-tight shell, the sealants and caulkings must be estimated. Where different materials meet—at expansion joints, at flashing penetrations, and at hundreds of other locations throughout a construction project—they provide another line of defense against water penetration. Often, an entire system is based on the proper location and placement of caulking or sealants. The detailed drawings that are included as part of a set of architectural plans show typical locations for these materials. When caulking or sealants are shown at typical locations, this means the estimator must include them for all the locations where this detail is applicable. Be careful to keep different types of sealants separate, and remember to consider backer rods and primers if necessary.

Reference Numbers

Reference numbers are shown in shaded boxes at the beginning of some major classifications. These numbers refer to related items in the Reference Section. The reference information may be an estimating procedure, an alternate pricing method, or technical information.

Note: Not all subdivisions listed here necessarily appear in this publication.

07 01 Operation and Maint. of Thermal and Moisture Protection

07 01 50 – Maintenance of Membrane Roofing

	07 01 50.10 Roof Coatings	Crew	Daily Output	Labor-Hours	Unit	Material	Labor (2012 Bare Costs)	Equipment	Total	Total Incl O&P
0010	**ROOF COATINGS**									
0012	Asphalt, brush grade, material only				Gal.	7.60			7.60	8.35
0200	Asphalt base, fibered aluminum coating G					10.35			10.35	11.40
0300	Asphalt primer, 5 gallon					6.60			6.60	7.30
0600	Coal tar pitch, 200 lb. barrels				Ton	1,200			1,200	1,325
0700	Tar roof cement, 5 gal. lots				Gal.	15			15	16.50
0800	Glass fibered roof & patching cement, 5 gallon				"	10.35			10.35	11.40
0900	Reinforcing glass membrane, 450 S.F./roll				Ea.	55.50			55.50	61
1000	Neoprene roof coating, 5 gal., 2 gal./sq.				Gal.	30			30	33
1100	Roof patch & flashing cement, 5 gallon					7.90			7.90	8.70
1200	Roof resaturant, glass fibered, 3 gal./sq.					7.40			7.40	8.15
1600	Reflective roof coating, white, elastomeric, approx. 50 S.F. per gal.					15.50			15.50	17.05

07 01 90 – Maintenance of Joint Protection

	07 01 90.81 Joint Sealant Replacement	Crew	Daily Output	Labor-Hours	Unit	Material	Labor	Equipment	Total	Total Incl O&P
0010	**JOINT SEALANT REPLACEMENT**									
0050	Control joints in concrete floors/slabs									
0100	Option 1 for joints with hard dry sealant									
0110	Step 1: Sawcut to remove 95% of old sealant									
0112	1/4" wide x 1/2" deep, with single saw blade	C-27	4800	.003	L.F.	.02	.14	.03	.19	.27
0114	3/8" wide x 3/4" deep, with single saw blade		4000	.004		.03	.17	.04	.24	.34
0116	1/2" wide x 1" deep, with double saw blades		3600	.004		.06	.19	.04	.29	.41
0118	3/4" wide x 1-1/2" deep, with double saw blades		3200	.005		.11	.21	.05	.37	.51
0120	Step 2: Water blast joint faces and edges	C-29	2500	.003			.11	.03	.14	.21
0130	Step 3: Air blast joint faces and edges	C-28	2000	.004			.17	.01	.18	.28
0140	Step 4: Sand blast joint faces and edges	E-11	2000	.016			.63	.11	.74	1.23
0150	Step 5: Air blast joint faces and edges	C-28	2000	.004			.17	.01	.18	.28
0200	Option 2 for joints with soft pliable sealant									
0210	Step 1: Plow joint with rectangular blade	B-62	2600	.009	L.F.		.35	.06	.41	.64
0220	Step 2: Sawcut to re-face joint faces									
0222	1/4" wide x 1/2" deep, with single saw blade	C-27	2400	.007	L.F.	.02	.28	.06	.36	.53
0224	3/8" wide x 3/4" deep, with single saw blade		2000	.008		.04	.34	.07	.45	.65
0226	1/2" wide x 1" deep, with double saw blades		1800	.009		.07	.38	.08	.53	.76
0228	3/4" wide x 1-1/2" deep, with double saw blades		1600	.010		.15	.42	.09	.66	.94
0230	Step 3: Water blast joint faces and edges	C-29	2500	.003			.11	.03	.14	.21
0240	Step 4: Air blast joint faces and edges	C-28	2000	.004			.17	.01	.18	.28
0250	Step 5: Sand blast joint faces and edges	E-11	2000	.016			.63	.11	.74	1.23
0260	Step 6: Air blast joint faces and edges	C-28	2000	.004			.17	.01	.18	.28
0290	For saw cutting new control joints, see Section 03 35 29.35									
8910	For backer rod, see Section 07 91 23.10									
8920	For joint sealant, see Sections 03 15 05.25 or 07 92 13.20									

07 05 05 – Selective Demolition

07 05 05.10 Selective Demo., Thermal and Moist. Protection		Crew	Daily Output	Labor-Hours	Unit	Material	2012 Bare Costs Labor	Equipment	Total	Total Incl O&P
0010	SELECTIVE DEMO., THERMAL AND MOISTURE PROTECTION									
0020	Caulking/sealant, to 1" x 1" joint R024119-10	1 Clab	600	.013	L.F.		.47		.47	.77
0120	Downspouts, including hangers		350	.023	"		.80		.80	1.32
0220	Flashing, sheet metal		290	.028	S.F.		.97		.97	1.59
0420	Gutters, aluminum or wood, edge hung		240	.033	L.F.		1.17		1.17	1.92
0520	Built-in		100	.080	"		2.81		2.81	4.60
0620	Insulation, air/vapor barrier		3500	.002	S.F.		.08		.08	.13
0670	Batts or blankets		1400	.006	C.F.		.20		.20	.33
0720	Foamed or sprayed in place	2 Clab	1000	.016	B.F.		.56		.56	.92
0770	Loose fitting	1 Clab	3000	.003	C.F.		.09		.09	.15
0870	Rigid board		3450	.002	B.F.		.08		.08	.13
1120	Roll roofing, cold adhesive		12	.667	Sq.		23.50		23.50	38.50
1170	Roof accessories, adjustable metal chimney flashing		9	.889	Ea.		31		31	51
1325	Plumbing vent flashing		32	.250	"		8.80		8.80	14.40
1375	Ridge vent strip, aluminum		310	.026	L.F.		.91		.91	1.49
1620	Skylight to 10 S.F.		8	1	Ea.		35		35	57.50
2120	Roof edge, aluminum soffit and fascia		570	.014	L.F.		.49		.49	.81
2170	Concrete coping, up to 12" wide	2 Clab	160	.100			3.51		3.51	5.75
2220	Drip edge	1 Clab	1650	.005			.17		.17	.28
2270	Gravel stop		1650	.005			.17		.17	.28
2370	Sheet metal coping, up to 12" wide		240	.033			1.17		1.17	1.92
2470	Roof insulation board, over 2" thick	B-2	7800	.005	B.F.		.18		.18	.30
2520	Up to 2" thick	"	3900	.010	S.F.		.36		.36	.60
2620	Roof ventilation, louvered gable vent	1 Clab	16	.500	Ea.		17.55		17.55	29
2670	Remove, roof hatch	G-3	15	2.133			93		93	149
2675	Rafter vents	1 Clab	960	.008			.29		.29	.48
2720	Soffit vent and/or fascia vent		575	.014	L.F.		.49		.49	.80
2775	Soffit vent strip, aluminum, 3" to 4" wide		160	.050			1.76		1.76	2.88
2820	Roofing accessories, shingle moulding, to 1" x 4"		1600	.005			.18		.18	.29
2870	Cant strip	B-2	2000	.020			.71		.71	1.16
2920	Concrete block walkway	1 Clab	230	.035			1.22		1.22	2
3070	Roofing, felt paper, 15#		70	.114	Sq.		4.01		4.01	6.60
3125	#30 felt		30	.267	"		9.35		9.35	15.35
3170	Asphalt shingles, 1 layer	B-2	3500	.011	S.F.		.41		.41	.67
3180	2 layers		1750	.023	"		.81		.81	1.33
3370	Modified bitumen		26	1.538	Sq.		54.50		54.50	89.50
3420	Built-up, no gravel, 3 ply		25	1.600			57		57	93
3470	4 ply		21	1.905			67.50		67.50	111
3620	5 ply		1600	.025	S.F.		.89		.89	1.46
3720	5 ply, with gravel		890	.045			1.60		1.60	2.62
3725	Gravel removal, minimum		5000	.008			.28		.28	.47
3730	Maximum		2000	.020			.71		.71	1.16
3870	Fiberglass sheet		1200	.033			1.18		1.18	1.94
4120	Slate shingles		1900	.021			.75		.75	1.23
4170	Ridge shingles, clay or slate		2000	.020	L.F.		.71		.71	1.16
4320	Single ply membrane, attached at seams		52	.769	Sq.		27.50		27.50	45
4370	Ballasted		75	.533			18.95		18.95	31
4420	Fully adhered		39	1.026			36.50		36.50	59.50
4550	Roof hatch, 2'-6" x 3'-0"	1 Clab	10	.800	Ea.		28		28	46
4670	Wood shingles	B-2	2200	.018	S.F.		.65		.65	1.06
4820	Sheet metal roofing	"	2150	.019			.66		.66	1.08
4970	Siding, horizontal wood clapboards	1 Clab	380	.021			.74		.74	1.21
5025	Exterior insulation finish system	"	120	.067			2.34		2.34	3.84

07 05 Common Work Results for Thermal and Moisture Protection

07 05 05 – Selective Demolition

07 05 05.10 Selective Demo., Thermal and Moist. Protection		Crew	Daily Output	Labor-Hours	Unit	Material	2012 Bare Costs Labor	Equipment	Total	Total Incl O&P
5070	Tempered hardboard, remove and reset	1 Carp	380	.021	S.F.		.93		.93	1.52
5120	Tempered hardboard sheet siding	"	375	.021			.94		.94	1.54
5170	Metal, corner strips	1 Clab	850	.009	L.F.		.33		.33	.54
5225	Horizontal strips		444	.018	S.F.		.63		.63	1.04
5320	Vertical strips		400	.020			.70		.70	1.15
5520	Wood shingles		350	.023			.80		.80	1.32
5620	Stucco siding		360	.022			.78		.78	1.28
5670	Textured plywood		725	.011			.39		.39	.63
5720	Vinyl siding		510	.016			.55		.55	.90
5770	Corner strips		900	.009	L.F.		.31		.31	.51
5870	Wood, boards, vertical		400	.020	S.F.		.70		.70	1.15
5920	Waterproofing, protection/drain board	2 Clab	3900	.004	B.F.		.14		.14	.24
5970	Over 1/2" thick		1750	.009	S.F.		.32		.32	.53
6020	To 1/2" thick		2000	.008	"		.28		.28	.46
9000	Minimum labor/equipment charge	1 Clab	2	4	Job		140		140	230

07 11 Dampproofing

07 11 13 – Bituminous Dampproofing

07 11 13.10 Bituminous Asphalt Coating

		Crew	Daily Output	Labor-Hours	Unit	Material	Labor	Equipment	Total	Total Incl O&P
0010	**BITUMINOUS ASPHALT COATING**									
0030	Brushed on, below grade, 1 coat	1 Rofc	665	.012	S.F.	.19	.45		.64	1.01
0100	2 coat		500	.016		.38	.60		.98	1.48
0300	Sprayed on, below grade, 1 coat		830	.010		.19	.36		.55	.85
0400	2 coat		500	.016		.37	.60		.97	1.47
0600	Troweled on, asphalt with fibers, 1/16" thick		500	.016		.45	.60		1.05	1.56
0700	1/8" thick		400	.020		.80	.75		1.55	2.21
1000	1/2" thick		350	.023		2.59	.86		3.45	4.37
9000	Minimum labor/equipment charge		3	2.667	Job		99.50		99.50	177

07 11 16 – Cementitious Dampproofing

07 11 16.20 Cementitious Parging

		Crew	Daily Output	Labor-Hours	Unit	Material	Labor	Equipment	Total	Total Incl O&P
0010	**CEMENTITIOUS PARGING**									
0020	Portland cement, 2 coats, 1/2" thick	D-1	250	.064	S.F.	.31	2.55		2.86	4.45
0100	Waterproofed Portland cement, 1/2" thick, 2 coats	"	250	.064	"	3.30	2.55		5.85	7.75

07 12 Built-up Bituminous Waterproofing

07 12 13 – Built-Up Asphalt Waterproofing

07 12 13.20 Membrane Waterproofing

		Crew	Daily Output	Labor-Hours	Unit	Material	Labor	Equipment	Total	Total Incl O&P
0010	**MEMBRANE WATERPROOFING**									
0012	On slabs, 1 ply, felt, mopped	G-1	3000	.019	S.F.	.36	.65	.15	1.16	1.73
0015	On walls, 1 ply, felt, mopped		3000	.019		.36	.65	.15	1.16	1.73
0100	On slabs, 1 ply, glass fiber fabric, mopped		2100	.027		.37	.93	.22	1.52	2.31
0105	On walls, 1 ply, glass fiber fabric, mopped		2100	.027		.37	.93	.22	1.52	2.31
0300	On slabs, 2 ply, felt, mopped		2500	.022		.72	.78	.18	1.68	2.39
0305	On walls, 2 ply, felt, mopped		2500	.022		.72	.78	.18	1.68	2.39
0400	On slabs, 2 ply, glass fiber fabric, mopped		1650	.034		.83	1.19	.27	2.29	3.33
0405	On walls, 2 ply, glass fiber fabric, mopped		1650	.034		.83	1.19	.27	2.29	3.33
0600	On slabs, 3 ply, felt, mopped		2100	.027		1.09	.93	.22	2.24	3.09
0605	On walls, 3 ply, felt, mopped		2100	.027		1.09	.93	.22	2.24	3.09
0700	On slabs, 3 ply, glass fiber fabric, mopped		1550	.036		1.12	1.26	.29	2.67	3.80

07 12 Built-up Bituminous Waterproofing

07 12 13 – Built-Up Asphalt Waterproofing

07 12 13.20 Membrane Waterproofing		Crew	Daily Output	Labor-Hours	Unit	2012 Bare Costs Material	Labor	Equipment	Total	Total Incl O&P
0705	On walls, 3 ply, glass fiber fabric, mopped	G-1	1550	.036	S.F.	1.12	1.26	.29	2.67	3.80
0710	Asphaltic hardboard protection board, 1/8" thick	2 Rofc	500	.032		.44	1.20		1.64	2.61
0715	1/4" thick		450	.036		.70	1.33		2.03	3.13
1000	1/4" EPS membrane protection board		3500	.005		.34	.17		.51	.67
1050	3/8" thick		3500	.005		.38	.17		.55	.71
1060	1/2" thick		3500	.005		.41	.17		.58	.75
1070	Fiberglass fabric, black, 20/10 mesh		116	.138	Sq.	18.95	5.15		24.10	30
9000	Minimum labor/equipment charge		2	8	Job		299		299	530

07 13 Sheet Waterproofing

07 13 53 – Elastomeric Sheet Waterproofing

07 13 53.10 Elastomeric Sheet Waterproofing and Access.

Line	Description	Crew	Daily Output	Labor-Hours	Unit	Material	Labor	Equipment	Total	Total Incl O&P
0010	**ELASTOMERIC SHEET WATERPROOFING AND ACCESS.**									
0090	EPDM, plain, 45 mils thick	2 Rofc	580	.028	S.F.	1.36	1.03		2.39	3.32
0100	60 mils thick		570	.028		1.45	1.05		2.50	3.46
0300	Nylon reinforced sheets, 45 mils thick		580	.028		1.37	1.03		2.40	3.33
0400	60 mils thick		570	.028		1.47	1.05		2.52	3.48
0600	Vulcanizing splicing tape for above, 2" wide				C.L.F.	50			50	55
0700	4" wide				"	113			113	124
0900	Adhesive, bonding, 60 S.F. per gal.				Gal.	22.50			22.50	24.50
1000	Splicing, 75 S.F. per gal.				"	32			32	35
1200	Neoprene sheets, plain, 45 mils thick	2 Rofc	580	.028	S.F.	1.67	1.03		2.70	3.67
1300	60 mils thick		570	.028		2.35	1.05		3.40	4.46
1500	Nylon reinforced, 45 mils thick		580	.028		1.93	1.03		2.96	3.95
1600	60 mils thick		570	.028		2.81	1.05		3.86	4.96
1800	120 mils thick		500	.032		5.60	1.20		6.80	8.30
1900	Adhesive, splicing, 150 S.F. per gal. per coat				Gal.	32			32	35
2100	Fiberglass reinforced, fluid applied, 1/8" thick	2 Rofc	500	.032	S.F.	1.59	1.20		2.79	3.88
2200	Polyethylene and rubberized asphalt sheets, 60 mils thick		550	.029		.63	1.09		1.72	2.62
2400	Polyvinyl chloride sheets, plain, 10 mils thick		580	.028		.17	1.03		1.20	2.02
2500	20 mils thick		570	.028		.26	1.05		1.31	2.16
2700	30 mils thick		560	.029		.36	1.07		1.43	2.30
3000	Adhesives, trowel grade, 40-100 S.F. per gal.				Gal.	23			23	25.50
3100	Brush grade, 100-250 S.F. per gal.				"	18.50			18.50	20.50
3300	Bitumen modified polyurethane, fluid applied, 55 mils thick	2 Rofc	665	.024	S.F.	.82	.90		1.72	2.50
9000	Minimum labor/equipment charge	"	2	8	Job		299		299	530

07 16 Cementitious and Reactive Waterproofing

07 16 16 – Crystalline Waterproofing

07 16 16.20 Cementitious Waterproofing

Line	Description	Crew	Daily Output	Labor-Hours	Unit	Material	Labor	Equipment	Total	Total Incl O&P
0010	**CEMENTITIOUS WATERPROOFING**									
0020	1/8" application, sprayed on	G-2A	1000	.024	S.F.	2	.80	.63	3.43	4.28
0050	4 coat cementitious metallic slurry	1 Cofi	1.20	6.667	C.S.F.	26.50	282		308.50	475

07 19 Water Repellents

07 19 19 – Silicone Water Repellents

07 19 19.10 Silicone Based Water Repellents		Crew	Daily Output	Labor-Hours	Unit	Material	2012 Bare Costs Labor	Equipment	Total	Total Incl O&P
0010	**SILICONE BASED WATER REPELLENTS**									
0020	Water base liquid, roller applied	2 Rofc	7000	.002	S.F.	.89	.09		.98	1.13
0200	Silicone or stearate, sprayed on CMU, 1 coat	1 Rofc	4000	.002		.33	.07		.40	.49
0300	2 coats		3000	.003		.66	.10		.76	.91
9000	Minimum labor/equipment charge		3	2.667	Job		99.50		99.50	177

07 21 Thermal Insulation

07 21 13 – Board Insulation

07 21 13.10 Rigid Insulation

			Crew	Daily Output	Labor-Hours	Unit	Material	Labor	Equipment	Total	Total Incl O&P
0010	**RIGID INSULATION**, for walls										
0040	Fiberglass, 1.5#/C.F., unfaced, 1" thick, R4.1	G	1 Carp	1000	.008	S.F.	.25	.35		.60	.86
0060	1-1/2" thick, R6.2	G		1000	.008		.37	.35		.72	.99
0080	2" thick, R8.3	G		1000	.008		.41	.35		.76	1.03
0120	3" thick, R12.4	G		800	.010		.51	.44		.95	1.28
0370	3#/C.F., unfaced, 1" thick, R4.3	G		1000	.008		.48	.35		.83	1.11
0390	1-1/2" thick, R6.5	G		1000	.008		.72	.35		1.07	1.37
0400	2" thick, R8.7	G		890	.009		.96	.40		1.36	1.71
0420	2-1/2" thick, R10.9	G		800	.010		1.01	.44		1.45	1.83
0440	3" thick, R13	G		800	.010		1.46	.44		1.90	2.33
0520	Foil faced, 1" thick, R4.3	G		1000	.008		.83	.35		1.18	1.49
0540	1-1/2" thick, R6.5	G		1000	.008		1.24	.35		1.59	1.94
0560	2" thick, R8.7	G		890	.009		1.55	.40		1.95	2.36
0580	2-1/2" thick, R10.9	G		800	.010		1.82	.44		2.26	2.72
0600	3" thick, R13	G		800	.010		2	.44		2.44	2.92
1600	Isocyanurate, 4' x 8' sheet, foil faced, both sides										
1610	1/2" thick	G	1 Carp	800	.010	S.F.	.29	.44		.73	1.04
1620	5/8" thick	G		800	.010		.32	.44		.76	1.07
1630	3/4" thick	G		800	.010		.33	.44		.77	1.08
1640	1" thick	G		800	.010		.51	.44		.95	1.28
1650	1-1/2" thick	G		730	.011		.64	.48		1.12	1.49
1660	2" thick	G		730	.011		.78	.48		1.26	1.65
1670	3" thick	G		730	.011		1.78	.48		2.26	2.75
1680	4" thick	G		730	.011		2	.48		2.48	2.99
1700	Perlite, 1" thick, R2.77	G		800	.010		.30	.44		.74	1.05
1750	2" thick, R5.55	G		730	.011		.60	.48		1.08	1.45
1900	Extruded polystyrene, 25 PSI compressive strength, 1" thick, R5	G		800	.010		.50	.44		.94	1.27
1940	2" thick R10	G		730	.011		1	.48		1.48	1.89
1960	3" thick, R15	G		730	.011		1.44	.48		1.92	2.37
2100	Expanded polystyrene, 1" thick, R3.85	G		800	.010		.28	.44		.72	1.03
2120	2" thick, R7.69	G		730	.011		.56	.48		1.04	1.41
2140	3" thick, R11.49	G		730	.011		.84	.48		1.32	1.71
9000	Minimum labor/equipment charge			4	2	Job		88		88	145

07 21 13.13 Foam Board Insulation

			Crew	Daily Output	Labor-Hours	Unit	Material	Labor	Equipment	Total	Total Incl O&P
0010	**FOAM BOARD INSULATION**										
0600	Polystyrene, expanded, 1" thick, R4	G	1 Carp	680	.012	S.F.	.28	.52		.80	1.16
0700	2" thick, R8	G		675	.012	"	.56	.52		1.08	1.48
9000	Minimum labor/equipment charge			4	2	Job		88		88	145

07 21 16 – Blanket Insulation

07 21 16.10 Blanket Insulation for Floors/Ceilings

	07 21 16.10 Blanket Insulation for Floors/Ceilings		Crew	Daily Output	Labor-Hours	Unit	2012 Bare Costs Material	Labor	Equipment	Total	Total Incl O&P
0010	**BLANKET INSULATION FOR FLOORS/CEILINGS**										
0020	Including spring type wire fasteners										
2000	Fiberglass, blankets or batts, paper or foil backing										
2100	3-1/2″ thick, R13	G	1 Carp	700	.011	S.F.	.38	.50		.88	1.25
2150	6-1/4″ thick, R19	G		600	.013		.43	.59		1.02	1.43
2210	9-1/2″ thick, R30	G		500	.016		.66	.71		1.37	1.89
2220	12″ thick, R38	G		475	.017		.72	.74		1.46	2.01
3000	Unfaced, 3-1/2″ thick, R13	G		600	.013		.31	.59		.90	1.30
3010	6-1/4″ thick, R19	G		500	.016		.35	.71		1.06	1.55
3020	9-1/2″ thick, R30	G		450	.018		.62	.78		1.40	1.97
3030	12″ thick, R38	G		425	.019		.71	.83		1.54	2.14
9000	Minimum labor/equipment charge			4	2	Job		88		88	145

07 21 16.20 Blanket Insulation for Walls

	07 21 16.20 Blanket Insulation for Walls		Crew	Daily Output	Labor-Hours	Unit	Material	Labor	Equipment	Total	Total Incl O&P
0010	**BLANKET INSULATION FOR WALLS**										
0020	Kraft faced fiberglass, 3-1/2″ thick, R11, 15″ wide	G	1 Carp	1350	.006	S.F.	.26	.26		.52	.72
0030	23″ wide	G		1600	.005		.26	.22		.48	.65
0060	R13, 11″ wide	G		1150	.007		.33	.31		.64	.86
0080	15″ wide	G		1350	.006		.33	.26		.59	.79
0100	23″ wide	G		1600	.005		.33	.22		.55	.72
0110	R15, 11″ wide	G		1150	.007		.43	.31		.74	.97
0120	15″ wide	G		1350	.006		.43	.26		.69	.90
0130	23″ wide	G		1600	.005		.43	.22		.65	.83
0140	6″ thick, R19, 11″ wide	G		1150	.007		.38	.31		.69	.92
0160	15″ wide	G		1350	.006		.38	.26		.64	.85
0180	23″ wide	G		1600	.005		.38	.22		.60	.78
0182	R21, 11″ wide	G		1150	.007		.52	.31		.83	1.07
0184	15″ wide	G		1350	.006		.52	.26		.78	1
0186	23″ wide	G		1600	.005		.52	.22		.74	.93
0188	9″ thick, R30, 11″ wide	G		985	.008		.62	.36		.98	1.27
0200	15″ wide	G		1150	.007		.62	.31		.93	1.18
0220	23″ wide	G		1350	.006		.62	.26		.88	1.11
0230	12″ thick, R38, 11″ wide	G		985	.008		.72	.36		1.08	1.38
0240	15″ wide	G		1150	.007		.72	.31		1.03	1.29
0260	23″ wide	G		1350	.006		.72	.26		.98	1.22
0410	Foil faced fiberglass, 3-1/2″ thick, R13, 11″ wide	G		1150	.007		.34	.31		.65	.87
0420	15″ wide	G		1350	.006		.34	.26		.60	.80
0440	23″ wide	G		1600	.005		.34	.22		.56	.73
0442	R15, 11″ wide	G		1150	.007		.44	.31		.75	.98
0444	15″ wide	G		1350	.006		.44	.26		.70	.91
0446	23″ wide	G		1600	.005		.44	.22		.66	.84
0448	6″ thick, R19, 11″ wide	G		1150	.007		.40	.31		.71	.94
0460	15″ wide	G		1350	.006		.40	.26		.66	.87
0480	23″ wide	G		1600	.005		.40	.22		.62	.80
0482	R21, 11″ wide	G		1150	.007		.55	.31		.86	1.11
0484	15″ wide	G		1350	.006		.55	.26		.81	1.04
0486	23″ wide	G		1600	.005		.55	.22		.77	.97
0488	9″ thick, R30, 11″ wide	G		985	.008		.65	.36		1.01	1.31
0500	15″ wide	G		1150	.007		.65	.31		.96	1.22
0550	23″ wide	G		1350	.006		.65	.26		.91	1.15
0560	12″ thick, R38, 11″ wide	G		985	.008		.75	.36		1.11	1.42
0570	15″ wide	G		1150	.007		.75	.31		1.06	1.33
0580	23″ wide	G		1350	.006		.75	.26		1.01	1.26

07 21 Thermal Insulation

07 21 16 – Blanket Insulation

07 21 16.20 Blanket Insulation for Walls			Crew	Daily Output	Labor-Hours	Unit	Material	2012 Bare Costs Labor	Equipment	Total	Total Incl O&P
0620	Unfaced fiberglass, 3-1/2" thick, R13, 11" wide	G	1 Carp	1150	.007	S.F.	.31	.31		.62	.84
0820	15" wide	G		1350	.006		.31	.26		.57	.77
0830	23" wide	G		1600	.005		.31	.22		.53	.70
0832	R15, 11" wide	G		1150	.007		.41	.31		.72	.95
0836	23" wide	G		1600	.005		.41	.22		.63	.81
0838	6" thick, R19, 11" wide	G		1150	.007		.35	.31		.66	.89
0860	15" wide	G		1150	.007		.35	.31		.66	.89
0880	23" wide	G		1350	.006		.35	.26		.61	.82
0882	R21, 11" wide	G		1150	.007		.50	.31		.81	1.05
0886	15" wide	G		1350	.006		.50	.26		.76	.98
0888	23" wide	G		1600	.005		.50	.22		.72	.91
0890	9" thick, R30, 11" wide	G		985	.008		.60	.36		.96	1.25
0900	15" wide	G		1150	.007		.60	.31		.91	1.16
0920	23" wide	G		1350	.006		.60	.26		.86	1.09
0930	12" thick, R38, 11" wide	G		985	.008		.71	.36		1.07	1.37
0940	15" wide	G		1000	.008		.71	.35		1.06	1.36
0960	23" wide	G		1150	.007		.71	.31		1.02	1.28
1300	Mineral fiber batts, kraft faced										
1320	3-1/2" thick, R12	G	1 Carp	1600	.005	S.F.	.38	.22		.60	.78
1340	6" thick, R19	G		1600	.005		.38	.22		.60	.78
1380	10" thick, R30	G		1350	.006		.62	.26		.88	1.11
1850	Friction fit wire insulation supports, 16" O.C.			960	.008	Ea.	.05	.37		.42	.66
9000	Minimum labor/equipment charge			4	2	Job		88		88	145

07 21 23 – Loose-Fill Insulation

07 21 23.10 Poured Loose-Fill Insulation			Crew	Daily Output	Labor-Hours	Unit	Material	Labor	Equipment	Total	Total Incl O&P
0010	**POURED LOOSE-FILL INSULATION**										
0020	Cellulose fiber, R3.8 per inch	G	1 Carp	200	.040	C.F.	.53	1.76		2.29	3.48
0040	Ceramic type (perlite), R3.2 per inch	G		200	.040		5	1.76		6.76	8.40
0080	Fiberglass wool, R4 per inch	G		200	.040		.10	1.76		1.86	3.01
0100	Mineral wool, R3 per inch	G		200	.040		.39	1.76		2.15	3.32
0300	Polystyrene, R4 per inch	G		200	.040		3.09	1.76		4.85	6.30
0400	Perlite, R2.7 per inch	G		200	.040		5	1.76		6.76	8.40
9000	Minimum labor/equipment charge			4	2	Job		88		88	145

07 21 23.20 Masonry Loose-Fill Insulation			Crew	Daily Output	Labor-Hours	Unit	Material	Labor	Equipment	Total	Total Incl O&P
0010	**MASONRY LOOSE-FILL INSULATION**, vermiculite or perlite										
0100	In cores of concrete block, 4" thick wall, .115 C.F./S.F.	G	D-1	4800	.003	S.F.	.58	.13		.71	.84
0200	6" thick wall, .175 C.F./S.F.	G		3000	.005		.88	.21		1.09	1.30
0300	8" thick wall, .258 C.F./S.F.	G		2400	.007		1.29	.27		1.56	1.85
0400	10" thick wall, .340 C.F./S.F.	G		1850	.009		1.70	.34		2.04	2.42
0500	12" thick wall, .422 C.F./S.F.	G		1200	.013		2.11	.53		2.64	3.18
0550	For sand fill, deduct from above	G					70%				
0600	Poured cavity wall, vermiculite or perlite, water repellant	G	D-1	250	.064	C.F.	5	2.55		7.55	9.60
0700	Foamed in place, urethane in 2-5/8" cavity	G	G-2A	1035	.023	S.F.	1.29	.78	.61	2.68	3.43
0800	For each 1" added thickness, add	G	"	2372	.010	"	.42	.34	.27	1.03	1.34

07 21 26 – Blown Insulation

07 21 26.10 Blown Insulation			Crew	Daily Output	Labor-Hours	Unit	Material	Labor	Equipment	Total	Total Incl O&P
0010	**BLOWN INSULATION** Ceilings, with open access										
0020	Cellulose, 3-1/2" thick, R13	G	G-4	5000	.005	S.F.	.18	.17	.07	.42	.56
0030	5-3/16" thick, R19	G		3800	.006		.27	.23	.09	.59	.77
0050	6-1/2" thick, R22	G		3000	.008		.35	.29	.12	.76	.98
0100	8-11/16" thick, R30	G		2600	.009		.47	.33	.13	.93	1.20

07 21 Thermal Insulation

07 21 26 – Blown Insulation

07 21 26.10 Blown Insulation			Crew	Daily Output	Labor-Hours	Unit	2012 Bare Costs Material	Labor	Equipment	Total	Total Incl O&P
1000	Fiberglass, 5.5" thick, R11	G	G-4	3800	.006	S.F.	.04	.23	.09	.36	.51
1050	6" thick, R12	G		3000	.008		.05	.29	.12	.46	.66
1100	8.8" thick, R19	G		2200	.011		.06	.39	.16	.61	.88
1200	10" thick, R22	G		1800	.013		.07	.48	.19	.74	1.07
1300	11.5" thick, R26	G		1500	.016		.09	.57	.23	.89	1.30
1350	13" thick, R30	G		1400	.017		.10	.61	.25	.96	1.38
1450	16" thick, R38	G		1145	.021		.13	.75	.30	1.18	1.70
1500	20" thick, R49	G		920	.026		.17	.93	.38	1.48	2.14
9000	Minimum labor/equipment charge			4	6	Job		215	87	302	445

07 21 27 – Reflective Insulation

07 21 27.10 Reflective Insulation Options			Crew	Daily Output	Labor-Hours	Unit	Material	Labor	Equipment	Total	Total Incl O&P
0010	**REFLECTIVE INSULATION OPTIONS**										
0020	Aluminum foil on reinforced scrim	G	1 Carp	19	.421	C.S.F.	14.20	18.55		32.75	46
0100	Reinforced with woven polyolefin	G		19	.421		18	18.55		36.55	50.50
0500	With single bubble air space, R8.8	G		15	.533		28	23.50		51.50	69.50
0600	With double bubble air space, R9.8	G		15	.533		29	23.50		52.50	70
9000	Minimum labor/equipment charge			4	2	Job		88		88	145

07 21 29 – Sprayed Insulation

07 21 29.10 Sprayed-On Insulation			Crew	Daily Output	Labor-Hours	Unit	Material	Labor	Equipment	Total	Total Incl O&P
0010	**SPRAYED-ON INSULATION**										
0020	Fibrous/cementitious, finished wall, 1" thick, R3.7	G	G-2	2050	.012	S.F.	.25	.43	.06	.74	1.05
0100	Attic, 5.2" thick, R19	G		1550	.015	"	.44	.57	.08	1.09	1.49
0200	Fiberglass, R4 per inch, vertical	G		2050	.012	B.F.	.15	.43	.06	.64	.94
0210	Horizontal	G		1550	.015	"	.15	.57	.08	.80	1.18
0300	Closed cell, spray polyurethane foam, 2 pounds per cubic foot density										
0310	1" thick	G	G-2A	6000	.004	S.F.	.42	.13	.11	.66	.81
0320	2" thick	G		3000	.008		.83	.27	.21	1.31	1.61
0330	3" thick	G		2000	.012		1.25	.40	.32	1.97	2.42
0335	3-1/2" thick	G		1715	.014		1.46	.47	.37	2.30	2.81
0340	4" thick	G		1500	.016		1.67	.54	.42	2.63	3.22
0350	5" thick	G		1200	.020		2.08	.67	.53	3.28	4.03
0355	5-1/2" thick	G		1090	.022		2.29	.74	.58	3.61	4.43
0360	6" thick	G		1000	.024		2.50	.80	.63	3.93	4.83
9000	Minimum labor/equipment charge		G-2	2	12	Job		445	63.50	508.50	785

07 22 Roof and Deck Insulation

07 22 16 – Roof Board Insulation

07 22 16.10 Roof Deck Insulation			Crew	Daily Output	Labor-Hours	Unit	Material	Labor	Equipment	Total	Total Incl O&P
0010	**ROOF DECK INSULATION**										
0020	Fiberboard low density, 1/2" thick R1.39	G	1 Rofc	1000	.008	S.F.	.26	.30		.56	.82
0030	1" thick R2.78	G		800	.010		.46	.37		.83	1.17
0080	1 1/2" thick R4.17	G		800	.010		.70	.37		1.07	1.43
0100	2" thick R5.56	G		800	.010		.93	.37		1.30	1.68
0110	Fiberboard high density, 1/2" thick R1.3	G		1000	.008		.24	.30		.54	.79
0120	1" thick R2.5	G		800	.010		.48	.37		.85	1.19
0130	1-1/2" thick R3.8	G		800	.010		.72	.37		1.09	1.45
0200	Fiberglass, 3/4" thick R2.78	G		1000	.008		.56	.30		.86	1.15
0400	15/16" thick R3.70	G		1000	.008		.74	.30		1.04	1.34
0460	1-1/16" thick R4.17	G		1000	.008		.93	.30		1.23	1.55
0600	1-5/16" thick R5.26	G		1000	.008		1.26	.30		1.56	1.92

07 22 Roof and Deck Insulation

07 22 16 – Roof Board Insulation

07 22 16.10 Roof Deck Insulation			Crew	Daily Output	Labor-Hours	Unit	2012 Bare Costs Material	Labor	Equipment	Total	Total Incl O&P
0650	2-1/16" thick R8.33	G	1 Rofc	800	.010	S.F.	1.33	.37		1.70	2.12
0700	2-7/16" thick R10	G		800	.010		1.54	.37		1.91	2.35
1650	Perlite, 1/2" thick R1.32	G		1050	.008		.34	.29		.63	.88
1655	3/4" thick R2.08	G		800	.010		.37	.37		.74	1.07
1660	1" thick R2.78	G		800	.010		.46	.37		.83	1.17
1670	1-1/2" thick R4.17	G		800	.010		.48	.37		.85	1.19
1680	2" thick R5.56	G		700	.011		.80	.43		1.23	1.64
1685	2-1/2" thick R6.67	G		700	.011		.95	.43		1.38	1.81
1690	Tapered for drainage	G		800	.010	B.F.	.73	.37		1.10	1.46
1700	Polyisocyanurate, 2#/C.F. density, 3/4" thick	G		1500	.005	S.F.	.44	.20		.64	.83
1705	1" thick	G		1400	.006		.51	.21		.72	.94
1715	1-1/2" thick	G		1250	.006		.66	.24		.90	1.16
1725	2" thick	G		1100	.007		.78	.27		1.05	1.34
1735	2-1/2" thick	G		1050	.008		1.01	.29		1.30	1.62
1745	3" thick	G		1000	.008		1.23	.30		1.53	1.88
1755	3-1/2" thick	G		1000	.008		1.95	.30		2.25	2.68
1765	Tapered for drainage	G		1400	.006	B.F.	.52	.21		.73	.95
1900	Extruded Polystyrene										
1910	15 PSI compressive strength, 1" thick, R5	G	1 Rofc	1500	.005	S.F.	.44	.20		.64	.83
1920	2" thick, R10	G		1250	.006		.56	.24		.80	1.05
1930	3" thick R15	G		1000	.008		1.15	.30		1.45	1.80
1932	4" thick R20	G		1000	.008		1.52	.30		1.82	2.20
1934	Tapered for drainage	G		1500	.005	B.F.	.51	.20		.71	.91
1940	25 PSI compressive strength, 1" thick R5	G		1500	.005	S.F.	.61	.20		.81	1.02
1942	2" thick R10	G		1250	.006		1.15	.24		1.39	1.70
1944	3" thick R15	G		1000	.008		1.75	.30		2.05	2.46
1946	4" thick R20	G		1000	.008		2.50	.30		2.80	3.28
1948	Tapered for drainage	G		1500	.005	B.F.	.53	.20		.73	.93
1950	40 psi compressive strength, 1" thick R5	G		1500	.005	S.F.	.47	.20		.67	.87
1952	2" thick R10	G		1250	.006		.88	.24		1.12	1.40
1954	3" thick R15	G		1000	.008		1.28	.30		1.58	1.94
1956	4" thick R20	G		1000	.008		1.70	.30		2	2.40
1958	Tapered for drainage	G		1400	.006	B.F.	.67	.21		.88	1.12
1960	60 PSI compressive strength, 1" thick R5	G		1450	.006	S.F.	.65	.21		.86	1.09
1962	2" thick R10	G		1200	.007		1.25	.25		1.50	1.82
1964	3" thick R15	G		975	.008		2	.31		2.31	2.75
1966	4" thick R20	G		950	.008		2.50	.31		2.81	3.31
1968	Tapered for drainage	G		1400	.006	B.F.	.85	.21		1.06	1.32
2010	Expanded polystyrene, 1#/C.F. density, 3/4" thick R2.89	G		1500	.005	S.F.	.21	.20		.41	.58
2020	1" thick R3.85	G		1500	.005		.28	.20		.48	.66
2100	2" thick R7.69	G		1250	.006		.56	.24		.80	1.05
2110	3" thick R11.49	G		1250	.006		.84	.24		1.08	1.35
2120	4" thick R15.38	G		1200	.007		1.12	.25		1.37	1.67
2130	5" thick R19.23	G		1150	.007		1.40	.26		1.66	2
2140	6" thick R23.26	G		1150	.007		1.68	.26		1.94	2.31
2150	Tapered for drainage	G		1500	.005	B.F.	.51	.20		.71	.91
2400	Composites with 2" EPS										
2410	1" fiberboard	G	1 Rofc	950	.008	S.F.	1.32	.31		1.63	2.01
2420	7/16" oriented strand board	G		800	.010		1.06	.37		1.43	1.83
2430	1/2" plywood	G		800	.010		1.31	.37		1.68	2.10
2440	1" perlite	G		800	.010		1.08	.37		1.45	1.85
2450	Composites with 1-1/2" polyisocyanurate										
2460	1" fiberboard	G	1 Rofc	800	.010	S.F.	1.12	.37		1.49	1.89

07 22 Roof and Deck Insulation

07 22 16 – Roof Board Insulation

07 22 16.10 Roof Deck Insulation			Crew	Daily Output	Labor-Hours	Unit	2012 Bare Costs Material	Labor	Equipment	Total	Total Incl O&P
2470	1″ perlite	G	1 Rofc	850	.009	S.F.	1	.35		1.35	1.73
2480	7/16″ oriented strand board	G	↓	800	.010	↓	.88	.37		1.25	1.63
9000	Minimum labor/equipment charge		↓	3.25	2.462	Job		92		92	164

07 24 Exterior Insulation and Finish Systems

07 24 13 – Polymer-Based Exterior Insulation and Finish System

07 24 13.10 Exterior Insulation and Finish Systems			Crew	Daily Output	Labor-Hours	Unit	Material	Labor	Equipment	Total	Total Incl O&P
0010	**EXTERIOR INSULATION AND FINISH SYSTEMS**										
0095	Field applied, 1″ EPS insulation	G	J-1	390	.103	S.F.	1.97	3.95	.34	6.26	8.85
0100	With 1/2″ cement board sheathing	G		268	.149		2.65	5.75	.49	8.89	12.65
0105	2″ EPS insulation	G		390	.103		2.25	3.95	.34	6.54	9.15
0110	With 1/2″ cement board sheathing	G		268	.149		2.93	5.75	.49	9.17	12.95
0115	3″ EPS insulation	G		390	.103		2.53	3.95	.34	6.82	9.45
0120	With 1/2″ cement board sheathing	G		268	.149		3.21	5.75	.49	9.45	13.25
0125	4″ EPS insulation	G		390	.103		2.81	3.95	.34	7.10	9.75
0130	With 1/2″ cement board sheathing	G		268	.149		4.17	5.75	.49	10.41	14.35
0140	Premium finish add			1265	.032		.31	1.22	.10	1.63	2.39
0150	Heavy duty reinforcement add		↓	914	.044	↓	.90	1.69	.14	2.73	3.84
0160	2.5#/S.Y. metal lath substrate add		1 Lath	75	.107	S.Y.	2.37	4.19		6.56	9.15
0170	3.4#/S.Y. metal lath substrate add		″	75	.107	″	2.58	4.19		6.77	9.40
0180	Color or texture change,		J-1	1265	.032	S.F.	.83	1.22	.10	2.15	2.96
0190	With substrate leveling base coat		1 Plas	530	.015		.83	.61		1.44	1.88
0210	With substrate sealing base coat		1 Pord	1224	.007	↓	.08	.25		.33	.49
0370	V groove shape in panel face					L.F.	.60			.60	.66
0380	U groove shape in panel face					″	.79			.79	.87
0433	Crack repair, acrylic rubber, fluid applied, 20 mils thick		1 Plas	350	.023	S.F.	1.43	.92		2.35	3.04
0437	50 mils thick, reinforced		″	200	.040	″	2.62	1.61		4.23	5.45
0440	For higher than one story, add							25%			

07 25 Weather Barriers

07 25 10 – Weather Barriers or Wraps

07 25 10.10 Weather Barriers		Crew	Daily Output	Labor-Hours	Unit	Material	Labor	Equipment	Total	Total Incl O&P
0010	**WEATHER BARRIERS**									
0400	Asphalt felt paper, 15#	1 Carp	37	.216	Sq.	5.65	9.55		15.20	22
0401	Per square foot	″	3700	.002	S.F.	.06	.10		.16	.22
0450	Housewrap, exterior, spun bonded polypropylene									
0470	Small roll	1 Carp	3800	.002	S.F.	.13	.09		.22	.30
0480	Large roll	″	4000	.002	″	.13	.09		.22	.28
2100	Asphalt felt roof deck vapor barrier, class 1 metal decks	1 Rofc	37	.216	Sq.	20.50	8.10		28.60	37.50
2200	For all other decks	″	37	.216		15	8.10		23.10	31
2800	Asphalt felt, 50% recycled content, 15 lb., 4 sq per roll	1 Carp	36	.222		5.65	9.80		15.45	22.50
2810	30 lb., 2 sq per roll	″	36	.222	↓	11.30	9.80		21.10	28.50
3000	Building wrap, spunbonded polyethylene	2 Carp	8000	.002	S.F.	.17	.09		.26	.33
9960	Minimum labor/equipment charge	1 Carp	2	4	Job		176		176	289

07 26 Vapor Retarders

07 26 10 – Above-Grade Vapor Retarders

07 26 10.10 Vapor Retarders		Crew	Daily Output	Labor-Hours	Unit	Material	2012 Bare Costs Labor	Equipment	Total	Total Incl O&P
0010	**VAPOR RETARDERS**									
0020	Aluminum and kraft laminated, foil 1 side G	1 Carp	37	.216	Sq.	11.25	9.55		20.80	28
0100	Foil 2 sides G		37	.216		12.50	9.55		22.05	29.50
0600	Polyethylene vapor barrier, standard, .002" thick G		37	.216		1.30	9.55		10.85	17.10
0700	.004" thick G		37	.216		3.23	9.55		12.78	19.20
0900	.006" thick G		37	.216		4.54	9.55		14.09	20.50
1200	.010" thick G		37	.216		7	9.55		16.55	23.50
1300	Clear reinforced, fire retardant, .008" thick G		37	.216		10.65	9.55		20.20	27.50
1350	Cross laminated type, .003" thick G		37	.216		7.45	9.55		17	24
1400	.004" thick G		37	.216		8.20	9.55		17.75	24.50
1800	Reinf. waterproof, .002" polyethylene backing, 1 side		37	.216		5.95	9.55		15.50	22
1900	2 sides		37	.216		7.80	9.55		17.35	24.50
2400	Waterproofed kraft with sisal or fiberglass fibers, minimum		37	.216		6.40	9.55		15.95	22.50
2500	Maximum		37	.216		15.85	9.55		25.40	33
9950	Minimum labor/equipment charge		4	2	Job		88		88	145

07 31 Shingles and Shakes

07 31 13 – Asphalt Shingles

07 31 13.10 Asphalt Roof Shingles		Crew	Daily Output	Labor-Hours	Unit	Material	Labor	Equipment	Total	Total Incl O&P
0010	**ASPHALT ROOF SHINGLES**									
0100	Standard strip shingles									
0150	Inorganic, class A, 210-235 lb./sq.	1 Rofc	5.50	1.455	Sq.	75.50	54.50		130	180
0155	Pneumatic nailed		7	1.143		75.50	42.50		118	159
0200	Organic, class C, 235-240 lb./sq.		5	1.600		82.50	60		142.50	197
0205	Pneumatic nailed		6.25	1.280		82.50	48		130.50	176
0250	Standard, laminated multi-layered shingles									
0300	Class A, 240-260 lb./sq.	1 Rofc	4.50	1.778	Sq.	92.50	66.50		159	220
0305	Pneumatic nailed		5.63	1.422		92.50	53		145.50	197
0350	Class C, 260-300 lb./square, 4 bundles/square		4	2		135	75		210	282
0355	Pneumatic nailed		5	1.600		135	60		195	255
0400	Premium, laminated multi-layered shingles									
0450	Class A, 260-300 lb., 4 bundles/sq	1 Rofc	3.50	2.286	Sq.	159	85.50		244.50	325
0455	Pneumatic nailed		4.37	1.831		159	68.50		227.50	297
0500	Class C, 300-385 lb./square, 5 bundles/square		3	2.667		165	99.50		264.50	360
0505	Pneumatic nailed		3.75	2.133		165	80		245	325
0800	#15 felt underlayment		64	.125		5.65	4.68		10.33	14.55
0825	#30 felt underlayment		58	.138		11.30	5.15		16.45	21.50
0850	Self adhering polyethylene and rubberized asphalt underlayment		22	.364		74	13.60		87.60	106
0900	Ridge shingles		330	.024	L.F.	1.90	.91		2.81	3.70
0905	Pneumatic nailed		412.50	.019	"	1.90	.73		2.63	3.38
1000	For steep roofs (7 to 12 pitch or greater), add						50%			
9000	Minimum labor/equipment charge	1 Rofc	3	2.667	Job		99.50		99.50	177

07 31 16 – Metal Shingles

07 31 16.10 Aluminum Shingles		Crew	Daily Output	Labor-Hours	Unit	Material	Labor	Equipment	Total	Total Incl O&P
0010	**ALUMINUM SHINGLES**									
0020	Mill finish, .019 thick	1 Carp	5	1.600	Sq.	191	70.50		261.50	325
0100	.020" thick	"	5	1.600		214	70.50		284.50	350
0300	For colors, add					20			20	22
0600	Ridge cap, .024" thick	1 Carp	170	.047	L.F.	2.18	2.08		4.26	5.80
0700	End wall flashing, .024" thick		170	.047		1.78	2.08		3.86	5.35

07 31 Shingles and Shakes

07 31 16 – Metal Shingles

07 31 16.10 Aluminum Shingles		Crew	Daily Output	Labor-Hours	Unit	Material	Labor	Equipment	Total	Total Incl O&P
						2012 Bare Costs				
0900	Valley section, .024" thick	1 Carp	170	.047	L.F.	2.90	2.08		4.98	6.60
1000	Starter strip, .024" thick		400	.020		1.41	.88		2.29	3
1200	Side wall flashing, .024" thick		170	.047		1.78	2.08		3.86	5.35
9000	Minimum labor/equipment charge		3	2.667	Job		118		118	193

07 31 16.20 Steel Shingles		Crew	Daily Output	Labor-Hours	Unit	Material	Labor	Equipment	Total	Total Incl O&P
0010	**STEEL SHINGLES**									
0012	Galvanized, 26 gauge	1 Rots	2.20	3.636	Sq.	299	137		436	575
0200	24 gauge	"	2.20	3.636		299	137		436	575
0300	For colored galvanized shingles, add					55			55	60.50
0500	For 1" factory applied polystyrene insulation, add					40			40	44
9000	Minimum labor/equipment charge	1 Rots	3	2.667	Job		100		100	178

07 31 26 – Slate Shingles

07 31 26.10 Slate Roof Shingles		Crew	Daily Output	Labor-Hours	Unit	Material	Labor	Equipment	Total	Total Incl O&P
0010	**SLATE ROOF SHINGLES**									
0100	Buckingham Virginia black, 3/16" - 1/4" thick G	1 Rots	1.75	4.571	Sq.	475	172		647	830
0200	1/4" thick G		1.75	4.571		475	172		647	830
0900	Pennsylvania black, Bangor, #1 clear G		1.75	4.571		490	172		662	845
1200	Vermont, unfading, green, mottled green G		1.75	4.571		480	172		652	830
1300	Semi-weathering green & gray G		1.75	4.571		350	172		522	685
1400	Purple G		1.75	4.571		425	172		597	770
1500	Black or gray G		1.75	4.571		460	172		632	810
2500	Slate roof repair, extensive replacement		1	8		590	300		890	1,175
2600	Repair individual pieces, scattered		19	.421	Ea.	5.90	15.80		21.70	34.50
9000	Minimum labor/equipment charge		3	2.667	Job		100		100	178

07 31 29 – Wood Shingles and Shakes

07 31 29.13 Wood Shingles		Crew	Daily Output	Labor-Hours	Unit	Material	Labor	Equipment	Total	Total Incl O&P
0010	**WOOD SHINGLES** R061110-30									
0012	16" No. 1 red cedar shingles, 5" exposure, on roof	1 Carp	2.50	3.200	Sq.	208	141		349	460
0015	Pneumatic nailed		3.25	2.462		208	109		317	405
0200	7-1/2" exposure, on walls		2.05	3.902		139	172		311	435
0205	Pneumatic nailed		2.67	2.996		139	132		271	370
0300	18" No. 1 red cedar perfections, 5-1/2" exposure, on roof		2.75	2.909		228	128		356	460
0305	Pneumatic nailed		3.57	2.241		228	99		327	415
0600	Resquared, and rebutted, 5-1/2" exposure, on roof		3	2.667		260	118		378	480
0605	Pneumatic nailed		3.90	2.051		260	90.50		350.50	435
0900	7-1/2" exposure, on walls		2.45	3.265		191	144		335	445
0905	Pneumatic nailed		3.18	2.516		191	111		302	390
1000	Add to above for fire retardant shingles					55			55	60.50
1060	Preformed ridge shingles	1 Carp	400	.020	L.F.	3.60	.88		4.48	5.40
2000	White cedar shingles, 16" long, extras, 5" exposure, on roof		2.40	3.333	Sq.	154	147		301	410
2005	Pneumatic nailed		3.12	2.564		154	113		267	355
2050	5" exposure on walls		2	4		154	176		330	460
2055	Pneumatic nailed		2.60	3.077		154	136		290	390
2100	7-1/2" exposure, on walls		2	4		110	176		286	410
2105	Pneumatic nailed		2.60	3.077		110	136		246	345
2150	"B" grade, 5" exposure on walls		2	4		136	176		312	440
2155	Pneumatic nailed		2.60	3.077		136	136		272	370
2300	For 15# organic felt underlayment on roof, 1 layer, add		64	.125		5.65	5.50		11.15	15.30
2400	2 layers, add		32	.250		11.30	11.05		22.35	30.50
2600	For steep roofs (7/12 pitch or greater), add to above						50%			
2700	Panelized systems, No.1 cedar shingles on 5/16" CDX plywood									

07 31 Shingles and Shakes

07 31 29 – Wood Shingles and Shakes

07 31 29.13 Wood Shingles		Crew	Daily Output	Labor-Hours	Unit	2012 Bare Costs Material	Labor	Equipment	Total	Total Incl O&P
2800	On walls, 8′ strips, 7″ or 14″ exposure	2 Carp	700	.023	S.F.	5.55	1.01		6.56	7.75
3500	On roofs, 8′ strips, 7″ or 14″ exposure	1 Carp	3	2.667	Sq.	555	118		673	805
3505	Pneumatic nailed		4	2	"	555	88		643	755
9000	Minimum labor/equipment charge		3	2.667	Job		118		118	193

07 31 29.16 Wood Shakes		Crew	Daily Output	Labor-Hours	Unit	Material	Labor	Equipment	Total	Total Incl O&P
0010	**WOOD SHAKES**									
1100	Hand-split red cedar shakes, 1/2″ thick x 24″ long, 10″ exp. on roof	1 Carp	2.50	3.200	Sq.	233	141		374	485
1105	Pneumatic nailed		3.25	2.462		233	109		342	435
1110	3/4″ thick x 24″ long, 10″ exp. on roof		2.25	3.556		233	157		390	515
1115	Pneumatic nailed		2.92	2.740		233	121		354	455
1200	1/2″ thick, 18″ long, 8-1/2″ exp. on roof		2	4		159	176		335	465
1205	Pneumatic nailed		2.60	3.077		159	136		295	395
1210	3/4″ thick x 18″ long, 8 1/2″ exp. on roof		1.80	4.444		159	196		355	495
1215	Pneumatic nailed		2.34	3.419		159	151		310	420
1255	10″ exp. on walls		2	4		154	176		330	460
1260	10″ exposure on walls, pneumatic nailed		2.60	3.077		154	136		290	390
1700	Add to above for fire retardant shakes, 24″ long					55			55	60.50
1800	18″ long					55			55	60.50
1810	Ridge shakes	1 Carp	350	.023	L.F.	3.60	1.01		4.61	5.60

07 32 Roof Tiles

07 32 13 – Clay Roof Tiles

07 32 13.10 Clay Tiles		Crew	Daily Output	Labor-Hours	Unit	Material	Labor	Equipment	Total	Total Incl O&P
0010	**CLAY TILES**, including accessories									
2500	Imported, English shingle, 555 pcs/sq., 6-1/2″ x 10-1/2″	3 Rots	5	4.800	Sq.	1,175	180		1,355	1,625
2510	Flat interlocking, 210 pcs/sq., 8″ x 14-1/4″		5.50	4.364		585	164		749	935
2520	192 pcs/sq., 9″ x 13″		5.50	4.364		610	164		774	960
2530	100 pcs/sq., 13″ x 16″		6	4		460	150		610	770
2540	94 pcs/sq., 12-1/2″ x 18-1/4″		6	4		485	150		635	800
2550	93 pcs/sq., 12″ x 18-1/2″		6	4		485	150		635	800
2560	French, 124 pcs/sq., 10-1/4″ x 17-1/2″		6	4		535	150		685	855
2570	120 pcs/sq., 9-3/4″ x 17-1/2″		6	4		385	150		535	690
2580	115 pcs/sq., 10-1/4″ x 17-1/4″		6	4		485	150		635	800
2590	91 pcs/sq., 12-1/2″ x 18-3/4″		6	4		510	150		660	825
2600	French shingle, 495 pcs/sq., 6-3/4″ x 12″		5	4.800		1,400	180		1,580	1,875
2610	Gaelic, 190 pcs/sq., 8-3/4″ x 12″		5.50	4.364		460	164		624	795
2620	120 pcs/sq., 10-3/4″ x 17-1/2″		6	4		560	150		710	880
2630	84 pcs/sq., 13-1/4″ x 19″		6	4		460	150		610	770
2640	Handcrafted shingle, 320 pcs/sq., 7-1/4″ x 14-3/4″		5	4.800		810	180		990	1,200
2650	Mediterranean S - Roman, 115 pcs/sq., 10″ x 17-3/4″		6	4		410	150		560	715
2660	112 pcs/sq., 10″ x 17-1/2″		6	4		355	150		505	655
2670	100 pcs/sq., 10-1/2″ x 18-1/2″		6	4		254	150		404	545
2680	Mission pan & cover, 260 pcs/sq., 6-1/4″ x 17-3/4″		5.50	4.364		435	164		599	770
2690	183 pcs/sq., 8″ x 18″		5.50	4.364		535	164		699	880
2700	Pantile, 180 pcs/sq., 9-3/4″ x 14″		5.50	4.364		685	164		849	1,050
2710	150 pcs/sq., 10-1/4″ x 14-1/2″		6	4		585	150		735	910
2720	115 pcs/sq., 11-1/4″ x 18″		6	4		535	150		685	855
3010	#15 felt underlayment	1 Rofc	64	.125		5.65	4.68		10.33	14.55
3020	#30 felt underlayment		58	.138		11.30	5.15		16.45	21.50
3040	Polyethylene and rubberized asph. underlayment		22	.364		74	13.60		87.60	106
9010	Minimum labor/equipment charge	1 Rots	2	4	Job		150		150	267

07 32 Roof Tiles

07 32 16 – Concrete Roof Tiles

07 32 16.10 Concrete Tiles		Crew	Daily Output	Labor-Hours	Unit	2012 Bare Costs Material	Labor	Equipment	Total	Total Incl O&P
0010	**CONCRETE TILES**									
0020	Corrugated, 13" x 16-1/2", 90 per sq., 950 lb. per sq.									
0050	Earthtone colors, nailed to wood deck	1 Rots	1.35	5.926	Sq.	97.50	223		320.50	500
0150	Blues		1.35	5.926		97.50	223		320.50	500
0200	Greens		1.35	5.926		101	223		324	505
0250	Premium colors	↓	1.35	5.926	↓	101	223		324	505
0500	Shakes, 13" x 16-1/2", 90 per sq., 950 lb. per sq.									
0600	All colors, nailed to wood deck	1 Rots	1.50	5.333	Sq.	83	200		283	445
1500	Accessory pieces, ridge & hip, 10" x 16-1/2", 8 lb. each	"	120	.067	Ea.	3.64	2.50		6.14	8.45
1700	Rake, 6-1/2" x 16-3/4", 9 lb. each					3.64			3.64	4
1800	Mansard hip, 10" x 16-1/2", 9.2 lb. each					3.64			3.64	4
1900	Hip starter, 10" x 16-1/2", 10.5 lb. each					11.45			11.45	12.60
2000	3 or 4 way apex, 10" each side, 11.5 lb. each				↓	13.25			13.25	14.55
9000	Minimum labor/equipment charge	1 Rots	3	2.667	Job		100		100	178

07 32 19 – Metal Roof Tiles

07 32 19.10 Metal Roof Tiles		Crew	Daily Output	Labor-Hours	Unit	Material	Labor	Equipment	Total	Total Incl O&P
0010	**METAL ROOF TILES**									
0020	Accessories included, .032" thick aluminum, mission tile	1 Carp	2.50	3.200	Sq.	815	141		956	1,125
0200	Spanish tiles		3	2.667	"	560	118		678	810
9000	Minimum labor/equipment charge	↓	3	2.667	Job		118		118	193

07 33 Natural Roof Coverings

07 33 63 – Vegetated Roofing

07 33 63.10 Green Roof Systems			Crew	Daily Output	Labor-Hours	Unit	Material	Labor	Equipment	Total	Total Incl O&P
0010	**GREEN ROOF SYSTEMS**										
0020	Soil mixture for green roof 30% sand, 55% gravel, 15% soil	G									
0100	Hoist and spread soil mixture 4 inch depth up to five stories tall roof	G	B-13B	4000	.014	S.F.	.25	.53	.28	1.06	1.43
0150	6 inch depth	G		2667	.021		.37	.80	.41	1.58	2.15
0200	8 inch depth	G		2000	.028		.50	1.07	.55	2.12	2.89
0250	10 inch depth	G		1600	.035		.62	1.33	.69	2.64	3.61
0300	12 inch depth	G		1335	.042		.75	1.60	.82	3.17	4.32
0310	Alt. man-made soil mix, hoist & spread, 4" deep up to 5 stories tall roof	G	↓	4000	.014	↓	1.50	.53	.28	2.31	2.81
0350	Mobilization 55 ton crane to site	G	1 Eqhv	3.60	2.222	Ea.		106		106	168
0355	Hoisting cost to five stories per day (Avg. 28 picks per day)	G	B-13B	1	56	Day		2,125	1,100	3,225	4,650
0360	Mobilization or demobilization, 100 ton crane to site driver & escort	G	A-3E	2.50	6.400	Ea.		266	55.50	321.50	485
0365	Hoisting cost six to ten stories per day (Avg. 21 picks per day)	G	B-13C	1	56	Day		2,125	1,625	3,750	5,250
0370	Hoist and spread soil mixture 4 inch depth six to ten stories tall roof	G		4000	.014	S.F.	.25	.53	.41	1.19	1.58
0375	6 inch depth	G		2667	.021		.37	.80	.61	1.78	2.37
0380	8 inch depth	G		2000	.028		.50	1.07	.81	2.38	3.18
0385	10 inch depth	G		1600	.035		.62	1.33	1.02	2.97	3.97
0390	12 inch depth	G	↓	1335	.042	↓	.75	1.60	1.22	3.57	4.75
0400	Green roof edging treated lumber 4" x 4" no hoisting included	G	2 Carp	400	.040	L.F.	.82	1.76		2.58	3.79
0410	4" x 6"	G		400	.040		2.11	1.76		3.07	5.20
0420	4" x 8"	G		360	.044		4.32	1.96		6.28	7.95
0430	4" x 6" double stacked	G		300	.053		4.21	2.35		6.56	8.50
0500	Green roof edging redwood lumber 4" x 4" no hoisting included	G		400	.040		6.40	1.76		8.16	9.95
0510	4" x 6"	G		400	.040		11.20	1.76		12.96	15.20
0520	4" x 8"	G		360	.044		20.50	1.96		22.46	25.50
0530	4" x 6" double stacked	G	↓	300	.053	↓	22.50	2.35		24.85	28.50
0550	Roof membrane and root barrier, not including insulation:	G									

07 33 Natural Roof Coverings

07 33 63 – Vegetated Roofing

07 33 63.10 Green Roof Systems			Crew	Daily Output	Labor-Hours	Unit	Material	Labor	Equipment	Total	Total Incl O&P
							2012 Bare Costs				
0560	Fluid applied rubber membrane, reinforced, 215 mil thick	G	G-5	350	.114	S.F.	.26	3.88	.51	4.65	7.75
0570	Root barrier	G	2 Rofc	775	.021		.70	.77		1.47	2.14
0580	Moisture retention barrier and reservoir	G	″	900	.018		2.70	.67		3.37	4.15
0600	Planting sedum, light soil, potted, 2-1/4″ diameter, two per S.F.	G	1 Clab	420	.019		5	.67		5.67	6.60
0610	one per S.F.	G	″	840	.010		2.50	.33		2.83	3.30
0630	Planting sedum mat per S.F. including shipping (4000 S.F. min)	G	4 Clab	4000	.008		6	.28		6.28	7.05
0640	Installation sedum mat system (no soil required) per S.F. (4000 S.F. min)	G	″	4000	.008		8.35	.28		8.63	9.65
0645	Note: pricing of sedum mats shipped in full truck loads (4000-5000 S.F.)	G									

07 41 Roof Panels

07 41 13 – Metal Roof Panels

07 41 13.10 Aluminum Roof Panels		Crew	Daily Output	Labor-Hours	Unit	Material	Labor	Equipment	Total	Total Incl O&P
0010	**ALUMINUM ROOF PANELS**									
0020	Corrugated or ribbed, .0155″ thick, natural	G-3	1200	.027	S.F.	.98	1.16		2.14	2.94
0300	Painted		1200	.027		1.43	1.16		2.59	3.43
0400	Corrugated, .018″ thick, on steel frame, natural finish		1200	.027		1.29	1.16		2.45	3.28
0600	Painted		1200	.027		1.58	1.16		2.74	3.60
0700	Corrugated, on steel frame, natural, .024″ thick		1200	.027		1.83	1.16		2.99	3.87
0800	Painted		1200	.027		2.23	1.16		3.39	4.31
0900	.032″ thick, natural		1200	.027		2.36	1.16		3.52	4.46
1200	Painted		1200	.027		3.03	1.16		4.19	5.20
9000	Minimum labor/equipment charge	1 Rofc	3	2.667	Job		99.50		99.50	177

07 41 13.20 Steel Roofing Panels		Crew	Daily Output	Labor-Hours	Unit	Material	Labor	Equipment	Total	Total Incl O&P
0010	**STEEL ROOFING PANELS**									
0012	Corrugated or ribbed, on steel framing, 30 ga galv	G-3	1100	.029	S.F.	1.75	1.27		3.02	3.96
0100	28 ga		1050	.030		1.86	1.33		3.19	4.18
0300	26 ga		1000	.032		1.92	1.40		3.32	4.34
0400	24 ga		950	.034		3.03	1.47		4.50	5.70
0600	Colored, 28 ga		1050	.030		1.75	1.33		3.08	4.06
0700	26 ga		1000	.032		2.07	1.40		3.47	4.51
0710	Flat profile, 1-3/4″ standing seams, 10″ wide, standard finish, 26 ga		1000	.032		4.05	1.40		5.45	6.70
0715	24 ga		950	.034		4.69	1.47		6.16	7.50
0720	22 ga		900	.036		5.80	1.55		7.35	8.85
0725	Zinc aluminum alloy finish, 26 ga		1000	.032		3.17	1.40		4.57	5.70
0730	24 ga		950	.034		3.79	1.47		5.26	6.50
0735	22 ga		900	.036		4.34	1.55		5.89	7.25
0740	12″ wide, standard finish, 26 ga		1000	.032		4.04	1.40		5.44	6.65
0745	24 ga		950	.034		5.30	1.47		6.77	8.20
0750	Zinc aluminum alloy finish, 26 ga		1000	.032		4.59	1.40		5.99	7.30
0755	24 ga		950	.034		3.78	1.47		5.25	6.50
0840	Flat profile, 1″ x 3/8″ batten, 12″ wide, standard finish, 26 ga		1000	.032		3.56	1.40		4.96	6.15
0845	24 ga		950	.034		4.17	1.47		5.64	6.95
0850	22 ga		900	.036		4.99	1.55		6.54	8
0855	Zinc aluminum alloy finish, 26 ga		1000	.032		3.41	1.40		4.81	6
0860	24 ga		950	.034		3.81	1.47		5.28	6.55
0865	22 ga		900	.036		4.40	1.55		5.95	7.30
0870	16-1/2″ wide, standard finish, 24 ga		950	.034		4.11	1.47		5.58	6.85
0875	22 ga		900	.036		4.60	1.55		6.15	7.55
0880	Zinc aluminum alloy finish, 24 ga		950	.034		3.59	1.47		5.06	6.30
0885	22 ga		900	.036		4	1.55		5.55	6.90
0890	Flat profile, 2″ x 2″ batten, 12″ wide, standard finish, 26 ga		1000	.032		4.09	1.40		5.49	6.75

07 41 Roof Panels

07 41 13 – Metal Roof Panels

07 41 13.20 Steel Roofing Panels		Crew	Daily Output	Labor-Hours	Unit	Material	2012 Bare Costs Labor	Equipment	Total	Total Incl O&P
0895	24 ga	G-3	950	.034	S.F.	4.88	1.47		6.35	7.70
0900	22 ga		900	.036		6	1.55		7.55	9.10
0905	Zinc aluminum alloy finish, 26 ga		1000	.032		3.81	1.40		5.21	6.40
0910	24 ga		950	.034		4.35	1.47		5.82	7.15
0915	22 ga		900	.036		5.05	1.55		6.60	8.05
0920	16-1/2" wide, standard finish, 24 ga		950	.034		4.49	1.47		5.96	7.30
0925	22 ga		900	.036		5.25	1.55		6.80	8.30
0930	Zinc aluminum alloy finish, 24 ga		950	.034		4.07	1.47		5.54	6.85
0935	22 ga	↓	900	.036	↓	4.63	1.55		6.18	7.60
9000	Minimum labor/equipment charge	1 Rofc	2	4	Job		150		150	266

07 41 33 – Plastic Roof Panels

07 41 33.10 Fiberglass Panels		Crew	Daily Output	Labor-Hours	Unit	Material	Labor	Equipment	Total	Total Incl O&P
0010	**FIBERGLASS PANELS**									
0012	Corrugated panels, roofing, 8 oz. per S.F.	G-3	1000	.032	S.F.	1.90	1.40		3.30	4.32
0100	12 oz. per S.F.		1000	.032		3.44	1.40		4.84	6
0300	Corrugated siding, 6 oz. per S.F.		880	.036		1.61	1.59		3.20	4.31
0400	8 oz. per S.F.		880	.036		1.90	1.59		3.49	4.63
0500	Fire retardant		880	.036		3.24	1.59		4.83	6.10
0600	12 oz. siding, textured		880	.036		3.41	1.59		5	6.30
0700	Fire retardant		880	.036		4.53	1.59		6.12	7.50
0900	Flat panels, 6 oz. per S.F., clear or colors		880	.036		2.01	1.59		3.60	4.75
1100	Fire retardant, class A		880	.036		3.28	1.59		4.87	6.15
1300	8 oz. per S.F., clear or colors	↓	880	.036	↓	2.21	1.59		3.80	4.97
9000	Minimum labor/equipment charge	1 Rofc	2	4	Job		150		150	266

07 42 Wall Panels

07 42 13 – Metal Wall Panels

07 42 13.10 Mansard Panels		Crew	Daily Output	Labor-Hours	Unit	Material	Labor	Equipment	Total	Total Incl O&P
0010	**MANSARD PANELS**									
0600	Aluminum, stock units, straight surfaces	1 Shee	115	.070	S.F.	3.41	3.63		7.04	9.45
0700	Concave or convex surfaces		75	.107	"	3.74	5.55		9.29	12.85
0800	For framing, to 5' high, add		115	.070	L.F.	3.74	3.63		7.37	9.80
0900	Soffits, to 1' wide		125	.064	S.F.	1.85	3.34		5.19	7.30
9000	Minimum labor/equipment charge	↓	2.50	3.200	Job		167		167	263

07 42 13.20 Aluminum Siding Panels		Crew	Daily Output	Labor-Hours	Unit	Material	Labor	Equipment	Total	Total Incl O&P
0010	**ALUMINUM SIDING PANELS**									
0012	Corrugated, on steel framing, .019 thick, natural finish	G-3	775	.041	S.F.	1.50	1.80		3.30	4.53
0100	Painted		775	.041		1.61	1.80		3.41	4.65
0400	Farm type, .021" thick on steel frame, natural		775	.041		1.51	1.80		3.31	4.54
0600	Painted		775	.041		1.61	1.80		3.41	4.65
0700	Industrial type, corrugated, on steel, .024" thick, mill		775	.041		2.10	1.80		3.90	5.20
0900	Painted		775	.041		2.26	1.80		4.06	5.35
1000	.032" thick, mill		775	.041		2.42	1.80		4.22	5.55
1200	Painted		775	.041		2.94	1.80		4.74	6.10
1300	V-Beam, on steel frame, .032" thick, mill		775	.041		2.72	1.80		4.52	5.85
1500	Painted		775	.041		3.12	1.80		4.92	6.30
1600	.040" thick, mill		775	.041		3.31	1.80		5.11	6.50
1800	Painted	↓	775	.041		3.87	1.80		5.67	7.15
3800	Horizontal, colored clapboard, 8" wide, plain	2 Carp	515	.031		2.21	1.37		3.58	4.68
3810	Insulated	↓	515	.031	↓	2.36	1.37		3.73	4.85

07 42 Wall Panels

07 42 13 – Metal Wall Panels

07 42 13.20 Aluminum Siding Panels		Crew	Daily Output	Labor-Hours	Unit	2012 Bare Costs Material	Labor	Equipment	Total	Total Incl O&P
3830	8″ embossed, painted	2 Carp	515	.031	S.F.	2.34	1.37		3.71	4.82
3840	Insulated		515	.031		2.50	1.37		3.87	5
3860	12″ painted, smooth		600	.027		2.21	1.18		3.39	4.36
3870	Insulated		600	.027		2.36	1.18		3.54	4.53
3890	12″ embossed, painted		600	.027		2.34	1.18		3.52	4.50
3900	Insulated		515	.031		2.21	1.37		3.58	4.68
4000	Vertical board & batten, colored, non-insulated		515	.031		1.68	1.37		3.05	4.10
4200	For simulated wood design, add					.10			.10	.11
4300	Corners for above, outside	2 Carp	515	.031	V.L.F.	3.25	1.37		4.62	5.85
4500	Inside corners	″	515	.031	″	1.42	1.37		2.79	3.81
9000	Minimum labor/equipment charge	1 Carp	3	2.667	Job		118		118	193

07 42 13.30 Steel Siding		Crew	Daily Output	Labor-Hours	Unit	Material	Labor	Equipment	Total	Total Incl O&P
0010	**STEEL SIDING**									
0020	Beveled, vinyl coated, 8″ wide	1 Carp	265	.030	S.F.	2.05	1.33		3.38	4.44
0050	10″ wide	″	275	.029		2.19	1.28		3.47	4.51
0080	Galv, corrugated or ribbed, on steel frame, 30 gauge	G-3	800	.040		1.37	1.75		3.12	4.30
0100	28 gauge		795	.040		1.44	1.76		3.20	4.39
0300	26 gauge		790	.041		2.02	1.77		3.79	5.05
0400	24 gauge		785	.041		2.03	1.78		3.81	5.10
0600	22 gauge		770	.042		2.34	1.81		4.15	5.45
0700	Colored, corrugated/ribbed, on steel frame, 10 yr. finish, 28 ga.		800	.040		2.13	1.75		3.88	5.15
0900	26 gauge		795	.040		2.23	1.76		3.99	5.25
1000	24 gauge		790	.041		2.59	1.77		4.36	5.70
1020	20 gauge		785	.041		3.29	1.78		5.07	6.45
9000	Minimum labor/equipment charge	1 Carp	3	2.667	Job		118		118	193

07 46 Siding

07 46 23 – Wood Siding

07 46 23.10 Wood Board Siding		Crew	Daily Output	Labor-Hours	Unit	Material	Labor	Equipment	Total	Total Incl O&P
0010	**WOOD BOARD SIDING**									
3200	Wood, cedar bevel, A grade, 1/2″ x 6″	1 Carp	295	.027	S.F.	4.35	1.20		5.55	6.75
3300	1/2″ x 8″		330	.024		6.30	1.07		7.37	8.70
3500	3/4″ x 10″, clear grade		375	.021		5.90	.94		6.84	8.05
3600	″B″ grade		375	.021		3.79	.94		4.73	5.70
3800	Cedar, rough sawn, 1″ x 4″, A grade, natural		220	.036		3.88	1.60		5.48	6.90
3900	Stained		220	.036		4.25	1.60		5.85	7.30
4100	1″ x 12″, board & batten, #3 & Btr., natural		420	.019		4.02	.84		4.86	5.80
4200	Stained		420	.019		4.25	.84		5.09	6.05
4400	1″ x 8″ channel siding, #3 & Btr., natural		330	.024		2.97	1.07		4.04	5
4500	Stained		330	.024		2.71	1.07		3.78	4.73
4700	Redwood, clear, beveled, vertical grain, 1/2″ x 4″		220	.036		3.85	1.60		5.45	6.85
4750	1/2″ x 6″		295	.027		3.85	1.20		5.05	6.20
4800	1/2″ x 8″		330	.024		4.10	1.07		5.17	6.25
5000	3/4″ x 10″		375	.021		4.12	.94		5.06	6.05
5200	Channel siding, 1″ x 10″, B grade		375	.021		4.12	.94		5.06	6.05
5250	Redwood, T&G boards, B grade, 1″ x 4″		220	.036		2.54	1.60		4.14	5.40
5270	1″ x 8″		330	.024		4.10	1.07		5.17	6.25
5400	White pine, rough sawn, 1″ x 8″, natural		330	.024		2	1.07		3.07	3.95
5500	Stained		330	.024		2.20	1.07		3.27	4.17
9000	Minimum labor/equipment charge		2	4	Job		176		176	289

07 46 Siding

07 46 29 – Plywood Siding

07 46 29.10 Plywood Siding Options		Crew	Daily Output	Labor-Hours	Unit	2012 Bare Costs Material	Labor	Equipment	Total	Total Incl O&P
0010	**PLYWOOD SIDING OPTIONS**									
0900	Plywood, medium density overlaid, 3/8" thick	2 Carp	750	.021	S.F.	1.15	.94		2.09	2.81
1000	1/2" thick		700	.023		1.26	1.01		2.27	3.04
1100	3/4" thick		650	.025		1.65	1.09		2.74	3.60
1600	Texture 1-11, cedar, 5/8" thick, natural		675	.024		2.43	1.05		3.48	4.38
1700	Factory stained		675	.024		2.69	1.05		3.74	4.67
1900	Texture 1-11, fir, 5/8" thick, natural		675	.024		1.06	1.05		2.11	2.88
2000	Factory stained		675	.024		1.32	1.05		2.37	3.16
2050	Texture 1-11, S.Y.P., 5/8" thick, natural		675	.024		1.19	1.05		2.24	3.02
2100	Factory stained		675	.024		1.24	1.05		2.29	3.07
2200	Rough sawn cedar, 3/8" thick, natural		675	.024		1.20	1.05		2.25	3.03
2300	Factory stained		675	.024		1.46	1.05		2.51	3.32
2500	Rough sawn fir, 3/8" thick, natural		675	.024		.76	1.05		1.81	2.55
2600	Factory stained		675	.024		1.02	1.05		2.07	2.83
2800	Redwood, textured siding, 5/8" thick		675	.024		1.92	1.05		2.97	3.82
3000	Polyvinyl chloride coated, 3/8" thick		750	.021		1.11	.94		2.05	2.76
9000	Minimum labor/equipment charge	1 Carp	2	4	Job		176		176	289

07 46 33 – Plastic Siding

07 46 33.10 Vinyl Siding		Crew	Daily Output	Labor-Hours	Unit	Material	Labor	Equipment	Total	Total Incl O&P
0010	**VINYL SIDING**									
3995	Clapboard profile, woodgrain texture, .048 thick, double 4	2 Carp	495	.032	S.F.	.85	1.43		2.28	3.28
4000	Double 5		550	.029		.85	1.28		2.13	3.04
4005	Single 8		495	.032		.98	1.43		2.41	3.42
4010	Single 10		550	.029		1.18	1.28		2.46	3.40
4015	.044 thick, double 4		495	.032		.82	1.43		2.25	3.25
4020	Double 5		550	.029		.82	1.28		2.10	3.01
4025	.042 thick, double 4		495	.032		.82	1.43		2.25	3.25
4030	Double 5		550	.029		.82	1.28		2.10	3.01
4035	Cross sawn texture, .040 thick, double 4		495	.032		.58	1.43		2.01	2.98
4040	Double 5		550	.029		.58	1.28		1.86	2.74
4045	Smooth texture, .042 thick, double 4		495	.032		.68	1.43		2.11	3.09
4050	Double 5		550	.029		.73	1.28		2.01	2.91
4055	Single 8		495	.032		.94	1.43		2.37	3.38
4060	Cedar texture, .044 thick, double 4		495	.032		.87	1.43		2.30	3.30
4065	Double 6		600	.027		.87	1.18		2.05	2.89
4070	Dutch lap profile, woodgrain texture, .048 thick, double 5		550	.029		.85	1.28		2.13	3.04
4075	.044 thick, double 4.5		525	.030		.87	1.34		2.21	3.16
4080	.042 thick, double 4.5		525	.030		.74	1.34		2.08	3.02
4085	.040 thick, double 4.5		525	.030		.58	1.34		1.92	2.84
4090	Shingle profile, random grooves, double 7		400	.040		2.73	1.76		4.49	5.90
4095	Triple 5		400	.040		2.72	1.76		4.48	5.90
4100	Shake profile, 10" wide		400	.040		2.91	1.76		4.67	6.10
4105	Vertical pattern, .046 thick, double 5		550	.029		1.22	1.28		2.50	3.45
4110	.044 thick, triple 3		550	.029		1.31	1.28		2.59	3.55
4115	.040 thick, triple 4		550	.029		1.25	1.28		2.53	3.48
4120	.040 thick, triple 2.66		550	.029		1.41	1.28		2.69	3.66
4125	Insulation, fan folded extruded polystyrene, 1/4"		2000	.008		.27	.35		.62	.88
4130	3/8"		2000	.008		.30	.35		.65	.91
4135	Accessories, J channel, 5/8" pocket		700	.023	L.F.	.38	1.01		1.39	2.06
4140	3/4" pocket		695	.023		.40	1.02		1.42	2.10
4145	1-1/4" pocket		680	.024		.59	1.04		1.63	2.35
4150	Flexible, 3/4" pocket		600	.027		1.92	1.18		3.10	4.04

07 46 Siding

07 46 33 – Plastic Siding

07 46 33.10 Vinyl Siding		Crew	Daily Output	Labor-Hours	Unit	2012 Bare Costs Material	Labor	Equipment	Total	Total Incl O&P
4155	Under sill finish trim	2 Carp	500	.032	L.F.	.41	1.41		1.82	2.76
4160	Vinyl starter strip		700	.023		.42	1.01		1.43	2.11
4165	Aluminum starter strip		700	.023		.34	1.01		1.35	2.02
4170	Window casing, 2-1/2" wide, 3/4" pocket		510	.031		1.22	1.38		2.60	3.61
4175	Outside corner, woodgrain finish, 4" face, 3/4" pocket		700	.023		1.58	1.01		2.59	3.39
4180	5/8" pocket		700	.023		1.66	1.01		2.67	3.48
4185	Smooth finish, 4" face, 3/4" pocket		700	.023		1.58	1.01		2.59	3.39
4190	7/8" pocket		690	.023		1.55	1.02		2.57	3.39
4195	1-1/4" pocket		700	.023		1.11	1.01		2.12	2.88
4200	Soffit and fascia, 1' overhang, solid		120	.133		3.35	5.90		9.25	13.35
4205	Vented		120	.133		3.35	5.90		9.25	13.35
4207	18" overhang, solid		110	.145		3.96	6.40		10.36	14.85
4208	Vented		110	.145		3.96	6.40		10.36	14.85
4210	2' overhang, solid		100	.160		4.57	7.05		11.62	16.60
4215	Vented		100	.160		4.57	7.05		11.62	16.60
4217	3' overhang, solid		100	.160		5.80	7.05		12.85	17.90
4218	Vented	↓	100	.160	↓	5.80	7.05		12.85	17.90
4220	Colors for siding and soffits, add				S.F.	.16			.16	.18
4225	Colors for accessories and trim, add				L.F.	.32			.32	.35
9000	Minimum labor/equipment charge	1 Carp	3	2.667	Job		118		118	193

07 46 46 – Fiber Cement Siding

07 46 46.10 Fiber Cement Siding		Crew	Daily Output	Labor-Hours	Unit	Material	Labor	Equipment	Total	Total Incl O&P
0010	**FIBER CEMENT SIDING**									
0020	Lap siding, 5/16" thick, 6" wide, 4-3/4" exposure, smooth texture	2 Carp	415	.039	S.F.	1.08	1.70		2.78	3.98
0025	Woodgrain texture		415	.039		1.08	1.70		2.78	3.98
0030	7-1/2" wide, 6-1/4" exposure, smooth texture		425	.038		1.04	1.66		2.70	3.86
0035	Woodgrain texture		425	.038		1.04	1.66		2.70	3.86
0040	8" wide, 6-3/4" exposure, smooth texture		425	.038		1.24	1.66		2.90	4.09
0045	Roughsawn texture		425	.038		1.24	1.66		2.90	4.09
0050	9-1/2" wide, 8-1/4" exposure, smooth texture		440	.036		.93	1.60		2.53	3.65
0055	Woodgrain texture		440	.036		.93	1.60		2.53	3.65
0060	12" wide, 10-3/8" exposure, smooth texture		455	.035		1.21	1.55		2.76	3.87
0065	Woodgrain texture		455	.035		1.21	1.55		2.76	3.87
0070	Panel siding, 5/16" thick, smooth texture		750	.021		1.17	.94		2.11	2.83
0075	Stucco texture		750	.021		1.17	.94		2.11	2.83
0080	Grooved woodgrain texture		750	.021		1.17	.94		2.11	2.83
0085	V - grooved woodgrain texture		750	.021	↓	1.17	.94		2.11	2.83
0090	Wood starter strip	↓	400	.040	L.F.	.43	1.76		2.19	3.36

07 46 73 – Soffit

07 46 73.10 Soffit Options		Crew	Daily Output	Labor-Hours	Unit	Material	Labor	Equipment	Total	Total Incl O&P
0010	**SOFFIT OPTIONS**									
0012	Aluminum, residential, .020" thick	1 Carp	210	.038	S.F.	1.96	1.68		3.64	4.91
0100	Baked enamel on steel, 16 or 18 gauge		105	.076		5.80	3.36		9.16	11.90
0300	Polyvinyl chloride, white, solid		230	.035		2.04	1.53		3.57	4.75
0400	Perforated	↓	230	.035		2.04	1.53		3.57	4.75
0500	For colors, add				↓	.12			.12	.13
9000	Minimum labor/equipment charge	1 Carp	3	2.667	Job		118		118	193

07 51 Built-Up Bituminous Roofing

07 51 13 – Built-Up Asphalt Roofing

07 51 13.10 Built-Up Roofing Components		Crew	Daily Output	Labor-Hours	Unit	2012 Bare Costs Material	Labor	Equipment	Total	Total Incl O&P
0010	**BUILT-UP ROOFING COMPONENTS**									
0012	Asphalt saturated felt, #30, 2 square per roll	1 Rofc	58	.138	Sq.	11.30	5.15		16.45	21.50
0200	#15, 4 sq per roll, plain or perforated, not mopped		58	.138		5.65	5.15		10.80	15.40
0300	Roll roofing, smooth, #65		15	.533		9.65	19.95		29.60	46
0500	#90		12	.667		34.50	25		59.50	82.50
0520	Mineralized		12	.667		34.50	25		59.50	82.50
0540	D.C. (Double coverage), 19" selvage edge		10	.800		44	30		74	101
0580	Adhesive (lap cement)				Gal.	5.50			5.50	6.05
0600	Steep, flat or dead level asphalt, 10 ton lots, bulk				Ton	455			455	500
0800	Packaged				"	780			780	860
9000	Minimum labor/equipment charge	1 Rofc	4	2	Job		75		75	133

07 51 13.13 Cold-Applied Built-Up Asphalt Roofing		Crew	Daily Output	Labor-Hours	Unit	Material	Labor	Equipment	Total	Total Incl O&P
0010	**COLD-APPLIED BUILT-UP ASPHALT ROOFING**									
0020	3 ply system, installation only (components listed below)	G-5	50	.800	Sq.		27	3.58	30.58	52.50
0100	Spunbond poly. fabric, 1.35 oz./S.Y., 36"W, 10.8 sq./roll				Ea.	175			175	193
0200	49" wide, 14.6 sq./roll					244			244	268
0300	2.10 oz./S.Y., 36" wide, 10.8 sq./roll					266			266	293
0400	49" wide, 14.6 sq./roll					360			360	395
0500	Base & finish coat, 3 gal./sq., 5 gal./can				Gal.	7.60			7.60	8.35
0600	Coating, ceramic granules, 1/2 sq./bag				Ea.	15.15			15.15	16.65
0700	Aluminum, 2 gal./sq.				Gal.	12.30			12.30	13.55
0800	Emulsion, fibered or non-fibered, 4 gal./sq.				"	5.20			5.20	5.70

07 51 13.20 Built-Up Roofing Systems		Crew	Daily Output	Labor-Hours	Unit	Material	Labor	Equipment	Total	Total Incl O&P
0010	**BUILT-UP ROOFING SYSTEMS**									
0120	Asphalt flood coat with gravel/slag surfacing, not including									
0140	Insulation, flashing or wood nailers									
0200	Asphalt base sheet, 3 plies #15 asphalt felt, mopped	G-1	22	2.545	Sq.	88	89	20.50	197.50	278
0350	On nailable decks		21	2.667		92.50	93	21.50	207	292
0500	4 plies #15 asphalt felt, mopped		20	2.800		122	98	22.50	242.50	335
0550	On nailable decks		19	2.947		109	103	24	236	330
0700	Coated glass base sheet, 2 plies glass (type IV), mopped		22	2.545		90.50	89	20.50	200	280
0850	3 plies glass, mopped		20	2.800		108	98	22.50	228.50	320
0950	On nailable decks		19	2.947		102	103	24	229	320
1100	4 plies glass fiber felt (type IV), mopped		20	2.800		133	98	22.50	253.50	345
1150	On nailable decks		19	2.947		119	103	24	246	340
1200	Coated & saturated base sheet, 3 plies #15 asph. felt, mopped		20	2.800		98	98	22.50	218.50	305
1250	On nailable decks		19	2.947		91.50	103	24	218.50	310
1300	4 plies #15 asphalt felt, mopped		22	2.545		114	89	20.50	223.50	305
2000	Asphalt flood coat, smooth surface									
2200	Asphalt base sheet & 3 plies #15 asphalt felt, mopped	G-1	24	2.333	Sq.	94.50	81.50	18.85	194.85	270
2400	On nailable decks		23	2.435		88	85	19.65	192.65	269
2600	4 plies #15 asphalt felt, mopped		24	2.333		111	81.50	18.85	211.35	288
2700	On nailable decks		23	2.435		104	85	19.65	208.65	288
2900	Coated glass fiber base sheet, mopped, and 2 plies of									
2910	glass fiber felt (type IV)	G-1	25	2.240	Sq.	86	78.50	18.10	182.60	253
3100	On nailable decks		24	2.333		81	81.50	18.85	181.35	255
3200	3 plies, mopped		23	2.435		104	85	19.65	208.65	287
3300	On nailable decks		22	2.545		97	89	20.50	206.50	288
3800	4 plies glass fiber felt (type IV), mopped		23	2.435		122	85	19.65	226.65	305
3900	On nailable decks		22	2.545		115	89	20.50	224.50	305
4000	Coated & saturated base sheet, 3 plies #15 asph. felt, mopped		24	2.333		93.50	81.50	18.85	193.85	269
4200	On nailable decks		23	2.435		87	85	19.65	191.65	268

07 51 Built-Up Bituminous Roofing

07 51 13 – Built-Up Asphalt Roofing

07 51 13.20 Built-Up Roofing Systems		Crew	Daily Output	Labor-Hours	Unit	Material	Labor	Equipment	Total	Total Incl O&P
						2012 Bare Costs				
4300	4 plies #15 organic felt, mopped	G-1	22	2.545	Sq.	110	89	20.50	219.50	300
4500	Coal tar pitch with gravel/slag surfacing									
4600	4 plies #15 tarred felt, mopped	G-1	21	2.667	Sq.	186	93	21.50	300.50	395
4800	3 plies glass fiber felt (type IV), mopped	"	19	2.947	"	154	103	24	281	380
5000	Coated glass fiber base sheet, and 2 plies of									
5010	glass fiber felt, (type IV), mopped	G-1	19	2.947	Sq.	159	103	24	286	385
5300	On nailable decks		18	3.111		138	109	25	272	375
5600	4 plies glass fiber felt (type IV), mopped		21	2.667		213	93	21.50	327.50	425
5800	On nailable decks		20	2.800		193	98	22.50	313.50	410

07 51 13.30 Cants		Crew	Daily Output	Labor-Hours	Unit	Material	Labor	Equipment	Total	Total Incl O&P
0010	**CANTS**									
0012	Lumber, treated, 4" x 4" cut diagonally	1 Rofc	325	.025	L.F.	2.02	.92		2.94	3.86
0100	Foamglass		325	.025		2.25	.92		3.17	4.12
0300	Mineral or fiber, trapezoidal, 1" x 4" x 48"		325	.025		.19	.92		1.11	1.85
0400	1-1/2" x 5-5/8" x 48"		325	.025		.31	.92		1.23	1.98
9000	Minimum labor/equipment charge		4	2	Job		75		75	133

07 51 13.40 Felts		Crew	Daily Output	Labor-Hours	Unit	Material	Labor	Equipment	Total	Total Incl O&P
0010	**FELTS**									
0012	Glass fibered roofing felt, #15, not mopped	1 Rofc	58	.138	Sq.	6.90	5.15		12.05	16.75
0300	Base sheet, #45, channel vented		58	.138		32.50	5.15		37.65	44.50
0400	#50, coated		58	.138		16.55	5.15		21.70	27.50
0500	Cap, mineral surfaced		58	.138		45	5.15		50.15	58.50
0600	Flashing membrane, #65		16	.500		9.65	18.70		28.35	43.50
0800	Coal tar fibered, #15, no mopping		58	.138		14.05	5.15		19.20	24.50
0900	Asphalt felt, #15, 4 sq per roll, no mopping		58	.138		5.65	5.15		10.80	15.40
1100	#30, 2 sq per roll		58	.138		11.30	5.15		16.45	21.50
1200	Double coated, #33		58	.138		9.95	5.15		15.10	20
1400	#40, base sheet		58	.138		10.95	5.15		16.10	21
1450	Coated and saturated		58	.138		10	5.15		15.15	20
1500	Tarred felt, organic, #15, 4 sq rolls		58	.138		11.55	5.15		16.70	22
1550	#30, 2 sq roll		58	.138		23	5.15		28.15	34.50
1700	Add for mopping above felts, per ply, asphalt, 24 lb. per sq.	G-1	192	.292		9.35	10.20	2.36	21.91	31
1800	Coal tar mopping, 30 lb. per sq.		186	.301		18	10.55	2.43	30.98	41
1900	Flood coat, with asphalt, 60 lb. per sq.		60	.933		23.50	32.50	7.55	63.55	92
2000	With coal tar, 75 lb. per sq.		56	1		45	35	8.10	88.10	120
9000	Minimum labor/equipment charge	1 Rofc	4	2	Job		75		75	133

07 51 13.50 Walkways for Built-Up Roofs		Crew	Daily Output	Labor-Hours	Unit	Material	Labor	Equipment	Total	Total Incl O&P
0010	**WALKWAYS FOR BUILT-UP ROOFS**									
0020	Asphalt impregnated, 3' x 6' x 1/2" thick	1 Rofc	400	.020	S.F.	1.73	.75		2.48	3.23
0100	3' x 3' x 3/4" thick		400	.020	"	4.99	.75		5.74	6.85
0600	100% recycled rubber, 3' x 4' x 3/8" G		400	.020	L.F.	6.35	.75		7.10	8.30
0610	3' x 4' x 1/2" G		400	.020		6.80	.75		7.55	8.80
0620	3' x 4' x 3/4" G		400	.020		8.45	.75		9.20	10.65
9000	Minimum labor/equipment charge		2.75	2.909	Job		109		109	193

07 52 Modified Bituminous Membrane Roofing

07 52 13 – Atactic-Polypropylene-Modified Bituminous Membrane Roofing

07 52 13.10 APP Modified Bituminous Membrane		Crew	Daily Output	Labor-Hours	Unit	2012 Bare Costs Material	Labor	Equipment	Total	Total Incl O&P
0010	**APP MODIFIED BITUMINOUS MEMBRANE** R075213-30									
0020	Base sheet, #15 glass fiber felt, nailed to deck	1 Rofc	58	.138	Sq.	7.85	5.15		13	17.75
0030	Spot mopped to deck	G-1	295	.190		11.60	6.65	1.53	19.78	26
0040	Fully mopped to deck	"	192	.292		16.25	10.20	2.36	28.81	38.50
0050	#15 organic felt, nailed to deck	1 Rofc	58	.138		6.60	5.15		11.75	16.40
0060	Spot mopped to deck	G-1	295	.190		10.35	6.65	1.53	18.53	25
0070	Fully mopped to deck	"	192	.292		15	10.20	2.36	27.56	37
2100	APP mod., smooth surf. cap sheet, poly. reinf., torched, 160 mils	G-5	2100	.019	S.F.	.52	.65	.09	1.26	1.81
2150	170 mils		2100	.019		.55	.65	.09	1.29	1.85
2200	Granule surface cap sheet, poly. reinf., torched, 180 mils		2000	.020		.62	.68	.09	1.39	1.99
2250	Smooth surface flashing, torched, 160 mils		1260	.032		.52	1.08	.14	1.74	2.65
2300	170 mils		1260	.032		.55	1.08	.14	1.77	2.69
2350	Granule surface flashing, torched, 180 mils		1260	.032		.62	1.08	.14	1.84	2.76
2400	Fibrated aluminum coating	1 Rofc	3800	.002		.10	.08		.18	.25

07 52 16 – Styrene-Butadiene-Styrene Modified Bituminous Membrane Roofing

07 52 16.10 SBS Modified Bituminous Membrane		Crew	Daily Output	Labor-Hours	Unit	Material	Labor	Equipment	Total	Total Incl O&P
0010	**SBS MODIFIED BITUMINOUS MEMBRANE**									
0080	SBS modified, granule surf cap sheet, polyester rein., mopped									
1500	Glass fiber reinforced, mopped, 160 mils	G-1	2000	.028	S.F.	.49	.98	.23	1.70	2.53
1600	Smooth surface cap sheet, mopped, 145 mils		2100	.027		.49	.93	.22	1.64	2.44
1700	Smooth surface flashing, 145 mils		1260	.044		.49	1.55	.36	2.40	3.70
1800	150 mils		1260	.044		.48	1.55	.36	2.39	3.69
1900	Granular surface flashing, 150 mils		1260	.044		.58	1.55	.36	2.49	3.80
2000	160 mils		1260	.044		.77	1.55	.36	2.68	4.01

07 53 Elastomeric Membrane Roofing

07 53 16 – Chlorosulfonate-Polyethylene Roofing

07 53 16.10 Chlorosulfonated Polyethylene Roofing		Crew	Daily Output	Labor-Hours	Unit	Material	Labor	Equipment	Total	Total Incl O&P
0010	**CHLOROSULFONATED POLYETHYLENE ROOFING**									
0800	Chlorosulfonated polyethylene-hypalon (CSPE), 45 mils,									
0900	0.29 P.S.F., fully adhered	G-5	26	1.538	Sq.	189	52.50	6.90	248.40	310
1200	Mechanically attached	"	35	1.143	"	190	39	5.10	234.10	284

07 53 23 – Ethylene-Propylene-Diene-Monomer Roofing

07 53 23.20 Ethylene-Propylene-Diene-Monomer Roofing		Crew	Daily Output	Labor-Hours	Unit	Material	Labor	Equipment	Total	Total Incl O&P
0010	**ETHYLENE-PROPYLENE-DIENE-MONOMER ROOFING (E.P.D.M.)**									
3500	Ethylene-propylene-diene-monomer (EPDM), 45 mils, 0.28 P.S.F.									
3600	Loose-laid & ballasted with stone (10 P.S.F.)	G-5	51	.784	Sq.	67.50	26.50	3.51	97.51	126
3700	Mechanically attached		35	1.143		63.50	39	5.10	107.60	144
3800	Fully adhered with adhesive		26	1.538		99.50	52.50	6.90	158.90	210
4500	60 mils, 0.40 P.S.F.									
4600	Loose-laid & ballasted with stone (10 P.S.F.)	G-5	51	.784	Sq.	78	26.50	3.51	108.01	137
4700	Mechanically attached		35	1.143		73	39	5.10	117.10	155
4800	Fully adhered with adhesive		26	1.538		109	52.50	6.90	168.40	221
4810	45 mil, .20 PSF, membrane only					39.50			39.50	43.50
4820	60 mil, .40 PSF, membrane only					48			48	53
4850	Seam tape for membrane, 3" x 100' roll				Ea.	50			50	55
4900	Batten strips, 10' sections					4.25			4.25	4.68
4910	Cover tape for batten strips, 6" x 100' roll					113			113	124
4930	Plate anchors				M	110			110	121
4970	Adhesive for fully adhered systems, 60 S.F./gal.				Gal.	22.50			22.50	24.50

07 53 Elastomeric Membrane Roofing

07 53 29 – Polyisobutylene Roofing

	07 53 29.10 Polyisobutylene Roofing	Crew	Daily Output	Labor-Hours	Unit	2012 Bare Costs Material	Labor	Equipment	Total	Total Incl O&P
0010	**POLYISOBUTYLENE ROOFING**									
7500	Polyisobutylene (PIB), 100 mils, 0.57 P.S.F.									
7600	Loose-laid & ballasted with stone/gravel (10 P.S.F.)	G-5	51	.784	Sq.	181	26.50	3.51	211.01	250
7700	Partially adhered with adhesive		35	1.143		227	39	5.10	271.10	325
7800	Hot asphalt attachment		35	1.143		217	39	5.10	261.10	315
7900	Fully adhered with contact cement		26	1.538		234	52.50	6.90	293.40	360

07 54 Thermoplastic Membrane Roofing

07 54 19 – Polyvinyl-Chloride Roofing

	07 54 19.10 Polyvinyl-Chloride Roofing (P.V.C.)	Crew	Daily Output	Labor-Hours	Unit	Material	Labor	Equipment	Total	Total Incl O&P
0010	**POLYVINYL-CHLORIDE ROOFING (P.V.C.)**									
8200	Heat welded seams									
8700	Reinforced, 48 mils, 0.33 P.S.F.									
8750	Loose-laid & ballasted with stone/gravel (12 P.S.F.)	G-5	51	.784	Sq.	100	26.50	3.51	130.01	161
8800	Mechanically attached		35	1.143		95.50	39	5.10	139.60	180
8850	Fully adhered with adhesive		26	1.538		125	52.50	6.90	184.40	239
8860	Reinforced, 60 mils, .40 P.S.F.									
8870	Loose-laid & ballasted with stone/gravel (12 P.S.F.)	G-5	51	.784	Sq.	102	26.50	3.51	132.01	164
8880	Mechanically attached		35	1.143		97.50	39	5.10	141.60	182
8890	Fully adhered with adhesive		26	1.538		127	52.50	6.90	186.40	241

07 54 23 – Thermoplastic-Polyolefin Roofing

	07 54 23.10 Thermoplastic Polyolefin Roofing (T.P.O)	Crew	Daily Output	Labor-Hours	Unit	Material	Labor	Equipment	Total	Total Incl O&P
0010	**THERMOPLASTIC POLYOLEFIN ROOFING (T.P.O.)**									
0100	45 mils, loose laid & ballasted with stone(1/2 ton/sq.)	G-5	51	.784	Sq.	82	26.50	3.51	112.01	141
0120	Fully adhered		25	1.600		73.50	54.50	7.15	135.15	185
0140	Mechanically attached		34	1.176		73.50	40	5.25	118.75	157
0160	Self adhered		35	1.143		88	39	5.10	132.10	172
0180	60 mil membrane, heat welded seams, ballasted		50	.800		88.50	27	3.58	119.08	149
0200	Fully adhered		25	1.600		79	54.50	7.15	140.65	191
0220	Mechanically attached		34	1.176		83.50	40	5.25	128.75	169
0240	Self adhered		35	1.143		107	39	5.10	151.10	192

07 56 Fluid-Applied Roofing

07 56 10 – Fluid-Applied Roofing Elastomers

	07 56 10.10 Elastomeric Roofing	Crew	Daily Output	Labor-Hours	Unit	Material	Labor	Equipment	Total	Total Incl O&P
0010	**ELASTOMERIC ROOFING**									
0110	Acrylic rubber, fluid applied, 20 mils thick	G-5	2000	.020	S.F.	2.04	.68	.09	2.81	3.55
0120	50 mils, reinforced		1200	.033		3.16	1.13	.15	4.44	5.65
0130	For walking surface, add		900	.044		.96	1.51	.20	2.67	3.96
0300	Hypalon neoprene, fluid applied, 20 mil thick, not-reinforced	G-1	1135	.049		2.57	1.72	.40	4.69	6.35
0600	Non-woven polyester, reinforced		960	.058		2.61	2.04	.47	5.12	7
0700	5 coat neoprene deck, 60 mil thick, under 10,000 S.F.		325	.172		5.75	6	1.39	13.14	18.55
0900	Over 10,000 S.F.		625	.090		5.35	3.13	.72	9.20	12.25
1300	Vinyl plastic traffic deck, sprayed, 2 to 4 mils thick		625	.090		1.68	3.13	.72	5.53	8.20
1500	Vinyl and neoprene membrane traffic deck		1550	.036		1.77	1.26	.29	3.32	4.51
9000	Minimum labor/equipment charge	1 Rofc	2	4	Job		150		150	266

07 57 Coated Foamed Roofing

07 57 13 – Sprayed Polyurethane Foam Roofing

07 57 13.10 Sprayed Polyurethane Foam Roofing (S.P.F.)		Crew	Daily Output	Labor-Hours	Unit	Material	2012 Bare Costs Labor	Equipment	Total	Total Incl O&P
0010	**SPRAYED POLYURETHANE FOAM ROOFING (S.P.F.)**									
0100	Primer for metal substrate (when required)	G-2A	3000	.008	S.F.	.42	.27	.21	.90	1.15
0200	Primer for non-metal substrate (when required)		3000	.008		.16	.27	.21	.64	.87
0300	Closed cell spray, polyurethane foam, 3 lb. per C.F. density, 1", R6.7		15000	.002		.57	.05	.04	.66	.76
0400	2", R13.4		13125	.002		1.14	.06	.05	1.25	1.41
0500	3", R18.6		11485	.002		1.70	.07	.05	1.82	2.05
0550	4", R24.8		10080	.002		2.27	.08	.06	2.41	2.71
0700	Spray-on silicone coating		2500	.010		.92	.32	.25	1.49	1.85
0800	Warranty 5-20 year manufacturer's									.15
0900	Warranty 20 year, no dollar limit									.20

07 58 Roll Roofing

07 58 10 – Asphalt Roll Roofing

07 58 10.10 Roll Roofing		Crew	Daily Output	Labor-Hours	Unit	Material	Labor	Equipment	Total	Total Incl O&P
0010	**ROLL ROOFING**									
0100	Asphalt, mineral surface									
0200	1 ply #15 organic felt, 1 ply mineral surfaced									
0300	Selvage roofing, lap 19", nailed & mopped	G-1	27	2.074	Sq.	62.50	72.50	16.75	151.75	216
0400	3 plies glass fiber felt (type IV), 1 ply mineral surfaced									
0500	Selvage roofing, lapped 19", mopped	G-1	25	2.240	Sq.	102	78.50	18.10	198.60	271
0600	Coated glass fiber base sheet, 2 plies of glass fiber									
0700	Felt (type IV), 1 ply mineral surfaced selvage									
0800	Roofing, lapped 19", mopped	G-1	25	2.240	Sq.	112	78.50	18.10	208.60	282
0900	On nailable decks	"	24	2.333	"	102	81.50	18.85	202.35	278
1000	3 plies glass fiber felt (type III), 1 ply mineral surfaced									
1100	Selvage roofing, lapped 19", mopped	G-1	25	2.240	Sq.	102	78.50	18.10	198.60	271

07 61 Sheet Metal Roofing

07 61 13 – Standing Seam Sheet Metal Roofing

07 61 13.10 Standing Seam Sheet Metal Roofing, Field Fab.		Crew	Daily Output	Labor-Hours	Unit	Material	Labor	Equipment	Total	Total Incl O&P
0010	**STANDING SEAM SHEET METAL ROOFING, FIELD FABRICATED**									
0400	Copper standing seam roofing, over 10 squares, 16 oz., 125 lb. per sq.	1 Shee	1.30	6.154	Sq.	985	320		1,305	1,575
0600	18 oz., 140 lb. per sq.		1.20	6.667		1,100	350		1,450	1,750
0700	20 oz., 150 lb. per sq.		1.10	7.273		1,225	380		1,605	1,950
1200	For abnormal conditions or small areas, add					25%	100%			
1300	For lead-coated copper, add					25%				
1400	Lead standing seam roofing, 5 lb. per S.F.	1 Shee	1.30	6.154		620	320		940	1,175
1500	Zinc standing seam roofing, .020" thick		1.20	6.667		1,100	350		1,450	1,775
1510	.027" thick		1.15	6.957		1,550	365		1,915	2,300
1520	.032" thick		1.10	7.273		1,875	380		2,255	2,675
1530	.040" thick		1.05	7.619		2,100	395		2,495	2,925
9000	Minimum labor/equipment charge		2	4	Job		209		209	330

07 61 16 – Batten Seam Sheet Metal Roofing

07 61 16.10 Batten Seam Sheet Metal Roofing, Field Fabricated		Crew	Daily Output	Labor-Hours	Unit	Material	Labor	Equipment	Total	Total Incl O&P
0009	**BATTEN SEAM SHEET METAL ROOFING, FIELD FABRICATED**									
0012	Copper batten seam roofing, over 10 sq., 16 oz., 130 lb. per sq.	1 Shee	1.10	7.273	Sq.	1,250	380		1,630	1,975
0020	Lead batten seam roofing, 5 lb. per S.F.		1.20	6.667		620	350		970	1,225
0100	Zinc/copper alloy batten seam roofing, .020 thick		1.20	6.667		1,100	350		1,450	1,775
0200	Copper roofing, batten seam, over 10 sq., 18 oz., 145 lb. per sq.		1	8		1,400	415		1,815	2,175

07 61 Sheet Metal Roofing

07 61 16 – Batten Seam Sheet Metal Roofing

07 61 16.10 Batten Seam Sheet Metal Roofing, Field Fabricated		Crew	Daily Output	Labor-Hours	Unit	2012 Bare Costs Material	Labor	Equipment	Total	Total Incl O&P
0300	20 oz., 160 lb. per sq.	1 Shee	1	8	Sq.	1,550	415		1,965	2,375
0500	Stainless steel batten seam roofing, type 304, 28 gauge		1.20	6.667		625	350		975	1,225
0600	26 gauge		1.15	6.957		630	365		995	1,275
0800	Zinc, copper alloy roofing, batten seam, .027" thick		1.15	6.957		1,550	365		1,915	2,300
0900	.032" thick		1.10	7.273		1,875	380		2,255	2,675
1000	.040" thick		1.05	7.619		2,100	395		2,495	2,925
9000	Minimum labor/equipment charge		2	4	Job		209		209	330

07 61 19 – Flat Seam Sheet Metal Roofing

07 61 19.10 Flat Seam Sheet Metal Roofing, Field Fabricated

Line	Description	Crew	Daily Output	Labor-Hours	Unit	Material	Labor	Equipment	Total	Total Incl O&P
0010	**FLAT SEAM SHEET METAL ROOFING, FIELD FABRICATED**									
0900	Copper flat seam roofing, over 10 squares, 16 oz., 115 lb./sq.	1 Shee	1.20	6.667	Sq.	915	350		1,265	1,550
0950	18 oz., 130 lb./sq.		1.15	6.957		1,025	365		1,390	1,700
1000	20 oz., 145 lb./sq.		1.10	7.273		1,150	380		1,530	1,850
1008	Zinc flat seam roofing, .020" thick		1.20	6.667		1,100	350		1,450	1,775
1010	.027" thick		1.15	6.957		1,550	365		1,915	2,300
1020	.032" thick		1.12	7.143		1,875	375		2,250	2,650
1100	Lead flat seam roofing, 5 lb. per S.F.		1.30	6.154		615	320		935	1,175
9000	Minimum labor/equipment charge		2.75	2.909	Job		152		152	239

07 62 Sheet Metal Flashing and Trim

07 62 10 – Sheet Metal Trim

07 62 10.10 Sheet Metal Cladding

Line	Description	Crew	Daily Output	Labor-Hours	Unit	Material	Labor	Equipment	Total	Total Incl O&P
0010	**SHEET METAL CLADDING**									
0100	Aluminum, up to 6 bends, .032" thick, window casing	1 Carp	180	.044	S.F.	1.31	1.96		3.27	4.65
0200	Window sill		72	.111	L.F.	1.31	4.90		6.21	9.50
0300	Door casing		180	.044	S.F.	1.31	1.96		3.27	4.65
0400	Fascia		250	.032		1.31	1.41		2.72	3.75
0500	Rake trim		225	.036		1.31	1.57		2.88	4.01
0700	.024" thick, window casing		180	.044		1.43	1.96		3.39	4.78
0800	Window sill		72	.111	L.F.	1.43	4.90		6.33	9.60
0900	Door casing		180	.044	S.F.	1.43	1.96		3.39	4.78
1000	Fascia		250	.032		1.43	1.41		2.84	3.88
1100	Rake trim		225	.036		1.43	1.57		3	4.14
1200	Vinyl coated aluminum, up to 6 bends, window casing		180	.044		1.11	1.96		3.07	4.43
1300	Window sill		72	.111	L.F.	1.11	4.90		6.01	9.25
1400	Door casing		180	.044	S.F.	1.11	1.96		3.07	4.43
1500	Fascia		250	.032		1.11	1.41		2.52	3.53
1600	Rake trim		225	.036		1.11	1.57		2.68	3.79

07 65 Flexible Flashing

07 65 10 – Sheet Metal Flashing

07 65 10.10 Sheet Metal Flashing and Counter Flashing		Crew	Daily Output	Labor-Hours	Unit	2012 Bare Costs Material	Labor	Equipment	Total	Total Incl O&P
0010	**SHEET METAL FLASHING AND COUNTER FLASHING**									
0011	Including up to 4 bends									
0020	Aluminum, mill finish, .013" thick	1 Rofc	145	.055	S.F.	.73	2.06		2.79	4.47
0030	.016" thick		145	.055		.84	2.06		2.90	4.59
0060	.019" thick		145	.055		1.07	2.06		3.13	4.85
0100	.032" thick		145	.055		1.25	2.06		3.31	5.05
0200	.040" thick		145	.055		2.14	2.06		4.20	6
0300	.050" thick		145	.055		2.25	2.06		4.31	6.15
0325	Mill finish 5" x 7" step flashing, .016" thick		1920	.004	Ea.	.18	.16		.34	.48
0350	Mill finish 12" x 12" step flashing, .016" thick		1600	.005	"	.62	.19		.81	1.01
0400	Painted finish, add				S.F.	.34			.34	.37
1000	Mastic-coated 2 sides, .005" thick	1 Rofc	330	.024		1.67	.91		2.58	3.45
1100	.016" thick		330	.024		1.84	.91		2.75	3.63
1600	Copper, 16 oz., sheets, under 1000 lb.		115	.070		7.85	2.60		10.45	13.25
1700	Over 4000 lb.		155	.052		7.85	1.93		9.78	12.10
1900	20 oz. sheets, under 1000 lb.		110	.073		9.85	2.72		12.57	15.70
2000	Over 4000 lb.		145	.055		9.35	2.06		11.41	13.95
2200	24 oz. sheets, under 1000 lb.		105	.076		12.10	2.85		14.95	18.40
2300	Over 4000 lb.		135	.059		11.50	2.22		13.72	16.60
2500	32 oz. sheets, under 1000 lb.		100	.080		16.15	2.99		19.14	23
2600	Over 4000 lb.		130	.062		15.35	2.30		17.65	21
2700	W shape for valleys, 16 oz., 24" wide		100	.080	L.F.	15.85	2.99		18.84	22.50
5800	Lead, 2.5 lb. per S.F., up to 12" wide		135	.059	S.F.	5.40	2.22		7.62	9.90
5900	Over 12" wide		135	.059		4.43	2.22		6.65	8.80
8900	Stainless steel sheets, 32 ga, .010" thick		155	.052		4	1.93		5.93	7.85
9000	28 ga, .015" thick		155	.052		5.65	1.93		7.58	9.65
9100	26 ga, .018" thick		155	.052		7.25	1.93		9.18	11.45
9200	24 ga, .025" thick		155	.052		8.90	1.93		10.83	13.25
9290	For mechanically keyed flashing, add					40%				
9320	Steel sheets, galvanized, 20 gauge	1 Rofc	130	.062	S.F.	1.43	2.30		3.73	5.65
9340	30 gauge		160	.050		1.33	1.87		3.20	4.78
9400	Terne coated stainless steel, .015" thick, 28 ga		155	.052		7.35	1.93		9.28	11.55
9500	.018" thick, 26 ga		155	.052		8.30	1.93		10.23	12.55
9600	Zinc and copper alloy (brass), .020" thick		155	.052		11.10	1.93		13.03	15.65
9700	.027" thick		155	.052		15.60	1.93		17.53	20.50
9800	.032" thick		155	.052		18.80	1.93		20.73	24
9900	.040" thick		155	.052		21	1.93		22.93	26.50
9950	Minimum labor/equipment charge		3	2.667	Job		99.50		99.50	177

07 65 12 – Fabric and Mastic Flashings

07 65 12.10 Fabric and Mastic Flashing and Counter Flashing		Crew	Daily Output	Labor-Hours	Unit	Material	Labor	Equipment	Total	Total Incl O&P
0010	**FABRIC AND MASTIC FLASHING AND COUNTER FLASHING**									
1300	Asphalt flashing cement, 5 gallon				Gal.	10.50			10.50	11.55
4900	Fabric, asphalt-saturated cotton, specification grade	1 Rofc	35	.229	S.Y.	2.96	8.55		11.51	18.45
5000	Utility grade		35	.229		1.42	8.55		9.97	16.75
5300	Close-mesh fabric, saturated, 17 oz. per S.Y.		35	.229		2.15	8.55		10.70	17.55
5500	Fiberglass, resin-coated		35	.229		1.05	8.55		9.60	16.35
8500	Shower pan, bituminous membrane, 7 oz.		155	.052	S.F.	2	1.93		3.93	5.65

07 65 13 – Laminated Sheet Flashing

07 65 13.10 Laminated Sheet Flashing		Crew	Daily Output	Labor-Hours	Unit	Material	Labor	Equipment	Total	Total Incl O&P
0010	**LAMINATED SHEET FLASHING**, Including up to 4 bends									
0500	Fabric-backed 2 sides, .004" thick	1 Rofc	330	.024	S.F.	1.41	.91		2.32	3.16
0700	.005" thick		330	.024		1.67	.91		2.58	3.45

07 65 Flexible Flashing

07 65 13 – Laminated Sheet Flashing

07 65 13.10 Laminated Sheet Flashing		Crew	Daily Output	Labor-Hours	Unit	Material	Labor	Equipment	Total	Total Incl O&P
						2012 Bare Costs				
0750	Mastic-backed, self adhesive	1 Rofc	460	.017	S.F.	3.46	.65		4.11	4.97
0800	Mastic-coated 2 sides, .004" thick		330	.024		1.41	.91		2.32	3.16
2800	Copper, paperbacked 1 side, 2 oz.		330	.024		1.67	.91		2.58	3.45
2900	3 oz.		330	.024		1.97	.91		2.88	3.78
3100	Paperbacked 2 sides, 2 oz.		330	.024		1.72	.91		2.63	3.50
3150	3 oz.		330	.024		2.07	.91		2.98	3.89
3200	5 oz.		330	.024		2.69	.91		3.60	4.57
3250	7 oz.		330	.024		4.81	.91		5.72	6.90
3400	Mastic-backed 2 sides, copper, 2 oz.		330	.024		1.65	.91		2.56	3.43
3500	3 oz.		330	.024		2	.91		2.91	3.81
3700	5 oz.		330	.024		3	.91		3.91	4.91
3800	Fabric-backed 2 sides, copper, 2 oz.		330	.024		1.93	.91		2.84	3.73
4000	3 oz.		330	.024		2.19	.91		3.10	4.02
4100	5 oz.		330	.024		3.13	.91		4.04	5.05
4300	Copper-clad stainless steel, .015" thick, under 500 lb.		115	.070		5.80	2.60		8.40	11
4400	Over 2000 lb.		155	.052		5.55	1.93		7.48	9.55
4600	.018" thick, under 500 lb.		100	.080		6.55	2.99		9.54	12.50
4700	Over 2000 lb.		145	.055		6.20	2.06		8.26	10.50
6100	Lead-coated copper, fabric-backed, 2 oz.		330	.024		2.64	.91		3.55	4.51
6200	5 oz.		330	.024		3.04	.91		3.95	4.95
6400	Mastic-backed 2 sides, 2 oz.		330	.024		2.06	.91		2.97	3.88
6500	5 oz.		330	.024		2.55	.91		3.46	4.42
6700	Paperbacked 1 side, 2 oz.		330	.024		1.78	.91		2.69	3.57
6800	3 oz.		330	.024		2.09	.91		3	3.91
7000	Paperbacked 2 sides, 2 oz.		330	.024		1.84	.91		2.75	3.63
7100	5 oz.		330	.024		3	.91		3.91	4.91
8550	3 ply copper and fabric, 3 oz.		155	.052		2.19	1.93		4.12	5.85
8600	7 oz.		155	.052		3.25	1.93		5.18	7
8700	Lead on copper and fabric, 5 oz.		155	.052		3.04	1.93		4.97	6.75
8800	7 oz.		155	.052		5.25	1.93		7.18	9.25
9300	Stainless steel, paperbacked 2 sides, .005" thick	↓	330	.024	↓	3.62	.91		4.53	5.60

07 65 19 – Plastic Sheet Flashing

07 65 19.10 Plastic Sheet Flashing and Counter Flashing		Crew	Daily Output	Labor-Hours	Unit	Material	Labor	Equipment	Total	Total Incl O&P
0010	**PLASTIC SHEET FLASHING AND COUNTER FLASHING**									
7300	Polyvinyl chloride, black, .010" thick	1 Rofc	285	.028	S.F.	.21	1.05		1.26	2.10
7400	.020" thick		285	.028		.30	1.05		1.35	2.20
7600	.030" thick		285	.028		.39	1.05		1.44	2.30
7700	.056" thick		285	.028		.94	1.05		1.99	2.90
7900	Black or white for exposed roofs, .060" thick	↓	285	.028	↓	2	1.05		3.05	4.07
8060	PVC tape, 5" x 45 mils, for joint covers, 100 L.F./roll				Ea.	136			136	150

07 65 23 – Rubber Sheet Flashing

07 65 23.10 Rubber Sheet Flashing and Counterflashing		Crew	Daily Output	Labor-Hours	Unit	Material	Labor	Equipment	Total	Total Incl O&P
0010	**RUBBER SHEET FLASHING AND COUNTERFLASHING**									
8100	Rubber, butyl, 1/32" thick	1 Rofc	285	.028	S.F.	1.75	1.05		2.80	3.80
8200	1/16" thick		285	.028		2.38	1.05		3.43	4.49
8300	Neoprene, cured, 1/16" thick		285	.028		1.92	1.05		2.97	3.98
8400	1/8" thick	↓	285	.028	↓	4	1.05		5.05	6.25

07 71 Roof Specialties

07 71 16 – Manufactured Counterflashing Systems

07 71 16.10 Roof Drain Boot		Crew	Daily Output	Labor-Hours	Unit	Material	2012 Bare Costs Labor	Equipment	Total	Total Incl O&P
0010	**ROOF DRAIN BOOT**									
0100	Cast iron, 2″ x 3″	1 Shee	125	.064	L.F.	114	3.34		117.34	130
0200	3″ x 4″		125	.064		227	3.34		230.34	255
0300	4″ x 5″		125	.064		238	3.34		241.34	267
0400	4″ x 6″		125	.064		286	3.34		289.34	320

07 71 19 – Manufactured Gravel Stops and Fascias

07 71 19.10 Gravel Stop

		Crew	Daily Output	Labor-Hours	Unit	Material	Labor	Equipment	Total	Total Incl O&P
0010	**GRAVEL STOP**									
0020	Aluminum, .050″ thick, 4″ face height, mill finish	1 Shee	145	.055	L.F.	6	2.88		8.88	11.15
0080	Duranodic finish		145	.055		6.50	2.88		9.38	11.70
0100	Painted		145	.055		6.85	2.88		9.73	12.05
0300	6″ face height		135	.059		6.45	3.09		9.54	11.95
0350	Duranodic finish		135	.059		6.95	3.09		10.04	12.50
0400	Painted		135	.059		8.10	3.09		11.19	13.75
0600	8″ face height		125	.064		7.30	3.34		10.64	13.25
0650	Duranodic finish		125	.064		8	3.34		11.34	14.05
0700	Painted		125	.064		8.15	3.34		11.49	14.20
0900	12″ face height, .080 thick, 2 piece		100	.080		10.15	4.17		14.32	17.75
0950	Duranodic finish		100	.080		10.05	4.17		14.22	17.60
1000	Painted		100	.080		11.90	4.17		16.07	19.65
1500	Polyvinyl chloride, 6″ face height		135	.059		4.90	3.09		7.99	10.25
1600	9″ face height		125	.064		5.80	3.34		9.14	11.60
1800	Stainless steel, 24 ga., 6″ face height		135	.059		9.85	3.09		12.94	15.65
1900	12″ face height		100	.080		19.70	4.17		23.87	28
2100	20 ga., 6″ face height		135	.059		11.90	3.09		14.99	17.90
2200	12″ face height		100	.080		24	4.17		28.17	32.50
9000	Minimum labor/equipment charge		3.50	2.286	Job		119		119	188

07 71 19.30 Fascia

		Crew	Daily Output	Labor-Hours	Unit	Material	Labor	Equipment	Total	Total Incl O&P
0010	**FASCIA**									
0100	Aluminum, reverse board and batten, .032″ thick, colored, no furring incl	1 Shee	145	.055	S.F.	6.25	2.88		9.13	11.45
0200	Residential type, aluminum	1 Carp	200	.040	L.F.	1.35	1.76		3.11	4.38
0220	Vinyl	"	200	.040	"	1.64	1.76		3.40	4.69
0300	Steel, galv and enameled, stock, no furring, long panels	1 Shee	145	.055	S.F.	4.20	2.88		7.08	9.15
0600	Short panels		115	.070	"	5.70	3.63		9.33	12
9000	Minimum labor/equipment charge		4	2	Job		104		104	164

07 71 23 – Manufactured Gutters and Downspouts

07 71 23.10 Downspouts

		Crew	Daily Output	Labor-Hours	Unit	Material	Labor	Equipment	Total	Total Incl O&P
0010	**DOWNSPOUTS**									
0020	Aluminum 2″ x 3″, .020″ thick, embossed	1 Shee	190	.042	L.F.	.97	2.20		3.17	4.53
0100	Enameled		190	.042		1.26	2.20		3.46	4.85
0300	Enameled, .024″ thick, 2″ x 3″		180	.044		2.05	2.32		4.37	5.90
0400	3″ x 4″		140	.057		2.75	2.98		5.73	7.70
0600	Round, corrugated aluminum, 3″ diameter, .020″ thick		190	.042		1.72	2.20		3.92	5.35
0700	4″ diameter, .025″ thick		140	.057		2.65	2.98		5.63	7.60
0900	Wire strainer, round, 2″ diameter		155	.052	Ea.	3.92	2.69		6.61	8.55
1000	4″ diameter		155	.052		7.20	2.69		9.89	12.20
1200	Rectangular, perforated, 2″ x 3″		145	.055		2.30	2.88		5.18	7.05
1300	3″ x 4″		145	.055		3.32	2.88		6.20	8.20
1500	Copper, round, 16 oz., stock, 2″ diameter		190	.042	L.F.	8.10	2.20		10.30	12.35
1600	3″ diameter		190	.042		8.10	2.20		10.30	12.35
1800	4″ diameter		145	.055		10.85	2.88		13.73	16.45

07 71 Roof Specialties

07 71 23 – Manufactured Gutters and Downspouts

07 71 23.10 Downspouts		Crew	Daily Output	Labor-Hours	Unit	Material	2012 Bare Costs Labor	Equipment	Total	Total Incl O&P
1900	5″ diameter	1 Shee	130	.062	L.F.	16.50	3.21		19.71	23
2100	Rectangular, corrugated copper, stock, 2″ x 3″		190	.042		8.45	2.20		10.65	12.70
2200	3″ x 4″		145	.055		11.10	2.88		13.98	16.75
2400	Rectangular, plain copper, stock, 2″ x 3″		190	.042		7.40	2.20		9.60	11.60
2500	3″ x 4″		145	.055		9.90	2.88		12.78	15.45
2700	Wire strainers, rectangular, 2″ x 3″		145	.055	Ea.	7.90	2.88		10.78	13.20
2800	3″ x 4″		145	.055		7.90	2.88		10.78	13.20
3000	Round, 2″ diameter		145	.055		7.90	2.88		10.78	13.20
3100	3″ diameter		145	.055		7.90	2.88		10.78	13.20
3300	4″ diameter		145	.055		14.25	2.88		17.13	20
3400	5″ diameter		115	.070		19.75	3.63		23.38	27
3600	Lead-coated copper, round, stock, 2″ diameter		190	.042	L.F.	34	2.20		36.20	41
3700	3″ diameter		190	.042		30	2.20		32.20	36.50
3900	4″ diameter		145	.055		31	2.88		33.88	39
4000	5″ diameter, corrugated		130	.062		24.50	3.21		27.71	32
4200	6″ diameter, corrugated		105	.076		30	3.97		33.97	39.50
4300	Rectangular, corrugated, stock, 2″ x 3″		190	.042		12.70	2.20		14.90	17.40
4500	Plain, stock, 2″ x 3″		190	.042		16.35	2.20		18.55	21.50
4600	3″ x 4″		145	.055		18.20	2.88		21.08	24.50
4800	Steel, galvanized, round, corrugated, 2″ or 3″ diameter, 28 gauge		190	.042		1.84	2.20		4.04	5.50
4900	4″ diameter, 28 gauge		145	.055		2.54	2.88		5.42	7.30
5100	5″ diameter, 28 gauge		130	.062		3.61	3.21		6.82	9
5200	26 gauge		130	.062		4	3.21		7.21	9.45
5400	6″ diameter, 28 gauge		105	.076		4	3.97		7.97	10.65
5500	26 gauge		105	.076		5.80	3.97		9.77	12.65
5700	Rectangular, corrugated, 28 gauge, 2″ x 3″		190	.042		1.28	2.20		3.48	4.87
5800	3″ x 4″		145	.055		1.82	2.88		4.70	6.55
6000	Rectangular, plain, 28 gauge, galvanized, 2″ x 3″		190	.042		1.28	2.20		3.48	4.87
6100	3″ x 4″		145	.055		2.44	2.88		5.32	7.20
6300	Epoxy painted, 24 gauge, corrugated, 2″ x 3″		190	.042		2	2.20		4.20	5.65
6400	3″ x 4″		145	.055		2.50	2.88		5.38	7.30
6600	Wire strainers, rectangular, 2″ x 3″		145	.055	Ea.	12	2.88		14.88	17.75
6700	3″ x 4″		145	.055		14	2.88		16.88	19.95
6900	Round strainers, 2″ or 3″ diameter		145	.055		3.12	2.88		6	7.95
7000	4″ diameter		145	.055		5.80	2.88		8.68	10.90
9000	Minimum labor/equipment charge		4	2	Job		104		104	164

07 71 23.20 Downspout Elbows		Crew	Daily Output	Labor-Hours	Unit	Material	Labor	Equipment	Total	Total Incl O&P
0010	**DOWNSPOUT ELBOWS**									
0020	Aluminum, 2″ x 3″, embossed	1 Shee	100	.080	Ea.	1.02	4.17		5.19	7.65
0100	Enameled		100	.080		1.02	4.17		5.19	7.65
0200	3″ x 4″, .025″ thick, embossed		100	.080		3.50	4.17		7.67	10.40
0300	Enameled		100	.080		2.02	4.17		6.19	8.75
0400	Round corrugated, 3″, embossed, .020″ thick		100	.080		2.74	4.17		6.91	9.55
0500	4″, .025″ thick		100	.080		5.80	4.17		9.97	12.90
0600	Copper, 16 oz. round, 2″ diameter		100	.080		8.50	4.17		12.67	15.90
0700	3″ diameter		100	.080		11.40	4.17		15.57	19.10
0800	4″ diameter		100	.080		12.90	4.17		17.07	21
1000	2″ x 3″ corrugated		100	.080		8.50	4.17		12.67	15.90
1100	3″ x 4″ corrugated		100	.080		14	4.17		18.17	22
9000	Minimum labor/equipment charge		4	2	Job		104		104	164

07 71 Roof Specialties

07 71 23 – Manufactured Gutters and Downspouts

07 71 23.30 Gutters

	07 71 23.30 Gutters	Crew	Daily Output	Labor-Hours	Unit	2012 Bare Costs Material	Labor	Equipment	Total	Total Incl O&P
0010	**GUTTERS**									
0012	Aluminum, stock units, 5" K type, .027" thick, plain	1 Shee	120	.067	L.F.	2.67	3.48		6.15	8.40
0100	Enameled		120	.067		2.58	3.48		6.06	8.30
0300	5" K type type, .032" thick, plain		120	.067		2.89	3.48		6.37	8.60
0400	Enameled		120	.067		3.26	3.48		6.74	9.05
0700	Copper, half round, 16 oz., stock units, 4" wide		120	.067		9.90	3.48		13.38	16.35
0900	5" wide		120	.067		9	3.48		12.48	15.35
1000	6" wide		115	.070		14.60	3.63		18.23	22
1200	K type, 16 oz., stock, 4" wide		120	.067		13.20	3.48		16.68	19.95
1300	5" wide		120	.067		11.70	3.48		15.18	18.30
1500	Lead coated copper, half round, stock, 4" wide		120	.067		10.40	3.48		13.88	16.90
1600	6" wide		115	.070		22	3.63		25.63	30
1800	K type, stock, 4" wide		120	.067		11.50	3.48		14.98	18.10
1900	5" wide		120	.067		14.45	3.48		17.93	21.50
2100	Stainless steel, half round or box, stock, 4" wide		120	.067		5.90	3.48		9.38	11.90
2200	5" wide		120	.067		6.35	3.48		9.83	12.40
2400	Steel, galv, half round or box, 28 ga, 5" wide, plain		120	.067		3.89	3.48		7.37	9.75
2500	Enameled		120	.067		2.02	3.48		5.50	7.65
2700	26 ga, stock, 5" wide		120	.067		2.18	3.48		5.66	7.85
2800	6" wide		120	.067		2.88	3.48		6.36	8.60
3000	Vinyl, O.G., 4" wide	1 Carp	110	.073		1.08	3.21		4.29	6.45
3100	5" wide		110	.073		1.38	3.21		4.59	6.75
3200	4" half round, stock units		110	.073		1.08	3.21		4.29	6.45
3250	Joint connectors				Ea.	2.88			2.88	3.17
3300	Wood, clear treated cedar, fir or hemlock, 3" x 4"	1 Carp	100	.080	L.F.	10	3.53		13.53	16.80
3400	4" x 5"	"	100	.080	"	13.50	3.53		17.03	20.50
9000	Minimum labor/equipment charge	1 Shee	3.75	2.133	Job		111		111	175

07 71 23.35 Gutter Guard

	07 71 23.35 Gutter Guard	Crew	Daily Output	Labor-Hours	Unit	Material	Labor	Equipment	Total	Total Incl O&P
0010	**GUTTER GUARD**									
0020	6" wide strip, aluminum mesh	1 Carp	500	.016	L.F.	2.85	.71		3.56	4.30
0100	Vinyl mesh		500	.016	"	2.58	.71		3.29	4
9000	Minimum labor/equipment charge		4	2	Job		88		88	145

07 71 26 – Reglets

07 71 26.10 Reglets and Accessories

	07 71 26.10 Reglets and Accessories	Crew	Daily Output	Labor-Hours	Unit	Material	Labor	Equipment	Total	Total Incl O&P
0010	**REGLETS AND ACCESSORIES**									
0020	Aluminum, .025" thick, in concrete parapet	1 Carp	225	.036	L.F.	1.77	1.57		3.34	4.52
0100	Copper, 10 oz.		225	.036		2.87	1.57		4.44	5.75
0300	16 oz.		225	.036		5.60	1.57		7.17	8.70
0400	Galvanized steel, 24 gauge		225	.036		1.03	1.57		2.60	3.70
0600	Stainless steel, .020" thick		225	.036		3.45	1.57		5.02	6.35
0700	Zinc and copper alloy, 20 oz.		225	.036		2.87	1.57		4.44	5.75
0900	Counter flashing for above, 12" wide, .032" aluminum	1 Shee	150	.053		1.78	2.78		4.56	6.35
1000	Copper, 10 oz.		150	.053		5.40	2.78		8.18	10.35
1200	16 oz.		150	.053		5.25	2.78		8.03	10.20
1300	Galvanized steel, .020" thick		150	.053		1.28	2.78		4.06	5.80
1500	Stainless steel, .020" thick		150	.053		5.15	2.78		7.93	10.05
1600	Zinc and copper alloy, 20 oz.		150	.053		5.25	2.78		8.03	10.15
9000	Minimum labor/equipment charge	1 Carp	3	2.667	Job		118		118	193

07 71 Roof Specialties

07 71 29 – Manufactured Roof Expansion Joints

	07 71 29.10 Expansion Joints	Crew	Daily Output	Labor-Hours	Unit	2012 Bare Costs Material	Labor	Equipment	Total	Total Incl O&P
0010	**EXPANSION JOINTS**									
0300	Butyl or neoprene center with foam insulation, metal flanges									
0400	Aluminum, .032″ thick for openings to 2-1/2″	1 Rofc	165	.048	L.F.	16	1.81		17.81	21
0600	For joint openings to 3-1/2″		165	.048		16	1.81		17.81	21
0610	For joint openings to 5″		165	.048		16.40	1.81		18.21	21.50
0620	For joint openings to 8″		165	.048		18.20	1.81		20.01	23
0700	Copper, 16 oz. for openings to 2-1/2″		165	.048		24.50	1.81		26.31	30
0900	For joint openings to 3-1/2″		165	.048		24.50	1.81		26.31	30
0910	For joint openings to 5″		165	.048		25.50	1.81		27.31	31
0920	For joint openings to 8″		165	.048		27	1.81		28.81	33
1000	Galvanized steel, 26 ga. for openings to 2-1/2″		165	.048		13	1.81		14.81	17.50
1200	For joint openings to 3-1/2″		165	.048		13	1.81		14.81	17.50
1210	For joint openings to 5″		165	.048		13.90	1.81		15.71	18.50
1220	For joint openings to 8″		165	.048		17.30	1.81		19.11	22.50
1300	Lead-coated copper, 16 oz. for openings to 2-1/2″		165	.048		33.50	1.81		35.31	40
1500	For joint openings to 3-1/2″		165	.048		40	1.81		41.81	47
1600	Stainless steel, .018″, for openings to 2-1/2″		165	.048		17.90	1.81		19.71	23
1800	For joint openings to 3-1/2″		165	.048		17.90	1.81		19.71	23
1810	For joint openings to 5″		165	.048		18.60	1.81		20.41	23.50
1820	For joint openings to 8″		165	.048		20.50	1.81		22.31	25.50
1900	Neoprene, double-seal type with thick center, 4-1/2″ wide		125	.064		14	2.39		16.39	19.65
1950	Polyethylene bellows, with galv steel flat flanges		100	.080		5.40	2.99		8.39	11.25
1960	With galvanized angle flanges		100	.080		6	2.99		8.99	11.90
2000	Roof joint with extruded aluminum cover, 2″	1 Shee	115	.070		37.50	3.63		41.13	47
2100	Roof joint, plastic curbs, foam center, standard	1 Rofc	100	.080		13.70	2.99		16.69	20.50
2200	Large		100	.080		18.25	2.99		21.24	25.50
2300	Transitions, regular, minimum		10	.800	Ea.	120	30		150	185
2350	Maximum		4	2		150	75		225	299
2400	Large, minimum		9	.889		173	33		206	250
2450	Maximum		3	2.667		182	99.50		281.50	375
2500	Roof to wall joint with extruded aluminum cover	1 Shee	115	.070	L.F.	31.50	3.63		35.13	40.50
2650										
2700	Wall joint, closed cell foam on PVC cover, 9″ wide	1 Rofc	125	.064	L.F.	4.90	2.39		7.29	9.65
2800	12″ wide	″	115	.070	″	5.50	2.60		8.10	10.65
9000	Minimum labor/equipment charge	1 Shee	3	2.667	Job		139		139	219

07 71 43 – Drip Edge

	07 71 43.10 Drip Edge, Rake Edge, Ice Belts	Crew	Daily Output	Labor-Hours	Unit	Material	Labor	Equipment	Total	Total Incl O&P
0010	**DRIP EDGE, RAKE EDGE, ICE BELTS**									
0020	Aluminum, .016″ thick, 5″ wide, mill finish	1 Carp	400	.020	L.F.	.38	.88		1.26	1.87
0100	White finish		400	.020		.38	.88		1.26	1.87
0200	8″ wide, mill finish		400	.020		.46	.88		1.34	1.96
0300	Ice belt, 28″ wide, mill finish		100	.080		3.85	3.53		7.38	10.05
0310	Vented, mill finish		400	.020		2.76	.88		3.64	4.49
0320	Painted finish		400	.020		2.86	.88		3.74	4.60
0400	Galvanized, 5″ wide		400	.020		.48	.88		1.36	1.98
0500	8″ wide, mill finish		400	.020		.69	.88		1.57	2.21
0510	Rake edge, aluminum, 1-1/2″ x 1-1/2″		400	.020		.28	.88		1.16	1.76
0520	3-1/2″ x 1-1/2″		400	.020		.38	.88		1.26	1.87
9000	Minimum labor/equipment charge		4	2	Job		88		88	145

07 72 Roof Accessories

07 72 23 – Relief Vents

07 72 23.10 Roof Vents		Crew	Daily Output	Labor-Hours	Unit	Material	Labor	Equipment	Total	Total Incl O&P
						2012 Bare Costs				
0010	ROOF VENTS									
0020	Mushroom shape, for built-up roofs, aluminum	1 Rofc	30	.267	Ea.	88.50	9.95		98.45	115
0100	PVC, 6" high		30	.267	"	28.50	9.95		38.45	49
9000	Minimum labor/equipment charge		2.75	2.909	Job		109		109	193

07 72 23.20 Vents

		Crew	Daily Output	Labor-Hours	Unit	Material	Labor	Equipment	Total	Total Incl O&P
0010	VENTS									
0100	Soffit or eave, aluminum, mill finish, strips, 2-1/2" wide	1 Carp	200	.040	L.F.	.40	1.76		2.16	3.33
0200	3" wide		200	.040		.45	1.76		2.21	3.39
0300	Enamel finish, 3" wide		200	.040		.50	1.76		2.26	3.44
0400	Mill finish, rectangular, 4" x 16"		72	.111	Ea.	1.37	4.90		6.27	9.55
0500	8" x 16"		72	.111		2.16	4.90		7.06	10.45
2420	Roof ventilator	Q-9	16	1		50.50	47		97.50	130

07 72 26 – Ridge Vents

07 72 26.10 Ridge Vents and Accessories

		Crew	Daily Output	Labor-Hours	Unit	Material	Labor	Equipment	Total	Total Incl O&P
0010	RIDGE VENTS AND ACCESSORIES									
0100	Aluminum strips, mill finish	1 Rofc	160	.050	L.F.	2.62	1.87		4.49	6.20
0150	Painted finish		160	.050	"	3.74	1.87		5.61	7.45
0200	Connectors		48	.167	Ea.	3.68	6.25		9.93	15.15
0300	End caps		48	.167	"	1.37	6.25		7.62	12.60
0400	Galvanized strips		160	.050	L.F.	2.19	1.87		4.06	5.75
0430	Molded polyethylene, shingles not included		160	.050	"	2.99	1.87		4.86	6.60
0440	End plugs		48	.167	Ea.	1.37	6.25		7.62	12.60
0450	Flexible roll, shingles not included		160	.050	L.F.	2.15	1.87		4.02	5.70
2300	Ridge vent strip, mill finish	1 Shee	155	.052	"	2.95	2.69		5.64	7.50

07 72 33 – Roof Hatches

07 72 33.10 Roof Hatch Options

		Crew	Daily Output	Labor-Hours	Unit	Material	Labor	Equipment	Total	Total Incl O&P
0010	ROOF HATCH OPTIONS									
0500	2'-6" x 3', aluminum curb and cover	G-3	10	3.200	Ea.	1,050	140		1,190	1,400
0520	Galvanized steel curb and aluminum cover		10	3.200		640	140		780	930
0540	Galvanized steel curb and cover		10	3.200		590	140		730	875
0600	2'-6" x 4'-6", aluminum curb and cover		9	3.556		920	155		1,075	1,250
0800	Galvanized steel curb and aluminum cover		9	3.556		650	155		805	965
0900	Galvanized steel curb and cover		9	3.556		890	155		1,045	1,225
1100	4' x 4' aluminum curb and cover		8	4		1,000	175		1,975	2,250
1120	Galvanized steel curb and aluminum cover		8	4		1,800	175		1,975	2,250
1140	Galvanized steel curb and cover		8	4		1,625	175		1,800	2,050
1200	2'-6" x 8'-0", aluminum curb and cover		6.60	4.848		1,800	212		2,012	2,325
1400	Galvanized steel curb and aluminum cover		6.60	4.848		1,800	212		2,012	2,325
1500	Galvanized steel curb and cover		6.60	4.848		1,300	212		1,512	1,775
1800	For plexiglass panels, 2'-6" x 3'-0", add to above					440			440	485
9000	Minimum labor/equipment charge	2 Carp	2	8	Job		355		355	580

07 72 36 – Smoke Vents

07 72 36.10 Smoke Hatches

		Crew	Daily Output	Labor-Hours	Unit	Material	Labor	Equipment	Total	Total Incl O&P
0010	SMOKE HATCHES									
0200	For 3'-0" long, add to roof hatches from Section 07 72 33.10				Ea.	25%	5%			
0250	For 4'-0" long, add to roof hatches from Section 07 72 33.10					20%	5%			
0300	For 8'-0" long, add to roof hatches from Section 07 72 33.10					10%	5%			

07 72 Roof Accessories

07 72 36 – Smoke Vents

07 72 36.20 Smoke Vent Options		Crew	Daily Output	Labor-Hours	Unit	Material	2012 Bare Costs Labor	Equipment	Total	Total Incl O&P
0010	**SMOKE VENT OPTIONS**									
0100	4' x 4' aluminum cover and frame	G-3	13	2.462	Ea.	2,050	107		2,157	2,425
0200	Galvanized steel cover and frame		13	2.462		1,800	107		1,907	2,150
0300	4' x 8' aluminum cover and frame		8	4		2,800	175		2,975	3,350
0400	Galvanized steel cover and frame	↓	8	4	↓	2,375	175		2,550	2,900
9000	Minimum labor/equipment charge	2 Carp	2	8	Job		355		355	580

07 72 53 – Snow Guards

07 72 53.10 Snow Guard Options		Crew	Daily Output	Labor-Hours	Unit	Material	Labor	Equipment	Total	Total Incl O&P
0010	**SNOW GUARD OPTIONS**									
0100	Slate & asphalt shingle roofs, fastened with nails	1 Rofc	160	.050	Ea.	9.40	1.87		11.27	13.65
0200	Standing seam metal roofs, fastened with set screws		48	.167		15.60	6.25		21.85	28.50
0300	Surface mount for metal roofs, fastened with solder		48	.167	↓	6.05	6.25		12.30	17.75
0400	Double rail pipe type, including pipe	↓	130	.062	L.F.	22.50	2.30		24.80	29

07 72 73 – Pitch Pockets

07 72 73.10 Pitch Pockets, Variable Sizes		Crew	Daily Output	Labor-Hours	Unit	Material	Labor	Equipment	Total	Total Incl O&P
0010	**PITCH POCKETS, VARIABLE SIZES**									
0100	Adjustable, 4" to 7", welded corners, 4" deep	1 Rofc	48	.167	Ea.	15	6.25		21.25	27.50
0200	Side extenders, 6"	"	240	.033	"	2	1.25		3.25	4.41

07 72 80 – Vents

07 72 80.30 Vent Options		Crew	Daily Output	Labor-Hours	Unit	Material	Labor	Equipment	Total	Total Incl O&P
0010	**VENT OPTIONS**									
0020	Plastic, for insulated decks, 1 per M.S.F., minimum	1 Rofc	40	.200	Ea.	19	7.50		26.50	34.50
0100	Maximum		20	.400		40	14.95		54.95	70.50
0300	Aluminum	↓	30	.267		19	9.95		28.95	38.50
0800	Polystyrene baffles, 12" wide for 16" O.C. rafter spacing	1 Carp	90	.089		.40	3.92		4.32	6.90
0900	For 24" O.C. rafter spacing		110	.073	↓	.75	3.21		3.96	6.10
9000	Minimum labor/equipment charge	↓	3	2.667	Job		118		118	193

07 81 Applied Fireproofing

07 81 16 – Cementitious Fireproofing

07 81 16.10 Sprayed Cementitious Fireproofing		Crew	Daily Output	Labor-Hours	Unit	Material	Labor	Equipment	Total	Total Incl O&P
0010	**SPRAYED CEMENTITIOUS FIREPROOFING**									
0050	Not including canvas protection, normal density									
0100	Per 1" thick, on flat plate steel	G-2	3000	.008	S.F.	.53	.30	.04	.87	1.11
0200	Flat decking		2400	.010		.53	.37	.05	.95	1.24
0400	Beams		1500	.016		.53	.59	.09	1.21	1.63
0500	Corrugated or fluted decks		1250	.019		.79	.71	.10	1.60	2.13
0700	Columns, 1-1/8" thick		1100	.022		.59	.81	.12	1.52	2.08
0800	2-3/16" thick		700	.034		1.20	1.27	.18	2.65	3.57
0900	For canvas protection, add		5000	.005	↓	.07	.18	.03	.28	.40
9000	Minimum labor/equipment charge	↓	3	8	Job		297	42.50	339.50	525

07 84 Firestopping

07 84 13 – Penetration Firestopping

07 84 13.10 Firestopping		Crew	Daily Output	Labor-Hours	Unit	2012 Bare Costs Material	Labor	Equipment	Total	Total Incl O&P
0010	FIRESTOPPING R078413-30									
0100	Metallic piping, non insulated									
0110	Through walls, 2″ diameter	1 Carp	16	.500	Ea.	14.65	22		36.65	52
0120	4″ diameter		14	.571		22.50	25		47.50	66
0130	6″ diameter		12	.667		30	29.50		59.50	81
0140	12″ diameter		10	.800		53	35.50		88.50	117
0150	Through floors, 2″ diameter		32	.250		8.85	11.05		19.90	28
0160	4″ diameter		28	.286		12.75	12.60		25.35	34.50
0170	6″ diameter		24	.333		16.75	14.70		31.45	42.50
0180	12″ diameter		20	.400		28.50	17.65		46.15	60
0190	Metallic piping, insulated									
0200	Through walls, 2″ diameter	1 Carp	16	.500	Ea.	20.50	22		42.50	59
0210	4″ diameter		14	.571		28.50	25		53.50	72.50
0220	6″ diameter		12	.667		36	29.50		65.50	87.50
0230	12″ diameter		10	.800		59	35.50		94.50	123
0240	Through floors, 2″ diameter		32	.250		15	11.05		26.05	34.50
0250	4″ diameter		28	.286		18.85	12.60		31.45	41
0260	6″ diameter		24	.333		23	14.70		37.70	49
0270	12″ diameter		20	.400		28.50	17.65		46.15	60
0280	Non metallic piping, non insulated									
0290	Through walls, 2″ diameter	1 Carp	12	.667	Ea.	60	29.50		89.50	115
0300	4″ diameter		10	.800		75.50	35.50		111	141
0310	6″ diameter		8	1		105	44		149	189
0330	Through floors, 2″ diameter		16	.500		47	22		69	88
0340	4″ diameter		6	1.333		58.50	59		117.50	161
0350	6″ diameter		6	1.333		70.50	59		129.50	174
0370	Ductwork, insulated & non insulated, round									
0380	Through walls, 6″ diameter	1 Carp	12	.667	Ea.	30.50	29.50		60	82
0390	12″ diameter		10	.800		61	35.50		96.50	125
0400	18″ diameter		8	1		99	44		143	182
0410	Through floors, 6″ diameter		16	.500		16.75	22		38.75	54.50
0420	12″ diameter		14	.571		30.50	25		55.50	75.50
0430	18″ diameter		12	.667		53	29.50		82.50	107
0440	Ductwork, insulated & non insulated, rectangular									
0450	With stiffener/closure angle, through walls, 6″ x 12″	1 Carp	8	1	Ea.	25.50	44		69.50	101
0460	12″ x 24″		6	1.333		33.50	59		92.50	134
0470	24″ x 48″		4	2		96.50	88		184.50	251
0480	With stiffener/closure angle, through floors, 6″ x 12″		10	.800		13.75	35.50		49.25	73
0490	12″ x 24″		8	1		25	44		69	100
0500	24″ x 48″		6	1.333		48.50	59		107.50	150
0510	Multi trade openings									
0520	Through walls, 6″ x 12″	1 Carp	2	4	Ea.	53	176		229	350
0530	12″ x 24″	″	1	8		215	355		570	815
0540	24″ x 48″	2 Carp	1	16		860	705		1,565	2,100
0550	48″ x 96″	″	.75	21.333		3,450	940		4,390	5,350
0560	Through floors, 6″ x 12″	1 Carp	2	4		53	176		229	350
0570	12″ x 24″	″	1	8		215	355		570	815
0580	24″ x 48″	2 Carp	.75	21.333		860	940		1,800	2,500
0590	48″ x 96″	″	.50	32		3,450	1,400		4,850	6,125
0600	Structural penetrations, through walls									
0610	Steel beams, W8 x 10	1 Carp	8	1	Ea.	33.50	44		77.50	110
0620	W12 x 14		6	1.333		53	59		112	155
0630	W21 x 44		5	1.600		107	70.50		177.50	234

07 84 Firestopping

07 84 13 – Penetration Firestopping

07 84 13.10 Firestopping		Crew	Daily Output	Labor-Hours	Unit	2012 Bare Costs Material	Labor	Equipment	Total	Total Incl O&P
0640	W36 x 135	1 Carp	3	2.667	Ea.	260	118		378	480
0650	Bar joists, 18″ deep		6	1.333		49	59		108	151
0660	24″ deep		6	1.333		61	59		120	164
0670	36″ deep		5	1.600		91.50	70.50		162	217
0680	48″ deep		4	2		107	88		195	263
0690	Construction joints, floor slab at exterior wall									
0700	Precast, brick, block or drywall exterior									
0710	2″ wide joint	1 Carp	125	.064	L.F.	7.60	2.82		10.42	13.05
0720	4″ wide joint	″	75	.107	″	15.25	4.70		19.95	24.50
0730	Metal panel, glass or curtain wall exterior									
0740	2″ wide joint	1 Carp	40	.200	L.F.	18.05	8.80		26.85	34.50
0750	4″ wide joint	″	25	.320	″	25	14.10		39.10	50.50
0760	Floor slab to drywall partition									
0770	Flat joint	1 Carp	100	.080	L.F.	7.50	3.53		11.03	14.05
0780	Fluted joint		50	.160		15.25	7.05		22.30	28.50
0790	Etched fluted joint		75	.107		9.90	4.70		14.60	18.60
0800	Floor slab to concrete/masonry partition									
0810	Flat joint	1 Carp	75	.107	L.F.	16.75	4.70		21.45	26
0820	Fluted joint	″	50	.160	″	19.85	7.05		26.90	33.50
0830	Concrete/CMU wall joints									
0840	1″ wide	1 Carp	100	.080	L.F.	9.15	3.53		12.68	15.85
0850	2″ wide		75	.107		16.75	4.70		21.45	26
0860	4″ wide		50	.160		32	7.05		39.05	46.50
0870	Concrete/CMU floor joints									
0880	1″ wide	1 Carp	200	.040	L.F.	4.59	1.76		6.35	7.95
0890	2″ wide		150	.053		8.40	2.35		10.75	13.05
0900	4″ wide		100	.080		16	3.53		19.53	23.50

07 91 Preformed Joint Seals

07 91 13 – Compression Seals

07 91 13.10 Compression Seals		Crew	Daily Output	Labor-Hours	Unit	Material	Labor	Equipment	Total	Total Incl O&P
0010	**COMPRESSION SEALS**									
4900	O-ring type cord, 1/4″	1 Bric	472	.017	L.F.	.48	.75		1.23	1.73
4910	1/2″		440	.018		1.79	.80		2.59	3.26
4920	3/4″		424	.019		3.62	.83		4.45	5.30
4930	1″		408	.020		6.95	.86		7.81	9.05
4940	1-1/4″		384	.021		13.75	.92		14.67	16.60
4950	1-1/2″		368	.022		17.40	.96		18.36	20.50
4960	1-3/4″		352	.023		36.50	1		37.50	41.50
4970	2″		344	.023		64.50	1.03		65.53	72

07 91 16 – Joint Gaskets

07 91 16.10 Joint Gaskets		Crew	Daily Output	Labor-Hours	Unit	Material	Labor	Equipment	Total	Total Incl O&P
0010	**JOINT GASKETS**									
4400	Joint gaskets, neoprene, closed cell w/adh, 1/8″ x 3/8″	1 Bric	240	.033	L.F.	.27	1.47		1.74	2.67
4500	1/4″ x 3/4″		215	.037		.59	1.64		2.23	3.29
4700	1/2″ x 1″		200	.040		1.31	1.76		3.07	4.28
4800	3/4″ x 1-1/2″		165	.048		1.45	2.14		3.59	5.05

07 91 Preformed Joint Seals

07 91 23 – Backer Rods

07 91 23.10 Backer Rods		Crew	Daily Output	Labor-Hours	Unit	2012 Bare Costs Material	Labor	Equipment	Total	Total Incl O&P
0010	**BACKER RODS**									
0032	Backer rod, polyethylene, 1/4" diameter	1 Bric	460	.017	L.F.	.03	.77		.80	1.26
0052	1/2" diameter		460	.017		.06	.77		.83	1.29
0072	3/4" diameter		460	.017		.07	.77		.84	1.31
0092	1" diameter		460	.017		.14	.77		.91	1.38

07 91 26 – Joint Fillers

07 91 26.10 Joint Fillers

Line	Description	Crew	Daily Output	Labor-Hours	Unit	Material	Labor	Equipment	Total	Total Incl O&P
0010	**JOINT FILLERS**									
4360	Butyl rubber filler, 1/4" x 1/4"	1 Bric	290	.028	L.F.	.14	1.22		1.36	2.12
4365	1/2" x 1/2"		250	.032		.57	1.41		1.98	2.90
4370	1/2" x 3/4"		210	.038		.86	1.68		2.54	3.65
4375	3/4" x 3/4"		230	.035		1.28	1.53		2.81	3.88
4380	1" x 1"		180	.044		1.71	1.96		3.67	5.05
4390	For coloring, add					12%				
4980	Polyethylene joint backing, 1/4" x 2"	1 Bric	2.08	3.846	C.L.F.	12	170		182	286
4990	1/4" x 6"		1.28	6.250	"	28	276		304	475
5600	Silicone, room temp vulcanizing foam seal, 1/4" x 1/2"		1312	.006	L.F.	.31	.27		.58	.78
5610	1/2" x 1/2"		656	.012		.63	.54		1.17	1.56
5620	1/2" x 3/4"		442	.018		.94	.80		1.74	2.33
5630	3/4" x 3/4"		328	.024		1.41	1.08		2.49	3.28
5640	1/8" x 1"		1312	.006		.31	.27		.58	.78
5650	1/8" x 3"		442	.018		.94	.80		1.74	2.33
5670	1/4" x 3"		295	.027		1.88	1.20		3.08	4
5680	1/4" x 6"		148	.054		3.77	2.38		6.15	8
5690	1/2" x 6"		82	.098		7.55	4.30		11.85	15.25
5700	1/2" x 9"		52.50	.152		11.30	6.70		18	23.50
5710	1/2" x 12"		33	.242		15.05	10.70		25.75	34

07 92 Joint Sealants

07 92 13 – Elastomeric Joint Sealants

07 92 13.10 Masonry Joint Sealants

Line	Description	Crew	Daily Output	Labor-Hours	Unit	Material	Labor	Equipment	Total	Total Incl O&P
0010	**MASONRY JOINT SEALANTS**, 1/2" x 1/2" joint									
0050	Re-caulk only, oil base	1 Bric	225	.036	L.F.	.41	1.57		1.98	2.97
0100	Acrylic latex		205	.039		.18	1.72		1.90	2.97
0200	Polyurethane		200	.040		4.25	1.76		6.01	7.50
0300	Silicone		195	.041		3.42	1.81		5.23	6.65
1000	Cut out and re-caulk, oil base		145	.055		.41	2.43		2.84	4.37
1050	Acrylic latex		130	.062		.18	2.71		2.89	4.57
1100	Polyurethane		125	.064		4.25	2.82		7.07	9.20
1150	Silicone		120	.067		3.42	2.94		6.36	8.50
9000	Minimum labor/equipment charge		4	2	Job		88		88	142

07 92 13.20 Caulking and Sealant Options

Line	Description	Crew	Daily Output	Labor-Hours	Unit	Material	Labor	Equipment	Total	Total Incl O&P
0010	**CAULKING AND SEALANT OPTIONS**									
0050	Latex acrylic based, bulk				Gal.	29.50			29.50	32.50
0055	Bulk in place 1/4" x 1/4" bead	1 Bric	300	.027	L.F.	.09	1.18		1.27	1.99
0060	1/4" x 3/8"		294	.027		.16	1.20		1.36	2.10
0065	1/4" x 1/2"		288	.028		.21	1.23		1.44	2.20
0075	3/8" x 3/8"		284	.028		.23	1.24		1.47	2.25
0080	3/8" x 1/2"		280	.029		.31	1.26		1.57	2.37
0085	3/8" x 5/8"		276	.029		.39	1.28		1.67	2.49

07 92 Joint Sealants

07 92 13 – Elastomeric Joint Sealants

07 92 13.20 Caulking and Sealant Options		Crew	Daily Output	Labor-Hours	Unit	Material	2012 Bare Costs Labor	Equipment	Total	Total Incl O&P
0095	3/8" x 3/4"	1 Bric	272	.029	L.F.	.46	1.30		1.76	2.60
0100	1/2" x 1/2"		275	.029		.41	1.28		1.69	2.53
0105	1/2" x 5/8"		269	.030		.52	1.31		1.83	2.68
0110	1/2" x 3/4"		263	.030		.62	1.34		1.96	2.84
0115	1/2" x 7/8"		256	.031		.72	1.38		2.10	3.02
0120	1/2" x 1"		250	.032		.83	1.41		2.24	3.18
0125	3/4" x 3/4"		244	.033		.93	1.45		2.38	3.36
0130	3/4" x 1"		225	.036		1.24	1.57		2.81	3.89
0135	1" x 1"		200	.040		1.66	1.76		3.42	4.66
0190	Cartridges				Gal.	25.50			25.50	28
0200	11 fl. oz. cartridge				Ea.	2.20			2.20	2.42
0500	1/4" x 1/2"	1 Bric	288	.028	L.F.	.18	1.23		1.41	2.17
0600	1/2" x 1/2"		275	.029		.36	1.28		1.64	2.47
0800	3/4" x 3/4"		244	.033		.81	1.45		2.26	3.22
0900	3/4" x 1"		225	.036		1.08	1.57		2.65	3.71
1000	1" x 1"		200	.040		1.35	1.76		3.11	4.32
1400	Butyl based, bulk				Gal.	28.50			28.50	31.50
1500	Cartridges				"	32			32	35
1700	1/4" x 1/2", 154 L.F./gal.	1 Bric	288	.028	L.F.	.19	1.23		1.42	2.17
1800	1/2" x 1/2", 77 L.F./gal.	"	275	.029	"	.37	1.28		1.65	2.48
2300	Polysulfide compounds, 1 component, bulk				Gal.	56			56	61.50
2400	Cartridges				"	54.50			54.50	60
2600	1 or 2 component, in place, 1/4" x 1/4", 308 L.F./gal.	1 Bric	300	.027	L.F.	.18	1.18		1.36	2.09
2700	1/2" x 1/4", 154 L.F./gal.		288	.028		.36	1.23		1.59	2.37
2900	3/4" x 3/8", 68 L.F./gal.		272	.029		.82	1.30		2.12	3
3000	1" x 1/2", 38 L.F./gal.		250	.032		1.47	1.41		2.88	3.89
3200	Polyurethane, 1 or 2 component				Gal.	51			51	56
3300	Cartridges				"	70			70	77
3500	Bulk, in place, 1/4" x 1/4"	1 Bric	300	.027	L.F.	.17	1.18		1.35	2.07
3655	1/2" x 1/4"		288	.028		.33	1.23		1.56	2.33
3800	3/4" x 3/8"		272	.029		.75	1.30		2.05	2.92
3900	1" x 1/2"		250	.032		1.33	1.41		2.74	3.73
4100	Silicone rubber, bulk				Gal.	42			42	46
4200	Cartridges				"	41			41	45
4300	Bulk in place, 1/4" x 1/2", 154 L.F./gal.	1 Bric	235	.034	L.F.	.27	1.50		1.77	2.72

07 92 16 – Rigid Joint Sealants

07 92 16.10 Rigid Joint Sealants		Crew	Daily Output	Labor-Hours	Unit	Material	Labor	Equipment	Total	Total Incl O&P
0010	**RIGID JOINT SEALANTS**									
5802	Tapes, sealant, P.V.C. foam adhesive, 1/16" x 1/4"				L.F.	.06			.06	.07
5902	1/16" x 1/2"					.08			.08	.09
5952	1/16" x 1"					.15			.15	.17
6002	1/8" x 1/2"					.09			.09	.10

07 92 19 – Acoustical Joint Sealants

07 92 19.10 Acoustical Sealant		Crew	Daily Output	Labor-Hours	Unit	Material	Labor	Equipment	Total	Total Incl O&P
0010	**ACOUSTICAL SEALANT**									
0020	Acoustical sealant, elastomeric, cartridges				Ea.	6.25			6.25	6.90
0025	In place, 1/4" x 1/4"	1 Bric	300	.027	L.F.	.26	1.18		1.44	2.17
0030	1/4" x 1/2"		288	.028		.51	1.23		1.74	2.53
0035	1/2" x 1/2"		275	.029		1.02	1.28		2.30	3.19
0040	1/2" x 3/4"		263	.030		1.53	1.34		2.87	3.85
0045	3/4" x 3/4"		244	.033		2.30	1.45		3.75	4.86
0050	1" x 1"		200	.040		4.09	1.76		5.85	7.35

07 95 Expansion Control

07 95 13 – Expansion Joint Cover Assemblies

07 95 13.50 Expansion Joint Assemblies		Crew	Daily Output	Labor-Hours	Unit	2012 Bare Costs Material	Labor	Equipment	Total	Total Incl O&P
0010	**EXPANSION JOINT ASSEMBLIES**									
0200	Floor cover assemblies, 1" space, aluminum	1 Sswk	38	.211	L.F.	19.60	10.40		30	41
0300	Bronze		38	.211		46.50	10.40		56.90	70.50
0500	2" space, aluminum		38	.211		24	10.40		34.40	46
0600	Bronze		38	.211		50	10.40		60.40	74.50
0800	Wall and ceiling assemblies, 1" space, aluminum		38	.211		11.70	10.40		22.10	32
0900	Bronze		38	.211		43.50	10.40		53.90	67
1100	2" space, aluminum		38	.211		19.90	10.40		30.30	41.50
1200	Bronze		38	.211		53.50	10.40		63.90	78
1400	Floor to wall assemblies, 1" space, aluminum		38	.211		17.85	10.40		28.25	39
1500	Bronze or stainless		38	.211		52.50	10.40		62.90	77
1700	Gym floor angle covers, aluminum, 3" x 3" angle		46	.174		15.45	8.60		24.05	33
1800	3" x 4" angle		46	.174		21.50	8.60		30.10	39.50
2000	Roof closures, aluminum, flat roof, low profile, 1" space		57	.140		35.50	6.95		42.45	52
2100	High profile		57	.140		51.50	6.95		58.45	69.50
2300	Roof to wall, low profile, 1" space		57	.140		19.75	6.95		26.70	34.50
2400	High profile		57	.140		25.50	6.95		32.45	41.50
9000	Minimum labor/equipment charge		2	4	Job		197		197	365

Division Notes

	CREW	DAILY OUTPUT	LABOR-HOURS	UNIT	2012 BARE COSTS				TOTAL INCL O&P
					MAT.	LABOR	EQUIP.	TOTAL	

Estimating Tips

08 10 00 Doors and Frames

All exterior doors should be addressed for their energy conservation (insulation and seals).

- Most metal doors and frames look alike, but there may be significant differences among them. When estimating these items, be sure to choose the line item that most closely compares to the specification or door schedule requirements regarding:
  - — type of metal
  - — metal gauge
  - — door core material
  - — fire rating
  - — finish
- Wood and plastic doors vary considerably in price. The primary determinant is the veneer material. Lauan, birch, and oak are the most common veneers. Other variables include the following:
  - — hollow or solid core
  - — fire rating
  - — flush or raised panel
  - — finish

08 30 00 Specialty Doors and Frames

- There are many varieties of special doors, and they are usually priced per each. Add frames, hardware, or operators required for a complete installation.

08 40 00 Entrances, Storefronts, and Curtain Walls

- Glazed curtain walls consist of the metal tube framing and the glazing material. The cost data in this subdivision is presented for the metal tube framing alone or the composite wall. If your estimate requires a detailed takeoff of the framing, be sure to add the glazing cost.

08 50 00 Windows

- Most metal windows are delivered preglazed. However, some metal windows are priced without glass. Refer to 08 80 00 Glazing for glass pricing. The grade C indicates commercial grade windows, usually ASTM C-35.
- All wood windows and vinyl are priced preglazed. The glazing is insulating glass. Some wood windows may have single pane float glass. Add the cost of screens and grills if required, and not already included.

08 70 00 Hardware

- Hardware costs add considerably to the cost of a door. The most efficient method to determine the hardware requirements for a project is to review the door schedule.
- Door hinges are priced by the pair, with most doors requiring 1-1/2 pairs per door. The hinge prices do not include installation labor because it is included in door installation. Hinges are classified according to the frequency of use.

08 80 00 Glazing

- Different openings require different types of glass. The most common types are:
  - — float
  - — tempered
  - — insulating
  - — impact-resistant
  - — ballistic-resistant
- Most exterior windows are glazed with insulating glass. Entrance doors and window walls, where the glass is less than 18″ from the floor, are generally glazed with tempered glass. Interior windows and some residential windows are glazed with float glass.
- Coastal communities require the use of impact-resistant glass.
- The insulation or 'u' value is a strong consideration, along with solar heat gain, to determine total energy efficiency.

Reference Numbers

Reference numbers are shown in shaded boxes at the beginning of some major classifications. These numbers refer to related items in the Reference Section. The reference information may be an estimating procedure, an alternate pricing method, or technical information.

Note: Not all subdivisions listed here necessarily appear in this publication.

08 01 Operation and Maintenance of Openings

08 01 53 – Operation and Maintenance of Plastic Windows

08 01 53.81 Solid Vinyl Replacement Windows		Crew	Daily Output	Labor-Hours	Unit	2012 Bare Costs Material	Labor	Equipment	Total	Total Incl O&P
0010	SOLID VINYL REPLACEMENT WINDOWS R085313-20									
0020	Double hung, insulated glass, up to 83 united inches G	2 Carp	8	2	Ea.	335	88		423	515
0040	84 to 93 G		8	2		370	88		458	550
0060	94 to 101 G		6	2.667		370	118		488	600
0080	102 to 111 G		6	2.667		390	118		508	625
0100	112 to 120 G		6	2.667		425	118		543	660
0120	For each united inch over 120 , add G		800	.020	Inch	5.35	.88		6.23	7.35
0140	Casement windows, one operating sash , 42 to 60 united inches G		8	2	Ea.	222	88		310	390
0160	61 to 70 G		8	2		248	88		336	420
0180	71 to 80 G		8	2		269	88		357	440
0200	81 to 96 G		8	2		286	88		374	460
0220	Two operating sash, 58 to 78 united inches G		8	2		440	88		528	630
0240	79 to 88 G		8	2		470	88		558	665
0260	89 to 98 G		8	2		515	88		603	710
0280	99 to 108 G		6	2.667		535	118		653	785
0300	109 to 121 G		6	2.667		575	118		693	830
0320	Two operating.one fixed sash, 73 to 108 united inches G		8	2		700	88		788	915
0340	109 to 118 G		8	2		735	88		823	955
0360	119 to 128 G		6	2.667		755	118		873	1,025
0380	129 to 138 G		6	2.667		805	118		923	1,075
0400	139 to 156 G		6	2.667		850	118		968	1,125
0420	Four operating sash, 98 to 118 united inches G		8	2		1,000	88		1,088	1,250
0440	119 to 128 G		8	2		1,075	88		1,163	1,325
0460	129 to 138 G		6	2.667		1,125	118		1,243	1,450
0480	139 to 148 G		6	2.667		1,175	118		1,293	1,500
0500	149 to 168 G		6	2.667		1,250	118		1,368	1,600
0520	169 to 178 G		6	2.667		1,350	118		1,468	1,700
0540	For venting unit to fixed unit, deduct					16			16	17.60
0560	Fixed picture window, up to 63 united inches G	2 Carp	8	2		180	88		268	345
0580	64 to 83 G		8	2		206	88		294	370
0600	84 to 101 G		8	2		257	88		345	430
0620	For each united inch over 101, add G		900	.018	Inch	3	.78		3.78	4.59
0760	Muntins, between glazing, square, per lite				Ea.	6			6	6.60
0780	Diamond shape, per full or partial diamond				"	8			8	8.80
0800	Cellulose fiber insulation, poured into sash balance cavity G	1 Carp	36	.222	C.F.	.53	9.80		10.33	16.65
0820	Silicone caulking at perimeter G	"	800	.010	L.F.	.13	.44		.57	.87

08 05 Common Work Results for Openings

08 05 05 – Selective Windows and Doors Demolition

08 05 05.10 Selective Demolition Doors		Crew	Daily Output	Labor-Hours	Unit	Material	Labor	Equipment	Total	Total Incl O&P
0010	SELECTIVE DEMOLITION DOORS R024119-10									
0200	Doors, exterior, 1-3/4″ thick, single, 3′ x 7′ high	1 Clab	16	.500	Ea.		17.55		17.55	29
0202	Doors, exterior, 3′ - 6′ wide x 7′ high		12	.667			23.50		23.50	38.50
0210	3′ x 8′ high		10	.800			28		28	46
0215	Double, 3′ x 8′ high		6	1.333			47		47	76.50
0220	Double, 6′ x 7′ high		12	.667			23.50		23.50	38.50
0500	Interior, 1-3/8″ thick, single, 3′ x 7′ high		20	.400			14.05		14.05	23
0520	Double, 6′ x 7′ high		16	.500			17.55		17.55	29
0700	Bi-folding, 3′ x 6′-8″ high	1 Clab	20	.400	Ea.		14.05		14.05	23
0720	6′ x 6′-8″ high		18	.444			15.60		15.60	25.50

08 05 05 – Selective Windows and Doors Demolition

08 05 05.10 Selective Demolition Doors		Crew	Daily Output	Labor-Hours	Unit	Material	2012 Bare Costs Labor	Equipment	Total	Total Incl O&P
0900	Bi-passing, 3′ x 6′-8″ high	1 Clab	16	.500	Ea.		17.55		17.55	29
0940	6′ x 6′-8″ high		14	.571			20		20	33
1500	Remove and reset, minimum	1 Carp	8	1			44		44	72.50
1520	Maximum		6	1.333			59		59	96.50
2000	Frames, including trim, metal		8	1			44		44	72.50
2200	Wood	2 Carp	32	.500			22		22	36
2201	Alternate pricing method	1 Carp	200	.040	L.F.		1.76		1.76	2.89
3000	Special doors, counter doors	2 Carp	6	2.667	Ea.		118		118	193
3100	Double acting		10	1.600			70.50		70.50	116
3200	Floor door (trap type), or access type		8	2			88		88	145
3300	Glass, sliding, including frames		12	1.333			59		59	96.50
3400	Overhead, commercial, 12′ x 12′ high		4	4			176		176	289
3440	up to 20′ x 16′ high		3	5.333			235		235	385
3445	up to 35′ x 30′ high		1	16			705		705	1,150
3500	Residential, 9′ x 7′ high		8	2			88		88	145
3540	16′ x 7′ high		7	2.286			101		101	165
3600	Remove and reset, minimum		4	4			176		176	289
3620	Maximum		2.50	6.400			282		282	465
3700	Roll-up grille		5	3.200			141		141	231
3800	Revolving door		2	8			355		355	580
3900	Storefront swing door		3	5.333			235		235	385
3902	Cafe/bar swing door	2 Clab	8	2			70		70	115
4000	Residential lockset, exterior	1 Carp	28	.286			12.60		12.60	20.50
4224	Pocket door	″	8	1			44		44	72.50
4654	Remove stool/apron	1 Clab	240	.033	L.F.		1.17		1.17	1.92
5585	Remove panic device		10	.800	Ea.		28		28	46
5590	Remove mail slot		45	.178	″		6.25		6.25	10.25
7100	Remove double swing pneumatic doors, openers and sensors	2 Skwk	.50	32	Opng.		1,450		1,450	2,350
7570	Remove shock absorbing door	2 Sswk	1.90	8.421	″		415		415	770
9000	Minimum labor/equipment charge	1 Carp	4	2	Job		88		88	145

08 05 05.20 Selective Demolition of Windows		Crew	Daily Output	Labor-Hours	Unit	Material	Labor	Equipment	Total	Total Incl O&P
0010	**SELECTIVE DEMOLITION OF WINDOWS** R024119-10									
0200	Aluminum, including trim, to 12 S.F.	1 Clab	16	.500	Ea.		17.55		17.55	29
0240	To 25 S.F.		11	.727			25.50		25.50	42
0280	To 50 S.F.		5	1.600			56		56	92
0320	Storm windows/screens, to 12 S.F.		27	.296			10.40		10.40	17.05
0360	To 25 S.F.		21	.381			13.35		13.35	22
0400	To 50 S.F.		16	.500			17.55		17.55	29
0600	Glass, minimum		200	.040	S.F.		1.40		1.40	2.30
0620	Maximum		150	.053	″		1.87		1.87	3.07
1000	Steel, including trim, to 12 S.F.		13	.615	Ea.		21.50		21.50	35.50
1020	To 25 S.F.		9	.889			31		31	51
1040	To 50 S.F.		4	2			70		70	115
2000	Wood, including trim, to 12 S.F.		22	.364			12.75		12.75	21
2020	To 25 S.F.		18	.444			15.60		15.60	25.50
2060	To 50 S.F.		13	.615			21.50		21.50	35.50
2065	To 180 S.F.		8	1			35		35	57.50
4400	Remove skylight, prefabricated glass block with metal frame	G-3	180	.178	S.F.		7.75		7.75	12.40
4410	Remove skylight, plstc domes, flush/curb mtd	″	395	.081	″		3.53		3.53	5.65
5020	Remove and reset window, minimum	1 Carp	6	1.333	Ea.		59		59	96.50
5040	Average		4	2			88		88	145
5080	Maximum		2	4			176		176	289

08 05 Common Work Results for Openings

08 05 05 – Selective Windows and Doors Demolition

08 05 05.20 Selective Demolition of Windows		Crew	Daily Output	Labor-Hours	Unit	Material	2012 Bare Costs Labor	Equipment	Total	Total Incl O&P
6000	Screening only	1 Clab	4000	.002	S.F.		.07		.07	.12
9000	Minimum labor/equipment charge	"	4	2	Job		70		70	115

08 11 Metal Doors and Frames

08 11 63 – Metal Screen and Storm Doors and Frames

08 11 63.23 Aluminum Screen and Storm Doors and Frames

		Crew	Daily Output	Labor-Hours	Unit	Material	Labor	Equipment	Total	Total Incl O&P
0010	**ALUMINUM SCREEN AND STORM DOORS AND FRAMES**									
0020	Combination storm and screen									
0420	Clear anodic coating, 2′-8″ wide	2 Carp	14	1.143	Ea.	161	50.50		211.50	260
0440	3′-0″ wide	"	14	1.143	"	161	50.50		211.50	260
0500	For 7′ door height, add					8%				
1020	Mill finish, 2′-8″ wide	2 Carp	14	1.143	Ea.	231	50.50		281.50	335
1040	3′-0″ wide	"	14	1.143		253	50.50		303.50	360
1100	For 7′-0″ door, add					8%				
1520	White painted, 2′-8″ wide	2 Carp	14	1.143		283	50.50		333.50	395
1540	3′-0″ wide	"	14	1.143		299	50.50		349.50	415
1600	For 7′-0″ door, add					8%				
2000	Wood door & screen, see Section 08 14 33.20									
9000	Minimum labor/equipment charge	1 Carp	4	2	Job		88		88	145

08 12 Metal Frames

08 12 13 – Hollow Metal Frames

08 12 13.13 Standard Hollow Metal Frames

			Crew	Daily Output	Labor-Hours	Unit	Material	Labor	Equipment	Total	Total Incl O&P
0010	**STANDARD HOLLOW METAL FRAMES**										
0020	16 ga., up to 5-3/4″ jamb depth										
0025	6′-8″ high, 3′-0″ wide, single	G	2 Carp	16	1	Ea.	143	44		187	230
0028	3′-6″ wide, single	G		16	1		150	44		194	237
0030	4′-0″ wide, single	G		16	1		150	44		194	237
0040	6′-0″ wide, double	G		14	1.143		204	50.50		254.50	305
0045	8′-0″ wide, double	G		14	1.143		213	50.50		263.50	315
0100	7′-0″ high, 3′-0″ wide, single	G		16	1		148	44		192	236
0110	3′-6″ wide, single			16	1		155	44		199	243
0112	4′-0″ wide, single			16	1		155	44		199	243
0140	6′-0″ wide, double			14	1.143		194	50.50		244.50	297
0145	8′-0″ wide, double			14	1.143		229	50.50		279.50	335
1000	16 ga., up to 4-7/8″ deep, 7′-0″ H, 3′-0″ W, single	G		16	1		163	44		207	252
1140	6′-0″ wide, double	G		14	1.143		184	50.50		234.50	285
1200	16 ga., 8-3/4″ deep, 7′-0″ H, 3′-0″ W, single	G		16	1		199	44		243	292
1240	6′-0″ wide, double	G		14	1.143		234	50.50		284.50	340
2800	14 ga., up to 3-7/8″ deep, 7′-0″ high, 3′-0″ wide, single	G		16	1		177	44		221	268
2840	6′-0″ wide, double	G		14	1.143		213	50.50		263.50	315
3000	14 ga., up to 5-3/4″ deep, 6′-8″ high, 3′-0″ wide, single	G		16	1		149	44		193	237
3002	3′-6″ wide, single	G		16	1		193	44		237	285
3005	4′-0″ wide, single	G		16	1		193	44		237	285
3600	up to 5-3/4″ jamb depth, 7′-0″ high, 4′-0″ wide, single			15	1.067		185	47		232	280
3620	6′-0″ wide, double			12	1.333		235	59		294	355
3640	8′-0″ wide, double			12	1.333		246	59		305	370
3700	8′-0″ high, 4′-0″ wide, single			15	1.067		235	47		282	335
3740	8′-0″ wide, double			12	1.333		290	59		349	415

08 12 Metal Frames

08 12 13 – Hollow Metal Frames

08 12 13.13 Standard Hollow Metal Frames

Line	08 12 13.13 Standard Hollow Metal Frames		Crew	Daily Output	Labor-Hours	Unit	Material	Labor (2012 Bare Costs)	Equipment (2012 Bare Costs)	Total (2012 Bare Costs)	Total Incl O&P
4000	6-3/4" deep, 7'-0" high, 4'-0" wide, single	G	2 Carp	15	1.067	Ea.	216	47		263	315
4020	6'-0" wide, double	G		12	1.333		268	59		327	390
4040	8'-0", wide double	G		12	1.333		276	59		335	400
4100	8'-0" high, 4'-0" wide, single	G		15	1.067		276	47		323	380
4140	8'-0" wide, double	G		12	1.333		320	59		379	450
4400	8-3/4" deep, 7'-0" high, 4'-0" wide, single			15	1.067		252	47		299	355
4440	8'-0" wide, double			12	1.333		315	59		374	440
4500	8'-0" high, 4'-0" wide, single			15	1.067		283	47		330	385
4540	8'-0" wide, double			12	1.333		350	59		409	480
4900	For welded frames, add						63			63	69.50
5380	Steel frames, KD, 14 ga, "B" label, to 5-3/4" throat, to 3'-0" x 7'-0"		2 Carp	15	1.067		210	47		257	310
5400	14 ga., "B" label, up to 5-3/4" deep, 7'-0" high, 4'-0" wide, single	G		15	1.067		204	47		251	300
5440	8'-0" wide, double	G		12	1.333		267	59		326	390
5800	6-3/4" deep, 7'-0" high, 4'-0" wide, single	G		15	1.067		218	47		265	315
5840	8'-0" wide, double	G		12	1.333		380	59		439	510
6200	8-3/4" deep, 7'-0" high, 4'-0" wide, single	G		15	1.067		305	47		352	410
6240	8'-0" wide, double	G		12	1.333		355	59		414	485
6300	For "A" label use same price as "B" label										
6400	For baked enamel finish, add						30%	15%			
6500	For galvanizing, add						20%				
6600	For hospital stop, add					Ea.	295			295	325
6620	For hospital stop, stainless steel add					"	380			380	420
7900	Transom lite frames, fixed, add		2 Carp	155	.103	S.F.	54	4.55		58.55	67
8000	Movable, add		"	130	.123	"	68	5.45		73.45	84
9000	Minimum labor/equipment charge		1 Carp	4	2	Job		88		88	145

08 12 13.25 Channel Metal Frames

Line	08 12 13.25 Channel Metal Frames		Crew	Daily Output	Labor-Hours	Unit	Material	Labor	Equipment	Total	Total Incl O&P
0010	**CHANNEL METAL FRAMES**										
0020	Steel channels with anchors and bar stops										
0100	6" channel @ 8.2#/L.F., 3' x 7' door, weighs 150#	G	E-4	13	2.462	Ea.	225	123	9.35	357.35	485
0200	8" channel @ 11.5#/L.F., 6' x 8' door, weighs 275#	G		9	3.556		415	177	13.50	605.50	800
0300	8' x 12' door, weighs 400#	G		6.50	4.923		600	245	18.70	863.70	1,125
0400	10" channel @ 15.3#/L.F., 10' x 10' door, weighs 500#	G		6	5.333		750	266	20.50	1,036.50	1,350
0500	12' x 12' door, weighs 600#	G		5.50	5.818		900	290	22	1,212	1,550
0600	12" channel @ 20.7#/L.F., 12' x 12' door, weighs 825#	G		4.50	7.111		1,250	355	27	1,632	2,050
0700	12' x 16' door, weighs 1000#	G		4	8		1,500	400	30.50	1,930.50	2,425
0800	For frames without bar stops, light sections, deduct						15%				
0900	Heavy sections, deduct						10%				
9000	Minimum labor/equipment charge		E-4	4	8	Job		400	30.50	430.50	775

08 13 Metal Doors

08 13 13 – Hollow Metal Doors

08 13 13.13 Standard Hollow Metal Doors

Line	08 13 13.13 Standard Hollow Metal Doors		Crew	Daily Output	Labor-Hours	Unit	Material	Labor	Equipment	Total	Total Incl O&P
0010	**STANDARD HOLLOW METAL DOORS**	R081313-20									
0015	Flush, full panel, hollow core										
0017	When noted doors are prepared but do not include glass or louvers										
0020	1-3/8" thick, 20 ga., 2'-0" x 6'-8"	G	2 Carp	20	.800	Ea.	335	35.50		370.50	430
0040	2'-8" x 6'-8"	G		18	.889		350	39		389	450
0060	3'-0" x 6'-8"	G		17	.941		350	41.50		391.50	455
0100	3'-0" x 7'-0"			17	.941		360	41.50		401.50	465
0120	For vision lite, add						93			93	102
0140	For narrow lite, add						101			101	111

08 13 13.13 Standard Hollow Metal Doors			Crew	Daily Output	Labor-Hours	Unit	2012 Bare Costs Material	Labor	Equipment	Total	Total Incl O&P
0320	Half glass, 20 ga., 2'-0" x 6'-8"	G	2 Carp	20	.800	Ea.	490	35.50		525.50	600
0340	2'-8" x 6'-8"	G		18	.889		515	39		554	630
0360	3'-0" x 6'-8"	G		17	.941		510	41.50		551.50	630
0400	3'-0" x 7'-0"	G		17	.941		615	41.50		656.50	750
0410	1-3/8" thick, 18 ga., 2'-0" x 6'-8"	G		20	.800		400	35.50		435.50	500
0420	3'-0" x 6'-8"	G		17	.941		405	41.50		446.50	515
0425	3'-0" x 7'-0"	G		17	.941		410	41.50		451.50	525
0450	For vision lite, add						93			93	102
0452	For narrow lite, add						101			101	111
0460	Half glass, 18 ga., 2'-0" x 6'-8"	G	2 Carp	20	.800		555	35.50		590.50	675
0465	2'-8" x 6'-8"	G		18	.889		575	39		614	700
0470	3'-0" x 6'-8"	G		17	.941		565	41.50		606.50	690
0475	3'-0" x 7'-0"	G		17	.941		565	41.50		606.50	690
0500	Hollow core, 1-3/4" thick, full panel, 20 ga., 2'-8" x 6'-8"	G		18	.889		415	39		454	520
0520	3'-0" x 6'-8"	G		17	.941		420	41.50		461.50	530
0640	3'-0" x 7'-0"	G		17	.941		435	41.50		476.50	545
0680	4'-0" x 7'-0"	G		15	1.067		630	47		677	770
0700	4'-0" x 8'-0"	G		13	1.231		735	54.50		789.50	900
1000	18 ga., 2'-8" x 6'-8"	G		17	.941		490	41.50		531.50	610
1020	3'-0" x 6'-8"	G		16	1		475	44		519	600
1120	3'-0" x 7'-0"	G		17	.941		595	41.50		636.50	720
1180	4'-0" x 7'-0"	G		14	1.143		630	50.50		680.50	780
1200	4'-0" x 8'-0"	G		17	.941		735	41.50		776.50	880
1212	For vision lite, add						93			93	102
1214	For narrow lite, add						101			101	111
1230	Half glass, 20 ga., 2'-8" x 6'-8"	G	2 Carp	20	.800		575	35.50		610.50	695
1240	3'-0" x 6'-8"	G		18	.889		575	39		614	700
1260	3'-0" x 7'-0"	G		18	.889		595	39		634	715
1320	18 ga., 2'-8" x 6'-8"	G		18	.889		640	39		679	770
1340	3'-0" x 6'-8"	G		17	.941		635	41.50		676.50	765
1360	3'-0" x 7'-0"	G		17	.941		645	41.50		686.50	780
1380	4'-0" x 7'-0"	G		15	1.067		785	47		832	940
1400	4'-0" x 8'-0"	G		14	1.143		890	50.50		940.50	1,075
1500	Flush full panel, 16 gauge, steel hollow core										
1505	1-3/8" same price as 1-3/4"										
1520	2'-0" x 6'-8"		2 Carp	20	.800	Ea.	555	35.50		590.50	670
1530	2'-8" x 6'-8"			20	.800		555	35.50		590.50	675
1540	3'-0" x 6'-8"			20	.800		550	35.50		585.50	665
1560	2'-8" x 7'-0"			18	.889		575	39		614	695
1570	3'-0" x 7'-0"			18	.889		560	39		599	680
1580	3'-6" x 7'-0"			18	.889		655	39		694	785
1590	4'-0" x 7'-0"			18	.889		725	39		764	860
1600	2'-8" x 8'-0"			18	.889		700	39		739	835
1620	3'-0" x 8'-0"			18	.889		720	39		759	855
1630	3'-6" x 8'-0"			18	.889		800	39		839	945
1640	4'-0" x 8'-0"			18	.889		850	39		889	995
1720	Insulated, 1-3/4" thick, full panel, 18 ga., 3'-0" x 6'-8"	G		15	1.067		475	47		522	600
1740	2'-8" x 7'-0"	G		16	1		505	44		549	630
1760	3'-0" x 7'-0"	G		15	1.067		490	47		537	615
1800	4'-0" x 8'-0"	G		13	1.231		740	54.50		794.50	900
1805	For vision lite, add						93			93	102
1810	For narrow lite, add						101			101	111
1820	Half glass, 18 ga., 3'-0" x 6'-8"	G	2 Carp	16	1		635	44		679	770

08 13 Metal Doors

08 13 13 – Hollow Metal Doors

08 13 13.13 Standard Hollow Metal Doors		Crew	Daily Output	Labor-Hours	Unit	Material	2012 Bare Costs Labor	Equipment	Total	Total Incl O&P
1840	2′-8″ x 7′-0″ G	2 Carp	17	.941	Ea.	660	41.50		701.50	795
1860	3′-0″ x 7′-0″ G		16	1		685	44		729	825
1900	4′-0″ x 8′-0″ G		14	1.143		670	50.50		720.50	820
8100	For bottom louver, add					274			274	300
8110	For baked enamel finish, add					30%	15%			
8120	For galvanizing, add					20%				
8170	For soundproofing STC 40, add				Ea.	1,725			1,725	1,900
8190	For 3 hour door, add					370			370	405
8270	For dutch door with shelf, add to standard door					300			300	330
9000	Minimum labor/equipment charge	1 Carp	4	2	Job		88		88	145

08 13 13.15 Metal Fire Doors		Crew	Daily Output	Labor-Hours	Unit	Material	Labor	Equipment	Total	Total Incl O&P
0010	**METAL FIRE DOORS** R081313-20									
0015	Steel, flush, "B" label, 90 minute									
0020	Full panel, 20 ga., 2′-0″ x 6′-8″	2 Carp	20	.800	Ea.	420	35.50		455.50	520
0040	2′-8″ x 6′-8″		18	.889		435	39		474	545
0060	3′-0″ x 6′-8″		17	.941		435	41.50		476.50	550
0080	3′-0″ x 7′-0″		17	.941		455	41.50		496.50	570
0140	18 ga., 3′-0″ x 6′-8″		16	1		495	44		539	620
0160	2′-8″ x 7′-0″		17	.941		520	41.50		561.50	640
0180	3′-0″ x 7′-0″		16	1		505	44		549	630
0200	4′-0″ x 7′-0″		15	1.067		650	47		697	790
0220	For "A" label, 3 hour, 18 ga., use same price as "B" label									
0240	For vision lite, add				Ea.	156			156	172
0520	Flush, "B" label 90 min., egress core, 20 ga., 2′-0″ x 6′-8″	2 Carp	18	.889		655	39		694	785
0540	2′-8″ x 6′-8″		17	.941		665	41.50		706.50	800
0560	3′-0″ x 6′-8″		16	1		665	44		709	805
0580	3′-0″ x 7′-0″		16	1		685	44		729	825
0640	Flush, "A" label 3 hour, egress core, 18 ga., 3′-0″ x 6′-8″		15	1.067		720	47		767	870
0660	2′-8″ x 7′-0″		16	1		750	44		794	895
0680	3′-0″ x 7′-0″		15	1.067		740	47		787	890
0700	4′-0″ x 7′-0″		14	1.143		885	50.50		935.50	1,050
9000	Minimum labor/equipment charge	1 Carp	4	2	Job		88		88	145

08 13 13.20 Residential Steel Doors		Crew	Daily Output	Labor-Hours	Unit	Material	Labor	Equipment	Total	Total Incl O&P
0010	**RESIDENTIAL STEEL DOORS**									
0020	Prehung, insulated, exterior									
0030	Embossed, full panel, 2′-8″ x 6′-8″ G	2 Carp	17	.941	Ea.	298	41.50		339.50	400
0040	3′-0″ x 6′-8″ G		15	1.067		243	47		290	345
0060	3′-0″ x 7′-0″ G		15	1.067		330	47		377	440
0070	5′-4″ x 6′-8″, double G		8	2		625	88		713	830
0220	Half glass, 2′-8″ x 6′-8″		17	.941		320	41.50		361.50	420
0240	3′-0″ x 6′-8″		16	1		320	44		364	425
0260	3′-0″ x 7′-0″		16	1		370	44		414	480
0270	5′-4″ x 6′-8″, double		8	2		650	88		738	860
1320	Flush face, full panel, 2′-8″ x 6′-8″		16	1		248	44		292	345
1340	3′-0″ x 6′-8″		15	1.067		248	47		295	350
1360	3′-0″ x 7′-0″		15	1.067		296	47		343	400
1380	5′-4″ x 6′-8″, double		8	2		495	88		583	690
1420	Half glass, 2′-8″ x 6′-8″		17	.941		325	41.50		366.50	425
1440	3′-0″ x 6′-8″		16	1		325	44		369	430
1460	3′-0″ x 7′-0″		16	1		350	44		394	460
1480	5′-4″ x 6′-8″, double		8	2		635	88		723	840
1500	Sidelight, full lite, 1′-0″ x 6′-8″ with grille					244			244	268

08 13 Metal Doors

08 13 13 – Hollow Metal Doors

08 13 13.20 Residential Steel Doors		Crew	Daily Output	Labor-Hours	Unit	Material	Labor	Equipment	Total	Total Incl O&P
						2012 Bare Costs				
1510	1′-0″ x 6′-8″, low e				Ea.	264			264	290
1520	1′-0″ x 6′-8″, half lite					254			254	280
1530	1′-0″ x 6′-8″, half lite, low e					262			262	289
2300	Interior, residential, closet, bi-fold, 6′-8″ x 2′-0″ wide	2 Carp	16	1		161	44		205	250
2330	3′-0″ wide		16	1		200	44		244	293
2360	4′-0″ wide		15	1.067		260	47		307	365
2400	5′-0″ wide		14	1.143		315	50.50		365.50	435
2420	6′-0″ wide		13	1.231		330	54.50		384.50	455
9000	Minimum labor/equipment charge	1 Carp	4	2	Job		88		88	145

08 13 13.25 Doors Hollow Metal		Crew	Daily Output	Labor-Hours	Unit	Material	Labor	Equipment	Total	Total Incl O&P
0010	**DOORS HOLLOW METAL**									
0500	Exterior, commercial, flush, 20 ga., 1-3/4″ x 7′-0″ x 2′-6″ wide	2 Carp	15	1.067	Ea.	390	47		437	505
0530	2′-8″ wide		15	1.067		435	47		482	550
0560	3′-0″ wide		14	1.143		435	50.50		485.50	560
1000	18 ga., 1-3/4″ x 7′-0″ x 2′-6″ wide		15	1.067		495	47		542	615
1030	2′-8″ wide		15	1.067		500	47		547	625
1060	3′-0″ wide		14	1.143		485	50.50		535.50	620
1500	16 ga., 1-3/4 x 7′-0″ x 2′-6″ wide		15	1.067		560	47		607	690
1530	2′-8″ wide		15	1.067		565	47		612	700
1560	3′-0″ wide		14	1.143		550	50.50		600.50	690
1590	3′-6″ wide		14	1.143		645	50.50		695.50	790
2900	Fire door, "A" label, 1-3/4″ x 7′-0″ x 2′-6″ wide		15	1.067		645	47		692	785
2930	2′-8″ wide		15	1.067		655	47		702	795
2960	3′-0″ wide		14	1.143		640	50.50		690.50	790
2990	3′-6″ wide		14	1.143		730	50.50		780.50	890
3100	"B" label, 2′-6″ wide		15	1.067		585	47		632	715
3130	2′-8″ wide		15	1.067		590	47		637	720
3160	3′-0″ wide		14	1.143		580	50.50		630.50	725

08 13 16 – Aluminum Doors

08 13 16.10 Commercial Aluminum Doors		Crew	Daily Output	Labor-Hours	Unit	Material	Labor	Equipment	Total	Total Incl O&P
0010	**COMMERCIAL ALUMINUM DOORS**, no glazing, flush									
5000	Flush panel doors, pair of 2′-6″ x 7′-0″	2 Sswk	2	8	Pr.	1,375	395		1,770	2,250
5050	3′-0″ x 7′-0″, single		2.50	6.400	Ea.	780	315		1,095	1,450
5100	Pair of 3′-0″ x 7′-0″		2	8	Pr.	1,475	395		1,870	2,350
5150	3′-6″ x 7′-0″, single		2.50	6.400	Ea.	935	315		1,250	1,600

08 14 Wood Doors

08 14 13 – Carved Wood Doors

08 14 13.10 Types of Wood Doors, Carved		Crew	Daily Output	Labor-Hours	Unit	Material	Labor	Equipment	Total	Total Incl O&P
0010	**TYPES OF WOOD DOORS, CARVED**									
3000	Solid wood, 1-3/4″ thick stile and rail									
3020	Mahogany, 3′-0″ x 7′-0″, minimum	2 Carp	14	1.143	Ea.	965	50.50		1,015.50	1,125
3030	Maximum		10	1.600		1,750	70.50		1,820.50	2,050
3040	3′-6″ x 8′-0″, minimum		10	1.600		1,250	70.50		1,320.50	1,500
3050	Maximum		8	2		2,400	88		2,488	2,800
3100	Pine, 3′-0″ x 7′-0″, minimum		14	1.143		490	50.50		540.50	620
3110	Maximum		10	1.600		825	70.50		895.50	1,025
3120	3′-6″ x 8′-0″, minimum		10	1.600		900	70.50		970.50	1,100
3130	Maximum		8	2		1,825	88		1,913	2,175
3200	Red oak, 3′-0″ x 7′-0″, minimum		14	1.143		1,650	50.50		1,700.50	1,900

08 14 Wood Doors

08 14 13 – Carved Wood Doors

08 14 13.10 Types of Wood Doors, Carved		Crew	Daily Output	Labor-Hours	Unit	Material	Labor (2012 Bare Costs)	Equipment	Total	Total Incl O&P
3210	Maximum	2 Carp	10	1.600	Ea.	2,200	70.50		2,270.50	2,550
3220	3'-6" x 8'-0", minimum		10	1.600		2,500	70.50		2,570.50	2,875
3230	Maximum		8	2		3,200	88		3,288	3,675
4000	Hand carved door, mahogany									
4020	3'-0" x 7'-0", minimum	2 Carp	14	1.143	Ea.	1,675	50.50		1,725.50	1,925
4030	Maximum		11	1.455		3,650	64		3,714	4,125
4040	3'-6" x 8'-0", minimum		10	1.600		2,500	70.50		2,570.50	2,875
4050	Maximum		8	2		3,600	88		3,688	4,100
4200	Rose wood, 3'-0" x 7'-0", minimum		14	1.143		5,100	50.50		5,150.50	5,675
4210	Maximum		11	1.455		13,600	64		13,664	15,100
4220	3'-6" x 8'-0", minimum		10	1.600		5,800	70.50		5,870.50	6,500
4225	Maximum		10	1.600		8,800	70.50		8,870.50	9,800
4280	For 6'-8" high door, deduct from 7'-0" door					34			34	37.50
4400	For custom finish, add					450			450	495
4600	Side light, mahogany, 7'-0" x 1'-6" wide, minimum	2 Carp	18	.889		1,000	39		1,039	1,175
4610	Maximum		14	1.143		2,625	50.50		2,675.50	2,975
4620	8'-0" x 1'-6" wide, minimum		14	1.143		1,700	50.50		1,750.50	1,950
4630	Maximum		10	1.600		2,000	70.50		2,070.50	2,325
4640	Side light, oak, 7'-0" x 1'-6" wide, minimum		18	.889		1,100	39		1,139	1,275
4650	Maximum		14	1.143		2,000	50.50		2,050.50	2,275
4660	8'-0" x 1'-6" wide, minimum		14	1.143		1,000	50.50		1,050.50	1,175
4670	Maximum		10	1.600		2,000	70.50		2,070.50	2,325

08 14 16 – Flush Wood Doors

08 14 16.09 Smooth Wood Doors

		Crew	Daily Output	Labor-Hours	Unit	Material	Labor	Equipment	Total	Total Incl O&P
0010	**SMOOTH WOOD DOORS**									
0015	Flush, interior, hollow core									
0025	Lauan face, 1-3/8", 3'-0" x 6'-8"	2 Carp	17	.941	Ea.	53.50	41.50		95	127
0030	4'-0" x 6'-8"		16	1		126	44		170	212
0080	1-3/4", 2'-0" x 6'-8"		17	.941		39	41.50		80.50	111
0085	3'-0" x 6'-8"	1 Carp	16	.500		53.50	22		75.50	95
0108	3'-0" x 7'-0"	2 Carp	16	1		99.50	44		143.50	182
0112	Pair of 3'-0" x 7'-0"		9	1.778	Pr.	192	78.50		270.50	340
0140	Birch face, 1-3/8", 2'-6" x 6'-8"		17	.941	Ea.	84	41.50		125.50	161
0180	3'-0" x 6'-8"		17	.941		94.50	41.50		136	172
0200	4'-0" x 6'-8"		16	1		158	44		202	247
0202	1-3/4", 2'-0" x 6'-8"		17	.941		48	41.50		89.50	121
0204	2'-4" x 7'-0"		16	1		93.50	44		137.50	176
0206	2'-6" x 7'-0"		16	1		96.50	44		140.50	179
0210	3'-0" x 7'-0"		16	1		108	44		152	192
0214	Pair of 3'-0" x 7'-0"		9	1.778	Pr.	207	78.50		285.50	355
0220	Oak face, 1-3/8", 2'-0" x 6'-8"		17	.941	Ea.	96	41.50		137.50	174
0280	3'-0" x 6'-8"		17	.941		110	41.50		151.50	189
0300	4'-0" x 6'-8"		16	1		134	44		178	221
0305	1-3/4", 2'-6" x 6'-8"		17	.941		106	41.50		147.50	184
0310	3'-0" x 7'-0"		16	1		196	44		240	288
0320	Walnut face, 1-3/8", 2'-0" x 6'-8"		17	941		178	41.50		219.50	263
0340	2'-6" x 6'-8"		17	.941		184	41.50		225.50	271
0380	3'-0" x 6'-8"		17	.941		192	41.50		233.50	279
0400	4'-0" x 6'-8"		16	1		213	44		257	305
0430	For 7'-0" high, add					25			25	27.50
0440	For 8'-0" high, add					35			35	38.50
0480	For prefinishing, clear, add					40			40	44

08 14 Wood Doors

08 14 16 – Flush Wood Doors

08 14 16.09 Smooth Wood Doors		Crew	Daily Output	Labor-Hours	Unit	Material	2012 Bare Costs Labor	Equipment	Total	Total Incl O&P
0500	For prefinishing, stain, add				Ea.	50			50	55
1320	M.D. overlay on hardboard, 1-3/8", 2'-0" x 6'-8"	2 Carp	17	.941		110	41.50		151.50	189
1340	2'-6" x 6'-8"		17	.941		110	41.50		151.50	189
1380	3'-0" x 6'-8"		17	.941		120	41.50		161.50	200
1400	4'-0" x 6'-8"		16	1		170	44		214	260
1420	For 7'-0" high, add					15			15	16.50
1440	For 8'-0" high, add					30			30	33
1720	H.P. plastic laminate, 1-3/8", 2'-0" x 6'-8"	2 Carp	16	1		232	44		276	330
1740	2'-6" x 6'-8"		16	1		232	44		276	330
1780	3'-0" x 6'-8"		15	1.067		281	47		328	385
1800	4'-0" x 6'-8"		14	1.143		365	50.50		415.50	490
1820	For 7'-0" high, add					15			15	16.50
1840	For 8'-0" high, add					30			30	33
2020	Particle core, lauan face, 1-3/8", 2'-6" x 6'-8"	2 Carp	15	1.067		90	47		137	176
2040	3'-0" x 6'-8"		14	1.143		92	50.50		142.50	184
2080	3'-0" x 7'-0"		13	1.231		96.50	54.50		151	195
2085	4'-0" x 7'-0"		12	1.333		120	59		179	229
2120	Birch face, 1-3/8", 2'-6" x 6'-8"		15	1.067		100	47		147	187
2140	3'-0" x 6'-8"		14	1.143		110	50.50		160.50	204
2180	3'-0" x 7'-0"		13	1.231		120	54.50		174.50	221
2200	4'-0" x 7'-0"		12	1.333		135	59		194	246
2205	1-3/4", 3'-0" x 7'-0"		13	1.231		120	54.50		174.50	221
2220	Oak face, 1-3/8", 2'-6" x 6'-8"		15	1.067		113	47		160	201
2240	3'-0" x 6'-8"		14	1.143		125	50.50		175.50	221
2280	3'-0" x 7'-0"		13	1.231		130	54.50		184.50	232
2300	4'-0" x 7'-0"		12	1.333		152	59		211	264
2305	1-3/4" 3'-0" x 7'-0"		13	1.231		159	54.50		213.50	264
2320	Walnut face, 1-3/8", 2'-0" x 6'-8"		15	1.067		120	47		167	209
2340	2'-6" x 6'-8"		14	1.143		135	50.50		185.50	232
2380	3'-0" x 6'-8"		13	1.231		152	54.50		206.50	256
2400	4'-0" x 6'-8"		12	1.333		200	59		259	315
2440	For 8'-0" high, add					40			40	44
2460	For 8'-0" high walnut, add					35			35	38.50
2480	For solid wood core, add					39			39	43
2720	For prefinishing, clear, add					35			35	38.50
2740	For prefinishing, stain, add					52			52	57
3320	M.D. overlay on hardboard, 1-3/8", 2'-6" x 6'-8"	2 Carp	14	1.143		106	50.50		156.50	200
3340	3'-0" x 6'-8"		13	1.231		112	54.50		166.50	212
3380	3'-0" x 7'-0"		12	1.333		114	59		173	222
3400	4'-0" x 7'-0"		10	1.600		150	70.50		220.50	281
3440	For 8'-0" height, add					35			35	38.50
3460	For solid wood core, add					40			40	44
3720	H.P. plastic laminate, 1-3/8", 2'-6" x 6'-8"	2 Carp	13	1.231		155	54.50		209.50	260
3740	3'-0" x 6'-8"		12	1.333		180	59		239	295
3780	3'-0" x 7'-0"		11	1.455		185	64		249	310
3800	4'-0" x 7'-0"		8	2		222	88		310	390
3840	For 8'-0" height, add					35			35	38.50
3860	For solid wood core, add					40			40	44
4000	Exterior, flush, solid core, birch, 1-3/4" x 2'-6" x 7'-0"	2 Carp	15	1.067		165	47		212	259
4020	2'-8" wide		15	1.067		153	47		200	245
4040	3'-0" wide		14	1.143		197	50.50		247.50	299
4045	3'-0" x 8'-0"	1 Carp	8	1		420	44		464	535
4100	Oak faced 1-3/4" x 2'-6" x 7'-0"	2 Carp	15	1.067	Ea.	185	47		232	281

08 14 Wood Doors

08 14 16 – Flush Wood Doors

08 14 16.09 Smooth Wood Doors		Crew	Daily Output	Labor-Hours	Unit	2012 Bare Costs Material	Labor	Equipment	Total	Total Incl O&P
4120	2′-8″ wide	2 Carp	15	1.067	Ea.	200	47		247	297
4140	3′-0″ wide		14	1.143		210	50.50		260.50	315
4160	Walnut faced, 1-3/4″ x 3′-0″ x 6′-8″	1 Carp	17	.471		281	21		302	345
4180	3′-6″ wide	″	17	.471		350	21		371	420
4200	Walnut faced, 1-3/4″ x 2′-6″ x 7′-0″	2 Carp	15	1.067		270	47		317	375
4220	2′-8″ wide		15	1.067		285	47		332	390
4240	3′-0″ wide		14	1.143		295	50.50		345.50	410
4300	For 6′-8″ high door, deduct from 7′-0″ door					16			16	17.60
5000	Wood doors, for vision lite, add					93			93	102
5010	Wood doors, for narrow lite, add					101			101	111
5015	Wood doors, for bottom (or top) louver, add					274			274	300
9000	Minimum labor/equipment charge	1 Carp	4	2	Job		88		88	145

08 14 16.10 Wood Doors Decorator		Crew	Daily Output	Labor-Hours	Unit	Material	Labor	Equipment	Total	Total Incl O&P
0010	**WOOD DOORS DECORATOR**									
1800	Exterior, flush, solid wood core, birch 1-3/4″ x 7′-0″ x 2′-6″ wide	2 Carp	15	1.067	Ea.	315	47		362	420
1820	2′-8″ wide		15	1.067		320	47		367	425
1840	3′-0″ wide		14	1.143		330	50.50		380.50	450
1900	Oak faced, 1-3/4″ x 7′-0″ x 2′-6″ wide		15	1.067		320	47		367	425
1920	2′-8″ wide		15	1.067		325	47		372	435
1940	3′-0″ wide		14	1.143		335	50.50		385.50	455
2100	Walnut faced, 1-3/4″ x 7′-0″ x 2′-6″ wide		15	1.067		360	47		407	470
2120	2′-8″ wide		15	1.067		375	47		422	490
2140	3′-0″ wide		14	1.143		405	50.50		455.50	530

08 14 16.20 Wood Fire Doors		Crew	Daily Output	Labor-Hours	Unit	Material	Labor	Equipment	Total	Total Incl O&P
0010	**WOOD FIRE DOORS**									
0020	Particle core, 7 face plys, "B" label,									
0040	1 hour, birch face, 1-3/4″ x 2′-6″ x 6′-8″	2 Carp	14	1.143	Ea.	405	50.50		455.50	530
0080	3′-0″ x 6′-8″		13	1.231		400	54.50		454.50	530
0090	3′-0″ x 7′-0″		12	1.333		435	59		494	575
0100	4′-0″ x 7′-0″		12	1.333		540	59		599	690
0140	Oak face, 2′-6″ x 6′-8″		14	1.143		340	50.50		390.50	460
0180	3′-0″ x 6′-8″		13	1.231		400	54.50		454.50	530
0190	3′-0″ x 7′-0″		12	1.333		450	59		509	590
0200	4′-0″ x 7′-0″		12	1.333		510	59		569	655
0240	Walnut face, 2′-6″ x 6′-8″		14	1.143		460	50.50		510.50	590
0280	3′-0″ x 6′-8″		13	1.231		480	54.50		534.50	620
0290	3′-0″ x 7′-0″		12	1.333		500	59		559	645
0300	4′-0″ x 7′-0″		12	1.333		620	59		679	775
0440	M.D. overlay on hardboard, 2′-6″ x 6′-8″		15	1.067		310	47		357	415
0480	3′-0″ x 6′-8″		14	1.143		370	50.50		420.50	490
0490	3′-0″ x 7′-0″		13	1.231		390	54.50		444.50	520
0500	4′-0″ x 7′-0″		12	1.333		410	59		469	545
0740	90 minutes, birch face, 1-3/4″ x 2′-6″ x 6′-8″		14	1.143		315	50.50		365.50	430
0780	3′-0″ x 6′-8″		13	1.231		310	54.50		364.50	430
0790	3′-0″ x 7′-0″		12	1.333		380	59		439	515
0800	4′-0″ x 7′-0″		12	1.333		490	59		549	635
0840	Oak face, 2′-6″ x 6′-8″		14	1.143		430	50.50		480.50	560
0880	3′-0″ x 6′-8″		13	1.231		440	54.50		494.50	575
0890	3′-0″ x 7′-0″		12	1.333		455	59		514	595
0900	4′-0″ x 7′-0″		12	1.333		590	59		649	745
0940	Walnut face, 2′-6″ x 6′-8″		14	1.143		400	50.50		450.50	525
0980	3′-0″ x 6′-8″		13	1.231		410	54.50		464.50	540

08 14 Wood Doors

08 14 16 – Flush Wood Doors

08 14 16.20 Wood Fire Doors		Crew	Daily Output	Labor-Hours	Unit	2012 Bare Costs Material	Labor	Equipment	Total	Total Incl O&P
0990	3'-0" x 7'-0"	2 Carp	12	1.333	Ea.	470	59		529	610
1000	4'-0" x 7'-0"		12	1.333		620	59		679	775
1140	M.D. overlay on hardboard, 2'-6" x 6'-8"		15	1.067		355	47		402	465
1180	3'-0" x 6'-8"		14	1.143		375	50.50		425.50	500
1190	3'-0" x 7'-0"		13	1.231		405	54.50		459.50	535
1200	4'-0" x 7'-0"		12	1.333		455	59		514	595
1240	For 8'-0" height, add					70			70	77
1260	For 8'-0" height walnut, add					85			85	93.50
2200	Custom architectural "B" label, flush, 1-3/4" thick, birch,									
2210	Solid core									
2220	2'-6" x 7'-0"	2 Carp	15	1.067	Ea.	300	47		347	405
2260	3'-0" x 7'-0"		14	1.143		310	50.50		360.50	425
2300	4'-0" x 7'-0"		13	1.231		400	54.50		454.50	530
2420	4'-0" x 8'-0"		11	1.455		430	64		494	580
2480	For oak veneer, add					50%				
2500	For walnut veneer, add					75%				
9000	Minimum labor/equipment charge	1 Carp	4	2	Job		88		88	145

08 14 23 – Clad Wood Doors

08 14 23.13 Metal-Faced Wood Doors		Crew	Daily Output	Labor-Hours	Unit	Material	Labor	Equipment	Total	Total Incl O&P
0010	**METAL-FACED WOOD DOORS**									
0020	Interior, flush type, 3' x 7'	2 Carp	4.30	3.721	Opng.	350	164		514	655
9000	Minimum labor/equipment charge	1 Carp	2	4	Job		176		176	289

08 14 23.20 Tin Clad Wood Doors		Crew	Daily Output	Labor-Hours	Unit	Material	Labor	Equipment	Total	Total Incl O&P
0010	**TIN CLAD WOOD DOORS**									
0020	3 ply, 6' x 7', double sliding, doors only	2 Carp	1	16	Opng.	2,000	705		2,705	3,350
1000	For electric operator, add	1 Elec	2	4	"	3,250	206		3,456	3,900
9000	Minimum labor/equipment charge	2 Carp	1	16	Job		705		705	1,150

08 14 33 – Stile and Rail Wood Doors

08 14 33.10 Wood Doors Paneled		Crew	Daily Output	Labor-Hours	Unit	Material	Labor	Equipment	Total	Total Incl O&P
0010	**WOOD DOORS PANELED**									
0020	Interior, six panel, hollow core, 1-3/8" thick									
0040	Molded hardboard, 2'-0" x 6'-8"	2 Carp	17	.941	Ea.	60	41.50		101.50	134
0060	2'-6" x 6'-8"		17	.941		62	41.50		103.50	136
0070	2'-8" x 6'-8"		17	.941		65	41.50		106.50	140
0080	3'-0" x 6'-8"		17	.941		68	41.50		109.50	143
0140	Embossed print, molded hardboard, 2'-0" x 6'-8"		17	.941		62	41.50		103.50	136
0160	2'-6" x 6'-8"		17	.941		62	41.50		103.50	136
0180	3'-0" x 6'-8"		17	.941		68	41.50		109.50	143
0540	Six panel, solid, 1-3/8" thick, pine, 2'-0" x 6'-8"		15	1.067		150	47		197	242
0560	2'-6" x 6'-8"		14	1.143		165	50.50		215.50	265
0580	3'-0" x 6'-8"		13	1.231		140	54.50		194.50	243
1020	Two panel, bored rail, solid, 1-3/8" thick, pine, 1'-6" x 6'-8"		16	1		265	44		309	365
1040	2'-0" x 6'-8"		15	1.067		350	47		397	460
1060	2'-6" x 6'-8"		14	1.143		395	50.50		445.50	520
1340	Two panel, solid, 1-3/8" thick, fir, 2'-0" x 6'-8"		15	1.067		155	47		202	248
1360	2'-6" x 6'-8"		14	1.143		205	50.50		255.50	310
1380	3'-0" x 6'-8"		13	1.231		410	54.50		464.50	540
1740	Five panel, solid, 1-3/8" thick, fir, 2'-0" x 6'-8"		15	1.067		275	47		322	380
1760	2'-6" x 6'-8"		14	1.143		415	50.50		465.50	540
1780	3'-0" x 6'-8"		13	1.231		415	54.50		469.50	545
4190	Exterior, Knotty pine, paneled, 1-3/4", 3'-0" x 6'-8"		16	1		750	44		794	900

08 14 33 – Stile and Rail Wood Doors

08 14 33.10 Wood Doors Paneled		Crew	Daily Output	Labor-Hours	Unit	Material	2012 Bare Costs Labor	Equipment	Total	Total Incl O&P
4195	Double 1-3/4", 3'-0" x 6'-8"	2 Carp	16	1	Ea.	1,500	44		1,544	1,725
4200	Ash, paneled, 1-3/4", 3'-0" x 6'-8"		16	1		850	44		894	1,000
4205	Double 1-3/4", 3'-0" x 6'-8"		16	1		1,700	44		1,744	1,950
4210	Cherry, paneled, 1-3/4", 3'-0" x 6'-8"		16	1		850	44		894	1,000
4215	Double 1-3/4", 3'-0" x 6'-8"		16	1		1,700	44		1,744	1,950
4230	Ash, paneled, 1-3/4", 3'-0" x 8'-0"		16	1		925	44		969	1,100
4235	Double 1-3/4", 3'-0" x 8'-0"		16	1		1,850	44		1,894	2,100
4240	Hard Maple, paneled, 1-3/4", 3'-0" x 8'-0"		16	1		925	44		969	1,100
4245	Double 1-3/4", 3'-0" x 8'-0"		16	1		1,850	44		1,894	2,100
4250	Cherry, paneled, 1-3/4", 3'-0" x 8'-0"		16	1		925	44		969	1,100
4255	Double 1-3/4", 3'-0" x 8'-0"		16	1		1,850	44		1,894	2,100
9000	Minimum labor/equipment charge	1 Carp	4	2	Job		88		88	145

08 14 33.20 Wood Doors Residential		Crew	Daily Output	Labor-Hours	Unit	Material	2012 Bare Costs Labor	Equipment	Total	Total Incl O&P
0010	**WOOD DOORS RESIDENTIAL**									
0200	Exterior, combination storm & screen, pine									
0260	2'-8" wide	2 Carp	10	1.600	Ea.	300	70.50		370.50	445
0280	3'-0" wide		9	1.778		310	78.50		388.50	470
0300	7'-1" x 3'-0" wide		9	1.778		330	78.50		408.50	495
0400	Full lite, 6'-9" x 2'-6" wide		11	1.455		320	64		384	455
0420	2'-8" wide		10	1.600		320	70.50		390.50	465
0440	3'-0" wide		9	1.778		325	78.50		403.50	485
0500	7'-1" x 3'-0" wide		9	1.778		355	78.50		433.50	520
0700	Dutch door, pine, 1-3/4" x 2'-8" x 6'-8", minimum		12	1.333		715	59		774	880
0720	Maximum		10	1.600		900	70.50		970.50	1,100
0800	3'-0" wide, minimum		12	1.333		715	59		774	880
0820	Maximum		10	1.600		900	70.50		970.50	1,100
1000	Entrance door, colonial, 1-3/4" x 6'-8" x 2'-8" wide		16	1		485	44		529	610
1020	6 panel pine, 3'-0" wide		15	1.067		445	47		492	565
1100	8 panel pine, 2'-8" wide		16	1		575	44		619	710
1120	3'-0" wide		15	1.067		560	47		607	690
1200	For tempered safety glass lites, (min of 2) add					76.50			76.50	84.50
1300	Flush, birch, solid core, 1-3/4" x 6'-8" x 2'-8" wide	2 Carp	16	1		110	44		154	194
1320	3'-0" wide		15	1.067		115	47		162	203
1350	7'-0" x 2'-8" wide		16	1		119	44		163	203
1360	3'-0" wide		15	1.067		130	47		177	220
1380	For tempered safety glass lites, add					103			103	113
1420	6'-8" x 3'-0" wide, fir	2 Carp	16	1		470	44		514	590
1720	Carved mahogany 3'-0" x 6'-8"		15	1.067		1,300	47		1,347	1,500
1800	Lauan, solid core, 1-3/4" x 7'-0" x 2'-4" wide		16	1		108	44		152	192
1810	2'-6" wide		15	1.067		111	47		158	199
1820	2'-8" wide		9	1.778		116	78.50		194.50	257
1830	3'-0" wide		16	1		128	44		172	213
1840	3'-4" wide		16	1		215	44		259	310
1850	Pair of 3'-0" wide		15	1.067	Pr.	254	47		301	355
2700	Interior, closet, bi-fold, w/hardware, no frame or trim incl.									
2720	Flush, birch, 6'-6" or 6'-8" x 2'-6" wide	2 Carp	13	1.231	Ea.	67.50	54.50		122	164
2740	3'-0" wide		13	1.231		71.50	54.50		126	168
2760	4'-0" wide		12	1.333		109	59		168	217
2780	5'-0" wide		11	1.455		105	64		169	221
2800	6'-0" wide		10	1.600		128	70.50		198.50	257
2820	Flush, hardboard, primed, 6'-8" x 2'-6" wide		13	1.231		52.50	54.50		107	147
2840	3'-0" wide		13	1.231		60	54.50		114.50	155

08 14 Wood Doors

08 14 33 – Stile and Rail Wood Doors

08 14 33.20 Wood Doors Residential		Crew	Daily Output	Labor-Hours	Unit	Material	2012 Bare Costs Labor	Equipment	Total	Total Incl O&P
2860	4'-0" wide	2 Carp	12	1.333	Ea.	90	59		149	196
2880	5'-0" wide		11	1.455		97.50	64		161.50	212
2900	6'-0" wide		10	1.600		138	70.50		208.50	268
3000	Raised panel pine, 6'-6" or 6'-8" x 2'-6" wide		13	1.231		190	54.50		244.50	298
3020	3'-0" wide		13	1.231		278	54.50		332.50	395
3040	4'-0" wide		12	1.333		305	59		364	430
3060	5'-0" wide		11	1.455		365	64		429	505
3080	6'-0" wide		10	1.600		400	70.50		470.50	555
3200	Louvered, pine 6'-6" or 6'-8" x 2'-6" wide		13	1.231		142	54.50		196.50	245
3220	3'-0" wide		13	1.231		206	54.50		260.50	315
3240	4'-0" wide		12	1.333		233	59		292	355
3260	5'-0" wide		11	1.455		260	64		324	390
3280	6'-0" wide		10	1.600		287	70.50		357.50	430
4400	Bi-passing closet, incl. hardware and frame, no trim incl.									
4420	Flush, lauan, 6'-8" x 4'-0" wide	2 Carp	12	1.333	Opng.	176	59		235	291
4440	5'-0" wide		11	1.455		187	64		251	310
4460	6'-0" wide		10	1.600		209	70.50		279.50	345
4600	Flush, birch, 6'-8" x 4'-0" wide		12	1.333		220	59		279	340
4620	5'-0" wide		11	1.455		212	64		276	340
4640	6'-0" wide		10	1.600		256	70.50		326.50	400
4800	Louvered, pine, 6'-8" x 4'-0" wide		12	1.333		405	59		464	540
4820	5'-0" wide		11	1.455		395	64		459	540
4840	6'-0" wide		10	1.600		525	70.50		595.50	690
5000	Paneled, pine, 6'-8" x 4'-0" wide		12	1.333		505	59		564	655
5020	5'-0" wide		11	1.455		410	64		474	555
5040	6'-0" wide		10	1.600		600	70.50		670.50	775
5042	8'-0" wide		12	1.333		1,025	59		1,084	1,225
6100	Folding accordion, closet, including track and frame									
6120	Vinyl, 2 layer, stock	2 Carp	10	1.600	Ea.	65.50	70.50		136	188
6200	Rigid PVC	"	10	1.600	"	52	70.50		122.50	173
7310	Passage doors, flush, no frame included									
7320	Hardboard, hollow core, 1-3/8" x 6'-8" x 1'-6" wide	2 Carp	18	.889	Ea.	43	39		82	112
7330	2'-0" wide		18	.889		43	39		82	112
7340	2'-6" wide		18	.889		47.50	39		86.50	117
7350	2'-8" wide		18	.889		49	39		88	118
7360	3'-0" wide		17	.941		52	41.50		93.50	125
7420	Lauan, hollow core, 1-3/8" x 6'-8" x 1'-6" wide		18	.889		32	39		71	99.50
7440	2'-0" wide		18	.889		34.50	39		73.50	103
7450	2'-4" wide		18	.889		38.50	39		77.50	107
7460	2'-6" wide		18	.889		38.50	39		77.50	107
7480	2'-8" wide		18	.889		40	39		79	108
7500	3'-0" wide		17	.941		42	41.50		83.50	114
7700	Birch, hollow core, 1-3/8" x 6'-8" x 1'-6" wide		18	.889		39.50	39		78.50	108
7720	2'-0" wide		18	.889		42	39		81	111
7740	2'-6" wide		18	.889		49.50	39		88.50	119
7760	2'-8" wide		18	.889		50.50	39		89.50	120
7780	3'-0" wide		17	.941		52	41.50		93.50	125
8000	Pine louvered, 1-3/8" x 6'-8" x 1'-6" wide		19	.842		102	37		139	173
8020	2'-0" wide		18	.889		128	39		167	206
8040	2'-6" wide		18	.889		140	39		179	218
8060	2'-8" wide		18	.889		147	39		186	227
8080	3'-0" wide		17	.941		173	41.50		214.50	259
8300	Pine paneled, 1-3/8" x 6'-8" x 1'-6" wide		19	.842		128	37		165	201

08 14 Wood Doors

08 14 33 – Stile and Rail Wood Doors

08 14 33.20 Wood Doors Residential		Crew	Daily Output	Labor-Hours	Unit	Material	Labor	Equipment	Total	Total Incl O&P
						2012 Bare Costs				
8320	2′-0″ wide	2 Carp	18	.889	Ea.	152	39		191	232
8330	2′-4″ wide		18	.889		180	39		219	263
8340	2′-6″ wide		18	.889		180	39		219	263
8360	2′-8″ wide		18	.889		184	39		223	267
8380	3′-0″ wide		17	.941		203	41.50		244.50	291
8490	3′-0″ wide		17	.941		252	41.50		293.50	345
8804	Pocket door,6 panel pine, 2′-6″ x 6′-8″		10.50	1.524		355	67		422	500
8814	2′-8″ x 6′-8″		10.50	1.524		360	67		427	505
8824	3′-0″ x 6′-8″		10.50	1.524		385	67		452	530
9000	Passage doors, flush, no frame, birch, solid core, 1-3/8″ x 7′-0″ x 2′-4″		16	1		114	44		158	198
9020	2′-8″ wide		16	1		122	44		166	207
9040	3′-0″ wide		16	1		131	44		175	217
9060	3′-4″ wide		15	1.067		151	47		198	243
9080	Pair of 3′-0″ wide		9	1.778	Pr.	261	78.50		339.50	415
9100	Lauan, solid core, 1-3/8″ x 7′-0″ x 2′-4″ wide		16	1	Ea.	106	44		150	189
9120	2′-8″ wide		16	1		113	44		157	197
9140	3′-0″ wide		16	1		119	44		163	204
9160	3′-4″ wide		15	1.067		127	47		174	217
9180	Pair of 3′-0″ wide		9	1.778	Pr.	218	78.50		296.50	370
9200	Hardboard, solid core, 1-3/8″ x 7′-0″ x 2′-4″ wide		16	1	Ea.	130	44		174	216
9220	2′-8″ wide		16	1		135	44		179	221
9240	3′-0″ wide		16	1		139	44		183	226
9260	3′-4″ wide		15	1.067		154	47		201	246
9900	Minimum labor/equipment charge	1 Carp	4	2	Job		88		88	145

08 14 40 – Interior Cafe Doors

08 14 40.10 Cafe Style Doors		Crew	Daily Output	Labor-Hours	Unit	Material	Labor	Equipment	Total	Total Incl O&P
0010	CAFE STYLE DOORS									
6520	Interior cafe doors, 2′-6″ opening, stock, panel pine	2 Carp	16	1	Ea.	192	44		236	284
6540	3′-0″ opening	″	16	1	″	199	44		243	292
6550	Louvered pine									
6560	2′-6″ opening	2 Carp	16	1	Ea.	166	44		210	255
8000	3′-0″ opening		16	1		180	44		224	271
8010	2′-6″ opening, hardwood		16	1		247	44		291	345
0020	3′-0″ opening		16	1		272	44		316	370
9000	Minimum labor/equipment charge	1 Carp	4	2	Job		88		88	145

08 16 Composite Doors

08 16 13 – Fiberglass Doors

08 16 13.10 Entrance Doors, Fiberous Glass			Crew	Daily Output	Labor-Hours	Unit	Material	Labor	Equipment	Total	Total Incl O&P
0010	ENTRANCE DOORS, FIBEROUS GLASS										
0020	Exterior, fiberglass, door, 2′-8″ wide x 6′-8″ high	G	2 Carp	15	1.067	Ea.	259	47		306	360
0040	3′-0″ wide x 6′-8″ high	G		15	1.067		259	47		306	360
0060	3′-0″ wide x 7′-0″ high	G		15	1.067		460	47		507	580
0080	3′-0″ wide x 6′-8″ high, with two lites	G		15	1.067		294	47		341	400
0100	3′-0″ wide x 8′-0″ high, with two lites	G		15	1.067		495	47		542	620
0110	Half glass, 3′-0″ wide x 6′-8″ high	G		15	1.067		305	47		352	410
0120	3′-0″ wide x 6′-8″ high, low e	G		15	1.067		330	47		377	440
0130	3′-0″ wide x 8′-0″ high	G		15	1.067		580	47		627	715
0140	3′-0″ wide x 8′-0″ high, low e	G		15	1.067		640	47		687	775
0150	Side lights, 1′-0″ wide x 6′-8″ high,	G					251			251	276

08 16 Composite Doors

08 16 13 – Fiberglass Doors

08 16 13.10 Entrance Doors, Fiberous Glass		Crew	Daily Output	Labor-Hours	Unit	Material	2012 Bare Costs Labor	Equipment	Total	Total Incl O&P
0160	1'-0" wide x 6'-8" high, low e G				Ea.	256			256	281
0180	1'-0" wide x 6'-8" high, full glass G					275			275	305
0190	1'-0" wide x 6'-8" high, low e G					284			284	315

08 16 14 – French Doors

08 16 14.10 Exterior Doors With Glass Lites

		Crew	Daily Output	Labor-Hours	Unit	Material	Labor	Equipment	Total	Total Incl O&P
0010	**EXTERIOR DOORS WITH GLASS LITES**									
0020	French, Fir, 1-3/4", 3'-0"wide x 6'-8" high	2 Carp	12	1.333	Ea.	600	59		659	755
0025	Double		12	1.333		1,200	59		1,259	1,425
0030	Maple, 1-3/4", 3'-0"wide x 6'-8" high		12	1.333		675	59		734	840
0035	Double		12	1.333		1,350	59		1,409	1,575
0040	Cherry, 1-3/4", 3'-0"wide x 6'-8" high		12	1.333		790	59		849	960
0045	Double		12	1.333		1,575	59		1,634	1,825
0100	Mahogany, 1-3/4", 3'-0"wide x 8'-0" high		10	1.600		700	70.50		770.50	885
0105	Double		10	1.600		1,400	70.50		1,470.50	1,675
0110	Fir, 1-3/4", 3'-0"wide x 8'-0" high		10	1.600		1,100	70.50		1,170.50	1,325
0115	Double		10	1.600		2,200	70.50		2,270.50	2,550
0120	Oak, 1-3/4", 3'-0"wide x 8'-0" high		10	1.600		1,825	70.50		1,895.50	2,125
0125	Double		10	1.600		3,650	70.50		3,720.50	4,125

08 17 Integrated Door Opening Assemblies

08 17 23 – Integrated Wood Door Opening Assemblies

08 17 23.10 Pre-Hung Doors

		Crew	Daily Output	Labor-Hours	Unit	Material	Labor	Equipment	Total	Total Incl O&P
0010	**PRE-HUNG DOORS**									
0300	Exterior, wood, comb. storm & screen, 6'-9" x 2'-6" wide	2 Carp	15	1.067	Ea.	296	47		343	400
0320	2'-8" wide		15	1.067		296	47		343	400
0340	3'-0" wide		15	1.067		305	47		352	410
0360	For 7'-0" high door, add					30			30	33
1600	Entrance door, flush, birch, solid core									
1620	4-5/8" solid jamb, 1-3/4" x 6'-8" x 2'-8" wide	2 Carp	16	1	Ea.	289	44		333	395
1640	3'-0" wide		16	1		330	44		374	440
1642	5-5/8" jamb		16	1		325	44		369	435
1680	For 7'-0" high door, add					25			25	27.50
2000	Entrance door, colonial, 6 panel pine									
2020	4-5/8" solid jamb, 1-3/4" x 6'-8" x 2'-8" wide	2 Carp	16	1	Ea.	635	44		679	775
2040	3'-0" wide	"	16	1		665	44		709	810
2060	For 7'-0" high door, add					53.50			53.50	58.50
2200	For 5-5/8" solid jamb, add					41.50			41.50	45.50
2230	French style, exterior, 1 lite, 1-3/4" x 3'-0" x 6'-8"	1 Carp	14	.571		590	25		615	690
2235	9 lites	"	14	.571		660	25		685	765
2245	15 lites	2 Carp	14	1.143		635	50.50		685.50	780
2250	Double, 15 lites, 2'-0" x 6'-8", 4'-0" opening		7	2.286	Pr.	1,200	101		1,301	1,500
2260	2'-6" x 6'-8", 5' -0" opening		7	2.286		1,325	101		1,426	1,625
2280	3'-0" x 6'-8", 6'-0" opening		7	2.286		1,325	101		1,426	1,650
2430	3'-0" x 7'-0", 15 lites		14	1.143	Ea.	940	50.50		990.50	1,100
2432	Two 3'-0" x 7'-0"		7	2.286	Pr.	1,950	101		2,051	2,325
2435	3'-0" x 8'-0"		14	1.143	Ea.	1,000	50.50		1,050.50	1,175
2437	Two, 3'-0" x 8'-0"		7	2.286	Pr.	2,075	101		2,176	2,450
4000	Interior, passage door, 4-5/8" solid jamb									
4400	Lauan, flush, solid core, 1-3/8" x 6'-8" x 2'-6" wide	2 Carp	17	.941	Ea.	185	41.50		226.50	271
4420	2'-8" wide		17	.941		185	41.50		226.50	271

08 17 Integrated Door Opening Assemblies

08 17 23 – Integrated Wood Door Opening Assemblies

08 17 23.10 Pre-Hung Doors		Crew	Daily Output	Labor-Hours	Unit	2012 Bare Costs Material	Labor	Equipment	Total	Total Incl O&P
4440	3′-0″ wide	2 Carp	16	1	Ea.	200	44		244	292
4600	Hollow core, 1-3/8″ x 6′-8″ x 2′-6″ wide		17	.941		125	41.50		166.50	205
4620	2′-8″ wide		17	.941		124	41.50		165.50	204
4640	3′-0″ wide		16	1		140	44		184	226
4700	For 7′-0″ high door, add					35			35	38.50
5000	Birch, flush, solid core, 1-3/8″ x 6′-8″ x 2′-6″ wide	2 Carp	17	.941		247	41.50		288.50	340
5020	2′-8″ wide		17	.941		200	41.50		241.50	287
5040	3′-0″ wide		16	1		277	44		321	380
5200	Hollow core, 1-3/8″ x 6′-8″ x 2′-6″ wide		17	.941		187	41.50		228.50	273
5220	2′-8″ wide		17	.941		161	41.50		202.50	245
5240	3′-0″ wide		16	1		204	44		248	297
5280	For 7′-0″ high door, add					30			30	33
5500	Hardboard paneled, 1-3/8″ x 6′-8″ x 2′-6″ wide	2 Carp	17	.941		144	41.50		185.50	226
5520	2′-8″ wide		17	.941		152	41.50		193.50	235
5540	3′-0″ wide		16	1		149	44		193	237
6000	Pine paneled, 1-3/8″ x 6′-8″ x 2′-6″ wide		17	.941		254	41.50		295.50	350
6020	2′-8″ wide		17	.941		273	41.50		314.50	370
6040	3′-0″ wide		16	1		281	44		325	385
6500	For 5-5/8″ solid jamb, add					22.50			22.50	25
6520	For split jamb, deduct					24			24	26.50
8200	Birch, flush, solid core, 1-3/4″ x 6′-8″ x 2′-4″ wide	1 Carp	17	.471		222	21		243	278
8220	2′-6″ wide		17	.471		218	21		239	273
8240	2′-8″ wide		17	.471		200	21		221	253
8260	3′-0″ wide		16	.500		208	22		230	265
8280	3′-6″ wide		15	.533		350	23.50		373.50	425
8500	Pocket door frame with lauan, flush, hollow core , 1-3/8″ x 3′-0″ x 6′-8″		17	.471		250	21		271	310
9000	Minimum labor/equipment charge		4	2	Job		88		88	145

08 31 Access Doors and Panels

08 31 13 – Access Doors and Frames

08 31 13.10 Types of Framed Access Doors		Crew	Daily Output	Labor-Hours	Unit	Material	Labor	Equipment	Total	Total Incl O&P
0010	**TYPES OF FRAMED ACCESS DOORS**									
1000	Fire rated door with lock									
1100	Metal, 12″ x 12″	1 Carp	10	.800	Ea.	165	35.50		200.50	240
1150	18″ x 18″		9	.889		220	39		259	305
1200	24″ x 24″		9	.889		325	39		364	425
1250	24″ x 36″		8	1		335	44		379	440
1300	24″ x 48″		8	1		430	44		474	550
1350	36″ x 36″		7.50	1.067		495	47		542	620
1400	48″ x 48″		7.50	1.067		635	47		682	770
1600	Stainless steel, 12″ x 12″		10	.800		282	35.50		317.50	370
1650	18″ x 18″		9	.889		410	39		449	515
1700	24″ x 24″		9	.889		500	39		539	615
1750	24″ x 36″		8	1		640	44		684	780
2000	Flush door for finishing									
2100	Metal 8″ x 8″	1 Carp	10	.800	Ea.	35	35.50		70.50	96.50
2150	12″ x 12″	″	10	.800	″	40	35.50		75.50	102
3000	Recessed door for acoustic tile									
3100	Metal, 12″ x 12″	1 Carp	4.50	1.778	Ea.	80	78.50		158.50	217
3150	12″ x 24″		4.50	1.778		90	78.50		168.50	228
3200	24″ x 24″		4	2		121	88		209	278

08 31 13.10 Types of Framed Access Doors		Crew	Daily Output	Labor-Hours	Unit	Material	2012 Bare Costs Labor	Equipment	Total	Total Incl O&P
3250	24" x 36"	1 Carp	4	2	Ea.	170	88		258	330
4000	Recessed door for drywall									
4100	Metal 12" x 12"	1 Carp	6	1.333	Ea.	80	59		139	185
4150	12" x 24"		5.50	1.455		114	64		178	230
4200	24" x 36"		5	1.600		182	70.50		252.50	315
6000	Standard door									
6100	Metal, 8" x 8"	1 Carp	10	.800	Ea.	35	35.50		70.50	96.50
6150	12" x 12"		10	.800		40	35.50		75.50	102
6200	18" x 18"		9	.889		62	39		101	133
6250	24" x 24"		9	.889		75	39		114	147
6300	24" x 36"		8	1		115	44		159	200
6350	36" x 36"		8	1		140	44		184	227
6500	Stainless steel, 8" x 8"		10	.800		80	35.50		115.50	146
6550	12" x 12"		10	.800		107	35.50		142.50	176
6600	18" x 18"		9	.889		198	39		237	283
6650	24" x 24"		9	.889		260	39		299	350
7000	Ceiling hatches, 2'-6" x 2'-6", swing up, single leaf, st fr & cover	G-3	11	2.909		580	127		707	845
7010	Aluminum cover	"	11	2.909		560	127		687	820
9000	Minimum labor/equipment charge	1 Carp	4	2	Job		88		88	145

08 31 13.20 Bulkhead/Cellar Doors		Crew	Daily Output	Labor-Hours	Unit	Material	Labor	Equipment	Total	Total Incl O&P
0010	**BULKHEAD/CELLAR DOORS**									
0020	Steel, not incl. sides, 44" x 62"	1 Carp	5.50	1.455	Ea.	560	64		624	720
0100	52" x 73"		5.10	1.569		770	69		839	960
0500	With sides and foundation plates, 57" x 45" x 24"		4.70	1.702		810	75		885	1,025
0600	42" x 49" x 51"		4.30	1.860		915	82		997	1,125
9000	Minimum labor/equipment charge		2	4	Job		176		176	289

08 31 13.30 Commercial Floor Doors		Crew	Daily Output	Labor-Hours	Unit	Material	Labor	Equipment	Total	Total Incl O&P
0010	**COMMERCIAL FLOOR DOORS**									
0020	Aluminum tile, steel frame, one leaf, 2' x 2' opng.	2 Sswk	3.50	4.571	Opng.	800	226		1,026	1,300
0050	3'-6" x 3'-6" opening		3.50	4.571		1,600	226		1,826	2,175
0500	Double leaf, 4' x 4' opening		3	5.333		1,675	263		1,938	2,350
0550	5' x 5' opening		3	5.333		3,000	263		3,263	3,800
9000	Minimum labor/equipment charge		2	8	Job		395		395	735

08 31 13.35 Industrial Floor Doors		Crew	Daily Output	Labor-Hours	Unit	Material	Labor	Equipment	Total	Total Incl O&P
0010	**INDUSTRIAL FLOOR DOORS**									
0020	Steel 300 psf L.L., single leaf, 2' x 2', 175#	2 Sswk	6	2.667	Opng.	760	132		892	1,075
0050	3' x 3' opening, 300#		5.50	2.909		1,075	144		1,219	1,450
0300	Double leaf, 4' x 4' opening, 455#		5	3.200		1,900	158		2,058	2,400
0350	5' x 5' opening, 645#		4.50	3.556		2,750	175		2,925	3,350
1000	Aluminum, 300 psf L.L., single leaf, 2' x 2', 60#		6	2.667		750	132		882	1,075
1050	3' x 3' opening, 100#		5.50	2.909		1,300	144		1,444	1,700
1500	Double leaf, 4' x 4' opening, 160#		5	3.200		2,100	158		2,258	2,600
1550	5' x 5' opening, 235#		4.50	3.556		2,800	175		2,975	3,400
9000	Minimum labor/equipment charge		2	8	Job		395		395	735

08 32 Sliding Glass Doors

08 32 13 – Sliding Aluminum-Framed Glass Doors

08 32 13.10 Sliding Aluminum Doors		Crew	Daily Output	Labor-Hours	Unit	2012 Bare Costs Material	Labor	Equipment	Total	Total Incl O&P
0010	**SLIDING ALUMINUM DOORS**									
0350	Aluminum, 5/8" tempered insulated glass, 6' wide									
0400	Premium	2 Carp	4	4	Ea.	1,575	176		1,751	2,025
0450	Economy		4	4		885	176		1,061	1,275
0500	8' wide, premium		3	5.333		1,650	235		1,885	2,200
0550	Economy		3	5.333		1,425	235		1,660	1,925
0600	12' wide, premium		2.50	6.400		2,975	282		3,257	3,750
0650	Economy		2.50	6.400		1,375	282		1,657	1,975
4000	Aluminum, baked on enamel, temp glass, 6'-8" x 10'-0" wide		4	4		1,050	176		1,226	1,450
4020	Insulating glass, 6'-8" x 6'-0" wide		4	4		915	176		1,091	1,300
4040	8'-0" wide		3	5.333		1,075	235		1,310	1,550
4060	10'-0" wide		2	8		1,325	355		1,680	2,025
4080	Anodized, temp glass, 6'-8" x 6'-0" wide		4	4		455	176		631	790
4100	8'-0" wide		3	5.333		575	235		810	1,025
4120	10'-0" wide		2	8		645	355		1,000	1,300
4200	Vinyl clad, anodized, temp glass, 6'-8" x 6'-0" wide		4	4		420	176		596	750
4240	8'-0" wide		3	5.333		585	235		820	1,025
4260	10'-0" wide		2	8		635	355		990	1,275
4280	Insulating glass, 6'-8" x 6'-0' wide		4	4		1,125	176		1,301	1,525
4300	8'-0" wide		3	5.333		935	235		1,170	1,400
4320	10'-0" wide		4	4		1,675	176		1,851	2,125

08 32 19 – Sliding Wood-Framed Glass Doors

08 32 19.10 Sliding Wood Doors		Crew	Daily Output	Labor-Hours	Unit	Material	Labor	Equipment	Total	Total Incl O&P
0010	**SLIDING WOOD DOORS**									
0020	Wood, tempered insul. glass, 6' wide, premium	2 Carp	4	4	Ea.	1,425	176		1,601	1,875
0100	Economy		4	4		1,125	176		1,301	1,550
0150	8' wide, wood, premium		3	5.333		1,825	235		2,060	2,375
0200	Economy		3	5.333		1,500	235		1,735	2,025
0235	10' wide, wood		2.50	6.400		2,600	282		2,882	3,325
0240	Economy		2.50	6.400		2,200	282		2,482	2,900
0250	12' wide, wood		2.50	6.400		3,000	282		3,282	3,775
0300	Economy		2.50	6.400		2,450	282		2,732	3,175
9000	Minimum labor/equipment charge	1 Carp	2	4	Job		176		176	289

08 32 19.15 Sliding Glass Vinyl-Clad Wood Doors		Crew	Daily Output	Labor-Hours	Unit	Material	Labor	Equipment	Total	Total Incl O&P
0010	**SLIDING GLASS VINYL-CLAD WOOD DOORS**									
0020	Glass, sliding vinyl clad, insul. glass, 6'-0" x 6'-8" G	2 Carp	4	4	Ea.	1,500	176		1,676	1,975
0025	6'-0" x 6'-10" high G		4	4	Opng.	1,650	176		1,826	2,125
0030	6'-0" x 8'-0" high G		4	4	Ea.	1,975	176		2,151	2,475
0050	5'-0" x 6'-8" high G		4	4	"	1,500	176		1,676	1,950
0100	8'-0" x 6'-10" high G		4	4	Opng.	2,025	176		2,201	2,525
0150	8'-0" x 8'-0" high G		4	4		2,225	176		2,401	2,750
0500	4 leaf, 9'-0" x 6'-10" high G		3	5.333		3,250	235		3,485	3,950
0550	9'-0" x 8'-0" high G		3	5.333	Ea.	3,775	235		4,010	4,525
0600	12'-0" x 6'-10" high G		3	5.333	Opng.	3,875	235		4,110	4,650
0650	12'-0" x 8'-0" high G		3	5.333	"	3,875	235		4,110	4,650
9000	Minimum labor/equipment charge	1 Carp	4	2	Job		88		88	145

08 33 Coiling Doors and Grilles

08 33 13 – Coiling Counter Doors

08 33 13.10 Counter Doors, Coiling Type		Crew	Daily Output	Labor-Hours	Unit	Material	Labor	Equipment	Total	Total Incl O&P
						2012 Bare Costs				
0010	**COUNTER DOORS, COILING TYPE**									
0020	Manual, incl. frm and hdwe, galv. stl., 4' roll-up, 6' long	2 Carp	2	8	Opng.	1,175	355		1,530	1,875
0300	Galvanized steel, UL label		1.80	8.889		1,225	390		1,615	2,000
0600	Stainless steel, 4' high roll-up, 6' long		2	8		2,100	355		2,455	2,875
0700	10' long		1.80	8.889		2,500	390		2,890	3,400
2000	Aluminum, 4' high, 4' long		2.20	7.273		1,375	320		1,695	2,025
2020	6' long		2	8		1,500	355		1,855	2,225
2040	8' long		1.90	8.421		1,725	370		2,095	2,500
2060	10' long		1.80	8.889		1,825	390		2,215	2,650
2080	14' long		1.40	11.429		2,575	505		3,080	3,650
2100	6' high, 4' long		2	8		1,550	355		1,905	2,275
2120	6' long		1.60	10		1,600	440		2,040	2,475
2140	10' long		1.40	11.429		2,100	505		2,605	3,150
9000	Minimum labor/equipment charge	1 Carp	2	4	Job		176		176	289

08 33 16 – Coiling Counter Grilles

08 33 16.10 Coiling Grilles

Line	Description	Crew	Daily Output	Labor-Hours	Unit	Material	Labor	Equipment	Total	Total Incl O&P
0010	**COILING GRILLES**									
2020	Aluminum, manual operated, mill finish	2 Sswk	82	.195	S.F.	28	9.65		37.65	48.50
2040	Bronze anodized		82	.195	"	44.50	9.65		54.15	67
2060	Steel, manual operated, 10' x 10' high		1	16	Opng.	2,500	790		3,290	4,225
2080	15' x 8' high		.80	20	"	2,900	985		3,885	5,000
3000	For safety edge bottom bar, electric, add				L.F.	49			49	54
8000	For motor operation, add	2 Sswk	5	3.200	Opng.	1,250	158		1,408	1,675
9000	Minimum labor/equipment charge	"	1	16	Job		790		790	1,475

08 33 23 – Overhead Coiling Doors

08 33 23.10 Rolling Service Doors

Line	Description	Crew	Daily Output	Labor-Hours	Unit	Material	Labor	Equipment	Total	Total Incl O&P
0010	**ROLLING SERVICE DOORS** Steel, manual, 20 ga., incl. hardware									
0050	8' x 8' high	2 Sswk	1.60	10	Ea.	1,100	495		1,595	2,125
0130	12' x 12' high, standard		1.20	13.333		2,025	660		2,685	3,450
0160	10' x 20' high, standard		.50	32		2,450	1,575		4,025	5,600
2000	Class A fire doors, manual, 20 ga., 8' x 8' high		1.40	11.429		1,475	565		2,040	2,675
2100	10' x 10' high		1.10	14.545		2,025	720		2,745	3,550
2200	20' x 10' high		.80	20		4,125	985		5,110	6,375
2300	12' x 12' high		1	16		3,175	790		3,965	4,975
2400	20' x 12' high		.80	20		4,475	985		5,460	6,750
2500	14' x 14' high		.60	26.667		3,475	1,325		4,800	6,275
2600	20' x 16' high		.50	32		5,625	1,575		7,200	9,100
2700	10' x 20' high		.40	40		4,350	1,975		6,325	8,450
3000	For 18 ga. doors, add				S.F.	1.45			1.45	1.60
3300	For enamel finish, add				"	1.70			1.70	1.87
3600	For safety edge bottom bar, pneumatic, add				L.F.	21			21	23
4000	For weatherstripping, extruded rubber, jambs, add					13.40			13.40	14.75
4100	Hood, add					8.25			8.25	9.05
4200	Sill, add					4.73			4.73	5.20
4500	Motor operators, to 14' x 14' opening	2 Sswk	5	3.200	Ea.	1,050	158		1,208	1,475
4700	For fire door, additional fusible link, add				"	26			26	28.50
9000	Minimum labor/equipment charge	2 Sswk	1	16	Job		790		790	1,475

08 34 Special Function Doors

08 34 13 – Cold Storage Doors

08 34 13.10 Doors for Cold Area Storage		Crew	Daily Output	Labor-Hours	Unit	Material	Labor	Equipment	Total	Total Incl O&P
						2012 Bare Costs				
0010	**DOORS FOR COLD AREA STORAGE**									
0020	Single, 20 ga. galvanized steel									
0300	Horizontal sliding, 5′ x 7′, manual operation, 3.5″ thick	2 Carp	2	8	Ea.	3,200	355		3,555	4,100
0400	4″ thick		2	8		3,225	355		3,580	4,125
0500	6″ thick		2	8		3,400	355		3,755	4,325
0800	5′ x 7′, power operation, 2″ thick		1.90	8.421		5,525	370		5,895	6,675
0900	4″ thick		1.90	8.421		5,650	370		6,020	6,825
1000	6″ thick		1.90	8.421		6,450	370		6,820	7,700
1300	9′ x 10′, manual operation, 2″ insulation		1.70	9.412		4,500	415		4,915	5,625
1400	4″ insulation		1.70	9.412		4,600	415		5,015	5,750
1500	6″ insulation		1.70	9.412		5,550	415		5,965	6,775
1800	Power operation, 2″ insulation		1.60	10		7,625	440		8,065	9,100
1900	4″ insulation		1.60	10		7,850	440		8,290	9,375
2000	6″ insulation		1.70	9.412		8,850	415		9,265	10,400
2300	For stainless steel face, add					25%				
3000	Hinged, lightweight, 3′ x 7′-0″, 2″ thick	2 Carp	2	8	Ea.	1,400	355		1,755	2,125
3050	4″ thick		1.90	8.421		1,750	370		2,120	2,525
3300	Aluminum doors, 3′ x 7′-0″, 4″ thick		1.90	8.421		1,325	370		1,695	2,075
3350	6″ thick		1.40	11.429		2,325	505		2,830	3,400
3600	Stainless steel, 3′ x 7′-0″, 4″ thick		1.90	8.421		1,725	370		2,095	2,500
3650	6″ thick		1.40	11.429		2,925	505		3,430	4,050
3900	Painted, 3′ x 7′-0″, 4″ thick		1.90	8.421		1,250	370		1,620	1,975
3950	6″ thick		1.40	11.429		2,325	505		2,830	3,400
5000	Bi-parting, electric operated									
5010	6′ x 8′ opening, galv. faces, 4″ thick for cooler	2 Carp	.80	20	Opng.	7,200	880		8,080	9,375
5050	For freezer, 4″ thick		.80	20		7,950	880		8,830	10,200
5300	For door buck framing and door protection, add		2.50	6.400		635	282		917	1,175
6000	Galvanized batten door, galvanized hinges, 4′ x 7′		2	8		1,800	355		2,155	2,550
6050	6′ x 8′		1.80	8.889		2,450	390		2,840	3,350
6500	Fire door, 3 hr., 6′ x 8′, single slide		.80	20		8,275	880		9,155	10,600
6550	Double, bi-parting		.70	22.857		12,200	1,000		13,200	15,100
9000	Minimum labor/equipment charge	1 Carp	2	4	Job		176		176	289

08 34 36 – Darkroom Doors

08 34 36.10 Various Types of Darkroom Doors		Crew	Daily Output	Labor-Hours	Unit	Material	Labor	Equipment	Total	Total Incl O&P
0010	**VARIOUS TYPES OF DARKROOM DOORS**									
0015	Revolving, standard, 2 way, 36″ diameter	2 Carp	3.10	5.161	Opng.	2,250	228		2,478	2,850
0020	41″ diameter		3.10	5.161		2,650	228		2,878	3,300
0050	3 way, 51″ diameter		1.40	11.429		3,075	505		3,580	4,200
1000	4 way, 49″ diameter		1.40	11.429		3,625	505		4,130	4,825
2000	Hinged safety, 2 way, 41″ diameter		2.30	6.957		2,825	305		3,130	3,625
2500	3 way, 51″ diameter		1.40	11.429		3,625	505		4,130	4,825
3000	Pop out safety, 2 way, 41″ diameter		3.10	5.161		3,600	228		3,828	4,325
4000	3 way, 51″ diameter		1.40	11.429		3,550	505		4,055	4,725
5000	Wheelchair-type, pop out, 51″ diameter		1.40	11.429		3,525	505		4,030	4,700
5020	72″ diameter		.90	17.778		7,225	785		8,010	9,225

08 34 53 – Security Doors and Frames

08 34 53.20 Steel Door		Crew	Daily Output	Labor-Hours	Unit	Material	Labor	Equipment	Total	Total Incl O&P
0010	**STEEL DOOR** with ballistic core and welded frame both 14 gauge									
0050	Flush, UL 752 Level 3, 1-3/4″, 3′-0″ x 6′-8″	2 Carp	1.50	10.667	Opng.	2,200	470		2,670	3,200
0055	1-3/4″, 3′-6″ x 6′-8″		1.50	10.667		2,300	470		2,770	3,300
0060	1-3/4″, 4′-0″ x 6′-8″		1.20	13.333		2,500	590		3,090	3,725
0100	UL 752 Level 8, 1-3/4″, 3′-0″ x 6′-8″		1.50	10.667		9,600	470		10,070	11,400

08 34 Special Function Doors

08 34 53 – Security Doors and Frames

08 34 53.20 Steel Door		Crew	Daily Output	Labor-Hours	Unit	Material	Labor	Equipment	Total	Total Incl O&P
						2012 Bare Costs				
0105	1-3/4", 3'-6" x 6'-8"	2 Carp	1.50	10.667	Opng.	9,800	470		10,270	11,600
0110	1-3/4", 4'-0" x 6'-8"		1.20	13.333		11,100	590		11,690	13,200
0120	UL 752 Level 3, 1-3/4", 3'-0" x 7'-0"		1.50	10.667		2,225	470		2,695	3,225
0125	1-3/4", 3'-6" x 7'-0"		1.50	10.667		2,325	470		2,795	3,350
0130	1-3/4", 4'-0" x 7'-0"		1.20	13.333		2,525	590		3,115	3,750
0150	UL 752 Level 8, 1-3/4", 3'-0" x 7'-0"		1.50	10.667		9,700	470		10,170	11,500
0155	1-3/4", 3'-6" x 7'-0"		1.50	10.667		9,800	470		10,270	11,600
0160	1-3/4", 4'-0" x 7'-0"		1.20	13.333		11,200	590		11,790	13,400
1000	Safe Room sliding door and hardware, 1-3/4", 3'-0" x 7'-0" UL 752 Level 3		.50	32		22,900	1,400		24,300	27,500
1050	Safe Room swinging door and hardware, 1-3/4", 3'-0" x 7'-0" UL 752 Level 3	↓	.50	32	↓	27,500	1,400		28,900	32,600

08 34 53.30 Wood Ballistic Doors		Crew	Daily Output	Labor-Hours	Unit	Material	Labor	Equipment	Total	Total Incl O&P
0010	**WOOD BALLISTIC DOORS** with frames and hardware									
0050	Wood, 1-3/4", 3'-0" x 7'-0" UL 752 Level 3	2 Carp	1.50	10.667	Opng.	2,125	470		2,595	3,100

08 34 56 – Security Gates

08 34 56.10 Gates		Crew	Daily Output	Labor-Hours	Unit	Material	Labor	Equipment	Total	Total Incl O&P
0010	**GATES**									
0015	Gates include mounting hardware									
0500	Wood, security gate, driveway, dual, 10' wide	H-4	.80	25	Opng.	3,900	1,025		4,925	5,975
0505	12' wide		.80	25		4,300	1,025		5,325	6,400
0510	15' wide		.80	25		5,100	1,025		6,125	7,275
0600	Steel, security gate, driveway, single, 10' wide		.80	25		2,150	1,025		3,175	4,025
0605	12' wide		.80	25		2,225	1,025		3,250	4,100
0620	Steel, security gate, driveway, dual, 12' wide		.80	25		2,050	1,025		3,075	3,950
0625	14' wide		.80	25		2,200	1,025		3,225	4,100
0630	16' wide		.80	25		2,525	1,025		3,550	4,450
0700	Aluminum, security gate, driveway, dual, 10' wide		.80	25		2,200	1,025		3,225	4,100
0705	12' wide		.80	25		2,600	1,025		3,625	4,525
0710	16' wide	↓	.80	25	↓	3,450	1,025		4,475	5,475
1000	Security gate, driveway, opener 12 VDC				Ea.	700			700	770
1010	Wireless					1,700			1,700	1,875
1020	Security gate, driveway, opener 24 VDC					1,100			1,100	1,200
1030	Wireless					1,300			1,300	1,425
1040	Security gate, driveway, opener 12 VDC, solar panel 10 watt	1 Elec	2	4		200	206		406	540
1050	20 watt	"	2	4	↓	320	206		526	670

08 34 73 – Sound Control Door Assemblies

08 34 73.10 Acoustical Doors		Crew	Daily Output	Labor-Hours	Unit	Material	Labor	Equipment	Total	Total Incl O&P
0010	**ACOUSTICAL DOORS**									
0020	Including framed seals, 3' x 7', wood, 27 STC rating	2 Carp	1.50	10.667	Ea.	950	470		1,420	1,825
0100	Steel, 40 STC rating		1.50	10.667		3,000	470		3,470	4,075
0200	45 STC rating		1.50	10.667		3,500	470		3,970	4,625
0300	48 STC rating		1.50	10.667		4,100	470		4,570	5,275
0400	52 STC rating	↓	1.50	10.667	↓	4,600	470		5,070	5,825
9000	Minimum labor/equipment charge	1 Carp	4	2	Job		88		88	145

08 36 13 – Sectional Doors

08 36 13.10 Overhead Commercial Doors		Crew	Daily Output	Labor-Hours	Unit	Material (2012 Bare Costs)	Labor (2012 Bare Costs)	Equipment (2012 Bare Costs)	Total (2012 Bare Costs)	Total Incl O&P
0010	**OVERHEAD COMMERCIAL DOORS**									
1000	Stock, sectional, heavy duty, wood, 1-3/4" thick, 8' x 8' high	2 Carp	2	8	Ea.	840	355		1,195	1,500
1200	12' x 12' high		1.50	10.667		1,750	470		2,220	2,700
1300	Chain hoist, 14' x 14' high		1.30	12.308		2,700	545		3,245	3,875
1600	20' x 16' high		.65	24.615		4,875	1,075		5,950	7,150
2100	For medium duty custom door, deduct					5%	5%			
2150	For medium duty stock doors, deduct					10%	5%			
2300	Fiberglass and aluminum, heavy duty, sectional, 12' x 12' high	2 Carp	1.50	10.667	Ea.	2,675	470		3,145	3,700
2450	Chain hoist, 20' x 20' high	"	.50	32		6,575	1,400		7,975	9,575
2900	For electric trolley operator, 1/3 H.P., to 12' x 12', add	1 Carp	2	4		925	176		1,101	1,325
2950	Over 12' x 12', 1/2 H.P., add	"	1	8		1,075	355		1,430	1,775
9000	Minimum labor/equipment charge	2 Carp	1.50	10.667	Job		470		470	770

08 36 13.20 Residential Garage Doors		Crew	Daily Output	Labor-Hours	Unit	Material	Labor	Equipment	Total	Total Incl O&P
0010	**RESIDENTIAL GARAGE DOORS**									
0050	Hinged, wood, custom, double door, 9' x 7'	2 Carp	4	4	Ea.	700	176		876	1,050
0070	16' x 7'		3	5.333		1,050	235		1,285	1,525
0200	Overhead, sectional, incl. hardware, fiberglass, 9' x 7', standard		5.28	3.030		865	134		999	1,175
0220	Deluxe		5.28	3.030		1,100	134		1,234	1,425
0300	16' x 7', standard		6	2.667		1,550	118		1,668	1,900
0320	Deluxe		6	2.667		2,100	118		2,218	2,500
0500	Hardboard, 9' x 7', standard		8	2		580	88		668	785
0520	Deluxe		8	2		800	88		888	1,025
0600	16' x 7', standard		6	2.667		1,200	118		1,318	1,525
0620	Deluxe		6	2.667		1,400	118		1,518	1,750
0700	Metal, 9' x 7', standard		5.28	3.030		640	134		774	925
0720	Deluxe		8	2		875	88		963	1,100
0800	16' x 7', standard		3	5.333		775	235		1,010	1,250
0820	Deluxe		6	2.667		1,375	118		1,493	1,700
0900	Wood, 9' x 7', standard		8	2		900	88		988	1,125
0920	Deluxe		8	2		2,100	88		2,188	2,450
1000	16' x 7', standard		6	2.667		1,600	118		1,718	1,950
1020	Deluxe		6	2.667		2,975	118		3,093	3,475
1800	Door hardware, sectional	1 Carp	4	2		320	88		408	495
1810	Door tracks only		4	2		159	88		247	320
1820	One side only		7	1.143		110	50.50		160.50	204
3000	Swing-up, including hardware, fiberglass, 9' x 7', standard	2 Carp	8	2		980	88		1,068	1,225
3020	Deluxe		8	2		1,000	88		1,088	1,250
3100	16' x 7', standard		6	2.667		1,225	118		1,343	1,550
3120	Deluxe		6	2.667		1,500	118		1,618	1,850
3200	Hardboard, 9' x 7', standard		8	2		500	88		588	695
3220	Deluxe		8	2		600	88		688	805
3300	16' x 7', standard		6	2.667		620	118		738	875
3320	Deluxe		6	2.667		800	118		918	1,075
3400	Metal, 9' x 7', standard		8	2		420	88		508	610
3420	Deluxe		8	2		950	88		1,038	1,200
3500	16' x 7', standard		6	2.667		685	118		803	950
3520	Deluxe		6	2.667		1,000	118		1,118	1,300
3600	Wood, 9' x 7', standard		8	2		600	88		688	805
3620	Deluxe		8	2		1,050	88		1,138	1,300
3700	16' x 7', standard		6	2.667		800	118		918	1,075
3720	Deluxe		6	2.667		2,000	118		2,118	2,400
3900	Door hardware only, swing up	1 Carp	4	2		163	88		251	325

08 36 Panel Doors

08 36 13 – Sectional Doors

08 36 13.20 Residential Garage Doors		Crew	Daily Output	Labor-Hours	Unit	Material	Labor	Equipment	Total	Total Incl O&P
						2012 Bare Costs				
3920	One side only	1 Carp	7	1.143	Ea.	80	50.50		130.50	171
4000	For electric operator, economy, add		8	1		410	44		454	525
4100	Deluxe, including remote control		8	1		600	44		644	735
4500	For transmitter/receiver control , add to operator				Total	110			110	121
4600	Transmitters, additional				"	55			55	60.50
6000	Replace section, on sectional door, fiberglass, 9' x 7'	1 Carp	4	2	Ea.	570	88		658	770
6020	16' x 7'		3.50	2.286		665	101		766	895
6200	Hardboard, 9' x 7'		4	2		115	88		203	271
6220	16' x 7'		3.50	2.286		218	101		319	405
6300	Metal, 9' x 7'		4	2		184	88		272	350
6320	16' x 7'		3.50	2.286		300	101		401	495
6500	Wood, 9' x 7'		4	2		116	88		204	273
6520	16' x 7'		3.50	2.286		225	101		326	415
9000	Minimum labor/equipment charge		2.50	3.200	Job		141		141	231

08 36 19 – Multi-Leaf Vertical Lift Doors

08 36 19.10 Sectional Vertical Lift Doors		Crew	Daily Output	Labor-Hours	Unit	Material	Labor	Equipment	Total	Total Incl O&P
0010	**SECTIONAL VERTICAL LIFT DOORS**									
0020	Motorized, 14 ga. stl, incl., frm and ctrl pnl									
0050	16' x 16' high	L-10	.50	48	Ea.	21,500	2,375	1,225	25,100	29,300
0100	10' x 20' high		1.30	18.462		31,800	915	470	33,185	37,100
0120	15' x 20' high		1.30	18.462		39,400	915	470	40,785	45,400
0140	20' x 20' high		1	24		45,900	1,200	610	47,710	53,500
0160	25' x 20' high		1	24		52,500	1,200	610	54,310	60,500
0170	32' x 24' high		.75	32		46,400	1,575	815	48,790	54,500
0180	20' x 25' high		1	24		54,000	1,200	610	55,810	62,000
0200	25' x 25' high		.70	34.286		62,500	1,700	875	65,075	73,000
0220	25' x 30' high		.70	34.286		68,000	1,700	875	70,575	78,500
0240	30' x 30' high		.70	34.286		79,000	1,700	875	81,575	90,500
0260	35' x 30' high		.70	34.286		89,000	1,700	875	91,575	102,000

08 38 Traffic Doors

08 38 19 – Rigid Traffic Doors

08 38 19.20 Double Acting Swing Doors		Crew	Daily Output	Labor-Hours	Unit	Material	Labor	Equipment	Total	Total Incl O&P
0010	**DOUBLE ACTING SWING DOORS**									
0020	Including frame, closer, hardware and vision panel									
1000	.063" aluminum, 7'-0" high, 4'-0" wide	2 Carp	4.20	3.810	Pr.	2,100	168		2,268	2,575
1025	6'-0" wide		4	4		2,200	176		2,376	2,725
1050	6'-8" wide		4	4		2,400	176		2,576	2,950
2000	Solid core wood, 3/4" thick, metal frame, stainless steel									
2010	base plate, 7' high opening, 4' wide	2 Carp	4	4	Pr.	2,600	176		2,776	3,150
2050	7' wide		3.80	4.211	"	2,800	186		2,986	3,375
9000	Minimum labor/equipment charge		2	8	Job		355		355	580

08 38 19.30 Shock Absorbing Doors		Crew	Daily Output	Labor-Hours	Unit	Material	Labor	Equipment	Total	Total Incl O&P
0010	**SHOCK ABSORBING DOORS**									
0020	Rigid, no frame, 1-1/2" thick, 5' x 7'	2 Sswk	1.90	8.421	Opng.	1,400	415		1,815	2,325
0100	8' x 8'		1.80	8.889		2,000	440		2,440	3,025
0500	Flexible, no frame, insulated, .16" thick, economy, 5' x 7'		2	8		1,750	395		2,145	2,650
0600	Deluxe		1.90	8.421		2,625	415		3,040	3,650
1000	8' x 8' opening, economy		2	8		2,750	395		3,145	3,750
1100	Deluxe		1.90	8.421		3,500	415		3,915	4,625

08 38 Traffic Doors

08 38 19 – Rigid Traffic Doors

	08 38 19.30 Shock Absorbing Doors	Crew	Daily Output	Labor-Hours	Unit	Material	2012 Bare Costs Labor	Equipment	Total	Total Incl O&P
9000	Minimum labor/equipment charge	2 Sswk	2	8	Job		395		395	735

08 41 Entrances and Storefronts

08 41 13 – Aluminum-Framed Entrances and Storefronts

	08 41 13.20 Tube Framing	Crew	Daily Output	Labor-Hours	Unit	Material	2012 Bare Costs Labor	Equipment	Total	Total Incl O&P
0010	**TUBE FRAMING**, For window walls and store fronts, aluminum stock									
0050	Plain tube frame, mill finish, 1-3/4" x 1-3/4"	2 Glaz	103	.155	L.F.	9.45	6.65		16.10	21
0150	1-3/4" x 4"		98	.163		12.95	7		19.95	25.50
0200	1-3/4" x 4-1/2"		95	.168		14.60	7.20		21.80	27.50
0250	2" x 6"		89	.180		22	7.70		29.70	36.50
0350	4" x 4"		87	.184		25	7.85		32.85	40
0400	4-1/2" x 4-1/2"		85	.188		26.50	8.05		34.55	42
0450	Glass bead		240	.067		2.80	2.85		5.65	7.70
1000	Flush tube frame, mill finish, 1/4" glass, 1-3/4" x 4", open header		80	.200		12.40	8.55		20.95	27.50
1050	Open sill		82	.195		10.10	8.35		18.45	24.50
1100	Closed back header		83	.193		17.75	8.25		26	33
1150	Closed back sill		85	.188		16.95	8.05		25	31.50
1160	Tube fmg, spandrel cover both sides, alum 1" wide	1 Sswk	85	.094	S.F.	96	4.64		100.64	114
1170	Tube fmg, spandrel cover both sides, alum 2" wide	"	85	.094	"	37	4.64		41.64	49
1200	Vertical mullion, one piece	2 Glaz	75	.213	L.F.	18.60	9.10		27.70	35.50
1250	Two piece		73	.219		19.75	9.35		29.10	36.50
1300	90° or 180° vertical corner post		75	.213		30	9.10		39.10	48
1400	1-3/4" x 4-1/2", open header		80	.200		15.10	8.55		23.65	30.50
1450	Open sill		82	.195		12.60	8.35		20.95	27.50
1500	Closed back header		83	.193		18	8.25		26.25	33
1550	Closed back sill		85	.188		17.75	8.05		25.80	32.50
1600	Vertical mullion, one piece		75	.213		19.80	9.10		28.90	37
1650	Two piece		73	.219		21	9.35		30.35	38
1700	90° or 180° vertical corner post		75	.213		21.50	9.10		30.60	38.50
2000	Flush tube frame, mil fin.,ins. glass w/thml brk, 2" x 4-1/2", open header		75	.213		15.55	9.10		24.65	32
2050	Open sill		77	.208		13.10	8.90		22	29
2100	Closed back header		78	.205		14.70	8.75		23.45	30.50
2150	Closed back sill		80	.200		15.75	8.55		24.30	31
2200	Vertical mullion, one piece		70	.229		16.30	9.75		26.05	33.50
2250	Two piece		68	.235		17.60	10.05		27.65	35.50
2300	90° or 180° vertical corner post		70	.229		16.75	9.75		26.50	34.50
5000	Flush tube frame, mill fin., thermal brk., 2-1/4" x 4-1/2", open header		74	.216		16.20	9.25		25.45	33
5050	Open sill		75	.213		14.40	9.10		23.50	30.50
5100	Vertical mullion, one piece		69	.232		16.75	9.90		26.65	34.50
5150	Two piece		67	.239		20.50	10.20		30.70	39
5200	90° or 180° vertical corner post		69	.232		18.05	9.90		27.95	36
6980	Door stop (snap in)		380	.042		3.09	1.80		4.89	6.30
7000	For joints, 90°, clip type, add				Ea.	24.50			24.50	27
7050	Screw spline joint, add					23			23	25
7100	For joint other than 90°, add					48			48	52.50
8000	For bronze anodized aluminum, add					15%				
8020	For black finish, add					30%				
8050	For stainless steel materials, add					350%				
8100	For monumental grade, add					53%				
8150	For steel stiffener, add	2 Glaz	200	.080	L.F.	11	3.42		14.42	17.60
8200	For 2 to 5 stories, add per story				Story		8%			

08 41 Entrances and Storefronts

08 41 13 – Aluminum-Framed Entrances and Storefronts

08 41 13.20 Tube Framing		Crew	Daily Output	Labor-Hours	Unit	Material	2012 Bare Costs Labor	Equipment	Total	Total Incl O&P
9000	Minimum labor/equipment charge	2 Glaz	2	8	Job		340		340	550

08 41 19 – Stainless-Steel-Framed Entrances and Storefronts

08 41 19.10 Stainless-Steel and Glass Entrance Unit		Crew	Daily Output	Labor-Hours	Unit	Material	Labor	Equipment	Total	Total Incl O&P
0010	**STAINLESS-STEEL AND GLASS ENTRANCE UNIT**, narrow stiles									
0020	3′ x 7′ opening, including hardware, minimum	2 Sswk	1.60	10	Opng.	6,475	495		6,970	8,025
0050	Average		1.40	11.429		6,875	565		7,440	8,600
0100	Maximum		1.20	13.333		7,425	660		8,085	9,375
1000	For solid bronze entrance units, statuary finish, add					64%				
1100	Without statuary finish, add					45%				
2000	Balanced doors, 3′ x 7′, economy	2 Sswk	.90	17.778	Ea.	8,675	875		9,550	11,200
2100	Premium		.70	22.857	"	14,900	1,125		16,025	18,500
9000	Minimum labor/equipment charge		2	8	Job		395		395	735

08 41 26 – All-Glass Entrances and Storefronts

08 41 26.10 Window Walls Aluminum, Stock		Crew	Daily Output	Labor-Hours	Unit	Material	Labor	Equipment	Total	Total Incl O&P
0010	**WINDOW WALLS ALUMINUM, STOCK**, including glazing									
0020	Minimum	H-2	160	.150	S.F.	43.50	6.05		49.55	58
0050	Average		140	.171		60	6.90		66.90	77
0100	Maximum		110	.218		165	8.75		173.75	195
0500	For translucent sandwich wall systems, see Section 07 41 33.10									
0850	Cost of the above walls depends on material,									
0860	finish, repetition, and size of units.									
0870	The larger the opening, the lower the S.F. cost									
1200	Double glazed acoustical window wall for airports,									
1220	including 1″ thick glass with 2″ x 4-1/2″ tube frame	H-2	40	.600	S.F.	113	24		137	163

08 42 Entrances

08 42 26 – All-Glass Entrances

08 42 26.10 Swinging Glass Doors		Crew	Daily Output	Labor-Hours	Unit	Material	Labor	Equipment	Total	Total Incl O&P
0010	**SWINGING GLASS DOORS**									
0020	Including hardware, 1/2″ thick, tempered, 3′ x 7′ opening	2 Glaz	2	8	Opng.	2,175	340		2,515	2,925
0100	6′ x 7′ opening		1.40	11.429	"	4,200	490		4,690	5,425
9000	Minimum labor/equipment charge		2	8	Job		340		340	550

08 42 29 – Automatic Entrances

08 42 29.23 Sliding Automatic Entrances		Crew	Daily Output	Labor-Hours	Unit	Material	Labor	Equipment	Total	Total Incl O&P
0010	**SLIDING AUTOMATIC ENTRANCES** 12′ x 7′-6″ opng., 5′ x 7′ door, 2 way traffic									
0020	Mat or electronic activated, panic pushout, incl. operator & hardware,									
0030	not including glass or glazing	2 Glaz	.70	22.857	Opng.	8,200	975		9,175	10,600
9000	Minimum labor/equipment charge	"	.70	22.857	Job		975		975	1,575

08 42 36 – Balanced Door Entrances

08 42 36.10 Balanced Entrance Doors		Crew	Daily Output	Labor-Hours	Unit	Material	Labor	Equipment	Total	Total Incl O&P
0010	**BALANCED ENTRANCE DOORS**									
0020	Hardware & frame, alum. & glass, 3′ x 7′, econ.	2 Sswk	.90	17.778	Ea.	6,300	875		7,175	8,550
0150	Premium		.70	22.857	"	7,700	1,125		8,825	10,600
9000	Minimum labor/equipment charge		1	16	Job		790		790	1,475

08 43 Storefronts

08 43 13 – Aluminum-Framed Storefronts

08 43 13.10 Aluminum-Framed Entrance Doors		Crew	Daily Output	Labor-Hours	Unit	2012 Bare Costs Material	Labor	Equipment	Total	Total Incl O&P
0010	**ALUMINUM-FRAMED ENTRANCE DOORS**									
0015	Standard hardware and glass stops but no glazing									
0020	Entrance door, 3′ x 7′ opening, clear anodized finish	2 Sswk	7	2.286	Opng.	365	113		478	610
0040	Bronze finish		7	2.286		380	113		493	630
0060	Black finish		7	2.286		425	113		538	680
0200	3′-6″ x 7′-0″, mill finish		7	2.286		525	113		638	785
0220	Bronze finish		7	2.286		540	113		653	805
0240	Black finish		7	2.286		580	113		693	845
0500	6′ x 7′ opening, clear finish		6	2.667		750	132		882	1,075
0520	Bronze finish		6	2.667		785	132		917	1,100
0600	Door Frame 3′-0″ x 7′-0″, mill finish		6	2.667		345	132		477	620
0620	Bronze finish		6	2.667		400	132		532	685
0640	Black finish		6	2.667		450	132		582	740
0700	3′-6″ x 7′-0″, mill finish		6	2.667		294	132		426	570
0720	Bronze finish		6	2.667		294	132		426	570
0740	Black finish		6	2.667		294	132		426	570
0800	6′-0″ x 7′-0″, mill finish		6	2.667		294	132		426	570
0820	Bronze finish		6	2.667		299	132		431	575
0840	Black finish		6	2.667		320	132		452	600
1000	With 3′ high transom above, 3′ x 7′ opening, clear finish		5.50	2.909		445	144		589	755
1050	Bronze finish		5.50	2.909		460	144		604	775
1100	Black finish		5.50	2.909		495	144		639	810
1300	3′-6″ x 7′-0″ opening, clear finish		5.50	2.909		330	144		474	625
1320	Bronze finish		5.50	2.909		335	144		479	635
1340	Black finish		5.50	2.909		355	144		499	655
1500	6′ x 7′ opening, clear finish		5.50	2.909		560	144		704	880
1550	Bronze finish		5.50	2.909		575	144		719	895
1600	Black finish		5.50	2.909		635	144		779	960
8000	For 8′ high doors add				Ea.	200			200	220
9000	Minimum labor/equipment charge	2 Sswk	4	4	Job		197		197	365

08 43 13.20 Storefront Systems		Crew	Daily Output	Labor-Hours	Unit	Material	Labor	Equipment	Total	Total Incl O&P
0010	**STOREFRONT SYSTEMS**, aluminum frame clear 3/8″ plate glass									
0020	incl. 3′ x 7′ door with hardware (400 sq. ft. max. wall)									
0500	Wall height to 12′ high, commercial grade	2 Glaz	150	.107	S.F.	20.50	4.56		25.06	30
0600	Institutional grade		130	.123		25.50	5.25		30.75	36.50
0700	Monumental grade		115	.139		38.50	5.95		44.45	51.50
1000	6′ x 7′ door with hardware, commercial grade		135	.119		20.50	5.05		25.55	30.50
1100	Institutional grade		115	.139		28	5.95		33.95	40.50
1200	Monumental grade		100	.160		52	6.85		58.85	68.50
1500	For bronze anodized finish, add					15%				
1600	For black anodized finish, add					36%				
1700	For stainless steel framing, add to monumental					78%				
9000	Minimum labor/equipment charge	2 Glaz	1	16	Job		685		685	1,100

08 43 29 – Sliding Storefronts

08 43 29.10 Sliding Panels		Crew	Daily Output	Labor-Hours	Unit	Material	Labor	Equipment	Total	Total Incl O&P
0010	**SLIDING PANELS**									
0020	Mall fronts, aluminum & glass, 15′ x 9′ high	2 Glaz	1.30	12.308	Opng.	3,275	525		3,800	4,450
0100	24′ x 9′ high		.70	22.857		4,650	975		5,625	6,700
0200	48′ x 9′ high, with fixed panels		.90	17.778		8,700	760		9,460	10,800
0500	For bronze finish, add					17%				
9000	Minimum labor/equipment charge	2 Glaz	1	16	Job		685		685	1,100

08 45 Translucent Wall and Roof Assemblies

08 45 10 – Translucent Roof Assemblies

08 45 10.10 Skyroofs		Crew	Daily Output	Labor-Hours	Unit	2012 Bare Costs Material	Labor	Equipment	Total	Total Incl O&P
0010	**SKYROOFS**, Translucent panels, 2-3/4" thick									
0020	Under 5000 S.F.	G-3	395	.081	SF Hor.	34	3.53		37.53	43
0100	Over 5000 S.F.		465	.069		31.50	3		34.50	40
0300	Continuous vaulted, semi-circular, to 8' wide, double glazed G		145	.221		52.50	9.65		62.15	73
0400	Single glazed		160	.200		35.50	8.75		44.25	53
0600	To 20' wide, single glazed		175	.183		46	8		54	63.50
0700	Over 20' wide, single glazed		200	.160		45	7		52	60.50
0900	Motorized opening type, single glazed, 1/3 opening		145	.221		46	9.65		55.65	66
1000	Full opening		130	.246		52.50	10.75		63.25	74.50
1200	Pyramid type units, self-supporting, to 30' clear opening,									
1300	square or circular, single glazed, minimum	G-3	200	.160	SF Hor.	27	7		34	40.50
1310	Average		165	.194		37	8.45		45.45	54
1400	Maximum		130	.246		53	10.75		63.75	75.50
1500	Grid type, 4' to 10' modules, single glass glazed, minimum		200	.160		36	7		43	50.50
1550	Maximum		128	.250		56	10.90		66.90	79
1600	Preformed acrylic, minimum		300	.107		43.50	4.65		48.15	55
1650	Maximum		175	.183		59.50	8		67.50	78.50
9000	Minimum labor/equipment charge	2 Carp	8	2	Job		88		88	145

08 51 Metal Windows

08 51 13 – Aluminum Windows

08 51 13.10 Aluminum Sash

Line	Description	Crew	Daily Output	Labor-Hours	Unit	Material	Labor	Equipment	Total	Total Incl O&P
0010	**ALUMINUM SASH**									
0020	Stock, grade C, glaze & trim not incl., casement	2 Sswk	200	.080	S.F.	38	3.95		41.95	49.50
0050	Double hung		200	.080		38.50	3.95		42.45	50
0100	Fixed casement		200	.080		17	3.95		20.95	26
0150	Picture window		200	.080		18.05	3.95		22	27
0200	Projected window		200	.080		35	3.95		38.95	46
0250	Single hung		200	.080		16.20	3.95		20.15	25
0300	Sliding		200	.080		21	3.95		24.95	30.50
1000	Mullions for above, tubular		240	.067	L.F.	6.05	3.29		9.34	12.75
2950	Single glazing for above, add	2 Glaz	200	.080	S.F.	7.15	3.42		10.57	13.35
3000	Double glazing for above, add		200	.080		11.95	3.42		15.37	18.65
3100	Triple glazing for above, add		85	.188		14.20	8.05		22.25	28.50
9000	Minimum labor/equipment charge	1 Sswk	2	4	Job		197		197	365

08 51 13.20 Aluminum Windows

Line	Description	Crew	Daily Output	Labor-Hours	Unit	Material	Labor	Equipment	Total	Total Incl O&P
0010	**ALUMINUM WINDOWS**, incl. frame and glazing, commercial grade									
1000	Stock units, casement, 3'-1" x 3'-2" opening	2 Sswk	10	1.600	Ea.	365	79		444	545
1050	Add for storms					118			118	130
1600	Projected, with screen, 3'-1" x 3'-2" opening	2 Sswk	10	1.600		345	79		424	525
1700	Add for storms					115			115	127
2000	4'-5" x 5'-3" opening	2 Sswk	8	2		390	98.50		488.50	615
2100	Add for storms					123			123	135
2500	Enamel finish windows, 3'-1" x 3'-2"	2 Sswk	10	1.600		350	79		429	530
2600	4'-5" x 5'-3"		8	2		395	98.50		493.50	620
3000	Single hung, 2' x 3' opening, enameled, standard glazed		10	1.600		203	79		282	370
3100	Insulating glass		10	1.600		247	79		326	420
3300	2'-8" x 6'-8" opening, standard glazed		8	2		360	98.50		458.50	580
3400	Insulating glass		8	2		460	98.50		558.50	695
3700	3'-4" x 5'-0" opening, standard glazed		9	1.778		294	87.50		381.50	490
3800	Insulating glass		9	1.778		325	87.50		412.50	525

08 51 Metal Windows

08 51 13 – Aluminum Windows

08 51 13.20 Aluminum Windows		Crew	Daily Output	Labor-Hours	Unit	Material	2012 Bare Costs Labor	Equipment	Total	Total Incl O&P
4000	Sliding aluminum, 3′ x 2′ opening, standard glazed	2 Sswk	10	1.600	Ea.	215	79		294	385
4100	Insulating glass		10	1.600		230	79		309	400
4300	5′ x 3′ opening, standard glazed		9	1.778		330	87.50		417.50	525
4400	Insulating glass		9	1.778		385	87.50		472.50	585
4600	8′ x 4′ opening, standard glazed		6	2.667		345	132		477	625
4700	Insulating glass		6	2.667		560	132		692	860
5000	9′ x 5′ opening, standard glazed		4	4		525	197		722	945
5100	Insulating glass		4	4		840	197		1,037	1,300
5500	Sliding, with thermal barrier and screen, 6′ x 4′, 2 track		8	2		715	98.50		813.50	975
5700	4 track		8	2		900	98.50		998.50	1,175
6000	For above units with bronze finish, add					15%				
6200	For installation in concrete openings, add					8%				
6400										

08 51 23 – Steel Windows

08 51 23.10 Steel Sash		Crew	Daily Output	Labor-Hours	Unit	Material	Labor	Equipment	Total	Total Incl O&P
0010	**STEEL SASH** Custom units, glazing and trim not included									
0100	Casement, 100% vented	2 Sswk	200	.080	S.F.	64.50	3.95		68.45	78
0200	50% vented		200	.080		52.50	3.95		56.45	65.50
0300	Fixed		200	.080		28	3.95		31.95	38.50
1000	Projected, commercial, 40% vented		200	.080		50	3.95		53.95	62.50
1100	Intermediate, 50% vented		200	.080		57	3.95		60.95	70
1500	Industrial, horizontally pivoted		200	.080		52	3.95		55.95	64.50
1600	Fixed		200	.080		30.50	3.95		34.45	41
2000	Industrial security sash, 50% vented		200	.080		56.50	3.95		60.45	69.50
2100	Fixed		200	.080		46	3.95		49.95	58
2500	Picture window		200	.080		29.50	3.95		33.45	40
3000	Double hung		200	.080		57.50	3.95		61.45	71
5000	Mullions for above, open interior face		240	.067	L.F.	10.10	3.29		13.39	17.25
5100	With interior cover		240	.067	"	16.85	3.29		20.14	24.50
5200	Single glazing for above, add	2 Glaz	200	.080	S.F.	6.80	3.42		10.22	13
6000	Double glazing for above, add		200	.080		12.80	3.42		16.22	19.60
6100	Triple glazing for above, add		85	.188		12.20	8.05		20.25	26.50
9000	Minimum labor/equipment charge	1 Sswk	2	4	Job		197		197	365

08 51 23.20 Steel Windows		Crew	Daily Output	Labor-Hours	Unit	Material	Labor	Equipment	Total	Total Incl O&P
0010	**STEEL WINDOWS** Stock, including frame, trim and insul. glass R085123-10									
1000	Custom units, double hung, 2′-8″ x 4′-6″ opening	2 Sswk	12	1.333	Ea.	700	66		766	890
1100	2′-4″ x 3′-9″ opening		12	1.333		580	66		646	755
1500	Commercial projected, 3′-9″ x 5′-5″ opening		10	1.600		1,200	79		1,279	1,475
1600	6′-9″ x 4′-1″ opening		7	2.286		1,600	113		1,713	1,950
2000	Intermediate projected, 2′-9″ x 4′-1″ opening		12	1.333		680	66		746	870
2100	4′-1″ x 5′-5″ opening		10	1.600		1,400	79		1,479	1,675
9000	Minimum labor/equipment charge	1 Sswk	3	2.667	Job		132		132	245

08 51 66 – Metal Window Screens

08 51 66.10 Screens		Crew	Daily Output	Labor-Hours	Unit	Material	Labor	Equipment	Total	Total Incl O&P
0010	**SCREENS**									
0020	For metal sash, aluminum or bronze mesh, flat screen	2 Sswk	1200	.013	S.F.	4.20	.66		4.86	5.85
0500	Wicket screen, inside window		1000	.016		6.50	.79		7.29	8.60
0800	Security screen, aluminum frame with stainless steel cloth		1200	.013		23.50	.66		24.16	26.50
0900	Steel grate, painted, on steel frame		1600	.010		12.80	.49		13.29	15
1000	For solar louvers, add		160	.100		24	4.94		28.94	35.50
4000	See Section 05 58 23.90									

08 52 Wood Windows

08 52 10 – Plain Wood Windows

08 52 10.20 Awning Window		Crew	Daily Output	Labor-Hours	Unit	Material	2012 Bare Costs Labor	Equipment	Total	Total Incl O&P
0010	**AWNING WINDOW**, Including frame, screens and grilles									
0100	Average quality, 34" x 22", double insulated glass	1 Carp	10	.800	Ea.	269	35.50		304.50	355
0200	Low E glass		10	.800		267	35.50		302.50	350
0300	40" x 28", double insulated glass		9	.889		310	39		349	410
0400	Low E Glass		9	.889		340	39		379	435
0500	48" x 36", double insulated glass		8	1		465	44		509	585
0600	Low E glass		8	1		490	44		534	610
4000	Impact windows, minimum, add					60%				
4010	Impact windows, maximum, add					160%				
9000	Minimum labor/equipment charge	1 Carp	4	2	Job		88		88	145

08 52 10.40 Casement Window			Crew	Daily Output	Labor-Hours	Unit	Material	Labor	Equipment	Total	Total Incl O&P
0010	**CASEMENT WINDOW**, including frame, screen and grilles										
0100	Avg. quality, bldrs. model, 2'-0" x 3'-0" H, dbl. insulated glass	G	1 Carp	10	.800	Ea.	284	35.50		319.50	375
0150	Low E glass	G		10	.800		259	35.50		294.50	345
0200	2'-0" x 4'-6" high, double insulated glass	G		9	.889		355	39		394	455
0250	Low E glass	G		9	.889		370	39		409	470
0260	Casement 4'-2" x 4'-2" double insulated glass	G		11	.727		875	32		907	1,025
0270	4'-0" x 4'-0" Low E glass	G		11	.727		535	32		567	645
0290	6'-4" x 5'-7" Low E glass	G		9	.889		1,125	39		1,164	1,300
0300	2'-4" x 6'-0" high, double insulated glass	G		8	1		430	44		474	550
0350	Low E glass	G		8	1		475	44		519	600
0522	Vinyl clad, premium, double insulated glass, 2'-0" x 3'-0"	G		10	.800		266	35.50		301.50	350
0524	2'-0" x 4'-0"	G		9	.889		310	39		349	405
0525	2'-0" x 5'-0"	G		8	1		355	44		399	465
0528	2'-0" x 6'-0"	G		8	1		375	44		419	485
0600	3'-0" x 5'-0"			8	1		650	44		694	790
0700	4'-0" x 3'-0"			8	1		715	44		759	865
0710	4'-0" x 4'-0"	G		8	1		615	44		659	750
0720	4'-8" x 4'-0"			8	1		675	44		719	820
0730	4'-8" x 5'-0"			6	1.333		775	59		834	945
0740	4'-8" x 6'-0"			6	1.333		870	59		929	1,050
0750	6'-0" x 4'-0"			6	1.333		790	59		849	965
0800	6'-0" x 5'-0"			6	1.333		885	59		944	1,075
0900	5'-6" x 5'-6"	G	2 Carp	15	1.067		1,400	47		1,447	1,600
8190	For installation, add per leaf							15%			
8200	For multiple leaf units, deduct for stationary sash										
8220	2' high					Ea.	23			23	25.50
8240	4'-6" high						26			26	28.50
8260	6' high						34.50			34.50	38
8300	Impact windows, minimum, add						60%				
8310	Impact windows, maximum, add						160%				
9000	Minimum labor/equipment charge		1 Carp	3	2.667	Job		118		118	193

08 52 10.50 Double Hung			Crew	Daily Output	Labor-Hours	Unit	Material	Labor	Equipment	Total	Total Incl O&P
0010	**DOUBLE HUNG**, Including frame, screens and grilles										
0100	Avg. quality, bldrs. model, 2'-0" x 3'-0" high, low E insul. glass	G	1 Carp	10	.800	Ea.	209	35.50		244.50	287
0200	3'-0" x 4'-0" high, double insulated glass	G		9	.889		270	39		309	360
0300	4'-0" x 4'-6" high, low E insulated glass	G		8	1		315	44		359	420
8000	Impact windows, minimum, add						60%				
8010	Impact windows, maximum, add						160%				
9000	Minimum labor/equipment charge		1 Carp	3	2.667	Job		118		118	193

08 52 10 – Plain Wood Windows

08 52 10.55 Picture Window

	08 52 10.55 Picture Window	Crew	Daily Output	Labor-Hours	Unit	Material	2012 Bare Costs Labor	Equipment	Total	Total Incl O&P
0010	**PICTURE WINDOW**, Including frame and grilles									
0100	Average quality, bldrs. model, 3'-6" x 4'-0" high, dbl. insulated glass	2 Carp	12	1.333	Ea.	400	59		459	530
0150	Low E glass		12	1.333		410	59		469	545
0200	4'-0" x 4'-6" high, double insulated glass		11	1.455		500	64		564	655
0250	Low E glass		11	1.455		530	64		594	690
0300	5'-0" x 4'-0" high, double insulated glass		11	1.455		580	64		644	745
0350	Low E glass		11	1.455		605	64		669	770
0400	6'-0" x 4'-6" high, double insulated glass		10	1.600		625	70.50		695.50	805
0450	Low E glass		10	1.600		635	70.50		705.50	815

08 52 10.65 Wood Sash

	08 52 10.65 Wood Sash	Crew	Daily Output	Labor-Hours	Unit	Material	Labor	Equipment	Total	Total Incl O&P
0010	**WOOD SASH**, Including glazing but not trim									
0050	Custom, 5'-0" x 4'-0", 1" dbl. glazed, 3/16" thick lites	2 Carp	3.20	5	Ea.	226	221		447	610
0100	1/4" thick lites		5	3.200		241	141		382	495
0200	1" thick, triple glazed		5	3.200		405	141		546	675
0300	7'-0" x 4'-6" high, 1" double glazed, 3/16" thick lites		4.30	3.721		415	164		579	725
0400	1/4" thick lites		4.30	3.721		465	164		629	780
0500	1" thick, triple glazed		4.30	3.721		530	164		694	855
0600	8'-6" x 5'-0" high, 1" double glazed, 3/16" thick lites		3.50	4.571		555	202		757	940
0700	1/4" thick lites		3.50	4.571		605	202		807	995
0800	1" thick, triple glazed		3.50	4.571		615	202		817	1,000
0900	Window frames only, based on perimeter length				L.F.	3.95			3.95	4.35
1200	Window sill, stock, per lineal foot					8.45			8.45	9.30
1250	Casing, stock					3.24			3.24	3.56

08 52 10.70 Sliding Windows

	08 52 10.70 Sliding Windows	Crew	Daily Output	Labor-Hours	Unit	Material	Labor	Equipment	Total	Total Incl O&P
0010	**SLIDING WINDOWS**									
0100	Average quality, bldrs. model, 3'-0" x 3'-0" high, double insulated G	1 Carp	10	.800	Ea.	280	35.50		315.50	370
0120	Low E glass G		10	.800		305	35.50		340.50	395
0200	4'-0" x 3'-6" high, double insulated G		9	.889		355	39		394	455
0220	Low E glass G		9	.889		355	39		394	455
0300	6'-0" x 5'-0" high, double insulated G		8	1		485	44		529	610
0320	Low E glass G		8	1		525	44		569	650
9000	Minimum labor/equipment charge		3	2.667	Job		118		118	193

08 52 13 – Metal-Clad Wood Windows

08 52 13.10 Awning Windows, Metal-Clad

	08 52 13.10 Awning Windows, Metal-Clad	Crew	Daily Output	Labor-Hours	Unit	Material	Labor	Equipment	Total	Total Incl O&P
0010	**AWNING WINDOWS, METAL-CLAD**									
2000	Metal clad, awning deluxe, double insulated glass, 34" x 22"	1 Carp	9	.889	Ea.	245	39		284	335
2050	36" x 25"		9	.889		272	39		311	365
2100	40" x 22"		9	.889		289	39		328	385
2150	40" x 30"		9	.889		340	39		379	440
2200	48" x 28"		8	1		345	44		389	455
2250	60" x 36"		8	1		370	44		414	485

08 52 13.20 Casement Windows, Metal-Clad

	08 52 13.20 Casement Windows, Metal-Clad	Crew	Daily Output	Labor-Hours	Unit	Material	Labor	Equipment	Total	Total Incl O&P
0010	**CASEMENT WINDOWS, METAL-CLAD**									
0100	Metal clad, deluxe, dbl. insul. glass, 2'-0" x 3'-0" high G	1 Carp	10	.800	Ea.	279	35.50		314.50	365
0120	2' 0" x 4'-0" high G		9	.889		315	39		354	410
0130	2'-0" x 5'-0" high G		8	1		325	44		369	435
0140	2'-0" x 6'-0" high G		8	1		365	44		409	475
0300	Metal clad, casement, bldrs mdl, 6'-0" x 4'-0", dbl. insltd gls, 3 panels	2 Carp	10	1.600		1,200	70.50		1,270.50	1,450
0310	9'-0" x 4'-0" , 4 panels		8	2		1,550	88		1,638	1,850
0320	10'-0" x 5'-0", 5 panels		7	2.286		2,100	101		2,201	2,500
0330	12'-0" x 6'-0", 6 panels		6	2.667		2,700	118		2,818	3,150

08 52 Wood Windows

08 52 13 – Metal-Clad Wood Windows

08 52 13.30 Double-Hung Windows, Metal-clad		Crew	Daily Output	Labor-Hours	Unit	2012 Bare Costs Material	Labor	Equipment	Total	Total Incl O&P
0010	**DOUBLE-HUNG WINDOWS, METAL-CLAD**									
0100	Metal clad, deluxe, dbl. insul. glass, 2'-6" x 3'-0" high G	1 Carp	10	.800	Ea.	272	35.50		307.50	355
0120	3'-0" x 3'-6" high G		10	.800		315	35.50		350.50	405
0140	3'-0" x 4'-0" high G		9	.889		325	39		364	425
0160	3'-0" x 4'-6" high G		9	.889		345	39		384	445
0180	3'-0" x 5'-0" high G		8	1		370	44		414	485
0200	3'-6" x 6'-0" high G		8	1		450	44		494	570

08 52 13.35 Picture and Sliding Windows Metal-Clad		Crew	Daily Output	Labor-Hours	Unit	Material	Labor	Equipment	Total	Total Incl O&P
0010	**PICTURE AND SLIDING WINDOWS METAL-CLAD**									
2000	Metal clad, dlx picture, dbl. insul. glass, 4'-0" x 4'-0" high	2 Carp	12	1.333	Ea.	375	59		434	505
2100	4'-0" x 6'-0" high		11	1.455		545	64		609	705
2200	5'-0" x 6'-0" high		10	1.600		610	70.50		680.50	785
2300	6'-0" x 6'-0" high		10	1.600		695	70.50		765.50	880
2400	Metal clad, dlx sliding, double insulated glass, 3'-0" x 3'-0" high G	1 Carp	10	.800		330	35.50		365.50	420
2420	4'-0" x 3'-6" high G		9	.889		400	39		439	505
2440	5'-0" x 4'-0" high G		9	.889		480	39		519	590
2460	6'-0" x 5'-0" high G		8	1		725	44		769	870
9000	Minimum labor/equipment charge	2 Carp	2.75	5.818	Job		257		257	420

08 52 13.40 Bow and Bay Windows, Metal-Clad		Crew	Daily Output	Labor-Hours	Unit	Material	Labor	Equipment	Total	Total Incl O&P
0010	**BOW AND BAY WINDOWS, METAL-CLAD**									
0100	Metal clad, deluxe, dbl. insul. glass, 8'-0" x 5'-0" high, 4 panels	2 Carp	10	1.600	Ea.	1,675	70.50		1,745.50	1,950
0120	10'-0" x 5'-0" high, 5 panels		8	2		1,800	88		1,888	2,125
0140	10'-0" x 6'-0" high, 5 panels		7	2.286		2,100	101		2,201	2,500
0160	12'-0" x 6'-0" high, 6 panels		6	2.667		2,925	118		3,043	3,425
0400	Double hung, bldrs. model, bay, 8' x 4' high, dbl. insulated glass		10	1.600		1,325	70.50		1,395.50	1,575
0440	Low E glass		10	1.600		1,425	70.50		1,495.50	1,700
0460	9'-0" x 5'-0" high, double insulated glass		6	2.667		1,425	118		1,543	1,775
0480	Low E glass		6	2.667		1,500	118		1,618	1,850
0500	Metal clad, deluxe, dbl. insul. glass, 7'-0" x 4'-0" high		10	1.600		1,275	70.50		1,345.50	1,525
0520	8'-0" x 4'-0" high		8	2		1,300	88		1,388	1,600
0540	8'-0" x 5'-0" high		7	2.286		1,350	101		1,451	1,675
0560	9'-0" x 5'-0" high		6	2.667		1,450	118		1,568	1,775

08 52 16 – Plastic-Clad Wood Windows

08 52 16.10 Bow Window		Crew	Daily Output	Labor-Hours	Unit	Material	Labor	Equipment	Total	Total Incl O&P
0010	**BOW WINDOW** Including frames, screens, and grilles									
0020	End panels operable									
1000	Bow type, casement, wood, bldrs mdl, 8' x 5' dbl. insltd glass, 4 panel	2 Carp	10	1.600	Ea.	1,500	70.50		1,570.50	1,775
1050	Low E glass		10	1.600		1,325	70.50		1,395.50	1,575
1100	10'-0" x 5'-0", double insulated glass, 6 panels		6	2.667		1,350	118		1,468	1,700
1200	Low E glass, 6 panels		6	2.667		1,450	118		1,568	1,800
1300	Vinyl clad, bldrs model, double insulated glass, 6'-0" x 4'-0", 3 panel		10	1.600		1,025	70.50		1,095.50	1,250
1340	9'-0" x 4'-0", 4 panel		8	2		1,350	88		1,438	1,650
1380	10'-0" x 6'-0", 5 panels		7	2.286		2,250	101		2,351	2,650
1420	12'-0" x 6'-0", 6 panels		6	2.667		2,925	118		3,043	3,425
2000	Bay window, 8' x 5', dbl. insul glass		10	1.600		1,875	70.50		1,945.50	2,200
2050	Low E glass		10	1.600		2,275	70.50		2,345.50	2,625
2100	12'-0" x 6'-0", double insulated glass, 6 panels		6	2.667		2,350	118		2,468	2,775
2200	Low E glass		6	2.667		2,350	118		2,468	2,800
2300	Vinyl clad, premium, double insulated glass, 8'-0" x 5'-0"		10	1.600		1,750	70.50		1,820.50	2,050
2340	10'-0" x 5'-0"		8	2		2,275	88		2,363	2,675
2380	10'-0" x 6'-0"		7	2.286		2,600	101		2,701	3,025

08 52 Wood Windows

08 52 16 – Plastic-Clad Wood Windows

08 52 16.10 Bow Window		Crew	Daily Output	Labor-Hours	Unit	Material (2012 Bare Costs)	Labor (2012 Bare Costs)	Equipment (2012 Bare Costs)	Total (2012 Bare Costs)	Total Incl O&P
2420	12′-0″ x 6′-0″	2 Carp	6	2.667	Ea.	3,200	118		3,318	3,725
3300	Vinyl clad, premium, double insulated glass, 7′-0″ x 4′-6″		10	1.600		1,375	70.50		1,445.50	1,625
3340	8′-0″ x 4′-6″		8	2		1,400	88		1,488	1,675
3380	8′-0″ x 5′-0″		7	2.286		1,450	101		1,551	1,775
3420	9′-0″ x 5′-0″		6	2.667		1,500	118		1,618	1,850
9000	Minimum labor/equipment charge		2.50	6.400	Job		282		282	465

08 52 16.15 Awning Window Vinyl-Clad		Crew	Daily Output	Labor-Hours	Unit	Material	Labor	Equipment	Total	Total Incl O&P
0010	**AWNING WINDOW VINYL-CLAD** Including frames, screens, and grilles									
0240	Vinyl clad, 34″ x 22″	1 Carp	10	.800	Ea.	262	35.50		297.50	345
0280	36″ x 28″		9	.889		305	39		344	400
0300	36″ x 36″		9	.889		340	39		379	440
0340	40″ x 22″		10	.800		287	35.50		322.50	375
0360	48″ x 28″		8	1		365	44		409	475
0380	60″ x 36″		8	1		515	44		559	640

08 52 16.20 Half Round, Vinyl Clad		Crew	Daily Output	Labor-Hours	Unit	Material	Labor	Equipment	Total	Total Incl O&P
0010	**HALF ROUND, VINYL CLAD**, double insulated glass, incl. grille									
0800	14″ height x 24″ base	2 Carp	9	1.778	Ea.	405	78.50		483.50	575
1040	15″ height x 25″ base		8	2		375	88		463	560
1060	16″ height x 28″ base		7	2.286		405	101		506	610
1080	17″ height x 29″ base		7	2.286		420	101		521	625
2000	19″ height x 33″ base	1 Carp	6	1.333		450	59		509	590
2100	20″ height x 35″ base		6	1.333		450	59		509	590
2200	21″ height x 37″ base		6	1.333		475	59		534	620
2250	23″ height x 41″ base	2 Carp	6	2.667		520	118		638	765
2300	26″ height x 48″ base		6	2.667		535	118		653	780
2350	30″ height x 56″ base		6	2.667		615	118		733	870
3000	36″ height x 67″base	1 Carp	4	2		1,050	88		1,138	1,300
3040	38″ height x 71″ base	2 Carp	5	3.200		975	141		1,116	1,300
3050	40″ height x 75″ base	″	5	3.200		1,275	141		1,416	1,625
5000	Elliptical, 71″ x 16″	1 Carp	11	.727		975	32		1,007	1,125
5100	95″ x 21″	″	10	.800		1,375	35.50		1,410.50	1,575

08 52 16.35 Double-Hung Window			Crew	Daily Output	Labor-Hours	Unit	Material	Labor	Equipment	Total	Total Incl O&P
0010	**DOUBLE-HUNG WINDOW** Including frames, screens, and grilles										
0300	Vinyl clad, premium, double insulated glass, 2′-6″ x 3′-0″	G	1 Carp	10	.800	Ea.	300	35.50		335.50	390
0305	2′-6″ x 4′-0″	G		10	.800		355	35.50		390.50	450
0400	3′-0″ x 3′-6″	G		10	.800		330	35.50		365.50	425
0500	3′-0″ x 4′-0″	G		9	.889		390	39		429	495
0600	3′-0″ x 4′-6″	G		9	.889		405	39		444	510
0700	3′-0″ x 5′-0″	G		8	1		440	44		484	560
0790	3′-4″ x 5′-0″	G		8	1		445	44		489	565
0800	3′-6″ x 6′-0″	G		8	1		485	44		529	605
0820	4′-0″ x 5′-0″	G		7	1.143		550	50.50		600.50	690
0830	4′-0″ x 6′-0″	G		7	1.143		685	50.50		735.50	840

08 52 16.40 Transom Windows		Crew	Daily Output	Labor-Hours	Unit	Material	Labor	Equipment	Total	Total Incl O&P
0010	**TRANSOM WINDOWS**									
0050	Vinyl clad, premium, double insulated glass, 32″ x 8″	1 Carp	16	.500	Ea.	184	22		206	238
0100	36″ x 8″		16	.500		195	22		217	251
0110	36″ x 12″		16	.500		209	22		231	265
2000	Custom sizes, up to 350 sq. in.		12	.667		252	29.50		281.50	325
2100	351 to 750 sq. in.		12	.667		320	29.50		349.50	400
2200	751 to 1150 sq. in.		11	.727		390	32		422	485
2300	1151 to 1450 sq. in.		11	.727		455	32		487	555

08 52 Wood Windows

08 52 16 – Plastic-Clad Wood Windows

08 52 16.40 Transom Windows		Crew	Daily Output	Labor-Hours	Unit	Material	Labor	Equipment	Total	Total Incl O&P
						2012 Bare Costs				
2400	1451 to 1850 sq. in.	2 Carp	12	1.333	Ea.	520	59		579	665
2500	1851 to 2250 sq. in.		12	1.333		595	59		654	750
2600	2251 to 2650 sq. in.		11	1.455		665	64		729	840
2700	2651 to 3050 sq. in.		11	1.455		670	64		734	845
2800	3051 to 3450 sq. in.		11	1.455		740	64		804	920
2900	3451 to 3850 sq. in.		10	1.600		800	70.50		870.50	990
3000	3851 to 4250 sq. in.		10	1.600		855	70.50		925.50	1,050
3100	4251 to 4650 sq. in.		10	1.600		870	70.50		940.50	1,075
3200	4651 to 5050 sq. in.		10	1.600		920	70.50		990.50	1,125
3300	5051 to 5450 sq. in.		9	1.778		1,000	78.50		1,078.50	1,225
3400	5451 to 5850 sq. in.		9	1.778		965	78.50		1,043.50	1,175
3600	6251 to 6650 sq. in.		8	2		1,150	88		1,238	1,425
3700	6651 to 7050 sq. in.		8	2		1,225	88		1,313	1,500

08 52 16.45 Trapezoid Windows		Crew	Daily Output	Labor-Hours	Unit	Material	Labor	Equipment	Total	Total Incl O&P
0010	**TRAPEZOID WINDOWS**									
0900	20" base x 44" leg x 53" leg	2 Carp	13	1.231	Ea.	390	54.50		444.50	520
1000	24" base x 90" leg x 102" leg		8	2		670	88		758	885
3000	36" base x 40" leg x 22" leg		12	1.333		430	59		489	565
3010	36" base x 44" leg x 25" leg		13	1.231		450	54.50		504.50	585
3050	36" base x 26" leg x 48" leg		9	1.778		455	78.50		533.50	630
3100	36" base x 42" legs, 50" peak		9	1.778		545	78.50		623.50	730
3200	36" base x 60" leg x 81" leg		11	1.455		710	64		774	890
4320	44" base x 23" leg x 56" leg		11	1.455		560	64		624	725
4350	44" base x 59" leg x 92" leg		10	1.600		840	70.50		910.50	1,050
4500	46" base x 15" leg x 46" leg		8	2		425	88		513	610
4550	46" base x 16" leg x 48" leg		8	2		450	88		538	640
4600	46" base x 50" leg x 80" leg		7	2.286		680	101		781	915
6600	66" base x 12" leg x 42" leg		8	2		575	88		663	780
6650	66" base x 12" legs, 28" peak		9	1.778		455	78.50		533.50	630
6700	68" base x 23" legs, 31" peak		8	2		575	88		663	780

08 52 16.70 Vinyl Clad, Premium, Dbl. Insulated Glass		Crew	Daily Output	Labor-Hours	Unit	Material	Labor	Equipment	Total	Total Incl O&P
0010	**VINYL CLAD, PREMIUM, DBL. INSULATED GLASS**									
1000	Sliding, 3'-0" x 3'-0" G	1 Carp	10	.800	Ea.	605	35.50		640.50	725
1050	4'-0" x 3'-6" G		9	.889		685	39		724	820
1100	5'-0" x 4'-0" G		9	.889		900	39		939	1,050
1150	6'-0" x 5'-0" G		8	1		1,125	44		1,169	1,325

08 52 50 – Window Accessories

08 52 50.10 Window Grille or Muntin		Crew	Daily Output	Labor-Hours	Unit	Material	Labor	Equipment	Total	Total Incl O&P
0010	**WINDOW GRILLE OR MUNTIN**, snap in type									
0020	Standard pattern interior grilles									
2000	Wood, awning window, glass size 28" x 16" high	1 Carp	30	.267	Ea.	26.50	11.75		38.25	48.50
2060	44" x 24" high		32	.250		37.50	11.05		48.55	59.50
2100	Casement, glass size, 20" x 36" high		30	.267		30.50	11.75		42.25	53
2180	20" x 56" high		32	.250		42	11.05		53.05	64
2200	Double hung, glass size, 16" x 24" high		24	.333	Set	50	14.70		64.70	79.50
2280	32" x 32" high		34	.235	"	131	10.40		141.40	161
2500	Picture, glass size, 48" x 48" high		30	.267	Ea.	117	11.75		128.75	147
2580	60" x 68" high		28	.286	"	180	12.60		192.60	219
2600	Sliding, glass size, 14" x 36" high		24	.333	Set	34.50	14.70		49.20	62
2680	36" x 36" high		22	.364	"	42.50	16.05		58.55	73
9000	Minimum labor/equipment charge		5	1.600	Job		70.50		70.50	116

08 52 Wood Windows

08 52 66 – Wood Window Screens

08 52 66.10 Wood Screens		Crew	Daily Output	Labor-Hours	Unit	2012 Bare Costs Material	Labor	Equipment	Total	Total Incl O&P
0010	**WOOD SCREENS**									
0020	Over 3 S.F., 3/4" frames	2 Carp	375	.043	S.F.	4.80	1.88		6.68	8.40
0100	1-1/8" frames		375	.043		8.10	1.88		9.98	12
0200	Rescreen wood frame		500	.032		1.58	1.41		2.99	4.05
9000	Minimum labor/equipment charge	1 Carp	4	2	Job		88		88	145

08 52 69 – Wood Storm Windows

08 52 69.10 Storm Windows		Crew	Daily Output	Labor-Hours	Unit	Material	Labor	Equipment	Total	Total Incl O&P
0010	**STORM WINDOWS**, aluminum residential									
0300	Basement, mill finish, incl. fiberglass screen									
0320	1'-10" x 1'-0" high G	2 Carp	30	.533	Ea.	35	23.50		58.50	77
0340	2'-9" x 1'-6" high G		30	.533		38	23.50		61.50	80.50
0360	3'-4" x 2'-0" high G		30	.533		44	23.50		67.50	87
1600	Double-hung, combination, storm & screen									
1700	Custom, clear anodic coating, 2'-0" x 3'-5" high	2 Carp	30	.533	Ea.	90	23.50		113.50	138
1720	2'-6" x 5'-0" high		28	.571		110	25		135	163
1740	4'-0" x 6'-0" high		25	.640		226	28		254	296
1800	White painted, 2'-0" x 3'-5" high		30	.533		100	23.50		123.50	149
1820	2'-6" x 5'-0" high		28	.571		160	25		185	218
1840	4'-0" x 6'-0" high		25	.640		273	28		301	345
2000	Average quality, clear anodic coating, 2'-0" x 3'-5" high G		30	.533		85	23.50		108.50	132
2020	2'-6" x 5'-0" high G		28	.571		113	25		138	166
2040	4'-0" x 6'-0" high G		25	.640		126	28		154	186
2400	White painted, 2'-0" x 3'-5" high G		30	.533		85	23.50		108.50	132
2420	2'-6" x 5'-0" high G		28	.571		90	25		115	141
2440	4'-0" x 6'-0" high G		25	.640		98	28		126	155
2600	Mill finish, 2'-0" x 3'-5" high G		30	.533		75	23.50		98.50	121
2620	2'-6" x 5'-0" high G		28	.571		85	25		110	135
2640	4'-0" x 6'-8" high G		25	.640		99	28		127	156
4000	Picture window, storm, 1 lite, white or bronze finish									
4020	4'-6" x 4'-6" high	2 Carp	25	.640	Ea.	120	28		148	179
4040	5'-8" x 4'-6" high		20	.800		144	35.50		179.50	216
4400	Mill finish, 4'-6" x 4'-6" high		25	.640		126	28		154	186
4420	5'-8" x 4'-6" high		20	.800		144	35.50		179.50	216
4600	3 lite, white or bronze finish									
4620	4'-6" x 4'-6" high	2 Carp	25	.640	Ea.	152	28		180	214
4640	5'-8" x 4'-6" high		20	.800		166	35.50		201.50	241
4800	Mill finish, 4'-6" x 4'-6" high		25	.640		144	28		172	205
4820	5'-8" x 4'-6" high		20	.800		152	35.50		187.50	225
6000	Sliding window, storm, 2 lite, white or bronze finish									
6020	3'-4" x 2'-7" high	2 Carp	28	.571	Ea.	126	25		151	181
6040	4'-4" x 3'-3" high		25	.640		150	28		178	212
6060	5'-4" x 6'-0" high		20	.800		240	35.50		275.50	320
8000	PVC framed									
8100	Double-hung, combination, storm & screen									
8120	2'-6" x 3'-5"	2 Carp	15	1.067	Ea.	131	47		178	221
8140	4' x 6'	"	12.50	1.280	"	158	56.50		214.50	266
8600	Single lite picture storm									
8620	4'-6" x 4'-6"	2 Carp	12.50	1.280	Ea.	158	56.50		214.50	266
8640	5'-8" x 4'-6"	"	10	1.600	"	167	70.50		237.50	299
9000	Magnetic interior storm window									
9100	3/16" plate glass	1 Glaz	107	.075	S.F.	5.40	3.20		8.60	11.10
9410	Minimum labor/equipment charge	1 Carp	4	2	Job		88		88	145

08 53 Plastic Windows

08 53 13 – Vinyl Windows

08 53 13.20 Vinyl Single Hung Windows		Crew	Daily Output	Labor-Hours	Unit	Material	2012 Bare Costs Labor	Equipment	Total	Total Incl O&P
0010	**VINYL SINGLE HUNG WINDOWS**, insulated glass									
0100	Grids, low E, J fin, ext. jambs, 21" x 53" G	2 Carp	18	.889	Ea.	198	39		237	283
0110	21" x 57" G		17	.941		195	41.50		236.50	283
0120	21" x 65" G		16	1		200	44		244	293
0130	25" x 41" G		20	.800		188	35.50		223.50	265
0140	25" x 49" G		18	.889		198	39		237	283
0150	25" x 57" G		17	.941		201	41.50		242.50	289
0160	25" x 65" G		16	1		207	44		251	300
0170	29" x 41" G		18	.889		192	39		231	276
0180	29" x 53" G		18	.889		202	39		241	287
0190	29" x 57" G		17	.941		206	41.50		247.50	295
0200	29" x 65" G		16	1		212	44		256	305
0210	33" x 41" G		20	.800		198	35.50		233.50	276
0220	33" x 53" G		18	.889		209	39		248	295
0230	33" x 57" G		17	.941		212	41.50		253.50	300
0240	33" x 65" G		16	1		218	44		262	315
0250	37" x 41" G		20	.800		206	35.50		241.50	285
0260	37" x 53" G		18	.889		218	39		257	305
0270	37" x 57" G		17	.941		221	41.50		262.50	310
0280	37" x 65" G	↓	16	1	↓	232	44		276	330

08 53 13.30 Vinyl Double Hung Windows		Crew	Daily Output	Labor-Hours	Unit	Material	Labor	Equipment	Total	Total Incl O&P
0010	**VINYL DOUBLE HUNG WINDOWS**, insulated glass									
0100	Grids, low E, J fin, ext. jambs, 21" x 53" G	2 Carp	18	.889	Ea.	215	39		254	300
0102	21" x 37" G		18	.889		197	39		236	282
0104	21" x 41" G		18	.889		201	39		240	286
0106	21" x 49" G		18	.889		221	39		260	310
0110	21" x 57" G		17	.941		218	41.50		259.50	310
0120	21" x 65" G		16	1		228	44		272	325
0128	25" x 37" G		20	.800		206	35.50		241.50	285
0130	25" x 41" G		20	.800		209	35.50		244.50	288
0140	25" x 49" G		18	.889		213	39		252	299
0145	25" x 53" G		18	.889		221	39		260	310
0150	25" x 57" G		17	.941		221	41.50		262.50	310
0160	25" x 65" G		16	1		231	44		275	325
0162	25" x 69" G		16	1		238	44		282	335
0164	25" x 77" G		16	1		248	44		292	345
0168	29" x 37" G		18	.889		211	39		250	297
0170	29" x 41" G		18	.889		214	39		253	300
0172	29" x 49" G		18	.889		221	39		260	310
0180	29" x 53" G		18	.889		225	39		264	315
0190	29" x 57" G		17	.941		229	41.50		270.50	320
0200	29" x 65" G		16	1		236	44		280	335
0202	29" x 69" G		16	1		243	44		287	340
0205	29" x 77" G		16	1		254	44		298	350
0208	33" x 37" G		20	.800		215	35.50		250.50	295
0210	33" x 41" G		20	.800		219	35.50		254.50	299
0215	33" x 49" G		20	.800		227	35.50		262.50	310
0220	33" x 53" G		18	.889		231	39		270	320
0230	33" x 57" G		17	.941		235	41.50		276.50	325
0240	33" x 65" G		16	1		239	44		283	335
0242	33" x 69" G		16	1		245	44		289	345
0246	33" x 77" G	↓	16	1	↓	251	44		295	350

08 53 Plastic Windows

08 53 13 – Vinyl Windows

08 53 13.30 Vinyl Double Hung Windows

			Crew	Daily Output	Labor-Hours	Unit	2012 Bare Costs Material	Labor	Equipment	Total	Total Incl O&P
0250	37" x 41"	G	2 Carp	20	.800	Ea.	223	35.50		258.50	305
0255	37" x 49"	G		20	.800		231	35.50		266.50	310
0260	37" x 53"	G		18	.889		239	39		278	330
0270	37" x 57"	G		17	.941		243	41.50		284.50	335
0280	37" x 65"	G		16	1		255	44		299	355
0282	37" x 69"	G		16	1		262	44		306	360
0286	37" x 77"	G		16	1		270	44		314	370
0300	Solid vinyl, average quality, double insulated glass, 2'-0" x 3'-0"	G	1 Carp	10	.800		287	35.50		322.50	375
0310	3'-0" x 4'-0"	G		9	.889		203	39		242	288
0320	4'-0" x 4'-6"	G		8	1		305	44		349	410
0330	Premium, double insulated glass, 2'-6" x 3'-0"	G		10	.800		239	35.50		274.50	320
0340	3'-0" x 3'-6"	G		9	.889		263	39		302	355
0350	3'-0" x 4"-0"	G		9	.889		288	39		327	380
0360	3'-0" x 4'-6"	G		9	.889		295	39		334	390
0370	3'-0" x 5'-0"	G		8	1		315	44		359	425
0380	3'-6" x 6"-0"	G		8	1		355	44		399	470

08 53 13.40 Vinyl Casement Windows

			Crew	Daily Output	Labor-Hours	Unit	Material	Labor	Equipment	Total	Total Incl O&P
0010	**VINYL CASEMENT WINDOWS**, insulated glass										
0015	Grids, low E, J fin, extension jambs, screens										
0100	One lite, 21" x 41"	G	2 Carp	20	.800	Ea.	271	35.50		306.50	355
0110	21" x 47"	G		20	.800		294	35.50		329.50	385
0120	21" x 53"	G		20	.800		315	35.50		350.50	410
0128	24" x 35"	G		19	.842		263	37		300	350
0130	24" x 41"	G		19	.842		283	37		320	370
0140	24" x 47"	G		19	.842		305	37		342	395
0150	24" x 53"	G		19	.842		325	37		362	420
0158	28" x 35"	G		19	.842		278	37		315	365
0160	28" x 41"	G		19	.842		298	37		335	390
0170	28" x 47"	G		19	.842		320	37		357	410
0180	28" x 53"	G		19	.842		350	37		387	445
0184	28" x 59"	G		19	.842		350	37		387	445
0188	Two lites, 33" x 35"	G		18	.889		430	39		469	535
0190	33" x 41"	G		18	.889		455	39		494	570
0200	33" x 47"	G		18	.889		490	39		529	600
0210	33" x 53"	G		18	.889		520	39		559	635
0212	33" x 59"	G		18	.889		550	39		589	670
0215	33" x 72"	G		18	.889		570	39		609	690
0220	41" x 41"	G		18	.889		495	39		534	610
0230	41" x 47"	G		18	.889		530	39		569	645
0240	41" x 53"	G		17	.941		560	41.50		601.50	685
0242	41" x 59"	G		17	.941		590	41.50		631.50	715
0246	41" x 72"	G		17	.941		615	41.50		656.50	745
0250	47" x 41"	G		17	.941		500	41.50		541.50	620
0260	47" x 47"	G		17	.941		530	41.50		571.50	655
0270	47" x 53"	G		17	.941		560	41.50		601.50	690
0272	47" x 59"	G		17	.941		615	41.50		656.50	745
0280	56" x 41"	G		15	1.067		535	47		582	665
0290	56" x 47"	G		15	1.067		560	47		607	695
0300	56" x 53"	G		15	1.067		610	47		657	745
0302	56" x 59"	G		15	1.067		635	47		682	775
0310	56" x 72"	G		15	1.067		690	47		737	835
0340	Solid vinyl, premium, double insulated glass, 2'-0" x 3'-0" high	G	1 Carp	10	.800		259	35.50		294.50	345

08 53 Plastic Windows

08 53 13 – Vinyl Windows

08 53 13.40 Vinyl Casement Windows		Crew	Daily Output	Labor-Hours	Unit	2012 Bare Costs Material	Labor	Equipment	Total	Total Incl O&P
0360	2′-0″ x 4′-0″ high G	1 Carp	9	.889	Ea.	289	39		328	385
0380	2′-0″ x 5′-0″ high G	↓	8	1	↓	310	44		354	415

08 53 13.50 Vinyl Picture Windows		Crew	Daily Output	Labor-Hours	Unit	Material	Labor	Equipment	Total	Total Incl O&P
0010	**VINYL PICTURE WINDOWS**, insulated glass									
0100	Grids, low E, J fin, ext. jambs, 33″ x 47″	2 Carp	12	1.333	Ea.	250	59		309	370
0110	35″ x 71″		12	1.333		335	59		394	465
0120	41″ x 47″		12	1.333		270	59		329	395
0130	41″ x 71″		12	1.333		350	59		409	480
0140	47″ x 47″		12	1.333		305	59		364	430
0150	47″ x 71″		11	1.455		380	64		444	525
0160	53″ x 47″		11	1.455		330	64		394	470
0170	53″ x 71″		11	1.455		345	64		409	485
0180	59″ x 47″		11	1.455		370	64		434	510
0190	59″ x 71″		11	1.455		380	64		444	525
0200	71″ x 47″		10	1.600		390	70.50		460.50	545
0210	71″ x 71″	↓	10	1.600	↓	405	70.50		475.50	560

08 53 13.60 Vinyl Half Round Windows		Crew	Daily Output	Labor-Hours	Unit	Material	Labor	Equipment	Total	Total Incl O&P
0010	**VINYL HALF ROUND WINDOWS**, Including grille, j fin low E, ext. jambs									
0100	10″ height x 20″ base	2 Carp	8	2	Ea.	475	88		563	665
0110	15″ height x 30″ base		8	2		440	88		528	630
0120	17″ height x 34″ base		7	2.286		315	101		416	515
0130	19″ height x 38″ base		7	2.286		340	101		441	540
0140	19″ height x 33″ base	↓	7	2.286		555	101		656	780
0150	24″ height x 48″ base	1 Carp	6	1.333		360	59		419	495
0160	25″ height x 50″ base	″	6	1.333		755	59		814	925
0170	30″ height x 60″ base	2 Carp	6	2.667	↓	660	118		778	920

08 54 Composite Windows

08 54 13 – Fiberglass Windows

08 54 13.10 Fiberglass Single Hung Windows		Crew	Daily Output	Labor-Hours	Unit	Material	Labor	Equipment	Total	Total Incl O&P
0010	**FIBERGLASS SINGLE HUNG WINDOWS**									
0100	Grids, low E, 18″ x 24″ G	2 Carp	18	.889	Ea.	335	39		374	435
0110	18″ x 40″ G		17	.941		340	41.50		381.50	445
0130	24″ x 40″ G		20	.800		360	35.50		395.50	455
0230	36″ x 36″ G		17	.941		370	41.50		411.50	475
0250	36″ x 48″ G		20	.800		405	35.50		440.50	505
0260	36″ x 60″ G		18	.889		445	39		484	555
0280	36″ x 72″ G		16	1		470	44		514	590
0290	48″ x 40″ G	↓	16	1	↓	470	44		514	590

08 56 Special Function Windows

08 56 63 – Detention Windows

08 56 63.13 Visitor Cubicle Windows		Crew	Daily Output	Labor-Hours	Unit	2012 Bare Costs Material	Labor	Equipment	Total	Total Incl O&P
0010	**VISITOR CUBICLE WINDOWS**									
4000	Visitor cubicle, vision panel, no intercom	E-4	2	16	Ea.	3,150	800	61	4,011	5,025

08 61 Roof Windows

08 61 13 – Metal Roof Windows

08 61 13.10 Roof Windows		Crew	Daily Output	Labor-Hours	Unit	Material	Labor	Equipment	Total	Total Incl O&P
0010	**ROOF WINDOWS**, fixed high perf tmpd glazing, metallic framed									
0020	46" x 21-1/2", Flashed for shingled roof	1 Carp	8	1	Ea.	245	44		289	345
0100	46" x 28"		8	1		272	44		316	370
0125	57" x 44"		6	1.333		335	59		394	465
0130	72" x 28"		7	1.143		335	50.50		385.50	455
0150	Fixed, laminated tempered glazing, 46" x 21-1/2"		8	1		415	44		459	530
0175	46" x 28"		8	1		460	44		504	580
0200	57" x 44"		6	1.333		430	59		489	570
0500	Vented flashing set for shingled roof, 46" x 21-1/2"		7	1.143		415	50.50		465.50	540
0525	46" x 28"		6	1.333		460	59		519	600
0550	57" x 44"		5	1.600		575	70.50		645.50	750
0560	72" x 28"		5	1.600		575	70.50		645.50	750
0575	Flashing set for low pitched roof, 46" x 21-1/2"		7	1.143		475	50.50		525.50	610
0600	46" x 28"		7	1.143		525	50.50		575.50	660
0625	57" x 44"		5	1.600		650	70.50		720.50	830
0650	Flashing set for curb 46" x 21-1/2"		7	1.143		550	50.50		600.50	690
0675	46" x 28"		7	1.143		600	50.50		650.50	745
0700	57" x 44"		5	1.600		740	70.50		810.50	925

08 62 Unit Skylights

08 62 13 – Domed Unit Skylights

08 62 13.10 Domed Skylights		Crew	Daily Output	Labor-Hours	Unit	Material	Labor	Equipment	Total	Total Incl O&P
0010	**DOMED SKYLIGHTS**									
0020	Skylight, fixed dome type, 22" x 22" G	G-3	12	2.667	Ea.	200	116		316	405
0030	22" x 46" G		10	3.200		229	140		369	475
0040	30" x 30" G		12	2.667		230	116		346	440
0050	30" x 46" G		10	3.200		290	140		430	545
0110	Fixed, double glazed, 22" x 27" G		12	2.667		213	116		329	420
0120	22" x 46" G		10	3.200		269	140		409	520
0130	44" x 46" G		10	3.200		400	140		540	665
0210	Operable, double glazed, 22" x 27" G		12	2.667		310	116		426	525
0220	22" x 46" G		10	3.200		365	140		505	625
0230	44" x 46" G		10	3.200		805	140		945	1,125
9000	Minimum labor/equipment charge		2	16	Job		700		700	1,125

08 62 13.20 Skylights		Crew	Daily Output	Labor-Hours	Unit	Material	Labor	Equipment	Total	Total Incl O&P
0010	**SKYLIGHTS**, Plastic domes, flush or curb mounted									
0300	Nominal size under 10 S.F., double G	G-3	130	.246	S.F.	29.50	10.75		40.25	49.50
0400	Single		160	.200		26	8.75		34.75	42.50
0600	10 S.F. to 20 S.F., double G		315	.102		25	4.43		29.43	34.50
0700	Single		395	.081		25	3.53		28.53	33
0900	20 S.F. to 30 S.F., double G		395	.081		23.50	3.53		27.03	31.50
1000	Single		465	.069		19.25	3		22.25	26
1200	30 S.F. to 65 S.F., double G		465	.069		24	3		27	31.50

08 62 Unit Skylights

08 62 13 – Domed Unit Skylights

08 62 13.20 Skylights		Crew	Daily Output	Labor-Hours	Unit	2012 Bare Costs Material	Labor	Equipment	Total	Total Incl O&P
1300	Single	G-3	610	.052	S.F.	17.60	2.29		19.89	23
1500	For insulated 4" curbs, double, add					27%				
1600	Single, add					30%				
1800	For integral insulated 9" curbs, double, add					30%				
1900	Single, add					40%				
2120	Ventilating insulated plexiglass dome with									
2130	curb mounting, 36" x 36" G	G-3	12	2.667	Ea.	460	116		576	690
2150	52" x 52" G		12	2.667		640	116		756	890
2160	28" x 52" G		10	3.200		470	140		610	740
2170	36" x 52" G		10	3.200		520	140		660	795
2180	For electric opening system, add G					300			300	330
2220	32" x 48"	G-3	9	3.556		590	155		745	895
2300	Insulated safety glass with aluminum frame G		160	.200	S.F.	90	8.75		98.75	113
2400	Sandwich panels, fiberglass, for walls, 1-9/16" thick, to 250 S.F. G		200	.160		17	7		24	30
2700	As above, but for roofs, 2-3/4" thick, to 250 S.F. G		295	.108		24.50	4.73		29.23	34.50
2800	250 S.F. and up G		330	.097		20	4.23		24.23	29
4000	Skylight, solar tube kit, incl dome, flashing, diffuser, 1 pipe, 9" dia.	1 Carp	2	4	Ea.	230	176		406	540
4010	13" dia.		2	4		310	176		486	630
4020	21" dia.		2	4		490	176		666	830
4030	Accessories for, 1' long x 9" dia pipe		24	.333		30	14.70		44.70	57
4040	2' long x 9" dia pipe		24	.333		44	14.70		58.70	72.50
4050	4' long x 9" dia pipe		20	.400		72	17.65		89.65	108
4060	1' long x 13" dia pipe		24	.333		44	14.70		58.70	72.50
4070	2' long x 13" dia pipe		24	.333		57	14.70		71.70	86.50
4080	4' long x 13" dia pipe		20	.400		99	17.65		116.65	138
4090	2' long x 21" dia pipe		16	.500		110	22		132	157
4100	4' long x 21" dia pipe		12	.667		178	29.50		207.50	244
4110	45 degree elbow, 9"		16	.500		100	22		122	146
4120	13"		16	.500		91	22		113	136
4130	Interior decorative ring, 9"		20	.400		22	17.65		39.65	53
4140	13"		20	.400		24	17.65		41.65	55.50

08 71 Door Hardware

08 71 13 – Automatic Door Operators

08 71 13.10 Automatic Openers Commercial		Crew	Daily Output	Labor-Hours	Unit	Material	Labor	Equipment	Total	Total Incl O&P
0010	**AUTOMATIC OPENERS COMMERCIAL**									
0020	Pneumatic, incl opener, motion sens, control box, tubing, compressor									
0050	For single swing door, per opening	2 Skwk	.80	20	Ea.	4,375	910		5,285	6,275
0100	Pair, per opening		.50	32	Opng.	7,100	1,450		8,550	10,200
1000	For single sliding door, per opening		.60	26.667		4,800	1,200		6,000	7,250
1300	Bi-parting pair		.50	32		7,200	1,450		8,650	10,300
1420	Electronic door opener incl motion sens, 12 V control box, motor									
1450	For single swing door, per opening	2 Skwk	.80	20	Opng.	3,550	910		4,460	5,375
1500	Pair, per opening		.50	32		5,800	1,450		7,250	8,725
1600	For single sliding door, per opening		.60	26.667		3,850	1,200		5,050	6,225
1700	Bi-parting pair		.50	32		5,900	1,450		7,350	8,850
1750	Handicap actuator buttons, 2, including 12 V DC wiring, add	1 Carp	1.50	5.333	Pr.	450	235		685	880

08 71 Door Hardware

08 71 20 – Hardware

08 71 20.15 Hardware

Line	08 71 20.15 Hardware	Crew	Daily Output	Labor-Hours	Unit	Material	Labor (2012 Bare Costs)	Equipment	Total	Total Incl O&P
0009	**HARDWARE**									
0010	Average percentage for hardware, total job cost									
0050	Maximum				Job					5%
0500	Total hardware for building, average distribution				″	85%	15%			
1000	Door hardware, apartment, interior	1 Carp	4	2	Door	515	88		603	715
1300	Average, door hardware, motel/hotel interior, with access card		4	2		635	88		723	840
1500	Hospital bedroom, minimum		4	2		605	88		693	810
2000	Maximum		3	2.667		720	118		838	985
2250	School, single exterior, incl. lever, incl. panic device		3	2.667		1,125	118		1,243	1,425
2500	Single interior, regular use, lever included		3	2.667		610	118		728	865
2600	Average, door hdwe, school, classroom, ANSI F88, incl. lever		3	2.667		820	118		938	1,100
2850	Stairway, single interior		3	2.667		535	118		653	785
3100	Double exterior, with panic device		2	4	Pr.	2,075	176		2,251	2,600
6020	Add for fire alarm door holder, electro-magnetic	1 Elec	4	2	Ea.	99.50	103		202.50	269

08 71 20.30 Door Closers

Line	08 71 20.30 Door Closers	Crew	Daily Output	Labor-Hours	Unit	Material	Labor	Equipment	Total	Total Incl O&P
0010	**DOOR CLOSERS**									
0020	Adjustable backcheck, 3 way mount, all sizes, regular arm	1 Carp	6	1.333	Ea.	184	59		243	299
0040	Hold open arm		6	1.333		234	59		293	355
0100	Fusible link		6.50	1.231		158	54.50		212.50	263
0200	Non sized, regular arm		6	1.333		168	59		227	282
0240	Hold open arm		6	1.333		204	59		263	320
0400	4 way mount, non sized, regular arm		6	1.333		224	59		283	345
0440	Hold open arm		6	1.333		234	59		293	355
2000	Backcheck and adjustable power, hinge face mount									
2010	All sizes, regular arm	1 Carp	6.50	1.231	Ea.	214	54.50		268.50	325
2040	Hold open arm		6.50	1.231		249	54.50		303.50	365
2400	Top jamb mount, all sizes, regular arm		6	1.333		204	59		263	320
2440	Hold open arm		6	1.333		214	59		273	330
2800	Top face mount, all sizes, regular arm		6.50	1.231		198	54.50		252.50	305
2840	Hold open arm		6.50	1.231		219	54.50		273.50	330
4000	Backcheck, overhead concealed, all sizes, regular arm		5.50	1.455		220	64		284	345
4040	Concealed arm		5	1.600		231	70.50		301.50	370
4400	Compact overhead, concealed, all sizes, regular arm		5.50	1.455		350	64		414	490
4440	Concealed arm		5	1.600		390	70.50		460.50	545
4800	Concealed in door, all sizes, regular arm		5.50	1.455		360	64		424	500
4840	Concealed arm		5	1.600		380	70.50		450.50	535
4900	Floor concealed, all sizes, single acting		2.20	3.636		385	160		545	690
4940	Double acting		2.20	3.636		325	160		485	625
5000	For cast aluminum cylinder, deduct					35			35	38.50
5040	For delayed action, add					46			46	50.50
5080	For fusible link arm, add					35			35	38.50
5120	For shock absorbing arm, add					50			50	55
5160	For spring power adjustment, add					40			40	44
6000	Closer-holder, hinge face mount, all sizes, exposed arm	1 Carp	6.50	1.231		168	54.50		222.50	274
7000	Electronic closer-holder, hinge facemount, concealed arm		5	1.600		295	70.50		365.50	440
7400	With built-in detector		5	1.600		730	70.50		800.50	920
9000	Minimum labor/equipment charge		4	2	Job		88		88	145

08 71 20.31 Door Closers

Line	08 71 20.31 Door Closers	Crew	Daily Output	Labor-Hours	Unit	Material	Labor	Equipment	Total	Total Incl O&P
0010	**DOOR CLOSERS**									
0015	Door closer, rack and pinion	1 Carp	6.50	1.231	Ea.	172	54.50		226.50	278

08 71 Door Hardware

08 71 20 – Hardware

08 71 20.35 Panic Devices		Crew	Daily Output	Labor-Hours	Unit	Material	2012 Bare Costs Labor	Equipment	Total	Total Incl O&P
0010	**PANIC DEVICES**									
0015	For rim locks, single door exit only	1 Carp	6	1.333	Ea.	530	59		589	680
0020	Outside key and pull		5	1.600		575	70.50		645.50	745
0200	Bar and vertical rod, exit only		5	1.600		735	70.50		805.50	925
0210	Outside key and pull		4	2		810	88		898	1,025
0400	Bar and concealed rod		4	2		600	88		688	805
0600	Touch bar, exit only		6	1.333		510	59		569	655
0610	Outside key and pull		5	1.600		605	70.50		675.50	780
0700	Touch bar and vertical rod, exit only		5	1.600		670	70.50		740.50	850
0710	Outside key and pull		4	2		780	88		868	1,000
1000	Mortise, bar, exit only		4	2		550	88		638	750
1600	Touch bar, exit only		4	2		625	88		713	830
2000	Narrow stile, rim mounted, bar, exit only		6	1.333		655	59		714	815
2010	Outside key and pull		5	1.600		700	70.50		770.50	885
2200	Bar and vertical rod, exit only		5	1.600		680	70.50		750.50	865
2210	Outside key and pull		4	2		680	88		768	895
2400	Bar and concealed rod, exit only		3	2.667		765	118		883	1,050
3000	Mortise, bar, exit only		4	2		610	88		698	815
3600	Touch bar, exit only		4	2		880	88		968	1,125
4000	Double doors, exit only		2	4	Pr.	955	176		1,131	1,350
4500	Exit & entrance		2	4	"	1,700	176		1,876	2,175
9000	Minimum labor/equipment charge		2.50	3.200	Job		141		141	231

08 71 20.40 Lockset		Crew	Daily Output	Labor-Hours	Unit	Material	Labor	Equipment	Total	Total Incl O&P
0010	**LOCKSET**, Standard duty									
0020	Non-keyed, passage, w/sect.trim	1 Carp	12	.667	Ea.	57.50	29.50		87	112
0100	Privacy		12	.667		62	29.50		91.50	116
0400	Keyed, single cylinder function		10	.800		92.50	35.50		128	160
0420	Hotel (see also Section 08 71 20.15)		8	1		135	44		179	222
0500	Lever handled, keyed, single cylinder function		10	.800		159	35.50		194.50	233
1000	Heavy duty with sectional trim, non-keyed, passages		12	.667		145	29.50		174.50	208
1100	Privacy		12	.667		185	29.50		214.50	252
1400	Keyed, single cylinder function		10	.800		220	35.50		255.50	300
1420	Hotel		8	1		325	44		369	435
1600	Communicating		10	.800		250	35.50		285.50	335
1690	For re-core cylinder, add					45			45	49.50
1700	Residential, interior door, minimum	1 Carp	16	.500		30	22		52	69
1720	Maximum		8	1		80	44		124	161
1800	Exterior, minimum		14	.571		45	25		70	91
1810	Average		8	1		72	44		116	152
1820	Maximum		8	1		180	44		224	271
3800	Cipher lockset w/key pad (security item)		13	.615		845	27		872	975
3900	Cipher lockset, with dial for swinging doors (security item)		13	.615		1,800	27		1,827	2,050
3920	with dial for swinging doors & drill resistant plate (security item)		12	.667		2,200	29.50		2,229.50	2,475
3950	Cipher lockset with dial for safe/vault door (security item)		12	.667		1,500	29.50		1,529.50	1,700
3980	Keyless, pushbutton type									
4000	Residential/light commercial, deadbolt, standard	1 Carp	9	.889	Ea.	150	39		189	230
4010	Heavy duty		9	.889		180	39		219	263
4020	Industrial, heavy duty, with deadbolt		9	.889		325	39		364	425
4030	Key override		9	.889		365	39		404	465
4040	Lever activated handle		9	.889		385	39		424	490
4050	Key override		9	.889		400	39		439	505
4060	Double sided pushbutton type		8	1		755	44		799	905

08 71 Door Hardware

08 71 20 – Hardware

08 71 20.40 Lockset		Crew	Daily Output	Labor-Hours	Unit	Material	2012 Bare Costs Labor	Equipment	Total	Total Incl O&P
4070	Key override	1 Carp	8	1	Ea.	790	44		834	945
9000	Minimum labor/equipment charge		6	1.333	Job		59		59	96.50

08 71 20.41 Dead Locks		Crew	Daily Output	Labor-Hours	Unit	Material	Labor	Equipment	Total	Total Incl O&P
0010	**DEAD LOCKS**									
0011	Mortise heavy duty outside key (security item)	1 Carp	9	.889	Ea.	275	39		314	370
0020	Double cylinder		9	.889		270	39		309	360
0100	Medium duty, outside key		10	.800		150	35.50		185.50	223
0110	Double cylinder		10	.800		165	35.50		200.50	240
1000	Tubular, standard duty, outside key		10	.800		66	35.50		101.50	131
1010	Double cylinder		10	.800		84	35.50		119.50	151
1200	Night latch, outside key		10	.800		83	35.50		118.50	150

08 71 20.42 Mortise Locksets		Crew	Daily Output	Labor-Hours	Unit	Material	Labor	Equipment	Total	Total Incl O&P
0010	**MORTISE LOCKSETS**, Comm., wrought knobs & full escutcheon trim									
0015	Assumes mortise is cut									
0020	Non-keyed, passage, minimum	1 Carp	9	.889	Ea.	194	39		233	278
0030	Maximum		8	1		300	44		344	405
0040	Privacy, minimum		9	.889		200	39		239	285
0050	Maximum		8	1		325	44		369	435
0100	Keyed, office/entrance/apartment, minimum		8	1		230	44		274	325
0110	Maximum		7	1.143		385	50.50		435.50	510
0120	Single cylinder, typical, minimum		8	1		198	44		242	291
0130	Maximum		7	1.143		360	50.50		410.50	480
0200	Hotel, minimum (see also Section 08 71 20.15)		7	1.143		250	50.50		300.50	360
0210	Maximum (see also Section 08 71 20.15)		6	1.333		400	59		459	535
0300	Communication, double cylinder, minimum		8	1		225	44		269	320
0310	Maximum		7	1.143		300	50.50		350.50	415
1000	Wrought knobs and sectional trim, non-keyed, passage, minimum		10	.800		130	35.50		165.50	201
1010	Maximum		9	.889		245	39		284	335
1040	Privacy, minimum		10	.800		145	35.50		180.50	218
1050	Maximum		9	.889		265	39		304	355
1100	Keyed, entrance, office/apartment, minimum		9	.889		220	39		259	305
1110	Maximum		8	1		315	44		359	425
1120	Single cylinder, typical, minimum		9	.889		225	39		264	315
1130	Maximum		8	1		305	44		349	410
2000	Cast knobs and full escutcheon trim									
2010	Non-keyed, passage, minimum	1 Carp	9	.889	Ea.	275	39		314	370
2020	Maximum		8	1		425	44		469	545
2040	Privacy, minimum		9	.889		320	39		359	415
2050	Maximum		8	1		450	44		494	570
2120	Keyed, single cylinder, typical, minimum		8	1		330	44		374	440
2130	Maximum		7	1.143		500	50.50		550.50	635
3000	Cast knob and sectional trim, non-keyed, passage, minimum		10	.800		210	35.50		245.50	289
3010	Maximum		10	.800		400	35.50		435.50	500
3040	Privacy, minimum		10	.800		225	35.50		260.50	305
3050	Maximum		10	.800		400	35.50		435.50	500
3100	Keyed, office/entrance/apartment, minimum		9	.889		255	39		294	345
3110	Maximum		9	.889		420	39		459	525
3120	Single cylinder, typical, minimum		9	.889		260	39		299	350
3130	Maximum		9	.889		510	39		549	625
3190	For re-core cylinder, add					45			45	49.50

08 71 Door Hardware

08 71 20 – Hardware

08 71 20.45 Peepholes		Crew	Daily Output	Labor-Hours	Unit	Material	Labor	Equipment	Total	Total Incl O&P
						2012 Bare Costs				
0010	PEEPHOLES									
2010	Peephole	1 Carp	32	.250	Ea.	15.80	11.05		26.85	35.50
2020	Peephole, wide view	"	32	.250	"	17	11.05		28.05	37

08 71 20.50 Door Stops		Crew	Daily Output	Labor-Hours	Unit	Material	Labor	Equipment	Total	Total Incl O&P
0010	DOOR STOPS									
0020	Holder & bumper, floor or wall	1 Carp	32	.250	Ea.	35.50	11.05		46.55	57
1300	Wall bumper, 4" diameter, with rubber pad, aluminum		32	.250		10.95	11.05		22	30
1600	Door bumper, floor type, aluminum		32	.250		7.10	11.05		18.15	26
1620	Brass		32	.250		6.75	11.05		17.80	25.50
1630	Bronze		32	.250		12.75	11.05		23.80	32
1900	Plunger type, door mounted		32	.250		27	11.05		38.05	47.50
2500	Holder, floor type, aluminum		32	.250		34	11.05		45.05	55.50
2520	Wall type, aluminum		32	.250		34	11.05		45.05	55.50
2530	Overhead type, bronze		32	.250		88.50	11.05		99.55	115
2540	Plunger type, aluminum		32	.250		27	11.05		38.05	47.50
2560	Brass		32	.250		41.50	11.05		52.55	64
3000	Electro-magnetic, wall mounted, US3		3	2.667		213	118		331	425
3020	Floor mounted, US3		3	2.667		287	118		405	510
4000	Doorstop, ceiling mounted		3	2.667		20	118		138	215
4030	Doorstop, header		3	2.667		56	118		174	255
9000	Minimum labor/equipment charge		6	1.333	Job		59		59	96.50

08 71 20.55 Push-Pull Plates		Crew	Daily Output	Labor-Hours	Unit	Material	Labor	Equipment	Total	Total Incl O&P
0010	PUSH-PULL PLATES									
0100	Push plate, .050 thick, 4" x 16", aluminum	1 Carp	12	.667	Ea.	10.20	29.50		39.70	59.50
0500	Bronze		12	.667		21.50	29.50		51	71.50
1500	Pull handle and push bar, aluminum		11	.727		131	32		163	197
2000	Bronze		10	.800		169	35.50		204.50	243
3000	Push plate both sides, aluminum		14	.571		19.95	25		44.95	63.50
3500	Bronze		13	.615		44	27		71	93
4000	Door pull, designer style, cast aluminum, minimum		12	.667		71.50	29.50		101	127
5000	Maximum		8	1		370	44		414	480
6000	Cast bronze, minimum		12	.667		87	29.50		116.50	144
7000	Maximum		8	1		405	44		449	520
8000	Walnut, minimum		12	.667		67	29.50		96.50	122
9000	Maximum		8	1		370	44		414	480
9800	Minimum labor/equipment charge		5	1.600	Job		70.50		70.50	116

08 71 20.60 Entrance Locks		Crew	Daily Output	Labor-Hours	Unit	Material	Labor	Equipment	Total	Total Incl O&P
0010	ENTRANCE LOCKS									
0015	Cylinder, grip handle deadlocking latch	1 Carp	9	.889	Ea.	132	39		171	210
0020	Deadbolt		8	1		160	44		204	249
0100	Push and pull plate, dead bolt		8	1		150	44		194	238
0900	For handicapped lever, add					170			170	187

08 71 20.65 Thresholds		Crew	Daily Output	Labor-Hours	Unit	Material	Labor	Equipment	Total	Total Incl O&P
0010	THRESHOLDS									
0011	Threshold 3' long saddles aluminum	1 Carp	48	.167	L.F.	5.60	7.35		12.95	18.25
0100	Aluminum, 8" wide, 1/2" thick		12	.667	Ea.	41.50	29.50		71	93.50
0500	Bronze		60	.133	L.F.	42	5.90		47.90	55.50
0600	Bronze, panic threshold, 5" wide, 1/2" thick		12	.667	Ea.	71.50	29.50		101	127
0700	Rubber, 1/2" thick, 5-1/2" wide		20	.400		42	17.65		59.65	75
0800	2-3/4" wide		20	.400		30.50	17.65		48.15	62.50
1950	ADA Compliant Thresholds									

08 71 20 – Hardware

08 71 20.65 Thresholds		Crew	Daily Output	Labor-Hours	Unit	Material	2012 Bare Costs Labor	Equipment	Total	Total Incl O&P
2000	Threshold, wood oak 3-1/2" wide x 24" long	1 Carp	12	.667	Ea.	9	29.50		38.50	58
2010	3-1/2" wide x 36" long		12	.667		13.50	29.50		43	63
2020	3-1/2" wide x 48" long		12	.667		18	29.50		47.50	68
2030	4-1/2" wide x 24" long		12	.667		12	29.50		41.50	61
2040	4-1/2" wide x 36" long		12	.667		17	29.50		46.50	66.50
2050	4-1/2" wide x 48" long		12	.667		23	29.50		52.50	73.50
2060	6-1/2" wide x 24" long		12	.667		16.50	29.50		46	66
2070	6-1/2" wide x 36" long		12	.667		25	29.50		54.50	75.50
2080	6-1/2" wide x 48" long		12	.667		33	29.50		62.50	84.50
2090	Threshold, wood cherry 3-1/2" wide x 24" long		12	.667		13.50	29.50		43	63
2100	3-1/2" wide x 36" long		12	.667		21	29.50		50.50	71
2110	3-1/2" wide x 48" long		12	.667		27	29.50		56.50	77.50
2120	4-1/2" wide x 24" long		12	.667		17.50	29.50		47	67.50
2130	4-1/2" wide x 36" long		12	.667		25.50	29.50		55	76
2140	4-1/2" wide x 48" long		12	.667		34	29.50		63.50	85.50
2150	6-1/2" wide x 24" long		12	.667		25.50	29.50		55	76
2160	6-1/2" wide x 36" long		12	.667		36.50	29.50		66	88
2170	6-1/2" wide x 48" long		12	.667		48.50	29.50		78	102
2180	Threshold, wood walnut 3-1/2" wide x 24" long		12	.667		15.50	29.50		45	65
2190	3-1/2" wide x 36" long		12	.667		22	29.50		51.50	72
2200	3-1/2" wide x 48" long		12	.667		30	29.50		59.50	81
2210	4-1/2" wide x 24" long		12	.667		19.50	29.50		49	69.50
2220	4-1/2" wide x 36" long		12	.667		28.50	29.50		58	79.50
2230	4-1/2" wide x 48" long		12	.667		38	29.50		67.50	90
2240	6-1/2" wide x 24" long		12	.667		28	29.50		57.50	79
2250	6-1/2" wide x 36" long		12	.667		41.50	29.50		71	93.50
2260	6-1/2" wide x 48" long		12	.667		55	29.50		84.50	109
2300	Threshold, aluminum 4" wide x 36" long		12	.667		33	29.50		62.50	84.50
2310	4" wide x 48" long		12	.667		41	29.50		70.50	93
2320	4" wide x 72" long		12	.667		66	29.50		95.50	121
2330	5" wide x 36" long		12	.667		45	29.50		74.50	97.50
2340	5" wide x 48" long		12	.667		56	29.50		85.50	110
2350	5" wide x 72" long		12	.667		89	29.50		118.50	146
2360	6" wide x 36" long		12	.667		53	29.50		82.50	107
2370	6" wide x 48" long		12	.667		68	29.50		97.50	123
2380	6" wide x 72" long		12	.667		106	29.50		135.50	165
2390	7" wide x 36" long		12	.667		69	29.50		98.50	124
2400	7" wide x 48" long		12	.667		90	29.50		119.50	147
2410	7" wide x 72" long		12	.667		137	29.50		166.50	199
2500	Threshold, ramp, aluminum or rubber 24" x 24"		12	.667		190	29.50		219.50	257
9000	Minimum labor/equipment charge		4	2	Job		88		88	145

08 71 20.75 Door Hardware Accessories		Crew	Daily Output	Labor-Hours	Unit	Material	Labor	Equipment	Total	Total Incl O&P
0010	**DOOR HARDWARE ACCESSORIES**									
0050	Door closing coordinator, 36" (for paired openings up to 56")	1 Carp	8	1	Ea.	94	44		138	176
0060	48" (for paired openings up to 84")		8	1		101	44		145	184
0070	56" (for paired openings up to 96")		8	1		111	44		155	195
1000	Knockers, brass, standard		16	.500		44	22		66	84.50
1100	Deluxe		10	.800		118	35.50		153.50	187
4100	Deluxe		18	.444		44	19.60		63.60	80.50
4500	Rubber door silencers		540	.015		.33	.65		.98	1.43
9000	Minimum labor/equipment charge		6	1.333	Job		59		59	96.50

08 71 Door Hardware

08 71 20 – Hardware

08 71 20.80 Hasps		Crew	Daily Output	Labor-Hours	Unit	Material	2012 Bare Costs Labor	Equipment	Total	Total Incl O&P
0010	**HASPS**, steel assembly									
0015	3"	1 Carp	26	.308	Ea.	4.77	13.55		18.32	28
0020	4-1/2"		13	.615		6.55	27		33.55	51.50
0040	6"		12.50	.640		9.30	28		37.30	57

08 71 20.90 Hinges

		Crew	Daily Output	Labor-Hours	Unit	Material	Labor	Equipment	Total	Total Incl O&P
0010	**HINGES** R087120-10									
0012	Full mortise, avg. freq., steel base, USP, 4-1/2" x 4-1/2"				Pr.	27.50			27.50	30
0100	5" x 5", USP					43.50			43.50	48
0200	6" x 6", USP					91.50			91.50	101
0400	Brass base, 4-1/2" x 4-1/2", US10					54.50			54.50	60
0500	5" x 5", US10					78			78	85.50
0600	6" x 6", US10					132			132	145
0800	Stainless steel base, 4-1/2" x 4-1/2", US32					82			82	90
0900	For non removable pin, add (security item)				Ea.	4.89			4.89	5.40
0910	For floating pin, driven tips, add					3.30			3.30	3.63
0930	For hospital type tip on pin, add					14.35			14.35	15.80
0940	For steeple type tip on pin, add					13.25			13.25	14.60
0950	Full mortise, high frequency, steel base, 3-1/2" x 3-1/2", US26D				Pr.	29.50			29.50	32.50
1000	4-1/2" x 4-1/2", USP					63			63	69.50
1100	5" x 5", USP					59.50			59.50	65.50
1200	6" x 6", USP					144			144	159
1400	Brass base, 3-1/2" x 3-1/2", US4					52			52	57.50
1430	4-1/2" x 4-1/2", US10					85.50			85.50	94
1500	5" x 5", US10					125			125	138
1600	6" x 6", US10					180			180	198
1800	Stainless steel base, 4-1/2" x 4-1/2", US32					127			127	140
1810	5" x 4-1/2", US32					182			182	200
1930	For hospital type tip on pin, add				Ea.	13.15			13.15	14.45
1950	Full mortise, low frequency, steel base, 3-1/2" x 3-1/2", US26D				Pr.	12.10			12.10	13.30
2000	4-1/2" x 4-1/2", USP					13.05			13.05	14.35
2100	5" x 5", USP					32			32	35.50
2200	6" x 6", USP					64.50			64.50	71
2300	4-1/2" x 4-1/2", US3					18.85			18.85	20.50
2310	5" x 5", US3					45			45	49.50
2400	Brass bass, 4-1/2" x 4-1/2", US10					45.50			45.50	50
2500	5" x 5", US10					68.50			68.50	75
2800	Stainless steel base, 4-1/2" x 4-1/2", US32					77			77	84.50

08 71 20.91 Special Hinges

		Crew	Daily Output	Labor-Hours	Unit	Material	Labor	Equipment	Total	Total Incl O&P
0010	**SPECIAL HINGES**									
0015	Paumelle, high frequency									
0020	Steel base, 6" x 4-1/2", US10				Pr.	145			145	160
0100	Brass base, 5" x 4-1/2", US10				Ea.	208			208	229
0200	Paumelle, average frequency, steel base, 4-1/2" x 3-1/2", US10				Pr.	98.50			98.50	108
0400	Olive knuckle, low frequency, brass base, 6" x 4-1/2", US10				Ea.	152			152	167
1000	Electric hinge with concealed conductor, average frequency									
1010	Steel base, 4-1/2" x 4-1/2", US26D				Pr.	315			315	345
1100	Bronze base, 4-1/2" x 4-1/2", US26D				"	320			320	350
1200	Electric hinge with concealed conductor, high frequency									
1210	Steel base, 4-1/2" x 4-1/2", US26D				Pr.	241			241	265
1600	Double weight, 800 lb., steel base, removable pin, 5" x 6", USP					132			132	145
1700	Steel base-welded pin, 5" x 6", USP					166			166	183
1800	Triple weight, 2000 lb., steel base, welded pin, 5" x 6", USP					151			151	166

08 71 20.91 Special Hinges

	08 71 20.91 Special Hinges	Crew	Daily Output	Labor-Hours	Unit	2012 Bare Costs Material	Labor	Equipment	Total	Total Incl O&P
2000	Pivot reinf., high frequency, steel base, 7-3/4" door plate, USP				Pr.	189			189	208
2200	Bronze base, 7-3/4" door plate, US10					232			232	255
3000	Swing clear, full mortise, full or half surface, high frequency,									
3010	Steel base, 5" high, USP				Pr.	159			159	175
3200	Swing clear, full mortise, average frequency									
3210	Steel base, 4-1/2" high, USP				Pr.	127			127	140
4000	Wide throw, average frequency, steel base, 4-1/2" x 6", USP					97			97	107
4200	High frequency, steel base, 4-1/2" x 6", USP					148			148	163
4440	US26D					89.50			89.50	98.50
4600	Spring hinge, single acting, 6" flange, steel				Ea.	54.50			54.50	60
4700	Brass					96			96	105
4900	Double acting, 6" flange, steel					97			97	107
4950	Brass					159			159	175
5000	T-strap, galvanized, 4"				Pr.	19.35			19.35	21.50
5010	6"					30.50			30.50	33.50
5020	8"					46			46	51
8000	Continuous hinges									
8010	Steel, piano, 2" x 72"	1 Carp	20	.400	Ea.	22	17.65		39.65	53
8020	Brass, piano, 1-1/16" x 30"		30	.267		8	11.75		19.75	28
8030	Acrylic, piano, 1-3/4" x 12"		40	.200		15	8.80		23.80	31
8040	Aluminum, door, standard duty, 7'		3	2.667		135	118		253	340
8050	Heavy duty, 7'		3	2.667		142	118		260	350
8060	8'		3	2.667		160	118		278	370
8070	Steel, door, heavy duty, 7'		3	2.667		200	118		318	415
8080	8'		3	2.667		250	118		368	470
8090	Stainless steel, door, heavy duty, 7'		3	2.667		260	118		378	480
8100	8'		3	2.667		280	118		398	505
9000	Continuous hinge, steel, full mortise, heavy duty, 96 inch		2	4		565	176		741	915

08 71 20.95 Kick Plates

	08 71 20.95 Kick Plates	Crew	Daily Output	Labor-Hours	Unit	Material	Labor	Equipment	Total	Total Incl O&P
0010	**KICK PLATES**									
0020	Stainless steel, .050, 16 ga, 8" x 28", US32	1 Carp	15	.533	Ea.	35	23.50		58.50	77
0030	8" x 30"		15	.533		36	23.50		59.50	78
0040	8" x 34"		15	.533		40	23.50		63.50	82.50
0050	10" x 28"		15	.533		67	23.50		90.50	112
0060	10" x 30"		15	.533		72	23.50		95.50	118
0070	10" x 34"		15	.533		81	23.50		104.50	128
0080	Mop/Kick, 4" x 28"		15	.533		30	23.50		53.50	71.50
0090	4" x 30"		15	.533		32	23.50		55.50	73.50
0100	4" x 34"		15	.533		36	23.50		59.50	78
0110	6" x 28"		15	.533		38	23.50		61.50	80.50
0120	6" x 30"		15	.533		41	23.50		64.50	83.50
0130	6" x 34"		15	.533		47	23.50		70.50	90
0500	Bronze, .050", 8" x 28"		15	.533		48.50	23.50		72	91.50
0510	8" x 30"		15	.533		63	23.50		86.50	108
0520	8" x 34"		15	.533		71	23.50		94.50	117
0530	10" x 28"		15	.533		73	23.50		96.50	119
0540	10" x 30"		15	.533		78	23.50		101.50	125
0550	10" x 34"		15	.533		89	23.50		112.50	137
0560	Mop/Kick, 4" x 28"		15	.533		31	23.50		54.50	72.50
0570	4" x 30"		15	.533		34	23.50		57.50	76
0580	4" x 34"		15	.533		35	23.50		58.50	77
0590	6" x 28"		15	.533		44	23.50		67.50	87

08 71 Door Hardware

08 71 20 – Hardware

08 71 20.95 Kick Plates		Crew	Daily Output	Labor-Hours	Unit	Material	Labor	Equipment	Total	Total Incl O&P
						2012 Bare Costs				
0600	6″ x 30″	1 Carp	15	.533	Ea.	50	23.50		73.50	93.50
0610	6″ x 34″		15	.533		54	23.50		77.50	98
1000	Acrylic, .125″, 8″ x 26″		15	.533		28	23.50		51.50	69.50
1010	8″ x 36″		15	.533		38	23.50		61.50	80.50
1020	8″ x 42″		15	.533		45	23.50		68.50	88
1030	10″ x 26″		15	.533		35	23.50		58.50	77
1040	10″ x 36″		15	.533		48	23.50		71.50	91.50
1050	10″ x 42″		15	.533		68.50	23.50		92	114
1060	Mop/Kick, 4″ x 26″		15	.533		17	23.50		40.50	57
1070	4″ x 36″		15	.533		24	23.50		47.50	65
1080	4″ x 42″		15	.533		27	23.50		50.50	68
1090	6″ x 26″		15	.533		23	23.50		46.50	64
1100	6″ x 36″		15	.533		34	23.50		57.50	76
1110	6″ x 42″		15	.533		39	23.50		62.50	81.50
1220	Brass, .050″, 8″ x 26″		15	.533		56	23.50		79.50	100
1230	8″ x 36″		15	.533		75	23.50		98.50	121
1240	8″ x 42″		15	.533		86	23.50		109.50	133
1250	10″ x 26″		15	.533		71	23.50		94.50	117
1260	10″ x 36″		15	.533		91	23.50		114.50	139
1270	10″ x 42″		15	.533		105	23.50		128.50	155
1320	Mop/Kick, 4″ x 26″		15	.533		28	23.50		51.50	69.50
1330	4″ x 36″		15	.533		39	23.50		62.50	81.50
1340	4″ x 42″		15	.533		44	23.50		67.50	87
1350	6″ x 26″		15	.533		38	23.50		61.50	80.50
1360	6″ x 36″		15	.533		48	23.50		71.50	91.50
1370	6″ x 42″		15	.533		55	23.50		78.50	99
1800	Aluminum, .050″, 8″ x 26″		15	.533		25	23.50		48.50	66
1810	8″ x 36″		15	.533		30.50	23.50		54	72
1820	8″ x 42″		15	.533		35.50	23.50		59	77.50
1830	10″ x 26″		15	.533		27.50	23.50		51	69
1840	10″ x 36″		15	.533		38	23.50		61.50	80.50
1850	10″ x 42″		15	.533		44.50	23.50		68	87.50
1860	Mop/Kick, 4″ x 26″		15	.533		11	23.50		34.50	50.50
1870	4″ x 36″		15	.533		15.50	23.50		39	55.50
1880	4″ x 42″		15	.533		17	23.50		40.50	57
1890	6″ x 26″		15	.533		16.50	23.50		40	56.50
1900	6″ x 36″		15	.533		23	23.50		46.50	64
1910	6″ x 42″		15	.533		27	23.50		50.50	68
9000	Minimum labor/equipment charge		6	1.333	Job		59		59	96.50

08 71 21 – Astragals

08 71 21.10 Exterior Mouldings, Astragals		Crew	Daily Output	Labor-Hours	Unit	Material	Labor	Equipment	Total	Total Incl O&P
0010	**EXTERIOR MOULDINGS, ASTRAGALS**									
0400	One piece, overlapping cadmium plated steel, flat, 3/16″ x 2″	1 Carp	90	.089	L.F.	3.78	3.92		7.70	10.60
0600	Prime coated steel, flat, 1/8″ x 3″		90	.089		4.95	3.92		8.87	11.90
0800	Stainless steel, flat, 3/32″ x 1-5/8″		90	.089		18.80	3.92		22.72	27
1000	Aluminum, flat, 1/8″ x 2″		90	.089		3.78	3.92		7.70	10.60
1200	Nail on, "T" extrusion		120	.067		1.05	2.94		3.99	6
1300	Vinyl bulb insert		105	.076		1.47	3.36		4.83	7.10
1600	Screw on, "T" extrusion		90	.089		5.45	3.92		9.37	12.45
1700	Vinyl insert		75	.107		3.65	4.70		8.35	11.70
2000	"L" extrusion, neoprene bulbs		75	.107		2.20	4.70		6.90	10.10
2100	Neoprene sponge insert		75	.107		6.35	4.70		11.05	14.70

08 71 Door Hardware

08 71 21 – Astragals

08 71 21.10 Exterior Mouldings, Astragals		Crew	Daily Output	Labor-Hours	Unit	Material	2012 Bare Costs Labor	Equipment	Total	Total Incl O&P
2200	Magnetic	1 Carp	75	.107	L.F.	10.50	4.70		15.20	19.25
2400	Spring hinged security seal, with cam		75	.107		6.75	4.70		11.45	15.15
2600	Spring loaded locking bolt, vinyl insert		45	.178		9.10	7.85		16.95	23
2800	Neoprene sponge strip, "Z" shaped, aluminum		60	.133		4.45	5.90		10.35	14.55
2900	Solid neoprene strip, nail on aluminum strip		90	.089		3.65	3.92		7.57	10.45
3000	One piece stile protection									
3020	Neoprene fabric loop, nail on aluminum strips	1 Carp	60	.133	L.F.	.95	5.90		6.85	10.70
3110	Flush mounted aluminum extrusion, 1/2" x 1-1/4"		60	.133		3.51	5.90		9.41	13.50
3140	3/4" x 1-3/8"		60	.133		4.46	5.90		10.36	14.55
3160	1-1/8" x 1-3/4"		60	.133		7.05	5.90		12.95	17.40
3300	Mortise, 9/16" x 3/4"		60	.133		3.80	5.90		9.70	13.85
3320	13/16" x 1-3/8"		60	.133		4.10	5.90		10	14.15
3600	Spring bronze strip, nail on type		105	.076		3.30	3.36		6.66	9.15
3620	Screw on, with retainer		75	.107		2.65	4.70		7.35	10.60
3800	Flexible stainless steel housing, pile insert, 1/2" door		105	.076		7.20	3.36		10.56	13.40
3820	3/4" door		105	.076		8.05	3.36		11.41	14.35
4000	Extruded aluminum retainer, flush mount, pile insert		105	.076		2.55	3.36		5.91	8.30
4080	Mortise, felt insert		90	.089		4.50	3.92		8.42	11.40
4160	Mortise with spring, pile insert		90	.089		3.57	3.92		7.49	10.40
4400	Rigid vinyl retainer, mortise, pile insert		105	.076		2.55	3.36		5.91	8.30
4600	Wool pile filler strip, aluminum backing		105	.076		2.60	3.36		5.96	8.35
5000	Two piece overlapping astragal, extruded aluminum retainer									
5010	Pile insert	1 Carp	60	.133	L.F.	3.30	5.90		9.20	13.30
5020	Vinyl bulb insert		60	.133		2.20	5.90		8.10	12.05
5040	Vinyl flap insert		60	.133		6.55	5.90		12.45	16.85
5060	Solid neoprene flap insert		60	.133		6.50	5.90		12.40	16.80
5080	Hypalon rubber flap insert		60	.133		6.60	5.90		12.50	16.90
5090	Snap on cover, pile insert		60	.133		7.55	5.90		13.45	17.95
5400	Magnetic aluminum, surface mounted		60	.133		25	5.90		30.90	37
5500	Interlocking aluminum, 5/8" x 1" neoprene bulb insert		45	.178		4.15	7.85		12	17.40
5600	Adjustable aluminum, 9/16" x 21/32", pile insert		45	.178		19.15	7.85		27	34
5790	For vinyl bulb, deduct					.70			.70	.77
5800	Magnetic, adjustable, 9/16" x 21/32"	1 Carp	45	.178		24.50	7.85		32.35	40
6000	Two piece stile protection									
6010	Cloth backed rubber loop, 1" gap, nail on aluminum strips	1 Carp	45	.178	L.F.	4.30	7.85		12.15	17.60
6040	Screw on aluminum strips		45	.178		6.50	7.85		14.35	20
6100	1-1/2" gap, screw on aluminum extrusion		45	.178		5.80	7.85		13.65	19.25
6240	Vinyl fabric loop, slotted aluminum extrusion, 1" gap		45	.178		2.15	7.85		10	15.20
6300	1-1/4" gap		45	.178		6.15	7.85		14	19.60

08 71 25 – Weatherstripping

08 71 25.10 Mechanical Seals, Weatherstripping		Crew	Daily Output	Labor-Hours	Unit	Material	Labor	Equipment	Total	Total Incl O&P
0010	**MECHANICAL SEALS, WEATHERSTRIPPING**									
1000	Doors, wood frame, interlocking, for 3' x 7' door, zinc	1 Carp	3	2.667	Opng.	20	118		138	215
1100	Bronze		3	2.667		29	118		147	225
1300	6' x 7' opening, zinc		2	4		20	176		196	310
1400	Bronze		2	4		36	176		212	330
1700	Wood frame, spring type, bronze									
1800	3' x 7' door	1 Carp	7.60	1.053	Opng.	19.70	46.50		66.20	97.50
1900	6' x 7' door	"	7	1.143	"	29	50.50		79.50	114
2200	Metal frame, spring type, bronze									
2300	3' x 7' door	1 Carp	3	2.667	Opng.	38.50	118		156.50	235
2400	6' x 7' door	"	2.50	3.200	"	49	141		190	285

08 71 Door Hardware

08 71 25 – Weatherstripping

08 71 25.10 Mechanical Seals, Weatherstripping		Crew	Daily Output	Labor-Hours	Unit	2012 Bare Costs Material	Labor	Equipment	Total	Total Incl O&P
2500	For stainless steel, spring type, add					133%				
2700	Metal frame, extruded sections, 3′ x 7′ door, aluminum	1 Carp	3	2.667	Opng.	46	118		164	244
2800	Bronze		3	2.667		120	118		238	325
3100	6′ x 7′ door, aluminum		1.50	5.333		58	235		293	450
3200	Bronze	↓	1.50	5.333	↓	137	235		372	535
3500	Threshold weatherstripping									
3650	Door sweep, flush mounted, aluminum	1 Carp	25	.320	Ea.	15	14.10		29.10	39.50
3700	Vinyl		25	.320		18	14.10		32.10	43
5000	Garage door bottom weatherstrip, 12′ aluminum, clear		14	.571		22	25		47	66
5010	Bronze		14	.571		86	25		111	136
5050	Bottom protection, Rubber		14	.571		31	25		56	75.50
5100	Threshold		14	.571	↓	92.50	25		117.50	144
9000	Minimum labor/equipment charge	↓	3	2.667	Job		118		118	193

08 74 Access Control Hardware

08 74 13 – Card Key Access Control Hardware

08 74 13.50 Card Key Access

		Crew	Daily Output	Labor-Hours	Unit	Material	Labor	Equipment	Total	Total Incl O&P
0010	**CARD KEY ACCESS**, does not include door hardware									
0020	Card type, 1 time zone, minimum				Ea.	665			665	730
0040	Maximum					1,775			1,775	1,950
0060	3 time zones, minimum					2,200			2,200	2,425
0080	Maximum				↓	2,450			2,450	2,700
0100	System with printer, and control console, 3 zones				Total	9,150			9,150	10,100
0120	6 zones				"	12,000			12,000	13,200
0140	For each door, minimum, add				Ea.	1,500			1,500	1,650
0160	Maximum, add				"	2,100			2,100	2,300

08 74 19 – Biometric Identity Access Control Hardware

08 74 19.50 Biometric Identity Access

		Crew	Daily Output	Labor-Hours	Unit	Material	Labor	Equipment	Total	Total Incl O&P
0010	**BIOMETRIC IDENTITY ACCESS**									
0220	Hand geometry scanner, mem of 512 users, excl striker/power	1 Elec	3	2.667	Ea.	2,100	137		2,237	2,500
0230	Memory upgrade for, adds 9,700 user profiles		8	1		300	51.50		351.50	410
0240	Adds 32,500 user profiles		8	1		600	51.50		651.50	740
0250	Prison type, memory of 256 users, excl striker, power		3	2.667		2,600	137		2,737	3,050
0260	Memory upgrade for, adds 3,300 user profiles		8	1		250	51.50		301.50	355
0270	Adds 9,700 user profiles		8	1		460	51.50		511.50	585
0280	Adds 27,900 user profiles		8	1		610	51.50		661.50	750
0290	All weather, mem of 512 users, excl striker/power		3	2.667		3,900	137		4,037	4,500
0300	Facial & fingerprint scanner, combination unit, excl striker/power		3	2.667		4,200	137		4,337	4,825
0310	Access for, for initial setup, excl striker/power	↓	3	2.667	↓	1,100	137		1,237	1,400

08 75 Window Hardware

08 75 10 – Window Handles and Latches

	08 75 10.10 Handles and Latches	Crew	Daily Output	Labor-Hours	Unit	Material	Labor	Equipment	Total	Total Incl O&P
						2012 Bare Costs				
0010	**HANDLES AND LATCHES**									
1000	Handles, surface mounted, aluminum	1 Carp	24	.333	Ea.	4.13	14.70		18.83	28.50
1020	Brass		24	.333		7.50	14.70		22.20	32.50
1040	Chrome		24	.333		6.30	14.70		21	31
1500	Recessed, aluminum		12	.667		2.29	29.50		31.79	50.50
1520	Brass		12	.667		2.78	29.50		32.28	51
1540	Chrome		12	.667		2.34	29.50		31.84	50.50
2000	Latches, aluminum		20	.400		3.06	17.65		20.71	32.50
2020	Brass		20	.400		4.07	17.65		21.72	33.50
2040	Chrome		20	.400		2.55	17.65		20.20	32
9000	Minimum labor/equipment charge		6	1.333	Job		59		59	96.50

08 75 30 – Weatherstripping

08 75 30.10 Mechanical Weather Seals

		Crew	Daily Output	Labor-Hours	Unit	Material	Labor	Equipment	Total	Total Incl O&P
0010	**MECHANICAL WEATHER SEALS**, Window, double hung, 3' X 5'									
0020	Zinc	1 Carp	7.20	1.111	Opng.	13	49		62	95
0100	Bronze		7.20	1.111		28	49		77	112
0200	Vinyl V strip		7	1.143		7	50.50		57.50	90
0500	As above but heavy duty, zinc		4.60	1.739		20	76.50		96.50	148
0600	Bronze		4.60	1.739		35	76.50		111.50	165
9000	Minimum labor/equipment charge	1 Clab	4.60	1.739	Job		61		61	100

08 79 Hardware Accessories

08 79 20 – Door Accessories

08 79 20.10 Door Hardware Accessories

		Crew	Daily Output	Labor-Hours	Unit	Material	Labor	Equipment	Total	Total Incl O&P
0010	**DOOR HARDWARE ACCESSORIES**									
0140	Door bolt, surface, 4"	1 Carp	32	.250	Ea.	11.70	11.05		22.75	31
0160	Door latch	"	12	.667	"	8.45	29.50		37.95	57.50
0200	Sliding closet door									
0220	Track and hanger, single	1 Carp	10	.800	Ea.	56	35.50		91.50	120
0240	Double		8	1		75	44		119	155
0260	Door guide, single		48	.167		25	7.35		32.35	39.50
0280	Double		48	.167		34	7.35		41.35	49.50
0600	Deadbolt and lock cover plate, brass or stainless steel		30	.267		28	11.75		39.75	50.50
0620	Hole cover plate, brass or chrome		35	.229		7.50	10.10		17.60	25
2240	Mortise lockset, passage, lever handle		9	.889		198	39		237	283
4000	Security chain, standard		18	.444		7.90	19.60		27.50	40.50

08 81 Glass Glazing

08 81 10 – Float Glass

08 81 10.10 Various Types and Thickness of Float Glass

		Crew	Daily Output	Labor-Hours	Unit	Material	Labor	Equipment	Total	Total Incl O&P
0010	**VARIOUS TYPES AND THICKNESS OF FLOAT GLASS**									
0020	3/16" Plain	2 Glaz	130	.123	S.F.	4.71	5.25		9.96	13.70
0200	Tempered, clear		130	.123		6.55	5.25		11.80	15.75
0300	Tinted		130	.123		7.45	5.25		12.70	16.70
0600	1/4" thick, clear, plain		120	.133		5.50	5.70		11.20	15.25
0700	Tinted		120	.133		8.80	5.70		14.50	18.90
0800	Tempered, clear		120	.133		8.25	5.70		13.95	18.30
0900	Tinted		120	.133		10.20	5.70		15.90	20.50
1600	3/8" thick, clear, plain		75	.213		9.60	9.10		18.70	25.50

08 81 Glass Glazing

08 81 10 – Float Glass

08 81 10.10 Various Types and Thickness of Float Glass		Crew	Daily Output	Labor-Hours	Unit	Material	Labor	Equipment	Total	Total Incl O&P
						2012 Bare Costs				
1700	Tinted	2 Glaz	75	.213	S.F.	15.10	9.10		24.20	31.50
1800	Tempered, clear		75	.213		15.85	9.10		24.95	32
1900	Tinted		75	.213		17.60	9.10		26.70	34
2200	1/2" thick, clear, plain		55	.291		16.25	12.45		28.70	38
2300	Tinted		55	.291		26.50	12.45		38.95	49
2400	Tempered, clear		55	.291		23.50	12.45		35.95	45.50
2500	Tinted		55	.291		24.50	12.45		36.95	46.50
2800	5/8" thick, clear, plain		45	.356		26.50	15.20		41.70	53.50
2900	Tempered, clear		45	.356		30	15.20		45.20	57.50
3200	3/4" thick, clear, plain		35	.457		34	19.55		53.55	69
3300	Tempered, clear		35	.457		39.50	19.55		59.05	75
3600	1" thick, clear, plain		30	.533		56.50	23		79.50	99.50
8900	For low emissivity coating for 3/16" & 1/4" only, add to above					18%				
9000	Minimum labor/equipment charge	1 Glaz	2	4	Job		171		171	276

08 81 17 – Fire Glass

08 81 17.10 Fire Resistant Glass

		Crew	Daily Output	Labor-Hours	Unit	Material	Labor	Equipment	Total	Total Incl O&P
0010	**FIRE RESISTANT GLASS**									
0020	Fire Glass Minimum	2 Glaz	40	.400	S.F.	26.50	17.10		43.60	56.50
0030	Mid Range		40	.400		67	17.10		84.10	102
0050	High End		40	.400		310	17.10		327.10	370

08 81 20 – Vision Panels

08 81 20.10 Full Vision

		Crew	Daily Output	Labor-Hours	Unit	Material	Labor	Equipment	Total	Total Incl O&P
0010	**FULL VISION**, window system with 3/4" glass mullions									
0020	Up to 10' high	H-2	130	.185	S.F.	60.50	7.40		67.90	78.50
0100	10' to 20' high, minimum		110	.218		64	8.75		72.75	85
0150	Average		100	.240		69	9.65		78.65	91.50
0200	Maximum		80	.300		78	12.05		90.05	105
9000	Minimum labor/equipment charge	1 Glaz	2	4	Job		171		171	276

08 81 25 – Glazing Variables

08 81 25.10 Applications of Glazing

		Crew	Daily Output	Labor-Hours	Unit	Material	Labor	Equipment	Total	Total Incl O&P
0010	**APPLICATIONS OF GLAZING** R088110-10									
0500	For high rise glazing, exterior, add per S.F. per story				S.F.					.49
0600	For glass replacement, add				"		100%			
0700	For gasket settings, add				L.F.	5.60			5.60	6.15
0900	For sloped glazing, add				S.F.		26%			
2000	Fabrication, polished edges, 1/4" thick				Inch	.52			.52	.57
2100	1/2" thick					1.25			1.25	1.38
2500	Mitered edges, 1/4" thick					1.25			1.25	1.38
2600	1/2" thick					2.05			2.05	2.26

08 81 30 – Insulating Glass

08 81 30.10 Reduce Heat Transfer Glass

		Crew	Daily Output	Labor-Hours	Unit	Material	Labor	Equipment	Total	Total Incl O&P
0010	**REDUCE HEAT TRANSFER GLASS**									
0015	2 lites 1/8" float, 1/2" thk under 15 S.F.									
0020	Clear R088110-10 G	2 Glaz	95	.168	S.F.	9.45	7.20		16.65	22
0100	Tinted G		95	.168		13.20	7.20		20.40	26
0200	2 lites 3/16" float, for 5/8" thk unit, 15 to 30 S.F., clear G		90	.178		13.20	7.60		20.80	27
0400	1" thk, dbl. glazed, 1/4" float, 30-70 S.F., clear G		75	.213		15.95	9.10		25.05	32.50
0500	Tinted G		75	.213		22.50	9.10		31.60	40
2000	Both lites, light & heat reflective G		85	.188		30	8.05		38.05	46
2500	Heat reflective, film inside, 1" thick unit, clear G		85	.188		26.50	8.05		34.55	42

08 81 Glass Glazing

08 81 30 – Insulating Glass

08 81 30.10 Reduce Heat Transfer Glass			Crew	Daily Output	Labor-Hours	Unit	2012 Bare Costs Material	Labor	Equipment	Total	Total Incl O&P
2600	Tinted	G	2 Glaz	85	.188	S.F.	27.50	8.05		35.55	43.50
3000	Film on weatherside, clear, 1/2" thick unit	G		95	.168		18.90	7.20		26.10	32.50
3100	5/8" thick unit	G		90	.178		18.70	7.60		26.30	33
3200	1" thick unit	G		85	.188		26	8.05		34.05	41.50
3350	Minimum	G	1 Glaz	50	.160		12.10	6.85		18.95	24.50
3360	Maximum	G	"	25	.320		12.85	13.70		26.55	36
3370	Reflective or tinted, add	G					4			4	4.40
5000	Spectrally selective film, on ext, blocks solar gain/allows 70% of light	G	2 Glaz	95	.168		12	7.20		19.20	25
9000	Minimum labor/equipment charge		1 Glaz	2	4	Job		171		171	276

08 81 40 – Plate Glass

08 81 40.10 Plate Glass		Crew	Daily Output	Labor-Hours	Unit	Material	Labor	Equipment	Total	Total Incl O&P
0010	**PLATE GLASS** Twin ground, polished,									
0020	3/16" thick	2 Glaz	100	.160	S.F.	4.98	6.85		11.83	16.55
0100	1/4" thick		94	.170		6.80	7.30		14.10	19.25
0200	3/8" thick		60	.267		11.80	11.40		23.20	31.50
0300	1/2" thick		40	.400		23	17.10		40.10	52.50

08 81 55 – Window Glass

08 81 55.10 Sheet Glass		Crew	Daily Output	Labor-Hours	Unit	Material	Labor	Equipment	Total	Total Incl O&P
0010	**SHEET GLASS** (window), clear float, stops, putty bed									
0015	1/8" thick, clear float	2 Glaz	480	.033	S.F.	3.47	1.42		4.89	6.10
0500	3/16" thick, clear		480	.033		5.50	1.42		6.92	8.35
0600	Tinted		480	.033		7.20	1.42		8.62	10.20
0700	Tempered		480	.033		8.80	1.42		10.22	12
2000	Replace broken window lite, 1/8" glass (9 S.F. maximum)	1 Glaz	48	.167		4.93	7.15		12.08	16.90
2100	1/4" plate (16 S.F. maximum)	"	48	.167		5.65	7.15		12.80	17.70
9000	Minimum labor/equipment charge	2 Glaz	5	3.200	Job		137		137	221

08 81 65 – Wire Glass

08 81 65.10 Glass Reinforced With Wire		Crew	Daily Output	Labor-Hours	Unit	Material	Labor	Equipment	Total	Total Incl O&P
0010	**GLASS REINFORCED WITH WIRE**									
0012	1/4" thick rough obscure	2 Glaz	135	.119	S.F.	21	5.05		26.05	31
1000	Polished wire, 1/4" thick, diamond, clear		135	.119		24.50	5.05		29.55	35
1500	Pinstripe, obscure		135	.119		30	5.05		35.05	41

08 83 Mirrors

08 83 13 – Mirrored Glass Glazing

08 83 13.10 Mirrors		Crew	Daily Output	Labor-Hours	Unit	Material	Labor	Equipment	Total	Total Incl O&P
0010	**MIRRORS**, No frames, wall type, 1/4" plate glass, polished edge									
0100	Up to 5 S.F.	2 Glaz	125	.128	S.F.	9.50	5.45		14.95	19.30
0200	Over 5 S.F.		160	.100		9.25	4.28		13.53	17.10
0500	Door type, 1/4" plate glass, up to 12 S.F.		160	.100		8.35	4.28		12.63	16.10
1000	Float glass, up to 10 S.F., 1/8" thick		160	.100		5.50	4.28		9.78	12.95
1100	3/16" thick		150	.107		6.75	4.56		11.31	14.80
1500	12" x 12" wall tiles, square edge, clear		195	.082		2	3.51		5.51	7.85
1600	Veined		195	.082		4.95	3.51		8.46	11.10
2000	1/4" thick, stock sizes, one way transparent		125	.128		18.45	5.45		23.90	29.50
2010	Bathroom, unframed, laminated		160	.100		13	4.28		17.28	21
2500	Tempered		160	.100		17.80	4.28		22.08	26.50

08 83 Mirrors

08 83 13 – Mirrored Glass Glazing

08 83 13.15 Reflective Glass		Crew	Daily Output	Labor-Hours	Unit	Material	2012 Bare Costs Labor	Equipment	Total	Total Incl O&P
0010	**REFLECTIVE GLASS**									
0100	1/4" float with fused metallic oxide fixed [G]	2 Glaz	115	.139	S.F.	16.30	5.95		22.25	27.50
0500	1/4" float glass with reflective applied coating [G]	"	115	.139	"	13.20	5.95		19.15	24

08 84 Plastic Glazing

08 84 10 – Plexiglass Glazing

08 84 10.10 Plexiglass Acrylic

Line	Description	Crew	Daily Output	Labor-Hours	Unit	Material	Labor	Equipment	Total	Total Incl O&P
0010	**PLEXIGLASS ACRYLIC**, clear, masked,									
0020	1/8" thick, cut sheets	2 Glaz	170	.094	S.F.	5.30	4.02		9.32	12.35
0200	Full sheets		195	.082		3.50	3.51		7.01	9.50
0500	1/4" thick, cut sheets		165	.097		9.20	4.15		13.35	16.80
0600	Full sheets		185	.086		6	3.70		9.70	12.55
0900	3/8" thick, cut sheets		155	.103		16.85	4.41		21.26	25.50
1000	Full sheets		180	.089		9.10	3.80		12.90	16.15
1300	1/2" thick, cut sheets		135	.119		19.40	5.05		24.45	29.50
1400	Full sheets		150	.107		18.80	4.56		23.36	28
1700	3/4" thick, cut sheets		115	.139		69	5.95		74.95	85.50
1800	Full sheets		130	.123		40	5.25		45.25	52.50
2100	1" thick, cut sheets		105	.152		77.50	6.50		84	96
2200	Full sheets		125	.128		48	5.45		53.45	62
3000	Colored, 1/8" thick, cut sheets		170	.094		16	4.02		20.02	24
3200	Full sheets		195	.082		10.30	3.51		13.81	17
3500	1/4" thick, cut sheets		165	.097		18	4.15		22.15	26.50
3600	Full sheets		185	.086		12.20	3.70		15.90	19.35
4000	Mirrors, untinted, cut sheets, 1/8" thick		185	.086		8.25	3.70		11.95	15.05
4200	1/4" thick		180	.089		13.20	3.80		17	20.50

08 84 20 – Polycarbonate

08 84 20.10 Thermoplastic

Line	Description	Crew	Daily Output	Labor-Hours	Unit	Material	Labor	Equipment	Total	Total Incl O&P
0010	**THERMOPLASTIC**, clear, masked, cut sheets									
0020	1/8" thick	2 Glaz	170	.094	S.F.	9.70	4.02		13.72	17.15
0500	3/16" thick		165	.097		11.10	4.15		15.25	18.90
1000	1/4" thick		155	.103		12.50	4.41		16.91	21
1500	3/8" thick		150	.107		22	4.56		26.56	32
9000	Minimum labor/equipment charge	1 Glaz	2	4	Job		171		171	276

08 85 Glazing Accessories

08 85 10 – Miscellaneous Glazing Accessories

08 85 10.10 Glazing Gaskets

Line	Description	Crew	Daily Output	Labor-Hours	Unit	Material	Labor	Equipment	Total	Total Incl O&P
0010	**GLAZING GASKETS**, Neoprene for glass tongued mullion									
0015	1/4" glass	2 Glaz	200	.080	L.F.	.68	3.42		4.10	6.25
0020	3/8"		200	.080		.87	3.42		4.29	6.45
0040	1/2"		200	.080		.98	3.42		4.40	6.60
0060	3/4"		180	.089		1.44	3.80		5.24	7.75
0080	1"		180	.089		1.16	3.80		4.96	7.45
1000	Glazing compound, wood	1 Glaz	58	.138		.90	5.90		6.80	10.50
1005	Glazing compound, per window, up to 30 L.F.					.90			.90	.99
1006	Glazing compound, per window, up to 30 L.F.				Ea.	27			27	29.50
1020	Metal	1 Glaz	58	.138	L.F.	.90	5.90		6.80	10.50

08 87 Glazing Surface Films

08 87 13 – Solar Control Films

08 87 13.10 Solar Films On Glass		Crew	Daily Output	Labor-Hours	Unit	2012 Bare Costs Material	Labor	Equipment	Total	Total Incl O&P
0010	**SOLAR FILMS ON GLASS** (glass not included)									
2000	Minimum G	2 Glaz	180	.089	S.F.	6.60	3.80		10.40	13.40
2050	Maximum G	"	225	.071	"	14.70	3.04		17.74	21

08 87 23 – Safety and Security Films

08 87 23.16 Security Films		Crew	Daily Output	Labor-Hours	Unit	2012 Bare Costs Material	Labor	Equipment	Total	Total Incl O&P
0010	**SECURITY FILMS**, clear, 32000 psi tensile strength, adhered to glass R088110-10									
0100	.002" thick, daylight installation	H-2	950	.025	S.F.	2.25	1.02		3.27	4.13
0150	.004" thick, daylight installation		800	.030		2.80	1.21		4.01	5.05
0200	.006" thick, daylight installation		700	.034		3	1.38		4.38	5.55
0210	Install for anchorage		600	.040		3.33	1.61		4.94	6.30
0400	.007" thick, daylight installation		600	.040		4	1.61		5.61	7
0410	Install for anchorage		500	.048		4.44	1.93		6.37	8
0500	.008" thick, daylight installation		500	.048		4.50	1.93		6.43	8.10
0510	Install for anchorage		500	.048		5	1.93		6.93	8.65
0600	.015" thick, daylight installation		400	.060		8	2.41		10.41	12.70
0610	Install for anchorage		400	.060		5	2.41		7.41	9.40
0900	Security film anchorage, mechanical attachment and cover plate	H-3	370	.043	L.F.	9.45	1.64		11.09	13.05
0950	Security film anchorage, wet glaze structural caulking	1 Glaz	225	.036	"	.84	1.52		2.36	3.38
1000	Adhered security film removal	1 Clab	275	.029	S.F.		1.02		1.02	1.67

08 88 Special Function Glazing

08 88 56 – Ballistics-Resistant Glazing

08 88 56.10 Laminated Glass		Crew	Daily Output	Labor-Hours	Unit	2012 Bare Costs Material	Labor	Equipment	Total	Total Incl O&P
0010	**LAMINATED GLASS**									
0020	Clear float .03" vinyl 1/4"	2 Glaz	90	.178	S.F.	11.75	7.60		19.35	25.50
0100	3/8" thick		78	.205		21.50	8.75		30.25	37.50
0200	.06" vinyl, 1/2" thick		65	.246		24.50	10.50		35	44
1000	5/8" thick		90	.178		28.50	7.60		36.10	44
2000	Bullet-resisting, 1-3/16" thick, to 15 S.F.		16	1		90	43		133	168
2100	Over 15 S.F.		16	1		104	43		147	183
2500	2-1/4" thick, to 15 S.F.		12	1.333		140	57		197	246
2600	Over 15 S.F.		12	1.333		130	57		187	235
2700	Level 2 (.357 magnum)		12	1.333		68	57		125	167
2750	Level 3A (.44 magnum)		12	1.333		73.50	57		130.50	173
2800	Level 4 (AK-47)		12	1.333		96	57		153	198
2850	Level 5 (M-16)		12	1.333		107	57		164	210
2900	Level 3 (7.62 Armor Piercing)		12	1.333		127	57		184	232

08 91 Louvers

08 91 19 – Fixed Louvers

08 91 19.10 Aluminum Louvers		Crew	Daily Output	Labor-Hours	Unit	2012 Bare Costs Material	Labor	Equipment	Total	Total Incl O&P
0010	**ALUMINUM LOUVERS**									
0020	Aluminum with screen, residential, 8" x 8"	1 Carp	30	.211	Ea.	13.50	9.30		22.80	30
0100	12" x 12"		38	.211		15	9.30		24.30	31.50
0200	12" x 18"		35	.229		19	10.10		29.10	37.50
0250	14" x 24"		30	.267		26	11.75		37.75	48
0300	18" x 24"		27	.296		29	13.05		42.05	53.50
0500	24" x 30"		24	.333		40	14.70		54.70	68
0700	Triangle, adjustable, small		20	.400		37.50	17.65		55.15	70.50

08 91 Louvers

08 91 19 – Fixed Louvers

08 91 19.10 Aluminum Louvers		Crew	Daily Output	Labor-Hours	Unit	Material	2012 Bare Costs Labor	Equipment	Total	Total Incl O&P
0800	Large	1 Carp	15	.533	Ea.	46.50	23.50		70	89.50
2100	Midget, aluminum, 3/4" deep, 1" diameter		85	.094		.72	4.15		4.87	7.60
2150	3" diameter		60	.133		2	5.90		7.90	11.85
2200	4" diameter		50	.160		3.64	7.05		10.69	15.55
2250	6" diameter		30	.267		3.85	11.75		15.60	23.50

08 91 26 – Door Louvers

08 91 26.10 Steel Louvers, 18 Gauge, Fixed Blade

Line	Description	Crew	Daily Output	Labor-Hours	Unit	Material	Labor	Equipment	Total	Total Incl O&P
0010	**STEEL LOUVERS, 18 GAUGE, FIXED BLADE**									
0050	12" x 12", with enamel or powder coat	1 Carp	20	.400	Ea.	69	17.65		86.65	105
0055	18" x 12"		20	.400		83	17.65		100.65	121
0060	18" x 18"		20	.400		84	17.65		101.65	122
0065	24" x 12"		20	.400		92	17.65		109.65	130
0070	24" x 18"		20	.400		126	17.65		143.65	168
0075	24" x 24"		20	.400		131	17.65		148.65	173
0100	12" x 12", galvanized		20	.400		60	17.65		77.65	95
0105	18" x 12"		20	.400		67	17.65		84.65	103
0115	24" x 12"		20	.400		83	17.65		100.65	121
0125	24" x 24"		20	.400		121	17.65		138.65	162

08 95 Vents

08 95 13 – Soffit Vents

08 95 13.10 Wall Louvers

Line	Description	Crew	Daily Output	Labor-Hours	Unit	Material	Labor	Equipment	Total	Total Incl O&P
0010	**WALL LOUVERS**									
2330	Soffit vent, continuous, 3" wide, aluminum, mill finish	1 Carp	200	.040	L.F.	.52	1.76		2.28	3.46
2340	Baked enamel finish		200	.040	"	.57	1.76		2.33	3.52
2400	Under eaves vent, aluminum, mill finish, 16" x 4"		48	.167	Ea.	1.84	7.35		9.19	14.05
2500	16" x 8"		48	.167	"	2.14	7.35		9.49	14.40

08 95 16 – Wall Vents

08 95 16.10 Louvers

Line	Description	Crew	Daily Output	Labor-Hours	Unit	Material	Labor	Equipment	Total	Total Incl O&P
0010	**LOUVERS**									
0020	Redwood, 2'-0" diameter, full circle	1 Carp	16	.500	Ea.	174	22		196	227
0100	Half circle		16	.500		192	22		214	247
0200	Octagonal		16	.500		132	22		154	181
0300	Triangular, 5/12 pitch, 5'-0" at base		16	.500		194	22		216	249
7000	Vinyl gable vent, 8" x 8"		38	.211		12.60	9.30		21.90	29
7020	12" x 12"		38	.211		24	9.30		33.30	41.50
7080	12" x 18"		35	.229		31	10.10		41.10	50.50
7200	18" x 24"		30	.267		38	11.75		49.75	61.50
9000	Minimum labor/equipment charge		3.50	2.286	Job		101		101	165

Estimating Tips

General

- Room Finish Schedule: A complete set of plans should contain a room finish schedule. If one is not available, it would be well worth the time and effort to obtain one.

09 20 00 Plaster and Gypsum Board

- Lath is estimated by the square yard plus a 5% allowance for waste. Furring, channels, and accessories are measured by the linear foot. An extra foot should be allowed for each accessory miter or stop.
- Plaster is also estimated by the square yard. Deductions for openings vary by preference, from zero deduction to 50% of all openings over 2 feet in width. The estimator should allow one extra square foot for each linear foot of horizontal interior or exterior angle located below the ceiling level. Also, double the areas of small radius work.
- Drywall accessories, studs, track, and acoustical caulking are all measured by the linear foot. Drywall taping is figured by the square foot. Gypsum wallboard is estimated by the square foot. No material deductions should be made for door or window openings under 32 S.F.

09 60 00 Flooring

- Tile and terrazzo areas are taken off on a square foot basis. Trim and base materials are measured by the linear foot. Accent tiles are listed per each. Two basic methods of installation are used. Mud set is approximately 30% more expensive than thin set. In terrazzo work, be sure to include the linear footage of embedded decorative strips, grounds, machine rubbing, and power cleanup.
- Wood flooring is available in strip, parquet, or block configuration. The latter two types are set in adhesives with quantities estimated by the square foot. The laying pattern will influence labor costs and material waste. In addition to the material and labor for laying wood floors, the estimator must make allowances for sanding and finishing these areas unless the flooring is prefinished.
- Sheet flooring is measured by the square yard. Roll widths vary, so consideration should be given to use the most economical width, as waste must be figured into the total quantity. Consider also the installation methods available, direct glue down or stretched.

09 70 00 Wall Finishes

- Wall coverings are estimated by the square foot. The area to be covered is measured, length by height of wall above baseboards, to calculate the square footage of each wall. This figure is divided by the number of square feet in the single roll which is being used. Deduct, in full, the areas of openings such as doors and windows. Where a pattern match is required allow 25%–30% waste.

09 80 00 Acoustic Treatment

- Acoustical systems fall into several categories. The takeoff of these materials should be by the square foot of area with a 5% allowance for waste. Do not forget about scaffolding, if applicable, when estimating these systems.

09 90 00 Painting and Coating

- A major portion of the work in painting involves surface preparation. Be sure to include cleaning, sanding, filling, and masking costs in the estimate.
- Protection of adjacent surfaces is not included in painting costs. When considering the method of paint application, an important factor is the amount of protection and masking required. These must be estimated separately and may be the determining factor in choosing the method of application.

Reference Numbers

Reference numbers are shown in shaded boxes at the beginning of some major classifications. These numbers refer to related items in the Reference Section. The reference information may be an estimating procedure, an alternate pricing method, or technical information.

Note: Not all subdivisions listed here necessarily appear in this publication.

09 01 Maintenance of Finishes

09 01 70 – Maintenance of Wall Finishes

09 01 70.10 Gypsum Wallboard Repairs		Crew	Daily Output	Labor-Hours	Unit	2012 Bare Costs Material	Labor	Equipment	Total	Total Incl O&P
0010	**GYPSUM WALLBOARD REPAIRS**									
0100	Fill and sand, pin/nail holes	1 Carp	960	.008	Ea.		.37		.37	.60
0110	Screw head pops		480	.017			.74		.74	1.21
0120	Dents, up to 2″ square		48	.167		.01	7.35		7.36	12.05
0130	2″ to 4″ square		24	.333		.03	14.70		14.73	24
0140	Cut square, patch, sand and finish, holes, up to 2″ square		12	.667		.04	29.50		29.54	48
0150	2″ to 4″ square		11	.727		.09	32		32.09	52.50
0160	4″ to 8″ square		10	.800		.21	35.50		35.71	58
0170	8″ to 12″ square		8	1		.40	44		44.40	73
0180	12″ to 32″ square		6	1.333		1.31	59		60.31	98
0210	16″ by 48″		5	1.600		2.17	70.50		72.67	118
0220	32″ by 48″		4	2		3.20	88		91.20	149
0230	48″ square		3.50	2.286		4.48	101		105.48	170
0240	60″ square		3.20	2.500		7.60	110		117.60	189
0500	Skim coat surface with joint compound		1600	.005	S.F.	.03	.22		.25	.39
0510	Prepare, retape and refinish joints		60	.133	L.F.	.59	5.90		6.49	10.30
9000	Minimum labor/equipment charge		2	4	Job		176		176	289

09 05 Common Work Results for Finishes

09 05 05 – Selective Finishes Demolition

09 05 05.10 Selective Demolition, Ceilings

Line	Description	Crew	Daily Output	Labor-Hours	Unit	Material	Labor	Equipment	Total	Total Incl O&P
0010	**SELECTIVE DEMOLITION, CEILINGS** R024119-10									
0200	Ceiling, drywall, furred and nailed or screwed	2 Clab	800	.020	S.F.		.70		.70	1.15
0220	On metal frame		760	.021			.74		.74	1.21
0240	On suspension system, including system		720	.022			.78		.78	1.28
1000	Plaster, lime and horse hair, on wood lath, incl. lath		700	.023			.80		.80	1.32
1020	On metal lath		570	.028			.99		.99	1.62
1100	Gypsum, on gypsum lath		720	.022			.78		.78	1.28
1120	On metal lath		500	.032			1.12		1.12	1.84
1200	Suspended ceiling, mineral fiber, 2′ x 2′ or 2′ x 4′		1500	.011			.37		.37	.61
1250	On suspension system, incl. system		1200	.013			.47		.47	.77
1500	Tile, wood fiber, 12″ x 12″, glued		900	.018			.62		.62	1.02
1540	Stapled		1500	.011			.37		.37	.61
1580	On suspension system, incl. system		760	.021			.74		.74	1.21
2000	Wood, tongue and groove, 1″ x 4″		1000	.016			.56		.56	.92
2040	1″ x 8″		1100	.015			.51		.51	.84
2400	Plywood or wood fiberboard, 4′ x 8′ sheets		1200	.013			.47		.47	.77
9000	Minimum labor/equipment charge	1 Clab	2	4	Job		140		140	230

09 05 05.20 Selective Demolition, Flooring

Line	Description	Crew	Daily Output	Labor-Hours	Unit	Material	Labor	Equipment	Total	Total Incl O&P
0010	**SELECTIVE DEMOLITION, FLOORING** R024119-10									
0200	Brick with mortar	2 Clab	475	.034	S.F.		1.18		1.18	1.94
0400	Carpet, bonded, including surface scraping		2000	.008			.28		.28	.46
0480	Tackless		9000	.002			.06		.06	.10
0550	Carpet tile, releasable adhesive		5000	.003			.11		.11	.18
0560	Permanent adhesive		1850	.009			.30		.30	.50
0600	Composition, acrylic or epoxy		400	.040			1.40		1.40	2.30
0700	Concrete, scarify skin	A-1A	225	.036			1.61	.94	2.55	3.65
0800	Resilient, sheet goods	2 Clab	1400	.011			.40		.40	.66
0820	For gym floors	″	900	.018			.62		.62	1.02
0850	Vinyl or rubber cove base	1 Clab	1000	.008	L.F.		.28		.28	.46
0860	Vinyl or rubber cove base, molded corner	″	1000	.008	Ea.		.28		.28	.46

09 05 05 – Selective Finishes Demolition

09 05 05.20 Selective Demolition, Flooring

	09 05 05.20 Selective Demolition, Flooring	Crew	Daily Output	Labor-Hours	Unit	Material	2012 Bare Costs Labor	Equipment	Total	Total Incl O&P
0870	For glued and caulked installation, add to labor						50%			
0900	Vinyl composition tile, 12" x 12"	2 Clab	1000	.016	S.F.		.56		.56	.92
2000	Tile, ceramic, thin set		675	.024			.83		.83	1.36
2020	Mud set		625	.026			.90		.90	1.47
2200	Marble, slate, thin set		675	.024			.83		.83	1.36
2220	Mud set		625	.026			.90		.90	1.47
2600	Terrazzo, thin set		450	.036			1.25		1.25	2.05
2620	Mud set		425	.038			1.32		1.32	2.17
2640	Terrazzo, cast in place		300	.053			1.87		1.87	3.07
3000	Wood, block, on end	1 Carp	400	.020			.88		.88	1.45
3200	Parquet		450	.018			.78		.78	1.29
3400	Strip flooring, interior, 2-1/4" x 25/32" thick		325	.025			1.09		1.09	1.78
3500	Exterior, porch flooring, 1" x 4"		220	.036			1.60		1.60	2.63
3800	Subfloor, tongue and groove, 1" x 6"		325	.025			1.09		1.09	1.78
3820	1" x 8"		430	.019			.82		.82	1.34
3840	1" x 10"		520	.015			.68		.68	1.11
4000	Plywood, nailed		600	.013			.59		.59	.96
4100	Glued and nailed		400	.020			.88		.88	1.45
4200	Hardboard, 1/4" thick		760	.011			.46		.46	.76
8000	Remove flooring, bead blast, minimum	A-1A	1000	.008			.36	.21	.57	.82
8100	Maximum		400	.020			.91	.53	1.44	2.06
8150	Mastic only		1500	.005			.24	.14	.38	.54
9000	Minimum labor/equipment charge	1 Clab	4	2	Job		70		70	115

09 05 05.30 Selective Demolition, Walls and Partitions

	09 05 05.30 Selective Demolition, Walls and Partitions	Crew	Daily Output	Labor-Hours	Unit	Material	2012 Bare Costs Labor	Equipment	Total	Total Incl O&P
0010	**SELECTIVE DEMOLITION, WALLS AND PARTITIONS** R024119-10									
0020	Walls, concrete, reinforced	B-39	120	.400	C.F.		14.80	1.74	16.54	26
0025	Plain	"	160	.300			11.10	1.31	12.41	19.55
0100	Brick, 4" to 12" thick	B-9	220	.182			6.45	.95	7.40	11.65
0200	Concrete block, 4" thick		1000	.040	S.F.		1.42	.21	1.63	2.56
0280	8" thick		810	.049			1.75	.26	2.01	3.15
0300	Exterior stucco 1" thick over mesh		3200	.013			.44	.07	.51	.80
1000	Drywall, nailed or screwed	1 Clab	1000	.008			.28		.28	.46
1010	2 layers		400	.020			.70		.70	1.15
1020	Glued and nailed		900	.009			.31		.31	.51
1500	Fiberboard, nailed		900	.009			.31		.31	.51
1520	Glued and nailed		800	.010			.35		.35	.58
1568	Plenum barrier, sheet lead		300	.027			.94		.94	1.53
2000	Movable walls, metal, 5' high		300	.027			.94		.94	1.53
2020	8' high		400	.020			.70		.70	1.15
2200	Metal or wood studs, finish 2 sides, fiberboard	B-1	520	.046			1.65		1.65	2.71
2250	Lath and plaster		260	.092			3.30		3.30	5.40
2300	Plasterboard (drywall)		520	.046			1.65		1.65	2.71
2350	Plywood		450	.053			1.91		1.91	3.13
2800	Paneling, 4' x 8' sheets	1 Clab	475	.017			.59		.59	.97
3000	Plaster, lime and horsehair, on wood lath		400	.020			.70		.70	1.15
3020	On metal lath		335	.024			.84		.84	1.37
3400	Gypsum or perlite, on gypsum lath		410	.020			.68		.68	1.12
3420	On metal lath		300	.027			.94		.94	1.53
3450	Plaster, interior gypsum, acoustic, or cement		60	.133	S.Y.		4.68		4.68	7.65
3500	Stucco, on masonry		145	.055			1.94		1.94	3.18
3510	Commercial 3-coat		80	.100			3.51		3.51	5.75
3520	Interior stucco		25	.320			11.25		11.25	18.40

09 05 Common Work Results for Finishes

09 05 05 – Selective Finishes Demolition

09 05 05.30 Selective Demolition, Walls and Partitions		Crew	Daily Output	Labor-Hours	Unit	Material	2012 Bare Costs Labor	Equipment	Total	Total Incl O&P
3760	Tile, ceramic, on walls, thin set	1 Clab	300	.027	S.F.		.94		.94	1.53
3765	Mud set		250	.032	"		1.12		1.12	1.84
3800	Toilet partitions, slate or marble		5	1.600	Ea.		56		56	92
3820	Metal or plastic	↓	8	1	"		35		35	57.50
5000	Wallcovering, vinyl	1 Pape	700	.011	S.F.		.44		.44	.70
5010	With release agent		1500	.005			.21		.21	.33
5025	Wallpaper, 2 layers or less, by hand		250	.032			1.23		1.23	1.97
5035	3 layers or more		165	.048			1.87		1.87	2.98
5040	Designer	↓	480	.017	↓		.64		.64	1.02
9000	Minimum labor/equipment charge	1 Clab	4	2	Job		70		70	115

09 21 Plaster and Gypsum Board Assemblies

09 21 16 – Gypsum Board Assemblies

09 21 16.23 Gypsum Board Shaft Wall Assemblies		Crew	Daily Output	Labor-Hours	Unit	Material	Labor	Equipment	Total	Total Incl O&P
0010	**GYPSUM BOARD SHAFT WALL ASSEMBLIES**									
0020	Cavity type on 25 ga. J-track & C-H studs, 24" O.C.									
0030	1" thick coreboard wall liner on shaft side									
0040	2-hour assembly with double layer									
0060	5/8" fire rated gypsum board on room side	2 Carp	220	.073	S.F.	1.54	3.21		4.75	6.95
0100	3-hour assembly with triple layer									
0300	5/8" fire rated gypsum board on room side	2 Carp	180	.089	S.F.	1.29	3.92		5.21	7.85
0400	4-hour assembly, 1" coreboard, 5/8" fire rated gypsum board									
0600	and 3/4" galv. metal furring channels, 24" O.C., with									
0700	Double layer 5/8" fire rated gypsum board on room side	2 Carp	110	.145	S.F.	1.25	6.40		7.65	11.90
0900	For taping & finishing, add per side	1 Carp	1050	.008	"	.05	.34		.39	.60
1000	For insulation, see Section 07 21									
5200	For work over 8' high, add	2 Carp	3060	.005	S.F.		.23		.23	.38
5300	For distribution cost over 3 stories high, add per story	"	6100	.003	"		.12		.12	.19

09 21 16.33 Partition Wall		Crew	Daily Output	Labor-Hours	Unit	Material	Labor	Equipment	Total	Total Incl O&P
0010	**PARTITION WALL** Stud wall, 8' to 12' high									
0050	1/2", interior, gypsum board, std, tape & finish 2 sides									
0500	Installed on and incl., 2" x 4" wood studs, 16" O.C.	2 Carp	310	.052	S.F.	.90	2.28		3.18	4.72
1000	Metal studs, NLB, 25 ga., 16" O.C., 3-5/8" wide		350	.046		.90	2.02		2.92	4.29
1200	6" wide		330	.048		1.08	2.14		3.22	4.70
1400	Water resistant, on 2" x 4" wood studs, 16" O.C.		310	.052		1.16	2.28		3.44	5
1600	Metal studs, NLB, 25 ga., 16" O.C., 3-5/8" wide		350	.046		1.16	2.02		3.18	4.58
1800	6" wide		330	.048		1.34	2.14		3.48	4.99
2000	Fire res., 2 layers, 1-1/2 hr., on 2" x 4" wood studs, 16" O.C.		210	.076		1.50	3.36		4.86	7.15
2200	Metal studs, NLB, 25 ga., 16" O.C., 3-5/8" wide		250	.064		1.50	2.82		4.32	6.30
2400	6" wide		230	.070		1.68	3.07		4.75	6.90
2600	Fire & water res., 2 layers, 1-1/2 hr., 2" x 4" studs, 16" O.C.		210	.076		1.50	3.36		4.86	7.15
2800	Metal studs, NLB, 25 ga., 16" O.C., 3-5/8" wide		250	.064		1.50	2.82		4.32	6.30
3000	6" wide	↓	230	.070	↓	1.68	3.07		4.75	6.90
3200	5/8", interior, gypsum board, standard, tape & finish 2 sides									
3400	Installed on and including 2" x 4" wood studs, 16" O.C.	2 Carp	300	.053	S.F.	.96	2.35		3.31	4.92
3600	24" O.C.		330	.048		.88	2.14		3.02	4.48
3800	Metal studs, NLB, 25 ga., 16" O.C., 3-5/8" wide		340	.047		.96	2.08		3.04	4.46
4000	6" wide		320	.050		1.14	2.21		3.35	4.88
4200	24" O.C., 3-5/8" wide		360	.044		.86	1.96		2.82	4.16
4400	6" wide		340	.047		1	2.08		3.08	4.50
4800	Water resistant, on 2" x 4" wood studs, 16" O.C.	↓	300	.053	↓	1.32	2.35		3.67	5.30

09 21 Plaster and Gypsum Board Assemblies

09 21 16 – Gypsum Board Assemblies

09 21 16.33 Partition Wall		Crew	Daily Output	Labor-Hours	Unit	Material	Labor	Equipment	Total	Total Incl O&P
						2012 Bare Costs				
5000	24″ O.C.	2 Carp	330	.048	S.F.	1.24	2.14		3.38	4.87
5200	Metal studs, NLB, 25 ga. 16″ O.C., 3-5/8″ wide		340	.047		1.32	2.08		3.40	4.85
5400	6″ wide		320	.050		1.50	2.21		3.71	5.25
5600	24″ O.C., 3-5/8″ wide		360	.044		1.22	1.96		3.18	4.56
5800	6″ wide		340	.047		1.36	2.08		3.44	4.89
6000	Fire resistant, 2 layers, 2 hr., on 2″ x 4″ wood studs, 16″ O.C.		205	.078		1.46	3.44		4.90	7.25
6200	24″ O.C.		235	.068		1.46	3		4.46	6.50
6400	Metal studs, NLB, 25 ga., 16″ O.C., 3-5/8″ wide		245	.065		1.59	2.88		4.47	6.45
6600	6″ wide		225	.071		1.72	3.14		4.86	7.05
6800	24″ O.C., 3-5/8″ wide		265	.060		1.44	2.66		4.10	5.95
7000	6″ wide		245	.065		1.58	2.88		4.46	6.45
7200	Fire & water resistant, 2 layers, 2 hr., 2″ x 4″ studs, 16″ O.C.		205	.078		1.54	3.44		4.98	7.35
7400	24″ O.C.		235	.068		1.46	3		4.46	6.50
7600	Metal studs, NLB, 25 ga., 16″ O.C., 3-5/8″ wide		245	.065		1.54	2.88		4.42	6.40
7800	6″ wide		225	.071		1.72	3.14		4.86	7.05
8000	24″ O.C., 3-5/8″ wide		265	.060		1.44	2.66		4.10	5.95
8200	6″ wide	↓	245	.065	↓	1.58	2.88		4.46	6.45
8600	1/2″ blueboard, mesh tape both sides									
8620	Installed on and including 2″ x 4″ wood studs, 16″ O.C.	2 Carp	300	.053	S.F.	.96	2.35		3.31	4.92
8640	Metal studs, NLB, 25 ga., 16″ O.C., 3-5/8″ wide		340	.047		.96	2.08		3.04	4.46
8660	6″ wide	↓	320	.050	↓	1.14	2.21		3.35	4.88
9000	Exterior, 1/2″ gypsum sheathing, 1/2″ gypsum finished, interior,									
9100	including foil faced insulation, metal studs, 20 ga.									
9200	16″ O.C., 3-5/8″ wide	2 Carp	290	.055	S.F.	1.54	2.43		3.97	5.70
9400	6″ wide		270	.059		1.77	2.61		4.38	6.25
9600	Partitions, for work over 8′ high, add	↓	1530	.010	↓		.46		.46	.76

09 22 Supports for Plaster and Gypsum Board

09 22 03 – Fastening Methods for Finishes

09 22 03.20 Drilling Plaster/Drywall

09 22 03.20 Drilling Plaster/Drywall		Crew	Daily Output	Labor-Hours	Unit	Material	Labor	Equipment	Total	Total Incl O&P
0010	**DRILLING PLASTER/DRYWALL**									
1100	Drilling & layout for drywall/plaster walls, up to 1″ deep, no anchor									
1200	Holes, 1/4″ diameter	1 Carp	150	.053	Ea.	.01	2.35		2.36	3.87
1300	3/8″ diameter		140	.057		.01	2.52		2.53	4.14
1400	1/2″ diameter		130	.062		.01	2.71		2.72	4.46
1500	3/4″ diameter		120	.067		.01	2.94		2.95	4.83
1600	1″ diameter		110	.073		.02	3.21		3.23	5.25
1700	1-1/4″ diameter		100	.080		.03	3.53		3.56	5.85
1800	1-1/2″ diameter	↓	90	.089		.04	3.92		3.96	6.50
1900	For ceiling installations, add				↓		40%			

09 22 13 – Metal Furring

09 22 13.13 Metal Channel Furring

09 22 13.13 Metal Channel Furring		Crew	Daily Output	Labor-Hours	Unit	Material	Labor	Equipment	Total	Total Incl O&P
0010	**METAL CHANNEL FURRING**									
0030	Beams and columns, 7/8″ channels, galvanized, 12″ O.C.	1 Lath	155	.052	S.F.	.37	2.03		2.40	3.58
0050	16″ O.C.		170	.047		.31	1.85		2.16	3.23
0070	24″ O.C.		185	.043		.20	1.70		1.90	2.88
0100	Ceilings, on steel, 7/8″ channels, galvanized, 12″ O.C.		210	.038		.34	1.50		1.84	2.71
0300	16″ O.C.		290	.028		.31	1.08		1.39	2.04
0400	24″ O.C.		420	.019		.20	.75		.95	1.39
0600	1-5/8″ channels, galvanized, 12″ O.C.	↓	190	.042	↓	.45	1.65		2.10	3.09

09 22 Supports for Plaster and Gypsum Board

09 22 13 – Metal Furring

09 22 13.13 Metal Channel Furring		Crew	Daily Output	Labor-Hours	Unit	Material	2012 Bare Costs Labor	Equipment	Total	Total Incl O&P
0700	16″ O.C.	1 Lath	260	.031	S.F.	.41	1.21		1.62	2.34
0900	24″ O.C.		390	.021		.27	.81		1.08	1.56
0930	7/8″ channels with sound isolation clips, 12″ O.C.		120	.067		1.40	2.62		4.02	5.65
0940	16″ O.C.		100	.080		1.93	3.14		5.07	7.05
0950	24″ O.C.		165	.048		1.27	1.90		3.17	4.37
0960	1-5/8″ channels, galvanized, 12″ O.C.		110	.073		1.52	2.85		4.37	6.15
0970	16″ O.C.		100	.080		2.04	3.14		5.18	7.15
0980	24″ O.C.		155	.052		1.34	2.03		3.37	4.64
1000	Walls, 7/8″ channels, galvanized, 12″ O.C.		235	.034		.34	1.34		1.68	2.46
1200	16″ O.C.		265	.030		.31	1.19		1.50	2.20
1300	24″ O.C.		350	.023		.20	.90		1.10	1.63
1500	1-5/8″ channels, galvanized, 12″ O.C.		210	.038		.45	1.50		1.95	2.84
1600	16″ O.C.		240	.033		.41	1.31		1.72	2.50
1800	24″ O.C.		305	.026		.27	1.03		1.30	1.91
1920	7/8″ channels with sound isolation clips, 12″ O.C.		125	.064		1.40	2.51		3.91	5.50
1940	16″ O.C.		100	.080		1.93	3.14		5.07	7.05
1950	24″ O.C.		150	.053		1.27	2.09		3.36	4.67
1960	1-5/8″ channels, galvanized, 12″ O.C.		115	.070		1.52	2.73		4.25	5.95
1970	16″ O.C.		95	.084		2.04	3.31		5.35	7.45
1980	24″ O.C.		140	.057		1.34	2.24		3.58	4.98
9000	Minimum labor/equipment charge		4	2	Job		78.50		78.50	123

09 22 16 – Non-Structural Metal Framing

09 22 16.13 Non-Structural Metal Stud Framing		Crew	Daily Output	Labor-Hours	Unit	Material	Labor	Equipment	Total	Total Incl O&P
0010	**NON-STRUCTURAL METAL STUD FRAMING**									
1600	Non-load bearing, galv., 8′ high, 25 ga. 1-5/8″ wide, 16″ O.C.	1 Carp	619	.013	S.F.	.29	.57		.86	1.24
1610	24″ O.C.		950	.008		.21	.37		.58	.85
1620	2-1/2″ wide, 16″ O.C.		613	.013		.33	.58		.91	1.30
1630	24″ O.C.		938	.009		.25	.38		.63	.89
1640	3-5/8″ wide, 16″ O.C.		600	.013		.39	.59		.98	1.39
1650	24″ O.C.		925	.009		.29	.38		.67	.95
1660	4″ wide, 16″ O.C.		594	.013		.45	.59		1.04	1.46
1670	24″ O.C.		925	.009		.34	.38		.72	1
1680	6″ wide, 16″ O.C.		588	.014		.58	.60		1.18	1.62
1690	24″ O.C.		906	.009		.43	.39		.82	1.12
1700	20 ga. studs, 1-5/8″ wide, 16″ O.C.		494	.016		.44	.71		1.15	1.65
1710	24″ O.C.		763	.010		.33	.46		.79	1.12
1720	2-1/2″ wide, 16″ O.C.		488	.016		.50	.72		1.22	1.74
1730	24″ O.C.		750	.011		.38	.47		.85	1.19
1740	3-5/8″ wide, 16″ O.C.		481	.017		.58	.73		1.31	1.84
1750	24″ O.C.		738	.011		.43	.48		.91	1.26
1760	4″ wide, 16″ O.C.		475	.017		.69	.74		1.43	1.97
1770	24″ O.C.		738	.011		.51	.48		.99	1.35
1780	6″ wide, 16″ O.C.		469	.017		.83	.75		1.58	2.14
1790	24″ O.C.		725	.011		.62	.49		1.11	1.48
2000	Non-load bearing, galv., 10′ high, 25 ga. 1-5/8″ wide, 16″ O.C.		495	.016		.27	.71		.98	1.47
2100	24″ O.C.		760	.011		.20	.46		.66	.98
2200	2-1/2″ wide, 16″ O.C.		490	.016		.31	.72		1.03	1.52
2250	24″ O.C.		750	.011		.23	.47		.70	1.02
2300	3-5/8″ wide, 16″ O.C.		480	.017		.37	.74		1.11	1.62
2350	24″ O.C.		740	.011		.27	.48		.75	1.08
2400	4″ wide, 16″ O.C.		475	.017		.43	.74		1.17	1.69
2450	24″ O.C.		740	.011		.31	.48		.79	1.12

09 22 16 – Non-Structural Metal Framing

09 22 16.13 Non-Structural Metal Stud Framing		Crew	Daily Output	Labor-Hours	Unit	Material	2012 Bare Costs Labor	Equipment	Total	Total Incl O&P
2500	6″ wide, 16″ O.C.	1 Carp	470	.017	S.F.	.55	.75		1.30	1.84
2550	24″ O.C.		725	.011		.40	.49		.89	1.24
2600	20 ga. studs, 1-5/8″ wide, 16″ O.C.		395	.020		.42	.89		1.31	1.92
2650	24″ O.C.		610	.013		.31	.58		.89	1.29
2700	2-1/2″ wide, 16″ O.C.		390	.021		.48	.90		1.38	2.01
2750	24″ O.C.		600	.013		.35	.59		.94	1.35
2800	3-5/8″ wide, 16″ OC		385	.021		.55	.92		1.47	2.10
2850	24″ O.C.		590	.014		.41	.60		1.01	1.43
2900	4″ wide, 16″ O.C.		380	.021		.65	.93		1.58	2.24
2950	24″ O.C.		590	.014		.48	.60		1.08	1.51
3000	6″ wide, 16″ O.C.		375	.021		.78	.94		1.72	2.40
3050	24″ O.C.		580	.014		.58	.61		1.19	1.63
3060	Non-load bearing, galv., 12′ high, 25 ga. 1-5/8″ wide, 16″ O.C.		413	.019		.26	.85		1.11	1.68
3070	24″ O.C.		633	.013		.19	.56		.75	1.12
3080	2-1/2″ wide, 16″ O.C.		408	.020		.30	.86		1.16	1.75
3090	24″ O.C.		625	.013		.22	.56		.78	1.17
3100	3-5/8″ wide, 16″ O.C.		400	.020		.35	.88		1.23	1.84
3110	24″ O.C.		617	.013		.26	.57		.83	1.22
3120	4″ wide, 16″ O.C.		396	.020		.41	.89		1.30	1.91
3130	24″ O.C.		617	.013		.30	.57		.87	1.27
3140	6″ wide, 16″ O.C.		392	.020		.53	.90		1.43	2.06
3150	24″ O.C.		604	.013		.38	.58		.96	1.38
3160	20 ga. studs, 1-5/8″ wide, 16″ O.C.		329	.024		.40	1.07		1.47	2.20
3170	24″ O.C.		508	.016		.29	.69		.98	1.46
3180	2-1/2″ wide, 16″ O.C.		325	.025		.46	1.09		1.55	2.28
3190	24″ O.C.		500	.016		.33	.71		1.04	1.53
3200	3-5/8″ wide, 16″ O.C.		321	.025		.53	1.10		1.63	2.38
3210	24″ O.C.		492	.016		.38	.72		1.10	1.60
3220	4″ wide, 16″ O.C.		317	.025		.62	1.11		1.73	2.50
3230	24″ O.C.		492	.016		.45	.72		1.17	1.68
3240	6″ wide, 16″ O.C.		313	.026		.75	1.13		1.88	2.67
3250	24″ O.C.	↓	483	.017	↓	.54	.73		1.27	1.80
5000	Load bearing studs, see Section 05 41 13.30									
9000	Minimum labor/equipment charge	1 Carp	4	2	Job		88		88	145

09 22 26 – Suspension Systems

09 22 26.13 Ceiling Suspension Systems		Crew	Daily Output	Labor-Hours	Unit	Material	Labor	Equipment	Total	Total Incl O&P
0010	**CEILING SUSPENSION SYSTEMS** for gypsum board or plaster									
8000	Suspended ceilings, including carriers									
8200	1-1/2″ carriers, 24″ O.C. with:									
8300	7/8″ channels, 16″ O.C.	1 Lath	275	.029	S.F.	.53	1.14		1.67	2.38
8320	24″ O.C.		310	.026		.43	1.01		1.44	2.07
8400	1-5/8″ channels, 16″ O.C.		205	.039		.64	1.53		2.17	3.10
8420	24″ O.C.	↓	250	.032	↓	.50	1.26		1.76	2.52
8600	2″ carriers, 24″ O.C. with:									
8700	7/8″ channels, 16″ O.C.	1 Lath	250	.032	S.F.	.56	1.26		1.82	2.58
8720	24″ O.C.		285	.028		.45	1.10		1.55	2.23
8800	1-5/8″ channels, 16″ O.C.		190	.042		.66	1.65		2.31	3.31
8820	24″ O.C.	↓	225	.036	↓	.52	1.40		1.92	2.76

09 22 36 – Lath

09 22 36.13 Gypsum Lath		Crew	Daily Output	Labor-Hours	Unit	Material	Labor	Equipment	Total	Total Incl O&P
0010	**GYPSUM LATH** R092000-50									
0020	Plain or perforated, nailed, 3/8″ thick	1 Lath	85	.094	S.Y.	3.87	3.69		7.56	10.05

09 22 Supports for Plaster and Gypsum Board

09 22 36 – Lath

09 22 36.13 Gypsum Lath		Crew	Daily Output	Labor-Hours	Unit	Material	2012 Bare Costs Labor	Equipment	Total	Total Incl O&P
0100	1/2″ thick	1 Lath	80	.100	S.Y.	4.50	3.93		8.43	11.10
0300	Clipped to steel studs, 3/8″ thick		75	.107		3.87	4.19		8.06	10.80
0400	1/2″ thick		70	.114		4.50	4.49		8.99	12
1500	For ceiling installations, add		216	.037			1.45		1.45	2.28
1600	For columns and beams, add		170	.047			1.85		1.85	2.89
9000	Minimum labor/equipment charge		4.25	1.882	Job		74		74	116

09 22 36.23 Metal Lath		Crew	Daily Output	Labor-Hours	Unit	Material	Labor	Equipment	Total	Total Incl O&P
0010	**METAL LATH** R092000-50									
0020	Diamond, expanded, 2.5 lb. per S.Y., painted				S.Y.	3.44			3.44	3.78
0100	Galvanized					3.02			3.02	3.32
0300	3.4 lb. per S.Y., painted					3.60			3.60	3.96
0400	Galvanized					4.46			4.46	4.91
0600	For 15# asphalt sheathing paper, add					.51			.51	.56
0900	Flat rib, 1/8″ high, 2.75 lb., painted					3.14			3.14	3.45
1000	Foil backed					2.83			2.83	3.11
1200	3.4 lb. per S.Y., painted					4			4	4.40
1300	Galvanized					3.50			3.50	3.85
1500	For 15# asphalt sheathing paper, add					.51			.51	.56
1800	High rib, 3/8″ high, 3.4 lb. per S.Y., painted					4.84			4.84	5.30
1900	Galvanized					4.84			4.84	5.30
2400	3/4″ high, painted, .60 lb. per S.F.				S.F.	.50			.50	.55
2500	.75 lb. per S.F.				″	1.08			1.08	1.19
2800	Stucco mesh, painted, 3.6 lb.				S.Y.	3.51			3.51	3.86
3000	K-lath, perforated, absorbent paper, regular					4.42			4.42	4.86
3100	Heavy duty					5.20			5.20	5.75
3300	Waterproof, heavy duty, grade B backing					5.10			5.10	5.60
3400	Fire resistant backing					5.65			5.65	6.20
3600	2.5 lb. diamond painted, on wood framing, on walls	1 Lath	85	.094		3.44	3.69		7.13	9.60
3700	On ceilings		75	.107		3.44	4.19		7.63	10.35
3900	3.4 lb. diamond painted, on wood framing, on walls		80	.100		4	3.93		7.93	10.55
4000	On ceilings		70	.114		4	4.49		8.49	11.45
4200	3.4 lb. diamond painted, wired to steel framing		75	.107		4	4.19		8.19	10.95
4300	On ceilings		60	.133		4	5.25		9.25	12.60
4600	Cornices, wired to steel		35	.229		4	8.95		12.95	18.45
4800	Screwed to steel studs, 2.5 lb.		80	.100		3.44	3.93		7.37	9.95
4900	3.4 lb.		75	.107		3.60	4.19		7.79	10.50
5100	Rib lath, painted, wired to steel, on walls, 2.5 lb.		75	.107		3.14	4.19		7.33	10
5200	3.4 lb.		70	.114		4.84	4.49		9.33	12.35
5400	4.0 lb.		65	.123		5.70	4.83		10.53	13.80
5500	For self-furring lath, add					.11			.11	.12
5700	Suspended ceiling system, incl. 3.4 lb. diamond lath, painted	1 Lath	15	.533		3.39	21		24.39	36.50
5800	Galvanized	″	15	.533		6.50	21		27.50	40
6000	Hollow metal stud partitions, 3.4 lb. painted lath both sides									
6010	Non-load bearing, 25 ga., w/rib lath 2-1/2″ studs, 12″ O.C.	1 Lath	20.30	.394	S.Y.	13.25	15.45		28.70	38.50
6300	16″ O.C.		21.10	.379		12.45	14.90		27.35	37
6350	24″ O.C.		22.70	.352		11.75	13.85		25.60	34.50
6400	3-5/8″ studs, 16″ O.C.		19.50	.410		13	16.10		29.10	39.50
6600	24″ O.C.		20.40	.392		12.10	15.40		27.50	37.50
6700	4″ studs, 16″ O.C.		20.40	.392		13.50	15.40		28.90	39
6900	24″ O.C.		21.60	.370		12.50	14.55		27.05	37
7000	6″ studs, 16″ O.C.		19.50	.410		14.65	16.10		30.75	41
7100	24″ O.C.		21.10	.379		13.30	14.90		28.20	38

09 22 Supports for Plaster and Gypsum Board

09 22 36 – Lath

09 22 36.23 Metal Lath		Crew	Daily Output	Labor-Hours	Unit	Material	Labor	Equipment	Total	Total Incl O&P
						2012 Bare Costs				
7200	L.B. partitions, 16 ga., w/rib lath, 2-1/2" studs, 16" O.C.	1 Lath	20	.400	S.Y.	14.50	15.70		30.20	40.50
7300	3-5/8" studs, 16 ga.		19.70	.406		16.35	15.95		32.30	43
7500	4" studs, 16 ga.		19.50	.410		17	16.10		33.10	43.50
7600	6" studs, 16 ga.		18.70	.428		19.95	16.80		36.75	48.50
9000	Minimum labor/equipment charge		4.25	1.882	Job		74		74	116

09 23 Gypsum Plastering

09 23 13 – Acoustical Gypsum Plastering

09 23 13.10 Perlite or Vermiculite Plaster

		Crew	Daily Output	Labor-Hours	Unit	Material	Labor	Equipment	Total	Total Incl O&P
0010	**PERLITE OR VERMICULITE PLASTER** R092000-50									
0020	In 100 lb. bags, under 200 bags				Bag	14.55			14.55	16.05
0300	2 coats, no lath included, on walls	J-1	92	.435	S.Y.	3.78	16.75	1.42	21.95	32
0400	On ceilings		79	.506		3.78	19.50	1.66	24.94	37
0900	3 coats, no lath included, on walls		74	.541		6.30	21	1.77	29.07	42.50
1000	On ceilings		63	.635		6.30	24.50	2.08	32.88	48
1700	For irregular or curved surfaces, add to above						30%			
1800	For columns and beams, add to above						50%			
1900	For soffits, add to ceiling prices						40%			
9000	Minimum labor/equipment charge	1 Plas	1	8	Job		325		325	515

09 23 20 – Gypsum Plaster

09 23 20.10 Gypsum Plaster On Walls and Ceilings

		Crew	Daily Output	Labor-Hours	Unit	Material	Labor	Equipment	Total	Total Incl O&P
0010	**GYPSUM PLASTER ON WALLS AND CEILINGS** R092000-50									
0020	80# bag, less than 1 ton				Bag	14.80			14.80	16.30
0300	2 coats, no lath included, on walls	J-1	105	.381	S.Y.	3.57	14.65	1.25	19.47	29
0400	On ceilings		92	.435		3.57	16.75	1.42	21.74	32
0900	3 coats, no lath included, on walls		87	.460		5.15	17.70	1.50	24.35	36
1000	On ceilings		78	.513		5.15	19.75	1.68	26.58	39
1600	For irregular or curved surfaces, add						30%			
1800	For columns & beams, add						50%			
9000	Minimum labor/equipment charge	1 Plas	1	8	Job		325		325	515

09 24 Cement Plastering

09 24 23 – Cement Stucco

09 24 23.40 Stucco

		Crew	Daily Output	Labor-Hours	Unit	Material	Labor	Equipment	Total	Total Incl O&P
0010	**STUCCO** R092000-50									
0015	3 coats 1" thick, float finish, with mesh, on wood frame	J-2	63	.762	S.Y.	5.80	29.50	2.08	37.38	55.50
0100	On masonry construction, no mesh incl.	J-1	67	.597		2.28	23	1.95	27.23	41
0300	For trowel finish, add	1 Plas	170	.047			1.90		1.90	3.03
0400	For 3/4" thick, on masonry, deduct	J-1	880	.045		.61	1.75	.15	2.51	3.63
0600	For coloring and special finish, add, minimum		685	.058		.39	2.25	.19	2.83	4.23
0700	Maximum		200	.200		1.36	7.70	.65	9.71	14.50
0900	For soffits, add	J-2	155	.310		2.11	11.95	.85	14.91	22.50
1000	Exterior stucco, with bonding agent, 3 coats, on walls, no mesh incl.	J-1	200	.200		3.54	7.70	.65	11.89	16.90
1200	Ceilings		180	.222		3.54	8.55	.73	12.82	18.35
1300	Beams		80	.500		3.54	19.25	1.64	24.43	36.50
1500	Columns		100	.400		3.54	15.40	1.31	20.25	30
1550	Minimum labor/equipment charge	1 Plas	1	8	Job		325		325	515
1600	Mesh, painted, nailed to wood, 1.8 lb.	1 Lath	60	.133	S.Y.	5	5.25		10.25	13.70
1800	3.6 lb.		55	.145		3.51	5.70		9.21	12.80

09 24 Cement Plastering

09 24 23 – Cement Stucco

09 24 23.40 Stucco		Crew	Daily Output	Labor-Hours	Unit	Material	Labor	Equipment	Total	Total Incl O&P
						2012 Bare Costs				
1900	Wired to steel, painted, 1.8 lb.	1 Lath	53	.151	S.Y.	5	5.90		10.90	14.80
2100	3.6 lb.		50	.160		3.51	6.30		9.81	13.70
9000	Minimum labor/equipment charge		4	2	Job		78.50		78.50	123

09 26 Veneer Plastering

09 26 13 – Gypsum Veneer Plastering

09 26 13.20 Blueboard

		Crew	Daily Output	Labor-Hours	Unit	Material	Labor	Equipment	Total	Total Incl O&P
0010	**BLUEBOARD** For use with thin coat									
0100	plaster application (see Section 09 26 13.80)									
1000	3/8" thick, on walls or ceilings, standard, no finish included	2 Carp	1900	.008	S.F.	.23	.37		.60	.86
1100	With thin coat plaster finish		875	.018		.33	.81		1.14	1.68
1400	On beams, columns, or soffits, standard, no finish included		675	.024		.26	1.05		1.31	2
1450	With thin coat plaster finish		475	.034		.36	1.49		1.85	2.84
3000	1/2" thick, on walls or ceilings, standard, no finish included		1900	.008		.25	.37		.62	.89
3100	With thin coat plaster finish		875	.018		.35	.81		1.16	1.70
3300	Fire resistant, no finish included		1900	.008		.25	.37		.62	.89
3400	With thin coat plaster finish		875	.018		.35	.81		1.16	1.70
3450	On beams, columns, or soffits, standard, no finish included		675	.024		.29	1.05		1.34	2.03
3500	With thin coat plaster finish		475	.034		.38	1.49		1.87	2.86
3700	Fire resistant, no finish included		675	.024		.29	1.05		1.34	2.03
3800	With thin coat plaster finish		475	.034		.38	1.49		1.87	2.86
5000	5/8" thick, on walls or ceilings, fire resistant, no finish included		1900	.008		.25	.37		.62	.89
5100	With thin coat plaster finish		875	.018		.35	.81		1.16	1.70
5500	On beams, columns, or soffits, no finish included		675	.024		.29	1.05		1.34	2.03
5600	With thin coat plaster finish		475	.034		.38	1.49		1.87	2.86
6000	For high ceilings, over 8' high, add		3060	.005			.23		.23	.38
6500	For over 3 stories high, add per story		6100	.003			.12		.12	.19
9000	Minimum labor/equipment charge	1 Carp	2	4	Job		176		176	289

09 26 13.80 Thin Coat Plaster

		Crew	Daily Output	Labor-Hours	Unit	Material	Labor	Equipment	Total	Total Incl O&P
0010	**THIN COAT PLASTER** R092000-50									
0012	1 coat veneer, not incl. lath	J-1	3600	.011	S.F.	.10	.43	.04	.57	.83
1000	In 50 lb. bags				Bag	12.95			12.95	14.25

09 28 Backing Boards and Underlayments

09 28 13 – Cementitious Backing Boards

09 28 13.10 Cementitious Backerboard

		Crew	Daily Output	Labor-Hours	Unit	Material	Labor	Equipment	Total	Total Incl O&P
0010	**CEMENTITIOUS BACKERBOARD**									
0070	Cementitious backerboard, on floor, 3' x 4' x 1/2" sheets	2 Carp	525	.030	S.F.	.73	1.34		2.07	3.01
0080	3' x 5' x 1/2" sheets		525	.030		.73	1.34		2.07	3
0090	3' x 6' x 1/2" sheets		525	.030		.68	1.34		2.02	2.95
0100	3' x 4' x 5/8" sheets		525	.030		.98	1.34		2.32	3.28
0110	3' x 5' x 5/8" sheets		525	.030		1	1.34		2.34	3.30
0120	3' x 6' x 5/8" sheets		525	.030		.90	1.34		2.24	3.19
0150	On wall, 3' x 4' x 1/2" sheets		350	.046		.73	2.02		2.75	4.11
0160	3' x 5' x 1/2" sheets		350	.046		.73	2.02		2.75	4.10
0170	3' x 6' x 1/2" sheets		350	.046		.68	2.02		2.70	4.05
0180	3' x 4' x 5/8" sheets		350	.046		.98	2.02		3	4.38
0190	3' x 5' x 5/8" sheets		350	.046		1	2.02		3.02	4.40
0200	3' x 6' x 5/8" sheets		350	.046		.90	2.02		2.92	4.29

09 28 Backing Boards and Underlayments

09 28 13 – Cementitious Backing Boards

09 28 13.10 Cementitious Backerboard		Crew	Daily Output	Labor-Hours	Unit	2012 Bare Costs Material	Labor	Equipment	Total	Total Incl O&P
0250	On counter, 3′ x 4′ x 1/2″ sheets	2 Carp	180	.089	S.F.	.73	3.92		4.65	7.25
0260	3′ x 5′ x 1/2″ sheets		180	.089		.73	3.92		4.65	7.25
0270	3′ x 6′ x 1/2″ sheets		180	.089		.68	3.92		4.60	7.20
0300	3′ x 4′ x 5/8″ sheets		180	.089		.98	3.92		4.90	7.55
0310	3′ x 5′ x 5/8″ sheets		180	.089		1	3.92		4.92	7.55
0320	3′ x 6′ x 5/8″ sheets		180	.089		.90	3.92		4.82	7.45

09 29 Gypsum Board

09 29 10 – Gypsum Board Panels

09 29 10.30 Gypsum Board		Crew	Daily Output	Labor-Hours	Unit	Material	Labor	Equipment	Total	Total Incl O&P
0010	**GYPSUM BOARD** on walls & ceilings R092910-10									
0100	Nailed or screwed to studs unless otherwise noted									
0150	3/8″ thick, on walls, standard, no finish included	2 Carp	2000	.008	S.F.	.22	.35		.57	.82
0200	On ceilings, standard, no finish included		1800	.009		.22	.39		.61	.88
0250	On beams, columns, or soffits, no finish included		675	.024		.22	1.05		1.27	1.95
0300	1/2″ thick, on walls, standard, no finish included		2000	.008		.22	.35		.57	.82
0350	Taped and finished (level 4 finish)		965	.017		.27	.73		1	1.49
0390	With compound skim coat (level 5 finish)		775	.021		.31	.91		1.22	1.84
0400	Fire resistant, no finish included		2000	.008		.26	.35		.61	.87
0450	Taped and finished (level 4 finish)		965	.017		.31	.73		1.04	1.54
0490	With compound skim coat (level 5 finish)		775	.021		.35	.91		1.26	1.88
0500	Water resistant, no finish included		2000	.008		.35	.35		.70	.97
0550	Taped and finished (level 4 finish)		965	.017		.40	.73		1.13	1.64
0590	With compound skim coat (level 5 finish)		775	.021		.44	.91		1.35	1.98
0600	Prefinished, vinyl, clipped to studs		900	.018		.47	.78		1.25	1.81
0700	Mold resistant, no finish included		2000	.008		.38	.35		.73	1
0710	Taped and finished (level 4 finish)		965	.017		.43	.73		1.16	1.67
0720	With compound skim coat (level 5 finish)		775	.021		.47	.91		1.38	2.01
1000	On ceilings, standard, no finish included		1800	.009		.22	.39		.61	.88
1050	Taped and finished (level 4 finish)		765	.021		.27	.92		1.19	1.80
1090	With compound skim coat (level 5 finish)		610	.026		.31	1.16		1.47	2.25
1100	Fire resistant, no finish included		1800	.009		.26	.39		.65	.93
1150	Taped and finished (level 4 finish)		765	.021		.31	.92		1.23	1.85
1195	With compound skim coat (level 5 finish)		610	.026		.35	1.16		1.51	2.29
1200	Water resistant, no finish included		1800	.009		.35	.39		.74	1.03
1250	Taped and finished (level 4 finish)		765	.021		.40	.92		1.32	1.95
1290	With compound skim coat (level 5 finish)		610	.026		.44	1.16		1.60	2.39
1310	Mold resistant, no finish included		1800	.009		.38	.39		.77	1.06
1320	Taped and finished (level 4 finish)		765	.021		.43	.92		1.35	1.98
1330	With compound skim coat (level 5 finish)		610	.026		.47	1.16		1.63	2.42
1500	On beams, columns, or soffits, standard, no finish included		675	.024		.25	1.05		1.30	1.99
1550	Taped and finished (level 4 finish)		475	.034		.27	1.49		1.76	2.73
1590	With compound skim coat (level 5 finish)		540	.030		.31	1.31		1.62	2.49
1600	Fire resistant, no finish included		675	.024		.26	1.05		1.31	2
1650	Taped and finished (level 4 finish)		475	.034		.31	1.49		1.80	2.78
1690	With compound skim coat (level 5 finish)		540	.030		.35	1.31		1.66	2.53
1700	Water resistant, no finish included		675	.024		.40	1.05		1.45	2.15
1750	Taped and finished (level 4 finish)		475	.034		.40	1.49		1.89	2.88
1790	With compound skim coat (level 5 finish)		540	.030		.44	1.31		1.75	2.63
1800	Mold resistant, no finish included		675	.024		.44	1.05		1.49	2.19
1810	Taped and finished (level 4 finish)		475	.034		.43	1.49		1.92	2.91

09 29 Gypsum Board

09 29 10 – Gypsum Board Panels

09 29 10.30 Gypsum Board		Crew	Daily Output	Labor-Hours	Unit	2012 Bare Costs Material	Labor	Equipment	Total	Total Incl O&P
1820	With compound skim coat (level 5 finish)	2 Carp	540	.030	S.F.	.47	1.31		1.78	2.66
2000	5/8" thick, on walls, standard, no finish included		2000	.008		.25	.35		.60	.86
2050	Taped and finished (level 4 finish)		965	.017		.30	.73		1.03	1.53
2090	With compound skim coat (level 5 finish)		775	.021		.34	.91		1.25	1.87
2100	Fire resistant, no finish included		2000	.008		.27	.35		.62	.88
2150	Taped and finished (level 4 finish)		965	.017		.32	.73		1.05	1.55
2195	With compound skim coat (level 5 finish)		775	.021		.36	.91		1.27	1.89
2200	Water resistant, no finish included		2000	.008		.43	.35		.78	1.05
2250	Taped and finished (level 4 finish)		965	.017		.48	.73		1.21	1.72
2290	With compound skim coat (level 5 finish)		775	.021		.52	.91		1.43	2.07
2300	Prefinished, vinyl, clipped to studs		900	.018		.87	.78		1.65	2.25
2510	Mold resistant, no finish included		2000	.008		.39	.35		.74	1.01
2520	Taped and finished (level 4 finish)		965	.017		.44	.73		1.17	1.68
2530	With compound skim coat (level 5 finish)		775	.021		.48	.91		1.39	2.02
3000	On ceilings, standard, no finish included		1800	.009		.25	.39		.64	.92
3050	Taped and finished (level 4 finish)		765	.021		.30	.92		1.22	1.84
3090	With compound skim coat (level 5 finish)		615	.026		.34	1.15		1.49	2.26
3100	Fire resistant, no finish included		1800	.009		.27	.39		.66	.94
3150	Taped and finished (level 4 finish)		765	.021		.32	.92		1.24	1.86
3190	With compound skim coat (level 5 finish)		615	.026		.36	1.15		1.51	2.28
3200	Water resistant, no finish included		1800	.009		.43	.39		.82	1.11
3250	Taped and finished (level 4 finish)		765	.021		.48	.92		1.40	2.03
3290	With compound skim coat (level 5 finish)		615	.026		.52	1.15		1.67	2.46
3300	Mold resistant, no finish included		1800	.009		.39	.39		.78	1.07
3310	Taped and finished (level 4 finish)		765	.021		.44	.92		1.36	1.99
3320	With compound skim coat (level 5 finish)		615	.026		.48	1.15		1.63	2.41
3500	On beams, columns, or soffits, no finish included		675	.024		.29	1.05		1.34	2.03
3550	Taped and finished (level 4 finish)		475	.034		.34	1.49		1.83	2.81
3590	With compound skim coat (level 5 finish)		380	.042		.40	1.86		2.26	3.48
3600	Fire resistant, no finish included		675	.024		.31	1.05		1.36	2.05
3650	Taped and finished (level 4 finish)		475	.034		.36	1.49		1.85	2.84
3690	With compound skim coat (level 5 finish)		380	.042		.36	1.86		2.22	3.44
3700	Water resistant, no finish included		675	.024		.49	1.05		1.54	2.25
3750	Taped and finished (level 4 finish)		475	.034		.55	1.49		2.04	3.04
3790	With compound skim coat (level 5 finish)		380	.042		.52	1.86		2.38	3.62
3800	Mold resistant, no finish included		675	.024		.45	1.05		1.50	2.20
3810	Taped and finished (level 4 finish)		475	.034		.50	1.49		1.99	2.99
3820	With compound skim coat (level 5 finish)		380	.042		.48	1.86		2.34	3.57
4000	Fireproofing, beams or columns, 2 layers, 1/2" thick, incl finish		330	.048		.57	2.14		2.71	4.13
4010	Mold resistant		330	.048		.81	2.14		2.95	4.40
4050	5/8" thick		300	.053		.63	2.35		2.98	4.55
4060	Mold resistant		300	.053		.87	2.35		3.22	4.82
4100	3 layers, 1/2" thick		225	.071		.83	3.14		3.97	6.05
4110	Mold resistant		225	.071		1.19	3.14		4.33	6.45
4150	5/8" thick		210	.076		.95	3.36		4.31	6.55
4160	Mold resistant		210	.076		1.31	3.36		4.67	6.95
5050	For 1" thick coreboard on columns		480	.033		.54	1.47		2.01	3
5100	For foil-backed board, add					.14			.14	.15
5200	For work over 8' high, add	2 Carp	3060	.005			.23		.23	.38
5270	For textured spray, add	2 Lath	1600	.010		.04	.39		.43	.66
5300	For distribution cost over 3 stories high, add per story	2 Carp	6100	.003			.12		.12	.19
5350	For finishing inner corners, add		950	.017	L.F.	.10	.74		.84	1.33
5355	For finishing outer corners, add		1250	.013		.22	.56		.78	1.17

09 29 Gypsum Board

09 29 10 – Gypsum Board Panels

09 29 10.30 Gypsum Board		Crew	Daily Output	Labor-Hours	Unit	Material	2012 Bare Costs Labor	Equipment	Total	Total Incl O&P
5500	For acoustical sealant, add per bead	1 Carp	500	.016	L.F.	.04	.71		.75	1.21
5550	Sealant, 1 quart tube				Ea.	7.05			7.05	7.80
6000	Gypsum sound dampening panels									
6010	1/2" thick on walls, multi-layer, light weight, no finish included	2 Carp	1500	.011	S.F.	1.36	.47		1.83	2.27
6015	Taped and finished (level 4 finish)		725	.022		1.41	.97		2.38	3.15
6020	With compound skim coat (level 5 finish)		580	.028		1.45	1.22		2.67	3.59
6025	5/8" thick on walls, for wood studs, no finish included		1500	.011		2.37	.47		2.84	3.38
6030	Taped and finished (level 4 finish)		725	.022		2.42	.97		3.39	4.26
6035	With compound skim coat (level 5 finish)		580	.028		2.46	1.22		3.68	4.70
6040	For metal stud, no finish included		1500	.011		2.08	.47		2.55	3.06
6045	Taped and finished (level 4 finish)		725	.022		2.13	.97		3.10	3.94
6050	With compound skim coat (level 5 finish)		580	.028		2.17	1.22		3.39	4.38
6055	Abuse resist, no finish included		1500	.011		2.54	.47		3.01	3.56
6060	Taped and finished (level 4 finish)		725	.022		2.59	.97		3.56	4.44
6065	With compound skim coat (level 5 finish)		580	.028		2.63	1.22		3.85	4.89
6070	Shear rated, no finish included		1500	.011		2.76	.47		3.23	3.81
6075	Taped and finished (level 4 finish)		725	.022		2.81	.97		3.78	4.69
6080	With compound skim coat (level 5 finish)		580	.028		2.85	1.22		4.07	5.15
6085	For SCIF applications, no finish included		1500	.011		3.05	.47		3.52	4.13
6090	Taped and finished (level 4 finish)		725	.022		3.10	.97		4.07	5
6095	With compound skim coat (level 5 finish)		580	.028		3.14	1.22		4.36	5.45
6100	1-3/8" thick on walls, THX Certified, no finish included		1500	.011		6.35	.47		6.82	7.75
6105	Taped and finished (level 4 finish)		725	.022		6.40	.97		7.37	8.65
6110	With compound skim coat (level 5 finish)		580	.028		6.45	1.22		7.67	9.10
6115	5/8" thick on walls, score & snap installation, no finish included		2000	.008		1.62	.35		1.97	2.36
6120	Taped and finished (level 4 finish)		965	.017		1.67	.73		2.40	3.03
6125	With compound skim coat (level 5 finish)		775	.021		1.71	.91		2.62	3.38
7020	5/8" thick on ceilings, for wood joists, no finish included		1200	.013		2.37	.59		2.96	3.57
7025	Taped and finished (level 4 finish)		510	.031		2.42	1.38		3.80	4.93
7030	With compound skim coat (level 5 finish)		410	.039		2.46	1.72		4.18	5.55
7035	For metal joists, no finish included		1200	.013		2.08	.59		2.67	3.25
7040	Taped and finished (level 4 finish)		510	.031		2.13	1.38		3.51	4.61
7045	With compound skim coat (level 5 finish)		410	.039		2.17	1.72		3.89	5.20
7050	Abuse resist, no finish included		1200	.013		2.54	.59		3.13	3.75
7055	Taped and finished (level 4 finish)		510	.031		2.59	1.38		3.97	5.10
7060	With compound skim coat (level 5 finish)		410	.039		2.63	1.72		4.35	5.70
7065	Shear rated, no finish included		1200	.013		2.76	.59		3.35	4
7070	Taped and finished (level 4 finish)		510	.031		2.81	1.38		4.19	5.35
7075	With compound skim coat (level 5 finish)		410	.039		2.85	1.72		4.57	5.95
7080	For SCIF applications, no finish included		1200	.013		3.05	.59		3.64	4.32
7085	Taped and finished (level 4 finish)		510	.031		3.10	1.38		4.48	5.70
7090	With compound skim coat (level 5 finish)		410	.039		3.14	1.72		4.86	6.30
8010	5/8" thick on ceilings, score & snap installation, no finish included		1600	.010		1.62	.44		2.06	2.50
8015	Taped and finished (level 4 finish)		680	.024		1.67	1.04		2.71	3.53
8020	With compound skim coat (level 5 finish)		545	.029		1.71	1.29		3	4.01
9000	Minimum labor/equipment charge	1 Carp	2	4	Job		176		176	289

09 29 15 – Gypsum Board Accessories

09 29 15.10 Accessories, Gypsum Board		Crew	Daily Output	Labor-Hours	Unit	Material	Labor	Equipment	Total	Total Incl O&P
0010	**ACCESSORIES, GYPSUM BOARD**									
0020	Casing bead, galvanized steel	1 Carp	2.90	2.759	C.L.F.	24	122		146	226
0100	Vinyl		3	2.667		21.50	118		139.50	217
0300	Corner bead, galvanized steel, 1" x 1"		4	2		14.50	88		102.50	161

09 29 Gypsum Board

09 29 15 – Gypsum Board Accessories

09 29 15.10 Accessories, Gypsum Board		Crew	Daily Output	Labor-Hours	Unit	2012 Bare Costs Material	Labor	Equipment	Total	Total Incl O&P
0400	1-1/4" x 1-1/4"	1 Carp	3.50	2.286	C.L.F.	15.85	101		116.85	182
0600	Vinyl		4	2		16.75	88		104.75	163
0900	Furring channel, galv. steel, 7/8" deep, standard		2.60	3.077		32.50	136		168.50	258
1000	Resilient		2.55	3.137		37.50	138		175.50	269
1100	J trim, galvanized steel, 1/2" wide		3	2.667		26.50	118		144.50	222
1120	5/8" wide		2.95	2.712		28	120		148	227
1500	Z stud, galvanized steel, 1-1/2" wide		2.60	3.077		37.50	136		173.50	264
9000	Minimum labor/equipment charge		3	2.667	Job		118		118	193

09 30 Tiling

09 30 13 – Ceramic Tiling

09 30 13.10 Ceramic Tile		Crew	Daily Output	Labor-Hours	Unit	Material	Labor	Equipment	Total	Total Incl O&P
0010	**CERAMIC TILE**									
0600	Cove base, 4-1/4" x 4-1/4" high, mud set	D-7	91	.176	L.F.	3.68	6.40		10.08	14.10
0700	Thin set		128	.125		3.71	4.55		8.26	11.25
0900	6" x 4-1/4" high, mud set		100	.160		3.40	5.85		9.25	12.90
1000	Thin set		137	.117		3.27	4.25		7.52	10.25
1200	Sanitary cove base, 6" x 4-1/4" high, mud set		93	.172		3.95	6.25		10.20	14.20
1300	Thin set		124	.129		3.82	4.70		8.52	11.55
1500	6" x 6" high, mud set		84	.190		3.72	6.95		10.67	15
1600	Thin set		117	.137		3.59	4.98		8.57	11.75
2400	Bullnose trim, 4-1/4" x 4-1/4", mud set		82	.195		3.52	7.10		10.62	15
2500	Thin set		128	.125		3.43	4.55		7.98	10.90
2700	6" x 4-1/4" bullnose trim, mud set		84	.190		4.84	6.95		11.79	16.25
2800	Thin set		124	.129		4.75	4.70		9.45	12.55
3000	Floors, natural clay, random or uniform, thin set, color group 1		183	.087	S.F.	3.83	3.19		7.02	9.20
3100	Color group 2		183	.087		5.55	3.19		8.74	11.10
3300	Porcelain type, 1 color, color group 2, 1" x 1"		183	.087		4.89	3.19		8.08	10.40
3310	2" x 2" or 2" x 1", thin set		190	.084		4.33	3.07		7.40	9.55
3350	For random blend, 2 colors, add					.95			.95	1.05
3360	4 colors, add					1.33			1.33	1.46
4300	Specialty tile, 4-1/4" x 4-1/4" x 1/2", decorator finish	D-7	183	.087		10.20	3.19		13.39	16.20
4500	Add for epoxy grout, 1/16" joint, 1" x 1" tile		800	.020		.64	.73		1.37	1.84
4600	2" x 2" tile		820	.020		.60	.71		1.31	1.78
4800	Pregrouted sheets, walls, 4-1/4" x 4-1/4", 6" x 4-1/4"									
4810	and 8-1/2" x 4-1/4", 4 S.F. sheets, silicone grout	D-7	240	.067	S.F.	4.87	2.43		7.30	9.15
5100	Floors, unglazed, 2 S.F. sheets,									
5110	Urethane adhesive	D-7	180	.089	S.F.	4.85	3.24		8.09	10.45
5400	Walls, interior, thin set, 4-1/4" x 4-1/4" tile		190	.084		2.20	3.07		5.27	7.25
5500	6" x 4-1/4" tile		190	.084		2.93	3.07		6	8.05
5700	8-1/2" x 4-1/4" tile		190	.084		4.09	3.07		7.16	9.30
5800	6" x 6" tile		175	.091		2.55	3.33		5.88	8.05
5810	8" x 8" tile		170	.094		3.64	3.43		7.07	9.40
5820	12" x 12" tile		160	.100		3.53	3.64		7.17	9.60
5830	16" x 16" tile		150	.107		4.53	3.89		8.42	11.10
6000	Decorated wall tile, 4-1/4" x 4-1/4", minimum		270	.059		2.80	2.16		4.96	6.45
6100	Maximum		180	.089		45.50	3.24		48.74	55
6600	Crystalline glazed, 4-1/4" x 4-1/4", mud set, plain		100	.160		3.58	5.85		9.43	13.10
6700	4-1/4" x 4-1/4", scored tile		100	.160		5.30	5.85		11.15	15
6900	6" x 6" plain		93	.172		5.65	6.25		11.90	16.05
7000	For epoxy grout, 1/16" joints, 4-1/4" tile, add		800	.020		.40	.73		1.13	1.58

09 30 Tiling

09 30 13 – Ceramic Tiling

09 30 13.10 Ceramic Tile		Crew	Daily Output	Labor-Hours	Unit	2012 Bare Costs Material	Labor	Equipment	Total	Total Incl O&P
7200	For tile set in dry mortar, add	D-7	1735	.009	S.F.		.34		.34	.53
7300	For tile set in Portland cement mortar, add		290	.055		.16	2.01		2.17	3.33
9300	Ceramic tiles, recycled glass, standard colors, 2″ x 2″ thru 6″ x 6″ G		190	.084		21	3.07		24.07	28
9310	6″ x 6″ G		175	.091		21	3.33		24.33	28.50
9320	8″ x 8″ G		170	.094		23	3.43		26.43	31
9330	12″ x 12″ G		160	.100		23	3.64		26.64	31
9340	Earthtones, 2″ x 2″ to 4″ x 8″ G		190	.084		26	3.07		29.07	33.50
9350	6″ x 6″ G		175	.091		26	3.33		29.33	34
9360	8″ x 8″ G		170	.094		27	3.43		30.43	35
9370	12″ x 12″ G		160	.100		27	3.64		30.64	35
9380	Deep colors, 2″ x 2″ to 4″ x 8″ G		190	.084		31	3.07		34.07	39
9390	6″ x 6″ G		175	.091		31	3.33		34.33	39.50
9400	8″ x 8″ G		170	.094		32	3.43		35.43	41
9410	12″ x 12″ G		160	.100		32	3.64		35.64	41
9500	Minimum labor/equipment charge		3.25	4.923	Job		179		179	281

09 30 13.20 Ceramic Tile Repairs		Crew	Daily Output	Labor-Hours	Unit	Material	Labor	Equipment	Total	Total Incl O&P
0010	**CERAMIC TILE REPAIRS**									
1000	Grout removal, carbide tipped, rotary grinder	1 Clab	180	.044	L.F.		1.56		1.56	2.56

09 30 13.45 Ceramic Tile Accessories		Crew	Daily Output	Labor-Hours	Unit	Material	Labor	Equipment	Total	Total Incl O&P
0010	**CERAMIC TILE ACCESSORIES**									
0100	Spacers, 1/8″				Ea.	1.44			1.44	1.58
1310	Sealer for natural stone tile, installed	1 Tilf	650	.012	S.F.	.04	.50		.54	.83

09 30 16 – Quarry Tiling

09 30 16.10 Quarry Tile		Crew	Daily Output	Labor-Hours	Unit	Material	Labor	Equipment	Total	Total Incl O&P
0010	**QUARRY TILE**									
0100	Base, cove or sanitary, mud set, to 5″ high, 1/2″ thick	D-7	110	.145	L.F.	5.60	5.30		10.90	14.45
0300	Bullnose trim, red, mud set, 6″ x 6″ x 1/2″ thick		120	.133		4.46	4.86		9.32	12.50
0400	4″ x 4″ x 1/2″ thick		110	.145		4.96	5.30		10.26	13.75
0600	4″ x 8″ x 1/2″ thick, using 8″ as edge		130	.123		4.38	4.48		8.86	11.85
0700	Floors, mud set, 1,000 S.F. lots, red, 4″ x 4″ x 1/2″ thick		120	.133	S.F.	6.45	4.86		11.31	14.70
0900	6″ x 6″ x 1/2″ thick		140	.114		6	4.16		10.16	13.15
1000	4″ x 8″ x 1/2″ thick		130	.123		6.45	4.48		10.93	14.15
1300	For waxed coating, add					.73			.73	.80
1500	For non-standard colors, add					.43			.43	.47
1600	For abrasive surface, add					.50			.50	.55
1800	Brown tile, imported, 6″ x 6″ x 3/4″	D-7	120	.133		7.45	4.86		12.31	15.80
1900	8″ x 8″ x 1″		110	.145		8.15	5.30		13.45	17.25
2100	For thin set mortar application, deduct		700	.023			.83		.83	1.31
2700	Stair tread, 6″ x 6″ x 3/4″, plain		50	.320		5.15	11.65		16.80	24
2800	Abrasive		47	.340		5.75	12.40		18.15	26
3000	Wainscot, 6″ x 6″ x 1/2″, thin set, red		105	.152		4.44	5.55		9.99	13.60
3100	Non-standard colors		105	.152		4.92	5.55		10.47	14.10
3300	Window sill, 6″ wide, 3/4″ thick		90	.178	L.F.	5.30	6.50		11.80	16
3400	Corners		80	.200	Ea.	6.20	7.30		13.50	18.25
9000	Minimum labor/equipment charge		3.25	4.923	Job		179		179	281

09 30 29 – Metal Tiling

09 30 29.10 Metal Tile		Crew	Daily Output	Labor-Hours	Unit	Material	Labor	Equipment	Total	Total Incl O&P
0010	**METAL TILE** 4′ x 4′ sheet, 24 ga., tile pattern, nailed									
0200	Stainless steel	2 Carp	512	.031	S.F.	27	1.38		28.38	32
0400	Aluminized steel	"	512	.031	"	14.50	1.38		15.88	18.20
9000	Minimum labor/equipment charge	1 Carp	4	2	Job		88		88	145

09 34 Waterproofing-Membrane Tiling

09 34 13 – Waterproofing-Membrane Ceramic Tiling

09 34 13.10 Ceramic Tile Waterproofing Membrane		Crew	Daily Output	Labor-Hours	Unit	2012 Bare Costs Material	Labor	Equipment	Total	Total Incl O&P
0010	**CERAMIC TILE WATERPROOFING MEMBRANE**									
0020	On floors, including thinset									
0030	Fleece laminated polyethylene grid, 1/8" thick	D-7	250	.064	S.F.	2.24	2.33		4.57	6.10
0040	5/16" thick	"	250	.064	"	2.60	2.33		4.93	6.50
0050	On walls, including thinset									
0060	Fleece laminated polyethylene sheet, 8 mil thick	D-7	480	.033	S.F.	2.24	1.21		3.45	4.36
0070	Accessories, including thinset									
0080	Joint and corner sheet, 4 mils thick, 5" wide	1 Tilf	240	.033	L.F.	1.45	1.36		2.81	3.72
0090	7-1/4" wide		180	.044		1.84	1.81		3.65	4.86
0100	10" wide		120	.067		2.24	2.72		4.96	6.70
0110	Pre-formed corners, inside		32	.250	Ea.	4.73	10.20		14.93	21
0120	Outside		32	.250		7.40	10.20		17.60	24
0130	2" flanged floor drain with 6" stainless steel grate		16	.500		355	20.50		375.50	420
0140	EPS, sloped shower floor		480	.017	S.F.	4.82	.68		5.50	6.35
0150	Curb		32	.250	L.F.	13.55	10.20		23.75	31

09 51 Acoustical Ceilings

09 51 23 – Acoustical Tile Ceilings

09 51 23.10 Suspended Acoustic Ceiling Tiles		Crew	Daily Output	Labor-Hours	Unit	Material	Labor	Equipment	Total	Total Incl O&P
0010	**SUSPENDED ACOUSTIC CEILING TILES**, not including									
0100	suspension system									
0300	Fiberglass boards, film faced, 2' x 2' or 2' x 4', 5/8" thick	1 Carp	625	.013	S.F.	.96	.56		1.52	1.99
0400	3/4" thick		600	.013		1.55	.59		2.14	2.67
0500	3" thick, thermal, R11		450	.018		1.57	.78		2.35	3.02
0600	Glass cloth faced fiberglass, 3/4" thick		500	.016		1.49	.71		2.20	2.80
0700	1" thick		485	.016		2.18	.73		2.91	3.59
0820	1-1/2" thick, nubby face		475	.017		2.66	.74		3.40	4.15
1110	Mineral fiber tile, lay-in, 2' x 2' or 2' x 4', 5/8" thick, fine texture		625	.013		.75	.56		1.31	1.76
1115	Rough textured		625	.013		1.36	.56		1.92	2.43
1125	3/4" thick, fine textured		600	.013		1.50	.59		2.09	2.61
1130	Rough textured		600	.013		1.79	.59		2.38	2.93
1135	Fissured		600	.013		2.45	.59		3.04	3.66
1150	Tegular, 5/8" thick, fine textured		470	.017		1.45	.75		2.20	2.83
1155	Rough textured		470	.017		1.90	.75		2.65	3.32
1165	3/4" thick, fine textured		450	.018		2.08	.78		2.86	3.58
1170	Rough textured		450	.018		2.35	.78		3.13	3.88
1175	Fissured		450	.018		3.63	.78		4.41	5.30
1180	For aluminum face, add					6.15			6.15	6.80
1185	For plastic film face, add					.94			.94	1.03
1190	For fire rating, add					.45			.45	.50
1300	Mirror tile, acrylic, 2' x 2'	1 Carp	500	.016		10.95	.71		11.66	13.20
1900	Eggcrate, acrylic, 1/2" x 1/2" x 1/2" cubes		500	.016		1.82	.71		2.53	3.16
2100	Polystyrene eggcrate, 3/8" x 3/8" x 1/2" cubes		510	.016		1.53	.69		2.22	2.81
2200	1/2" x 1/2" x 1/2" cubes		500	.016		1.85	.71		2.56	3.20
2400	Luminous panels, prismatic, acrylic		400	.020		2.22	.88		3.10	3.89
2500	Polystyrene		400	.020		1.14	.88		2.02	2.70
2700	Flat white acrylic		400	.020		3.86	.88		4.74	5.70
2800	Polystyrene		400	.020		2.65	.88		3.53	4.37
3000	Drop pan, white, acrylic		400	.020		5.65	.88		6.53	7.70
3100	Polystyrene		400	.020		4.73	.88		5.61	6.65
3600	Perforated aluminum sheets, .024" thick, corrugated, painted		490	.016		2.29	.72		3.01	3.70

09 51 Acoustical Ceilings

09 51 23 – Acoustical Tile Ceilings

09 51 23.10 Suspended Acoustic Ceiling Tiles		Crew	Daily Output	Labor-Hours	Unit	2012 Bare Costs Material	Labor	Equipment	Total	Total Incl O&P
3700	Plain	1 Carp	500	.016	S.F.	3.78	.71		4.49	5.30
3720	Mineral fiber, 24" x 24" or 48", reveal edge, painted, 5/8" thick		600	.013		1.38	.59		1.97	2.48
3740	3/4" thick		575	.014		2.24	.61		2.85	3.47
5020	66 – 78% recycled content, 3/4" thick G		600	.013		1.82	.59		2.41	2.96
5040	Mylar, 42% recycled content, 3/4" thick G		600	.013		4.28	.59		4.87	5.65
9000	Minimum labor/equipment charge		4	2	Job		88		88	145

09 51 23.30 Suspended Ceilings, Complete		Crew	Daily Output	Labor-Hours	Unit	Material	Labor	Equipment	Total	Total Incl O&P
0010	**SUSPENDED CEILINGS, COMPLETE** including standard									
0100	suspension system but not incl. 1-1/2" carrier channels									
0600	Fiberglass ceiling board, 2' x 4' x 5/8", plain faced	1 Carp	500	.016	S.F.	1.66	.71		2.37	2.98
0700	Offices, 2' x 4' x 3/4"		380	.021		2.25	.93		3.18	3.99
0800	Mineral fiber, on 15/16" T bar susp. 2' x 2' x 3/4" lay-in board		345	.023		2.37	1.02		3.39	4.28
0810	2' x 4' x 5/8" tile		380	.021		1.45	.93		2.38	3.11
0820	Tegular, 2' x 2' x 5/8" tile on 9/16" grid		250	.032		2.48	1.41		3.89	5.05
0830	2' x 4' x 3/4" tile		275	.029		2.94	1.28		4.22	5.35
0900	Luminous panels, prismatic, acrylic		255	.031		2.92	1.38		4.30	5.50
1200	Metal pan with acoustic pad, steel		75	.107		4.41	4.70		9.11	12.55
1300	Painted aluminum		75	.107		2.99	4.70		7.69	11
1500	Aluminum, degreased finish		75	.107		5.10	4.70		9.80	13.30
1600	Stainless steel		75	.107		9.50	4.70		14.20	18.15
1800	Tile, Z bar suspension, 5/8" mineral fiber tile		150	.053		2.27	2.35		4.62	6.35
1900	3/4" mineral fiber tile		150	.053		2.43	2.35		4.78	6.55
2402	For strip lighting, see Section 26 51 13.50									
2500	For rooms under 500 S.F., add				S.F.		25%			
9000	Minimum labor/equipment charge	1 Carp	2	4	Job		176		176	289

09 51 53 – Direct-Applied Acoustical Ceilings

09 51 53.10 Ceiling Tile		Crew	Daily Output	Labor-Hours	Unit	Material	Labor	Equipment	Total	Total Incl O&P
0010	**CEILING TILE**, stapled or cemented									
0100	12" x 12" or 12" x 24", not including furring									
0600	Mineral fiber, vinyl coated, 5/8" thick	1 Carp	300	.027	S.F.	1.72	1.18		2.90	3.82
0700	3/4" thick		300	.027		2.12	1.18		3.30	4.26
0900	Fire rated, 3/4" thick, plain faced		300	.027		1.46	1.18		2.64	3.54
1000	Plastic coated face		300	.027		1.24	1.18		2.42	3.29
1200	Aluminum faced, 5/8" thick, plain		300	.027		1.59	1.18		2.77	3.68
3000	Wood fiber tile, 1/2" thick		400	.020		.97	.88		1.85	2.52
3100	3/4" thick		400	.020		1.17	.88		2.05	2.74
3300	For flameproofing, add					.10			.10	.11
3400	For sculptured 3 dimensional, add					.29			.29	.32
3900	For ceiling primer, add					.13			.13	.14
4000	For ceiling cement, add					.37			.37	.41
9000	Minimum labor/equipment charge	1 Carp	4	2	Job		88		88	145

09 53 Acoustical Ceiling Suspension Assemblies

09 53 23 – Metal Acoustical Ceiling Suspension Assemblies

09 53 23.30 Ceiling Suspension Systems		Crew	Daily Output	Labor-Hours	Unit	2012 Bare Costs Material	Labor	Equipment	Total	Total Incl O&P
0010	**CEILING SUSPENSION SYSTEMS** for boards and tile									
0050	Class A suspension system, 15/16" T bar, 2' x 4' grid	1 Carp	800	.010	S.F.	.70	.44		1.14	1.49
0300	2' x 2' grid		650	.012		.87	.54		1.41	1.84
0310	25% recycled steel, 2' x 4' grid G		800	.010		.76	.44		1.20	1.56
0320	2' x 2' grid G	↓	650	.012		.96	.54		1.50	1.94
0350	For 9/16" grid, add					.16			.16	.18
0360	For fire rated grid, add					.09			.09	.10
0370	For colored grid, add					.21			.21	.23
0400	Concealed Z bar suspension system, 12" module	1 Carp	520	.015		.73	.68		1.41	1.91
0600	1-1/2" carrier channels, 4' O.C., add	"	470	.017	↓	.13	.75		.88	1.37
0700	Carrier channels for ceilings with									
0900	recessed lighting fixtures, add	1 Carp	460	.017	S.F.	.23	.77		1	1.51
1200	Seismic bracing, 2-1/2" compression post with four bracing wires		50	.160	Ea.	2.88	7.05		9.93	14.70
9000	Minimum labor/equipment charge	↓	5	1.600	Job		70.50		70.50	116

09 54 Specialty Ceilings

09 54 33 – Decorative Panel Ceilings

09 54 33.20 Metal Panel Ceilings		Crew	Daily Output	Labor-Hours	Unit	Material	Labor	Equipment	Total	Total Incl O&P
0010	**METAL PANEL CEILINGS**									
0020	Lay-in or screwed to furring, not including grid									
0100	Tin ceilings, 2' x 2' or 2' x 4', bare steel finish	2 Carp	300	.053	S.F.	2.36	2.35		4.71	6.45
0120	Painted white finish		300	.053	"	3.64	2.35		5.99	7.85
0140	Copper, chrome or brass finish		300	.053	L.F.	6.65	2.35		9	11.15
0200	Cornice molding, 2-1/2" to 3-1/2" wide, 4' long, bare steel finish		200	.080	S.F.	2.25	3.53		5.78	8.25
0220	Painted white finish		200	.080		2.94	3.53		6.47	9.05
0240	Copper, chrome or brass finish		200	.080		3.96	3.53		7.49	10.15
0320	5" to 6-1/2" wide, 4' long, bare steel finish		150	.107		3.19	4.70		7.89	11.20
0340	Painted white finish		150	.107		4.20	4.70		8.90	12.30
0360	Copper, chrome or brass finish		150	.107		6.50	4.70		11.20	14.85
0420	Flat molding, 3-1/2" to 5" wide, 4' long, bare steel finish		250	.064		3.69	2.82		6.51	8.70
0440	Painted white finish		250	.064		4.09	2.82		6.91	9.15
0460	Copper, chrome or brass finish	↓	250	.064	↓	7.35	2.82		10.17	12.75

09 62 Specialty Flooring

09 62 19 – Laminate Flooring

09 62 19.10 Floating Floor		Crew	Daily Output	Labor-Hours	Unit	Material	Labor	Equipment	Total	Total Incl O&P
0010	**FLOATING FLOOR**									
8300	Floating floor, laminate, wood pattern strip, complete	1 Clab	133	.060	S.F.	4.18	2.11		6.29	8.05
8310	Components, T & G wood composite strips					3.27			3.27	3.59
8320	Film					.13			.13	.14
8330	Foam					.25			.25	.28
8340	Adhesive					.50			.50	.55
8350	Installation kit				↓	.17			.17	.19
8360	Trim, 2" wide x 3' long				L.F.	3.85			3.85	4.24
8370	Reducer moulding				"	5.20			5.20	5.70

09 62 23 – Bamboo Flooring

09 62 23.10 Flooring, Bamboo		Crew	Daily Output	Labor-Hours	Unit	Material	Labor	Equipment	Total	Total Incl O&P
0010	**FLOORING, BAMBOO**									
8600	Flooring, wood, bamboo strips, unfinished, 5/8" x 4" x 3' G	1 Carp	255	.031	S.F.	4.44	1.38		5.82	7.15

09 62 Specialty Flooring

09 62 23 – Bamboo Flooring

09 62 23.10 Flooring, Bamboo			Crew	Daily Output	Labor-Hours	Unit	Material (2012 Bare Costs)	Labor (2012 Bare Costs)	Equipment (2012 Bare Costs)	Total (2012 Bare Costs)	Total Incl O&P
8610	5/8" x 4" x 4'	G	1 Carp	275	.029	S.F.	4.61	1.28		5.89	7.15
8620	5/8" x 4" x 6'	G		295	.027		5.05	1.20		6.25	7.50
8630	Finished, 5/8" x 4" x 3'	G		255	.031		4.89	1.38		6.27	7.65
8640	5/8" x 4" x 4'	G		275	.029		5.10	1.28		6.38	7.75
8650	5/8" x 4" x 6'	G		295	.027		4.28	1.20		5.48	6.65
8660	Stair treads, unfinished, 1-1/16" x 11-1/2" x 4'	G		18	.444	Ea.	42.50	19.60		62.10	79
8670	Finished, 1-1/16" x 11-1/2" x 4'	G		18	.444		71.50	19.60		91.10	111
8680	Stair risers, unfinished, 5/8" x 7-1/2" x 4'	G		18	.444		15.75	19.60		35.35	49.50
8690	Finished, 5/8" x 7-1/2" x 4'	G		18	.444		30	19.60		49.60	65
8700	Stair nosing, unfinished, 6' long	G		16	.500		35	22		57	74.50
8710	Finished, 6' long	G		16	.500		39.50	22		61.50	79.50

09 63 Masonry Flooring

09 63 13 – Brick Flooring

09 63 13.10 Miscellaneous Brick Flooring			Crew	Daily Output	Labor-Hours	Unit	Material	Labor	Equipment	Total	Total Incl O&P
0010	**MISCELLANEOUS BRICK FLOORING**										
0020	Acid-proof shales, red, 8" x 3-3/4" x 1-1/4" thick		D-7	.43	37.209	M	760	1,350		2,110	2,950
0050	2-1/4" thick		D-1	.40	40		950	1,600		2,550	3,625
0200	Acid-proof clay brick, 8" x 3-3/4" x 2-1/4" thick	G		.40	40		890	1,600		2,490	3,550
0250	9" x 4-1/2" x 3"	G		95	.168	S.F.	4.14	6.70		10.84	15.35
0260	Cast ceramic, pressed, 4" x 8" x 1/2", unglazed		D-7	100	.160		6.25	5.85		12.10	16.05
0270	Glazed			100	.160		8.35	5.85		14.20	18.30
0280	Hand molded flooring, 4" x 8" x 3/4", unglazed			95	.168		8.25	6.15		14.40	18.75
0290	Glazed			95	.168		10.35	6.15		16.50	21
0300	8" hexagonal, 3/4" thick, unglazed			85	.188		9.05	6.85		15.90	20.50
0310	Glazed			85	.188		16.35	6.85		23.20	28.50
0400	Heavy duty industrial, cement mortar bed, 2" thick, not incl. brick		D-1	80	.200		.83	7.95		8.78	13.75
0450	Acid-proof joints, 1/4" wide		"	65	.246		1.44	9.80		11.24	17.40
0500	Pavers, 8" x 4", 1" to 1-1/4" thick, red		D-7	95	.168		3.64	6.15		9.79	13.65
0510	Ironspot		"	95	.168		5.15	6.15		11.30	15.30
0540	1-3/8" to 1-3/4" thick, red		D-1	95	.168		3.51	6.70		10.21	14.65
0560	Ironspot			95	.168		5.10	6.70		11.80	16.40
0580	2-1/4" thick, red			90	.178		3.57	7.10		10.67	15.35
0590	Ironspot			90	.178		5.55	7.10		12.65	17.50
0700	Paver, adobe brick, 6" x 12", 1/2" joint	G		42	.381		1.38	15.20		16.58	26
0710	Mexican red, 12" x 12"	G	1 Tilf	48	.167		1.64	6.80		8.44	12.45
0720	Saltillo, 12" x 12"	G	"	48	.167		1.64	6.80		8.44	12.45
0800	For sidewalks and patios with pavers, see Section 32 14 16.10										
0870	For epoxy joints, add		D-1	600	.027	S.F.	2.72	1.06		3.78	4.70
0880	For Furan underlayment, add		"	600	.027		2.25	1.06		3.31	4.19
0890	For waxed surface, steam cleaned, add		A-1H	1000	.008		.20	.28	.07	.55	.75
9000	Minimum labor/equipment charge		1 Bric	2	4	Job		176		176	284

09 63 40 – Stone Flooring

09 63 40.10 Marble		Crew	Daily Output	Labor-Hours	Unit	Material	Labor	Equipment	Total	Total Incl O&P
0010	**MARBLE**									
0020	Thin gauge tile, 12" x 6", 3/8", white Carara	D-7	60	.267	S.F.	11.45	9.70		21.15	28
0100	Travertine		60	.267		13.15	9.70		22.85	30
0200	12" x 12" x 3/8", thin set, floors		60	.267		6.25	9.70		15.95	22
0300	On walls		52	.308		9.85	11.20		21.05	28.50
1000	Marble threshold, 4" wide x 36" long x 5/8" thick, white		60	.267	Ea.	6.25	9.70		15.95	22

09 63 Masonry Flooring

09 63 40 – Stone Flooring

09 63 40.10 Marble		Crew	Daily Output	Labor-Hours	Unit	Material	2012 Bare Costs Labor	Equipment	Total	Total Incl O&P
9000	Minimum labor/equipment charge	D-7	3	5.333	Job		194		194	305
09 63 40.20 Slate Tile										
0010	**SLATE TILE**									
0020	Vermont, 6" x 6" x 1/4" thick, thin set	D-7	180	.089	S.F.	7.50	3.24		10.74	13.35
9000	Minimum labor/equipment charge	"	3	5.333	Job		194		194	305

09 64 Wood Flooring

09 64 16 – Wood Block Flooring

09 64 16.10 End Grain Block Flooring		Crew	Daily Output	Labor-Hours	Unit	Material	Labor	Equipment	Total	Total Incl O&P
0010	**END GRAIN BLOCK FLOORING**									
0020	End grain flooring, coated, 2" thick	1 Carp	295	.027	S.F.	3.52	1.20		4.72	5.85
0400	Natural finish, 1" thick, fir		125	.064		3.64	2.82		6.46	8.65
0600	1-1/2" thick, pine		125	.064		3.57	2.82		6.39	8.55
0700	2" thick, pine		125	.064		5	2.82		7.82	10.15
9000	Minimum labor/equipment charge		2	4	Job		176		176	289

09 64 23 – Wood Parquet Flooring

09 64 23.10 Wood Parquet		Crew	Daily Output	Labor-Hours	Unit	Material	Labor	Equipment	Total	Total Incl O&P
0010	**WOOD PARQUET** flooring									
5200	Parquetry, standard, 5/16" thick, not incl. finish, oak, minimum	1 Carp	160	.050	S.F.	4.56	2.21		6.77	8.60
5300	Maximum		100	.080		5.80	3.53		9.33	12.15
5500	Teak, minimum		160	.050		4.86	2.21		7.07	8.95
5600	Maximum		100	.080		8.50	3.53		12.03	15.15
5650	13/16" thick, select grade oak, minimum		160	.050		9.50	2.21		11.71	14.05
5700	Maximum		100	.080		14.40	3.53		17.93	21.50
5800	Custom parquetry, including finish, minimum		100	.080		15.65	3.53		19.18	23
5900	Maximum		50	.160		21	7.05		28.05	34.50
6700	Parquetry, prefinished white oak, 5/16" thick, minimum		160	.050		3.30	2.21		5.51	7.25
6800	Maximum		100	.080		4.47	3.53		8	10.70
7000	Walnut or teak, parquetry, minimum		160	.050		5.35	2.21		7.56	9.50
7100	Maximum		100	.080		9.35	3.53		12.88	16.10
7200	Acrylic wood parquet blocks, 12" x 12" x 5/16",									
7210	Irradiated, set in epoxy	1 Carp	160	.050	S.F.	7.80	2.21		10.01	12.20

09 64 29 – Wood Strip and Plank Flooring

09 64 29.10 Wood		Crew	Daily Output	Labor-Hours	Unit	Material	Labor	Equipment	Total	Total Incl O&P
0010	**WOOD** R061110-30									
0020	Fir, vertical grain, 1" x 4", not incl. finish, grade B & better	1 Carp	255	.031	S.F.	2.74	1.38		4.12	5.30
0100	C grade & better		255	.031		2.58	1.38		3.96	5.10
0300	Flat grain, 1" x 4", not incl. finish, B & better		255	.031		3.13	1.38		4.51	5.70
0400	C & better		255	.031		3.01	1.38		4.39	5.60
4000	Maple, strip, 25/32" x 2-1/4", not incl. finish, select		170	.047		5.20	2.08		7.28	9.10
4100	#2 & better		170	.047		3.99	2.08		6.07	7.80
4300	33/32" x 3-1/4", not incl. finish, #1 grade		170	.047		4.40	2.08		6.48	8.25
4400	#2 & better		170	.047		3.92	2.08		6	7.70
4600	Oak, white or red, 25/32" x 2-1/4", not incl. finish									
4700	#1 common	1 Carp	170	.047	S.F.	1.99	2.08		4.07	5.60
4900	Select quartered, 2-1/4" wide		170	.047		2.99	2.08		5.07	6.70
5000	Clear		170	.047		3.75	2.08		5.83	7.55
6100	Prefinished, white oak, prime grade, 2-1/4" wide		170	.047		4.79	2.08		6.87	8.65
6200	3-1/4" wide		185	.043		5.05	1.91		6.96	8.70
6400	Ranch plank		145	.055		6.50	2.43		8.93	11.15

09 64 Wood Flooring

09 64 29 – Wood Strip and Plank Flooring

09 64 29.10 Wood		Crew	Daily Output	Labor-Hours	Unit	Material	2012 Bare Costs Labor	Equipment	Total	Total Incl O&P
6500	Hardwood blocks, 9" x 9", 25/32" thick	1 Carp	160	.050	S.F.	5.75	2.21		7.96	9.95
7400	Yellow pine, 3/4" x 3-1/8", T & G, C & better, not incl. finish	↓	200	.040		1.49	1.76		3.25	4.53
7500	Refinish wood floor, sand, 2 coats poly, wax, soft wood, min.	1 Clab	400	.020		.86	.70		1.56	2.10
7600	Hard wood, max.		130	.062		1.28	2.16		3.44	4.95
7800	Sanding and finishing, 2 coats polyurethane	↓	295	.027	↓	.86	.95		1.81	2.51
7900	Subfloor and underlayment, see Section 06 16									
8015	Transition molding, 2 1/4" wide, 5' long	1 Carp	19.20	.417	Ea.	15	18.40		33.40	46.50
9000	Minimum labor/equipment charge	"	2	4	Job		176		176	289

09 65 Resilient Flooring

09 65 10 – Resilient Tile Underlayment

09 65 10.10 Latex Underlayment		Crew	Daily Output	Labor-Hours	Unit	Material	Labor	Equipment	Total	Total Incl O&P
0010	**LATEX UNDERLAYMENT**									
3600	Latex underlayment, 1/8" thk., cementitious for resilient flooring	1 Tilf	160	.050	S.F.	.97	2.04		3.01	4.27
4000	Liquid, fortified				Gal.	29.50			29.50	32.50

09 65 13 – Resilient Base and Accessories

09 65 13.13 Resilient Base		Crew	Daily Output	Labor-Hours	Unit	Material	Labor	Equipment	Total	Total Incl O&P
0010	**RESILIENT BASE**									
0800	Base, cove, rubber or vinyl									
1100	Standard colors, 0.080" thick, 2-1/2" high	1 Tilf	315	.025	L.F.	1.05	1.04		2.09	2.78
1150	4" high		315	.025		1.01	1.04		2.05	2.73
1200	6" high		315	.025		1.55	1.04		2.59	3.33
1450	1/8" thick, 2-1/2" high		315	.025		1.05	1.04		2.09	2.78
1500	4" high		315	.025		.72	1.04		1.76	2.41
1550	6" high		315	.025	↓	1.50	1.04		2.54	3.27
1600	Corners, 2-1/2" high		315	.025	Ea.	1.70	1.04		2.74	3.49
1630	4" high		315	.025		3.15	1.04		4.19	5.10
1660	6" high	↓	315	.025	↓	3.15	1.04		4.19	5.10

09 65 13.23 Resilient Stair Treads and Risers		Crew	Daily Output	Labor-Hours	Unit	Material	Labor	Equipment	Total	Total Incl O&P
0010	**RESILIENT STAIR TREADS AND RISERS**									
0300	Rubber, molded tread, 12" wide, 5/16" thick, black	1 Tilf	115	.070	L.F.	13.50	2.84		16.34	19.30
0400	Colors		115	.070		14	2.84		16.84	19.85
0600	1/4" thick, black		115	.070		13	2.84		15.84	18.75
0700	Colors		115	.070		13.50	2.84		16.34	19.30
0900	Grip strip safety tread, colors, 5/16" thick		115	.070		17.60	2.84		20.44	24
1000	3/16" thick		120	.067	↓	12.75	2.72		15.47	18.30
1200	Landings, smooth sheet rubber, 1/8" thick		120	.067	S.F.	6.65	2.72		9.37	11.55
1300	3/16" thick		120	.067	"	8.35	2.72		11.07	13.45
1500	Nosings, 3" wide, 3/16" thick, black		140	.057	L.F.	3.69	2.33		6.02	7.70
1600	Colors		140	.057		4.09	2.33		6.42	8.15
1800	Risers, 7" high, 1/8" thick, flat		250	.032		6	1.30		7.30	8.65
1900	Coved		250	.032		8.15	1.30		9.45	11
2100	Vinyl, molded tread, 12" wide, colors, 1/8" thick		115	.070		6.65	2.84		9.49	11.75
2200	1/4" thick		115	.070	↓	8.35	2.84		11.19	13.65
2300	Landing material, 1/8" thick		200	.040	S.F.	6.65	1.63		8.28	9.85
2400	Riser, 7" high, 1/8" thick, coved		175	.046	L.F.	3.75	1.86		5.61	7.05
2500	Tread and riser combined, 1/8" thick		80	.100	"	10	4.08		14.08	17.40
9000	Minimum labor/equipment charge	↓	3	2.667	Job		109		109	171

09 65 Resilient Flooring

09 65 16 – Resilient Sheet Flooring

09 65 16.10 Rubber and Vinyl Sheet Flooring			Crew	Daily Output	Labor-Hours	Unit	2012 Bare Costs Material	Labor	Equipment	Total	Total Incl O&P
0010	**RUBBER AND VINYL SHEET FLOORING**										
5500	Linoleum, sheet goods	G	1 Tilf	360	.022	S.F.	4.28	.91		5.19	6.15
5900	Rubber, sheet goods, 36" wide, 1/8" thick			120	.067		7.30	2.72		10.02	12.30
5950	3/16" thick			100	.080		9.90	3.26		13.16	16
6000	1/4" thick			90	.089		11.70	3.62		15.32	18.55
8000	Vinyl sheet goods, backed, .065" thick, minimum			250	.032		3.95	1.30		5.25	6.40
8050	Maximum			200	.040		3.83	1.63		5.46	6.75
8100	.080" thick, minimum			230	.035		3.95	1.42		5.37	6.55
8150	Maximum			200	.040		5.65	1.63		7.28	8.75
8200	.125" thick, minimum			230	.035		4.57	1.42		5.99	7.25
8250	Maximum		↓	200	.040	↓	6.85	1.63		8.48	10.10
8700	Adhesive cement, 1 gallon per 200 to 300 S.F.					Gal.	26			26	28.50
8800	Asphalt primer, 1 gallon per 300 S.F.						13.40			13.40	14.75
8900	Emulsion, 1 gallon per 140 S.F.					↓	17.25			17.25	19

09 65 19 – Resilient Tile Flooring

09 65 19.10 Miscellaneous Resilient Tile Flooring			Crew	Daily Output	Labor-Hours	Unit	Material	Labor	Equipment	Total	Total Incl O&P
0010	**MISCELLANEOUS RESILIENT TILE FLOORING**										
2200	Cork tile, standard finish, 1/8" thick	G	1 Tilf	315	.025	S.F.	7.35	1.04		8.39	9.70
2250	3/16" thick	G		315	.025		6.65	1.04		7.69	8.90
2300	5/16" thick	G		315	.025		8.60	1.04		9.64	11.10
2350	1/2" thick	G		315	.025		10.50	1.04		11.54	13.15
2500	Urethane finish, 1/8" thick	G		315	.025		9.50	1.04		10.54	12.05
2550	3/16" thick	G		315	.025		10.20	1.04		11.24	12.80
2600	5/16" thick	G		315	.025		11.50	1.04		12.54	14.25
2650	1/2" thick	G		315	.025		15.60	1.04		16.64	18.75
6050	Rubber tile, marbleized colors, 12" x 12", 1/8" thick			400	.020		7	.82		7.82	9
6100	3/16" thick			400	.020		10.70	.82		11.52	13.05
6300	Special tile, plain colors, 1/8" thick			400	.020		10.70	.82		11.52	13.05
6350	3/16" thick			400	.020		10.70	.82		11.52	13.05
7000	Vinyl composition tile, 12" x 12", 1/16" thick			500	.016		.94	.65		1.59	2.05
7050	Embossed			500	.016		2.17	.65		2.82	3.41
7100	Marbleized			500	.016		2.17	.65		2.82	3.41
7150	Solid			500	.016		2.79	.65		3.44	4.09
7200	3/32" thick, embossed			500	.016		1.34	.65		1.99	2.49
7250	Marbleized			500	.016		2.49	.65		3.14	3.76
7300	Solid			500	.016		2.31	.65		2.96	3.56
7350	1/8" thick, marbleized			500	.016		1.85	.65		2.50	3.06
7400	Solid			500	.016		1.58	.65		2.23	2.76
7450	Conductive			500	.016		6.40	.65		7.05	8.05
7500	Vinyl tile, 12" x 12", .050" thick, minimum			500	.016		3.25	.65		3.90	4.60
7550	Maximum			500	.016		6.50	.65		7.15	8.15
7600	1/8" thick, minimum			500	.016		4.80	.65		5.45	6.30
7650	Solid colors			500	.016		5.85	.65		6.50	7.45
7700	Marbleized or Travertine pattern			500	.016		5.90	.65		6.55	7.50
7750	Florentine pattern			500	.016		6.30	.65		6.95	7.95
7800	Maximum			500	.016	↓	14	.65		14.65	16.40
9500	Minimum labor/equipment charge		↓	4	2	Job		81.50		81.50	128

09 65 Resilient Flooring

09 65 33 – Conductive Resilient Flooring

09 65 33.10 Conductive Rubber and Vinyl Flooring		Crew	Daily Output	Labor-Hours	Unit	Material	Labor	Equipment	Total	Total Incl O&P
						2012 Bare Costs				
0010	**CONDUCTIVE RUBBER AND VINYL FLOORING**									
1700	Conductive flooring, rubber tile, 1/8″ thick	1 Tilf	315	.025	S.F.	6.30	1.04		7.34	8.55
1800	Homogeneous vinyl tile, 1/8″ thick	″	315	.025	″	8.40	1.04		9.44	10.85

09 66 Terrazzo Flooring

09 66 13 – Portland Cement Terrazzo Flooring

09 66 13.10 Portland Cement Terrazzo

Line	Description	Crew	Daily Output	Labor-Hours	Unit	Material	Labor	Equipment	Total	Total Incl O&P
0010	**PORTLAND CEMENT TERRAZZO**, cast-in-place									
0020	Cove base, 6″ high, 16 ga. zinc	1 Mstz	20	.400	L.F.	3.04	16.15		19.19	29
0100	Curb, 6″ high and 6″ wide		6	1.333		5.40	54		59.40	90.50
0300	Divider strip for floors, 14 ga., 1-1/4″ deep, zinc		375	.021		1.43	.86		2.29	2.92
0400	Brass		375	.021		2.48	.86		3.34	4.08
0600	Heavy top strip 1/4″ thick, 1-1/4″ deep, zinc		300	.027		1.87	1.08		2.95	3.75
1200	For thin set floors, 16 ga., 1/2″ x 1/2″, zinc		350	.023		.85	.92		1.77	2.39
1500	Floor, bonded to concrete, 1-3/4″ thick, gray cement	J-3	75	.213	S.F.	3.15	7.90	3.49	14.54	19.70
1600	White cement, mud set		75	.213		3.51	7.90	3.49	14.90	20
1800	Not bonded, 3″ total thickness, gray cement		70	.229		3.93	8.50	3.74	16.17	21.50
1900	White cement, mud set		70	.229		4.58	8.50	3.74	16.82	22.50
9000	Minimum labor/equipment charge	1 Mstz	1	8	Job		325		325	505

09 66 13.30 Terrazzo, Precast

Line	Description	Crew	Daily Output	Labor-Hours	Unit	Material	Labor	Equipment	Total	Total Incl O&P
0010	**TERRAZZO, PRECAST**									
0020	Base, 6″ high, straight	1 Mstz	70	.114	L.F.	11.50	4.62		16.12	19.90
0100	Cove		60	.133		12.50	5.40		17.90	22
0300	8″ high, straight		60	.133		11.15	5.40		16.55	21
0400	Cove		50	.160		16.45	6.45		22.90	28.50
0600	For white cement, add					.44			.44	.48
0700	For 16 ga. zinc toe strip, add					1.68			1.68	1.85
0900	Curbs, 4″ x 4″ high	1 Mstz	40	.200		31.50	8.10		39.60	47
1000	8″ x 8″ high	″	30	.267		36	10.75		46.75	56.50
1200	Floor tiles, non-slip, 1″ thick, 12″ x 12″	D-1	60	.267	S.F.	18.90	10.65		29.55	38
1300	1-1/4″ thick, 12″ x 12″		60	.267		20	10.65		30.65	39
1500	16″ x 16″		50	.320		21.50	12.75		34.25	44.50
1600	1-1/2″ thick, 16″ x 16″		45	.356		19.85	14.15		34	45
1800	For Venetian terrazzo, add					5.95			5.95	6.55
1900	For white cement, add					.56			.56	.62
2400	Stair treads, 1-1/2″ thick, non-slip, three line pattern	2 Mstz	70	.229	L.F.	40	9.25		49.25	58.50
2500	Nosing and two lines		70	.229		40	9.25		49.25	58.50
2700	2″ thick treads, straight		60	.267		44	10.75		54.75	65.50
2800	Curved		50	.320		57.50	12.95		70.45	84
3000	Stair risers, 1″ thick, to 6″ high, straight sections		60	.267		10.05	10.75		20.80	28
3100	Cove		50	.320		14.55	12.95		27.50	36.50
3300	Curved, 1″ thick, to 6″ high, vertical		48	.333		19.90	13.45		33.35	43
3400	Cove		38	.421		37.50	17		54.50	68
3600	Stair tread and riser, single piece, straight, minimum		60	.267		51	10.75		61.75	73
3700	Maximum		40	.400		66	16.15		82.15	98
3900	Curved tread and riser, minimum		40	.400		71.50	16.15		87.65	104
4000	Maximum		32	.500		90	20		110	131
4200	Stair stringers, notched, 1″ thick		25	.640		29	26		55	72.50
4300	2″ thick		22	.727		34.50	29.50		64	84
4500	Stair landings, structural, non-slip, 1-1/2″ thick		85	.188	S.F.	32	7.60		39.60	47

09 66 Terrazzo Flooring

09 66 13 – Portland Cement Terrazzo Flooring

09 66 13.30 Terrazzo, Precast		Crew	Daily Output	Labor-Hours	Unit	Material	2012 Bare Costs Labor	Equipment	Total	Total Incl O&P
4600	3" thick	2 Mstz	75	.213	S.F.	45	8.60		53.60	63
4800	Wainscot, 12" x 12" x 1" tiles	1 Mstz	12	.667		6.70	27		33.70	50
4900	16" x 16" x 1-1/2" tiles	"	8	1		13.95	40.50		54.45	79
9500	Minimum labor/equipment charge	1 Tilf	2	4	Job		163		163	256

09 66 16 – Terrazzo Floor Tile

09 66 16.10 Tile or Terrazzo Base		Crew	Daily Output	Labor-Hours	Unit	Material	Labor	Equipment	Total	Total Incl O&P
0010	**TILE OR TERRAZZO BASE**									
0020	Scratch coat only	1 Mstz	150	.053	S.F.	.42	2.15		2.57	3.84
0500	Scratch and brown coat only		75	.107	"	.80	4.31		5.11	7.65
9000	Minimum labor/equipment charge		1	8	Job		325		325	505

09 67 Fluid-Applied Flooring

09 67 20 – Epoxy-Marble Chip Flooring

09 67 20.13 Elastomeric Liquid Flooring		Crew	Daily Output	Labor-Hours	Unit	Material	Labor	Equipment	Total	Total Incl O&P
0010	**ELASTOMERIC LIQUID FLOORING**									
0020	Cementitious acrylic, 1/4" thick	C-6	520	.092	S.F.	1.58	3.38	.11	5.07	7.35
0200	Methyl methachrylate, 1/4" thick	C-8A	3000	.016	"	5.70	.61		6.31	7.25

09 67 20.16 Epoxy Terrazzo		Crew	Daily Output	Labor-Hours	Unit	Material	Labor	Equipment	Total	Total Incl O&P
0010	**EPOXY TERRAZZO**									
1800	Epoxy terrazzo, 1/4" thick, chemical resistant, minimum	J-3	200	.080	S.F.	5.65	2.97	1.31	9.93	12.30
1900	Maximum	"	150	.107	"	8.55	3.96	1.75	14.26	17.55

09 67 20.26 Quartz Flooring		Crew	Daily Output	Labor-Hours	Unit	Material	Labor	Equipment	Total	Total Incl O&P
0010	**QUARTZ FLOORING**									
0600	Epoxy, with colored quartz chips, broadcast, minimum	C-6	675	.071	S.F.	2.67	2.61	.08	5.36	7.25
0700	Maximum		490	.098		3.85	3.59	.11	7.55	10.20
0900	Trowelled, minimum		560	.086		3.40	3.14	.10	6.64	8.95
1000	Maximum		480	.100		5.10	3.66	.11	8.87	11.75
1200	Heavy duty epoxy topping, 1/4" thick,									
1300	500 to 1,000 S.F.	C-6	420	.114	S.F.	5.25	4.19	.13	9.57	12.75
1500	1,000 to 2,000 S.F.		450	.107		4.76	3.91	.12	8.79	11.75
1600	Over 10,000 S.F.		480	.100		4.39	3.66	.11	8.16	10.90

09 68 Carpeting

09 68 05 – Carpet Accessories

09 68 05.11 Flooring Transition Strip		Crew	Daily Output	Labor-Hours	Unit	Material	Labor	Equipment	Total	Total Incl O&P
0010	**FLOORING TRANSITION STRIP**									
0107	Clamp down brass divider, 12' strip, vinyl to carpet	1 Tilf	31.25	.256	Ea.	11.45	10.45		21.90	29
0117	Vinyl to hard surface	"	31.25	.256	"	11.45	10.45		21.90	29

09 68 10 – Carpet Pad

09 68 10.10 Commercial Grade Carpet Pad		Crew	Daily Output	Labor-Hours	Unit	Material	Labor	Equipment	Total	Total Incl O&P
0010	**COMMERCIAL GRADE CARPET PAD**									
9000	Sponge rubber pad, minimum	1 Tilf	150	.053	S.Y.	4.18	2.17		6.35	8
9100	Maximum		150	.053		10.40	2.17		12.57	14.85
9200	Felt pad, minimum		150	.053		3.84	2.17		6.01	7.65
9300	Maximum		150	.053		7.45	2.17		9.62	11.60
9400	Bonded urethane pad, minimum		150	.053		4.57	2.17		6.74	8.45
9500	Maximum		150	.053		8.20	2.17		10.37	12.40
9600	Prime urethane pad, minimum		150	.053		3	2.17		5.17	6.70

09 68 Carpeting

09 68 10 – Carpet Pad

09 68 10.10 Commercial Grade Carpet Pad		Crew	Daily Output	Labor-Hours	Unit	Material	Labor	Equipment	Total	Total Incl O&P
						2012 Bare Costs				
9700	Maximum	1 Tilf	150	.053	S.Y.	5.25	2.17		7.42	9.20

09 68 13 – Tile Carpeting

09 68 13.10 Carpet Tile		Crew	Daily Output	Labor-Hours	Unit	Material	Labor	Equipment	Total	Total Incl O&P
0010	**CARPET TILE**									
0100	Tufted nylon, 18" x 18", hard back, 20 oz.	1 Tilf	80	.100	S.Y.	22	4.08		26.08	31
0110	26 oz.		80	.100		38	4.08		42.08	48
0200	Cushion back, 20 oz.		80	.100		28	4.08		32.08	37
0210	26 oz.		80	.100		43.50	4.08		47.58	54.50
1100	Tufted, 24" x 24", hard back, 24 oz. nylon		80	.100		28	4.08		32.08	37.50
1180	35 oz.		80	.100		34.50	4.08		38.58	44.50
5060	42 oz.	↓	80	.100	↓	45.50	4.08		49.58	56.50

09 68 16 – Sheet Carpeting

09 68 16.10 Sheet Carpet		Crew	Daily Output	Labor-Hours	Unit	Material	Labor	Equipment	Total	Total Incl O&P
0010	**SHEET CARPET**									
0700	Nylon, level loop, 26 oz., light to medium traffic	1 Tilf	75	.107	S.Y.	29	4.35		33.35	39
0720	28 oz., light to medium traffic		75	.107		30	4.35		34.35	40
0900	32 oz., medium traffic		75	.107		36.50	4.35		40.85	47
1100	40 oz., medium to heavy traffic		75	.107		53.50	4.35		57.85	66
2920	Nylon plush, 30 oz., medium traffic		57	.140		28.50	5.70		34.20	40
3000	36 oz., medium traffic		75	.107		37.50	4.35		41.85	48.50
3100	42 oz., medium to heavy traffic		70	.114		42.50	4.66		47.16	54
3200	46 oz., medium to heavy traffic		70	.114		49.50	4.66		54.16	62
3300	54 oz., heavy traffic		70	.114		54	4.66		58.66	67
4500	50 oz., medium to heavy traffic		75	.107		127	4.35		131.35	147
4700	Patterned, 32 oz., medium to heavy traffic		70	.114		126	4.66		130.66	146
4900	48 oz., heavy traffic	↓	70	.114	↓	129	4.66		133.66	149
5000	For less than full roll (approx. 1500 S.F.), add					25%				
5100	For small rooms, less than 12' wide, add						25%			
5200	For large open areas (no cuts), deduct						25%			
5600	For bound carpet baseboard, add	1 Tilf	300	.027	L.F.	2.75	1.09		3.84	4.74
5610	For stairs, not incl. price of carpet, add	"	30	.267	Riser		10.85		10.85	17.05
5620	For borders and patterns, add to labor						18%			
8950	For tackless, stretched installation, add padding from 09 68 10.10 to above									
9850	For brand-named specific fiber, add				S.Y.	25%				
9900	Carpet cleaning machine, rent				Day				38	42
9910	Minimum labor/equipment charge	1 Tilf	3	2.667	Job		109		109	171

09 68 20 – Athletic Carpet

09 68 20.10 Indoor Athletic Carpet		Crew	Daily Output	Labor-Hours	Unit	Material	Labor	Equipment	Total	Total Incl O&P
0010	**INDOOR ATHLETIC CARPET**									
3700	Polyethylene, in rolls, no base incl., landscape surfaces	1 Tilf	275	.029	S.F.	3.47	1.19		4.66	5.70
3800	Nylon action surface, 1/8" thick		275	.029		3.63	1.19		4.82	5.85
3900	1/4" thick		275	.029		5.25	1.19		6.44	7.60
4000	3/8" thick		275	.029		6.55	1.19		7.74	9.10
5500	Polyvinyl chloride, sheet goods for gyms, 1/4" thick		80	.100		9.20	4.08		13.28	16.50
5600	3/8" thick	↓	60	.133	↓	10.60	5.45		16.05	20

09 69 Access Flooring

09 69 13 – Rigid-Grid Access Flooring

09 69 13.10 Access Floors		Crew	Daily Output	Labor-Hours	Unit	Material	Labor (2012 Bare Costs)	Equipment	Total	Total Incl O&P
0010	**ACCESS FLOORS**									
0015	Access floor package including panel, pedestal, stringers & laminate cover									
0100	Computer room, greater than 6,000 S.F.	4 Carp	750	.043	S.F.	7.50	1.88		9.38	11.35
0110	Less than 6,000 S.F.	2 Carp	375	.043		8.15	1.88		10.03	12.10
0120	Office, greater than 6,000 S.F.	4 Carp	1050	.030		5.35	1.34		6.69	8.10
0250	Panels, particle board or steel, 1250# load, no covering, under 6,000 S.F.	2 Carp	600	.027		3.75	1.18		4.93	6.05
0300	Over 6,000 S.F.		640	.025		3.21	1.10		4.31	5.35
0400	Aluminum, 24" panels		500	.032		33	1.41		34.41	39
0600	For carpet covering, add					8.75			8.75	9.65
0700	For vinyl floor covering, add					7			7	7.70
0900	For high pressure laminate covering, add					5.45			5.45	6
0910	For snap on stringer system, add	2 Carp	1000	.016		1.44	.71		2.15	2.74
0950	Office applications, steel or concrete panels,									
0960	no covering, over 6,000 S.F.	2 Carp	960	.017	S.F.	10.65	.74		11.39	12.90
1050	Pedestals, 6" to 12"	"	85	.188	Ea.	8.35	8.30		16.65	23

09 72 Wall Coverings

09 72 23 – Wallpapering

09 72 23.10 Wallpaper		Crew	Daily Output	Labor-Hours	Unit	Material	Labor (2012 Bare Costs)	Equipment	Total	Total Incl O&P
0010	**WALLPAPER** including sizing; add 10-30 percent waste @ takeoff									
0050	Aluminum foil	1 Pape	275	.029	S.F.	1.02	1.12		2.14	2.91
0100	Copper sheets, .025" thick, vinyl backing		240	.033		5.45	1.28		6.73	8.05
0300	Phenolic backing		240	.033		7.05	1.28		8.33	9.85
0600	Cork tiles, light or dark, 12" x 12" x 3/16"		240	.033		4.47	1.28		5.75	6.95
0700	5/16" thick		235	.034		3.80	1.31		5.11	6.25
0900	1/4" basketweave		240	.033		5.85	1.28		7.13	8.50
1000	1/2" natural, non-directional pattern		240	.033		7.50	1.28		8.78	10.30
1100	3/4" natural, non-directional pattern		240	.033		10.45	1.28		11.73	13.55
1200	Granular surface, 12" x 36", 1/2" thick		385	.021		1.26	.80		2.06	2.67
1300	1" thick		370	.022		1.63	.83		2.46	3.12
1500	Polyurethane coated, 12" x 12" x 3/16" thick		240	.033		3.95	1.28		5.23	6.40
1600	5/16" thick		235	.034		5.65	1.31		6.96	8.30
1800	Cork wallpaper, paperbacked, natural		480	.017		1.99	.64		2.63	3.21
1900	Colors		480	.017		2.78	.64		3.42	4.08
2100	Flexible wood veneer, 1/32" thick, plain woods		100	.080		2.36	3.08		5.44	7.50
2200	Exotic woods		95	.084		3.58	3.24		6.82	9.10
2400	Gypsum-based, fabric-backed, fire resistant									
2500	for masonry walls, minimum, 21 oz./S.Y.	1 Pape	800	.010	S.F.	.87	.39		1.26	1.57
2600	Average		720	.011		1.23	.43		1.66	2.03
2700	Maximum (small quantities)		640	.013		1.37	.48		1.85	2.28
2750	Acrylic, modified, semi-rigid PVC, .028" thick	2 Carp	330	.048		1.13	2.14		3.27	4.75
2800	.040" thick	"	320	.050		1.49	2.21		3.70	5.25
3000	Vinyl wall covering, fabric-backed, lightweight, type 1 (12-15 oz./S.Y.)	1 Pape	640	.013		.82	.48		1.30	1.67
3300	Medium weight, type 2 (20-24 oz./S.Y.)		480	.017		.90	.64		1.54	2.01
3400	Heavy weight, type 3 (28 oz./S.Y.)		435	.018		1.52	.71		2.23	2.80
3600	Adhesive, 5 gal. lots (18 S.Y./gal.)				Gal.	11.10			11.10	12.20
3700	Wallpaper, average workmanship, solid pattern, low cost paper	1 Pape	640	.013	S.F.	.37	.48		.85	1.18
3900	basic patterns (matching required), avg. cost paper		535	.015		.97	.58		1.55	1.99
4000	Paper at $85 per double roll, quality workmanship		435	.018		2.47	.71		3.18	3.85
4190	Grass cloth, natural fabric G		400	.020		2.99	.77		3.76	4.52
4200	Grass cloths with lining paper, minimum G		400	.020		.78	.77		1.55	2.09

09 72 Wall Coverings

09 72 23 – Wallpapering

09 72 23.10 Wallpaper		Crew	Daily Output	Labor-Hours	Unit	Material	Labor	Equipment	Total	Total Incl O&P
						2012 Bare Costs				
4300	Maximum G	1 Pape	350	.023	S.F.	2.49	.88		3.37	4.14
5990	Wallpaper removal, 1 layer		800	.010		.06	.39		.45	.68
6000	Wallpaper removal, 3 layer		400	.020		.14	.77		.91	1.38
9000	Minimum labor/equipment charge		2	4	Job		154		154	246

09 77 Special Wall Surfacing

09 77 33 – Fiberglass Reinforced Panels

09 77 33.10 Fiberglass Reinforced Plastic Panels

		Crew	Daily Output	Labor-Hours	Unit	Material	Labor	Equipment	Total	Total Incl O&P
0010	**FIBERGLASS REINFORCED PLASTIC PANELS**, .090" thick									
0020	On walls, adhesive mounted, embossed surface	2 Carp	640	.025	S.F.	1.15	1.10		2.25	3.08
0030	Smooth surface		640	.025		1.41	1.10		2.51	3.36
0040	Fire rated, embossed surface		640	.025		2.05	1.10		3.15	4.07
0050	Nylon rivet mounted, on drywall, embossed surface		480	.033		1.15	1.47		2.62	3.68
0060	Smooth surface		480	.033		1.41	1.47		2.88	3.96
0070	Fire rated, embossed surface		480	.033		2.05	1.47		3.52	4.67
0080	On masonry, embossed surface		320	.050		1.15	2.21		3.36	4.89
0090	Smooth surface		320	.050		1.41	2.21		3.62	5.15
0100	Fire rated, embossed surface		320	.050		2.05	2.21		4.26	5.90
0110	Nylon rivet and adhesive mounted, on drywall, embossed surface		240	.067		1.30	2.94		4.24	6.25
0120	Smooth surface		240	.067		1.30	2.94		4.24	6.25
0130	Fire rated, embossed surface		240	.067		2.26	2.94		5.20	7.30
0140	On masonry, embossed surface		190	.084		1.30	3.71		5.01	7.55
0150	Smooth surface		190	.084		1.30	3.71		5.01	7.55
0160	Fire rated, embossed surface		190	.084		2.26	3.71		5.97	8.60
0170	For moldings add	1 Carp	250	.032	L.F.	.32	1.41		1.73	2.66
0180	On ceilings, for lay in grid system, embossed surface		400	.020	S.F.	1.15	.88		2.03	2.72
0190	Smooth surface		400	.020		1.41	.88		2.29	3
0200	Fire rated, embossed surface		400	.020		2.05	.88		2.93	3.71

09 77 43 – Panel Systems

09 77 43.20 Slatwall Panels and Accessories

		Crew	Daily Output	Labor-Hours	Unit	Material	Labor	Equipment	Total	Total Incl O&P
0010	**SLATWALL PANELS AND ACCESSORIES**									
0100	Slatwall panel, 4' x 8' x 3/4" T, MDF, paint grade	1 Carp	500	.016	S.F.	1.32	.71		2.03	2.61
0110	Melamine finish		500	.016		1.87	.71		2.58	3.22
0120	High pressure plastic laminate finish		500	.016		3.25	.71		3.96	4.74
0130	Aluminum channel inserts, add					2.94			2.94	3.23
0200	Accessories, corner forms, 8' L				L.F.	4.28			4.28	4.71
0210	T-connector, 8' L					5.15			5.15	5.70
0220	J-mold, 8' L					1.33			1.33	1.46
0230	Edge cap, 8' L					2.05			2.05	2.26
0240	Finish end cap, 8' L					3.18			3.18	3.50
0300	Display hook, metal, 4" L				Ea.	.44			.44	.48
0310	6" L					.49			.49	.54
0320	8" L					.54			.54	.59
0330	10" L					.58			.58	.64
0340	12" L					.63			.63	.69
0350	Acrylic, 4" L					.91			.91	1
0360	6" L					1.05			1.05	1.16
0370	8" L					1.10			1.10	1.21
0380	10" L					1.12			1.12	1.23
0400	Waterfall hanger, metal, 12" - 16"					4.75			4.75	5.25

09 77 Special Wall Surfacing

09 77 43 – Panel Systems

09 77 43.20 Slatwall Panels and Accessories		Crew	Daily Output	Labor-Hours	Unit	Material (2012 Bare Costs)	Labor (2012 Bare Costs)	Equipment (2012 Bare Costs)	Total (2012 Bare Costs)	Total Incl O&P
0410	Acrylic				Ea.	9.85			9.85	10.85
0500	Shelf bracket, metal, 8″				↓	4.50			4.50	4.95
0510	10″					4.62			4.62	5.10
0520	12″					4.52			4.52	4.97
0530	14″					4.86			4.86	5.35
0540	16″					6.10			6.10	6.75
0550	Acrylic, 8″					3.04			3.04	3.34
0560	10″					3.37			3.37	3.71
0570	12″					3.72			3.72	4.09
0580	14″					3.99			3.99	4.39
0600	Shelf, acrylic, 12″ x 16″ x 1/4″					19.90			19.90	22
0610	12″ x 24″ x 1/4″				↓	30			30	33

09 81 Acoustic Insulation

09 81 16 – Acoustic Blanket Insulation

09 81 16.10 Sound Attenuation Blanket

		Crew	Daily Output	Labor-Hours	Unit	Material	Labor	Equipment	Total	Total Incl O&P
0010	**SOUND ATTENUATION BLANKET**									
0020	Blanket, 1″ thick	1 Carp	925	.009	S.F.	.25	.38		.63	.91
0500	1-1/2″ thick	↓	920	.009	↓	.25	.38		.63	.91
1000	2″ thick		915	.009		.34	.39		.73	1
1500	3″ thick	↓	910	.009		.50	.39		.89	1.19
3400	Urethane plastic foam, open cell, on wall, 2″ thick	2 Carp	2050	.008		3.05	.34		3.39	3.92
3500	3″ thick	↓	1550	.010		4.05	.46		4.51	5.20
3600	4″ thick		1050	.015		5.70	.67		6.37	7.35
3700	On ceiling, 2″ thick		1700	.009		3.04	.42		3.46	4.02
3800	3″ thick		1300	.012		4.05	.54		4.59	5.35
3900	4″ thick	↓	900	.018	↓	5.70	.78		6.48	7.55
4000	Nylon matting 0.4″ thick, with carbon black spinerette									
4010	plus polyester fabric, on floor	D-7	1600	.010	S.F.	1.30	.36		1.66	2
4200	Fiberglass reinf. backer board underlayment, 7/16″ thick, on floor	″	1500	.011	″	1.90	.39		2.29	2.70
9000	Minimum labor/equipment charge	1 Carp	5	1.600	Job		70.50		70.50	116

09 84 Acoustic Room Components

09 84 13 – Fixed Sound-Absorptive Panels

09 84 13.10 Fixed Panels

		Crew	Daily Output	Labor-Hours	Unit	Material	Labor	Equipment	Total	Total Incl O&P
0010	**FIXED PANELS** Perforated steel facing, painted with									
0100	Fiberglass or mineral filler, no backs, 2-1/4″ thick, modular									
0200	space units, ceiling or wall hung, white or colored	1 Carp	100	.080	S.F.	8.55	3.53		12.08	15.20
0300	Fiberboard sound deadening panels, 1/2″ thick	″	600	.013	″	.32	.59		.91	1.31
0500	Fiberglass panels, 4′ x 8′ x 1″ thick, with									
0600	glass cloth face for walls, cemented	1 Carp	155	.052	S.F.	7.90	2.28		10.18	12.45
0700	1-1/2″ thick, dacron covered, inner aluminum frame,									
0710	wall mounted	1 Carp	300	.027	S.F.	8.75	1.18		9.93	11.55
0900	Mineral fiberboard panels, fabric covered, 30″ x 108″,									
1000	3/4″ thick, concealed spline, wall mounted	1 Carp	150	.053	S.F.	6.25	2.35		8.60	10.75
9000	Minimum labor/equipment charge	″	4	2	Job		88		88	145

09 91 03 – Paint Restoration

09 91 03.20 Sanding		Crew	Daily Output	Labor-Hours	Unit	Material	2012 Bare Costs Labor	Equipment	Total	Total Incl O&P
0010	**SANDING** and puttying interior trim, compared to									
0100	Painting 1 coat, on quality work				L.F.		100%			
0300	Medium work						50%			
0400	Industrial grade						25%			
0500	Surface protection, placement and removal									
0510	Basic drop cloths	1 Pord	6400	.001	S.F.		.05		.05	.08
0520	Masking with paper		800	.010		.05	.38		.43	.67
0530	Volume cover up (using plastic sheathing, or building paper)		16000	.001			.02		.02	.03

09 91 03.30 Exterior Surface Preparation

		Crew	Daily Output	Labor-Hours	Unit	Material	Labor	Equipment	Total	Total Incl O&P
0010	**EXTERIOR SURFACE PREPARATION**									
0015	Doors, per side, not incl. frames or trim									
0020	Scrape & sand									
0030	Wood, flush	1 Pord	616	.013	S.F.		.50		.50	.79
0040	Wood, detail		496	.016			.62		.62	.98
0050	Wood, louvered		280	.029			1.09		1.09	1.74
0060	Wood, overhead		616	.013			.50		.50	.79
0070	Wire brush									
0080	Metal, flush	1 Pord	640	.013	S.F.		.48		.48	.76
0090	Metal, detail		520	.015			.59		.59	.94
0100	Metal, louvered		360	.022			.85		.85	1.36
0110	Metal or fibr., overhead		640	.013			.48		.48	.76
0120	Metal, roll up		560	.014			.55		.55	.87
0130	Metal, bulkhead		640	.013			.48		.48	.76
0140	Power wash, based on 2500 lb. operating pressure									
0150	Metal, flush	A-1H	2240	.004	S.F.		.13	.03	.16	.24
0160	Metal, detail		2120	.004			.13	.03	.16	.25
0170	Metal, louvered		2000	.004			.14	.03	.17	.27
0180	Metal or fibr., overhead		2400	.003			.12	.03	.15	.22
0190	Metal, roll up		2400	.003			.12	.03	.15	.22
0200	Metal, bulkhead		2200	.004			.13	.03	.16	.24
0400	Windows, per side, not incl. trim									
0410	Scrape & sand									
0420	Wood, 1-2 lite	1 Pord	320	.025	S.F.		.96		.96	1.53
0430	Wood, 3-6 lite		280	.029			1.09		1.09	1.74
0440	Wood, 7-10 lite		240	.033			1.27		1.27	2.03
0450	Wood, 12 lite		200	.040			1.53		1.53	2.44
0460	Wood, Bay/Bow		320	.025			.96		.96	1.53
0470	Wire brush									
0480	Metal, 1-2 lite	1 Pord	480	.017	S.F.		.64		.64	1.02
0490	Metal, 3-6 lite		400	.020			.77		.77	1.22
0500	Metal, Bay/Bow		480	.017			.64		.64	1.02
0510	Power wash, based on 2500 lb. operating pressure									
0520	1-2 lite	A-1H	4400	.002	S.F.		.06	.02	.08	.12
0530	3-6 lite		4320	.002			.06	.02	.08	.13
0540	7-10 lite		4240	.002			.07	.02	.09	.13
0550	12 lite		4160	.002			.07	.02	.09	.13
0560	Bay/Bow		4400	.002			.06	.02	.08	.12
0600	Siding, scrape and sand, light=10-30%, med.=30-70%									
0610	Heavy=70-100% of surface to sand									
0650	Texture 1-11, light	1 Pord	480	.017	S.F.		.64		.64	1.02
0660	Med.		440	.018			.70		.70	1.11
0670	Heavy		360	.022			.85		.85	1.36

09 91 Painting

09 91 03 – Paint Restoration

09 91 03.30 Exterior Surface Preparation

	09 91 03.30 Exterior Surface Preparation	Crew	Daily Output	Labor-Hours	Unit	Material	2012 Bare Costs Labor	Equipment	Total	Total Incl O&P
0680	Wood shingles, shakes, light	1 Pord	440	.018	S.F.		.70		.70	1.11
0690	Med.		360	.022			.85		.85	1.36
0700	Heavy		280	.029			1.09		1.09	1.74
0710	Clapboard, light		520	.015			.59		.59	.94
0720	Med.		480	.017			.64		.64	1.02
0730	Heavy		400	.020			.77		.77	1.22
0740	Wire brush									
0750	Aluminum, light	1 Pord	600	.013	S.F.		.51		.51	.81
0760	Med.		520	.015			.59		.59	.94
0770	Heavy		440	.018			.70		.70	1.11
0780	Pressure wash, based on 2500 lb. operating pressure									
0790	Stucco	A-1H	3080	.003	S.F.		.09	.02	.11	.17
0800	Aluminum or vinyl		3200	.003			.09	.02	.11	.16
0810	Siding, masonry, brick & block		2400	.003			.12	.03	.15	.22
1300	Miscellaneous, wire brush									
1310	Metal, pedestrian gate	1 Pord	100	.080	S.F.		3.06		3.06	4.88
8000	For chemical washing, see Section 04 01 30									
8010	For steam cleaning, see Section 04 01 30.20									

09 91 03.40 Interior Surface Preparation

	09 91 03.40 Interior Surface Preparation	Crew	Daily Output	Labor-Hours	Unit	Material	2012 Bare Costs Labor	Equipment	Total	Total Incl O&P
0010	**INTERIOR SURFACE PREPARATION**									
0020	Doors, per side, not incl. frames or trim									
0030	Scrape & sand									
0040	Wood, flush	1 Pord	616	.013	S.F.		.50		.50	.79
0050	Wood, detail		496	.016			.62		.62	.98
0060	Wood, louvered		280	.029			1.09		1.09	1.74
0070	Wire brush									
0080	Metal, flush	1 Pord	640	.013	S.F.		.48		.48	.76
0090	Metal, detail		520	.015			.59		.59	.94
0100	Metal, louvered		360	.022			.85		.85	1.36
0110	Hand wash									
0120	Wood, flush	1 Pord	2160	.004	S.F.		.14		.14	.23
0130	Wood, detailed		2000	.004			.15		.15	.24
0140	Wood, louvered		1360	.006			.22		.22	.36
0150	Metal, flush		2160	.004			.14		.14	.23
0160	Metal, detail		2000	.004			.15		.15	.24
0170	Metal, louvered		1360	.006			.22		.22	.36
0400	Windows, per side, not incl. trim									
0410	Scrape & sand									
0420	Wood, 1-2 lite	1 Pord	360	.022	S.F.		.85		.85	1.36
0430	Wood, 3-6 lite		320	.025			.96		.96	1.53
0440	Wood, 7-10 lite		280	.029			1.09		1.09	1.74
0450	Wood, 12 lite		240	.033			1.27		1.27	2.03
0460	Wood, Bay/Bow		360	.022			.85		.85	1.36
0470	Wire brush									
0480	Metal, 1-2 lite	1 Pord	520	.015	S.F.		.59		.59	.94
0490	Metal, 3-6 lite		440	.018			.70		.70	1.11
0500	Metal, Bay/Bow		520	.015			.59		.59	.94
0600	Walls, sanding, light=10-30%, medium - 30-70%,									
0610	heavy=70-100% of surface to sand									
0650	Walls, sand									
0660	Gypsum board or plaster, light	1 Pord	3077	.003	S.F.		.10		.10	.16
0670	Gypsum board or plaster, medium		2160	.004			.14		.14	.23

09 91 Painting

09 91 03 – Paint Restoration

09 91 03.40 Interior Surface Preparation

	09 91 03.40 Interior Surface Preparation	Crew	Daily Output	Labor-Hours	Unit	Material	2012 Bare Costs Labor	Equipment	Total	Total Incl O&P
0680	Gypsum board or plaster, heavy	1 Pord	923	.009	S.F.		.33		.33	.53
0690	Wood, T&G, light		2400	.003			.13		.13	.20
0700	Wood, T&G, med.		1600	.005			.19		.19	.31
0710	Wood, T&G, heavy		800	.010			.38		.38	.61
0720	Walls, wash									
0730	Gypsum board or plaster	1 Pord	3200	.003	S.F.		.10		.10	.15
0740	Wood, T&G		3200	.003			.10		.10	.15
0750	Masonry, brick & block, smooth		2800	.003			.11		.11	.17
0760	Masonry, brick & block, coarse		2000	.004			.15		.15	.24
8000	For chemical washing, see Section 04 01 30									
8010	For steam cleaning, see Section 04 01 30.20									
9010	Minimum labor/equipment charge	1 Pord	3	2.667	Job		102		102	163

09 91 03.41 Scrape After Fire Damage

	09 91 03.41 Scrape After Fire Damage	Crew	Daily Output	Labor-Hours	Unit	Material	Labor	Equipment	Total	Total Incl O&P
0010	**SCRAPE AFTER FIRE DAMAGE**									
0050	Boards, 1″ x 4″	1 Pord	336	.024	L.F.		.91		.91	1.45
0060	1″ x 6″		260	.031			1.18		1.18	1.88
0070	1″ x 8″		207	.039			1.48		1.48	2.36
0080	1″ x 10″		174	.046			1.76		1.76	2.81
0500	Framing, 2″ x 4″		265	.030			1.15		1.15	1.84
0510	2″ x 6″		221	.036			1.38		1.38	2.21
0520	2″ x 8″		190	.042			1.61		1.61	2.57
0530	2″ x 10″		165	.048			1.85		1.85	2.96
0540	2″ x 12″		144	.056			2.13		2.13	3.39
1000	Heavy framing, 3″ x 4″		226	.035			1.35		1.35	2.16
1010	4″ x 4″		210	.038			1.46		1.46	2.33
1020	4″ x 6″		191	.042			1.60		1.60	2.56
1030	4″ x 8″		165	.048			1.85		1.85	2.96
1040	4″ x 10″		144	.056			2.13		2.13	3.39
1060	4″ x 12″		131	.061			2.34		2.34	3.73
2900	For sealing, minimum		825	.010	S.F.	.14	.37		.51	.74
2920	Maximum		460	.017	″	.29	.67		.96	1.38
3000	For sandblasting, see Section 04 01 30.20									
3020										
9000	Minimum labor/equipment charge	1 Pord	3	2.667	Job		102		102	163

09 91 13 – Exterior Painting

09 91 13.30 Fences

	09 91 13.30 Fences	Crew	Daily Output	Labor-Hours	Unit	Material	Labor	Equipment	Total	Total Incl O&P
0010	**FENCES**									
0100	Chain link or wire metal, one side, water base									
0110	Roll & brush, first coat	1 Pord	960	.008	S.F.	.07	.32		.39	.58
0120	Second coat		1280	.006		.06	.24		.30	.45
0130	Spray, first coat		2275	.004		.07	.13		.20	.28
0140	Second coat		2600	.003		.07	.12		.19	.26
0150	Picket, water base									
0160	Roll & brush, first coat	1 Pord	865	.009	S.F.	.07	.35		.42	.64
0170	Second coat		1050	.008		.07	.29		.36	.55
0180	Spray, first coat		2275	.004		.07	.13		.20	.29
0190	Second coat		2600	.003		.07	.12		.19	.27
0200	Stockade, water base									
0210	Roll & brush, first coat	1 Pord	1040	.008	S.F.	.07	.29		.36	.55
0220	Second coat		1200	.007		.07	.26		.33	.49
0230	Spray, first coat		2275	.004		.07	.13		.20	.29
0240	Second coat		2600	.003		.07	.12		.19	.27

09 91 Painting

09 91 13 – Exterior Painting

09 91 13.30 Fences		Crew	Daily Output	Labor-Hours	Unit	Material	2012 Bare Costs Labor	Equipment	Total	Total Incl O&P
9000	Minimum labor/equipment charge	1 Pord	2	4	Job		153		153	244

09 91 13.42 Miscellaneous, Exterior

		Crew	Daily Output	Labor-Hours	Unit	Material	Labor	Equipment	Total	Total Incl O&P
0010	**MISCELLANEOUS, EXTERIOR**									
0100	Railing, ext., decorative wood, incl. cap & baluster									
0110	Newels & spindles @ 12″ O.C.									
0120	Brushwork, stain, sand, seal & varnish									
0130	First coat	1 Pord	90	.089	L.F.	.73	3.40		4.13	6.25
0140	Second coat	″	120	.067	″	.73	2.55		3.28	4.88
0150	Rough sawn wood, 42″ high, 2″ x 2″ verticals, 6″ O.C.									
0160	Brushwork, stain, each coat	1 Pord	90	.089	L.F.	.24	3.40		3.64	5.70
0170	Wrought iron, 1″ rail, 1/2″ sq. verticals									
0180	Brushwork, zinc chromate, 60″ high, bars 6″ O.C.									
0190	Primer	1 Pord	130	.062	L.F.	.83	2.35		3.18	4.67
0200	Finish coat		130	.062		1.09	2.35		3.44	4.96
0210	Additional coat		190	.042		1.27	1.61		2.88	3.97
0220	Shutters or blinds, single panel, 2′ x 4′, paint all sides									
0230	Brushwork, primer	1 Pord	20	.400	Ea.	.62	15.30		15.92	25
0240	Finish coat, exterior latex		20	.400		.54	15.30		15.84	25
0250	Primer & 1 coat, exterior latex		13	.615		1.02	23.50		24.52	38.50
0260	Spray, primer		35	.229		.90	8.75		9.65	14.95
0270	Finish coat, exterior latex		35	.229		1.14	8.75		9.89	15.20
0280	Primer & 1 coat, exterior latex		20	.400		.97	15.30		16.27	25.50
0290	For louvered shutters, add				S.F.	10%				
0300	Stair stringers, exterior, metal									
0310	Roll & brush, zinc chromate, to 14″, each coat	1 Pord	320	.025	L.F.	.36	.96		1.32	1.93
0320	Rough sawn wood, 4″ x 12″									
0330	Roll & brush, exterior latex, each coat	1 Pord	215	.037	L.F.	.08	1.42		1.50	2.36
0340	Trellis/lattice, 2″ x 2″ @ 3″ O.C. with 2″ x 8″ supports									
0350	Spray, latex, per side, each coat	1 Pord	475	.017	S.F.	.08	.64		.72	1.12
0450	Decking, ext., sealer, alkyd, brushwork, sealer coat		1140	.007		.09	.27		.36	.53
0460	1st coat		1140	.007		.09	.27		.36	.53
0470	2nd coat		1300	.006		.07	.24		.31	.45
0500	Paint, alkyd, brushwork, primer coat		1140	.007		.09	.27		.36	.53
0510	1st coat		1140	.007		.16	.27		.43	.60
0520	2nd coat		1300	.006		.12	.24		.36	.51
0600	Sand paint, alkyd, brushwork, 1 coat		150	.053		.16	2.04		2.20	3.43
9000	Minimum labor/equipment charge		2	4	Job		153		153	244

09 91 13.60 Siding Exterior

		Crew	Daily Output	Labor-Hours	Unit	Material	Labor	Equipment	Total	Total Incl O&P
0010	**SIDING EXTERIOR**, Alkyd (oil base)									
0450	Steel siding, oil base, paint 1 coat, brushwork	2 Pord	2015	.008	S.F.	.13	.30		.43	.62
0500	Spray		4550	.004		.19	.13		.32	.42
0800	Paint 2 coats, brushwork		1300	.012		.25	.47		.72	1.03
1000	Spray		2750	.006		.17	.22		.39	.55
1200	Stucco, rough, oil base, paint 2 coats, brushwork		1300	.012		.25	.47		.72	1.03
1400	Roller		1625	.010		.27	.38		.65	.89
1600	Spray		2925	.005		.28	.21		.49	.64
1800	Texture 1-11 or clapboard, oil base, primer coat, brushwork		1300	.012		.12	.47		.59	.88
2000	Spray		4550	.004		.12	.13		.25	.34
2400	Paint 2 coats, brushwork		810	.020		.37	.76		1.13	1.61
2600	Spray		2600	.006		.41	.24		.65	.83
3400	Stain 2 coats, brushwork		950	.017		.16	.64		.80	1.21
4000	Spray		3050	.005		.18	.20		.38	.52

09 91 13 – Exterior Painting

09 91 13.60 Siding Exterior		Crew	Daily Output	Labor-Hours	Unit	Material	Labor	Equipment	Total	Total Incl O&P
						2012 Bare Costs				
4200	Wood shingles, oil base primer coat, brushwork	2 Pord	1300	.012	S.F.	.11	.47		.58	.87
4400	Spray		3900	.004		.10	.16		.26	.36
5000	Paint 2 coats, brushwork		810	.020		.31	.76		1.07	1.55
5200	Spray		2275	.007		.29	.27		.56	.75
6500	Stain 2 coats, brushwork		950	.017		.16	.64		.80	1.21
7000	Spray		2660	.006		.22	.23		.45	.61
8000	For latex paint, deduct					10%				
8100	For work over 12′ H, from pipe scaffolding, add						15%			
8200	For work over 12′ H, from extension ladder, add						25%			
8300	For work over 12′ H, from swing staging, add						35%			
9000	Minimum labor/equipment charge	1 Pord	2	4	Job		153		153	244

09 91 13.62 Siding, Misc.		Crew	Daily Output	Labor-Hours	Unit	Material	Labor	Equipment	Total	Total Incl O&P
0010	**SIDING, MISC.**, latex paint									
0100	Aluminum siding									
0110	Brushwork, primer	2 Pord	2275	.007	S.F.	.06	.27		.33	.49
0120	Finish coat, exterior latex		2275	.007		.05	.27		.32	.49
0130	Primer & 1 coat exterior latex		1300	.012		.12	.47		.59	.88
0140	Primer & 2 coats exterior latex		975	.016		.17	.63		.80	1.18
0150	Mineral fiber shingles									
0160	Brushwork, primer	2 Pord	1495	.011	S.F.	.12	.41		.53	.78
0170	Finish coat, industrial enamel		1495	.011		.15	.41		.56	.82
0180	Primer & 1 coat enamel		810	.020		.27	.76		1.03	1.50
0190	Primer & 2 coats enamel		540	.030		.42	1.13		1.55	2.27
0200	Roll, primer		1625	.010		.13	.38		.51	.74
0210	Finish coat, industrial enamel		1625	.010		.16	.38		.54	.78
0220	Primer & 1 coat enamel		975	.016		.29	.63		.92	1.32
0230	Primer & 2 coats enamel		650	.025		.45	.94		1.39	2
0240	Spray, primer		3900	.004		.10	.16		.26	.36
0250	Finish coat, industrial enamel		3900	.004		.14	.16		.30	.40
0260	Primer & 1 coat enamel		2275	.007		.24	.27		.51	.69
0270	Primer & 2 coats enamel		1625	.010		.37	.38		.75	1.01
0280	Waterproof sealer, first coat		4485	.004		.08	.14		.22	.31
0290	Second coat		5235	.003		.07	.12		.19	.27
0300	Rough wood incl. shingles, shakes or rough sawn siding									
0310	Brushwork, primer	2 Pord	1280	.013	S.F.	.12	.48		.60	.90
0320	Finish coat, exterior latex		1280	.013		.09	.48		.57	.86
0330	Primer & 1 coat exterior latex		960	.017		.21	.64		.85	1.26
0340	Primer & 2 coats exterior latex		700	.023		.30	.87		1.17	1.73
0350	Roll, primer		2925	.005		.17	.21		.38	.51
0360	Finish coat, exterior latex		2925	.005		.11	.21		.32	.45
0370	Primer & 1 coat exterior latex		1790	.009		.27	.34		.61	.85
0380	Primer & 2 coats exterior latex		1300	.012		.38	.47		.85	1.17
0390	Spray, primer		3900	.004		.14	.16		.30	.40
0400	Finish coat, exterior latex		3900	.004		.08	.16		.24	.34
0410	Primer & 1 coat exterior latex		2600	.006		.22	.24		.46	.63
0420	Primer & 2 coats exterior latex		2080	.008		.31	.29		.60	.81
0430	Waterproof sealer, first coat		4485	.004		.14	.14		.28	.37
0440	Second coat		4485	.004		.08	.14		.22	.31
0450	Smooth wood incl. butt, T&G, beveled, drop or B&B siding									
0460	Brushwork, primer	2 Pord	2325	.007	S.F.	.09	.26		.35	.52
0470	Finish coat, exterior latex		1280	.013		.09	.48		.57	.86
0480	Primer & 1 coat exterior latex		800	.020		.18	.77		.95	1.42

09 91 Painting

09 91 13 – Exterior Painting

09 91 13.62 Siding, Misc.		Crew	Daily Output	Labor-Hours	Unit	2012 Bare Costs Material	Labor	Equipment	Total	Total Incl O&P
0490	Primer & 2 coats exterior latex	2 Pord	630	.025	S.F.	.27	.97		1.24	1.85
0500	Roll, primer		2275	.007		.10	.27		.37	.54
0510	Finish coat, exterior latex		2275	.007		.10	.27		.37	.54
0520	Primer & 1 coat exterior latex		1300	.012		.20	.47		.67	.97
0530	Primer & 2 coats exterior latex		975	.016		.29	.63		.92	1.32
0540	Spray, primer		4550	.004		.08	.13		.21	.30
0550	Finish coat, exterior latex		4550	.004		.08	.13		.21	.30
0560	Primer & 1 coat exterior latex		2600	.006		.16	.24		.40	.56
0570	Primer & 2 coats exterior latex		1950	.008		.24	.31		.55	.77
0580	Waterproof sealer, first coat		5230	.003		.08	.12		.20	.28
0590	Second coat		5980	.003		.08	.10		.18	.25
0600	For oil base paint, add					10%				
9000	Minimum labor/equipment charge	1 Pord	2	4	Job		153		153	244

09 91 13.70 Doors and Windows, Exterior		Crew	Daily Output	Labor-Hours	Unit	Material	Labor	Equipment	Total	Total Incl O&P
0010	**DOORS AND WINDOWS, EXTERIOR**									
0100	Door frames & trim, only									
0110	Brushwork, primer	1 Pord	512	.016	L.F.	.06	.60		.66	1.01
0120	Finish coat, exterior latex		512	.016		.07	.60		.67	1.02
0130	Primer & 1 coat, exterior latex		300	.027		.12	1.02		1.14	1.77
0135	2 coats, exterior latex, both sides		15	.533	Ea.	6	20.50		26.50	39
0140	Primer & 2 coats, exterior latex		265	.030	L.F.	.19	1.15		1.34	2.05
0150	Doors, flush, both sides, incl. frame & trim									
0160	Roll & brush, primer	1 Pord	10	.800	Ea.	4.26	30.50		34.76	53.50
0170	Finish coat, exterior latex		10	.800		5.10	30.50		35.60	54.50
0180	Primer & 1 coat, exterior latex		7	1.143		9.35	43.50		52.85	80.50
0190	Primer & 2 coats, exterior latex		5	1.600		14.45	61		75.45	113
0200	Brushwork, stain, sealer & 2 coats polyurethane		4	2		25.50	76.50		102	150
0210	Doors, French, both sides, 10-15 lite, incl. frame & trim									
0220	Brushwork, primer	1 Pord	6	1.333	Ea.	2.13	51		53.13	84
0230	Finish coat, exterior latex		6	1.333		2.55	51		53.55	84.50
0240	Primer & 1 coat, exterior latex		3	2.667		4.67	102		106.67	168
0250	Primer & 2 coats, exterior latex		2	4		7.05	153		160.05	252
0260	Brushwork, stain, sealer & 2 coats polyurethane		2.50	3.200		9.30	122		131.30	205
0270	Doors, louvered, both sides, incl. frame & trim									
0280	Brushwork, primer	1 Pord	7	1.143	Ea.	4.26	43.50		47.76	74.50
0290	Finish coat, exterior latex		7	1.143		5.10	43.50		48.60	75.50
0300	Primer & 1 coat, exterior latex		4	2		9.35	76.50		85.85	132
0310	Primer & 2 coats, exterior latex		3	2.667		14.15	102		116.15	179
0320	Brushwork, stain, sealer & 2 coats polyurethane		4.50	1.778		25.50	68		93.50	137
0330	Doors, panel, both sides, incl. frame & trim									
0340	Roll & brush, primer	1 Pord	6	1.333	Ea.	4.26	51		55.26	86
0350	Finish coat, exterior latex		6	1.333		5.10	51		56.10	87
0360	Primer & 1 coat, exterior latex		3	2.667		9.35	102		111.35	173
0370	Primer & 2 coats, exterior latex		2.50	3.200		14.15	122		136.15	211
0380	Brushwork, stain, sealer & 2 coats polyurethane		3	2.667		25.50	102		127.50	191
0400	Windows, per ext. side, based on 15 S.F.									
0410	1 to 6 lite									
0420	Brushwork, primer	1 Pord	13	.615	Ea.	.84	23.50		24.34	38.50
0430	Finish coat, exterior latex		13	.615		1	23.50		24.50	38.50
0440	Primer & 1 coat, exterior latex		8	1		1.85	38.50		40.35	63
0450	Primer & 2 coats, exterior latex		6	1.333		2.79	51		53.79	84.50
0460	Stain, sealer & 1 coat varnish		7	1.143		3.67	43.50		47.17	74

09 91 13 – Exterior Painting

09 91 13.70 Doors and Windows, Exterior		Crew	Daily Output	Labor-Hours	Unit	Material	2012 Bare Costs Labor	Equipment	Total	Total Incl O&P
0470	7 to 10 lite									
0480	Brushwork, primer	1 Pord	11	.727	Ea.	.84	28		28.84	45.50
0490	Finish coat, exterior latex		11	.727		1	28		29	45.50
0500	Primer & 1 coat, exterior latex		7	1.143		1.85	43.50		45.35	72
0510	Primer & 2 coats, exterior latex		5	1.600		2.79	61		63.79	101
0520	Stain, sealer & 1 coat varnish		6	1.333		3.67	51		54.67	85.50
0530	12 lite									
0540	Brushwork, primer	1 Pord	10	.800	Ea.	.84	30.50		31.34	50
0550	Finish coat, exterior latex		10	.800		1	30.50		31.50	50
0560	Primer & 1 coat, exterior latex		6	1.333		1.85	51		52.85	83.50
0570	Primer & 2 coats, exterior latex		5	1.600		2.79	61		63.79	101
0580	Stain, sealer & 1 coat varnish		6	1.333		3.68	51		54.68	85.50
0590	For oil base paint, add					10%				
9000	Minimum labor/equipment charge	1 Pord	2	4	Job		153		153	244

09 91 13.80 Trim, Exterior		Crew	Daily Output	Labor-Hours	Unit	Material	Labor	Equipment	Total	Total Incl O&P
0010	**TRIM, EXTERIOR**									
0100	Door frames & trim (see Doors, interior or exterior)									
0110	Fascia, latex paint, one coat coverage									
0120	1" x 4", brushwork	1 Pord	640	.013	L.F.	.02	.48		.50	.78
0130	Roll		1280	.006		.02	.24		.26	.40
0140	Spray		2080	.004		.02	.15		.17	.26
0150	1" x 6" to 1" x 10", brushwork		640	.013		.07	.48		.55	.84
0160	Roll		1230	.007		.07	.25		.32	.48
0170	Spray		2100	.004		.06	.15		.21	.29
0180	1" x 12", brushwork		640	.013		.07	.48		.55	.84
0190	Roll		1050	.008		.07	.29		.36	.55
0200	Spray		2200	.004		.06	.14		.20	.28
0210	Gutters & downspouts, metal, zinc chromate paint									
0220	Brushwork, gutters, 5", first coat	1 Pord	640	.013	L.F.	.38	.48		.86	1.18
0230	Second coat		960	.008		.36	.32		.68	.91
0240	Third coat		1280	.006		.29	.24		.53	.70
0250	Downspouts, 4", first coat		640	.013		.38	.48		.86	1.18
0260	Second coat		960	.008		.36	.32		.68	.91
0270	Third coat		1280	.006		.29	.24		.53	.70
0280	Gutters & downspouts, wood									
0290	Brushwork, gutters, 5", primer	1 Pord	640	.013	L.F.	.06	.48		.54	.82
0300	Finish coat, exterior latex		640	.013		.06	.48		.54	.83
0310	Primer & 1 coat exterior latex		400	.020		.12	.77		.89	1.36
0320	Primer & 2 coats exterior latex		325	.025		.19	.94		1.13	1.71
0330	Downspouts, 4", primer		640	.013		.06	.48		.54	.82
0340	Finish coat, exterior latex		640	.013		.06	.48		.54	.83
0350	Primer & 1 coat exterior latex		400	.020		.12	.77		.89	1.36
0360	Primer & 2 coats exterior latex		325	.025		.10	.94		1.04	1.60
0370	Molding, exterior, up to 14" wide									
0380	Brushwork, primer	1 Pord	640	.013	L.F.	.07	.48		.55	.83
0390	Finish coat, exterior latex		640	.013		.07	.48		.55	.84
0400	Primer & 1 coat exterior latex		400	.020		.15	.77		.92	1.38
0410	Primer & 2 coats exterior latex		315	.025		.15	.97		1.12	1.71
0420	Stain & fill		1050	.008		.10	.29		.39	.58
0430	Shellac		1850	.004		.10	.17		.27	.37
0440	Varnish		1275	.006		.11	.24		.35	.51
9000	Minimum labor/equipment charge		2	4	Job		153		153	244

09 91 Painting

09 91 13 – Exterior Painting

09 91 13.90 Walls, Masonry (CMU), Exterior		Crew	Daily Output	Labor-Hours	Unit	Material	2012 Bare Costs Labor	Equipment	Total	Total Incl O&P
0350	**WALLS, MASONRY (CMU), EXTERIOR**									
0360	Concrete masonry units (CMU), smooth surface									
0370	Brushwork, latex, first coat	1 Pord	640	.013	S.F.	.06	.48		.54	.83
0380	Second coat		960	.008		.05	.32		.37	.56
0390	Waterproof sealer, first coat		736	.011		.24	.42		.66	.93
0400	Second coat		1104	.007		.24	.28		.52	.71
0410	Roll, latex, paint, first coat		1465	.005		.07	.21		.28	.41
0420	Second coat		1790	.004		.05	.17		.22	.33
0430	Waterproof sealer, first coat		1680	.005		.24	.18		.42	.56
0440	Second coat		2060	.004		.24	.15		.39	.51
0450	Spray, latex, paint, first coat		1950	.004		.06	.16		.22	.31
0460	Second coat		2600	.003		.05	.12		.17	.24
0470	Waterproof sealer, first coat		2245	.004		.24	.14		.38	.49
0480	Second coat		2990	.003		.24	.10		.34	.43
0490	Concrete masonry unit (CMU), porous									
0500	Brushwork, latex, first coat	1 Pord	640	.013	S.F.	.12	.48		.60	.89
0510	Second coat		960	.008		.06	.32		.38	.58
0520	Waterproof sealer, first coat		736	.011		.24	.42		.66	.93
0530	Second coat		1104	.007		.24	.28		.52	.71
0540	Roll latex, first coat		1465	.005		.09	.21		.30	.43
0550	Second coat		1790	.004		.06	.17		.23	.33
0560	Waterproof sealer, first coat		1680	.005		.24	.18		.42	.56
0570	Second coat		2060	.004		.24	.15		.39	.51
0580	Spray latex, first coat		1950	.004		.07	.16		.23	.32
0590	Second coat		2600	.003		.05	.12		.17	.24
0600	Waterproof sealer, first coat		2245	.004		.24	.14		.38	.49
0610	Second coat		2990	.003		.24	.10		.34	.43
9000	Minimum labor/equipment charge		2	4	Job		153		153	244

09 91 23 – Interior Painting

09 91 23.20 Cabinets and Casework

Code	Description	Crew	Daily Output	Labor-Hours	Unit	Material	Labor	Equipment	Total	Total Incl O&P
0010	**CABINETS AND CASEWORK**									
1000	Primer coat, oil base, brushwork	1 Pord	650	.012	S.F.	.06	.47		.53	.82
2000	Paint, oil base, brushwork, 1 coat		650	.012		.11	.47		.58	.87
3000	Stain, brushwork, wipe off		650	.012		.08	.47		.55	.84
4000	Shellac, 1 coat, brushwork		650	.012		.08	.47		.55	.84
4500	Varnish, 3 coats, brushwork, sand after 1st coat		325	.025		.29	.94		1.23	1.82
5000	For latex paint, deduct					10%				
9010	Minimum labor/equipment charge	1 Pord	2	4	Job		153		153	244

09 91 23.33 Doors and Windows, Interior Alkyd (Oil Base)

Code	Description	Crew	Daily Output	Labor-Hours	Unit	Material	Labor	Equipment	Total	Total Incl O&P
0010	**DOORS AND WINDOWS, INTERIOR ALKYD (OIL BASE)**									
0500	Flush door & frame, 3' x 7', oil, primer, brushwork	1 Pord	10	.800	Ea.	3.08	30.50		33.58	52.50
1000	Paint, 1 coat		10	.800		4.11	30.50		34.61	53.50
1400	Stain, brushwork, wipe off		18	.444		1.68	17		18.68	29
1600	Shellac, 1 coat, brushwork		25	.320		1.70	12.25		13.95	21.50
1800	Varnish, 3 coats, brushwork, sand after 1st coat		9	.889		6.05	34		40.05	61
2000	Panel door & frame, 3' x 7', oil, primer, brushwork		6	1.333		2.28	51		53.28	84
2200	Paint, 1 coat		6	1.333		4.11	51		55.11	86
2600	Stain, brushwork, panel door, 3' x 7', not incl. frame		16	.500		1.68	19.15		20.83	32.50
2800	Shellac, 1 coat, brushwork		22	.364		1.70	13.90		15.60	24
3000	Varnish, 3 coats, brushwork, sand after 1st coat		7.50	1.067		6.05	41		47.05	71.50
4400	Windows, including frame and trim, per side									

09 91 23 – Interior Painting

09 91 23.33 Doors and Windows, Interior Alkyd (Oil Base)		Crew	Daily Output	Labor-Hours	Unit	Material	Labor	Equipment	Total	Total Incl O&P
						2012 Bare Costs				
4600	Colonial type, 6/6 lites, 2' x 3', oil, primer, brushwork	1 Pord	14	.571	Ea.	.36	22		22.36	35.50
5800	Paint, 1 coat		14	.571		.65	22		22.65	35.50
6200	3' x 5' opening, 6/6 lites, primer coat, brushwork		12	.667		.90	25.50		26.40	41.50
6400	Paint, 1 coat		12	.667		1.62	25.50		27.12	42.50
6800	4' x 8' opening, 6/6 lites, primer coat, brushwork		8	1		1.92	38.50		40.42	63
7000	Paint, 1 coat		8	1		3.46	38.50		41.96	65
8000	Single lite type, 2' x 3', oil base, primer coat, brushwork		33	.242		.36	9.25		9.61	15.20
8200	Paint, 1 coat		33	.242		.65	9.25		9.90	15.50
8600	3' x 5' opening, primer coat, brushwork		20	.400		.90	15.30		16.20	25.50
8800	Paint, 1 coat		20	.400		1.62	15.30		16.92	26.50
9200	4' x 8' opening, primer coat, brushwork		14	.571		1.92	22		23.92	37
9400	Paint, 1 coat		14	.571		3.46	22		25.46	39
9900	Minimum labor/equipment charge		2	4	Job		153		153	244

09 91 23.35 Doors and Windows, Interior Latex		Crew	Daily Output	Labor-Hours	Unit	Material	Labor	Equipment	Total	Total Incl O&P
0010	**DOORS & WINDOWS, INTERIOR LATEX**									
0100	Doors, flush, both sides, incl. frame & trim									
0110	Roll & brush, primer	1 Pord	10	.800	Ea.	3.94	30.50		34.44	53.50
0120	Finish coat, latex		10	.800		4.43	30.50		34.93	54
0130	Primer & 1 coat latex		7	1.143		8.35	43.50		51.85	79
0140	Primer & 2 coats latex		5	1.600		12.55	61		73.55	111
0160	Spray, both sides, primer		20	.400		4.15	15.30		19.45	29
0170	Finish coat, latex		20	.400		4.64	15.30		19.94	29.50
0180	Primer & 1 coat latex		11	.727		8.85	28		36.85	54.50
0190	Primer & 2 coats latex		8	1		13.25	38.50		51.75	75.50
0200	Doors, French, both sides, 10-15 lite, incl. frame & trim									
0210	Roll & brush, primer	1 Pord	6	1.333	Ea.	1.97	51		52.97	83.50
0220	Finish coat, latex		6	1.333		2.21	51		53.21	84
0230	Primer & 1 coat latex		3	2.667		4.18	102		106.18	168
0240	Primer & 2 coats latex		2	4		6.25	153		159.25	251
0260	Doors, louvered, both sides, incl. frame & trim									
0270	Roll & brush, primer	1 Pord	7	1.143	Ea.	3.94	43.50		47.44	74.50
0280	Finish coat, latex		7	1.143		4.43	43.50		47.93	75
0290	Primer & 1 coat, latex		4	2		8.15	76.50		84.65	131
0300	Primer & 2 coats, latex		3	2.667		12.80	102		114.80	177
0320	Spray, both sides, primer		20	.400		4.15	15.30		19.45	29
0330	Finish coat, latex		20	.400		4.64	15.30		19.94	29.50
0340	Primer & 1 coat, latex		11	.727		8.85	28		36.85	54.50
0350	Primer & 2 coats, latex		8	1		13.55	38.50		52.05	76
0360	Doors, panel, both sides, incl. frame & trim									
0370	Roll & brush, primer	1 Pord	6	1.333	Ea.	4.15	51		55.15	86
0380	Finish coat, latex		6	1.333		4.43	51		55.43	86.50
0390	Primer & 1 coat, latex		3	2.667		8.35	102		110.35	172
0400	Primer & 2 coats, latex		2.50	3.200		12.80	122		134.80	209
0420	Spray, both sides, primer		10	.800		4.15	30.50		34.65	53.50
0430	Finish coat, latex		10	.800		4.64	30.50		35.14	54
0440	Primer & 1 coat, latex		5	1.600		8.85	61		69.85	107
0450	Primer & 2 coats, latex		4	2		13.55	76.50		90.05	137
0460	Windows, per interior side, based on 15 S.F.									
0470	1 to 6 lite									
0480	Brushwork, primer	1 Pord	13	.615	Ea.	.78	23.50		24.28	38.50
0490	Finish coat, enamel		13	.615		.87	23.50		24.37	38.50
0500	Primer & 1 coat enamel		8	1		1.65	38.50		40.15	63

09 91 23.35 Doors and Windows, Interior Latex			Crew	Daily Output	Labor-Hours	Unit	Material	Labor	Equipment	Total	Total Incl O&P
0510	Primer & 2 coats enamel		1 Pord	6	1.333	Ea.	2.52	51		53.52	84.50
0530	7 to 10 lite										
0540	Brushwork, primer		1 Pord	11	.727	Ea.	.78	28		28.78	45.50
0550	Finish coat, enamel			11	.727		.87	28		28.87	45.50
0560	Primer & 1 coat enamel			7	1.143		1.65	43.50		45.15	72
0570	Primer & 2 coats enamel		↓	5	1.600	↓	2.52	61		63.52	100
0590	12 lite										
0600	Brushwork, primer		1 Pord	10	.800	Ea.	.78	30.50		31.28	50
0610	Finish coat, enamel			10	.800		.87	30.50		31.37	50
0620	Primer & 1 coat enamel			6	1.333		1.65	51		52.65	83.50
0630	Primer & 2 coats enamel		↓	5	1.600		2.52	61		63.52	100
0650	For oil base paint, add					↓	10%				
9000	Minimum labor/equipment charge		1 Pord	2	4	Job		153		153	244

09 91 23.39 Doors and Windows, Interior Latex, Zero Voc			Crew	Daily Output	Labor-Hours	Unit	Material	Labor	Equipment	Total	Total Incl O&P
0010	**DOORS & WINDOWS, INTERIOR LATEX, ZERO VOC**										
0100	Doors flush, both sides, incl. frame & trim										
0110	Roll & brush, primer	G	1 Pord	10	.800	Ea.	5.10	30.50		35.60	54.50
0120	Finish coat, latex	G		10	.800		5.80	30.50		36.30	55.50
0130	Primer & 1 coat latex	G		7	1.143		10.90	43.50		54.40	82
0140	Primer & 2 coats latex	G		5	1.600		16.35	61		77.35	116
0160	Spray, both sides, primer	G		20	.400		5.35	15.30		20.65	30.50
0170	Finish coat, latex	G		20	.400		6.10	15.30		21.40	31
0180	Primer & 1 coat latex	G		11	.727		11.50	28		39.50	57
0190	Primer & 2 coats latex	G	↓	8	1	↓	17.30	38.50		55.80	80
0200	Doors, French, both sides, 10-15 lite, incl. frame & trim										
0210	Roll & brush, primer	G	1 Pord	6	1.333	Ea.	2.55	51		53.55	84.50
0220	Finish coat, latex	G		6	1.333		2.90	51		53.90	84.50
0230	Primer & 1 coat latex	G		3	2.667		5.45	102		107.45	169
0240	Primer & 2 coats latex	G	↓	2	4	↓	8.20	153		161.20	253
0360	Doors, panel, both sides, incl. frame & trim										
0370	Roll & brush, primer	G	1 Pord	6	1.333	Ea.	5.35	51		56.35	87.50
0380	Finish coat, latex	G		6	1.333		5.80	51		56.80	88
0390	Primer & 1 coat, latex	G		3	2.667		10.90	102		112.90	175
0400	Primer & 2 coats, latex	G		2.50	3.200		16.70	122		138.70	213
0420	Spray, both sides, primer	G		10	.800		5.35	30.50		35.85	55
0430	Finish coat, latex	G		10	.800		6.10	30.50		36.60	55.50
0440	Primer & 1 coat, latex	G		5	1.600		11.50	61		72.50	110
0450	Primer & 2 coats, latex	G	↓	4	2	↓	17.70	76.50		94.20	141
0460	Windows, per interior side, based on 15 S.F.										
0470	1 to 6 lite										
0480	Brushwork, primer	G	1 Pord	13	.615	Ea.	1	23.50		24.50	38.50
0490	Finish coat, enamel	G		13	.615		1.15	23.50		24.65	39
0500	Primer & 1 coat enamel	G		8	1		2.15	38.50		40.65	63.50
0510	Primer & 2 coats enamel	G		6	1.333	↓	3.30	51		54.30	85
9000	Minimum labor/equipment charge		↓	2	4	Job		153		153	244

09 91 23.40 Floors, Interior			Crew	Daily Output	Labor-Hours	Unit	Material	Labor	Equipment	Total	Total Incl O&P
0010	**FLOORS, INTERIOR**										
0100	Concrete paint, latex										
0110	Brushwork										
0120	1st coat		1 Pord	975	.008	S.F.	.16	.31		.47	.67
0130	2nd coat			1150	.007		.10	.27		.37	.53
0140	3rd coat		↓	1300	.006	↓	.08	.24		.32	.47

09 91 23.40 Floors, Interior		Crew	Daily Output	Labor-Hours	Unit	2012 Bare Costs Material	Labor	Equipment	Total	Total Incl O&P
0150	Roll									
0160	1st coat	1 Pord	2600	.003	S.F.	.21	.12		.33	.42
0170	2nd coat		3250	.002		.13	.09		.22	.29
0180	3rd coat		3900	.002		.09	.08		.17	.23
0190	Spray									
0200	1st coat	1 Pord	2600	.003	S.F.	.18	.12		.30	.39
0210	2nd coat		3250	.002		.10	.09		.19	.26
0220	3rd coat		3900	.002		.08	.08		.16	.22
0300	Acid stain and sealer									
0310	Stain, one coat	1 Pord	650	.012	S.F.	.11	.47		.58	.87
0320	Two coats		570	.014		.22	.54		.76	1.10
0330	Acrylic sealer, one coat		2600	.003		.18	.12		.30	.39
0340	Two coats		1400	.006		.37	.22		.59	.76

09 91 23.52 Miscellaneous, Interior		Crew	Daily Output	Labor-Hours	Unit	Material	Labor	Equipment	Total	Total Incl O&P
0010	**MISCELLANEOUS, INTERIOR**									
2400	Floors, conc./wood, oil base, primer/sealer coat, brushwork	2 Pord	1950	.008	S.F.	.07	.31		.38	.58
2450	Roller		5200	.003		.08	.12		.20	.27
2600	Spray		6000	.003		.08	.10		.18	.24
2650	Paint 1 coat, brushwork		1950	.008		.09	.31		.40	.60
2800	Roller		5200	.003		.09	.12		.21	.29
2850	Spray		6000	.003		.10	.10		.20	.27
3000	Stain, wood floor, brushwork, 1 coat		4550	.004		.08	.13		.21	.30
3200	Roller		5200	.003		.08	.12		.20	.28
3250	Spray		6000	.003		.08	.10		.18	.25
3400	Varnish, wood floor, brushwork		4550	.004		.10	.13		.23	.32
3450	Roller		5200	.003		.10	.12		.22	.30
3600	Spray		6000	.003		.11	.10		.21	.28
3800	Grilles, per side, oil base, primer coat, brushwork	1 Pord	520	.015		.12	.59		.71	1.07
3850	Spray		1140	.007		.13	.27		.40	.57
3920	Paint 2 coats, brushwork		325	.025		.42	.94		1.36	1.96
3940	Spray		650	.012		.48	.47		.95	1.28
5000	Pipe, 1" - 4" diameter, primer or sealer coat, oil base, brushwork	2 Pord	1250	.013	L.F.	.08	.49		.57	.86
5100	Spray		2165	.007		.07	.28		.35	.53
5350	Paint 2 coats, brushwork		775	.021		.20	.79		.99	1.48
5400	Spray		1240	.013		.23	.49		.72	1.04
6300	13" - 16" diameter, primer or sealer coat, brushwork		310	.052		.31	1.97		2.28	3.49
6450	Spray		540	.030		.47	1.13		1.60	2.32
6500	Paint 2 coats, brushwork		195	.082		.82	3.14		3.96	5.90
6550	Spray		310	.052		.91	1.97		2.88	4.15
7000	Trim, wood, incl. puttying, under 6" wide									
7200	Primer coat, oil base, brushwork	1 Pord	650	.012	L.F.	.03	.47		.50	.78
7250	Paint, 1 coat, brushwork		650	.012		.05	.47		.52	.81
7450	3 coats		325	.025		.16	.94		1.10	1.67
7500	Over 6" wide, primer coat, brushwork		650	.012		.06	.47		.53	.82
7550	Paint, 1 coat, brushwork		650	.012		.11	.47		.58	.87
7650	3 coats		325	.025		.31	.94		1.25	1.84
8000	Cornice, simple design, primer coat, oil base, brushwork		650	.012	S.F.	.06	.47		.53	.82
8250	Paint, 1 coat		650	.012		.11	.47		.58	.87
8350	Ornate design, primer coat		350	.023		.06	.87		.93	1.47
8400	Paint, 1 coat		350	.023		.11	.87		.98	1.52
8600	Balustrades, primer coat, oil base, brushwork		520	.015		.06	.59		.65	1.01
8800										

09 91 Painting

09 91 23 – Interior Painting

09 91 23.52 Miscellaneous, Interior

	09 91 23.52 Miscellaneous, Interior	Crew	Daily Output	Labor-Hours	Unit	2012 Bare Costs Material	Labor	Equipment	Total	Total Incl O&P
8900	Trusses and wood frames, primer coat, oil base, brushwork	1 Pord	800	.010	S.F.	.06	.38		.44	.68
8950	Spray		1200	.007		.06	.26		.32	.48
9220	Paint 2 coats, brushwork		500	.016		.21	.61		.82	1.21
9240	Spray		600	.013		.23	.51		.74	1.07
9260	Stain, brushwork, wipe off		600	.013		.08	.51		.59	.90
9280	Varnish, 3 coats, brushwork		275	.029		.29	1.11		1.40	2.10
9350	For latex paint, deduct					10%				
9900	Minimum labor/equipment charge	1 Pord	2	4	Job		153		153	244

09 91 23.62 Electrostatic Painting

	09 91 23.62 Electrostatic Painting	Crew	Daily Output	Labor-Hours	Unit	Material	Labor	Equipment	Total	Total Incl O&P
0010	**ELECTROSTATIC PAINTING**									
0100	In shop									
0200	Flat surfaces (lockers, casework, elevator doors. etc.)									
0300	One coat	1 Pord	200	.040	S.F.	.46	1.53		1.99	2.95
0400	Two coats	"	120	.067	"	.67	2.55		3.22	4.81
0500	Irregular surfaces (furniture, door frames, etc)									
0600	One coat	1 Pord	150	.053	S.F.	.46	2.04		2.50	3.77
0700	Two coats	"	100	.080	"	.67	3.06		3.73	5.60
0800	On site									
0900	Flat surfaces (lockers, casework, elev doors. etc)									
1000	One coat	1 Pord	150	.053	S.F.	.46	2.04		2.50	3.77
1100	Two coats	"	100	.080	"	.67	3.06		3.73	5.60
1200	Irregular surfaces (furniture, door frames, etc)									
1300	One coat	1 Pord	115	.070	S.F.	.46	2.66		3.12	4.76
1400	Two coats	"	70	.114	"	.67	4.37		5.04	7.75

09 91 23.72 Walls and Ceilings, Interior

	09 91 23.72 Walls and Ceilings, Interior	Crew	Daily Output	Labor-Hours	Unit	Material	Labor	Equipment	Total	Total Incl O&P
0010	**WALLS AND CEILINGS, INTERIOR**									
0100	Concrete, drywall or plaster, latex , primer or sealer coat									
0200	Smooth finish, brushwork	1 Pord	1150	.007	S.F.	.06	.27		.33	.49
0240	Roller		1350	.006		.06	.23		.29	.43
0280	Spray		2750	.003		.05	.11		.16	.24
0300	Sand finish, brushwork		975	.008		.06	.31		.37	.57
0340	Roller		1150	.007		.06	.27		.33	.49
0380	Spray		2275	.004		.05	.13		.18	.27
0800	Paint 2 coats, smooth finish, brushwork		680	.012		.12	.45		.57	.86
0840	Roller		800	.010		.12	.38		.50	.75
0880	Spray		1625	.005		.12	.19		.31	.43
0900	Sand finish, brushwork		605	.013		.12	.51		.63	.95
0940	Roller		1020	.008		.12	.30		.42	.62
0980	Spray		1700	.005		.12	.18		.30	.42
1600	Glaze coating, 2 coats, spray, clear		1200	.007		.49	.26		.75	.95
1640	Multicolor		1200	.007		.98	.26		1.24	1.49
1660	Painting walls, complete, including surface prep, primer &									
1670	2 coats finish, on drywall or plaster, with roller	1 Pord	325	.025	S.F.	.19	.94		1.13	1.71
1700	For oil base paint, add					10%				
1800	For ceiling installations, add						25%			
2000	Masonry or concrete block, primer/sealer, latex paint									
2100	Primer, smooth finish, brushwork	1 Pord	1000	.008	S.F.	.10	.31		.41	.60
2110	Roller		1150	.007		.10	.27		.37	.53
2180	Spray		2400	.003		.09	.13		.22	.30
2200	Sand finish, brushwork		850	.009		.10	.36		.46	.68
2210	Roller		975	.008		.10	.31		.41	.61
2280	Spray		2050	.004		.09	.15		.24	.34

09 91 Painting

09 91 23 – Interior Painting

09 91 23.72 Walls and Ceilings, Interior		Crew	Daily Output	Labor-Hours	Unit	2012 Bare Costs Material	Labor	Equipment	Total	Total Incl O&P
2800	Primer plus one finish coat, smooth brush	1 Pord	525	.015	S.F.	.29	.58		.87	1.25
2810	Roller		615	.013		.18	.50		.68	.99
2880	Spray		1200	.007		.16	.26		.42	.59
2900	Sand finish, brushwork		450	.018		.18	.68		.86	1.29
2910	Roller		515	.016		.18	.59		.77	1.15
2980	Spray		1025	.008		.16	.30		.46	.66
3600	Glaze coating, 3 coats, spray, clear		900	.009		.70	.34		1.04	1.31
3620	Multicolor		900	.009		1.13	.34		1.47	1.79
4000	Block filler, 1 coat, brushwork		425	.019		.13	.72		.85	1.29
4100	Silicone, water repellent, 2 coats, spray		2000	.004		.30	.15		.45	.57
4120	For oil base paint, add					10%				
8200	For work 8' - 15' H, add						10%			
8300	For work over 15' H, add						20%			
8400	For light textured surfaces, add						10%			
8410	Heavy textured, add						25%			
9900	Minimum labor/equipment charge	1 Pord	2	4	Job		153		153	244

09 91 23.74 Walls and Ceilings, Interior, Zero VOC Latex		Crew	Daily Output	Labor-Hours	Unit	Material	Labor	Equipment	Total	Total Incl O&P
0010	**WALLS AND CEILINGS, INTERIOR, ZERO VOC LATEX**									
0100	Concrete, dry wall or plaster, latex, primer or sealer coat									
0200	Smooth finish, brushwork G	1 Pord	1150	.007	S.F.	.07	.27		.34	.49
0240	Roller G		1350	.006		.06	.23		.29	.43
0280	Spray G		2750	.003		.05	.11		.16	.23
0300	Sand finish, brushwork G		975	.008		.06	.31		.37	.57
0340	Roller G		1150	.007		.07	.27		.34	.50
0380	Spray G		2275	.004		.05	.13		.18	.27
0800	Paint 2 coats, smooth finish, brushwork G		680	.012		.17	.45		.62	.90
0840	Roller G		800	.010		.17	.38		.55	.80
0880	Spray G		1625	.005		.15	.19		.34	.46
0900	Sand finish, brushwork G		605	.013		.16	.51		.67	.99
0940	Roller G		1020	.008		.17	.30		.47	.67
0980	Spray G		1700	.005		.15	.18		.33	.45
1800	For ceiling installations, add G						25%			
8200	For work 8' - 15' H, add						10%			
8300	For work over 15' H, add						20%			
9900	Minimum labor/equipment charge	1 Pord	2	4	Job		153		153	244

09 91 23.75 Dry Fall Painting		Crew	Daily Output	Labor-Hours	Unit	Material	Labor	Equipment	Total	Total Incl O&P
0010	**DRY FALL PAINTING**									
0100	Sprayed on walls, gypsum board or plaster									
0220	One coat	1 Pord	2600	.003	S.F.	.05	.12		.17	.25
0250	Two coats		1560	.005		.11	.20		.31	.43
0280	Concrete or textured plaster, one coat		1560	.005		.05	.20		.25	.37
0310	Two coats		1300	.006		.11	.24		.35	.50
0340	Concrete block, one coat		1560	.005		.05	.20		.25	.37
0370	Two coats		1300	.006		.11	.24		.35	.50
0400	Wood, one coat		877	.009		.05	.35		.40	.62
0430	Two coats		650	.012		.11	.47		.58	.87
0440	On ceilings, gypsum board or plaster									
0470	One coat	1 Pord	1560	.005	S.F.	.05	.20		.25	.37
0500	Two coats		1300	.006		.11	.24		.35	.50
0530	Concrete or textured plaster, one coat		1560	.005		.05	.20		.25	.37
0560	Two coats		1300	.006		.11	.24		.35	.50
0570	Structural steel, bar joists or metal deck, one coat		1560	.005		.05	.20		.25	.37

09 91 Painting

09 91 23 – Interior Painting

09 91 23.75 Dry Fall Painting		Crew	Daily Output	Labor-Hours	Unit	Material	Labor	Equipment	Total	Total Incl O&P
						2012 Bare Costs				
0580	Two coats	1 Pord	1040	.008	S.F.	.11	.29		.40	.59
9900	Minimum labor/equipment charge	↓	2	4	Job		153		153	244

09 93 Staining and Transparent Finishing

09 93 23 – Interior Staining and Finishing

09 93 23.10 Varnish

		Crew	Daily Output	Labor-Hours	Unit	Material	Labor	Equipment	Total	Total Incl O&P
0010	**VARNISH**									
0012	1 coat + sealer, on wood trim, brush, no sanding included	1 Pord	400	.020	S.F.	.07	.77		.84	1.30
0100	Hardwood floors, 2 coats, no sanding included, roller		1890	.004	"	.15	.16		.31	.43
9000	Minimum labor/equipment charge	↓	4	2	Job		76.50		76.50	122

09 96 High-Performance Coatings

09 96 23 – Graffiti-Resistant Coatings

09 96 23.10 Graffiti Resistant Treatments

		Crew	Daily Output	Labor-Hours	Unit	Material	Labor	Equipment	Total	Total Incl O&P
0010	**GRAFFITI RESISTANT TREATMENTS**, sprayed on walls									
0100	Non-sacrificial, permanent non-stick coating, clear, on metals	1 Pord	2000	.004	S.F.	2.06	.15		2.21	2.51
0200	Concrete		2000	.004		2.35	.15		2.50	2.82
0300	Concrete block		2000	.004		3.03	.15		3.18	3.58
0400	Brick		2000	.004		3.44	.15		3.59	4.02
0500	Stone		2000	.004		3.44	.15		3.59	4.02
0600	Unpainted wood		2000	.004		3.97	.15		4.12	4.61
2000	Semi-permanent cross linking polymer primer, on metals		2000	.004		.40	.15		.55	.68
2100	Concrete		2000	.004		.48	.15		.63	.77
2200	Concrete block		2000	.004		.60	.15		.75	.90
2300	Brick		2000	.004		.48	.15		.63	.77
2400	Stone		2000	.004		.48	.15		.63	.77
2500	Unpainted wood		2000	.004		.67	.15		.82	.97
3000	Top coat, on metals		2000	.004		.43	.15		.58	.71
3100	Concrete		2000	.004		.49	.15		.64	.77
3200	Concrete block		2000	.004		.68	.15		.83	.99
3300	Brick		2000	.004		.57	.15		.72	.86
3400	Stone		2000	.004		.57	.15		.72	.86
3500	Unpainted wood		2000	.004		.68	.15		.83	.99
5000	Sacrificial, water based, on metal		2000	.004		.24	.15		.39	.51
5100	Concrete		2000	.004		.24	.15		.39	.51
5200	Concrete block		2000	.004		.24	.15		.39	.51
5300	Brick		2000	.004		.24	.15		.39	.51
5400	Stone		2000	.004		.24	.15		.39	.51
5500	Unpainted wood	↓	2000	.004	↓	.24	.15		.39	.51
8000	Cleaner for use after treatment									
8100	Towels or wipes, per package of 30				Ea.	.63			.63	.70
8200	Aerosol spray, 24 oz. can				"	18			18	19.80

09 96 53 – Elastomeric Coatings

09 96 53.10 Coatings, Elastomeric

		Crew	Daily Output	Labor-Hours	Unit	Material	Labor	Equipment	Total	Total Incl O&P
0010	**COATINGS, ELASTOMERIC**									
0020	High build, water proof, one coat system									
0100	Concrete, brush	1 Pord	650	.012	S.F.	.27	.47		.74	1.05

09 96 High-Performance Coatings

09 96 56 – Epoxy Coatings

09 96 56.20 Wall Coatings		Crew	Daily Output	Labor-Hours	Unit	Material	2012 Bare Costs Labor	Equipment	Total	Total Incl O&P
0010	**WALL COATINGS**									
0100	Acrylic glazed coatings, minimum	1 Pord	525	.015	S.F.	.30	.58		.88	1.26
0200	Maximum		305	.026		.64	1		1.64	2.30
0300	Epoxy coatings, minimum		525	.015		.39	.58		.97	1.36
0400	Maximum		170	.047		1.18	1.80		2.98	4.17
2400	Sprayed perlite or vermiculite, 1/16" thick, minimum		2935	.003		.26	.10		.36	.46
2500	Maximum		640	.013		.72	.48		1.20	1.55
2700	Vinyl plastic wall coating, minimum		735	.011		.32	.42		.74	1.01
2800	Maximum		240	.033		.79	1.27		2.06	2.90
3000	Urethane on smooth surface, 2 coats, minimum		1135	.007		.26	.27		.53	.72
3100	Maximum	▼	665	.012	▼	.57	.46		1.03	1.36

Division Notes

		CREW	DAILY OUTPUT	LABOR-HOURS	UNIT	2012 BARE COSTS MAT.	LABOR	EQUIP.	TOTAL	TOTAL INCL O&P

Estimating Tips

General

- The items in this division are usually priced per square foot or each.
- Many items in Division 10 require some type of support system or special anchors that are not usually furnished with the item. The required anchors must be added to the estimate in the appropriate division.
- Some items in Division 10, such as lockers, may require assembly before installation. Verify the amount of assembly required. Assembly can often exceed installation time.

10 20 00 Interior Specialties

- Support angles and blocking are not included in the installation of toilet compartments, shower/dressing compartments, or cubicles. Appropriate line items from Divisions 5 or 6 may need to be added to support the installations.
- Toilet partitions are priced by the stall. A stall consists of a side wall, pilaster, and door with hardware. Toilet tissue holders and grab bars are extra.
- The required acoustical rating of a folding partition can have a significant impact on costs. Verify the sound transmission coefficient rating of the panel priced to the specification requirements.
- Grab bar installation does not include supplemental blocking or backing to support the required load. When grab bars are installed at an existing facility, provisions must be made to attach the grab bars to solid structure.

Reference Numbers

Reference numbers are shown in shaded boxes at the beginning of some major classifications. These numbers refer to related items in the Reference Section. The reference information may be an estimating procedure, an alternate pricing method, or technical information.

Note: Not all subdivisions listed here necessarily appear in this publication.

10 11 Visual Display Units

10 11 13 – Chalkboards

10 11 13.13 Fixed Chalkboards		Crew	Daily Output	Labor-Hours	Unit	2012 Bare Costs Material	Labor	Equipment	Total	Total Incl O&P
0010	**FIXED CHALKBOARDS** Porcelain enamel steel									
3900	Wall hung									
4000	Aluminum frame and chalktrough									
4300	3′ x 5′	2 Carp	15	1.067	Ea.	299	47		346	405
4600	4′ x 12′	"	13	1.231	"	575	54.50		629.50	720
4700	Wood frame and chalktrough									
4800	3′ x 4′	2 Carp	16	1	Ea.	217	44		261	310
5300	4′ x 8′	"	13	1.231	"	355	54.50		409.50	480
5400	Liquid chalk, white porcelain enamel, wall hung									
5420	Deluxe units, aluminum trim and chalktrough									
5450	4′ x 4′	2 Carp	16	1	Ea.	250	44		294	350
5550	4′ x 12′	"	12	1.333	"	530	59		589	680
5700	Wood trim and chalktrough									
5900	4′ x 4′	2 Carp	16	1	Ea.	715	44		759	860
6200	4′ x 8′	↓	14	1.143	"	970	50.50		1,020.50	1,150
9000	Minimum labor/equipment charge	↓	3	5.333	Job		235		235	385

10 11 13.43 Portable Chalkboards		Crew	Daily Output	Labor-Hours	Unit	Material	Labor	Equipment	Total	Total Incl O&P
0010	**PORTABLE CHALKBOARDS**									
0100	Freestanding, reversible									
0120	Economy, wood frame, 4′ x 6′									
0140	Chalkboard both sides				Ea.	595			595	655
0160	Chalkboard one side, cork other side				"	565			565	620
0200	Standard, lightweight satin finished aluminum, 4′ x 6′									
0220	Chalkboard both sides				Ea.	675			675	745
0240	Chalkboard one side, cork other side				"	695			695	765
0300	Deluxe, heavy duty extruded aluminum, 4′ x 6′									
0320	Chalkboard both sides				Ea.	1,025			1,025	1,125
0340	Chalkboard one side, cork other side				"	930			930	1,025

10 11 16 – Markerboards

10 11 16.53 Electronic Markerboards		Crew	Daily Output	Labor-Hours	Unit	Material	Labor	Equipment	Total	Total Incl O&P
0010	**ELECTRONIC MARKERBOARDS**									

10 11 23 – Tackboards

10 11 23.10 Fixed Tackboards		Crew	Daily Output	Labor-Hours	Unit	Material	Labor	Equipment	Total	Total Incl O&P
0010	**FIXED TACKBOARDS**									
2120	Prefabricated, 1/4″ cork, 3′ x 5′ with aluminum frame	2 Carp	16	1	Ea.	128	44		172	214
2140	Wood frame	"	16	1	"	153	44		197	242
2300	Glass enclosed cabinets, alum., cork panel, hinged doors									
2600	4′ x 7′, 3 door	2 Carp	10	1.600	Ea.	1,775	70.50		1,845.50	2,075
9000	Minimum labor/equipment charge	"	4	4	Job		176		176	289

10 13 Directories

10 13 10 – Building Directories

	10 13 10.10 Directory Boards	Crew	Daily Output	Labor-Hours	Unit	Material	2012 Bare Costs Labor	Equipment	Total	Total Incl O&P
0010	**DIRECTORY BOARDS**									
0050	Plastic, glass covered, 30" x 20"	2 Carp	3	5.333	Ea.	260	235		495	670
0100	36" x 48"		2	8		845	355		1,200	1,500
0900	Outdoor, weatherproof, black plastic, 36" x 24"		2	8		760	355		1,115	1,425
1000	36" x 36"		1.50	10.667		880	470		1,350	1,750
9000	Minimum labor/equipment charge	1 Carp	1	8	Job		355		355	580

10 14 Signage

10 14 19 – Dimensional Letter Signage

	10 14 19.10 Exterior Signs	Crew	Daily Output	Labor-Hours	Unit	Material	2012 Bare Costs Labor	Equipment	Total	Total Incl O&P
0010	**EXTERIOR SIGNS**									
3900	Plaques, custom, 20" x 30", for up to 450 letters, cast aluminum	2 Carp	4	4	Ea.	1,125	176		1,301	1,525
4000	Cast bronze		4	4		1,675	176		1,851	2,125
4200	30" x 36", up to 900 letters cast aluminum		3	5.333		2,000	235		2,235	2,575
4300	Cast bronze		3	5.333		3,250	235		3,485	3,950
5100	Exit signs, 24 ga. alum., 14" x 12" surface mounted	1 Carp	30	.267		43	11.75		54.75	66.50
5200	10" x 7"	"	20	.400		29.50	17.65		47.15	61.50
6400	Replacement sign faces, 6" or 8"	1 Clab	50	.160		62.50	5.60		68.10	77.50
9000	Minimum labor/equipment charge	1 Carp	4	2	Job		88		88	145

10 14 23 – Panel Signage

	10 14 23.13 Engraved Panel Signage	Crew	Daily Output	Labor-Hours	Unit	Material	2012 Bare Costs Labor	Equipment	Total	Total Incl O&P
0010	**ENGRAVED PANEL SIGNAGE**, interior									
1010	Flexible door sign, adhesive back, w/Braille, 5/8" letters, 4" x 4"	1 Clab	32	.250	Ea.	35	8.80		43.80	53
1050	6" x 6"		32	.250		52.50	8.80		61.30	72
1100	8" x 2"		32	.250		36	8.80		44.80	54
1150	8" x 4"		32	.250		45.50	8.80		54.30	64.50
1200	8" x 8"		32	.250		71.50	8.80		80.30	93
1250	12" x 2"		32	.250		35	8.80		43.80	53
1300	12" x 6"		32	.250		50	8.80		58.80	69.50
1350	12" x 12"		32	.250		143	8.80		151.80	171
1500	Graphic symbols, 2" x 2"		32	.250		12	8.80		20.80	27.50
1550	6" x 6"		32	.250		31	8.80		39.80	48.50
1600	8" x 8"		32	.250		31	8.80		39.80	48.50
2010	Corridor, stock acrylic, 2-sided, with mounting bracket, 2" x 8"	1 Carp	24	.333		26	14.70		40.70	52.50
2020	2" x 10"		24	.333		35.50	14.70		50.20	63
2050	3" x 8"		24	.333		31.50	14.70		46.20	58.50
2060	3" x 10"		24	.333		44.50	14.70		59.20	73
2070	3" x 12"		24	.333		65	14.70		79.70	95.50
2100	4" x 8"		24	.333		35.50	14.70		50.20	63
2110	4" x 10"		24	.333		42	14.70		56.70	70
2120	4" x 12"		24	.333		56.50	14.70		71.20	86

10 14 53 – Traffic Signage

	10 14 53.20 Traffic Signs	Crew	Daily Output	Labor-Hours	Unit	Material	2012 Bare Costs Labor	Equipment	Total	Total Incl O&P
0010	**TRAFFIC SIGNS**									
0012	Stock, 24" x 24", no posts, .080" alum. reflectorized	B-80	70	.457	Ea.	81.50	17.30	9.55	108.35	128
0100	High intensity		70	.457		81.50	17.30	9.55	108.35	128
0300	30" x 30", reflectorized		70	.457		248	17.30	9.55	274.85	310
0400	High intensity		70	.457		275	17.30	9.55	301.85	345
0600	Guide and directional signs, 12" x 18", reflectorized		70	.457		34.50	17.30	9.55	61.35	76.50
0700	High intensity		70	.457		57	17.30	9.55	83.85	102

10 14 Signage

10 14 53 – Traffic Signage

10 14 53.20 Traffic Signs		Crew	Daily Output	Labor-Hours	Unit	Material	2012 Bare Costs Labor	Equipment	Total	Total Incl O&P
0900	18" x 24", stock signs, reflectorized	B-80	70	.457	Ea.	45.50	17.30	9.55	72.35	88.50
1000	High intensity		70	.457		50.50	17.30	9.55	77.35	94
1200	24" x 24", stock signs, reflectorized		70	.457		55	17.30	9.55	81.85	99
1300	High intensity		70	.457		60	17.30	9.55	86.85	105
1500	Add to above for steel posts, galvanized, 10'-0" upright, bolted		200	.160		42	6.05	3.34	51.39	60
1600	12'-0" upright, bolted		140	.229		39	8.65	4.77	52.42	62
1800	Highway road signs, aluminum, over 20 S. F., reflectorized		350	.091	S.F.	29	3.46	1.91	34.37	39
2000	High intensity		350	.091		29	3.46	1.91	34.37	39
2200	Highway, suspended over road, 80 S.F. min., reflectorized		165	.194		29	7.35	4.05	40.40	48
2300	High intensity		165	.194		29	7.35	4.05	40.40	48

10 17 Telephone Specialties

10 17 16 – Telephone Enclosures

10 17 16.10 Commercial Telephone Enclosures		Crew	Daily Output	Labor-Hours	Unit	Material	Labor	Equipment	Total	Total Incl O&P
0010	**COMMERCIAL TELEPHONE ENCLOSURES**									
0300	Shelf type, wall hung, minimum	2 Carp	5	3.200	Ea.	995	141		1,136	1,325
0400	Maximum		5	3.200	"	3,225	141		3,366	3,775
9000	Minimum labor/equipment charge		4	4	Job		176		176	289

10 21 Compartments and Cubicles

10 21 13 – Toilet Compartments

10 21 13.13 Metal Toilet Compartments		Crew	Daily Output	Labor-Hours	Unit	Material	Labor	Equipment	Total	Total Incl O&P
0010	**METAL TOILET COMPARTMENTS**									
0110	Cubicles, ceiling hung									
0200	Powder coated steel	2 Carp	4	4	Ea.	500	176		676	840
0500	Stainless steel	"	4	4		945	176		1,121	1,350
0600	For handicap units, incl. 52" grab bars, add					420			420	460
0900	Floor and ceiling anchored									
1000	Powder coated steel	2 Carp	5	3.200	Ea.	615	141		756	910
1300	Stainless steel	"	5	3.200		1,075	141		1,216	1,425
1400	For handicap units, incl. 52" grab bars, add					300			300	330
1610	Floor anchored									
1700	Powder coated steel	2 Carp	7	2.286	Ea.	590	101		691	815
2000	Stainless steel	"	7	2.286		1,375	101		1,476	1,700
2100	For handicap units, incl. 52" grab bars, add					299			299	330
2200	For juvenile units, deduct					41.50			41.50	45.50
2450	Floor anchored, headrail braced									
2500	Powder coated steel	2 Carp	6	2.667	Ea.	385	118		503	615
2804	Stainless steel	"	4.60	3.478		1,025	153		1,178	1,375
2900	For handicap units, incl. 52" grab bars, add					350			350	385
3000	Wall hung partitions, powder coated steel	2 Carp	7	2.286		625	101		726	850
3300	Stainless steel	"	7	2.286		1,650	101		1,751	2,000
3400	For handicap units, incl. 52" grab bars, add					350			350	385
4000	Screens, entrance, floor mounted, 58" high, 48" wide									
4200	Powder coated steel	2 Carp	15	1.067	Ea.	238	47		285	340
4500	Stainless steel	"	15	1.067	"	895	47		942	1,050
4650	Urinal screen, 18" wide									
4704	Powder coated steel	2 Carp	6.15	2.602	Ea.	220	115		335	430
5004	Stainless steel	"	6.15	2.602	"	580	115		695	830

10 21 Compartments and Cubicles

10 21 13 – Toilet Compartments

10 21 13.13 Metal Toilet Compartments		Crew	Daily Output	Labor-Hours	Unit	2012 Bare Costs Material	Labor	Equipment	Total	Total Incl O&P
5100	Floor mounted, head rail braced									
5300	Powder coated steel	2 Carp	8	2	Ea.	228	88		316	395
5600	Stainless steel	"	8	2	"	595	88		683	800
5750	Pilaster, flush									
5800	Powder coated steel	2 Carp	10	1.600	Ea.	276	70.50		346.50	420
6100	Stainless steel		10	1.600		670	70.50		740.50	850
6300	Post braced, powder coated steel		10	1.600		153	70.50		223.50	284
6600	Stainless steel	↓	10	1.600	↓	450	70.50		520.50	610
6700	Wall hung, bracket supported									
6800	Powder coated steel	2 Carp	10	1.600	Ea.	153	70.50		223.50	284
7100	Stainless steel		10	1.600		219	70.50		289.50	355
7400	Flange supported, powder coated steel		10	1.600		106	70.50		176.50	232
7700	Stainless steel		10	1.600		310	70.50		380.50	455
7800	Wedge type, powder coated steel		10	1.600		134	70.50		204.50	264
8100	Stainless steel	↓	10	1.600	↓	565	70.50		635.50	735
9000	Minimum labor/equipment charge	1 Carp	2.50	3.200	Job		141		141	231

10 21 13.14 Metal Toilet Compartment Components		Crew	Daily Output	Labor-Hours	Unit	Material	Labor	Equipment	Total	Total Incl O&P
0010	**METAL TOILET COMPARTMENT COMPONENTS**									
0100	Pilasters									
0110	Overhead braced, powder coated steel, 7" wide x 82" high	2 Carp	22.20	.721	Ea.	72.50	32		104.50	132
0120	Stainless steel		22.20	.721		121	32		153	185
0130	Floor braced, powder coated steel, 7" wide x 70" high		23.30	.687		122	30.50		152.50	184
0140	Stainless steel		23.30	.687		236	30.50		266.50	310
0150	Ceiling hung, powder coated steel, 7" wide x 83" high		13.30	1.203		129	53		182	229
0160	Stainless steel		13.30	1.203		264	53		317	375
0170	Wall hung, powder coated steel, 3" wide x 58" high		18.90	.847		129	37.50		166.50	203
0180	Stainless steel	↓	18.90	.847	↓	183	37.50		220.50	262
0200	Panels									
0210	Powder coated steel, 31" wide x 58" high	2 Carp	18.90	.847	Ea.	128	37.50		165.50	202
0220	Stainless steel		18.90	.847		350	37.50		387.50	445
0230	Powder coated steel, 53" wide x 58" high		18.90	.847		158	37.50		195.50	235
0240	Stainless steel		18.90	.847		450	37.50		487.50	555
0250	Powder coated steel, 63" wide x 58" high		18.90	.847		186	37.50		223.50	266
0260	Stainless steel	↓	18.90	.847	↓	495	37.50		532.50	605
0300	Doors									
0310	Powder coated steel, 24" wide x 58" high	2 Carp	14.10	1.135	Ea.	133	50		183	229
0320	Stainless steel		14.10	1.135		275	50		325	385
0330	Powder coated steel, 26" wide x 58" high		14.10	1.135		136	50		186	231
0340	Stainless steel		14.10	1.135		284	50		334	390
0350	Powder coated steel, 28" wide x 58" high		14.10	1.135		156	50		206	253
0360	Stainless steel		14.10	1.135		315	50		365	430
0370	Powder coated steel, 36" wide x 58" high		14.10	1.135		167	50		217	266
0380	Stainless steel	↓	14.10	1.135	↓	355	50		405	470
0400	Headrails									
0410	For powder coated steel, 62" long	2 Carp	65	.246	Ea.	21	10.85		31.85	41.50
0420	Stainless steel		65	.246		21	10.85		31.85	41.50
0430	For powder coated steel, 84" long		50	.320		32	14.10		46.10	58
0440	Stainless steel		50	.320		32	14.10		46.10	58
0450	For powder coated steel, 120" long		30	.533		42	23.50		65.50	84.50
0460	Stainless steel	↓	30	.533	↓	42	23.50		65.50	84.50
9000	Minimum labor/equipment charge	1 Carp	4	2	Job		88		88	145

10 21 Compartments and Cubicles

10 21 13 – Toilet Compartments

10 21 13.16 Plastic-Laminate-Clad Toilet Compartments		Crew	Daily Output	Labor-Hours	Unit	2012 Bare Costs Material	Labor	Equipment	Total	Total Incl O&P
0010	**PLASTIC-LAMINATE-CLAD TOILET COMPARTMENTS**									
0110	Cubicles, ceiling hung									
0300	Plastic laminate on particle board	2 Carp	4	4	Ea.	540	176		716	885
0600	For handicap units, incl. 52" grab bars, add				"	420			420	460
0900	Floor and ceiling anchored									
1100	Plastic laminate on particle board	2 Carp	5	3.200	Ea.	790	141		931	1,100
1400	For handicap units, incl. 52" grab bars, add				"	300			300	330
1610	Floor mounted									
1800	Plastic laminate on particle board	2 Carp	7	2.286	Ea.	495	101		596	710
2450	Floor mounted, headrail braced									
2600	Plastic laminate on particle board	2 Carp	6	2.667	Ea.	750	118		868	1,025
3400	For handicap units, incl. 52" grab bars, add					350			350	385
4300	Entrance screen, floor mtd., plas. lam., 58" high, 48" wide	2 Carp	15	1.067		610	47		657	745
4800	Urinal screen, 18" wide, ceiling braced, plastic laminate		8	2		174	88		262	335
5400	Floor mounted, headrail braced		8	2		180	88		268	345
5900	Pilaster, flush, plastic laminate		10	1.600		505	70.50		575.50	675
6400	Post braced, plastic laminate		10	1.600		295	70.50		365.50	440
6700	Wall hung, bracket supported									
6900	Plastic laminate on particle board	2 Carp	10	1.600	Ea.	90.50	70.50		161	216
7450	Flange supported									
7500	Plastic laminate on particle board	2 Carp	10	1.600	Ea.	229	70.50		299.50	370
9000	Minimum labor/equipment charge	1 Carp	2.50	3.200	Job		141		141	231

10 21 13.17 Plastic-Laminate Clad Toilet Compartment Components		Crew	Daily Output	Labor-Hours	Unit	Material	Labor	Equipment	Total	Total Incl O&P
0010	**PLASTIC-LAMINATE CLAD TOILET COMPARTMENT COMPONENTS**									
0100	Pilasters									
0110	Overhead braced, 7" wide x 82" high	2 Carp	22.20	.721	Ea.	92	32		124	153
0130	Floor anchored, 7" wide x 70" high		23.30	.687		92.50	30.50		123	152
0150	Ceiling hung, 7" wide x 83" high		13.30	1.203		98	53		151	195
0180	Wall hung, 3" wide x 58" high		18.90	.847		90.50	37.50		128	161
0200	Panels									
0210	31" wide x 58" high	2 Carp	18.90	.847	Ea.	138	37.50		175.50	213
0230	51" wide x 58" high		18.90	.847		190	37.50		227.50	270
0250	63" wide x 58" high		18.90	.847		221	37.50		258.50	305
0300	Doors									
0310	24" wide x 58" high	2 Carp	14.10	1.135	Ea.	135	50		185	231
0330	26" wide x 58" high		14.10	1.135		140	50		190	236
0350	28" wide x 58" high		14.10	1.135		145	50		195	242
0370	36" wide x 58" high		14.10	1.135		174	50		224	273
0400	Headrails									
0410	62" long	2 Carp	65	.246	Ea.	22.50	10.85		33.35	42.50
0430	84" long		60	.267		32	11.75		43.75	54.50
0450	120" long		30	.533		42	23.50		65.50	84.50
9000	Minimum labor/equipment charge	1 Carp	4	2	Job		88		88	145

10 21 13.19 Plastic Toilet Compartments		Crew	Daily Output	Labor-Hours	Unit	Material	Labor	Equipment	Total	Total Incl O&P
0010	**PLASTIC TOILET COMPARTMENTS**									
0110	Cubicles, ceiling hung									
0250	Phenolic	2 Carp	4	4	Ea.	850	176		1,026	1,225
0600	For handicap units, incl. 52" grab bars, add				"	420			420	460
0900	Floor and ceiling anchored									
1050	Phenolic	2 Carp	5	3.200	Ea.	860	141		1,001	1,175
1400	For handicap units, incl. 52" grab bars, add				"	300			300	330
1610	Floor mounted									

10 21 13.19 Plastic Toilet Compartments

Line	10 21 13.19 Plastic Toilet Compartments	Crew	Daily Output	Labor-Hours	Unit	2012 Bare Costs Material	Labor	Equipment	Total	Total Incl O&P
1750	Phenolic	2 Carp	7	2.286	Ea.	750	101		851	990
2100	For handicap units, incl. 52″ grab bars, add					299			299	330
2200	For juvenile units, deduct					41.50			41.50	45.50
2450	Floor mounted, headrail braced									
2550	Phenolic	2 Carp	6	2.667	Ea.	750	118		868	1,025
9000	Minimum labor/equipment charge	1 Carp	2.50	3.200	Job		141		141	231

10 21 13.20 Plastic Toilet Compartment Components

Line	10 21 13.20 Plastic Toilet Compartment Components	Crew	Daily Output	Labor-Hours	Unit	2012 Bare Costs Material	Labor	Equipment	Total	Total Incl O&P
0010	**PLASTIC TOILET COMPARTMENT COMPONENTS**									
0100	Pilasters									
0110	Overhead braced, polymer plastic, 7″ wide x 82″ high	2 Carp	22.20	.721	Ea.	103	32		135	166
0120	Phenolic		22.20	.721		146	32		178	213
0130	Floor braced, polymer plastic, 7″ wide x 70″ high		23.30	.687		169	30.50		199.50	236
0140	Phenolic		23.30	.687		138	30.50		168.50	202
0150	Ceiling hung, polymer plastic, 7″ wide x 83″ high		13.30	1.203		166	53		219	270
0160	Phenolic		13.30	1.203		156	53		209	259
0180	Wall hung, phenolic, 3″ wide x 58″ high		18.90	.847		93	37.50		130.50	163
0200	Panels									
0210	Polymer plastic, 31″ high x 55″ high	2 Carp	18.90	.847	Ea.	295	37.50		332.50	385
0220	Phenolic, 31″ wide x 58″ high		18.90	.847		249	37.50		286.50	335
0230	Polymer plastic, 51″ wide x 55″ high		18.90	.847		395	37.50		432.50	495
0240	Phenolic, 51″ wide x 58″ high		18.90	.847		380	37.50		417.50	480
0250	Polymer plastic, 63″ wide x 55″ high		18.90	.847		540	37.50		577.50	655
0260	Phenolic, 63″ wide x 58″ high		18.90	.847		400	37.50		437.50	500
0300	Doors									
0310	Polymer plastic, 24″ wide x 55″ high	2 Carp	14.10	1.135	Ea.	202	50		252	305
0320	Phenolic, 24″ wide x 58″ high		14.10	1.135		283	50		333	390
0330	Polymer plastic, 26″ high x 55″ high		14.10	1.135		215	50		265	320
0340	Phenolic, 26″ wide x 58″ high		14.10	1.135		300	50		350	410
0350	Polymer plastic, 28″ wide x 55″ high		14.10	1.135		240	50		290	345
0360	Phenolic, 28″ wide x 58″ high		14.10	1.135		315	50		365	430
0370	Polymer plastic, 36″ wide x 55″ high		14.10	1.135		280	50		330	390
0380	Phenolic, 36″ wide x 58″ high		14.10	1.135		395	50		445	515
0400	Headrails									
0410	For polymer plastic, 62″ long	2 Carp	65	.246	Ea.	22.50	10.85		33.35	42.50
0420	Phenolic		65	.246		22.50	10.85		33.35	42.50
0430	For polymer plastic, 84″ long		50	.320		32	14.10		46.10	58
0440	Phenolic		50	.320		32	14.10		46.10	58
0450	For polymer plastic, 120″ long		30	.533		42	23.50		65.50	84.50
0460	Phenolic		30	.533		42	23.50		65.50	84.50
9000	Minimum labor/equipment charge	1 Carp	4	2	Job		88		88	145

10 21 13.40 Stone Toilet Compartments

Line	10 21 13.40 Stone Toilet Compartments	Crew	Daily Output	Labor-Hours	Unit	2012 Bare Costs Material	Labor	Equipment	Total	Total Incl O&P
0010	**STONE TOILET COMPARTMENTS**									
0100	Cubicles, ceiling hung, marble	2 Marb	2	8	Ea.	1,525	330		1,855	2,200
0600	For handicap units, incl. 52″ grab bars, add					420			420	460
0800	Floor & ceiling anchored, marble	2 Marb	2.50	6.400		1,925	263		2,188	2,525
1400	For handicap units, incl. 52″ grab bars, add					300			300	330
1600	Floor mounted, marble	2 Marb	3	5.333		1,125	219		1,344	1,600
2400	Floor mounted, headrail braced, marble	″	3	5.333		1,100	219		1,319	1,550
2900	For handicap units, incl. 52″ grab bars, add					350			350	385
4100	Entrance screen, floor mounted marble, 58″ high, 48″ wide	2 Marb	9	1.778		710	73		783	900
4600	Urinal screen, 18″ wide, ceiling braced, marble	D-1	6	2.667		725	106		831	970
5100	Floor mounted, head rail braced									

10 21 Compartments and Cubicles

10 21 13 – Toilet Compartments

10 21 13.40 Stone Toilet Compartments		Crew	Daily Output	Labor-Hours	Unit	Material	Labor	Equipment	Total	Total Incl O&P
						2012 Bare Costs				
5200	Marble	D-1	6	2.667	Ea.	615	106		721	850
5700	Pilaster, flush, marble		9	1.778		810	71		881	1,000
6200	Post braced, marble		9	1.778		800	71		871	995
9000	Minimum labor/equipment charge	1 Carp	2.50	3.200	Job		141		141	231

10 21 16 – Shower and Dressing Compartments

10 21 16.10 Partitions, Shower

		Crew	Daily Output	Labor-Hours	Unit	Material	Labor	Equipment	Total	Total Incl O&P
0010	**PARTITIONS, SHOWER** floor mounted, no plumbing									
0400	Cabinet, one piece, fiberglass, 32" x 32"	2 Carp	5	3.200	Ea.	505	141		646	785
0420	36" x 36"		5	3.200		570	141		711	855
0440	36" x 48"		5	3.200		1,450	141		1,591	1,825
0460	Acrylic, 32" x 32"		5	3.200		325	141		466	585
0480	36" x 36"		5	3.200		1,025	141		1,166	1,350
0500	36" x 48"		5	3.200		1,475	141		1,616	1,850
0520	Shower door for above, clear plastic, 24" wide	1 Carp	8	1		177	44		221	268
0540	28" wide		8	1		201	44		245	295
0560	Tempered glass, 24" wide		8	1		191	44		235	283
0580	28" wide		8	1		216	44		260	310
4100	Shower doors, economy plastic, 24" wide	1 Shee	9	.889		134	46.50		180.50	221
4200	Tempered glass door, economy		8	1		252	52		304	360
4700	Deluxe, tempered glass, chrome on brass frame, minimum		8	1		385	52		437	500
4800	Maximum		1	8		625	415		1,040	1,350
9990	Minimum labor/equipment charge		2.50	3.200	Job		167		167	263

10 22 Partitions

10 22 16 – Folding Gates

10 22 16.10 Security Gates

		Crew	Daily Output	Labor-Hours	Unit	Material	Labor	Equipment	Total	Total Incl O&P
0010	**SECURITY GATES** for roll up type, see Section 08 33 13.10									
0300	Scissors type folding gate, ptd. steel, single, 6-1/2' high, 5-1/2' wide	2 Sswk	4	4	Opng.	229	197		426	615
0350	6-1/2' wide		4	4		242	197		439	630
0400	7-1/2' wide		4	4		215	197		412	600
0600	Double gate, 8' high, 8' wide		2.50	6.400		355	315		670	975
0650	10' wide		2.50	6.400		390	315		705	1,025
0700	12' wide		2	8		590	395		985	1,375
0750	14' wide		2	8		675	395		1,070	1,475
0900	Door gate, folding steel, 4' wide, 61" high		4	4		129	197		326	505
1000	71" high		4	4		161	197		358	540
1200	81" high		4	4		175	197		372	555
1300	Window gates, 2' to 4' wide, 31" high		4	4		62.50	197		259.50	435
1500	55" high		3.75	4.267		97.50	211		308.50	495
1600	79" high		3.50	4.571		113	226		339	545

10 22 19 – Demountable Partitions

10 22 19.43 Demountable Composite Partitions

		Crew	Daily Output	Labor-Hours	Unit	Material	Labor	Equipment	Total	Total Incl O&P
0010	**DEMOUNTABLE COMPOSITE PARTITIONS**, add for doors									
0100	Do not deduct door openings from total L.F.									
0900	Demountable gypsum system on 2" to 2-1/2"									
1000	steel studs, 9' high, 3" to 3-3/4" thick									
1200	Vinyl clad gypsum	2 Carp	48	.333	L.F.	58	14.70		72.70	87.50
1300	Fabric clad gypsum		44	.364		144	16.05		160.05	185
1500	Steel clad gypsum		40	.400		156	17.65		173.65	201
1600	1.75 system, aluminum framing, vinyl clad hardboard,									

10 22 19 – Demountable Partitions

10 22 19.43 Demountable Composite Partitions		Crew	Daily Output	Labor-Hours	Unit	2012 Bare Costs Material	Labor	Equipment	Total	Total Incl O&P
1800	paper honeycomb core panel, 1-3/4″ to 2-1/2″ thick									
1900	9′ high	2 Carp	48	.333	L.F.	97	14.70		111.70	131
2100	7′ high		60	.267		87	11.75		98.75	115
2200	5′ high		80	.200		73.50	8.80		82.30	95.50
2250	Unitized gypsum system									
2300	Unitized panel, 9′ high, 2″ to 2-1/2″ thick									
2350	Vinyl clad gypsum	2 Carp	48	.333	L.F.	125	14.70		139.70	161
2400	Fabric clad gypsum	″	44	.364	″	206	16.05		222.05	253
2500	Unitized mineral fiber system									
2510	Unitized panel, 9′ high, 2-1/4″ thick, aluminum frame									
2550	Vinyl clad mineral fiber	2 Carp	48	.333	L.F.	124	14.70		138.70	161
2600	Fabric clad mineral fiber	″	44	.364	″	185	16.05		201.05	231
2800	Movable steel walls, modular system									
2900	Unitized panels, 9′ high, 48″ wide									
3100	Baked enamel, pre-finished	2 Carp	60	.267	L.F.	141	11.75		152.75	174
3200	Fabric clad steel	″	56	.286	″	203	12.60		215.60	245
5500	For acoustical partitions, add, minimum				S.F.	2.22			2.22	2.44
5550	Maximum				″	10.35			10.35	11.40
5700	For doors, see Sections 08 11 & 08 16									
5800	For door hardware, see Section 08 71									
6100	In-plant modular office system, w/prehung hollow core door									
6200	3″ thick polystyrene core panels									
6250	12′ x 12′, 2 wall	2 Clab	3.80	4.211	Ea.	3,900	148		4,048	4,525
6300	4 wall		1.90	8.421		6,800	296		7,096	7,950
6350	16′ x 16′, 2 wall		3.60	4.444		5,850	156		6,006	6,675
6400	4 wall		1.80	8.889		8,225	310		8,535	9,525
9000	Minimum labor/equipment charge	2 Carp	3	5.333	Job		235		235	385

10 22 23 – Portable Partitions, Screens, and Panels

10 22 23.13 Wall Screens		Crew	Daily Output	Labor-Hours	Unit	Material	Labor	Equipment	Total	Total Incl O&P
0010	**WALL SCREENS**, divider panels, free standing, fiber core									
0020	Fabric face straight									
0100	3′-0″ long, 4′-0″ high	2 Carp	100	.160	L.F.	146	7.05		153.05	173
0200	5′-0″ high		90	.178		103	7.85		110.85	126
0500	6′-0″ high		75	.213		120	9.40		129.40	147
0900	5′-0″ long, 4′-0″ high		175	.091		90.50	4.03		94.53	106
1000	5′-0″ high		150	.107		98.50	4.70		103.20	116
1500	6″-0″ high		125	.128		114	5.65		119.65	135
1600	6′-0″ long, 5′-0″ high		162	.099		95.50	4.36		99.86	112
3200	Economical panels, fabric face, 4′-0″ long, 5′-0″ high		132	.121		50.50	5.35		55.85	64.50
3250	6′-0″ high		112	.143		54	6.30		60.30	69.50
3300	5′-0″ long, 5′-0″ high		150	.107		43	4.70		47.70	54.50
3350	6′-0″ high		125	.128		44	5.65		49.65	57.50
3450	Acoustical panels, 60 to 90 NRC, 3′-0″ long, 5′-0″ high		90	.178		89.50	7.85		97.35	111
3550	6′-0″ high		75	.213		99.50	9.40		108.90	124
3600	5′-0″ long, 5′-0″ high		150	.107		56.50	4.70		61.20	69.50
3650	6′-0″ high		125	.128		65	5.65		70.65	81.50
3700	6′-0″ long, 5′-0″ high		162	.099		48	4.36		52.36	59.50
3750	6′-0″ high		138	.116		56	5.10		61.10	70
3800	Economy acoustical panels, 40 N.R.C., 4′-0″ long, 5′-0″ high		132	.121		50.50	5.35		55.85	64.50
3850	6′-0″ high		112	.143		54	6.30		60.30	69.50
3900	5′-0″ long, 6′-0″ high		125	.128		44	5.65		49.65	57.50
3950	6′-0″ long, 5′-0″ high		162	.099		41	4.36		45.36	52

10 22 Partitions

10 22 23 – Portable Partitions, Screens, and Panels

10 22 23.13 Wall Screens		Crew	Daily Output	Labor-Hours	Unit	Material	2012 Bare Costs Labor	Equipment	Total	Total Incl O&P
9000	Minimum labor/equipment charge	2 Carp	3	5.333	Job		235		235	385

10 22 33 – Accordion Folding Partitions

10 22 33.10 Partitions, Accordion Folding

		Crew	Daily Output	Labor-Hours	Unit	Material	Labor	Equipment	Total	Total Incl O&P
0010	**PARTITIONS, ACCORDION FOLDING**									
0100	Vinyl covered, over 150 S.F., frame not included									
0300	Residential, 1.25 lb. per S.F., 8′ maximum height	2 Carp	300	.053	S.F.	20.50	2.35		22.85	27
0400	Commercial, 1.75 lb. per S.F., 8′ maximum height		225	.071		23.50	3.14		26.64	31
0900	Acoustical, 3 lb. per S.F., 17′ maximum height		100	.160		27.50	7.05		34.55	41.50
1200	5 lb. per S.F., 20′ maximum height		95	.168		38	7.45		45.45	54
1500	Vinyl clad wood or steel, electric operation, 5.0 psf		160	.100		52	4.41		56.41	64.50
1900	Wood, non-acoustic, birch or mahogany, to 10′ high		300	.053		28	2.35		30.35	34.50
9000	Minimum labor/equipment charge		4	4	Job		176		176	289

10 22 39 – Folding Panel Partitions

10 22 39.10 Partitions, Folding Panel

		Crew	Daily Output	Labor-Hours	Unit	Material	Labor	Equipment	Total	Total Incl O&P
0010	**PARTITIONS, FOLDING PANEL**, acoustic, wood									
0100	Vinyl faced, to 18′ high, 6 psf, minimum	2 Carp	60	.267	S.F.	54	11.75		65.75	79
0150	Average		45	.356		64.50	15.70		80.20	96.50
0200	Maximum		30	.533		83.50	23.50		107	131
0400	Formica or hardwood finish, minimum		60	.267		55.50	11.75		67.25	81
0500	Maximum		30	.533		59.50	23.50		83	104
0600	Wood, low acoustical type, 4.5 psf, to 14′ high		50	.320		40.50	14.10		54.60	67.50
9000	Minimum labor/equipment charge		4	4	Job		176		176	289

10 26 Wall and Door Protection

10 26 13 – Corner Guards

10 26 13.10 Metal Corner Guards

		Crew	Daily Output	Labor-Hours	Unit	Material	Labor	Equipment	Total	Total Incl O&P
0010	**METAL CORNER GUARDS**									
0020	Steel angle w/anchors, 1″ x 1″ x 1/4″, 1.5#/L.F.	2 Carp	160	.100	L.F.	6.85	4.41		11.26	14.80
0100	2″ x 2″ x 1/4″ angles, 3.2#/L.F.	″	150	.107	″	9.60	4.70		14.30	18.25
9000	Minimum labor/equipment charge	1 Carp	2	4	Job		176		176	289

10 28 Toilet, Bath, and Laundry Accessories

10 28 13 – Toilet Accessories

10 28 13.13 Commercial Toilet Accessories

		Crew	Daily Output	Labor-Hours	Unit	Material	Labor	Equipment	Total	Total Incl O&P
0010	**COMMERCIAL TOILET ACCESSORIES**									
0200	Curtain rod, stainless steel, 5′ long, 1″ diameter	1 Carp	13	.615	Ea.	36.50	27		63.50	84.50
0300	1-1/4″ diameter		13	.615		27.50	27		54.50	74.50
0350	Chrome, 1″ diameter		13	.615		28	27		55	75
0360	For vinyl curtain, add		1950	.004	S.F.	.84	.18		1.02	1.22
0400	Diaper changing station, horizontal, wall mounted, plastic		10	.800	Ea.	218	35.50		253.50	297
0500	Dispenser units, combined soap & towel dispensers,									
0510	mirror and shelf, flush mounted	1 Carp	10	.800	Ea.	310	35.50		345.50	400
0600	Towel dispenser and waste receptacle,									
0610	18 gallon capacity	1 Carp	10	.800	Ea.	325	35.50		360.50	420
0800	Grab bar, straight, 1-1/4″ diameter, stainless steel, 18″ long		24	.333		29	14.70		43.70	56
0900	24″ long		23	.348		29.50	15.35		44.85	57.50
1000	30″ long		22	.364		31.50	16.05		47.55	61
1100	36″ long		20	.400		37	17.65		54.65	69.50

	10 28 13.13 Commercial Toilet Accessories	Crew	Daily Output	Labor-Hours	Unit	Material	2012 Bare Costs Labor	Equipment	Total	Total Incl O&P
1105	42" long	1 Carp	20	.400	Ea.	37.50	17.65		55.15	70
1200	1-1/2" diameter, 24" long		23	.348		30	15.35		45.35	58
1300	36" long		20	.400		34.50	17.65		52.15	66.50
1310	42" long		18	.444		43	19.60		62.60	79.50
1500	Tub bar, 1-1/4" diameter, 24" x 36"		14	.571		87	25		112	138
1600	Plus vertical arm		12	.667		85.50	29.50		115	142
1900	End tub bar, 1" diameter, 90° angle, 16" x 32"		12	.667		106	29.50		135.50	164
2010	Tub/shower/toilet, 2-wall, 36" x 24"		12	.667		78	29.50		107.50	134
2300	Hand dryer, surface mounted, electric, 115 volt, 20 amp		4	2		475	88		563	665
2400	230 volt, 10 amp		4	2		720	88		808	940
2600	Hat and coat strip, stainless steel, 4 hook, 36" long		24	.333		71	14.70		85.70	103
2700	6 hook, 60" long		20	.400		112	17.65		129.65	152
3000	Mirror, with stainless steel 3/4" square frame, 18" x 24"		20	.400		42.50	17.65		60.15	76
3100	36" x 24"		15	.533		96.50	23.50		120	145
3200	48" x 24"		10	.800		144	35.50		179.50	217
3300	72" x 24"		6	1.333		206	59		265	325
3500	With 5" stainless steel shelf, 18" x 24"		20	.400		181	17.65		198.65	228
3600	36" x 24"		15	.533		200	23.50		223.50	259
3700	48" x 24"		10	.800		236	35.50		271.50	320
3800	72" x 24"		6	1.333		291	59		350	415
4100	Mop holder strip, stainless steel, 5 holders, 48" long		20	.400		86	17.65		103.65	124
4200	Napkin/tampon dispenser, recessed		15	.533		520	23.50		543.50	615
4250	Napkin receptacle, recessed		6.50	1.231		157	54.50		211.50	262
4300	Robe hook, single, regular		36	.222		21	9.80		30.80	39
4400	Heavy duty, concealed mounting		36	.222		15.20	9.80		25	33
4600	Soap dispenser, chrome, surface mounted, liquid		20	.400		42	17.65		59.65	75.50
4700	Powder		20	.400		54.50	17.65		72.15	89
5000	Recessed stainless steel, liquid		10	.800		123	35.50		158.50	193
5100	Powder		10	.800		180	35.50		215.50	256
5300	Soap tank, stainless steel, 1 gallon		10	.800		200	35.50		235.50	278
5400	5 gallon		5	1.600		275	70.50		345.50	415
5600	Shelf, stainless steel, 5" wide, 18 ga., 24" long		24	.333		64.50	14.70		79.20	95
5700	48" long		16	.500		106	22		128	153
5800	8" wide shelf, 18 ga., 24" long		22	.364		68.50	16.05		84.55	102
5900	48" long		14	.571		121	25		146	175
6000	Toilet seat cover dispenser, stainless steel, recessed		20	.400		158	17.65		175.65	203
6050	Surface mounted		15	.533		32	23.50		55.50	74
6100	Toilet tissue dispenser, surface mounted, SS, single roll		30	.267		17.05	11.75		28.80	38
6200	Double roll		24	.333		22.50	14.70		37.20	49
6240	Plastic, twin/jumbo dbl. roll		24	.333		25.50	14.70		40.20	52
6290	Toilet seat	1 Plum	40	.200		30	10.90		40.90	50
6400	Towel bar, stainless steel, 18" long	1 Carp	23	.348		39.50	15.35		54.85	68.50
6500	30" long		21	.381		111	16.80		127.80	151
6700	Towel dispenser, stainless steel, surface mounted		16	.500		42.50	22		64.50	82.50
6800	Flush mounted, recessed		10	.800		231	35.50		266.50	310
6900	Plastic, touchless, battery operated		16	.500		78.50	22		100.50	123
7000	Towel holder, hotel type, 2 guest size		20	.400		44.50	17.65		62.15	78
7200	Towel shelf, stainless steel, 24" long, 8" wide		20	.400		56.50	17.65		74.15	91
7400	Tumbler holder, tumbler only		30	.267		42	11.75		53.75	66
7500	Soap, tumbler & toothbrush		30	.267		19.25	11.75		31	40.50
7700	Wall urn ash receiver, surface mount, 11" long		12	.667		94.50	29.50		124	152
8000	Waste receptacles, stainless steel, with top, 13 gallon		10	.800		276	35.50		311.50	365
8100	36 gallon		8	1		365	44		409	475

10 28 Toilet, Bath, and Laundry Accessories

10 28 13 – Toilet Accessories

10 28 13.13 Commercial Toilet Accessories		Crew	Daily Output	Labor-Hours	Unit	Material	2012 Bare Costs Labor	Equipment	Total	Total Incl O&P
9000	Minimum labor/equipment charge	1 Carp	5	1.600	Job		70.50		70.50	116

10 28 16 – Bath Accessories

10 28 16.20 Medicine Cabinets		Crew	Daily Output	Labor-Hours	Unit	Material	Labor	Equipment	Total	Total Incl O&P
0010	**MEDICINE CABINETS**									
0020	With mirror, st. st. frame, 16" x 22", unlighted	1 Carp	14	.571	Ea.	78	25		103	128
0100	Wood frame		14	.571		121	25		146	176
0300	Sliding mirror doors, 20" x 16" x 4-3/4", unlighted		7	1.143		114	50.50		164.50	208
0400	24" x 19" x 8-1/2", lighted		5	1.600		177	70.50		247.50	310
0600	Triple door, 30" x 32", unlighted, plywood body		7	1.143		320	50.50		370.50	435
0700	Steel body		7	1.143		370	50.50		420.50	490
0900	Oak door, wood body, beveled mirror, single door		7	1.143		175	50.50		225.50	275
1000	Double door		6	1.333		390	59		449	525
1200	Hotel cabinets, stainless, with lower shelf, unlighted		10	.800		200	35.50		235.50	277
1300	Lighted		5	1.600	↓	300	70.50		370.50	445
9000	Minimum labor/equipment charge	↓	4	2	Job		88		88	145

10 31 Manufactured Fireplaces

10 31 13 – Manufactured Fireplace Chimneys

10 31 13.10 Fireplace Chimneys		Crew	Daily Output	Labor-Hours	Unit	Material	Labor	Equipment	Total	Total Incl O&P
0010	**FIREPLACE CHIMNEYS**									
0500	Chimney dbl. wall, all stainless, over 8'-6", 7" diam., add to fireplace	1 Carp	33	.242	V.L.F.	81	10.70		91.70	107
0600	10" diameter, add to fireplace		32	.250		119	11.05		130.05	148
0700	12" diameter, add to fireplace		31	.258		210	11.40		221.40	250
0800	14" diameter, add to fireplace		30	.267	↓	233	11.75		244.75	275
1000	Simulated brick chimney top, 4' high, 16" x 16"		10	.800	Ea.	425	35.50		460.50	530
1100	24" x 24"	↓	7	1.143	"	520	50.50		570.50	655

10 31 13.20 Chimney Accessories		Crew	Daily Output	Labor-Hours	Unit	Material	Labor	Equipment	Total	Total Incl O&P
0010	**CHIMNEY ACCESSORIES**									
0020	Chimney screens, galv., 13" x 13" flue	1 Bric	8	1	Ea.	58.50	44		102.50	135
0050	24" x 24" flue		5	1.600		118	70.50		188.50	244
0200	Stainless steel, 13" x 13" flue		8	1		97.50	44		141.50	178
0250	20" x 20" flue		5	1.600		175	70.50		245.50	305
2400	Squirrel and bird screens, galvanized, 8" x 8" flue		16	.500		46.50	22		68.50	86.50
2450	13" x 13" flue		12	.667	↓	50.50	29.50		80	103
9000	Minimum labor/equipment charge	↓	3.50	2.286	Job		101		101	162

10 31 16 – Manufactured Fireplace Forms

10 31 16.10 Fireplace Forms		Crew	Daily Output	Labor-Hours	Unit	Material	Labor	Equipment	Total	Total Incl O&P
0010	**FIREPLACE FORMS**									
1800	Fireplace forms, no accessories, 32" opening	1 Bric	3	2.667	Ea.	670	118		788	930
1900	36" opening		2.50	3.200		855	141		996	1,175
2000	40" opening		2	4		1,125	176		1,301	1,525
2100	78" opening	↓	1.50	5.333	↓	1,650	235		1,885	2,200

10 31 23 – Prefabricated Fireplaces

10 31 23.10 Fireplace, Prefabricated		Crew	Daily Output	Labor-Hours	Unit	Material	Labor	Equipment	Total	Total Incl O&P
0010	**FIREPLACE, PREFABRICATED**, free standing or wall hung									
0100	With hood & screen, minimum	1 Carp	1.30	6.154	Ea.	1,200	271		1,471	1,775
0150	Average		1	8		1,425	355		1,780	2,125
0200	Maximum		.90	8.889	↓	2,725	390		3,115	3,650
1500	Simulated logs, gas fired, 40,000 BTU, 2' long, minimum	↓	7	1.143	Set	500	50.50		550.50	635

10 31 Manufactured Fireplaces

10 31 23 – Prefabricated Fireplaces

10 31 23.10 Fireplace, Prefabricated		Crew	Daily Output	Labor-Hours	Unit	Material	2012 Bare Costs Labor	Equipment	Total	Total Incl O&P
1600	Maximum	1 Carp	6	1.333	Set	1,100	59		1,159	1,325
1700	Electric, 1,500 BTU, 1'-6" long, minimum		7	1.143		200	50.50		250.50	305
1800	11,500 BTU, maximum		6	1.333		450	59		509	590
2000	Fireplace, built-in, 36" hearth, radiant		1.30	6.154	Ea.	650	271		921	1,150
2100	Recirculating, small fan		1	8		1,025	355		1,380	1,725
2150	Large fan		.90	8.889		2,050	390		2,440	2,900
2200	42" hearth, radiant		1.20	6.667		920	294		1,214	1,500
2300	Recirculating, small fan		.90	8.889		1,250	390		1,640	2,025
2350	Large fan		.80	10		2,050	440		2,490	2,975
2400	48" hearth, radiant		1.10	7.273		1,925	320		2,245	2,625
2500	Recirculating, small fan		.80	10		2,400	440		2,840	3,350
2550	Large fan		.70	11.429		3,700	505		4,205	4,875
3000	See through, including doors		.80	10		3,125	440		3,565	4,175
3200	Corner (2 wall)		1	8		3,175	355		3,530	4,050

10 32 Fireplace Specialties

10 32 13 – Fireplace Dampers

10 32 13.10 Dampers

		Crew	Daily Output	Labor-Hours	Unit	Material	Labor	Equipment	Total	Total Incl O&P
0010	**DAMPERS**									
0800	Damper, rotary control, steel, 30" opening	1 Bric	6	1.333	Ea.	89	59		148	193
0850	Cast iron, 30" opening		6	1.333		97	59		156	202
0880	36" opening		6	1.333		105	59		164	211
0900	48" opening		6	1.333		150	59		209	260
0920	60" opening		6	1.333		335	59		394	465
1000	84" opening, special order		5	1.600		865	70.50		935.50	1,075
1050	96" opening, special order		4	2		880	88		968	1,100
1200	Steel plate, poker control, 60" opening		8	1		305	44		349	405
1250	84" opening, special order		5	1.600		560	70.50		630.50	730
1400	"Universal" type, chain operated, 32" x 20" opening		8	1		239	44		283	335
1450	48" x 24" opening		5	1.600		355	70.50		425.50	505

10 32 23 – Fireplace Doors

10 32 23.10 Doors

		Crew	Daily Output	Labor-Hours	Unit	Material	Labor	Equipment	Total	Total Incl O&P
0010	**DOORS**									
0400	Cleanout doors and frames, cast iron, 8" x 8"	1 Bric	12	.667	Ea.	39	29.50		68.50	90.50
0450	12" x 12"		10	.800		108	35.50		143.50	176
0500	18" x 24"		8	1		143	44		187	228
0550	Cast iron frame, steel door, 24" x 30"		5	1.600		310	70.50		380.50	455
1600	Dutch Oven door and frame, cast iron, 12" x 15" opening		13	.615		125	27		152	182
1650	Copper plated, 12" x 15" opening		13	.615		245	27		272	315

10 35 Stoves

10 35 13 – Heating Stoves

10 35 13.10 Woodburning Stoves		Crew	Daily Output	Labor-Hours	Unit	2012 Bare Costs Material	Labor	Equipment	Total	Total Incl O&P
0010	**WOODBURNING STOVES**									
0015	Cast iron, minimum	2 Carp	1.30	12.308	Ea.	905	545		1,450	1,900
0020	Average		1	16		1,675	705		2,380	3,000
0030	Maximum		.80	20		2,450	880		3,330	4,150
0050	For gas log lighter, add					42.50			42.50	46.50

10 44 Fire Protection Specialties

10 44 13 – Fire Protection Cabinets

10 44 13.53 Fire Equipment Cabinets

		Crew	Daily Output	Labor-Hours	Unit	Material	Labor	Equipment	Total	Total Incl O&P
0010	**FIRE EQUIPMENT CABINETS**, not equipped, 20 ga. steel box									
0040	recessed, D.S. glass in door, box size given									
1000	Portable extinguisher, single, 8" x 12" x 27", alum. door & frame	Q-12	8	2	Ea.	131	97.50		228.50	296
1100	Steel door and frame	"	8	2	"	90.50	97.50		188	252
3000	Hose rack assy., 1-1/2" valve & 100' hose, 24" x 40" x 5-1/2"									
3200	Steel door and frame	Q-12	6	2.667	Ea.	360	130		490	595
4000	Hose rack assy., 2-1/2" x 1-1/2" valve, 100' hose, 24" x 40" x 8"									
4200	Steel door and frame	Q-12	6	2.667	Ea.	208	130		338	430
5000	Hose rack assy., 2-1/2" x 1-1/2" valve, 100' hose									
5010	and extinguisher, 30" x 40" x 8"									
5200	Steel door and frame	Q-12	5	3.200	Ea.	260	156		416	530

10 44 16 – Fire Extinguishers

10 44 16.13 Portable Fire Extinguishers

		Crew	Daily Output	Labor-Hours	Unit	Material	Labor	Equipment	Total	Total Incl O&P
0010	**PORTABLE FIRE EXTINGUISHERS**									
0140	CO_2, with hose and "H" horn, 10 lb.				Ea.	248			248	273
1000	Dry chemical, pressurized									
1040	Standard type, portable, painted, 2-1/2 lb.				Ea.	34.50			34.50	38
1080	10 lb.					74.50			74.50	81.50
1100	20 lb.					126			126	138
1120	30 lb.					385			385	420
2000	ABC all purpose type, portable, 2-1/2 lb.					21			21	23.50
2080	9-1/2 lb.					44			44	48

10 51 Lockers

10 51 13 – Metal Lockers

10 51 13.10 Lockers

		Crew	Daily Output	Labor-Hours	Unit	Material	Labor	Equipment	Total	Total Incl O&P
0011	**LOCKERS** steel, baked enamel, pre-assembled									
0110	Single tier box locker, 12" x 15" x 72"	1 Shee	20	.400	Ea.	220	21		241	274
0120	18" x 15" x 72"		20	.400		241	21		262	298
0130	12" x 18" x 72"		20	.400		255	21		276	315
0140	18" x 18" x 72"		20	.400		265	21		286	325
0410	Double tier, 12" x 15" x 36"		30	.267		225	13.90		238.90	269
0420	18" x 15" x 36"		30	.267		235	13.90		248.90	281
0430	12" x 18" x 36"		30	.267		236	13.90		249.90	282
0440	18" x 18" x 36"		30	.267		236	13.90		249.90	282
0500	Two person, 18" x 15" x 72"		20	.400		268	21		289	325
0510	18" x 18" x 72"		20	.400		279	21		300	340
0520	Duplex, 15" x 15" x 72"		20	.400		268	21		289	325
0530	15" x 21" x 72"		20	.400		315	21		336	385
1100	Wire meshed wardrobe, floor. mtd., open front varsity type		7.50	1.067		247	55.50		302.50	360

10 51 Lockers

10 51 13 – Metal Lockers

10 51 13.10 Lockers		Crew	Daily Output	Labor-Hours	Unit	2012 Bare Costs Material	Labor	Equipment	Total	Total Incl O&P
2400	16-person locker unit with clothing rack									
2500	72 wide x 15" deep x 72" high	1 Shee	15	.533	Ea.	445	28		473	535
2550	18" deep		15	.533	"	600	28		628	705
3250	Rack w/24 wire mesh baskets		1.50	5.333	Set	390	278		668	865
3260	30 baskets		1.25	6.400		365	335		700	925
3270	36 baskets		.95	8.421		460	440		900	1,200
3280	42 baskets		.80	10		435	520		955	1,300
3600	For hanger rods, add				Ea.	1.86			1.86	2.05
3650	For number plate kit, 100 plates #1 - #100, add	1 Shee	4	2		66	104		170	237
3700	For locker base, closed front panel		90	.089		6.90	4.64		11.54	14.85
3710	End panel, bolted		36	.222		8.25	11.60		19.85	27.50
3800	For sloping top, 12" wide		24	.333		28.50	17.40		45.90	59
3810	15" wide		24	.333		31	17.40		48.40	61.50
3820	18" wide		24	.333		28	17.40		45.40	58.50
3850	Sloping top end panel, 12" deep		72	.111		10.95	5.80		16.75	21
3860	15" deep		72	.111		11.10	5.80		16.90	21.50
3870	18" deep		72	.111		11.60	5.80		17.40	22
3900	For finish end panels, steel, 60" high, 15" deep		12	.667		50	35		85	110
3910	72" high, 12" deep		12	.667		50	35		85	110
3920	18" deep		12	.667		60	35		95	121
5000	For 'ready to assemble' lockers,									
5010	Add to labor						75%			
5020	Deduct from material					20%				
9000	Minimum labor/equipment charge	1 Shee	2.50	3.200	Job		167		167	263

10 55 Postal Specialties

10 55 23 – Mail Boxes

10 55 23.10 Commercial Mail Boxes		Crew	Daily Output	Labor-Hours	Unit	Material	Labor	Equipment	Total	Total Incl O&P
0010	**COMMERCIAL MAIL BOXES**									
0020	Horiz., key lock, 5"H x 6"W x 15"D, alum., rear load	1 Carp	34	.235	Ea.	35	10.40		45.40	55.50
0100	Front loading		34	.235		35	10.40		45.40	55.50
0200	Double, 5"H x 12"W x 15"D, rear loading		26	.308		67.50	13.55		81.05	97
0300	Front loading		26	.308		67	13.55		80.55	96
0500	Quadruple, 10"H x 12"W x 15"D, rear loading		20	.400		81.50	17.65		99.15	119
0600	Front loading		20	.400		95.50	17.65		113.15	134
1600	Vault type, horizontal, for apartments, 4" x 5"		34	.235		18.75	10.40		29.15	37.50
1700	Alphabetical directories, 120 names		10	.800		130	35.50		165.50	201
1830	Lobby collection boxes, aluminum	2 Shee	5	3.200		2,025	167		2,192	2,500
1840	Bronze or stainless	"	4.50	3.556		2,500	185		2,685	3,050
1900	Letter slot, residential	1 Carp	20	.400		89.50	17.65		107.15	128
2000	Post office type		8	1		105	44		149	189
9000	Minimum labor/equipment charge		5	1.600	Job		70.50		70.50	116

10 56 Storage Assemblies

10 56 13 – Metal Storage Shelving

10 56 13.10 Shelving		Crew	Daily Output	Labor-Hours	Unit	2012 Bare Costs Material	Labor	Equipment	Total	Total Incl O&P
0010	**SHELVING**									
0020	Metal, industrial, cross-braced, 3′ wide, 12″ deep	1 Sswk	175	.046	SF Shlf	7.15	2.26		9.41	12.10
0100	24″ deep		330	.024		4.61	1.20		5.81	7.25
2200	Wide span, 1600 lb. capacity per shelf, 6′ wide, 24″ deep		380	.021		6.30	1.04		7.34	8.90
2400	36″ deep		440	.018		6.40	.90		7.30	8.70
9000	Minimum labor/equipment charge	1 Carp	4	2	Job		88		88	145

10 57 Wardrobe and Closet Specialties

10 57 23 – Closet and Utility Shelving

10 57 23.19 Wood Closet and Utility Shelving

Line	Description	Crew	Daily Output	Labor-Hours	Unit	Material	Labor	Equipment	Total	Total Incl O&P
0010	**WOOD CLOSET AND UTILITY SHELVING**									
0020	Pine, clear grade, no edge band, 1″ x 8″	1 Carp	115	.070	L.F.	3.01	3.07		6.08	8.35
0100	1″ x 10″		110	.073		3.75	3.21		6.96	9.40
0200	1″ x 12″		105	.076		4.52	3.36		7.88	10.45
0600	Plywood, 3/4″ thick with lumber edge, 12″ wide		75	.107		1.55	4.70		6.25	9.40
0700	24″ wide		70	.114		2.72	5.05		7.77	11.25
0900	Bookcase, clear grade pine, shelves 12″ O.C., 8″ deep, per S.F. shelf		70	.114	S.F.	9.80	5.05		14.85	19
1000	12″ deep shelves		65	.123	″	14.70	5.45		20.15	25
1200	Adjustable closet rod and shelf, 12″ wide, 3′ long		20	.400	Ea.	10.45	17.65		28.10	40.50
1300	8′ long		15	.533	″	19.35	23.50		42.85	60
1500	Prefinished shelves with supports, stock, 8″ wide		75	.107	L.F.	3.17	4.70		7.87	11.20
1600	10″ wide		70	.114	″	4.32	5.05		9.37	13
9000	Minimum labor/equipment charge		4	2	Job		88		88	145

10 73 Protective Covers

10 73 13 – Awnings

10 73 13.10 Awnings, Fabric

Line	Description	Crew	Daily Output	Labor-Hours	Unit	Material	Labor	Equipment	Total	Total Incl O&P
0010	**AWNINGS, FABRIC**									
0020	Including acrylic canvas and frame, standard design									
0100	Door and window, slope, 3′ high, 4′ wide	1 Carp	4.50	1.778	Ea.	705	78.50		783.50	905
0110	6′ wide		3.50	2.286		910	101		1,011	1,175
0120	8′ wide		3	2.667		1,100	118		1,218	1,425
0200	Quarter round convex, 4′ wide		3	2.667		1,100	118		1,218	1,400
0210	6′ wide		2.25	3.556		1,425	157		1,582	1,825
0220	8′ wide		1.80	4.444		1,750	196		1,946	2,250
0300	Dome, 4′ wide		7.50	1.067		425	47		472	540
0310	6′ wide		3.50	2.286		950	101		1,051	1,225
0320	8′ wide		2	4		1,700	176		1,876	2,150
0350	Elongated dome, 4′ wide		1.33	6.015		1,575	265		1,840	2,175
0360	6′ wide		1.11	7.207		1,900	320		2,220	2,625
0370	8′ wide		1	8		2,225	355		2,580	3,025
1000	Entry or walkway, peak, 12′ long, 4′ wide	2 Carp	.90	17.778		5,025	785		5,810	6,800
1010	6′ wide		.60	26.667		7,725	1,175		8,900	10,400
1020	8′ wide		.40	40		10,700	1,775		12,475	14,600
1100	Radius with dome end, 4′ wide		1.10	14.545		3,800	640		4,440	5,225
1110	6′ wide		.70	22.857		6,125	1,000		7,125	8,375
1120	8′ wide		.50	32		8,700	1,400		10,100	11,900
2000	Retractable lateral arm awning, manual									
2010	To 12′ wide, 8′ - 6″ projection	2 Carp	1.70	9.412	Ea.	1,125	415		1,540	1,925

10 73 Protective Covers

10 73 13 – Awnings

10 73 13.10 Awnings, Fabric		Crew	Daily Output	Labor-Hours	Unit	Material	2012 Bare Costs Labor	Equipment	Total	Total Incl O&P
2020	To 14′ wide, 8′ - 6″ projection	2 Carp	1.10	14.545	Ea.	1,325	640		1,965	2,500
2030	To 19′ wide, 8′ - 6″ projection		.85	18.824		1,800	830		2,630	3,325
2040	To 24′ wide, 8′ - 6″ projection		.67	23.881		2,275	1,050		3,325	4,225
2050	Motor for above, add	1 Carp	2.67	3		970	132		1,102	1,300
3000	Patio/deck canopy with frame									
3010	12′ wide, 12′ projection	2 Carp	2	8	Ea.	1,600	355		1,955	2,325
3020	16′ wide, 14′ projection	"	1.20	13.333		2,500	590		3,090	3,725
9000	For fire retardant canvas, add					7%				
9010	For lettering or graphics, add					35%				
9020	For painted or coated acrylic canvas, deduct					8%				
9030	For translucent or opaque vinyl canvas, add					10%				
9040	For 6 or more units, deduct					20%	15%			

10 73 16 – Canopies

10 73 16.10 Canopies, Residential		Crew	Daily Output	Labor-Hours	Unit	Material	Labor	Equipment	Total	Total Incl O&P
0010	**CANOPIES, RESIDENTIAL** Prefabricated									
0500	Carport, free standing, baked enamel, alum., .032″, 40 psf									
0520	16′ x 8′, 4 posts	2 Carp	3	5.333	Ea.	4,275	235		4,510	5,075
0600	20′ x 10′, 6 posts		2	8		4,450	355		4,805	5,475
0605	30′ x 10′, 8 posts		2	8		6,700	355		7,055	7,925
1000	Door canopies, extruded alum., .032″, 42″ projection, 4′ wide	1 Carp	8	1		202	44		246	296
1020	6′ wide	"	6	1.333		286	59		345	410
1040	8′ wide	2 Carp	9	1.778		375	78.50		453.50	540
1060	10′ wide		7	2.286		345	101		446	540
1080	12′ wide		5	3.200		560	141		701	845
1200	54″ projection, 4′ wide	1 Carp	8	1		249	44		293	345
1220	6′ wide	"	6	1.333		300	59		359	425
1240	8′ wide	2 Carp	9	1.778		355	78.50		433.50	520
1260	10′ wide		7	2.286		515	101		616	735
1280	12′ wide		5	3.200		555	141		696	845
1300	Painted, add					20%				
1310	Bronze anodized, add					50%				
3000	Window awnings, aluminum, window 3′ high, 4′ wide	1 Carp	10	.800		144	35.50		179.50	216
3020	6′ wide	"	8	1		195	44		239	288
3040	9′ wide	2 Carp	9	1.778		292	78.50		370.50	450
3060	12′ wide	"	5	3.200		325	141		466	590
3100	Window, 4′ high, 4′ wide	1 Carp	10	.800		163	35.50		198.50	237
3120	6′ wide	"	8	1		227	44		271	325
3140	9′ wide	2 Carp	9	1.778		325	78.50		403.50	490
3160	12′ wide	"	5	3.200		405	141		546	675
3200	Window, 6′ high, 4′ wide	1 Carp	10	.800		295	35.50		330.50	385
3220	6′ wide	"	8	1		340	44		384	450
3240	9′ wide	2 Carp	9	1.778		900	78.50		978.50	1,125
3260	12′ wide	"	5	3.200		1,375	141		1,516	1,750
3400	Roll-up aluminum, 2′-6″ wide	1 Carp	14	.571		190	25		215	251
3420	3′ wide		12	.667		195	29.50		224.50	262
3440	4′ wide		10	.800		234	35.50		269.50	315
3460	6′ wide		8	1		261	44		305	360
3480	9′ wide	2 Carp	9	1.778		370	78.50		448.50	540
3500	12′ wide	"	5	3.200		475	141		616	755
3600	Window awnings, canvas, 24″ drop, 3′ wide	1 Carp	30	.267	L.F.	48.50	11.75		60.25	73
3620	4′ wide		40	.200		42	8.80		50.80	60.50
3700	30″ drop, 3′ wide		30	.267		51.50	11.75		63.25	76.50

10 73 Protective Covers

10 73 16 – Canopies

10 73 16.10 Canopies, Residential		Crew	Daily Output	Labor-Hours	Unit	2012 Bare Costs Material	Labor	Equipment	Total	Total Incl O&P
3720	4′ wide	1 Carp	40	.200	L.F.	44.50	8.80		53.30	63.50
3740	5′ wide		45	.178		40	7.85		47.85	57
3760	6′ wide		48	.167		37	7.35		44.35	52.50
3780	8′ wide		48	.167		33.50	7.35		40.85	48.50
3800	10′ wide		50	.160		41	7.05		48.05	57
3900	Repair canvas, minimum		8	1	Ea.		44		44	72.50
3920	Maximum		4	2	"		88		88	145
9000	Minimum labor/equipment charge		3	2.667	Job		118		118	193

10 73 16.20 Metal Canopies		Crew	Daily Output	Labor-Hours	Unit	Material	Labor	Equipment	Total	Total Incl O&P
0010	**METAL CANOPIES**									
0020	Wall hung, .032", aluminum, prefinished, 8′ x 10′	K-2	1.30	18.462	Ea.	2,150	830	203	3,183	4,075
0300	8′ x 20′		1.10	21.818		4,275	980	240	5,495	6,750
1000	12′ x 20′		1	24		5,475	1,075	264	6,814	8,275
1050										
2300	Aluminum entrance canopies, flat soffit, .032"									
2500	3′-6" x 4′-0", clear anodized	2 Carp	4	4	Ea.	900	176		1,076	1,275
9000	Minimum labor/equipment charge	"	2	8	Job		355		355	580

10 74 Manufactured Exterior Specialties

10 74 23 – Cupolas

10 74 23.10 Wood Cupolas		Crew	Daily Output	Labor-Hours	Unit	Material	Labor	Equipment	Total	Total Incl O&P
0010	**WOOD CUPOLAS**									
0020	Stock units, pine, painted, 18" sq., 28" high, alum. roof	1 Carp	4.10	1.951	Ea.	167	86		253	325
0100	Copper roof		3.80	2.105		233	93		326	410
0300	23" square, 33" high, aluminum roof		3.70	2.162		385	95.50		480.50	580
0400	Copper roof		3.30	2.424		480	107		587	705
0600	30" square, 37" high, aluminum roof		3.70	2.162		510	95.50		605.50	720
0700	Copper roof		3.30	2.424		610	107		717	845
0900	Hexagonal, 31" wide, 46" high, copper roof		4	2		755	88		843	975
1000	36" wide, 50" high, copper roof		3.50	2.286		1,375	101		1,476	1,700
1200	For deluxe stock units, add to above					25%				
1400	For custom built units, add to above					50%	50%			
9000	Minimum labor/equipment charge	1 Carp	2.75	2.909	Job		128		128	210

10 74 29 – Steeples

10 74 29.10 Prefabricated Steeples		Crew	Daily Output	Labor-Hours	Unit	Material	Labor	Equipment	Total	Total Incl O&P
0010	**PREFABRICATED STEEPLES**									
4000	Steeples, translucent fiberglass, 30" square, 15′ high	F-3	2	20	Ea.	6,700	895	305	7,900	9,150
4150	25′ high	"	1.80	22.222		7,825	995	340	9,160	10,600
4600	Aluminum, baked finish, 16" square, 14′ high					4,850			4,850	5,325
4640	35′ high, 8′ base					31,400			31,400	34,500
4680	152′ high, custom					553,000			553,000	608,500

10 74 33 – Weathervanes

10 74 33.10 Residential Weathervanes		Crew	Daily Output	Labor-Hours	Unit	Material	Labor	Equipment	Total	Total Incl O&P
0010	**RESIDENTIAL WEATHERVANES**									
0020	Residential types, minimum	1 Carp	8	1	Ea.	151	44		195	239
0100	Maximum		2	4	"	1,700	176		1,876	2,175
9000	Minimum labor/equipment charge		4	2	Job		88		88	145

10 74 Manufactured Exterior Specialties

10 74 46 – Window Wells

10 74 46.10 Area Window Wells		Crew	Daily Output	Labor-Hours	Unit	Material	Labor	Equipment	Total	Total Incl O&P
						2012 Bare Costs				
0010	**AREA WINDOW WELLS**, Galvanized steel									
0020	20 ga., 3'-2" wide, 1' deep	1 Sswk	29	.276	Ea.	14.45	13.60		28.05	41.50
0100	2' deep		23	.348		24	17.15		41.15	58
0300	16 ga., 3'-2" wide, 1' deep		29	.276		19.85	13.60		33.45	47.50
0400	3' deep		23	.348		37	17.15		54.15	73
0600	Welded grating for above, 15 lb., painted		45	.178		85	8.75		93.75	110
0700	Galvanized		45	.178		116	8.75		124.75	143
0900	Translucent plastic cap for above	↓	60	.133	↓	19	6.60		25.60	33.50

Division Notes

	CREW	DAILY OUTPUT	LABOR-HOURS	UNIT	2012 BARE COSTS MAT.	LABOR	EQUIP.	TOTAL	TOTAL INCL O&P

Estimating Tips

General

- The items in this division are usually priced per square foot or each. Many of these items are purchased by the owner for installation by the contractor. Check the specifications for responsibilities and include time for receiving, storage, installation, and mechanical and electrical hookups in the appropriate divisions.
- Many items in Division 11 require some type of support system that is not usually furnished with the item. Examples of these systems include blocking for the attachment of casework and support angles for ceiling-hung projection screens. The required blocking or supports must be added to the estimate in the appropriate division.
- Some items in Division 11 may require assembly or electrical hookups. Verify the amount of assembly required or the need for a hard electrical connection and add the appropriate costs.

Reference Numbers

Reference numbers are shown in shaded boxes at the beginning of some major classifications. These numbers refer to related items in the Reference Section. The reference information may be an estimating procedure, an alternate pricing method, or technical information.

Note: Not all subdivisions listed here necessarily appear in this publication.

11 11 Vehicle Service Equipment

11 11 13 – Compressed-Air Vehicle Service Equipment

11 11 13.10 Compressed Air Equipment		Crew	Daily Output	Labor-Hours	Unit	2012 Bare Costs Material	Labor	Equipment	Total	Total Incl O&P
0010	**COMPRESSED AIR EQUIPMENT**									
0030	Compressors, electric, 1-1/2 H.P., standard controls	L-4	1.50	16	Ea.	420	660		1,080	1,525
0550	Dual controls		1.50	16		765	660		1,425	1,925
0600	5 H.P., 115/230 volt, standard controls		1	24		2,025	995		3,020	3,850
0650	Dual controls		1	24		2,575	995		3,570	4,450

11 11 19 – Vehicle Lubrication Equipment

11 11 19.10 Lubrication Equipment

Code	Description	Crew	Daily Output	Labor-Hours	Unit	Material	Labor	Equipment	Total	Total Incl O&P
0010	**LUBRICATION EQUIPMENT**									
3000	Lube equipment, 3 reel type, with pumps, not including piping	L-4	.50	48	Set	8,700	1,975		10,675	12,800

11 11 33 – Vehicle Spray Painting Equipment

11 11 33.10 Spray Painting Equipment

Code	Description	Crew	Daily Output	Labor-Hours	Unit	Material	Labor	Equipment	Total	Total Incl O&P
0010	**SPRAY PAINTING EQUIPMENT**									
4000	Spray painting booth, 26′ long, complete	L-4	.40	60	Ea.	17,600	2,475		20,075	23,500

11 12 Parking Control Equipment

11 12 13 – Parking Key and Card Control Units

11 12 13.10 Parking Control Units

Code	Description	Crew	Daily Output	Labor-Hours	Unit	Material	Labor	Equipment	Total	Total Incl O&P
0010	**PARKING CONTROL UNITS**									
5100	Card reader	1 Elec	2	4	Ea.	1,900	206		2,106	2,425
5120	Proximity with customer display	2 Elec	1	16		5,500	825		6,325	7,325
6000	Parking control software, minimum	1 Elec	.50	16		23,600	825		24,425	27,200
6020	Maximum	″	.20	40		103,000	2,050		105,050	116,500

11 12 16 – Parking Ticket Dispensers

11 12 16.10 Ticket Dispensers

Code	Description	Crew	Daily Output	Labor-Hours	Unit	Material	Labor	Equipment	Total	Total Incl O&P
0010	**TICKET DISPENSERS**									
5900	Ticket spitter with time/date stamp, standard	2 Elec	2	8	Ea.	6,400	410		6,810	7,650
5920	Mag stripe encoding	″	2	8	″	18,600	410		19,010	21,000

11 12 26 – Parking Fee Collection Equipment

11 12 26.13 Parking Fee Coin Collection Equipment

Code	Description	Crew	Daily Output	Labor-Hours	Unit	Material	Labor	Equipment	Total	Total Incl O&P
0010	**PARKING FEE COIN COLLECTION EQUIPMENT**									
5200	Cashier booth, average	B-22	1	30	Ea.	10,100	1,225	199	11,524	13,300
5300	Collector station, pay on foot	2 Elec	.20	80		109,500	4,125		113,625	127,000
5320	Credit card only	″	.50	32		20,200	1,650		21,850	24,900

11 12 26.23 Fee Equipment

Code	Description	Crew	Daily Output	Labor-Hours	Unit	Material	Labor	Equipment	Total	Total Incl O&P
0010	**FEE EQUIPMENT**									
5600	Fee computer	1 Elec	1.50	5.333	Ea.	13,900	275		14,175	15,700

11 12 33 – Parking Gates

11 12 33.13 Parking Gate Equipment

Code	Description	Crew	Daily Output	Labor-Hours	Unit	Material	Labor	Equipment	Total	Total Incl O&P
0010	**PARKING GATE EQUIPMENT**									
5000	Barrier gate with programmable controller	2 Elec	3	5.333	Ea.	3,525	275		3,800	4,325
5020	Industrial		3	5.333		4,875	275		5,150	5,800
5500	Exit verifier		1	16		17,400	825		18,225	20,500
5700	Full sign, 4″ letters	1 Elec	2	4		1,225	206		1,431	1,650
5800	Inductive loop	2 Elec	4	4		169	206		375	505
5950	Vehicle detector, microprocessor based	1 Elec	3	2.667		405	137		542	655

11 13 Loading Dock Equipment

11 13 13 – Loading Dock Bumpers

11 13 13.10 Dock Bumpers		Crew	Daily Output	Labor-Hours	Unit	2012 Bare Costs Material	Labor	Equipment	Total	Total Incl O&P
0010	**DOCK BUMPERS** Bolts not included									
0012	2" x 6" to 4" x 8", average	1 Carp	300	.027	B.F.	.69	1.18		1.87	2.69
0050	Bumpers, rubber blocks 4-1/2" thick, 10" high, 14" long		26	.308	Ea.	51	13.55		64.55	78.50
0200	24" long		22	.364		95	16.05		111.05	131
0300	36" long		17	.471		104	21		125	149
0500	12" high, 14" long		25	.320		100	14.10		114.10	133

11 13 16 – Loading Dock Seals and Shelters

11 13 16.10 Dock Seals and Shelters		Crew	Daily Output	Labor-Hours	Unit	Material	Labor	Equipment	Total	Total Incl O&P
0010	**DOCK SEALS AND SHELTERS**									
6200	Shelters, fabric, for truck or train, scissor arms, minimum	1 Carp	1	8	Ea.	1,650	355		2,005	2,400
6300	Maximum	"	.50	16	"	2,200	705		2,905	3,550

11 13 19 – Stationary Loading Dock Equipment

11 13 19.10 Dock Equipment		Crew	Daily Output	Labor-Hours	Unit	Material	Labor	Equipment	Total	Total Incl O&P
0010	**DOCK EQUIPMENT**									
2200	Dock boards, heavy duty, 60" x 60", aluminum, 5,000 lb. capacity				Ea.	1,250			1,250	1,375
4200	Platform lifter, 6' x 6', portable, 3,000 lb. capacity				"	9,150			9,150	10,100

11 16 Vault Equipment

11 16 16 – Safes

11 16 16.10 Commercial Safes		Crew	Daily Output	Labor-Hours	Unit	Material	Labor	Equipment	Total	Total Incl O&P
0010	**COMMERCIAL SAFES**									
0800	Money, "B" label, 9" x 14" x 14"				Ea.	525			525	580
0900	Tool resistive, 24" x 24" x 20"					3,875			3,875	4,250
1050	Tool and torch resistive, 24" x 24" x 20"					7,050			7,050	7,750

11 17 Teller and Service Equipment

11 17 13 – Teller Equipment Systems

11 17 13.10 Bank Equipment		Crew	Daily Output	Labor-Hours	Unit	Material	Labor	Equipment	Total	Total Incl O&P
0010	**BANK EQUIPMENT**									
0020	Alarm system, police	2 Elec	1.60	10	Ea.	4,675	515		5,190	5,950
0400	Bullet resistant teller window, 44" x 60"	1 Glaz	.60	13.333		3,600	570		4,170	4,875
0500	48" x 60"	"	.60	13.333		4,775	570		5,345	6,200
3000	Counters for banks, frontal only	2 Carp	1	16	Station	1,750	705		2,455	3,075
3100	Complete with steel undercounter		.50	32	"	3,400	1,400		4,800	6,075
5400	Partitions, bullet-resistant, 1-3/16" glass, 8' high		10	1.600	L.F.	189	70.50		259.50	325
5600	Pass thru, bullet-res. window, painted steel, 24" x 36"	2 Sswk	1.60	10	Ea.	2,450	495		2,945	3,625

11 19 Detention Equipment

11 19 30 – Detention Cell Equipment

11 19 30.10 Cell Equipment		Crew	Daily Output	Labor-Hours	Unit	2012 Bare Costs Material	Labor	Equipment	Total	Total Incl O&P
0010	**CELL EQUIPMENT**									
3000	Toilet apparatus including wash basin, average	L-8	1.50	13.333	Ea.	3,350	615		3,965	4,700

11 21 Mercantile and Service Equipment

11 21 13 – Cash Registers and Checking Equipment

11 21 13.10 Checkout Counter

		Crew	Daily Output	Labor-Hours	Unit	Material	Labor	Equipment	Total	Total Incl O&P
0010	**CHECKOUT COUNTER**									
0020	Supermarket conveyor, single belt	2 Clab	10	1.600	Ea.	2,975	56		3,031	3,375
0100	Double belt, power take-away		9	1.778		4,300	62.50		4,362.50	4,825
0400	Double belt, power take-away, incl. side scanning		7	2.286		5,050	80		5,130	5,675
0800	Warehouse or bulk type		6	2.667		5,975	93.50		6,068.50	6,725
1000	Scanning system, 2 lanes, w/registers, scan gun & memory				System	15,600			15,600	17,200
1100	10 lanes, single processor, full scan, with scales				″	148,500			148,500	163,500
2000	Register, restaurant, minimum				Ea.	695			695	765
2100	Maximum					2,925			2,925	3,225
2150	Store, minimum					695			695	765
2200	Maximum					2,925			2,925	3,225

11 21 53 – Barber and Beauty Shop Equipment

11 21 53.10 Barber Equipment

		Crew	Daily Output	Labor-Hours	Unit	Material	Labor	Equipment	Total	Total Incl O&P
0010	**BARBER EQUIPMENT**									
0020	Chair, hydraulic, movable, minimum	1 Carp	24	.333	Ea.	465	14.70		479.70	535
0050	Maximum	″	16	.500		2,975	22		2,997	3,300
0200	Wall hung styling station with mirrors, minimum	L-2	8	2		395	77.50		472.50	560
0300	Maximum	″	4	4		1,975	155		2,130	2,425
0500	Sink, hair washing basin, rough plumbing not incl.	1 Plum	8	1		425	54.50		479.50	550
1000	Sterilizer, liquid solution for tools					138			138	151
1100	Total equipment, rule of thumb, per chair, minimum	L-8	1	20		1,725	925		2,650	3,400
1150	Maximum	″	1	20		4,400	925		5,325	6,350

11 23 Commercial Laundry and Dry Cleaning Equipment

11 23 13 – Dry Cleaning Equipment

11 23 13.13 Dry Cleaners

		Crew	Daily Output	Labor-Hours	Unit	Material	Labor	Equipment	Total	Total Incl O&P
0010	**DRY CLEANERS** Not incl. rough-in									
2000	Dry cleaners, electric, 20 lb. capacity	L-1	.20	80	Ea.	33,300	4,250		37,550	43,200
2050	25 lb. capacity		.17	94.118		46,500	5,000		51,500	59,000
2100	30 lb. capacity		.15	106		49,000	5,650		54,650	63,000
2150	60 lb. capacity		.09	177		78,000	9,425		87,425	100,500

11 23 16 – Drying and Conditioning Equipment

11 23 16.13 Dryers

		Crew	Daily Output	Labor-Hours	Unit	Material	Labor	Equipment	Total	Total Incl O&P
0010	**DRYERS**, Not including rough-in									
1500	Industrial, 30 lb. capacity	1 Plum	2	4	Ea.	3,025	218		3,243	3,675
1600	50 lb. capacity	″	1.70	4.706		3,250	257		3,507	3,975
4700	Lint collector, ductwork not included, 8,000 to 10,000 CFM	Q-10	.30	80		8,800	3,900		12,700	15,800

11 23 19 – Finishing Equipment

11 23 19.13 Folders and Spreaders

		Crew	Daily Output	Labor-Hours	Unit	Material	Labor	Equipment	Total	Total Incl O&P
0010	**FOLDERS AND SPREADERS**									
3500	Folders, blankets & sheets, minimum	1 Elec	.17	47.059	Ea.	31,600	2,425		34,025	38,600
3700	King size with automatic stacker		.10	80		57,000	4,125		61,125	69,500

11 23 Commercial Laundry and Dry Cleaning Equipment

11 23 19 – Finishing Equipment

11 23 19.13 Folders and Spreaders		Crew	Daily Output	Labor-Hours	Unit	Material	2012 Bare Costs Labor	Equipment	Total	Total Incl O&P
3800	For conveyor delivery, add	1 Elec	.45	17.778	Ea.	13,500	915		14,415	16,300

11 23 23 – Commercial Ironing Equipment

11 23 23.13 Irons and Pressers		Crew	Daily Output	Labor-Hours	Unit	Material	Labor	Equipment	Total	Total Incl O&P
0010	**IRONS AND PRESSERS**									
4500	Ironers, institutional, 110", single roll	1 Elec	.20	40	Ea.	30,700	2,050		32,750	37,000

11 23 26 – Commercial Washers and Extractors

11 23 26.13 Washers and Extractors		Crew	Daily Output	Labor-Hours	Unit	Material	Labor	Equipment	Total	Total Incl O&P
0010	**WASHERS AND EXTRACTORS**, not including rough-in									
6000	Combination washer/extractor, 20 lb. capacity	L-6	1.50	8	Ea.	5,700	430		6,130	6,950
6100	30 lb. capacity		.80	15		8,975	805		9,780	11,100
6200	50 lb. capacity		.68	17.647		10,500	945		11,445	13,100
6300	75 lb. capacity		.30	40		19,900	2,150		22,050	25,200
6350	125 lb. capacity	↓	.16	75	↓	27,300	4,025		31,325	36,400

11 23 33 – Coin-Operated Laundry Equipment

11 23 33.13 Coin Operated Washers and Dryers		Crew	Daily Output	Labor-Hours	Unit	Material	Labor	Equipment	Total	Total Incl O&P
0010	**COIN OPERATED WASHERS AND DRYERS**									
0990	Dryer, gas fired									
1000	Commercial, 30 lb. capacity, coin operated, single	1 Plum	3	2.667	Ea.	3,150	146		3,296	3,700
1100	Double stacked	"	2	4	"	6,825	218		7,043	7,850
5290	Clothes washer									
5300	Commercial, coin operated, average	1 Plum	3	2.667	Ea.	1,250	146		1,396	1,600

11 24 Maintenance Equipment

11 24 19 – Vacuum Cleaning Systems

11 24 19.10 Vacuum Cleaning		Crew	Daily Output	Labor-Hours	Unit	Material	Labor	Equipment	Total	Total Incl O&P
0010	**VACUUM CLEANING**									
0020	Central, 3 inlet, residential	1 Skwk	.90	8.889	Total	1,025	405		1,430	1,800
0200	Commercial		.70	11.429		1,225	520		1,745	2,200
0400	5 inlet system, residential		.50	16		1,450	725		2,175	2,775
0600	7 inlet system, commercial	↓	.40	20		1,675	910		2,585	3,325
4010	Rule of thumb: First 1200 S.F., installed				↓				1,425	1,575
4020	For each additional S.F., add				S.F.					.26

11 26 Unit Kitchens

11 26 13 – Metal Unit Kitchens

11 26 13.10 Commercial Unit Kitchens		Crew	Daily Output	Labor-Hours	Unit	Material	Labor	Equipment	Total	Total Incl O&P
0010	**COMMERCIAL UNIT KITCHENS**									
1500	Combination range, refrigerator and sink, 30" wide, minimum	L-1	2	8	Ea.	1,125	425		1,550	1,875
1550	Maximum	"	1	16	"	3,875	850		4,725	5,575
1640	Combination range, refrigerator, sink, microwave									
1660	Oven and ice maker	L-1	.80	20	Ea.	4,675	1,050		5,725	6,775

11 27 Photographic Processing Equipment

11 27 13 – Darkroom Processing Equipment

	11 27 13.10 Darkroom Equipment	Crew	Daily Output	Labor-Hours	Unit	2012 Bare Costs Material	Labor	Equipment	Total	Total Incl O&P
0010	**DARKROOM EQUIPMENT**									
0020	Developing sink, 5" deep, 24" x 48"	Q-1	2	8	Ea.	5,625	395		6,020	6,800
0050	48" x 52"		1.70	9.412		5,675	465		6,140	6,950
0200	10" deep, 24" x 48"		1.70	9.412		7,775	465		8,240	9,275
0250	24" x 108"		1.50	10.667		10,200	525		10,725	12,100
0500	Dryers, dehumidified filtered air, 36" x 25" x 68" high	L-7	6	4.667		5,075	199		5,274	5,900
0550	48" x 25" x 68" high		5	5.600		10,000	239		10,239	11,400
2000	Processors, automatic, color print, minimum		4	7		15,700	298		15,998	17,800
2050	Maximum		.60	46.667		24,700	2,000		26,700	30,400
2300	Black and white print, minimum		2	14		11,900	595		12,495	14,100
2350	Maximum		.80	35		61,500	1,500		63,000	70,000
2600	Manual processor, 16" x 20" maximum print size		2	14		9,450	595		10,045	11,400
2650	20" x 24" maximum print size		1	28		8,975	1,200		10,175	11,800
3000	Viewing lights, 20" x 24"		6	4.667		380	199		579	740
3100	20" x 24" with color correction		6	4.667		535	199		734	915
3500	Washers, round, minimum sheet 11" x 14"	Q-1	2	8		3,400	395		3,795	4,375
3550	Maximum sheet 20" x 24"		1	16		3,475	785		4,260	5,050
3800	Square, minimum sheet 20" x 24"		1	16		3,125	785		3,910	4,650
3900	Maximum sheet 50" x 56"		.80	20		4,850	985		5,835	6,850
4500	Combination tank sink, tray sink, washers, with									
4510	Dry side tables, average	Q-1	.45	35.556	Ea.	10,600	1,750		12,350	14,400

11 31 Residential Appliances

11 31 13 – Residential Kitchen Appliances

	11 31 13.13 Cooking Equipment	Crew	Daily Output	Labor-Hours	Unit	Material	Labor	Equipment	Total	Total Incl O&P
0010	**COOKING EQUIPMENT**									
0020	Cooking range, 30" free standing, 1 oven, minimum	2 Clab	10	1.600	Ea.	405	56		461	535
0050	Maximum	"	4	4		1,775	140		1,915	2,175
0350	Built-in, 30" wide, 1 oven, minimum	1 Elec	6	1.333		650	68.50		718.50	820
0400	Maximum	2 Carp	2	8		1,250	355		1,605	1,950
0900	Countertop cooktops, 4 burner, standard, minimum	1 Elec	6	1.333		275	68.50		343.50	410
0950	Maximum		3	2.667		1,100	137		1,237	1,400
1250	Microwave oven, minimum		4	2		110	103		213	280
1300	Maximum		2	4		420	206		626	780
5380	Oven, built in, standard		4	2		550	103		653	765
5390	Deluxe		2	4		1,850	206		2,056	2,350

	11 31 13.23 Refrigeration Equipment	Crew	Daily Output	Labor-Hours	Unit	Material	Labor	Equipment	Total	Total Incl O&P
0010	**REFRIGERATION EQUIPMENT**									
5450	Refrigerator, no frost, 6 C.F.	2 Clab	15	1.067	Ea.	207	37.50		244.50	290
5500	Refrigerator, no frost, 10 C.F. to 12 C.F., minimum		10	1.600		425	56		481	560
5600	Maximum		6	2.667		510	93.50		603.50	715
6790	Energy-star qualified, 18 C.F., minimum G	2 Carp	4	4		450	176		626	785
6795	Maximum G		2	8		1,150	355		1,505	1,825
6797	21.7 C.F., minimum G		4	4		955	176		1,131	1,350
6799	Maximum G		4	4		1,875	176		2,051	2,375

	11 31 13.33 Kitchen Cleaning Equipment	Crew	Daily Output	Labor-Hours	Unit	Material	Labor	Equipment	Total	Total Incl O&P
0010	**KITCHEN CLEANING EQUIPMENT**									
2750	Dishwasher, built-in, 2 cycles, minimum	L-1	4	4	Ea.	212	212		424	565
2800	Maximum		2	8		400	425		825	1,100
3100	Energy-star qualified, minimum G		4	4		355	212		567	720

11 31 Residential Appliances

11 31 13 – Residential Kitchen Appliances

11 31 13.33 Kitchen Cleaning Equipment			Crew	Daily Output	Labor-Hours	Unit	2012 Bare Costs Material	Labor	Equipment	Total	Total Incl O&P
3110	Maximum	G	L-1	2	8	Ea.	1,300	425		1,725	2,075

11 31 13.43 Waste Disposal Equipment

			Crew	Daily Output	Labor-Hours	Unit	Material	Labor	Equipment	Total	Total Incl O&P
0010	**WASTE DISPOSAL EQUIPMENT**										
1750	Compactor, residential size, 4 to 1 compaction, minimum		1 Carp	5	1.600	Ea.	580	70.50		650.50	750
1800	Maximum		"	3	2.667		860	118		978	1,150
3300	Garbage disposal, sink type, minimum		L-1	10	1.600		83	85		168	223
3350	Maximum		"	10	1.600		197	85		282	350

11 31 13.53 Kitchen Ventilation Equipment

		Crew	Daily Output	Labor-Hours	Unit	Material	Labor	Equipment	Total	Total Incl O&P
0010	**KITCHEN VENTILATION EQUIPMENT**									
4150	Hood for range, 2 speed, vented, 30" wide, minimum	L-3	5	3.200	Ea.	57.50	154		211.50	310
4200	Maximum		3	5.333		750	256		1,006	1,225
4300	42" wide, minimum		5	3.200		181	154		335	445
4330	Custom		5	3.200		1,650	154		1,804	2,050
4350	Maximum		3	5.333		2,000	256		2,256	2,600
4400	Ventless hood, 2 speed, 30" wide, minimum	1 Elec	8	1		73.50	51.50		125	161
4450	Maximum		7	1.143		555	59		614	700
4510	36" wide, minimum		7	1.143		77	59		136	176
4550	Maximum		6	1.333		555	68.50		623.50	715
4580	42" wide, minimum		6	1.333		198	68.50		266.50	325
4600	Maximum		5	1.600		595	82.50		677.50	780
4650	For vented 1 speed, deduct from maximum					47.50			47.50	52.50
4700										

11 31 23 – Residential Laundry Appliances

11 31 23.13 Washers

			Crew	Daily Output	Labor-Hours	Unit	Material	Labor	Equipment	Total	Total Incl O&P
0010	**WASHERS**										
5000	Residential, 4 cycle, average		1 Plum	3	2.667	Ea.	850	146		996	1,150
6750	Energy star qualified, front loading, minimum	G		3	2.667		550	146		696	835
6760	Maximum	G		1	8		1,425	435		1,860	2,250
6764	Top loading, minimum	G		3	2.667		370	146		516	635
6766	Maximum	G		3	2.667		1,825	146		1,971	2,225

11 31 23.23 Dryers

			Crew	Daily Output	Labor-Hours	Unit	Material	Labor	Equipment	Total	Total Incl O&P
0010	**DRYERS**										
0500	Gas fired residential, 16 lb. capacity, average		1 Plum	3	2.667	Ea.	645	146		791	935
6770	Electric, front loading, energy-star qualified, minimum	G	L-2	3	5.333		395	206		601	775
6780	Maximum	G	"	2	8		1,700	310		2,010	2,375

11 31 33 – Miscellaneous Residential Appliances

11 31 33.13 Sump Pumps

		Crew	Daily Output	Labor-Hours	Unit	Material	Labor	Equipment	Total	Total Incl O&P
0010	**SUMP PUMPS**									
6400	Cellar drainer, pedestal, 1/3 H.P., molded PVC base	1 Plum	3	2.667	Ea.	128	146		274	365
6450	Solid brass	"	2	4	"	272	218		490	640
6460	Sump pump, see also Section 22 14 29.16									

11 33 Retractable Stairs

11 33 10 – Disappearing Stairs

11 33 10.10 Disappearing Stairway		Crew	Daily Output	Labor-Hours	Unit	Material	Labor (2012 Bare Costs)	Equipment	Total	Total Incl O&P
0010	**DISAPPEARING STAIRWAY** No trim included									
0020	One piece, yellow pine, 8'-0" ceiling	2 Carp	4	4	Ea.	260	176		436	575
0030	9'-0" ceiling		4	4		330	176		506	650
0040	10'-0" ceiling		3	5.333		260	235		495	670
0050	11'-0" ceiling		3	5.333		360	235		595	780
0060	12'-0" ceiling		3	5.333		385	235		620	805
0100	Custom grade, pine, 8'-6" ceiling, minimum	1 Carp	4	2		155	88		243	315
0150	Average		3.50	2.286		239	101		340	430
0200	Maximum		3	2.667		310	118		428	535
0250										
0500	Heavy duty, pivoted, from 7'-7" to 12'-10" floor to floor	1 Carp	3	2.667	Ea.	735	118		853	1,000
0600	16'-0" ceiling		2	4		1,500	176		1,676	1,950
0800	Economy folding, pine, 8'-6" ceiling		4	2		132	88		220	290
0900	9'-6" ceiling		4	2		155	88		243	315
1100	Automatic electric, aluminum, floor to floor height, 8' to 9'	2 Carp	1	16		8,575	705		9,280	10,600
1400	11' to 12'		.90	17.778		9,275	785		10,060	11,500
1700	14' to 15'		.70	22.857		9,875	1,000		10,875	12,600
9000	Minimum labor/equipment charge	1 Carp	2	4	Job		176		176	289

11 41 Foodservice Storage Equipment

11 41 13 – Refrigerated Food Storage Cases

11 41 13.10 Refrigerated Food Cases		Crew	Daily Output	Labor-Hours	Unit	Material	Labor (2012 Bare Costs)	Equipment	Total	Total Incl O&P
0010	**REFRIGERATED FOOD CASES**									
0030	Dairy, multi-deck, 12' long	Q-5	3	5.333	Ea.	11,200	267		11,467	12,700
0100	For rear sliding doors, add					1,725			1,725	1,900
0200	Delicatessen case, service deli, 12' long, single deck	Q-5	3.90	4.103		6,575	206		6,781	7,575
0300	Multi-deck, 18 S.F. shelf display		3	5.333		6,225	267		6,492	7,275
0400	Freezer, self-contained, chest-type, 30 C.F.		3.90	4.103		7,425	206		7,631	8,500
0500	Glass door, upright, 78 C.F.		3.30	4.848		11,900	243		12,143	13,500
0600	Frozen food, chest type, 12' long		3.30	4.848		7,975	243		8,218	9,150
0700	Glass door, reach-in, 5 door		3	5.333		13,900	267		14,167	15,700
0800	Island case, 12' long, single deck		3.30	4.848		7,200	243		7,443	8,300
0900	Multi-deck		3	5.333		8,400	267		8,667	9,675
1000	Meat case, 12' long, single deck		3.30	4.848		6,000	243		6,243	6,975
1050	Multi-deck		3.10	5.161		10,300	259		10,559	11,700
1100	Produce, 12' long, single deck		3.30	4.848		6,775	243		7,018	7,825
1200	Multi-deck		3.10	5.161		8,625	259		8,884	9,875

11 41 13.20 Refrigerated Food Storage Equipment		Crew	Daily Output	Labor-Hours	Unit	Material	Labor (2012 Bare Costs)	Equipment	Total	Total Incl O&P
0010	**REFRIGERATED FOOD STORAGE EQUIPMENT**									
2350	Cooler, reach-in, beverage, 6' long	Q-1	6	2.667	Ea.	3,625	131		3,756	4,175
4300	Freezers, reach-in, 44 C.F.		4	4		2,625	197		2,822	3,200
4500	68 C.F.		3	5.333		4,550	262		4,812	5,425
4600	Freezer, pre-fab, 8' x 8' w/refrigeration	2 Carp	.45	35.556		11,000	1,575		12,575	14,700
4620	8' x 12'		.35	45.714		11,800	2,025		13,825	16,200
4640	8' x 16'		.25	64		16,600	2,825		19,425	22,800
4660	8' x 20'		.17	94.118		19,900	4,150		24,050	28,700
4680	Reach-in, 1 compartment	Q-1	4	4		2,275	197		2,472	2,800
4685	Energy star rated G	R-18	7.80	3.333		2,325	135		2,460	2,775
4700	2 compartment	Q-1	3	5.333		3,650	262		3,912	4,400
4705	Energy star rated G	R-18	6.20	4.194		3,250	169		3,419	3,850
4710	3 compartment	Q-1	3	5.333		4,825	262		5,087	5,725

11 41 Foodservice Storage Equipment

11 41 13 – Refrigerated Food Storage Cases

11 41 13.20 Refrigerated Food Storage Equipment

	11 41 13.20 Refrigerated Food Storage Equipment		Crew	Daily Output	Labor-Hours	Unit	2012 Bare Costs Material	Labor	Equipment	Total	Total Incl O&P
4715	Energy star rated	G	R-18	5.60	4.643	Ea.	4,025	187		4,212	4,725
8320	Refrigerator, reach-in, 1 compartment			7.80	3.333		2,175	135		2,310	2,600
8325	Energy star rated	G		7.80	3.333		2,325	135		2,460	2,775
8330	2 compartment			6.20	4.194		3,450	169		3,619	4,050
8335	Energy star rated	G		6.20	4.194		3,250	169		3,419	3,850
8340	3 compartment			5.60	4.643		4,275	187		4,462	5,000
8345	Energy star rated	G		5.60	4.643		4,025	187		4,212	4,725
8350	Pre-fab, with refrigeration, 8′ x 8′		2 Carp	.45	35.556		6,575	1,575		8,150	9,800
8360	8′ x 12′			.35	45.714		7,025	2,025		9,050	11,100
8370	8′ x 16′			.25	64		11,800	2,825		14,625	17,600
8380	8′ x 20′			.17	94.118		14,900	4,150		19,050	23,200
8390	Pass-thru/roll-in, 1 compartment		R-18	7.80	3.333		4,300	135		4,435	4,975
8400	2 compartment			6.24	4.167		6,075	168		6,243	6,975
8410	3 compartment			5.60	4.643		8,375	187		8,562	9,500
8420	Walk-in, alum, door & floor only, no refrig, 6′ x 6′ x 7′-6″		2 Carp	1.40	11.429		9,175	505		9,680	10,900
8430	10′ x 6′ x 7′-6″			.55	29.091		13,100	1,275		14,375	16,500
8440	12′ x 14′ x 7′-6″			.25	64		18,400	2,825		21,225	24,800
8450	12′ x 20′ x 7′-6″			.17	94.118		22,400	4,150		26,550	31,500
8460	Refrigerated cabinets, mobile						3,725			3,725	4,075
8470	Refrigerator/freezer, reach-in, 1 compartment		R-18	5.60	4.643		5,050	187		5,237	5,875
8480	2 compartment		″	4.80	5.417		7,125	219		7,344	8,175

11 41 13.30 Wine Cellar

	11 41 13.30 Wine Cellar	Crew	Daily Output	Labor-Hours	Unit	Material	Labor	Equipment	Total	Total Incl O&P
0010	**WINE CELLAR**, refrigerated, Redwood interior, carpeted, walk-in type									
0020	6′-8″ high, including racks									
0200	80″ W x 48″ D for 900 bottles	2 Carp	1.50	10.667	Ea.	3,275	470		3,745	4,375
0250	80″ W x 72″ D for 1300 bottles		1.33	12.030		4,125	530		4,655	5,425
0300	80″ W x 94″ D for 1900 bottles		1.17	13.675		4,900	605		5,505	6,375
0400	80″ W x 124″ D for 2500 bottles		1	16		6,000	705		6,705	7,750

11 41 33 – Foodservice Shelving

11 41 33.20 Metal Food Storage Shelving

	11 41 33.20 Metal Food Storage Shelving	Crew	Daily Output	Labor-Hours	Unit	Material	Labor	Equipment	Total	Total Incl O&P
0010	**METAL FOOD STORAGE SHELVING**									
8600	Stainless steel shelving, louvered 4-tier, 20″ x 3′	1 Clab	6	1.333	Ea.	1,300	47		1,347	1,525
8605	20″ x 4′		6	1.333		1,875	47		1,922	2,150
8610	20″ x 6′		6	1.333		2,725	47		2,772	3,075
8615	24″ x 3′		6	1.333		1,725	47		1,772	1,975
8620	24″ x 4′		6	1.333		2,000	47		2,047	2,275
8625	24″ x 6′		6	1.333		2,900	47		2,947	3,275
8630	Flat 4-tier, 20″ x 3′		6	1.333		1,175	47		1,222	1,375
8635	20″ x 4′		6	1.333		1,425	47		1,472	1,625
8640	20″ x 5′		6	1.333		1,650	47		1,697	1,875
8645	24″ x 3′		6	1.333		1,325	47		1,372	1,550
8650	24″ x 4′		6	1.333		2,200	47		2,247	2,475
8655	24″ x 6′		6	1.333		2,625	47		2,672	2,950
8700	Galvanized shelving, louvered 4-tier, 20″ x 3′		6	1.333		670	47		717	810
8705	20″ x 4′		6	1.333		745	47		792	895
8710	20″ x 6′		6	1.333		915	47		962	1,075
8715	24″ x 3′		6	1.333		580	47		627	715
8720	24″ x 4′		6	1.333		805	47		852	960
8725	24″ x 6′		6	1.333		1,150	47		1,197	1,350
8730	Flat 4-tier, 20″ x 3′		6	1.333		595	47		642	730
8735	20″ x 4′		6	1.333		655	47		702	800
8740	20″ x 6′		6	1.333		940	47		987	1,100

11 41 Foodservice Storage Equipment

11 41 33 – Foodservice Shelving

11 41 33.20 Metal Food Storage Shelving		Crew	Daily Output	Labor-Hours	Unit	Material	2012 Bare Costs Labor	Equipment	Total	Total Incl O&P
8745	24" x 3'	1 Clab	6	1.333	Ea.	660	47		707	805
8750	24" x 4'		6	1.333		670	47		717	815
8755	24" x 6'		6	1.333		980	47		1,027	1,150
8760	Stainless steel dunnage rack, 24" x 3'		8	1		330	35		365	420
8765	24" x 4'		8	1		420	35		455	520
8770	Galvanized dunnage rack, 24" x 3'		8	1		178	35		213	254
8775	24" x 4'		8	1		198	35		233	276

11 42 Food Preparation Equipment

11 42 10 – Commercial Food Preparation Equipment

11 42 10.10 Choppers, Mixers and Misc. Equipment

		Crew	Daily Output	Labor-Hours	Unit	Material	Labor	Equipment	Total	Total Incl O&P
0010	**CHOPPERS, MIXERS AND MISC. EQUIPMENT**									
1700	Choppers, 5 pounds	R-18	7	3.714	Ea.	1,950	150		2,100	2,400
1720	16 pounds		5	5.200		2,300	210		2,510	2,875
1740	35 to 40 pounds		4	6.500		3,425	262		3,687	4,200
1840	Coffee brewer, 5 burners	1 Plum	3	2.667		1,275	146		1,421	1,625
1850	Coffee urn, twin 6 gallon urns		2	4		2,375	218		2,593	2,975
1860	Single, 3 gallon		3	2.667		1,725	146		1,871	2,125
3000	Fast food equipment, total package, minimum	6 Skwk	.08	600		201,000	27,200		228,200	266,000
3100	Maximum	"	.07	685		274,000	31,100		305,100	352,000
3800	Food mixers, bench type, 20 quarts	L-7	7	4		2,475	170		2,645	3,000
3850	40 quarts		5.40	5.185		6,425	221		6,646	7,425
3900	60 quarts		5	5.600		10,600	239		10,839	12,100
4040	80 quarts		3.90	7.179		11,600	305		11,905	13,300
4080	130 quarts		2.20	12.727		16,700	540		17,240	19,300
4100	Floor type, 20 quarts		15	1.867		3,075	79.50		3,154.50	3,500
4120	60 quarts		14	2		9,475	85		9,560	10,500
4140	80 quarts		12	2.333		11,200	99.50		11,299.50	12,600
4160	140 quarts		8.60	3.256		18,600	139		18,739	20,600
6700	Peelers, small	R-18	8	3.250		1,650	131		1,781	2,025
6720	Large	"	6	4.333		4,975	175		5,150	5,750
6800	Pulper/extractor, close coupled, 5 HP	1 Plum	1.90	4.211		3,650	230		3,880	4,375
8580	Slicer with table	R-18	9	2.889		5,075	117		5,192	5,750

11 43 Food Delivery Carts and Conveyors

11 43 13 – Food Delivery Carts

11 43 13.10 Mobile Carts, Racks and Trays

		Crew	Daily Output	Labor-Hours	Unit	Material	Labor	Equipment	Total	Total Incl O&P
0010	**MOBILE CARTS, RACKS AND TRAYS**									
1650	Cabinet, heated, 1 compartment, reach-in	R-18	5.60	4.643	Ea.	3,850	187		4,037	4,525
1655	Pass-thru roll-in		5.60	4.643		5,200	187		5,387	6,025
1660	2 compartment, reach-in		4.80	5.417		8,275	219		8,494	9,450
1670	Mobile					2,875			2,875	3,175
6850	Mobile rack w/pan slide					1,350			1,350	1,475
9180	Tray and silver dispenser, mobile	1 Clab	16	.500		890	17.55		907.55	1,000

11 44 Food Cooking Equipment

11 44 13 – Commercial Ranges

11 44 13.10 Cooking Equipment		Crew	Daily Output	Labor-Hours	Unit	2012 Bare Costs Material	Labor	Equipment	Total	Total Incl O&P
0010	**COOKING EQUIPMENT**									
0020	Bake oven, gas, one section	Q-1	8	2	Ea.	4,800	98.50		4,898.50	5,425
0300	Two sections		7	2.286		9,175	112		9,287	10,300
0600	Three sections		6	2.667		14,900	131		15,031	16,600
0900	Electric convection, single deck	L-7	4	7		6,425	298		6,723	7,550
1300	Broiler, without oven, standard	Q-1	8	2		3,500	98.50		3,598.50	4,000
1550	Infrared	L-7	4	7		9,350	298		9,648	10,800
4750	Fryer, with twin baskets, modular model	Q-1	7	2.286		1,125	112		1,237	1,400
5000	Floor model, on 6" legs	"	5	3.200		2,125	157		2,282	2,600
5100	Extra single basket, large					97			97	107
5170	Energy star rated, 50 lb. capacity [G]	R-18	4	6.500		4,100	262		4,362	4,925
5175	85 lb. capacity [G]	"	4	6.500		6,775	262		7,037	7,875
5300	Griddle, SS, 24" plate, w/4" legs, elec, 208 V, 3 phase, 3' long	Q-1	7	2.286		1,250	112		1,362	1,550
5550	4' long	"	6	2.667		2,200	131		2,331	2,625
6200	Iced tea brewer	1 Plum	3.44	2.326		710	127		837	980
6350	Kettle, w/steam jacket, tilting, w/positive lock, SS, 20 gallons	L-7	7	4		7,375	170		7,545	8,375
6600	60 gallons	"	6	4.667		10,400	199		10,599	11,700
6900	Range, restaurant type, 6 burners and 1 standard oven, 36" wide	Q-1	7	2.286		2,875	112		2,987	3,325
6950	Convection		7	2.286		4,400	112		4,512	5,025
7150	2 standard ovens, 24" griddle, 60" wide		6	2.667		3,775	131		3,906	4,375
7200	1 standard, 1 convection oven		6	2.667		8,800	131		8,931	9,875
7450	Heavy duty, single 34" standard oven, open top		5	3.200		4,650	157		4,807	5,350
7500	Convection oven		5	3.200		4,750	157		4,907	5,475
7700	Griddle top		6	2.667		2,600	131		2,731	3,075
7750	Convection oven		6	2.667		7,375	131		7,506	8,300
8850	Steamer, electric 27 KW	L-7	7	4		8,500	170		8,670	9,625
9100	Electric, 10 KW or gas 100,000 BTU	"	5	5.600		6,250	239		6,489	7,250
9150	Toaster, conveyor type, 16-22 slices per minute					990			990	1,100
9160	Pop-up, 2 slot					610			610	670
9200	For deluxe models of above equipment, add					75%				
9400	Rule of thumb: Equipment cost based									
9410	on kitchen work area									
9420	Office buildings, minimum	L-7	77	.364	S.F.	89.50	15.50		105	124
9450	Maximum		58	.483		151	20.50		171.50	200
9550	Public eating facilities, minimum		77	.364		117	15.50		132.50	154
9600	Maximum		46	.609		191	26		217	252
9750	Hospitals, minimum		58	.483		121	20.50		141.50	167
9800	Maximum		39	.718		222	30.50		252.50	294

11 46 Food Dispensing Equipment

11 46 16 – Service Line Equipment

11 46 16.10 Commercial Food Dispensing Equipment		Crew	Daily Output	Labor-Hours	Unit	Material	Labor	Equipment	Total	Total Incl O&P
0010	**COMMERCIAL FOOD DISPENSING EQUIPMENT**									
1050	Butter pat dispenser	1 Clab	13	.615	Ea.	945	21.50		966.50	1,075
1100	Bread dispenser, counter top		13	.615		920	21.50		941.50	1,050
1900	Cup and glass dispenser, drop in		4	2		1,050	70		1,120	1,275
1920	Disposable cup, drop in		16	.500		470	17.55		487.55	550
2650	Dish dispenser, drop in, 12"		11	.727		2,275	25.50		2,300.50	2,550
2660	Mobile		10	.800		2,700	28		2,728	3,025
3300	Food warmer, counter, 1.2 KW					510			510	560
3550	1.6 KW					2,050			2,050	2,275

11 46 Food Dispensing Equipment

11 46 16 – Service Line Equipment

11 46 16.10 Commercial Food Dispensing Equipment		Crew	Daily Output	Labor-Hours	Unit	2012 Bare Costs Material	Labor	Equipment	Total	Total Incl O&P
3600	Well, hot food, built-in, rectangular, 12" x 20"	R-30	10	2.600	Ea.	710	108		818	955
3610	Circular, 7 qt.		10	2.600		375	108		483	580
3620	Refrigerated, 2 compartments		10	2.600		2,875	108		2,983	3,350
3630	3 compartments		9	2.889		3,450	120		3,570	4,000
3640	4 compartments		8	3.250		4,225	135		4,360	4,875
4720	Frost cold plate		9	2.889		18,200	120		18,320	20,200
5700	Hot chocolate dispenser	1 Plum	4	2		1,175	109		1,284	1,475
5750	Ice dispenser 567 pound	Q-1	6	2.667		5,000	131		5,131	5,700
6250	Jet spray dispenser	R-18	4.50	5.778		3,225	233		3,458	3,925
6300	Juice dispenser, concentrate	"	4.50	5.778		2,125	233		2,358	2,725
6690	Milk dispenser, bulk, 2 flavor	R-30	8	3.250		1,700	135		1,835	2,100
6695	3 flavor	"	8	3.250		2,300	135		2,435	2,750
8800	Serving counter, straight	1 Carp	40	.200	L.F.	840	8.80		848.80	940
8820	Curved section	"	30	.267	"	1,025	11.75		1,036.75	1,150
8825	Solid surface, see Section 12 36 61.16									
8860	Sneeze guard with lights, 60" L	1 Clab	16	.500	Ea.	375	17.55		392.55	440
9100	Soft serve ice cream machine, medium	R-18	11	2.364		11,900	95.50		11,995.50	13,300
9110	Large	"	9	2.889		19,200	117		19,317	21,300

11 47 Ice Machines

11 47 10 – Commercial Ice Machines

11 47 10.10 Commercial Ice Equipment		Crew	Daily Output	Labor-Hours	Unit	Material	Labor	Equipment	Total	Total Incl O&P
0010	**COMMERCIAL ICE EQUIPMENT**									
5800	Ice cube maker, 50 pounds per day	Q-1	6	2.667	Ea.	1,475	131		1,606	1,825
5900	250 pounds per day		1.20	13.333		2,625	655		3,280	3,900
6050	500 pounds per day		4	4		2,750	197		2,947	3,325
6060	With bin		1.20	13.333		3,100	655		3,755	4,425
6090	1000 pounds per day, with bin		1	16		4,700	785		5,485	6,400
6100	Ice flakers, 300 pounds per day		1.60	10		2,575	490		3,065	3,625
6120	600 pounds per day		.95	16.842		3,750	830		4,580	5,425
6130	1000 pounds per day		.75	21.333		4,325	1,050		5,375	6,375
6140	2000 pounds per day		.65	24.615		21,300	1,200		22,500	25,300
6160	Ice storage bin, 500 pound capacity	Q-5	1	16		975	800		1,775	2,325
6180	1000 pound	"	.56	28.571		2,500	1,425		3,925	4,950

11 48 Cleaning and Disposal Equipment

11 48 13 – Commercial Dishwashers

11 48 13.10 Dishwashers		Crew	Daily Output	Labor-Hours	Unit	Material	Labor	Equipment	Total	Total Incl O&P
0010	**DISHWASHERS**									
2700	Dishwasher, commercial, rack type									
2720	10 to 12 racks per hour	Q-1	3.20	5	Ea.	4,300	246		4,546	5,100
2730	Energy star rated, 35 to 40 racks/hour G		1.30	12.308		4,725	605		5,330	6,150
2740	50 to 60 racks/hour G		1.30	12.308		10,200	605		10,805	12,100
2800	Automatic, 190 to 230 racks per hour	L-6	.35	34.286		13,500	1,825		15,325	17,700
2820	235 to 275 racks per hour		.25	48		31,900	2,575		34,475	39,100
2840	8,750 to 12,500 dishes per hour		.10	120		49,500	6,425		55,925	64,500
2950	Dishwasher hood, canopy type	L-3A	10	1.200	L.F.	800	57.50		857.50	975
2960	Pant leg type	"	2.50	4.800	Ea.	8,600	231		8,831	9,850
5200	Garbage disposal 1.5 HP, 100 GPH	L-1	4.80	3.333		2,000	177		2,177	2,475

11 48 Cleaning and Disposal Equipment

11 48 13 – Commercial Dishwashers

11 48 13.10 Dishwashers		Crew	Daily Output	Labor-Hours	Unit	Material	Labor	Equipment	Total	Total Incl O&P
						2012 Bare Costs				
5210	3 HP, 120 GPH	L-1	4.60	3.478	Ea.	2,100	185		2,285	2,575
5220	5 HP, 250 GPH	↓	4.50	3.556	↓	3,375	189		3,564	4,025
6750	Pot sink, 3 compartment	1 Plum	7.25	1.103	L.F.	1,000	60.50		1,060.50	1,200
6760	Pot washer, low temp wash/rinse		1.60	5	Ea.	3,800	273		4,073	4,625
6770	High pressure wash, high temperature rinse	↓	1.20	6.667		36,500	365		36,865	40,700
9170	Trash compactor, small, up to 125 lb. compacted weight	L-4	4	6		23,900	248		24,148	26,700
9175	Large, up to 175 lb. compacted weight	"	3	8	↓	29,200	330		29,530	32,600

11 52 Audio-Visual Equipment

11 52 13 – Projection Screens

11 52 13.10 Projection Screens, Wall or Ceiling Hung

Line	Description	Crew	Daily Output	Labor-Hours	Unit	Material	Labor	Equipment	Total	Total Incl O&P
0010	**PROJECTION SCREENS, WALL OR CEILING HUNG**, matte white									
0100	Manually operated, economy	2 Carp	500	.032	S.F.	5.85	1.41		7.26	8.75
0300	Intermediate		450	.036		6.80	1.57		8.37	10.05
0400	Deluxe		400	.040	↓	9.45	1.76		11.21	13.25
9000	Minimum labor/equipment charge	↓	3	5.333	Job		235		235	385

11 52 16 – Projectors

11 52 16.10 Movie Equipment

Line	Description	Crew	Daily Output	Labor-Hours	Unit	Material	Labor	Equipment	Total	Total Incl O&P
0011	**MOVIE EQUIPMENT**									
0020	Changeover, minimum				Ea.	465			465	510
3000	Projection screens, rigid, in wall, acrylic, 1/4" thick	2 Glaz	195	.082	S.F.	42	3.51		45.51	51.50
3100	1/2" thick	"	130	.123	"	48.50	5.25		53.75	62
3700	Sound systems, incl. amplifier, mono, minimum	1 Elec	.90	8.889	Ea.	3,300	460		3,760	4,325
3800	Dolby/Super Sound, maximum		.40	20		18,100	1,025		19,125	21,500
4100	Dual system, 2 channel, front surround, minimum		.70	11.429		4,625	590		5,215	6,000
4200	Dolby/Super Sound, 4 channel, maximum	↓	.40	20		16,500	1,025		17,525	19,800
5700	Seating, painted steel, upholstered, minimum	2 Carp	35	.457		132	20		152	178
5800	Maximum	"	28	.571	↓	420	25		445	505

11 53 Laboratory Equipment

11 53 33 – Emergency Safety Appliances

11 53 33.13 Emergency Equipment

Line	Description	Crew	Daily Output	Labor-Hours	Unit	Material	Labor	Equipment	Total	Total Incl O&P
0010	**EMERGENCY EQUIPMENT**									
1400	Safety equipment, eye wash, hand held				Ea.	405			405	445
1450	Deluge shower				"	740			740	815

11 53 43 – Service Fittings and Accessories

11 53 43.13 Fittings

Line	Description	Crew	Daily Output	Labor-Hours	Unit	Material	Labor	Equipment	Total	Total Incl O&P
0010	**FITTINGS**									
1600	Sink, one piece plastic, flask wash, hose, free standing	1 Plum	1.60	5	Ea.	1,850	273		2,123	2,450
1610	Epoxy resin sink, 25" x 16" x 10"	"	2	4	"	204	218		422	565
1630	Steel table, open underneath				L.F.	465			465	515
1640	Doors underneath				"	700			700	770
1900	Utility cabinet, stainless steel				Ea.	740			740	815
1950	Utility table, acid resistant top with drawers	2 Carp	30	.533	L.F.	151	23.50		174.50	205
8000	Alternate pricing method: as percent of lab furniture									
8050	Installation, not incl. plumbing & duct work				% Furn.				20%	22%
8100	Plumbing, final connections, simple system								9.09%	10%
8110	Moderately complex system				↓				13.64%	15%

11 53 Laboratory Equipment

11 53 43 – Service Fittings and Accessories

11 53 43.13 Fittings		Crew	Daily Output	Labor-Hours	Unit	Material	Labor	Equipment	Total	Total Incl O&P
						2012 Bare Costs				
8120	Complex system				% Furn.				18.18%	20%
8150	Electrical, simple system								9.09%	10%
8160	Moderately complex system								18.18%	20%
8170	Complex system								31.90%	35%

11 57 Vocational Shop Equipment

11 57 10 – Shop Equipment

11 57 10.10 Vocational School Shop Equipment

		Crew	Daily Output	Labor-Hours	Unit	Material	Labor	Equipment	Total	Total Incl O&P
0010	**VOCATIONAL SCHOOL SHOP EQUIPMENT**									
0020	Benches, work, wood, average	2 Carp	5	3.200	Ea.	530	141		671	810
0100	Metal, average		5	3.200		490	141		631	770
0400	Combination belt & disc sander, 6"		4	4		1,425	176		1,601	1,850
0700	Drill press, floor mounted, 12", 1/2 H.P.		4	4		385	176		561	715
0800	Dust collector, not incl. ductwork, 6" diameter	1 Shee	1.10	7.273		4,175	380		4,555	5,175
0810	Dust collector bag, 20" diameter	"	5	1.600		405	83.50		488.50	575
1000	Grinders, double wheel, 1/2 H.P.	2 Carp	5	3.200		203	141		344	455
1300	Jointer, 4", 3/4 H.P.		4	4		1,275	176		1,451	1,700
1600	Kilns, 16 C.F., to 2000°		4	4		1,500	176		1,676	1,925
1900	Lathe, woodworking, 10", 1/2 H.P.		4	4		830	176		1,006	1,200
2200	Planer, 13" x 6"		4	4		950	176		1,126	1,350
2500	Potter's wheel, motorized		4	4		1,050	176		1,226	1,450
2800	Saws, band, 14", 3/4 H.P.		4	4		860	176		1,036	1,225
3100	Metal cutting band saw, 14"		4	4		2,275	176		2,451	2,800
3400	Radial arm saw, 10", 2 H.P.		4	4		1,300	176		1,476	1,725
3700	Scroll saw, 24"		4	4		590	176		766	940
4000	Table saw, 10", 3 H.P.		4	4		2,275	176		2,451	2,800
4300	Welder AC arc, 30 amp capacity		4	4		3,100	176		3,276	3,725

11 61 Theater and Stage Equipment

11 61 23 – Folding and Portable Stages

11 61 23.10 Portable Stages

		Crew	Daily Output	Labor-Hours	Unit	Material	Labor	Equipment	Total	Total Incl O&P
0010	**PORTABLE STAGES**									
5000	Stages, portable with steps, folding legs, stock, 8" high				SF Stg.	31			31	34.50
5100	16" high					24.50			24.50	27
5200	32" high					43			43	47
5300	40" high					59.50			59.50	65

11 61 33 – Rigging Systems and Controls

11 61 33.10 Controls

		Crew	Daily Output	Labor-Hours	Unit	Material	Labor	Equipment	Total	Total Incl O&P
0010	**CONTROLS**									
0050	Control boards with dimmers and breakers, minimum	1 Elec	1	8	Ea.	12,300	410		12,710	14,100
0150	Maximum	"	.20	40	"	126,000	2,050		128,050	141,500
0160										

11 61 43 – Stage Curtains

11 61 43.10 Curtains

		Crew	Daily Output	Labor-Hours	Unit	Material	Labor	Equipment	Total	Total Incl O&P
0010	**CURTAINS**									
0500	Curtain track, straight, light duty	2 Carp	20	.800	L.F.	27	35.50		62.50	87.50
0700	Curved sections		12	1.333	"	173	59		232	287
1000	Curtains, velour, medium weight		600	.027	S.F.	7.90	1.18		9.08	10.65

11 66 Athletic Equipment

11 66 13 – Exercise Equipment

11 66 13.10 Physical Training Equipment

		Crew	Daily Output	Labor-Hours	Unit	2012 Bare Costs Material	Labor	Equipment	Total	Total Incl O&P
0010	**PHYSICAL TRAINING EQUIPMENT**									
0020	Abdominal rack, 2 board capacity				Ea.	510			510	565
0050	Abdominal board, upholstered					685			685	750
0200	Bicycle trainer, minimum					430			430	470
0300	Deluxe, electric					4,300			4,300	4,725
0400	Barbell set, chrome plated steel, 25 lb.					249			249	274
0420	100 lb.					400			400	440
0450	200 lb.					665			665	730
0500	Weight plates, cast iron, per lb.				Lb.	5.20			5.20	5.70
0520	Storage rack, 10 station				Ea.	900			900	990
0600	Circuit training apparatus, 12 machines minimum	2 Clab	1.25	12.800	Set	28,700	450		29,150	32,200
0700	Average		1	16		35,200	560		35,760	39,600
0800	Maximum		.75	21.333		41,700	750		42,450	47,100
0820	Dumbbell set, cast iron, with rack and 5 pair					615			615	675
0900	Squat racks	2 Clab	5	3.200	Ea.	890	112		1,002	1,175
4150	Exercise equipment, bicycle trainer					750			750	825
4180	Chinning bar, adjustable, wall mounted	1 Carp	5	1.600		300	70.50		370.50	445
4200	Exercise ladder, 16′ x 1′-7″, suspended	L-2	3	5.333		1,700	206		1,906	2,200
4210	High bar, floor plate attached	1 Carp	4	2		1,250	88		1,338	1,525
4240	Parallel bars, adjustable		4	2		2,250	88		2,338	2,650
4270	Uneven parallel bars, adjustable		4	2		2,950	88		3,038	3,400
4280	Wall mounted, adjustable	L-2	1.50	10.667	Set	820	415		1,235	1,575
4300	Rope, ceiling mounted, 18′ long	1 Carp	3.66	2.186	Ea.	286	96.50		382.50	475
4330	Side horse, vaulting		5	1.600		1,775	70.50		1,845.50	2,075
4360	Treadmill, motorized, deluxe, training type		5	1.600		3,000	70.50		3,070.50	3,425
4390	Weight lifting multi-station, minimum	2 Clab	1	16		450	560		1,010	1,425

11 66 23 – Gymnasium Equipment

11 66 23.13 Basketball Equipment

		Crew	Daily Output	Labor-Hours	Unit	Material	Labor	Equipment	Total	Total Incl O&P
0010	**BASKETBALL EQUIPMENT**									
1000	Backstops, wall mtd., 6′ extended, fixed, minimum	L-2	1	16	Ea.	1,300	620		1,920	2,450
1100	Maximum		1	16		1,925	620		2,545	3,150
1200	Swing up, minimum		1	16		1,500	620		2,120	2,675
1250	Maximum		1	16		6,125	620		6,745	7,775
1300	Portable, manual, heavy duty, spring operated		1.90	8.421		15,200	325		15,525	17,200
1400	Ceiling suspended, stationary, minimum		.78	20.513		3,650	795		4,445	5,325
1450	Fold up, with accessories, maximum		.40	40		6,975	1,550		8,525	10,200
1600	For electrically operated, add	1 Elec	1	8		2,300	410		2,710	3,150
5800	Wall pads, 1-1/2″ thick, standard (not fire rated)	2 Carp	640	.025	S.F.	7.30	1.10		8.40	9.85

11 66 23.19 Boxing Ring

		Crew	Daily Output	Labor-Hours	Unit	Material	Labor	Equipment	Total	Total Incl O&P
0010	**BOXING RING**									
4100	Elevated, 22′ x 22′	L-4	.10	240	Ea.	7,725	9,925		17,650	24,700
4110	For cellular plastic foam padding, add		.10	240		975	9,925		10,900	17,300
4120	Floor level, including posts and ropes only, 20′ x 20′		.80	30		3,750	1,250		5,000	6,150
4130	Canvas, 30′ x 30′		5	4.800		1,250	199		1,449	1,700

11 66 23.47 Gym Mats

		Crew	Daily Output	Labor-Hours	Unit	Material	Labor	Equipment	Total	Total Incl O&P
0010	**GYM MATS**									
5500	2″ thick, naugahyde covered				S.F.	3.81			3.81	4.19
5600	Vinyl/nylon covered					6.60			6.60	7.30
6000	Wrestling mats, 1″ thick, heavy duty					5.60			5.60	6.15

11 66 Athletic Equipment

11 66 43 – Interior Scoreboards

11 66 43.10 Scoreboards		Crew	Daily Output	Labor-Hours	Unit	Material	2012 Bare Costs Labor	Equipment	Total	Total Incl O&P
0010	**SCOREBOARDS**									
7000	Baseball, minimum	R-3	1.30	15.385	Ea.	4,025	785	102	4,912	5,775
7200	Maximum		.05	400		17,700	20,400	2,650	40,750	54,000
7300	Football, minimum		.86	23.256		5,225	1,175	154	6,554	7,775
7400	Maximum		.20	100		15,600	5,100	660	21,360	25,700
7500	Basketball (one side), minimum		2.07	9.662		2,350	495	64	2,909	3,425
7600	Maximum		.30	66.667		3,450	3,400	440	7,290	9,525
7700	Hockey-basketball (four sides), minimum		.25	80		5,550	4,075	530	10,155	13,000
7800	Maximum		.15	133		5,625	6,800	885	13,310	17,600

11 66 53 – Gymnasium Dividers

11 66 53.10 Divider Curtains		Crew	Daily Output	Labor-Hours	Unit	Material	Labor	Equipment	Total	Total Incl O&P
0010	**DIVIDER CURTAINS**									
4500	Gym divider curtain, mesh top, vinyl bottom, manual	L-4	500	.048	S.F.	9.15	1.99		11.14	13.35
4700	Electric roll up	L-7	400	.070	"	11.15	2.98		14.13	17.10

11 82 Solid Waste Handling Equipment

11 82 26 – Facility Waste Compactors

11 82 26.10 Compactors		Crew	Daily Output	Labor-Hours	Unit	Material	Labor	Equipment	Total	Total Incl O&P
0010	**COMPACTORS**									
0020	Compactors, 115 volt, 250#/hr., chute fed	L-4	1	24	Ea.	11,900	995		12,895	14,700
0100	Hand fed		2.40	10		8,725	415		9,140	10,300
1000	Heavy duty industrial compactor, 0.5 C.Y. capacity		1	24		7,750	995		8,745	10,200
1050	1.0 C.Y. capacity		1	24		11,800	995		12,795	14,600
1400	For handling hazardous waste materials, 55 gallon drum packer, std.					18,300			18,300	20,200
1410	55 gallon drum packer w/HEPA filter					22,900			22,900	25,200
1420	55 gallon drum packer w/charcoal & HEPA filter					30,600			30,600	33,600
1430	All of the above made explosion proof, add					13,900			13,900	15,300

11 91 Religious Equipment

11 91 13 – Baptisteries

11 91 13.10 Baptistry		Crew	Daily Output	Labor-Hours	Unit	Material	Labor	Equipment	Total	Total Incl O&P
0010	**BAPTISTRY**									
0150	Fiberglass, 3'-6" deep, x 13'-7" long,									
0160	steps at both ends, incl. plumbing, minimum	L-8	1	20	Ea.	4,600	925		5,525	6,550
0200	Maximum	"	.70	28.571		7,500	1,325		8,825	10,400
0250	Add for filter, heater and lights					1,650			1,650	1,800

11 91 23 – Sanctuary Equipment

11 91 23.10 Sanctuary Furnishings		Crew	Daily Output	Labor-Hours	Unit	Material	Labor	Equipment	Total	Total Incl O&P
0010	**SANCTUARY FURNISHINGS**									
0020	Altar, wood, custom design, plain	1 Carp	1.40	5.714	Ea.	2,150	252		2,402	2,800
0050	Deluxe		.20	40	"	10,400	1,775		12,175	14,300
5000	Wall cross, aluminum, extruded, 2" x 2" section		34	.235	L.F.	160	10.40		170.40	193
5150	4" x 4" section		29	.276		230	12.15		242.15	273
5300	Bronze, extruded, 1" x 2" section		31	.258		315	11.40		326.40	365
5350	2-1/2" x 2-1/2" section		34	.235		480	10.40		490.40	540

Estimating Tips

General

- The items in this division are usually priced per square foot or each. Most of these items are purchased by the owner and installed by the contractor. Do not assume the items in Division 12 will be purchased and installed by the contractor. Check the specifications for responsibilities and include receiving, storage, installation, and mechanical and electrical hookups in the appropriate divisions.
- Some items in this division require some type of support system that is not usually furnished with the item. Examples of these systems include blocking for the attachment of casework and heavy drapery rods. The required blocking must be added to the estimate in the appropriate division.

Reference Numbers

Reference numbers are shown in shaded boxes at the beginning of some major classifications. These numbers refer to related items in the Reference Section. The reference information may be an estimating procedure, an alternate pricing method, or technical information.

Note: Not all subdivisions listed here necessarily appear in this publication.

12 21 Window Blinds

12 21 13 – Horizontal Louver Blinds

12 21 13.13 Metal Horizontal Louver Blinds		Crew	Daily Output	Labor-Hours	Unit	Material	2012 Bare Costs Labor	Equipment	Total	Total Incl O&P
0010	**METAL HORIZONTAL LOUVER BLINDS**									
0020	Horizontal, 1" aluminum slats, solid color, stock	1 Carp	590	.014	S.F.	4.99	.60		5.59	6.50

12 21 13.33 Vinyl Horizontal Louver Blinds		Crew	Daily Output	Labor-Hours	Unit	Material	Labor	Equipment	Total	Total Incl O&P
0010	**VINYL HORIZONTAL LOUVER BLINDS**									
0100	2" custom composite, 48" wide, 48" high	1 Carp	30	.267	Ea.	89	11.75		100.75	117
0120	72" high		29	.276		133	12.15		145.15	166
0140	96" high		28	.286		167	12.60		179.60	205
0200	60" wide, 60" high		27	.296		109	13.05		122.05	142
0220	72" high		25	.320		127	14.10		141.10	162
0240	96" high		24	.333		176	14.70		190.70	218
0300	72" wide, 72" high		25	.320		183	14.10		197.10	224
0320	96" high		23	.348		255	15.35		270.35	305
0400	96" wide, 96" high		20	.400		310	17.65		327.65	375
1000	2" custom faux wood, 48" wide, 48" high		30	.267		57.50	11.75		69.25	82.50
1020	72" high		29	.276		75.50	12.15		87.65	103
1040	96" high		28	.286		94	12.60		106.60	124
1300	72" wide, 72" high		25	.320		117	14.10		131.10	152
1320	96" high		23	.348		186	15.35		201.35	229
1400	96" wide, 96" high		20	.400		207	17.65		224.65	256

12 23 Interior Shutters

12 23 10 – Wood Interior Shutters

12 23 10.10 Wood Interior Shutters		Crew	Daily Output	Labor-Hours	Unit	Material	Labor	Equipment	Total	Total Incl O&P
0010	**WOOD INTERIOR SHUTTERS**, louvered									
0200	Two panel, 27" wide, 36" high	1 Carp	5	1.600	Set	146	70.50		216.50	277
0300	33" wide, 36" high		5	1.600		192	70.50		262.50	325
0500	47" wide, 36" high		5	1.600		253	70.50		323.50	395
1000	Four panel, 27" wide, 36" high		5	1.600		179	70.50		249.50	310
1100	33" wide, 36" high		5	1.600		230	70.50		300.50	370
1300	47" wide, 36" high		5	1.600		305	70.50		375.50	450

12 23 10.13 Wood Panels		Crew	Daily Output	Labor-Hours	Unit	Material	Labor	Equipment	Total	Total Incl O&P
0010	**WOOD PANELS**									
3000	Wood folding panels with movable louvers, 7" x 20" each	1 Carp	17	.471	Pr.	78	21		99	120
4000	Fixed louver type, stock units, 8" x 20" each		17	.471		92	21		113	135
4450	18" x 40" each		17	.471		132	21		153	179

12 24 Window Shades

12 24 13 – Roller Window Shades

12 24 13.10 Shades		Crew	Daily Output	Labor-Hours	Unit	Material	Labor	Equipment	Total	Total Incl O&P
0010	**SHADES**									
0020	Basswood, roll-up, stain finish, 3/8" slats	1 Carp	300	.027	S.F.	13.95	1.18		15.13	17.30
0030	Double layered, heat reflective		685	.012		9	.52		9.52	10.75
0200	7/8" slats		300	.027		13.15	1.18		14.33	16.45
0900	Mylar, single layer, non-heat reflective		685	.012		6.90	.52		7.42	8.40
1000	Double layered, heat reflective		685	.012		9.55	.52		10.07	11.35
1100	Triple layered, heat reflective		685	.012		11.10	.52		11.62	13.05
5000	Thermal, roll up, R4		44	.182		11.60	8		19.60	26
5030	R10.7		44	.182		13.30	8		21.30	28
5050	Magnetic clips, set of 20				Set	25			25	27.50

12 32 Manufactured Wood Casework

12 32 16 – Manufactured Plastic-Laminate-Clad Casework

12 32 16.20 Plastic Laminate Casework Doors		Crew	Daily Output	Labor-Hours	Unit	Material	2012 Bare Costs Labor	Equipment	Total	Total Incl O&P
0010	**PLASTIC LAMINATE CASEWORK DOORS**									
1000	For casework frames, see Section 12 32 23.15									
1100	For casework hardware, see Section 12 32 23.35									
6000	Plastic laminate on particle board									
6100	12" wide, 18" high	1 Carp	25	.320	Ea.	12.20	14.10		26.30	36.50
6140	30" high		23	.348		20.50	15.35		35.85	47.50
6500	18" wide, 18" high		24	.333		18.35	14.70		33.05	44
6600	30" high		22	.364		30.50	16.05		46.55	60

12 32 16.25 Plastic Laminate Drawer Fronts		Crew	Daily Output	Labor-Hours	Unit	Material	Labor	Equipment	Total	Total Incl O&P
0010	**PLASTIC LAMINATE DRAWER FRONTS**									
2800	Plastic laminate on particle board front									
3000	4" high, 12" wide	1 Carp	17	.471	Ea.	5.20	21		26.20	39.50
3200	18" wide	"	16	.500	"	7.75	22		29.75	44.50

12 32 23 – Hardwood Casework

12 32 23.10 Manufactured Wood Casework, Stock Units		Crew	Daily Output	Labor-Hours	Unit	Material	Labor	Equipment	Total	Total Incl O&P
0010	**MANUFACTURED WOOD CASEWORK, STOCK UNITS**									
0300	Built-in drawer units, pine, 18" deep, 32" high, unfinished									
0400	Minimum	2 Carp	53	.302	L.F.	149	13.30		162.30	185
0500	Maximum	"	40	.400	"	185	17.65		202.65	232
0700	Kitchen base cabinets, hardwood, not incl. counter tops,									
0710	24" deep, 35" high, prefinished									
0800	One top drawer, one door below, 12" wide	2 Carp	24.80	.645	Ea.	234	28.50		262.50	305
0820	15" wide		24	.667		240	29.50		269.50	310
0840	18" wide		23.30	.687		263	30.50		293.50	340
0860	21" wide		22.70	.705		279	31		310	355
0880	24" wide		22.30	.717		310	31.50		341.50	390
1000	Four drawers, 12" wide		24.80	.645		246	28.50		274.50	320
1020	15" wide		24	.667		249	29.50		278.50	320
1040	18" wide		23.30	.687		275	30.50		305.50	355
1060	24" wide		22.30	.717		310	31.50		341.50	390
1200	Two top drawers, two doors below, 27" wide		22	.727		340	32		372	430
1220	30" wide		21.40	.748		375	33		408	465
1240	33" wide		20.90	.766		395	34		429	485
1260	36" wide		20.30	.788		410	35		445	505
1280	42" wide		19.80	.808		435	35.50		470.50	540
1300	48" wide		18.90	.847		470	37.50		507.50	575
1500	Range or sink base, two doors below, 30" wide		21.40	.748		315	33		348	400
1520	33" wide		20.90	.766		335	34		369	425
1540	36" wide		20.30	.788		355	35		390	445
1560	42" wide		19.80	.808		375	35.50		410.50	475
1580	48" wide		18.90	.847		395	37.50		432.50	495
1800	For sink front units, deduct					140			140	154
2000	Corner base cabinets, 36" wide, standard	2 Carp	18	.889		550	39		589	675
2100	Lazy Susan with revolving door	"	16.50	.970		740	43		783	885
4000	Kitchen wall cabinets, hardwood, 12" deep with two doors									
4050	12" high, 30" wide	2 Carp	24.80	.645	Ea.	210	28.50		238.50	278
4100	36" wide		24	.667		246	29.50		275.50	320
4400	15" high, 30" wide		24	.667		210	29.50		239.50	279
4420	33" wide		23.30	.687		232	30.50		262.50	305
4440	36" wide		22.70	.705		246	31		277	320
4450	42" wide		22.70	.705		296	31		327	375
4700	24" high, 30" wide		23.30	.687		282	30.50		312.50	360

12 32 Manufactured Wood Casework

12 32 23 – Hardwood Casework

12 32 23.10 Manufactured Wood Casework, Stock Units

	12 32 23.10 Manufactured Wood Casework, Stock Units	Crew	Daily Output	Labor-Hours	Unit	2012 Bare Costs Material	Labor	Equipment	Total	Total Incl O&P
4720	36″ wide	2 Carp	22.70	.705	Ea.	310	31		341	390
4740	42″ wide		22.30	.717		360	31.50		391.50	445
5000	30″ high, one door, 12″ wide		22	.727		200	32		232	273
5020	15″ wide		21.40	.748		213	33		246	288
5040	18″ wide		20.90	.766		234	34		268	315
5060	24″ wide		20.30	.788		273	35		308	355
5300	Two doors, 27″ wide		19.80	.808		300	35.50		335.50	390
5320	30″ wide		19.30	.829		310	36.50		346.50	405
5340	36″ wide		18.80	.851		350	37.50		387.50	445
5360	42″ wide		18.50	.865		390	38		428	495
5380	48″ wide		18.40	.870		440	38.50		478.50	550
6000	Corner wall, 30″ high, 24″ wide		18	.889		271	39		310	365
6050	30″ wide		17.20	.930		280	41		321	380
6100	36″ wide		16.50	.970		296	43		339	395
6500	Revolving Lazy Susan		15.20	1.053		425	46.50		471.50	540
7000	Broom cabinet, 84″ high, 24″ deep, 18″ wide		10	1.600		590	70.50		660.50	765
7500	Oven cabinets, 84″ high, 24″ deep, 27″ wide		8	2		860	88		948	1,100
7750	Valance board trim		396	.040	L.F.	12.40	1.78		14.18	16.55
9000	For deluxe models of all cabinets, add					40%				
9500	For custom built in place, add					25%	10%			
9558	Rule of thumb, kitchen cabinets not including									
9560	appliances & counter top, minimum	2 Carp	30	.533	L.F.	150	23.50		173.50	204
9600	Maximum	″	25	.640	″	320	28		348	395
9700	Minimum labor/equipment charge	1 Carp	3	2.667	Job		118		118	193

12 32 23.15 Manufactured Wood Casework Frames

	12 32 23.15 Manufactured Wood Casework Frames	Crew	Daily Output	Labor-Hours	Unit	Material	Labor	Equipment	Total	Total Incl O&P
0010	**MANUFACTURED WOOD CASEWORK FRAMES**									
0050	Base cabinets, counter storage, 36″ high									
0100	One bay, 18″ wide	1 Carp	2.70	2.963	Ea.	137	131		268	365
0400	Two bay, 36″ wide		2.20	3.636		209	160		369	495
1100	Three bay, 54″ wide		1.50	5.333		249	235		484	660
2800	Bookcases, one bay, 7′ high, 18″ wide		2.40	3.333		161	147		308	420
3500	Two bay, 36″ wide		1.60	5		234	221		455	615
4100	Three bay, 54″ wide		1.20	6.667		385	294		679	905
5100	Coat racks, one bay, 7′ high, 24″ wide		4.50	1.778		161	78.50		239.50	305
5300	Two bay, 48″ wide		2.75	2.909		224	128		352	455
5800	Three bay, 72″ wide		2.10	3.810		330	168		498	635
6100	Wall mounted cabinet, one bay, 24″ high, 18″ wide		3.60	2.222		88.50	98		186.50	258
6800	Two bay, 36″ wide		2.20	3.636		129	160		289	405
7400	Three bay, 54″ wide		1.70	4.706		161	208		369	515
8400	30″ high, one bay, 18″ wide		3.60	2.222		96	98		194	267
9000	Two bay, 36″ wide		2.15	3.721		128	164		292	410
9400	Three bay, 54″ wide		1.60	5		160	221		381	535
9800	Wardrobe, 7′ high, single, 24″ wide		2.70	2.963		177	131		308	410
9880	Partition & adjustable shelves, 48″ wide		1.70	4.706		226	208		434	590
9950	Partition, adjustable shelves & drawers, 48″ wide		1.40	5.714		340	252		592	785
9970	Minimum labor/equipment charge		4	2	Job		88		88	145

12 32 23.20 Manufactured Hardwood Casework Doors

	12 32 23.20 Manufactured Hardwood Casework Doors	Crew	Daily Output	Labor-Hours	Unit	Material	Labor	Equipment	Total	Total Incl O&P
0010	**MANUFACTURED HARDWOOD CASEWORK DOORS**									
2000	Glass panel, hardwood frame									
2200	12″ wide, 18″ high	1 Carp	34	.235	Ea.	25	10.40		35.40	44.50
2600	30″ high		32	.250		41.50	11.05		52.55	64
4450	18″ wide, 18″ high		32	.250		37.50	11.05		48.55	59.50

12 32 Manufactured Wood Casework

12 32 23 – Hardwood Casework

12 32 23.20 Manufactured Hardwood Casework Doors		Crew	Daily Output	Labor-Hours	Unit	2012 Bare Costs Material	Labor	Equipment	Total	Total Incl O&P
4550	30″ high	1 Carp	29	.276	Ea.	62.50	12.15		74.65	89
5000	Hardwood, raised panel									
5100	12″ wide, 18″ high	1 Carp	16	.500	Ea.	28	22		50	67
5200	30″ high		15	.533		47	23.50		70.50	90
5500	18″ wide, 18″ high		15	.533		42	23.50		65.50	85
5600	30″ high		14	.571		70.50	25		95.50	119
9000	Minimum labor/equipment charge		4	2	Job		88		88	145

12 32 23.25 Manufactured Wood Casework Drawer Fronts

		Crew	Daily Output	Labor-Hours	Unit	Material	Labor	Equipment	Total	Total Incl O&P
0010	**MANUFACTURED WOOD CASEWORK DRAWER FRONTS**									
0100	Solid hardwood front									
1000	4″ high, 12″ wide	1 Carp	17	.471	Ea.	3.41	21		24.41	38
1200	18″ wide		16	.500	″	5.05	22		27.05	41.50
9000	Minimum labor/equipment charge		4	2	Job		88		88	145

12 32 23.30 Manufactured Wood Casework Vanities

		Crew	Daily Output	Labor-Hours	Unit	Material	Labor	Equipment	Total	Total Incl O&P
0010	**MANUFACTURED WOOD CASEWORK VANITIES**									
8000	Vanity bases, 2 doors, 30″ high, 21″ deep, 24″ wide	2 Carp	20	.800	Ea.	233	35.50		268.50	315
8050	30″ wide		16	1		268	44		312	370
8100	36″ wide		13.33	1.200		355	53		408	475
8150	48″ wide		11.43	1.400		425	61.50		486.50	570
9000	For deluxe models of all vanities, add to above					40%				
9500	For custom built in place, add to above					25%	10%			

12 32 23.35 Manufactured Wood Casework Hardware

		Crew	Daily Output	Labor-Hours	Unit	Material	Labor	Equipment	Total	Total Incl O&P
0010	**MANUFACTURED WOOD CASEWORK HARDWARE**									
1000	Catches, minimum	1 Carp	235	.034	Ea.	1.15	1.50		2.65	3.73
1040	Maximum	″	80	.100	″	7.20	4.41		11.61	15.15
2000	Door/drawer pulls, handles									
2200	Handles and pulls, projecting, metal, minimum	1 Carp	48	.167	Ea.	4.66	7.35		12.01	17.20
2240	Maximum		24	.333		9.85	14.70		24.55	35
2300	Wood, minimum		48	.167		4.90	7.35		12.25	17.45
2340	Maximum		24	.333		9	14.70		23.70	34
2600	Flush, metal, minimum		48	.167		4.90	7.35		12.25	17.45
2640	Maximum		24	.333		9	14.70		23.70	34
3000	Drawer tracks/glides, minimum		48	.167	Pr.	8.30	7.35		15.65	21
3040	Maximum		24	.333		24	14.70		38.70	50.50
4000	Cabinet hinges, minimum		160	.050		2.82	2.21		5.03	6.70
4040	Maximum		68	.118		10.70	5.20		15.90	20.50

12 34 Manufactured Plastic Casework

12 34 16 – Manufactured Solid-Plastic Casework

12 34 16.10 Outdoor Casework

		Crew	Daily Output	Labor-Hours	Unit	Material	Labor	Equipment	Total	Total Incl O&P
0010	**OUTDOOR CASEWORK**									
0020	Cabinet, base, sink/range, 36″	2 Carp	20.30	.788	Ea.	1,775	35		1,810	2,025
0100	Base, 36″		20.30	.788		2,000	35		2,035	2,250
0200	Filler strip, 1″ x 30″		158	.101		31	4.47		35.47	41.50
0210	Filler strip, 2″ x 30″		158	.101		41	4.47		45.47	52.50

12 35 Specialty Casework

12 35 53 – Laboratory Casework

	12 35 53.13 Metal Laboratory Casework	Crew	Daily Output	Labor-Hours	Unit	2012 Bare Costs Material	Labor	Equipment	Total	Total Incl O&P
0010	**METAL LABORATORY CASEWORK**									
0020	Cabinets, base, door units, metal	2 Carp	18	.889	L.F.	212	39		251	298
0300	Drawer units		18	.889		475	39		514	585
0700	Tall storage cabinets, open, 7′ high		20	.800		455	35.50		490.50	560
0900	With glazed doors		20	.800		695	35.50		730.50	825
1300	Wall cabinets, metal, 12-1/2″ deep, open		20	.800		149	35.50		184.50	222
1500	With doors		20	.800		310	35.50		345.50	400

12 36 Countertops

12 36 16 – Metal Countertops

12 36 16.10 Stainless Steel Countertops

		Crew	Daily Output	Labor-Hours	Unit	Material	Labor	Equipment	Total	Total Incl O&P
0010	**STAINLESS STEEL COUNTERTOPS**									
3200	Stainless steel, custom	1 Carp	24	.333	S.F.	145	14.70		159.70	184

12 36 19 – Wood Countertops

12 36 19.10 Maple Countertops

		Crew	Daily Output	Labor-Hours	Unit	Material	Labor	Equipment	Total	Total Incl O&P
0010	**MAPLE COUNTERTOPS**									
2900	Solid, laminated, 1-1/2″ thick, no splash	1 Carp	28	.286	L.F.	74	12.60		86.60	102
3000	With square splash		28	.286	″	88	12.60		100.60	118
3400	Recessed cutting block with trim, 16″ x 20″ x 1″		8	1	Ea.	74	44		118	154

12 36 23 – Plastic Countertops

12 36 23.13 Plastic-Laminate-Clad Countertops

		Crew	Daily Output	Labor-Hours	Unit	Material	Labor	Equipment	Total	Total Incl O&P
0010	**PLASTIC-LAMINATE-CLAD COUNTERTOPS**									
0020	Stock, 24″ wide w/backsplash, minimum	1 Carp	30	.267	L.F.	9.85	11.75		21.60	30
0100	Maximum		25	.320		21	14.10		35.10	46
0300	Custom plastic, 7/8″ thick, aluminum molding, no splash		30	.267		22.50	11.75		34.25	44.50
0400	Cove splash		30	.267		29	11.75		40.75	51.50
0600	1-1/4″ thick, no splash		28	.286		31.50	12.60		44.10	55
0700	Square splash		28	.286		32	12.60		44.60	55.50
0900	Square edge, plastic face, 7/8″ thick, no splash		30	.267		28	11.75		39.75	50.50
1000	With splash		30	.267		36	11.75		47.75	59
1200	For stainless channel edge, 7/8″ thick, add					3			3	3.30
1300	1-1/4″ thick, add					3.50			3.50	3.85
1500	For solid color suede finish, add					3			3	3.30
1700	For end splash, add				Ea.	18			18	19.80
1900	For cut outs, standard, add, minimum	1 Carp	32	.250		4	11.05		15.05	22.50
2000	Maximum		8	1		6	44		50	79
2100	Postformed, including backsplash and front edge		30	.267	L.F.	11	11.75		22.75	31.50
2110	Mitred, add		12	.667	Ea.		29.50		29.50	48
2200	Built-in place, 25″ wide, plastic laminate		25	.320	L.F.	30	14.10		44.10	56
9000	Minimum labor/equipment charge		3.75	2.133	Job		94		94	154

12 36 33 – Tile Countertops

12 36 33.10 Ceramic Tile Countertops

		Crew	Daily Output	Labor-Hours	Unit	Material	Labor	Equipment	Total	Total Incl O&P
0010	**CERAMIC TILE COUNTERTOPS**									
2300	Ceramic tile mosaic	1 Carp	25	.320	L.F.	33	14.10		47.10	59.50

12 36 40 – Stone Countertops

12 36 40.10 Natural Stone Countertops

		Crew	Daily Output	Labor-Hours	Unit	Material	Labor	Equipment	Total	Total Incl O&P
0010	**NATURAL STONE COUNTERTOPS**									
2500	Marble, stock, with splash, 1/2″ thick, minimum	1 Bric	17	.471	L.F.	41	21		62	78.50
2700	3/4″ thick, maximum	″	13	.615	″	103	27		130	157

12 36 53 – Laboratory Countertops

12 36 53.10 Laboratory Countertops and Sinks		Crew	Daily Output	Labor-Hours	Unit	Material	Labor (2012 Bare Costs)	Equipment (2012 Bare Costs)	Total (2012 Bare Costs)	Total Incl O&P
0010	**LABORATORY COUNTERTOPS AND SINKS**									
0020	Countertops, not incl. base cabinets, acid-proof, minimum	2 Carp	82	.195	S.F.	36.50	8.60		45.10	54
0030	Maximum		70	.229		46.50	10.10		56.60	68
0040	Stainless steel		82	.195		135	8.60		143.60	162

12 36 61 – Simulated Stone Countertops

12 36 61.16 Solid Surface Countertops		Crew	Daily Output	Labor-Hours	Unit	Material	Labor	Equipment	Total	Total Incl O&P
0010	**SOLID SURFACE COUNTERTOPS**, Acrylic polymer									
0020	Pricing for orders of 100 L.F. or greater									
0100	25″ wide, solid colors	2 Carp	28	.571	L.F.	44.50	25		69.50	90.50
0200	Patterned colors		28	.571		56.50	25		81.50	104
0300	Premium patterned colors		28	.571		70.50	25		95.50	120
0400	With silicone attached 4″ backsplash, solid colors		27	.593		49	26		75	97
0500	Patterned colors		27	.593		62	26		88	111
0600	Premium patterned colors		27	.593		77	26		103	128
0700	With hard seam attached 4″ backsplash, solid colors		23	.696		49	30.50		79.50	105
0800	Patterned colors		23	.696		62	30.50		92.50	119
0900	Premium patterned colors		23	.696		77	30.50		107.50	136
1000	Pricing for order of 51 – 99 L.F.									
1100	25″ wide, solid colors	2 Carp	24	.667	L.F.	51.50	29.50		81	105
1200	Patterned colors		24	.667		65	29.50		94.50	120
1300	Premium patterned colors		24	.667		81.50	29.50		111	138
1400	With silicone attached 4″ backsplash, solid colors		23	.696		56.50	30.50		87	113
1500	Patterned colors		23	.696		71.50	30.50		102	129
1600	Premium patterned colors		23	.696		89	30.50		119.50	148
1700	With hard seam attached 4″ backsplash, solid colors		20	.800		56.50	35.50		92	120
1800	Patterned colors		20	.800		71.50	35.50		107	137
1900	Premium patterned colors		20	.800		89	35.50		124.50	156
2000	Pricing for order of 1 – 50 L.F.									
2100	25″ wide, solid colors	2 Carp	20	.800	L.F.	60	35.50		95.50	124
2200	Patterned colors		20	.800		76.50	35.50		112	142
2300	Premium patterned colors		20	.800		95.50	35.50		131	163
2400	With silicone attached 4″ backsplash, solid colors		19	.842		66	37		103	134
2500	Patterned colors		19	.842		83.50	37		120.50	153
2600	Premium patterned colors		19	.842		104	37		141	176
2700	With hard seam attached 4″ backsplash, solid colors		15	1.067		66	47		113	150
2800	Patterned colors		15	1.067		83.50	47		130.50	169
2900	Premium patterned colors		15	1.067		104	47		151	192
3000	Sinks, pricing for order of 100 or greater units									
3100	Single bowl, hard seamed, solid colors, 13″ x 17″	1 Carp	3	2.667	Ea.	300	118		418	525
3200	10″ x 15″		7	1.143		139	50.50		189.50	236
3300	Cutouts for sinks		8	1			44		44	72.50
3400	Sinks, pricing for order of 51 – 99 units									
3500	Single bowl, hard seamed, solid colors, 13″ x 17″	1 Carp	2.55	3.137	Ea.	345	138		483	605
3600	10″ x 15″		6	1.333		160	59		219	273
3700	Cutouts for sinks		7	1.143			50.50		50.50	82.50
3800	Sinks, pricing for order of 1 – 50 units									
3900	Single bowl, hard seamed, solid colors, 13″ x 17″	1 Carp	2	4	Ea.	405	176		581	735
4000	10″ x 15″		4.55	1.758		188	77.50		265.50	335
4100	Cutouts for sinks		5.25	1.524			67		67	110
4200	Cooktop cutouts, pricing for 100 or greater units		4	2		22	88		110	170
4300	51 – 99 units		3.40	2.353		25.50	104		129.50	198
4400	1 – 50 units		3	2.667		30	118		148	226

12 36 Countertops

12 36 61 – Simulated Stone Countertops

12 36 61.17 Solid Surface Vanity Tops		Crew	Daily Output	Labor-Hours	Unit	Material	Labor	Equipment	Total	Total Incl O&P
						2012 Bare Costs				
0010	**SOLID SURFACE VANITY TOPS**									
0015	Solid surface, center bowl, 17″ x 19″	1 Carp	12	.667	Ea.	175	29.50		204.50	241
0020	19″ x 25″		12	.667		180	29.50		209.50	246
0030	19″ x 31″		12	.667		210	29.50		239.50	279
0040	19″ x 37″		12	.667		244	29.50		273.50	315
0050	22″ x 25″		10	.800		320	35.50		355.50	415
0060	22″ x 31″		10	.800		375	35.50		410.50	470
0070	22″ x 37″		10	.800		435	35.50		470.50	540
0080	22″ x 43″		10	.800		495	35.50		530.50	605
0090	22″ x 49″		10	.800		550	35.50		585.50	665
0110	22″ x 55″		8	1		625	44		669	760
0120	22″ x 61″		8	1		715	44		759	860
0220	Double bowl, 22″ x 61″		8	1		805	44		849	960
0230	Double bowl, 22″ x 73″		8	1		1,400	44		1,444	1,600
0240	For aggregate colors, add					35%				
0250	For faucets and fittings, see Section 22 41 39.10									

12 36 61.19 Quartz Agglomerate Countertops		Crew	Daily Output	Labor-Hours	Unit	Material	Labor	Equipment	Total	Total Incl O&P
0010	**QUARTZ AGGLOMERATE COUNTERTOPS**									
0100	25″ wide, 4″ backsplash, color group A, minimum	2 Carp	15	1.067	L.F.	58	47		105	141
0110	Maximum		15	1.067		81	47		128	166
0120	Color group B, minimum		15	1.067		60	47		107	143
0130	Maximum		15	1.067		85	47		132	171
0140	Color group C, minimum		15	1.067		70	47		117	154
0150	Maximum		15	1.067		96	47		143	183
0160	Color group D, minimum		15	1.067		76	47		123	161
0170	Maximum		15	1.067		103	47		150	190

12 48 Rugs and Mats

12 48 13 – Entrance Floor Mats and Frames

12 48 13.13 Entrance Floor Mats		Crew	Daily Output	Labor-Hours	Unit	Material	Labor	Equipment	Total	Total Incl O&P
0010	**ENTRANCE FLOOR MATS**									
2000	Recycled rubber tire tile, 12″ x 12″ x 3/8″ thick G	1 Clab	125	.064	S.F.	8.65	2.25		10.90	13.25
2510	Natural cocoa fiber, 1/2″ thick G		125	.064		7.65	2.25		9.90	12.10
2520	3/4″ thick G		125	.064		6.40	2.25		8.65	10.70
2530	1″ thick G		125	.064		7.05	2.25		9.30	11.45

12 51 Office Furniture

12 51 16 – Case Goods

12 51 16.13 Metal Case Goods		Crew	Daily Output	Labor-Hours	Unit	Material	Labor	Equipment	Total	Total Incl O&P
0010	**METAL CASE GOODS**									
0020	Desks, 29″ high, double pedestal, 30″ x 60″, metal, minimum				Ea.	540			540	595
0030	Maximum				″	1,325			1,325	1,475

12 54 Hospitality Furniture

12 54 13 – Hotel and Motel Furniture

12 54 13.10 Hotel Furniture		Crew	Daily Output	Labor-Hours	Unit	Material	2012 Bare Costs Labor	Equipment	Total	Total Incl O&P
0010	**HOTEL FURNITURE**									
0020	Standard quality set, minimum				Room	2,500			2,500	2,750
0200	Maximum				"	8,875			8,875	9,750

12 54 16 – Restaurant Furniture

12 54 16.20 Furniture, Restaurant		Crew	Daily Output	Labor-Hours	Unit	Material	Labor	Equipment	Total	Total Incl O&P
0010	**FURNITURE, RESTAURANT**									
0020	Bars, built-in, front bar	1 Carp	5	1.600	L.F.	262	70.50		332.50	405
0200	Back bar		5	1.600	"	191	70.50		261.50	325
0500	Booth unit, molded plastic, stub wall and 2 seats, minimum		2	4	Set	350	176		526	675
0600	Maximum		1.50	5.333	"	1,275	235		1,510	1,775
0800	Booth seat, upholstered, foursome, single (end) minimum		5	1.600	Ea.	685	70.50		755.50	870
0900	Maximum		4	2		915	88		1,003	1,150
1000	Foursome, double, minimum		4	2		1,125	88		1,213	1,375
1100	Maximum		3	2.667		1,575	118		1,693	1,950
1300	Circle booth, upholstered, 1/4 circle, minimum		3	2.667		1,250	118		1,368	1,575
1400	Maximum		2	4		1,700	176		1,876	2,175
1500	3/4 circle, minimum		1.50	5.333		4,850	235		5,085	5,700
1600	Maximum		1	8		5,875	355		6,230	7,025

12 55 Detention Furniture

12 55 13 – Detention Bunks

12 55 13.13 Cots		Crew	Daily Output	Labor-Hours	Unit	Material	Labor	Equipment	Total	Total Incl O&P
0010	**COTS**									
2500	Bolted, single, painted steel	E-4	20	1.600	Ea.	330	80	6.10	416.10	520
2700	Stainless steel	"	20	1.600	"	960	80	6.10	1,046.10	1,200

12 56 Institutional Furniture

12 56 43 – Dormitory Furniture

12 56 43.10 Dormitory Furnishings		Crew	Daily Output	Labor-Hours	Unit	Material	Labor	Equipment	Total	Total Incl O&P
0010	**DORMITORY FURNISHINGS**									
0300	Bunkable bed, twin, minimum				Ea.	390			390	430
0320	Maximum				"	765			765	845

12 56 51 – Library Furniture

12 56 51.10 Library Furnishings		Crew	Daily Output	Labor-Hours	Unit	Material	Labor	Equipment	Total	Total Incl O&P
0010	**LIBRARY FURNISHINGS**									
1710	Carrels, hardwood, 36" x 24", minimum	1 Carp	5	1.600	Ea.	580	70.50		650.50	755
1720	Maximum		4	2		1,325	88		1,413	1,600
1730	Metal, minimum		5	1.600		251	70.50		321.50	390
1740	Maximum		4	2		595	88		683	795
6010	Bookshelf, metal, 90" high, 10" shelf, double face		11.50	.696	L.F.	122	30.50		152.50	185
6020	Single face		12	.667	"	105	29.50		134.50	163
6050	For 8" shelving, subtract from above					10%				
6070	For 42" high with countertop, subtract from above					20%				

12 56 70 – Healthcare Furniture

12 56 70.10 Furniture, Hospital		Crew	Daily Output	Labor-Hours	Unit	Material	Labor	Equipment	Total	Total Incl O&P
0010	**FURNITURE, HOSPITAL**									
0020	Beds, manual, minimum				Ea.	770			770	850
0100	Maximum				"	2,600			2,600	2,875

12 56 Institutional Furniture

12 56 70 – Healthcare Furniture

12 56 70.10 Furniture, Hospital		Crew	Daily Output	Labor-Hours	Unit	Material	Labor	Equipment	Total	Total Incl O&P
						2012 Bare Costs				
1100	Patient wall systems, not incl. plumbing, minimum				Room	1,100			1,100	1,225
1200	Maximum				"	2,050			2,050	2,250

12 63 Stadium and Arena Seating

12 63 13 – Stadium and Arena Bench Seating

12 63 13.13 Bleachers

		Crew	Daily Output	Labor-Hours	Unit	Material	Labor	Equipment	Total	Total Incl O&P
0010	**BLEACHERS**									
3000	Telescoping, manual to 15 tier, minimum	F-5	65	.492	Seat	86	22		108	131
3100	Maximum		60	.533		129	24		153	181
3300	16 to 20 tier, minimum		60	.533		207	24		231	267
3400	Maximum		55	.582		259	26		285	325
3600	21 to 30 tier, minimum		50	.640		215	28.50		243.50	284
3700	Maximum	↓	40	.800		259	35.50		294.50	345
3900	For integral power operation, add, minimum	2 Elec	300	.053		43	2.75		45.75	52
4000	Maximum	"	250	.064	↓	69	3.30		72.30	81
5000	Benches, folding, in wall, 14′ table, 2 benches	L-4	2	12	Set	730	495		1,225	1,600

12 67 Pews and Benches

12 67 13 – Pews

12 67 13.13 Sanctuary Pews

		Crew	Daily Output	Labor-Hours	Unit	Material	Labor	Equipment	Total	Total Incl O&P
0010	**SANCTUARY PEWS**									
1500	Bench type, hardwood, minimum	1 Carp	20	.400	L.F.	83	17.65		100.65	121
1550	Maximum	"	15	.533		165	23.50		188.50	220
1570	For kneeler, add				↓	20.50			20.50	22.50

12 93 Site Furnishings

12 93 23 – Trash and Litter Receptors

12 93 23.10 Trash Receptacles

		Crew	Daily Output	Labor-Hours	Unit	Material	Labor	Equipment	Total	Total Incl O&P
0010	**TRASH RECEPTACLES**									
0500	Recycled plastic, var colors, round, 32 gal., 28″ x 38″ H [G]	2 Clab	5	3.200	Ea.	445	112		557	675
0510	32 gal., 31″ x 32″ H [G]	"	5	3.200	"	490	112		602	725

12 93 33 – Manufactured Planters

12 93 33.10 Planters

		Crew	Daily Output	Labor-Hours	Unit	Material	Labor	Equipment	Total	Total Incl O&P
0010	**PLANTERS**									
0012	Concrete, sandblasted, precast, 48″ diameter, 24″ high	2 Clab	15	1.067	Ea.	695	37.50		732.50	820
0100	Fluted, precast, 7′ diameter, 36″ high		10	1.600		1,425	56		1,481	1,675
0300	Fiberglass, circular, 36″ diameter, 24″ high		15	1.067		660	37.50		697.50	785
0400	60″ diameter, 24″ high	↓	10	1.600	↓	775	56		831	940
9000	Minimum labor/equipment charge	1 Clab	2	4	Job		140		140	230

12 93 43 – Site Seating and Tables

12 93 43.13 Site Seating

		Crew	Daily Output	Labor-Hours	Unit	Material	Labor	Equipment	Total	Total Incl O&P
0010	**SITE SEATING**									
0012	Seating, benches, park, precast conc, w/backs, wood rails, 4′ long	2 Clab	5	3.200	Ea.	535	112		647	775
0100	8′ long		4	4		1,050	140		1,190	1,375
0300	Fiberglass, without back, one piece, 4′ long		10	1.600		585	56		641	735
0400	8′ long	↓	7	2.286	↓	800	80		880	1,000

12 93 Site Furnishings

12 93 43 – Site Seating and Tables

12 93 43.13 Site Seating		Crew	Daily Output	Labor-Hours	Unit	2012 Bare Costs Material	Labor	Equipment	Total	Total Incl O&P
0500	Steel barstock pedestals w/backs, 2" x 3" wood rails, 4' long	2 Clab	10	1.600	Ea.	1,175	56		1,231	1,400
0510	8' long		7	2.286		1,425	80		1,505	1,700
0520	3" x 8" wood plank, 4' long		10	1.600		1,175	56		1,231	1,400
0530	8' long		7	2.286		1,425	80		1,505	1,700
0540	Backless, 4" x 4" wood plank, 4' square		10	1.600		1,200	56		1,256	1,400
0550	8' long		7	2.286		920	80		1,000	1,125
0600	Aluminum pedestals, with backs, aluminum slats, 8' long		8	2		440	70		510	600
0610	15' long		5	3.200		440	112		552	670
0620	Portable, aluminum slats, 8' long		8	2		440	70		510	600
0630	15' long		5	3.200		480	112		592	715
0800	Cast iron pedestals, back & arms, wood slats, 4' long		8	2		370	70		440	520
0820	8' long		5	3.200		1,025	112		1,137	1,300
0840	Backless, wood slats, 4' long		8	2		455	70		525	615
0860	8' long		5	3.200		1,075	112		1,187	1,350
1700	Steel frame, fir seat, 10' long		10	1.600		325	56		381	450
9000	Minimum labor/equipment charge		2	8	Job		281		281	460

Division Notes

	CREW	DAILY OUTPUT	LABOR-HOURS	UNIT	2012 BARE COSTS MAT.	LABOR	EQUIP.	TOTAL	TOTAL INCL O&P

Estimating Tips

General

- The items and systems in this division are usually estimated, purchased, supplied, and installed as a unit by one or more subcontractors. The estimator must ensure that all parties are operating from the same set of specifications and assumptions, and that all necessary items are estimated and will be provided. Many times the complex items and systems are covered, but the more common ones, such as excavation or a crane, are overlooked for the very reason that everyone assumes nobody could miss them. The estimator should be the central focus and be able to ensure that all systems are complete.
- Another area where problems can develop in this division is at the interface between systems. The estimator must ensure, for instance, that anchor bolts, nuts, and washers are estimated and included for the air-supported structures and pre-engineered buildings to be bolted to their foundations. Utility supply is a common area where essential items or pieces of equipment can be missed or overlooked due to the fact that each subcontractor may feel it is another's responsibility. The estimator should also be aware of certain items which may be supplied as part of a package but installed by others, and ensure that the installing contractor's estimate includes the cost of installation. Conversely, the estimator must also ensure that items are not costed by two different subcontractors, resulting in an inflated overall estimate.

13 30 00 Special Structures

- The foundations and floor slab, as well as rough mechanical and electrical, should be estimated, as this work is required for the assembly and erection of the structure. Generally, as noted in the book, the pre-engineered building comes as a shell. Pricing is based on the size and structural design parameters stated in the reference section. Additional features, such as windows and doors with their related structural framing, must also be included by the estimator. Here again, the estimator must have a clear understanding of the scope of each portion of the work and all the necessary interfaces.

Reference Numbers

Reference numbers are shown in shaded boxes at the beginning of some major classifications. These numbers refer to related items in the Reference Section. The reference information may be an estimating procedure, an alternate pricing method, or technical information.

Note: Not all subdivisions listed here necessarily appear in this publication.

13 05 05 – Selective Special Construction Demolition

13 05 05.10 Selective Demolition, Air Supported Structures

	13 05 05.10 Selective Demolition, Air Supported Structures	Crew	Daily Output	Labor-Hours	Unit	Material	2012 Bare Costs Labor	Equipment	Total	Total Incl O&P
0010	**SELECTIVE DEMOLITION, AIR SUPPORTED STRUCTURES**									
0020	Tank covers, scrim, dbl. layer, vinyl poly w/hdw., blower & controls									
0050	Round and rectangular R024119-10	B-2	9000	.004	S.F.		.16		.16	.26
0100	Warehouse structures									
0120	Poly/vinyl fabric, 28 oz., incl. tension cables & inflation system	4 Clab	9000	.004	SF Flr.		.13		.13	.20
0150	Reinforced vinyl, 12 oz., 3000 S.F.	"	5000	.006			.22		.22	.37
0200	12,000 to 24,000 S.F.	8 Clab	20000	.003			.11		.11	.18
0250	Tedlar vinyl fabric, 28 oz. w/liner, to 3000 S.F.	4 Clab	5000	.006			.22		.22	.37
0300	12,000 to 24,000 S.F.	8 Clab	20000	.003			.11		.11	.18
0350	Greenhouse/shelter, woven polyethylene with liner									
0400	3000 S.F.	4 Clab	5000	.006	SF Flr.		.22		.22	.37
0450	12,000 to 24,000 S.F.	8 Clab	20000	.003			.11		.11	.18
0500	Tennis/gymnasium, poly/vinyl fabric, 28 oz., incl. thermal liner	4 Clab	9000	.004			.13		.13	.20
0600	Stadium/convention center, teflon coated fiberglass, incl. thermal liner	9 Clab	40000	.002			.06		.06	.10
0700	Doors, air lock, 15′ long, 10′ x 10′	2 Carp	1.50	10.667	Ea.		470		470	770
0720	15′ x 15′		.80	20			880		880	1,450
0750	Revolving personnel door, 6′ dia. x 6′-6″ high		1.50	10.667			470		470	770

13 05 05.20 Selective Demolition, Garden Houses

	Description	Crew	Daily Output	Labor-Hours	Unit	Material	Labor	Equipment	Total	Total Incl O&P
0010	**SELECTIVE DEMOLITION, GARDEN HOUSES**									
0020	Prefab, wood, excl foundation, average	2 Clab	400	.040	SF Flr.		1.40		1.40	2.30

13 05 05.25 Selective Demolition, Geodesic Domes

	Description	Crew	Daily Output	Labor-Hours	Unit	Material	Labor	Equipment	Total	Total Incl O&P
0010	**SELECTIVE DEMOLITION, GEODESIC DOMES**									
0050	Shell only, interlocking plywood panels, 30′ diameter	F-5	3.20	10	Ea.		445		445	730
0060	34′ diameter		2.30	13.913			620		620	1,025
0070	39′ diameter		2	16			715		715	1,175
0080	45′ diameter	F-3	2.20	18.182			815	278	1,093	1,625
0090	55′ diameter		2	20			895	305	1,200	1,775
0100	60′ diameter		2	20			895	305	1,200	1,775
0110	65′ diameter		1.60	25			1,125	385	1,510	2,250

13 05 05.30 Selective Demolition, Greenhouses

	Description	Crew	Daily Output	Labor-Hours	Unit	Material	Labor	Equipment	Total	Total Incl O&P
0010	**SELECTIVE DEMOLITION, GREENHOUSES**									
0020	Resi-type, free standing, excl. foundations, 9′ long x 8′ wide	2 Clab	160	.100	SF Flr.		3.51		3.51	5.75
0030	9′ long x 11′ wide		170	.094			3.30		3.30	5.40
0040	9′ long x 14′ wide		220	.073			2.55		2.55	4.19
0050	9′ long x 17′ wide		320	.050			1.76		1.76	2.88
0060	Lean-to type, 4′ wide		64	.250			8.80		8.80	14.40
0070	7′ wide		120	.133			4.68		4.68	7.65
0080	Geodesic hemisphere, 1/8″ plexiglass glazing, 8′ diam.		4	4	Ea.		140		140	230
0090	24′ diam.		.80	20			700		700	1,150
0100	48′ diam.		.40	40			1,400		1,400	2,300

13 05 05.35 Selective Demolition, Hangars

	Description	Crew	Daily Output	Labor-Hours	Unit	Material	Labor	Equipment	Total	Total Incl O&P
0010	**SELECTIVE DEMOLITION, HANGARS**									
0020	T type hangars, prefab, steel , galv roof & walls, incl doors, excl fndtn	E-2	2550	.022	SF Flr.		1.06	.59	1.65	2.54
0030	Circular type, prefab, steel frame, plastic skin, incl foundation, 80′ diam	"	.50	112	Total		5,400	3,000	8,400	13,000

13 05 05.45 Selective Demolition, Lightning Protection

	Description	Crew	Daily Output	Labor-Hours	Unit	Material	Labor	Equipment	Total	Total Incl O&P
0010	**SELECTIVE DEMOLITION, LIGHTNING PROTECTION**									
0020	Air terminal & base, copper, 3/8″ diam. x 10″, to 75′ h	1 Clab	16	.500	Ea.		17.55		17.55	29
0030	1/2″ diam. x 12″, over 75′ h		16	.500			17.55		17.55	29
0050	Aluminum, 1/2″ diam. x 12″, to 75′ h		16	.500			17.55		17.55	29
0060	5/8″ diam. x 12″, over 75′ h		16	.500			17.55		17.55	29
0070	Cable, copper, 220 lb. per thousand feet, to 75′ high		640	.013	L.F.		.44		.44	.72

13 05 05 – Selective Special Construction Demolition

13 05 05.45 Selective Demolition, Lightning Protection		Crew	Daily Output	Labor-Hours	Unit	Material	2012 Bare Costs Labor	Equipment	Total	Total Incl O&P
0080	375 lb. per thousand feet, over 75′ high	1 Clab	460	.017	L.F.		.61		.61	1
0090	Aluminum, 101 lb. per thousand feet, to 75′ high		560	.014			.50		.50	.82
0100	199 lb. per thousand feet, over 75′ high		480	.017			.59		.59	.96
0110	Arrester, 175 V AC, to ground		16	.500	Ea.		17.55		17.55	29
0120	650 V AC, to ground		13	.615	″		21.50		21.50	35.50

13 05 05.50 Selective Demolition, Pre-Engineered Steel Buildings		Crew	Daily Output	Labor-Hours	Unit	Material	Labor	Equipment	Total	Total Incl O&P
0010	**SELECTIVE DEMOLITION, PRE-ENGINEERED STEEL BUILDINGS**									
0500	Pre-engd. steel bldgs., rigid frame, clear span & multi post, excl. salvage									
0550	3,500 to 7,500 S.F.	L-10	1000	.024	SF Flr.		1.19	.61	1.80	2.77
0600	7,501 to 12,500 S.F.		1500	.016			.79	.41	1.20	1.85
0650	12,500 S.F. or greater		1650	.015			.72	.37	1.09	1.68
0700	Pre-engd. steel building components									
0710	Entrance canopy, including frame 4′ x 4′	E-24	8	4	Ea.		195	87	282	445
0720	4′ x 8′	″	7	4.571			222	99.50	321.50	510
0730	H.M. doors, self framing, single leaf	2 Skwk	8	2			91		91	148
0740	Double leaf		5	3.200			145		145	236
0760	Gutter, eave type		600	.027	L.F.		1.21		1.21	1.97
0770	Sash, single slide, double slide or fixed		24	.667	Ea.		30.50		30.50	49
0780	Skylight, fiberglass, to 30 S.F.		16	1			45.50		45.50	74
0785	Roof vents, circular, 12″ to 24″ diameter		12	1.333			60.50		60.50	98.50
0790	Continuous, 10′ long		8	2			91		91	148
0900	Shelters, aluminum frame									
0910	Acrylic glazing, 3′ x 9′ x 8′ high	2 Skwk	2	8	Ea.		365		365	590
0920	9′ x 12′ x 8′ high	″	1.50	10.667	″		485		485	785

13 05 05.60 Selective Demolition, Silos		Crew	Daily Output	Labor-Hours	Unit	Material	Labor	Equipment	Total	Total Incl O&P
0010	**SELECTIVE DEMOLITION, SILOS**									
0020	Conc stave, indstrl, conical/sloping bott, excl fndtn, 12′ diam., 35′ h	E-24	.18	177	Ea.		8,650	3,875	12,525	19,800
0030	16′ diam., 45′ h		.12	266			13,000	5,800	18,800	29,600
0040	25′ diam., 75′ h		.08	400			19,500	8,725	28,225	44,500
0050	Steel, factory fabricated, 30,000 gal. cap, painted or epoxy lined	L-5	2	28			1,375	350	1,725	2,900

13 05 05.65 Selective Demolition, Sound Control		Crew	Daily Output	Labor-Hours	Unit	Material	Labor	Equipment	Total	Total Incl O&P
0010	**SELECTIVE DEMOLITION, SOUND CONTROL**									
0120	Acoustical enclosure, 4″ thick walls & ceiling panels, 8 lb./S.F.	3 Carp	144	.167	SF Surf		7.35		7.35	12.05
0130	10.5 lb./S.F.		128	.188			8.25		8.25	13.55
0140	Reverb chamber, parallel walls, 4″ thick		120	.200			0.00		8.80	14.45
0150	Skewed walls, parallel roof, 4″ thick		110	.218			9.60		9.60	15.75
0160	Skewed walls/roof, 4″ layer/air space		96	.250			11.05		11.05	18.10
0170	Sound-absorbing panels, painted metal, 2′-6″ x 8′, under 1,000 S.F.		430	.056			2.46		2.46	4.04
0180	Over 1,000 S.F.		480	.050			2.21		2.21	3.62
0190	Flexible transparent curtain, clear	3 Shee	430	.056			2.91		2.91	4.58
0192	50% clear, 50% foam		430	.056			2.91		2.91	4.58
0194	25% clear, 75% foam		430	.056			2.91		2.91	4.58
0196	100% foam		430	.056			2.91		2.91	4.58
0200	Audio-masking sys., incl. speakers, amplfr., signal gnrtr.									
0205	Ceiling mounted, 5,000 S.F.	2 Elec	4800	.003	S.F.		.17		.17	.27
0210	10,000 S.F.		5600	.003			.15		.15	.23
0220	Plenum mounted, 5,000 S.F.		7600	.002			.11		.11	.17
0230	10,000 S.F.		8800	.002			.09		.09	.15

13 05 05.70 Selective Demolition, Special Purpose Rooms		Crew	Daily Output	Labor-Hours	Unit	Material	Labor	Equipment	Total	Total Incl O&P
0010	**SELECTIVE DEMOLITION, SPECIAL PURPOSE ROOMS**									
0100	Audiometric rooms, under 500 S.F. surface	4 Carp	200	.160	SF Surf		7.05		7.05	11.55
0110	Over 500 S.F. surface	″	240	.133	″		5.90		5.90	9.65

13 05 Common Work Results for Special Construction

13 05 05 – Selective Special Construction Demolition

13 05 05.70 Selective Demolition, Special Purpose Rooms

	13 05 05.70 Selective Demolition, Special Purpose Rooms	Crew	Daily Output	Labor-Hours	Unit	Material (2012 Bare Costs)	Labor (2012 Bare Costs)	Equipment (2012 Bare Costs)	Total (2012 Bare Costs)	Total Incl O&P
0200	Clean rooms, 12′ x 12′ soft wall, class 100	1 Carp	.30	26.667	Ea.		1,175		1,175	1,925
0210	Class 1000		.30	26.667			1,175		1,175	1,925
0220	Class 10,000		.35	22.857			1,000		1,000	1,650
0230	Class 100,000		.35	22.857			1,000		1,000	1,650
0300	Darkrooms, shell complete, 8′ high	2 Carp	220	.073	SF Flr.		3.21		3.21	5.25
0310	12′ high		110	.145	"		6.40		6.40	10.50
0350	Darkrooms doors, mini-cylindrical, revolving		4	4	Ea.		176		176	289
0400	Music room, practice modular		140	.114	SF Surf		5.05		5.05	8.25
0500	Refrigeration structures and finishes									
0510	Wall finish, 2 coat portland cement plaster, 1/2″ thick	1 Clab	200	.040	S.F.		1.40		1.40	2.30
0520	Fiberglass panels, 1/8″ thick		400	.020			.70		.70	1.15
0530	Ceiling finish, polystyrene plastic, 1″ to 2″ thick		500	.016			.56		.56	.92
0540	4″ thick		450	.018			.62		.62	1.02
0550	Refrigerator, prefab aluminum walk-in, 7′-6″ high, 6′ x 6′ OD	2 Carp	100	.160	SF Flr.		7.05		7.05	11.55
0560	10′ x 10′ OD		160	.100			4.41		4.41	7.25
0570	Over 150 S.F.		200	.080			3.53		3.53	5.80
0600	Sauna, prefabricated, including heater & controls, 7′ high, to 30 S.F.		120	.133			5.90		5.90	9.65
0610	To 40 S.F.		140	.114			5.05		5.05	8.25
0620	To 60 S.F.		175	.091			4.03		4.03	6.60
0630	To 100 S.F.		220	.073			3.21		3.21	5.25
0640	To 130 S.F.		250	.064			2.82		2.82	4.63

13 05 05.75 Selective Demolition, Storage Tanks

	13 05 05.75 Selective Demolition, Storage Tanks	Crew	Daily Output	Labor-Hours	Unit	Material	Labor	Equipment	Total	Total Incl O&P
0010	**SELECTIVE DEMOLITION, STORAGE TANKS**									
0500	Steel tank, single wall, above ground, not incl. fdn., pumps or piping									
0510	Single wall, 275 gallon	Q-1	3	5.333	Ea.		262		262	410
0520	550 thru 2,000 gallon	B-34P	2	12			545	299	844	1,200
0530	5,000 thru 10,000 gallon	B-34Q	2	12			550	580	1,130	1,500
0540	15,000 thru 30,000 gallon	B-34S	2	16			780	1,750	2,530	3,125
0600	Steel tank, double wall, above ground not incl. fdn., pumps & piping									
0620	500 thru 2,000 gallon	B-34P	2	12	Ea.		545	299	844	1,200

13 05 05.85 Selective Demolition, Swimming Pool Equip

	13 05 05.85 Selective Demolition, Swimming Pool Equip	Crew	Daily Output	Labor-Hours	Unit	Material	Labor	Equipment	Total	Total Incl O&P
0010	**SELECTIVE DEMOLITION, SWIMMING POOL EQUIP**									
0020	Diving stand, stainless steel, 3 meter	2 Clab	3	5.333	Ea.		187		187	305
0030	1 meter		5	3.200			112		112	184
0040	Diving board, 16′ long, aluminum		5.40	2.963			104		104	171
0050	Fiberglass		5.40	2.963			104		104	171
0070	Ladders, heavy duty, stainless steel, 2 tread		14	1.143			40		40	66
0080	4 tread		12	1.333			47		47	76.50
0090	Lifeguard chair, stainless steel, fixed		5	3.200			112		112	184
0100	Slide, tubular, fiberglass, aluminum handrails & ladder, 5′, straight		4	4			140		140	230
0110	8′, curved		6	2.667			93.50		93.50	153
0120	10′, curved		3	5.333			187		187	305
0130	12′ straight, with platform		2.50	6.400			225		225	370
0140	Removable access ramp, stainless steel		4	4			140		140	230
0150	Removable stairs, stainless steel, collapsible		4	4			140		140	230

13 05 05.90 Selective Demolition, Tension Structures

	13 05 05.90 Selective Demolition, Tension Structures	Crew	Daily Output	Labor-Hours	Unit	Material	Labor	Equipment	Total	Total Incl O&P
0010	**SELECTIVE DEMOLITION, TENSION STRUCTURES**									
0020	Steel/alum. frame, fabric shell, 60′ clear span, 6,000 S.F.	B-41	2000	.022	SF Flr.		.80	.14	.94	1.45
0030	12,000 S.F.		2200	.020			.73	.13	.86	1.33
0040	80′ clear span, 20,800 S.F.		2440	.018			.65	.11	.76	1.20
0050	100′ clear span, 10,000 S.F.	L-5	4350	.013			.64	.16	.80	1.34
0060	26,000 S.F.	"	4600	.012			.60	.15	.75	1.26

13 05 Common Work Results for Special Construction

13 05 05 – Selective Special Construction Demolition

13 05 05.95 Selective Demo, X-Ray/Radio Freq Protection		Crew	Daily Output	Labor-Hours	Unit	2012 Bare Costs Material	Labor	Equipment	Total	Total Incl O&P
0010	**SELECTIVE DEMO, X-RAY/RADIO FREQ PROTECTION**									
0020	Shielding lead, lined door frame, excl. hdwe., 1/16" thick	1 Clab	4.80	1.667	Ea.		58.50		58.50	96
0030	Lead sheets, 1/16" thick	2 Clab	270	.059	S.F.		2.08		2.08	3.41
0050	Lead shielding, 1/4" thick		270	.059			2.08		2.08	3.41
0060	1/2" thick		240	.067			2.34		2.34	3.84
0070	Lead glass, 1/4" thick, 2.0 mm LE, 12" x 16"	2 Glaz	16	1	Ea.		43		43	69
0080	24" x 36"		8	2			85.50		85.50	138
0090	36" x 60"		4	4			171		171	276
0100	Lead glass window frame, with 1/16" lead & voice passage, 36" x 60"		4	4			171		171	276
0110	Lead glass window frame, 24" x 36"		8	2			85.50		85.50	138
0120	Lead gypsum board, 5/8" thick with 1/16" lead	2 Clab	320	.050	S.F.		1.76		1.76	2.88
0130	1/8" lead		280	.057			2.01		2.01	3.29
0140	1/32" lead		400	.040			1.40		1.40	2.30
0150	Butt joints, 1/8" lead or thicker, 2" x 7' long batten strip		480	.033	Ea.		1.17		1.17	1.92
0160	X-ray protection, average radiography room, up to 300 S.F., 1/16" lead, min		.50	32	Total		1,125		1,125	1,850
0170	Maximum		.30	53.333			1,875		1,875	3,075
0180	Deep therapy X-ray room, 250 kV cap, up to 300 S.F., 1/4" lead, min		.20	80			2,800		2,800	4,600
0190	Maximum		.12	133			4,675		4,675	7,675
0880	Radio frequency shielding, prefab or screen-type copper or steel, minimum		360	.044	SF Surf		1.56		1.56	2.56
0890	Average		310	.052			1.81		1.81	2.97
0895	Maximum		290	.055			1.94		1.94	3.18

13 11 Swimming Pools

13 11 13 – Below-Grade Swimming Pools

13 11 13.50 Swimming Pools		Crew	Daily Output	Labor-Hours	Unit	Material	Labor	Equipment	Total	Total Incl O&P
0011	**SWIMMING POOLS**, Outdoor, incl. equip. & houses, minimum								32.50	36
0300	Maximum								62	68
0400	Residential, incl. equipment, permanent type, minimum								10.90	12
0700	Maximum								22.50	25
0900	Municipal, including equipment only, over 5000 S.F., minimum								22.50	25
1000	Maximum								46.50	51
1300	Motel or apt., incl. equipment only, under 5000 S.F., minimum								20	22
1400	Maximum								31	34

13 12 Fountains

13 12 13 – Exterior Fountains

13 12 13.10 Yard Fountains		Crew	Daily Output	Labor-Hours	Unit	Material	Labor	Equipment	Total	Total Incl O&P
0010	**YARD FOUNTAINS**									
0100	Outdoor fountain, 48" high with bowl and figures	2 Clab	2	8	Ea.	550	281		831	1,075

13 17 Tubs and Pools

13 17 13 – Hot Tubs

13 17 13.10 Redwood Hot Tub System		Crew	Daily Output	Labor-Hours	Unit	Material	2012 Bare Costs Labor	Equipment	Total	Total Incl O&P
0010	**REDWOOD HOT TUB SYSTEM**									
7050	4' diameter x 4' deep	Q-1	1	16	Ea.	2,925	785		3,710	4,450
7150	6' diameter x 4' deep		.80	20		4,500	985		5,485	6,475
7200	8' diameter x 4' deep		.80	20		6,600	985		7,585	8,775

13 17 33 – Whirlpool Tubs

13 17 33.10 Whirlpool Bath		Crew	Daily Output	Labor-Hours	Unit	Material	Labor	Equipment	Total	Total Incl O&P
0010	**WHIRLPOOL BATH**									
6000	Whirlpool, bath with vented overflow, molded fiberglass									
6100	66" x 36" x 24"	Q-1	1	16	Ea.	3,100	785		3,885	4,650
6400	72" x 36" x 21"		1	16		1,750	785		2,535	3,150
6500	60" x 34" x 21"		1	16		1,650	785		2,435	3,025
6600	72" x 42" x 23"		1	16		1,975	785		2,760	3,400
6710	For color add					10%				
6711	For designer colors and trim add					25%				

13 21 Controlled Environment Rooms

13 21 26 – Cold Storage Rooms

13 21 26.50 Refrigeration		Crew	Daily Output	Labor-Hours	Unit	Material	Labor	Equipment	Total	Total Incl O&P
0010	**REFRIGERATION**									
0020	Curbs, 12" high, 4" thick, concrete	2 Carp	58	.276	L.F.	4.61	12.15		16.76	25
6300	Rule of thumb for complete units, w/o doors & refrigeration, cooler		146	.110	SF Flr.	134	4.83		138.83	155
6400	Freezer		109.60	.146	"	158	6.45		164.45	185

13 24 Special Activity Rooms

13 24 16 – Saunas

13 24 16.50 Saunas and Heaters		Crew	Daily Output	Labor-Hours	Unit	Material	Labor	Equipment	Total	Total Incl O&P
0010	**SAUNAS AND HEATERS**									
0020	Prefabricated, incl. heater & controls, 7' high, 6' x 4', C/C	L-7	2.20	12.727	Ea.	5,125	540		5,665	6,525
1700	Door only, cedar, 2' x 6', with tempered insulated glass window	2 Carp	3.40	4.706		620	208		828	1,025
1800	Prehung, incl. jambs, pulls & hardware	"	12	1.333		620	59		679	780
2500	Heaters only (incl. above), wall mounted, to 200 C.F.					615			615	675
4480	For additional equipment, see Section 11 66 13.10									

13 24 26 – Steam Baths

13 24 26.50 Steam Baths and Components		Crew	Daily Output	Labor-Hours	Unit	Material	Labor	Equipment	Total	Total Incl O&P
0010	**STEAM BATHS AND COMPONENTS**									
0020	Heater, timer & head, single, to 140 C.F.	1 Plum	1.20	6.667	Ea.	2,100	365		2,465	2,875
0500	To 300 C.F.	"	1.10	7.273		2,425	395		2,820	3,300
2000	Multiple, motels, apts., 2 baths, w/blow-down assm., 500 C.F.	Q-1	1.30	12.308		5,900	605		6,505	7,450
2500	4 baths	"	.70	22.857		7,450	1,125		8,575	9,950

13 28 Athletic and Recreational Special Construction

13 28 33 – Athletic and Recreational Court Walls

13 28 33.50 Sport Court		Crew	Daily Output	Labor-Hours	Unit	Material	2012 Bare Costs Labor	Equipment	Total	Total Incl O&P
0010	**SPORT COURT**									
0020	Floors, No. 2 & better maple, 25/32" thick				SF Flr.				6.05	6.65
0300	Squash, regulation court in existing building, minimum				Court	36,800			36,800	40,400
0400	Maximum				"	41,000			41,000	45,000
0450	Rule of thumb for components:									
0470	Walls	3 Carp	.15	160	Court	11,000	7,050		18,050	23,700
0500	Floor	"	.25	96		8,725	4,225		12,950	16,500
0550	Lighting	2 Elec	.60	26.667	↓	2,100	1,375		3,475	4,425

13 34 Fabricated Engineered Structures

13 34 13 – Glazed Structures

13 34 13.13 Greenhouses

		Crew	Daily Output	Labor-Hours	Unit	Material	Labor	Equipment	Total	Total Incl O&P
0010	**GREENHOUSES**, Shell only, stock units, not incl. 2' stub walls,									
0020	foundation, floors, heat or compartments									
0300	Residential type, free standing, 8'-6" long x 7'-6" wide	2 Carp	59	.271	SF Flr.	33	11.95		44.95	55.50
0400	10'-6" wide		85	.188		28	8.30		36.30	44.50
0600	13'-6" wide		108	.148		24.50	6.55		31.05	37.50
0700	17'-0" wide		160	.100		23.50	4.41		27.91	33.50
0900	Lean-to type, 3'-10" wide		34	.471		32.50	21		53.50	70
1000	6'-10" wide		58	.276		43.50	12.15		55.65	68
1050	8'-0" wide	↓	60	.267	↓	38.50	11.75		50.25	61.50
1060										
1100	Wall mounted to existing window, 3' x 3'	1 Carp	4	2	Ea.	1,475	88		1,563	1,750
1120	4' x 5'	"	3	2.667		1,850	118		1,968	2,250
3900	Cooling, 1200cfm exhaust fan, add					239			239	263
4000	7850 cfm					900			900	990
4200	For heaters, 10 MBH, add					195			195	215
4300	60 MBH, add					645			645	710
4500	For benches, 2' x 8', add					150			150	165
4600	4' x 10', add				↓	220			220	242
4800	For controls, add, minimum				Total	210			210	231
4900	Maximum				"	700			700	770
5100	For humidification equipment, add				Ea.	560			560	615
5200	For vinyl shading, add				S.F.	.22			.22	.24

13 34 13.19 Swimming Pool Enclosures

		Crew	Daily Output	Labor-Hours	Unit	Material	Labor	Equipment	Total	Total Incl O&P
0010	**SWIMMING POOL ENCLOSURES** Translucent, free standing									
0020	not including foundations, heat or light									
0200	Economy, minimum	2 Carp	200	.080	SF Hor.	14.80	3.53		18.33	22
0300	Maximum		100	.160		45.50	7.05		52.55	61.50
0400	Deluxe, minimum		100	.160		50.50	7.05		57.55	67
0600	Maximum	↓	70	.229	↓	227	10.10		237.10	267

13 34 23 – Fabricated Structures

13 34 23.16 Fabricated Control Booths

		Crew	Daily Output	Labor-Hours	Unit	Material	Labor	Equipment	Total	Total Incl O&P
0010	**FABRICATED CONTROL BOOTHS**									
0100	Guard House, prefab conc w/bullet resistant doors & windows, roof & wiring									
0110	8' x 8', Level III	L-10	1	24	Ea.	24,200	1,200	610	26,010	29,400
0120	8' x 8', Level IV	"	1	24	"	30,300	1,200	610	32,110	36,100

13 34 Fabricated Engineered Structures

13 34 23 – Fabricated Structures

13 34 23.45 Kiosks		Crew	Daily Output	Labor-Hours	Unit	Material	2012 Bare Costs Labor	Equipment	Total	Total Incl O&P
0010	**KIOSKS**									
0020	Round, 5′ diameter, 8′ high, 1/4″ fiberglass wall				Ea.	5,800			5,800	6,375
0100	1″ insulated double wall, fiberglass					6,600			6,600	7,250
0500	Rectangular, 5′ x 9′, 7′-6″ high, 1/4″ fiberglass wall					8,450			8,450	9,300
0600	1″ insulated double wall, fiberglass				↓	10,000			10,000	11,000

13 34 63 – Natural Fiber Construction

13 34 63.50 Straw Bale Construction		Crew	Daily Output	Labor-Hours	Unit	Material	Labor	Equipment	Total	Total Incl O&P
0010	**STRAW BALE CONSTRUCTION**									
2000	Straw bales wall, incl. labor & material complete, minimum G				SF Flr.					130
2010	Maximum G				″					160
2020	Straw bales in walls w/modified post and beam frame G	2 Carp	320	.050	S.F.	6.60	2.21		8.81	10.85

13 42 Building Modules

13 42 63 – Detention Cell Modules

13 42 63.16 Steel Detention Cell Modules		Crew	Daily Output	Labor-Hours	Unit	Material	Labor	Equipment	Total	Total Incl O&P
0010	**STEEL DETENTION CELL MODULES**									
2000	Cells, prefab., 5′ to 6′ wide, 7′ to 8′ high, 7′ to 8′ deep,									
2010	bar front, cot, not incl. plumbing	E-4	1.50	21.333	Ea.	9,650	1,075	81	10,806	12,700

13 48 Sound, Vibration, and Seismic Control

13 48 13 – Manufactured Sound and Vibration Control Components

13 48 13.50 Audio Masking		Crew	Daily Output	Labor-Hours	Unit	Material	Labor	Equipment	Total	Total Incl O&P
0010	**AUDIO MASKING**, acoustical enclosure, 4″ thick wall and ceiling									
0020	8# per S.F., up to 12′ span	3 Carp	72	.333	SF Surf	32	14.70		46.70	59.50
0300	Better quality panels, 10.5# per S.F.		64	.375		36.50	16.55		53.05	67
0400	Reverb-chamber, 4″ thick, parallel walls	↓	60	.400	↓	45.50	17.65		63.15	79

13 49 Radiation Protection

13 49 13 – Integrated X-Ray Shielding Assemblies

13 49 13.50 Lead Sheets		Crew	Daily Output	Labor-Hours	Unit	Material	Labor	Equipment	Total	Total Incl O&P
0010	**LEAD SHEETS**									
0300	Lead sheets, 1/16″ thick	2 Lath	135	.119	S.F.	10.05	4.65		14.70	18.35
0400	1/8″ thick		120	.133		20	5.25		25.25	30
0500	Lead shielding, 1/4″ thick		135	.119		49	4.65		53.65	61.50
0550	1/2″ thick	↓	120	.133	↓	83.50	5.25		88.75	99.50
1200	X-ray protection, average radiography or fluoroscopy									
1210	room, up to 300 S.F. floor, 1/16″ lead, minimum	2 Lath	.25	64	Total	9,200	2,500		11,700	14,000
1500	Maximum, 7′-0″ walls	″	.15	106	″	11,100	4,175		15,275	18,800
1600	Deep therapy X-ray room, 250 kV capacity,									
1800	up to 300 S.F. floor, 1/4″ lead, minimum	2 Lath	.08	200	Total	25,700	7,850		33,550	40,600
1900	Maximum, 7′-0″ walls		.06	266	″	31,700	10,500		42,200	51,500
1999	Minimum labor/equipment charge	↓	4.50	3.556	Job		140		140	219

13 49 21 – Lead Glazing

13 49 21.50 Lead Glazing		Crew	Daily Output	Labor-Hours	Unit	Material	Labor	Equipment	Total	Total Incl O&P
0010	**LEAD GLAZING**									
0600	Lead glass, 1/4″ thick, 2.0 mm LE, 12″ x 16″	2 Glaz	13	1.231	Ea.	305	52.50		357.50	420

13 49 Radiation Protection

13 49 21 – Lead Glazing

13 49 21.50 Lead Glazing		Crew	Daily Output	Labor-Hours	Unit	2012 Bare Costs Material	Labor	Equipment	Total	Total Incl O&P
0700	24″ x 36″	2 Glaz	8	2	Ea.	1,100	85.50		1,185.50	1,375
2000	X-ray viewing panels, clear lead plastic									
2010	7 mm thick, 0.3 mm LE, 2.3 lb./S.F.	H-3	139	.115	S.F.	224	4.38		228.38	253
2020	12 mm thick, 0.5 mm LE, 3.9 lb./S.F.		82	.195		305	7.40		312.40	345
2030	18 mm thick, 0.8 mm LE, 5.9 lb./S.F.		54	.296		330	11.25		341.25	380
2040	22 mm thick, 1.0 mm LE, 7.2 lb./S.F.		44	.364		335	13.85		348.85	395
2050	35 mm thick, 1.5 mm LE, 11.5 lb./S.F.		28	.571		375	21.50		396.50	450
2060	46 mm thick, 2.0 mm LE, 15.0 lb./S.F.		21	.762		495	29		524	590
2090	For panels 12 S.F. to 48 S.F., add crating charge				Ea.					50

13 49 23 – Integrated RFI/EMI Shielding Assemblies

13 49 23.50 Modular Shielding Partitions		Crew	Daily Output	Labor-Hours	Unit	Material	Labor	Equipment	Total	Total Incl O&P
0010	**MODULAR SHIELDING PARTITIONS**									
4000	X-ray barriers, modular, panels mounted within framework for									
4002	attaching to floor, wall or ceiling, upper portion is clear lead									
4005	plastic window panels 48″H, lower portion is opaque leaded									
4008	steel panels 36″H, structural supports not incl.									
4010	1-section barrier, 36″W x 84″H overall									
4020	0.5 mm LE panels	H-3	6.40	2.500	Ea.	7,075	95		7,170	7,925
4030	0.8 mm LE panels		6.40	2.500		7,525	95		7,620	8,450
4040	1.0 mm LE panels		5.33	3.002		7,675	114		7,789	8,625
4050	1.5 mm LE panels		5.33	3.002		8,175	114		8,289	9,175
4060	2-section barrier, 72″W x 84″H overall									
4070	0.5 mm LE panels	H-3	4	4	Ea.	14,400	152		14,552	16,100
4080	0.8 mm LE panels		4	4		15,300	152		15,452	17,000
4090	1.0 mm LE panels		3.56	4.494		15,600	171		15,771	17,500
5000	1.5 mm LE panels		3.20	5		17,700	190		17,890	19,800
5010	3-section barrier, 108″W x 84″H overall									
5020	0.5 mm LE panels	H-3	3.20	5	Ea.	21,700	190		21,890	24,200
5030	0.8 mm LE panels		3.20	5		22,900	190		23,090	25,500
5040	1.0 mm LE panels		2.67	5.993		23,300	228		23,528	26,000
5050	1.5 mm LE panels		2.46	6.504		25,300	247		25,547	28,200
7000	X-ray barriers, mobile, mounted within framework w/casters on									
7005	bottom, clear lead plastic window panels on upper portion,									
7010	opaque on lower, 30″W x 75″H overall, incl. framework									
7020	24″H upper w/0.5 mm LE, 48″H lower w/0.8 mm LE	1 Carp	16	.500	Ea.	3,775	22		3,797	4,175
7030	48″W x 75″H overall, incl. framework									
7040	36″H upper w/0.5 mm LE, 36″H lower w/0.8 mm LE	1 Carp	16	.500	Ea.	6,900	22		6,922	7,625
7050	36″H upper w/1.0 mm LE, 36″H lower w/1.5 mm LE	″	16	.500	″	8,200	22		8,222	9,050
7060	72″W x 75″H overall, incl. framework									
7070	36″H upper w/0.5 mm LE, 36″H lower w/0.8 mm LE	1 Carp	16	.500	Ea.	8,175	22		8,197	9,025
7080	36″H upper w/1.0 mm LE, 36″H lower w/1.5 mm LE	″	16	.500	″	10,200	22		10,222	11,300

Division Notes

		CREW	DAILY OUTPUT	LABOR-HOURS	UNIT	2012 BARE COSTS MAT.	LABOR	EQUIP.	TOTAL	TOTAL INCL O&P

Estimating Tips

General

- Many products in Division 14 will require some type of support or blocking for installation not included with the item itself. Examples are supports for conveyors or tube systems, attachment points for lifts, and footings for hoists or cranes. Add these supports in the appropriate division.

14 10 00 Dumbwaiters
14 20 00 Elevators

- Dumbwaiters and elevators are estimated and purchased in a method similar to buying a car. The manufacturer has a base unit with standard features. Added to this base unit price will be whatever options the owner or specifications require. Increased load capacity, additional vertical travel, additional stops, higher speed, and cab finish options are items to be considered. When developing an estimate for dumbwaiters and elevators, remember that some items needed by the installers may have to be included as part of the general contract.

  Examples are:
  - — shaftway
  - — rail support brackets
  - — machine room
  - — electrical supply
  - — sill angles
  - — electrical connections
  - — pits
  - — roof penthouses
  - — pit ladders

Check the job specifications and drawings before pricing.

- Installation of elevators and handicapped lifts in historic structures can require significant additional costs. The associated structural requirements may involve cutting into and repairing finishes, mouldings, flooring, etc. The estimator must account for these special conditions.

14 30 00 Escalators and Moving Walks

- Escalators and moving walks are specialty items installed by specialty contractors. There are numerous options associated with these items. For specific options, contact a manufacturer or contractor. In a method similar to estimating dumbwaiters and elevators, you should verify the extent of general contract work and add items as necessary.

14 40 00 Lifts
14 90 00 Other Conveying Equipment

- Products such as correspondence lifts, chutes, and pneumatic tube systems, as well as other items specified in this subdivision, may require trained installers. The general contractor might not have any choice as to who will perform the installation or when it will be performed. Long lead times are often required for these products, making early decisions in scheduling necessary.

Reference Numbers

Reference numbers are shown in shaded boxes at the beginning of some major classifications. These numbers refer to related items in the Reference Section. The reference information may be an estimating procedure, an alternate pricing method, or technical information.

Note: Not all subdivisions listed here necessarily appear in this publication.

14 11 Manual Dumbwaiters

14 11 10 – Hand Operated Dumbwaiters

14 11 10.20 Manual Dumbwaiters		Crew	Daily Output	Labor-Hours	Unit	Material	2012 Bare Costs Labor	Equipment	Total	Total Incl O&P
0010	MANUAL DUMBWAITERS									
0020	2 stop, minimum	2 Elev	.75	21.333	Ea.	2,925	1,500		4,425	5,575
0100	Maximum		.50	32	"	6,650	2,275		8,925	10,800
0300	For each additional stop, add	↓	.75	21.333	Stop	1,050	1,500		2,550	3,525

14 12 Electric Dumbwaiters

14 12 10 – Dumbwaiters

14 12 10.10 Electric Dumbwaiters		Crew	Daily Output	Labor-Hours	Unit	Material	Labor	Equipment	Total	Total Incl O&P
0010	ELECTRIC DUMBWAITERS									
0020	2 stop, minimum	2 Elev	.13	123	Ea.	7,375	8,725		16,100	21,600
0100	Maximum		.11	145	"	22,200	10,300		32,500	40,400
0600	For each additional stop, add	↓	.54	29.630	Stop	3,250	2,100		5,350	6,825
0750	Correspondence lift, 1 floor, 2 stop, 45 lb. capacity	2 Elec	.20	80	Ea.	8,350	4,125		12,475	15,600

14 21 Electric Traction Elevators

14 21 13 – Electric Traction Freight Elevators

14 21 13.10 Electric Traction Freight Elevators and Options		Crew	Daily Output	Labor-Hours	Unit	Material	Labor	Equipment	Total	Total Incl O&P
0010	ELECTRIC TRACTION FREIGHT ELEVATORS AND OPTIONS R142000-10									
0425	Electric freight, base unit, 4000 lb., 200 fpm, 4 stop, std. fin.	2 Elev	.05	320	Ea.	99,000	22,700		121,700	144,000
0450	For 5000 lb. capacity, add R142000-40					5,500			5,500	6,050
0500	For 6000 lb. capacity, add					16,000			16,000	17,600
0525	For 7000 lb. capacity, add					19,800			19,800	21,700
0550	For 8000 lb. capacity, add					43,400			43,400	47,700
0575	For 10000 lb. capacity, add					30,000			30,000	33,000
0600	For 12000 lb. capacity, add					36,500			36,500	40,200
0625	For 16000 lb. capacity, add					46,500			46,500	51,000
0650	For 20000 lb. capacity, add					55,000			55,000	60,500
0675	For increased speed, 250 fpm, add					20,400			20,400	22,400
0700	300 fpm, geared electric, add					25,800			25,800	28,300
0725	350 fpm, geared electric, add					31,200			31,200	34,300
0750	400 fpm, geared electric, add					36,300			36,300	39,900
0775	500 fpm, gearless electric, add					43,800			43,800	48,100
0800	600 fpm, gearless electric, add					53,000			53,000	58,000
0825	700 fpm, gearless electric, add					63,000			63,000	69,500
0850	800 fpm, gearless electric, add					72,500			72,500	80,000
0875	For class "B" loading, add					5,875			5,875	6,450
0900	For class "C-1" loading, add					8,300			8,300	9,125
0925	For class "C-2" loading, add					9,475			9,475	10,400
0950	For class "C-3" loading, add				↓	12,000			12,000	13,200
0975	For travel over 40 V.L.F., add	2 Elev	7.25	2.207	V.L.F.	775	156		931	1,100
1000	For number of stops over 4, add	"	.27	59.259	Stop	3,925	4,200		8,125	10,800

14 21 23 – Electric Traction Passenger Elevators

14 21 23.10 Electric Traction Passenger Elevators and Options		Crew	Daily Output	Labor-Hours	Unit	Material	Labor	Equipment	Total	Total Incl O&P
0010	ELECTRIC TRACTION PASSENGER ELEVATORS AND OPTIONS									
1625	Electric pass., base unit, 2000 lb., 200 fpm, 4 stop, std. fin.	2 Elev	.05	320	Ea.	84,000	22,700		106,700	127,500
1650	For 2500 lb. capacity, add					3,525			3,525	3,875
1675	For 3000 lb. capacity, add					4,200			4,200	4,600
1700	For 3500 lb. capacity, add					7,025			7,025	7,725
1725	For 4000 lb. capacity, add				↓	6,600			6,600	7,250

14 21 Electric Traction Elevators

14 21 23 – Electric Traction Passenger Elevators

14 21 23.10 Electric Traction Passenger Elevators and Options		Crew	Daily Output	Labor-Hours	Unit	2012 Bare Costs Material	Labor	Equipment	Total	Total Incl O&P
1750	For 4500 lb. capacity, add				Ea.	8,325			8,325	9,175
1775	For 5000 lb. capacity, add					10,100			10,100	11,100
1800	For increased speed, 250 fpm, geared electric, add					6,100			6,100	6,700
1825	300 fpm, geared electric, add					9,325			9,325	10,200
1850	350 fpm, geared electric, add					10,600			10,600	11,700
1875	400 fpm, geared electric, add					13,600			13,600	14,900
1900	500 fpm, gearless electric, add					32,000			32,000	35,200
1925	600 fpm, gearless electric, add					39,600			39,600	43,600
1950	700 fpm, gearless electric, add					41,800			41,800	46,000
1975	800 fpm, gearless electric, add					48,000			48,000	53,000
2000	For travel over 40 V.L.F., add	2 Elev	7.25	2.207	V.L.F.	910	156		1,066	1,250
2025	For number of stops over 4, add		.27	59.259	Stop	6,825	4,200		11,025	14,000
2400	Electric hospital, base unit, 4000 lb., 200 fpm, 4 stop, std fin.		.05	320	Ea.	72,500	22,700		95,200	114,500
2425	For 4500 lb. capacity, add					5,175			5,175	5,675
2450	For 5000 lb. capacity, add					6,225			6,225	6,825
2475	For increased speed, 250 fpm, geared electric, add					6,150			6,150	6,775
2500	300 fpm, geared electric, add					9,200			9,200	10,100
2525	350 fpm, geared electric, add					10,500			10,500	11,600
2550	400 fpm, geared electric, add					13,400			13,400	14,700
2575	500 fpm, gearless electric, add					33,200			33,200	36,500
2600	600 fpm, gearless electric, add					38,300			38,300	42,100
2625	700 fpm, gearless electric, add					43,700			43,700	48,100
2650	800 fpm, gearless electric, add					50,500			50,500	55,500
2675	For travel over 40 V.L.F., add	2 Elev	7.25	2.207	V.L.F.	580	156		736	880
2700	For number of stops over 4, add	"	.27	59.259	Stop	3,875	4,200		8,075	10,800

14 21 33 – Electric Traction Residential Elevators

14 21 33.20 Residential Elevators		Crew	Daily Output	Labor-Hours	Unit	Material	Labor	Equipment	Total	Total Incl O&P
0010	**RESIDENTIAL ELEVATORS**									
7000	Residential, cab type, 1 floor, 2 stop, minimum	2 Elev	.20	80	Ea.	11,000	5,675		16,675	20,900
7100	Maximum		.10	160		18,600	11,300		29,900	38,100
7200	2 floor, 3 stop, minimum		.12	133		16,400	9,450		25,850	32,600
7300	Maximum		.06	266		26,700	18,900		45,600	58,500

14 24 Hydraulic Elevators

14 24 13 – Hydraulic Freight Elevators

14 24 13.10 Hydraulic Freight Elevators and Options		Crew	Daily Output	Labor-Hours	Unit	Material	Labor	Equipment	Total	Total Incl O&P
0010	**HYDRAULIC FREIGHT ELEVATORS AND OPTIONS**									
1025	Hydraulic freight, base unit, 2000 lb., 50 fpm, 2 stop, std. fin.	2 Elev	.10	160	Ea.	76,000	11,300		87,300	101,500
1050	For 2500 lb. capacity, add					4,450			4,450	4,900
1075	For 3000 lb. capacity, add					7,825			7,825	8,600
1100	For 3500 lb. capacity, add					14,100			14,100	15,500
1125	For 4000 lb. capacity, add					15,100			15,100	16,600
1150	For 4500 lb. capacity, add					17,800			17,800	19,600
1175	For 5000 lb. capacity, add					22,400			22,400	24,700
1200	For 6000 lb. capacity, add					27,800			27,800	30,600
1225	For 7000 lb. capacity, add					33,800			33,800	37,100
1250	For 8000 lb. capacity, add					37,200			37,200	40,900
1275	For 10000 lb. capacity, add					46,600			46,600	51,500
1300	For 12000 lb. capacity, add					61,000			61,000	67,000
1325	For 16000 lb. capacity, add					83,500			83,500	92,000

14 24 Hydraulic Elevators

14 24 13 – Hydraulic Freight Elevators

14 24 13.10 Hydraulic Freight Elevators and Options		Crew	Daily Output	Labor-Hours	Unit	2012 Bare Costs Material	Labor	Equipment	Total	Total Incl O&P
1350	For 20000 lb. capacity, add				Ea.	101,500			101,500	112,000
1375	For increased speed, 100 fpm, add					1,950			1,950	2,150
1400	125 fpm, add					4,250			4,250	4,675
1425	150 fpm, add					5,100			5,100	5,625
1450	175 fpm, add					8,650			8,650	9,525
1475	For class "B" loading, add					5,850			5,850	6,425
1500	For class "C-1" loading, add					8,275			8,275	9,100
1525	For class "C-2" loading, add					9,450			9,450	10,400
1550	For class "C-3" loading, add				↓	12,000			12,000	13,200
1575	For travel over 20 V.L.F., add	2 Elev	7.25	2.207	V.L.F.	1,525	156		1,681	1,925
1600	For number of stops over 2, add	"	.27	59.259	Stop	4,125	4,200		8,325	11,100

14 24 23 – Hydraulic Passenger Elevators

14 24 23.10 Hydraulic Passenger Elevators and Options

		Crew	Daily Output	Labor-Hours	Unit	Material	Labor	Equipment	Total	Total Incl O&P
0010	**HYDRAULIC PASSENGER ELEVATORS AND OPTIONS**									
2050	Hyd. pass., base unit, 1500 lb., 100 fpm, 2 stop, std. fin.	2 Elev	.10	160	Ea.	44,400	11,300		55,700	66,500
2075	For 2000 lb. capacity, add					1,750			1,750	1,925
2100	For 2500 lb. capacity, add					2,800			2,800	3,075
2125	For 3000 lb. capacity, add					3,725			3,725	4,075
2150	For 3500 lb. capacity, add					5,675			5,675	6,250
2175	For 4000 lb. capacity, add					7,000			7,000	7,700
2200	For 4500 lb. capacity, add					7,900			7,900	8,700
2225	For 5000 lb. capacity, add					10,500			10,500	11,600
2250	For increased speed, 125 fpm, add					2,575			2,575	2,825
2275	150 fpm, add					3,300			3,300	3,625
2300	175 fpm, add					5,000			5,000	5,500
2325	200 fpm, add				↓	7,025			7,025	7,725
2350	For travel over 12 V.L.F., add	2 Elev	7.25	2.207	V.L.F.	835	156		991	1,150
2375	For number of stops over 2, add		.27	59.259	Stop	2,250	4,200		6,450	8,975
2725	Hydraulic hospital, base unit, 4000 lb., 100 fpm, 2 stop, std. fin.	↓	.10	160	Ea.	56,500	11,300		67,800	80,000
2775	For 4500 lb. capacity, add					6,425			6,425	7,050
2800	For 5000 lb. capacity, add					8,500			8,500	9,350
2825	For increased speed, 125 fpm, add					3,000			3,000	3,325
2850	150 fpm, add					3,900			3,900	4,275
2875	175 fpm, add					5,675			5,675	6,250
2900	200 fpm, add				↓	6,850			6,850	7,550
2925	For travel over 12 V.L.F., add	2 Elev	7.25	2.207	V.L.F.	665	156		821	970
2950	For number of stops over 2, add	"	.27	59.259	Stop	3,875	4,200		8,075	10,800

14 27 Custom Elevator Cabs and Doors

14 27 13 – Custom Elevator Cab Finishes

14 27 13.10 Cab Finishes

		Crew	Daily Output	Labor-Hours	Unit	Material	Labor	Equipment	Total	Total Incl O&P
0010	**CAB FINISHES**									
3325	Passenger elevator cab finishes (based on 3500 lb. cab size)									
3350	Acrylic panel ceiling				Ea.	715			715	785
3375	Aluminum eggcrate ceiling					1,075			1,075	1,175
3400	Stainless steel doors					3,075			3,075	3,375
3425	Carpet flooring					580			580	640
3450	Epoxy flooring					440			440	485
3475	Quarry tile flooring					825			825	905
3500	Slate flooring				↓	1,450			1,450	1,600

14 27 Custom Elevator Cabs and Doors

14 27 13 – Custom Elevator Cab Finishes

14 27 13.10 Cab Finishes		Crew	Daily Output	Labor-Hours	Unit	Material	2012 Bare Costs Labor	Equipment	Total	Total Incl O&P
3525	Textured rubber flooring				Ea.	190			190	208
3550	Stainless steel walls					4,725			4,725	5,200
3575	Stainless steel returns at door				↓	1,075			1,075	1,200
4450	Hospital elevator cab finishes (based on 3500 lb. cab size)									
4475	Aluminum eggcrate ceiling				Ea.	1,050			1,050	1,175
4500	Stainless steel doors					2,575			2,575	2,825
4525	Epoxy flooring					355			355	390
4550	Quarry tile flooring					830			830	910
4575	Textured rubber flooring					630			630	690
4600	Stainless steel walls					4,700			4,700	5,175
4625	Stainless steel returns at door				↓	1,075			1,075	1,200

14 28 Elevator Equipment and Controls

14 28 10 – Elevator Equipment and Control Options

14 28 10.10 Elevator Controls and Doors		Crew	Daily Output	Labor-Hours	Unit	Material	Labor	Equipment	Total	Total Incl O&P
0010	**ELEVATOR CONTROLS AND DOORS**									
2975	Passenger elevator options									
3000	2 car group automatic controls	2 Elev	.66	24.242	Ea.	5,825	1,725		7,550	9,075
3025	3 car group automatic controls		.44	36.364		11,400	2,575		13,975	16,600
3050	4 car group automatic controls		.33	48.485		23,200	3,425		26,625	30,800
3075	5 car group automatic controls		.26	61.538		42,300	4,350		46,650	53,500
3100	6 car group automatic controls		.22	72.727		86,500	5,150		91,650	103,000
3125	Intercom service		3	5.333		1,225	380		1,605	1,925
3150	Duplex car selective collective		.66	24.242		10,700	1,725		12,425	14,500
3175	Center opening 1 speed doors		2	8		1,750	565		2,315	2,800
3200	Center opening 2 speed doors		2	8		2,450	565		3,015	3,575
3225	Rear opening doors (opposite front)		2	8		2,825	565		3,390	3,975
3250	Side opening 2 speed doors		2	8		4,475	565		5,040	5,800
3275	Automatic emergency power switching		.66	24.242		1,100	1,725		2,825	3,875
3300	Manual emergency power switching	↓	8	2		485	142		627	755
3625	Hall finishes, stainless steel doors					1,825			1,825	2,000
3650	Stainless steel frames					1,475			1,425	1,575
3675	12 month maintenance contract								3,600	3,950
3700	Signal devices, hall lanterns	2 Elev	8	2		530	142		672	805
3725	Position indicators, up to 3		9.40	1.702		570	121		691	815
3750	Position indicators, per each over 3	↓	32	.500		540	35.50		575.50	650
3775	High speed heavy duty door opener					3,775			3,775	4,150
3800	Variable voltage, O.H. gearless machine, min.	2 Elev	.16	100		61,500	7,075		68,575	78,500
3815	Maximum		.07	228		92,000	16,200		108,200	126,500
3825	Basement installed geared machine	↓	.33	48.485	↓	63,000	3,425		66,425	74,500
3850	Freight elevator options									
3875	Doors, bi-parting	2 Elev	.66	24.242	Ea.	8,900	1,725		10,625	12,500
3900	Power operated door and gate	"	.66	24.242		22,300	1,725		24,025	27,300
3925	Finishes, steel plate floor					2,200			2,200	2,425
3950	14 ga. 1/4" x 4' steel plate walls					1,825			1,825	2,000
3975	12 month maintenance contract								3,600	3,950
4000	Signal devices, hall lanterns	2 Elev	8	2		525	142		667	795
4025	Position indicators, up to 3		9.40	1.702		580	121		701	825
4050	Position indicators, per each over 3		32	.500		545	35.50		580.50	655
4075	Variable voltage basement installed geared machine	↓	.66	24.242	↓	18,600	1,725		20,325	23,200
4100	Hospital elevator options									

14 28 Elevator Equipment and Controls

14 28 10 – Elevator Equipment and Control Options

14 28 10.10 Elevator Controls and Doors		Crew	Daily Output	Labor-Hours	Unit	2012 Bare Costs Material	Labor	Equipment	Total	Total Incl O&P
4125	2 car group automatic controls	2 Elev	.66	24.242	Ea.	5,825	1,725		7,550	9,075
4150	3 car group automatic controls		.44	36.364		11,400	2,575		13,975	16,500
4175	4 car group automatic controls		.33	48.485		23,100	3,425		26,525	30,700
4200	5 car group automatic controls		.26	61.538		42,100	4,350		46,450	53,000
4225	6 car group automatic controls		.22	72.727		86,000	5,150		91,150	102,500
4250	Intercom service		3	5.333		1,225	380		1,605	1,925
4275	Duplex car selective collective		.66	24.242		10,700	1,725		12,425	14,400
4300	Center opening 1 speed doors		2	8		1,750	565		2,315	2,800
4325	Center opening 2 speed doors		2	8		2,375	565		2,940	3,500
4350	Rear opening doors (opposite front)		2	8		3,400	565		3,965	4,600
4375	Side opening 2 speed doors		2	8		4,950	565		5,515	6,325
4400	Automatic emergency power switching		.66	24.242		1,075	1,725		2,800	3,825
4425	Manual emergency power switching		8	2		475	142		617	740
4675	Hall finishes, stainless steel doors					1,825			1,825	2,000
4700	Stainless steel frames					1,425			1,425	1,575
4725	12 month maintenance contract								3,600	3,950
4750	Signal devices, hall lanterns	2 Elev	8	2		520	142		662	790
4775	Position indicators, up to 3		9.40	1.702		570	121		691	810
4800	Position indicators, per each over 3		32	.500		540	35.50		575.50	650
4825	High speed heavy duty door opener					3,775			3,775	4,150
4850	Variable voltage, O.H. gearless machine, min.	2 Elev	.16	100		61,000	7,075		68,075	78,500
4865	Maximum		.07	228		90,500	16,200		106,700	124,500
4875	Basement installed geared machine		.33	48.485		23,400	3,425		26,825	31,100
5000	Drilling for piston, casing included, 18" diameter	B-48	80	.700	V.L.F.	49.50	27.50	34.50	111.50	137

14 42 Wheelchair Lifts

14 42 13 – Inclined Wheelchair Lifts

14 42 13.10 Inclined Wheelchair Lifts and Stairclimbers		Crew	Daily Output	Labor-Hours	Unit	Material	Labor	Equipment	Total	Total Incl O&P
0010	**INCLINED WHEELCHAIR LIFTS AND STAIRCLIMBERS**									
7700	Stair climber (chair lift), single seat, minimum	2 Elev	1	16	Ea.	5,325	1,125		6,450	7,625
7800	Maximum		.20	80		7,350	5,675		13,025	16,900
8700	Stair lift, minimum		1	16		14,500	1,125		15,625	17,700
8900	Maximum		.20	80		22,900	5,675		28,575	34,000

14 42 16 – Vertical Wheelchair Lifts

14 42 16.10 Wheelchair Lifts		Crew	Daily Output	Labor-Hours	Unit	Material	Labor	Equipment	Total	Total Incl O&P
0010	**WHEELCHAIR LIFTS**									
8000	Wheelchair lift, minimum	2 Elev	1	16	Ea.	7,325	1,125		8,450	9,800
8500	Maximum	"	.50	32	"	17,300	2,275		19,575	22,500

14 45 Vehicle Lifts

14 45 10 – Hydraulic Vehicle Lifts

14 45 10.10 Hydraulic Lifts		Crew	Daily Output	Labor-Hours	Unit	Material	2012 Bare Costs Labor	Equipment	Total	Total Incl O&P
0010	**HYDRAULIC LIFTS**									
0200	Double post, 9000 lb. frame	L-4	1.15	20.870	Ea.	3,350	865		4,215	5,075
0500	Four post, 50,000 lb. frame		.18	133		49,200	5,525		54,725	63,000
2200	Hoists, single post, 8000 lb. capacity, swivel arms		.40	60		5,325	2,475		7,800	9,925
2400	Two posts, adjustable frames, 11,000 lb. capacity		.25	96		6,850	3,975		10,825	14,000
2500	24,000 lb. capacity		.15	160		9,150	6,625		15,775	20,900
2700	7500 lb. capacity, frame supports		.50	48		7,625	1,975		9,600	11,600
2800	Four post, roll on ramp		.50	48		6,850	1,975		8,825	10,800
2810	Hydraulic lifts, above ground, 2 post, clear floor, 6000 lb. cap		2.67	8.989		7,525	370		7,895	8,875
2815	9000 lb. capacity		2.29	10.480		17,800	435		18,235	20,300
2820	15,000 lb. capacity		2	12		20,200	495		20,695	23,000
2825	30,000 lb. capacity		1.60	15		44,600	620		45,220	50,000
2830	4 post, ramp style, 25,000 lb. capacity		2	12		17,400	495		17,895	20,000
2835	35,000 lb. capacity		1	24		81,500	995		82,495	91,000
2840	50,000 lb. capacity		1	24		91,000	995		91,995	101,500
2845	75,000 lb. capacity	↓	1	24		105,500	995		106,495	117,500
2850	For drive thru tracks, add, minimum					1,125			1,125	1,225
2855	Maximum					1,925			1,925	2,100
2860	Ramp extensions, 3′ (set of 2)					920			920	1,000
2865	Rolling jack platform					3,200			3,200	3,525
2870	Elec/hyd jacking beam					8,550			8,550	9,425
2880	Scissor lift, portable, 6000 lb. capacity				↓	8,400			8,400	9,250

14 51 Correspondence and Parcel Lifts

14 51 10 – Electric Correspondence and Parcel Lifts

14 51 10.10 Correspondence Lifts		Crew	Daily Output	Labor-Hours	Unit	Material	Labor	Equipment	Total	Total Incl O&P
0010	**CORRESPONDENCE LIFTS**									
0020	1 floor, 2 stop, 25 lb. capacity, electric	2 Elev	.20	80	Ea.	5,975	5,675		11,650	15,400
0100	Hand, 5 lb. capacity	"	.20	80	"	2,250	5,675		7,925	11,300

14 91 Facility Chutes

14 91 33 – Laundry and Linen Chutes

14 91 33.10 Chutes		Crew	Daily Output	Labor-Hours	Unit	Material	Labor	Equipment	Total	Total Incl O&P
0011	**CHUTES**, linen, trash or refuse									
0050	Aluminized steel, 16 ga., 18″ diameter	2 Shee	3.50	4.571	Floor	1,700	238		1,938	2,250
0100	24″ diameter		3.20	5		1,775	261		2,036	2,350
0400	Galvanized steel, 16 ga., 18″ diameter		3.50	4.571		1,000	238		1,238	1,475
0500	24″ diameter		3.20	5		1,125	261		1,386	1,650
0800	Stainless steel, 18″ diameter		3.50	4.571		3,000	238		3,238	3,675
0900	24″ diameter	↓	3.20	5	↓	3,150	261		3,411	3,850

14 91 82 – Trash Chutes

14 91 82.10 Trash Chutes and Accessories		Crew	Daily Output	Labor-Hours	Unit	Material	Labor	Equipment	Total	Total Incl O&P
0010	**TRASH CHUTES AND ACCESSORIES**									
9000	Minimum labor/equipment charge	1 Shee	1	8	Job		415		415	655

14 92 Pneumatic Tube Systems

14 92 10 – Conventional, Automatic and Computer Controlled Pneumatic Tube Systems

	14 92 10.10 Pneumatic Tube Systems	Crew	Daily Output	Labor-Hours	Unit	2012 Bare Costs Material	Labor	Equipment	Total	Total Incl O&P
0010	**PNEUMATIC TUBE SYSTEMS**									
0020	100′ long, single tube, 2 stations, stock									
0100	3″ diameter	2 Stpi	.12	133	Total	6,600	7,425		14,025	18,900
0300	4″ diameter	″	.09	177	″	7,450	9,900		17,350	23,600
0400	Twin tube, two stations or more, conventional system									
0700	3″ round	2 Stpi	46	.348	L.F.	15.25	19.35		34.60	47
1050	Add for blower		2	8	System	5,200	445		5,645	6,425
1110	Plus for each round station, add		7.50	2.133	Ea.	585	119		704	830
1200	Alternate pricing method: base cost, minimum		.75	21.333	Total	6,375	1,200		7,575	8,875
1300	Maximum		.25	64	″	12,700	3,575		16,275	19,500
1500	Plus total system length, add, minimum		93.40	.171	L.F.	8.20	9.55		17.75	24
1600	Maximum	↓	37.60	.426	″	24.50	23.50		48	63.50

Estimating Tips

Pipe for fire protection and all uses is located in Subdivisions 21 11 13 and 22 11 13.

The labor adjustment factors listed in Subdivision 22 01 02.20 also apply to Division 21.

Many, but not all, areas in the U.S. require backflow protection in the fire system. It is advisable to check local building codes for specific requirements.

For your reference, the following is a list of the most applicable Fire Codes and Standards which may be purchased from the NFPA, 1 Batterymarch Park, Quincy, MA 02169-7471.

NFPA 1: Uniform Fire Code
NFPA 10: Portable Fire Extinguishers
NFPA 11: Low-, Medium-, and High-Expansion Foam
NFPA 12: Carbon Dioxide Extinguishing Systems (Also companion 12A)
NFPA 13: Installation of Sprinkler Systems (Also companion 13D, 13E, and 13R)
NFPA 14: Installation of Standpipe and Hose Systems
NFPA 15: Water Spray Fixed Systems for Fire Protection
NFPA 16: Installation of Foam-Water Sprinkler and Foam-Water Spray Systems
NFPA 17: Dry Chemical Extinguishing Systems (Also companion 17A)
NFPA 18: Wetting Agents
NFPA 20: Installation of Stationary Pumps for Fire Protection
NFPA 22: Water Tanks for Private Fire Protection
NFPA 24: Installation of Private Fire Service Mains and their Appurtenances
NFPA 25: Inspection, Testing and Maintenance of Water-Based Fire Protection

Reference Numbers

Reference numbers are shown in shaded boxes at the beginning of some major classifications. These numbers refer to related items in the Reference Section. The reference information may be an estimating procedure, an alternate pricing method, or technical information.

Note: Not all subdivisions listed here necessarily appear in this publication.

Note: **Trade Service,** *in part, has been used as a reference source for some of the material prices used in Division 21.*

21 05 Common Work Results for Fire Suppression

21 05 23 – General-Duty Valves for Water-Based Fire-Suppression Piping

21 05 23.50 General-Duty Valves		Crew	Daily Output	Labor-Hours	Unit	2012 Bare Costs Material	Labor	Equipment	Total	Total Incl O&P
0010	**GENERAL-DUTY VALVES**, for water-based fire suppression									
6500	Check, swing, C.I. body, brass fittings, auto. ball drip									
6520	4" size	Q-12	3	5.333	Ea.	291	259		550	725
9990	Minimum labor/equipment charge	1 Spri	3	2.667	Job		144		144	225

21 11 Facility Fire-Suppression Water-Service Piping

21 11 16 – Facility Fire Hydrants

21 11 16.50 Fire Hydrants for Buildings

Line	Description	Crew	Daily Output	Labor-Hours	Unit	Material	Labor	Equipment	Total	Total Incl O&P
0010	**FIRE HYDRANTS FOR BUILDINGS**									
3750	Hydrants, wall, w/caps, single, flush, polished brass									
3800	2-1/2" x 2-1/2"	Q-12	5	3.200	Ea.	175	156		331	435
4350	Double, projecting, polished brass									
4400	2-1/2" x 2-1/2" x 4"	Q-12	5	3.200	Ea.	209	156		365	475

21 11 19 – Fire-Department Connections

21 11 19.50 Connections for the Fire-Department

Line	Description	Crew	Daily Output	Labor-Hours	Unit	Material	Labor	Equipment	Total	Total Incl O&P
0010	**CONNECTIONS FOR THE FIRE-DEPARTMENT**									
7140	Standpipe connections, wall, w/plugs & chains									
7280	Double, flush, polished brass									
7300	2-1/2" x 2-1/2" x 4"	Q-12	5	3.200	Ea.	425	156		581	715
9990	Minimum labor/equipment charge	1 Spri	4	2	Job		108		108	169

21 12 Fire-Suppression Standpipes

21 12 13 – Fire-Suppression Hoses and Nozzles

21 12 13.50 Fire Hoses and Nozzles

Line	Description	Crew	Daily Output	Labor-Hours	Unit	Material	Labor	Equipment	Total	Total Incl O&P
0010	**FIRE HOSES AND NOZZLES**									
2200	Hose, less couplings									
2260	Synthetic jacket, lined, 300 lb. test, 1-1/2" diameter	Q-12	2600	.006	L.F.	2.39	.30		2.69	3.10
2280	2-1/2" diameter	"	2200	.007	"	3.98	.35		4.33	4.93
5600	Nozzles, brass									
5620	Adjustable fog, 3/4" booster line				Ea.	87.50			87.50	96
5640	1-1/2" leader line					94			94	104
5660	2-1/2" direct connection				↓	123			123	135
9900	Minimum labor/equipment charge	1 Plum	2	4	Job		218		218	340

21 12 19 – Fire-Suppression Hose Racks

21 12 19.50 Fire Hose Racks

Line	Description	Crew	Daily Output	Labor-Hours	Unit	Material	Labor	Equipment	Total	Total Incl O&P
0010	**FIRE HOSE RACKS**									
2600	Hose rack, swinging, for 1-1/2" diameter hose,									
2620	Enameled steel, 50' & 75' lengths of hose	Q-12	20	.800	Ea.	50	39		89	116

21 12 23 – Fire-Suppression Hose Valves

21 12 23.70 Fire Hose Valves

Line	Description	Crew	Daily Output	Labor-Hours	Unit	Material	Labor	Equipment	Total	Total Incl O&P
0010	**FIRE HOSE VALVES**									
3000	Gate, hose, wheel handle, N.R.S., rough brass, 1-1/2"	1 Spri	12	.667	Ea.	111	36		147	178
3040	2-1/2", 300 lb.	"	7	1.143	"	156	62		218	268

21 13 Fire-Suppression Sprinkler Systems

21 13 13 – Wet-Pipe Sprinkler Systems

21 13 13.50 Wet-Pipe Sprinkler System Components		Crew	Daily Output	Labor-Hours	Unit	2012 Bare Costs Material	Labor	Equipment	Total	Total Incl O&P
0010	**WET-PIPE SPRINKLER SYSTEM COMPONENTS**									
1220	Water motor, complete with gong	1 Spri	4	2	Ea.	350	108		458	555
1800	Firecycle system, controls, includes panel,									
1820	batteries, solenoid valves and pressure switches	Q-13	1	32	Ea.	18,100	1,650		19,750	22,500
1860	Detector	1 Spri	16	.500	"	600	27		627	700
1900	Flexible sprinkler head connectors									
1910	Braided stainless steel hose with mounting bracket									
1920	1/2" and 1" outlet size									
1930	24" length	1 Spri	32	.250	Ea.	25.50	13.50		39	49
1940	36" length		30	.267		29.50	14.40		43.90	55
1950	48" length		26	.308		33.50	16.65		50.15	63
1960	60" length		22	.364		37.50	19.65		57.15	72
1970	72" length	↓	18	.444	↓	41.50	24		65.50	83
1982	May replace hard-pipe armovers									
1984	For wet, pre-action, deluge or dry pipe systems									
2600	Sprinkler heads, not including supply piping									
3700	Standard spray, pendent or upright, brass, 135°F to 286°F									
3740	1/2" NPT, 1/2" orifice	1 Spri	16	.500	Ea.	8.55	27		35.55	51.50
3790	For riser & feeder piping, see Section 22 11 13.00									

21 13 16 – Dry-Pipe Sprinkler Systems

21 13 16.50 Dry-Pipe Sprinkler System Components		Crew	Daily Output	Labor-Hours	Unit	Material	Labor	Equipment	Total	Total Incl O&P
0010	**DRY-PIPE SPRINKLER SYSTEM COMPONENTS**									
0800	Air compressor for dry pipe system, automatic, complete									
0820	280 gal. system capacity, 3/4 HP	1 Spri	1.30	6.154	Ea.	685	335		1,020	1,275
2600	Sprinkler heads, not including supply piping									
2640	Dry, pendent, 1/2" orifice, 3/4" or 1" NPT									
2660	1/2" to 6" length	1 Spri	14	.571	Ea.	105	31		136	163
2700	15-1/4" to 18" length	"	14	.571		123	31		154	183
8200	Dry pipe valve, incl. trim and gauges, 3" size	Q-12	2	8		2,250	390		2,640	3,075
8220	4" size	"	1	16	↓	2,450	780		3,230	3,900

21 13 39 – Foam-Water Systems

21 13 39.50 Foam-Water System Components		Crew	Daily Output	Labor-Hours	Unit	Material	Labor	Equipment	Total	Total Incl O&P
0010	**FOAM-WATER SYSTEM COMPONENTS**									
2600	Sprinkler heads, not including supply piping									
3600	Foam-water, pendent or upright, 1/2" NPT	1 Spri	12	.667	Ea.	157	36		193	228

Division Notes

		DAILY	LABOR-		2012 BARE COSTS				TOTAL
	CREW	OUTPUT	HOURS	UNIT	MAT.	LABOR	EQUIP.	TOTAL	INCL O&P

Estimating Tips

22 10 00 Plumbing Piping and Pumps

This subdivision is primarily basic pipe and related materials. The pipe may be used by any of the mechanical disciplines, i.e., plumbing, fire protection, heating, and air conditioning.

Note: CPVC plastic piping approved for tire protection is located in 21 11 13.

- The labor adjustment factors listed in Subdivision 22 01 02.20 apply throughout Divisions 21, 22, and 23. CAUTION: the correct percentage may vary for the same items. For example, the percentage add for the basic pipe installation should be based on the maximum height that the craftsman must install for that particular section. If the pipe is to be located 14′ above the floor but it is suspended on threaded rod from beams, the bottom flange of which is 18′ high (4′ rods), then the height is actually 18′ and the add is 20%. The pipe coverer, however, does not have to go above the 14′, and so his or her add should be 10%.
- Most pipe is priced first as straight pipe with a joint (coupling, weld, etc.) every 10′ and a hanger usually every 10′. There are exceptions with hanger spacing such as for cast iron pipe (5′) and plastic pipe (3 per 10′). Following each type of pipe there are several lines listing sizes and the amount to be subtracted to delete couplings and hangers. This is for pipe that is to be buried or supported together on trapeze hangers. The reason that the couplings are deleted is that these runs are usually long, and frequently longer lengths of pipe are used. By deleting the couplings, the estimator is expected to look up and add back the correct reduced number of couplings.
- When preparing an estimate, it may be necessary to approximate the fittings. Fittings usually run between 25% and 50% of the cost of the pipe. The lower percentage is for simpler runs, and the higher number is for complex areas, such as mechanical rooms.
- For historic restoration projects, the systems must be as invisible as possible, and pathways must be sought for pipes, conduit, and ductwork. While installations in accessible spaces (such as basements and attics) are relatively straightforward to estimate, labor costs may be more difficult to determine when delivery systems must be concealed.

22 40 00 Plumbing Fixtures

- Plumbing fixture costs usually require two lines: the fixture itself and its "rough-in, supply, and waste."
- In the Assemblies Section (Plumbing D2010) for the desired fixture, the System Components Group at the center of the page shows the fixture on the first line. The rest of the list (fittings, pipe, tubing, etc.) will total up to what we refer to in the Unit Price section as "Rough-in, supply, waste, and vent." Note that for most fixtures we allow a nominal 5′ of tubing to reach from the fixture to a main or riser.
- Remember that gas- and oil-fired units need venting.

Reference Numbers

Reference numbers are shown in shaded boxes at the beginning of some major classifications. These numbers refer to related items in the Reference Section. The reference information may be an estimating procedure, an alternate pricing method, or technical information.

Note: Not all subdivisions listed here necessarily appear in this publication.

Note: **Trade Service,** *in part, has been used as a reference source for some of the material prices used in Division 22.*

22 01 Operation and Maintenance of Plumbing

22 01 02 – Labor Adjustments

22 01 02.20 Labor Adjustment Factors			Crew	Daily Output	Labor-Hours	Unit	Material	2012 Bare Costs Labor	Equipment	Total	Total Incl O&P
0010	**LABOR ADJUSTMENT FACTORS**, (For Div. 21, 22 and 23)	R220102-20									
1000	Add to labor for elevated installation (Above floor level)										
1080	10′ to 14.5′ high							10%			
1100	15′ to 19.5′ high							20%			
1120	20′ to 24.5′ high							25%			
1140	25′ to 29.5′ high							35%			
1160	30′ to 34.5′ high							40%			
1180	35′ to 39.5′ high							50%			
1200	40′ and higher							55%			
2000	Add to labor for crawl space										
2100	3′ high							40%			
2140	4′ high							30%			
3000	Add to labor for multi-story building										
3100	Add per floor for floors 3 thru 19							2%			
3140	Add per floor for floors 20 and up							4%			
4000	Add to labor for working in existing occupied buildings										
4100	Hospital							35%			
4140	Office building							25%			
4180	School							20%			
4220	Factory or warehouse							15%			
4260	Multi dwelling							15%			
5000	Add to labor, miscellaneous										
5100	Cramped shaft							35%			
5140	Congested area							15%			
5180	Excessive heat or cold							30%			
9000	Labor factors, The above are reasonable suggestions, however										
9010	each project should be evaluated for its own peculiarities.										
9100	Other factors to be considered are:										
9140	Movement of material and equipment through finished areas										
9180	Equipment room										
9220	Attic space										
9260	No service road										
9300	Poor unloading/storage area										
9340	Congested site area/heavy traffic										

22 05 Common Work Results for Plumbing

22 05 05 – Selective Plumbing Demolition

22 05 05.10 Plumbing Demolition			Crew	Daily Output	Labor-Hours	Unit	Material	Labor	Equipment	Total	Total Incl O&P
0010	**PLUMBING DEMOLITION**	R220105-10									
1020	Fixtures, including 10′ piping										
1100	Bathtubs, cast iron	R024119-10	1 Plum	4	2	Ea.		109		109	170
1120	Fiberglass			6	1.333			73		73	113
1140	Steel			5	1.600			87.50		87.50	136
1200	Lavatory, wall hung			10	.800			43.50		43.50	68
1220	Counter top			8	1			54.50		54.50	85
1300	Sink, single compartment			8	1			54.50		54.50	85
1320	Double compartment			7	1.143			62.50		62.50	97
1400	Water closet, floor mounted			8	1			54.50		54.50	85
1420	Wall mounted			7	1.143			62.50		62.50	97
1500	Urinal, floor mounted			4	2			109		109	170
1520	Wall mounted		↓	7	1.143	↓		62.50		62.50	97

22 05 Common Work Results for Plumbing

22 05 05 – Selective Plumbing Demolition

22 05 05.10 Plumbing Demolition		Crew	Daily Output	Labor-Hours	Unit	Material	2012 Bare Costs Labor	Equipment	Total	Total Incl O&P
1600	Water fountains, free standing	1 Plum	8	1	Ea.		54.50		54.50	85
1620	Wall or deck mounted		6	1.333	↓		73		73	113
2000	Piping, metal, up thru 1-1/2" diameter		200	.040	L.F.		2.18		2.18	3.40
2050	2" thru 3-1/2" diameter	↓	150	.053			2.91		2.91	4.54
2100	4" thru 6" diameter	2 Plum	100	.160			8.75		8.75	13.60
2150	8" thru 14" diameter	"	60	.267			14.55		14.55	22.50
2153	16" thru 20" diameter	Q-18	70	.343			17.80	.53	18.33	28.50
2155	24" thru 26" diameter		55	.436			22.50	.67	23.17	36
2156	30" thru 36" diameter	↓	40	.600	↓		31	.92	31.92	49.50
2250	Water heater, 40 gal.	1 Plum	6	1.333	Ea.		73		73	113
6000	Remove and reset fixtures, minimum		6	1.333			73		73	113
6100	Maximum		4	2	↓		109		109	170
9000	Minimum labor/equipment charge	↓	2	4	Job		218		218	340

22 05 23 – General-Duty Valves for Plumbing Piping

22 05 23.20 Valves, Bronze		Crew	Daily Output	Labor-Hours	Unit	Material	Labor	Equipment	Total	Total Incl O&P
0010	**VALVES, BRONZE**									
1020	Angle, 150 lb., rising stem, threaded									
1070	3/4"	1 Plum	20	.400	Ea.	162	22		184	212
1080	1"		19	.421		234	23		257	293
1100	1-1/2"		13	.615		360	33.50		393.50	450
1110	2"	↓	11	.727	↓	630	39.50		669.50	755
1300	Ball									
1398	150 psi, threaded									
1460	3/4"	1 Plum	20	.400	Ea.	21	22		43	57
1750	Check, swing, class 150, regrinding disc, threaded									
1850	1/2"	1 Plum	24	.333	Ea.	107	18.20		125.20	147
1860	3/4"		20	.400		142	22		164	190
1870	1"		19	.421		145	23		168	195
1890	1-1/2"		13	.615		292	33.50		325.50	375
1900	2"	↓	11	.727	↓	400	39.50		439.50	495
2850	Gate, N.R.S., soldered, 125 psi									
2920	1/2"	1 Plum	24	.333	Ea.	41	18.20		59.20	73.50
2940	3/4"		20	.400		47	22		69	85.50
2950	1"		19	.421		66.50	23		89.50	109
2970	1-1/2"		13	.615		114	33.50		147.50	178
2980	2"	↓	11	.727	↓	161	39.50		200.50	239
4850	Globe, class 150, rising stem, threaded									
4950	1/2"	1 Plum	24	.333	Ea.	88.50	18.20		106.70	126
4960	3/4"	"	20	.400	"	119	22		141	165
5600	Relief, pressure & temperature, self-closing, ASME, threaded									
5640	3/4"	1 Plum	28	.286	Ea.	142	15.60		157.60	181
5650	1"		24	.333		217	18.20		235.20	267
5660	1-1/4"		20	.400		420	22		442	495
5670	1-1/2"		18	.444		800	24.50		824.50	920
5680	2"	↓	16	.500	↓	870	27.50		897.50	1,000
6400	Pressure, water, ASME, threaded									
6440	3/4"	1 Plum	28	.286	Ea.	80	15.60		95.60	113
6450	1"		24	.333		170	18.20		188.20	216
6470	1-1/2"	↓	18	.444	↓	375	24.50		399.50	450
6900	Reducing, water pressure									
6940	1/2"	1 Plum	24	.333	Ea.	257	18.20		275.20	310
6960	1"	"	19	.421	"	400	23		423	475

22 05 Common Work Results for Plumbing

22 05 23 – General-Duty Valves for Plumbing Piping

22 05 23.20 Valves, Bronze		Crew	Daily Output	Labor-Hours	Unit	2012 Bare Costs Material	Labor	Equipment	Total	Total Incl O&P
8350	Tempering, water, sweat connections									
8400	1/2"	1 Plum	24	.333	Ea.	85	18.20		103.20	122
8440	3/4"	"	20	.400	"	104	22		126	149
8650	Threaded connections									
8700	1/2"	1 Plum	24	.333	Ea.	128	18.20		146.20	170
8740	3/4"		20	.400	"	475	22		497	560
9000	Minimum labor/equipment charge	↓	4	2	Job		109		109	170

22 05 23.60 Valves, Plastic

		Crew	Daily Output	Labor-Hours	Unit	Material	Labor	Equipment	Total	Total Incl O&P
0010	**VALVES, PLASTIC**									
0020										
1150	Ball, PVC, socket or threaded, single union									
1280	2"	1 Plum	17	.471	Ea.	89	25.50		114.50	138
1650	CPVC, socket or threaded, single union									
1770	2"	1 Plum	17	.471	Ea.	176	25.50		201.50	233
3150	Ball check, PVC, socket or threaded									
3290	2"	1 Plum	17	.471	Ea.	133	25.50		158.50	187
9000	Minimum labor/equipment charge	"	3.50	2.286	Job		125		125	194

22 05 76 – Facility Drainage Piping Cleanouts

22 05 76.10 Cleanouts

		Crew	Daily Output	Labor-Hours	Unit	Material	Labor	Equipment	Total	Total Incl O&P
0010	**CLEANOUTS**									
0060	Floor type									
0080	Round or square, scoriated nickel bronze top									
0100	2" pipe size	1 Plum	10	.800	Ea.	167	43.50		210.50	252
0140	4" pipe size	"	6	1.333	"	251	73		324	390
0980	Round top, recessed for terrazzo									
1000	2" pipe size	1 Plum	9	.889	Ea.	167	48.50		215.50	260
1100	4" pipe size	"	4	2		251	109		360	445
1120	5" pipe size	Q-1	6	2.667	↓	320	131		451	555
9000	Minimum labor/equipment charge	1 Plum	3	2.667	Job		146		146	227

22 05 76.20 Cleanout Tees

		Crew	Daily Output	Labor-Hours	Unit	Material	Labor	Equipment	Total	Total Incl O&P
0010	**CLEANOUT TEES**									
0100	Cast iron, B&S, with countersunk plug									
0220	3" pipe size	1 Plum	3.60	2.222	Ea.	248	121		369	460
0240	4" pipe size	"	3.30	2.424	"	310	132		442	545
0500	For round smooth access cover, same price									
4000	Plastic, tees and adapters. Add plugs									
4010	ABS, DWV									
4020	Cleanout tee, 1-1/2" pipe size	1 Plum	15	.533	Ea.	10.40	29		39.40	57
9000	Minimum labor/equipment charge	"	2.75	2.909	Job		159		159	247

22 07 Plumbing Insulation

22 07 16 – Plumbing Equipment Insulation

22 07 16.10 Insulation for Plumbing Equipment			Crew	Daily Output	Labor-Hours	Unit	2012 Bare Costs Material	Labor	Equipment	Total	Total Incl O&P
0010	**INSULATION FOR PLUMBING EQUIPMENT**										
2900	Domestic water heater wrap kit										
2920	1-1/2" with vinyl jacket, 20-60 gal.	G	1 Plum	8	1	Ea.	17.60	54.50		72.10	104
2925	50 to 80 gallons	G	"	8	1	"	21	54.50		75.50	108

22 07 19 – Plumbing Piping Insulation

22 07 19.10 Piping Insulation

22 07 19.10 Piping Insulation			Crew	Daily Output	Labor-Hours	Unit	Material	Labor	Equipment	Total	Total Incl O&P
0010	**PIPING INSULATION**										
0100	Rule of thumb, as a percentage of total mechanical costs					Job				10%	
0110	Insulation req'd is based on the surface size/area to be covered										
0230	Insulated protectors, (ADA)										
0235	For exposed piping under sinks or lavatories										
0240	Vinyl coated foam, velcro tabs										
0245	P Trap, 1-1/4" or 1-1/2"		1 Plum	32	.250	Ea.	20.50	13.65		34.15	44
0260	Valve and supply cover										
0265	1/2", 3/8", and 7/16" pipe size		1 Plum	32	.250	Ea.	19.35	13.65		33	43
0280	Tailpiece offset (wheelchair)										
0285	1-1/4" pipe size		1 Plum	32	.250	Ea.	16.55	13.65		30.20	39.50
0600	Pipe covering (price copper tube one size less than IPS)										
6120	2" wall, 1/2" iron pipe size	G	Q-14	145	.110	L.F.	4.67	4.85		9.52	12.95
6210	4" iron pipe size	G	"	125	.128	"	8.50	5.65		14.15	18.40
6600	Fiberglass, with all service jacket										
6840	1" wall, 1/2" iron pipe size	G	Q-14	240	.067	L.F.	.89	2.93		3.82	5.70
6860	3/4" iron pipe size	G		230	.070		.97	3.06		4.03	6
6870	1" iron pipe size	G		220	.073		1.04	3.20		4.24	6.30
6880	1-1/4" iron pipe size	G		210	.076		1.13	3.35		4.48	6.65
6890	1-1/2" iron pipe size	G		210	.076		1.22	3.35		4.57	6.75
6900	2" iron pipe size	G		200	.080		1.32	3.52		4.84	7.10
7080	1-1/2" wall, 1/2" iron pipe size	G		230	.070		1.69	3.06		4.75	6.80
7140	2" iron pipe size	G		190	.084		2.28	3.70		5.98	8.45
7160	3" iron pipe size	G		170	.094		2.56	4.14		6.70	9.45
7879	Rubber tubing, flexible closed cell foam										
8100	1/2" wall, 1/4" iron pipe size	G	1 Asbe	90	.089	L.F.	.57	4.34		4.91	7.65
8120	3/8" iron pipe size	G		90	.089		.64	4.34		4.98	7.70
8130	1/2" iron pipe size	G		89	.090		.71	4.39		5.10	7.85
8140	3/4" iron pipe size	G		89	.090		.79	4.39		5.18	7.90
8150	1" iron pipe size	G		88	.091		.87	4.44		5.31	8.10
8170	1-1/2" iron pipe size	G		87	.092		1.21	4.49		5.70	8.55
8180	2" iron pipe size	G		86	.093		1.55	4.54		6.09	9
8200	3" iron pipe size	G		85	.094		2.18	4.60		6.78	9.80
8220	4" iron pipe size	G		80	.100		3.24	4.89		8.13	11.40
8300	3/4" wall, 1/4" iron pipe size	G		90	.089		.90	4.34		5.24	8
8330	1/2" iron pipe size	G		89	.090		1.17	4.39		5.56	8.35
8340	3/4" iron pipe size	G		89	.090		1.43	4.39		5.82	8.60
8350	1" iron pipe size	G		88	.091		1.63	4.44		6.07	8.95
8360	1-1/4" iron pipe size	G		87	.092		2.17	4.49		6.66	9.60
8370	1-1/2" iron pipe size	G		87	.092		2.46	4.49		6.95	9.90
8380	2" iron pipe size	G		86	.093		2.84	4.54		7.38	10.40
8444	1" wall, 1/2" iron pipe size	G		86	.093		2.17	4.54		6.71	9.70
8445	3/4" iron pipe size	G		84	.095		2.63	4.65		7.28	10.40
8446	1" iron pipe size	G		84	.095		3.06	4.65		7.71	10.85
8447	1-1/4" iron pipe size	G		82	.098		3.42	4.77		8.19	11.40
8448	1-1/2" iron pipe size	G		82	.098		3.98	4.77		8.75	12.05

22 07 Plumbing Insulation

22 07 19 – Plumbing Piping Insulation

22 07 19.10 Piping Insulation		Crew	Daily Output	Labor-Hours	Unit	Material	Labor (2012 Bare Costs)	Equipment	Total	Total Incl O&P
8449	2" iron pipe size [G]	1 Asbe	80	.100	L.F.	5.25	4.89		10.14	13.60
8450	2-1/2" iron pipe size [G]		80	.100		6.85	4.89		11.74	15.35
8456	Rubber insulation tape, 1/8" x 2" x 30' [G]				Ea.	12.05			12.05	13.25
9600	Minimum labor/equipment charge	1 Plum	4	2	Job		109		109	170

22 11 Facility Water Distribution

22 11 13 – Facility Water Distribution Piping

22 11 13.14 Pipe, Brass

Line	Description	Crew	Daily Output	Labor-Hours	Unit	Material	Labor	Equipment	Total	Total Incl O&P
0010	**PIPE, BRASS**, Plain end									
0900	Field threaded, coupling & clevis hanger assembly 10' O.C.									
0920	Regular weight									
1120	1/2" diameter	1 Plum	48	.167	L.F.	6.35	9.10		15.45	21
1140	3/4" diameter		46	.174		8.45	9.50		17.95	24
1160	1" diameter		43	.186		12.25	10.15		22.40	29.50
1180	1-1/4" diameter	Q-1	72	.222		18.65	10.90		29.55	37.50
1200	1-1/2" diameter		65	.246		22.50	12.10		34.60	43.50
1220	2" diameter		53	.302		31	14.85		45.85	57.50
9000	Minimum labor/equipment charge	1 Plum	4	2	Job		109		109	170

22 11 13.16 Pipe Fittings, Brass

Line	Description	Crew	Daily Output	Labor-Hours	Unit	Material	Labor	Equipment	Total	Total Incl O&P
0010	**PIPE FITTINGS, BRASS**, Rough bronze, threaded.									
0020										
1000	Standard wt., 90° Elbow									
1100	1/2"	1 Plum	12	.667	Ea.	16.60	36.50		53.10	75
1120	3/4"		11	.727		22	39.50		61.50	86.50
1140	1"		10	.800		36	43.50		79.50	108
1160	1-1/4"	Q-1	17	.941		58	46.50		104.50	136
1180	1-1/2"	"	16	1		72	49		121	156
1500	45° Elbow, 1/8"	1 Plum	13	.615		20.50	33.50		54	75
1580	1/2"		12	.667		20.50	36.50		57	79
1600	3/4"		11	.727		29	39.50		68.50	94
1620	1"		10	.800		49.50	43.50		93	123
1640	1-1/4"	Q-1	17	.941		78.50	46.50		125	158
1660	1-1/2"	"	16	1		98.50	49		147.50	185
2000	Tee, 1/8"	1 Plum	9	.889		22	48.50		70.50	99.50
2080	1/2"		8	1		22	54.50		76.50	109
2100	3/4"		7	1.143		27.50	62.50		90	128
2120	1"		6	1.333		50	73		123	168
2140	1-1/4"	Q-1	10	1.600		86	78.50		164.50	218
2160	1-1/2"	"	9	1.778		97	87.50		184.50	243
2500	Coupling, 1/8"	1 Plum	26	.308		13.85	16.80		30.65	41.50
2580	1/2"		15	.533		13.85	29		42.85	61
2600	3/4"		14	.571		19.45	31		50.45	70
2620	1"		13	.615		33	33.50		66.50	89
2640	1-1/4"	Q-1	22	.727		55.50	36		91.50	117
2660	1-1/2"		20	.800		72	39.50		111.50	141
2680	2"		18	.889		119	43.50		162.50	199
9000	Minimum labor/equipment charge	1 Plum	4	2	Job		109		109	170

22 11 13.23 Pipe/Tube, Copper

Line	Description	Crew	Daily Output	Labor-Hours	Unit	Material	Labor	Equipment	Total	Total Incl O&P
0010	**PIPE/TUBE, COPPER**, Solder joints									
1000	Type K tubing, couplings & clevis hanger assemblies 10' O.C.									

22 11 13 – Facility Water Distribution Piping

22 11 13.23 Pipe/Tube, Copper		Crew	Daily Output	Labor-Hours	Unit	2012 Bare Costs Material	Labor	Equipment	Total	Total Incl O&P
1100	1/4" diameter	1 Plum	84	.095	L.F.	4.94	5.20		10.14	13.55
1200	1" diameter		66	.121		12.85	6.60		19.45	24.50
1260	2" diameter		40	.200		32	10.90		42.90	52
2000	Type L tubing, couplings & clevis hanger assemblies 10' O.C.									
2100	1/4" diameter	1 Plum	88	.091	L.F.	2.69	4.96		7.65	10.70
2120	3/8" diameter		84	.095		3.73	5.20		8.93	12.20
2140	1/2" diameter		81	.099		4.30	5.40		9.70	13.15
2160	5/8" diameter		79	.101		6.25	5.55		11.80	15.50
2180	3/4" diameter		76	.105		6.60	5.75		12.35	16.20
2200	1" diameter		68	.118		9.85	6.40		16.25	21
2220	1-1/4" diameter		58	.138		13.45	7.55		21	26.50
2240	1-1/2" diameter		52	.154		17.25	8.40		25.65	32
2260	2" diameter		42	.190		27	10.40		37.40	46
2280	2-1/2" diameter	Q-1	62	.258		42	12.70		54.70	66
2300	3" diameter		56	.286		57	14.05		71.05	85
2320	3-1/2" diameter		43	.372		81	18.30		99.30	118
2340	4" diameter		39	.410		99	20		119	141
2360	5" diameter		34	.471		211	23		234	268
2380	6" diameter	Q-2	40	.600		300	30.50		330.50	380
2410	For other than full hard temper, add					21%				
2590	For silver solder, add						15%			
3000	Type M tubing, couplings & clevis hanger assemblies 10' O.C.									
3140	1/2" diameter	1 Plum	84	.095	L.F.	3.31	5.20		8.51	11.75
3180	3/4" diameter		78	.103		5.05	5.60		10.65	14.25
3200	1" diameter		70	.114		7.65	6.25		13.90	18.15
4000	Type DWV tubing, couplings & clevis hanger assemblies 10' O.C.									
4100	1-1/4" diameter	1 Plum	60	.133	L.F.	11.95	7.30		19.25	24.50
4120	1-1/2" diameter		54	.148		14.95	8.10		23.05	29
4140	2" diameter		44	.182		21	9.95		30.95	38.50
4160	3" diameter	Q-1	58	.276		39	13.55		52.55	63.50
4180	4" diameter		40	.400		70	19.65		89.65	108
4200	5" diameter		36	.444		191	22		213	244
4220	6" diameter	Q-2	42	.571		280	29		309	355
9000	Minimum labor/equipment charge	1 Plum	4	2	Job		109		109	170

22 11 13.25 Pipe/Tube Fittings, Copper		Crew	Daily Output	Labor-Hours	Unit	Material	Labor	Equipment	Total	Total Incl O&P
0010	**PIPE/TUBE FITTINGS, COPPER**, Wrought unless otherwise noted									
0040	Solder joints, copper x copper									
0070	90° elbow, 1/4"	1 Plum	22	.364	Ea.	8.95	19.85		28.80	41
0100	1/2"		20	.400		2.84	22		24.84	37
0120	3/4"		19	.421		6.40	23		29.40	43
0130	1"		16	.500		15.70	27.50		43.20	60
0250	45° elbow, 1/4"		22	.364		15.95	19.85		35.80	48.50
0270	3/8"		22	.364		13.70	19.85		33.55	46
0280	1/2"		20	.400		5.25	22		27.25	40
0290	5/8"		19	.421		26	23		49	64.50
0300	3/4"		19	.421		9.15	23		32.15	46
0310	1"		16	.500		23	27.50		50.50	68
0320	1-1/4"		15	.533		31	29		60	79.50
0330	1-1/2"		13	.615		37.50	33.50		71	93.50
0340	2"		11	.727		62.50	39.50		102	131
0350	2-1/2"	Q-1	13	1.231		134	60.50		194.50	242
0360	3"		13	1.231		197	60.50		257.50	310

22 11 Facility Water Distribution

22 11 13 – Facility Water Distribution Piping

22 11 13.25 Pipe/Tube Fittings, Copper		Crew	Daily Output	Labor-Hours	Unit	2012 Bare Costs Material	Labor	Equipment	Total	Total Incl O&P
0370	3-1/2"	Q-1	10	1.600	Ea.	320	78.50		398.50	475
0380	4"		9	1.778		420	87.50		507.50	600
0390	5"		6	2.667		1,625	131		1,756	2,000
0400	6"	Q-2	9	2.667		2,575	136		2,711	3,025
0450	Tee, 1/4"	1 Plum	14	.571		17.90	31		48.90	68
0470	3/8"		14	.571		14.40	31		45.40	64.50
0480	1/2"		13	.615		4.88	33.50		38.38	58
0490	5/8"		12	.667		31.50	36.50		68	91
0500	3/4"		12	.667		11.70	36.50		48.20	69.50
0510	1"		10	.800		36	43.50		79.50	108
0520	1-1/4"		9	.889		49.50	48.50		98	130
0530	1-1/2"		8	1		76.50	54.50		131	169
0540	2"		7	1.143		119	62.50		181.50	228
0550	2-1/2"	Q-1	8	2		240	98.50		338.50	415
0560	3"		7	2.286		365	112		477	580
0570	3-1/2"		6	2.667		1,050	131		1,181	1,375
0580	4"		5	3.200		885	157		1,042	1,225
0590	5"		4	4		2,900	197		3,097	3,500
0600	6"	Q-2	6	4		3,975	204		4,179	4,675
0612	Tee, reducing on the outlet, 1/4"	1 Plum	15	.533		30.50	29		59.50	79
0613	3/8"		15	.533		28.50	29		57.50	77
0614	1/2"		14	.571		25	31		56	76
0615	5/8"		13	.615		48	33.50		81.50	106
0616	3/4"		12	.667		13.20	36.50		49.70	71
0617	1"		11	.727		44.50	39.50		84	111
0618	1-1/4"		10	.800		53.50	43.50		97	127
0619	1-1/2"		9	.889		57	48.50		105.50	138
0620	2"		8	1		92.50	54.50		147	187
0621	2-1/2"	Q-1	9	1.778		280	87.50		367.50	445
0622	3"		8	2		310	98.50		408.50	495
0623	4"		6	2.667		590	131		721	855
0624	5"		5	3.200		2,900	157		3,057	3,450
0625	6"	Q-2	7	3.429		3,775	175		3,950	4,425
0626	8"	"	6	4		15,300	204		15,504	17,100
0630	Tee, reducing on the run, 1/4"	1 Plum	15	.533		38	29		67	87.50
0631	3/8"		15	.533		51	29		80	102
0632	1/2"		14	.571		34.50	31		65.50	86.50
0633	5/8"		13	.615		51	33.50		84.50	109
0634	3/4"		12	.667		13.80	36.50		50.30	71.50
0635	1"		11	.727		43	39.50		82.50	110
0636	1-1/4"		10	.800		68.50	43.50		112	144
0637	1-1/2"		9	.889		114	48.50		162.50	201
0638	2"		8	1		161	54.50		215.50	262
0639	2-1/2"	Q-1	9	1.778		360	87.50		447.50	535
0640	3"		8	2		535	98.50		633.50	740
0641	4"		6	2.667		1,175	131		1,306	1,500
0642	5"		5	3.200		2,750	157		2,907	3,275
0643	6"	Q-2	7	3.429		4,175	175		4,350	4,875
0644	8"	"	6	4		15,300	204		15,504	17,100
0650	Coupling, 1/4"	1 Plum	24	.333		2.07	18.20		20.27	31
0670	3/8"		24	.333		2.75	18.20		20.95	31.50
0680	1/2"		22	.364		2.15	19.85		22	33.50
0690	5/8"		21	.381		6.50	21		27.50	39.50

22 11 13.25 Pipe/Tube Fittings, Copper		Crew	Daily Output	Labor-Hours	Unit	Material	Labor	Equipment	Total	Total Incl O&P
						2012 Bare Costs				
0700	3/4″	1 Plum	21	.381	Ea.	4.33	21		25.33	37.50
0710	1″		18	.444		8.65	24.50		33.15	47.50
0715	1-1/4″		17	.471		15.30	25.50		40.80	57
0716	1-1/2″		15	.533		20.50	29		49.50	68
0718	2″		13	.615		34	33.50		67.50	89.50
0721	2-1/2″	Q-1	15	1.067		72	52.50		124.50	161
0722	3″		13	1.231		108	60.50		168.50	214
0724	3-1/2″		8	2		207	98.50		305.50	380
0726	4″		7	2.286		227	112		339	425
0728	5″		6	2.667		560	131		691	820
0731	6″	Q-2	8	3		925	153		1,078	1,275
2000	DWV, solder joints, copper x copper									
2030	90° Elbow, 1-1/4″	1 Plum	13	.615	Ea.	27.50	33.50		61	83
2050	1-1/2″		12	.667		41	36.50		77.50	102
2070	2″		10	.800		54	43.50		97.50	127
2090	3″	Q-1	10	1.600		133	78.50		211.50	269
2100	4″	″	9	1.778		715	87.50		802.50	920
2250	Tee, Sanitary, 1-1/4″	1 Plum	9	.889		54.50	48.50		103	135
2270	1-1/2″		8	1		67.50	54.50		122	160
2290	2″		7	1.143		78.50	62.50		141	183
2310	3″	Q-1	7	2.286		320	112		432	530
2330	4″	″	6	2.667		815	131		946	1,100
2400	Coupling, 1-1/4″	1 Plum	14	.571		12.90	31		43.90	62.50
2420	1-1/2″		13	.615		16	33.50		49.50	70
2440	2″		11	.727		22	39.50		61.50	86.50
2460	3″	Q-1	11	1.455		43	71.50		114.50	159
2480	4″	″	10	1.600		137	78.50		215.50	274
9000	Minimum labor/equipment charge	1 Plum	4	2	Job		109		109	170

22 11 13.44 Pipe, Steel		Crew	Daily Output	Labor-Hours	Unit	Material	Labor	Equipment	Total	Total Incl O&P
0010	**PIPE, STEEL** R221113-50									
0020	All pipe sizes are to Spec. A-53 unless noted otherwise									
0050	Schedule 40, threaded, with couplings, and clevis hanger									
0060	assemblies sized for covering, 10′ O.C.									
0540	Black, 1/4″ diameter	1 Plum	66	.121	L.F.	6	6.60		12.60	16.90
0560	1/2″ diameter		63	.127		4.38	6.95		11.33	15.60
0570	3/4″ diameter		61	.131		5.15	7.15		12.30	16.80
0580	1″ diameter		53	.151		7.55	8.25		15.80	21
0590	1-1/4″ diameter	Q-1	89	.180		9.50	8.85		18.35	24
0600	1-1/2″ diameter		80	.200		11.15	9.85		21	27.50
0610	2″ diameter		64	.250		14.90	12.30		27.20	35.50
0620	2-1/2″ diameter		50	.320		23	15.75		38.75	49.50
0630	3″ diameter		43	.372		29.50	18.30		47.80	61
0640	3-1/2″ diameter		40	.400		40	19.65		59.65	74.50
0650	4″ diameter		36	.444		43.50	22		65.50	81.50
1280	All pipe sizes are to Spec. A-53 unless noted otherwise									
1281	Schedule 40, threaded, with couplings and clevis hanger									
1282	assemblies sized for covering, 10′ O. C.									
1290	Galvanized, 1/4″ diameter	1 Plum	66	.121	L.F.	8.10	6.60		14.70	19.20
1310	1/2″ diameter		63	.127		6.35	6.95		13.30	17.80
1320	3/4″ diameter		61	.131		7.60	7.15		14.75	19.50
1330	1″ diameter		53	.151		10.40	8.25		18.65	24.50
1350	1-1/2″ diameter	Q-1	80	.200		15.65	9.85		25.50	32.50

22 11 13.44 Pipe, Steel

	22 11 13.44 Pipe, Steel	Crew	Daily Output	Labor-Hours	Unit	Material	2012 Bare Costs Labor	Equipment	Total	Total Incl O&P
1360	2″ diameter	Q-1	64	.250	L.F.	21	12.30		33.30	42
1370	2-1/2″ diameter		50	.320		34.50	15.75		50.25	62.50
1380	3″ diameter		43	.372		44	18.30		62.30	77
1400	4″ diameter		36	.444		63.50	22		85.50	104
2000	Welded, sch. 40, on yoke & roll hanger assy's, sized for covering, 10′ O.C.									
2040	Black, 1″ diameter	Q-15	93	.172	L.F.	6.80	8.45	.40	15.65	21
2070	2″ diameter		61	.262		12.35	12.90	.61	25.86	34.50
2090	3″ diameter		43	.372		25.50	18.30	.86	44.66	58
2110	4″ diameter		37	.432		35	21.50	1	57.50	72.50
2120	5″ diameter		32	.500		38	24.50	1.16	63.66	81.50
2130	6″ diameter	Q-16	36	.667		50.50	34	1.03	85.53	110
9990	Minimum labor/equipment charge	1 Plum	3	2.667	Job		146		146	227

22 11 13.45 Pipe Fittings, Steel, Threaded

	22 11 13.45 Pipe Fittings, Steel, Threaded	Crew	Daily Output	Labor-Hours	Unit	Material	Labor	Equipment	Total	Total Incl O&P
0010	**PIPE FITTINGS, STEEL, THREADED** R221113-50									
0020	Cast Iron									
0040	Standard weight, black									
0060	90° Elbow, straight									
0070	1/4″	1 Plum	16	.500	Ea.	8.50	27.50		36	52
0080	3/8″		16	.500		12.25	27.50		39.75	56
0090	1/2″		15	.533		5.40	29		34.40	51.50
0100	3/4″		14	.571		5.60	31		36.60	54.50
0110	1″		13	.615		6.65	33.50		40.15	60
0130	1-1/2″	Q-1	20	.800		13	39.50		52.50	76
0140	2″		18	.889		20.50	43.50		64	90.50
0150	2-1/2″		14	1.143		48.50	56		104.50	141
0160	3″		10	1.600		80	78.50		158.50	211
0180	4″		6	2.667		148	131		279	365
0500	Tee, straight									
0510	1/4″	1 Plum	10	.800	Ea.	13.25	43.50		56.75	82.50
0520	3/8″		10	.800		12.90	43.50		56.40	82
0540	3/4″		9	.889		9.80	48.50		58.30	86.50
0550	1″		8	1		8.70	54.50		63.20	94.50
0570	1-1/2″	Q-1	13	1.231		20.50	60.50		81	117
0580	2″		11	1.455		28.50	71.50		100	143
0590	2-1/2″		9	1.778		74.50	87.50		162	218
0600	3″		6	2.667		114	131		245	330
0620	4″		4	4		223	197		420	550
0700	Standard weight, galvanized cast iron									
0720	90° Elbow, straight									
0730	1/4″	1 Plum	16	.500	Ea.	13.15	27.50		40.65	57
0740	3/8″		16	.500		13.15	27.50		40.65	57
0750	1/2″		15	.533		15	29		44	62
0760	3/4″		14	.571		14.70	31		45.70	64.50
0770	1″		13	.615		17	33.50		50.50	71
0790	1-1/2″	Q-1	20	.800		36.50	39.50		76	102
0800	2″		18	.889		53.50	43.50		97	127
0810	2-1/2″		14	1.143		110	56		166	209
0820	3″		10	1.600		167	78.50		245.50	305
0840	4″		6	2.667		305	131		436	545
1100	Tee, straight									
1110	1/4″	1 Plum	10	.800	Ea.	15.85	43.50		59.35	85.50
1120	3/8″		10	.800		15.85	43.50		59.35	85.50

22 11 13.45 Pipe Fittings, Steel, Threaded		Crew	Daily Output	Labor-Hours	Unit	2012 Bare Costs Material	Labor	Equipment	Total	Total Incl O&P
1140	3/4"	1 Plum	9	.889	Ea.	21	48.50		69.50	99
1150	1"		8	1		23	54.50		77.50	111
1170	1-1/2"	Q-1	13	1.231		53	60.50		113.50	153
1180	2"		11	1.455		66.50	71.50		138	184
1190	2-1/2"		9	1.778		135	87.50		222.50	284
1200	3"		6	2.667		350	131		481	590
1220	4"		4	4		425	197		622	775
5000	Malleable iron, 150 lb.									
5020	Black									
5040	90° elbow, straight									
5060	1/4"	1 Plum	16	.500	Ea.	4.17	27.50		31.67	47
5070	3/8"		16	.500		4.17	27.50		31.67	47
5090	3/4"		14	.571		3.49	31		34.49	52.50
5100	1"		13	.615		6.05	33.50		39.55	59
5120	1-1/2"	Q-1	20	.800		13.15	39.50		52.65	76
5130	2"		18	.889		22.50	43.50		66	93
5140	2-1/2"		14	1.143		50.50	56		106.50	143
5150	3"		10	1.600		74	78.50		152.50	204
5170	4"		6	2.667		159	131		290	380
5450	Tee, straight									
5470	1/4"	1 Plum	10	.800	Ea.	6.05	43.50		49.55	74.50
5480	3/8"		10	.800		6.05	43.50		49.55	74.50
5500	3/4"		9	.889		5.55	48.50		54.05	81.50
5510	1"		8	1		9.50	54.50		64	95.50
5520	1-1/4"	Q-1	14	1.143		15.40	56		71.40	104
5530	1-1/2"		13	1.231		19.10	60.50		79.60	116
5540	2"		11	1.455		32.50	71.50		104	147
5550	2-1/2"		9	1.778		70.50	87.50		158	214
5560	3"		6	2.667		104	131		235	320
5570	3-1/2"		5	3.200		240	157		397	510
5580	4"		4	4		250	197		447	580
5650	Coupling									
5670	1/4"	1 Plum	19	.421	Ea.	5.20	23		28.20	41.50
5680	3/8"		19	.421		5.20	23		28.20	41.50
5690	1/2"		19	.421		3.99	23		26.99	40.50
5700	3/4"		18	.444		4.68	24.50		29.18	43
5710	1"		15	.533		7	29		36	53
5730	1-1/2"	Q-1	24	.667		12.25	33		45.25	64.50
5740	2"		21	.762		18.15	37.50		55.65	78.50
5750	2-1/2"		18	.889		50	43.50		93.50	123
5760	3"		14	1.143		68	56		124	162
5780	4"		10	1.600		137	78.50		215.50	273
6000	For galvanized elbows, tees, and couplings add					20%				
9990	Minimum labor/equipment charge	1 Plum	4	2	Job		109		109	170

22 11 13.47 Pipe Fittings, Steel		Crew	Daily Output	Labor-Hours	Unit	Material	Labor	Equipment	Total	Total Incl O&P
0010	**PIPE FITTINGS, STEEL**, Flanged, Welded & Special									
3000	Weld joint, butt, carbon steel, standard weight									
3040	90° elbow, long radius									
3050	1/2" pipe size	Q-15	16	1	Ea.	45	49	2.31	96.31	129
3060	3/4" pipe size		16	1		45	49	2.31	96.31	129
3070	1" pipe size		16	1		21	49	2.31	72.31	102
3100	2" pipe size		10	1.600		22.50	78.50	3.70	104.70	152

22 11 Facility Water Distribution

22 11 13 – Facility Water Distribution Piping

22 11 13.47 Pipe Fittings, Steel		Crew	Daily Output	Labor-Hours	Unit	2012 Bare Costs Material	Labor	Equipment	Total	Total Incl O&P
3120	3" pipe size	Q-15	7	2.286	Ea.	29	112	5.30	146.30	213
3130	4" pipe size		5	3.200		48	157	7.40	212.40	305
3136	5" pipe size		4	4		103	197	9.25	309.25	430
3140	6" pipe size	Q-16	5	4.800		106	245	7.40	358.40	505
3350	Tee, straight									
3360	1/2" pipe size	Q-15	10	1.600	Ea.	110	78.50	3.70	192.20	248
3370	3/4" pipe size		10	1.600		110	78.50	3.70	192.20	248
3380	1" pipe size		10	1.600		54.50	78.50	3.70	136.70	187
3410	2" pipe size		6	2.667		54.50	131	6.15	191.65	271
3430	3" pipe size		4	4		84	197	9.25	290.25	410
3440	4" pipe size		3	5.333		118	262	12.30	392.30	555
3446	5" pipe size		2.50	6.400		195	315	14.80	524.80	720
3450	6" pipe size	Q-16	3	8		203	410	12.30	625.30	870
9990	Minimum labor/equipment charge	Q-15	3	5.333	Job		262	12.30	274.30	425

22 11 13.48 Pipe, Fittings and Valves, Steel, Grooved-Joint		Crew	Daily Output	Labor-Hours	Unit	Material	Labor	Equipment	Total	Total Incl O&P
0010	**PIPE, FITTINGS AND VALVES, STEEL, GROOVED-JOINT**									
0012	Fittings are ductile iron. Steel fittings noted.									
0020	Pipe includes coupling & clevis type hanger assemblies, 10' O.C.									
1000	Schedule 40, black									
1040	3/4" diameter	1 Plum	71	.113	L.F.	6.30	6.15		12.45	16.50
1050	1" diameter		63	.127		6.30	6.95		13.25	17.75
1060	1-1/4" diameter		58	.138		8	7.55		15.55	20.50
1070	1-1/2" diameter		51	.157		9.25	8.55		17.80	23.50
1080	2" diameter		40	.200		11.85	10.90		22.75	30
1090	2-1/2" diameter	Q-1	57	.281		20.50	13.80		34.30	44
1100	3" diameter		50	.320		26	15.75		41.75	53
1110	4" diameter		45	.356		35	17.50		52.50	65.50
1120	5" diameter		37	.432		38.50	21.50		60	75.50
4000	Elbow, 90° or 45°, painted									
4030	3/4" diameter	1 Plum	50	.160	Ea.	46	8.75		54.75	64
4040	1" diameter		50	.160		24.50	8.75		33.25	40.50
4050	1-1/4" diameter		40	.200		24.50	10.90		35.40	44
4060	1-1/2" diameter		33	.242		24.50	13.25		37.75	47.50
4070	2" diameter		25	.320		24.50	17.45		41.95	54
4080	2-1/2" diameter	Q-1	40	.400		24.50	19.65		44.15	57.50
4100	4" diameter		25	.640		47	31.50		78.50	101
4110	5" diameter		20	.800		112	39.50		151.50	186
4250	For galvanized elbows, add					26%				
4690	Tee, painted									
4700	3/4" diameter	1 Plum	38	.211	Ea.	49	11.50		60.50	72
4740	1" diameter		33	.242		38	13.25		51.25	62
4750	1-1/4" diameter		27	.296		38	16.20		54.20	66.50
4760	1-1/2" diameter		22	.364		38	19.85		57.85	72.50
4770	2" diameter		17	.471		38	25.50		63.50	81.50
4780	2-1/2" diameter	Q-1	27	.593		38	29		67	87
4800	4" diameter		17	.941		79.50	46.50		126	160
4810	5" diameter		13	1.231		185	60.50		245.50	299
4900	For galvanized tees, add					24%				
9990	Minimum labor/equipment charge	1 Plum	4	2	Job		109		109	170

22 11 13.74 Pipe, Plastic		Crew	Daily Output	Labor-Hours	Unit	Material	Labor	Equipment	Total	Total Incl O&P
0010	**PIPE, PLASTIC**									
1800	PVC, couplings 10' O.C., clevis hanger assemblies, 3 per 10'									

22 11 13 – Facility Water Distribution Piping

22 11 13.74 Pipe, Plastic		Crew	Daily Output	Labor-Hours	Unit	2012 Bare Costs Material	Labor	Equipment	Total	Total Incl O&P
1820	Schedule 40									
1860	1/2" diameter	1 Plum	54	.148	L.F.	2.83	8.10		10.93	15.70
1870	3/4" diameter		51	.157		3.17	8.55		11.72	16.85
1880	1" diameter		46	.174		3.59	9.50		13.09	18.75
1890	1-1/4" diameter		42	.190		4.33	10.40		14.73	21
1900	1-1/2" diameter		36	.222		4.59	12.15		16.74	24
1910	2" diameter	Q-1	59	.271		5.75	13.35		19.10	27.50
1920	2-1/2" diameter		56	.286		8.40	14.05		22.45	31.50
1930	3" diameter		53	.302		10.70	14.85		25.55	35
1940	4" diameter		48	.333		14.35	16.40		30.75	41.50
1950	5" diameter		43	.372		20	18.30		38.30	50.50
1960	6" diameter		39	.410		26	20		46	60.50
4100	DWV type, schedule 40, couplings 10' O.C., clevis hanger assy's, 3 per 10'									
4210	ABS, schedule 40, foam core type									
4212	Plain end black									
4214	1-1/2" diameter	1 Plum	39	.205	L.F.	2.97	11.20		14.17	20.50
4216	2" diameter	Q-1	62	.258		3.44	12.70		16.14	23.50
4218	3" diameter		56	.286		6.15	14.05		20.20	29
4220	4" diameter		51	.314		8.55	15.40		23.95	33.50
4222	6" diameter		42	.381		17.90	18.70		36.60	48.50
4240	To delete coupling & hangers, subtract									
4244	1-1/2" diam. to 6" diam.					43%	48%			
4400	PVC									
4410	1-1/4" diameter	1 Plum	42	.190	L.F.	3.21	10.40		13.61	19.75
4460	2" diameter	Q-1	59	.271		3.46	13.35		16.81	25
4470	3" diameter		53	.302		6.15	14.85		21	30
4480	4" diameter		48	.333		8.30	16.40		24.70	34.50
4490	6" diameter		39	.410		15.65	20		35.65	48.50
5300	CPVC, socket joint, couplings 10' O.C., clevis hanger assemblies, 3 per 10'									
5302	Schedule 40									
5304	1/2" diameter	1 Plum	54	.148	L.F.	3.58	8.10		11.68	16.55
5305	3/4" diameter		51	.157		4.17	8.55		12.72	17.95
5306	1" diameter		46	.174		5.25	9.50		14.75	20.50
5307	1-1/4" diameter		42	.190		6.60	10.40		17	23.50
5308	1-1/2" diameter		36	.222		7.60	12.15		19.75	27.50
5309	2" diameter	Q-1	59	.271		9.35	13.35		22.70	31.50
5310	2-1/2" diameter		56	.286		15.65	14.05		29.70	39
5311	3" diameter		53	.302		18.85	14.85		33.70	43.50
5360	CPVC, threaded, couplings 10' O.C., clevis hanger assemblies, 3 per 10'									
5380	Schedule 40									
5460	1/2" diameter	1 Plum	54	.148	L.F.	4.34	8.10		12.44	17.35
5470	3/4" diameter		51	.157		5.50	8.55		14.05	19.40
5480	1" diameter		46	.174		6.60	9.50		16.10	22
5490	1-1/4" diameter		42	.190		7.65	10.40		18.05	24.50
5500	1-1/2" diameter		36	.222		8.50	12.15		20.65	28.50
5510	2" diameter	Q-1	59	.271		10.45	13.35		23.80	32.50
5520	2-1/2" diameter		56	.286		16.80	14.05		30.85	40.50
5530	3" diameter		53	.302		20.50	14.85		35.35	45.50
5560										
9900	Minimum labor/equipment charge	1 Plum	4	2	Job		109		109	170

22 11 Facility Water Distribution

22 11 13 – Facility Water Distribution Piping

22 11 13.76 Pipe Fittings, Plastic		Crew	Daily Output	Labor-Hours	Unit	2012 Bare Costs Material	Labor	Equipment	Total	Total Incl O&P
0010	**PIPE FITTINGS, PLASTIC**									
2700	PVC (white), schedule 40, socket joints									
2760	90° elbow, 1/2″	1 Plum	33.30	.240	Ea.	.46	13.10		13.56	21
2810	2″	Q-1	36.40	.440	″	2.75	21.50		24.25	36.50
4500	DWV, ABS, non pressure, socket joints									
4540	1/4 Bend, 1-1/4″	1 Plum	20.20	.396	Ea.	3.73	21.50		25.23	37.50
4560	1-1/2″	″	18.20	.440		2.84	24		26.84	40.50
4570	2″	Q-1	33.10	.483		4.10	24		28.10	41.50
4800	Tee, sanitary									
4820	1-1/4″	1 Plum	13.50	.593	Ea.	4.95	32.50		37.45	56
4830	1-1/2″	″	12.10	.661		4.33	36		40.33	61
4840	2″	Q-1	20	.800		6.70	39.50		46.20	69
5000	DWV, PVC, schedule 40, socket joints									
5040	1/4 bend, 1-1/4″	1 Plum	20.20	.396	Ea.	7.45	21.50		28.95	41.50
5060	1-1/2″	″	18.20	.440		2.05	24		26.05	40
5070	2″	Q-1	33.10	.483		3.23	24		27.23	40.50
5080	3″		20.80	.769		9.55	38		47.55	69.50
5090	4″		16.50	.970		18.80	47.50		66.30	95
5100	6″		10.10	1.584		66	78		144	194
5105	8″	Q-2	9.30	2.581		133	132		265	350
5110	1/4 bend, long sweep, 1-1/2″	1 Plum	18.20	.440		4.92	24		28.92	43
5112	2″	Q-1	33.10	.483		5.30	24		29.30	43
5114	3″		20.80	.769		12.55	38		50.55	73
5116	4″		16.50	.970		24	47.50		71.50	101
5215	8″	Q-2	9.30	2.581		78.50	132		210.50	292
5250	Tee, sanitary 1-1/4″	1 Plum	13.50	.593		7.50	32.50		40	59
5254	1-1/2″	″	12.10	.661		3.72	36		39.72	60
5255	2″	Q-1	20	.800		5.45	39.50		44.95	67.50
5256	3″		13.90	1.151		14	56.50		70.50	103
5257	4″		11	1.455		24.50	71.50		96	138
5259	6″		6.70	2.388		106	117		223	300
5261	8″	Q-2	6.20	3.871		320	197		517	655
5264	2″ x 1-1/2″	Q-1	22	.727		4.83	36		40.83	61
5266	3″ x 1-1/2″		15.50	1.032		10.15	50.50		60.65	90
5268	4″ x 3″		12.10	1.322		31	65		96	135
5271	6″ x 4″		6.90	2.319		103	114		217	291
5314	Combination Y & 1/8 bend, 1-1/2″	1 Plum	12.10	.661		8.95	36		44.95	66
5315	2″	Q-1	20	.800		9.60	39.50		49.10	72
5317	3″		13.90	1.151		24	56.50		80.50	115
5318	4″		11	1.455		47.50	71.50		119	163
5324	Combination Y & 1/8 bend, reducing									
5325	2″ x 2″ x 1-1/2″	Q-1	22	.727	Ea.	12.60	36		48.60	69.50
5327	3″ x 3″ x 1-1/2″		15.50	1.032		22	50.50		72.50	103
5328	3″ x 3″ x 2″		15.30	1.046		15.80	51.50		67.30	97.50
5329	4″ x 4″ x 2″		12.20	1.311		25.50	64.50		90	129
5331	Wye, 1-1/4″	1 Plum	13.50	.593		9.50	32.50		42	61
5332	1-1/2″	″	12.10	.661		7	36		43	63.50
5333	2″	Q-1	20	.800		6.65	39.50		46.15	69
5334	3″		13.90	1.151		17.90	56.50		74.40	108
5335	4″		11	1.455		32.50	71.50		104	147
5336	6″		6.70	2.388		98.50	117		215.50	291
5337	8″	Q-2	6.20	3.871		229	197		426	555

22 11 13.76 Pipe Fittings, Plastic		Crew	Daily Output	Labor-Hours	Unit	2012 Bare Costs Material	Labor	Equipment	Total	Total Incl O&P
5341	2" x 1-1/2"	Q-1	22	.727	Ea.	8.15	36		44.15	64.50
5342	3" x 1-1/2"		15.50	1.032		11.75	50.50		62.25	92
5343	4" x 3"		12.10	1.322		26.50	65		91.50	130
5344	6" x 4"		6.90	2.319		72.50	114		186.50	258
5345	8" x 6"	Q-2	6.40	3.750		173	191		364	490
5347	Double wye, 1-1/2"	1 Plum	9.10	.879		15.15	48		63.15	91.50
5348	2"	Q-1	16.60	.964		17.05	47.50		64.55	93
5349	3"		10.40	1.538		35.50	75.50		111	157
5350	4"		8.25	1.939		72.50	95.50		168	229
5353	Double wye, reducing									
5354	2" x 2" x 1-1/2" x 1-1/2"	Q-1	16.80	.952	Ea.	15.45	47		62.45	90
5355	3" x 3" x 2" x 2"		10.60	1.509		26.50	74		100.50	145
5356	4" x 4" x 3" x 3"		8.45	1.893		57	93		150	208
5357	6" x 6" x 4" x 4"		7.25	2.207		187	108		295	375
5410	Reducer bushing, 2" x 1-1/4"		36.50	.438		4.20	21.50		25.70	38
5412	3" x 1-1/2"		27.30	.586		8.40	29		37.40	54.50
5414	4" x 2"		18.20	.879		14.65	43		57.65	83.50
5416	6" x 4"		11.10	1.441		39	71		110	153
5418	8" x 6"	Q-2	10.20	2.353		84.50	120		204.50	280
5500	CPVC, Schedule 80, threaded joints									
5540	90° Elbow, 1/4"	1 Plum	32	.250	Ea.	11.55	13.65		25.20	34
5560	1/2"		30.30	.264		6.70	14.40		21.10	30
5570	3/4"		26	.308		10	16.80		26.80	37
5580	1"		22.70	.352		14.05	19.25		33.30	45.50
5590	1-1/4"		20.20	.396		27	21.50		48.50	63.50
5600	1-1/2"		18.20	.440		29	24		53	69.50
5610	2"	Q-1	33.10	.483		39	24		63	80
5620	2-1/2"		24.20	.661		121	32.50		153.50	184
5630	3"		20.80	.769		130	38		168	202
5730	Coupling, 1/4"	1 Plum	32	.250		14.70	13.65		28.35	37.50
5732	1/2"		30.30	.264		12.10	14.40		26.50	36
5734	3/4"		26	.308		19.55	16.80		36.35	47.50
5736	1"		22.70	.352		22	19.25		41.25	54.50
5738	1-1/4"		20.20	.396		23.50	21.50		45	59.50
5740	1-1/2"		18.20	.440		25.50	24		49.50	65.50
5742	2"	Q-1	33.10	.483		30	24		54	70
5744	2-1/2"		24.20	.661		53.50	32.50		86	110
5746	3"		20.80	.769		62	38		100	128
5900	CPVC, Schedule 80, socket joints									
5904	90° Elbow, 1/4"	1 Plum	32	.250	Ea.	11	13.65		24.65	33.50
5906	1/2"		30.30	.264		4.31	14.40		18.71	27
5908	3/4"		26	.308		5.50	16.80		22.30	32
5910	1"		22.70	.352		8.70	19.25		27.95	39.50
5912	1-1/4"		20.20	.396		18.85	21.50		40.35	54.50
5914	1-1/2"		18.20	.440		21	24		45	60.50
5916	2"	Q-1	33.10	.483		25.50	24		49.50	65
5918	2 1/2"		24.20	.661		58.50	32.50		91	115
5920	3"		20.80	.769		66	38		104	132
5930	45° Elbow, 1/4"	1 Plum	32	.250		16.35	13.65		30	39.50
5932	1/2"		30.30	.264		5.25	14.40		19.65	28.50
5934	3/4"		26	.308		7.60	16.80		24.40	34.50
5936	1"		22.70	.352		12.10	19.25		31.35	43.50
5938	1-1/4"		20.20	.396		24	21.50		45.50	59.50

22 11 Facility Water Distribution

22 11 13 – Facility Water Distribution Piping

22 11 13.76 Pipe Fittings, Plastic		Crew	Daily Output	Labor-Hours	Unit	2012 Bare Costs Material	Labor	Equipment	Total	Total Incl O&P
5940	1-1/2"	1 Plum	18.20	.440	Ea.	24.50	24		48.50	64.50
5942	2"	Q-1	33.10	.483		27.50	24		51.50	67
5944	2-1/2"		24.20	.661		56	32.50		88.50	112
5946	3"		20.80	.769		72	38		110	138
5990	Coupling, 1/4"	1 Plum	32	.250		11.70	13.65		25.35	34.50
5992	1/2"		30.30	.264		4.55	14.40		18.95	27.50
5994	3/4"		26	.308		6.35	16.80		23.15	33
5996	1"		22.70	.352		8.55	19.25		27.80	39.50
5998	1-1/4"		20.20	.396		12.80	21.50		34.30	47.50
6000	1-1/2"		18.20	.440		16.15	24		40.15	55.50
6002	2"	Q-1	33.10	.483		18.75	24		42.75	57.50
6004	2-1/2"		24.20	.661		41.50	32.50		74	96.50
6006	3"		20.80	.769		45.50	38		83.50	109
9990	Minimum labor/equipment charge	1 Plum	4	2	Job		109		109	170

22 11 19 – Domestic Water Piping Specialties

22 11 19.10 Flexible Connectors

		Crew	Daily Output	Labor-Hours	Unit	Material	Labor	Equipment	Total	Total Incl O&P
0010	**FLEXIBLE CONNECTORS**, Corrugated, 7/8" O.D., 1/2" I.D.									
0050	Gas, seamless brass, steel fittings									
0200	12" long	1 Plum	36	.222	Ea.	16.30	12.15		28.45	37
0220	18" long		36	.222		20	12.15		32.15	41
0240	24" long		34	.235		24	12.85		36.85	46.50
0260	30" long		34	.235		25.50	12.85		38.35	48
9000	Minimum labor/equipment charge		4	2	Job		109		109	170

22 11 19.18 Mixing Valve

		Crew	Daily Output	Labor-Hours	Unit	Material	Labor	Equipment	Total	Total Incl O&P
0010	**MIXING VALVE**, Automatic, water tempering.									
0040	1/2" size	1 Stpi	19	.421	Ea.	505	23.50		528.50	590
0050	3/4" size		18	.444	"	505	25		530	595
9000	Minimum labor/equipment charge		5	1.600	Job		89		89	139

22 11 19.38 Water Supply Meters

		Crew	Daily Output	Labor-Hours	Unit	Material	Labor	Equipment	Total	Total Incl O&P
0010	**WATER SUPPLY METERS**									
2000	Domestic/commercial, bronze									
2020	Threaded									
2060	5/8" diameter, to 20 GPM	1 Plum	16	.500	Ea.	46	27.50		73.50	93
2080	3/4" diameter, to 30 GPM		14	.571		83.50	31		114.50	141
2100	1" diameter, to 50 GPM		12	.667		127	36.50		163.50	197
2300	Threaded/flanged									
2340	1-1/2" diameter, to 100 GPM	1 Plum	8	1	Ea.	310	54.50		364.50	425
9000	Minimum labor/equipment charge	"	3.25	2.462	Job		134		134	209

22 11 19.42 Backflow Preventers

		Crew	Daily Output	Labor-Hours	Unit	Material	Labor	Equipment	Total	Total Incl O&P
0010	**BACKFLOW PREVENTERS**, Includes valves									
0020	and four test cocks, corrosion resistant, automatic operation									
1000	Double check principle									
1010	Threaded, with ball valves									
1020	3/4" pipe size	1 Plum	16	.500	Ea.	220	27.50		247.50	285
1030	1" pipe size		14	.571		252	31		283	325
1040	1-1/2" pipe size		10	.800		535	43.50		578.50	660
1050	2" pipe size		7	1.143		615	62.50		677.50	775
1080	Threaded, with gate valves									
1100	3/4" pipe size	1 Plum	16	.500	Ea.	1,025	27.50		1,052.50	1,175
1120	1" pipe size		14	.571		1,050	31		1,081	1,200
1140	1-1/2" pipe size		10	.800		1,350	43.50		1,393.50	1,550

22 11 19 – Domestic Water Piping Specialties

22 11 19.42 Backflow Preventers		Crew	Daily Output	Labor-Hours	Unit	Material	2012 Bare Costs Labor	Equipment	Total	Total Incl O&P
1160	2" pipe size	1 Plum	7	1.143	Ea.	1,650	62.50		1,712.50	1,900
4000	Reduced pressure principle									
4100	Threaded, bronze, valves are ball									
4120	3/4" pipe size	1 Plum	16	.500	Ea.	420	27.50		447.50	510
4140	1" pipe size		14	.571		455	31		486	550
4160	1-1/2" pipe size		10	.800		915	43.50		958.50	1,075
4180	2" pipe size		7	1.143		1,025	62.50		1,087.50	1,225
5000	Flanged, bronze, valves are OS&Y									
5060	2-1/2" pipe size	Q-1	5	3.200	Ea.	6,100	157		6,257	6,975
5080	3" pipe size		4.50	3.556		7,100	175		7,275	8,100
5100	4" pipe size		3	5.333		8,325	262		8,587	9,575
5120	6" pipe size	Q-2	3	8		13,300	410		13,710	15,200
5600	Flanged, iron, valves are OS&Y									
5660	2-1/2" pipe size	Q-1	5	3.200	Ea.	3,150	157		3,307	3,725
5680	3" pipe size		4.50	3.556		3,425	175		3,600	4,050
5700	4" pipe size		3	5.333		3,800	262		4,062	4,575
5720	6" pipe size	Q-2	3	8		5,500	410		5,910	6,675
5800	Rebuild 4" diameter reduced pressure BFP	1 Plum	4	2		310	109		419	510
5810	6" diameter		2.66	3.008		330	164		494	615
5820	8" diameter		2	4		405	218		623	790
5830	10" diameter		1.60	5		475	273		748	950
9010	Minimum labor/equipment charge		2	4	Job		218		218	340

22 11 19.50 Vacuum Breakers		Crew	Daily Output	Labor-Hours	Unit	Material	Labor	Equipment	Total	Total Incl O&P
0010	**VACUUM BREAKERS**									
0013	See also backflow preventers Section 22 11 19.42									
1000	Anti-siphon continuous pressure type									
1010	Max. 150 PSI - 210°F									
1020	Bronze body									
1030	1/2" size	1 Stpi	24	.333	Ea.	171	18.55		189.55	217
1040	3/4" size		20	.400		171	22.50		193.50	223
1050	1" size		19	.421		177	23.50		200.50	231
1060	1-1/4" size		15	.533		350	29.50		379.50	430
1070	1-1/2" size		13	.615		430	34.50		464.50	525
1080	2" size		11	.727		445	40.50		485.50	550
1200	Max. 125 PSI with atmospheric vent									
1210	Brass, in-line construction									
1220	1/4" size	1 Stpi	24	.333	Ea.	68	18.55		86.55	104
1230	3/8" size	"	24	.333		68	18.55		86.55	104
1260	For polished chrome finish, add					13%				
2000	Anti-siphon, non-continuous pressure type									
2010	Hot or cold water 125 PSI - 210°F									
2020	Bronze body									
2030	1/4" size	1 Stpi	24	.333	Ea.	43.50	18.55		62.05	76.50
2040	3/8" size		24	.333		43.50	18.55		62.05	76.50
2050	1/2" size		24	.333		49	18.55		67.55	83
2060	3/4" size		20	.400		58	22.50		80.50	98.50
2070	1" size		19	.421		91	23.50		114.50	137
2080	1-1/4" size		15	.533		159	29.50		188.50	222
2090	1-1/2" size		13	.615		187	34.50		221.50	260
2100	2" size		11	.727		291	40.50		331.50	385
2110	2-1/2" size		8	1		840	55.50		895.50	1,000
2120	3" size		6	1.333		1,125	74.50		1,199.50	1,350

22 11 Facility Water Distribution

22 11 19 – Domestic Water Piping Specialties

	22 11 19.50 Vacuum Breakers	Crew	Daily Output	Labor-Hours	Unit	2012 Bare Costs Material	Labor	Equipment	Total	Total Incl O&P
2150	For polished chrome finish, add					50%				

22 11 19.54 Water Hammer Arresters/Shock Absorbers

		Crew	Daily Output	Labor-Hours	Unit	Material	Labor	Equipment	Total	Total Incl O&P
0010	**WATER HAMMER ARRESTERS/SHOCK ABSORBERS**									
0490	Copper									
0500	3/4" male I.P.S. For 1 to 11 fixtures	1 Plum	12	.667	Ea.	21.50	36.50		58	80.50
0600	1" male I.P.S. For 12 to 32 fixtures		8	1		34	54.50		88.50	122
0700	1-1/4" male I.P.S. For 33 to 60 fixtures		8	1		44	54.50		98.50	134
0800	1-1/2" male I.P.S. For 61 to 113 fixtures		8	1		63	54.50		117.50	155
0900	2" male I.P.S. For 114 to 154 fixtures		8	1		92	54.50		146.50	186
1000	2-1/2" male I.P.S. For 155 to 330 fixtures		4	2		285	109		394	485
9000	Minimum labor/equipment charge		3.50	2.286	Job		125		125	194

22 11 19.64 Hydrants

		Crew	Daily Output	Labor-Hours	Unit	Material	Labor	Equipment	Total	Total Incl O&P
0010	**HYDRANTS**									
0050	Wall type, moderate climate, bronze, encased									
0200	3/4" IPS connection	1 Plum	16	.500	Ea.	650	27.50		677.50	760
1000	Non-freeze, bronze, exposed									
1100	3/4" IPS connection, 4" to 9" thick wall	1 Plum	14	.571	Ea.	440	31		471	535
1120	10" to 14" thick wall	"	12	.667		475	36.50		511.50	575
1280	For anti-siphon type, add					115			115	126
9000	Minimum labor/equipment charge	1 Plum	3	2.667	Job		146		146	227

22 13 Facility Sanitary Sewerage

22 13 16 – Sanitary Waste and Vent Piping

22 13 16.20 Pipe, Cast Iron

		Crew	Daily Output	Labor-Hours	Unit	Material	Labor	Equipment	Total	Total Incl O&P
0010	**PIPE, CAST IRON**, Soil, on clevis hanger assemblies, 5' O.C. R221113-50									
0020	Single hub, service wt., lead & oakum joints 10' O.C.									
2120	2" diameter	Q-1	63	.254	L.F.	8.50	12.50		21	29
2140	3" diameter		60	.267		11.95	13.10		25.05	33.50
2160	4" diameter		55	.291		15.45	14.30		29.75	39.50
2180	5" diameter	Q-2	76	.316		21	16.10		37.10	48.50
2200	6" diameter	"	73	.329		26.50	16.75		43.25	55
2220	8" diameter	Q-3	59	.542		41	28		69	89
4000	No hub, couplings 10' O.C.									
4100	1-1/2" diameter	Q-1	71	.225	L.F.	8.45	11.10		19.55	26.50
4120	2" diameter		67	.239		8.65	11.75		20.40	28
4140	3" diameter		64	.250		12.05	12.30		24.35	32.50
4160	4" diameter		58	.276		15.70	13.55		29.25	38.50
4180	5" diameter	Q-2	83	.289		21	14.75		35.75	46
9000	Minimum labor/equipment charge	1 Plum	4	2	Job		109		109	170

22 13 16.30 Pipe Fittings, Cast Iron

		Crew	Daily Output	Labor-Hours	Unit	Material	Labor	Equipment	Total	Total Incl O&P
0010	**PIPE FITTINGS, CAST IRON**, Soil R221113-50									
0040	Hub and spigot, service weight, lead & oakum joints									
0080	1/4 bend, 2"	Q-1	16	1	Ea.	16.95	49		65.95	95
0120	3"		14	1.143		22.50	56		78.50	113
0140	4"		13	1.231		35.50	60.50		96	134
0160	5"	Q-2	18	1.333		49.50	68		117.50	161
0180	6"	"	17	1.412		61.50	72		133.50	180
0200	8"	Q-3	11	2.909		185	151		336	440
0340	1/8 bend, 2"	Q-1	16	1		12.05	49		61.05	90
0350	3"		14	1.143		18.90	56		74.90	109

22 13 16 – Sanitary Waste and Vent Piping

22 13 16.30 Pipe Fittings, Cast Iron		Crew	Daily Output	Labor-Hours	Unit	Material	Labor	Equipment	Total	Total Incl O&P
						2012 Bare Costs				
0360	4"	Q-1	13	1.231	Ea.	27.50	60.50		88	125
0380	5"	Q-2	18	1.333		38	68		106	148
0400	6"	"	17	1.412		47	72		119	164
0420	8"	Q-3	11	2.909		140	151		291	390
0500	Sanitary tee, 2"	Q-1	10	1.600		23.50	78.50		102	149
0540	3"		9	1.778		38	87.50		125.50	178
0620	4"		8	2		50.50	98.50		149	209
0700	5"	Q-2	12	2		93	102		195	261
0800	6"	"	11	2.182		105	111		216	289
0880	8"	Q-3	7	4.571		280	238		518	680
0954	10" x 6"	"	8	4		405	208		613	770
5990	No hub									
6000	Cplg. & labor required at joints not incl. in fitting									
6010	price. Add 1 coupling per joint for installed price									
6020	1/4 Bend, 1-1/2"				Ea.	8.75			8.75	9.65
6060	2"					9.45			9.45	10.40
6080	3"					13.25			13.25	14.55
6120	4"					19.60			19.60	21.50
6184	1/4 Bend, long sweep, 1-1/2"					20.50			20.50	23
6186	2"					20.50			20.50	23
6188	3"					25			25	27.50
6189	4"					40			40	44
6190	5"					77.50			77.50	85
6191	6"					88			88	97
6192	8"					239			239	263
6193	10"					410			410	450
6200	1/8 Bend, 1-1/2"					7.35			7.35	8.10
6210	2"					8.10			8.10	8.95
6212	3"					10.90			10.90	12
6214	4"					14.35			14.35	15.80
6380	Sanitary Tee, tapped, 1-1/2"					17.35			17.35	19.05
6382	2" x 1-1/2"					15.30			15.30	16.85
6384	2"					16.40			16.40	18
6386	3" x 2"					24.50			24.50	27
6388	3"					42			42	46.50
6390	4" x 1-1/2"					21.50			21.50	24
6392	4" x 2"					24.50			24.50	27
6393	4"					24.50			24.50	27
6394	6" x 1-1/2"					56.50			56.50	62
6396	6" x 2"					57.50			57.50	63.50
6459	Sanitary Tee, 1-1/2"					12.25			12.25	13.50
6460	2"					13			13	14.30
6470	3"					16.15			16.15	17.80
6472	4"					30.50			30.50	33.50
8000	Coupling, standard (by CISPI Mfrs.)									
8020	1-1/2"	Q-1	48	.333	Ea.	13	16.40		29.40	40
8040	2"		44	.364		13	17.85		30.85	42.50
8080	3"		38	.421		15.55	20.50		36.05	49
8120	4"		33	.485		18.35	24		42.35	57
9000	Minimum labor/equipment charge	1 Plum	4	2	Job		109		109	170

22 13 Facility Sanitary Sewerage

22 13 16 – Sanitary Waste and Vent Piping

22 13 16.60 Traps

	22 13 16.60 Traps	Crew	Daily Output	Labor-Hours	Unit	2012 Bare Costs Material	Labor	Equipment	Total	Total Incl O&P
0010	**TRAPS**									
0030	Cast iron, service weight									
0050	Running P trap, without vent									
1100	2″	Q-1	16	1	Ea.	122	49		171	211
1150	4″	″	13	1.231		122	60.50		182.50	229
1160	6″	Q-2	17	1.412		530	72		602	695
3000	P trap, B&S, 2″ pipe size	Q-1	16	1		29	49		78	109
3040	3″ pipe size	″	14	1.143		43	56		99	135
3900										
4700	Copper, drainage, drum trap									
4840	3″ x 6″ swivel, 1-1/2″ pipe size	1 Plum	16	.500	Ea.	405	27.50		432.50	490
5100	P trap, standard pattern									
5200	1-1/4″ pipe size	1 Plum	18	.444	Ea.	189	24.50		213.50	246
5240	1-1/2″ pipe size		17	.471		183	25.50		208.50	241
5260	2″ pipe size		15	.533		282	29		311	355
5280	3″ pipe size		11	.727		675	39.50		714.50	805
9000	Minimum labor/equipment charge		3	2.667	Job		146		146	227

22 13 16.80 Vent Flashing and Caps

	22 13 16.80 Vent Flashing and Caps	Crew	Daily Output	Labor-Hours	Unit	Material	Labor	Equipment	Total	Total Incl O&P
0010	**VENT FLASHING AND CAPS**									
0120	Vent caps									
0140	Cast iron									
0180	2-1/2″ - 3-5/8″ pipe	1 Plum	21	.381	Ea.	40	21		61	76.50
0190	4″ - 4-1/8″ pipe	″	19	.421	″	48	23		71	89
0900	Vent flashing									
1000	Aluminum with lead ring									
1050	3″ pipe	1 Plum	17	.471	Ea.	13.30	25.50		38.80	54.50
1060	4″ pipe	″	16	.500	″	16.05	27.50		43.55	60
1350	Copper with neoprene ring									
1440	2″ pipe	1 Plum	18	.444	Ea.	21	24.50		45.50	61
1450	3″ pipe		17	.471		24.50	25.50		50	67
1460	4″ pipe		16	.500		27	27.50		54.50	72.50
9000	Minimum labor/equipment charge		4	2	Job		109		109	170

22 13 19 – Sanitary Waste Piping Specialties

22 13 19.13 Sanitary Drains

	22 13 19.13 Sanitary Drains	Crew	Daily Output	Labor-Hours	Unit	Material	Labor	Equipment	Total	Total Incl O&P
0010	**SANITARY DRAINS**									
0400	Deck, auto park, C.I., 13″ top									
0440	3″, 4″, 5″, and 6″ pipe size	Q-1	8	2	Ea.	1,225	98.50		1,323.50	1,500
0480	For galvanized body, add				″	660			660	725
0500										
2000	Floor, medium duty, C.I., deep flange, 7″ dia top									
2040	2″ and 3″ pipe size	Q-1	12	1.333	Ea.	171	65.50		236.50	290
2080	For galvanized body, add					82			82	90
2120	With polished bronze top					267			267	293
2500	Heavy duty, cleanout & trap w/bucket, C.I., 15″ top									
2540	2″, 3″, and 4″ pipe size	Q-1	6	2.667	Ea.	5,575	131		5,706	6,350
2560	For galvanized body, add					1,425			1,425	1,575
2580	With polished bronze top					6,200			6,200	6,825

22 13 19.14 Floor Receptors

	22 13 19.14 Floor Receptors	Crew	Daily Output	Labor-Hours	Unit	Material	Labor	Equipment	Total	Total Incl O&P
0010	**FLOOR RECEPTORS**, For connection to 2″, 3″ & 4″ diameter pipe									
0200	12-1/2″ square top, 25 sq. in. open area	Q-1	10	1.600	Ea.	825	78.50		903.50	1,025
0300	For grate with 4″ diam. x 3-3/4″ high funnel, add					148			148	162

22 13 Facility Sanitary Sewerage

22 13 19 – Sanitary Waste Piping Specialties

22 13 19.14 Floor Receptors		Crew	Daily Output	Labor-Hours	Unit	2012 Bare Costs Material	Labor	Equipment	Total	Total Incl O&P
0400	For grate with 6" diameter x 6" high funnel, add				Ea.	190			190	209
0700	For acid-resisting bucket, add					232			232	256
0900	For stainless steel mesh bucket liner, add					183			183	202
2000	12-5/8" diameter top, 40 sq. in. open area	Q-1	10	1.600		655	78.50		733.50	845
2100	For options, add same prices as square top									
3000	8" x 4" rectangular top, 7.5 sq. in. open area	Q-1	14	1.143	Ea.	635	56		691	790
3100	For trap primer connections, add				"	75			75	82.50
9000	Minimum labor/equipment charge	Q-1	3	5.333	Job		262		262	410

22 13 23 – Sanitary Waste Interceptors

22 13 23.10 Interceptors

		Crew	Daily Output	Labor-Hours	Unit	Material	Labor	Equipment	Total	Total Incl O&P
0010	**INTERCEPTORS**									
0150	Grease, fabricated steel, 4 GPM, 8 lb. fat capacity	1 Plum	4	2	Ea.	1,000	109		1,109	1,275
0200	7 GPM, 14 lb. fat capacity		4	2		1,400	109		1,509	1,725
1000	10 GPM, 20 lb. fat capacity		4	2		1,650	109		1,759	2,000
1160	100 GPM, 200 lb. fat capacity	Q-1	2	8		12,500	395		12,895	14,300
1240	300 GPM, 600 lb. fat capacity	"	1	16		27,500	785		28,285	31,500
9000	Minimum labor/equipment charge	1 Plum	3	2.667	Job		146		146	227

22 13 29 – Sanitary Sewerage Pumps

22 13 29.14 Sewage Ejector Pumps

		Crew	Daily Output	Labor-Hours	Unit	Material	Labor	Equipment	Total	Total Incl O&P
0010	**SEWAGE EJECTOR PUMPS**, With operating and level controls									
0100	Simplex system incl. tank, cover, pump 15' head									
0500	37 gal. PE tank, 12 GPM, 1/2 HP, 2" discharge	Q-1	3.20	5	Ea.	455	246		701	885
0510	3" discharge		3.10	5.161		495	254		749	940
0530	87 GPM, .7 HP, 2" discharge		3.20	5		690	246		936	1,150
0540	3" discharge		3.10	5.161		750	254		1,004	1,225
0600	45 gal. coated stl tank, 12 GPM, 1/2 HP, 2" discharge		3	5.333		805	262		1,067	1,300
0610	3" discharge		2.90	5.517		840	271		1,111	1,350
0630	87 GPM, .7 HP, 2" discharge		3	5.333		1,025	262		1,287	1,525
0640	3" discharge		2.90	5.517		1,100	271		1,371	1,625
0660	134 GPM, 1 HP, 2" discharge		2.80	5.714		1,125	281		1,406	1,675
0680	3" discharge		2.70	5.926		1,175	291		1,466	1,750
0700	70 gal. PE tank, 12 GPM, 1/2 HP, 2" discharge		2.60	6.154		880	300		1,180	1,425
0710	3" discharge		2.40	6.667		935	330		1,265	1,525
0730	87 GPM, 0.7 HP, 2" discharge		2.50	6.400		1,125	315		1,440	1,725
0740	3" discharge		2.30	6.957		1,200	340		1,540	1,850
0760	134 GPM, 1 HP, 2" discharge		2.20	7.273		1,225	355		1,580	1,900
0770	3" discharge		2	8		1,300	395		1,695	2,050
9000	Minimum labor/equipment charge		2.50	6.400	Job		315		315	490

22 14 Facility Storm Drainage

22 14 26 – Facility Storm Drains

22 14 26.13 Roof Drains

		Crew	Daily Output	Labor-Hours	Unit	Material	Labor	Equipment	Total	Total Incl O&P
0010	**ROOF DRAINS**									
0140	Cornice, C.I., 45° or 90° outlet									
0200	3" and 4" pipe size	Q-1	12	1.333	Ea.	360	65.50		425.50	495
0260	For galvanized body, add					64.50			64.50	71
0280	For polished bronze dome, add					50			50	55
3860	Roof, flat metal deck, C.I. body, 12" C.I. dome									
3890	3" pipe size	Q-1	14	1.143	Ea.	330	56		386	455
3900	4" pipe size	"	13	1.231	"	330	60.50		390.50	460

22 14 Facility Storm Drainage

22 14 26 – Facility Storm Drains

22 14 26.13 Roof Drains		Crew	Daily Output	Labor-Hours	Unit	Material	2012 Bare Costs Labor	Equipment	Total	Total Incl O&P
4620	Main, all aluminum, 12" low profile dome									
4640	2", 3" and 4" pipe size	Q-1	14	1.143	Ea.	370	56		426	500
9000	Minimum labor/equipment charge	1 Plum	4	2	Job		109		109	170

22 14 29 – Sump Pumps

22 14 29.16 Submersible Sump Pumps		Crew	Daily Output	Labor-Hours	Unit	Material	Labor	Equipment	Total	Total Incl O&P
0010	**SUBMERSIBLE SUMP PUMPS**									
7000	Sump pump, automatic									
7100	Plastic, 1-1/4" discharge, 1/4 HP	1 Plum	6	1.333	Ea.	126	73		199	252
7140	1/3 HP		5	1.600		189	87.50		276.50	345
7180	1-1/2" discharge, 1/2 HP		4	2		256	109		365	450
7500	Cast iron, 1-1/4" discharge, 1/4 HP		6	1.333		177	73		250	310
7540	1/3 HP		6	1.333		209	73		282	345
7560	1/2 HP		5	1.600		252	87.50		339.50	415
9000	Minimum labor/equipment charge		4	2	Job		109		109	170

22 15 General Service Compressed-Air Systems

22 15 19 – General Service Packaged Air Compressors and Receivers

22 15 19.10 Air Compressors		Crew	Daily Output	Labor-Hours	Unit	Material	Labor	Equipment	Total	Total Incl O&P
0010	**AIR COMPRESSORS**									
5250	Air, reciprocating air cooled, splash lubricated, tank mounted									
5300	Single stage, 1 phase, 140 psi									
5303	1/2 HP, 30 gal. tank	1 Stpi	3	2.667	Ea.	1,750	149		1,899	2,150
5305	3/4 HP, 30 gal. tank		2.60	3.077		1,775	171		1,946	2,225
5307	1 HP, 30 gal. tank		2.20	3.636		2,175	203		2,378	2,725

22 31 Domestic Water Softeners

22 31 13 – Residential Domestic Water Softeners

22 31 13.10 Residential Water Softeners		Crew	Daily Output	Labor-Hours	Unit	Material	Labor	Equipment	Total	Total Incl O&P
0010	**RESIDENTIAL WATER SOFTENERS**									
7350	Water softener, automatic, to 30 grains per gallon	2 Plum	5	3.200	Ea.	340	175		515	645
7400	To 100 grains per gallon	"	4	4	"	680	218		898	1,075

22 33 Electric Domestic Water Heaters

22 33 13 – Instantaneous Electrical Domestic Water Heaters

22 33 13.20 Instantaneous Elec. Point-of-Use Water Heaters		Crew	Daily Output	Labor-Hours	Unit	Material	Labor	Equipment	Total	Total Incl O&P
0010	**INSTANTANEOUS ELECTRIC POINT-OF-USE WATER HEATERS**									
8965	Point of use, electric, glass lined									
8969	Energy saver									
8970	2.5 gal. single element	1 Plum	2.80	2.857	Ea.	320	156		476	600
8971	4 gal. single element		2.80	2.857		340	156		496	620
8974	6 gal. single element		2.50	3.200		365	175		540	670
8975	10 gal. single element		2.50	3.200		385	175		560	695
8976	15 gal. single element		2.40	3.333		405	182		587	730
8977	20 gal. single element		2.40	3.333		440	182		622	770
8978	30 gal. single element		2.30	3.478		490	190		680	835
8979	40 gal. single element		2.20	3.636		615	199		814	985
8988	Commercial (ASHRAE energy std. 90)									

22 33 Electric Domestic Water Heaters

22 33 13 – Instantaneous Electrical Domestic Water Heaters

22 33 13.20	Instantaneous Elec. Point-of-Use Water Heaters	Crew	Daily Output	Labor-Hours	Unit	Material	2012 Bare Costs Labor	Equipment	Total	Total Incl O&P
8989	6 gallon	1 Plum	2.50	3.200	Ea.	650	175		825	985
8990	10 gallon		2.50	3.200		690	175		865	1,025
8991	15 gallon		2.40	3.333		735	182		917	1,100
8992	20 gallon		2.40	3.333		785	182		967	1,150
8993	30 gallon		2.30	3.478		1,650	190		1,840	2,125
8995	Under the sink, copper, w/bracket									
8996	2.5 gallon	1 Plum	4	2	Ea.	335	109		444	535
9000	Minimum labor/equipment charge	"	1.75	4.571	Job		250		250	390

22 33 30 – Residential, Electric Domestic Water Heaters

22 33 30.13	Residential, Small-Capacity Elec. Water Heaters	Crew	Daily Output	Labor-Hours	Unit	Material	Labor	Equipment	Total	Total Incl O&P
0010	RESIDENTIAL, SMALL-CAPACITY ELECTRIC DOMESTIC WATER HEATERS									
0050										
1000	Residential, electric, glass lined tank, 5 yr., 10 gal., single element	1 Plum	2.30	3.478	Ea.	385	190		575	720
1060	30 gallon, double element		2.20	3.636		575	199		774	945
1080	40 gallon, double element		2	4		615	218		833	1,025
1100	52 gallon, double element		2	4		690	218		908	1,100
1120	66 gallon, double element		1.80	4.444		930	243		1,173	1,400
1140	80 gallon, double element		1.60	5		1,025	273		1,298	1,575

22 33 33 – Light-Commercial Electric Domestic Water Heaters

22 33 33.10	Commercial Electric Water Heaters	Crew	Daily Output	Labor-Hours	Unit	Material	Labor	Equipment	Total	Total Incl O&P
0010	COMMERCIAL ELECTRIC WATER HEATERS									
4000	Commercial, 100° rise. NOTE: for each size tank, a range of									
4010	heaters between the ones shown are available									
4020	Electric									
4100	5 gal., 3 kW, 12 GPH, 208 volt	1 Plum	2	4	Ea.	2,625	218		2,843	3,225
4160	50 gal., 36 kW, 148 GPH, 208 volt	"	1.80	4.444		6,075	243		6,318	7,075
4480	400 gal., 210 kW, 860 GPH, 480 volt	Q-1	1	16		36,600	785		37,385	41,500

22 34 Fuel-Fired Domestic Water Heaters

22 34 13 – Instantaneous, Tankless, Gas Domestic Water Heaters

22 34 13.10	Instantaneous, Tankless, Gas Water Heaters	Crew	Daily Output	Labor-Hours	Unit	Material	Labor	Equipment	Total	Total Incl O&P
0010	INSTANTANEOUS, TANKLESS, GAS WATER HEATERS									
9410	Natural gas/propane, 3.2 GPM C	1 Plum	2	4	Ea.	535	218		753	930
9420	6.4 GPM		1.90	4.211		630	230		860	1,050
9430	8.4 GPM		1.80	4.444		755	243		998	1,225
9440	9.5 GPM		1.60	5		930	273		1,203	1,450

22 34 30 – Residential Gas Domestic Water Heaters

22 34 30.13	Residential, Atmos, Gas Domestic Wtr Heaters	Crew	Daily Output	Labor-Hours	Unit	Material	Labor	Equipment	Total	Total Incl O&P
0010	RESIDENTIAL, ATMOSPHERIC, GAS DOMESTIC WATER HEATERS									
2000	Gas fired, foam lined tank, 10 yr., vent not incl.									
2040	30 gallon	1 Plum	2	4	Ea.	895	218		1,113	1,325
2060	40 gallon		1.90	4.211		895	230		1,125	1,350
2100	75 gallon		1.50	5.333		1,225	291		1,516	1,800

22 34 36 – Commercial Gas Domestic Water Heaters

22 34 36.13	Commercial, Atmos., Gas Domestic Water Htrs.	Crew	Daily Output	Labor-Hours	Unit	Material	Labor	Equipment	Total	Total Incl O&P
0010	COMMERCIAL, ATMOSPHERIC, GAS DOMESTIC WATER HEATERS									
6000	Gas fired, flush jacket, std. controls, vent not incl.									
6040	75 MBH input, 73 GPH	1 Plum	1.40	5.714	Ea.	1,575	310		1,885	2,200
6060	98 MBH input, 95 GPH		1.40	5.714		4,575	310		4,885	5,525

22 34 Fuel-Fired Domestic Water Heaters

22 34 36 – Commercial Gas Domestic Water Heaters

22 34 36.13 Commercial, Atmos., Gas Domestic Water Htrs.		Crew	Daily Output	Labor-Hours	Unit	Material	Labor	Equipment	Total	Total Incl O&P
						2012 Bare Costs				
6180	200 MBH input, 192 GPH	1 Plum	.60	13.333	Ea.	8,025	730		8,755	9,950

22 34 46 – Oil-Fired Domestic Water Heaters

22 34 46.10 Residential Oil-Fired Water Heaters

Line	Description	Crew	Daily Output	Labor-Hours	Unit	Material	Labor	Equipment	Total	Total Incl O&P
0010	**RESIDENTIAL OIL-FIRED WATER HEATERS**									
3000	Oil fired, glass lined tank, 5 yr., vent not included, 30 gallon	1 Plum	2	4	Ea.	665	218		883	1,075
3040	50 gallon	"	1.80	4.444	"	1,050	243		1,293	1,525

22 34 46.20 Commercial Oil-Fired Water Heaters

Line	Description	Crew	Daily Output	Labor-Hours	Unit	Material	Labor	Equipment	Total	Total Incl O&P
0010	**COMMERCIAL OIL-FIRED WATER HEATERS**									
8000	Oil fired, glass lined, UL listed, std. controls, vent not incl.									
8060	140 gal., 140 MBH input, 134 GPH	Q-1	2.13	7.512	Ea.	14,900	370		15,270	17,000
8080	140 gal., 199 MBH input, 191 GPH		2	8		15,400	395		15,795	17,500
8160	140 gal., 540 MBH input, 519 GPH		.96	16.667		21,000	820		21,820	24,400
8280	201 gal., 1250 MBH input, 1200 GPH	Q-2	1.22	19.672		32,900	1,000		33,900	37,700

22 41 Residential Plumbing Fixtures

22 41 13 – Residential Water Closets, Urinals, and Bidets

22 41 13.40 Water Closets

Line	Description	Crew	Daily Output	Labor-Hours	Unit	Material	Labor	Equipment	Total	Total Incl O&P
0010	**WATER CLOSETS** R224000-40									
0032	For automatic flush, see Line 22 42 39.10 0972									
0150	Tank type, vitreous china, incl. seat, supply pipe w/stop, 1.6 gpf or noted									
0200	Wall hung									
0400	Two piece, close coupled	Q-1	5.30	3.019	Ea.	605	148		753	895
0960	For rough-in, supply, waste, vent and carrier	"	2.73	5.861	"	1,000	288		1,288	1,550
0999	Floor mounted									
1100	Two piece, close coupled	Q-1	5.30	3.019	Ea.	247	148		395	505
1102	Economy		5.30	3.019		137	148		285	380
1110	Two piece, close coupled, dual flush		5.30	3.019		273	148		421	530
1140	Two piece, close coupled, 1.28 gpf, ADA G		5.30	3.019		295	148		443	555
1960	For color, add					30%				
1980	For rough-in, supply, waste and vent	Q-1	3.05	5.246	Ea.	340	258		598	775

22 41 16 – Residential Lavatories and Sinks

22 41 16.10 Lavatories

Line	Description	Crew	Daily Output	Labor-Hours	Unit	Material	Labor	Equipment	Total	Total Incl O&P
0010	**LAVATORIES**, With trim, white unless noted otherwise R224000-40									
0020										
0500	Vanity top, porcelain enamel on cast iron									
0600	20" x 18"	Q-1	6.40	2.500	Ea.	305	123		428	525
0640	33" x 19" oval	"	6.40	2.500	"	625	123		748	875
0860	For color, add					25%				
1000	Cultured marble, 19" x 17", single bowl	Q-1	6.40	2.500	Ea.	179	123		302	390
1120	25" x 22", single bowl		6.40	2.500		220	123		343	435
1900	Stainless steel, self-rimming, 25" x 22", single bowl, ledge		6.40	2.500		360	123		483	585
1960	17" x 22", single bowl		6.40	2.500		350	123		473	575
2600	Steel, enameled, 20" x 17", single bowl		5.80	2.759		188	136		324	420
2900	Vitreous china, 20" x 16", single bowl		5.40	2.963		260	146		406	515
2960	20" x 17", single bowl		5.40	2.963		176	146		322	420
3580	Rough-in, supply, waste and vent for all above lavatories		2.30	6.957		420	340		760	995
4000	Wall hung									
4040	Porcelain enamel on cast iron, 16" x 14", single bowl	Q-1	8	2	Ea.	460	98.50		558.50	660
4180	20" x 18", single bowl	"	8	2	"	355	98.50		453.50	550
4580	For color, add					30%				

22 41 16 – Residential Lavatories and Sinks

22 41 16.10 Lavatories		Crew	Daily Output	Labor-Hours	Unit	2012 Bare Costs Material	Labor	Equipment	Total	Total Incl O&P
6000	Vitreous china, 18" x 15", single bowl with backsplash	Q-1	7	2.286	Ea.	231	112		343	430
6960	Rough-in, supply, waste and vent for above lavatories	"	1.66	9.639	"	480	475		955	1,275
7000	Pedestal type									
7600	Vitreous china, 27" x 21", white	Q-1	6.60	2.424	Ea.	555	119		674	800
7610	27" x 21", colored		6.60	2.424		670	119		789	920
7620	27" x 21", premium color		6.60	2.424		785	119		904	1,050
7660	26" x 20", white		6.60	2.424		505	119		624	740
7670	26" x 20", colored		6.60	2.424		600	119		719	845
7680	26" x 20", premium color		6.60	2.424		635	119		754	885
7700	24" x 20", white		6.60	2.424		490	119		609	720
7710	24" x 20", colored		6.60	2.424		585	119		704	825
7720	24" x 20", premium color		6.60	2.424		680	119		799	930
7760	21" x 18", white		6.60	2.424		235	119		354	445
7770	21" x 18", colored		6.60	2.424		267	119		386	480
7990	Rough-in, supply, waste and vent for pedestal lavatories		1.66	9.639		480	475		955	1,275
9000	Minimum labor/equipment charge	1 Plum	3	2.667	Job		146		146	227

22 41 16.30 Sinks		Crew	Daily Output	Labor-Hours	Unit	Material	Labor	Equipment	Total	Total Incl O&P
0010	**SINKS**, With faucets and drain									
2000	Kitchen, counter top style, P.E. on C.I., 24" x 21" single bowl	Q-1	5.60	2.857	Ea.	257	140		397	500
2100	31" x 22" single bowl		5.60	2.857		525	140		665	795
2200	32" x 21" double bowl		4.80	3.333		286	164		450	570
3000	Stainless steel, self rimming, 19" x 18" single bowl		5.60	2.857		580	140		720	855
3100	25" x 22" single bowl		5.60	2.857		645	140		785	930
3200	33" x 22" double bowl		4.80	3.333		940	164		1,104	1,275
3300	43" x 22" double bowl		4.80	3.333		1,100	164		1,264	1,450
4000	Steel, enameled, with ledge, 24" x 21" single bowl		5.60	2.857		440	140		580	705
4100	32" x 21" double bowl		4.80	3.333		400	164		564	695
4960	For color sinks except stainless steel, add					10%				
4980	For rough-in, supply, waste and vent, counter top sinks	Q-1	2.14	7.477		470	365		835	1,100
5790	For rough-in, supply, waste & vent, sinks	"	1.85	8.649		470	425		895	1,175

22 41 19 – Residential Bathtubs

22 41 19.10 Baths		Crew	Daily Output	Labor-Hours	Unit	Material	Labor	Equipment	Total	Total Incl O&P
0010	**BATHS** R224000-40									
0100	Tubs, recessed porcelain enamel on cast iron, with trim									
0180	48" x 42"	Q-1	4	4	Ea.	2,225	197		2,422	2,750
0220	72" x 36"	"	3	5.333	"	2,500	262		2,762	3,150
0300	Mat bottom									
0380	5' long	Q-1	4.40	3.636	Ea.	1,025	179		1,204	1,400
0560	Corner 48" x 44"		4.40	3.636		2,375	179		2,554	2,875
2000	Enameled formed steel, 4'-6" long		5.80	2.759		490	136		626	750
4600	Module tub & showerwall surround, molded fiberglass									
4610	5' long x 34" wide x 76" high	Q-1	4	4	Ea.	790	197		987	1,175
4750	Handicap with 1-1/2" OD grab bar, antiskid bottom									
4760	60" x 32-3/4" x 72" high	Q-1	4	4	Ea.	785	197		982	1,175
4770	60" x 30" x 71" high with molded seat		3.50	4.571		775	225		1,000	1,200
9600	Rough-in, supply, waste and vent, for all above tubs, add		2.07	7.729		485	380		865	1,125
9900	Minimum labor/equipment charge		3	5.333	Job		262		262	410

22 41 Residential Plumbing Fixtures

22 41 23 – Residential Showers

22 41 23.20 Showers		Crew	Daily Output	Labor-Hours	Unit	Material	Labor	Equipment	Total	Total Incl O&P
						2012 Bare Costs				
0011	**SHOWERS**, Stall, with drain only									
0020										
1520	32" square	Q-1	5	3.200	Ea.	745	157		902	1,075
1530	36" square		4.80	3.333		2,175	164		2,339	2,650
1540	Terrazzo receptor, 32" square		5	3.200		1,075	157		1,232	1,450
1580	36" corner angle		4.80	3.333		1,275	164		1,439	1,650
3000	Fiberglass, one piece, with 3 walls, 32" x 32" square		5.50	2.909		545	143		688	820
3100	36" x 36" square		5.50	2.909		555	143		698	840
4200	Rough-in, supply, waste and vent for above showers		2.05	7.805		590	385		975	1,250

22 41 23.40 Shower System Components

Line	Description	Crew	Daily Output	Labor-Hours	Unit	Material	Labor	Equipment	Total	Total Incl O&P
0010	**SHOWER SYSTEM COMPONENTS**									
5500	Head, water economizer, 1.6 GPM G	1 Plum	24	.333	Ea.	52	18.20		70.20	85.50

22 41 36 – Residential Laundry Trays

22 41 36.10 Laundry Sinks

Line	Description	Crew	Daily Output	Labor-Hours	Unit	Material	Labor	Equipment	Total	Total Incl O&P
0010	**LAUNDRY SINKS**, With trim									
0020	Porcelain enamel on cast iron, black iron frame									
0050	24" x 21", single compartment	Q-1	6	2.667	Ea.	465	131		596	715
0100	26" x 21", single compartment	"	6	2.667	"	450	131		581	700
2000	Molded stone, on wall hanger or legs									
2020	22" x 23", single compartment	Q-1	6	2.667	Ea.	138	131		269	355
2100	45" x 21", double compartment	"	5	3.200	"	259	157		416	530
3000	Plastic, on wall hanger or legs									
3020	18" x 23", single compartment	Q-1	6.50	2.462	Ea.	149	121		270	355
3100	20" x 24", single compartment		6.50	2.462		149	121		270	355
3200	36" x 23", double compartment		5.50	2.909		176	143		319	415
3300	40" x 24", double compartment		5.50	2.909		267	143		410	515
5000	Stainless steel, counter top, 22" x 17" single compartment		6	2.667		61.50	131		192.50	272
5200	33" x 22", double compartment		5	3.200		73.50	157		230.50	325
9600	Rough-in, supply, waste and vent, for all laundry sinks		2.14	7.477		470	365		835	1,100
9810	Minimum labor/equipment charge	1 Plum	3	2.667	Job		146		146	227

22 41 39 – Residential Faucets, Supplies and Trim

22 41 39.10 Faucets and Fittings

Line	Description	Crew	Daily Output	Labor-Hours	Unit	Material	Labor	Equipment	Total	Total Incl O&P
0010	**FAUCETS AND FITTINGS**									
0150	Bath, faucets, diverter spout combination, sweat	1 Plum	8	1	Ea.	86.50	54.50		141	180
0200	For integral stops, IPS unions, add					109			109	120
0420	Bath, press-bal mix valve w/diverter, spout, shower hd, arm/flange	1 Plum	8	1		163	54.50		217.50	264
0500	Drain, central lift, 1-1/2" IPS male		20	.400		71	22		93	113
0600	Trip lever, 1-1/2" IPS male		20	.400		45	22		67	83.50
0810	Bidet									
0812	Fitting, over the rim, swivel spray/pop-up drain	1 Plum	8	1	Ea.	172	54.50		226.50	274
1000	Kitchen sink faucets, top mount, cast spout		10	.800		61.50	43.50		105	136
1100	For spray, add		24	.333		16.15	18.20		34.35	46.50
1200	Wall type, swing tube spout		10	.800		74.50	43.50		118	150
1300	Single control lever handle									
1310	With pull out spray									
1320	Polished chrome	1 Plum	10	.800	Ea.	171	43.50		214.50	257
2000	Laundry faucets, shelf type, IPS or copper unions		12	.667		49.50	36.50		86	111
2100	Lavatory faucet, centerset, without drain		10	.800		44.50	43.50		88	117
2120	With pop-up drain		6.66	1.201		62.50	65.50		128	171
2210	Porcelain cross handles and pop-up drain									
2220	Polished chrome	1 Plum	6.66	1.201	Ea.	128	65.50		193.50	243

22 41 39 – Residential Faucets, Supplies and Trim

22 41 39.10 Faucets and Fittings		Crew	Daily Output	Labor-Hours	Unit	Material	Labor	Equipment	Total	Total Incl O&P
						2012 Bare Costs				
2230	Polished brass	1 Plum	6.66	1.201	Ea.	193	65.50		258.50	315
2260	Single lever handle and pop-up drain									
2280	Satin nickel	1 Plum	6.66	1.201	Ea.	274	65.50		339.50	400
2290	Polished chrome		6.66	1.201		182	65.50		247.50	300
2800	Self-closing, center set		10	.800		131	43.50		174.50	212
2810	Automatic sensor and operator, with faucet head G		6.15	1.301		370	71		441	520
4000	Shower by-pass valve with union		18	.444		68.50	24.50		93	113
4100	Shower arm with flange and head		22	.364		71.50	19.85		91.35	110
4140	Shower, hand held, pin mount, massage action, chrome		22	.364		70	19.85		89.85	108
4142	Polished brass		22	.364		134	19.85		153.85	178
4144	Shower, hand held, wall mtd, adj. spray, 2 wall mounts, chrome		20	.400		94	22		116	137
4146	Polished brass		20	.400		182	22		204	234
4148	Shower, hand held head, bar mounted 24", adj. spray, chrome		20	.400		151	22		173	200
4150	Polished brass		20	.400		320	22		342	385
4200	Shower thermostatic mixing valve, concealed, with shower head trim kit	↓	8	1		320	54.50		374.50	435
4204	For inlet strainer, check, and stops, add				↓	49.50			49.50	54.50
4220	Shower pressure balancing mixing valve,									
4230	With shower head, arm, flange and diverter tub spout									
4240	Chrome	1 Plum	6.14	1.303	Ea.	222	71		293	355
4250	Polished brass		6.14	1.303		279	71		350	415
4260	Satin		6.14	1.303		279	71		350	415
4270	Polished chrome/brass		6.14	1.303		229	71		300	365
5000	Sillcock, compact, brass, IPS or copper to hose	↓	24	.333	↓	13.30	18.20		31.50	43
6000	Stop and waste valves, bronze									
6100	Angle, solder end 1/2"	1 Plum	24	.333	Ea.	11.35	18.20		29.55	41
6110	3/4"		20	.400		13.30	22		35.30	48.50
6300	Straightway, solder end 3/8"		24	.333		11.35	18.20		29.55	41
6310	1/2"		24	.333		11.60	18.20		29.80	41.50
6410	Straightway, threaded 1/2"		24	.333		15	18.20		33.20	45
6420	3/4"		20	.400		21	22		43	57
6430	1"		19	.421		12.60	23		35.60	50
7800	Water closet, wax gasket		96	.083		1.83	4.55		6.38	9.10
7820	Gasket toilet tank to bowl		32	.250		2.38	13.65		16.03	24
7830	Replacement diaphragm washer assy for ballcock valve		12	.667		2.38	36.50		38.88	59
7850	Dual flush valve	↓	12	.667	↓	30	36.50		66.50	89.50
8000	Water supply stops, polished chrome plate									
8200	Angle, 3/8"	1 Plum	24	.333	Ea.	9.10	18.20		27.30	38.50
8300	1/2"		22	.364		9.10	19.85		28.95	41
8400	Straight, 3/8"		26	.308		9.55	16.80		26.35	36.50
8500	1/2"		24	.333		9.55	18.20		27.75	39
8600	Water closet, angle, w/flex riser, 3/8"		24	.333	↓	35	18.20		53.20	67
9000	Minimum labor/equipment charge	↓	4	2	Job		109		109	170

22 42 Commercial Plumbing Fixtures

22 42 13 – Commercial Water Closets, Urinals, and Bidets

22 42 13.30 Urinals		Crew	Daily Output	Labor-Hours	Unit	2012 Bare Costs Material	Labor	Equipment	Total	Total Incl O&P
0010	**URINALS**									
3000	Wall hung, vitreous china, with hanger & self-closing valve									
3100	Siphon jet type	Q-1	3	5.333	Ea.	293	262		555	735
3120	Blowout type		3	5.333		285	262		547	725
3140	Water saving .5 gpf G		3	5.333		655	262		917	1,125
3300	Rough-in, supply, waste & vent		2.83	5.654		585	278		863	1,075
5000	Stall type, vitreous china, includes valve		2.50	6.400		655	315		970	1,200
6980	Rough-in, supply, waste and vent		1.99	8.040		590	395		985	1,275
9000	Minimum labor/equipment charge		4	4	Job		197		197	305

22 42 13.40 Water Closets

Code	Description	Crew	Daily Output	Labor-Hours	Unit	Material	Labor	Equipment	Total	Total Incl O&P
0010	**WATER CLOSETS**									
3000	Bowl only, with flush valve, seat, 1.6 gpf unless noted									
3100	Wall hung	Q-1	5.80	2.759	Ea.	455	136		591	710
3200	For rough-in, supply, waste and vent, single WC		2.56	6.250		1,075	305		1,380	1,650
3360	With floor outlet, 1.28 gpf G		5.80	2.759		505	136		641	765
3362	With floor outlet, 1.28 gpf, ADA G		5.80	2.759		595	136		731	865
9000	Minimum labor/equipment charge		4	4	Job		197		197	305

22 42 16 – Commercial Lavatories and Sinks

22 42 16.10 Handwasher-Dryer Module

Code	Description	Crew	Daily Output	Labor-Hours	Unit	Material	Labor	Equipment	Total	Total Incl O&P
0010	**HANDWASHER-DRYER MODULE**									
0030	Wall mounted									
0040	With electric dryer									
0050	Sensor operated	Q-1	8	2	Ea.	3,650	98.50		3,748.50	4,150
0110	Sensor operated (ADA)		8	2		3,650	98.50		3,748.50	4,175
0140	Sensor operated (ADA), surface mounted		8	2		4,400	98.50		4,498.50	5,000
0150	With paper towels									
0180	Sensor operated	Q-1	8	2	Ea.	3,550	98.50		3,648.50	4,050

22 42 16.14 Lavatories

Code	Description	Crew	Daily Output	Labor-Hours	Unit	Material	Labor	Equipment	Total	Total Incl O&P
0010	**LAVATORIES**, With trim, white unless noted otherwise									
0020	Commercial lavatories same as residential. See Section 22 41 16.10									

22 42 16.40 Service Sinks

Code	Description	Crew	Daily Output	Labor-Hours	Unit	Material	Labor	Equipment	Total	Total Incl O&P
0010	**SERVICE SINKS**									
6650	Service, floor, corner, P.E. on C.I., 28" x 28"	Q-1	4.40	3.636	Ea.	880	179		1,059	1,250
6790	For rough-in, supply, waste & vent, floor service sinks		1.64	9.756	"	1,400	480		1,880	2,275
9000	Minimum labor/equipment charge		4	4	Job		197		197	305

22 42 23 – Commercial Showers

22 42 23.30 Group Showers

Code	Description	Crew	Daily Output	Labor-Hours	Unit	Material	Labor	Equipment	Total	Total Incl O&P
0010	**GROUP SHOWERS**									
9000	Minimum labor/equipment charge	Q-1	4	4	Job		197		197	305

22 42 33 – Wash Fountains

22 42 33.20 Commercial Wash Fountains

Code	Description	Crew	Daily Output	Labor-Hours	Unit	Material	Labor	Equipment	Total	Total Incl O&P
0010	**COMMERCIAL WASH FOUNTAINS**									
1900	Group, foot control									
2000	Precast terrazzo, circular, 36" diam., 5 or 6 persons	Q-2	3	8	Ea.	5,450	410		5,860	6,625
2100	54" diameter for 8 or 10 persons		2.50	9.600		6,800	490		7,290	8,225
2400	Semi-circular, 36" diam. for 3 persons		3	8		4,775	410		5,185	5,875
2500	54" diam. for 4 or 5 persons		2.50	9.600		6,050	490		6,540	7,400
5610	Group, infrared control, barrier free									
5614	Precast terrazzo									
5620	Semi-circular 36" diam. for 3 persons	Q-2	3	8	Ea.	7,150	410		7,560	8,500

22 42 33 – Wash Fountains

22 42 33.20 Commercial Wash Fountains		Crew	Daily Output	Labor-Hours	Unit	Material	2012 Bare Costs Labor	Equipment	Total	Total Incl O&P
5630	46″ diam. for 4 persons	Q-2	2.80	8.571	Ea.	7,725	435		8,160	9,175
5640	Circular, 54″ diam. for 8 persons, button control		2.50	9.600		9,125	490		9,615	10,800
5700	Rough-in, supply, waste and vent for above wash fountains	Q-1	1.82	8.791		675	430		1,105	1,425
9000	Minimum labor/equipment charge	Q-2	3	8	Job		410		410	635

22 42 39 – Commercial Faucets, Supplies, and Trim

22 42 39.10 Faucets and Fittings

Line	Description	Crew	Daily Output	Labor-Hours	Unit	Material	Labor	Equipment	Total	Total Incl O&P
0010	**FAUCETS AND FITTINGS**									
0840	Flush valves, with vacuum breaker									
0850	Water closet									
0860	Exposed, rear spud	1 Plum	8	1	Ea.	137	54.50		191.50	236
0870	Top spud		8	1		126	54.50		180.50	224
0880	Concealed, rear spud		8	1		180	54.50		234.50	283
0890	Top spud		8	1		149	54.50		203.50	249
0900	Wall hung		8	1		159	54.50		213.50	260
0910	Dual flush flushometer		12	.667		197	36.50		233.50	273
0912	Flushometer retrofit kit		18	.444		14.75	24.50		39.25	54.50
0920	Urinal									
0930	Exposed, stall	1 Plum	8	1	Ea.	126	54.50		180.50	224
0940	Wall, (washout)		8	1		126	54.50		180.50	224
0950	Pedestal, top spud		8	1		129	54.50		183.50	227
0960	Concealed, stall		8	1		145	54.50		199.50	244
0970	Wall (washout)		8	1		156	54.50		210.50	257
0971	Automatic flush sensor and operator for G									
0972	urinals or water closets, standard G	1 Plum	8	1	Ea.	415	54.50		469.50	545
0980	High efficiency water saving									
0984	Water closets, 1.28 gpf G	1 Plum	8	1	Ea.	395	54.50		449.50	520
0988	Urinals, .5 gpf G	″	8	1	″	395	54.50		449.50	520
2790	Faucets for lavatories									
2800	Self-closing, center set	1 Plum	10	.800	Ea.	131	43.50		174.50	212
2810	Automatic sensor and operator, with faucet head		6.15	1.301		370	71		441	520
3000	Service sink faucet, cast spout, pail hook, hose end		14	.571		80	31		111	137

22 42 39.30 Carriers and Supports

Line	Description	Crew	Daily Output	Labor-Hours	Unit	Material	Labor	Equipment	Total	Total Incl O&P
0010	**CARRIERS AND SUPPORTS**, For plumbing fixtures									
0020										
3000	Lavatory, concealed arm									
3050	Floor mounted, single									
3100	High back fixture	1 Plum	6	1.333	Ea.	425	73		498	580
3200	Flat slab fixture		6	1.333		370	73		443	520
3220	Paraplegic		6	1.333		475	73		548	635
6980	Water closet, siphon jet									
7000	Horizontal, adjustable, caulk									
7040	Single, 4″ pipe size	1 Plum	5.33	1.501	Ea.	815	82		897	1,025
7060	5″ pipe size		5.33	1.501		1,025	82		1,107	1,250
7100	Double, 4″ pipe size		5	1.600		1,425	87.50		1,512.50	1,675
7120	5″ pipe size		5	1.600		1,725	87.50		1,812.50	2,000
8200	Water closet, residential									
8220	Vertical centerline, floor mount									
8240	Single, 3″ caulk, 2″ or 3″ vent	1 Plum	6	1.333	Ea.	490	73		563	655
8260	4″ caulk, 2″ or 4″ vent		6	1.333	″	630	73		703	810
9990	Minimum labor/equipment charge		3.50	2.286	Job		125		125	194

22 43 Healthcare Plumbing Fixtures

22 43 39 – Healthcare Faucets

22 43 39.10 Faucets and Fittings		Crew	Daily Output	Labor-Hours	Unit	Material	2012 Bare Costs Labor	Equipment	Total	Total Incl O&P
0010	**FAUCETS AND FITTINGS**									
2850	Medical, bedpan cleanser, with pedal valve,	1 Plum	12	.667	Ea.	760	36.50		796.50	895
2860	With screwdriver stop valve		12	.667		395	36.50		431.50	490
2870	With self-closing spray valve		12	.667		243	36.50		279.50	325
2900	Faucet, gooseneck spout, wrist handles, grid drain		10	.800		150	43.50		193.50	233
2940	Mixing valve, knee action, screwdriver stops		4	2		425	109		534	640

22 45 Emergency Plumbing Fixtures

22 45 13 – Emergency Showers

22 45 13.10 Emergency Showers

		Crew	Daily Output	Labor-Hours	Unit	Material	Labor	Equipment	Total	Total Incl O&P
0010	**EMERGENCY SHOWERS**, Rough-in not included									
0020										
5000	Shower, single head, drench, ball valve, pull, freestanding	Q-1	4	4	Ea.	350	197		547	690
5200	Horizontal or vertical supply		4	4		520	197		717	875
6000	Multi-nozzle, eye/face wash combination		4	4		615	197		812	980
6400	Multi-nozzle, 12 spray, shower only		4	4		1,950	197		2,147	2,450
6600	For freeze-proof, add		6	2.667		440	131		571	690
9000	Minimum labor/equipment charge		3	5.333	Job		262		262	410

22 45 16 – Eyewash Equipment

22 45 16.10 Eyewash Safety Equipment

		Crew	Daily Output	Labor-Hours	Unit	Material	Labor	Equipment	Total	Total Incl O&P
0010	**EYEWASH SAFETY EQUIPMENT**, Rough-in not included									
1000	Eye wash fountain									
1400	Plastic bowl, pedestal mounted	Q-1	4	4	Ea.	262	197		459	595

22 47 Drinking Fountains and Water Coolers

22 47 13 – Drinking Fountains

22 47 13.10 Drinking Water Fountains

		Crew	Daily Output	Labor-Hours	Unit	Material	Labor	Equipment	Total	Total Incl O&P
0010	**DRINKING WATER FOUNTAINS**, For connection to cold water supply									
1000	Wall mounted, non-recessed									
2700	Stainless steel, single bubbler, no back	1 Plum	4	2	Ea.	920	109		1,029	1,175
2740	With back		4	2		635	109		744	870
2780	Dual handle & wheelchair projection type		4	2		685	109		794	925
2820	Dual level for handicapped type		3.20	2.500		1,400	137		1,537	1,775
3980	For rough-in, supply and waste, add		2.21	3.620		345	198		543	690
4000	Wall mounted, semi-recessed									
4200	Poly-marble, single bubbler	1 Plum	4	2	Ea.	925	109		1,034	1,200
4600	Stainless steel, satin finish, single bubbler	"	4	2	"	995	109		1,104	1,275
6000	Wall mounted, fully recessed									
6400	Poly-marble, single bubbler	1 Plum	4	2	Ea.	1,525	109		1,634	1,850
6800	Stainless steel, single bubbler		4	2		1,525	109		1,634	1,850
7580	For rough-in, supply and waste, add		1.83	4.372		345	239		584	750
7590										
7600	Floor mounted, pedestal type									
8600	Enameled iron, heavy duty service, 2 bubblers	1 Plum	2	4	Ea.	2,500	218		2,718	3,125
8880	For freeze-proof valve system, add		2	4		610	218		828	1,000
8900	For rough-in, supply and waste, add		1.83	4.372		345	239		584	750
9000	Minimum labor/equipment charge		2	4	Job		218		218	340

22 47 Drinking Fountains and Water Coolers

22 47 16 – Pressure Water Coolers

22 47 16.10 Electric Water Coolers		Crew	Daily Output	Labor-Hours	Unit	Material	2012 Bare Costs Labor	Equipment	Total	Total Incl O&P
0010	**ELECTRIC WATER COOLERS**									
0100	Wall mounted, non-recessed									
0140	4 GPH	Q-1	4	4	Ea.	680	197		877	1,050
0160	8 GPH, barrier free, sensor operated		4	4		890	197		1,087	1,275
1000	Dual height, 8.2 GPH		3.80	4.211		1,975	207		2,182	2,500
1040	14.3 GPH		3.80	4.211		1,125	207		1,332	1,575
3300	Semi-recessed, 8.1 GPH		4	4		630	197		827	995
4600	Floor mounted, flush-to-wall									
4640	4 GPH	1 Plum	3	2.667	Ea.	620	146		766	910
4980	For stainless steel cabinet, add					128			128	141
5000	Dual height, 8.2 GPH	1 Plum	2	4		1,100	218		1,318	1,575
9000	Minimum labor/equipment charge	"	2	4	Job		218		218	340

22 66 Chemical-Waste Systems for Lab. and Healthcare Facilities

22 66 53 – Laboratory Chemical-Waste and Vent Piping

22 66 53.30 Glass Pipe

Line	Description	Crew	Daily Output	Labor-Hours	Unit	Material	Labor	Equipment	Total	Total Incl O&P
0010	**GLASS PIPE**, Borosilicate, couplings & clevis hanger assemblies, 10′ O.C.									
0020	Drainage									
1100	1-1/2″ diameter	Q-1	52	.308	L.F.	10	15.10		25.10	34.50
1120	2″ diameter		44	.364		13.05	17.85		30.90	42.50
1140	3″ diameter		39	.410		17.65	20		37.65	51
1160	4″ diameter		30	.533		31.50	26		57.50	75.50
1180	6″ diameter		26	.615		58.50	30.50		89	111
9000	Minimum labor/equipment charge	1 Plum	4	2	Job		109		109	170

22 66 53.40 Pipe Fittings, Glass

Line	Description	Crew	Daily Output	Labor-Hours	Unit	Material	Labor	Equipment	Total	Total Incl O&P
0010	**PIPE FITTINGS, GLASS**									
0020	Drainage, beaded ends									
0040	Coupling & labor required at joints not incl. in fitting									
0050	price. Add 1 per joint for installed price									
0070	90° Bend or sweep, 1-1/2″				Ea.	30			30	33
0090	2″					38			38	41.50
0100	3″					62.50			62.50	68.50
0110	4″					100			100	110
0120	6″ (sweep only)					305			305	335
0350	Tee, single sanitary, 1-1/2″					49			49	53.50
0370	2″					49			49	53.50
0380	3″					72.50			72.50	79.50
0390	4″					130			130	143
0400	6″					345			345	380
0500	Coupling, stainless steel, TFE seal ring									
0520	1-1/2″	Q-1	32	.500	Ea.	21	24.50		45.50	61.50
0530	2″		30	.533		26.50	26		52.50	70
0540	3″		25	.640		36	31.50		67.50	88.50
0550	4″		23	.696		61.50	34		95.50	122
0560	6″		20	.800		139	39.50		178.50	214
9000	Minimum labor/equipment charge	1 Plum	4	2	Job		109		109	170

22 66 53.60 Corrosion Resistant Pipe

Line	Description	Crew	Daily Output	Labor-Hours	Unit	Material	Labor	Equipment	Total	Total Incl O&P
0010	**CORROSION RESISTANT PIPE**, No couplings or hangers									
0020	Iron alloy, drain, mechanical joint									
1000	1-1/2″ diameter	Q-1	70	.229	L.F.	46	11.25		57.25	68

22 66 Chemical-Waste Systems for Lab. and Healthcare Facilities

22 66 53 – Laboratory Chemical-Waste and Vent Piping

	22 66 53.60 Corrosion Resistant Pipe	Crew	Daily Output	Labor-Hours	Unit	Material	2012 Bare Costs Labor	Equipment	Total	Total Incl O&P
1100	2" diameter	Q-1	66	.242	L.F.	47	11.90		58.90	70
1120	3" diameter		60	.267		60.50	13.10		73.60	87
1140	4" diameter		52	.308		77.50	15.10		92.60	109
2980	Plastic, epoxy, fiberglass filament wound, B&S joint									
3000	2" diameter	Q-1	62	.258	L.F.	12	12.70		24.70	33
3100	3" diameter		51	.314		14	15.40		29.40	39.50
3120	4" diameter		45	.356		20	17.50		37.50	49
3160	8" diameter	Q-2	38	.632		44	32		76	98.50
3200	12" diameter	"	28	.857		72	43.50		115.50	148
9800	Minimum labor/equipment charge	1 Plum	4	2	Job		109		109	170

	22 66 53.70 Pipe Fittings, Corrosion Resistant	Crew	Daily Output	Labor-Hours	Unit	Material	Labor	Equipment	Total	Total Incl O&P
0010	**PIPE FITTINGS, CORROSION RESISTANT**									
0030	Iron alloy									
0050	Mechanical joint									
0060	1/4 Bend, 1-1/2"	Q-1	12	1.333	Ea.	77	65.50		142.50	187
0080	2"		10	1.600		126	78.50		204.50	262
0090	3"		9	1.778		151	87.50		238.50	300
0100	4"		8	2		174	98.50		272.50	345
0160	Tee and Y, sanitary, straight									
0170	1-1/2"	Q-1	8	2	Ea.	84	98.50		182.50	246
0180	2"		7	2.286		112	112		224	298
0190	3"		6	2.667		174	131		305	395
0200	4"		5	3.200		320	157		477	595
0360	Coupling, 1-1/2"		14	1.143		47.50	56		103.50	140
0380	2"		12	1.333		54	65.50		119.50	162
0390	3"		11	1.455		56.50	71.50		128	174
0400	4"		10	1.600		64.50	78.50		143	194
3000	Epoxy, filament wound									
3030	Quick-lock joint									
3040	90° Elbow, 2"	Q-1	28	.571	Ea.	97.50	28		125.50	151
3060	3"		16	1		112	49		161	200
3070	4"		13	1.231		153	60.50		213.50	263
3190	Tee, 2"		19	.842		233	41.50		274.50	320
3200	3"		11	1.455		280	71.50		351.50	420
3210	4"		9	1.778		335	87.50		422.50	505
9000	Minimum labor/equipment charge	1 Plum	4	2	Job		109		109	170

Estimating Tips

The labor adjustment factors listed in Subdivision 22 01 02.20 also apply to Division 23.

23 10 00 Facility Fuel Systems

- The prices in this subdivision for above- and below-ground storage tanks do not include foundations or hold-down slabs, unless noted. The estimator should refer to Divisions 3 and 31 for foundation system pricing. In addition to the foundations, required tank accessories, such as tank gauges, leak detection devices, and additional manholes and piping, must be added to the tank prices.

23 50 00 Central Heating Equipment

- When estimating the cost of an HVAC system, check to see who is responsible for providing and installing the temperature control system. It is possible to overlook controls, assuming that they would be included in the electrical estimate.
- When looking up a boiler, be careful on specified capacity. Some manufacturers rate their products on output while others use input.
- Include HVAC insulation for pipe, boiler, and duct (wrap and liner).
- Be careful when looking up mechanical items to get the correct pressure rating and connection type (thread, weld, flange).

23 70 00 Central HVAC Equipment

- Combination heating and cooling units are sized by the air conditioning requirements. (See Reference No. R236000-20 for preliminary sizing guide.)
- A ton of air conditioning is nominally 400 CFM.
- Rectangular duct is taken off by the linear foot for each size, but its cost is usually estimated by the pound. Remember that SMACNA standards now base duct on internal pressure.
- Prefabricated duct is estimated and purchased like pipe: straight sections and fittings.
- Note that cranes or other lifting equipment are not included on any lines in Division 23. For example, if a crane is required to lift a heavy piece of pipe into place high above a gym floor, or to put a rooftop unit on the roof of a four-story building, etc., it must be added. Due to the potential for extreme variation—from nothing additional required to a major crane or helicopter—we feel that including a nominal amount for "lifting contingency" would be useless and detract from the accuracy of the estimate. When using equipment rental cost data from RSMeans, do not forget to include the cost of the operator(s).

Reference Numbers

Reference numbers are shown in shaded boxes at the beginning of some major classifications. These numbers refer to related items in the Reference Section. The reference information may be an estimating procedure, an alternate pricing method, or technical information.

Note: Not all subdivisions listed here necessarily appear in this publication.

Note: **Trade Service,** *in part, has been used as a reference source for some of the material prices used in Division 23.*

23 05 Common Work Results for HVAC

23 05 05 – Selective HVAC Demolition

23 05 05.10 HVAC Demolition		Crew	Daily Output	Labor-Hours	Unit	Material	2012 Bare Costs Labor	Equipment	Total	Total Incl O&P
0010	**HVAC DEMOLITION** R220105-10									
0100	Air conditioner, split unit, 3 ton	Q-5	2	8	Ea.		400		400	625
0150	Package unit, 3 ton R024119-10	Q-6	3	8	"		415		415	650
0298	Boilers									
0300	Electric, up thru 148 kW	Q-19	2	12	Ea.		605		605	945
0310	150 thru 518 kW	"	1	24			1,225		1,225	1,875
0320	550 thru 2000 kW	Q-21	.40	80			4,150		4,150	6,450
0330	2070 kW and up	"	.30	106			5,525		5,525	8,600
0340	Gas and/or oil, up thru 150 MBH	Q-7	2.20	14.545			770		770	1,200
0350	160 thru 2000 MBH		.80	40			2,125		2,125	3,300
0360	2100 thru 4500 MBH		.50	64			3,400		3,400	5,300
0370	4600 thru 7000 MBH		.30	106			5,650		5,650	8,825
0380	7100 thru 12,000 MBH		.16	200			10,600		10,600	16,500
0390	12,200 thru 25,000 MBH		.12	266			14,100		14,100	22,000
1000	Ductwork, 4" high, 8" wide	1 Clab	200	.040	L.F.		1.40		1.40	2.30
1020	10" wide		190	.042			1.48		1.48	2.42
1040	14" wide		180	.044			1.56		1.56	2.56
1100	6" high, 8" wide		165	.048			1.70		1.70	2.79
1120	12" wide		150	.053			1.87		1.87	3.07
1140	18" wide		135	.059			2.08		2.08	3.41
1200	10" high, 12" wide		125	.064			2.25		2.25	3.68
1220	18" wide		115	.070			2.44		2.44	4
1240	24" wide		110	.073			2.55		2.55	4.19
1300	12"-14" high, 16"-18" wide		85	.094			3.30		3.30	5.40
1320	24" wide		75	.107			3.74		3.74	6.15
1340	48" wide		71	.113			3.96		3.96	6.50
1400	18" high, 24" wide		67	.119			4.19		4.19	6.85
1420	36" wide		63	.127			4.46		4.46	7.30
1440	48" wide		59	.136			4.76		4.76	7.80
1500	30" high, 36" wide		56	.143			5		5	8.20
1520	48" wide		53	.151			5.30		5.30	8.70
1540	72" wide		50	.160			5.60		5.60	9.20
1550	Duct heater, electric strip	1 Elec	8	1	Ea.		51.50		51.50	79.50
1850	Minimum labor/equipment charge	1 Clab	3	2.667	Job		93.50		93.50	153
2200	Furnace, electric	Q-20	2	10	Ea.		480		480	750
2300	Gas or oil, under 120 MBH	Q-9	4	4			188		188	296
2340	Over 120 MBH	"	3	5.333			250		250	395
2800	Heat pump, package unit, 3 ton	Q-5	2.40	6.667			335		335	520
2840	Split unit, 3 ton		2	8			400		400	625
3000	Mechanical equipment, light items. Unit is weight, not cooling.		.90	17.778	Ton		890		890	1,400
3600	Heavy items		1.10	14.545	"		730		730	1,125
9000	Minimum labor/equipment charge	Q-6	3	8	Job		415		415	650

23 05 23 – General-Duty Valves for HVAC Piping

23 05 23.30 Valves, Iron Body		Crew	Daily Output	Labor-Hours	Unit	Material	Labor	Equipment	Total	Total Incl O&P
0010	**VALVES, IRON BODY**									
1020	Butterfly, wafer type, gear actuator, 200 lb.									
1030	2"	1 Plum	14	.571	Ea.	172	31		203	239
1060	4"	Q-1	5	3.200	"	209	157		366	475
1650	Gate, 125 lb., N.R.S.									
2150	Flanged									
2240	2-1/2"	Q-1	5	3.200	Ea.	625	157		782	930
2260	3"		4.50	3.556		700	175		875	1,050

23 05 Common Work Results for HVAC

23 05 23 – General-Duty Valves for HVAC Piping

23 05 23.30 Valves, Iron Body		Crew	Daily Output	Labor-Hours	Unit	2012 Bare Costs Material	Labor	Equipment	Total	Total Incl O&P
2280	4″	Q-1	3	5.333	Ea.	1,000	262		1,262	1,500
3550	OS&Y, 125 lb., flanged									
3680	4″	Q-1	3	5.333	Ea.	635	262		897	1,100
3690	5″	Q-2	3.40	7.059		1,050	360		1,410	1,700
3700	6″	″	3	8		1,050	410		1,460	1,775
9000	Minimum labor/equipment charge	1 Plum	3	2.667	Job		146		146	227

23 07 HVAC Insulation

23 07 13 – Duct Insulation

23 07 13.10 Duct Thermal Insulation			Crew	Daily Output	Labor-Hours	Unit	Material	Labor	Equipment	Total	Total Incl O&P
0010	**DUCT THERMAL INSULATION**										
0100	Rule of thumb, as a percentage of total mechanical costs					Job				10%	
0110	Insulation req'd is based on the surface size/area to be covered										
3000	Ductwork										
3020	Blanket type, fiberglass, flexible										
3140	FSK vapor barrier wrap, .75 lb. density										
3160	1″ thick	G	Q-14	350	.046	S.F.	.18	2.01		2.19	3.43
3170	1-1/2″ thick	G		320	.050		.21	2.20		2.41	3.77
3180	2″ thick	G		300	.053		.26	2.35		2.61	4.06
3190	3″ thick	G		260	.062		.37	2.71		3.08	4.76
3200	4″ thick	G		242	.066		.60	2.91		3.51	5.35
3210	Vinyl jacket, same as FSK										
9600	Minimum labor/equipment charge		1 Stpi	4	2	Job		111		111	174

23 07 16 – HVAC Equipment Insulation

23 07 16.10 HVAC Equipment Thermal Insulation		Crew	Daily Output	Labor-Hours	Unit	Material	Labor	Equipment	Total	Total Incl O&P
0010	**HVAC EQUIPMENT THERMAL INSULATION**									
0100	Rule of thumb, as a percentage of total mechanical costs				Job				10%	
0110	Insulation req'd is based on the surface size/area to be covered									

23 09 Instrumentation and Control for HVAC

23 09 33 – Electric and Electronic Control System for HVAC

23 09 33.10 Electronic Control Systems		Crew	Daily Output	Labor-Hours	Unit	Material	Labor	Equipment	Total	Total Incl O&P
0010	**ELECTRONIC CONTROL SYSTEMS**									
9000	Minimum labor/equipment charge	1 Plum	8	1	Job		54.50		54.50	85

23 09 53 – Pneumatic and Electric Control System for HVAC

23 09 53.10 Control Components			Crew	Daily Output	Labor-Hours	Unit	Material	Labor	Equipment	Total	Total Incl O&P
0010	**CONTROL COMPONENTS**										
5000	Thermostats										
5030	Manual		1 Stpi	8	1	Ea.	45	55.50		100.50	136
5040	1 set back, electric, timed	G		8	1		35	55.50		90.50	126
5050	2 set back, electric, timed	G		8	1		161	55.50		216.50	264
5200	24 hour, automatic, clock	G	1 Shee	8	1		142	52		194	238
6000	Valves, motorized zone										
6100	Sweat connections, 1/2″ C x C		1 Stpi	20	.400	Ea.	159	22.50		181.50	210
6110	3/4″ C x C			20	.400		163	22.50		185.50	215
6120	1″ C x C			19	.421		212	23.50		235.50	271
9000	Minimum labor/equipment charge		1 Plum	4	2	Job		109		109	170

23 21 20.34 Dielectric Unions

	23 21 20.34 Dielectric Unions	Crew	Daily Output	Labor-Hours	Unit	Material	2012 Bare Costs Labor	Equipment	Total	Total Incl O&P
0010	**DIELECTRIC UNIONS**, Standard gaskets for water and air									
0020	250 psi maximum pressure									
0280	Female IPT to sweat, straight									
0340	3/4" pipe size	1 Plum	20	.400	Ea.	6.10	22		28.10	40.50
0780	Female IPT to female IPT, straight									
0800	1/2" pipe size	1 Plum	24	.333	Ea.	10.35	18.20		28.55	40
0840	3/4" pipe size		20	.400	"	11.70	22		33.70	47
9000	Minimum labor/equipment charge		4	2	Job		109		109	170

23 21 20.46 Expansion Tanks

	23 21 20.46 Expansion Tanks	Crew	Daily Output	Labor-Hours	Unit	Material	Labor	Equipment	Total	Total Incl O&P
0010	**EXPANSION TANKS**									
2000	Steel, liquid expansion, ASME, painted, 15 gallon capacity	Q-5	17	.941	Ea.	595	47		642	730
2040	30 gallon capacity		12	1.333		665	67		732	840
2080	60 gallon capacity		8	2		935	100		1,035	1,175
2120	100 gallon capacity		6	2.667		1,350	134		1,484	1,700
3000	Steel ASME expansion, rubber diaphragm, 19 gal. cap. accept.		12	1.333		2,450	67		2,517	2,800
3020	31 gallon capacity		8	2		2,600	100		2,700	3,000
3040	61 gallon capacity		6	2.667		3,650	134		3,784	4,225
3080	119 gallon capacity		4	4		3,925	201		4,126	4,625
9000	Minimum labor/equipment charge		4	4	Job		201		201	310

23 21 20.70 Steam Traps

	23 21 20.70 Steam Traps	Crew	Daily Output	Labor-Hours	Unit	Material	Labor	Equipment	Total	Total Incl O&P
0010	**STEAM TRAPS**									
0030	Cast iron body, threaded									
0040	Inverted bucket									
0050	1/2" pipe size	1 Stpi	12	.667	Ea.	136	37		173	207
0100	1" pipe size		9	.889		365	49.50		414.50	475
0120	1-1/4" pipe size		8	1		550	55.50		605.50	690
1000	Float & thermostatic, 15 psi									
1010	3/4" pipe size	1 Stpi	16	.500	Ea.	109	28		137	164
1020	1" pipe size		15	.533		131	29.50		160.50	191
1040	1-1/2" pipe size		9	.889		231	49.50		280.50	330
1060	2" pipe size		6	1.333		425	74.50		499.50	585
9000	Minimum labor/equipment charge		4	2	Job		111		111	174

23 21 20.76 Strainers, Y Type, Bronze Body

	23 21 20.76 Strainers, Y Type, Bronze Body	Crew	Daily Output	Labor-Hours	Unit	Material	Labor	Equipment	Total	Total Incl O&P
0010	**STRAINERS, Y TYPE, BRONZE BODY**									
0050	Screwed, 125 lb., 1/4" pipe size	1 Stpi	24	.333	Ea.	22	18.55		40.55	53
0100	1/2" pipe size		20	.400		29	22.50		51.50	66
0120	3/4" pipe size		19	.421		32.50	23.50		56	72.50
0140	1" pipe size		17	.471		38.50	26		64.50	83
0160	1-1/2" pipe size		14	.571		83	32		115	141
0180	2" pipe size		13	.615		110	34.50		144.50	175
0182	3" pipe size		12	.667		835	37		872	980
0220	3" pipe size	Q-5	16	1		960	50		1,010	1,125
0240	4" pipe size	"	15	1.067		2,200	53.50		2,253.50	2,475
1000	Flanged, 150 lb., 1-1/2" pipe size	1 Stpi	11	.727		450	40.50		490.50	560
1020	2" pipe size	"	8	1		610	55.50		665.50	755
1030	2-1/2" pipe size	Q-5	5	3.200		905	160		1,065	1,250
1040	3" pipe size		4.50	3.556		1,125	178		1,303	1,500
1060	4" pipe size		3	5.333		1,700	267		1,967	2,275
1100	6" pipe size	Q-6	3	8		3,225	415		3,640	4,200
1106	8" pipe size	"	2.60	9.231		3,550	480		4,030	4,650
1500	For 300 lb. rating, add					40%				
9000	Minimum labor/equipment charge	1 Stpi	3.75	2.133	Job		119		119	185

23 21 Hydronic Piping and Pumps

23 21 23 – Hydronic Pumps

23 21 23.13 In-Line Centrifugal Hydronic Pumps		Crew	Daily Output	Labor-Hours	Unit	Material	Labor	Equipment	Total	Total Incl O&P
						2012 Bare Costs				
0010	**IN-LINE CENTRIFUGAL HYDRONIC PUMPS**									
0600	Bronze, sweat connections, 1/40 HP, in line									
0640	3/4" size	Q-1	16	1	Ea.	206	49		255	305
1000	Flange connection, 3/4" to 1-1/2" size									
1040	1/12 HP	Q-1	6	2.667	Ea.	530	131		661	790
1060	1/8 HP		6	2.667		915	131		1,046	1,200
1100	1/3 HP		6	2.667		1,025	131		1,156	1,325
1140	2" size, 1/6 HP		5	3.200		1,325	157		1,482	1,700
1180	2-1/2" size, 1/4 HP		5	3.200		1,675	157		1,832	2,100
1220	3" size, 1/4 HP		4	4		1,750	197		1,947	2,225
9000	Minimum labor/equipment charge		3.25	4.923	Job		242		242	375

23 31 HVAC Ducts and Casings

23 31 13 – Metal Ducts

23 31 13.13 Rectangular Metal Ducts

		Crew	Daily Output	Labor-Hours	Unit	Material	Labor	Equipment	Total	Total Incl O&P
0010	**RECTANGULAR METAL DUCTS** R233100-20									
0020	Fabricated rectangular, includes fittings, joints, supports,									
0100	Aluminum, alloy 3003-H14, under 100 lb.	Q-10	75	.320	Lb.	3.58	15.55		19.13	28.50
0110	100 to 500 lb.		80	.300		2.34	14.60		16.94	25.50
0120	500 to 1,000 lb.		95	.253		2.21	12.30		14.51	22
0500	Galvanized steel, under 200 lb.		235	.102		.70	4.97		5.67	8.60
0520	200 to 500 lb.		245	.098		.69	4.77		5.46	8.25
0540	500 to 1,000 lb.		255	.094		.68	4.58		5.26	7.95
0560	1,000 to 2,000 lb.		265	.091		.67	4.41		5.08	7.70
0580	Over 5,000 lb.		285	.084		.66	4.10		4.76	7.20
0590										
1000	Stainless steel, type 304, under 100 lb.	Q-10	165	.145	Lb.	6.35	7.10		13.45	18.10
1020	100 to 500 lb.		175	.137		4.05	6.65		10.70	14.95
1030	500 to 1,000 lb.		190	.126		2.95	6.15		9.10	12.95

23 31 13.16 Round and Flat-Oval Spiral Ducts

		Crew	Daily Output	Labor-Hours	Unit	Material	Labor	Equipment	Total	Total Incl O&P
0010	**ROUND AND FLAT-OVAL SPIRAL DUCTS**									
9990	Minimum labor/equipment charge	1 Shee	3	2.667	Job		139		139	219

23 31 13.19 Metal Duct Fittings

		Crew	Daily Output	Labor-Hours	Unit	Material	Labor	Equipment	Total	Total Incl O&P
0010	**METAL DUCT FITTINGS**									

23 31 16 – Nonmetal Ducts

23 31 16.13 Fibrous-Glass Ducts

		Crew	Daily Output	Labor-Hours	Unit	Material	Labor	Equipment	Total	Total Incl O&P
0010	**FIBROUS-GLASS DUCTS** R233100-20									
9990	Minimum labor/equipment charge	1 Shee	3	2.667	Job		139		139	219

23 31 16.19 PVC Ducts

		Crew	Daily Output	Labor-Hours	Unit	Material	Labor	Equipment	Total	Total Incl O&P
0010	**PVC DUCTS**									
4000	Rigid plastic, corrosive fume resistant PVC									
4020	Straight, 6" diameter	Q-9	220	.073	L.F.	10.05	3.41		13.46	16.40
4040	8" diameter		160	.100		15.35	4.69		20.04	24.50
4060	10" diameter		120	.133		19.40	6.25		25.65	31.50
4070	12" diameter		100	.160		22.50	7.50		30	37
4080	14" diameter		70	.229		28	10.75		38.75	48
4090	16" diameter		62	.258		32.50	12.10		44.60	55
4100	18" diameter		58	.276		44.50	12.95		57.45	69.50
4110	20" diameter	Q-10	75	.320		54.50	15.55		70.05	84.50

23 31 HVAC Ducts and Casings

23 31 16 – Nonmetal Ducts

23 31 16.19 PVC Ducts		Crew	Daily Output	Labor-Hours	Unit	Material	2012 Bare Costs Labor	Equipment	Total	Total Incl O&P
4130	24" diameter	Q-10	55	.436	L.F.	57.50	21		78.50	96.50
4250	Coupling, 6" diameter	Q-9	88	.182	Ea.	22	8.55		30.55	37.50
4270	8" diameter		78	.205		28.50	9.65		38.15	46
4290	10" diameter		70	.229		32.50	10.75		43.25	53
4300	12" diameter		55	.291		35.50	13.65		49.15	60.50
4310	14" diameter		44	.364		46	17.05		63.05	77.50
4320	16" diameter		40	.400		48.50	18.75		67.25	83
4330	18" diameter	↓	38	.421		63	19.75		82.75	100
4340	20" diameter	Q-10	55	.436		75.50	21		96.50	117
4360	24" diameter	"	45	.533		94.50	26		120.50	145
4470	Elbow, 90°, 6" diameter	Q-9	44	.364		99	17.05		116.05	136
4490	8" diameter		28	.571		111	27		138	165
4510	10" diameter		18	.889		135	41.50		176.50	215
4520	12" diameter		15	1.067		154	50		204	248
4540	16" diameter	↓	10	1.600		231	75		306	375
4550	18" diameter	Q-10	15	1.600		320	78		398	475
4560	20" diameter		14	1.714		480	83.50		563.50	655
4580	24" diameter	↓	12	2	↓	590	97.50		687.50	800
4750	Elbow 45°, use 90° and deduct					25%				
9990	Minimum labor/equipment charge	1 Shee	3	2.667	Job		139		139	219

23 33 Air Duct Accessories

23 33 13 – Dampers

23 33 13.13 Volume-Control Dampers		Crew	Daily Output	Labor-Hours	Unit	Material	Labor	Equipment	Total	Total Incl O&P
0010	**VOLUME-CONTROL DAMPERS**									
5990	Multi-blade dampers, opposed blade, 8" x 6"	1 Shee	24	.333	Ea.	21	17.40		38.40	51
5994	8" x 8"		22	.364		22	18.95		40.95	54
5996	10" x 10"		21	.381		25.50	19.85		45.35	60
6000	12" x 12"		21	.381		28.50	19.85		48.35	63
6020	12" x 18"		18	.444		38.50	23		61.50	79
6030	14" x 10"		20	.400		28	21		49	64
6031	14" x 14"		17	.471		34	24.50		58.50	76
6033	16" x 12"		17	.471		34	24.50		58.50	76
6035	16" x 16"		16	.500		42.50	26		68.50	87.50
6037	18" x 16"		15	.533		47	28		75	95.50
6038	18" x 18"		15	.533		50.50	28		78.50	99.50
6070	20" x 16"		14	.571		50.50	30		80.50	103
6072	20" x 20"		13	.615		60.50	32		92.50	117
6074	22" x 18"		14	.571		60.50	30		90.50	114
6076	24" x 16"		11	.727		59.50	38		97.50	125
6078	24" x 20"		8	1		70.50	52		122.50	160
6080	24" x 24"		8	1		82.50	52		134.50	173
6110	26" x 26"	↓	6	1.333		90.50	69.50		160	209
6133	30" x 30"	Q-9	6.60	2.424		130	114		244	320
6135	32" x 32"		6.40	2.500		151	117		268	350
6180	48" x 36"	↓	5.60	2.857	↓	249	134		383	485
7500	Variable volume modulating motorized damper, incl. elect. mtr.									
7504	8" x 6"	1 Shee	15	.533	Ea.	115	28		143	170
7506	10" x 6"		14	.571		115	30		145	173
7510	10" x 10"		13	.615		118	32		150	181
7520	12" x 12"	↓	12	.667	↓	123	35		158	191

23 33 Air Duct Accessories

23 33 13 – Dampers

23 33 13.13 Volume-Control Dampers		Crew	Daily Output	Labor-Hours	Unit	Material	2012 Bare Costs Labor	Equipment	Total	Total Incl O&P
7522	12″ x 16″	1 Shee	11	.727	Ea.	124	38		162	196
7524	16″ x 10″		12	.667		121	35		156	188
7526	16″ x 14″		10	.800		126	41.50		167.50	204
7528	16″ x 18″		9	.889		131	46.50		177.50	217
7542	18″ x 18″		8	1		134	52		186	229
7544	20″ x 14″		8	1		137	52		189	233
7546	20″ x 18″		7	1.143		143	59.50		202.50	251
7560	24″ x 12″		8	1		138	52		190	234
7562	24″ x 18″		7	1.143		150	59.50		209.50	259
7564	24″ x 24″		6	1.333		155	69.50		224.50	279
7568	28″ x 10″		7	1.143		138	59.50		197.50	246
7590	30″ x 14″		5	1.600		147	83.50		230.50	292
7600	30″ x 18″		4	2		189	104		293	370
7610	30″ x 24″		3.80	2.105		251	110		361	450
7700	For thermostat, add		8	1		42.50	52		94.50	129
8000	Multi-blade dampers, parallel blade									
8100	8″ x 8″	1 Shee	24	.333	Ea.	82	17.40		99.40	118
8140	16″ x 10″		20	.400		103	21		124	147
8160	18″ x 12″		18	.444		113	23		136	161
8220	28″ x 16″		10	.800		150	41.50		191.50	231

23 33 13.16 Fire Dampers		Crew	Daily Output	Labor-Hours	Unit	Material	Labor	Equipment	Total	Total Incl O&P
0010	**FIRE DAMPERS**									
3000	Fire damper, curtain type, 1-1/2 hr. rated, vertical, 6″ x 6″	1 Shee	24	.333	Ea.	24	17.40		41.40	54
3020	8″ x 6″		22	.364		24	18.95		42.95	56.50
3240	16″ x 14″		18	.444		44	23		67	85
3400	24″ x 20″		8	1		54	52		106	142

23 33 19 – Duct Silencers

23 33 19.10 Duct Silencers		Crew	Daily Output	Labor-Hours	Unit	Material	Labor	Equipment	Total	Total Incl O&P
0009	**DUCT SILENCERS**									
0010	Silencers, noise control for air flow, duct				MCFM	57			57	63

23 33 23 – Turning Vanes

23 33 23.13 Air Turning Vanes		Crew	Daily Output	Labor-Hours	Unit	Material	Labor	Equipment	Total	Total Incl O&P
0010	**AIR TURNING VANES**									
9900	Minimum labor/equipment charge	1 Shee	4	2	Job		104		104	164

23 33 46 – Flexible Ducts

23 33 46.10 Flexible Air Ducts		Crew	Daily Output	Labor-Hours	Unit	Material	Labor	Equipment	Total	Total Incl O&P
0010	**FLEXIBLE AIR DUCTS** R233100-20									
1280	Add to labor for elevated installation									
1282	of prefabricated (purchased) ductwork									
1283	10′ to 15′ high						10%			
1284	15′ to 20′ high						20%			
1285	20′ to 25′ high						25%			
1286	25′ to 30′ high						35%			
1287	30′ to 35′ high						40%			
1288	35′ to 40′ high						50%			
1289	Over 40′ high						55%			
1300	Flexible, coated fiberglass fabric on corr. resist. metal helix									
1400	pressure to 12″ (WG) UL-181									
1500	Non-insulated, 3″ diameter	Q-9	400	.040	L.F.	1.12	1.88		3	4.19
1540	5″ diameter		320	.050		1.12	2.35		3.47	4.92
1560	6″ diameter		280	.057		1.40	2.68		4.08	5.75

23 33 Air Duct Accessories

23 33 46 – Flexible Ducts

23 33 46.10 Flexible Air Ducts			Crew	Daily Output	Labor-Hours	Unit	2012 Bare Costs Material	Labor	Equipment	Total	Total Incl O&P
1580	7″ diameter		Q-9	240	.067	L.F.	1.46	3.13		4.59	6.55
1600	8″ diameter			200	.080		1.76	3.75		5.51	7.85
1640	10″ diameter			160	.100		2.28	4.69		6.97	9.90
1660	12″ diameter			120	.133		2.73	6.25		8.98	12.85
1900	Insulated, 1″ thick, PE jacket, 3″ diameter	G		380	.042		2.22	1.98		4.20	5.55
1910	4″ diameter	G		340	.047		2.22	2.21		4.43	5.90
1920	5″ diameter	G		300	.053		2.23	2.50		4.73	6.40
1940	6″ diameter	G		260	.062		2.50	2.89		5.39	7.30
1960	7″ diameter	G		220	.073		2.73	3.41		6.14	8.35
1980	8″ diameter	G		180	.089		3.01	4.17		7.18	9.85
2020	10″ diameter	G		140	.114		3.66	5.35		9.01	12.50
2040	12″ diameter	G		100	.160		4.20	7.50		11.70	16.40
2060	14″ diameter	G	↓	80	.200	↓	5.10	9.40		14.50	20.50
9990	Minimum labor/equipment charge		1 Shee	3	2.667	Job		139		139	219

23 33 53 – Duct Liners

23 33 53.10 Duct Liner Board			Crew	Daily Output	Labor-Hours	Unit	Material	Labor	Equipment	Total	Total Incl O&P
0010	**DUCT LINER BOARD**										
3490	Board type, fiberglass liner, 3 lb. density										
3940	Board type, non-fibrous foam										
3950	Temperature, bacteria and fungi resistant										
3960	1″ thick	G	Q-14	150	.107	S.F.	2.25	4.69		6.94	10.05
3970	1-1/2″ thick	G		130	.123		2.95	5.40		8.35	11.95
3980	2″ thick	G	↓	120	.133	↓	3.60	5.85		9.45	13.40

23 34 HVAC Fans

23 34 13 – Axial HVAC Fans

23 34 13.10 Axial Flow HVAC Fans		Crew	Daily Output	Labor-Hours	Unit	Material	Labor	Equipment	Total	Total Incl O&P
0010	**AXIAL FLOW HVAC FANS**									
0020	Air conditioning and process air handling									
0030	Axial flow, compact, low sound, 2.5″ S.P.									
0050	3,800 CFM, 5 HP	Q-20	3.40	5.882	Ea.	4,800	281		5,081	5,725
0120	15,600 CFM, 10 HP	″	1.60	12.500	″	8,400	600		9,000	10,200

23 34 14 – Blower HVAC Fans

23 34 14.10 Blower Type HVAC Fans		Crew	Daily Output	Labor-Hours	Unit	Material	Labor	Equipment	Total	Total Incl O&P
0010	**BLOWER TYPE HVAC FANS**									
2500	Ceiling fan, right angle, extra quiet, 0.10″ S.P.									
2540	210 CFM	Q-20	19	1.053	Ea.	310	50.50		360.50	425
2580	885 CFM		16	1.250		780	60		840	955
2620	2,960 CFM	↓	11	1.818		1,450	87		1,537	1,700
2640	For wall or roof cap, add	1 Shee	16	.500	↓	264	26		290	330

23 34 16 – Centrifugal HVAC Fans

23 34 16.10 Centrifugal Type HVAC Fans		Crew	Daily Output	Labor-Hours	Unit	Material	Labor	Equipment	Total	Total Incl O&P
0010	**CENTRIFUGAL TYPE HVAC FANS**									
0200	In-line centrifugal, supply/exhaust booster									
0220	aluminum wheel/hub, disconnect switch, 1/4″ S.P.									
0240	500 CFM, 10″ diameter connection	Q-20	3	6.667	Ea.	1,150	320		1,470	1,775
0280	1,520 CFM, 16″ diameter connection		2	10		1,325	480		1,805	2,200
0320	3,480 CFM, 20″ diameter connection		.80	25		1,725	1,200		2,925	3,750
0326	5,080 CFM, 20″ diameter connection	↓	.75	26.667	↓	1,875	1,275		3,150	4,050

23 34 16 – Centrifugal HVAC Fans

23 34 16.10 Centrifugal Type HVAC Fans		Crew	Daily Output	Labor-Hours	Unit	2012 Bare Costs Material	Labor	Equipment	Total	Total Incl O&P
4500	Corrosive fume resistant, plastic									
4600	roof ventilators, centrifugal, V belt drive, motor									
4620	1/4" S.P., 250 CFM, 1/4 HP	Q-20	6	3.333	Ea.	4,150	160		4,310	4,825
4640	895 CFM, 1/3 HP		5	4		4,525	191		4,716	5,275
4660	1630 CFM, 1/2 HP		4	5		5,350	239		5,589	6,250
4680	2240 CFM, 1 HP		3	6.667		5,575	320		5,895	6,625
5000	Utility set, centrifugal, V belt drive, motor									
5020	1/4" S.P., 1200 CFM, 1/4 HP	Q-20	6	3.333	Ea.	1,450	160		1,610	1,825
5040	1520 CFM, 1/3 HP		5	4		1,775	191		1,966	2,275
5060	1850 CFM, 1/2 HP		4	5		1,775	239		2,014	2,350
5080	2180 CFM, 3/4 HP		3	6.667		2,125	320		2,445	2,825
7100	Direct drive, 320 CFM, 11" sq. damper		7	2.857		625	137		762	905
7120	600 CFM, 11" sq. damper		6	3.333		785	160		945	1,100
9900	Minimum labor/equipment charge	1 Elec	4	2	Job		103		103	159

23 34 23 – HVAC Power Ventilators

23 34 23.10 HVAC Power Circulators and Ventilators		Crew	Daily Output	Labor-Hours	Unit	Material	Labor	Equipment	Total	Total Incl O&P
0010	**HVAC POWER CIRCULATORS AND VENTILATORS**									
6650	Residential, bath exhaust, grille, back draft damper									
6660	50 CFM	Q-20	24	.833	Ea.	49	40		89	117
6670	110 CFM		22	.909		82	43.50		125.50	158
6680	Light combination, squirrel cage, 100 watt, 70 CFM		24	.833		92.50	40		132.50	165
6700	Light/heater combination, ceiling mounted									
6710	70 CFM, 1450 watt	Q-20	24	.833	Ea.	174	40		214	254
6800	Heater combination, recessed, 70 CFM		24	.833		55	40		95	123
6820	With 2 infrared bulbs		23	.870		86	41.50		127.50	160
6900	Kitchen exhaust, grille, complete, 160 CFM		22	.909		92	43.50		135.50	169
6910	180 CFM		20	1		78	48		126	161
6920	270 CFM		18	1.111		141	53		194	239
6940	Residential roof jacks and wall caps									
6944	Wall cap with back draft damper									
6946	3" & 4" dia. round duct	1 Shee	11	.727	Ea.	22.50	38		60.50	84
6948	6" dia. round duct	"	11	.727	"	55.50	38		93.50	121
6958	Roof jack with bird screen and back draft damper									
6960	3" & 4" dia. round duct	1 Shee	11	.727	Ea.	22.50	38		60.50	84
6962	3-1/4" x 10" rectangular duct	"	10	.800	"	42	41.50		83.50	112
6980	Transition									
6982	3-1/4" x 10" to 6" dia. round	1 Shee	20	.400	Ea.	25.50	21		46.50	61
8020	Attic, roof type									
8030	Aluminum dome, damper & curb									
8040	6" diameter, 300 CFM	1 Elec	16	.500	Ea.	460	26		486	550
8050	7" diameter, 450 CFM		15	.533		505	27.50		532.50	600
8060	9" diameter, 900 CFM		14	.571		555	29.50		584.50	655
8080	12" diameter, 1000 CFM (gravity)		10	.800		520	41		561	635
8090	16" diameter, 1500 CFM (gravity)		9	.889		625	46		671	760
8100	20" diameter, 2500 CFM (gravity)		8	1		770	51.50		821.50	925
8110	26" diameter, 4000 CFM (gravity)		7	1.143		930	59		989	1,125
8120	32" diameter, 6500 CFM (gravity)		6	1.333		1,275	68.50		1,343.50	1,500
8130	38" diameter, 8000 CFM (gravity)		5	1.600		1,900	82.50		1,982.50	2,200
8140	50" diameter, 13,000 CFM (gravity)		4	2		2,750	103		2,853	3,175
8160	Plastic, ABS dome									
8180	1050 CFM	1 Elec	14	.571	Ea.	153	29.50		182.50	214
8200	1600 CFM	"	12	.667	"	229	34.50		263.50	305

23 34 HVAC Fans

23 34 23 – HVAC Power Ventilators

23 34 23.10 HVAC Power Circulators and Ventilators		Crew	Daily Output	Labor-Hours	Unit	Material	2012 Bare Costs Labor	Equipment	Total	Total Incl O&P
8240	Attic, wall type, with shutter, one speed									
8250	12" diameter, 1000 CFM	1 Elec	14	.571	Ea.	365	29.50		394.50	445
8260	14" diameter, 1500 CFM		12	.667		395	34.50		429.50	490
8270	16" diameter, 2000 CFM		9	.889		445	46		491	560
8290	Whole house, wall type, with shutter, one speed									
8300	30" diameter, 4800 CFM	1 Elec	7	1.143	Ea.	955	59		1,014	1,150
8310	36" diameter, 7000 CFM		6	1.333		1,050	68.50		1,118.50	1,250
8320	42" diameter, 10,000 CFM		5	1.600		1,175	82.50		1,257.50	1,400
8330	48" diameter, 16,000 CFM		4	2		1,450	103		1,553	1,750
8340	For two speed, add					87			87	96
8350	Whole house, lay-down type, with shutter, one speed									
8360	30" diameter, 4500 CFM	1 Elec	8	1	Ea.	1,025	51.50		1,076.50	1,200
8370	36" diameter, 6500 CFM		7	1.143		1,100	59		1,159	1,300
8380	42" diameter, 9000 CFM		6	1.333		1,200	68.50		1,268.50	1,425
8390	48" diameter, 12,000 CFM		5	1.600		1,375	82.50		1,457.50	1,625
8440	For two speed, add					65.50			65.50	72
8450	For 12 hour timer switch, add	1 Elec	32	.250		65.50	12.90		78.40	92

23 36 Air Terminal Units

23 36 13 – Constant-Air-Volume Units

23 36 13.10 Constant Volume Mixing Boxes

		Crew	Daily Output	Labor-Hours	Unit	Material	Labor	Equipment	Total	Total Incl O&P
0010	**CONSTANT VOLUME MIXING BOXES**									
5180	Mixing box, includes electric or pneumatic motor									
5200	Constant volume, 150 to 270 CFM	Q-9	12	1.333	Ea.	670	62.50		732.50	835
5210	270 to 600 CFM	"	11	1.455	"	685	68.50		753.50	860

23 36 16 – Variable-Air-Volume Units

23 36 16.10 Variable Volume Mixing Boxes

		Crew	Daily Output	Labor-Hours	Unit	Material	Labor	Equipment	Total	Total Incl O&P
0010	**VARIABLE VOLUME MIXING BOXES**									
5180	Mixing box, includes electric or pneumatic motor									
5500	VAV Cool only, pneumatic, pressure independent 300 to 600 CFM	Q-9	11	1.455	Ea.	385	68.50		453.50	525
5510	500 to 1000 CFM		9	1.778		395	83.50		478.50	560
5520	800 to 1600 CFM		9	1.778		405	83.50		488.50	575
5530	1100 to 2000 CFM		8	2		415	94		509	605
5540	1500 to 3000 CFM		7	2.286		440	107		547	655
5550	2000 to 4000 CFM		6	2.667		445	125		570	685
5560	For electric, w/thermostat, pressure dependent, add					23			23	25.50

23 37 Air Outlets and Inlets

23 37 13 – Diffusers, Registers, and Grilles

23 37 13.10 Diffusers

		Crew	Daily Output	Labor-Hours	Unit	Material	Labor	Equipment	Total	Total Incl O&P
0010	**DIFFUSERS**, Aluminum, opposed blade damper unless noted									
0100	Ceiling, linear, also for sidewall									
0120	2" wide	1 Shee	32	.250	L.F.	44.50	13.05		57.55	69.50
0160	4" wide		26	.308		60	16.05		76.05	91.50
0180	6" wide		24	.333		74	17.40		91.40	109
0200	8" wide		22	.364		86.50	18.95		105.45	126
0500	Perforated, 24" x 24" lay-in panel size, 6" x 6"		16	.500	Ea.	151	26		177	207
0520	8" x 8"		15	.533		159	28		187	219
0530	9" x 9"		14	.571		161	30		191	224

23 37 13.10 Diffusers		Crew	Daily Output	Labor-Hours	Unit	2012 Bare Costs Material	Labor	Equipment	Total	Total Incl O&P
0540	10″ x 10″	1 Shee	14	.571	Ea.	162	30		192	225
0560	12″ x 12″		12	.667		168	35		203	240
0590	16″ x 16″		11	.727		189	38		227	267
0600	18″ x 18″		10	.800		202	41.50		243.50	288
0610	20″ x 20″		10	.800		218	41.50		259.50	305
0620	24″ x 24″		9	.889		239	46.50		285.50	335
1000	Rectangular, 1 to 4 way blow, 6″ x 6″		16	.500		50	26		76	96
1010	8″ x 8″		15	.533		58	28		86	108
1014	9″ x 9″		15	.533		64.50	28		92.50	115
1016	10″ x 10″		15	.533		80	28		108	132
1020	12″ x 6″		15	.533		71.50	28		99.50	123
1040	12″ x 9″		14	.571		75.50	30		105.50	131
1060	12″ x 12″		12	.667		84	35		119	147
1070	14″ x 6″		13	.615		77.50	32		109.50	136
1074	14″ x 14″		12	.667		128	35		163	195
1150	18″ x 18″		9	.889		136	46.50		182.50	222
1170	24″ x 12″		10	.800		163	41.50		204.50	246
2000	T bar mounting, 24″ x 24″ lay-in frame, 6″ x 6″		16	.500		45.50	26		71.50	91
2020	8″ x 8″		14	.571		47	30		77	99
2040	12″ x 12″		12	.667		55.50	35		90.50	116
2060	16″ x 16″		11	.727		75	38		113	142
2080	18″ x 18″		10	.800		83	41.50		124.50	157
6000	For steel diffusers instead of aluminum, deduct					10%				
9000	Minimum labor/equipment charge	1 Shee	4	2	Job		104		104	164

23 37 13.30 Grilles		Crew	Daily Output	Labor-Hours	Unit	Material	Labor	Equipment	Total	Total Incl O&P
0010	**GRILLES**									
0020	Aluminum									
1000	Air return, steel, 6″ x 6″	1 Shee	26	.308	Ea.	18.15	16.05		34.20	45.50
1020	10″ x 6″		24	.333		18.15	17.40		35.55	47.50
1080	16″ x 8″		22	.364		26.50	18.95		45.45	59
1100	12″ x 12″		22	.364		26.50	18.95		45.45	59
1120	24″ x 12″		18	.444		34.50	23		57.50	74.50
1300	40″ x 24″		12	.667		125	35		160	193
9000	Minimum labor/equipment charge		4	2	Job		104		104	164

23 37 13.60 Registers		Crew	Daily Output	Labor-Hours	Unit	Material	Labor	Equipment	Total	Total Incl O&P
0010	**REGISTERS**									
0020										
0980	Air supply									
1000	Ceiling/wall, O.B. damper, anodized aluminum									
1010	One or two way deflection, adj. curved face bars									
1140	14″ x 8″	1 Shee	17	.471	Ea.	55.50	24.50		80	99.50
3000	Baseboard, hand adj. damper, enameled steel									
3012	8″ x 6″	1 Shee	26	.308	Ea.	14.85	16.05		30.90	42
3020	10″ x 6″		24	.333		16.15	17.40		33.55	45.50
3040	12″ x 5″		23	.348		17.50	18.15		35.65	48
3060	12″ x 6″		23	.340		17.50	18.15		35.65	48
3080	12″ x 8″		22	.364		25.50	18.95		44.45	58
3100	14″ x 6″		20	.400		19	21		40	54
9000	Minimum labor/equipment charge		4	2	Job		104		104	164

23 37 Air Outlets and Inlets

23 37 23 – HVAC Gravity Ventilators

23 37 23.10 HVAC Gravity Air Ventilators		Crew	Daily Output	Labor-Hours	Unit	Material	2012 Bare Costs Labor	Equipment	Total	Total Incl O&P
0010	**HVAC GRAVITY AIR VENTILATORS**, Incl. base and damper									
9000	Minimum labor/equipment charge	1 Plum	2	4	Job		218		218	340

23 38 Ventilation Hoods

23 38 13 – Commercial-Kitchen Hoods

23 38 13.10 Hood and Ventilation Equipment		Crew	Daily Output	Labor-Hours	Unit	Material	Labor	Equipment	Total	Total Incl O&P
0010	**HOOD AND VENTILATION EQUIPMENT**									
2970	Exhaust hood, sst, gutter on all sides, 4' x 4' x 2'	1 Carp	1.80	4.444	Ea.	4,675	196		4,871	5,450
2980	4' x 4' x 7'	"	1.60	5		7,425	221		7,646	8,525
7950	Hood fire protection system, minimum	Q-1	3	5.333		4,400	262		4,662	5,250
8050	Maximum	"	1	16	↓	35,300	785		36,085	40,100

23 41 Particulate Air Filtration

23 41 13 – Panel Air Filters

23 41 13.10 Panel Type Air Filters		Crew	Daily Output	Labor-Hours	Unit	Material	Labor	Equipment	Total	Total Incl O&P
0010	**PANEL TYPE AIR FILTERS**									
0020										
2950	Mechanical media filtration units									
3000	High efficiency type, with frame, non-supported G				MCFM	45			45	49.50
3100	Supported type G				"	60			60	66
5500	Throwaway glass or paper media type				Ea.	3.35			3.35	3.69

23 41 16 – Renewable-Media Air Filters

23 41 16.10 Disposable Media Air Filters		Crew	Daily Output	Labor-Hours	Unit	Material	Labor	Equipment	Total	Total Incl O&P
0010	**DISPOSABLE MEDIA AIR FILTERS**									
5000	Renewable disposable roll				MCFM	11.50			11.50	12.65

23 41 19 – Washable Air Filters

23 41 19.10 Permanent Air Filters		Crew	Daily Output	Labor-Hours	Unit	Material	Labor	Equipment	Total	Total Incl O&P
0010	**PERMANENT AIR FILTERS**									
4500	Permanent washable G				MCFM	20			20	22

23 41 23 – Extended Surface Filters

23 41 23.10 Expanded Surface Filters		Crew	Daily Output	Labor-Hours	Unit	Material	Labor	Equipment	Total	Total Incl O&P
0010	**EXPANDED SURFACE FILTERS**									
4000	Medium efficiency, extended surface G				MCFM	5.50			5.50	6.05

23 42 Gas-Phase Air Filtration

23 42 13 – Activated-Carbon Air Filtration

23 42 13.10 Charcoal Type Air Filtration		Crew	Daily Output	Labor-Hours	Unit	Material	Labor	Equipment	Total	Total Incl O&P
0010	**CHARCOAL TYPE AIR FILTRATION**									
0050	Activated charcoal type, full flow				MCFM	600			600	660
0060	Full flow, impregnated media 12" deep					225			225	248
0070	HEPA filter & frame for field erection					300			300	330
0080	HEPA filter-diffuser, ceiling install.				↓	300			300	330

23 42 Gas-Phase Air Filtration

23 42 16 – Chemically-Impregnated Adsorption Air Filtration

23 42 16.10 Chemical Adsorption Air Filtration		Crew	Daily Output	Labor-Hours	Unit	2012 Bare Costs Material	Labor	Equipment	Total	Total Incl O&P
0010	**CHEMICAL ADSORPTION AIR FILTRATION**									
9000	Minimum labor/equipment charge	1 Shee	2.25	3.556	Job		185		185	292

23 43 Electronic Air Cleaners

23 43 13 – Washable Electronic Air Cleaners

23 43 13.10 Electronic Air Cleaners

		Crew	Daily Output	Labor-Hours	Unit	Material	Labor	Equipment	Total	Total Incl O&P
0010	**ELECTRONIC AIR CLEANERS**									
2000	Electronic air cleaner, duct mounted									
2150	1000 CFM	1 Shee	4	2	Ea.	380	104		484	585
2200	1200 CFM		3.80	2.105		430	110		540	650
2250	1400 CFM		3.60	2.222		435	116		551	660

23 51 Breechings, Chimneys, and Stacks

23 51 13 – Draft Control Devices

23 51 13.16 Vent Dampers

		Crew	Daily Output	Labor-Hours	Unit	Material	Labor	Equipment	Total	Total Incl O&P
0010	**VENT DAMPERS**									
5000	Vent damper, bi-metal, gas, 3" diameter	Q-9	24	.667	Ea.	44	31.50		75.50	98
5010	4" diameter		24	.667		44	31.50		75.50	98
5020	5" diameter		23	.696		44	32.50		76.50	100
5030	6" diameter		22	.727		44	34		78	102
5040	7" diameter		21	.762		45.50	36		81.50	107
5050	8" diameter		20	.800		49	37.50		86.50	113
9000	Minimum labor/equipment charge	1 Shee	4	2	Job		104		104	164

23 51 13.19 Barometric Dampers

		Crew	Daily Output	Labor-Hours	Unit	Material	Labor	Equipment	Total	Total Incl O&P
0010	**BAROMETRIC DAMPERS**									
1000	Barometric, gas fired system only, 6" size for 5" and 6" pipes	1 Shee	20	.400	Ea.	63	21		84	103
1040	8" size, for 7" and 8" pipes	"	18	.444	"	87	23		110	133
2000	All fuel, oil, oil/gas, coal									
2020	10" for 9" and 10" pipes	1 Shee	15	.533	Ea.	146	28		174	205
3260	For thermal switch for above, add	"	24	.333	"	60	17.40		77.40	94

23 51 23 – Gas Vents

23 51 23.10 Gas Chimney Vents

		Crew	Daily Output	Labor-Hours	Unit	Material	Labor	Equipment	Total	Total Incl O&P
0010	**GAS CHIMNEY VENTS**, Prefab metal, U.L. listed									
0020	Gas, double wall, galvanized steel									
0080	3" diameter	Q-9	72	.222	V.L.F.	4.65	10.45		15.10	21.50
0100	4" diameter		68	.235		6.35	11.05		17.40	24.50
0120	5" diameter		64	.250		7.20	11.75		18.95	26.50
0140	6" diameter		60	.267		9.05	12.50		21.55	29.50
0160	7" diameter		56	.286		10.40	13.40		23.80	32.50
0180	8" diameter		52	.308		15.80	14.45		30.25	40
0200	10" diameter		48	.333		30	15.65		45.65	57.50
0220	12" diameter		44	.364		36	17.05		53.05	66.50
0260	16" diameter		40	.400		86	18.75		104.75	125
0300	20" diameter	Q-10	36	.667		125	32.50		157.50	189
0340	24" diameter	"	32	.750		197	36.50		233.50	274

23 51 Breechings, Chimneys, and Stacks

23 51 26 – All-Fuel Vent Chimneys

23 51 26.30 All-Fuel Vent Chimneys, Double Wall, St. Stl.		Crew	Daily Output	Labor-Hours	Unit	Material	2012 Bare Costs Labor	Equipment	Total	Total Incl O&P
0010	**ALL-FUEL VENT CHIMNEYS, DOUBLE WALL, STAINLESS STEEL**									
7800	6" diameter	Q-9	60	.267	V.L.F.	62	12.50		74.50	87.50
7804	8" diameter		52	.308		71	14.45		85.45	101
7806	10" diameter		48	.333		79	15.65		94.65	112
7808	12" diameter		44	.364		90.50	17.05		107.55	127
7810	14" diameter		42	.381		102	17.90		119.90	140
8000	All fuel, double wall, stainless steel fittings									
8010	Roof support 6" diameter	Q-9	30	.533	Ea.	109	25		134	160
8030	8" diameter		26	.615		127	29		156	186
8040	10" diameter		24	.667		135	31.50		166.50	198
8050	12" diameter		22	.727		143	34		177	211
8060	14" diameter		21	.762		150	36		186	222
8100	Elbow 45°, 6" diameter		30	.533		233	25		258	296
8140	8" diameter		26	.615		262	29		291	335
8160	10" diameter		24	.667		298	31.50		329.50	380
8180	12" diameter		22	.727		340	34		374	430
8200	14" diameter		21	.762		380	36		416	475
8300	Insulated tee, 6" diameter		30	.533		274	25		299	340
8360	8" diameter		26	.615		300	29		329	375
8380	10" diameter		24	.667		335	31.50		366.50	420
8400	12" diameter		22	.727		380	34		414	475
8420	14" diameter		20	.800		445	37.50		482.50	550
8500	Boot tee, 6" diameter		28	.571		525	27		552	615
8520	8" diameter		24	.667		570	31.50		601.50	675
8530	10" diameter		22	.727		665	34		699	785
8540	12" diameter		20	.800		785	37.50		822.50	925
8550	14" diameter		18	.889		895	41.50		936.50	1,050
8600	Rain cap with bird screen, 6" diameter		30	.533		280	25		305	350
8640	8" diameter		26	.615		325	29		354	400
8660	10" diameter		24	.667		380	31.50		411.50	465
8680	12" diameter		22	.727		440	34		474	535
8700	14" diameter		21	.762		495	36		531	600
8800	Flat roof flashing, 6" diameter		30	.533		103	25		128	153
8840	8" diameter		26	.615		112	29		141	169
8860	10" diameter		24	.667		120	31.50		151.50	182
8880	12" diameter		22	.727		132	34		166	199
8900	14" diameter		21	.762		135	36		171	205

23 51 33 – Insulated Sectional Chimneys

23 51 33.10 Prefabricated Insulated Sectional Chimneys		Crew	Daily Output	Labor-Hours	Unit	Material	Labor	Equipment	Total	Total Incl O&P
0010	**PREFABRICATED INSULATED SECTIONAL CHIMNEYS**									
9990	Minimum labor/equipment charge	Q-9	3	5.333	Job		250		250	395

23 52 Heating Boilers

23 52 13 – Electric Boilers

23 52 13.10 Electric Boilers, ASME		Crew	Daily Output	Labor-Hours	Unit	2012 Bare Costs Material	Labor	Equipment	Total	Total Incl O&P
0010	**ELECTRIC BOILERS, ASME**, Standard controls and trim									
1000	Steam, 6 KW, 20.5 MBH	Q-19	1.20	20	Ea.	3,625	1,000		4,625	5,575
1160	60 KW, 205 MBH		1	24		6,250	1,225		7,475	8,750
1300	296 KW, 1010 MBH		.45	53.333		13,600	2,700		16,300	19,200
1400	592 KW, 2020 MBH	Q-21	.34	94.118		25,800	4,875		30,675	36,000
1540	1110 KW, 3788 MBH	"	.19	168		36,100	8,750		44,850	53,500
2000	Hot water, 7.5 KW, 25.6 MBH	Q-19	1.30	18.462		3,825	935		4,760	5,650
2040	30 KW, 102 MBH		1.20	20		4,075	1,000		5,075	6,050
2060	45 KW, 164 MBH		1.20	20		4,550	1,000		5,550	6,575
2070	60 KW, 205 MBH		1.20	20		4,800	1,000		5,800	6,875
2080	75 KW, 256 MBH		1.10	21.818		5,175	1,100		6,275	7,425
2100	90 KW, 307 MBH		1.10	21.818		5,575	1,100		6,675	7,850
9000	Minimum labor/equipment charge	Q-20	1	20	Job		955		955	1,500

23 52 23 – Cast-Iron Boilers

23 52 23.20 Gas-Fired Boilers

Line	Description	Crew	Daily Output	Labor-Hours	Unit	Material	Labor	Equipment	Total	Total Incl O&P
0010	**GAS-FIRED BOILERS**, Natural or propane, standard controls, packaged									
1000	Cast iron, with insulated jacket									
2000	Steam, gross output, 81 MBH	Q-7	1.40	22.857	Ea.	1,950	1,200		3,150	4,050
2060	163 MBH		.90	35.556		2,375	1,875		4,250	5,575
2200	440 MBH		.51	62.500		5,525	3,325		8,850	11,300
2320	1,875 MBH		.30	106		16,400	5,650		22,050	26,800
2400	3570 MBH		.18	181		24,600	9,650		34,250	42,100
2540	6,970 MBH		.10	320		88,500	17,000		105,500	124,000
2800										
3000	Hot water, gross output, 80 MBH	Q-7	1.46	21.918	Ea.	1,775	1,175		2,950	3,775
3020	100 MBH		1.35	23.704		2,050	1,250		3,300	4,200
3040	122 MBH		1.10	29.091		2,200	1,550		3,750	4,825
3060	163 MBH		1	32		2,675	1,700		4,375	5,600
3200	440 MBH		.58	54.983		5,300	2,925		8,225	10,400
3320	2,000 MBH		.26	125		18,100	6,625		24,725	30,200
3540	6,970 MBH		.09	359		88,500	19,100		107,600	127,000
7000	For tankless water heater, add					10%				
7050	For additional zone valves up to 312 MBH add					157			157	173
9900	Minimum labor/equipment charge	Q-6	1	24	Job		1,250		1,250	1,950

23 52 23.30 Gas/Oil Fired Boilers

Line	Description	Crew	Daily Output	Labor-Hours	Unit	Material	Labor	Equipment	Total	Total Incl O&P
0010	**GAS/OIL FIRED BOILERS**, Combination with burners and controls, packaged									
1000	Cast iron with insulated jacket									
2000	Steam, gross output, 720 MBH	Q-7	.43	74.074	Ea.	13,300	3,925		17,225	20,800
2140	2,700 MBH		.19	165		25,400	8,800		34,200	41,700
2380	6,970 MBH		.09	372		93,000	19,700		112,700	133,000
2900	Hot water, gross output									
2910	200 MBH	Q-6	.62	39.024	Ea.	8,925	2,025		10,950	13,000
2920	300 MBH		.49	49.080		8,925	2,550		11,475	13,800
2930	400 MBH		.41	57.971		10,400	3,025		13,425	16,200
2940	500 MBH		.36	67.039		11,300	3,475		14,775	17,800
3000	584 MBH	Q-7	.44	72.072		12,400	3,825		16,225	19,700
3060	1,460 MBH		.28	113		24,800	6,000		30,800	36,600
3300	13,500 MBH, 403.3 BHP		.04	727		161,500	38,600		200,100	237,500

23 52 23.40 Oil-Fired Boilers

Line	Description	Crew	Daily Output	Labor-Hours	Unit	Material	Labor	Equipment	Total	Total Incl O&P
0010	**OIL-FIRED BOILERS**, Standard controls, flame retention burner, packaged									
1000	Cast iron, with insulated flush jacket									
2000	Steam, gross output, 109 MBH	Q-7	1.20	26.667	Ea.	2,025	1,425		3,450	4,425

23 52 Heating Boilers

23 52 23 – Cast-Iron Boilers

23 52 23.40 Oil-Fired Boilers		Crew	Daily Output	Labor-Hours	Unit	Material	Labor	Equipment	Total	Total Incl O&P
						2012 Bare Costs				
2060	207 MBH	Q-7	.90	35.556	Ea.	2,775	1,875		4,650	6,000
2180	1,084 MBH		.38	85.106		9,400	4,525		13,925	17,300
2200	1,360 MBH		.33	98.160		10,800	5,200		16,000	20,000
2240	2,175 MBH		.24	133		17,300	7,100		24,400	30,200
2340	4,360 MBH		.15	214		37,300	11,400		48,700	58,500
2460	6,970 MBH		.09	363		88,000	19,300		107,300	127,000
3000	Hot water, same price as steam									

23 52 26 – Steel Boilers

23 52 26.40 Oil-Fired Boilers		Crew	Daily Output	Labor-Hours	Unit	Material	Labor	Equipment	Total	Total Incl O&P
0010	**OIL-FIRED BOILERS**, Standard controls, flame retention burner									
5000	Steel, with insulated flush jacket									
7000	Hot water, gross output, 103 MBH	Q-6	1.60	15	Ea.	1,800	780		2,580	3,200
7020	122 MBH		1.45	16.506		1,900	860		2,760	3,425
7040	137 MBH		1.36	17.595		2,025	915		2,940	3,650
7060	168 MBH		1.30	18.405		2,125	955		3,080	3,850
7080	225 MBH		1.22	19.704		2,700	1,025		3,725	4,550
7100	315 MBH		.96	25.105		5,225	1,300		6,525	7,775
7120	420 MBH		.70	34.483		5,500	1,800		7,300	8,850
9000	Minimum labor/equipment charge		1.75	13.714	Job		715		715	1,100

23 52 88 – Burners

23 52 88.10 Replacement Type Burners		Crew	Daily Output	Labor-Hours	Unit	Material	Labor	Equipment	Total	Total Incl O&P
0010	**REPLACEMENT TYPE BURNERS**									
0990	Residential, conversion, gas fired, LP or natural									
1000	Gun type, atmospheric input 50 to 225 MBH	Q-1	2.50	6.400	Ea.	720	315		1,035	1,275
1020	100 to 400 MBH		2	8		1,200	395		1,595	1,950
1040	300 to 1000 MBH		1.70	9.412		3,625	465		4,090	4,700
3000	Flame retention oil fired assembly, input									
3040	2.0 to 5.0 GPH	Q-1	2	8	Ea.	345	395		740	995

23 54 Furnaces

23 54 13 – Electric-Resistance Furnaces

23 54 13.10 Electric Furnaces		Crew	Daily Output	Labor-Hours	Unit	Material	Labor	Equipment	Total	Total Incl O&P
0010	**ELECTRIC FURNACES**, Hot air, blowers, std. controls									
0011	not including gas, oil or flue piping									
1000	Electric, UL listed									
1020	10.2 MBH	Q-20	5	4	Ea.	475	191		666	820
1100	34.1 MBH	"	4.40	4.545	"	635	218		853	1,050

23 54 16 – Fuel-Fired Furnaces

23 54 16.13 Gas-Fired Furnaces		Crew	Daily Output	Labor-Hours	Unit	Material	Labor	Equipment	Total	Total Incl O&P
0010	**GAS-FIRED FURNACES**									
3000	Gas, AGA certified, upflow, direct drive models									
3020	45 MBH input	Q-9	4	4	Ea.	620	188		808	975
3040	60 MBH input		3.80	4.211		645	198		843	1,025
3060	75 MBH input		3.60	4.444		665	209		874	1,050
3100	100 MBH input		3.20	5		720	235		955	1,150
3120	125 MBH input		3	5.333		750	250		1,000	1,225
3130	150 MBH input		2.80	5.714		770	268		1,038	1,275
3140	200 MBH input		2.60	6.154		2,800	289		3,089	3,525

23 54 16 – Fuel-Fired Furnaces

	23 54 16.16 Oil-Fired Furnaces	Crew	Daily Output	Labor-Hours	Unit	2012 Bare Costs Material	Labor	Equipment	Total	Total Incl O&P
0010	**OIL-FIRED FURNACES**									
6000	Oil, UL listed, atomizing gun type burner									
6020	56 MBH output	Q-9	3.60	4.444	Ea.	1,475	209		1,684	1,950
6040	95 MBH output		3.40	4.706		1,875	221		2,096	2,425
6060	134 MBH output		3.20	5		2,175	235		2,410	2,775
6080	151 MBH output		3	5.333		2,250	250		2,500	2,875
6100	200 MBH input		2.60	6.154		2,450	289		2,739	3,150
9000	Minimum labor/equipment charge		2.75	5.818	Job		273		273	430

23 54 24 – Furnace Components for Cooling

	23 54 24.10 Furnace Components and Combinations	Crew	Daily Output	Labor-Hours	Unit	Material	Labor	Equipment	Total	Total Incl O&P
0010	**FURNACE COMPONENTS AND COMBINATIONS**									
0080	Coils, A/C evaporator, for gas or oil furnaces									
0090	Add-on, with holding charge									
0100	Upflow									
0120	1-1/2 ton cooling	Q-5	4	4	Ea.	261	201		462	595
0130	2 ton cooling		3.70	4.324		292	217		509	660
0140	3 ton cooling		3.30	4.848		455	243		698	880
0150	4 ton cooling		3	5.333		565	267		832	1,025
0160	5 ton cooling		2.70	5.926		575	297		872	1,100
0300	Downflow									
0330	2-1/2 ton cooling	Q-5	3	5.333	Ea.	345	267		612	795
0340	3-1/2 ton cooling		2.60	6.154		535	310		845	1,075
0350	5 ton cooling		2.20	7.273		575	365		940	1,200
0600	Horizontal									
0630	2 ton cooling	Q-5	3.90	4.103	Ea.	390	206		596	745
0640	3 ton cooling		3.50	4.571		430	229		659	830
0650	4 ton cooling		3.20	5		550	251		801	995
0660	5 ton cooling		2.90	5.517		550	277		827	1,025
2000	Cased evaporator coils for air handlers									
2100	1-1/2 ton cooling	Q-5	4.40	3.636	Ea.	320	182		502	635
2110	2 ton cooling		4.10	3.902		365	196		561	710
2120	2-1/2 ton cooling		3.90	4.103		385	206		591	745
2130	3 ton cooling		3.70	4.324		430	217		647	815
2140	3-1/2 ton cooling		3.50	4.571		550	229		779	960
2150	4 ton cooling		3.20	5		625	251		876	1,075
2160	5 ton cooling		2.90	5.517		705	277		982	1,200
3010	Air handler, modular									
3100	With cased evaporator cooling coil									
3120	1-1/2 ton cooling	Q-5	3.80	4.211	Ea.	1,275	211		1,486	1,725
3130	2 ton cooling		3.50	4.571		1,325	229		1,554	1,800
3140	2-1/2 ton cooling		3.30	4.848		1,425	243		1,668	1,950
3150	3 ton cooling		3.10	5.161		1,525	259		1,784	2,100
3160	3-1/2 ton cooling		2.90	5.517		1,575	277		1,852	2,175
3170	4 ton cooling		2.50	6.400		1,775	320		2,095	2,450
3180	5 ton cooling		2.10	7.619		2,025	380		2,405	2,825
3500	With no cooling coil									
3520	1-1/2 ton coil size	Q-5	12	1.333	Ea.	685	67		752	855
3530	2 ton coil size		10	1.600		685	80		765	875
3540	2-1/2 ton coil size		10	1.600		815	80		895	1,025
3554	3 ton coil size		9	1.778		925	89		1,014	1,175
3560	3-1/2 ton coil size		9	1.778		935	89		1,024	1,175
3570	4 ton coil size		8.50	1.882		1,125	94.50		1,219.50	1,400

23 54 Furnaces

23 54 24 – Furnace Components for Cooling

23 54 24.10 Furnace Components and Combinations		Crew	Daily Output	Labor-Hours	Unit	2012 Bare Costs Material	Labor	Equipment	Total	Total Incl O&P
3580	5 ton coil size	Q-5	8	2	Ea.	1,425	100		1,525	1,725
4000	With heater									
4120	5 kW, 17.1 MBH	Q-5	16	1	Ea.	670	50		720	815
4130	7.5 kW, 25.6 MBH		15.60	1.026		825	51.50		876.50	985
4140	10 kW, 34.2 MBH		15.20	1.053		875	53		928	1,050
4150	12.5 KW, 42.7 MBH		14.80	1.081		940	54		994	1,100

23 55 Fuel-Fired Heaters

23 55 13 – Fuel-Fired Duct Heaters

23 55 13.16 Gas-Fired Duct Heaters

		Crew	Daily Output	Labor-Hours	Unit	Material	Labor	Equipment	Total	Total Incl O&P
0010	**GAS-FIRED DUCT HEATERS**, Includes burner, controls, stainless steel									
0020	heat exchanger. Gas fired, electric ignition									
1000	Outdoor installation, with power venter									
1120	225 MBH output	Q-5	2.50	6.400	Ea.	5,225	320		5,545	6,250
1160	375 MBH output		1.60	10		6,425	500		6,925	7,850
1180	450 MBH output		1.40	11.429		7,450	575		8,025	9,100

23 55 33 – Fuel-Fired Unit Heaters

23 55 33.16 Gas-Fired Unit Heaters

		Crew	Daily Output	Labor-Hours	Unit	Material	Labor	Equipment	Total	Total Incl O&P
0010	**GAS-FIRED UNIT HEATERS**, Cabinet, grilles, fan, ctrls., burner, no piping									
0022	thermostat, no piping. For flue see Section 23 51 23.10									
1000	Gas fired, floor mounted									
1100	60 MBH output	Q-5	10	1.600	Ea.	830	80		910	1,025
1180	180 MBH output		6	2.667		1,300	134		1,434	1,650
2000	Suspension mounted, propeller fan, 20 MBH output		8.50	1.882		985	94.50		1,079.50	1,225
2040	60 MBH output		7	2.286		1,200	115		1,315	1,500
2100	130 MBH output		5	3.200		1,500	160		1,660	1,900
2240	320 MBH output		2	8		2,650	400		3,050	3,525
2500	For powered venter and adapter, add					355			355	390
5000	Wall furnace, 17.5 MBH output	Q-5	6	2.667		715	134		849	995
5020	24 MBH output		5	3.200		835	160		995	1,175
5040	35 MBH output		4	4		960	201		1,161	1,350
9000	Minimum labor/equipment charge		3.50	4.571	Job		229		229	355

23 56 Solar Energy Heating Equipment

23 56 16 – Packaged Solar Heating Equipment

23 56 16.40 Solar Heating Systems

		Crew	Daily Output	Labor-Hours	Unit	Material	Labor	Equipment	Total	Total Incl O&P
0010	**SOLAR HEATING SYSTEMS**									
0020	System/Package prices, not including connecting									
0030	pipe, insulation, or special heating/plumbing fixtures									
0500	Hot water, standard package, low temperature									
0580	2 collectors, circulator, fittings, 120 gal. tank G	Q-1	.40	40	Ea.	2,375	1,975		4,350	5,675

23 56 19 – Solar Heating Components

23 56 19.50 Solar Heating Ancillary

		Crew	Daily Output	Labor-Hours	Unit	Material	Labor	Equipment	Total	Total Incl O&P
0010	**SOLAR HEATING ANCILLARY**									
2300	Circulators, air									
2310	Blowers									
2580	8" diameter, 150 CFM G	Q-9	16	1	Ea.	41	47		88	120
2660	Shutter/damper G		12	1.333		52	62.50		114.50	156

23 56 19 – Solar Heating Components

23 56 19.50 Solar Heating Ancillary			Crew	Daily Output	Labor-Hours	Unit	2012 Bare Costs Material	Labor	Equipment	Total	Total Incl O&P
2670	Shutter motor	G	Q-9	16	1	Ea.	129	47		176	216
2800	Circulators, liquid, 1/25 HP, 5.3 GPM	G	Q-1	14	1.143		268	56		324	385
2820	1/20 HP, 17 GPM	G	″	12	1.333		161	65.50		226.50	279
3000	Collector panels, air with aluminum absorber plate										
3010	Wall or roof mount										
3040	Flat black, plastic glazing										
3080	4′ x 8′	G	Q-9	6	2.667	Ea.	660	125		785	925
3300	Collector panels, liquid with copper absorber plate										
3320	Black chrome, tempered glass glazing										
3330	Alum. frame, 4′ x 8′, 5/32″ single glazing	G	Q-1	9.50	1.684	Ea.	995	83		1,078	1,225
3450	Flat black, alum. frame, 3.5′ x 7.5′	G		9	1.778		690	87.50		777.50	895
3500	4′ x 8′	G		5.50	2.909		865	143		1,008	1,175
3600	Liquid, full wetted, plastic, alum. frame, 4′ x 10′	G		5	3.200		234	157		391	500
3650	Collector panel mounting, flat roof or ground rack	G		7	2.286		228	112		340	425
3670	Roof clamps	G		70	.229	Set	2.46	11.25		13.71	20
3700	Roof strap, teflon	G	1 Plum	205	.039	L.F.	19.95	2.13		22.08	25.50
3900	Differential controller with two sensors										
3930	Thermostat, hard wired	G	1 Plum	8	1	Ea.	83	54.50		137.50	177
4050	Pool valve system	G	″	2.50	3.200	″	278	175		453	575
4150	Sensors										
4220	Freeze prevention	G	1 Plum	32	.250	Ea.	23	13.65		36.65	46.50
4300	Heat exchanger										
4315	includes coil, blower, circulator										
4316	and controller for DHW and space hot air										
4380	70 MBH	G	Q-1	3.50	4.571	Ea.	350	225		575	730
4400	80 MBH	G	″	3	5.333	″	465	262		727	920
4580	Fluid to fluid package includes two circulating pumps										
4590	expansion tank, check valve, relief valve										
4600	controller, high temperature cutoff and sensors	G	Q-1	2.50	6.400	Ea.	725	315		1,040	1,275
4650	Heat transfer fluid										
4700	Propylene glycol, inhibited anti-freeze	G	1 Plum	28	.286	Gal.	14.50	15.60		30.10	40.50
4800	Solar storage tanks, knocked down										
5020	7′ x 10′-6″ – 459 C.F./3000 gallons	G	Q-10	.80	30	Ea.	15,500	1,450		16,950	19,300
5210	7′ x 14′ = 613 C.F./4000 gallons	G		.60	40		18,600	1,950		20,550	23,600
5230	10′-6″ x 14′ = 919 C.F./6000 gallons	G		.40	60		22,500	2,925		25,425	29,400
5250	14′ x 17′-6″ = 1531 C.F./10,000 gallons	G	Q-11	.30	106		29,500	5,300		34,800	40,900
7000	Solar control valves and vents										
7070	Air eliminator, automatic 3/4″ size	G	1 Plum	32	.250	Ea.	30	13.65		43.65	54.50
7090	Air vent, automatic, 1/8″ fitting	G		32	.250		17.25	13.65		30.90	40.50
7100	Manual, 1/8″ NPT	G		32	.250		2.73	13.65		16.38	24.50
7120	Backflow preventer, 1/2″ pipe size	G		16	.500		120	27.50		147.50	175
7130	3/4″ pipe size	G		16	.500		121	27.50		148.50	177
7150	Balancing valve, 3/4″ pipe size	G		20	.400		49.50	22		71.50	88.50
7180	Draindown valve, 1/2″ copper tube	G		9	.889		212	48.50		260.50	310
7200	Flow control valve, 1/2″ pipe size	G		22	.364		151	19.85		170.85	198
7220	Expansion tank, up to 5 gal.	G		32	.250		64	13.65		77.65	92
7400	Pressure gauge, 2″ dial	G		32	.250		23	13.65		36.65	47
7450	Relief valve, temp. and pressure 3/4″ pipe size	G		30	.267		18.65	14.55		33.20	43
7500	Solenoid valve, normally closed										
7750	Vacuum relief valve, 3/4″ pipe size	G	1 Plum	32	.250	Ea.	27.50	13.65		41.15	52
7800	Thermometers										
8250	Water storage tank with heat exchanger and electric element										
8270	66 gal. with 2″ x 2 lb. density insulation	G	1 Plum	1.60	5	Ea.	1,100	273		1,373	1,625

23 56 Solar Energy Heating Equipment

23 56 19 – Solar Heating Components

23 56 19.50 Solar Heating Ancillary			Crew	Daily Output	Labor-Hours	Unit	Material	2012 Bare Costs Labor	Equipment	Total	Total Incl O&P
8300	80 gal. with 2" x 2 lb. density insulation	G	1 Plum	1.60	5	Ea.	1,100	273		1,373	1,625
8500	Water storage module, plastic										
8600	Tubular, 12" diameter, 4' high	G	1 Carp	48	.167	Ea.	103	7.35		110.35	125
8610	12" diameter, 8' high	G	"	40	.200	↓	159	8.80		167.80	189
8650	Cap, 12" diameter	G				↓	19			19	21
9000	Minimum labor/equipment charge		1 Plum	2	4	Job		218		218	340

23 61 Refrigerant Compressors

23 61 16 – Reciprocating Refrigerant Compressors

23 61 16.10 Reciprocating Compressors		Crew	Daily Output	Labor-Hours	Unit	Material	Labor	Equipment	Total	Total Incl O&P
0010	**RECIPROCATING COMPRESSORS**									
0990	Refrigeration, recip. hermetic, switches & protective devices									
1000	10 ton	Q-5	1	16	Ea.	8,850	800		9,650	11,000
1100	20 ton	Q-6	.72	33.333	↓	13,400	1,725		15,125	17,400
1400	50 ton	"	.20	120	↓	17,500	6,250		23,750	28,900
1600	130 ton	Q-7	.21	152	↓	22,100	8,075		30,175	36,900

23 62 Packaged Compressor and Condenser Units

23 62 13 – Packaged Air-Cooled Refrigerant Compressor and Condenser Units

23 62 13.10 Packaged Air-Cooled Refrig. Condensing Units		Crew	Daily Output	Labor-Hours	Unit	Material	Labor	Equipment	Total	Total Incl O&P
0010	**PACKAGED AIR-COOLED REFRIGERANT CONDENSING UNITS**									
0020	Condensing unit									
0030	Air cooled, compressor, standard controls									
0050	1.5 ton	Q-5	2.50	6.400	Ea.	1,125	320		1,445	1,725
0200	2.5 ton	↓	1.70	9.412	↓	1,400	470		1,870	2,250
0300	3 ton	↓	1.30	12.308	↓	1,425	615		2,040	2,525
0400	4 ton	↓	.90	17.778	↓	1,875	890		2,765	3,450
0500	5 ton	↓	.60	26.667	↓	2,250	1,325		3,575	4,575

23 64 Packaged Water Chillers

23 64 19 – Reciprocating Water Chillers

23 64 19.10 Reciprocating Type Water Chillers		Crew	Daily Output	Labor-Hours	Unit	Material	Labor	Equipment	Total	Total Incl O&P
0010	**RECIPROCATING TYPE WATER CHILLERS**, With standard controls									
0494	Water chillers, integral air cooled condenser									
0980	Water cooled, multiple compressor, semi-hermetic, tower not incl.									
1100	50 ton cooling	Q-7	.28	113	Ea.	32,300	6,050		38,350	44,900
1160	100 ton cooling	↓	.18	179	↓	58,500	9,525		68,025	79,500
1200	145 ton cooling	↓	.16	202	↓	77,500	10,700		88,200	102,000

23 64 23 – Scroll Water Chillers

23 64 23.10 Scroll Water Chillers		Crew	Daily Output	Labor-Hours	Unit	Material	Labor	Equipment	Total	Total Incl O&P
0010	**SCROLL WATER CHILLERS**, With standard controls									
0490	Packaged w/integral air cooled condenser, 15 ton cool	Q-7	.37	86.486	Ea.	20,400	4,575		24,975	29,700
0500	20 ton cooling	↓	.34	94.118	↓	26,400	5,000		31,400	36,800
0520	40 ton cooling	↓	.30	108	↓	34,100	5,725		39,825	46,400
0680	Scroll water cooled, single compressor, hermetic, tower not incl.									
0760	10 ton cooling	Q-6	.36	67.039	Ea.	6,425	3,475		9,900	12,500

23 64 Packaged Water Chillers

23 64 26 – Rotary-Screw Water Chillers

23 64 26.10 Rotary-Screw Type Water Chillers		Crew	Daily Output	Labor-Hours	Unit	Material	2012 Bare Costs Labor	Equipment	Total	Total Incl O&P
0010	**ROTARY-SCREW TYPE WATER CHILLERS**, With standard controls									
0110	Screw, liquid chiller, air cooled, insulated evaporator									
0120	130 ton	Q-7	.14	228	Ea.	86,500	12,100		98,600	114,000
0124	160 ton	"	.13	246	"	106,500	13,100		119,600	137,500

23 64 33 – Direct Expansion Water Chillers

23 64 33.10 Direct Expansion Type Water Chillers		Crew	Daily Output	Labor-Hours	Unit	Material	Labor	Equipment	Total	Total Incl O&P
0010	**DIRECT EXPANSION TYPE WATER CHILLERS**, With standard controls									
9000	Minimum labor/equipment charge	Q-6	1	24	Job		1,250		1,250	1,950

23 65 Cooling Towers

23 65 13 – Forced-Draft Cooling Towers

23 65 13.10 Forced-Draft Type Cooling Towers		Crew	Daily Output	Labor-Hours	Unit	Material	Labor	Equipment	Total	Total Incl O&P
0010	**FORCED-DRAFT TYPE COOLING TOWERS**, Packaged units									
0070	Galvanized steel									
0080	Induced draft, crossflow									
0100	Vertical, belt drive, 61 tons	Q-6	90	.267	TonAC	197	13.85		210.85	238
0150	100 ton		100	.240		190	12.50		202.50	228
0200	115 ton		109	.220		165	11.45		176.45	200
0250	131 ton		120	.200		198	10.40		208.40	234
0260	162 ton		132	.182		160	9.45		169.45	192
1000	For higher capacities, use multiples									
1500	Induced air, double flow									
1900	Vertical, gear drive, 167 ton	Q-6	126	.190	TonAC	156	9.90		165.90	186
2000	297 ton		129	.186		96	9.65		105.65	121
2100	582 ton		132	.182		53.50	9.45		62.95	73.50
2200	1016 ton		150	.160		72.50	8.30		80.80	93
3500	For pumps and piping, add		38	.632		98.50	33		131.50	159
4000	For absorption systems, add					75%	75%			
5000	Fiberglass tower on galvanized steel support structure									
5010	Draw thru									
5100	100 ton	Q-6	1.40	17.143	Ea.	10,100	890		10,990	12,500
5120	120 ton		1.20	20		11,700	1,050		12,750	14,500
5140	140 ton		1	24		12,700	1,250		13,950	16,000
5160	160 ton		.80	30		14,100	1,550		15,650	17,900
5180	180 ton		.65	36.923		16,000	1,925		17,925	20,600
5200	200 ton		.48	50		18,100	2,600		20,700	24,000
5300	For stainless steel support structure, add					30%				
5360	For higher capacities, use multiples of each size									
6000	Stainless steel									
6010	Induced draft, crossflow, horizontal, belt drive									
6100	57 ton	Q-6	1.50	16	Ea.	25,300	830		26,130	29,100
6120	91 ton		.99	24.242		30,600	1,250		31,850	35,600
6140	111 ton		.43	55.814		38,900	2,900		41,800	47,200
6160	126 ton		.22	109		38,900	5,675		44,575	51,500
6170	Induced draft, crossflow, vertical, gear drive									
6172	167 ton	Q-6	.75	32	Ea.	49,600	1,675		51,275	57,000
6174	297 ton		.43	55.814		55,000	2,900		57,900	65,000
6176	582 ton		.23	104		87,500	5,425		92,925	104,500
6180	1016 ton		.15	160		148,000	8,325		156,325	176,000
9000	Minimum labor/equipment charge		1	24	Job		1,250		1,250	1,950

23 72 Air-to-Air Energy Recovery Equipment

23 72 13 – Heat-Wheel Air-to-Air Energy-Recovery Equipment

23 72 13.10 Heat-Wheel Air-to-Air Energy Recovery Equip.		Crew	Daily Output	Labor-Hours	Unit	Material	2012 Bare Costs Labor	Equipment	Total	Total Incl O&P
0010	**HEAT-WHEEL AIR-TO-AIR ENERGY RECOVERY EQUIPMENT**									
0100	Air to air									
9000	Minimum labor/equipment charge	1 Shee	4	2	Job		104		104	164

23 73 Indoor Central-Station Air-Handling Units

23 73 39 – Indoor, Direct Gas-Fired Heating and Ventilating Units

23 73 39.10 Make-Up Air Unit

		Crew	Daily Output	Labor-Hours	Unit	Material	Labor	Equipment	Total	Total Incl O&P
0010	**MAKE-UP AIR UNIT**									
0020	Indoor suspension, natural/LP gas, direct fired,									
0032	standard control. For flue see Section 23 51 23.10									
0040	70°F temperature rise, MBH is input									
0100	75 MBH input	Q-6	3.60	6.667	Ea.	2,400	345		2,745	3,200
0220	225 MBH input		2.40	10		3,300	520		3,820	4,425
0300	400 MBH input		1.60	15		7,450	780		8,230	9,400

23 74 Packaged Outdoor HVAC Equipment

23 74 23 – Packaged, Outdoor, Heating-Only Makeup-Air Units

23 74 23.16 Packaged, Ind.-Fired, In/Outdoor

		Crew	Daily Output	Labor-Hours	Unit	Material	Labor	Equipment	Total	Total Incl O&P
0010	**PACKAGED, IND.-FIRED, IN/OUTDOOR**, Heat-Only Makeup-Air Units									
9000	Minimum labor/equipment charge	Q-6	2.75	8.727	Job		455		455	705

23 74 33 – Dedicated Outdoor-Air Units

23 74 33.10 Roof Top Air Conditioners

		Crew	Daily Output	Labor-Hours	Unit	Material	Labor	Equipment	Total	Total Incl O&P
0010	**ROOF TOP AIR CONDITIONERS**, Standard controls, curb, economizer									
0020										
1000	Single zone, electric cool, gas heat									
1100	3 ton cooling, 60 MBH heating	Q-5	.70	22.857	Ea.	3,600	1,150		4,750	5,725
1120	4 ton cooling, 95 MBH heating	"	.61	26.403		4,250	1,325		5,575	6,725
1160	10 ton cooling, 200 MBH heating	Q-6	.67	35.982		10,600	1,875		12,475	14,500
1220	30 ton cooling, 540 MBH heating	Q-7	.47	68.376		25,500	3,625		29,125	33,800
1240	40 ton cooling, 675 MBH heating	"	.35	91.168		33,400	4,825		38,225	44,200
2000	Multizone, electric cool, gas heat, economizer									
2120	20 ton cooling, 360 MBH heating	Q-7	.53	60.038	Ea.	64,000	3,175		67,175	75,500
2180	30 ton cooling, 540 MBH heating		.37	85.562		98,500	4,550		103,050	115,500
2210	50 ton cooling, 540 MBH heating		.23	142		140,000	7,550		147,550	166,000
2220	70 ton cooling, 1500 MBH heating		.16	198		151,500	10,500		162,000	183,000
2240	80 ton cooling, 1500 MBH heating		.14	228		173,000	12,100		185,100	209,500
2280	105 ton cooling, 1500 MBH heating		.11	290		200,000	15,400		215,400	244,000
2400	For hot water heat coil, deduct					5%				
2500	For steam heat coil, deduct					2%				
2600	For electric heat, deduct					3%	5%			
9000	Minimum labor/equipment charge	Q-5	1.50	10.667	Job		535		535	835

23 81 19 – Self-Contained Air-Conditioners

23 81 19.10 Window Unit Air Conditioners

Line	Description	Crew	Daily Output	Labor-Hours	Unit	2012 Bare Costs Material	Labor	Equipment	Total	Total Incl O&P
0010	**WINDOW UNIT AIR CONDITIONERS**									
4500	10,000 BTUH	1 Carp	6	1.333	Ea.	525	59		584	675
4520	12,000 BTUH	L-2	8	2	"	645	77.50		722.50	830
9000	Minimum labor/equipment charge	1 Carp	2	4	Job		176		176	289

23 81 19.20 Self-Contained Single Package

Line	Description	Crew	Daily Output	Labor-Hours	Unit	Material	Labor	Equipment	Total	Total Incl O&P
0010	**SELF-CONTAINED SINGLE PACKAGE**									
0100	Air cooled, for free blow or duct, not incl. remote condenser									
0200	3 ton cooling	Q-5	1	16	Ea.	3,475	800		4,275	5,075
0210	4 ton cooling	"	.80	20		3,800	1,000		4,800	5,725
0240	10 ton cooling	Q-7	1	32		7,000	1,700		8,700	10,400
0280	30 ton cooling	"	.80	40		23,600	2,125		25,725	29,300
1000	Water cooled for free blow or duct, not including tower									
1010	Constant volume									
1100	3 ton cooling	Q-6	1	24	Ea.	3,475	1,250		4,725	5,775
1120	5 ton cooling	"	1	24		4,550	1,250		5,800	6,975
1140	10 ton cooling	Q-7	.90	35.556		8,550	1,875		10,425	12,400
1180	30 ton cooling	"	.70	45.714		35,500	2,425		37,925	42,900
1240	60 ton cooling	Q-8	.30	106		65,500	5,650	123	71,273	81,000
9000	Minimum labor/equipment charge	Q-5	1	16	Job		800		800	1,250

23 81 23 – Computer-Room Air-Conditioners

23 81 23.10 Computer Room Units

Line	Description	Crew	Daily Output	Labor-Hours	Unit	Material	Labor	Equipment	Total	Total Incl O&P
0010	**COMPUTER ROOM UNITS**									
1000	Air cooled, includes remote condenser but not									
1020	interconnecting tubing or refrigerant									
1160	6 ton	Q-5	.30	53.333	Ea.	33,600	2,675		36,275	41,100
1240	10 ton		.25	64		37,200	3,200		40,400	45,900
1260	12 ton		.24	66.667		38,600	3,350		41,950	47,600
1300	20 ton	Q-6	.26	92.308		49,000	4,800		53,800	61,500
1320	22 ton	"	.24	100		49,700	5,200		54,900	62,500

23 81 26 – Split-System Air-Conditioners

23 81 26.10 Split Ductless Systems

Line	Description	Crew	Daily Output	Labor-Hours	Unit	Material	Labor	Equipment	Total	Total Incl O&P
0010	**SPLIT DUCTLESS SYSTEMS**									
0100	Cooling only, single zone									
0110	Wall mount									
0120	3/4 ton cooling	Q-5	2	8	Ea.	690	400		1,090	1,375
0130	1 ton cooling		1.80	8.889		765	445		1,210	1,550
0140	1-1/2 ton cooling		1.60	10		1,225	500		1,725	2,125
0150	2 ton cooling		1.40	11.429		1,400	575		1,975	2,450
1000	Ceiling mount									
1020	2 ton cooling	Q-5	1.40	11.429	Ea.	1,250	575		1,825	2,275
1030	3 ton cooling	"	1.20	13.333	"	1,600	670		2,270	2,800
3000	Multizone									
3010	Wall mount									
3020	2 @ 3/4 ton cooling	Q-5	1.80	8.889	Ea.	2,100	445		2,545	3,025
5000	Cooling/Heating									
5010	Wall mount									
5110	1 ton cooling	Q-5	1.70	9.412	Ea.	765	470		1,235	1,575
5120	1-1/2 ton cooling	"	1.50	10.667	"	1,225	535		1,760	2,175
7000	Accessories for all split ductless systems									
7010	Add for ambient frost control	Q-5	8	2	Ea.	79	100		179	243
7020	Add for tube/wiring kit, (Line sets)									

23 81 Decentralized Unitary HVAC Equipment

23 81 26 – Split-System Air-Conditioners

23 81 26.10 Split Ductless Systems		Crew	Daily Output	Labor-Hours	Unit	Material	Labor	Equipment	Total	Total Incl O&P
						2012 Bare Costs				
7030	15′ kit	Q-5	32	.500	Ea.	96	25		121	145
7036	25′ kit		28	.571		116	28.50		144.50	173
7040	35′ kit		24	.667		169	33.50		202.50	238
7050	50′ kit		20	.800		207	40		247	291

23 81 43 – Air-Source Unitary Heat Pumps

23 81 43.10 Air-Source Heat Pumps

		Crew	Daily Output	Labor-Hours	Unit	Material	Labor	Equipment	Total	Total Incl O&P
0010	**AIR-SOURCE HEAT PUMPS**, Not including interconnecting tubing									
1000	Air to air, split system, not including curbs, pads, fan coil and ductwork									
1012	Outside condensing unit only, for fan coil see Section 23 82 19.10									
1020	2 ton cooling, 8.5 MBH heat @ 0°F	Q-5	2	8	Ea.	2,125	400		2,525	2,950
1040	3 ton cooling, 13 MBH heat @ 0°F		1.20	13.333		2,650	670		3,320	3,975
1080	7.5 ton cooling, 33 MBH heat @ 0°F		.45	35.556		5,925	1,775		7,700	9,275
1120	15 ton cooling, 64 MBH heat @ 0°F	Q-6	.50	48		12,200	2,500		14,700	17,300
1130	20 ton cooling, 85 MBH heat @ 0°F		.35	68.571		16,000	3,575		19,575	23,200
1140	25 ton cooling, 119 MBH heat @ 0°F		.25	96		19,300	5,000		24,300	29,000
1500	Single package, not including curbs, pads, or plenums									
1520	2 ton cooling, 6.5 MBH heat @ 0°F	Q-5	1.50	10.667	Ea.	2,750	535		3,285	3,850
1560	3 ton cooling, 10 MBH heat @ 0°F	"	1.20	13.333	"	3,275	670		3,945	4,650

23 81 46 – Water-Source Unitary Heat Pumps

23 81 46.10 Water Source Heat Pumps

		Crew	Daily Output	Labor-Hours	Unit	Material	Labor	Equipment	Total	Total Incl O&P
0010	**WATER SOURCE HEAT PUMPS**, Not incl. connecting tubing or water source									
2000	Water source to air, single package									
2100	1 ton cooling, 13 MBH heat @ 75°F	Q-5	2	8	Ea.	1,700	400		2,100	2,475
2140	2 ton cooling, 19 MBH heat @ 75°F		1.70	9.412		1,875	470		2,345	2,800
2220	5 ton cooling, 29 MBH heat @ 75°F		.90	17.778		2,825	890		3,715	4,500
9000	Minimum labor/equipment charge		1.75	9.143	Job		460		460	715

23 82 Convection Heating and Cooling Units

23 82 16 – Air Coils

23 82 16.10 Flanged Coils

		Crew	Daily Output	Labor-Hours	Unit	Material	Labor	Equipment	Total	Total Incl O&P
0010	**FLANGED COILS**									
1500	Hot water heating, 1 row, 24″ x 48″	Q-5	4	4	Ea.	1,525	201		1,726	2,000
9000	Minimum labor/equipment charge	1 Plum	1	8	Job		435		435	680

23 82 16.20 Duct Heaters

		Crew	Daily Output	Labor-Hours	Unit	Material	Labor	Equipment	Total	Total Incl O&P
0010	**DUCT HEATERS**, Electric, 480 V, 3 Ph.									
0020	Finned tubular insert, 500°F									
0100	8″ wide x 6″ high, 4.0 kW	Q-20	16	1.250	Ea.	740	60		800	910
0120	12″ high, 8.0 kW		15	1.333		1,225	64		1,289	1,450
0140	18″ high, 12.0 kW		14	1.429		1,725	68.50		1,793.50	2,000
0160	24″ high, 16.0 kW		13	1.538		2,225	73.50		2,298.50	2,550
0180	30″ high, 20.0 kW		12	1.667		2,700	80		2,780	3,100
0300	12″ wide x 6″ high, 6.7 kW		15	1.333		785	64		849	965
0320	12″ high, 13.3 kW		14	1.429		1,275	68.50		1,343.50	1,500
0340	18″ high, 20.0 kW		13	1.538		1,775	73.50		1,848.50	2,075
8000	To obtain BTU multiply kW by 3413									
9900	Minimum labor/equipment charge	1 Shee	4	2	Job		104		104	164

23 82 Convection Heating and Cooling Units

23 82 19 – Fan Coil Units

23 82 19.10 Fan Coil Air Conditioning		Crew	Daily Output	Labor-Hours	Unit	Material	Labor	Equipment	Total	Total Incl O&P
						2012 Bare Costs				
0010	**FAN COIL AIR CONDITIONING** Cabinet mounted, filters, controls									
0020										
0100	Chilled water, 1/2 ton cooling	Q-5	8	2	Ea.	710	100		810	935
0120	1 ton cooling		6	2.667		815	134		949	1,100
0160	2.5 ton cooling		5	3.200		1,525	160		1,685	1,925
0180	3 ton cooling	↓	4	4	↓	1,725	201		1,926	2,200
0262	For hot water coil, add					40%	10%			
0940	Direct expansion, for use w/air cooled condensing unit, 1.5 ton cooling	Q-5	5	3.200	Ea.	690	160		850	1,000
1000	5 ton cooling	"	3	5.333		1,325	267		1,592	1,875
1040	10 ton cooling	Q-6	2.60	9.231		2,975	480		3,455	4,025
1060	20 ton cooling	"	.70	34.286	↓	5,575	1,775		7,350	8,925
1512	For condensing unit add see Section 23 62									
3000	Chilled water, horizontal unit, housing, 2 pipe, fan control, no valves									
3100	1/2 ton cooling	Q-5	8	2	Ea.	1,150	100		1,250	1,400
3110	1 ton cooling		6	2.667		1,375	134		1,509	1,725
3120	1.5 ton cooling		5.50	2.909		1,650	146		1,796	2,050
3130	2 ton cooling		5.25	3.048		1,975	153		2,128	2,425
3140	3 ton cooling		4	4		2,300	201		2,501	2,825
3150	3.5 ton cooling		3.80	4.211		2,525	211		2,736	3,100
3160	4 ton cooling	↓	3.80	4.211	↓	2,525	211		2,736	3,100
4000	With electric heat, 2 pipe									
4100	1/2 ton cooling	Q-5	8	2	Ea.	1,400	100		1,500	1,675
4105	3/4 ton cooling		7	2.286		1,550	115		1,665	1,875
4110	1 ton cooling		6	2.667		1,700	134		1,834	2,050
4120	1.5 ton cooling		5.50	2.909		1,875	146		2,021	2,300
4130	2 ton cooling		5.25	3.048		2,375	153		2,528	2,850
4135	2.5 ton cooling		4.80	3.333		3,275	167		3,442	3,850
4140	3 ton cooling		4	4		4,500	201		4,701	5,250
4150	3.5 ton cooling		3.80	4.211		5,375	211		5,586	6,250
4160	4 ton cooling		3.60	4.444	↓	6,225	223		6,448	7,200
9000	Minimum labor/equipment charge	↓	3.75	4.267	Job		214		214	335

23 82 29 – Radiators

23 82 29.10 Hydronic Heating		Crew	Daily Output	Labor-Hours	Unit	Material	Labor	Equipment	Total	Total Incl O&P
0010	**HYDRONIC HEATING**, Terminal units, not incl. main supply pipe									
1000	Radiation									
3000	Radiators, cast iron									
3100	Free standing or wall hung, 6 tube, 25" high	Q-5	96	.167	Section	37.50	8.35		45.85	54.50
9000	Minimum labor/equipment charge	"	3.20	5	Job		251		251	390

23 82 36 – Finned-Tube Radiation Heaters

23 82 36.10 Finned Tube Radiation		Crew	Daily Output	Labor-Hours	Unit	Material	Labor	Equipment	Total	Total Incl O&P
0010	**FINNED TUBE RADIATION**, Terminal units, not incl. main supply pipe									
1150	Fin tube, wall hung, 14" slope top cover, with damper									
1200	1-1/4" copper tube, 4-1/4" alum. fin	Q-5	38	.421	L.F.	42.50	21		63.50	79.50
1250	1-1/4" steel tube, 4-1/4" steel fin		36	.444		39	22.50		61.50	77.50
1255	2" steel tube, 4-1/4" steel fin		32	.500		42	25		67	85.50
1310	Baseboard, pkgd, 1/2" copper tube, alum. fin, 7" high		60	.267		7.65	13.35		21	29.50
1320	3/4" copper tube, alum. fin, 7" high		58	.276		8.40	13.85		22.25	31
1340	1" copper tube, alum. fin, 8-7/8" high		56	.286		18.30	14.30		32.60	42.50
1360	1-1/4" copper tube, alum. fin, 8-7/8" high	↓	54	.296	↓	30	14.85		44.85	56
1500	Note: fin tube may also require corners, caps, etc.									

23 82 Convection Heating and Cooling Units

23 82 39 – Unit Heaters

23 82 39.16 Propeller Unit Heaters		Crew	Daily Output	Labor-Hours	Unit	Material	2012 Bare Costs Labor	Equipment	Total	Total Incl O&P
0010	**PROPELLER UNIT HEATERS**									
3950	Unit heaters, propeller, 115 V 2 psi steam, 60°F entering air									
4000	Horizontal, 12 MBH	Q-5	12	1.333	Ea.	345	67		412	485
4060	43.9 MBH		8	2		520	100		620	730
4240	286.9 MBH		2	8		1,575	400		1,975	2,350
4250	326.0 MBH		1.90	8.421		1,825	420		2,245	2,650
4260	364 MBH		1.80	8.889		1,975	445		2,420	2,875
4270	404 MBH		1.60	10		2,500	500		3,000	3,525
4300	For vertical diffuser, add					174			174	191
4310	Vertical flow, 40 MBH	Q-5	11	1.455		735	73		808	925
4326	131.0 MBH	"	4	4		1,150	201		1,351	1,575
4354	420 MBH,(460 V)	Q-6	1.80	13.333		3,375	695		4,070	4,800
4358	500 MBH, (460 V)		1.71	14.035		4,700	730		5,430	6,300
4362	570 MBH, (460 V)		1.40	17.143		5,975	890		6,865	7,975
4366	620 MBH, (460 V)		1.30	18.462		6,525	960		7,485	8,675
4370	960 MBH, (460 V)		1.10	21.818		7,200	1,125		8,325	9,675

23 83 Radiant Heating Units

23 83 33 – Electric Radiant Heaters

23 83 33.10 Electric Heating		Crew	Daily Output	Labor-Hours	Unit	Material	Labor	Equipment	Total	Total Incl O&P
0010	**ELECTRIC HEATING**, not incl. conduit or feed wiring									
1300	Baseboard heaters, 2' long, 375 watt	1 Elec	8	1	Ea.	40	51.50		91.50	124
1400	3' long, 500 watt		8	1		46	51.50		97.50	130
1600	4' long, 750 watt		6.70	1.194		64	61.50		125.50	166
1800	5' long, 935 watt		5.70	1.404		68	72.50		140.50	187
2000	6' long, 1125 watt		5	1.600		74	82.50		156.50	209
2400	8' long, 1500 watt		4	2		92	103		195	260
2600	9' long, 1680 watt		3.60	2.222		104	115		219	291
2800	10' long, 1875 watt		3.30	2.424		193	125		318	405
2950	Wall heaters with fan, 120 to 277 volt									
3190	1500 watt	1 Elec	4	2	Ea.	134	103		237	305
3230	2500 watt		3.50	2.286		335	118		453	550
3250	4000 watt		2.70	2.963		495	153		648	780
3600	Thermostats, integral		16	.500		30	26		56	73
3800	Line voltage, 1 pole		8	1		31	51.50		82.50	114
3810	2 pole		8	1		29.50	51.50		81	112
9990	Minimum labor/equipment charge		4	2	Job		103		103	159

23 84 Humidity Control Equipment

23 84 13 – Humidifiers

23 84 13.10 Humidifier Units		Crew	Daily Output	Labor-Hours	Unit	Material	Labor	Equipment	Total	Total Incl O&P
0010	**HUMIDIFIER UNITS**									
0520	Steam, room or duct, filter, regulators, auto. controls, 220 V									
0560	22 lb. per hour	Q-5	5	3.200	Ea.	2,700	160		2,860	3,225
0580	33 lb. per hour		4	4		2,775	201		2,976	3,350
0620	100 lb. per hour		3	5.333		4,075	267		4,342	4,900
9000	Minimum labor/equipment charge		3.50	4.571	Job		229		229	355

23 91 Prefabricated Equipment Supports

23 91 10 – Prefabricated Curbs, Pads and Stands

23 91 10.10 Prefabricated Pads and Stands		Crew	Daily Output	Labor-Hours	Unit	2012 Bare Costs Material	Labor	Equipment	Total	Total Incl O&P
0010	**PREFABRICATED PADS AND STANDS**									
6000	Pad, fiberglass reinforced concrete with polystyrene foam core									
6050	Condenser, 2" thick, 20" x 38"	1 Shee	8	1	Ea.	25.50	52		77.50	110
6090	24" x 36"	"	12	.667		32.50	35		67.50	90
6280	36" x 36"	Q-9	8	2		46.50	94		140.50	199
6300	36" x 40"		7	2.286		53	107		160	227
6320	36" x 48"		7	2.286		62	107		169	237

Division Notes

		CREW	DAILY OUTPUT	LABOR-HOURS	UNIT	2012 BARE COSTS MAT.	LABOR	EQUIP.	TOTAL	TOTAL INCL O&P

Estimating Tips

26 05 00 Common Work Results for Electrical

- Conduit should be taken off in three main categories—power distribution, branch power, and branch lighting—so the estimator can concentrate on systems and components, therefore making it easier to ensure all items have been accounted for.
- For cost modifications for elevated conduit installation, add the percentages to labor according to the height of installation, and only to the quantities exceeding the different height levels, not to the total conduit quantities.
- Remember that aluminum wiring of equal ampacity is larger in diameter than copper and may require larger conduit.
- If more than three wires at a time are being pulled, deduct percentages from the labor hours of that grouping of wires.
- When taking off grounding systems, identify separately the type and size of wire, and list each unique type of ground connection.
- The estimator should take the weights of materials into consideration when completing a takeoff. Topics to consider include: How will the materials be supported? What methods of support are available? How high will the support structure have to reach? Will the final support structure be able to withstand the total burden? Is the support material included or separate from the fixture, equipment, and material specified?
- Do not overlook the costs for equipment used in the installation. If scaffolding or highlifts are available in the field, contractors may use them in lieu of the proposed ladders and rolling staging.

26 20 00 Low-Voltage Electrical Transmission

- Supports and concrete pads may be shown on drawings for the larger equipment, or the support system may be only a piece of plywood for the back of a panelboard. In either case, it must be included in the costs.

26 40 00 Electrical and Cathodic Protection

- When taking off cathodic protections systems, identify the type and size of cable, and list each unique type of anode connection.

26 50 00 Lighting

- Fixtures should be taken off room by room, using the fixture schedule, specifications, and the ceiling plan. For large concentrations of lighting fixtures in the same area, deduct the percentages from labor hours.

Reference Numbers

Reference numbers are shown in shaded boxes at the beginning of some major classifications. These numbers refer to related items in the Reference Section. The reference information may be an estimating procedure, an alternate pricing method, or technical information.

Note: Not all subdivisions listed here necessarily appear in this publication.

Note: **Trade Service,** *in part, has been used as a reference source for some of the material prices used in Division 26.*

26 01 Operation and Maintenance of Electrical Systems

26 01 50 – Operation and Maintenance of Lighting

26 01 50.81 Luminaire Replacement		Crew	Daily Output	Labor-Hours	Unit	Material	2012 Bare Costs Labor	Equipment	Total	Total Incl O&P
0010	**LUMINAIRE REPLACEMENT**									
0962	Remove and install new replacement lens, 2′ x 2′ troffer	1 Elec	16	.500	Ea.	16.40	26		42.40	58
0964	2′ x 4′ troffer	↓	16	.500	↓	16.10	26		42.10	57.50
3200	Remove and replace (reinstall), lighting fixture	↓	4	2	↓		103		103	159

26 05 Common Work Results for Electrical

26 05 05 – Selective Electrical Demolition

26 05 05.10 Electrical Demolition

		Crew	Daily Output	Labor-Hours	Unit	Material	Labor	Equipment	Total	Total Incl O&P
0010	**ELECTRICAL DEMOLITION** R260105-30									
0020	Conduit to 15′ high, including fittings & hangers									
0100	Rigid galvanized steel, 1/2″ to 1″ diameter R024119-10	1 Elec	242	.033	L.F.		1.70		1.70	2.63
0120	1-1/4″ to 2″	″	200	.040	↓		2.06		2.06	3.19
0140	2-1/2″ to 3-1/2″	2 Elec	302	.053	↓		2.73		2.73	4.22
0200	Electric metallic tubing (EMT), 1/2″ to 1″	1 Elec	394	.020	↓		1.05		1.05	1.62
0220	1-1/4″ to 1-1/2″	↓	326	.025	↓		1.27		1.27	1.95
0240	2″ to 3″	↓	236	.034	↓		1.75		1.75	2.70
0400	Wiremold raceway, including fittings & hangers									
0420	No. 3000	1 Elec	250	.032	L.F.		1.65		1.65	2.55
0440	No. 4000	″	217	.037	″		1.90		1.90	2.94
0500	Channels, steel, including fittings & hangers									
0520	3/4″ x 1-1/2″	1 Elec	308	.026	L.F.		1.34		1.34	2.07
0540	1-1/2″ x 1-1/2″	″	269	.030	″		1.53		1.53	2.37
0600	Copper bus duct, indoor, 3 phase									
0610	Including hangers & supports									
0620	225 amp	2 Elec	135	.119	L.F.		6.10		6.10	9.45
0640	400 amp	↓	106	.151	↓		7.80		7.80	12
0660	600 amp	↓	86	.186	↓		9.60		9.60	14.80
0680	1000 amp	↓	60	.267	↓		13.75		13.75	21
0700	1600 amp	↓	40	.400	↓		20.50		20.50	32
0720	3000 amp	↓	10	1.600	↓		82.50		82.50	127
0800	Plug-in switches, 600V 3 ph, incl. disconnecting									
0820	wire, conduit terminations, 30 amp	1 Elec	15.50	.516	Ea.		26.50		26.50	41
0840	60 amp	↓	13.90	.576	↓		29.50		29.50	46
0850	100 amp	↓	10.40	.769	↓		39.50		39.50	61.50
0860	200 amp	↓	6.20	1.290	↓		66.50		66.50	103
0940	1200 amp	2 Elec	2	8	↓		410		410	635
0960	1600 amp	″	1.70	9.412	↓		485		485	750
1010	Safety switches, 250 or 600V, incl. disconnection									
1050	of wire & conduit terminations									
1100	30 amp	1 Elec	12.30	.650	Ea.		33.50		33.50	52
1120	60 amp	↓	8.80	.909	↓		47		47	72.50
1140	100 amp	↓	7.30	1.096	↓		56.50		56.50	87.50
1160	200 amp	↓	5	1.600	↓		82.50		82.50	127
1210	Panel boards, incl. removal of all breakers,									
1220	conduit terminations & wire connections									
1230	3 wire, 120/240 V, 100A, to 20 circuits	1 Elec	2.60	3.077	Ea.		159		159	245
1240	200 amps, to 42 circuits	2 Elec	2.60	6.154	↓		315		315	490
1260	4 wire, 120/208 V, 125A, to 20 circuits	1 Elec	2.40	3.333	↓		172		172	266
1270	200 amps, to 42 circuits	2 Elec	2.40	6.667	↓		345		345	530
1300	Transformer, dry type, 1 phase, incl. removal of									
1320	supports, wire & conduit terminations									

26 05 05.10 Electrical Demolition		Crew	Daily Output	Labor-Hours	Unit	Material	2012 Bare Costs Labor	Equipment	Total	Total Incl O&P
1340	1 kVA	1 Elec	7.70	1.039	Ea.		53.50		53.50	83
1360	5 kVA	″	4.70	1.702			87.50		87.50	136
1420	75 kVA	2 Elec	2.50	6.400			330		330	510
1440	3 phase to 600V, primary									
1460	3 kVA	1 Elec	3.85	2.078	Ea.		107		107	166
1480	15 kVA	2 Elec	4.20	3.810			196		196	305
1500	30 kVA	″	3.50	4.571			236		236	365
1530	112.5 kVA	R-3	2.90	6.897			350	45.50	395.50	595
1560	500 kVA		1.40	14.286			730	94.50	824.50	1,225
1570	750 kVA		1.10	18.182			925	120	1,045	1,575
1600	Pull boxes & cabinets, sheet metal, incl. removal									
1620	of supports and conduit terminations									
1640	6″ x 6″ x 4″	1 Elec	31.10	.257	Ea.		13.25		13.25	20.50
1660	12″ x 12″ x 4″		23.30	.343			17.70		17.70	27.50
1720	Junction boxes, 4″ sq. & oct.		80	.100			5.15		5.15	7.95
1740	Handy box		107	.075			3.85		3.85	5.95
1760	Switch box		107	.075			3.85		3.85	5.95
1780	Receptacle & switch plates		257	.031			1.60		1.60	2.48
1800	Wire, THW-THWN-THHN, removed from									
1810	in place conduit, to 15′ high									
1830	#14	1 Elec	65	.123	C.L.F.		6.35		6.35	9.80
1840	#12		55	.145			7.50		7.50	11.60
1850	#10		45.50	.176			9.05		9.05	14
1880	#4	2 Elec	53	.302			15.55		15.55	24
1890	#3		50	.320			16.50		16.50	25.50
1910	1/0		33.20	.482			25		25	38.50
1920	2/0		29.20	.548			28.50		28.50	43.50
1930	3/0		25	.640			33		33	51
1980	400 kcmil		17	.941			48.50		48.50	75
1990	500 kcmil		16.20	.988			51		51	78.50
2000	Interior fluorescent fixtures, incl. supports									
2010	& whips, to 15′ high									
2100	Recessed drop-in 2′ x 2′, 2 lamp	2 Elec	35	.457	Ea.		23.50		23.50	36.50
2120	2′ x 4′, 2 lamp		33	.485			25		25	38.50
2140	2′ x 4′, 4 lamp		30	.533			27.50		27.50	42.50
2160	4′ x 4′, 4 lamp		20	.800			41		41	63.50
2180	Surface mount, acrylic lens & hinged frame									
2200	1′ x 4′, 2 lamp	2 Elec	44	.364	Ea.		18.75		18.75	29
2220	2′ x 2′, 2 lamp		44	.364			18.75		18.75	29
2260	2′ x 4′, 4 lamp		33	.485			25		25	38.50
2280	4′ x 4′, 4 lamp		23	.696			36		36	55.50
2300	Strip fixtures, surface mount									
2320	4′ long, 1 lamp	2 Elec	53	.302	Ea.		15.55		15.55	24
2340	4′ long, 2 lamp		50	.320			16.50		16.50	25.50
2360	8′ long, 1 lamp		42	.381			19.65		19.65	30.50
2380	8′ long, 2 lamp		40	.400			20.50		20.50	32
2400	Pendant mount, industrial, incl. removal									
2410	of chain or rod hangers, to 15′ high									
2420	4′ long, 2 lamp	2 Elec	35	.457	Ea.		23.50		23.50	36.50
2440	8′ long, 2 lamp	″	27	.593	″		30.50		30.50	47
2460	Interior incandescent, surface, ceiling									
2470	or wall mount, to 12′ high									
2480	Metal cylinder type, 75 Watt	2 Elec	62	.258	Ea.		13.30		13.30	20.50

26 05 05 – Selective Electrical Demolition

26 05 05.10 Electrical Demolition		Crew	Daily Output	Labor-Hours	Unit	Material	2012 Bare Costs Labor	Equipment	Total	Total Incl O&P
2500	150 Watt	2 Elec	62	.258	Ea.		13.30		13.30	20.50
2520	Metal halide, high bay									
2540	400 Watt	2 Elec	15	1.067	Ea.		55		55	85
2560	1000 Watt		12	1.333			68.50		68.50	106
2580	150 Watt, low bay		20	.800			41		41	63.50
2600	Exterior fixtures, incandescent, wall mount									
2620	100 Watt	2 Elec	50	.320	Ea.		16.50		16.50	25.50
2640	Quartz, 500 Watt		33	.485			25		25	38.50
2660	1500 Watt		27	.593			30.50		30.50	47
2680	Wall pack, mercury vapor									
2700	175 Watt	2 Elec	25	.640	Ea.		33		33	51
2720	250 Watt	"	25	.640	"		33		33	51
9000	Minimum labor/equipment charge	1 Elec	4	2	Job		103		103	159
9900	Add to labor for higher elevated installation									
9910	15′ to 20′ high, add						10%			
9920	20′ to 25′ high, add						20%			
9930	25′ to 30′ high, add						25%			
9940	30′ to 35′ high, add						30%			
9950	35′ to 40′ high, add						35%			
9960	Over 40′ high, add						40%			

26 05 19 – Low-Voltage Electrical Power Conductors and Cables

26 05 19.20 Armored Cable		Crew	Daily Output	Labor-Hours	Unit	Material	Labor	Equipment	Total	Total Incl O&P
0010	**ARMORED CABLE**									
0050	600 volt, copper (BX), #14, 2 conductor, solid	1 Elec	2.40	3.333	C.L.F.	71.50	172		243.50	345
0100	3 conductor, solid		2.20	3.636		111	187		298	410
0150	#12, 2 conductor, solid		2.30	3.478		72.50	179		251.50	355
0200	3 conductor, solid		2	4		117	206		323	450
0250	#10, 2 conductor, solid		2	4		133	206		339	465
0300	3 conductor, solid		1.60	5		184	258		442	600
0350	#8, 3 conductor, solid		1.30	6.154		345	315		660	870
9010	600 volt, copper (MC) steel clad, #14, 2 wire		2.40	3.333		72.50	172		244.50	345
9020	3 wire		2.20	3.636		116	187		303	415
9040	#12, 2 wire		2.30	3.478		73	179		252	360
9050	3 wire		2	4		128	206		334	460
9070	#10, 2 wire		2	4		158	206		364	495
9080	3 wire		1.60	5		221	258		479	645
9100	#8, 2 wire, stranded		1.80	4.444		296	229		525	680
9110	3 wire, stranded		1.30	6.154		415	315		730	945
9900	Minimum labor/equipment charge		4	2	Job		103		103	159

26 05 19.50 Mineral Insulated Cable		Crew	Daily Output	Labor-Hours	Unit	Material	Labor	Equipment	Total	Total Incl O&P
0010	**MINERAL INSULATED CABLE** 600 volt									
0100	1 conductor, #12	1 Elec	1.60	5	C.L.F.	385	258		643	820
1500	2 conductor, #12		1.40	5.714		800	295		1,095	1,325
1600	#10		1.20	6.667		985	345		1,330	1,600
1800	#8		1.10	7.273		1,125	375		1,500	1,825
2000	#6		1.05	7.619		1,550	395		1,945	2,300
2100	#4	2 Elec	2	8		2,100	410		2,510	2,925
2200	3 conductor, #12	1 Elec	1.20	6.667		1,000	345		1,345	1,625
2400	#10		1.10	7.273		1,125	375		1,500	1,825
2600	#8		1.05	7.619		1,425	395		1,820	2,150
2800	#6		1	8		1,875	410		2,285	2,675
3000	#4	2 Elec	1.80	8.889		2,325	460		2,785	3,275

26 05 19.50 Mineral Insulated Cable		Crew	Daily Output	Labor-Hours	Unit	2012 Bare Costs Material	Labor	Equipment	Total	Total Incl O&P
3100	4 conductor, #12	1 Elec	1.20	6.667	C.L.F.	1,050	345		1,395	1,675
3200	#10		1.10	7.273		1,275	375		1,650	1,975
3400	#8		1	8		1,650	410		2,060	2,450
3600	#6		.90	8.889		2,150	460		2,610	3,075
3620	7 conductor, #12		1.10	7.273		1,350	375		1,725	2,050
3640	#10		1	8		1,775	410		2,185	2,575
3800	Terminations, 600 volt, 1 conductor, #12		8	1	Ea.	16.90	51.50		68.40	98
5500	2 conductor, #12		6.70	1.194		16.90	61.50		78.40	114
5600	#10		6.40	1.250		25.50	64.50		90	128
5800	#8		6.20	1.290		25.50	66.50		92	131
6000	#6		5.70	1.404		25.50	72.50		98	140
6200	#4		5.30	1.509		56.50	78		134.50	182
6400	3 conductor, #12		5.70	1.404		25.50	72.50		98	140
6500	#10		5.50	1.455		25.50	75		100.50	144
6600	#8		5.20	1.538		25.50	79.50		105	151
6800	#6		4.80	1.667		25.50	86		111.50	161
7200	#4		4.60	1.739		56.50	89.50		146	201
7400	4 conductor, #12		4.60	1.739		28.50	89.50		118	170
7500	#10		4.40	1.818		28.50	93.50		122	176
7600	#8		4.20	1.905		28.50	98		126.50	183
8400	#6		4	2		59.50	103		162.50	225
8500	7 conductor, #12		3.50	2.286		28.50	118		146.50	213
8600	#10		3	2.667		61.50	137		198.50	280
8800	Crimping tool, plier type					58.50			58.50	64
9000	Stripping tool					220			220	242
9200	Hand vise					65.50			65.50	72
9500	Minimum labor/equipment charge	1 Elec	4	2	Job		103		103	159

26 05 19.55 Non-Metallic Sheathed Cable		Crew	Daily Output	Labor-Hours	Unit	Material	Labor	Equipment	Total	Total Incl O&P
0010	**NON-METALLIC SHEATHED CABLE** 600 volt									
0100	Copper with ground wire, (Romex)									
0150	#14, 2 conductor	1 Elec	2.70	2.963	C.L.F.	29.50	153		182.50	269
0200	3 conductor		2.40	3.333		41	172		213	310
0250	#12, 2 conductor		2.50	3.200		45	165		210	305
0300	3 conductor		2.20	3.636		63	187		250	360
0350	#10, 2 conductor		2.20	3.636		71	187		258	370
0400	3 conductor		1.80	4.444		100	229		329	465
0450	#8, 3 conductor		1.50	5.333		160	275		435	600
0500	#6, 3 conductor		1.40	5.714		259	295		554	740
0550	SE type SER aluminum cable, 3 RHW and									
0600	1 bare neutral, 3 #8 & 1 #8	1 Elec	1.60	5	C.L.F.	140	258		398	555
0650	3 #6 & 1 #6	"	1.40	5.714		159	295		454	630
0700	3 #4 & 1 #6	2 Elec	2.40	6.667		148	345		493	695
0750	3 #2 & 1 #4		2.20	7.273		263	375		638	870
0800	3 #1/0 & 1 #2		2	8		395	410		805	1,075
0850	3 #2/0 & 1 #1		1.80	8.889		470	460		930	1,225
0900	3 #4/0 & 1 #2/0		1.60	10		665	515		1,180	1,525
6500	Service entrance cap for copper SEU									
6700	150 amp	1 Elec	10	.800	Ea.	24.50	41		65.50	90.50
6800	200 amp		8	1	"	37.50	51.50		89	121
9000	Minimum labor/equipment charge		4	2	Job		103		103	159

26 05 19 – Low-Voltage Electrical Power Conductors and Cables

26 05 19.90 Wire

	26 05 19.90 Wire	Crew	Daily Output	Labor-Hours	Unit	Material	2012 Bare Costs Labor	Equipment	Total	Total Incl O&P
0010	**WIRE** R260533-22									
0020	600 volt, copper type THW, solid, #14	1 Elec	13	.615	C.L.F.	8.20	31.50		39.70	58
0030	#12		11	.727		12.40	37.50		49.90	71.50
0040	#10		10	.800		19.55	41		60.55	85
0140	#8		8	1		37	51.50		88.50	120
0160	#6		6.50	1.231		63	63.50		126.50	167
0180	#4	2 Elec	10.60	1.509		98.50	78		176.50	229
0200	#3		10	1.600		124	82.50		206.50	263
0220	#2		9	1.778		157	91.50		248.50	315
0240	#1		8	2		198	103		301	375
0260	1/0		6.60	2.424		247	125		372	465
0280	2/0		5.80	2.759		310	142		452	560
0300	3/0		5	3.200		390	165		555	685
0350	4/0		4.40	3.636		495	187		682	835
0400	250 kcmil	3 Elec	6	4		580	206		786	955
0420	300 kcmil		5.70	4.211		690	217		907	1,100
0450	350 kcmil		5.40	4.444		845	229		1,074	1,275
0480	400 kcmil		5.10	4.706		970	243		1,213	1,450
0490	500 kcmil		4.80	5		1,150	258		1,408	1,650
0510	750 kcmil		3.30	7.273		1,950	375		2,325	2,700
0540	600 volt, aluminum type THHN, stranded, #6	1 Elec	8	1		28	51.50		79.50	111
0560	#4	2 Elec	13	1.231		34.50	63.50		98	136
0580	#2		10.60	1.509		47	78		125	172
0600	#1		9	1.778		68.50	91.50		160	218
0620	1/0		8	2		82.50	103		185.50	250
0640	2/0		7.20	2.222		97.50	115		212.50	284
0680	3/0		6.60	2.424		121	125		246	325
0700	4/0		6.20	2.581		134	133		267	355
0720	250 kcmil	3 Elec	8.70	2.759		164	142		306	400
0740	300 kcmil		8.10	2.963		227	153		380	485
0760	350 kcmil		7.50	3.200		231	165		396	510
0780	400 kcmil		6.90	3.478		270	179		449	575
0800	500 kcmil		6	4		297	206		503	645
0850	600 kcmil		5.70	4.211		375	217		592	750
0880	700 kcmil		5.10	4.706		435	243		678	855
9000	Minimum labor/equipment charge	1 Elec	4	2	Job		103		103	159

26 05 26 – Grounding and Bonding for Electrical Systems

26 05 26.80 Grounding

	26 05 26.80 Grounding	Crew	Daily Output	Labor-Hours	Unit	Material	Labor	Equipment	Total	Total Incl O&P
0010	**GROUNDING**									
0030	Rod, copper clad, 8′ long, 1/2″ diameter	1 Elec	5.50	1.455	Ea.	18	75		93	136
0040	5/8″ diameter		5.50	1.455		19.20	75		94.20	137
0050	3/4″ diameter		5.30	1.509		32.50	78		110.50	156
0080	10′ long, 1/2″ diameter		4.80	1.667		21.50	86		107.50	157
0090	5/8″ diameter		4.60	1.739		23	89.50		112.50	164
0100	3/4″ diameter		4.40	1.818		38.50	93.50		132	188
0130	15′ long, 3/4″ diameter		4	2		83.50	103		186.50	251
0260	Wire ground bare armored, #8-1 conductor		2	4	C.L.F.	124	206		330	455
0280	#4-1 conductor		1.60	5		202	258		460	625
0390	Bare copper wire, stranded, #8		11	.727		32	37.50		69.50	93.50
0400	#6		10	.800		56.50	41		97.50	126
0600	#2	2 Elec	10	1.600		151	82.50		233.50	293
0800	3/0		6.60	2.424		335	125		460	560

26 05 Common Work Results for Electrical

26 05 26 – Grounding and Bonding for Electrical Systems

26 05 26.80 Grounding		Crew	Daily Output	Labor-Hours	Unit	2012 Bare Costs Material	Labor	Equipment	Total	Total Incl O&P
1000	4/0	2 Elec	5.70	2.807	C.L.F.	420	145		565	690
1200	250 kcmil	3 Elec	7.20	3.333		500	172		672	815
1800	Water pipe ground clamps, heavy duty									
2000	Bronze, 1/2" to 1" diameter	1 Elec	8	1	Ea.	29.50	51.50		81	112
2100	1-1/4" to 2" diameter		8	1		38.50	51.50		90	122
2200	2-1/2" to 3" diameter		6	1.333		67.50	68.50		136	181
2800	Brazed connections, #6 wire		12	.667		15.85	34.50		50.35	70.50
3000	#2 wire		10	.800		21	41		62	87
3100	3/0 wire		8	1		32	51.50		83.50	115
3200	4/0 wire		7	1.143		36.50	59		95.50	131
3400	250 kcmil wire		5	1.600		42.50	82.50		125	174
3600	500 kcmil wire		4	2		52.50	103		155.50	217
9000	Minimum labor/equipment charge		4	2	Job		103		103	159

26 05 33 – Raceway and Boxes for Electrical Systems

26 05 33.13 Conduit		Crew	Daily Output	Labor-Hours	Unit	Material	Labor	Equipment	Total	Total Incl O&P
0010	**CONDUIT** To 15' high, includes 2 terminations, 2 elbows, R260533-22									
0020	11 beam clamps, and 11 couplings per 100 L.F.									
1161	Field bends, 45° to 90°, 1/2" diameter	1 Elec	53	.151	Ea.		7.80		7.80	12
1162	3/4" diameter		47	.170			8.75		8.75	13.55
1163	1" diameter		44	.182			9.35		9.35	14.50
1164	1-1/4" diameter		23	.348			17.95		17.95	27.50
1165	1-1/2" diameter		21	.381			19.65		19.65	30.50
1166	2" diameter		16	.500			26		26	40
1991	Field bends, 45° to 90°, 1/2" diameter		44	.182			9.35		9.35	14.50
1992	3/4" diameter		40	.200			10.30		10.30	15.95
1993	1" diameter		36	.222			11.45		11.45	17.70
1994	1-1/4" diameter		19	.421			21.50		21.50	33.50
1995	1-1/2" diameter		18	.444			23		23	35.50
1996	2" diameter		13	.615			31.50		31.50	49
2500	Steel, intermediate conduit (IMC), 1/2" diameter		100	.080	L.F.	1.99	4.12		6.11	8.55
2530	3/4" diameter		90	.089		2.38	4.58		6.96	9.70
2550	1" diameter		70	.114		3.32	5.90		9.22	12.75
2570	1-1/4" diameter		65	.123		4.42	6.35		10.77	14.65
2600	1-1/2" diameter		60	.133		5.35	6.85		12.20	16.50
2630	2" diameter		50	.160		6.80	8.25		15.05	20.50
2650	2-1/2" diameter		40	.200		13.60	10.30		23.90	31
2670	3" diameter	2 Elec	60	.267		18.25	13.75		32	41
2700	3-1/2" diameter		54	.296		23	15.25		38.25	49
2730	4" diameter		50	.320		25.50	16.50		42	54
2731	Field bends, 45° to 90°, 1/2" diameter	1 Elec	44	.182	Ea.		9.35		9.35	14.50
2732	3/4" diameter		40	.200			10.30		10.30	15.95
2733	1" diameter		36	.222			11.45		11.45	17.70
2734	1-1/4" diameter		19	.421			21.50		21.50	33.50
2735	1-1/2" diameter		18	.444			23		23	35.50
2736	2" diameter		13	.615			31.50		31.50	49
5000	Electric metallic tubing (EMT), 1/2" diameter		170	.047	L.F.	.70	2.43		3.13	4.52
5020	3/4" diameter		130	.062		1.12	3.17		4.29	6.15
5040	1" diameter		115	.070		1.94	3.59		5.53	7.70
5060	1-1/4" diameter		100	.080		3.22	4.12		7.34	9.90
5080	1-1/2" diameter		90	.089		4.21	4.58		8.79	11.75
5100	2" diameter		80	.100		5.45	5.15		10.60	13.95
5120	2-1/2" diameter		60	.133		13.05	6.85		19.90	25

26 05 33.13 Conduit

	26 05 33.13 Conduit	Crew	Daily Output	Labor-Hours	Unit	2012 Bare Costs Material	Labor	Equipment	Total	Total Incl O&P
5140	3″ diameter	2 Elec	100	.160	L.F.	15.45	8.25		23.70	30
5160	3-1/2″ diameter		90	.178		19.70	9.15		28.85	35.50
5180	4″ diameter		80	.200		21.50	10.30		31.80	40
5200	Field bends, 45° to 90°, 1/2″ diameter	1 Elec	89	.090	Ea.		4.63		4.63	7.15
5220	3/4″ diameter		80	.100			5.15		5.15	7.95
5240	1″ diameter		73	.110			5.65		5.65	8.75
5260	1-1/4″ diameter		38	.211			10.85		10.85	16.75
5280	1-1/2″ diameter		36	.222			11.45		11.45	17.70
5300	2″ diameter		26	.308			15.85		15.85	24.50
5320	Offsets, 1/2″ diameter		65	.123			6.35		6.35	9.80
5340	3/4″ diameter		62	.129			6.65		6.65	10.30
5360	1″ diameter		53	.151			7.80		7.80	12
5380	1-1/4″ diameter		30	.267			13.75		13.75	21
5400	1-1/2″ diameter		28	.286			14.75		14.75	23
7600	EMT, "T" fittings with covers, 1/2″ diameter, set screw		16	.500		12.85	26		38.85	54
9990	Minimum labor/equipment charge		4	2	Job		103		103	159

26 05 33.16 Outlet Boxes

	26 05 33.16 Outlet Boxes	Crew	Daily Output	Labor-Hours	Unit	Material	Labor	Equipment	Total	Total Incl O&P
0010	**OUTLET BOXES**									
0020	Pressed steel, octagon, 4″	1 Elec	20	.400	Ea.	4.24	20.50		24.74	36.50
0060	Covers, blank		64	.125		1.77	6.45		8.22	11.90
0100	Extension rings		40	.200		7.05	10.30		17.35	23.50
0150	Square, 4″		20	.400		2.90	20.50		23.40	35
0200	Extension rings		40	.200		7.10	10.30		17.40	24
0250	Covers, blank		64	.125		1.93	6.45		8.38	12.05
0300	Plaster rings		64	.125		3.90	6.45		10.35	14.25
0652	Switchbox		24	.333		6.80	17.20		24	34
1100	Concrete, floor, 1 gang		5.30	1.509		124	78		202	256
9000	Minimum labor/equipment charge		4	2	Job		103		103	159

26 05 33.17 Outlet Boxes, Plastic

	26 05 33.17 Outlet Boxes, Plastic	Crew	Daily Output	Labor-Hours	Unit	Material	Labor	Equipment	Total	Total Incl O&P
0010	**OUTLET BOXES, PLASTIC**									
0050	4″ diameter, round with 2 mounting nails	1 Elec	25	.320	Ea.	3.58	16.50		20.08	29.50
0100	Bar hanger mounted		25	.320		7.35	16.50		23.85	33.50
0200	4″, square with 2 mounting nails		25	.320		6.05	16.50		22.55	32
0300	Plaster ring		64	.125		2.78	6.45		9.23	13
0400	Switch box with 2 mounting nails, 1 gang		30	.267		4.73	13.75		18.48	26
0500	2 gang		25	.320		5.10	16.50		21.60	31
0600	3 gang		20	.400		7.95	20.50		28.45	41
0700	Old work box		30	.267		9.30	13.75		23.05	31
9000	Minimum labor/equipment charge		4	2	Job		103		103	159

26 05 33.18 Pull Boxes

	26 05 33.18 Pull Boxes	Crew	Daily Output	Labor-Hours	Unit	Material	Labor	Equipment	Total	Total Incl O&P
0010	**PULL BOXES**									
0100	Sheet metal, pull box, NEMA 1, type SC, 6″ W x 6″ H x 4″ D	1 Elec	8	1	Ea.	15.50	51.50		67	96.50
0200	8″ W x 8″ H x 4″ D		8	1		21	51.50		72.50	103
0300	10″ W x 12″ H x 6″ D		5.30	1.509		25.50	78		103.50	148
0400	16″ W x 20″ H x 8″ D		4	2		86.50	103		189.50	255
0500	20″ W x 24″ H x 8″ D		3.20	2.500		94	129		223	300
0600	24″ W x 36″ H x 8″ D		2.70	2.963		145	153		298	395
0650	Pull box, hinged , NEMA 1, 6″ W x 6″ H x 4″ D		8	1		9.95	51.50		61.45	90.50
0800	12″ W x 16″ H x 6″ D		4.70	1.702		32.50	87.50		120	172
1000	20″ W x 20″ H x 6″ D		3.60	2.222		65	115		180	249
1200	20″ W x 20″ H x 8″ D		3.20	2.500		130	129		259	340
1400	24″ W x 36″ H x 8″ D		2.70	2.963		208	153		361	465

26 05 Common Work Results for Electrical

26 05 33 – Raceway and Boxes for Electrical Systems

26 05 33.18 Pull Boxes		Crew	Daily Output	Labor-Hours	Unit	Material	2012 Bare Costs Labor	Equipment	Total	Total Incl O&P
1600	24" W x 42" H x 8" D	1 Elec	2	4	Ea.	315	206		521	665

26 05 33.23 Wireway		Crew	Daily Output	Labor-Hours	Unit	Material	Labor	Equipment	Total	Total Incl O&P
0010	**WIREWAY** to 15' high									
0100	Screw cover, NEMA 1 w/fittings and supports, 2-1/2" x 2-1/2"	1 Elec	45	.178	L.F.	11.45	9.15		20.60	27
0200	4" x 4"	"	40	.200		12.65	10.30		22.95	30
0400	6" x 6"	2 Elec	60	.267		19.75	13.75		33.50	42.50
0600	8" x 8"	"	40	.400		33	20.50		53.50	68.50
4475	Screw cover, NEMA 3R w/fittings and supports, 4" x 4"	1 Elec	36	.222		27	11.45		38.45	47
4480	6" x 6"	2 Elec	55	.291		34	15		49	60.50
4485	8" x 8"		36	.444		55	23		78	96
4490	12" x 12"		18	.889		77	46		123	156

26 05 33.95 Cutting and Drilling		Crew	Daily Output	Labor-Hours	Unit	Material	Labor	Equipment	Total	Total Incl O&P
0010	**CUTTING AND DRILLING**									
0100	Hole drilling to 10' high, concrete wall									
0110	8" thick, 1/2" pipe size	R-31	12	.667	Ea.	.25	34.50	3.92	38.67	57.50
0120	3/4" pipe size		12	.667		.25	34.50	3.92	38.67	57.50
0130	1" pipe size		9.50	.842		.39	43.50	4.95	48.84	73
0140	1-1/4" pipe size		9.50	.842		.39	43.50	4.95	48.84	73
0150	1-1/2" pipe size		9.50	.842		.39	43.50	4.95	48.84	73
0160	2" pipe size		4.40	1.818		.54	93.50	10.70	104.74	157
0170	2-1/2" pipe size		4.40	1.818		.54	93.50	10.70	104.74	157
0180	3" pipe size		4.40	1.818		.54	93.50	10.70	104.74	157
0190	3-1/2" pipe size		3.30	2.424		.65	125	14.25	139.90	209
0200	4" pipe size		3.30	2.424		.65	125	14.25	139.90	209
0500	12" thick, 1/2" pipe size		9.40	.851		.37	44	5	49.37	74
0520	3/4" pipe size		9.40	.851		.37	44	5	49.37	74
0540	1" pipe size		7.30	1.096		.58	56.50	6.45	63.53	95
0560	1-1/4" pipe size		7.30	1.096		.58	56.50	6.45	63.53	95
0570	1-1/2" pipe size		7.30	1.096		.58	56.50	6.45	63.53	95
0580	2" pipe size		3.60	2.222		.81	115	13.05	128.86	192
0590	2-1/2" pipe size		3.60	2.222		.81	115	13.05	128.86	192
0600	3" pipe size		3.60	2.222		.81	115	13.05	128.86	192
0610	3-1/2" pipe size		2.80	2.857		.98	147	16.80	164.78	248
0630	4" pipe size		2.50	3.200		.98	165	18.80	184.78	277
0650	16" thick, 1/2" pipe size		7.60	1.053		.49	54.50	6.20	61.19	91.50
0670	3/4" pipe size		7	1.143		.49	59	6.70	66.19	99
0690	1" pipe size		6	1.333		.78	68.50	7.85	77.13	115
0710	1-1/4" pipe size		5.50	1.455		.78	75	8.55	84.33	126
0730	1-1/2" pipe size		5.50	1.455		.78	75	8.55	84.33	126
0750	2" pipe size		3	2.667		1.08	137	15.70	153.78	230
0770	2-1/2" pipe size		2.70	2.963		1.08	153	17.40	171.48	256
0790	3" pipe size		2.50	3.200		1.08	165	18.80	184.88	277
0810	3-1/2" pipe size		2.30	3.478		1.30	179	20.50	200.80	300
0830	4" pipe size		2	4		1.30	206	23.50	230.80	345
0850	20" thick, 1/2" pipe size		6.40	1.250		.61	64.50	7.35	72.46	108
0870	3/4" pipe size		6	1.333		.61	68.50	7.85	76.96	115
0890	1" pipe size		5	1.600		.97	82.50	9.40	92.87	138
0910	1-1/4" pipe size		4.80	1.667		.97	86	9.80	96.77	145
0930	1-1/2" pipe size		4.60	1.739		.97	89.50	10.25	100.72	151
0950	2" pipe size		2.70	2.963		1.35	153	17.40	171.75	257
0970	2-1/2" pipe size		2.40	3.333		1.35	172	19.60	192.95	289
0990	3" pipe size		2.20	3.636		1.35	187	21.50	209.85	315

26 05 33.95 Cutting and Drilling		Crew	Daily Output	Labor-Hours	Unit	2012 Bare Costs Material	Labor	Equipment	Total	Total Incl O&P
1010	3-1/2" pipe size	R-31	2	4	Ea.	1.63	206	23.50	231.13	350
1030	4" pipe size		1.70	4.706		1.63	243	27.50	272.13	405
1050	24" thick, 1/2" pipe size		5.50	1.455		.74	75	8.55	84.29	126
1070	3/4" pipe size		5.10	1.569		.74	81	9.20	90.94	136
1090	1" pipe size		4.30	1.860		1.16	96	10.95	108.11	161
1110	1-1/4" pipe size		4	2		1.16	103	11.75	115.91	173
1130	1-1/2" pipe size		4	2		1.16	103	11.75	115.91	173
1150	2" pipe size		2.40	3.333		1.63	172	19.60	193.23	289
1170	2-1/2" pipe size		2.20	3.636		1.63	187	21.50	210.13	315
1190	3" pipe size		2	4		1.63	206	23.50	231.13	350
1210	3-1/2" pipe size		1.80	4.444		1.95	229	26	256.95	385
1230	4" pipe size		1.50	5.333		1.95	275	31.50	308.45	460
1500	Brick wall, 8" thick, 1/2" pipe size		18	.444		.25	23	2.61	25.86	38.50
1520	3/4" pipe size		18	.444		.25	23	2.61	25.86	38.50
1540	1" pipe size		13.30	.602		.39	31	3.54	34.93	52.50
1560	1-1/4" pipe size		13.30	.602		.39	31	3.54	34.93	52.50
1580	1-1/2" pipe size		13.30	.602		.39	31	3.54	34.93	52.50
1600	2" pipe size		5.70	1.404		.54	72.50	8.25	81.29	122
1620	2-1/2" pipe size		5.70	1.404		.54	72.50	8.25	81.29	122
1640	3" pipe size		5.70	1.404		.54	72.50	8.25	81.29	122
1660	3-1/2" pipe size		4.40	1.818		.65	93.50	10.70	104.85	157
1680	4" pipe size		4	2		.65	103	11.75	115.40	173
1700	12" thick, 1/2" pipe size		14.50	.552		.37	28.50	3.24	32.11	48
1720	3/4" pipe size		14.50	.552		.37	28.50	3.24	32.11	48
1740	1" pipe size		11	.727		.58	37.50	4.28	42.36	63.50
1760	1-1/4" pipe size		11	.727		.58	37.50	4.28	42.36	63.50
1780	1-1/2" pipe size		11	.727		.58	37.50	4.28	42.36	63.50
1800	2" pipe size		5	1.600		.81	82.50	9.40	92.71	138
1820	2-1/2" pipe size		5	1.600		.81	82.50	9.40	92.71	138
1840	3" pipe size		5	1.600		.81	82.50	9.40	92.71	138
1860	3-1/2" pipe size		3.80	2.105		.98	109	12.40	122.38	183
1880	4" pipe size		3.30	2.424		.98	125	14.25	140.23	210
1900	16" thick, 1/2" pipe size		12.30	.650		.49	33.50	3.82	37.81	56.50
1920	3/4" pipe size		12.30	.650		.49	33.50	3.82	37.81	56.50
1940	1" pipe size		9.30	.860		.78	44.50	5.05	50.33	75
1960	1-1/4" pipe size		9.30	.860		.78	44.50	5.05	50.33	75
1980	1-1/2" pipe size		9.30	.860		.78	44.50	5.05	50.33	75
2000	2" pipe size		4.40	1.818		1.08	93.50	10.70	105.28	158
2010	2-1/2" pipe size		4.40	1.818		1.08	93.50	10.70	105.28	158
2030	3" pipe size		4.40	1.818		1.08	93.50	10.70	105.28	158
2050	3-1/2" pipe size		3.30	2.424		1.30	125	14.25	140.55	210
2070	4" pipe size		3	2.667		1.30	137	15.70	154	231
2090	20" thick, 1/2" pipe size		10.70	.748		.61	38.50	4.40	43.51	65
2110	3/4" pipe size		10.70	.748		.61	38.50	4.40	43.51	65
2130	1" pipe size		8	1		.97	51.50	5.90	58.37	87
2150	1-1/4" pipe size		8	1		.97	51.50	5.90	58.37	87
2170	1-1/2" pipe size		8	1		.97	51.50	5.90	58.37	87
2190	2" pipe size		4	2		1.35	103	11.75	116.10	173
2210	2-1/2" pipe size		4	2		1.35	103	11.75	116.10	173
2230	3" pipe size		4	2		1.35	103	11.75	116.10	173
2250	3-1/2" pipe size		3	2.667		1.63	137	15.70	154.33	231
2270	4" pipe size		2.70	2.963		1.63	153	17.40	172.03	257
2290	24" thick, 1/2" pipe size		9.40	.851		.74	44	5	49.74	74.50

26 05 33 – Raceway and Boxes for Electrical Systems

26 05 33.95 Cutting and Drilling		Crew	Daily Output	Labor-Hours	Unit	2012 Bare Costs Material	Labor	Equipment	Total	Total Incl O&P
2310	3/4" pipe size	R-31	9.40	.851	Ea.	.74	44	5	49.74	74.50
2330	1" pipe size		7.10	1.127		1.16	58	6.65	65.81	98.50
2350	1-1/4" pipe size		7.10	1.127		1.16	58	6.65	65.81	98.50
2370	1-1/2" pipe size		7.10	1.127		1.16	58	6.65	65.81	98.50
2390	2" pipe size		3.60	2.222		1.63	115	13.05	129.68	193
2410	2-1/2" pipe size		3.60	2.222		1.63	115	13.05	129.68	193
2430	3" pipe size		3.60	2.222		1.63	115	13.05	129.68	193
2450	3-1/2" pipe size		2.80	2.857		1.95	147	16.80	165.75	249
2470	4" pipe size		2.50	3.200		1.95	165	18.80	185.75	278
2480										
3000	Knockouts to 8' high, metal boxes & enclosures									
3020	With hole saw, 1/2" pipe size	1 Elec	53	.151	Ea.		7.80		7.80	12
3040	3/4" pipe size		47	.170			8.75		8.75	13.55
3050	1" pipe size		40	.200			10.30		10.30	15.95
3060	1-1/4" pipe size		36	.222			11.45		11.45	17.70
3070	1-1/2" pipe size		32	.250			12.90		12.90	19.90
3080	2" pipe size		27	.296			15.25		15.25	23.50
3090	2-1/2" pipe size		20	.400			20.50		20.50	32
4010	3" pipe size		16	.500			26		26	40
4030	3-1/2" pipe size		13	.615			31.50		31.50	49
4050	4" pipe size		11	.727			37.50		37.50	58
4070	With hand punch set, 1/2" pipe size		40	.200			10.30		10.30	15.95
4090	3/4" pipe size		32	.250			12.90		12.90	19.90
4110	1" pipe size		30	.267			13.75		13.75	21
4130	1-1/4" pipe size		28	.286			14.75		14.75	23
4150	1-1/2" pipe size		26	.308			15.85		15.85	24.50
4170	2" pipe size		20	.400			20.50		20.50	32
4190	2-1/2" pipe size		17	.471			24.50		24.50	37.50
4200	3" pipe size		15	.533			27.50		27.50	42.50
4220	3-1/2" pipe size		12	.667			34.50		34.50	53
4240	4" pipe size		10	.800			41		41	63.50
4260	With hydraulic punch, 1/2" pipe size		44	.182			9.35		9.35	14.50
4280	3/4" pipe size		38	.211			10.85		10.85	16.75
4300	1" pipe size		38	.211			10.85		10.85	16.75
4320	1-1/4" pipe size		38	.211			10.85		10.85	16.75
4340	1-1/2" pipe size		38	.211			10.85		10.85	16.75
4360	2" pipe size		32	.250			12.90		12.90	19.90
4380	2-1/2" pipe size		27	.296			15.25		15.25	23.50
4400	3" pipe size		23	.348			17.95		17.95	27.50
4420	3-1/2" pipe size		20	.400			20.50		20.50	32

26 05 39 – Underfloor Raceways for Electrical Systems

26 05 39.30 Conduit In Concrete Slab

Code	Description	Crew	Daily Output	Labor-Hours	Unit	Material	Labor	Equipment	Total	Total Incl O&P
0010	**CONDUIT IN CONCRETE SLAB** Including terminations,									
0020	fittings and supports									
3230	PVC, schedule 40, 1/2" diameter	1 Elec	270	.030	L.F.	.80	1.53		2.33	3.24
3250	3/4" diameter		230	.035		97	1.79		2.76	3.83
3270	1" diameter		200	.040		1.30	2.06		3.36	4.62
3300	1-1/4" diameter		170	.047		1.82	2.43		4.25	5.75
3330	1-1/2" diameter		140	.057		2.19	2.95		5.14	6.95
3350	2" diameter		120	.067		2.80	3.44		6.24	8.40
4350	Rigid galvanized steel, 1/2" diameter		200	.040		2.65	2.06		4.71	6.10
4400	3/4" diameter		170	.047		2.83	2.43		5.26	6.85

26 05 Common Work Results for Electrical

26 05 39 – Underfloor Raceways for Electrical Systems

26 05 39.30 Conduit In Concrete Slab		Crew	Daily Output	Labor-Hours	Unit	Material	2012 Bare Costs Labor	Equipment	Total	Total Incl O&P
4450	1" diameter	1 Elec	130	.062	L.F.	3.91	3.17		7.08	9.20
4500	1-1/4" diameter		110	.073		5.35	3.75		9.10	11.70
4600	1-1/2" diameter		100	.080		6.25	4.12		10.37	13.25
4800	2" diameter		90	.089		7.95	4.58		12.53	15.85
9000	Minimum labor/equipment charge		4	2	Job		103		103	159

26 05 39.40 Conduit In Trench

		Crew	Daily Output	Labor-Hours	Unit	Material	Labor	Equipment	Total	Total Incl O&P
0010	**CONDUIT IN TRENCH** Includes terminations and fittings									
0020	Does not include excavation or backfill, see Section 31 23 16.00									
0200	Rigid galvanized steel, 2" diameter	1 Elec	150	.053	L.F.	7.65	2.75		10.40	12.65
0400	2-1/2" diameter	"	100	.080		14.90	4.12		19.02	23
0600	3" diameter	2 Elec	160	.100		19.25	5.15		24.40	29
0800	3-1/2" diameter		140	.114		25	5.90		30.90	36.50
1000	4" diameter		100	.160		28	8.25		36.25	44
1200	5" diameter		80	.200		57	10.30		67.30	79
1400	6" diameter		60	.267		84	13.75		97.75	114
9000	Minimum labor/equipment charge	1 Elec	4	2	Job		103		103	159

26 05 80 – Wiring Connections

26 05 80.10 Motor Connections

		Crew	Daily Output	Labor-Hours	Unit	Material	Labor	Equipment	Total	Total Incl O&P
0010	**MOTOR CONNECTIONS**									
0020	Flexible conduit and fittings, 115 volt, 1 phase, up to 1 HP motor	1 Elec	8	1	Ea.	10.30	51.50		61.80	91
9000	Minimum labor/equipment charge	"	4	2	Job		103		103	159

26 05 90 – Residential Applications

26 05 90.10 Residential Wiring

		Crew	Daily Output	Labor-Hours	Unit	Material	Labor	Equipment	Total	Total Incl O&P
0010	**RESIDENTIAL WIRING**									
0020	20' avg. runs and #14/2 wiring incl. unless otherwise noted									
1000	Service & panel, includes 24' SE-AL cable, service eye, meter,									
1010	Socket, panel board, main bkr., ground rod, 15 or 20 amp									
1020	1-pole circuit breakers, and misc. hardware									
1100	100 amp, with 10 branch breakers	1 Elec	1.19	6.723	Ea.	590	345		935	1,175
1110	With PVC conduit and wire		.92	8.696		670	450		1,120	1,425
1120	With RGS conduit and wire		.73	10.959		845	565		1,410	1,800
1150	150 amp, with 14 branch breakers		1.03	7.767		915	400		1,315	1,625
1170	With PVC conduit and wire		.82	9.756		1,075	505		1,580	1,950
1180	With RGS conduit and wire		.67	11.940		1,400	615		2,015	2,500
1200	200 amp, with 18 branch breakers	2 Elec	1.80	8.889		1,200	460		1,660	2,025
1220	With PVC conduit and wire		1.46	10.959		1,375	565		1,940	2,375
1230	With RGS conduit and wire		1.24	12.903		1,825	665		2,490	3,050
1800	Lightning surge suppressor for above services, add	1 Elec	32	.250		59.50	12.90		72.40	85
2000	Switch devices									
2100	Single pole, 15 amp, Ivory, with a 1-gang box, cover plate,									
2110	Type NM (Romex) cable	1 Elec	17.10	.468	Ea.	13.95	24		37.95	53
2120	Type MC (BX) cable		14.30	.559		30	29		59	77.50
2130	EMT & wire		5.71	1.401		38	72		110	154
2150	3-way, #14/3, type NM cable		14.55	.550		18.60	28.50		47.10	64.50
2170	Type MC cable		12.31	.650		40.50	33.50		74	96.50
2180	EMT & wire		5	1.600		42	82.50		124.50	174
2200	4-way, #14/3, type NM cable		14.55	.550		28	28.50		56.50	75
2220	Type MC cable		12.31	.650		50	33.50		83.50	107
2230	EMT & wire		5	1.600		51.50	82.50		134	184
2250	S.P., 20 amp, #12/2, type NM cable		13.33	.600		21.50	31		52.50	72
2270	Type MC cable		11.43	.700		35	36		71	94.50

26 05 90.10 Residential Wiring		Crew	Daily Output	Labor-Hours	Unit	Material	2012 Bare Costs Labor	Equipment	Total	Total Incl O&P
2280	EMT & wire	1 Elec	4.85	1.649	Ea.	47	85		132	183
2290	S.P. rotary dimmer, 600W, no wiring		17	.471		24	24.50		48.50	64
2300	S.P. rotary dimmer, 600W, type NM cable		14.55	.550		30	28.50		58.50	77
2320	Type MC cable		12.31	.650		46	33.50		79.50	103
2330	EMT & wire		5	1.600		55.50	82.50		138	189
2350	3-way rotary dimmer, type NM cable		13.33	.600		37.50	31		68.50	89
2370	Type MC cable		11.43	.700		53.50	36		89.50	115
2380	EMT & wire		4.85	1.649		63	85		148	201
2400	Interval timer wall switch, 20 amp, 1-30 min., #12/2									
2410	Type NM cable	1 Elec	14.55	.550	Ea.	59	28.50		87.50	109
2420	Type MC cable		12.31	.650		65.50	33.50		99	124
2430	EMT & wire		5	1.600		84	82.50		166.50	220
2500	Decorator style									
2510	S.P., 15 amp, type NM cable	1 Elec	17.10	.468	Ea.	18.85	24		42.85	58
2520	Type MC cable		14.30	.559		35	29		64	83
2530	EMT & wire		5.71	1.401		43	72		115	160
2550	3-way, #14/3, type NM cable		14.55	.550		23.50	28.50		52	70
2570	Type MC cable		12.31	.650		45	33.50		78.50	102
2580	EMT & wire		5	1.600		47	82.50		129.50	179
2600	4-way, #14/3, type NM cable		14.55	.550		33	28.50		61.50	80.50
2620	Type MC cable		12.31	.650		54.50	33.50		88	112
2630	EMT & wire		5	1.600		56.50	82.50		139	189
2650	S.P., 20 amp, #12/2, type NM cable		13.33	.600		26.50	31		57.50	77
2670	Type MC cable		11.43	.700		40	36		76	100
2680	EMT & wire		4.85	1.649		52	85		137	188
2700	S.P., slide dimmer, type NM cable		17.10	.468		37.50	24		61.50	79
2720	Type MC cable		14.30	.559		54	29		83	104
2730	EMT & wire		5.71	1.401		63.50	72		135.50	182
2770	Type MC cable		14.30	.559		48.50	29		77.50	98
2780	EMT & wire		5.71	1.401		58.50	72		130.50	176
2800	3-way touch dimmer, type NM cable		13.33	.600		52.50	31		83.50	106
2820	Type MC cable		11.43	.700		68.50	36		104.50	132
2830	EMT & wire		4.85	1.649		78	85		163	217
3100	S.P. switch/15 amp recpt., Ivory, 1-gang box, plate									
3110	Type NM cable	1 Elec	11.43	.700	Ea.	26.50	36		62.50	85.50
3120	Type MC cable		10	.800		43	41		84	111
3130	EMT & wire		4.40	1.818		52.50	93.50		146	203
3150	S.P. switch/pilot light, type NM cable		11.43	.700		27.50	36		63.50	86.50
3170	Type MC cable		10	.800		44	41		85	112
3180	EMT & wire		4.43	1.806		53.50	93		146.50	203
3190	2-S.P. switches, 2-#14/2, no wiring		14	.571		8.90	29.50		38.40	55.50
3200	2-S.P. switches, 2-#14/2, type NM cables		10	.800		31.50	41		72.50	98
3220	Type MC cable		8.89	.900		56	46.50		102.50	134
3230	EMT & wire		4.10	1.951		56.50	101		157.50	217
3250	3-way switch/15 amp recpt., #14/3, type NM cable		10	.800		35.50	41		76.50	103
3270	Type MC cable		8.89	.900		57	46.50		103.50	135
3280	EMT & wire		4.10	1.951		59	101		160	220
3300	2-3 way switches, 2-#14/3, type NM cables		8.89	.900		46.50	46.50		93	123
3320	Type MC cable		8	1		82.50	51.50		134	170
3330	EMT & wire		4	2		67	103		170	233
3350	S.P. switch/20 amp recpt., #12/2, type NM cable		10	.800		39.50	41		80.50	107
3370	Type MC cable		8.89	.900		46.50	46.50		93	123
3380	EMT & wire		4.10	1.951		65	101		166	227

26 05 90.10 Residential Wiring		Crew	Daily Output	Labor-Hours	Unit	2012 Bare Costs Material	Labor	Equipment	Total	Total Incl O&P
3400	Decorator style									
3410	S.P. switch/15 amp recpt., type NM cable	1 Elec	11.43	.700	Ea.	31.50	36		67.50	91
3420	Type MC cable		10	.800		48	41		89	116
3430	EMT & wire		4.40	1.818		57.50	93.50		151	208
3450	S.P. switch/pilot light, type NM cable		11.43	.700		32.50	36		68.50	92
3470	Type MC cable		10	.800		48.50	41		89.50	117
3480	EMT & wire		4.40	1.818		58.50	93.50		152	209
3500	2-S.P. switches, 2-#14/2, type NM cables		10	.800		36.50	41		77.50	104
3520	Type MC cable		8.89	.900		61	46.50		107.50	139
3530	EMT & wire		4.10	1.951		61.50	101		162.50	223
3550	3-way/15 amp recpt., #14/3, type NM cable		10	.800		40.50	41		81.50	108
3580	EMT & wire		4.10	1.951		64	101		165	226
3650	2-3 way switches, 2-#14/3, type NM cables		8.89	.900		51.50	46.50		98	129
3670	Type MC cable		8	1		87	51.50		138.50	176
3680	EMT & wire		4	2		72	103		175	238
3700	S.P. switch/20 amp recpt., #12/2, type NM cable		10	.800		44.50	41		85.50	113
3720	Type MC cable		8.89	.900		51.50	46.50		98	128
3730	EMT & wire	↓	4.10	1.951	↓	70	101		171	232
4000	Receptacle devices									
4010	Duplex outlet, 15 amp recpt., Ivory, 1-gang box, plate									
4015	Type NM cable	1 Elec	14.55	.550	Ea.	12.45	28.50		40.95	57.50
4020	Type MC cable		12.31	.650		28.50	33.50		62	83.50
4030	EMT & wire		5.33	1.501		36.50	77.50		114	161
4050	With #12/2, type NM cable		12.31	.650		15.55	33.50		49.05	69
4070	Type MC cable		10.67	.750		29	38.50		67.50	91
4080	EMT & wire		4.71	1.699		41	87.50		128.50	180
4100	20 amp recpt., #12/2, type NM cable		12.31	.650		25	33.50		58.50	79.50
4120	Type MC cable		10.67	.750		38.50	38.50		77	102
4130	EMT & wire	↓	4.71	1.699	↓	50.50	87.50		138	191
4140	For GFI see Section 26 05 90.10 line 4300 below									
4150	Decorator style, 15 amp recpt., type NM cable	1 Elec	14.55	.550	Ea.	17.30	28.50		45.80	63
4170	Type MC cable		12.31	.650		33.50	33.50		67	89
4180	EMT & wire		5.33	1.501		41.50	77.50		119	166
4200	With #12/2, type NM cable		12.31	.650		20.50	33.50		54	74.50
4220	Type MC cable		10.67	.750		33.50	38.50		72	96.50
4230	EMT & wire		4.71	1.699		45.50	87.50		133	185
4250	20 amp recpt. #12/2, type NM cable		12.31	.650		30	33.50		63.50	85
4270	Type MC cable		10.67	.750		43	38.50		81.50	107
4280	EMT & wire		4.71	1.699		55	87.50		142.50	196
4300	GFI, 15 amp recpt., type NM cable		12.31	.650		48	33.50		81.50	105
4320	Type MC cable		10.67	.750		64.50	38.50		103	131
4330	EMT & wire		4.71	1.699		72.50	87.50		160	215
4350	GFI with #12/2, type NM cable		10.67	.750		51.50	38.50		90	116
4370	Type MC cable		9.20	.870		64.50	45		109.50	141
4380	EMT & wire		4.21	1.900		76.50	98		174.50	235
4400	20 amp recpt., #12/2 type NM cable		10.67	.750		53.50	38.50		92	119
4420	Type MC cable		9.20	.870		66.50	45		111.50	143
4430	EMT & wire		4.21	1.900		78.50	98		176.50	238
4500	Weather-proof cover for above receptacles, add	↓	32	.250	↓	5.65	12.90		18.55	26
4550	Air conditioner outlet, 20 amp-240 volt recpt.									
4560	30' of #12/2, 2 pole circuit breaker									
4570	Type NM cable	1 Elec	10	.800	Ea.	65.50	41		106.50	136
4580	Type MC cable	↓	9	.889	↓	82	46		128	161

26 05 90.10 Residential Wiring		Crew	Daily Output	Labor-Hours	Unit	Material	Labor	Equipment	Total	Total Incl O&P
						2012 Bare Costs				
4590	EMT & wire	1 Elec	4	2	Ea.	90	103		193	258
4600	Decorator style, type NM cable		10	.800		70.50	41		111.50	141
4620	Type MC cable		9	.889		86.50	46		132.50	166
4630	EMT & wire		4	2		95	103		198	263
4650	Dryer outlet, 30 amp-240 volt recpt., 20' of #10/3									
4660	2 pole circuit breaker									
4670	Type NM cable	1 Elec	6.41	1.248	Ea.	73	64.50		137.50	180
4680	Type MC cable		5.71	1.401		81.50	72		153.50	202
4690	EMT & wire		3.48	2.299		86.50	119		205.50	278
4700	Range outlet, 50 amp-240 volt recpt., 30' of #8/3									
4710	Type NM cable	1 Elec	4.21	1.900	Ea.	105	98		203	267
4720	Type MC cable		4	2		167	103		270	345
4730	EMT & wire		2.96	2.703		119	139		258	345
4750	Central vacuum outlet, Type NM cable		6.40	1.250		71	64.50		135.50	178
4770	Type MC cable		5.71	1.401		94	72		166	215
4780	EMT & wire		3.48	2.299		98	119		217	291
4800	30 amp-110 volt locking recpt., #10/2 circ. bkr.									
4810	Type NM cable	1 Elec	6.20	1.290	Ea.	82.50	66.50		149	194
4820	Type MC cable		5.40	1.481		108	76.50		184.50	237
4830	EMT & wire		3.20	2.500		109	129		238	320
4900	Low voltage outlets									
4910	Telephone recpt., 20' of 4/C phone wire	1 Elec	26	.308	Ea.	12.15	15.85		28	38
4920	TV recpt., 20' of RG59U coax wire, F type connector	″	16	.500	″	20.50	26		46.50	62.50
4950	Door bell chime, transformer, 2 buttons, 60' of bellwire									
4970	Economy model	1 Elec	11.50	.696	Ea.	76.50	36		112.50	140
4980	Custom model		11.50	.696		119	36		155	187
4990	Luxury model, 3 buttons		9.50	.842		320	43.50		363.50	420
6000	Lighting outlets									
6050	Wire only (for fixture), type NM cable	1 Elec	32	.250	Ea.	8.05	12.90		20.95	29
6070	Type MC cable		24	.333		17.30	17.20		34.50	45.50
6080	EMT & wire		10	.800		24	41		65	90
6100	Box (4"), and wire (for fixture), type NM cable		25	.320		20.50	16.50		37	48
6120	Type MC cable		20	.400		29.50	20.50		50	64.50
6130	EMT & wire		11	.727		36	37.50		73.50	98
6200	Fixtures (use with lines 6050 or 6100 above)									
6210	Canopy style, economy grade	1 Elec	40	.200	Ea.	37	10.30		47.30	57
6220	Custom grade		40	.200		57	10.30		67.30	78.50
6250	Dining room chandelier, economy grade		19	.421		92.50	21.50		114	136
6270	Luxury grade		15	.533		610	27.50		637.50	720
6310	Kitchen fixture (fluorescent), economy grade		30	.267		68.50	13.75		82.25	96.50
6320	Custom grade		25	.320		188	16.50		204.50	233
6350	Outdoor, wall mounted, economy grade		30	.267		35	13.75		48.75	59.50
6360	Custom grade		30	.267		120	13.75		133.75	153
6370	Luxury grade		25	.320		272	16.50		288.50	325
6410	Outdoor PAR floodlights, 1 lamp, 150 watt		20	.400		37.50	20.50		58	73.50
6420	2 lamp, 150 watt each		20	.400		62.50	20.50		83	101
6425	Motion sensing, 2 lamp, 150 watt each		20	.400		75	20.50		95.50	115
6430	For infrared security sensor, add		32	.250		126	12.90		138.90	158
6450	Outdoor, quartz-halogen, 300 watt flood		20	.400		45.50	20.50		66	82
6600	Recessed downlight, round, pre-wired, 50 or 75 watt trim		30	.267		46.50	13.75		60.25	72
6610	With shower light trim		30	.267		57.50	13.75		71.25	84
6620	With wall washer trim		28	.286		69.50	14.75		84.25	99
6630	With eye-ball trim		28	.286		69.50	14.75		84.25	99

26 05 90.10 Residential Wiring		Crew	Daily Output	Labor-Hours	Unit	Material	2012 Bare Costs Labor	Equipment	Total	Total Incl O&P
6640	For direct contact with insulation, add				Ea.	2.47			2.47	2.72
6700	Porcelain lamp holder	1 Elec	40	.200		4.83	10.30		15.13	21.50
6710	With pull switch		40	.200		5.65	10.30		15.95	22
6750	Fluorescent strip, 1-20 watt tube, wrap around diffuser, 24"		24	.333		58	17.20		75.20	90.50
6760	1-34 watt tube, 48"		24	.333		74.50	17.20		91.70	109
6770	2-34 watt tubes, 48"		20	.400		88.50	20.50		109	130
6780	With residential ballast		20	.400		99.50	20.50		120	141
6800	Bathroom heat lamp, 1-250 watt		28	.286		51	14.75		65.75	79
6810	2-250 watt lamps		28	.286		81.50	14.75		96.25	113
6820	For timer switch, see Section 26 05 90.10 line 2400									
6900	Outdoor post lamp, incl. post, fixture, 35' of #14/2									
6910	Type NMC cable	1 Elec	3.50	2.286	Ea.	207	118		325	410
6920	Photo-eye, add		27	.296		37.50	15.25		52.75	65
6950	Clock dial time switch, 24 hr., w/enclosure, type NM cable		11.43	.700		72	36		108	135
6970	Type MC cable		11	.727		88	37.50		125.50	155
6980	EMT & wire		4.85	1.649		96	85		181	237
7000	Alarm systems									
7050	Smoke detectors, box, #14/3, type NM cable	1 Elec	14.55	.550	Ea.	37	28.50		65.50	84.50
7070	Type MC cable		12.31	.650		52.50	33.50		86	110
7080	EMT & wire		5	1.600		54	82.50		136.50	187
7090	For relay output to security system, add					14			14	15.40
8000	Residential equipment									
8050	Disposal hook-up, incl. switch, outlet box, 3' of flex									
8060	20 amp-1 pole circ. bkr., and 25' of #12/2									
8070	Type NM cable	1 Elec	10	.800	Ea.	36.50	41		77.50	104
8080	Type MC cable		8	1		51	51.50		102.50	136
8090	EMT & wire		5	1.600		64	82.50		146.50	197
8100	Trash compactor or dishwasher hook-up, incl. outlet box,									
8110	3' of flex, 15 amp-1 pole circ. bkr., and 25' of #14/2									
8120	Type NM cable	1 Elec	10	.800	Ea.	27	41		68	93
8130	Type MC cable		8	1		45.50	51.50		97	130
8140	EMT & wire		5	1.600		55	82.50		137.50	188
8150	Hot water sink dispensor hook-up, use line 8100									
8200	Vent/exhaust fan hook-up, type NM cable	1 Elec	32	.250	Ea.	8.05	12.90		20.95	29
8220	Type MC cable		24	.333		17.30	17.20		34.50	45.50
8230	EMT & wire		10	.800		24	41		65	90
8250	Bathroom vent fan, 50 CFM (use with above hook-up)									
8260	Economy model	1 Elec	15	.533	Ea.	27	27.50		54.50	72
8270	Low noise model		15	.533		36.50	27.50		64	82.50
8280	Custom model		12	.667		126	34.50		160.50	191
8300	Bathroom or kitchen vent fan, 110 CFM									
8310	Economy model	1 Elec	15	.533	Ea.	74.50	27.50		102	125
8320	Low noise model	"	15	.533	"	92.50	27.50		120	145
8350	Paddle fan, variable speed (w/o lights)									
8360	Economy model (AC motor)	1 Elec	10	.800	Ea.	124	41		165	200
8362	With light kit		10	.800		162	41		203	242
8370	Custom model (AC motor)		10	.800		195	41		236	279
8372	With light kit		10	.800		234	41		275	320
8380	Luxury model (DC motor)		8	1		385	51.50		436.50	505
8382	With light kit		8	1		425	51.50		476.50	550
8390	Remote speed switch for above, add		12	.667		32	34.50		66.50	88.50
8500	Whole house exhaust fan, ceiling mount, 36", variable speed									
8510	Remote switch, incl. shutters, 20 amp-1 pole circ. bkr.									

26 05 Common Work Results for Electrical

26 05 90 – Residential Applications

26 05 90.10 Residential Wiring		Crew	Daily Output	Labor-Hours	Unit	Material	Labor	Equipment	Total	Total Incl O&P
						2012 Bare Costs				
8520	30′ of #12/2, type NM cable	1 Elec	4	2	Ea.	1,200	103		1,303	1,450
8530	Type MC cable		3.50	2.286		1,200	118		1,318	1,500
8540	EMT & wire		3	2.667		1,225	137		1,362	1,550
8600	Whirlpool tub hook-up, incl. timer switch, outlet box									
8610	3′ of flex, 20 amp-1 pole GFI circ. bkr.									
8620	30′ of #12/2, type NM cable	1 Elec	5	1.600	Ea.	133	82.50		215.50	273
8630	Type MC cable		4.20	1.905		142	98		240	310
8640	EMT & wire		3.40	2.353		154	121		275	355
8650	Hot water heater hook-up, incl. 1-2 pole circ. bkr., box;									
8660	3′ of flex, 20′ of #10/2, type NM cable	1 Elec	5	1.600	Ea.	37.50	82.50		120	169
8670	Type MC cable		4.20	1.905		58	98		156	216
8680	EMT & wire		3.40	2.353		58	121		179	251
9000	Heating/air conditioning									
9050	Furnace/boiler hook-up, incl. firestat, local on-off switch									
9060	Emergency switch, and 40′ of type NM cable	1 Elec	4	2	Ea.	62	103		165	228
9070	Type MC cable		3.50	2.286		87	118		205	278
9080	EMT & wire		1.50	5.333		101	275		376	535
9100	Air conditioner hook-up, incl. local 60 amp disc. switch									
9110	3′ sealtite, 40 amp, 2 pole circuit breaker									
9130	40′ of #8/2, type NM cable	1 Elec	3.50	2.286	Ea.	234	118		352	440
9140	Type MC cable		3	2.667		330	137		467	570
9150	EMT & wire		1.30	6.154		273	315		588	790
9200	Heat pump hook-up, 1-40 & 1-100 amp 2 pole circ. bkr.									
9210	Local disconnect switch, 3′ sealtite									
9220	40′ of #8/2 & 30′ of #3/2									
9230	Type NM cable	1 Elec	1.30	6.154	Ea.	615	315		930	1,175
9240	Type MC cable		1.08	7.407		760	380		1,140	1,425
9250	EMT & wire		.94	8.511		670	440		1,110	1,425
9500	Thermostat hook-up, using low voltage wire									
9520	Heating only, 25′ of #18-3	1 Elec	24	.333	Ea.	10.95	17.20		28.15	38.50
9530	Heating/cooling, 25′ of #18-4	"	20	.400	"	13.20	20.50		33.70	46.50

26 22 Low-Voltage Transformers

26 22 13 – Low-Voltage Distribution Transformers

26 22 13.10 Transformer, Dry-Type		Crew	Daily Output	Labor-Hours	Unit	Material	Labor	Equipment	Total	Total Incl O&P
0010	**TRANSFORMER, DRY-TYPE**									
0050	Single phase, 240/480 volt primary, 120/240 volt secondary									
0100	1 kVA	1 Elec	2	4	Ea.	257	206		463	600
0300	2 kVA		1.60	5		385	258		643	825
0500	3 kVA		1.40	5.714		475	295		770	980
0700	5 kVA		1.20	6.667		660	345		1,005	1,250
0900	7.5 kVA	2 Elec	2.20	7.273		915	375		1,290	1,575
1100	10 kVA		1.60	10		1,175	515		1,690	2,075
1300	15 kVA		1.20	13.333		1,275	685		1,960	2,450
2300	3 phase, 480 volt primary 120/208 volt secondary									
2310	Ventilated, 3 kVA	1 Elec	1	8	Ea.	805	410		1,215	1,525
2700	6 kVA		.80	10		1,100	515		1,615	2,000
2900	9 kVA		.70	11.429		1,250	590		1,840	2,275
3100	15 kVA	2 Elec	1.10	14.545		1,275	750		2,025	2,550
3300	30 kVA		.90	17.778		1,300	915		2,215	2,850
3500	45 kVA		.80	20		1,550	1,025		2,575	3,300

26 22 Low-Voltage Transformers

26 22 13 – Low-Voltage Distribution Transformers

26 22 13.10 Transformer, Dry-Type		Crew	Daily Output	Labor-Hours	Unit	2012 Bare Costs Material	Labor	Equipment	Total	Total Incl O&P
9000	Minimum labor/equipment charge	1 Elec	1	8	Job		410		410	635

26 24 Switchboards and Panelboards

26 24 16 – Panelboards

26 24 16.20 Panelboard and Load Center Circuit Breakers

		Crew	Daily Output	Labor-Hours	Unit	Material	Labor	Equipment	Total	Total Incl O&P
0010	**PANELBOARD AND LOAD CENTER CIRCUIT BREAKERS**									
0050	Bolt-on, 10,000 amp IC, 120 volt, 1 pole									
0100	15 to 50 amp	1 Elec	10	.800	Ea.	14.80	41		55.80	80
0200	60 amp		8	1		21.50	51.50		73	103
0300	70 amp		8	1		28	51.50		79.50	111
0350	240 volt, 2 pole									
0400	15 to 50 amp	1 Elec	8	1	Ea.	32.50	51.50		84	116
0500	60 amp		7.50	1.067		58	55		113	149
0600	80 to 100 amp		5	1.600		83.50	82.50		166	219
0700	3 pole, 15 to 60 amp		6.20	1.290		103	66.50		169.50	216
0800	70 amp		5	1.600		130	82.50		212.50	269
0900	80 to 100 amp		3.60	2.222		147	115		262	340
1000	22,000 amp I.C., 240 volt, 2 pole, 70 – 225 amp		2.70	2.963		635	153		788	930
1100	3 pole, 70 – 225 amp		2.30	3.478		705	179		884	1,050
1200	14,000 amp I.C., 277 volts, 1 pole, 15 – 30 amp		8	1		39	51.50		90.50	123
1300	22,000 amp I.C., 480 volts, 2 pole, 70 – 225 amp		2.70	2.963		635	153		788	930
1400	3 pole, 70 – 225 amp		2.30	3.478		780	179		959	1,125
2060	Plug-in tandem, 120/240 V, 2-15 A, 1 pole		11	.727		33.50	37.50		71	95
2070	1-15 A & 1-20 A		11	.727		33.50	37.50		71	95
2080	2-20 A		11	.727		33.50	37.50		71	95
2082	Arc fault circuit interrupter, 120/240 V, 1-15 A & 1-20 A, 1 pole		11	.727		89	37.50		126.50	156
9000	Minimum labor/equipment charge		3	2.667	Job		137		137	212

26 24 16.30 Panelboards Commercial Applications

		Crew	Daily Output	Labor-Hours	Unit	Material	Labor	Equipment	Total	Total Incl O&P
0010	**PANELBOARDS COMMERCIAL APPLICATIONS**									
0050	NQOD, w/20 amp 1 pole bolt-on circuit breakers									
0100	3 wire, 120/240 volts, 100 amp main lugs									
0150	10 circuits	1 Elec	1	8	Ea.	545	410		955	1,225
0200	14 circuits		.88	9.091		665	470		1,135	1,450
0250	18 circuits		.75	10.667		720	550		1,270	1,650
0300	20 circuits		.65	12.308		810	635		1,445	1,875
0600	4 wire, 120/208 volts, 100 amp main lugs, 12 circuits		1	8		640	410		1,050	1,350
0650	16 circuits		.75	10.667		735	550		1,285	1,650
0700	20 circuits		.65	12.308		850	635		1,485	1,925
0750	24 circuits		.60	13.333		915	685		1,600	2,050
0800	30 circuits		.53	15.094		1,050	780		1,830	2,350
0850	225 amp main lugs, 32 circuits	2 Elec	.90	17.778		1,175	915		2,090	2,725
0900	34 circuits		.84	19.048		1,200	980		2,180	2,850
0950	36 circuits		.80	20		1,225	1,025		2,250	2,950
1000	42 circuits		.68	23.529		1,375	1,225		2,600	3,400
1200	NEHB,w/20 amp, 1 pole bolt-on circuit breakers									
1250	4 wire, 277/480 volts, 100 amp main lugs, 12 circuits	1 Elec	.88	9.091	Ea.	1,200	470		1,670	2,025
1300	20 circuits	"	.60	13.333		1,775	685		2,460	3,000
1350	225 amp main lugs, 24 circuits	2 Elec	.90	17.778		2,000	915		2,915	3,625
1400	30 circuits		.80	20		2,400	1,025		3,425	4,250
1450	36 circuits		.72	22.222		2,800	1,150		3,950	4,850
1600	NQOD panel, w/20 amp, 1 pole, circuit breakers									

26 24 Switchboards and Panelboards

26 24 16 – Panelboards

26 24 16.30 Panelboards Commercial Applications		Crew	Daily Output	Labor-Hours	Unit	2012 Bare Costs Material	Labor	Equipment	Total	Total Incl O&P
2000	4 wire, 120/208 volts with main circuit breaker									
2050	100 amp main, 24 circuits	1 Elec	.47	17.021	Ea.	1,150	875		2,025	2,600
2100	30 circuits	"	.40	20		1,300	1,025		2,325	3,025
2200	225 amp main, 32 circuits	2 Elec	.72	22.222		2,150	1,150		3,300	4,150
2250	42 circuits		.56	28.571		2,350	1,475		3,825	4,850
2300	400 amp main, 42 circuits		.48	33.333		3,150	1,725		4,875	6,125
2350	600 amp main, 42 circuits		.40	40		4,675	2,050		6,725	8,325
2400	NEHB, with 20 amp, 1 pole circuit breaker									
2450	4 wire, 277/480 volts with main circuit breaker									
2500	100 amp main, 24 circuits	1 Elec	.42	19.048	Ea.	2,300	980		3,280	4,075
2550	30 circuits	"	.38	21.053		2,700	1,075		3,775	4,650
2600	225 amp main, 30 circuits	2 Elec	.72	22.222		3,400	1,150		4,550	5,500
2650	42 circuits	"	.56	28.571		4,175	1,475		5,650	6,875
9000	Minimum labor/equipment charge	1 Elec	1	8	Job		410		410	635

26 27 Low-Voltage Distribution Equipment

26 27 13 – Electricity Metering

26 27 13.10 Meter Centers and Sockets

		Crew	Daily Output	Labor-Hours	Unit	Material	Labor	Equipment	Total	Total Incl O&P
0010	**METER CENTERS AND SOCKETS**									
0100	Sockets, single position, 4 terminal, 100 amp	1 Elec	3.20	2.500	Ea.	46.50	129		175.50	251
0200	150 amp		2.30	3.478		55.50	179		234.50	340
0300	200 amp		1.90	4.211		78.50	217		295.50	420
0400	Transformer rated, 20 amp		3.20	2.500		138	129		267	350
0500	Double position, 4 terminal, 100 amp		2.80	2.857		191	147		338	440
0600	150 amp		2.10	3.810		228	196		424	555
0700	200 amp		1.70	4.706		460	243		703	885
9000	Minimum labor/equipment charge		3	2.667	Job		137		137	212

26 27 16 – Electrical Cabinets and Enclosures

26 27 16.10 Cabinets

		Crew	Daily Output	Labor-Hours	Unit	Material	Labor	Equipment	Total	Total Incl O&P
0010	**CABINETS**									
7000	Cabinets, current transformer									
7050	Single door, 24" H x 24" W x 10" D	1 Elec	1.60	5	Ea.	134	258		392	545
7100	30" H x 24" W x 10" D		1.30	6.154		137	315		452	640
7150	36" H x 24" W x 10" D		1.10	7.273		140	375		515	735
7200	30" H x 30" W x 10" D		1	8		146	410		556	795
7250	36" H x 30" W x 10" D		.90	8.889		198	460		658	930
7300	36" H x 36" W x 10" D		.80	10		211	515		726	1,025
7500	Double door, 48" H x 36" W x 10" D		.60	13.333		435	685		1,120	1,525
7550	24" H x 24" W x 12" D		1	8		178	410		588	830
9990	Minimum labor/equipment charge		2	4	Job		206		206	320

26 27 23 – Indoor Service Poles

26 27 23.40 Surface Raceway

		Crew	Daily Output	Labor-Hours	Unit	Material	Labor	Equipment	Total	Total Incl O&P
0010	**SURFACE RACEWAY**									
0090	Metal, straight section									
0100	No. 500	1 Elec	100	.080	L.F.	1.03	4.12		5.15	7.50
0110	No. 700		100	.080		1.16	4.12		5.28	7.65
0400	No. 1500, small pancake		90	.089		2.08	4.58		6.66	9.40
0600	No. 2000, base & cover, blank		90	.089		2.13	4.58		6.71	9.45
0800	No. 3000, base & cover, blank		75	.107		4.10	5.50		9.60	13
1000	No. 4000, base & cover, blank		65	.123		6.70	6.35		13.05	17.15

26 27 23 – Indoor Service Poles

26 27 23.40 Surface Raceway		Crew	Daily Output	Labor-Hours	Unit	Material	2012 Bare Costs Labor	Equipment	Total	Total Incl O&P
1200	No. 6000, base & cover, blank	1 Elec	50	.160	L.F.	11.10	8.25		19.35	25
2400	Fittings, elbows, No. 500		40	.200	Ea.	1.85	10.30		12.15	18
2800	Elbow cover, No. 2000		40	.200		3.50	10.30		13.80	19.80
2880	Tee, No. 500		42	.190		3.56	9.80		13.36	19.05
2900	No. 2000		27	.296		11.70	15.25		26.95	36.50
3000	Switch box, No. 500		16	.500		12.40	26		38.40	53.50
3400	Telephone outlet, No. 1500		16	.500		14.55	26		40.55	56
3600	Junction box, No. 1500		16	.500		9.95	26		35.95	51
3800	Plugmold wired sections, No. 2000									
4000	1 circuit, 6 outlets, 3 ft. long	1 Elec	8	1	Ea.	36	51.50		87.50	119
4100	2 circuits, 8 outlets, 6 ft. long	"	5.30	1.509	"	59.50	78		137.50	186
9300	Non-metallic, straight section									
9310	7/16" x 7/8", base & cover, blank	1 Elec	160	.050	L.F.	1.68	2.58		4.26	5.85
9320	Base & cover w/adhesive		160	.050		1.84	2.58		4.42	6
9340	7/16" x 1-5/16", base & cover, blank		145	.055		1.84	2.84		4.68	6.40
9350	Base & cover w/adhesive		145	.055		2.15	2.84		4.99	6.75
9370	11/16" x 2-1/4", base & cover, blank		130	.062		2.60	3.17		5.77	7.75
9380	Base & cover w/adhesive		130	.062		2.98	3.17		6.15	8.20
9385	1-11/16" x 5-1/4", two compartment base & cover w/screws		80	.100		7.35	5.15		12.50	16.05
9400	Fittings, elbows, 7/16" x 7/8"		50	.160	Ea.	1.84	8.25		10.09	14.75
9410	7/16" x 1-5/16"		45	.178		1.91	9.15		11.06	16.25
9420	11/16" x 2-1/4"		40	.200		2.08	10.30		12.38	18.25
9425	1-11/16" x 5-1/4"		28	.286		11.20	14.75		25.95	35.50
9430	Tees, 7/16" x 7/8"		35	.229		2.42	11.80		14.22	21
9440	7/16" x 1-5/16"		32	.250		2.49	12.90		15.39	22.50
9450	11/16" x 2-1/4"		30	.267		2.52	13.75		16.27	24
9455	1-11/16" x 5-1/4"		24	.333		17.90	17.20		35.10	46
9460	Cover clip, 7/16" x 7/8"		80	.100		.48	5.15		5.63	8.50
9470	7/16" x 1-5/16"		72	.111		.43	5.75		6.18	9.30
9480	11/16" x 2-1/4"		64	.125		.74	6.45		7.19	10.75
9484	1-11/16" x 5-1/4"		42	.190		2.63	9.80		12.43	18.05
9486	Wire clip, 1-11/16" x 5-1/4"		68	.118		.45	6.05		6.50	9.85
9490	Blank end, 7/16" x 7/8"		50	.160		.69	8.25		8.94	13.50
9500	7/16" x 1-5/16"		45	.178		.77	9.15		9.92	15
9510	11/16" x 2-1/4"		40	.200		1.17	10.30		11.47	17.25
9515	1-11/16" x 5-1/4"		38	.211		5.35	10.85		16.20	22.50
9520	Round fixture box, 5.5" dia x 1"		25	.320		11.10	16.50		27.60	37.50
9530	Device box, 1 gang		30	.267		4.91	13.75		18.66	26.50
9540	2 gang		25	.320		7.30	16.50		23.80	33.50
9990	Minimum labor/equipment charge		5	1.600	Job		82.50		82.50	127

26 27 26 – Wiring Devices

26 27 26.20 Wiring Devices Elements

		Crew	Daily Output	Labor-Hours	Unit	Material	Labor	Equipment	Total	Total Incl O&P
0010	**WIRING DEVICES ELEMENTS**									
0200	Toggle switch, quiet type, single pole, 15 amp	1 Elec	40	.200	Ea.	5.40	10.30		15.70	22
0500	20 amp		27	.296		7.60	15.25		22.85	32
0550	Rocker, 15 amp		40	.200		6.70	10.30		17	23.50
0560	20 amp		27	.296		15.45	15.25		30.70	40.50
0600	3 way, 15 amp		23	.348		8.20	17.95		26.15	36.50
0850	Rocker, 15 amp		23	.348		9.40	17.95		27.35	38
0860	20 amp		18	.444		22.50	23		45.50	60
0900	4 way, 15 amp		15	.533		14.80	27.50		42.30	59
1030	Rocker, 15 amp		15	.533		37	27.50		64.50	83

26 27 Low-Voltage Distribution Equipment

26 27 26 – Wiring Devices

26 27 26.20 Wiring Devices Elements		Crew	Daily Output	Labor-Hours	Unit	2012 Bare Costs Material	Labor	Equipment	Total	Total Incl O&P
1040	20 amp	1 Elec	11	.727	Ea.	55	37.50		92.50	119
1650	Dimmer switch, 120 volt, incandescent, 600 watt, 1 pole G		16	.500		14	26		40	55.50
2460	Receptacle, duplex, 120 volt, grounded, 15 amp		40	.200		1.42	10.30		11.72	17.50
2470	20 amp		27	.296		10.95	15.25		26.20	35.50
2480	Ground fault interrupting, 15 amp		27	.296		37	15.25		52.25	64.50
2482	20 amp		27	.296		39.50	15.25		54.75	67
2490	Dryer, 30 amp		15	.533		6.75	27.50		34.25	50
2500	Range, 50 amp		11	.727		15	37.50		52.50	74.50
2600	Wall plates, stainless steel, 1 gang		80	.100		2.50	5.15		7.65	10.70
2800	2 gang		53	.151		4.25	7.80		12.05	16.70
3200	Lampholder, keyless		26	.308		13.60	15.85		29.45	39.50
3400	Pullchain with receptacle		22	.364		20	18.75		38.75	51
9000	Minimum labor/equipment charge		4	2	Job		103		103	159

26 27 73 – Door Chimes

26 27 73.10 Doorbell System		Crew	Daily Output	Labor-Hours	Unit	Material	Labor	Equipment	Total	Total Incl O&P
0010	**DOORBELL SYSTEM**, incl. transformer, button & signal									
1000	Door chimes, 2 notes, minimum	1 Elec	16	.500	Ea.	30	26		56	73
1020	Maximum		12	.667		158	34.50		192.50	227
1100	Tube type, 3 tube system		12	.667		234	34.50		268.50	310
1180	4 tube system		10	.800		375	41		416	475
1900	For transformer & button, minimum add		5	1.600		17	82.50		99.50	146
1960	Maximum, add		4.50	1.778		51	91.50		142.50	198
3000	For push button only, minimum		24	.333		3.44	17.20		20.64	30.50
3100	Maximum		20	.400		27	20.50		47.50	61.50
9000	Minimum labor/equipment charge		4	2	Job		103		103	159

26 28 Low-Voltage Circuit Protective Devices

26 28 16 – Enclosed Switches and Circuit Breakers

26 28 16.10 Circuit Breakers		Crew	Daily Output	Labor-Hours	Unit	Material	Labor	Equipment	Total	Total Incl O&P
0010	**CIRCUIT BREAKERS** (in enclosure)									
0100	Enclosed (NEMA 1), 600 volt, 3 pole, 30 amp	1 Elec	3.20	2.500	Ea.	485	129		614	735
0200	60 amp		2.80	2.857		600	147		747	885
0400	100 amp		2.30	3.478		685	179		864	1,025
0600	225 amp		1.50	5.333		1,575	275		1,850	2,175
0700	400 amp	2 Elec	1.60	10		2,700	515		3,215	3,775
9000	Minimum labor/equipment charge	1 Elec	4	2	Job		103		103	159

26 28 16.20 Safety Switches		Crew	Daily Output	Labor-Hours	Unit	Material	Labor	Equipment	Total	Total Incl O&P
0010	**SAFETY SWITCHES**									
0100	General duty 240 volt, 3 pole NEMA 1, fusible, 30 amp	1 Elec	3.20	2.500	Ea.	119	129		248	330
0200	60 amp		2.30	3.478		202	179		381	500
0300	100 amp		1.90	4.211		345	217		562	715
0400	200 amp		1.30	6.154		745	315		1,060	1,300
0500	400 amp	2 Elec	1.80	8.889		1,875	460		2,335	2,775
9990	Minimum labor/equipment charge	1 Elec	3	2.667	Job		137		137	212

26 28 16.40 Time Switches		Crew	Daily Output	Labor-Hours	Unit	Material	Labor	Equipment	Total	Total Incl O&P
0010	**TIME SWITCHES**									
0100	Single pole, single throw, 24 hour dial	1 Elec	4	2	Ea.	119	103		222	289
0200	24 hour dial with reserve power		3.60	2.222		505	115		620	730
0300	Astronomic dial		3.60	2.222		183	115		298	380
0400	Astronomic dial with reserve power		3.30	2.424		655	125		780	915

26 28 Low-Voltage Circuit Protective Devices

26 28 16 – Enclosed Switches and Circuit Breakers

26 28 16.40 Time Switches		Crew	Daily Output	Labor-Hours	Unit	2012 Bare Costs Material	Labor	Equipment	Total	Total Incl O&P
0500	7 day calendar dial	1 Elec	3.30	2.424	Ea.	156	125		281	365
0600	7 day calendar dial with reserve power		3.20	2.500		545	129		674	800
0700	Photo cell 2000 watt		8	1		21	51.50		72.50	103
1080	Load management device, 4 loads		2	4		670	206		876	1,050
1100	8 loads		1	8		1,675	410		2,085	2,475
9000	Minimum labor/equipment charge		3.50	2.286	Job		118		118	182

26 29 Low-Voltage Controllers

26 29 13 – Enclosed Controllers

26 29 13.40 Relays

		Crew	Daily Output	Labor-Hours	Unit	Material	Labor	Equipment	Total	Total Incl O&P
0010	**RELAYS** Enclosed (NEMA 1)									
0100	2 pole, 12 amp	1 Elec	5	1.600	Ea.	88	82.50		170.50	224
0200	4 pole, 10 amp	"	4.50	1.778	"	118	91.50		209.50	271

26 29 23 – Variable-Frequency Motor Controllers

26 29 23.10 Variable Frequency Drives/Adj. Frequency Drives

			Crew	Daily Output	Labor-Hours	Unit	Material	Labor	Equipment	Total	Total Incl O&P
0010	**VARIABLE FREQUENCY DRIVES/ADJ. FREQUENCY DRIVES**										
0100	Enclosed (NEMA 1), 460 volt, for 3 HP motor size	G	1 Elec	.80	10	Ea.	1,100	515		1,615	2,025
0110	5 HP motor size	G		.80	10		1,250	515		1,765	2,175
0120	7.5 HP motor size	G		.67	11.940		1,500	615		2,115	2,600
0130	10 HP motor size	G		.67	11.940		1,775	615		2,390	2,900
0140	15 HP motor size	G	2 Elec	.89	17.978		2,125	925		3,050	3,750
0150	20 HP motor size	G		.89	17.978		2,525	925		3,450	4,200
0160	25 HP motor size	G		.67	23.881		3,125	1,225		4,350	5,350
0170	30 HP motor size	G		.67	23.881		3,850	1,225		5,075	6,150
0180	40 HP motor size	G		.67	23.881		4,625	1,225		5,850	6,975
0190	50 HP motor size	G		.53	30.189		5,775	1,550		7,325	8,750
0200	60 HP motor size	G	R-3	.56	35.714		6,875	1,825	236	8,936	10,700
0210	75 HP motor size	G		.56	35.714		7,950	1,825	236	10,011	11,800
0220	100 HP motor size	G		.50	40		9,275	2,050	265	11,590	13,700
0230	125 HP motor size	G		.50	40		10,300	2,050	265	12,615	14,900
0240	150 HP motor size	G		.50	40		11,900	2,050	265	14,215	16,500
0250	200 HP motor size	G		.42	47.619		15,400	2,425	315	18,140	21,000
1100	Custom-engineered, 460 volt, for 3 HP motor size	G	1 Elec	.56	14.286		2,575	735		3,310	3,975
1110	5 HP motor size	G		.56	14.286		2,575	735		3,310	3,975
1120	7.5 HP motor size	G		.47	17.021		2,700	875		3,575	4,325
1130	10 HP motor size	G		.47	17.021		2,825	875		3,700	4,450
1140	15 HP motor size	G	2 Elec	.62	25.806		3,525	1,325		4,850	5,950
1150	20 HP motor size	G		.62	25.806		3,975	1,325		5,300	6,425
1160	25 HP motor size	G		.47	34.043		4,650	1,750		6,400	7,825
1170	30 HP motor size	G		.47	34.043		5,825	1,750		7,575	9,100
1180	40 HP motor size	G		.47	34.043		6,875	1,750		8,625	10,300
1190	50 HP motor size	G		.37	43.243		7,900	2,225		10,125	12,100
1200	60 HP motor size	G	R-3	.39	51.282		11,900	2,625	340	14,865	17,500
1210	75 HP motor size	G		.39	51.282		12,700	2,625	340	15,665	18,300
1220	100 HP motor size	G		.35	57.143		13,900	2,925	380	17,205	20,200
1230	125 HP motor size	G		.35	57.143		14,800	2,925	380	18,105	21,200
1240	150 HP motor size	G		.35	57.143		17,000	2,925	380	20,305	23,600
1250	200 HP motor size	G		.29	68.966		22,200	3,525	455	26,180	30,400
2000	For complex & special design systems to meet specific										
2010	requirements, obtain quote from vendor.										

26 32 Packaged Generator Assemblies

26 32 13 – Engine Generators

26 32 13.16 Gas-Engine-Driven Generator Sets		Crew	Daily Output	Labor-Hours	Unit	Material	2012 Bare Costs Labor	Equipment	Total	Total Incl O&P
0010	**GAS-ENGINE-DRIVEN GENERATOR SETS**									
0020	Gas or gasoline operated, includes battery,									
0050	charger, muffler & transfer switch									
0200	3 phase 4 wire, 277/480 volt, 7.5 kW	R-3	.83	24.096	Ea.	8,550	1,225	160	9,935	11,500
0300	11.5 kW		.71	28.169		12,100	1,425	186	13,711	15,700
0400	20 kW		.63	31.746		14,300	1,625	210	16,135	18,400
0500	35 kW		.55	36.364		17,000	1,850	241	19,091	21,800

26 51 Interior Lighting

26 51 13 – Interior Lighting Fixtures, Lamps, and Ballasts

26 51 13.50 Interior Lighting Fixtures		Crew	Daily Output	Labor-Hours	Unit	Material	2012 Bare Costs Labor	Equipment	Total	Total Incl O&P
0010	**INTERIOR LIGHTING FIXTURES** Including lamps, mounting									
0030	hardware and connections									
0100	Fluorescent, C.W. lamps, troffer, recess mounted in grid, RS									
0130	Grid ceiling mount									
0200	Acrylic lens, 1'W x 4'L, two 40 watt	1 Elec	5.70	1.404	Ea.	54	72.50		126.50	172
0300	2'W x 2'L, two U40 watt		5.70	1.404		59.50	72.50		132	178
0600	2'W x 4'L, four 40 watt		4.70	1.702		67	87.50		154.50	210
0910	Acrylic lens, 1'W x 4'L, two 32 watt T8 G		5.70	1.404		67	72.50		139.50	186
0930	2'W x 2'L, two U32 watt T8 G		5.70	1.404		80	72.50		152.50	200
0940	2'W x 4'L, two 32 watt T8 G		5.30	1.509		72.50	78		150.50	200
0950	2'W x 4'L, three 32 watt T8 G		5	1.600		76.50	82.50		159	211
0960	2'W x 4'L, four 32 watt T8 G		4.70	1.702		78.50	87.50		166	222
1000	Surface mounted, RS									
1030	Acrylic lens with hinged & latched door frame									
1100	1'W x 4'L, two 40 watt	1 Elec	7	1.143	Ea.	74	59		133	173
1200	2'W x 2'L, two U40 watt		7	1.143		79.50	59		138.50	178
1500	2'W x 4'L, four 40 watt		5.30	1.509		94.50	78		172.50	224
2100	Strip fixture									
2130	Surface mounted									
2200	4' long, one 40 watt, RS	1 Elec	8.50	.941	Ea.	27	48.50		75.50	105
2300	4' long, two 40 watt, RS	″	8	1		38	51.50		89.50	121
2600	8' long, one 75 watt, SL	2 Elec	13.40	1.194		47	61.50		108.50	147
2700	8' long, two 75 watt, SL	″	12.40	1.290		57	66.50		123.50	166
3000	Strip, pendent mounted, industrial, white porcelain enamel									
3100	4' long, two 40 watt, RS	1 Elec	5.70	1.404	Ea.	52	72.50		124.50	169
3200	4' long, two 60 watt, HO	″	5	1.600		81	82.50		163.50	216
3300	8' long, two 75 watt, SL	2 Elec	8.80	1.818		96	93.50		189.50	251
4220	Metal halide, integral ballast, ceiling, recess mounted									
4230	prismatic glass lens, floating door									
4240	2'W x 2'L, 250 watt	1 Elec	3.20	2.500	Ea.	310	129		439	540
4250	2'W x 2'L, 400 watt	2 Elec	5.80	2.759		360	142		502	615
4260	Surface mounted, 2'W x 2'L, 250 watt	1 Elec	2.70	2.963		335	153		488	605
4270	400 watt	2 Elec	4.80	3.333		395	172		567	700
4280	High bay, aluminum reflector,									
4290	Single unit, 400 watt	2 Elec	4.60	3.478	Ea.	405	179		584	720
4300	Single unit, 1000 watt		4	4		580	206		786	960
4310	Twin unit, 400 watt		3.20	5		805	258		1,063	1,275
4320	Low bay, aluminum reflector, 250W DX lamp	1 Elec	3.20	2.500		355	129		484	590
4340	High pressure sodium integral ballast ceiling, recess mounted									
4350	prismatic glass lens, floating door									

26 51 Interior Lighting

26 51 13 – Interior Lighting Fixtures, Lamps, and Ballasts

26 51 13.50 Interior Lighting Fixtures		Crew	Daily Output	Labor-Hours	Unit	Material	2012 Bare Costs Labor	Equipment	Total	Total Incl O&P
4360	2'W x 2'L, 150 watt lamp	1 Elec	3.20	2.500	Ea.	380	129		509	615
4370	2'W x 2'L, 400 watt lamp	2 Elec	5.80	2.759		450	142		592	715
4380	Surface mounted, 2'W x 2'L, 150 watt lamp	1 Elec	2.70	2.963		460	153		613	745
4390	400 watt lamp	2 Elec	4.80	3.333		515	172		687	835
4400	High bay, aluminum reflector,									
4410	Single unit, 400 watt lamp	2 Elec	4.60	3.478	Ea.	370	179		549	685
4430	Single unit, 1000 watt lamp	"	4	4		535	206		741	910
4440	Low bay, aluminum reflector, 150 watt lamp	1 Elec	3.20	2.500		320	129		449	555
4450	Incandescent, high hat can, round alzak reflector, prewired									
4470	100 watt	1 Elec	8	1	Ea.	70	51.50		121.50	157
4480	150 watt		8	1		102	51.50		153.50	192
4500	300 watt		6.70	1.194		236	61.50		297.50	355
4600	Square glass lens with metal trim, prewired									
4630	100 watt	1 Elec	6.70	1.194	Ea.	54	61.50		115.50	155
4700	200 watt		6.70	1.194		95	61.50		156.50	200
6010	Vapor tight, incandescent, ceiling mounted, 200 watt		6.20	1.290		81	66.50		147.50	192
6100	Fluorescent, surface mounted, 2 lamps, 4'L, RS, 40 watt		3.20	2.500		113	129		242	325
6850	Vandalproof, surface mounted, fluorescent, two 32 watt T8 G		3.20	2.500		248	129		377	470
6860	Incandescent, one 150 watt		8	1		84	51.50		135.50	172
6900	Mirror light, fluorescent, RS, acrylic enclosure, two 40 watt		8	1		125	51.50		176.50	218
6910	One 40 watt		8	1		104	51.50		155.50	194
6920	One 20 watt		12	.667		80	34.50		114.50	141
7500	Ballast replacement, by weight of ballast, to 15' high									
7520	Indoor fluorescent, less than 2 lb.	1 Elec	10	.800	Ea.	25.50	41		66.50	91.50
7540	Two 40W, watt reducer, 2 to 5 lb.		9.40	.851		40.50	44		84.50	113
7560	Two F96 slimline, over 5 lb.		8	1		75.50	51.50		127	163
7580	Vaportite ballast, less than 2 lb.		9.40	.851		25.50	44		69.50	96
7600	2 lb. to 5 lb.		8.90	.899		40.50	46.50		87	116
7620	Over 5 lb.		7.60	1.053		75.50	54.50		130	167
7630	Electronic ballast for two tubes		8	1		36.50	51.50		88	120
7640	Dimmable ballast one lamp G		8	1		104	51.50		155.50	194
7650	Dimmable ballast two-lamp G		7.60	1.053		102	54.50		156.50	196
9000	Minimum labor/equipment charge		3	2.667	Job		137		137	212

26 51 13.70 Residential Fixtures

		Crew	Daily Output	Labor-Hours	Unit	Material	Labor	Equipment	Total	Total Incl O&P
0010	**RESIDENTIAL FIXTURES**									
0400	Fluorescent, interior, surface, circline, 32 watt & 40 watt	1 Elec	20	.400	Ea.	103	20.50		123.50	145
0700	Shallow under cabinet, two 20 watt		16	.500		47.50	26		73.50	92.50
0900	Wall mounted, 4'L, two 32 watt T8, with baffle		10	.800		161	41		202	241
2000	Incandescent, exterior lantern, wall mounted, 60 watt		16	.500		42	26		68	86
2100	Post light, 150W, with 7' post		4	2		117	103		220	288
2500	Lamp holder, weatherproof with 150W PAR		16	.500		32	26		58	75
2550	With reflector and guard		12	.667		64	34.50		98.50	124
2600	Interior pendent, globe with shade, 150 watt		20	.400		209	20.50		229.50	262
9000	Minimum labor/equipment charge		4	2	Job		103		103	159

26 52 Emergency Lighting

26 52 13 – Emergency Lighting Equipments

26 52 13.10 Emergency Lighting and Battery Units		Crew	Daily Output	Labor-Hours	Unit	Material	2012 Bare Costs Labor	Equipment	Total	Total Incl O&P
0010	**EMERGENCY LIGHTING AND BATTERY UNITS**									
0300	Emergency light units, battery operated									
0350	Twin sealed beam light, 25 watt, 6 volt each									
0500	Lead battery operated	1 Elec	4	2	Ea.	124	103		227	295
0700	Nickel cadmium battery operated		4	2	"	560	103		663	775
9000	Minimum labor/equipment charge	↓	4	2	Job		103		103	159

26 53 Exit Signs

26 53 13 – Exit Lighting

26 53 13.10 Exit Lighting Fixtures		Crew	Daily Output	Labor-Hours	Unit	Material	Labor	Equipment	Total	Total Incl O&P
0010	**EXIT LIGHTING FIXTURES**									
0080	Exit light ceiling or wall mount, incandescent, single face	1 Elec	8	1	Ea.	38	51.50		89.50	122
0100	Double face		6.70	1.194	"	38	61.50		99.50	137
9000	Minimum labor/equipment charge	↓	4	2	Job		103		103	159

26 55 Special Purpose Lighting

26 55 59 – Display Lighting

26 55 59.10 Track Lighting		Crew	Daily Output	Labor-Hours	Unit	Material	Labor	Equipment	Total	Total Incl O&P
0010	**TRACK LIGHTING**									
0080	Track, 1 circuit, 4′ section	1 Elec	6.70	1.194	Ea.	49.50	61.50		111	149
0100	8′ section	2 Elec	10.60	1.509		81	78		159	210
0300	3 circuits, 4′ section	1 Elec	6.70	1.194		68	61.50		129.50	170
0400	8′ section	2 Elec	10.60	1.509	↓	106	78		184	236
9000	Minimum labor/equipment charge	1 Elec	3	2.667	Job		137		137	212

26 56 Exterior Lighting

26 56 23 – Area Lighting

26 56 23.10 Exterior Fixtures		Crew	Daily Output	Labor-Hours	Unit	Material	Labor	Equipment	Total	Total Incl O&P
0010	**EXTERIOR FIXTURES** With lamps									
0200	Wall mounted, incandescent, 100 watt	1 Elec	8	1	Ea.	33	51.50		84.50	116
0400	Quartz, 500 watt		5.30	1.509		57.50	78		135.50	184
1100	Wall pack, low pressure sodium, 35 watt		4	2		221	103		324	400
1150	55 watt		4	2		262	103		365	445
1160	High pressure sodium, 70 watt		4	2		217	103		320	400
1170	150 watt		4	2		237	103		340	420
1180	Metal Halide, 175 watt		4	2		240	103		343	425
1190	250 watt		4	2		273	103		376	460
1195	400 watt		4	2		325	103		428	515
1250	Induction lamp, 40 watt		4	2		330	103		433	525
1260	80 watt	↓	4	2	↓	565	103		668	780

26 56 33 – Walkway Lighting

26 56 33.10 Walkway Luminaire		Crew	Daily Output	Labor-Hours	Unit	Material	Labor	Equipment	Total	Total Incl O&P
0010	**WALKWAY LUMINAIRE**									
9000	Minimum labor/equipment charge	1 Elec	3.75	2.133	Job		110		110	170

26 56 Exterior Lighting

26 56 36 – Flood Lighting

26 56 36.20 Floodlights		Crew	Daily Output	Labor-Hours	Unit	Material	2012 Bare Costs Labor	Equipment	Total	Total Incl O&P
0010	**FLOODLIGHTS** with ballast and lamp,									
1290	floor mtd, mount with swivel bracket									
1400	Pole mounted, pole not included									
2250	Low pressure sodium, 55 watt	1 Elec	2.70	2.963	Ea.	565	153		718	855
2270	90 watt	"	2	4	"	620	206		826	1,000

26 61 Lighting Systems and Accessories

26 61 23 – Lamps Applications

26 61 23.10 Lamps

		Crew	Daily Output	Labor-Hours	Unit	Material	Labor	Equipment	Total	Total Incl O&P
0010	**LAMPS**									
0080	Fluorescent, rapid start, cool white, 2′ long, 20 watt	1 Elec	1	8	C	370	410		780	1,050
0100	4′ long, 40 watt		.90	8.889		315	460		775	1,050
0170	4′ long, 34 watt energy saver G		.90	8.889		425	460		885	1,175
0176	2′ long, T8, 17 W energy saver G		1	8		320	410		730	985
0178	3′ long, T8, 25 W energy saver G		.90	8.889		490	460		950	1,250
0180	4′ long, T8, 32 watt energy saver G		.90	8.889		500	460		960	1,250
0560	Twin tube compact lamp G		.90	8.889		585	460		1,045	1,350
0570	Double twin tube compact lamp G		.80	10		1,300	515		1,815	2,250
0600	Mercury vapor, mogul base, deluxe white, 100 watt		.30	26.667		3,175	1,375		4,550	5,600
0800	400 watt		.30	26.667		7,050	1,375		8,425	9,875
1000	Metal halide, mogul base, 175 watt		.30	26.667		4,175	1,375		5,550	6,725
1100	250 watt		.30	26.667		4,725	1,375		6,100	7,325
1350	High pressure sodium, 70 watt		.30	26.667		5,050	1,375		6,425	7,675
1370	150 watt		.30	26.667		5,400	1,375		6,775	8,075
1500	Low pressure sodium, 35 watt		.30	26.667		11,300	1,375		12,675	14,500
1600	90 watt		.30	26.667		13,200	1,375		14,575	16,600
1800	Incandescent, interior, A21, 100 watt		1.60	5		200	258		458	620
1900	A21, 150 watt		1.60	5		192	258		450	610
2300	R30, 75 watt		1.30	6.154		605	315		920	1,150
2500	Exterior, PAR 38, 75 watt		1.30	6.154		1,475	315		1,790	2,100
2600	PAR 38, 150 watt		1.30	6.154	↓	1,725	315		2,040	2,400
9000	Minimum labor/equipment charge	↓	4	2	Job		103		103	159

Estimating Tips

27 20 00 Data Communications
27 30 00 Voice Communications
27 40 00 Audio-Video Communications

When estimating material costs for special systems, it is always prudent to obtain manufacturers' quotations for equipment prices and special installation requirements which will affect the total costs.

Reference Numbers

Reference numbers are shown in shaded boxes at the beginning of some major classifications. These numbers refer to related items in the Reference Section. The reference information may be an estimating procedure, an alternate pricing method, or technical information.

Note: Not all subdivisions listed here necessarily appear in this publication.

Note: **Trade Service,** *in part, has been used as a reference source for some of the material prices used in Division 27.*

27 15 Communications Horizontal Cabling

27 15 10 – Special Communications Cabling

27 15 10.23 Sound and Video Cables and Fittings		Crew	Daily Output	Labor-Hours	Unit	2012 Bare Costs Material	Labor	Equipment	Total	Total Incl O&P
0010	**SOUND AND VIDEO CABLES & FITTINGS**									
1250	Nonshielded, #22-2 conductor	1 Elec	10	.800	C.L.F.	14.65	41		55.65	79.50

27 15 13 – Communications Copper Horizontal Cabling

27 15 13.13 Communication Cables

Line	Description	Crew	Daily Output	Labor-Hours	Unit	Material	Labor	Equipment	Total	Total Incl O&P
0010	**COMMUNICATION CABLES**									
5000	High performance unshielded twisted pair (UTP)									
7000	Category 5, #24, 4 pair solid, PVC jacket	1 Elec	7	1.143	C.L.F.	14.50	59		73.50	107
7100	4 pair solid, plenum		7	1.143		44.50	59		103.50	140
7200	4 pair stranded, PVC jacket		7	1.143		21.50	59		80.50	115
7210	Category 5e, #24, 4 pair solid, PVC jacket		7	1.143		13.90	59		72.90	106
7212	4 pair solid, plenum		7	1.143		23	59		82	117
7214	4 pair stranded, PVC jacket		7	1.143		25	59		84	119
7240	Category 6, #24, 4 pair solid, PVC jacket		7	1.143		20	59		79	113
7242	4 pair solid, plenum		7	1.143		24	59		83	118
7244	4 pair stranded, PVC jacket		7	1.143		25	59		84	119
7300	Connector, RJ-45, category 5		80	.100	Ea.	1.30	5.15		6.45	9.40
7302	Shielded RJ-45, category 5		72	.111		3.30	5.75		9.05	12.50
7310	Jack, UTP RJ-45, category 3		72	.111		2.29	5.75		8.04	11.35
7312	Category 5		65	.123		4.90	6.35		11.25	15.20
7314	Category 5e		65	.123		3.20	6.35		9.55	13.30
7316	Category 6		65	.123		3.25	6.35		9.60	13.40

27 41 Audio-Video Systems

27 41 19 – Portable Audio-Video Equipment

27 41 19.10 T.V. Systems

Line	Description	Crew	Daily Output	Labor-Hours	Unit	Material	Labor	Equipment	Total	Total Incl O&P
0010	**T.V. SYSTEMS**, not including rough-in wires, cables & conduits									
5000	T.V. Antenna only, minimum	1 Elec	6	1.333	Ea.	45	68.50		113.50	156
5100	Maximum	"	4	2	"	190	103		293	370

27 51 Distributed Audio-Video Communications Systems

27 51 16 – Public Address and Mass Notification Systems

27 51 16.10 Public Address System

Line	Description	Crew	Daily Output	Labor-Hours	Unit	Material	Labor	Equipment	Total	Total Incl O&P
0010	**PUBLIC ADDRESS SYSTEM**									
0100	Conventional, office	1 Elec	5.33	1.501	Speaker	131	77.50		208.50	265
0200	Industrial		2.70	2.963	"	254	153		407	515
9000	Minimum labor/equipment charge		3.50	2.286	Job		118		118	182

27 51 19 – Sound Masking Systems

27 51 19.10 Sound System

Line	Description	Crew	Daily Output	Labor-Hours	Unit	Material	Labor	Equipment	Total	Total Incl O&P
0010	**SOUND SYSTEM**, not including rough-in wires, cables & conduits									
2000	Intercom, 30 station capacity, master station	2 Elec	2	8	Ea.	2,250	410		2,660	3,100
2020	10 station capacity	"	4	4		1,250	206		1,456	1,700
3600	House telephone, talking station	1 Elec	1.60	5		455	258		713	900
3800	Press to talk, release to listen	"	5.30	1.509		106	78		184	237
4000	System-on button					63.50			63.50	70
4200	Door release	1 Elec	4	2		113	103		216	284
4400	Combination speaker and microphone		8	1		193	51.50		244.50	292
4600	Termination box		3.20	2.500		61	129		190	266
4800	Amplifier or power supply		5.30	1.509		700	78		778	890

27 51 Distributed Audio-Video Communications Systems

27 51 19 – Sound Masking Systems

27 51 19.10 Sound System		Crew	Daily Output	Labor-Hours	Unit	2012 Bare Costs Material	Labor	Equipment	Total	Total Incl O&P
5000	Vestibule door unit	1 Elec	16	.500	Name	128	26		154	181
5200	Strip cabinet		27	.296	Ea.	242	15.25		257.25	291
5400	Directory		16	.500	"	114	26		140	165
9000	Minimum labor/equipment charge	↓	3.50	2.286	Job		118		118	182

Division Notes

		CREW	DAILY OUTPUT	LABOR-HOURS	UNIT	2012 BARE COSTS MAT.	LABOR	EQUIP.	TOTAL	TOTAL INCL O&P

Division 28 - Electronic Safety and Security

Estimating Tips

- When estimating material costs for electronic safety and security systems, it is always prudent to obtain manufacturers' quotations for equipment prices and special installation requirements that affect the total cost.
- Fire alarm systems consist of control panels, annunciator panels, battery with rack, charger, and fire alarm actuating and indicating devices. Some fire alarm systems include speakers, telephone lines, door closer controls, and other components. Be careful not to overlook the costs related to installation for these items. Also be aware of costs for integrated automation instrumentation and terminal devices, control equipment, control wiring, and programming.
- Security equipment includes items such as CCTV, access control, and other detection and identification systems to perform alert and alarm functions. Be sure to consider the costs related to installation for this security equipment, such as for integrated automation instrumentation and terminal devices, control equipment, control wiring, and programming.

Reference Numbers

Reference numbers are shown in shaded boxes at the beginning of some major classifications. These numbers refer to related items in the Reference Section. The reference information may be an estimating procedure, an alternate pricing method, or technical information.

Note: Not all subdivisions listed here necessarily appear in this publication.

28 13 Access Control

28 13 53 – Security Access Detection

28 13 53.13 Security Access Metal Detectors		Crew	Daily Output	Labor-Hours	Unit	2012 Bare Costs Material	Labor	Equipment	Total	Total Incl O&P
0010	**SECURITY ACCESS METAL DETECTORS**									
0240	Metal detector, hand-held, wand type, unit only				Ea.	89			89	98
0250	Metal detector, walk through portal type, single zone	1 Elec	2	4		3,150	206		3,356	3,775
0260	Multi-zone	"	2	4	↓	4,000	206		4,206	4,725

28 13 53.16 Security Access X-Ray Equipment		Crew	Daily Output	Labor-Hours	Unit	Material	Labor	Equipment	Total	Total Incl O&P
0010	**SECURITY ACCESS X-RAY EQUIPMENT**									
0290	X-ray machine, desk top, for mail/small packages/letters	1 Elec	4	2	Ea.	3,350	103		3,453	3,850
0300	Conveyor type, incl monitor, minimum		2	4		15,700	206		15,906	17,500
0310	Maximum	↓	2	4		28,000	206		28,206	31,100
0320	X-ray machine, large unit, for airports, incl monitor, min	2 Elec	1	16		39,200	825		40,025	44,400
0330	Maximum	"	.50	32	↓	67,000	1,650		68,650	76,500

28 13 53.23 Security Access Explosive Detection Equipment		Crew	Daily Output	Labor-Hours	Unit	Material	Labor	Equipment	Total	Total Incl O&P
0010	**SECURITY ACCESS EXPLOSIVE DETECTION EQUIPMENT**									
0270	Explosives detector, walk through portal type	1 Elec	2	4	Ea.	43,300	206		43,506	47,900
0280	Hand-held, battery operated				"				25,500	28,100

28 16 Intrusion Detection

28 16 16 – Intrusion Detection Systems Infrastructure

28 16 16.50 Intrusion Detection		Crew	Daily Output	Labor-Hours	Unit	Material	Labor	Equipment	Total	Total Incl O&P
0010	**INTRUSION DETECTION**, not including wires & conduits									
0100	Burglar alarm, battery operated, mechanical trigger	1 Elec	4	2	Ea.	272	103		375	460
0200	Electrical trigger		4	2		325	103		428	515
0400	For outside key control, add		8	1		82	51.50		133.50	170
0600	For remote signaling circuitry, add		8	1		122	51.50		173.50	215
0800	Card reader, flush type, standard		2.70	2.963		910	153		1,063	1,225
1000	Multi-code		2.70	2.963		1,175	153		1,328	1,525
1200	Door switches, hinge switch		5.30	1.509		57.50	78		135.50	183
1400	Magnetic switch		5.30	1.509		67.50	78		145.50	195
1600	Exit control locks, horn alarm		4	2		270	103		373	455
1800	Flashing light alarm		4	2		300	103		403	490
2000	Indicating panels, 1 channel	↓	2.70	2.963		360	153		513	630
2200	10 channel	2 Elec	3.20	5		1,225	258		1,483	1,750
2400	20 channel		2	8		2,400	410		2,810	3,275
2600	40 channel	↓	1.14	14.035		4,375	725		5,100	5,925
2800	Ultrasonic motion detector, 12 volt	1 Elec	2.30	3.478		225	179		404	525
3000	Infrared photoelectric detector		4	2		186	103		289	365
3200	Passive infrared detector		4	2		278	103		381	465
3400	Glass break alarm switch		8	1		91.50	51.50		143	180
3420	Switchmats, 30" x 5'		5.30	1.509		83	78		161	212
3440	30" x 25'		4	2		199	103		302	380
3460	Police connect panel		4	2		239	103		342	420
3480	Telephone dialer		5.30	1.509		375	78		453	535
3500	Alarm bell		4	2		102	103		205	272
3520	Siren		4	2		143	103		246	315
3540	Microwave detector, 10' to 200'		2	4		655	206		861	1,050
3560	10' to 350'	↓	2	4	↓	1,900	206		2,106	2,425

28 23 Video Surveillance

28 23 23 – Video Surveillance Systems Infrastructure

28 23 23.50 Video Surveillance Equipments		Crew	Daily Output	Labor-Hours	Unit	Material	2012 Bare Costs Labor	Equipment	Total	Total Incl O&P
0010	**VIDEO SURVEILLANCE EQUIPMENTS**									
0200	Video cameras, wireless, hidden in exit signs, clocks, etc, incl receiver	1 Elec	3	2.667	Ea.	106	137		243	330
0210	Accessories for video recorder, single camera		3	2.667		179	137		316	410
0220	For multiple cameras		3	2.667		1,725	137		1,862	2,100
0230	Video cameras, wireless, for under vehicle searching, complete		2	4		9,975	206		10,181	11,300

28 31 Fire Detection and Alarm

28 31 23 – Fire Detection and Alarm Annunciation Panels and Fire Stations

28 31 23.50 Alarm Panels and Devices

Line	Description	Crew	Daily Output	Labor-Hours	Unit	Material	Labor	Equipment	Total	Total Incl O&P
0010	**ALARM PANELS AND DEVICES**, not including wires & conduits									
3594	Fire, alarm control panel									
3600	4 zone	2 Elec	2	8	Ea.	390	410		800	1,050
3800	8 zone		1	16		760	825		1,585	2,100
4000	12 zone		.67	23.988		2,325	1,225		3,550	4,450
4200	Battery and rack	1 Elec	4	2		400	103		503	600
4400	Automatic charger		8	1		565	51.50		616.50	705
4600	Signal bell		8	1		71	51.50		122.50	158
4800	Trouble buzzer or manual station		8	1		81	51.50		132.50	169
5600	Strobe and horn		5.30	1.509		148	78		226	282
5800	Fire alarm horn		6.70	1.194		59	61.50		120.50	160
6000	Door holder, electro-magnetic		4	2		99.50	103		202.50	269
6200	Combination holder and closer		3.20	2.500		120	129		249	330
6600	Drill switch		8	1		360	51.50		411.50	475
6800	Master box		2.70	2.963		6,225	153		6,378	7,075
7000	Break glass station		8	1		65.50	51.50		117	152
7800	Remote annunciator, 8 zone lamp		1.80	4.444		203	229		432	580
8000	12 zone lamp	2 Elec	2.60	6.154		325	315		640	850
8200	16 zone lamp	"	2.20	7.273		405	375		780	1,025

28 31 43 – Fire Detection Sensors

28 31 43.50 Fire and Heat Detectors

Line	Description	Crew	Daily Output	Labor-Hours	Unit	Material	Labor	Equipment	Total	Total Incl O&P
0010	**FIRE & HEAT DETECTORS**									
5000	Detector, rate of rise	1 Elec	8	1	Ea.	49.50	51.50		101	134
5100	Fixed temperature	"	8	1	"	49.50	51.50		101	134

28 31 46 – Smoke Detection Sensors

28 31 46.50 Smoke Detectors

Line	Description	Crew	Daily Output	Labor-Hours	Unit	Material	Labor	Equipment	Total	Total Incl O&P
0010	**SMOKE DETECTORS**									
5200	Smoke detector, ceiling type	1 Elec	6.20	1.290	Ea.	127	66.50		193.50	243
5400	Duct type	"	3.20	2.500	"	320	129		449	550

28 39 Mass Notification Systems

28 39 10 – Notification Systems

28 39 10.10 Mass Notification System		Crew	Daily Output	Labor-Hours	Unit	Material	2012 Bare Costs Labor	Equipment	Total	Total Incl O&P
0010	**MASS NOTIFICATION SYSTEM**									
0100	Wireless command center, 10,000 devices	2 Elec	1.33	12.030	Ea.	3,500	620		4,120	4,800
0200	Option, email notification					1,700			1,700	1,875
0210	Remote device supervision & monitor					2,375			2,375	2,600
0300	Antenna VHF or UHF, for medium range	1 Elec	4	2		125	103		228	297
0310	For high-power transmitter		2	4		650	206		856	1,050
0400	Transmitter, 25 watt		4	2		1,975	103		2,078	2,325
0410	40 watt		2.66	3.008		2,625	155		2,780	3,150
0420	100 watt		1.33	6.015		6,575	310		6,885	7,700
0500	Wireless receiver/control module for speaker		8	1		257	51.50		308.50	365
0600	Desktop paging controller, stand alone	↓	4	2	↓	360	103		463	555

Estimating Tips

31 05 00 Common Work Results for Earthwork

- Estimating the actual cost of performing earthwork requires careful consideration of the variables involved. This includes items such as type of soil, whether water will be encountered, dewatering, whether banks need bracing, disposal of excavated earth, and length of haul to fill or spoil sites, etc. If the project has large quantities of cut or fill, consider raising or lowering the site to reduce costs, while paying close attention to the effect on site drainage and utilities.
- If the project has large quantities of fill, creating a borrow pit on the site can significantly lower the costs.
- It is very important to consider what time of year the project is scheduled for completion. Bad weather can create large cost overruns from dewatering, site repair, and lost productivity from cold weather.
- New lines have been added for marine piling.

Reference Numbers

Reference numbers are shown in shaded boxes at the beginning of some major classifications. These numbers refer to related items in the Reference Section. The reference information may be an estimating procedure, an alternate pricing method, or technical information.

Note: Not all subdivisions listed here necessarily appear in this publication.

31 05 Common Work Results for Earthwork

31 05 13 – Soils for Earthwork

31 05 13.10 Borrow		Crew	Daily Output	Labor-Hours	Unit	2012 Bare Costs Material	Labor	Equipment	Total	Total Incl O&P
0010	**BORROW**									
0020	Spread, 200 H.P. dozer, no compaction, 2 mi. RT haul									
0200	Common borrow	B-15	600	.047	C.Y.	14.50	1.80	4.02	20.32	23.50
0700	Screened loam		600	.047		33.50	1.80	4.02	39.32	44.50
0800	Topsoil, weed free		600	.047		23	1.80	4.02	28.82	33
0900	For 5 mile haul, add	B-34B	200	.040			1.41	2.92	4.33	5.50

31 05 16 – Aggregates for Earthwork

31 05 16.10 Borrow		Crew	Daily Output	Labor-Hours	Unit	Material	Labor	Equipment	Total	Total Incl O&P
0010	**BORROW**									
0020	Spread, with 200 H.P. dozer, no compaction, 2 mi. RT haul									
0100	Bank run gravel	B-15	600	.047	C.Y.	23	1.80	4.02	28.82	32.50
0300	Crushed stone (1.40 tons per CY) , 1-1/2″		600	.047		31.50	1.80	4.02	37.32	42
0320	3/4″		600	.047		31.50	1.80	4.02	37.32	42
0340	1/2″		600	.047		44	1.80	4.02	49.82	56
0360	3/8″		600	.047		32.50	1.80	4.02	38.32	43.50
0400	Sand, washed, concrete		600	.047		36.50	1.80	4.02	42.32	47.50
0500	Dead or bank sand		600	.047		17	1.80	4.02	22.82	26
0600	Select structural fill		600	.047		21.50	1.80	4.02	27.32	31
0900	For 5 mile haul, add	B-34B	200	.040			1.41	2.92	4.33	5.50

31 06 Schedules for Earthwork

31 06 60 – Schedules for Special Foundations and Load Bearing Elements

31 06 60.14 Piling Special Costs		Crew	Daily Output	Labor-Hours	Unit	Material	Labor	Equipment	Total	Total Incl O&P
0010	**PILING SPECIAL COSTS**									
0011	Piling special costs, pile caps, see Section 03 30 53.40									
0500	Cutoffs, concrete piles, plain	1 Pile	5.50	1.455	Ea.		62.50		62.50	103
0600	With steel thin shell, add		38	.211			9		9	14.95
0700	Steel pile or "H" piles		19	.421			18		18	30
0800	Wood piles		38	.211			9		9	14.95
1000	Testing, any type piles, test load is twice the design load									
1050	50 ton design load, 100 ton test				Ea.				14,000	15,500
1100	100 ton design load, 200 ton test								20,000	22,000
1200	200 ton design load, 400 ton test								28,000	31,000

31 06 60.15 Mobilization		Crew	Daily Output	Labor-Hours	Unit	Material	Labor	Equipment	Total	Total Incl O&P
0010	**MOBILIZATION**									
0020	Set up & remove, air compressor, 600 CFM	A-5	3.30	5.455	Ea.		191	16	207	335
0100	1200 CFM	″	2.20	8.182			287	24	311	495
0200	Crane, with pile leads and pile hammer, 75 ton	B-19	.60	106			4,700	3,050	7,750	11,000
0300	150 ton	″	.36	177			7,850	5,100	12,950	18,400

31 23 Excavation and Fill

31 23 16 – Excavation

31 23 16.13 Excavating, Trench

31 23 16.13 Excavating, Trench		Crew	Daily Output	Labor-Hours	Unit	2012 Bare Costs Material	Labor	Equipment	Total	Total Incl O&P
0010	**EXCAVATING, TRENCH**									
0011	Or continuous footing									
0020	Common earth with no sheeting or dewatering included									
1400	By hand with pick and shovel 2′ to 6′ deep, light soil	1 Clab	8	1	B.C.Y.		35		35	57.50
1500	Heavy soil	″	4	2	″		70		70	115
1700	For tamping backfilled trenches, air tamp, add	A-1G	100	.080	E.C.Y.		2.81	.50	3.31	5.15
1900	Vibrating plate, add	B-18	180	.133	″		4.77	.23	5	8.05
2100	Trim sides and bottom for concrete pours, common earth		1500	.016	S.F.		.57	.03	.60	.97
2300	Hardpan		600	.040	″		1.43	.07	1.50	2.42
9000	Minimum labor/equipment charge	1 Clab	4	2	Job		70		70	115

31 23 16.14 Excavating, Utility Trench

31 23 16.14 Excavating, Utility Trench		Crew	Daily Output	Labor-Hours	Unit	Material	Labor	Equipment	Total	Total Incl O&P
0010	**EXCAVATING, UTILITY TRENCH**									
0011	Common earth									
0050	Trenching with chain trencher, 12 H.P., operator walking									
0100	4″ wide trench, 12″ deep	B-53	800	.010	L.F.		.45	.08	.53	.80
0150	18″ deep		750	.011			.48	.08	.56	.84
0200	24″ deep		700	.011			.51	.09	.60	.91
0300	6″ wide trench, 12″ deep		650	.012			.55	.10	.65	.98
0350	18″ deep		600	.013			.60	.10	.70	1.06
0400	24″ deep		550	.015			.65	.11	.76	1.16
0450	36″ deep		450	.018			.80	.14	.94	1.41
0600	8″ wide trench, 12″ deep		475	.017			.75	.13	.88	1.34
0650	18″ deep		400	.020			.90	.16	1.06	1.58
0700	24″ deep		350	.023			1.02	.18	1.20	1.82
0750	36″ deep		300	.027			1.19	.21	1.40	2.11
0900	Minimum labor/equipment charge		2	4	Job		179	31.50	210.50	320
1000	Backfill by hand including compaction, add									
1050	4″ wide trench, 12″ deep	A-1G	800	.010	L.F.		.35	.06	.41	.65
1100	18″ deep		530	.015			.53	.09	.62	.97
1150	24″ deep		400	.020			.70	.12	.82	1.29
1300	6″ wide trench, 12″ deep		540	.015			.52	.09	.61	.95
1350	18″ deep		405	.020			.69	.12	.81	1.28
1400	24″ deep		270	.030			1.04	.18	1.22	1.91
1450	36″ deep		180	.044			1.56	.28	1.84	2.86
1600	8″ wide trench, 12″ deep		400	.020			.70	.12	.82	1.29
1650	18″ deep		265	.030			1.06	.19	1.25	1.95
1700	24″ deep		200	.040			1.40	.25	1.65	2.57
1750	36″ deep		135	.059			2.08	.37	2.45	3.82
2000	Chain trencher, 40 H.P. operator riding									
2050	6″ wide trench and backfill, 12″ deep	B-54	1200	.007	L.F.		.30	.26	.56	.76
2100	18″ deep		1000	.008			.36	.31	.67	.91
2150	24″ deep		975	.008			.37	.32	.69	.93
2200	36″ deep		900	.009			.40	.35	.75	1.01
2250	48″ deep		750	.011			.48	.42	.90	1.21
2300	60″ deep		650	.012			.55	.48	1.03	1.40
2400	8″ wide trench and backfill, 12″ deep		1000	.008			.36	.31	.67	.91
2450	18″ deep		950	.008			.38	.33	.71	.95
2500	24″ deep		900	.009			.40	.35	.75	1.01
2550	36″ deep		800	.010			.45	.39	.84	1.14
2600	48″ deep		650	.012			.55	.48	1.03	1.40
2700	12″ wide trench and backfill, 12″ deep		975	.008			.37	.32	.69	.93
2750	18″ deep		860	.009			.42	.36	.78	1.06

31 23 Excavation and Fill

31 23 16 – Excavation

31 23 16.14 Excavating, Utility Trench

	31 23 16.14 Excavating, Utility Trench	Crew	Daily Output	Labor-Hours	Unit	Material	2012 Bare Costs Labor	Equipment	Total	Total Incl O&P
2800	24" deep	B-54	800	.010	L.F.		.45	.39	.84	1.14
2850	36" deep		725	.011			.49	.43	.92	1.25
3000	16" wide trench and backfill, 12" deep		835	.010			.43	.37	.80	1.09
3050	18" deep		750	.011			.48	.42	.90	1.21
3100	24" deep		700	.011			.51	.45	.96	1.30
3200	Compaction with vibratory plate, add								35%	35%
5100	Hand excavate and trim for pipe bells after trench excavation									
5200	8" pipe	1 Clab	155	.052	L.F.		1.81		1.81	2.97
5300	18" pipe	"	130	.062	"		2.16		2.16	3.54
9000	Minimum labor/equipment charge	A-1G	4	2	Job		70	12.45	82.45	129

31 23 16.16 Structural Excavation for Minor Structures

	31 23 16.16 Structural Excavation for Minor Structures	Crew	Daily Output	Labor-Hours	Unit	Material	2012 Bare Costs Labor	Equipment	Total	Total Incl O&P
0010	**STRUCTURAL EXCAVATION FOR MINOR STRUCTURES**									
0015	Hand, pits to 6' deep, sandy soil	1 Clab	8	1	B.C.Y.		35		35	57.50
0100	Heavy soil or clay		4	2			70		70	115
0300	Pits 6' to 12' deep, sandy soil		5	1.600			56		56	92
0500	Heavy soil or clay		3	2.667			93.50		93.50	153
0700	Pits 12' to 18' deep, sandy soil		4	2			70		70	115
0900	Heavy soil or clay		2	4			140		140	230
1100	Hand loading trucks from stock pile, sandy soil		12	.667			23.50		23.50	38.50
1300	Heavy soil or clay		8	1			35		35	57.50
1500	For wet or muck hand excavation, add to above								50%	50%
6000	Machine excavation, for spread and mat footings, elevator pits,									
6001	and small building foundations									
6030	Common earth, hydraulic backhoe, 1/2 C.Y. bucket	B-12E	55	.291	B.C.Y.		12.05	6.75	18.80	27
6035	3/4 C.Y. bucket	B-12F	90	.178			7.35	7.20	14.55	19.75
6040	1 C.Y. bucket	B-12A	108	.148			6.15	6.95	13.10	17.50
6050	1-1/2 C.Y. bucket	B-12B	144	.111			4.61	6.65	11.26	14.75
6060	2 C.Y. bucket	B-12C	200	.080			3.32	6.25	9.57	12.20
6070	Sand and gravel, 3/4 C.Y. bucket	B-12F	100	.160			6.65	6.45	13.10	17.75
6080	1 C.Y. bucket	B-12A	120	.133			5.55	6.25	11.80	15.70
6090	1-1/2 C.Y. bucket	B-12B	160	.100			4.15	6	10.15	13.25
6100	2 C.Y. bucket	B-12C	220	.073			3.01	5.70	8.71	11.10
6110	Clay, till, or blasted rock, 3/4 C.Y. bucket	B-12F	80	.200			8.30	8.10	16.40	22
6120	1 C.Y. bucket	B-12A	95	.168			7	7.90	14.90	19.85
6130	1-1/2 C.Y. bucket	B-12B	130	.123			5.10	7.40	12.50	16.35
6140	2 C.Y. bucket	B-12C	175	.091			3.79	7.15	10.94	14
6230	Sandy clay & loam, hydraulic backhoe, 1/2 C.Y. bucket	B-12E	60	.267			11.05	6.20	17.25	24.50
6235	3/4 C.Y. bucket	B-12F	98	.163			6.75	6.60	13.35	18.10
6240	1 C.Y. bucket	B-12A	116	.138			5.70	6.45	12.15	16.30
6250	1-1/2 C.Y. bucket	B-12B	156	.103			4.25	6.15	10.40	13.55
9000	Minimum labor/equipment charge	1 Clab	4	2	Job		70		70	115

31 23 16.42 Excavating, Bulk Bank Measure

	31 23 16.42 Excavating, Bulk Bank Measure	Crew	Daily Output	Labor-Hours	Unit	Material	2012 Bare Costs Labor	Equipment	Total	Total Incl O&P
0010	**EXCAVATING, BULK BANK MEASURE** R312316-40									
0011	Common earth piled									
0020	For loading onto trucks, add								15%	15%
0200	Excavator, hydraulic, crawler mtd., 1 C.Y. cap. = 100 C.Y./hr. R312316-45	B-12A	800	.020	B.C.Y.		.83	.94	1.77	2.36
1200	Front end loader, track mtd., 1-1/2 C.Y. cap. = 70 C.Y./hr.	B-10N	560	.021			.92	.87	1.79	2.42
1500	Wheel mounted, 3/4 C.Y. cap. = 45 C.Y./hr.	B-10R	360	.033			1.42	.74	2.16	3.09
5000	Excavating, bulk bank measure, sandy clay & loam piled									
5020	For loading onto trucks, add								15%	15%
5100	Excavator, hydraulic, crawler mtd., 1 C.Y. cap. = 120 C.Y./hr.	B-12A	960	.017	B.C.Y.		.69	.78	1.47	1.97
9000	Minimum labor/equipment charge	B-10L	2	6	Job		256	221	477	655

31 23 16 – Excavation

31 23 16.46 Excavating, Bulk, Dozer

	31 23 16.46 Excavating, Bulk, Dozer	Crew	Daily Output	Labor-Hours	Unit	Material	2012 Bare Costs Labor	Equipment	Total	Total Incl O&P
0010	**EXCAVATING, BULK, DOZER**									
0011	Open site									
2000	80 H.P., 50′ haul, sand & gravel	B-10L	460	.026	B.C.Y.		1.11	.96	2.07	2.84
2200	150′ haul, sand & gravel		230	.052			2.23	1.92	4.15	5.70
2400	300′ haul, sand & gravel	↓	120	.100			4.27	3.69	7.96	10.85
3000	105 H.P., 50′ haul, sand & gravel	B-10W	700	.017			.73	.79	1.52	2.04
3200	150′ haul, sand & gravel		310	.039			1.65	1.78	3.43	4.59
3300	300′ haul, sand & gravel	↓	140	.086			3.66	3.93	7.59	10.20
4000	200 H.P., 50′ haul, sand & gravel	B-10B	1400	.009			.37	.89	1.26	1.56
4200	150′ haul, sand & gravel		595	.020			.86	2.09	2.95	3.68
4400	300′ haul, sand & gravel	↓	310	.039			1.65	4.01	5.66	7.05
5040	Clay	B-10M	1025	.012			.50	1.68	2.18	2.64
5400	300′ haul, sand & gravel	″	470	.026	↓		1.09	3.66	4.75	5.75

31 23 23 – Fill

31 23 23.13 Backfill

	31 23 23.13 Backfill	Crew	Daily Output	Labor-Hours	Unit	Material	Labor	Equipment	Total	Total Incl O&P
0010	**BACKFILL** R312323-30									
0015	By hand, no compaction, light soil	1 Clab	14	.571	L.C.Y.		20		20	33
0100	Heavy soil		11	.727	″		25.50		25.50	42
0300	Compaction in 6″ layers, hand tamp, add to above	↓	20.60	.388	E.C.Y.		13.65		13.65	22.50
0400	Roller compaction operator walking, add	B-10A	100	.120			5.15	1.72	6.87	10.10
0500	Air tamp, add	B-9D	190	.211			7.45	1.27	8.72	13.65
0600	Vibrating plate, add	A-1D	60	.133			4.68	.54	5.22	8.25
0800	Compaction in 12″ layers, hand tamp, add to above	1 Clab	34	.235			8.25		8.25	13.55
1000	Air tamp, add	B-9	285	.140			4.98	.73	5.71	8.95
1100	Vibrating plate, add	A-1E	90	.089	↓		3.12	.45	3.57	5.60
1200	Trench, dozer, no compaction, 60 HP	B-10L	425	.028	L.C.Y.		1.21	1.04	2.25	3.08
1300	Dozer backfilling, bulk, up to 300′ haul, no compaction	B-10B	1200	.010	″		.43	1.04	1.47	1.82
1400	Air tamped, add	B-11B	80	.200	E.C.Y.		8	3.35	11.35	16.50
1900	Dozer backfilling, trench, up to 300′ haul, no compaction	B-10B	900	.013	L.C.Y.		.57	1.38	1.95	2.43
2000	Air tamped, add	B-11B	80	.200	E.C.Y.		8	3.35	11.35	16.50
2350	Spreading in 8″ layers, small dozer	B-10B	1060	.011	L.C.Y.		.48	1.17	1.65	2.06
2450	Compacting with vibrating plate, 8″ lifts	A-1D	73	.110	E.C.Y.		3.85	.44	4.29	6.80

31 23 23.14 Backfill, Structural

	31 23 23.14 Backfill, Structural	Crew	Daily Output	Labor-Hours	Unit	Material	Labor	Equipment	Total	Total Incl O&P
0010	**BACKFILL, STRUCTURAL**									
0011	Dozer or F.E. loader									
0020	From existing stockpile, no compaction									
2000	80 H.P., 50′ haul, sand & gravel	B-10L	1100	.011	L.C.Y.		.47	.40	.87	1.18
2010	Sandy clay & loam		1070	.011			.48	.41	.89	1.21
2020	Common earth		975	.012			.53	.45	.98	1.34
2040	Clay		850	.014			.60	.52	1.12	1.53
2400	300′ haul, sand & gravel		370	.032			1.39	1.20	2.59	3.53
2410	Sandy clay & loam		360	.033			1.42	1.23	2.65	3.62
2420	Common earth		330	.036			1.55	1.34	2.89	3.95
2440	Clay	↓	290	.041			1.77	1.53	3.30	4.50
3000	105 H.P., 50′ haul, sand & gravel	B-10W	1350	.009			.38	.41	.79	1.06
3010	Sandy clay & loam		1325	.009			.39	.42	.81	1.08
3020	Common earth		1225	.010			.42	.45	.87	1.16
3040	Clay		1100	.011			.47	.50	.97	1.29
3300	300′ haul, sand & gravel		465	.026			1.10	1.18	2.28	3.06
3310	Sandy clay & loam		455	.026			1.13	1.21	2.34	3.13
3320	Common earth	↓	415	.029	↓		1.24	1.33	2.57	3.43

31 23 Excavation and Fill

31 23 23 – Fill

31 23 23.14 Backfill, Structural		Crew	Daily Output	Labor-Hours	Unit	Material	2012 Bare Costs Labor	Equipment	Total	Total Incl O&P
3340	Clay	B-10W	370	.032	L.C.Y.		1.39	1.49	2.88	3.85

31 23 23.16 Fill By Borrow and Utility Bedding

		Crew	Daily Output	Labor-Hours	Unit	Material	Labor	Equipment	Total	Total Incl O&P
0010	**FILL BY BORROW AND UTILITY BEDDING**									
0049	Utility bedding, for pipe & conduit, not incl. compaction									
0050	Crushed or screened bank run gravel	B-6	150	.160	L.C.Y.	26	6.15	2.23	34.38	41
0100	Crushed stone 3/4" to 1/2"		150	.160		31.50	6.15	2.23	39.88	47
0200	Sand, dead or bank		150	.160		17	6.15	2.23	25.38	31
0500	Compacting bedding in trench	A-1D	90	.089	E.C.Y.		3.12	.36	3.48	5.50
0600	If material source exceeds 2 miles, add for extra mileage.									

31 23 23.17 General Fill

		Crew	Daily Output	Labor-Hours	Unit	Material	Labor	Equipment	Total	Total Incl O&P
0010	**GENERAL FILL**									
0011	Spread dumped material, no compaction									
0020	By dozer, no compaction	B-10B	1000	.012	L.C.Y.		.51	1.24	1.75	2.19
0100	By hand	1 Clab	12	.667	"		23.50		23.50	38.50
9000	Minimum labor/equipment charge	"	4	2	Job		70		70	115

31 23 23.20 Hauling

		Crew	Daily Output	Labor-Hours	Unit	Material	Labor	Equipment	Total	Total Incl O&P
0010	**HAULING**									
0011	Excavated or borrow, loose cubic yards									
0012	no loading equipment, including hauling,waiting, loading/dumping									
0013	time per cycle (wait, load, travel, unload or dump & return)									
0014	8 C.Y. truck, 15 MPH ave, cycle 0.5 miles, 10 min. wait/Ld./Uld.	B-34A	320	.025	L.C.Y.		.88	1.12	2	2.66
0016	cycle 1 mile		272	.029			1.04	1.31	2.35	3.13
0018	cycle 2 miles		208	.038			1.36	1.72	3.08	4.09
0020	cycle 4 miles		144	.056			1.96	2.48	4.44	5.90
0022	cycle 6 miles		112	.071			2.52	3.19	5.71	7.60
0024	cycle 8 miles		88	.091			3.21	4.06	7.27	9.65
0026	20 MPH ave,cycle 0.5 mile		336	.024			.84	1.06	1.90	2.53
0028	cycle 1 mile		296	.027			.95	1.21	2.16	2.87
0030	cycle 2 miles		240	.033			1.18	1.49	2.67	3.54
0032	cycle 4 miles		176	.045			1.60	2.03	3.63	4.83
0034	cycle 6 miles		136	.059			2.08	2.63	4.71	6.25
0036	cycle 8 miles		112	.071			2.52	3.19	5.71	7.60
0044	25 MPH ave, cycle 4 miles		192	.042			1.47	1.86	3.33	4.43
0046	cycle 6 miles		160	.050			1.77	2.23	4	5.30
0048	cycle 8 miles		128	.063			2.21	2.79	5	6.65
0050	30 MPH ave, cycle 4 miles		216	.037			1.31	1.65	2.96	3.94
0052	cycle 6 miles		176	.045			1.60	2.03	3.63	4.83
0054	cycle 8 miles		144	.056			1.96	2.48	4.44	5.90
0114	15 MPH ave, cycle 0.5 mile, 15 min. wait/Ld./Uld.		224	.036			1.26	1.60	2.86	3.79
0116	cycle 1 mile		200	.040			1.41	1.79	3.20	4.25
0118	cycle 2 miles		168	.048			1.68	2.13	3.81	5.05
0120	cycle 4 miles		120	.067			2.35	2.98	5.33	7.10
0122	cycle 6 miles		96	.083			2.94	3.72	6.66	8.85
0124	cycle 8 miles		80	.100			3.53	4.47	8	10.60
0126	20 MPH ave, cycle 0.5 mile		232	.034			1.22	1.54	2.76	3.66
0128	cycle 1 mile		208	.038			1.36	1.72	3.08	4.09
0130	cycle 2 miles		184	.043			1.53	1.94	3.47	4.62
0132	cycle 4 miles		144	.056			1.96	2.48	4.44	5.90
0134	cycle 6 miles		112	.071			2.52	3.19	5.71	7.60
0136	cycle 8 miles		96	.083			2.94	3.72	6.66	8.85
0144	25 MPH ave, cycle 4 miles		152	.053			1.86	2.35	4.21	5.60
0146	cycle 6 miles		128	.063			2.21	2.79	5	6.65

31 23 23.20 Hauling		Crew	Daily Output	Labor-Hours	Unit	Material	2012 Bare Costs Labor	Equipment	Total	Total Incl O&P
0148	cycle 8 miles	B-34A	112	.071	L.C.Y.		2.52	3.19	5.71	7.60
0150	30 MPH ave, cycle 4 miles		168	.048			1.68	2.13	3.81	5.05
0152	cycle 6 miles		144	.056			1.96	2.48	4.44	5.90
0154	cycle 8 miles		120	.067			2.35	2.98	5.33	7.10
0214	15 MPH ave, cycle 0.5 mile, 20 min wait/Ld./Uld.		176	.045			1.60	2.03	3.63	4.83
0216	cycle 1 mile		160	.050			1.77	2.23	4	5.30
0218	cycle 2 miles		136	.059			2.08	2.63	4.71	6.25
0220	cycle 4 miles		104	.077			2.72	3.44	6.16	8.15
0222	cycle 6 miles		88	.091			3.21	4.06	7.27	9.65
0224	cycle 8 miles		72	.111			3.92	4.96	8.88	11.80
0226	20 MPH ave, cycle 0.5 mile		176	.045			1.60	2.03	3.63	4.83
0228	cycle 1 mile		168	.048			1.68	2.13	3.81	5.05
0230	cycle 2 miles		144	.056			1.96	2.48	4.44	5.90
0232	cycle 4 miles		120	.067			2.35	2.98	5.33	7.10
0234	cycle 6 miles		96	.083			2.94	3.72	6.66	8.85
0236	cycle 8 miles		88	.091			3.21	4.06	7.27	9.65
0244	25 MPH ave, cycle 4 miles		128	.063			2.21	2.79	5	6.65
0246	cycle 6 miles		112	.071			2.52	3.19	5.71	7.60
0248	cycle 8 miles		96	.083			2.94	3.72	6.66	8.85
0250	30 MPH ave, cycle 4 miles		136	.059			2.08	2.63	4.71	6.25
0252	cycle 6 miles		120	.067			2.35	2.98	5.33	7.10
0254	cycle 8 miles		104	.077			2.72	3.44	6.16	8.15
0314	15 MPH ave, cycle 0.5 mile, 25 min wait/Ld./Uld.		144	.056			1.96	2.48	4.44	5.90
0316	cycle 1 mile		128	.063			2.21	2.79	5	6.65
0318	cycle 2 miles		112	.071			2.52	3.19	5.71	7.60
0320	cycle 4 miles		96	.083			2.94	3.72	6.66	8.85
0322	cycle 6 miles		80	.100			3.53	4.47	8	10.60
0324	cycle 8 miles		64	.125			4.41	5.60	10.01	13.30
0326	20 MPH ave, cycle 0.5 mile		144	.056			1.96	2.48	4.44	5.90
0328	cycle 1 mile		136	.059			2.08	2.63	4.71	6.25
0330	cycle 2 miles		120	.067			2.35	2.98	5.33	7.10
0332	cycle 4 miles		104	.077			2.72	3.44	6.16	8.15
0334	cycle 6 miles		88	.091			3.21	4.06	7.27	9.65
0336	cycle 8 miles		80	.100			3.53	4.47	8	10.60
0344	25 MPH ave, cycle 4 miles		112	.071			2.52	3.19	5.71	7.60
0346	cycle 6 miles		96	.083			2.94	3.72	6.66	8.85
0348	cycle 8 miles		88	.091			3.21	4.06	7.27	9.65
0350	30 MPH ave, cycle 4 miles		112	.071			2.52	3.19	5.71	7.60
0352	cycle 6 miles		104	.077			2.72	3.44	6.16	8.15
0354	cycle 8 miles		96	.083			2.94	3.72	6.66	8.85
0414	15 MPH ave, cycle 0.5 mile, 30 min wait/Ld./Uld.		120	.067			2.35	2.98	5.33	7.10
0416	cycle 1 mile		112	.071			2.52	3.19	5.71	7.60
0418	cycle 2 miles		96	.083			2.94	3.72	6.66	8.85
0420	cycle 4 miles		80	.100			3.53	4.47	8	10.60
0422	cycle 6 miles		72	.111			3.92	4.96	8.88	11.80
0424	cycle 8 miles		64	.125			4.41	5.60	10.01	13.30
0426	20 MPH ave, cycle 0.5 mile		120	.067			2.35	2.98	5.33	7.10
0428	cycle 1 mile		112	.071			2.52	3.19	5.71	7.60
0430	cycle 2 miles		104	.077			2.72	3.44	6.16	8.15
0432	cycle 4 miles		88	.091			3.21	4.06	7.27	9.65
0434	cycle 6 miles		80	.100			3.53	4.47	8	10.60
0436	cycle 8 miles		72	.111			3.92	4.96	8.88	11.80
0444	25 MPH ave, cycle 4 miles		96	.083			2.94	3.72	6.66	8.85

31 23 23.20 Hauling		Crew	Daily Output	Labor-Hours	Unit	Material	2012 Bare Costs Labor	Equipment	Total	Total Incl O&P
0446	cycle 6 miles	B-34A	88	.091	L.C.Y.		3.21	4.06	7.27	9.65
0448	cycle 8 miles		80	.100			3.53	4.47	8	10.60
0450	30 MPH ave, cycle 4 miles		96	.083			2.94	3.72	6.66	8.85
0452	cycle 6 miles		88	.091			3.21	4.06	7.27	9.65
0454	cycle 8 miles		80	.100			3.53	4.47	8	10.60
0514	15 MPH ave, cycle 0.5 mile, 35 min wait/Ld./Uld.		104	.077			2.72	3.44	6.16	8.15
0516	cycle 1 mile		96	.083			2.94	3.72	6.66	8.85
0518	cycle 2 miles		88	.091			3.21	4.06	7.27	9.65
0520	cycle 4 miles		72	.111			3.92	4.96	8.88	11.80
0522	cycle 6 miles		64	.125			4.41	5.60	10.01	13.30
0524	cycle 8 miles		56	.143			5.05	6.40	11.45	15.15
0526	20 MPH ave, cycle 0.5 mile		104	.077			2.72	3.44	6.16	8.15
0528	cycle 1 mile		96	.083			2.94	3.72	6.66	8.85
0530	cycle 2 miles		96	.083			2.94	3.72	6.66	8.85
0532	cycle 4 miles		80	.100			3.53	4.47	8	10.60
0534	cycle 6 miles		72	.111			3.92	4.96	8.88	11.80
0536	cycle 8 miles		64	.125			4.41	5.60	10.01	13.30
0544	25 MPH ave, cycle 4 miles		88	.091			3.21	4.06	7.27	9.65
0546	cycle 6 miles		80	.100			3.53	4.47	8	10.60
0548	cycle 8 miles		72	.111			3.92	4.96	8.88	11.80
0550	30 MPH ave, cycle 4 miles		88	.091			3.21	4.06	7.27	9.65
0552	cycle 6 miles		80	.100			3.53	4.47	8	10.60
0554	cycle 8 miles		72	.111			3.92	4.96	8.88	11.80
1014	12 C.Y. truck, cycle 0.5 mile, 15 MPH ave, 15 min. wait/Ld./Uld.	B-34B	336	.024			.84	1.74	2.58	3.27
1016	cycle 1 mile		300	.027			.94	1.95	2.89	3.66
1018	cycle 2 miles		252	.032			1.12	2.32	3.44	4.36
1020	cycle 4 miles		180	.044			1.57	3.25	4.82	6.10
1022	cycle 6 miles		144	.056			1.96	4.06	6.02	7.65
1024	cycle 8 miles		120	.067			2.35	4.87	7.22	9.15
1025	cycle 10 miles		96	.083			2.94	6.10	9.04	11.45
1026	20 MPH ave, cycle 0.5 mile		348	.023			.81	1.68	2.49	3.16
1028	cycle 1 mile		312	.026			.91	1.87	2.78	3.52
1030	cycle 2 miles		276	.029			1.02	2.12	3.14	3.99
1032	cycle 4 miles		216	.037			1.31	2.71	4.02	5.10
1034	cycle 6 miles		168	.048			1.68	3.48	5.16	6.55
1036	cycle 8 miles		144	.056			1.96	4.06	6.02	7.65
1038	cycle 10 miles		120	.067			2.35	4.87	7.22	9.15
1040	25 MPH ave, cycle 4 miles		228	.035			1.24	2.56	3.80	4.82
1042	cycle 6 miles		192	.042			1.47	3.05	4.52	5.75
1044	cycle 8 miles		168	.048			1.68	3.48	5.16	6.55
1046	cycle 10 miles		144	.056			1.96	4.06	6.02	7.65
1050	30 MPH ave, cycle 4 miles		252	.032			1.12	2.32	3.44	4.36
1052	cycle 6 miles		216	.037			1.31	2.71	4.02	5.10
1054	cycle 8 miles		180	.044			1.57	3.25	4.82	6.10
1056	cycle 10 miles		156	.051			1.81	3.75	5.56	7.05
1060	35 MPH ave, cycle 4 miles		264	.030			1.07	2.21	3.28	4.17
1062	cycle 6 miles		228	.035			1.24	2.56	3.80	4.82
1064	cycle 8 miles		204	.039			1.38	2.87	4.25	5.40
1066	cycle 10 miles		180	.044			1.57	3.25	4.82	6.10
1068	cycle 20 miles		120	.067			2.35	4.87	7.22	9.15
1069	cycle 30 miles		84	.095			3.36	6.95	10.31	13.10
1070	cycle 40 miles		72	.111			3.92	8.10	12.02	15.30
1072	40 MPH ave, cycle 6 miles		240	.033			1.18	2.44	3.62	4.58

31 23 23.20 Hauling		Crew	Daily Output	Labor-Hours	Unit	Material	2012 Bare Costs Labor	Equipment	Total	Total Incl O&P
1074	cycle 8 miles	B-34B	216	.037	L.C.Y.		1.31	2.71	4.02	5.10
1076	cycle 10 miles		192	.042			1.47	3.05	4.52	5.75
1078	cycle 20 miles		120	.067			2.35	4.87	7.22	9.15
1080	cycle 30 miles		96	.083			2.94	6.10	9.04	11.45
1082	cycle 40 miles		72	.111			3.92	8.10	12.02	15.30
1084	cycle 50 miles		60	.133			4.71	9.75	14.46	18.30
1094	45 MPH ave, cycle 8 miles		216	.037			1.31	2.71	4.02	5.10
1096	cycle 10 miles		204	.039			1.38	2.87	4.25	5.40
1098	cycle 20 miles		132	.061			2.14	4.43	6.57	8.35
1100	cycle 30 miles		108	.074			2.61	5.40	8.01	10.20
1102	cycle 40 miles		84	.095			3.36	6.95	10.31	13.10
1104	cycle 50 miles		72	.111			3.92	8.10	12.02	15.30
1106	50 MPH ave, cycle 10 miles		216	.037			1.31	2.71	4.02	5.10
1108	cycle 20 miles		144	.056			1.96	4.06	6.02	7.65
1110	cycle 30 miles		108	.074			2.61	5.40	8.01	10.20
1112	cycle 40 miles		84	.095			3.36	6.95	10.31	13.10
1114	cycle 50 miles		72	.111			3.92	8.10	12.02	15.30
1214	15 MPH ave, cycle 0.5 mile, 20 min. wait/Ld./Uld.		264	.030			1.07	2.21	3.28	4.17
1216	cycle 1 mile		240	.033			1.18	2.44	3.62	4.58
1218	cycle 2 miles		204	.039			1.38	2.87	4.25	5.40
1220	cycle 4 miles		156	.051			1.81	3.75	5.56	7.05
1222	cycle 6 miles		132	.061			2.14	4.43	6.57	8.35
1224	cycle 8 miles		108	.074			2.61	5.40	8.01	10.20
1225	cycle 10 miles		96	.083			2.94	6.10	9.04	11.45
1226	20 MPH ave, cycle 0.5 mile		264	.030			1.07	2.21	3.28	4.17
1228	cycle 1 mile		252	.032			1.12	2.32	3.44	4.36
1230	cycle 2 miles		216	.037			1.31	2.71	4.02	5.10
1232	cycle 4 miles		180	.044			1.57	3.25	4.82	6.10
1234	cycle 6 miles		144	.056			1.96	4.06	6.02	7.65
1236	cycle 8 miles		132	.061			2.14	4.43	6.57	8.35
1238	cycle 10 miles		108	.074			2.61	5.40	8.01	10.20
1240	25 MPH ave, cycle 4 miles		192	.042			1.47	3.05	4.52	5.75
1242	cycle 6 miles		168	.048			1.68	3.48	5.16	6.55
1244	cycle 8 miles		144	.056			1.96	4.06	6.02	7.65
1246	cycle 10 miles		132	.061			2.14	4.43	6.57	8.35
1250	30 MPH ave, cycle 4 miles		204	.039			1.38	2.87	4.25	5.40
1252	cycle 6 miles		180	.044			1.57	3.25	4.82	6.10
1254	cycle 8 miles		156	.051			1.81	3.75	5.56	7.05
1256	cycle 10 miles		144	.056			1.96	4.06	6.02	7.65
1260	35 MPH ave, cycle 4 miles		216	.037			1.31	2.71	4.02	5.10
1262	cycle 6 miles		192	.042			1.47	3.05	4.52	5.75
1264	cycle 8 miles		168	.048			1.68	3.48	5.16	6.55
1266	cycle 10 miles		156	.051			1.81	3.75	5.56	7.05
1268	cycle 20 miles		108	.074			2.61	5.40	8.01	10.20
1269	cycle 30 miles		72	.111			3.92	8.10	12.02	15.30
1270	cycle 40 miles		60	.133			4.71	9.75	14.46	18.30
1272	40 MPH ave, cycle 6 miles		192	.042			1.47	3.05	4.52	5.75
1274	cycle 8 miles		180	.044			1.57	3.25	4.82	6.10
1276	cycle 10 miles		156	.051			1.81	3.75	5.56	7.05
1278	cycle 20 miles		108	.074			2.61	5.40	8.01	10.20
1280	cycle 30 miles		84	.095			3.36	6.95	10.31	13.10
1282	cycle 40 miles		72	.111			3.92	8.10	12.02	15.30
1284	cycle 50 miles	↓	60	.133	↓		4.71	9.75	14.46	18.30

31 23 23.20 Hauling		Crew	Daily Output	Labor-Hours	Unit	Material	2012 Bare Costs Labor	Equipment	Total	Total Incl O&P
1294	45 MPH ave, cycle 8 miles	B-34B	180	.044	L.C.Y.		1.57	3.25	4.82	6.10
1296	cycle 10 miles		168	.048			1.68	3.48	5.16	6.55
1298	cycle 20 miles		120	.067			2.35	4.87	7.22	9.15
1300	cycle 30 miles		96	.083			2.94	6.10	9.04	11.45
1302	cycle 40 miles		72	.111			3.92	8.10	12.02	15.30
1304	cycle 50 miles		60	.133			4.71	9.75	14.46	18.30
1306	50 MPH ave, cycle 10 miles		180	.044			1.57	3.25	4.82	6.10
1308	cycle 20 miles		132	.061			2.14	4.43	6.57	8.35
1310	cycle 30 miles		96	.083			2.94	6.10	9.04	11.45
1312	cycle 40 miles		84	.095			3.36	6.95	10.31	13.10
1314	cycle 50 miles		72	.111			3.92	8.10	12.02	15.30
1414	15 MPH ave, cycle 0.5 mile, 25 min. wait/Ld./Uld.		204	.039			1.38	2.87	4.25	5.40
1416	cycle 1 mile		192	.042			1.47	3.05	4.52	5.75
1418	cycle 2 miles		168	.048			1.68	3.48	5.16	6.55
1420	cycle 4 miles		132	.061			2.14	4.43	6.57	8.35
1422	cycle 6 miles		120	.067			2.35	4.87	7.22	9.15
1424	cycle 8 miles		96	.083			2.94	6.10	9.04	11.45
1425	cycle 10 miles		84	.095			3.36	6.95	10.31	13.10
1426	20 MPH ave, cycle 0.5 mile		216	.037			1.31	2.71	4.02	5.10
1428	cycle 1 mile		204	.039			1.38	2.87	4.25	5.40
1430	cycle 2 miles		180	.044			1.57	3.25	4.82	6.10
1432	cycle 4 miles		156	.051			1.81	3.75	5.56	7.05
1434	cycle 6 miles		132	.061			2.14	4.43	6.57	8.35
1436	cycle 8 miles		120	.067			2.35	4.87	7.22	9.15
1438	cycle 10 miles		96	.083			2.94	6.10	9.04	11.45
1440	25 MPH ave, cycle 4 miles		168	.048			1.68	3.48	5.16	6.55
1442	cycle 6 miles		144	.056			1.96	4.06	6.02	7.65
1444	cycle 8 miles		132	.061			2.14	4.43	6.57	8.35
1446	cycle 10 miles		108	.074			2.61	5.40	8.01	10.20
1450	30 MPH ave, cycle 4 miles		168	.048			1.68	3.48	5.16	6.55
1452	cycle 6 miles		156	.051			1.81	3.75	5.56	7.05
1454	cycle 8 miles		132	.061			2.14	4.43	6.57	8.35
1456	cycle 10 miles		120	.067			2.35	4.87	7.22	9.15
1460	35 MPH ave, cycle 4 miles		180	.044			1.57	3.25	4.82	6.10
1462	cycle 6 miles		156	.051			1.81	3.75	5.56	7.05
1464	cycle 8 miles		144	.056			1.96	4.06	6.02	7.65
1466	cycle 10 miles		132	.061			2.14	4.43	6.57	8.35
1468	cycle 20 miles		96	.083			2.94	6.10	9.04	11.45
1469	cycle 30 miles		72	.111			3.92	8.10	12.02	15.30
1470	cycle 40 miles		60	.133			4.71	9.75	14.46	18.30
1472	40 MPH ave, cycle 6 miles		168	.048			1.68	3.48	5.16	6.55
1474	cycle 8 miles		156	.051			1.81	3.75	5.56	7.05
1476	cycle 10 miles		144	.056			1.96	4.06	6.02	7.65
1478	cycle 20 miles		96	.083			2.94	6.10	9.04	11.45
1480	cycle 30 miles		84	.095			3.36	6.95	10.31	13.10
1482	cycle 40 miles		60	.133			4.71	9.75	14.46	18.30
1484	cycle 50 miles		60	.133			4.71	9.75	14.46	18.30
1494	45 MPH ave, cycle 8 miles		156	.051			1.81	3.75	5.56	7.05
1496	cycle 10 miles		144	.056			1.96	4.06	6.02	7.65
1498	cycle 20 miles		108	.074			2.61	5.40	8.01	10.20
1500	cycle 30 miles		84	.095			3.36	6.95	10.31	13.10
1502	cycle 40 miles		72	.111			3.92	8.10	12.02	15.30
1504	cycle 50 miles		60	.133			4.71	9.75	14.46	18.30

31 23 23.20 Hauling		Crew	Daily Output	Labor-Hours	Unit	Material	2012 Bare Costs Labor	Equipment	Total	Total Incl O&P
1506	50 MPH ave, cycle 10 miles	B-34B	156	.051	L.C.Y.		1.81	3.75	5.56	7.05
1508	cycle 20 miles		120	.067			2.35	4.87	7.22	9.15
1510	cycle 30 miles		96	.083			2.94	6.10	9.04	11.45
1512	cycle 40 miles		72	.111			3.92	8.10	12.02	15.30
1514	cycle 50 miles		60	.133			4.71	9.75	14.46	18.30
1614	15 MPH, cycle 0.5 mile, 30 min. wait/Ld./Uld.		180	.044			1.57	3.25	4.82	6.10
1616	cycle 1 mile		168	.048			1.68	3.48	5.16	6.55
1618	cycle 2 miles		144	.056			1.96	4.06	6.02	7.65
1620	cycle 4 miles		120	.067			2.35	4.87	7.22	9.15
1622	cycle 6 miles		108	.074			2.61	5.40	8.01	10.20
1624	cycle 8 miles		84	.095			3.36	6.95	10.31	13.10
1625	cycle 10 miles		84	.095			3.36	6.95	10.31	13.10
1626	20 MPH ave, cycle 0.5 mile		180	.044			1.57	3.25	4.82	6.10
1628	cycle 1 mile		168	.048			1.68	3.48	5.16	6.55
1630	cycle 2 miles		156	.051			1.81	3.75	5.56	7.05
1632	cycle 4 miles		132	.061			2.14	4.43	6.57	8.35
1634	cycle 6 miles		120	.067			2.35	4.87	7.22	9.15
1636	cycle 8 miles		108	.074			2.61	5.40	8.01	10.20
1638	cycle 10 miles		96	.083			2.94	6.10	9.04	11.45
1640	25 MPH ave, cycle 4 miles		144	.056			1.96	4.06	6.02	7.65
1642	cycle 6 miles		132	.061			2.14	4.43	6.57	8.35
1644	cycle 8 miles		108	.074			2.61	5.40	8.01	10.20
1646	cycle 10 miles		108	.074			2.61	5.40	8.01	10.20
1650	30 MPH ave, cycle 4 miles		144	.056			1.96	4.06	6.02	7.65
1652	cycle 6 miles		132	.061			2.14	4.43	6.57	8.35
1654	cycle 8 miles		120	.067			2.35	4.87	7.22	9.15
1656	cycle 10 miles		108	.074			2.61	5.40	8.01	10.20
1660	35 MPH ave, cycle 4 miles		156	.051			1.81	3.75	5.56	7.05
1662	cycle 6 miles		144	.056			1.96	4.06	6.02	7.65
1664	cycle 8 miles		132	.061			2.14	4.43	6.57	8.35
1666	cycle 10 miles		120	.067			2.35	4.87	7.22	9.15
1668	cycle 20 miles		84	.095			3.36	6.95	10.31	13.10
1669	cycle 30 miles		72	.111			3.92	8.10	12.02	15.30
1670	cycle 40 miles		60	.133			4.71	9.75	14.46	18.30
1672	40 MPH, cycle 6 miles		144	.056			1.96	4.06	6.02	7.65
1674	cycle 8 miles		132	.061			2.14	4.43	6.57	8.35
1676	cycle 10 miles		120	.067			2.35	4.87	7.22	9.15
1678	cycle 20 miles		96	.083			2.94	6.10	9.04	11.45
1680	cycle 30 miles		72	.111			3.92	8.10	12.02	15.30
1682	cycle 40 miles		60	.133			4.71	9.75	14.46	18.30
1684	cycle 50 miles		48	.167			5.90	12.20	18.10	23
1694	45 MPH ave, cycle 8 miles		144	.056			1.96	4.06	6.02	7.65
1696	cycle 10 miles		132	.061			2.14	4.43	6.57	8.35
1698	cycle 20 miles		96	.083			2.94	6.10	9.04	11.45
1700	cycle 30 miles		84	.095			3.36	6.95	10.31	13.10
1702	cycle 40 miles		60	.133			4.71	9.75	14.46	18.30
1704	cycle 50 miles		60	.133			4.71	9.75	14.46	18.30
1706	50 MPH ave, cycle 10 miles		132	.061			2.14	4.43	6.57	8.35
1708	cycle 20 miles		108	.074			2.61	5.40	8.01	10.20
1710	cycle 30 miles		84	.095			3.36	6.95	10.31	13.10
1712	cycle 40 miles		72	.111			3.92	8.10	12.02	15.30
1714	cycle 50 miles	↓	60	.133	↓		4.71	9.75	14.46	18.30
2000	Hauling, 8 C.Y. truck, small project cost per hour	B-34A	8	1	Hr.		35.50	44.50	80	106

31 23 23.20 Hauling		Crew	Daily Output	Labor-Hours	Unit	Material	2012 Bare Costs Labor	Equipment	Total	Total Incl O&P
2100	12 C.Y. Truck	B-34B	8	1	Hr.		35.50	73	108.50	138
2150	16.5 C.Y. Truck	B-34C	8	1			35.50	80.50	116	146
2175	18 C.Y. 8 wheel Truck	B-34I	8	1			35.50	93	128.50	159
2200	20 C.Y. Truck	B-34D	8	1			35.50	82.50	118	148
2300	Grading at dump, or embankment if required, by dozer	B-10B	1000	.012	L.C.Y.		.51	1.24	1.75	2.19
2310	Spotter at fill or cut, if required	1 Clab	8	1	Hr.		35		35	57.50
9014	18 C.Y. truck, 8 wheels,15 min. wait/Ld./Uld.,15 MPH, cycle 0.5 mi.	B-34I	504	.016	L.C.Y.		.56	1.48	2.04	2.54
9016	cycle 1 mile		450	.018			.63	1.66	2.29	2.84
9018	cycle 2 miles		378	.021			.75	1.97	2.72	3.38
9020	cycle 4 miles		270	.030			1.05	2.76	3.81	4.73
9022	cycle 6 miles		216	.037			1.31	3.45	4.76	5.90
9024	cycle 8 miles		180	.044			1.57	4.14	5.71	7.10
9025	cycle 10 miles		144	.056			1.96	5.15	7.11	8.85
9026	20 MPH ave, cycle 0.5 mile		522	.015			.54	1.43	1.97	2.45
9028	cycle 1 mile		468	.017			.60	1.59	2.19	2.73
9030	cycle 2 miles		414	.019			.68	1.80	2.48	3.08
9032	cycle 4 miles		324	.025			.87	2.30	3.17	3.94
9034	cycle 6 miles		252	.032			1.12	2.96	4.08	5.05
9036	cycle 8 miles		216	.037			1.31	3.45	4.76	5.90
9038	cycle 10 miles		180	.044			1.57	4.14	5.71	7.10
9040	25 MPH ave, cycle 4 miles		342	.023			.83	2.18	3.01	3.74
9042	cycle 6 miles		288	.028			.98	2.59	3.57	4.44
9044	cycle 8 miles		252	.032			1.12	2.96	4.08	5.05
9046	cycle 10 miles		216	.037			1.31	3.45	4.76	5.90
9050	30 MPH ave, cycle 4 miles		378	.021			.75	1.97	2.72	3.38
9052	cycle 6 miles		324	.025			.87	2.30	3.17	3.94
9054	cycle 8 miles		270	.030			1.05	2.76	3.81	4.73
9056	cycle 10 miles		234	.034			1.21	3.18	4.39	5.45
9060	35 MPH ave, cycle 4 miles		396	.020			.71	1.88	2.59	3.22
9062	cycle 6 miles		342	.023			.83	2.18	3.01	3.74
9064	cycle 8 miles		288	.028			.98	2.59	3.57	4.44
9066	cycle 10 miles		270	.030			1.05	2.76	3.81	4.73
9068	cycle 20 miles		162	.049			1.74	4.60	6.34	7.85
9070	cycle 30 miles		126	.063			2.24	5.90	8.14	10.15
9072	cycle 40 miles		90	.089			3.14	8.30	11.44	14.20
9074	40 MPH ave, cycle 6 miles		360	.022			.78	2.07	2.85	3.55
9076	cycle 8 miles		324	.025			.87	2.30	3.17	3.94
9078	cycle 10 miles		288	.028			.98	2.59	3.57	4.44
9080	cycle 20 miles		180	.044			1.57	4.14	5.71	7.10
9082	cycle 30 miles		144	.056			1.96	5.15	7.11	8.85
9084	cycle 40 miles		108	.074			2.61	6.90	9.51	11.85
9086	cycle 50 miles		90	.089			3.14	8.30	11.44	14.20
9094	45 MPH ave, cycle 8 miles		324	.025			.87	2.30	3.17	3.94
9096	cycle 10 miles		306	.026			.92	2.43	3.35	4.17
9098	cycle 20 miles		198	.040			1.43	3.76	5.19	6.45
9100	cycle 30 miles		144	.056			1.96	5.15	7.11	8.85
9102	cycle 40 miles		126	.063			2.24	5.90	8.14	10.15
9104	cycle 50 miles		108	.074			2.61	6.90	9.51	11.85
9106	50 MPH ave, cycle 10 miles		324	.025			.87	2.30	3.17	3.94
9108	cycle 20 miles		216	.037			1.31	3.45	4.76	5.90
9110	cycle 30 miles		162	.049			1.74	4.60	6.34	7.85
9112	cycle 40 miles		126	.063			2.24	5.90	8.14	10.15
9114	cycle 50 miles		108	.074			2.61	6.90	9.51	11.85

31 23 23.20 Hauling		Crew	Daily Output	Labor-Hours	Unit	Material	2012 Bare Costs Labor	Equipment	Total	Total Incl O&P
9214	20 min. wait/Ld./Uld.,15 MPH, cycle 0.5 mi.	B-34I	396	.020	L.C.Y.		.71	1.88	2.59	3.22
9216	cycle 1 mile		360	.022			.78	2.07	2.85	3.55
9218	cycle 2 miles		306	.026			.92	2.43	3.35	4.17
9220	cycle 4 miles		234	.034			1.21	3.18	4.39	5.45
9222	cycle 6 miles		198	.040			1.43	3.76	5.19	6.45
9224	cycle 8 miles		162	.049			1.74	4.60	6.34	7.85
9225	cycle 10 miles		144	.056			1.96	5.15	7.11	8.85
9226	20 MPH ave, cycle 0.5 mile		396	.020			.71	1.88	2.59	3.22
9228	cycle 1 mile		378	.021			.75	1.97	2.72	3.38
9230	cycle 2 miles		324	.025			.87	2.30	3.17	3.94
9232	cycle 4 miles		270	.030			1.05	2.76	3.81	4.73
9234	cycle 6 miles		216	.037			1.31	3.45	4.76	5.90
9236	cycle 8 miles		198	.040			1.43	3.76	5.19	6.45
9238	cycle 10 miles		162	.049			1.74	4.60	6.34	7.85
9240	25 MPH ave, cycle 4 miles		288	.028			.98	2.59	3.57	4.44
9242	cycle 6 miles		252	.032			1.12	2.96	4.08	5.05
9244	cycle 8 miles		216	.037			1.31	3.45	4.76	5.90
9246	cycle 10 miles		198	.040			1.43	3.76	5.19	6.45
9250	30 MPH ave, cycle 4 miles		306	.026			.92	2.43	3.35	4.17
9252	cycle 6 miles		270	.030			1.05	2.76	3.81	4.73
9254	cycle 8 miles		234	.034			1.21	3.18	4.39	5.45
9256	cycle 10 miles		216	.037			1.31	3.45	4.76	5.90
9260	35 MPH ave, cycle 4 miles		324	.025			.87	2.30	3.17	3.94
9262	cycle 6 miles		288	.028			.98	2.59	3.57	4.44
9264	cycle 8 miles		252	.032			1.12	2.96	4.08	5.05
9266	cycle 10 miles		234	.034			1.21	3.18	4.39	5.45
9268	cycle 20 miles		162	.049			1.74	4.60	6.34	7.85
9270	cycle 30 miles		108	.074			2.61	6.90	9.51	11.85
9272	cycle 40 miles		90	.089			3.14	8.30	11.44	14.20
9274	40 MPH ave, cycle 6 miles		288	.028			.98	2.59	3.57	4.44
9276	cycle 8 miles		270	.030			1.05	2.76	3.81	4.73
9278	cycle 10 miles		234	.034			1.21	3.18	4.39	5.45
9280	cycle 20 miles		162	.049			1.74	4.60	6.34	7.85
9282	cycle 30 miles		126	.063			2.24	5.90	8.14	10.15
9284	cycle 40 miles		108	.074			2.61	6.90	9.51	11.85
9286	cycle 50 miles		90	.089			3.14	8.30	11.44	14.20
9294	45 MPH ave, cycle 8 miles		270	.030			1.05	2.76	3.81	4.73
9296	cycle 10 miles		252	.032			1.12	2.96	4.08	5.05
9298	cycle 20 miles		180	.044			1.57	4.14	5.71	7.10
9300	cycle 30 miles		144	.056			1.96	5.15	7.11	8.85
9302	cycle 40 miles		108	.074			2.61	6.90	9.51	11.85
9304	cycle 50 miles		90	.089			3.14	8.30	11.44	14.20
9306	50 MPH ave, cycle 10 miles		270	.030			1.05	2.76	3.81	4.73
9308	cycle 20 miles		198	.040			1.43	3.76	5.19	6.45
9310	cycle 30 miles		144	.056			1.96	5.15	7.11	8.85
9312	cycle 40 miles		126	.063			2.24	5.90	8.14	10.15
9314	cycle 50 miles		108	.074			2.61	6.90	9.51	11.85
9414	25 min. wait/Ld./Uld.,15 MPH, cycle 0.5 mi.		306	.026			.92	2.43	3.35	4.17
9416	cycle 1 mile		288	.028			.98	2.59	3.57	4.44
9418	cycle 2 miles		252	.032			1.12	2.96	4.08	5.05
9420	cycle 4 miles		198	.040			1.43	3.76	5.19	6.45
9422	cycle 6 miles		180	.044			1.57	4.14	5.71	7.10
9424	cycle 8 miles	B-34I	144	.056	L.C.Y.		1.96	5.15	7.11	8.85

31 23 23 – Fill

31 23 23.20 Hauling		Crew	Daily Output	Labor-Hours	Unit	Material	2012 Bare Costs Labor	Equipment	Total	Total Incl O&P
9425	cycle 10 miles	B-34I	126	.063	L.C.Y.		2.24	5.90	8.14	10.15
9426	20 MPH ave, cycle 0.5 mile		324	.025			.87	2.30	3.17	3.94
9428	cycle 1 mile		306	.026			.92	2.43	3.35	4.17
9430	cycle 2 miles		270	.030			1.05	2.76	3.81	4.73
9432	cycle 4 miles		234	.034			1.21	3.18	4.39	5.45
9434	cycle 6 miles		198	.040			1.43	3.76	5.19	6.45
9436	cycle 8 miles		180	.044			1.57	4.14	5.71	7.10
9438	cycle 10 miles		144	.056			1.96	5.15	7.11	8.85
9440	25 MPH ave, cycle 4 miles		252	.032			1.12	2.96	4.08	5.05
9442	cycle 6 miles		216	.037			1.31	3.45	4.76	5.90
9444	cycle 8 miles		198	.040			1.43	3.76	5.19	6.45
9446	cycle 10 miles		180	.044			1.57	4.14	5.71	7.10
9450	30 MPH ave, cycle 4 miles		252	.032			1.12	2.96	4.08	5.05
9452	cycle 6 miles		234	.034			1.21	3.18	4.39	5.45
9454	cycle 8 miles		198	.040			1.43	3.76	5.19	6.45
9456	cycle 10 miles		180	.044			1.57	4.14	5.71	7.10
9460	35 MPH ave, cycle 4 miles		270	.030			1.05	2.76	3.81	4.73
9462	cycle 6 miles		234	.034			1.21	3.18	4.39	5.45
9464	cycle 8 miles		216	.037			1.31	3.45	4.76	5.90
9466	cycle 10 miles		198	.040			1.43	3.76	5.19	6.45
9468	cycle 20 miles		144	.056			1.96	5.15	7.11	8.85
9470	cycle 30 miles		108	.074			2.61	6.90	9.51	11.85
9472	cycle 40 miles		90	.089			3.14	8.30	11.44	14.20
9474	40 MPH ave, cycle 6 miles		252	.032			1.12	2.96	4.08	5.05
9476	cycle 8 miles		234	.034			1.21	3.18	4.39	5.45
9478	cycle 10 miles		216	.037			1.31	3.45	4.76	5.90
9480	cycle 20 miles		144	.056			1.96	5.15	7.11	8.85
9482	cycle 30 miles		126	.063			2.24	5.90	8.14	10.15
9484	cycle 40 miles		90	.089			3.14	8.30	11.44	14.20
9486	cycle 50 miles		90	.089			3.14	8.30	11.44	14.20
9494	45 MPH ave, cycle 8 miles		234	.034			1.21	3.18	4.39	5.45
9496	cycle 10 miles		216	.037			1.31	3.45	4.76	5.90
9498	cycle 20 miles		162	.049			1.74	4.60	6.34	7.85
9500	cycle 30 miles		126	.063			2.24	5.90	8.14	10.15
9502	cycle 40 miles		108	.074			2.61	6.90	9.51	11.85
9504	cycle 50 miles		90	.089			3.14	8.30	11.44	14.20
9506	50 MPH ave, cycle 10 miles		234	.034			1.21	3.18	4.39	5.45
9508	cycle 20 miles		180	.044			1.57	4.14	5.71	7.10
9510	cycle 30 miles		144	.056			1.96	5.15	7.11	8.85
9512	cycle 40 miles		108	.074			2.61	6.90	9.51	11.85
9514	cycle 50 miles		90	.089			3.14	8.30	11.44	14.20
9614	30 min. wait/Ld./Uld.,15 MPH, cycle 0.5 mi.		270	.030			1.05	2.76	3.81	4.73
9616	cycle 1 mile		252	.032			1.12	2.96	4.08	5.05
9618	cycle 2 miles		216	.037			1.31	3.45	4.76	5.90
9620	cycle 4 miles		180	.044			1.57	4.14	5.71	7.10
9622	cycle 6 miles		162	.049			1.74	4.60	6.34	7.85
9624	cycle 8 miles		126	.063			2.24	5.90	8.14	10.15
9625	cycle 10 miles		126	.063			2.24	5.90	8.14	10.15
9626	20 MPH ave, cycle 0.5 mile		270	.030			1.05	2.76	3.81	4.73
9628	cycle 1 mile		252	.032			1.12	2.96	4.08	5.05
9630	cycle 2 miles		234	.034			1.21	3.18	4.39	5.45
9632	cycle 4 miles		198	.040			1.43	3.76	5.19	6.45
9634	cycle 6 miles	↓	180	.044	↓		1.57	4.14	5.71	7.10

31 23 23.20 Hauling

	31 23 23.20 Hauling	Crew	Daily Output	Labor-Hours	Unit	Material	2012 Bare Costs Labor	Equipment	Total	Total Incl O&P
9636	cycle 8 miles	B-34I	162	.049	L.C.Y.		1.74	4.60	6.34	7.85
9638	cycle 10 miles		144	.056			1.96	5.15	7.11	8.85
9640	25 MPH ave, cycle 4 miles		216	.037			1.31	3.45	4.76	5.90
9642	cycle 6 miles		198	.040			1.43	3.76	5.19	6.45
9644	cycle 8 miles		180	.044			1.57	4.14	5.71	7.10
9646	cycle 10 miles		162	.049			1.74	4.60	6.34	7.85
9650	30 MPH ave, cycle 4 miles		216	.037			1.31	3.45	4.76	5.90
9652	cycle 6 miles		198	.040			1.43	3.76	5.19	6.45
9654	cycle 8 miles		180	.044			1.57	4.14	5.71	7.10
9656	cycle 10 miles		162	.049			1.74	4.60	6.34	7.85
9660	35 MPH ave, cycle 4 miles		234	.034			1.21	3.18	4.39	5.45
9662	cycle 6 miles		216	.037			1.31	3.45	4.76	5.90
9664	cycle 8 miles		198	.040			1.43	3.76	5.19	6.45
9666	cycle 10 miles		180	.044			1.57	4.14	5.71	7.10
9668	cycle 20 miles		126	.063			2.24	5.90	8.14	10.15
9670	cycle 30 miles		108	.074			2.61	6.90	9.51	11.85
9672	cycle 40 miles		90	.089			3.14	8.30	11.44	14.20
9674	40 MPH ave, cycle 6 miles		216	.037			1.31	3.45	4.76	5.90
9676	cycle 8 miles		198	.040			1.43	3.76	5.19	6.45
9678	cycle 10 miles		180	.044			1.57	4.14	5.71	7.10
9680	cycle 20 miles		144	.056			1.96	5.15	7.11	8.85
9682	cycle 30 miles		108	.074			2.61	6.90	9.51	11.85
9684	cycle 40 miles		90	.089			3.14	8.30	11.44	14.20
9686	cycle 50 miles		72	.111			3.92	10.35	14.27	17.75
9694	45 MPH ave, cycle 8 miles		216	.037			1.31	3.45	4.76	5.90
9696	cycle 10 miles		198	.040			1.43	3.76	5.19	6.45
9698	cycle 20 miles		144	.056			1.96	5.15	7.11	8.85
9700	cycle 30 miles		126	.063			2.24	5.90	8.14	10.15
9702	cycle 40 miles		108	.074			2.61	6.90	9.51	11.85
9704	cycle 50 miles		90	.089			3.14	8.30	11.44	14.20
9706	50 MPH ave, cycle 10 miles		198	.040			1.43	3.76	5.19	6.45
9708	cycle 20 miles		162	.049			1.74	4.60	6.34	7.85
9710	cycle 30 miles		126	.063			2.24	5.90	8.14	10.15
9712	cycle 40 miles		108	.074			2.61	6.90	9.51	11.85
9714	cycle 50 miles		90	.089			3.14	8.30	11.44	14.20

31 23 23.23 Compaction

	31 23 23.23 Compaction	Crew	Daily Output	Labor-Hours	Unit	Material	Labor	Equipment	Total	Total Incl O&P
0010	**COMPACTION**									
5000	Riding, vibrating roller, 6" lifts, 2 passes	B-10Y	3000	.004	E.C.Y.		.17	.17	.34	.46
5020	3 passes		2300	.005			.22	.22	.44	.61
5040	4 passes		1900	.006			.27	.27	.54	.73
5050	8" lifts, 2 passes		4100	.003			.13	.13	.26	.34
5060	12" lifts, 2 passes		5200	.002			.10	.10	.20	.27
5080	3 passes		3500	.003			.15	.15	.30	.39
5100	4 passes		2600	.005			.20	.20	.40	.54
5600	Sheepsfoot or wobbly wheel roller, 6" lifts, 2 passes	B-10G	2400	.005			.21	.46	.67	.84
5620	3 passes		1735	.007			.30	.63	.93	1.17
5640	4 passes		1300	.009			.39	.85	1.24	1.56
5680	12" lifts, 2 passes		5200	.002			.10	.21	.31	.39
5700	3 passes		3500	.003			.15	.31	.46	.58
5720	4 passes		2600	.005			.20	.42	.62	.79
7000	Walk behind, vibrating plate 18" wide, 6" lifts, 2 passes	A-1D	200	.040			1.40	.16	1.56	2.48
7020	3 passes		185	.043			1.52	.17	1.69	2.68

31 23 Excavation and Fill

31 23 23 – Fill

31 23 23.23 Compaction

	31 23 23.23 Compaction	Crew	Daily Output	Labor-Hours	Unit	Material	2012 Bare Costs Labor	Equipment	Total	Total Incl O&P
7040	4 passes	A-1D	140	.057	E.C.Y.		2.01	.23	2.24	3.54
7200	12″ lifts, 2 passes	A-1E	560	.014			.50	.07	.57	.90
7220	3 passes		375	.021			.75	.11	.86	1.35
7240	4 passes		280	.029			1	.15	1.15	1.80
7500	Vibrating roller 24″ wide, 6″ lifts, 2 passes	B-10A	420	.029			1.22	.41	1.63	2.40
7520	3 passes		280	.043			1.83	.61	2.44	3.59
7540	4 passes		210	.057			2.44	.82	3.26	4.80
7600	12″ lifts, 2 passes		840	.014			.61	.20	.81	1.19
7620	3 passes		560	.021			.92	.31	1.23	1.80
7640	4 passes		420	.029			1.22	.41	1.63	2.40
8000	Rammer tamper, 6″ to 11″, 4″ lifts, 2 passes	A-1F	130	.062			2.16	.34	2.50	3.91
8050	3 passes		97	.082			2.89	.46	3.35	5.25
8100	4 passes		65	.123			4.32	.68	5	7.85
8200	8″ lifts, 2 passes		260	.031			1.08	.17	1.25	1.96
8250	3 passes		195	.041			1.44	.23	1.67	2.61
8300	4 passes		130	.062			2.16	.34	2.50	3.91
8400	13″ to 18″, 4″ lifts, 2 passes	A-1G	390	.021			.72	.13	.85	1.32
8450	3 passes		290	.028			.97	.17	1.14	1.78
8500	4 passes		195	.041			1.44	.26	1.70	2.64
8600	8″ lifts, 2 passes		780	.010			.36	.06	.42	.66
8650	3 passes		585	.014			.48	.09	.57	.88
8700	4 passes		390	.021			.72	.13	.85	1.32
9000	Water, 3000 gal. truck, 3 mile haul	B-45	1888	.008		1.18	.35	.44	1.97	2.33
9010	6 mile haul		1444	.011		1.18	.45	.57	2.20	2.65
9020	12 mile haul		1000	.016		1.18	.65	.83	2.66	3.25
9030	6000 gal. wagon, 3 mile haul	B-59	2000	.004		1.18	.14	.23	1.55	1.78
9040	6 mile haul	″	1600	.005		1.18	.18	.28	1.64	1.90

31 23 23.24 Compaction, Structural

		Crew	Daily Output	Labor-Hours	Unit	Material	Labor	Equipment	Total	Total Incl O&P
0010	**COMPACTION, STRUCTURAL** R312323-30									
0020	Steel wheel tandem roller, 5 tons	B-10E	8	1.500	Hr.		64	18.65	82.65	123
0050	Air tamp, 6″ to 8″ lifts, common fill	B-9	250	.160	E.C.Y.		5.70	.84	6.54	10.20
0060	Select fill	″	300	.133			4.73	.70	5.43	8.50
0600	Vibratory plate, 8″ lifts, common fill	A-1D	200	.040			1.40	.16	1.56	2.48
0700	Select fill	″	216	.037			1.30	.15	1.45	2.29
9000	Minimum labor/equipment charge	1 Clab	4	2	Job		70		70	115

31 25 Erosion and Sedimentation Controls

31 25 14 – Stabilization Measures for Erosion and Sedimentation Control

31 25 14.16 Rolled Erosion Control Mats and Blankets

		Crew	Daily Output	Labor-Hours	Unit	Material	Labor	Equipment	Total	Total Incl O&P
0010	**ROLLED EROSION CONTROL MATS AND BLANKETS**									
0020	Jute mesh, 100 S.Y. per roll, 4′ wide, stapled G	B-80A	2400	.010	S.Y.	.60	.35	.11	1.06	1.36
0100	Plastic netting, stapled, 2″ x 1″ mesh, 20 mil G	B-1	2500	.010		.34	.34		.68	.93
0200	Polypropylene mesh, stapled, 6.5 oz./S.Y. G		2500	.010		1.32	.34		1.66	2.01
0300	Tobacco netting, or jute mesh #2, stapled G		2500	.010		.12	.34		.46	.69
1000	Silt fence, polypropylene, 3′ high, ideal conditions G	2 Clab	1600	.010	L.F.	.25	.35		.60	.86
1100	Adverse conditions G	″	950	.017	″	.25	.59		.84	1.25
1200	Place and remove hay bales G	A-2	3	8	Ton	265	279	70.50	614.50	825
1250	Hay bales, staked G	″	2500	.010	L.F.	10.60	.33	.08	11.01	12.30

31 31 Soil Treatment

31 31 16 – Termite Control

31 31 16.13 Chemical Termite Control		Crew	Daily Output	Labor-Hours	Unit	Material	Labor	Equipment	Total	Total Incl O&P
						2012 Bare Costs				
0010	**CHEMICAL TERMITE CONTROL**									
0020	Slab and walls, residential	1 Skwk	1200	.007	SF Flr.	.33	.30		.63	.85
0100	Commercial, minimum		2496	.003		.33	.15		.48	.60
0200	Maximum		1645	.005		.50	.22		.72	.90
0390	Minimum labor/equipment charge		4	2	Job		91		91	148
0400	Insecticides for termite control, minimum		14.20	.563	Gal.	59	25.50		84.50	107
0500	Maximum		11	.727	"	101	33		134	165

31 41 Shoring

31 41 13 – Timber Shoring

31 41 13.10 Building Shoring

		Crew	Daily Output	Labor-Hours	Unit	Material	Labor	Equipment	Total	Total Incl O&P
0010	**BUILDING SHORING**									
0020	Shoring, existing building, with timber, no salvage allowance	B-51	2.20	21.818	M.B.F.	640	770	96	1,506	2,050
1000	On cribbing with 35 ton screw jacks, per box and jack		3.60	13.333	Jack	66	470	58.50	594.50	905
1090	Minimum labor/equipment charge		2	24	Ea.		845	106	951	1,500
1100	Masonry openings in walls, see Section 02 41 19.16									

31 41 16 – Sheet Piling

31 41 16.10 Sheet Piling Systems

		Crew	Daily Output	Labor-Hours	Unit	Material	Labor	Equipment	Total	Total Incl O&P
0010	**SHEET PILING SYSTEMS**									
0020	Sheet piling steel, not incl. wales, 22 psf, 15' excav., left in place	B-40	10.81	5.920	Ton	1,525	261	320	2,106	2,450
0100	Drive, extract & salvage R314116-40		6	10.667	"	530	470	575	1,575	1,975
1200	15' deep excavation, 22 psf, left in place		983	.065	S.F.	17.65	2.87	3.51	24.03	28
1300	Drive, extract & salvage		545	.117	"	5.95	5.20	6.35	17.50	22
2100	Rent steel sheet piling and wales, first month				Ton	305			305	335
2200	Per added month				"	30.50			30.50	33.50
3900	Wood, solid sheeting, incl. wales, braces and spacers,									
3910	drive, extract & salvage, 8' deep excavation	B-31	330	.121	S.F.	1.56	4.52	.59	6.67	9.75
4520	Left in place, 8' deep, 55 S.F./hr.		440	.091	"	2.82	3.39	.44	6.65	9.15
4990	Minimum labor/equipment charge		2	20	Job		745	97.50	842.50	1,325
5000	For treated lumber add cost of treatment to lumber									

31 43 Concrete Raising

31 43 13 – Pressure Grouting

31 43 13.13 Concrete Pressure Grouting

		Crew	Daily Output	Labor-Hours	Unit	Material	Labor	Equipment	Total	Total Incl O&P
0010	**CONCRETE PRESSURE GROUTING**									
0020	Grouting, pressure, cement & sand, 1:1 mix, minimum	B-61	124	.323	Bag	14.10	12.05	2.64	28.79	38
0100	Maximum		51	.784	"	14.10	29.50	6.40	50	70
0200	Cement and sand, 1:1 mix, minimum		250	.160	C.F.	28	6	1.31	35.31	42
0300	Maximum		100	.400		42.50	14.95	3.27	60.72	74.50
0400	Epoxy cement grout, minimum		137	.292		560	10.95	2.39	573.34	635
0500	Maximum		57	.702		560	26.50	5.75	592.25	665
0700	Alternate pricing method: (Add for materials)									
0710	5 person crew and equipment	B-61	1	40	Day		1,500	325	1,825	2,775

31 46 Needle Beams

31 46 13 – Cantilever Needle Beams

31 46 13.10 Needle Beams		Crew	Daily Output	Labor-Hours	Unit	2012 Bare Costs Material	Labor	Equipment	Total	Total Incl O&P
0010	**NEEDLE BEAMS**									
0011	Incl. wood shoring 10′ x 10′ opening									
0400	Block, concrete, 8″ thick	B-9	7.10	5.634	Ea.	48	200	29.50	277.50	415
0420	12″ thick		6.70	5.970		56	212	31	299	440
0800	Brick, 4″ thick with 8″ backup block		5.70	7.018		56	249	36.50	341.50	510
1000	Brick, solid, 8″ thick		6.20	6.452		48	229	33.50	310.50	465
1040	12″ thick		4.90	8.163		56	290	42.50	388.50	585
1080	16″ thick		4.50	8.889		71.50	315	46.50	433	645
2000	Add for additional floors of shoring	B-1	6	4		48	143		191	288
9000	Minimum labor/equipment charge	″	2	12	Job		430		430	705

31 48 Underpinning

31 48 13 – Underpinning Piers

31 48 13.10 Underpinning Foundations

		Crew	Daily Output	Labor-Hours	Unit	Material	Labor	Equipment	Total	Total Incl O&P
0010	**UNDERPINNING FOUNDATIONS**									
0011	Including excavation,									
0020	forming, reinforcing, concrete and equipment									
0100	5′ to 16′ below grade, 100 to 500 C.Y.	B-52	2.30	24.348	C.Y.	272	995	241	1,508	2,200
0200	Over 500 C.Y.		2.50	22.400		245	915	221	1,381	1,975
0400	16′ to 25′ below grade, 100 to 500 C.Y.		2	28		299	1,150	277	1,726	2,475
0500	Over 500 C.Y.		2.10	26.667		283	1,100	263	1,646	2,375
0700	26′ to 40′ below grade, 100 to 500 C.Y.		1.60	35		325	1,425	345	2,095	3,075
0800	Over 500 C.Y.		1.80	31.111		299	1,275	305	1,879	2,750
0900	For under 50 C.Y., add					10%	40%			

31 62 Driven Piles

31 62 13 – Concrete Piles

31 62 13.23 Prestressed Concrete Piles

		Crew	Daily Output	Labor-Hours	Unit	Material	Labor	Equipment	Total	Total Incl O&P
0010	**PRESTRESSED CONCRETE PILES**, 200 piles									
0020	Unless specified otherwise, not incl. pile caps or mobilization									
3100	Precast, prestressed, 40′ long, 10″ thick, square	B-19	700	.091	V.L.F.	15.85	4.03	2.62	22.50	27
3200	12″ thick, square		680	.094		20	4.15	2.69	26.84	31.50
3400	14″ thick, square		600	.107		23.50	4.71	3.05	31.26	36.50
4000	18″ thick, square	B-19A	520	.123		44.50	5.45	4.30	54.25	62.50

31 62 16 – Steel Piles

31 62 16.13 Steel Piles

		Crew	Daily Output	Labor-Hours	Unit	Material	Labor	Equipment	Total	Total Incl O&P
0010	**STEEL PILES**									
0100	Step tapered, round, concrete filled									
0110	8″ tip, 60 ton capacity, 30′ depth	B-19	760	.084	V.L.F.	10.50	3.71	2.41	16.62	20.50
0120	60′ depth		740	.086		11.90	3.82	2.48	18.20	22
0250	"H" Sections, 50′ long, HP8 x 36		640	.100		19.30	4.41	2.86	26.57	31.50
1300	HP14 X 102	B-19A	510	.125		56	5.55	4.38	65.93	75.50

31 62 19 – Timber Piles

31 62 19.10 Wood Piles

		Crew	Daily Output	Labor-Hours	Unit	Material	Labor	Equipment	Total	Total Incl O&P
0010	**WOOD PILES**									
0011	Friction or end bearing, not including									
0050	mobilization or demobilization									
0100	Untreated piles, up to 30′ long, 12″ butts, 8″ points	B-19	625	.102	V.L.F.	20	4.52	2.93	27.45	32.50

31 62 Driven Piles

31 62 19 – Timber Piles

31 62 19.10 Wood Piles		Crew	Daily Output	Labor-Hours	Unit	Material	2012 Bare Costs Labor	Equipment	Total	Total Incl O&P
0200	30′ to 39′ long, 12″ butts, 8″ points	B-19	700	.091	V.L.F.	20	4.03	2.62	26.65	31.50
0800	Treated piles, 12 lb. per C.F.,									
0810	friction or end bearing, ASTM class B									
1000	Up to 30′ long, 12″ butts, 8″ points	B-19	625	.102	V.L.F.	18.60	4.52	2.93	26.05	31
1100	30′ to 39′ long, 12″ butts, 8″ points		700	.091		20	4.03	2.62	26.65	31.50
2700	Mobilization for 10,000 L.F. pile job, add		3300	.019			.86	.55	1.41	2
2800	25,000 L.F. pile job, add		8500	.008			.33	.22	.55	.78

31 62 23 – Composite Piles

31 62 23.13 Concrete-Filled Steel Piles		Crew	Daily Output	Labor-Hours	Unit	Material	Labor	Equipment	Total	Total Incl O&P
0010	**CONCRETE-FILLED STEEL PILES** no mobilization or demobilization									
2600	Pipe piles, 50′ lg. 8″ diam., 29 lb. per L.F., no concrete	B-19	500	.128	V.L.F.	20.50	5.65	3.66	29.81	35.50
2700	Concrete filled		460	.139		21.50	6.15	3.98	31.63	38
3500	14″ diameter, 46 lb. per L.F., no concrete		430	.149		33.50	6.55	4.26	44.31	52.50
3600	Concrete filled		355	.180		34.50	7.95	5.15	47.60	56.50
4100	18″ diameter, 59 lb. per L.F., no concrete		355	.180		49	7.95	5.15	62.10	72
4200	Concrete filled		310	.206		55	9.10	5.90	70	82

31 63 Bored Piles

31 63 26 – Drilled Caissons

31 63 26.13 Fixed End Caisson Piles		Crew	Daily Output	Labor-Hours	Unit	Material	Labor	Equipment	Total	Total Incl O&P
0010	**FIXED END CAISSON PILES**									
0015	Including excavation, concrete, 50 lb. reinforcing									
0020	per C.Y., not incl. mobilization, boulder removal, disposal									
0100	Open style, machine drilled, to 50′ deep, in stable ground, no R316326-60									
0110	casings or ground water, 18″ diam., 0.065 C.Y./L.F.	B-43	200	.240	V.L.F.	8.20	9.25	12.20	29.65	37.50
0200	24″ diameter, 0.116 C.Y./L.F.		190	.253	"	14.65	9.75	12.85	37.25	46
0210	4′ bell diameter , add		20	2.400	Ea.	246	92.50	122	460.50	555
0500	48″ diameter, 0.465 C.Y./L.F.		100	.480	V.L.F.	59	18.50	24.50	102	122
0510	9′ bell diameter , add		2	24	Ea.	1,475	925	1,225	3,625	4,475
1200	Open style, machine drilled, to 50′ deep, in wet ground, pulled									
1220	pulled casing and pumping									
1400	24″ diameter, 0.116 C.Y./L.F.	B-48	125	.448	V.L.F.	14.65	17.70	22	54.35	69
1410	4′ bell diameter , add	"	19.80	2.828	Ea.	246	112	140	498	605
1700	48″ diameter, 0.465 C.Y./L.F.	B-49	55	1.600	V.L.F.	59	65.50	63	187.50	240
1710	9′ bell diameter , add	"	3.30	26.667	Ea.	1,475	1,100	1,050	3,625	4,550
2300	Open style, machine drilled, to 50′ deep, in soft rocks and									
2320	medium hard shales									
2500	24″ diameter, 0.116 C.Y./L.F.	B-49	30	2.933	V.L.F.	14.65	120	116	250.65	335
2510	4′ bell diameter, add		10.90	8.073	Ea.	246	330	320	896	1,150
2800	48″ diameter, 0.465 C.Y./L.F.		10	8.800	V.L.F.	59	360	345	764	1,025
2810	9′ bell diameter , add		1.10	80	Ea.	1,100	3,275	3,150	7,525	9,975
3600	For rock excavation, sockets, add, minimum		120	.733	C.F.		30	29	59	80.50
3650	Average		95	.926			38	36.50	74.50	101
3700	Maximum		48	1.833			75	72.50	147.50	201
3900	For 50′ to 100′ deep, add				V.L.F.				7%	7%
4000	For 100′ to 150′ deep, add								25%	25%
4100	For 150′ to 200′ deep, add								30%	30%
4200	For casings left in place, add				Lb.	1.04			1.04	1.14
4300	For other than 50 lb. reinf. per C.Y., add or deduct				"	1.03			1.03	1.13
4400	For steel "I" beam cores, add	B-49	8.30	10.602	Ton	2,000	435	420	2,855	3,350

31 63 Bored Piles

31 63 26 – Drilled Caissons

31 63 26.13 Fixed End Caisson Piles		Crew	Daily Output	Labor-Hours	Unit	Material	2012 Bare Costs Labor	Equipment	Total	Total Incl O&P
4500	Load and haul excess excavation, 2 miles	B-34B	178	.045	L.C.Y.		1.59	3.28	4.87	6.20
4600	For mobilization, 50 mile radius, rig to 36″	B-43	2	24	Ea.		925	1,225	2,150	2,850
4650	Rig to 84″	B-48	1.75	32			1,275	1,575	2,850	3,775
4700	For low headroom, add								50%	
4750	For difficult access, add				↓				25%	

31 63 29 – Drilled Concrete Piers and Shafts

31 63 29.13 Uncased Drilled Concrete Piers

		Crew	Daily Output	Labor-Hours	Unit	Material	Labor	Equipment	Total	Total Incl O&P
0010	**UNCASED DRILLED CONCRETE PIERS**									
0020	Unless specified otherwise, not incl. pile caps or mobilization									
0800	Cast in place friction pile, 50′ long, fluted,									
0810	tapered steel, 4000 psi concrete, no reinforcing									
0900	12″ diameter, 7 ga.	B-19	600	.107	V.L.F.	28.50	4.71	3.05	36.26	42.50
1200	18″ diameter, 7 ga.	″	480	.133	″	43	5.90	3.82	52.72	61.50
1300	End bearing, fluted, constant diameter,									
1320	4000 psi concrete, no reinforcing									
1340	12″ diameter, 7 ga.	B-19	600	.107	V.L.F.	30	4.71	3.05	37.76	44
1400	18″ diameter, 7 ga.	″	480	.133	″	48	5.90	3.82	57.72	66.50

Estimating Tips

32 01 00 Operations and Maintenance of Exterior Improvements

- Recycling of asphalt pavement is becoming very popular and is an alternative to removal and replacement. It can be a good value engineering proposal if removed pavement can be recycled, either at the project site or at another site that is reasonably close to the project site. Two new sections on repair of flexible and rigid pavement have been added.

32 10 00 Bases, Ballasts, and Paving

- When estimating paving, keep in mind the project schedule. If an asphaltic paving project is in a colder climate and runs through to the spring, consider placing the base course in the autumn and then topping it in the spring, just prior to completion. This could save considerable costs in spring repair. Keep in mind that prices for asphalt and concrete are generally higher in the cold seasons. New lines have been added for pavement markings including tactile warning systems and new fence lines.

32 90 00 Planting

- The timing of planting and guarantee specifications often dictate the costs for establishing tree and shrub growth and a stand of grass or ground cover. Establish the work performance schedule to coincide with the local planting season. Maintenance and growth guarantees can add from 20%–100% to the total landscaping cost. The cost to replace trees and shrubs can be as high as 5% of the total cost, depending on the planting zone, soil conditions, and time of year.

Reference Numbers

Reference numbers are shown in shaded boxes at the beginning of some major classifications. These numbers refer to related items in the Reference Section. The reference information may be an estimating procedure, an alternate pricing method, or technical information.

Note: Not all subdivisions listed here necessarily appear in this publication.

32 01 13 – Flexible Paving Surface Treatment

32 01 13.61 Slurry Seal (Latex Modified)		Crew	Daily Output	Labor-Hours	Unit	2012 Bare Costs Material	Labor	Equipment	Total	Total Incl O&P
0010	**SLURRY SEAL (LATEX MODIFIED)**									
3600	Waterproofing, membrane, tar and fabric, small area	B-63	233	.172	S.Y.	13.50	6.35	.69	20.54	26
3640	Large area		1435	.028		12.45	1.03	.11	13.59	15.45
3680	Preformed rubberized asphalt, small area		100	.400		18.40	14.80	1.60	34.80	46
3720	Large area		367	.109		16.75	4.04	.43	21.22	25.50
3780	Rubberized asphalt (latex) seal	B-45	5000	.003		2.70	.13	.17	3	3.36
9000	Minimum labor/equipment charge	1 Clab	2	4	Job		140		140	230

32 01 13.62 Asphalt Surface Treatment		Crew	Daily Output	Labor-Hours	Unit	Material	Labor	Equipment	Total	Total Incl O&P
0010	**ASPHALT SURFACE TREATMENT**									
3000	Pavement overlay, polypropylene									
3040	6 oz. per S.Y., ideal conditions	B-63	10000	.004	S.Y.	2.13	.15	.02	2.30	2.60
3080	Adverse conditions		1000	.040		2.85	1.48	.16	4.49	5.75
3120	4 oz. per S.Y., ideal conditions		10000	.004		1.42	.15	.02	1.59	1.82
3160	Adverse conditions		1000	.040		1.83	1.48	.16	3.47	4.60
3200	Tack coat, emulsion, .05 gal. per S.Y., 1000 S.Y.	B-45	2500	.006		.37	.26	.33	.96	1.19
3240	10,000 S.Y.		10000	.002		.30	.07	.08	.45	.52
3270	.10 gal. per S.Y., 1000 S.Y.		2500	.006		.69	.26	.33	1.28	1.54
3275	10,000 S.Y.		10000	.002		.56	.07	.08	.71	.80
3280	.15 gal. per S.Y., 1000 S.Y.		2500	.006		1.02	.26	.33	1.61	1.90
3320	10,000 S.Y.		10000	.002		.81	.07	.08	.96	1.09

32 01 13.64 Sand Seal		Crew	Daily Output	Labor-Hours	Unit	Material	Labor	Equipment	Total	Total Incl O&P
0010	**SAND SEAL**									
2080	Sand sealing, sharp sand, asphalt emulsion, small area	B-91	10000	.006	S.Y.	1.40	.26	.22	1.88	2.20
2120	Roadway or large area	"	18000	.004	"	1.20	.15	.12	1.47	1.68
3000	Sealing random cracks, min 1/2" wide, to 1-1/2", 1,000 L.F.	B-77	2800	.014	L.F.	1.05	.51	.19	1.75	2.19
3040	10,000 L.F.		4000	.010	"	.73	.35	.13	1.21	1.52
3080	Alternate method, 1,000 L.F.		200	.200	Gal.	30	7.05	2.59	39.64	47.50
3120	10,000 L.F.		325	.123	"	24	4.35	1.59	29.94	35.50
3200	Multi-cracks (flooding), 1 coat, small area	B-92	460	.070	S.Y.	2.42	2.48	1.13	6.03	7.95
3240	Large area		2850	.011		2.15	.40	.18	2.73	3.22
3280	2 coat, small area		230	.139		13	4.95	2.26	20.21	25
3320	Large area		1425	.022		11.80	.80	.36	12.96	14.70
3360	Alternate method, small area		115	.278	Gal.	13	9.90	4.52	27.42	35.50
3400	Large area		715	.045	"	12.20	1.59	.73	14.52	16.85

32 01 13.66 Fog Seal		Crew	Daily Output	Labor-Hours	Unit	Material	Labor	Equipment	Total	Total Incl O&P
0010	**FOG SEAL**									
0012	Sealcoating, 2 coat coal tar pitch emulsion over 10,000 S.Y.	B-45	5000	.003	S.Y.	1	.13	.17	1.30	1.49
0030	1000 to 10,000 S.Y.	"	3000	.005		1	.22	.28	1.50	1.75
0100	Under 1000 S.Y.	B-1	1050	.023		1	.82		1.82	2.44
0300	Petroleum resistant, over 10,000 S.Y.	B-45	5000	.003		1.28	.13	.17	1.58	1.80
0320	1000 to 10,000 S.Y.	"	3000	.005		1.28	.22	.28	1.78	2.06
0400	Under 1000 S.Y.	B-1	1050	.023		1.28	.82		2.10	2.75
0600	Non-skid pavement renewal, over 10,000 S.Y.	B-45	5000	.003		1.40	.13	.17	1.70	1.93
0620	1000 to 10,000 S.Y.	"	3000	.005		1.40	.22	.28	1.90	2.19
0700	Under 1000 S.Y.	B-1	1050	.023		1.40	.82		2.22	2.88
0800	Prepare and clean surface for above	A-2	8545	.003			.10	.02	.12	.19
1000	Hand seal asphalt curbing	B-1	4420	.005	L.F.	.84	.19		1.03	1.24
1900	Asphalt surface treatment, single course, small area									
1901	0.30 gal/S.Y. asphalt material, 20#/S.Y. aggregate	B-91	5000	.013	S.Y.	1.92	.53	.43	2.88	3.42
1910	Roadway or large area		10000	.006		1.77	.26	.22	2.25	2.60
1950	Asphalt surface treatment, dbl. course for small area		3000	.021		3.84	.88	.72	5.44	6.40
1960	Roadway or large area		6000	.011		3.46	.44	.36	4.26	4.89

32 01 13 – Flexible Paving Surface Treatment

32 01 13.66 Fog Seal

	32 01 13.66 Fog Seal	Crew	Daily Output	Labor-Hours	Unit	Material	Labor	Equipment	Total	Total Incl O&P
						2012 Bare Costs				
1980	Asphalt surface treatment, single course, for shoulders	B-91	7500	.009	S.Y.	1.68	.35	.29	2.32	2.73

32 01 13.68 Slurry Seal

	32 01 13.68 Slurry Seal	Crew	Daily Output	Labor-Hours	Unit	Material	Labor	Equipment	Total	Total Incl O&P
0010	**SLURRY SEAL**									
0100	Slurry seal, type I, 8 lb. agg./S.Y., 1 coat, small or irregular area	B-90	2800	.023	S.Y.	1.78	.86	.77	3.41	4.21
0150	Roadway or large area		10000	.006		1.78	.24	.22	2.24	2.59
0200	Type II, 12 lb. aggregate/S.Y., 2 coats,small or irregular area		2000	.032		3.56	1.21	1.08	5.85	7.05
0250	Roadway or large area		8000	.008		3.56	.30	.27	4.13	4.71
0300	Type III, 20 lb. aggregate/S.Y., 2 coats, small or irregular area		1800	.036		4.20	1.34	1.20	6.74	8.10
0350	Roadway or large area		6000	.011		4.20	.40	.36	4.96	5.65
0400	Slurry seal, thermoplastic coal-tar, type I, small or irregular area		2400	.027		3.62	1.01	.90	5.53	6.60
0450	Roadway or large area		8000	.008		3.62	.30	.27	4.19	4.77
0500	Type II, small or irregular area		2400	.027		4.63	1.01	.90	6.54	7.70
0550	Roadway or large area		7800	.008		4.63	.31	.28	5.22	5.90

32 01 16 – Flexible Paving Rehabilitation

32 01 16.71 Cold Milling Asphalt Paving

	32 01 16.71 Cold Milling Asphalt Paving	Crew	Daily Output	Labor-Hours	Unit	Material	Labor	Equipment	Total	Total Incl O&P
0010	**COLD MILLING ASPHALT PAVING**									
5200	Cold planing & cleaning, 1″ to 3″ asphalt pavmt., over 25,000 S.Y.	B-71	6000	.009	S.Y.		.38	1.07	1.45	1.79
5280	5,000 S.Y. to 10,000 S.Y.	″	4000	.014	″		.56	1.60	2.16	2.67
5300	Asphalt pavement removal from conc. base, no haul									
5320	Rip, load & sweep 1″ to 3″	B-70	8000	.007	S.Y.		.28	.21	.49	.68
5330	3″ to 6″ deep	″	5000	.011			.45	.33	.78	1.09
5340	Profile grooving, asphalt pavement load & sweep, 1″ deep	B-71	12500	.004			.18	.51	.69	.85
5350	3″ deep		9000	.006			.25	.71	.96	1.18
5360	6″ deep		5000	.011			.45	1.28	1.73	2.14

32 01 16.73 In Place Cold Reused Asphalt Paving

	32 01 16.73 In Place Cold Reused Asphalt Paving	Crew	Daily Output	Labor-Hours	Unit	Material	Labor	Equipment	Total	Total Incl O&P
0010	**IN PLACE COLD REUSED ASPHALT PAVING**									
5000	Reclamation, pulverizing and blending with existing base									
5040	Aggregate base, 4″ thick pavement, over 15,000 S.Y.	B-73	2400	.027	S.Y.		1.13	2.02	3.15	4.03
5080	5,000 S.Y. to 15,000 S.Y.		2200	.029			1.24	2.20	3.44	4.40
5120	8″ thick pavement, over 15,000 S.Y.		2200	.029			1.24	2.20	3.44	4.40
5160	5,000 S.Y. to 15,000 S.Y.		2000	.032			1.36	2.42	3.78	4.83

32 01 16.74 In Place Hot Reused Asphalt Paving

	32 01 16.74 In Place Hot Reused Asphalt Paving	Crew	Daily Output	Labor-Hours	Unit	Material	Labor	Equipment	Total	Total Incl O&P
0010	**IN PLACE HOT REUSED ASPHALT PAVING**									
5500	Recycle asphalt pavement at site									
5520	Remove, rejuvenate and spread 4″ deep G	B-72	2500	.026	S.Y.	4.19	1.05	4.28	9.52	11
5521	6″ deep G	″	2000	.032	″	6.15	1.31	5.35	12.81	14.75

32 01 17 – Flexible Paving Repair

32 01 17.10 Repair of Asphalt Pavement Holes

	32 01 17.10 Repair of Asphalt Pavement Holes	Crew	Daily Output	Labor-Hours	Unit	Material	Labor	Equipment	Total	Total Incl O&P
0010	**REPAIR OF ASPHALT PAVEMENT HOLES** (cold patch)									
0100	Flexible pavement repair holes, roadway,light traffic, 1 C.F. size	B-37A	24	1	Ea.	9.85	35	12	56.85	81
0150	Group of two, 1 C.F. size each		16	1.500	Set	19.70	52.50	18	90.20	127
0200	Group of three, 1 C.F. size each		12	2	″	29.50	69.50	24	123	173
0300	Medium traffic, 1 C.F. size each	B-37B	24	1.333	Ea.	9.85	46.50	12	68.35	100
0350	Group of two, 1 C.F. size each		16	2	Set	19.70	70	18	107.70	155
0400	Group of three, 1 C.F. size each		12	2.667	″	29.50	93	24	146.50	211
0500	Highway/heavy traffic, 1 C.F. size each	B-37C	24	1.333	Ea.	9.85	46.50	21	77.35	109
0550	Group of two, 1 C.F. size each		16	2	Set	19.70	69.50	31	120.20	169
0600	Group of three, 1 C.F. size each		12	2.667	″	29.50	92.50	41.50	163.50	230
0700	Add police officer and car for traffic control				Hr.	42			42	46
1000	Flexible pavement repair holes,parking lot, bag material, 1 C.F. size each	B-37D	18	.889	Ea.	65.50	31	7.05	103.55	130
1010	Economy bag material, 1 C.F. each		18	.889	″	42.50	31	7.05	80.55	105

32 01 Operation and Maintenance of Exterior Improvements

32 01 17 – Flexible Paving Repair

32 01 17.10 Repair of Asphalt Pavement Holes		Crew	Daily Output	Labor-Hours	Unit	Material	Labor	Equipment	Total	Total Incl O&P
						2012 Bare Costs				
1100	Group of two, 1 C.F. size each	B-37D	12	1.333	Set	131	46.50	10.60	188.10	231
1110	Group of two, economy bag, 1 C.F. size each		12	1.333		85	46.50	10.60	142.10	181
1200	Group of three 1 C.F. size each		10	1.600		196	55.50	12.70	264.20	320
1210	Economy bag, group of three, 1 C.F. size each		10	1.600		127	55.50	12.70	195.20	244
1300	Flexible pavement repair holes, parking lot, 1 C.F. single hole	A-3A	4	2	Ea.	65.50	68.50	35	169	221
1310	Economy material, 1 C.F. single hole		4	2	"	42.50	68.50	35	146	196
1400	Flexible pavement repair holes, parking lot, 1 C.F. four holes		2	4	Set	262	137	69.50	468.50	585
1410	Economy material, 1 C.F. four holes		2	4	"	170	137	69.50	376.50	485
1500	Flexible pavement repair holes,large parking lot, bulk matl , 1 C.F. size	B-37A	32	.750	Ea.	9.85	26	9	44.85	63.50

32 01 17.20 Repair of Asphalt Pavement Patches		Crew	Daily Output	Labor-Hours	Unit	Material	Labor	Equipment	Total	Total Incl O&P
0010	REPAIR OF ASPHALT PAVEMENT PATCHES									
0100	Flexible pavement patches, roadway,light traffic,sawcut 10-25 S.F.	B-89	12	1.333	Ea.		52.50	36	88.50	124
0150	Sawcut 26-60 S.F.		10	1.600			63.50	43	106.50	149
0200	Sawcut 61-100 S.F.		8	2			79	53.50	132.50	185
0210	Sawcut groups of small size patches		24	.667			26.50	17.90	44.40	61.50
0220	Large size patches		16	1			39.50	27	66.50	92.50
0300	Flexible pavement patches, roadway,light traffic,digout 10-25 S.F.	B-6	16	1.500			57.50	21	78.50	116
0350	Digout 26-60 S.F.		12	2			76.50	28	104.50	155
0400	Digout 61-100 S.F.		8	3			115	42	157	232
0450	Add 8 C.Y. truck, small project debris haulaway	B-34A	8	1	Hr.		35.50	44.50	80	106
0460	Add 12 C.Y. truck, small project debris haulaway	B-34B	8	1			35.50	73	108.50	138
0480	Add flagger for non-intersection medium traffic	1 Clab	8	1			35		35	57.50
0490	Add flasher truck for intersection medium traffic or heavy traffic	A-2B	8	1			34.50	26.50	61	84.50
0500	Flexible pavement patches, roadway, repave,cold, 15 S.F., 4″ D	B-37	8	6	Ea.	46.50	222	18.65	287.15	430
0510	6″ depth		8	6		70.50	222	18.65	311.15	460
0520	Repave,cold, 20 S.F., 4″ depth		8	6		62	222	18.65	302.65	450
0530	6″ depth		8	6		94	222	18.65	334.65	485
0540	Repave,cold, 25 S.F., 4″ depth		8	6		77.50	222	18.65	318.15	465
0550	6″ depth		8	6		117	222	18.65	357.65	510
0600	Repave,cold, 30 S.F., 4″ depth		8	6		93	222	18.65	333.65	485
0610	6″ depth		8	6		141	222	18.65	381.65	535
0640	Repave,cold, 40 S.F., 4″ depth		8	6		124	222	18.65	364.65	515
0650	6″ depth		8	6		187	222	18.65	427.65	585
0680	Repave,cold, 50 S.F., 4″ depth		8	6		155	222	18.65	395.65	550
0690	6″ depth		8	6		234	222	18.65	474.65	640
0720	Repave,cold, 60 S.F., 4″ depth		8	6		186	222	18.65	426.65	585
0730	6″ depth		8	6		281	222	18.65	521.65	690
0800	Repave,cold, 70 S.F., 4″ depth		7	6.857		217	254	21.50	492.50	675
0810	6″ depth		7	6.857		330	254	21.50	605.50	800
0820	Repave,cold, 75 S.F., 4″ depth		7	6.857		231	254	21.50	506.50	695
0830	6″ depth		7	6.857		350	254	21.50	625.50	825
0900	Repave,cold, 80 S.F., 4″ depth		6	8		247	296	25	568	780
0910	6″ depth		6	8		375	296	25	696	920
0940	Repave,cold, 90 S.F., 4″ depth		6	8		278	296	25	599	815
0950	6″ depth		6	8		425	296	25	746	975
0980	Repave,cold, 100 S.F., 4″ depth		6	8		310	296	25	631	850
0990	6″ depth		6	8		470	296	25	791	1,025
1000	Add flasher truck for paving operations in medium or heavy traffic	A-2B	8	1	Hr.		34.50	26.50	61	84.50
1100	Prime coat for repair 15-40 S.F., 25% overspray	B-37A	48	.500	Ea.	9.25	17.45	6	32.70	45.50
1150	41-60 S.F.		40	.600		18.50	21	7.20	46.70	62.50
1175	61-80 S.F.		32	.750		25	26	9	60	80
1200	81-100 S.F.		32	.750		31	26	9	66	86.50

32 01 Operation and Maintenance of Exterior Improvements

32 01 17 – Flexible Paving Repair

32 01 17.20 Repair of Asphalt Pavement Patches		Crew	Daily Output	Labor-Hours	Unit	2012 Bare Costs Material	Labor	Equipment	Total	Total Incl O&P
1210	Groups of patches w/25% overspray	B-37A	3600	.007	S.F.	.31	.23	.08	.62	.81
1300	Flexible pavement repair patches,street repave,hot, 60 S.F., 4" D	B-37	8	6	Ea.	100	222	18.65	340.65	490
1310	6" depth		8	6		151	222	18.65	391.65	545
1320	Repave,hot, 70 S.F., 4" depth		7	6.857		116	254	21.50	391.50	565
1330	6" depth		7	6.857		176	254	21.50	451.50	635
1340	Repave,hot, 75 S.F., 4" depth		7	6.857		124	254	21.50	399.50	575
1350	6" depth		7	6.857		189	254	21.50	464.50	645
1360	Repave,hot, 80 S.F., 4" depth		6	8		133	296	25	454	655
1370	6" depth		6	8		202	296	25	523	730
1380	Repave,hot, 90 S.F., 4" depth		6	8		150	296	25	471	675
1390	6" depth		6	8		227	296	25	548	760
1400	Repave,hot, 100 S.F., 4" depth		6	8		166	296	25	487	690
1410	6" depth		6	8		252	296	25	573	785
1420	Pave hot groups of patches, 4" depth		900	.053	S.F.	1.66	1.98	.17	3.81	5.20
1430	6" depth		900	.053	"	2.52	1.98	.17	4.67	6.15
1500	Add 8 C.Y. truck for hot asphalt paving operations	B-34A	1	8	Day		282	355	637	850
1550	Add flasher truck for hot paving operations in med/heavy traffic	A-2B	8	1	Hr.		34.50	26.50	61	84.50
2000	Add police officer and car for traffic control				"	42			42	46

32 01 17.61 Sealing Cracks In Asphalt Paving		Crew	Daily Output	Labor-Hours	Unit	Material	Labor	Equipment	Total	Total Incl O&P
0010	**SEALING CRACKS IN ASPHALT PAVING**									
0100	Sealing cracks in asphalt paving, 1/8" wide x 1/2" depth, slow set	B-37A	2000	.012	L.F.	.18	.42	.14	.74	1.03
0110	1/4" wide x 1/2" depth		2000	.012		.20	.42	.14	.76	1.06
0130	3/8" wide x 1/2" depth		2000	.012		.22	.42	.14	.78	1.09
0140	1/2" wide x 1/2" depth		1800	.013		.25	.46	.16	.87	1.22
0150	3/4" wide x 1/2" depth		1600	.015		.30	.52	.18	1	1.38
0160	1" wide x 1/2" depth		1600	.015		.36	.52	.18	1.06	1.44
0165	1/8" wide x 1" depth, rapid set	B-37F	2000	.016		.26	.56	.26	1.08	1.47
0170	1/4" wide x 1" depth		2000	.016		.50	.56	.26	1.32	1.74
0175	3/8" wide x 1" depth		1800	.018		.75	.62	.29	1.66	2.16
0180	1/2" wide x 1" depth		1800	.018		1.01	.62	.29	1.92	2.44
0185	5/8" wide x 1" depth		1800	.018		1.26	.62	.29	2.17	2.71
0190	3/4" wide x 1" depth		1600	.020		1.51	.70	.32	2.53	3.16
0195	1" wide x 1" depth		1600	.020		2.01	.70	.32	3.03	3.71
0200	1/8" wide x 2" depth, rapid set		2000	.016		.50	.56	.26	1.32	1.74
0210	1/4" wide x 2" depth		2000	.016		1.01	.56	.26	1.83	2.30
0220	3/8" wide x 2" depth		1800	.018		1.51	.62	.29	2.42	2.99
0230	1/2" wide x 2" depth		1800	.018		2.01	.62	.29	2.92	3.54
0240	5/8" wide x 2" depth		1800	.018		2.51	.62	.29	3.42	4.09
0250	3/4" wide x 2" depth		1600	.020		3.01	.70	.32	4.03	4.82
0260	1" wide x 2" depth		1600	.020		4.02	.70	.32	5.04	5.95
0300	Add flagger for non-intersection medium traffic	1 Clab	8	1	Hr.		35		35	57.50
0400	Add flasher truck for intersection medium traffic or heavy traffic	A-2B	8	1	"		34.50	26.50	61	84.50

32 01 29 – Rigid Paving Repair

32 01 29.61 Partial Depth Patching of Rigid Pavement		Crew	Daily Output	Labor-Hours	Unit	Material	Labor	Equipment	Total	Total Incl O&P
0010	**PARTIAL DEPTH PATCHING OF RIGID PAVEMENT**									
0100	Rigid pavement repair, roadway,light traffic, 25% pitting 15 S.F.	B-37F	16	2	Ea.	28.50	70	32.50	131	181
0110	Pitting 20 S.F.		16	2		38.50	70	32.50	141	192
0120	Pitting 25 S.F.		16	2		48	70	32.50	150.50	202
0130	Pitting 30 S.F.		16	2		57.50	70	32.50	160	213
0140	Pitting 40 S.F.		16	2		76.50	70	32.50	179	234
0150	Pitting 50 S.F.		12	2.667		95.50	93	43	231.50	305
0160	Pitting 60 S.F.		12	2.667		115	93	43	251	325

32 01 29.61 Partial Depth Patching of Rigid Pavement		Crew	Daily Output	Labor-Hours	Unit	Material	2012 Bare Costs Labor	Equipment	Total	Total Incl O&P
0170	Pitting 70 S.F.	B-37F	12	2.667	Ea.	134	93	43	270	345
0180	Pitting 75 S.F.		12	2.667		144	93	43	280	360
0190	Pitting 80 S.F.		8	4		153	140	64.50	357.50	465
0200	Pitting 90 S.F.		8	4		172	140	64.50	376.50	490
0210	Pitting 100 S.F.		8	4		191	140	64.50	395.50	510
0300	Roadway,light traffic, 50% pitting 15 S.F.		12	2.667		57.50	93	43	193.50	263
0310	Pitting 20 S.F.		12	2.667		103	93	43	239	315
0320	Pitting 25 S.F.		12	2.667		95.50	93	43	231.50	305
0330	Pitting 30 S.F.		12	2.667		115	93	43	251	325
0340	Pitting 40 S.F.		12	2.667		153	93	43	289	370
0350	Pitting 50 S.F.		8	4		191	140	64.50	395.50	510
0360	Pitting 60 S.F.		8	4		230	140	64.50	434.50	550
0370	Pitting 70 S.F.		8	4		268	140	64.50	472.50	595
0380	Pitting 75 S.F.		8	4		287	140	64.50	491.50	615
0390	Pitting 80 S.F.		6	5.333		305	186	86	577	735
0400	Pitting 90 S.F.		6	5.333		345	186	86	617	780
0410	Pitting 100 S.F.		6	5.333		385	186	86	657	820
1000	Rigid pavement repair,light traffic,surface patch, 2" deep, 15 S.F.		8	4		115	140	64.50	319.50	425
1010	Surface patch, 2" deep, 20 S.F.		8	4		153	140	64.50	357.50	465
1020	Surface patch, 2" deep, 25 S.F.		8	4		191	140	64.50	395.50	510
1030	Surface patch, 2" deep, 30 S.F.		8	4		230	140	64.50	434.50	550
1040	Surface patch, 2" deep, 40 S.F.		8	4		305	140	64.50	509.50	635
1050	Surface patch, 2" deep, 50 S.F.		6	5.333		385	186	86	657	820
1060	Surface patch, 2" deep, 60 S.F.		6	5.333		460	186	86	732	905
1070	Surface patch, 2" deep, 70 S.F.		6	5.333		535	186	86	807	990
1080	Surface patch, 2" deep, 75 S.F.		6	5.333		575	186	86	847	1,025
1090	Surface patch, 2" deep, 80 S.F.		5	6.400		615	223	103	941	1,150
1100	Surface patch, 2" deep, 90 S.F.		5	6.400		690	223	103	1,016	1,250
1200	Surface patch, 2" deep, 100 S.F.		5	6.400		765	223	103	1,091	1,325
2000	Add flagger for non-intersection medium traffic	1 Clab	8	1	Hr.		35		35	57.50
2100	Add flasher truck for intersection medium traffic or heavy traffic	A-2B	8	1			34.50	26.50	61	84.50
2200	Add police officer and car for traffic control					42			42	46

32 01 29.70 Full Depth Patching of Rigid Pavement		Crew	Daily Output	Labor-Hours	Unit	Material	Labor	Equipment	Total	Total Incl O&P
0010	**FULL DEPTH PATCHING OF RIGID PAVEMENT**									
0015	Pavement preparation includes sawcut, remove pavement and replace									
0020	6 inches of subbase and one layer of reinforcement									
0030	Trucking and haulaway of debris excluded									
0100	Rigid pavement replace, light traffic,prep, 15 S.F.,6" depth	B-37E	16	3.500	Ea.	22	133	57.50	212.50	300
0110	Replacement preparation 20 S.F., 6" depth		16	3.500		29	133	57.50	219.50	310
0115	25 S.F., 6" depth		16	3.500		36.50	133	57.50	227	315
0120	30 S.F., 6" depth		14	4		44	152	65.50	261.50	365
0125	35 S.F., 6" depth		14	4		51	152	65.50	268.50	375
0130	40 S.F., 6" depth		14	4		58.50	152	65.50	276	380
0135	45 S.F., 6" depth		14	4		66	152	65.50	283.50	390
0140	50 S.F., 6" depth		12	4.667		73	177	76.50	326.50	450
0145	55 S.F., 6" depth		12	4.667		80.50	177	76.50	334	460
0150	60 S.F., 6" depth		10	5.600		87.50	212	91.50	391	540
0155	65 S.F., 6" depth		10	5.600		95	212	91.50	398.50	545
0160	70 S.F., 6" depth		10	5.600		102	212	91.50	405.50	555
0165	75 S.F., 6" depth		10	5.600		110	212	91.50	413.50	560
0170	80 S.F., 6" depth		8	7		117	265	115	497	685
0175	85 S.F., 6" depth		8	7		124	265	115	504	695

32 01 29.70 Full Depth Patching of Rigid Pavement		Crew	Daily Output	Labor-Hours	Unit	2012 Bare Costs Material	Labor	Equipment	Total	Total Incl O&P
0180	90 S.F., 6" depth	B-37E	8	7	Ea.	132	265	115	512	700
0185	95 S.F., 6" depth		8	7		139	265	115	519	710
0190	100 S.F., 6" depth		8	7		146	265	115	526	715
0200	Pavement preparation for 8 inch rigid paving same as 6 inch									
0290	Pavement preparation for 9 inch rigid paving same as 10 inch									
0300	Rigid pavement replace, light traffic,prep, 15 S.F.,10" depth	B-37E	16	3.500	Ea.	35	133	57.50	225.50	315
0302	Pavement preparation 10 inch includes sawcut, remove pavement and									
0304	replace 6 inches of subbase and two layers of reinforcement									
0306	Trucking and haulaway of debris excluded									
0310	Replacement preparation 20 S.F., 10" depth	B-37E	16	3.500	Ea.	47	133	57.50	237.50	330
0315	25 S.F., 10" depth		16	3.500		58.50	133	57.50	249	340
0320	30 S.F., 10" depth		14	4		70.50	152	65.50	288	395
0325	35 S.F., 10" depth		14	4		55.50	152	65.50	273	380
0330	40 S.F., 10" depth		14	4		93.50	152	65.50	311	420
0335	45 S.F., 10" depth		12	4.667		105	177	76.50	358.50	485
0340	50 S.F., 10" depth		12	4.667		117	177	76.50	370.50	500
0345	55 S.F., 10" depth		12	4.667		129	177	76.50	382.50	510
0350	60 S.F., 10" depth		10	5.600		141	212	91.50	444.50	595
0355	65 S.F., 10" depth		10	5.600		152	212	91.50	455.50	610
0360	70 S.F., 10" depth		10	5.600		164	212	91.50	467.50	620
0365	75 S.F., 10" depth		8	7		176	265	115	556	750
0370	80 S.F., 10" depth		8	7		187	265	115	567	760
0375	85 S.F., 10" depth		8	7		199	265	115	579	775
0380	90 S.F., 10" depth		8	7		211	265	115	591	790
0385	95 S.F., 10" depth		8	7		223	265	115	603	800
0390	100 S.F., 10" depth		8	7		234	265	115	614	815
0500	Add 8 C.Y. truck, small project debris haulaway	B-34A	8	1	Hr.		35.50	44.50	80	106
0550	Add 12 C.Y. truck, small project debris haulaway	B-34B	8	1			35.50	73	108.50	138
0600	Add flagger for non-intersection medium traffic	1 Clab	8	1			35		35	57.50
0650	Add flasher truck for intersection medium traffic or heavy traffic	A-2B	8	1			34.50	26.50	61	84.50
0700	Add police officer and car for traffic control					42			42	46
0990	Concrete will replaced using quick set mix with aggregate									
1000	Rigid pavement replace, light traffic,repour, 15 S.F.,6" depth	B-37F	18	1.778	Ea.	171	62	28.50	261.50	320
1010	20 S.F., 6" depth		18	1.778		229	62	28.50	319.50	385
1015	25 S.F., 6" depth		18	1.778		286	62	28.50	376.50	450
1020	30 S.F., 6" depth		16	2		345	70	32.50	447.50	525
1025	35 S.F., 6" depth		16	2		400	70	32.50	502.50	590
1030	40 S.F., 6" depth		16	2		455	70	32.50	557.50	655
1035	45 S.F., 6" depth		16	2		515	70	32.50	617.50	715
1040	50 S.F., 6" depth		16	2		570	70	32.50	672.50	780
1045	55 S.F., 6" depth		12	2.667		630	93	43	766	890
1050	60 S.F., 6" depth		12	2.667		685	93	43	821	955
1055	65 S.F., 6" depth		12	2.667		745	93	43	881	1,025
1060	70 S.F., 6" depth		12	2.667		800	93	43	936	1,075
1065	75 S.F., 6" depth		10	3.200		855	112	51.50	1,018.50	1,175
1070	80 S.F., 6" depth		10	3.200		915	112	51.50	1,078.50	1,250
1075	85 S.F., 6" depth		8	4		970	140	64.50	1,174.50	1,375
1080	90 S.F., 6" depth		8	4		1,025	140	64.50	1,229.50	1,425
1085	95 S.F., 6" depth		8	4		1,075	140	64.50	1,279.50	1,500
1090	100 S.F., 6" depth		8	4		1,150	140	64.50	1,354.50	1,550
1099	Concrete will be replaced using 4500 PSI Concrete ready mix									
1100	Rigid pavement replace, light traffic,repour, 15 S.F.,6" depth	A-2	12	2	Ea.	325	69.50	17.60	412.10	490
1110	20-50 S.F., 6" depth		12	2		325	69.50	17.60	412.10	490

32 01 Operation and Maintenance of Exterior Improvements

32 01 29 – Rigid Paving Repair

32 01 29.70 Full Depth Patching of Rigid Pavement		Crew	Daily Output	Labor-Hours	Unit	2012 Bare Costs Material	Labor	Equipment	Total	Total Incl O&P
1120	55-65 S.F., 6" depth	A-2	10	2.400	Ea.	355	83.50	21	459.50	550
1130	70-80 S.F., 6" depth		10	2.400		380	83.50	21	484.50	580
1140	85-100 S.F., 6" depth		8	3		440	105	26.50	571.50	680
1200	Repour 15-40 S.F.,8" depth		12	2		325	69.50	17.60	412.10	490
1210	45-50 S.F.,8" depth		12	2		355	69.50	17.60	442.10	525
1220	55-60 S.F.,8" depth		10	2.400		380	83.50	21	484.50	580
1230	65-80 S.F.,8" depth		10	2.400		440	83.50	21	544.50	640
1240	85-90 S.F.,8" depth		8	3		470	105	26.50	601.50	715
1250	95-100 S.F.,8" depth		8	3		495	105	26.50	626.50	745
1300	Repour 15-30 S.F.,10" depth		12	2		325	69.50	17.60	412.10	490
1310	35-40 S.F.,10" depth		12	2		355	69.50	17.60	442.10	525
1320	45 S.F.,10" depth		12	2		380	69.50	17.60	467.10	555
1330	50-65 S.F.,10" depth		10	2.400		440	83.50	21	544.50	640
1340	70 S.F.,10" depth		10	2.400		470	83.50	21	574.50	675
1350	75-80 S.F.,10" depth		10	2.400		495	83.50	21	599.50	705
1360	85-95 S.F.,10" depth		8	3		555	105	26.50	686.50	810
1370	100 S.F.,10" depth		8	3		585	105	26.50	716.50	845

32 06 Schedules for Exterior Improvements

32 06 10 – Schedules for Bases, Ballasts, and Paving

32 06 10.10 Sidewalks, Driveways and Patios

		Crew	Daily Output	Labor-Hours	Unit	Material	Labor	Equipment	Total	Total Incl O&P
0010	**SIDEWALKS, DRIVEWAYS AND PATIOS** No base									
0020	Asphaltic concrete, 2" thick	B-37	720	.067	S.Y.	6.55	2.47	.21	9.23	11.45
0100	2-1/2" thick	"	660	.073	"	8.35	2.69	.23	11.27	13.80
0110	Bedding for brick or stone, mortar, 1" thick	D-1	300	.053	S.F.	.79	2.13		2.92	4.29
0120	2" thick	"	200	.080		1.97	3.19		5.16	7.30
0130	Sand, 2" thick	B-18	8000	.003		.32	.11	.01	.44	.54
0140	4" thick	"	4000	.006		.64	.21	.01	.86	1.07
0300	Concrete, 3000 psi, CIP, 6 x 6 - W1.4 x W1.4 mesh,									
0310	broomed finish, no base, 4" thick	B-24	600	.040	S.F.	1.70	1.62		3.32	4.49
0350	5" thick		545	.044		2.27	1.78		4.05	5.35
0400	6" thick		510	.047		2.64	1.91		4.55	6
0450	For bank run gravel base, 4" thick, add	B-18	2500	.010		.53	.34	.02	.89	1.16
0520	8" thick, add	"	1600	.015		1.07	.54	.03	1.64	2.09
0550	Exposed aggregate finish, add to above, minimum	B-24	1875	.013		.13	.52		.65	.98
0600	Maximum	"	455	.053		.43	2.14		2.57	3.93
0950	Concrete tree grate, 5' square	B-6	25	.960	Ea.	400	37	13.35	450.35	515
0955	Tree well & cover, concrete, 3' square		25	.960		300	37	13.35	350.35	405
0960	Cast iron tree grate with frame, 2 piece, round, 5' diameter		25	.960		1,000	37	13.35	1,050.35	1,175
0980	Square, 5' side		25	.960		1,000	37	13.35	1,050.35	1,175
1700	Redwood, prefabricated, 4' x 4' sections	2 Carp	316	.051	S.F.	4.74	2.23		6.97	8.85
1750	Redwood planks, 1" thick, on sleepers	"	240	.067	"	4.74	2.94		7.68	10
2250	Stone dust, 4" thick	B-62	900	.027	S.Y.	3.66	1.02	.18	4.86	5.85
9000	Minimum labor/equipment charge	D-1	2	8	Job		320		320	515

32 06 10.20 Steps

		Crew	Daily Output	Labor-Hours	Unit	Material	Labor	Equipment	Total	Total Incl O&P
0010	**STEPS**									
0011	Incl. excav., borrow & concrete base as required									
0100	Brick steps	B-24	35	.686	LF Riser	11.70	28		39.70	58
0200	Railroad ties	2 Clab	25	.640		3.10	22.50		25.60	40.50
0300	Bluestone treads, 12" x 2" or 12" x 1-1/2"	B-24	30	.800		28	32.50		60.50	83
0490	Minimum labor/equipment charge	D-1	2	8	Job		320		320	515

32 06 Schedules for Exterior Improvements

32 06 10 – Schedules for Bases, Ballasts, and Paving

32 06 10.20 Steps		Crew	Daily Output	Labor-Hours	Unit	Material	Labor	Equipment	Total	Total Incl O&P
						2012 Bare Costs				
0500	Concrete, cast in place, see Section 03 30 53.40									
0600	Precast concrete, see Section 03 41 23.50									

32 11 Base Courses

32 11 23 – Aggregate Base Courses

32 11 23.23 Base Course Drainage Layers

		Crew	Daily Output	Labor-Hours	Unit	Material	Labor	Equipment	Total	Total Incl O&P
0010	**BASE COURSE DRAINAGE LAYERS**									
0011	For roadways and large areas									
0050	Crushed 3/4" stone base, compacted, 3" deep	B-36C	5200	.008	S.Y.	3.04	.33	.72	4.09	4.66
0100	6" deep		5000	.008		6.10	.34	.75	7.19	8.05
0200	9" deep		4600	.009		9.15	.37	.81	10.33	11.55
0300	12" deep		4200	.010		12.20	.40	.89	13.49	15
0301	Crushed 1-1/2" stone base, compacted to 4" deep	B-36B	6000	.011		3.98	.44	.72	5.14	5.90
0302	6" deep		5400	.012		5.95	.49	.80	7.24	8.20
0303	8" deep		4500	.014		7.95	.58	.97	9.50	10.75
0304	12" deep		3800	.017		11.95	.69	1.14	13.78	15.50
0310	Minimum labor/equipment charge	B-36	1	40	Job		1,600	1,500	3,100	4,225
0350	Bank run gravel, spread and compacted									
0370	6" deep	B-32	6000	.005	S.Y.	4.44	.23	.35	5.02	5.65
0390	9" deep		4900	.007		6.65	.29	.42	7.36	8.25
0400	12" deep		4200	.008		8.90	.33	.50	9.73	10.90
6900	For small and irregular areas, add						50%	50%		
6990	Minimum labor/equipment charge	1 Clab	3	2.667	Job		93.50		93.50	153
7000	Prepare and roll sub-base, small areas to 2500 S.Y.	B-32A	1500	.016	S.Y.		.68	.83	1.51	2.01
8000	Large areas over 2500 S.Y.	"	3500	.007			.29	.36	.65	.86
8050	For roadways	B-32	4000	.008			.35	.52	.87	1.13
9000	Minimum labor/equipment charge	1 Clab	4	2	Job		70		70	115

32 12 Flexible Paving

32 12 16 – Asphalt Paving

32 12 16.13 Plant-Mix Asphalt Paving

		Crew	Daily Output	Labor-Hours	Unit	Material	Labor	Equipment	Total	Total Incl O&P
0010	**PLANT-MIX ASPHALT PAVING**									
0080	Binder course, 1-1/2" thick	B-25	7725	.011	S.Y.	5.05	.44	.33	5.82	6.65
0120	2" thick		6345	.014		6.75	.53	.40	7.68	8.70
0130	2-1/2" thick		5620	.016		8.45	.60	.45	9.50	10.70
0160	3" thick		4905	.018		10.10	.69	.52	11.31	12.85
0170	3-1/2" thick		4520	.019		11.80	.75	.56	13.11	14.85
0300	Wearing course, 1" thick	B-25B	10575	.009		3.34	.35	.26	3.95	4.54
0340	1-1/2" thick		7725	.012		5.60	.49	.36	6.45	7.35
0380	2" thick		6345	.015		7.55	.59	.44	8.58	9.75
0420	2-1/2" thick		5480	.018		9.30	.68	.50	10.48	11.90
0460	3" thick		4900	.020		11.10	.77	.56	12.43	14.05
0470	3-1/2" thick		4520	.021		13.05	.83	.61	14.49	16.35
0480	4" thick		4140	.023		14.90	.91	.67	16.48	18.55
0500	Open graded friction course	B-25C	5000	.010		2.70	.38	.44	3.52	4.07
0800	Alternate method of figuring paving costs									
0810	Binder course, 1-1/2" thick	B-25	630	.140	Ton	62	5.35	4.03	71.38	81
0811	2" thick		690	.128		62	4.90	3.68	70.58	80
0812	3" thick		800	.110		62	4.22	3.17	69.39	78.50

32 12 Flexible Paving

32 12 16 – Asphalt Paving

32 12 16.13 Plant-Mix Asphalt Paving		Crew	Daily Output	Labor-Hours	Unit	Material	2012 Bare Costs Labor	Equipment	Total	Total Incl O&P
0813	4″ thick	B-25	900	.098	Ton	62	3.75	2.82	68.57	77
0850	Wearing course, 1″ thick	B-25B	575	.167		68	6.50	4.81	79.31	91
0851	1-1/2″ thick		630	.152		68	5.95	4.39	78.34	89.50
0852	2″ thick		690	.139		68	5.45	4.01	77.46	88
0853	2-1/2″ thick		765	.125		68	4.90	3.61	76.51	87
0854	3″ thick		800	.120		68	4.69	3.46	76.15	86.50
1000	Pavement replacement over trench, 2″ thick	B-37	90	.533	S.Y.	6.95	19.75	1.66	28.36	41.50
1050	4″ thick		70	.686		13.80	25.50	2.13	41.43	59
1080	6″ thick		55	.873		22	32.50	2.71	57.21	79.50
3000	Prime coat, emulsion, .30 gal. per S.Y., 1000 S.Y.	B-45	2500	.006		2.22	.26	.33	2.81	3.22
3100	Tack coat, emulsion, .10 gal. per S.Y., 1000 S.Y.	″	2500	.006		.74	.26	.33	1.33	1.59

32 12 16.14 Asphaltic Concrete Paving		Crew	Daily Output	Labor-Hours	Unit	Material	Labor	Equipment	Total	Total Incl O&P
0011	**ASPHALTIC CONCRETE PAVING**, parking lots & driveways									
0015	No asphalt hauling included									
0020	6″ stone base, 2″ binder course, 1″ topping	B-25C	9000	.005	S.F.	1.70	.21	.25	2.16	2.48
0025	2″ binder course, 2″ topping		9000	.005		2.11	.21	.25	2.57	2.93
0030	3″ binder course, 2″ topping		9000	.005		2.49	.21	.25	2.95	3.34
0035	4″ binder course, 2″ topping		9000	.005		2.86	.21	.25	3.32	3.75
0040	1.5″ binder course, 1″ topping		9000	.005		1.51	.21	.25	1.97	2.27
0042	3″ binder course, 1″ topping		9000	.005		2.07	.21	.25	2.53	2.89
0045	3″ binder course, 3″ topping		9000	.005		2.90	.21	.25	3.36	3.80
0050	4″ binder course, 3″ topping		9000	.005		3.27	.21	.25	3.73	4.21
0055	4″ binder course, 4″ topping		9000	.005		3.68	.21	.25	4.14	4.66
0300	Binder course, 1-1/2″ thick		35000	.001		.56	.05	.06	.67	.78
0400	2″ thick		25000	.002		.73	.08	.09	.90	1.02
0500	3″ thick		15000	.003		1.13	.13	.15	1.41	1.60
0600	4″ thick		10800	.004		1.48	.17	.21	1.86	2.13
0800	Sand finish course, 3/4″ thick		41000	.001		.32	.05	.05	.42	.48
0900	1″ thick		34000	.001		.39	.06	.07	.52	.59
1000	Fill pot holes, hot mix, 2″ thick	B-16	4200	.008		.84	.27	.14	1.25	1.51
1100	4″ thick		3500	.009		1.23	.33	.17	1.73	2.06
1120	6″ thick		3100	.010		1.65	.37	.19	2.21	2.63
1140	Cold patch, 2″ thick	B-51	3000	.016		.85	.57	.07	1.49	1.94
1160	4″ thick		2700	.018		1.62	.63	.08	2.33	2.90
1180	6″ thick		1900	.025		2.52	.89	.11	3.52	4.35
3000	Prime coat, emulsion, .30 gal. per S.Y., 1000 S.Y.	B-45	2500	.006	S.Y.	2.22	.26	.33	2.81	3.22
3100	Tack coat, emulsion, .10 gal. per S.Y., 1000 S.Y.	″	2500	.006	″	.74	.26	.33	1.33	1.59

32 13 Rigid Paving

32 13 13 – Concrete Paving

32 13 13.23 Concrete Paving Surface Treatment		Crew	Daily Output	Labor-Hours	Unit	Material	Labor	Equipment	Total	Total Incl O&P
0010	**CONCRETE PAVING SURFACE TREATMENT**									
0015	Including joints, finishing and curing									
0020	Fixed form, 12′ pass, unreinforced, 6″ thick	B-26	3000	.029	S.Y.	20.50	1.15	1.05	22.70	25.50
0100	8″ thick		2750	.032		28	1.26	1.14	30.40	34
0400	12″ thick		1800	.049		40.50	1.92	1.75	44.17	49.50
0410	Conc pavement, w/jt,fnsh&curing,fix form,24′ pass,unreinforced,6″T		6000	.015		19.50	.58	.52	20.60	23
0430	8″ thick		5500	.016		26.50	.63	.57	27.70	31
0460	12″ thick		3600	.024		39	.96	.87	40.83	45.50
0505	Minimum labor/equipment charge	1 Cefi	1	8	Job		340		340	530
0520	Welded wire fabric, sheets for rigid paving 2.33 lb./S.Y.	2 Rodm	389	.041	S.Y.	1.24	2.02		3.26	4.74

32 13 Rigid Paving

32 13 13 – Concrete Paving

32 13 13.23 Concrete Paving Surface Treatment		Crew	Daily Output	Labor-Hours	Unit	Material	2012 Bare Costs Labor	Equipment	Total	Total Incl O&P
0530	Reinforcing steel for rigid paving 12 lb./S.Y.	2 Rodm	666	.024	S.Y.	6.20	1.18		7.38	8.75
0540	Reinforcing steel for rigid paving 18 lb./S.Y.		444	.036		9.25	1.77		11.02	13.15
0620	Slip form, 12′ pass, unreinforced, 6″ thick	B-26A	5600	.016		19.80	.62	.59	21.01	23.50
0624	8″ thick		5300	.017		27	.65	.62	28.27	31
0630	12″ thick		3470	.025		39.50	1	.95	41.45	46
0632	15″ thick		2890	.030		50	1.20	1.14	52.34	57.50
0640	Slip form, 24′ pass, unreinforced, 6″ thick		11200	.008		19.45	.31	.29	20.05	22.50
0644	8″ thick		10600	.008		26	.33	.31	26.64	30
0650	12″ thick		6940	.013		38.50	.50	.47	39.47	44
0652	15″ thick		5780	.015		48.50	.60	.57	49.67	55
0700	Finishing, broom finish small areas	2 Cefi	120	.133			5.65		5.65	8.85
1000	Curing, with sprayed membrane by hand	2 Clab	1500	.011		.67	.37		1.04	1.35

32 14 Unit Paving

32 14 13 – Precast Concrete Unit Paving

32 14 13.18 Precast Concrete Plantable Pavers

		Crew	Daily Output	Labor-Hours	Unit	Material	Labor	Equipment	Total	Total Incl O&P
0010	**PRECAST CONCRETE PLANTABLE PAVERS** (50% grass)									
0300	3/4″ crushed stone base for plantable pavers, 6 inch depth	B-62	1000	.024	S.Y.	6.10	.92	.16	7.18	8.35
0400	8 inch depth		900	.027		8.15	1.02	.18	9.35	10.85
0500	10 inch depth		800	.030		10.15	1.15	.20	11.50	13.25
0600	12 inch depth		700	.034		12.20	1.31	.23	13.74	15.75
0700	Hydro seeding plantable pavers	B-81A	20	.800	M.S.F.	19.60	28	19.25	66.85	87.50
0800	Apply fertilizer and seed to plantable pavers	1 Clab	8	1	"	34.50	35		69.50	95.50

32 14 16 – Brick Unit Paving

32 14 16.10 Brick Paving

		Crew	Daily Output	Labor-Hours	Unit	Material	Labor	Equipment	Total	Total Incl O&P
0010	**BRICK PAVING**									
0012	4″ x 8″ x 1-1/2″, without joints (4.5 brick/S.F.)	D-1	110	.145	S.F.	2.86	5.80		8.66	12.50
0100	Grouted, 3/8″ joint (3.9 brick/S.F.)		90	.178		3.38	7.10		10.48	15.10
0200	4″ x 8″ x 2-1/4″, without joints (4.5 bricks/S.F.)		110	.145		3.74	5.80		9.54	13.45
0300	Grouted, 3/8″ joint (3.9 brick/S.F.)		90	.178		3.45	7.10		10.55	15.20
0500	Bedding, asphalt, 3/4″ thick	B-25	5130	.017		.62	.66	.49	1.77	2.29
0540	Course washed sand bed, 1″ thick	B-18	5000	.005		.36	.17	.01	.54	.69
0580	Mortar, 1″ thick	D-1	300	.053		.66	2.13		2.79	4.14
0620	2″ thick		200	.080		1.31	3.19		4.50	6.60
1500	Brick on 1″ thick sand bed laid flat, 4.5 per S.F.		100	.160		3.03	6.40		9.43	13.60
2000	Brick pavers, laid on edge, 7.2 per S.F.		70	.229		2.80	9.10		11.90	17.75
2500	For 4″ thick concrete bed and joints, add		595	.027		1.12	1.07		2.19	2.97
2800	For steam cleaning, add	A-1H	950	.008		.08	.30	.07	.45	.65
9000	Minimum labor/equipment charge	1 Bric	2	4	Job		176		176	284

32 14 23 – Asphalt Unit Paving

32 14 23.10 Asphalt Blocks

		Crew	Daily Output	Labor-Hours	Unit	Material	Labor	Equipment	Total	Total Incl O&P
0010	**ASPHALT BLOCKS**									
0020	Rectangular, 6″ x 12″ x 1-1/4″, w/bed & neopr. adhesive	D-1	135	.119	S.F.	9.05	4.72		13.77	17.55
0100	3″ thick		130	.123		12.65	4.90		17.55	22
0300	Hexagonal tile, 8″ wide, 1-1/4″ thick		135	.119		9.05	4.72		13.77	17.55
0400	2″ thick		130	.123		12.65	4.90		17.55	22
0500	Square, 8″ x 8″, 1-1/4″ thick		135	.119		9.05	4.72		13.77	17.55
0600	2″ thick		130	.123		12.65	4.90		17.55	22
9000	Minimum labor/equipment charge	1 Bric	2	4	Job		176		176	284

32 14 Unit Paving

32 14 40 – Stone Paving

32 14 40.10 Stone Pavers		Crew	Daily Output	Labor-Hours	Unit	2012 Bare Costs Material	Labor	Equipment	Total	Total Incl O&P
0010	**STONE PAVERS**									
1300	Slate, natural cleft, irregular, 3/4" thick	D-1	92	.174	S.F.	9.60	6.95		16.55	21.50
1350	Random rectangular, gauged, 1/2" thick		105	.152		21	6.05		27.05	33
1400	Random rectangular, butt joint, gauged, 1/4" thick		150	.107		22.50	4.25		26.75	31.50
1450	For sand rubbed finish, add					9.60			9.60	10.55
1550	Granite blocks, 3-1/2" x 3-1/2" x 3-1/2"	D-1	92	.174		10	6.95		16.95	22
1600	4" to 12" long, 3" to 5" wide, 3" to 5" thick	"	98	.163		8.35	6.50		14.85	19.60

32 16 Curbs, Gutters, Sidewalks, and Driveways

32 16 13 – Curbs and Gutters

		Crew	Daily Output	Labor-Hours	Unit	Material	Labor	Equipment	Total	Total Incl O&P
	32 16 13.13 Cast-in-Place Concrete Curbs and Gutters									
0010	**CAST-IN-PLACE CONCRETE CURBS AND GUTTERS**									
0290	Forms only, no concrete									
0300	Concrete, wood forms, 6" x 18", straight	C-2	500	.096	L.F.	2.15	4.12		6.27	9.10
0400	6" x 18", radius	"	200	.240		2.25	10.30		12.55	19.35
0404	Concrete, wood forms, 6" x 18", straight & concrete	C-2A	500	.096		5	4.09		9.09	12.15
0406	6" x 18", radius	"	200	.240		5.10	10.25		15.35	22.50
0415	Machine formed, 6" x 18", straight	B-69A	2000	.024		6.15	.93	.38	7.46	8.65
0416	6" x 18", radius	"	900	.053		6.25	2.06	.85	9.16	11.15
	32 16 13.23 Precast Concrete Curbs and Gutters									
0010	**PRECAST CONCRETE CURBS AND GUTTERS**									
0550	Precast, 6" x 18", straight	B-29	700	.080	L.F.	8.75	3.05	1.21	13.01	15.90
0600	6" x 18", radius	"	325	.172	"	19.50	6.55	2.60	28.65	35
	32 16 13.33 Asphalt Curbs									
0010	**ASPHALT CURBS**									
0012	Curbs, asphaltic, machine formed, 8" wide, 6" high, 40 L.F./ton	B-27	1000	.032	L.F.	1.65	1.14	.28	3.07	4
0100	8" wide, 8" high, 30 L.F. per ton		900	.036		2.20	1.27	.31	3.78	4.84
0150	Asphaltic berm, 12" W, 3"-6" H, 35 L.F./ton, before pavement		700	.046		.04	1.63	.40	2.07	3.15
0200	12" W, 1-1/2" to 4" H, 60 L.F. per ton, laid with pavement	B-2	1050	.038		.02	1.35		1.37	2.24
	32 16 13.43 Stone Curbs									
0010	**STONE CURBS**									
1000	Granite, split face, straight, 5" x 16"	D-13	500	.096	L.F.	12.75	4.05	.90	17.70	21.50
1100	6" x 18"	"	450	.107		16.75	4.50	1	22.25	27
1300	Radius curbing, 6" x 18", over 10' radius	B-29	260	.215		20.50	8.20	3.25	31.95	39.50
1400	Corners, 2' radius		80	.700	Ea.	69	26.50	10.55	106.05	130
1600	Edging, 4-1/2" x 12", straight		300	.187	L.F.	6.40	7.10	2.81	16.31	21.50
1800	Curb inlets, (guttermouth) straight		41	1.366	Ea.	153	52	20.50	225.50	275
2000	Indian granite (belgian block)									
2100	Jumbo, 10-1/2" x 7-1/2" x 4", grey	D-1	150	.107	L.F.	6.50	4.25		10.75	14
2150	Pink		150	.107		6.70	4.25		10.95	14.25
2200	Regular, 9" x 4-1/2" x 4-1/2", grey		160	.100		6.60	3.99		10.59	13.65
2250	Pink		160	.100		5.95	3.99		9.94	12.90
2300	Cubes, 4" x 4" x 4", grey		175	.091		3.83	3.64		7.47	10.05
2350	Pink		175	.091		3.23	3.64		6.87	9.40
2400	6" x 6" x 6", pink		155	.103		11.20	4.11		15.31	18.90
2500	Alternate pricing method for indian granite									
2550	Jumbo, 10-1/2" x 7-1/2" x 4" (30 lb), grey				Ton	370			370	405
2600	Pink					390			390	430
2650	Regular, 9" x 4-1/2" x 4-1/2" (20 lb), grey					465			465	510
2700	Pink					415			415	455

32 16 Curbs, Gutters, Sidewalks, and Driveways

32 16 13 – Curbs and Gutters

32 16 13.43 Stone Curbs		Crew	Daily Output	Labor-Hours	Unit	2012 Bare Costs Material	Labor	Equipment	Total	Total Incl O&P
2750	Cubes, 4" x 4" x 4" (5 lb), grey				Ton	465			465	510
2800	Pink					415			415	455
2850	6" x 6" x 6" (25 lb), pink					435			435	480
2900	For pallets, add					20			20	22

32 17 Paving Specialties

32 17 13 – Parking Bumpers

32 17 13.13 Metal Parking Bumpers

		Crew	Daily Output	Labor-Hours	Unit	Material	Labor	Equipment	Total	Total Incl O&P
0010	**METAL PARKING BUMPERS**									
0015	Bumper rails for garages, 12 Ga. rail, 6" wide, with steel									
0020	posts 12'-6" O.C., minimum	E-4	190	.168	L.F.	16.15	8.40	.64	25.19	34
0030	Average		165	.194		20	9.65	.74	30.39	41
0100	Maximum		140	.229		24	11.40	.87	36.27	48.50
1300	Pipe bollards, conc filled/paint, 8' L x 4' D hole, 6" diam.	B-6	20	1.200	Ea.	705	46	16.70	767.70	870
1400	8" diam.		15	1.600		1,025	61.50	22.50	1,109	1,250
1500	12" diam.		12	2		1,150	76.50	28	1,254.50	1,425
1592	Bollards, steel, 3' H, retractable, incl hydraulic controls, min		4	6		44,000	230	83.50	44,313.50	48,900
1594	Max		2	12		46,000	460	167	46,627	51,500
8900	Security bollards, SS, lighted, hyd, incl. controls, group of 3	L-7	.06	509		47,000	21,700		68,700	86,500
8910	Group of 5	"	.04	682		63,000	29,100		92,100	116,500
9000	Minimum labor/equipment charge	E-4	2	16	Job		800	61	861	1,550

32 17 13.16 Plastic Parking Bumpers

		Crew	Daily Output	Labor-Hours	Unit	Material	Labor	Equipment	Total	Total Incl O&P
0010	**PLASTIC PARKING BUMPERS**									
1600	Bollards, recycled plastic, 5" x 5" x 5' L [G]	B-6	20	1.200	Ea.	73	46	16.70	135.70	173
1610	Wheel stops, recycled plastic, yellow, 4" x 6" x 6' L [G]		20	1.200		73	46	16.70	135.70	173
1620	Speed bumps, recycled plastic, yellow, 3"H x 10" W x 6' L [G]		20	1.200		163	46	16.70	225.70	272
1630	3"H x 10" W x 9' L [G]		20	1.200		244	46	16.70	306.70	360

32 17 13.19 Precast Concrete Parking Bumpers

		Crew	Daily Output	Labor-Hours	Unit	Material	Labor	Equipment	Total	Total Incl O&P
0010	**PRECAST CONCRETE PARKING BUMPERS**									
1000	Wheel stops, precast concrete incl. dowels, 6" x 10" x 6'-0"	B-2	120	.333	Ea.	59.50	11.85		71.35	85
1100	8" x 13" x 6'-0"	"	120	.333		87.50	11.85		99.35	116
1540	Bollards, precast concrete, free standing, round, 15" Dia x 36" H	B-6	14	1.714		305	65.50	24	394.50	470
1550	16" Dia x 30" H		14	1.714		390	65.50	24	479.50	565
1560	18" Dia x 34" H		14	1.714		475	65.50	24	564.50	655
1570	18" Dia x 72" H		14	1.714		510	65.50	24	599.50	695
1572	26" Dia x 18" H		12	2		425	76.50	28	529.50	625
1574	26" Dia x 30" H		12	2		710	76.50	28	814.50	935
1578	26" Dia x 50" H		12	2		1,175	76.50	28	1,279.50	1,450
1580	Square, 22" x 32" H		12	2		490	76.50	28	594.50	690
1585	22" x 41" H		12	2		715	76.50	28	819.50	945
1590	Hexagon, 26" x 41" H		10	2.400		1,200	92	33.50	1,325.50	1,500

32 17 13.26 Wood Parking Bumpers

		Crew	Daily Output	Labor-Hours	Unit	Material	Labor	Equipment	Total	Total Incl O&P
0010	**WOOD PARKING BUMPERS**									
0020	Parking barriers, timber w/saddles, treated type									
0100	4" x 4" for cars	B-2	520	.077	L.F.	2.26	2.73		4.99	6.95
0200	6" x 6" for trucks	"	520	.077	"	4.61	2.73		7.34	9.55

32 17 23 – Pavement Markings

32 17 23.13 Painted Pavement Markings

		Crew	Daily Output	Labor-Hours	Unit	Material	Labor	Equipment	Total	Total Incl O&P
0010	**PAINTED PAVEMENT MARKINGS**									
0020	Acrylic waterborne, white or yellow, 4" wide, less than 3000 L.F.	B-78	20000	.002	L.F.	.12	.08	.03	.23	.30

32 17 Paving Specialties

32 17 23 – Pavement Markings

32 17 23.13 Painted Pavement Markings		Crew	Daily Output	Labor-Hours	Unit	Material	2012 Bare Costs Labor	Equipment	Total	Total Incl O&P
0200	6″ wide, less than 3000 L.F.	B-78	11000	.004	L.F.	.18	.15	.05	.38	.50
0500	8″ wide, less than 3000 L.F.		10000	.005		.24	.17	.05	.46	.61
0600	12″ wide, less than 3000 L.F.		4000	.012		.36	.42	.14	.92	1.24
0620	Arrows or gore lines		2300	.021	S.F.	.15	.74	.23	1.12	1.64
0640	Temporary paint, white or yellow, less than 3000 L.F.		15000	.003	L.F.	.05	.11	.04	.20	.28
0660	Removal	1 Clab	300	.027			.94		.94	1.53
0680	Temporary tape	2 Clab	1500	.011		.42	.37		.79	1.07
0710	Thermoplastic, white or yellow, 4″ wide, less than 6000 L.F.	B-79	15000	.003		.28	.09	.09	.46	.56
0730	6″ wide, less than 6000 L.F.		14000	.003		.42	.10	.09	.61	.73
0740	8″ wide, less than 6000 L.F.		12000	.003		.56	.12	.11	.79	.93
0750	12″ wide, less than 6000 L.F.		6000	.007		.83	.24	.22	1.29	1.54
0760	Arrows		660	.061	S.F.	.56	2.14	1.97	4.67	6.30
0770	Gore lines		2500	.016		.56	.57	.52	1.65	2.11
0780	Letters		660	.061		.56	2.14	1.97	4.67	6.30
32 17 23.14 Pavement Parking Markings										
0010	**PAVEMENT PARKING MARKINGS**									
0790	Layout of pavement marking	A-2	25000	.001	L.F.		.03	.01	.04	.06
0800	Lines on pvmt, parking stall, paint, white, 4″ wide	B-78B	400	.045	Stall	4.40	1.63	.86	6.89	8.45
0825	Parking stall, small quantities	2 Pord	80	.200		8.80	7.65		16.45	22
0830	Lines on pvmt, parking stall, thermoplastic, white, 4″ wide	B-79	300	.133		12	4.71	4.33	21.04	25.50
1000	Street letters and numbers	B-78B	1600	.011	S.F.	.64	.41	.21	1.26	1.60

32 31 Fences and Gates

32 31 13 – Chain Link Fences and Gates

32 31 13.20 Fence, Chain Link Industrial		Crew	Daily Output	Labor-Hours	Unit	Material	Labor	Equipment	Total	Total Incl O&P
0010	**FENCE, CHAIN LINK INDUSTRIAL**									
0011	Schedule 40, including concrete									
0020	3 strands barb wire, 2″ post @ 10′ O.C., set in concrete, 6′ H									
0200	9 ga. wire, galv. steel, in concrete	B-80C	240	.100	L.F.	18.50	3.49	.91	22.90	27
0248	Fence, add for vinyl coated fabric				S.F.	.68			.68	.75
0300	Aluminized steel	B-80C	240	.100	L.F.	18.90	3.49	.91	23.30	27.50
0500	6 ga. wire, galv. steel		240	.100		20	3.49	.91	24.40	28.50
0600	Aluminized steel		240	.100		23	3.49	.91	27.40	32
0800	6 ga. wire, 6′ high but omit barbed wire, galv. steel		250	.096		18.55	3.35	.87	22.77	27
0900	Aluminized steel, in concrete		250	.096		22.50	3.35	.87	26.72	31
0920	8′ H, 6 ga. wire, 2-1/2″ line post, galv. steel, in concrete		180	.133		29.50	4.65	1.21	35.36	41.50
0940	Aluminized steel, in concrete		180	.133		36	4.65	1.21	41.86	48.50
1100	Add for corner posts, 3″ diam., galv. steel, in concrete		40	.600	Ea.	86	21	5.45	112.45	135
1200	Aluminized steel, in concrete		40	.600		99	21	5.45	125.45	149
1300	Add for braces, galv. steel		80	.300		30	10.45	2.73	43.18	53
1350	Aluminized steel		80	.300		39	10.45	2.73	52.18	63
1400	Gate for 6′ high fence, 1-5/8″ frame, 3′ wide, galv. steel		10	2.400		188	83.50	22	293.50	370
1500	Aluminized steel, in concrete		10	2.400		188	83.50	22	293.50	370
2000	5′-0″ high fence, 9 ga., no barbed wire, 2″ line post, in concrete									
2010	10′ O.C., 1-5/8″ top rail, in concrete									
2100	Galvanized steel, in concrete	B-80C	300	.080	L.F.	18.20	2.79	.73	21.72	25.50
2200	Aluminized steel, in concrete		300	.080	″	18.90	2.79	.73	22.42	26.50
2400	Gate, 4′ wide, 5′ high, 2″ frame, galv. steel, in concrete		10	2.400	Ea.	82.50	83.50	22	188	252
2500	Aluminized steel, in concrete		10	2.400	″	96.50	83.50	22	202	267
3100	Overhead slide gate, chain link, 6′ high, to 18′ wide, in concrete		38	.632	L.F.	92.50	22	5.75	120.25	144
3110	Cantilever type, in concrete	B-80	48	.667		65.50	25	13.90	104.40	128

32 31 Fences and Gates

32 31 13 – Chain Link Fences and Gates

32 31 13.20 Fence, Chain Link Industrial		Crew	Daily Output	Labor-Hours	Unit	2012 Bare Costs Material	Labor	Equipment	Total	Total Incl O&P
3120	8′ high, in concrete	B-80	24	1.333	L.F.	147	50.50	28	225.50	274
3130	10′ high, in concrete		18	1.778		180	67.50	37	284.50	350
9000	Minimum labor/equipment charge		2	16	Job		605	335	940	1,350

32 31 13.25 Fence, Chain Link Residential

		Crew	Daily Output	Labor-Hours	Unit	Material	Labor	Equipment	Total	Total Incl O&P
0010	**FENCE, CHAIN LINK RESIDENTIAL**									
0011	Schedule 20, 11 gauge wire, 1-5/8″ post									
3300	Residential, 11 ga. wire, 1-5/8″ line post @ 10′ O.C.									
3310	1-3/8″ top rail									
3350	4′ high, galvanized steel	B-80C	475	.051	L.F.	10.85	1.76	.46	13.07	15.35
3400	Aluminized		475	.051	″	12.55	1.76	.46	14.77	17.20
3600	Gate, 3′ wide, 1-3/8″ frame, galv. steel		10	2.400	Ea.	137	83.50	22	242.50	310
3700	Aluminized		10	2.400	″	143	83.50	22	248.50	320
3900	3′ high, galvanized steel		620	.039	L.F.	29.50	1.35	.35	31.20	35
4000	Aluminized		620	.039	″	33.50	1.35	.35	35.20	39
4200	Gate, 3′ wide, 1-3/8″ frame, galv. steel		12	2	Ea.	88	69.50	18.20	175.70	231
4300	Aluminized		12	2	″	90	69.50	18.20	177.70	233
9000	Minimum labor/equipment charge	B-1	2	12	Job		430		430	705

32 31 13.26 Tennis Court Fences and Gates

		Crew	Daily Output	Labor-Hours	Unit	Material	Labor	Equipment	Total	Total Incl O&P
0010	**TENNIS COURT FENCES AND GATES**									
3150	Tennis courts, 11 ga. wire, 1-3/4″ mesh, 2-1/2″ line posts									
3170	1-5/8″ top rail, 3″ corner and gate posts									
3190	10′ high, galvanized steel	B-80	190	.168	L.F.	22.50	6.35	3.51	32.36	38.50
3210	Vinyl covered 9 ga. wire		190	.168	″	26.50	6.35	3.51	36.36	43
3240	Corner posts for above, 3″ diameter, 10′ high		30	1.067	Ea.	73	40.50	22.50	136	170

32 31 13.40 Fence, Fabric and Accessories

		Crew	Daily Output	Labor-Hours	Unit	Material	Labor	Equipment	Total	Total Incl O&P
0010	**FENCE, FABRIC & ACCESSORIES**									
8000	Components only, fabric, galvanized, 4′ high	B-80C	585	.041	L.F.	3.10	1.43	.37	4.90	6.15
8040	6′ high		430	.056	″	4.60	1.95	.51	7.06	8.80
8200	Posts, line post, 4′ high		34	.706	Ea.	14.60	24.50	6.45	45.55	63
8240	6′ high		26	.923	″	14.65	32	8.40	55.05	78
8400	Top rails, 1-3/8″ O.D.		1000	.024	L.F.	.77	.84	.22	1.83	2.46
8440	1-5/8″ O.D.		1000	.024	″	1.36	.84	.22	2.42	3.11

32 31 19 – Decorative Metal Fences and Gates

32 31 19.10 Decorative Fence

		Crew	Daily Output	Labor-Hours	Unit	Material	Labor	Equipment	Total	Total Incl O&P
0010	**DECORATIVE FENCE**									
5300	Tubular picket, steel, 6′ sections, 1-9/16″ posts, 4′ high	B-80C	300	.080	L.F.	29	2.79	.73	32.52	37.50
5400	2″ posts, 5′ high		240	.100		44	3.49	.91	48.40	54.50
5600	2″ posts, 6′ high		200	.120		50.50	4.18	1.09	55.77	64
5700	Staggered picket 1-9/16″ posts, 4′ high		300	.080		29	2.79	.73	32.52	37.50
5800	2″ posts, 5′ high		240	.100		44	3.49	.91	48.40	54.50
5900	2″ posts, 6′ high		200	.120		50.50	4.18	1.09	55.77	64
6200	Gates, 4′ high, 3′ wide	B-1	10	2.400	Ea.	250	86		336	415
6300	5′ high, 3′ wide		10	2.400		264	86		350	430
6400	6′ high, 3′ wide		10	2.400		315	86		401	490
6500	4′ wide		10	2.400		345	86		431	520

32 31 23 – Plastic Fences and Gates

32 31 23.20 Fence, Recycled Plastic

		Crew	Daily Output	Labor-Hours	Unit	Material	Labor	Equipment	Total	Total Incl O&P
0010	**FENCE, RECYCLED PLASTIC**									
9015	Fence rail, made from recycled plastic, various colors, 2 rail G	B-1	150	.160	L.F.	10.90	5.70		16.60	21.50
9018	3 rail G		150	.160		13.35	5.70		19.05	24
9020	4 rail G		150	.160		15.80	5.70		21.50	27

32 31 Fences and Gates

32 31 23 – Plastic Fences and Gates

32 31 23.20 Fence, Recycled Plastic			Crew	Daily Output	Labor-Hours	Unit	2012 Bare Costs Material	Labor	Equipment	Total	Total Incl O&P
9030	Fence pole, made from recycled plastic, various colors, 7′	G	B-1	96	.250	Ea.	55.50	8.95		64.45	75.50
9040	Stockade fence, made from recycled plastic, various colors, 4′ high	G	B-80C	160	.150	L.F.	37	5.25	1.37	43.62	51
9050	6′ high	G		160	.150	"	38.50	5.25	1.37	45.12	52.50
9060	6′ pole	G		96	.250	Ea.	40.50	8.70	2.28	51.48	61.50
9070	9′ pole	G		96	.250	"	47.50	8.70	2.28	58.48	69
9080	Picket fence, made from recycled plastic, various colors, 3′ high	G		160	.150	L.F.	25	5.25	1.37	31.62	37.50
9090	4′ high	G		160	.150	"	31	5.25	1.37	37.62	44.50
9100	3′ high gate	G		8	3	Ea.	24	105	27.50	156.50	228
9110	4′ high gate	G		8	3		30.50	105	27.50	163	235
9120	5′ high pole	G		96	.250		40	8.70	2.28	50.98	60
9130	6′ high pole	G		96	.250		47.50	8.70	2.28	58.48	69
9140	Pole cap only	G					7.15			7.15	7.85
9150	Keeper pins only	G					.42			.42	.46

32 31 26 – Wire Fences and Gates

32 31 26.10 Fences, Misc. Metal		Crew	Daily Output	Labor-Hours	Unit	Material	Labor	Equipment	Total	Total Incl O&P
0010	**FENCES, MISC. METAL**									
0012	Chicken wire, posts @ 4′, 1″ mesh, 4′ high	B-80C	410	.059	L.F.	4.30	2.04	.53	6.87	8.65
0100	2″ mesh, 6′ high		350	.069		2.32	2.39	.62	5.33	7.15
0200	Galv. steel, 12 ga., 2″ x 4″ mesh, posts 5′ O.C., 3′ high		300	.080		2.59	2.79	.73	6.11	8.20
0300	5′ high		300	.080		3.96	2.79	.73	7.48	9.70
0400	14 ga., 1″ x 2″ mesh, 3′ high		300	.080		2.93	2.79	.73	6.45	8.55
0500	5′ high		300	.080		4.60	2.79	.73	8.12	10.40
1000	Kennel fencing, 1-1/2″ mesh, 6′ long, 3′-6″ wide, 6′-2″ high	2 Clab	4	4	Ea.	490	140		630	765
1050	12′ long		4	4		635	140		775	930
1200	Top covers, 1-1/2″ mesh, 6′ long		15	1.067		163	37.50		200.50	241
1250	12′ long		12	1.333		237	47		284	340
4500	Security fence, prison grade, set in concrete, 12′ high	B-80	25	1.280	L.F.	54	48.50	26.50	129	168
4600	16′ high	"	20	1.600	"	73.50	60.50	33.50	167.50	216

32 31 29 – Wood Fences and Gates

32 31 29.20 Fence, Wood Rail		Crew	Daily Output	Labor-Hours	Unit	Material	Labor	Equipment	Total	Total Incl O&P
0010	**FENCE, WOOD RAIL**									
0012	Picket, No. 2 cedar, Gothic, 2 rail, 3′ high	B-1	160	.150	L.F.	6.85	5.35		12.20	16.35
0050	Gate, 3′-6″ wide	B-80C	9	2.667	Ea.	59	93	24.50	176.50	243
0600	Open rail, rustic, No. 1 cedar, 2 rail, 3′ high		160	.150	L.F.	6.10	5.25	1.37	12.72	16.75
0650	Gate, 3′ wide		9	2.667	Ea.	68.50	93	24.50	186	254
1200	Stockade, No. 2 cedar, treated wood rails, 6′ high		160	.150	L.F.	8.55	5.25	1.37	15.17	19.45
1250	Gate, 3′ wide		9	2.667	Ea.	70	93	24.50	187.50	256
3300	Board, shadow box, 1″ x 6″, treated pine, 6′ high		160	.150	L.F.	11.60	5.25	1.37	18.22	23
3400	No. 1 cedar, 6′ high		150	.160		23	5.60	1.46	30.06	35.50
3900	Basket weave, No. 1 cedar, 6′ high		160	.150		35	5.25	1.37	41.62	48.50
3950	Gate, 3′-6″ wide	B-1	8	3	Ea.	180	107		287	375
4000	Treated pine, 6′ high		150	.160	L.F.	17.55	5.70		23.25	28.50
4200	Gate, 3′-6″ wide		9	2.667	Ea.	92.50	95.50		188	258
8000	Posts only, 4′ high		40	.600		16.65	21.50		38.15	53.50
8040	6′ high		30	.800		23	28.50		51.50	72.50
9000	Minimum labor/equipment charge	1 Clab	2	4	Job		140		140	230

32 32 Retaining Walls

32 32 60 – Stone Retaining Walls

32 32 60.10 Retaining Walls, Stone		Crew	Daily Output	Labor-Hours	Unit	2012 Bare Costs Material	Labor	Equipment	Total	Total Incl O&P
0010	**RETAINING WALLS, STONE**									
0015	Including excavation, concrete footing and									
0020	stone 3′ below grade. Price is exposed face area.									
0200	Decorative random stone, to 6′ high, 1′-6″ thick, dry set	D-1	35	.457	S.F.	43	18.20		61.20	77
0300	Mortar set		40	.400		48	15.95		63.95	78.50
0500	Cut stone, to 6′ high, 1′-6″ thick, dry set		35	.457		43	18.20		61.20	77
0600	Mortar set		40	.400		46	15.95		61.95	76
0800	Random stone, 6′ to 10′ high, 2′ thick, dry set		45	.356		46	14.15		60.15	73.50
0900	Mortar set		50	.320		51	12.75		63.75	76.50
1100	Cut stone, 6′ to 10′ high, 2′ thick, dry set		45	.356		46	14.15		60.15	73.50
1200	Mortar set		50	.320		51	12.75		63.75	76.50
9000	Minimum labor/equipment charge		2	8	Job		320		320	515

32 84 Planting Irrigation

32 84 23 – Underground Sprinklers

32 84 23.10 Sprinkler Irrigation System		Crew	Daily Output	Labor-Hours	Unit	Material	Labor	Equipment	Total	Total Incl O&P
0010	**SPRINKLER IRRIGATION SYSTEM**									
0011	For lawns									
0100	Golf course with fully automatic system	C-17	.05	1600	9 holes	98,000	73,500		171,500	227,000
0200	24′ diam. head at 15′ O.C incl. piping, auto oper., minimum	B-20	70	.343	Head	22	13.45		35.45	46.50
0300	Maximum		40	.600		54.50	23.50		78	98.50
0500	60′ diam. head at 40′ O.C. incl. piping, auto oper., minimum		28	.857		69.50	33.50		103	131
0600	Maximum		23	1.043		192	41		233	278
0800	Residential system, custom, 1″ supply		2000	.012	S.F.	.34	.47		.81	1.14
0900	1-1/2″ supply		1800	.013	″	.39	.52		.91	1.28

32 91 Planting Preparation

32 91 13 – Soil Preparation

32 91 13.23 Structural Soil Mixing		Crew	Daily Output	Labor-Hours	Unit	Material	Labor	Equipment	Total	Total Incl O&P
0010	**STRUCTURAL SOIL MIXING**									
0100	Rake topsoil, site material, harley rock rake, ideal	B-6	33	.727	M.S.F.		28	10.10	38.10	56
0200	Adverse	″	7	3.429			131	47.50	178.50	265
0300	Screened loam, york rake and finish, ideal	B-62	24	1			38.50	6.65	45.15	69.50
0400	Adverse	″	20	1.200			46	8	54	83.50
1000	Remove topsoil & stock pile on site, 75 HP dozer, 6″ deep, 50′ haul	B-10L	30	.400			17.10	14.75	31.85	43.50
1050	300′ haul		6.10	1.967			84	72.50	156.50	214
1100	12″ deep, 50′ haul		15.50	.774			33	28.50	61.50	84.50
1150	300′ haul		3.10	3.871			165	143	308	420
1200	200 HP dozer, 6″ deep, 50′ haul	B-10B	125	.096			4.10	9.95	14.05	17.50
1250	300′ haul		30.70	.391			16.70	40.50	57.20	71
1300	12″ deep, 50′ haul		62	.194			8.25	20	28.25	35
1350	300′ haul		15.40	.779			33.50	80.50	114	142
1400	Alternate method, 75 HP dozer, 50′ haul	B-10L	860	.014	C.Y.		.60	.51	1.11	1.52
1450	300′ haul	″	114	.105			4.50	3.88	8.38	11.45
1500	200 HP dozer, 50′ haul	B-10B	2660	.005			.19	.47	.66	.82
1600	300′ haul	″	570	.021			.90	2.18	3.08	3.84
1800	Rolling topsoil, hand push roller	1 Clab	3200	.003	S.F.		.09		.09	.14
1850	Tractor drawn roller	B-66	10666	.001	″		.03	.02	.05	.08
2000	Root raking and loading, residential, no boulders	B-6	53.30	.450	M.S.F.		17.25	6.25	23.50	35

32 91 Planting Preparation

32 91 13 – Soil Preparation

32 91 13.23 Structural Soil Mixing		Crew	Daily Output	Labor-Hours	Unit	Material	2012 Bare Costs Labor	Equipment	Total	Total Incl O&P
2100	With boulders	B-6	32	.750	M.S.F.		28.50	10.45	38.95	58
2200	Municipal, no boulders		200	.120			4.60	1.67	6.27	9.30
2300	With boulders	↓	120	.200			7.65	2.78	10.43	15.45
2400	Large commercial, no boulders	B-10B	400	.030			1.28	3.11	4.39	5.45
2500	With boulders	"	240	.050			2.14	5.20	7.34	9.10
3000	Scarify subsoil, residential, skid steer loader w/scarifiers, 50 HP	B-66	32	.250			11.20	7.75	18.95	26
3050	Municipal, skid steer loader w/scarifiers, 50 HP	"	120	.067			2.98	2.06	5.04	7
3100	Large commercial, 75 HP, dozer w/scarifier	B-10L	240	.050	↓		2.14	1.84	3.98	5.45
3500	Screen topsoil from stockpile, vibrating screen, wet material (organic)	B-10P	200	.060	C.Y.		2.56	5.55	8.11	10.20
3550	Dry material	"	300	.040			1.71	3.69	5.40	6.80
3600	Mixing with conditioners, manure and peat	B-10R	550	.022	↓		.93	.49	1.42	2.03
3650	Mobilization add for 2 days or less operation	B-34K	3	2.667	Job		94	283	377	460
3800	Spread conditioned topsoil, 6" deep, by hand	B-1	360	.067	S.Y.	5.10	2.38		7.48	9.50
3850	300 HP dozer	B-10M	27	.444	M.S.F.	550	19	63.50	632.50	705
4000	Spread soil conditioners, alum. sulfate, 1#/S.Y., hand push spreader	1 Clab	17500	.001	S.Y.	17.75	.02		17.77	19.55
4050	Tractor spreader	B-66	700	.011	M.S.F.	1,975	.51	.35	1,975.86	2,175
4100	Fertilizer, 0.2#/S.Y., push spreader	1 Clab	17500	.001	S.Y.	.07	.02		.09	.11
4150	Tractor spreader	B-66	700	.011	M.S.F.	7.80	.51	.35	8.66	9.75
4200	Ground limestone, 1#/S.Y., push spreader	1 Clab	17500	.001	S.Y.	.16	.02		.18	.21
4250	Tractor spreader	B-66	700	.011	M.S.F.	17.80	.51	.35	18.66	21
4300	Lusoil, 3#/S.Y., push spreader	1 Clab	17500	.001	S.Y.	.63	.02		.65	.72
4350	Tractor spreader	B-66	700	.011	M.S.F.	70	.51	.35	70.86	78
4400	Manure, 18#/S.Y., push spreader	1 Clab	2500	.003	S.Y.	8.60	.11		8.71	9.70
4450	Tractor spreader	B-66	280	.029	M.S.F.	960	1.28	.88	962.16	1,050
4500	Perlite, 1" deep, push spreader	1 Clab	17500	.001	S.Y.	9.20	.02		9.22	10.15
4550	Tractor spreader	B-66	700	.011	M.S.F.	1,025	.51	.35	1,025.86	1,125
4600	Vermiculite, push spreader	1 Clab	17500	.001	S.Y.	9.20	.02		9.22	10.15
4650	Tractor spreader	B-66	700	.011	M.S.F.	1,025	.51	.35	1,025.86	1,125
5000	Spread topsoil, skid steer loader and hand dress	B-62	270	.089	C.Y.	23	3.41	.59	27	31.50
5100	Articulated loader and hand dress	B-100	320	.038		23	1.60	2.80	27.40	31
5200	Articulated loader and 75HP dozer	B-10M	500	.024		23	1.03	3.44	27.47	31
5300	Road grader and hand dress	B-11L	1000	.016	↓	23	.65	.61	24.26	27
6000	Tilling topsoil, 20 HP tractor, disk harrow, 2" deep	B-66	450	.018	M.S.F.		.80	.55	1.35	1.86
6050	4" deep		360	.022			.99	.69	1.68	2.33
6100	6" deep	↓	270	.030	↓		1.33	.92	2.25	3.10
6150	26" rototiller, 2" deep	A-1J	1250	.006	S.Y.		.22	.05	.27	.42
6200	4" deep		1000	.008			.28	.06	.34	.52
6250	6" deep	↓	750	.011	↓		.37	.08	.45	.70

32 91 13.26 Planting Beds		Crew	Daily Output	Labor-Hours	Unit	Material	Labor	Equipment	Total	Total Incl O&P
0010	**PLANTING BEDS**									
0100	Backfill planting pit, by hand, on site topsoil	2 Clab	18	.889	C.Y.		31		31	51
0200	Prepared planting mix, by hand	"	24	.667			23.50		23.50	38.50
0300	Skid steer loader, on site topsoil	B-62	340	.071			2.71	.47	3.18	4.89
0400	Prepared planting mix	"	410	.059			2.24	.39	2.63	4.05
1000	Excavate planting pit, by hand, sandy soil	2 Clab	16	1			35		35	57.50
1100	Heavy soil or clay	"	8	2			70		70	115
1200	1/2 C.Y. backhoe, sandy soil	B-11C	150	.107			4.36	2.23	6.59	9.45
1300	Heavy soil or clay	"	115	.139			5.70	2.91	8.61	12.30
2000	Mix planting soil, incl. loam, manure, peat, by hand	2 Clab	60	.267		40	9.35		49.35	59.50
2100	Skid steer loader	B-62	150	.160	↓	40	6.15	1.06	47.21	55
3000	Pile sod, skid steer loader	"	2800	.009	S.Y.		.33	.06	.39	.59
3100	By hand	2 Clab	400	.040	↓		1.40		1.40	2.30

32 91 Planting Preparation

32 91 13 – Soil Preparation

32 91 13.26 Planting Beds		Crew	Daily Output	Labor-Hours	Unit	Material	2012 Bare Costs Labor	Equipment	Total	Total Incl O&P
4000	Remove sod, F.E. loader	B-10S	2000	.006	S.Y.		.26	.17	.43	.59
4100	Sod cutter	B-12K	3200	.005			.21	.30	.51	.66
4200	By hand	2 Clab	240	.067			2.34		2.34	3.84

32 91 19 – Landscape Grading

32 91 19.13 Topsoil Placement and Grading		Crew	Daily Output	Labor-Hours	Unit	Material	Labor	Equipment	Total	Total Incl O&P
0010	**TOPSOIL PLACEMENT AND GRADING**									
0700	Furnish and place, truck dumped, screened, 4" deep	B-10S	1300	.009	S.Y.	4.21	.39	.26	4.86	5.55
0800	6" deep		820	.015	″	5.40	.63	.41	6.44	7.40
0810	Minimum labor/equipment charge		2	6	Ea.		256	167	423	595
0900	Fine grading and seeding, incl. lime, fertilizer & seed,									
1000	With equipment	B-14	1000	.048	S.Y.	.56	1.78	.33	2.67	3.88
2000	Minimum labor/equipment charge	1 Clab	4	2	Job		70		70	115

32 92 Turf and Grasses

32 92 19 – Seeding

32 92 19.13 Mechanical Seeding		Crew	Daily Output	Labor-Hours	Unit	Material	Labor	Equipment	Total	Total Incl O&P
0010	**MECHANICAL SEEDING** R329219-50									
0020	Mechanical seeding, 215 lb./acre	B-66	1.50	5.333	Acre	605	239	165	1,009	1,225
0100	44 lb./M.S.Y.	″	2500	.003	S.Y.	.16	.14	.10	.40	.52
0600	Limestone hand push spreader, 50 lb. per M.S.F.	1 Clab	180	.044	M.S.F.	3.30	1.56		4.86	6.20
9000	Minimum labor/equipment charge	″	4	2	Job		70		70	115

32 92 23 – Sodding

32 92 23.10 Sodding Systems		Crew	Daily Output	Labor-Hours	Unit	Material	Labor	Equipment	Total	Total Incl O&P
0010	**SODDING SYSTEMS**									
0020	Sodding, 1" deep, bluegrass sod, on level ground, over 8 M.S.F.	B-63	22	1.818	M.S.F.	175	67.50	7.25	249.75	310
0200	4 M.S.F.		17	2.353		240	87	9.40	336.40	415
0300	1000 S.F.		13.50	2.963		295	110	11.80	416.80	515
0500	Sloped ground, over 8 M.S.F.		6	6.667		175	247	26.50	448.50	625
0600	4 M.S.F.		5	8		240	296	32	568	780
0700	1000 S.F.		4	10		295	370	40	705	970
1000	Bent grass sod, on level ground, over 6 M.S.F.		20	2		600	74	8	682	790
1100	3 M.S.F.		18	2.222		665	82.50	8.85	756.35	875
1200	Sodding 1000 S.F. or less		14	2.857		875	106	11.40	992.40	1,150
1500	Sloped ground, over 6 M.S.F.		15	2.667		600	99	10.65	709.65	830
1600	3 M.S.F.		13.50	2.963		665	110	11.80	786.80	920
1700	1000 S.F.		12	3.333		875	123	13.30	1,011.30	1,175

32 93 Plants

32 93 13 – Ground Covers

32 93 13.10 Ground Cover Plants		Crew	Daily Output	Labor-Hours	Unit	Material	Labor	Equipment	Total	Total Incl O&P
0010	**GROUND COVER PLANTS**									
0012	Plants, pachysandra, in prepared beds	B-1	15	1.600	C	31	57		88	128
0200	Vinca minor, 1 yr., bare root, in prepared beds		12	2	″	39	71.50		110.50	160
0600	Stone chips, in 50 lb. bags, Georgia marble		520	.046	Bag	5	1.65		6.65	8.20
0700	Onyx gemstone		260	.092		18	3.30		21.30	25
0800	Quartz		260	.092		6.90	3.30		10.20	12.95
0900	Pea gravel, truckload lots		28	.857	Ton	30.50	30.50		61	84

32 93 33 – Shrubs

32 93 33.10 Shrubs and Trees

	32 93 33.10 Shrubs and Trees	Crew	Daily Output	Labor-Hours	Unit	2012 Bare Costs Material	Labor	Equipment	Total	Total Incl O&P
0010	**SHRUBS AND TREES**									
0011	Evergreen, in prepared beds, B & B									
0100	Arborvitae pyramidal, 4′-5′	B-17	30	1.067	Ea.	52	40	23	115	148
0150	Globe, 12″-15″	B-1	96	.250		21.50	8.95		30.45	38
0300	Cedar, blue, 8′-10′	B-17	18	1.778		220	67	38.50	325.50	395
0500	Hemlock, Canadian, 2-1/2′-3′	B-1	36	.667		36	24		60	78.50
0550	Holly, Savannah, 8′ - 10′ H		9.68	2.479		163	88.50		251.50	325
0600	Juniper, andorra, 18″-24″		80	.300		20	10.75		30.75	39.50
0620	Wiltoni, 15″-18″		80	.300		24.50	10.75		35.25	44.50
0640	Skyrocket, 4-1/2′-5′	B-17	55	.582		139	22	12.55	173.55	202
0660	Blue pfitzer, 2′-2-1/2′	B-1	44	.545		39	19.50		58.50	75
0680	Ketleerie, 2-1/2′-3′		50	.480		40	17.15		57.15	72
0700	Pine, black, 2-1/2′-3′		50	.480		53	17.15		70.15	86.50
0720	Mugo, 18″-24″		60	.400		45	14.30		59.30	73
0740	White, 4′-5′	B-17	75	.427		40	16.05	9.20	65.25	80
0800	Spruce, blue, 18″-24″	B-1	60	.400		49	14.30		63.30	77.50
0840	Norway, 4′-5′	B-17	75	.427		75	16.05	9.20	100.25	119
0900	Yew, denisforma, 12″-15″	B-1	60	.400		34.50	14.30		48.80	61.50
1000	Capitata, 18″-24″		30	.800		18	28.50		46.50	67
1100	Hicksi, 2′-2-1/2′		30	.800		32	28.50		60.50	82.50

32 93 33.20 Shrubs

	32 93 33.20 Shrubs	Crew	Daily Output	Labor-Hours	Unit	Material	Labor	Equipment	Total	Total Incl O&P
0010	**SHRUBS**									
0011	Broadleaf Evergreen, planted in prepared beds									
0100	Andromeda, 15″-18″, container	B-1	96	.250	Ea.	18	8.95		26.95	34.50
0200	Azalea, 15″ - 18″, container		96	.250		22	8.95		30.95	38.50
0300	Barberry, 9″-12″, container		130	.185		13	6.60		19.60	25
0400	Boxwood, 15″-18″, B&B		96	.250		30	8.95		38.95	47.50
0500	Euonymus, emerald gaiety, 12″ to 15″, container		115	.209		22	7.45		29.45	36.50
0600	Holly, 15″-18″, B & B		96	.250		17	8.95		25.95	33.50
0900	Mount laurel, 18″ - 24″, B & B		80	.300		55	10.75		65.75	78
1000	Paxistema, 9 – 12″ high		130	.185		20	6.60		26.60	33
1100	Rhododendron, 18″-24″, container		48	.500		28.50	17.90		46.40	61
1200	Rosemary, 1 gal. container		600	.040		10.60	1.43		12.03	14
2000	Deciduous, planted in prepared beds, amelanchier, 2′-3′, B & B		57	.421		119	15.05		134.05	156
2100	Azalea, 15″-18″, B & B		96	.250		17.50	8.95		26.45	34
2300	Bayberry, 2′-3′, B & B		57	.421		22	15.05		37.05	48.50
2600	Cotoneaster, 15″-18″, B & B		80	.300		10.60	10.75		21.35	29.50
2800	Dogwood, 3′-4′, B & B	B-17	40	.800		27	30	17.30	74.30	97
2900	Euonymus, alatus compacta, 15″ to 18″, container	B-1	80	.300		22	10.75		32.75	41.50
3200	Forsythia, 2′-3′, container	″	60	.400		22	14.30		36.30	47.50
3300	Hibiscus, 3′-4′, B & B	B-17	75	.427		41.50	16.05	9.20	66.75	81.50
3400	Honeysuckle, 3′-4′, B & B	B-1	60	.400		20.50	14.30		34.80	46
3500	Hydrangea, 2′-3′, B & B	″	57	.421		27	15.05		42.05	54
3600	Lilac, 3′-4′, B & B	B-17	40	.800		19.65	30	17.30	66.95	89
3900	Privet, bare root, 18″-24″	B-1	80	.300		22	10.75		32.75	41.50
4100	Quince, 2′-3′, B & B	″	57	.421		23	15.05		38.05	50
4200	Russian olive, 3′-4′, B & B	B-17	75	.427		25	16.05	9.20	50.25	63.50
4400	Spirea, 3′-4′, B & B	B-1	70	.343		21.50	12.25		33.75	44
4500	Viburnum, 3′-4′, B & B	B-17	40	.800		21.50	30	17.30	68.80	91

32 93 Plants

32 93 43 – Trees

32 93 43.20 Trees			Crew	Daily Output	Labor-Hours	Unit	Material	2012 Bare Costs Labor	Equipment	Total	Total Incl O&P
0010	**TREES**										
0011	Deciduous, in prep. beds, balled & burlapped (B&B)										
0100	Ash, 2″ caliper	G	B-17	8	4	Ea.	200	150	86.50	436.50	560
0200	Beech, 5′-6′	G		50	.640		180	24	13.85	217.85	252
0300	Birch, 6′-8′, 3 stems	G		20	1.600		210	60	34.50	304.50	365
0500	Crabapple, 6′-8′	G		20	1.600		99	60	34.50	193.50	244
0600	Dogwood, 4′-5′	G		40	.800		138	30	17.30	185.30	219
0700	Eastern redbud 4′-5′	G		40	.800		117	30	17.30	164.30	196
0800	Elm, 8′-10′	G		20	1.600		137	60	34.50	231.50	286
0900	Ginkgo, 6′-7′	G		24	1.333		177	50	29	256	310
1000	Hawthorn, 8′-10′, 1″ caliper	G		20	1.600		173	60	34.50	267.50	325
1100	Honeylocust, 10′-12′, 1-1/2″ caliper	G		10	3.200		162	120	69	351	450
1300	Larch, 8′	G		32	1		106	37.50	21.50	165	202
1400	Linden, 8′-10′, 1″ caliper	G		20	1.600		129	60	34.50	223.50	277
1500	Magnolia, 4′-5′	G		20	1.600		94	60	34.50	188.50	238
1600	Maple, red, 8′-10′, 1-1/2″ caliper	G		10	3.200		191	120	69	380	480
1700	Mountain ash, 8′-10′, 1″ caliper	G		16	2		184	75	43	302	370
1800	Oak, 2-1/2″-3″ caliper	G		6	5.333		264	200	115	579	740
2100	Planetree, 9′-11′, 1-1/4″ caliper	G		10	3.200		122	120	69	311	405
2200	Plum, 6′-8′, 1″ caliper	G		20	1.600		98	60	34.50	192.50	243
2300	Poplar, 9′-11′, 1-1/4″ caliper	G		10	3.200		101	120	69	290	380
2500	Sumac, 2′-3′	G		75	.427		23	16.05	9.20	48.25	61.50
2700	Tulip, 5′-6′	G		40	.800		59	30	17.30	106.30	133
2800	Willow, 6′-8′, 1″ caliper	G		20	1.600		67.50	60	34.50	162	210
9000	Minimum labor/equipment charge		1 Clab	4	2	Job		70		70	115

32 94 Planting Accessories

32 94 13 – Landscape Edging

32 94 13.20 Edging		Crew	Daily Output	Labor-Hours	Unit	Material	Labor	Equipment	Total	Total Incl O&P
0010	**EDGING**									
0050	Aluminum alloy, including stakes, 1/8″ x 4″, mill finish	B-1	390	.062	L.F.	3.94	2.20		6.14	7.95
0051	Black paint		390	.062		4.57	2.20		6.77	8.65
0052	Black anodized		390	.062		5.30	2.20		7.50	9.40
0100	Brick, set horizontally, 1-1/2 bricks per L.F.	D-1	370	.043		1.35	1.72		3.07	4.26
0150	Set vertically, 3 bricks per L.F.	″	135	.119		3	4.72		7.72	10.90
0200	Corrugated aluminum, roll, 4″ wide	1 Carp	650	.012		1.84	.54		2.38	2.91
0250	6″ wide	″	550	.015		2.30	.64		2.94	3.58
0600	Railroad ties, 6″ x 8″	2 Carp	170	.094		3.97	4.15		8.12	11.15
0650	7″ x 9″		136	.118		4.41	5.20		9.61	13.35
0750	Redwood 2″ x 4″		330	.048		2.26	2.14		4.40	6
0800	Steel edge strips, incl. stakes, 1/4″ x 5″	B-1	390	.062		2.50	2.20		4.70	6.35
0850	3/16″ x 4″	″	390	.062		1.98	2.20		4.18	5.80
9000	Minimum labor/equipment charge	1 Carp	4	2	Job		88		88	145

32 94 50 – Tree Guying

32 94 50.10 Tree Guying Systems		Crew	Daily Output	Labor-Hours	Unit	Material	Labor	Equipment	Total	Total Incl O&P
0010	**TREE GUYING SYSTEMS**									
0015	Tree guying Including stakes, guy wire and wrap									
0100	Less than 3″ caliper, 2 stakes	2 Clab	35	.457	Ea.	14	16.05		30.05	42
0200	3″ to 4″ caliper, 3 stakes	″	21	.762	″	18.95	26.50		45.45	65
1000	Including arrowhead anchor, cable, turnbuckles and wrap									

32 94 Planting Accessories

32 94 50 – Tree Guying

32 94 50.10 Tree Guying Systems		Crew	Daily Output	Labor-Hours	Unit	Material	2012 Bare Costs Labor	Equipment	Total	Total Incl O&P
1100	Less than 3″ caliper, 3″ anchors	2 Clab	20	.800	Ea.	44	28		72	94.50
1200	3″ to 6″ caliper, 4″ anchors		15	1.067		37.50	37.50		75	103
1300	6″ caliper, 6″ anchors		12	1.333		44	47		91	125
1400	8″ caliper, 8″ anchors		9	1.778		126	62.50		188.50	240

32 96 Transplanting

32 96 23 – Plant and Bulb Transplanting

32 96 23.23 Planting

		Crew	Daily Output	Labor-Hours	Unit	Material	Labor	Equipment	Total	Total Incl O&P
0010	**PLANTING**									
0012	Moving shrubs on site, 12″ ball	B-62	28	.857	Ea.		33	5.70	38.70	59.50
0100	24″ ball	″	22	1.091	″		42	7.25	49.25	75.50

32 96 23.43 Moving Trees

		Crew	Daily Output	Labor-Hours	Unit	Material	Labor	Equipment	Total	Total Incl O&P
0290	**MOVING TREES**, On site									
0300	Moving trees on site, 36″ ball	B-6	3.75	6.400	Ea.		245	89	334	495
0400	60″ ball	″	1	24	″		920	335	1,255	1,850

32 96 43 – Tree Transplanting

32 96 43.20 Tree Removal

		Crew	Daily Output	Labor-Hours	Unit	Material	Labor	Equipment	Total	Total Incl O&P
0010	**TREE REMOVAL**									
0100	Dig & lace, shrubs, broadleaf evergreen, 18″-24″ high	B-1	55	.436	Ea.		15.60		15.60	25.50
0200	2′-3′	″	35	.686			24.50		24.50	40
0300	3′-4′	B-6	30	.800			30.50	11.15	41.65	62
0400	4′-5′	″	20	1.200			46	16.70	62.70	93
1000	Deciduous, 12″-15″	B-1	110	.218			7.80		7.80	12.80
1100	18″-24″		65	.369			13.20		13.20	21.50
1200	2′-3′		55	.436			15.60		15.60	25.50
1300	3′-4′	B-6	50	.480			18.40	6.70	25.10	37
2000	Evergreen, 18″-24″	B-1	55	.436			15.60		15.60	25.50
2100	2′-0″ to 2′-6″		50	.480			17.15		17.15	28
2200	2′-6″ to 3′-0″		35	.686			24.50		24.50	40
2300	3′-0″ to 3′-6″		20	1.200			43		43	70.50
3000	Trees, deciduous, small, 2′-3′		55	.436			15.60		15.60	25.50
3100	3′-4′	B-6	50	.480			18.40	6.70	25.10	37
3200	4′-5′		35	.686			26.50	9.55	36.05	53
3300	5′-6′		30	.800			30.50	11.15	41.65	62
4000	Shade, 5′-6′		50	.480			18.40	6.70	25.10	37
4100	6′-8′		35	.686			26.50	9.55	36.05	53
4200	8′-10′		25	.960			37	13.35	50.35	74
4300	2″ caliper		12	2			76.50	28	104.50	155
5000	Evergreen, 4′-5′		35	.686			26.50	9.55	36.05	53
5100	5′-6′		25	.960			37	13.35	50.35	74
5200	6′-7′		19	1.263			48.50	17.60	66.10	97.50
5300	7′-8′		15	1.600			61.50	22.50	84	124
5400	8′-10′		11	2.182			83.50	30.50	114	169

Estimating Tips

33 10 00 Water Utilities
33 30 00 Sanitary Sewerage Utilities
33 40 00 Storm Drainage Utilities

- Never assume that the water, sewer, and drainage lines will go in at the early stages of the project. Consider the site access needs before dividing the site in half with open trenches, loose pipe, and machinery obstructions. Always inspect the site to establish that the site drawings are complete. Check off all existing utilities on your drawings as you locate them. Be especially careful with underground utilities because appurtenances are sometimes buried during regrading or repaving operations. If you find any discrepancies, mark up the site plan for further research. Differing site conditions can be very costly if discovered later in the project.
- See also Section 33 01 00 for restoration of pipe where removal/replacement may be undesirable. Use of new types of piping materials can reduce the overall project cost. Owners/design engineers should consider the installing contractor as a valuable source of current information on utility products and local conditions that could lead to significant cost savings.

Reference Numbers

Reference numbers are shown in shaded boxes at the beginning of some major classifications. These numbers refer to related items in the Reference Section. The reference information may be an estimating procedure, an alternate pricing method, or technical information.

Note: Not all subdivisions listed here necessarily appear in this publication.

Note: **Trade Service,** *in part, has been used as a reference source for some of the material prices used in Division 33.*

33 01 Operation and Maintenance of Utilities

33 01 30 – Operation and Maintenance of Sewer Utilities

33 01 30.72 Relining Sewers		Crew	Daily Output	Labor-Hours	Unit	Material	2012 Bare Costs Labor	Equipment	Total	Total Incl O&P
0010	**RELINING SEWERS**									
0011	With cement incl. bypass & cleaning									
0020	Less than 10,000 L.F., urban, 6" to 10"	C-17E	130	.615	L.F.	9.20	28	.70	37.90	57
0050	10" to 12"		125	.640		11.30	29.50	.72	41.52	60.50
0070	12" to 16"		115	.696		11.60	32	.79	44.39	65.50
0100	16" to 20"		95	.842		13.60	38.50	.95	53.05	78.50
0200	24" to 36"		90	.889		14.65	40.50	1	56.15	83
0300	48" to 72"		80	1		23.50	46	1.13	70.63	101
0500	Rural, 6" to 10"		180	.444		9.20	20.50	.50	30.20	43.50
0550	10" to 12"		175	.457		11.30	21	.52	32.82	47
0570	12" to 16"		160	.500		11.60	23	.57	35.17	50.50
0600	16" to 20"		135	.593		13.60	27	.67	41.27	59.50
0700	24" to 36"		125	.640		14.65	29.50	.72	44.87	64.50
0800	48" to 72"		100	.800		23.50	36.50	.90	60.90	86
1000	Greater than 10,000 L.F., urban, 6" to 10"		160	.500		9.20	23	.57	32.77	47.50
1050	10" to 12"		155	.516		11.30	23.50	.58	35.38	51.50
1070	12" to 16"		140	.571		11.60	26	.65	38.25	56
1100	16" to 20"		120	.667		13.60	30.50	.75	44.85	65.50
1200	24" to 36"		115	.696		14.65	32	.79	47.44	69
1300	48" to 72"		95	.842		23.50	38.50	.95	62.95	89
1500	Rural, 6" to 10"		215	.372		9.20	17.05	.42	26.67	38
1550	10" to 12"		210	.381		11.30	17.45	.43	29.18	41.50
1570	12" to 16"		185	.432		11.60	19.80	.49	31.89	45.50
1600	16" to 20"		150	.533		13.60	24.50	.60	38.70	55
1700	24" to 36"		140	.571		14.65	26	.65	41.30	59.50
1800	48" to 72"	▼	120	.667	▼	23.50	30.50	.75	54.75	76

33 05 Common Work Results for Utilities

33 05 16 – Utility Structures

33 05 16.13 Precast Concrete Utility Boxes

		Crew	Daily Output	Labor-Hours	Unit	Material	Labor	Equipment	Total	Total Incl O&P
0010	**PRECAST CONCRETE UTILITY BOXES**, 6" thick									
0050	5' x 10' x 6' high, I.D.	B-13	2	28	Ea.	2,475	1,075	350	3,900	4,825
0350	Hand hole, precast concrete, 1-1/2" thick									
0400	1'-0" x 2'-0" x 1'-9", I.D., light duty	B-1	4	6	Ea.	390	215		605	780
0450	4'-6" x 3'-2" x 2'-0", O.D., heavy duty	B-6	3	8	"	1,400	305	111	1,816	2,175

33 05 23 – Trenchless Utility Installation

33 05 23.19 Microtunneling

		Crew	Daily Output	Labor-Hours	Unit	Material	Labor	Equipment	Total	Total Incl O&P
0010	**MICROTUNNELING**									
0011	Not including excavation, backfill, shoring,									
0020	or dewatering, average 50'/day, slurry method									
0100	24" to 48" outside diameter, minimum				L.F.				830	915
0110	Adverse conditions, add				%				50%	50%
1000	Rent microtunneling machine, average monthly lease				Month				92,000	101,000
1010	Operating technician				Day				610	680
1100	Mobilization and demobilization, minimum				Job				40,000	45,000
1110	Maximum				"				420,000	462,000

33 05 23.20 Horizontal Boring

		Crew	Daily Output	Labor-Hours	Unit	Material	Labor	Equipment	Total	Total Incl O&P
0010	**HORIZONTAL BORING**									
0011	Casing only, 100' minimum,									
0020	not incl. jacking pits or dewatering									

33 05 Common Work Results for Utilities

33 05 23 – Trenchless Utility Installation

33 05 23.20 Horizontal Boring		Crew	Daily Output	Labor-Hours	Unit	2012 Bare Costs Material	Labor	Equipment	Total	Total Incl O&P
0100	Roadwork, 1/2″ thick wall, 24″ diameter casing	B-42	20	3.200	L.F.	146	126	63.50	335.50	440
0200	36″ diameter		16	4		220	158	79	457	590
0300	48″ diameter		15	4.267		294	168	84.50	546.50	695
0500	Railroad work, 24″ diameter		15	4.267		146	168	84.50	398.50	535
0600	36″ diameter		14	4.571		220	181	90.50	491.50	640
0700	48″ diameter		12	5.333		294	211	106	611	790
0900	For ledge, add									20%
1000	Small diameter boring, 3″, sandy soil	B-82	900	.018		23	.71	.09	23.80	26
1040	Rocky soil	″	500	.032		23	1.28	.15	24.43	27

33 05 26 – Utility Identification

33 05 26.10 Utility Accessories		Crew	Daily Output	Labor-Hours	Unit	Material	Labor	Equipment	Total	Total Incl O&P
0010	**UTILITY ACCESSORIES** R312316-40									
0400	Underground tape, detectable, reinforced, alum. foil core, 2″	1 Clab	150	.053	C.L.F.	2.05	1.87		3.92	5.35
0500	6″		140	.057	″	5.60	2.01		7.61	9.45
9000	Minimum labor/equipment charge		4	2	Job		70		70	115

33 11 Water Utility Distribution Piping

33 11 13 – Public Water Utility Distribution Piping

33 11 13.15 Water Supply, Ductile Iron Pipe		Crew	Daily Output	Labor-Hours	Unit	Material	Labor	Equipment	Total	Total Incl O&P
0010	**WATER SUPPLY, DUCTILE IRON PIPE** R331113-80									
0020	Not including excavation or backfill									
2000	Pipe, class 50 water piping, 18′ lengths									
2020	Mechanical joint, 4″ diameter	B-21A	200	.200	L.F.	12.60	8.75	2.24	23.59	30
3000	Tyton, push-on joint, 4″ diameter		400	.100		11.40	4.37	1.12	16.89	20.50
3020	6″ diameter		333.33	.120		15.80	5.25	1.35	22.40	27
8000	Fittings, mechanical joint									
8006	90° bend, 4″ diameter	B-20A	16	2	Ea.	251	85.50		336.50	410
8020	6″ diameter		12.80	2.500		375	107		482	580
8200	Wye or tee, 4″ diameter		10.67	2.999		405	128		533	650
8220	6″ diameter		8.53	3.751		615	160		775	930
8398	45° bends, 4″ diameter		16	2		234	85.50		319.50	395
8450	Decreaser, 6″ x 4″ diameter		14.22	2.250		330	96		426	520
8460	8″ x 6″ diameter		11.64	2.749		485	117		602	715
8550	Piping, butterfly valves, cast iron									
8560	4″ diameter	B-20	6	4	Ea.	800	157		957	1,125
9600	Steel sleeve with tap, 4″ diameter		3	8		410	315		725	960
9620	6″ diameter		2	12		465	470		935	1,275

33 12 Water Utility Distribution Equipment

33 12 13 – Water Service Connections

33 12 13.15 Tapping, Crosses and Sleeves		Crew	Daily Output	Labor-Hours	Unit	Material	Labor	Equipment	Total	Total Incl O&P
0010	**TAPPING, CROSSES AND SLEEVES**									
4000	Drill and tap pressurized main (labor only)									
4100	6″ main, 1″ to 2″ service	Q-1	3	5.333	Ea.		262		262	410
4150	8″ main, 1″ to 2″ service	″	2.75	5.818	″		286		286	445
4500	Tap and insert gate valve									
4600	8″ main, 4″ branch	B-21	3.20	8.750	Ea.		355	41.50	396.50	620
4650	6″ branch		2.70	10.370			420	49	469	735
4800	12″ main, 6″ branch		2.35	11.915			480	56.50	536.50	845

33 21 Water Supply Wells

33 21 13 – Public Water Supply Wells

33 21 13.10 Wells and Accessories		Crew	Daily Output	Labor-Hours	Unit	2012 Bare Costs Material	Labor	Equipment	Total	Total Incl O&P
0010	**WELLS & ACCESSORIES** R221113-50									
0011	Domestic									
0100	Drilled, 4″ to 6″ diameter	B-23	120	.333	L.F.		11.85	22.50	34.35	44.50
0200	8″ diameter	″	95.20	.420	″		14.90	28.50	43.40	56
1400	Remove & reset pump, minimum	B-21	4	7	Ea.		283	33	316	495
1420	Maximum	″	2	14	″		565	66	631	995
1500	Pumps, installed in wells to 100′ deep, 4″ submersible									
1510	1/2 H.P.	Q-1	3.22	4.969	Ea.	420	244		664	840
1520	3/4 H.P.		2.66	6.015		625	296		921	1,150
1600	1 H.P.		2.29	6.987		680	345		1,025	1,275
1700	1-1/2 H.P.	Q-22	1.60	10		780	490	385	1,655	2,050
1800	2 H.P.		1.33	12.030		850	590	460	1,900	2,350
1900	3 H.P.		1.14	14.035		1,325	690	535	2,550	3,150
2000	5 H.P.		1.14	14.035		2,225	690	535	3,450	4,125
2050	Remove and install motor only, 4 H.P.		1.14	14.035		1,200	690	535	2,425	3,000
5000	Wells to 180 ft. deep, 4″ submersible, 1 HP	B-21	1.10	25.455		2,150	1,025	120	3,295	4,175
5500	2 HP		1.10	25.455		3,025	1,025	120	4,170	5,125
6000	3 HP		1	28		4,075	1,125	132	5,332	6,475
7000	5 HP		.90	31.111		5,625	1,250	147	7,022	8,400
9000	Minimum labor/equipment charge		1.80	15.556	Job		630	73.50	703.50	1,100

33 31 Sanitary Utility Sewerage Piping

33 31 13 – Public Sanitary Utility Sewerage Piping

33 31 13.15 Sewage Collection, Concrete Pipe

		Crew	Daily Output	Labor-Hours	Unit	Material	Labor	Equipment	Total	Total Incl O&P
0010	**SEWAGE COLLECTION, CONCRETE PIPE**									
0020	See Section 33 41 13.60 for sewage/drainage collection, concrete pipe									

33 31 13.25 Sewage Collection, Polyvinyl Chloride Pipe

		Crew	Daily Output	Labor-Hours	Unit	Material	Labor	Equipment	Total	Total Incl O&P
0010	**SEWAGE COLLECTION, POLYVINYL CHLORIDE PIPE**									
0020	Not including excavation or backfill									
2000	20′ lengths, S.D.R. 35, B&S, 4″ diameter	B-20	375	.064	L.F.	1.64	2.51		4.15	5.90
2040	6″ diameter		350	.069		3.45	2.69		6.14	8.20
2080	13′ lengths , S.D.R. 35, B&S, 8″ diameter		335	.072		7.85	2.81		10.66	13.20
2120	10″ diameter	B-21	330	.085		12.70	3.43	.40	16.53	19.95
4000	Piping, DWV PVC, no exc./bkfill., 10′ L, Sch 40, 4″ diameter	B-20	375	.064		4.09	2.51		6.60	8.60
4010	6″ diameter		350	.069		8.45	2.69		11.14	13.70
4020	8″ diameter		335	.072		15.30	2.81		18.11	21.50

33 36 Utility Septic Tanks

33 36 13 – Utility Septic Tank and Effluent Wet Wells

33 36 13.10 Septic Tanks

		Crew	Daily Output	Labor-Hours	Unit	Material	Labor	Equipment	Total	Total Incl O&P
0010	**SEPTIC TANKS**									
0011	Not including excavation or piping									
0015	Septic tanks, precast, 1,000 gallon	B-21	8	3.500	Ea.	1,025	142	16.55	1,183.55	1,375
0020	1,250 gallon		8	3.500		1,025	142	16.55	1,183.55	1,375
0060	1,500 gallon		7	4		1,575	162	18.90	1,755.90	2,025
0100	2,000 gallon		5	5.600		2,000	226	26.50	2,252.50	2,600
0140	2,500 gallon		5	5.600		2,425	226	26.50	2,677.50	3,050
0180	4,000 gallon		4	7		5,500	283	33	5,816	6,550
0220	5,000 gal., 4 piece	B-13	3	18.667		10,000	710	232	10,942	12,400

33 36 Utility Septic Tanks

33 36 13 – Utility Septic Tank and Effluent Wet Wells

33 36 13.10 Septic Tanks		Crew	Daily Output	Labor-Hours	Unit	2012 Bare Costs Material	Labor	Equipment	Total	Total Incl O&P
0300	15,000 gallon, 4 piece	B-13B	1.70	32.941	Ea.	20,900	1,250	645	22,795	25,700
0400	25,000 gallon, 4 piece		1.10	50.909		40,500	1,950	1,000	43,450	48,900
0500	40,000 gallon, 4 piece		.80	70		52,000	2,675	1,375	56,050	63,000
0520	50,000 gallon, 5 piece	B-13C	.60	93.333		60,000	3,550	2,725	66,275	74,500
0540	75,000 gallon, cast in place	C-14C	.25	448		73,000	18,900	108	92,008	111,000
0560	100,000 gallon	"	.15	746		90,000	31,600	179	121,779	150,500
0600	High density polyethylene, 1,000 gallon	B-21	6	4.667		1,100	189	22	1,311	1,525
0700	1,500 gallon		4	7		1,400	283	33	1,716	2,025
0900	Galley, 4′ x 4′ x 4′		16	1.750		260	71	8.30	339.30	410
1000	Distribution boxes, concrete, 7 outlets	2 Clab	16	1		75	35		110	140
1100	9 outlets		8	2		460	70		530	620
1120	Distribution boxes, HDPE, 5 outlets		20	.800		45	28		73	95.50
1125	8 outlets		10	1.600		56	56		112	154
1150	Leaching field chambers, 13′ x 3′-7″ x 1′-4″, standard	B-13	16	3.500		470	133	43.50	646.50	785
1190	Distribution boxes, Outlet Flow Leveler	1 Clab	50	.160		2.95	5.60		8.55	12.45
1200	Heavy duty, 8′ x 4′ x 1′-6″	B-13	14	4		278	152	50	480	605
1300	13′ x 3′-9″ x 1′-6″		12	4.667		1,050	178	58	1,286	1,500
1350	20′ x 4′ x 1′-6″		5	11.200		1,100	425	139	1,664	2,075
1420	Leaching pit, 6′, dia, 3′ deep complete					815			815	895
1600	Leaching pit, 6′-6″ diameter, 6′ deep	B-21	5	5.600		1,200	226	26.50	1,452.50	1,725
1620	8′ deep		4	7		1,375	283	33	1,691	2,000
1700	8′ diameter, H-20 load, 6′ deep		4	7		2,300	283	33	2,616	3,025
1720	8′ deep		3	9.333		2,750	375	44	3,169	3,700
2000	Velocity reducing pit, precast conc., 6′ diameter, 3′ deep		4.70	5.957		1,550	241	28	1,819	2,125
2200	Excavation for septic tank, 3/4 C.Y. backhoe	B-12F	145	.110	C.Y.		4.57	4.46	9.03	12.25
2400	4′ trench for disposal field, 3/4 C.Y. backhoe	"	335	.048	L.F.		1.98	1.93	3.91	5.30
2600	Gravel fill, run of bank	B-6	150	.160	C.Y.	20.50	6.15	2.23	28.88	35.50
2800	Crushed stone, 3/4″	"	150	.160	"	39.50	6.15	2.23	47.88	55.50
3000	Effluent filter, 4″ diameter	1 Skwk	8	1	Ea.	50	45.50		95.50	129
3020	Effluent filter, 6″ diameter	"	7	1.143	"	200	52		252	305

33 41 Storm Utility Drainage Piping

33 41 13 – Public Storm Utility Drainage Piping

33 41 13.40 Piping, Storm Drainage, Corrugated Metal		Crew	Daily Output	Labor-Hours	Unit	Material	Labor	Equipment	Total	Total Incl O&P
0010	**PIPING, STORM DRAINAGE, CORRUGATED METAL** R221113-50									
0020	Not including excavation or backfill									
2000	Corrugated metal pipe, galvanized									
2020	Bituminous coated with paved invert, 20′ lengths									
2040	8″ diameter, 16 ga.	B-14	330	.145	L.F.	11.10	5.40	1.01	17.51	22
2080	12″ diameter, 16 ga.	"	210	.229	"	14	8.45	1.59	24.04	31

33 41 13.60 Sewage/Drainage Collection, Concrete Pipe		Crew	Daily Output	Labor-Hours	Unit	Material	Labor	Equipment	Total	Total Incl O&P
0010	**SEWAGE/DRAINAGE COLLECTION, CONCRETE PIPE**									
0020	Not including excavation or backfill									
1000	Non-reinforced pipe, extra strength, B&S or T&G joints									
1010	6″ diameter	B-14	265.04	.181	L.F.	6.40	6.70	1.26	14.36	19.35
1020	8″ diameter		224	.214		7.05	7.95	1.49	16.49	22.50
1040	12″ diameter		200	.240		9.55	8.90	1.67	20.12	27
2000	Reinforced culvert, class 3, no gaskets									
2010	12″ diameter	B-14	150	.320	L.F.	11.75	11.85	2.23	25.83	34.50
2020	15″ diameter	"	150	.320	"	16	11.85	2.23	30.08	39.50

33 44 Storm Utility Water Drains

33 44 13 – Utility Area Drains

33 44 13.13 Catchbasins		Crew	Daily Output	Labor-Hours	Unit	2012 Bare Costs Material	2012 Bare Costs Labor	2012 Bare Costs Equipment	2012 Bare Costs Total	Total Incl O&P
0010	**CATCHBASINS**									
0011	Not including footing & excavation									
1600	Frames & covers, C.I., 24″ square, 500 lb.	B-6	7.80	3.077	Ea.	320	118	43	481	590
1700	26″ D shape, 600 lb.	″	7	3.429	″	480	131	47.50	658.50	795
3320	Frames and covers, existing, raised for paving, 2″, including									
3340	row of brick, concrete collar, up to 12″ wide frame	B-6	18	1.333	Ea.	46.50	51	18.55	116.05	154

33 44 13.15 Remove and Replace Catch Basin Cover

		Crew	Daily Output	Labor-Hours	Unit	Material	Labor	Equipment	Total	Total Incl O&P
0010	**REMOVE AND REPLACE CATCH BASIN COVER**									
0023	Remove catch basin cover	1 Clab	80	.100	Ea.		3.51		3.51	5.75
0033	Replace catch basin cover	″	80	.100	″		3.51		3.51	5.75

33 49 Storm Drainage Structures

33 49 13 – Storm Drainage Manholes, Frames, and Covers

33 49 13.10 Storm Drainage Manholes, Frames and Covers

		Crew	Daily Output	Labor-Hours	Unit	Material	Labor	Equipment	Total	Total Incl O&P
0010	**STORM DRAINAGE MANHOLES, FRAMES & COVERS**									
0020	Excludes footing, excavation, backfill (See line items for frame & cover)									
0050	Brick, 4′ inside diameter, 4′ deep	D-1	1	16	Ea.	410	640		1,050	1,475
0100	6′ deep		.70	22.857		570	910		1,480	2,100
0150	8′ deep		.50	32		730	1,275		2,005	2,850
0200	For depths over 8′, add		4	4	V.L.F.	85	159		244	350
1110	Precast, 4′ I.D., 4′ deep	B-22	4.10	7.317	Ea.	775	299	48.50	1,122.50	1,400
1120	6′ deep		3	10		900	410	66	1,376	1,725
1130	8′ deep		2	15		1,025	615	99.50	1,739.50	2,225
1140	For depths over 8′, add		16	1.875	V.L.F.	121	76.50	12.40	209.90	271

33 51 Natural-Gas Distribution

33 51 13 – Natural-Gas Piping

33 51 13.10 Piping, Gas Service and Distribution, P.E.

		Crew	Daily Output	Labor-Hours	Unit	Material	Labor	Equipment	Total	Total Incl O&P
0010	**PIPING, GAS SERVICE AND DISTRIBUTION, POLYETHYLENE**									
0020	Not including excavation or backfill									
1000	60 psi coils, compression coupling @ 100′, 1/2″ diameter, SDR 11	B-20A	608	.053	L.F.	.91	2.24		3.15	4.57
1010	1″ diameter, SDR 11		544	.059		2.03	2.51		4.54	6.20
1040	1-1/4″ diameter, SDR 11		544	.059		2.04	2.51		4.55	6.25
1100	2″ diameter, SDR 11		488	.066		3.36	2.80		6.16	8.15
1160	3″ diameter, SDR 11		408	.078		4.37	3.34		7.71	10.10
1500	60 PSI 40′ joints with coupling, 3″ diameter, SDR 11	B-21A	408	.098		4.37	4.28	1.10	9.75	12.80
1540	4″ diameter, SDR 11		352	.114		9.15	4.96	1.27	15.38	19.35
1600	6″ diameter, SDR 11		328	.122		26.50	5.30	1.37	33.17	39
1640	8″ diameter, SDR 11		272	.147		32.50	6.40	1.65	40.55	47.50
9000	Minimum labor/equipment charge	B-20	2	12	Job		470		470	770

33 52 16 – Gasoline Distribution

33 52 16.13 Gasoline Piping		Crew	Daily Output	Labor-Hours	Unit	2012 Bare Costs Material	Labor	Equipment	Total	Total Incl O&P
0010	**GASOLINE PIPING**									
0020	Primary containment pipe, fiberglass-reinforced									
0030	Plastic pipe 15′ & 30′ lengths									
0040	2″ diameter	Q-6	425	.056	L.F.	5.60	2.94		8.54	10.75
0050	3″ diameter		400	.060		8.60	3.12		11.72	14.30
0060	4″ diameter		375	.064		11.25	3.33		14.58	17.60
0100	Fittings									
0110	Elbows, 90° & 45°, bell-ends, 2″	Q-6	24	1	Ea.	41	52		93	126
0120	3″ diameter		22	1.091		51.50	56.50		108	146
0130	4″ diameter		20	1.200		66	62.50		128.50	170
0200	Tees, bell ends, 2″		21	1.143		57	59.50		116.50	155
0210	3″ diameter		18	1.333		60.50	69.50		130	175
0220	4″ diameter		15	1.600		79.50	83		162.50	217
0230	Flanges bell ends, 2″		24	1		31.50	52		83.50	116
0240	3″ diameter		22	1.091		37	56.50		93.50	129
0250	4″ diameter		20	1.200		42.50	62.50		105	144
0260	Sleeve couplings, 2″		21	1.143		13.50	59.50		73	107
0270	3″ diameter		18	1.333		17.10	69.50		86.60	127
0280	4″ diameter		15	1.600		24.50	83		107.50	157
0290	Threaded adapters 2″		21	1.143		17.10	59.50		76.60	111
0300	3″ diameter		18	1.333		32.50	69.50		102	144
0310	4″ diameter		15	1.600		36	83		119	170
0320	Reducers, 2″		27	.889		26	46		72	101
0330	3″ diameter		22	1.091		26	56.50		82.50	117
0340	4″ diameter		20	1.200		35	62.50		97.50	136
1010	Gas station product line for secondary containment (double wall)									
1100	Fiberglass reinforced plastic pipe 25′ lengths									
1120	Pipe, plain end, 3″ diameter	Q-6	375	.064	L.F.	6.75	3.33		10.08	12.65
1130	4″ diameter		350	.069		7.90	3.56		11.46	14.25
1140	5″ diameter		325	.074		13.80	3.84		17.64	21
1150	6″ diameter		300	.080		14.20	4.16		18.36	22
1200	Fittings									
1230	Elbows, 90° & 45°, 3″ diameter	Q-6	18	1.333	Ea.	56.50	69.50		126	170
1240	4″ diameter		16	1.500		105	78		183	238
1250	5″ diameter		14	1.714		280	89		369	450
1260	6″ diameter		12	2		320	104		424	510
1270	Tees, 3″ diameter		15	1.600		86	83		169	225
1280	4″ diameter		12	2		131	104		235	305
1290	5″ diameter		9	2.667		300	139		439	545
1300	6″ diameter		6	4		365	208		573	730
1310	Couplings, 3″ diameter		18	1.333		55	69.50		124.50	169
1320	4″ diameter		16	1.500		115	78		193	249
1330	5″ diameter		14	1.714		250	89		339	415
1340	6″ diameter		12	2		310	104		414	500
1350	Cross-over nipples, 3″ diameter		18	1.333		10	69.50		79.50	119
1360	4″ diameter		16	1.500		12	78		90	135
1370	5″ diameter		14	1.714		15	89		104	156
1380	6″ diameter		12	2		18	104		122	182
1400	Telescoping, reducers, concentric 4″ x 3″		18	1.333		45	69.50		114.50	158
1410	5″ x 4″		17	1.412		90	73.50		163.50	213
1420	6″ x 5″		16	1.500		220	78		298	365

33 71 19 – Electrical Underground Ducts and Manholes

33 71 19.17 Electric and Telephone Underground		Crew	Daily Output	Labor-Hours	Unit	Material	Labor	Equipment	Total	Total Incl O&P
						2012 Bare Costs				
0010	**ELECTRIC AND TELEPHONE UNDERGROUND**									
0011	Not including excavation									
0200	backfill and cast in place concrete									
4200	Underground duct, banks ready for concrete fill, min. of 7.5"									
4400	between conduits, center to center									
4600	2 @ 2" diameter	2 Elec	240	.067	L.F.	.74	3.44		4.18	6.10
4800	4 @ 2" diameter		120	.133		1.48	6.85		8.33	12.25
5600	4 @ 4" diameter		80	.200		3.02	10.30		13.32	19.30
6200	Rigid galvanized steel, 2 @ 2" diameter		180	.089		15.15	4.58		19.73	24
6400	4 @ 2" diameter		90	.178		30.50	9.15		39.65	47.50
7400	4 @ 4" diameter	▼	34	.471	▼	110	24.50		134.50	159
9990	Minimum labor/equipment charge	1 Elec	3.50	2.286	Job		118		118	182

Estimating Tips

34 11 00 Rail Tracks

This subdivision includes items that may involve either repair of existing, or construction of new, railroad tracks. Additional preparation work, such as the roadbed earthwork, would be found in Division 31. Additional new construction siding and turnouts are found in Subdivision 34 72. Maintenance of railroads is found under 34 01 23 Operation and Maintenance of Railways.

34 41 13 Traffic Signals

This subdivision includes traffic signal systems. Other traffic control devices such as traffic signs are found in Subdivision 10 14 53 Traffic Signage.

34 71 13 Vehicle Barriers

This subdivision includes security vehicle barriers, guide and guard rails, crash barriers, and delineators. The actual maintenance and construction of concrete and asphalt pavement is found in Division 32.

Reference Numbers

Reference numbers are shown in shaded boxes at the beginning of some major classifications. These numbers refer to related items in the Reference Section. The reference information may be an estimating procedure, an alternate pricing method, or technical information.

Note: Not all subdivisions listed here necessarily appear in this publication.

34 71 Roadway Construction

34 71 13 – Vehicle Barriers

34 71 13.26 Vehicle Guide Rails		Crew	Daily Output	Labor-Hours	Unit	Material	Labor	Equipment	Total	Total Incl O&P
						2012 Bare Costs				
0010	**VEHICLE GUIDE RAILS**									
0012	Corrugated stl, galv. stl posts, 6′-3″ O.C.	B-80	850	.038	L.F.	20	1.42	.79	22.21	25
0200	End sections, galvanized, flared		50	.640	Ea.	88	24	13.35	125.35	151
0300	Wrap around end		50	.640	″	128	24	13.35	165.35	195
0400	Timber guide rail, 4″ x 8″ with 6″ x 8″ wood posts, treated	↓	960	.033	L.F.	10.35	1.26	.70	12.31	14.15

Division 41 - Mat'l. Processing and Handling Equip.

Estimating Tips

Products such as conveyors, material handling cranes and hoists, as well as other items specified in this division, require trained installers. The general contractor may not have any choice as to who will perform the installation or when it will be performed. Long lead times are often required for these products, making early decisions in purchasing and scheduling necessary.

The installation of this type of equipment may require the embedment of mounting hardware during construction of floors, structural walls, or interior walls/partitions. Electrical connections will require coordination with the electrical contractor.

Reference Numbers

Reference numbers are shown in shaded boxes at the beginning of some major classifications. These numbers refer to related items in the Reference Section. The reference information may be an estimating procedure, an alternate pricing method, or technical information.

Note: Not all subdivisions listed here necessarily appear in this publication.

41 22 Cranes and Hoists

41 22 13 – Cranes

41 22 13.10 Crane Rail		Crew	Daily Output	Labor-Hours	Unit	Material	2012 Bare Costs Labor	Equipment	Total	Total Incl O&P
0010	**CRANE RAIL**									
0020	Box beam bridge, no equipment included	E-4	3400	.009	Lb.	1.19	.47	.04	1.70	2.22
0210	Running track only, 104 lb. per yard, 20′ piece	"	160	.200	L.F.	21.50	9.95	.76	32.21	43.50

41 22 13.13 Bridge Cranes		Crew	Daily Output	Labor-Hours	Unit	Material	Labor	Equipment	Total	Total Incl O&P
0010	**BRIDGE CRANES**									
0100	1 girder, 20′ span, 3 ton	M-3	1	34	Ea.	23,800	1,825	130	25,755	29,200
0125	5 ton		1	34		26,000	1,825	130	27,955	31,600
0150	7.5 ton		1	34		31,000	1,825	130	32,955	37,100
0175	10 ton		.80	42.500		41,000	2,275	162	43,437	48,700
0200	15 ton		.80	42.500		52,500	2,275	162	54,937	61,500
0225	30′ span, 3 ton		1	34		24,700	1,825	130	26,655	30,200
0250	5 ton		1	34		27,200	1,825	130	29,155	32,900
0275	7.5 ton		1	34		32,600	1,825	130	34,555	38,800
0300	10 ton		.80	42.500		42,400	2,275	162	44,837	50,500
0325	15 ton		.80	42.500		55,000	2,275	162	57,437	64,000
0350	2 girder, 40′ span, 3 ton	M-4	.50	72		40,900	3,825	330	45,055	51,000
0375	5 ton		.50	72		42,800	3,825	330	46,955	53,500
0400	7.5 ton		.50	72		46,900	3,825	330	51,055	58,000
0425	10 ton		.40	90		55,000	4,775	415	60,190	68,500
0450	15 ton		.40	90		74,500	4,775	415	79,690	90,000
0475	25 ton		.30	120		88,000	6,350	555	94,905	107,500
0500	50′ span, 3 ton		.50	72		46,500	3,825	330	50,655	57,500
0525	5 ton		.50	72		48,400	3,825	330	52,555	59,500
0550	7.5 ton		.50	72		52,000	3,825	330	56,155	63,500
0575	10 ton		.40	90		59,500	4,775	415	64,690	73,500
0600	15 ton	M-4	.40	90	Ea.	78,000	4,775	415	83,190	94,000
0625	25 ton	M-4	.30	120	Ea.	92,500	6,350	555	99,405	112,000

Assemblies Section

Table of Contents

How to Use the Assemblies Cost Tables

The following is a detailed explanation of a sample Assemblies Cost Table. Most Assembly Tables are separated into three parts: 1) an illustration of the system to be estimated; 2) the components and related costs of a typical system; and 3) the costs for similar systems with dimensional and/or size variations. For costs of the components that comprise these systems or "assemblies," refer to the Unit Price Section. Next to each bold number below is the item being described with the appropriate component of the sample entry following in parenthesis. In most cases, if the work is to be subcontracted, the general contractor will need to add an additional markup (RSMeans suggests using 10%) to the "Total" figures.

1 *System/Line Numbers (B1010 263 1700)*

Each Assemblies Cost Line has been assigned a unique identification number based on the UNIFORMAT II classification system.

UNIFORMAT II Major Group

B1010 263 1700

UNIFORMAT II Level 3

Means Major Classification

Means Individual Line Number

B10 Superstructure

B1010 Floor Construction 1

2

This page illustrates and describes a wood joist floor system including wood joist, oak floor, sub-floor, bridging, sand and finish floor, furring, plasterboard, taped, finished and painted ceiling. Lines within System Components give the unit price and total price per square foot for this system. Prices for wood joist floor systems are on Line Items B1010 263 1700 thru 2300. Both material quantities and labor costs have been adjusted for the system listed.

Factors: To adjust for job conditions other than normal working situations use Lines B1010 263 2700 thru 4000.

Example: You are to install the system during off peak hours, 6 P.M. to 2 A.M. Go to Line B1010 263 3800 and apply this percentage to the appropriate INST. costs.

6

System Components	QUANTITY	UNIT	COST PER S.F. MAT.	INST.	TOTAL
Wood joists, 2" x 8", 16" O.C.,oak floor (sanded & finished),1/2" sub Floor, 1"x 3" bridging, furring, 5/8" drywall, taped, finished and painted. 3					
Wood joists, 2" x 8", 16" O.C.	1.000	L.F.	.83	1.05	1.88
Subfloor plywood CDX 1/2"	1.000	SF Flr.	.52	.77	1.29
Bridging 1" x 3"	.150	Pr.	.08	.67	.75
Oak flooring, No. 1 common	1.000	S.F.	4.13	3.40	7.53
Sand and finish floor	1.000	S.F.	.95	1.56	2.51
Wood furring, 1" x 3", 16" O.C.	1.000	[illegible]	.38	1.65	2.03
Gypsum drywall, 5/8" thick	1.000	S.F.	.28	.64	.92
Tape and finishing	1.000	S.F.	.05	.58	.63
Paint ceiling	1.000	S.F.	.19	.64	.83
TOTAL	4	5	7.41 (7)	10.96 (8)	18.37 (9)

B1010 263	Floor-Ceiling, Wood Joists	COST PER S.F. MAT.	INST.	TOTAL
1700	16" on center, 2" x 6" joists	7.15	10.85	18
1750	2" x 8"	7.40	10.95	18.35
1800	2" x 10"	7.70	11.20	18.90
1900	2" x 12"	8	11.25	19.25
2000	2" x 14"	8.45	11.40	19.85
2100	12" on center, 2" x 10" joists	7.95	11.50	19.45
2200	2" x 12"	8.35	11.55	19.90
2300	2" x 14"	8.90	11.80	20.70

2 Illustration

At the top of most assembly tables are an illustration, a brief description, and the design criteria used to develop the cost.

3 System Components

The components of a typical system are listed separately to show what has been included in the development of the total system price. The table below contains prices for other similar systems with dimensional and/or size variations.

4 Quantity

This is the number of line item units required for one system unit. For example, we assume that it will take .15 pair of 1″ × 3″ bridging on a square foot basis.

5 Unit of Measure for Each Item

The abbreviated designation indicates the unit of measure, as defined by industry standards, upon which the price of the component is based. For example, wood joists are priced by L.F. (linear foot) while plywood is priced by S.F. (square foot).

6 Unit of Measure for Each System (Cost per S.F.)

Costs shown in the three right-hand columns have been adjusted by the component quantity and unit of measure for the entire system. In this example, "Cost per S.F." is the unit of measure for this system, or assembly.

7 Materials (7.41)

This column contains the Materials Cost of each component. These cost figures are bare costs plus 10% for profit.

8 Installation (10.96)

Installation includes labor and equipment plus the installing contractor's overhead and profit. Equipment costs are the bare rental costs plus 10% for profit. The labor overhead and profit are defined on the inside back cover of this book.

9 Total (18.37)

The figure in this column is the sum of the material and installation costs.

Material Cost	+	Installation Cost	=	Total
$7.41	+	$10.96	=	$18.37

A Substructure

A1010 Standard Foundations

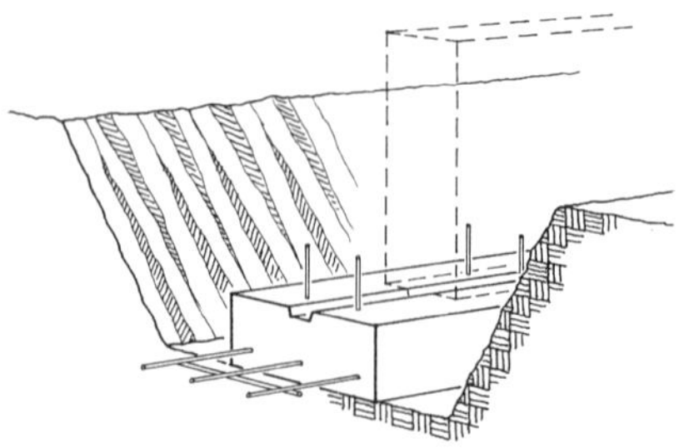

This page illustrates and describes a strip footing system including concrete, forms, reinforcing, keyway and dowels. Lines within System Components give the unit price and total price per linear foot for this system. Prices for strip footing systems are on Line Items A1010 120 1400 thru 2500. Both material quantities and labor costs have been adjusted for the system listed.

Factors: To adjust for job conditions other than normal working situations use Lines A1010 120 2900 thru 4000.

Example: You are to install this footing, and due to a lack of accessibility, only hand tools can be used. Material handling is also a problem. Go to Lines A1010 120 3400 and 3600 and apply these percentages to the appropriate MAT. and INST. costs.

System Components	QUANTITY	UNIT	COST PER L.F. MAT.	COST PER L.F. INST.	COST PER L.F. TOTAL
Strip footing, 2′-0″ wide x 1′-0″ thick, 2000 psi concrete including forms Reinforcing, keyway, and dowels.					
Concrete, 2000 psi	.074	C.Y.	7.47		7.47
Placing concrete	.074	C.Y.		1.82	1.82
Forms, footing, 4 uses	2.000	S.F.	4.24	9.06	13.30
Reinforcing	3.170	Lb.	1.78	1.97	3.75
Keyway, 2″ x 4″, 4 uses	1.000	L.F.	.19	1.09	1.28
Dowels, #4 bars, 2′ long, 24″ O.C.	.500	Ea.	.40	1.37	1.77
TOTAL			14.08	15.31	29.39

A1010 120	Strip Footing	COST PER L.F. MAT.	COST PER L.F. INST.	COST PER L.F. TOTAL
1400	2′-0″ wide x 1′ thick, 2000 psi concrete	14.10	15.30	29.40
1500	3000 psi concrete	14.90	15.30	30.20
1600	4000 psi concrete	14.95	15.30	30.25
1800	For alternate footing systems:			
1900	2′-6″ wide x 1′ thick, 2000 psi concrete	16.45	16.30	32.75
2000	3000 psi concrete	17.50	16.30	33.80
2100	4000 psi concrete	17.60	16.30	33.90
2300	3′-0″ wide x 1′ thick, 2000 psi concrete	18.65	17.20	35.85
2400	3000 psi concrete	19.85	17.20	37.05
2500	4000 psi concrete	19.95	17.20	37.15
2700				
2800				
2900	Cut & patch to match existing construction, add, minimum	2%	3%	
3000	Maximum	5%	9%	
3100	Dust protection, add, minimum	1%	2%	
3200	Maximum	4%	11%	
3300	Equipment usage curtailment, add, minimum	1%	1%	
3400	Maximum	3%	10%	
3500	Material handling & storage limitation, add, minimum	1%	1%	
3600	Maximum	6%	7%	
3700	Protection of existing work, add, minimum	2%	2%	
3800	Maximum	5%	7%	
3900	Shift work requirements, add, minimum		5%	
4000	Maximum		30%	

A1010 Standard Foundations

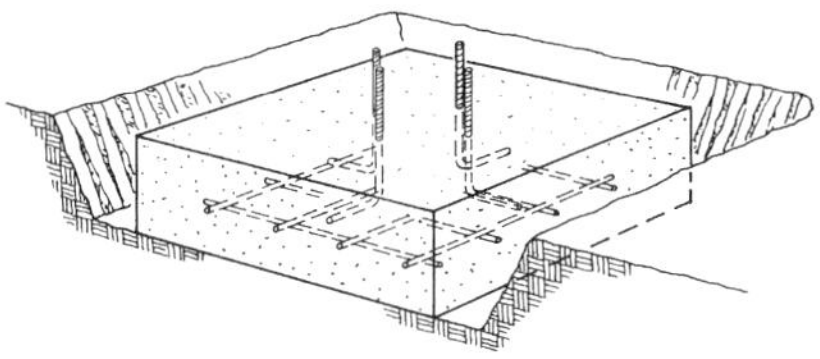

This page illustrates and describes a spread footing system including concrete, forms, reinforcing and anchor bolts. Lines within System Components give the unit price and total price on a cost each basis for this system. Prices for spread footing systems are on Line Items A1010 230 1200 thru 2300. Both material quantities and labor costs have been adjusted for the system listed.

Factors: To adjust for job conditions other than normal working situations use Lines A1010 230 2900 thru 4000.

Example: You are to install the system in an existing occupied building. Access to the site and protection to the building are mandatory. Go to Lines A1010 230 3600 and 3800 and apply these percentages to the appropriate MAT. and INST. costs.

System Components	QUANTITY	UNIT	COST EACH		
			MAT.	INST.	TOTAL
Interior column footing, 3′ square 1′ thick, 2000 psi concrete, Including forms, reinforcing, and anchor bolts.					
Concrete, 2000 psi	.330	C.Y.	33.33		33.33
Placing concrete	.330	C.Y.		17.52	17.52
Forms, footing, 4 uses	12.000	SFCA	8.16	63.60	71.76
Reinforcing	11.000	Lb.	6.16	6.82	12.98
Anchor bolts, 3/4″ diameter	2.000	Ea.	3.04	8.60	11.64
TOTAL			50.69	96.54	147.23

A1010 230	Spread Footing	COST EACH		
		MAT.	INST.	TOTAL
1200	3′ square x 1′ thick, 2000 psi concrete	50.50	96.50	147
1300	3000 psi concrete	54.50	96.50	151
1400	4000 psi concrete	54.50	96.50	151
1600	For alternate footing systems:			
1700	4′ square x 1′ thick, 2000 psi concrete	84.50	137	221.50
1800	3000 psi concrete	91	137	228
1900	4000 psi concrete	91.50	137	228.50
2100	5′ square x 1′-3″ thick, 2000 psi concrete	158	226	384
2200	3000 psi concrete	171	226	397
2300	4000 psi concrete	172	226	398
2600				
2700				
2900	Cut & patch to match existing construction, add, minimum	2%	3%	
3000	Maximum	5%	9%	
3100	Dust protection, add, minimum	1%	2%	
3200	Maximum	4%	11%	
3300	Equipment usage curtailment, add, minimum	1%	1%	
3400	Maximum	3%	10%	
3500	Material handling & storage limitation, add, minimum	1%	1%	
3600	Maximum	6%	7%	
3700	Protection of existing work, add, minimum	2%	2%	
3800	Maximum	5%	7%	
3900	Shift work requirements, add, minimum		5%	
4000	Maximum		30%	

A1010 Standard Foundations

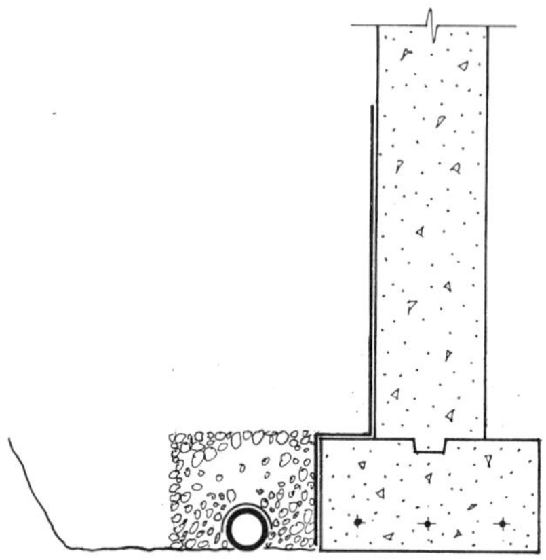

General: Footing drains can be placed either inside or outside of foundation walls depending upon the source of water to be intercepted. If the source of subsurface water is principally from grade or a subsurface stream above the bottom of the footing, outside drains should be used. For high water tables, use inside drains or both inside and outside.

The effectiveness of underdrains depends on good waterproofing. This must be carefully installed and protected during construction.

Costs below include the labor and materials for the pipe and 6″ of gravel or crushed stone around pipe. Excavation and backfill are not included.

System Components	QUANTITY	UNIT	COST PER L.F. MAT.	INST.	TOTAL
Foundation underdrain, outside, PVC 4″ diameter.					
PVC pipe 4″ diam. S.D.R. 35	1.000	L.F.	1.80	4.10	5.90
Crushed stone 3/4″ to 1/2″	.070	C.Y.	2.42	.86	3.28
TOTAL			4.22	4.96	9.18

A1010 310	Foundation Underdrain	COST PER L.F. MAT.	INST.	TOTAL
1000	Foundation underdrain, outside only, PVC, 4″ diameter	4.22	4.96	9.18
1100	6″ diameter	6.90	5.50	12.40
1400	Perforated HDPE, 6″ diameter	5.30	2.13	7.43
1450	8″ diameter	6.90	2.68	9.58
1500	12″ diameter	12.85	6.50	19.35
1600	Corrugated metal, 16 ga. asphalt coated, 6″ diameter	9.70	9.70	19.40
1650	8″ diameter	16	10.15	26.15
1700	10″ diameter	19.75	10.70	30.45
3000	Outside and inside, PVC, 4″ diameter	8.45	9.95	18.40
3100	6″ diameter	13.80	11	24.80
3400	Perforated HDPE, 6″ diameter	10.60	4.26	14.86
3450	8″ diameter	13.75	5.35	19.10
3500	12″ diameter	25.50	13	38.50
3600	Corrugated metal, 16 ga., asphalt coated, 6″ diameter	19.40	19.40	38.80
3650	8″ diameter	32	20.50	52.50
3700	10″ diameter	39.50	21.50	61
4700	Cut & patch to match existing construction, add, minimum	2%	3%	
4800	Maximum	5%	9%	
4900	Dust protection, add, minimum	1%	2%	
5000	Maximum	4%	11%	
5100	Equipment usage curtailment, add, minimum	1%	1%	
5200	Maximum	3%	10%	
5300	Material handling & storage limitation, add, minimum	1%	1%	
5400	Maximum	6%	7%	
5500	Protection of existing work, add, minimum	2%	2%	
5600	Maximum	5%	7%	
5700	Shift work requirements, add, minimum		5%	
5800	Maximum		30%	
5900	Temporary shoring and bracing, add, minimum	2%	5%	
6000	Maximum	5%	12%	

A1030 Slab on Grade

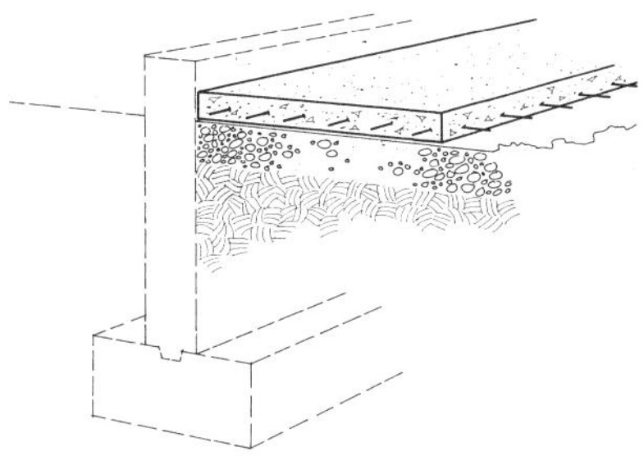

This page illustrates and describes a slab on grade system including slab, bank run gravel, bulkhead forms, placing concrete, welded wire fabric, vapor barrier, steel trowel finish and curing paper. Lines within System Components give the unit price and total price per square foot for this system. Prices for slab on grade systems are on Line Items A1030 110 1400 thru 2600. Both material quantities and labor costs have been adjusted for the system listed.

Factors: To adjust for job conditions other than normal working situations use Lines A1030 110 2900 thru 4000.

Example: You are to install the system at a site where protection of the existing building is required. Go to Line A1030 110 3800 and apply these percentages to the appropriate MAT. and INST. costs.

System Components	QUANTITY	UNIT	COST PER S.F.		
			MAT.	INST.	TOTAL
Ground slab, 4″ thick, 3000 psi concrete, 4″ granular base, vapor barrier					
Welded wire fabric, screed and steel trowel finish.					
Concrete, 4″ thick, 3000 psi concrete	.012	C.Y.	1.34		1.34
Bank run gravel, 4″ deep	.074	C.Y.	.30	.08	.38
Polyethylene vapor barrier, 10 mil.	.011	C.S.F.	.08	.17	.25
Bulkhead forms, expansion material	.100	L.F.	.03	.37	.40
Welded wire fabric, 6 x 6 - #10/10	.011	C.S.F.	.17	.41	.58
Place concrete	.012	C.Y.		.32	.32
Screed & steel trowel finish	1.000	S.F.		.92	.92
TOTAL			1.92	2.27	4.19

A1030 110	Interior Slab on Grade	COST PER S.F.		
		MAT.	INST.	TOTAL
1400	4″ thick slab, 3000 PSI concrete, 4″ deep bank run gravel	1.92	2.27	4.19
1500	6″ deep bank run gravel	2.16	2.27	4.43
1600	12″ deep bank run gravel	2.70	2.31	5.01
1700				
1800				
1900				
2000	For alternate slab systems:			
2100	5″ thick slab, 3000 psi concrete, 6″ deep bank run gravel	2.50	2.35	4.85
2200	12″ deep bank run gravel	3.04	2.39	5.43
2300				
2400				
2500	6″ thick slab, 3000 psi concrete, 6″ deep bank run gravel	2.95	2.45	5.40
2600	12″ deep bank run gravel	3.49	2.49	5.98
2700				
2900	Cut & patch to match existing construction, add, minimum	2%	3%	
3000	Maximum	5%	9%	
3100	Dust protection, add, minimum	1%	2%	
3200	Maximum	4%	11%	
3300	Equipment usage curtailment, add, minimum	1%	1%	
3400	Maximum	3%	10%	
3500	Material handling & storage limitation, add, minimum	1%	1%	
3600	Maximum	6%	7%	
3700	Protection of existing work, add, minimum	2%	2%	
3800	Maximum	5%	7%	
3900	Shift work requirements, add, minimum		5%	
4000	Maximum		30%	

A2010 Basement Excavation

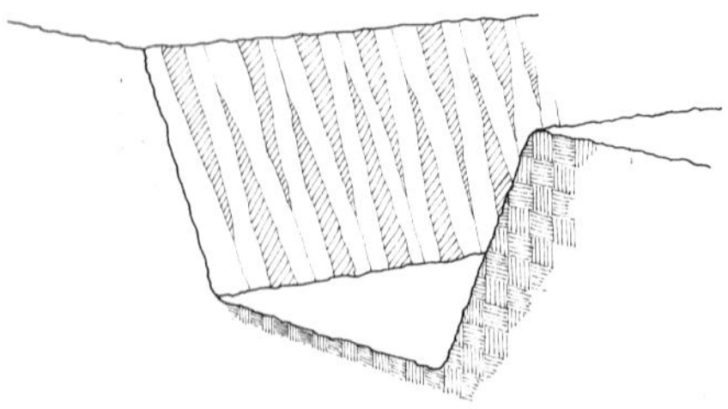

This page illustrates and describes continuous footing and trench excavation systems including a wheel mounted backhoe, operator, equipment rental, fuel, oil, mobilization, no hauling or backfill. Lines within System Components give the unit price and total price per linear foot for this system. Prices for alternate continuous footing and trench excavation systems are on Line Items A2010 130 1200 thru 2700. Both material quantities and labor costs have been adjusted for the system listed.

Factors: To adjust for job conditions other than normal working situations use Lines A2010 130 2900 thru 4000.

Example: You are to install the system and supply dust protection. Go to Line A2010 130 3000 and apply this percentage to the appropriate TOTAL costs.

System Components	QUANTITY	UNIT	COST PER L.F. EQUIP.	LABOR	TOTAL
Continuous footing or trench excav. w/ 3/4 C.Y. wheel mntd backhoe Including operator, equipment rental, fuel, oil and mobilization. Trench Is 4′ wide at bottom, 3′ deep w/sides sloped 1 to 2 and 200′ long. Costs Are based on a production rate of 240 C.Y./day. No hauling/backfill incl.					
Equipment operator	4.000	Hr.		295	295
3/4 C.Y. backhoe, wheel mounted	.500	Day	96.80		96.80
Operating expense (fuel, oil)	4.000	Hr.	86.92		86.92
Mobilization	1.000	Ea.	259	127	386
TOTAL			442.72	422	864.72
COST PER L.F.		L.F.	2.21	2.11	4.32

A2010 130	Excavation, Footings or Trench	COST PER L.F. EQUIP.	LABOR	TOTAL
1200	For alternate trench sizes, 4′ bottom with sloped sides:			
1300	2′ deep, 50′ long	7.45	8.45	15.90
1400	100′ long	3.89	4.22	8.11
1500	300′ long	1.45	1.40	2.85
1600	4′ deep, 50′ long	7.75	8.45	16.20
1700	100′ long	4.21	4.22	8.43
1800	300′ long	2.14	2.51	4.65
1900	6′ deep, 50′ long	8.25	8.45	16.70
2000	100′ long	4.43	5.35	9.78
2100	300′ long	3.31	4.35	7.66
2200	8′ deep, 50′ long	8.85	8.45	17.30
2300	100′ long	6.25	7.15	13.40
2400	300′ long	4.54	4.35	8.89
2500	10′ deep, 50′ long	10.70	11.40	22.10
2600	100′ long	7.90	9.75	17.65
2700	300′ long	6	8.65	14.65
2800				
2900	Dust protection, add, minimum		2%	2%
3000	Maximum		11%	11%
3100	Equipment usage curtailment, add, minimum		1%	1%
3200	Maximum		10%	10%
3300	Material handling & storage limitation, add, minimum		1%	1%
3400	Maximum		7%	7%
3500	Protection of existing work, add, minimum		2%	2%
3600	Maximum		7%	7%
3700	Shift work requirements, add, minimum		5%	5%
3800	Maximum		30%	30%
3900	Temporary shoring and bracing, add, minimum		5%	5%
4000	Maximum		12%	12%

A2010 Basement Excavation

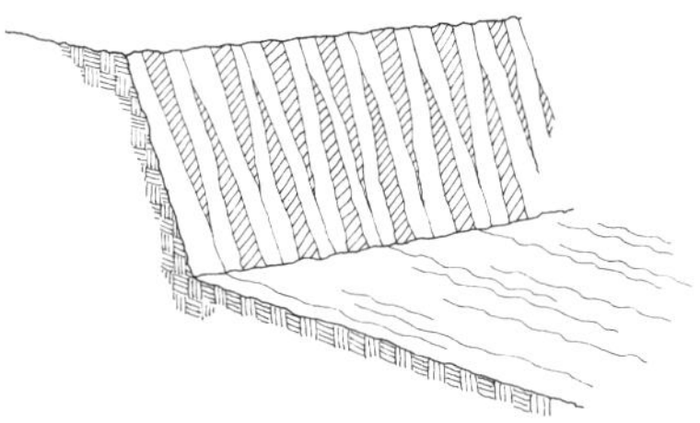

This page illustrates and describes foundation excavation systems including a backhoe-loader, operator, equipment rental, fuel, oil, mobilization, hauling material, no backfilling. Lines within System Components give the unit price and total price per cubic yard for this system. Prices for foundation excavation systems are on Line Items A2010 140 1600 thru 2200. Both material quantities and labor costs have been adjusted for the system listed.

Factors: To adjust for job conditions other than normal working situations use Lines A2010 140 2900 thru 4000.

Example: You are to install the system with the use of temporary shoring and bracing. Go to Line A2010 140 3900 and apply these percentages to the appropriate TOTAL costs.

System Components	QUANTITY	UNIT	COST PER C.Y. EQUIP.	COST PER C.Y. LABOR	COST PER C.Y. TOTAL
Foundation excav w/ 3/4 C.Y. backhoe-loader, incl. operator, equip Rental, fuel, oil, and mobilization. Hauling of excavated material is Included. Prices based on one day production of 360 C.Y. in medium soil Without backfilling.					
Equipment operator	8.000	Hr.		590	590
Backhoe-loader, 3/4 C.Y.	1.000	Day	193.60		193.60
Operating expense (fuel, oil)	8.000	Hr.	173.84		173.84
Hauling, 12 C.Y. trucks, 1 mile round trip	360.000	L.C.Y.	918	651.60	1,569.60
Mobilization	1.000	Ea.	259	127	386
TOTAL			1,544.44	1,368.60	2,913.04
COST PER C.Y.		C.Y.	4.28	3.79	8.09

A2010 140	Excavation, Foundation	COST PER C.Y. EQUIP.	COST PER C.Y. LABOR	COST PER C.Y. TOTAL
1600	100 C.Y.	5.65	6.40	12.05
1700	200 C.Y.	4.88	4.11	8.99
1800	300 C.Y.	4.46	3.88	8.34
1850	360 C.Y.	4.31	3.82	8.13
1900	400 C.Y.	4.21	3.75	7.96
2000	500 C.Y.	4.10	3.71	7.81
2100	600 C.Y.	3.98	3.66	7.64
2200	700 C.Y.	3.93	3.59	7.52
2300				
2400				
2900	Dust protection, add, minimum		2%	2%
3000	Maximum		11%	11%
3100	Equipment usage curtailment, add, minimum		1%	1%
3200	Maximum		10%	10%
3300	Material handling & storage limitation, add, minimum		1%	1%
3400	Maximum		7%	7%
3500	Protection of existing work, add, minimum		2%	2%
3600	Maximum		7%	7%
3700	Shift work requirements, add, minimum		5%	5%
3800	Maximum		30%	30%
3900	Temporary shoring and bracing, add, minimum		5%	5%
4000	Maximum		12%	12%

A2020 Basement Walls

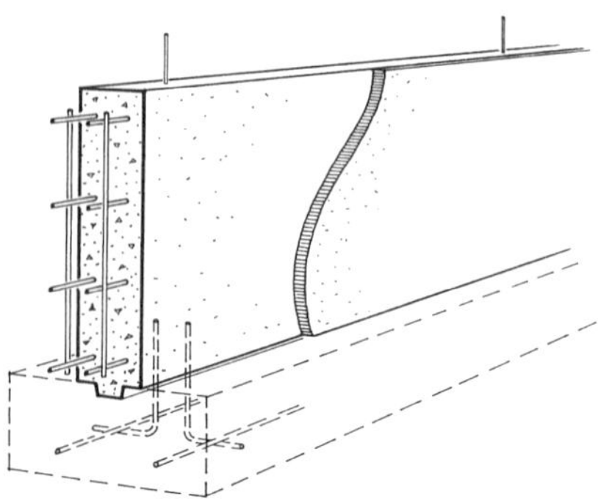

This page illustrates and describes a concrete wall system including concrete, placing concrete, forms, reinforcing, insulation, waterproofing and anchor bolts. Lines within System Components give the unit price and total price per linear foot for this system. Prices for concrete wall systems are on Line Items A2020 120 1400 thru 2600. Both material quantities and labor costs have been adjusted for the system listed.

Factors: To adjust for job conditions other than normal working situations use Lines A2020 120 2900 thru 4000.

Example: You are to install this wall system where delivery of material is difficult. Go to Line A2020 120 3600 and apply these percentages to the appropriate MAT. and INST. costs.

System Components	QUANTITY	UNIT	COST PER L.F. MAT.	INST.	TOTAL
Cast in place concrete foundation wall, 8″ thick, 3′ high, 2500 psi					
Concrete including forms, reinforcing, waterproofing, and anchor bolts.					
Concrete, 2500 psi, 8″ thick, 3′ high	.070	C.Y.	7.21		7.21
Forms, wall, 4 uses	6.000	S.F.	5.40	40.20	45.60
Reinforcing	6.000	Lb.	3.36	2.64	6
Placing concrete	.070	C.Y.		3.32	3.32
Waterproofing	3.000	S.F.	1.26	3.18	4.44
Rigid insulaton, 1″ polystyrene	3.000	S.F.	.93	2.55	3.48
Anchor bolts, 1/2″ diameter, 4′ O.C.	.250	Ea.	.42	1.09	1.51
TOTAL			18.58	52.98	71.56

A2020 120	Concrete Wall	COST PER L.F. MAT.	INST.	TOTAL
1400	8″ thick, 2500 psi concrete, 3′ high	18.60	53	71.60
1500	4′ high	25.50	71	96.50
1600	6′ high	38	105	143
1700	8′ high	50	140	190
1800	3500 psi concrete, 4′ high	26	71	97
1900	6′ high	39	105	144
2000	8′ high	51.50	140	191.50
2100	12″ thick, 2500 psi concrete, 4′ high	32.50	75	107.50
2200	6′ high	47	111	158
2300	8′ high	63	147	210
2400	3500 psi concrete, 4′ high	34	75	109
2500	8′ high	65	147	212
2600	10′ high	80.50	183	263.50
2700				
2900	Cut & patch to match existing construction, add, minimum	2%	3%	
3000	Maximum	5%	9%	
3100	Dust protection, add, minimum	1%	2%	
3200	Maximum	4%	11%	
3300	Equipment usage curtailment, add, minimum	1%	1%	
3400	Maximum	3%	10%	
3500	Material handling & storage limitation, add, minimum	1%	1%	
3600	Maximum	6%	7%	
3700	Protection of existing work, add, minimum	2%	2%	
3800	Maximum	5%	7%	
3900	Shift work requirements, add, minimum		5%	
4000	Maximum		30%	

A2020 Basement Walls

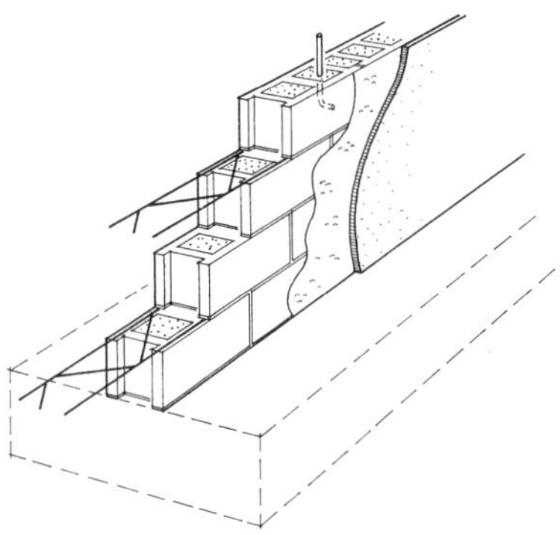

This page illustrates and describes a concrete block wall system including concrete block, masonry reinforcing, parging, waterproofing insulation and anchor bolts. Lines within System Components give the unit price and total price per linear foot for this system. Prices for concrete block wall systems are on Line Items A2020 140 1200 thru 2600. Both material quantities and labor costs have been adjusted for the system listed.

Factors: To adjust for job conditions other than normal working situations use Lines A2020 140 2900 thru 4000.

Example: You are to install the system to match an existing foundation wall. Go to Line A2020 140 3000 and apply these percentages to the appropriate MAT. and INST. costs.

System Components	QUANTITY	UNIT	COST PER L.F. MAT.	INST.	TOTAL
Concrete block, 8″ thick, masonry reinforcing, parged and Waterproofed, insulation and anchor bolts, wall 2′-8″ high.					
Concrete block, 8″ x 8″ x 16″	2.670	S.F.	8.70	16.42	25.12
Masonry reinforcing	2.000	L.F.	.26	.38	.64
Parging	.193	S.Y.	.75	2.51	3.26
Waterproofing	2.670	S.F.	1.12	2.83	3.95
Insulation, 1″ rigid polystyrene	2.670	S.F.	.83	2.27	3.10
Anchor bolts, 1/2″ diameter, 4′ O.C.	.250	Ea.	.42	1.09	1.51
TOTAL			12.08	25.50	37.58

A2020 140	Concrete Block Wall	COST PER L.F. MAT.	INST.	TOTAL
1200	8″ thick block, 2′-8″ high	12.10	25.50	37.60
1300	4′ high	17.90	37.50	55.40
1400	6′ high	26.50	56	82.50
1500	8′ high	35.50	74.50	110
1600	Grouted solid, 4′ high	22.50	50.50	73
1700	6′ high	34	75	109
1800	8′ high	45	100	145
2100	12″ thick block, 4′ high	22.50	54	76.50
2200	6′ high	33.50	81	114.50
2300	8′ high	45	107	152
2400	Grouted solid, 4′ high	30.50	68	98.50
2500	6′ high	45.50	102	147.50
2600	8′ high	60.50	135	195.50
2700				
2900	Cut & patch to match existing construction, add, minimum	2%	3%	
3000	Maximum	5%	9%	
3100	Dust protection, add, minimum	1%	2%	
3200	Maximum	4%	11%	
3300	Equipment usage curtailment, add, minimum	1%	1%	
3400	Maximum	3%	10%	
3500	Material handling & storage limitation, add, minimum	1%	1%	
3600	Maximum	6%	7%	
3700	Protection of existing work, add, minimum	2%	2%	
3800	Maximum	5%	7%	
3900	Shift work requirements, add, minimum		5%	
4000	Maximum		30%	

B1010 Floor Construction

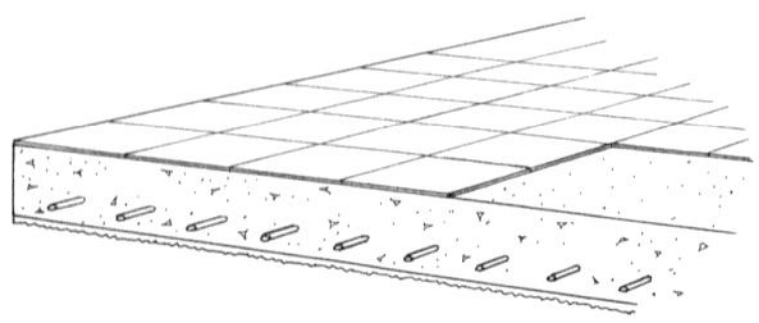

This page illustrates and describes a reinforced concrete slab system including concrete, placing concrete, formwork, reinforcing, steel trowel finish, V.A. floor tile and acoustical spray ceiling finish. Lines within System Components give the unit price per square foot for this system. Prices for reinforced concrete slab systems are on Line Items B1010 225 1400 thru 1800. Both material quantities and labor costs have been adjusted for the system listed.

Factors: To adjust for job conditions other than normal working situations use Lines B1010 225 2700 thru 4000.

Example: You are to install the system to match an existing floor system. Go to Line B1010 225 2800 and apply these percentages to the appropriate MAT. and INST. costs.

System Components	QUANTITY	UNIT	COST PER S.F. MAT.	COST PER S.F. INST.	COST PER S.F. TOTAL
Flat slab system, reinforced concrete with vinyl tile floor, sprayed Acoustical ceiling finish, not including columns.					
Concrete, 4000 psi, 6″ thick	.020	C.Y.	2.26		2.26
Placing concrete	.020	C.Y.		.68	.68
Formwork, 4 use	1.000	S.F.	1.14	6.05	7.19
Edge form	.120	S.F.	.06	1.17	1.23
Reinforcing steel	3.000	Lb.	1.68	1.86	3.54
Steel trowel finish	1.000	S.F.		1.21	1.21
Vinyl floor tile	1.000	S.F.	5.30	1.02	6.32
Acoustical spray ceiling finish	1.000	S.F.	.29	.17	.46
TOTAL			10.73	12.16	22.89

B1010 225	Floor - Ceiling, Concrete Slab	COST PER S.F. MAT.	COST PER S.F. INST.	COST PER S.F. TOTAL
1400	Concrete, 4000 psi, 6″ thick	10.75	12.15	22.90
1500	7″ thick	11.20	12.70	23.90
1600	8″ thick	11.85	13.35	25.20
1700	9″ thick	12.45	13.85	26.30
1800	10″ thick	13.10	14.55	27.65
1900				
2000				
2100				
2200				
2300				
2400				
2500				
2700	Cut & patch to match existing construction, add, minimum	2%	3%	
2800	Maximum	5%	9%	
2900	Dust protection, add, minimum	1%	2%	
3000	Maximum	4%	11%	
3100	Equipment usage curtailment, add, minimum	1%	1%	
3200	Maximum	3%	10%	
3300	Material handling & storage limitation, add, minimum	1%	1%	
3400	Maximum	6%	7%	
3500	Protection of existing work, add, minimum	2%	2%	
3600	Maximum	5%	7%	
3700	Shift work requirements, add, minimum		5%	
3800	Maximum		30%	
3900	Temporary shoring and bracing, add, minimum	2%	5%	
4000	Maximum	5%	12%	

B1010 Floor Construction

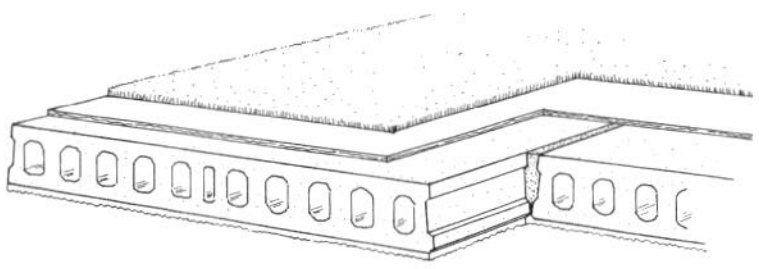

This page illustrates and describes a hollow core prestressed concrete panel system including hollow core slab, grout, carpet, carpet padding, and sprayed ceiling. Lines within System Components give the unit price and total price per square foot for this system. Prices for hollow core prestressed concrete panel systems are on Line Items B1010 232 1200 thru 1600. Both material and labor costs have been adjusted for the system listed.

Factors: To adjust for job conditions other than normal working situations use Lines B1010 232 2700 thru 3800.

Example: You are to install the system where dust control is a major concern. Go to Line B1010 232 2800 and apply these percentages to the appropriate MAT. and INST. costs.

System Components	QUANTITY	UNIT	COST PER S.F. MAT.	COST PER S.F. INST.	COST PER S.F. TOTAL
Precast hollow core plank with carpeted floors, padding					
And sprayed textured ceiling.					
Hollow core plank, 4″ thick with grout topping	1.000	S.F.	6.65	3.46	10.11
Nylon carpet, 26 oz. medium traffic	.110	S.Y.	4.40	.75	5.15
Carpet padding, minimum quality	.110	S.Y.	.51	.38	.89
Sprayed texture ceiling	1.000	S.F.	.04	.62	.66
TOTAL			11.60	5.21	16.81

B1010 232	Floor - Ceiling, Conc. Panel	COST PER S.F. MAT.	COST PER S.F. INST.	COST PER S.F. TOTAL
1200	Hollow core concrete plank, with grout, 4″ thick	11.60	5.20	16.80
1300	6″ thick	12.10	4.72	16.82
1400	8″ thick	12.80	4.35	17.15
1500	10″ thick	13.10	4.06	17.16
1600	12″ thick	13.65	3.82	17.47
1700				
1800				
1900				
2000				
2100				
2200				
2300				
2400				
2500				
2700	Dust protection, add, minimum	1%	2%	
2800	Maximum	4%	11%	
2900	Equipment usage curtailment, add, minimum	1%	1%	
3000	Maximum	3%	10%	
3100	Material handling & storage limitation, add, minimum	1%	1%	
3200	Maximum	6%	7%	
3300	Protection of existing work, add, minimum	2%	2%	
3400	Maximum	5%	7%	
3500	Shift work requirements, add, minimum		5%	
3600	Maximum		30%	
3700	Temporary shoring and bracing, add, minimum	2%	5%	
3800	Maximum	5%	12%	

B1010 Floor Construction

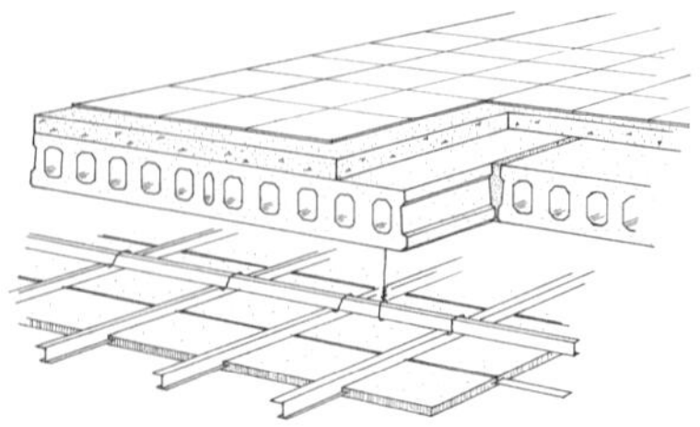

This page illustrates and describes a hollow core plank system including precast hollow core plank, concrete topping, V.A. tile, concealed suspension system and ceiling tile. Lines within System Components give the unit price and total price per square foot for this system. Prices for hollow core plank systems are on Line Items B1010 233 1400 thru 1800. Both material quantities and labor costs have been adjusted for the system listed.

Factors: To adjust for job conditions other than normal working situations use Lines B1010 233 2900 thru 4000.

Example: You are to install the system where equipment usage is a problem. Go to Line B1010 233 3200 and apply these percentages to the appropriate MAT. and INST. costs.

System Components	QUANTITY	UNIT	COST PER S.F.		
			MAT.	INST.	TOTAL
Precast hollow core plank with 2″ topping, vinyl composition floor tile And suspended acoustical tile ceiling.					
Hollow core concrete plank, 4″ thick	1.000	S.F.	6.65	3.46	10.11
Concrete topping, 2″ thick	1.000	S.F.	1.43	5.44	6.87
Vinyl composition tile, .08″ thick	1.000	S.F.	1.47	1.02	2.49
Concealed Z bar suspension system, 12″ module	1.000	S.F.	.80	1.11	1.91
Ceiling tile, mineral fiber, 3/4″ thick	1.000	S.F.	2.33	1.93	4.26
TOTAL			12.68	12.96	25.64

B1010 233	Floor - Ceiling, Conc. Plank	COST PER S.F.		
		MAT.	INST.	TOTAL
1400	Hollow core concrete plank, 4″ thick	12.70	12.95	25.65
1500	6″ thick	13.20	12.45	25.65
1600	8″ thick	13.90	12.10	26
1700	10″ thick	14.20	11.80	26
1800	12″ thick	14.75	11.55	26.30
1900				
2000				
2100				
2200				
2300				
2400				
2500				
2600				
2700				
2900	Dust protection, add, minimum	1%	2%	
3000	Maximum	4%	11%	
3100	Equipment usage curtailment, add, minimum	1%	1%	
3200	Maximum	3%	10%	
3300	Material handling & storage limitation, add, minimum	1%	1%	
3400	Maximum	6%	7%	
3500	Protection of existing work, add, minimum	2%	2%	
3600	Maximum	5%	7%	
3700	Shift work requirements, add, minimum		5%	
3800	Maximum		30%	
3900	Temporary shoring and bracing, add, minimum	2%	5%	
4000	Maximum	5%	12%	

B1010 Floor Construction

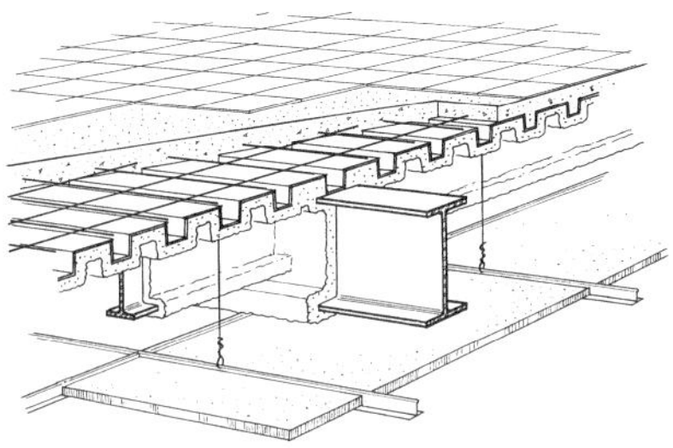

This page illustrates and describes a structural steel w/metal decking and concrete system including steel beams, steel decking, shear studs, concrete, placing concrete, edge form, steel trowel finish, curing, wire fabric, fireproofing, beams and decking, tile floor, and suspended ceiling. Lines within System Components give the unit price and total price per square foot for this system. Prices for alternate structural steel w/metal decking and concrete systems are on Line Items B1010 243 1900 thru 2400. Both material quantities and labor costs have been adjusted for the system listed.

Factors: To adjust for job conditions other than normal working situations use Lines B1010 243 2900 thru 4000.

Example: You are to install the system where material handling and storage are a problem. Go to Line B1010 243 3400 and apply these percentages to the appropriate MAT. and INST. costs.

System Components	QUANTITY	UNIT	COST PER S.F. MAT.	COST PER S.F. INST.	COST PER S.F. TOTAL
Composite structural beams, 20 ga. 3″ deep steel decking, shear studs, 3000 psi, concrete, placing concrete, edge forms, steel trowel finish, Welded wire fabric fireproofing, tile floor and suspended ceiling.					
Steel deck, 20 gage, 3″ deep, galvanized	1.000	S.F.	2.21	1.03	3.24
Structural steel framing	.004	Ton	11	2.73	13.73
Shear studs, 3/4″	.250	Ea.	.13	.46	.59
Concrete, 3000 psi, 5″ thick	.015	C.Y.	1.68		1.68
Place concrete	.015	C.Y.		.51	.51
Edge form	.140	L.F.	.03	.61	.64
Steel trowel finish	1.000	S.F.		1.21	1.21
Welded wire fabric 6 x 6 - #10/10	.011	C.S.F.	.17	.41	.58
Fireproofing, sprayed	1.880	S.F.	1.09	1.97	3.06
Vinyl composition floor tile	1.000	S.F.	1.03	1.02	2.05
Suspended acoustical ceiling	1.000	S.F.	2.47	1.52	3.99
TOTAL			19.81	11.47	31.28

B1010 243	Floor - Ceiling, Struc. Steel	COST PER S.F. MAT.	COST PER S.F. INST.	COST PER S.F. TOTAL
1900	Composite deck, galvanized, 3″ deep, 22 ga.	19.60	11.40	31
1950	20 ga.	19.80	11.45	31.25
2000	18 ga.	20.50	11.50	32
2100	16 ga.	21	11.60	32.60
2200	Composite deck, galvanized, 2″ deep, 22 ga.	19.40	11.25	30.65
2300	20 ga.	19.60	11.30	30.90
2400	16 ga.	21	11.40	32.40
2500				
2900	Dust protection, add, minimum	1%	2%	
3000	Maximum	4%	11%	
3100	Equipment usage curtailment, add, minimum	1%	1%	
3200	Maximum	3%	10%	
3300	Material handling & storage limitation, add, minimum	1%	1%	
3400	Maximum	6%	7%	
3500	Protection of existing work, add, minimum	2%	2%	
3600	Maximum	5%	7%	
3700	Shift work requirements, add, minimum		5%	
3800	Maximum		30%	
3900	Temporary shoring and bracing, add, minimum	2%	5%	
4000	Maximum	5%	12%	

B1010 Floor Construction

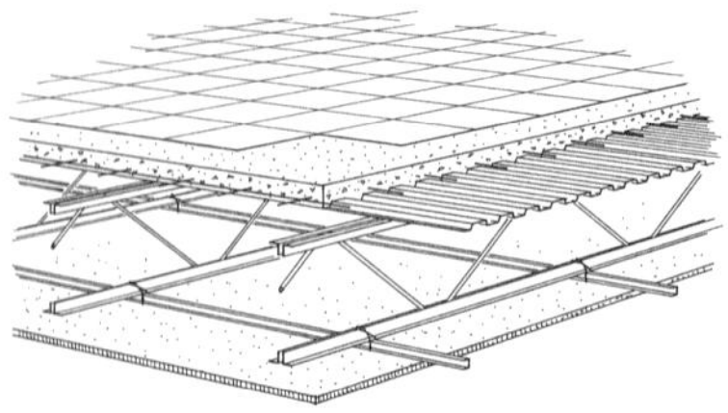

This page illustrates and describes an open web joist and steel slab-form system including open web steel joist, slab form, concrete, placing concrete, wire fabric, steel trowel finish, tile floor and plasterboard. Lines within System Components give the unit price and total price per square foot for this system. Prices for open web joists and steel slab-form systems are on Line Items B1010 245 1800 thru 2200. Both material quantities and labor costs have been adjusted for the systems listed.

Factors: To adjust for job conditions other than normal working situations use Lines B1010 245 2900 thru 4000.

Example: You are to install the system in a congested commercial area and most work will be done at night. Go to Line B1010 245 3800 and apply this percentage to the appropriate INST. cost.

System Components	QUANTITY	UNIT	COST PER S.F.		
			MAT.	INST.	TOTAL
Open web steel joists, 24″ O.C., slab form, 3000 psi concrete, welded Wire fabric, steel trowel finish, floor tile, 5/8″ drywall ceiling.					
Open web steel joists, 12″ deep, 5.2#/L.F., 24″ O.C.	2.630	Lb.	2.39	1.18	3.57
Slab form, 28 gage, 9/16″ deep, galvanized	1.000	S.F.	1.29	.77	2.06
Concrete, 3000 psi, 2-1/2″ thick	.008	C.Y.	.90		.90
Placing concrete	.008	C.Y.		.27	.27
Welded wire fabric 6 x 6 - #10/10	.011	C.S.F.	.17	.41	.58
Steel trowel finish	1.000	S.F.		1.21	1.21
Vinyl composition floor tile	1.000	S.F.	1.03	1.02	2.05
Ceiling furring, 3/4″ channels, 24″ O.C.	1.000	S.F.	.22	1.17	1.39
Gypsum drywall, 5/8″ thick, finished	1.000	S.F.	.33	1.51	1.84
Paint ceiling	1.000	S.F.	.19	.64	.83
TOTAL			6.52	8.18	14.70

B1010 245	Floor - Ceiling, Steel Joists	COST PER S.F.		
		MAT.	INST.	TOTAL
1800	Open web joists, 12″ deep, 5.2#/L.F.	6.50	8.20	14.70
1900	16″ deep, 6.6#/L.F.	7.15	8.50	15.65
2000	20″ deep, 8.4# L.F.	7.95	8.90	16.85
2100	24″ deep, 11.5# L.F.	9.35	9.60	18.95
2200	26″ deep, 12.8# L.F.	9.95	9.85	19.80
2300				
2400				
2500				
2600				
2700				
2900	Dust protection, add, minimum	1%	2%	
3000	Maximum	4%	11%	
3100	Equipment usage curtailment, add, minimum	1%	1%	
3200	Maximum	3%	10%	
3300	Material handling & storage limitation, add, minimum	1%	1%	
3400	Maximum	6%	7%	
3500	Protection of existing work, add, minimum	2%	2%	
3600	Maximum	5%	7%	
3700	Shift work requirements, add, minimum		5%	
3800	Maximum		30%	
3900	Temporary shoring and bracing, add, minimum	2%	5%	
4000	Maximum	5%	12%	

B1010 Floor Construction

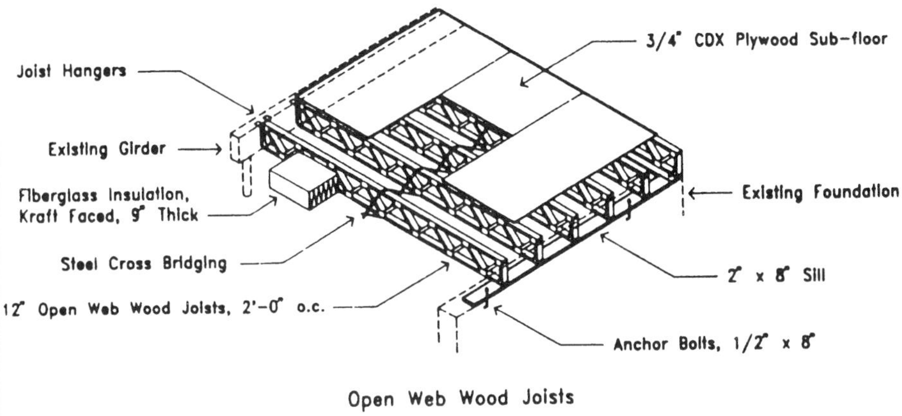

Open Web Wood Joists

This page illustrates and describes floor framing systems including sills, joist hangers, joists, bridging, subfloor and insulation. Lines within systems components give the unit price and total price per square foot for this system. Prices for alternate joist types are on line items B1010 262 0100 through 0700. Both material quantities and labor costs have been adjusted for the system listed.

Factors: To adjust for job conditions other than normal working situations use lines B1010 262 2700 through 4000.

Example: You are to install the system where delivery of material is difficult. Go to line B1010 262 3400 and apply these percentages to the appropriate MAT. and INST. costs.

System Components	QUANTITY	UNIT	COST PER S.F.		
			MAT.	INST.	TOTAL
Open web wood joists, sill, butt to girder, 5/8″ subfloor, insul.					
Anchor bolt, hook type, with nut and washer, 1/2″ x 8″ long	.250	Ea.	.04	.11	.15
Joist & beam hanger, 18 ga. galvanized	.750	Ea.	.11	.26	.37
Bridging, steel, compression, for 2″ x 12″ joists	.010	C.Pr.	.02	.03	.05
Framing, open web wood joists, 12″ deep	.001	M.L.F.	3.20	1.33	4.53
Framing, sills, 2″ x 8″	.001	M.B.F.	.62	1.73	2.35
Sub-floor, plywood, CDX, 3/4″ thick	1.000	S.F.	.68	.86	1.54
Insulation, fiberglass, kraft face, 9″ thick x 23″ wide	1.000	S.F.	.68	.43	1.11
TOTAL			5.35	4.75	10.10

B1010 262	Open Web Wood Joists	COST PER S.F.		
		MAT.	INST.	TOTAL
0100	12″ open web joists, sill, butt to girder, 5/8″ sub-floor	5.55	5	10.55
0200	14″ joists	5.80	5.10	10.90
0300	16″ joists	5.85	5.15	11
0400	18″ joists	6	5.25	11.25
0450	Wood struc "I" joists, 24″ O.C., to 24′ span, 50 PSF LL, butt to girder	4.36	4.65	9.01
0500	55 PSF LL	4.62	4.72	9.34
0600	to 30′span, 45 PSF LL	5.50	4.58	10.08
0700	55 PSF LL	6.25	4.65	10.90
2700	Cut & patch to match existing construction, add, minimum	2%	3%	
2800	Maximum	5%	9%	
2900	Dust protection, add, minimum	1%	2%	
3000	Maximum	4%	11%	
3100	Equipment usage curtailment, add, minimum	1%	1%	
3200	Maximum	3%	10%	
3300	Material handling & storage limitation, add, minimum	1%	1%	
3400	Maximum	6%	7%	
3500	Protection of existing work, add, minimum	2%	2%	
3600	Maximum	5%	7%	
3700	Shift work requirements, add, minimum		5%	
3800	Maximum		30%	
3900	Temporary shoring and bracing, add, minimum	2%	5%	
4000	Maximum	5%	12%	

B1010 Floor Construction

This page illustrates and describes a wood joist floor system including wood joist, oak floor, sub-floor, bridging, sand and finish floor, furring, plasterboard, taped, finished and painted ceiling. Lines within System Components give the unit price and total price per square foot for this system. Prices for wood joist floor systems are on Line Items B1010 263 1700 thru 2300. Both material quantities and labor costs have been adjusted for the system listed.

Factors: To adjust for job conditions other than normal working situations use Lines B1010 263 2700 thru 4000.

Example: You are to install the system during off peak hours, 6 P.M. to 2 A.M. Go to Line B1010 263 3800 and apply this percentage to the appropriate INST. costs.

System Components	QUANTITY	UNIT	COST PER S.F.		
			MAT.	INST.	TOTAL
Wood joists, 2″ x 8″, 16″ O.C.,oak floor (sanded & finished),1/2″ sub Floor, 1″x 3″ bridging, furring, 5/8″ drywall, taped, finished and painted.					
Wood joists, 2″ x 8″, 16″ O.C.	1.000	L.F.	.83	1.05	1.88
Subfloor plywood CDX 1/2″	1.000	SF Flr.	.52	.77	1.29
Bridging 1″ x 3″	.150	Pr.	.08	.67	.75
Oak flooring, No. 1 common	1.000	S.F.	4.13	3.40	7.53
Sand and finish floor	1.000	S.F.	.95	1.56	2.51
Wood furring, 1″ x 3″, 16″ O.C.	1.000	L.F.	.38	1.65	2.03
Gypsum drywall, 5/8″ thick	1.000	S.F.	.28	.64	.92
Tape and finishing	1.000	S.F.	.05	.58	.63
Paint ceiling	1.000	S.F.	.19	.64	.83
TOTAL			7.41	10.96	18.37

B1010 263	Floor-Ceiling, Wood Joists	COST PER S.F.		
		MAT.	INST.	TOTAL
1700	16″ on center, 2″ x 6″ joists	7.15	10.85	18
1750	2″ x 8″	7.40	10.95	18.35
1800	2″ x 10″	7.70	11.20	18.90
1900	2″ x 12″	8	11.25	19.25
2000	2″ x 14″	8.45	11.40	19.85
2100	12″ on center, 2″ x 10″ joists	7.95	11.50	19.45
2200	2″ x 12″	8.35	11.55	19.90
2300	2″ x 14″	8.90	11.80	20.70
2400				
2500				
2700	Cut & patch to match existing construction, add, minimum	2%	3%	
2800	Maximum	5%	9%	
2900	Dust protection, add, minimum	1%	2%	
3000	Maximum	4%	11%	
3100	Equipment usage curtailment, add, minimum	1%	1%	
3200	Maximum	3%	10%	
3300	Material handling & storage limitation, add, minimum	1%	1%	
3400	Maximum	6%	7%	
3500	Protection of existing work, add, minimum	2%	2%	
3600	Maximum	5%	7%	
3700	Shift work requirements, add, minimum		5%	
3800	Maximum		30%	
3900	Temporary shoring and bracing, add, minimum	2%	5%	
4000	Maximum	5%	12%	

B1020 Roof Coverings

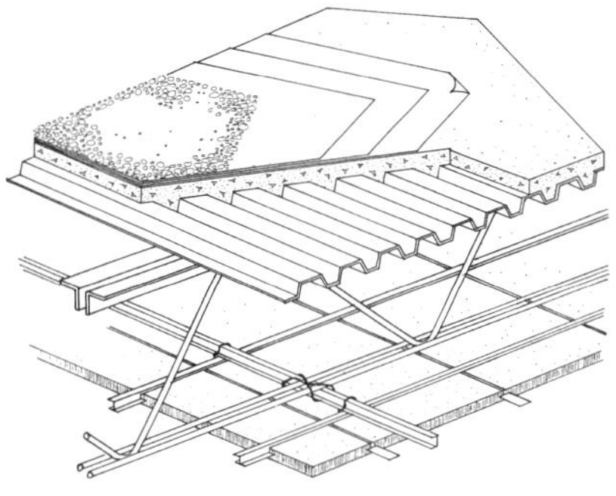

This page illustrates and describes a flat roof bar joist system including asphalt and gravel roof, lightweight concrete, metal decking, open web joists, and suspended acoustic ceiling. Lines within System Components give the unit price and total price per square foot for this system. Prices for flat roof bar joist systems are on Line Items B1020 106 1200 thru 1600. Both material quantities and labor costs have been adjusted for the system listed.

Factors: To adjust for job conditions other than normal working situations use Lines B1020 106 2700 thru 4000.

Example: You are to install the system where material handling and storage present some problem. Go to Line B1020 106 3300 and apply these percentages to the appropriate MAT. and INST. costs.

System Components	QUANTITY	UNIT	COST PER S.F.		
			MAT.	INST.	TOTAL
Three ply asphalt and gravel roof on lightweight concrete, metal decking on Web joist 4′ O.C., with suspended acoustic ceiling.					
Open web joist, 10″ deep, 4′ O.C.	1.250	Lb.	1.14	.57	1.71
Metal decking, 1-1/2″ deep, 22 Ga., galvanized	1.000	S.F.	1.91	.69	2.60
Gravel roofing, 3 ply asphalt	.010	C.S.F.	1.26	1.81	3.07
Perlite roof fill, 3″ thick	1.000	S.F.	2.25	.56	2.81
Suspended ceiling, 3/4″ mineral fiber on "Z" bar suspension	1.000	S.F.	2.67	3.86	6.53
TOTAL			9.23	7.49	16.72

B1020 106	Steel Joist Roof & Ceiling	COST PER S.F.		
		MAT.	INST.	TOTAL
1200	Open web bar joist, 10″ deep, 5#/L.F.	9.25	7.50	16.75
1300	16″ deep, 5#/L.F.	9.60	7.70	17.30
1400	18″ deep, 6.6#/L.F.	9.90	7.80	17.70
1500	20″ deep, 9.6#/L.F.	10.25	8	18.25
1600	24″ deep, 11.52#/L.F.	10.70	8.20	18.90
1700				
1800				
1900				
2000				
2100				
2200				
2300				
2400				
2500				
2700	Cut & patch to match existing construction, add, minimum	2%	3%	
2800	Maximum	5%	9%	
2900	Dust protection, add, minimum	1%	2%	
3000	Maximum	4%	11%	
3100	Equipment usage curtailment, add, minimum	1%	1%	
3200	Maximum	3%	10%	
3300	Material handling & storage limitation, add, minimum	1%	1%	
3400	Maximum	6%	7%	
3500	Protection of existing work, add, minimum	2%	2%	
3600	Maximum	5%	7%	
3700	Shift work requirements, add, minimum		5%	
3800	Maximum		30%	
3900	Temporary shoring and bracing, add, minimum	2%	5%	
4000	Maximum	5%	12%	

B1020 Roof Coverings

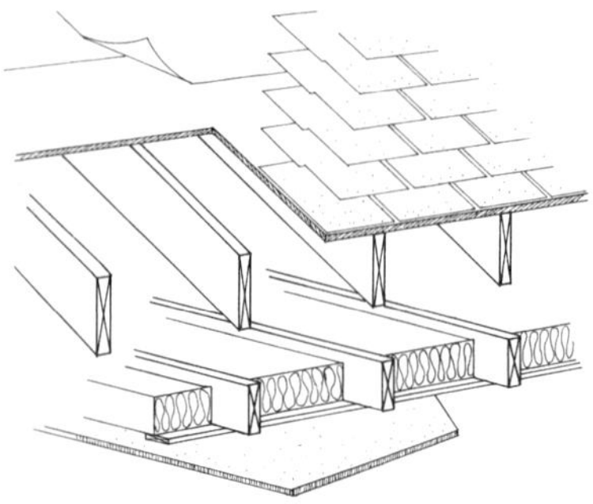

This page illustrates and describes a wood frame roof system including rafters, ceiling joists, sheathing, building paper, asphalt shingles, roof trim, furring, insulation, plaster and paint. Lines within System Components give the unit price and total price per square foot for this system. Prices for wood frame roof systems are on Line Items B1020 202 1800 thru 2700. Both material quantities and labor costs have been adjusted for the system listed.

Factors: To adjust for job conditions other than normal working situations use Lines B1020 202 3300 thru 4000.

Example: You are to install the system while protecting existing work. Go to Line B1020 202 3800 and apply these percentages to the appropriate MAT. and INST. costs.

System Components	QUANTITY	UNIT	COST PER S.F. MAT.	COST PER S.F. INST.	COST PER S.F. TOTAL
Wood frame roof system, 4 in 12 pitch, including rafters, sheathing, Shingles, insulation, drywall, thin coat plaster, and painting.					
Rafters, 2″ x 6″, 16″ O.C., 4 in 12 pitch	1.080	L.F.	.64	1.25	1.89
Ceiling joists, 2″ x 6″, 16 O.C.	1.000	L.F.	.59	.93	1.52
Sheathing, 1/2″ CDX	1.080	S.F.	.56	.90	1.46
Building paper, 15# felt	.011	C.S.F.	.07	.17	.24
Asphalt shingles, 240#	.011	C.S.F.	1	1.17	2.17
Roof trim	.100	L.F.	.16	.26	.42
Furring, 1″ x 3″, 16″ O.C.	1.000	L.F.	.38	1.65	2.03
Fiberglass insulation, 6″ batts	1.000	S.F.	.44	.43	.87
Gypsum board, 1/2″ thick	1.000	S.F.	.24	.64	.88
Thin coat plaster	1.000	S.F.	.11	.72	.83
Paint, roller, 2 coats	1.000	S.F.	.19	.64	.83
TOTAL			4.38	8.76	13.14

B1020 202	Wood Frame Roof & Ceiling	COST PER S.F. MAT.	COST PER S.F. INST.	COST PER S.F. TOTAL
1800	Rafters 16″ O.C., 2″ x 6″	4.38	8.75	13.13
1900	2″ x 8″	4.64	8.85	13.49
2000	2″ x 10″	4.94	9.50	14.44
2100	2″ x 12″	5.25	9.70	14.95
2200	Rafters 24″ O.C., 2″ x 6″	4.06	8.20	12.26
2300	2″ x 8″	4.25	8.25	12.50
2400	2″ x 10″	4.48	8.75	13.23
2500	2″ x 12″	4.71	8.90	13.61
2600	Roof pitch, 6 in 12, add	3%	10%	
2700	8 in 12, add	5%	12%	
2800				
2900				
3000				
3100				
3300	Cut & patch to match existing construction, add, minimum	2%	3%	
3400	Maximum	5%	9%	
3500	Material handling & storage limitation, add, minimum	1%	1%	
3600	Maximum	6%	7%	
3700	Protection of existing work, add, minimum	2%	2%	
3800	Maximum	5%	7%	
3900	Shift work requirements, add, minimum		5%	
4000	Maximum		30%	

B1020 Roof Construction

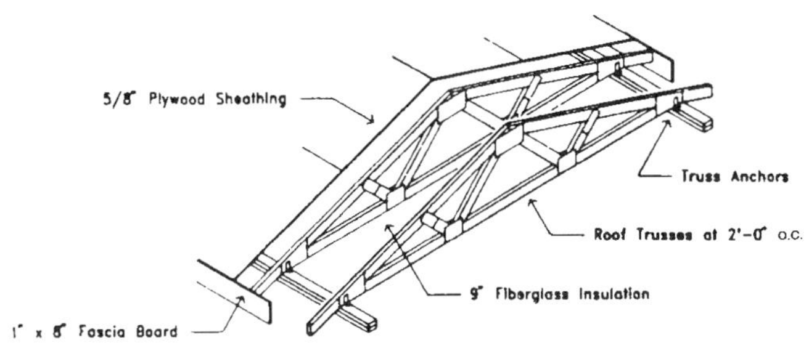

Wood Fabricated Roof Truss System

This page illustrates and describes a sloped roof truss framing system including trusses, truss anchors, sheathing, fascia and insulation. Lines B1020 210 1500 through 1800 are for a flat roof framing system including flat roof trusses, truss anchors, sheathing, fascia and insulation. Lines within system components give the unit price and total price per square foot for this system. Prices for alternate sizes and systems are given in line items B1020 210 0100 through 1800. Both material quantities and labor costs have been adjusted for the system listed.

Factors: To adjust for job conditions other than normal working situations use lines B1020 210 2700 through 4000.

Example: You are to install the system where crane placement and movement will be difficult. Go to line B1020 210 3200 and apply these percentages to the appropriate MAT. and INST. costs.

System Components	QUANTITY	UNIT	COST PER S.F. MAT.	COST PER S.F. INST.	COST PER S.F. TOTAL
Roof truss, 2' O.C., 4/12 slope, 12' span, 5/8 sheath., fascia, insul.					
Timber connectors, rafter anchors, galv., 1 1/2" x 5 1/4"	.084	Ea.	.04	.34	.38
Plywood sheathing, CDX, 5/8" thick	1.054	S.F.	.72	.94	1.66
Truss, 2' O.C., metal plate connectors, 12' span	.042	Ea.	2	1.79	3.79
Molding, fascia trim, 1" x 8"	.167	L.F.	.26	.43	.69
Insulation, fiberglass, kraft face, 9" thick x 23" wide	1.000	S.F.	.68	.43	1.11
TOTAL			3.70	3.93	7.63

B1020 210	Roof Truss, Wood	COST PER S.F. MAT.	COST PER S.F. INST.	COST PER S.F. TOTAL
0100	Roof truss, 2'O.C., 4/12p, 1'ovhg, 12'span, 5/8"sheath, fascia, insul.	3.70	3.93	7.63
0200	20' span	3.48	3.33	6.81
0300	24' span	3.34	3.10	6.44
0400	26' span	4.90	4.15	9.05
0500	28' span	3.39	3.03	6.42
0600	30' span	3.20	2.99	6.19
0700	32' span	3.44	2.95	6.39
0800	34' span	3.45	2.91	6.36
0890	60' span	4.68	3.30	7.98
0900	8/12 pitch, 20' span	3.80	3.58	7.38
1000	24' span	3.92	3.35	7.27
1100	26' span	3.77	3.27	7.04
1200	28' span	3.86	3.25	7.11
1300	32' span	3.99	3.19	7.18
1400	36' span	4.19	3.13	7.32
1500	Flat roof frame, fab.strl.joists, 2' O.C., 15'-24' span, 50 PSF LL	3.58	2.65	6.23
1600	55 PSF LL	3.84	2.72	6.56
1700	24'-30'span, 45 PSF LL	4.68	2.54	7.22
1800	55 PSF LL	5.45	2.61	8.06
1900				
2700	Cut & patch to match existing construction, add, minimum	2%	3%	
2800	Maximum	5%	9%	
2900	Dust protection, add, minimum	1%	2%	
3000	Maximum	4%	11%	
3100	Equipment usage curtailment, add, minimum	1%	1%	
3200	Maximum	3%	10%	

B1020 Roof Construction

B1020 210	Roof Truss, Wood	COST PER S.F.		
		MAT.	INST.	TOTAL
3300	Material handling & storage limitation, add, minimum	1%	1%	
3400	Maximum	6%	7%	
3500	Protection of existing work, add, minimum	2%	2%	
3600	Maximum	5%	7%	
3700	Shift work requirements, add, minimum		5%	
3800	Maximum		30%	
3900	Temporary shoring and bracing, add, minimum	2%	5%	
4000	Maximum	5%	12%	

B2010 Exterior Walls

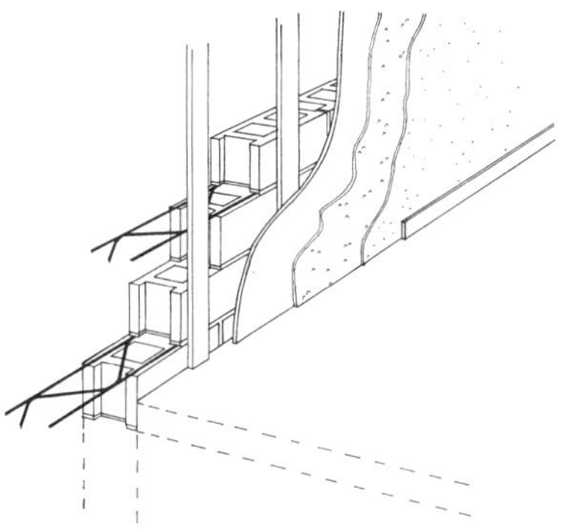

This page illustrates and describes a masonry concrete block system including concrete block wall, pointed, reinforcing, waterproofing, gypsum plaster on gypsum lath, and furring. Lines within System Components give the unit price and total price per square foot for this system. Prices for masonry concrete block wall systems are on Line Items B2010 108 1400 thru 2200. Both material quantities and labor costs have been adjusted for the system listed.

Factors: To adjust for job conditions other than normal working situations use Lines B2010 108 2700 thru 4000.

Example: You are to install the system and match existing construction at several locations. Go to Line B2010 108 2800 and apply these percentages to the appropriate MAT. and INST. costs.

System Components	QUANTITY	UNIT	COST PER S.F.		
			MAT.	INST.	TOTAL
Concrete block, 8″ thick, reinforced every 2 courses, waterproofing gypsum Plaster over gypsum lath on 1″ x 3″ furring, interior painting & baseboard.					
Concrete block, 8″ x 8″ x 16″, reinforced	1.000	S.F.	4.57	6.30	10.87
Silicone waterproofing, 2 coats	1.000	S.F.	.73	.18	.91
Bituminous coating, 1/16″ thick	1.000	S.F.	.50	1.06	1.56
Furring, 1″ x 3″, 16″ O.C.	1.000	L.F.	.38	2.22	2.60
Gypsum lath, 3/8″ thick	.110	S.Y.	.47	.64	1.11
Gypsum plaster, 2 coats	.110	S.Y.	.43	2.74	3.17
Painting, 2 coats	1.000	S.F.	.12	.50	.62
Baseboard wood, 9/16″ x 2-5/8″	.100	L.F.	.14	.24	.38
Paint baseboard, primer + 1 coat enamel	.100	L.F.	.01	.06	.07
TOTAL			7.35	13.94	21.29

B2010 108	Masonry Wall, Concrete Block	COST PER S.F.		
		MAT.	INST.	TOTAL
1400	8″ thick block, regular	7.35	13.95	21.30
1500	Fluted 2 sides	10.05	16.40	26.45
1600	Deep grooved	7.95	16.40	24.35
1700	Slump block	8.35	14.25	22.60
1800	Split rib	7.95	16.40	24.35
1900				
2000	12″ thick block, regular	6.60	16.70	23.30
2100	Slump block	14.05	15.55	29.60
2200	Split rib	8.80	18.85	27.65
2300				
2700	Cut & patch to match existing construction, add, minimum	2%	3%	
2800	Maximum	5%	9%	
2900	Dust protection, add, minimum	1%	2%	
3000	Maximum	4%	11%	
3100	Equipment usage curtailment, add, minimum	1%	1%	
3200	Maximum	3%	10%	
3300	Material handling & storage limitation, add, minimum	1%	1%	
3400	Maximum	6%	7%	
3500	Protection of existing work, add, minimum	2%	2%	
3600	Maximum	5%	7%	
3700	Shift work requirements, add, minimum		5%	
3800	Maximum		30%	
3900	Temporary shoring and bracing, add, minimum	2%	5%	
4000	Maximum	5%	12%	

B2010 Exterior Walls

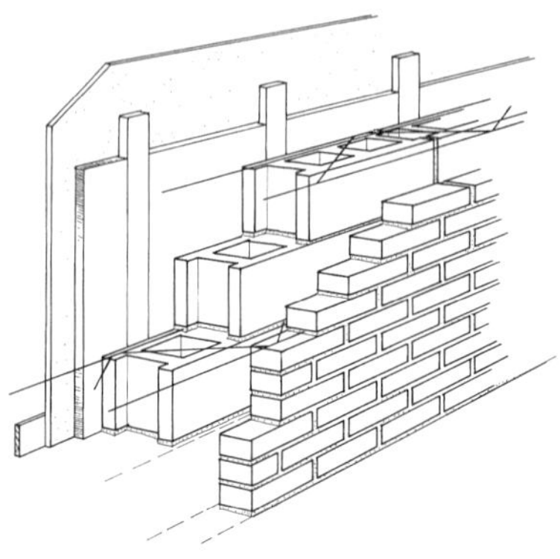

This page illustrates and describes a masonry wall, brick-stone system including brick, concrete block, durawall, insulation, plasterboard, taped and finished, furring, baseboard and painting interior. Lines within System Components give the unit price and total price per square foot for this system. Prices for masonry wall, brick-stone systems are on Line Item B2010 129 1400 thru 2500. Both material quantities and labor costs have been adjusted for the system listed.

Factors: To adjust for job conditions other than normal working situations use Lines B2010 129 3100 thru 4200.

Example: You are to install the system without damaging the existing work. Go to Line B2010 129 3900 and apply these percentages to the appropriate MAT. and INST. costs.

System Components	QUANTITY	UNIT	COST PER S.F.		
			MAT.	INST.	TOTAL
Face brick, 4″thick, concrete block back-up, reinforce every second course, 3/4″insulation, furring, 1/2″drywall, taped, finish, and painted, baseboard					
Face brick, 4″ brick	1.000	S.F.	4.61	11.90	16.51
Concrete back-up block, reinforced 8″ thick	1.000	S.F.	2.91	6.65	9.56
3/4″ rigid polystyrene insulation	1.000	S.F.	.55	.72	1.27
Furring, 1″ x 3″, wood, 16″ O.C.	1.000	L.F.	.38	1.17	1.55
Drywall, 1/2″ thick	1.000	S.F.	.24	.58	.82
Taping & finishing	1.000	S.F.	.05	.58	.63
Painting, 2 coats	1.000	S.F.	.19	.64	.83
Baseboard, wood, 9/16″ x 2-5/8″	.100	L.F.	.14	.24	.38
Paint baseboard, primer + 1 coat enamel	.100	L.F.	.01	.06	.07
TOTAL			9.08	22.54	31.62

B2010 131	Masonry Wall, Brick - Stone	COST PER S.F.		
		MAT.	INST.	TOTAL
1400	Face brick, standard, red, 4″ x 2-2/3″ x 8″ (6.75/S.F.)	9.10	22.50	31.60
1500	Norman, 4″ x 2-2/3″ x 12″ (4.5 per S.F.)	10.75	18.85	29.60
1600	Roman, 4″ x 2″ x 12″ (6.0 per S.F.)	10.80	21	31.80
1700	Engineer, 4″ x 3-1/5″ x 8″ (5.63 per S.F.)	8.35	20.50	28.85
1800	S.C.R., 6″ x 2-2/3″ x 12″ (4.5 per S.F.)	10.20	19.10	29.30
1900	Jumbo, 6″ x 4″ x 12″ (3.0 per S.F.)	9.85	16.65	26.50
2000	Norwegian, 6″ x 3-1/5″ x 12″ (3.75 per S.F.)	9.05	17.65	26.70
2100				
2200				
2300	Stone, veneer, fieldstone, 6″ thick	10.90	19.20	30.10
2400	Marble, 2″ thick	50	33.50	83.50
2500	Limestone, 2″ thick	43	20.50	63.50
3100	Cut & patch to match existing construction, add, minimum	2%	3%	
3200	Maximum	5%	9%	
3300	Dust protection, add, minimum	1%	2%	
3400	Maximum	4%	11%	
3500	Equipment usage curtailment, add, minimum	1%	1%	
3600	Maximum	3%	10%	
3700	Material handling & storage limitation, add, minimum	1%	1%	
3800	Maximum	6%	7%	
3900	Protection of existing work, add, minimum	2%	2%	
4000	Maximum	5%	7%	
4100	Shift work requirements, add, minimum		5%	
4200	Maximum		30%	

B2010 Exterior Walls

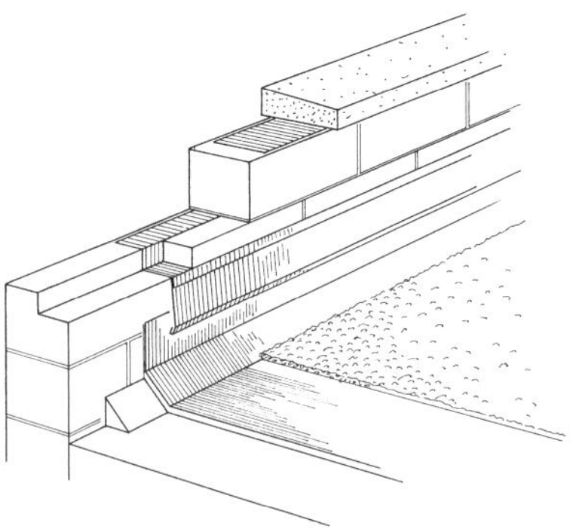

This page illustrates and describes a parapet wall system including wall, coping, flashing and cant strip. Lines within System Components give the unit price and cost per lineal foot for this system. Prices for parapet wall systems are on Lines B2010 142 1300 thru 2400. Both material quantities and labor costs have been adjusted for the system listed.

Factors: To adjust for job conditions other than normal working situations, use Lines B2010 142 2900 thru 4000.

Example: You are to install the system without damaging the adjacent property. Go to Line B2010 142 3600 and apply these percentages to the appropriate MAT. and INST. costs.

System Components	QUANTITY	UNIT	COST PER L.F. MAT.	INST.	TOTAL
Concrete block parapet, incl. reinf., coping, flashing & cant strip, 2'high					
8" concrete block	2.000	S.F.	9.14	12.60	21.74
Masonry reinforcing	1.000	L.F.	.13	.19	.32
Coping, precast	1.000	L.F.	21.50	14.65	36.15
Roof cant	1.000	L.F.	2.48	1.64	4.12
Through wall flashing	1.000	L.F.	3.16	2.57	5.73
Cap flashing	1.000	L.F.	8.65	4.62	13.27
TOTAL			45.06	36.27	81.33

B2010 142	Parapet Wall	COST PER L.F. MAT.	INST.	TOTAL
1300	Concrete block, 8" thick	45	36.50	81.50
1400	12" thick	52.50	47.50	100
1500	Split rib block, 8" thick	46	41	87
1600	12" thick	51.50	48.50	100
1700	Brick, common brick, 8" wall	49.50	62.50	112
1800	12" wall	56.50	78.50	135
1900	4" brick, 4" backup block	49	59.50	108.50
2000	8" backup block	51	61.50	112.50
2100	Stucco on masonry	45.50	40.50	86
2200	On wood frame	18.85	33	51.85
2300	Wood, T 1-11 siding	29	40	69
2400	Boards, 1" x 6" cedar	32	44	76
2500				
2600				
2900	Dust protection, add, minimum	1%	2%	
3000	Maximum	4%	11%	
3100	Equipment usage curtailment, add, minimum	1%	1%	
3200	Maximum	3%	10%	
3300	Material handling & storage limitation, add, minimum	1%	1%	
3400	Maximum	6%	7%	
3500	Protection of existing work, add, minimum	2%	2%	
3600	Maximum	5%	7%	
3700	Shift work requirements, add, minimum		5%	
3800	Maximum		30%	
3900	Temporary shoring and bracing, add, minimum	2%	5%	
4000	Maximum	5%	12%	

B2010 Exterior Walls

This page illustrates and describes a masonry cleaning and restoration system, including staging, cleaning, repointing. Lines within System Components give the unit price and cost per square foot for this system. Prices for systems are on Lines B2010 143 1300 thru 2500. Both material quantities and labor costs have been adjusted for the system listed.

Factors: To adjust for conditions other than normal working situations, use Lines B2010 143 2900 thru 4200.

Example: You are to clean a wall and be concerned about dust control. Go to line B2010 143 3100 and apply these percentages to the appropriate MAT. and INST. costs.

System Components	QUANTITY	UNIT	COST PER S.F. MAT.	COST PER S.F. INST.	COST PER S.F. TOTAL
Repoint existing building, brick, common bond, high pressure cleaning, Water only, soft old mortar.					
Scaffolding,steel tubular,bldg ext wall face, labor only, 1 to 5 stories	.010	C.S.F.		2.17	2.17
Cleaning, high pressure water only	1.100	S.F.		1.55	1.55
Repoint, brick common bond	1.000	S.F.	.58	5.90	6.48
TOTAL			.58	9.62	10.20

B2010 143	Masonry Restoration - Cleaning	COST PER S.F. MAT.	COST PER S.F. INST.	COST PER S.F. TOTAL
1300	Brick, common bond	.58	9.60	10.18
1400	Flemish bond	.61	10	10.61
1500	English bond	.61	10.65	11.26
1700	Stone, 2′ x 2′ blocks	.77	7.25	8.02
1800	2′ x 4′ blocks	.58	6.40	6.98
2000	Add to above prices for alternate cleaning systems:			
2100	Chemical brush and wash	.05	.64	.69
2200	High pressure chemical and water	.07	.53	.60
2300	Sandblasting, wet system	.62	1.53	2.15
2400	Dry system	.57	.89	1.46
2500	Steam cleaning		.43	.43
2600				
2900	Cut & patch to match existing construction, add, minimum	2%	3%	
3000	Maximum	5%	9%	
3100	Dust protection, add, minimum	1%	2%	
3200	Maximum	4%	11%	
3300	Equipment usage curtailment, add, minimum	1%	1%	
3400	Maximum	3%	10%	
3500	Material handling & storage limitation, add, minimum	1%	1%	
3600	Maximum	6%	7%	
3900	Protection of existing work, add, minimum	2%	2%	
4000	Maximum	5%	7%	
4100	Shift work requirements, add, minimum		5%	
4200	Maximum		30%	

B2010 Exterior Walls

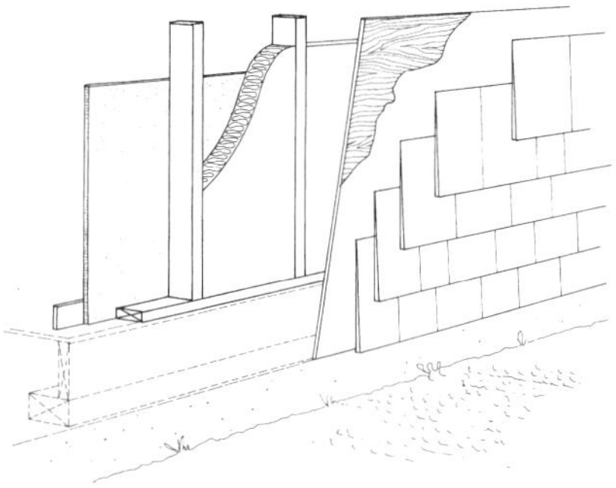

This page illustrates and describes a wood frame exterior wall system including wood studs, sheathing, felt, insulation, plasterboard, taped and finished, baseboard and painted interior. Lines within System Components give the unit price and total price per square foot for this system. Prices for wood frame exterior wall systems are on Line Items B2010 150 1600 thru 2700. Both material quantities and labor costs have been adjusted for the system listed.

Factors: To adjust for job conditions other than normal working situations use Lines B2010 150 3100 thru 4200.

Example: You are to install the system with need for complete temporary bracing. Go to Line B2010 150 4200 and apply these percentages to the appropriate MAT. and INST. costs.

System Components	QUANTITY	UNIT	COST PER S.F. MAT.	COST PER S.F. INST.	COST PER S.F. TOTAL
Wood stud wall, cedar shingle siding, building paper, plywood sheathing, Insulation, 5/8″ drywall, taped, finished and painted, baseboard.					
2″ x 4″ wood studs, 16″ O.C.	.100	L.F.	.42	1.16	1.58
1/2″ CDX sheathing	1.000	S.F.	.52	.83	1.35
18″ No. 1 red cedar shingles, 7-1/2″ exposure	.008	C.S.F.	1.47	2.06	3.53
15# felt paper	.010	C.S.F.	.06	.16	.22
3-1/2″ fiberglass insulation	1.000	S.F.	.37	.43	.80
5/8″ drywall	1.000	S.F.	.28	.58	.86
Drywall finishing adder	1.000	S.F.	.05	.58	.63
Baseboard trim, stock pine, 9/16″ x 3-1/2″, painted	.100	L.F.	.14	.24	.38
Paint baseboard, primer + 1 coat enamel	.100	L.F.	.01	.06	.07
Paint, 2 coats, interior	1.000	S.F.	.19	.64	.83
TOTAL			3.51	6.74	10.25

B2010 150	Wood Frame Exterior Wall	COST PER S.F. MAT.	COST PER S.F. INST.	COST PER S.F. TOTAL
1600	Cedar shingle siding, painted	3.51	6.75	10.26
1700	Aluminum siding, horizontal clapboard	4.47	6.95	11.42
1800	Cedar bevel siding, 1/2″ x 6″, vertical , painted	6.85	6.65	13.50
1900	Redwood siding 1″ x 4″ to 1″ x 6″ vertical, T & G	6.30	7.30	13.60
2000	Board and batten	6.45	6.05	12.50
2100	Ship lap siding	5.30	6.45	11.75
2200	Plywood, grooved (T1-11) fir	3.21	6.40	9.61
2300	Redwood	4.71	6.40	11.11
2400	Southern yellow pine	3.35	6.40	9.75
2500	Masonry on stud wall, stucco, wire and plaster	2.74	10.10	12.84
2600	Stone veneer	8.45	13.25	21.70
2700	Brick veneer, brick $275 per M	6.65	16.60	23.25
3100	Cut & patch to match existing construction, add, minimum	2%	3%	
3200	Maximum	5%	9%	
3300	Dust protection, add, minimum	1%	2%	
3400	Maximum	4%	11%	
3500	Material handling & storage limitation, add, minimum	1%	1%	
3600	Maximum	6%	7%	
3700	Protection of existing work, add, minimum	2%	2%	
3800	Maximum	5%	7%	
3900	Shift work requirements, add, minimum		5%	
4000	Maximum		30%	
4100	Temporary shoring and bracing, add, minimum	2%	5%	
4200	Maximum	5%	12%	

B2010 Exterior Walls

B2010 190	Selective Price Sheet	COST PER S.F. MAT.	INST.	TOTAL
0100	Exterior surface, masonry, concrete block, standard 4" thick	1.63	6.10	7.73
0200	6" thick	2.31	6.55	8.86
0300	8" thick	2.64	7	9.64
0400	12" thick	3.82	9.05	12.87
0500	Split rib, 4" thick	3.84	7.60	11.44
0600	8" thick	5.15	8.75	13.90
0700	Brick running bond, standard size, 6.75/S.F.	4.61	11.90	16.51
0800	Buff, 6.75/S.F.	4.87	11.90	16.77
0900	Stucco, on frame	.70	5.45	6.15
1000	On masonry	.28	4.29	4.57
1100	Metal, aluminum, horizontal, plain	2.43	2.25	4.68
1200	Insulated	2.60	2.25	4.85
1300	Vertical, plain	1.85	2.25	4.10
1400	Insulated	1.93	2.35	4.28
1500	Wood, beveled siding, "A" grade cedar, 1/2" x 6"	4.79	1.96	6.75
1600	1/2" x 8"	6.95	1.75	8.70
1700	Shingles, 16" #1 red, 7-1/2" exposure	1.53	2.82	4.35
1800	18" perfections, 7-1/2" exposure	2.10	2.36	4.46
1900	Handsplit, 10" exposure	2.56	2.31	4.87
2000	White cedar, 7-1/2" exposure	1.21	2.89	4.10
2100	Vertical, board & batten, redwood	4.24	2.63	6.87
2200	White pine	2.20	1.75	3.95
2300	T. & G. boards, redwood, 1" x 4"	2.79	2.63	5.42
2400	1' x 8"	4.51	1.75	6.26
2500				
2600	Interior surface, drywall, taped & finished, standard, 1/2"	.29	1.20	1.49
2700	5/8" thick	.33	1.20	1.53
2800	Fire resistant, 1/2" thick	.34	1.20	1.54
2900	5/8" thick	.35	1.20	1.55
3000	Moisture resistant, 1/2" thick	.44	1.20	1.64
3100	5/8" thick	.52	1.20	1.72
3200	Core board, 1" thick	.59	2.41	3
3300	Plaster, gypsum, 2 coats	.44	2.76	3.20
3400	3 coats	.63	3.34	3.97
3500	Perlite or vermiculite, 2 coats	.46	3.11	3.57
3600	3 coats	.77	3.94	4.71
3700	Gypsum lath, standard, 3/8" thick	.47	.64	1.11
3800	1/2" thick	.55	.68	1.23
3900	Fire resistant, 3/8" thick	.28	.78	1.06
4000	1/2" thick	.30	.84	1.14
4100	Metal lath, diamond, 2.5 lb.	.42	.64	1.06
4200	Rib, 3.4 lb.	.59	.78	1.37
4300	Framing metal studs including top and bottom			
4400	Runners, walls 10' high			
4500	24" O.C., non load bearing 20 gauge, 2-1/2" wide	.39	.96	1.35
4600	3-5/8" wide	.45	.98	1.43
4700	4" wide	.53	.98	1.51
4800	6" wide	.63	1	1.63
4900	Load bearing 18 gauge, 2-1/2" wide	.73	1.08	1.81
5000	3-5/8" wide	.87	1.10	1.97
5100	4" wide	.98	1.41	2.39
5200	6" wide	1.16	1.15	2.31
5300	16" O.C., non load bearing 20 gauge, 2-1/2" wide	.53	1.48	2.01
5400	3-5/8" wide	.60	1.50	2.10
5500	4" wide	.72	1.52	2.24
5600	6" wide	.86	1.54	2.40
5700	Load bearing 18 gauge, 2-1/2" wide	.99	1.50	2.49
5800	3-5/8" wide	1.18	1.52	2.70

B2010 Exterior Walls

B2010 190	Selective Price Sheet	COST PER S.F.		
		MAT.	INST.	TOTAL
5900	4″ wide	1.23	1.57	2.80
6000	6″ wide	1.56	1.59	3.15
6100	Framing wood studs incl. double top plate and			
6200	Single bottom plate, walls 10′ high			
6300	24″ O.C., 2″ x 4″	.32	.93	1.25
6400	2″ x 6″	.52	1.01	1.53
6500	16″ O.C., 2″ x 4″	.42	1.16	1.58
6600	2″ x 6″	.68	1.29	1.97
6700	Sheathing, boards, 1″ x 6″	1.73	1.78	3.51
6800	1″x 8″	1.76	1.51	3.27
6900	Plywood, 3/8″ thick	.52	.96	1.48
7000	1/2″ thick	.52	1.03	1.55
7100	5/8″ thick	.68	1.10	1.78
7200	3/4″ thick	.77	1.19	1.96
7300	Wood fiber, 5/8″ thick	.78	.96	1.74
7400	Gypsum weatherproof, 1/2″ thick	.42	1.10	1.52
7500	Insulation, fiberglass batts, 3-1/2″ thick, R13	.42	.83	1.25
7600	6″ thick, R19	.47	.96	1.43
7700	Poured 4″ thick, fiberglass wool, R4/inch	.12	2.89	3.01
7800	Mineral wool, R3/inch	.43	2.89	3.32
7900	Polystyrene, R4/inch	3.40	2.89	6.29
8000	Perlite or vermiculite, R2.7/inch	5.50	2.89	8.39
8100	Rigid, fiberglass, R4.3/inch, 1″ thick	.53	.58	1.11
8200	R8.7/inch, 2″ thick	1.06	.65	1.71

B2020 Exterior Windows

This page illustrates and describes an aluminum window system including double hung aluminum window, exterior and interior trim, hardware and insulating glass. Lines within System Components give the unit price and total price on a cost each basis for this system. Prices for aluminum window systems are on Line Items B2020 110 1000 thru 2400. Both material quantities and labor costs have been adjusted for the system listed.

Factors: To adjust for job conditions other than normal working situations use Lines B2020 110 3100 thru 4000.

Example: You are to install the system and cut and patch to match existing construction. Go to Line B2020 110 3200 and apply these percentages to the appropriate MAT. and INST. costs.

System Components	QUANTITY	UNIT	COST EACH		
			MAT.	INST.	TOTAL
Double hung aluminum window, 2′-4″ x 2′-6″ exterior and interior trim, Hardware glazed with insulating glass.					
Double hung aluminum window, 2′-4″ x 2′ 6″	1.000	Ea.	247.78	42.85	290.63
Exterior and interior trim	1.000	Set	37.50	44.50	82
Hardware	1.000	Set	4.54	24	28.54
Insulating glass	5.830	S.F.	76.66	32.07	108.73
TOTAL			366.48	143.42	509.90

B2020 110	Windows - Aluminum	COST EACH		
		MAT.	INST.	TOTAL
1000	Aluminum, double hung, 2′-4″ x 2′-6″	365	143	508
1100	2′-8″ x 4′-6″	715	236	951
1200	3′- 4″ x 5′-6″	1,100	355	1,455
1300	Casement, 3′-6″ x 2′-4″	490	173	663
1400	4′-6″ x 2′-4″	630	217	847
1500	5′-6″ x 2′-4″	785	285	1,070
1600	Projected window, 2′-1″ x 3′-0″	365	149	514
1700	3′-5″ x 3′-0″	580	214	794
1800	4′-0″ x 4′-0″	900	325	1,225
1900	Horizontal sliding 3′-0″ x 3′-0″	365	184	549
2000	3′-6″ x 4′-0″	555	262	817
2100	5′-0″ x 6′-0″	1,150	505	1,655
2200	Picture window, 3′-8″ x 3′-1″	415	213	628
2300	4′-4″ x 4′-5″	680	330	1,010
2400	5′-8″ x 4′-9″	965	465	1,430
2500				
2600				
2700				
2800				
2900				
3100	Cut & patch to match existing construction, add, minimum	2%	3%	
3200	Maximum	5%	9%	
3300	Dust protection, add, minimum	1%	2%	
3400	Maximum	4%	11%	
3500	Material handling & storage limitation, add, minimum	1%	1%	
3600	Maximum	6%	7%	
3700	Protection of existing work, add, minimum	2%	2%	
3800	Maximum	5%	7%	
3900	Shift work requirements, add, minimum		5%	
4000	Maximum		30%	

B2020 Exterior Windows

This page illustrates and describes a wood window system including wood picture window, exterior and interior trim, hardware and insulating glass. Lines within System Components give the unit price and total price on a cost each basis for this system. Prices for wood window systems are on Line Items B2020 112 1200 thru 2900. Both material quantities and labor costs have been adjusted for the system listed.

Factors: To adjust for job conditions other than normal working situations use Lines B2020 112 3100 thru 4000.

Example: You are to install the above system where dust control is vital. Go to Line B2020 112 3400 and apply these percentages to the appropriate MAT. and INST. costs.

System Components	QUANTITY	UNIT	COST EACH MAT.	COST EACH INST.	COST EACH TOTAL
Wood picture window 4′-0″ x 4′-6″, exterior and interior trim, hardware, Glazed with insulating glass.					
4′-0″ x 4′-6″ wood picture window with insulating glass	1.000	Ea.	585	105	690
Exterior and interior trim	1.000	Set	71	96.50	167.50
Hardware	1.000	Set	4.54	24	28.54
TOTAL			660.54	225.50	886.04

B2020 112	Windows - Wood	COST EACH MAT.	COST EACH INST.	COST EACH TOTAL
1200	Picture window, 4′-0″ x 4′-6″	660	226	886
1300	5′-0″ x 4′-0″	740	226	966
1400	6′-0″ x 4′-6″	775	237	1,012
1500	Bow, bay window, 8′-0″ x 5′-0″, standard	2,575	237	2,812
1600	Deluxe	1,900	237	2,137
1700	Bow, bay window, 12′-0″ x 6′-0″ standard	3,300	315	3,615
1800	Deluxe	2,650	315	2,965
1900	Double hung, 3′-0″ x 4′-0″	365	147	512
2000	4′-0″ x 4′-6″	450	193	643
2100	Casement 2′-0″ x 3′-0″	325	127	452
2200	2 leaf, 4′-0″ x 4′-0″	860	211	1,071
2300	3 leaf, 6′-0″ x 6′-0″	1,650	340	1,990
2400	Awning, 2′-10″ x 1′-10″	335	127	462
2500	3′-6″ x 2′-4″	420	147	567
2600	4′-0″ x 3′-0″	610	193	803
2700	Horizontal sliding 3′-0″ x 2′-0″	375	127	502
2800	4′-0″ x 3′-6″	440	147	587
2900	6′-0″ x 5′-0″	650	193	843
3100	Cut & patch to match existing construction, add, minimum	2%	3%	
3200	Maximum	5%	9%	
3300	Dust protection, add, minimum	1%	2%	
3400	Maximum	4%	11%	
3500	Material handling & storage limitation, add, minimum	1%	1%	
3600	Maximum	6%	7%	
3700	Protection of existing work, add, minimum	2%	2%	
3800	Maximum	5%	7%	
3900	Shift work requirements, add, minimum		5%	
4000	Maximum		30%	

B2020 Exterior Windows

This page illustrates and describes storm window and door systems based on a cost each price. Prices for storm window and door systems are on Line Items B2020 116 0400 thru 2400. Both material quantities and labor costs have been adjusted for the system listed.

Factors: To adjust for job conditions other than normal working situations use Lines B2020 116 3100 thru 4000 and apply these percentages to the appropriate MAT and INST. costs.

Example: You are to install the system and protect all existing construction. Go to Line B2020 116 3800 and apply these percentages to the appropriate MAT. and INST. costs.

B2020 116	Storm Windows & Doors	COST EACH		
		MAT.	INST.	TOTAL
0350	Storm Window and door systems, single glazed			
0400	Window, custom aluminum anodized, 2'-0" x 3'-5"	99	38.50	137.50
0500	2'-6" x 5'-0"	121	41.50	162.50
0600	4'-0" x 6'-0"	249	46.50	295.50
0700	White painted aluminum, 2'-0" x 3'-5"	110	38.50	148.50
0800	2'-6" x 5'-0"	176	41.50	217.50
0900	4'-0" x 6'-0"	300	46.50	346.50
1000	Average quality aluminum, anodized, 2'-0" x 3'-5"	93.50	38.50	132
1100	2'-6" x 5'-0"	124	41.50	165.50
1200	4'-0" x 6'-0"	139	46.50	185.50
1300	White painted aluminum, 2'-0" x 3'-5"	93.50	38.50	132
1400	2'-6" x 5'-0"	99	41.50	140.50
1500	4'-0" x 6'-0"	108	46.50	154.50
1600	Mill finish, 2'-0" x 3'-5"	82.50	38.50	121
1700	2'-6" x 5'-0"	93.50	41.50	135
1800	4'-0" x 6'-0"	109	46.50	155.50
1900				
2000	Door, aluminum anodized 3'-0" x 6'-8"	177	82.50	259.50
2100	White painted aluminum, 3'-0" x 6'-8"	330	82.50	412.50
2200	Mill finish, 3'-0" x 6'-8"	278	82.50	360.50
2300				
2400	Wood, storm and screen, painted	340	129	469
2500				
2600				
2700				
2800				
3100	Cut & patch to match existing construction, add, minimum	2%	3%	
3200	Maximum	5%	9%	
3300	Dust protection, add, minimum	1%	2%	
3400	Maximum	4%	11%	
3500	Material handling & storage limitation, add, minimum	1%	1%	
3600	Maximum	6%	7%	
3700	Protection of existing work, add, minimum	2%	2%	
3800	Maximum	5%	7%	
3900	Shift work requirements, add, minimum		5%	
4000	Maximum		30%	

B2020 Exterior Windows

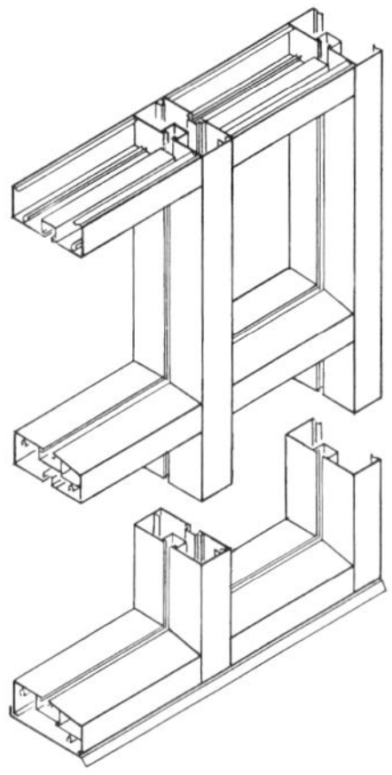

This page illustrates and describes a window wall system including aluminum tube framing, caulking, and glass. Lines within System Components give the unit price and total price per square foot for this system. Prices for window wall systems are on Line Items B2020 124 1300 thru 2600. Both material quantities and labor costs have been adjusted for the system listed.

Factors: To adjust for job conditions other than normal working situations use Lines B2020 124 3100 thru 4000.

Example: You are to install the system with need for complete dust protection. Go to Line B2020 124 3400 and apply these percentages to the appropriate MAT. and INST. costs.

System Components	QUANTITY	UNIT	COST PER S.F.		
			MAT.	INST.	TOTAL
Window wall, including aluminum header, sill, mullions, caulking					
And glass					
Header, mill finish, 1-3/4" x 4-1/2" deep	.167	L.F.	3.31	2.22	5.53
Sill, mill finish, 1-3/4" x 4-1/2" deep	.167	L.F.	3.26	2.17	5.43
Vertical mullion, 1-3/4" x 4-1/2" deep, 6' O.C.	.191	L.F.	8.23	5.42	13.65
Caulking	.381	L.F.	.08	.75	.83
Glass, 1/4" plate	1.000	S.F.	6.05	9.20	15.25
TOTAL			20.93	19.76	40.69

B2020 124	Aluminum Frame, Window Wall	COST PER S.F.		
		MAT.	INST.	TOTAL
1300	Mill finish, 1-3/4" x 4-1/2" deep, 1/4" plate glass	21	19.75	40.75
1400	2" x 4-1/2" deep, insulating glass	23	23	46
1500	Thermo-break, 2-1/4" x 4-1/2" deep, insulating glass	24	23.50	47.50
1600	Bronze finish, 1-3/4" x 4-1/2" deep, 1/4" plate glass	23.50	21.50	45
1700	2" x 4-1/2" deep, insulating glass	25	25	50
1800	Thermo-break, 2-1/4" x 4-1/2" deep, insulating glass	26.50	25	51.50
2000	Black finish, 1-3/4" x 4-1/2" deep, 1/4" plate glass	25	22.50	47.50
2100	2" x 4-1/2" deep, insulating glass	26.50	25.50	52
2200	Thermo-break, 2-1/4" x 4-1/2" deep, insulating glass	28	26	54
2300				
2400	Stainless steel, 1-3/4" x 4-1/2" deep, 1/4" plate glass	32	27	59
2500	2" x 4-1/2" deep, insulating glass	32.50	30.50	63
2600	Thermo-break, 2-1/4" x 4-1/2" deep, insulating glass	34.50	31.50	66
2700				
3100	Cut & patch to match existing construction, add, minimum	2%	3%	
3200	Maximum	5%	9%	
3300	Dust protection, add, minimum	1%	2%	
3400	Maximum	4%	11%	
3500	Material handling, & storage limitation, add, minimum	1%	1%	
3600	Maximum	6%	7%	
3700	Protection of existing work, add, minimum	2%	2%	
3800	Maximum	5%	7%	
3900	Shift work requirements, add, minimum		5%	
4000	Maximum		30%	

B2030 Exterior Doors

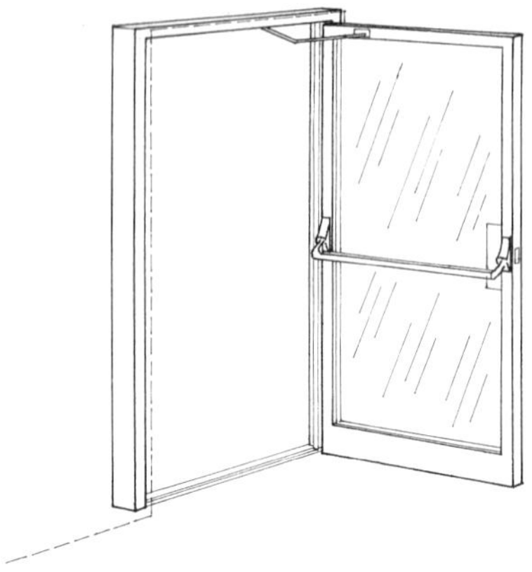

This page illustrates and describes a commercial metal door system, including a single aluminum and glass door, narrow stiles, jamb, hardware weatherstripping, panic hardware and closer. Lines within System Components give the unit price and total price on a cost each basis for this system. Prices for commercial metal door systems are on Line Items B2030 125 1200 thru 2500. Both material quantities and labor costs have been adjusted for the system listed.

Factors: To adjust for job conditions other than normal working situations, use Lines B2030 125 3100 thru 4000.

Example: You are to install the system and cut and patch to match existing construction. Go to Line B2030 125 3200 and apply these percentages to the appropriate MAT. and INST. costs.

System Components	QUANTITY	UNIT	COST EACH MAT.	COST EACH INST.	COST EACH TOTAL
Single aluminum and glass door, 3′-0″x7′-0″, with narrow stiles, ext. jamb. Weatherstripping, 1/2″ tempered insul. glass, panic hardware, and closer.					
Aluminum door, 3′-0″ x 7′-0″ x 1-3/4″, narrow stiles	1.000	Ea.	850	735	1,585
Tempered insulating glass, 1/2″ thick	20.000	S.F.	510	400	910
Panic hardware	1.000	Set	585	96.50	681.50
Automatic closer	1.000	Ea.	144	89	233
TOTAL			2,089	1,320.50	3,409.50

B2030 125	Doors, Metal - Commercial	COST EACH MAT.	COST EACH INST.	COST EACH TOTAL
1200	Single aluminum and glass, 3′-0″ x 7′-0″	2,100	1,325	3,425
1300	With transom, 3′-0″ x 10′-0″	2,775	1,550	4,325
1400	Anodized aluminum and glass, 3′-0″ x 7′-0″	2,375	1,575	3,950
1500	With transom, 3′-0″ x 10′-0″	3,225	1,850	5,075
1600	Steel, deluxe, hollow metal 3′-0″ x 7′-0″	1,700	500	2,200
1700	With transom 3′-0″ x 10′-0″	2,225	565	2,790
1800	Fire door, "A" label, 3′-0″ x 7′-0″	1,850	500	2,350
1900	Double, aluminum and glass, 6′-0″ x 7′-0″	3,200	2,250	5,450
2000	With transom, 6′-0″ x 10′-0″	4,050	2,775	6,825
2100	Anodized aluminum and glass 6′-0″ x 7′-0″	3,625	2,650	6,275
2200	With transom, 6′-0″ x 10′-0″	4,650	3,225	7,875
2300	Steel, deluxe, hollow metal, 6′-0″ x 7′-0″	2,700	875	3,575
2400	With transom, 6′-0″ x 10′-0″	3,750	1,000	4,750
2500	Fire door, "A" label, 6′-0″ x 7′-0″	2,900	865	3,765
2800				
2900				
3100	Cut & patch to match existing construction, add, minimum	2%	3%	
3200	Maximum	5%	9%	
3300	Dust protection, add, minimum	1%	2%	
3400	Maximum	4%	11%	
3500	Material handling & storage limitation, add, minimum	1%	1%	
3600	Maximum	6%	7%	
3700	Protection of existing work, add, minimum	2%	2%	
3800	Maximum	5%	7%	
3900	Shift work requirements, add, minimum		5%	
4000	Maximum		30%	

B2030 Exterior Doors

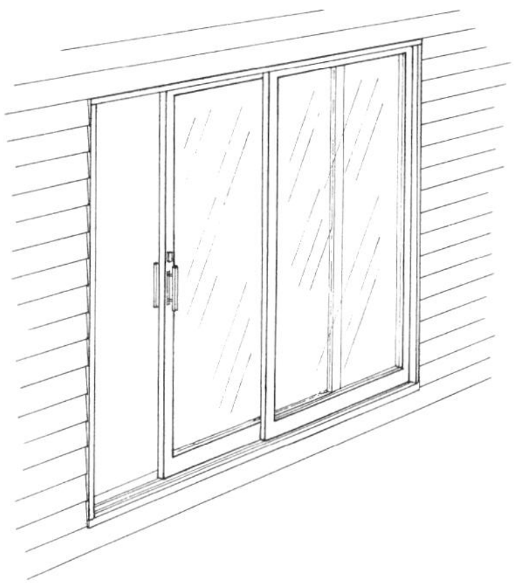

This page illustrates and describes sliding door systems including a sliding door, frame, interior and exterior trim with exterior staining. Lines within System Components give the unit price and total price on a cost each basis for this system. Prices for sliding door systems are on Line Items B2030 150 1000 thru 2400. Both material quantities and labor costs have been adjusted for the system listed.

Factors: To adjust for job conditions other than normal working situations use Lines B2030 150 2700 thru 4000.

Example: You are to install the system with temporary shoring and bracing. Go to Line B2030 150 3900 and apply these percentages to the appropriate MAT. and INST. costs.

System Components	QUANTITY	UNIT	COST EACH		
			MAT.	INST.	TOTAL
Sliding wood door, 6′-0″ x 6′-8″, with wood frame, interior and exterior Trim and exterior staining.					
Sliding wood door, standard, 6′-0″ x 6′-8″, insulated glass	1.000	Ea.	1,250	289	1,539
Interior & exterior trim	1.000	Set	29	46.20	75.20
Stain door	1.000	Ea.	2.06	28.50	30.56
Stain trim	1.000	Ea.	2.60	12.20	14.80
TOTAL			1,283.66	375.90	1,659.56

B2030 150	Doors, Sliding - Patio	COST EACH		
		MAT.	INST.	TOTAL
1000	Wood, standard, 6′-0″ x 6′-8″, insulated glass	1,275	375	1,650
1100	8′-0″ x 6′-8″	1,675	480	2,155
1200	12′-0″ x 6′-8″	2,750	585	3,335
1300	Vinyl coated, 6′-0″ x 6′-8″	1,600	375	1,975
1400	8′-0″ x 6′-8″	2,025	480	2,505
1500	12′-0″ x 6′-8″	3,350	585	3,935
1700	Aluminum, standard, 6′-0″ x 6′-8″, insulated glass	1,025	405	1,430
1800	8′-0″ x 6′-8″	1,600	515	2,115
1900	12′-0″ x 6′-8″	1,575	615	2,190
2000	Anodized, 6′-0″ x 6′-8″	1,775	405	2,180
2100	8′-0″ x 6′-8″	1,875	515	2,390
2200	12′-0″ x 6′-8″	3,350	615	3,965
2300				
2400	Deduct for single glazing	89.50		89.50
2500				
2700	Cut & patch to match existing construction, add, minimum	2%	3%	
2800	Maximum	5%	9%	
2900	Dust protection, add, minimum	1%	2%	
3000	Maximum	4%	11%	
3100	Equipment usage curtailment, add, minimum	1%	1%	
3200	Maximum	3%	10%	
3300	Material handling & storage limitation, add, minimum	1%	1%	
3400	Maximum	6%	7%	
3500	Protection of existing, work, add, minimum	2%	2%	
3600	Maximum	5%	7%	
3700	Shift work requirements, add, minimum		5%	
3800	Maximum		30%	
3900	Temporary shoring and bracing, add, minimum	2%	5%	
4000	Maximum	5%	12%	

B2030 Exterior Doors

This page illustrates and describes residential door systems including a door, frame, trim, hardware, weatherstripping, stained and finished. Lines within System Components give the unit price and total price on a cost each basis for this system. Prices for residential door systems are on Line Items B2030 240 1500 thru 2200. Both material quantities and labor costs have been adjusted for the system listed.

Factors: To adjust for job conditions other than normal working situations use Lines B2030 240 3100 thru 4000.

Example: You are to install the system with a material handling and storage limitation. Go to Line B2030 240 3600 and apply these percentages to the appropriate MAT. and INST. costs.

System Components	QUANTITY	UNIT	COST EACH		
			MAT.	INST.	TOTAL
Single solid wood colonial door, 3′ x 6′-8″ with wood frame and trim, Stained and finished, including hardware and weatherstripping.					
Solid wood colonial door, fir 3′ x 6′-8″ x 1-3/4″, hinges	1.000	Ea.	490	77	567
Hinges, ball bearing	1.000	Ea.	36.50		36.50
Exterior frame, with trim	1.000	Set	143.65	52.53	196.18
Interior trim	1.000	Set	24.50	48	72.50
Lockset	1.000	Set	49.50	41.50	91
Sill, oak 8″ deep	3.500	Ea.	65.80	40.43	106.23
Weatherstripping	1.000	Set	21.50	76	97.50
Stained and finished	1.000	Ea.	4.41	61	65.41
TOTAL			835.86	396.46	1,232.32

B2030 240	Doors, Residential - Exterior	COST EACH		
		MAT.	INST.	TOTAL
1400	Single doors, 3′ x 6′-8″			
1500	Solid colonial door, fir	835	395	1,230
1600	Hollow metal exterior door, plain	805	370	1,175
1700	Solid core wood door, plain	470	380	850
1800				
1900	Double doors, 6′ x 6′-8″			
2000	Solid colonial double doors, fir	1,425	525	1,950
2100	Hollow metal exterior doors, plain	1,350	490	1,840
2200	Hollow core wood doors, plain	685	510	1,195
2300				
2800				
2900				
3100	Cut & patch to match existing construction, add, minimum	2%	3%	
3200	Maximum	5%	9%	
3300	Dust protection, add, minimum	1%	2%	
3400	Maximum	4%	11%	
3500	Material handling & storage limitation, add, minimum	1%	1%	
3600	Maximum	6%	7%	
3700	Protection of existing work, add, minimum	2%	2%	
3800	Maximum	5%	7%	
3900	Shift work requirements, add, minimum		5%	
4000	Maximum		30%	

B2030 Exterior Doors

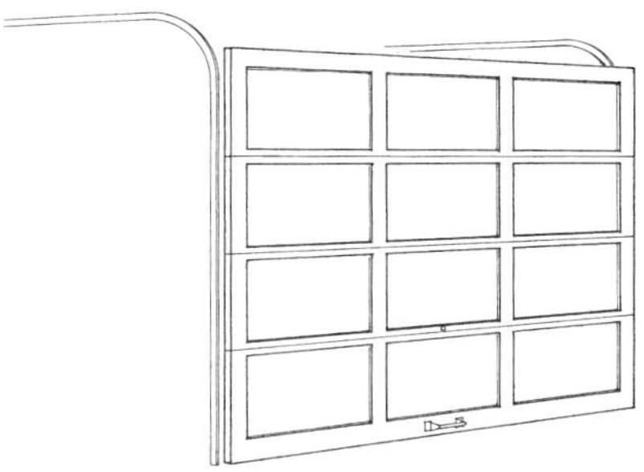

This page illustrates and describes overhead door systems including an overhead door, track, hardware, trim, and electric door opener. Lines within System Components give the unit price and total price on a cost each basis for this system. Prices for overhead door systems are on Line Items B2030 410 1200 thru 2300. Both material quantities and labor costs have been adjusted for the system listed.

Factors: To adjust for job conditions other than normal working situations use Lines B2030 410 3100 thru 4000.

Example: You are to install the system and match existing construction. Go to Line B2030 410 3200 and apply these percentages to the appropriate MAT. and INST. costs.

System Components	QUANTITY	UNIT	COST EACH		
			MAT.	INST.	TOTAL
Wood overhead door, commercial sectional, including track, door, hardware And trim, electrically operated.					
Commercial, heavy duty wood door, 8′ x 8′ x 1-3/4″ thick	1.000	Ea.	925	580	1,505
Frame, 2 x 8, pressure treated	1.000	Set	23.28	72.96	96.24
Wood trim	1.000	Set	32.40	57.84	90.24
Painting, two coats	1.000	Ea.	18.72	195.84	214.56
Electric trolley operator	1.000	Ea.	1,025	289	1,314
TOTAL			2,024.40	1,195.64	3,220.04

B2030 410	Doors, Overhead	COST EACH		
		MAT.	INST.	TOTAL
1200	Commercial, wood, 1-3/4″ thick, 8′ x 8′	2,025	1,200	3,225
1300	12′ x 12′	3,075	1,675	4,750
1400	14′ x 14′	4,150	1,975	6,125
1500	Fiberglass & aluminum 12′ x 12′	4,075	1,700	5,775
1600	20′ x 20′	8,675	4,375	13,050
1700	Residential, wood, 9′ x 7′	1,050	455	1,505
1800	16′ x 7′	1,850	675	2,525
1900	Hardboard faced, 9′ x 7′	705	455	1,160
2000	16′ x 7′	1,425	675	2,100
2100	Fiberglass & aluminum, 9′ x 7′	1,025	530	1,555
2200	16′ x 7′	1,800	675	2,475
2300	For residential electric opener, add	660	72.50	732.50
2400				
2500				
2600				
2700				
2800				
2900				
3100	Cut & patch to match existing construction, add, minumum	2%	3%	
3200	Maximum	5%	9%	
3300	Dust protection, add, minimum	1%	2%	
3400	Maximum	4%	11%	
3500	Material handling & storage limitation, add, minimum	1%	1%	
3600	Maximum	6%	7%	
3700	Protection of existing work, add, minimum	2%	2%	
3800	Maximum	5%	7%	
3900	Shift work requirements, add, minimum		5%	
4000	Maximum		30%	

B2030 Exterior Doors

B2030 810	Selective Price Sheet	COST EACH		
		MAT.	INST.	TOTAL
0100	Door closer, rack and pinion	189	89	278
0200	Backcheck and adjustable power	202	96.50	298.50
0300	Regular, hinge face mount, all sizes, regular arm	235	89	324
0400	Hold open arm	274	89	363
0500	Top jamb mount, all sizes, regular arm	224	96.50	320.50
0600	Hold open arm	235	96.50	331.50
0700	Stop face mount, all sizes, regular arm	218	89	307
0800	Hold open arm	241	89	330
0900	Fusible link, hinge face mount, all sizes, regular arm	274	89	363
1000	Hold open arm	315	89	404
1100	Top jamb mount, all sizes, regular arm	263	96.50	359.50
1200	Hold open arm	274	96.50	370.50
1300	Stop face mount, all sizes, regular arm	257	89	346
1400	Hold open arm	280	89	369
1500				
1600				
1700	Door stops			
1800				
1900	Holder & bumper, floor or wall	39	18.10	57.10
2000	Wall bumper	12.05	18.10	30.15
2100	Floor bumper	7.80	18.10	25.90
2200	Plunger type, door mounted	29.50	18.10	47.60
2300	Hinges, full mortise, material only, per pair			
2400	Low frequency, 4-1/2" x 4-1/2", steel base, USP	14.35		14.35
2500	Brass base, US10	50		50
2600	Stainless steel base, US32	84.50		84.50
2700	Average frequency, 4-1/2" x 4-1/2", steel base, USP	30		30
2800	Brass base, US10	60		60
2900	Stainless steel base, US32	90		90
3000	High frequency, 4-1/2" x 4-1/2", steel base, USP	69.50		69.50
3100	Brass base, US10	94		94
3200	Stainless steel base, US32	140		140
3300	Kick plate			
3400				
3500	6" high, for 3'-0" door, aluminum	33.50	38.50	72
3600	Bronze	59.50	38.50	98
3700	Panic device			
3800				
3900	For rim locks, single door, exit	585	96.50	681.50
4000	Outside key and pull	630	116	746
4100	Bar and vertical rod, exit only	810	116	926
4200	Outside key and pull	890	145	1,035
4300	Lockset			
4400				
4500	Heavy duty, cylindrical, passage doors	160	48	208
4600	Classroom	360	72.50	432.50
4700	Bedroom, bathroom, and inner office doors	204	48	252
4800	Apartment, office, and corridor doors	275	58	333
4900	Standard duty, cylindrical, exit doors	102	58	160
5100	Passage doors	68	48	116
5200	Public restroom, classroom, & office doors	149	72.50	221.50
5300	Deadlock, mortise, heavy duty	305	64.50	369.50
5400	Double cylinder	297	64.50	361.50
5500	Entrance lock, cylinder, deadlocking latch	145	64.50	209.50
5600	Deadbolt	176	72.50	248.50
5700	Commercial, mortise, wrought knob, keyed, minimum	218	72.50	290.50
5800	Maximum	395	82.50	477.50
5900	Cast knob, keyed, minimum	286	64.50	350.50

B2030 Exterior Doors

B2030 810	Selective Price Sheet	COST EACH		
		MAT.	INST.	TOTAL
6000	Maximum	560	64.50	624.50
6100	Push-pull			
6200				
6300	Aluminum	11.25	48	59.25
6400	Bronze	23.50	48	71.50
6500	Door pull, designer style, minimum	78.50	48	126.50
6600	Maximum	405	72.50	477.50
6700	Threshold			
6800				
6900	3′-0″ long door saddles, aluminum, minimum	6.20	12.05	18.25
7000	Maximum	45.50	48	93.50
7100	Bronze, minimum	46	9.65	55.65
7200	Maximum	78.50	48	126.50
7300	Rubber, 1/2″ thick, 5-1/2″ wide	46	29	75
7400	2-3/4″ wide	33.50	29	62.50
7500	Weatherstripping, per set			
7600				
7700	Doors, wood frame, interlocking for 3′ x 7′ door, zinc	22	193	215
7800	Bronze	32	193	225
7900	Wood frame, spring type for 3′ x 7′ door, bronze	21.50	76	97.50
8000	Metal frame, spring type for 3′ x 7′ door, bronze	42	193	235
8100	For stainless steel, spring type, add	133%		
8200				
8300	Metal frame, extruded sections, 3′ x 7′ door, aluminum	50.50	193	243.50
8400	Bronze	132	193	325

B3010 Roof Coverings

B3010 160	Selective Price Sheet	COST PER S.F.		
		MAT.	INST.	TOTAL
0100	Roofing, built-up, asphalt roll roof, 3 ply organic/mineral surface	.69	1.47	2.16
0200	3 plies glass fiber felt type iv, 1 ply mineral surface	1.12	1.59	2.71
0300	Cold applied, 3 ply		.53	.53
0400	Coal tar pitch, 4 ply tarred felt	2.05	1.90	3.95
0500	Mopped, 3 ply glass fiber	1.69	2.09	3.78
0600	4 ply organic felt	2.05	1.90	3.95
0700	Elastomeric, hypalon, neoprene unreinforced	2.83	3.50	6.33
0800	Polyester reinforced	2.87	4.14	7.01
0900	Neoprene, 5 coats 60 mils	6.30	12.25	18.55
1000	Over 10,000 S.F.	5.90	6.35	12.25
1100	PVC, traffic deck sprayed	1.85	6.35	8.20
1200	With neoprene	1.95	2.56	4.51
1600				
1700	Asphalt, strip, 210-235#/sq.	.83	.97	1.80
1800	235-240#/sq.	.91	1.06	1.97
1900	Class A laminated	1.02	1.18	2.20
2000	Class C laminated	1.49	1.33	2.82
2100	Slate, buckingham, 3/16" thick	5.25	3.05	8.30
2200	Black, 1/4" thick	5.25	3.05	8.30
2300	Wood, shingles, 16" no. 1, 5" exp.	2.29	2.31	4.60
2400	Red cedar, 18" perfections	2.51	2.10	4.61
2500	Shakes, 24", 10" exposure	2.56	2.31	4.87
2600	18", 8-1/2" exposure	1.75	2.89	4.64
2700	Insulation, ceiling batts, fiberglass, 3-1/2" thick, R13	.36	.43	.79
2800	6" thick, R19	.42	.43	.85
2900	9" thick, R30	.68	.50	1.18
3000	12" thick, R38	.79	.50	1.29
3100	Mineral, 3-1/2" thick, R13	.42	.36	.78
3200	Fiber, 6" thick, R19	.42	.36	.78
3300	Roof deck, fiberboard, 1" thick, R2.78	.51	.66	1.17
3400	Mineral, 2" thick, R5.26	1.02	.66	1.68
3500	Perlite boards, 3/4" thick, R2.08	.41	.66	1.07
3600	2" thick, R5.26	.88	.76	1.64
3700	Polystyrene extruded, R5.26, 1" thick,	.48	.35	.83
3800	2" thick R10	.62	.43	1.05
3900	40 PSI compressive strength, 1" thick R5	.52	.35	.87
4000	Tapered for drainage	.74	.38	1.12
4300	Ceiling, plaster, gypsum, 2 coats	.44	3.11	3.55
4400	3 coats	.63	3.71	4.34
4500	Perlite or vermiculite, 2 coats	.46	3.64	4.10
4600	3 coats	.77	4.58	5.35
4700	Gypsum lath, plain 3/8" thick	.47	.64	1.11
4800	1/2" thick	.55	.68	1.23
4900	Firestop, 3/8" thick	.28	.78	1.06
5000	1/2" thick	.30	.84	1.14
5100	Metal lath, rib, 2.75 lb.	.38	.73	1.11
5200	3.40 lb.	.59	.78	1.37
5300	Diamond, 2.50 lb.	.42	.73	1.15
5400	3.40 lb.	.49	.91	1.40
5500	Drywall, taped and finished standard, 1/2" thick	.29	1.51	1.80
5600	5/8" thick	.33	1.51	1.84
5700	Fire resistant, 1/2" thick	.34	1.51	1.85
5800	5/8" thick	.35	1.51	1.86
5900	Water resist., 1/2" thick	.44	1.51	1.95
6000	5/8" thick	.52	1.51	2.03
6100	Finish, instead of taping			
6200	For thin coat plaster, add	.11	.72	.83
6300	Finish textured spray, add	.04	.62	.66

B3010 Roof Coverings

B3010 160	Selective Price Sheet	COST PER S.F.		
		MAT.	INST.	TOTAL
6500	Tile, stapled glued, mineral fiber plastic coated, 5/8″ thick	1.89	1.93	3.82
6600	3/4″ thick	2.33	1.93	4.26
6700	Wood fiber, 1/2″ thick	1.07	1.45	2.52
6800	3/4″ thick	1.29	1.45	2.74
6900	Suspended, fiberglass film faced, 5/8″ thick	1.06	.93	1.99
7000	3″ thick	1.73	1.29	3.02
7100	Mineral fiber 5/8″ thick, standard face	.86	.86	1.72
7200	Aluminum faced	2.04	.96	3
7300	Wood fiber reveal edge, 1″ thick			
7400	3″ thick			
7500	Ceiling suspension systems, for tile, "T" bar, class "A", 2′ x 4′ grid	.77	.72	1.49
7600	2′ x 2′ grid	.95	.89	1.84
7700	Concealed "Z" bar, 12″ module	.80	1.11	1.91
7800				
7900	Plaster/drywall, 3/4″ channels, steel furring, 16″ O.C.	.34	1.70	2.04
8000	24″ O.C.	.22	1.17	1.39
8100	1-1/2″ channels, 16″ O.C.	.45	1.89	2.34
8200	24″ O.C.	.30	1.26	1.56
8300	Ceiling framing, 2″ x 4″ studs, 16″ O.C.	.27	.97	1.24
8400	24″ O.C.	.18	.65	.83

B3020 Roof Openings

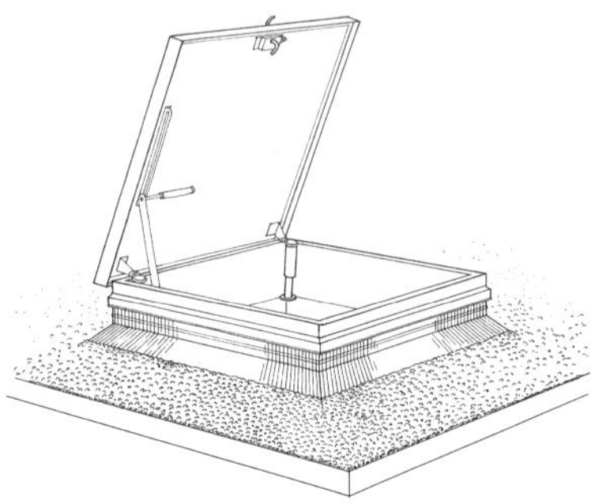

This page illustrates and describes a roof hatch system. Lines within System Components give the unit price and total cost each for this system. Prices for systems are on Lines B3020 230 1400 thru 2400. Both material quantities and labor costs have been adjusted for the system listed.

Factors: To adjust for job conditions other than normal working situtations use Lines B3020 230 2900 thru 4000.

Example: You are to install the system and cut and patch to match existing construction. Use line B3020 230 3000 and apply these percentages to the appropriate MAT. and INST. costs.

System Components	QUANTITY	UNIT	COST EACH MAT.	COST EACH INST.	COST EACH TOTAL
Roof hatch, 2'-6" x 3'-0", aluminum, curb and cover included.					
Through steel construction.					
Roof hatch, 2'-6" x 3'-0", aluminum	1.000	Ea.	1,175	223	1,398
Cutout decking	11.000	L.F.		29.92	29.92
Frame opening	44.000	L.F.	36.52	391.60	428.12
Flashing, 16 oz. copper	16.000	S.F.	138.40	73.92	212.32
Cant strip, 4 x 4, foamglass	12.000	L.F.	29.76	19.68	49.44
TOTAL			1,379.68	738.12	2,117.80

B3020 230	Roof Hatches, Skylights	COST EACH MAT.	COST EACH INST.	COST EACH TOTAL
1400	Roof hatch, aluminum, curb and cover, 2'-6" x 3'-0"	1,375	735	2,110
1500	2'-6" x 4'-6"	1,250	905	2,155
1600	2'-6" x 8'-0"	2,325	1,300	3,625
1700	Skylight, plexiglass dome, with curb mounting, 30" x 32"	710	700	1,410
1800	30" x 45"	770	880	1,650
1900	40" x 45"	930	1,200	2,130
2000	Smoke hatch, unlabeled, 2'-6" x 3'-0"	1,675	795	2,470
2100	2'-6" x 4'-6"	1,500	965	2,465
2200	2'-6" x 8'-0"	2,525	1,350	3,875
2300				
2400	For steel ladder, add per vertical linear foot	39.50	36.50	76
2500				
2600				
2700				
2900	Cut & patch to match existing construction, add, minimum	2%	3%	
3000	Maximum	5%	9%	
3100	Dust protection, add, minimum	1%	2%	
3200	Maximum	4%	11%	
3300	Equipment usage curtailment, add, minimum	1%	1%	
3400	Maximum	3%	10%	
3500	Material handling & storage limitation, add, minimum	1%	1%	
3600	Maximum	6%	7%	
3700	Protection of existing work, add, minimum	2%	2%	
3800	Maximum	5%	7%	
3900	Shift work requirements, add, minimum		5%	
4000	Maximum		30%	

B3020 Roof Openings

B3020 240	Selective Price Sheet	COST PER L.F. MAT.	INST.	TOTAL
0100	Downspouts per L.F., aluminum, enameled .024" thick, 2" x 3"	2.26	3.65	5.91
0200	3" x 4"	3.03	4.69	7.72
0300	Round .025" thick, 3" diam.	1.89	3.46	5.35
0400	4" diam.	2.92	4.69	7.61
0500	Copper, round 16 oz. stock, 2" diam.	8.90	3.46	12.36
0600	3" diam.	8.90	3.46	12.36
0700	4" diam.	11.90	4.53	16.43
0800	5" diam.	18.15	5.05	23.20
0900	Rectangular, 2" x 3"	8.15	3.46	11.61
1000	3" x 4"	10.90	4.53	15.43
1100	Lead coated copper, round, 2" diam.	37.50	3.46	40.96
1200	3" diam.	33	3.46	36.46
1300	4" diam.	34.50	4.53	39.03
1400	5" diam.	27	5.05	32.05
1500	Rectangular, 2" x 3"	18	3.46	21.46
1600	3" x 4"	20	4.53	24.53
1700	Steel galvanized, round 28 gauge, 3" diam.	2.02	3.46	5.48
1800	4" diam.	2.79	4.53	7.32
1900	5" diam.	3.97	5.05	9.02
2000	6" diam.	4.40	6.25	10.65
2100	Rectangular, 2" x 3"	1.41	3.46	4.87
2200	3" x 4"	2.68	4.53	7.21
2300	Elbows, aluminum, round, 3" diam.	3.01	6.55	9.56
2400	4" diam.	6.35	6.55	12.90
2500	Rectangular, 2" x 3"	1.12	6.55	7.67
2600	3" x 4"	3.85	6.55	10.40
2700	Copper, round 16 oz., 2" diam.	9.35	6.55	15.90
2800	3" diam.	12.55	6.55	19.10
2900	4" diam.	14.20	6.55	20.75
3100	Rectangular, 2" x 3"	9.35	6.55	15.90
3200	3" x 4"	15.40	6.55	21.95
3300	Drip edge per L.F., aluminum, 5" wide	.42	1.45	1.87
3400	8" wide	.51	1.45	1.96
3500	28" wide	4.24	5.80	10.04
3600				
3700	Steel galvanized, 5" wide	.53	1.45	1.98
3800	8" wide	.76	1.45	2.21
3900				
4000				
4100				
4200				
4300	Flashing 12" wide per S.F., aluminum, mill finish, .013" thick	.80	3.67	4.47
4400	.019" thick	1.18	3.67	4.85
4500	.040" thick	2.35	3.67	6.02
4600	.050" thick	2.48	3.67	6.15
4700	Copper, mill finish, 16 oz.	8.65	4.62	13.27
4800	20 oz.	10.85	4.83	15.68
4900	24 oz.	13.35	5.05	18.40
5000	32 oz.	17.80	5.30	23.10
5100	Lead, 2.5 lb./S.F., 12" wide	5.95	3.94	9.89
5200	Over 12" wide	4.87	3.94	8.81
5300	Lead-coated copper, fabric backed, 2 oz.	2.90	1.61	4.51
5400	5 oz.	3.34	1.61	4.95
5500	Mastic backed, 2 oz.	2.27	1.61	3.88
5600	5 oz.	2.81	1.61	4.42
5700	Paper backed, 2 oz.	1.96	1.61	3.57
5800	3 oz.	2.30	1.61	3.91
5900	Polyvinyl chloride, black, .010" thick	.23	1.87	2.10

B3020 Roof Openings

B3020 240	Selective Price Sheet	COST PER L.F. MAT.	INST.	TOTAL
6000	.020″ thick	.33	1.87	2.20
6100	.030″ thick	.43	1.87	2.30
6200	.056″ thick	1.03	1.87	2.90
6300	Steel, galvanized, 20 gauge	1.57	4.09	5.66
6400	30 gauge	1.46	3.32	4.78
6500	Stainless, 32 gauge, .010″ thick	4.40	3.43	7.83
6600	28 gauge, .015″ thick	6.20	3.43	9.63
6700	26 gauge, .018″ thick	8	3.43	11.43
6800	24 gauge, .025″ thick	9.80	3.43	13.23
6900	Gutters per L.F., aluminum, 5″ box, .027″ thick	2.93	5.45	8.38
7000	.032″ thick	3.17	5.45	8.62
7100	Copper, half round, 4″ wide	10.90	5.45	16.35
7200	6″ wide	16.05	5.70	21.75
7300	Steel, 26 gauge galvanized, 5″ wide	2.40	5.45	7.85
7400	6″ wide	3.17	5.45	8.62
7500	Wood, treated hem-fir, 3″ x 4″	11	5.80	16.80
7600	4″ x 5″	14.85	5.80	20.65
7700	Reglet per L.F., aluminum, .025″ thick	1.95	2.57	4.52
7800	Copper, 10 oz.	3.16	2.57	5.73
7900	Steel, galvanized, 24 gauge	1.13	2.57	3.70
8000	Stainless, .020″ thick	3.80	2.57	6.37
8100	Counter flashing 12″ wide per L.F., aluminum, .032″ thick	1.96	4.38	6.34
8200	Copper, 10 oz.	5.95	4.38	10.33
8300	Steel, galvanized, 24 gauge	1.41	4.38	5.79
8400	Stainless, .020″ thick	5.65	4.38	10.03

C Interiors

C1010 Partitions

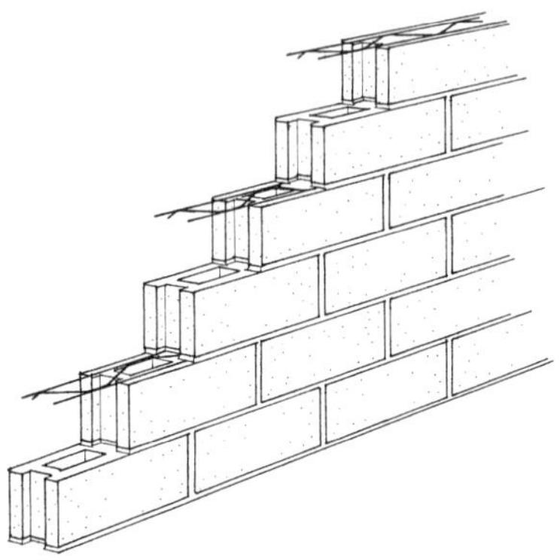

This page illustrates and describes a concrete block wall system including concrete block, horizontal reinforcing alternate courses, mortar, and tooled joints both sides. Lines within System Components give the unit price and total price per square foot for this system. Prices for concrete block wall systems are on Line Items C1010 106 0800 thru 2100. Both material quantities and labor costs have been adjusted for the system listed.

Factors: To adjust for job conditions other than normal working situations use Lines C1010 106 2900 thru 4000.

Example: You are to install the system and protect all existing work. Go to Line C1010 106 3600 and apply these percentages to the appropriate MAT. and INST. costs.

System Components	QUANTITY	UNIT	COST PER S.F.		
			MAT.	INST.	TOTAL
Concrete block partition, including horizontal reinforcing every second Course, mortar, tooled joints both sides.					
8″ x 16″ concrete block, normal weight, 4″ thick	1.000	S.F.	1.63	6.10	7.73
Horizontal reinforcing every second course	.750	L.F.	.10	.14	.24
TOTAL			1.73	6.24	7.97

C1010 106	Partitions, Concrete Block	COST PER S.F.		
		MAT.	INST.	TOTAL
0800	8″ x 16″ concrete block, normal weight, 4″ thick	1.73	6.25	7.98
0900	6″ thick	2.41	6.70	9.11
1000	8″ thick	2.74	7.15	9.89
1100	10″ thick	3.48	7.45	10.93
1200	12″ thick	3.92	9.20	13.12
1300	8″ x 16″ concrete block, lightweight, 4″ thick	1.91	6.10	8.01
1400	6″ thick	2.71	6.55	9.26
1500	8″ thick	3.27	6.95	10.22
1600	10″ thick	3.93	7.25	11.18
1700	12″ thick	4.86	8.95	13.81
1800	8″ x 16″ glazed concrete block, 4″ thick	10.50	7.75	18.25
1900	8″ thick	11.05	8.60	19.65
2000	Structural facing tile, 6T series, glazed 2 sides, 4″ thick	16.50	13.60	30.10
2100	6″ thick	22.50	14.30	36.80
2200				
2300				
2400				
2500				
2600				
2900	Cut & patch to match existing construction, add, minimum	2%	3%	
3000	Maximum	5%	9%	
3100	Dust protection, add, minimum	1%	2%	
3200	Maximum	4%	11%	
3300	Material handling & storage limitation, add, minimum	1%	1%	
3400	Maximum	6%	7%	
3500	Protection of existing work, add, minimum	2%	2%	
3600	Maximum	5%	7%	
3700	Shift work requirements, add, minimum		5%	
3800	Maximum		30%	
3900	Temporary shoring and bracing, add, minimum	2%	5%	
4000	Maximum	5%	12%	

C1010 Partitions

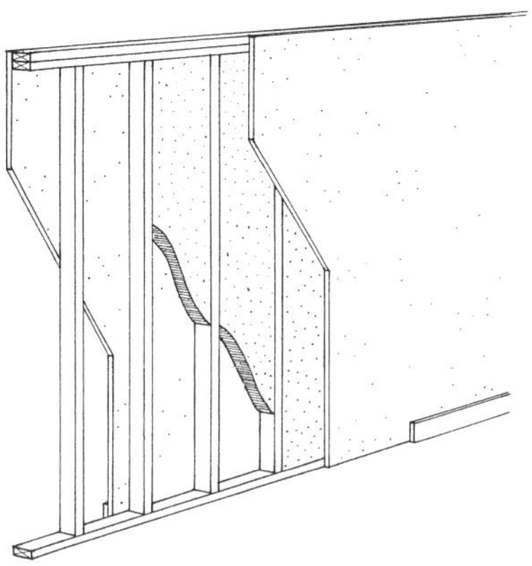

This page illustrates and describes a wood stud partition system including wood studs with plates, gypsum plasterboard – taped and finished, insulation, baseboard and painting. Lines within System Components give the unit price and total price per square foot for this system. Prices for wood stud partition systems are on Line Items C1010 132 1300 thru 2700. Both material quantities and labor costs have been adjusted for the system listed.

Factors: To adjust for job conditions other than normal working situations use Lines C1010 132 2900 thru 4000.

Example: You are to install the system where material handling and storage present a serious problem. Go to Line C1010 132 3400 and apply these percentages to the appropriate MAT. and INST. costs.

System Components	QUANTITY	UNIT	COST PER S.F. MAT.	INST.	TOTAL
Wood stud wall, 2″x4″, 16″ O.C., dbl. top plate, sngl bot. plate, 5/8″ dwl. Taped, finished and painted on 2 faces, insulation, baseboard, wall 8′ high					
Wood studs, 2″ x 4″, 16″ O.C., 8′ high	1.000	S.F.	.45	1.44	1.89
Gypsum drywall, 5/8″ thick	2.000	S.F.	.56	1.16	1.72
Taping and finishing	2.000	S.F.	.10	1.16	1.26
Insulation, 3-1/2″ fiberglass batts	1.000	S.F.	.37	.43	.80
Baseboard	.200	L.F.	.34	.60	.94
Paint baseboard, primer + 2 coats	.200	L.F.	.05	.36	.41
Painting, roller, 2 coats	2.000	S.F.	.38	1.28	1.66
TOTAL			2.25	6.43	8.68

C1010 132	Partitions, Wood Stud	COST PER S.F. MAT.	INST.	TOTAL
1300	2″ x 3″ studs, 8′ high, 16″ O.C.	2.22	6.35	8.57
1400	24″ O.C.	2.13	6.10	8.23
1500	10′ high, 16″ O.C.	2.20	6.10	8.30
1600	24″ O.C.	2.10	5.85	7.95
1650	2″ x 4″ studs, 8′ high, 16″ O.C.	2.25	6.45	8.70
1700	24″ O.C.	2.15	6.15	8.30
1800	10′ high, 16″ O.C.	2.22	6.15	8.37
1900	24″ O.C.	2.12	5.90	8.02
2000	12′ high, 16″ O.C.	2.18	6.15	8.33
2100	24″ O.C.	2.06	5.90	7.96
2200	2″ x 6″ studs, 8′ high, 16″ O.C.	2.53	6.60	9.13
2300	24″ O.C.	2.36	6.25	8.61
2400	10′ high, 16″ O.C.	2.48	6.30	8.78
2500	24″ O.C.	2.32	6	8.32
2600	12′ high, 16″ O.C.	2.42	6.30	8.72
2700	24″ O.C.	2.26	6	8.26
2900	Cut & patch to match existing construction, add, minimum	2%	3%	
3000	Maximum	5%	9%	
3100	Dust protection, add, minimum	1%	2%	
3200	Maximum	4%	11%	
3300	Material handling & storage limitation, add, minimum	1%	1%	
3400	Maximum	6%	7%	
3500	Protection of existing work, add, minimum	2%	2%	
3600	Maximum	5%	7%	
3700	Shift work requirements, add, minimum		5%	
3800	Maximum		30%	
3900	Temporary shoring and bracing, add, minimum	2%	5%	
4000	Maximum	5%	12%	

C1010 Partitions

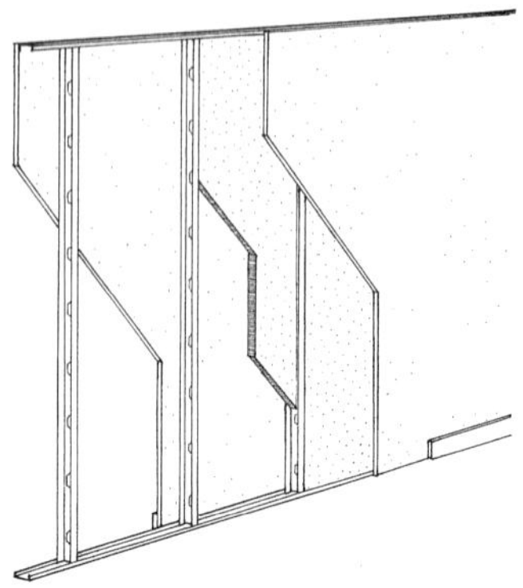

This page illustrates and describes a non-load bearing metal stud partition system including metal studs with runners, gypsum plasterboard, taped and finished, insulation, baseboard and painting. Lines within System Components give the unit price and total price per square foot for this system. Prices for non-load bearing metal stud partition systems are on Line Items C1010 134 1300 thru 2300. Both material quantities and labor costs have been adjusted for the system listed.

Factors: To adjust for job conditions other than normal working situations use Lines C1010 134 2900 thru 4000.

Example: You are to install the system and cut and patch to match existing construction. Go to Line C1010 134 3000 and apply these percentages to the appropriate MAT. and INST. costs.

System Components	QUANTITY	UNIT	COST PER S.F. MAT.	COST PER S.F. INST.	COST PER S.F. TOTAL
Non-load bearing metal studs, including top & bottom runners, 5/8″ drywall, Taped, finished and painted 2 faces, insulation, painted baseboard.					
Metal studs, 25 ga., 3-5/8″ wide, 24″ O.C.	1.000	S.F.	.30	.78	1.08
Gypsum drywall, 5/8″ thick	2.000	S.F.	.56	1.16	1.72
Taping & finishing	2.000	S.F.	.10	1.16	1.26
Insulation, 3-1/2″ fiberglass batts	1.000	S.F.	.37	.43	.80
Baseboard	.200	L.F.	.27	.48	.75
Paint baseboard, primer + 2 coats	.200	L.F.	.04	.29	.33
Painting, roller work, 2 coats	2.000	S.F.	.38	1.28	1.66
TOTAL			2.02	5.58	7.60

C1010 134	Partitions, Metal Stud, NLB	COST PER S.F. MAT.	COST PER S.F. INST.	COST PER S.F. TOTAL
1300	Non-load bearing, 25 ga., 24″ O.C., 2-1/2″ wide	1.97	5.55	7.52
1350	3-5/8″ wide	2.02	5.60	7.62
1400	6″ wide	2.16	5.60	7.76
1500	16″ O.C., 2-1/2″ wide	2.03	5.75	7.78
1600	3-5/8″ wide	2.10	5.80	7.90
1700	6″ wide	2.27	5.80	8.07
1800	20 ga., 24″ O.C., 2-1/2″ wide	2.11	5.75	7.86
1900	3-5/8″ wide	2.25	5.80	8.05
2000	6″ wide	2.35	5.80	8.15
2100	16″ O.C., 2-1/2″ wide	2.21	6	8.21
2200	3-5/8″ wide	2.38	6.05	8.43
2300	6″ wide	2.51	6.05	8.56
2400				
2500				
2600				
2700				
2900	Cut & patch to match existing construction, add, minimum	2%	3%	
3000	Maximum	5%	9%	
3100	Dust protection, add, minimum	1%	2%	
3200	Maximum	4%	11%	
3300	Material handling & storage limitation, add, minimum	1%	1%	
3400	Maximum	6%	7%	
3500	Protection of existing work, add, minimum	2%	2%	
3600	Maximum	5%	7%	
3700	Shift work requirements, add, minimum		5%	
3800	Maximum		30%	
3900	Temporary shoring and bracing, add, minimum	2%	5%	
4000	Maximum	5%	12%	

C1010 Partitions

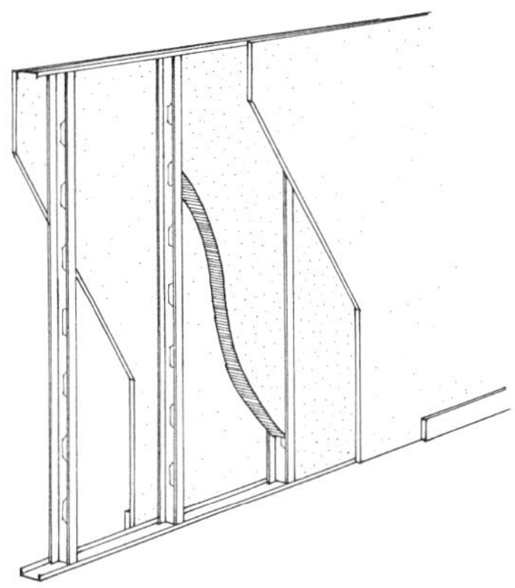

This page illustrates and describes a drywall system including gypsum plasterboard, taped and finished, metal studs with runners, insulation, baseboard and painting. Lines within System Components give the unit price and total price per square foot for this system. Prices for drywall systems are on Line Items C1010 136 1200 thru 1900. Both material quantities and labor costs have been adjusted for the system listed.

Factors: To adjust for job conditions other than normal working situations use Lines C1010 136 2900 thru 4000.

Example: You are to install the system and control dust in the work area. Go to Line C1010 136 3100 and apply these percentages to the appropriate MAT. and INST. costs.

System Components	QUANTITY	UNIT	COST PER S.F.		
			MAT.	INST.	TOTAL
Gypsum drywall, taped, finished and painted 2 faces, galvanized metal studs					
Including top & bottom runners, insulation, painted baseboard, wall 10'high					
Gypsum drywall, 5/8" thick, standard	2.000	S.F.	.56	1.16	1.72
Taping and finishing	2.000	S.F.	.10	1.16	1.26
Metal studs, 20 ga., 3-5/8" wide, 24" O.C.	1.000	S.F.	.45	.98	1.43
Insulation, 3-1/2" fiberglass batts	1.000	S.F.	.37	.43	.80
Baseboard	.200	L.F.	.27	.48	.75
Paint baseboard, primer + 2 coats	.200	L.F.	.03	.12	.15
Painting, roller 2 coats	2.000	S.F.	.28	1.22	1.50
TOTAL			2.06	5.55	7.61

C1010 136	Partitions, Drywall	COST PER S.F.		
		MAT.	INST.	TOTAL
1200	Gypsum drywall, 5/8" thick, standard	2.06	5.55	7.61
1300	Fire resistant	1.95	5.35	7.30
1400	Water resistant	2.29	5.35	7.64
1500	1/2" thick, standard	1.83	5.35	7.18
1600	Fire resistant	1.93	5.35	7.28
1700	Water resistant	2.13	5.35	7.48
1800	3/8" thick, vinyl faced, standard	2.39	6.75	9.14
1900	5/8" thick, vinyl faced, fire resistant	3.27	6.75	10.02
2000				
2100				
2200				
2300				
2400				
2500				
2600				
2700				
2900	Cut & patch to match existing construction, add, minimum	2%	3%	
3000	Maximum	5%	9%	
3100	Dust protection, add, minimum	1%	2%	
3200	Maximum	4%	11%	
3300	Material handling & storage limitation, add, minimum	1%	1%	
3400	Maximum	6%	7%	
3500	Protection of existing work, add, minimum	2%	2%	
3600	Maximum	5%	7%	
3700	Shift work requirements, add, minimum		5%	
3800	Maximum		30%	
3900	Temporary shoring and bracing, add, minimum	2%	5%	
4000	Maximum	5%	12%	

C1010 Partitions

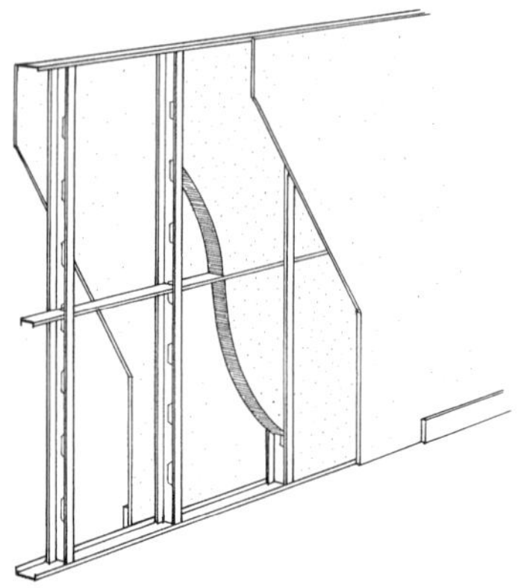

This page illustrates and describes a load bearing metal stud wall system including metal studs, sheetrock–taped and finished, insulation, baseboard and painting. Lines within System Components give the unit price and total price per square foot for this system. Prices for load bearing metal stud wall systems are on Line Items C1010 138 1500 thru 2500. Both material quantities and labor costs have been adjusted for the system listed.

Factors: To adjust for job conditions other than normal working situations use Lines C1010 138 2900 thru 4000.

Example: You are to install the system using temporary shoring and bracing. Go to Line C1010 138 3900 and apply these percentages to the appropriate MAT. and INST. costs.

System Components	QUANTITY	UNIT	COST PER S.F.		
			MAT.	INST.	TOTAL
Load bearing, 18 ga., 3-5/8″, galvanized metal studs, 24″ O.C.,including Top and bottom runners, 1/2″ drywall, taped, finished and painted 2 Faces, 3″ insulation, and painted baseboard, wall 10′ high.					
Metal studs, 24″ O.C., 18 ga., 3-5/8″ wide, galvanized	1.000	S.F.	.87	1.10	1.97
Gypsum drywall 1/2″ thick	2.000	S.F.	.48	1.16	1.64
Taping and finishing	2.000	S.F.	.10	1.16	1.26
Insulation, 3-1/2″ fiberglass batts	1.000	S.F.	.37	.43	.80
Baseboard	.200	L.F.	.27	.48	.75
Paint baseboard, primer + 2 coats	.200	L.F.	.04	.29	.33
Painting, roller 2 coats	2.000	S.F.	.38	1.28	1.66
TOTAL			2.51	5.90	8.41

C1010 138	Partitions, Metal Stud, LB	COST PER S.F.		
		MAT.	INST.	TOTAL
1500	Load bearing, 18 ga., 24″ O.C., 2-1/2″ wide	2.37	5.90	8.27
1550	3-5/8″ wide	2.51	5.90	8.41
1600	6″ wide	2.80	5.95	8.75
1700	16″ O.C. 2-1/2″ wide	2.55	6.15	8.70
1800	3-5/8″ wide	2.72	6.20	8.92
1900	6″ wide	3.08	6.25	9.33
2000	16 ga., 24″ O.C., 2-1/2″ wide	2.49	6.05	8.54
2100	3-5/8″ wide	2.65	6.05	8.70
2200	6″ wide	2.98	6.10	9.08
2300	16″ O.C., 2-1/2″ wide	2.70	6.35	9.05
2400	3-5/8″ wide	2.90	6.35	9.25
2500	6″ wide	3.31	6.45	9.76
2600				
2700				
2900	Cut & patch to match existing construction, add, minimum	2%	3%	
3000	Maximum	5%	9%	
3100	Dust protection, add, minumum	1%	2%	
3200	Maximum	4%	11%	
3300	Material handling & storage limitation, add, minimum	1%	1%	
3400	Maximum	6%	7%	
3500	Protection of existing work, add, minimum	2%	2%	
3600	Maximum	5%	7%	
3700	Shift work requirements, add, minimum		5%	
3800	Maximum		30%	
3900	Temporary shoring and bracing, add, minimum	2%	5%	
4000	Maximum	5%	12%	

C1010 Partitions

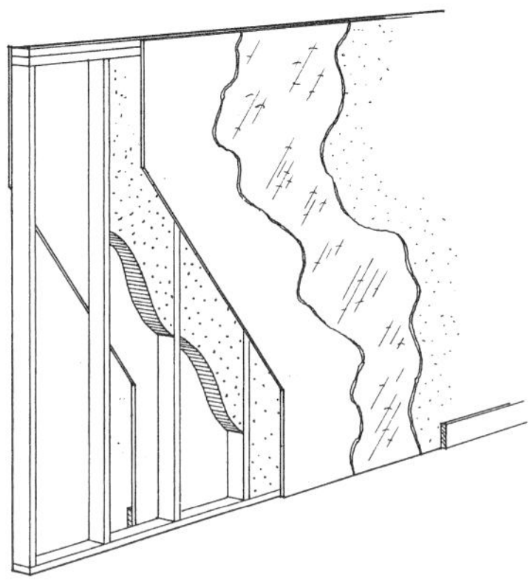

This page illustrates and describes a plaster and lath system including gypsum plaster, gypsum lath, wood studs with plates, insulation, baseboard and painting. Lines within System Components give the unit price and price per square foot for this system. Prices for plaster and lath systems are on Line Items C1010 148 1400 thru 2500. Both material quantities and labor costs have been adjusted for the system listed.

Factors: To adjust for job conditions other than normal working situations use Lines C1010 148 2900 thru 4000.

Example: You are to install the system during evening hours only. Go to Line C1010 148 3800 and apply this percentage to the appropriate INST. costs.

System Components	QUANTITY	UNIT	COST PER S.F.		
			MAT.	INST.	TOTAL
Gypsum plaster, 2 coats, over 3/8″ lath, 2 faces, 2″ x 4″ wood Stud partition, 24″O.C. incl. double top plate,single bottom plate,3-1/2″ Insulation, baseboard and painting.					
Gypsum plaster, 2 coats	.220	S.Y.	.86	5.47	6.33
Lath, gypsum, 3/8″ thick	.220	S.Y.	.94	1.28	2.22
Wood studs, 2″ x 4″, 24″ O.C.	1.000	S.F.	.32	.93	1.25
Insulation, 3-1/2″ fiberglass batts	1.000	S.F.	.37	.43	.80
Baseboard, 9/16″ x 3-1/2″	.200	L.F.	.27	.48	.75
Paint baseboard, primer + 2 coats	.200	L.F.	.03	.12	.15
Paint, 2 coats	2.000	S.F.	.38	1.28	1.66
TOTAL			3.17	9.99	13.16

C1010 148	Partitions, Plaster & Lath	COST PER S.F.		
		MAT.	INST.	TOTAL
1400	Gypsum plaster, 2 coats	3.17	10	13.17
1500	3 coats	3.55	11.15	14.70
1600	Perlite plaster, 2 coats	3.23	10.70	13.93
1700	3 coats	3.84	12.35	16.19
1800				
1900				
2000	For alternate lath systems:			
2100	Gypsum lath, 1/2″ thick	3.38	10.75	14.13
2200	Foil back, 3/8″ thick	2.92	10.85	13.77
2300	1/2″ thick	2.94	10.95	13.89
2400	Metal lath, 2.5 Lb. diamond	3.12	10.70	13.82
2500	3.4 Lb. diamond	3.26	10.85	14.11
2600				
2700				
2900	Cut & patch to match existing construction, add, minimum	2%	3%	
3000	Maximum	5%	9%	
3100	Dust protection, add, minimum	1%	2%	
3200	Maximum	4%	11%	
3300	Material handling & storage limitation, add, minimum	1%	1%	
3400	Maximum	6%	7%	
3500	Protection of existing work, add, minimum	2%	2%	
3600	Maximum	5%	7%	
3700	Shift work requirements, add, minimum		5%	
3800	Maximum		30%	
3900	Temporary shoring and bracing, add, minimum	2%	5%	
4000	Maximum	5%	12%	

C1010 Partitions

C1010 170	Selective Price Sheet	COST PER S.F. MAT.	COST PER S.F. INST.	COST PER S.F. TOTAL
0100	Studs			
0200				
0300	24" O.C. metal, 10' high wall, including			
0400	Top and bottom runners			
0500	Non load bearing, galvanized 25 Ga., 1-5/8" wide	.22	.76	.98
0600	2-1/2" wide	.25	.77	1.02
0700	3-5/8" wide	.30	.78	1.08
0800	4" wide	.34	.78	1.12
0900	6" wide	.44	.80	1.24
1000				
1100	Galvanized 20 Ga., 2-1/2" wide	.39	.96	1.35
1200	3-5/8" wide	.45	.98	1.43
1300	4" wide	.53	.98	1.51
1400	6" wide	.63	1	1.63
1500	Load bearing, painted 18 Ga., 2-1/2" wide	.73	1.08	1.81
1600	3-5/8" wide	.87	1.10	1.97
1700	4" wide	.78	1.13	1.91
1800	6" wide	1.16	1.15	2.31
1900	Galvanized 18 Ga., 2-1/2" wide	.73	1.08	1.81
2000	3-5/8" wide	.87	1.10	1.97
2100	4" wide	.78	1.13	1.91
2200	6" wide	1.16	1.15	2.31
2300	Galvanized 16 Ga., 2-1/2" wide	.85	1.23	2.08
2400	3-5/8" wide	1.01	1.26	2.27
2500	4" wide	1.07	1.29	2.36
2600	6" wide	1.34	1.32	2.66
2700				
2800				
2900	24" O.C. wood, 10' high wall, including			
3000	Double top plate and shoe			
3100	2" x 4"	.37	1.09	1.46
3200	2" x 6"	.60	1.17	1.77
3300				
3400				
3500	Furring, 24" O.C. 10' high wall, metal, 3/4" channels	.22	1.17	1.39
3600	1-1/2" channels	.30	1.26	1.56
3700	Wood, on wood, 1" x 2" strips	.14	.53	.67
3800	1" x 3" strips	.19	.53	.72
3900	On masonry, 1" x 2" strips	.14	.59	.73
4000	1" x 3" strips	.19	.59	.78
4100	On concrete, 1" x 2" strips	.14	1.11	1.25
4200	1" x 3" strips	.19	1.11	1.30
4300	Studs			
4400				
4500	16" O.C., metal, 10' high wall, including			
4600	Top and bottom runners			
4700	Non load bearing, galvanized 25 Ga., 1-5/8" wide	.30	1.17	1.47
4800	2-1/2" wide	.34	1.18	1.52
4900	3-5/8" wide	.41	1.21	1.62
5000	4" wide	.47	1.22	1.69
5100	6" wide	.61	1.23	1.84
5200				
5300	Galvanized 20 Ga., 2-1/2" wide	.53	1.48	2.01
5400	3-5/8" wide	.60	1.50	2.10
5500	4" wide	.72	1.52	2.24
5600	6" wide	.86	1.54	2.40

C1010 Partitions

C1010 170	Selective Price Sheet	COST PER S.F.		
		MAT.	INST.	TOTAL
5700	Load bearing, painted 18 Ga., 2-1/2″ wide	.99	1.50	2.49
5800	3-5/8″ wide	1.18	1.52	2.70
5900	4″ wide	1.23	1.57	2.80
6000	6″ wide	1.56	1.59	3.15
6100	Galvanized 18 Ga., 2-1/2″ wide	.99	1.50	2.49
6200	3-5/8″ wide	1.18	1.52	2.70
6300	4″ wide	1.23	1.57	2.80
6400	6″ wide	1.56	1.59	3.15
6500	Galvanized 16 Ga., 2-1/2″ wide	1.17	1.70	2.87
6600	3-5/8″ wide	1.39	1.76	3.15
6700	4″ wide	1.47	1.78	3.25
6800	6″ wide	1.83	1.81	3.64
6900				
7000				
7100	16″ O.C., wood, 10′ high wall, including			
7200	Double top plate and shoe			
7300	2″ x 4″	.45	1.26	1.71
7400	2″ x 6″	.59	1.13	1.72
7500				
7600				
7700	Furring, 16″ O.C. 10′ high wall, metal, 3/4″ channels	.34	1.70	2.04
7800	1-1/2″ channels	.45	1.89	2.34
7900	Wood, on wood, 1″ x 2″ strips	.21	.79	1
8000	1″ x 3″ strips	.29	.79	1.08
8100	On masonry, 1″ x 2″ strips	.21	.88	1.09
8200	1″ x 3″ strips	.29	.88	1.17
8300	On concrete, 1″ x 2″ strips	.21	1.67	1.88
8400	1″ x 3″ strips	.29	1.67	1.96

C1010 Partitions

C1010 180	Selective Price Sheet, Drywall & Plaster	COST PER S.F.		
		MAT.	INST.	TOTAL
0100	Lath, gypsum perforated			
0200				
0300	Lath, gypsum, perforated, regular, 3/8" thick	.47	.64	1.11
0400	1/2" thick	.55	.78	1.33
0500	Fire resistant, 3/8" thick	.28	.78	1.06
0600	1/2" thick	.30	.84	1.14
0700	Foil back, 3/8" thick	.32	.73	1.05
0800	1/2" thick	.33	.78	1.11
0900				
1000	Metal lath			
1100	Metal lath, diamond, painted, 2.5 lb.	.42	.64	1.06
1200	3.4 lb.	.49	.73	1.22
1300	Rib painted, 2.75 lb	.38	.73	1.11
1400	3.40 lb	.59	.78	1.37
1500				
1600				
1700	Plaster, gypsum, 2 coats	.44	2.76	3.20
1800	3 coats	.63	3.34	3.97
1900	Perlite/vermiculite, 2 coats	.46	3.11	3.57
2000	3 coats	.77	3.94	4.71
2100	Bondcrete, 1 coat	.43	1.45	1.88
2200				
2500				
2600				
2700	Drywall, standard, 3/8" thick, no finish included	.24	.58	.82
2800	1/2" thick, no finish included	.24	.58	.82
2900	Taped and finished	.29	1.20	1.49
3000	5/8" thick, no finish included	.28	.58	.86
3100	Taped and finished	.33	1.20	1.53
3200	Fire resistant, 1/2" thick, no finish included	.29	.58	.87
3300	Taped and finished	.34	1.20	1.54
3400	5/8" thick, no finish included	.30	.58	.88
3500	Taped and finished	.35	1.20	1.55
3600	Water resistant, 1/2" thick, no finish included	.39	.58	.97
3700	Taped and finished	.44	1.20	1.64
3800	5/8" thick, no finish included	.47	.58	1.05
3900	Taped and finished	.52	1.20	1.72
4000	Finish, instead of taping			
4100	For thin coat plaster, add	.11	.72	.83
4200	Finish, textured spray, add	.04	.62	.66

C1010 Partitions

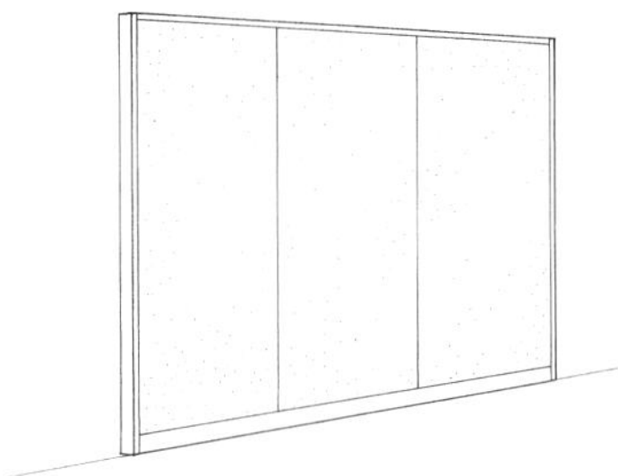

This page illustrates and describes a movable office partition system including demountable office partitions of various styles and sizes based on a cost per square foot basis. Prices for movable office partition systems are on Line Items C1010 210 0300 thru 3600.

C1010 210	Partitions, Movable Office	COST PER S.F. MAT.	COST PER S.F. INST.	COST PER S.F. TOTAL
0250	Office partition, demountable, no deduction for door opening, add for doors			
0300	Air wall, cork finish, semi acoustic, 1-5/8" thick, minimum	40.50	3.56	44.06
0400	Maximum	45	6.10	51.10
0500	Acoustic, 2" thick, minimum	38.50	3.79	42.29
0600	Maximum	59	5.15	64.15
0700				
0800				
0900	In-plant modular office system, w/prehung steel door			
1000	3" thick honeycomb core panels			
1100	12' x 12', 2 wall	8.55	.48	9.03
1200	4 wall	22.50	1.46	23.96
1300				
1400				
1500	Gypsum, demountable, 3" to 3-3/4" thick x 9' high, vinyl clad	7.05	2.66	9.71
1600	Fabric clad	17.55	2.94	20.49
1700	1.75 system, vinyl clad hardboard, paper honeycomb core panel			
1800	1-3/4" to 2-1/2" thick x 9' high	11.90	2.66	14.56
1900	Unitized gypsum panel, 2" to 2-1/2" thick x 9' high, vinyl clad	15.20	2.66	17.86
2000	Fabric clad	25	2.94	27.94
2100				
2200				
2300	Unitized mineral fiber panel system, 2-1/4" thick x 9' high			
2400	Vinyl clad mineral fiber	15.20	2.66	17.86
2500	Fabric clad mineral fiber	22.50	2.94	25.44
2600				
2700	Movable steel walls, modular system			
2800	Unitized panels, 48" wide x 9' high			
2900	Baked enamel, pre-finished	17.20	2.14	19.34
3000	Fabric clad	25	2.28	27.28
3100	For acoustical partitions, add, minimum	2.44		2.44
3200	Maximum	11.40		11.40
3300				
3400				
3500	Note: For door prices, see Div. 08			
3600	For door hardware prices, see Div. 08			
3700				
3800				
3900				
4000				

C1020 Interior Doors

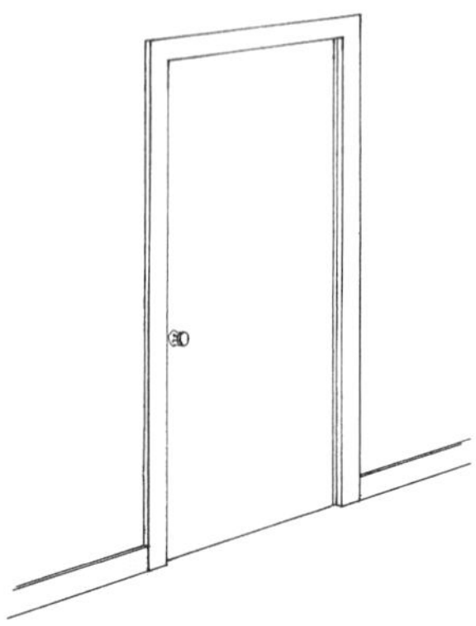

This page illustrates and describes flush interior door systems including hollow core door, jamb, header and trim with hardware. Lines within System Components give the unit price and total price on a cost each basis for this system. Prices for flush interior door systems are on Line items C1020 106 1000 thru 2400. Both material quantities and labor costs have been adjusted for the system listed.

Factors: To adjust for job conditions other than normal working situations use Lines C1020 106 2700 thru 3800.

Example: You are to install the system in an area where dust protection is vital. Go to Line C1020 106 3000 and apply these percentages to the appropriate MAT. and INST. costs.

System Components	QUANTITY	UNIT	COST EACH		
			MAT.	INST.	TOTAL
Single hollow core door, include jamb, header, trim and hardware, painted.					
Hollow core Lauan, 1-3/8" thick, 2'-0" x 6'-8", painted	1.000	Ea.	38	64.50	102.50
Paint door and frame, 1 coat	1.000	Ea.	2.06	28.50	30.56
Wood jamb, 4-9/16" deep	1.000	Set	104.80	49.44	154.24
Trim, casing	1.000	Set	46.40	73.92	120.32
Hardware, hinges	1.000	Set	14.35		14.35
Hardware, lockset	1.000	Set	33	36	69
TOTAL			238.61	252.36	490.97

C1020 106	Doors, Interior Flush, Wood	COST EACH		
		MAT.	INST.	TOTAL
1000	Lauan (Mahogany) hollow core, 1-3/8" x 2'-0" x 6'-8"	239	252	491
1100	2'-6" x 6'-8"	247	256	503
1200	2'-8" x 6'-8"	252	259	511
1300	3'-0" x 6'-8"	259	268	527
1500	Birch, hollow core, 1-3/8" x 2'-0" x 6'-8"	247	252	499
1600	2'-6" x 6'-8"	260	256	516
1700	2'-8" x 6'-8"	264	259	523
1800	3'-0" x 6'-8"	270	268	538
1900	Solid core, pre-hung, 1-3/8" x 2'-6" x 6'-8"	475	260	735
2000	2'-8" x 6'-8"	425	263	688
2100	3'-0" x 6'-8"	520	273	793
2200				
2400	For metal frame instead of wood, add	50%	20%	
2600				
2700	Cut & patch to match existing construction, add, minimum	2%	3%	
2800	Maximum	5%	9%	
2900	Dust protection, add, minimum	1%	2%	
3000	Maximum	4%	11%	
3100	Equipment usage curtailment, add, minimum	1%	1%	
3200	Maximum	3%	10%	
3300	Material handling & storage limitation, add, minimum	1%	1%	
3400	Maximum	6%	7%	
3500	Protection of existing work, add, minimum	2%	2%	
3600	Maximum	5%	7%	
3700	Shift work requirements, add, minimum		5%	
3800	Maximum		30%	
3900				

C1020 Interior Doors

This page illustrates and describes interior, solid and louvered door systems including a pine panel door, wood jambs, header, and trim with hardware. Lines within System Components give the unit price and total price on a cost each basis for this system. Prices for interior, solid and louvered systems are on Line Items C1020 108 1100 thru 2400. Both material quantities and labor costs have been adjusted for the system listed.

Factors: To adjust for job conditions other than normal working situations use Lines C1020 108 2900 thru 4000.

Example: You are to install the system during night hours only. Go to Line C1020 108 4000 and apply these percentages to the appropriate INST. costs.

System Components	QUANTITY	UNIT	COST EACH		
			MAT.	INST.	TOTAL
Single interior door, including jamb, header, trim and hardware, painted.					
Solid pine panel door, 1-3/8" thick, 2'-0" x 6'-8"	1.000	Ea.	167	64.50	231.50
Paint door and frame, 1 coat	1.000	Ea.	2.06	28.50	30.56
Wooden jamb, 4-5/8" deep	1.000	Set	104.80	49.44	154.24
Trim, casing	1.000	Set	46.40	73.92	120.32
Hardware, hinges	1.000	Set	14.35		14.35
Hardware, lockset	1.000	Set	33	36	69
TOTAL			367.61	252.36	619.97

C1020 108	Doors, Interior Solid & Louvered	COST EACH		
		MAT.	INST.	TOTAL
1000				
1100	Solid pine, painted raised panel, 1-3/8" x 2'-0" x 6'-8"	370	252	622
1200	2'-6" x 6'-8"	405	256	661
1300	2'-8" x 6'-8"	410	259	669
1400	3'-0" x 6'-8"	435	268	703
1500				
1600	Louvered pine, painted 1'-6" x 6'-8"	315	249	564
1700	2'-0" x 6'-8"	345	256	601
1800	2'-6" x 6'-8"	360	259	619
1900	3'-0" x 6'-8"	405	268	673
2200	For prehung door, deduct	5%	30%	
2300				
2400	For metal frame instead of wood, add	50%	20%	
2500				
2900	Cut & patch to match existing construction, add, minimum	2%	3%	
3000	Maximum	5%	9%	
3100	Dust protection, add, minimum	1%	2%	
3200	Maximum	4%	11%	
3300	Equipment usage curtailment, add, minimum	1%	1%	
3400	Maximum	3%	10%	
3500	Material handling & storage limitation, add, minimum	1%	1%	
3600	Maximum	6%	7%	
3700	Protection of existing work, add, minimum	2%	2%	
3800	Maximum	5%	7%	
3900	Shift work requirements, add, minimum		5%	
4000	Maximum		30%	

C1020 Interior Doors

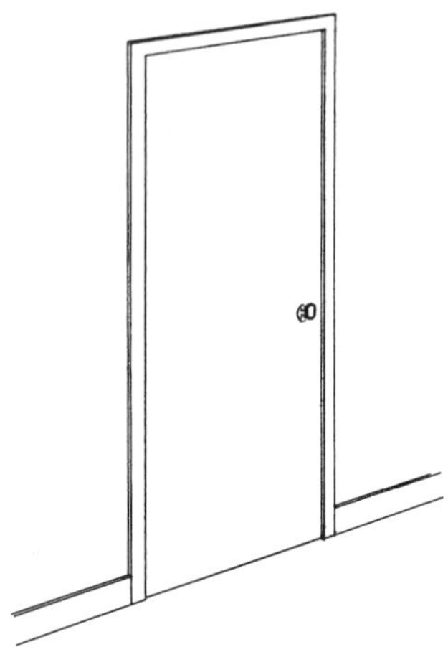

This page illustrates and describes interior metal door systems including a metal door, metal frame and hardware. Lines within System Components give the unit price and total price on a cost each for this system. Prices for interior metal door systems are on Line Items C1020 110 1000 thru 2000. Both material quantities and labor costs have been adjusted for the system listed.

Factors: To adjust for job conditions other than normal working situations use Lines C1020 110 2900 thru 4000.

Example: You are to install the system while protecting existing construction. Go to Line C1020 110 3700 and apply these percentages to the appropriate MAT. and INST. costs.

System Components	QUANTITY	UNIT	COST EACH		
			MAT.	INST.	TOTAL
Single metal door, including frame and hardware.					
Hollow metal door, 1-3/8″ thick, 2′-6″ x 6′-8″, painted	1.000	Ea.	540	77	617
Metal frame, 6-1/4″ deep	1.000	Set	183	72.50	255.50
Paint door and frame, 1 coat	1.000	Ea.	2.06	28.50	30.56
Hardware, hinges	1.000	Set	85.50		85.50
Hardware, passage lockset	1.000	Set	33	36	69
TOTAL			843.56	214	1,057.56

C1020 110	Doors, Interior Flush, Metal	COST EACH		
		MAT.	INST.	TOTAL
1000	Hollow metal doors, 1-3/8″ thick, 2′-6″ x 6′-8″	845	214	1,059
1100	1-3/8″ thick, 2′-8″ x 6′-8″	855	214	1,069
1200	3′-0″ x 7′-0″	840	224	1,064
1300				
1400	Interior fire door, 1-3/8″ thick, 2′-6″ x 6′-8″	1,050	214	1,264
1500	2′-8″ x 6′-8″	1,050	214	1,264
1600	3′-0″ x 7′-0″	1,050	224	1,274
1700				
1800	Add to fire doors:			
1900	Baked enamel finish	30%	15%	
2000	Galvanizing	15%		
2200				
2300				
2400				
2900	Cut & patch to match existing construction, add, minimum	2%	3%	
3000	Maximum	5%	9%	
3100	Dust protection, add, minimum	1%	2%	
3200	Maximum	4%	11%	
3300	Equipment usage curtailment, add, minimum	1%	1%	
3400	Maximum	3%	10%	
3500	Material handling & storage limitation, add, minimum	1%	1%	
3600	Maximum	6%	7%	
3700	Protection of existing work, add, minimum	2%	2%	
3800	Maximum	5%	7%	
3900	Shift work requirements, add, minimum		5%	
4000	Maximum		30%	

C1020 Interior Doors

This page illustrates and describes an interior closet door system including an interior closet door, painted, with trim and hardware. Prices for interior closet door systems are on Line Items C1020 112 0500 thru 2200. Both material quantities and labor costs have been adjusted for the system listed.

Factors: To adjust for job conditions other than normal working situations use Lines C1020 112 2900 thru 4000.

Example: You are to install the system and match the existing construction. Go to Line C1020 112 2900 and apply these percentages to the appropriate MAT. and INST. costs.

C1020 112	Doors, Closet	COST PER SET		
		MAT.	INST.	TOTAL
0350	Interior closet door painted, including frame, trim and hardware, prehung.			
0400	Bi-fold doors			
0500	Pine paneled, 3′-0″ x 6′-8″	425	209	634
0600	6′-0″ x 6′-8″	585	281	866
0700	Birch, hollow core, 3′-0″ x 6′-8″	216	227	443
0800	6′-0″ x 6′-8″	315	305	620
0900	Lauan, hollow core, 3′-0″ x 6′-8″	200	209	409
1000	6′-0″ x 6′-8″	286	281	567
1100	Louvered pine, 3′-0″ x 6′-8″	350	209	559
1200	6′-0″ x 6′-8″	460	281	741
1300				
1400	Sliding, bi-passing closet doors			
1500	Pine paneled, 4′-0″ x 6′-8″	690	222	912
1600	6′-0″ x 6′-8″	805	281	1,086
1700	Birch, hollow core, 4′-0″ x 6′-8″	370	222	592
1800	6′-0″ x 6′-8″	425	281	706
1900	Lauan, hollow corc, 4′ 0″ x 6′ 8″	325	222	547
2000	6′-0″ x 6′-8″	375	281	656
2100	Louvered pinc, 4′ 0″ x 6′ 8″	575	222	797
2200	6′-0″ x 6′-8″	720	281	1,001
2300				
2400				
2500				
2600				
2700				
2800				
2900	Cut & patch to match existing construction, add, minimum	2%	3%	
3000	Maximum	5%	9%	
3100	Dust protection, add, minimum	1%	2%	
3200	Maximum	4%	11%	
3300	Equipment usage curtailment, add, minimum	1%	1%	
3400	Maximum	3%	10%	
3500	Material handling & storage limitation, add, minimum	1%	1%	
3600	Maximum	6%	7%	
3700	Protection of existing work, add, minimum	2%	2%	
3800	Maximum	5%	7%	
3900	Shift work requirements, add, minimum		5%	
4000	Maximum		30%	

C2010 Stair Construction

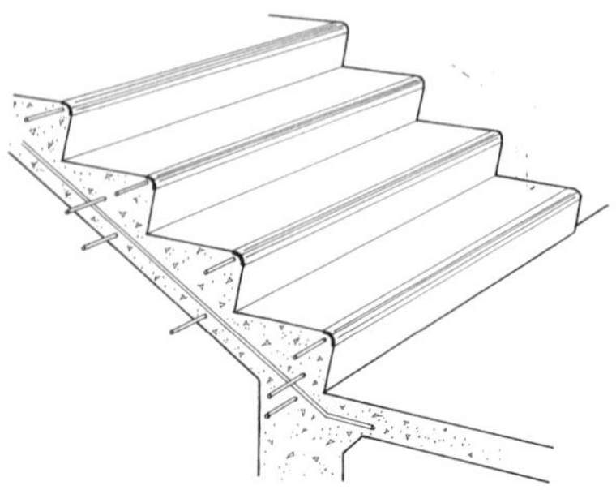

This page illustrates and describes a stair system based on a cost per flight price. Prices for various stair systems are on Line Items C2010 130 0700 thru 3200. Both material quantities and labor costs have been adjusted for the system listed.

Factors: To adjust for job conditions other than normal working situations use Lines C2010 130 3500 thru 4200.

Example: You are to install the system during evenings only. Go to Line C2010 130 4200 and apply this percentage to the appropriate MAT. and INST. costs.

System Components	QUANTITY	UNIT	COST PER FLIGHT MAT.	COST PER FLIGHT INST.	COST PER FLIGHT TOTAL
Below are various stair systems based on cost per flight of stairs, no side Walls. Stairs are 4′-0″ wide, railings are included unless otherwise noted					

C2010 130	Stairs	COST PER FLIGHT MAT.	COST PER FLIGHT INST.	COST PER FLIGHT TOTAL
0700	Concrete, cast in place, no nosings, no railings, 12 risers	269	2,000	2,269
0800	24 risers	540	3,950	4,490
0900	Add for 1 intermediate landing	145	550	695
1000	Concrete, cast in place, with nosings, no railings, 12 risers	970	2,350	3,320
1100	24 risers	1,950	4,650	6,600
1200	Add for 1 intermediate landing	204	580	784
1300	Steel, grating tread, safety nosing, 12 risers	4,500	1,175	5,675
1400	24 risers	8,975	2,325	11,300
1500	Add for intermediate landing	1,100	310	1,410
1600	Steel, cement fill pan tread, 12 risers	6,725	1,175	7,900
1700	24 risers	13,500	2,325	15,825
1800	Add for intermediate landing	1,100	310	1,410
1900	Spiral, industrial, 4′ - 6″ diameter, 12 risers	6,650	825	7,475
2000	24 risers	13,300	1,650	14,950
2100	Wood, box stairs, oak treads, 12 risers	2,325	655	2,980
2200	24 risers	4,675	1,300	5,975
2300	Add for 1 intermediate landing	146	148	294
2400	Wood, basement stairs, no risers, 12 steps	785	270	1,055
2500	24 steps	1,575	540	2,115
2600	Add for 1 intermediate landing	30	28.50	58.50
2700	Wood, open, rough sawn cedar, 12 steps	2,650	530	3,180
2800	24 steps	5,300	1,050	6,350
2900	Add for 1 intermediate landing	23.50	29	52.50
3000	Wood, residential, oak treads, 12 risers	2,175	2,900	5,075
3100	24 risers	4,350	5,775	10,125
3200	Add for 1 intermediate landing	131	134	265
3500	Dust protection, add, minimum	1%	2%	
3600	Maximum	4%	11%	
3700	Material handling & storage limitation, add, minimum	1%	1%	
3800	Maximum	6%	7%	
3900	Protection of existing work, add, minimum	2%	2%	
4000	Maximum	5%	7%	
4100	Shift work requirements, add, minimum		5%	
4200	Maximum		30%	

C3010 Wall Finishes

C3010 210	Selective Price Sheet	COST PER S.F. MAT.	COST PER S.F. INST.	COST PER S.F. TOTAL
0100	Painting, on plaster or drywall, brushwork, primer and 1 ct.	.14	.72	.86
0200	Primer and 2 ct.	.21	1.14	1.35
0300	Rollerwork, primer and 1 ct.	.14	.61	.75
0400	Primer and 2 ct.	.21	.97	1.18
0500	Woodwork incl. puttying, brushwork, primer and 1 ct.	.12	1.09	1.21
0600	Primer and 2 ct.	.19	1.44	1.63
0700	Wood trim to 6″ wide, enamel, primer and 1 ct.	.13	.61	.74
0800	Primer and 2 ct.	.19	.78	.97
0900	Cabinets and casework, enamel, primer and 1 ct.	.13	1.22	1.35
1000	Primer and 2 ct.	.19	1.50	1.69
1100	On masonry or concrete, latex, brushwork, primer and 1 ct.	.29	1.02	1.31
1200	Primer and 2 ct.	.37	1.46	1.83
1300	For block filler, add	.20	1.25	1.45
1400				
1500	Varnish, wood trim, sealer 1 ct., sanding, puttying, quality work	.14	2.20	2.34
1600	Medium work	.11	1.71	1.82
1700	Without sanding	.17	.26	.43
1800				
1900	Wall coverings, wall paper, at low price per double roll, avg workmanship	.41	.77	1.18
2000	At medium price per double roll, average workmanship	1.07	.92	1.99
2100	At high price per double roll, quality workmanship	2.72	1.13	3.85
2200				
2300	Grass cloths with lining paper, minimum	.86	1.23	2.09
2400	Maximum	2.74	1.40	4.14
2500	Vinyl, fabric backed, light weight	.90	.77	1.67
2600	Medium weight	.99	1.02	2.01
2700	Heavy weight	1.67	1.13	2.80
2800				
2900	Cork tiles, 12″ x 12″, 3/16″ thick	4.92	2.05	6.97
3000	5/16″ thick	4.18	2.09	6.27
3100	Granular surface, 12″ x 36″, 1/2″ thick	1.39	1.28	2.67
3200	1″ thick	1.79	1.33	3.12
3300	Aluminum foil	1.12	1.79	2.91
3400				
3500	Tile, ceramic, adhesive set, 4-1/4″ x 4-1/4″	2.42	4.81	7.23
3600	6″ x 6″	2.80	5.25	8.05
3700	Decorated, 4-1/4″ x 4-1/4″, minimum	3.08	3.39	6.47
3800	Maximum	50	5.10	55.10
3900	For epoxy grout, add	.44	1.14	1.58
4000	Pregrouted sheets	5.35	3.81	9.16
4100	Glass mosaics, 3/4″ tile on 12″ sheets, minimum	18.90	12.55	31.45
4200	Color group 8	62.50	14.30	76.80
4300	Metal, tile pattern, 4′ x 4′ sheet, 24 ga., nailed			
4400	Stainless steel	29.50	2.26	31.76
4500	Aluminized steel	15.95	2.26	18.21
4600				
4700	Brick, interior veneer, 4″ face brick, running bond, minimum	4.53	12.20	16.73
4800	Maximum	4.53	12.20	16.73
4900	Simulated, urethane pieces, set in mastic	7.35	3.86	11.21
5000	Fiberglass panels	4.26	2.89	7.15
5100	Wall coating, on drywall, thin coat, plain	.11	.72	.83
5200	Stipple	.11	.72	.83
5300	Textured spray	.04	.62	.66
5400				
5500	Paneling not incl. furring or trim, hardboard, tempered, 1/8″ thick	.42	2.31	2.73
5600	1/4″ thick	.61	2.31	2.92
5700	Plastic faced, 1/8″ thick	.75	2.31	3.06
5800	1/4″ thick	.92	2.31	3.23

C3010 Wall Finishes

C3010 210	Selective Price Sheet	COST PER S.F.		
		MAT.	INST.	TOTAL
5900	Woodgrained, 1/4" thick, minimum	.64	2.31	2.95
6000	Maximum	1.23	2.72	3.95
6100	Plywood, 4' x 8' shts. 1/4" thick, prefin., birch faced, min.	.99	2.31	3.30
6200	Maximum	2.19	3.30	5.49
6300	Walnut, minimum	3.06	2.31	5.37
6400	Maximum	5.80	2.89	8.69
6500	Mahogany, African	2.79	2.89	5.68
6600	Philippine	1.20	2.31	3.51
6700	Chestnut	5.30	3.09	8.39
6800	Pecan	2.29	2.89	5.18
6900	Rosewood	5.10	3.62	8.72
7000	Teak	3.58	2.89	6.47
7100	Aromatic cedar, plywood	2.31	2.89	5.20
7200	Particle board	1.12	2.89	4.01
7300	Wood board, 3/4" thick, knotty pine	2.08	3.86	5.94
7400	Rough sawn cedar	3.15	3.86	7.01
7500	Redwood, clear	5.45	3.86	9.31
7600	Aromatic cedar	2.20	4.21	6.41

C3020 Floor Finishes

C3020 430	Selective Price Sheet	COST PER S.F.		
		MAT.	INST.	TOTAL
0100	Flooring, carpet, acrylic, 26 oz. light traffic	3.55	.75	4.30
0200	35 oz. heavy traffic	6.55	.75	7.30
0300	Nylon anti-static, 15 oz. light traffic	2.33	1	3.33
0400	22 oz. medium traffic	4.61	.75	5.36
0500	26 oz. heavy traffic	6.05	.81	6.86
0600	28 oz. heavy traffic	6.60	.81	7.41
0700	Tile, foamed back, needle punch	4.66	.90	5.56
0800	Tufted loop	3.25	.90	4.15
0900	Wool, 36 oz. medium traffic	15.45	.81	16.26
1000	42 oz. heavy traffic	15.75	.81	16.56
1100	Composition, epoxy, with colored chips, minimum	2.94	4.33	7.27
1200	Maximum	4.24	5.95	10.19
1300	Trowelled, minimum	3.74	5.20	8.94
1400	Maximum	5.65	6.10	11.75
1500	Terrazzo, 1/4" thick, chemical resistant, minimum	6.20	6.10	12.30
1600	Maximum	9.45	8.10	17.55
1700	Resilient, asphalt tile, 1/8" thick	1.31	1.28	2.59
1800	Conductive flooring, rubber, 1/8" thick	6.95	1.62	8.57
1900	Cork tile 1/8" thick, standard finish	8.10	1.62	9.72
2000	Urethane finish	10.45	1.62	12.07
2100	PVC sheet goods for gyms, 1/4" thick	10.10	6.40	16.50
2200	3/8" thick	11.65	8.55	20.20
2300	Vinyl composition 12" x 12" tile, plain, 1/16" thick	1.03	1.02	2.05
2400	1/8" thick	1.74	1.02	2.76
2500	Vinyl tile, 12" x 12" x 1/8" thick, minimum	5.30	1.02	6.32
2600	Maximum	15.40	1.02	16.42
2700	Vinyl sheet goods, backed, .093" thick	4.35	2.22	6.57
2900	Slate, random rectangular, 1/4" thick	24.50	6.85	31.35
3000	1/2" thick	23	9.80	32.80
3100	Natural cleft, irregular, 3/4" thick	10.55	11.15	21.70
3200	For sand rubbed finish, add	10.55		10.55
3300	Terrazzo, cast in place, bonded 1-3/4" thick, gray cement	3.47	16.25	19.72
3400	White cement	3.86	16.25	20.11
3500	Not bonded 3" thick, gray cement	4.32	17.40	21.72
3600	White cement	5.05	17.40	22.45
3700	Precast, 12" x 12" x 1" thick	21	17.10	38.10
3800	1-1/4" thick	22	17.10	39.10
3900	16" x 16" x 1-1/4" thick	24	20.50	44.50
4000	1-1/2" thick	22	23	45
4100	Marble travertine, standard, 12" x 12" x 3/4" thick	14.50	15.25	29.75
4200				
4300	Tile, ceramic, natural clay, thin set	4.21	5	9.21
4400	Porcelain, thin set	4.76	4.81	9.57
4500	Specialty, decorator finish	11.20	5	16.20
4600				
4700	Quarry, red, mud set, 4" x 4" x 1/2" thick	7.10	7.60	14.70
4800	6" x 6" x 1/2" thick	6.60	6.55	13.15
4900	Brown, imported, 6" x 6" x 3/4" thick	8.20	7.60	15.80
5000	8" x 8" x 1" thick	8.95	8.30	17.25
5100	Slate, Vermont, thin set, 6" x 6" x 1/4" thick	8.25	5.10	13.35
5200				
5300	Wood, maple strip, 25/32" x 2-1/4", finished, select	5.85	3.66	9.51
5400	2nd and better	4.56	3.66	8.22
5500	Oak, 25/32" x 2-1/4" finished, clear	2.36	3.66	6.02
5600	No. 1 common	4.30	3.66	7.96
5700	Parquet, standard, 5/16", finished, minimum	5.15	3.88	9.03
5800	Maximum	6.50	6.05	12.55
5900	Custom, finished, minimum	17.25	5.80	23.05

C3020 Floor Finishes

C3020 430	Selective Price Sheet	COST PER S.F.		
		MAT.	INST.	TOTAL
6000	Maximum	23	11.55	34.55
6100	Prefinished, oak, 2-1/4" wide	5.25	3.40	8.65
6200	Ranch plank	7.15	3.99	11.14
6300	Sleepers on concrete, treated, 24" O.C., 1" x 2"	.25	.38	.63
6400	1" x 3"	.40	.49	.89
6500	2" x 4"	.33	.68	1.01
6600	2" x 6"	.61	.78	1.39
6700	Refinish old floors, minimum	.95	1.15	2.10
6800	Maximum	1.41	3.54	4.95
6900	Subflooring, plywood, CDX, 1/2" thick	.52	.77	1.29
7000	5/8" thick	.68	.86	1.54
7100	3/4" thick	.77	.93	1.70
7200				
7300	1" x 10" boards, S4S, laid regular	1.80	1.05	2.85
7400	Laid diagonal	1.80	1.29	3.09
7500	1" x 8" boards, S4S, laid regular	1.76	1.16	2.92
7600	Laid diagonal	1.76	1.36	3.12
7700	Underlayment, plywood, underlayment grade, 3/8" thick	.80	.77	1.57
7800	1/2" thick	.85	.80	1.65
7900	5/8" thick	1.06	.83	1.89
8000	3/4" thick	1.29	.89	2.18
8100	Particle board, 3/8" thick	.39	.77	1.16
8200	1/2" thick	.40	.80	1.20
8300	5/8" thick	.51	.83	1.34
8400	3/4" thick	.72	.89	1.61

C3030 Ceiling Finishes

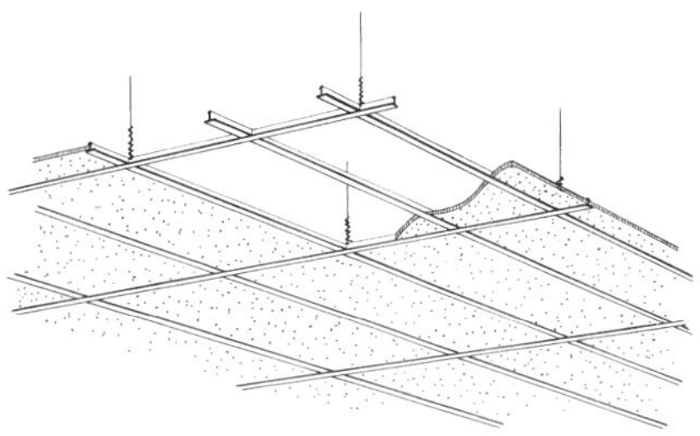

This page illustrates suspended acoustical board systems including acoustic ceiling board, hangers, and T bar suspension. Lines within System Components give the unit price and total price per square foot for this system. Prices for suspended acoustical board systems are on Line Items C3030 210 1000 thru 2400. Both material quantities and labor costs have been adjusted for the system listed.

Factors: To adjust for job conditions other than normal working situations use Lines C3030 210 2900 thru 4000.

Example: You are to install the system and protect existing construction. Go to Line C3030 210 3800 and apply these percentages to the appropriate MAT. and INST. costs.

System Components	QUANTITY	UNIT	COST PER S.F. MAT.	COST PER S.F. INST.	COST PER S.F. TOTAL
Suspended acoustical ceiling board installed on exposed grid system.					
Fiberglass boards, film faced, 2′ x 4′, 5/8″ thick	1.000	S.F.	1.06	.93	1.99
Hangers, #12 wire	1.000	S.F.	.51	.56	1.07
T bar suspension system, 2′ x 4′ grid	1.000	S.F.	.77	.72	1.49
TOTAL			2.34	2.21	4.55

C3030 210	Ceiling, Suspended Acoustical	COST PER S.F. MAT.	COST PER S.F. INST.	COST PER S.F. TOTAL
1000	2′ x 4′ grid, fiberglass board, film faced, 5/8″ thick	2.34	2.21	4.55
1100	Mineral fiber board, aluminum faced	2.95	2.17	5.12
1200	Standard faced	2.14	2.14	4.28
1300	Plastic faced	2.63	2.73	5.36
1400	Fiberglass, film faced, 3″ thick, R11	3.01	2.57	5.58
1500	Grass cloth faced, 3/4″ thick	2.92	2.44	5.36
1600	1″ thick	3.68	2.47	6.15
1700	1-1/2″ thick, nubby face	4.21	2.50	6.71
2200				
2300				
2400	Add for 2′ x 2′ grid system	.19	.18	.37
2500				
2600				
2700				
2900	Cut & patch to match existing construction, add, minimum	2%	3%	
3000	Maximum	5%	9%	
3100	Dust protection, add, minimum	1%	2%	
3200	Maximum	4%	11%	
3300	Equipment usage curtailment, add, minimum	1%	1%	
3400	Maximum	3%	10%	
3500	Material handling & storage limitation, add, minimum	1%	1%	
3600	Maximum	6%	7%	
3700	Protection of existing work, add, minimum	2%	2%	
3800	Maximum	5%	7%	
3900	Shift work requirements, add, minimum		5%	
4000	Maximum		30%	

C3030 Ceiling Finishes

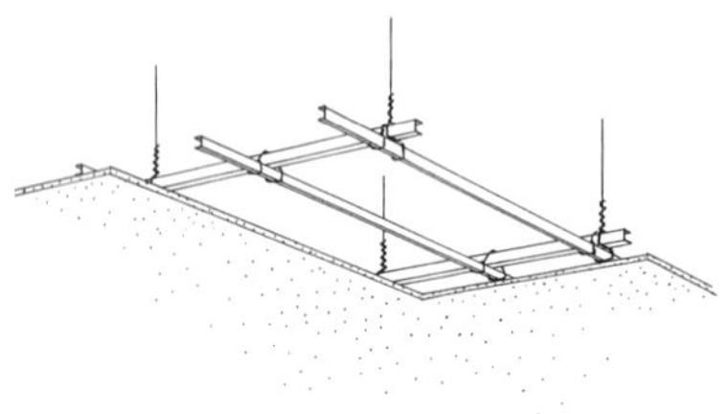

This page illustrates and describes suspended gypsum board systems including gypsum board, metal furring, taping, finished and painted. Lines within System Components give the unit price and total price per square foot for this system. Prices for suspended gypsum board systems are on Line Items C3030 220 1400 thru 1700. Both material quantities and labor costs have been adjusted for the system listed.

Factors: To adjust for job conditions other than normal working situations use Lines C3030 220 2900 thru 4000.

Example: You are to install the system and control dust in the working area. Go to Line C3030 220 3200 and apply these percentages to the appropriate MAT. and INST. costs.

System Components	QUANTITY	UNIT	COST PER S.F. MAT.	COST PER S.F. INST.	COST PER S.F. TOTAL
Suspended ceiling gypsum board,4′ x 8′ x 5/8″ thick, On metal furring, taped, finished and painted.					
Gypsum drywall, 4′ x 8′, 5/8″ thick, screwed	1.000	S.F.	.28	.64	.92
Main runners, 1-1/2″ C.R.C., 4′ O.C.	.500	S.F.	.15	.63	.78
25 ga., channels, 2′ O.C.	1.000	S.F.	.22	1.17	1.39
Taped and finished	1.000	S.F.	.05	.58	.63
Paint, 2 coats, roller work	1.000	S.F.	.19	.64	.83
TOTAL			.89	3.66	4.55

C3030 220	Ceilings, Suspended Gypsum Board	COST PER S.F. MAT.	COST PER S.F. INST.	COST PER S.F. TOTAL
1400	4′ x 8′ x 5/8″, gypsum drywall, 2 coats paint	.89	3.66	4.55
1500	Thin coat plaster, 2 coats paint	.76	3.16	3.92
1600	Spray-on sand finish, no paint	.69	3.06	3.75
1700	12″ x 12″ x 3/4″ acoustical wood fiber tile	1.94	3.89	5.83
1800				
1900				
2000				
2100				
2200				
2300				
2400				
2500				
2600				
2700				
2900	Cut & patch to match existing construction, add, minimum	2%	3%	
3000	Maximum	5%	9%	
3100	Dust protection, add, minimum	1%	2%	
3200	Maximum	4%	11%	
3300	Equipment usage curtailment, add, minimum	1%	1%	
3400	Maximum	3%	10%	
3500	Material handling & storage limitation, add, minimum	1%	1%	
3600	Maximum	6%	7%	
3700	Protection of existing work, add, minimum	2%	2%	
3800	Maximum	5%	7%	
3900	Shift work requirements, add, minimum		5%	
4000	Maximum		30%	

C3030 Ceiling Finishes

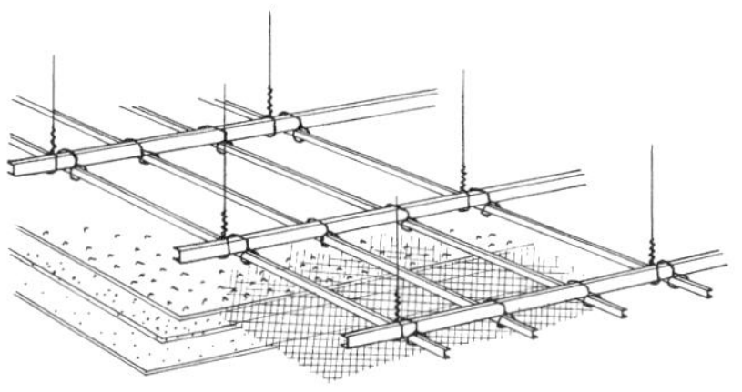

This page illustrates and describes suspended plaster and lath systems including gypsum plaster, lath, furring and runners with ceiling painted. Lines within System Components give the unit price and total price per square foot for this system. Prices for suspended plaster and lath systems are on Line Items C3030 230 1200 thru 2300. Both material quantities and labor costs have been adjusted for the system listed.

Factors: To adjust for job conditions other than normal working situations use Lines C3030 230 2900 thru 4000.

Example: You are to install the system to match existing construction. Go to Line C3030 230 2900 and apply these percentages to the appropriate MAT. and INST. costs.

System Components	QUANTITY	UNIT	COST PER S.F. MAT.	COST PER S.F. INST.	COST PER S.F. TOTAL
Gypsum plaster, 3 coats, on 3.4# rib lath, on 3/4″ C.R.C. furring on 1-1/2″ main runners, ceiling painted.					
Gypsum plaster, 3 coats	.110	S.Y.	.62	3.67	4.29
3.4# rib lath	.110	S.Y.	.58	.78	1.36
Main runners, 1-1/2″ C.R.C. 24″ O.C.	.333	S.F.	.30	1.26	1.56
Furring 3/4″ C.R.C. 16″ O.C.	1.000	S.F.	.34	1.70	2.04
Painting, 2 coats, roller work	1.000	S.F.	.19	.64	.83
TOTAL			2.03	8.05	10.08

C3030 230	Ceiling, Suspended Plaster	COST PER S.F. MAT.	COST PER S.F. INST.	COST PER S.F. TOTAL
1200	Gypsum plaster, 3 coats, on 3.4# rib lath	2.03	8.05	10.08
1300	On 2.5# diamond lath	1.87	8	9.87
1400	On 3/8″ gypsum lath	1.92	8	9.92
1500	2 coats, on 3.4# rib lath	1.99	8.05	10.04
1600	On 2.5# diamond lath	1.68	7.40	9.08
1700	On 3/8″ gypsum lath	1.73	7.40	9.13
1800	Perlite plaster, 3 coats, on 3.4# rib lath	2.17	8.90	11.07
1900	On 2.5# diamond lath	2.01	8.85	10.86
2000	On 3/8″ gypsum lath	2.06	8.85	10.91
2100	2 coats, on 3.4# rib lath	1.87	8	9.87
2200	On 2.5# diamond lath	1.71	7.95	9.66
2300	On 3/8″ gypsum lath	1.76	7.95	9.71
2400				
2500				
2600				
2700				
2900	Cut & patch to match existing construction, add, minimum	2%	3%	
3000	Maximum	5%	9%	
3100	Dust protection, add, minimum	1%	2%	
3200	Maximum	4%	11%	
3300	Equipment usage curtailment, add, minimum	1%	1%	
3400	Maximum	3%	10%	
3500	Material handling & storage limitation, add, minimum	1%	1%	
3600	Maximum	6%	7%	
3700	Protection of existing work, add, minimum	2%	2%	
3800	Maximum	5%	7%	
3900	Shift work requirements, add, minimum		5%	
4000	Maximum		30%	

C3030 Ceiling Finishes

C3030 250	Selective Price Sheet	COST PER S.F.		
		MAT.	INST.	TOTAL
0100	Ceiling, plaster, gypsum, 2 coats	.44	3.11	3.55
0200	3 coats	.63	3.71	4.34
0300	Perlite or vermiculite, 2 coats	.46	3.64	4.10
0400	3 coats	.77	4.58	5.35
0500	Gypsum lath, plain, 3/8" thick	.47	.64	1.11
0600	1/2" thick	.55	.68	1.23
0700	Firestop, 3/8" thick	.28	.78	1.06
0800	1/2" thick	.30	.84	1.14
0900	Metal lath, rib, 2.75 lb.	.38	.73	1.11
1000	3.40 lb.	.59	.78	1.37
1100	Diamond, 2.50 lb.	.42	.73	1.15
1200	3.40 lb.	.49	.91	1.40
1500	Drywall, standard, 1/2" thick, no finish included	.24	.64	.88
1600	Taped and finished	.29	1.51	1.80
1700	5/8" thick, no finish included	.28	.64	.92
1800	Taped and finished	.33	1.51	1.84
1900	Fire resistant, 1/2" thick, no finish included	.29	.64	.93
2000	Taped and finished	.34	1.51	1.85
2100	5/8" thick, no finish included	.30	.64	.94
2200	Taped and finished	.35	1.51	1.86
2300	Water resistant, 1/2" thick, no finish included	.39	.64	1.03
2400	Taped and finished	.44	1.51	1.95
2500	5/8" thick, no finish included	.47	.64	1.11
2600	Taped and finished	.52	1.51	2.03
2700	Finish, instead of taping			
2800	For thin coat plaster, add	.11	.72	.83
2900	Finish, textured spray, add	.04	.62	.66
3000				
3100				
3200				
3300	Tile, stapled or glued, mineral fiber plastic coated, 5/8" thick	1.89	1.93	3.82
3400	3/4" thick	2.33	1.93	4.26
3500	Wood fiber, 1/2" thick	1.07	1.45	2.52
3600	3/4" thick	1.29	1.45	2.74
3700	Suspended, fiberglass, film faced, 5/8" thick	1.06	.93	1.99
3800	3" thick	1.73	1.29	3.02
3900	Mineral fiber, 5/8" thick, standard	.86	.86	1.72
4000	Aluminum	2.04	.96	3
4100	Wood fiber, reveal edge, 1" thick			
4200	3" thick			
4300	Framing, metal furring, 3/4" channels, 12" O.C.	.37	2.34	2.71
4400	16" O.C.	.34	1.70	2.04
4500	24" O.C.	.22	1.17	1.39
4600				
4700	1-1/2" channels, 12" O.C.	.50	2.59	3.09
4800	16" O.C.	.45	1.89	2.34
4900	24" O.C.	.30	1.26	1.56
5000				
5100				
5200				
5300	Ceiling suspension systems, for tile,			
5400	Concealed "Z" bar, 12" module	.80	1.11	1.91
5500	Class A, "T" bar 2'-0" x 4'-0" grid	.77	.72	1.49
5600	2'-0" x 2'-0" grid	.95	.89	1.84
5700	Carrier channels for lighting fixtures, add	.25	1.26	1.51
5800				
5900				
6000				

C3030 Ceiling Finishes

C3030 250	Selective Price Sheet	COST PER S.F.		
		MAT.	INST.	TOTAL
6100	Wood, furring 1" x 3", on wood, 12" O.C.	.38	1.65	2.03
6200	16" O.C..	.29	1.24	1.53
6300	24" O.C.	.19	.83	1.02
6400				
6500	On concrete, 12" O.C.	.38	2.75	3.13
6600	16" O.C.	.29	2.06	2.35
6700	24" O.C.	.19	1.38	1.57
6800				
6900	Joists, 2" x 4", 12" O.C.	.46	1.20	1.66
7000	16" O.C.	.39	1	1.39
7100	24" O.C.	.32	.81	1.13
7200	32" O.C.	.24	.62	.86
7300	2" x 6", 12" O.C.	.77	1.21	1.98
7400	16" O.C.	.64	1.01	1.65
7500	24" O.C.	.51	.80	1.31
7600	32" O.C.	.38	.59	.97

D Services

D1010 Elevators and Lifts

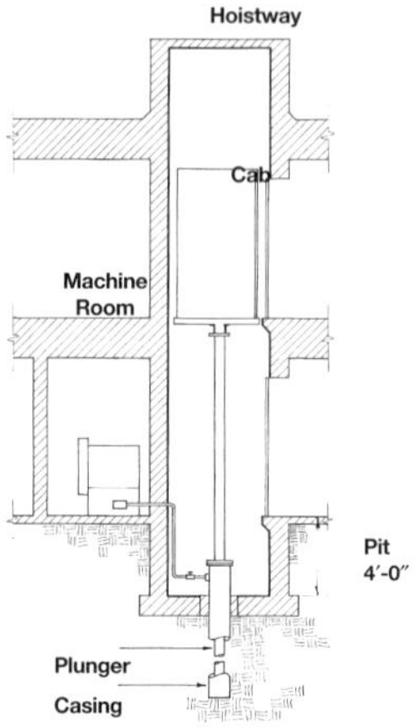

This page illustrates and describes oil hydraulic elevator systems. Prices for various oil hydraulic elevator systems are on Line Items D1010 130 0400 thru 2000. Both material quantities and labor costs have been adjusted for the system listed.

Factors: To adjust for job conditions other than normal working situations use Lines D1010 130 3100 thru 4000.

Example: You are to install the system with limited equipment usage. Go to Line D1010 130 3400 and apply this percentage to the appropriate TOTAL costs.

D1010 130	Oil Hydraulic Elevators	COST EACH		
		MAT.	INST.	TOTAL
0310	Oil-hydraulic elevator systems			
0320	Including piston and piston shaft			
0330				
0400	1500 Lb. passenger, 2 floors	48,800	17,600	66,400
0500	3 floors	62,000	27,200	89,200
0600	4 floors	97,500	38,200	135,700
0700	5 floors	110,000	47,600	157,600
0800	6 floors	122,500	56,500	179,000
0900				
1000	2500 Lb. passenger, 2 floors	52,000	17,600	69,600
1100	3 floors	65,500	27,200	92,700
1200	4 floors	96,500	38,200	134,700
1300	5 floors	109,500	47,600	157,100
1400	6 floors	125,500	56,500	182,000
1500				
1600	4000 Lb. passenger, 2 floors	56,500	17,600	74,100
1700	3 floors	72,500	27,200	99,700
1800	4 floors	105,000	38,200	143,200
1900	5 floors	118,000	47,600	165,600
2000	6 floors	130,000	56,500	186,500
2100				
2200				
2300				
2400				
2500				
2600				
2700				
2800				
2900				
3000				
3100	Dust protection, add, minimum	1%	2%	
3200	Maximum	4%	11%	
3300	Equipment usage curtailment, add, minimum	1%	1%	
3400	Maximum	3%	10%	
3500	Material handling & storage limitation, add, minimum	1%	1%	
3600	Maximum	6%	7%	
3700	Protection of existing work, add, minimum	2%	2%	
3800	Maximum	5%	7%	
3900	Shift work requirements, add, minimum		5%	
4000	Maximum		30%	

D2010 Plumbing Fixtures

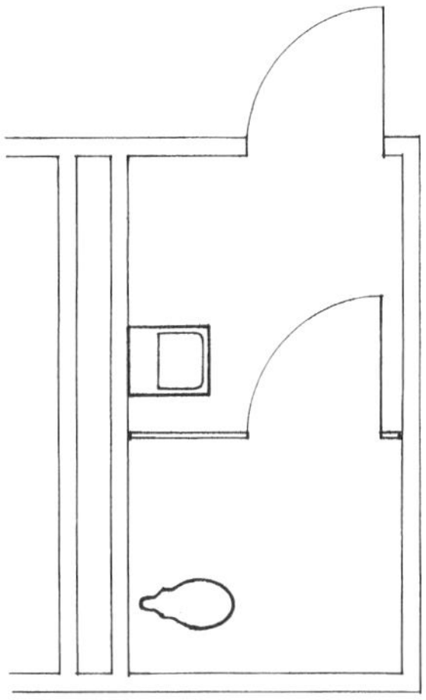

This page illustrates and describes a women's public restroom system including a water closet, lavatory, accessories, and service piping. Lines within System Components give the unit price and total price on a cost each basis for this system. Prices for women's public restroom systems are on Line Items D2010 956 1300 thru 1700. Both material quantities and labor costs have been adjusted for the system listed.

Factors: To adjust for job conditions other than normal working situations use Lines D2010 956 2900 thru 4000.

Example: You are to install the system and protect surrounding area from dust. Go to Line D2010 956 3100 and apply these percentages to the MAT. and INST. costs.

System Components	QUANTITY	UNIT	COST EACH		
			MAT.	INST.	TOTAL
Public women's restroom incl. water closet, lavatory, accessories and Necessary service piping to install this system in one wall.					
Water closet, wall mounted, one piece	1.000	Ea.	500	211	711
Rough-in waste and vent for water closet	1.000	Set	1,175	480	1,655
Lavatory, 20" x 18" P.E. cast iron with accessories	1.000	Ea.	395	153	548
Rough-in waste and vent for lavatory	1.000	Set	525	740	1,265
Toilet partition, painted metal between walls, floor mounted	1.000	Ea.	420	193	613
For handicap unit, add	1.000	Ea.	385		385
Grab bar, 36" long	1.000	Ea.	81	58	139
Mirror, 18" x 24", with stainless steel shelf	1.000	Ea.	199	29	228
Napkin/tampon dispenser, recessed	1.000	Ea.	575	38.50	613.50
Soap Dispenser, chrome, surface mounted, liquid	1.000	Ea.	46.50	29	75.50
Toilet tissue dispenser, surface mounted, stainless steel	1.000	Ea.	18.75	19.30	38.05
Towel dispenser, surface mounted, stainless steel	1.000	Ea.	46.50	36	82.50
TOTAL			4,366.75	1,986.80	6,353.55

D2010 956	Plumbing - Public Restroom	COST EACH		
		MAT.	INST.	TOTAL
1200				
1300	Public women's restroom, one water closet, one lavatory	4,375	1,975	6,350
1400	Two water closets, two lavatories	8,100	3,775	11,875
1500				
1600	For each additional water closet over 2, add	2,350	875	3,225
1700	For each additional lavatory over 2, add	1,175	950	2,125
1800				
1900				
2400	NOTE: PLUMBING APPROXIMATIONS			
2500	WATER CONTROL: water meter, backflow preventer,			
2600	Shock absorbers, vacuum breakers, mixer....10 to 15% of fixtures			
2700	PIPE AND FITTINGS: .... 30 to 60% of fixtures			
2800				
2900	Cut & patch to match existing construction, add, minimum	2%	3%	
3000	Maximum	5%	9%	
3100	Dust protection, add, minimum	1%	2%	
3200	Maximum	4%	11%	
3300	Equipment usage curtailment, add, minimum	1%	1%	
3400	Maximum	3%	10%	
3500	Material handling & storage limitation, add, minimum	1%	1%	

D2010 Plumbing Fixtures

D2010 956	Plumbing - Public Restroom	COST EACH		
		MAT.	INST.	TOTAL
3600	Maximum	6%	7%	
3700	Protection of existing work, add, minimum	2%	2%	
3800	Maximum	5%	7%	
3900	Shift work requirements, add, minimum		5%	
4000	Maximum		30%	

D2010 Plumbing Fixtures

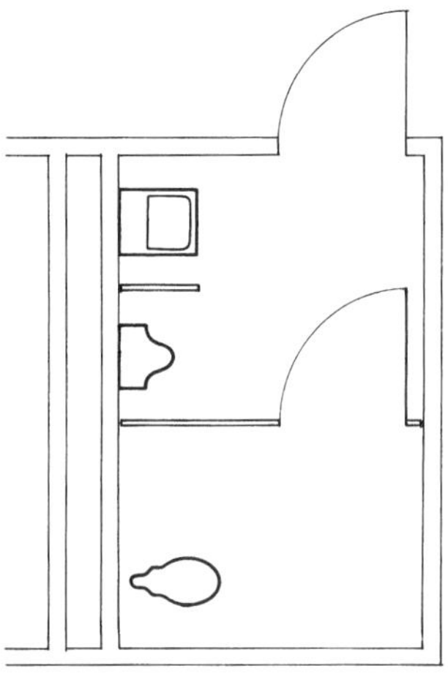

This page illustrates and describes a men's public restroom system including a water closet, urinal, lavatory, accessories and service piping. Lines within System Components give the unit price and total price on a cost each basis for this system. Prices for men's public restroom systems are on Line Items D2010 957 1700 thru 2200. Both material quantities and labor costs have been adjusted for the system listed.

Factors: To adjust for job conditions other than normal working situations use Lines D2010 957 2900 thru 4000.

Example: You are to install the system and match existing construction. Go to Line D2010 957 3000 and apply these percentages to the appropriate MAT. and INST. costs.

System Components	QUANTITY	UNIT	COST EACH		
			MAT.	INST.	TOTAL
Public men's restroom incl. water closet, urinal, lavatory, accessories, And necessary service piping to install this system in one wall.					
Water closet, wall mounted, one piece	1.000	Ea.	500	211	711
Rough-in waste & vent for water closet	1.000	Set	1,175	480	1,655
Urinal, wall hung	1.000	Ea.	325	410	735
Rough-in waste & vent for urinal	1.000	Set	640	435	1,075
Lavatory, 20" x 18", P.E. cast iron with accessories	1.000	Ea.	395	153	548
Rough-in waste & vent for lavatory	1.000	Set	525	740	1,265
Partition, painted mtl., between walls, floor mntd.	1.000	Ea.	420	193	613
For handicap unit, add	1.000	Ea.	385		385
Grab bars	1.000	Ea.	81	58	139
Urinal screen, powder coated steel, wall mounted	1.000	Ea.	168	116	284
Mirror, 18" x 24" with stainless steel shelf	1.000	Ea.	199	29	228
Soap dispenser, surface mounted, liquid	1.000	Ea.	46.50	29	75.50
Toilet tissue dispenser, surface mounted, stainless steel	1.000	Ea.	18.75	19.30	38.05
Towel dispenser, surface mounted, stainless steel	1.000	Ea.	46.50	36	82.50
TOTAL			4,924.75	2,909.30	7,834.05

D2010 957	Plumbing - Public Restroom	COST EACH		
		MAT.	INST.	TOTAL
1600				
1700	Public men's restroom, one water closet, one urinal, one lavatory	4,925	2,900	7,825
1800	Two water closets, two urinals, two lavatories	9,800	5,675	15,475
1900				
2000	For each additional water closet over 2, add	2,350	875	3,225
2100	For each additional urinal over 2, add	1,125	960	2,085
2200	For each additional lavatory over 2, add	1,175	950	2,125
2300				
2400	NOTE: PLUMBING APPROXIMATIONS			
2500	WATER CONTROL: water meter, backflow preventer,			
2600	Shock absorbers, vacuum breakers, mixer....10 to 15% of fixtures			
2700	PIPE AND FITTINGS: .... 30 to 60% of fixtures			
2800				
2900	Cut & patch to match existing construction, add, minimum	2%	3%	
3000	Maximum	5%	9%	
3100	Dust protection, add, minimum	1%	2%	
3200	Maximum	4%	11%	
3300	Equipment usage curtailment, add, minimum	1%	1%	

D2010 Plumbing Fixtures

D2010 957	Plumbing - Public Restroom	COST EACH		
		MAT.	INST.	TOTAL
3400	Maximum	3%	10%	
3500	Material handling, & storage limitation, add, minimum	1%	1%	
3600	Maximum	6%	7%	
3700	Protection of existing work, add, minimum	2%	2%	
3800	Maximum	5%	7%	
3900	Shift work requirements, add, minimum		5%	
4000	Maximum		30%	

D2010 Plumbing Fixtures

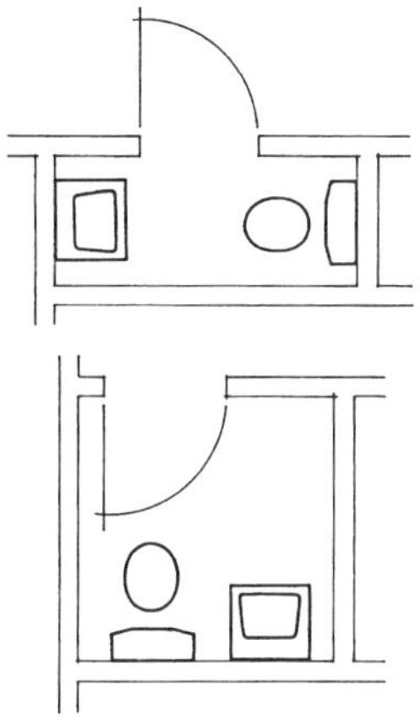

This page illustrates and describes a two fixture lavatory system including a water closet, lavatory, accessories and all service piping. Lines within System Components give the unit price and total price on a cost each basis for this system. Prices for two fixture lavatory systems are on Line Items D2010 958 1800 and 1900. Both material quantities and labor costs have been adjusted for the system listed.

Factors: To adjust for job conditions other than normal working situations use Lines D2010 958 2900 thru 4000.

Example: You are to install the system while controlling dust in the work area. Go to Line D2010 958 3200 and apply these percentages to the appropriate MAT. and INST. costs.

System Components	QUANTITY	UNIT	COST EACH MAT.	COST EACH INST.	COST EACH TOTAL
Two fixture bathroom incl. water closet, lavatory, accessories and Necessary service piping to install this system in 2 walls.					
Water closet, floor mounted, 2 piece, close coupled	1.000	Ea.	272	231	503
Rough in waste & vent for water closet	1.000	Set	375	400	775
Lavatory, 20" x 18", P.E. cast iron with accessories	1.000	Ea.	395	153	548
Rough in waste & vent for lavatory	1.000	Set	525	740	1,265
Additional service piping					
1/2" copper pipe with sweat solder joints	10.000	L.F.	47.30	84	131.30
2" black schedule 40 steel pipe with threaded couplings	12.000	L.F.	196.80	229.80	426.60
4" cast iron soil pipe with lead and oakum joints	7.000	L.F.	119	157.50	276.50
Accessories					
Toilet tissue dispenser, chrome, single roll	1.000	Ea.	18.75	19.30	38.05
18" long stainless steel towel bar	1.000	Ea.	43.50	25	68.50
Medicine cabinet with mirror, 20" x 16", unlighted	1.000	Ea.	86	41.50	127.50
TOTAL			2,078.35	2,081.10	4,159.45

D2010 958	Plumbing - Two Fixture Bathroom	COST EACH MAT.	COST EACH INST.	COST EACH TOTAL
1800	Water closet, lavavatory & accessories, in 2 walls with all service piping	2,075	2,075	4,150
1900	Above system installed in 1 wall	1,875	1,800	3,675
2000				
2400	NOTE: PLUMBING APPROXIMATIONS			
2500	WATER CONTROL: water meter, backflow preventer,			
2600	Shock absorbers, vacuum breakers, mixer....10 to 15% of fixtures			
2700	PIPE AND FITTINGS: .... 30 to 60% of fixtures			
2800				
2900	Cut & patch to match existing construction, add, minimum	2%	3%	
3000	Maximum	5%	9%	
3100	Dust protection, add, minimum	1%	2%	
3200	Maximum	4%	11%	
3300	Equipment usage curtailment, add, minimum	1%	1%	
3400	Maximum	3%	10%	
3500	Material handling & storage limitation, add, minimum	1%	1%	
3600	Maximum	6%	7%	
3700	Protection of existing work, add, minimum	2%	2%	
3800	Maximum	5%	7%	
3900	Shift work requirements, add, minimum		5%	
4000	Maximum		30%	

D2010 Plumbing Fixtures

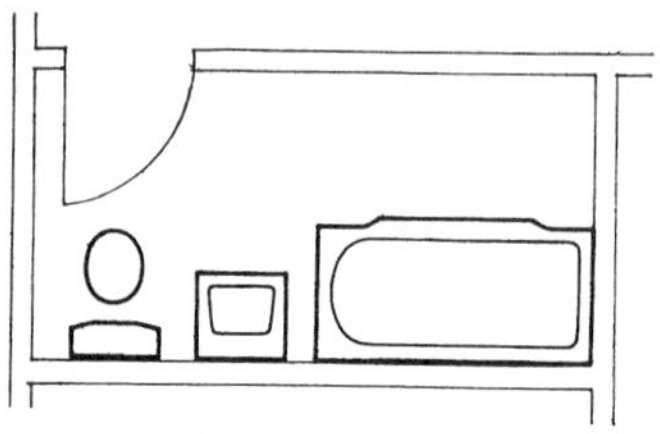

This page illustrates and describes a three fixture bathroom system including a water closet, tub, lavatory, accessories and service piping. Lines within System Components give the unit price and total price on a cost each basis for this system. Prices for a three fixture bathroom system is on Line Item D2010 959 1700. Both material quantities and labor costs have been adjusted for the system listed.

Factors: To adjust for job conditions other than normal working situations use Lines D2010 959 2900 thru 4000.

Example: You are to install the system and protect all existing work. Go to Line D2010 959 3800 and apply these percentages to the appropriate MAT. and INST. costs.

System Components	QUANTITY	UNIT	COST EACH		
			MAT.	INST.	TOTAL
Three fixture bathroom incl. water closet, bathtub, lavatory, accessories, And necessary service piping to install this system in 1 wall.					
Two piece, close coupled	1.000	Ea.	272	231	503
For rough-in, supply, waste and vent	.930	Set	348.75	372	720.75
5′ long	1.000	Ea.	1,125	278	1,403
Rough-in, supply, waste and vent, for all above tubs, add	.950	Set	508.25	560.50	1,068.75
20″ x 18″, single bowl	1.000	Ea.	395	153	548
Rough-in, supply, waste and vent for above lavatories	1.000	Set	525	740	1,265
Accessories					
Toilet tissue dispenser, surface mounted, SS, single roll	1.000	Ea.	18.75	19.30	38.05
Towel bar, stainless steel, 18″ long	2.000	Ea.	87	50	137
Medicine cabinets with mirror, st. st. frame, 16″ x 22″, unlighted	1.000	Ea.	86	41.50	127.50
TOTAL			3,365.75	2,445.30	5,811.05

D2010 959	Plumbing - Three Fixture Bathroom	COST EACH		
		MAT.	INST.	TOTAL
1600				
1700	Water closet, lavatory, bathtub, in 1 wall with all service piping	3,375	2,450	5,825
2400	NOTE: PLUMBING APPROXIMATIONS			
2500	WATER CONTROL: water meter, backflow preventer,			
2600	Shock absorbers, vacuum breakers, mixer....10 to 15% of fixtures			
2700	PIPE AND FITTINGS: .... 30 to 60% of fixtures			
2800				
2900	Cut & patch to match existing construction, add, minimum	2%	3%	
3000	Maximum	5%	9%	
3100	Dust protection, add, minimum	1%	2%	
3200	Maximum	4%	11%	
3300	Equipment usage curtailment, add, minimum	1%	1%	
3400	Maximum	3%	10%	
3500	Material handling & storage limitation, add, minimum	1%	1%	
3600	Maximum	6%	7%	
3700	Protection of existing work, add, minimum	2%	2%	
3800	Maximum	5%	7%	
3900	Shift work requirements, add, minimum		5%	
4000	Maximum		30%	

D2010 Plumbing Fixtures

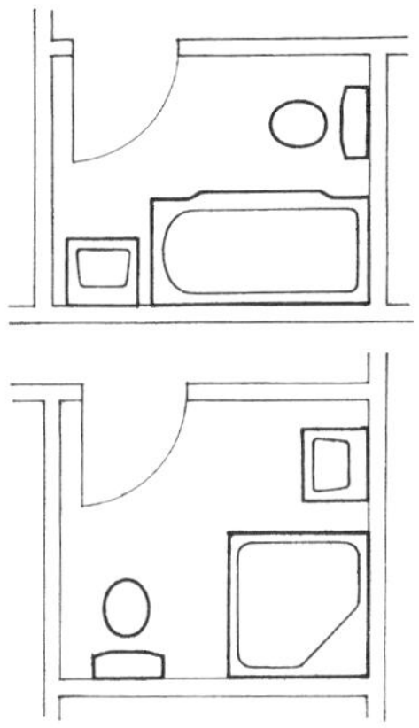

This page illustrates and describes a three fixture bathroom system including a water closet, tub, lavatory, accessories, and service piping. Lines within System Components give the unit price and total price on a cost each basis for this system. Prices for a three fixture bathroom system are on Line Item D2010 960 1900 thru 2000. Both material quantities and labor costs have been adjusted for the system listed.

Factors: To adjust for job conditions other than normal working situations use Lines D2010 960 2900 thru 4000.

Example: You are to install the system and protect the surrounding area from dust. Go to Line D2010 960 3100 and apply these percentages to the appropriate MAT. and INST. costs.

System Components	QUANTITY	UNIT	COST EACH		
			MAT.	INST.	TOTAL
Three fixture bathroom incl. water closet, bathtub, lavatory, accessories, And necessary service piping to install this system in 2 walls.					
Water closet, floor mounted, 2 piece, close coupled	1.000	Ea.	272	231	503
Rough-in waste & vent for water closet	1.000	Set	375	400	775
Bathtub, P.E. cast iron, 5′ long with accessories	1.000	Ea.	1,125	278	1,403
Rough-in waste & vent for bathtub	1.000	Set	535	590	1,125
Lavatory, 20″ x 18″ P.E. cast iron with accessories	1.000	Ea.	395	153	548
Rough-in waste & vent for lavatory	1.000	Set	525	740	1,265
Additional service piping					
1-1/4″ copper DWV type tubing, sweat solder joints	6.000	L.F.	78.90	68.10	147
2″ black schedule 40 steel pipe with threaded couplings	12.000	L.F.	196.80	229.80	426.60
Accessories					
Toilet tissue dispenser, chrome, single roll	1.000	Ea.	18.75	19.30	38.05
18″ long stainless steel towel bar	2.000	Ea.	87	50	137
Medicine cabinet with mirror, 20″ x 16″, unlighted	1.000	Ea.	86	41.50	127.50
TOTAL			3,694.45	2,800.70	6,495.15

D2010 960	Plumbing - Three Fixture Bathroom	COST EACH		
		MAT.	INST.	TOTAL
1900	Water closet, lavatory, bathtub, in 2 walls with all service piping	3,700	2,800	6,500
2000	Above system with corner contour tub, P.E. cast iron	5,175	2,800	7,975
2300				
2400	NOTE: PLUMBING APPROXIMATIONS			
2500	WATER CONTROL: water meter, backflow preventer,			
2600	Shock absorbers, vacuum breakers, mixer....10 to 15% of fixtures			
2700	PIPE AND FITTINGS: .... 30 to 60% of fixtures			
2800				
2900	Cut & patch to match existing construction, add, minimum	2%	3%	
3000	Maximum	5%	9%	
3100	Dust protection, add, minimum	1%	2%	
3200	Maximum	4%	11%	
3300	Equipment usage curtailment, add, minimum	1%	1%	
3400	Maximum	3%	10%	
3500	Material handling & storage limitation, add, minimum	1%	1%	
3600	Maximum	6%	7%	
3700	Protection of existing work, add, minimum	2%	2%	
3800	Maximum	5%	7%	
3900	Shift work requirements, add, minimum		5%	
4000	Maximum		30%	

D2010 Plumbing Fixtures

This page illustrates and describes a three fixture bathroom system including a water closet, shower, lavatory, accessories, and service piping, Lines within System Components give the unit price and total price on a cost each basis for this system. Prices for three fixture bathroom systems are on Line Items D2010 961 2100 thru 2200. Both material quantities and labor costs have been adjusted for the system listed.

Factors: To adjust for job conditions other than normal working situations use Lines D2010 961 2900 thru 4000.

Example: You are to install the system and protect existing construction. Go to Line D2010 961 3700 and apply these percentages to the appropriate MAT. and INST. costs.

System Components	QUANTITY	UNIT	COST EACH MAT.	COST EACH INST.	COST EACH TOTAL
Three fixture bathroom incl. water closet, shower, lavatory, accessories,					
And necessary service piping to install this system in 2 walls.					
Water closet, floor mounted, 2 piece, close coupled	1.000	Ea.	272	231	503
Rough-in waste & vent for water closet	1.000	Set	375	400	775
32″ shower, enameled stall, molded stone receptor	1.000	Ea.	820	245	1,065
Rough-in waste & vent for shower	1.000	Set	650	600	1,250
Lavatory, 20″ x 18″ P.E. cast iron with accessories	1.000	Ea.	395	153	548
Rough-in waste & vent for lavatory	1.000	Set	525	740	1,265
Additional service piping					
1-1/4″ Copper DWV type tubing, sweat solder joints	4.000	L.F.	52.60	45.40	98
2″ black schedule 40 steel pipe with threaded couplings	6.000	L.F.	98.40	114.90	213.30
4″ cast iron soil with lead and oakum joints	7.000	L.F.	119	157.50	276.50
Accessories					
Toilet tissue dispenser, chrome, single roll	1.000	Ea.	18.75	19.30	38.05
18″ long stainless steel towel bar	2.000	Ea.	87	50	137
Medicine cabinet with mirror, 20″ x 16″, unlighted	1.000	Ea.	86	41.50	127.50
TOTAL			3,498.75	2,797.60	6,296.35

D2010 961	Plumbing - Three Fixture Bathroom	COST EACH MAT.	COST EACH INST.	COST EACH TOTAL
2100	Water closet, 32″ shower, lavavatory, in 2 walls with all service piping	3,500	2,800	6,300
2200	Above system with 36″ corner angle shower	4,075	2,800	6,875
2300				
2400	NOTE: PLUMBING APPROXIMATIONS			
2500	WATER CONTROL: water meter, backflow preventer,			
2600	Shock absorbers, vacuum breakers, mixer....10 to 15% of fixtures			
2700	PIPE AND FITTINGS: .... 30 to 60% of fixtures			
2800				
2900	Cut & patch to match existing construction, add, minimum	2%	3%	
3000	Maximum	5%	9%	
3100	Dust protection, add, minimum	1%	2%	
3200	Maximum	4%	11%	
3300	Equipment usage curtailment, add, minimum	1%	1%	
3400	Maximum	3%	10%	
3500	Material handling & storage limitation, add, minimum	1%	1%	
3600	Maximum	6%	7%	
3700	Protection of existing work, add, minimum	2%	2%	
3800	Maximum	5%	7%	
3900	Shift work requirements, add, minimum		5%	
4000	Maximum		30%	

D2010 Plumbing Fixtures

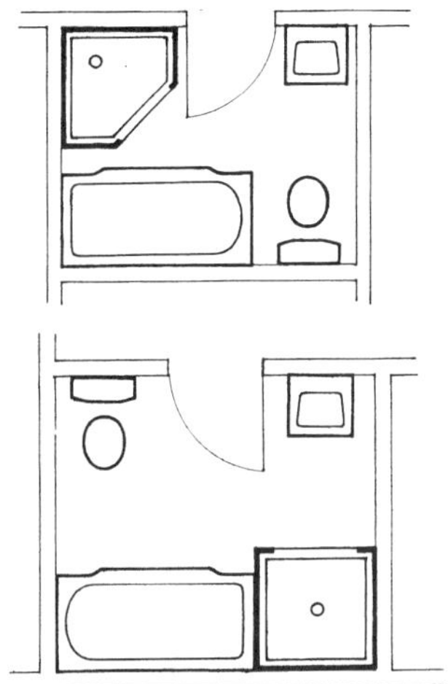

This page illustrates and describes a four fixture bathroom system including a water closet, shower, bathtub, lavatory, accessories, and service piping. Lines within System Components give the unit price and total price on a cost each basis for this system. Prices for four fixture bathroom systems are on Line Items D2010 962 1800 thru 1900. Both material quantities and labor costs have been adjusted for the system listed.

Factors: To adjust for job conditions other than normal working situations use Lines D2010 962 2900 thru 4000.

Example: You are to install the system during weekends and evenings. Go to Line D2010 962 4000 and apply these percentages to the appropriate MAT. and INST. costs.

System Components	QUANTITY	UNIT	COST EACH MAT.	COST EACH INST.	COST EACH TOTAL
Four fixture bathroom incl. water closet, shower, bathtub, lavatory					
Accessories and necessary service piping to install this system in 2 walls.					
Water closet, floor mounted, 2 piece, close coupled	1.000	Ea.	272	231	503
Rough-in waste & vent for water closet	1.000	Set	375	400	775
32″ shower, enameled steel stall, molded stone receptor	1.000	Ea.	820	245	1,065
Rough-in waste & vent for shower	1.000	Set	650	600	1,250
Bathtub, P.E. cast iron, 5′ long with accessories	1.000	Ea.	1,125	278	1,403
Rough-in waste & vent for bathtub	1.000	Set	535	590	1,125
Lavatory, 20″ x 18″ P.E. cast iron with accessories	1.000	Ea.	395	153	548
Rough-in waste & vent for lavatory	1.000	Set	525	740	1,265
Accessories					
Toilet tissue dispenser, chrome, single roll	1.000	Ea.	18.75	19.30	38.05
18″ long stainless steel towel bar	2.000	Ea.	87	50	137
Medicine cabinet with mirror, 20″ x 16″ unlighted	1.000	Ea.	86	41.50	127.50
TOTAL			4,888.75	3,347.80	8,236.55

D2010 962	Plumbing - Four Fixture Bathroom	COST EACH MAT.	COST EACH INST.	COST EACH TOTAL
1800	Water closet, 32″ shower, bathtub, lavavatory, in 2 walls	4,900	3,350	8,250
1900	Above system with 36″ corner angle shower, plumbing in 3 walls	5,725	3,675	9,400
2000				
2300				
2400	NOTE: PLUMBING APPROXIMATIONS			
2500	WATER CONTROL: water meter, backflow preventer,			
2600	Shock absorbers, vacuum breakers, mixer....10 to 15% of fixtures			
2700	PIPE AND FITTINGS: .... 30 to 60% of fixtures			
2800				
2900	Cut & patch to match existing construction, add, minimum	2%	3%	
3000	Maximum	5%	9%	
3100	Dust protection, add, minimum	1%	2%	
3200	Maximum	4%	11%	
3300	Equipment usage curtailment, add, minimum	1%	1%	
3400	Maximum	3%	10%	
3500	Material handling & storage limitation, add, minimum	1%	1%	
3600	Maximum	6%	7%	
3700	Protection of existing work, add, minimum	2%	2%	
3800	Maximum	5%	7%	
3900	Shift work requirements, add, minimum		5%	
4000	Maximum		30%	

D2010 Plumbing Fixtures

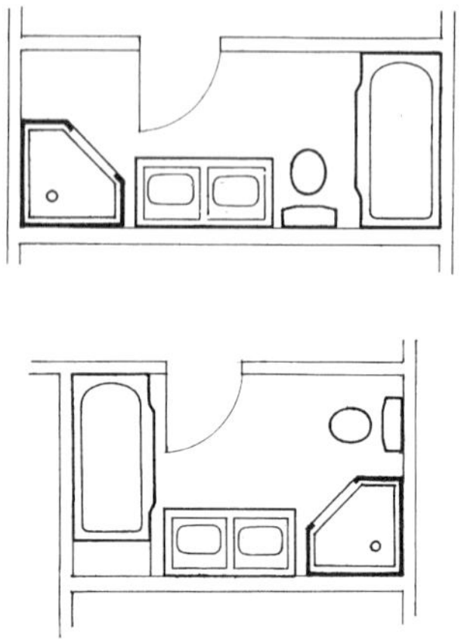

This page illustrates and describes a five fixture bathroom system including a water closet, shower, bathtub, lavatories, accessories, and service piping. Lines within System Components give the unit price and total price on a cost each basis for this system. Prices for five fixture bathroom systems are on Line Items D2010 963 1800 thru 1900. Both material quantities and labor costs have been adjusted for the system listed.

Factors: To adjust for job conditions other than normal working situations use Lines D2010 963 2900 thru 4000.

Example: You are to install and match any existing construction. Go to Line D2010 963 2900 and apply these percentages to the appropriate MAT. and INST. costs.

System Components	QUANTITY	UNIT	COST EACH MAT.	COST EACH INST.	COST EACH TOTAL
Five fixture bathroom incl. water closet, shower, bathtub, 2 lavatories, Accessories and necessary service piping to install this system in 1 wall.					
Water closet, floor mounted, 2 piece, close coupled	1.000	Ea.	272	231	503
Rough-in waste & vent for water closet	1.000	Set	375	400	775
36″ corner angle shower, enameled steel stall, molded stone receptor	1.000	Ea.	1,400	255	1,655
Rough-in waste & vent for shower	1.000	Set	650	600	1,250
Bathtub, P.E. cast iron, 5′ long with accessories	1.000	Ea.	1,125	278	1,403
Rough-in waste & vent for bathtub	1.000	Set	535	590	1,125
Countertop, plastic laminate, with backsplash	2.000	Ea.	158	77.20	235.20
Vanity base, 2 doors, 24″ wide	2.000	Ea.	259	105	364
Lavatories, 20″ x 18″ cabinet mntd, PECI	2.000	Ea.	670	382	1,052
Rough-in waste & vent for lavatories	1.600	Set	736	856	1,592
Accessories					
Toilet tissue dispenser, chrome, single roll	1.000	Ea.	18.75	19.30	38.05
18″ long stainless steel towel bars	2.000	Ea.	87	50	137
Medicine cabinet with mirror, 20″ x 16″, unlighted	2.000	Ea.	172	83	255
TOTAL			6,457.75	3,926.50	10,384.25

D2010 963	Plumbing - Five Fixture Bathroom	COST EACH MAT.	COST EACH INST.	COST EACH TOTAL
1800	Water closet, shower, bathtub, 2 lavatories, on 1 wall	6,450	3,925	10,375
1900	Above system installed in 2 walls with all necessary service piping	6,575	4,075	10,650
2000				
2100				
2200				
2300				
2400	NOTE: PLUMBING APPROXIMATIONS			
2500	WATER CONTROL: water meter, backflow preventer,			
2600	Shock absorbers, vacuum breakers, mixer....10 to 15% of fixtures			
2700	PIPE AND FITTINGS: .... 30 to 60% of fixtures			
2800				
2900	Cut & patch to match existing construction, add, minimum	2%	3%	
3000	Maximum	5%	9%	
3100	Dust protection, add, minimum	1%	2%	
3200	Maximum	4%	11%	
3300	Equipment usage curtailment, add, minimum	1%	1%	
3400	Maximum	3%	10%	
3500	Material handling & storage limitation, add, minimum	1%	1%	

D2010 Plumbing Fixtures

D2010 963	Plumbing - Five Fixture Bathroom	COST EACH		
		MAT.	INST.	TOTAL
3600	Maximum	6%	7%	
3700	Protection of existing work, add, minimum	2%	2%	
3800	Maximum	5%	7%	
3900	Shift work requirements, add, minimum		5%	
4000	Maximum		30%	

D3010 Energy Supply

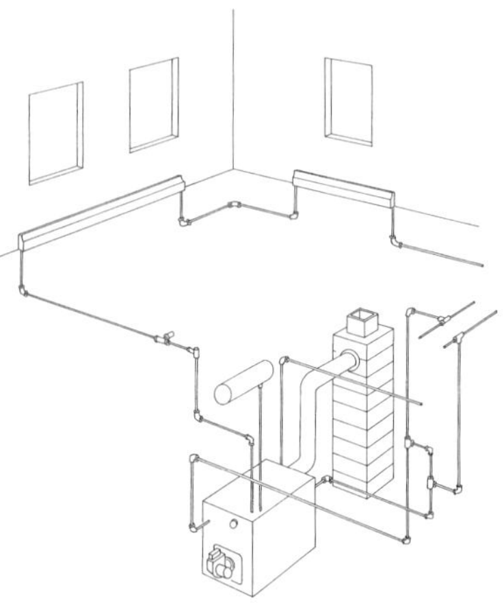

This page illustrates and describes an oil fired hot water baseboard system including an oil fired boiler, fin tube radiation and all fittings and piping. Lines within System Components give the unit price and total price per square foot for this system. Prices for oil fired hot water baseboard systems are on Line Items D3010 540 1600 thru 2100. Both material quantities and labor costs have been adjusted for the system listed.

Factors: To adjust for job conditions other than normal working situations use Lines D3010 540 2900 thru 4000.

Example: You are to install the system while protecting all existing work. Go to Line D3010 540 3700 and apply these percentages to the appropriate MAT. and INST. costs.

System Components	QUANTITY	UNIT	COST PER S.F.		
			MAT.	INST.	TOTAL
Oil fired hot water baseboard system including boiler, fin tube radiation And all necessary fittings and piping.					
Area to 800 S.F.					
Boiler, cast iron, w/oil piping, 97 MBH	1.000	Ea.	2,468.75	1,531.25	4,000
Copper piping	130.000	L.F.	1,924	1,527.50	3,451.50
Fin tube radiation	68.000	L.F.	629	1,462	2,091
Circulator	1.000	Ea.	585	204	789
Oil tank	1.000	Ea.	490	250	740
Expansion tank, ASME	1.000	Ea.	700	89.50	789.50
TOTAL			6,796.75	5,064.25	11,861
COST PER S.F.		S.F.	8.50	6.33	14.83

D3010 540	Heating - Oil Fired Hot Water	COST PER S.F.		
		MAT.	INST.	TOTAL
1600	Cast iron boiler, area to 800 S.F.	8.50	6.34	14.84
1700	To 1000 S.F.	7.40	5.45	12.85
1800	To 1200 S.F.	6.40	4.95	11.35
1900	To 1600 S.F.	7.60	4.93	12.53
2000	To 2000 S.F.	6.40	4.37	10.77
2100	To 3000 S.F.	4.79	3.68	8.47
2200				
2300				
2700				
2800				
2900	Cut & patch to match existing construction, add, minimum	2%	3%	
3000	Maximum	5%	9%	
3100	Dust protection, add, minimum	1%	2%	
3200	Maximum	4%	11%	
3300	Equipment usage curtailment, add, minimum	1%	1%	
3400	Maximum	3%	10%	
3500	Material handling & storage limitation, add, minimum	1%	1%	
3600	Maximum	6%	7%	
3700	Protection of existing work, add, minimum	2%	2%	
3800	Maximum	5%	7%	
3900	Shift work requirements, add, minimum		5%	
4000	Maximum		30%	

D3010 Energy Supply

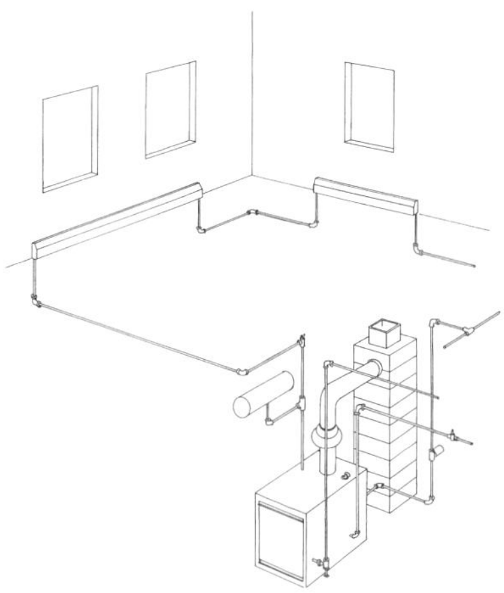

This page illustrates and describes a gas fired hot water baseboard system including a gas fired boiler, fin tube radiation, fittings and piping. Lines within System Components give the unit price and total price per square foot for this system. Prices for gas fired hot water baseboard systems are on Line Items D3010 550 1400 thru 1900. Both material quantities and labor costs have been adjusted for the system listed.

Factors: To adjust for job conditions other than normal working situations use Lines D3010 550 2900 thru 4000.

Example: You are to install the system with minimal equipment usage. Go to Line D3010 550 3400 and apply these percentages to the appropriate MAT. and INST. costs.

System Components	QUANTITY	UNIT	COST PER S.F. MAT.	COST PER S.F. INST.	COST PER S.F. TOTAL
Gas fired hot water baseboard system including boiler, fin tube radiation, All necessary fittings and piping.					
Area to 800 S.F.					
Cast iron boiler, insulating jacket, gas piping, 80 MBH	1.000	Ea.	2,468.75	2,250	4,718.75
Copper piping	130.000	L.F.	1,924	1,527.50	3,451.50
Fin tube radiation	68.000	L.F.	629	1,462	2,091
Circulator, flange connection	1.000	Ea.	585	204	789
Expansion tank, ASME	1.000	Ea.	700	89.50	789.50
TOTAL			6,306.75	5,533	11,839.75
COST PER S.F.		S.F.	7.88	6.92	14.80

D3010 550	Heating - Gas Fired Hot Water	COST PER S.F. MAT.	COST PER S.F. INST.	COST PER S.F. TOTAL
1400	Cast iron boiler, area to 800 S.F.	7.88	6.91	14.79
1500	To 1000 S.F.	6.70	6.05	12.75
1600	To 1200 S.F.	5.85	5.45	11.30
1700	To 1600 S.F.	7.20	5.30	12.50
1800	To 2000 S.F.	6.20	4.77	10.97
1900	To 3000 S.F.	4.88	4.17	9.05
2000				
2100				
2900	Cut & patch to match existing construction, add, minimum	2%	3%	
3000	Maximum	5%	9%	
3100	Dust protection, add, minimum	1%	2%	
3200	Maximum	4%	11%	
3300	Equipment usage curtailment, add, minimum	1%	1%	
3400	Maximum	3%	10%	
3500	Material handling & storage limitation, add, minimum	1%	1%	
3600	Maximum	6%	7%	
3700	Protection of existing work, add, minimum	2%	2%	
3800	Maximum	5%	7%	
3900	Shift work requirements, add, minimum		5%	
4000	Maximum		30%	

D3010 Energy Supply

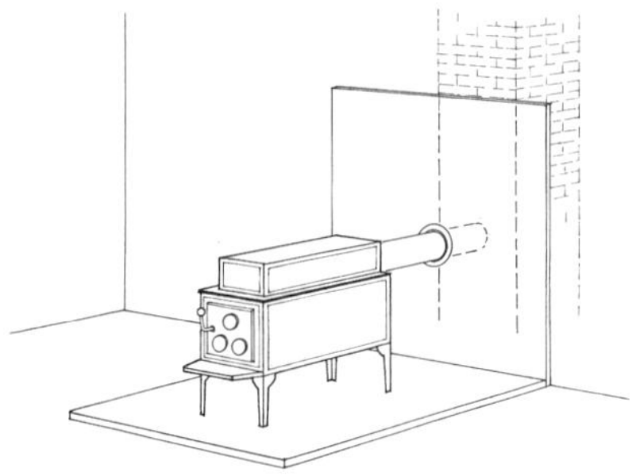

This page illustrates and describes a wood burning stove system including a free standing stove, preformed hearth, masonry chimney, and necessary piping and fittings. Lines within System Components give the unit price and total price on a cost each basis for this system. Prices for wood burning stove systems are on Line Items D3010 991 1400 thru 2000. Both material quantities and labor costs have been adjusted for the system listed.

Factors: To adjust for job conditions other than normal working situations use Lines D3010 991 2900 thru 4000.

Example: You are to install the system and protect existing construction. Go to Line D3010 991 3500 and apply these percentages to the appropriate MAT., and INST. costs.

System Components	QUANTITY	UNIT	COST EACH		
			MAT.	INST.	TOTAL
Cast iron, free standing,wood burning stove with preformed hearth, masonry Chimney, and all necessary piping and fittings to install in chimney.					
Cast iron wood burning stove, stove pipe	1.000	Ea.	1,000	890	1,890
Wall panel or hearth, non-combustible	1.000	Ea.	570	162.45	732.45
16″ x 16″ brick chimney, 8″ x 8″ flue	20.000	V.L.F.	587.50	1,412.50	2,000
Foundation	.500	C.Y.	79	120.97	199.97
TOTAL			2,236.50	2,585.92	4,822.42

D3010 991	Wood Burning Stoves	COST EACH		
		MAT.	INST.	TOTAL
1400	Cast iron, free standing with preformed hearth	2,225	2,575	4,800
1500				
1600	Installed in existing fireplace	1,300	975	2,275
1700				
1800	Installed with insulated metal chimney			
1900	System including ceiling package,			
2000	Metal chimney to 10 L.F.	2,825	1,375	4,200
2100				
2200				
2300				
2400				
2500				
2600				
2700				
2900	Dust protection, add, minimum	1%	2%	
3000	Maximum	4%	11%	
3100	Equipment usage curtailment, add, minimum	1%	1%	
3200	Maximum	3%	10%	
3300	Material handling & storage limitation, add, minimum	1%	1%	
3400	Maximum	6%	7%	
3500	Protection of existing work, add, minimum	2%	2%	
3600	Maximum	5%	7%	
3700	Shift work requirements, add, minimum		5%	
3800	Maximum		30%	
3900	Temporary shoring and bracing, add, minimum	2%	5%	
4000	Maximum	5%	12%	

D3010 Energy Supply

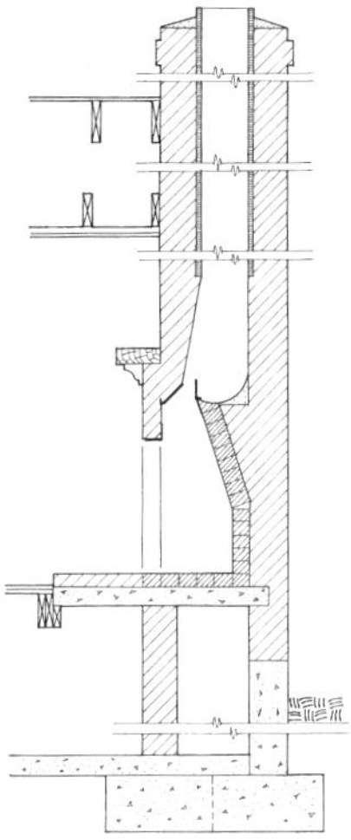

This page illustrates and describes masonry fireplace systems including a brick fireplace, footing, foundation, hearth, firebox, chimney and flue. Lines within System Components give the unit price and total price on a cost each basis for this system. Prices for masonry fireplace systems are on Line Items D3010 992 1200 thru 1500. Both material quantities and labor costs have been adjusted for the system listed.

Factors: To adjust for job conditions other than normal working situations use Lines D3010 992 2700 thru 4000.

Example: You are to install the system with some temporary shoring and bracing. Go to Line D3010 992 3900 and apply these percentages to the appropriate MAT. and INST. costs.

System Components	QUANTITY	UNIT	COST EACH MAT.	COST EACH INST.	COST EACH TOTAL
Brick masonry fireplace, including footing, foundation, hearth, firebox, Chimney and flue, chimney 12′ above firebox.					
Footing, 4′ x 7′ x 12″ thick, 3000 psi concrete	1.040	C.Y.	190.32	289.20	479.52
Foundation, 12″ concrete block	180.000	S.F.	799.20	1,845	2,644.20
Fireplace, brick faced, 6′-0″ wide x 5′-0″ high	1.000	Ea.	590	2,575	3,165
Hearth	1.000	Ea.	217	515	732
Chimney, 20″ x 20″, one 12″ x 12″ flue, 12 V.L.F.	12.000	V.L.F.	510	900	1,410
Mantle, wood	7.000	L.F.	63.70	112.35	176.05
TOTAL			2,370.22	6,236.55	8,606.77

D3010 992	Masonry Fireplace	COST EACH MAT.	COST EACH INST.	COST EACH TOTAL
1200	Chimney, 20″ x 20″, one 12″ x 12″ flue, 12 V.L.F.	2,375	6,225	8,600
1300	18 V.L.F.	2,625	6,675	9,300
1400	24 V.L.F.	2,875	7,125	10,000
1500	Fieldstone face instead of brick	2,825	6,225	9,050
1600				
1700				
1800				
1900				
2000				
2100				
2200				
2300				
2700	Cut & patch to match existing construction, add, minimum	2%	3%	
2800	Maximum	5%	9%	
2900	Dust protection, add, minimum	1%	2%	
3000	Maximum	4%	11%	
3100	Equipment usage curtailment, add, minimum	1%	1%	
3200	Maximum	3%	10%	
3300	Material handling & storage limitation, add, minimum	1%	1%	
3400	Maximum	6%	7%	
3500	Protection of existing work, add, minimum	2%	2%	
3600	Maximum	5%	7%	
3700	Shift work requirements, add, minimum		5%	
3800	Maximum		30%	
3900	Temporary shoring and bracing, add, minimum	2%	5%	
4000	Maximum	5%	12%	

D3020 Heat Generating Systems

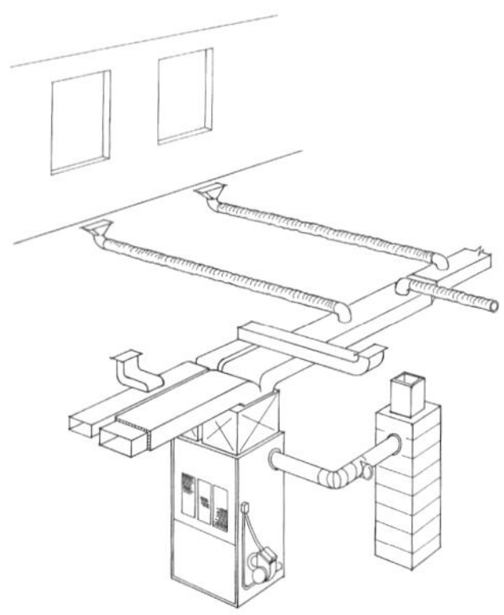

This page illustrates and describes an oil fired forced air system including an oil fired furnace, ductwork, registers and hookups. Lines within System Components give the unit price and total price per square foot for this system. Prices for oil fired forced air systems are on Line Items D3020 122 1600 thru 2800. Both material quantities and labor costs have been adjusted for the system listed.

Factors: To adjust for job conditions other than normal working situations use Lines D3020 122 3100 thru 4200.

Example: You are to install the system during evenings and weekends. Go to Line D3020 122 4200 and apply this percentage to the appropriate INST. cost.

System Components	QUANTITY	UNIT	COST PER S.F.		
			MAT.	INST.	TOTAL
Oil fired hot air heating system including furnace, ductwork, registers And all necessary hookups.					
Area to 800 S.F., heat only					
Furnace, oil, atomizing gun type burner, w/oil piping	1.000	Ea.	2,031.25	412.50	2,443.75
Oil tank, steel, 275 gallon	1.000	Ea.	490	250	740
Duct, galvanized, steel	312.000	Lb.	240.24	2,449.20	2,689.44
Insulation, blanket type, duct work	270.000	S.F.	243	872.10	1,115.10
Flexible duct, 6″ diameter, insulated	100.000	L.F.	275	455	730
Registers, baseboard, gravity, 12″ x 6″	8.000	Ea.	154	228	382
Return, damper, 36″ x 18″	1.000	Ea.	149	65.50	214.50
TOTAL			3,582.49	4,732.30	8,314.79
COST PER S.F.		S.F.	4.48	5.92	10.40

D3020 122	Heating-Cooling, Oil, Forced Air	COST PER S.F.		
		MAT.	INST.	TOTAL
1600	Oil fired, area to 800 S.F.	4.47	5.91	10.38
1700	To 1000 S.F.	3.62	4.80	8.42
1800	To 1200 S.F.	3.15	4.23	7.38
1900	To 1600 S.F.	2.59	3.54	6.13
2000	To 2000 S.F.	2.79	4.54	7.33
2100	To 3000 S.F.	2.19	3.60	5.79
2200	For combined heating and cooling systems:			
2300	Oil fired, heating and cooling, area to 800 S.F.	7	7.05	14.05
2400	To 1000 S.F.	5.75	5.75	11.50
2500	To 1200 S.F.	4.93	5	9.93
2600	To 1600 S.F.	3.96	4.16	8.12
2700	To 2000 S.F.	3.98	5.10	9.08
2800	To 3000 S.F.	3.02	4	7.02
3100	Cut & patch to match existing construction, add, minimum	2%	3%	
3200	Maximum	5%	9%	
3300	Dust protection, add, minimum	1%	2%	
3400	Maximum	4%	11%	
3500	Equipment usage curtailment, add, minimum	1%	1%	
3600	Maximum	3%	10%	
3700	Material handling & storage limitation, add, minimum	1%	1%	
3800	Maximum	6%	7%	
3900	Protection of existing work, add, minimum	2%	2%	
4000	Maximum	5%	7%	
4100	Shift work requirements, add, minimum		5%	
4200	Maximum		30%	

D3020 Heat Generating Systems

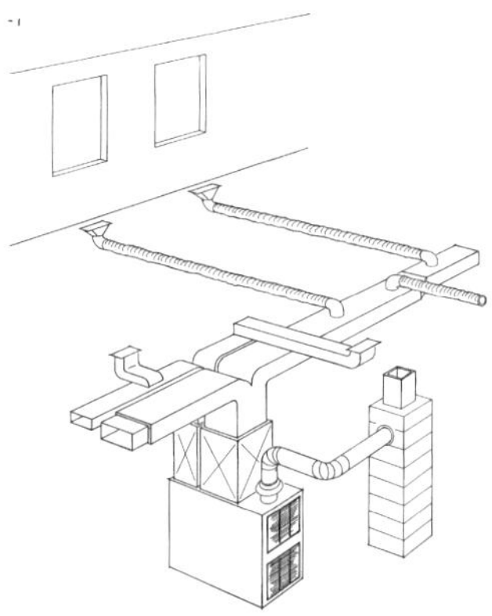

This page illustrates and describes a gas fired forced air system including a gas fired furnace, ductwork, registers and hookups. Lines within System Components give the unit price and total price per square foot for this system. Prices for gas fired forced air systems are on Line Items D3020 124 1400 thru 2600. Both material quantities and labor costs have been adjusted for the system listed.

Factors: To adjust for job conditions other than normal working situations use Lines D3020 124 2900 thru 4000.

Example: You are to install the system with material handling and storage limitations. Go to Line D3020 124 3500 and apply these percentages to the appropriate MAT. and INST. costs.

System Components	QUANTITY	UNIT	COST PER S.F.		
			MAT.	INST.	TOTAL
Gas fired hot air heating system including furnace, ductwork, registers And all necessary hookups.					
Area to 800 S.F., heat only					
Furnace, gas, AGA certified, direct drive, w/gas piping, 44 MBH	1.000	Ea.	850	370	1,220
Duct, galvanized steel	312.000	Lb.	240.24	2,449.20	2,689.44
Insulation, blanket type, ductwork	270.000	S.F.	243	872.10	1,115.10
Flexible duct, 6″ diameter, insulated	100.000	L.F.	275	455	730
Registers, baseboard, gravity, 12″ x 6″	8.000	Ea.	154	228	382
Return, damper, 36″ x 18″	1.000	Ea.	149	65.50	214.50
TOTAL			1,911.24	4,439.80	6,351.04
COST PER S.F.		S.F.	2.39	5.55	7.94

D3020 124	Heating-Cooling, Gas, Forced Air	COST PER S.F.		
		MAT.	INST.	TOTAL
1400	Gas fired, area to 800 S.F.	2.38	5.56	7.94
1500	To 1000 S.F.	1.98	4.52	6.50
1600	To 1200 S.F.	1.78	4	5.78
1700	To 1600 S.F.	1.73	3.75	5.48
1800	To 2000 S.F.	1.79	4.42	6.21
1900	To 3000 S.F.	1.40	3.51	4.91
2000	For combined heating and cooling systems:			
2100	Gas fired, heating and cooling, area to 800 S.F.	4.93	6.65	11.58
2200	To 1000 S.F.	4.11	5.45	9.56
2300	To 1200 S.F.	3.56	4.77	8.33
2400	To 1600 S.F.	3.10	4.37	7.47
2500	To 2000 S.F.	2.97	4.97	7.94
2600	To 3000 S.F.	2.23	3.91	6.14
2900	Cut & patch to match existing construction, add, minimum	2%	3%	
3000	Maximum	5%	9%	
3100	Dust protection, add, minimum	1%	2%	
3200	Maximum	4%	11%	
3300	Equipment usage curtailment, add, minimum	1%	1%	
3400	Maximum	3%	10%	
3500	Material handling & storage limitation, add, minimum	1%	1%	
3600	Maximum	6%	7%	
3700	Protection of existing work, add, minimum	2%	2%	
3800	Maximum	5%	7%	
3900	Shift work requirements, add, minimum		5%	
4000	Maximum		30%	

D3020 Heat Generating Systems

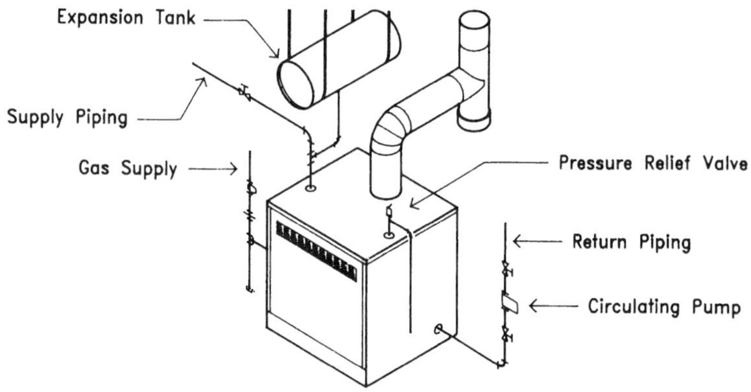

Cast Iron Boiler, Hot Water, Gas Fired

This page illustrates and describes boilers including expansion tank, circulating pump and all service piping. Lines within Systems Components give the material and installation price on a cost each basis for the components. Prices for boiler systems are on Line Items D3020 130 1010 thru D3020 138 1040. Material quantities and labor costs have been adjusted for the system listed.

Factors: To adjust for job conditions other than normal working situations use Line D3020 138 2700 thru 4000.

System Components	QUANTITY	UNIT	COST EACH MAT.	COST EACH INST.	COST EACH TOTAL
Cast iron boiler, including piping, chimney and accessories					
Boilers, gas fired, std controls, CI, insulated, HW, gross output 100 MBH	1.000	Ea.	2,250	1,950	4,200
Pipe, black steel, Sch 40, threaded, W/coupling & hangers, 10′ OC, 3/4″ dia	20.000	L.F.	127.13	250.88	378.01
Pipe, black steel, Sch 40, threaded, W/coupling & hangers, 10′ OC, 1″ dia.	20.000	L.F.	178.45	276.28	454.73
Elbow, 90°, black, straight, 3/4″ dia.	9.000	Ea.	34.56	436.50	471.06
Elbow, 90°, black, straight, 1″ dia.	6.000	Ea.	39.90	315	354.90
Tee, black, straight, 3/4″ dia.	2.000	Ea.	12.20	151	163.20
Tee, black, straight, 1″ dia.	2.000	Ea.	20.90	170	190.90
Tee, black, reducing, 1″ dia.	2.000	Ea.	33.50	170	203.50
Pipe cap, black, 3/4″ dia.	1.000	Ea.	4.36	21.50	25.86
Union, black with brass seat, 3/4″ dia.	2.000	Ea.	32.30	105	137.30
Union, black with brass seat, 1″ dia.	2.000	Ea.	42	113	155
Pipe nipples, black, 3/4″ dia.	5.000	Ea.	14.13	27.88	42.01
Pipe nipples, black, 1″ dia.	3.000	Ea.	12.45	19.28	31.73
Valves, bronze, gate, N.R.S., threaded, class 150, 3/4″ size	2.000	Ea.	150	68	218
Valves, bronze, gate, N.R.S., threaded, class 150, 1″ size	2.000	Ea.	190	72	262
Gas cock, brass, 3/4″ size	1.000	Ea.	17.05	31	48.05
Thermometer, stem type, 9″ case, 8″ stem, 3/4″ NPT	2.000	Ea.	372	50	422
Tank, steel, liquid expansion, ASME, painted, 15 gallon capacity	1.000	Ea.	655	73.50	728.50
Pump, circulating, bronze, flange connection, 3/4″ to 1-1/2″ size, 1/8 HP	1.000	Ea.	1,000	204	1,204
Vent chimney, all fuel, pressure tight, double wall, SS, 6″ dia.	20.000	L.F.	1,180	394	1,574
Vent chimney, elbow, 90° fixed, 6″ dia.	2.000	Ea.	860	79	939
Vent chimney, Tee, 6″ dia.	2.000	Ea.	504	99	603
Vent chimney, ventilated roof thimble, 6″ dia.	1.000	Ea.	263	45.50	308.50
Vent chimney, flat roof flashing, 6″ dia.	1.000	Ea.	113	39.50	152.50
Vent chimney, stack cap, 6″ diameter	1.000	Ea.	310	25.50	335.50
Insulation, fiberglass pipe covering, 1″ wall, 1″ IPS	20.000	L.F.	22.80	103	125.80
TOTAL			8,438.73	5,290.32	13,729.05

D3020 130	Boiler, Cast Iron, Hot Water, Gas	COST EACH MAT.	COST EACH INST.	COST EACH TOTAL
1000	For alternate boilers:			
1010	Boiler, cast iron, gas, hot water, 100 MBH	8,450	5,300	13,750
1020	200 MBH	9,650	6,075	15,725
1030	320 MBH	10,100	6,600	16,700
1040	440 MBH	13,600	8,525	22,125
1050	544 MBH	22,400	13,200	35,600
1060	765 MBH	27,000	14,200	41,200
1070	1088 MBH	29,900	15,200	45,100

D3020 Heat Generating Systems

D3020 130	Boiler, Cast Iron, Hot Water, Gas	COST EACH		
		MAT.	INST.	TOTAL
1080	1530 MBH	35,000	20,100	55,100
1090	2312 MBH	40,800	23,600	64,400

D3020 134	Boiler, Cast Iron, Steam, Gas	COST EACH		
		MAT.	INST.	TOTAL
1010	Boiler, cast iron, gas, steam, 100 MBH	7,075	5,125	12,200
1020	200 MBH	14,100	9,675	23,775
1030	320 MBH	16,000	10,800	26,800
1040	544 MBH	23,800	16,900	40,700
1050	765 MBH	27,100	17,600	44,700
1060	1275 MBH	32,000	19,300	51,300
1070	2675 MBH	41,300	28,200	69,500

D3020 136	Boiler, Cast Iron, Hot Water, Gas/Oil	COST EACH		
		MAT.	INST.	TOTAL
1010	Boiler, cast iron, gas & oil, hot water, 584 MBH	27,300	14,700	42,000
1020	876 MBH	34,700	15,800	50,500
1030	1168 MBH	38,500	17,800	56,300
1040	1460 MBH	45,500	21,400	66,900
1050	2044 MBH	48,400	22,200	70,600
1060	2628 MBH	62,000	26,900	88,900

D3020 138	Boiler, Cast Iron, Steam, Gas/Oil	COST EACH		
		MAT.	INST.	TOTAL
1010	Boiler, cast iron, gas & oil, steam, 810 MBH	30,800	19,100	49,900
1020	1360 MBH	35,700	20,400	56,100
1030	2040 MBH	44,000	26,100	70,100
1040	2700 MBH	47,300	29,100	76,400
2700	Cut & patch to match existing construction, add, minimum	2%	3%	
2800	Maximum	5%	9%	
2900	Dust protection, add, minimum	1%	2%	
3000	Maximum	4%	11%	
3100	Equipment usage curtailment, add, minimum	1%	1%	
3200	Maximum	3%	10%	
3300	Material handling & storage limitation, add, minimum	1%	1%	
3400	Maximum	6%	7%	
3500	Protection of existing work, add, minimum	2%	2%	
3600	Maximum	5%	7%	
3700	Shift work requirements, add, minimum		5%	
3800	Maximum		30%	
3900	Temporary shoring and bracing, add, minimum	2%	5%	
4000	Maximum	5%	12%	

D3020 Heat Generating Systems

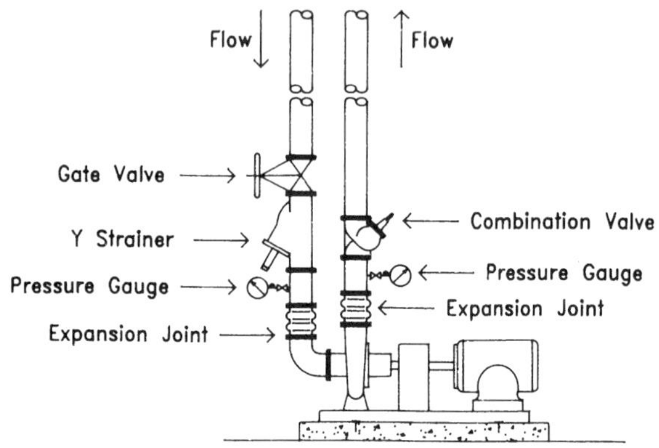

Base Mounted End-Suction Pump

This page illustrates and describes end suction base mounted circulating pumps, including strainer and piping. Lines within system components give the material and installation price on a cost each basis for the components. Prices for circulating pumps are on Line items D3020 330 1010 thru 1050. Material quantities and labor costs have been adjusting for the system listed.

Factors: To adjust for job conditions other than normal working situations uses Lines D3020 330 2700 thru 4000.

System Components	QUANTITY	UNIT	COST EACH MAT.	COST EACH INST.	COST EACH TOTAL
Circulator pump, end suction, including pipe,					
Valves, fittings and guages.					
Pump, circulating, CI, base mounted, 2-1/2" size, 3 HP, to 150 GPM	1.000	Ea.	7,850	680	8,530
Pipe, black steel, Sch. 40, on yoke & roll hangers, 10′ O.C., 2-1/2" dia	12.000	L.F.	276	322.32	598.32
Elbow, 90°, weld joint, steel, 2-1/2" pipe size	1.000	Ea.	30.50	158.10	188.60
Flange, weld neck, 150 LB, 2-1/2" pipe size	8.000	Ea.	280	632.32	912.32
Valve, iron body, gate, 125 lb., N.R.S., flanged, 2-1/2" size	1.000	Ea.	685	245	930
Strainer, Y type, iron body, flanged, 125 lb., 2-1/2" pipe size	1.000	Ea.	121	250	371
Multipurpose valve, CI body, 2" size	1.000	Ea.	480	87	567
Expansion joint, flanged spool, 6" F to F, 2-1/2" dia	2.000	Ea.	588	202	790
T-O-L, weld joint, socket, 1/4" pipe size, nozzle	2.000	Ea.	15.60	110.54	126.14
T-O-L, weld joint, socket, 1/2" pipe size, nozzle	2.000	Ea.	15.60	114.70	130.30
Control gauges, pressure or vacuum, 3-1/2" diameter dial	2.000	Ea.	79	43	122
Pressure/temperature relief plug, 316 SS, 3/4" OD, 7-1/2" insertion	2.000	Ea.	188	43	231
Insulation, fiberglass pipe covering, 1-1/2" wall, 2-1/2" IPS	12.000	L.F.	32.16	75.60	107.76
Pump control system	1.000	Ea.	1,275	630	1,905
Pump balancing	1.000	Ea.		305	305
TOTAL			11,915.86	3,898.58	15,814.44

D3020 330	Circulating Pump Systems, End Suction	COST EACH MAT.	COST EACH INST.	COST EACH TOTAL
1000	For alternate pump sizes:			
1010	Pump, base mtd with motor, end-suction, 2-1/2" size, 3 HP, to 150 GPM	11,900	3,900	15,800
1020	3" size, 5 HP, to 225 GPM	13,400	4,475	17,875
1030	4" size, 7-1/2 HP, to 350 GPM	15,000	5,375	20,375
1040	5" size, 15 HP, to 1000 GPM	21,500	8,175	29,675
1050	6" size, 25 HP, to 1550 GPM	27,200	9,900	37,100
2700	Cut & patch to match existing construction, add, minimum	2%	3%	
2800	Maximum	5%	9%	
2900	Dust protection, add, minimum	1%	2%	
3000	Maximum	4%	11%	
3100	Equipment usage curtailment, add, minimum	1%	1%	
3200	Maximum	3%	10%	
3300	Material handling & storage limitation, add, minimum	1%	1%	
3400	Maximum	6%	7%	
3500	Protection of existing work, add, minimum	2%	2%	
3600	Maximum	5%	7%	

D3020 Heat Generating Systems

D3020 330	Circulating Pump Systems, End Suction	COST EACH		
		MAT.	INST.	TOTAL
3700	Shift work requirements, add, minimum		5%	
3800	Maximum		30%	
3900	Temporary shoring and bracing, add, minimum	2%	5%	
4000	Maximum	5%	12%	

D3040 Distribution Systems

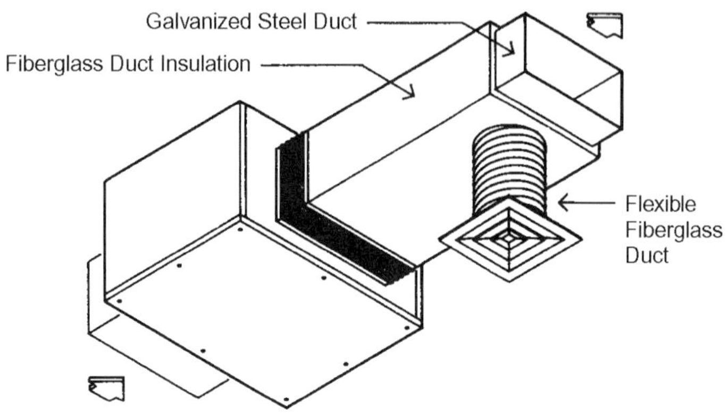

Horizontal Fan Coil Air Conditioning System

This page illustrates and describes fan coil air conditioners including duct work, duct installation, piping and diffusers. Lines within Systems Components give the material and installation price on a cost each basis for the components. Prices for fan coil A/C unit systems are on Line Items D3040 118 1010 thru D3040 126 1120. Material quantities and labor costs have been adjusted for the system listed.

Factors: To adjust for job conditions other than normal working situations use Lines D3040 126 2700 thru 4000.

System Components	QUANTITY	UNIT	COST EACH		
			MAT.	INST.	TOTAL
Fan coil A/C system, horizontal, with housing, controls, 2 pipe, 1/2 ton					
Fan coil A/C, horizontal housing, filters, chilled water, 1/2 ton cooling	1.000	Ea.	1,250	156	1,406
Pipe, black steel, Sch 40, threaded, W/cplgs & hangers, 10′ OC, 3/4″ dia.	20.000	L.F.	121.48	239.73	361.21
Elbow, 90°, black, straight, 3/4″ dia.	6.000	Ea.	23.04	291	314.04
Tee, black, straight, 3/4″ dia.	2.000	Ea.	12.20	151	163.20
Union, black with brass seat, 3/4″ dia.	2.000	Ea.	32.30	105	137.30
Pipe nipples, 3/4″ diam	3.000	Ea.	8.48	16.73	25.21
Valves, bronze, gate, N.R.S., threaded, class 150, 3/4″ size	2.000	Ea.	150	68	218
Circuit setter, balance valve, bronze body, threaded, 3/4″ pipe size	1.000	Ea.	83	34.50	117.50
Insulation, fiberglass pipe covering, 1″ wall, 3/4″ IPS	20.000	L.F.	21.40	98.40	119.80
Ductwork, 12″ x 8″ fabricated, galvanized steel, 12 LF	55.000	Lb.	42.35	431.75	474.10
Insulation, ductwork, blanket type, fiberglass, 1″ thk, 1-1/2 LB density	40.000	S.F.	36	129.20	165.20
Diffusers, aluminum, OB damper, ceiling, perf, 24″x24″ panel size, 6″x6″	2.000	Ea.	332	82	414
Ductwork, flexible, fiberglass fabric, insulated, 1″thk, PE jacket, 6″ dia	16.000	L.F.	44	72.80	116.80
Round volume control damper 6″ dia.	2.000	Ea.	53	60	113
Fan coil unit balancing	1.000	Ea.		85.50	85.50
Re-heat coil balancing	1.000	Ea.		128	128
Diffuser/register, high, balancing	2.000	Ea.		236	236
TOTAL			2,209.25	2,385.61	4,594.86

D3040 118	Fan Coil A/C Unit, Two Pipe	COST EACH		
		MAT.	INST.	TOTAL
1000	For alternate A/C units:			
1010	Fan coil A/C system, cabinet mounted, controls, 2 pipe, 1/2 Ton	1,225	1,275	2,500
1020	1 Ton	1,500	1,450	2,950
1030	1-1/2 Ton	1,625	1,450	3,075
1040	2 Ton	2,550	1,775	4,325
1050	3 Ton	3,175	1,850	5,025

D3040 120	Fan Coil A/C Unit, Two Pipe, Electric Heat	COST EACH		
		MAT.	INST.	TOTAL
1010	Fan coil A/C system, cabinet mntd, elect. ht, controls, 2 pipe, 1/2 Ton	1,800	1,275	3,075
1020	1 Ton	2,225	1,450	3,675
1030	1-1/2 Ton	2,575	1,450	4,025
1040	2 Ton	3,875	1,800	5,675
1050	3 Ton	6,450	1,850	8,300

D3040 Distribution Systems

D3040 122	Fan Coil A/C Unit, Four Pipe	COST EACH		
		MAT.	INST.	TOTAL
1010	Fan coil A/C system, cabinet mounted, controls, 4 pipe, 1/2 Ton	2,025	2,500	4,525
1020	1 Ton	2,325	2,650	4,975
1030	1-1/2 Ton	2,475	2,675	5,150
1040	2 Ton	3,075	2,775	5,850
1050	3 Ton	4,100	2,975	7,075

D3040 124	Fan Coil A/C, Horizontal, Duct Mount, 2 Pipe	COST EACH		
		MAT.	INST.	TOTAL
1010	Fan coil A/C system, horizontal w/housing, controls, 2 pipe, 1/2 Ton	2,200	2,375	4,575
1020	1 Ton	2,950	3,475	6,425
1030	1-1/2 Ton	3,625	4,500	8,125
1040	2 Ton	4,725	5,400	10,125
1050	3 Ton	5,750	7,125	12,875
1060	3-1/2 Ton	6,000	7,400	13,400
1070	4 Ton	6,150	8,375	14,525
1080	5 Ton	7,300	10,600	17,900
1090	6 Ton	7,400	11,400	18,800
1100	7 Ton	7,500	11,800	19,300
1110	8 Ton	7,500	12,100	19,600
1120	10 Ton	8,450	12,500	20,950

D3040 126	Fan Coil A/C, Horiz., Duct Mount, 2 Pipe, Elec. Ht.	COST EACH		
		MAT.	INST.	TOTAL
1010	Fan coil A/C system, horiz. hsng, elect. ht, ctrls, 2 pipe, 1/2 Ton	2,475	2,375	4,850
1020	1 Ton	3,275	3,475	6,750
1030	1-1/2 Ton	3,875	4,500	8,375
1040	2 Ton	5,150	5,400	10,550
1050	3 Ton	8,175	7,375	15,550
1060	3-1/2 Ton	9,150	7,400	16,550
1070	4 Ton	10,200	8,400	18,600
1080	5 Ton	12,200	10,600	22,800
1090	6 Ton	12,300	11,400	23,700
1100	7 Ton	12,500	12,100	24,600
1110	8 Ton	13,200	12,600	25,800
1120	10 Ton	14,000	11,800	25,800
2700	Cut & patch to match existing construction, add, minimum	2%	3%	
2800	Maximum	5%	9%	
2900	Dust protection, add, minimum	1%	2%	
3000	Maximum	4%	11%	
3100	Equipment usage curtailment, add, minimum	1%	1%	
3200	Maximum	3%	10%	
3300	Material handling & storage limitation, add, minimum	1%	1%	
3400	Maximum	6%	7%	
3500	Protection of existing work, add, minimum	2%	2%	
3600	Maximum	5%	7%	
3700	Shift work requirements, add, minimum		5%	
3800	Maximum		30%	
3900	Temporary shoring and bracing, add, minimum	2%	5%	
4000	Maximum	5%	12%	

D3040 Distribution Systems

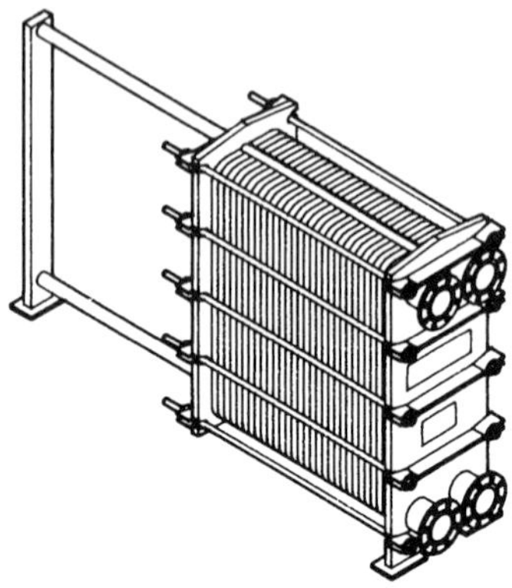

Plate Heat Exchanger

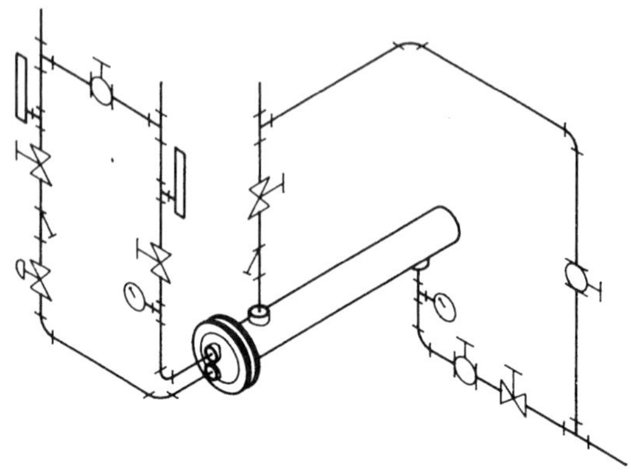

Shell and Tube Heat Exchanger

This page illustrates and describes plate and shell and tube type heat exchangers including pressure gauge and exchanger control system. Lines within Systems Components give the material and installation price on a cost each basis for the components. Prices for heat exchanger systems are on Line Items D3040 610 1010 thru D3040 620 1040. Material quantities and labor costs have been adjusted for the system listed.

Factors To adjust for job conditions other then normal working situations use Lines D3040 620 2700 thru 4000.

System Components	QUANTITY	UNIT	COST EACH		
			MAT.	INST.	TOTAL
Shell and tube type heat exchanger with related valves and piping.					
Heat exchanger, 4 pass, 3/4" O.D. copper tubes, by steam at 10 psi, 40 GPM	1.000	Ea.	4,950	310	5,260
Pipe, black steel, Sch 40, threaded, W/couplings & hangers, 10' OC, 2" dia.	40.000	L.F.	828.20	967.08	1,795.28
Elbow, 90°, straight, 2" dia.	16.000	Ea.	400	1,088	1,488
Reducer, black steel, concentric, 2" dia.	2.000	Ea.	56	117	173
Valves, bronze, gate, rising stem, threaded, class 150, 2" size	6.000	Ea.	1,374	372	1,746
Valves, bronze, globe, class 150, rising stem, threaded, 2" size	2.000	Ea.	1,190	124	1,314
Strainers, Y type, bronze body, screwed, 125 lb, 2" pipe size	2.000	Ea.	242	107	349
Valve, electric motor actuated, brass, 2 way, screwed, 1-1/2" pipe size	1.000	Ea.	660	52.50	712.50
Union, black with brass seat, 2" dia.	8.000	Ea.	352	576	928
Tee, black, straight, 2" dia.	9.000	Ea.	283.50	999	1,282.50
Tee, black, reducing run and outlet, 2" dia.	4.000	Ea.	230	444	674
Pipe nipple, black, 2" dia.	21.000	Ea.	8.20	9.58	17.78
Thermometers, stem type, 9" case, 8" stem, 3/4" NPT	2.000	Ea.	372	50	422
Gauges, pressure or vacuum, 3-1/2" diameter dial	2.000	Ea.	79	43	122
Insulation, fiberglass pipe covering, 1" wall, 2" IPS	40.000	L.F.	58	226	284
Heat exchanger control system	1.000	Ea.	2,675	2,200	4,875
Coil balancing	1.000	Ea.		128	128
TOTAL			13,757.90	7,813.16	21,571.06

D3040 610	Heat Exchanger, Plate Type	COST EACH		
		MAT.	INST.	TOTAL
1000	For alternate heat exchangers:			
1010	Plate heat exchanger, 400 GPM	52,000	15,700	67,700
1020	800 GPM	83,000	20,000	103,000
1030	1200 GPM	124,000	27,100	151,100
1040	1800 GPM	168,000	33,700	201,700

D3040 620	Heat Exchanger, Shell & Tube	COST EACH		
		MAT.	INST.	TOTAL
1010	Shell & tube heat exchanger, 40 GPM	13,800	7,825	21,625
1020	96 GPM	24,200	13,800	38,000
1030	240 GPM	43,900	18,700	62,600
1040	600 GPM	84,500	26,300	110,800
2700	Cut & patch to match existing construction, add, minimum	2%	3%	
2800	Maximum	5%	9%	
2900	Dust protection, add, minimum	1%	2%	
3000	Maximum	4%	11%	

D3040 Distribution Systems

D3040 620	Heat Exchanger, Shell & Tube	COST EACH		
		MAT.	INST.	TOTAL
3100	Equipment usage curtailment, add, minimum	1%	1%	
3200	Maximum	3%	10%	
3300	Material handling & storage limitation, add, minimum	1%	1%	
3400	Maximum	6%	7%	
3500	Protection of existing work, add, minimum	2%	2%	
3600	Maximum	5%	7%	
3700	Shift work requirements, add, minimum		5%	
3800	Maximum		30%	
3900	Temporary shoring and bracing, add, minimum	2%	5%	
4000	Maximum	5%	12%	

D3050 Terminal & Package Units

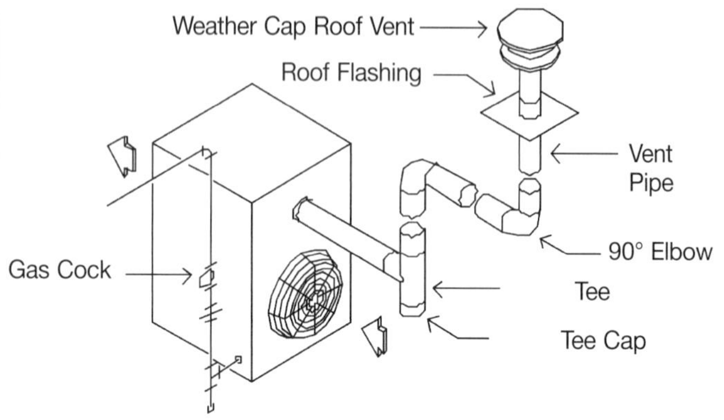

Gas Unit Heater

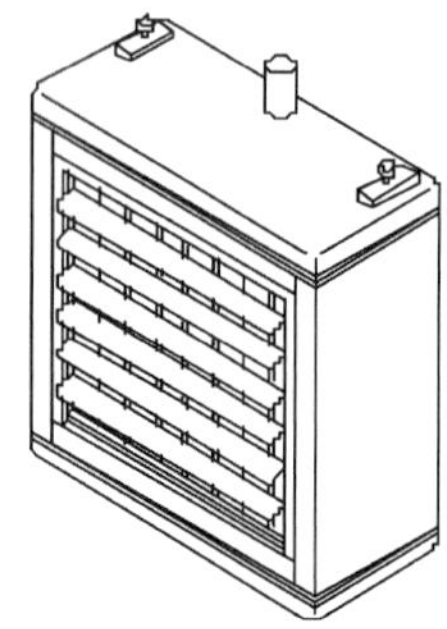

Hydronic Unit Heater

This page illustrates and describes unit heaters including piping and vents. Lines within Systems Components give the material and installation price on a cost each basis for this system. Prices for unit heater systems are on Line items D3050 120 1010 thru D3050 140 1040. Material quantities and labor costs have been adjusted for the system listed.

Factors: To adjust for job conditions other than normal working situations use Lines D3050 140 2700 thru 4000.

System Components	QUANTITY	UNIT	COST EACH MAT.	COST EACH INST.	COST EACH TOTAL
Unit heaters complete with piping, thermostats and chimney as required					
Space heater, propeller fan, 20 MBH output	1.000	Ea.	1,075	147	1,222
Pipe, black steel, Sch 40, threaded, W/coupling & hangers, 10′ OC, 3/4″ dia	20.000	L.F.	118.65	234.15	352.80
Elbow, 90°, black steel, straight, 1/2″ dia.	3.000	Ea.	9.54	136.50	146.04
Elbow, 90°, black steel, straight, 3/4″ dia.	3.000	Ea.	11.52	145.50	157.02
Tee, black steel, reducing, 3/4″ dia.	1.000	Ea.	10.10	75.50	85.60
Union, black with brass seat, 3/4″ dia.	1.000	Ea.	16.15	52.50	68.65
Pipe nipples, black, 1/2″ dia	2.000	Ea.	9.64	21.60	31.24
Pipe nipples, black, 3/4″ dia	2.000	Ea.	5.65	11.15	16.80
Pipe cap, black, 3/4″ dia.	1.000	Ea.	4.36	21.50	25.86
Gas cock, brass, 3/4″ size	1.000	Ea.	17.05	31	48.05
Thermostat, 1 set back, electric, timed	1.000	Ea.	38.50	87	125.50
Wiring, thermostat hook-up,25′ of #18-3	1.000	Ea.	12.05	26.50	38.55
Vent chimney, prefab metal, U.L. listed, gas, double wall, galv. st, 4″ dia	12.000	L.F.	84	208.80	292.80
Vent chimney, gas, double wall, galv. steel, elbow 90°, 4″ dia.	2.000	Ea.	52	70	122
Vent chimney, gas, double wall, galv. steel, Tee, 4″ dia.	1.000	Ea.	37	45.50	82.50
Vent chimney, gas, double wall, galv. steel, T cap, 4″ dia.	1.000	Ea.	3.85	28	31.85
Vent chimney, gas, double wall, galv. steel, roof flashing, 4″ dia.	1.000	Ea.	9.60	35	44.60
Vent chimney, gas, double wall, galv. steel, top, 4″ dia.	1.000	Ea.	23.50	27	50.50
TOTAL			1,538.16	1,404.20	2,942.36

D3050 120	Unit Heaters, Gas	COST EACH MAT.	COST EACH INST.	COST EACH TOTAL
1000	For alternate unit heaters:			
1010	Space heater, suspended, gas fired, propeller fan, 20 MBH	1,550	1,400	2,950
1020	60 MBH	1,825	1,450	3,275
1030	100 MBH	2,050	1,550	3,600
1040	160 MBH	2,400	1,700	4,100
1050	200 MBH	2,850	1,875	4,725
1060	280 MBH	3,550	2,000	5,550
1070	320 MBH	4,500	2,275	6,775

D3050 Terminal & Package Units

D3050 130	Unit Heaters, Hydronic	COST EACH		
		MAT.	INST.	TOTAL
1010	Space heater, suspended, horiz. mount, HW, prop. fan, 20 MBH	1,975	1,500	3,475
1020	60 MBH	2,725	1,700	4,425
1030	100 MBH	3,275	1,850	5,125
1040	150 MBH	3,725	2,050	5,775
1050	200 MBH	4,175	2,300	6,475
1060	300 MBH	5,200	2,700	7,900

D3050 140	Cabinet Unit Heaters, Hydronic	COST EACH		
		MAT.	INST.	TOTAL
1010	Unit heater, cabinet type, horizontal blower, hot water, 20 MBH	2,300	1,475	3,775
1020	60 MBH	3,400	1,625	5,025
1030	100 MBH	3,800	1,800	5,600
1040	120 MBH	3,875	1,850	5,725
2700	Cut & patch to match existing construction, add, minimum	2%	3%	
2800	Maximum	5%	9%	
2900	Dust protection, add, minimum	1%	2%	
3000	Maximum	4%	11%	
3100	Equipment usage curtailment, add, minimum	1%	1%	
3200	Maximum	3%	10%	
3300	Material handling & storage limitation, add, minimum	1%	1%	
3400	Maximum	6%	7%	
3500	Protection of existing work, add, minimum	2%	2%	
3600	Maximum	5%	7%	
3700	Shift work requirements, add, minimum		5%	
3800	Maximum		30%	
3900	Temporary shoring and bracing, add, minimum	2%	5%	
4000	Maximum	5%	12%	

D3050 Terminal & Package Units

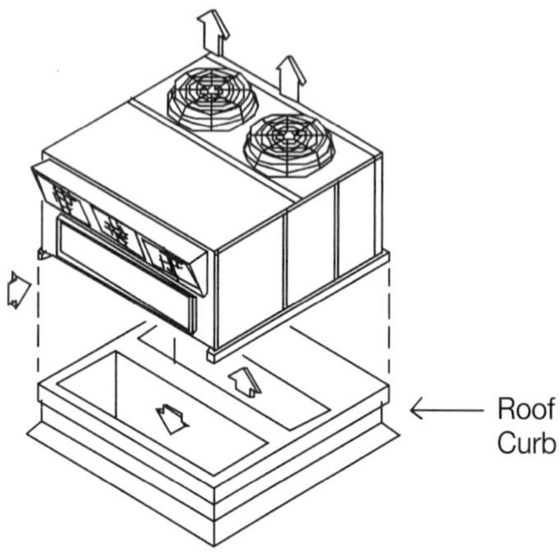

Rooftop Air Conditioner Unit

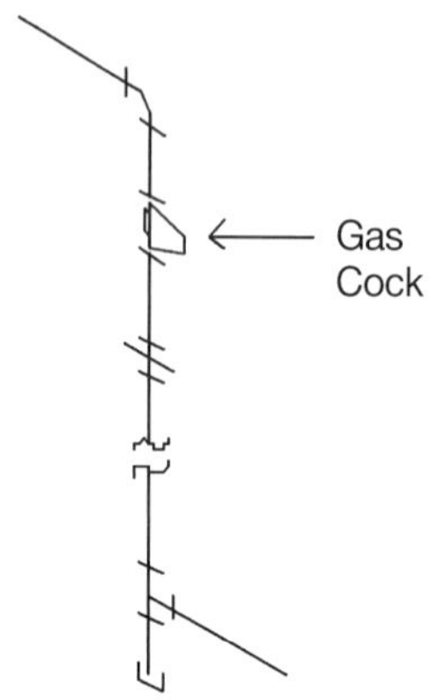

Gas Piping

This page illustrates and describes packaged air conditioners including rooftop DX unit, pipes and fittings. Lines within Systems Components give the material and installation price on a cost each basis for the components. Prices for A/C systems are on Line Items D3050 175 1010 thru D3050 180 1060. Material quantities and labor costs have been adjusted for the system listed.

Factors: To adjust for job conditions other than normal working situations use Lines D3050 180 2700 thru 4000.

System Components	QUANTITY	UNIT	COST EACH MAT.	COST EACH INST.	COST EACH TOTAL
Packaged rooftop DX unit, gas, with related piping and valves.					
Roof top A/C, curb, economizer, sgl zone, elec cool, gas ht, 5 ton, 112 MBH	1.000	Ea.	5,475	2,225	7,700
Pipe, black steel, Sch 40, threaded, W/cplgs & hangers, 10′ OC, 1″ dia	20.000	L.F.	174.30	269.85	444.15
Elbow, 90°, black, straight, 3/4″ dia.	3.000	Ea.	11.52	145.50	157.02
Elbow, 90°, black, straight, 1″ dia.	3.000	Ea.	19.95	157.50	177.45
Tee, black, reducing, 1″ dia.	1.000	Ea.	16.75	85	101.75
Union, black with brass seat, 1″ dia.	1.000	Ea.	21	56.50	77.50
Pipe nipple, black, 3/4″ dia	2.000	Ea.	11.30	22.30	33.60
Pipe nipple, black, 1″ dia	2.000	Ea.	8.30	12.85	21.15
Cap, black, 1″ dia.	1.000	Ea.	5.25	22.50	27.75
Gas cock, brass, 1″ size	1.000	Ea.	34	36	70
Control system, pneumatic, Rooftop A/C unit	1.000	Ea.	2,750	2,450	5,200
Rooftop unit heat/cool balancing	1.000	Ea.		460	460
TOTAL			8,527.37	5,943	14,470.37

D3050 175	Rooftop Air Conditioner, Const. Volume	COST EACH MAT.	COST EACH INST.	COST EACH TOTAL
1000	For alternate A/C systems:			
1010	A/C, Rooftop, DX cool, gas heat, curb, economizer, fltrs, 5 Ton	8,525	5,950	14,475
1020	7-1/2 Ton	11,500	6,200	17,700
1030	12-1/2 Ton	17,900	6,800	24,700
1040	18 Ton	23,900	7,450	31,350
1050	25 Ton	30,700	8,475	39,175
1060	40 Ton	39,900	11,300	51,200

D3050 180	Rooftop Air Conditioner, Variable Air Volume	COST EACH MAT.	COST EACH INST.	COST EACH TOTAL
1010	A/C, Rooftop, DX cool, gas heat, curb, ecmizr, fltrs, VAV, 12-1/2 Ton	19,900	7,275	27,175
1020	18 Ton	26,200	8,000	34,200
1030	25 Ton	33,500	9,200	42,700
1040	40 Ton	45,000	12,500	57,500
1050	60 Ton	65,000	17,000	82,000
1060	80 Ton	100,500	21,300	121,800
2700	Cut & patch to match existing construction, add, minimum	2%	3%	
2800	Maximum	5%	9%	
2900	Dust protection, add, minimum	1%	2%	
3000	Maximum	4%	11%	

D3050 Terminal & Package Units

D3050 180	Rooftop Air Conditioner, Variable Air Volume	COST EACH		
		MAT.	INST.	TOTAL
3100	Equipment usage curtailment, add, minimum	1%	1%	
3200	Maximum	3%	10%	
3300	Material handling & storage limitation, add, minimum	1%	1%	
3400	Maximum	6%	7%	
3500	Protection of existing work, add, minimum	2%	2%	
3600	Maximum	5%	7%	
3700	Shift work requirements, add, minimum		5%	
3800	Maximum		30%	
3900	Temporary shoring and bracing, add, minimum	2%	5%	
4000	Maximum	5%	12%	

D3050 Terminal & Package Units

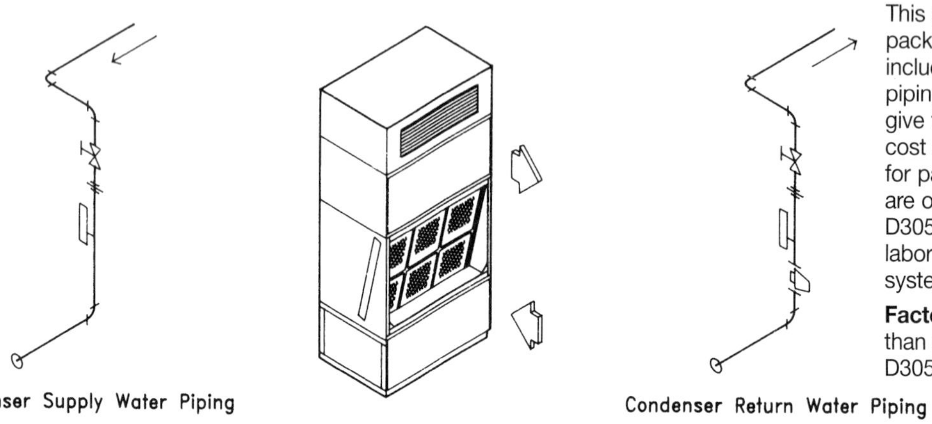

Self–Contained Air Conditioner Unit, Water Cooled

This page illustrates and describes packaged electrical air conditioning units including multizone control and related piping. Lines within Systems Components give the material and installation price on a cost each basis for the components. Prices for packaged electrical A/C unit systems are on Line Items D3050 210 1010 thru D3050 225 1060. Material quantities and labor costs have been adjusted for the system listed.

Factors: To adjust for job conditions other than normal working situations use Lines D3050 225 2700 thru 4000.

System Components	QUANTITY	UNIT	COST EACH MAT.	COST EACH INST.	COST EACH TOTAL
Packaged water cooled electric air conditioning unit with related piping and valves.					
Self-contained, water cooled, elect. heat, not inc tower, 5 ton, const vol	1.000	Ea.	8,750	1,625	10,375
Pipe, black steel, Sch 40, threaded, W/cplg & hangers, 10′ OC, 1-1/4″ dia.	20.000	L.F.	224.68	295.63	520.31
Elbow, 90°, black, straight, 1-1/4″ dia.	6.000	Ea.	66	333	399
Tee, black, straight, 1-1/4″ dia.	2.000	Ea.	33.80	175	208.80
Tee, black, reducing, 1-1/4″ dia.	2.000	Ea.	59	175	234
Thermometers, stem type, 9″ case, 8″ stem, 3/4″ NPT	2.000	Ea.	372	50	422
Union, black with brass seat, 1-1/4″ dia.	2.000	Ea.	60	117	177
Pipe nipple, black, 1-1/4″ dia	3.000	Ea.	15.68	20.63	36.31
Valves, bronze, gate, N.R.S., threaded, class 150, 1-1/4″ size	2.000	Ea.	256	91	347
Circuit setter, bal valve, bronze body, threaded, 1-1/4″ pipe size	1.000	Ea.	157	46.50	203.50
Insulation, fiberglass pipe covering, 1″ wall, 1-1/4″ IPS	20.000	L.F.	24.80	108	132.80
Control system, pneumatic, A/C Unit with heat	1.000	Ea.	2,750	2,450	5,200
Re-heat coil balancing	1.000	Ea.		128	128
Rooftop unit heat/cool balancing	1.000	Ea.		460	460
TOTAL			12,768.96	6,074.76	18,843.72

D3050 210	AC Unit, Package, Elec. Ht., Water Cooled	COST EACH MAT.	COST EACH INST.	COST EACH TOTAL
1000	For alternate A/C systems:			
1010	A/C, Self contained, single pkg., water cooled, elect. heat, 5 Ton	12,800	6,075	18,875
1020	10 Ton	19,700	7,475	27,175
1030	20 Ton	28,800	9,450	38,250
1040	30 Ton	45,000	10,300	55,300
1050	40 Ton	58,000	11,400	69,400
1060	50 Ton	75,500	13,600	89,100

D3050 215	AC Unit, Package, Elec. Ht., Water Cooled, VAV	COST EACH MAT.	COST EACH INST.	COST EACH TOTAL
1010	A/C, Self contained, single pkg., water cooled, elect. ht, VAV, 10 Ton	21,400	7,475	28,875
1020	20 Ton	31,600	9,450	41,050
1030	30 Ton	49,200	10,300	59,500
1040	40 Ton	63,000	11,400	74,400
1050	50 Ton	81,500	13,600	95,100
1060	60 Ton	93,500	16,500	110,000

D3050 Terminal & Package Units

D3050 220	AC Unit, Package, Hot Water Coil, Water Cooled	COST EACH		
		MAT.	INST.	TOTAL
1010	A/C, Self contn'd, single pkg., water cool, H/W ht, const. vol, 5 Ton	11,300	8,425	19,725
1020	10 Ton	16,500	10,000	26,500
1030	20 Ton	40,800	12,500	53,300
1040	30 Ton	53,000	15,700	68,700
1050	40 Ton	65,000	19,100	84,100
1060	50 Ton	80,000	23,600	103,600

D3050 225	AC Unit, Package, HW Coil, Water Cooled, VAV	COST EACH		
		MAT.	INST.	TOTAL
1010	A/C, Self contn'd, single pkg., water cool, H/W ht, VAV, 10 Ton	18,200	9,450	27,650
1020	20 Ton	28,300	12,200	40,500
1030	30 Ton	42,900	15,000	57,900
1040	40 Ton	57,000	16,400	73,400
1050	50 Ton	72,000	19,400	91,400
1060	60 Ton	85,500	21,000	106,500
2700	Cut & patch to match existing construction, add, minimum	2%	3%	
2800	Maximum	5%	9%	
2900	Dust protection, add, minimum	1%	2%	
3000	Maximum	4%	11%	
3100	Equipment usage curtailment, add, minimum	1%	1%	
3200	Maximum	3%	10%	
3300	Material handling & storage limitation, add, minimum	1%	1%	
3400	Maximum	6%	7%	
3500	Protection of existing work, add, minimum	2%	2%	
3600	Maximum	5%	7%	
3700	Shift work requirements, add, minimum		5%	
3800	Maximum		30%	
3900	Temporary shoring and bracing, add, minimum	2%	5%	
4000	Maximum	5%	12%	

D3050 Terminal & Package Units

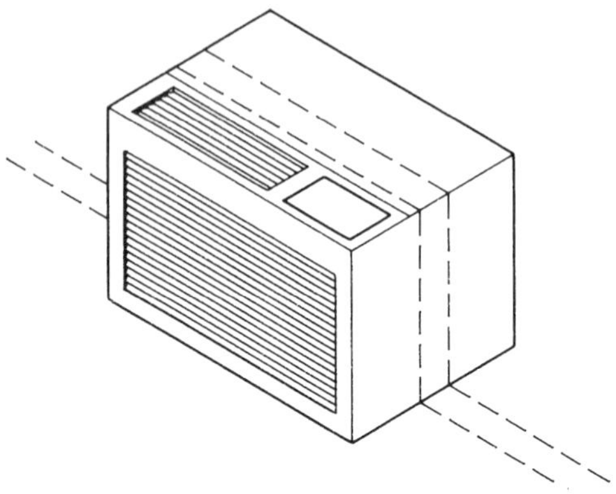

Thru–Wall Heat Pump

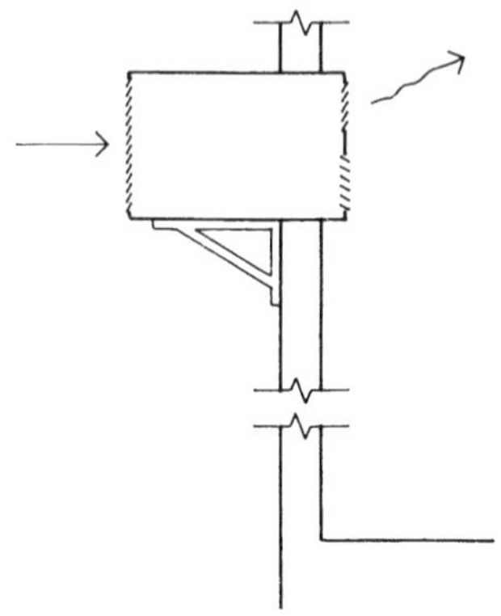

Thru–Wall Air Conditioner

This page illustrates and describes thru-wall air conditioning and heat pump units. Lines within Systems Components give the material and installation price on a cost each basis for the components. Prices for thru-wall unit systems are on Line Items D3050 255 1010 thru D3050 260 1050. Material quantities and labor costs have been adjusted for the system listed.

Factors: To adjust for job conditions other than normal working conditions use Lines D3050 260 2700 thru 4000.

System Components	QUANTITY	UNIT	COST EACH MAT.	COST EACH INST.	COST EACH TOTAL
Electric thru-wall air conditioning unit.					
A/C unit, thru-wall, electric heat., cabinet, louver, 1/2 ton	1.000	Ea.	790	208	998
TOTAL			790	208	998

D3050 255	Thru-Wall A/C Unit	COST EACH MAT.	COST EACH INST.	COST EACH TOTAL
1000	For alternate A/C units:			
1010	A/C Unit, thru-the-wall, supplemental elect. heat, cabinet, louver, 1/2 Ton	790	208	998
1020	3/4 Ton	865	250	1,115
1030	1 Ton	935	310	1,245
1040	1-1/2 Ton	1,950	520	2,470
1050	2 Ton	1,975	660	2,635

D3050 260	Thru-Wall Heat Pump	COST EACH MAT.	COST EACH INST.	COST EACH TOTAL
1010	Heat pump, thru-the-wall, cabinet, louver, 1/2 Ton	2,425	156	2,581
1020	3/4 Ton	2,650	208	2,858
1030	1 Ton	2,725	310	3,035
1040	Supplemental. elect. heat, 1-1/2 Ton	2,800	805	3,605
1050	2 Ton	3,025	835	3,860
2700	Cut & patch to match existing construction, add, minimum	2%	3%	
2800	Maximum	5%	9%	
2900	Dust protection, add, minimum	1%	2%	
3000	Maximum	4%	11%	
3100	Equipment usage curtailment, add, minimum	1%	1%	
3200	Maximum	3%	10%	
3300	Material handling & storage limitation, add, minimum	1%	1%	
3400	Maximum	6%	7%	
3500	Protection of existing work, add, minimum	2%	2%	
3600	Maximum	5%	7%	
3700	Shift work requirements, add, minimum		5%	
3800	Maximum		30%	
3900	Temporary shoring and bracing, add, minimum	2%	5%	
4000	Maximum	5%	12%	

D4010 Sprinklers

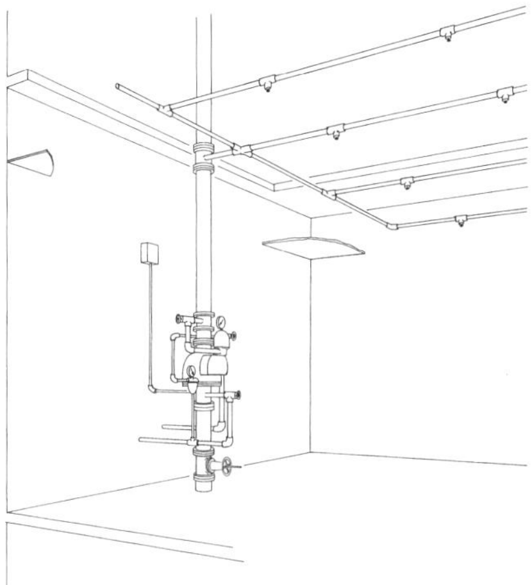

This page illustrates and describes a dry type fire sprinkler system. Lines within System Components give the unit cost of a system for a 2,000 square foot building. Lines D4010 320 1700 thru 2400 give the square foot costs for alternate systems. Both material quantities and labor costs have been adjusted for the system listed.

Factors: To adjust for conditions other than normal working conditions, use Lines D4010 320 2900 thru 4000.

System Components	QUANTITY	UNIT	COST EACH		
			MAT.	INST.	TOTAL
Pre-action fire sprinkler system, ordinary hazard, open area to 2000 S.F. On one floor.					
Air compressor	1.000	Ea.	755	520	1,275
4″ OS & Y valve	1.000	Ea.	695	410	1,105
Pipe riser 4″ diameter	10.000	L.F.	385	341	726
Sprinkler head supply piping, 1″	163.000	L.F.	622.50	963.75	1,586.25
Sprinkler head supply piping, 1-1/4″	163.000	L.F.	574.75	756.25	1,331
Sprinkler head supply piping, 2-1/2″	163.000	L.F.	950	931	1,881
Pipe fittings, 1″	8.000	Ea.	83.60	680	763.60
Pipe fittings, 1-1/4″	8.000	Ea.	135.20	700	835.20
Pipe fittings, 2-1/2″	4.000	Ea.	310	544	854
Pipe fittings, 4″	2.000	Ea.	550	610	1,160
Detectors	2.000	Ea.	1,320	84	1,404
Sprinkler heads	16.000	Ea.	150.40	672	822.40
Fire department connection	1.000	Ea.	470	243	713
TOTAL			7,001.45	7,455	14,456.45

D4010 320	Fire Sprinkler Systems, Dry	COST PER S.F.		
		MAT.	INST.	TOTAL
1700	Ordinary hazard, one floor, area to 2000 S.F./floor	3.53	3.75	7.28
1800	For each additional floor, add per floor	1.90	3.12	5.02
1900	Area to 3200 S.F./ floor	2.98	3.70	6.68
2000	For each additional floor, add per floor	1.96	3.30	5.26
2100	Area to 5000 S.F./ floor	3.41	3.76	7.17
2200	For each additional floor, add per floor	2.50	3.50	6
2300	Area to 8000 S.F./ floor	2.68	3.18	5.86
2400	For each additional floor, add per floor	2.03	3	5.03
2500				
2600				
2700				
2800				
2900	Cut & patch to match existing construction, add, minimum	2%	3%	
3000	Maximum	5%	9%	
3100	Dust protection, add, minimum	1%	2%	
3200	Maximum	4%	11%	
3300	Equipment usage curtailment, add, minimum	1%	1%	
3400	Maximum	3%	10%	

D40 Fire Protection

D4010 Sprinklers

D4010 320	Fire Sprinkler Systems, Dry	COST PER S.F.		
		MAT.	INST.	TOTAL
3500	Material handling & storage limitation, add, minimum	1%	1%	
3600	Maximum	6%	7%	
3700	Protection of existing work, add, minimum	2%	2%	
3800	Maximum	5%	7%	
3900	Shift work requirements, add, minimum		5%	
4000	Maximum		30%	

D4010 Sprinklers

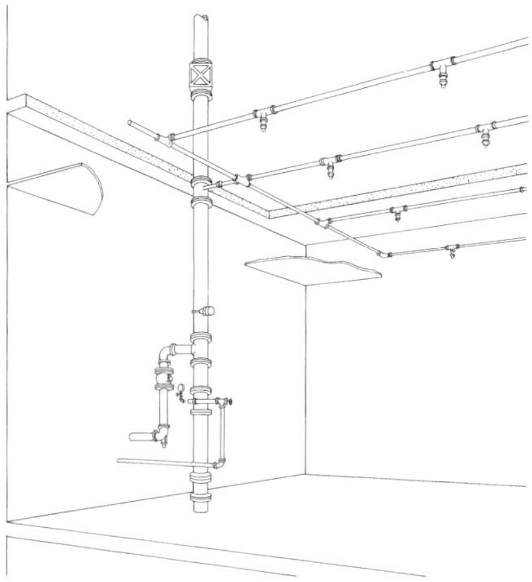

This page illustrates and describes a wet type fire sprinkler system. Lines within System Components give the unit cost of a system for a 2,000 square foot building. Lines D4010 420 1900 thru 2600 give the square foot costs for alternate systems. Both material quantities and labor costs have been adjusted for the system listed.

Factors: To adjust for conditions other than normal working conditions, use Lines D4010 420 2900 thru 3600.

System Components	QUANTITY	UNIT	COST EACH MAT.	COST EACH INST.	COST EACH TOTAL
Wet pipe fire sprinkler system, ordinary hazard, open area to 2000 S.F.					
On one floor.					
4″ OS & Y valve	1.000	Ea.	695	410	1,105
Wet pipe alarm valve	1.000	Ea.	1,625	605	2,230
Water motor alarm	1.000	Ea.	385	169	554
3″ check valve	1.000	Ea.	320	405	725
Pipe riser, 4″ diameter	10.000	L.F.	385	341	726
Water gauges and trim	1.000	Set	2,675	1,225	3,900
Electric fire horn	1.000	Ea.	65	95	160
Sprinkler head supply piping, 1″ diameter	168.000	L.F.	622.50	963.75	1,586.25
Sprinkler head supply piping, 1-1/2″ diameter	168.000	L.F.	673.75	841.50	1,515.25
Sprinkler head supply piping, 2-1/2″ diameter	168.000	L.F.	950	931	1,881
Pipe fittings, 1″	8.000	Ea.	83.60	680	763.60
Pipe fittings, 1-1/2″	8.000	Ea.	168	756	924
Pipe fittings, 2-1/2″	4.000	Ea.	310	544	854
Pipe fittings, 4″	2.000	Ea.	550	610	1,160
Sprinkler heads	16.000	Ea.	150.40	672	822.40
Fire department connection	1.000	Ea.	470	243	713
TOTAL			10,128.25	9,491.25	19,619.50

D4010 420	Fire Sprinkler Systems, Wet	COST PER S.F. MAT.	COST PER S.F. INST.	COST PER S.F. TOTAL
1900	Ordinary hazard, one floor, area to 2000 S.F./floor	5.10	4.76	9.86
2000	For each additional floor, add per floor	1.96	3.19	5.15
2100	Area to 3200 S.F./floor	3.92	4.29	8.21
2200	For each additional floor, add per floor	1.96	3.30	5.26
2300	Area to 5000 S.F./ floor	3.75	4.13	7.88
2400	For each additional floor, add per floor	2.50	3.50	6
2500	Area to 8000 S.F./ floor	2.81	3.39	6.20
2600	For each additional floor, add per floor	2.19	3.19	5.38
2700				
2900	Cut & patch to match existing construction, add minimum	2%	3%	
3000	Maximum	5%	9%	
3100	Dust protection, add, minimum	1%	2%	
3200	Maximum	4%	11%	
3300	Equipment usage curtailment, add, minimum	1%	1%	
3400	Maximum	3%	10%	
3500	Material handling & storage limitation, add, minimum	1%	1%	
3600	Maximum	6%	7%	

D5010 Electrical Service/Distribution

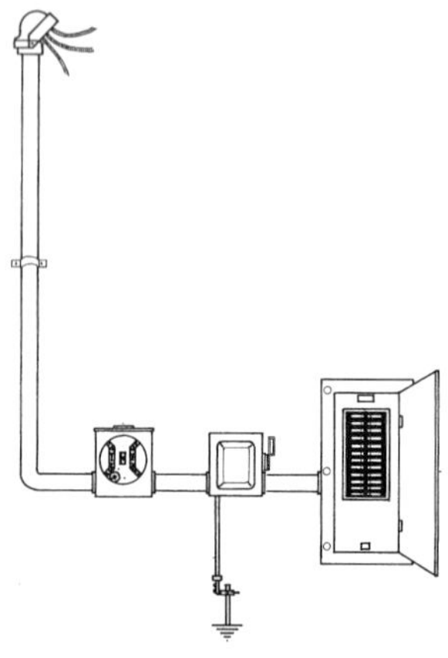

This page illustrates and describes commercial service systems including a meter socket, service head and cable, entrance switch, steel conduit, copper wire, panel board, ground rod, wire, and conduit. Lines within System Components give the unit price and total price on a cost each basis for this system. Prices for commercial service systems are on Line Items D5010 210 1500 thru 1700. Both material quantities and labor costs have been adjusted for the system listed.

Factors: To adjust for job conditions other than normal working situations use Lines D5010 210 2900 thru 4000.

Example: You are to install the system with maximum equipment usage curtailment. Go to Line D5010 210 3400 and apply these percentages to the appropriate MAT. and INST. costs.

System Components	QUANTITY	UNIT	COST EACH		
			MAT.	INST.	TOTAL
Commercial electric service including service breakers, metering 120/208 Volt, 3 phase, 4 wire, feeder, and panel board.					
100 Amp Service					
Meter socket	1.000	Ea.	51.50	199	250.50
Service head	.200	C.L.F.	57.80	127	184.80
Service entrance switch	1.000	Ea.	380	335	715
Rigid steel conduit	20.000	L.F.	97.20	196	293.20
600 volt copper wire #3	1.000	C.L.F.	136	127	263
Panel board, 24 circuits, 20 Amp breakers	1.000	Ea.	1,000	1,050	2,050
Ground rod	1.000	Ea.	19.80	116	135.80
TOTAL			1,742.30	2,150	3,892.30

D5010 210	Commercial Service - 3 Phase	COST EACH		
		MAT.	INST.	TOTAL
1500	120/208 Volt, 3 phase, 4 wire service, 60 Amp	1,175	1,675	2,850
1550	100 Amp	1,750	2,150	3,900
1600	200 Amp	3,575	3,800	7,375
1700	400 Amp	8,425	4,775	13,200
1800				
1900				
2000				
2100				
2200				
2300				
2400				
2500				
2900	Cut & patch to match existing construction, add, minimum	2%	3%	
3000	Maximum	5%	9%	
3100	Dust protection, add, minimum	1%	2%	
3200	Maximum	4%	11%	
3300	Equipment usage curtailment, add, minimum	1%	1%	
3400	Maximum	3%	10%	
3500	Material handling & storage limitation, add, minimum	1%	1%	
3600	Maximum	6%	7%	
3700	Protection of existing work, add, minimum	2%	2%	
3800	Maximum	5%	7%	
3900	Shift work requirements, add, minimum		5%	
4000	Maximum		30%	

D5010 Electrical Service/Distribution

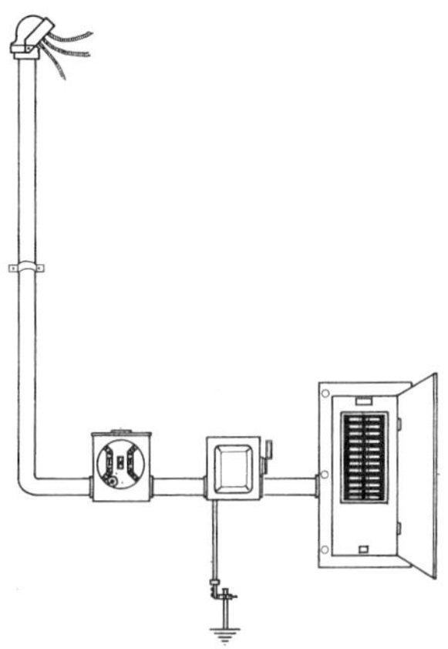

This page illustrates and describes a residential, single phase system including a weather cap, service entrance cable, meter socket, entrance switch, ground rod, ground cable, EMT, and panelboard. Lines within System Components give the unit price and total price on a cost each basis for this system. Prices for a residential, single phase system are also given. Both material quantities and labor costs have been adjusted for the system listed.

Factors: To adjust for job conditions other than normal working situations use Lines D5010 220 2900 thru 4000.

Example: You are to install the system with a minimum equipment usage curtailment. Go to Line D5010 220 3300 and apply these percentages to the appropriate MAT. and INST. costs.

System Components	QUANTITY	UNIT	COST EACH		
			MAT.	INST.	TOTAL
100 Amp Service, single phase					
Weathercap	1.000	Ea.	14.90	53	67.90
Service entrance cable	.200	C.L.F.	57.80	127	184.80
Meter socket	1.000	Ea.	51.50	199	250.50
Entrance disconnect switch	1.000	Ea.	380	335	715
Ground rod, with clamp	1.000	Ea.	23.50	133	156.50
Ground cable	.100	C.L.F.	22.30	40	62.30
Panelboard, 12 circuit	1.000	Ea.	295	530	825
TOTAL			845	1,417	2,262
200 Amp Service , single phase					
Weathercap	1.000	Ea.	41	79.50	120.50
Service entrance cable	.200	C.L.F.	147	182	329
Meter socket	1.000	Ea.	86	335	421
Entrance disconnect switch	1.000	Ea.	820	490	1,310
Ground rod, with clamp	1.000	Ea.	42.50	145	187.50
Ground cable	.100	C.L.F.	34	22	56
3/4" EMT	10.000	L.F.	12.30	49	61.30
Panelboard, 24 circuit	1.000	Ea.	665	825	1,490
TOTAL			1,847.80	2,127.50	3,975.30

D5010 220	Residential Service - Single Phase	COST EACH		
		MAT.	INST.	TOTAL
2800				
2810	100 Amp Service, single phase	845	1,425	2,270
2820	200 Amp Service, single phase	1,850	2,125	3,975
2900	Cut & patch to match existing construction, add, minimum	2%	3%	
3000	Maximum	5%	9%	
3100	Dust protection, add, minimum	1%	2%	
3200	Maximum	4%	11%	
3300	Equipment usage curtailment, add, minimum	1%	1%	
3400	Maximum	3%	10%	
3500	Material handling & storage limitation, add, minimum	1%	1%	
3600	Maximum	6%	7%	
3700	Protection of existing work, add, minimum	2%	2%	
3800	Maximum	5%	7%	
3900	Shift work requirements, add, minimum		5%	
4000	Maximum		30%	

D5020 Lighting and Branch Wiring

D5020 180	Selective Price Sheet	COST EACH		
		MAT.	INST.	TOTAL
0100	Using non-metallic sheathed, cable, air conditioning receptacle	27.50	63.50	91
0200	Disposal wiring	22.50	71	93.50
0300	Dryer circuit	52	116	168
0400	Duplex receptacle	27.50	49	76.50
0500	Fire alarm or smoke detector	101	63.50	164.50
0600	Furnace circuit & switch	31.50	106	137.50
0700	Ground fault receptacle	62.50	79.50	142
0800	Heater circuit	35.50	79.50	115
0900	Lighting wiring	37	40	77
1000	Range circuit	98	159	257
1100	Switches single pole	24	40	64
1200	3-way	29.50	53	82.50
1300	Water heater circuit	37	127	164
1400	Weatherproof receptacle	152	106	258
1500	Using BX cable, air conditioning receptacle	42	77	119
1600	Disposal wiring	36	85	121
1700	Dryer circuit	61	139	200
1800	Duplex receptacle	42	59	101
1900	Fire alarm or smoke detector	101	63.50	164.50
2000	Furnace circuit & switch	49.50	127	176.50
2100	Ground fault receptacle	76	96.50	172.50
2200	Heater circuit	43	96.50	139.50
2300	Lighting wiring	44.50	48	92.50
2400	Range circuit	146	193	339
2500	Switches, single pole	38.50	48	86.50
2600	3-way	41.50	63.50	105
2700	Water heater circuit	59	152	211
2800	Weatherproof receptacle	160	127	287
2900	Using EMT conduit, air conditioning receptacle	55.50	95	150.50
3000	Disposal wiring	51	106	157
3100	Dryer circuit	66.50	172	238.50
3200	Duplex receptacle	55.50	73	128.50
3300	Fire alarm or smoke detector	122	95	217
3400	Furnace circuit & switch	58.50	159	217.50
3500	Ground fault receptacle	99	118	217
3600	Heater circuit	55	118	173
3700	Lighting wiring	53	59.50	112.50
3800	Range circuit	110	236	346
3900	Switches, single pole	51.50	59.50	111
4000	3-way	46	79.50	125.50
4100	Water heater circuit	59	187	246
4200	Weatherproof receptacle	173	159	332
4300	Using aluminum conduit, air conditioning receptacle	62	127	189
4400	Disposal wiring	58	142	200
4500	Dryer circuit	87	228	315
4600	Duplex receptacle	59.50	98	157.50
4700	Fire alarm or smoke detector	136	127	263
4800	Furnace circuit & switch	71	212	283
4900	Ground fault receptacle	106	159	265
5000	Heater circuit	60	159	219
5100	Lighting wiring	71.50	79.50	151
5200	Range circuit	136	320	456
5300	Switches, single pole	70	79.50	149.50
5400	3-way	64.50	106	170.50
5500	Water heater circuit	73	255	328
5600	Weatherproof receptacle	199	212	411
5700	Using galvanized steel conduit	63.50	136	199.50
5800	Disposal wiring	52.50	152	204.50

D5020 Lighting and Branch Wiring

D5020 180	Selective Price Sheet	COST EACH		
		MAT.	INST.	TOTAL
5900	Dryer circuit	90	245	335
6000	Duplex receptacle	63.50	104	167.50
6100	Fire alarm or smoke detector	130	136	266
6200	Furnace circuit & switch	65.50	228	293.50
6300	Ground fault receptacle	106	168	274
6400	Heater circuit	62	168	230
6500	Lighting wiring	69.50	85	154.50
6600	Range circuit	136	335	471
6700	Switches, single pole	69.50	85	154.50
6800	3-way	64	110	174
6900	Water heater circuit	73.50	266	339.50
7000	Weatherproof receptacle	191	228	419
7100				
7200				
7300				
7400				
7500				
7600				
7700				
7800				
7900				
8000				
8100				
8200				
8300				
8400				

D5020 Lighting and Branch Wiring

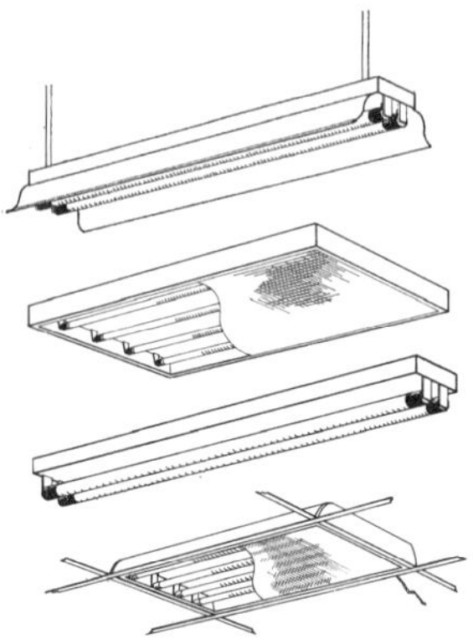

This page illustrates and describes fluorescent lighting systems including a fixture, lamp, outlet box and wiring. Lines within System Components give the unit price and total price on a cost each basis for this system. Prices for fluorescent lighting systems are on Line Items D5020 248 1200 thru 1500. Both material quantities and labor costs have been adjusted for the system listed.

Factors: To adjust for job conditions other than normal working situations use Lines D5020 248 2900 thru 4000.

Example: You are to install the system during evening hours. Go to Line D5020 248 3900 and apply this percentage to the appropriate INST. cost.

System Components	QUANTITY	UNIT	COST EACH		
			MAT.	INST.	TOTAL
Fluorescent lighting, including fixture, lamp, outlet box and wiring.					
Recessed lighting fixture, on suspended system	1.000	Ea.	73.50	136	209.50
Outlet box	1.000	Ea.	3.19	35.50	38.69
#12 wire	.660	C.L.F.	8.98	38.28	47.26
Conduit, EMT, 1/2″ conduit	20.000	L.F.	15.40	75	90.40
TOTAL			101.07	284.78	385.85

D5020 248	Lighting, Fluorescent	COST EACH		
		MAT.	INST.	TOTAL
1200	Recessed lighting fixture, on suspended system	101	285	386
1300	Surface mounted, 2′ x 4′, acrylic prismatic diffuser	135	298	433
1400	Strip fixture, 8′ long, two 8′ lamps	93	281	374
1500	Pendant mounted, industrial, 8′ long, with reflectors	137	325	462
1600				
1700				
1800				
1900				
2000				
2100				
2200				
2300				
2900	Cut & patch to match existing construction, add, minimum	2%	3%	
3000	Maximum	5%	9%	
3100	Dust protection, add, minimum	1%	2%	
3200	Maximum	4%	11%	
3300	Equipment usage curtailment, add, minimum	1%	1%	
3400	Maximum	3%	10%	
3500	Material handling & storage limitation, add, minimum	1%	1%	
3600	Maximum	6%	7%	
3700	Protection of existing work, add, minimum	2%	2%	
3800	Maximum	5%	7%	
3900	Shift work requirements, add, minimum		5%	
4000	Maximum		30%	

D5020 Lighting and Branch Wiring

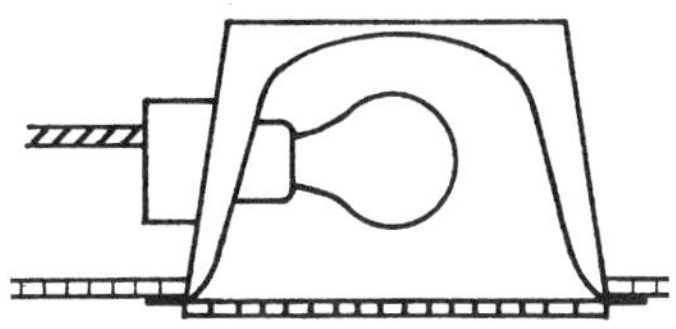

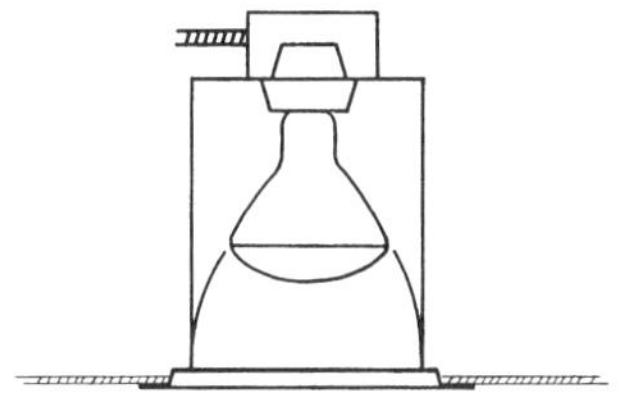

This page illustrates and describes incandescent lighting systems including a fixture, lamp, outlet box, conduit and wiring. Lines within System Components give the unit price and total price on a cost each basis for this system. Prices for an alternate incandescent lighting system are also given. Both material quantities and labor costs have been adjusted for the system listed.

Factors: To adjust for conditions other than normal working situations use Lines D5020 250 2900 thru 4000.

Example: You are to install the system and cut and match existing construction. Go to Line D5020 250 3000 and apply this percentage to the appropriate INST. costs.

System Components	QUANTITY	UNIT	COST EACH MAT.	COST EACH INST.	COST EACH TOTAL
Incandescent light fixture, including lamp, outlet box, conduit and wiring.					
Recessed wide reflector with flat glass lens	1.000	Ea.	105	95	200
Outlet box	1.000	Ea.	3.19	35.50	38.69
Armored cable, 3 wire	.200	C.L.F.	24.40	58	82.40
TOTAL			132.59	188.50	321.09
Recessed flood light fixture, including lamp, outlet box, conduit & wire					
Recessed, R-40 flood lamp with refractor	1.000	Ea.	112	79.50	191.50
150 watt R-40 flood lamp	.010	Ea.	6.65	4.90	11.55
Outlet box	1.000	Ea.	3.19	35.50	38.69
Outlet box cover	1.000	Ea.	2.12	9.95	12.07
Romex, 12-2 with ground	.200	C.L.F.	9.90	58	67.90
Conduit, 1/2" EMT	20.000	L.F.	15.40	75	90.40
TOTAL			149.26	262.85	412.11

D5020 250	Lighting, Incandescent	COST EACH MAT.	COST EACH INST.	COST EACH TOTAL
2810	Incandescent light fixture, including lamp, outlet box, conduit and wiring.	133	189	322
2820	Recessed flood light fixture, including lamp, outlet box, conduit & wire	149	263	412
2900	Cut & patch to match existing construction, add, minimum	2%	3%	
3000	Maximum	5%	9%	
3100	Dust protection, add, minimum	1%	2%	
3200	Maximum	4%	11%	
3300	Equipment usage curtailment, add, minimum	1%	1%	
3400	Maximum	3%	10%	
3500	Material handling & storage limitation, add, minimum	1%	1%	
3600	Maximum	6%	7%	
3700	Protection of existing work, add, minimum	2%	2%	
3800	Maximum	5%	7%	
3900	Shift work requirements, add, minimum		5%	
4000	Maximum		30%	

D5020 Lighting and Branch Wiring

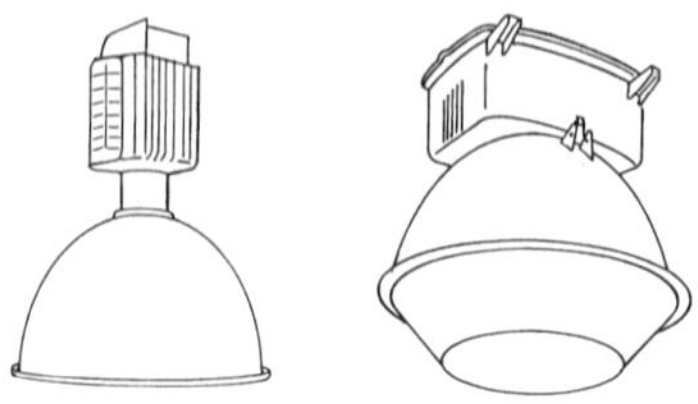

This page illustrates and describes high intensity lighting systems including a lamp, EMT conduit, EMT "T" fitting with cover, and wire. Lines within System Components give the unit price and total price on a cost each basis for this system. Prices for high intensity lighting systems are on Line Items D5020 252 1100 thru 1800. Both material quantities and labor costs have been adjusted for the system listed.

Factors: To adjust for job conditions other than normal working situations use Lines D5020 252 2900 thru 4000.

Example: You are to install the system and protect existing construction. Go to Line D5020 252 3700 and apply these percentages to the appropriate MAT. and INST. costs.

System Components	QUANTITY	UNIT	COST EACH		
			MAT.	INST.	TOTAL
High intensity lighting system consisting of 400 watt metal halide fixture And lamp with 1/2″ EMT conduit and fittings using #12 wire, high bay.					
Single unit, 400 watt	1.000	Ea.	445	277	722
Electric metallic tubing (EMT), 1/2″ diameter	30.000	Ea.	23.10	112.50	135.60
1/2″ EMT "T" fitting with cover	1.000	Ea.	14.10	40	54.10
Wire, 600 volt, type THWN-THHN, copper, solid, #12	.600	Ea.	8.16	34.80	42.96
TOTAL			490.36	464.30	954.66

D5020 252	Lighting, High Intensity	COST EACH		
		MAT.	INST.	TOTAL
1100	High bay: 400 watt, metal halide fixture and lamp	490	465	955
1200	High pressure sodium and lamp	455	465	920
1400	1000 watt, metal halide	685	505	1,190
1500	High pressure sodium	635	505	1,140
1700	Low bay: 250 watt, metal halide	435	385	820
1800	150 watt, high pressure sodium	400	385	785
1900				
2000				
2100				
2200				
2300				
2400				
2500				
2600				
2700				
2900	Cut & patch to match existing construction, add, minimum	2%	3%	
3000	Maximum	5%	9%	
3100	Dust protection, add, minimum	1%	2%	
3200	Maximum	4%	11%	
3300	Equipment usage curtailment, add, minimum	1%	1%	
3400	Maximum	3%	10%	
3500	Material handling & storage limitation, add, minimum	1%	1%	
3600	Maximum	6%	7%	
3700	Protection of existing work, add, minimum	2%	2%	
3800	Maximum	5%	7%	
3900	Shift work requirements, add, minimum		5%	
4000	Maximum		30%	

D5090 Other Electrical Systems

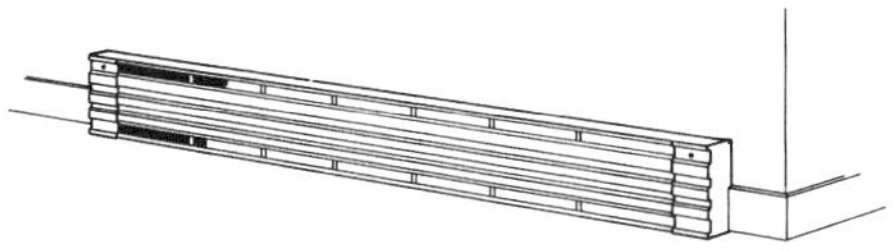

This page illustrates and describes baseboard heat systems including a thermostat, outlet box, breaker, and feed. Lines within System Components give the unit price and total price on a cost each basis for this system. Prices for baseboard heat systems are on Line Items D5090 530 1500 thru 1800. Both material quantities and labor costs have been adjusted for the system listed.

Factors: To adjust for job conditions other than normal working situations use Lines D5090 530 2900 thru 4000.

Example: You are to install the system during evenings and weekends only. Go to Line D5090 530 4000 and apply this percentage to the appropriate INST. cost.

System Components	QUANTITY	UNIT	COST EACH		
			MAT.	INST.	TOTAL
Baseboard heat including thermostat, outlet box, breaker and feed.					
Electric baseboard heater, 4′ long	1.000	Ea.	70.50	95	165.50
Thermostat, integral	1.000	Ea.	33	40	73
Romex, 12-3 with ground	.400	C.L.F.	27.80	128	155.80
Panel board breaker, 20 Amp	1.000	Ea.	36	79.50	115.50
Outlet box	1.000	Ea.	3.19	35.50	38.69
TOTAL			170.49	378	548.49

D5090 530	Heat, Baseboard	COST EACH		
		MAT.	INST.	TOTAL
1400	Heat, electric baseboard, 4′ long			
1500	Electric baseboard, 2′ long	144	365	509
1550	4′ long	170	380	550
1600	6′ long	181	410	591
1700	8′ long	201	440	641
1800	10′ long	310	475	785
1900				
2000				
2100				
2200				
2300				
2400				
2500				
2600				
2700				
2900	Cut & patch to match existing construction, add, minimum	2%	3%	
3000	Maximum	5%	9%	
3100	Dust protection, add, minimum	1%	2%	
3200	Maximum	4%	11%	
3300	Equipment usage curtailment, add, minimum	1%	1%	
3400	Maximum	3%	10%	
3500	Material handling & storage limitation, add, minimum	1%	1%	
3600	Maximum	6%	7%	
3700	Protection of existing work, add, minimum	2%	2%	
3800	Maximum	5%	7%	
3900	Shift work requirements, add, minimum		5%	
4000	Maximum		30%	

E Equipment & Furnishings

E1090 Other Equipment

This page illustrates and describes kitchen systems including top and bottom cabinets, custom laminated plastic top, single bowl sink, and appliances. Lines within System Components give the unit price and total price on a cost each basis for this system. Prices for kitchen systems are on Line Items E1090 310 1400 thru 1600. Both material quantities and labor costs have been adjusted for the system listed.

Factors: To adjust for job conditions other than normal working situations use Lines E1090 310 2900 thru 4000.

Example: You are to install the system and protect the work area from dust. Go to Line E1090 310 3200 and apply these percentages to the appropriate MAT. and INST. costs.

System Components	QUANTITY	UNIT	COST EACH		
			MAT.	INST.	TOTAL
Kitchen cabinets including wall and base cabinets, custom laminated Plastic top,sink & appliances,no plumbing or electrical rough-in included.					
Prefinished wood cabinets, average quality, wall and base	20.000	L.F.	3,300	770	4,070
Custom laminated plastic counter top	20.000	L.F.	216	386	602
Stainless steel sink, 22" x 25"	1.000	Ea.	710	219	929
Faucet, top mount	1.000	Ea.	67.50	68	135.50
Dishwasher, built-in	1.000	Ea.	233	330	563
Compactor, built-in	1.000	Ea.	635	116	751
Range hood, vented, 30" wide	1.000	Ea.	63	245	308
TOTAL			5,224.50	2,134	7,358.50

E1090 310	Kitchens	COST EACH		
		MAT.	INST.	TOTAL
1400	Prefinished wood cabinets, average quality	5,225	2,125	7,350
1500	Prefinished wood cabinets, high quality	9,250	2,450	11,700
1600	Custom cabinets, built in place, high quality	12,200	2,825	15,025
1700				
1800				
1900				
2000				
2100				
2200	NOTE: No plumbing or electric rough-ins are included in the above			
2300	Prices, for plumbing see Division 22, for electric see Division 26.			
2400				
2500				
2600				
2700				
2900	Cut & patch to match existing construction, add, minimum	2%	3%	
3000	Maximum	5%	9%	
3100	Dust protection, add, minimum	1%	2%	
3200	Maximum	4%	11%	
3300	Equipment usage curtailment, add, minimum	1%	1%	
3400	Maximum	3%	10%	
3500	Material handling & storage limitation, add, minimum	1%	1%	
3600	Maximum	6%	7%	
3700	Protection of existing work, add, minimum	2%	2%	
3800	Maximum	5%	7%	
3900	Shift work requirements, add, minimum		5%	
4000	Maximum		30%	

E1090 Other Equipment

E1090 315	Selective Price Sheet	COST EACH		
		MAT.	INST.	TOTAL
0100	Cabinets standard wood, base, one drawer one door, 12″ wide	257	46.50	303.50
0200	15″ wide	264	48	312
0300	18″ wide	289	49.50	338.50
0400	21″ wide	305	51	356
0500	24″ wide	340	52	392
0600				
0700	Two drawers two doors, 27″ wide	375	52.50	427.50
0800	30″ wide	410	54	464
0900	33″ wide	430	55.50	485.50
1000	36″ wide	450	57	507
1100	42″ wide	480	58.50	538.50
1200	48″ wide	515	61	576
1300	Drawer base (4 drawers), 12″ wide	271	46.50	317.50
1400	15″ wide	274	48	322
1500	18″ wide	305	49.50	354.50
1600	24″ wide	340	52	392
1700	Sink or range base, 30″ wide	345	54	399
1800	33″ wide	370	55.50	425.50
1900	36″ wide	390	57	447
2000	42″ wide	415	58.50	473.50
2100	Corner base, 36″ wide	610	64.50	674.50
2200	Lazy susan with revolving door	815	70	885
2300	Cabinets standard wood, wall two doors, 12″ high, 30″ wide	231	46.50	277.50
2400	36″ wide	271	48	319
2500	15″ high, 30″ high	231	48	279
2600	36″ wide	271	51	322
2700	24″ high, 30″ wide	310	49.50	359.50
2800	36″ wide	340	51	391
2900	30″ high, 30″ wide	345	60	405
3000	36″ wide	385	61.50	446.50
3100	42″ wide	430	62.50	492.50
3200	48″ wide	485	63	548
3300	One door, 30″ high, 12″ wide	220	52.50	272.50
3400	15″ wide	234	54	288
3500	18″ wide	258	55.50	313.50
3600	24″ wide	300	57	357
3700	Corner, 30″ high, 24″ wide	299	64.50	363.50
3800	36″ wide	325	70	395
3900	Broom, 84″ high, 24″ deep, 18″ wide	650	116	766
4000	Oven, 84″ high, 24″ deep, 27″ wide	945	145	1,090
4100	Valance board, 4′ long	54.50	11.70	66.20
4200	6′ long	82	17.50	99.50
4300	Counter tops, laminated plastic, stock 25″ wide w/backsplash, min.	10.80	19.30	30.10
4400	Maximum	23	23	46
4500	Custom, 7/8″ thick, no splash	25	19.30	44.30
4600	Cove splash	32	19.30	51.30
4700	1-1/4″ thick, no splash	34.50	20.50	55
4800	Square splash	35	20.50	55.50
4900	Post formed	12.10	19.30	31.40
5000				
5100	Maple laminated 1-1/2″ thick, no splash	81.50	20.50	102
5200	Square splash	97	20.50	117.50
5300				
5400				
5500	Appliances, range, free standing, minimum	445	92	537
5600	Maximum	1,950	230	2,180
5700	Built-in, minimum	715	106	821
5800	Maximum	1,375	580	1,955

E1090 Other Equipment

E1090 315	Selective Price Sheet	COST EACH		
		MAT.	INST.	TOTAL
5900	Counter top range 4 burner, maximum	305	106	411
6000	Maximum	1,200	212	1,412
6100	Compactor, built-in, minimum	635	116	751
6200	Maximum	945	193	1,138
6300	Dishwasher, built-in, minimum	233	330	563
6400	Maximum	440	660	1,100
6500	Garbage disposer, sink-pipe, minimum	91	132	223
6600	Maximum	216	132	348
6700	Range hood, 30" wide, 2 speed, minimum	63	245	308
6800	Maximum	825	410	1,235
6900	Refrigerator, no frost, 12 cu. ft.	470	92	562
7000	20 cu. ft.	560	153	713
7100	Plumb. not incl. rough-ins, sinks porc. C.I., single bowl, 21" x 24"	283	219	502
7200	21" x 30"	575	219	794
7300	Double bowl, 20" x 32"	315	255	570
7400				
7500	Stainless steel, single bowl, 19" x 18"	635	219	854
7600	22" x 25"	710	219	929
7700				
7800				
7900				
8000				
8100				
8200				
8300				
8400				

G Building Sitework

G1030 Site Clearing

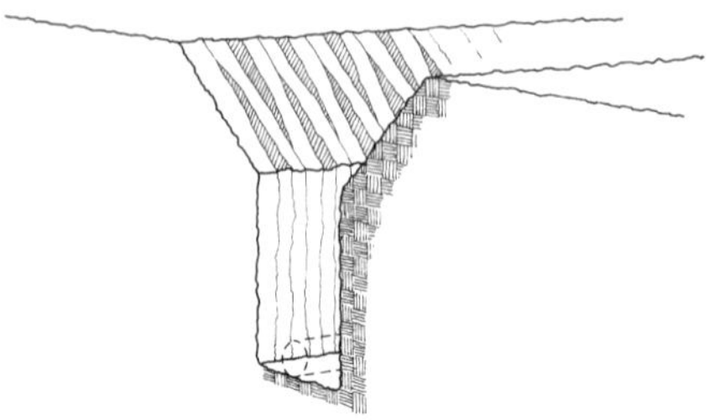

This page illustrates and describes utility trench excavation with backfill systems including a trench, backfill, excavated material, utility pipe or ductwork not included. Lines within System Components give the unit price and total price on a linear foot basis for this system. Prices for utility trench excavation with backfill systems are on Line Items G1030 810 1100 thru 2500. Both material quantities and labor costs have been adjusted for the system listed.

Factors: To adjust for job conditions other than normal working situations use Lines G1030 810 3100 thru 4000.

Example: You are to install the system and protect existing construction. Go to Line G1030 810 3600 and apply these percentages to the appropriate TOTAL cost.

System Components	QUANTITY	UNIT	COST PER L.F. EQUIP.	LABOR	TOTAL
Cont. utility trench excav. with 3/8 C.Y. wheel mtd. backhoe, Backfilling with excav. mat. and comp. in 12″ lifts. Trench is 2′ wide by 4′ dp. Cost of utility piping or ductwork is not incl. Cost based on an Excavation production rate of 150 C.Y. daily no hauling included.					
Machine excavate trench, 2′ wide by 4′ deep	.296	C.Y.	.73	2.07	2.80
Dozer backfill with excavated material	.296	C.Y.	.34	.57	.91
Compact in 12″ lifts, vibrating plate	.296	C.Y.	.07	2.31	2.38
TOTAL			1.14	4.95	6.09

G1030 810	Excavation, Utility Trench	COST PER L.F. EQUIP.	LABOR	TOTAL
1100	2′ wide x 2′ deep	.57	2.48	3.05
1200	3′ deep	.86	3.71	4.57
1250	4′ deep	1.14	4.95	6.10
1300	With sloping sides, 2′ wide x 5′ deep	3.21	13.95	17.15
1400	6′ deep	4.28	18.60	23
1500	7′ deep	5.50	24	29.50
1600	8′ deep	6.85	30	37
1700	9′ deep	8.35	36.50	45
1800	10′ deep	10	43.50	53.50
2000	For hauling excavated material up to 2 miles & backfilling w/gravel			
2100	Gravel, 2′ wide by 2′ deep, add	4.22	3.40	7.62
2200	4′ deep, add	8.45	6.80	15.25
2300	6′ deep, add	31.50	25.50	57
2400	8′ deep, add	50.50	40.50	91
2500	10′ deep, add	74	59.50	133.50
2600				
2700				
2800	For shallow, hand excavated trenches, no backfill, to 6′ dp, light soil		57.50	57.50
2900	Heavy soil		115	115
3000				
3100	Equipment usage curtailment, add, minimum	1%	1%	
3200	Maximum	3%	10%	
3300	Material handling & storage limitation, add, minimum	1%	1%	
3400	Maximum	6%	7%	
3500	Protection of existing work, add, minimum	2%	2%	
3600	Maximum	5%	7%	
3700	Shift work requirements, add, minimum		5%	
3800	Maximum		30%	
3900	Temporary shoring and bracing, add, minimum	2%	5%	
4000	Maximum	5%	12%	

G2010 Roadways

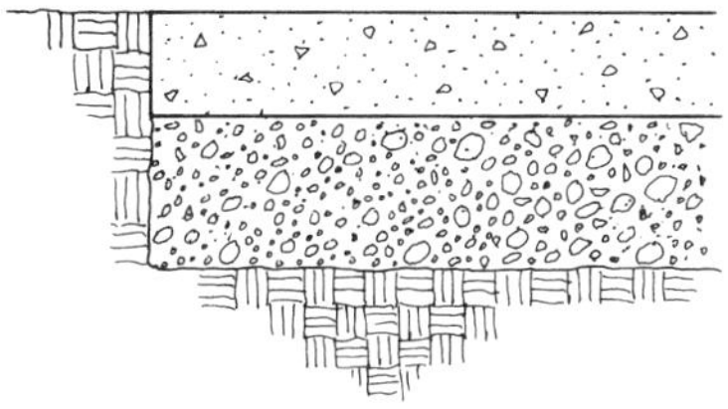

This page illustrates and describes driveway systems including concrete slab, gravel base, broom finish, compaction, and joists. Lines within System Components give the unit price and total price on a cost each basis for this system. Prices for driveway systems are on Line Items G2010 240 1200 thru 2700. Both material quantities and labor costs have been adjusted for the system listed.

Factors: To adjust for job conditions other than normal working situations use Lines G2010 240 2900 thru 4000.

Example: You are to install the system and match existing construction. Go to Line G2010 240 3000 and apply these percentages to the appropriate MAT. and INST. costs.

System Components	QUANTITY	UNIT	COST EACH		
			MAT.	INST.	TOTAL
Complete driveway, 10′ x 30′, 4″ concrete slab, 3000 psi, on 6″ compacted Gravel base, concrete broom finished and cured with 10′ x 10′ joints.					
Grade and compact subgrade	33.000	S.Y.		66.33	66.33
Place and compact 6″ crushed stone base	33.000	S.Y.	221.10	44.88	265.98
Place and remove edgeforms (4 use)	80.000	L.F.	25.60	292.80	318.40
Concrete for slab, 3000 psi	4.000	C.Y.	448		448
Place concrete, 4″ thick slab, direct chute	4.000	C.Y.		106.20	106.20
Screed, float and broom finish slab	4.000	C.Y.		246	246
Spray on membrane curing compound	3.000	S.F.	24.60	29.10	53.70
Saw cut 3″ deep joints	60.000	L.F.	9	90.60	99.60
TOTAL			728.30	875.91	1,604.21

G2010 240	Driveways	COST EACH		
		MAT.	INST.	TOTAL
1200	Concrete: 10′ x 30′ with 4″ concrete on 6″ crushed stone	730	875	1,605
1300	With 6″ concrete	950	930	1,880
1400	20′ x 30′ with 4″ concrete	1,450	1,675	3,125
1500	With 6″ concrete	1,900	1,775	3,675
1600	10′ wide, for each additional 10′ length over 30′, 4″ thick, add	233	281	514
1700	6″ thick, add	315	300	615
1800	20′ wide, for each additional 10′ length over 30′, 4″ thick, add	465	535	1,000
1900	6″ thick, add	605	570	1,175
2000	Asphalt: 10′ x 30′ with 2″ binder, 1″ topping on 6″ cr. st., sealed	625	620	1,245
2100	3″ binder, 1″ topping	745	635	1,380
2200	20′ x 30′ with 2″ binder, 1″ topping	1,250	860	2,110
2300	3″ binding, 1″ topping	1,500	885	2,385
2400	10′ wide for each add'l. 10′ length over 30′, 2″ b + 1″ t, add	208	78.50	286.50
2500	3″ binder, 1″ topping	249	83	332
2600	20′ wide for each add'l. 10′ length over 30′, 2″ b + 1″ t, add	415	158	573
2700	3″ binder, 1″ topping	500	167	667
2800				
2900	Cut & patch to match existing construction, add, minimum	2%	3%	
3000	Maximum	5%	9%	
3100	Dust protection, add, minimum	1%	2%	
3200	Maximum	4%	11%	
3300	Equipment usage curtailment, add, minimum	1%	1%	
3400	Maximum	3%	10%	
3500	Material handling & storage limitation, add, minimum	1%	1%	
3600	Maximum	6%	7%	
3700	Protection of existing work, add, minimum	2%	2%	
3800	Maximum	5%	7%	
3900	Shift work requirements, add, minimum		5%	
4000	Maximum		30%	

G2020 Parking Lots

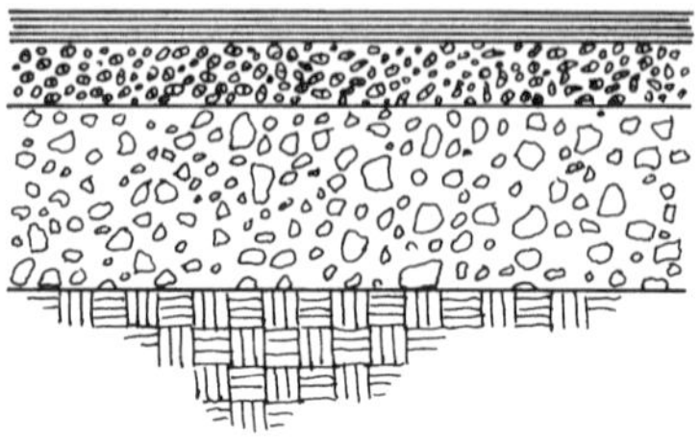

This page illustrates and describes asphalt parking lot systems including asphalt binder, topping, crushed stone base, painted parking stripes and concrete parking blocks. Lines within System Components give the unit price and total price per square yard for this system. Prices for asphalt parking lot systems are on Line Items G2020 230 1400 thru 2500. Both material quantities and labor costs have been adjusted for the system listed.

Factors: To adjust for job conditions other than normal working situations use Lines G2020 230 2900 thru 4000.

Example: You are to install the system and match existing construction. Go to Line G2020 230 2900 and apply these percentages to the appropriate MAT. and INST. costs.

System Components	QUANTITY	UNIT	COST PER S.Y. MAT.	INST.	TOTAL
Parking lot consisting of 2″ asphalt binder and 1″ topping on 6″ Crushed stone base with painted parking stripes and concrete parking blocks.					
Fine grade and compact subgrade	1.000	S.Y.		2.01	2.01
6″ crushed stone base, stone	.320	Ton	6.70	1.36	8.06
2″ asphalt binder	1.000	S.Y.	7.40	1.30	8.70
1″ asphalt topping	1.000	S.Y.	3.68	.86	4.54
Paint parking stripes	.500	L.F.	.07	.09	.16
6″ x 10″ x 6′ precast concrete parking blocks	.020	Ea.	1.31	.39	1.70
Mobilization of equipment	.005	Ea.		1.94	1.94
TOTAL			19.16	7.95	27.11

G2020 230	Parking Lots, Asphalt	COST PER S.Y. MAT.	INST.	TOTAL
1400	2″ binder plus 1″ topping on 6″ crushed stone	19.15	7.95	27.10
1500	9″ crushed stone	22.50	8.10	30.60
1600	12″ crushed stone	26	8.20	34.20
1700	On bank run gravel, 6″ deep	17.35	7.35	24.70
1800	9″ deep	19.80	7.50	27.30
1900	12″ deep	22.50	7.65	30.15
2000	3″ binder plus 1″ topping on 6″ crushed stone	23	8.35	31.35
2100	9″ deep crushed stone	26.50	8.45	34.95
2200	12″ deep crushed stone	29.50	8.60	38.10
2300	On bank run gravel, 6″ deep	21	7.75	28.75
2400	9″ deep	23.50	7.90	31.40
2500	12″ deep	26	8.05	34.05
2600				
2700				
2900	Cut & patch to match existing construction, add, minimum	2%	3%	
3000	Maximum	5%	9%	
3100	Dust protection, add, minimum	1%	2%	
3200	Maximum	4%	11%	
3300	Equipment usage curtailment, add, minimum	1%	1%	
3400	Maximum	3%	10%	
3500	Material handling & storage limitation, add, minimum	1%	1%	
3600	Maximum	6%	7%	
3700	Protection of existing work, add, minimum	2%	2%	
3800	Maximum	5%	7%	
3900	Shift work requirements, add, minimum		5%	
4000	Maximum		30%	

G2020 Parking Lots

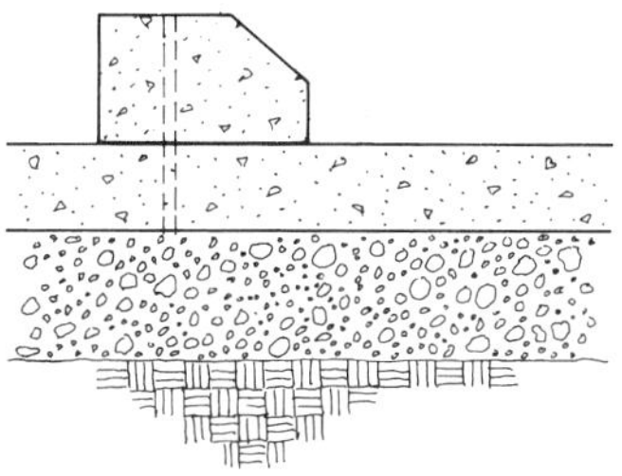

This page illustrates and describes concrete parking lot systems including a concrete slab, unreinforced, compacted gravel base, broom finished, cured, joints, painted parking stripes, and parking blocks. Lines within System Components give the unit price and total price per square yard for this system. Prices for concrete parking lot systems are on Line Items G2020 240 1500 thru 2500. All materials have been adjusted according to the system listed.

Factors: To adjust for job conditions other than normal working situations use Lines G2020 240 2900 thru 4000.

Example: You are to install the system and protect existing construction. Go to Line G2020 240 3800 and apply these percentages to the appropriate MAT. and INST. costs.

System Components	QUANTITY	UNIT	COST PER S.Y. MAT.	INST.	TOTAL
Parking lot, 4″ slab, 3000 psi concrete on 6″ compacted gravel base					
With 12′ x 12′ joints, painted parking strips, conc. parking blocks.					
Grade and compact subgrade	1.000	S.Y.		2.01	2.01
6″ crushed stone base, stone	.320	Ton	6.70	1.36	8.06
Place and remove edge forms, 4 use	.250	L.F.	.08	.92	1
Concrete for slab, 3000 psi	1.000	S.F.	12.32		12.32
Place concrete, 4″ thick slab, direct chute	1.000	S.F.		2.92	2.92
Spray on membrane curing compound	1.000	S.F.	.74	.61	1.35
Saw cut 3″ deep joints	2.250	L.F.	.34	3.40	3.74
Paint parking stripes	.500	L.F.	.07	.09	.16
6″ x 10″ x 6′ precast concrete parking blocks	.020	Ea.	1.31	.39	1.70
TOTAL			21.56	11.70	33.26

G2020 240	Parking Lots, Concrete	COST PER S.Y. MAT.	INST.	TOTAL
1500	4″ concrete slab, 3000 psi on 3″ crushed stone	18.20	11.65	29.85
1550	6″ crushed stone	21.50	11.70	33.20
1600	9″ crushed stone	25	11.85	36.85
1700	On bank run gravel, 6″ deep	19.75	11.10	30.85
1800	9″ deep	22	11.25	33.25
1900	On compacted subgrade only	14.85	10.35	25.20
2000	6″ concrete slab, 3000 psi on 3″ crushed stone	24	13.05	37.05
2100	6″ crushed stone	27.50	13.15	40.65
2200	9″ crushed stone	31	13.25	44.25
2300	On bank run gravel, 6″ deep	25.50	12.50	38
2400	9″ deep	28	12.70	40.70
2500	On compacted subgrade only	27.50	13.15	40.65
2600				
2700				
2900	Cut & patch to match existing construction, add, minimum	2%	3%	
3000	Maximum	5%	9%	
3100	Dust protection, add, minimum	1%	2%	
3200	Maximum	4%	11%	
3300	Equipment usage curtailment, add, minimum	1%	1%	
3400	Maximum	3%	10%	
3500	Material handling & storage limitation, add, minimum	1%	1%	
3600	Maximum	6%	7%	
3700	Protection of existing work, add, minimum	2%	2%	
3800	Maximum	5%	7%	
3900	Shift work requirements, add, minimum		5%	
4000	Maximum		30%	

G2030 Pedestrian Paving

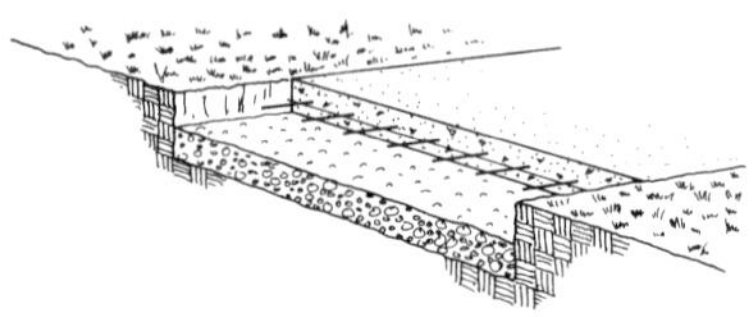

This page illustrates and describes sidewalk systems including concrete, welded wire and broom finish. Lines within System Components give unit price and total price per square foot for this system. Prices for sidewalk systems are on Line Items G2030 105 1700 thru 1900. Both material quantities and labor costs have been adjusted for the system listed.

Factors: To adjust for job conditions other than normal working situations use Lines G2030 105 2900 thru 4000.

Example: You are to install the system and match existing construction. Go to Line G2030 105 2900 and apply these percentages to the appropriate MAT. and INST. costs.

System Components	QUANTITY	UNIT	COST PER S.F. MAT.	COST PER S.F. INST.	COST PER S.F. TOTAL
4″ thick concrete sidewalk with welded wire fabric 3000 psi air entrained concrete, broom finish.					
Gravel fill, 4″ deep	.012	C.Y.	.30	.08	.38
Compact fill	.012	C.Y.		.03	.03
Hand grade	1.000	S.F.		2.07	2.07
Edge form	.250	L.F.	.08	.92	1
Welded wire fabric	.011	S.F.	.17	.41	.58
Concrete, 3000 psi air entrained	.012	C.Y.	1.34		1.34
Place concrete	.012	C.Y.		.32	.32
Broom finish	1.000	S.F.		.82	.82
TOTAL			1.89	4.65	6.54

G2030 105	Sidewalks	COST PER S.F. MAT.	COST PER S.F. INST.	COST PER S.F. TOTAL
1700	Asphalt (bituminous), 2″ thick	1.09	2.65	3.74
1750	4″ thick	1.89	4.65	6.54
1800	Brick, on sand, bed, 4.5 brick per S.F.	3.33	10.25	13.58
1900	Flagstone, slate, 1″ thick, rectangular	9.60	11.15	20.75
2000				
2100				
2200				
2300				
2400				
2500				
2600				
2700				
2900	Cut & patch to match existing construction, add, minimum	2%	3%	
3000	Maximum	5%	9%	
3100	Dust protection, add, minimum	1%	2%	
3200	Maximum	4%	11%	
3300	Equipment usage curtailment, add, minimum	1%	1%	
3400	Maximum	3%	10%	
3500	Material handling & storage limitation, add, minimum	1%	1%	
3600	Maximum	6%	7%	
3700	Protection of existing work, add, minimum	2%	2%	
3800	Maximum	5%	7%	
3900	Shift work requirements, add, minimum		5%	
4000	Maximum		30%	

G2050 Landscaping

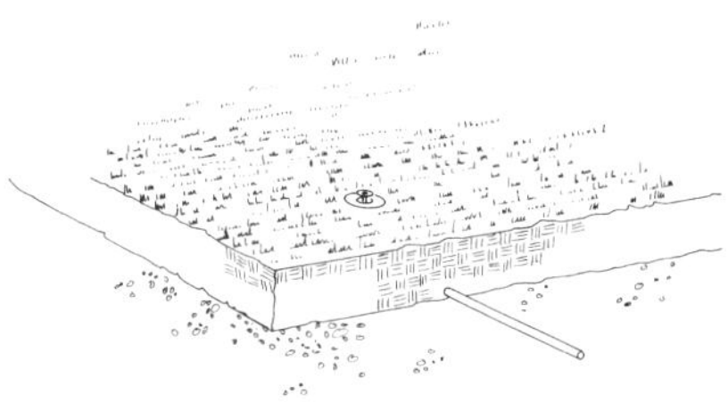

This page describes landscaping—lawn establishment systems including loam, lime, fertilizer, side and top mulching. Lines within system components give the unit price and total price per square yard for this system. Prices for landscaping—lawn establishment systems are on Line Items G2050 420 1000 thru 1200. Both material quantities and labor costs have been adjusted for the system listed.

Factors: To adjust for job conditions other than normal working situations use Lines G2050 420 2900 thru 4000.

Example: You are to install the system and provide dust protection. Go to Line G2050 420 3100 and apply these percentages to the appropriate MAT. and INST. costs.

System Components	QUANTITY	UNIT	COST PER S.Y.		
			MAT.	INST.	TOTAL
Establishing lawns with loam, lime, fertilizer, seed and top mulching					
On rough graded areas.					
Furnish and place loam 4" deep	.110	C.Y.	4.63	.91	5.54
Fine grade, lime, fertilize and seed	1.000	S.Y.	.62	3.26	3.88
Hay mulch, 1 bale/M.S.F.	1.000	S.Y.	.13	.56	.69
Rolling with hand roller	1.000	S.Y.		1.01	1.01
TOTAL			5.38	5.74	11.12

G2050 420	Landscaping - Lawn Establishment	COST PER S.Y.		
		MAT.	INST.	TOTAL
1000	Establishing lawns with loam, lime, fertilizer, seed and hay mulch	5.40	5.75	11.15
1100	Above system with jute mesh in place of hay mulch	5.90	5.90	11.80
1200	Above system with sod in place of seed	6.55	2	8.55
1300				
1400				
1500				
1600				
1700				
1800				
1900				
2000				
2100				
2200				
2300				
2400				
2500				
2600				
2700				
2900	Cut & patch to match existing construction, add, minimum	2%	3%	
3000	Maximum	5%	9%	
3100	Dust protection, add, minimum	1%	2%	
3200	Maximum	4%	11%	
3300	Equipment usage curtailment, add, minimum	1%	1%	
3400	Maximum	3%	10%	
3500	Material handling & storage limitation, add, minimum	1%	1%	
3600	Maximum	6%	7%	
3700	Protection of existing work, add, minimum	2%	2%	
3800	Maximum	5%	7%	
3900	Shift work requirements, add, minimum		5%	
4000	Maximum		30%	

G2050 Landscaping

G2050 510	Selective Price Sheet	COST PER UNIT		
		MAT.	INST.	TOTAL
0100	Shrubs and trees, Evergreen, in prepared beds			
0200				
0300	Arborvitae pyramidal, 4′-5′	57	90.50	147.50
0400	Globe, 12″-15″	23.50	14.65	38.15
0500	Cedar, blue, 8′-10′	242	151	393
0600	Hemlock, Canadian, 2-1/2′-3′	39.50	39	78.50
0700	Juniper, andora, 18″-24″	22	17.60	39.60
0800	Wiltoni, 15″-18″	27	17.60	44.60
0900	Skyrocket, 4-1/2′-5′	153	49.50	202.50
1000	Blue pfitzer, 2′-2-1/2′	43	32	75
1100	Ketleerie, 2-1/2′-3′	44	28	72
1200	Pine, black, 2-1/2′-3′	58.50	28	86.50
1300	Mugo, 18″-24″	49.50	23.50	73
1400	White, 4′-5′	44	36	80
1500	Spruce, blue, 18″-24″	54	23.50	77.50
1600	Norway, 4′-5′	82.50	36	118.50
1700	Yew, denisforma, 12″-15″	38	23.50	61.50
1800	Capitata, 18″-24″	19.80	47	66.80
1900	Hicksi, 2′-2-1/2′	35.50	47	82.50
2000				
2100	Trees, Deciduous, in prepared beds			
2200				
2300	Beech, 5′-6′	198	54	252
2400	Dogwood, 4′-5′	151	67.50	218.50
2500	Elm, 8′-10′	151	135	286
2600	Magnolia, 4′-5′	103	135	238
2700	Maple, red, 8′-10′, 1-1/2″ caliper	210	270	480
2800	Oak, 2-1/2″-3″ caliper	290	450	740
2900	Willow, 6′-8′, 1″ caliper	74.50	135	209.50
3000				
3100	Shrubs, Broadleaf Evergreen, in prepared beds			
3200				
3300	Andromeda, 15″-18″, container	19.80	14.65	34.45
3400	Azalea, 15″-18″, container	24	14.65	38.65
3500	Barberry, 9″-12″, container	14.30	10.80	25.10
3600	Boxwood, 15″-18″, B & B	33	14.65	47.65
3700	Euonymus, emerald gaiety, 12″-15″, container	24	12.25	36.25
3800	Holly, 15″-18″, B & B	18.70	14.65	33.35
3900	Mount laurel, 18″-24″, B & B	60.50	17.60	78.10
4000	Privet, 18″-24″, B & B	22	10.80	32.80
4100	Rhododendron, 18″-24″, container	31.50	29.50	61
4200	Rosemary, 1 gal. container	11.65	2.35	14
4300	Deciduous, amalanchier, 2′-3′, B & B	131	24.50	155.50
4400	Azalea, 15″-18″, B & B	19.25	14.65	33.90
4500	Bayberry, 2′-3′, B & B	24	24.50	48.50
4600	Cotoneaster, 15″-18″, B & B	11.65	17.60	29.25
4700	Dogwood, 3′-4′, B & B	29.50	67.50	97
4800	Euonymus, alatus compacta, 15″-18″, container	24	17.60	41.60
4900	Forsythia, 2′-3′, container	24	23.50	47.50
5000	Honeysuckle, 3′-4′, B & B	22.50	23.50	46
5100	Hydrangea, 2′-3′, B & B	29.50	24.50	54
5200	Lilac, 3′-4′, B & B	21.50	67.50	89
5300	Quince, 2′-3′, B & B	25.50	24.50	50
5400				
5500	Ground Cover			
5600				
5700	Plants, Pachysandra, prepared beds, per hundred	34	94	128
5800	Vinca minor or English ivy, per hundred	42.50	117	159.50

G2050 Landscaping

G2050 510	Selective Price Sheet	COST PER UNIT		
		MAT.	INST.	TOTAL
5900	Plant bed prep. 18" dp., mach., per square foot	2.47	.41	2.88
6000	By hand, per square foot	2.47	2.97	5.44
6100	Stone chips, Georgia marble, 50# bags, per bag	5.50	2.71	8.21
6200	Onyx gemstone, per bag	19.80	5.40	25.20
6300	Quartz, per bag	7.55	5.40	12.95
6400	Pea gravel, truck load lots, per cubic yard	33.50	50.50	84
6500	Mulch, polyethylene mulch, per square yard	.26	.46	.72
6600	Wood chips, 2" deep, per square yard	1.93	2.09	4.02
6700	Peat moss, 1" deep, per square yard	3.08	.51	3.59
6800	Erosion control per square yard, Jute mesh stapled	.66	.70	1.36
6900	Plastic netting, stapled	.37	.56	.93
7000	Polypropylene mesh, stapled	1.45	.56	2.01
7100	Tobacco netting, stapled	.13	.56	.69
7200	Lawns per square yard, seeding incl. fine grade, limestone, fert. and seed	.62	3.26	3.88
7300	Sodding, incl. fine grade, level ground	1.74	1.05	2.79
7500	Edging, per linear foot, Redwood, untreated, 1" x 4"	1.28	2.31	3.59
7600	2" x 4"	2.49	3.51	6
7700	Stl. edge strips, 1/4" x 5" inc. stakes	3.70	3.51	7.21
7800	3/16" x 4"	2.33	3.51	5.84
7900	Brick edging, set on edge	2.97	7.60	10.57
8000	Set flat	1.49	2.77	4.26
8100				
8200				
8300				
8400				

G3020 Sanitary Sewer

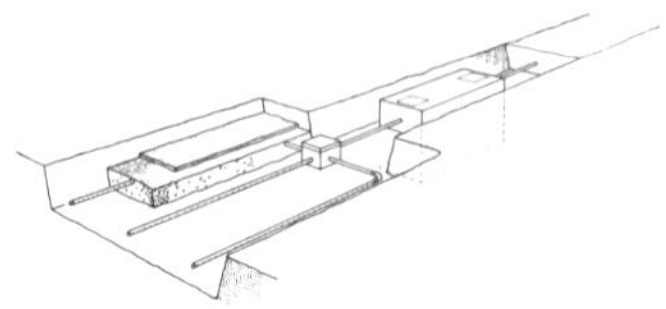

This page illustrates and describes drainage and utilities – septic system including a tank, distribution box, excavation, piping, crushed stone and backfill. Lines within System Components give the unit price and total price on a cost each basis for this system. Prices for drainage and utilities – septic systems are on Line Items G3020 610 1600 thru 2000. Both material quantities and labor costs have been adjusted for the system listed.

Factors: To adjust for job conditions other than normal working situations use Lines G3020 610 3100 thru 4000.

Example: You are to install the system with a material handling limitation. Go to Line G3020 610 3500 and apply these percentages to the appropriate MAT. and INST. costs.

System Components	QUANTITY	UNIT	COST EACH		
			MAT.	INST.	TOTAL
Septic system including tank, distribution box, excavation, piping, crushed Stone and backfill for a 1000 S.F. leaching field.					
Precast concrete septic tank, 1000 gal. capacity	1.000	Ea.	1,125	248.20	1,373.20
Concrete distribution box	1.000	Ea.	82.50	57.50	140
Sch. 40 sewer pipe, PVC, 4″ diameter	25.000	L.F.	306	697	1,003
Sch. 40 sewer pipe fittings, PVC, 4″ diameter	8.000	Ea.	116.80	788	904.80
Excavation for septic tank	120.000	C.Y.		220.68	220.68
Excavation for disposal field	120.000	C.Y.		536.31	536.31
Crushed stone backfill	76.000	C.Y.	1,748	938.60	2,686.60
Backfill with excavated material	26.000	C.Y.		47.32	47.32
Asphalt felt paper, 15#	6.000	Sq.	37.50	93.90	131.40
TOTAL			3,415.80	3,627.51	7,043.31

G3020 610	Septic Systems	COST EACH		
		MAT.	INST.	TOTAL
1600	1000 gal. tank with 1000 S.F. field	3,425	3,650	7,075
1700	1000 gal. tank with 2000 S.F. field	4,875	5,275	10,150
1800	With leaching pits	3,125	1,175	4,300
1900	2000 gal. tank with 2000 S.F. field	5,950	5,425	11,375
2000	With leaching pits	4,300	1,750	6,050
2100				
2200				
2300				
2400				
2500				
2600				
2700				
2800				
2900				
3100	Dust protection, add, minimum	1%	2%	
3200	Maximum	4%	11%	
3300	Equipment usage curtailment, add, minimum	1%	1%	
3400	Maximum	3%	10%	
3500	Material handling & storage limitation, add, minimum	1%	1%	
3600	Maximum	6%	7%	
3700	Protection of existing work, add, minimum	2%	2%	
3800	Maximum	5%	7%	
3900	Shift work requirements, add, minimum		5%	
4000	Maximum		30%	

G3060 Fuel Distribution

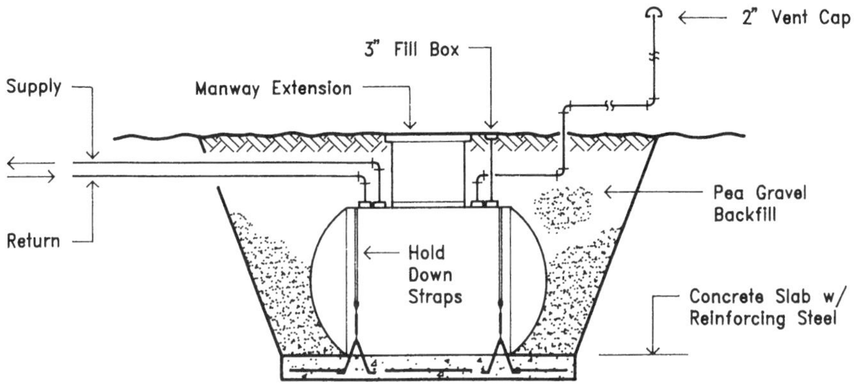

Fiberglass Underground Fuel Storage Tank, Single Wall

This page illustrates and describes fiberglass tanks including hold down slab, reinforcing steel and peastone gravel. Lines within Systems Components give the material and installation price on a cost each basis for the components. Prices for fiberglass tank systems are on Line Items G3060 310 1010 thru 1100. Material quantities and labor costs have been adjusted for the system listed.

Factors: To adjust for job conditions other than normal working situations use Lines G3060 310 2700 thru 4000.

System Components	QUANTITY	UNIT	COST EACH MAT.	COST EACH INST.	COST EACH TOTAL
Double wall underground fiberglass storage tanks including, excavation, hold down slab, backfill, manway extension and related piping.					
Excavating trench, no sheeting or dewatering, 6'-10 ' D, 3/4 CY hyd backhoe	18.000	Ea.		142.02	142.02
Corrosion resistance wrap & coat, small diam. pipe, 1" diam., add	120.000	Ea.	193.20		193.20
Fiberglass-reinforced elbows 90 & 45 degree, 2 inch	7.000	Ea.	315	567	882
Fiberglass-reinforced threaded adapters, 2 inch	2.000	Ea.	37.60	185	222.60
Remote tank gauging system, 30', 5" pointer travel	1.000	Ea.	5,725	345	6,070
Corrosion resistance wrap & coat, small diam. pipe. 4 " diam. add	30.000	Ea.	73.80		73.80
Pea gravel	9.000	Ea.	711	454.50	1,165.50
Reinforcing in hold down pad, #3 to #7	120.000	Ea.	64.80	74.40	139.20
Tank leak detector system, 8 channel, external monitoring	1.000	Ea.	980		980
Tubbing copper, type L, 3/4" diam.	120.000	Ea.	870	1,074	1,944
Elbow 90 degree ,wrought cu x cu 3/4 " diam.	4.000	Ea.	28.20	144	172.20
Pipe, black steel welded, sch 40, 3" diam.	30.000	Ea.	855	883.50	1,738.50
Elbow 90 degree, steel welded butt joint, 3 " diam.	3.000	Ea.	96	542.40	638.40
Fiberglass-reinforced plastic pipe, 2 inch	156.000	Ea.	967.20	712.92	1,680.12
Fiberglass-reinforced plastic pipe, 3 inch	36.000	Ea.	268.20	187.20	455.40
Fiberglass-reinforced elbows 90 & 45 degree, 3 inch	3.000	Ea.	186	324	510
Tank leak detection system, probes, well monitoring, liquid phase detection	2.000	Ea.	1,010		1,010
Double wall fiberglass annular space	1.000	Ea.	345		345
Flange, weld neck, 150 lb, 3 " pipe size	1.000	Ea.	38.50	90.40	128.90
Vent protector/breather, 2" diam.	1.000	Ea.	31.50	21.50	53
Corrosion resistance wrap & coat, small diam. pipe, 2" diam., add	36.000	Ea.	63.36		63.36
Concrete hold down pad, 6' x 5' x 8" thick	30.000	Ea.	89.10	45.90	135
600 gallon capacity	1.000	Ea.	8,650	515	9,165
Tank, for hold-downs 500-4000 gal., add	1.000	Ea.	445	165	610
Foot valve, single poppet, 3/4" diam.	1.000	Ea.	78.50	38.50	117
Fuel fill box, locking inner cover, 3" diam.	1.000	Ea.	130	139	269
Fuel oil specialties, valve,ball chk, globe type fusible, 3/4" diam.	2.000	Ea.	154	69	223
TOTAL			22,405.96	6,720.24	29,126.20

G3060 310	Fiberglass Fuel Tank, Double Wall	COST EACH MAT.	COST EACH INST.	COST EACH TOTAL
1010	Storage tank, fuel, underground, double wall fiberglass, 600 Gal.	22,400	6,725	29,125
1020	2500 Gal.	34,000	9,325	43,325
1030	4000 Gal.	39,400	11,500	50,900
1040	6000 Gal.	46,300	13,100	59,400
1050	8000 Gal.	54,000	15,700	69,700
1060	10,000 Gal.	58,500	18,100	76,600

G3060 Fuel Distribution

G3060 310	Fiberglass Fuel Tank, Double Wall	COST EACH		
		MAT.	INST.	TOTAL
1070	15,000 Gal.	79,000	20,900	99,900
1080	20,000 Gal.	98,500	25,300	123,800
1090	25,000 Gal.	122,000	29,500	151,500
1100	30,000 Gal.	140,000	31,200	171,200
2700	Cut & patch to match existing construction, add, minimum	2%	3%	
2800	Maximum	5%	9%	
2900	Dust protection, add, minimum	1%	2%	
3000	Maximum	4%	11%	
3100	Equipment usage curtailment, add, minimum	1%	1%	
3200	Maximum	3%	10%	
3300	Material handling & storage limitation, add, minimum	1%	1%	
3400	Maximum	6%	7%	
3500	Protection of existing work, add, minimum	2%	2%	
3600	Maximum	5%	7%	
3700	Shift work requirements, add, minimum		5%	
3800	Maximum		30%	
3900	Temporary shoring and bracing, add, minimum	2%	5%	
4000	Maximum	5%	12%	

G3060 Fuel Distribution

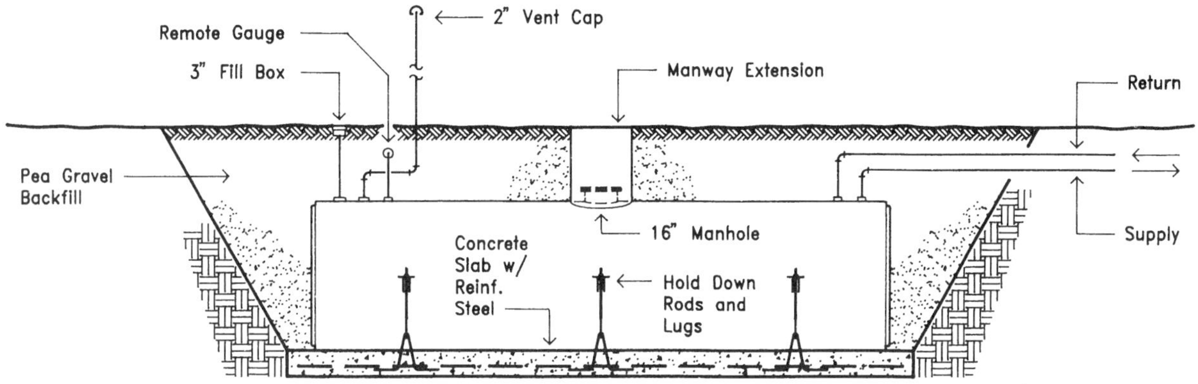

Steel Underground Fuel Storage Tank, Double Wall

System Components	QUANTITY	UNIT	COST EACH MAT.	COST EACH INST.	COST EACH TOTAL
Double wall underground steel storage tank including, excavation, hold down slab, backfill, manway extension and related piping.					
Fiberglass-reinforced plastic pipe, 2 inch	156.000	Ea.	967.20	712.92	1,680.12
Fiberglass-reinforced plastic pipe, 3 inch	36.000	Ea.	268.20	187.20	455.40
Fiberglass-reinforced elbows 90 & 45 degree, 3 inch	3.000	Ea.	186	324	510
Tank leak detection system, probes, well monitoring, liquid phase detection	2.000	Ea.	1,010		1,010
Flange, weld neck, 150 lb, 3″ pipe size	1.000	Ea.	38.50	90.40	128.90
Vent protector/breather, 2″ diam.	1.000	Ea.	31.50	21.50	53
Concrete hold down pad, 6′ x 5′ x 8″ thick	30.000	Ea.	89.10	45.90	135
Tanks, for hold-downs, 500-2000 gal, add	1.000	Ea.	183	165	348
Double wall steel tank annular space	1.000	Ea.	360		360
Foot valve, single poppet, 3/4″ diam.	1.000	Ea.	78.50	38.50	117
Fuel fill box, locking inner cover, 3″ diam.	1.000	Ea.	130	139	269
Fuel oil specialties, valve, ball chk, globe type fusible, 3/4″ diam.	2.000	Ea.	154	69	223
Pea gravel	9.000	Ea.	711	454.50	1,165.50
Reinforcing in hold down pad, #3 to #7	120.000	Ea.	64.80	74.40	139.20
Tank, for manways, add	1.000	Ea.	1,800		1,800
Tank leak detector system, 8 channel, external monitoring	1.000	Ea.	980		980
Tubbing copper, type L, 3/4″ diam.	120.000	Ea.	870	1,074	1,944
Elbow 90 degree, wrought cu x cu 3/4″ diam.	8.000	Ea.	56.40	288	344.40
Pipe, black steel welded, sch 40, 3″ diam.	30.000	Ea.	855	883.50	1,738.50
Elbow 90 degree, steel welded butt joint, 3″ diam.	3.000	Ea.	96	542.40	638.40
Excavating trench, no sheeting or dewatering, 6′-10′ D, 3/4 CY hyd backhoe	18.000	Ea.		142.02	142.02
Fiberglass-reinforced elbows 90 & 45 degree, 2 inch	7.000	Ea.	315	567	882
Fiberglass-reinforced threaded adapters, 2 inch	2.000	Ea.	37.60	185	222.60
500 gallon capacity	1.000	Ea.	4,725	520	5,245
Remote tank gauging system, 30′, 5″ pointer travel	1.000	Ea.	5,725	345	6,070
TOTAL			19,731.80	6,869.24	26,601.04

G3060 320	Steel Fuel Tank, Double Wall	COST EACH MAT.	COST EACH INST.	COST EACH TOTAL
1010	Storage tank, fuel, underground, double wall steel, 500 Gal.	19,700	6,875	26,575
1020	2000 Gal.	27,100	9,000	36,100
1030	4000 Gal.	37,200	11,700	48,900
1040	6000 Gal.	47,200	13,000	60,200
1050	8000 Gal.	53,000	16,000	69,000
1060	10,000 Gal.	58,000	18,300	76,300
1070	15,000 Gal.	65,000	21,400	86,400
1080	20,000 Gal.	78,500	25,700	104,200

G3060 Fuel Distribution

G3060 320	Steel Fuel Tank, Double Wall	COST EACH		
		MAT.	INST.	TOTAL
1090	25,000 Gal.	117,500	29,400	146,900
1100	30,000 Gal.	135,000	31,400	166,400
1110	40,000 Gal.	169,000	39,400	208,400
1120	50,000 Gal.	202,500	47,300	249,800

G3060 325	Steel Tank, Above Ground, Single Wall	COST EACH		
		MAT.	INST.	TOTAL
1010	Storage tank, fuel, above ground, single wall steel, 550 Gal.	13,800	4,725	18,525
1020	2000 Gal.	20,100	4,825	24,925
1030	5000 Gal.	33,600	7,425	41,025
1040	10,000 Gal.	44,500	7,925	52,425
1050	15,000 Gal.	53,500	8,150	61,650
1060	20,000 Gal.	65,500	8,425	73,925
1070	25,000 Gal.	74,500	8,625	83,125
1080	30,000 Gal.	86,500	9,000	95,500

G3060 330	Steel Tank, Above Ground, Double Wall	COST EACH		
		MAT.	INST.	TOTAL
1010	Storage tank, fuel, above ground, double wall steel, 500 Gal.	14,400	4,775	19,175
1020	2000 Gal.	21,700	4,900	26,600
1030	4000 Gal.	29,900	5,000	34,900
1040	6000 Gal.	35,300	7,700	43,000
1050	8000 Gal.	41,700	7,925	49,625
1060	10,000 Gal.	44,900	8,075	52,975
1070	15,000 Gal.	60,500	8,375	68,875
1080	20,000 Gal.	67,000	8,625	75,625
1090	25,000 Gal.	78,000	8,900	86,900
1100	30,000 Gal.	84,000	9,250	93,250
2700	Cut & patch to match existing construction, add, minimum	2%	3%	
2800	Maximum	5%	9%	
2900	Dust protection, add, minimum	1%	2%	
3000	Maximum	4%	11%	
3100	Equipment usage curtailment, add, minimum	1%	1%	
3200	Maximum	3%	10%	
3300	Material handling & storage limitation, add, minimum	1%	1%	
3400	Maximum	6%	7%	
3500	Protection of existing work, add, minimum	2%	2%	
3600	Maximum	5%	7%	
3700	Shift work requirements, add, minimum		5%	
3800	Maximum		30%	
3900	Temporary shoring and bracing, add, minimum	2%	5%	
4000	Maximum	5%	12%	

G4020 Site Lighting

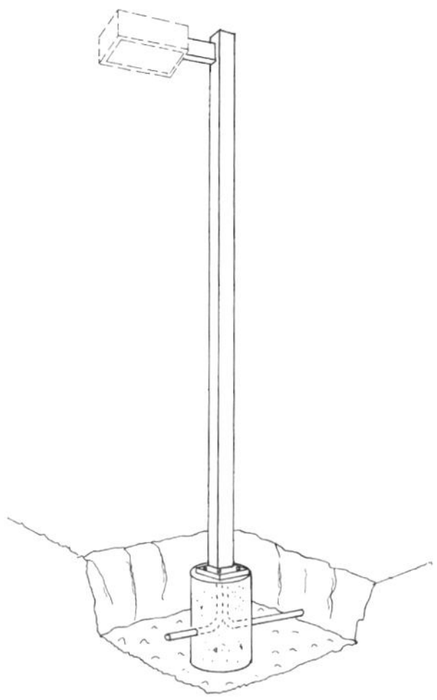

This page illustrates and describes light poles for parking or walkway area lighting. Included are aluminum or steel light poles, single or multiple fixture bracket arms, excavation, concrete footing, backfill and compaction. Lines within system components give the unit price and total price on a cost per each basis. Prices for systems are shown on lines G4020 210 0200 through 1440. Both material quantities and labor costs have been adjusted for the system listed.

Factors: To adjust for job conditions other than normal working situations use lines G4020 210 2700 through 3600.

Example: You are to install the system where you must be careful not to damage existing walks or parking areas. Go to line G4020 210 3300 and apply these percentages to the appropriate MAT. and INST. costs.

System Components	QUANTITY	UNIT	COST EACH MAT.	COST EACH INST.	COST EACH TOTAL
Light poles, aluminum, 20′ high, 1 arm bracket					
Aluminum light pole, 20′, no concrete base	1.000	Ea.	990	595	1,585
Bracket arm for Aluminum light pole	1.000	Ea.	134	79.50	213.50
Excavation by hand, pits to 6′ deep, heavy soil or clay	2.368	C.Y.		272.32	272.32
Footing, concrete incl forms, reinforcing, spread, under 1 C.Y.	.465	C.Y.	85.10	129.31	214.41
Backfill by hand	1.903	C.Y.		79.93	79.93
Compaction vibrating plate	1.903	C.Y.		10.66	10.66
TOTAL			1,209.10	1,166.72	2,375.82

G4020 210	Light Pole (Installed)	COST EACH MAT.	COST EACH INST.	COST EACH TOTAL
0200	Light pole, aluminum, 20′ high, 1 arm bracket	1,200	1,175	2,375
0240	2 arm brackets	1,350	1,175	2,525
0280	3 arm brackets	1,500	1,200	2,700
0320	4 arm brackets	1,625	1,200	2,825
0360	30′ high, 1 arm bracket	2,050	1,475	3,525
0400	2 arm brackets	2,175	1,475	3,650
0440	3 arm brackets	2,325	1,525	3,850
0480	4 arm brackets	2,450	1,525	3,975
0680	40′ high, 1 arm bracket	2,600	1,975	4,575
0720	2 arm brackets	2,750	1,975	4,725
0760	3 arm brackets	2,900	2,000	4,900
0800	4 arm brackets	3,025	2,000	5,025
0840	Steel, 20′ high, 1 arm bracket	1,475	1,225	2,700
0880	2 arm brackets	1,575	1,225	2,800
0920	3 arm brackets	1,600	1,275	2,875
0960	4 arm brackets	1,725	1,275	3,000
1000	30′ high, 1 arm bracket	1,575	1,575	3,150
1040	2 arm brackets	1,675	1,575	3,250
1080	3 arm brackets	1,700	1,625	3,325
1120	4 arm brackets	1,825	1,625	3,450
1320	40′ high, 1 arm bracket	2,200	2,125	4,325
1360	2 arm brackets	2,300	2,125	4,425
1400	3 arm brackets	2,325	2,175	4,500
1440	4 arm brackets	2,450	2,175	4,625
2700	Cut & patch to match existing construction, add, minimum	2%	3%	
2800	Maximum	5%	9%	
2900	Equipment usage curtailment, add, minimum	1%	1%	
3000	Maximum	3%	10%	

G4020 Site Lighting

G4020 210	Light Pole (Installed)	COST EACH		
		MAT.	INST.	TOTAL
3100	Material handling & storage limitation, add, minimum	1%	1%	
3200	Maximum	6%	7%	
3300	Protection of existing work, add, minimum	2%	2%	
3400	Maximum	5%	7%	
3500	Shift work requirements, add, minimum		5%	
3600	Maximum		30%	

Reference Section

All the reference information is in one section, making it easy to find what you need to know . . . and easy to use the book on a daily basis. This section is visually identified by a vertical gray bar on the page edges.

In this Reference Section, we've included Equipment Rental Costs, a listing of rental and operating costs; Crew Listings, a full listing of all crews and equipment, and their costs; Historical Cost Indexes for cost comparisons over time; City Cost Indexes and Location Factors for adjusting costs to the region you are in; Reference Tables, where you will find explanations, estimating information and procedures, or technical data; Change Orders, information on pricing changes to contract documents; Square Foot Costs that allow you to make a rough estimate for the overall cost of a project; and an explanation of all the Abbreviations in the book.

Table of Contents

Reference Tables (cont.)

Estimating Tips

- This section contains the average costs to rent and operate hundreds of pieces of construction equipment. This is useful information when estimating the time and material requirements of any particular operation in order to establish a unit or total cost. Equipment costs include not only rental, but also operating costs for equipment under normal use.

Rental Costs

- Equipment rental rates are obtained from industry sources throughout North America–contractors, suppliers, dealers, manufacturers, and distributors.
- Rental rates vary throughout the country, with larger cities generally having lower rates. Lease plans for new equipment are available for periods in excess of six months, with a percentage of payments applying toward purchase.
- Monthly rental rates vary from 2% to 5% of the purchase price of the equipment depending on the anticipated life of the equipment and its wearing parts.
- Weekly rental rates are about 1/3 the monthly rates, and daily rental rates are about 1/3 the weekly rate.
- Rental rates can also be treated as reimbursement costs for contractor-owned equipment. Owned equipment costs include depreciation, loan payments, interest, taxes, insurance, storage, and major repairs.

Operating Costs

- The operating costs include parts and labor for routine servicing, such as repair and replacement of pumps, filters and worn lines. Normal operating expendables, such as fuel, lubricants, tires and electricity (where applicable), are also included.
- Extraordinary operating expendables with highly variable wear patterns, such as diamond bits and blades, are excluded. These costs can be found as material costs in the Unit Price section.
- The hourly operating costs listed do not include the operator's wages.

Equipment Cost/Day

- Any power equipment required by a crew is shown in the Crew Listings with a daily cost.
- The daily cost of equipment needed by a crew is based on dividing the weekly rental rate by 5 (number of working days in the week), and then adding the hourly operating cost times 8 (the number of hours in a day). This "Equipment Cost/Day" is shown in the far right column of the Equipment Rental pages.
- If equipment is needed for only one or two days, it is best to develop your own cost by including components for daily rent and hourly operating cost. This is important when the listed Crew for a task does not contain the equipment needed, such as a crane for lifting mechanical heating/cooling equipment up onto a roof.
- If the quantity of work is less than the crew's Daily Output shown for a Unit Price line item that includes a bare unit equipment cost, it is recommended to estimate one day's rental cost and operating cost for equipment shown in the Crew Listing for that line item.

Mobilization/ Demobilization

- The cost to move construction equipment from an equipment yard or rental company to the jobsite and back again is not included in equipment rental costs listed in the Reference section, nor in the bare equipment cost of any Unit Price line item, nor in any equipment costs shown in the Crew listings.
- Mobilization (to the site) and demobilization (from the site) costs can be found in the Unit Price section.
- If a piece of equipment is already at the jobsite, it is not appropriate to utilize mobil./demob. costs again in an estimate.

01 54 33 | Equipment Rental

		01 54 33 Equipment Rental	UNIT	HOURLY OPER. COST	RENT PER DAY	RENT PER WEEK	RENT PER MONTH	EQUIPMENT COST/DAY	
10	0010	**CONCRETE EQUIPMENT RENTAL** without operators (R015433-10)							10
	0200	Bucket, concrete lightweight, 1/2 C.Y.	Ea.	.75	22	66	198	19.20	
	0300	1 C.Y.		.80	26.50	80	240	22.40	
	0400	1-1/2 C.Y.		1.05	36.50	110	330	30.40	
	0500	2 C.Y.		1.15	43.50	130	390	35.20	
	0580	8 C.Y.		5.80	255	765	2,300	199.40	
	0600	Cart, concrete, self-propelled, operator walking, 10 C.F.		2.85	56.50	170	510	56.80	
	0700	Operator riding, 18 C.F.		4.80	93.50	280	840	94.40	
	0800	Conveyer for concrete, portable, gas, 16″ wide, 26′ long		11.00	122	365	1,100	161	
	0900	46′ long		11.40	147	440	1,325	179.20	
	1000	56′ long		11.50	157	470	1,400	186	
	1100	Core drill, electric, 2-1/2 H.P., 1″ to 8″ bit diameter		1.40	59.50	179	535	47	
	1150	11 H.P., 8″ to 18″ cores		5.40	112	335	1,000	110.20	
	1200	Finisher, concrete floor, gas, riding trowel, 96″ wide		10.90	145	435	1,300	174.20	
	1300	Gas, walk-behind, 3 blade, 36″ trowel		1.80	20	60	180	26.40	
	1400	4 blade, 48″ trowel		3.55	27	81	243	44.60	
	1500	Float, hand-operated (Bull float) 48″ wide		.08	13.65	41	123	8.85	
	1570	Curb builder, 14 H.P., gas, single screw		13.10	245	735	2,200	251.80	
	1590	Double screw		13.75	283	850	2,550	280	
	1600	Floor grinder, concrete and terrazzo, electric, 22″ path		1.84	117	350	1,050	84.70	
	1700	Edger, concrete, electric, 7″ path		.90	51.50	155	465	38.20	
	1750	Vacuum pick-up system for floor grinders, wet/dry		1.35	81.50	245	735	59.80	
	1800	Mixer, powered, mortar and concrete, gas, 6 C.F., 18 H.P.		7.60	117	350	1,050	130.80	
	1900	10 C.F., 25 H.P.		9.45	142	425	1,275	160.60	
	2000	16 C.F.		9.75	165	495	1,475	177	
	2100	Concrete, stationary, tilt drum, 2 C.Y.		6.55	230	690	2,075	190.40	
	2120	Pump, concrete, truck mounted 4″ line 80′ boom		23.95	875	2,620	7,850	715.60	
	2140	5″ line, 110′ boom		31.05	1,150	3,465	10,400	941.40	
	2160	Mud jack, 50 C.F. per hr.		6.55	125	375	1,125	127.40	
	2180	225 C.F. per hr.		8.70	142	425	1,275	154.60	
	2190	Shotcrete pump rig, 12 C.Y./hr.		13.75	220	660	1,975	242	
	2200	35 C.Y./hr.		15.85	237	710	2,125	268.80	
	2600	Saw, concrete, manual, gas, 18 H.P.		5.00	43.50	130	390	66	
	2650	Self-propelled, gas, 30 H.P.		11.00	100	300	900	148	
	2700	Vibrators, concrete, electric, 60 cycle, 2 H.P.		.30	8.65	26	78	7.60	
	2800	3 H.P.		.43	11.65	35	105	10.45	
	2900	Gas engine, 5 H.P.		1.70	16	48	144	23.20	
	3000	8 H.P.		2.30	15	45	135	27.40	
	3050	Vibrating screed, gas engine, 8 H.P.		2.81	71.50	215	645	65.50	
	3120	Concrete transit mixer, 6 x 4, 250 H.P., 8 C.Y., rear discharge		52.20	570	1,705	5,125	758.60	
	3200	Front discharge		61.30	695	2,090	6,275	908.40	
	3300	6 x 6, 285 H.P., 12 C.Y., rear discharge		60.40	655	1,970	5,900	877.20	
	3400	Front discharge	↓	63.10	705	2,115	6,350	927.80	
20	0010	**EARTHWORK EQUIPMENT RENTAL** without operators (R015433-10)							20
	0040	Aggregate spreader, push type 8′ to 12′ wide	Ea.	2.70	25.50	76	228	36.80	
	0045	Tailgate type, 8′ wide		2.55	33	99	297	40.20	
	0055	Earth auger, truck-mounted, for fence & sign posts, utility poles		15.30	470	1,405	4,225	403.40	
	0060	For borings and monitoring wells		41.30	645	1,940	5,825	718.40	
	0070	Portable, trailer mounted		2.75	31.50	95	285	41	
	0075	Truck-mounted, for caissons, water wells		89.05	2,875	8,645	25,900	2,441	
	0080	Horizontal boring machine, 12″ to 36″ diameter, 45 H.P.		22.20	192	575	1,725	292.60	
	0090	12″ to 48″ diameter, 65 H.P.		30.95	335	1,000	3,000	447.60	
	0095	Auger, for fence posts, gas engine, hand held		.50	6	18	54	7.60	
	0100	Excavator, diesel hydraulic, crawler mounted, 1/2 C.Y. capacity		18.80	370	1,105	3,325	371.40	
	0120	5/8 C.Y. capacity		28.60	540	1,615	4,850	551.80	
	0140	3/4 C.Y. capacity		34.90	615	1,840	5,525	647.20	
	0150	1 C.Y. capacity	↓	42.70	680	2,035	6,100	748.60	

01 54 33 | Equipment Rental

			UNIT	HOURLY OPER. COST	RENT PER DAY	RENT PER WEEK	RENT PER MONTH	EQUIPMENT COST/DAY	
20	0200	1-1/2 C.Y. capacity	Ea.	51.80	910	2,730	8,200	960.40	20
	0300	2 C.Y. capacity		68.70	1,175	3,515	10,500	1,253	
	0320	2-1/2 C.Y. capacity		92.40	1,525	4,565	13,700	1,652	
	0325	3-1/2 C.Y. capacity		124.00	2,025	6,095	18,300	2,211	
	0330	4-1/2 C.Y. capacity		149.80	2,575	7,750	23,300	2,748	
	0335	6 C.Y. capacity		188.25	2,875	8,615	25,800	3,229	
	0340	7 C.Y. capacity		190.45	2,975	8,900	26,700	3,304	
	0342	Excavator attachments, bucket thumbs		3.05	240	720	2,150	168.40	
	0345	Grapples		2.75	208	625	1,875	147	
	0347	Hydraulic hammer for boom mounting, 5000 ft lb.		13.70	425	1,275	3,825	364.60	
	0349	11,000 ft lb.		22.35	755	2,260	6,775	630.80	
	0350	Gradall type, truck mounted, 3 ton @ 15′ radius, 5/8 C.Y.		39.75	875	2,630	7,900	844	
	0370	1 C.Y. capacity		43.65	1,025	3,060	9,175	961.20	
	0400	Backhoe-loader, 40 to 45 H.P., 5/8 C.Y. capacity		12.90	240	720	2,150	247.20	
	0450	45 H.P. to 60 H.P., 3/4 C.Y. capacity		19.75	293	880	2,650	334	
	0460	80 H.P., 1-1/4 C.Y. capacity		21.60	320	955	2,875	363.80	
	0470	112 H.P., 1-1/2 C.Y. capacity		35.65	655	1,960	5,875	677.20	
	0482	Backhoe-loader attachment, compactor, 20,000 lb.		5.50	142	425	1,275	129	
	0485	Hydraulic hammer, 750 ft lb.		3.10	95	285	855	81.80	
	0486	Hydraulic hammer, 1200 ft lb.		6.05	217	650	1,950	178.40	
	0500	Brush chipper, gas engine, 6″ cutter head, 35 H.P.		9.75	107	320	960	142	
	0550	Diesel engine, 12″ cutter head, 130 H.P.		18.60	225	675	2,025	283.80	
	0600	15″ cutter head, 165 H.P.		23.35	265	795	2,375	345.80	
	0750	Bucket, clamshell, general purpose, 3/8 C.Y.		1.20	38.50	115	345	32.60	
	0800	1/2 C.Y.		1.30	45	135	405	37.40	
	0850	3/4 C.Y.		1.45	53.50	160	480	43.60	
	0900	1 C.Y.		1.50	58.50	175	525	47	
	0950	1-1/2 C.Y.		2.45	80	240	720	67.60	
	1000	2 C.Y.		2.55	88.50	265	795	73.40	
	1010	Bucket, dragline, medium duty, 1/2 C.Y.		.70	23.50	70	210	19.60	
	1020	3/4 C.Y.		.70	24.50	74	222	20.40	
	1030	1 C.Y.		.75	26.50	80	240	22	
	1040	1-1/2 C.Y.		1.15	40	120	360	33.20	
	1050	2 C.Y.		1.25	45	135	405	37	
	1070	3 C.Y.		1.85	61.50	185	555	51.80	
	1200	Compactor, manually guided 2-drum vibratory smooth roller, 7.5 H.P.		6.70	197	590	1,775	171.60	
	1250	Rammer/tamper, gas, 8″		2.40	41.50	125	375	44.20	
	1260	15″		2.60	48.50	145	435	49.80	
	1300	Vibratory plate, gas, 18″ plate, 3000 lb. blow		2.30	23	69	207	32.20	
	1350	21″ plate, 5000 lb. blow		2.85	30	90	270	40.80	
	1370	Curb builder/extruder, 14 H.P., gas, single screw		13.10	245	735	2,200	251.80	
	1390	Double screw		13.75	283	850	2,550	280	
	1500	Disc harrow attachment, for tractor		.42	69.50	209	625	45.15	
	1810	Feller buncher, shearing & accumulating trees, 100 H.P.		31.15	545	1,640	4,925	577.20	
	1860	Grader, self-propelled, 25,000 lb.		27.85	525	1,580	4,750	538.80	
	1910	30,000 lb.		31.10	605	1,810	5,425	610.80	
	1920	40,000 lb.		48.75	1,025	3,105	9,325	1,011	
	1930	55,000 lb.		64.25	1,500	4,530	13,600	1,420	
	1950	Hammer, pavement breaker, self-propelled, diesel, 1000 to 1250 lb.		26.10	345	1,040	3,125	416.80	
	2000	1300 to 1500 lb.		39.15	695	2,080	6,250	729.20	
	2050	Pile driving hammer, steam or air, 4150 ft lb. @ 225 bpm		10.00	480	1,435	4,300	367	
	2100	8750 ft lb. @ 145 bpm		12.00	665	2,000	6,000	496	
	2150	15,000 ft lb. @ 60 bpm		13.55	805	2,415	7,250	591.40	
	2200	24,450 ft lb. @ 111 bpm		14.60	895	2,680	8,050	652.80	
	2250	Leads, 60′ high for pile driving hammers up to 20,000 ft lb.		3.15	80.50	242	725	73.60	
	2300	90′ high for hammers over 20,000 ft lb.		4.75	143	430	1,300	124	
	2350	Diesel type hammer, 22,400 ft lb.		27.60	620	1,860	5,575	592.80	
	2400	41,300 ft lb.	▼	35.35	675	2,020	6,050	686.80	

01 54 33	Equipment Rental		UNIT	HOURLY OPER. COST	RENT PER DAY	RENT PER WEEK	RENT PER MONTH	EQUIPMENT COST/DAY
20	2450	141,000 ft lb.	Ea.	52.60	1,150	3,440	10,300	1,109
	2500	Vib. elec. hammer/extractor, 200 kW diesel generator, 34 H.P.		45.65	645	1,940	5,825	753.20
	2550	80 H.P.		82.40	945	2,840	8,525	1,227
	2600	150 H.P.		155.30	1,825	5,465	16,400	2,335
	2800	Log chipper, up to 22" diameter, 600 H.P.		50.70	510	1,525	4,575	710.60
	2850	Logger, for skidding & stacking logs, 150 H.P.		46.90	815	2,450	7,350	865.20
	2860	Mulcher, diesel powered, trailer mounted		19.70	215	645	1,925	286.60
	2900	Rake, spring tooth, with tractor		15.41	335	1,007	3,025	324.70
	3000	Roller, vibratory, tandem, smooth drum, 20 H.P.		7.90	143	430	1,300	149.20
	3050	35 H.P.		9.95	247	740	2,225	227.60
	3100	Towed type vibratory compactor, smooth drum, 50 H.P.		23.60	335	1,000	3,000	388.80
	3150	Sheepsfoot, 50 H.P.		24.90	370	1,105	3,325	420.20
	3170	Landfill compactor, 220 H.P.		77.35	1,425	4,305	12,900	1,480
	3200	Pneumatic tire roller, 80 H.P.		14.15	335	1,000	3,000	313.20
	3250	120 H.P.		21.25	560	1,680	5,050	506
	3300	Sheepsfoot vibratory roller, 240 H.P.		58.10	1,050	3,180	9,550	1,101
	3320	340 H.P.		85.45	1,525	4,575	13,700	1,599
	3350	Smooth drum vibratory roller, 75 H.P.		22.10	565	1,690	5,075	514.80
	3400	125 H.P.		28.40	690	2,065	6,200	640.20
	3410	Rotary mower, brush, 60", with tractor		19.80	310	925	2,775	343.40
	3420	Rototiller, walk-behind, gas, 5 H.P.		2.23	67.50	203	610	58.45
	3422	8 H.P.		3.47	103	310	930	89.75
	3440	Scrapers, towed type, 7 C.Y. capacity		5.40	108	325	975	108.20
	3450	10 C.Y. capacity		6.15	150	450	1,350	139.20
	3500	15 C.Y. capacity		6.60	173	520	1,550	156.80
	3525	Self-propelled, single engine, 14 C.Y. capacity		91.35	1,350	4,030	12,100	1,537
	3550	Dual engine, 21 C.Y. capacity		131.60	1,525	4,605	13,800	1,974
	3600	31 C.Y. capacity		181.80	2,350	7,070	21,200	2,868
	3640	44 C.Y. capacity		230.40	3,200	9,625	28,900	3,768
	3650	Elevating type, single engine, 11 C.Y. capacity		61.60	1,025	3,060	9,175	1,105
	3700	22 C.Y. capacity		113.95	1,975	5,935	17,800	2,099
	3710	Screening plant 110 H.P. w/5' x 10' screen		33.90	435	1,305	3,925	532.20
	3720	5' x 16' screen		36.15	540	1,625	4,875	614.20
	3850	Shovel, crawler-mounted, front-loading, 7 C.Y. capacity		200.60	2,750	8,270	24,800	3,259
	3855	12 C.Y. capacity		268.10	3,500	10,520	31,600	4,249
	3860	Shovel/backhoe bucket, 1/2 C.Y.		2.30	65	195	585	57.40
	3870	3/4 C.Y.		2.40	71.50	215	645	62.20
	3880	1 C.Y.		2.50	80	240	720	68
	3890	1-1/2 C.Y.		2.60	93.50	280	840	76.80
	3910	3 C.Y.		2.95	127	380	1,150	99.60
	3950	Stump chipper, 18" deep, 30 H.P.		7.21	188	565	1,700	170.70
	4110	Dozer, crawler, torque converter, diesel 80 H.P.		25.30	400	1,200	3,600	442.40
	4150	105 H.P.		30.85	505	1,520	4,550	550.80
	4200	140 H.P.		44.30	770	2,315	6,950	817.40
	4260	200 H.P.		65.75	1,200	3,585	10,800	1,243
	4310	300 H.P.		86.45	1,700	5,130	15,400	1,718
	4360	410 H.P.		117.20	2,225	6,670	20,000	2,272
	4370	500 H.P.		150.10	2,900	8,680	26,000	2,937
	4380	700 H.P.		222.15	4,100	12,290	36,900	4,235
	4400	Loader, crawler, torque conv., diesel, 1-1/2 C.Y., 80 H.P.		26.70	460	1,375	4,125	488.60
	4450	1-1/2 to 1-3/4 C.Y., 95 H.P.		30.40	590	1,775	5,325	598.20
	4510	1-3/4 to 2-1/4 C.Y., 130 H.P.		47.40	860	2,580	7,750	895.20
	4530	2-1/2 to 3-1/4 C.Y., 190 H.P.		57.75	1,075	3,220	9,650	1,106
	4560	3-1/2 to 5 C.Y., 275 H.P.		73.90	1,425	4,260	12,800	1,443
	4610	Front end loader, 4WD, articulated frame, diesel, 1 to 1-1/4 C.Y., 70 H.P.		16.65	225	675	2,025	268.20
	4620	1-1/2 to 1-3/4 C.Y., 95 H.P.		20.85	278	835	2,500	333.80
	4650	1-3/4 to 2 C.Y., 130 H.P.		24.40	365	1,100	3,300	415.20
	4710	2-1/2 to 3-1/2 C.Y., 145 H.P.	↓	27.55	420	1,260	3,775	472.40

01 54 | Construction Aids

01 54 33 | Equipment Rental

			UNIT	HOURLY OPER. COST	RENT PER DAY	RENT PER WEEK	RENT PER MONTH	EQUIPMENT COST/DAY	
20	4730	3 to 4-1/2 C.Y., 185 H.P.	Ea.	35.25	550	1,645	4,925	611	20
	4760	5-1/4 to 5-3/4 C.Y., 270 H.P.		56.50	845	2,540	7,625	960	
	4810	7 to 9 C.Y., 475 H.P.		98.35	1,775	5,345	16,000	1,856	
	4870	9 - 11 C.Y., 620 H.P.		132.15	2,500	7,495	22,500	2,556	
	4880	Skid steer loader, wheeled, 10 C.F., 30 H.P. gas		8.95	147	440	1,325	159.60	
	4890	1 C.Y., 78 H.P., diesel		16.90	243	730	2,200	281.20	
	4892	Skid-steer attachment, auger		.55	92	276	830	59.60	
	4893	Backhoe		.70	116	348	1,050	75.20	
	4894	Broom		.74	124	371	1,125	80.10	
	4895	Forks		.25	41	123	370	26.60	
	4896	Grapple		.57	94.50	283	850	61.15	
	4897	Concrete hammer		1.08	179	538	1,625	116.25	
	4898	Tree spade		.94	156	468	1,400	101.10	
	4899	Trencher		.76	126	378	1,125	81.70	
	4900	Trencher, chain, boom type, gas, operator walking, 12 H.P.		4.35	46.50	140	420	62.80	
	4910	Operator riding, 40 H.P.		17.10	293	880	2,650	312.80	
	5000	Wheel type, diesel, 4′ deep, 12″ wide		73.05	835	2,500	7,500	1,084	
	5100	6′ deep, 20″ wide		84.15	1,900	5,715	17,100	1,816	
	5150	Chain type, diesel, 5′ deep, 8″ wide		32.30	560	1,685	5,050	595.40	
	5200	Diesel, 8′ deep, 16″ wide		134.05	2,950	8,880	26,600	2,848	
	5202	Rock trencher, wheel type, 6″ wide x 18″ deep		19.05	320	960	2,875	344.40	
	5206	Chain type, 18″ wide x 7′ deep		103.05	2,825	8,465	25,400	2,517	
	5210	Tree spade, self-propelled		13.73	267	800	2,400	269.85	
	5250	Truck, dump, 2-axle, 12 ton, 8 C.Y. payload, 220 H.P.		27.80	225	675	2,025	357.40	
	5300	Three axle dump, 16 ton, 12 C.Y. payload, 400 H.P.		48.95	320	965	2,900	584.60	
	5310	Four axle dump, 25 ton, 18 C.Y. payload, 450 H.P.		58.00	470	1,405	4,225	745	
	5350	Dump trailer only, rear dump, 16-1/2 C.Y.		5.15	137	410	1,225	123.20	
	5400	20 C.Y.		5.60	155	465	1,400	137.80	
	5450	Flatbed, single axle, 1-1/2 ton rating		21.25	68.50	205	615	211	
	5500	3 ton rating		25.80	96.50	290	870	264.40	
	5550	Off highway rear dump, 25 ton capacity		67.70	1,225	3,645	10,900	1,271	
	5600	35 ton capacity		63.10	1,075	3,260	9,775	1,157	
	5610	50 ton capacity		85.40	1,475	4,425	13,300	1,568	
	5620	65 ton capacity		88.20	1,450	4,330	13,000	1,572	
	5630	100 ton capacity		113.50	1,950	5,815	17,400	2,071	
	6000	Vibratory plow, 25 H.P., walking		7.35	60	180	540	94.80	
40	0010	**GENERAL EQUIPMENT RENTAL** without operators (R015433 -10)							40
	0150	Aerial lift, scissor type, to 15′ high, 1000 lb. capacity, electric	Ea.	3.00	53.50	160	480	56	
	0160	To 25′ high, 2000 lb. capacity		3.40	68.50	205	615	68.20	
	0170	Telescoping boom to 40′ high, 500 lb. capacity, gas		17.40	310	925	2,775	324.20	
	0180	To 45′ high, 500 lb. capacity		18.50	355	1,070	3,200	362	
	0190	To 60′ high, 600 lb. capacity		20.80	470	1,405	4,225	447.40	
	0195	Air compressor, portable, 6.5 CFM, electric		.31	12.65	38	114	10.10	
	0196	Gasoline		.77	19	57	171	17.55	
	0200	Towed type, gas engine, 60 CFM		11.55	46.50	140	420	120.40	
	0300	160 CFM		13.35	50	150	450	136.80	
	0400	Diesel engine, rotary screw, 250 CFM		14.00	108	325	975	177	
	0500	365 CFM		18.75	132	395	1,175	229	
	0550	450 CFM		23.70	165	495	1,475	288.60	
	0600	600 CFM		41.45	227	680	2,050	467.60	
	0700	750 CFM		41.60	237	710	2,125	474.80	
	0800	For silenced models, small sizes, add to rent		3%	5%	5%	5%		
	0900	Large sizes, add to rent		5%	7%	7%	7%		
	0930	Air tools, breaker, pavement, 60 lb.	Ea.	.50	9.65	29	87	9.80	
	0940	80 lb.		.50	10.35	31	93	10.20	
	0950	Drills, hand (jackhammer) 65 lb.		.60	16.65	50	150	14.80	
	0960	Track or wagon, swing boom, 4″ drifter		55.90	815	2,450	7,350	937.20	
	0970	5″ drifter		67.20	915	2,750	8,250	1,088	

01 54 33 | Equipment Rental

			UNIT	HOURLY OPER. COST	RENT PER DAY	RENT PER WEEK	RENT PER MONTH	EQUIPMENT COST/DAY	
40	0975	Track mounted quarry drill, 6" diameter drill	Ea.	102.85	1,350	4,060	12,200	1,635	40
	0980	Dust control per drill		.63	22	66	198	18.25	
	0990	Hammer, chipping, 12 lb.		.55	25.50	77	231	19.80	
	1000	Hose, air with couplings, 50' long, 3/4" diameter		.03	4.67	14	42	3.05	
	1100	1" diameter		.04	6.35	19	57	4.10	
	1200	1-1/2" diameter		.06	9.65	29	87	6.30	
	1300	2" diameter		.07	12	36	108	7.75	
	1400	2-1/2" diameter		.11	19	57	171	12.30	
	1410	3" diameter		.14	23	69	207	14.90	
	1450	Drill, steel, 7/8" x 2'		.05	8.65	26	78	5.60	
	1460	7/8" x 6'		.05	9	27	81	5.80	
	1520	Moil points		.02	4	12	36	2.55	
	1525	Pneumatic nailer w/accessories		.48	31.50	95	285	22.85	
	1530	Sheeting driver for 60 lb. breaker		.04	6	18	54	3.90	
	1540	For 90 lb. breaker		.12	8	24	72	5.75	
	1550	Spade, 25 lb.		.45	6.65	20	60	7.60	
	1560	Tamper, single, 35 lb.		.55	36.50	109	325	26.20	
	1570	Triple, 140 lb.		.82	54.50	164	490	39.35	
	1580	Wrenches, impact, air powered, up to 3/4" bolt		.40	11.35	34	102	10	
	1590	Up to 1-1/4" bolt		.50	23.50	70	210	18	
	1600	Barricades, barrels, reflectorized, 1 to 99 barrels		.03	4.60	13.80	41.50	3	
	1610	100 to 200 barrels		.02	3.53	10.60	32	2.30	
	1620	Barrels with flashers, 1 to 99 barrels		.03	5.25	15.80	47.50	3.40	
	1630	100 to 200 barrels		.03	4.20	12.60	38	2.75	
	1640	Barrels with steady burn type C lights		.04	7	21	63	4.50	
	1650	Illuminated board, trailer mounted, with generator		3.25	133	400	1,200	106	
	1670	Portable barricade, stock, with flashers, 1 to 6 units		.03	5.25	15.80	47.50	3.40	
	1680	25 to 50 units		.03	4.90	14.70	44	3.20	
	1685	Butt fusion machine, wheeled, 1.5 HP electric, 2" - 8" diameter pipe		2.56	167	500	1,500	120.50	
	1690	Tracked, 20 HP diesel, 4"-12" diameter pipe		10.43	490	1,465	4,400	376.45	
	1695	83 HP diesel, 8" - 24" diameter pipe		27.52	975	2,930	8,800	806.15	
	1700	Carts, brick, hand powered, 1000 lb. capacity		.44	72.50	218	655	47.10	
	1800	Gas engine, 1500 lb., 7-1/2' lift		4.04	115	345	1,025	101.30	
	1822	Dehumidifier, medium, 6 lb./hr., 150 CFM		.93	60	180	540	43.45	
	1824	Large, 18 lb./hr., 600 CFM		1.89	122	366	1,100	88.30	
	1830	Distributor, asphalt, trailer mounted, 2000 gal., 38 H.P. diesel		9.80	330	990	2,975	276.40	
	1840	3000 gal., 38 H.P. diesel		11.25	360	1,080	3,250	306	
	1850	Drill, rotary hammer, electric		.82	26.50	80	240	22.55	
	1860	Carbide bit, 1-1/2" diameter, add to electric rotary hammer		.02	4.97	14.90	44.50	3.15	
	1865	Rotary, crawler, 250 H.P.		131.70	2,050	6,175	18,500	2,289	
	1870	Emulsion sprayer, 65 gal., 5 H.P. gas engine		2.85	97	291	875	81	
	1880	200 gal., 5 H.P. engine		7.25	162	485	1,450	155	
	1900	Floor auto-scrubbing machine, walk-behind, 28" path		3.76	247	740	2,225	178.10	
	1930	Floodlight, mercury vapor, or quartz, on tripod, 1000 watt		.30	16	48	144	12	
	1940	2000 watt		.55	30.50	91	273	22.60	
	1950	Floodlights, trailer mounted with generator, 1 - 300 watt light		3.35	71.50	215	645	69.80	
	1960	2 - 1000 watt lights		4.35	96.50	290	870	92.80	
	2000	4 - 300 watt lights		4.10	91.50	275	825	87.80	
	2005	Foam spray rig, incl. box trailer, compressor, generator, proportioner		27.23	475	1,430	4,300	503.85	
	2020	Forklift, straight mast, 12' lift, 5000 lb., 2 wheel drive, gas		22.00	198	595	1,775	295	
	2040	21' lift, 5000 lb., 4 wheel drive, diesel		17.90	243	730	2,200	289.20	
	2050	For rough terrain, 42' lift, 35' reach, 9000 lb., 110 H.P.		25.10	470	1,405	4,225	481.80	
	2060	For plant, 4 ton capacity, 80 H.P., 2 wheel drive, gas		13.40	98.50	295	885	166.20	
	2080	10 ton capacity, 120 H.P., 2 wheel drive, diesel		19.95	172	515	1,550	262.60	
	2100	Generator, electric, gas engine, 1.5 kW to 3 kW		3.10	11.35	34	102	31.60	
	2200	5 kW		4.00	15	45	135	41	
	2300	10 kW		7.50	36.50	110	330	82	
	2400	25 kW		8.80	85	255	765	121.40	

01 54 | Construction Aids

01 54 33 | Equipment Rental

		01 54 33 Equipment Rental	Unit	Hourly Oper. Cost	Rent Per Day	Rent Per Week	Rent Per Month	Equipment Cost/Day	
40	2500	Diesel engine, 20 kW	Ea.	10.10	71.50	215	645	123.80	40
	2600	50 kW		18.15	98.50	295	885	204.20	
	2700	100 kW		35.05	138	415	1,250	363.40	
	2800	250 kW		69.05	243	730	2,200	698.40	
	2850	Hammer, hydraulic, for mounting on boom, to 500 ft lb.		2.40	75	225	675	64.20	
	2860	1000 ft lb.		4.15	125	375	1,125	108.20	
	2900	Heaters, space, oil or electric, 50 MBH		1.84	8	24	72	19.50	
	3000	100 MBH		3.32	11.35	34	102	33.35	
	3100	300 MBH		10.62	36.50	110	330	106.95	
	3150	500 MBH		21.42	48.50	145	435	200.35	
	3200	Hose, water, suction with coupling, 20′ long, 2″ diameter		.02	3	9	27	1.95	
	3210	3″ diameter		.03	4.33	13	39	2.85	
	3220	4″ diameter		.03	5	15	45	3.25	
	3230	6″ diameter		.11	17.65	53	159	11.50	
	3240	8″ diameter		.20	33.50	100	300	21.60	
	3250	Discharge hose with coupling, 50′ long, 2″ diameter		.02	2.67	8	24	1.75	
	3260	3″ diameter		.03	4.67	14	42	3.05	
	3270	4″ diameter		.04	7.35	22	66	4.70	
	3280	6″ diameter		.11	18.65	56	168	12.10	
	3290	8″ diameter		.20	33.50	100	300	21.60	
	3295	Insulation blower		.67	6	18	54	8.95	
	3300	Ladders, extension type, 16′ to 36′ long		.14	24	72	216	15.50	
	3400	40′ to 60′ long		.20	32.50	98	294	21.20	
	3405	Lance for cutting concrete		2.17	104	313	940	79.95	
	3407	Lawn mower, rotary, 22″, 5 H.P.		2.14	63.50	190	570	55.10	
	3408	48″ self propelled		2.88	75	225	675	68.05	
	3410	Level, electronic, automatic, with tripod and leveling rod		1.64	109	328	985	78.70	
	3430	Laser type, for pipe and sewer line and grade		.69	46	138	415	33.10	
	3440	Rotating beam for interior control		1.15	76.50	230	690	55.20	
	3460	Builder's optical transit, with tripod and rod		.09	15.35	46	138	9.90	
	3500	Light towers, towable, with diesel generator, 2000 watt		4.10	91.50	275	825	87.80	
	3600	4000 watt		4.35	96.50	290	870	92.80	
	3700	Mixer, powered, plaster and mortar, 6 C.F., 7 H.P.		2.30	19.65	59	177	30.20	
	3800	10 C.F., 9 H.P.		2.45	30.50	92	276	38	
	3850	Nailer, pneumatic		.49	32.50	98	294	23.50	
	3900	Paint sprayers complete, 8 CFM		.73	48.50	145	435	34.85	
	4000	17 CFM		1.35	89.50	269	805	64.60	
	4020	Pavers, bituminous, rubber tires, 8′ wide, 50 H.P., diesel		29.45	490	1,465	4,400	528.60	
	4030	10′ wide, 150 H.P.		96.65	1,800	5,395	16,200	1,852	
	4050	Crawler, 8′ wide, 100 H.P., diesel		83.45	1,800	5,410	16,200	1,750	
	4060	10′ wide, 150 H.P.		103.55	2,225	6,645	19,900	2,157	
	4070	Concrete paver, 12′ to 24′ wide, 250 H.P.		88.70	1,550	4,660	14,000	1,642	
	4080	Placer-spreader-trimmer, 24′ wide, 300 H.P.		127.05	2,525	7,575	22,700	2,531	
	4100	Pump, centrifugal gas pump, 1-1/2″ diameter, 65 GPM		3.70	48.50	145	435	58.60	
	4200	2″ diameter, 130 GPM		4.85	53.50	160	480	70.80	
	4300	3″ diameter, 250 GPM		5.10	55	165	495	73.80	
	4400	6″ diameter, 1500 GPM		25.55	173	520	1,550	308.40	
	4500	Submersible electric pump, 1-1/4″ diameter, 55 GPM		.31	16.35	49	147	12.30	
	4600	1-1/2″ diameter, 83 GPM		.35	18.65	56	168	14	
	4700	2″ diameter, 120 GPM		1.35	23.50	70	210	24.80	
	4800	3″ diameter, 300 GPM		2.25	41.50	125	375	43	
	4900	4″ diameter, 560 GPM		9.65	158	475	1,425	172.20	
	5000	6″ diameter, 1590 GPM		14.15	213	640	1,925	241.20	
	5100	Diaphragm pump, gas, single, 1-1/2″ diameter		1.15	49.50	148	445	38.80	
	5200	2″ diameter		3.95	61.50	185	555	68.60	
	5300	3″ diameter		3.95	61.50	185	555	68.60	
	5400	Double, 4″ diameter		5.90	107	320	960	111.20	
	5450	Pressure washer 5 GPM, 3000 psi		4.25	48.50	145	435	63	

01 54 33 | Equipment Rental

			UNIT	HOURLY OPER. COST	RENT PER DAY	RENT PER WEEK	RENT PER MONTH	EQUIPMENT COST/DAY	
40	5500	Trash pump, self-priming, gas, 2″ diameter	Ea.	4.05	21	63	189	45	40
	5600	Diesel, 4″ diameter		7.75	73.50	220	660	106	
	5650	Diesel, 6″ diameter		20.50	140	420	1,250	248	
	5655	Grout Pump		21.80	177	530	1,600	280.40	
	5700	Salamanders, L.P. gas fired, 100,000 Btu		3.33	12.65	38	114	34.25	
	5705	50,000 Btu		2.50	10.35	31	93	26.20	
	5720	Sandblaster, portable, open top, 3 C.F. capacity		.55	26.50	80	240	20.40	
	5730	6 C.F. capacity		.90	40	120	360	31.20	
	5740	Accessories for above		.12	20.50	62	186	13.35	
	5750	Sander, floor		.59	19.65	59	177	16.50	
	5760	Edger		.54	21.50	64	192	17.10	
	5800	Saw, chain, gas engine, 18″ long		1.95	19.65	59	177	27.40	
	5900	Hydraulic powered, 36″ long		.70	63.50	190	570	43.60	
	5950	60″ long		.70	65	195	585	44.60	
	6000	Masonry, table mounted, 14″ diameter, 5 H.P.		1.25	56.50	170	510	44	
	6050	Portable cut-off, 8 H.P.		2.10	32.50	98	294	36.40	
	6100	Circular, hand held, electric, 7-1/4″ diameter		.14	4.67	14	42	3.90	
	6200	12″ diameter		.21	7.65	23	69	6.30	
	6250	Wall saw, w/hydraulic power, 10 H.P.		7.00	60	180	540	92	
	6275	Shot blaster, walk-behind, 20″ wide		4.65	290	870	2,600	211.20	
	6280	Sidewalk broom, walk-behind		2.09	60.50	182	545	53.10	
	6300	Steam cleaner, 100 gallons per hour		3.35	66.50	200	600	66.80	
	6310	200 gallons per hour		4.65	81.50	245	735	86.20	
	6340	Tar Kettle/Pot, 400 gallons		5.50	78.50	235	705	91	
	6350	Torch, cutting, acetylene-oxygen, 150′ hose, excludes gases		.30	14	42	126	10.80	
	6360	Hourly operating cost includes tips and gas		10.80				86.40	
	6410	Toilet, portable chemical		.12	20	60	180	12.95	
	6420	Recycle flush type		.14	24	72	216	15.50	
	6430	Toilet, fresh water flush, garden hose,		.17	28.50	85	255	18.35	
	6440	Hoisted, non-flush, for high rise		.14	23.50	71	213	15.30	
	6465	Tractor, farm with attachment		18.25	275	825	2,475	311	
	6500	Trailers, platform, flush deck, 2 axle, 25 ton capacity		5.20	112	335	1,000	108.60	
	6600	40 ton capacity		6.75	155	465	1,400	147	
	6700	3 axle, 50 ton capacity		7.25	172	515	1,550	161	
	6800	75 ton capacity		9.05	225	675	2,025	207.40	
	6810	Trailer mounted cable reel for high voltage line work		5.14	245	734	2,200	187.90	
	6820	Trailer mounted cable tensioning rig		10.22	485	1,460	4,375	373.75	
	6830	Cable pulling rig		68.64	2,725	8,180	24,500	2,185	
	6900	Water tank trailer, engine driven discharge, 5000 gallons		6.75	145	435	1,300	141	
	6925	10,000 gallons		9.20	202	605	1,825	194.60	
	6950	Water truck, off highway, 6000 gallons		75.60	800	2,400	7,200	1,085	
	7010	Tram car for high voltage line work, powered, 2 conductor		4.44	133	399	1,200	115.30	
	7020	Transit (builder's level) with tripod		.09	15.35	46	138	9.90	
	7030	Trench box, 3000 lb., 6′ x 8′		.56	93	279	835	60.30	
	7040	7200 lb., 6′ x 20′		1.05	175	525	1,575	113.40	
	7050	8000 lb., 8′ x 16′		1.05	175	525	1,575	113.40	
	7060	9500 lb., 8′ x 20′		1.19	199	597	1,800	128.90	
	7065	11,000 lb., 8′ x 24′		1.25	209	627	1,875	135.40	
	7070	12,000 lb., 10′ x 20′		1.36	227	680	2,050	146.90	
	7100	Truck, pickup, 3/4 ton, 2 wheel drive		11.50	58.50	175	525	127	
	7200	4 wheel drive		11.75	75	225	675	139	
	7250	Crew carrier, 9 passenger		16.30	86.50	260	780	182.40	
	7290	Flat bed truck, 20,000 lb. GVW		17.60	125	375	1,125	215.80	
	7300	Tractor, 4 x 2, 220 H.P.		24.60	197	590	1,775	314.80	
	7410	330 H.P.		36.35	270	810	2,425	452.80	
	7500	6 x 4, 380 H.P.		41.65	315	945	2,825	522.20	
	7600	450 H.P.		50.75	380	1,145	3,425	635	
	7610	Tractor, with A frame, boom and winch, 225 H.P.		27.40	272	815	2,450	382.20	

01 54 33 | Equipment Rental

		01 54 33 | Equipment Rental	UNIT	HOURLY OPER. COST	RENT PER DAY	RENT PER WEEK	RENT PER MONTH	EQUIPMENT COST/DAY	
40	7620	Vacuum truck, hazardous material, 2500 gallons	Ea.	11.55	298	895	2,675	271.40	40
	7625	5,000 gallons		13.40	415	1,250	3,750	357.20	
	7650	Vacuum, HEPA, 16 gallon, wet/dry		.74	18	54	162	16.70	
	7655	55 gallon, wet/dry		.59	27	81	243	20.90	
	7660	Water tank, portable		.16	27	80.70	242	17.40	
	7690	Sewer/catch basin vacuum, 14 C.Y., 1500 gallons		17.81	625	1,880	5,650	518.50	
	7700	Welder, electric, 200 amp		2.14	16.35	49	147	26.90	
	7800	300 amp		3.11	20	60	180	36.90	
	7900	Gas engine, 200 amp		11.70	23.50	70	210	107.60	
	8000	300 amp		13.35	24.50	74	222	121.60	
	8100	Wheelbarrow, any size		.08	12.65	38	114	8.25	
	8200	Wrecking ball, 4000 lb.	↓	2.25	70	210	630	60	
50	0010	**HIGHWAY EQUIPMENT RENTAL** without operators (R015433-10)							50
	0050	Asphalt batch plant, portable drum mixer, 100 ton/hr.	Ea.	70.30	1,425	4,310	12,900	1,424	
	0060	200 ton/hr.		78.75	1,525	4,570	13,700	1,544	
	0070	300 ton/hr.		91.25	1,800	5,375	16,100	1,805	
	0100	Backhoe attachment, long stick, up to 185 H.P., 10.5′ long		.34	22.50	67	201	16.10	
	0140	Up to 250 H.P., 12′ long		.36	24	72	216	17.30	
	0180	Over 250 H.P., 15′ long		.50	33	99	297	23.80	
	0200	Special dipper arm, up to 100 H.P., 32′ long		1.02	67.50	203	610	48.75	
	0240	Over 100 H.P., 33′ long		1.27	84.50	253	760	60.75	
	0280	Catch basin/sewer cleaning truck, 3 ton, 9 C.Y., 1000 gal.		37.35	395	1,185	3,550	535.80	
	0300	Concrete batch plant, portable, electric, 200 C.Y./hr.		27.69	510	1,530	4,600	527.50	
	0520	Grader/dozer attachment, ripper/scarifier, rear mounted, up to 135 H.P.		3.35	65	195	585	65.80	
	0540	Up to 180 H.P.		3.95	83.50	250	750	81.60	
	0580	Up to 250 H.P.		4.35	95	285	855	91.80	
	0700	Pavement removal bucket, for hydraulic excavator, up to 90 H.P.		1.80	51.50	155	465	45.40	
	0740	Up to 200 H.P.		2.00	71.50	215	645	59	
	0780	Over 200 H.P.		2.15	85	255	765	68.20	
	0900	Aggregate spreader, self-propelled, 187 H.P.		46.10	690	2,065	6,200	781.80	
	1000	Chemical spreader, 3 C.Y.		3.00	43.50	130	390	50	
	1900	Hammermill, traveling, 250 H.P.		68.15	1,950	5,880	17,600	1,721	
	2000	Horizontal borer, 3″ diameter, 13 H.P. gas driven		5.65	53.50	160	480	77.20	
	2150	Horizontal directional drill, 20,000 lb. thrust, 78 H.P. diesel		28.00	665	1,990	5,975	622	
	2160	30,000 lb. thrust, 115 H.P.		34.65	1,025	3,045	9,125	886.20	
	2170	50,000 lb. thrust, 170 H.P.		49.60	1,300	3,895	11,700	1,176	
	2190	Mud trailer for HDD, 1500 gallons, 175 H.P., gas		27.15	153	460	1,375	309.20	
	2200	Hydromulcher, diesel, 3000 gallon, for truck mounting		19.10	252	755	2,275	303.80	
	2300	Gas, 600 gallon		7.55	100	300	900	120.40	
	2400	Joint & crack cleaner, walk behind, 25 H.P.		3.40	50	150	450	57.20	
	2500	Filler, trailer mounted, 400 gallons, 20 H.P.		8.55	213	640	1,925	196.40	
	3000	Paint striper, self-propelled, 40 gallon, 22 H.P.		6.70	158	475	1,425	148.60	
	3100	120 gallon, 120 H.P.		20.45	410	1,225	3,675	408.60	
	3200	Post drivers, 6″ I-Beam frame, for truck mounting		15.15	405	1,210	3,625	363.20	
	3400	Road sweeper, self-propelled, 8′ wide, 90 H.P.		35.10	565	1,700	5,100	620.80	
	3450	Road sweeper, vacuum assisted, 4 C.Y., 220 gallons		63.20	620	1,855	5,575	876.60	
	4000	Road mixer, self-propelled, 130 H.P.		43.95	770	2,305	6,925	812.60	
	4100	310 H.P.		76.25	2,125	6,360	19,100	1,882	
	4220	Cold mix paver, incl. pug mill and bitumen tank, 165 H.P.		90.00	2,150	6,460	19,400	2,012	
	4240	Pavement brush, towed		3.10	95	285	855	81.80	
	4250	Paver, asphalt, wheel or crawler, 130 H.P., diesel		89.70	2,125	6,390	19,200	1,996	
	4300	Paver, road widener, gas 1′ to 6′, 67 H.P.		44.75	875	2,620	7,850	882	
	4400	Diesel, 2′ to 14′, 88 H.P.		56.90	1,050	3,155	9,475	1,086	
	4600	Slipform pavers, curb and gutter, 2 track, 75 H.P.		40.75	725	2,175	6,525	761	
	4700	4 track, 165 H.P.		36.65	735	2,200	6,600	733.20	
	4800	Median barrier, 215 H.P.		41.40	755	2,260	6,775	783.20	
	4901	Trailer, low bed, 75 ton capacity		9.75	225	675	2,025	213	
	5000	Road planer, walk behind, 10″ cutting width, 10 H.P.	↓	3.20	32.50	98	294	45.20	

01 54 33 | Equipment Rental

		01 54 33 Equipment Rental		UNIT	HOURLY OPER. COST	RENT PER DAY	RENT PER WEEK	RENT PER MONTH	EQUIPMENT COST/DAY	
50	5100	Self-propelled, 12" cutting width, 64 H.P.		Ea.	8.90	112	335	1,000	138.20	50
	5120	Traffic line remover, metal ball blaster, truck mounted, 115 H.P.			44.60	740	2,220	6,650	800.80	
	5140	Grinder, truck mounted, 115 H.P.			50.10	805	2,410	7,225	882.80	
	5160	Walk-behind, 11 H.P.			3.65	53.50	160	480	61.20	
	5200	Pavement profiler, 4' to 6' wide, 450 H.P.			227.55	3,350	10,035	30,100	3,827	
	5300	8' to 10' wide, 750 H.P.			353.65	4,375	13,150	39,500	5,459	
	5400	Roadway plate, steel, 1" x 8' x 20'			.07	12.35	37	111	7.95	
	5600	Stabilizer, self-propelled, 150 H.P.			43.25	610	1,825	5,475	711	
	5700	310 H.P.			84.85	1,500	4,530	13,600	1,585	
	5800	Striper, thermal, truck mounted, 120 gallon paint, 460 H.P.			54.85	495	1,485	4,450	735.80	
	5900	Thermal paint heating kettle, 115 gallons			3.64	25.50	77	231	44.50	
	6000	Tar kettle, 330 gallon, trailer mounted			5.15	60	180	540	77.20	
	7000	Tunnel locomotive, diesel, 8 to 12 ton			29.20	570	1,710	5,125	575.60	
	7005	Electric, 10 ton			25.40	650	1,955	5,875	594.20	
	7010	Muck cars, 1/2 C.Y. capacity			2.00	23.50	71	213	30.20	
	7020	1 C.Y. capacity			2.20	31.50	95	285	36.60	
	7030	2 C.Y. capacity			2.30	36.50	110	330	40.40	
	7040	Side dump, 2 C.Y. capacity			2.50	43.50	130	390	46	
	7050	3 C.Y. capacity			3.40	50	150	450	57.20	
	7060	5 C.Y. capacity			4.85	63.50	190	570	76.80	
	7100	Ventilating blower for tunnel, 7-1/2 H.P.			1.75	50	150	450	44	
	7110	10 H.P.			1.88	51.50	155	465	46.05	
	7120	20 H.P.			2.68	67.50	202	605	61.85	
	7140	40 H.P.			4.17	95	285	855	90.35	
	7160	60 H.P.			6.38	147	440	1,325	139.05	
	7175	75 H.P.			8.35	196	587	1,750	184.20	
	7180	200 H.P.			15.40	293	880	2,650	299.20	
	7800	Windrow loader, elevating		↓	51.85	1,325	3,960	11,900	1,207	
60	0010	**LIFTING AND HOISTING EQUIPMENT RENTAL** without operators	R015433 -10							60
	0120	Aerial lift truck, 2 person, to 80'		Ea.	26.70	705	2,120	6,350	637.60	
	0140	Boom work platform, 40' snorkel	R015433 -15		15.80	270	810	2,425	288.40	
	0150	Crane, flatbed mounted, 3 ton capacity			14.55	193	580	1,750	232.40	
	0200	Crane, climbing, 106' jib, 6000 lb. capacity, 410 fpm	R312316 -45		29.95	1,550	4,670	14,000	1,174	
	0300	101' jib, 10,250 lb. capacity, 270 fpm			36.20	1,975	5,920	17,800	1,474	
	0500	Tower, static, 130' high, 106' jib, 6200 lb. capacity at 400 fpm			33.60	1,800	5,400	16,200	1,349	
	0600	Crawler mounted, lattice boom, 1/2 C.Y., 15 tons at 12' radius			33.19	635	1,910	5,725	647.50	
	0700	3/4 C.Y., 20 tons at 12' radius			44.25	795	2,390	7,175	832	
	0800	1 C.Y., 25 tons at 12' radius			59.00	1,075	3,190	9,575	1,110	
	0900	1-1/2 C.Y., 40 tons at 12' radius			59.00	1,075	3,215	9,650	1,115	
	1000	2 C.Y., 50 tons at 12' radius			62.90	1,250	3,770	11,300	1,257	
	1100	3 C.Y., 75 tons at 12' radius			67.55	1,475	4,415	13,200	1,423	
	1200	100 ton capacity, 60' boom			76.50	1,700	5,080	15,200	1,628	
	1300	165 ton capacity, 60' boom			96.25	1,975	5,940	17,800	1,958	
	1400	200 ton capacity, 70' boom			118.45	2,475	7,440	22,300	2,436	
	1500	350 ton capacity, 80' boom			166.45	3,750	11,245	33,700	3,581	
	1600	Truck mounted, lattice boom, 6 x 4, 20 tons at 10' radius			35.08	1,125	3,350	10,100	950.65	
	1700	25 tons at 10' radius			37.95	1,225	3,650	11,000	1,034	
	1800	8 x 4, 30 tons at 10' radius			41.11	1,300	3,880	11,600	1,105	
	1900	40 tons at 12' radius			43.92	1,350	4,060	12,200	1,163	
	2000	60 tons at 15' radius			49.48	1,425	4,300	12,900	1,256	
	2050	82 tons at 15' radius			55.38	1,525	4,590	13,800	1,361	
	2100	90 tons at 15' radius			62.13	1,675	5,000	15,000	1,497	
	2200	115 tons at 15' radius			70.13	1,875	5,590	16,800	1,679	
	2300	150 tons at 18' radius			81.45	1,950	5,885	17,700	1,829	
	2350	165 tons at 18' radius			82.43	2,075	6,240	18,700	1,907	
	2400	Truck mounted, hydraulic, 12 ton capacity			36.75	530	1,590	4,775	612	
	2500	25 ton capacity			39.15	640	1,920	5,750	697.20	
	2550	33 ton capacity		↓	39.70	655	1,970	5,900	711.60	

01 54 33 | Equipment Rental

			UNIT	HOURLY OPER. COST	RENT PER DAY	RENT PER WEEK	RENT PER MONTH	EQUIPMENT COST/DAY	
60	2560	40 ton capacity	Ea.	51.70	770	2,305	6,925	874.60	60
	2600	55 ton capacity		67.65	930	2,795	8,375	1,100	
	2700	80 ton capacity		88.65	1,400	4,225	12,700	1,554	
	2720	100 ton capacity		84.25	1,450	4,365	13,100	1,547	
	2740	120 ton capacity		96.15	1,575	4,690	14,100	1,707	
	2760	150 ton capacity		114.30	2,075	6,190	18,600	2,152	
	2800	Self-propelled, 4 x 4, with telescoping boom, 5 ton		15.70	232	695	2,075	264.60	
	2900	12-1/2 ton capacity		28.15	370	1,115	3,350	448.20	
	3000	15 ton capacity		28.85	390	1,175	3,525	465.80	
	3050	20 ton capacity		31.05	450	1,355	4,075	519.40	
	3100	25 ton capacity		32.65	505	1,520	4,550	565.20	
	3150	40 ton capacity		40.25	570	1,710	5,125	664	
	3200	Derricks, guy, 20 ton capacity, 60′ boom, 75′ mast		23.51	380	1,144	3,425	416.90	
	3300	100′ boom, 115′ mast		37.04	655	1,970	5,900	690.30	
	3400	Stiffleg, 20 ton capacity, 70′ boom, 37′ mast		25.93	495	1,490	4,475	505.45	
	3500	100′ boom, 47′ mast		39.91	795	2,380	7,150	795.30	
	3550	Helicopter, small, lift to 1250 lb. maximum, w/pilot		97.18	3,075	9,240	27,700	2,625	
	3600	Hoists, chain type, overhead, manual, 3/4 ton		.10	.33	1	3	1	
	3900	10 ton		.70	6	18	54	9.20	
	4000	Hoist and tower, 5000 lb. capacity, portable electric, 40′ high		3.96	220	659	1,975	163.50	
	4100	For each added 10′ section, add		.10	17.35	52	156	11.20	
	4200	Hoist and single tubular tower, 5000 lb. electric, 100′ high		5.43	305	920	2,750	227.45	
	4300	For each added 6′-6″ section, add		.18	29.50	89	267	19.25	
	4400	Hoist and double tubular tower, 5000 lb., 100′ high		5.89	340	1,013	3,050	249.70	
	4500	For each added 6′-6″ section, add		.20	33	99	297	21.40	
	4550	Hoist and tower, mast type, 6000 lb., 100′ high		6.25	350	1,051	3,150	260.20	
	4570	For each added 10′ section, add		.12	20.50	62	186	13.35	
	4600	Hoist and tower, personnel, electric, 2000 lb., 100′ @ 125 fpm		14.61	930	2,790	8,375	674.90	
	4700	3000 lb., 100′ @ 200 fpm		16.68	1,050	3,170	9,500	767.45	
	4800	3000 lb., 150′ @ 300 fpm		18.58	1,175	3,550	10,700	858.65	
	4900	4000 lb., 100′ @ 300 fpm		19.09	1,200	3,620	10,900	876.70	
	5000	6000 lb., 100′ @ 275 fpm		20.27	1,275	3,790	11,400	920.15	
	5100	For added heights up to 500′, add	L.F.	.01	1.67	5	15	1.10	
	5200	Jacks, hydraulic, 20 ton	Ea.	.05	2	6	18	1.60	
	5500	100 ton		.40	11.65	35	105	10.20	
	6100	Jacks, hydraulic, climbing w/50′ jackrods, control console, 30 ton capacity		1.89	126	378	1,125	90.70	
	6150	For each added 10′ jackrod section, add		.05	3.33	10	30	2.40	
	6300	50 ton capacity		3.05	203	609	1,825	146.20	
	6350	For each added 10′ jackrod section, add		.06	4	12	36	2.90	
	6500	125 ton capacity		8.00	535	1,600	4,800	384	
	6550	For each added 10′ jackrod section, add		.55	36.50	109	325	26.20	
	6600	Cable jack, 10 ton capacity with 200′ cable		1.59	106	317	950	76.10	
	6650	For each added 50′ of cable, add		.19	12.35	37	111	8.90	
70	0010	**WELLPOINT EQUIPMENT RENTAL** without operators R015433-10							70
	0020	Based on 2 months rental							
	0100	Combination jetting & wellpoint pump, 60 H.P. diesel	Ea.	15.87	315	938	2,825	314.55	
	0200	High pressure gas jet pump, 200 H.P., 300 psi	″	36.61	267	802	2,400	453.30	
	0300	Discharge pipe, 8″ diameter	L.F.	.01	.51	1.53	4.59	.40	
	0350	12″ diameter		.01	.75	2.24	6.70	.55	
	0400	Header pipe, flows up to 150 GPM, 4″ diameter		.01	.46	1.38	4.14	.35	
	0500	400 GPM, 6″ diameter		.01	.54	1.63	4.89	.40	
	0600	800 GPM, 8″ diameter		.01	.75	2.24	6.70	.55	
	0700	1500 GPM, 10″ diameter		.01	.79	2.36	7.10	.55	
	0800	2500 GPM, 12″ diameter		.02	1.49	4.47	13.40	1.05	
	0900	4500 GPM, 16″ diameter		.03	1.91	5.72	17.15	1.40	
	0950	For quick coupling aluminum and plastic pipe, add		.03	1.97	5.92	17.75	1.40	
	1100	Wellpoint, 25′ long, with fittings & riser pipe, 1-1/2″ or 2″ diameter	Ea.	.06	3.94	11.81	35.50	2.85	

01 54 | Construction Aids

01 54 33 | Equipment Rental

			UNIT	HOURLY OPER. COST	RENT PER DAY	RENT PER WEEK	RENT PER MONTH	EQUIPMENT COST/DAY	
70	1200	Wellpoint pump, diesel powered, 4″ suction, 20 H.P.	Ea.	6.89	180	541	1,625	163.30	70
	1300	6″ suction, 30 H.P.		9.35	224	671	2,025	209	
	1400	8″ suction, 40 H.P.		12.64	305	920	2,750	285.10	
	1500	10″ suction, 75 H.P.		19.15	360	1,075	3,225	368.20	
	1600	12″ suction, 100 H.P.		27.47	570	1,710	5,125	561.75	
	1700	12″ suction, 175 H.P.	↓	.36	630	1,890	5,675	700.90	
80	0010	**MARINE EQUIPMENT RENTAL** without operators R015433-10							80
	0200	Barge, 400 Ton, 30′ wide x 90′ long	Ea.	16.65	1,075	3,230	9,700	779.20	
	0240	800 Ton, 45′ wide x 90′ long		20.15	1,300	3,915	11,700	944.20	
	2000	Tugboat, diesel, 100 H.P.		31.60	215	645	1,925	381.80	
	2040	250 H.P.		64.00	395	1,180	3,550	748	
	2080	380 H.P.	↓	131.50	1,175	3,505	10,500	1,753	

Crew No.	Bare Costs		Incl. Subs O&P		Cost Per Labor-Hour	

Crew A-1	Hr.	Daily	Hr.	Daily	Bare Costs	Incl. O&P
1 Building Laborer	$35.10	$280.80	$57.55	$460.40	$35.10	$57.55
1 Concrete saw, gas manual		66.00		72.60	8.25	9.07
8 L.H., Daily Totals		$346.80		$533.00	$43.35	$66.63

Crew A-1A	Hr.	Daily	Hr.	Daily	Bare Costs	Incl. O&P
1 Skilled Worker	$45.40	$363.20	$73.80	$590.40	$45.40	$73.80
1 Shot Blaster, 20"		211.20		232.32	26.40	29.04
8 L.H., Daily Totals		$574.40		$822.72	$71.80	$102.84

Crew A-1B	Hr.	Daily	Hr.	Daily	Bare Costs	Incl. O&P
1 Building Laborer	$35.10	$280.80	$57.55	$460.40	$35.10	$57.55
1 Concrete Saw		148.00		162.80	18.50	20.35
8 L.H., Daily Totals		$428.80		$623.20	$53.60	$77.90

Crew A-1C	Hr.	Daily	Hr.	Daily	Bare Costs	Incl. O&P
1 Building Laborer	$35.10	$280.80	$57.55	$460.40	$35.10	$57.55
1 Chain Saw, gas, 18"		27.40		30.14	3.42	3.77
8 L.H., Daily Totals		$308.20		$490.54	$38.52	$61.32

Crew A-1D	Hr.	Daily	Hr.	Daily	Bare Costs	Incl. O&P
1 Building Laborer	$35.10	$280.80	$57.55	$460.40	$35.10	$57.55
1 Vibrating plate, gas, 18"		32.20		35.42	4.03	4.43
8 L.H., Daily Totals		$313.00		$495.82	$39.13	$61.98

Crew A-1E	Hr.	Daily	Hr.	Daily	Bare Costs	Incl. O&P
1 Building Laborer	$35.10	$280.80	$57.55	$460.40	$35.10	$57.55
1 Vibratory Plate, gas, 21"		40.80		44.88	5.10	5.61
8 L.H., Daily Totals		$321.60		$505.28	$40.20	$63.16

Crew A-1F	Hr.	Daily	Hr.	Daily	Bare Costs	Incl. O&P
1 Building Laborer	$35.10	$280.80	$57.55	$460.40	$35.10	$57.55
1 Rammer/tamper, gas, 8"		44.20		48.62	5.53	6.08
8 L.H., Daily Totals		$325.00		$509.02	$40.63	$63.63

Crew A-1G	Hr.	Daily	Hr.	Daily	Bare Costs	Incl. O&P
1 Building Laborer	$35.10	$280.80	$57.55	$460.40	$35.10	$57.55
1 Rammer/tamper, gas, 15"		49.80		54.78	6.22	6.85
8 L.H., Daily Totals		$330.60		$515.18	$41.33	$64.40

Crew A-1H	Hr.	Daily	Hr.	Daily	Bare Costs	Incl. O&P
1 Building Laborer	$35.10	$280.80	$57.55	$460.40	$35.10	$57.55
1 Exterior Steam Cleaner		66.80		73.48	8.35	9.19
8 L.H., Daily Totals		$347.60		$533.88	$43.45	$66.73

Crew A-1J	Hr.	Daily	Hr.	Daily	Bare Costs	Incl. O&P
1 Building Laborer	$35.10	$280.80	$57.55	$460.40	$35.10	$57.55
1 Cultivator, Walk-Behind, 5 H.P.		58.45		64.30	7.31	8.04
8 L.H., Daily Totals		$339.25		$524.70	$42.41	$65.59

Crew A-1K	Hr.	Daily	Hr.	Daily	Bare Costs	Incl. O&P
1 Building Laborer	$35.10	$280.80	$57.55	$460.40	$35.10	$57.55
1 Cultivator, Walk-Behind, 8 H.P.		89.75		98.72	11.22	12.34
8 L.H., Daily Totals		$370.55		$559.13	$46.32	$69.89

Crew A-1M	Hr.	Daily	Hr.	Daily	Bare Costs	Incl. O&P
1 Building Laborer	$35.10	$280.80	$57.55	$460.40	$35.10	$57.55
1 Snow Blower, Walk-Behind		53.10		58.41	6.64	7.30
8 L.H., Daily Totals		$333.90		$518.81	$41.74	$64.85

Crew A-2	Hr.	Daily	Hr.	Daily	Bare Costs	Incl. O&P
2 Laborers	$35.10	$561.60	$57.55	$920.80	$34.85	$56.88
1 Truck Driver (light)	34.35	274.80	55.55	444.40		
1 Flatbed Truck, Gas, 1.5 Ton		211.00		232.10	8.79	9.67
24 L.H., Daily Totals		$1047.40		$1597.30	$43.64	$66.55

Crew A-2A	Hr.	Daily	Hr.	Daily	Bare Costs	Incl. O&P
2 Laborers	$35.10	$561.60	$57.55	$920.80	$34.85	$56.88
1 Truck Driver (light)	34.35	274.80	55.55	444.40		
1 Flatbed Truck, Gas, 1.5 Ton		211.00		232.10		
1 Concrete Saw		148.00		162.80	14.96	16.45
24 L.H., Daily Totals		$1195.40		$1760.10	$49.81	$73.34

Crew A-2B	Hr.	Daily	Hr.	Daily	Bare Costs	Incl. O&P
1 Truck Driver (light)	$34.35	$274.80	$55.55	$444.40	$34.35	$55.55
1 Flatbed Truck, Gas, 1.5 Ton		211.00		232.10	26.38	29.01
8 L.H., Daily Totals		$485.80		$676.50	$60.73	$84.56

Crew A-3A	Hr.	Daily	Hr.	Daily	Bare Costs	Incl. O&P
1 Truck Driver (light)	$34.35	$274.80	$55.55	$444.40	$34.35	$55.55
1 Pickup truck, 4 x 4, 3/4 ton		139.00		152.90	17.38	19.11
8 L.H., Daily Totals		$413.80		$597.30	$51.73	$74.66

Crew A-3B	Hr.	Daily	Hr.	Daily	Bare Costs	Incl. O&P
1 Equip. Oper. (med.)	$46.55	$372.40	$73.50	$588.00	$40.92	$65.30
1 Truck Driver (heav.)	35.30	282.40	57.10	456.80		
1 Dump Truck, 12 C.Y., 400 H.P.		584.60		643.06		
1 F.E. Loader, W.M.,2.5 C.Y.		472.40		519.64	66.06	72.67
16 L.H., Daily Totals		$1711.80		$2207.50	$106.99	$137.97

Crew A-3C	Hr.	Daily	Hr.	Daily	Bare Costs	Incl. O&P
1 Equip. Oper. (light)	$44.75	$358.00	$70.65	$565.20	$44.75	$70.65
1 Loader, Skid Steer, 78 H.P.		281.20		309.32	35.15	38.66
8 L.H., Daily Totals		$639.20		$874.52	$79.90	$109.32

Crew A-3D	Hr.	Daily	Hr.	Daily	Bare Costs	Incl. O&P
1 Truck Driver, Light	$34.35	$274.80	$55.55	$444.40	$34.35	$55.55
1 Pickup truck, 4 x 4, 3/4 ton		139.00		152.90		
1 Flatbed Trailer, 25 Ton		108.60		119.46	30.95	34.05
8 L.H., Daily Totals		$522.40		$716.76	$65.30	$89.59

Crew A-3E	Hr.	Daily	Hr.	Daily	Bare Costs	Incl. O&P
1 Equip. Oper. (crane)	$47.80	$382.40	$75.50	$604.00	$41.55	$66.30
1 Truck Driver (heavy)	35.30	282.40	57.10	456.80		
1 Pickup truck, 4 x 4, 3/4 ton		139.00		152.90	8.69	9.56
16 L.H., Daily Totals		$803.80		$1213.70	$50.24	$75.86

Crew A-3F	Hr.	Daily	Hr.	Daily	Bare Costs	Incl. O&P
1 Equip. Oper. (crane)	$47.80	$382.40	$75.50	$604.00	$41.55	$66.30
1 Truck Driver (heavy)	35.30	282.40	57.10	456.80		
1 Pickup truck, 4 x 4, 3/4 ton		139.00		152.90		
1 Truck Tractor, 6x4, 380 H.P.		522.20		574.42		
1 Lowbed Trailer, 75 Ton		213.00		234.30	54.64	60.10
16 L.H., Daily Totals		$1539.00		$2022.42	$96.19	$126.40

Crew No.	Bare Costs		Incl. Subs O&P		Cost Per Labor-Hour	
Crew A-3G	**Hr.**	**Daily**	**Hr.**	**Daily**	**Bare Costs**	**Incl. O&P**
1 Equip. Oper. (crane)	$47.80	$382.40	$75.50	$604.00	$41.55	$66.30
1 Truck Driver (heavy)	35.30	282.40	57.10	456.80		
1 Pickup truck, 4 x 4, 3/4 ton		139.00		152.90		
1 Truck Tractor, 6x4, 450 H.P.		635.00		698.50		
1 Lowbed Trailer, 75 Ton		213.00		234.30	61.69	67.86
16 L.H., Daily Totals		$1651.80		$2146.50	$103.24	$134.16
Crew A-3H	**Hr.**	**Daily**	**Hr.**	**Daily**	**Bare Costs**	**Incl. O&P**
1 Equip. Oper. (crane)	$47.80	$382.40	$75.50	$604.00	$47.80	$75.50
1 Hyd. crane, 12 Ton (daily)		824.00		906.40	103.00	113.30
8 L.H., Daily Totals		$1206.40		$1510.40	$150.80	$188.80
Crew A-3I	**Hr.**	**Daily**	**Hr.**	**Daily**	**Bare Costs**	**Incl. O&P**
1 Equip. Oper. (crane)	$47.80	$382.40	$75.50	$604.00	$47.80	$75.50
1 Hyd. crane, 25 Ton (daily)		953.20		1048.52	119.15	131.07
8 L.H., Daily Totals		$1335.60		$1652.52	$166.95	$206.57
Crew A-3J	**Hr.**	**Daily**	**Hr.**	**Daily**	**Bare Costs**	**Incl. O&P**
1 Equip. Oper. (crane)	$47.80	$382.40	$75.50	$604.00	$47.80	$75.50
1 Hyd. crane, 40 Ton (daily)		1184.00		1302.40	148.00	162.80
8 L.H., Daily Totals		$1566.40		$1906.40	$195.80	$238.30
Crew A-3K	**Hr.**	**Daily**	**Hr.**	**Daily**	**Bare Costs**	**Incl. O&P**
1 Equip. Oper. (crane)	$47.80	$382.40	$75.50	$604.00	$44.52	$70.33
1 Equip. Oper. Oiler	41.25	330.00	65.15	521.20		
1 Hyd. crane, 55 Ton (daily)		1471.00		1618.10		
1 P/U Truck, 3/4 Ton (daily)		152.00		167.20	101.44	111.58
16 L.H., Daily Totals		$2335.40		$2910.50	$145.96	$181.91
Crew A-3L	**Hr.**	**Daily**	**Hr.**	**Daily**	**Bare Costs**	**Incl. O&P**
1 Equip. Oper. (crane)	$47.80	$382.40	$75.50	$604.00	$44.52	$70.33
1 Equip. Oper. Oiler	41.25	330.00	65.15	521.20		
1 Hyd. crane, 80 Ton (daily)		2119.00		2330.90		
1 P/U Truck, 3/4 Ton (daily)		152.00		167.20	141.94	156.13
16 L.H., Daily Totals		$2983.40		$3623.30	$186.46	$226.46
Crew A-3M	**Hr.**	**Daily**	**Hr.**	**Daily**	**Bare Costs**	**Incl. O&P**
1 Equip. Oper. (crane)	$47.80	$382.40	$75.50	$604.00	$44.52	$70.33
1 Equip. Oper. Oiler	41.25	330.00	65.15	521.20		
1 Hyd. crane, 100 Ton (daily)		2129.00		2341.90		
1 P/U Truck, 3/4 Ton (daily)		152.00		167.20	142.56	156.82
16 L.H., Daily Totals		$2993.40		$3634.30	$187.09	$227.14
Crew A-3N	**Hr.**	**Daily**	**Hr.**	**Daily**	**Bare Costs**	**Incl. O&P**
1 Equip. Oper. (crane)	$47.80	$382.40	$75.50	$604.00	$47.80	$75.50
1 Tower crane (monthly)		1005.00		1105.50	125.63	138.19
8 L.H., Daily Totals		$1387.40		$1709.50	$173.43	$213.69
Crew A-3P	**Hr.**	**Daily**	**Hr.**	**Daily**	**Bare Costs**	**Incl. O&P**
1 Equip. Oper., Light	$44.75	$358.00	$70.65	$565.20	$44.75	$70.65
1 A.T. Forklift, 42' lift		481.80		529.98	60.23	66.25
8 L.H., Daily Totals		$839.80		$1095.18	$104.97	$136.90
Crew A-4	**Hr.**	**Daily**	**Hr.**	**Daily**	**Bare Costs**	**Incl. O&P**
2 Carpenters	$44.10	$705.60	$72.30	$1156.80	$42.15	$68.55
1 Painter, Ordinary	38.25	306.00	61.05	488.40		
24 L.H., Daily Totals		$1011.60		$1645.20	$42.15	$68.55

Crew No.	Bare Costs		Incl. Subs O&P		Cost Per Labor-Hour	
Crew A-5	**Hr.**	**Daily**	**Hr.**	**Daily**	**Bare Costs**	**Incl. O&P**
2 Laborers	$35.10	$561.60	$57.55	$920.80	$35.02	$57.33
.25 Truck Driver (light)	34.35	68.70	55.55	111.10		
.25 Flatbed Truck, Gas, 1.5 Ton		52.75		58.02	2.93	3.22
18 L.H., Daily Totals		$683.05		$1089.93	$37.95	$60.55
Crew A-6	**Hr.**	**Daily**	**Hr.**	**Daily**	**Bare Costs**	**Incl. O&P**
1 Instrument Man	$45.40	$363.20	$73.80	$590.40	$44.08	$71.42
1 Rodman/Chainman	42.75	342.00	69.05	552.40		
1 Level, electronic		78.70		86.57	4.92	5.41
16 L.H., Daily Totals		$783.90		$1229.37	$48.99	$76.84
Crew A-7	**Hr.**	**Daily**	**Hr.**	**Daily**	**Bare Costs**	**Incl. O&P**
1 Chief Of Party	$56.40	$451.20	$89.95	$719.60	$48.18	$77.60
1 Instrument Man	45.40	363.20	73.80	590.40		
1 Rodman/Chainman	42.75	342.00	69.05	552.40		
1 Level, electronic		78.70		86.57	3.28	3.61
24 L.H., Daily Totals		$1235.10		$1948.97	$51.46	$81.21
Crew A-8	**Hr.**	**Daily**	**Hr.**	**Daily**	**Bare Costs**	**Incl. O&P**
1 Chief of Party	$56.40	$451.20	$89.95	$719.60	$46.83	$75.46
1 Instrument Man	45.40	363.20	73.80	590.40		
2 Rodmen/Chainmen	42.75	684.00	69.05	1104.80		
1 Level, electronic		78.70		86.57	2.46	2.71
32 L.H., Daily Totals		$1577.10		$2501.37	$49.28	$78.17
Crew A-9	**Hr.**	**Daily**	**Hr.**	**Daily**	**Bare Costs**	**Incl. O&P**
1 Asbestos Foreman	$49.35	$394.80	$79.35	$634.80	$48.91	$78.65
7 Asbestos Workers	48.85	2735.60	78.55	4398.80		
64 L.H., Daily Totals		$3130.40		$5033.60	$48.91	$78.65
Crew A-10A	**Hr.**	**Daily**	**Hr.**	**Daily**	**Bare Costs**	**Incl. O&P**
1 Asbestos Foreman	$49.35	$394.80	$79.35	$634.80	$49.02	$78.82
2 Asbestos Workers	48.85	781.60	78.55	1256.80		
24 L.H., Daily Totals		$1176.40		$1891.60	$49.02	$78.82
Crew A-10B	**Hr.**	**Daily**	**Hr.**	**Daily**	**Bare Costs**	**Incl. O&P**
1 Asbestos Foreman	$49.35	$394.80	$79.35	$634.80	$48.98	$78.75
3 Asbestos Workers	48.85	1172.40	78.55	1885.20		
32 L.H., Daily Totals		$1567.20		$2520.00	$48.98	$78.75
Crew A-10C	**Hr.**	**Daily**	**Hr.**	**Daily**	**Bare Costs**	**Incl. O&P**
3 Asbestos Workers	$48.85	$1172.40	$78.55	$1885.20	$48.85	$78.55
1 Flatbed Truck, Gas, 1.5 Ton		211.00		232.10	8.79	9.67
24 L.H., Daily Totals		$1383.40		$2117.30	$57.64	$88.22
Crew A-10D	**Hr.**	**Daily**	**Hr.**	**Daily**	**Bare Costs**	**Incl. O&P**
2 Asbestos Workers	$48.85	$781.60	$78.55	$1256.80	$46.69	$74.44
1 Equip. Oper. (crane)	47.80	382.40	75.50	604.00		
1 Equip. Oper. Oiler	41.25	330.00	65.15	521.20		
1 Hydraulic Crane, 33 Ton		711.60		782.76	22.24	24.46
32 L.H., Daily Totals		$2205.60		$3164.76	$68.92	$98.90
Crew A-11	**Hr.**	**Daily**	**Hr.**	**Daily**	**Bare Costs**	**Incl. O&P**
1 Asbestos Foreman	$49.35	$394.80	$79.35	$634.80	$48.91	$78.65
7 Asbestos Workers	48.85	2735.60	78.55	4398.80		
2 Chip. Hammers, 12 Lb., Elec.		39.60		43.56	.62	.68
64 L.H., Daily Totals		$3170.00		$5077.16	$49.53	$79.33

Crew No.	Bare Costs		Incl. Subs O&P		Cost Per Labor-Hour	
Crew A-12	**Hr.**	**Daily**	**Hr.**	**Daily**	**Bare Costs**	**Incl. O&P**
1 Asbestos Foreman	$49.35	$394.80	$79.35	$634.80	$48.91	$78.65
7 Asbestos Workers	48.85	2735.60	78.55	4398.80		
1 Trk-mtd vac, 14 CY, 1500 Gal.		518.50		570.35		
1 Flatbed Truck, 20,000 GVW		215.80		237.38	11.47	12.62
64 L.H., Daily Totals		$3864.70		$5841.33	$60.39	$91.27
Crew A-13	**Hr.**	**Daily**	**Hr.**	**Daily**	**Bare Costs**	**Incl. O&P**
1 Equip. Oper. (light)	$44.75	$358.00	$70.65	$565.20	$44.75	$70.65
1 Trk-mtd vac, 14 CY, 1500 Gal.		518.50		570.35		
1 Flatbed Truck, 20,000 GVW		215.80		237.38	91.79	100.97
8 L.H., Daily Totals		$1092.30		$1372.93	$136.54	$171.62
Crew B-1	**Hr.**	**Daily**	**Hr.**	**Daily**	**Bare Costs**	**Incl. O&P**
1 Labor Foreman (outside)	$37.10	$296.80	$60.80	$486.40	$35.77	$58.63
2 Laborers	35.10	561.60	57.55	920.80		
24 L.H., Daily Totals		$858.40		$1407.20	$35.77	$58.63
Crew B-1A	**Hr.**	**Daily**	**Hr.**	**Daily**	**Bare Costs**	**Incl. O&P**
1 Labor Foreman (outside)	$37.10	$296.80	$60.80	$486.40	$35.77	$58.63
2 Laborers	35.10	561.60	57.55	920.80		
2 Cutting Torches		21.60		23.76		
2 Sets of Gases		172.80		190.08	8.10	8.91
24 L.H., Daily Totals		$1052.80		$1621.04	$43.87	$67.54
Crew B-1B	**Hr.**	**Daily**	**Hr.**	**Daily**	**Bare Costs**	**Incl. O&P**
1 Labor Foreman (outside)	$37.10	$296.80	$60.80	$486.40	$38.77	$62.85
2 Laborers	35.10	561.60	57.55	920.80		
1 Equip. Oper. (crane)	47.80	382.40	75.50	604.00		
2 Cutting Torches		21.60		23.76		
2 Sets of Gases		172.80		190.08		
1 Hyd. Crane, 12 Ton		612.00		673.20	25.20	27.72
32 L.H., Daily Totals		$2047.20		$2898.24	$63.98	$90.57
Crew B-2	**Hr.**	**Daily**	**Hr.**	**Daily**	**Bare Costs**	**Incl. O&P**
1 Labor Foreman (outside)	$37.10	$296.80	$60.80	$486.40	$35.50	$58.20
4 Laborers	35.10	1123.20	57.55	1841.60		
40 L.H., Daily Totals		$1420.00		$2328.00	$35.50	$58.20
Crew B-3	**Hr.**	**Daily**	**Hr.**	**Daily**	**Bare Costs**	**Incl. O&P**
1 Labor Foreman (outside)	$37.10	$296.80	$60.80	$486.40	$37.41	$60.60
2 Laborers	35.10	561.60	57.55	920.80		
1 Equip. Oper. (med.)	46.55	372.40	73.50	588.00		
2 Truck Drivers (heavy)	35.30	564.80	57.10	913.60		
1 Crawler Loader, 3 C.Y.		1106.00		1216.60		
2 Dump Trucks, 12 C.Y., 400 H.P.		1169.20		1286.12	47.40	52.14
48 L.H., Daily Totals		$4070.80		$5411.52	$84.81	$112.74
Crew B-3A	**Hr.**	**Daily**	**Hr.**	**Daily**	**Bare Costs**	**Incl. O&P**
4 Laborers	$35.10	$1123.20	$57.55	$1841.60	$37.39	$60.74
1 Equip. Oper. (med.)	46.55	372.40	73.50	588.00		
1 Hyd. Excavator, 1.5 C.Y.		960.40		1056.44	24.01	26.41
40 L.H., Daily Totals		$2456.00		$3486.04	$61.40	$87.15
Crew B-3B	**Hr.**	**Daily**	**Hr.**	**Daily**	**Bare Costs**	**Incl. O&P**
2 Laborers	$35.10	$561.60	$57.55	$920.80	$38.01	$61.42
1 Equip. Oper. (med.)	46.55	372.40	73.50	588.00		
1 Truck Driver (heavy)	35.30	282.40	57.10	456.80		
1 Backhoe Loader, 80 H.P.		363.80		400.18		
1 Dump Truck, 12 C.Y., 400 H.P.		584.60		643.06	29.64	32.60
32 L.H., Daily Totals		$2164.80		$3008.84	$67.65	$94.03

Crew No.	Bare Costs		Incl. Subs O&P		Cost Per Labor-Hour	
Crew B-3C	**Hr.**	**Daily**	**Hr.**	**Daily**	**Bare Costs**	**Incl. O&P**
3 Laborers	$35.10	$842.40	$57.55	$1381.20	$37.96	$61.54
1 Equip. Oper. (med.)	46.55	372.40	73.50	588.00		
1 Crawler Loader, 4 C.Y.		1443.00		1587.30	45.09	49.60
32 L.H., Daily Totals		$2657.80		$3556.50	$83.06	$111.14
Crew B-4	**Hr.**	**Daily**	**Hr.**	**Daily**	**Bare Costs**	**Incl. O&P**
1 Labor Foreman (outside)	$37.10	$296.80	$60.80	$486.40	$35.47	$58.02
4 Laborers	35.10	1123.20	57.55	1841.60		
1 Truck Driver (heavy)	35.30	282.40	57.10	456.80		
1 Truck Tractor, 220 H.P.		314.80		346.28		
1 Flatbed Trailer, 40 Ton		147.00		161.70	9.62	10.58
48 L.H., Daily Totals		$2164.20		$3292.78	$45.09	$68.60
Crew B-5	**Hr.**	**Daily**	**Hr.**	**Daily**	**Bare Costs**	**Incl. O&P**
1 Labor Foreman (outside)	$37.10	$296.80	$60.80	$486.40	$38.66	$62.57
4 Laborers	35.10	1123.20	57.55	1841.60		
2 Equip. Oper. (med.)	46.55	744.80	73.50	1176.00		
1 Air Compressor, 250 cfm		177.00		194.70		
2 Breakers, Pavement, 60 lb.		19.60		21.56		
2 -50' Air Hoses, 1.5"		12.60		13.86		
1 Crawler Loader, 3 C.Y.		1106.00		1216.60	23.49	25.83
56 L.H., Daily Totals		$3480.00		$4950.72	$62.14	$88.41
Crew B-5D	**Hr.**	**Daily**	**Hr.**	**Daily**	**Bare Costs**	**Incl. O&P**
1 Labor Foreman (outside)	$37.10	$296.80	$60.80	$486.40	$38.24	$61.89
4 Laborers	35.10	1123.20	57.55	1841.60		
2 Equip. Oper. (med.)	46.55	744.80	73.50	1176.00		
1 Truck Driver (heavy)	35.30	282.40	57.10	456.80		
1 Air Compressor, 250 cfm		177.00		194.70		
2 Breakers, Pavement, 60 lb.		19.60		21.56		
2 -50' Air Hoses, 1.5"		12.60		13.86		
1 Crawler Loader, 3 C.Y.		1106.00		1216.60		
1 Dump Truck, 12 C.Y., 400 H.P.		584.60		643.06	29.68	32.65
64 L.H., Daily Totals		$4347.00		$6050.58	$67.92	$94.54
Crew B-6	**Hr.**	**Daily**	**Hr.**	**Daily**	**Bare Costs**	**Incl. O&P**
2 Laborers	$35.10	$561.60	$57.55	$920.80	$38.32	$61.92
1 Equip. Oper. (light)	44.75	358.00	70.65	565.20		
1 Backhoe Loader, 48 H.P.		334.00		367.40	13.92	15.31
24 L.H., Daily Totals		$1253.60		$1853.40	$52.23	$77.22
Crew B-6B	**Hr.**	**Daily**	**Hr.**	**Daily**	**Bare Costs**	**Incl. O&P**
2 Labor Foremen (outside)	$37.10	$593.60	$60.80	$972.80	$35.77	$58.63
4 Laborers	35.10	1123.20	57.55	1841.60		
1 S.P. Crane, 4x4, 5 Ton		264.60		291.06		
1 Flatbed Truck, Gas, 1.5 Ton		211.00		232.10		
1 Butt Fusion Mach., 4"-12" diam.		376.45		414.10	17.75	19.53
48 L.H., Daily Totals		$2568.85		$3751.66	$53.52	$78.16
Crew B-6C	**Hr.**	**Daily**	**Hr.**	**Daily**	**Bare Costs**	**Incl. O&P**
2 Labor Foremen (outside)	$37.10	$593.60	$60.80	$972.80	$35.77	$58.63
4 Laborers	35.10	1123.20	57.55	1841.60		
1 S.P. Crane, 4x4, 12 Ton		448.20		493.02		
1 Flatbed Truck, Gas, 3 Ton		264.40		290.84		
1 Butt Fusion Mach., 8"-24" diam.		806.15		886.76	31.64	34.80
48 L.H., Daily Totals		$3235.55		$4485.02	$67.41	$93.44

Crew No.	Bare Costs		Incl. Subs O&P		Cost Per Labor-Hour	
Crew B-7	**Hr.**	**Daily**	**Hr.**	**Daily**	**Bare Costs**	**Incl. O&P**
1 Labor Foreman (outside)	$37.10	$296.80	$60.80	$486.40	$37.34	$60.75
4 Laborers	35.10	1123.20	57.55	1841.60		
1 Equip. Oper. (med.)	46.55	372.40	73.50	588.00		
1 Brush Chipper, 12", 130 H.P.		283.80		312.18		
1 Crawler Loader, 3 C.Y.		1106.00		1216.60		
2 Chain Saws, gas, 36" Long		87.20		95.92	30.77	33.85
48 L.H., Daily Totals		$3269.40		$4540.70	$68.11	$94.60
Crew B-7A	**Hr.**	**Daily**	**Hr.**	**Daily**	**Bare Costs**	**Incl. O&P**
2 Laborers	$35.10	$561.60	$57.55	$920.80	$38.32	$61.92
1 Equip. Oper. (light)	44.75	358.00	70.65	565.20		
1 Rake w/Tractor		324.70		357.17		
2 Chain Saws, gas, 18"		54.80		60.28	15.81	17.39
24 L.H., Daily Totals		$1299.10		$1903.45	$54.13	$79.31
Crew B-7B	**Hr.**	**Daily**	**Hr.**	**Daily**	**Bare Costs**	**Incl. O&P**
1 Labor Foreman (outside)	$37.10	$296.80	$60.80	$486.40	$37.05	$60.23
4 Laborers	35.10	1123.20	57.55	1841.60		
1 Equip. Oper. (med.)	46.55	372.40	73.50	588.00		
1 Truck Driver (heavy)	35.30	282.40	57.10	456.80		
1 Brush Chipper, 12", 130 H.P.		283.80		312.18		
1 Crawler Loader, 3 C.Y.		1106.00		1216.60		
2 Chain Saws, gas, 36" Long		87.20		95.92		
1 Dump Truck, 8 C.Y., 220 H.P.		357.40		393.14	32.76	36.03
56 L.H., Daily Totals		$3909.20		$5390.64	$69.81	$96.26
Crew B-7C	**Hr.**	**Daily**	**Hr.**	**Daily**	**Bare Costs**	**Incl. O&P**
1 Labor Foreman (outside)	$37.10	$296.80	$60.80	$486.40	$37.05	$60.23
4 Laborers	35.10	1123.20	57.55	1841.60		
1 Equip. Oper. (med.)	46.55	372.40	73.50	588.00		
1 Truck Driver (heavy)	35.30	282.40	57.10	456.80		
1 Brush Chipper, 12", 130 H.P.		283.80		312.18		
1 Crawler Loader, 3 C.Y.		1106.00		1216.60		
2 Chain Saws, gas, 36" Long		87.20		95.92		
1 Dump Truck, 12 C.Y., 400 H.P.		584.60		643.06	36.81	40.50
56 L.H., Daily Totals		$4136.40		$5640.56	$73.86	$100.72
Crew B-8	**Hr.**	**Daily**	**Hr.**	**Daily**	**Bare Costs**	**Incl. O&P**
1 Labor Foreman (outside)	$37.10	$296.80	$60.80	$486.40	$39.03	$62.78
2 Laborers	35.10	561.60	57.55	920.80		
2 Equip. Oper. (med.)	46.55	744.80	73.50	1176.00		
1 Equip. Oper. Oiler	41.25	330.00	65.15	521.20		
2 Truck Drivers (heavy)	35.30	564.80	57.10	913.60		
1 Hyd. Crane, 25 Ton		697.20		766.92		
1 Crawler Loader, 3 C.Y.		1106.00		1216.60		
2 Dump Trucks, 12 C.Y., 400 H.P.		1169.20		1286.12	46.44	51.09
64 L.H., Daily Totals		$5470.40		$7287.64	$85.48	$113.87
Crew B-9	**Hr.**	**Daily**	**Hr.**	**Daily**	**Bare Costs**	**Incl. O&P**
1 Labor Foreman (outside)	$37.10	$296.80	$60.80	$486.40	$35.50	$58.20
4 Laborers	35.10	1123.20	57.55	1841.60		
1 Air Compressor, 250 cfm		177.00		194.70		
2 Breakers, Pavement, 60 lb.		19.60		21.56		
2 -50' Air Hoses, 1.5"		12.60		13.86	5.23	5.75
40 L.H., Daily Totals		$1629.20		$2558.12	$40.73	$63.95

Crew No.	Bare Costs		Incl. Subs O&P		Cost Per Labor-Hour	
Crew B-9A	**Hr.**	**Daily**	**Hr.**	**Daily**	**Bare Costs**	**Incl. O&P**
2 Laborers	$35.10	$561.60	$57.55	$920.80	$35.17	$57.40
1 Truck Driver (heavy)	35.30	282.40	57.10	456.80		
1 Water Tank Trailer, 5000 Gal.		141.00		155.10		
1 Truck Tractor, 220 H.P.		314.80		346.28		
2 -50' Discharge Hoses, 3"		6.10		6.71	19.25	21.17
24 L.H., Daily Totals		$1305.90		$1885.69	$54.41	$78.57
Crew B-9B	**Hr.**	**Daily**	**Hr.**	**Daily**	**Bare Costs**	**Incl. O&P**
2 Laborers	$35.10	$561.60	$57.55	$920.80	$35.17	$57.40
1 Truck Driver (heavy)	35.30	282.40	57.10	456.80		
2 -50' Discharge Hoses, 3"		6.10		6.71		
1 Water Tank Trailer, 5000 Gal.		141.00		155.10		
1 Truck Tractor, 220 H.P.		314.80		346.28		
1 Pressure Washer		63.00		69.30	21.87	24.06
24 L.H., Daily Totals		$1368.90		$1954.99	$57.04	$81.46
Crew B-9D	**Hr.**	**Daily**	**Hr.**	**Daily**	**Bare Costs**	**Incl. O&P**
1 Labor Foreman (Outside)	$37.10	$296.80	$60.80	$486.40	$35.50	$58.20
4 Common Laborers	35.10	1123.20	57.55	1841.60		
1 Air Compressor, 250 cfm		177.00		194.70		
2 -50' Air Hoses, 1.5"		12.60		13.86		
2 Air Powered Tampers		52.40		57.64	6.05	6.66
40 L.H., Daily Totals		$1662.00		$2594.20	$41.55	$64.86
Crew B-10	**Hr.**	**Daily**	**Hr.**	**Daily**	**Bare Costs**	**Incl. O&P**
1 Equip. Oper. (med.)	$46.55	$372.40	$73.50	$588.00	$42.73	$68.18
.5 Laborer	35.10	140.40	57.55	230.20		
12 L.H., Daily Totals		$512.80		$818.20	$42.73	$68.18
Crew B-10A	**Hr.**	**Daily**	**Hr.**	**Daily**	**Bare Costs**	**Incl. O&P**
1 Equip. Oper. (med.)	$46.55	$372.40	$73.50	$588.00	$42.73	$68.18
.5 Laborer	35.10	140.40	57.55	230.20		
1 Roller, 2-Drum, W.B., 7.5 H.P.		171.60		188.76	14.30	15.73
12 L.H., Daily Totals		$684.40		$1006.96	$57.03	$83.91
Crew B-10B	**Hr.**	**Daily**	**Hr.**	**Daily**	**Bare Costs**	**Incl. O&P**
1 Equip. Oper. (med.)	$46.55	$372.40	$73.50	$588.00	$42.73	$68.18
.5 Laborer	35.10	140.40	57.55	230.20		
1 Dozer, 200 H.P.		1243.00		1367.30	103.58	113.94
12 L.H., Daily Totals		$1755.80		$2185.50	$146.32	$182.13
Crew B-10C	**Hr.**	**Daily**	**Hr.**	**Daily**	**Bare Costs**	**Incl. O&P**
1 Equip. Oper. (med.)	$46.55	$372.40	$73.50	$588.00	$42.73	$68.18
.5 Laborer	35.10	140.40	57.55	230.20		
1 Dozer, 200 H.P.		1243.00		1367.30		
1 Vibratory Roller, Towed, 23 Ton		388.80		427.68	135.98	149.58
12 L.H., Daily Totals		$2144.60		$2613.18	$178.72	$217.76
Crew B-10D	**Hr.**	**Daily**	**Hr.**	**Daily**	**Bare Costs**	**Incl. O&P**
1 Equip. Oper. (med.)	$46.55	$372.40	$73.50	$588.00	$42.73	$68.18
.5 Laborer	35.10	140.40	57.55	230.20		
1 Dozer, 200 H.P.		1243.00		1367.30		
1 Sheepsft. Roller, Towed		420.20		462.22	138.60	152.46
12 L.H., Daily Totals		$2176.00		$2647.72	$181.33	$220.64
Crew B-10E	**Hr.**	**Daily**	**Hr.**	**Daily**	**Bare Costs**	**Incl. O&P**
1 Equip. Oper. (med.)	$46.55	$372.40	$73.50	$588.00	$42.73	$68.18
.5 Laborer	35.10	140.40	57.55	230.20		
1 Tandem Roller, 5 Ton		149.20		164.12	12.43	13.68
12 L.H., Daily Totals		$662.00		$982.32	$55.17	$81.86

Crew No.	Bare Costs		Incl. Subs O&P		Cost Per Labor-Hour	
Crew B-10F	**Hr.**	**Daily**	**Hr.**	**Daily**	**Bare Costs**	**Incl. O&P**
1 Equip. Oper. (med.)	$46.55	$372.40	$73.50	$588.00	$42.73	$68.18
.5 Laborer	35.10	140.40	57.55	230.20		
1 Tandem Roller, 10 Ton		227.60		250.36	18.97	20.86
12 L.H., Daily Totals		$740.40		$1068.56	$61.70	$89.05
Crew B-10G	**Hr.**	**Daily**	**Hr.**	**Daily**	**Bare Costs**	**Incl. O&P**
1 Equip. Oper. (med.)	$46.55	$372.40	$73.50	$588.00	$42.73	$68.18
.5 Laborer	35.10	140.40	57.55	230.20		
1 Sheepsft. Roll., 240 H.P.		1101.00		1211.10	91.75	100.93
12 L.H., Daily Totals		$1613.80		$2029.30	$134.48	$169.11
Crew B-10H	**Hr.**	**Daily**	**Hr.**	**Daily**	**Bare Costs**	**Incl. O&P**
1 Equip. Oper. (med.)	$46.55	$372.40	$73.50	$588.00	$42.73	$68.18
.5 Laborer	35.10	140.40	57.55	230.20		
1 Diaphragm Water Pump, 2"		68.60		75.46		
1 -20' Suction Hose, 2"		1.95		2.15		
2 -50' Discharge Hoses, 2"		3.50		3.85	6.17	6.79
12 L.H., Daily Totals		$586.85		$899.65	$48.90	$74.97
Crew B-10I	**Hr.**	**Daily**	**Hr.**	**Daily**	**Bare Costs**	**Incl. O&P**
1 Equip. Oper. (med.)	$46.55	$372.40	$73.50	$588.00	$42.73	$68.18
.5 Laborer	35.10	140.40	57.55	230.20		
1 Diaphragm Water Pump, 4"		111.20		122.32		
1 -20' Suction Hose, 4"		3.25		3.58		
2 -50' Discharge Hoses, 4"		9.40		10.34	10.32	11.35
12 L.H., Daily Totals		$636.65		$954.43	$53.05	$79.54
Crew B-10J	**Hr.**	**Daily**	**Hr.**	**Daily**	**Bare Costs**	**Incl. O&P**
1 Equip. Oper. (med.)	$46.55	$372.40	$73.50	$588.00	$42.73	$68.18
.5 Laborer	35.10	140.40	57.55	230.20		
1 Centrifugal Water Pump, 3"		73.80		81.18		
1 -20' Suction Hose, 3"		2.85		3.13		
2 -50' Discharge Hoses, 3"		6.10		6.71	6.90	7.59
12 L.H., Daily Totals		$595.55		$909.23	$49.63	$75.77
Crew B-10K	**Hr.**	**Daily**	**Hr.**	**Daily**	**Bare Costs**	**Incl. O&P**
1 Equip. Oper. (med.)	$46.55	$372.40	$73.50	$588.00	$42.73	$68.18
.5 Laborer	35.10	140.40	57.55	230.20		
1 Centr. Water Pump, 6"		308.40		339.24		
1 -20' Suction Hose, 6"		11.50		12.65		
2 -50' Discharge Hoses, 6"		24.20		26.62	28.68	31.54
12 L.H., Daily Totals		$856.90		$1196.71	$71.41	$99.73
Crew B-10L	**Hr.**	**Daily**	**Hr.**	**Daily**	**Bare Costs**	**Incl. O&P**
1 Equip. Oper. (med.)	$46.55	$372.40	$73.50	$588.00	$42.73	$68.18
.5 Laborer	35.10	140.40	57.55	230.20		
1 Dozer, 80 H.P.		442.40		486.64	36.87	40.55
12 L.H., Daily Totals		$955.20		$1304.84	$79.60	$108.74
Crew B-10M	**Hr.**	**Daily**	**Hr.**	**Daily**	**Bare Costs**	**Incl. O&P**
1 Equip. Oper. (med.)	$46.55	$372.40	$73.50	$588.00	$42.73	$68.18
.5 Laborer	35.10	140.40	57.55	230.20		
1 Dozer, 300 H.P.		1718.00		1889.80	143.17	157.48
12 L.H., Daily Totals		$2230.80		$2708.00	$185.90	$225.67
Crew B-10N	**Hr.**	**Daily**	**Hr.**	**Daily**	**Bare Costs**	**Incl. O&P**
1 Equip. Oper. (med.)	$46.55	$372.40	$73.50	$588.00	$42.73	$68.18
.5 Laborer	35.10	140.40	57.55	230.20		
1 F.E. Loader, T.M., 1.5 C.Y		488.60		537.46	40.72	44.79
12 L.H., Daily Totals		$1001.40		$1355.66	$83.45	$112.97
Crew B-10O	**Hr.**	**Daily**	**Hr.**	**Daily**	**Bare Costs**	**Incl. O&P**
1 Equip. Oper. (med.)	$46.55	$372.40	$73.50	$588.00	$42.73	$68.18
.5 Laborer	35.10	140.40	57.55	230.20		
1 F.E. Loader, T.M., 2.25 C.Y.		895.20		984.72	74.60	82.06
12 L.H., Daily Totals		$1408.00		$1802.92	$117.33	$150.24
Crew B-10P	**Hr.**	**Daily**	**Hr.**	**Daily**	**Bare Costs**	**Incl. O&P**
1 Equip. Oper. (med.)	$46.55	$372.40	$73.50	$588.00	$42.73	$68.18
.5 Laborer	35.10	140.40	57.55	230.20		
1 Crawler Loader, 3 C.Y.		1106.00		1216.60	92.17	101.38
12 L.H., Daily Totals		$1618.80		$2034.80	$134.90	$169.57
Crew B-10Q	**Hr.**	**Daily**	**Hr.**	**Daily**	**Bare Costs**	**Incl. O&P**
1 Equip. Oper. (med.)	$46.55	$372.40	$73.50	$588.00	$42.73	$68.18
.5 Laborer	35.10	140.40	57.55	230.20		
1 Crawler Loader, 4 C.Y.		1443.00		1587.30	120.25	132.28
12 L.H., Daily Totals		$1955.80		$2405.50	$162.98	$200.46
Crew B-10R	**Hr.**	**Daily**	**Hr.**	**Daily**	**Bare Costs**	**Incl. O&P**
1 Equip. Oper. (med.)	$46.55	$372.40	$73.50	$588.00	$42.73	$68.18
.5 Laborer	35.10	140.40	57.55	230.20		
1 F.E. Loader, W.M., 1 C.Y.		268.20		295.02	22.35	24.59
12 L.H., Daily Totals		$781.00		$1113.22	$65.08	$92.77
Crew B-10S	**Hr.**	**Daily**	**Hr.**	**Daily**	**Bare Costs**	**Incl. O&P**
1 Equip. Oper. (med.)	$46.55	$372.40	$73.50	$588.00	$42.73	$68.18
.5 Laborer	35.10	140.40	57.55	230.20		
1 F.E. Loader, W.M., 1.5 C.Y.		333.80		367.18	27.82	30.60
12 L.H., Daily Totals		$846.60		$1185.38	$70.55	$98.78
Crew B-10T	**Hr.**	**Daily**	**Hr.**	**Daily**	**Bare Costs**	**Incl. O&P**
1 Equip. Oper. (med.)	$46.55	$372.40	$73.50	$588.00	$42.73	$68.18
.5 Laborer	35.10	140.40	57.55	230.20		
1 F.E. Loader, W.M.,2.5 C.Y.		472.40		519.64	39.37	43.30
12 L.H., Daily Totals		$985.20		$1337.84	$82.10	$111.49
Crew B-10U	**Hr.**	**Daily**	**Hr.**	**Daily**	**Bare Costs**	**Incl. O&P**
1 Equip. Oper. (med.)	$46.55	$372.40	$73.50	$588.00	$42.73	$68.18
.5 Laborer	35.10	140.40	57.55	230.20		
1 F.E. Loader, W.M., 5.5 C.Y.		960.00		1056.00	80.00	88.00
12 L.H., Daily Totals		$1472.80		$1874.20	$122.73	$156.18
Crew B-10V	**Hr.**	**Daily**	**Hr.**	**Daily**	**Bare Costs**	**Incl. O&P**
1 Equip. Oper. (med.)	$46.55	$372.40	$73.50	$588.00	$42.73	$68.18
.5 Laborer	35.10	140.40	57.55	230.20		
1 Dozer, 700 H.P.		4235.00		4658.50	352.92	388.21
12 L.H., Daily Totals		$4747.80		$5476.70	$395.65	$456.39
Crew B-10W	**Hr.**	**Daily**	**Hr.**	**Daily**	**Bare Costs**	**Incl. O&P**
1 Equip. Oper. (med.)	$46.55	$372.40	$73.50	$588.00	$42.73	$68.18
.5 Laborer	35.10	140.40	57.55	230.20		
1 Dozer, 105 H.P.		550.80		605.88	45.90	50.49
12 L.H., Daily Totals		$1063.60		$1424.08	$88.63	$118.67
Crew B-10X	**Hr.**	**Daily**	**Hr.**	**Daily**	**Bare Costs**	**Incl. O&P**
1 Equip. Oper. (med.)	$46.55	$372.40	$73.50	$588.00	$42.73	$68.18
.5 Laborer	35.10	140.40	57.55	230.20		
1 Dozer, 410 H.P.		2272.00		2499.20	189.33	208.27
12 L.H., Daily Totals		$2784.80		$3317.40	$232.07	$276.45

Crew No.	Bare Costs		Incl. Subs O&P		Cost Per Labor-Hour	

Crew B-10Y	Hr.	Daily	Hr.	Daily	Bare Costs	Incl. O&P
1 Equip. Oper. (med.)	$46.55	$372.40	$73.50	$588.00	$42.73	$68.18
.5 Laborer	35.10	140.40	57.55	230.20		
1 Vibr. Roller, Towed, 12 Ton		514.80		566.28	42.90	47.19
12 L.H., Daily Totals		$1027.60		$1384.48	$85.63	$115.37

Crew B-11A	Hr.	Daily	Hr.	Daily	Bare Costs	Incl. O&P
1 Equipment Oper. (med.)	$46.55	$372.40	$73.50	$588.00	$40.83	$65.53
1 Laborer	35.10	280.80	57.55	460.40		
1 Dozer, 200 H.P.		1243.00		1367.30	77.69	85.46
16 L.H., Daily Totals		$1896.20		$2415.70	$118.51	$150.98

Crew B-11B	Hr.	Daily	Hr.	Daily	Bare Costs	Incl. O&P
1 Equipment Oper. (light)	$44.75	$358.00	$70.65	$565.20	$39.92	$64.10
1 Laborer	35.10	280.80	57.55	460.40		
1 Air Powered Tamper		26.20		28.82		
1 Air Compressor, 365 cfm		229.00		251.90		
2 -50' Air Hoses, 1.5"		12.60		13.86	16.74	18.41
16 L.H., Daily Totals		$906.60		$1320.18	$56.66	$82.51

Crew B-11C	Hr.	Daily	Hr.	Daily	Bare Costs	Incl. O&P
1 Equipment Oper. (med.)	$46.55	$372.40	$73.50	$588.00	$40.83	$65.53
1 Laborer	35.10	280.80	57.55	460.40		
1 Backhoe Loader, 48 H.P.		334.00		367.40	20.88	22.96
16 L.H., Daily Totals		$987.20		$1415.80	$61.70	$88.49

Crew B-11K	Hr.	Daily	Hr.	Daily	Bare Costs	Incl. O&P
1 Equipment Oper. (med.)	$46.55	$372.40	$73.50	$588.00	$40.83	$65.53
1 Laborer	35.10	280.80	57.55	460.40		
1 Trencher, Chain Type, 8' D		2848.00		3132.80	178.00	195.80
16 L.H., Daily Totals		$3501.20		$4181.20	$218.82	$261.32

Crew B-11L	Hr.	Daily	Hr.	Daily	Bare Costs	Incl. O&P
1 Equipment Oper. (med.)	$46.55	$372.40	$73.50	$588.00	$40.83	$65.53
1 Laborer	35.10	280.80	57.55	460.40		
1 Grader, 30,000 Lbs.		610.80		671.88	38.17	41.99
16 L.H., Daily Totals		$1264.00		$1720.28	$79.00	$107.52

Crew B-11M	Hr.	Daily	Hr.	Daily	Bare Costs	Incl. O&P
1 Equipment Oper. (med.)	$46.55	$372.40	$73.50	$588.00	$40.83	$65.53
1 Laborer	35.10	280.80	57.55	460.40		
1 Backhoe Loader, 80 H.P.		363.80		400.18	22.74	25.01
16 L.H., Daily Totals		$1017.00		$1448.58	$63.56	$90.54

Crew B-11S	Hr.	Daily	Hr.	Daily	Bare Costs	Incl. O&P
1 Equipment Operator (med.)	$46.55	$372.40	$73.50	$588.00	$42.73	$68.18
.5 Laborer	35.10	140.40	57.55	230.20		
1 Dozer, 300 H.P.		1718.00		1889.80		
1 Ripper, beam & 1 shank		81.60		89.76	149.97	164.96
12 L.H., Daily Totals		$2312.40		$2797.76	$192.70	$233.15

Crew B-11T	Hr.	Daily	Hr.	Daily	Bare Costs	Incl. O&P
1 Equipment Operator (med.)	$46.55	$372.40	$73.50	$588.00	$42.73	$68.18
.5 Laborer	35.10	140.40	57.55	230.20		
1 Dozer, 410 H.P.		2272.00		2499.20		
1 Ripper, beam & 2 shanks		91.80		100.98	196.98	216.68
12 L.H., Daily Totals		$2876.60		$3418.38	$239.72	$284.87

Crew B-11W	Hr.	Daily	Hr.	Daily	Bare Costs	Incl. O&P
1 Equipment Operator (med.)	$46.55	$372.40	$73.50	$588.00	$36.22	$58.50
1 Common Laborer	35.10	280.80	57.55	460.40		
10 Truck Drivers (hvy.)	35.30	2824.00	57.10	4568.00		
1 Dozer, 200 H.P.		1243.00		1367.30		
1 Vibratory Roller, Towed, 23 Ton		388.80		427.68		
10 Dump Trucks, 8 C.Y., 220 H.P.		3574.00		3931.40	54.23	59.65
96 L.H., Daily Totals		$8683.00		$11342.78	$90.45	$118.15

Crew B-11Y	Hr.	Daily	Hr.	Daily	Bare Costs	Incl. O&P
1 Labor Foreman (Outside)	$37.10	$296.80	$60.80	$486.40	$39.14	$63.23
5 Common Laborers	35.10	1404.00	57.55	2302.00		
3 Equipment Operators (med.)	46.55	1117.20	73.50	1764.00		
1 Dozer, 80 H.P.		442.40		486.64		
2 Roller, 2-Drum, W.B., 7.5 H.P.		343.20		377.52		
4 Vibratory Plates, gas, 21"		163.20		179.52	13.18	14.50
72 L.H., Daily Totals		$3766.80		$5596.08	$52.32	$77.72

Crew B-12A	Hr.	Daily	Hr.	Daily	Bare Costs	Incl. O&P
1 Equip. Oper. (crane)	$47.80	$382.40	$75.50	$604.00	$41.45	$66.53
1 Laborer	35.10	280.80	57.55	460.40		
1 Hyd. Excavator, 1 C.Y.		748.60		823.46	46.79	51.47
16 L.H., Daily Totals		$1411.80		$1887.86	$88.24	$117.99

Crew B-12B	Hr.	Daily	Hr.	Daily	Bare Costs	Incl. O&P
1 Equip. Oper. (crane)	$47.80	$382.40	$75.50	$604.00	$41.45	$66.53
1 Laborer	35.10	280.80	57.55	460.40		
1 Hyd. Excavator, 1.5 C.Y.		960.40		1056.44	60.02	66.03
16 L.H., Daily Totals		$1623.60		$2120.84	$101.47	$132.55

Crew B-12C	Hr.	Daily	Hr.	Daily	Bare Costs	Incl. O&P
1 Equip. Oper. (crane)	$47.80	$382.40	$75.50	$604.00	$41.45	$66.53
1 Laborer	35.10	280.80	57.55	460.40		
1 Hyd. Excavator, 2 C.Y.		1253.00		1378.30	78.31	86.14
16 L.H., Daily Totals		$1916.20		$2442.70	$119.76	$152.67

Crew B-12D	Hr.	Daily	Hr.	Daily	Bare Costs	Incl. O&P
1 Equip. Oper. (crane)	$47.80	$382.40	$75.50	$604.00	$41.45	$66.53
1 Laborer	35.10	280.80	57.55	460.40		
1 Hyd. Excavator, 3.5 C.Y.		2211.00		2432.10	138.19	152.01
16 L.H., Daily Totals		$2874.20		$3496.50	$179.64	$218.53

Crew B-12E	Hr.	Daily	Hr.	Daily	Bare Costs	Incl. O&P
1 Equip. Oper. (crane)	$47.80	$382.40	$75.50	$604.00	$41.45	$66.53
1 Laborer	35.10	280.80	57.55	460.40		
1 Hyd. Excavator, .5 C.Y.		371.40		408.54	23.21	25.53
16 L.H., Daily Totals		$1034.60		$1472.94	$64.66	$92.06

Crew B-12F	Hr.	Daily	Hr.	Daily	Bare Costs	Incl. O&P
1 Equip. Oper. (crane)	$47.80	$382.40	$75.50	$604.00	$41.45	$66.53
1 Laborer	35.10	280.80	57.55	460.40		
1 Hyd. Excavator, .75 C.Y.		647.20		711.92	40.45	44.49
16 L.H., Daily Totals		$1310.40		$1776.32	$81.90	$111.02

Crew B-12G	Hr.	Daily	Hr.	Daily	Bare Costs	Incl. O&P
1 Equip. Oper. (crane)	$47.80	$382.40	$75.50	$604.00	$41.45	$66.53
1 Laborer	35.10	280.80	57.55	460.40		
1 Crawler Crane, 15 Ton		647.50		712.25		
1 Clamshell Bucket, .5 C.Y.		37.40		41.14	42.81	47.09
16 L.H., Daily Totals		$1348.10		$1817.79	$84.26	$113.61

Crews

Crew No.	Bare Costs		Incl. Subs O&P		Cost Per Labor-Hour	
Crew B-12H	**Hr.**	**Daily**	**Hr.**	**Daily**	**Bare Costs**	**Incl. O&P**
1 Equip. Oper. (crane)	$47.80	$382.40	$75.50	$604.00	$41.45	$66.53
1 Laborer	35.10	280.80	57.55	460.40		
1 Crawler Crane, 25 Ton		1110.00		1221.00		
1 Clamshell Bucket, 1 C.Y.		47.00		51.70	72.31	79.54
16 L.H., Daily Totals		$1820.20		$2337.10	$113.76	$146.07

Crew B-12I	Hr.	Daily	Hr.	Daily	Bare Costs	Incl. O&P
1 Equip. Oper. (crane)	$47.80	$382.40	$75.50	$604.00	$41.45	$66.53
1 Laborer	35.10	280.80	57.55	460.40		
1 Crawler Crane, 20 Ton		832.00		915.20		
1 Dragline Bucket, .75 C.Y.		20.40		22.44	53.27	58.60
16 L.H., Daily Totals		$1515.60		$2002.04	$94.72	$125.13

Crew B-12J	Hr.	Daily	Hr.	Daily	Bare Costs	Incl. O&P
1 Equip. Oper. (crane)	$47.80	$382.40	$75.50	$604.00	$41.45	$66.53
1 Laborer	35.10	280.80	57.55	460.40		
1 Gradall, 5/8 C.Y.		844.00		928.40	52.75	58.02
16 L.H., Daily Totals		$1507.20		$1992.80	$94.20	$124.55

Crew B-12K	Hr.	Daily	Hr.	Daily	Bare Costs	Incl. O&P
1 Equip. Oper. (crane)	$47.80	$382.40	$75.50	$604.00	$41.45	$66.53
1 Laborer	35.10	280.80	57.55	460.40		
1 Gradall, 3 Ton, 1 C.Y.		961.20		1057.32	60.08	66.08
16 L.H., Daily Totals		$1624.40		$2121.72	$101.53	$132.61

Crew B-12L	Hr.	Daily	Hr.	Daily	Bare Costs	Incl. O&P
1 Equip. Oper. (crane)	$47.80	$382.40	$75.50	$604.00	$41.45	$66.53
1 Laborer	35.10	280.80	57.55	460.40		
1 Crawler Crane, 15 Ton		647.50		712.25		
1 F.E. Attachment, .5 C.Y.		57.40		63.14	44.06	48.46
16 L.H., Daily Totals		$1368.10		$1839.79	$85.51	$114.99

Crew B-12M	Hr.	Daily	Hr.	Daily	Bare Costs	Incl. O&P
1 Equip. Oper. (crane)	$47.80	$382.40	$75.50	$604.00	$41.45	$66.53
1 Laborer	35.10	280.80	57.55	460.40		
1 Crawler Crane, 20 Ton		832.00		915.20		
1 F.E. Attachment, .75 C.Y.		62.20		68.42	55.89	61.48
16 L.H., Daily Totals		$1557.40		$2048.02	$97.34	$128.00

Crew B-12N	Hr.	Daily	Hr.	Daily	Bare Costs	Incl. O&P
1 Equip. Oper. (crane)	$47.80	$382.40	$75.50	$604.00	$41.45	$66.53
1 Laborer	35.10	280.80	57.55	460.40		
1 Crawler Crane, 25 Ton		1110.00		1221.00		
1 F.E. Attachment, 1 C.Y.		68.00		74.80	73.63	80.99
16 L.H., Daily Totals		$1841.20		$2360.20	$115.08	$147.51

Crew B-12O	Hr.	Daily	Hr.	Daily	Bare Costs	Incl. O&P
1 Equip. Oper. (crane)	$47.80	$382.40	$75.50	$604.00	$41.45	$66.53
1 Laborer	35.10	280.80	57.55	460.40		
1 Crawler Crane, 40 Ton		1115.00		1226.50		
1 F.E. Attachment, 1.5 C.Y.		76.80		84.48	74.49	81.94
16 L.H., Daily Totals		$1855.00		$2375.38	$115.94	$148.46

Crew B-12P	Hr.	Daily	Hr.	Daily	Bare Costs	Incl. O&P
1 Equip. Oper. (crane)	$47.80	$382.40	$75.50	$604.00	$41.45	$66.53
1 Laborer	35.10	280.80	57.55	460.40		
1 Crawler Crane, 40 Ton		1115.00		1226.50		
1 Dragline Bucket, 1.5 C.Y.		33.20		36.52	71.76	78.94
16 L.H., Daily Totals		$1811.40		$2327.42	$113.21	$145.46

Crew B-12Q	Hr.	Daily	Hr.	Daily	Bare Costs	Incl. O&P
1 Equip. Oper. (crane)	$47.80	$382.40	$75.50	$604.00	$41.45	$66.53
1 Laborer	35.10	280.80	57.55	460.40		
1 Hyd. Excavator, 5/8 C.Y.		551.80		606.98	34.49	37.94
16 L.H., Daily Totals		$1215.00		$1671.38	$75.94	$104.46

Crew B-12S	Hr.	Daily	Hr.	Daily	Bare Costs	Incl. O&P
1 Equip. Oper. (crane)	$47.80	$382.40	$75.50	$604.00	$41.45	$66.53
1 Laborer	35.10	280.80	57.55	460.40		
1 Hyd. Excavator, 2.5 C.Y.		1652.00		1817.20	103.25	113.58
16 L.H., Daily Totals		$2315.20		$2881.60	$144.70	$180.10

Crew B-12T	Hr.	Daily	Hr.	Daily	Bare Costs	Incl. O&P
1 Equip. Oper. (crane)	$47.80	$382.40	$75.50	$604.00	$41.45	$66.53
1 Laborer	35.10	280.80	57.55	460.40		
1 Crawler Crane, 75 Ton		1423.00		1565.30		
1 F.E. Attachment, 3 C.Y.		99.60		109.56	95.16	104.68
16 L.H., Daily Totals		$2185.80		$2739.26	$136.61	$171.20

Crew B-12V	Hr.	Daily	Hr.	Daily	Bare Costs	Incl. O&P
1 Equip. Oper. (crane)	$47.80	$382.40	$75.50	$604.00	$41.45	$66.53
1 Laborer	35.10	280.80	57.55	460.40		
1 Crawler Crane, 75 Ton		1423.00		1565.30		
1 Dragline Bucket, 3 C.Y.		51.80		56.98	92.17	101.39
16 L.H., Daily Totals		$2138.00		$2686.68	$133.63	$167.92

Crew B-12Y	Hr.	Daily	Hr.	Daily	Bare Costs	Incl. O&P
1 Equip. Oper. (crane)	$47.80	$382.40	$75.50	$604.00	$39.33	$63.53
2 Laborers	35.10	561.60	57.55	920.80		
1 Hyd. Excavator, 3.5 C.Y.		2211.00		2432.10	92.13	101.34
24 L.H., Daily Totals		$3155.00		$3956.90	$131.46	$164.87

Crew B-12Z	Hr.	Daily	Hr.	Daily	Bare Costs	Incl. O&P
1 Equip. Oper. (crane)	$47.80	$382.40	$75.50	$604.00	$39.33	$63.53
2 Laborers	35.10	561.60	57.55	920.80		
1 Hyd. Excavator, 2.5 C.Y.		1652.00		1817.20	68.83	75.72
24 L.H., Daily Totals		$2596.00		$3342.00	$108.17	$139.25

Crew B-13	Hr.	Daily	Hr.	Daily	Bare Costs	Incl. O&P
1 Labor Foreman (outside)	$37.10	$296.80	$60.80	$486.40	$38.08	$61.66
4 Laborers	35.10	1123.20	57.55	1841.60		
1 Equip. Oper. (crane)	47.80	382.40	75.50	604.00		
1 Equip. Oper. Oiler	41.25	330.00	65.15	521.20		
1 Hyd. Crane, 25 Ton		697.20		766.92	12.45	13.70
56 L.H., Daily Totals		$2829.60		$4220.12	$50.53	$75.36

Crew B-13A	Hr.	Daily	Hr.	Daily	Bare Costs	Incl. O&P
1 Labor Foreman (outside)	$37.10	$296.80	$60.80	$486.40	$38.71	$62.44
2 Laborers	35.10	561.60	57.55	920.80		
2 Equipment Operators (med.)	46.55	744.80	73.50	1176.00		
2 Truck Drivers (heavy)	35.30	564.80	57.10	913.60		
1 Crawler Crane, 75 Ton		1423.00		1565.30		
1 Crawler Loader, 4 C.Y.		1443.00		1587.30		
2 Dump Trucks, 8 C.Y., 220 H.P.		714.80		786.28	63.94	70.34
56 L.H., Daily Totals		$5748.80		$7435.68	$102.66	$132.78

Crew No.	Bare Costs		Incl. Subs O&P		Cost Per Labor-Hour	
Crew B-13B	**Hr.**	**Daily**	**Hr.**	**Daily**	**Bare Costs**	**Incl. O&P**
1 Labor Foreman (outside)	$37.10	$296.80	$60.80	$486.40	$38.08	$61.66
4 Laborers	35.10	1123.20	57.55	1841.60		
1 Equip. Oper. (crane)	47.80	382.40	75.50	604.00		
1 Equip. Oper. Oiler	41.25	330.00	65.15	521.20		
1 Hyd. Crane, 55 Ton		1100.00		1210.00	19.64	21.61
56 L.H., Daily Totals		$3232.40		$4663.20	$57.72	$83.27
Crew B-13C	**Hr.**	**Daily**	**Hr.**	**Daily**	**Bare Costs**	**Incl. O&P**
1 Labor Foreman (outside)	$37.10	$296.80	$60.80	$486.40	$38.08	$61.66
4 Laborers	35.10	1123.20	57.55	1841.60		
1 Equip. Oper. (crane)	47.80	382.40	75.50	604.00		
1 Equip. Oper. Oiler	41.25	330.00	65.15	521.20		
1 Crawler Crane, 100 Ton		1628.00		1790.80	29.07	31.98
56 L.H., Daily Totals		$3760.40		$5244.00	$67.15	$93.64
Crew B-13D	**Hr.**	**Daily**	**Hr.**	**Daily**	**Bare Costs**	**Incl. O&P**
1 Laborer	$35.10	$280.80	$57.55	$460.40	$41.45	$66.53
1 Equip. Oper. (crane)	47.80	382.40	75.50	604.00		
1 Hyd. Excavator, 1 C.Y.		748.60		823.46		
1 Trench Box		113.40		124.74	53.88	59.26
16 L.H., Daily Totals		$1525.20		$2012.60	$95.33	$125.79
Crew B-13E	**Hr.**	**Daily**	**Hr.**	**Daily**	**Bare Costs**	**Incl. O&P**
1 Laborer	$35.10	$280.80	$57.55	$460.40	$41.45	$66.53
1 Equip. Oper. (crane)	47.80	382.40	75.50	604.00		
1 Hyd. Excavator, 1.5 C.Y.		960.40		1056.44		
1 Trench Box		113.40		124.74	67.11	73.82
16 L.H., Daily Totals		$1737.00		$2245.58	$108.56	$140.35
Crew B-13F	**Hr.**	**Daily**	**Hr.**	**Daily**	**Bare Costs**	**Incl. O&P**
1 Laborer	$35.10	$280.80	$57.55	$460.40	$41.45	$66.53
1 Equip. Oper. (crane)	47.80	382.40	75.50	604.00		
1 Hyd. Excavator, 3.5 C.Y.		2211.00		2432.10		
1 Trench Box		113.40		124.74	145.28	159.80
16 L.H., Daily Totals		$2987.60		$3621.24	$186.72	$226.33
Crew B-13G	**Hr.**	**Daily**	**Hr.**	**Daily**	**Bare Costs**	**Incl. O&P**
1 Laborer	$35.10	$280.80	$57.55	$460.40	$41.45	$66.53
1 Equip. Oper. (crane)	47.80	382.40	75.50	604.00		
1 Hyd. Excavator, .75 C.Y.		647.20		711.92		
1 Trench Box		113.40		124.74	47.54	52.29
16 L.H., Daily Totals		$1423.80		$1901.06	$88.99	$118.82
Crew B-13H	**Hr.**	**Daily**	**Hr.**	**Daily**	**Bare Costs**	**Incl. O&P**
1 Laborer	$35.10	$280.80	$57.55	$460.40	$41.45	$66.53
1 Equip. Oper. (crane)	47.80	382.40	75.50	604.00		
1 Gradall, 5/8 C.Y.		844.00		928.40		
1 Trench Box		113.40		124.74	59.84	65.82
16 L.H., Daily Totals		$1620.60		$2117.54	$101.29	$132.35
Crew B-13I	**Hr.**	**Daily**	**Hr.**	**Daily**	**Bare Costs**	**Incl. O&P**
1 Laborer	$35.10	$280.80	$57.55	$460.40	$41.45	$66.53
1 Equip. Oper. (crane)	47.80	382.40	75.50	604.00		
1 Gradall, 3 Ton, 1 C.Y.		961.20		1057.32		
1 Trench Box		113.40		124.74	67.16	73.88
16 L.H., Daily Totals		$1737.80		$2246.46	$108.61	$140.40

Crew No.	Bare Costs		Incl. Subs O&P		Cost Per Labor-Hour	
Crew B-13J	**Hr.**	**Daily**	**Hr.**	**Daily**	**Bare Costs**	**Incl. O&P**
1 Laborer	$35.10	$280.80	$57.55	$460.40	$41.45	$66.53
1 Equip. Oper. (crane)	47.80	382.40	75.50	604.00		
1 Hyd. Excavator, 2.5 C.Y.		1652.00		1817.20		
1 Trench Box		113.40		124.74	110.34	121.37
16 L.H., Daily Totals		$2428.60		$3006.34	$151.79	$187.90
Crew B-14	**Hr.**	**Daily**	**Hr.**	**Daily**	**Bare Costs**	**Incl. O&P**
1 Labor Foreman (outside)	$37.10	$296.80	$60.80	$486.40	$37.04	$60.27
4 Laborers	35.10	1123.20	57.55	1841.60		
1 Equip. Oper. (light)	44.75	358.00	70.65	565.20		
1 Backhoe Loader, 48 H.P.		334.00		367.40	6.96	7.65
48 L.H., Daily Totals		$2112.00		$3260.60	$44.00	$67.93
Crew B-14A	**Hr.**	**Daily**	**Hr.**	**Daily**	**Bare Costs**	**Incl. O&P**
1 Equip. Oper. (crane)	$47.80	$382.40	$75.50	$604.00	$43.57	$69.52
.5 Laborer	35.10	140.40	57.55	230.20		
1 Hyd. Excavator, 4.5 C.Y.		2748.00		3022.80	229.00	251.90
12 L.H., Daily Totals		$3270.80		$3857.00	$272.57	$321.42
Crew B-14B	**Hr.**	**Daily**	**Hr.**	**Daily**	**Bare Costs**	**Incl. O&P**
1 Equip. Oper. (crane)	$47.80	$382.40	$75.50	$604.00	$43.57	$69.52
.5 Laborer	35.10	140.40	57.55	230.20		
1 Hyd. Excavator, 6 C.Y.		3229.00		3551.90	269.08	295.99
12 L.H., Daily Totals		$3751.80		$4386.10	$312.65	$365.51
Crew B-14C	**Hr.**	**Daily**	**Hr.**	**Daily**	**Bare Costs**	**Incl. O&P**
1 Equip. Oper. (crane)	$47.80	$382.40	$75.50	$604.00	$43.57	$69.52
.5 Laborer	35.10	140.40	57.55	230.20		
1 Hyd. Excavator, 7 C.Y.		3304.00		3634.40	275.33	302.87
12 L.H., Daily Totals		$3826.80		$4468.60	$318.90	$372.38
Crew B-14F	**Hr.**	**Daily**	**Hr.**	**Daily**	**Bare Costs**	**Incl. O&P**
1 Equip. Oper. (crane)	$47.80	$382.40	$75.50	$604.00	$43.57	$69.52
.5 Laborer	35.10	140.40	57.55	230.20		
1 Hyd. Shovel, 7 C.Y.		3259.00		3584.90	271.58	298.74
12 L.H., Daily Totals		$3781.80		$4419.10	$315.15	$368.26
Crew B-14G	**Hr.**	**Daily**	**Hr.**	**Daily**	**Bare Costs**	**Incl. O&P**
1 Equip. Oper. (crane)	$47.80	$382.40	$75.50	$604.00	$43.57	$69.52
.5 Laborer	35.10	140.40	57.55	230.20		
1 Hyd. Shovel, 12 C.Y.		4249.00		4673.90	354.08	389.49
12 L.H., Daily Totals		$4771.80		$5508.10	$397.65	$459.01
Crew B-14J	**Hr.**	**Daily**	**Hr.**	**Daily**	**Bare Costs**	**Incl. O&P**
1 Equip. Oper. (med.)	$46.55	$372.40	$73.50	$588.00	$42.73	$68.18
.5 Laborer	35.10	140.40	57.55	230.20		
1 F.E. Loader, 8 C.Y.		1856.00		2041.60	154.67	170.13
12 L.H., Daily Totals		$2368.80		$2859.80	$197.40	$238.32
Crew B-14K	**Hr.**	**Daily**	**Hr.**	**Daily**	**Bare Costs**	**Incl. O&P**
1 Equip. Oper. (med.)	$46.55	$372.40	$73.50	$588.00	$42.73	$68.18
.5 Laborer	35.10	140.40	57.55	230.20		
1 F.E. Loader, 10 C.Y.		2556.00		2811.60	213.00	234.30
12 L.H., Daily Totals		$3068.80		$3629.80	$255.73	$302.48

Crew No.	Bare Costs		Incl. Subs O&P		Cost Per Labor-Hour	
Crew B-15	**Hr.**	**Daily**	**Hr.**	**Daily**	**Bare Costs**	**Incl. O&P**
1 Equipment Oper. (med.)	$46.55	$372.40	$73.50	$588.00	$38.49	$61.85
.5 Laborer	35.10	140.40	57.55	230.20		
2 Truck Drivers (heavy)	35.30	564.80	57.10	913.60		
2 Dump Trucks, 12 C.Y., 400 H.P.		1169.20		1286.12		
1 Dozer, 200 H.P.		1243.00		1367.30	86.15	94.77
28 L.H., Daily Totals		$3489.80		$4385.22	$124.64	$156.62
Crew B-16	**Hr.**	**Daily**	**Hr.**	**Daily**	**Bare Costs**	**Incl. O&P**
1 Labor Foreman (outside)	$37.10	$296.80	$60.80	$486.40	$35.65	$58.25
2 Laborers	35.10	561.60	57.55	920.80		
1 Truck Driver (heavy)	35.30	282.40	57.10	456.80		
1 Dump Truck, 12 C.Y., 400 H.P.		584.60		643.06	18.27	20.10
32 L.H., Daily Totals		$1725.40		$2507.06	$53.92	$78.35
Crew B-17	**Hr.**	**Daily**	**Hr.**	**Daily**	**Bare Costs**	**Incl. O&P**
2 Laborers	$35.10	$561.60	$57.55	$920.80	$37.56	$60.71
1 Equip. Oper. (light)	44.75	358.00	70.65	565.20		
1 Truck Driver (heavy)	35.30	282.40	57.10	456.80		
1 Backhoe Loader, 48 H.P.		334.00		367.40		
1 Dump Truck, 8 C.Y., 220 H.P.		357.40		393.14	21.61	23.77
32 L.H., Daily Totals		$1893.40		$2703.34	$59.17	$84.48
Crew B-17A	**Hr.**	**Daily**	**Hr.**	**Daily**	**Bare Costs**	**Incl. O&P**
2 Labor Foremen (outside)	$37.10	$593.60	$60.80	$972.80	$37.76	$61.77
6 Laborers	35.10	1684.80	57.55	2762.40		
1 Skilled Worker Foreman (out)	47.40	379.20	77.05	616.40		
1 Skilled Worker	45.40	363.20	73.80	590.40		
80 L.H., Daily Totals		$3020.80		$4942.00	$37.76	$61.77
Crew B-17B	**Hr.**	**Daily**	**Hr.**	**Daily**	**Bare Costs**	**Incl. O&P**
2 Laborers	$35.10	$561.60	$57.55	$920.80	$37.56	$60.71
1 Equip. Oper. (light)	44.75	358.00	70.65	565.20		
1 Truck Driver (heavy)	35.30	282.40	57.10	456.80		
1 Backhoe Loader, 48 H.P.		334.00		367.40		
1 Dump Truck, 12 C.Y., 400 H.P.		584.60		643.06	28.71	31.58
32 L.H., Daily Totals		$2120.60		$2953.26	$66.27	$92.29
Crew B-18	**Hr.**	**Daily**	**Hr.**	**Daily**	**Bare Costs**	**Incl. O&P**
1 Labor Foreman (outside)	$37.10	$296.80	$60.80	$486.40	$35.77	$58.63
2 Laborers	35.10	561.60	57.55	920.80		
1 Vibratory Plate, gas, 21"		40.80		44.88	1.70	1.87
24 L.H., Daily Totals		$899.20		$1452.08	$37.47	$60.50
Crew B-19	**Hr.**	**Daily**	**Hr.**	**Daily**	**Bare Costs**	**Incl. O&P**
1 Pile Driver Foreman (outside)	$44.80	$358.40	$74.25	$594.00	$44.11	$71.75
4 Pile Drivers	42.80	1369.60	70.90	2268.80		
2 Equip. Oper. (crane)	47.80	764.80	75.50	1208.00		
1 Equip. Oper. Oiler	41.25	330.00	65.15	521.20		
1 Crawler Crane, 40 Ton		1115.00		1226.50		
1 Lead, 90' high		124.00		136.40		
1 Hammer, Diesel, 22k ft-lb		592.80		652.08	28.62	31.48
64 L.H., Daily Totals		$4654.60		$6606.98	$72.73	$103.23

Crew No.	Bare Costs		Incl. Subs O&P		Cost Per Labor-Hour	
Crew B-19A	**Hr.**	**Daily**	**Hr.**	**Daily**	**Bare Costs**	**Incl. O&P**
1 Pile Driver Foreman (outside)	$44.80	$358.40	$74.25	$594.00	$44.11	$71.75
4 Pile Drivers	42.80	1369.60	70.90	2268.80		
2 Equip. Oper. (crane)	47.80	764.80	75.50	1208.00		
1 Equip. Oper. Oiler	41.25	330.00	65.15	521.20		
1 Crawler Crane, 75 Ton		1423.00		1565.30		
1 Lead, 90' high		124.00		136.40		
1 Hammer, Diesel, 41k ft-lb		686.80		755.48	34.90	38.39
64 L.H., Daily Totals		$5056.60		$7049.18	$79.01	$110.14
Crew B-19B	**Hr.**	**Daily**	**Hr.**	**Daily**	**Bare Costs**	**Incl. O&P**
1 Pile Driver Foreman (outside)	$44.80	$358.40	$74.25	$594.00	$44.11	$71.75
4 Pile Drivers	42.80	1369.60	70.90	2268.80		
2 Equip. Oper. (crane)	47.80	764.80	75.50	1208.00		
1 Equip. Oper. Oiler	41.25	330.00	65.15	521.20		
1 Crawler Crane, 40 Ton		1115.00		1226.50		
1 Lead, 90' high		124.00		136.40		
1 Hammer, Diesel, 22k ft-lb		592.80		652.08		
1 Barge, 400 Ton		779.20		857.12	40.80	44.88
64 L.H., Daily Totals		$5433.80		$7464.10	$84.90	$116.63
Crew B-19C	**Hr.**	**Daily**	**Hr.**	**Daily**	**Bare Costs**	**Incl. O&P**
1 Pile Driver Foreman (outside)	$44.80	$358.40	$74.25	$594.00	$44.11	$71.75
4 Pile Drivers	42.80	1369.60	70.90	2268.80		
2 Equip. Oper. (crane)	47.80	764.80	75.50	1208.00		
1 Equip. Oper. Oiler	41.25	330.00	65.15	521.20		
1 Crawler Crane, 75 Ton		1423.00		1565.30		
1 Lead, 90' high		124.00		136.40		
1 Hammer, Diesel, 41k ft-lb		686.80		755.48		
1 Barge, 400 Ton		779.20		857.12	47.08	51.79
64 L.H., Daily Totals		$5835.80		$7906.30	$91.18	$123.54
Crew B-20	**Hr.**	**Daily**	**Hr.**	**Daily**	**Bare Costs**	**Incl. O&P**
1 Labor Foreman (outside)	$37.10	$296.80	$60.80	$486.40	$39.20	$64.05
1 Skilled Worker	45.40	363.20	73.80	590.40		
1 Laborer	35.10	280.80	57.55	460.40		
24 L.H., Daily Totals		$940.80		$1537.20	$39.20	$64.05
Crew B-20A	**Hr.**	**Daily**	**Hr.**	**Daily**	**Bare Costs**	**Incl. O&P**
1 Labor Foreman (outside)	$37.10	$296.80	$60.80	$486.40	$42.63	$67.88
1 Laborer	35.10	280.80	57.55	460.40		
1 Plumber	54.60	436.80	85.05	680.40		
1 Plumber Apprentice	43.70	349.60	68.10	544.80		
32 L.H., Daily Totals		$1364.00		$2172.00	$42.63	$67.88
Crew B-21	**Hr.**	**Daily**	**Hr.**	**Daily**	**Bare Costs**	**Incl. O&P**
1 Labor Foreman (outside)	$37.10	$296.80	$60.80	$486.40	$40.43	$65.69
1 Skilled Worker	45.40	363.20	73.80	590.40		
1 Laborer	35.10	280.80	57.55	460.40		
.5 Equip. Oper. (crane)	47.80	191.20	75.50	302.00		
.5 S.P. Crane, 4x4, 5 Ton		132.30		145.53	4.72	5.20
28 L.H., Daily Totals		$1264.30		$1984.73	$45.15	$70.88
Crew B-21A	**Hr.**	**Daily**	**Hr.**	**Daily**	**Bare Costs**	**Incl. O&P**
1 Labor Foreman (outside)	$37.10	$296.80	$60.80	$486.40	$43.66	$69.40
1 Laborer	35.10	280.80	57.55	460.40		
1 Plumber	54.60	436.80	85.05	680.40		
1 Plumber Apprentice	43.70	349.60	68.10	544.80		
1 Equip. Oper. (crane)	47.80	382.40	75.50	604.00		
1 S.P. Crane, 4x4, 12 Ton		448.20		493.02	11.21	12.33
40 L.H., Daily Totals		$2194.60		$3269.02	$54.87	$81.73

Crew B-21B	Bare Costs Hr.	Bare Costs Daily	Incl. Subs O&P Hr.	Incl. Subs O&P Daily	Cost Per Labor-Hour Bare Costs	Cost Per Labor-Hour Incl. O&P
1 Labor Foreman (outside)	$37.10	$296.80	$60.80	$486.40	$38.04	$61.79
3 Laborers	35.10	842.40	57.55	1381.20		
1 Equip. Oper. (crane)	47.80	382.40	75.50	604.00		
1 Hyd. Crane, 12 Ton		612.00		673.20	15.30	16.83
40 L.H., Daily Totals		$2133.60		$3144.80	$53.34	$78.62

Crew B-21C	Hr.	Daily	Hr.	Daily	Bare Costs	Incl. O&P
1 Labor Foreman (outside)	$37.10	$296.80	$60.80	$486.40	$38.08	$61.66
4 Laborers	35.10	1123.20	57.55	1841.60		
1 Equip. Oper. (crane)	47.80	382.40	75.50	604.00		
1 Equip. Oper. Oiler	41.25	330.00	65.15	521.20		
2 Cutting Torches		21.60		23.76		
2 Sets of Gases		172.80		190.08		
1 Lattice Boom Crane, 90 Ton		1497.00		1646.70	30.20	33.22
56 L.H., Daily Totals		$3823.80		$5313.74	$68.28	$94.89

Crew B-22	Hr.	Daily	Hr.	Daily	Bare Costs	Incl. O&P
1 Labor Foreman (outside)	$37.10	$296.80	$60.80	$486.40	$40.92	$66.34
1 Skilled Worker	45.40	363.20	73.80	590.40		
1 Laborer	35.10	280.80	57.55	460.40		
.75 Equip. Oper. (crane)	47.80	286.80	75.50	453.00		
.75 S.P. Crane, 4x4, 5 Ton		198.45		218.29	6.62	7.28
30 L.H., Daily Totals		$1426.05		$2208.49	$47.53	$73.62

Crew B-22A	Hr.	Daily	Hr.	Daily	Bare Costs	Incl. O&P
1 Labor Foreman (outside)	$37.10	$296.80	$60.80	$486.40	$40.10	$65.04
1 Skilled Worker	45.40	363.20	73.80	590.40		
2 Laborers	35.10	561.60	57.55	920.80		
1 Equipment Oper. (crane)	47.80	382.40	75.50	604.00		
1 S.P. Crane, 4x4, 5 Ton		264.60		291.06		
1 Butt Fusion Mach., 4"-12" diam.		376.45		414.10	16.03	17.63
40 L.H., Daily Totals		$2245.05		$3306.76	$56.13	$82.67

Crew B-22B	Hr.	Daily	Hr.	Daily	Bare Costs	Incl. O&P
1 Labor Foreman (outside)	$37.10	$296.80	$60.80	$486.40	$40.10	$65.04
1 Skilled Worker	45.40	363.20	73.80	590.40		
2 Laborers	35.10	561.60	57.55	920.80		
1 Equip. Oper. (crane)	47.80	382.40	75.50	604.00		
1 S.P. Crane, 4x4, 5 Ton		264.60		291.06		
1 Butt Fusion Mach., 8"-24" diam.		806.15		886.76	26.77	29.45
40 L.H., Daily Totals		$2674.75		$3779.43	$66.87	$94.49

Crew B-22C	Hr.	Daily	Hr.	Daily	Bare Costs	Incl. O&P
1 Skilled Worker	$45.40	$363.20	$73.80	$590.40	$40.25	$65.67
1 Laborer	35.10	280.80	57.55	460.40		
1 Butt Fusion Mach., 2"-8" diam.		120.50		132.55	7.53	8.28
16 L.H., Daily Totals		$764.50		$1183.35	$47.78	$73.96

Crew B-23	Hr.	Daily	Hr.	Daily	Bare Costs	Incl. O&P
1 Labor Foreman (outside)	$37.10	$296.80	$60.80	$486.40	$35.50	$58.20
4 Laborers	35.10	1123.20	57.55	1841.60		
1 Drill Rig, Truck-Mounted		2441.00		2685.10		
1 Flatbed Truck, Gas, 3 Ton		264.40		290.84	67.64	74.40
40 L.H., Daily Totals		$4125.40		$5303.94	$103.14	$132.60

Crew B-23A	Hr.	Daily	Hr.	Daily	Bare Costs	Incl. O&P
1 Labor Foreman (outside)	$37.10	$296.80	$60.80	$486.40	$39.58	$63.95
1 Laborer	35.10	280.80	57.55	460.40		
1 Equip. Operator (med.)	46.55	372.40	73.50	588.00		
1 Drill Rig, Truck-Mounted		2441.00		2685.10		
1 Pickup Truck, 3/4 Ton		127.00		139.70	107.00	117.70
24 L.H., Daily Totals		$3518.00		$4359.60	$146.58	$181.65

Crew B-23B	Hr.	Daily	Hr.	Daily	Bare Costs	Incl. O&P
1 Labor Foreman (outside)	$37.10	$296.80	$60.80	$486.40	$39.58	$63.95
1 Laborer	35.10	280.80	57.55	460.40		
1 Equip. Operator (med.)	46.55	372.40	73.50	588.00		
1 Drill Rig, Truck-Mounted		2441.00		2685.10		
1 Pickup Truck, 3/4 Ton		127.00		139.70		
1 Centr. Water Pump, 6"		308.40		339.24	119.85	131.84
24 L.H., Daily Totals		$3826.40		$4698.84	$159.43	$195.79

Crew B-24	Hr.	Daily	Hr.	Daily	Bare Costs	Incl. O&P
1 Cement Finisher	$42.35	$338.80	$66.55	$532.40	$40.52	$65.47
1 Laborer	35.10	280.80	57.55	460.40		
1 Carpenter	44.10	352.80	72.30	578.40		
24 L.H., Daily Totals		$972.40		$1571.20	$40.52	$65.47

Crew B-25	Hr.	Daily	Hr.	Daily	Bare Costs	Incl. O&P
1 Labor Foreman (outside)	$37.10	$296.80	$60.80	$486.40	$38.40	$62.20
7 Laborers	35.10	1965.60	57.55	3222.80		
3 Equip. Oper. (med.)	46.55	1117.20	73.50	1764.00		
1 Asphalt Paver, 130 H.P.		1996.00		2195.60		
1 Tandem Roller, 10 Ton		227.60		250.36		
1 Roller, Pneum. Whl, 12 Ton		313.20		344.52	28.83	31.71
88 L.H., Daily Totals		$5916.40		$8263.68	$67.23	$93.91

Crew B-25B	Hr.	Daily	Hr.	Daily	Bare Costs	Incl. O&P
1 Labor Foreman (outside)	$37.10	$296.80	$60.80	$486.40	$39.08	$63.14
7 Laborers	35.10	1965.60	57.55	3222.80		
4 Equip. Oper. (med.)	46.55	1489.60	73.50	2352.00		
1 Asphalt Paver, 130 H.P.		1996.00		2195.60		
2 Tandem Rollers, 10 Ton		455.20		500.72		
1 Roller, Pneum. Whl, 12 Ton		313.20		344.52	28.80	31.68
96 L.H., Daily Totals		$6516.40		$9102.04	$67.88	$94.81

Crew B-25C	Hr.	Daily	Hr.	Daily	Bare Costs	Incl. O&P
1 Labor Foreman (outside)	$37.10	$296.80	$60.80	$486.40	$39.25	$63.41
3 Laborers	35.10	842.40	57.55	1381.20		
2 Equip. Oper. (med.)	46.55	744.80	73.50	1176.00		
1 Asphalt Paver, 130 H.P.		1996.00		2195.60		
1 Tandem Roller, 10 Ton		227.60		250.36	46.33	50.96
48 L.H., Daily Totals		$4107.60		$5489.56	$85.58	$114.37

Crew B-26	Hr.	Daily	Hr.	Daily	Bare Costs	Incl. O&P
1 Labor Foreman (outside)	$37.10	$296.80	$60.80	$486.40	$39.30	$63.77
6 Laborers	35.10	1684.80	57.55	2762.40		
2 Equip. Oper. (med.)	46.55	744.80	73.50	1176.00		
1 Rodman (reinf.)	49.10	392.80	81.85	654.80		
1 Cement Finisher	42.35	338.80	66.55	532.40		
1 Grader, 30,000 Lbs.		610.80		671.88		
1 Paving Mach. & Equip.		2531.00		2784.10	35.70	39.27
88 L.H., Daily Totals		$6599.80		$9067.98	$75.00	$103.05

Crew No.	Bare Costs		Incl. Subs O&P		Cost Per Labor-Hour	
Crew B-26A	**Hr.**	**Daily**	**Hr.**	**Daily**	**Bare Costs**	**Incl. O&P**
1 Labor Foreman (outside)	$37.10	$296.80	$60.80	$486.40	$39.30	$63.77
6 Laborers	35.10	1684.80	57.55	2762.40		
2 Equip. Oper. (med.)	46.55	744.80	73.50	1176.00		
1 Rodman (reinf.)	49.10	392.80	81.85	654.80		
1 Cement Finisher	42.35	338.80	66.55	532.40		
1 Grader, 30,000 Lbs.		610.80		671.88		
1 Paving Mach. & Equip.		2531.00		2784.10		
1 Concrete Saw		148.00		162.80	37.38	41.12
88 L.H., Daily Totals		$6747.80		$9230.78	$76.68	$104.90
Crew B-26B	**Hr.**	**Daily**	**Hr.**	**Daily**	**Bare Costs**	**Incl. O&P**
1 Labor Foreman (outside)	$37.10	$296.80	$60.80	$486.40	$39.90	$64.58
6 Laborers	35.10	1684.80	57.55	2762.40		
3 Equip. Oper. (med.)	46.55	1117.20	73.50	1764.00		
1 Rodman (reinf.)	49.10	392.80	81.85	654.80		
1 Cement Finisher	42.35	338.80	66.55	532.40		
1 Grader, 30,000 Lbs.		610.80		671.88		
1 Paving Mach. & Equip.		2531.00		2784.10		
1 Concrete Pump, 110' Boom		941.40		1035.54	42.53	46.79
96 L.H., Daily Totals		$7913.60		$10691.52	$82.43	$111.37
Crew B-27	**Hr.**	**Daily**	**Hr.**	**Daily**	**Bare Costs**	**Incl. O&P**
1 Labor Foreman (outside)	$37.10	$296.80	$60.80	$486.40	$35.60	$58.36
3 Laborers	35.10	842.40	57.55	1381.20		
1 Berm Machine		280.00		308.00	8.75	9.63
32 L.H., Daily Totals		$1419.20		$2175.60	$44.35	$67.99
Crew B-28	**Hr.**	**Daily**	**Hr.**	**Daily**	**Bare Costs**	**Incl. O&P**
2 Carpenters	$44.10	$705.60	$72.30	$1156.80	$41.10	$67.38
1 Laborer	35.10	280.80	57.55	460.40		
24 L.H., Daily Totals		$986.40		$1617.20	$41.10	$67.38
Crew B-29	**Hr.**	**Daily**	**Hr.**	**Daily**	**Bare Costs**	**Incl. O&P**
1 Labor Foreman (outside)	$37.10	$296.80	$60.80	$486.40	$38.08	$61.66
4 Laborers	35.10	1123.20	57.55	1841.60		
1 Equip. Oper. (crane)	47.80	382.40	75.50	604.00		
1 Equip. Oper. Oiler	41.25	330.00	65.15	521.20		
1 Gradall, 5/8 C.Y.		844.00		928.40	15.07	16.58
56 L.H., Daily Totals		$2976.40		$4381.60	$53.15	$78.24
Crew B-30	**Hr.**	**Daily**	**Hr.**	**Daily**	**Bare Costs**	**Incl. O&P**
1 Equip. Oper. (med.)	$46.55	$372.40	$73.50	$588.00	$39.05	$62.57
2 Truck Drivers (heavy)	35.30	564.80	57.10	913.60		
1 Hyd. Excavator, 1.5 C.Y.		960.40		1056.44		
2 Dump Trucks, 12 C.Y., 400 H.P.		1169.20		1286.12	88.73	97.61
24 L.H., Daily Totals		$3066.80		$3844.16	$127.78	$160.17
Crew B-31	**Hr.**	**Daily**	**Hr.**	**Daily**	**Bare Costs**	**Incl. O&P**
1 Labor Foreman (outside)	$37.10	$296.80	$60.80	$486.40	$37.30	$61.15
3 Laborers	35.10	842.40	57.55	1381.20		
1 Carpenter	44.10	352.80	72.30	578.40		
1 Air Compressor, 250 cfm		177.00		194.70		
1 Sheeting Driver		5.75		6.33		
2 -50' Air Hoses, 1.5"		12.60		13.86	4.88	5.37
40 L.H., Daily Totals		$1687.35		$2660.89	$42.18	$66.52

Crew No.	Bare Costs		Incl. Subs O&P		Cost Per Labor-Hour	
Crew B-32	**Hr.**	**Daily**	**Hr.**	**Daily**	**Bare Costs**	**Incl. O&P**
1 Laborer	$35.10	$280.80	$57.55	$460.40	$43.69	$69.51
3 Equip. Oper. (med.)	46.55	1117.20	73.50	1764.00		
1 Grader, 30,000 Lbs.		610.80		671.88		
1 Tandem Roller, 10 Ton		227.60		250.36		
1 Dozer, 200 H.P.		1243.00		1367.30	65.04	71.55
32 L.H., Daily Totals		$3479.40		$4513.94	$108.73	$141.06
Crew B-32A	**Hr.**	**Daily**	**Hr.**	**Daily**	**Bare Costs**	**Incl. O&P**
1 Laborer	$35.10	$280.80	$57.55	$460.40	$42.73	$68.18
2 Equip. Oper. (med.)	46.55	744.80	73.50	1176.00		
1 Grader, 30,000 Lbs.		610.80		671.88		
1 Roller, Vibratory, 25 Ton		640.20		704.22	52.13	57.34
24 L.H., Daily Totals		$2276.60		$3012.50	$94.86	$125.52
Crew B-32B	**Hr.**	**Daily**	**Hr.**	**Daily**	**Bare Costs**	**Incl. O&P**
1 Laborer	$35.10	$280.80	$57.55	$460.40	$42.73	$68.18
2 Equip. Oper. (med.)	46.55	744.80	73.50	1176.00		
1 Dozer, 200 H.P.		1243.00		1367.30		
1 Roller, Vibratory, 25 Ton		640.20		704.22	78.47	86.31
24 L.H., Daily Totals		$2908.80		$3707.92	$121.20	$154.50
Crew B-32C	**Hr.**	**Daily**	**Hr.**	**Daily**	**Bare Costs**	**Incl. O&P**
1 Labor Foreman (outside)	$37.10	$296.80	$60.80	$486.40	$41.16	$66.07
2 Laborers	35.10	561.60	57.55	920.80		
3 Equip. Oper. (med.)	46.55	1117.20	73.50	1764.00		
1 Grader, 30,000 Lbs.		610.80		671.88		
1 Tandem Roller, 10 Ton		227.60		250.36		
1 Dozer, 200 H.P.		1243.00		1367.30	43.36	47.70
48 L.H., Daily Totals		$4057.00		$5460.74	$84.52	$113.77
Crew B-33A	**Hr.**	**Daily**	**Hr.**	**Daily**	**Bare Costs**	**Incl. O&P**
1 Equip. Oper. (med.)	$46.55	$372.40	$73.50	$588.00	$43.28	$68.94
.5 Laborer	35.10	140.40	57.55	230.20		
.25 Equip. Oper. (med.)	46.55	93.10	73.50	147.00		
1 Scraper, Towed, 7 C.Y.		108.20		119.02		
1.25 Dozers, 300 H.P.		2147.50		2362.25	161.12	177.23
14 L.H., Daily Totals		$2861.60		$3446.47	$204.40	$246.18
Crew B-33B	**Hr.**	**Daily**	**Hr.**	**Daily**	**Bare Costs**	**Incl. O&P**
1 Equip. Oper. (med.)	$46.55	$372.40	$73.50	$588.00	$43.28	$68.94
.5 Laborer	35.10	140.40	57.55	230.20		
.25 Equip. Oper. (med.)	46.55	93.10	73.50	147.00		
1 Scraper, Towed, 10 C.Y.		139.20		153.12		
1.25 Dozers, 300 H.P.		2147.50		2362.25	163.34	179.67
14 L.H., Daily Totals		$2892.60		$3480.57	$206.61	$248.61
Crew B-33C	**Hr.**	**Daily**	**Hr.**	**Daily**	**Bare Costs**	**Incl. O&P**
1 Equip. Oper. (med.)	$46.55	$372.40	$73.50	$588.00	$43.28	$68.94
.5 Laborer	35.10	140.40	57.55	230.20		
.25 Equip. Oper. (med.)	46.55	93.10	73.50	147.00		
1 Scraper, Towed, 15 C.Y.		156.80		172.48		
1.25 Dozers, 300 H.P.		2147.50		2362.25	164.59	181.05
14 L.H., Daily Totals		$2910.20		$3499.93	$207.87	$250.00
Crew B-33D	**Hr.**	**Daily**	**Hr.**	**Daily**	**Bare Costs**	**Incl. O&P**
1 Equip. Oper. (med.)	$46.55	$372.40	$73.50	$588.00	$43.28	$68.94
.5 Laborer	35.10	140.40	57.55	230.20		
.25 Equip. Oper. (med.)	46.55	93.10	73.50	147.00		
1 S.P. Scraper, 14 C.Y.		1537.00		1690.70		
.25 Dozer, 300 H.P.		429.50		472.45	140.46	154.51
14 L.H., Daily Totals		$2572.40		$3128.35	$183.74	$223.45

Crew No.	Bare Costs		Incl. Subs O&P		Cost Per Labor-Hour	
Crew B-33E	**Hr.**	**Daily**	**Hr.**	**Daily**	**Bare Costs**	**Incl. O&P**
1 Equip. Oper. (med.)	$46.55	$372.40	$73.50	$588.00	$43.28	$68.94
.5 Laborer	35.10	140.40	57.55	230.20		
.25 Equip. Oper. (med.)	46.55	93.10	73.50	147.00		
1 S.P. Scraper, 21 C.Y.		1974.00		2171.40		
.25 Dozer, 300 H.P.		429.50		472.45	171.68	188.85
14 L.H., Daily Totals		$3009.40		$3609.05	$214.96	$257.79

Crew B-33F	Hr.	Daily	Hr.	Daily	Bare Costs	Incl. O&P
1 Equip. Oper. (med.)	$46.55	$372.40	$73.50	$588.00	$43.28	$68.94
.5 Laborer	35.10	140.40	57.55	230.20		
.25 Equip. Oper. (med.)	46.55	93.10	73.50	147.00		
1 Elev. Scraper, 11 C.Y.		1105.00		1215.50		
.25 Dozer, 300 H.P.		429.50		472.45	109.61	120.57
14 L.H., Daily Totals		$2140.40		$2653.15	$152.89	$189.51

Crew B-33G	Hr.	Daily	Hr.	Daily	Bare Costs	Incl. O&P
1 Equip. Oper. (med.)	$46.55	$372.40	$73.50	$588.00	$43.28	$68.94
.5 Laborer	35.10	140.40	57.55	230.20		
.25 Equip. Oper. (med.)	46.55	93.10	73.50	147.00		
1 Elev. Scraper, 22 C.Y.		2099.00		2308.90		
.25 Dozer, 300 H.P.		429.50		472.45	180.61	198.67
14 L.H., Daily Totals		$3134.40		$3746.55	$223.89	$267.61

Crew B-33K	Hr.	Daily	Hr.	Daily	Bare Costs	Incl. O&P
1 Equipment Operator (med.)	$46.55	$372.40	$73.50	$588.00	$43.28	$68.94
.25 Equipment Operator (med.)	46.55	93.10	73.50	147.00		
.5 Laborer	35.10	140.40	57.55	230.20		
1 S.P. Scraper, 31 C.Y.		2868.00		3154.80		
.25 Dozer, 410 H.P.		568.00		624.80	245.43	269.97
14 L.H., Daily Totals		$4041.90		$4744.80	$288.71	$338.91

Crew B-34A	Hr.	Daily	Hr.	Daily	Bare Costs	Incl. O&P
1 Truck Driver (heavy)	$35.30	$282.40	$57.10	$456.80	$35.30	$57.10
1 Dump Truck, 8 C.Y., 220 H.P.		357.40		393.14	44.67	49.14
8 L.H., Daily Totals		$639.80		$849.94	$79.97	$106.24

Crew B-34B	Hr.	Daily	Hr.	Daily	Bare Costs	Incl. O&P
1 Truck Driver (heavy)	$35.30	$282.40	$57.10	$456.80	$35.30	$57.10
1 Dump Truck, 12 C.Y., 400 H.P.		584.60		643.06	73.08	80.38
8 L.H., Daily Totals		$867.00		$1099.86	$108.38	$137.48

Crew B-34C	Hr.	Daily	Hr.	Daily	Bare Costs	Incl. O&P
1 Truck Driver (heavy)	$35.30	$282.40	$57.10	$456.80	$35.30	$57.10
1 Truck Tractor, 6x4, 380 H.P.		522.20		574.42		
1 Dump Trailer, 16.5 C.Y.		123.20		135.52	80.67	88.74
8 L.H., Daily Totals		$927.80		$1166.74	$115.97	$145.84

Crew B-34D	Hr.	Daily	Hr.	Daily	Bare Costs	Incl. O&P
1 Truck Driver (heavy)	$35.30	$282.40	$57.10	$456.80	$35.30	$57.10
1 Truck Tractor, 6x4, 380 H.P.		522.20		574.42		
1 Dump Trailer, 20 C.Y.		137.80		151.58	82.50	90.75
8 L.H., Daily Totals		$942.40		$1182.80	$117.80	$147.85

Crew B-34E	Hr.	Daily	Hr.	Daily	Bare Costs	Incl. O&P
1 Truck Driver (heavy)	$35.30	$282.40	$57.10	$456.80	$35.30	$57.10
1 Dump Truck, Off Hwy., 25 Ton		1271.00		1398.10	158.88	174.76
8 L.H., Daily Totals		$1553.40		$1854.90	$194.18	$231.86

Crew B-34F	Hr.	Daily	Hr.	Daily	Bare Costs	Incl. O&P
1 Truck Driver (heavy)	$35.30	$282.40	$57.10	$456.80	$35.30	$57.10
1 Dump Truck, Off Hwy., 35 Ton		1157.00		1272.70	144.63	159.09
8 L.H., Daily Totals		$1439.40		$1729.50	$179.93	$216.19

Crew B-34G	Hr.	Daily	Hr.	Daily	Bare Costs	Incl. O&P
1 Truck Driver (heavy)	$35.30	$282.40	$57.10	$456.80	$35.30	$57.10
1 Dump Truck, Off Hwy., 50 Ton		1568.00		1724.80	196.00	215.60
8 L.H., Daily Totals		$1850.40		$2181.60	$231.30	$272.70

Crew B-34H	Hr.	Daily	Hr.	Daily	Bare Costs	Incl. O&P
1 Truck Driver (heavy)	$35.30	$282.40	$57.10	$456.80	$35.30	$57.10
1 Dump Truck, Off Hwy., 65 Ton		1572.00		1729.20	196.50	216.15
8 L.H., Daily Totals		$1854.40		$2186.00	$231.80	$273.25

Crew B-34I	Hr.	Daily	Hr.	Daily	Bare Costs	Incl. O&P
1 Truck Driver (heavy)	$35.30	$282.40	$57.10	$456.80	$35.30	$57.10
1 Dump Truck, 18 C.Y., 450 H.P.		745.00		819.50	93.13	102.44
8 L.H., Daily Totals		$1027.40		$1276.30	$128.43	$159.54

Crew B-34J	Hr.	Daily	Hr.	Daily	Bare Costs	Incl. O&P
1 Truck Driver (heavy)	$35.30	$282.40	$57.10	$456.80	$35.30	$57.10
1 Dump Truck, Off Hwy., 100 Ton		2071.00		2278.10	258.88	284.76
8 L.H., Daily Totals		$2353.40		$2734.90	$294.18	$341.86

Crew B-34K	Hr.	Daily	Hr.	Daily	Bare Costs	Incl. O&P
1 Truck Driver (heavy)	$35.30	$282.40	$57.10	$456.80	$35.30	$57.10
1 Truck Tractor, 6x4, 450 H.P.		635.00		698.50		
1 Lowbed Trailer, 75 Ton		213.00		234.30	106.00	116.60
8 L.H., Daily Totals		$1130.40		$1389.60	$141.30	$173.70

Crew B-34L	Hr.	Daily	Hr.	Daily	Bare Costs	Incl. O&P
1 Equip. Oper. (light)	$44.75	$358.00	$70.65	$565.20	$44.75	$70.65
1 Flatbed Truck, Gas, 1.5 Ton		211.00		232.10	26.38	29.01
8 L.H., Daily Totals		$569.00		$797.30	$71.13	$99.66

Crew B-34N	Hr.	Daily	Hr.	Daily	Bare Costs	Incl. O&P
1 Truck Driver (heavy)	$35.30	$282.40	$57.10	$456.80	$35.30	$57.10
1 Dump Truck, 8 C.Y., 220 H.P.		357.40		393.14		
1 Flatbed Trailer, 40 Ton		147.00		161.70	63.05	69.36
8 L.H., Daily Totals		$786.80		$1011.64	$98.35	$126.46

Crew B-34P	Hr.	Daily	Hr.	Daily	Bare Costs	Incl. O&P
1 Pipe Fitter	$55.70	$445.60	$86.80	$694.40	$45.53	$71.95
1 Truck Driver (light)	34.35	274.80	55.55	444.40		
1 Equip. Oper. (med.)	46.55	372.40	73.50	588.00		
1 Flatbed Truck, Gas, 3 Ton		264.40		290.84		
1 Backhoe Loader, 48 H.P.		334.00		367.40	24.93	27.43
24 L.H., Daily Totals		$1691.20		$2385.04	$70.47	$99.38

Crew B-34Q	Hr.	Daily	Hr.	Daily	Bare Costs	Incl. O&P
1 Pipe Fitter	$55.70	$445.60	$86.80	$694.40	$45.95	$72.62
1 Truck Driver (light)	34.35	274.80	55.55	444.40		
1 Eqip. Oper. (crane)	47.80	382.40	75.50	604.00		
1 Flatbed Trailer, 25 Ton		108.60		119.46		
1 Dump Truck, 8 C.Y., 220 H.P.		357.40		393.14		
1 Hyd. Crane, 25 Ton		697.20		766.92	48.47	53.31
24 L.H., Daily Totals		$2266.00		$3022.32	$94.42	$125.93

Crew No.	Bare Costs		Incl. Subs O&P		Cost Per Labor-Hour	
Crew B-34R	**Hr.**	**Daily**	**Hr.**	**Daily**	**Bare Costs**	**Incl. O&P**
1 Pipe Fitter	$55.70	$445.60	$86.80	$694.40	$45.95	$72.62
1 Truck Driver (light)	34.35	274.80	55.55	444.40		
1 Eqip. Oper. (crane)	47.80	382.40	75.50	604.00		
1 Flatbed Trailer, 25 Ton		108.60		119.46		
1 Dump Truck, 8 C.Y., 220 H.P.		357.40		393.14		
1 Hyd. Crane, 25 Ton		697.20		766.92		
1 Hyd. Excavator, 1 C.Y.		748.60		823.46	79.66	87.62
24 L.H., Daily Totals		$3014.60		$3845.78	$125.61	$160.24
Crew B-34S	**Hr.**	**Daily**	**Hr.**	**Daily**	**Bare Costs**	**Incl. O&P**
2 Pipe Fitters	$55.70	$891.20	$86.80	$1388.80	$48.63	$76.55
1 Truck Driver (heavy)	35.30	282.40	57.10	456.80		
1 Eqip. Oper. (crane)	47.80	382.40	75.50	604.00		
1 Flatbed Trailer, 40 Ton		147.00		161.70		
1 Truck Tractor, 6x4, 380 H.P.		522.20		574.42		
1 Hyd. Crane, 80 Ton		1554.00		1709.40		
1 Hyd. Excavator, 2 C.Y.		1253.00		1378.30	108.63	119.49
32 L.H., Daily Totals		$5032.20		$6273.42	$157.26	$196.04
Crew B-34T	**Hr.**	**Daily**	**Hr.**	**Daily**	**Bare Costs**	**Incl. O&P**
2 Pipe Fitters	$55.70	$891.20	$86.80	$1388.80	$48.63	$76.55
1 Truck Driver (heavy)	35.30	282.40	57.10	456.80		
1 Eqip. Oper. (crane)	47.80	382.40	75.50	604.00		
1 Flatbed Trailer, 40 Ton		147.00		161.70		
1 Truck Tractor, 6x4, 380 H.P.		522.20		574.42		
1 Hyd. Crane, 80 Ton		1554.00		1709.40	69.47	76.42
32 L.H., Daily Totals		$3779.20		$4895.12	$118.10	$152.97
Crew B-35	**Hr.**	**Daily**	**Hr.**	**Daily**	**Bare Costs**	**Incl. O&P**
1 Labor Foreman (outside)	$37.10	$296.80	$60.80	$486.40	$43.54	$69.64
1 Skilled Worker	45.40	363.20	73.80	590.40		
1 Welder (plumber)	54.60	436.80	85.05	680.40		
1 Laborer	35.10	280.80	57.55	460.40		
1 Equip. Oper. (crane)	47.80	382.40	75.50	604.00		
1 Equip. Oper. Oiler	41.25	330.00	65.15	521.20		
1 Welder, electric, 300 amp		36.90		40.59		
1 Hyd. Excavator, .75 C.Y.		647.20		711.92	14.25	15.68
48 L.H., Daily Totals		$2774.10		$4095.31	$57.79	$85.32
Crew B-35A	**Hr.**	**Daily**	**Hr.**	**Daily**	**Bare Costs**	**Incl. O&P**
1 Labor Foreman (outside)	$37.10	$296.80	$60.80	$486.40	$42.34	$67.91
2 Laborers	35.10	561.60	57.55	920.80		
1 Skilled Worker	45.40	363.20	73.80	590.40		
1 Welder (plumber)	54.60	436.80	85.05	680.40		
1 Equip. Oper. (crane)	47.80	382.40	75.50	604.00		
1 Equip. Oper. Oiler	41.25	330.00	65.15	521.20		
1 Welder, gas engine, 300 amp		121.60		133.76		
1 Crawler Crane, 75 Ton		1423.00		1565.30	27.58	30.34
56 L.H., Daily Totals		$3915.40		$5502.26	$69.92	$98.25
Crew B-36	**Hr.**	**Daily**	**Hr.**	**Daily**	**Bare Costs**	**Incl. O&P**
1 Labor Foreman (outside)	$37.10	$296.80	$60.80	$486.40	$40.08	$64.58
2 Laborers	35.10	561.60	57.55	920.80		
2 Equip. Oper. (med.)	46.55	744.80	73.50	1176.00		
1 Dozer, 200 H.P.		1243.00		1367.30		
1 Aggregate Spreader		36.80		40.48		
1 Tandem Roller, 10 Ton		227.60		250.36	37.69	41.45
40 L.H., Daily Totals		$3110.60		$4241.34	$77.77	$106.03

Crew No.	Bare Costs		Incl. Subs O&P		Cost Per Labor-Hour	
Crew B-36A	**Hr.**	**Daily**	**Hr.**	**Daily**	**Bare Costs**	**Incl. O&P**
1 Labor Foreman (outside)	$37.10	$296.80	$60.80	$486.40	$41.93	$67.13
2 Laborers	35.10	561.60	57.55	920.80		
4 Equip. Oper. (med.)	46.55	1489.60	73.50	2352.00		
1 Dozer, 200 H.P.		1243.00		1367.30		
1 Aggregate Spreader		36.80		40.48		
1 Tandem Roller, 10 Ton		227.60		250.36		
1 Roller, Pneum. Whl, 12 Ton		313.20		344.52	32.51	35.76
56 L.H., Daily Totals		$4168.60		$5761.86	$74.44	$102.89
Crew B-36B	**Hr.**	**Daily**	**Hr.**	**Daily**	**Bare Costs**	**Incl. O&P**
1 Labor Foreman (outside)	$37.10	$296.80	$60.80	$486.40	$41.10	$65.88
2 Laborers	35.10	561.60	57.55	920.80		
4 Equip. Oper. (med.)	46.55	1489.60	73.50	2352.00		
1 Truck Driver, Heavy	35.30	282.40	57.10	456.80		
1 Grader, 30,000 Lbs.		610.80		671.88		
1 F.E. Loader, crl, 1.5 C.Y.		598.20		658.02		
1 Dozer, 300 H.P.		1718.00		1889.80		
1 Roller, Vibratory, 25 Ton		640.20		704.22		
1 Truck Tractor, 6x4, 450 H.P.		635.00		698.50		
1 Water Tank Trailer, 5000 Gal.		141.00		155.10	67.86	74.65
64 L.H., Daily Totals		$6973.60		$8993.52	$108.96	$140.52
Crew B-36C	**Hr.**	**Daily**	**Hr.**	**Daily**	**Bare Costs**	**Incl. O&P**
1 Labor Foreman (outside)	$37.10	$296.80	$60.80	$486.40	$42.41	$67.68
3 Equip. Oper. (med.)	46.55	1117.20	73.50	1764.00		
1 Truck Driver, Heavy	35.30	282.40	57.10	456.80		
1 Grader, 30,000 Lbs.		610.80		671.88		
1 Dozer, 300 H.P.		1718.00		1889.80		
1 Roller, Vibratory, 25 Ton		640.20		704.22		
1 Truck Tractor, 6x4, 450 H.P.		635.00		698.50		
1 Water Tank Trailer, 5000 Gal.		141.00		155.10	93.63	102.99
40 L.H., Daily Totals		$5441.40		$6826.70	$136.04	$170.67
Crew B-37	**Hr.**	**Daily**	**Hr.**	**Daily**	**Bare Costs**	**Incl. O&P**
1 Labor Foreman (outside)	$37.10	$296.80	$60.80	$486.40	$37.04	$60.27
4 Laborers	35.10	1123.20	57.55	1841.60		
1 Equip. Oper. (light)	44.75	358.00	70.65	565.20		
1 Tandem Roller, 5 Ton		149.20		164.12	3.11	3.42
48 L.H., Daily Totals		$1927.20		$3057.32	$40.15	$63.69
Crew B-37A	**Hr.**	**Daily**	**Hr.**	**Daily**	**Bare Costs**	**Incl. O&P**
2 Laborers	$35.10	$561.60	$57.55	$920.80	$34.85	$56.88
1 Truck Driver (light)	34.35	274.80	55.55	444.40		
1 Flatbed Truck, Gas, 1.5 Ton		211.00		232.10		
1 Tar Kettle, T.M.		77.20		84.92	12.01	13.21
24 L.H., Daily Totals		$1124.60		$1682.22	$46.86	$70.09
Crew B-37B	**Hr.**	**Daily**	**Hr.**	**Daily**	**Bare Costs**	**Incl. O&P**
3 Laborers	$35.10	$842.40	$57.55	$1381.20	$34.91	$57.05
1 Truck Driver (light)	34.35	274.80	55.55	444.40		
1 Flatbed Truck, Gas, 1.5 Ton		211.00		232.10		
1 Tar Kettle, T.M.		77.20		84.92	9.01	9.91
32 L.H., Daily Totals		$1405.40		$2142.62	$43.92	$66.96
Crew B-37C	**Hr.**	**Daily**	**Hr.**	**Daily**	**Bare Costs**	**Incl. O&P**
2 Laborers	$35.10	$561.60	$57.55	$920.80	$34.73	$56.55
2 Truck Drivers (light)	34.35	549.60	55.55	888.80		
2 Flatbed Trucks, Gas, 1.5 Ton		422.00		464.20		
1 Tar Kettle, T.M.		77.20		84.92	15.60	17.16
32 L.H., Daily Totals		$1610.40		$2358.72	$50.33	$73.71

Crew No.	Bare Costs		Incl. Subs O&P		Cost Per Labor-Hour	
Crew B-37D	**Hr.**	**Daily**	**Hr.**	**Daily**	**Bare Costs**	**Incl. O&P**
1 Laborer	$35.10	$280.80	$57.55	$460.40	$34.73	$56.55
1 Truck Driver (light)	34.35	274.80	55.55	444.40		
1 Pickup Truck, 3/4 Ton		127.00		139.70	7.94	8.73
16 L.H., Daily Totals		$682.60		$1044.50	$42.66	$65.28
Crew B-37E	**Hr.**	**Daily**	**Hr.**	**Daily**	**Bare Costs**	**Incl. O&P**
3 Laborers	$35.10	$842.40	$57.55	$1381.20	$37.90	$61.13
1 Equip. Oper. (light)	44.75	358.00	70.65	565.20		
1 Equip. Oper. (medium)	46.55	372.40	73.50	588.00		
2 Truck Drivers (light)	34.35	549.60	55.55	888.80		
4 Barrels w/ Flasher		13.60		14.96		
1 Concrete Saw		148.00		162.80		
1 Rotary Hammer Drill		22.55		24.81		
1 Hammer Drill Bit		3.15		3.46		
1 Loader, Skid Steer, 30 H.P., gas		159.60		175.56		
1 Conc. Hammer Attach.		116.25		127.88		
1 Vibrating plate, gas, 18"		32.20		35.42		
2 Flatbed Trucks, Gas, 1.5 Ton		422.00		464.20	16.38	18.02
56 L.H., Daily Totals		$3039.75		$4432.28	$54.28	$79.15
Crew B-37F	**Hr.**	**Daily**	**Hr.**	**Daily**	**Bare Costs**	**Incl. O&P**
3 Laborers	$35.10	$842.40	$57.55	$1381.20	$34.91	$57.05
1 Truck Driver (light)	34.35	274.80	55.55	444.40		
4 Barrels w/ Flasher		13.60		14.96		
1 Concrete Mixer, 10 C.F.		160.60		176.66		
1 Air Compressor, 60 cfm		120.40		132.44		
1 -50' Air Hose, 3/4"		3.05		3.36		
1 Spade (chipper)		7.60		8.36		
1 Flatbed Truck, Gas, 1.5 Ton		211.00		232.10	16.13	17.75
32 L.H., Daily Totals		$1633.45		$2393.47	$51.05	$74.80
Crew B-37G	**Hr.**	**Daily**	**Hr.**	**Daily**	**Bare Costs**	**Incl. O&P**
1 Labor Foreman (outside)	$37.10	$296.80	$60.80	$486.40	$37.04	$60.27
4 Laborers	35.10	1123.20	57.55	1841.60		
1 Equip. Oper. (light)	44.75	358.00	70.65	565.20		
1 Berm Machine		280.00		308.00		
1 Tandem Roller, 5 Ton		149.20		164.12	8.94	9.84
48 L.H., Daily Totals		$2207.20		$3365.32	$45.98	$70.11
Crew B-38	**Hr.**	**Daily**	**Hr.**	**Daily**	**Bare Costs**	**Incl. O&P**
1 Labor Foreman (outside)	$37.10	$296.80	$60.80	$486.40	$39.72	$64.01
2 Laborers	35.10	561.60	57.55	920.80		
1 Equip. Oper. (light)	44.75	358.00	70.65	565.20		
1 Equip. Oper. (med.)	46.55	372.40	73.50	588.00		
1 Backhoe Loader, 48 H.P.		334.00		367.40		
1 Hyd.Hammer, (1200 lb.)		178.40		196.24		
1 F.E. Loader, W.M., 4 C.Y.		611.00		672.10		
1 Pvmt. Rem. Bucket		59.00		64.90	29.56	32.52
40 L.H., Daily Totals		$2771.20		$3861.04	$69.28	$96.53
Crew B-39	**Hr.**	**Daily**	**Hr.**	**Daily**	**Bare Costs**	**Incl. O&P**
1 Labor Foreman (outside)	$37.10	$296.80	$60.80	$486.40	$37.04	$60.27
4 Laborers	35.10	1123.20	57.55	1841.60		
1 Equip. Oper. (light)	44.75	358.00	70.65	565.20		
1 Air Compressor, 250 cfm		177.00		194.70		
2 Breakers, Pavement, 60 lb.		19.60		21.56		
2 -50' Air Hoses, 1.5"		12.60		13.86	4.36	4.79
48 L.H., Daily Totals		$1987.20		$3123.32	$41.40	$65.07

Crew No.	Bare Costs		Incl. Subs O&P		Cost Per Labor-Hour	
Crew B-40	**Hr.**	**Daily**	**Hr.**	**Daily**	**Bare Costs**	**Incl. O&P**
1 Pile Driver Foreman (outside)	$44.80	$358.40	$74.25	$594.00	$44.11	$71.75
4 Pile Drivers	42.80	1369.60	70.90	2268.80		
2 Equip. Oper. (crane)	47.80	764.80	75.50	1208.00		
1 Equip. Oper. Oiler	41.25	330.00	65.15	521.20		
1 Crawler Crane, 40 Ton		1115.00		1226.50		
1 Vibratory Hammer & Gen.		2335.00		2568.50	53.91	59.30
64 L.H., Daily Totals		$6272.80		$8387.00	$98.01	$131.05
Crew B-40B	**Hr.**	**Daily**	**Hr.**	**Daily**	**Bare Costs**	**Incl. O&P**
1 Labor Foreman (outside)	$37.10	$296.80	$60.80	$486.40	$38.58	$62.35
3 Laborers	35.10	842.40	57.55	1381.20		
1 Equip. Oper. (crane)	47.80	382.40	75.50	604.00		
1 Equip. Oper. Oiler	41.25	330.00	65.15	521.20		
1 Lattice Boom Crane, 40 Ton		1163.00		1279.30	24.23	26.65
48 L.H., Daily Totals		$3014.60		$4272.10	$62.80	$89.00
Crew B-41	**Hr.**	**Daily**	**Hr.**	**Daily**	**Bare Costs**	**Incl. O&P**
1 Labor Foreman (outside)	$37.10	$296.80	$60.80	$486.40	$36.32	$59.30
4 Laborers	35.10	1123.20	57.55	1841.60		
.25 Equip. Oper. (crane)	47.80	95.60	75.50	151.00		
.25 Equip. Oper. Oiler	41.25	82.50	65.15	130.30		
.25 Crawler Crane, 40 Ton		278.75		306.63	6.34	6.97
44 L.H., Daily Totals		$1876.85		$2915.93	$42.66	$66.27
Crew B-42	**Hr.**	**Daily**	**Hr.**	**Daily**	**Bare Costs**	**Incl. O&P**
1 Labor Foreman (outside)	$37.10	$296.80	$60.80	$486.40	$39.49	$65.42
4 Laborers	35.10	1123.20	57.55	1841.60		
1 Equip. Oper. (crane)	47.80	382.40	75.50	604.00		
1 Equip. Oper. Oiler	41.25	330.00	65.15	521.20		
1 Welder	49.35	394.80	91.70	733.60		
1 Hyd. Crane, 25 Ton		697.20		766.92		
1 Welder, gas engine, 300 amp		121.60		133.76		
1 Horz. Boring Csg. Mch.		447.60		492.36	19.79	21.77
64 L.H., Daily Totals		$3793.60		$5579.84	$59.27	$87.19
Crew B-43	**Hr.**	**Daily**	**Hr.**	**Daily**	**Bare Costs**	**Incl. O&P**
1 Labor Foreman (outside)	$37.10	$296.80	$60.80	$486.40	$38.58	$62.35
3 Laborers	35.10	842.40	57.55	1381.20		
1 Equip. Oper. (crane)	47.80	382.40	75.50	604.00		
1 Equip. Oper. Oiler	41.25	330.00	65.15	521.20		
1 Drill Rig, Truck-Mounted		2441.00		2685.10	50.85	55.94
48 L.H., Daily Totals		$4292.60		$5677.90	$89.43	$118.29
Crew B-44	**Hr.**	**Daily**	**Hr.**	**Daily**	**Bare Costs**	**Incl. O&P**
1 Pile Driver Foreman (outside)	$44.80	$358.40	$74.25	$594.00	$43.34	$70.80
4 Pile Drivers	42.80	1369.60	70.90	2268.80		
2 Equip. Oper. (crane)	47.80	764.80	75.50	1208.00		
1 Laborer	35.10	280.80	57.55	460.40		
1 Crawler Crane, 40 Ton		1115.00		1226.50		
1 Lead, 60' high		73.60		80.96		
1 Hammer, diesel, 15K ft.-lbs.		591.40		650.54	27.81	30.59
64 L.H., Daily Totals		$4553.60		$6489.20	$71.15	$101.39
Crew B-45	**Hr.**	**Daily**	**Hr.**	**Daily**	**Bare Costs**	**Incl. O&P**
1 Equip. Oper. (med.)	$46.55	$372.40	$73.50	$588.00	$40.92	$65.30
1 Truck Driver (heavy)	35.30	282.40	57.10	456.80		
1 Dist. Tanker, 3000 Gallon		306.00		336.60		
1 Truck Tractor, 6x4, 380 H.P.		522.20		574.42	51.76	56.94
16 L.H., Daily Totals		$1483.00		$1955.82	$92.69	$122.24

Crew No.	Bare Costs		Incl. Subs O&P		Cost Per Labor-Hour	
Crew B-46	**Hr.**	**Daily**	**Hr.**	**Daily**	**Bare Costs**	**Incl. O&P**
1 Pile Driver Foreman (outside)	$44.80	$358.40	$74.25	$594.00	$39.28	$64.78
2 Pile Drivers	42.80	684.80	70.90	1134.40		
3 Laborers	35.10	842.40	57.55	1381.20		
1 Chain Saw, gas, 36" Long		43.60		47.96	.91	1.00
48 L.H., Daily Totals		$1929.20		$3157.56	$40.19	$65.78

Crew B-47	Hr.	Daily	Hr.	Daily	Bare Costs	Incl. O&P
1 Blast Foreman (outside)	$37.10	$296.80	$60.80	$486.40	$38.98	$63.00
1 Driller	35.10	280.80	57.55	460.40		
1 Equip. Oper. (light)	44.75	358.00	70.65	565.20		
1 Air Track Drill, 4"		937.20		1030.92		
1 Air Compressor, 600 cfm		467.60		514.36		
2 -50' Air Hoses, 3"		29.80		32.78	59.77	65.75
24 L.H., Daily Totals		$2370.20		$3090.06	$98.76	$128.75

Crew B-47A	Hr.	Daily	Hr.	Daily	Bare Costs	Incl. O&P
1 Drilling Foreman (outside)	$37.10	$296.80	$60.80	$486.40	$42.05	$67.15
1 Equip. Oper. (heavy)	47.80	382.40	75.50	604.00		
1 Oiler	41.25	330.00	65.15	521.20		
1 Air Track Drill, 5"		1088.00		1196.80	45.33	49.87
24 L.H., Daily Totals		$2097.20		$2808.40	$87.38	$117.02

Crew B-47C	Hr.	Daily	Hr.	Daily	Bare Costs	Incl. O&P
1 Laborer	$35.10	$280.80	$57.55	$460.40	$39.92	$64.10
1 Equip. Oper. (light)	44.75	358.00	70.65	565.20		
1 Air Compressor, 750 cfm		474.80		522.28		
2 -50' Air Hoses, 3"		29.80		32.78		
1 Air Track Drill, 4"		937.20		1030.92	90.11	99.12
16 L.H., Daily Totals		$2080.60		$2611.58	$130.04	$163.22

Crew B-47E	Hr.	Daily	Hr.	Daily	Bare Costs	Incl. O&P
1 Labor Foreman (outside)	$37.10	$296.80	$60.80	$486.40	$35.60	$58.36
3 Laborers	35.10	842.40	57.55	1381.20		
1 Flatbed Truck, Gas, 3 Ton		264.40		290.84	8.26	9.09
32 L.H., Daily Totals		$1403.60		$2158.44	$43.86	$67.45

Crew B-47G	Hr.	Daily	Hr.	Daily	Bare Costs	Incl. O&P
1 Labor Foreman (outside)	$37.10	$296.80	$60.80	$486.40	$38.01	$61.64
2 Laborers	35.10	561.60	57.55	920.80		
1 Equip. Oper. (light)	44.75	358.00	70.65	565.20		
1 Air Track Drill, 4"		937.20		1030.92		
1 Air Compressor, 600 cfm		467.60		514.36		
2 -50' Air Hoses, 3"		29.80		32.78		
1 Gunite Pump Rig		280.40		308.44	53.59	58.95
32 L.H., Daily Totals		$2931.40		$3858.90	$91.61	$120.59

Crew B-47H	Hr.	Daily	Hr.	Daily	Bare Costs	Incl. O&P
1 Skilled Worker Foreman (out)	$47.40	$379.20	$77.05	$616.40	$45.90	$74.61
3 Skilled Workers	45.40	1089.60	73.80	1771.20		
1 Flatbed Truck, Gas, 3 Ton		264.40		290.84	8.26	9.09
32 L.H., Daily Totals		$1733.20		$2678.44	$54.16	$83.70

Crew B-48	Hr.	Daily	Hr.	Daily	Bare Costs	Incl. O&P
1 Labor Foreman (outside)	$37.10	$296.80	$60.80	$486.40	$39.46	$63.54
3 Laborers	35.10	842.40	57.55	1381.20		
1 Equip. Oper. (crane)	47.80	382.40	75.50	604.00		
1 Equip. Oper. Oiler	41.25	330.00	65.15	521.20		
1 Equip. Oper. (light)	44.75	358.00	70.65	565.20		
1 Centr. Water Pump, 6"		308.40		339.24		
1 -20' Suction Hose, 6"		11.50		12.65		
1 -50' Discharge Hose, 6"		12.10		13.31		
1 Drill Rig, Truck-Mounted		2441.00		2685.10	49.52	54.47
56 L.H., Daily Totals		$4982.60		$6608.30	$88.97	$118.01

Crew B-49	Hr.	Daily	Hr.	Daily	Bare Costs	Incl. O&P
1 Labor Foreman (outside)	$37.10	$296.80	$60.80	$486.40	$40.99	$66.11
3 Laborers	35.10	842.40	57.55	1381.20		
2 Equip. Oper. (crane)	47.80	764.80	75.50	1208.00		
2 Equip. Oper. Oilers	41.25	660.00	65.15	1042.40		
1 Equip. Oper. (light)	44.75	358.00	70.65	565.20		
2 Pile Drivers	42.80	684.80	70.90	1134.40		
1 Hyd. Crane, 25 Ton		697.20		766.92		
1 Centr. Water Pump, 6"		308.40		339.24		
1 -20' Suction Hose, 6"		11.50		12.65		
1 -50' Discharge Hose, 6"		12.10		13.31		
1 Drill Rig, Truck-Mounted		2441.00		2685.10	39.43	43.38
88 L.H., Daily Totals		$7077.00		$9634.82	$80.42	$109.49

Crew B-50	Hr.	Daily	Hr.	Daily	Bare Costs	Incl. O&P
2 Pile Driver Foremen (outside)	$44.80	$716.80	$74.25	$1188.00	$42.04	$68.76
6 Pile Drivers	42.80	2054.40	70.90	3403.20		
2 Equip. Oper. (crane)	47.80	764.80	75.50	1208.00		
1 Equip. Oper. Oiler	41.25	330.00	65.15	521.20		
3 Laborers	35.10	842.40	57.55	1381.20		
1 Crawler Crane, 40 Ton		1115.00		1226.50		
1 Lead, 60' high		73.60		80.96		
1 Hammer, diesel, 15K ft.-lbs.		591.40		650.54		
1 Air Compressor, 600 cfm		467.60		514.36		
2 -50' Air Hoses, 3"		29.80		32.78		
1 Chain Saw, gas, 36" Long		43.60		47.96	20.72	22.80
112 L.H., Daily Totals		$7029.40		$10254.70	$62.76	$91.56

Crew B-51	Hr.	Daily	Hr.	Daily	Bare Costs	Incl. O&P
1 Labor Foreman (outside)	$37.10	$296.80	$60.80	$486.40	$35.31	$57.76
4 Laborers	35.10	1123.20	57.55	1841.60		
1 Truck Driver (light)	34.35	274.80	55.55	444.40		
1 Flatbed Truck, Gas, 1.5 Ton		211.00		232.10	4.40	4.84
48 L.H., Daily Totals		$1905.80		$3004.50	$39.70	$62.59

Crew B-52	Hr.	Daily	Hr.	Daily	Bare Costs	Incl. O&P
1 Carpenter Foreman (outside)	$46.10	$368.80	$75.55	$604.40	$40.81	$66.39
1 Carpenter	44.10	352.80	72.30	578.40		
3 Laborers	35.10	842.40	57.55	1381.20		
1 Cement Finisher	42.35	338.80	66.55	532.40		
.5 Rodman (reinf.)	49.10	196.40	81.85	327.40		
.5 Equip. Oper. (med.)	46.55	186.20	73.50	294.00		
.5 Crawler Loader, 3 C.Y.		553.00		608.30	9.88	10.86
56 L.H., Daily Totals		$2838.40		$4326.10	$50.69	$77.25

Crew B-53	Hr.	Daily	Hr.	Daily	Bare Costs	Incl. O&P
1 Equip. Oper. (light)	$44.75	$358.00	$70.65	$565.20	$44.75	$70.65
1 Trencher, Chain, 12 H.P.		62.80		69.08	7.85	8.63
8 L.H., Daily Totals		$420.80		$634.28	$52.60	$79.28

Crew No.	Bare Costs		Incl. Subs O&P		Cost Per Labor-Hour	

Crew B-54	Hr.	Daily	Hr.	Daily	Bare Costs	Incl. O&P
1 Equip. Oper. (light)	$44.75	$358.00	$70.65	$565.20	$44.75	$70.65
1 Trencher, Chain, 40 H.P.		312.80		344.08	39.10	43.01
8 L.H., Daily Totals		$670.80		$909.28	$83.85	$113.66

Crew B-54A	Hr.	Daily	Hr.	Daily	Bare Costs	Incl. O&P
.17 Labor Foreman (outside)	$37.10	$50.46	$60.80	$82.69	$45.18	$71.65
1 Equipment Operator (med.)	46.55	372.40	73.50	588.00		
1 Wheel Trencher, 67 H.P.		1084.00		1192.40	115.81	127.39
9.36 L.H., Daily Totals		$1506.86		$1863.09	$160.99	$199.05

Crew B-54B	Hr.	Daily	Hr.	Daily	Bare Costs	Incl. O&P
.25 Labor Foreman (outside)	$37.10	$74.20	$60.80	$121.60	$44.66	$70.96
1 Equipment Operator (med.)	46.55	372.40	73.50	588.00		
1 Wheel Trencher, 150 H.P.		1816.00		1997.60	181.60	199.76
10 L.H., Daily Totals		$2262.60		$2707.20	$226.26	$270.72

Crew B-54D	Hr.	Daily	Hr.	Daily	Bare Costs	Incl. O&P
1 Laborer	$35.10	$280.80	$57.55	$460.40	$40.83	$65.53
1 Equipment Operator (med.)	46.55	372.40	73.50	588.00		
1 Rock trencher, 6" width		344.40		378.84	21.52	23.68
16 L.H., Daily Totals		$997.60		$1427.24	$62.35	$89.20

Crew B-54E	Hr.	Daily	Hr.	Daily	Bare Costs	Incl. O&P
1 Laborer	$35.10	$280.80	$57.55	$460.40	$40.83	$65.53
1 Equipment Operator (med.)	46.55	372.40	73.50	588.00		
1 Rock trencher, 18" width		2517.00		2768.70	157.31	173.04
16 L.H., Daily Totals		$3170.20		$3817.10	$198.14	$238.57

Crew B-55	Hr.	Daily	Hr.	Daily	Bare Costs	Incl. O&P
2 Laborers	$35.10	$561.60	$57.55	$920.80	$34.85	$56.88
1 Truck Driver (light)	34.35	274.80	55.55	444.40		
1 Truck-mounted earth auger		718.40		790.24		
1 Flatbed Truck, Gas, 3 Ton		264.40		290.84	40.95	45.05
24 L.H., Daily Totals		$1819.20		$2446.28	$75.80	$101.93

Crew B-56	Hr.	Daily	Hr.	Daily	Bare Costs	Incl. O&P
1 Laborer	$35.10	$280.80	$57.55	$460.40	$39.92	$64.10
1 Equip. Oper. (light)	44.75	358.00	70.65	565.20		
1 Air Track Drill, 4"		937.20		1030.92		
1 Air Compressor, 600 cfm		467.60		514.36		
1 -50' Air Hose, 3"		14.90		16.39	88.73	97.60
16 L.H., Daily Totals		$2058.50		$2587.27	$128.66	$161.70

Crew B-57	Hr.	Daily	Hr.	Daily	Bare Costs	Incl. O&P
1 Labor Foreman (outside)	$37.10	$296.80	$60.80	$486.40	$40.18	$64.53
2 Laborers	35.10	561.60	57.55	920.80		
1 Equip. Oper. (crane)	47.80	382.40	75.50	604.00		
1 Equip. Oper. (light)	44.75	358.00	70.65	565.20		
1 Equip. Oper. Oiler	41.25	330.00	65.15	521.20		
1 Crawler Crane, 25 Ton		1110.00		1221.00		
1 Clamshell Bucket, 1 C.Y.		47.00		51.70		
1 Centr. Water Pump, 6"		308.40		339.24		
1 -20' Suction Hose, 6"		11.50		12.65		
20 -50' Discharge Hoses, 6"		242.00		266.20	35.81	39.39
48 L.H., Daily Totals		$3647.70		$4988.39	$75.99	$103.92

Crew B-58	Hr.	Daily	Hr.	Daily	Bare Costs	Incl. O&P
2 Laborers	$35.10	$561.60	$57.55	$920.80	$38.32	$61.92
1 Equip. Oper. (light)	44.75	358.00	70.65	565.20		
1 Backhoe Loader, 48 H.P.		334.00		367.40		
1 Small Helicopter, w/pilot		2625.00		2887.50	123.29	135.62
24 L.H., Daily Totals		$3878.60		$4740.90	$161.61	$197.54

Crew B-59	Hr.	Daily	Hr.	Daily	Bare Costs	Incl. O&P
1 Truck Driver (heavy)	$35.30	$282.40	$57.10	$456.80	$35.30	$57.10
1 Truck Tractor, 220 H.P.		314.80		346.28		
1 Water Tank Trailer, 5000 Gal.		141.00		155.10	56.98	62.67
8 L.H., Daily Totals		$738.20		$958.18	$92.28	$119.77

Crew B-59A	Hr.	Daily	Hr.	Daily	Bare Costs	Incl. O&P
2 Laborers	$35.10	$561.60	$57.55	$920.80	$35.17	$57.40
1 Truck Driver (heavy)	35.30	282.40	57.10	456.80		
1 Water Tank Trailer, 5000 Gal.		141.00		155.10		
1 Truck Tractor, 220 H.P.		314.80		346.28	18.99	20.89
24 L.H., Daily Totals		$1299.80		$1878.98	$54.16	$78.29

Crew B-60	Hr.	Daily	Hr.	Daily	Bare Costs	Incl. O&P
1 Labor Foreman (outside)	$37.10	$296.80	$60.80	$486.40	$40.84	$65.41
2 Laborers	35.10	561.60	57.55	920.80		
1 Equip. Oper. (crane)	47.80	382.40	75.50	604.00		
2 Equip. Oper. (light)	44.75	716.00	70.65	1130.40		
1 Equip. Oper. Oiler	41.25	330.00	65.15	521.20		
1 Crawler Crane, 40 Ton		1115.00		1226.50		
1 Lead, 60' high		73.60		80.96		
1 Hammer, diesel, 15K ft.-lbs.		591.40		650.54		
1 Backhoe Loader, 48 H.P.		334.00		367.40	37.75	41.52
56 L.H., Daily Totals		$4400.80		$5988.20	$78.59	$106.93

Crew B-61	Hr.	Daily	Hr.	Daily	Bare Costs	Incl. O&P
1 Labor Foreman (outside)	$37.10	$296.80	$60.80	$486.40	$37.43	$60.82
3 Laborers	35.10	842.40	57.55	1381.20		
1 Equip. Oper. (light)	44.75	358.00	70.65	565.20		
1 Cement Mixer, 2 C.Y.		190.40		209.44		
1 Air Compressor, 160 cfm		136.80		150.48	8.18	9.00
40 L.H., Daily Totals		$1824.40		$2792.72	$45.61	$69.82

Crew B-62	Hr.	Daily	Hr.	Daily	Bare Costs	Incl. O&P
2 Laborers	$35.10	$561.60	$57.55	$920.80	$38.32	$61.92
1 Equip. Oper. (light)	44.75	358.00	70.65	565.20		
1 Loader, Skid Steer, 30 H.P.,gas		159.60		175.56	6.65	7.32
24 L.H., Daily Totals		$1079.20		$1661.56	$44.97	$69.23

Crew B-63	Hr.	Daily	Hr.	Daily	Bare Costs	Incl. O&P
4 Laborers	$35.10	$1123.20	$57.55	$1841.60	$37.03	$60.17
1 Equip. Oper. (light)	44.75	358.00	70.65	565.20		
1 Loader, Skid Steer, 30 H.P.,gas		159.60		175.56	3.99	4.39
40 L.H., Daily Totals		$1640.80		$2582.36	$41.02	$64.56

Crew B-64	Hr.	Daily	Hr.	Daily	Bare Costs	Incl. O&P
1 Laborer	$35.10	$280.80	$57.55	$460.40	$34.73	$56.55
1 Truck Driver (light)	34.35	274.80	55.55	444.40		
1 Power Mulcher (small)		142.00		156.20		
1 Flatbed Truck, Gas, 1.5 Ton		211.00		232.10	22.06	24.27
16 L.H., Daily Totals		$908.60		$1293.10	$56.79	$80.82

Crew No.	Bare Costs		Incl. Subs O&P		Cost Per Labor-Hour	
Crew B-65	**Hr.**	**Daily**	**Hr.**	**Daily**	**Bare Costs**	**Incl. O&P**
1 Laborer	$35.10	$280.80	$57.55	$460.40	$34.73	$56.55
1 Truck Driver (light)	34.35	274.80	55.55	444.40		
1 Power Mulcher (large)		286.60		315.26		
1 Flatbed Truck, Gas, 1.5 Ton		211.00		232.10	31.10	34.21
16 L.H., Daily Totals		$1053.20		$1452.16	$65.83	$90.76

Crew B-66	Hr.	Daily	Hr.	Daily	Bare Costs	Incl. O&P
1 Equip. Oper. (light)	$44.75	$358.00	$70.65	$565.20	$44.75	$70.65
1 Loader-Backhoe		247.20		271.92	30.90	33.99
8 L.H., Daily Totals		$605.20		$837.12	$75.65	$104.64

Crew B-67	Hr.	Daily	Hr.	Daily	Bare Costs	Incl. O&P
1 Millwright	$46.05	$368.40	$72.45	$579.60	$45.40	$71.55
1 Equip. Oper. (light)	44.75	358.00	70.65	565.20		
1 Forklift, R/T, 4,000 Lb.		289.20		318.12	18.07	19.88
16 L.H., Daily Totals		$1015.60		$1462.92	$63.48	$91.43

Crew B-68	Hr.	Daily	Hr.	Daily	Bare Costs	Incl. O&P
2 Millwrights	$46.05	$736.80	$72.45	$1159.20	$45.62	$71.85
1 Equip. Oper. (light)	44.75	358.00	70.65	565.20		
1 Forklift, R/T, 4,000 Lb.		289.20		318.12	12.05	13.26
24 L.H., Daily Totals		$1384.00		$2042.52	$57.67	$85.11

Crew B-69	Hr.	Daily	Hr.	Daily	Bare Costs	Incl. O&P
1 Labor Foreman (outside)	$37.10	$296.80	$60.80	$486.40	$38.58	$62.35
3 Laborers	35.10	842.40	57.55	1381.20		
1 Equip Oper. (crane)	47.80	382.40	75.50	604.00		
1 Equip Oper. Oiler	41.25	330.00	65.15	521.20		
1 Hyd. Crane, 80 Ton		1554.00		1709.40	32.38	35.61
48 L.H., Daily Totals		$3405.60		$4702.20	$70.95	$97.96

Crew B-69A	Hr.	Daily	Hr.	Daily	Bare Costs	Incl. O&P
1 Labor Foreman (outside)	$37.10	$296.80	$60.80	$486.40	$38.55	$62.25
3 Laborers	35.10	842.40	57.55	1381.20		
1 Equip. Oper. (med.)	46.55	372.40	73.50	588.00		
1 Concrete Finisher	42.35	338.80	66.55	532.40		
1 Curb/Gutter Paver, 2-Track		761.00		837.10	15.85	17.44
48 L.H., Daily Totals		$2611.40		$3825.10	$54.40	$79.69

Crew B 69B	Hr.	Daily	Hr.	Daily	Bare Costs	Incl. O&P
1 Labor Foreman (outside)	$37.10	$296.80	$60.80	$486.40	$38.55	$62.25
3 Laborers	35.10	842.40	57.55	1381.20		
1 Equip. Oper. (med.)	46.55	372.40	73.50	588.00		
1 Cement Finisher	42.35	338.80	66.55	532.40		
1 Curb/Gutter Paver, 4-Track		733.20		806.52	15.28	16.80
48 L.H., Daily Totals		$2583.60		$3794.52	$53.83	$79.05

Crew B-70	Hr.	Daily	Hr.	Daily	Bare Costs	Incl. O&P
1 Labor Foreman (outside)	$37.10	$296.80	$60.80	$486.40	$40.29	$64.85
3 Laborers	35.10	842.40	57.55	1381.20		
3 Equip. Oper. (med.)	46.55	1117.20	73.50	1764.00		
1 Grader, 30,000 Lbs.		610.80		671.88		
1 Ripper, beam & 1 shank		81.60		89.76		
1 Road Sweeper, S.P., 8' wide		620.80		682.88		
1 F.E. Loader, W.M., 1.5 C.Y.		333.80		367.18	29.41	32.35
56 L.H., Daily Totals		$3903.40		$5443.30	$69.70	$97.20

Crew B-71	Hr.	Daily	Hr.	Daily	Bare Costs	Incl. O&P
1 Labor Foreman (outside)	$37.10	$296.80	$60.80	$486.40	$40.29	$64.85
3 Laborers	35.10	842.40	57.55	1381.20		
3 Equip. Oper. (med.)	46.55	1117.20	73.50	1764.00		
1 Pvmt. Profiler, 750 H.P.		5459.00		6004.90		
1 Road Sweeper, S.P., 8' wide		620.80		682.88		
1 F.E. Loader, W.M., 1.5 C.Y.		333.80		367.18	114.53	125.98
56 L.H., Daily Totals		$8670.00		$10686.56	$154.82	$190.83

Crew B-72	Hr.	Daily	Hr.	Daily	Bare Costs	Incl. O&P
1 Labor Foreman (outside)	$37.10	$296.80	$60.80	$486.40	$41.08	$65.93
3 Laborers	35.10	842.40	57.55	1381.20		
4 Equip. Oper. (med.)	46.55	1489.60	73.50	2352.00		
1 Pvmt. Profiler, 750 H.P.		5459.00		6004.90		
1 Hammermill, 250 H.P.		1721.00		1893.10		
1 Windrow Loader		1207.00		1327.70		
1 Mix Paver 165 H.P.		2012.00		2213.20		
1 Roller, Pneum. Whl, 12 Ton		313.20		344.52	167.38	184.12
64 L.H., Daily Totals		$13341.00		$16003.02	$208.45	$250.05

Crew B-73	Hr.	Daily	Hr.	Daily	Bare Costs	Incl. O&P
1 Labor Foreman (outside)	$37.10	$296.80	$60.80	$486.40	$42.51	$67.92
2 Laborers	35.10	561.60	57.55	920.80		
5 Equip. Oper. (med.)	46.55	1862.00	73.50	2940.00		
1 Road Mixer, 310 H.P.		1882.00		2070.20		
1 Tandem Roller, 10 Ton		227.60		250.36		
1 Hammermill, 250 H.P.		1721.00		1893.10		
1 Grader, 30,000 Lbs.		610.80		671.88		
.5 F.E. Loader, W.M., 1.5 C.Y.		166.90		183.59		
.5 Truck Tractor, 220 H.P.		157.40		173.14		
.5 Water Tank Trailer, 5000 Gal.		70.50		77.55	75.57	83.12
64 L.H., Daily Totals		$7556.60		$9667.02	$118.07	$151.05

Crew B-74	Hr.	Daily	Hr.	Daily	Bare Costs	Incl. O&P
1 Labor Foreman (outside)	$37.10	$296.80	$60.80	$486.40	$41.13	$65.82
1 Laborer	35.10	280.80	57.55	460.40		
4 Equip. Oper. (med.)	46.55	1489.60	73.50	2352.00		
2 Truck Drivers (heavy)	35.30	564.80	57.10	913.60		
1 Grader, 30,000 Lbs.		610.80		671.88		
1 Ripper, beam & 1 shank		81.60		89.76		
2 Stabilizers, 310 H.P.		3170.00		3487.00		
1 Flatbed Truck, Gas, 3 Ton		264.40		290.84		
1 Chem. Spreader, Towed		50.00		55.00		
1 Roller, Vibratory, 25 Ton		640.20		704.22		
1 Water Tank Trailer, 5000 Gal.		141.00		155.10		
1 Truck Tractor, 220 H.P.		314.80		346.28	82.39	90.63
64 L.H., Daily Totals		$7904.80		$10012.48	$123.51	$156.45

Crew B-75	Hr.	Daily	Hr.	Daily	Bare Costs	Incl. O&P
1 Labor Foreman (outside)	$37.10	$296.80	$60.80	$486.40	$41.96	$67.06
1 Laborer	35.10	280.80	57.55	460.40		
4 Equip. Oper. (med.)	46.55	1489.60	73.50	2352.00		
1 Truck Driver (heavy)	35.30	282.40	57.10	456.80		
1 Grader, 30,000 Lbs.		610.80		671.88		
1 Ripper, beam & 1 shank		81.60		89.76		
2 Stabilizers, 310 H.P.		3170.00		3487.00		
1 Dist. Tanker, 3000 Gallon		306.00		336.60		
1 Truck Tractor, 6x4, 380 H.P.		522.20		574.42		
1 Roller, Vibratory, 25 Ton		640.20		704.22	95.19	104.71
56 L.H., Daily Totals		$7680.40		$9619.48	$137.15	$171.78

Crew No.	Bare Costs		Incl. Subs O&P		Cost Per Labor-Hour	
Crew B-76	**Hr.**	**Daily**	**Hr.**	**Daily**	**Bare Costs**	**Incl. O&P**
1 Dock Builder Foreman (outside)	$44.80	$358.40	$74.25	$594.00	$43.96	$71.66
5 Dock Builders	42.80	1712.00	70.90	2836.00		
2 Equip. Oper. (crane)	47.80	764.80	75.50	1208.00		
1 Equip. Oper. Oiler	41.25	330.00	65.15	521.20		
1 Crawler Crane, 50 Ton		1257.00		1382.70		
1 Barge, 400 Ton		779.20		857.12		
1 Hammer, diesel, 15K ft.-lbs.		591.40		650.54		
1 Lead, 60' high		73.60		80.96		
1 Air Compressor, 600 cfm		467.60		514.36		
2 -50' Air Hoses, 3"		29.80		32.78	44.42	48.87
72 L.H., Daily Totals		$6363.80		$8677.66	$88.39	$120.52

Crew B-76A	Hr.	Daily	Hr.	Daily	Bare Costs	Incl. O&P
1 Labor Foreman (outside)	$37.10	$296.80	$60.80	$486.40	$37.71	$61.15
5 Laborers	35.10	1404.00	57.55	2302.00		
1 Equip. Oper. (crane)	47.80	382.40	75.50	604.00		
1 Equip. Oper. Oiler	41.25	330.00	65.15	521.20		
1 Crawler Crane, 50 Ton		1257.00		1382.70		
1 Barge, 400 Ton		779.20		857.12	31.82	35.00
64 L.H., Daily Totals		$4449.40		$6153.42	$69.52	$96.15

Crew B-77	Hr.	Daily	Hr.	Daily	Bare Costs	Incl. O&P
1 Labor Foreman (outside)	$37.10	$296.80	$60.80	$486.40	$35.35	$57.80
3 Laborers	35.10	842.40	57.55	1381.20		
1 Truck Driver (light)	34.35	274.80	55.55	444.40		
1 Crack Cleaner, 25 H.P.		57.20		62.92		
1 Crack Filler, Trailer Mtd.		196.40		216.04		
1 Flatbed Truck, Gas, 3 Ton		264.40		290.84	12.95	14.24
40 L.H., Daily Totals		$1932.00		$2881.80	$48.30	$72.05

Crew B-78	Hr.	Daily	Hr.	Daily	Bare Costs	Incl. O&P
1 Labor Foreman (outside)	$37.10	$296.80	$60.80	$486.40	$35.31	$57.76
4 Laborers	35.10	1123.20	57.55	1841.60		
1 Truck Driver (light)	34.35	274.80	55.55	444.40		
1 Paint Striper, S.P., 40 Gallon		148.60		163.46		
1 Flatbed Truck, Gas, 3 Ton		264.40		290.84		
1 Pickup Truck, 3/4 Ton		127.00		139.70	11.25	12.38
48 L.H., Daily Totals		$2234.80		$3366.40	$46.56	$70.13

Crew B-78A	Hr.	Daily	Hr.	Daily	Bare Costs	Incl. O&P
1 Equip. Oper. (light)	$44.75	$358.00	$70.65	$565.20	$44.75	$70.65
1 Line Rem. (metal balls) 115 H.P.		800.80		880.88	100.10	110.11
8 L.H., Daily Totals		$1158.80		$1446.08	$144.85	$180.76

Crew B-78B	Hr.	Daily	Hr.	Daily	Bare Costs	Incl. O&P
2 Laborers	$35.10	$561.60	$57.55	$920.80	$36.17	$59.01
.25 Equip. Oper. (light)	44.75	89.50	70.65	141.30		
1 Pickup Truck, 3/4 Ton		127.00		139.70		
1 Line Rem., 11 H.P.,walk behind		61.20		67.32		
.25 Road Sweeper, S.P., 8' wide		155.20		170.72	19.08	20.99
18 L.H., Daily Totals		$994.50		$1439.84	$55.25	$79.99

Crew B-78C	Hr.	Daily	Hr.	Daily	Bare Costs	Incl. O&P
1 Labor Foreman (outside)	$37.10	$296.80	$60.80	$486.40	$35.31	$57.76
4 Laborers	35.10	1123.20	57.55	1841.60		
1 Truck Driver (light)	34.35	274.80	55.55	444.40		
1 Paint Striper, T.M., 120 Gal.		735.80		809.38		
1 Flatbed Truck, Gas, 3 Ton		264.40		290.84		
1 Pickup Truck, 3/4 Ton		127.00		139.70	23.48	25.83
48 L.H., Daily Totals		$2822.00		$4012.32	$58.79	$83.59

Crew No.	Bare Costs		Incl. Subs O&P		Cost Per Labor-Hour	
Crew B-78D	**Hr.**	**Daily**	**Hr.**	**Daily**	**Bare Costs**	**Incl. O&P**
2 Labor Foremen (Outside)	$37.10	$593.60	$60.80	$972.80	$35.42	$58.00
7 Laborers	35.10	1965.60	57.55	3222.80		
1 Truck Driver (light)	34.35	274.80	55.55	444.40		
1 Paint Striper, T.M., 120 Gal.		735.80		809.38		
1 Flatbed Truck, Gas, 3 Ton		264.40		290.84		
3 Pickup Trucks, 3/4 Ton		381.00		419.10		
1 Air Compressor, 60 cfm		120.40		132.44		
1 -50' Air Hose, 3/4"		3.05		3.36		
1 Breakers, Pavement, 60 lb.		9.80		10.78	18.93	20.82
80 L.H., Daily Totals		$4348.45		$6305.90	$54.36	$78.82

Crew B-78E	Hr.	Daily	Hr.	Daily	Bare Costs	Incl. O&P
2 Labor Foremen (Outside)	$37.10	$593.60	$60.80	$972.80	$35.37	$57.92
9 Laborers	35.10	2527.20	57.55	4143.60		
1 Truck Driver (light)	34.35	274.80	55.55	444.40		
1 Paint Striper, T.M., 120 Gal.		735.80		809.38		
1 Flatbed Truck, Gas, 3 Ton		264.40		290.84		
4 Pickup Trucks, 3/4 Ton		508.00		558.80		
2 Air Compressor, 60 cfm		240.80		264.88		
2 -50' Air Hose, 3/4"		6.10		6.71		
2 Breakers, Pavement, 60 lb.		19.60		21.56	18.49	20.34
96 L.H., Daily Totals		$5170.30		$7512.97	$53.86	$78.26

Crew B-78F	Hr.	Daily	Hr.	Daily	Bare Costs	Incl. O&P
2 Labor Foremen (Outside)	$37.10	$593.60	$60.80	$972.80	$35.33	$57.87
11 Laborers	35.10	3088.80	57.55	5064.40		
1 Truck Driver (light)	34.35	274.80	55.55	444.40		
1 Paint Striper, T.M., 120 Gal.		735.80		809.38		
1 Flatbed Truck, Gas, 3 Ton		264.40		290.84		
7 Pickup Trucks, 3/4 Ton		889.00		977.90		
3 Air Compressor, 60 cfm		361.20		397.32		
3 -50' Air Hose, 3/4"		9.15		10.07		
3 Breakers, Pavement, 60 lb.		29.40		32.34	20.44	22.48
112 L.H., Daily Totals		$6246.15		$8999.44	$55.77	$80.35

Crew B-79	Hr.	Daily	Hr.	Daily	Bare Costs	Incl. O&P
1 Labor Foreman (outside)	$37.10	$296.80	$60.80	$486.40	$35.35	$57.80
3 Laborers	35.10	842.40	57.55	1381.20		
1 Truck Driver (light)	34.35	274.80	55.55	444.40		
1 Paint Striper, T.M., 120 Gal.		735.80		809.38		
1 Heating Kettle, 115 Gallon		44.50		48.95		
1 Flatbed Truck, Gas, 3 Ton		264.40		290.84		
2 Pickup Trucks, 3/4 Ton		254.00		279.40	32.47	35.71
40 L.H., Daily Totals		$2712.70		$3740.57	$67.82	$93.51

Crew B-79A	Hr.	Daily	Hr.	Daily	Bare Costs	Incl. O&P
1.5 Equip. Oper. (light)	$44.75	$537.00	$70.65	$847.80	$44.75	$70.65
.5 Line Remov. (grinder) 115 H.P.		441.40		485.54		
1 Line Rem. (metal balls) 115 H.P.		800.80		880.88	103.52	113.87
12 L.H., Daily Totals		$1779.20		$2214.22	$148.27	$184.52

Crew B-79B	Hr.	Daily	Hr.	Daily	Bare Costs	Incl. O&P
1 Laborer	$35.10	$280.80	$57.55	$460.40	$35.10	$57.55
1 Set of Gases		86.40		95.04	10.80	11.88
8 L.H., Daily Totals		$367.20		$555.44	$45.90	$69.43

Crew No.	Bare Costs		Incl. Subs O&P		Cost Per Labor-Hour	
Crew B-79C	**Hr.**	**Daily**	**Hr.**	**Daily**	**Bare Costs**	**Incl. O&P**
1 Labor Foreman (outside)	$37.10	$296.80	$60.80	$486.40	$35.28	$57.73
5 Laborers	35.10	1404.00	57.55	2302.00		
1 Truck Driver (light)	34.35	274.80	55.55	444.40		
1 Paint Striper, T.M., 120 Gal.		735.80		809.38		
1 Heating Kettle, 115 Gallon		44.50		48.95		
1 Flatbed Truck, Gas, 3 Ton		264.40		290.84		
3 Pickup Trucks, 3/4 Ton		381.00		419.10		
1 Air Compressor, 60 cfm		120.40		132.44		
1 -50' Air Hose, 3/4"		3.05		3.36		
1 Breakers, Pavement, 60 lb.		9.80		10.78	27.84	30.62
56 L.H., Daily Totals		$3534.55		$4947.65	$63.12	$88.35
Crew B-79D	**Hr.**	**Daily**	**Hr.**	**Daily**	**Bare Costs**	**Incl. O&P**
2 Labor Foremen (Outside)	$37.10	$593.60	$60.80	$972.80	$35.51	$58.11
5 Laborers	35.10	1404.00	57.55	2302.00		
1 Truck Driver (light)	34.35	274.80	55.55	444.40		
1 Paint Striper, T.M., 120 Gal.		735.80		809.38		
1 Heating Kettle, 115 Gallon		44.50		48.95		
1 Flatbed Truck, Gas, 3 Ton		264.40		290.84		
4 Pickup Trucks, 3/4 Ton		508.00		558.80		
1 Air Compressor, 60 cfm		120.40		132.44		
1 -50' Air Hose, 3/4"		3.05		3.36		
1 Breakers, Pavement, 60 lb.		9.80		10.78	26.34	28.98
64 L.H., Daily Totals		$3958.35		$5573.74	$61.85	$87.09
Crew B-79E	**Hr.**	**Daily**	**Hr.**	**Daily**	**Bare Costs**	**Incl. O&P**
2 Labor Foremen (Outside)	$37.10	$593.60	$60.80	$972.80	$35.42	$58.00
7 Laborers	35.10	1965.60	57.55	3222.80		
1 Truck Driver (light)	34.35	274.80	55.55	444.40		
1 Paint Striper, T.M., 120 Gal.		735.80		809.38		
1 Heating Kettle, 115 Gallon		44.50		48.95		
1 Flatbed Truck, Gas, 3 Ton		264.40		290.84		
5 Pickup Trucks, 3/4 Ton		635.00		698.50		
2 Air Compressors, 60 cfm		240.80		264.88		
2 -50' Air Hoses, 3/4"		6.10		6.71		
2 Breakers, Pavement, 60 lb.		19.60		21.56	24.33	26.76
80 L.H., Daily Totals		$4780.20		$6780.82	$59.75	$84.76
Crew B-80	**Hr.**	**Daily**	**Hr.**	**Daily**	**Bare Costs**	**Incl. O&P**
1 Labor Foreman (outside)	$37.10	$296.80	$60.80	$486.40	$37.83	$61.14
1 Laborer	35.10	280.80	57.55	460.40		
1 Truck Driver (light)	34.35	274.80	55.55	444.40		
1 Equip. Oper. (light)	44.75	358.00	70.65	565.20		
1 Flatbed Truck, Gas, 3 Ton		264.40		290.84		
1 Earth Auger, Truck-Mtd.		403.40		443.74	20.87	22.96
32 L.H., Daily Totals		$1878.20		$2690.98	$58.69	$84.09
Crew B-80A	**Hr.**	**Daily**	**Hr.**	**Daily**	**Bare Costs**	**Incl. O&P**
3 Laborers	$35.10	$842.40	$57.55	$1381.20	$35.10	$57.55
1 Flatbed Truck, Gas, 3 Ton		264.40		290.84	11.02	12.12
24 L.H., Daily Totals		$1106.80		$1672.04	$46.12	$69.67
Crew B-80B	**Hr.**	**Daily**	**Hr.**	**Daily**	**Bare Costs**	**Incl. O&P**
3 Laborers	$35.10	$842.40	$57.55	$1381.20	$37.51	$60.83
1 Equip. Oper. (light)	44.75	358.00	70.65	565.20		
1 Crane, Flatbed Mounted, 3 Ton		232.40		255.64	7.26	7.99
32 L.H., Daily Totals		$1432.80		$2202.04	$44.77	$68.81

Crew No.	Bare Costs		Incl. Subs O&P		Cost Per Labor-Hour	
Crew B-80C	**Hr.**	**Daily**	**Hr.**	**Daily**	**Bare Costs**	**Incl. O&P**
2 Laborers	$35.10	$561.60	$57.55	$920.80	$34.85	$56.88
1 Truck Driver (light)	34.35	274.80	55.55	444.40		
1 Flatbed Truck, Gas, 1.5 Ton		211.00		232.10		
1 Manual fence post auger, gas		7.60		8.36	9.11	10.02
24 L.H., Daily Totals		$1055.00		$1605.66	$43.96	$66.90
Crew B-81	**Hr.**	**Daily**	**Hr.**	**Daily**	**Bare Costs**	**Incl. O&P**
1 Laborer	$35.10	$280.80	$57.55	$460.40	$38.98	$62.72
1 Equip. Oper. (med.)	46.55	372.40	73.50	588.00		
1 Truck Driver (heavy)	35.30	282.40	57.10	456.80		
1 Hydromulcher, T.M., 3000 Gal.		303.80		334.18		
1 Truck Tractor, 220 H.P.		314.80		346.28	25.77	28.35
24 L.H., Daily Totals		$1554.20		$2185.66	$64.76	$91.07
Crew B-81A	**Hr.**	**Daily**	**Hr.**	**Daily**	**Bare Costs**	**Incl. O&P**
1 Laborer	$35.10	$280.80	$57.55	$460.40	$34.73	$56.55
1 Truck Driver (light)	34.35	274.80	55.55	444.40		
1 Hydromulcher, T.M., 600 Gal.		120.40		132.44		
1 Flatbed Truck, Gas, 3 Ton		264.40		290.84	24.05	26.45
16 L.H., Daily Totals		$940.40		$1328.08	$58.77	$83.00
Crew B-82	**Hr.**	**Daily**	**Hr.**	**Daily**	**Bare Costs**	**Incl. O&P**
1 Laborer	$35.10	$280.80	$57.55	$460.40	$39.92	$64.10
1 Equip. Oper. (light)	44.75	358.00	70.65	565.20		
1 Horiz. Borer, 6 H.P.		77.20		84.92	4.83	5.31
16 L.H., Daily Totals		$716.00		$1110.52	$44.75	$69.41
Crew B-82A	**Hr.**	**Daily**	**Hr.**	**Daily**	**Bare Costs**	**Incl. O&P**
1 Laborer	$35.10	$280.80	$57.55	$460.40	$39.92	$64.10
1 Equip. Oper. (light)	44.75	358.00	70.65	565.20		
1 Flatbed Truck, Gas, 3 Ton		264.40		290.84		
1 Flatbed Trailer, 25 Ton		108.60		119.46		
1 Horiz. Dir. Drill, 20k lb. thrust		622.00		684.20	62.19	68.41
16 L.H., Daily Totals		$1633.80		$2120.10	$102.11	$132.51
Crew B-82B	**Hr.**	**Daily**	**Hr.**	**Daily**	**Bare Costs**	**Incl. O&P**
2 Laborers	$35.10	$561.60	$57.55	$920.80	$38.32	$61.92
1 Equip. Oper. (light)	44.75	358.00	70.65	565.20		
1 Flatbed Truck, Gas, 3 Ton		264.40		290.84		
1 Flatbed Trailer, 25 Ton		108.60		119.46		
1 Horiz. Dir. Drill, 30k lb. thrust		886.20		974.82	52.47	57.71
24 L.H., Daily Totals		$2178.80		$2871.12	$90.78	$119.63
Crew B-82C	**Hr.**	**Daily**	**Hr.**	**Daily**	**Bare Costs**	**Incl. O&P**
2 Laborers	$35.10	$561.60	$57.55	$920.80	$38.32	$61.92
1 Equip. Oper. (light)	44.75	358.00	70.65	565.20		
1 Flatbed Truck, Gas, 3 Ton		264.40		290.84		
1 Flatbed Trailer, 25 Ton		108.60		119.46		
1 Horiz. Dir. Drill, 50k lb. thrust		1176.00		1293.60	64.54	71.00
24 L.H., Daily Totals		$2468.60		$3189.90	$102.86	$132.91
Crew B-82D	**Hr.**	**Daily**	**Hr.**	**Daily**	**Bare Costs**	**Incl. O&P**
1 Equip. Oper. (light)	$44.75	$358.00	$70.65	$565.20	$44.75	$70.65
1 Mud Trailer for HDD, 1500 gallon		309.20		340.12	38.65	42.52
8 L.H., Daily Totals		$667.20		$905.32	$83.40	$113.17

Crew No.	Bare Costs		Incl. Subs O&P		Cost Per Labor-Hour	
Crew B-83	**Hr.**	**Daily**	**Hr.**	**Daily**	**Bare Costs**	**Incl. O&P**
1 Tugboat Captain	$46.55	$372.40	$73.50	$588.00	$40.83	$65.53
1 Tugboat Hand	35.10	280.80	57.55	460.40		
1 Tugboat, 250 H.P.		748.00		822.80	46.75	51.42
16 L.H., Daily Totals		$1401.20		$1871.20	$87.58	$116.95

Crew B-84	Hr.	Daily	Hr.	Daily	Bare Costs	Incl. O&P
1 Equip. Oper. (med.)	$46.55	$372.40	$73.50	$588.00	$46.55	$73.50
1 Rotary Mower/Tractor		343.40		377.74	42.92	47.22
8 L.H., Daily Totals		$715.80		$965.74	$89.47	$120.72

Crew B-85	Hr.	Daily	Hr.	Daily	Bare Costs	Incl. O&P
3 Laborers	$35.10	$842.40	$57.55	$1381.20	$37.43	$60.65
1 Equip. Oper. (med.)	46.55	372.40	73.50	588.00		
1 Truck Driver (heavy)	35.30	282.40	57.10	456.80		
1 Aerial Lift Truck, 80′		637.60		701.36		
1 Brush Chipper, 12″, 130 H.P.		283.80		312.18		
1 Pruning Saw, Rotary		6.30		6.93	23.19	25.51
40 L.H., Daily Totals		$2424.90		$3446.47	$60.62	$86.16

Crew B-86	Hr.	Daily	Hr.	Daily	Bare Costs	Incl. O&P
1 Equip. Oper. (med.)	$46.55	$372.40	$73.50	$588.00	$46.55	$73.50
1 Stump Chipper, S.P.		170.70		187.77	21.34	23.47
8 L.H., Daily Totals		$543.10		$775.77	$67.89	$96.97

Crew B-86A	Hr.	Daily	Hr.	Daily	Bare Costs	Incl. O&P
1 Equip. Oper. (med.)	$46.55	$372.40	$73.50	$588.00	$46.55	$73.50
1 Grader, 30,000 Lbs.		610.80		671.88	76.35	83.98
8 L.H., Daily Totals		$983.20		$1259.88	$122.90	$157.49

Crew B-86B	Hr.	Daily	Hr.	Daily	Bare Costs	Incl. O&P
1 Equip. Oper. (med.)	$46.55	$372.40	$73.50	$588.00	$46.55	$73.50
1 Dozer, 200 H.P.		1243.00		1367.30	155.38	170.91
8 L.H., Daily Totals		$1615.40		$1955.30	$201.93	$244.41

Crew B-87	Hr.	Daily	Hr.	Daily	Bare Costs	Incl. O&P
1 Laborer	$35.10	$280.80	$57.55	$460.40	$44.26	$70.31
4 Equip. Oper. (med.)	46.55	1489.60	73.50	2352.00		
2 Feller Bunchers, 100 H.P.		1154.40		1269.84		
1 Log Chipper, 22″ Tree		710.60		781.66		
1 Dozer, 105 H.P.		550.80		605.88		
1 Chain Saw, gas, 36″ Long		43.60		47.96	61.48	67.63
40 L.H., Daily Totals		$4229.80		$5517.74	$105.75	$137.94

Crew B-88	Hr.	Daily	Hr.	Daily	Bare Costs	Incl. O&P
1 Laborer	$35.10	$280.80	$57.55	$460.40	$44.91	$71.22
6 Equip. Oper. (med.)	46.55	2234.40	73.50	3528.00		
2 Feller Bunchers, 100 H.P.		1154.40		1269.84		
1 Log Chipper, 22″ Tree		710.60		781.66		
2 Log Skidders, 50 H.P.		1730.40		1903.44		
1 Dozer, 105 H.P.		550.80		605.88		
1 Chain Saw, gas, 36″ Long		43.60		47.96	74.82	82.30
56 L.H., Daily Totals		$6705.00		$8597.18	$119.73	$153.52

Crew B-89	Hr.	Daily	Hr.	Daily	Bare Costs	Incl. O&P
1 Equip. Oper. (light)	$44.75	$358.00	$70.65	$565.20	$39.55	$63.10
1 Truck Driver (light)	34.35	274.80	55.55	444.40		
1 Flatbed Truck, Gas, 3 Ton		264.40		290.84		
1 Concrete Saw		148.00		162.80		
1 Water Tank, 65 Gal.		17.40		19.14	26.86	29.55
16 L.H., Daily Totals		$1062.60		$1482.38	$66.41	$92.65

Crew B-89A	Hr.	Daily	Hr.	Daily	Bare Costs	Incl. O&P
1 Skilled Worker	$45.40	$363.20	$73.80	$590.40	$40.25	$65.67
1 Laborer	35.10	280.80	57.55	460.40		
1 Core Drill (large)		110.20		121.22	6.89	7.58
16 L.H., Daily Totals		$754.20		$1172.02	$47.14	$73.25

Crew B-89B	Hr.	Daily	Hr.	Daily	Bare Costs	Incl. O&P
1 Equip. Oper. (light)	$44.75	$358.00	$70.65	$565.20	$39.55	$63.10
1 Truck Driver, Light	34.35	274.80	55.55	444.40		
1 Wall Saw, Hydraulic, 10 H.P.		92.00		101.20		
1 Generator, Diesel, 100 kW		363.40		399.74		
1 Water Tank, 65 Gal.		17.40		19.14		
1 Flatbed Truck, Gas, 3 Ton		264.40		290.84	46.08	50.68
16 L.H., Daily Totals		$1370.00		$1820.52	$85.63	$113.78

Crew B-90	Hr.	Daily	Hr.	Daily	Bare Costs	Incl. O&P
1 Labor Foreman (outside)	$37.10	$296.80	$60.80	$486.40	$37.81	$61.12
3 Laborers	35.10	842.40	57.55	1381.20		
2 Equip. Oper. (light)	44.75	716.00	70.65	1130.40		
2 Truck Drivers (heavy)	35.30	564.80	57.10	913.60		
1 Road Mixer, 310 H.P.		1882.00		2070.20		
1 Dist. Truck, 2000 Gal.		276.40		304.04	33.73	37.10
64 L.H., Daily Totals		$4578.40		$6285.84	$71.54	$98.22

Crew B-90A	Hr.	Daily	Hr.	Daily	Bare Costs	Incl. O&P
1 Labor Foreman (outside)	$37.10	$296.80	$60.80	$486.40	$41.93	$67.13
2 Laborers	35.10	561.60	57.55	920.80		
4 Equip. Oper. (med.)	46.55	1489.60	73.50	2352.00		
2 Graders, 30,000 Lbs.		1221.60		1343.76		
1 Tandem Roller, 10 Ton		227.60		250.36		
1 Roller, Pneum. Whl, 12 Ton		313.20		344.52	31.47	34.62
56 L.H., Daily Totals		$4110.40		$5697.84	$73.40	$101.75

Crew B-90B	Hr.	Daily	Hr.	Daily	Bare Costs	Incl. O&P
1 Labor Foreman (outside)	$37.10	$296.80	$60.80	$486.40	$41.16	$66.07
2 Laborers	35.10	561.60	57.55	920.80		
3 Equip. Oper. (med.)	46.55	1117.20	73.50	1764.00		
1 Roller, Pneum. Whl, 12 Ton		313.20		344.52		
1 Road Mixer, 310 H.P.		1882.00		2070.20	45.73	50.31
48 L.H., Daily Totals		$4170.80		$5585.92	$86.89	$116.37

Crew B-90C	Hr.	Daily	Hr.	Daily	Bare Costs	Incl. O&P
1 Labor Foreman (outside)	$37.10	$296.80	$60.80	$486.40	$38.46	$62.07
4 Laborers	35.10	1123.20	57.55	1841.60		
3 Equip. Oper. (med.)	46.55	1117.20	73.50	1764.00		
3 Truck Drivers (heavy)	35.30	847.20	57.10	1370.40		
3 Road Mixers, 310 H.P.		5646.00		6210.60	64.16	70.58
88 L.H., Daily Totals		$9030.40		$11673.00	$102.62	$132.65

Crew B-90D	Hr.	Daily	Hr.	Daily	Bare Costs	Incl. O&P
1 Labor Foreman (outside)	$37.10	$296.80	$60.80	$486.40	$37.94	$61.38
6 Laborers	35.10	1684.80	57.55	2762.40		
3 Equip. Oper. (med.)	46.55	1117.20	73.50	1764.00		
3 Truck Drivers (heavy)	35.30	847.20	57.10	1370.40		
3 Road Mixers, 310 H.P.		5646.00		6210.60	54.29	59.72
104 L.H., Daily Totals		$9592.00		$12593.80	$92.23	$121.09

Crews

Crew No.	Bare Costs		Incl. Subs O&P		Cost Per Labor-Hour	
Crew B-91	**Hr.**	**Daily**	**Hr.**	**Daily**	**Bare Costs**	**Incl. O&P**
1 Labor Foreman (outside)	$37.10	$296.80	$60.80	$486.40	$41.10	$65.88
2 Laborers	35.10	561.60	57.55	920.80		
4 Equip. Oper. (med.)	46.55	1489.60	73.50	2352.00		
1 Truck Driver (heavy)	35.30	282.40	57.10	456.80		
1 Dist. Tanker, 3000 Gallon		306.00		336.60		
1 Truck Tractor, 6x4, 380 H.P.		522.20		574.42		
1 Aggreg. Spreader, S.P.		781.80		859.98		
1 Roller, Pneum. Whl, 12 Ton		313.20		344.52		
1 Tandem Roller, 10 Ton		227.60		250.36	33.61	36.97
64 L.H., Daily Totals		$4781.20		$6581.88	$74.71	$102.84
Crew B-91B	**Hr.**	**Daily**	**Hr.**	**Daily**	**Bare Costs**	**Incl. O&P**
1 Laborer	$35.10	$280.80	$57.55	$460.40	$40.83	$65.53
1 Equipment Oper. (med.)	46.55	372.40	73.50	588.00		
1 Road Sweeper, Vac. Assist.		876.60		964.26	54.79	60.27
16 L.H., Daily Totals		$1529.80		$2012.66	$95.61	$125.79
Crew B-91C	**Hr.**	**Daily**	**Hr.**	**Daily**	**Bare Costs**	**Incl. O&P**
1 Laborer	$35.10	$280.80	$57.55	$460.40	$34.73	$56.55
1 Truck Driver (light)	34.35	274.80	55.55	444.40		
1 Catch Basin Cleaning Truck		535.80		589.38	33.49	36.84
16 L.H., Daily Totals		$1091.40		$1494.18	$68.21	$93.39
Crew B-91D	**Hr.**	**Daily**	**Hr.**	**Daily**	**Bare Costs**	**Incl. O&P**
1 Labor Foreman (outside)	$37.10	$296.80	$60.80	$486.40	$39.69	$63.87
5 Laborers	35.10	1404.00	57.55	2302.00		
5 Equip. Oper. (med.)	46.55	1862.00	73.50	2940.00		
2 Truck Drivers (heavy)	35.30	564.80	57.10	913.60		
1 Aggreg. Spreader, S.P.		781.80		859.98		
2 Truck Tractor, 6x4, 380 H.P.		1044.40		1148.84		
2 Dist. Tanker, 3000 Gallon		612.00		673.20		
2 Pavement brush, towed		163.60		179.96		
2 Roller, Pneum. Whl, 12 Ton		626.40		689.04	31.04	34.14
104 L.H., Daily Totals		$7355.80		$10193.02	$70.73	$98.01
Crew B-92	**Hr.**	**Daily**	**Hr.**	**Daily**	**Bare Costs**	**Incl. O&P**
1 Labor Foreman (outside)	$37.10	$296.80	$60.80	$486.40	$35.60	$58.36
3 Laborers	35.10	842.40	57.55	1381.20		
1 Crack Cleaner, 25 H.P.		57.20		62.92		
1 Air Compressor, 60 cfm		120.40		132.44		
1 Tar Kettle, T.M.		77.20		84.92		
1 Flatbed Truck, Gas, 3 Ton		264.40		290.84	16.23	17.85
32 L.H., Daily Totals		$1658.40		$2438.72	$51.83	$76.21
Crew B-93	**Hr.**	**Daily**	**Hr.**	**Daily**	**Bare Costs**	**Incl. O&P**
1 Equip. Oper. (med.)	$46.55	$372.40	$73.50	$588.00	$46.55	$73.50
1 Feller Buncher, 100 H.P.		577.20		634.92	72.15	79.36
8 L.H., Daily Totals		$949.60		$1222.92	$118.70	$152.87
Crew B-94A	**Hr.**	**Daily**	**Hr.**	**Daily**	**Bare Costs**	**Incl. O&P**
1 Laborer	$35.10	$280.80	$57.55	$460.40	$35.10	$57.55
1 Diaphragm Water Pump, 2″		68.60		75.46		
1 -20′ Suction Hose, 2″		1.95		2.15		
2 -50′ Discharge Hoses, 2″		3.50		3.85	9.26	10.18
8 L.H., Daily Totals		$354.85		$541.86	$44.36	$67.73

Crew No.	Bare Costs		Incl. Subs O&P		Cost Per Labor-Hour	
Crew B-94B	**Hr.**	**Daily**	**Hr.**	**Daily**	**Bare Costs**	**Incl. O&P**
1 Laborer	$35.10	$280.80	$57.55	$460.40	$35.10	$57.55
1 Diaphragm Water Pump, 4″		111.20		122.32		
1 -20′ Suction Hose, 4″		3.25		3.58		
2 -50′ Discharge Hoses, 4″		9.40		10.34	15.48	17.03
8 L.H., Daily Totals		$404.65		$596.63	$50.58	$74.58
Crew B-94C	**Hr.**	**Daily**	**Hr.**	**Daily**	**Bare Costs**	**Incl. O&P**
1 Laborer	$35.10	$280.80	$57.55	$460.40	$35.10	$57.55
1 Centrifugal Water Pump, 3″		73.80		81.18		
1 -20′ Suction Hose, 3″		2.85		3.13		
2 -50′ Discharge Hoses, 3″		6.10		6.71	10.34	11.38
8 L.H., Daily Totals		$363.55		$551.42	$45.44	$68.93
Crew B-94D	**Hr.**	**Daily**	**Hr.**	**Daily**	**Bare Costs**	**Incl. O&P**
1 Laborer	$35.10	$280.80	$57.55	$460.40	$35.10	$57.55
1 Centr. Water Pump, 6″		308.40		339.24		
1 -20′ Suction Hose, 6″		11.50		12.65		
2 -50′ Discharge Hoses, 6″		24.20		26.62	43.01	47.31
8 L.H., Daily Totals		$624.90		$838.91	$78.11	$104.86
Crew C-1	**Hr.**	**Daily**	**Hr.**	**Daily**	**Bare Costs**	**Incl. O&P**
3 Carpenters	$44.10	$1058.40	$72.30	$1735.20	$41.85	$68.61
1 Laborer	35.10	280.80	57.55	460.40		
32 L.H., Daily Totals		$1339.20		$2195.60	$41.85	$68.61
Crew C-2	**Hr.**	**Daily**	**Hr.**	**Daily**	**Bare Costs**	**Incl. O&P**
1 Carpenter Foreman (outside)	$46.10	$368.80	$75.55	$604.40	$42.93	$70.38
4 Carpenters	44.10	1411.20	72.30	2313.60		
1 Laborer	35.10	280.80	57.55	460.40		
48 L.H., Daily Totals		$2060.80		$3378.40	$42.93	$70.38
Crew C-2A	**Hr.**	**Daily**	**Hr.**	**Daily**	**Bare Costs**	**Incl. O&P**
1 Carpenter Foreman (outside)	$46.10	$368.80	$75.55	$604.40	$42.64	$69.42
3 Carpenters	44.10	1058.40	72.30	1735.20		
1 Cement Finisher	42.35	338.80	66.55	532.40		
1 Laborer	35.10	280.80	57.55	460.40		
48 L.H., Daily Totals		$2046.80		$3332.40	$42.64	$69.42
Crew C-3	**Hr.**	**Daily**	**Hr.**	**Daily**	**Bare Costs**	**Incl. O&P**
1 Rodman Foreman (outside)	$51.10	$408.80	$85.20	$681.60	$45.31	$74.79
4 Rodmen (reinf.)	49.10	1571.20	81.85	2619.20		
1 Equip. Oper. (light)	44.75	358.00	70.65	565.20		
2 Laborers	35.10	561.60	57.55	920.80		
3 Stressing Equipment		30.60		33.66		
.5 Grouting Equipment		77.30		85.03	1.69	1.85
64 L.H., Daily Totals		$3007.50		$4905.49	$46.99	$76.65
Crew C-4	**Hr.**	**Daily**	**Hr.**	**Daily**	**Bare Costs**	**Incl. O&P**
1 Rodman Foreman (outside)	$51.10	$408.80	$85.20	$681.60	$49.60	$82.69
3 Rodmen (reinf.)	49.10	1178.40	81.85	1964.40		
3 Stressing Equipment		30.60		33.66	.96	1.05
32 L.H., Daily Totals		$1617.80		$2679.66	$50.56	$83.74
Crew C-4A	**Hr.**	**Daily**	**Hr.**	**Daily**	**Bare Costs**	**Incl. O&P**
2 Rodmen (reinf.)	$49.10	$785.60	$81.85	$1309.60	$49.10	$81.85
4 Stressing Equipment		40.80		44.88	2.55	2.81
16 L.H., Daily Totals		$826.40		$1354.48	$51.65	$84.66

Crew No.	Bare Costs		Incl. Subs O&P		Cost Per Labor-Hour	
Crew C-5	**Hr.**	**Daily**	**Hr.**	**Daily**	**Bare Costs**	**Incl. O&P**
1 Rodman Foreman (outside)	$51.10	$408.80	$85.20	$681.60	$48.08	$79.04
4 Rodmen (reinf.)	49.10	1571.20	81.85	2619.20		
1 Equip. Oper. (crane)	47.80	382.40	75.50	604.00		
1 Equip. Oper. Oiler	41.25	330.00	65.15	521.20		
1 Hyd. Crane, 25 Ton		697.20		766.92	12.45	13.70
56 L.H., Daily Totals		$3389.60		$5192.92	$60.53	$92.73

Crew C-6	Hr.	Daily	Hr.	Daily	Bare Costs	Incl. O&P
1 Labor Foreman (outside)	$37.10	$296.80	$60.80	$486.40	$36.64	$59.59
4 Laborers	35.10	1123.20	57.55	1841.60		
1 Cement Finisher	42.35	338.80	66.55	532.40		
2 Gas Engine Vibrators		54.80		60.28	1.14	1.26
48 L.H., Daily Totals		$1813.60		$2920.68	$37.78	$60.85

Crew C-7	Hr.	Daily	Hr.	Daily	Bare Costs	Incl. O&P
1 Labor Foreman (outside)	$37.10	$296.80	$60.80	$486.40	$38.08	$61.53
5 Laborers	35.10	1404.00	57.55	2302.00		
1 Cement Finisher	42.35	338.80	66.55	532.40		
1 Equip. Oper. (med.)	46.55	372.40	73.50	588.00		
1 Equip. Oper. (oiler)	41.25	330.00	65.15	521.20		
2 Gas Engine Vibrators		54.80		60.28		
1 Concrete Bucket, 1 C.Y.		22.40		24.64		
1 Hyd. Crane, 55 Ton		1100.00		1210.00	16.35	17.98
72 L.H., Daily Totals		$3919.20		$5724.92	$54.43	$79.51

Crew C-8	Hr.	Daily	Hr.	Daily	Bare Costs	Incl. O&P
1 Labor Foreman (outside)	$37.10	$296.80	$60.80	$486.40	$39.09	$62.86
3 Laborers	35.10	842.40	57.55	1381.20		
2 Cement Finishers	42.35	677.60	66.55	1064.80		
1 Equip. Oper. (med.)	46.55	372.40	73.50	588.00		
1 Concrete Pump (small)		715.60		787.16	12.78	14.06
56 L.H., Daily Totals		$2904.80		$4307.56	$51.87	$76.92

Crew C-8A	Hr.	Daily	Hr.	Daily	Bare Costs	Incl. O&P
1 Labor Foreman (outside)	$37.10	$296.80	$60.80	$486.40	$37.85	$61.09
3 Laborers	35.10	842.40	57.55	1381.20		
2 Cement Finishers	42.35	677.60	66.55	1064.80		
48 L.H., Daily Totals		$1816.80		$2932.40	$37.85	$61.09

Crew C-8B	Hr.	Daily	Hr.	Daily	Bare Costs	Incl. O&P
1 Labor Foreman (outside)	$37.10	$296.80	$60.80	$486.40	$37.79	$61.39
3 Laborers	35.10	842.40	57.55	1381.20		
1 Equip. Oper. (med.)	46.55	372.40	73.50	588.00		
1 Vibrating Power Screed		65.50		72.05		
1 Roller, Vibratory, 25 Ton		640.20		704.22		
1 Dozer, 200 H.P.		1243.00		1367.30	48.72	53.59
40 L.H., Daily Totals		$3460.30		$4599.17	$86.51	$114.98

Crew C-8C	Hr.	Daily	Hr.	Daily	Bare Costs	Incl. O&P
1 Labor Foreman (outside)	$37.10	$296.80	$60.80	$486.40	$38.55	$62.25
3 Laborers	35.10	842.40	57.55	1381.20		
1 Cement Finisher	42.35	338.80	66.55	532.40		
1 Equip. Oper. (med.)	46.55	372.40	73.50	588.00		
1 Shotcrete Rig, 12 C.Y./hr		242.00		266.20		
1 Air Compressor, 160 cfm		136.80		150.48		
4 -50' Air Hoses, 1"		16.40		18.04		
4 -50' Air Hoses, 2"		31.00		34.10	8.88	9.77
48 L.H., Daily Totals		$2276.60		$3456.82	$47.43	$72.02

Crew C-8D	Hr.	Daily	Hr.	Daily	Bare Costs	Incl. O&P
1 Labor Foreman (outside)	$37.10	$296.80	$60.80	$486.40	$39.83	$63.89
1 Laborer	35.10	280.80	57.55	460.40		
1 Cement Finisher	42.35	338.80	66.55	532.40		
1 Equipment Oper. (light)	44.75	358.00	70.65	565.20		
1 Air Compressor, 250 cfm		177.00		194.70		
2 -50' Air Hoses, 1"		8.20		9.02	5.79	6.37
32 L.H., Daily Totals		$1459.60		$2248.12	$45.61	$70.25

Crew C-8E	Hr.	Daily	Hr.	Daily	Bare Costs	Incl. O&P
1 Labor Foreman (outside)	$37.10	$296.80	$60.80	$486.40	$38.25	$61.77
3 Laborers	35.10	842.40	57.55	1381.20		
1 Cement Finisher	42.35	338.80	66.55	532.40		
1 Equipment Oper. (light)	44.75	358.00	70.65	565.20		
1 Shotcrete Rig, 35 C.Y./hr		268.80		295.68		
1 Air Compressor, 250 cfm		177.00		194.70		
4 -50' Air Hoses, 1"		16.40		18.04		
4 -50' Air Hoses, 2"		31.00		34.10	10.28	11.30
48 L.H., Daily Totals		$2329.20		$3507.72	$48.52	$73.08

Crew C-10	Hr.	Daily	Hr.	Daily	Bare Costs	Incl. O&P
1 Laborer	$35.10	$280.80	$57.55	$460.40	$39.93	$63.55
2 Cement Finishers	42.35	677.60	66.55	1064.80		
24 L.H., Daily Totals		$958.40		$1525.20	$39.93	$63.55

Crew C-10B	Hr.	Daily	Hr.	Daily	Bare Costs	Incl. O&P
3 Laborers	$35.10	$842.40	$57.55	$1381.20	$38.00	$61.15
2 Cement Finishers	42.35	677.60	66.55	1064.80		
1 Concrete Mixer, 10 C.F.		160.60		176.66		
2 Trowels, 48" Walk-Behind		89.20		98.12	6.25	6.87
40 L.H., Daily Totals		$1769.80		$2720.78	$44.24	$68.02

Crew C-10C	Hr.	Daily	Hr.	Daily	Bare Costs	Incl. O&P
1 Laborer	$35.10	$280.80	$57.55	$460.40	$39.93	$63.55
2 Cement Finishers	42.35	677.60	66.55	1064.80		
1 Trowel, 48" Walk-Behind		44.60		49.06	1.86	2.04
24 L.H., Daily Totals		$1003.00		$1574.26	$41.79	$65.59

Crew C-10D	Hr.	Daily	Hr.	Daily	Bare Costs	Incl. O&P
1 Laborer	$35.10	$280.80	$57.55	$460.40	$39.93	$63.55
2 Cement Finishers	42.35	677.60	66.55	1064.80		
1 Vibrating Power Screed		65.50		72.05		
1 Trowel, 48" Walk-Behind		44.60		49.06	4.59	5.05
24 L.H., Daily Totals		$1068.50		$1646.31	$44.52	$68.60

Crew C-10E	Hr.	Daily	Hr.	Daily	Bare Costs	Incl. O&P
1 Laborer	$35.10	$280.80	$57.55	$460.40	$39.93	$63.55
2 Cement Finishers	42.35	677.60	66.55	1064.80		
1 Vibrating Power Screed		65.50		72.05		
1 Cement Trowel, 96" Ride-On		174.20		191.62	9.99	10.99
24 L.H., Daily Totals		$1198.10		$1788.87	$49.92	$74.54

Crew C-11	Hr.	Daily	Hr.	Daily	Bare Costs	Incl. O&P
1 Struc. Steel Foreman (outside)	$51.35	$410.80	$95.40	$763.20	$48.50	$87.36
6 Struc. Steel Workers	49.35	2368.80	91.70	4401.60		
1 Equip. Oper. (crane)	47.80	382.40	75.50	604.00		
1 Equip. Oper. Oiler	41.25	330.00	65.15	521.20		
1 Lattice Boom Crane, 150 Ton		1829.00		2011.90	25.40	27.94
72 L.H., Daily Totals		$5321.00		$8301.90	$73.90	$115.30

Crew No.	Bare Costs		Incl. Subs O&P		Cost Per Labor-Hour	
Crew C-12	**Hr.**	**Daily**	**Hr.**	**Daily**	**Bare Costs**	**Incl. O&P**
1 Carpenter Foreman (outside)	$46.10	$368.80	$75.55	$604.40	$43.55	$70.92
3 Carpenters	44.10	1058.40	72.30	1735.20		
1 Laborer	35.10	280.80	57.55	460.40		
1 Equip. Oper. (crane)	47.80	382.40	75.50	604.00		
1 Hyd. Crane, 12 Ton		612.00		673.20	12.75	14.03
48 L.H., Daily Totals		$2702.40		$4077.20	$56.30	$84.94

Crew C-13	Hr.	Daily	Hr.	Daily	Bare Costs	Incl. O&P
1 Struc. Steel Worker	$49.35	$394.80	$91.70	$733.60	$47.60	$85.23
1 Welder	49.35	394.80	91.70	733.60		
1 Carpenter	44.10	352.80	72.30	578.40		
1 Welder, gas engine, 300 amp		121.60		133.76	5.07	5.57
24 L.H., Daily Totals		$1264.00		$2179.36	$52.67	$90.81

Crew C-14	Hr.	Daily	Hr.	Daily	Bare Costs	Incl. O&P
1 Carpenter Foreman (outside)	$46.10	$368.80	$75.55	$604.40	$43.17	$70.47
5 Carpenters	44.10	1764.00	72.30	2892.00		
4 Laborers	35.10	1123.20	57.55	1841.60		
4 Rodmen (reinf.)	49.10	1571.20	81.85	2619.20		
2 Cement Finishers	42.35	677.60	66.55	1064.80		
1 Equip. Oper. (crane)	47.80	382.40	75.50	604.00		
1 Equip. Oper. Oiler	41.25	330.00	65.15	521.20		
1 Hyd. Crane, 80 Ton		1554.00		1709.40	10.79	11.87
144 L.H., Daily Totals		$7771.20		$11856.60	$53.97	$82.34

Crew C-14A	Hr.	Daily	Hr.	Daily	Bare Costs	Incl. O&P
1 Carpenter Foreman (outside)	$46.10	$368.80	$75.55	$604.40	$44.29	$72.60
16 Carpenters	44.10	5644.80	72.30	9254.40		
4 Rodmen (reinf.)	49.10	1571.20	81.85	2619.20		
2 Laborers	35.10	561.60	57.55	920.80		
1 Cement Finisher	42.35	338.80	66.55	532.40		
1 Equip. Oper. (med.)	46.55	372.40	73.50	588.00		
1 Gas Engine Vibrator		27.40		30.14		
1 Concrete Pump (small)		715.60		787.16	3.71	4.09
200 L.H., Daily Totals		$9600.60		$15336.50	$48.00	$76.68

Crew C-14B	Hr.	Daily	Hr.	Daily	Bare Costs	Incl. O&P
1 Carpenter Foreman (outside)	$46.10	$368.80	$75.55	$604.40	$44.21	$72.36
16 Carpenters	44.10	5644.80	72.30	9254.40		
4 Rodmen (reinf.)	49.10	1571.20	81.85	2619.20		
2 Laborers	35.10	561.60	57.55	920.80		
2 Cement Finishers	42.35	677.60	66.55	1064.80		
1 Equip. Oper. (med.)	46.55	372.40	73.50	588.00		
1 Gas Engine Vibrator		27.40		30.14		
1 Concrete Pump (small)		715.60		787.16	3.57	3.93
208 L.H., Daily Totals		$9939.40		$15868.90	$47.79	$76.29

Crew C-14C	Hr.	Daily	Hr.	Daily	Bare Costs	Incl. O&P
1 Carpenter Foreman (outside)	$46.10	$368.80	$75.55	$604.40	$42.26	$69.27
6 Carpenters	44.10	2116.80	72.30	3470.40		
2 Rodmen (reinf.)	49.10	785.60	81.85	1309.60		
4 Laborers	35.10	1123.20	57.55	1841.60		
1 Cement Finisher	42.35	338.80	66.55	532.40		
1 Gas Engine Vibrator		27.40		30.14	.24	.27
112 L.H., Daily Totals		$4760.60		$7788.54	$42.51	$69.54

Crew No.	Bare Costs		Incl. Subs O&P		Cost Per Labor-Hour	
Crew C-14D	**Hr.**	**Daily**	**Hr.**	**Daily**	**Bare Costs**	**Incl. O&P**
1 Carpenter Foreman (outside)	$46.10	$368.80	$75.55	$604.40	$43.89	$71.83
18 Carpenters	44.10	6350.40	72.30	10411.20		
2 Rodmen (reinf.)	49.10	785.60	81.85	1309.60		
2 Laborers	35.10	561.60	57.55	920.80		
1 Cement Finisher	42.35	338.80	66.55	532.40		
1 Equip. Oper. (med.)	46.55	372.40	73.50	588.00		
1 Gas Engine Vibrator		27.40		30.14		
1 Concrete Pump (small)		715.60		787.16	3.71	4.09
200 L.H., Daily Totals		$9520.60		$15183.70	$47.60	$75.92

Crew C-14E	Hr.	Daily	Hr.	Daily	Bare Costs	Incl. O&P
1 Carpenter Foreman (outside)	$46.10	$368.80	$75.55	$604.40	$43.49	$71.52
2 Carpenters	44.10	705.60	72.30	1156.80		
4 Rodmen (reinf.)	49.10	1571.20	81.85	2619.20		
3 Laborers	35.10	842.40	57.55	1381.20		
1 Cement Finisher	42.35	338.80	66.55	532.40		
1 Gas Engine Vibrator		27.40		30.14	.31	.34
88 L.H., Daily Totals		$3854.20		$6324.14	$43.80	$71.87

Crew C-14F	Hr.	Daily	Hr.	Daily	Bare Costs	Incl. O&P
1 Labor Foreman (outside)	$37.10	$296.80	$60.80	$486.40	$40.16	$63.91
2 Laborers	35.10	561.60	57.55	920.80		
6 Cement Finishers	42.35	2032.80	66.55	3194.40		
1 Gas Engine Vibrator		27.40		30.14	.38	.42
72 L.H., Daily Totals		$2918.60		$4631.74	$40.54	$64.33

Crew C-14G	Hr.	Daily	Hr.	Daily	Bare Costs	Incl. O&P
1 Labor Foreman (outside)	$37.10	$296.80	$60.80	$486.40	$39.53	$63.16
2 Laborers	35.10	561.60	57.55	920.80		
4 Cement Finishers	42.35	1355.20	66.55	2129.60		
1 Gas Engine Vibrator		27.40		30.14	.49	.54
56 L.H., Daily Totals		$2241.00		$3566.94	$40.02	$63.70

Crew C-14H	Hr.	Daily	Hr.	Daily	Bare Costs	Incl. O&P
1 Carpenter Foreman (outside)	$46.10	$368.80	$75.55	$604.40	$43.48	$71.02
2 Carpenters	44.10	705.60	72.30	1156.80		
1 Rodman (reinf.)	49.10	392.80	81.85	654.80		
1 Laborer	35.10	280.80	57.55	460.40		
1 Cement Finisher	42.35	338.80	66.55	532.40		
1 Gas Engine Vibrator		27.40		30.14	.57	.63
48 L.H., Daily Totals		$2114.20		$3438.94	$44.05	$71.64

Crew C-14L	Hr.	Daily	Hr.	Daily	Bare Costs	Incl. O&P
1 Carpenter Foreman (outside)	$46.10	$368.80	$75.55	$604.40	$41.12	$67.17
6 Carpenters	44.10	2116.80	72.30	3470.40		
4 Laborers	35.10	1123.20	57.55	1841.60		
1 Cement Finisher	42.35	338.80	66.55	532.40		
1 Gas Engine Vibrator		27.40		30.14	.29	.31
96 L.H., Daily Totals		$3975.00		$6478.94	$41.41	$67.49

Crew C-15	Hr.	Daily	Hr.	Daily	Bare Costs	Incl. O&P
1 Carpenter Foreman (outside)	$46.10	$368.80	$75.55	$604.40	$41.49	$67.53
2 Carpenters	44.10	705.60	72.30	1156.80		
3 Laborers	35.10	842.40	57.55	1381.20		
2 Cement Finishers	42.35	677.60	66.55	1064.80		
1 Rodman (reinf.)	49.10	392.80	81.85	654.80		
72 L.H., Daily Totals		$2987.20		$4862.00	$41.49	$67.53

Crew No.	Bare Costs		Incl. Subs O&P		Cost Per Labor-Hour	
Crew C-16	**Hr.**	**Daily**	**Hr.**	**Daily**	**Bare Costs**	**Incl. O&P**
1 Labor Foreman (outside)	$37.10	$296.80	$60.80	$486.40	$39.09	$62.86
3 Laborers	35.10	842.40	57.55	1381.20		
2 Cement Finishers	42.35	677.60	66.55	1064.80		
1 Equip. Oper. (med.)	46.55	372.40	73.50	588.00		
1 Gunite Pump Rig		280.40		308.44		
2 -50' Air Hoses, 3/4"		6.10		6.71		
2 -50' Air Hoses, 2"		15.50		17.05	5.39	5.93
56 L.H., Daily Totals		$2491.20		$3852.60	$44.49	$68.80

Crew C-17	Hr.	Daily	Hr.	Daily	Bare Costs	Incl. O&P
2 Skilled Worker Foremen (out)	$47.40	$758.40	$77.05	$1232.80	$45.80	$74.45
8 Skilled Workers	45.40	2905.60	73.80	4723.20		
80 L.H., Daily Totals		$3664.00		$5956.00	$45.80	$74.45

Crew C-17A	Hr.	Daily	Hr.	Daily	Bare Costs	Incl. O&P
2 Skilled Worker Foremen (out)	$47.40	$758.40	$77.05	$1232.80	$45.82	$74.46
8 Skilled Workers	45.40	2905.60	73.80	4723.20		
.125 Equip. Oper. (crane)	47.80	47.80	75.50	75.50		
.125 Hyd. Crane, 80 Ton		194.25		213.68	2.40	2.64
81 L.H., Daily Totals		$3906.05		$6245.18	$48.22	$77.10

Crew C-17B	Hr.	Daily	Hr.	Daily	Bare Costs	Incl. O&P
2 Skilled Worker Foremen (out)	$47.40	$758.40	$77.05	$1232.80	$45.85	$74.48
8 Skilled Workers	45.40	2905.60	73.80	4723.20		
.25 Equip. Oper. (crane)	47.80	95.60	75.50	151.00		
.25 Hyd. Crane, 80 Ton		388.50		427.35		
.25 Trowel, 48" Walk-Behind		11.15		12.27	4.87	5.36
82 L.H., Daily Totals		$4159.25		$6546.61	$50.72	$79.84

Crew C-17C	Hr.	Daily	Hr.	Daily	Bare Costs	Incl. O&P
2 Skilled Worker Foremen (out)	$47.40	$758.40	$77.05	$1232.80	$45.87	$74.49
8 Skilled Workers	45.40	2905.60	73.80	4723.20		
.375 Equip. Oper. (crane)	47.80	143.40	75.50	226.50		
.375 Hyd. Crane, 80 Ton		582.75		641.02	7.02	7.72
83 L.H., Daily Totals		$4390.15		$6823.52	$52.89	$82.21

Crew C-17D	Hr.	Daily	Hr.	Daily	Bare Costs	Incl. O&P
2 Skilled Worker Foremen (out)	$47.40	$758.40	$77.05	$1232.80	$45.90	$74.50
8 Skilled Workers	45.40	2905.60	73.80	4723.20		
.5 Equip. Oper. (crane)	47.80	191.20	75.50	302.00		
.5 Hyd. Crane, 80 Ton		777.00		854.70	9.25	10.18
84 L.H., Daily Totals		$4632.20		$7112.70	$55.15	$84.67

Crew C-17E	Hr.	Daily	Hr.	Daily	Bare Costs	Incl. O&P
2 Skilled Worker Foremen (out)	$47.40	$758.40	$77.05	$1232.80	$45.80	$74.45
8 Skilled Workers	45.40	2905.60	73.80	4723.20		
1 Hyd. Jack with Rods		90.70		99.77	1.13	1.25
80 L.H., Daily Totals		$3754.70		$6055.77	$46.93	$75.70

Crew C-18	Hr.	Daily	Hr.	Daily	Bare Costs	Incl. O&P
.125 Labor Foreman (outside)	$37.10	$37.10	$60.80	$60.80	$35.32	$57.91
1 Laborer	35.10	280.80	57.55	460.40		
1 Concrete Cart, 10 C.F.		56.80		62.48	6.31	6.94
9 L.H., Daily Totals		$374.70		$583.68	$41.63	$64.85

Crew C-19	Hr.	Daily	Hr.	Daily	Bare Costs	Incl. O&P
.125 Labor Foreman (outside)	$37.10	$37.10	$60.80	$60.80	$35.32	$57.91
1 Laborer	35.10	280.80	57.55	460.40		
1 Concrete Cart, 18 C.F.		94.40		103.84	10.49	11.54
9 L.H., Daily Totals		$412.30		$625.04	$45.81	$69.45

Crew No.	Bare Costs		Incl. Subs O&P		Cost Per Labor-Hour	
Crew C-20	**Hr.**	**Daily**	**Hr.**	**Daily**	**Bare Costs**	**Incl. O&P**
1 Labor Foreman (outside)	$37.10	$296.80	$60.80	$486.40	$37.69	$61.08
5 Laborers	35.10	1404.00	57.55	2302.00		
1 Cement Finisher	42.35	338.80	66.55	532.40		
1 Equip. Oper. (med.)	46.55	372.40	73.50	588.00		
2 Gas Engine Vibrators		54.80		60.28		
1 Concrete Pump (small)		715.60		787.16	12.04	13.24
64 L.H., Daily Totals		$3182.40		$4756.24	$49.73	$74.32

Crew C-21	Hr.	Daily	Hr.	Daily	Bare Costs	Incl. O&P
1 Labor Foreman (outside)	$37.10	$296.80	$60.80	$486.40	$37.69	$61.08
5 Laborers	35.10	1404.00	57.55	2302.00		
1 Cement Finisher	42.35	338.80	66.55	532.40		
1 Equip. Oper. (med.)	46.55	372.40	73.50	588.00		
2 Gas Engine Vibrators		54.80		60.28		
1 Concrete Conveyer		186.00		204.60	3.76	4.14
64 L.H., Daily Totals		$2652.80		$4173.68	$41.45	$65.21

Crew C-22	Hr.	Daily	Hr.	Daily	Bare Costs	Incl. O&P
1 Rodman Foreman (outside)	$51.10	$408.80	$85.20	$681.60	$49.26	$81.94
4 Rodmen (reinf.)	49.10	1571.20	81.85	2619.20		
.125 Equip. Oper. (crane)	47.80	47.80	75.50	75.50		
.125 Equip. Oper. Oiler	41.25	41.25	65.15	65.15		
.125 Hyd. Crane, 25 Ton		87.15		95.86	2.08	2.28
42 L.H., Daily Totals		$2156.20		$3537.32	$51.34	$84.22

Crew C-23	Hr.	Daily	Hr.	Daily	Bare Costs	Incl. O&P
2 Skilled Worker Foremen (out)	$47.40	$758.40	$77.05	$1232.80	$45.63	$73.75
6 Skilled Workers	45.40	2179.20	73.80	3542.40		
1 Equip. Oper. (crane)	47.80	382.40	75.50	604.00		
1 Equip. Oper. Oiler	41.25	330.00	65.15	521.20		
1 Lattice Boom Crane, 90 Ton		1497.00		1646.70	18.71	20.58
80 L.H., Daily Totals		$5147.00		$7547.10	$64.34	$94.34

Crew C-24	Hr.	Daily	Hr.	Daily	Bare Costs	Incl. O&P
2 Skilled Worker Foremen (out)	$47.40	$758.40	$77.05	$1232.80	$45.63	$73.75
6 Skilled Workers	45.40	2179.20	73.80	3542.40		
1 Equip. Oper. (crane)	47.80	382.40	75.50	604.00		
1 Equip. Oper. Oiler	41.25	330.00	65.15	521.20		
1 Lattice Boom Crane, 150 Ton		1829.00		2011.90	22.86	25.15
80 L.H., Daily Totals		$5479.00		$7912.30	$68.49	$98.90

Crew C-25	Hr.	Daily	Hr.	Daily	Bare Costs	Incl. O&P
2 Rodmen (reinf.)	$49.10	$785.60	$81.85	$1309.60	$38.48	$65.67
2 Rodmen Helpers	27.85	445.60	49.50	792.00		
32 L.H., Daily Totals		$1231.20		$2101.60	$38.48	$65.67

Crew C-27	Hr.	Daily	Hr.	Daily	Bare Costs	Incl. O&P
2 Cement Finishers	$42.35	$677.60	$66.55	$1064.80	$42.35	$66.55
1 Concrete Saw		148.00		162.80	9.25	10.18
16 L.H., Daily Totals		$825.60		$1227.60	$51.60	$76.72

Crew C-28	Hr.	Daily	Hr.	Daily	Bare Costs	Incl. O&P
1 Cement Finisher	$42.35	$338.80	$66.55	$532.40	$42.35	$66.55
1 Portable Air Compressor, Gas		17.55		19.31	2.19	2.41
8 L.H., Daily Totals		$356.35		$551.71	$44.54	$68.96

Crew C-29	Hr.	Daily	Hr.	Daily	Bare Costs	Incl. O&P
1 Laborer	$35.10	$280.80	$57.55	$460.40	$35.10	$57.55
1 Pressure Washer		63.00		69.30	7.88	8.66
8 L.H., Daily Totals		$343.80		$529.70	$42.98	$66.21

Crew No.	Bare Costs		Incl. Subs O&P		Cost Per Labor-Hour	
Crew C-30	**Hr.**	**Daily**	**Hr.**	**Daily**	**Bare Costs**	**Incl. O&P**
1 Laborer	$35.10	$280.80	$57.55	$460.40	$35.10	$57.55
1 Concrete Mixer, 10 C.F.		160.60		176.66	20.07	22.08
8 L.H., Daily Totals		$441.40		$637.06	$55.17	$79.63
Crew D-1	**Hr.**	**Daily**	**Hr.**	**Daily**	**Bare Costs**	**Incl. O&P**
1 Bricklayer	$44.10	$352.80	$71.00	$568.00	$39.85	$64.15
1 Bricklayer Helper	35.60	284.80	57.30	458.40		
16 L.H., Daily Totals		$637.60		$1026.40	$39.85	$64.15
Crew D-2	**Hr.**	**Daily**	**Hr.**	**Daily**	**Bare Costs**	**Incl. O&P**
3 Bricklayers	$44.10	$1058.40	$71.00	$1704.00	$41.01	$66.14
2 Bricklayer Helpers	35.60	569.60	57.30	916.80		
.5 Carpenter	44.10	176.40	72.30	289.20		
44 L.H., Daily Totals		$1804.40		$2910.00	$41.01	$66.14
Crew D-3	**Hr.**	**Daily**	**Hr.**	**Daily**	**Bare Costs**	**Incl. O&P**
3 Bricklayers	$44.10	$1058.40	$71.00	$1704.00	$40.86	$65.84
2 Bricklayer Helpers	35.60	569.60	57.30	916.80		
.25 Carpenter	44.10	88.20	72.30	144.60		
42 L.H., Daily Totals		$1716.20		$2765.40	$40.86	$65.84
Crew D-4	**Hr.**	**Daily**	**Hr.**	**Daily**	**Bare Costs**	**Incl. O&P**
1 Bricklayer	$44.10	$352.80	$71.00	$568.00	$40.01	$64.06
2 Bricklayer Helpers	35.60	569.60	57.30	916.80		
1 Equip. Oper. (light)	44.75	358.00	70.65	565.20		
1 Grout Pump, 50 C.F./hr.		127.40		140.14	3.98	4.38
32 L.H., Daily Totals		$1407.80		$2190.14	$43.99	$68.44
Crew D-5	**Hr.**	**Daily**	**Hr.**	**Daily**	**Bare Costs**	**Incl. O&P**
1 Bricklayer	$44.10	$352.80	$71.00	$568.00	$44.10	$71.00
8 L.H., Daily Totals		$352.80		$568.00	$44.10	$71.00
Crew D-6	**Hr.**	**Daily**	**Hr.**	**Daily**	**Bare Costs**	**Incl. O&P**
3 Bricklayers	$44.10	$1058.40	$71.00	$1704.00	$40.02	$64.48
3 Bricklayer Helpers	35.60	854.40	57.30	1375.20		
.25 Carpenter	44.10	88.20	72.30	144.60		
50 L.H., Daily Totals		$2001.00		$3223.80	$40.02	$64.48
Crew D-7	**Hr.**	**Daily**	**Hr.**	**Daily**	**Bare Costs**	**Incl. O&P**
1 Tile Layer	$40.75	$326.00	$63.95	$511.60	$36.42	$57.15
1 Tile Layer Helper	32.10	256.80	50.35	402.80		
16 L.H., Daily Totals		$582.80		$914.40	$36.42	$57.15
Crew D-8	**Hr.**	**Daily**	**Hr.**	**Daily**	**Bare Costs**	**Incl. O&P**
3 Bricklayers	$44.10	$1058.40	$71.00	$1704.00	$40.70	$65.52
2 Bricklayer Helpers	35.60	569.60	57.30	916.80		
40 L.H., Daily Totals		$1628.00		$2620.80	$40.70	$65.52
Crew D-9	**Hr.**	**Daily**	**Hr.**	**Daily**	**Bare Costs**	**Incl. O&P**
3 Bricklayers	$44.10	$1058.40	$71.00	$1704.00	$39.85	$64.15
3 Bricklayer Helpers	35.60	854.40	57.30	1375.20		
48 L.H., Daily Totals		$1912.80		$3079.20	$39.85	$64.15

Crew No.	Bare Costs		Incl. Subs O&P		Cost Per Labor-Hour	
Crew D-10	**Hr.**	**Daily**	**Hr.**	**Daily**	**Bare Costs**	**Incl. O&P**
1 Bricklayer Foreman (outside)	$46.10	$368.80	$74.20	$593.60	$43.40	$69.50
1 Bricklayer	44.10	352.80	71.00	568.00		
1 Bricklayer Helper	35.60	284.80	57.30	458.40		
1 Equip. Oper. (crane)	47.80	382.40	75.50	604.00		
1 S.P. Crane, 4x4, 12 Ton		448.20		493.02	14.01	15.41
32 L.H., Daily Totals		$1837.00		$2717.02	$57.41	$84.91
Crew D-11	**Hr.**	**Daily**	**Hr.**	**Daily**	**Bare Costs**	**Incl. O&P**
1 Bricklayer Foreman (outside)	$46.10	$368.80	$74.20	$593.60	$41.93	$67.50
1 Bricklayer	44.10	352.80	71.00	568.00		
1 Bricklayer Helper	35.60	284.80	57.30	458.40		
24 L.H., Daily Totals		$1006.40		$1620.00	$41.93	$67.50
Crew D-12	**Hr.**	**Daily**	**Hr.**	**Daily**	**Bare Costs**	**Incl. O&P**
1 Bricklayer Foreman (outside)	$46.10	$368.80	$74.20	$593.60	$40.35	$64.95
1 Bricklayer	44.10	352.80	71.00	568.00		
2 Bricklayer Helpers	35.60	569.60	57.30	916.80		
32 L.H., Daily Totals		$1291.20		$2078.40	$40.35	$64.95
Crew D-13	**Hr.**	**Daily**	**Hr.**	**Daily**	**Bare Costs**	**Incl. O&P**
1 Bricklayer Foreman (outside)	$46.10	$368.80	$74.20	$593.60	$42.22	$67.93
1 Bricklayer	44.10	352.80	71.00	568.00		
2 Bricklayer Helpers	35.60	569.60	57.30	916.80		
1 Carpenter	44.10	352.80	72.30	578.40		
1 Equip. Oper. (crane)	47.80	382.40	75.50	604.00		
1 S.P. Crane, 4x4, 12 Ton		448.20		493.02	9.34	10.27
48 L.H., Daily Totals		$2474.60		$3753.82	$51.55	$78.20
Crew E-1	**Hr.**	**Daily**	**Hr.**	**Daily**	**Bare Costs**	**Incl. O&P**
1 Welder Foreman (outside)	$51.35	$410.80	$95.40	$763.20	$48.48	$85.92
1 Welder	49.35	394.80	91.70	733.60		
1 Equip. Oper. (light)	44.75	358.00	70.65	565.20		
1 Welder, gas engine, 300 amp		121.60		133.76	5.07	5.57
24 L.H., Daily Totals		$1285.20		$2195.76	$53.55	$91.49
Crew E-2	**Hr.**	**Daily**	**Hr.**	**Daily**	**Bare Costs**	**Incl. O&P**
1 Struc. Steel Foreman (outside)	$51.35	$410.80	$95.40	$763.20	$48.26	$86.12
4 Struc. Steel Workers	49.35	1579.20	91.70	2934.40		
1 Equip. Oper. (crane)	47.80	382.40	75.50	604.00		
1 Equip. Oper. Oiler	41.25	330.00	65.15	521.20		
1 Lattice Boom Crane, 90 Ton		1497.00		1646.70	26.73	29.41
56 L.H., Daily Totals		$4199.40		$6469.50	$74.99	$115.53
Crew E-3	**Hr.**	**Daily**	**Hr.**	**Daily**	**Bare Costs**	**Incl. O&P**
1 Struc. Steel Foreman (outside)	$51.35	$410.80	$95.40	$763.20	$50.02	$92.93
1 Struc. Steel Worker	49.35	394.80	91.70	733.60		
1 Welder	49.35	394.80	91.70	733.60		
1 Welder, gas engine, 300 amp		121.60		133.76	5.07	5.57
24 L.H., Daily Totals		$1322.00		$2364.16	$55.08	$98.51
Crew E-4	**Hr.**	**Daily**	**Hr.**	**Daily**	**Bare Costs**	**Incl. O&P**
1 Struc. Steel Foreman (outside)	$51.35	$410.80	$95.40	$763.20	$49.85	$92.63
3 Struc. Steel Workers	49.35	1184.40	91.70	2200.80		
1 Welder, gas engine, 300 amp		121.60		133.76	3.80	4.18
32 L.H., Daily Totals		$1716.80		$3097.76	$53.65	$96.81

Crew No.	Bare Costs		Incl. Subs O&P		Cost Per Labor-Hour	

Crew E-5	Hr.	Daily	Hr.	Daily	Bare Costs	Incl. O&P
2 Struc. Steel Foremen (outside)	$51.35	$821.60	$95.40	$1526.40	$48.78	$88.17
5 Struc. Steel Workers	49.35	1974.00	91.70	3668.00		
1 Equip. Oper. (crane)	47.80	382.40	75.50	604.00		
1 Welder	49.35	394.80	91.70	733.60		
1 Equip. Oper. Oiler	41.25	330.00	65.15	521.20		
1 Lattice Boom Crane, 90 Ton		1497.00		1646.70		
1 Welder, gas engine, 300 amp		121.60		133.76	20.23	22.26
80 L.H., Daily Totals		$5521.40		$8833.66	$69.02	$110.42

Crew E-6	Hr.	Daily	Hr.	Daily	Bare Costs	Incl. O&P
3 Struc. Steel Foremen (outside)	$51.35	$1232.40	$95.40	$2289.60	$48.83	$88.41
9 Struc. Steel Workers	49.35	3553.20	91.70	6602.40		
1 Equip. Oper. (crane)	47.80	382.40	75.50	604.00		
1 Welder	49.35	394.80	91.70	733.60		
1 Equip. Oper. Oiler	41.25	330.00	65.15	521.20		
1 Equip. Oper. (light)	44.75	358.00	70.65	565.20		
1 Lattice Boom Crane, 90 Ton		1497.00		1646.70		
1 Welder, gas engine, 300 amp		121.60		133.76		
1 Air Compressor, 160 cfm		136.80		150.48		
2 Impact Wrenches		36.00		39.60	14.00	15.39
128 L.H., Daily Totals		$8042.20		$13286.54	$62.83	$103.80

Crew E-7	Hr.	Daily	Hr.	Daily	Bare Costs	Incl. O&P
1 Struc. Steel Foreman (outside)	$51.35	$410.80	$95.40	$763.20	$48.78	$88.17
4 Struc. Steel Workers	49.35	1579.20	91.70	2934.40		
1 Equip. Oper. (crane)	47.80	382.40	75.50	604.00		
1 Equip. Oper. Oiler	41.25	330.00	65.15	521.20		
1 Welder Foreman (outside)	51.35	410.80	95.40	763.20		
2 Welders	49.35	789.60	91.70	1467.20		
1 Lattice Boom Crane, 90 Ton		1497.00		1646.70		
2 Welders, gas engine, 300 amp		243.20		267.52	21.75	23.93
80 L.H., Daily Totals		$5643.00		$8967.42	$70.54	$112.09

Crew E-8	Hr.	Daily	Hr.	Daily	Bare Costs	Incl. O&P
1 Struc. Steel Foreman (outside)	$51.35	$410.80	$95.40	$763.20	$48.56	$87.36
4 Struc. Steel Workers	49.35	1579.20	91.70	2934.40		
1 Welder Foreman (outside)	51.35	410.80	95.40	763.20		
4 Welders	49.35	1579.20	91.70	2934.40		
1 Equip. Oper. (crane)	47.80	382.40	75.50	604.00		
1 Equip. Oper. Oiler	41.25	330.00	65.15	521.20		
1 Equip. Oper. (light)	44.75	358.00	70.65	565.20		
1 Lattice Boom Crane, 90 Ton		1497.00		1646.70		
4 Welders, gas engine, 300 amp		486.40		535.04	19.07	20.98
104 L.H., Daily Totals		$7033.80		$11267.34	$67.63	$108.34

Crew E-9	Hr.	Daily	Hr.	Daily	Bare Costs	Incl. O&P
2 Struc. Steel Foremen (outside)	$51.35	$821.60	$95.40	$1526.40	$48.83	$88.41
5 Struc. Steel Workers	49.35	1974.00	91.70	3668.00		
1 Welder Foreman (outside)	51.35	410.80	95.40	763.20		
5 Welders	49.35	1974.00	91.70	3668.00		
1 Equip. Oper. (crane)	47.80	382.40	75.50	604.00		
1 Equip. Oper. Oiler	41.25	330.00	65.15	521.20		
1 Equip. Oper. (light)	44.75	358.00	70.65	565.20		
1 Lattice Boom Crane, 90 Ton		1497.00		1646.70		
5 Welders, gas engine, 300 amp		608.00		668.80	16.45	18.09
128 L.H., Daily Totals		$8355.80		$13631.50	$65.28	$106.50

Crew E-10	Hr.	Daily	Hr.	Daily	Bare Costs	Incl. O&P
1 Welder Foreman (outside)	$51.35	$410.80	$95.40	$763.20	$50.35	$93.55
1 Welder	49.35	394.80	91.70	733.60		
1 Welder, gas engine, 300 amp		121.60		133.76		
1 Flatbed Truck, Gas, 3 Ton		264.40		290.84	24.13	26.54
16 L.H., Daily Totals		$1191.60		$1921.40	$74.47	$120.09

Crew E-11	Hr.	Daily	Hr.	Daily	Bare Costs	Incl. O&P
2 Painters, Struc. Steel	$39.30	$628.80	$75.10	$1201.60	$39.61	$69.60
1 Building Laborer	35.10	280.80	57.55	460.40		
1 Equip. Oper. (light)	44.75	358.00	70.65	565.20		
1 Air Compressor, 250 cfm		177.00		194.70		
1 Sandblaster, portable, 3 C.F.		20.40		22.44		
1 Set Sand Blasting Accessories		13.35		14.69	6.59	7.24
32 L.H., Daily Totals		$1478.35		$2459.03	$46.20	$76.84

Crew E-12	Hr.	Daily	Hr.	Daily	Bare Costs	Incl. O&P
1 Welder Foreman (outside)	$51.35	$410.80	$95.40	$763.20	$48.05	$83.03
1 Equip. Oper. (light)	44.75	358.00	70.65	565.20		
1 Welder, gas engine, 300 amp		121.60		133.76	7.60	8.36
16 L.H., Daily Totals		$890.40		$1462.16	$55.65	$91.39

Crew E-13	Hr.	Daily	Hr.	Daily	Bare Costs	Incl. O&P
1 Welder Foreman (outside)	$51.35	$410.80	$95.40	$763.20	$49.15	$87.15
.5 Equip. Oper. (light)	44.75	179.00	70.65	282.60		
1 Welder, gas engine, 300 amp		121.60		133.76	10.13	11.15
12 L.H., Daily Totals		$711.40		$1179.56	$59.28	$98.30

Crew E-14	Hr.	Daily	Hr.	Daily	Bare Costs	Incl. O&P
1 Welder Foreman (outside)	$51.35	$410.80	$95.40	$763.20	$51.35	$95.40
1 Welder, gas engine, 300 amp		121.60		133.76	15.20	16.72
8 L.H., Daily Totals		$532.40		$896.96	$66.55	$112.12

Crew E-16	Hr.	Daily	Hr.	Daily	Bare Costs	Incl. O&P
1 Welder Foreman (outside)	$51.35	$410.80	$95.40	$763.20	$50.35	$93.55
1 Welder	49.35	394.80	91.70	733.60		
1 Welder, gas engine, 300 amp		121.60		133.76	7.60	8.36
16 L.H., Daily Totals		$927.20		$1630.56	$57.95	$101.91

Crew E-17	Hr.	Daily	Hr.	Daily	Bare Costs	Incl. O&P
1 Struc. Steel Foreman (outside)	$51.35	$410.80	$95.40	$763.20	$50.35	$93.55
1 Structural Steel Worker	49.35	394.80	91.70	733.60		
16 L.H., Daily Totals		$805.60		$1496.80	$50.35	$93.55

Crew E-18	Hr.	Daily	Hr.	Daily	Bare Costs	Incl. O&P
1 Struc. Steel Foreman (outside)	$51.35	$410.80	$95.40	$763.20	$49.19	$88.80
3 Structural Steel Workers	49.35	1184.40	91.70	2200.80		
1 Equipment Operator (med.)	46.55	372.40	73.50	588.00		
1 Lattice Boom Crane, 20 Ton		950.65		1045.71	23.77	26.14
40 L.H., Daily Totals		$2918.25		$4597.72	$72.96	$114.94

Crew E-19	Hr.	Daily	Hr.	Daily	Bare Costs	Incl. O&P
1 Structural Steel Worker	$49.35	$394.80	$91.70	$733.60	$48.48	$85.92
1 Struc. Steel Foreman (outside)	51.35	410.80	95.40	763.20		
1 Equip. Oper. (light)	44.75	358.00	70.65	565.20		
1 Lattice Boom Crane, 20 Ton		950.65		1045.71	39.61	43.57
24 L.H., Daily Totals		$2114.25		$3107.72	$88.09	$129.49

Crew No.	Bare Costs		Incl. Subs O&P		Cost Per Labor-Hour	
Crew E-20	**Hr.**	**Daily**	**Hr.**	**Daily**	**Bare Costs**	**Incl. O&P**
1 Struc. Steel Foreman (outside)	$51.35	$410.80	$95.40	$763.20	$48.39	$86.82
5 Structural Steel Workers	49.35	1974.00	91.70	3668.00		
1 Equip. Oper. (crane)	47.80	382.40	75.50	604.00		
1 Oiler	41.25	330.00	65.15	521.20		
1 Lattice Boom Crane, 40 Ton		1163.00		1279.30	18.17	19.99
64 L.H., Daily Totals		$4260.20		$6835.70	$66.57	$106.81

Crew E-22	Hr.	Daily	Hr.	Daily	Bare Costs	Incl. O&P
1 Skilled Worker Foreman (out)	$47.40	$379.20	$77.05	$616.40	$46.07	$74.88
2 Skilled Workers	45.40	726.40	73.80	1180.80		
24 L.H., Daily Totals		$1105.60		$1797.20	$46.07	$74.88

Crew E-24	Hr.	Daily	Hr.	Daily	Bare Costs	Incl. O&P
3 Structural Steel Workers	$49.35	$1184.40	$91.70	$2200.80	$48.65	$87.15
1 Equipment Operator (med.)	46.55	372.40	73.50	588.00		
1 Hyd. Crane, 25 Ton		697.20		766.92	21.79	23.97
32 L.H., Daily Totals		$2254.00		$3555.72	$70.44	$111.12

Crew E-25	Hr.	Daily	Hr.	Daily	Bare Costs	Incl. O&P
1 Welder Foreman (outside)	$51.35	$410.80	$95.40	$763.20	$51.35	$95.40
1 Cutting Torch		10.80		11.88		
1 Set of Gases		86.40		95.04	12.15	13.37
8 L.H., Daily Totals		$508.00		$870.12	$63.50	$108.77

Crew F-3	Hr.	Daily	Hr.	Daily	Bare Costs	Incl. O&P
4 Carpenters	$44.10	$1411.20	$72.30	$2313.60	$44.84	$72.94
1 Equip. Oper. (crane)	47.80	382.40	75.50	604.00		
1 Hyd. Crane, 12 Ton		612.00		673.20	15.30	16.83
40 L.H., Daily Totals		$2405.60		$3590.80	$60.14	$89.77

Crew F-4	Hr.	Daily	Hr.	Daily	Bare Costs	Incl. O&P
4 Carpenters	$44.10	$1411.20	$72.30	$2313.60	$44.24	$71.64
1 Equip. Oper. (crane)	47.80	382.40	75.50	604.00		
1 Equip. Oper. Oiler	41.25	330.00	65.15	521.20		
1 Hyd. Crane, 55 Ton		1100.00		1210.00	22.92	25.21
48 L.H., Daily Totals		$3223.60		$4648.80	$67.16	$96.85

Crew F-5	Hr.	Daily	Hr.	Daily	Bare Costs	Incl. O&P
1 Carpenter Foreman (outside)	$46.10	$368.80	$75.55	$604.40	$44.60	$73.11
3 Carpenters	44.10	1058.40	72.30	1735.20		
32 L.H., Daily Totals		$1427.20		$2339.60	$44.60	$73.11

Crew F-6	Hr.	Daily	Hr.	Daily	Bare Costs	Incl. O&P
2 Carpenters	$44.10	$705.60	$72.30	$1156.80	$41.24	$67.04
2 Building Laborers	35.10	561.60	57.55	920.80		
1 Equip. Oper. (crane)	47.80	382.40	75.50	604.00		
1 Hyd. Crane, 12 Ton		612.00		673.20	15.30	16.83
40 L.H., Daily Totals		$2261.60		$3354.80	$56.54	$83.87

Crew F-7	Hr.	Daily	Hr.	Daily	Bare Costs	Incl. O&P
2 Carpenters	$44.10	$705.60	$72.30	$1156.80	$39.60	$64.92
2 Building Laborers	35.10	561.60	57.55	920.80		
32 L.H., Daily Totals		$1267.20		$2077.60	$39.60	$64.92

Crew No.	Bare Costs		Incl. Subs O&P		Cost Per Labor-Hour	
Crew G-1	**Hr.**	**Daily**	**Hr.**	**Daily**	**Bare Costs**	**Incl. O&P**
1 Roofer Foreman (outside)	$39.40	$315.20	$70.00	$560.00	$34.96	$62.11
4 Roofers, Composition	37.40	1196.80	66.45	2126.40		
2 Roofer Helpers	27.85	445.60	49.50	792.00		
1 Application Equipment		179.20		197.12		
1 Tar Kettle/Pot		91.00		100.10		
1 Crew Truck		182.40		200.64	8.08	8.89
56 L.H., Daily Totals		$2410.20		$3976.26	$43.04	$71.00

Crew G-2	Hr.	Daily	Hr.	Daily	Bare Costs	Incl. O&P
1 Plasterer	$40.35	$322.80	$64.45	$515.60	$37.07	$59.70
1 Plasterer Helper	35.75	286.00	57.10	456.80		
1 Building Laborer	35.10	280.80	57.55	460.40		
1 Grout Pump, 50 C.F./hr.		127.40		140.14	5.31	5.84
24 L.H., Daily Totals		$1017.00		$1572.94	$42.38	$65.54

Crew G-2A	Hr.	Daily	Hr.	Daily	Bare Costs	Incl. O&P
1 Roofer, composition	$37.40	$299.20	$66.45	$531.60	$33.45	$57.83
1 Roofer Helper	27.85	222.80	49.50	396.00		
1 Building Laborer	35.10	280.80	57.55	460.40		
1 Foam spray rig, trailer-mtd.		503.85		554.24		
1 Pickup Truck, 3/4 Ton		127.00		139.70	26.29	28.91
24 L.H., Daily Totals		$1433.65		$2081.93	$59.74	$86.75

Crew G-3	Hr.	Daily	Hr.	Daily	Bare Costs	Incl. O&P
2 Sheet Metal Workers	$52.15	$834.40	$82.10	$1313.60	$43.63	$69.83
2 Building Laborers	35.10	561.60	57.55	920.80		
32 L.H., Daily Totals		$1396.00		$2234.40	$43.63	$69.83

Crew G-4	Hr.	Daily	Hr.	Daily	Bare Costs	Incl. O&P
1 Labor Foreman (outside)	$37.10	$296.80	$60.80	$486.40	$35.77	$58.63
2 Building Laborers	35.10	561.60	57.55	920.80		
1 Flatbed Truck, Gas, 1.5 Ton		211.00		232.10		
1 Air Compressor, 160 cfm		136.80		150.48	14.49	15.94
24 L.H., Daily Totals		$1206.20		$1789.78	$50.26	$74.57

Crew G-5	Hr.	Daily	Hr.	Daily	Bare Costs	Incl. O&P
1 Roofer Foreman (outside)	$39.40	$315.20	$70.00	$560.00	$33.98	$60.38
2 Roofers, Composition	37.40	598.40	66.45	1063.20		
2 Roofer Helpers	27.85	445.60	49.50	792.00		
1 Application Equipment		179.20		197.12	4.48	4.93
40 L.H., Daily Totals		$1538.40		$2612.32	$38.46	$65.31

Crew G-6A	Hr.	Daily	Hr.	Daily	Bare Costs	Incl. O&P
2 Roofers Composition	$37.40	$598.40	$66.45	$1063.20	$37.40	$66.45
1 Small Compressor, Electric		10.10		11.11		
2 Pneumatic Nailers		45.70		50.27	3.49	3.84
16 L.H., Daily Totals		$654.20		$1124.58	$40.89	$70.29

Crew G-7	Hr.	Daily	Hr.	Daily	Bare Costs	Incl. O&P
1 Carpenter	$44.10	$352.80	$72.30	$578.40	$44.10	$72.30
1 Small Compressor, Electric		10.10		11.11		
1 Pneumatic Nailer		22.85		25.14	4.12	4.53
8 L.H., Daily Totals		$385.75		$614.64	$48.22	$76.83

Crew H-1	Hr.	Daily	Hr.	Daily	Bare Costs	Incl. O&P
2 Glaziers	$42.75	$684.00	$69.05	$1104.80	$46.05	$80.38
2 Struc. Steel Workers	49.35	789.60	91.70	1467.20		
32 L.H., Daily Totals		$1473.60		$2572.00	$46.05	$80.38

Crew H-2	Bare Costs Hr.	Bare Costs Daily	Incl. Subs O&P Hr.	Incl. Subs O&P Daily	Cost Per Labor-Hour Bare Costs	Cost Per Labor-Hour Incl. O&P
2 Glaziers	$42.75	$684.00	$69.05	$1104.80	$40.20	$65.22
1 Building Laborer	35.10	280.80	57.55	460.40		
24 L.H., Daily Totals		$964.80		$1565.20	$40.20	$65.22

Crew H-3	Hr.	Daily	Hr.	Daily	Bare Costs	Incl. O&P
1 Glazier	$42.75	$342.00	$69.05	$552.40	$38.02	$61.80
1 Helper	33.30	266.40	54.55	436.40		
16 L.H., Daily Totals		$608.40		$988.80	$38.02	$61.80

Crew H-4	Hr.	Daily	Hr.	Daily	Bare Costs	Incl. O&P
1 Carpenter	$44.10	$352.80	$72.30	$578.40	$41.27	$66.67
1 Carpenter Helper	33.30	266.40	54.55	436.40		
.5 Electrician	51.55	206.20	79.65	318.60		
20 L.H., Daily Totals		$825.40		$1333.40	$41.27	$66.67

Crew J-1	Hr.	Daily	Hr.	Daily	Bare Costs	Incl. O&P
3 Plasterers	$40.35	$968.40	$64.45	$1546.80	$38.51	$61.51
2 Plasterer Helpers	35.75	572.00	57.10	913.60		
1 Mixing Machine, 6 C.F.		130.80		143.88	3.27	3.60
40 L.H., Daily Totals		$1671.20		$2604.28	$41.78	$65.11

Crew J-2	Hr.	Daily	Hr.	Daily	Bare Costs	Incl. O&P
3 Plasterers	$40.35	$968.40	$64.45	$1546.80	$38.63	$61.51
2 Plasterer Helpers	35.75	572.00	57.10	913.60		
1 Lather	39.25	314.00	61.50	492.00		
1 Mixing Machine, 6 C.F.		130.80		143.88	2.73	3.00
48 L.H., Daily Totals		$1985.20		$3096.28	$41.36	$64.51

Crew J-3	Hr.	Daily	Hr.	Daily	Bare Costs	Incl. O&P
1 Terrazzo Worker	$40.40	$323.20	$63.40	$507.20	$37.08	$58.17
1 Terrazzo Helper	33.75	270.00	52.95	423.60		
1 Floor Grinder, 22" path		84.70		93.17		
1 Terrazzo Mixer		177.00		194.70	16.36	17.99
16 L.H., Daily Totals		$854.90		$1218.67	$53.43	$76.17

Crew J-4	Hr.	Daily	Hr.	Daily	Bare Costs	Incl. O&P
2 Cement Finishers	$42.35	$677.60	$66.55	$1064.80	$39.93	$63.55
1 Laborer	35.10	280.80	57.55	460.40		
1 Floor Grinder, 22" path		84.70		93.17		
1 Floor Edger, 7" path		38.20		42.02		
1 Vacuum Pick-Up System		59.80		65.78	7.61	8.37
24 L.H., Daily Totals		$1141.10		$1726.17	$47.55	$71.92

Crew J-4A	Hr.	Daily	Hr.	Daily	Bare Costs	Incl. O&P
2 Cement Finishers	$42.35	$677.60	$66.55	$1064.80	$38.73	$62.05
2 Laborers	35.10	561.60	57.55	920.80		
1 Floor Grinder, 22" path		84.70		93.17		
1 Floor Edger, 7" path		38.20		42.02		
1 Vacuum Pick-Up System		59.80		65.78		
1 Floor Auto Scrubber		178.10		195.91	11.28	12.40
32 L.H., Daily Totals		$1600.00		$2382.48	$50.00	$74.45

Crew J-4B	Hr.	Daily	Hr.	Daily	Bare Costs	Incl. O&P
1 Laborer	$35.10	$280.80	$57.55	$460.40	$35.10	$57.55
1 Floor Auto Scrubber		178.10		195.91	22.26	24.49
8 L.H., Daily Totals		$458.90		$656.31	$57.36	$82.04

Crew J-6	Hr.	Daily	Hr.	Daily	Bare Costs	Incl. O&P
2 Painters	$38.25	$612.00	$61.05	$976.80	$39.09	$62.58
1 Building Laborer	35.10	280.80	57.55	460.40		
1 Equip. Oper. (light)	44.75	358.00	70.65	565.20		
1 Air Compressor, 250 cfm		177.00		194.70		
1 Sandblaster, portable, 3 C.F.		20.40		22.44		
1 Set Sand Blasting Accessories		13.35		14.69	6.59	7.24
32 L.H., Daily Totals		$1461.55		$2234.22	$45.67	$69.82

Crew K-1	Hr.	Daily	Hr.	Daily	Bare Costs	Incl. O&P
1 Carpenter	$44.10	$352.80	$72.30	$578.40	$39.23	$63.92
1 Truck Driver (light)	34.35	274.80	55.55	444.40		
1 Flatbed Truck, Gas, 3 Ton		264.40		290.84	16.52	18.18
16 L.H., Daily Totals		$892.00		$1313.64	$55.75	$82.10

Crew K-2	Hr.	Daily	Hr.	Daily	Bare Costs	Incl. O&P
1 Struc. Steel Foreman (outside)	$51.35	$410.80	$95.40	$763.20	$45.02	$80.88
1 Struc. Steel Worker	49.35	394.80	91.70	733.60		
1 Truck Driver (light)	34.35	274.80	55.55	444.40		
1 Flatbed Truck, Gas, 3 Ton		264.40		290.84	11.02	12.12
24 L.H., Daily Totals		$1344.80		$2232.04	$56.03	$93.00

Crew L-1	Hr.	Daily	Hr.	Daily	Bare Costs	Incl. O&P
1 Electrician	$51.55	$412.40	$79.65	$637.20	$53.08	$82.35
1 Plumber	54.60	436.80	85.05	680.40		
16 L.H., Daily Totals		$849.20		$1317.60	$53.08	$82.35

Crew L-2	Hr.	Daily	Hr.	Daily	Bare Costs	Incl. O&P
1 Carpenter	$44.10	$352.80	$72.30	$578.40	$38.70	$63.42
1 Carpenter Helper	33.30	266.40	54.55	436.40		
16 L.H., Daily Totals		$619.20		$1014.80	$38.70	$63.42

Crew L-3	Hr.	Daily	Hr.	Daily	Bare Costs	Incl. O&P
1 Carpenter	$44.10	$352.80	$72.30	$578.40	$47.98	$76.59
.5 Electrician	51.55	206.20	79.65	318.60		
.5 Sheet Metal Worker	52.15	208.60	82.10	328.40		
16 L.H., Daily Totals		$767.60		$1225.40	$47.98	$76.59

Crew L-3A	Hr.	Daily	Hr.	Daily	Bare Costs	Incl. O&P
1 Carpenter Foreman (outside)	$46.10	$368.80	$75.55	$604.40	$48.12	$77.73
.5 Sheet Metal Worker	52.15	208.60	82.10	328.40		
12 L.H., Daily Totals		$577.40		$932.80	$48.12	$77.73

Crew L-4	Hr.	Daily	Hr.	Daily	Bare Costs	Incl. O&P
2 Skilled Workers	$45.40	$726.40	$73.80	$1180.80	$41.37	$67.38
1 Helper	33.30	266.40	54.55	436.40		
24 L.H., Daily Totals		$992.80		$1617.20	$41.37	$67.38

Crew L-5	Hr.	Daily	Hr.	Daily	Bare Costs	Incl. O&P
1 Struc. Steel Foreman (outside)	$51.35	$410.80	$95.40	$763.20	$49.41	$89.91
5 Struc. Steel Workers	49.35	1974.00	91.70	3668.00		
1 Equip. Oper. (crane)	47.80	382.40	75.50	604.00		
1 Hyd. Crane, 25 Ton		697.20		766.92	12.45	13.70
56 L.H., Daily Totals		$3464.40		$5802.12	$61.86	$103.61

Crew L-5A	Hr.	Daily	Hr.	Daily	Bare Costs	Incl. O&P
1 Struc. Steel Foreman (outside)	$51.35	$410.80	$95.40	$763.20	$49.46	$88.58
2 Structural Steel Workers	49.35	789.60	91.70	1467.20		
1 Equip. Oper. (crane)	47.80	382.40	75.50	604.00		
1 S.P. Crane, 4x4, 25 Ton		565.20		621.72	17.66	19.43
32 L.H., Daily Totals		$2148.00		$3456.12	$67.13	$108.00

Crew L-5B	Hr.	Daily	Hr.	Daily	Bare Costs	Incl. O&P
1 Struc. Steel Foreman (outside)	$51.35	$410.80	$95.40	$763.20	$50.40	$83.59
2 Structural Steel Workers	49.35	789.60	91.70	1467.20		
2 Electricians	51.55	824.80	79.65	1274.40		
2 Steamfitters/Pipefitters	55.70	891.20	86.80	1388.80		
1 Equip. Oper. (crane)	47.80	382.40	75.50	604.00		
1 Equip. Oper. Oiler	41.25	330.00	65.15	521.20		
1 Hyd. Crane, 80 Ton		1554.00		1709.40	21.58	23.74
72 L.H., Daily Totals		$5182.80		$7728.20	$71.98	$107.34

Crew L-6	Hr.	Daily	Hr.	Daily	Bare Costs	Incl. O&P
1 Plumber	$54.60	$436.80	$85.05	$680.40	$53.58	$83.25
.5 Electrician	51.55	206.20	79.65	318.60		
12 L.H., Daily Totals		$643.00		$999.00	$53.58	$83.25

Crew L-7	Hr.	Daily	Hr.	Daily	Bare Costs	Incl. O&P
2 Carpenters	$44.10	$705.60	$72.30	$1156.80	$42.59	$69.14
1 Building Laborer	35.10	280.80	57.55	460.40		
.5 Electrician	51.55	206.20	79.65	318.60		
28 L.H., Daily Totals		$1192.60		$1935.80	$42.59	$69.14

Crew L-8	Hr.	Daily	Hr.	Daily	Bare Costs	Incl. O&P
2 Carpenters	$44.10	$705.60	$72.30	$1156.80	$46.20	$74.85
.5 Plumber	54.60	218.40	85.05	340.20		
20 L.H., Daily Totals		$924.00		$1497.00	$46.20	$74.85

Crew L-9	Hr.	Daily	Hr.	Daily	Bare Costs	Incl. O&P
1 Labor Foreman (inside)	$35.60	$284.80	$58.35	$466.80	$40.21	$67.77
2 Building Laborers	35.10	561.60	57.55	920.80		
1 Struc. Steel Worker	49.35	394.80	91.70	733.60		
.5 Electrician	51.55	206.20	79.65	318.60		
36 L.H., Daily Totals		$1447.40		$2439.80	$40.21	$67.77

Crew L-10	Hr.	Daily	Hr.	Daily	Bare Costs	Incl. O&P
1 Struc. Steel Foreman (outside)	$51.35	$410.80	$95.40	$763.20	$49.50	$87.53
1 Structural Steel Worker	49.35	394.80	91.70	733.60		
1 Equip. Oper. (crane)	47.80	382.40	75.50	604.00		
1 Hyd. Crane, 12 Ton		612.00		673.20	25.50	28.05
24 L.H., Daily Totals		$1800.00		$2774.00	$75.00	$115.58

Crew L-11	Hr.	Daily	Hr.	Daily	Bare Costs	Incl. O&P
2 Wreckers	$35.10	$561.60	$62.95	$1007.20	$40.69	$68.01
1 Equip. Oper. (crane)	47.80	382.40	75.50	604.00		
1 Equip. Oper. (light)	44.75	358.00	70.65	565.20		
1 Hyd. Excavator, 2.5 C.Y.		1652.00		1817.20		
1 Loader, Skid Steer, 78 H.P.		281.20		309.32	60.41	66.45
32 L.H., Daily Totals		$3235.20		$4302.92	$101.10	$134.47

Crew M-1	Hr.	Daily	Hr.	Daily	Bare Costs	Incl. O&P
3 Elevator Constructors	$70.85	$1700.40	$109.70	$2632.80	$67.31	$104.21
1 Elevator Apprentice	56.70	453.60	87.75	702.00		
5 Hand Tools		46.00		50.60	1.44	1.58
32 L.H., Daily Totals		$2200.00		$3385.40	$68.75	$105.79

Crew M-3	Hr.	Daily	Hr.	Daily	Bare Costs	Incl. O&P
1 Electrician Foreman (outside)	$53.55	$428.40	$82.75	$662.00	$53.61	$83.79
1 Common Laborer	35.10	280.80	57.55	460.40		
.25 Equipment Operator, Med.	46.55	93.10	73.50	147.00		
1 Elevator Constructor	70.85	566.80	109.70	877.60		
1 Elevator Apprentice	56.70	453.60	87.75	702.00		
.25 S.P. Crane, 4x4, 20 Ton		129.85		142.84	3.82	4.20
34 L.H., Daily Totals		$1952.55		$2991.84	$57.43	$88.00

Crew M-4	Hr.	Daily	Hr.	Daily	Bare Costs	Incl. O&P
1 Electrician Foreman (outside)	$53.55	$428.40	$82.75	$662.00	$52.99	$82.87
1 Common Laborer	35.10	280.80	57.55	460.40		
.25 Equipment Operator, Crane	47.80	95.60	75.50	151.00		
.25 Equipment Operator, Oiler	41.25	82.50	65.15	130.30		
1 Elevator Constructor	70.85	566.80	109.70	877.60		
1 Elevator Apprentice	56.70	453.60	87.75	702.00		
.25 S.P. Crane, 4x4, 40 Ton		166.00		182.60	4.61	5.07
36 L.H., Daily Totals		$2073.70		$3165.90	$57.60	$87.94

Crew Q-1	Hr.	Daily	Hr.	Daily	Bare Costs	Incl. O&P
1 Plumber	$54.60	$436.80	$85.05	$680.40	$49.15	$76.58
1 Plumber Apprentice	43.70	349.60	68.10	544.80		
16 L.H., Daily Totals		$786.40		$1225.20	$49.15	$76.58

Crew Q-1A	Hr.	Daily	Hr.	Daily	Bare Costs	Incl. O&P
.25 Plumber Foreman (outside)	$56.60	$113.20	$88.20	$176.40	$55.00	$85.68
1 Plumber	54.60	436.80	85.05	680.40		
10 L.H., Daily Totals		$550.00		$856.80	$55.00	$85.68

Crew Q-1C	Hr.	Daily	Hr.	Daily	Bare Costs	Incl. O&P
1 Plumber	$54.60	$436.80	$85.05	$680.40	$48.28	$75.55
1 Plumber Apprentice	43.70	349.60	68.10	544.80		
1 Equip. Oper. (medium)	46.55	372.40	73.50	588.00		
1 Trencher, Chain Type, 8' D		2848.00		3132.80	118.67	130.53
24 L.H., Daily Totals		$4006.80		$4946.00	$166.95	$206.08

Crew Q-2	Hr.	Daily	Hr.	Daily	Bare Costs	Incl. O&P
2 Plumbers	$54.60	$873.60	$85.05	$1360.80	$50.97	$79.40
1 Plumber Apprentice	43.70	349.60	68.10	544.80	50.97	79.40
24 L.H., Daily Totals		$1223.20		$1905.60	$50.97	$79.40

Crew Q-3	Hr.	Daily	Hr.	Daily	Bare Costs	Incl. O&P
1 Plumber Foreman (inside)	$55.10	$440.80	$85.85	$686.80	$52.00	$81.01
2 Plumbers	54.60	873.60	85.05	1360.80		
1 Plumber Apprentice	43.70	349.60	68.10	544.80		
32 L.H., Daily Totals		$1664.00		$2592.40	$52.00	$81.01

Crew Q-4	Hr.	Daily	Hr.	Daily	Bare Costs	Incl. O&P
1 Plumber Foreman (inside)	$55.10	$440.80	$85.85	$686.80	$52.00	$81.01
1 Plumber	54.60	436.80	85.05	680.40		
1 Welder (plumber)	54.60	436.80	85.05	680.40		
1 Plumber Apprentice	43.70	349.60	68.10	544.80		
1 Welder, electric, 300 amp		36.90		40.59	1.15	1.27
32 L.H., Daily Totals		$1700.90		$2632.99	$53.15	$82.28

Crew Q-5	Hr.	Daily	Hr.	Daily	Bare Costs	Incl. O&P
1 Steamfitter	$55.70	$445.60	$86.80	$694.40	$50.13	$78.10
1 Steamfitter Apprentice	44.55	356.40	69.40	555.20		
16 L.H., Daily Totals		$802.00		$1249.60	$50.13	$78.10

Crew No.	Bare Costs		Incl. Subs O&P		Cost Per Labor-Hour	
Crew Q-6	**Hr.**	**Daily**	**Hr.**	**Daily**	**Bare Costs**	**Incl. O&P**
2 Steamfitters	$55.70	$891.20	$86.80	$1388.80	$51.98	$81.00
1 Steamfitter Apprentice	44.55	356.40	69.40	555.20	51.98	81.00
24 L.H., Daily Totals		$1247.60		$1944.00	$51.98	$81.00
Crew Q-7	**Hr.**	**Daily**	**Hr.**	**Daily**	**Bare Costs**	**Incl. O&P**
1 Steamfitter Foreman (inside)	$56.20	$449.60	$87.55	$700.40	$53.04	$82.64
2 Steamfitters	55.70	891.20	86.80	1388.80		
1 Steamfitter Apprentice	44.55	356.40	69.40	555.20		
32 L.H., Daily Totals		$1697.20		$2644.40	$53.04	$82.64
Crew Q-8	**Hr.**	**Daily**	**Hr.**	**Daily**	**Bare Costs**	**Incl. O&P**
1 Steamfitter Foreman (inside)	$56.20	$449.60	$87.55	$700.40	$53.04	$82.64
1 Steamfitter	55.70	445.60	86.80	694.40		
1 Welder (steamfitter)	55.70	445.60	86.80	694.40		
1 Steamfitter Apprentice	44.55	356.40	69.40	555.20		
1 Welder, electric, 300 amp		36.90		40.59	1.15	1.27
32 L.H., Daily Totals		$1734.10		$2684.99	$54.19	$83.91
Crew Q-9	**Hr.**	**Daily**	**Hr.**	**Daily**	**Bare Costs**	**Incl. O&P**
1 Sheet Metal Worker	$52.15	$417.20	$82.10	$656.80	$46.92	$73.88
1 Sheet Metal Apprentice	41.70	333.60	65.65	525.20		
16 L.H., Daily Totals		$750.80		$1182.00	$46.92	$73.88
Crew Q-10	**Hr.**	**Daily**	**Hr.**	**Daily**	**Bare Costs**	**Incl. O&P**
2 Sheet Metal Workers	$52.15	$834.40	$82.10	$1313.60	$48.67	$76.62
1 Sheet Metal Apprentice	41.70	333.60	65.65	525.20		
24 L.H., Daily Totals		$1168.00		$1838.80	$48.67	$76.62
Crew Q-11	**Hr.**	**Daily**	**Hr.**	**Daily**	**Bare Costs**	**Incl. O&P**
1 Sheet Metal Foreman (inside)	$52.65	$421.20	$82.85	$662.80	$49.66	$78.17
2 Sheet Metal Workers	52.15	834.40	82.10	1313.60		
1 Sheet Metal Apprentice	41.70	333.60	65.65	525.20		
32 L.H., Daily Totals		$1589.20		$2501.60	$49.66	$78.17
Crew Q-12	**Hr.**	**Daily**	**Hr.**	**Daily**	**Bare Costs**	**Incl. O&P**
1 Sprinkler Installer	$54.05	$432.40	$84.25	$674.00	$48.65	$75.85
1 Sprinkler Apprentice	43.25	346.00	67.45	539.60		
16 L.H., Daily Totals		$778.40		$1213.60	$48.65	$75.85
Crew Q-13	**Hr.**	**Daily**	**Hr.**	**Daily**	**Bare Costs**	**Incl. O&P**
1 Sprinkler Foreman (inside)	$54.55	$436.40	$85.05	$680.40	$51.48	$80.25
2 Sprinkler Installers	54.05	864.80	84.25	1348.00		
1 Sprinkler Apprentice	43.25	346.00	67.45	539.60		
32 L.H., Daily Totals		$1647.20		$2568.00	$51.48	$80.25
Crew Q-14	**Hr.**	**Daily**	**Hr.**	**Daily**	**Bare Costs**	**Incl. O&P**
1 Asbestos Worker	$48.85	$390.80	$78.55	$628.40	$43.98	$70.70
1 Asbestos Apprentice	39.10	312.80	62.85	502.80		
16 L.H., Daily Totals		$703.60		$1131.20	$43.98	$70.70
Crew Q-15	**Hr.**	**Daily**	**Hr.**	**Daily**	**Bare Costs**	**Incl. O&P**
1 Plumber	$54.60	$436.80	$85.05	$680.40	$49.15	$76.58
1 Plumber Apprentice	43.70	349.60	68.10	544.80		
1 Welder, electric, 300 amp		36.90		40.59	2.31	2.54
16 L.H., Daily Totals		$823.30		$1265.79	$51.46	$79.11

Crew No.	Bare Costs		Incl. Subs O&P		Cost Per Labor-Hour	
Crew Q-16	**Hr.**	**Daily**	**Hr.**	**Daily**	**Bare Costs**	**Incl. O&P**
2 Plumbers	$54.60	$873.60	$85.05	$1360.80	$50.97	$79.40
1 Plumber Apprentice	43.70	349.60	68.10	544.80		
1 Welder, electric, 300 amp		36.90		40.59	1.54	1.69
24 L.H., Daily Totals		$1260.10		$1946.19	$52.50	$81.09
Crew Q-17	**Hr.**	**Daily**	**Hr.**	**Daily**	**Bare Costs**	**Incl. O&P**
1 Steamfitter	$55.70	$445.60	$86.80	$694.40	$50.13	$78.10
1 Steamfitter Apprentice	44.55	356.40	69.40	555.20		
1 Welder, electric, 300 amp		36.90		40.59	2.31	2.54
16 L.H., Daily Totals		$838.90		$1290.19	$52.43	$80.64
Crew Q-17A	**Hr.**	**Daily**	**Hr.**	**Daily**	**Bare Costs**	**Incl. O&P**
1 Steamfitter	$55.70	$445.60	$86.80	$694.40	$49.35	$77.23
1 Steamfitter Apprentice	44.55	356.40	69.40	555.20		
1 Equip. Oper. (crane)	47.80	382.40	75.50	604.00		
1 Hyd. Crane, 12 Ton		612.00		673.20		
1 Welder, electric, 300 amp		36.90		40.59	27.04	29.74
24 L.H., Daily Totals		$1833.30		$2567.39	$76.39	$106.97
Crew Q-18	**Hr.**	**Daily**	**Hr.**	**Daily**	**Bare Costs**	**Incl. O&P**
2 Steamfitters	$55.70	$891.20	$86.80	$1388.80	$51.98	$81.00
1 Steamfitter Apprentice	44.55	356.40	69.40	555.20		
1 Welder, electric, 300 amp		36.90		40.59	1.54	1.69
24 L.H., Daily Totals		$1284.50		$1984.59	$53.52	$82.69
Crew Q-19	**Hr.**	**Daily**	**Hr.**	**Daily**	**Bare Costs**	**Incl. O&P**
1 Steamfitter	$55.70	$445.60	$86.80	$694.40	$50.60	$78.62
1 Steamfitter Apprentice	44.55	356.40	69.40	555.20		
1 Electrician	51.55	412.40	79.65	637.20		
24 L.H., Daily Totals		$1214.40		$1886.80	$50.60	$78.62
Crew Q-20	**Hr.**	**Daily**	**Hr.**	**Daily**	**Bare Costs**	**Incl. O&P**
1 Sheet Metal Worker	$52.15	$417.20	$82.10	$656.80	$47.85	$75.03
1 Sheet Metal Apprentice	41.70	333.60	65.65	525.20		
.5 Electrician	51.55	206.20	79.65	318.60		
20 L.H., Daily Totals		$957.00		$1500.60	$47.85	$75.03
Crew Q-21	**Hr.**	**Daily**	**Hr.**	**Daily**	**Bare Costs**	**Incl. O&P**
2 Steamfitters	$55.70	$891.20	$86.80	$1388.80	$51.88	$80.66
1 Steamfitter Apprentice	44.55	356.40	69.40	555.20		
1 Electrician	51.55	412.40	79.65	637.20		
32 L.H., Daily Totals		$1660.00		$2581.20	$51.88	$80.66
Crew Q-22	**Hr.**	**Daily**	**Hr.**	**Daily**	**Bare Costs**	**Incl. O&P**
1 Plumber	$54.60	$436.80	$85.05	$680.40	$49.15	$76.58
1 Plumber Apprentice	43.70	349.60	68.10	544.80		
1 Hyd. Crane, 12 Ton		612.00		673.20	38.25	42.08
16 L.H., Daily Totals		$1398.40		$1898.40	$87.40	$118.65
Crew Q-22A	**Hr.**	**Daily**	**Hr.**	**Daily**	**Bare Costs**	**Incl. O&P**
1 Plumber	$54.60	$436.80	$85.05	$680.40	$45.30	$71.55
1 Plumber Apprentice	43.70	349.60	68.10	544.80		
1 Laborer	35.10	280.80	57.55	460.40		
1 Equip. Oper. (crane)	47.80	382.40	75.50	604.00		
1 Hyd. Crane, 12 Ton		612.00		673.20	19.13	21.04
32 L.H., Daily Totals		$2061.60		$2962.80	$64.42	$92.59

Crew No.	Bare Costs		Incl. Subs O&P		Cost Per Labor-Hour	
Crew Q-23	**Hr.**	**Daily**	**Hr.**	**Daily**	**Bare Costs**	**Incl. O&P**
1 Plumber Foreman (outside)	$56.60	$452.80	$88.20	$705.60	$52.58	$82.25
1 Plumber	54.60	436.80	85.05	680.40		
1 Equip. Oper. (med.)	46.55	372.40	73.50	588.00		
1 Lattice Boom Crane, 20 Ton		950.65		1045.71	39.61	43.57
24 L.H., Daily Totals		$2212.65		$3019.72	$92.19	$125.82
Crew R-1	**Hr.**	**Daily**	**Hr.**	**Daily**	**Bare Costs**	**Incl. O&P**
1 Electrician Foreman	$52.05	$416.40	$80.40	$643.20	$45.55	$71.41
3 Electricians	51.55	1237.20	79.65	1911.60		
2 Helpers	33.30	532.80	54.55	872.80		
48 L.H., Daily Totals		$2186.40		$3427.60	$45.55	$71.41
Crew R-1A	**Hr.**	**Daily**	**Hr.**	**Daily**	**Bare Costs**	**Incl. O&P**
1 Electrician	$51.55	$412.40	$79.65	$637.20	$42.42	$67.10
1 Helper	33.30	266.40	54.55	436.40		
16 L.H., Daily Totals		$678.80		$1073.60	$42.42	$67.10
Crew R-2	**Hr.**	**Daily**	**Hr.**	**Daily**	**Bare Costs**	**Incl. O&P**
1 Electrician Foreman	$52.05	$416.40	$80.40	$643.20	$45.87	$71.99
3 Electricians	51.55	1237.20	79.65	1911.60		
2 Helpers	33.30	532.80	54.55	872.80		
1 Equip. Oper. (crane)	47.80	382.40	75.50	604.00		
1 S.P. Crane, 4x4, 5 Ton		264.60		291.06	4.72	5.20
56 L.H., Daily Totals		$2833.40		$4322.66	$50.60	$77.19
Crew R-3	**Hr.**	**Daily**	**Hr.**	**Daily**	**Bare Costs**	**Incl. O&P**
1 Electrician Foreman	$52.05	$416.40	$80.40	$643.20	$51.00	$79.12
1 Electrician	51.55	412.40	79.65	637.20		
.5 Equip. Oper. (crane)	47.80	191.20	75.50	302.00		
.5 S.P. Crane, 4x4, 5 Ton		132.30		145.53	6.62	7.28
20 L.H., Daily Totals		$1152.30		$1727.93	$57.62	$86.40
Crew R-4	**Hr.**	**Daily**	**Hr.**	**Daily**	**Bare Costs**	**Incl. O&P**
1 Struc. Steel Foreman (outside)	$51.35	$410.80	$95.40	$763.20	$50.19	$90.03
3 Struc. Steel Workers	49.35	1184.40	91.70	2200.80		
1 Electrician	51.55	412.40	79.65	637.20		
1 Welder, gas engine, 300 amp		121.60		133.76	3.04	3.34
40 L.H., Daily Totals		$2129.20		$3734.96	$53.23	$93.37
Crew R-5	**Hr.**	**Daily**	**Hr.**	**Daily**	**Bare Costs**	**Incl. O&P**
1 Electrician Foreman	$52.05	$416.40	$80.40	$643.20	$44.96	$70.59
4 Electrician Linemen	51.55	1649.60	79.65	2548.80		
2 Electrician Operators	51.55	824.80	79.65	1274.40		
4 Electrician Groundmen	33.30	1065.60	54.55	1745.60		
1 Crew Truck		182.40		200.64		
1 Flatbed Truck, 20,000 GVW		215.80		237.38		
1 Pickup Truck, 3/4 Ton		127.00		139.70		
.2 Hyd. Crane, 55 Ton		220.00		242.00		
.2 Hyd. Crane, 12 Ton		122.40		134.64		
.2 Earth Auger, Truck-Mtd.		80.68		88.75		
1 Tractor w/Winch		382.20		420.42	15.12	16.63
88 L.H., Daily Totals		$5286.88		$7675.53	$60.08	$87.22

Crew No.	Bare Costs		Incl. Subs O&P		Cost Per Labor-Hour	
Crew R-6	**Hr.**	**Daily**	**Hr.**	**Daily**	**Bare Costs**	**Incl. O&P**
1 Electrician Foreman	$52.05	$416.40	$80.40	$643.20	$44.96	$70.59
4 Electrician Linemen	51.55	1649.60	79.65	2548.80		
2 Electrician Operators	51.55	824.80	79.65	1274.40		
4 Electrician Groundmen	33.30	1065.60	54.55	1745.60		
1 Crew Truck		182.40		200.64		
1 Flatbed Truck, 20,000 GVW		215.80		237.38		
1 Pickup Truck, 3/4 Ton		127.00		139.70		
.2 Hyd. Crane, 55 Ton		220.00		242.00		
.2 Hyd. Crane, 12 Ton		122.40		134.64		
.2 Earth Auger, Truck-Mtd.		80.68		88.75		
1 Tractor w/Winch		382.20		420.42		
3 Cable Trailers		563.70		620.07		
.5 Tensioning Rig		186.88		205.56		
.5 Cable Pulling Rig		1092.50		1201.75	36.06	39.67
88 L.H., Daily Totals		$7129.95		$9702.91	$81.02	$110.26
Crew R-7	**Hr.**	**Daily**	**Hr.**	**Daily**	**Bare Costs**	**Incl. O&P**
1 Electrician Foreman	$52.05	$416.40	$80.40	$643.20	$36.42	$58.86
5 Electrician Groundmen	33.30	1332.00	54.55	2182.00		
1 Crew Truck		182.40		200.64	3.80	4.18
48 L.H., Daily Totals		$1930.80		$3025.84	$40.23	$63.04
Crew R-8	**Hr.**	**Daily**	**Hr.**	**Daily**	**Bare Costs**	**Incl. O&P**
1 Electrician Foreman	$52.05	$416.40	$80.40	$643.20	$45.55	$71.41
3 Electrician Linemen	51.55	1237.20	79.65	1911.60		
2 Electrician Groundmen	33.30	532.80	54.55	872.80		
1 Pickup Truck, 3/4 Ton		127.00		139.70		
1 Crew Truck		182.40		200.64	6.45	7.09
48 L.H., Daily Totals		$2495.80		$3767.94	$52.00	$78.50
Crew R-9	**Hr.**	**Daily**	**Hr.**	**Daily**	**Bare Costs**	**Incl. O&P**
1 Electrician Foreman	$52.05	$416.40	$80.40	$643.20	$42.49	$67.19
1 Electrician Lineman	51.55	412.40	79.65	637.20		
2 Electrician Operators	51.55	824.80	79.65	1274.40		
4 Electrician Groundmen	33.30	1065.60	54.55	1745.60		
1 Pickup Truck, 3/4 Ton		127.00		139.70		
1 Crew Truck		182.40		200.64	4.83	5.32
64 L.H., Daily Totals		$3028.60		$4640.74	$47.32	$72.51
Crew R-10	**Hr.**	**Daily**	**Hr.**	**Daily**	**Bare Costs**	**Incl. O&P**
1 Electrician Foreman	$52.05	$416.40	$80.40	$643.20	$48.59	$75.59
4 Electrician Linemen	51.55	1649.60	79.65	2548.80		
1 Electrician Groundman	33.30	266.40	54.55	436.40		
1 Crew Truck		182.40		200.64		
3 Tram Cars		345.90		380.49	11.01	12.11
48 L.H., Daily Totals		$2860.70		$4209.53	$59.60	$87.70
Crew R-11	**Hr.**	**Daily**	**Hr.**	**Daily**	**Bare Costs**	**Incl. O&P**
1 Electrician Foreman	$52.05	$416.40	$80.40	$643.20	$48.74	$76.01
4 Electricians	51.55	1649.60	79.65	2548.80		
1 Equip. Oper. (crane)	47.80	382.40	75.50	604.00		
1 Common Laborer	35.10	280.80	57.55	460.40		
1 Crew Truck		182.40		200.64		
1 Hyd. Crane, 12 Ton		612.00		673.20	14.19	15.60
56 L.H., Daily Totals		$3523.60		$5130.24	$62.92	$91.61

Crew No.	Bare Costs		Incl. Subs O & P		Cost Per Labor-Hour	
Crew R-12	**Hr.**	**Daily**	**Hr.**	**Daily**	**Bare Costs**	**Incl. O&P**
1 Carpenter Foreman (inside)	$44.60	$356.80	$73.10	$584.80	$41.57	$68.88
4 Carpenters	44.10	1411.20	72.30	2313.60		
4 Common Laborers	35.10	1123.20	57.55	1841.60		
1 Equip. Oper. (med.)	46.55	372.40	73.50	588.00		
1 Steel Worker	49.35	394.80	91.70	733.60		
1 Dozer, 200 H.P.		1243.00		1367.30		
1 Pickup Truck, 3/4 Ton		127.00		139.70	15.57	17.13
88 L.H., Daily Totals		$5028.40		$7568.60	$57.14	$86.01

Crew R-13	Hr.	Daily	Hr.	Daily	Bare Costs	Incl. O&P
1 Electrician Foreman	$52.05	$416.40	$80.40	$643.20	$49.50	$76.83
3 Electricians	51.55	1237.20	79.65	1911.60		
.25 Equip. Oper. (crane)	47.80	95.60	75.50	151.00		
1 Equipment Oiler	41.25	330.00	65.15	521.20		
.25 Hydraulic Crane, 33 Ton		177.90		195.69	4.24	4.66
42 L.H., Daily Totals		$2257.10		$3422.69	$53.74	$81.49

Crew R-15	Hr.	Daily	Hr.	Daily	Bare Costs	Incl. O&P
1 Electrician Foreman	$52.05	$416.40	$80.40	$643.20	$50.50	$78.28
4 Electricians	51.55	1649.60	79.65	2548.80		
1 Equipment Oper. (light)	44.75	358.00	70.65	565.20		
1 Aerial Lift Truck		324.20		356.62	6.75	7.43
48 L.H., Daily Totals		$2748.20		$4113.82	$57.25	$85.70

Crew R-18	Hr.	Daily	Hr.	Daily	Bare Costs	Incl. O&P
.25 Electrician Foreman	$52.05	$104.10	$80.40	$160.80	$40.36	$64.26
1 Electrician	51.55	412.40	79.65	637.20		
2 Helpers	33.30	532.80	54.55	872.80		
26 L.H., Daily Totals		$1049.30		$1670.80	$40.36	$64.26

Crew R-19	Hr.	Daily	Hr.	Daily	Bare Costs	Incl. O&P
.5 Electrician Foreman	$52.05	$208.20	$80.40	$321.60	$51.65	$79.80
2 Electricians	51.55	824.80	79.65	1274.40		
20 L.H., Daily Totals		$1033.00		$1596.00	$51.65	$79.80

Crew R-21	Hr.	Daily	Hr.	Daily	Bare Costs	Incl. O&P
1 Electrician Foreman	$52.05	$416.40	$80.40	$643.20	$51.55	$79.68
3 Electricians	51.55	1237.20	79.65	1911.60		
.1 Equip. Oper. (med.)	46.55	37.24	73.50	58.80		
.1 S.P. Crane, 4x4, 25 Ton		56.52		62.17	1.72	1.90
32.8 L.H., Daily Totals		$1747.36		$2675.77	$53.27	$81.58

Crew R-22	Hr.	Daily	Hr.	Daily	Bare Costs	Incl. O&P
.66 Electrician Foreman	$52.05	$274.82	$80.40	$424.51	$43.79	$68.98
2 Helpers	33.30	532.80	54.55	872.80		
2 Electricians	51.55	824.80	79.65	1274.40		
37.28 L.H., Daily Totals		$1632.42		$2571.71	$43.79	$68.98

Crew R-30	Hr.	Daily	Hr.	Daily	Bare Costs	Incl. O&P
.25 Electrician Foreman (outside)	$53.55	$107.10	$82.75	$165.50	$41.58	$66.29
1 Electrician	51.55	412.40	79.65	637.20		
2 Laborers, (Semi-Skilled)	35.10	561.60	57.55	920.80		
26 L.H., Daily Totals		$1081.10		$1723.50	$41.58	$66.29

Crew R-31	Hr.	Daily	Hr.	Daily	Bare Costs	Incl. O&P
1 Electrician	$51.55	$412.40	$79.65	$637.20	$51.55	$79.65
1 Core Drill, Electric, 2.5 H.P.		47.00		51.70	5.88	6.46
8 L.H., Daily Totals		$459.40		$688.90	$57.42	$86.11

Historical Cost Indexes

The table below lists both the RSMeans Historical Cost Index based on Jan. 1, 1993 = 100 as well as the computed value of an index based on Jan. 1, 2012 costs. Since the Jan. 1, 2012 figure is estimated, space is left to write in the actual index figures as they become available through either the quarterly *Means Construction Cost Indexes* or as printed in the *Engineering News-Record*. To compute the actual index based on Jan. 1, 2012 = 100, divide the Historical Cost Index for a particular year by the actual Jan. 1, 2012 Construction Cost Index. Space has been left to advance the index figures as the year progresses.

Year	Historical Cost Index Jan. 1, 1993 = 100		Current Index Based on Jan. 1, 2012 = 100	
	Est.	Actual	Est.	Actual
Oct 2012*				
July 2012*				
April 2012*				
Jan 2012*	192.8		100.0	100.0
July 2011		185.7	96.3	
2010		183.5	95.2	
2009		180.1	93.4	
2008		180.4	93.6	
2007		169.4	87.9	
2006		162.0	84.0	
2005		151.6	78.6	
2004		143.7	74.5	
2003		132.0	68.5	
2002		128.7	66.8	
2001		125.1	64.9	
2000		120.9	62.7	
1999		117.6	61.0	
1998		115.1	59.7	

Year	Historical Cost Index Jan. 1, 1993 = 100	Current Index Based on Jan. 1, 2012 = 100	
	Actual	Est.	Actual
July 1997	112.8	58.5	
1996	110.2	57.2	
1995	107.6	55.8	
1994	104.4	54.1	
1993	101.7	52.7	
1992	99.4	51.6	
1991	96.8	50.2	
1990	94.3	48.9	
1989	92.1	47.8	
1988	89.9	46.6	
1987	87.7	45.5	
1986	84.2	43.7	
1985	82.6	42.8	
1984	82.0	42.5	
1983	80.2	41.6	
1982	76.1	39.5	
1981	70.0	36.3	
1980	62.9	32.6	

Year	Historical Cost Index Jan. 1, 1993 = 100	Current Index Based on Jan. 1, 2012 = 100	
	Actual	Est.	Actual
July 1979	57.8	30.0	
1978	53.5	27.7	
1977	49.5	25.7	
1976	46.9	24.3	
1975	44.8	23.2	
1974	41.4	21.5	
1973	37.7	19.6	
1972	34.8	18.0	
1971	32.1	16.6	
1970	28.7	14.9	
1969	26.9	14.0	
1968	24.9	12.9	
1967	23.5	12.2	
1966	22.7	11.8	
1965	21.7	11.3	
1964	21.2	11.0	
1963	20.7	10.7	
1962	20.2	10.5	

Adjustments to Costs

The Historical Cost Index can be used to convert National Average building costs at a particular time to the approximate building costs for some other time.

Time Adjustment using the Historical Cost Indexes:

$$\frac{\text{Index for Year A}}{\text{Index for Year B}} \times \text{Cost in Year B} = \text{Cost in Year A}$$

Example:

Estimate and compare construction costs for different years in the same city.

To estimate the National Average construction cost of a building in 1970, knowing that it cost $900,000 in 2012:

INDEX in 1970 = 28.7

INDEX in 2012 = 192.8

$$\frac{\text{INDEX 1970}}{\text{INDEX 2012}} \times \text{Cost 2012} = \text{Cost 1970}$$

$$\frac{28.7}{192.8} \times \$900{,}000 = .149 \times \$900{,}000 = \$134{,}100$$

The construction cost of the building in 1970 is $134,100.

Note: The City Cost Indexes for Canada can be used to convert U.S. National averages to local costs in Canadian dollars.

*Historical Cost Index updates and other resources are provided on the following website.
http://www.reedconstructiondata.com/rsmeans/chgnotice/176028

How to Use the City Cost Indexes

What you should know before you begin

RSMeans City Cost Indexes (CCI) are an extremely useful tool to use when you want to compare costs from city to city and region to region.

This publication contains average construction cost indexes for 731 U.S. and Canadian cities covering over 930 three-digit zip code locations, as listed directly under each city.

Keep in mind that a City Cost Index number is a percentage ratio of a specific city's cost to the national average cost of the same item at a stated time period.

In other words, these index figures represent relative construction factors (or, if you prefer, multipliers) for Material and Installation costs, as well as the weighted average for Total In Place costs for each CSI MasterFormat division. Installation costs include both labor and equipment rental costs. When estimating equipment rental rates only, for a specific location, use 015433 CONTRACTOR EQUIPMENT index.

The 30 City Average Index is the average of 30 major U.S. cities and serves as a National Average.

Index figures for both material and installation are based on the 30 major city average of 100 and represent the cost relationship as of July 1, 2011. The index for each division is computed from representative material and labor quantities for that division. The weighted average for each city is a weighted total of the components listed above it, but does not include relative productivity between trades or cities.

As changes occur in local material prices, labor rates, and equipment rental rates, (including fuel costs) the impact of these changes should be accurately measured by the change in the City Cost Index for each particular city (as compared to the 30 City Average).

Therefore, if you know (or have estimated) building costs in one city today, you can easily convert those costs to expected building costs in another city.

In addition, by using the Historical Cost Index, you can easily convert National Average building costs at a particular time to the approximate building costs for some other time. The City Cost Indexes can then be applied to calculate the costs for a particular city.

Quick Calculations

Location Adjustment Using the City Cost Indexes:

$$\frac{\text{Index for City A}}{\text{Index for City B}} \times \text{Cost in City B} = \text{Cost in City A}$$

Time Adjustment for the National Average Using the Historical Cost Index:

$$\frac{\text{Index for Year A}}{\text{Index for Year B}} \times \text{Cost in Year B} = \text{Cost in Year A}$$

Adjustment from the National Average:

$$\frac{\text{Index for City A}}{100} \times \text{National Average Cost} = \text{Cost in City A}$$

Since each of the other RSMeans publications contains many different items, any *one* item multiplied by the particular city index may give incorrect results. However, the larger the number of items compiled, the closer the results should be to actual costs for that particular city.

The City Cost Indexes for Canadian cities are calculated using Canadian material and equipment prices and labor rates, in Canadian dollars. Therefore, indexes for Canadian cities can be used to convert U.S. National Average prices to local costs in Canadian dollars.

How to use this section

1. Compare costs from city to city.

In using the RSMeans Indexes, remember that an index number is not a fixed number but a ratio: It's a percentage ratio of a building component's cost at any stated time to the National Average cost of that same component at the same time period. Put in the form of an equation:

$$\frac{\text{Specific City Cost}}{\text{National Average Cost}} \times 100 = \text{City Index Number}$$

Therefore, when making cost comparisons between cities, do not subtract one city's index number from the index number of another city and read the result as a percentage difference. Instead, divide one city's index number by that of the other city. The resulting number may then be used as a multiplier to calculate cost differences from city to city.

The formula used to find cost differences between cities for the purpose of comparison is as follows:

$$\frac{\text{City A Index}}{\text{City B Index}} \times \text{City B Cost (Known)} = \text{City A Cost (Unknown)}$$

In addition, you can use RSMeans CCI to calculate and compare costs division by division between cities using the same basic formula. (Just be sure that you're comparing similar divisions.)

2. Compare a specific city's construction costs with the National Average.

When you're studying construction location feasibility, it's advisable to compare a prospective project's cost index with an index of the National Average cost.

For example, divide the weighted average index of construction costs of a specific city by that of the 30 City Average, which = 100.

$$\frac{\text{City Index}}{100} = \text{\% of National Average}$$

As a result, you get a ratio that indicates the relative cost of construction in that city in comparison with the National Average.

3. Convert U.S. National Average to actual costs in Canadian City.

$$\frac{\text{Index for Canadian City}}{100} \times \text{National Average Cost} = \text{Cost in Canadian City in \$ CAN}$$

4. **Adjust construction cost data based on a National Average.**

When you use a source of construction cost data which is based on a National Average (such as RSMeans cost data publications), it is necessary to adjust those costs to a specific location.

$$\frac{\text{City Index}}{100} \times \text{"Book" Cost Based on National Average Costs} = \text{City Cost (Unknown)}$$

5. **When applying the City Cost Indexes to demolition projects, use the appropriate division installation index.** For example, for removal of existing doors and windows, use Division 8 (Openings) index.

What you might like to know about how we developed the Indexes

The information presented in the CCI is organized according to the Construction Specifications Institute (CSI) MasterFormat 2010.

To create a reliable index, RSMeans researched the building type most often constructed in the United States and Canada. Because it was concluded that no one type of building completely represented the building construction industry, nine different types of buildings were combined to create a composite model.

The exact material, labor, and equipment quantities are based on detailed analyses of these nine building types, then each quantity is weighted in proportion to expected usage. These various material items, labor hours, and equipment rental rates are thus combined to form a composite building representing as closely as possible the actual usage of materials, labor and equipment used in the North American Building Construction Industry.

The following structures were chosen to make up that composite model:

1. Factory, 1 story
2. Office, 2–4 story
3. Store, Retail
4. Town Hall, 2–3 story
5. High School, 2–3 story
6. Hospital, 4–8 story
7. Garage, Parking
8. Apartment, 1–3 story
9. Hotel/Motel, 2–3 story

For the purposes of ensuring the timeliness of the data, the components of the index for the composite model have been streamlined. They currently consist of:

- specific quantities of 66 commonly used construction materials;
- specific labor-hours for 21 building construction trades; and
- specific days of equipment rental for 6 types of construction equipment (normally used to install the 66 material items by the 21 trades.) Fuel costs and routine maintenance costs are included in the equipment cost.

A sophisticated computer program handles the updating of all costs for each city on a quarterly basis. Material and equipment price quotations are gathered quarterly from 731 cities in the United States and Canada. These prices and the latest negotiated labor wage rates for 21 different building trades are used to compile the quarterly update of the City Cost Index.

The 30 major U.S. cities used to calculate the National Average are:

Atlanta, GA	Memphis, TN
Baltimore, MD	Milwaukee, WI
Boston, MA	Minneapolis, MN
Buffalo, NY	Nashville, TN
Chicago, IL	New Orleans, LA
Cincinnati, OH	New York, NY
Cleveland, OH	Philadelphia, PA
Columbus, OH	Phoenix, AZ
Dallas, TX	Pittsburgh, PA
Denver, CO	St. Louis, MO
Detroit, MI	San Antonio, TX
Houston, TX	San Diego, CA
Indianapolis, IN	San Francisco, CA
Kansas City, MO	Seattle, WA
Los Angeles, CA	Washington, DC

What the CCI does not indicate

The weighted average for each city is a total of the divisional components weighted to reflect typical usage, but it does not include the productivity variations between trades or cities.

In addition, the CCI does not take into consideration factors such as the following:

- managerial efficiency
- competitive conditions
- automation
- restrictive union practices
- unique local requirements
- regional variations due to specific building codes

	DIVISION	UNITED STATES 30 CITY AVERAGE			ALABAMA ANNISTON 362			BIRMINGHAM 350 - 352			BUTLER 369			DECATUR 356			DOTHAN 363		
		MAT.	INST.	TOTAL	MAT.	INST.	TOTAL	MAT.	INST.	TOTAL	MAT.	INST.	TOTAL	MAT.	INST.	TOTAL	MAT.	INST.	TOTAL
015433	**CONTRACTOR EQUIPMENT**		100.0	100.0		100.7	100.7		100.8	100.8		98.0	98.0		100.7	100.7		98.0	98.0
0241, 31 - 34	**SITE & INFRASTRUCTURE, DEMOLITION**	100.0	100.0	100.0	87.6	93.4	91.6	85.6	94.0	91.4	103.5	87.5	92.4	80.6	92.0	88.5	100.8	87.5	91.6
0310	Concrete Forming & Accessories	100.0	100.0	100.0	91.9	49.9	55.4	92.3	72.4	75.0	88.6	45.2	50.8	96.0	48.1	54.3	96.8	44.6	51.4
0320	Concrete Reinforcing	100.0	100.0	100.0	93.3	83.3	88.3	93.3	83.7	88.5	98.6	48.8	73.5	93.3	78.1	85.7	98.6	48.2	73.2
0330	Cast-in-Place Concrete	100.0	100.0	100.0	92.5	51.8	76.5	98.7	66.6	86.0	90.2	57.3	77.2	91.9	64.9	81.3	90.2	48.2	73.7
03	**CONCRETE**	100.0	100.0	100.0	95.3	58.4	77.3	94.3	73.5	84.2	96.1	51.9	74.6	91.1	61.2	76.5	95.5	48.4	72.6
04	**MASONRY**	100.0	100.0	100.0	92.0	70.5	78.9	94.8	79.1	85.2	96.9	51.3	69.1	93.0	54.2	69.3	98.1	42.0	63.8
05	**METALS**	100.0	100.0	100.0	93.6	91.3	92.9	94.5	93.1	94.0	92.6	78.9	88.1	95.8	87.9	93.2	92.7	77.7	87.7
06	**WOOD, PLASTICS & COMPOSITES**	100.0	100.0	100.0	94.1	45.4	66.0	96.0	71.0	81.5	89.3	45.0	63.7	99.0	45.6	68.1	100.4	45.0	68.4
07	**THERMAL & MOISTURE PROTECTION**	100.0	100.0	100.0	93.1	55.8	78.0	95.1	82.0	89.8	93.2	59.5	79.5	92.9	61.3	80.1	93.1	51.8	76.3
08	**OPENINGS**	100.0	100.0	100.0	93.8	51.1	83.1	97.6	73.3	91.4	93.9	46.8	82.0	97.9	52.6	86.4	93.9	46.8	82.1
0920	Plaster & Gypsum Board	100.0	100.0	100.0	88.2	44.1	56.7	97.1	70.5	78.1	86.0	43.7	55.8	93.2	44.4	58.3	96.0	43.7	58.7
0950, 0980	Ceilings & Acoustic Treatment	100.0	100.0	100.0	89.3	44.1	59.4	94.9	70.5	78.8	89.3	43.7	59.2	93.8	44.4	61.2	89.3	43.7	59.2
0960	Flooring	100.0	100.0	100.0	101.8	42.1	84.3	104.5	63.1	92.4	107.6	58.6	93.2	104.5	53.0	89.4	114.6	28.8	89.4
0970, 0990	Wall Finishes & Painting/Coating	100.0	100.0	100.0	96.6	33.2	57.5	96.6	58.6	73.1	96.6	56.2	71.7	96.6	59.8	73.9	96.6	56.2	71.7
09	**FINISHES**	100.0	100.0	100.0	95.4	45.5	67.6	98.5	68.8	81.9	99.4	48.4	70.9	97.1	49.4	70.5	102.3	42.4	68.9
COVERS	**DIVS. 10 - 14, 25, 28, 41, 43, 44, 46**	100.0	100.0	100.0	100.0	73.9	94.8	100.0	88.7	97.8	100.0	44.7	89.0	100.0	46.7	89.4	100.0	44.8	89.0
21, 22, 23	**FIRE SUPPRESSION, PLUMBING & HVAC**	100.0	100.0	100.0	99.2	59.4	83.0	99.9	65.9	86.1	96.7	35.7	72.0	99.9	43.7	77.2	96.7	35.6	72.0
26, 27, 3370	**ELECTRICAL, COMMUNICATIONS & UTIL.**	100.0	100.0	100.0	96.6	60.9	78.2	103.0	62.5	82.1	98.5	42.7	69.7	98.2	65.6	81.4	97.3	61.0	78.6
MF2010	**WEIGHTED AVERAGE**	100.0	100.0	100.0	96.0	64.5	82.1	97.6	74.6	87.4	96.5	52.0	76.9	96.5	60.1	80.5	96.6	52.0	76.9

	DIVISION	ALABAMA EVERGREEN 364			GADSDEN 359			HUNTSVILLE 357 - 358			JASPER 355			MOBILE 365 - 366			MONTGOMERY 360 - 361		
		MAT.	INST.	TOTAL	MAT.	INST.	TOTAL	MAT.	INST.	TOTAL	MAT.	INST.	TOTAL	MAT.	INST.	TOTAL	MAT.	INST.	TOTAL
015433	**CONTRACTOR EQUIPMENT**		98.0	98.0		100.7	100.7		100.7	100.7		100.7	100.7		98.0	98.0		98.0	98.0
0241, 31 - 34	**SITE & INFRASTRUCTURE, DEMOLITION**	104.0	87.8	92.8	86.3	92.9	90.9	80.4	93.1	89.1	85.7	92.7	90.6	95.1	88.5	90.5	93.6	88.4	90.0
0310	Concrete Forming & Accessories	85.4	46.9	51.9	88.7	43.9	49.7	96.0	64.3	68.4	93.6	36.4	43.9	96.1	56.9	62.0	94.0	45.9	52.2
0320	Concrete Reinforcing	98.7	48.8	73.5	98.9	82.5	90.6	93.3	78.6	85.9	93.3	82.4	87.8	96.3	81.7	89.0	96.3	82.3	89.3
0330	Cast-in-Place Concrete	90.2	59.9	78.3	92.0	51.0	75.8	89.6	63.3	79.2	101.9	45.9	79.8	94.5	69.3	84.6	96.5	48.6	77.6
03	**CONCRETE**	96.3	53.6	75.5	95.1	55.3	75.7	89.9	67.9	79.2	98.6	50.3	75.0	92.9	67.3	80.4	93.7	55.4	75.1
04	**MASONRY**	96.9	56.2	72.0	91.4	45.4	63.3	94.5	60.3	73.6	88.0	43.6	60.9	95.2	55.7	71.1	93.3	41.6	61.7
05	**METALS**	92.7	78.9	88.1	93.7	90.5	92.6	95.8	89.5	93.7	93.6	89.8	92.4	94.6	91.6	93.6	94.5	90.1	93.0
06	**WOOD, PLASTICS & COMPOSITES**	85.9	45.0	62.3	89.9	41.1	61.7	99.0	65.2	79.4	96.0	33.5	59.9	99.1	56.1	74.3	95.6	45.0	66.4
07	**THERMAL & MOISTURE PROTECTION**	93.1	60.1	79.7	93.0	66.1	82.1	92.8	74.7	85.4	93.0	47.8	74.7	92.8	73.3	84.9	92.4	64.8	81.2
08	**OPENINGS**	93.9	46.8	82.0	93.8	49.9	82.8	97.9	63.2	89.1	93.8	48.8	82.5	97.9	60.7	88.5	97.9	54.3	86.9
0920	Plaster & Gypsum Board	85.1	43.7	55.5	85.5	39.7	52.8	93.2	64.5	72.7	88.7	31.9	48.1	93.2	55.2	66.1	97.1	43.7	59.0
0950, 0980	Ceilings & Acoustic Treatment	89.3	43.7	59.2	89.3	39.7	56.6	95.8	64.5	75.2	89.3	31.9	51.4	95.8	55.2	69.0	94.9	43.7	61.1
0960	Flooring	105.0	58.6	91.4	100.0	47.1	84.5	104.5	56.6	90.4	102.6	42.1	84.9	113.6	58.6	97.5	112.0	31.0	88.2
0970, 0990	Wall Finishes & Painting/Coating	96.6	56.2	71.7	96.6	64.0	76.5	96.6	61.6	75.0	96.6	40.7	62.1	100.1	57.2	73.6	96.6	56.2	71.7
09	**FINISHES**	98.6	49.6	71.2	94.4	44.9	66.8	97.6	62.2	77.8	95.4	36.5	62.5	102.6	56.1	76.7	101.8	42.8	68.9
COVERS	**DIVS. 10 - 14, 25, 28, 41, 43, 44, 46**	100.0	46.4	89.3	100.0	80.4	96.1	100.0	85.1	97.0	100.0	42.9	88.6	100.0	81.6	96.3	100.0	80.3	96.1
21, 22, 23	**FIRE SUPPRESSION, PLUMBING & HVAC**	96.7	38.2	73.0	101.8	38.6	76.2	99.9	55.5	81.9	101.8	56.4	83.4	99.9	64.6	85.6	100.0	36.4	74.2
26, 27, 3370	**ELECTRICAL, COMMUNICATIONS & UTIL.**	95.9	42.7	68.5	98.3	62.5	79.9	99.4	65.6	82.0	97.7	62.5	79.6	99.4	60.7	79.5	98.5	64.7	81.0
MF2010	**WEIGHTED AVERAGE**	96.2	53.4	77.3	96.6	57.6	79.4	96.6	68.1	84.1	96.8	57.6	79.5	97.7	67.5	84.4	97.4	56.6	79.4

	DIVISION	ALABAMA PHENIX CITY 368			SELMA 367			TUSCALOOSA 354			ALASKA ANCHORAGE 995 - 996			FAIRBANKS 997			JUNEAU 998		
		MAT.	INST.	TOTAL	MAT.	INST.	TOTAL	MAT.	INST.	TOTAL	MAT.	INST.	TOTAL	MAT.	INST.	TOTAL	MAT.	INST.	TOTAL
015433	**CONTRACTOR EQUIPMENT**		98.0	98.0		98.0	98.0		100.7	100.7		114.5	114.5		114.5	114.5		114.5	114.5
0241, 31 - 34	**SITE & INFRASTRUCTURE, DEMOLITION**	108.1	88.4	94.5	100.6	88.3	92.2	80.8	93.1	89.3	125.9	129.7	128.5	118.0	129.7	126.1	125.5	129.7	128.4
0310	Concrete Forming & Accessories	91.9	40.7	47.4	89.8	45.1	50.9	95.9	49.5	55.6	139.3	117.2	120.1	142.3	117.3	120.6	138.5	117.2	119.9
0320	Concrete Reinforcing	98.6	65.1	81.7	98.6	82.1	90.3	93.3	82.4	87.9	148.6	108.8	128.5	143.5	108.9	126.0	111.7	108.8	110.3
0330	Cast-in-Place Concrete	90.2	49.2	74.0	90.2	48.4	73.7	93.3	51.0	76.6	144.5	116.5	133.5	136.6	116.9	128.8	144.4	116.5	133.4
03	**CONCRETE**	98.7	50.1	75.1	95.0	54.9	75.5	91.7	57.9	75.2	131.5	114.7	123.3	119.6	114.9	117.3	122.1	114.7	118.5
04	**MASONRY**	96.9	43.4	64.2	100.2	40.8	64.0	93.3	48.5	66.0	193.3	121.7	149.6	195.8	121.7	150.6	178.8	121.7	144.0
05	**METALS**	92.6	83.8	89.7	92.6	89.0	91.4	94.9	90.5	93.4	120.1	101.7	114.0	117.1	101.8	112.1	117.1	101.6	112.0
06	**WOOD, PLASTICS & COMPOSITES**	93.8	37.6	61.4	90.9	45.0	64.4	99.0	47.2	69.0	116.7	116.0	116.3	121.5	116.0	118.4	116.7	116.0	116.3
07	**THERMAL & MOISTURE PROTECTION**	93.5	64.7	81.8	93.0	54.7	77.4	92.9	68.2	82.9	191.0	116.3	160.6	179.7	117.6	154.4	181.2	116.3	154.8
08	**OPENINGS**	93.8	46.1	81.8	93.8	54.3	83.9	97.9	60.4	88.4	119.9	112.9	118.1	116.4	113.0	115.5	115.7	112.9	115.0
0920	Plaster & Gypsum Board	89.6	36.1	51.4	87.3	43.7	56.2	93.2	46.0	59.5	143.4	116.3	124.1	150.2	116.3	126.0	130.9	116.3	120.5
0950, 0980	Ceilings & Acoustic Treatment	89.3	36.1	54.2	89.3	43.7	59.2	95.8	46.0	62.9	136.0	116.3	123.0	122.5	116.3	118.4	116.2	116.3	116.3
0960	Flooring	110.1	31.0	86.9	108.3	29.8	85.3	104.5	39.3	85.4	160.1	128.4	150.8	150.4	128.4	143.9	150.6	128.4	144.1
0970, 0990	Wall Finishes & Painting/Coating	96.6	56.2	71.7	96.6	56.2	71.7	96.6	48.4	66.9	150.9	116.9	129.9	150.4	125.8	135.2	150.9	116.9	129.9
09	**FINISHES**	101.1	38.7	66.3	99.4	42.6	67.7	97.5	46.3	69.0	150.9	119.6	133.4	143.6	120.5	130.8	140.5	119.6	128.8
COVERS	**DIVS. 10 - 14, 25, 28, 41, 43, 44, 46**	100.0	79.9	96.0	100.0	44.7	89.0	100.0	82.3	96.5	100.0	108.7	101.7	100.0	108.7	101.7	100.0	108.7	101.7
21, 22, 23	**FIRE SUPPRESSION, PLUMBING & HVAC**	96.7	36.8	72.4	96.7	35.9	72.1	99.9	36.2	74.1	100.3	104.5	102.0	100.3	108.2	103.5	100.3	104.0	101.8
26, 27, 3370	**ELECTRICAL, COMMUNICATIONS & UTIL.**	98.0	73.9	85.5	97.0	42.7	69.0	98.9	62.5	80.1	133.3	116.7	124.7	143.5	116.7	129.6	129.3	116.7	122.8
MF2010	**WEIGHTED AVERAGE**	97.1	55.9	78.9	96.3	51.9	76.7	96.6	58.6	79.9	124.5	114.0	119.9	122.4	115.0	119.1	120.3	113.9	117.5

	DIVISION	ALASKA			ARIZONA															
		KETCHIKAN			CHAMBERS			FLAGSTAFF			GLOBE			KINGMAN			MESA/TEMPE			
		999			865			860			855			864			852			
		MAT.	INST.	TOTAL	MAT.	INST.	TOTAL	MAT.	INST.	TOTAL	MAT.	INST.	TOTAL	MAT.	INST.	TOTAL	MAT.	INST.	TOTAL	
015433	CONTRACTOR EQUIPMENT		114.5	114.5		94.0	94.0		94.0	94.0		95.9	95.9		94.0	94.0		95.9	95.9	
0241, 31 - 34	SITE & INFRASTRUCTURE, DEMOLITION	176.4	129.7	144.2	64.0	96.6	86.5	82.4	98.5	93.5	96.6	98.8	98.1	64.0	98.6	87.9	86.8	100.4	96.2	
0310	Concrete Forming & Accessories	134.3	117.2	119.4	96.5	45.5	52.2	101.3	68.1	72.4	99.2	45.4	52.4	94.8	67.3	70.9	101.8	67.2	71.7	
0320	Concrete Reinforcing	112.2	108.8	110.5	98.0	78.0	87.9	97.8	85.2	91.5	102.6	77.9	90.2	98.1	85.2	91.6	103.3	85.2	94.1	
0330	Cast-in-Place Concrete	276.4	116.5	213.4	89.3	54.1	75.4	89.4	75.5	83.9	86.2	54.2	73.6	89.0	71.9	82.3	86.9	71.9	81.0	
03	CONCRETE	203.0	114.7	159.9	94.8	55.0	75.4	112.9	73.8	93.9	105.5	55.0	80.9	94.4	72.3	83.6	97.5	72.3	85.2	
04	MASONRY	204.8	121.7	154.1	92.9	41.2	61.3	93.0	62.8	74.6	107.4	41.1	66.9	92.9	61.3	73.6	107.5	61.3	79.3	
05	METALS	117.3	101.6	112.1	94.0	68.6	85.6	94.5	76.8	88.7	90.6	68.8	83.4	94.7	76.6	88.7	90.9	76.8	86.3	
06	WOOD, PLASTICS & COMPOSITES	112.9	116.0	114.7	94.6	43.4	65.0	99.9	68.3	81.6	88.8	43.5	62.6	90.6	68.3	77.7	92.0	68.4	78.3	
07	THERMAL & MOISTURE PROTECTION	182.8	116.3	155.8	96.4	50.9	77.9	98.2	66.3	85.2	102.6	48.4	80.6	96.4	62.1	82.5	101.8	62.8	86.0	
08	OPENINGS	118.6	112.9	117.2	104.8	50.8	91.2	105.0	73.2	97.0	97.7	50.9	85.9	105.0	72.0	96.7	97.8	72.0	91.3	
0920	Plaster & Gypsum Board	137.9	116.3	122.5	91.3	41.7	55.9	94.8	67.3	75.2	97.8	41.7	57.7	85.2	67.3	72.4	100.6	67.3	76.8	
0950, 0980	Ceilings & Acoustic Treatment	117.4	116.3	116.7	112.4	41.7	65.7	113.3	67.3	82.9	98.9	41.7	61.1	113.3	67.3	82.9	98.9	67.3	78.0	
0960	Flooring	150.4	128.4	143.9	92.7	41.4	77.7	95.1	41.6	79.4	98.1	41.4	81.4	91.6	56.2	81.2	99.5	51.9	85.5	
0970, 0990	Wall Finishes & Painting/Coating	150.4	116.9	129.7	93.2	38.9	59.7	93.2	57.7	71.3	97.8	38.9	61.5	93.2	57.7	71.3	97.8	57.7	73.1	
09	FINISHES	146.4	119.6	131.4	97.4	42.8	67.0	100.7	61.4	78.8	98.1	42.9	67.3	96.5	63.5	78.1	97.7	62.7	78.2	
COVERS	DIVS. 10 - 14, 25, 28, 41, 43, 44, 46	100.0	108.7	101.7	100.0	77.7	95.6	100.0	82.9	96.6	100.0	77.9	95.6	100.0	82.3	96.5	100.0	82.6	96.5	
21, 22, 23	FIRE SUPPRESSION, PLUMBING & HVAC	98.6	104.0	100.8	96.4	74.0	87.3	100.2	83.0	93.2	94.3	73.7	86.0	96.4	82.0	90.6	100.2	76.9	90.8	
26, 27, 3370	ELECTRICAL, COMMUNICATIONS & UTIL.	143.4	116.7	129.6	101.4	68.5	84.4	100.4	62.3	80.7	96.6	63.7	79.7	101.4	62.3	81.2	93.6	62.3	77.4	
MF2010	WEIGHTED AVERAGE	134.1	113.9	125.2	96.5	63.0	81.7	100.3	73.9	88.7	97.4	62.5	82.0	96.4	73.4	86.3	97.4	72.4	86.3	

	DIVISION	ARIZONA												ARKANSAS					
		PHOENIX			PRESCOTT			SHOW LOW			TUCSON			BATESVILLE			CAMDEN		
		850,853			863			859			856 - 857			725			717		
		MAT.	INST.	TOTAL	MAT.	INST.	TOTAL	MAT.	INST.	TOTAL	MAT.	INST.	TOTAL	MAT.	INST.	TOTAL	MAT.	INST.	TOTAL
015433	CONTRACTOR EQUIPMENT		96.4	96.4		94.0	94.0		95.9	95.9		95.9	95.9		88.3	88.3		88.3	88.3
0241, 31 - 34	SITE & INFRASTRUCTURE, DEMOLITION	87.3	100.9	96.7	70.3	96.6	88.4	98.8	98.8	98.8	82.3	100.7	95.0	71.0	84.4	80.2	71.7	83.9	80.1
0310	Concrete Forming & Accessories	102.7	70.4	74.6	97.8	45.3	52.1	105.3	45.7	53.4	102.0	70.0	74.2	83.5	47.6	52.3	83.3	32.4	39.0
0320	Concrete Reinforcing	101.5	85.4	93.4	97.8	78.2	87.9	103.3	78.1	90.6	85.5	85.2	85.3	95.4	64.5	79.8	96.7	64.3	80.4
0330	Cast-in-Place Concrete	87.0	75.7	82.5	89.3	54.0	75.4	86.2	54.3	73.6	89.4	75.5	84.0	80.2	47.4	67.3	82.2	40.0	65.6
03	CONCRETE	97.1	75.1	86.3	99.7	54.9	77.9	107.6	55.2	82.0	95.8	74.7	85.5	85.2	51.5	68.8	86.9	42.2	65.1
04	MASONRY	94.3	65.9	77.0	93.0	46.4	64.5	107.4	42.7	67.9	92.4	62.8	74.3	105.5	43.1	67.4	113.6	33.2	64.5
05	METALS	92.5	78.2	87.7	94.6	68.7	86.0	90.4	69.2	83.4	91.7	77.0	86.8	96.9	66.9	87.0	97.0	66.4	86.9
06	WOOD, PLASTICS & COMPOSITES	92.9	71.2	80.3	95.8	43.4	65.5	96.2	43.5	65.8	92.3	71.2	80.1	86.9	48.3	64.6	86.7	31.5	54.8
07	THERMAL & MOISTURE PROTECTION	101.5	67.0	87.4	96.9	50.2	77.9	102.8	50.8	81.7	102.9	62.7	86.6	96.2	46.4	76.0	96.0	37.7	72.3
08	OPENINGS	98.9	74.8	92.9	105.0	50.8	91.3	96.7	50.9	85.2	94.0	74.8	89.2	95.1	46.9	83.0	91.3	41.3	78.7
0920	Plaster & Gypsum Board	103.2	70.2	79.6	91.5	41.7	55.9	103.3	41.7	59.3	107.3	70.2	80.8	76.7	47.2	55.6	76.7	29.9	43.3
0950, 0980	Ceilings & Acoustic Treatment	106.1	70.2	82.4	111.5	41.7	65.4	98.9	41.7	61.1	99.8	70.2	80.2	88.4	47.2	61.2	88.4	29.9	49.7
0960	Flooring	99.8	54.2	86.4	93.6	41.4	78.3	101.3	47.2	85.4	90.5	41.6	76.1	102.7	61.9	90.7	102.6	41.5	84.6
0970, 0990	Wall Finishes & Painting/Coating	97.8	64.0	76.9	93.2	38.9	59.7	97.8	38.9	61.5	98.4	57.7	73.3	96.2	43.4	63.6	96.2	49.9	67.6
09	FINISHES	99.7	66.1	81.0	98.0	42.8	67.2	99.9	43.9	68.7	95.9	63.2	77.6	89.7	49.5	67.3	89.8	35.1	59.3
COVERS	DIVS. 10 - 14, 25, 28, 41, 43, 44, 46	100.0	83.5	96.7	100.0	77.7	95.6	100.0	77.9	95.6	100.0	83.4	96.7	100.0	42.7	88.6	100.0	38.4	87.7
21, 22, 23	FIRE SUPPRESSION, PLUMBING & HVAC	100.2	83.1	93.2	100.2	71.2	88.4	94.3	74.1	86.1	100.2	73.5	89.4	94.3	53.6	77.8	94.3	50.9	76.7
26, 27, 3370	ELECTRICAL, COMMUNICATIONS & UTIL.	100.7	70.9	85.3	100.1	62.2	80.5	93.9	63.6	78.3	95.6	59.9	77.2	98.4	63.4	80.3	94.9	60.4	77.1
MF2010	WEIGHTED AVERAGE	97.9	76.6	88.5	98.2	62.1	82.2	97.5	62.9	82.2	96.1	72.1	85.5	94.2	56.1	77.4	94.1	50.1	74.7

	DIVISION	ARKANSAS																	
		FAYETTEVILLE			FORT SMITH			HARRISON			HOT SPRINGS			JONESBORO			LITTLE ROCK		
		727			729			726			719			724			720 - 722		
		MAT.	INST.	TOTAL	MAT.	INST.	TOTAL	MAT.	INST.	TOTAL	MAT.	INST.	TOTAL	MAT.	INST.	TOTAL	MAT.	INST.	TOTAL
015433	CONTRACTOR EQUIPMENT		88.3	88.3		88.3	88.3		88.3	88.3		88.3	88.3		106.3	106.3		88.3	88.3
0241, 31 - 34	SITE & INFRASTRUCTURE, DEMOLITION	70.4	85.9	81.1	75.1	85.8	82.5	75.5	84.4	81.6	74.9	85.3	82.1	94.8	99.4	97.9	82.1	85.8	84.6
0310	Concrete Forming & Accessories	79.5	40.9	45.9	96.7	58.2	63.2	87.6	47.6	52.8	81.0	37.7	43.4	86.9	51.1	55.7	89.3	62.6	66.1
0320	Concrete Reinforcing	95.4	69.4	82.3	96.6	70.1	83.2	95.0	64.4	79.6	94.9	64.5	79.6	92.3	66.8	79.4	96.7	65.2	80.9
0330	Cast-in-Place Concrete	80.2	51.7	69.0	91.7	66.6	81.8	89.0	45.4	71.8	84.1	40.5	66.9	87.4	57.5	75.6	92.1	66.6	82.1
03	CONCRETE	85.0	50.9	68.4	93.1	63.8	78.8	92.5	50.7	72.1	90.4	44.7	68.2	89.8	57.6	74.1	92.9	64.9	79.2
04	MASONRY	95.0	43.7	63.7	100.3	52.0	70.8	105.9	41.3	66.4	85.2	35.4	54.8	96.4	44.3	64.6	97.1	52.0	69.6
05	METALS	96.9	69.0	87.7	99.3	70.7	89.8	98.1	66.6	87.7	96.9	66.9	87.0	93.8	80.0	89.2	95.4	69.1	86.7
06	WOOD, PLASTICS & COMPOSITES	82.9	37.7	56.8	102.5	60.6	78.3	92.0	48.3	66.8	84.0	37.9	57.4	90.3	51.3	67.8	97.2	66.4	79.4
07	THERMAL & MOISTURE PROTECTION	97.1	47.9	77.1	97.4	56.0	80.6	96.5	45.1	75.6	96.2	40.0	73.4	100.9	51.9	81.0	96.5	56.6	80.3
08	OPENINGS	95.1	46.4	82.8	96.0	57.0	86.2	96.0	47.6	83.8	91.3	43.2	79.1	98.2	55.4	87.4	96.0	60.2	87.0
0920	Plaster & Gypsum Board	74.5	36.3	47.2	82.8	59.9	66.5	81.4	47.2	57.0	75.4	36.5	47.6	90.7	50.0	61.6	88.4	65.9	72.3
0950, 0980	Ceilings & Acoustic Treatment	88.4	36.3	54.0	92.0	59.9	70.8	92.0	47.2	62.4	88.4	36.5	54.1	92.5	50.0	64.4	93.0	65.9	75.1
0960	Flooring	99.6	61.9	88.5	110.3	63.7	96.6	105.4	61.9	92.6	101.3	61.9	89.7	73.6	55.9	68.4	111.7	63.7	97.6
0970, 0990	Wall Finishes & Painting/Coating	96.2	31.0	55.9	96.2	56.7	71.8	96.2	43.4	63.6	96.2	49.9	67.6	86.3	51.0	64.5	96.2	58.2	72.8
09	FINISHES	88.6	42.4	62.8	93.7	59.3	74.5	92.2	49.5	68.4	89.6	42.5	63.3	86.8	51.4	67.0	95.4	62.9	77.3
COVERS	DIVS. 10 - 14, 25, 28, 41, 43, 44, 46	100.0	52.6	90.6	100.0	77.1	95.4	100.0	53.1	90.7	100.0	39.4	87.9	100.0	48.7	89.8	100.0	77.7	95.6
21, 22, 23	FIRE SUPPRESSION, PLUMBING & HVAC	94.3	52.2	77.2	100.3	51.6	80.5	94.3	51.3	76.9	94.3	50.4	76.5	100.5	54.5	81.9	100.2	59.3	83.6
26, 27, 3370	ELECTRICAL, COMMUNICATIONS & UTIL.	92.3	54.4	72.7	95.7	68.6	81.7	97.0	39.4	67.3	97.1	67.2	81.6	103.4	63.4	82.8	100.5	69.7	84.6
MF2010	WEIGHTED AVERAGE	92.9	54.1	75.8	97.1	62.5	81.8	95.5	52.3	76.5	93.4	52.9	75.5	96.8	60.7	80.9	97.0	65.0	82.9

	DIVISION	ARKANSAS												CALIFORNIA					
		PINE BLUFF			RUSSELLVILLE			TEXARKANA			WEST MEMPHIS			ALHAMBRA			ANAHEIM		
		716			728			718			723			917 - 918			928		
		MAT.	INST.	TOTAL	MAT.	INST.	TOTAL	MAT.	INST.	TOTAL	MAT.	INST.	TOTAL	MAT.	INST.	TOTAL	MAT.	INST.	TOTAL
015433	CONTRACTOR EQUIPMENT		88.3	88.3		88.3	88.3		88.9	88.9		106.3	106.3		99.3	99.3		100.8	100.8
0241, 31 - 34	SITE & INFRASTRUCTURE, DEMOLITION	76.8	85.8	83.0	72.2	84.3	80.5	89.7	86.3	87.4	101.5	99.4	100.0	97.9	109.3	105.8	96.8	108.4	104.8
0310	Concrete Forming & Accessories	80.7	62.5	64.9	84.1	56.5	60.1	86.5	42.6	48.3	91.7	51.3	56.5	121.6	119.9	120.1	107.0	121.9	119.9
0320	Concrete Reinforcing	96.6	65.2	80.8	96.1	64.3	80.1	96.2	64.5	80.2	92.3	66.8	79.4	109.7	118.2	114.0	95.9	118.1	107.1
0330	Cast-in-Place Concrete	84.1	66.6	77.2	84.1	47.3	69.6	91.7	45.3	73.4	91.6	57.5	78.2	89.2	118.3	100.7	92.9	120.8	103.9
03	CONCRETE	91.2	64.8	78.3	88.4	55.4	72.3	89.6	48.5	69.6	96.9	57.8	77.8	100.0	118.0	108.8	99.7	119.8	109.5
04	MASONRY	122.3	52.0	79.4	99.5	39.5	62.9	100.3	35.0	60.5	81.7	44.3	58.9	116.0	118.2	117.4	80.3	116.4	102.4
05	METALS	97.8	69.0	88.3	96.9	66.5	86.9	90.9	67.0	83.0	92.8	80.2	88.6	83.5	101.6	89.5	100.7	101.8	101.1
06	WOOD, PLASTICS & COMPOSITES	83.5	66.4	73.6	88.1	60.6	72.2	91.5	44.6	64.4	96.0	51.3	70.2	96.8	117.7	108.9	97.0	120.3	110.5
07	THERMAL & MOISTURE PROTECTION	96.3	56.6	80.2	97.3	46.1	76.5	97.1	46.7	76.6	101.4	51.9	81.2	101.2	117.4	107.8	101.6	121.3	109.6
08	OPENINGS	91.3	60.2	83.4	95.1	56.2	85.3	96.5	47.6	84.2	98.2	55.4	87.4	91.5	119.7	98.6	103.4	121.3	107.9
0920	Plaster & Gypsum Board	74.9	65.9	68.5	76.7	59.9	64.7	79.7	43.4	53.8	93.4	50.0	62.4	98.5	118.3	112.7	103.2	120.9	115.8
0950, 0980	Ceilings & Acoustic Treatment	88.4	65.9	73.5	88.4	59.9	69.6	93.8	43.4	60.5	90.5	50.0	63.7	92.8	118.3	109.7	107.5	120.9	116.3
0960	Flooring	101.0	63.7	90.0	102.3	61.9	90.4	103.4	53.4	88.7	76.2	55.9	70.2	97.2	105.0	99.4	108.9	111.4	109.6
0970, 0990	Wall Finishes & Painting/Coating	96.2	58.2	72.8	96.2	36.9	59.6	96.2	32.0	56.6	86.3	51.0	64.5	108.1	108.5	108.4	102.8	106.7	105.2
09	FINISHES	89.5	62.9	74.7	89.9	56.0	71.0	92.6	42.7	64.8	88.2	51.4	67.7	97.3	115.9	107.7	104.4	118.3	112.2
COVERS	DIVS. 10 - 14, 25, 28, 41, 43, 44, 46	100.0	77.7	95.6	100.0	44.0	88.9	100.0	35.2	87.1	100.0	48.7	89.8	100.0	109.9	102.0	100.0	110.5	102.1
21, 22, 23	FIRE SUPPRESSION, PLUMBING & HVAC	100.3	54.0	81.5	94.3	53.4	77.8	100.3	57.3	82.9	94.5	66.3	83.1	94.1	111.1	101.0	100.1	112.0	104.9
26, 27, 3370	ELECTRICAL, COMMUNICATIONS & UTIL.	95.0	69.7	82.0	95.7	47.5	70.8	97.0	38.2	66.7	105.0	60.3	81.9	112.7	115.8	114.3	92.2	105.4	99.0
MF2010	WEIGHTED AVERAGE	96.7	63.8	82.2	94.1	55.4	77.0	95.9	51.3	76.2	95.8	62.8	81.2	97.1	113.5	104.3	98.9	112.8	105.1

	DIVISION	CALIFORNIA																	
		BAKERSFIELD			BERKELEY			EUREKA			FRESNO			INGLEWOOD			LONG BEACH		
		932 - 933			947			955			936 - 938			903 - 905			906 - 908		
		MAT.	INST.	TOTAL	MAT.	INST.	TOTAL	MAT.	INST.	TOTAL	MAT.	INST.	TOTAL	MAT.	INST.	TOTAL	MAT.	INST.	TOTAL
015433	CONTRACTOR EQUIPMENT		98.7	98.7		98.5	98.5		98.4	98.4		98.7	98.7		96.5	96.5		96.5	96.5
0241, 31 - 34	SITE & INFRASTRUCTURE, DEMOLITION	99.1	106.0	103.9	120.8	105.3	110.1	110.3	102.8	105.2	102.0	105.5	104.4	91.4	103.8	100.0	98.8	103.8	102.3
0310	Concrete Forming & Accessories	105.5	121.4	119.3	119.6	148.1	144.4	117.0	132.1	130.1	103.0	133.1	129.2	111.2	119.8	118.7	106.4	119.8	118.1
0320	Concrete Reinforcing	108.0	118.0	113.0	95.0	119.4	107.3	104.7	118.8	111.8	91.5	118.7	105.2	107.6	118.1	112.9	106.6	118.1	112.4
0330	Cast-in-Place Concrete	94.6	119.7	104.5	128.5	121.8	125.9	100.7	116.2	106.8	97.9	117.0	105.4	88.4	118.7	100.3	100.7	118.7	107.8
03	CONCRETE	100.2	119.1	109.4	114.5	131.6	122.8	110.2	122.7	116.3	101.3	123.4	112.1	97.2	118.2	107.4	107.0	118.2	112.5
04	MASONRY	99.4	115.9	109.4	130.4	132.7	131.8	103.8	127.3	118.1	102.6	122.1	114.5	76.5	118.3	102.0	85.4	118.3	105.5
05	METALS	96.1	100.9	97.7	105.9	101.0	104.3	100.6	100.0	100.4	100.7	101.6	101.0	91.1	102.0	94.7	90.9	102.0	94.6
06	WOOD, PLASTICS & COMPOSITES	92.1	120.4	108.4	118.6	153.8	138.9	110.4	137.2	125.9	103.5	137.2	123.0	99.9	117.6	110.1	93.6	117.6	107.5
07	THERMAL & MOISTURE PROTECTION	100.5	111.9	105.1	114.4	135.6	123.0	105.7	115.6	109.7	95.3	115.0	103.3	99.4	118.7	107.2	99.6	119.0	107.5
08	OPENINGS	100.9	116.8	104.9	101.4	141.2	111.4	102.2	114.6	105.3	102.7	126.0	108.6	93.5	119.6	100.1	93.5	119.6	100.1
0920	Plaster & Gypsum Board	94.5	120.9	113.3	110.4	154.9	142.2	108.4	138.2	129.7	97.4	138.2	126.6	103.7	118.3	114.2	99.2	118.3	112.9
0950, 0980	Ceilings & Acoustic Treatment	108.0	120.9	116.5	109.4	154.9	139.4	112.9	138.2	129.6	110.5	138.2	128.8	100.7	118.3	112.3	100.7	118.3	112.3
0960	Flooring	109.1	97.4	105.7	115.8	126.8	119.1	113.5	122.0	116.0	115.5	139.8	122.6	100.0	105.0	101.5	97.3	105.0	99.6
0970, 0990	Wall Finishes & Painting/Coating	104.9	93.8	98.0	103.9	141.5	127.1	104.2	48.7	70.0	120.2	111.0	114.5	99.7	108.5	105.2	99.7	108.5	105.2
09	FINISHES	105.4	114.7	110.6	111.0	145.5	130.2	110.9	123.9	118.1	109.8	133.8	123.2	102.5	115.8	109.9	101.8	115.8	109.6
COVERS	DIVS. 10 - 14, 25, 28, 41, 43, 44, 46	100.0	107.6	101.5	100.0	124.0	104.8	100.0	120.3	104.0	100.0	120.3	104.0	100.0	109.6	101.9	100.0	109.6	101.9
21, 22, 23	FIRE SUPPRESSION, PLUMBING & HVAC	100.1	105.5	102.3	94.3	147.5	115.9	94.1	103.5	97.9	100.2	113.2	105.5	93.7	111.0	100.7	93.7	111.0	100.7
26, 27, 3370	ELECTRICAL, COMMUNICATIONS & UTIL.	101.6	101.2	101.4	106.6	138.5	123.1	99.8	116.8	108.6	91.7	100.4	96.2	103.4	115.9	109.8	103.1	115.9	109.7
MF2010	WEIGHTED AVERAGE	100.0	109.5	104.2	105.5	133.2	117.7	101.6	114.2	107.2	100.5	115.7	107.2	95.3	113.1	103.2	96.9	113.2	104.1

	DIVISION	CALIFORNIA																	
		LOS ANGELES			MARYSVILLE			MODESTO			MOJAVE			OAKLAND			OXNARD		
		900 - 902			959			953			935			946			930		
		MAT.	INST.	TOTAL	MAT.	INST.	TOTAL	MAT.	INST.	TOTAL	MAT.	INST.	TOTAL	MAT.	INST.	TOTAL	MAT.	INST.	TOTAL
015433	CONTRACTOR EQUIPMENT		100.1	100.1		98.4	98.4		98.4	98.4		98.7	98.7		98.5	98.5		97.6	97.6
0241, 31 - 34	SITE & INFRASTRUCTURE, DEMOLITION	98.8	107.5	104.8	106.2	105.0	105.4	100.5	105.0	103.6	95.5	106.0	102.7	127.4	105.3	112.2	102.3	104.1	103.5
0310	Concrete Forming & Accessories	108.2	121.9	120.1	107.6	133.0	129.7	104.2	133.2	129.5	116.2	116.2	116.2	109.1	148.1	143.0	108.1	121.9	120.1
0320	Concrete Reinforcing	108.4	118.3	113.4	104.7	118.5	111.7	108.5	118.7	113.7	110.0	117.9	114.0	97.1	119.4	108.4	108.0	118.0	113.0
0330	Cast-in-Place Concrete	95.5	119.4	104.9	112.6	117.0	114.3	100.8	117.1	107.2	84.7	119.6	98.5	121.8	121.8	121.8	98.6	120.0	107.0
03	CONCRETE	102.8	119.4	110.9	112.2	123.4	117.7	103.3	123.6	113.2	93.8	116.8	105.0	113.9	131.7	122.6	102.3	119.4	110.7
04	MASONRY	90.2	120.1	108.5	104.8	119.1	113.5	103.8	123.0	115.5	100.7	115.1	109.5	138.3	132.7	134.9	104.1	114.5	110.5
05	METALS	98.1	103.5	99.9	100.1	101.6	100.6	97.3	102.1	98.9	97.7	100.7	98.7	99.9	101.0	100.3	95.9	101.3	97.7
06	WOOD, PLASTICS & COMPOSITES	99.6	120.2	111.5	98.2	137.2	120.7	94.4	137.2	119.1	102.8	113.6	109.0	106.6	153.8	133.9	98.1	120.4	111.0
07	THERMAL & MOISTURE PROTECTION	98.8	121.0	107.8	105.2	118.0	110.4	104.8	117.7	110.0	101.2	110.4	105.0	112.1	135.6	121.6	104.7	116.9	109.7
08	OPENINGS	100.6	121.3	105.8	101.4	125.9	107.6	100.2	126.6	106.8	96.7	113.1	100.9	101.4	141.2	111.4	99.8	121.4	105.2
0920	Plaster & Gypsum Board	103.7	120.9	115.9	100.4	138.2	127.4	103.6	138.2	128.3	109.9	113.9	112.7	103.8	154.9	140.3	103.2	120.9	115.8
0950, 0980	Ceilings & Acoustic Treatment	110.8	120.9	117.4	112.0	138.2	129.3	107.5	138.2	127.8	112.0	113.9	113.2	112.3	154.9	140.4	112.5	120.9	118.0
0960	Flooring	97.7	111.4	101.7	108.6	110.5	109.2	109.1	116.7	111.3	115.9	97.4	110.5	110.2	126.8	115.1	107.5	111.4	108.7
0970, 0990	Wall Finishes & Painting/Coating	98.8	108.5	104.8	104.2	115.2	111.0	104.2	118.4	112.9	104.2	93.8	97.7	103.9	141.5	127.1	104.2	99.7	101.4
09	FINISHES	104.2	118.4	112.1	107.7	129.4	119.8	106.6	130.8	120.1	110.0	110.6	110.3	109.9	145.5	129.7	107.2	117.6	113.0
COVERS	DIVS. 10 - 14, 25, 28, 41, 43, 44, 46	100.0	110.2	102.0	100.0	120.3	104.0	100.0	120.3	104.0	100.0	106.8	101.3	100.0	124.0	104.8	100.0	110.7	102.1
21, 22, 23	FIRE SUPPRESSION, PLUMBING & HVAC	100.2	111.9	104.9	94.1	108.9	100.1	100.1	113.2	105.4	94.1	100.9	96.9	100.2	147.5	119.4	100.1	112.0	104.9
26, 27, 3370	ELECTRICAL, COMMUNICATIONS & UTIL.	101.7	122.2	112.3	96.4	103.8	100.2	98.9	110.5	104.9	90.3	98.2	94.4	105.8	143.6	125.3	96.2	107.5	102.0
MF2010	WEIGHTED AVERAGE	100.1	115.5	106.9	100.9	114.4	106.8	100.7	116.9	107.9	97.0	106.9	101.3	106.3	133.9	118.5	100.2	112.3	105.5

	DIVISION	CALIFORNIA																	
		PALM SPRINGS			PALO ALTO			PASADENA			REDDING			RICHMOND			RIVERSIDE		
		922			943			910 - 912			960			948			925		
		MAT.	INST.	TOTAL	MAT.	INST.	TOTAL	MAT.	INST.	TOTAL	MAT.	INST.	TOTAL	MAT.	INST.	TOTAL	MAT.	INST.	TOTAL
015433	CONTRACTOR EQUIPMENT		99.6	99.6		98.5	98.5		99.3	99.3		98.4	98.4		98.5	98.5		99.6	99.6
0241, 31 - 34	SITE & INFRASTRUCTURE, DEMOLITION	88.3	106.5	100.8	116.5	105.3	108.8	94.7	109.3	104.8	106.2	105.0	105.4	126.8	105.3	112.0	95.1	106.5	102.9
0310	Concrete Forming & Accessories	103.5	119.9	117.7	107.3	140.1	135.9	111.8	119.9	118.9	107.4	133.0	129.7	122.0	147.7	144.4	107.1	121.9	119.9
0320	Concrete Reinforcing	110.2	118.0	114.1	95.0	119.2	107.2	110.6	118.2	114.4	104.7	118.5	111.7	95.0	119.2	107.2	107.0	118.0	112.6
0330	Cast-in-Place Concrete	88.5	120.7	101.2	108.7	121.6	113.8	84.7	118.3	97.9	111.6	117.0	113.7	125.0	121.6	123.7	96.3	120.8	105.9
03	CONCRETE	95.0	118.8	106.6	102.8	128.0	115.1	95.7	118.1	106.6	111.2	123.4	117.1	116.0	131.4	123.5	101.0	119.7	110.1
04	MASONRY	78.1	115.4	100.8	109.6	128.8	121.3	100.8	118.2	111.4	104.8	119.1	113.5	130.1	128.8	129.3	79.1	115.4	101.2
05	METALS	101.3	101.5	101.4	97.4	100.3	98.3	83.6	101.6	89.5	100.4	101.6	100.8	97.4	100.4	98.4	100.9	101.7	101.2
06	WOOD, PLASTICS & COMPOSITES	91.9	117.8	106.9	104.0	143.6	126.9	84.4	117.7	103.7	97.9	137.2	120.6	122.2	153.8	140.4	97.0	120.3	110.5
07	THERMAL & MOISTURE PROTECTION	101.3	119.3	108.6	111.6	128.7	118.6	100.9	117.9	107.8	105.2	117.5	110.2	112.2	133.9	121.0	101.8	120.7	109.5
08	OPENINGS	98.8	118.2	103.7	101.4	134.7	109.8	91.5	119.7	98.6	102.4	125.9	108.3	101.4	140.2	111.2	101.9	121.3	106.8
0920	Plaster & Gypsum Board	97.3	118.3	112.3	101.5	144.3	132.1	91.2	118.3	110.6	100.7	138.2	127.5	112.4	154.9	142.8	102.3	120.9	115.6
0950, 0980	Ceilings & Acoustic Treatment	104.8	118.3	113.7	110.3	144.3	132.8	92.8	118.3	109.7	118.0	138.2	131.3	110.3	154.9	139.7	109.8	120.9	117.1
0960	Flooring	111.1	105.0	109.3	109.3	120.8	112.7	91.7	105.0	95.6	108.6	110.5	109.1	117.7	120.8	118.6	112.9	111.4	112.5
0970, 0990	Wall Finishes & Painting/Coating	100.9	117.3	111.0	103.9	139.2	125.6	108.1	108.5	108.4	104.2	115.2	111.0	103.9	139.2	125.6	100.9	106.7	104.5
09	FINISHES	102.9	116.9	110.7	108.0	138.2	124.8	94.6	115.9	106.5	108.9	129.4	120.3	112.6	144.2	130.2	105.6	118.3	112.7
COVERS	DIVS. 10 - 14, 25, 28, 41, 43, 44, 46	100.0	110.2	102.0	100.0	122.9	104.5	100.0	109.9	102.0	100.0	120.3	104.0	100.0	124.0	104.8	100.0	110.5	102.1
21, 22, 23	FIRE SUPPRESSION, PLUMBING & HVAC	94.1	107.3	99.5	94.3	141.2	113.3	94.1	111.1	101.0	100.1	108.9	103.6	94.3	136.5	111.4	100.1	112.0	104.9
26, 27, 3370	ELECTRICAL, COMMUNICATIONS & UTIL.	95.9	104.4	100.3	105.7	142.0	124.4	109.7	115.8	112.9	99.3	103.8	101.6	106.2	135.0	121.1	92.4	104.4	98.6
MF2010	WEIGHTED AVERAGE	96.4	110.9	102.8	101.3	129.9	113.9	95.2	113.6	103.3	102.8	114.4	107.9	104.7	129.7	115.7	98.9	112.4	104.9

	DIVISION	CALIFORNIA																	
		SACRAMENTO			SALINAS			SAN BERNARDINO			SAN DIEGO			SAN FRANCISCO			SAN JOSE		
		942,956 - 958			939			923 - 924			919 - 921			940 - 941			951		
		MAT.	INST.	TOTAL	MAT.	INST.	TOTAL	MAT.	INST.	TOTAL	MAT.	INST.	TOTAL	MAT.	INST.	TOTAL	MAT.	INST.	TOTAL
015433	CONTRACTOR EQUIPMENT		98.0	98.0		98.7	98.7		99.6	99.6		99.3	99.3		109.1	109.1		99.1	99.1
0241, 31 - 34	SITE & INFRASTRUCTURE, DEMOLITION	100.9	110.7	107.6	116.7	105.7	109.1	74.3	106.5	96.5	100.8	103.0	102.3	129.7	111.0	116.8	135.8	100.2	111.2
0310	Concrete Forming & Accessories	107.4	136.1	132.3	110.7	136.4	133.0	110.5	119.8	118.6	108.2	110.6	110.3	108.7	148.9	143.7	110.0	147.9	142.9
0320	Concrete Reinforcing	89.7	118.8	104.4	108.5	119.1	113.9	107.0	118.0	112.5	107.2	118.0	112.6	110.6	119.7	115.2	95.5	119.4	107.6
0330	Cast-in-Place Concrete	103.0	118.1	109.0	97.4	117.4	105.3	66.5	120.7	87.8	92.8	106.6	98.2	124.9	123.1	124.2	117.1	121.0	118.6
03	CONCRETE	103.2	124.7	113.7	111.5	125.2	118.1	75.6	118.8	96.7	100.5	110.0	105.1	117.5	133.1	125.1	109.8	131.7	120.5
04	MASONRY	111.6	123.0	118.5	100.4	128.2	117.4	85.8	115.4	103.8	92.0	113.7	105.2	138.8	135.9	137.0	132.7	132.8	132.7
05	METALS	95.6	96.0	95.7	100.3	103.2	101.2	100.9	101.4	101.0	98.7	102.7	100.0	106.5	110.8	107.9	95.7	107.7	99.6
06	WOOD, PLASTICS & COMPOSITES	99.3	140.7	123.2	102.3	140.4	124.3	100.6	117.8	110.6	97.9	107.2	103.3	106.6	153.9	134.0	106.5	153.5	133.6
07	THERMAL & MOISTURE PROTECTION	118.7	122.1	120.1	102.0	125.8	111.7	100.4	119.3	108.1	106.5	105.1	106.0	114.2	137.1	123.5	101.2	136.7	115.6
08	OPENINGS	114.1	128.5	117.7	101.2	133.9	109.4	98.9	118.2	103.7	99.5	111.9	102.6	105.9	141.3	114.8	93.1	141.0	105.1
0920	Plaster & Gypsum Board	97.8	141.5	129.0	104.5	141.5	130.9	104.1	118.3	114.3	98.8	107.2	104.8	106.8	154.9	141.1	97.5	154.9	138.5
0950, 0980	Ceilings & Acoustic Treatment	110.3	141.5	130.9	112.0	141.5	131.5	107.5	118.3	114.6	100.0	107.2	104.8	120.2	154.9	143.1	102.6	154.9	137.1
0960	Flooring	109.4	116.7	111.5	109.9	126.8	114.9	115.1	105.0	112.1	100.1	111.4	103.4	110.2	126.8	115.1	101.4	126.8	108.9
0970, 0990	Wall Finishes & Painting/Coating	101.1	118.4	111.8	105.3	141.5	127.7	100.9	104.9	103.4	104.5	109.4	107.5	103.9	151.4	133.2	104.7	141.5	127.4
09	FINISHES	106.3	132.9	121.1	110.0	137.0	125.0	104.1	115.6	110.5	100.3	110.2	105.9	112.1	146.7	131.4	104.7	145.3	127.3
COVERS	DIVS. 10 - 14, 25, 28, 41, 43, 44, 46	100.0	121.3	104.2	100.0	120.7	104.1	100.0	107.1	101.4	100.0	108.4	101.7	100.0	124.5	104.9	100.0	123.4	104.7
21, 22, 23	FIRE SUPPRESSION, PLUMBING & HVAC	100.1	116.8	106.9	94.1	120.0	104.6	94.1	107.3	99.5	100.1	110.4	104.3	100.2	169.6	128.3	100.1	146.7	118.9
26, 27, 3370	ELECTRICAL, COMMUNICATIONS & UTIL.	101.1	113.3	107.4	91.7	125.8	109.3	95.9	103.6	99.9	96.0	101.0	98.6	105.9	155.0	131.2	102.1	152.2	128.0
MF2010	WEIGHTED AVERAGE	102.9	118.6	109.9	100.5	122.7	110.3	94.3	110.5	101.4	99.3	107.8	103.0	108.5	142.3	123.4	103.2	135.1	117.3

	DIVISION	CALIFORNIA																	
		SAN LUIS OBISPO			SAN MATEO			SAN RAFAEL			SANTA ANA			SANTA BARBARA			SANTA CRUZ		
		934			944			949			926 - 927			931			950		
		MAT.	INST.	TOTAL	MAT.	INST.	TOTAL	MAT.	INST.	TOTAL	MAT.	INST.	TOTAL	MAT.	INST.	TOTAL	MAT.	INST.	TOTAL
015433	CONTRACTOR EQUIPMENT		98.7	98.7		98.5	98.5		98.5	98.5		99.6	99.6		98.7	98.7		99.1	99.1
0241, 31 - 34	SITE & INFRASTRUCTURE, DEMOLITION	108.1	106.0	106.7	124.1	105.3	111.1	114.7	110.7	112.0	86.7	106.5	100.3	102.3	106.0	104.9	135.1	100.0	110.9
0310	Concrete Forming & Accessories	117.8	119.6	119.4	112.9	147.8	143.2	115.2	147.5	143.3	110.8	119.8	118.6	109.1	121.8	120.1	110.1	136.5	133.1
0320	Concrete Reinforcing	110.0	117.9	114.0	95.0	119.4	107.3	95.5	119.3	107.5	110.7	118.0	114.4	108.0	118.0	113.0	118.3	119.2	118.7
0330	Cast-in-Place Concrete	104.7	119.7	110.6	121.0	121.7	121.3	140.9	120.6	132.9	85.0	120.7	99.0	98.2	119.8	106.7	116.3	119.3	117.5
03	CONCRETE	110.0	118.3	114.1	112.3	131.4	121.6	134.4	130.8	132.6	92.6	118.8	105.4	102.2	119.4	110.6	112.9	126.1	119.4
04	MASONRY	102.3	115.1	110.1	129.8	132.2	131.3	106.7	132.2	122.2	75.1	115.7	99.9	100.9	115.1	109.6	136.8	128.4	131.7
05	METALS	98.3	100.9	99.2	97.2	100.6	98.4	98.8	97.7	98.4	101.0	101.4	101.1	96.3	101.3	97.9	102.6	106.6	103.9
06	WOOD, PLASTICS & COMPOSITES	105.0	117.9	112.5	111.6	153.8	135.9	106.4	153.5	133.6	102.2	117.8	111.2	98.1	120.4	111.0	106.5	140.5	126.1
07	THERMAL & MOISTURE PROTECTION	102.1	115.9	107.7	112.0	134.8	121.3	116.6	131.7	122.7	101.6	119.3	108.8	101.4	116.2	107.4	101.1	128.6	112.3
08	OPENINGS	99.0	115.5	103.2	101.4	140.2	111.2	111.3	140.1	118.5	98.1	118.2	103.1	100.9	121.4	106.1	94.3	134.0	104.3
0920	Plaster & Gypsum Board	111.8	118.3	116.5	107.4	154.9	141.3	109.6	154.9	141.9	105.0	118.3	114.5	103.2	120.9	115.8	104.0	141.5	130.8
0950, 0980	Ceilings & Acoustic Treatment	112.0	118.3	116.2	110.3	154.9	139.7	118.4	154.9	142.5	107.5	118.3	114.6	112.5	120.9	118.0	109.3	141.5	130.6
0960	Flooring	116.9	97.4	111.1	112.3	120.8	114.8	123.2	120.8	122.5	115.7	105.0	112.5	109.1	97.4	105.7	105.9	126.8	112.0
0970, 0990	Wall Finishes & Painting/Coating	104.2	99.7	101.4	103.9	139.2	125.6	99.9	139.2	124.1	100.9	104.9	103.4	104.2	99.7	101.4	104.9	141.5	127.5
09	FINISHES	111.7	112.9	112.4	110.2	143.9	129.0	112.8	144.0	130.2	105.7	115.6	111.2	107.9	115.3	112.0	108.3	137.0	124.3
COVERS	DIVS. 10 - 14, 25, 28, 41, 43, 44, 46	100.0	117.5	103.5	100.0	124.0	104.8	100.0	123.3	104.6	100.0	110.2	102.0	100.0	110.7	102.1	100.0	121.0	104.2
21, 22, 23	FIRE SUPPRESSION, PLUMBING & HVAC	94.1	107.3	99.5	94.3	138.4	112.2	94.2	146.1	115.3	94.1	107.4	99.5	100.1	112.0	104.9	100.1	120.0	108.2
26, 27, 3370	ELECTRICAL, COMMUNICATIONS & UTIL.	90.3	98.1	94.3	105.7	141.1	124.0	102.3	116.2	109.5	95.9	105.0	100.6	89.3	108.2	99.0	101.0	125.8	113.8
MF2010	WEIGHTED AVERAGE	99.7	109.4	104.0	103.8	131.3	115.9	106.0	129.5	116.4	96.1	110.8	102.6	99.5	112.3	105.1	105.0	122.8	112.9

	DIVISION	CALIFORNIA															COLORADO		
		SANTA ROSA			STOCKTON			SUSANVILLE			VALLEJO			VAN NUYS			ALAMOSA		
		954			952			961			945			913 - 916			811		
		MAT.	INST.	TOTAL	MAT.	INST.	TOTAL	MAT.	INST.	TOTAL	MAT.	INST.	TOTAL	MAT.	INST.	TOTAL	MAT.	INST.	TOTAL
015433	CONTRACTOR EQUIPMENT		99.0	99.0		98.4	98.4		98.4	98.4		98.5	98.5		99.3	99.3		95.2	95.2
0241, 31 - 34	SITE & INFRASTRUCTURE, DEMOLITION	100.9	105.1	103.8	100.2	105.0	103.5	113.2	105.0	107.5	100.3	110.6	107.4	112.4	109.3	110.3	131.0	91.6	103.8
0310	Concrete Forming & Accessories	104.5	146.9	141.4	107.9	133.3	130.0	108.7	133.0	129.8	106.1	146.6	141.4	117.6	119.9	119.6	105.9	71.5	76.0
0320	Concrete Reinforcing	105.7	119.6	112.7	108.5	118.8	113.7	104.7	118.5	111.7	96.7	119.5	108.2	110.6	118.2	114.4	105.8	80.7	93.2
0330	Cast-in-Place Concrete	110.6	118.7	113.8	98.2	117.1	105.6	101.6	117.0	107.6	112.2	119.3	115.0	89.3	118.3	100.7	99.6	82.0	92.7
03	CONCRETE	111.7	130.4	120.8	102.2	123.6	112.6	112.1	123.4	117.6	108.8	130.0	119.1	108.8	118.0	113.3	113.2	77.2	95.6
04	MASONRY	102.8	133.5	121.5	103.8	123.0	115.5	103.2	119.1	112.9	80.2	133.5	112.7	116.0	118.2	117.4	126.0	75.6	95.2
05	METALS	101.3	104.9	102.5	97.5	102.2	99.0	99.5	101.6	100.2	98.8	98.0	98.5	82.6	101.6	88.9	96.0	82.7	91.6
06	WOOD, PLASTICS & COMPOSITES	93.2	153.3	127.9	99.7	137.2	121.4	99.8	137.2	121.4	96.0	153.5	129.2	92.1	117.7	106.9	93.2	71.8	80.8
07	THERMAL & MOISTURE PROTECTION	102.1	133.9	115.0	104.9	119.8	111.0	105.8	117.5	110.5	114.2	133.6	122.1	101.8	117.4	108.2	106.1	80.8	95.8
08	OPENINGS	99.8	140.9	110.2	100.2	126.6	106.8	102.3	125.9	108.3	113.7	141.0	120.6	91.3	119.7	98.5	95.6	78.0	91.2
0920	Plaster & Gypsum Board	99.1	154.9	138.9	103.6	138.2	128.3	101.8	138.2	127.8	103.7	154.9	140.3	96.1	118.3	112.0	81.6	70.7	73.8
0950, 0980	Ceilings & Acoustic Treatment	107.5	154.9	138.8	112.5	138.2	129.5	112.0	138.2	129.3	120.4	154.9	143.2	91.0	118.3	109.0	108.8	70.7	83.6
0960	Flooring	111.7	120.4	114.3	109.1	116.7	111.3	109.1	110.5	109.5	118.0	126.8	120.6	94.9	105.0	97.8	111.5	57.4	95.6
0970, 0990	Wall Finishes & Painting/Coating	100.9	141.5	126.0	104.2	118.4	112.9	104.2	115.2	111.0	100.8	141.5	125.9	108.1	108.5	108.4	107.9	24.9	56.7
09	FINISHES	105.0	143.5	126.5	107.7	130.8	120.6	108.9	129.4	120.3	109.3	144.7	129.0	97.2	115.9	107.6	106.9	63.5	82.7
COVERS	DIVS. 10 - 14, 25, 28, 41, 43, 44, 46	100.0	122.0	104.4	100.0	120.3	104.0	100.0	120.3	104.0	100.0	122.6	104.5	100.0	109.9	102.0	100.0	90.2	98.0
21, 22, 23	FIRE SUPPRESSION, PLUMBING & HVAC	94.1	168.2	124.1	100.1	113.2	105.4	94.1	108.9	100.1	100.2	130.4	112.4	94.1	111.1	101.0	94.1	74.6	86.2
26, 27, 3370	ELECTRICAL, COMMUNICATIONS & UTIL.	96.2	120.9	108.9	98.9	112.2	105.8	99.7	103.8	101.8	98.1	129.6	114.4	109.7	115.8	112.9	95.7	75.1	85.1
MF2010	WEIGHTED AVERAGE	100.3	135.0	115.6	100.7	117.2	108.0	101.5	114.4	107.2	102.2	128.2	113.7	98.1	113.5	104.9	101.5	76.7	90.6

	DIVISION	COLORADO																	
		BOULDER			COLORADO SPRINGS			DENVER			DURANGO			FORT COLLINS			FORT MORGAN		
		803			808 - 809			800 - 802			813			805			807		
		MAT.	INST.	TOTAL	MAT.	INST.	TOTAL	MAT.	INST.	TOTAL	MAT.	INST.	TOTAL	MAT.	INST.	TOTAL	MAT.	INST.	TOTAL
015433	CONTRACTOR EQUIPMENT		96.2	96.2		94.7	94.7		99.1	99.1		95.2	95.2		96.2	96.2		96.2	96.2
0241, 31 - 34	SITE & INFRASTRUCTURE, DEMOLITION	91.9	97.2	95.6	93.3	95.4	94.7	92.4	103.6	100.1	123.9	91.6	101.6	103.6	96.8	98.9	94.1	96.8	96.0
0310	Concrete Forming & Accessories	104.0	80.0	83.1	95.1	80.4	82.3	100.4	80.7	83.2	111.3	71.8	77.0	101.4	79.1	82.0	104.3	79.5	82.7
0320	Concrete Reinforcing	98.9	80.8	89.8	98.1	85.2	91.6	98.1	85.2	91.6	105.8	80.7	93.2	99.0	79.6	89.2	99.1	80.7	89.8
0330	Cast-in-Place Concrete	103.2	82.4	95.0	106.0	85.8	98.0	100.3	85.5	94.5	114.5	82.1	101.8	116.5	81.2	102.6	101.3	81.3	93.4
03	CONCRETE	102.3	81.3	92.0	105.5	83.4	94.8	100.0	83.5	91.9	116.2	77.4	97.3	112.8	80.1	96.9	100.9	80.5	90.9
04	MASONRY	92.8	76.1	82.6	93.5	76.1	82.9	95.5	76.1	83.7	113.4	75.6	90.3	110.8	55.2	76.9	107.6	75.8	88.2
05	METALS	99.5	86.0	95.0	102.6	88.5	98.0	105.2	88.7	99.7	96.0	83.0	91.7	100.7	81.1	94.2	99.2	82.9	93.8
06	WOOD, PLASTICS & COMPOSITES	97.6	82.1	88.7	88.7	82.2	85.0	95.6	82.2	87.8	101.0	71.8	84.1	95.4	82.1	87.7	97.8	82.1	88.7
07	THERMAL & MOISTURE PROTECTION	105.0	78.8	94.4	105.9	83.1	96.6	104.2	77.0	93.2	106.0	80.8	95.7	105.4	67.9	90.2	105.0	82.4	95.8
08	OPENINGS	101.5	83.6	96.9	106.1	84.9	100.8	106.3	84.8	100.9	103.4	78.0	97.0	101.4	83.6	96.9	101.4	83.6	96.9
0920	Plaster & Gypsum Board	108.6	81.8	89.4	89.1	81.8	83.9	104.6	81.8	88.3	93.4	70.7	77.2	102.7	81.8	87.8	109.1	81.8	89.6
0950, 0980	Ceilings & Acoustic Treatment	99.5	81.8	87.8	106.8	81.8	90.3	105.9	81.8	90.0	108.8	70.7	83.6	99.5	81.8	87.8	99.5	81.8	87.8
0960	Flooring	109.4	87.8	103.1	100.7	71.5	92.1	105.8	88.8	100.8	116.4	57.4	99.1	105.7	57.4	91.5	109.9	57.4	94.5
0970, 0990	Wall Finishes & Painting/Coating	103.0	70.2	82.8	102.7	42.7	65.7	103.0	79.8	88.7	107.9	24.9	56.7	103.0	42.4	65.6	103.0	56.4	74.3
09	FINISHES	104.2	80.5	91.0	101.0	74.7	86.3	103.9	81.9	91.6	108.8	63.5	83.5	103.0	71.4	85.4	104.3	73.0	86.8
COVERS	DIVS. 10 - 14, 25, 28, 41, 43, 44, 46	100.0	90.4	98.1	100.0	90.8	98.2	100.0	90.4	98.1	100.0	90.1	98.0	100.0	90.4	98.1	100.0	90.4	98.1
21, 22, 23	FIRE SUPPRESSION, PLUMBING & HVAC	94.1	80.2	88.4	100.2	74.8	89.9	99.9	80.5	92.1	94.1	89.1	92.1	100.0	80.1	91.9	94.1	80.1	88.4
26, 27, 3370	ELECTRICAL, COMMUNICATIONS & UTIL.	98.9	84.0	91.2	102.4	84.7	93.3	104.1	87.0	95.3	95.1	73.7	84.1	98.9	86.8	92.6	99.3	86.8	92.8
MF2010	WEIGHTED AVERAGE	98.6	82.9	91.7	101.6	81.7	92.8	101.7	84.7	94.2	101.9	79.6	92.1	102.6	79.1	92.2	99.3	82.0	91.7

	DIVISION	COLORADO																	
		GLENWOOD SPRINGS			GOLDEN			GRAND JUNCTION			GREELEY			MONTROSE			PUEBLO		
		816			804			815			806			814			810		
		MAT.	INST.	TOTAL	MAT.	INST.	TOTAL	MAT.	INST.	TOTAL	MAT.	INST.	TOTAL	MAT.	INST.	TOTAL	MAT.	INST.	TOTAL
015433	CONTRACTOR EQUIPMENT		97.9	97.9		96.2	96.2		97.9	97.9		96.2	96.2		96.5	96.5		95.2	95.2
0241, 31 - 34	SITE & INFRASTRUCTURE, DEMOLITION	140.5	98.5	111.5	105.6	97.1	99.7	123.1	98.2	105.9	90.5	96.2	94.5	133.7	94.7	106.8	114.8	94.1	100.5
0310	Concrete Forming & Accessories	103.3	79.6	82.7	97.4	79.5	81.8	110.1	79.2	83.2	99.6	83.0	85.2	102.9	79.4	82.4	107.8	80.7	84.2
0320	Concrete Reinforcing	104.7	80.7	92.6	99.1	80.7	89.8	105.0	80.7	92.7	98.9	79.3	89.0	104.6	80.7	92.5	101.2	85.2	93.1
0330	Cast-in-Place Concrete	99.5	81.5	92.4	101.3	81.4	93.5	110.2	80.8	98.6	97.4	60.3	82.8	99.6	81.4	92.4	98.8	86.6	94.0
03	CONCRETE	117.9	80.7	99.8	110.1	80.5	95.7	112.6	80.1	96.8	97.6	74.6	86.4	109.6	80.4	95.4	102.7	83.9	93.5
04	MASONRY	100.2	75.8	85.3	110.4	75.8	89.3	133.2	75.6	98.0	104.7	51.6	72.3	106.6	75.6	87.7	96.7	76.1	84.1
05	METALS	95.7	83.5	91.7	99.4	83.0	94.0	97.3	82.2	92.3	100.7	80.0	93.8	94.9	82.5	90.8	98.9	89.3	95.7
06	WOOD, PLASTICS & COMPOSITES	88.7	82.2	85.0	90.7	82.1	85.7	98.8	82.2	89.2	92.9	88.2	90.2	89.9	82.4	85.6	95.5	82.5	88.0
07	THERMAL & MOISTURE PROTECTION	105.9	81.9	96.2	105.9	76.4	93.9	104.9	70.8	91.0	104.8	63.3	87.9	106.1	81.9	96.3	104.4	82.6	95.6
08	OPENINGS	102.5	83.6	97.8	101.4	83.6	96.9	103.4	83.6	98.4	101.4	86.9	97.8	103.6	83.7	98.6	97.5	85.0	94.4
0920	Plaster & Gypsum Board	113.3	81.8	90.8	100.0	81.8	87.0	125.1	81.8	94.2	100.9	88.1	91.8	80.7	81.8	81.5	86.6	81.8	83.1
0950, 0980	Ceilings & Acoustic Treatment	107.9	81.8	90.6	99.5	81.8	87.8	107.9	81.8	90.6	99.5	88.1	92.0	108.8	81.8	90.9	117.8	81.8	94.0
0960	Flooring	110.6	52.8	93.6	103.3	57.4	89.8	115.8	57.4	98.7	104.5	57.4	90.7	113.9	47.5	94.4	112.6	88.8	105.6
0970, 0990	Wall Finishes & Painting/Coating	107.9	56.4	76.2	103.0	70.2	82.8	107.9	70.2	84.6	103.0	26.4	55.8	107.9	24.9	56.7	107.9	39.6	65.8
09	FINISHES	110.7	72.3	89.3	102.9	75.3	87.5	111.4	74.6	90.9	101.5	73.3	85.8	107.5	67.7	85.3	107.7	77.7	91.0
COVERS	DIVS. 10 - 14, 25, 28, 41, 43, 44, 46	100.0	90.6	98.1	100.0	90.4	98.1	100.0	90.5	98.1	100.0	91.1	98.2	100.0	90.9	98.2	100.0	91.4	98.3
21, 22, 23	FIRE SUPPRESSION, PLUMBING & HVAC	94.1	88.9	92.0	94.1	79.7	88.3	100.0	88.6	95.4	100.0	80.0	91.9	94.1	88.9	92.0	100.0	74.8	89.8
26, 27, 3370	ELECTRICAL, COMMUNICATIONS & UTIL.	92.7	73.8	82.9	99.3	83.9	91.4	94.8	57.2	75.4	98.9	86.8	92.6	94.8	59.9	76.8	95.7	75.1	85.1
MF2010	WEIGHTED AVERAGE	101.6	82.2	93.1	100.7	81.7	92.3	104.2	79.6	93.4	100.0	78.2	90.4	100.8	79.3	91.3	100.4	80.8	91.8

City Cost Indexes

	DIVISION	COLORADO			CONNECTICUT														
		SALIDA			BRIDGEPORT			BRISTOL			HARTFORD			MERIDEN			NEW BRITAIN		
		812			066			060			061			064			060		
		MAT.	INST.	TOTAL	MAT.	INST.	TOTAL	MAT.	INST.	TOTAL	MAT.	INST.	TOTAL	MAT.	INST.	TOTAL	MAT.	INST.	TOTAL
015433	CONTRACTOR EQUIPMENT		96.5	96.5		100.4	100.4		100.4	100.4		100.4	100.4		100.9	100.9		100.4	100.4
0241, 31 - 34	SITE & INFRASTRUCTURE, DEMOLITION	123.6	95.0	103.9	111.8	103.3	105.9	111.0	103.3	105.7	103.6	103.3	103.4	109.2	104.0	105.6	111.2	103.3	105.7
0310	Concrete Forming & Accessories	110.1	79.2	83.3	99.3	121.7	118.8	99.3	121.6	118.7	97.3	121.6	118.4	99.0	121.6	118.6	99.6	121.6	118.7
0320	Concrete Reinforcing	104.3	80.7	92.4	103.6	128.8	116.3	103.6	128.8	116.3	103.6	128.8	116.3	103.6	128.8	116.3	103.6	128.8	116.3
0330	Cast-in-Place Concrete	114.1	81.4	101.2	107.8	125.0	114.6	101.0	125.0	110.4	107.0	125.0	114.1	97.1	125.0	108.1	102.6	125.0	111.4
03	CONCRETE	111.5	80.4	96.3	110.6	124.0	117.1	107.1	123.9	115.3	110.0	123.9	116.8	105.2	123.9	114.3	108.0	123.9	115.7
04	MASONRY	134.8	75.6	98.7	104.1	129.8	119.8	96.7	129.8	116.9	97.2	129.8	117.1	96.3	129.8	116.7	98.0	129.8	117.4
05	METALS	94.6	82.4	90.6	99.1	125.2	107.7	99.1	125.0	107.7	103.9	125.0	110.9	96.5	125.0	105.9	95.7	125.0	105.4
06	WOOD, PLASTICS & COMPOSITES	96.5	82.4	88.3	98.6	120.3	111.1	98.6	120.3	111.1	97.1	120.3	110.5	98.6	120.3	111.1	98.6	120.3	111.1
07	THERMAL & MOISTURE PROTECTION	105.0	81.9	95.6	101.2	126.6	111.6	101.3	123.7	110.4	102.6	123.7	111.2	101.3	123.7	110.4	101.3	123.7	110.4
08	OPENINGS	95.7	83.7	92.7	102.4	130.1	109.4	102.4	130.1	109.4	103.1	130.1	109.9	105.1	130.1	111.4	102.4	130.1	109.4
0920	Plaster & Gypsum Board	81.2	81.8	81.6	97.8	120.3	113.8	97.8	120.3	113.8	95.9	120.3	113.3	99.6	120.3	114.4	97.8	120.3	113.8
0950, 0980	Ceilings & Acoustic Treatment	108.8	81.8	90.9	102.0	120.3	114.1	102.0	120.3	114.1	100.2	120.3	113.5	106.3	120.3	115.5	102.0	120.3	114.1
0960	Flooring	119.1	47.5	98.1	94.5	134.4	106.2	94.5	134.4	106.2	94.5	134.4	106.2	94.5	134.4	106.2	94.5	134.4	106.2
0970, 0990	Wall Finishes & Painting/Coating	107.9	24.9	56.7	90.1	117.5	107.0	90.1	117.5	107.0	90.1	117.5	107.0	90.1	117.5	107.0	90.1	117.5	107.0
09	FINISHES	107.9	67.7	85.5	101.7	123.2	113.7	101.8	123.2	113.7	99.9	123.2	112.9	102.9	123.2	114.2	101.8	123.2	113.7
COVERS	DIVS. 10 - 14, 25, 28, 41, 43, 44, 46	100.0	91.0	98.2	100.0	108.6	101.7	100.0	108.6	101.7	100.0	108.6	101.7	100.0	108.6	101.7	100.0	108.6	101.7
21, 22, 23	FIRE SUPPRESSION, PLUMBING & HVAC	94.1	74.6	86.2	100.0	114.7	106.0	100.0	114.7	106.0	100.1	114.7	106.0	94.1	114.7	102.4	100.0	114.7	106.0
26, 27, 3370	ELECTRICAL, COMMUNICATIONS & UTIL.	95.1	75.1	84.8	102.2	109.9	106.2	102.2	109.6	106.0	99.2	110.5	105.1	102.1	109.6	106.0	102.3	109.6	106.1
MF2010	WEIGHTED AVERAGE	101.3	78.3	91.2	102.2	118.7	109.5	101.5	118.6	109.0	102.0	118.7	109.4	99.7	118.6	108.0	101.1	118.6	108.8

	DIVISION	CONNECTICUT																	
		NEW HAVEN			NEW LONDON			NORWALK			STAMFORD			WATERBURY			WILLIMANTIC		
		065			063			068			069			067			062		
		MAT.	INST.	TOTAL	MAT.	INST.	TOTAL	MAT.	INST.	TOTAL	MAT.	INST.	TOTAL	MAT.	INST.	TOTAL	MAT.	INST.	TOTAL
015433	CONTRACTOR EQUIPMENT		100.9	100.9		100.9	100.9		100.4	100.4		100.4	100.4		100.4	100.4		100.4	100.4
0241, 31 - 34	SITE & INFRASTRUCTURE, DEMOLITION	111.1	104.0	106.2	102.8	104.0	103.6	111.6	103.3	105.9	112.3	103.3	106.1	111.4	103.3	105.8	111.9	103.3	105.9
0310	Concrete Forming & Accessories	99.0	121.6	118.6	99.0	121.6	118.6	99.3	122.1	119.1	99.3	122.1	119.1	99.3	121.6	118.7	99.3	121.5	118.6
0320	Concrete Reinforcing	103.6	128.8	116.3	81.2	128.8	105.2	103.6	129.0	116.4	103.6	129.0	116.4	103.6	128.8	116.3	103.6	128.8	116.3
0330	Cast-in-Place Concrete	104.4	125.0	112.5	88.9	125.0	103.1	106.0	126.5	114.1	107.8	126.5	115.2	107.8	125.0	114.6	100.7	124.9	110.2
03	CONCRETE	123.2	123.9	123.5	94.7	123.9	108.9	109.7	124.7	117.0	110.6	124.7	117.4	110.6	123.9	117.0	107.0	123.8	115.2
04	MASONRY	96.9	129.8	117.0	95.3	129.8	116.3	96.4	131.3	117.7	97.2	131.3	118.0	97.2	129.8	117.1	96.5	129.8	116.8
05	METALS	95.9	125.0	105.5	95.6	125.0	105.3	99.1	125.7	107.9	99.1	125.7	107.9	99.1	125.0	107.7	98.9	124.9	107.5
06	WOOD, PLASTICS & COMPOSITES	98.6	120.3	111.1	98.6	120.3	111.1	98.6	120.3	111.1	98.6	120.3	111.1	98.6	120.3	111.1	98.6	120.3	111.1
07	THERMAL & MOISTURE PROTECTION	101.4	123.6	110.4	101.3	123.7	110.4	101.4	127.3	111.9	101.3	127.3	111.9	101.3	123.6	110.4	101.5	123.3	110.4
08	OPENINGS	102.4	130.1	109.4	105.6	130.1	111.8	102.4	130.1	109.4	102.4	130.1	109.4	102.4	130.1	109.4	105.6	130.1	111.8
0920	Plaster & Gypsum Board	97.8	120.3	113.8	97.8	120.3	113.8	97.8	120.3	113.8	97.8	120.3	113.8	97.8	120.3	113.8	97.8	120.3	113.8
0950, 0980	Ceilings & Acoustic Treatment	102.0	120.3	114.1	100.0	120.3	113.4	102.0	120.3	114.1	102.0	120.3	114.1	102.0	120.3	114.1	100.0	120.3	113.4
0960	Flooring	94.5	134.4	106.2	94.5	134.4	106.2	94.5	134.4	106.2	94.5	134.4	106.2	94.5	134.4	106.2	94.5	119.7	101.9
0970, 0990	Wall Finishes & Painting/Coating	90.1	117.5	107.0	90.1	117.5	107.0	90.1	117.5	107.0	90.1	117.5	107.0	90.1	117.5	107.0	90.1	117.5	107.0
09	FINISHES	101.8	123.2	113.7	100.6	123.2	113.2	101.8	123.2	113.7	101.9	123.2	113.7	101.6	123.2	113.6	101.5	120.7	112.2
COVERS	DIVS. 10 - 14, 25, 28, 41, 43, 44, 46	100.0	108.6	101.7	100.0	108.6	101.7	100.0	108.8	101.7	100.0	108.8	101.7	100.0	108.6	101.7	100.0	108.6	101.7
21, 22, 23	FIRE SUPPRESSION, PLUMBING & HVAC	100.0	114.7	106.0	94.1	114.7	102.4	100.0	114.8	106.0	100.0	114.8	106.0	100.0	114.7	106.0	100.0	114.7	106.0
26, 27, 3370	ELECTRICAL, COMMUNICATIONS & UTIL.	102.1	109.6	106.0	98.7	109.6	104.3	102.2	162.3	133.2	102.2	162.3	133.2	101.6	109.9	105.9	102.2	110.5	106.5
MF2010	WEIGHTED AVERAGE	102.8	118.6	109.8	97.6	118.6	106.9	101.8	126.3	112.6	101.9	126.3	112.7	101.8	118.6	109.2	101.7	118.4	109.1

	DIVISION	D.C.			DELAWARE									FLORIDA					
		WASHINGTON			DOVER			NEWARK			WILMINGTON			DAYTONA BEACH			FORT LAUDERDALE		
		200 - 205			199			197			198			321			333		
		MAT.	INST.	TOTAL	MAT.	INST.	TOTAL	MAT.	INST.	TOTAL	MAT.	INST.	TOTAL	MAT.	INST.	TOTAL	MAT.	INST.	TOTAL
015433	CONTRACTOR EQUIPMENT		103.2	103.2		116.9	116.9		116.9	116.9		117.0	117.0		98.0	98.0		90.9	90.9
0241, 31 - 34	SITE & INFRASTRUCTURE, DEMOLITION	109.6	92.6	97.9	98.2	111.9	107.6	98.6	111.9	107.8	88.8	112.2	104.9	116.5	89.6	97.9	99.7	77.9	84.6
0310	Concrete Forming & Accessories	97.4	81.1	83.2	98.3	102.7	102.1	100.4	102.7	102.4	101.4	102.7	102.5	96.7	72.1	75.3	95.3	72.8	75.7
0320	Concrete Reinforcing	98.5	89.7	94.1	96.0	102.1	99.1	96.8	102.1	99.5	96.8	102.1	99.5	96.3	77.0	86.6	96.3	76.7	86.4
0330	Cast-in-Place Concrete	129.8	90.0	114.1	95.5	101.9	98.0	85.8	101.9	92.2	92.6	101.9	96.3	91.8	73.9	84.7	96.2	81.0	90.2
03	CONCRETE	115.7	87.0	101.7	97.3	103.2	100.2	92.7	103.2	97.8	96.2	103.2	99.6	91.6	74.8	83.4	93.8	77.5	85.8
04	MASONRY	101.4	81.4	89.2	103.8	96.8	99.5	102.1	96.8	98.9	108.2	96.8	101.2	93.6	69.4	78.8	93.9	72.2	80.6
05	METALS	96.5	106.6	99.8	103.9	116.9	108.2	103.9	116.9	108.2	100.8	116.9	106.1	96.1	92.2	94.8	96.0	93.1	95.1
06	WOOD, PLASTICS & COMPOSITES	101.9	79.9	89.2	97.2	102.6	100.3	100.1	102.6	101.6	102.5	102.6	102.5	100.0	73.6	84.7	95.9	71.4	81.8
07	THERMAL & MOISTURE PROTECTION	100.2	84.9	94.0	97.9	112.4	103.8	97.9	112.4	103.8	98.4	112.4	104.1	95.1	77.8	88.1	95.1	85.4	91.2
08	OPENINGS	103.0	88.8	99.4	95.6	110.4	99.3	95.6	110.4	99.3	95.3	110.4	99.1	100.2	70.5	92.7	97.9	69.5	90.7
0920	Plaster & Gypsum Board	108.6	79.3	87.7	105.9	102.5	103.5	107.3	102.5	103.9	107.8	102.5	104.0	93.2	73.2	78.9	92.3	71.0	77.1
0950, 0980	Ceilings & Acoustic Treatment	105.8	79.3	88.3	104.2	102.5	103.1	104.2	102.5	103.1	99.5	102.5	101.5	95.8	73.2	80.9	95.8	71.0	79.4
0960	Flooring	115.0	97.2	109.8	97.1	107.6	100.1	97.3	107.6	100.3	96.9	107.6	100.0	117.7	77.1	105.8	117.7	72.6	104.5
0970, 0990	Wall Finishes & Painting/Coating	121.0	86.6	99.8	98.1	102.8	101.0	98.1	102.8	101.0	98.1	102.8	101.0	111.2	78.1	90.8	105.7	74.0	86.2
09	FINISHES	107.0	83.9	94.1	100.1	103.1	101.8	100.6	103.1	102.0	99.4	103.1	101.5	107.1	73.4	88.3	103.9	72.1	86.2
COVERS	DIVS. 10 - 14, 25, 28, 41, 43, 44, 46	100.0	99.4	99.9	100.0	93.7	98.7	100.0	93.7	98.7	100.0	93.7	98.7	100.0	78.3	95.7	100.0	84.8	97.0
21, 22, 23	FIRE SUPPRESSION, PLUMBING & HVAC	100.2	94.2	97.7	99.9	113.4	105.4	100.0	113.4	105.5	100.1	113.4	105.5	99.9	77.0	90.6	99.9	70.6	88.1
26, 27, 3370	ELECTRICAL, COMMUNICATIONS & UTIL.	98.8	106.0	102.5	96.4	110.1	103.5	100.3	110.1	105.3	98.6	110.1	104.5	98.0	58.4	77.6	98.0	74.3	85.8
MF2010	WEIGHTED AVERAGE	102.4	92.8	98.2	99.5	107.9	103.2	99.4	107.9	103.2	99.1	108.0	103.0	98.8	75.2	88.4	98.1	75.9	88.3

	DIVISION	FLORIDA																	
		FORT MYERS			GAINESVILLE			JACKSONVILLE			LAKELAND			MELBOURNE			MIAMI		
		339,341			326,344			320,322			338			329			330 - 332,340		
		MAT.	INST.	TOTAL	MAT.	INST.	TOTAL	MAT.	INST.	TOTAL	MAT.	INST.	TOTAL	MAT.	INST.	TOTAL	MAT.	INST.	TOTAL
015433	CONTRACTOR EQUIPMENT		98.0	98.0		98.0	98.0		98.0	98.0		98.0	98.0		98.0	98.0		90.9	90.9
0241, 31 - 34	SITE & INFRASTRUCTURE, DEMOLITION	113.5	88.9	96.5	127.3	88.6	100.6	116.5	88.9	97.5	115.7	89.1	97.3	124.9	88.8	100.0	101.6	77.6	85.1
0310	Concrete Forming & Accessories	91.1	79.5	81.1	92.0	57.5	62.0	96.5	57.9	62.9	87.6	80.1	81.1	93.1	73.8	76.3	99.7	73.2	76.6
0320	Concrete Reinforcing	97.4	98.6	98.0	102.3	67.6	84.8	96.3	67.7	81.9	99.9	99.5	99.7	97.4	77.0	87.1	96.3	76.7	86.5
0330	Cast-in-Place Concrete	100.4	71.6	89.1	105.3	65.3	89.6	92.7	71.4	84.3	102.7	72.9	90.9	110.6	76.4	97.1	98.1	82.0	91.8
03	CONCRETE	94.7	81.3	88.2	102.6	63.6	83.6	92.0	65.9	79.3	96.6	82.1	89.6	103.3	76.4	90.2	95.0	78.0	86.7
04	MASONRY	87.5	65.4	74.0	109.0	64.6	81.9	93.6	64.7	76.0	105.5	80.1	90.0	92.7	73.7	81.1	94.4	76.0	83.2
05	METALS	98.2	100.5	99.0	95.1	87.1	92.4	94.7	87.5	92.3	98.1	101.2	99.1	104.2	92.3	100.3	100.9	91.9	97.9
06	WOOD, PLASTICS & COMPOSITES	92.8	82.3	86.7	94.0	55.0	71.5	100.0	55.0	74.0	88.6	82.3	85.0	95.6	73.6	82.9	101.6	71.4	84.2
07	THERMAL & MOISTURE PROTECTION	94.9	84.5	90.7	95.4	64.7	82.9	95.3	65.3	83.1	94.9	89.3	92.6	95.5	80.3	89.3	103.6	79.5	93.8
08	OPENINGS	98.8	79.0	93.8	98.2	55.6	87.4	100.2	58.9	89.8	98.7	79.7	93.9	99.3	73.9	92.9	97.9	69.5	90.7
0920	Plaster & Gypsum Board	87.3	82.2	83.7	88.2	54.1	63.8	93.2	54.1	65.3	85.1	82.2	83.0	88.2	73.2	77.5	90.6	71.0	76.6
0950, 0980	Ceilings & Acoustic Treatment	89.3	82.2	84.6	89.3	54.1	66.0	95.8	54.1	68.2	89.3	82.2	84.6	91.3	73.2	79.3	93.1	71.0	78.5
0960	Flooring	114.1	57.7	97.5	114.6	45.2	94.2	117.7	67.6	103.0	111.4	59.2	96.1	114.8	77.1	103.7	125.3	76.4	111.0
0970, 0990	Wall Finishes & Painting/Coating	111.2	64.1	82.1	111.2	64.1	82.1	111.2	65.3	82.9	111.2	64.1	82.1	111.2	96.9	102.4	105.7	74.0	86.2
09	FINISHES	103.7	74.1	87.2	105.5	54.8	77.2	107.1	59.4	80.5	102.8	74.4	86.9	105.5	76.6	89.4	105.5	73.3	87.6
COVERS	DIVS. 10 - 14, 25, 28, 41, 43, 44, 46	100.0	78.6	95.7	100.0	76.6	95.3	100.0	74.6	95.0	100.0	78.6	95.7	100.0	79.8	96.0	100.0	85.4	97.1
21, 22, 23	FIRE SUPPRESSION, PLUMBING & HVAC	96.7	66.2	84.4	98.8	68.6	86.6	99.9	68.6	87.3	96.7	86.7	92.7	99.9	79.2	91.5	99.9	71.3	88.3
26, 27, 3370	ELECTRICAL, COMMUNICATIONS & UTIL.	100.3	65.5	82.3	98.3	76.3	87.0	97.7	66.1	81.4	98.3	65.4	81.3	99.4	68.0	83.2	99.9	76.9	88.0
MF2010	WEIGHTED AVERAGE	98.1	75.8	88.3	100.4	69.7	86.9	98.7	69.4	85.7	99.0	82.0	91.5	101.5	78.2	91.2	99.6	76.7	89.6

	DIVISION	FLORIDA																	
		ORLANDO			PANAMA CITY			PENSACOLA			SARASOTA			ST. PETERSBURG			TALLAHASSEE		
		327 - 328,347			324			325			342			337			323		
		MAT.	INST.	TOTAL	MAT.	INST.	TOTAL	MAT.	INST.	TOTAL	MAT.	INST.	TOTAL	MAT.	INST.	TOTAL	MAT.	INST.	TOTAL
015433	CONTRACTOR EQUIPMENT		98.0	98.0		98.0	98.0		98.0	98.0		98.0	98.0		98.0	98.0		98.0	98.0
0241, 31 - 34	SITE & INFRASTRUCTURE, DEMOLITION	116.7	88.7	97.4	131.7	87.8	101.4	130.9	88.3	101.5	117.9	88.8	97.8	117.3	88.4	97.4	112.6	88.0	95.6
0310	Concrete Forming & Accessories	96.3	76.4	79.0	95.8	46.6	53.0	93.8	55.0	60.0	96.7	79.6	81.9	94.8	54.1	59.4	98.6	46.8	53.5
0320	Concrete Reinforcing	96.3	76.3	86.2	100.6	77.1	88.7	103.2	77.6	90.3	96.3	99.4	97.9	99.9	86.2	93.0	96.3	67.5	81.8
0330	Cast-in-Place Concrete	115.0	73.8	98.7	97.4	59.3	82.4	120.3	68.9	100.0	104.9	72.7	92.2	103.8	67.2	89.4	100.2	58.8	83.9
03	CONCRETE	104.5	76.5	90.8	100.2	58.4	79.8	111.1	65.6	88.9	98.2	81.8	90.2	98.1	66.2	82.6	95.9	56.6	76.8
04	MASONRY	99.2	69.4	81.0	98.6	49.9	68.9	116.3	57.5	80.4	94.5	80.1	85.7	148.1	51.7	89.3	96.8	56.8	72.4
05	METALS	101.6	91.5	98.3	95.9	89.3	93.7	97.0	90.8	95.0	98.8	100.7	99.4	99.0	93.7	97.3	89.8	86.8	88.8
06	WOOD, PLASTICS & COMPOSITES	99.5	79.9	88.2	98.8	44.4	67.3	96.9	54.8	72.6	100.0	82.3	89.8	97.3	53.2	71.9	97.3	43.9	66.5
07	THERMAL & MOISTURE PROTECTION	96.3	78.4	89.0	95.6	59.7	81.0	95.5	65.8	83.4	95.1	89.3	92.8	95.1	61.0	81.2	101.0	75.5	90.6
08	OPENINGS	100.2	72.9	93.3	97.7	49.3	85.5	97.7	60.4	88.3	100.2	78.7	94.8	98.7	63.3	89.8	100.7	49.9	87.9
0920	Plaster & Gypsum Board	104.6	79.8	86.9	91.0	43.1	56.8	91.5	53.8	64.6	93.2	82.2	85.4	90.1	52.2	63.0	107.2	42.6	61.0
0950, 0980	Ceilings & Acoustic Treatment	98.5	79.8	86.1	91.3	43.1	59.4	91.3	53.8	66.5	93.8	82.2	86.1	91.3	52.2	65.5	98.5	42.6	61.6
0960	Flooring	117.9	77.1	105.9	117.2	45.0	96.0	112.0	65.9	98.5	117.7	60.6	100.9	116.3	57.8	99.2	117.7	64.9	102.2
0970, 0990	Wall Finishes & Painting/Coating	111.2	74.0	88.3	111.2	66.3	83.5	111.2	66.3	83.5	111.2	64.1	82.1	111.2	64.1	82.1	111.2	66.3	83.5
09	FINISHES	108.8	76.8	90.9	107.5	47.2	73.9	105.6	57.5	78.8	106.7	74.6	88.8	105.4	54.8	77.2	108.7	50.8	76.4
COVERS	DIVS. 10 - 14, 25, 28, 41, 43, 44, 46	100.0	79.0	95.8	100.0	49.9	90.0	100.0	50.0	90.1	100.0	78.6	95.7	100.0	60.3	92.1	100.0	71.2	94.3
21, 22, 23	FIRE SUPPRESSION, PLUMBING & HVAC	99.9	60.6	84.0	99.9	55.8	82.1	99.9	56.2	82.2	99.8	64.8	85.6	99.9	60.1	83.8	99.9	41.8	76.4
26, 27, 3370	ELECTRICAL, COMMUNICATIONS & UTIL.	100.1	62.8	80.8	96.7	60.1	77.8	100.7	57.2	78.3	98.0	65.4	81.2	98.3	65.4	81.3	103.8	60.1	81.3
MF2010	WEIGHTED AVERAGE	101.8	73.0	89.1	100.1	60.5	82.7	102.6	64.1	85.6	100.0	77.3	90.0	102.5	65.8	86.3	99.4	59.3	81.7

	DIVISION	FLORIDA						GEORGIA											
		TAMPA			WEST PALM BEACH			ALBANY			ATHENS			ATLANTA			AUGUSTA		
		335 - 336,346			334,349			317,398			306			300 - 303,399			308 - 309		
		MAT.	INST.	TOTAL	MAT.	INST.	TOTAL	MAT.	INST.	TOTAL	MAT.	INST.	TOTAL	MAT.	INST.	TOTAL	MAT.	INST.	TOTAL
015433	CONTRACTOR EQUIPMENT		98.0	98.0		90.9	90.9		91.8	91.8		93.2	93.2		93.7	93.7		93.2	93.2
0241, 31 - 34	SITE & INFRASTRUCTURE, DEMOLITION	117.6	89.0	97.9	96.4	77.9	83.6	99.9	79.2	85.6	103.6	92.3	95.8	99.4	94.1	95.7	95.8	92.0	93.2
0310	Concrete Forming & Accessories	97.6	80.2	82.4	98.6	72.4	75.8	96.0	45.8	52.3	93.6	43.5	50.0	96.9	75.3	78.1	94.5	67.0	70.6
0320	Concrete Reinforcing	96.3	99.5	97.9	99.1	76.4	87.6	96.0	83.5	89.7	102.7	82.7	92.6	102.0	84.9	93.4	103.0	75.5	89.1
0330	Cast-in-Place Concrete	101.4	72.9	90.2	91.5	78.0	86.2	98.0	47.8	78.2	111.6	53.1	88.5	111.6	71.7	95.9	105.4	50.0	83.6
03	CONCRETE	96.5	82.2	89.5	90.7	76.2	83.7	94.5	55.1	75.3	103.1	55.0	79.7	101.1	75.9	88.8	96.6	63.0	80.2
04	MASONRY	94.9	80.1	85.9	93.4	70.4	79.4	96.0	44.6	64.7	82.4	56.9	66.8	96.8	67.7	79.1	97.1	41.4	63.1
05	METALS	98.1	101.2	99.1	95.2	92.6	94.3	95.4	86.9	92.6	91.2	75.4	86.0	92.1	79.5	87.9	90.9	72.9	84.9
06	WOOD, PLASTICS & COMPOSITES	101.2	82.3	90.3	100.6	71.4	83.7	99.0	40.3	65.1	93.8	37.3	61.2	97.7	78.0	86.3	95.1	72.7	82.1
07	THERMAL & MOISTURE PROTECTION	95.3	89.3	92.9	94.9	74.1	86.4	92.8	61.4	80.0	94.5	54.7	78.3	94.4	74.2	86.2	94.1	58.5	79.6
08	OPENINGS	100.2	82.1	95.6	97.0	69.5	90.0	97.9	46.6	84.9	91.2	45.2	79.6	97.0	73.7	91.1	91.3	64.4	84.5
0920	Plaster & Gypsum Board	93.2	82.2	85.4	96.0	71.0	78.1	93.0	38.9	54.4	96.5	35.7	53.1	98.7	77.7	83.7	97.4	72.2	79.4
0950, 0980	Ceilings & Acoustic Treatment	95.8	82.2	86.8	89.3	71.0	77.2	94.9	38.9	57.9	95.8	35.7	56.1	95.8	77.7	83.8	96.8	72.2	80.6
0960	Flooring	117.7	59.2	100.5	120.1	69.6	105.3	117.7	37.5	94.2	89.2	56.4	79.5	90.6	68.3	84.1	89.6	41.1	75.4
0970, 0990	Wall Finishes & Painting/Coating	111.2	64.1	82.1	105.7	74.0	86.2	105.7	55.0	74.4	96.6	36.9	59.8	96.6	84.6	89.2	96.6	38.5	60.8
09	FINISHES	107.2	74.4	88.9	103.2	71.5	85.6	103.7	43.2	70.0	94.6	43.5	66.1	94.8	75.0	83.8	94.2	59.9	75.1
COVERS	DIVS. 10 - 14, 25, 28, 41, 43, 44, 46	100.0	79.1	95.9	100.0	84.8	97.0	100.0	80.6	96.1	100.0	42.2	88.5	100.0	85.3	97.1	100.0	53.6	90.8
21, 22, 23	FIRE SUPPRESSION, PLUMBING & HVAC	99.9	86.7	94.6	96.7	67.3	84.8	99.9	68.2	87.1	94.0	72.9	85.4	99.9	73.4	89.2	100.0	65.7	86.1
26, 27, 3370	ELECTRICAL, COMMUNICATIONS & UTIL.	98.0	65.4	81.2	99.4	74.3	86.5	93.5	60.1	76.3	102.3	73.9	87.7	101.7	75.5	88.2	103.0	65.6	83.7
MF2010	WEIGHTED AVERAGE	99.8	82.1	92.0	96.7	74.4	86.9	97.6	61.5	81.7	95.5	64.2	81.7	98.0	76.5	88.5	96.7	64.5	82.5

	DIVISION	GEORGIA																	
		COLUMBUS			DALTON			GAINESVILLE			MACON			SAVANNAH			STATESBORO		
		318 - 319			307			305			310 - 312			313 - 314			304		
		MAT.	INST.	TOTAL	MAT.	INST.	TOTAL	MAT.	INST.	TOTAL	MAT.	INST.	TOTAL	MAT.	INST.	TOTAL	MAT.	INST.	TOTAL
015433	CONTRACTOR EQUIPMENT		91.8	91.8		104.7	104.7		93.2	93.2		101.9	101.9		92.7	92.7		92.6	92.6
0241, 31 - 34	SITE & INFRASTRUCTURE, DEMOLITION	99.8	79.3	85.7	104.0	96.3	98.7	103.0	92.3	95.6	101.1	94.0	96.2	100.8	80.0	86.5	104.5	77.0	85.6
0310	Concrete Forming & Accessories	95.9	57.1	62.2	87.3	49.0	54.0	96.6	45.3	52.0	95.2	54.3	59.6	96.0	50.7	56.6	82.3	42.5	47.6
0320	Concrete Reinforcing	96.3	84.2	90.2	102.2	78.3	90.1	102.5	82.7	92.5	97.6	83.5	90.5	97.2	75.7	86.4	101.8	44.0	72.6
0330	Cast-in-Place Concrete	97.6	56.3	81.3	108.4	47.0	84.2	117.3	52.9	91.9	96.3	50.2	78.1	107.4	49.2	84.5	111.4	61.7	91.8
03	CONCRETE	94.4	63.2	79.2	102.0	55.4	79.3	105.3	55.8	81.2	93.9	59.8	77.3	99.4	56.3	78.4	102.6	51.2	77.5
04	MASONRY	96.0	58.4	73.1	84.2	33.9	53.5	91.4	57.3	70.6	108.2	39.4	66.3	95.1	51.5	68.5	85.6	41.6	58.8
05	METALS	94.9	88.6	92.8	91.9	85.0	89.6	90.4	75.1	85.3	91.1	87.7	90.0	91.4	84.2	89.1	95.4	72.4	87.8
06	WOOD, PLASTICS & COMPOSITES	99.0	55.4	73.8	76.7	50.4	61.5	97.5	40.2	64.4	108.9	54.4	77.4	115.1	47.5	76.1	70.7	41.8	54.0
07	THERMAL & MOISTURE PROTECTION	92.7	65.7	81.7	96.3	52.7	78.6	94.4	57.6	79.4	91.3	63.1	79.9	95.0	58.2	80.1	95.0	51.0	77.1
08	OPENINGS	97.9	56.7	87.5	92.9	47.1	81.3	91.2	42.4	78.9	96.2	55.2	85.9	99.1	48.5	86.4	93.6	33.4	78.4
0920	Plaster & Gypsum Board	93.0	54.5	65.4	80.3	49.2	58.1	98.7	38.7	55.9	101.8	53.4	67.2	92.2	46.4	59.5	82.2	40.4	52.3
0950, 0980	Ceilings & Acoustic Treatment	94.9	54.5	68.2	107.8	49.2	69.1	95.8	38.7	58.1	89.7	53.4	65.7	98.5	46.4	64.1	104.0	40.4	62.0
0960	Flooring	117.7	52.8	98.6	90.0	37.5	74.6	90.6	37.5	75.0	91.6	38.0	75.9	118.3	48.6	97.8	106.1	47.0	88.8
0970, 0990	Wall Finishes & Painting/Coating	105.7	68.2	82.5	87.3	56.3	68.2	96.6	36.9	59.8	108.1	55.0	75.3	107.1	57.2	76.3	94.6	34.6	57.6
09	FINISHES	103.6	56.4	77.3	103.6	47.0	72.0	95.0	40.9	64.8	91.3	50.2	68.4	104.7	50.1	74.2	106.4	41.6	70.3
COVERS	DIVS. 10 - 14, 25, 28, 41, 43, 44, 46	100.0	82.3	96.5	100.0	24.2	84.9	100.0	42.5	88.6	100.0	80.3	96.1	100.0	52.6	90.6	100.0	39.2	87.9
21, 22, 23	FIRE SUPPRESSION, PLUMBING & HVAC	99.9	49.9	79.7	94.1	59.2	79.9	94.0	72.8	85.4	99.9	66.3	86.3	99.9	61.6	84.4	94.5	56.7	79.2
26, 27, 3370	ELECTRICAL, COMMUNICATIONS & UTIL.	93.5	74.0	83.4	113.8	72.0	92.2	102.3	73.9	87.7	92.5	62.1	76.8	95.0	57.4	75.6	102.6	57.4	79.3
MF2010	WEIGHTED AVERAGE	97.5	64.6	83.0	97.6	60.1	81.1	96.1	64.0	82.0	96.4	64.2	82.2	98.1	60.5	81.5	97.4	54.0	78.3

	DIVISION	GEORGIA						HAWAII									IDAHO		
		VALDOSTA			WAYCROSS			HILO			HONOLULU			STATES & POSS., GUAM			BOISE		
		316			315			967			968			969			836 - 837		
		MAT.	INST.	TOTAL	MAT.	INST.	TOTAL	MAT.	INST.	TOTAL	MAT.	INST.	TOTAL	MAT.	INST.	TOTAL	MAT.	INST.	TOTAL
015433	CONTRACTOR EQUIPMENT		91.8	91.8		91.8	91.8		99.4	99.4		99.4	99.4		162.1	162.1		99.3	99.3
0241, 31 - 34	SITE & INFRASTRUCTURE, DEMOLITION	110.3	79.2	88.9	106.8	78.0	86.9	124.4	107.4	112.6	127.4	107.4	113.6	166.7	104.8	124.0	77.8	98.3	91.9
0310	Concrete Forming & Accessories	86.0	44.9	50.2	87.8	68.6	71.1	110.2	135.0	131.8	117.2	135.0	132.7	112.4	67.0	72.9	100.7	73.9	77.4
0320	Concrete Reinforcing	98.3	71.4	84.8	98.3	77.2	87.7	109.0	122.1	115.6	108.0	122.1	115.1	192.4	32.0	111.5	99.8	74.9	87.2
0330	Cast-in-Place Concrete	95.9	51.7	78.5	108.3	50.4	85.5	187.8	125.4	163.2	158.4	125.4	145.4	162.8	108.9	141.6	90.1	74.5	84.0
03	CONCRETE	98.4	53.9	76.7	102.6	64.7	84.1	147.4	128.1	138.0	132.9	128.1	130.5	148.0	75.4	112.6	98.8	74.4	86.9
04	MASONRY	102.1	51.1	71.0	103.0	41.5	65.5	121.3	127.0	124.8	114.1	127.0	122.0	173.3	46.0	95.6	122.1	62.5	85.8
05	METALS	94.6	83.1	90.8	93.6	79.9	89.1	107.4	108.1	107.6	128.1	108.1	121.5	138.5	77.6	118.4	99.9	74.8	91.6
06	WOOD, PLASTICS & COMPOSITES	87.0	39.3	59.5	88.8	77.0	82.0	101.4	137.6	122.4	111.7	137.6	126.7	112.0	70.9	88.3	87.9	76.2	81.2
07	THERMAL & MOISTURE PROTECTION	92.9	64.4	81.3	92.7	53.1	76.6	104.7	124.7	112.8	114.5	124.7	118.6	120.8	69.9	100.1	96.0	66.6	84.1
08	OPENINGS	93.2	42.7	80.4	93.2	60.7	85.0	99.7	131.3	107.7	105.0	131.3	111.6	104.5	57.0	92.5	96.3	70.8	89.9
0920	Plaster & Gypsum Board	85.1	37.9	51.4	85.5	76.7	79.2	103.7	138.6	128.6	134.7	138.6	137.5	202.8	59.2	100.2	86.8	75.4	78.7
0950, 0980	Ceilings & Acoustic Treatment	91.3	37.9	56.0	89.3	76.7	81.0	109.3	138.6	128.7	114.4	138.6	130.4	220.9	59.2	114.0	111.7	75.4	87.7
0960	Flooring	110.4	48.7	92.3	111.8	31.3	88.2	133.3	133.6	133.4	157.0	133.6	150.1	156.7	48.6	125.0	96.5	66.5	87.7
0970, 0990	Wall Finishes & Painting/Coating	105.7	55.0	74.4	105.7	42.0	66.4	98.7	144.2	126.8	103.1	144.2	128.5	104.2	37.9	63.3	98.8	41.6	63.5
09	FINISHES	100.9	45.0	69.7	100.4	60.1	77.9	115.8	136.8	127.5	128.5	136.8	133.1	190.9	61.9	118.9	99.3	70.1	83.0
COVERS	DIVS. 10 - 14, 25, 28, 41, 43, 44, 46	100.0	51.7	90.4	100.0	43.0	88.7	100.0	115.6	103.1	100.0	115.6	103.1	100.0	79.7	96.0	100.0	62.2	92.5
21, 22, 23	FIRE SUPPRESSION, PLUMBING & HVAC	99.9	68.1	87.0	96.1	58.1	80.7	100.1	109.7	104.0	100.1	109.7	104.0	102.1	40.0	76.9	100.0	67.6	86.9
26, 27, 3370	ELECTRICAL, COMMUNICATIONS & UTIL.	91.6	59.8	75.2	95.8	60.5	77.6	107.0	127.2	117.4	105.9	127.2	116.9	154.1	43.8	97.2	96.4	73.0	84.3
MF2010	WEIGHTED AVERAGE	97.6	60.8	81.4	97.3	61.4	81.5	110.3	121.2	115.1	113.3	121.2	116.8	131.1	60.8	100.1	99.3	72.4	87.4

	DIVISION	IDAHO															ILLINOIS		
		COEUR D'ALENE			IDAHO FALLS			LEWISTON			POCATELLO			TWIN FALLS			BLOOMINGTON		
		838			834			835			832			833			617		
		MAT.	INST.	TOTAL	MAT.	INST.	TOTAL	MAT.	INST.	TOTAL	MAT.	INST.	TOTAL	MAT.	INST.	TOTAL	MAT.	INST.	TOTAL
015433	CONTRACTOR EQUIPMENT		93.8	93.8		99.3	99.3		93.8	93.8		99.3	99.3		99.3	99.3		100.8	100.8
0241, 31 - 34	SITE & INFRASTRUCTURE, DEMOLITION	76.2	93.9	88.4	76.3	97.4	90.8	82.2	94.4	90.6	79.6	98.2	92.5	86.4	96.6	93.4	97.8	96.8	97.1
0310	Concrete Forming & Accessories	117.9	57.8	65.6	95.4	41.4	48.4	122.5	59.1	67.4	100.9	73.6	77.2	101.6	29.6	39.0	86.0	120.1	115.6
0320	Concrete Reinforcing	107.4	93.5	100.4	101.8	77.7	89.6	107.4	93.6	100.5	100.2	74.9	87.5	102.1	77.9	89.9	98.6	107.6	103.1
0330	Cast-in-Place Concrete	97.4	86.9	93.3	85.8	52.0	72.5	101.3	87.8	96.0	92.7	74.3	85.4	95.1	39.8	73.3	97.9	114.2	104.3
03	CONCRETE	105.9	75.0	90.9	91.4	52.8	72.6	109.5	76.0	93.1	98.3	74.2	86.6	105.5	43.4	75.2	96.3	115.6	105.7
04	MASONRY	123.7	82.4	98.5	117.2	34.6	66.8	124.1	83.3	99.2	119.8	64.4	86.0	122.3	32.6	67.6	113.8	119.2	117.1
05	METALS	94.2	84.0	90.8	107.5	73.7	96.4	93.6	85.0	90.8	107.6	74.7	96.7	107.6	73.1	96.2	96.8	111.3	101.6
06	WOOD, PLASTICS & COMPOSITES	93.4	50.5	68.6	82.6	39.8	57.9	98.0	50.5	70.6	88.0	76.2	81.2	88.7	28.3	53.8	85.6	119.8	105.4
07	THERMAL & MOISTURE PROTECTION	147.8	73.7	117.7	95.6	44.6	74.9	148.0	73.7	117.8	96.1	61.3	81.9	96.9	38.7	73.2	97.9	113.9	104.4
08	OPENINGS	115.3	56.0	100.4	100.5	46.8	87.0	115.3	58.7	101.0	97.1	65.5	89.1	100.6	38.4	84.9	89.4	107.0	93.8
0920	Plaster & Gypsum Board	154.5	49.0	79.1	81.0	37.8	50.1	155.4	49.0	79.4	83.8	75.4	77.8	84.5	25.9	42.6	97.2	120.3	113.7
0950, 0980	Ceilings & Acoustic Treatment	147.1	49.0	82.3	109.7	37.8	62.2	147.1	49.0	82.3	117.8	75.4	89.8	112.4	25.9	55.3	94.9	120.3	111.7
0960	Flooring	135.3	48.0	109.7	96.2	45.1	81.2	138.5	89.5	124.1	99.7	66.5	90.0	100.9	45.1	84.6	97.4	112.2	101.8
0970, 0990	Wall Finishes & Painting/Coating	116.6	61.9	82.8	98.8	34.2	59.0	116.6	61.9	82.8	98.7	43.3	64.5	98.8	25.5	53.6	96.0	124.2	113.4
09	FINISHES	167.7	54.8	104.7	97.5	40.6	65.8	168.9	63.3	110.0	101.2	70.3	84.0	101.2	30.9	62.0	97.5	119.8	109.9
COVERS	DIVS. 10 - 14, 25, 28, 41, 43, 44, 46	100.0	52.4	90.5	100.0	39.5	88.0	100.0	65.0	93.0	100.0	62.2	92.5	100.0	35.5	87.2	100.0	105.0	101.0
21, 22, 23	FIRE SUPPRESSION, PLUMBING & HVAC	99.6	83.2	93.0	100.9	61.5	84.9	100.8	90.2	96.5	100.0	67.5	86.8	100.0	55.6	82.0	94.0	106.7	99.1
26, 27, 3370	ELECTRICAL, COMMUNICATIONS & UTIL.	87.4	80.3	83.7	87.6	73.0	80.1	85.5	80.8	83.1	93.9	73.0	83.1	88.9	62.9	75.5	94.3	96.7	95.6
MF2010	WEIGHTED AVERAGE	107.2	76.3	93.6	98.9	58.7	81.2	107.9	79.7	95.5	100.3	72.2	87.9	101.3	52.5	79.8	96.3	109.3	102.0

DIVISION		ILLINOIS																	
		CARBONDALE			CENTRALIA			CHAMPAIGN			CHICAGO			DECATUR			EAST ST. LOUIS		
		629			628			618 - 619			606 - 608			625			620 - 622		
		MAT.	INST.	TOTAL	MAT.	INST.	TOTAL	MAT.	INST.	TOTAL	MAT.	INST.	TOTAL	MAT.	INST.	TOTAL	MAT.	INST.	TOTAL
015433	CONTRACTOR EQUIPMENT		107.0	107.0		107.0	107.0		101.6	101.6		92.7	92.7		101.6	101.6		107.0	107.0
0241, 31 - 34	SITE & INFRASTRUCTURE, DEMOLITION	104.8	97.4	99.7	105.2	98.7	100.7	107.5	97.7	100.7	111.8	93.7	99.3	91.6	97.7	95.8	107.2	98.6	101.3
0310	Concrete Forming & Accessories	94.0	104.1	102.8	95.9	112.4	110.2	91.7	118.5	115.0	98.0	156.1	148.5	95.8	118.0	115.1	91.7	112.7	110.0
0320	Concrete Reinforcing	97.6	94.6	96.1	97.6	104.5	101.1	98.6	102.5	100.6	99.0	161.1	130.3	95.6	102.9	99.3	97.5	108.5	103.1
0330	Cast-in-Place Concrete	91.6	99.1	94.6	92.1	116.1	101.5	113.5	109.9	112.1	107.8	147.0	123.2	100.0	112.9	105.1	93.6	116.2	102.5
03	CONCRETE	87.4	101.4	94.2	87.9	112.8	100.0	108.7	112.4	110.5	101.0	152.5	126.1	97.5	113.3	105.2	88.8	113.7	101.0
04	MASONRY	75.5	102.6	92.0	75.6	116.7	100.6	137.4	116.3	124.5	98.4	153.1	131.8	72.3	115.4	98.6	75.8	116.6	100.7
05	METALS	96.0	111.1	101.0	96.0	117.9	103.3	96.8	107.3	100.3	95.3	133.9	108.1	99.5	107.8	102.3	97.2	119.8	104.6
06	WOOD, PLASTICS & COMPOSITES	99.2	103.4	101.6	101.6	109.8	106.4	92.0	118.1	107.1	101.5	155.5	132.7	98.8	118.1	110.0	96.8	109.8	104.3
07	THERMAL & MOISTURE PROTECTION	93.1	98.9	95.5	93.2	112.9	101.2	98.6	113.3	104.5	99.7	143.6	117.6	98.7	108.0	102.5	93.2	112.4	101.0
08	OPENINGS	85.7	108.5	91.4	85.7	113.9	92.8	90.1	113.0	95.9	97.5	158.0	112.7	96.0	113.1	100.3	85.7	115.7	93.3
0920	Plaster & Gypsum Board	100.8	103.4	102.7	102.2	110.1	107.8	99.9	118.6	113.3	101.0	157.2	141.1	104.1	118.6	114.5	99.9	110.1	107.2
0950, 0980	Ceilings & Acoustic Treatment	94.9	103.4	100.5	94.9	110.1	104.9	94.9	118.6	110.6	109.9	157.2	141.1	101.2	118.6	112.7	94.9	110.1	104.9
0960	Flooring	116.3	125.8	119.1	117.4	118.6	117.8	100.5	125.8	107.9	97.0	148.5	112.1	104.0	117.6	108.0	115.1	118.6	116.2
0970, 0990	Wall Finishes & Painting/Coating	104.0	101.0	102.1	104.0	108.3	106.6	96.0	112.3	106.1	95.1	161.0	135.8	96.0	110.6	105.0	104.0	108.3	106.6
09	FINISHES	100.9	107.0	104.3	101.4	113.0	107.9	99.6	119.8	110.9	101.2	156.2	131.9	101.7	118.2	110.9	100.5	113.2	107.6
COVERS	DIVS. 10 - 14, 25, 28, 41, 43, 44, 46	100.0	100.5	100.1	100.0	103.9	100.8	100.0	104.7	100.9	100.0	125.2	105.0	100.0	104.4	100.9	100.0	103.9	100.8
21, 22, 23	FIRE SUPPRESSION, PLUMBING & HVAC	93.9	96.7	95.1	93.9	92.8	93.5	94.0	105.5	98.6	100.0	133.9	113.7	99.9	101.5	100.6	99.9	97.1	98.8
26, 27, 3370	ELECTRICAL, COMMUNICATIONS & UTIL.	93.0	101.5	97.4	94.4	105.4	100.1	97.6	95.8	96.7	95.4	133.4	115.0	97.1	90.3	93.6	94.0	105.4	99.8
MF2010	WEIGHTED AVERAGE	93.1	102.0	97.1	93.4	107.0	99.4	99.8	108.1	103.5	99.1	138.9	116.6	97.4	106.3	101.3	95.1	108.3	100.9

DIVISION		ILLINOIS																	
		EFFINGHAM			GALESBURG			JOLIET			KANKAKEE			LA SALLE			NORTH SUBURBAN		
		624			614			604			609			613			600 - 603		
		MAT.	INST.	TOTAL	MAT.	INST.	TOTAL	MAT.	INST.	TOTAL	MAT.	INST.	TOTAL	MAT.	INST.	TOTAL	MAT.	INST.	TOTAL
015433	CONTRACTOR EQUIPMENT		101.6	101.6		100.8	100.8		90.9	90.9		90.9	90.9		100.8	100.8		90.9	90.9
0241, 31 - 34	SITE & INFRASTRUCTURE, DEMOLITION	99.0	97.2	97.7	100.3	96.7	97.8	111.7	92.6	98.5	104.8	92.0	96.0	99.7	97.6	98.2	110.7	91.8	97.7
0310	Concrete Forming & Accessories	100.1	111.5	110.0	91.6	115.7	112.6	100.1	154.6	147.5	94.2	135.5	130.1	104.0	121.0	118.8	99.5	148.9	142.4
0320	Concrete Reinforcing	98.6	99.9	99.3	98.1	107.2	102.7	99.0	146.2	122.8	99.8	140.4	120.3	98.3	133.4	116.0	99.0	155.2	127.3
0330	Cast-in-Place Concrete	99.7	105.3	101.9	100.8	105.2	102.5	107.7	142.3	121.3	100.4	127.1	110.9	100.6	114.4	106.1	107.8	139.8	120.4
03	CONCRETE	98.3	107.2	102.7	99.0	110.5	104.6	101.1	147.4	123.7	95.3	132.6	113.5	99.8	121.1	110.2	101.1	145.7	122.8
04	MASONRY	79.3	104.4	94.6	113.9	113.7	113.8	101.0	146.3	128.6	97.8	132.5	119.0	113.9	117.0	115.8	98.3	142.4	125.2
05	METALS	96.7	105.6	99.6	96.8	110.1	101.2	93.2	125.8	104.0	93.2	121.4	102.5	96.9	125.6	106.4	94.4	128.7	105.7
06	WOOD, PLASTICS & COMPOSITES	101.2	113.3	108.2	91.8	115.0	105.2	103.2	156.1	133.8	96.5	134.4	118.4	105.5	120.3	114.1	101.5	150.6	129.9
07	THERMAL & MOISTURE PROTECTION	98.2	106.0	101.4	98.1	106.6	101.5	99.6	140.1	116.0	98.8	131.9	112.2	98.2	114.9	105.0	100.0	138.5	115.7
08	OPENINGS	89.9	109.7	94.9	89.4	108.8	94.3	95.8	154.3	110.5	87.5	139.4	100.5	89.4	122.7	97.8	95.8	152.1	110.0
0920	Plaster & Gypsum Board	104.0	113.6	110.9	99.9	115.4	111.0	97.8	157.8	140.7	95.5	135.5	124.1	106.3	120.9	116.7	101.0	152.2	137.6
0950, 0980	Ceilings & Acoustic Treatment	94.9	113.6	107.3	94.9	115.4	108.4	109.9	157.8	141.6	109.9	135.5	126.8	94.9	120.9	112.1	109.9	152.2	137.8
0960	Flooring	105.3	125.8	111.3	100.4	112.2	103.9	96.6	138.1	108.8	93.7	138.1	106.7	107.4	116.3	110.0	97.0	138.1	109.1
0970, 0990	Wall Finishes & Painting/Coating	96.0	105.9	102.1	96.0	93.9	94.7	93.2	151.2	129.0	93.2	124.2	112.4	96.0	124.2	113.4	95.1	151.2	129.7
09	FINISHES	100.7	114.7	108.5	98.9	113.6	107.1	100.7	152.6	129.6	99.0	133.0	118.0	101.7	119.6	111.7	101.2	148.4	127.5
COVERS	DIVS. 10 - 14, 25, 28, 41, 43, 44, 46	100.0	72.5	94.5	100.0	100.9	100.2	100.0	123.8	104.7	100.0	114.8	102.9	100.0	101.0	100.2	100.0	117.0	103.4
21, 22, 23	FIRE SUPPRESSION, PLUMBING & HVAC	94.0	101.1	96.9	94.0	102.7	97.5	100.1	133.2	113.5	94.1	127.3	107.5	94.0	124.7	106.4	100.0	123.2	109.4
26, 27, 3370	ELECTRICAL, COMMUNICATIONS & UTIL.	95.0	101.5	98.4	95.3	87.7	91.4	94.7	137.9	117.0	89.5	137.1	114.0	92.4	134.8	114.2	94.5	133.3	114.6
MF2010	WEIGHTED AVERAGE	95.3	103.9	99.1	96.9	104.8	100.4	98.6	136.4	115.3	94.6	127.4	109.1	97.0	120.8	107.5	98.7	132.4	113.5

DIVISION		ILLINOIS																	
		PEORIA			QUINCY			ROCK ISLAND			ROCKFORD			SOUTH SUBURBAN			SPRINGFIELD		
		615 - 616			623			612			610 - 611			605			626 - 627		
		MAT.	INST.	TOTAL	MAT.	INST.	TOTAL	MAT.	INST.	TOTAL	MAT.	INST.	TOTAL	MAT.	INST.	TOTAL	MAT.	INST.	TOTAL
015433	CONTRACTOR EQUIPMENT		100.8	100.8		101.6	101.6		100.8	100.8		100.8	100.8		90.9	90.9		101.6	101.6
0241, 31 - 34	SITE & INFRASTRUCTURE, DEMOLITION	100.5	96.8	97.9	97.8	97.3	97.4	98.4	96.1	96.8	99.6	98.0	98.5	110.7	91.8	97.7	99.2	97.8	98.2
0310	Concrete Forming & Accessories	94.2	116.9	113.9	98.0	114.6	112.4	93.1	100.0	99.1	98.2	132.3	127.9	99.5	148.9	142.5	96.5	118.5	115.6
0320	Concrete Reinforcing	95.6	107.3	101.5	98.3	99.8	99.0	98.1	99.8	99.0	90.4	140.0	115.4	99.0	155.2	127.3	95.6	102.9	99.3
0330	Cast-in-Place Concrete	97.7	115.2	104.6	99.8	97.6	99.0	98.6	96.0	97.6	100.1	125.9	110.2	107.8	139.8	120.4	89.9	108.1	97.1
03	CONCRETE	96.3	114.5	105.2	98.0	106.0	101.9	97.1	99.0	98.0	96.9	131.2	113.6	101.1	145.7	122.8	92.5	111.9	101.9
04	MASONRY	114.0	114.4	114.3	99.5	101.1	100.5	113.8	95.6	102.7	88.9	137.3	118.4	98.3	142.4	125.2	77.2	116.5	101.2
05	METALS	99.5	111.0	103.3	96.7	105.3	99.6	96.8	106.1	99.9	99.5	128.8	109.2	94.4	128.7	105.7	102.0	107.9	103.9
06	WOOD, PLASTICS & COMPOSITES	98.8	115.0	108.2	98.7	118.1	109.9	93.3	99.3	96.7	98.8	131.0	117.4	101.5	150.6	129.9	100.1	118.1	110.5
07	THERMAL & MOISTURE PROTECTION	98.8	112.1	104.2	98.2	103.5	100.4	98.1	97.2	97.7	101.3	129.6	112.8	100.0	138.7	115.8	98.4	112.1	104.0
08	OPENINGS	96.1	112.8	100.3	90.8	112.4	96.2	89.4	98.4	91.7	96.1	132.0	105.1	95.8	152.1	110.0	97.1	113.1	101.1
0920	Plaster & Gypsum Board	104.1	115.4	112.2	102.2	118.6	113.9	99.9	99.1	99.4	104.1	131.9	124.0	101.0	152.2	137.6	105.9	118.6	115.0
0950, 0980	Ceilings & Acoustic Treatment	101.2	115.4	110.6	94.9	118.6	110.6	94.9	99.1	97.7	101.2	131.9	121.5	109.9	152.2	137.8	100.3	118.6	112.4
0960	Flooring	104.0	112.2	106.4	104.0	104.0	104.0	101.6	94.7	99.6	104.0	116.3	107.6	97.0	138.1	109.1	104.0	109.0	105.5
0970, 0990	Wall Finishes & Painting/Coating	96.0	124.2	113.4	96.0	108.2	103.5	96.0	97.0	96.6	96.0	131.3	117.8	95.1	151.2	129.7	96.0	108.2	103.5
09	FINISHES	101.8	117.1	110.4	100.1	113.5	107.6	99.1	98.6	98.9	101.8	129.7	117.4	101.2	148.4	127.5	101.5	116.5	109.9
COVERS	DIVS. 10 - 14, 25, 28, 41, 43, 44, 46	100.0	104.7	100.9	100.0	75.5	95.1	100.0	98.1	99.6	100.0	94.4	98.9	100.0	117.0	103.4	100.0	104.7	100.9
21, 22, 23	FIRE SUPPRESSION, PLUMBING & HVAC	99.9	105.7	102.3	94.0	98.3	95.7	94.0	99.4	96.2	100.0	115.6	106.3	100.0	123.2	109.4	99.8	104.8	101.8
26, 27, 3370	ELECTRICAL, COMMUNICATIONS & UTIL.	96.2	97.2	96.7	92.5	81.6	86.9	87.4	98.4	93.1	96.6	131.8	114.8	94.5	133.3	114.6	97.5	92.4	94.9
MF2010	WEIGHTED AVERAGE	99.5	108.3	103.4	96.0	100.1	97.8	95.9	98.9	97.2	98.4	124.3	109.8	98.7	132.4	113.5	97.8	107.1	101.9

DIVISION		INDIANA																	
		ANDERSON 460			BLOOMINGTON 474			COLUMBUS 472			EVANSVILLE 476 - 477			FORT WAYNE 467 - 468			GARY 463 - 464		
		MAT.	INST.	TOTAL	MAT.	INST.	TOTAL	MAT.	INST.	TOTAL	MAT.	INST.	TOTAL	MAT.	INST.	TOTAL	MAT.	INST.	TOTAL
015433	CONTRACTOR EQUIPMENT		96.6	96.6		88.1	88.1		88.1	88.1		116.9	116.9		96.6	96.6		96.6	96.6
0241, 31 - 34	SITE & INFRASTRUCTURE, DEMOLITION	94.9	94.9	94.9	79.4	95.4	90.4	76.4	95.1	89.3	84.4	124.7	112.2	95.4	94.6	94.8	95.4	98.2	97.3
0310	Concrete Forming & Accessories	95.0	81.5	83.3	100.5	82.4	84.8	95.0	79.3	81.4	94.6	80.5	82.3	93.4	74.7	77.1	95.1	110.3	108.3
0320	Concrete Reinforcing	97.9	82.7	90.3	89.6	82.5	86.0	90.0	82.5	86.2	98.1	76.6	87.3	97.9	77.8	87.8	97.9	103.3	100.6
0330	Cast-in-Place Concrete	98.2	79.7	90.9	98.2	80.8	91.3	97.8	74.5	88.6	93.8	87.5	91.3	104.4	75.3	93.0	102.7	107.9	104.7
03	CONCRETE	94.7	81.6	88.3	100.8	81.6	91.4	100.2	78.1	89.4	101.0	82.4	91.9	97.7	76.2	87.2	96.9	108.0	102.3
04	MASONRY	87.4	81.7	83.9	88.8	79.1	82.9	88.6	77.9	82.1	84.5	83.0	83.6	91.2	79.2	83.9	88.7	113.2	103.7
05	METALS	94.2	89.5	92.7	94.6	78.8	89.4	94.7	78.1	89.2	88.3	84.7	87.1	94.2	86.9	91.8	94.2	104.8	97.7
06	WOOD, PLASTICS & COMPOSITES	107.1	81.7	92.4	119.4	82.1	97.9	114.6	78.8	93.9	98.2	79.1	87.1	106.8	74.0	87.8	104.7	108.2	106.8
07	THERMAL & MOISTURE PROTECTION	98.2	80.2	90.9	93.9	82.8	89.3	93.5	82.0	88.8	97.6	84.4	92.2	98.0	82.6	91.7	97.1	109.5	102.1
08	OPENINGS	98.3	82.0	94.2	106.5	82.3	100.4	101.5	80.4	96.2	98.2	78.9	93.3	98.3	74.2	92.2	98.3	111.6	101.7
0920	Plaster & Gypsum Board	99.8	81.6	86.8	97.7	82.2	86.6	94.5	78.8	83.3	93.0	77.8	82.1	98.8	73.6	80.8	92.9	108.9	104.4
0950, 0980	Ceilings & Acoustic Treatment	91.1	81.6	84.8	83.4	82.2	82.6	83.4	78.8	80.3	88.4	77.8	81.4	91.1	73.6	79.5	91.1	108.9	102.9
0960	Flooring	98.3	81.3	93.3	105.9	78.5	97.9	100.9	78.5	94.3	99.1	84.5	94.8	98.3	78.5	92.4	98.3	121.4	105.1
0970, 0990	Wall Finishes & Painting/Coating	98.8	70.0	81.0	91.4	83.3	86.4	91.4	83.3	86.4	96.0	85.8	89.7	98.8	75.2	84.2	98.8	122.6	113.5
09	FINISHES	96.5	80.3	87.5	96.7	81.8	88.4	94.9	79.5	86.3	95.2	81.6	87.6	96.2	75.5	84.7	95.6	113.6	105.7
COVERS	DIVS. 10 - 14, 25, 28, 41, 43, 44, 46	100.0	92.5	98.5	100.0	87.8	97.6	100.0	87.0	97.4	100.0	92.0	98.4	100.0	87.5	97.5	100.0	103.4	100.7
21, 22, 23	FIRE SUPPRESSION, PLUMBING & HVAC	100.0	77.4	90.8	99.8	80.8	92.1	93.8	78.7	87.7	99.9	81.1	92.3	100.0	74.5	89.7	100.0	105.7	102.3
26, 27, 3370	ELECTRICAL, COMMUNICATIONS & UTIL.	88.2	86.3	87.2	99.7	85.7	92.5	98.9	86.3	92.4	95.8	85.6	90.6	89.0	74.1	81.3	100.4	114.3	107.6
MF2010	WEIGHTED AVERAGE	96.1	83.4	90.5	98.4	82.9	91.6	96.0	81.5	89.6	96.0	86.5	91.8	96.7	78.7	88.8	97.6	108.5	102.4

DIVISION		INDIANA																	
		INDIANAPOLIS 461 - 462			KOKOMO 469			LAFAYETTE 479			LAWRENCEBURG 470			MUNCIE 473			NEW ALBANY 471		
		MAT.	INST.	TOTAL	MAT.	INST.	TOTAL	MAT.	INST.	TOTAL	MAT.	INST.	TOTAL	MAT.	INST.	TOTAL	MAT.	INST.	TOTAL
015433	CONTRACTOR EQUIPMENT		92.4	92.4		96.6	96.6		88.1	88.1		105.8	105.8		96.4	96.4		96.0	96.0
0241, 31 - 34	SITE & INFRASTRUCTURE, DEMOLITION	95.1	97.9	97.0	91.5	94.8	93.8	77.3	95.2	89.6	74.4	111.8	100.2	79.2	95.5	90.5	71.8	98.1	90.0
0310	Concrete Forming & Accessories	95.7	85.2	86.6	98.2	77.1	79.9	92.9	82.0	83.5	91.6	75.9	78.0	93.2	81.3	82.9	88.8	73.7	75.7
0320	Concrete Reinforcing	97.3	82.8	90.0	88.6	82.6	85.6	89.6	82.7	86.1	88.9	78.8	83.8	99.1	82.7	90.8	90.2	76.5	83.3
0330	Cast-in-Place Concrete	94.5	86.4	91.3	97.3	84.0	92.0	98.3	83.1	92.3	91.9	73.6	84.7	103.2	79.9	94.0	94.9	72.5	86.1
03	CONCRETE	96.5	84.9	90.8	91.6	81.1	86.5	100.3	82.3	91.6	93.6	76.3	85.1	100.1	81.6	91.1	98.6	74.3	86.8
04	MASONRY	94.6	82.9	87.4	87.1	79.6	82.5	93.5	81.5	86.2	73.6	77.8	76.2	90.7	81.6	85.2	80.7	71.0	74.8
05	METALS	96.3	80.9	91.2	90.8	89.2	90.3	93.1	78.9	88.4	90.0	86.5	88.8	96.5	89.8	94.3	91.8	81.8	88.5
06	WOOD, PLASTICS & COMPOSITES	103.8	85.5	93.2	110.1	75.5	90.1	111.9	81.9	94.5	97.0	75.2	84.4	114.3	81.6	95.4	98.7	74.2	84.6
07	THERMAL & MOISTURE PROTECTION	96.2	84.6	91.5	97.8	79.9	90.5	93.5	83.3	89.3	96.8	79.3	89.6	96.4	80.5	90.0	85.6	70.4	79.4
08	OPENINGS	106.5	84.1	100.8	92.4	78.7	89.0	99.8	82.1	95.4	99.9	76.5	94.0	101.3	81.9	96.4	97.7	76.1	92.3
0920	Plaster & Gypsum Board	97.7	85.3	88.8	105.2	75.2	83.8	91.4	81.9	84.6	72.1	74.9	74.1	93.0	81.6	84.9	91.2	73.7	78.7
0950, 0980	Ceilings & Acoustic Treatment	95.6	85.3	88.8	91.1	75.2	80.6	78.9	81.9	80.9	93.8	74.9	81.3	84.3	81.6	82.5	88.4	73.7	78.7
0960	Flooring	98.1	87.5	94.9	102.4	88.8	98.4	99.7	89.1	96.6	73.7	87.5	77.7	99.4	81.3	94.1	97.3	65.8	88.1
0970, 0990	Wall Finishes & Painting/Coating	98.8	83.3	89.3	98.8	75.6	84.5	91.4	98.1	95.6	92.5	74.4	81.3	91.4	70.0	78.2	96.0	87.2	90.6
09	FINISHES	97.5	85.6	90.9	98.1	78.8	87.3	93.2	85.1	88.7	86.4	78.2	81.8	94.1	80.2	86.3	94.7	74.0	83.1
COVERS	DIVS. 10 - 14, 25, 28, 41, 43, 44, 46	100.0	93.8	98.8	100.0	87.1	97.4	100.0	92.2	98.4	100.0	46.4	89.3	100.0	92.0	98.4	100.0	45.8	89.2
21, 22, 23	FIRE SUPPRESSION, PLUMBING & HVAC	99.9	83.3	93.2	94.0	80.7	88.6	93.8	79.6	88.1	94.5	75.6	86.8	99.8	77.3	90.7	94.0	77.0	87.1
26, 27, 3370	ELECTRICAL, COMMUNICATIONS & UTIL.	102.4	86.3	94.1	92.7	79.2	85.7	98.4	83.4	90.7	93.7	75.6	84.4	91.5	80.3	85.7	94.3	76.4	85.0
MF2010	WEIGHTED AVERAGE	99.2	85.6	93.2	93.7	82.2	88.6	95.7	83.2	90.2	92.5	79.6	86.8	97.1	82.6	90.7	93.6	76.6	86.1

DIVISION		INDIANA									IOWA								
		SOUTH BEND 465 - 466			TERRE HAUTE 478			WASHINGTON 475			BURLINGTON 526			CARROLL 514			CEDAR RAPIDS 522 - 524		
		MAT.	INST.	TOTAL	MAT.	INST.	TOTAL	MAT.	INST.	TOTAL	MAT.	INST.	TOTAL	MAT.	INST.	TOTAL	MAT.	INST.	TOTAL
015433	CONTRACTOR EQUIPMENT		104.8	104.8		116.9	116.9		116.9	116.9		99.8	99.8		99.8	99.8		96.5	96.5
0241, 31 - 34	SITE & INFRASTRUCTURE, DEMOLITION	97.8	95.1	95.9	86.1	125.0	113.0	86.3	122.8	111.5	100.9	96.9	98.2	88.6	97.1	94.4	103.1	96.4	98.5
0310	Concrete Forming & Accessories	98.2	78.8	81.3	95.5	80.5	82.4	96.2	79.3	81.5	99.1	77.1	79.9	88.1	52.7	57.3	104.3	81.6	84.6
0320	Concrete Reinforcing	97.9	77.9	87.8	98.1	82.9	90.5	90.7	51.8	71.1	100.6	82.3	91.4	101.5	81.8	91.5	101.4	84.8	93.1
0330	Cast-in-Place Concrete	95.6	79.5	89.3	90.8	88.9	90.1	98.8	88.2	94.7	111.3	56.9	89.8	111.3	61.8	91.8	111.6	80.4	99.3
03	CONCRETE	89.6	80.2	85.0	103.4	84.1	94.0	108.9	77.3	93.5	104.3	71.9	88.5	103.3	62.7	83.5	104.4	82.4	93.7
04	MASONRY	87.2	80.0	82.8	91.9	82.2	86.0	84.6	83.2	83.7	100.0	68.3	80.7	102.2	74.9	85.6	106.1	80.3	90.3
05	METALS	94.2	99.3	95.9	89.0	88.1	88.7	83.7	70.8	79.5	89.2	91.4	90.0	89.3	91.1	89.9	91.6	93.4	92.2
06	WOOD, PLASTICS & COMPOSITES	104.4	78.5	89.5	100.1	79.4	88.1	100.5	78.9	88.0	97.1	77.9	86.0	85.2	47.7	63.5	102.9	80.6	90.1
07	THERMAL & MOISTURE PROTECTION	92.8	83.8	89.2	97.6	82.7	91.6	97.8	84.5	92.4	106.8	76.9	94.6	107.1	71.5	92.7	107.7	80.3	96.6
08	OPENINGS	91.4	78.1	88.1	98.8	80.8	94.3	95.2	69.1	88.6	91.9	71.2	86.7	97.0	56.1	86.7	98.0	81.7	93.9
0920	Plaster & Gypsum Board	90.8	78.3	81.9	93.0	78.1	82.4	92.5	77.7	81.9	104.2	77.2	84.9	100.1	46.1	61.5	108.9	80.3	88.5
0950, 0980	Ceilings & Acoustic Treatment	85.6	78.3	80.8	88.4	78.1	81.6	82.0	77.7	79.1	109.7	77.2	88.2	109.7	46.1	67.7	113.3	80.3	91.5
0960	Flooring	96.5	83.4	92.7	99.1	87.5	95.7	100.2	80.0	94.3	113.3	39.6	91.7	106.4	34.8	85.4	131.2	82.0	116.8
0970, 0990	Wall Finishes & Painting/Coating	98.3	87.2	91.4	96.0	85.2	89.3	96.0	85.8	89.7	99.0	82.8	89.0	99.0	78.3	86.2	101.0	72.4	83.3
09	FINISHES	93.7	80.5	86.4	95.2	82.1	87.9	94.3	80.7	86.8	107.8	70.8	87.2	103.3	50.3	73.7	114.3	80.6	95.5
COVERS	DIVS. 10 - 14, 25, 28, 41, 43, 44, 46	100.0	71.1	94.2	100.0	94.6	98.9	100.0	92.0	98.4	100.0	87.3	97.5	100.0	69.8	94.0	100.0	88.3	97.7
21, 22, 23	FIRE SUPPRESSION, PLUMBING & HVAC	99.9	77.4	90.8	99.9	79.9	91.8	94.0	80.3	88.4	94.3	77.9	87.7	94.3	71.1	84.9	100.3	81.9	92.8
26, 27, 3370	ELECTRICAL, COMMUNICATIONS & UTIL.	99.7	84.2	91.7	94.0	85.7	89.7	94.5	85.6	89.9	100.8	71.9	85.9	101.5	82.4	91.7	98.4	81.0	89.4
MF2010	WEIGHTED AVERAGE	95.7	82.9	90.1	96.7	86.9	92.4	94.4	83.7	89.7	97.6	77.3	88.6	97.4	72.4	86.4	100.7	84.0	93.3

DIVISION		IOWA																	
		COUNCIL BLUFFS			CRESTON			DAVENPORT			DECORAH			DES MOINES			DUBUQUE		
		515			508			527 - 528			521			500 - 503,509			520		
		MAT.	INST.	TOTAL	MAT.	INST.	TOTAL	MAT.	INST.	TOTAL	MAT.	INST.	TOTAL	MAT.	INST.	TOTAL	MAT.	INST.	TOTAL
015433	CONTRACTOR EQUIPMENT		96.0	96.0		99.8	99.8		99.8	99.8		99.8	99.8		101.4	101.4		95.3	95.3
0241, 31 - 34	SITE & INFRASTRUCTURE, DEMOLITION	111.3	92.7	98.5	88.5	96.6	94.1	100.8	99.4	99.8	99.4	96.5	97.4	89.8	100.2	96.9	100.9	93.6	95.9
0310	Concrete Forming & Accessories	87.6	72.0	74.0	88.6	68.7	71.3	103.7	88.0	90.1	96.8	46.5	53.0	105.2	78.3	81.8	88.8	70.9	73.2
0320	Concrete Reinforcing	103.5	77.8	90.5	100.7	81.8	91.2	101.4	95.8	98.6	100.6	81.3	90.9	101.4	82.0	91.7	100.0	84.7	92.3
0330	Cast-in-Place Concrete	116.0	76.6	100.5	110.9	65.8	93.2	107.4	86.9	99.3	108.3	58.5	88.7	93.5	90.5	92.3	109.2	87.7	100.8
03	CONCRETE	107.0	75.5	91.7	103.0	71.1	87.5	102.3	89.7	96.2	102.2	58.6	81.0	94.4	84.0	89.3	101.3	80.2	91.0
04	MASONRY	107.6	75.8	88.2	102.2	82.0	89.8	102.7	85.8	92.4	121.3	71.1	90.7	96.6	83.1	88.3	107.0	76.6	88.5
05	METALS	96.9	89.9	94.6	89.2	91.4	89.9	91.6	101.5	94.9	89.4	89.3	89.3	91.7	93.6	92.4	90.3	92.9	91.1
06	WOOD, PLASTICS & COMPOSITES	84.2	71.4	76.8	85.7	65.5	74.0	102.9	87.0	93.8	94.3	39.3	62.6	102.7	76.1	87.4	85.7	67.2	75.0
07	THERMAL & MOISTURE PROTECTION	107.2	69.8	92.0	107.9	78.9	96.1	107.2	85.9	98.6	107.0	54.9	85.8	108.6	80.1	97.0	107.5	72.1	93.1
08	OPENINGS	97.0	76.6	91.9	109.5	66.2	98.6	98.0	89.0	95.7	95.3	51.6	84.3	98.0	82.1	94.0	97.0	76.1	91.7
0920	Plaster & Gypsum Board	100.1	70.8	79.2	100.1	64.5	74.7	108.9	86.7	93.0	102.8	37.5	56.2	97.7	75.4	81.8	100.1	66.4	76.1
0950, 0980	Ceilings & Acoustic Treatment	109.7	70.8	84.0	109.7	64.5	79.8	113.3	86.7	95.7	109.7	37.5	62.0	110.6	75.4	87.4	109.7	66.4	81.1
0960	Flooring	105.4	84.5	99.3	106.6	34.8	85.5	116.0	86.9	107.5	112.8	49.4	94.2	115.6	84.5	106.5	120.5	79.7	108.5
0970, 0990	Wall Finishes & Painting/Coating	96.0	67.5	78.4	99.0	78.3	86.2	99.0	88.6	92.6	99.0	34.7	59.3	99.0	82.5	88.8	100.0	67.2	79.8
09	FINISHES	104.6	73.5	87.3	103.3	62.6	80.6	109.8	87.7	97.5	107.4	44.0	72.0	106.3	79.5	91.3	109.1	71.2	88.0
COVERS	DIVS. 10 - 14, 25, 28, 41, 43, 44, 46	100.0	81.7	96.4	100.0	74.2	94.9	100.0	90.2	98.1	100.0	81.9	96.4	100.0	88.2	97.7	100.0	86.3	97.3
21, 22, 23	FIRE SUPPRESSION, PLUMBING & HVAC	100.3	74.6	89.9	94.3	80.7	88.8	100.3	89.6	96.0	94.3	73.3	85.8	100.2	81.3	92.5	100.3	76.4	90.6
26, 27, 3370	ELECTRICAL, COMMUNICATIONS & UTIL.	103.6	82.0	92.5	96.3	82.4	89.1	95.3	86.7	90.8	98.4	47.1	71.9	107.5	82.5	94.6	102.2	79.1	90.3
MF2010	WEIGHTED AVERAGE	101.7	78.8	91.6	98.1	78.8	89.6	99.6	90.4	95.5	98.5	65.7	84.0	99.1	84.7	92.7	99.9	79.7	91.0

DIVISION		IOWA																	
		FORT DODGE			MASON CITY			OTTUMWA			SHENANDOAH			SIBLEY			SIOUX CITY		
		505			504			525			516			512			510 - 511		
		MAT.	INST.	TOTAL	MAT.	INST.	TOTAL	MAT.	INST.	TOTAL	MAT.	INST.	TOTAL	MAT.	INST.	TOTAL	MAT.	INST.	TOTAL
015433	CONTRACTOR EQUIPMENT		99.8	99.8		99.8	99.8		95.3	95.3		96.0	96.0		99.8	99.8		99.8	99.8
0241, 31 - 34	SITE & INFRASTRUCTURE, DEMOLITION	96.4	95.4	95.7	96.4	96.5	96.5	101.7	91.7	94.8	109.3	91.8	97.2	112.0	95.3	100.4	113.2	96.1	101.4
0310	Concrete Forming & Accessories	89.2	47.5	52.9	93.5	47.2	53.2	95.0	59.3	64.0	89.1	59.4	63.2	89.4	39.7	46.1	104.3	65.8	70.8
0320	Concrete Reinforcing	100.7	64.4	82.4	100.6	81.3	90.9	100.6	82.6	91.5	103.5	68.7	85.9	103.5	64.3	83.7	101.4	64.7	82.9
0330	Cast-in-Place Concrete	104.1	46.1	81.2	104.1	59.1	86.4	112.0	56.3	90.1	112.2	62.1	92.4	109.9	46.5	84.9	110.6	57.3	89.6
03	CONCRETE	98.4	51.7	75.6	98.6	59.2	79.4	104.0	63.9	84.4	104.3	63.2	84.3	103.2	48.3	76.5	103.8	63.7	84.2
04	MASONRY	100.7	40.0	63.7	114.1	69.9	87.1	103.5	59.4	76.6	107.1	75.8	88.0	126.0	40.3	73.7	99.9	56.3	73.3
05	METALS	89.3	81.0	86.6	89.3	89.6	89.4	89.2	91.1	89.8	95.9	84.7	92.2	89.5	79.8	86.3	91.6	82.5	88.6
06	WOOD, PLASTICS & COMPOSITES	86.1	48.0	64.1	90.1	39.3	60.8	92.0	60.8	74.0	85.7	56.1	68.6	86.4	38.0	58.5	102.9	65.9	81.5
07	THERMAL & MOISTURE PROTECTION	107.3	59.7	87.9	106.8	65.1	89.8	107.6	64.5	90.1	106.5	68.1	90.9	106.8	50.6	84.0	107.2	65.2	90.2
08	OPENINGS	102.2	50.4	89.2	92.8	51.5	82.4	97.0	66.2	89.2	86.8	56.9	79.3	93.1	45.1	81.0	98.0	62.0	88.9
0920	Plaster & Gypsum Board	100.1	46.4	61.7	100.1	37.5	55.4	101.0	59.9	71.6	100.1	55.0	67.9	100.1	36.2	54.4	108.9	64.9	77.4
0950, 0980	Ceilings & Acoustic Treatment	109.7	46.4	67.9	109.7	37.5	62.0	109.7	59.9	76.8	109.7	55.0	73.6	109.7	36.2	61.1	113.3	64.9	81.3
0960	Flooring	108.5	49.4	91.1	111.1	49.4	93.0	124.0	52.5	103.1	106.3	35.9	85.7	107.4	35.6	86.3	116.3	56.2	98.7
0970, 0990	Wall Finishes & Painting/Coating	99.0	62.0	76.2	99.0	31.7	57.5	100.0	81.2	88.4	96.0	67.5	78.4	99.0	62.0	76.2	99.5	62.0	76.4
09	FINISHES	105.5	48.4	73.7	106.3	43.9	71.5	110.4	59.5	82.0	104.7	54.8	76.9	107.4	39.9	69.8	111.9	63.5	84.9
COVERS	DIVS. 10 - 14, 25, 28, 41, 43, 44, 46	100.0	77.8	95.6	100.0	82.3	96.5	100.0	80.1	96.0	100.0	71.0	94.2	100.0	76.8	95.4	100.0	84.2	96.9
21, 22, 23	FIRE SUPPRESSION, PLUMBING & HVAC	94.3	66.8	83.2	94.3	74.4	86.2	94.3	72.6	85.5	94.3	72.1	85.3	94.3	62.7	81.5	100.3	69.3	87.7
26, 27, 3370	ELECTRICAL, COMMUNICATIONS & UTIL.	103.2	44.5	72.9	102.2	59.1	80.0	100.6	67.9	83.7	98.4	82.0	89.9	98.4	44.0	70.3	98.4	74.4	86.0
MF2010	WEIGHTED AVERAGE	97.9	59.7	81.0	97.6	67.8	84.5	98.4	70.7	86.2	98.2	72.3	86.8	98.9	56.5	80.2	100.5	70.7	87.3

DIVISION		IOWA						KANSAS											
		SPENCER			WATERLOO			BELLEVILLE			COLBY			DODGE CITY			EMPORIA		
		513			506 - 507			669			677			678			668		
		MAT.	INST.	TOTAL	MAT.	INST.	TOTAL	MAT.	INST.	TOTAL	MAT.	INST.	TOTAL	MAT.	INST.	TOTAL	MAT.	INST.	TOTAL
015433	CONTRACTOR EQUIPMENT		99.8	99.8		99.8	99.8		102.5	102.5		102.5	102.5		102.5	102.5		100.8	100.8
0241, 31 - 34	SITE & INFRASTRUCTURE, DEMOLITION	112.0	95.3	100.5	101.2	96.5	98.0	115.8	93.8	100.6	113.1	93.8	99.8	115.5	93.4	100.3	106.5	91.6	96.2
0310	Concrete Forming & Accessories	94.9	39.7	46.9	104.6	49.7	56.8	94.1	56.6	61.5	100.0	61.1	66.2	93.6	61.0	65.3	86.5	68.5	70.8
0320	Concrete Reinforcing	103.5	64.3	83.7	101.4	84.5	92.9	103.5	55.5	79.3	106.1	55.5	80.6	103.5	55.6	79.4	102.0	56.1	78.9
0330	Cast-in-Place Concrete	109.9	46.5	84.9	111.6	56.5	89.9	118.6	57.9	94.7	114.0	57.9	91.9	116.0	57.6	93.0	114.6	48.5	88.6
03	CONCRETE	103.6	48.3	76.6	104.5	60.0	82.8	116.2	58.2	88.0	113.4	60.2	87.5	114.6	60.0	88.0	109.0	60.4	85.3
04	MASONRY	126.0	40.3	73.7	101.4	75.9	85.8	97.3	58.4	73.6	98.4	58.4	74.0	106.5	60.9	78.7	102.8	68.7	82.0
05	METALS	89.4	79.8	86.3	91.6	92.4	91.9	95.0	78.6	89.5	95.3	78.5	89.8	96.8	77.5	90.4	94.7	79.8	89.8
06	WOOD, PLASTICS & COMPOSITES	91.9	38.0	60.8	103.5	39.3	66.4	93.4	55.4	71.4	100.1	61.5	77.8	92.7	61.5	74.7	85.8	70.8	77.2
07	THERMAL & MOISTURE PROTECTION	107.7	50.6	84.5	107.0	72.1	92.8	99.5	59.3	83.1	99.4	59.9	83.4	99.4	60.9	83.8	98.0	74.3	88.4
08	OPENINGS	106.2	45.1	90.8	93.8	54.1	83.8	95.8	50.6	84.5	95.9	53.9	85.4	95.9	53.9	85.3	93.4	59.0	84.8
0920	Plaster & Gypsum Board	101.0	36.2	54.7	108.9	37.5	57.9	101.7	53.9	67.6	106.3	60.2	73.4	100.8	60.2	71.8	98.5	69.8	78.0
0950, 0980	Ceilings & Acoustic Treatment	109.7	36.2	61.1	113.3	37.5	63.2	94.9	53.9	67.8	94.9	60.2	72.0	94.9	60.2	72.0	94.9	69.8	78.4
0960	Flooring	110.6	35.6	88.6	117.7	79.0	106.4	103.6	41.0	85.2	107.3	41.0	87.8	103.4	41.0	85.1	98.6	38.1	80.8
0970, 0990	Wall Finishes & Painting/Coating	99.0	62.0	76.2	99.5	82.0	88.7	96.0	41.0	62.1	96.0	41.0	62.1	96.0	41.0	62.1	96.0	41.0	62.1
09	FINISHES	108.5	39.9	70.2	110.4	56.3	80.2	101.5	51.9	73.8	102.9	55.5	76.5	101.3	55.5	75.7	98.5	60.4	77.3
COVERS	DIVS. 10 - 14, 25, 28, 41, 43, 44, 46	100.0	76.8	95.4	100.0	83.6	96.7	100.0	43.6	88.8	100.0	44.2	88.9	100.0	44.3	88.9	100.0	43.3	88.7
21, 22, 23	FIRE SUPPRESSION, PLUMBING & HVAC	94.3	62.7	81.5	100.3	81.2	92.6	94.0	70.8	84.6	94.0	67.4	83.2	99.9	67.4	86.8	94.0	71.0	84.7
26, 27, 3370	ELECTRICAL, COMMUNICATIONS & UTIL.	100.1	44.0	71.2	98.4	59.1	78.1	99.9	67.3	83.1	99.9	67.3	83.1	96.7	73.7	84.8	97.1	73.6	84.9
MF2010	WEIGHTED AVERAGE	100.6	56.5	81.2	99.7	72.2	87.6	99.6	65.5	84.6	99.5	65.8	84.6	101.2	66.8	86.0	97.9	69.7	85.4

		KANSAS																	
DIVISION		FORT SCOTT			HAYS			HUTCHINSON			INDEPENDENCE			KANSAS CITY			LIBERAL		
		667			676			675			673			660 - 662			679		
		MAT.	INST.	TOTAL	MAT.	INST.	TOTAL	MAT.	INST.	TOTAL	MAT.	INST.	TOTAL	MAT.	INST.	TOTAL	MAT.	INST.	TOTAL
015433	CONTRACTOR EQUIPMENT		101.7	101.7		102.5	102.5		102.5	102.5		102.5	102.5		99.2	99.2		102.5	102.5
0241, 31 - 34	SITE & INFRASTRUCTURE, DEMOLITION	102.8	92.0	95.4	119.6	93.8	101.8	95.2	93.9	94.3	116.8	94.0	101.0	95.0	91.8	92.8	119.8	93.9	102.0
0310	Concrete Forming & Accessories	100.7	80.7	83.3	97.7	61.1	65.9	88.8	61.0	64.6	107.8	69.5	74.5	97.6	97.1	97.2	94.1	61.0	65.3
0320	Concrete Reinforcing	101.4	89.1	95.2	103.5	55.5	79.3	103.5	55.6	79.3	102.9	60.5	81.5	98.3	94.4	96.3	105.0	55.6	80.1
0330	Cast-in-Place Concrete	106.3	56.7	86.7	89.9	57.9	77.3	83.3	47.9	69.4	116.5	48.2	89.6	91.1	96.1	93.0	89.9	47.9	73.4
03	CONCRETE	103.9	74.6	89.7	104.1	60.2	82.7	87.8	56.7	72.6	115.8	61.4	89.3	95.6	96.5	96.0	106.1	56.7	82.0
04	MASONRY	104.0	59.4	76.8	106.7	58.4	77.3	97.7	60.6	75.1	95.7	63.2	75.9	106.0	97.5	100.8	104.3	60.6	77.6
05	METALS	94.6	91.1	93.5	94.9	78.5	89.5	94.7	77.4	89.0	94.6	79.9	89.8	102.1	98.7	101.0	95.2	77.4	89.3
06	WOOD, PLASTICS & COMPOSITES	102.0	88.8	94.4	97.5	61.5	76.7	88.1	61.5	72.7	109.2	72.3	87.9	98.3	97.7	97.9	93.4	61.5	75.0
07	THERMAL & MOISTURE PROTECTION	98.8	71.0	87.5	99.7	59.9	83.6	98.4	60.8	83.1	99.5	72.9	88.7	98.2	97.3	97.8	99.9	60.8	84.0
08	OPENINGS	93.4	80.6	90.2	95.8	53.9	85.3	95.8	53.9	85.3	93.4	61.0	85.3	94.8	90.8	93.8	95.9	53.9	85.3
0920	Plaster & Gypsum Board	104.5	88.4	93.0	104.5	60.2	72.9	99.9	60.2	71.6	112.2	71.4	83.0	96.2	97.5	97.1	101.7	60.2	72.1
0950, 0980	Ceilings & Acoustic Treatment	94.9	88.4	90.6	94.9	60.2	72.0	94.9	60.2	72.0	94.9	71.4	79.4	94.9	97.5	96.6	94.9	60.2	72.0
0960	Flooring	115.0	41.3	93.4	106.1	41.0	87.0	100.4	41.0	83.0	112.0	41.0	91.2	91.6	98.3	93.6	103.6	41.0	85.2
0970, 0990	Wall Finishes & Painting/Coating	98.0	41.0	62.8	96.0	41.0	62.1	96.0	41.0	62.1	96.0	41.0	62.1	104.0	71.4	83.9	96.0	41.0	62.1
09	FINISHES	104.2	70.7	85.5	103.0	55.5	76.5	98.4	55.5	74.5	105.2	61.9	81.0	98.4	93.7	95.8	102.3	55.5	76.2
COVERS	DIVS. 10 - 14, 25, 28, 41, 43, 44, 46	100.0	49.7	90.0	100.0	44.2	88.9	100.0	44.3	88.9	100.0	45.5	89.2	100.0	66.4	93.3	100.0	44.3	88.9
21, 22, 23	FIRE SUPPRESSION, PLUMBING & HVAC	94.0	69.1	83.9	94.0	67.4	83.2	94.0	67.4	83.2	94.0	69.4	84.0	99.9	99.0	99.5	94.0	67.4	83.2
26, 27, 3370	ELECTRICAL, COMMUNICATIONS & UTIL.	96.3	73.6	84.6	98.8	67.3	82.5	93.7	67.3	80.1	95.9	73.6	84.4	102.3	95.9	99.0	96.7	69.6	82.7
MF2010	WEIGHTED AVERAGE	97.7	73.9	87.2	98.8	65.8	84.3	94.8	65.4	81.8	99.2	69.5	86.1	99.5	95.4	97.7	98.7	65.8	84.2

		KANSAS									KENTUCKY								
DIVISION		SALINA			TOPEKA			WICHITA			ASHLAND			BOWLING GREEN			CAMPTON		
		674			664 - 666			670 - 672			411 - 412			421 - 422			413 - 414		
		MAT.	INST.	TOTAL	MAT.	INST.	TOTAL	MAT.	INST.	TOTAL	MAT.	INST.	TOTAL	MAT.	INST.	TOTAL	MAT.	INST.	TOTAL
015433	CONTRACTOR EQUIPMENT		102.5	102.5		100.8	100.8		102.5	102.5		98.5	98.5		96.0	96.0		102.2	102.2
0241, 31 - 34	SITE & INFRASTRUCTURE, DEMOLITION	105.0	93.8	97.3	97.0	90.7	92.6	96.9	92.6	93.9	106.0	87.7	93.4	71.9	98.8	90.5	80.4	100.0	93.9
0310	Concrete Forming & Accessories	90.7	57.0	61.4	97.4	42.8	49.9	96.6	46.4	53.0	89.5	100.8	99.4	85.3	80.1	80.8	90.0	87.7	88.0
0320	Concrete Reinforcing	102.9	64.3	83.5	95.6	96.6	96.1	95.6	75.8	85.6	91.7	106.6	99.2	88.9	89.0	88.9	89.8	105.6	97.8
0330	Cast-in-Place Concrete	100.7	51.5	81.3	90.6	47.7	73.7	88.6	50.6	73.6	86.4	101.2	92.2	85.7	99.8	91.2	95.6	63.7	83.0
03	CONCRETE	101.7	57.8	80.3	92.9	56.2	75.0	91.8	54.8	73.8	94.5	102.6	98.4	92.7	88.7	90.8	96.8	82.8	90.0
04	MASONRY	121.5	58.4	83.0	100.2	56.8	73.7	93.8	53.1	69.0	90.6	104.8	99.3	92.4	73.6	80.9	90.3	73.0	79.7
05	METALS	96.6	82.2	91.8	102.0	96.1	100.0	102.0	84.1	96.1	90.9	110.1	97.3	92.5	86.6	90.6	91.8	94.2	92.6
06	WOOD, PLASTICS & COMPOSITES	89.5	55.4	69.8	94.5	39.6	62.8	96.0	45.2	66.6	80.0	99.3	91.2	93.9	78.7	85.1	92.6	94.7	93.8
07	THERMAL & MOISTURE PROTECTION	99.0	59.3	82.8	100.0	63.3	85.0	98.3	54.6	80.6	89.9	98.9	93.6	85.5	81.3	83.8	97.9	72.3	87.5
08	OPENINGS	95.8	51.5	84.6	96.1	56.5	86.1	96.1	53.0	85.2	96.9	96.7	96.9	97.8	77.0	92.5	99.5	92.4	97.7
0920	Plaster & Gypsum Board	99.9	53.9	67.1	98.6	37.6	55.0	98.3	43.3	59.0	59.3	99.3	87.9	87.1	78.4	80.8	87.1	93.9	91.9
0950, 0980	Ceilings & Acoustic Treatment	94.9	53.9	67.8	95.8	37.6	57.4	94.9	43.3	60.9	83.3	99.3	93.9	88.4	78.4	81.7	88.4	93.9	92.0
0960	Flooring	101.7	64.9	90.9	105.0	77.2	96.8	103.8	66.0	92.7	78.5	104.9	86.3	95.1	90.1	93.6	96.5	43.2	80.9
0970, 0990	Wall Finishes & Painting/Coating	96.0	41.0	62.1	96.0	55.4	70.9	96.0	50.6	67.9	98.3	95.6	96.6	96.0	69.9	79.9	96.0	69.0	79.3
09	FINISHES	99.8	56.5	75.6	100.0	48.6	71.4	99.6	49.9	71.9	82.5	101.1	92.9	93.4	80.8	86.4	94.1	78.4	85.4
COVERS	DIVS. 10 - 14, 25, 28, 41, 43, 44, 46	100.0	54.1	90.9	100.0	51.0	90.2	100.0	50.3	90.1	100.0	99.4	99.9	100.0	61.4	92.3	100.0	60.3	92.1
21, 22, 23	FIRE SUPPRESSION, PLUMBING & HVAC	99.9	67.5	86.8	99.9	57.4	82.7	99.7	63.6	85.1	93.9	91.3	92.9	99.9	83.9	93.5	94.1	80.9	88.7
26, 27, 3370	ELECTRICAL, COMMUNICATIONS & UTIL.	96.4	73.7	84.7	102.0	73.6	87.4	99.9	73.7	86.4	92.2	97.3	94.8	94.8	83.7	89.1	92.2	70.3	80.9
MF2010	WEIGHTED AVERAGE	100.0	66.9	85.4	99.2	64.5	83.9	98.4	64.1	83.3	93.3	98.5	95.6	95.0	83.5	90.0	94.5	81.2	88.6

		KENTUCKY																	
DIVISION		CORBIN			COVINGTON			ELIZABETHTOWN			FRANKFORT			HAZARD			HENDERSON		
		407 - 409			410			427			406			417 - 418			424		
		MAT.	INST.	TOTAL	MAT.	INST.	TOTAL	MAT.	INST.	TOTAL	MAT.	INST.	TOTAL	MAT.	INST.	TOTAL	MAT.	INST.	TOTAL
015433	CONTRACTOR EQUIPMENT		102.2	102.2		105.8	105.8		96.0	96.0		102.2	102.2		102.2	102.2		116.9	116.9
0241, 31 - 34	SITE & INFRASTRUCTURE, DEMOLITION	77.8	99.7	92.9	75.9	113.8	102.1	67.0	98.7	88.8	82.9	100.7	95.2	78.3	101.4	94.2	75.4	125.3	109.8
0310	Concrete Forming & Accessories	87.3	69.2	71.5	86.1	96.4	95.1	81.0	79.1	79.3	97.1	75.2	78.1	87.3	88.4	88.3	92.9	83.1	84.4
0320	Concrete Reinforcing	89.4	68.8	79.0	88.4	95.8	92.2	89.4	92.5	91.0	90.7	92.5	91.6	90.2	105.6	98.0	89.1	90.0	89.5
0330	Cast-in-Place Concrete	90.5	61.2	78.9	91.5	94.0	92.5	77.5	76.0	76.9	87.5	71.7	81.3	91.9	87.2	90.1	75.7	89.4	81.1
03	CONCRETE	92.1	66.9	79.8	95.1	95.5	95.3	85.2	80.8	83.0	91.4	77.6	84.7	93.7	91.2	92.5	89.7	86.7	88.2
04	MASONRY	90.5	65.5	75.2	103.6	107.4	105.9	77.0	79.6	78.6	87.4	81.8	84.0	87.7	75.7	80.4	96.3	92.5	93.9
05	METALS	91.7	78.8	87.4	89.9	93.5	91.1	91.7	88.5	90.6	98.7	89.4	95.6	91.8	94.4	92.7	83.5	89.6	85.5
06	WOOD, PLASTICS & COMPOSITES	89.8	73.1	80.1	90.9	89.5	90.1	89.5	78.7	83.3	101.2	73.1	84.9	89.8	94.7	92.6	96.2	80.6	87.2
07	THERMAL & MOISTURE PROTECTION	97.7	67.5	85.4	97.0	97.7	97.3	85.2	78.2	82.4	97.6	76.9	89.2	97.8	71.0	86.9	97.3	93.0	95.5
08	OPENINGS	98.7	64.1	90.0	101.0	88.3	97.8	97.8	82.7	94.0	97.9	79.5	93.3	99.9	82.6	95.5	95.7	83.0	92.5
0920	Plaster & Gypsum Board	86.2	71.6	75.8	69.3	89.6	83.8	86.2	78.4	80.6	97.9	71.6	79.1	86.2	93.9	91.7	89.3	79.4	82.2
0950, 0980	Ceilings & Acoustic Treatment	88.4	71.6	77.3	92.0	89.6	90.4	88.1	78.4	81.7	91.1	71.6	78.2	88.4	93.9	92.0	82.0	79.4	80.3
0960	Flooring	94.9	43.2	79.7	71.3	88.1	76.2	92.5	90.1	91.8	99.5	57.0	87.0	94.9	44.0	79.9	98.3	90.1	95.9
0970, 0990	Wall Finishes & Painting/Coating	96.0	52.1	68.9	92.5	85.5	88.2	96.0	78.3	85.1	96.0	84.1	88.7	96.0	69.0	79.3	96.0	93.8	94.6
09	FINISHES	93.2	62.8	76.3	85.2	93.5	89.9	92.2	80.3	85.6	96.9	72.3	83.2	93.3	79.3	85.5	92.6	85.4	88.6
COVERS	DIVS. 10 - 14, 25, 28, 41, 43, 44, 46	100.0	51.5	90.3	100.0	104.5	100.9	100.0	90.4	98.1	100.0	70.1	94.1	100.0	61.2	92.3	100.0	69.6	93.9
21, 22, 23	FIRE SUPPRESSION, PLUMBING & HVAC	94.0	69.6	84.1	94.5	93.1	93.9	94.4	82.4	89.5	100.0	83.6	93.4	94.1	81.8	89.1	94.4	78.2	87.8
26, 27, 3370	ELECTRICAL, COMMUNICATIONS & UTIL.	92.4	83.7	87.9	96.0	80.3	87.9	92.0	83.7	87.7	100.0	92.6	96.2	92.1	62.0	76.6	94.2	83.9	88.9
MF2010	WEIGHTED AVERAGE	93.7	72.5	84.4	94.4	95.2	94.8	91.4	83.9	88.1	97.2	83.5	91.2	93.9	81.5	88.4	92.4	87.9	90.4

DIVISION		KENTUCKY																	
		LEXINGTON			LOUISVILLE			OWENSBORO			PADUCAH			PIKEVILLE			SOMERSET		
		403 - 405			400 - 402			423			420			415 - 416			425 - 426		
		MAT.	INST.	TOTAL	MAT.	INST.	TOTAL	MAT.	INST.	TOTAL	MAT.	INST.	TOTAL	MAT.	INST.	TOTAL	MAT.	INST.	TOTAL
015433	CONTRACTOR EQUIPMENT		102.2	102.2		96.0	96.0		116.9	116.9		116.9	116.9		98.5	98.5		102.2	102.2
0241, 31 - 34	SITE & INFRASTRUCTURE, DEMOLITION	79.5	101.7	94.8	69.8	98.7	89.7	84.6	125.2	112.6	78.1	124.8	110.3	116.7	87.0	96.2	71.7	100.3	91.4
0310	Concrete Forming & Accessories	98.1	76.1	79.0	93.5	79.1	80.9	91.5	82.6	83.8	89.6	79.3	80.6	97.3	94.3	94.7	87.9	73.9	75.7
0320	Concrete Reinforcing	98.1	92.0	95.0	98.1	92.5	95.3	89.1	89.7	89.4	89.6	86.1	87.8	92.2	106.5	99.4	89.4	91.8	90.6
0330	Cast-in-Place Concrete	92.5	86.7	90.3	94.5	75.9	87.2	88.3	98.0	92.1	80.7	87.5	83.4	94.9	98.7	96.4	75.7	102.8	86.4
03	CONCRETE	95.2	83.1	89.3	95.9	80.7	88.5	101.3	89.4	95.5	94.0	83.6	89.0	107.4	98.7	103.2	80.6	87.5	84.0
04	MASONRY	88.7	70.1	77.4	89.9	79.4	83.5	88.8	83.7	85.7	91.8	91.6	91.7	88.1	101.0	96.0	83.0	66.6	73.0
05	METALS	94.1	89.2	92.5	100.9	88.6	96.9	85.0	89.3	86.4	82.0	87.6	83.8	90.9	108.9	96.8	91.7	88.0	90.5
06	WOOD, PLASTICS & COMPOSITES	103.7	73.1	86.0	101.2	78.7	88.2	94.4	80.6	86.4	92.1	77.7	83.8	88.4	94.0	91.6	90.3	73.1	80.3
07	THERMAL & MOISTURE PROTECTION	97.9	86.3	93.2	92.0	78.2	86.4	97.6	88.8	94.0	97.3	81.8	91.0	90.5	84.0	87.9	97.3	67.5	85.2
08	OPENINGS	98.8	77.0	93.3	98.8	82.7	94.8	95.7	83.8	92.7	94.8	77.4	90.4	97.8	90.9	96.0	98.7	77.3	93.3
0920	Plaster & Gypsum Board	95.9	71.6	78.5	89.0	78.4	81.4	87.9	79.4	81.8	86.5	76.4	79.3	62.0	93.9	84.8	86.2	71.6	75.8
0950, 0980	Ceilings & Acoustic Treatment	93.8	71.6	79.1	91.1	78.4	82.7	82.0	79.4	80.3	82.0	76.4	78.3	83.3	93.9	90.3	88.4	71.6	77.3
0960	Flooring	99.9	73.6	92.2	98.3	90.1	95.9	97.7	90.1	95.4	96.3	62.2	86.3	82.9	104.9	89.3	95.1	43.2	79.9
0970, 0990	Wall Finishes & Painting/Coating	96.0	84.1	88.7	96.0	78.3	85.1	96.0	97.0	96.6	96.0	81.0	86.7	98.3	85.6	90.5	96.0	78.3	85.1
09	FINISHES	97.1	75.7	85.2	94.7	80.9	87.0	92.9	85.2	88.6	91.9	76.0	83.1	85.0	95.9	91.1	92.6	68.3	79.1
COVERS	DIVS. 10 - 14, 25, 28, 41, 43, 44, 46	100.0	92.6	98.5	100.0	90.3	98.1	100.0	98.7	99.7	100.0	65.5	93.1	100.0	61.0	92.2	100.0	61.8	92.4
21, 22, 23	FIRE SUPPRESSION, PLUMBING & HVAC	99.9	76.7	90.5	100.0	82.3	92.8	99.9	77.2	90.7	94.4	82.5	89.6	93.9	88.8	91.8	94.4	75.1	86.6
26, 27, 3370	ELECTRICAL, COMMUNICATIONS & UTIL.	95.3	83.7	89.3	102.7	80.8	91.4	94.3	83.9	88.9	96.6	80.5	88.3	95.2	78.0	86.3	92.6	83.7	88.0
MF2010	WEIGHTED AVERAGE	96.5	81.8	90.0	97.8	83.5	91.5	95.2	87.9	92.0	92.6	85.7	89.6	95.6	91.7	93.9	91.9	79.1	86.2

DIVISION		LOUISIANA																	
		ALEXANDRIA			BATON ROUGE			HAMMOND			LAFAYETTE			LAKE CHARLES			MONROE		
		713 - 714			707 - 708			704			705			706			712		
		MAT.	INST.	TOTAL	MAT.	INST.	TOTAL	MAT.	INST.	TOTAL	MAT.	INST.	TOTAL	MAT.	INST.	TOTAL	MAT.	INST.	TOTAL
015433	CONTRACTOR EQUIPMENT		88.9	88.9		88.7	88.7		89.4	89.4		89.4	89.4		88.7	88.7		88.9	88.9
0241, 31 - 34	SITE & INFRASTRUCTURE, DEMOLITION	96.1	85.7	88.9	104.0	85.2	91.1	105.3	86.5	92.3	106.0	86.4	92.5	106.7	85.0	91.8	96.1	85.0	88.5
0310	Concrete Forming & Accessories	82.5	36.4	42.4	93.8	62.0	66.1	79.3	48.7	52.7	95.1	53.7	59.1	96.6	57.3	62.4	82.1	36.7	42.6
0320	Concrete Reinforcing	98.0	57.5	77.6	99.0	59.3	79.0	97.6	61.1	79.2	99.0	59.3	79.0	99.0	58.2	78.5	96.9	57.3	76.9
0330	Cast-in-Place Concrete	95.8	39.4	73.6	94.8	57.7	80.2	97.6	45.1	76.9	97.1	45.7	76.9	102.2	66.5	88.2	95.8	55.8	80.1
03	CONCRETE	95.1	42.8	69.6	95.1	60.5	78.2	95.4	50.7	73.6	96.3	52.8	75.1	99.0	61.3	80.6	94.9	48.4	72.3
04	MASONRY	118.1	49.3	76.2	106.1	51.1	72.5	105.6	52.8	73.4	105.6	50.3	71.8	104.6	56.7	75.4	113.4	45.8	72.1
05	METALS	89.5	71.8	83.7	101.3	70.2	91.1	97.0	70.9	88.4	96.1	70.4	87.6	96.1	70.1	87.5	89.5	69.5	82.9
06	WOOD, PLASTICS & COMPOSITES	86.7	35.1	56.9	100.8	67.8	81.8	86.7	49.9	65.5	107.0	57.0	78.1	105.2	61.1	79.7	86.0	35.5	56.8
07	THERMAL & MOISTURE PROTECTION	97.5	50.9	78.6	95.5	57.9	80.2	97.2	58.8	81.6	97.7	56.4	80.9	96.9	58.9	81.5	97.5	50.1	78.2
08	OPENINGS	98.2	41.5	83.9	101.1	59.6	90.7	96.8	54.3	86.1	101.1	53.8	89.2	101.1	56.8	90.0	98.2	45.4	84.9
0920	Plaster & Gypsum Board	76.2	33.6	45.8	95.1	67.2	75.2	89.4	48.7	60.3	98.8	56.0	68.3	98.8	60.2	71.3	75.7	34.0	45.9
0950, 0980	Ceilings & Acoustic Treatment	91.1	33.6	53.1	102.2	67.2	79.1	99.3	48.7	65.9	97.5	56.0	70.1	98.5	60.2	73.2	91.1	34.0	53.3
0960	Flooring	101.3	66.9	91.2	102.1	66.9	91.8	93.4	66.9	85.6	102.3	66.9	91.9	102.3	76.4	94.7	100.8	59.8	88.8
0970, 0990	Wall Finishes & Painting/Coating	96.2	33.6	57.6	99.9	38.5	62.0	99.9	50.9	69.7	99.9	38.5	62.0	99.9	35.4	60.1	96.2	40.1	61.6
09	FINISHES	91.6	40.7	63.2	100.4	61.1	78.5	96.5	52.3	71.8	99.8	54.5	74.5	100.0	58.7	77.0	91.4	40.3	62.9
COVERS	DIVS. 10 - 14, 25, 28, 41, 43, 44, 46	100.0	51.0	90.2	100.0	79.6	95.9	100.0	49.0	89.8	100.0	78.1	95.6	100.0	78.9	95.8	100.0	46.8	89.4
21, 22, 23	FIRE SUPPRESSION, PLUMBING & HVAC	100.3	51.2	80.4	100.1	59.1	83.5	94.2	45.9	74.6	100.2	59.4	83.6	100.2	59.8	83.8	100.3	50.1	80.0
26, 27, 3370	ELECTRICAL, COMMUNICATIONS & UTIL.	93.9	56.5	74.6	110.9	61.3	85.3	102.0	56.6	78.6	103.3	67.0	84.5	102.7	67.0	84.3	95.8	57.7	76.1
MF2010	WEIGHTED AVERAGE	97.1	53.5	77.9	101.2	63.0	84.4	97.5	56.1	79.3	99.9	61.5	83.0	100.1	64.1	84.2	97.0	53.6	77.9

DIVISION		LOUISIANA									MAINE								
		NEW ORLEANS			SHREVEPORT			THIBODAUX			AUGUSTA			BANGOR			BATH		
		700 - 701			710 - 711			703			043			044			045		
		MAT.	INST.	TOTAL	MAT.	INST.	TOTAL	MAT.	INST.	TOTAL	MAT.	INST.	TOTAL	MAT.	INST.	TOTAL	MAT.	INST.	TOTAL
015433	CONTRACTOR EQUIPMENT		89.6	89.6		88.9	88.9		89.4	89.4		100.4	100.4		100.4	100.4		100.4	100.4
0241, 31 - 34	SITE & INFRASTRUCTURE, DEMOLITION	109.6	88.0	94.7	98.6	84.9	89.1	107.9	86.4	93.1	87.9	100.2	96.4	90.9	97.9	95.7	89.1	100.2	96.7
0310	Concrete Forming & Accessories	98.5	68.9	72.7	96.1	38.8	46.2	91.7	61.6	65.5	98.1	100.3	100.0	93.7	91.0	91.3	90.1	100.3	98.9
0320	Concrete Reinforcing	99.0	61.8	80.3	96.6	57.7	77.0	97.6	61.0	79.1	84.6	101.0	92.9	84.3	101.1	92.8	83.4	100.9	92.2
0330	Cast-in-Place Concrete	101.2	69.9	88.9	95.3	44.9	75.5	104.9	48.0	82.5	95.3	61.9	82.2	83.5	104.4	91.8	83.5	61.9	75.0
03	CONCRETE	98.6	68.4	83.9	94.9	45.7	70.9	100.8	57.4	79.6	105.7	86.6	96.4	99.1	96.8	98.0	99.1	86.5	93.0
04	MASONRY	105.9	62.0	79.1	103.4	47.4	69.2	133.9	48.5	81.8	95.4	60.0	73.8	109.0	59.7	78.9	115.5	56.9	79.7
05	METALS	109.1	74.3	97.6	86.3	69.6	80.7	97.0	70.6	88.3	86.7	85.0	86.1	86.3	80.8	84.5	84.7	84.7	84.7
06	WOOD, PLASTICS & COMPOSITES	101.7	70.5	83.7	101.8	38.5	65.2	94.6	67.0	78.7	97.2	112.0	105.8	92.2	98.8	96.0	87.9	112.0	101.8
07	THERMAL & MOISTURE PROTECTION	97.4	70.4	86.5	96.6	49.1	77.3	97.5	54.5	80.0	100.9	66.1	86.8	101.5	65.7	86.9	101.5	65.3	86.8
08	OPENINGS	101.9	68.4	93.5	96.0	43.5	82.8	102.1	63.6	92.4	99.3	90.9	97.2	99.2	83.5	95.3	99.2	90.7	97.1
0920	Plaster & Gypsum Board	97.0	70.0	77.7	82.8	37.1	50.1	92.6	66.4	73.9	95.2	111.7	107.0	92.4	98.1	96.5	90.2	111.7	105.6
0950, 0980	Ceilings & Acoustic Treatment	97.5	70.0	79.3	92.0	37.1	55.7	99.3	66.4	77.6	94.8	111.7	106.0	94.8	98.1	97.0	92.7	111.7	105.3
0960	Flooring	102.7	66.9	92.2	109.5	64.2	96.2	99.6	46.2	83.9	94.5	57.9	83.7	92.3	64.4	84.1	90.5	57.9	80.9
0970, 0990	Wall Finishes & Painting/Coating	101.3	67.6	80.5	96.2	33.6	57.6	101.3	50.9	70.2	90.1	36.4	57.0	90.1	37.8	57.8	90.1	36.4	57.0
09	FINISHES	100.3	68.6	82.6	95.1	42.1	65.6	99.0	58.3	76.3	97.4	87.6	91.9	96.9	81.6	88.4	95.6	87.6	91.1
COVERS	DIVS. 10 - 14, 25, 28, 41, 43, 44, 46	100.0	83.6	96.7	100.0	64.9	93.0	100.0	80.1	96.0	100.0	93.7	98.7	100.0	106.1	101.2	100.0	93.7	98.7
21, 22, 23	FIRE SUPPRESSION, PLUMBING & HVAC	100.1	65.8	86.2	100.2	55.1	81.9	94.2	59.6	80.2	99.9	65.4	85.9	100.0	61.4	84.4	94.1	65.4	82.4
26, 27, 3370	ELECTRICAL, COMMUNICATIONS & UTIL.	103.4	71.7	87.0	100.0	66.6	82.8	100.4	69.3	84.4	98.5	80.1	89.0	100.1	80.1	89.8	98.0	80.1	88.8
MF2010	WEIGHTED AVERAGE	102.3	70.5	88.3	96.7	56.3	78.9	100.2	63.4	84.0	97.6	79.8	89.8	97.7	78.9	89.4	96.0	79.5	88.7

DIVISION		MAINE																	
		HOULTON			KITTERY			LEWISTON			MACHIAS			PORTLAND			ROCKLAND		
		047			039			042			046			040 - 041			048		
		MAT.	INST.	TOTAL	MAT.	INST.	TOTAL	MAT.	INST.	TOTAL	MAT.	INST.	TOTAL	MAT.	INST.	TOTAL	MAT.	INST.	TOTAL
015433	CONTRACTOR EQUIPMENT		100.4	100.4		100.4	100.4		100.4	100.4		100.4	100.4		100.4	100.4		100.4	100.4
0241, 31 - 34	SITE & INFRASTRUCTURE, DEMOLITION	90.8	100.0	97.2	84.7	100.2	95.4	87.5	97.9	94.7	90.1	100.0	97.0	85.0	97.9	93.9	86.0	100.1	95.7
0310	Concrete Forming & Accessories	96.7	99.4	99.0	90.7	100.7	99.4	98.3	91.0	91.9	94.3	99.5	98.8	97.2	91.0	91.8	95.2	99.7	99.1
0320	Concrete Reinforcing	84.3	101.0	92.7	82.9	101.0	92.0	103.6	101.1	102.3	84.3	101.0	92.7	103.6	101.1	102.3	84.3	101.1	92.8
0330	Cast-in-Place Concrete	83.5	60.1	74.3	83.5	62.8	75.4	85.2	104.4	92.8	83.5	60.2	74.4	105.7	104.4	105.2	85.3	61.3	75.8
03	CONCRETE	100.0	85.5	93.0	94.6	87.0	90.9	99.2	96.8	98.0	99.5	85.6	92.7	109.4	96.8	103.2	96.7	86.1	91.5
04	MASONRY	93.4	58.0	71.8	106.2	59.5	77.7	94.0	59.7	73.1	93.4	58.3	72.0	94.3	59.7	73.2	88.0	59.7	70.8
05	METALS	84.9	84.9	84.9	84.7	84.8	84.7	89.3	80.8	86.5	85.0	84.9	84.9	90.6	80.8	87.4	84.8	84.9	84.9
06	WOOD, PLASTICS & COMPOSITES	96.0	112.0	105.3	88.3	112.0	102.0	97.9	98.8	98.4	93.1	112.0	104.0	97.1	98.8	98.1	93.9	112.0	104.4
07	THERMAL & MOISTURE PROTECTION	101.6	63.6	86.1	101.1	66.3	86.9	101.3	65.7	86.8	101.5	63.6	86.1	104.6	65.7	88.8	101.3	64.5	86.3
08	OPENINGS	99.3	89.0	96.7	99.2	90.2	96.9	102.4	83.5	97.7	99.3	89.0	96.7	102.4	83.5	97.7	99.2	89.0	96.7
0920	Plaster & Gypsum Board	95.2	111.7	107.0	90.2	111.7	105.6	98.7	98.1	98.3	93.3	111.7	106.5	99.5	98.1	98.5	93.3	111.7	106.5
0950, 0980	Ceilings & Acoustic Treatment	92.7	111.7	105.3	92.7	111.7	105.3	102.0	98.1	99.4	92.7	111.7	105.3	99.3	98.1	98.5	92.7	111.7	105.3
0960	Flooring	93.5	54.2	82.0	90.6	60.4	81.7	95.1	64.4	86.1	92.6	54.2	81.3	94.5	64.4	85.6	92.9	54.2	81.5
0970, 0990	Wall Finishes & Painting/Coating	90.1	37.8	57.8	90.1	36.4	57.0	90.1	37.8	57.8	90.1	37.8	57.8	90.1	37.8	57.8	90.1	37.8	57.8
09	FINISHES	97.2	86.6	91.3	95.2	88.3	91.3	99.6	81.6	89.6	96.6	86.6	91.0	98.3	81.6	89.0	96.3	86.8	91.0
COVERS	DIVS. 10 - 14, 25, 28, 41, 43, 44, 46	100.0	92.9	98.6	100.0	95.2	99.1	100.0	106.1	101.2	100.0	93.0	98.6	100.0	106.1	101.2	100.0	93.2	98.7
21, 22, 23	FIRE SUPPRESSION, PLUMBING & HVAC	94.1	64.3	82.0	94.1	66.0	82.7	100.0	61.4	84.4	94.1	64.4	82.0	99.9	61.4	84.3	94.1	64.7	82.2
26, 27, 3370	ELECTRICAL, COMMUNICATIONS & UTIL.	102.3	80.1	90.9	98.0	80.1	88.8	102.3	80.1	90.9	102.3	80.1	90.9	103.9	80.1	91.6	102.2	80.1	90.8
MF2010	WEIGHTED AVERAGE	95.7	78.9	88.3	94.8	80.1	88.3	98.1	78.9	89.7	95.5	79.0	88.2	99.5	78.9	90.4	94.7	79.4	88.0

DIVISION		MAINE			MARYLAND														
		WATERVILLE			ANNAPOLIS			BALTIMORE			COLLEGE PARK			CUMBERLAND			EASTON		
		049			214			210 - 212			207 - 208			215			216		
		MAT.	INST.	TOTAL	MAT.	INST.	TOTAL	MAT.	INST.	TOTAL	MAT.	INST.	TOTAL	MAT.	INST.	TOTAL	MAT.	INST.	TOTAL
015433	CONTRACTOR EQUIPMENT		100.4	100.4		98.6	98.6		102.0	102.0		103.1	103.1		98.6	98.6		98.6	98.6
0241, 31 - 34	SITE & INFRASTRUCTURE, DEMOLITION	91.0	100.2	97.3	99.7	90.5	93.4	100.2	94.4	96.2	106.9	92.8	97.2	91.2	90.2	90.5	98.2	87.6	90.9
0310	Concrete Forming & Accessories	89.7	100.3	98.9	94.1	71.0	74.0	96.9	75.3	78.1	85.5	71.8	73.6	88.5	77.8	79.2	86.7	72.4	74.3
0320	Concrete Reinforcing	84.3	101.0	92.7	99.8	82.5	91.1	99.8	82.5	91.1	98.5	80.9	89.6	88.2	74.7	81.4	87.4	80.7	84.0
0330	Cast-in-Place Concrete	83.5	61.9	75.0	140.3	79.5	116.4	124.9	80.3	107.3	131.0	80.5	111.1	111.5	84.1	100.7	123.8	50.6	95.0
03	CONCRETE	100.8	86.6	93.8	120.9	77.2	99.6	113.4	79.4	96.8	115.7	77.8	97.2	101.2	80.3	91.0	110.1	67.4	89.3
04	MASONRY	103.1	60.0	76.8	101.1	75.8	85.7	102.2	75.8	86.1	115.1	70.5	87.9	100.2	78.5	86.9	114.3	45.4	72.3
05	METALS	84.9	85.0	84.9	94.9	94.7	94.9	96.1	95.3	95.8	83.9	96.9	88.2	92.5	91.3	92.1	92.8	88.8	91.4
06	WOOD, PLASTICS & COMPOSITES	87.3	112.0	101.6	90.0	70.8	78.9	92.8	76.5	83.4	87.6	71.1	78.1	83.3	76.3	79.2	81.3	79.5	80.2
07	THERMAL & MOISTURE PROTECTION	101.6	66.1	87.1	97.1	79.7	90.0	97.4	80.5	90.6	100.5	77.8	91.3	96.8	77.3	88.9	96.9	61.6	82.5
08	OPENINGS	99.3	90.9	97.2	93.9	78.5	90.0	96.1	81.6	92.4	97.2	75.0	91.6	95.1	77.5	90.7	93.1	73.4	88.2
0920	Plaster & Gypsum Board	90.2	111.7	105.6	98.3	70.3	78.3	99.9	76.0	82.9	99.7	70.3	78.7	94.5	75.9	81.2	94.0	79.2	83.4
0950, 0980	Ceilings & Acoustic Treatment	92.7	111.7	105.3	98.9	70.3	80.0	99.8	76.0	84.1	95.8	70.3	78.9	99.8	75.9	84.0	99.8	79.2	86.2
0960	Flooring	90.2	57.9	80.7	93.8	80.9	90.0	93.8	80.9	90.0	107.1	80.9	99.4	89.8	82.2	87.6	89.1	54.1	78.8
0970, 0990	Wall Finishes & Painting/Coating	90.1	36.4	57.0	96.8	83.5	88.6	96.8	83.5	88.6	121.0	78.7	94.9	96.8	77.3	84.8	96.8	78.7	85.6
09	FINISHES	95.8	87.6	91.2	96.0	73.2	83.3	96.6	76.6	85.4	101.4	73.5	85.8	94.1	78.3	85.3	94.3	71.3	81.4
COVERS	DIVS. 10 - 14, 25, 28, 41, 43, 44, 46	100.0	93.7	98.7	100.0	86.1	97.2	100.0	87.1	97.4	100.0	82.2	96.5	100.0	89.8	98.0	100.0	81.1	96.2
21, 22, 23	FIRE SUPPRESSION, PLUMBING & HVAC	94.1	65.4	82.4	99.9	83.7	93.3	99.9	83.8	93.4	94.3	86.1	91.0	93.9	72.0	85.0	93.9	69.5	84.0
26, 27, 3370	ELECTRICAL, COMMUNICATIONS & UTIL.	102.3	80.1	90.9	96.2	93.2	94.7	99.4	93.8	96.5	98.3	98.9	98.6	97.1	82.0	89.3	96.6	66.1	80.9
MF2010	WEIGHTED AVERAGE	96.1	79.8	88.9	100.2	83.2	92.7	100.2	84.7	93.3	98.5	84.2	92.2	95.8	80.2	89.0	97.5	70.3	85.5

DIVISION		MARYLAND															MASSACHUSETTS		
		ELKTON			HAGERSTOWN			SALISBURY			SILVER SPRING			WALDORF			BOSTON		
		219			217			218			209			206			020 - 022, 024		
		MAT.	INST.	TOTAL	MAT.	INST.	TOTAL	MAT.	INST.	TOTAL	MAT.	INST.	TOTAL	MAT.	INST.	TOTAL	MAT.	INST.	TOTAL
015433	CONTRACTOR EQUIPMENT		98.6	98.6		98.6	98.6		98.6	98.6		96.1	96.1		96.1	96.1		106.0	106.0
0241, 31 - 34	SITE & INFRASTRUCTURE, DEMOLITION	85.3	88.3	87.4	89.2	91.1	90.5	98.2	87.7	90.9	94.3	86.0	88.6	101.6	86.0	90.9	95.2	107.4	103.6
0310	Concrete Forming & Accessories	92.1	85.8	86.6	87.8	78.1	79.3	99.2	54.9	60.7	93.9	70.1	73.2	100.6	70.2	74.1	101.8	142.2	136.9
0320	Concrete Reinforcing	87.4	104.7	96.1	88.2	74.7	81.4	87.4	67.5	77.4	97.4	76.0	86.6	98.0	76.0	86.9	102.5	149.5	126.2
0330	Cast-in-Place Concrete	100.2	78.9	91.8	106.2	84.2	97.5	123.8	49.2	94.5	134.1	79.4	112.5	150.3	79.4	122.4	107.8	149.9	124.4
03	CONCRETE	92.9	87.6	90.3	97.1	80.5	89.0	110.9	56.8	84.5	114.3	75.6	95.5	126.1	75.7	101.5	113.4	145.1	128.9
04	MASONRY	100.1	62.3	77.0	106.0	78.5	89.2	114.1	50.0	75.0	114.2	68.7	86.4	97.9	68.7	80.1	112.0	162.5	142.8
05	METALS	92.8	101.2	95.6	92.7	91.5	92.3	92.8	83.9	89.9	87.9	93.3	89.7	87.9	93.4	89.7	101.2	129.1	110.4
06	WOOD, PLASTICS & COMPOSITES	87.5	92.3	90.3	82.4	76.4	78.9	96.6	58.5	74.6	93.8	70.4	80.3	101.0	70.4	83.3	101.1	141.6	124.5
07	THERMAL & MOISTURE PROTECTION	96.5	75.1	87.8	96.5	76.1	88.2	97.1	66.8	84.8	103.1	81.4	94.3	103.6	81.4	94.6	101.6	150.1	121.3
08	OPENINGS	93.1	84.4	90.9	93.1	76.3	88.9	93.4	64.7	86.2	90.3	74.6	86.4	91.1	74.6	87.0	98.5	143.3	109.8
0920	Plaster & Gypsum Board	96.7	92.4	93.7	94.0	76.0	81.2	102.7	57.5	70.4	105.4	70.3	80.3	109.0	70.3	81.4	101.7	142.2	130.7
0950, 0980	Ceilings & Acoustic Treatment	99.8	92.4	94.9	100.8	76.0	84.4	99.8	57.5	71.9	105.8	70.3	82.3	105.8	70.3	82.3	101.0	142.2	128.2
0960	Flooring	91.6	61.2	82.7	89.5	82.2	87.4	95.3	67.8	87.2	113.9	80.9	104.2	118.0	80.9	107.1	95.0	178.2	119.4
0970, 0990	Wall Finishes & Painting/Coating	96.8	78.7	85.6	96.8	78.7	85.6	96.8	78.7	85.6	128.5	78.7	97.8	128.5	78.7	97.8	92.8	153.8	130.4
09	FINISHES	94.5	81.1	87.0	94.0	78.5	85.3	97.1	59.2	76.0	101.6	71.7	84.9	103.6	72.3	86.1	99.4	150.6	128.0
COVERS	DIVS. 10 - 14, 25, 28, 41, 43, 44, 46	100.0	64.7	93.0	100.0	89.8	98.0	100.0	41.7	88.4	100.0	79.7	96.0	100.0	79.7	96.0	100.0	120.2	104.0
21, 22, 23	FIRE SUPPRESSION, PLUMBING & HVAC	93.9	83.3	89.6	99.9	84.4	93.6	93.9	64.3	81.9	94.3	84.8	90.4	94.3	84.8	90.4	100.1	128.6	111.6
26, 27, 3370	ELECTRICAL, COMMUNICATIONS & UTIL.	98.4	93.2	95.7	96.9	82.0	89.2	95.4	71.4	83.0	95.4	98.9	97.2	93.0	98.9	96.0	101.5	132.0	117.2
MF2010	WEIGHTED AVERAGE	94.7	84.3	90.1	96.8	82.9	90.7	97.9	65.3	83.5	97.7	82.3	90.9	98.5	82.4	91.4	102.2	136.8	117.5

	DIVISION	MASSACHUSETTS																	
		BROCKTON			BUZZARDS BAY			FALL RIVER			FITCHBURG			FRAMINGHAM			GREENFIELD		
		023			025			027			014			017			013		
		MAT.	INST.	TOTAL	MAT.	INST.	TOTAL	MAT.	INST.	TOTAL	MAT.	INST.	TOTAL	MAT.	INST.	TOTAL	MAT.	INST.	TOTAL
015433	CONTRACTOR EQUIPMENT		102.2	102.2		102.2	102.2		103.1	103.1		100.4	100.4		101.6	101.6		100.4	100.4
0241, 31 - 34	SITE & INFRASTRUCTURE, DEMOLITION	92.5	104.0	100.4	82.7	103.9	97.3	91.6	104.0	100.2	84.5	103.9	97.9	81.0	103.6	96.6	88.4	102.4	98.1
0310	Concrete Forming & Accessories	102.4	127.9	124.6	100.1	127.2	123.7	102.4	127.5	124.2	94.8	127.1	122.9	101.1	127.7	124.3	93.3	108.3	106.3
0320	Concrete Reinforcing	103.6	149.4	126.7	83.0	121.5	102.4	103.6	121.5	112.6	83.0	142.0	112.7	83.0	149.0	116.3	86.5	125.7	106.2
0330	Cast-in-Place Concrete	102.6	152.3	122.2	85.3	152.0	111.6	99.3	152.6	120.3	89.4	145.9	111.6	89.4	139.7	109.2	91.9	124.2	104.6
03	CONCRETE	110.5	139.5	124.7	93.3	133.9	113.1	108.9	134.2	121.2	91.7	135.4	113.0	94.6	135.0	114.3	95.7	116.5	105.8
04	MASONRY	107.9	158.8	139.0	99.3	158.8	135.6	108.4	158.8	139.1	95.5	151.9	129.9	101.7	152.0	132.4	99.8	125.9	115.7
05	METALS	98.0	126.9	107.6	93.0	115.2	100.3	98.0	115.6	103.8	92.9	120.7	102.1	93.0	126.1	103.9	95.1	109.4	99.8
06	WOOD, PLASTICS & COMPOSITES	100.0	124.7	114.3	97.2	124.7	113.1	100.0	125.0	114.4	93.7	124.5	111.5	99.7	124.5	114.0	92.0	106.0	100.1
07	THERMAL & MOISTURE PROTECTION	101.6	145.9	119.6	101.0	141.2	117.3	101.6	140.4	117.4	101.0	137.8	116.0	101.1	142.8	118.1	101.1	118.2	108.0
08	OPENINGS	96.6	133.8	105.9	92.1	120.5	99.3	96.6	120.6	102.6	101.1	131.7	108.8	92.6	133.6	102.9	101.2	111.9	103.9
0920	Plaster & Gypsum Board	95.7	124.8	116.5	91.3	124.8	115.2	95.7	124.8	116.5	92.4	124.6	115.4	95.6	124.8	116.5	93.7	105.5	102.1
0950, 0980	Ceilings & Acoustic Treatment	103.0	124.8	117.4	93.7	124.8	114.3	103.0	124.8	117.4	92.7	124.6	113.8	92.7	124.8	113.9	100.0	105.5	103.6
0960	Flooring	95.5	178.2	119.7	94.2	178.2	118.8	95.3	178.2	119.6	92.3	178.2	117.5	93.9	178.2	118.7	91.6	140.6	106.0
0970, 0990	Wall Finishes & Painting/Coating	92.3	139.1	121.2	92.3	139.1	121.2	92.3	139.1	121.2	90.1	139.1	120.4	91.2	139.1	120.8	90.1	109.4	102.0
09	FINISHES	99.4	138.1	121.0	95.6	138.1	119.3	99.3	138.3	121.1	95.0	138.0	119.0	95.6	137.9	119.2	96.9	114.2	106.5
COVERS	DIVS. 10 - 14, 25, 28, 41, 43, 44, 46	100.0	117.1	103.4	100.0	117.1	103.4	100.0	117.7	103.5	100.0	107.0	101.4	100.0	116.6	103.3	100.0	105.3	101.1
21, 22, 23	FIRE SUPPRESSION, PLUMBING & HVAC	100.0	111.9	104.8	94.1	111.6	101.2	100.0	111.7	104.7	95.0	115.0	103.1	95.0	124.6	107.0	95.0	102.0	97.8
26, 27, 3370	ELECTRICAL, COMMUNICATIONS & UTIL.	100.6	98.8	99.7	97.6	98.8	98.3	100.5	98.8	99.7	102.8	101.8	102.3	99.1	128.4	114.2	102.8	93.0	97.7
MF2010	WEIGHTED AVERAGE	100.8	124.7	111.4	94.9	122.1	106.9	100.6	122.2	110.1	96.1	123.3	108.1	95.5	129.9	110.7	97.3	108.4	102.2

	DIVISION	MASSACHUSETTS																	
		HYANNIS			LAWRENCE			LOWELL			NEW BEDFORD			PITTSFIELD			SPRINGFIELD		
		026			019			018			027			012			010 - 011		
		MAT.	INST.	TOTAL	MAT.	INST.	TOTAL	MAT.	INST.	TOTAL	MAT.	INST.	TOTAL	MAT.	INST.	TOTAL	MAT.	INST.	TOTAL
015433	CONTRACTOR EQUIPMENT		102.2	102.2		102.2	102.2		100.4	100.4		103.1	103.1		100.4	100.4		100.4	100.4
0241, 31 - 34	SITE & INFRASTRUCTURE, DEMOLITION	88.8	103.9	99.2	93.1	104.0	100.6	92.4	103.9	100.3	90.0	104.0	99.7	93.4	102.2	99.4	92.8	102.4	99.5
0310	Concrete Forming & Accessories	95.0	127.2	123.1	102.1	128.0	124.6	99.4	128.0	124.3	102.4	127.5	124.2	99.3	106.5	105.5	99.6	108.4	107.2
0320	Concrete Reinforcing	83.0	121.5	102.4	102.7	135.5	119.2	103.6	135.2	119.5	103.6	121.5	112.6	85.8	108.6	97.3	103.6	125.7	114.7
0330	Cast-in-Place Concrete	93.6	152.0	116.6	103.5	146.8	120.5	94.1	146.8	114.8	87.4	152.6	113.1	102.6	121.8	110.2	97.9	124.3	108.3
03	CONCRETE	100.4	133.9	116.7	110.8	135.0	122.6	101.4	134.8	117.7	102.9	134.2	118.2	102.9	111.7	107.2	103.3	116.6	109.8
04	MASONRY	106.9	158.8	138.6	107.1	158.8	138.7	94.7	151.9	129.6	106.8	158.8	138.5	95.4	121.6	111.4	95.0	126.1	114.0
05	METALS	94.5	115.2	101.3	95.5	121.4	104.0	95.4	118.6	103.1	98.0	115.6	103.8	95.3	102.4	97.6	98.0	109.4	101.7
06	WOOD, PLASTICS & COMPOSITES	91.2	124.7	110.6	100.0	124.7	114.3	99.5	124.7	114.1	100.0	125.0	114.4	99.5	106.0	103.3	99.5	106.0	103.3
07	THERMAL & MOISTURE PROTECTION	101.2	141.2	117.5	101.5	145.1	119.2	101.4	143.2	118.4	101.5	140.4	117.3	101.4	116.3	107.5	101.4	118.3	108.2
08	OPENINGS	92.8	120.5	99.8	96.6	130.0	105.0	102.4	130.0	109.4	96.6	120.6	102.6	102.4	107.3	103.6	102.4	111.9	104.8
0920	Plaster & Gypsum Board	87.6	124.8	114.2	98.7	124.8	117.4	98.7	124.8	117.4	95.7	124.8	116.5	98.7	105.5	103.6	98.7	105.5	103.6
0950, 0980	Ceilings & Acoustic Treatment	95.8	124.8	115.0	102.0	124.8	117.1	102.0	124.8	117.1	103.0	124.8	117.4	102.0	105.5	104.3	102.0	105.5	104.3
0960	Flooring	92.0	178.2	117.3	94.5	178.2	119.0	94.5	178.2	119.0	95.3	178.2	119.6	94.7	140.6	108.1	94.3	140.6	107.9
0970, 0990	Wall Finishes & Painting/Coating	92.3	139.1	121.2	90.2	139.1	120.4	90.1	139.1	120.4	92.3	139.1	121.2	90.1	109.4	102.0	91.7	109.4	102.6
09	FINISHES	95.5	138.1	119.3	99.0	138.1	120.8	99.0	138.1	120.8	99.2	138.3	121.0	99.1	113.1	106.9	99.1	114.2	107.5
COVERS	DIVS. 10 - 14, 25, 28, 41, 43, 44, 46	100.0	117.1	103.4	100.0	117.2	103.4	100.0	117.2	103.4	100.0	117.7	103.5	100.0	103.8	100.8	100.0	105.4	101.1
21, 22, 23	FIRE SUPPRESSION, PLUMBING & HVAC	100.0	111.6	104.7	100.0	119.0	107.7	100.0	124.5	109.9	100.0	111.7	104.7	100.0	97.6	99.1	100.0	102.1	100.9
26, 27, 3370	ELECTRICAL, COMMUNICATIONS & UTIL.	98.1	98.8	98.5	101.8	128.4	115.5	102.3	124.2	113.6	101.4	98.8	100.1	102.3	93.0	97.5	102.3	94.9	98.5
MF2010	WEIGHTED AVERAGE	98.0	122.1	108.6	100.5	129.0	113.1	99.4	128.5	112.3	99.9	122.2	109.7	99.6	105.3	102.1	100.0	108.7	103.9

	DIVISION	MASSACHUSETTS			MICHIGAN														
		WORCESTER			ANN ARBOR			BATTLE CREEK			BAY CITY			DEARBORN			DETROIT		
		015 - 016			481			490			487			481			482		
		MAT.	INST.	TOTAL	MAT.	INST.	TOTAL	MAT.	INST.	TOTAL	MAT.	INST.	TOTAL	MAT.	INST.	TOTAL	MAT.	INST.	TOTAL
015433	CONTRACTOR EQUIPMENT		100.4	100.4		110.6	110.6		103.2	103.2		110.6	110.6		110.6	110.6		98.9	98.9
0241, 31 - 34	SITE & INFRASTRUCTURE, DEMOLITION	92.8	103.9	100.4	74.7	96.5	89.8	84.8	88.5	87.4	66.5	95.7	86.6	74.5	96.6	89.7	89.3	98.2	95.4
0310	Concrete Forming & Accessories	99.8	127.1	123.6	100.7	110.6	109.3	98.6	89.8	90.9	100.8	89.2	90.7	100.5	114.6	112.8	101.3	114.6	112.9
0320	Concrete Reinforcing	103.6	144.3	124.1	99.7	123.6	111.8	98.1	91.1	94.6	99.7	122.8	111.4	99.7	123.8	111.9	99.0	123.8	111.5
0330	Cast-in-Place Concrete	97.3	145.9	116.5	81.9	112.4	93.9	92.4	102.5	96.4	78.4	93.3	84.2	80.1	111.5	92.5	87.9	111.6	97.2
03	CONCRETE	103.0	135.8	119.0	90.3	113.9	101.8	95.2	93.8	94.5	88.5	97.6	92.9	89.3	115.4	102.0	93.2	114.3	103.5
04	MASONRY	94.5	151.9	129.5	97.9	108.5	104.4	96.0	86.8	90.3	97.5	88.5	92.0	97.8	111.9	106.4	96.9	111.9	106.1
05	METALS	98.0	121.6	105.8	93.8	119.0	102.1	93.0	87.3	91.1	94.5	115.9	101.5	93.9	119.4	102.3	97.9	102.2	99.3
06	WOOD, PLASTICS & COMPOSITES	99.9	124.5	114.1	97.8	111.1	105.5	100.1	89.6	94.1	97.8	88.4	92.4	97.8	115.7	108.1	98.2	115.7	108.3
07	THERMAL & MOISTURE PROTECTION	101.4	137.8	116.2	101.3	110.0	104.8	93.7	87.4	91.2	99.1	96.0	97.8	99.9	114.7	105.9	98.3	114.8	105.0
08	OPENINGS	102.4	132.4	110.0	95.1	109.8	98.8	92.7	83.8	90.5	95.1	95.6	95.2	95.1	112.3	99.4	96.6	112.3	100.6
0920	Plaster & Gypsum Board	98.7	124.6	117.2	104.7	110.4	108.8	94.6	85.4	88.0	104.7	87.0	92.1	104.7	115.1	112.2	104.7	115.1	112.2
0950, 0980	Ceilings & Acoustic Treatment	102.0	124.6	116.9	93.7	110.4	104.7	93.8	85.4	88.2	94.7	87.0	89.6	93.7	115.1	107.9	94.7	115.1	108.2
0960	Flooring	94.5	170.9	116.9	94.2	117.1	100.9	98.7	96.2	98.0	94.0	80.5	90.0	93.8	117.8	100.8	93.9	117.8	100.9
0970, 0990	Wall Finishes & Painting/Coating	90.1	139.1	120.4	92.8	102.0	98.5	96.0	82.6	87.7	92.8	83.3	87.0	92.8	102.1	98.5	94.3	102.1	99.1
09	FINISHES	99.0	136.5	119.9	94.8	111.1	103.9	96.5	90.1	92.9	94.5	86.2	89.9	94.7	114.3	105.6	96.2	114.3	106.3
COVERS	DIVS. 10 - 14, 25, 28, 41, 43, 44, 46	100.0	110.4	102.1	100.0	106.8	101.4	100.0	103.0	100.6	100.0	97.1	99.4	100.0	107.6	101.5	100.0	107.6	101.5
21, 22, 23	FIRE SUPPRESSION, PLUMBING & HVAC	100.0	115.1	106.1	100.0	100.2	100.1	99.9	90.9	96.3	100.0	87.9	95.1	100.0	110.5	104.3	100.1	111.8	104.8
26, 27, 3370	ELECTRICAL, COMMUNICATIONS & UTIL.	102.3	102.6	102.4	95.6	110.4	103.2	94.8	83.4	88.9	94.4	91.3	92.8	95.6	109.2	102.6	96.7	109.1	103.1
MF2010	WEIGHTED AVERAGE	100.0	123.5	110.4	95.8	108.1	101.2	96.0	89.2	93.0	95.2	93.6	94.5	95.7	111.4	102.6	97.4	110.1	103.0

DIVISION		MICHIGAN																	
		FLINT			GAYLORD			GRAND RAPIDS			IRON MOUNTAIN			JACKSON			KALAMAZOO		
		484 - 485			497			493,495			498 - 499			492			491		
		MAT.	INST.	TOTAL	MAT.	INST.	TOTAL	MAT.	INST.	TOTAL	MAT.	INST.	TOTAL	MAT.	INST.	TOTAL	MAT.	INST.	TOTAL
015433	**CONTRACTOR EQUIPMENT**		110.6	110.6		105.6	105.6		103.2	103.2		94.9	94.9		105.6	105.6		103.2	103.2
0241, 31 - 34	**SITE & INFRASTRUCTURE, DEMOLITION**	64.3	95.9	86.1	79.0	86.7	84.3	86.3	88.4	87.7	87.5	94.9	92.6	101.3	88.4	92.4	85.1	88.5	87.4
0310	Concrete Forming & Accessories	103.5	91.4	93.0	95.9	56.1	61.2	98.9	81.0	83.4	88.0	86.1	86.4	93.4	88.0	88.7	98.6	89.5	90.7
0320	Concrete Reinforcing	99.7	123.1	111.5	91.5	92.7	92.1	98.1	90.8	94.4	91.3	94.4	92.8	88.9	122.9	106.0	98.1	91.0	94.6
0330	Cast-in-Place Concrete	82.5	95.4	87.6	92.2	71.4	84.0	98.4	95.5	97.3	109.2	89.4	101.4	92.1	94.5	93.0	94.3	102.4	97.5
03	**CONCRETE**	90.7	99.4	95.0	91.6	69.8	81.0	98.2	87.5	93.0	101.2	88.9	95.2	86.8	97.4	91.9	98.4	93.6	96.1
04	**MASONRY**	98.0	96.2	96.9	106.1	82.9	91.9	91.8	81.9	85.8	92.5	88.8	90.3	86.7	93.2	90.7	94.6	86.8	89.8
05	**METALS**	93.9	116.5	101.4	94.1	100.6	96.3	94.2	86.3	91.6	93.5	90.7	92.6	94.3	113.8	100.8	93.0	87.0	91.0
06	**WOOD, PLASTICS & COMPOSITES**	100.9	89.7	94.5	92.5	48.8	67.3	97.2	79.1	86.8	87.3	86.1	86.6	91.3	86.3	88.4	100.1	89.6	94.1
07	**THERMAL & MOISTURE PROTECTION**	99.1	99.3	99.2	92.2	70.8	83.5	95.5	77.1	88.0	95.4	84.1	90.8	91.6	96.6	93.6	93.7	87.4	91.2
08	**OPENINGS**	95.1	96.3	95.4	95.0	53.9	84.7	96.6	82.2	92.9	100.6	73.4	93.7	94.0	87.2	92.3	92.7	83.8	90.5
0920	Plaster & Gypsum Board	106.5	88.4	93.5	94.0	46.0	59.7	95.5	74.6	80.6	58.1	86.1	78.1	92.5	84.6	86.8	94.6	85.4	88.0
0950, 0980	Ceilings & Acoustic Treatment	93.7	88.4	90.2	92.0	46.0	61.6	92.9	74.6	80.8	91.8	86.1	88.0	92.9	84.6	87.4	93.8	85.4	88.2
0960	Flooring	94.0	83.2	90.8	91.8	50.4	79.7	98.9	70.5	90.5	112.8	96.8	108.1	90.5	69.4	84.3	98.7	75.8	92.0
0970, 0990	Wall Finishes & Painting/Coating	92.8	90.0	91.1	92.5	45.6	63.5	96.0	81.3	86.9	111.5	58.8	79.0	92.5	75.7	82.1	96.0	82.6	87.7
09	**FINISHES**	94.0	89.0	91.2	96.7	51.8	71.7	96.6	78.4	86.5	99.3	85.3	91.5	98.0	82.4	89.3	96.5	86.2	90.7
COVERS	**DIVS. 10 - 14, 25, 28, 41, 43, 44, 46**	100.0	98.2	99.6	100.0	91.8	98.4	100.0	101.6	100.3	100.0	93.7	98.7	100.0	98.0	99.6	100.0	103.0	100.6
21, 22, 23	**FIRE SUPPRESSION, PLUMBING & HVAC**	100.0	94.1	97.6	94.4	82.8	89.7	99.9	83.2	93.1	94.3	86.8	91.3	94.4	92.0	93.4	99.9	83.0	93.1
26, 27, 3370	**ELECTRICAL, COMMUNICATIONS & UTIL.**	95.6	95.0	95.3	92.9	49.0	70.3	99.1	75.1	86.7	99.2	87.6	93.2	96.8	110.3	103.8	94.7	86.9	90.7
MF2010	**WEIGHTED AVERAGE**	95.5	97.1	96.2	94.7	72.8	85.1	97.2	83.0	90.9	96.7	87.8	92.8	94.3	95.9	95.0	96.3	87.4	92.4

DIVISION		MICHIGAN															MINNESOTA		
		LANSING			MUSKEGON			ROYAL OAK			SAGINAW			TRAVERSE CITY			BEMIDJI		
		488 - 489			494			480,483			486			496			566		
		MAT.	INST.	TOTAL	MAT.	INST.	TOTAL	MAT.	INST.	TOTAL	MAT.	INST.	TOTAL	MAT.	INST.	TOTAL	MAT.	INST.	TOTAL
015433	**CONTRACTOR EQUIPMENT**		110.6	110.6		103.2	103.2		96.4	96.4		110.6	110.6		94.9	94.9		98.7	98.7
0241, 31 - 34	**SITE & INFRASTRUCTURE, DEMOLITION**	83.2	95.8	91.9	82.9	88.4	86.7	79.0	96.0	90.7	67.4	95.7	86.9	74.3	94.6	88.3	98.4	98.6	98.6
0310	Concrete Forming & Accessories	104.2	88.4	90.5	99.1	88.1	89.5	97.1	111.7	109.8	100.7	89.0	90.5	88.0	71.1	73.3	90.8	94.2	93.7
0320	Concrete Reinforcing	99.7	122.9	111.4	98.8	90.5	94.6	90.4	120.5	105.6	99.7	122.8	111.3	92.7	90.2	91.4	103.5	109.4	106.5
0330	Cast-in-Place Concrete	87.0	93.7	89.6	92.1	96.5	93.8	71.7	110.4	87.0	80.9	93.2	85.8	85.2	63.7	76.7	105.6	106.6	106.0
03	**CONCRETE**	93.1	97.5	95.2	93.6	90.9	92.3	78.0	111.9	94.5	89.8	97.5	93.6	83.4	72.7	78.2	99.5	102.4	100.9
04	**MASONRY**	91.7	93.8	92.9	93.1	76.2	82.8	93.0	111.9	104.5	99.5	88.5	92.8	90.7	82.8	85.8	100.3	107.1	104.4
05	**METALS**	92.5	115.9	100.2	90.7	85.9	89.1	97.0	99.9	98.0	93.9	115.7	101.1	93.5	88.8	91.9	92.6	122.5	102.5
06	**WOOD, PLASTICS & COMPOSITES**	100.3	86.5	92.3	96.1	88.5	91.7	93.2	112.2	104.2	93.6	88.4	90.6	87.3	69.5	77.0	72.8	90.9	83.3
07	**THERMAL & MOISTURE PROTECTION**	100.2	92.0	96.9	92.9	78.2	86.9	97.7	114.0	104.4	99.8	96.0	98.3	94.5	71.4	85.1	108.5	99.7	104.9
08	**OPENINGS**	95.1	94.5	94.9	92.0	87.7	90.9	95.3	110.1	99.0	92.9	95.6	93.6	100.6	64.9	91.6	97.7	111.9	101.3
0920	Plaster & Gypsum Board	97.4	85.0	88.6	75.1	84.2	81.6	101.5	111.5	108.6	104.7	87.0	92.1	58.1	69.0	65.9	105.6	91.0	95.1
0950, 0980	Ceilings & Acoustic Treatment	93.7	85.0	88.0	94.7	84.2	87.8	92.9	111.5	105.2	93.7	87.0	89.3	91.8	69.0	76.7	137.6	91.0	106.8
0960	Flooring	104.0	83.2	97.9	97.9	65.3	88.3	91.8	116.4	99.0	94.2	80.5	90.2	112.8	50.4	94.5	109.8	126.5	114.7
0970, 0990	Wall Finishes & Painting/Coating	103.9	84.1	91.7	95.3	81.3	86.7	94.3	102.0	99.0	92.8	83.3	87.0	111.5	45.6	70.8	96.0	101.0	99.1
09	**FINISHES**	98.5	86.3	91.7	93.4	83.0	87.6	93.8	111.9	103.9	94.4	86.2	89.9	98.2	63.8	79.0	112.4	100.2	105.6
COVERS	**DIVS. 10 - 14, 25, 28, 41, 43, 44, 46**	100.0	97.6	99.5	100.0	102.8	100.6	100.0	102.7	100.5	100.0	97.1	99.4	100.0	90.3	98.1	100.0	95.7	99.1
21, 22, 23	**FIRE SUPPRESSION, PLUMBING & HVAC**	99.9	92.2	96.8	99.8	85.2	93.9	94.5	109.6	100.6	100.0	87.5	95.0	94.3	82.4	89.5	94.5	88.1	91.9
26, 27, 3370	**ELECTRICAL, COMMUNICATIONS & UTIL.**	98.7	93.0	95.8	95.2	80.1	87.4	97.6	94.0	95.8	92.7	91.3	92.0	94.6	76.7	85.3	104.5	104.0	104.2
MF2010	**WEIGHTED AVERAGE**	96.4	95.1	95.9	94.9	84.9	90.5	93.4	106.2	99.1	95.0	93.5	94.4	93.6	78.7	87.0	98.7	101.3	99.9

DIVISION		MINNESOTA																	
		BRAINERD			DETROIT LAKES			DULUTH			MANKATO			MINNEAPOLIS			ROCHESTER		
		564			565			556 - 558			560			553 - 555			559		
		MAT.	INST.	TOTAL	MAT.	INST.	TOTAL	MAT.	INST.	TOTAL	MAT.	INST.	TOTAL	MAT.	INST.	TOTAL	MAT.	INST.	TOTAL
015433	**CONTRACTOR EQUIPMENT**		101.4	101.4		98.7	98.7		101.2	101.2		101.4	101.4		104.7	104.7		101.2	101.2
0241, 31 - 34	**SITE & INFRASTRUCTURE, DEMOLITION**	96.6	103.5	101.3	96.4	99.1	98.2	98.8	102.3	101.2	92.9	104.1	100.6	101.3	107.6	105.7	100.3	102.1	101.6
0310	Concrete Forming & Accessories	94.0	97.0	96.6	88.1	96.9	95.7	101.1	116.4	114.4	101.0	104.7	104.2	103.2	129.6	126.1	103.8	110.6	109.7
0320	Concrete Reinforcing	102.3	109.6	106.0	103.5	109.4	106.5	98.8	109.9	104.4	102.1	119.8	111.0	99.0	120.2	109.7	98.8	119.9	109.4
0330	Cast-in-Place Concrete	114.8	111.6	113.6	102.6	110.8	105.8	107.4	109.8	108.4	105.8	110.7	107.7	120.3	119.0	119.8	117.9	103.0	112.0
03	**CONCRETE**	103.9	105.4	104.6	97.0	105.0	100.9	102.3	113.5	107.7	99.0	110.4	104.6	110.5	124.5	117.3	107.7	110.5	109.1
04	**MASONRY**	123.6	117.6	119.9	123.6	114.8	118.2	108.6	121.3	116.3	111.9	114.4	113.4	108.1	129.1	120.9	107.3	117.0	113.2
05	**METALS**	93.8	122.5	103.3	92.6	122.3	102.4	99.6	125.1	108.0	93.6	128.2	105.1	102.4	131.9	112.1	99.5	130.1	109.6
06	**WOOD, PLASTICS & COMPOSITES**	93.7	90.7	92.0	70.2	90.9	82.2	110.4	116.3	113.8	101.9	102.6	102.3	113.6	127.8	121.8	113.6	109.4	111.2
07	**THERMAL & MOISTURE PROTECTION**	107.6	112.3	109.5	108.3	112.5	110.0	109.0	119.9	113.5	108.0	103.2	106.0	108.8	132.6	118.5	109.1	104.6	107.3
08	**OPENINGS**	85.7	111.8	92.2	97.6	111.9	101.2	93.9	121.1	100.7	90.4	121.3	98.2	97.2	135.7	106.9	93.9	125.7	101.9
0920	Plaster & Gypsum Board	86.5	91.0	89.7	104.7	91.0	94.9	94.3	117.2	110.7	90.6	103.2	99.6	102.2	129.0	121.3	101.7	110.1	107.7
0950, 0980	Ceilings & Acoustic Treatment	67.9	91.0	83.1	137.6	91.0	106.8	92.0	117.2	108.7	67.9	103.2	91.2	93.8	129.0	117.1	92.0	110.1	103.9
0960	Flooring	108.6	126.5	113.9	108.4	126.5	113.7	105.9	127.8	112.4	110.7	126.5	115.4	103.7	126.5	110.4	106.1	126.5	112.1
0970, 0990	Wall Finishes & Painting/Coating	90.2	101.0	96.9	96.0	101.0	99.1	88.5	111.6	102.7	101.7	105.1	103.8	97.6	132.1	118.9	92.0	105.1	100.1
09	**FINISHES**	94.1	101.9	98.5	111.7	102.1	106.4	97.0	118.4	108.9	95.7	108.6	102.9	98.4	129.7	115.8	98.2	113.1	106.5
COVERS	**DIVS. 10 - 14, 25, 28, 41, 43, 44, 46**	100.0	97.9	99.6	100.0	98.2	99.6	100.0	105.4	101.1	100.0	98.1	99.6	100.0	110.2	102.0	100.0	104.1	100.8
21, 22, 23	**FIRE SUPPRESSION, PLUMBING & HVAC**	93.7	92.0	93.0	94.5	91.7	93.3	100.1	101.0	100.5	93.7	91.7	92.9	100.0	121.6	108.8	100.0	99.4	99.8
26, 27, 3370	**ELECTRICAL, COMMUNICATIONS & UTIL.**	102.1	108.9	105.6	104.1	70.5	86.8	99.5	108.9	104.3	108.7	89.0	98.5	101.6	113.1	107.6	100.6	89.1	94.7
MF2010	**WEIGHTED AVERAGE**	97.6	105.3	101.0	99.4	99.3	99.4	100.1	112.0	105.4	97.7	104.6	100.7	102.1	123.0	111.3	100.9	107.6	103.8

	DIVISION	MINNESOTA															MISSISSIPPI		
		SAINT PAUL			ST. CLOUD			THIEF RIVER FALLS			WILLMAR			WINDOM			BILOXI		
		550 - 551			563			567			562			561			395		
		MAT.	INST.	TOTAL	MAT.	INST.	TOTAL	MAT.	INST.	TOTAL	MAT.	INST.	TOTAL	MAT.	INST.	TOTAL	MAT.	INST.	TOTAL
015433	CONTRACTOR EQUIPMENT		101.2	101.2		101.4	101.4		98.7	98.7		101.4	101.4		101.4	101.4		98.5	98.5
0241, 31 - 34	SITE & INFRASTRUCTURE, DEMOLITION	101.0	102.9	102.3	91.1	105.4	101.0	97.1	98.6	98.1	90.8	103.2	99.4	84.1	101.9	96.4	101.8	88.9	92.9
0310	Concrete Forming & Accessories	97.8	129.3	125.2	91.3	124.9	120.5	91.4	93.4	93.2	91.2	96.2	95.5	95.2	87.9	88.9	94.7	56.1	61.2
0320	Concrete Reinforcing	95.4	120.1	107.8	102.3	120.0	111.2	103.9	109.3	106.7	101.9	119.6	110.8	101.9	118.7	110.4	96.3	60.1	78.1
0330	Cast-in-Place Concrete	120.3	118.4	119.5	101.3	117.9	107.8	104.7	91.3	99.4	102.9	89.3	97.5	89.0	67.0	80.4	106.6	55.2	86.4
03	CONCRETE	112.0	124.1	117.9	94.9	121.9	108.0	98.2	96.8	97.5	95.0	99.3	97.1	85.6	87.8	86.6	98.8	58.2	79.0
04	MASONRY	117.2	129.1	124.4	108.6	126.5	119.5	100.2	107.1	104.4	113.0	117.6	115.8	123.1	90.1	102.9	95.6	45.7	65.1
05	METALS	99.0	131.6	109.8	94.5	129.5	106.0	92.8	121.8	102.3	93.6	127.0	104.6	93.5	124.6	103.8	92.5	83.7	89.6
06	WOOD, PLASTICS & COMPOSITES	106.6	127.6	118.7	91.3	122.2	109.2	73.6	90.9	83.6	91.1	91.0	91.0	94.9	88.1	91.0	99.3	59.6	76.4
07	THERMAL & MOISTURE PROTECTION	110.3	132.4	119.3	107.8	122.8	113.9	109.4	101.5	106.2	107.6	109.6	108.4	107.6	86.6	99.0	92.8	57.0	78.3
08	OPENINGS	91.7	135.6	102.7	90.4	132.7	101.1	97.7	111.9	101.3	88.0	96.2	90.0	92.1	94.6	92.8	97.9	60.7	88.5
0920	Plaster & Gypsum Board	91.2	128.9	118.1	86.5	123.4	112.9	105.2	91.0	95.0	86.5	91.2	89.9	86.5	88.2	87.8	97.3	58.8	69.8
0950, 0980	Ceilings & Acoustic Treatment	89.3	128.9	115.4	67.9	123.4	104.6	137.6	91.0	106.8	67.9	91.2	83.3	67.9	88.2	81.3	95.8	58.8	71.4
0960	Flooring	100.0	126.5	107.8	104.6	126.5	111.1	109.4	126.5	114.4	106.5	126.5	112.4	109.2	126.5	114.3	117.7	55.3	99.4
0970, 0990	Wall Finishes & Painting/Coating	97.6	127.6	116.1	101.7	132.1	120.5	96.0	101.0	99.1	96.0	101.0	99.1	96.0	105.1	101.6	105.7	45.8	68.7
09	FINISHES	95.0	129.0	114.0	93.2	126.2	111.6	112.2	100.2	105.5	93.4	99.6	96.8	93.5	95.0	94.3	104.4	55.1	76.9
COVERS	DIVS. 10 - 14, 25, 28, 41, 43, 44, 46	100.0	109.9	102.0	100.0	104.5	100.9	100.0	95.6	99.1	100.0	97.8	99.6	100.0	92.9	98.6	100.0	59.9	92.0
21, 22, 23	FIRE SUPPRESSION, PLUMBING & HVAC	100.0	117.1	106.9	99.7	121.3	108.4	94.5	87.7	91.7	93.7	110.0	100.3	93.7	86.1	90.6	99.9	50.4	79.9
26, 27, 3370	ELECTRICAL, COMMUNICATIONS & UTIL.	98.7	113.1	106.1	102.1	113.1	107.8	101.7	70.5	85.6	102.1	90.5	96.1	108.7	89.0	98.5	98.8	59.0	78.3
MF2010	WEIGHTED AVERAGE	101.1	121.5	110.1	97.6	120.8	107.9	98.3	95.8	97.2	96.0	105.1	100.0	96.4	93.8	95.2	98.3	60.2	81.5

	DIVISION	MISSISSIPPI																	
		CLARKSDALE			COLUMBUS			GREENVILLE			GREENWOOD			JACKSON			LAUREL		
		386			397			387			389			390 - 392			394		
		MAT.	INST.	TOTAL	MAT.	INST.	TOTAL	MAT.	INST.	TOTAL	MAT.	INST.	TOTAL	MAT.	INST.	TOTAL	MAT.	INST.	TOTAL
015433	CONTRACTOR EQUIPMENT		98.5	98.5		98.5	98.5		98.5	98.5		98.5	98.5		98.5	98.5		98.5	98.5
0241, 31 - 34	SITE & INFRASTRUCTURE, DEMOLITION	99.2	88.7	91.9	100.5	89.0	92.6	105.7	88.9	94.1	102.4	88.4	92.8	97.3	88.9	91.5	107.1	88.8	94.5
0310	Concrete Forming & Accessories	86.1	52.4	56.8	84.3	54.5	58.4	83.0	67.7	69.7	94.6	52.3	57.8	94.0	61.1	65.4	84.4	56.3	59.9
0320	Concrete Reinforcing	103.8	44.2	73.8	103.2	45.2	74.0	104.4	64.0	84.0	103.8	55.8	79.6	96.3	59.1	77.6	103.8	41.5	72.4
0330	Cast-in-Place Concrete	103.9	56.1	85.1	108.7	61.4	90.1	107.0	58.3	87.8	111.8	55.7	89.7	90.7	59.1	78.2	106.3	57.8	87.2
03	CONCRETE	97.0	53.8	76.0	100.6	56.8	79.2	102.0	65.1	84.0	103.3	55.7	80.1	90.8	61.5	76.5	102.6	55.7	79.8
04	MASONRY	95.4	47.6	66.2	122.5	54.6	81.1	141.8	51.1	86.4	96.0	47.4	66.4	96.6	51.1	68.8	118.2	50.9	77.1
05	METALS	89.6	76.2	85.2	89.5	77.1	85.4	90.7	85.6	89.0	89.6	77.7	85.7	92.4	83.5	89.5	89.6	76.1	85.1
06	WOOD, PLASTICS & COMPOSITES	87.3	54.4	68.3	85.4	55.5	68.1	84.1	72.4	77.4	99.3	54.4	73.4	98.0	63.3	78.0	86.4	58.0	70.0
07	THERMAL & MOISTURE PROTECTION	92.7	49.8	75.2	92.8	54.3	77.1	93.0	63.0	80.8	93.0	49.7	75.4	94.6	62.0	81.4	92.9	56.1	77.9
08	OPENINGS	97.1	52.4	85.9	97.1	52.5	85.9	97.1	65.1	89.1	97.1	53.8	86.2	98.3	58.8	88.3	93.7	53.3	83.5
0920	Plaster & Gypsum Board	85.5	53.5	62.6	85.1	54.5	63.3	85.1	72.0	75.7	96.0	53.5	65.6	91.2	62.7	70.8	85.1	57.2	65.2
0950, 0980	Ceilings & Acoustic Treatment	89.3	53.5	65.6	89.3	54.5	66.3	91.3	72.0	78.6	89.3	53.5	65.6	93.1	62.7	73.0	89.3	57.2	68.1
0960	Flooring	111.4	55.3	95.0	110.2	62.1	96.1	109.4	55.3	93.5	117.7	55.3	99.4	118.1	55.3	99.7	108.4	55.3	92.8
0970, 0990	Wall Finishes & Painting/Coating	105.7	48.6	70.5	105.7	55.8	74.9	105.7	69.4	83.3	105.7	48.6	70.5	105.7	69.4	83.3	105.7	64.2	80.1
09	FINISHES	99.5	52.8	73.5	99.4	56.4	75.4	100.0	66.6	81.4	103.0	52.8	75.0	103.4	61.3	79.9	99.5	57.5	76.1
COVERS	DIVS. 10 - 14, 25, 28, 41, 43, 44, 46	100.0	60.3	92.1	100.0	61.5	92.3	100.0	63.2	92.7	100.0	60.2	92.1	100.0	62.2	92.5	100.0	40.4	88.1
21, 22, 23	FIRE SUPPRESSION, PLUMBING & HVAC	97.4	44.3	75.9	96.7	40.3	73.8	99.9	51.1	80.1	97.4	42.2	75.0	99.9	56.5	82.3	96.7	49.2	77.5
26, 27, 3370	ELECTRICAL, COMMUNICATIONS & UTIL.	98.1	49.3	72.9	96.3	52.2	73.5	98.1	66.6	81.9	98.1	46.2	71.3	99.5	66.6	82.5	97.8	59.5	78.1
MF2010	WEIGHTED AVERAGE	96.4	55.5	78.4	97.8	56.9	79.8	100.3	65.1	84.7	97.6	55.1	78.8	97.4	64.4	82.9	97.9	58.8	80.6

	DIVISION	MISSISSIPPI									MISSOURI								
		MCCOMB			MERIDIAN			TUPELO			BOWLING GREEN			CAPE GIRARDEAU			CHILLICOTHE		
		396			393			388			633			637			646		
		MAT.	INST.	TOTAL	MAT.	INST.	TOTAL	MAT.	INST.	TOTAL	MAT.	INST.	TOTAL	MAT.	INST.	TOTAL	MAT.	INST.	TOTAL
015433	CONTRACTOR EQUIPMENT		98.5	98.5		98.5	98.5		98.5	98.5		106.1	106.1		106.1	106.1		102.2	102.2
0241, 31 - 34	SITE & INFRASTRUCTURE, DEMOLITION	92.8	88.6	89.9	97.0	89.2	91.6	96.8	88.6	91.1	91.4	93.6	92.9	92.7	93.5	93.3	100.3	90.9	93.8
0310	Concrete Forming & Accessories	84.3	54.2	58.1	81.9	67.2	69.1	83.5	54.5	58.3	91.8	87.1	87.7	85.7	75.0	76.4	90.3	90.0	90.0
0320	Concrete Reinforcing	104.4	43.6	73.8	103.2	59.1	81.0	101.6	55.8	78.5	100.9	96.9	98.9	102.1	83.3	92.7	97.5	100.1	98.8
0330	Cast-in-Place Concrete	94.4	55.6	79.1	101.0	61.6	85.4	103.9	57.6	85.7	94.6	75.6	87.1	93.7	82.5	89.2	93.0	65.8	82.3
03	CONCRETE	90.5	54.2	72.8	94.8	65.1	80.3	96.5	57.3	77.4	96.3	86.1	91.4	95.5	80.6	88.3	99.8	83.9	92.1
04	MASONRY	123.9	46.9	76.9	95.3	55.3	70.9	131.6	50.6	82.2	124.4	95.9	107.0	121.5	77.3	94.5	98.6	90.8	93.8
05	METALS	89.7	74.3	84.6	90.6	83.7	88.3	89.5	77.9	85.7	97.9	110.8	102.1	98.9	104.4	100.8	95.5	96.5	95.8
06	WOOD, PLASTICS & COMPOSITES	85.4	57.4	69.2	83.0	69.3	75.1	84.7	55.5	67.8	88.9	87.4	88.0	82.6	72.3	76.6	91.9	96.0	94.3
07	THERMAL & MOISTURE PROTECTION	92.4	48.9	74.7	92.6	64.5	81.1	92.6	55.8	77.7	101.0	98.4	99.9	100.9	81.1	92.8	97.6	84.9	92.4
08	OPENINGS	97.2	54.1	86.3	97.1	67.9	89.8	97.1	53.7	86.2	99.0	83.1	95.0	99.0	74.3	92.7	90.8	84.5	89.2
0920	Plaster & Gypsum Board	85.1	56.5	64.7	85.1	68.8	73.4	85.1	54.5	63.3	97.3	87.0	89.9	95.5	71.3	78.2	103.1	95.7	97.8
0950, 0980	Ceilings & Acoustic Treatment	89.3	56.5	67.6	91.3	68.8	76.4	89.3	54.5	66.3	90.1	87.0	88.0	90.1	71.3	77.7	89.5	95.7	93.6
0960	Flooring	110.2	55.3	94.1	108.4	55.3	92.8	109.7	55.3	93.7	101.2	95.8	99.6	97.5	88.0	94.7	93.5	45.0	79.3
0970, 0990	Wall Finishes & Painting/Coating	105.7	44.6	68.0	105.7	80.3	90.0	105.7	61.9	78.7	102.0	103.4	102.8	102.0	72.7	83.9	94.3	47.3	65.3
09	FINISHES	98.6	54.0	73.7	98.9	67.0	81.1	98.9	55.7	74.8	98.5	89.4	93.4	97.1	75.3	84.9	96.4	79.5	87.0
COVERS	DIVS. 10 - 14, 25, 28, 41, 43, 44, 46	100.0	63.9	92.8	100.0	64.3	92.9	100.0	61.5	92.3	100.0	89.5	97.9	100.0	90.4	98.1	100.0	55.7	91.2
21, 22, 23	FIRE SUPPRESSION, PLUMBING & HVAC	96.7	37.1	72.6	99.9	58.7	83.2	97.7	46.4	76.9	93.9	86.7	91.0	99.9	94.9	97.8	94.2	44.0	73.9
26, 27, 3370	ELECTRICAL, COMMUNICATIONS & UTIL.	94.8	52.5	73.0	97.8	59.2	77.9	97.8	52.3	74.4	99.7	82.9	91.0	99.7	107.9	103.9	95.9	34.9	64.5
MF2010	WEIGHTED AVERAGE	96.3	54.6	77.9	96.7	66.1	83.2	98.0	57.9	80.4	98.4	90.5	94.9	99.7	89.6	95.3	96.1	70.0	84.6

	DIVISION	MISSOURI																	
		COLUMBIA			FLAT RIVER			HANNIBAL			HARRISONVILLE			JEFFERSON CITY			JOPLIN		
		652			636			634			647			650 - 651			648		
		MAT.	INST.	TOTAL	MAT.	INST.	TOTAL	MAT.	INST.	TOTAL	MAT.	INST.	TOTAL	MAT.	INST.	TOTAL	MAT.	INST.	TOTAL
015433	CONTRACTOR EQUIPMENT		107.0	107.0		106.1	106.1		106.1	106.1		102.2	102.2		107.0	107.0		105.7	105.7
0241, 31 - 34	SITE & INFRASTRUCTURE, DEMOLITION	105.3	95.9	98.8	94.1	93.6	93.8	89.0	93.4	92.0	91.7	95.8	94.5	101.7	95.8	97.6	100.4	98.5	99.1
0310	Concrete Forming & Accessories	87.3	79.5	80.5	97.1	87.1	88.4	90.2	77.3	79.0	87.6	104.1	101.9	96.7	79.3	81.6	101.1	73.5	77.1
0320	Concrete Reinforcing	99.3	115.6	107.5	102.1	105.8	104.0	100.3	96.8	98.5	97.1	111.9	104.5	98.1	106.7	102.4	100.6	72.3	86.3
0330	Cast-in-Place Concrete	89.5	82.6	86.8	97.7	94.2	96.3	89.6	75.1	83.9	95.3	106.7	99.8	90.0	82.5	87.1	100.7	66.1	87.1
03	CONCRETE	86.1	88.8	87.4	99.3	94.2	96.8	92.5	81.6	87.2	96.4	106.7	101.4	87.5	87.0	87.2	99.6	71.6	86.0
04	MASONRY	133.3	85.6	104.2	122.0	78.8	95.6	115.2	95.9	103.4	93.1	107.9	102.1	96.0	85.6	89.7	92.4	84.8	87.7
05	METALS	94.5	119.5	102.7	97.8	113.8	103.1	97.9	109.6	101.7	95.9	112.3	101.4	93.9	115.8	101.1	98.6	88.4	95.2
06	WOOD, PLASTICS & COMPOSITES	92.4	77.1	83.6	95.6	87.4	90.9	87.1	75.1	80.2	88.6	103.7	97.3	102.4	77.1	87.8	102.8	74.5	86.4
07	THERMAL & MOISTURE PROTECTION	93.3	86.0	90.3	101.2	93.8	98.2	100.8	92.3	97.4	96.9	107.7	101.3	93.4	86.0	90.4	97.4	72.2	87.1
08	OPENINGS	90.6	94.5	91.6	99.0	85.7	95.6	99.0	76.4	93.3	91.2	106.0	94.9	87.8	92.1	88.9	92.0	72.4	87.1
0920	Plaster & Gypsum Board	99.9	76.3	83.1	102.7	87.0	91.5	95.9	74.3	80.5	97.1	103.6	101.8	104.2	76.3	84.3	110.2	73.5	84.0
0950, 0980	Ceilings & Acoustic Treatment	94.9	76.3	82.6	90.1	87.0	88.0	90.1	74.3	79.7	89.5	103.6	98.8	94.0	76.3	82.3	90.4	73.5	79.2
0960	Flooring	112.4	95.9	107.5	104.5	88.0	99.6	100.3	95.8	99.0	89.0	107.3	94.4	120.0	95.9	112.9	118.8	42.2	96.3
0970, 0990	Wall Finishes & Painting/Coating	104.0	79.4	88.8	102.0	62.6	77.7	102.0	103.4	102.8	99.2	107.0	104.0	104.0	79.4	88.8	93.8	37.0	58.7
09	FINISHES	99.5	81.1	89.2	100.2	83.4	90.9	97.8	82.1	89.1	93.9	104.8	100.0	102.4	81.1	90.5	103.0	64.4	81.5
COVERS	DIVS. 10 - 14, 25, 28, 41, 43, 44, 46	100.0	95.3	99.1	100.0	92.6	98.5	100.0	54.7	91.0	100.0	97.6	99.5	100.0	95.3	99.1	100.0	62.0	92.4
21, 22, 23	FIRE SUPPRESSION, PLUMBING & HVAC	99.9	95.2	98.0	93.9	86.7	91.0	93.9	86.6	91.0	94.1	103.3	97.8	99.9	89.5	95.7	100.2	48.0	79.0
26, 27, 3370	ELECTRICAL, COMMUNICATIONS & UTIL.	95.6	88.0	91.7	104.2	107.9	106.1	98.4	82.9	90.4	102.6	101.7	102.1	99.3	88.0	93.4	93.7	69.0	81.0
MF2010	WEIGHTED AVERAGE	97.7	92.4	95.4	99.4	92.9	96.5	97.3	87.1	92.8	95.7	104.5	99.6	96.2	90.5	93.7	98.2	70.4	85.9

	DIVISION	MISSOURI																	
		KANSAS CITY			KIRKSVILLE			POPLAR BLUFF			ROLLA			SEDALIA			SIKESTON		
		640 - 641			635			639			654 - 655			653			638		
		MAT.	INST.	TOTAL	MAT.	INST.	TOTAL	MAT.	INST.	TOTAL	MAT.	INST.	TOTAL	MAT.	INST.	TOTAL	MAT.	INST.	TOTAL
015433	CONTRACTOR EQUIPMENT		103.4	103.4		98.5	98.5		100.9	100.9		107.0	107.0		99.2	99.2		100.9	100.9
0241, 31 - 34	SITE & INFRASTRUCTURE, DEMOLITION	93.2	98.2	96.7	90.3	89.5	89.8	77.2	92.9	88.1	104.8	94.0	97.3	96.7	91.8	93.3	80.7	93.2	89.3
0310	Concrete Forming & Accessories	100.1	107.9	106.9	83.5	77.9	78.6	83.7	72.6	74.0	93.9	90.4	90.8	91.3	79.8	81.3	84.7	73.6	75.0
0320	Concrete Reinforcing	95.6	116.5	106.1	101.2	86.0	93.5	104.2	83.3	93.7	99.8	67.8	83.7	98.3	104.7	101.5	103.6	83.2	93.3
0330	Cast-in-Place Concrete	93.7	108.0	99.3	97.6	75.1	88.8	75.1	79.7	77.0	91.6	85.5	89.2	95.7	83.2	90.8	80.3	82.0	81.0
03	CONCRETE	95.6	109.7	102.5	110.1	79.3	95.1	85.7	78.1	82.0	87.8	85.8	86.8	99.2	86.5	93.0	89.4	79.3	84.5
04	MASONRY	95.1	109.7	104.0	128.1	87.4	103.2	119.9	71.6	90.4	109.8	78.8	90.8	115.7	86.0	97.5	119.6	76.2	93.1
05	METALS	106.5	114.9	109.3	97.4	96.4	97.1	98.0	96.5	97.5	93.8	102.3	96.6	92.4	106.7	97.1	98.3	95.7	97.5
06	WOOD, PLASTICS & COMPOSITES	102.0	107.6	105.2	75.9	77.1	76.6	75.1	72.3	73.5	99.1	93.2	95.7	92.0	77.1	83.4	76.5	72.3	74.1
07	THERMAL & MOISTURE PROTECTION	97.0	109.6	102.1	107.3	85.8	98.6	105.9	80.9	95.7	93.4	91.0	92.4	98.8	93.0	96.4	106.1	79.4	95.2
08	OPENINGS	98.2	109.8	101.2	103.3	74.9	96.2	104.3	73.6	96.6	90.6	87.3	89.8	93.9	87.1	92.2	104.3	67.1	94.9
0920	Plaster & Gypsum Board	106.1	107.6	107.2	90.4	76.3	80.3	90.9	71.3	76.9	100.8	92.9	95.1	93.9	76.3	81.4	92.7	71.3	77.5
0950, 0980	Ceilings & Acoustic Treatment	96.7	107.6	103.9	88.3	76.3	80.4	90.1	71.3	77.7	94.9	92.9	93.6	94.9	76.3	82.6	90.1	71.3	77.7
0960	Flooring	94.3	109.0	98.6	77.5	95.9	82.9	92.2	81.4	89.1	116.3	95.9	110.3	88.4	97.1	90.9	92.7	81.4	89.4
0970, 0990	Wall Finishes & Painting/Coating	99.2	114.7	108.7	97.2	79.4	86.2	96.7	72.1	81.5	104.0	86.4	93.1	104.0	105.7	105.0	96.7	57.6	72.6
09	FINISHES	98.1	108.6	104.0	96.7	80.2	87.5	97.0	72.7	83.4	100.9	90.4	95.0	97.4	84.5	90.2	97.7	71.6	83.1
COVERS	DIVS. 10 - 14, 25, 28, 41, 43, 44, 46	100.0	101.5	100.3	100.0	53.1	90.7	100.0	88.4	97.7	100.0	57.1	91.5	100.0	89.7	97.9	100.0	90.0	98.0
21, 22, 23	FIRE SUPPRESSION, PLUMBING & HVAC	100.0	110.9	104.4	93.9	86.1	90.7	93.9	91.3	92.9	93.9	86.9	91.1	93.9	96.0	94.7	93.9	84.4	90.1
26, 27, 3370	ELECTRICAL, COMMUNICATIONS & UTIL.	104.3	102.0	103.1	98.6	82.8	90.5	98.9	107.8	103.5	94.1	87.9	90.9	95.7	101.6	98.8	97.9	107.8	103.0
MF2010	WEIGHTED AVERAGE	100.1	108.0	103.6	100.4	83.8	93.1	97.0	86.8	92.5	95.2	87.9	92.0	96.6	92.8	94.9	97.5	85.5	92.2

	DIVISION	MISSOURI									MONTANA								
		SPRINGFIELD			ST. JOSEPH			ST. LOUIS			BILLINGS			BUTTE			GREAT FALLS		
		656 - 658			644 - 645			630 - 631			590 - 591			597			594		
		MAT.	INST.	TOTAL	MAT.	INST.	TOTAL	MAT.	INST.	TOTAL	MAT.	INST.	TOTAL	MAT.	INST.	TOTAL	MAT.	INST.	TOTAL
015433	CONTRACTOR EQUIPMENT		101.7	101.7		102.2	102.2		107.1	107.1		98.9	98.9		98.7	98.7		98.7	98.7
0241, 31 - 34	SITE & INFRASTRUCTURE, DEMOLITION	99.2	93.6	95.4	94.7	94.0	94.2	92.6	97.5	96.0	98.6	97.0	97.5	109.4	96.3	100.4	113.3	97.1	102.1
0310	Concrete Forming & Accessories	98.8	74.1	77.4	99.9	87.6	89.2	97.1	103.4	102.6	101.9	65.9	70.6	90.3	64.0	67.5	101.7	67.4	71.8
0320	Concrete Reinforcing	95.6	115.1	105.4	94.4	111.1	102.8	93.4	107.6	100.5	101.4	80.2	90.7	110.0	80.3	95.0	101.4	80.2	90.7
0330	Cast-in-Place Concrete	97.2	72.4	87.5	93.6	104.1	97.7	93.7	106.3	98.7	115.2	66.3	95.9	126.8	67.9	103.6	134.0	64.4	106.5
03	CONCRETE	96.3	82.1	89.4	95.4	98.4	96.9	94.9	106.0	100.3	106.3	69.7	88.5	110.4	69.5	90.5	115.5	69.7	93.2
04	MASONRY	89.5	84.5	86.4	94.9	93.9	94.3	101.5	109.4	106.3	125.4	69.9	91.5	120.7	71.7	90.8	125.3	73.9	93.9
05	METALS	98.4	106.2	101.0	102.5	110.5	105.1	103.8	118.1	108.5	110.2	89.8	103.5	103.0	89.9	98.7	106.8	89.8	101.2
06	WOOD, PLASTICS & COMPOSITES	99.0	73.3	84.2	102.8	85.1	92.5	95.5	101.7	99.1	97.8	66.2	79.6	86.9	63.6	73.4	98.9	66.2	80.0
07	THERMAL & MOISTURE PROTECTION	97.5	80.1	90.5	97.5	95.5	96.7	100.9	106.7	103.3	107.4	69.0	91.8	107.3	69.5	91.9	107.8	68.4	91.8
08	OPENINGS	96.1	81.5	92.4	96.8	96.4	96.7	98.2	108.6	100.9	96.6	66.3	89.0	94.6	66.1	87.4	98.0	65.6	89.8
0920	Plaster & Gypsum Board	102.2	72.3	80.9	111.8	84.4	92.2	103.9	101.7	102.3	96.9	65.5	74.4	100.6	62.7	73.6	108.9	65.5	77.9
0950, 0980	Ceilings & Acoustic Treatment	94.9	72.3	80.0	95.8	84.4	88.3	95.5	101.7	99.6	106.2	65.5	79.3	111.5	62.7	79.3	113.3	65.5	81.7
0960	Flooring	113.8	42.2	92.8	97.5	97.1	97.4	104.2	99.3	102.8	109.2	73.6	98.7	106.4	77.7	98.0	114.8	77.7	103.9
0970, 0990	Wall Finishes & Painting/Coating	98.0	52.9	70.2	94.3	88.8	90.9	102.0	106.7	104.9	97.5	54.5	71.0	96.0	54.0	70.1	96.0	51.3	68.4
09	FINISHES	103.2	65.4	82.1	99.2	88.7	93.4	101.4	102.4	101.9	104.5	65.7	82.8	105.0	65.0	82.7	109.2	67.0	85.7
COVERS	DIVS. 10 - 14, 25, 28, 41, 43, 44, 46	100.0	93.4	98.7	100.0	96.8	99.4	100.0	102.3	100.5	100.0	74.9	95.0	100.0	74.7	95.0	100.0	76.3	95.3
21, 22, 23	FIRE SUPPRESSION, PLUMBING & HVAC	99.9	70.8	88.1	100.2	87.3	94.9	99.9	101.7	100.6	100.2	71.4	88.5	100.3	68.4	87.4	100.3	73.5	89.4
26, 27, 3370	ELECTRICAL, COMMUNICATIONS & UTIL.	99.8	72.4	85.7	102.7	76.5	89.2	102.2	108.3	105.3	97.3	76.7	86.7	104.9	72.6	88.2	97.5	73.5	85.1
MF2010	WEIGHTED AVERAGE	98.5	79.9	90.3	99.3	91.8	96.0	100.0	105.7	102.5	103.5	74.7	90.8	103.5	73.5	90.3	105.0	75.3	91.9

DIVISION		MONTANA																	
		HAVRE			HELENA			KALISPELL			MILES CITY			MISSOULA			WOLF POINT		
		595			596			599			593			598			592		
		MAT.	INST.	TOTAL	MAT.	INST.	TOTAL	MAT.	INST.	TOTAL	MAT.	INST.	TOTAL	MAT.	INST.	TOTAL	MAT.	INST.	TOTAL
015433	CONTRACTOR EQUIPMENT		98.7	98.7		98.7	98.7		98.7	98.7		98.7	98.7		98.7	98.7		98.7	98.7
0241, 31 - 34	SITE & INFRASTRUCTURE, DEMOLITION	118.8	96.7	103.6	101.9	96.2	98.0	98.2	96.0	96.7	105.9	96.4	99.3	89.2	96.2	94.0	127.4	96.5	106.1
0310	Concrete Forming & Accessories	84.2	57.2	60.7	101.9	66.6	71.2	93.2	55.2	60.1	100.4	55.7	61.5	93.2	62.3	66.4	93.7	50.7	56.3
0320	Concrete Reinforcing	110.8	64.3	87.4	104.9	64.0	84.3	112.9	80.0	96.3	110.4	64.3	87.2	111.8	80.2	95.9	112.0	64.3	88.0
0330	Cast-in-Place Concrete	136.5	62.9	107.5	104.0	63.3	88.0	110.1	60.6	90.6	120.5	62.6	97.7	93.4	62.6	81.3	135.0	63.0	106.6
03	CONCRETE	118.9	61.7	91.0	101.0	65.9	83.9	99.8	62.9	81.8	107.2	60.9	84.6	87.9	66.8	77.7	122.5	58.8	91.5
04	MASONRY	121.6	72.0	91.3	116.6	68.5	87.2	119.6	66.9	87.4	127.2	67.2	90.6	145.2	70.2	99.4	128.3	67.9	91.5
05	METALS	99.4	82.0	93.7	106.0	81.3	97.8	99.3	88.2	95.6	98.6	81.9	93.1	99.8	89.0	96.3	98.7	81.8	93.1
06	WOOD, PLASTICS & COMPOSITES	79.6	54.1	64.9	98.9	66.2	80.0	89.8	52.4	68.2	96.4	54.5	72.2	89.8	62.4	74.0	88.9	47.5	65.0
07	THERMAL & MOISTURE PROTECTION	107.6	65.6	90.5	107.6	70.4	92.5	106.9	66.5	90.5	107.2	63.8	89.5	106.5	68.9	91.2	108.0	62.8	89.6
08	OPENINGS	94.7	54.5	84.6	97.4	60.0	88.0	94.7	58.1	85.5	94.2	54.7	84.2	94.7	63.3	86.8	94.2	50.7	83.2
0920	Plaster & Gypsum Board	96.1	52.9	65.3	97.7	65.5	74.7	100.6	51.2	65.3	107.8	53.4	68.9	100.6	61.5	72.7	102.3	46.2	62.2
0950, 0980	Ceilings & Acoustic Treatment	111.5	52.9	72.8	110.6	65.5	80.8	111.5	51.2	71.7	109.7	53.4	72.5	111.5	61.5	78.5	109.7	46.2	67.7
0960	Flooring	103.1	77.7	95.6	114.8	61.9	99.3	108.5	77.7	99.4	114.6	73.6	102.5	108.5	77.7	99.4	110.2	73.6	99.4
0970, 0990	Wall Finishes & Painting/Coating	96.0	51.3	68.4	96.0	45.6	64.9	96.0	39.4	61.1	96.0	47.3	66.0	96.0	50.2	67.7	96.0	51.3	68.4
09	FINISHES	104.3	59.4	79.2	106.6	63.7	82.7	104.9	56.7	78.0	107.6	57.3	79.6	104.2	63.5	81.5	107.4	53.7	77.4
COVERS	DIVS. 10 - 14, 25, 28, 41, 43, 44, 46	100.0	74.3	94.9	100.0	76.3	95.3	100.0	73.4	94.7	100.0	72.7	94.6	100.0	74.1	94.9	100.0	72.1	94.5
21, 22, 23	FIRE SUPPRESSION, PLUMBING & HVAC	94.3	72.1	85.3	100.2	69.9	87.9	94.3	68.1	83.7	94.3	69.6	84.3	100.3	67.6	87.0	94.3	70.0	84.5
26, 27, 3370	ELECTRICAL, COMMUNICATIONS & UTIL.	97.5	66.0	81.2	101.9	73.5	87.3	101.7	71.9	86.3	97.5	68.8	82.7	102.7	71.9	86.8	97.5	64.0	80.2
MF2010	WEIGHTED AVERAGE	102.0	70.2	88.0	102.6	72.0	89.1	99.6	70.1	86.6	100.7	69.1	86.8	100.8	72.2	88.2	103.1	67.5	87.4

DIVISION		NEBRASKA																	
		ALLIANCE			COLUMBUS			GRAND ISLAND			HASTINGS			LINCOLN			MCCOOK		
		693			686			688			689			683 - 685			690		
		MAT.	INST.	TOTAL	MAT.	INST.	TOTAL	MAT.	INST.	TOTAL	MAT.	INST.	TOTAL	MAT.	INST.	TOTAL	MAT.	INST.	TOTAL
015433	CONTRACTOR EQUIPMENT		97.0	97.0		100.8	100.8		100.8	100.8		100.8	100.8		100.8	100.8		100.8	100.8
0241, 31 - 34	SITE & INFRASTRUCTURE, DEMOLITION	99.4	97.5	98.1	98.7	90.9	93.3	103.4	92.1	95.6	102.3	91.3	94.7	91.0	92.1	91.7	103.2	91.1	94.9
0310	Concrete Forming & Accessories	88.2	56.3	60.5	96.3	59.1	64.0	95.9	76.0	78.6	98.8	74.3	77.4	100.2	60.3	65.5	93.7	56.6	61.5
0320	Concrete Reinforcing	113.6	47.6	80.3	104.7	46.4	75.3	104.1	75.5	89.7	104.1	47.1	75.4	95.6	74.9	85.2	106.1	75.0	90.4
0330	Cast-in-Place Concrete	108.2	58.5	88.6	110.2	55.4	88.6	116.7	69.8	98.2	116.7	64.1	96.0	87.2	72.3	81.3	116.7	58.2	93.7
03	CONCRETE	117.8	56.1	87.7	104.1	56.6	81.0	108.7	74.4	92.0	108.8	66.4	88.2	91.3	68.3	80.1	108.8	61.8	85.9
04	MASONRY	110.1	66.8	83.7	115.3	71.6	88.7	108.6	79.0	90.6	117.7	88.6	99.9	96.8	69.7	80.2	105.1	66.9	81.8
05	METALS	100.6	65.3	88.9	92.7	74.6	86.8	94.5	86.3	91.8	95.2	76.2	88.9	99.6	85.3	94.9	95.5	84.2	91.7
06	WOOD, PLASTICS & COMPOSITES	85.2	55.5	68.0	96.4	60.3	75.6	95.8	76.8	84.8	99.3	76.8	86.3	98.5	55.8	73.8	93.9	55.5	71.7
07	THERMAL & MOISTURE PROTECTION	104.5	67.8	89.6	98.3	64.8	84.7	98.4	75.3	89.0	98.5	86.7	93.7	97.9	70.6	86.8	98.4	65.0	84.8
08	OPENINGS	89.9	51.7	80.3	89.8	56.2	81.4	89.8	72.7	85.5	89.8	65.2	83.6	95.3	57.3	85.7	89.9	58.7	82.0
0920	Plaster & Gypsum Board	91.3	54.1	64.7	100.4	59.0	70.8	99.9	76.0	82.8	101.7	76.0	83.4	101.6	54.4	67.9	99.9	54.1	67.2
0950, 0980	Ceilings & Acoustic Treatment	99.2	54.1	69.4	94.9	59.0	71.2	94.9	76.0	82.4	94.9	76.0	82.4	97.6	54.4	69.0	94.9	54.1	67.9
0960	Flooring	103.9	86.0	98.6	102.6	105.1	103.3	102.4	112.3	105.3	103.8	105.1	104.2	104.0	87.5	99.2	101.5	86.0	96.9
0970, 0990	Wall Finishes & Painting/Coating	173.8	57.7	102.2	96.0	67.2	78.2	96.0	71.3	80.7	96.0	67.2	78.2	96.0	82.1	87.4	96.0	50.3	67.8
09	FINISHES	103.3	61.1	79.7	100.1	67.4	81.8	100.3	81.4	89.8	100.9	78.6	88.5	100.9	66.7	81.9	100.0	60.2	77.8
COVERS	DIVS. 10 - 14, 25, 28, 41, 43, 44, 46	100.0	68.0	93.6	100.0	77.2	95.5	100.0	82.6	96.5	100.0	81.2	96.3	100.0	80.2	96.1	100.0	67.3	93.5
21, 22, 23	FIRE SUPPRESSION, PLUMBING & HVAC	94.2	59.4	80.1	94.0	72.3	85.2	99.9	78.2	91.1	94.0	75.4	86.5	99.8	78.2	91.1	94.0	72.8	85.4
26, 27, 3370	ELECTRICAL, COMMUNICATIONS & UTIL.	92.7	73.7	82.9	92.3	81.9	86.9	91.0	67.3	78.8	90.4	88.8	89.6	99.8	67.3	83.0	94.8	73.7	83.9
MF2010	WEIGHTED AVERAGE	99.7	65.8	84.8	96.8	71.7	85.7	98.7	78.4	89.8	97.8	79.3	89.6	98.0	73.6	87.2	97.6	70.7	85.7

DIVISION		NEBRASKA												NEVADA					
		NORFOLK			NORTH PLATTE			OMAHA			VALENTINE			CARSON CITY			ELKO		
		687			691			680 - 681			692			897			898		
		MAT.	INST.	TOTAL	MAT.	INST.	TOTAL	MAT.	INST.	TOTAL	MAT.	INST.	TOTAL	MAT.	INST.	TOTAL	MAT.	INST.	TOTAL
015433	CONTRACTOR EQUIPMENT		92.2	92.2		100.8	100.8		92.2	92.2		95.1	95.1		99.3	99.3		99.3	99.3
0241, 31 - 34	SITE & INFRASTRUCTURE, DEMOLITION	81.1	90.5	87.6	104.3	91.3	95.3	81.5	91.1	88.1	86.2	95.0	92.3	69.7	101.3	91.5	56.0	101.0	87.0
0310	Concrete Forming & Accessories	83.1	73.4	74.7	95.8	74.1	76.9	91.5	73.1	75.5	84.3	55.9	59.6	101.5	97.6	98.1	108.0	84.0	87.1
0320	Concrete Reinforcing	104.8	49.1	76.7	105.5	76.1	90.7	100.1	75.5	87.7	106.1	46.1	75.9	105.0	117.5	111.3	106.7	91.8	99.2
0330	Cast-in-Place Concrete	111.4	63.3	92.4	116.7	65.5	96.5	93.6	78.7	87.7	103.2	55.9	84.6	94.6	87.8	91.9	95.4	83.5	90.7
03	CONCRETE	103.0	65.7	84.9	108.9	72.2	91.0	93.9	75.7	85.0	105.5	54.9	80.8	100.1	97.9	99.0	97.9	85.5	91.9
04	MASONRY	122.3	69.5	90.1	92.9	88.6	90.3	102.3	79.9	88.6	105.4	67.0	81.9	116.9	76.6	92.3	119.2	69.9	89.1
05	METALS	96.0	68.0	86.8	94.7	86.2	91.9	98.9	78.6	92.2	106.5	66.3	93.3	94.1	103.2	97.1	93.2	90.7	92.4
06	WOOD, PLASTICS & COMPOSITES	80.0	76.3	77.9	95.7	76.8	84.8	89.6	72.5	79.8	79.7	55.0	65.5	87.7	100.4	95.1	96.7	84.4	89.6
07	THERMAL & MOISTURE PROTECTION	97.9	73.7	88.1	98.4	80.6	91.2	94.1	80.5	88.6	98.8	67.6	86.1	99.6	87.0	94.5	100.9	75.9	90.7
08	OPENINGS	91.3	65.7	84.9	89.1	72.0	84.8	98.0	72.4	91.6	91.4	52.9	81.7	97.0	112.6	100.9	99.8	82.2	95.4
0920	Plaster & Gypsum Board	100.7	76.0	83.1	99.9	76.0	82.8	104.9	72.1	81.5	102.6	54.1	67.9	90.4	100.3	97.5	94.5	83.8	86.9
0950, 0980	Ceilings & Acoustic Treatment	109.8	76.0	87.5	94.9	76.0	82.4	111.6	72.1	85.5	113.1	54.1	74.1	109.7	100.3	103.5	112.4	83.8	93.5
0960	Flooring	130.7	105.1	123.2	102.4	105.1	103.2	135.4	85.8	120.8	133.2	83.6	118.6	101.1	82.0	95.5	106.6	53.5	91.0
0970, 0990	Wall Finishes & Painting/Coating	172.0	67.2	107.3	96.0	67.2	78.2	172.0	68.9	108.4	172.0	69.6	108.8	99.8	83.3	89.6	98.8	83.3	89.2
09	FINISHES	121.5	78.6	97.6	100.3	78.6	88.2	123.9	74.7	96.5	123.8	61.5	89.0	100.8	93.4	96.7	101.9	78.3	88.7
COVERS	DIVS. 10 - 14, 25, 28, 41, 43, 44, 46	100.0	80.0	96.0	100.0	70.8	94.2	100.0	81.0	96.2	100.0	65.2	93.1	100.0	97.1	99.4	100.0	67.8	93.6
21, 22, 23	FIRE SUPPRESSION, PLUMBING & HVAC	93.8	75.3	86.3	99.9	73.7	89.3	99.7	78.0	90.9	93.7	57.5	79.0	99.9	84.2	93.5	97.2	83.8	91.8
26, 27, 3370	ELECTRICAL, COMMUNICATIONS & UTIL.	91.2	82.5	86.7	93.0	81.9	87.3	99.2	82.8	90.8	89.9	92.0	91.0	91.1	98.5	94.9	93.9	98.4	96.3
MF2010	WEIGHTED AVERAGE	98.5	75.1	88.2	98.1	79.4	89.9	100.0	79.2	90.8	99.7	67.6	85.6	97.8	93.5	95.9	97.2	85.3	92.0

City Cost Indexes

DIVISION		NEVADA									NEW HAMPSHIRE								
		ELY			LAS VEGAS			RENO			CHARLESTON			CLAREMONT			CONCORD		
		893			889 - 891			894 - 895			036			037			032 - 033		
		MAT.	INST.	TOTAL	MAT.	INST.	TOTAL	MAT.	INST.	TOTAL	MAT.	INST.	TOTAL	MAT.	INST.	TOTAL	MAT.	INST.	TOTAL
015433	CONTRACTOR EQUIPMENT		99.3	99.3		99.3	99.3		99.3	99.3		100.4	100.4		100.4	100.4		100.4	100.4
0241, 31 - 34	SITE & INFRASTRUCTURE, DEMOLITION	60.8	101.2	88.6	63.1	103.8	91.1	60.7	101.3	88.7	85.6	96.0	92.8	79.1	96.0	90.7	91.9	98.0	96.1
0310	Concrete Forming & Accessories	101.8	104.9	104.5	103.6	114.8	113.3	98.9	97.6	97.7	88.6	43.4	49.3	93.8	43.4	50.0	95.5	87.1	88.2
0320	Concrete Reinforcing	105.5	85.0	95.1	97.2	125.4	111.4	99.7	124.1	112.0	82.9	46.7	64.6	82.9	46.7	64.6	82.9	91.9	87.4
0330	Cast-in-Place Concrete	102.5	103.1	102.7	99.2	110.8	103.8	108.4	87.8	100.3	100.0	50.2	80.4	91.9	50.2	75.5	108.8	86.5	100.0
03	CONCRETE	106.0	100.3	103.3	101.3	114.9	107.9	106.0	99.0	102.6	104.8	47.3	76.8	96.2	47.3	72.4	107.6	87.7	97.9
04	MASONRY	124.0	79.1	96.6	111.6	106.7	108.6	118.3	76.6	92.8	90.1	37.8	58.2	90.1	37.8	58.2	93.0	80.2	85.2
05	METALS	93.1	90.2	92.2	100.3	110.4	103.6	94.6	105.8	98.3	92.3	62.2	82.3	92.3	62.2	82.3	93.0	88.7	91.6
06	WOOD, PLASTICS & COMPOSITES	88.6	106.8	99.1	87.8	114.8	103.4	84.3	100.4	93.6	86.7	45.0	62.6	92.5	45.0	65.0	96.2	91.2	93.3
07	THERMAL & MOISTURE PROTECTION	101.5	87.3	95.7	115.1	101.8	109.7	101.0	87.0	95.3	100.8	41.5	76.7	100.6	41.5	76.6	99.4	84.9	93.5
08	OPENINGS	99.7	92.5	97.9	98.6	122.2	104.6	97.4	108.0	100.1	99.1	40.7	84.4	100.3	40.7	85.3	100.3	81.9	95.7
0920	Plaster & Gypsum Board	90.9	106.9	102.3	90.9	115.2	108.3	83.8	100.3	95.6	90.2	42.5	56.1	91.5	42.5	56.5	94.2	90.3	91.4
0950, 0980	Ceilings & Acoustic Treatment	112.4	106.9	108.8	121.4	115.2	117.3	117.8	100.3	106.3	92.7	42.5	59.6	92.7	42.5	59.6	91.8	90.3	90.8
0960	Flooring	104.0	59.9	91.0	94.9	103.4	97.4	100.7	82.0	95.2	89.2	33.6	72.9	91.6	33.6	74.6	89.2	93.7	90.5
0970, 0990	Wall Finishes & Painting/Coating	98.8	118.3	110.8	101.6	122.1	114.2	98.8	83.3	89.2	90.1	48.2	64.3	90.1	48.2	64.3	90.1	100.6	96.6
09	FINISHES	101.3	98.9	100.0	100.7	113.8	108.0	100.5	93.4	96.5	93.7	42.6	65.2	93.9	42.6	65.3	94.9	90.6	92.5
COVERS	DIVS. 10 - 14, 25, 28, 41, 43, 44, 46	100.0	66.9	93.4	100.0	93.2	98.7	100.0	97.1	99.4	100.0	38.6	87.8	100.0	38.6	87.8	100.0	62.1	92.5
21, 22, 23	FIRE SUPPRESSION, PLUMBING & HVAC	97.2	104.8	100.3	100.1	108.5	103.5	100.0	84.2	93.6	94.1	39.3	71.9	94.1	39.4	71.9	99.9	83.5	93.2
26, 27, 3370	ELECTRICAL, COMMUNICATIONS & UTIL.	94.2	112.0	103.4	98.5	116.6	107.8	94.5	98.5	96.5	99.5	34.1	65.8	99.5	34.1	65.8	98.3	59.4	78.2
MF2010	WEIGHTED AVERAGE	98.5	98.1	98.3	99.8	110.7	104.6	98.7	93.7	96.5	96.3	47.1	74.6	95.3	47.1	74.1	98.6	82.4	91.5

DIVISION		NEW HAMPSHIRE															NEW JERSEY		
		KEENE			LITTLETON			MANCHESTER			NASHUA			PORTSMOUTH			ATLANTIC CITY		
		034			035			031			030			038			082,084		
		MAT.	INST.	TOTAL	MAT.	INST.	TOTAL	MAT.	INST.	TOTAL	MAT.	INST.	TOTAL	MAT.	INST.	TOTAL	MAT.	INST.	TOTAL
015433	CONTRACTOR EQUIPMENT		100.4	100.4		100.4	100.4		100.4	100.4		100.4	100.4		100.4	100.4		98.7	98.7
0241, 31 - 34	SITE & INFRASTRUCTURE, DEMOLITION	94.3	96.2	95.6	79.0	95.7	90.5	89.6	98.1	95.4	94.9	98.1	97.1	88.6	99.5	96.1	98.9	103.0	101.7
0310	Concrete Forming & Accessories	92.6	41.5	48.1	102.2	46.3	53.6	97.6	87.7	89.0	99.6	87.7	89.3	89.7	92.6	92.3	107.0	124.2	121.9
0320	Concrete Reinforcing	82.9	55.6	69.1	83.6	33.8	58.5	103.6	92.0	97.7	103.6	92.0	97.7	82.9	92.1	87.5	79.0	114.8	97.0
0330	Cast-in-Place Concrete	100.4	51.7	81.2	90.1	52.4	75.3	112.0	108.4	110.5	95.1	108.4	100.3	90.2	115.8	100.3	85.3	130.0	102.9
03	CONCRETE	104.5	48.6	77.3	95.4	46.9	71.8	112.5	95.5	104.2	104.2	95.5	100.0	95.5	100.4	97.9	97.1	123.4	109.9
04	MASONRY	93.9	40.0	61.0	101.2	43.4	65.9	95.6	80.2	86.2	94.7	80.2	85.9	90.6	92.3	91.6	97.3	127.6	115.8
05	METALS	93.0	66.3	84.1	93.0	57.0	81.1	98.0	89.3	95.1	98.0	89.3	95.1	94.5	92.1	93.7	92.7	104.6	96.6
06	WOOD, PLASTICS & COMPOSITES	90.9	40.1	61.6	100.8	45.9	69.1	96.2	91.2	93.3	99.5	91.2	94.7	87.9	91.2	89.8	109.4	123.9	117.8
07	THERMAL & MOISTURE PROTECTION	101.2	43.5	77.7	100.7	47.3	79.0	100.7	88.2	95.6	101.5	88.2	96.1	101.2	112.4	105.7	101.2	124.8	110.8
08	OPENINGS	97.6	45.4	84.5	101.4	38.5	85.6	102.4	81.9	97.2	102.4	81.9	97.2	103.1	77.2	96.6	100.0	121.6	105.4
0920	Plaster & Gypsum Board	90.6	37.6	52.7	103.3	43.5	60.6	102.2	90.3	93.7	98.7	90.3	92.7	90.2	90.3	90.2	102.9	124.0	118.0
0950, 0980	Ceilings & Acoustic Treatment	92.7	37.6	56.3	92.7	43.5	60.2	98.4	90.3	93.0	102.0	90.3	94.2	94.8	90.3	91.8	92.7	124.0	113.4
0960	Flooring	91.1	55.1	80.6	100.4	33.6	80.8	95.4	93.7	94.9	94.5	93.7	94.2	89.6	93.7	90.8	98.1	133.1	108.4
0970, 0990	Wall Finishes & Painting/Coating	90.1	48.2	64.3	90.1	63.2	73.5	90.1	100.6	96.6	90.1	100.6	96.6	90.1	100.6	96.6	90.1	121.1	109.2
09	FINISHES	95.8	44.4	67.1	97.8	46.0	68.9	98.5	90.6	94.1	99.7	90.6	94.6	95.1	93.5	94.2	100.0	125.9	114.4
COVERS	DIVS. 10 - 14, 25, 28, 41, 43, 44, 46	100.0	52.8	90.6	100.0	40.4	88.1	100.0	62.2	92.5	100.0	62.2	92.5	100.0	66.2	93.3	100.0	110.4	102.1
21, 22, 23	FIRE SUPPRESSION, PLUMBING & HVAC	94.1	42.6	73.2	94.1	65.6	82.5	99.9	83.5	93.2	100.0	83.5	93.3	100.0	89.6	95.8	99.6	123.3	109.2
26, 27, 3370	ELECTRICAL, COMMUNICATIONS & UTIL.	99.5	34.1	65.8	100.7	59.3	79.3	100.3	81.4	90.6	102.1	81.4	91.5	100.0	81.5	90.5	97.2	136.9	117.7
MF2010	WEIGHTED AVERAGE	96.9	49.4	76.0	96.5	56.6	78.9	100.7	86.7	94.5	100.2	86.7	94.3	97.8	91.2	94.9	98.1	122.1	108.7

DIVISION		NEW JERSEY																	
		CAMDEN			DOVER			ELIZABETH			HACKENSACK			JERSEY CITY			LONG BRANCH		
		081			078			072			076			073			077		
		MAT.	INST.	TOTAL	MAT.	INST.	TOTAL	MAT.	INST.	TOTAL	MAT.	INST.	TOTAL	MAT.	INST.	TOTAL	MAT.	INST.	TOTAL
015433	CONTRACTOR EQUIPMENT		98.7	98.7		100.4	100.4		100.4	100.4		100.4	100.4		98.7	98.7		98.3	98.3
0241, 31 - 34	SITE & INFRASTRUCTURE, DEMOLITION	99.4	103.5	102.3	111.1	104.1	106.3	115.9	104.1	107.8	111.3	104.6	106.7	99.4	104.5	102.9	105.2	104.4	104.7
0310	Concrete Forming & Accessories	99.6	124.3	121.1	96.2	125.0	121.2	106.6	125.0	122.6	96.2	125.0	121.2	99.6	125.0	121.7	100.1	124.7	121.5
0320	Concrete Reinforcing	103.6	116.4	110.1	79.8	123.6	101.9	79.8	123.6	101.9	79.8	123.6	101.9	103.6	123.6	113.7	79.8	123.6	101.9
0330	Cast-in-Place Concrete	82.7	131.7	102.0	106.0	130.9	115.8	91.0	130.9	106.8	103.6	131.0	114.4	82.7	131.0	101.7	91.9	129.9	106.9
03	CONCRETE	98.0	124.3	110.8	105.8	125.9	115.6	100.6	125.9	112.9	103.8	125.9	114.6	98.0	125.8	111.5	102.8	125.3	113.7
04	MASONRY	89.2	127.6	112.6	93.0	128.1	114.4	108.0	128.1	120.2	96.7	128.1	115.8	86.5	128.1	111.9	100.9	127.6	117.1
05	METALS	97.8	104.7	100.1	92.7	111.9	99.1	94.2	111.9	100.1	92.8	111.9	99.1	97.9	109.6	101.8	92.8	109.4	98.3
06	WOOD, PLASTICS & COMPOSITES	99.5	123.9	113.6	98.9	123.9	113.4	112.4	123.9	119.1	98.9	123.9	113.4	99.5	123.9	113.6	100.5	123.9	114.0
07	THERMAL & MOISTURE PROTECTION	101.1	124.5	110.6	101.6	126.4	111.6	101.7	126.4	111.7	101.3	123.9	110.5	101.1	126.4	111.4	101.2	125.0	110.9
08	OPENINGS	102.4	121.6	107.2	105.4	123.6	110.0	103.4	123.6	108.5	102.6	123.6	107.9	102.4	123.6	107.8	98.4	123.6	104.8
0920	Plaster & Gypsum Board	98.7	124.0	116.8	95.2	124.0	115.8	102.9	124.0	118.0	95.2	124.0	115.8	98.7	124.0	116.8	97.4	124.0	116.4
0950, 0980	Ceilings & Acoustic Treatment	102.0	124.0	116.5	92.7	124.0	113.4	94.8	124.0	114.1	92.7	124.0	113.4	102.0	124.0	116.5	92.7	124.0	113.4
0960	Flooring	94.5	133.1	105.8	93.4	164.3	114.2	98.7	164.3	117.9	93.4	164.3	114.2	94.5	164.3	114.9	94.7	164.3	115.1
0970, 0990	Wall Finishes & Painting/Coating	90.1	121.1	109.2	90.0	123.1	110.4	90.0	123.1	110.4	90.0	123.1	110.4	90.1	123.1	110.5	90.1	121.1	109.2
09	FINISHES	100.2	125.9	114.6	97.8	131.5	116.6	101.4	131.5	118.2	97.6	131.5	116.5	100.2	131.5	117.7	98.8	131.1	116.9
COVERS	DIVS. 10 - 14, 25, 28, 41, 43, 44, 46	100.0	110.5	102.1	100.0	110.3	102.1	100.0	110.3	102.1	100.0	110.4	102.1	100.0	110.4	102.1	100.0	110.3	102.1
21, 22, 23	FIRE SUPPRESSION, PLUMBING & HVAC	100.0	124.9	110.1	99.6	125.7	110.2	100.0	125.6	110.4	99.6	125.7	110.2	100.0	125.7	110.4	99.6	124.7	109.8
26, 27, 3370	ELECTRICAL, COMMUNICATIONS & UTIL.	102.6	136.9	120.3	98.6	137.0	118.4	99.3	137.0	118.8	98.6	136.9	118.4	103.8	136.9	120.9	98.3	132.7	116.0
MF2010	WEIGHTED AVERAGE	99.4	122.6	109.6	99.7	124.6	110.7	100.6	124.5	111.2	99.4	124.5	110.5	99.4	124.4	110.4	99.0	123.3	109.7

DIVISION		NEW JERSEY																	
		NEW BRUNSWICK			NEWARK			PATERSON			POINT PLEASANT			SUMMIT			TRENTON		
		088 - 089			070 - 071			074 - 075			087			079			085 - 086		
		MAT.	INST.	TOTAL	MAT.	INST.	TOTAL	MAT.	INST.	TOTAL	MAT.	INST.	TOTAL	MAT.	INST.	TOTAL	MAT.	INST.	TOTAL
015433	CONTRACTOR EQUIPMENT		98.3	98.3		100.4	100.4		100.4	100.4		98.3	98.3		100.4	100.4		98.3	98.3
0241, 31 - 34	SITE & INFRASTRUCTURE, DEMOLITION	114.2	104.4	107.5	115.2	104.1	107.5	113.0	104.6	107.2	115.7	104.4	107.9	113.1	104.1	106.9	96.1	104.4	101.8
0310	Concrete Forming & Accessories	102.1	124.9	121.9	96.0	125.0	121.2	98.1	125.0	121.5	97.5	124.4	120.9	98.8	125.0	121.6	96.7	124.6	121.0
0320	Concrete Reinforcing	79.8	123.6	101.9	103.6	123.6	113.7	103.6	123.6	113.7	79.8	123.5	101.8	79.8	123.6	101.9	103.6	119.0	111.3
0330	Cast-in-Place Concrete	105.3	130.2	115.1	109.9	131.0	118.2	105.4	131.0	115.5	105.3	129.8	115.0	88.0	131.0	104.9	101.5	129.9	112.7
03	CONCRETE	115.3	125.5	120.3	111.4	125.9	118.5	109.3	125.9	117.4	115.0	125.1	119.9	97.4	125.9	111.3	107.3	124.4	115.6
04	MASONRY	95.5	128.0	115.3	98.2	128.1	116.4	93.5	128.1	114.6	86.9	127.6	111.7	95.2	128.1	115.3	85.8	127.6	111.3
05	METALS	92.8	109.5	98.3	97.9	111.9	102.5	93.0	111.9	99.3	92.8	109.2	98.2	92.7	111.9	99.1	93.0	107.2	97.7
06	WOOD, PLASTICS & COMPOSITES	103.6	123.9	115.3	98.2	123.9	113.1	101.3	123.9	114.4	97.2	123.9	112.6	102.4	123.9	114.8	96.6	123.9	112.4
07	THERMAL & MOISTURE PROTECTION	101.4	125.2	111.1	101.3	126.4	111.5	101.6	123.9	110.7	101.4	125.0	111.0	102.0	126.4	111.9	97.2	124.6	108.3
08	OPENINGS	94.5	123.6	101.8	108.3	123.6	112.1	108.3	123.6	112.1	96.5	124.5	103.6	110.6	123.6	113.9	103.0	122.1	107.8
0920	Plaster & Gypsum Board	99.7	124.0	117.1	96.2	124.0	116.1	98.7	124.0	116.8	95.2	124.0	115.8	97.4	124.0	116.4	96.5	124.0	116.1
0950, 0980	Ceilings & Acoustic Treatment	92.7	124.0	113.4	98.4	124.0	115.3	102.0	124.0	116.5	92.7	124.0	113.4	92.7	124.0	113.4	99.3	124.0	115.6
0960	Flooring	95.9	164.3	116.0	94.7	164.3	115.1	94.5	164.3	114.9	93.4	133.1	105.1	94.8	164.3	115.2	94.8	164.3	115.2
0970, 0990	Wall Finishes & Painting/Coating	90.1	123.1	110.5	90.0	123.1	110.4	90.0	123.1	110.4	90.1	121.1	109.2	90.0	123.1	110.4	90.1	121.1	109.2
09	FINISHES	100.5	131.5	117.8	99.0	131.5	117.2	100.4	131.5	117.8	99.3	125.9	114.1	98.9	131.5	117.1	98.8	131.1	116.9
COVERS	DIVS. 10 - 14, 25, 28, 41, 43, 44, 46	100.0	110.2	102.0	100.0	110.3	102.1	100.0	110.4	102.1	100.0	107.1	101.4	100.0	110.3	102.1	100.0	110.3	102.1
21, 22, 23	FIRE SUPPRESSION, PLUMBING & HVAC	99.6	125.0	109.9	100.1	126.1	110.6	100.0	125.7	110.4	99.6	124.5	109.7	99.6	126.1	110.3	100.2	124.7	110.1
26, 27, 3370	ELECTRICAL, COMMUNICATIONS & UTIL.	98.0	138.0	118.7	105.5	136.9	121.7	103.8	137.0	120.9	97.2	132.7	115.5	99.3	137.0	118.8	105.1	137.6	121.9
MF2010	WEIGHTED AVERAGE	100.1	124.3	110.8	102.7	124.6	112.4	101.4	124.5	111.6	99.7	122.5	109.7	99.6	124.7	110.7	99.6	123.6	110.2

DIVISION		NEW JERSEY			NEW MEXICO														
		VINELAND			ALBUQUERQUE			CARRIZOZO			CLOVIS			FARMINGTON			GALLUP		
		080,083			870 - 872			883			881			874			873		
		MAT.	INST.	TOTAL	MAT.	INST.	TOTAL	MAT.	INST.	TOTAL	MAT.	INST.	TOTAL	MAT.	INST.	TOTAL	MAT.	INST.	TOTAL
015433	CONTRACTOR EQUIPMENT		98.7	98.7		111.1	111.1		111.1	111.1		111.1	111.1		111.1	111.1		111.1	111.1
0241, 31 - 34	SITE & INFRASTRUCTURE, DEMOLITION	104.1	103.5	103.7	78.9	106.9	98.2	100.5	106.9	104.9	87.9	106.9	101.0	85.1	106.9	100.1	93.6	106.9	102.8
0310	Concrete Forming & Accessories	95.3	124.3	120.5	99.4	65.8	70.2	99.4	65.8	70.2	99.4	65.6	70.0	99.4	65.8	70.2	99.4	65.8	70.2
0320	Concrete Reinforcing	79.0	114.8	97.0	100.4	71.1	85.6	107.7	71.1	89.2	108.9	71.0	89.8	109.7	71.1	90.2	105.0	71.1	87.9
0330	Cast-in-Place Concrete	91.9	130.0	106.9	101.4	73.4	90.4	94.3	73.4	86.1	94.3	73.4	86.0	102.3	73.4	90.9	96.3	73.4	87.3
03	CONCRETE	102.3	123.5	112.6	102.6	70.5	87.0	116.2	70.5	93.9	105.5	70.4	88.4	106.2	70.5	88.8	111.7	70.5	91.6
04	MASONRY	85.9	127.6	111.3	107.7	62.9	80.4	104.4	62.9	79.1	104.5	62.9	79.1	114.7	62.9	83.1	104.5	62.9	79.2
05	METALS	92.7	104.7	96.6	100.4	88.2	96.4	97.3	88.2	94.3	96.9	88.1	94.0	98.2	88.2	94.9	97.3	88.2	94.3
06	WOOD, PLASTICS & COMPOSITES	94.4	123.9	111.5	87.9	66.1	75.3	88.0	66.1	75.4	88.0	66.1	75.4	88.0	66.1	75.4	88.0	66.1	75.4
07	THERMAL & MOISTURE PROTECTION	101.0	124.8	110.7	100.6	73.4	89.5	102.2	73.4	90.5	101.0	73.4	89.8	100.9	73.4	89.7	101.9	73.4	90.3
08	OPENINGS	96.0	121.6	102.5	97.1	68.8	90.0	95.8	68.8	89.0	96.0	68.8	89.2	100.1	68.8	92.2	100.1	68.8	92.2
0920	Plaster & Gypsum Board	93.3	124.0	115.2	89.3	64.7	71.7	81.2	64.7	69.4	81.2	64.7	69.4	81.2	64.7	69.4	81.2	64.7	69.4
0950, 0980	Ceilings & Acoustic Treatment	92.7	124.0	113.4	112.0	64.7	80.7	108.8	64.7	79.7	108.8	64.7	79.7	108.8	64.7	79.7	108.8	64.7	79.7
0960	Flooring	92.6	133.1	104.4	98.4	69.1	89.8	100.7	69.1	91.4	100.7	69.1	91.4	100.7	69.1	91.4	100.7	69.1	91.4
0970, 0990	Wall Finishes & Painting/Coating	90.1	121.1	109.2	103.7	51.7	71.6	98.8	51.7	69.8	98.8	51.7	69.8	98.8	51.7	69.8	98.8	51.7	69.8
09	FINISHES	97.7	125.9	113.5	100.3	64.6	80.4	101.2	64.6	80.8	99.7	64.6	80.2	99.3	64.6	80.0	100.6	64.6	80.5
COVERS	DIVS. 10 - 14, 25, 28, 41, 43, 44, 46	100.0	110.5	102.1	100.0	80.2	96.1	100.0	80.2	96.1	100.0	80.2	96.1	100.0	80.2	96.1	100.0	80.2	96.1
21, 22, 23	FIRE SUPPRESSION, PLUMBING & HVAC	99.6	124.8	109.8	100.1	68.2	87.2	96.6	68.2	85.1	96.6	67.8	84.9	100.0	68.2	87.1	96.4	68.2	85.0
26, 27, 3370	ELECTRICAL, COMMUNICATIONS & UTIL.	97.2	136.9	117.7	89.5	71.1	80.0	89.5	71.1	80.0	87.2	71.1	78.9	87.8	71.1	79.2	87.1	71.1	78.8
MF2010	WEIGHTED AVERAGE	97.6	122.4	108.6	98.7	73.6	87.7	99.4	73.6	88.1	97.4	73.5	86.9	99.4	73.6	88.0	98.8	73.6	87.7

DIVISION		NEW MEXICO																	
		LAS CRUCES			LAS VEGAS			ROSWELL			SANTA FE			SOCORRO			TRUTH/CONSEQUENCES		
		880			877			882			875			878			879		
		MAT.	INST.	TOTAL	MAT.	INST.	TOTAL	MAT.	INST.	TOTAL	MAT.	INST.	TOTAL	MAT.	INST.	TOTAL	MAT.	INST.	TOTAL
015433	CONTRACTOR EQUIPMENT		87.3	87.3		111.1	111.1		111.1	111.1		111.1	111.1		111.1	111.1		87.3	87.3
0241, 31 - 34	SITE & INFRASTRUCTURE, DEMOLITION	90.5	86.0	87.4	84.4	106.9	99.9	90.0	106.9	101.7	83.4	106.9	99.6	81.0	106.9	98.9	101.7	86.0	90.9
0310	Concrete Forming & Accessories	96.3	64.6	68.7	99.4	65.8	70.2	99.4	65.8	70.2	98.6	65.8	70.1	99.4	65.8	70.2	96.3	64.6	68.7
0320	Concrete Reinforcing	103.6	70.9	87.1	106.7	71.1	88.8	108.9	71.1	89.8	107.7	71.1	89.2	108.9	71.1	89.8	102.4	70.9	86.5
0330	Cast-in-Place Concrete	89.0	65.8	79.9	99.5	73.4	89.2	94.3	73.4	86.1	103.6	73.4	91.7	97.5	73.4	88.0	106.9	65.8	90.7
03	CONCRETE	85.1	67.0	76.3	103.5	70.5	87.4	106.2	70.5	88.8	104.8	70.5	88.1	102.3	70.5	86.8	96.9	67.0	82.4
04	MASONRY	100.5	62.6	77.4	104.8	62.9	79.3	115.3	62.9	83.3	108.8	62.9	80.8	104.7	62.9	79.2	102.1	62.6	78.0
05	METALS	98.1	81.7	92.7	97.0	88.2	94.1	98.2	88.2	94.9	98.2	88.2	94.9	97.3	88.2	94.3	96.9	81.7	91.9
06	WOOD, PLASTICS & COMPOSITES	79.3	65.0	71.0	88.0	66.1	75.4	88.0	66.1	75.4	87.5	66.1	75.2	88.0	66.1	75.4	79.3	65.0	71.0
07	THERMAL & MOISTURE PROTECTION	87.1	68.7	79.6	100.5	73.4	89.5	101.1	73.4	89.8	100.2	73.4	89.3	100.4	73.4	89.4	87.6	68.7	79.9
08	OPENINGS	90.1	68.2	84.6	96.0	68.8	89.2	95.8	68.8	89.0	96.0	68.8	89.1	95.8	68.8	89.0	90.0	68.2	84.5
0920	Plaster & Gypsum Board	83.3	64.7	70.0	81.2	64.7	69.4	81.2	64.7	69.4	89.2	64.7	71.7	81.2	64.7	69.4	83.3	64.7	70.0
0950, 0980	Ceilings & Acoustic Treatment	97.8	64.7	75.9	108.8	64.7	79.7	108.8	64.7	79.7	109.7	64.7	80.0	108.8	64.7	79.7	97.8	64.7	75.9
0960	Flooring	133.3	69.1	114.4	100.7	69.1	91.4	100.7	69.1	91.4	100.7	69.1	91.4	100.7	69.1	91.4	133.3	69.1	114.4
0970, 0990	Wall Finishes & Painting/Coating	90.0	51.7	66.4	98.8	51.7	69.8	98.8	51.7	69.8	98.8	51.7	69.8	98.8	51.7	69.8	90.0	51.7	66.4
09	FINISHES	112.8	63.8	85.5	99.1	64.6	79.9	99.8	64.6	80.2	100.6	64.6	80.5	99.0	64.6	79.9	113.6	63.8	85.8
COVERS	DIVS. 10 - 14, 25, 28, 41, 43, 44, 46	100.0	77.4	95.5	100.0	80.2	96.1	100.0	80.2	96.1	100.0	80.2	96.1	100.0	80.2	96.1	100.0	77.4	95.5
21, 22, 23	FIRE SUPPRESSION, PLUMBING & HVAC	100.3	67.9	87.2	96.4	68.2	85.0	100.0	68.2	87.1	99.9	68.2	87.1	96.4	68.2	85.0	96.3	67.9	84.8
26, 27, 3370	ELECTRICAL, COMMUNICATIONS & UTIL.	88.4	66.3	77.0	89.5	71.1	80.0	88.6	71.1	79.6	96.6	71.1	83.4	87.6	71.1	79.1	91.2	71.1	80.8
MF2010	WEIGHTED AVERAGE	96.1	69.6	84.4	97.3	73.6	86.8	99.3	73.6	88.0	99.4	73.6	88.0	96.8	73.6	86.6	97.0	70.3	85.3

City Cost Indexes

	DIVISION	NEW MEXICO			NEW YORK															
		TUCUMCARI			ALBANY			BINGHAMTON			BRONX			BROOKLYN			BUFFALO		
		884			120 - 122			137 - 139			104			112			140 - 142		
		MAT.	INST.	TOTAL	MAT.	INST.	TOTAL	MAT.	INST.	TOTAL	MAT.	INST.	TOTAL	MAT.	INST.	TOTAL	MAT.	INST.	TOTAL
015433	CONTRACTOR EQUIPMENT		111.1	111.1		113.6	113.6		114.2	114.2		113.9	113.9		114.5	114.5		96.5	96.5
0241, 31 - 34	SITE & INFRASTRUCTURE, DEMOLITION	87.6	106.9	100.9	70.5	106.9	95.6	93.4	92.7	92.9	114.2	124.5	121.3	119.7	128.0	125.4	97.5	97.3	97.3
0310	Concrete Forming & Accessories	99.4	65.6	70.0	97.8	95.5	95.8	102.1	82.7	85.2	101.7	176.5	166.8	108.5	176.2	167.4	102.0	115.7	113.9
0320	Concrete Reinforcing	106.7	71.0	88.7	98.3	91.8	95.0	97.3	95.1	96.2	95.9	179.8	138.2	98.8	187.6	143.5	102.5	101.8	102.1
0330	Cast-in-Place Concrete	94.3	73.4	86.0	76.6	102.0	86.6	98.4	94.5	96.8	89.2	164.2	118.8	100.4	162.5	124.9	103.1	117.8	108.9
03	CONCRETE	104.7	70.4	88.0	93.9	97.8	95.8	96.7	90.8	93.8	96.2	171.0	132.6	108.5	171.5	139.2	106.4	113.0	109.6
04	MASONRY	116.0	62.9	83.6	92.4	100.8	97.6	107.9	89.5	96.7	93.8	175.1	143.4	117.3	175.0	152.5	106.8	116.8	112.9
05	METALS	96.9	88.1	94.0	100.3	104.0	101.5	96.9	114.2	102.6	104.9	142.9	117.5	105.2	141.8	117.3	101.5	94.7	99.2
06	WOOD, PLASTICS & COMPOSITES	88.0	66.1	75.4	96.1	94.3	95.1	106.3	81.2	91.8	103.1	178.3	146.6	107.9	177.9	148.3	99.6	116.5	109.3
07	THERMAL & MOISTURE PROTECTION	101.0	73.4	89.7	88.6	93.7	90.7	103.4	83.3	95.2	104.1	164.3	128.6	105.8	163.1	129.1	100.9	108.4	104.0
08	OPENINGS	95.8	68.8	89.0	94.3	87.7	92.6	89.3	80.3	87.0	85.0	172.8	107.1	88.7	172.1	109.7	90.7	103.7	94.0
0920	Plaster & Gypsum Board	81.2	64.7	69.4	101.2	93.9	96.0	112.3	80.2	89.3	103.6	180.4	158.4	109.8	180.4	160.2	109.2	116.7	114.6
0950, 0980	Ceilings & Acoustic Treatment	108.8	64.7	79.7	94.4	93.9	94.1	97.1	80.2	85.9	83.5	180.4	147.5	93.6	180.4	150.9	105.9	116.7	113.1
0960	Flooring	100.7	69.1	91.4	91.7	103.9	95.3	103.7	89.6	99.6	99.0	177.0	121.9	110.9	177.0	130.3	102.3	116.7	106.5
0970, 0990	Wall Finishes & Painting/Coating	98.8	51.7	69.8	89.3	78.4	82.6	90.8	81.8	85.2	106.1	152.7	134.8	121.2	152.7	140.6	98.1	112.6	107.1
09	FINISHES	99.7	64.6	80.1	95.0	95.3	95.2	97.4	83.5	89.7	101.2	175.0	142.3	113.7	174.7	147.7	102.0	116.7	110.2
COVERS	DIVS. 10 - 14, 25, 28, 41, 43, 44, 46	100.0	80.2	96.1	100.0	93.6	98.7	100.0	94.3	98.9	100.0	134.3	106.8	100.0	133.5	106.7	100.0	106.5	101.3
21, 22, 23	FIRE SUPPRESSION, PLUMBING & HVAC	96.6	67.8	84.9	100.0	92.1	96.8	100.4	80.7	92.4	100.1	166.8	127.1	99.7	166.7	126.9	99.9	92.8	97.0
26, 27, 3370	ELECTRICAL, COMMUNICATIONS & UTIL.	89.5	71.1	80.0	99.2	91.6	95.3	100.7	90.2	95.3	96.9	173.3	136.3	100.9	173.3	138.3	98.6	99.7	99.2
MF2010	WEIGHTED AVERAGE	98.1	73.5	87.3	96.7	96.4	96.5	98.4	89.2	94.4	98.9	163.6	127.4	103.4	163.8	130.0	100.3	104.0	101.9

	DIVISION	NEW YORK																	
		ELMIRA			FAR ROCKAWAY			FLUSHING			GLENS FALLS			HICKSVILLE			JAMAICA		
		148 - 149			116			113			128			115,117,118			114		
		MAT.	INST.	TOTAL	MAT.	INST.	TOTAL	MAT.	INST.	TOTAL	MAT.	INST.	TOTAL	MAT.	INST.	TOTAL	MAT.	INST.	TOTAL
015433	CONTRACTOR EQUIPMENT		115.2	115.2		114.5	114.5		114.5	114.5		113.6	113.6		114.5	114.5		114.5	114.5
0241, 31 - 34	SITE & INFRASTRUCTURE, DEMOLITION	93.1	93.6	93.4	123.6	128.0	126.7	123.6	128.0	126.7	62.4	106.2	92.6	112.3	126.6	122.2	117.5	128.0	124.8
0310	Concrete Forming & Accessories	87.4	85.3	85.6	95.4	176.2	165.7	99.1	176.2	166.2	86.7	88.8	88.5	92.1	146.0	139.0	99.1	176.2	166.2
0320	Concrete Reinforcing	103.7	85.9	94.8	98.8	187.6	143.5	100.5	187.6	144.4	97.9	88.6	93.2	98.8	179.5	139.5	98.8	187.6	143.5
0330	Cast-in-Place Concrete	96.4	94.5	95.6	108.5	162.5	129.8	108.5	162.5	129.8	74.8	99.7	84.6	92.2	153.5	116.4	100.4	162.5	124.9
03	CONCRETE	100.2	90.4	95.4	114.7	171.5	142.4	115.2	171.5	142.6	86.9	93.4	90.0	100.4	153.5	126.3	107.9	171.5	138.9
04	MASONRY	103.2	91.0	95.7	121.4	175.0	154.1	115.4	175.0	151.8	93.8	96.8	95.6	111.9	160.6	141.6	119.3	175.0	153.3
05	METALS	96.8	113.1	102.2	105.2	141.8	117.3	105.2	141.8	117.3	96.4	102.7	98.5	106.7	136.5	116.5	105.2	141.8	117.3
06	WOOD, PLASTICS & COMPOSITES	83.9	84.4	84.2	92.0	177.9	141.7	96.6	177.9	143.6	83.6	87.5	85.9	88.6	145.9	121.7	96.6	177.9	143.6
07	THERMAL & MOISTURE PROTECTION	101.8	84.8	94.9	105.8	163.1	129.1	105.8	163.1	129.1	88.0	90.6	89.0	105.4	149.6	123.3	105.6	163.1	129.0
08	OPENINGS	91.6	79.9	88.7	87.1	172.1	108.5	87.1	172.1	108.5	88.0	83.3	86.8	87.1	152.9	103.7	87.1	172.1	108.5
0920	Plaster & Gypsum Board	103.6	83.7	89.4	98.4	180.4	156.9	101.1	180.4	157.7	95.8	86.9	89.5	97.9	147.3	133.2	101.1	180.4	157.7
0950, 0980	Ceilings & Acoustic Treatment	101.4	83.7	89.7	82.7	180.4	147.2	82.7	180.4	147.2	91.8	86.9	88.6	81.7	147.3	125.0	82.7	180.4	147.2
0960	Flooring	91.5	89.6	90.9	105.5	177.0	126.5	107.0	177.0	127.5	83.3	103.9	89.4	104.4	149.4	117.6	107.0	177.0	127.5
0970, 0990	Wall Finishes & Painting/Coating	96.6	77.9	85.0	121.2	152.7	140.6	121.2	152.7	140.6	89.3	81.9	84.7	121.2	152.7	140.6	121.2	152.7	140.6
09	FINISHES	96.8	85.1	90.3	108.9	174.7	145.6	109.6	174.7	145.9	90.6	90.7	90.7	107.2	146.7	129.2	109.1	174.7	145.7
COVERS	DIVS. 10 - 14, 25, 28, 41, 43, 44, 46	100.0	87.8	97.6	100.0	133.5	106.7	100.0	133.5	106.7	100.0	91.5	98.3	100.0	123.0	104.6	100.0	133.5	106.7
21, 22, 23	FIRE SUPPRESSION, PLUMBING & HVAC	94.2	86.3	91.0	93.8	166.7	123.3	93.8	166.7	123.3	94.3	84.6	90.4	99.7	148.7	119.6	93.8	166.7	123.3
26, 27, 3370	ELECTRICAL, COMMUNICATIONS & UTIL.	97.3	93.5	95.3	108.1	173.3	141.7	108.1	173.3	141.7	96.2	91.6	93.9	100.1	141.0	121.2	99.0	173.3	137.4
MF2010	WEIGHTED AVERAGE	96.7	91.0	94.2	103.1	163.8	129.8	102.9	163.8	129.7	92.3	92.6	92.5	101.3	145.6	120.8	101.1	163.8	128.7

	DIVISION	NEW YORK																	
		JAMESTOWN			KINGSTON			LONG ISLAND CITY			MONTICELLO			MOUNT VERNON			NEW ROCHELLE		
		147			124			111			127			105			108		
		MAT.	INST.	TOTAL	MAT.	INST.	TOTAL	MAT.	INST.	TOTAL	MAT.	INST.	TOTAL	MAT.	INST.	TOTAL	MAT.	INST.	TOTAL
015433	CONTRACTOR EQUIPMENT		93.5	93.5		114.5	114.5		114.5	114.5		114.5	114.5		113.9	113.9		113.9	113.9
0241, 31 - 34	SITE & INFRASTRUCTURE, DEMOLITION	94.3	92.9	93.3	119.2	124.7	123.0	121.0	128.0	125.9	114.5	124.7	121.5	122.3	119.2	120.2	121.1	119.2	119.8
0310	Concrete Forming & Accessories	87.5	81.0	81.8	88.4	108.3	105.7	103.2	176.2	166.7	95.0	108.4	106.6	92.6	129.3	124.5	106.7	129.3	126.4
0320	Concrete Reinforcing	103.9	97.6	100.7	98.2	134.7	116.6	98.8	187.6	143.5	97.5	134.7	116.2	94.6	178.4	136.9	94.7	178.4	136.9
0330	Cast-in-Place Concrete	100.0	93.5	97.5	101.1	130.7	112.8	103.5	162.5	126.8	94.7	130.7	108.9	99.7	129.7	111.5	99.7	129.7	111.5
03	CONCRETE	103.4	88.4	96.1	108.8	120.6	114.6	111.0	171.5	140.5	103.6	120.7	111.9	105.7	137.6	121.3	105.1	137.6	121.0
04	MASONRY	111.8	90.8	99.0	109.1	135.1	124.9	114.0	175.0	151.2	102.1	135.1	122.2	99.7	127.3	116.5	99.7	127.3	116.5
05	METALS	94.1	89.2	92.5	105.2	115.1	108.4	105.2	141.8	117.3	105.1	115.2	108.4	104.7	132.6	113.9	105.0	132.6	114.1
06	WOOD, PLASTICS & COMPOSITES	82.5	78.0	79.9	84.9	102.6	95.1	102.0	177.9	145.9	92.5	102.6	98.3	93.2	132.4	115.9	110.3	132.4	123.1
07	THERMAL & MOISTURE PROTECTION	101.2	85.8	95.0	107.5	136.0	119.1	105.7	163.1	129.1	107.1	136.0	118.9	105.0	137.9	118.3	105.1	137.9	118.4
08	OPENINGS	91.5	81.4	88.9	92.3	119.6	99.2	87.1	172.1	108.5	87.2	119.6	95.3	85.0	145.6	100.2	85.0	145.6	100.3
0920	Plaster & Gypsum Board	98.0	77.1	83.0	94.2	102.6	100.2	104.8	180.4	158.8	96.5	102.6	100.9	98.5	133.0	123.2	111.2	133.0	126.8
0950, 0980	Ceilings & Acoustic Treatment	98.8	77.1	84.4	70.4	102.6	93.7	82.7	180.4	147.2	76.4	102.6	93.7	81.7	133.0	115.6	81.7	133.0	115.6
0960	Flooring	94.5	92.3	93.9	97.7	75.9	91.3	108.8	177.0	128.8	100.4	75.9	93.2	91.1	172.3	114.9	97.9	172.3	119.8
0970, 0990	Wall Finishes & Painting/Coating	98.1	78.2	85.9	118.7	106.3	111.0	121.2	152.7	140.6	118.7	108.8	112.6	104.1	152.7	134.1	104.1	152.7	134.1
09	FINISHES	96.7	82.1	88.5	104.2	99.9	101.8	110.3	174.7	146.2	104.8	100.2	102.2	98.7	139.1	121.2	102.0	139.1	122.7
COVERS	DIVS. 10 - 14, 25, 28, 41, 43, 44, 46	100.0	93.3	98.7	100.0	108.4	101.7	100.0	133.5	106.7	100.0	108.4	101.7	100.0	117.2	103.4	100.0	117.2	103.4
21, 22, 23	FIRE SUPPRESSION, PLUMBING & HVAC	94.1	82.2	89.3	94.2	115.7	102.9	99.7	166.7	126.9	94.2	118.1	103.9	94.3	117.9	103.8	94.3	117.9	103.8
26, 27, 3370	ELECTRICAL, COMMUNICATIONS & UTIL.	96.0	86.4	91.1	98.0	115.7	107.1	99.6	173.3	137.6	98.0	115.7	107.1	95.2	157.5	127.4	95.2	157.5	127.4
MF2010	WEIGHTED AVERAGE	96.9	86.4	92.3	100.9	117.4	108.2	103.0	163.8	129.8	99.4	118.0	107.6	98.6	133.0	113.8	99.0	133.0	114.0

	DIVISION	NEW YORK																	
		NEW YORK			NIAGARA FALLS			PLATTSBURGH			POUGHKEEPSIE			QUEENS			RIVERHEAD		
		100 - 102			143			129			125 - 126			110			119		
		MAT.	INST.	TOTAL	MAT.	INST.	TOTAL	MAT.	INST.	TOTAL	MAT.	INST.	TOTAL	MAT.	INST.	TOTAL	MAT.	INST.	TOTAL
015433	CONTRACTOR EQUIPMENT		114.3	114.3		93.5	93.5		99.5	99.5		114.5	114.5		114.5	114.5		114.5	114.5
0241, 31 - 34	SITE & INFRASTRUCTURE, DEMOLITION	123.0	125.1	124.5	96.5	94.7	95.3	85.4	102.9	97.5	115.3	124.3	121.5	115.7	128.0	124.2	113.2	126.8	122.6
0310	Concrete Forming & Accessories	106.0	184.9	174.6	87.4	117.3	113.5	91.0	86.4	87.0	88.4	153.4	144.9	92.3	176.2	165.3	96.5	146.8	140.3
0320	Concrete Reinforcing	101.5	179.8	141.0	102.5	102.3	102.4	102.6	88.2	95.4	98.2	134.9	116.7	100.5	187.6	144.4	100.6	176.6	138.9
0330	Cast-in-Place Concrete	100.4	180.1	131.8	103.4	122.6	111.0	91.6	95.4	93.1	97.9	128.2	109.9	95.4	162.5	121.8	93.7	154.7	117.8
03	CONCRETE	106.4	180.1	142.3	105.7	115.4	110.4	100.5	89.5	95.2	105.9	139.8	122.4	103.6	171.5	136.7	101.1	153.7	126.7
04	MASONRY	103.9	178.2	149.2	119.6	123.4	121.9	89.5	90.7	90.2	102.2	129.4	118.8	108.3	175.0	149.0	117.2	162.8	145.1
05	METALS	119.0	143.1	126.9	96.9	93.5	95.8	101.0	83.5	95.2	105.2	116.8	109.0	105.2	141.8	117.3	107.1	135.4	116.5
06	WOOD, PLASTICS & COMPOSITES	107.5	187.8	153.9	82.4	115.3	101.4	90.5	85.1	87.4	84.9	164.5	130.9	88.7	177.9	140.3	93.7	145.9	123.8
07	THERMAL & MOISTURE PROTECTION	104.1	166.8	129.6	101.3	110.7	105.1	101.1	88.2	95.9	107.4	140.3	120.8	105.4	163.1	128.9	106.0	150.5	124.1
08	OPENINGS	90.2	178.0	112.3	91.5	103.2	94.4	96.7	81.2	92.8	92.3	151.4	107.2	87.1	172.1	108.5	87.1	152.2	103.5
0920	Plaster & Gypsum Board	112.0	190.1	167.8	98.0	115.5	110.5	117.0	83.9	93.4	94.2	166.5	145.9	97.9	180.4	156.8	99.6	147.3	133.6
0950, 0980	Ceilings & Acoustic Treatment	102.5	190.1	160.4	98.6	115.5	109.8	104.9	83.9	91.0	76.4	166.5	135.9	82.7	180.4	147.2	82.6	147.3	125.3
0960	Flooring	100.3	177.0	122.8	94.4	117.0	101.0	107.8	103.9	106.7	97.7	145.8	111.8	104.4	177.0	125.7	105.5	149.4	118.4
0970, 0990	Wall Finishes & Painting/Coating	106.1	155.0	136.2	98.1	113.1	107.4	116.5	78.8	93.2	118.7	106.3	111.0	121.2	152.7	140.6	121.2	152.7	140.6
09	FINISHES	107.5	181.6	148.8	96.8	117.8	108.5	101.2	88.9	94.3	103.9	149.2	129.2	107.6	174.7	145.0	107.8	147.2	129.8
COVERS	DIVS. 10 - 14, 25, 28, 41, 43, 44, 46	100.0	137.9	107.5	100.0	104.2	100.8	100.0	51.4	90.3	100.0	113.5	102.7	100.0	133.5	106.7	100.0	123.8	104.7
21, 22, 23	FIRE SUPPRESSION, PLUMBING & HVAC	100.0	168.4	127.7	94.1	99.4	96.2	94.2	84.6	90.3	94.2	117.2	103.5	99.7	166.7	126.9	99.9	149.8	120.1
26, 27, 3370	ELECTRICAL, COMMUNICATIONS & UTIL.	105.0	173.3	140.3	94.7	98.8	96.8	93.9	85.7	89.7	98.0	115.7	107.1	100.1	173.3	137.9	101.7	141.0	122.0
MF2010	WEIGHTED AVERAGE	104.8	166.9	132.2	97.9	106.1	101.5	96.9	87.0	92.6	100.1	128.4	112.6	101.4	163.8	128.9	102.1	146.1	121.5

	DIVISION	NEW YORK																	
		ROCHESTER			SCHENECTADY			STATEN ISLAND			SUFFERN			SYRACUSE			UTICA		
		144 - 146			123			103			109			130 - 132			133 - 135		
		MAT.	INST.	TOTAL	MAT.	INST.	TOTAL	MAT.	INST.	TOTAL	MAT.	INST.	TOTAL	MAT.	INST.	TOTAL	MAT.	INST.	TOTAL
015433	CONTRACTOR EQUIPMENT		115.6	115.6		113.6	113.6		113.9	113.9		113.9	113.9		113.6	113.6		113.6	113.6
0241, 31 - 34	SITE & INFRASTRUCTURE, DEMOLITION	74.5	107.6	97.3	71.9	106.9	96.0	127.1	124.5	125.3	117.7	120.3	119.5	92.4	106.6	102.2	70.2	104.2	93.7
0310	Concrete Forming & Accessories	103.5	91.7	93.3	102.0	95.1	96.0	92.1	176.5	165.6	100.6	126.4	123.1	101.0	88.0	89.7	102.5	80.6	83.5
0320	Concrete Reinforcing	103.4	89.2	96.2	96.9	91.8	94.3	95.9	179.8	138.2	94.7	135.0	115.0	98.3	95.5	96.9	98.3	87.1	92.6
0330	Cast-in-Place Concrete	96.8	96.6	96.7	90.4	101.4	94.7	99.7	164.2	125.1	96.7	126.8	108.5	91.3	95.8	93.1	83.6	87.5	85.1
03	CONCRETE	112.7	93.9	103.5	100.8	97.4	99.2	107.4	171.0	138.4	102.1	127.4	114.4	99.7	93.0	96.5	97.7	85.3	91.7
04	MASONRY	103.3	96.3	99.0	91.9	99.8	96.7	106.6	175.1	148.4	99.0	123.0	113.6	98.9	101.8	100.6	90.6	88.2	89.1
05	METALS	98.0	104.2	100.1	100.4	104.0	101.6	102.8	142.9	116.0	102.8	118.0	107.8	100.3	105.0	101.8	98.3	101.6	99.4
06	WOOD, PLASTICS & COMPOSITES	96.6	91.7	93.8	102.5	94.3	97.8	91.7	178.3	141.7	102.8	132.6	120.0	102.5	85.6	92.7	102.5	82.4	90.9
07	THERMAL & MOISTURE PROTECTION	96.7	93.9	95.5	89.1	93.2	90.8	104.4	164.3	128.8	104.9	134.4	116.9	99.5	96.2	98.2	88.9	90.0	89.4
08	OPENINGS	94.2	86.1	92.2	94.3	87.7	92.6	85.0	172.8	107.1	85.0	136.9	98.1	92.3	79.9	89.2	94.3	78.6	90.4
0920	Plaster & Gypsum Board	114.8	91.4	98.1	106.3	93.9	97.5	98.6	180.4	157.0	102.6	133.3	124.5	106.3	84.8	91.0	106.3	81.6	88.7
0950, 0980	Ceilings & Acoustic Treatment	106.9	91.4	96.7	97.1	93.9	95.0	83.5	180.4	147.5	81.7	133.3	115.7	97.1	84.8	89.0	97.1	81.6	86.9
0960	Flooring	90.3	103.0	94.1	91.0	103.9	94.8	95.2	177.0	119.2	94.3	149.4	110.4	93.1	95.0	93.7	91.0	90.4	90.8
0970, 0990	Wall Finishes & Painting/Coating	97.9	95.5	96.4	89.3	78.4	82.6	106.1	152.7	134.8	104.1	114.4	110.5	96.6	98.5	97.8	89.3	98.5	95.0
09	FINISHES	102.9	94.4	98.2	96.0	95.1	95.5	100.8	175.0	142.2	99.6	126.9	114.9	97.9	90.0	93.5	96.0	84.2	89.4
COVERS	DIVS. 10 - 14, 25, 28, 41, 43, 44, 46	100.0	94.1	98.8	100.0	93.3	98.7	100.0	134.3	106.8	100.0	115.6	103.1	100.0	95.7	99.2	100.0	87.2	97.4
21, 22, 23	FIRE SUPPRESSION, PLUMBING & HVAC	100.0	87.1	94.8	100.2	91.5	96.7	100.1	166.8	127.1	94.3	112.0	101.4	100.2	88.3	95.3	100.2	78.4	91.3
26, 27, 3370	ELECTRICAL, COMMUNICATIONS & UTIL.	99.2	89.6	94.2	100.9	91.6	96.1	96.9	173.3	136.3	103.2	123.8	113.8	100.9	94.3	97.5	99.0	94.3	96.6
MF2010	WEIGHTED AVERAGE	100.0	94.0	97.3	97.8	96.1	97.0	100.8	163.6	128.5	98.7	122.0	109.0	98.9	94.5	97.0	96.8	88.2	93.0

	DIVISION	NEW YORK									NORTH CAROLINA								
		WATERTOWN			WHITE PLAINS			YONKERS			ASHEVILLE			CHARLOTTE			DURHAM		
		136			106			107			287 - 288			281 - 282			277		
		MAT.	INST.	TOTAL	MAT.	INST.	TOTAL	MAT.	INST.	TOTAL	MAT.	INST.	TOTAL	MAT.	INST.	TOTAL	MAT.	INST.	TOTAL
015433	CONTRACTOR EQUIPMENT		113.6	113.6		113.9	113.9		113.9	113.9		94.7	94.7		94.7	94.7		99.9	99.9
0241, 31 - 34	SITE & INFRASTRUCTURE, DEMOLITION	78.2	106.5	97.7	113.8	119.2	117.5	122.5	119.1	120.1	108.4	75.5	85.7	109.7	75.6	86.2	108.6	84.0	91.6
0310	Concrete Forming & Accessories	88.2	90.3	90.1	105.3	129.3	126.2	105.7	129.3	126.2	92.5	37.4	44.6	96.6	39.2	46.7	94.8	38.0	45.4
0320	Concrete Reinforcing	98.9	95.6	97.2	94.7	178.4	136.9	98.5	178.4	138.8	93.6	41.0	67.1	94.0	38.3	65.9	95.2	47.9	71.4
0330	Cast-in-Place Concrete	97.3	94.7	96.3	88.5	129.7	104.8	99.0	129.8	111.1	118.0	45.0	89.3	124.5	47.0	93.9	119.9	45.3	90.5
03	CONCRETE	110.4	93.6	102.2	95.6	137.6	116.1	105.3	137.6	121.1	109.0	42.7	76.7	111.9	43.6	78.6	109.7	44.3	77.8
04	MASONRY	91.6	101.4	97.6	98.5	127.3	116.1	103.4	127.3	118.0	87.1	38.2	57.3	93.2	41.8	61.8	92.4	34.4	57.0
05	METALS	98.3	105.0	100.5	104.3	132.6	113.7	114.4	132.6	120.4	87.2	72.4	82.3	91.2	71.6	84.7	100.4	75.5	92.2
06	WOOD, PLASTICS & COMPOSITES	85.2	89.4	87.6	108.6	132.4	122.4	108.4	132.4	122.3	94.5	36.9	61.3	99.7	39.0	64.6	96.9	38.3	63.1
07	THERMAL & MOISTURE PROTECTION	89.2	96.3	92.1	104.8	137.9	118.2	105.1	137.9	118.4	99.9	41.2	76.1	97.3	42.7	75.1	100.7	43.2	77.3
08	OPENINGS	94.3	85.7	92.1	85.1	145.6	100.3	88.2	145.6	102.7	93.2	36.7	79.0	97.3	38.1	82.4	97.3	40.5	83.0
0920	Plaster & Gypsum Board	97.7	88.8	91.3	106.2	133.0	125.4	111.4	133.0	126.9	104.3	34.8	54.7	100.7	36.9	55.1	108.3	36.2	56.8
0950, 0980	Ceilings & Acoustic Treatment	97.1	88.8	91.6	81.7	133.0	115.6	100.7	133.0	122.1	93.7	34.8	54.8	96.4	36.9	57.1	97.3	36.2	57.0
0960	Flooring	83.8	91.2	85.9	96.3	172.3	118.6	95.9	172.3	118.3	103.7	43.6	86.1	106.7	43.8	88.2	107.3	43.6	88.6
0970, 0990	Wall Finishes & Painting/Coating	89.3	88.9	89.1	104.1	152.7	134.1	104.1	152.7	134.1	114.8	33.7	64.8	114.8	38.5	67.7	114.8	35.4	65.8
09	FINISHES	93.7	90.1	91.7	100.1	139.1	121.8	105.6	139.1	124.3	100.8	37.9	65.7	101.5	39.8	67.1	103.2	38.4	67.0
COVERS	DIVS. 10 - 14, 25, 28, 41, 43, 44, 46	100.0	95.7	99.2	100.0	117.2	103.4	100.0	120.4	104.1	100.0	64.5	92.9	100.0	65.0	93.0	100.0	72.4	94.5
21, 22, 23	FIRE SUPPRESSION, PLUMBING & HVAC	100.2	83.8	93.5	100.3	117.9	107.4	100.3	117.9	107.4	100.2	36.9	74.6	99.9	37.1	74.5	100.3	35.8	74.2
26, 27, 3370	ELECTRICAL, COMMUNICATIONS & UTIL.	100.9	91.7	96.2	95.2	157.5	127.4	103.3	157.5	131.3	96.6	36.6	65.7	97.2	47.6	71.6	97.5	46.3	71.1
MF2010	WEIGHTED AVERAGE	98.5	93.5	96.3	98.8	133.0	113.9	103.5	133.1	116.5	97.8	45.5	74.7	99.4	47.8	76.7	100.8	47.9	77.5

	DIVISION	NORTH CAROLINA																	
		ELIZABETH CITY			FAYETTEVILLE			GASTONIA			GREENSBORO			HICKORY			KINSTON		
		279			283			280			270,272 - 274			286			285		
		MAT.	INST.	TOTAL	MAT.	INST.	TOTAL	MAT.	INST.	TOTAL	MAT.	INST.	TOTAL	MAT.	INST.	TOTAL	MAT.	INST.	TOTAL
015433	CONTRACTOR EQUIPMENT		103.9	103.9		99.9	99.9		94.7	94.7		99.9	99.9		99.9	99.9		99.9	99.9
0241, 31 - 34	SITE & INFRASTRUCTURE, DEMOLITION	114.3	85.0	94.1	107.4	84.1	91.3	108.0	75.5	85.6	108.5	84.1	91.7	107.6	84.0	91.4	106.2	84.1	90.9
0310	Concrete Forming & Accessories	82.0	33.0	39.4	92.1	38.8	45.7	98.4	38.3	46.1	94.5	38.6	45.9	88.9	36.6	43.4	85.5	38.8	44.8
0320	Concrete Reinforcing	93.1	39.9	66.3	97.3	48.1	72.5	94.0	40.9	67.3	94.0	56.2	75.0	93.6	18.6	55.8	93.1	18.1	55.3
0330	Cast-in-Place Concrete	120.1	36.9	87.4	123.7	48.2	94.0	115.3	52.1	90.4	119.0	46.5	90.5	117.9	46.1	89.6	113.8	39.3	84.5
03	CONCRETE	109.9	37.7	74.7	111.4	45.7	79.4	107.4	45.5	77.2	109.0	46.6	78.6	108.7	38.6	74.6	105.4	37.2	72.2
04	MASONRY	105.0	25.7	56.6	90.6	39.1	59.2	91.7	36.1	57.8	90.5	35.0	56.6	76.8	39.3	53.9	82.8	31.5	51.5
05	METALS	89.2	71.9	83.5	104.3	75.6	94.9	88.1	72.4	82.9	94.5	78.6	89.2	87.3	66.1	80.3	86.3	65.2	79.3
06	WOOD, PLASTICS & COMPOSITES	81.9	34.2	54.3	93.5	38.3	61.6	102.4	38.3	65.3	96.6	38.3	62.9	89.3	36.7	58.9	86.3	39.6	59.3
07	THERMAL & MOISTURE PROTECTION	100.0	32.5	72.6	99.8	41.7	76.2	100.2	41.5	76.3	100.5	40.8	76.2	100.3	40.2	75.8	100.1	33.8	73.2
08	OPENINGS	94.2	32.1	78.5	93.3	40.5	80.0	97.3	38.2	82.4	97.3	42.8	83.6	93.3	29.8	77.3	93.4	31.4	77.7
0920	Plaster & Gypsum Board	101.8	31.3	51.4	109.0	36.2	57.0	110.4	36.2	57.4	110.4	36.2	57.4	104.3	34.6	54.5	104.5	37.6	56.7
0950, 0980	Ceilings & Acoustic Treatment	97.3	31.3	53.7	94.6	36.2	56.0	97.3	36.2	56.9	97.3	36.2	56.9	93.7	34.6	54.6	97.3	37.6	57.9
0960	Flooring	98.0	23.7	76.2	103.8	43.6	86.2	107.3	43.6	88.6	107.3	40.3	87.6	103.5	33.6	83.0	100.7	19.2	76.8
0970, 0990	Wall Finishes & Painting/Coating	114.8	43.6	70.9	114.8	33.7	64.8	114.8	38.7	67.9	114.8	32.8	64.2	114.8	32.8	64.2	114.8	29.9	62.4
09	FINISHES	100.1	32.2	62.2	101.5	38.7	66.5	103.3	39.1	67.5	103.4	38.1	67.0	101.0	35.3	64.3	100.7	34.6	63.8
COVERS	DIVS. 10 - 14, 25, 28, 41, 43, 44, 46	100.0	72.8	94.6	100.0	73.1	94.7	100.0	64.5	92.9	100.0	64.7	93.0	100.0	64.5	92.9	100.0	73.6	94.7
21, 22, 23	FIRE SUPPRESSION, PLUMBING & HVAC	94.3	29.0	67.8	100.2	36.9	74.5	100.2	36.6	74.4	100.2	37.0	74.6	94.3	30.4	68.4	94.3	31.0	68.6
26, 27, 3370	ELECTRICAL, COMMUNICATIONS & UTIL.	97.1	29.0	62.0	95.6	50.5	72.3	96.1	34.0	64.1	96.4	36.8	65.7	94.2	30.8	61.5	94.0	47.9	70.2
MF2010	WEIGHTED AVERAGE	97.8	40.6	72.6	100.7	49.4	78.1	98.6	45.4	75.1	99.7	47.3	76.6	95.5	42.3	72.1	95.2	43.8	72.6

	DIVISION	NORTH CAROLINA															NORTH DAKOTA		
		MURPHY			RALEIGH			ROCKY MOUNT			WILMINGTON			WINSTON-SALEM			BISMARCK		
		289			275 - 276			278			284			271			585		
		MAT.	INST.	TOTAL	MAT.	INST.	TOTAL	MAT.	INST.	TOTAL	MAT.	INST.	TOTAL	MAT.	INST.	TOTAL	MAT.	INST.	TOTAL
015433	CONTRACTOR EQUIPMENT		94.7	94.7		99.9	99.9		99.9	99.9		94.7	94.7		99.9	99.9		98.7	98.7
0241, 31 - 34	SITE & INFRASTRUCTURE, DEMOLITION	110.1	75.3	86.1	109.1	84.1	91.9	112.0	84.2	92.8	109.5	75.7	86.2	108.8	84.0	91.7	100.1	97.4	98.2
0310	Concrete Forming & Accessories	99.0	23.9	33.7	94.9	39.5	46.7	87.7	31.6	38.9	93.7	44.1	50.5	95.8	37.7	45.2	99.8	42.5	49.9
0320	Concrete Reinforcing	93.2	17.9	55.2	94.0	48.1	70.9	93.1	26.3	59.5	94.3	46.4	70.2	94.0	38.0	65.8	110.0	85.6	97.7
0330	Cast-in-Place Concrete	122.0	35.2	87.8	127.6	51.5	97.6	117.5	38.5	86.4	117.5	48.0	90.1	122.0	44.0	91.3	101.1	49.3	80.7
03	CONCRETE	112.2	29.2	71.8	113.3	47.1	81.1	110.4	35.1	73.7	108.8	47.7	79.0	110.6	41.9	77.1	100.3	54.3	77.9
04	MASONRY	79.4	25.4	46.4	91.7	40.0	60.2	83.2	33.1	52.6	77.1	40.0	54.5	90.7	36.0	57.3	104.3	59.6	77.0
05	METALS	85.3	64.5	78.4	92.2	75.5	86.7	88.4	65.4	80.8	87.0	74.9	83.0	92.2	71.1	85.2	96.5	86.1	93.0
06	WOOD, PLASTICS & COMPOSITES	103.1	22.1	56.3	96.9	38.9	63.4	88.6	28.9	54.1	96.2	44.1	66.1	96.6	37.6	62.5	82.3	36.9	56.0
07	THERMAL & MOISTURE PROTECTION	100.1	30.6	71.9	99.5	42.0	76.1	100.5	31.8	72.6	99.9	46.1	78.0	100.5	40.5	76.1	107.4	52.6	85.1
08	OPENINGS	93.2	24.5	75.9	94.1	40.6	80.6	93.4	25.6	76.3	93.4	43.9	80.9	97.3	37.5	82.2	98.0	48.4	85.5
0920	Plaster & Gypsum Board	109.3	19.5	45.1	100.7	36.9	55.1	103.7	26.5	48.6	106.0	42.2	60.4	110.4	35.5	56.9	98.3	35.2	53.2
0950, 0980	Ceilings & Acoustic Treatment	93.7	19.5	44.7	96.4	36.9	57.1	94.6	26.5	49.6	94.6	42.2	60.0	97.3	35.5	56.5	144.8	35.2	72.4
0960	Flooring	107.6	24.6	83.3	107.3	43.6	88.6	102.0	22.3	78.6	104.5	45.5	87.2	107.3	43.6	88.6	115.3	75.5	103.6
0970, 0990	Wall Finishes & Painting/Coating	114.8	18.1	55.1	114.8	37.2	66.9	114.8	40.0	68.6	114.8	40.2	68.8	114.8	34.4	65.2	96.0	31.0	55.9
09	FINISHES	102.8	22.7	58.1	101.8	39.7	67.2	101.1	30.3	61.6	101.4	43.9	69.3	103.4	38.2	67.1	114.9	45.3	76.1
COVERS	DIVS. 10 - 14, 25, 28, 41, 43, 44, 46	100.0	61.8	92.4	100.0	73.5	94.7	100.0	72.9	94.6	100.0	74.7	95.0	100.0	64.4	92.9	100.0	81.2	96.3
21, 22, 23	FIRE SUPPRESSION, PLUMBING & HVAC	94.3	18.5	63.6	100.0	35.3	73.8	94.3	35.7	70.6	100.2	38.3	75.1	100.2	35.7	74.1	100.3	60.7	84.3
26, 27, 3370	ELECTRICAL, COMMUNICATIONS & UTIL.	97.5	25.7	60.5	100.6	36.5	67.5	99.1	38.8	68.0	96.8	36.9	65.9	96.4	40.1	67.4	98.1	71.7	84.5
MF2010	WEIGHTED AVERAGE	96.4	33.2	68.5	99.8	47.6	76.8	96.8	42.5	72.9	97.4	48.5	75.8	99.5	46.0	75.9	100.7	64.4	84.7

	DIVISION	NORTH DAKOTA																	
		DEVILS LAKE			DICKINSON			FARGO			GRAND FORKS			JAMESTOWN			MINOT		
		583			586			580 - 581			582			584			587		
		MAT.	INST.	TOTAL	MAT.	INST.	TOTAL	MAT.	INST.	TOTAL	MAT.	INST.	TOTAL	MAT.	INST.	TOTAL	MAT.	INST.	TOTAL
015433	CONTRACTOR EQUIPMENT		98.7	98.7		98.7	98.7		98.7	98.7		98.7	98.7		98.7	98.7		98.7	98.7
0241, 31 - 34	SITE & INFRASTRUCTURE, DEMOLITION	109.6	95.4	99.8	119.7	94.1	102.1	100.7	97.4	98.4	114.2	94.1	100.4	108.5	94.1	98.6	111.2	97.4	101.7
0310	Concrete Forming & Accessories	103.3	36.9	45.6	93.3	36.1	43.6	101.0	43.2	50.7	96.9	36.1	44.0	94.6	35.8	43.4	93.0	60.8	65.0
0320	Concrete Reinforcing	110.0	85.9	97.9	111.0	42.0	76.2	101.4	85.7	93.5	108.3	85.8	96.9	110.6	49.8	80.0	112.0	85.9	98.9
0330	Cast-in-Place Concrete	122.8	47.9	93.3	111.1	46.3	85.6	100.2	51.2	80.9	111.1	46.3	85.6	121.3	46.2	91.7	111.1	48.3	86.4
03	CONCRETE	114.0	51.4	83.5	113.0	42.2	78.5	107.8	55.3	82.2	110.1	50.2	80.9	112.6	43.4	78.9	108.8	62.1	86.1
04	MASONRY	112.8	65.2	83.8	114.8	59.0	80.8	106.0	59.6	77.7	107.3	64.8	81.3	125.1	34.7	69.9	106.5	64.2	80.7
05	METALS	96.5	85.1	92.8	96.4	64.8	86.0	98.8	86.4	94.7	96.5	80.5	91.2	96.5	65.9	86.4	96.8	86.7	93.4
06	WOOD, PLASTICS & COMPOSITES	85.5	34.0	55.7	75.2	34.0	51.4	82.1	37.4	56.3	78.9	34.0	53.0	76.9	34.0	52.1	74.9	61.3	67.1
07	THERMAL & MOISTURE PROTECTION	108.2	50.8	84.9	108.8	48.4	84.2	105.7	53.7	84.6	108.5	50.8	85.0	108.1	43.0	81.6	108.2	56.4	87.2
08	OPENINGS	98.0	42.3	84.0	97.9	31.9	81.3	97.9	48.7	85.5	98.0	42.3	83.9	98.0	34.1	81.9	98.1	61.7	88.9
0920	Plaster & Gypsum Board	118.9	32.2	57.0	109.9	32.2	54.4	98.3	35.7	53.6	111.7	32.2	54.9	111.2	32.2	54.8	109.9	60.4	74.5
0950, 0980	Ceilings & Acoustic Treatment	145.7	32.2	70.7	145.7	32.2	70.7	144.8	35.7	72.8	145.7	32.2	70.7	145.7	32.2	70.7	145.7	60.4	89.4
0960	Flooring	119.5	37.6	95.4	111.2	37.6	89.6	114.9	75.5	103.3	113.6	37.6	91.3	112.2	37.6	90.3	111.0	84.2	103.1
0970, 0990	Wall Finishes & Painting/Coating	96.0	23.7	51.4	96.0	23.7	51.4	96.0	73.8	82.3	96.0	23.7	51.4	96.0	23.7	51.4	96.0	26.0	52.8
09	FINISHES	119.5	34.1	71.9	117.1	34.1	70.8	114.9	50.4	78.9	117.4	34.1	70.9	116.4	34.1	70.5	116.1	60.9	85.3
COVERS	DIVS. 10 - 14, 25, 28, 41, 43, 44, 46	100.0	34.9	87.1	100.0	35.1	87.1	100.0	81.3	96.3	100.0	35.0	87.1	100.0	79.1	95.8	100.0	84.0	96.8
21, 22, 23	FIRE SUPPRESSION, PLUMBING & HVAC	94.5	63.0	81.7	94.5	70.5	84.7	100.3	66.0	86.4	100.4	35.6	74.2	94.5	37.5	71.4	100.4	60.7	84.3
26, 27, 3370	ELECTRICAL, COMMUNICATIONS & UTIL.	96.3	38.0	66.3	105.1	71.7	87.8	99.7	66.9	82.8	99.9	56.9	77.7	96.3	38.0	66.2	103.1	80.4	91.4
MF2010	WEIGHTED AVERAGE	101.7	57.0	82.0	102.7	58.8	83.3	102.1	65.7	86.0	102.8	53.1	80.9	101.9	46.4	77.4	102.7	70.2	88.4

	DIVISION	NORTH DAKOTA			OHIO														
		WILLISTON			AKRON			ATHENS			CANTON			CHILLICOTHE			CINCINNATI		
		588			442 - 443			457			446 - 447			456			451 - 452		
		MAT.	INST.	TOTAL	MAT.	INST.	TOTAL	MAT.	INST.	TOTAL	MAT.	INST.	TOTAL	MAT.	INST.	TOTAL	MAT.	INST.	TOTAL
015433	CONTRACTOR EQUIPMENT		98.7	98.7		97.0	97.0		91.7	91.7		97.0	97.0		101.6	101.6		101.4	101.4
0241, 31 - 34	SITE & INFRASTRUCTURE, DEMOLITION	112.2	94.1	99.7	91.7	104.6	100.6	103.8	94.8	97.6	91.8	104.1	100.3	89.0	106.6	101.2	86.1	106.1	99.9
0310	Concrete Forming & Accessories	98.4	36.1	44.2	100.3	92.6	93.6	91.7	89.0	89.4	100.3	82.7	85.0	93.3	95.4	95.1	95.0	82.2	83.8
0320	Concrete Reinforcing	113.2	42.0	77.3	93.4	92.0	92.7	93.4	87.6	90.4	93.4	77.3	85.3	90.6	81.0	85.8	96.0	83.2	89.6
0330	Cast-in-Place Concrete	111.1	46.3	85.6	95.5	98.4	96.6	108.5	95.2	103.3	96.4	96.0	96.2	98.5	94.3	96.8	90.7	86.6	89.1
03	CONCRETE	110.2	42.2	77.1	95.1	93.8	94.5	102.6	90.3	96.6	95.6	85.9	90.9	96.9	92.1	94.6	91.6	84.0	87.9
04	MASONRY	101.2	59.0	75.5	90.4	95.6	93.6	76.1	91.1	85.2	91.1	86.5	88.3	82.9	100.4	93.6	82.5	87.3	85.4
05	METALS	96.6	64.8	86.1	92.7	82.5	89.3	100.5	79.3	93.5	92.7	76.2	87.2	92.8	85.1	90.3	95.2	84.7	91.7
06	WOOD, PLASTICS & COMPOSITES	80.2	34.0	53.5	102.2	91.4	96.0	83.8	90.3	87.5	102.7	80.9	90.1	96.2	94.1	95.0	98.5	80.0	87.8
07	THERMAL & MOISTURE PROTECTION	108.4	48.4	84.0	107.2	96.9	103.0	99.8	93.0	97.0	108.2	92.7	101.9	100.9	96.4	99.1	99.3	91.1	95.9
08	OPENINGS	98.0	31.9	81.4	112.2	91.5	107.0	101.0	86.3	97.3	105.9	78.0	98.9	92.9	86.1	91.2	101.7	79.7	96.1
0920	Plaster & Gypsum Board	111.7	32.2	54.9	91.0	90.8	90.8	91.1	89.6	90.1	92.3	80.0	83.5	91.7	94.2	93.5	93.8	79.6	83.7
0950, 0980	Ceilings & Acoustic Treatment	145.7	32.2	70.7	94.7	90.8	92.1	103.6	89.6	94.4	94.7	80.0	85.0	97.7	94.2	95.4	98.6	79.6	86.1
0960	Flooring	114.7	37.6	92.1	100.4	92.1	98.0	127.7	100.1	119.6	100.6	82.1	95.2	103.9	103.6	103.8	104.8	90.6	100.7
0970, 0990	Wall Finishes & Painting/Coating	96.0	23.7	51.4	100.8	102.3	101.7	103.1	96.1	98.8	100.8	83.3	90.0	100.1	92.5	95.4	100.1	86.1	91.5
09	FINISHES	117.6	34.1	71.0	98.1	93.3	95.4	103.0	92.7	97.3	98.3	82.1	89.3	99.6	97.1	98.2	100.1	83.7	90.9
COVERS	DIVS. 10 - 14, 25, 28, 41, 43, 44, 46	100.0	35.1	87.1	100.0	94.0	98.8	100.0	56.1	91.3	100.0	91.5	98.3	100.0	92.0	98.4	100.0	92.4	98.5
21, 22, 23	FIRE SUPPRESSION, PLUMBING & HVAC	94.5	70.5	84.7	100.0	97.0	98.8	94.0	55.0	78.2	100.0	83.9	93.5	94.4	93.4	94.0	99.9	83.0	93.1
26, 27, 3370	ELECTRICAL, COMMUNICATIONS & UTIL.	100.2	71.7	85.5	97.9	94.5	96.1	99.3	102.6	101.0	97.2	90.9	93.9	98.9	85.2	91.8	97.7	80.4	88.8
MF2010	WEIGHTED AVERAGE	101.1	58.8	82.4	98.7	94.5	96.9	98.0	83.1	91.4	98.2	86.3	93.0	95.2	93.4	94.4	96.9	85.8	92.0

	DIVISION	OHIO																	
		CLEVELAND			COLUMBUS			DAYTON			HAMILTON			LIMA			LORAIN		
		441			430 - 432			453 - 454			450			458			440		
		MAT.	INST.	TOTAL	MAT.	INST.	TOTAL	MAT.	INST.	TOTAL	MAT.	INST.	TOTAL	MAT.	INST.	TOTAL	MAT.	INST.	TOTAL
015433	CONTRACTOR EQUIPMENT		97.3	97.3		95.9	95.9		96.7	96.7		101.6	101.6		94.2	94.2		97.0	97.0
0241, 31 - 34	SITE & INFRASTRUCTURE, DEMOLITION	91.7	104.9	100.8	93.2	100.9	98.5	84.9	106.0	99.4	84.8	106.2	99.5	97.6	94.5	95.5	91.1	103.4	99.6
0310	Concrete Forming & Accessories	100.4	97.9	98.2	100.6	87.4	89.1	95.0	81.6	83.4	95.0	81.7	83.4	91.6	85.9	86.6	100.4	87.9	89.5
0320	Concrete Reinforcing	93.8	92.3	93.0	103.7	83.0	93.3	96.0	81.2	88.6	96.0	83.2	89.6	93.4	81.4	87.4	93.4	92.1	92.8
0330	Cast-in-Place Concrete	93.6	106.4	98.7	88.9	94.0	90.9	84.4	87.6	85.6	90.4	86.0	88.7	99.7	97.4	98.8	90.9	104.2	96.1
03	CONCRETE	94.3	99.0	96.6	93.5	88.5	91.1	88.4	83.4	85.9	91.5	83.6	87.6	95.8	88.7	92.4	92.8	93.8	93.3
04	MASONRY	95.0	103.4	100.1	91.4	95.0	93.6	82.0	83.7	83.1	82.4	86.0	84.6	104.8	84.9	92.6	87.1	97.6	93.5
05	METALS	94.3	84.9	91.2	94.3	81.2	90.0	94.4	77.9	88.9	94.5	84.7	91.2	100.5	81.1	94.1	93.3	83.0	89.9
06	WOOD, PLASTICS & COMPOSITES	101.2	94.9	97.6	98.6	84.9	90.7	99.9	80.3	88.6	98.5	80.0	87.8	83.7	85.2	84.6	102.2	85.3	92.5
07	THERMAL & MOISTURE PROTECTION	105.9	107.7	106.6	100.7	95.1	98.5	104.2	89.0	98.0	101.0	90.7	96.8	99.3	95.8	97.9	108.1	101.0	105.3
08	OPENINGS	101.7	93.4	99.6	103.4	81.6	97.9	101.8	79.9	96.3	99.1	79.7	94.2	101.0	80.0	95.7	105.9	88.2	101.4
0920	Plaster & Gypsum Board	90.0	94.4	93.1	96.1	84.5	87.8	93.8	80.0	83.9	93.8	79.6	83.7	91.1	84.5	86.4	91.0	84.5	86.3
0950, 0980	Ceilings & Acoustic Treatment	92.9	94.4	93.9	95.5	84.5	88.2	99.7	80.0	86.7	98.6	79.6	86.1	102.6	84.5	90.6	94.7	84.5	88.0
0960	Flooring	100.2	105.0	101.6	95.2	90.0	93.7	107.7	83.0	100.5	104.8	90.6	100.7	126.8	90.8	116.3	100.6	105.0	101.9
0970, 0990	Wall Finishes & Painting/Coating	100.8	106.5	104.3	99.0	92.5	95.0	100.1	85.9	91.4	100.1	86.1	91.5	103.2	84.9	91.9	100.8	106.5	104.3
09	FINISHES	97.6	99.7	98.8	97.3	88.0	92.1	101.1	82.0	90.4	100.0	83.4	90.7	102.1	86.4	93.3	98.1	92.6	95.1
COVERS	DIVS. 10 - 14, 25, 28, 41, 43, 44, 46	100.0	98.3	99.7	100.0	94.7	98.9	100.0	91.9	98.4	100.0	92.0	98.4	100.0	92.9	98.6	100.0	95.2	99.0
21, 22, 23	FIRE SUPPRESSION, PLUMBING & HVAC	99.9	98.8	99.5	99.9	93.9	97.5	100.6	84.8	94.2	100.4	82.4	93.1	94.0	86.2	90.9	100.0	84.7	93.8
26, 27, 3370	ELECTRICAL, COMMUNICATIONS & UTIL.	97.5	107.4	102.6	98.7	88.1	93.2	96.3	83.8	89.9	96.7	81.9	89.1	99.7	80.8	89.9	97.4	89.9	93.5
MF2010	WEIGHTED AVERAGE	97.9	99.8	98.7	97.8	90.6	94.6	96.7	85.2	91.6	96.6	85.6	91.7	98.4	86.1	93.0	97.7	91.3	94.9

	DIVISION	OHIO																	
		MANSFIELD			MARION			SPRINGFIELD			STEUBENVILLE			TOLEDO			YOUNGSTOWN		
		448 - 449			433			455			439			434 - 436			444 - 445		
		MAT.	INST.	TOTAL	MAT.	INST.	TOTAL	MAT.	INST.	TOTAL	MAT.	INST.	TOTAL	MAT.	INST.	TOTAL	MAT.	INST.	TOTAL
015433	CONTRACTOR EQUIPMENT		97.0	97.0		95.6	95.6		96.7	96.7		100.3	100.3		98.3	98.3		97.0	97.0
0241, 31 - 34	SITE & INFRASTRUCTURE, DEMOLITION	87.9	104.4	99.3	89.2	99.8	96.5	85.2	104.6	98.6	134.7	109.2	117.1	92.4	101.8	98.9	91.6	105.0	100.8
0310	Concrete Forming & Accessories	91.1	86.6	87.2	97.5	84.4	86.1	95.0	84.7	86.1	100.5	86.7	88.5	100.5	95.9	96.5	100.3	86.0	87.9
0320	Concrete Reinforcing	85.0	77.9	81.4	95.6	83.0	89.3	96.0	81.2	88.6	93.1	85.4	89.2	103.7	89.8	96.7	93.4	86.9	90.1
0330	Cast-in-Place Concrete	88.4	93.0	90.2	81.2	91.1	85.1	86.6	87.7	87.1	88.2	93.1	90.1	88.9	103.4	94.6	94.6	98.5	96.1
03	CONCRETE	87.8	86.7	87.2	85.8	86.1	86.0	89.6	84.7	87.2	89.7	88.1	88.9	93.5	97.0	95.2	94.7	90.0	92.4
04	MASONRY	89.6	97.4	94.3	92.9	96.1	94.8	82.2	84.0	83.3	79.8	91.2	86.7	99.5	100.1	99.8	90.7	92.7	91.9
05	METALS	93.5	77.1	88.1	93.4	79.2	88.7	94.3	77.6	88.8	90.1	79.6	86.6	94.1	87.6	91.9	92.7	80.3	88.6
06	WOOD, PLASTICS & COMPOSITES	91.2	85.3	87.8	94.9	84.9	89.2	101.5	84.7	91.8	90.9	85.1	87.6	98.6	95.1	96.6	102.2	83.8	91.6
07	THERMAL & MOISTURE PROTECTION	107.6	95.8	102.8	100.3	84.8	94.0	104.0	89.6	98.1	112.3	94.3	105.0	102.5	103.6	102.9	108.3	94.9	102.9
08	OPENINGS	105.1	79.6	98.7	97.3	80.2	93.0	99.5	80.4	94.7	98.3	82.9	94.4	101.2	91.4	98.7	105.9	84.7	100.6
0920	Plaster & Gypsum Board	84.9	84.5	84.6	93.8	84.5	87.1	93.8	84.5	87.1	94.4	84.1	87.1	96.1	94.9	95.2	91.0	83.0	85.3
0950, 0980	Ceilings & Acoustic Treatment	95.6	84.5	88.3	95.5	84.5	88.2	99.7	84.5	89.6	92.3	84.1	86.9	95.5	94.9	95.1	94.7	83.0	87.0
0960	Flooring	95.7	104.5	98.3	93.8	73.0	87.7	107.7	83.0	100.5	122.3	95.1	114.3	94.5	101.8	96.6	100.6	93.7	98.6
0970, 0990	Wall Finishes & Painting/Coating	100.8	91.2	94.9	99.0	50.1	68.8	100.1	85.9	91.4	113.1	95.5	102.2	99.0	102.9	101.4	100.8	93.5	96.3
09	FINISHES	95.9	90.3	92.8	96.1	78.9	86.5	101.1	84.6	91.9	114.7	88.6	100.2	97.0	97.5	97.3	98.2	87.6	92.3
COVERS	DIVS. 10 - 14, 25, 28, 41, 43, 44, 46	100.0	92.4	98.5	100.0	57.9	91.6	100.0	92.5	98.5	100.0	96.3	99.3	100.0	98.5	99.7	100.0	92.3	98.5
21, 22, 23	FIRE SUPPRESSION, PLUMBING & HVAC	94.0	83.4	89.7	94.0	89.0	92.0	100.6	85.0	94.3	94.3	91.2	93.0	100.0	99.2	99.7	100.0	92.1	96.8
26, 27, 3370	ELECTRICAL, COMMUNICATIONS & UTIL.	95.0	80.6	87.6	92.6	80.6	86.4	96.3	84.0	89.9	87.8	111.5	100.0	98.7	104.1	101.5	97.4	86.1	91.5
MF2010	WEIGHTED AVERAGE	95.2	87.4	91.7	93.9	85.4	90.2	96.6	85.8	91.8	96.1	93.5	95.0	97.9	98.3	98.1	98.1	90.1	94.6

	DIVISION	OHIO			OKLAHOMA															
		ZANESVILLE			ARDMORE			CLINTON			DURANT			ENID			GUYMON			
		437 - 438			734			736			747			737			739			
		MAT.	INST.	TOTAL	MAT.	INST.	TOTAL	MAT.	INST.	TOTAL	MAT.	INST.	TOTAL	MAT.	INST.	TOTAL	MAT.	INST.	TOTAL	
015433	CONTRACTOR EQUIPMENT		95.6	95.6		81.8	81.8		81.0	81.0		81.0	81.0		81.0	81.0		81.0	81.0	
0241, 31 - 34	SITE & INFRASTRUCTURE, DEMOLITION	92.0	101.1	98.2	98.9	90.4	93.0	100.5	89.0	92.5	93.8	88.8	90.3	101.9	89.0	93.0	105.1	87.6	93.0	
0310	Concrete Forming & Accessories	94.8	84.5	85.9	91.3	39.6	46.3	90.2	34.3	41.5	83.8	39.5	45.3	93.4	33.6	41.4	96.7	21.9	31.6	
0320	Concrete Reinforcing	95.0	84.9	89.9	96.4	77.7	87.0	96.9	77.7	87.2	97.3	61.6	79.3	96.3	77.7	86.9	96.9	27.3	61.8	
0330	Cast-in-Place Concrete	85.6	90.1	87.3	97.9	43.1	76.3	94.6	43.1	74.3	91.5	43.0	72.4	94.6	45.4	75.2	94.6	31.2	69.6	
03	CONCRETE	89.2	86.2	87.7	96.9	48.5	73.3	96.3	46.1	71.8	91.5	45.4	69.1	96.7	46.7	72.3	99.5	27.0	64.1	
04	MASONRY	90.3	82.2	85.4	97.2	55.8	72.0	125.5	55.8	83.0	95.8	63.6	76.2	106.1	55.8	75.4	100.3	22.2	52.6	
05	METALS	94.8	80.4	90.0	91.0	69.6	84.0	91.1	69.6	84.0	91.1	63.2	81.8	92.5	69.6	85.0	91.6	40.7	74.8	
06	WOOD, PLASTICS & COMPOSITES	91.3	84.9	87.6	97.2	38.2	63.1	96.2	31.0	58.6	88.4	38.2	59.4	99.9	30.2	59.6	103.5	20.0	55.3	
07	THERMAL & MOISTURE PROTECTION	100.5	89.1	95.8	97.5	59.2	81.9	97.6	58.5	81.7	97.2	57.0	80.9	97.6	58.4	81.7	97.9	28.2	69.6	
08	OPENINGS	97.3	82.1	93.5	94.3	48.7	82.8	94.4	44.8	81.9	94.4	44.6	81.8	94.3	44.1	81.7	94.5	21.4	76.1	
0920	Plaster & Gypsum Board	91.1	84.5	86.4	79.5	36.9	49.1	79.0	29.5	43.7	74.9	36.9	47.8	80.4	28.6	43.4	81.2	18.2	36.2	
0950, 0980	Ceilings & Acoustic Treatment	95.5	84.5	88.2	83.8	36.9	52.9	83.8	29.5	48.0	83.8	36.9	52.9	83.8	28.6	47.4	86.5	18.2	41.4	
0960	Flooring	92.0	87.4	90.7	107.2	45.2	89.0	105.8	43.1	87.4	102.0	64.8	91.1	108.0	43.1	88.9	110.0	25.5	85.2	
0970, 0990	Wall Finishes & Painting/Coating	99.0	78.4	86.3	96.2	54.1	70.2	96.2	54.1	70.2	96.2	54.1	70.2	96.2	54.1	70.2	96.2	19.3	48.7	
09	FINISHES	95.6	84.2	89.2	92.7	40.3	63.5	92.5	35.7	60.8	90.3	44.4	64.7	93.4	35.2	60.9	95.1	21.9	54.3	
COVERS	DIVS. 10 - 14, 25, 28, 41, 43, 44, 46	100.0	88.0	97.6	100.0	72.6	94.5	100.0	71.8	94.4	100.0	72.7	94.6	100.0	71.7	94.4	100.0	69.8	94.0	
21, 22, 23	FIRE SUPPRESSION, PLUMBING & HVAC	94.0	88.9	92.0	94.3	61.3	80.9	94.3	61.3	80.9	94.3	61.4	81.0	100.3	61.3	84.5	94.3	23.1	65.5	
26, 27, 3370	ELECTRICAL, COMMUNICATIONS & UTIL.	93.1	81.1	86.9	94.4	68.6	81.1	95.5	68.6	81.6	97.1	51.7	73.7	95.5	68.6	81.6	97.1	16.4	55.5	
MF2010	WEIGHTED AVERAGE	94.5	86.1	90.8	95.0	60.1	79.6	96.5	58.8	79.8	94.1	57.7	78.1	97.3	58.7	80.3	96.2	31.0	67.5	

	DIVISION	OKLAHOMA																	
		LAWTON			MCALESTER			MIAMI			MUSKOGEE			OKLAHOMA CITY			PONCA CITY		
		735			745			743			744			730 - 731			746		
		MAT.	INST.	TOTAL	MAT.	INST.	TOTAL	MAT.	INST.	TOTAL	MAT.	INST.	TOTAL	MAT.	INST.	TOTAL	MAT.	INST.	TOTAL
015433	CONTRACTOR EQUIPMENT		81.8	81.8		81.0	81.0		88.9	88.9		88.9	88.9		82.1	82.1		81.0	81.0
0241, 31 - 34	SITE & INFRASTRUCTURE, DEMOLITION	97.6	90.4	92.6	86.9	89.1	88.4	88.4	86.3	87.0	88.6	85.0	86.1	96.1	90.9	92.5	94.4	89.0	90.6
0310	Concrete Forming & Accessories	96.5	45.2	51.9	82.0	40.9	46.2	93.6	69.2	72.4	97.6	30.1	38.9	95.7	39.3	46.6	89.6	39.9	46.4
0320	Concrete Reinforcing	96.6	77.7	87.1	96.9	40.9	68.7	95.4	77.8	86.5	96.3	31.1	63.5	96.6	77.7	87.1	96.3	77.7	86.9
0330	Cast-in-Place Concrete	91.5	45.5	73.3	80.2	45.5	66.5	84.0	44.9	68.6	85.0	35.9	65.7	92.6	48.1	75.0	93.9	38.4	72.0
03	CONCRETE	93.0	51.8	72.9	82.3	43.0	63.2	86.7	63.0	75.1	88.4	33.7	61.7	93.5	50.1	72.3	93.5	47.0	70.9
04	MASONRY	100.2	55.8	73.1	114.9	56.4	79.2	98.6	57.0	73.2	117.7	45.3	73.5	99.8	57.5	74.0	91.1	56.4	69.9
05	METALS	96.1	69.7	87.4	91.0	53.9	78.8	91.0	81.7	87.9	92.5	56.2	80.5	98.0	69.6	88.6	91.0	69.3	83.8
06	WOOD, PLASTICS & COMPOSITES	102.5	45.8	69.8	86.0	41.4	60.3	99.8	77.5	86.9	103.8	29.9	61.1	101.8	36.9	64.3	95.6	38.8	62.8
07	THERMAL & MOISTURE PROTECTION	97.4	60.0	82.2	96.9	55.9	80.2	97.2	60.5	82.3	97.3	41.5	74.6	94.2	59.8	80.2	97.4	58.4	81.5
08	OPENINGS	96.0	52.6	85.1	94.3	40.2	80.7	94.3	70.5	88.3	94.3	29.0	77.9	96.0	47.8	83.9	94.3	49.0	82.9
0920	Plaster & Gypsum Board	82.8	44.8	55.6	73.6	40.2	49.8	80.4	77.3	78.2	82.8	28.2	43.8	84.9	35.6	49.7	79.0	37.5	49.4
0950, 0980	Ceilings & Acoustic Treatment	92.0	44.8	60.8	83.8	40.2	55.0	83.8	77.3	79.5	92.0	28.2	49.8	92.0	35.6	54.7	83.8	37.5	53.2
0960	Flooring	110.3	43.1	90.6	100.8	43.1	83.9	108.8	64.8	95.9	111.4	41.7	90.9	110.3	43.1	90.6	105.4	43.1	87.1
0970, 0990	Wall Finishes & Painting/Coating	96.2	54.1	70.2	96.2	39.8	61.4	96.2	81.7	87.2	96.2	28.9	54.7	96.2	54.1	70.2	96.2	54.1	70.2
09	FINISHES	95.5	44.4	67.0	89.1	40.4	61.9	92.1	70.8	80.2	95.1	32.0	59.9	95.5	39.6	64.3	91.9	40.4	63.2
COVERS	DIVS. 10 - 14, 25, 28, 41, 43, 44, 46	100.0	73.5	94.7	100.0	73.1	94.7	100.0	77.8	95.6	100.0	71.3	94.3	100.0	73.0	94.6	100.0	72.8	94.6
21, 22, 23	FIRE SUPPRESSION, PLUMBING & HVAC	100.3	61.3	84.5	94.3	30.2	68.3	94.3	60.2	80.5	100.3	23.4	69.1	100.2	62.1	84.7	94.3	59.9	80.4
26, 27, 3370	ELECTRICAL, COMMUNICATIONS & UTIL.	97.1	68.6	82.4	95.5	66.4	80.5	97.0	66.5	81.2	95.1	28.0	60.5	98.7	68.6	83.2	95.1	62.6	78.3
MF2010	WEIGHTED AVERAGE	97.5	61.4	81.6	93.6	50.6	74.6	93.8	67.9	82.3	96.7	39.3	71.4	97.8	60.6	81.4	94.1	58.7	78.5

	DIVISION	OKLAHOMA												OREGON					
		POTEAU			SHAWNEE			TULSA			WOODWARD			BEND			EUGENE		
		749			748			740 - 741			738			977			974		
		MAT.	INST.	TOTAL	MAT.	INST.	TOTAL	MAT.	INST.	TOTAL	MAT.	INST.	TOTAL	MAT.	INST.	TOTAL	MAT.	INST.	TOTAL
015433	CONTRACTOR EQUIPMENT		88.3	88.3		81.0	81.0		88.9	88.9		81.0	81.0		98.7	98.7		98.7	98.7
0241, 31 - 34	SITE & INFRASTRUCTURE, DEMOLITION	73.2	85.1	81.4	97.5	88.9	91.6	94.7	86.4	89.0	100.9	89.0	92.7	113.0	103.8	106.6	101.4	103.8	103.0
0310	Concrete Forming & Accessories	87.7	42.1	48.1	83.7	38.6	44.5	97.5	39.1	46.7	90.3	34.2	41.5	111.3	94.4	96.6	108.0	94.3	96.1
0320	Concrete Reinforcing	97.4	77.7	87.5	96.3	45.0	70.4	96.6	77.6	87.0	96.3	77.7	86.9	100.8	104.5	102.7	105.2	104.5	104.8
0330	Cast-in-Place Concrete	84.1	44.4	68.4	97.0	42.6	75.5	92.7	44.4	73.7	94.6	43.1	74.3	99.4	104.7	101.5	96.2	104.7	99.6
03	CONCRETE	88.8	50.8	70.3	95.4	41.8	69.2	93.6	49.5	72.1	96.5	46.1	72.0	109.1	99.7	104.5	100.7	99.6	100.2
04	MASONRY	99.5	56.4	73.2	116.2	55.2	79.0	99.4	57.6	73.9	93.9	55.8	70.7	106.6	108.5	107.8	103.9	108.5	106.7
05	METALS	91.0	81.3	87.8	90.9	55.7	79.3	95.7	80.4	90.6	91.2	69.6	84.0	86.8	98.1	90.5	87.5	98.0	90.9
06	WOOD, PLASTICS & COMPOSITES	92.7	41.5	63.1	88.3	38.0	59.2	103.1	38.0	65.5	96.4	30.9	58.6	102.0	91.3	95.8	98.0	91.3	94.1
07	THERMAL & MOISTURE PROTECTION	97.3	56.3	80.6	97.4	59.0	81.8	97.3	54.8	80.0	97.6	58.4	81.7	105.3	95.4	101.2	104.5	92.4	99.6
08	OPENINGS	94.3	51.0	83.4	94.3	37.4	80.0	96.0	48.0	83.9	94.4	44.7	81.9	97.8	99.5	98.2	98.1	99.5	98.5
0920	Plaster & Gypsum Board	77.7	40.2	50.9	74.9	36.7	47.6	82.8	36.6	49.8	79.8	29.4	43.8	102.2	90.9	94.1	99.9	90.9	93.5
0950, 0980	Ceilings & Acoustic Treatment	83.8	40.2	55.0	83.8	36.7	52.7	92.0	36.6	55.4	86.5	29.4	48.8	96.6	90.9	92.8	97.6	90.9	93.2
0960	Flooring	104.7	64.8	93.0	102.0	34.3	82.1	110.1	44.8	91.0	105.9	45.2	88.1	109.2	108.2	108.9	107.4	108.2	107.6
0970, 0990	Wall Finishes & Painting/Coating	96.2	54.1	70.2	96.2	36.0	59.0	96.2	42.7	63.2	96.2	54.1	70.2	108.7	69.0	84.2	108.7	69.0	84.2
09	FINISHES	89.7	46.4	65.5	90.6	36.1	60.2	94.9	39.3	63.9	93.3	36.0	61.4	105.7	93.4	98.8	103.8	93.4	98.0
COVERS	DIVS. 10 - 14, 25, 28, 41, 43, 44, 46	100.0	73.4	94.7	100.0	72.4	94.5	100.0	73.4	94.7	100.0	71.8	94.4	100.0	102.9	100.6	100.0	102.9	100.6
21, 22, 23	FIRE SUPPRESSION, PLUMBING & HVAC	94.3	59.9	80.4	94.3	60.6	80.6	100.3	47.3	78.8	94.3	61.3	80.9	94.1	97.6	95.5	100.1	97.6	99.1
26, 27, 3370	ELECTRICAL, COMMUNICATIONS & UTIL.	95.2	66.5	80.4	97.2	68.6	82.4	97.1	36.5	65.9	97.0	68.6	82.3	100.5	93.9	97.1	99.1	93.9	96.4
MF2010	WEIGHTED AVERAGE	93.2	61.4	79.2	95.7	56.5	78.4	97.4	53.5	78.0	95.1	58.8	79.1	98.8	98.6	98.7	98.6	98.5	98.6

DIVISION		OREGON																	
		KLAMATH FALLS			MEDFORD			PENDLETON			PORTLAND			SALEM			VALE		
		976			975			978			970 - 972			973			979		
		MAT.	INST.	TOTAL	MAT.	INST.	TOTAL	MAT.	INST.	TOTAL	MAT.	INST.	TOTAL	MAT.	INST.	TOTAL	MAT.	INST.	TOTAL
015433	CONTRACTOR EQUIPMENT		98.7	98.7		98.7	98.7		96.2	96.2		98.7	98.7		98.7	98.7		96.2	96.2
0241, 31 - 34	SITE & INFRASTRUCTURE, DEMOLITION	117.6	103.8	108.1	110.0	103.8	105.7	107.8	97.3	100.6	104.0	103.8	103.9	96.6	103.8	101.5	94.7	97.3	96.5
0310	Concrete Forming & Accessories	104.9	94.3	95.6	104.1	94.3	95.5	104.4	94.5	95.8	109.1	94.6	96.5	106.9	94.4	96.1	110.3	93.7	95.8
0320	Concrete Reinforcing	100.8	104.5	102.7	102.6	104.5	103.5	100.0	104.5	102.3	106.0	104.5	105.2	106.2	104.5	105.3	97.5	104.4	101.0
0330	Cast-in-Place Concrete	99.4	104.7	101.5	99.3	104.7	101.4	100.1	106.1	102.5	98.9	104.8	101.2	88.2	104.7	94.7	78.9	105.7	89.4
03	CONCRETE	111.9	99.6	105.9	106.7	99.6	103.2	93.1	100.2	96.6	102.2	99.8	101.0	96.8	99.7	98.2	78.2	99.7	88.6
04	MASONRY	120.1	108.5	113.0	100.9	108.5	105.5	110.4	108.6	109.3	104.5	108.5	107.0	108.7	108.5	108.6	108.7	108.6	108.6
05	METALS	86.8	97.9	90.5	87.1	97.9	90.6	92.6	98.5	94.5	88.3	98.2	91.6	87.8	98.1	91.2	92.5	97.6	94.2
06	WOOD, PLASTICS & COMPOSITES	93.4	91.3	92.2	92.3	91.3	91.7	94.8	91.4	92.9	98.9	91.3	94.5	95.2	91.3	93.0	102.9	91.4	96.3
07	THERMAL & MOISTURE PROTECTION	105.5	93.9	100.8	105.1	93.9	100.6	97.3	94.2	96.0	104.5	98.0	101.8	102.5	95.4	99.6	96.7	91.9	94.7
08	OPENINGS	97.8	99.5	98.2	100.7	99.5	100.4	93.2	99.6	94.8	95.7	99.5	96.7	97.6	99.5	98.1	93.1	92.3	92.9
0920	Plaster & Gypsum Board	97.3	90.9	92.7	96.4	90.9	92.5	80.4	90.9	87.9	100.2	90.9	93.5	94.4	90.9	91.9	87.2	90.9	89.8
0950, 0980	Ceilings & Acoustic Treatment	104.8	90.9	95.6	109.8	90.9	97.3	67.6	90.9	83.0	99.7	90.9	93.9	99.9	90.9	93.9	67.6	90.9	83.0
0960	Flooring	105.9	108.2	106.6	105.2	108.2	106.1	74.1	108.2	84.1	104.9	108.2	105.9	107.6	108.2	107.8	76.1	108.2	85.5
0970, 0990	Wall Finishes & Painting/Coating	108.7	61.3	79.4	108.7	61.3	79.4	98.3	69.0	80.2	108.3	69.0	84.0	108.7	69.0	84.2	98.3	69.0	80.2
09	FINISHES	106.5	92.6	98.7	106.3	92.6	98.7	72.6	93.5	84.3	103.6	93.4	97.9	103.2	93.4	97.7	72.8	93.5	84.3
COVERS	DIVS. 10 - 14, 25, 28, 41, 43, 44, 46	100.0	102.9	100.6	100.0	102.9	100.6	100.0	95.8	99.2	100.0	103.0	100.6	100.0	102.9	100.6	100.0	103.2	100.6
21, 22, 23	FIRE SUPPRESSION, PLUMBING & HVAC	94.1	97.6	95.5	100.1	97.6	99.1	95.8	106.9	100.3	100.1	100.9	100.4	100.1	97.6	99.1	95.8	74.9	87.3
26, 27, 3370	ELECTRICAL, COMMUNICATIONS & UTIL.	99.2	83.4	91.1	102.6	83.4	92.7	91.4	94.3	92.9	99.4	100.2	99.8	103.5	93.9	98.6	91.4	76.7	83.8
MF2010	WEIGHTED AVERAGE	99.8	97.0	98.6	100.1	97.0	98.7	94.1	100.0	96.7	98.8	100.3	99.5	98.6	98.6	98.6	91.9	90.5	91.3

DIVISION		PENNSYLVANIA																	
		ALLENTOWN			ALTOONA			BEDFORD			BRADFORD			BUTLER			CHAMBERSBURG		
		181			166			155			167			160			172		
		MAT.	INST.	TOTAL	MAT.	INST.	TOTAL	MAT.	INST.	TOTAL	MAT.	INST.	TOTAL	MAT.	INST.	TOTAL	MAT.	INST.	TOTAL
015433	CONTRACTOR EQUIPMENT		113.6	113.6		113.6	113.6		110.9	110.9		113.6	113.6		113.6	113.6		112.9	112.9
0241, 31 - 34	SITE & INFRASTRUCTURE, DEMOLITION	91.4	105.8	101.4	95.6	105.6	102.5	109.6	102.5	104.7	90.6	105.0	100.5	86.7	107.1	100.8	84.9	103.1	97.4
0310	Concrete Forming & Accessories	100.4	113.5	111.8	86.4	79.5	80.4	85.6	82.0	82.4	88.4	80.2	81.2	87.8	93.7	92.9	86.1	81.1	81.8
0320	Concrete Reinforcing	98.3	111.7	105.0	95.3	90.1	92.7	96.3	81.9	89.1	97.3	90.3	93.8	96.0	110.0	103.0	95.7	102.2	99.0
0330	Cast-in-Place Concrete	82.8	104.8	91.5	92.2	84.8	89.3	107.1	72.8	93.6	88.1	89.2	88.6	81.5	96.8	87.5	89.5	72.5	82.8
03	CONCRETE	94.4	110.9	102.4	90.1	84.8	87.5	101.7	80.1	91.2	96.3	86.7	91.6	82.1	99.0	90.4	102.0	83.6	93.0
04	MASONRY	95.7	104.8	101.3	97.8	68.1	79.7	107.4	85.2	93.9	95.0	83.6	88.1	100.0	96.2	97.7	96.0	90.3	92.5
05	METALS	100.6	123.5	108.2	94.4	110.6	99.7	96.9	105.0	99.6	98.5	110.2	102.4	94.1	121.9	103.3	98.5	114.9	103.9
06	WOOD, PLASTICS & COMPOSITES	101.9	115.4	109.7	81.3	81.7	81.5	82.7	81.7	82.1	87.6	78.4	82.3	82.6	91.5	87.7	91.0	81.7	85.6
07	THERMAL & MOISTURE PROTECTION	99.5	117.2	106.7	98.9	90.5	95.4	104.8	88.5	98.2	99.5	91.2	96.1	98.5	100.1	99.2	98.7	75.5	89.3
08	OPENINGS	92.3	114.6	97.9	86.8	85.4	86.4	94.6	83.2	91.8	92.7	87.3	91.3	86.8	103.0	90.9	89.1	83.4	87.7
0920	Plaster & Gypsum Board	103.9	115.6	112.3	96.6	80.9	85.4	95.4	80.9	85.0	97.2	77.5	83.1	96.6	91.0	92.6	100.8	80.9	86.6
0950, 0980	Ceilings & Acoustic Treatment	88.9	115.6	106.6	93.5	80.9	85.2	89.7	80.9	83.9	91.8	77.5	82.4	94.5	91.0	92.2	91.8	80.9	84.6
0960	Flooring	93.1	98.0	94.6	86.9	92.2	88.4	88.4	92.3	89.5	87.0	100.1	90.8	87.7	84.8	86.9	90.8	48.3	78.3
0970, 0990	Wall Finishes & Painting/Coating	96.6	74.4	82.9	92.6	106.9	101.4	97.3	86.0	90.3	96.6	92.3	93.9	92.6	106.9	101.4	96.6	86.0	90.1
09	FINISHES	95.8	106.8	101.9	94.6	84.0	88.7	94.2	82.9	87.9	93.9	83.3	88.0	94.4	92.2	93.2	94.4	75.3	83.7
COVERS	DIVS. 10 - 14, 25, 28, 41, 43, 44, 46	100.0	106.2	101.2	100.0	95.9	99.2	100.0	99.0	99.8	100.0	98.9	99.8	100.0	102.9	100.6	100.0	96.2	99.2
21, 22, 23	FIRE SUPPRESSION, PLUMBING & HVAC	100.2	112.2	105.0	99.7	83.7	93.2	94.1	88.4	91.7	94.3	90.3	92.7	93.8	95.0	94.3	94.3	89.9	92.5
26, 27, 3370	ELECTRICAL, COMMUNICATIONS & UTIL.	99.9	97.7	98.7	89.8	107.8	99.1	92.6	107.8	100.5	93.5	107.8	100.9	90.4	102.0	96.4	92.1	85.8	88.8
MF2010	WEIGHTED AVERAGE	97.9	109.2	102.9	94.8	90.6	93.0	97.2	91.7	94.8	95.5	93.8	94.7	92.3	100.5	95.9	95.5	89.5	92.9

DIVISION		PENNSYLVANIA																	
		DOYLESTOWN			DUBOIS			ERIE			GREENSBURG			HARRISBURG			HAZLETON		
		189			158			164 - 165			156			170 - 171			182		
		MAT.	INST.	TOTAL	MAT.	INST.	TOTAL	MAT.	INST.	TOTAL	MAT.	INST.	TOTAL	MAT.	INST.	TOTAL	MAT.	INST.	TOTAL
015433	CONTRACTOR EQUIPMENT		94.9	94.9		110.9	110.9		113.6	113.6		110.9	110.9		112.9	112.9		113.6	113.6
0241, 31 - 34	SITE & INFRASTRUCTURE, DEMOLITION	104.8	91.7	95.8	115.4	102.6	106.6	92.2	106.2	101.8	105.1	104.7	104.8	81.2	104.7	97.4	84.5	106.0	99.3
0310	Concrete Forming & Accessories	85.3	128.2	122.6	85.0	83.0	83.2	100.0	87.4	89.0	91.4	93.7	93.4	93.7	90.7	91.1	83.2	88.3	87.6
0320	Concrete Reinforcing	95.1	138.9	117.2	95.7	90.5	93.1	97.3	92.0	94.6	95.7	110.0	102.9	98.3	102.6	100.5	95.4	106.5	101.0
0330	Cast-in-Place Concrete	78.3	89.1	82.5	103.2	91.4	98.6	90.7	84.2	88.1	99.4	96.4	98.2	87.2	96.3	90.8	78.2	91.6	83.5
03	CONCRETE	90.2	116.4	103.0	102.6	88.6	95.8	89.0	88.4	88.7	96.6	98.8	97.7	94.5	96.1	95.3	87.3	94.2	90.6
04	MASONRY	99.0	128.9	117.3	107.4	86.0	94.4	87.8	90.1	89.2	118.0	96.2	104.7	95.5	93.5	94.3	108.2	96.8	101.2
05	METALS	98.1	127.7	107.9	96.9	109.8	101.1	94.6	110.9	100.0	96.8	120.7	104.7	102.7	118.6	107.9	100.3	120.3	106.9
06	WOOD, PLASTICS & COMPOSITES	83.3	130.2	110.4	81.6	81.7	81.6	97.9	85.9	91.0	89.0	91.4	90.4	95.9	90.0	92.5	82.0	86.8	84.8
07	THERMAL & MOISTURE PROTECTION	97.8	132.0	111.7	105.1	94.8	100.9	99.2	93.5	96.9	104.7	99.8	102.7	105.2	108.9	106.7	99.1	105.7	101.8
08	OPENINGS	94.1	139.9	105.6	94.6	89.0	93.2	87.0	88.0	87.2	94.6	102.9	96.7	92.3	94.4	92.8	93.0	92.9	93.0
0920	Plaster & Gypsum Board	94.8	130.9	120.6	94.0	80.9	84.6	103.9	85.2	90.5	96.5	91.0	92.5	100.9	89.4	92.7	95.3	86.1	88.7
0950, 0980	Ceilings & Acoustic Treatment	88.1	130.9	116.4	89.7	80.9	83.9	88.9	85.2	86.5	88.8	91.0	90.2	93.5	89.4	90.8	90.0	86.1	87.4
0960	Flooring	76.9	133.3	93.4	88.1	100.1	91.7	92.3	96.1	93.4	91.8	71.2	85.7	93.7	93.7	93.7	84.3	96.7	87.9
0970, 0990	Wall Finishes & Painting/Coating	96.1	72.7	81.7	97.3	106.9	103.2	100.7	90.2	94.2	97.3	106.9	103.2	96.6	89.0	91.9	96.6	100.5	99.0
09	FINISHES	86.8	122.4	106.6	94.6	87.9	90.8	96.4	88.9	92.2	94.7	90.5	92.4	95.1	90.8	92.7	92.0	89.1	90.4
COVERS	DIVS. 10 - 14, 25, 28, 41, 43, 44, 46	100.0	74.2	94.9	100.0	99.0	99.8	100.0	101.1	100.2	100.0	102.8	100.6	100.0	98.2	99.6	100.0	96.5	99.3
21, 22, 23	FIRE SUPPRESSION, PLUMBING & HVAC	93.8	126.9	107.2	94.1	88.6	91.8	99.7	93.0	97.0	94.1	92.8	93.6	100.0	92.2	96.9	94.3	98.3	95.9
26, 27, 3370	ELECTRICAL, COMMUNICATIONS & UTIL.	92.7	117.9	105.7	93.2	107.8	100.8	91.7	87.4	89.5	93.2	107.8	100.8	96.7	85.8	91.1	94.3	85.8	89.9
MF2010	WEIGHTED AVERAGE	94.7	120.1	105.9	97.6	94.5	96.2	94.6	93.6	94.1	97.2	100.3	98.5	97.7	96.1	97.0	95.1	97.2	96.0

DIVISION		PENNSYLVANIA																	
		INDIANA			JOHNSTOWN			KITTANNING			LANCASTER			LEHIGH VALLEY			MONTROSE		
		157			159			162			175 - 176			180			188		
		MAT.	INST.	TOTAL	MAT.	INST.	TOTAL	MAT.	INST.	TOTAL	MAT.	INST.	TOTAL	MAT.	INST.	TOTAL	MAT.	INST.	TOTAL
015433	CONTRACTOR EQUIPMENT		110.9	110.9		110.9	110.9		113.6	113.6		112.9	112.9		113.6	113.6		113.6	113.6
0241, 31 - 34	SITE & INFRASTRUCTURE, DEMOLITION	103.1	103.6	103.4	110.2	103.7	105.8	89.4	107.1	101.6	77.0	104.6	96.1	88.6	105.6	100.3	87.5	103.9	98.8
0310	Concrete Forming & Accessories	86.1	86.2	86.2	85.0	83.3	83.5	87.8	93.8	93.0	87.9	89.8	89.6	94.3	112.4	110.1	84.0	87.6	87.2
0320	Concrete Reinforcing	94.9	110.0	102.6	96.3	109.8	103.1	96.0	110.1	103.1	95.4	102.5	99.0	95.4	111.7	103.6	100.0	106.3	103.1
0330	Cast-in-Place Concrete	97.4	96.3	97.0	108.0	89.9	100.9	84.7	96.8	89.5	76.2	95.5	83.8	84.7	103.5	92.1	83.0	90.6	86.0
03	CONCRETE	94.3	95.5	94.9	102.5	91.9	97.3	84.6	99.1	91.7	89.6	95.5	92.5	93.5	109.9	101.5	92.2	93.3	92.7
04	MASONRY	103.3	97.9	100.0	104.1	85.6	92.8	103.7	97.9	100.2	101.3	92.4	95.9	95.7	102.7	100.0	95.7	97.8	97.0
05	METALS	96.9	120.5	104.7	96.9	118.7	104.1	94.2	122.2	103.4	98.5	118.1	105.0	100.3	122.7	107.7	98.6	115.0	104.0
06	WOOD, PLASTICS & COMPOSITES	83.4	81.7	82.4	81.6	81.7	81.6	82.6	91.5	87.7	93.5	90.0	91.5	94.2	115.4	106.5	82.7	86.9	85.1
07	THERMAL & MOISTURE PROTECTION	104.6	97.8	101.8	104.8	93.3	100.2	98.6	100.6	99.4	98.3	97.3	97.9	99.4	105.2	101.7	99.1	94.3	97.1
08	OPENINGS	94.6	94.9	94.7	94.6	91.2	93.8	86.8	100.2	90.2	89.1	98.1	91.4	93.0	114.6	98.5	89.3	92.9	90.2
0920	Plaster & Gypsum Board	95.8	80.9	85.2	93.7	80.9	84.6	96.6	91.0	92.6	103.1	89.4	93.3	98.5	115.6	110.7	95.8	86.2	88.9
0950, 0980	Ceilings & Acoustic Treatment	89.7	80.9	83.9	88.8	80.9	83.6	94.5	91.0	92.2	91.8	89.4	90.2	90.0	115.6	106.9	91.8	86.2	88.1
0960	Flooring	88.9	100.1	92.2	88.1	92.3	89.4	87.7	102.4	92.0	91.8	93.7	92.3	89.8	98.0	92.2	84.9	60.6	77.8
0970, 0990	Wall Finishes & Painting/Coating	97.3	106.9	103.2	97.3	106.9	103.2	92.6	106.9	101.4	96.6	60.2	74.2	96.6	70.0	80.2	96.6	100.5	99.0
09	FINISHES	93.8	89.1	91.2	93.8	86.4	89.7	94.6	95.3	95.0	94.3	87.3	90.4	94.3	104.8	100.1	92.9	83.7	87.8
COVERS	DIVS. 10 - 14, 25, 28, 41, 43, 44, 46	100.0	101.7	100.3	100.0	99.2	99.8	100.0	102.9	100.6	100.0	97.6	99.5	100.0	102.1	100.4	100.0	97.1	99.4
21, 22, 23	FIRE SUPPRESSION, PLUMBING & HVAC	94.1	92.8	93.5	94.1	89.0	92.0	93.8	97.7	95.4	94.3	91.6	93.2	94.3	111.1	101.1	94.3	99.6	96.4
26, 27, 3370	ELECTRICAL, COMMUNICATIONS & UTIL.	93.2	107.8	100.8	93.2	107.8	100.8	89.8	107.9	99.1	93.5	46.8	69.4	94.3	138.5	117.1	93.5	94.4	93.9
MF2010	WEIGHTED AVERAGE	96.0	99.1	97.4	97.2	95.8	96.6	92.8	102.4	97.0	94.3	89.7	92.3	95.6	113.5	103.5	94.5	97.0	95.6

DIVISION		PENNSYLVANIA																	
		NEW CASTLE			NORRISTOWN			OIL CITY			PHILADELPHIA			PITTSBURGH			POTTSVILLE		
		161			194			163			190 - 191			150 - 152			179		
		MAT.	INST.	TOTAL	MAT.	INST.	TOTAL	MAT.	INST.	TOTAL	MAT.	INST.	TOTAL	MAT.	INST.	TOTAL	MAT.	INST.	TOTAL
015433	CONTRACTOR EQUIPMENT		113.6	113.6		98.2	98.2		113.6	113.6		97.6	97.6		112.1	112.1		112.9	112.9
0241, 31 - 34	SITE & INFRASTRUCTURE, DEMOLITION	87.2	107.0	100.9	91.9	99.2	96.9	85.6	105.2	99.1	98.6	98.3	98.4	108.7	106.8	107.4	79.9	104.8	97.1
0310	Concrete Forming & Accessories	87.8	96.2	95.1	87.0	132.2	126.3	87.8	81.4	82.3	102.8	134.7	130.6	98.5	97.4	97.5	80.2	88.4	87.3
0320	Concrete Reinforcing	94.7	93.7	94.2	95.5	146.3	121.1	96.0	93.5	94.7	98.5	146.3	122.6	96.8	110.5	103.7	94.6	102.0	98.3
0330	Cast-in-Place Concrete	82.3	95.9	87.6	81.1	128.6	99.8	79.9	93.7	85.4	99.4	128.6	110.9	103.2	96.6	100.6	81.1	78.3	80.0
03	CONCRETE	82.5	96.8	89.4	87.6	133.2	109.8	80.9	89.4	85.0	100.0	134.3	116.7	99.9	100.6	100.3	93.4	88.8	91.2
04	MASONRY	99.3	96.2	97.4	109.3	129.0	121.3	99.8	89.9	93.8	96.2	129.1	116.3	97.1	103.6	101.0	95.4	94.4	94.8
05	METALS	94.2	114.0	100.7	99.7	133.0	110.7	94.2	111.8	100.0	103.5	133.2	113.3	98.3	121.7	106.0	98.7	118.1	105.1
06	WOOD, PLASTICS & COMPOSITES	82.6	95.8	90.3	84.3	132.2	112.0	82.6	78.4	80.2	102.7	135.5	121.7	98.2	95.8	96.8	84.3	86.8	85.7
07	THERMAL & MOISTURE PROTECTION	98.6	97.7	98.2	97.5	132.5	111.7	98.5	93.8	96.6	98.4	132.9	112.4	104.9	102.2	103.8	98.3	105.9	101.4
08	OPENINGS	86.8	97.5	89.5	89.6	143.1	103.1	86.8	80.3	85.2	98.1	145.0	109.9	97.6	106.0	99.7	89.2	91.5	89.8
0920	Plaster & Gypsum Board	96.6	95.4	95.8	90.8	133.0	121.0	96.6	77.5	83.0	98.6	136.5	125.7	103.1	95.4	97.6	97.6	86.1	89.4
0950, 0980	Ceilings & Acoustic Treatment	94.5	95.4	95.1	99.7	133.0	121.7	94.5	77.5	83.3	99.7	136.5	124.0	89.7	95.4	93.5	91.8	86.1	88.0
0960	Flooring	87.7	60.2	79.6	91.9	139.3	105.8	87.7	100.1	91.4	99.8	141.2	112.0	95.3	107.4	98.9	87.8	93.7	89.6
0970, 0990	Wall Finishes & Painting/Coating	92.6	106.9	101.4	99.6	146.7	128.7	92.6	106.9	101.4	99.6	146.7	128.7	97.3	114.0	107.6	96.6	100.5	99.0
09	FINISHES	94.5	90.3	92.1	95.6	134.9	117.5	94.3	85.7	89.5	99.2	137.2	120.4	96.9	100.3	98.8	92.8	90.4	91.4
COVERS	DIVS. 10 - 14, 25, 28, 41, 43, 44, 46	100.0	103.3	100.7	100.0	115.0	103.0	100.0	100.0	100.0	100.0	119.0	103.8	100.0	103.3	100.7	100.0	94.5	98.9
21, 22, 23	FIRE SUPPRESSION, PLUMBING & HVAC	93.8	94.7	94.2	93.9	128.3	107.9	93.8	92.3	93.2	99.9	128.4	111.5	100.0	100.3	100.1	94.3	99.5	96.4
26, 27, 3370	ELECTRICAL, COMMUNICATIONS & UTIL.	90.4	91.4	90.9	94.7	142.8	119.5	92.4	102.0	97.4	99.1	142.8	121.7	95.5	107.9	101.9	91.7	90.2	90.9
MF2010	WEIGHTED AVERAGE	92.4	97.5	94.6	95.2	130.2	110.6	92.4	94.7	93.4	99.9	130.9	113.6	99.1	104.6	101.5	94.2	96.8	95.4

DIVISION		PENNSYLVANIA																	
		READING			SCRANTON			STATE COLLEGE			STROUDSBURG			SUNBURY			UNIONTOWN		
		195 - 196			184 - 185			168			183			178			154		
		MAT.	INST.	TOTAL	MAT.	INST.	TOTAL	MAT.	INST.	TOTAL	MAT.	INST.	TOTAL	MAT.	INST.	TOTAL	MAT.	INST.	TOTAL
015433	CONTRACTOR EQUIPMENT		116.9	116.9		113.6	113.6		112.9	112.9		113.6	113.6		113.6	113.6		110.9	110.9
0241, 31 - 34	SITE & INFRASTRUCTURE, DEMOLITION	96.3	110.9	106.4	91.8	106.1	101.7	82.1	104.4	97.4	86.3	104.0	98.5	90.3	105.8	101.0	103.6	104.7	104.4
0310	Concrete Forming & Accessories	102.3	86.0	88.1	100.6	89.2	90.7	85.2	79.4	80.2	89.1	88.8	88.8	92.3	87.5	88.1	79.8	93.7	91.9
0320	Concrete Reinforcing	96.8	102.1	99.5	98.3	106.5	102.4	96.5	90.8	93.6	98.6	118.6	108.7	97.3	102.6	100.0	95.7	110.0	102.9
0330	Cast-in-Place Concrete	72.7	96.7	82.1	86.5	92.6	88.9	83.4	68.3	77.4	81.5	75.1	79.0	88.7	95.3	91.3	97.4	96.4	97.0
03	CONCRETE	86.3	94.2	90.1	96.2	95.0	95.6	96.3	79.2	88.0	91.0	90.8	90.9	96.9	94.4	95.7	94.0	98.8	96.4
04	MASONRY	97.8	95.5	96.4	96.1	98.3	97.4	101.4	81.5	89.3	92.5	102.3	98.5	95.0	92.0	93.1	120.8	96.2	105.8
05	METALS	99.9	118.9	106.2	102.7	120.6	108.6	98.3	111.5	102.7	100.4	120.2	106.9	98.4	118.4	105.0	96.6	120.7	104.6
06	WOOD, PLASTICS & COMPOSITES	102.8	82.8	91.3	101.9	86.9	93.2	89.7	81.7	85.1	88.5	86.9	87.6	91.8	86.8	88.9	75.5	91.4	84.7
07	THERMAL & MOISTURE PROTECTION	97.9	108.9	102.4	99.4	98.0	98.9	98.5	91.5	95.7	99.3	87.1	94.3	99.4	104.9	101.6	104.5	99.8	102.6
08	OPENINGS	95.3	94.0	95.0	92.3	92.9	92.5	89.1	85.4	88.2	93.0	89.7	92.2	89.3	91.7	89.9	94.6	102.9	96.7
0920	Plaster & Gypsum Board	103.0	82.1	88.0	106.3	86.2	92.0	98.7	80.9	86.0	96.6	86.2	89.2	97.3	86.1	89.3	91.0	91.0	91.0
0950, 0980	Ceilings & Acoustic Treatment	88.7	82.1	84.3	97.1	86.2	89.9	89.0	80.9	83.7	88.1	86.2	86.9	89.0	86.1	87.1	88.8	91.0	90.2
0960	Flooring	97.3	93.7	96.2	93.1	110.2	98.1	90.3	102.1	93.8	87.5	53.8	77.6	88.7	91.8	89.6	85.4	100.1	89.7
0970, 0990	Wall Finishes & Painting/Coating	98.1	100.9	99.8	96.6	100.5	99.0	96.6	106.9	103.0	96.6	68.1	79.0	96.6	100.5	99.0	97.3	106.9	103.2
09	FINISHES	96.5	87.6	91.5	97.9	93.6	95.5	93.1	86.0	89.1	92.8	79.9	85.6	93.9	89.5	91.4	92.0	94.6	93.5
COVERS	DIVS. 10 - 14, 25, 28, 41, 43, 44, 46	100.0	99.6	99.9	100.0	97.1	99.4	100.0	93.1	98.6	100.0	65.9	93.2	100.0	93.6	98.7	100.0	102.8	100.6
21, 22, 23	FIRE SUPPRESSION, PLUMBING & HVAC	100.0	107.9	103.2	100.2	100.5	100.3	94.3	85.7	90.8	94.3	102.2	97.5	94.3	91.3	93.1	94.1	92.8	93.6
26, 27, 3370	ELECTRICAL, COMMUNICATIONS & UTIL.	101.3	90.2	95.6	99.9	94.5	97.1	92.6	107.8	100.5	94.3	142.6	119.2	92.0	86.7	89.3	90.2	107.8	99.3
MF2010	WEIGHTED AVERAGE	97.6	100.0	98.6	98.6	99.5	99.0	95.0	91.8	93.6	94.9	103.0	98.5	95.1	95.1	95.1	96.4	100.8	98.3

DIVISION		PENNSYLVANIA																	
		WASHINGTON			WELLSBORO			WESTCHESTER			WILKES-BARRE			WILLIAMSPORT			YORK		
		153			169			193			186 - 187			177			173 - 174		
		MAT.	INST.	TOTAL	MAT.	INST.	TOTAL	MAT.	INST.	TOTAL	MAT.	INST.	TOTAL	MAT.	INST.	TOTAL	MAT.	INST.	TOTAL
015433	CONTRACTOR EQUIPMENT		110.9	110.9		113.6	113.6		98.2	98.2		113.6	113.6		113.6	113.6		112.9	112.9
0241, 31 - 34	SITE & INFRASTRUCTURE, DEMOLITION	103.7	104.7	104.4	94.7	103.8	101.0	98.0	96.9	97.2	84.3	106.1	99.3	81.6	104.4	97.3	80.5	104.7	97.2
0310	Concrete Forming & Accessories	86.2	93.8	92.9	87.8	85.9	86.1	93.5	129.2	124.6	91.5	89.2	89.5	89.1	61.9	65.4	83.3	87.9	87.3
0320	Concrete Reinforcing	95.7	110.1	103.0	96.5	106.1	101.4	94.6	112.9	103.8	97.3	106.5	102.0	96.5	68.5	82.4	97.3	102.6	100.0
0330	Cast-in-Place Concrete	97.4	96.5	97.1	87.4	88.5	87.8	89.7	128.0	104.8	78.2	92.6	83.9	75.0	74.0	74.6	81.5	95.7	87.1
03	CONCRETE	94.4	99.0	96.6	98.7	91.8	95.3	95.0	125.4	109.8	88.1	95.0	91.4	84.9	69.3	77.3	94.6	94.7	94.7
04	MASONRY	103.1	97.9	99.9	101.1	92.0	95.6	103.4	129.0	119.0	108.8	98.3	102.4	86.9	92.3	90.2	96.1	92.4	93.9
05	METALS	96.6	121.0	104.6	98.4	115.2	104.0	99.7	117.0	105.4	98.6	120.7	105.9	98.5	101.5	99.5	100.1	118.6	106.2
06	WOOD, PLASTICS & COMPOSITES	83.5	91.4	88.1	87.1	86.8	86.9	91.6	130.1	113.8	90.9	86.7	88.5	88.5	54.2	68.7	87.8	86.8	87.2
07	THERMAL & MOISTURE PROTECTION	104.6	100.2	102.8	99.7	92.0	96.6	97.8	131.9	111.7	99.0	106.3	102.0	98.9	101.4	99.9	98.4	108.1	102.4
08	OPENINGS	94.5	102.9	96.7	92.6	92.9	92.7	89.6	126.7	99.0	89.3	97.5	91.4	89.3	56.8	81.1	89.1	92.6	90.0
0920	Plaster & Gypsum Board	95.6	91.0	92.3	96.4	86.1	89.1	92.7	130.9	120.0	97.6	86.0	89.4	97.6	52.5	65.4	99.0	86.1	89.8
0950, 0980	Ceilings & Acoustic Treatment	88.8	91.0	90.2	89.0	86.1	87.1	99.7	130.9	120.3	91.8	86.0	88.0	91.8	52.5	65.9	90.8	86.1	87.7
0960	Flooring	88.9	100.1	92.2	86.7	53.2	76.9	95.4	141.2	108.9	88.4	96.7	90.9	87.5	52.1	77.1	89.2	93.7	90.5
0970, 0990	Wall Finishes & Painting/Coating	97.3	106.9	103.2	96.6	100.5	99.0	99.6	146.7	128.7	96.6	103.2	100.7	96.6	103.2	100.7	96.6	89.0	91.9
09	FINISHES	93.6	95.0	94.4	93.6	81.2	86.7	97.4	131.9	116.6	93.8	91.1	92.3	93.4	63.5	76.7	93.1	88.7	90.6
COVERS	DIVS. 10 - 14, 25, 28, 41, 43, 44, 46	100.0	102.8	100.6	100.0	93.6	98.7	100.0	114.7	102.9	100.0	97.1	99.4	100.0	93.3	98.7	100.0	97.5	99.5
21, 22, 23	FIRE SUPPRESSION, PLUMBING & HVAC	94.1	97.9	95.6	94.3	90.6	92.8	93.9	125.8	106.8	94.3	99.2	96.3	94.3	90.7	92.8	100.2	91.7	96.7
26, 27, 3370	ELECTRICAL, COMMUNICATIONS & UTIL.	92.6	107.8	100.5	93.5	84.0	88.6	94.6	120.8	108.1	94.3	85.8	89.9	92.5	49.5	70.3	93.5	85.8	89.5
MF2010	WEIGHTED AVERAGE	95.9	102.1	98.6	96.1	92.4	94.5	96.1	122.8	107.9	94.8	98.2	96.3	93.0	79.6	87.1	96.3	95.2	95.8

DIVISION		PUERTO RICO			RHODE ISLAND						SOUTH CAROLINA								
		SAN JUAN			NEWPORT			PROVIDENCE			AIKEN			BEAUFORT			CHARLESTON		
		009			028			029			298			299			294		
		MAT.	INST.	TOTAL	MAT.	INST.	TOTAL	MAT.	INST.	TOTAL	MAT.	INST.	TOTAL	MAT.	INST.	TOTAL	MAT.	INST.	TOTAL
015433	CONTRACTOR EQUIPMENT		90.2	90.2		101.9	101.9		101.9	101.9		99.5	99.5		99.5	99.5		99.5	99.5
0241, 31 - 34	SITE & INFRASTRUCTURE, DEMOLITION	131.5	90.1	103.0	87.5	102.9	98.1	88.6	102.9	98.5	128.4	86.4	99.4	122.8	84.5	96.4	103.7	84.1	90.2
0310	Concrete Forming & Accessories	95.4	17.4	27.6	102.3	117.2	115.2	101.0	117.2	115.0	96.3	71.7	74.9	95.2	39.4	46.6	94.3	41.1	48.1
0320	Concrete Reinforcing	186.3	12.2	98.6	103.6	112.7	108.2	103.6	112.7	108.2	95.1	69.3	82.1	94.2	27.3	60.5	94.0	58.9	76.3
0330	Cast-in-Place Concrete	107.8	31.0	77.6	83.5	117.3	96.8	102.6	117.3	108.4	93.1	73.1	85.2	93.1	48.6	75.6	109.2	49.3	85.6
03	CONCRETE	115.0	22.2	69.8	100.9	115.9	108.2	110.4	115.9	113.1	110.3	72.8	92.0	107.6	42.4	75.8	104.1	49.0	77.2
04	MASONRY	84.4	16.3	42.8	101.1	122.3	114.0	104.6	122.3	115.4	81.2	65.0	71.3	97.2	34.8	59.1	98.6	38.8	62.1
05	METALS	111.9	35.1	86.5	98.0	108.9	101.6	98.0	108.9	101.6	86.1	86.6	86.2	86.1	69.5	80.6	87.9	77.0	84.3
06	WOOD, PLASTICS & COMPOSITES	93.4	17.2	49.4	99.9	117.8	110.2	99.1	117.8	109.9	99.8	72.9	84.2	97.8	40.1	64.5	96.6	40.1	64.0
07	THERMAL & MOISTURE PROTECTION	129.6	21.4	85.6	101.5	114.5	106.8	99.1	114.5	105.3	101.3	70.2	88.7	101.0	37.8	75.3	100.1	42.3	76.6
08	OPENINGS	146.8	14.6	113.5	96.6	115.7	101.4	97.2	115.7	101.9	93.2	68.7	87.1	93.3	38.8	79.6	97.3	43.8	83.8
0920	Plaster & Gypsum Board	115.3	14.4	43.3	95.1	117.7	111.2	93.0	117.7	110.6	108.4	71.9	82.3	111.3	38.1	59.0	113.4	38.1	59.6
0950, 0980	Ceilings & Acoustic Treatment	243.9	14.4	92.3	99.2	117.7	111.4	98.5	117.7	111.2	93.7	71.9	79.3	97.3	38.1	58.2	97.3	38.1	58.2
0960	Flooring	197.1	13.7	143.3	95.3	134.8	106.9	94.9	134.8	106.6	106.2	71.6	96.0	107.8	53.6	91.9	107.3	61.0	93.7
0970, 0990	Wall Finishes & Painting/Coating	193.0	12.6	81.7	92.3	120.8	109.9	92.3	120.8	109.9	114.8	74.1	89.7	114.8	36.6	66.5	114.8	41.3	69.4
09	FINISHES	207.2	16.4	100.8	98.5	121.4	111.2	97.7	121.4	110.9	105.4	72.0	86.7	106.5	42.0	70.5	104.2	44.3	70.8
COVERS	DIVS. 10 - 14, 25, 28, 41, 43, 44, 46	100.0	17.0	83.5	100.0	103.1	100.6	100.0	103.1	100.6	100.0	77.6	95.5	100.0	46.7	89.4	100.0	70.3	94.1
21, 22, 23	FIRE SUPPRESSION, PLUMBING & HVAC	102.8	13.3	66.5	100.0	108.3	103.4	99.9	108.3	103.3	94.3	67.7	83.5	94.3	25.4	66.4	100.2	39.9	75.8
26, 27, 3370	ELECTRICAL, COMMUNICATIONS & UTIL.	130.6	12.6	69.7	101.4	99.0	100.2	99.7	99.0	99.3	94.3	70.0	81.8	97.7	32.2	63.9	96.2	94.3	95.2
MF2010	WEIGHTED AVERAGE	121.3	24.1	78.4	99.2	111.1	104.5	100.2	111.1	105.0	96.8	72.8	86.2	97.5	42.5	73.3	98.4	57.4	80.3

DIVISION		SOUTH CAROLINA															SOUTH DAKOTA		
		COLUMBIA			FLORENCE			GREENVILLE			ROCK HILL			SPARTANBURG			ABERDEEN		
		290 - 292			295			296			297			293			574		
		MAT.	INST.	TOTAL	MAT.	INST.	TOTAL	MAT.	INST.	TOTAL	MAT.	INST.	TOTAL	MAT.	INST.	TOTAL	MAT.	INST.	TOTAL
015433	CONTRACTOR EQUIPMENT		99.5	99.5		99.5	99.5		99.5	99.5		99.5	99.5		99.5	99.5		98.7	98.7
0241, 31 - 34	SITE & INFRASTRUCTURE, DEMOLITION	101.5	84.0	89.5	115.4	84.1	93.8	109.7	83.7	91.8	107.1	83.6	90.9	109.5	83.7	91.7	98.8	94.9	96.1
0310	Concrete Forming & Accessories	96.0	40.2	47.5	83.0	40.3	45.9	93.9	40.2	47.2	92.5	37.9	45.0	96.8	40.2	47.6	100.2	38.9	46.9
0320	Concrete Reinforcing	94.0	58.7	76.2	93.7	58.8	76.1	93.6	46.9	70.1	94.4	46.7	70.4	93.6	46.9	70.1	108.5	40.4	74.1
0330	Cast-in-Place Concrete	112.1	49.9	87.6	93.0	49.1	75.7	93.0	49.0	75.7	93.0	44.1	73.7	93.0	49.0	75.7	105.9	45.7	82.2
03	CONCRETE	105.6	48.7	77.9	102.2	48.5	76.1	100.8	46.3	74.2	98.9	43.5	71.9	101.0	46.3	74.3	103.1	43.0	73.8
04	MASONRY	92.9	36.8	58.6	81.3	38.8	55.4	79.1	38.8	54.5	105.1	34.1	61.8	81.3	38.8	55.4	108.2	56.7	76.8
05	METALS	87.9	76.3	84.1	86.9	76.6	83.5	86.9	72.3	82.1	86.1	71.8	81.4	86.9	72.3	82.1	102.9	66.1	90.8
06	WOOD, PLASTICS & COMPOSITES	99.0	39.7	64.7	83.1	39.7	58.0	96.6	39.7	63.7	94.9	37.7	61.8	100.4	39.7	65.3	96.9	38.2	63.0
07	THERMAL & MOISTURE PROTECTION	98.8	41.0	75.3	100.4	42.2	76.8	100.3	42.2	76.7	100.2	39.9	75.7	100.4	42.2	76.7	107.2	50.3	84.0
08	OPENINGS	97.3	43.5	83.7	93.3	43.5	80.8	93.2	40.7	80.0	93.3	39.8	79.8	93.3	40.7	80.0	93.8	39.0	80.0
0920	Plaster & Gypsum Board	100.7	37.6	55.6	101.0	37.6	55.7	106.6	37.6	57.3	105.7	35.6	55.6	108.9	37.6	58.0	111.3	36.6	57.9
0950, 0980	Ceilings & Acoustic Treatment	96.4	37.6	57.6	94.6	37.6	57.0	93.7	37.6	56.6	93.7	35.6	55.3	93.7	37.6	56.6	112.9	36.6	62.5
0960	Flooring	104.0	45.7	86.8	98.5	45.7	83.0	104.9	59.9	91.7	104.0	46.5	87.1	106.4	59.9	92.7	115.7	52.7	97.2
0970, 0990	Wall Finishes & Painting/Coating	112.2	41.3	68.5	114.8	41.3	69.4	114.8	41.3	69.4	114.8	41.3	69.4	114.8	41.3	69.4	96.0	39.1	60.9
09	FINISHES	100.8	41.0	67.5	101.0	41.5	67.8	102.9	43.9	70.0	102.2	39.5	67.2	103.6	43.9	70.3	109.1	41.1	71.1
COVERS	DIVS. 10 - 14, 25, 28, 41, 43, 44, 46	100.0	70.1	94.0	100.0	70.1	94.0	100.0	70.1	94.0	100.0	69.3	93.9	100.0	70.1	94.0	100.0	43.8	88.8
21, 22, 23	FIRE SUPPRESSION, PLUMBING & HVAC	99.9	35.3	73.7	100.2	35.3	73.9	100.2	33.9	73.3	94.3	32.9	69.4	100.2	33.9	73.3	100.2	42.7	76.9
26, 27, 3370	ELECTRICAL, COMMUNICATIONS & UTIL.	95.2	48.2	71.0	94.2	45.9	69.3	96.2	49.6	72.2	96.2	49.6	72.2	96.2	49.6	72.2	100.3	54.4	76.6
MF2010	WEIGHTED AVERAGE	97.8	49.3	76.4	96.6	49.3	75.8	96.6	49.0	75.6	96.0	47.2	74.5	96.8	49.0	75.7	101.6	52.2	79.8

DIVISION		SOUTH DAKOTA																	
		MITCHELL			MOBRIDGE			PIERRE			RAPID CITY			SIOUX FALLS			WATERTOWN		
		573			576			575			577			570 - 571			572		
		MAT.	INST.	TOTAL	MAT.	INST.	TOTAL	MAT.	INST.	TOTAL	MAT.	INST.	TOTAL	MAT.	INST.	TOTAL	MAT.	INST.	TOTAL
015433	**CONTRACTOR EQUIPMENT**		98.7	98.7		98.7	98.7		98.7	98.7		98.7	98.7		99.6	99.6		98.7	98.7
0241, 31 - 34	**SITE & INFRASTRUCTURE, DEMOLITION**	95.7	94.9	95.1	95.6	94.9	95.1	94.0	94.9	94.7	96.9	95.0	95.6	94.6	96.6	96.0	95.5	94.9	95.1
0310	Concrete Forming & Accessories	99.5	39.3	47.1	91.1	39.1	45.8	98.9	40.0	47.7	106.9	38.1	47.0	99.6	42.6	50.0	88.1	38.9	45.3
0320	Concrete Reinforcing	107.8	49.2	78.2	110.6	40.4	75.2	108.0	59.1	83.3	101.4	59.2	80.1	101.4	59.3	80.2	104.9	40.5	72.4
0330	Cast-in-Place Concrete	102.9	44.4	79.9	102.9	45.7	80.4	88.8	44.0	71.2	102.1	44.6	79.5	94.5	45.0	75.0	102.9	45.7	80.4
03	**CONCRETE**	100.8	44.2	73.2	100.7	43.1	72.6	93.7	46.5	70.7	99.9	45.8	73.5	94.5	48.0	71.8	99.6	43.0	72.0
04	**MASONRY**	96.8	54.5	71.0	105.7	56.7	75.8	103.8	54.8	73.9	105.7	55.9	75.3	101.3	56.6	74.0	131.0	58.9	87.0
05	**METALS**	101.9	66.3	90.2	102.0	66.1	90.1	102.9	74.6	93.6	104.7	75.1	94.9	104.9	75.4	95.1	101.9	66.6	90.3
06	**WOOD, PLASTICS & COMPOSITES**	96.0	38.8	62.9	86.3	38.5	58.7	94.8	39.0	62.6	100.2	35.4	62.7	95.6	41.7	64.5	83.3	38.2	57.2
07	**THERMAL & MOISTURE PROTECTION**	107.0	48.0	83.0	107.1	50.3	84.0	106.9	48.3	83.1	107.5	49.8	84.1	109.4	51.2	85.7	106.8	50.9	84.1
08	**OPENINGS**	92.3	39.5	79.0	95.2	38.5	80.9	96.9	44.9	83.8	98.0	43.0	84.1	98.2	46.4	85.2	92.3	38.7	78.8
0920	Plaster & Gypsum Board	109.3	37.1	57.8	103.1	36.8	55.8	98.3	37.4	54.8	110.5	33.7	55.6	108.3	40.2	59.6	101.1	36.6	55.0
0950, 0980	Ceilings & Acoustic Treatment	109.2	37.1	61.6	112.9	36.8	62.6	108.3	37.4	61.5	114.7	33.7	61.2	112.0	40.2	64.6	109.2	36.6	61.2
0960	Flooring	115.1	52.7	96.8	109.6	52.7	92.9	114.8	37.5	92.1	114.8	75.4	103.2	114.8	78.2	104.1	108.1	52.7	91.8
0970, 0990	Wall Finishes & Painting/Coating	96.0	42.8	63.2	96.0	43.9	63.9	96.0	47.0	65.8	96.0	47.0	65.8	96.0	47.0	65.8	96.0	39.1	60.9
09	**FINISHES**	107.7	41.8	70.9	106.2	41.8	70.2	105.9	39.5	68.9	109.0	45.0	73.3	108.0	49.1	75.2	104.6	41.1	69.2
COVERS	**DIVS. 10 - 14, 25, 28, 41, 43, 44, 46**	100.0	40.6	88.2	100.0	43.8	88.8	100.0	74.2	94.9	100.0	74.0	94.8	100.0	74.5	94.9	100.0	43.7	88.8
21, 22, 23	**FIRE SUPPRESSION, PLUMBING & HVAC**	94.2	42.3	73.2	94.2	42.8	73.4	100.1	42.4	76.7	100.2	42.9	77.0	100.2	41.2	76.3	94.2	42.8	73.4
26, 27, 3370	**ELECTRICAL, COMMUNICATIONS & UTIL.**	98.7	46.7	71.9	100.3	47.6	73.2	100.1	52.1	75.3	97.0	52.1	73.8	99.2	63.1	80.6	97.9	47.6	72.0
MF2010	**WEIGHTED AVERAGE**	98.6	51.0	77.6	99.3	51.4	78.2	100.1	53.8	79.7	101.4	54.6	80.7	100.7	57.1	81.5	99.8	51.6	78.5

DIVISION		TENNESSEE																	
		CHATTANOOGA			COLUMBIA			COOKEVILLE			JACKSON			JOHNSON CITY			KNOXVILLE		
		373 - 374			384			385			383			376			377 - 379		
		MAT.	INST.	TOTAL	MAT.	INST.	TOTAL	MAT.	INST.	TOTAL	MAT.	INST.	TOTAL	MAT.	INST.	TOTAL	MAT.	INST.	TOTAL
015433	**CONTRACTOR EQUIPMENT**		103.8	103.8		98.0	98.0		98.0	98.0		104.0	104.0		97.8	97.8		97.8	97.8
0241, 31 - 34	**SITE & INFRASTRUCTURE, DEMOLITION**	102.1	98.6	99.7	89.3	87.3	87.9	95.5	86.7	89.4	98.8	97.1	97.6	110.0	87.5	94.5	87.2	88.0	87.7
0310	Concrete Forming & Accessories	96.6	57.1	62.3	83.4	61.3	64.1	83.5	37.4	43.4	89.7	47.5	53.0	84.5	40.7	46.4	95.3	43.8	50.5
0320	Concrete Reinforcing	88.5	62.6	75.5	90.9	62.6	76.7	90.9	62.6	76.6	90.9	62.9	76.8	89.1	61.0	75.0	88.5	61.1	74.7
0330	Cast-in-Place Concrete	97.3	62.1	83.4	92.7	53.3	77.2	105.0	45.2	81.5	102.6	51.5	82.4	78.3	62.7	72.1	91.4	50.1	75.1
03	**CONCRETE**	93.3	61.6	77.9	93.9	60.4	77.6	103.7	47.0	76.1	95.7	53.7	75.3	98.9	54.2	77.1	90.6	51.3	71.4
04	**MASONRY**	103.9	51.6	72.0	121.4	56.6	81.8	115.6	46.2	73.3	121.6	49.9	77.8	118.8	47.3	75.2	81.4	55.0	65.3
05	**METALS**	96.2	87.4	93.3	90.6	88.0	89.7	90.7	86.9	89.4	92.9	87.6	91.1	93.5	86.1	91.0	96.8	86.3	93.3
06	**WOOD, PLASTICS & COMPOSITES**	106.2	58.8	78.8	74.0	62.8	67.6	74.2	35.7	52.0	90.5	48.2	66.0	79.2	39.2	56.1	92.4	39.2	61.7
07	**THERMAL & MOISTURE PROTECTION**	95.6	59.6	81.0	90.4	61.1	78.5	90.8	48.2	73.5	92.6	51.5	75.8	91.1	55.5	76.6	89.2	59.3	77.0
08	**OPENINGS**	102.1	57.3	90.8	92.0	57.6	83.4	92.0	44.0	79.9	101.5	53.9	89.5	97.5	46.9	84.8	94.3	47.0	82.4
0920	Plaster & Gypsum Board	76.1	58.0	63.2	82.3	62.1	67.9	82.3	34.2	47.9	83.9	47.0	57.6	98.3	37.8	55.1	105.8	37.8	57.3
0950, 0980	Ceilings & Acoustic Treatment	99.7	58.0	72.2	82.9	62.1	69.2	82.9	34.2	50.7	91.8	47.0	62.2	95.8	37.8	57.5	96.7	37.8	57.8
0960	Flooring	98.1	59.3	86.7	92.8	21.5	71.9	92.9	61.4	83.7	91.7	42.7	77.3	94.7	42.6	79.4	99.5	58.4	87.4
0970, 0990	Wall Finishes & Painting/Coating	104.4	55.7	74.3	94.6	35.2	57.9	94.6	38.4	59.9	96.0	52.3	69.0	101.6	44.4	66.3	101.6	51.2	70.5
09	**FINISHES**	96.1	57.4	74.5	94.7	51.5	70.6	95.3	40.7	64.8	95.0	46.3	67.8	100.1	40.3	66.8	92.6	46.1	66.7
COVERS	**DIVS. 10 - 14, 25, 28, 41, 43, 44, 46**	100.0	44.4	88.9	100.0	50.7	90.2	100.0	44.2	88.9	100.0	46.5	89.3	100.0	46.6	89.4	100.0	49.3	89.9
21, 22, 23	**FIRE SUPPRESSION, PLUMBING & HVAC**	100.2	63.3	85.2	96.9	77.4	89.0	96.9	72.1	86.9	100.0	71.6	88.5	99.9	56.3	82.2	99.9	66.7	86.5
26, 27, 3370	**ELECTRICAL, COMMUNICATIONS & UTIL.**	102.4	69.3	85.3	95.1	61.0	77.5	96.8	61.9	78.8	102.5	62.1	81.7	92.3	49.2	70.1	97.9	58.1	77.3
MF2010	**WEIGHTED AVERAGE**	99.0	66.2	84.6	95.7	67.1	83.1	97.0	60.2	80.8	99.2	63.8	83.6	98.7	56.7	80.2	95.4	61.4	80.4

DIVISION		TENNESSEE									TEXAS								
		MCKENZIE			MEMPHIS			NASHVILLE			ABILENE			AMARILLO			AUSTIN		
		382			375,380 - 381			370 - 372			795 - 796			790 - 791			786 - 787		
		MAT.	INST.	TOTAL	MAT.	INST.	TOTAL	MAT.	INST.	TOTAL	MAT.	INST.	TOTAL	MAT.	INST.	TOTAL	MAT.	INST.	TOTAL
015433	**CONTRACTOR EQUIPMENT**		98.0	98.0		102.2	102.2		104.1	104.1		88.9	88.9		88.9	88.9		88.6	88.6
0241, 31 - 34	**SITE & INFRASTRUCTURE, DEMOLITION**	95.0	87.0	89.5	90.5	93.3	92.5	96.6	99.5	98.6	95.1	86.0	88.8	95.3	87.1	89.6	97.0	86.8	90.0
0310	Concrete Forming & Accessories	90.8	40.3	46.9	95.4	60.4	64.9	97.0	64.0	68.3	95.1	40.2	47.4	95.6	54.1	59.5	90.2	53.4	58.2
0320	Concrete Reinforcing	91.1	62.9	76.9	97.3	63.5	80.3	92.6	63.0	77.7	93.8	51.8	72.6	93.8	50.0	71.7	88.8	48.8	68.7
0330	Cast-in-Place Concrete	102.8	60.0	85.9	90.8	62.7	79.8	86.1	66.3	78.3	96.1	48.7	77.4	95.3	55.4	79.6	92.4	49.3	75.4
03	**CONCRETE**	102.3	53.5	78.5	91.9	63.4	78.0	88.9	66.1	77.8	94.8	46.5	71.3	94.4	54.7	75.0	87.5	51.9	70.2
04	**MASONRY**	119.9	48.8	76.5	96.6	63.7	76.6	86.2	59.2	69.7	103.1	57.5	75.3	103.1	58.3	75.8	99.7	47.6	67.9
05	**METALS**	90.7	87.9	89.8	97.8	89.5	95.0	97.8	88.7	94.8	99.3	68.9	89.2	99.3	68.6	89.2	95.5	64.8	85.4
06	**WOOD, PLASTICS & COMPOSITES**	82.6	38.5	57.1	100.8	61.2	78.0	99.9	65.0	79.8	99.6	38.7	64.4	98.1	56.5	74.1	94.3	56.6	72.5
07	**THERMAL & MOISTURE PROTECTION**	90.8	49.4	74.0	90.2	64.8	79.9	93.9	61.9	80.9	97.4	51.0	78.5	97.4	54.2	79.8	86.7	51.1	72.2
08	**OPENINGS**	92.0	45.8	80.4	101.4	61.7	91.4	99.6	64.1	90.6	92.6	42.1	79.9	92.6	53.7	82.8	98.2	54.2	87.1
0920	Plaster & Gypsum Board	85.5	37.0	50.9	91.7	60.4	69.3	92.1	64.3	72.3	82.8	37.3	50.3	84.4	55.7	63.9	81.5	55.8	63.1
0950, 0980	Ceilings & Acoustic Treatment	82.9	37.0	52.6	99.8	60.4	73.7	94.7	64.3	74.6	92.0	37.3	55.9	92.0	55.7	68.0	93.7	55.8	68.6
0960	Flooring	96.1	41.2	80.0	94.8	37.8	78.0	99.8	67.3	90.2	110.3	77.2	100.6	110.1	77.2	100.5	95.9	41.4	79.9
0970, 0990	Wall Finishes & Painting/Coating	94.6	52.3	68.5	99.3	55.3	72.2	102.2	68.9	81.7	94.7	51.8	68.2	94.7	51.1	67.8	92.2	42.1	61.3
09	**FINISHES**	96.5	40.5	65.2	96.3	55.1	73.3	98.3	64.9	79.7	94.9	47.5	68.5	95.2	58.1	74.5	90.4	49.9	67.8
COVERS	**DIVS. 10 - 14, 25, 28, 41, 43, 44, 46**	100.0	29.5	86.0	100.0	81.1	96.2	100.0	81.8	96.4	100.0	76.2	95.3	100.0	71.3	94.3	100.0	78.0	95.6
21, 22, 23	**FIRE SUPPRESSION, PLUMBING & HVAC**	96.9	71.8	86.8	100.0	70.2	87.9	100.0	77.3	90.8	100.3	45.2	78.0	100.2	51.2	80.3	100.1	55.3	82.0
26, 27, 3370	**ELECTRICAL, COMMUNICATIONS & UTIL.**	96.5	66.8	81.2	103.7	65.9	84.2	95.6	61.9	78.2	97.3	50.5	73.1	99.5	67.1	82.8	96.4	64.5	79.9
MF2010	**WEIGHTED AVERAGE**	97.1	61.8	81.6	98.2	69.6	85.6	96.8	72.2	86.0	97.9	54.2	78.6	98.1	61.0	81.7	96.1	58.8	79.6

DIVISION		TEXAS																	
		BEAUMONT			BROWNWOOD			BRYAN			CHILDRESS			CORPUS CHRISTI			DALLAS		
		776 - 777			768			778			792			783 - 784			752 - 753		
		MAT.	INST.	TOTAL	MAT.	INST.	TOTAL	MAT.	INST.	TOTAL	MAT.	INST.	TOTAL	MAT.	INST.	TOTAL	MAT.	INST.	TOTAL
015433	CONTRACTOR EQUIPMENT		89.8	89.8		88.9	88.9		89.8	89.8		88.9	88.9		95.7	95.7		98.0	98.0
0241, 31 - 34	SITE & INFRASTRUCTURE, DEMOLITION	86.6	86.3	86.4	101.1	85.5	90.3	78.2	87.9	84.9	106.1	85.9	92.1	137.0	82.6	99.5	109.8	87.3	94.2
0310	Concrete Forming & Accessories	101.5	50.1	56.8	93.6	34.1	41.8	81.2	41.6	46.8	93.7	61.5	65.7	96.4	38.1	45.7	93.9	64.3	68.1
0320	Concrete Reinforcing	91.2	42.7	66.7	94.1	49.1	71.4	93.7	48.9	71.1	93.9	49.3	71.4	88.2	47.3	67.6	101.0	52.3	76.5
0330	Cast-in-Place Concrete	86.3	53.1	73.2	103.7	43.8	80.1	69.1	66.0	67.9	98.6	50.1	79.5	102.5	46.2	80.3	96.5	56.2	80.6
03	CONCRETE	93.0	50.7	72.4	102.2	41.5	72.6	78.1	52.6	65.7	103.2	55.9	80.2	95.3	44.5	70.5	97.3	60.6	79.4
04	MASONRY	104.5	58.5	76.5	132.2	50.2	82.2	143.7	57.2	90.9	107.8	51.5	73.4	88.8	49.3	64.7	102.6	58.4	75.6
05	METALS	97.0	65.3	86.5	96.9	62.5	85.6	96.5	68.2	87.1	96.7	64.4	86.0	95.1	75.2	88.5	100.4	79.9	93.6
06	WOOD, PLASTICS & COMPOSITES	111.8	50.2	76.2	98.3	30.2	59.0	78.2	34.9	53.2	99.0	66.6	80.3	113.9	37.3	69.7	98.2	67.2	80.3
07	THERMAL & MOISTURE PROTECTION	103.6	56.8	84.6	97.5	45.7	76.4	96.1	56.9	80.1	98.0	50.3	78.6	95.4	46.1	75.4	92.6	64.0	81.0
08	OPENINGS	95.9	46.8	83.5	88.6	37.5	75.7	97.6	43.0	83.9	89.6	55.9	81.1	105.0	38.6	88.3	100.7	58.4	90.1
0920	Plaster & Gypsum Board	95.1	49.3	62.4	81.7	28.6	43.8	83.6	33.4	47.8	81.7	66.1	70.5	86.3	35.8	50.2	88.7	66.6	72.9
0950, 0980	Ceilings & Acoustic Treatment	99.7	49.3	66.4	88.4	28.6	48.9	93.0	33.4	53.7	88.4	66.1	73.6	93.7	35.8	55.4	94.6	66.6	76.1
0960	Flooring	109.1	74.6	99.0	109.2	46.8	90.9	82.1	55.2	74.2	108.2	42.8	89.0	109.0	66.0	96.4	99.3	50.0	84.8
0970, 0990	Wall Finishes & Painting/Coating	91.8	50.2	66.1	94.7	30.2	54.9	89.0	62.4	72.6	94.7	39.9	60.9	104.7	59.6	76.8	103.0	53.2	72.3
09	FINISHES	91.2	54.3	70.6	94.3	35.2	61.3	79.8	44.5	60.1	94.8	56.8	73.6	98.7	44.9	68.7	97.7	60.8	77.1
COVERS	DIVS. 10 - 14, 25, 28, 41, 43, 44, 46	100.0	79.6	95.9	100.0	39.2	87.9	100.0	78.8	95.8	100.0	73.4	94.7	100.0	77.0	95.4	100.0	82.1	96.4
21, 22, 23	FIRE SUPPRESSION, PLUMBING & HVAC	100.1	59.3	83.6	94.3	48.4	75.7	94.1	66.4	82.9	94.3	52.7	77.5	100.1	40.1	75.8	100.0	63.7	85.3
26, 27, 3370	ELECTRICAL, COMMUNICATIONS & UTIL.	94.8	72.6	83.4	93.3	45.4	68.6	92.9	67.4	79.8	97.3	45.1	70.3	92.7	54.7	73.1	95.5	67.6	81.1
MF2010	WEIGHTED AVERAGE	97.2	62.0	81.7	97.7	49.0	76.2	94.3	61.7	79.9	97.3	57.2	79.6	98.9	52.3	78.4	99.4	66.8	85.0

DIVISION		TEXAS																	
		DEL RIO			DENTON			EASTLAND			EL PASO			FORT WORTH			GALVESTON		
		788			762			764			798 - 799,885			760 - 761			775		
		MAT.	INST.	TOTAL	MAT.	INST.	TOTAL	MAT.	INST.	TOTAL	MAT.	INST.	TOTAL	MAT.	INST.	TOTAL	MAT.	INST.	TOTAL
015433	CONTRACTOR EQUIPMENT		88.6	88.6		94.7	94.7		88.9	88.9		88.9	88.9		88.9	88.9		98.5	98.5
0241, 31 - 34	SITE & INFRASTRUCTURE, DEMOLITION	117.8	85.1	95.2	101.7	78.7	85.9	104.0	85.4	91.2	92.4	85.4	87.6	97.1	86.4	89.7	107.8	85.7	92.6
0310	Concrete Forming & Accessories	91.2	37.7	44.7	99.0	33.5	42.0	94.2	37.1	44.5	94.4	45.0	51.4	91.3	63.7	67.3	89.8	61.9	65.5
0320	Concrete Reinforcing	88.8	44.3	66.4	95.5	50.3	72.7	94.3	49.0	71.5	93.8	47.2	70.3	93.8	52.2	72.8	93.1	64.0	78.4
0330	Cast-in-Place Concrete	110.4	43.7	84.1	80.2	45.7	66.6	109.6	44.5	83.9	89.8	41.8	70.9	99.0	52.1	80.5	91.7	67.1	82.0
03	CONCRETE	112.3	42.1	78.1	80.7	42.9	62.3	106.9	43.0	75.8	91.6	45.3	69.0	96.0	58.2	77.6	94.4	65.6	80.4
04	MASONRY	103.6	51.9	72.0	140.8	48.6	84.6	101.6	50.1	70.2	100.8	47.9	68.5	97.3	58.3	73.5	99.9	57.2	73.9
05	METALS	94.0	59.6	82.6	96.5	75.6	89.6	96.7	62.9	85.6	99.0	63.4	87.3	96.3	68.7	87.2	98.0	87.1	94.4
06	WOOD, PLASTICS & COMPOSITES	93.7	35.9	60.3	107.9	33.1	64.7	105.3	34.0	64.1	99.2	47.5	69.3	100.9	67.1	81.4	95.3	61.8	76.0
07	THERMAL & MOISTURE PROTECTION	92.3	45.2	73.2	96.0	45.6	75.5	97.9	44.9	76.3	93.6	53.8	77.4	98.3	54.2	80.3	95.3	65.2	83.1
08	OPENINGS	99.0	34.2	82.6	102.7	38.2	86.4	76.9	39.5	67.5	92.6	43.6	80.3	86.7	58.3	79.5	102.3	61.6	92.0
0920	Plaster & Gypsum Board	80.9	34.4	47.7	85.3	31.4	46.8	81.7	32.4	46.5	87.4	46.4	58.1	85.4	66.6	72.0	89.9	61.1	69.3
0950, 0980	Ceilings & Acoustic Treatment	89.2	34.4	53.0	92.9	31.4	52.3	88.4	32.4	51.4	92.0	46.4	61.9	93.8	66.6	75.8	95.7	61.1	72.8
0960	Flooring	92.8	35.7	76.0	101.7	44.8	85.0	143.0	46.8	114.7	110.3	65.3	97.1	144.3	42.6	114.5	95.7	55.2	83.8
0970, 0990	Wall Finishes & Painting/Coating	92.2	26.5	51.7	104.7	30.5	59.0	96.2	31.5	56.3	94.7	36.0	58.5	96.2	52.7	69.3	99.2	48.5	67.9
09	FINISHES	90.6	35.8	60.0	90.7	34.6	59.4	104.7	37.5	67.2	95.1	47.9	68.8	106.0	59.2	79.9	89.3	58.8	72.3
COVERS	DIVS. 10 - 14, 25, 28, 41, 43, 44, 46	100.0	39.0	87.9	100.0	37.8	87.6	100.0	40.3	88.1	100.0	77.9	95.6	100.0	81.8	96.4	100.0	83.6	96.7
21, 22, 23	FIRE SUPPRESSION, PLUMBING & HVAC	94.1	46.3	74.7	94.3	39.8	72.2	94.3	48.7	75.8	100.2	35.4	73.9	100.2	59.1	83.5	94.1	65.9	82.7
26, 27, 3370	ELECTRICAL, COMMUNICATIONS & UTIL.	94.1	31.2	61.7	95.8	45.5	69.9	93.2	45.4	68.5	93.9	53.3	72.9	93.4	62.4	77.4	94.5	67.3	80.4
MF2010	WEIGHTED AVERAGE	98.0	46.5	75.3	97.1	47.8	75.4	96.5	49.7	75.9	96.8	51.2	76.7	97.2	63.2	82.2	96.5	68.2	84.0

DIVISION		TEXAS																	
		GIDDINGS			GREENVILLE			HOUSTON			HUNTSVILLE			LAREDO			LONGVIEW		
		789			754			770 - 772			773			780			756		
		MAT.	INST.	TOTAL	MAT.	INST.	TOTAL	MAT.	INST.	TOTAL	MAT.	INST.	TOTAL	MAT.	INST.	TOTAL	MAT.	INST.	TOTAL
015433	CONTRACTOR EQUIPMENT		88.6	88.6		95.6	95.6		98.4	98.4		89.8	89.8		88.6	88.6		90.5	90.5
0241, 31 - 34	SITE & INFRASTRUCTURE, DEMOLITION	103.3	85.4	90.9	102.5	82.2	88.5	106.0	85.5	91.9	94.1	86.0	88.5	96.9	86.6	89.8	97.8	89.3	91.9
0310	Concrete Forming & Accessories	89.4	57.6	61.7	86.6	28.2	35.8	91.7	62.0	65.9	87.7	57.1	61.1	91.2	38.3	45.2	83.7	29.0	36.2
0320	Concrete Reinforcing	89.4	46.2	67.6	101.6	25.6	63.3	92.9	64.5	78.6	94.0	48.3	70.9	88.8	47.2	67.9	100.4	23.4	61.6
0330	Cast-in-Place Concrete	93.5	44.1	74.1	87.9	38.2	68.3	88.9	67.1	80.3	95.0	48.5	76.7	78.9	59.2	71.2	102.2	35.8	76.0
03	CONCRETE	91.2	51.4	71.8	90.1	33.2	62.4	92.1	65.8	79.3	100.4	53.3	77.4	85.8	48.3	67.5	105.5	31.7	69.5
04	MASONRY	111.6	52.0	75.2	160.5	47.4	91.5	99.7	67.2	79.9	142.0	43.8	82.1	97.3	47.4	66.9	156.1	43.7	87.5
05	METALS	93.4	61.9	83.0	97.7	63.6	86.4	101.0	88.0	96.7	96.4	63.8	85.6	96.5	63.8	85.7	91.2	51.2	78.0
06	WOOD, PLASTICS & COMPOSITES	92.7	61.8	74.8	89.6	27.3	53.6	97.5	61.8	76.9	86.1	60.4	71.2	93.7	37.2	61.1	83.9	29.5	52.5
07	THERMAL & MOISTURE PROTECTION	92.9	49.4	75.2	92.7	40.8	71.6	94.9	67.5	83.8	97.0	47.4	76.9	91.3	49.4	74.3	93.9	36.2	70.5
08	OPENINGS	98.2	50.2	86.1	99.5	26.9	81.2	104.5	61.6	93.7	97.6	49.3	85.5	98.8	39.2	83.8	90.5	26.9	74.5
0920	Plaster & Gypsum Board	79.5	61.1	66.4	82.8	25.5	41.8	92.8	61.1	70.2	87.2	59.7	67.6	82.2	35.8	49.1	80.6	27.9	42.9
0950, 0980	Ceilings & Acoustic Treatment	89.2	61.1	70.6	90.1	25.5	47.4	101.1	61.1	74.7	93.0	59.7	71.0	93.7	35.8	55.4	87.3	27.9	48.1
0960	Flooring	92.8	35.7	76.0	95.0	49.1	81.5	97.0	55.2	84.7	85.6	35.7	70.9	92.7	66.0	84.9	100.6	36.1	81.6
0970, 0990	Wall Finishes & Painting/Coating	92.2	31.5	54.8	103.0	25.9	55.5	99.2	62.4	76.5	89.0	32.3	54.0	92.2	56.1	69.9	94.6	30.3	54.9
09	FINISHES	89.2	52.0	68.4	94.1	31.4	59.2	97.1	60.4	76.6	82.3	51.7	65.2	89.9	44.4	64.5	99.7	30.4	61.0
COVERS	DIVS. 10 - 14, 25, 28, 41, 43, 44, 46	100.0	45.2	89.1	100.0	36.1	87.3	100.0	83.6	96.7	100.0	45.4	89.1	100.0	74.9	95.0	100.0	22.9	84.7
21, 22, 23	FIRE SUPPRESSION, PLUMBING & HVAC	94.1	60.1	80.3	94.1	45.7	74.5	100.0	66.6	86.5	94.1	61.0	80.7	100.0	38.8	75.2	94.1	28.9	67.6
26, 27, 3370	ELECTRICAL, COMMUNICATIONS & UTIL.	91.1	45.1	67.3	92.2	28.1	59.1	96.5	67.5	81.5	92.9	45.1	68.2	94.3	62.2	77.7	92.4	45.6	68.3
MF2010	WEIGHTED AVERAGE	95.1	56.2	77.9	98.6	43.3	74.2	99.1	69.8	86.2	97.5	55.9	79.2	95.8	52.7	76.8	98.6	40.4	72.9

	DIVISION	TEXAS																	
		LUBBOCK			LUFKIN			MCALLEN			MCKINNEY			MIDLAND			ODESSA		
		793 - 794			759			785			750			797			797		
		MAT.	INST.	TOTAL	MAT.	INST.	TOTAL	MAT.	INST.	TOTAL	MAT.	INST.	TOTAL	MAT.	INST.	TOTAL	MAT.	INST.	TOTAL
015433	CONTRACTOR EQUIPMENT		97.2	97.2		90.5	90.5		95.7	95.7		95.6	95.6		97.2	97.2		88.9	88.9
0241, 31 - 34	SITE & INFRASTRUCTURE, DEMOLITION	120.9	84.3	95.7	92.5	90.3	91.0	142.0	82.5	101.0	98.5	82.8	87.6	124.3	84.3	96.7	95.2	86.3	89.1
0310	Concrete Forming & Accessories	93.2	52.6	57.9	86.6	58.7	62.3	95.8	35.7	43.5	85.7	37.6	43.9	96.6	39.6	47.0	95.0	39.5	46.7
0320	Concrete Reinforcing	95.0	51.8	73.2	102.1	60.8	81.3	88.3	47.2	67.6	101.6	50.3	75.7	96.1	51.1	73.4	93.8	51.0	72.2
0330	Cast-in-Place Concrete	96.3	49.0	77.6	91.4	45.2	73.2	111.4	44.5	85.1	82.4	46.4	68.2	102.3	46.3	80.3	96.1	45.3	76.1
03	CONCRETE	93.6	52.9	73.7	97.9	55.0	77.0	101.2	42.9	72.8	85.6	44.9	65.8	98.1	46.1	72.8	94.8	44.9	70.5
04	MASONRY	102.4	55.9	74.0	119.7	43.8	73.4	103.4	49.3	70.4	172.7	49.9	97.8	120.6	43.6	73.7	103.1	43.6	66.8
05	METALS	102.6	80.4	95.3	97.8	67.5	87.8	93.9	74.8	87.6	97.6	74.6	90.0	100.8	79.2	93.7	98.6	67.4	88.3
06	WOOD, PLASTICS & COMPOSITES	98.7	56.7	74.4	92.0	61.7	74.5	112.6	34.8	67.7	88.3	37.7	59.1	103.0	39.3	66.2	99.6	39.1	64.7
07	THERMAL & MOISTURE PROTECTION	88.0	53.3	73.9	93.7	47.1	74.8	95.4	44.9	74.9	92.6	47.5	74.2	88.2	44.9	70.6	97.4	44.0	75.7
08	OPENINGS	102.2	50.7	89.2	73.0	54.6	68.4	103.3	37.7	86.8	99.5	40.6	84.7	101.1	40.9	85.9	92.6	40.8	79.6
0920	Plaster & Gypsum Board	83.3	55.7	63.6	79.8	61.1	66.5	86.3	33.2	48.4	81.4	36.3	49.2	84.8	37.7	51.2	82.8	37.7	50.6
0950, 0980	Ceilings & Acoustic Treatment	93.8	55.7	68.6	80.1	61.1	67.5	93.7	33.2	53.7	90.1	36.3	54.5	91.1	37.7	55.9	92.0	37.7	56.2
0960	Flooring	102.8	40.5	84.5	135.1	39.2	107.0	108.5	66.0	96.0	94.6	36.6	77.6	104.3	50.0	88.3	110.3	50.0	92.6
0970, 0990	Wall Finishes & Painting/Coating	105.2	51.1	71.8	94.6	39.0	60.3	104.7	29.8	58.5	103.0	42.5	65.7	105.2	51.1	71.8	94.7	51.1	67.8
09	FINISHES	97.7	50.3	71.2	107.8	53.8	77.7	99.1	40.1	66.2	93.5	37.9	62.5	98.1	42.0	66.8	94.9	41.9	65.3
COVERS	DIVS. 10 - 14, 25, 28, 41, 43, 44, 46	100.0	77.9	95.6	100.0	74.6	94.9	100.0	74.8	95.0	100.0	41.7	88.4	100.0	77.6	95.5	100.0	77.3	95.5
21, 22, 23	FIRE SUPPRESSION, PLUMBING & HVAC	99.9	44.3	77.3	94.1	61.0	80.7	94.1	30.4	68.3	94.1	43.4	73.6	93.9	44.5	73.9	100.3	44.4	77.6
26, 27, 3370	ELECTRICAL, COMMUNICATIONS & UTIL.	95.9	51.5	73.0	93.7	45.1	68.6	91.9	35.5	62.8	92.3	46.9	68.9	95.9	41.9	68.0	97.4	41.8	68.8
MF2010	WEIGHTED AVERAGE	99.6	56.8	80.7	95.8	58.3	79.3	98.6	46.6	75.7	98.5	50.2	77.2	99.4	51.4	78.2	97.8	50.2	76.8

	DIVISION	TEXAS																	
		PALESTINE			SAN ANGELO			SAN ANTONIO			TEMPLE			TEXARKANA			TYLER		
		758			769			781 - 782			765			755			757		
		MAT.	INST.	TOTAL	MAT.	INST.	TOTAL	MAT.	INST.	TOTAL	MAT.	INST.	TOTAL	MAT.	INST.	TOTAL	MAT.	INST.	TOTAL
015433	CONTRACTOR EQUIPMENT		90.5	90.5		88.9	88.9		91.0	91.0		88.9	88.9		90.5	90.5		90.5	90.5
0241, 31 - 34	SITE & INFRASTRUCTURE, DEMOLITION	99.6	90.1	93.1	97.3	86.8	90.1	96.6	91.3	92.9	85.6	85.2	85.3	86.6	90.1	89.0	97.2	90.1	92.3
0310	Concrete Forming & Accessories	79.0	57.8	60.6	93.9	37.4	44.7	91.2	56.8	61.3	96.8	35.2	43.2	92.5	35.2	42.6	88.0	34.1	41.1
0320	Concrete Reinforcing	99.6	49.1	74.2	94.0	51.0	72.3	95.1	48.1	71.4	94.1	48.8	71.3	99.5	52.0	75.6	100.4	52.2	76.1
0330	Cast-in-Place Concrete	83.5	43.8	67.9	97.8	43.5	76.4	77.4	67.5	73.5	80.2	48.9	67.8	84.2	44.6	68.6	100.4	43.5	78.0
03	CONCRETE	99.0	51.9	76.1	97.6	43.2	71.1	86.0	59.7	73.2	83.4	43.7	64.1	90.9	42.9	67.5	105.0	42.1	74.3
04	MASONRY	115.0	50.4	75.6	128.8	43.0	76.4	97.2	59.3	74.1	140.2	48.8	84.4	177.1	46.4	97.3	166.5	49.3	94.9
05	METALS	97.5	62.4	86.0	97.1	64.6	86.4	97.2	68.3	87.7	96.8	63.5	85.8	91.1	67.2	83.2	97.4	67.7	87.6
06	WOOD, PLASTICS & COMPOSITES	83.1	61.7	70.8	98.5	37.4	63.2	93.7	54.9	71.3	107.6	34.0	65.1	94.2	33.5	59.1	93.5	31.6	57.7
07	THERMAL & MOISTURE PROTECTION	94.2	48.2	75.5	97.3	43.7	75.5	91.0	63.9	80.0	96.9	45.3	75.9	93.5	48.8	75.4	94.0	53.2	77.4
08	OPENINGS	72.9	51.6	67.6	88.6	39.8	76.3	100.7	53.0	88.7	74.8	36.2	65.1	90.5	38.7	77.4	72.9	37.7	64.0
0920	Plaster & Gypsum Board	75.3	61.1	65.2	81.7	36.0	49.1	82.2	53.9	62.0	81.7	32.4	46.5	84.7	32.0	47.1	79.8	30.0	44.3
0950, 0980	Ceilings & Acoustic Treatment	80.1	61.1	67.5	88.4	36.0	53.8	93.7	53.9	67.4	88.4	32.4	51.4	87.3	32.0	50.8	80.1	30.0	47.0
0960	Flooring	126.5	60.6	107.1	109.3	35.7	87.7	92.7	66.2	84.9	145.6	42.9	115.5	108.5	61.0	94.6	137.0	38.1	108.0
0970, 0990	Wall Finishes & Painting/Coating	94.6	31.5	55.7	94.7	51.1	67.8	92.2	56.1	69.9	96.2	34.7	58.3	94.6	40.6	61.3	94.6	40.6	61.3
09	FINISHES	105.5	56.4	78.1	94.0	37.7	62.6	89.9	57.4	71.8	104.0	36.0	66.1	101.8	39.5	67.0	108.7	33.8	66.9
COVERS	DIVS. 10 - 14, 25, 28, 41, 43, 44, 46	100.0	42.8	88.6	100.0	74.8	95.0	100.0	81.0	96.2	100.0	37.3	87.5	100.0	74.9	95.0	100.0	74.6	94.9
21, 22, 23	FIRE SUPPRESSION, PLUMBING & HVAC	94.1	57.7	79.3	94.3	43.7	73.8	99.9	66.1	86.2	94.3	34.9	70.3	94.1	34.4	69.9	94.1	59.2	79.9
26, 27, 3370	ELECTRICAL, COMMUNICATIONS & UTIL.	89.9	45.1	66.8	97.3	47.5	71.6	94.7	62.3	78.0	94.4	67.4	80.5	93.6	63.5	78.1	92.4	54.0	72.6
MF2010	WEIGHTED AVERAGE	95.2	56.6	78.2	97.3	49.6	76.3	96.2	65.0	82.4	95.1	49.4	75.0	98.0	51.0	77.2	99.0	54.4	79.4

	DIVISION	TEXAS															UTAH		
		VICTORIA			WACO			WAXAHACKIE			WHARTON			WICHITA FALLS			LOGAN		
		779			766 - 767			751			774			763			843		
		MAT.	INST.	TOTAL	MAT.	INST.	TOTAL	MAT.	INST.	TOTAL	MAT.	INST.	TOTAL	MAT.	INST.	TOTAL	MAT.	INST.	TOTAL
015433	CONTRACTOR EQUIPMENT		97.4	97.4		88.9	88.9		95.6	95.6		98.5	98.5		88.9	88.9		98.6	98.6
0241, 31 - 34	SITE & INFRASTRUCTURE, DEMOLITION	112.8	81.1	90.9	94.0	86.4	88.7	100.6	83.3	88.7	118.9	83.9	94.8	94.7	86.0	88.7	90.7	98.4	96.0
0310	Concrete Forming & Accessories	90.1	36.4	43.4	95.5	40.5	47.7	85.7	59.2	62.6	85.1	37.9	44.0	95.4	38.4	45.8	105.4	60.2	66.1
0320	Concrete Reinforcing	89.4	31.2	60.1	93.8	48.8	71.1	101.6	50.4	75.8	93.0	53.1	72.9	93.8	51.0	72.2	100.2	75.9	87.9
0330	Cast-in-Place Concrete	103.0	42.0	79.0	86.7	53.2	73.5	87.0	42.4	69.4	105.8	52.1	84.7	92.7	45.8	74.2	86.0	72.1	80.5
03	CONCRETE	101.7	39.6	71.4	90.1	47.7	69.4	89.2	53.1	71.7	105.7	47.7	77.4	93.1	44.6	69.5	105.3	67.8	87.0
04	MASONRY	116.7	35.4	67.1	99.9	55.8	73.0	161.1	50.0	93.3	101.0	46.2	67.5	100.3	57.5	74.2	106.8	62.0	79.4
05	METALS	96.6	70.9	88.1	99.2	66.7	88.5	97.7	76.6	90.7	98.0	78.0	91.4	99.1	68.9	89.1	99.9	76.8	92.3
06	WOOD, PLASTICS & COMPOSITES	99.2	37.2	63.4	106.1	36.4	65.8	88.3	66.6	75.8	89.5	33.8	57.3	106.1	36.3	65.8	79.9	58.4	67.5
07	THERMAL & MOISTURE PROTECTION	98.9	40.9	75.3	97.9	48.7	77.9	92.7	47.5	74.3	95.6	50.7	77.4	97.9	52.0	79.2	99.6	63.7	85.0
08	OPENINGS	102.1	36.4	85.6	86.7	37.5	74.3	99.5	56.3	88.7	102.3	38.3	86.2	86.7	41.3	75.3	91.7	57.6	83.1
0920	Plaster & Gypsum Board	87.8	35.7	50.6	82.8	34.9	48.6	82.0	66.1	70.6	85.8	32.1	47.5	82.8	34.8	48.5	81.4	57.0	64.0
0950, 0980	Ceilings & Acoustic Treatment	96.6	35.7	58.4	92.0	34.9	54.3	91.0	66.1	71.8	95.7	32.1	53.7	92.0	34.8	54.2	109.7	57.0	74.9
0960	Flooring	95.1	39.2	78.7	144.3	46.4	115.6	94.6	39.2	78.3	93.2	46.8	79.6	145.2	77.2	125.3	100.7	50.0	85.8
0970, 0990	Wall Finishes & Painting/Coating	99.4	39.0	62.1	96.2	34.7	58.3	103.0	39.0	63.5	99.2	32.3	57.9	98.7	51.1	69.3	98.8	62.0	76.1
09	FINISHES	87.5	37.3	59.5	105.1	39.8	68.7	94.2	54.7	72.2	89.0	38.0	60.5	105.5	46.0	72.3	100.6	57.8	76.8
COVERS	DIVS. 10 - 14, 25, 28, 41, 43, 44, 46	100.0	38.4	87.7	100.0	78.3	95.7	100.0	42.2	88.5	100.0	40.6	88.2	100.0	68.7	93.8	100.0	57.0	91.5
21, 22, 23	FIRE SUPPRESSION, PLUMBING & HVAC	94.1	32.8	69.3	100.3	51.3	80.4	94.1	42.5	73.2	94.1	61.0	80.7	100.3	46.6	78.5	100.0	71.3	88.4
26, 27, 3370	ELECTRICAL, COMMUNICATIONS & UTIL.	99.4	41.2	69.4	97.5	67.6	82.0	92.3	67.6	79.5	98.1	45.5	71.0	99.2	57.9	77.9	94.5	73.7	83.8
MF2010	WEIGHTED AVERAGE	98.6	44.0	74.5	97.4	56.4	79.4	98.5	57.3	80.3	98.5	54.0	78.9	98.0	54.8	79.0	99.2	70.0	86.3

City Cost Indexes

	DIVISION	UTAH OGDEN 842,844			UTAH PRICE 845			UTAH PROVO 846 - 847			UTAH SALT LAKE CITY 840 - 841			VERMONT BELLOWS FALLS 051			VERMONT BENNINGTON 052		
		MAT.	INST.	TOTAL	MAT.	INST.	TOTAL	MAT.	INST.	TOTAL	MAT.	INST.	TOTAL	MAT.	INST.	TOTAL	MAT.	INST.	TOTAL
015433	**CONTRACTOR EQUIPMENT**		98.6	98.6		97.6	97.6		97.6	97.6		98.6	98.6		100.4	100.4		100.4	100.4
0241, 31 - 34	**SITE & INFRASTRUCTURE, DEMOLITION**	78.7	98.4	92.3	88.4	96.7	94.1	87.1	96.7	93.7	78.3	98.3	92.1	81.9	97.8	92.8	81.3	97.5	92.5
0310	Concrete Forming & Accessories	105.5	60.2	66.1	107.4	51.0	58.4	106.6	60.1	66.1	107.9	60.1	66.3	99.3	48.8	55.4	97.2	53.0	58.8
0320	Concrete Reinforcing	99.9	75.9	87.8	107.5	75.7	91.5	108.4	75.8	92.0	102.1	75.8	88.9	82.9	63.7	73.2	82.9	63.6	73.2
0330	Cast-in-Place Concrete	87.3	72.1	81.3	86.0	61.7	76.4	86.0	72.0	80.5	95.4	72.0	86.2	96.9	71.2	86.8	96.9	67.9	85.5
03	**CONCRETE**	95.8	67.8	82.2	106.6	60.1	84.0	105.2	67.7	86.9	114.2	67.7	91.5	100.9	60.0	81.0	100.7	60.7	81.2
04	**MASONRY**	100.3	62.0	76.9	112.2	62.0	81.6	112.3	62.0	81.6	114.0	62.0	82.3	93.6	60.6	73.5	102.3	60.5	76.8
05	**METALS**	100.4	76.8	92.6	97.1	76.1	90.2	98.1	76.7	91.1	105.0	76.7	95.6	92.5	69.6	85.0	92.5	68.7	84.7
06	**WOOD, PLASTICS & COMPOSITES**	79.9	58.4	67.5	82.5	47.1	62.1	81.2	58.4	68.0	81.5	58.4	68.2	99.5	38.4	64.2	96.7	44.9	66.8
07	**THERMAL & MOISTURE PROTECTION**	98.5	63.7	84.4	101.4	59.4	84.4	101.4	63.7	86.1	105.1	63.7	88.3	100.7	59.6	84.0	100.7	54.7	82.0
08	**OPENINGS**	91.7	57.6	83.1	96.2	50.3	84.6	96.2	57.6	86.5	93.7	57.6	84.6	99.2	41.1	84.5	99.2	43.4	85.1
0920	Plaster & Gypsum Board	81.4	57.0	64.0	83.7	45.3	56.3	81.9	57.0	64.1	84.4	57.0	64.8	96.5	35.8	53.1	94.2	42.5	57.3
0950, 0980	Ceilings & Acoustic Treatment	109.7	57.0	74.9	109.7	45.3	67.2	109.7	57.0	74.9	103.6	57.0	72.8	92.7	35.8	55.1	92.7	42.5	59.5
0960	Flooring	98.5	50.0	84.2	101.8	37.1	82.8	101.4	50.0	86.3	102.1	50.0	86.8	94.5	31.9	76.1	93.4	31.9	75.3
0970, 0990	Wall Finishes & Painting/Coating	98.8	62.0	76.1	98.8	39.6	62.3	98.8	50.6	69.1	101.9	50.6	70.2	90.1	28.1	51.9	90.1	28.1	51.9
09	**FINISHES**	98.6	57.8	75.9	101.8	46.1	70.7	101.3	56.6	76.4	99.4	56.6	75.5	95.6	42.0	65.7	95.0	45.9	67.6
COVERS	**DIVS. 10 - 14, 25, 28, 41, 43, 44, 46**	100.0	57.0	91.5	100.0	49.5	90.0	100.0	57.0	91.4	100.0	57.0	91.4	100.0	45.9	89.2	100.0	46.5	89.4
21, 22, 23	**FIRE SUPPRESSION, PLUMBING & HVAC**	100.0	71.3	88.4	97.0	68.8	85.6	100.0	71.3	88.4	100.1	71.3	88.4	94.1	95.7	94.7	94.1	89.5	92.2
26, 27, 3370	**ELECTRICAL, COMMUNICATIONS & UTIL.**	94.8	73.7	83.9	100.0	73.7	86.4	95.4	63.5	78.9	97.8	63.5	80.1	102.0	94.8	98.3	102.0	64.9	82.8
MF2010	**WEIGHTED AVERAGE**	97.3	70.0	85.3	99.5	65.9	84.7	99.7	68.3	85.8	101.6	68.4	86.9	96.5	72.7	86.0	96.8	67.8	84.0

	DIVISION	VERMONT BRATTLEBORO 053			VERMONT BURLINGTON 054			VERMONT GUILDHALL 059			VERMONT MONTPELIER 056			VERMONT RUTLAND 057			VERMONT ST. JOHNSBURY 058		
		MAT.	INST.	TOTAL	MAT.	INST.	TOTAL	MAT.	INST.	TOTAL	MAT.	INST.	TOTAL	MAT.	INST.	TOTAL	MAT.	INST.	TOTAL
015433	**CONTRACTOR EQUIPMENT**		100.4	100.4		100.4	100.4		100.4	100.4		100.4	100.4		100.4	100.4		100.4	100.4
0241, 31 - 34	**SITE & INFRASTRUCTURE, DEMOLITION**	82.9	97.8	93.1	83.8	96.5	92.5	80.9	95.8	91.1	84.8	96.5	92.9	84.8	96.5	92.8	80.9	96.0	91.3
0310	Concrete Forming & Accessories	99.6	48.7	55.3	96.8	51.3	57.3	97.6	43.8	50.8	101.5	51.4	57.9	99.8	51.4	57.7	96.5	43.9	50.7
0320	Concrete Reinforcing	82.0	63.7	72.8	103.6	82.7	93.1	83.6	37.7	60.4	82.5	82.7	82.6	103.6	82.7	93.1	82.0	42.9	62.3
0330	Cast-in-Place Concrete	100.0	68.3	87.5	105.7	98.9	103.0	94.0	54.9	78.6	111.9	98.9	106.8	95.0	98.9	96.5	94.0	55.2	78.7
03	**CONCRETE**	103.3	59.0	81.7	109.4	74.2	92.2	98.3	47.4	73.5	109.5	74.2	92.3	104.1	74.2	89.6	98.0	48.4	73.8
04	**MASONRY**	101.3	60.6	76.4	104.9	49.1	70.8	101.5	39.0	63.4	89.6	49.1	64.9	85.0	49.1	63.1	126.1	39.1	73.0
05	**METALS**	92.5	69.5	84.9	99.7	78.9	92.8	92.5	58.5	81.3	92.5	78.9	88.0	98.0	78.9	91.7	92.5	59.0	81.5
06	**WOOD, PLASTICS & COMPOSITES**	99.8	38.4	64.3	94.7	50.7	69.3	96.2	44.9	66.6	98.8	50.7	71.0	99.9	50.7	71.5	92.3	44.9	64.9
07	**THERMAL & MOISTURE PROTECTION**	100.8	53.6	81.6	100.7	59.7	84.0	100.6	41.4	76.5	100.1	60.1	83.9	100.9	60.1	84.3	100.5	41.4	76.5
08	**OPENINGS**	99.2	41.1	84.5	102.4	52.5	89.9	99.2	38.2	83.8	99.2	52.5	87.4	102.4	52.5	89.9	99.2	38.2	83.8
0920	Plaster & Gypsum Board	96.5	35.8	53.1	98.1	48.5	62.7	102.4	42.5	59.6	105.6	48.5	64.8	97.3	48.5	62.5	105.2	42.5	60.4
0950, 0980	Ceilings & Acoustic Treatment	92.7	35.8	55.1	94.8	48.5	64.2	92.7	42.5	59.5	91.8	48.5	63.2	97.5	48.5	65.1	92.7	42.5	59.5
0960	Flooring	94.6	31.9	76.2	94.7	57.7	83.8	97.7	31.9	78.4	101.3	57.7	88.5	94.5	57.7	83.7	101.3	31.9	80.9
0970, 0990	Wall Finishes & Painting/Coating	90.1	28.1	51.9	90.1	32.7	54.7	90.1	21.8	48.0	90.1	32.7	54.7	90.1	32.7	54.7	90.1	22.9	48.7
09	**FINISHES**	95.7	42.0	65.8	96.4	49.3	70.1	97.1	39.4	64.9	98.5	49.3	71.1	96.8	49.3	70.3	98.5	39.6	65.6
COVERS	**DIVS. 10 - 14, 25, 28, 41, 43, 44, 46**	100.0	45.9	89.2	100.0	70.6	94.1	100.0	38.6	87.8	100.0	70.6	94.1	100.0	70.6	94.1	100.0	38.6	87.8
21, 22, 23	**FIRE SUPPRESSION, PLUMBING & HVAC**	94.1	97.8	95.6	99.9	62.1	84.5	94.1	60.4	80.4	93.9	62.1	81.0	100.0	62.1	84.6	94.1	60.4	80.4
26, 27, 3370	**ELECTRICAL, COMMUNICATIONS & UTIL.**	102.0	94.8	98.3	99.4	64.9	81.6	102.0	64.9	82.8	100.7	64.9	82.2	102.0	64.9	82.9	102.0	64.9	82.8
MF2010	**WEIGHTED AVERAGE**	97.2	72.9	86.5	100.6	65.4	85.1	96.6	55.0	78.3	97.4	65.4	83.3	99.2	65.4	84.3	97.9	55.2	79.1

	DIVISION	VERMONT WHITE RIVER JCT. 050			VIRGINIA ALEXANDRIA 223			VIRGINIA ARLINGTON 222			VIRGINIA BRISTOL 242			VIRGINIA CHARLOTTESVILLE 229			VIRGINIA CULPEPER 227		
		MAT.	INST.	TOTAL	MAT.	INST.	TOTAL	MAT.	INST.	TOTAL	MAT.	INST.	TOTAL	MAT.	INST.	TOTAL	MAT.	INST.	TOTAL
015433	**CONTRACTOR EQUIPMENT**		100.4	100.4		101.1	101.1		99.9	99.9		99.9	99.9		104.0	104.0		99.9	99.9
0241, 31 - 34	**SITE & INFRASTRUCTURE, DEMOLITION**	85.4	95.9	92.7	118.6	88.3	97.7	130.3	85.1	99.2	113.8	84.7	93.7	119.2	86.7	96.8	116.8	85.5	95.2
0310	Concrete Forming & Accessories	94.8	42.1	49.0	89.8	75.3	77.2	89.3	71.8	74.1	85.7	44.1	49.5	84.2	50.2	54.6	81.7	70.4	71.9
0320	Concrete Reinforcing	82.9	37.6	60.1	83.6	79.9	81.7	94.4	71.6	82.9	94.3	68.5	81.3	93.7	69.4	81.5	94.4	65.7	79.9
0330	Cast-in-Place Concrete	100.0	54.7	82.1	122.1	83.8	107.0	118.8	79.2	103.2	118.3	48.9	91.0	123.1	57.0	97.0	121.6	66.3	99.8
03	**CONCRETE**	105.2	46.6	76.6	111.6	80.1	96.2	115.8	75.0	95.9	112.2	52.2	83.0	113.3	57.9	86.3	110.3	69.0	90.2
04	**MASONRY**	113.5	37.8	67.3	93.7	72.5	80.7	107.7	66.1	82.3	95.8	41.5	62.7	121.9	48.0	76.8	109.7	61.8	80.5
05	**METALS**	92.5	58.7	81.4	95.8	94.2	95.3	94.5	81.8	90.3	93.3	85.4	90.7	93.6	88.0	91.8	93.7	80.6	89.4
06	**WOOD, PLASTICS & COMPOSITES**	93.7	43.2	64.5	95.7	74.4	83.4	92.9	74.4	82.2	85.9	42.4	60.8	84.5	48.3	63.6	83.2	74.4	78.1
07	**THERMAL & MOISTURE PROTECTION**	100.9	36.6	74.7	99.0	80.9	91.6	101.0	68.6	87.8	100.5	48.1	79.2	100.1	53.5	81.2	100.3	62.0	84.7
08	**OPENINGS**	99.2	39.5	84.1	97.3	75.4	91.8	95.2	69.4	88.7	98.6	48.4	86.0	96.6	53.3	85.7	96.9	65.0	88.9
0920	Plaster & Gypsum Board	92.4	40.7	55.5	110.4	73.5	84.0	107.8	73.5	83.3	103.4	40.5	58.4	103.4	45.9	62.3	103.7	73.5	82.1
0950, 0980	Ceilings & Acoustic Treatment	92.7	40.7	58.4	97.3	73.5	81.6	94.6	73.5	80.6	93.7	40.5	58.5	93.7	45.9	62.1	94.6	73.5	80.6
0960	Flooring	92.3	31.9	74.6	107.3	86.1	101.1	105.6	63.4	93.2	101.8	39.2	83.4	100.3	46.2	84.4	100.3	63.4	89.5
0970, 0990	Wall Finishes & Painting/Coating	90.1	19.1	46.3	130.0	81.1	99.8	130.0	81.1	99.8	114.8	43.8	71.0	114.8	62.0	82.2	130.0	55.1	83.8
09	**FINISHES**	94.8	37.9	63.1	104.4	77.4	89.3	104.5	71.9	86.3	100.8	42.4	68.2	100.4	49.9	72.2	101.1	68.8	83.1
COVERS	**DIVS. 10 - 14, 25, 28, 41, 43, 44, 46**	100.0	38.2	87.7	100.0	89.3	97.9	100.0	70.2	94.1	100.0	73.2	94.7	100.0	73.1	94.6	100.0	59.7	92.0
21, 22, 23	**FIRE SUPPRESSION, PLUMBING & HVAC**	94.1	60.0	80.2	100.2	88.8	95.6	100.2	84.6	93.9	94.3	53.8	77.9	94.3	56.9	79.1	94.3	62.8	81.5
26, 27, 3370	**ELECTRICAL, COMMUNICATIONS & UTIL.**	102.0	64.9	82.8	96.3	102.5	99.5	94.0	98.3	96.2	96.0	37.6	65.9	95.9	71.0	83.1	98.4	55.4	76.2
MF2010	**WEIGHTED AVERAGE**	98.0	54.4	78.8	100.6	86.0	94.2	101.6	79.9	92.0	98.6	54.3	79.1	100.0	62.8	83.6	99.3	67.0	85.1

	DIVISION	VIRGINIA																	
		FAIRFAX			FARMVILLE			FREDERICKSBURG			GRUNDY			HARRISONBURG			LYNCHBURG		
		220 - 221			239			224 - 225			246			228			245		
		MAT.	INST.	TOTAL	MAT.	INST.	TOTAL	MAT.	INST.	TOTAL	MAT.	INST.	TOTAL	MAT.	INST.	TOTAL	MAT.	INST.	TOTAL
015433	**CONTRACTOR EQUIPMENT**		99.9	99.9		104.0	104.0		99.9	99.9		99.9	99.9		99.9	99.9		99.9	99.9
0241, 31 - 34	**SITE & INFRASTRUCTURE, DEMOLITION**	129.2	85.9	99.4	114.8	86.0	94.9	116.2	85.8	95.3	111.3	83.6	92.2	125.8	84.8	97.5	112.2	84.9	93.3
0310	Concrete Forming & Accessories	84.3	72.3	73.9	95.1	45.7	52.1	84.3	50.0	54.5	88.6	38.1	44.7	80.7	44.2	49.0	85.7	63.1	66.1
0320	Concrete Reinforcing	94.4	79.8	87.0	94.3	52.9	73.5	95.1	78.8	86.9	93.0	53.5	73.1	94.4	48.1	71.1	93.7	69.1	81.3
0330	Cast-in-Place Concrete	118.8	80.1	103.5	123.0	54.9	96.2	120.6	53.8	94.3	118.3	49.3	91.1	118.7	60.0	95.6	118.3	59.8	95.3
03	**CONCRETE**	115.5	77.6	97.0	112.4	51.9	82.9	109.9	58.4	84.8	111.0	46.4	79.5	113.3	52.5	83.7	110.9	64.5	88.3
04	**MASONRY**	107.6	66.4	82.4	104.7	47.3	69.7	108.8	46.0	70.5	99.2	50.0	69.2	105.8	52.1	73.0	114.0	54.2	77.5
05	**METALS**	93.7	95.1	94.2	93.6	77.5	88.3	93.7	91.6	93.0	93.3	71.5	86.1	93.6	83.7	90.4	93.5	86.7	91.3
06	**WOOD, PLASTICS & COMPOSITES**	85.9	74.4	79.2	97.7	46.7	68.2	85.9	47.7	63.8	88.6	37.2	58.9	82.3	41.4	58.6	85.9	66.5	74.7
07	**THERMAL & MOISTURE PROTECTION**	100.9	68.9	87.9	100.1	48.5	79.1	100.3	51.6	80.5	100.4	47.8	79.0	100.7	61.4	84.7	100.3	55.7	82.2
08	**OPENINGS**	95.2	76.6	90.5	96.9	45.6	84.0	96.6	55.2	86.2	98.6	36.5	83.0	96.9	51.9	85.6	96.9	61.5	88.0
0920	Plaster & Gypsum Board	103.7	73.5	82.1	109.3	44.2	62.8	103.7	45.9	62.4	103.4	35.1	54.6	103.4	39.4	57.7	103.4	65.3	76.2
0950, 0980	Ceilings & Acoustic Treatment	94.6	73.5	80.6	93.7	44.2	61.0	94.6	45.9	62.4	93.7	35.1	55.0	93.7	39.4	57.8	93.7	65.3	74.9
0960	Flooring	102.1	66.2	91.6	107.3	63.1	94.3	102.1	77.5	94.9	103.3	35.3	83.3	100.0	77.5	93.4	101.8	43.7	84.7
0970, 0990	Wall Finishes & Painting/Coating	130.0	81.1	99.8	114.8	40.8	69.2	130.0	67.7	91.6	114.8	37.0	66.8	130.0	53.5	82.8	114.8	43.8	71.0
09	**FINISHES**	102.9	72.4	85.9	102.8	48.7	72.6	101.6	55.7	76.0	101.0	37.1	65.4	101.7	50.2	73.0	100.6	57.9	76.8
COVERS	**DIVS. 10 - 14, 25, 28, 41, 43, 44, 46**	100.0	70.2	94.1	100.0	49.1	89.9	100.0	74.0	94.8	100.0	47.1	89.5	100.0	72.9	94.6	100.0	76.6	95.3
21, 22, 23	**FIRE SUPPRESSION, PLUMBING & HVAC**	94.3	84.9	90.5	94.3	36.7	70.9	94.3	82.3	89.4	94.3	41.1	72.7	94.3	53.7	77.8	94.3	56.4	78.9
26, 27, 3370	**ELECTRICAL, COMMUNICATIONS & UTIL.**	97.1	98.3	97.7	93.3	49.9	70.9	94.2	91.3	92.7	96.0	47.4	70.9	96.2	83.0	89.4	97.0	55.9	75.8
MF2010	**WEIGHTED AVERAGE**	100.1	82.0	92.1	98.9	52.3	78.4	98.8	71.8	86.9	98.6	49.7	77.0	99.6	63.0	83.4	99.3	63.6	83.5

	DIVISION	VIRGINIA																	
		NEWPORT NEWS			NORFOLK			PETERSBURG			PORTSMOUTH			PULASKI			RICHMOND		
		236			233 - 235			238			237			243			230 - 232		
		MAT.	INST.	TOTAL	MAT.	INST.	TOTAL	MAT.	INST.	TOTAL	MAT.	INST.	TOTAL	MAT.	INST.	TOTAL	MAT.	INST.	TOTAL
015433	**CONTRACTOR EQUIPMENT**		104.0	104.0		104.6	104.6		104.0	104.0		103.9	103.9		99.9	99.9		104.0	104.0
0241, 31 - 34	**SITE & INFRASTRUCTURE, DEMOLITION**	113.0	87.4	95.3	111.3	88.3	95.4	117.7	87.7	97.0	111.5	86.7	94.4	110.7	83.6	92.0	110.4	87.7	94.7
0310	Concrete Forming & Accessories	94.5	66.1	69.8	95.9	66.2	70.1	88.5	59.6	63.3	84.9	54.3	58.3	88.6	38.3	44.8	94.0	59.6	64.1
0320	Concrete Reinforcing	94.0	72.3	83.1	94.0	72.3	83.1	93.7	70.1	81.8	93.7	69.8	81.6	93.0	53.5	73.1	94.0	70.1	82.0
0330	Cast-in-Place Concrete	119.6	60.2	96.2	130.7	60.3	102.9	127.0	55.2	98.7	118.4	59.7	95.3	118.3	49.3	91.1	115.2	55.2	91.6
03	**CONCRETE**	109.3	66.5	88.4	114.9	66.6	91.4	115.2	61.5	89.1	108.0	60.6	84.9	111.0	46.5	79.6	107.1	61.5	84.9
04	**MASONRY**	99.5	54.1	71.8	105.6	54.1	74.2	113.2	55.6	78.1	105.0	53.9	73.8	92.5	50.0	66.6	97.3	55.6	71.9
05	**METALS**	95.8	89.9	93.9	94.8	90.0	93.2	93.7	89.5	92.3	94.8	87.2	92.3	93.4	71.6	86.2	98.0	89.5	95.2
06	**WOOD, PLASTICS & COMPOSITES**	96.6	70.3	81.4	98.6	70.3	82.2	89.0	62.1	73.5	85.7	54.5	67.7	88.6	37.2	58.9	95.7	62.1	76.3
07	**THERMAL & MOISTURE PROTECTION**	100.1	56.1	82.2	97.5	56.1	80.6	100.1	56.3	82.3	100.1	53.8	81.3	100.4	47.8	79.0	98.7	56.3	81.5
08	**OPENINGS**	97.3	64.9	89.1	97.3	64.9	89.1	96.6	60.8	87.6	97.4	55.7	86.9	98.6	36.5	83.0	97.3	60.8	88.1
0920	Plaster & Gypsum Board	110.4	68.5	80.5	101.0	68.5	77.8	103.7	60.1	72.6	104.1	52.3	67.1	103.4	35.1	54.6	114.0	60.1	75.5
0950, 0980	Ceilings & Acoustic Treatment	97.3	68.5	78.3	97.3	68.5	78.3	94.6	60.1	71.8	97.3	52.3	67.6	93.7	35.1	55.0	100.9	60.1	74.0
0960	Flooring	107.3	52.9	91.3	106.7	52.9	90.9	102.8	74.6	94.5	99.5	66.9	90.0	103.3	35.3	83.3	106.7	74.6	97.3
0970, 0990	Wall Finishes & Painting/Coating	114.8	44.3	71.3	114.8	67.8	85.8	114.8	62.0	82.2	114.8	67.8	85.8	114.8	37.0	66.8	114.8	62.0	82.2
09	**FINISHES**	103.5	61.7	80.2	101.8	64.4	80.9	101.3	62.7	79.8	100.5	57.4	76.5	101.0	37.1	65.4	104.0	62.7	81.0
COVERS	**DIVS. 10 - 14, 25, 28, 41, 43, 44, 46**	100.0	78.3	95.7	100.0	78.3	95.7	100.0	73.3	94.7	100.0	76.6	95.3	100.0	47.1	89.5	100.0	73.3	94.7
21, 22, 23	**FIRE SUPPRESSION, PLUMBING & HVAC**	100.2	63.9	85.5	100.0	63.7	85.3	94.3	55.0	78.4	100.2	62.1	84.8	94.3	41.4	72.9	100.0	55.0	81.8
26, 27, 3370	**ELECTRICAL, COMMUNICATIONS & UTIL.**	96.0	60.4	77.6	101.2	59.9	79.9	96.2	71.0	83.2	94.3	59.8	76.5	96.0	51.2	72.9	99.3	71.0	84.7
MF2010	**WEIGHTED AVERAGE**	100.4	67.3	85.8	101.4	67.6	86.5	99.9	66.0	85.0	99.9	64.4	84.3	98.3	50.3	77.1	100.6	66.0	85.3

	DIVISION	VIRGINIA									WASHINGTON								
		ROANOKE			STAUNTON			WINCHESTER			CLARKSTON			EVERETT			OLYMPIA		
		240 - 241			244			226			994			982			985		
		MAT.	INST.	TOTAL	MAT.	INST.	TOTAL	MAT.	INST.	TOTAL	MAT.	INST.	TOTAL	MAT.	INST.	TOTAL	MAT.	INST.	TOTAL
015433	**CONTRACTOR EQUIPMENT**		99.9	99.9		104.0	104.0		99.9	99.9		91.0	91.0		104.1	104.1		104.1	104.1
0241, 31 - 34	**SITE & INFRASTRUCTURE, DEMOLITION**	110.3	84.8	92.7	115.4	86.5	95.5	124.3	84.9	97.1	96.4	90.4	92.2	92.9	112.9	106.7	94.4	112.9	107.1
0310	Concrete Forming & Accessories	94.3	62.9	67.0	88.1	52.4	57.0	82.9	49.2	53.6	124.9	70.5	77.6	113.7	99.4	101.2	103.5	99.5	100.0
0320	Concrete Reinforcing	94.0	69.2	81.5	93.7	51.9	72.6	93.7	83.8	88.7	106.6	84.5	95.5	116.6	103.0	109.7	117.2	103.0	110.1
0330	Cast-in-Place Concrete	134.4	59.7	105.0	123.0	53.2	95.5	118.7	56.3	94.1	102.5	82.4	94.6	101.0	105.8	102.9	91.9	106.0	97.5
03	**CONCRETE**	116.7	64.3	91.2	112.6	54.3	84.2	112.7	59.7	86.9	99.5	77.4	88.7	98.4	101.8	100.0	94.4	101.9	98.0
04	**MASONRY**	100.6	55.8	73.3	110.2	54.3	76.1	103.4	49.9	70.8	104.1	79.6	89.2	122.3	99.2	108.2	112.2	99.3	104.3
05	**METALS**	95.6	86.6	92.6	93.6	82.3	89.9	93.7	90.2	92.6	87.3	80.3	84.9	100.8	92.9	98.2	95.8	93.0	94.9
06	**WOOD, PLASTICS & COMPOSITES**	96.6	66.5	79.2	88.6	52.9	68.0	84.5	47.7	63.2	101.0	67.8	81.8	106.4	99.0	102.1	93.4	99.0	96.6
07	**THERMAL & MOISTURE PROTECTION**	100.0	54.7	81.6	100.1	52.3	80.6	100.8	58.7	83.7	151.2	76.5	120.8	102.4	99.0	101.1	102.1	96.8	99.9
08	**OPENINGS**	97.3	61.5	88.3	96.9	51.0	85.4	98.7	56.5	88.1	106.4	71.0	97.5	101.2	100.2	101.0	101.0	100.2	100.8
0920	Plaster & Gypsum Board	110.4	65.3	78.2	103.4	50.6	65.7	103.7	45.9	62.4	130.2	66.6	84.8	115.5	98.9	103.6	105.1	98.9	100.7
0950, 0980	Ceilings & Acoustic Treatment	97.3	65.3	76.2	93.7	50.6	65.2	94.6	45.9	62.4	99.3	66.6	77.7	107.8	98.9	101.9	101.2	98.9	99.7
0960	Flooring	107.3	47.1	89.6	102.7	41.3	84.7	101.4	77.5	94.4	103.2	49.4	87.4	116.5	95.9	110.5	109.2	96.1	105.3
0970, 0990	Wall Finishes & Painting/Coating	114.8	43.8	71.0	114.8	30.6	62.8	130.0	47.1	78.8	110.1	61.7	80.2	105.5	74.8	86.6	105.5	84.1	92.3
09	**FINISHES**	103.3	58.7	78.4	100.8	48.6	71.7	102.2	53.4	75.0	116.9	64.5	87.7	108.7	96.0	101.6	104.0	97.1	100.1
COVERS	**DIVS. 10 - 14, 25, 28, 41, 43, 44, 46**	100.0	76.5	95.3	100.0	76.3	95.3	100.0	55.1	91.1	100.0	81.0	96.2	100.0	96.6	99.3	100.0	101.0	100.2
21, 22, 23	**FIRE SUPPRESSION, PLUMBING & HVAC**	100.2	47.5	78.9	94.3	55.4	78.5	94.3	67.4	83.4	94.4	87.9	91.8	100.0	90.9	96.3	99.9	90.9	96.2
26, 27, 3370	**ELECTRICAL, COMMUNICATIONS & UTIL.**	96.0	40.4	67.3	94.9	43.9	68.6	94.6	45.6	69.3	99.9	92.2	96.0	103.8	82.3	92.7	98.1	97.0	97.6
MF2010	**WEIGHTED AVERAGE**	101.2	59.8	82.9	99.2	58.2	81.1	99.4	62.2	83.0	100.4	81.3	92.0	102.2	95.6	99.3	99.4	97.9	98.7

DIVISION		WASHINGTON																	
		RICHLAND			SEATTLE			SPOKANE			TACOMA			VANCOUVER			WENATCHEE		
		993			980 - 981,987			990 - 992			983 - 984			986			988		
		MAT.	INST.	TOTAL	MAT.	INST.	TOTAL	MAT.	INST.	TOTAL	MAT.	INST.	TOTAL	MAT.	INST.	TOTAL	MAT.	INST.	TOTAL
015433	CONTRACTOR EQUIPMENT		91.0	91.0		103.9	103.9		91.0	91.0		104.1	104.1		98.8	98.8		104.1	104.1
0241, 31 - 34	SITE & INFRASTRUCTURE, DEMOLITION	98.0	91.3	93.4	97.0	111.1	106.7	97.4	91.3	93.1	96.1	112.9	107.7	107.4	100.2	102.5	107.0	111.5	110.1
0310	Concrete Forming & Accessories	125.0	79.9	85.7	105.2	100.3	100.9	129.2	79.7	86.1	105.2	99.5	100.2	106.3	86.6	89.2	106.6	78.0	81.7
0320	Concrete Reinforcing	102.3	84.6	93.4	115.1	103.1	109.1	103.0	84.6	93.7	115.1	103.0	109.0	116.1	102.7	109.4	116.1	91.2	103.6
0330	Cast-in-Place Concrete	102.7	84.5	95.5	106.0	106.3	106.1	106.7	84.4	97.9	103.9	106.0	104.7	116.0	100.3	109.8	106.0	79.4	95.5
03	CONCRETE	99.0	82.4	90.9	101.2	102.5	101.8	101.4	82.2	92.1	100.1	101.9	101.0	110.1	94.3	102.4	107.3	81.1	94.5
04	MASONRY	105.0	81.9	90.9	116.4	99.3	106.0	105.6	81.9	91.2	116.2	99.3	105.9	117.3	95.8	104.1	119.0	81.3	96.0
05	METALS	87.6	82.9	86.1	102.5	95.1	100.1	89.9	82.7	87.6	102.5	93.0	99.4	100.0	93.5	97.8	100.0	86.7	95.6
06	WOOD, PLASTICS & COMPOSITES	101.2	78.7	88.2	97.9	99.0	98.5	109.7	78.7	91.8	96.5	99.0	98.0	88.4	83.7	85.7	97.6	76.5	85.4
07	THERMAL & MOISTURE PROTECTION	152.4	78.6	122.4	102.3	99.1	101.0	148.9	79.5	120.6	102.2	96.8	100.0	102.7	91.2	98.0	102.8	79.8	93.5
08	OPENINGS	109.0	72.5	99.8	103.4	100.2	102.6	109.6	72.9	100.3	101.8	100.2	101.4	99.5	90.0	97.1	101.3	72.9	94.1
0920	Plaster & Gypsum Board	130.2	77.9	92.8	109.6	98.9	101.9	130.6	77.9	92.9	112.5	98.9	102.8	110.2	83.4	91.0	114.1	75.7	86.7
0950, 0980	Ceilings & Acoustic Treatment	104.4	77.9	86.9	110.3	98.9	102.7	101.4	77.9	85.8	110.5	98.9	102.8	105.2	83.4	90.8	103.0	75.7	85.0
0960	Flooring	103.4	83.9	97.7	109.1	103.9	107.6	103.5	89.3	99.4	109.6	96.1	105.6	115.2	107.0	112.8	112.7	65.6	98.9
0970, 0990	Wall Finishes & Painting/Coating	110.1	61.7	80.2	105.5	84.1	92.3	110.1	61.7	80.3	105.5	84.1	92.3	108.9	65.2	82.0	105.5	61.7	78.5
09	FINISHES	118.3	78.3	96.0	106.6	98.7	102.2	117.9	79.4	96.4	107.0	97.1	101.5	104.9	87.7	95.3	107.6	73.2	88.4
COVERS	DIVS. 10 - 14, 25, 28, 41, 43, 44, 46	100.0	66.9	93.4	100.0	101.0	100.2	100.0	66.8	93.4	100.0	101.0	100.2	100.0	64.1	92.9	100.0	82.3	96.5
21, 22, 23	FIRE SUPPRESSION, PLUMBING & HVAC	100.4	100.1	100.3	100.0	122.1	109.0	100.3	89.0	95.7	100.1	90.9	96.4	100.2	93.0	97.3	94.1	85.4	90.6
26, 27, 3370	ELECTRICAL, COMMUNICATIONS & UTIL.	97.1	93.1	95.1	103.6	107.9	105.8	95.1	79.1	86.8	103.6	97.0	100.2	108.5	98.0	103.1	104.4	80.2	91.9
MF2010	WEIGHTED AVERAGE	102.1	86.8	95.4	102.5	106.3	104.2	102.5	82.7	93.7	102.2	97.9	100.3	103.5	93.0	98.8	101.8	83.6	93.8

DIVISION		WASHINGTON			WEST VIRGINIA														
		YAKIMA			BECKLEY			BLUEFIELD			BUCKHANNON			CHARLESTON			CLARKSBURG		
		989			258 - 259			247 - 248			262			250 - 253			263 - 264		
		MAT.	INST.	TOTAL	MAT.	INST.	TOTAL	MAT.	INST.	TOTAL	MAT.	INST.	TOTAL	MAT.	INST.	TOTAL	MAT.	INST.	TOTAL
015433	CONTRACTOR EQUIPMENT		104.1	104.1		99.9	99.9		99.9	99.9		99.9	99.9		99.9	99.9		99.9	99.9
0241, 31 - 34	SITE & INFRASTRUCTURE, DEMOLITION	98.9	112.3	108.2	104.9	86.9	92.5	104.7	86.9	92.4	111.5	86.7	94.4	102.7	88.4	92.9	112.1	86.7	94.6
0310	Concrete Forming & Accessories	105.5	95.1	96.5	84.6	86.5	86.3	85.8	86.2	86.1	85.2	84.5	84.6	99.5	88.5	89.9	82.9	84.4	84.2
0320	Concrete Reinforcing	115.7	91.3	103.4	92.7	84.2	88.4	92.7	66.3	79.4	93.3	83.1	88.2	94.0	84.4	89.2	93.3	83.7	88.4
0330	Cast-in-Place Concrete	111.0	84.2	100.4	115.7	102.7	110.6	115.7	102.6	110.5	115.4	100.0	109.3	107.0	100.9	104.6	126.4	100.0	116.0
03	CONCRETE	104.9	90.3	97.8	106.5	92.3	99.6	106.6	88.8	97.9	109.5	90.3	100.2	103.3	92.6	98.1	114.6	90.4	102.8
04	MASONRY	109.5	81.3	92.3	99.4	89.6	93.5	94.2	89.6	91.4	105.6	86.9	94.2	96.2	91.5	93.3	108.1	86.9	95.2
05	METALS	100.7	87.0	96.2	93.6	95.2	94.1	93.6	88.7	91.9	93.7	95.7	94.4	95.8	96.4	96.0	93.7	95.8	94.4
06	WOOD, PLASTICS & COMPOSITES	97.0	99.0	98.2	86.0	87.8	87.1	87.7	87.8	87.8	87.0	83.3	84.9	102.6	87.8	94.1	83.9	83.3	83.6
07	THERMAL & MOISTURE PROTECTION	102.3	81.4	93.8	100.2	87.2	94.9	100.2	87.2	94.9	100.5	86.7	94.9	97.3	87.7	93.4	100.4	86.7	94.8
08	OPENINGS	101.3	81.7	96.4	98.0	81.2	93.7	98.8	77.1	93.4	98.9	78.5	93.8	98.5	81.2	94.2	98.9	80.5	94.2
0920	Plaster & Gypsum Board	112.3	98.9	102.7	101.5	87.3	91.4	102.4	87.3	91.6	103.0	82.7	88.5	100.5	87.3	91.1	100.7	82.7	87.8
0950, 0980	Ceilings & Acoustic Treatment	105.0	98.9	101.0	91.9	87.3	88.9	91.9	87.3	88.9	93.7	82.7	86.4	95.5	87.3	90.1	93.7	82.7	86.4
0960	Flooring	110.8	60.7	96.1	98.9	110.4	102.3	99.5	110.4	102.7	99.2	104.1	100.6	107.3	110.4	108.2	98.0	104.1	99.8
0970, 0990	Wall Finishes & Painting/Coating	105.5	61.7	78.5	114.8	36.9	66.7	114.8	36.9	66.7	114.8	89.3	99.0	114.8	89.5	99.2	114.8	89.3	99.0
09	FINISHES	106.4	85.5	94.8	98.6	86.4	91.8	98.9	86.4	91.9	99.9	88.4	93.5	101.2	92.7	96.5	99.2	88.4	93.2
COVERS	DIVS. 10 - 14, 25, 28, 41, 43, 44, 46	100.0	98.2	99.6	100.0	57.3	91.5	100.0	57.2	91.5	100.0	94.6	98.9	100.0	95.6	99.1	100.0	94.6	98.9
21, 22, 23	FIRE SUPPRESSION, PLUMBING & HVAC	100.1	100.2	100.1	94.3	80.1	88.6	94.3	79.7	88.4	94.3	88.0	91.7	100.0	82.2	92.8	94.3	87.7	91.6
26, 27, 3370	ELECTRICAL, COMMUNICATIONS & UTIL.	106.6	93.1	99.6	92.4	90.9	91.6	95.0	71.5	82.9	96.3	91.5	93.8	97.2	90.9	93.9	96.3	91.5	93.8
MF2010	WEIGHTED AVERAGE	102.5	92.6	98.1	97.3	86.6	92.6	97.4	82.6	90.9	98.8	89.0	94.5	99.2	89.5	94.9	99.4	89.1	94.9

DIVISION		WEST VIRGINIA																	
		GASSAWAY			HUNTINGTON			LEWISBURG			MARTINSBURG			MORGANTOWN			PARKERSBURG		
		266			255 - 257			249			254			265			261		
		MAT.	INST.	TOTAL	MAT.	INST.	TOTAL	MAT.	INST.	TOTAL	MAT.	INST.	TOTAL	MAT.	INST.	TOTAL	MAT.	INST.	TOTAL
015433	CONTRACTOR EQUIPMENT		99.9	99.9		99.9	99.9		99.9	99.9		99.9	99.9		99.9	99.9		99.9	99.9
0241, 31 - 34	SITE & INFRASTRUCTURE, DEMOLITION	108.5	87.2	93.8	109.3	89.4	95.5	122.2	86.9	97.9	108.8	84.9	92.3	105.6	87.0	92.8	114.4	88.4	96.5
0310	Concrete Forming & Accessories	84.7	84.6	84.6	95.4	91.3	91.8	83.2	86.2	85.8	84.6	76.9	77.9	83.2	84.1	84.0	86.9	86.0	86.1
0320	Concrete Reinforcing	93.3	84.4	88.8	94.0	90.6	92.3	93.3	66.3	79.7	92.7	74.1	83.3	93.3	83.7	88.4	92.7	83.0	87.8
0330	Cast-in-Place Concrete	120.8	100.1	112.6	126.2	105.1	117.9	115.8	102.6	110.6	121.2	89.7	108.8	115.3	99.9	109.2	117.9	91.5	107.5
03	CONCRETE	110.4	90.6	100.8	112.7	96.4	104.7	115.9	88.8	102.7	110.5	81.5	96.4	106.3	90.2	98.5	111.7	88.0	100.1
04	MASONRY	109.2	90.1	97.6	99.9	91.8	95.0	98.6	89.6	93.1	101.2	79.5	87.9	128.6	86.9	103.2	84.3	84.7	84.5
05	METALS	93.7	96.4	94.6	95.9	98.6	96.8	93.7	88.7	92.0	93.9	85.3	91.1	93.8	95.5	94.4	94.4	95.5	94.8
06	WOOD, PLASTICS & COMPOSITES	86.2	83.3	84.5	96.6	89.9	92.7	84.2	87.8	86.3	86.0	78.1	81.5	84.2	83.3	83.7	87.1	84.9	85.8
07	THERMAL & MOISTURE PROTECTION	100.2	87.5	95.1	100.3	88.4	95.5	101.1	87.2	95.5	100.4	72.7	89.2	100.3	86.7	94.8	100.1	86.0	94.4
08	OPENINGS	96.8	78.8	92.2	97.3	83.8	93.9	98.9	77.1	93.4	100.2	66.9	91.8	100.3	80.5	95.3	97.9	79.4	93.2
0920	Plaster & Gypsum Board	102.1	82.7	88.2	109.3	89.5	95.2	100.7	87.3	91.1	102.1	77.3	84.4	100.7	82.7	87.8	103.4	84.3	89.7
0950, 0980	Ceilings & Acoustic Treatment	93.7	82.7	86.4	93.7	89.5	90.9	93.7	87.3	89.5	93.7	77.3	82.9	93.7	82.7	86.4	93.7	84.3	87.5
0960	Flooring	99.0	110.4	102.3	107.1	105.1	106.5	98.1	110.4	101.7	98.9	56.2	86.4	98.1	104.1	99.9	102.3	96.6	100.6
0970, 0990	Wall Finishes & Painting/Coating	114.8	89.5	99.2	114.8	83.2	95.3	114.8	36.9	66.7	114.8	43.6	70.9	114.8	89.3	99.0	114.8	88.4	98.5
09	FINISHES	99.4	89.7	94.0	102.3	92.9	97.1	100.5	86.4	92.6	99.4	70.1	83.0	98.7	88.4	93.0	100.9	88.0	93.7
COVERS	DIVS. 10 - 14, 25, 28, 41, 43, 44, 46	100.0	94.6	98.9	100.0	96.9	99.4	100.0	57.2	91.5	100.0	55.4	91.1	100.0	66.6	93.4	100.0	95.1	99.0
21, 22, 23	FIRE SUPPRESSION, PLUMBING & HVAC	94.3	88.8	92.1	100.2	84.4	93.8	94.3	80.0	88.5	94.3	82.6	89.6	94.3	87.6	91.6	100.2	87.8	95.1
26, 27, 3370	ELECTRICAL, COMMUNICATIONS & UTIL.	96.3	90.9	93.5	96.0	93.9	94.9	92.4	71.5	81.6	98.0	81.6	89.6	96.5	91.5	93.9	96.4	90.3	93.3
MF2010	WEIGHTED AVERAGE	98.7	89.8	94.8	100.6	91.4	96.6	99.1	82.7	91.9	98.9	79.1	90.2	99.5	88.2	94.5	99.6	88.4	94.7

DIVISION		WEST VIRGINIA									WISCONSIN								
		PETERSBURG			ROMNEY			WHEELING			BELOIT			EAU CLAIRE			GREEN BAY		
		268			267			260			535			547			541 - 543		
		MAT.	INST.	TOTAL	MAT.	INST.	TOTAL	MAT.	INST.	TOTAL	MAT.	INST.	TOTAL	MAT.	INST.	TOTAL	MAT.	INST.	TOTAL
015433	**CONTRACTOR EQUIPMENT**		99.9	99.9		99.9	99.9		99.9	99.9		100.9	100.9		100.8	100.8		98.7	98.7
0241, 31 - 34	**SITE & INFRASTRUCTURE, DEMOLITION**	104.9	87.3	92.8	108.0	87.3	93.7	115.1	87.6	96.1	97.8	106.2	103.6	98.6	103.4	101.9	102.8	99.5	100.5
0310	Concrete Forming & Accessories	86.0	79.7	80.5	82.6	79.8	80.1	88.6	84.5	85.1	103.3	94.6	95.7	101.9	96.3	97.0	107.3	95.6	97.1
0320	Concrete Reinforcing	92.7	74.6	83.6	93.3	71.8	82.5	92.1	83.7	87.9	102.0	118.4	110.3	100.1	101.7	100.9	98.1	88.4	93.2
0330	Cast-in-Place Concrete	115.3	94.0	106.9	120.8	94.0	110.2	117.9	100.1	110.9	103.5	92.8	99.3	103.6	93.5	99.6	107.2	98.0	103.6
03	**CONCRETE**	106.4	84.5	95.7	110.3	84.1	97.5	111.7	90.5	101.3	99.7	98.6	99.2	101.1	96.6	98.9	104.0	95.4	99.8
04	**MASONRY**	101.2	86.9	92.5	99.0	83.7	89.7	108.9	86.9	95.5	104.6	93.9	98.1	92.6	97.9	95.8	124.4	97.2	107.8
05	**METALS**	93.8	91.8	93.2	93.9	90.8	92.9	94.6	95.9	95.0	95.9	108.3	100.0	95.7	102.8	98.1	97.9	96.9	97.6
06	**WOOD, PLASTICS & COMPOSITES**	88.0	78.1	82.3	83.3	78.1	80.3	88.6	83.3	85.5	101.5	97.0	98.9	108.3	97.1	101.8	109.3	97.1	102.2
07	**THERMAL & MOISTURE PROTECTION**	100.3	75.1	90.1	100.4	74.2	89.8	100.5	87.3	95.2	105.8	91.4	99.9	106.3	84.9	97.6	107.8	85.2	98.6
08	**OPENINGS**	100.3	73.8	93.6	100.2	73.1	93.4	98.8	80.5	94.2	102.1	105.3	102.9	99.6	95.9	98.7	97.8	94.6	97.0
0920	Plaster & Gypsum Board	103.0	77.3	84.6	100.3	77.3	83.8	103.4	82.7	88.6	100.3	97.3	98.1	105.2	97.3	99.5	96.6	97.3	97.1
0950, 0980	Ceilings & Acoustic Treatment	93.7	77.3	82.9	93.7	77.3	82.9	93.7	82.7	86.4	87.5	97.3	93.9	101.5	97.3	98.7	93.4	97.3	96.0
0960	Flooring	100.0	104.1	101.2	97.9	97.8	97.9	103.3	104.1	103.5	104.5	117.5	108.3	96.3	113.3	101.3	115.9	113.3	115.1
0970, 0990	Wall Finishes & Painting/Coating	114.8	89.3	99.0	114.8	89.3	99.0	114.8	89.3	99.0	91.9	97.4	95.3	88.4	91.1	90.1	98.0	80.3	87.1
09	**FINISHES**	99.5	85.4	91.6	98.9	84.1	90.6	101.2	88.4	94.1	100.3	99.3	99.7	99.7	99.3	99.5	104.0	97.9	100.6
COVERS	**DIVS. 10 - 14, 25, 28, 41, 43, 44, 46**	100.0	66.0	93.2	100.0	94.0	98.8	100.0	96.2	99.2	100.0	89.4	97.9	100.0	91.8	98.4	100.0	91.5	98.3
21, 22, 23	**FIRE SUPPRESSION, PLUMBING & HVAC**	94.3	88.7	92.0	94.3	88.6	92.0	100.2	87.7	95.2	99.9	89.1	95.5	100.2	84.8	93.9	100.4	82.9	93.3
26, 27, 3370	**ELECTRICAL, COMMUNICATIONS & UTIL.**	99.6	81.7	90.4	98.8	81.7	90.0	93.7	91.5	92.6	101.1	87.5	94.1	103.7	90.2	96.7	98.4	82.8	90.3
MF2010	**WEIGHTED AVERAGE**	98.5	84.9	92.5	98.7	85.1	92.7	100.7	89.2	95.7	100.0	96.0	98.3	99.7	94.3	97.3	101.8	91.6	97.3

DIVISION		WISCONSIN																	
		KENOSHA			LA CROSSE			LANCASTER			MADISON			MILWAUKEE			NEW RICHMOND		
		531			546			538			537			530,532			540		
		MAT.	INST.	TOTAL	MAT.	INST.	TOTAL	MAT.	INST.	TOTAL	MAT.	INST.	TOTAL	MAT.	INST.	TOTAL	MAT.	INST.	TOTAL
015433	**CONTRACTOR EQUIPMENT**		98.8	98.8		100.8	100.8		100.9	100.9		100.9	100.9		89.3	89.3		101.4	101.4
0241, 31 - 34	**SITE & INFRASTRUCTURE, DEMOLITION**	104.7	103.6	103.9	91.8	103.5	99.9	97.2	106.6	103.7	93.3	106.7	102.5	98.6	97.2	97.6	93.7	103.7	100.6
0310	Concrete Forming & Accessories	107.0	103.3	103.7	90.2	96.4	95.6	102.7	95.4	96.4	102.8	97.2	97.9	105.4	120.2	118.3	98.0	103.1	102.4
0320	Concrete Reinforcing	101.8	101.1	101.5	99.7	90.6	95.1	103.4	90.4	96.9	102.0	90.8	96.4	102.0	101.7	101.8	97.1	101.6	99.3
0330	Cast-in-Place Concrete	112.7	97.4	106.7	93.2	91.0	92.3	102.9	96.5	100.4	94.6	97.2	95.6	101.1	113.3	105.9	107.9	81.0	97.3
03	**CONCRETE**	104.5	100.9	102.7	92.3	93.7	93.0	99.5	95.0	97.3	95.2	96.1	95.6	98.8	113.4	105.9	99.3	95.3	97.4
04	**MASONRY**	101.9	107.0	105.0	91.7	99.8	96.7	104.7	99.6	101.6	102.0	102.1	102.1	104.5	122.3	115.4	119.1	94.3	104.0
05	**METALS**	96.6	102.4	98.5	95.6	97.6	96.3	93.4	94.5	93.8	98.5	96.5	97.8	99.2	96.4	98.3	95.8	101.9	97.8
06	**WOOD, PLASTICS & COMPOSITES**	102.0	102.5	102.3	95.1	97.1	96.2	100.9	97.0	98.6	98.8	97.0	97.7	103.3	120.5	113.2	97.7	109.0	104.2
07	**THERMAL & MOISTURE PROTECTION**	105.4	96.3	101.7	105.9	84.5	97.2	105.6	79.8	95.1	102.4	90.8	97.7	103.7	115.0	108.3	107.8	92.1	101.4
08	**OPENINGS**	97.5	105.6	99.5	99.6	83.8	95.6	97.2	84.6	94.0	102.1	97.5	100.9	104.3	115.3	107.1	87.7	102.4	91.4
0920	Plaster & Gypsum Board	88.1	103.0	98.8	100.1	97.3	98.1	99.0	97.3	97.8	102.7	97.3	98.8	101.7	121.4	115.7	87.2	109.8	103.3
0950, 0980	Ceilings & Acoustic Treatment	87.5	103.0	97.7	99.7	97.3	98.1	82.9	97.3	92.4	92.2	97.3	95.5	92.0	121.4	111.4	67.0	109.8	95.3
0960	Flooring	123.0	112.2	119.8	89.2	113.3	96.2	103.7	106.2	104.5	98.6	106.2	100.8	107.5	119.4	111.0	108.5	113.3	109.9
0970, 0990	Wall Finishes & Painting/Coating	100.5	103.6	102.4	88.4	72.5	78.6	91.9	70.0	78.4	91.9	97.4	95.3	93.9	117.5	108.5	101.7	91.1	95.2
09	**FINISHES**	105.2	105.4	105.3	96.1	97.7	97.0	98.9	94.4	96.4	99.4	99.2	99.3	102.5	120.6	112.6	94.5	103.7	99.6
COVERS	**DIVS. 10 - 14, 25, 28, 41, 43, 44, 46**	100.0	99.3	99.9	100.0	92.4	98.5	100.0	51.0	90.3	100.0	92.2	98.5	100.0	104.2	100.8	100.0	89.5	97.9
21, 22, 23	**FIRE SUPPRESSION, PLUMBING & HVAC**	100.1	94.8	97.9	100.2	85.5	94.3	93.9	85.3	90.4	99.8	94.6	97.7	99.9	103.7	101.4	93.7	82.2	89.1
26, 27, 3370	**ELECTRICAL, COMMUNICATIONS & UTIL.**	101.7	99.4	100.5	104.0	90.4	97.0	100.9	90.4	95.5	99.4	94.1	96.7	102.6	100.1	101.3	102.0	90.4	96.0
MF2010	**WEIGHTED AVERAGE**	100.8	101.0	100.9	98.1	93.2	95.9	97.6	91.5	94.9	99.3	97.2	98.4	100.9	108.3	104.2	97.3	94.4	96.0

DIVISION		WISCONSIN																	
		OSHKOSH			PORTAGE			RACINE			RHINELANDER			SUPERIOR			WAUSAU		
		549			539			534			545			548			544		
		MAT.	INST.	TOTAL	MAT.	INST.	TOTAL	MAT.	INST.	TOTAL	MAT.	INST.	TOTAL	MAT.	INST.	TOTAL	MAT.	INST.	TOTAL
015433	**CONTRACTOR EQUIPMENT**		98.7	98.7		100.9	100.9		100.9	100.9		98.7	98.7		101.4	101.4		98.7	98.7
0241, 31 - 34	**SITE & INFRASTRUCTURE, DEMOLITION**	93.9	98.5	97.0	87.1	106.6	100.6	97.6	107.3	104.3	107.9	99.6	102.2	90.2	104.1	99.7	89.3	99.5	96.3
0310	Concrete Forming & Accessories	92.2	84.6	85.6	96.3	96.3	96.3	103.6	103.3	103.3	89.9	95.5	94.8	96.5	92.7	93.2	91.6	95.6	95.1
0320	Concrete Reinforcing	98.2	77.7	87.9	103.5	90.5	97.0	102.0	101.1	101.6	98.4	92.7	95.5	97.1	93.7	95.4	98.4	90.6	94.4
0330	Cast-in-Place Concrete	99.4	88.3	95.0	88.3	98.7	92.4	101.5	97.2	99.8	112.7	98.7	107.2	101.6	96.9	99.8	92.6	98.0	94.7
03	**CONCRETE**	94.0	85.1	89.7	87.6	96.2	91.8	98.7	100.8	99.7	106.2	96.3	101.4	93.8	94.7	94.2	89.0	95.8	92.3
04	**MASONRY**	105.6	87.9	94.8	103.6	99.5	101.1	104.6	107.0	106.0	122.5	99.1	108.2	118.4	104.0	109.6	105.1	97.2	100.3
05	**METALS**	96.0	89.8	94.0	94.1	95.4	94.5	97.5	102.4	99.1	95.9	97.9	96.5	96.7	99.7	97.7	95.8	97.8	96.4
06	**WOOD, PLASTICS & COMPOSITES**	91.8	86.2	88.5	93.0	97.0	95.3	101.7	102.5	102.2	89.6	97.1	93.9	96.2	91.2	93.3	91.3	97.1	94.6
07	**THERMAL & MOISTURE PROTECTION**	106.9	78.3	95.3	105.1	89.3	98.7	105.8	95.9	101.8	107.7	83.5	97.9	107.5	95.1	102.5	106.8	84.2	97.6
08	**OPENINGS**	93.2	83.2	90.7	97.4	91.6	95.9	102.1	105.6	103.0	93.2	86.8	91.6	87.5	92.3	88.7	93.4	95.2	93.9
0920	Plaster & Gypsum Board	85.7	86.0	85.9	93.6	97.3	96.2	100.3	103.0	102.2	85.7	97.3	94.0	87.0	91.5	90.2	85.7	97.3	94.0
0950, 0980	Ceilings & Acoustic Treatment	93.4	86.0	88.5	84.7	97.3	93.0	87.5	103.0	97.7	93.4	97.3	96.0	67.9	91.5	83.5	93.4	97.3	96.0
0960	Flooring	106.5	99.7	104.5	99.6	112.7	103.5	104.5	112.2	106.7	105.6	113.3	107.9	110.0	122.5	113.6	106.3	113.3	108.4
0970, 0990	Wall Finishes & Painting/Coating	96.0	68.3	78.9	91.9	70.0	78.4	91.9	108.7	102.2	96.0	63.6	76.0	90.2	100.4	96.5	96.0	80.3	86.3
09	**FINISHES**	99.1	85.2	91.4	96.7	96.4	96.5	100.3	105.9	103.4	100.1	95.5	97.5	94.0	98.9	96.7	98.7	97.9	98.3
COVERS	**DIVS. 10 - 14, 25, 28, 41, 43, 44, 46**	100.0	64.1	92.8	100.0	65.9	93.2	100.0	99.3	99.9	100.0	56.9	91.4	100.0	89.2	97.8	100.0	91.5	98.3
21, 22, 23	**FIRE SUPPRESSION, PLUMBING & HVAC**	94.5	74.6	86.4	93.9	90.2	92.4	99.9	94.8	97.8	94.5	84.8	90.5	93.7	92.3	93.1	94.5	83.6	90.0
26, 27, 3370	**ELECTRICAL, COMMUNICATIONS & UTIL.**	102.6	76.2	89.0	104.8	93.7	99.0	100.9	97.3	99.1	101.9	81.4	91.3	107.0	96.9	101.8	103.7	81.4	92.2
MF2010	**WEIGHTED AVERAGE**	97.2	82.6	90.8	96.1	94.5	95.4	100.1	101.0	100.5	99.8	90.5	95.7	97.1	96.9	97.0	96.5	91.7	94.4

	DIVISION	WYOMING																	
		CASPER			CHEYENNE			NEWCASTLE			RAWLINS			RIVERTON			ROCK SPRINGS		
		826			820			827			823			825			829 - 831		
		MAT.	INST.	TOTAL	MAT.	INST.	TOTAL	MAT.	INST.	TOTAL	MAT.	INST.	TOTAL	MAT.	INST.	TOTAL	MAT.	INST.	TOTAL
015433	CONTRACTOR EQUIPMENT		99.3	99.3		99.3	99.3		99.3	99.3		99.3	99.3		99.3	99.3		99.3	99.3
0241, 31 - 34	SITE & INFRASTRUCTURE, DEMOLITION	96.2	97.4	97.0	88.5	97.4	94.7	79.6	96.4	91.2	94.4	96.4	95.8	87.6	96.4	93.6	82.9	96.4	92.2
0310	Concrete Forming & Accessories	103.0	41.3	49.3	104.6	56.0	62.3	96.0	43.8	50.6	99.8	43.8	51.1	95.0	43.8	50.5	101.4	43.8	51.3
0320	Concrete Reinforcing	106.8	64.1	85.3	100.8	64.9	82.7	108.3	64.7	86.3	107.9	64.7	86.1	108.9	64.7	86.6	108.9	66.3	87.4
0330	Cast-in-Place Concrete	92.9	74.7	85.7	91.3	74.9	84.8	92.2	51.9	76.3	92.3	51.9	76.4	92.3	51.9	76.4	92.2	51.9	76.3
03	CONCRETE	99.6	58.0	79.3	98.3	64.7	81.9	99.0	51.3	75.8	111.4	51.3	82.2	106.5	51.3	79.6	99.5	51.6	76.2
04	MASONRY	101.3	35.8	61.4	98.1	49.4	68.4	94.8	44.3	64.0	94.8	44.3	64.0	94.8	44.3	64.0	152.6	44.3	86.5
05	METALS	97.6	68.4	87.9	98.8	69.8	89.2	93.9	67.7	85.2	93.9	67.7	85.2	94.0	67.7	85.3	94.7	68.3	86.0
06	WOOD, PLASTICS & COMPOSITES	91.8	38.4	60.9	93.5	57.8	72.9	84.3	44.8	61.5	87.9	44.8	63.0	83.3	44.8	61.0	91.2	44.8	64.4
07	THERMAL & MOISTURE PROTECTION	97.8	47.5	77.3	99.5	53.1	80.6	100.8	46.6	78.8	102.5	46.6	79.8	101.9	46.6	79.4	100.9	46.6	78.8
08	OPENINGS	96.3	43.2	83.0	98.6	53.8	87.3	102.8	46.7	88.7	102.5	46.7	88.4	102.7	46.7	88.6	103.2	47.1	89.1
0920	Plaster & Gypsum Board	100.7	36.4	54.8	89.0	56.4	65.7	83.7	42.9	54.6	84.1	42.9	54.7	83.7	42.9	54.6	92.8	42.9	57.2
0950, 0980	Ceilings & Acoustic Treatment	114.2	36.4	62.8	106.7	56.4	73.5	108.8	42.9	65.3	108.8	42.9	65.3	108.8	42.9	65.3	108.8	42.9	65.3
0960	Flooring	99.6	34.6	80.5	104.1	55.5	89.8	97.6	48.8	83.3	100.6	48.8	85.4	97.0	48.8	82.8	102.9	48.8	87.1
0970, 0990	Wall Finishes & Painting/Coating	98.8	45.4	65.9	102.3	45.4	67.2	98.1	29.1	55.5	98.1	29.1	55.5	98.1	29.1	55.5	98.1	29.1	55.5
09	FINISHES	105.1	39.1	68.3	104.1	54.8	76.6	98.5	42.4	67.2	101.2	42.4	68.4	99.4	42.4	67.6	101.1	42.4	68.4
COVERS	DIVS. 10 - 14, 25, 28, 41, 43, 44, 46	100.0	74.3	94.9	100.0	76.5	95.3	100.0	57.2	91.5	100.0	57.2	91.5	100.0	57.2	91.5	100.0	57.2	91.5
21, 22, 23	FIRE SUPPRESSION, PLUMBING & HVAC	100.0	55.0	81.8	100.0	55.0	81.8	97.0	52.6	79.0	97.0	52.6	79.0	97.0	52.6	79.0	100.0	52.7	80.8
26, 27, 3370	ELECTRICAL, COMMUNICATIONS & UTIL.	94.9	66.9	80.5	97.6	72.9	84.9	95.1	66.8	80.5	95.1	66.8	80.5	95.1	66.9	80.5	93.4	66.9	79.7
MF2010	WEIGHTED AVERAGE	98.9	57.7	80.7	99.1	63.8	83.5	96.9	57.0	79.3	99.0	57.0	80.5	98.1	57.0	80.0	101.0	57.2	81.7

	DIVISION	WYOMING												CANADA					
		SHERIDAN			WHEATLAND			WORLAND			YELLOWSTONE NAT'L PA			BARRIE, ONTARIO			BATHURST, NEW BRUNSWICK		
		828			822			824			821								
		MAT.	INST.	TOTAL	MAT.	INST.	TOTAL	MAT.	INST.	TOTAL	MAT.	INST.	TOTAL	MAT.	INST.	TOTAL	MAT.	INST.	TOTAL
015433	CONTRACTOR EQUIPMENT		99.3	99.3		99.3	99.3		99.3	99.3		99.3	99.3		101.7	101.7		102.0	102.0
0241, 31 - 34	SITE & INFRASTRUCTURE, DEMOLITION	88.1	97.4	94.5	83.8	96.4	92.5	81.2	96.4	91.7	81.3	96.7	91.9	114.9	101.4	105.6	94.6	97.7	96.7
0310	Concrete Forming & Accessories	102.6	46.2	53.5	97.9	44.2	51.2	97.9	43.8	50.9	97.9	46.1	52.9	123.2	93.6	97.5	103.1	66.6	71.4
0320	Concrete Reinforcing	108.9	64.8	86.7	108.3	64.7	86.3	108.9	64.7	86.6	110.8	69.6	90.0	180.2	87.6	133.5	148.8	59.1	103.6
0330	Cast-in-Place Concrete	95.4	74.8	87.3	96.2	52.4	79.0	92.2	51.9	76.3	92.2	55.2	77.6	167.6	92.3	137.9	138.2	64.1	109.0
03	CONCRETE	106.6	60.3	84.1	103.7	51.7	78.3	99.2	51.3	75.9	99.5	54.4	77.5	153.4	92.1	123.5	135.8	65.0	101.3
04	MASONRY	95.1	48.8	66.9	95.1	45.2	64.7	94.8	44.3	64.0	94.9	50.3	67.7	167.0	102.0	127.3	158.0	67.6	102.8
05	METALS	97.5	69.6	88.3	93.8	67.7	85.2	94.0	67.7	85.3	94.6	69.5	86.3	107.3	92.6	102.5	104.7	73.5	94.4
06	WOOD, PLASTICS & COMPOSITES	94.7	44.8	65.9	85.9	44.8	62.2	86.0	44.8	62.2	86.0	44.8	62.2	117.4	92.2	102.8	94.7	66.9	78.6
07	THERMAL & MOISTURE PROTECTION	101.9	51.5	81.4	101.1	47.5	79.3	100.9	46.6	78.8	100.4	49.7	79.8	111.6	93.1	104.0	105.0	65.1	88.8
08	OPENINGS	103.5	47.6	89.4	101.4	46.7	87.6	103.0	46.7	88.9	95.9	47.9	83.8	92.4	90.3	91.8	87.4	58.3	80.1
0920	Plaster & Gypsum Board	111.2	42.9	62.5	83.7	42.9	54.6	83.7	42.9	54.6	84.0	42.9	54.7	155.0	91.9	109.9	162.1	65.7	93.3
0950, 0980	Ceilings & Acoustic Treatment	111.8	42.9	66.3	108.8	42.9	65.3	108.8	42.9	65.3	109.7	42.9	65.6	95.8	91.9	93.2	96.7	65.7	76.2
0960	Flooring	100.8	48.8	85.5	99.2	48.8	84.4	99.2	48.8	84.4	99.2	48.8	84.4	132.5	101.8	123.5	115.8	48.6	96.1
0970, 0990	Wall Finishes & Painting/Coating	100.4	45.4	66.5	98.1	29.1	55.5	98.1	29.1	55.5	98.1	29.1	55.5	112.8	93.1	100.7	112.0	53.5	75.9
09	FINISHES	106.5	45.4	72.4	99.4	42.6	67.7	99.0	42.4	67.4	99.2	43.9	68.3	118.4	95.4	105.5	112.6	62.1	84.5
COVERS	DIVS. 10 - 14, 25, 28, 41, 43, 44, 46	100.0	75.0	95.0	100.0	57.5	91.5	100.0	57.2	91.5	100.0	59.2	91.9	140.0	79.2	127.9	140.5	69.5	126.4
21, 22, 23	FIRE SUPPRESSION, PLUMBING & HVAC	97.0	55.0	80.0	97.0	53.1	79.2	97.0	52.6	79.0	97.0	55.7	80.3	103.0	104.9	103.8	103.2	73.7	91.3
26, 27, 3370	ELECTRICAL, COMMUNICATIONS & UTIL.	97.8	66.9	81.9	95.1	72.8	83.6	95.1	66.9	80.5	94.1	66.9	80.1	119.3	94.5	106.5	119.9	64.6	91.4
MF2010	WEIGHTED AVERAGE	99.7	60.6	82.4	97.5	58.2	80.2	97.1	57.0	79.4	96.4	59.3	80.1	118.6	96.9	109.1	114.1	70.1	94.7

	DIVISION	CANADA																	
		BRANDON, MANITOBA			BRANTFORD, ONTARIO			BRIDGEWATER, NOVA SCOTIA			CALGARY, ALBERTA			CAP-DE-LA-MADELEINE, QUEBEC			CHARLESBOURG, QUEBEC		
		MAT.	INST.	TOTAL	MAT.	INST.	TOTAL	MAT.	INST.	TOTAL	MAT.	INST.	TOTAL	MAT.	INST.	TOTAL	MAT.	INST.	TOTAL
015433	CONTRACTOR EQUIPMENT		103.9	103.9		101.7	101.7		101.7	101.7		106.4	106.4		102.6	102.6		102.6	102.6
0241, 31 - 34	SITE & INFRASTRUCTURE, DEMOLITION	110.7	100.3	103.5	112.7	101.7	105.1	97.0	99.3	98.5	125.9	105.2	111.6	92.7	100.9	98.4	92.7	100.9	98.4
0310	Concrete Forming & Accessories	122.5	75.8	81.9	123.0	101.4	104.2	97.4	75.0	77.9	127.5	96.2	100.3	128.1	92.1	96.8	128.1	92.1	96.8
0320	Concrete Reinforcing	161.8	55.6	108.3	175.1	86.3	130.3	148.8	48.8	98.4	143.8	72.5	107.9	148.8	82.4	115.3	148.8	82.4	115.3
0330	Cast-in-Place Concrete	151.1	79.4	122.8	163.5	114.5	144.2	168.1	74.8	131.3	201.8	106.1	164.1	131.9	101.5	119.9	131.9	101.5	119.9
03	CONCRETE	135.9	73.9	105.7	150.5	103.0	127.4	148.4	70.7	110.5	167.2	95.3	132.2	130.9	93.6	112.7	130.9	93.6	112.7
04	MASONRY	161.7	70.4	106.0	163.3	106.7	128.8	158.8	74.8	107.6	212.7	90.7	138.3	159.0	90.6	117.3	159.0	90.6	117.3
05	METALS	107.5	78.8	98.0	106.7	93.4	102.3	105.8	76.3	96.1	143.7	89.0	125.6	105.0	90.0	100.1	105.0	90.0	100.1
06	WOOD, PLASTICS & COMPOSITES	112.9	76.8	92.0	115.2	100.1	106.5	87.4	74.3	79.8	113.5	96.0	103.4	126.1	92.0	106.4	126.1	92.0	106.4
07	THERMAL & MOISTURE PROTECTION	105.2	75.0	93.0	111.5	98.9	106.3	106.5	73.1	92.9	123.7	94.0	111.6	105.4	93.9	100.7	105.4	93.9	100.7
08	OPENINGS	93.4	67.6	86.9	92.2	96.7	93.4	83.9	67.7	79.9	93.4	85.7	91.4	93.4	84.6	91.2	93.4	84.6	91.2
0920	Plaster & Gypsum Board	130.0	75.6	91.2	155.0	100.1	115.8	155.9	73.4	96.9	171.9	95.4	117.2	185.9	91.5	118.5	185.9	91.5	118.5
0950, 0980	Ceilings & Acoustic Treatment	95.8	75.6	82.5	95.8	100.1	98.6	95.8	73.4	81.0	151.3	95.4	114.4	95.8	91.5	93.0	95.8	91.5	93.0
0960	Flooring	132.5	72.3	114.8	132.5	101.8	123.5	112.1	69.3	99.5	134.6	95.4	123.1	132.5	102.5	123.7	132.5	102.5	123.7
0970, 0990	Wall Finishes & Painting/Coating	112.0	60.8	80.5	112.0	102.2	106.0	112.0	66.5	83.9	111.9	109.7	110.5	112.0	95.0	101.5	112.0	95.0	101.5
09	FINISHES	114.5	74.0	91.9	117.1	102.0	108.7	110.3	73.2	89.6	133.8	98.0	113.8	119.7	94.6	105.7	119.7	94.6	105.7
COVERS	DIVS. 10 - 14, 25, 28, 41, 43, 44, 46	140.5	71.8	126.9	140.5	81.5	128.8	140.5	70.8	126.7	140.5	101.4	132.8	140.5	90.7	130.6	140.5	90.7	130.6
21, 22, 23	FIRE SUPPRESSION, PLUMBING & HVAC	103.2	88.8	97.4	103.2	108.0	105.2	103.2	89.2	97.5	100.5	92.6	97.3	103.8	96.1	100.7	103.8	96.1	100.7
26, 27, 3370	ELECTRICAL, COMMUNICATIONS & UTIL.	122.0	72.8	96.6	117.2	94.0	105.2	123.2	67.3	94.4	123.5	84.2	103.2	116.9	75.9	95.8	116.9	75.9	95.8
MF2010	WEIGHTED AVERAGE	116.2	79.0	99.8	117.7	101.0	110.4	115.6	77.8	98.9	129.8	93.1	113.6	114.8	91.3	104.4	114.8	91.3	104.4

	DIVISION	CANADA																	
		CHARLOTTETOWN, PRINCE EDWARD ISLAND			CHICOUTIMI, QUEBEC			CORNER BROOK, NEWFOUNDLAND			CORNWALL, ONTARIO			DALHOUSIE, NEW BRUNSWICK			DARTMOUTH, NOVA SCOTIA		
		MAT.	INST.	TOTAL	MAT.	INST.	TOTAL	MAT.	INST.	TOTAL	MAT.	INST.	TOTAL	MAT.	INST.	TOTAL	MAT.	INST.	TOTAL
015433	CONTRACTOR EQUIPMENT		101.6	101.6		102.6	102.6		103.3	103.3		101.7	101.7		102.0	102.0		101.7	101.7
0241, 31 - 34	SITE & INFRASTRUCTURE, DEMOLITION	108.2	96.8	100.3	92.2	99.7	97.4	112.8	98.3	102.8	110.8	101.2	104.2	94.5	97.7	96.7	104.7	99.3	100.9
0310	Concrete Forming & Accessories	97.8	60.9	65.7	128.1	89.7	94.7	104.5	64.5	69.7	121.0	93.8	97.3	103.1	66.9	71.6	97.5	75.0	77.9
0320	Concrete Reinforcing	155.0	48.9	101.5	112.2	92.1	102.1	148.8	50.8	99.4	175.1	86.0	130.2	148.8	59.1	103.6	155.2	48.8	101.6
0330	Cast-in-Place Concrete	160.5	63.3	122.2	130.1	96.8	117.0	172.3	73.2	133.2	147.0	104.4	130.2	138.2	64.2	109.1	169.8	74.8	132.4
03	CONCRETE	153.8	60.3	108.3	124.3	92.7	108.9	162.3	65.9	115.3	142.1	96.0	119.7	135.8	65.1	101.4	150.3	70.7	111.5
04	MASONRY	164.1	65.2	103.7	159.5	90.6	117.5	159.2	66.7	102.8	162.0	97.8	122.8	155.8	67.6	101.9	170.6	74.8	112.2
05	METALS	111.6	69.4	97.7	104.5	91.1	100.1	107.9	75.0	97.1	106.7	92.0	101.8	104.7	73.8	94.5	108.0	76.3	97.5
06	WOOD, PLASTICS & COMPOSITES	87.7	60.4	71.9	126.1	89.8	105.1	95.7	63.5	77.1	114.1	93.1	102.0	94.7	66.9	78.6	87.4	74.3	79.8
07	THERMAL & MOISTURE PROTECTION	106.6	62.5	88.6	105.4	93.0	100.3	108.6	64.7	90.7	111.3	93.2	103.9	108.9	65.1	91.1	106.5	73.1	92.9
08	OPENINGS	95.7	52.1	84.7	92.9	77.9	89.1	99.7	59.6	89.6	93.4	90.0	92.5	87.4	58.3	80.1	83.9	67.7	79.9
0920	Plaster & Gypsum Board	157.7	59.1	87.2	185.9	89.3	116.9	162.1	62.2	90.7	224.8	92.8	130.5	162.1	65.7	93.3	158.9	73.4	97.8
0950, 0980	Ceilings & Acoustic Treatment	110.3	59.1	76.5	95.8	89.3	91.5	96.7	62.2	73.9	98.5	92.8	94.8	96.7	65.7	76.2	105.8	73.4	84.4
0960	Flooring	112.2	64.8	98.2	132.5	102.5	123.7	116.5	59.0	99.6	132.5	100.3	123.1	115.8	74.3	103.6	112.1	69.3	99.5
0970, 0990	Wall Finishes & Painting/Coating	112.0	44.5	70.4	112.0	101.1	105.3	112.0	64.0	82.4	112.0	95.3	101.7	112.0	53.5	75.9	112.0	66.5	83.9
09	FINISHES	115.2	60.2	84.5	119.7	93.7	105.2	114.0	63.5	85.8	125.8	95.4	108.8	112.6	67.2	87.3	112.9	73.2	90.7
COVERS	DIVS. 10 - 14, 25, 28, 41, 43, 44, 46	140.5	68.9	126.3	140.5	90.1	130.5	140.5	69.8	126.5	140.5	78.8	128.2	140.5	69.4	126.4	140.5	70.8	126.7
21, 22, 23	FIRE SUPPRESSION, PLUMBING & HVAC	103.3	67.4	88.8	103.2	95.6	100.1	103.2	75.9	92.1	103.8	105.8	104.6	103.2	73.7	91.3	103.2	89.2	97.5
26, 27, 3370	ELECTRICAL, COMMUNICATIONS & UTIL.	113.6	54.7	83.2	116.5	83.0	99.2	119.1	60.9	89.1	118.3	95.0	106.3	120.0	61.1	89.6	123.7	67.3	94.6
MF2010	WEIGHTED AVERAGE	118.2	65.4	95.0	113.7	91.6	104.0	119.5	70.5	97.9	117.7	97.2	108.7	114.1	70.3	94.8	117.2	77.8	99.8

	DIVISION	CANADA																	
		EDMONTON, ALBERTA			FORT MCMURRAY, ALBERTA			FREDERICTON, NEW BRUNSWICK			GATINEAU, QUEBEC			GRANBY, QUEBEC			HALIFAX, NOVA SCOTIA		
		MAT.	INST.	TOTAL	MAT.	INST.	TOTAL	MAT.	INST.	TOTAL	MAT.	INST.	TOTAL	MAT.	INST.	TOTAL	MAT.	INST.	TOTAL
015433	CONTRACTOR EQUIPMENT		106.4	106.4		103.7	103.7		102.0	102.0		102.6	102.6		102.6	102.6		101.7	101.7
0241, 31 - 34	SITE & INFRASTRUCTURE, DEMOLITION	151.7	105.2	119.6	118.6	102.2	107.3	96.4	97.7	97.3	92.5	100.9	98.3	93.0	100.9	98.5	99.9	99.0	99.3
0310	Concrete Forming & Accessories	129.7	96.2	100.6	121.2	96.4	99.6	120.7	67.2	74.1	128.1	92.0	96.7	128.1	91.9	96.6	97.8	77.3	80.0
0320	Concrete Reinforcing	143.8	72.5	107.9	161.8	72.5	116.8	145.2	59.3	101.9	157.4	82.4	119.6	157.4	82.4	119.6	153.0	61.2	106.7
0330	Cast-in-Place Concrete	210.4	106.1	169.3	217.2	105.3	173.1	132.5	64.3	105.6	130.1	101.4	118.8	134.5	101.4	121.4	172.4	80.6	136.2
03	CONCRETE	171.7	95.3	134.5	175.3	95.1	136.2	133.6	65.3	100.4	131.3	93.5	112.9	133.5	93.5	114.0	151.2	76.1	114.6
04	MASONRY	200.5	90.7	133.5	202.4	90.7	134.2	175.9	69.2	110.8	158.9	90.6	117.2	159.2	90.6	117.3	171.4	82.4	117.1
05	METALS	144.2	89.0	126.0	132.2	89.1	118.0	110.3	74.3	98.4	105.0	89.9	100.0	105.0	89.8	100.0	126.5	81.6	111.7
06	WOOD, PLASTICS & COMPOSITES	110.0	96.0	101.9	109.8	96.0	101.8	115.9	66.9	87.6	126.1	92.0	106.4	126.1	92.0	106.4	87.5	76.2	81.0
07	THERMAL & MOISTURE PROTECTION	125.1	94.0	112.5	118.9	93.5	108.5	109.8	66.1	92.0	105.4	93.9	100.7	105.4	92.3	100.0	107.3	77.1	95.0
08	OPENINGS	93.4	85.7	91.4	93.4	85.7	91.4	87.3	57.1	79.7	93.4	79.7	89.9	93.4	79.7	89.9	83.9	71.4	80.8
0920	Plaster & Gypsum Board	166.3	95.4	115.6	154.6	95.5	112.4	177.8	65.7	97.7	147.1	91.5	107.4	150.1	91.5	108.3	158.0	75.4	99.0
0950, 0980	Ceilings & Acoustic Treatment	158.0	95.4	116.6	105.8	95.5	99.0	107.6	65.7	79.9	95.8	91.5	93.0	95.8	91.5	93.0	111.2	75.4	87.5
0960	Flooring	135.8	95.4	123.9	132.5	95.4	121.6	128.0	77.5	113.2	132.5	102.5	123.7	132.5	102.5	123.7	112.1	91.0	105.9
0970, 0990	Wall Finishes & Painting/Coating	112.0	109.7	110.6	112.0	99.0	104.0	112.0	68.2	84.9	112.0	95.0	101.5	112.0	95.0	101.5	112.0	79.8	92.1
09	FINISHES	137.0	98.0	115.2	120.8	96.8	107.4	120.8	69.4	92.2	115.2	94.6	103.7	115.5	94.6	103.9	114.2	80.4	95.4
COVERS	DIVS. 10 - 14, 25, 28, 41, 43, 44, 46	140.5	101.4	132.8	140.5	101.3	132.7	140.5	69.5	126.4	140.5	90.7	130.6	140.5	90.7	130.6	140.5	71.5	126.8
21, 22, 23	FIRE SUPPRESSION, PLUMBING & HVAC	100.6	92.6	97.4	103.5	104.4	103.9	103.3	83.8	95.4	103.8	96.1	100.7	103.2	96.1	100.4	99.8	78.0	91.0
26, 27, 3370	ELECTRICAL, COMMUNICATIONS & UTIL.	115.6	84.2	99.4	111.0	89.3	99.8	124.5	80.5	101.8	116.9	75.9	95.8	117.6	75.9	96.1	120.9	79.8	99.7
MF2010	WEIGHTED AVERAGE	130.0	93.1	113.7	126.5	95.8	113.0	117.0	75.6	98.8	114.4	91.1	104.2	114.7	91.0	104.3	119.0	80.3	101.9

	DIVISION	CANADA																	
		HAMILTON, ONTARIO			HULL, QUEBEC			JOLIETTE, QUEBEC			KAMLOOPS, BRITISH COLUMBIA			KINGSTON, ONTARIO			KITCHENER, ONTARIO		
		MAT.	INST.	TOTAL	MAT.	INST.	TOTAL	MAT.	INST.	TOTAL	MAT.	INST.	TOTAL	MAT.	INST.	TOTAL	MAT.	INST.	TOTAL
015433	CONTRACTOR EQUIPMENT		108.4	108.4		102.6	102.6		102.6	102.6		105.7	105.7		104.0	104.0		103.9	103.9
0241, 31 - 34	SITE & INFRASTRUCTURE, DEMOLITION	116.3	112.8	113.9	92.5	100.9	98.3	93.2	100.9	98.5	117.7	104.3	108.5	110.8	104.9	106.7	100.3	105.1	103.6
0310	Concrete Forming & Accessories	130.8	94.3	99.0	128.1	92.0	96.7	128.1	92.1	96.8	123.5	95.0	98.7	121.2	93.8	97.4	117.6	85.6	89.8
0320	Concrete Reinforcing	162.7	85.9	124.0	157.4	82.4	119.6	148.8	82.4	115.3	116.8	80.3	98.4	175.1	86.0	130.2	110.5	85.8	98.1
0330	Cast-in-Place Concrete	153.0	95.8	130.4	130.1	101.4	118.8	135.6	101.5	122.1	117.3	105.3	112.5	147.0	104.4	130.2	140.8	86.9	119.5
03	CONCRETE	143.8	93.3	119.2	131.3	93.5	112.9	132.7	93.6	113.7	137.7	95.8	117.3	142.1	96.1	119.7	122.6	86.4	105.0
04	MASONRY	179.1	100.4	131.1	158.9	90.6	117.2	159.3	90.6	117.4	165.6	98.8	124.8	168.6	97.9	125.4	159.8	92.6	118.8
05	METALS	125.5	92.3	114.6	105.0	89.9	100.0	105.0	90.0	100.1	107.3	89.9	101.6	108.4	91.9	102.9	115.4	92.1	107.7
06	WOOD, PLASTICS & COMPOSITES	114.5	93.4	102.3	126.1	92.0	106.4	126.1	92.0	106.4	98.3	93.4	95.5	114.1	93.2	102.0	108.7	84.9	94.9
07	THERMAL & MOISTURE PROTECTION	122.5	95.7	111.6	105.4	93.9	100.7	105.4	93.9	100.7	120.7	90.7	108.5	111.3	94.4	104.4	110.8	90.8	102.7
08	OPENINGS	93.4	91.9	93.0	93.4	79.7	89.9	93.4	84.6	91.2	89.1	89.4	89.2	93.4	89.6	92.4	84.5	85.5	84.7
0920	Plaster & Gypsum Board	191.9	93.2	121.4	147.1	91.5	107.4	185.9	91.5	118.5	136.1	92.7	105.1	229.1	93.0	131.9	153.0	84.4	104.0
0950, 0980	Ceilings & Acoustic Treatment	130.9	93.2	106.0	95.8	91.5	93.0	95.8	91.5	93.0	95.8	92.7	93.8	113.0	93.0	99.8	105.8	84.4	91.6
0960	Flooring	134.6	105.3	126.0	132.5	102.5	123.7	132.5	102.5	123.7	130.3	57.3	108.8	132.5	100.3	123.1	128.5	105.3	121.7
0970, 0990	Wall Finishes & Painting/Coating	112.0	100.9	105.2	112.0	95.0	101.5	112.0	95.0	101.5	112.0	87.0	96.6	112.0	88.0	97.2	112.0	91.1	99.1
09	FINISHES	130.9	97.2	112.1	115.2	94.6	103.7	119.7	94.6	105.7	116.6	87.7	100.5	129.5	94.7	110.1	116.9	89.5	101.6
COVERS	DIVS. 10 - 14, 25, 28, 41, 43, 44, 46	140.5	104.0	133.3	140.5	90.7	130.6	140.5	90.7	130.6	140.5	101.7	132.8	140.5	78.8	128.2	140.5	100.9	132.7
21, 22, 23	FIRE SUPPRESSION, PLUMBING & HVAC	100.7	84.0	93.9	103.2	96.1	100.4	103.2	96.1	100.4	103.2	99.7	101.8	103.8	105.9	104.7	99.8	89.1	95.4
26, 27, 3370	ELECTRICAL, COMMUNICATIONS & UTIL.	119.7	87.5	103.1	119.0	75.9	96.7	117.6	75.9	96.1	121.3	85.6	102.9	118.3	93.6	105.6	117.6	85.1	100.8
MF2010	WEIGHTED AVERAGE	121.9	93.7	109.4	114.5	91.1	104.2	114.9	91.3	104.5	116.9	94.4	107.0	118.6	97.3	109.2	113.7	90.4	103.4

	DIVISION	CANADA																	
		LAVAL, QUEBEC			LETHBRIDGE, ALBERTA			LLOYDMINSTER, ALBERTA			LONDON, ONTARIO			MEDICINE HAT, ALBERTA			MONCTON, NEW BRUNSWICK		
		MAT.	INST.	TOTAL	MAT.	INST.	TOTAL	MAT.	INST.	TOTAL	MAT.	INST.	TOTAL	MAT.	INST.	TOTAL	MAT.	INST.	TOTAL
015433	**CONTRACTOR EQUIPMENT**		102.6	102.6		103.7	103.7		103.7	103.7		104.0	104.0		103.7	103.7		102.0	102.0
0241, 31 - 34	**SITE & INFRASTRUCTURE, DEMOLITION**	93.0	100.9	98.5	110.9	102.8	105.3	110.9	102.2	104.9	115.1	105.3	108.4	109.5	102.3	104.5	94.0	97.8	96.7
0310	Concrete Forming & Accessories	128.2	92.0	96.7	122.4	96.5	99.9	120.7	86.2	90.7	130.9	88.2	93.8	122.3	86.2	90.9	103.1	67.1	71.8
0320	Concrete Reinforcing	157.4	82.4	119.6	161.8	72.5	116.8	161.8	72.5	116.8	142.3	84.7	113.2	161.8	72.5	116.8	148.8	62.3	105.2
0330	Cast-in-Place Concrete	134.5	101.4	121.4	162.9	105.3	140.2	151.1	101.5	131.5	164.1	93.7	136.3	151.1	101.5	131.5	133.1	64.5	106.1
03	**CONCRETE**	133.5	93.5	114.0	148.1	95.2	122.3	142.1	89.3	116.4	146.2	89.7	118.6	142.2	89.3	116.4	133.2	66.1	100.5
04	**MASONRY**	159.2	90.6	117.3	176.5	90.7	124.1	158.7	83.8	113.0	183.3	97.7	131.1	158.7	83.8	113.0	157.6	67.6	102.7
05	**METALS**	105.1	89.9	100.1	125.8	89.1	113.7	107.5	89.0	101.4	124.2	91.6	113.5	107.5	88.9	101.4	104.7	79.4	96.4
06	**WOOD, PLASTICS & COMPOSITES**	126.3	92.0	106.5	113.0	96.0	103.2	109.8	85.7	95.9	114.5	86.5	98.3	113.0	85.7	97.2	94.7	66.9	78.6
07	**THERMAL & MOISTURE PROTECTION**	105.9	93.9	101.0	116.4	93.5	107.1	113.6	88.8	103.5	123.3	93.4	111.1	119.6	88.8	107.1	108.9	66.6	91.7
08	**OPENINGS**	93.4	79.7	89.9	93.4	85.7	91.4	93.4	80.1	90.0	94.4	87.0	92.5	93.4	80.1	90.0	87.4	61.6	80.9
0920	Plaster & Gypsum Board	150.5	91.5	108.4	139.1	95.5	107.9	133.7	84.9	98.8	193.7	86.0	116.8	136.4	84.9	99.6	162.1	65.7	93.3
0950, 0980	Ceilings & Acoustic Treatment	95.8	91.5	93.0	104.9	95.5	98.7	95.8	84.9	88.6	137.2	86.0	103.4	95.8	84.9	88.6	96.7	65.7	76.2
0960	Flooring	132.5	102.5	123.7	132.5	95.4	121.6	132.5	95.4	121.6	134.8	105.3	126.2	132.5	95.4	121.6	115.8	74.3	103.6
0970, 0990	Wall Finishes & Painting/Coating	112.0	95.0	101.5	111.9	108.0	109.5	112.0	84.2	94.8	112.0	97.7	103.2	111.9	84.2	94.8	112.0	53.5	75.9
09	**FINISHES**	115.6	94.6	103.9	117.8	97.8	106.6	115.3	87.4	99.7	132.1	92.3	109.9	115.4	87.4	99.8	112.6	67.2	87.3
COVERS	**DIVS. 10 - 14, 25, 28, 41, 43, 44, 46**	140.5	90.7	130.6	140.5	101.3	132.7	140.5	97.8	132.0	140.5	102.6	133.0	140.5	97.8	132.0	140.5	69.5	126.4
21, 22, 23	**FIRE SUPPRESSION, PLUMBING & HVAC**	99.8	96.1	98.3	103.4	100.8	102.4	103.8	100.9	102.6	100.7	81.2	92.8	103.2	97.3	100.8	103.2	73.8	91.3
26, 27, 3370	**ELECTRICAL, COMMUNICATIONS & UTIL.**	118.9	75.9	96.7	112.5	89.3	100.5	109.9	89.3	99.3	112.9	85.1	98.6	109.9	89.3	99.3	124.6	64.6	93.7
MF2010	**WEIGHTED AVERAGE**	114.0	91.1	103.9	120.8	95.3	109.5	116.0	91.8	105.4	121.6	90.2	107.8	116.1	91.1	105.1	114.3	71.7	95.5

	DIVISION	CANADA																	
		MONTREAL, QUEBEC			MOOSE JAW, SASKATCHEWAN			NEW GLASGOW, NOVA SCOTIA			NEWCASTLE, NEW BRUNSWICK			NORTH BAY, ONTARIO			OSHAWA, ONTARIO		
		MAT.	INST.	TOTAL	MAT.	INST.	TOTAL	MAT.	INST.	TOTAL	MAT.	INST.	TOTAL	MAT.	INST.	TOTAL	MAT.	INST.	TOTAL
015433	**CONTRACTOR EQUIPMENT**		104.2	104.2		100.2	100.2		101.7	101.7		102.0	102.0		101.7	101.7		103.9	103.9
0241, 31 - 34	**SITE & INFRASTRUCTURE, DEMOLITION**	102.6	100.0	100.8	110.6	96.8	101.1	97.2	99.3	98.6	94.6	97.7	96.7	112.7	100.8	104.5	112.6	104.9	107.3
0310	Concrete Forming & Accessories	135.8	90.0	96.0	106.7	63.3	69.0	97.4	75.0	77.9	103.1	66.9	71.6	123.0	91.1	95.2	123.2	93.2	97.1
0320	Concrete Reinforcing	150.6	92.1	121.1	114.2	64.7	89.2	148.8	48.8	98.4	148.8	59.1	103.6	175.1	85.5	129.9	175.1	88.2	131.3
0330	Cast-in-Place Concrete	182.2	98.1	149.1	146.6	73.7	117.9	169.8	74.8	132.4	138.2	64.2	109.1	163.5	90.3	134.7	162.8	91.2	134.6
03	**CONCRETE**	156.9	93.2	125.9	125.4	67.8	97.4	149.3	70.7	111.0	135.8	65.2	101.4	150.5	89.9	121.0	150.2	91.7	121.7
04	**MASONRY**	166.1	90.6	120.1	157.3	67.0	102.2	159.0	74.8	107.6	158.0	67.6	102.8	163.3	93.5	120.7	162.9	98.2	123.4
05	**METALS**	124.2	91.6	113.4	103.9	75.9	94.7	105.8	76.3	96.1	104.7	73.8	94.5	106.7	91.6	101.7	106.8	93.8	102.5
06	**WOOD, PLASTICS & COMPOSITES**	126.3	90.0	105.3	96.2	61.8	76.3	87.4	74.3	79.8	94.7	66.9	78.6	115.2	91.6	101.6	115.4	92.2	102.0
07	**THERMAL & MOISTURE PROTECTION**	119.4	93.5	108.9	104.8	67.1	89.5	106.5	73.1	92.9	108.9	65.1	91.1	111.5	89.4	102.5	111.8	91.2	103.4
08	**OPENINGS**	93.4	79.8	89.9	88.2	59.2	80.9	83.9	67.7	79.9	87.4	58.3	80.1	92.2	87.4	91.0	92.2	91.7	92.1
0920	Plaster & Gypsum Board	171.6	89.3	112.8	131.9	60.5	80.9	155.9	73.4	96.9	162.1	65.7	93.3	155.0	91.3	109.5	156.8	91.9	110.4
0950, 0980	Ceilings & Acoustic Treatment	115.6	89.3	98.2	95.8	60.5	72.5	95.8	73.4	81.0	96.7	65.7	76.2	95.8	91.3	92.8	100.3	91.9	94.8
0960	Flooring	134.1	102.5	124.8	121.1	65.2	104.7	112.1	69.3	99.5	115.8	74.3	103.6	132.5	100.3	123.1	132.5	108.0	125.3
0970, 0990	Wall Finishes & Painting/Coating	112.0	101.1	105.3	112.0	69.2	85.6	112.0	66.5	83.9	112.0	53.5	75.9	112.0	94.5	101.2	112.0	104.9	107.6
09	**FINISHES**	123.5	93.9	107.0	111.5	64.0	85.0	110.3	73.2	89.6	112.6	67.2	87.3	117.1	93.6	104.0	118.2	97.2	106.5
COVERS	**DIVS. 10 - 14, 25, 28, 41, 43, 44, 46**	140.5	90.6	130.6	140.5	68.6	126.2	140.5	70.8	126.7	140.5	69.5	126.4	140.5	77.4	128.0	140.5	103.5	133.2
21, 22, 23	**FIRE SUPPRESSION, PLUMBING & HVAC**	100.7	95.6	98.6	103.8	80.4	94.3	103.2	89.2	97.5	103.2	73.7	91.3	103.2	103.6	103.4	99.8	106.3	102.4
26, 27, 3370	**ELECTRICAL, COMMUNICATIONS & UTIL.**	123.4	83.0	102.5	120.0	65.6	91.9	119.4	67.3	92.5	119.2	64.6	91.0	119.4	94.9	106.8	118.7	94.5	106.2
MF2010	**WEIGHTED AVERAGE**	121.9	91.9	108.7	113.4	72.4	95.3	115.3	77.8	98.8	114.1	70.8	95.0	118.0	94.9	107.8	117.1	98.2	108.7

	DIVISION	CANADA																	
		OTTAWA, ONTARIO			OWEN SOUND, ONTARIO			PETERBOROUGH, ONTARIO			PORTAGE LA PRAIRIE, MANITOBA			PRINCE ALBERT, SASKATCHEWAN			PRINCE GEORGE, BRITISH COLUMBIA		
		MAT.	INST.	TOTAL	MAT.	INST.	TOTAL	MAT.	INST.	TOTAL	MAT.	INST.	TOTAL	MAT.	INST.	TOTAL	MAT.	INST.	TOTAL
015433	**CONTRACTOR EQUIPMENT**		103.9	103.9		101.7	101.7		101.7	101.7		103.9	103.9		100.2	100.2		105.7	105.7
0241, 31 - 34	**SITE & INFRASTRUCTURE, DEMOLITION**	112.3	105.1	107.3	114.9	101.2	105.5	112.7	101.1	104.7	111.2	100.2	103.6	105.5	97.0	99.7	121.2	104.3	109.6
0310	Concrete Forming & Accessories	131.7	90.8	96.1	123.2	89.5	93.9	123.0	92.2	96.2	122.5	75.3	81.5	106.7	63.1	68.8	111.9	89.5	92.4
0320	Concrete Reinforcing	155.4	84.6	119.7	180.2	87.6	133.5	175.1	86.0	130.2	161.8	55.6	108.3	119.4	64.6	91.8	116.8	80.3	98.4
0330	Cast-in-Place Concrete	168.5	92.9	138.7	167.6	86.2	135.5	163.5	92.0	135.4	151.1	78.7	122.6	132.8	73.6	109.5	147.1	105.3	130.6
03	**CONCRETE**	150.5	90.5	121.3	153.4	88.2	121.6	150.5	91.1	121.6	135.9	73.5	105.5	119.4	67.7	94.2	151.9	93.4	123.4
04	**MASONRY**	180.1	97.6	129.8	167.0	99.5	125.8	163.3	100.6	125.0	161.7	69.3	105.3	156.3	67.0	101.8	167.9	98.8	125.8
05	**METALS**	129.3	91.5	116.8	107.3	92.4	102.4	106.7	92.1	101.9	107.5	78.7	98.0	104.0	75.7	94.6	107.3	89.9	101.6
06	**WOOD, PLASTICS & COMPOSITES**	114.4	90.0	100.3	117.4	88.1	100.5	115.2	90.2	100.8	112.9	76.8	92.0	96.2	61.8	76.3	98.3	85.8	91.1
07	**THERMAL & MOISTURE PROTECTION**	125.1	92.9	112.0	111.6	90.3	102.9	111.5	95.2	104.9	105.2	74.5	92.8	104.7	65.9	88.9	114.9	89.9	104.8
08	**OPENINGS**	93.4	88.9	92.2	92.4	86.7	90.9	92.2	89.2	91.5	93.4	67.6	86.9	86.7	59.2	79.8	89.1	85.3	88.2
0920	Plaster & Gypsum Board	230.4	89.7	129.9	155.0	87.7	107.0	155.0	89.9	108.5	130.0	75.6	91.2	131.9	60.5	80.9	136.1	84.9	99.5
0950, 0980	Ceilings & Acoustic Treatment	140.8	89.7	107.0	95.8	87.7	90.5	95.8	89.9	91.9	95.8	75.6	82.5	95.8	60.5	72.5	95.8	84.9	88.6
0960	Flooring	134.6	100.3	124.6	132.5	101.8	123.5	132.5	100.3	123.1	132.5	72.3	114.8	121.1	65.2	104.7	127.3	78.5	113.0
0970, 0990	Wall Finishes & Painting/Coating	112.0	92.6	100.0	112.8	93.1	100.7	112.0	96.9	102.7	112.0	60.8	80.5	112.0	59.0	79.3	112.0	87.0	96.6
09	**FINISHES**	136.5	92.9	112.2	118.4	92.3	103.8	117.1	94.3	104.4	114.6	73.7	91.8	111.5	62.9	84.4	115.7	86.8	99.6
COVERS	**DIVS. 10 - 14, 25, 28, 41, 43, 44, 46**	140.5	100.6	132.6	140.0	77.9	127.7	140.5	79.0	128.3	140.5	71.4	126.8	140.5	68.7	126.2	140.5	100.8	132.6
21, 22, 23	**FIRE SUPPRESSION, PLUMBING & HVAC**	100.7	81.7	93.0	103.0	103.5	103.2	103.2	107.4	104.9	103.2	88.2	97.2	103.8	72.4	91.1	103.2	99.7	101.8
26, 27, 3370	**ELECTRICAL, COMMUNICATIONS & UTIL.**	112.3	85.9	98.7	120.9	93.6	106.8	117.2	94.5	105.5	119.7	62.8	90.3	120.0	65.6	91.9	117.4	85.6	101.0
MF2010	**WEIGHTED AVERAGE**	122.9	90.7	108.7	118.8	95.0	108.3	117.7	97.0	108.6	116.0	77.2	98.9	112.3	70.5	93.9	118.1	93.7	107.3

	DIVISION	CANADA																	
		QUEBEC, QUEBEC			RED DEER, ALBERTA			REGINA, SASKATCHEWAN			RIMOUSKI, QUEBEC			ROUYN-NORANDA, QUEBEC			SAINT HYACINTHE, QUEBEC		
		MAT.	INST.	TOTAL	MAT.	INST.	TOTAL	MAT.	INST.	TOTAL	MAT.	INST.	TOTAL	MAT.	INST.	TOTAL	MAT.	INST.	TOTAL
015433	CONTRACTOR EQUIPMENT		104.5	104.5		103.7	103.7		100.2	100.2		102.6	102.6		102.6	102.6		102.6	102.6
0241, 31 - 34	SITE & INFRASTRUCTURE, DEMOLITION	100.7	100.1	100.3	109.5	102.3	104.5	115.5	98.5	103.8	93.0	99.7	97.7	92.5	100.9	98.3	93.0	100.9	98.5
0310	Concrete Forming & Accessories	137.2	90.3	96.4	139.2	86.2	93.1	107.0	88.1	90.6	128.1	89.7	94.7	128.1	92.0	96.7	128.1	92.0	96.7
0320	Concrete Reinforcing	141.0	92.2	116.4	161.8	72.5	116.8	128.8	77.8	103.1	112.2	92.1	102.1	157.4	82.4	119.6	157.4	82.4	119.6
0330	Cast-in-Place Concrete	149.3	98.5	129.3	151.1	101.5	131.5	170.8	93.2	140.2	136.9	96.8	121.1	130.1	101.4	118.8	134.5	101.4	121.4
03	CONCRETE	139.0	93.5	116.8	143.2	89.3	116.9	139.9	88.0	114.6	127.7	92.7	110.6	131.3	93.5	112.9	133.5	93.5	114.0
04	MASONRY	173.5	90.7	122.9	158.7	83.8	113.0	167.7	91.1	121.0	158.7	90.6	117.1	158.9	90.6	117.2	159.1	90.6	117.3
05	METALS	124.2	91.8	113.5	107.5	88.9	101.4	110.6	83.7	101.7	104.5	91.1	100.1	105.0	89.9	100.0	105.0	89.9	100.0
06	WOOD, PLASTICS & COMPOSITES	126.9	90.1	105.6	113.0	85.7	97.2	96.3	88.8	92.0	126.1	89.8	105.1	126.1	92.0	106.4	126.1	92.0	106.4
07	THERMAL & MOISTURE PROTECTION	121.4	93.7	110.1	129.1	88.8	112.8	106.7	84.3	97.6	105.4	93.0	100.3	105.4	93.9	100.7	105.6	93.9	100.9
08	OPENINGS	93.4	87.8	92.0	93.4	80.1	90.0	88.2	77.7	85.5	92.9	77.9	89.1	93.4	79.7	89.9	93.4	79.7	89.9
0920	Plaster & Gypsum Board	191.4	89.3	118.4	136.4	84.9	99.6	168.2	88.5	111.2	185.7	89.3	116.8	146.8	91.5	107.3	149.8	91.5	108.2
0950, 0980	Ceilings & Acoustic Treatment	120.7	89.3	99.9	95.8	84.9	88.6	125.7	88.5	101.1	94.9	89.3	91.2	94.9	91.5	92.7	94.9	91.5	92.7
0960	Flooring	134.6	102.5	125.2	134.6	95.4	123.1	121.1	67.4	105.4	132.5	102.5	123.7	132.5	102.5	123.7	132.5	102.5	123.7
0970, 0990	Wall Finishes & Painting/Coating	112.5	101.1	105.4	111.9	84.2	94.8	112.0	87.3	96.8	112.0	101.1	105.3	112.0	95.0	101.5	112.0	95.0	101.5
09	FINISHES	127.5	93.9	108.8	116.0	87.4	100.0	123.0	84.8	101.7	119.5	93.7	105.1	114.9	94.6	103.6	115.3	94.6	103.8
COVERS	DIVS. 10 - 14, 25, 28, 41, 43, 44, 46	140.5	90.7	130.6	140.5	97.8	132.0	140.5	75.1	127.5	140.5	90.1	130.5	140.5	90.7	130.6	140.5	90.7	130.6
21, 22, 23	FIRE SUPPRESSION, PLUMBING & HVAC	100.6	95.7	98.6	103.2	97.3	100.8	99.9	82.3	92.8	103.2	95.6	100.1	103.2	96.1	100.4	98.6	96.1	97.6
26, 27, 3370	ELECTRICAL, COMMUNICATIONS & UTIL.	116.5	83.0	99.2	109.9	89.3	99.3	119.9	87.3	103.1	117.6	83.0	99.7	117.6	75.9	96.1	118.5	75.9	96.5
MF2010	WEIGHTED AVERAGE	119.8	92.3	107.7	116.6	91.1	105.3	116.7	86.2	103.2	114.2	91.6	104.2	114.4	91.1	104.1	113.6	91.1	103.7

	DIVISION	CANADA																	
		SAINT JOHN, NEW BRUNSWICK			SARNIA, ONTARIO			SASKATOON, SASKATCHEWAN			SAULT STE MARIE, ONTARIO			SHERBROOKE, QUEBEC			SOREL, QUEBEC		
		MAT.	INST.	TOTAL	MAT.	INST.	TOTAL	MAT.	INST.	TOTAL	MAT.	INST.	TOTAL	MAT.	INST.	TOTAL	MAT.	INST.	TOTAL
015433	CONTRACTOR EQUIPMENT		102.0	102.0		101.7	101.7		100.2	100.2		101.7	101.7		102.6	102.6		102.6	102.6
0241, 31 - 34	SITE & INFRASTRUCTURE, DEMOLITION	94.8	98.6	97.4	111.2	101.2	104.3	106.6	98.5	101.0	101.2	100.8	100.9	93.0	100.9	98.5	93.2	100.9	98.5
0310	Concrete Forming & Accessories	120.5	63.9	71.3	121.9	99.7	102.6	106.9	88.1	90.5	112.0	88.0	91.1	128.1	92.0	96.7	128.1	92.1	96.8
0320	Concrete Reinforcing	148.8	62.5	105.3	124.0	87.4	105.6	119.4	77.8	98.4	112.4	85.6	98.9	157.4	82.4	119.6	148.8	82.4	115.3
0330	Cast-in-Place Concrete	136.2	64.3	107.8	150.9	106.0	133.2	142.0	93.2	122.8	135.5	90.4	117.8	134.5	101.4	121.4	135.6	101.5	122.1
03	CONCRETE	136.0	64.7	101.3	136.1	99.5	118.3	124.0	88.0	106.4	120.0	88.6	104.7	133.5	93.5	114.0	132.7	93.6	113.7
04	MASONRY	176.3	69.8	111.3	174.1	103.5	131.0	164.2	91.1	119.6	159.7	95.1	120.3	159.2	90.6	117.3	159.3	90.6	117.4
05	METALS	104.6	79.9	96.5	106.6	92.5	102.0	104.9	83.7	97.9	105.9	90.8	100.9	105.0	89.9	100.0	105.0	90.0	100.1
06	WOOD, PLASTICS & COMPOSITES	115.9	62.4	85.0	114.3	99.0	105.5	94.6	88.8	91.3	102.6	87.0	93.6	126.1	92.0	106.4	126.1	92.0	106.4
07	THERMAL & MOISTURE PROTECTION	109.2	65.5	91.4	111.5	98.9	106.4	105.4	84.3	96.8	110.3	91.0	102.4	105.4	93.9	100.7	105.4	93.9	100.7
08	OPENINGS	87.3	58.0	79.9	94.4	93.3	94.1	87.2	77.7	84.8	84.9	85.4	85.0	93.4	79.7	89.9	93.4	84.6	91.2
0920	Plaster & Gypsum Board	179.4	61.1	94.9	178.2	98.9	121.6	144.8	88.5	104.6	143.4	86.6	102.8	149.8	91.5	108.2	185.7	91.5	118.4
0950, 0980	Ceilings & Acoustic Treatment	103.1	61.1	75.3	100.3	98.9	99.4	117.5	88.5	98.3	95.8	86.6	89.7	94.9	91.5	92.7	94.9	91.5	92.7
0960	Flooring	128.0	74.3	112.2	132.5	109.7	125.8	121.1	67.4	105.4	124.3	103.8	118.3	132.5	102.5	123.7	132.5	102.5	123.7
0970, 0990	Wall Finishes & Painting/Coating	112.0	71.6	87.1	112.0	109.5	110.5	112.0	87.3	96.8	112.0	101.5	105.5	112.0	95.0	101.5	112.0	95.0	101.5
09	FINISHES	119.7	66.4	90.0	120.8	103.1	110.9	117.8	84.8	99.4	112.4	92.2	101.2	115.3	94.6	103.8	119.5	94.6	105.6
COVERS	DIVS. 10 - 14, 25, 28, 41, 43, 44, 46	140.5	68.9	126.3	140.5	80.5	128.6	140.5	75.1	127.5	140.5	102.1	132.9	140.5	90.7	130.6	140.5	90.7	130.6
21, 22, 23	FIRE SUPPRESSION, PLUMBING & HVAC	103.2	78.9	93.4	103.2	113.8	107.5	99.8	82.3	92.7	103.2	100.0	101.9	103.8	96.1	100.7	103.2	96.1	100.4
26, 27, 3370	ELECTRICAL, COMMUNICATIONS & UTIL.	127.7	84.5	105.4	120.2	96.9	108.2	121.9	87.3	104.1	118.9	94.9	106.5	117.6	75.9	96.1	117.6	75.9	96.1
MF2010	WEIGHTED AVERAGE	116.6	75.3	98.4	117.4	101.6	110.5	113.2	86.2	101.3	112.6	94.5	104.6	114.8	91.1	104.4	114.9	91.3	104.5

	DIVISION	CANADA																	
		ST CATHARINES, ONTARIO			ST JEROME, QUEBEC			ST JOHNS, NEWFOUNDLAND			SUDBURY, ONTARIO			SUMMERSIDE, PRINCE EDWARD ISLAND			SYDNEY, NOVA SCOTIA		
		MAT.	INST.	TOTAL	MAT.	INST.	TOTAL	MAT.	INST.	TOTAL	MAT.	INST.	TOTAL	MAT.	INST.	TOTAL	MAT.	INST.	TOTAL
015433	CONTRACTOR EQUIPMENT		101.7	101.7		102.6	102.6		103.3	103.3		101.7	101.7		101.6	101.6		101.7	101.7
0241, 31 - 34	SITE & INFRASTRUCTURE, DEMOLITION	101.1	101.9	101.7	92.5	100.9	98.3	113.3	99.7	103.9	101.2	101.4	101.3	105.2	96.8	99.4	92.7	99.3	97.2
0310	Concrete Forming & Accessories	115.6	94.0	96.8	128.1	92.0	96.7	104.7	72.5	76.7	112.0	88.0	91.1	97.7	61.0	65.8	97.4	75.0	77.9
0320	Concrete Reinforcing	111.5	85.8	98.6	157.4	82.4	119.6	160.7	64.0	112.0	112.4	84.3	98.2	146.6	48.9	97.4	148.8	48.8	98.4
0330	Cast-in-Place Concrete	134.4	95.6	119.1	130.1	101.4	118.8	173.5	85.2	138.7	135.5	92.0	118.4	161.2	63.4	122.6	130.9	74.8	108.8
03	CONCRETE	119.5	93.1	106.6	131.3	93.5	112.9	164.8	75.9	121.5	120.0	89.0	104.9	152.9	60.4	107.8	129.7	70.7	101.0
04	MASONRY	159.3	100.4	123.4	158.9	90.6	117.2	164.8	77.3	111.4	159.4	95.6	120.4	158.1	65.2	101.4	155.9	74.8	106.5
05	METALS	105.9	92.0	101.3	105.0	89.9	100.0	113.8	79.9	102.6	105.9	91.6	101.1	105.8	69.5	93.8	105.8	76.3	96.1
06	WOOD, PLASTICS & COMPOSITES	106.4	93.4	98.9	126.1	92.0	106.4	95.8	71.2	81.6	102.6	87.0	93.6	87.7	60.4	71.9	87.4	74.3	79.8
07	THERMAL & MOISTURE PROTECTION	110.8	96.4	105.0	105.4	93.9	100.7	110.3	75.5	96.2	110.3	90.9	102.4	106.0	63.9	88.9	106.5	73.1	92.9
08	OPENINGS	84.0	90.5	85.6	93.4	79.7	89.9	99.7	66.2	91.2	84.9	85.4	85.0	95.7	52.1	84.7	83.9	67.7	79.9
0920	Plaster & Gypsum Board	137.6	93.2	105.9	146.8	91.5	107.3	172.6	70.1	99.4	143.4	86.6	102.8	156.4	59.1	86.9	155.9	73.4	96.9
0950, 0980	Ceilings & Acoustic Treatment	100.3	93.2	95.6	94.9	91.5	92.7	110.3	70.1	83.8	95.8	86.6	89.7	95.8	59.1	71.5	95.8	73.4	81.0
0960	Flooring	126.9	101.8	119.5	132.5	102.5	123.7	116.5	61.0	100.2	124.3	103.8	118.3	112.2	64.8	98.2	112.1	69.3	99.5
0970, 0990	Wall Finishes & Painting/Coating	112.0	100.9	105.2	112.0	95.0	101.5	112.0	74.7	89.0	112.0	92.1	99.7	112.0	44.5	70.4	112.0	66.5	83.9
09	FINISHES	113.5	96.6	104.1	114.9	94.6	103.6	118.2	70.6	91.6	112.4	91.3	100.6	111.4	60.2	82.9	110.3	73.2	89.6
COVERS	DIVS. 10 - 14, 25, 28, 41, 43, 44, 46	140.5	78.9	128.3	140.5	90.7	130.6	140.5	72.2	126.9	140.5	102.2	132.9	140.5	68.9	126.3	140.5	70.8	126.7
21, 22, 23	FIRE SUPPRESSION, PLUMBING & HVAC	99.8	82.9	92.9	103.2	96.1	100.4	100.1	71.3	88.4	103.2	85.9	96.2	103.2	67.4	88.7	103.2	89.2	97.5
26, 27, 3370	ELECTRICAL, COMMUNICATIONS & UTIL.	119.2	86.0	102.1	118.3	75.9	96.4	116.0	72.3	93.4	116.2	86.0	100.6	118.4	54.7	85.5	119.4	67.3	92.5
MF2010	WEIGHTED AVERAGE	111.7	91.3	102.8	114.4	91.1	104.1	120.3	75.7	100.6	112.3	90.4	102.7	117.0	65.5	94.3	112.8	77.8	97.4

DIVISION		CANADA																	
		THUNDER BAY, ONTARIO			TIMMINS, ONTARIO			TORONTO, ONTARIO			TROIS RIVIERES, QUEBEC			TRURO, NOVA SCOTIA			VANCOUVER, BRITISH COLUMBIA		
		MAT.	INST.	TOTAL	MAT.	INST.	TOTAL	MAT.	INST.	TOTAL	MAT.	INST.	TOTAL	MAT.	INST.	TOTAL	MAT.	INST.	TOTAL
015433	**CONTRACTOR EQUIPMENT**		101.7	101.7		101.7	101.7		104.0	104.0		102.6	102.6		101.7	101.7		112.1	112.1
0241, 31 - 34	**SITE & INFRASTRUCTURE, DEMOLITION**	106.5	101.9	103.3	112.7	100.8	104.5	139.0	105.9	116.2	93.2	100.9	98.5	97.2	99.3	98.6	121.1	108.5	112.4
0310	Concrete Forming & Accessories	123.1	91.6	95.7	123.0	91.1	95.2	132.2	99.5	103.7	128.1	92.1	96.8	97.4	75.0	77.9	126.6	87.0	92.2
0320	Concrete Reinforcing	99.7	85.3	92.4	175.1	85.5	129.9	162.1	86.5	124.0	148.8	82.4	115.3	148.8	48.8	98.4	152.8	79.7	116.0
0330	Cast-in-Place Concrete	148.0	94.0	126.7	163.5	90.3	134.7	154.3	114.8	138.8	135.6	101.5	122.1	169.8	74.8	132.4	164.0	101.0	139.2
03	**CONCRETE**	128.2	91.4	110.3	150.5	89.9	121.0	144.5	102.2	123.9	132.7	93.6	113.7	149.3	70.7	111.0	153.3	90.8	122.9
04	**MASONRY**	160.1	99.7	123.2	163.3	93.5	120.7	193.0	105.0	139.3	159.3	90.6	117.4	159.0	74.8	107.6	177.9	93.9	126.6
05	**METALS**	105.7	90.9	100.8	106.7	91.6	101.7	127.1	92.9	115.8	105.0	90.0	100.1	105.8	76.3	96.1	157.9	91.2	135.9
06	**WOOD, PLASTICS & COMPOSITES**	115.4	90.3	100.9	115.2	91.6	101.6	115.4	98.1	105.4	126.1	92.0	106.4	87.4	74.3	79.8	114.9	85.2	97.7
07	**THERMAL & MOISTURE PROTECTION**	111.0	93.1	103.8	111.5	89.4	102.5	123.5	100.4	114.1	105.4	93.9	100.7	106.5	73.1	92.9	133.7	89.0	115.5
08	**OPENINGS**	82.9	88.1	84.2	92.2	87.4	91.0	92.2	96.5	93.3	93.4	84.6	91.2	83.9	67.7	79.9	95.1	84.3	92.4
0920	Plaster & Gypsum Board	170.4	89.9	112.9	155.0	91.3	109.5	176.3	98.0	120.3	185.7	91.5	118.4	155.9	73.4	96.9	162.8	84.1	106.6
0950, 0980	Ceilings & Acoustic Treatment	95.8	89.9	91.9	95.8	91.3	92.8	132.8	98.0	109.8	94.9	91.5	92.7	95.8	73.4	81.0	157.5	84.1	109.0
0960	Flooring	132.5	109.0	125.6	132.5	100.3	123.1	135.8	111.6	128.7	132.5	102.5	123.7	112.1	69.3	99.5	135.0	98.8	124.4
0970, 0990	Wall Finishes & Painting/Coating	112.0	89.5	98.1	112.0	94.5	101.2	114.0	104.9	108.4	112.0	95.0	101.5	112.0	66.5	83.9	111.9	81.4	93.1
09	**FINISHES**	118.4	94.8	105.3	117.1	93.6	104.0	131.6	102.4	115.4	119.5	94.6	105.6	110.3	73.2	89.6	133.9	88.2	108.4
COVERS	**DIVS. 10 - 14, 25, 28, 41, 43, 44, 46**	140.5	78.7	128.2	140.5	77.4	128.0	140.5	105.7	133.6	140.5	90.7	130.6	140.5	70.8	126.7	140.5	99.5	132.4
21, 22, 23	**FIRE SUPPRESSION, PLUMBING & HVAC**	99.8	82.7	92.8	103.2	103.6	103.4	100.6	97.7	99.4	103.2	96.1	100.4	103.2	89.2	97.5	100.6	84.7	94.2
26, 27, 3370	**ELECTRICAL, COMMUNICATIONS & UTIL.**	117.5	84.5	100.5	118.9	94.9	106.5	114.0	88.1	100.7	119.8	75.9	97.1	117.3	67.3	91.5	115.1	77.7	95.8
MF2010	**WEIGHTED AVERAGE**	113.1	90.2	103.0	117.9	94.9	107.8	123.0	98.8	112.3	115.1	91.3	104.6	115.1	77.8	98.6	128.1	89.1	110.9

DIVISION		CANADA																	
		VICTORIA, BRITISH COLUMBIA			WHITEHORSE, YUKON			WINDSOR, ONTARIO			WINNIPEG, MANITOBA			YARMOUTH, NOVA SCOTIA			YELLOWKNIFE, NWT		
		MAT.	INST.	TOTAL	MAT.	INST.	TOTAL	MAT.	INST.	TOTAL	MAT.	INST.	TOTAL	MAT.	INST.	TOTAL	MAT.	INST.	TOTAL
015433	**CONTRACTOR EQUIPMENT**		108.8	108.8		101.8	101.8		101.7	101.7		105.9	105.9		101.7	101.7		101.7	101.7
0241, 31 - 34	**SITE & INFRASTRUCTURE, DEMOLITION**	121.1	108.0	112.1	108.8	97.7	101.2	96.3	101.8	100.1	121.5	101.6	107.8	97.0	99.3	98.5	128.6	102.3	110.5
0310	Concrete Forming & Accessories	111.9	86.4	89.7	107.0	62.3	68.1	123.1	89.9	94.3	131.6	69.8	77.8	97.4	75.0	77.9	106.9	85.3	88.2
0320	Concrete Reinforcing	116.8	79.6	98.0	115.8	62.7	89.0	109.3	84.6	96.9	143.8	58.2	100.7	148.8	48.8	98.4	118.4	66.6	92.3
0330	Cast-in-Place Concrete	147.1	100.3	128.6	162.2	72.9	127.0	137.7	94.9	120.8	192.3	75.4	146.2	168.1	74.8	131.3	245.0	92.6	185.0
03	**CONCRETE**	151.9	90.1	121.8	133.5	66.8	101.0	121.3	90.8	106.4	162.7	70.1	117.6	148.4	70.7	110.5	175.5	84.6	131.2
04	**MASONRY**	167.0	93.9	122.4	170.1	65.3	106.2	159.5	98.5	122.3	188.0	70.9	116.5	158.8	74.8	107.6	177.2	79.2	117.4
05	**METALS**	107.3	87.2	100.7	110.9	75.6	99.2	105.7	91.4	101.0	154.6	75.5	128.5	105.8	76.3	96.1	109.7	81.6	100.4
06	**WOOD, PLASTICS & COMPOSITES**	98.3	85.1	90.7	96.2	60.9	75.8	115.4	88.8	100.0	113.4	70.6	88.7	87.4	74.3	79.8	96.2	87.0	90.9
07	**THERMAL & MOISTURE PROTECTION**	114.9	88.8	104.3	105.3	64.9	88.9	110.8	92.2	103.3	116.3	72.4	98.5	106.5	73.1	92.9	108.2	83.1	98.0
08	**OPENINGS**	89.1	80.4	86.9	86.7	57.9	79.5	82.6	87.8	84.0	93.4	62.8	85.7	83.9	67.7	79.9	86.7	73.1	83.3
0920	Plaster & Gypsum Board	136.1	84.1	99.0	124.3	59.5	78.0	158.4	88.4	108.4	161.4	69.1	95.5	155.9	73.4	96.9	175.0	86.5	111.8
0950, 0980	Ceilings & Acoustic Treatment	95.8	84.1	88.1	100.3	59.5	73.4	95.8	88.4	90.9	142.8	69.1	94.1	95.8	73.4	81.0	130.2	86.5	101.3
0960	Flooring	127.3	78.5	113.0	121.1	63.2	104.1	132.5	106.1	124.8	133.7	74.8	116.4	112.1	69.3	99.5	121.1	96.9	114.0
0970, 0990	Wall Finishes & Painting/Coating	112.0	81.4	93.1	112.0	58.1	78.7	112.0	93.9	100.8	112.0	56.2	77.6	112.0	66.5	83.9	112.0	84.6	95.1
09	**FINISHES**	115.7	84.7	98.4	111.7	61.7	83.9	116.6	93.6	103.8	129.7	69.6	96.2	110.3	73.2	89.6	124.3	86.8	103.4
COVERS	**DIVS. 10 - 14, 25, 28, 41, 43, 44, 46**	140.5	74.5	127.4	140.5	67.9	126.1	140.5	78.1	128.1	140.5	70.5	126.6	140.5	70.8	126.7	140.5	72.5	127.0
21, 22, 23	**FIRE SUPPRESSION, PLUMBING & HVAC**	103.2	84.6	95.7	103.3	79.0	93.5	99.8	82.3	92.7	100.5	67.9	87.3	103.2	89.2	97.5	103.6	100.9	102.5
26, 27, 3370	**ELECTRICAL, COMMUNICATIONS & UTIL.**	118.3	77.7	97.4	127.6	65.0	95.3	122.0	86.0	103.4	117.9	68.3	92.3	119.4	67.3	92.5	131.3	90.6	110.3
MF2010	**WEIGHTED AVERAGE**	118.1	87.2	104.5	116.5	71.3	96.6	112.2	90.0	102.4	128.4	72.4	103.7	115.2	77.8	98.7	123.5	88.9	108.3

Location Factors

Costs shown in RSMeans cost data publications are based on national averages for materials and installation. To adjust these costs to a specific location, simply multiply the base cost by the factor and divide by 100 for that city. The data is arranged alphabetically by state and postal zip code numbers. For a city not listed, use the factor for a nearby city with similar economic characteristics.

STATE/ZIP	CITY	MAT.	INST.	TOTAL
ALABAMA				
350-352	Birmingham	97.6	74.6	87.4
354	Tuscaloosa	96.6	58.6	79.9
355	Jasper	96.8	57.6	79.5
356	Decatur	96.5	60.1	80.5
357-358	Huntsville	96.6	68.1	84.1
359	Gadsden	96.6	57.6	79.4
360-361	Montgomery	97.4	56.6	79.4
362	Anniston	96.0	64.5	82.1
363	Dothan	96.6	52.0	76.9
364	Evergreen	96.2	53.4	77.3
365-366	Mobile	97.7	67.5	84.4
367	Selma	96.3	51.9	76.7
368	Phenix City	97.1	55.9	78.9
369	Butler	96.5	52.0	76.9
ALASKA				
995-996	Anchorage	124.5	114.0	119.9
997	Fairbanks	122.4	115.0	119.1
998	Juneau	120.3	113.9	117.5
999	Ketchikan	134.1	113.9	125.2
ARIZONA				
850,853	Phoenix	97.9	76.6	88.5
851,852	Mesa/Tempe	97.4	72.4	86.3
855	Globe	97.4	62.5	82.0
856-857	Tucson	96.1	72.1	85.5
859	Show Low	97.5	62.9	82.2
860	Flagstaff	100.3	73.9	88.7
863	Prescott	98.2	62.1	82.2
864	Kingman	96.4	73.4	86.3
865	Chambers	96.5	63.0	81.7
ARKANSAS				
716	Pine Bluff	96.7	63.8	82.2
717	Camden	94.1	50.1	74.7
718	Texarkana	95.9	51.3	76.2
719	Hot Springs	93.4	52.9	75.5
720-722	Little Rock	97.0	65.0	82.9
723	West Memphis	95.8	62.8	81.2
724	Jonesboro	96.8	60.7	80.9
725	Batesville	94.2	56.1	77.4
726	Harrison	95.5	52.3	76.5
727	Fayetteville	92.9	54.1	75.8
728	Russellville	94.1	55.4	77.0
729	Fort Smith	97.1	62.5	81.8
CALIFORNIA				
900-902	Los Angeles	100.1	115.5	106.9
903-905	Inglewood	95.3	113.1	103.2
906-908	Long Beach	96.9	113.2	104.1
910-912	Pasadena	95.2	113.6	103.3
913-916	Van Nuys	98.1	113.5	104.9
917-918	Alhambra	97.1	113.5	104.3
919-921	San Diego	99.3	107.8	103.0
922	Palm Springs	96.4	110.9	102.8
923-924	San Bernardino	94.3	110.5	101.4
925	Riverside	98.9	112.4	104.9
926-927	Santa Ana	96.1	110.8	102.6
928	Anaheim	98.9	112.8	105.1
930	Oxnard	100.2	112.3	105.5
931	Santa Barbara	99.5	112.3	105.1
932-933	Bakersfield	100.0	109.5	104.2
934	San Luis Obispo	99.7	109.4	104.0
935	Mojave	97.0	106.9	101.3
936-938	Fresno	100.5	115.7	107.2
939	Salinas	100.5	122.7	110.3
940-941	San Francisco	108.5	142.3	123.4
942,956-958	Sacramento	102.9	118.6	109.9
943	Palo Alto	101.3	129.9	113.9
944	San Mateo	103.8	131.3	115.9
945	Vallejo	102.2	128.2	113.7
946	Oakland	106.3	133.9	118.5
947	Berkeley	105.5	133.2	117.7
948	Richmond	104.7	129.7	115.7
949	San Rafael	106.0	129.5	116.4
950	Santa Cruz	105.0	122.8	112.9

STATE/ZIP	CITY	MAT.	INST.	TOTAL
CALIFORNIA (CONT'D)				
951	San Jose	103.2	135.1	117.3
952	Stockton	100.7	117.2	108.0
953	Modesto	100.7	116.9	107.9
954	Santa Rosa	100.3	135.0	115.6
955	Eureka	101.6	114.2	107.2
959	Marysville	100.9	114.4	106.8
960	Redding	102.8	114.4	107.9
961	Susanville	101.5	114.4	107.2
COLORADO				
800-802	Denver	101.7	84.7	94.2
803	Boulder	98.6	82.9	91.7
804	Golden	100.7	81.7	92.3
805	Fort Collins	102.6	79.1	92.2
806	Greeley	100.0	78.2	90.4
807	Fort Morgan	99.3	82.0	91.7
808-809	Colorado Springs	101.6	81.7	92.8
810	Pueblo	100.4	80.8	91.8
811	Alamosa	101.5	76.7	90.6
812	Salida	101.3	78.3	91.2
813	Durango	101.9	79.6	92.1
814	Montrose	100.8	79.3	91.3
815	Grand Junction	104.2	79.6	93.4
816	Glenwood Springs	101.6	82.2	93.1
CONNECTICUT				
060	New Britain	101.1	118.6	108.8
061	Hartford	102.0	118.7	109.4
062	Willimantic	101.7	118.4	109.1
063	New London	97.6	118.6	106.9
064	Meriden	99.7	118.6	108.0
065	New Haven	102.8	118.6	109.8
066	Bridgeport	102.2	118.7	109.5
067	Waterbury	101.8	118.6	109.2
068	Norwalk	101.8	126.3	112.6
069	Stamford	101.9	126.3	112.7
D.C.				
200-205	Washington	102.4	92.8	98.2
DELAWARE				
197	Newark	99.4	107.9	103.2
198	Wilmington	99.1	108.0	103.0
199	Dover	99.5	107.9	103.2
FLORIDA				
320,322	Jacksonville	98.7	69.4	85.7
321	Daytona Beach	98.8	75.2	88.4
323	Tallahassee	99.4	59.3	81.7
324	Panama City	100.1	60.5	82.7
325	Pensacola	102.6	64.1	85.6
326,344	Gainesville	100.4	69.7	86.9
327-328,347	Orlando	101.8	73.0	89.1
329	Melbourne	101.5	78.2	91.2
330-332,340	Miami	99.6	76.7	89.6
333	Fort Lauderdale	98.1	75.9	88.3
334,349	West Palm Beach	96.7	74.4	86.9
335-336,346	Tampa	99.8	82.1	92.0
337	St. Petersburg	102.5	65.8	86.3
338	Lakeland	99.0	82.0	91.5
339,341	Fort Myers	98.1	75.8	88.3
342	Sarasota	100.0	77.3	90.0
GEORGIA				
300-303,399	Atlanta	98.0	76.5	88.5
304	Statesboro	97.4	54.0	78.3
305	Gainesville	96.1	64.0	82.0
306	Athens	95.5	64.2	81.7
307	Dalton	97.6	60.1	81.1
308-309	Augusta	96.7	64.5	82.5
310-312	Macon	96.4	64.2	82.2
313-314	Savannah	98.1	60.5	81.5
315	Waycross	97.3	61.4	81.5
316	Valdosta	97.6	60.8	81.4
317,398	Albany	97.6	61.5	81.7
318-319	Columbus	97.5	64.6	83.0

Location Factors

STATE/ZIP	CITY	MAT.	INST.	TOTAL
HAWAII				
967	Hilo	110.3	121.2	115.1
968	Honolulu	113.3	121.2	116.8
STATES & POSS.				
969	Guam	131.1	60.8	100.1
IDAHO				
832	Pocatello	100.3	72.2	87.9
833	Twin Falls	101.3	52.5	79.8
834	Idaho Falls	98.9	58.7	81.2
835	Lewiston	107.9	79.7	95.5
836-837	Boise	99.3	72.4	87.4
838	Coeur d'Alene	107.2	76.3	93.6
ILLINOIS				
600-603	North Suburban	98.7	132.4	113.5
604	Joliet	98.6	136.4	115.3
605	South Suburban	98.7	132.4	113.5
606-608	Chicago	99.1	138.9	116.6
609	Kankakee	94.6	127.4	109.1
610-611	Rockford	98.4	124.3	109.8
612	Rock Island	95.9	98.9	97.2
613	La Salle	97.0	120.8	107.5
614	Galesburg	96.9	104.8	100.4
615-616	Peoria	99.5	108.3	103.4
617	Bloomington	96.3	109.3	102.0
618-619	Champaign	99.8	108.1	103.5
620-622	East St. Louis	95.1	108.3	100.9
623	Quincy	96.0	100.1	97.8
624	Effingham	95.3	103.9	99.1
625	Decatur	97.4	106.3	101.3
626-627	Springfield	97.8	107.1	101.9
628	Centralia	93.4	107.0	99.4
629	Carbondale	93.1	102.0	97.1
INDIANA				
460	Anderson	96.1	83.4	90.5
461-462	Indianapolis	99.2	85.6	93.2
463-464	Gary	97.6	108.5	102.4
465-466	South Bend	95.7	82.9	90.1
467-468	Fort Wayne	96.7	78.7	88.8
469	Kokomo	93.7	82.2	88.6
470	Lawrenceburg	92.5	79.6	86.8
471	New Albany	93.6	76.6	86.1
472	Columbus	96.0	81.5	89.6
473	Muncie	97.1	82.6	90.7
474	Bloomington	98.4	82.9	91.6
475	Washington	94.4	83.7	89.7
476-477	Evansville	96.0	86.5	91.8
478	Terre Haute	96.7	86.9	92.4
479	Lafayette	95.7	83.2	90.2
IOWA				
500-503,509	Des Moines	99.1	84.7	92.7
504	Mason City	97.6	67.8	84.5
505	Fort Dodge	97.9	59.7	81.0
506-507	Waterloo	99.7	72.2	87.6
508	Creston	98.1	78.8	89.6
510-511	Sioux City	100.5	70.7	87.3
512	Sibley	98.9	56.5	80.2
513	Spencer	100.6	56.5	81.2
514	Carroll	97.4	72.4	86.4
515	Council Bluffs	101.7	78.8	91.6
516	Shenandoah	98.2	72.3	86.8
520	Dubuque	99.9	79.7	91.0
521	Decorah	98.5	65.7	84.0
522-524	Cedar Rapids	100.7	84.0	93.3
525	Ottumwa	98.4	70.7	86.2
526	Burlington	97.6	77.3	88.6
527-528	Davenport	99.6	90.4	95.5
KANSAS				
660-662	Kansas City	99.5	95.4	97.7
664-666	Topeka	99.2	64.5	83.9
667	Fort Scott	97.7	73.9	87.2
668	Emporia	97.9	69.7	85.4
669	Belleville	99.6	65.5	84.6
670-672	Wichita	98.4	64.1	83.3
673	Independence	99.2	69.5	86.1
674	Salina	100.0	66.9	85.4
675	Hutchinson	94.8	65.4	81.8
676	Hays	98.8	65.8	84.3
677	Colby	99.5	65.8	84.6

STATE/ZIP	CITY	MAT.	INST.	TOTAL
KANSAS (CONT'D)				
678	Dodge City	101.2	66.8	86.0
679	Liberal	98.7	65.8	84.2
KENTUCKY				
400-402	Louisville	97.8	83.5	91.5
403-405	Lexington	96.5	81.8	90.0
406	Frankfort	97.2	83.5	91.2
407-409	Corbin	93.7	72.5	84.4
410	Covington	94.4	95.2	94.8
411-412	Ashland	93.3	98.5	95.6
413-414	Campton	94.5	81.2	88.6
415-416	Pikeville	95.6	91.7	93.9
417-418	Hazard	93.9	81.5	88.4
420	Paducah	92.6	85.7	89.6
421-422	Bowling Green	95.0	83.5	90.0
423	Owensboro	95.2	87.9	92.0
424	Henderson	92.4	87.9	90.4
425-426	Somerset	91.9	79.1	86.2
427	Elizabethtown	91.4	83.9	88.1
LOUISIANA				
700-701	New Orleans	102.3	70.5	88.3
703	Thibodaux	100.2	63.4	84.0
704	Hammond	97.5	56.1	79.3
705	Lafayette	99.9	61.5	83.0
706	Lake Charles	100.1	64.1	84.2
707-708	Baton Rouge	101.2	63.0	84.4
710-711	Shreveport	96.7	56.3	78.9
712	Monroe	97.0	53.6	77.9
713-714	Alexandria	97.1	53.5	77.9
MAINE				
039	Kittery	94.8	80.1	88.3
040-041	Portland	99.5	78.9	90.4
042	Lewiston	98.1	78.9	89.7
043	Augusta	97.6	79.8	89.8
044	Bangor	97.7	78.9	89.4
045	Bath	96.0	79.5	88.7
046	Machias	95.5	79.0	88.2
047	Houlton	95.7	78.9	88.3
048	Rockland	94.7	79.4	88.0
049	Waterville	96.1	79.8	88.9
MARYLAND				
206	Waldorf	98.5	82.4	91.4
207-208	College Park	98.5	84.2	92.2
209	Silver Spring	97.7	82.3	90.9
210-212	Baltimore	100.2	84.7	93.3
214	Annapolis	100.2	83.2	92.7
215	Cumberland	95.8	80.2	89.0
216	Easton	97.5	70.3	85.5
217	Hagerstown	96.8	82.9	90.7
218	Salisbury	97.9	65.3	83.5
219	Elkton	94.7	84.3	90.1
MASSACHUSETTS				
010-011	Springfield	100.0	108.7	103.9
012	Pittsfield	99.6	105.3	102.1
013	Greenfield	97.3	108.4	102.2
014	Fitchburg	96.1	123.3	108.1
015-016	Worcester	100.0	123.5	110.4
017	Framingham	95.5	129.9	110.7
018	Lowell	99.4	128.5	112.3
019	Lawrence	100.5	129.0	113.1
020-022, 024	Boston	102.2	136.8	117.5
023	Brockton	100.8	124.7	111.4
025	Buzzards Bay	94.9	122.1	106.9
026	Hyannis	98.0	122.1	108.6
027	New Bedford	99.9	122.2	109.7
MICHIGAN				
480,483	Royal Oak	93.4	106.2	99.1
481	Ann Arbor	95.8	108.1	101.2
482	Detroit	97.4	110.1	103.0
484-485	Flint	95.5	97.1	96.2
486	Saginaw	95.0	93.5	94.4
487	Bay City	95.2	93.6	94.5
488-489	Lansing	96.4	95.1	95.9
490	Battle Creek	96.0	89.2	93.0
491	Kalamazoo	96.3	87.4	92.4
492	Jackson	94.3	95.9	95.0
493,495	Grand Rapids	97.2	83.0	90.9
494	Muskegon	94.9	84.9	90.5

STATE/ZIP	CITY	MAT.	INST.	TOTAL
MICHIGAN (CONT'D)				
496	Traverse City	93.6	78.7	87.0
497	Gaylord	94.7	72.8	85.1
498-499	Iron mountain	96.7	87.8	92.8
MINNESOTA				
550-551	Saint Paul	101.1	121.5	110.1
553-555	Minneapolis	102.1	123.0	111.3
556-558	Duluth	100.1	112.0	105.4
559	Rochester	100.9	107.6	103.8
560	Mankato	97.7	104.6	100.7
561	Windom	96.4	93.8	95.2
562	Willmar	96.0	105.1	100.0
563	St. Cloud	97.6	120.8	107.9
564	Brainerd	97.6	105.3	101.0
565	Detroit Lakes	99.4	99.3	99.4
566	Bemidji	98.7	101.3	99.9
567	Thief River Falls	98.3	95.8	97.2
MISSISSIPPI				
386	Clarksdale	96.4	55.5	78.4
387	Greenville	100.3	65.1	84.7
388	Tupelo	98.0	57.9	80.4
389	Greenwood	97.6	55.1	78.8
390-392	Jackson	97.4	64.4	82.9
393	Meridian	96.7	66.1	83.2
394	Laurel	97.9	58.8	80.6
395	Biloxi	98.3	60.2	81.5
396	Mccomb	96.3	54.6	77.9
397	Columbus	97.8	56.9	79.8
MISSOURI				
630-631	St. Louis	100.0	105.7	102.5
633	Bowling Green	98.4	90.5	94.9
634	Hannibal	97.3	87.1	92.8
635	Kirksville	100.4	83.8	93.1
636	Flat River	99.4	92.9	96.5
637	Cape Girardeau	99.7	89.6	95.3
638	Sikeston	97.5	85.5	92.2
639	Poplar Bluff	97.0	86.8	92.5
640-641	Kansas City	100.1	108.0	103.6
644-645	St. Joseph	99.3	91.8	96.0
646	Chillicothe	96.1	70.0	84.6
647	Harrisonville	95.7	104.5	99.6
648	Joplin	98.2	70.4	85.9
650-651	Jefferson City	96.2	90.5	93.7
652	Columbia	97.7	92.4	95.4
653	Sedalia	96.6	92.8	94.9
654-655	Rolla	95.2	87.9	92.0
656-658	Springfield	98.5	79.9	90.3
MONTANA				
590-591	Billings	103.5	74.7	90.8
592	Wolf Point	103.1	67.5	87.4
593	Miles City	100.7	69.1	86.8
594	Great Falls	105.0	75.3	91.9
595	Havre	102.0	70.2	88.0
596	Helena	102.6	72.0	89.1
597	Butte	103.5	73.5	90.3
598	Missoula	100.8	72.2	88.2
599	Kalispell	99.6	70.1	86.6
NEBRASKA				
680-681	Omaha	100.0	79.2	90.8
683-685	Lincoln	98.0	73.6	87.2
686	Columbus	96.8	71.7	85.7
687	Norfolk	98.5	75.1	88.2
688	Grand Island	98.7	78.4	89.8
689	Hastings	97.8	79.3	89.6
690	Mccook	97.6	70.7	85.7
691	North Platte	98.1	79.4	89.9
692	Valentine	99.7	67.6	85.6
693	Alliance	99.7	65.8	84.8
NEVADA				
889-891	Las Vegas	99.8	110.7	104.6
893	Ely	98.5	98.1	98.3
894-895	Reno	98.7	93.7	96.5
897	Carson City	97.8	93.5	95.9
898	Elko	97.2	85.3	92.0
NEW HAMPSHIRE				
030	Nashua	100.2	86.7	94.3
031	Manchester	100.7	86.7	94.5

STATE/ZIP	CITY	MAT.	INST.	TOTAL
NEW HAMPSHIRE (CONT'D)				
032-033	Concord	98.6	82.4	91.5
034	Keene	96.9	49.4	76.0
035	Littleton	96.5	56.6	78.9
036	Charleston	96.3	47.1	74.6
037	Claremont	95.3	47.1	74.1
038	Portsmouth	97.8	91.2	94.9
NEW JERSEY				
070-071	Newark	102.7	124.6	112.4
072	Elizabeth	100.6	124.5	111.2
073	Jersey City	99.4	124.4	110.4
074-075	Paterson	101.4	124.5	111.6
076	Hackensack	99.4	124.5	110.5
077	Long Branch	99.0	123.3	109.7
078	Dover	99.7	124.6	110.7
079	Summit	99.6	124.7	110.7
080,083	Vineland	97.6	122.4	108.6
081	Camden	99.4	122.6	109.6
082,084	Atlantic City	98.1	122.1	108.7
085-086	Trenton	99.6	123.6	110.2
087	Point Pleasant	99.7	122.5	109.7
088-089	New Brunswick	100.1	124.3	110.8
NEW MEXICO				
870-872	Albuquerque	98.7	73.6	87.7
873	Gallup	98.8	73.6	87.7
874	Farmington	99.4	73.6	88.0
875	Santa Fe	99.4	73.6	88.0
877	Las Vegas	97.3	73.6	86.8
878	Socorro	96.8	73.6	86.6
879	Truth/Consequences	97.0	70.3	85.3
880	Las Cruces	96.1	69.6	84.4
881	Clovis	97.4	73.5	86.9
882	Roswell	99.3	73.6	88.0
883	Carrizozo	99.4	73.6	88.1
884	Tucumcari	98.1	73.5	87.3
NEW YORK				
100-102	New York	104.8	166.9	132.2
103	Staten Island	100.8	163.6	128.5
104	Bronx	98.9	163.6	127.4
105	Mount Vernon	98.6	133.0	113.8
106	White Plains	98.8	133.0	113.9
107	Yonkers	103.5	133.1	116.5
108	New Rochelle	99.0	133.0	114.0
109	Suffern	98.7	122.0	109.0
110	Queens	101.4	163.8	128.9
111	Long Island City	103.0	163.8	129.8
112	Brooklyn	103.4	163.8	130.0
113	Flushing	102.9	163.8	129.7
114	Jamaica	101.1	163.8	128.7
115,117,118	Hicksville	101.3	145.6	120.8
116	Far Rockaway	103.1	163.8	129.8
119	Riverhead	102.1	146.1	121.5
120-122	Albany	96.7	96.4	96.5
123	Schenectady	97.8	96.1	97.0
124	Kingston	100.9	117.4	108.2
125-126	Poughkeepsie	100.1	128.4	112.6
127	Monticello	99.4	118.0	107.6
128	Glens Falls	92.3	92.6	92.5
129	Plattsburgh	96.9	87.0	92.6
130-132	Syracuse	98.9	94.5	97.0
133-135	Utica	96.8	88.2	93.0
136	Watertown	98.5	93.5	96.3
137-139	Binghamton	98.4	89.2	94.4
140-142	Buffalo	100.3	104.0	101.9
143	Niagara Falls	97.9	106.1	101.5
144-146	Rochester	100.0	94.0	97.3
147	Jamestown	96.9	86.4	92.3
148-149	Elmira	96.7	91.0	94.2
NORTH CAROLINA				
270,272-274	Greensboro	99.7	47.3	76.6
271	Winston-Salem	99.5	46.0	75.9
275-276	Raleigh	99.8	47.6	76.8
277	Durham	100.8	47.9	77.5
278	Rocky Mount	96.8	42.5	72.9
279	Elizabeth City	97.8	40.6	72.6
280	Gastonia	98.6	45.4	75.1
281-282	Charlotte	99.4	47.8	76.7
283	Fayetteville	100.7	49.4	78.1
284	Wilmington	97.4	48.5	75.8
285	Kinston	95.2	43.8	72.6

STATE/ZIP	CITY	MAT.	INST.	TOTAL
NORTH CAROLINA (CONT'D)				
286	Hickory	95.5	42.3	72.1
287-288	Asheville	97.8	45.5	74.7
289	Murphy	96.4	33.2	68.5
NORTH DAKOTA				
580-581	Fargo	102.1	65.7	86.0
582	Grand Forks	102.8	53.1	80.9
583	Devils Lake	101.7	57.0	82.0
584	Jamestown	101.9	46.4	77.4
585	Bismarck	100.7	64.4	84.7
586	Dickinson	102.7	58.8	83.3
587	Minot	102.7	70.2	88.4
588	Williston	101.1	58.8	82.4
OHIO				
430-432	Columbus	97.8	90.6	94.6
433	Marion	93.9	85.4	90.2
434-436	Toledo	97.9	98.3	98.1
437-438	Zanesville	94.5	86.1	90.8
439	Steubenville	96.1	93.5	95.0
440	Lorain	97.7	91.3	94.9
441	Cleveland	97.9	99.8	98.7
442-443	Akron	98.7	94.5	96.9
444-445	Youngstown	98.1	90.1	94.6
446-447	Canton	98.2	86.3	93.0
448-449	Mansfield	95.2	87.4	91.7
450	Hamilton	96.6	85.6	91.7
451-452	Cincinnati	96.9	85.8	92.0
453-454	Dayton	96.7	85.2	91.6
455	Springfield	96.6	85.8	91.8
456	Chillicothe	95.2	93.4	94.4
457	Athens	98.0	83.1	91.4
458	Lima	98.4	86.1	93.0
OKLAHOMA				
730-731	Oklahoma City	97.8	60.6	81.4
734	Ardmore	95.0	60.1	79.6
735	Lawton	97.5	61.4	81.6
736	Clinton	96.5	58.8	79.8
737	Enid	97.3	58.7	80.3
738	Woodward	95.1	58.8	79.1
739	Guymon	96.2	31.0	67.5
740-741	Tulsa	97.4	53.5	78.0
743	Miami	93.8	67.9	82.3
744	Muskogee	96.7	39.3	71.4
745	Mcalester	93.6	50.6	74.6
746	Ponca City	94.1	58.7	78.5
747	Durant	94.1	57.7	78.1
748	Shawnee	95.7	56.5	78.4
749	Poteau	93.2	61.4	79.2
OREGON				
970-972	Portland	98.8	100.3	99.5
973	Salem	98.6	98.6	98.6
974	Eugene	98.6	98.5	98.6
975	Medford	100.1	97.0	98.7
976	Klamath Falls	99.8	97.0	98.6
977	Bend	98.8	98.6	98.7
978	Pendleton	94.1	100.0	96.7
979	Vale	91.9	90.5	91.3
PENNSYLVANIA				
150-152	Pittsburgh	99.1	104.6	101.5
153	Washington	95.9	102.1	98.6
154	Uniontown	96.4	100.8	98.3
155	Bedford	97.2	91.7	94.8
156	Greensburg	97.2	100.3	98.5
157	Indiana	96.0	99.1	97.4
158	Dubois	97.6	94.5	96.2
159	Johnstown	97.2	95.8	96.6
160	Butler	92.3	100.5	95.9
161	New Castle	92.4	97.5	94.6
162	Kittanning	92.8	102.4	97.0
163	Oil City	92.4	94.7	93.4
164-165	Erie	94.6	93.6	94.1
166	Altoona	94.8	90.6	93.0
167	Bradford	95.5	93.8	94.7
168	State College	95.0	91.8	93.6
169	Wellsboro	96.1	92.4	94.5
170-171	Harrisburg	97.7	96.1	97.0
172	Chambersburg	95.5	89.5	92.9
173-174	York	96.3	95.2	95.8
175-176	Lancaster	94.3	89.7	92.3

STATE/ZIP	CITY	MAT.	INST.	TOTAL
PENNSYLVANIA (CONT'D)				
177	Williamsport	93.0	79.6	87.1
178	Sunbury	95.1	95.1	95.1
179	Pottsville	94.2	96.8	95.4
180	Lehigh Valley	95.6	113.5	103.5
181	Allentown	97.9	109.2	102.9
182	Hazleton	95.1	97.2	96.0
183	Stroudsburg	94.9	103.0	98.5
184-185	Scranton	98.6	99.5	99.0
186-187	Wilkes-Barre	94.8	98.2	96.3
188	Montrose	94.5	97.0	95.6
189	Doylestown	94.7	120.1	105.9
190-191	Philadelphia	99.9	130.9	113.6
193	Westchester	96.1	122.8	107.9
194	Norristown	95.2	130.2	110.6
195-196	Reading	97.6	100.0	98.6
PUERTO RICO				
009	San Juan	121.3	24.1	78.4
RHODE ISLAND				
028	Newport	99.2	111.1	104.5
029	Providence	100.2	111.1	105.0
SOUTH CAROLINA				
290-292	Columbia	97.8	49.3	76.4
293	Spartanburg	96.8	49.0	75.7
294	Charleston	98.4	57.4	80.3
295	Florence	96.6	49.3	75.8
296	Greenville	96.6	49.0	75.6
297	Rock Hill	96.0	47.2	74.5
298	Aiken	96.8	72.8	86.2
299	Beaufort	97.5	42.5	73.3
SOUTH DAKOTA				
570-571	Sioux Falls	100.7	57.1	81.5
572	Watertown	99.8	51.6	78.5
573	Mitchell	98.6	51.0	77.6
574	Aberdeen	101.6	52.2	79.8
575	Pierre	100.1	53.8	79.7
576	Mobridge	99.3	51.4	78.2
577	Rapid City	101.4	54.6	80.7
TENNESSEE				
370-372	Nashville	96.8	72.2	86.0
373-374	Chattanooga	99.0	66.2	84.6
375,380-381	Memphis	98.2	69.6	85.6
376	Johnson City	98.7	56.7	80.2
377-379	Knoxville	95.4	61.4	80.4
382	Mckenzie	97.1	61.8	81.6
383	Jackson	99.2	63.8	83.6
384	Columbia	95.7	67.1	83.1
385	Cookeville	97.0	60.2	80.8
TEXAS				
750	Mckinney	98.5	50.2	77.2
751	Waxahackie	98.5	57.3	80.3
752-753	Dallas	99.4	66.8	85.0
754	Greenville	98.6	43.3	74.2
755	Texarkana	98.0	51.0	77.2
756	Longview	98.6	40.4	72.9
757	Tyler	99.0	54.4	79.4
758	Palestine	95.2	56.6	78.2
759	Lufkin	95.8	58.3	79.3
760-761	Fort Worth	97.2	63.2	82.2
762	Denton	97.1	47.8	75.4
763	Wichita Falls	98.0	54.8	79.0
764	Eastland	96.5	49.7	75.9
765	Temple	95.1	49.4	75.0
766-767	Waco	97.4	56.4	79.4
768	Brownwood	97.7	49.0	76.2
769	San Angelo	97.3	49.6	76.3
770-772	Houston	99.1	69.8	86.2
773	Huntsville	97.5	55.9	79.2
774	Wharton	98.5	54.0	78.9
775	Galveston	96.5	68.2	84.0
776-777	Beaumont	97.2	62.0	81.7
778	Bryan	94.3	61.7	79.9
779	Victoria	98.6	44.0	74.5
780	Laredo	95.8	52.7	76.8
781-782	San Antonio	96.2	65.0	82.4
783-784	Corpus Christi	98.9	52.3	78.4
785	Mc Allen	98.6	46.6	75.7
786-787	Austin	96.1	58.8	79.6

Location Factors

STATE/ZIP	CITY	MAT.	INST.	TOTAL
TEXAS (CONT'D)				
788	Del Rio	98.0	46.5	75.3
789	Giddings	95.1	56.2	77.9
790-791	Amarillo	98.1	61.0	81.7
792	Childress	97.3	57.2	79.6
793-794	Lubbock	99.6	56.8	80.7
795-796	Abilene	97.9	54.2	78.6
797	Midland	99.4	51.4	78.2
798-799,885	El Paso	96.8	51.2	76.7
UTAH				
840-841	Salt Lake City	101.6	68.4	86.9
842,844	Ogden	97.3	70.0	85.3
843	Logan	99.2	70.0	86.3
845	Price	99.5	65.9	84.7
846-847	Provo	99.7	68.3	85.8
VERMONT				
050	White River Jct.	98.0	54.4	78.8
051	Bellows Falls	96.5	72.7	86.0
052	Bennington	96.8	67.8	84.0
053	Brattleboro	97.2	72.9	86.5
054	Burlington	100.6	65.4	85.1
056	Montpelier	97.4	65.4	83.3
057	Rutland	99.2	65.4	84.3
058	St. Johnsbury	97.9	55.2	79.1
059	Guildhall	96.6	55.0	78.3
VIRGINIA				
220-221	Fairfax	100.1	82.0	92.1
222	Arlington	101.6	79.9	92.0
223	Alexandria	100.6	86.0	94.2
224-225	Fredericksburg	98.8	71.8	86.9
226	Winchester	99.4	62.2	83.0
227	Culpeper	99.3	67.0	85.1
228	Harrisonburg	99.6	63.0	83.4
229	Charlottesville	100.0	62.8	83.6
230-232	Richmond	100.6	66.0	85.3
233-235	Norfolk	101.4	67.6	86.5
236	Newport News	100.4	67.3	85.8
237	Portsmouth	99.9	64.4	84.3
238	Petersburg	99.9	66.0	85.0
239	Farmville	98.9	52.3	78.4
240-241	Roanoke	101.2	59.8	82.9
242	Bristol	98.6	54.3	79.1
243	Pulaski	98.3	50.3	77.1
244	Staunton	99.2	58.2	81.1
245	Lynchburg	99.3	63.6	83.5
246	Grundy	98.6	49.7	77.0
WASHINGTON				
980-981,987	Seattle	102.5	106.3	104.2
982	Everett	102.2	95.6	99.3
983-984	Tacoma	102.2	97.9	100.3
985	Olympia	99.4	97.9	98.7
986	Vancouver	103.5	93.0	98.8
988	Wenatchee	101.8	83.6	93.8
989	Yakima	102.5	92.6	98.1
990-992	Spokane	102.5	82.7	93.7
993	Richland	102.1	86.8	95.4
994	Clarkston	100.4	81.3	92.0
WEST VIRGINIA				
247-248	Bluefield	97.4	82.6	90.9
249	Lewisburg	99.1	82.7	91.9
250-253	Charleston	99.2	89.5	94.9
254	Martinsburg	98.9	79.1	90.2
255-257	Huntington	100.6	91.4	96.6
258-259	Beckley	97.3	86.6	92.6
260	Wheeling	100.7	89.2	95.7
261	Parkersburg	99.6	88.4	94.7
262	Buckhannon	98.8	89.0	94.5
263-264	Clarksburg	99.4	89.1	94.9
265	Morgantown	99.5	88.2	94.5
266	Gassaway	98.7	89.8	94.8
267	Romney	98.7	85.1	92.7
268	Petersburg	98.5	84.9	92.5
WISCONSIN				
530,532	Milwaukee	100.9	108.3	104.2
531	Kenosha	100.8	101.0	100.9
534	Racine	100.1	101.0	100.5
535	Beloit	100.0	96.0	98.3
537	Madison	99.3	97.2	98.4

STATE/ZIP	CITY	MAT.	INST.	TOTAL
WISCONSIN (CONT'D)				
538	Lancaster	97.6	91.5	94.9
539	Portage	96.1	94.5	95.4
540	New Richmond	97.3	94.4	96.0
541-543	Green Bay	101.8	91.6	97.3
544	Wausau	96.5	91.7	94.4
545	Rhinelander	99.8	90.5	95.7
546	La Crosse	98.1	93.2	95.9
547	Eau Claire	99.7	94.3	97.3
548	Superior	97.1	96.9	97.0
549	Oshkosh	97.2	82.6	90.8
WYOMING				
820	Cheyenne	99.1	63.8	83.5
821	Yellowstone Nat'l Park	96.4	59.3	80.1
822	Wheatland	97.5	58.2	80.2
823	Rawlins	99.0	57.0	80.5
824	Worland	97.1	57.0	79.4
825	Riverton	98.1	57.0	80.0
826	Casper	98.9	57.7	80.7
827	Newcastle	96.9	57.0	79.3
828	Sheridan	99.7	60.6	82.4
829-831	Rock Springs	101.0	57.2	81.7
CANADIAN FACTORS (reflect Canadian currency)				
ALBERTA				
	Calgary	129.8	93.1	113.6
	Edmonton	130.0	93.1	113.7
	Fort McMurray	126.5	95.8	113.0
	Lethbridge	120.8	95.3	109.5
	Lloydminster	116.0	91.8	105.4
	Medicine Hat	116.1	91.1	105.1
	Red Deer	116.6	91.1	105.3
BRITISH COLUMBIA				
	Kamloops	116.9	94.4	107.0
	Prince George	118.1	93.7	107.3
	Vancouver	128.1	89.1	110.9
	Victoria	118.1	87.2	104.5
MANITOBA				
	Brandon	116.2	79.0	99.8
	Portage la Prairie	116.0	77.2	98.9
	Winnipeg	128.4	72.4	103.7
NEW BRUNSWICK				
	Bathurst	114.1	70.1	94.7
	Dalhousie	114.1	70.3	94.8
	Fredericton	117.0	75.6	98.8
	Moncton	114.3	71.7	95.5
	Newcastle	114.1	70.8	95.0
	St. John	116.6	75.3	98.4
NEWFOUNDLAND				
	Corner Brook	119.5	70.5	97.9
	St Johns	120.3	75.7	100.6
NORTHWEST TERRITORIES				
	Yellowknife	123.5	88.9	108.3
NOVA SCOTIA				
	Bridgewater	115.6	77.8	98.9
	Dartmouth	117.2	77.8	99.8
	Halifax	119.0	80.3	101.9
	New Glasgow	115.3	77.8	98.8
	Sydney	112.8	77.8	97.4
	Truro	115.1	77.8	98.6
	Yarmouth	115.2	77.8	98.7
ONTARIO				
	Barrie	118.6	96.9	109.1
	Brantford	117.7	101.0	110.4
	Cornwall	117.7	97.2	108.7
	Hamilton	121.9	93.7	109.4
	Kingston	118.6	97.3	109.2
	Kitchener	113.7	90.4	103.4
	London	121.6	90.2	107.8
	North Bay	118.0	94.9	107.8
	Oshawa	117.1	98.2	108.7
	Ottawa	122.9	90.7	108.7
	Owen Sound	118.8	95.0	108.3
	Peterborough	117.7	97.0	108.6
	Sarnia	117.4	101.6	110.5

Location Factors

STATE/ZIP	CITY	MAT.	INST.	TOTAL
ONTARIO (CONT'D)	Sault Ste Marie	112.6	94.5	104.6
	St. Catharines	111.7	91.3	102.8
	Sudbury	112.3	90.4	102.7
	Thunder Bay	113.1	90.2	103.0
	Timmins	117.9	94.9	107.8
	Toronto	123.0	98.8	112.3
	Windsor	112.2	90.0	102.4
PRINCE EDWARD ISLAND				
	Charlottetown	118.2	65.4	95.0
	Summerside	117.0	65.5	94.3
QUEBEC				
	Cap-de-la-Madeleine	114.8	91.3	104.4
	Charlesbourg	114.8	91.3	104.4
	Chicoutimi	113.7	91.6	104.0
	Gatineau	114.4	91.1	104.2
	Granby	114.7	91.0	104.3
	Hull	114.5	91.1	104.2
	Joliette	114.9	91.3	104.5
	Laval	114.0	91.1	103.9
	Montreal	121.9	91.9	108.7
	Quebec	119.8	92.3	107.7
	Rimouski	114.2	91.6	104.2
	Rouyn-Noranda	114.4	91.1	104.1
	Saint Hyacinthe	113.6	91.1	103.7
	Sherbrooke	114.8	91.1	104.4
	Sorel	114.9	91.3	104.5
	St Jerome	114.4	91.1	104.1
	Trois Rivieres	115.1	91.3	104.6
SASKATCHEWAN				
	Moose Jaw	113.4	72.4	95.3
	Prince Albert	112.3	70.5	93.9
	Regina	116.7	86.2	103.2
	Saskatoon	113.2	86.2	101.3
YUKON				
	Whitehorse	116.5	71.3	96.6

R011105-05 Tips for Accurate Estimating

1. Use pre-printed or columnar forms for orderly sequence of dimensions and locations and for recording telephone quotations.
2. Use only the front side of each paper or form except for certain pre-printed summary forms.
3. Be consistent in listing dimensions: For example, length x width x height. This helps in rechecking to ensure that, the total length of partitions is appropriate for the building area.
4. Use printed (rather than measured) dimensions where given.
5. Add up multiple printed dimensions for a single entry where possible.
6. Measure all other dimensions carefully.
7. Use each set of dimensions to calculate multiple related quantities.
8. Convert foot and inch measurements to decimal feet when listing. Memorize decimal equivalents to .01 parts of a foot (1/8″ equals approximately .01′).
9. Do not "round off" quantities until the final summary.
10. Mark drawings with different colors as items are taken off.
11. Keep similar items together, different items separate.
12. Identify location and drawing numbers to aid in future checking for completeness.
13. Measure or list everything on the drawings or mentioned in the specifications.
14. It may be necessary to list items not called for to make the job complete.
15. Be alert for: Notes on plans such as N.T.S. (not to scale); changes in scale throughout the drawings; reduced size drawings; discrepancies between the specifications and the drawings.
16. Develop a consistent pattern of performing an estimate. For example:
    a. Start the quantity takeoff at the lower floor and move to the next higher floor.
    b. Proceed from the main section of the building to the wings.
    c. Proceed from south to north or vice versa, clockwise or counterclockwise.
    d. Take off floor plan quantities first, elevations next, then detail drawings.
17. List all gross dimensions that can be either used again for different quantities, or used as a rough check of other quantities for verification (exterior perimeter, gross floor area, individual floor areas, etc.).
18. Utilize design symmetry or repetition (repetitive floors, repetitive wings, symmetrical design around a center line, similar room layouts, etc.). Note: Extreme caution is needed here so as not to omit or duplicate an area.
19. Do not convert units until the final total is obtained. For instance, when estimating concrete work, keep all units to the nearest cubic foot, then summarize and convert to cubic yards.
20. When figuring alternatives, it is best to total all items involved in the basic system, then total all items involved in the alternates. Therefore you work with positive numbers in all cases. When adds and deducts are used, it is often confusing whether to add or subtract a portion of an item; especially on a complicated or involved alternate.

R011110-10 Architectural Fees

Tabulated below are typical percentage fees by project size, for good professional architectural service. Fees may vary from those listed depending upon degree of design difficulty and economic conditions in any particular area.

Rates can be interpolated horizontally and vertically. Various portions of the same project requiring different rates should be adjusted proportionately. For alterations, add 50% to the fee for the first $500,000 of project cost and add 25% to the fee for project cost over $500,000.

Architectural fees tabulated below include Structural, Mechanical and Electrical Engineering Fees. They do not include the fees for special consultants such as kitchen planning, security, acoustical, interior design, etc.

Civil Engineering fees are included in the Architectural fee for project sites requiring minimal design such as city sites. However, separate Civil Engineering fees must be added when utility connections require design, drainage calculations are needed, stepped foundations are required, or provisions are required to protect adjacent wetlands.

	Total Project Size in Thousands of Dollars						
Building Types	100	250	500	1,000	5,000	10,000	50,000
Factories, garages, warehouses, repetitive housing	9.0%	8.0%	7.0%	6.2%	5.3%	4.9%	4.5%
Apartments, banks, schools, libraries, offices, municipal buildings	12.2	12.3	9.2	8.0	7.0	6.6	6.2
Churches, hospitals, homes, laboratories, museums, research	15.0	13.6	12.7	11.9	9.5	8.8	8.0
Memorials, monumental work, decorative furnishings	—	16.0	14.5	13.1	10.0	9.0	8.3

General Requirements — R0111 Summary of Work

R011110-30 Engineering Fees

Typical **Structural Engineering Fees** based on type of construction and total project size. These fees are included in Architectural Fees.

Type of Construction	Total Project Size (in thousands of dollars)			
	$500	$500-$1,000	$1,000-$5,000	Over $5000
Industrial buildings, factories & warehouses	Technical payroll times 2.0 to 2.5	1.60%	1.25%	1.00%
Hotels, apartments, offices, dormitories, hospitals, public buildings, food stores	↓	2.00%	1.70%	1.20%
Museums, banks, churches and cathedrals	↓	2.00%	1.75%	1.25%
Thin shells, prestressed concrete, earthquake resistive	↓	2.00%	1.75%	1.50%
Parking ramps, auditoriums, stadiums, convention halls, hangars & boiler houses	↓	2.50%	2.00%	1.75%
Special buildings, major alterations, underpinning & future expansion	↓	Add to above 0.5%	Add to above 0.5%	Add to above 0.5%

For complex reinforced concrete or unusually complicated structures, add 20% to 50%.

Typical **Mechanical and Electrical Engineering Fees** are based on the size of the subcontract. The fee structure for both are shown below. These fees are included in Architectural Fees.

Type of Construction	Subcontract Size							
	$25,000	$50,000	$100,000	$225,000	$350,000	$500,000	$750,000	$1,000,000
Simple structures	6.4%	5.7%	4.8%	4.5%	4.4%	4.3%	4.2%	4.1%
Intermediate structures	8.0	7.3	6.5	5.6	5.1	5.0	4.9	4.8
Complex structures	10.1	9.0	9.0	8.0	7.5	7.5	7.0	7.0

For renovations, add 15% to 25% to applicable fee.

General Requirements — R0121 Allowances

R012153-10 Repair and Remodeling

Cost figures are based on new construction utilizing the most cost-effective combination of labor, equipment and material with the work scheduled in proper sequence to allow the various trades to accomplish their work in an efficient manner.

The costs for repair and remodeling work must be modified due to the following factors that may be present in any given repair and remodeling project.

1. Equipment usage curtailment due to the physical limitations of the project, with only hand-operated equipment being used.
2. Increased requirement for shoring and bracing to hold up the building while structural changes are being made and to allow for temporary storage of construction materials on above-grade floors.
3. Material handling becomes more costly due to having to move within the confines of an enclosed building. For multi-story construction, low capacity elevators and stairwells may be the only access to the upper floors.
4. Large amount of cutting and patching and attempting to match the existing construction is required. It is often more economical to remove entire walls rather than create many new door and window openings. This sort of trade-off has to be carefully analyzed.
5. Cost of protection of completed work is increased since the usual sequence of construction usually cannot be accomplished.
6. Economies of scale usually associated with new construction may not be present. If small quantities of components must be custom fabricated due to job requirements, unit costs will naturally increase. Also, if only small work areas are available at a given time, job scheduling between trades becomes difficult and subcontractor quotations may reflect the excessive start-up and shut-down phases of the job.
7. Work may have to be done on other than normal shifts and may have to be done around an existing production facility which has to stay in production during the course of the repair and remodeling.
8. Dust and noise protection of adjoining non-construction areas can involve substantial special protection and alter usual construction methods.
9. Job may be delayed due to unexpected conditions discovered during demolition or removal. These delays ultimately increase construction costs.
10. Piping and ductwork runs may not be as simple as for new construction. Wiring may have to be snaked through walls and floors.
11. Matching "existing construction" may be impossible because materials may no longer be manufactured. Substitutions may be expensive.
12. Weather protection of existing structure requires additional temporary structures to protect building at openings.
13. On small projects, because of local conditions, it may be necessary to pay a tradesman for a minimum of four hours for a task that is completed in one hour.

All of the above areas can contribute to increased costs for a repair and remodeling project. Each of the above factors should be considered in the planning, bidding and construction stage in order to minimize the increased costs associated with repair and remodeling jobs.

R012909-80 Sales Tax by State

State sales tax on materials is tabulated below (5 states have no sales tax). Many states allow local jurisdictions, such as a county or city, to levy additional sales tax.

Some projects may be sales tax exempt, particularly those constructed with public funds.

State	Tax (%)	State	Tax (%)	State	Tax (%)	State	Tax (%)
Alabama	4	Illinois	6.25	Montana	0	Rhode Island	7
Alaska	0	Indiana	7	Nebraska	5.5	South Carolina	6
Arizona	6.6	Iowa	6	Nevada	6.85	South Dakota	4
Arkansas	6	Kansas	5.3	New Hampshire	0	Tennessee	7
California	9.25	Kentucky	6	New Jersey	7	Texas	6.25
Colorado	2.9	Louisiana	4	New Mexico	5.13	Utah	4.75
Connecticut	6	Maine	5	New York	4	Vermont	6
Delaware	0	Maryland	6	North Carolina	5.75	Virginia	4
District of Columbia	6	Massachusetts	6.25	North Dakota	5	Washington	6.5
Florida	6	Michigan	6	Ohio	5.5	West Virginia	6
Georgia	4	Minnesota	6.875	Oklahoma	4.5	Wisconsin	5
Hawaii	4	Mississippi	7	Oregon	0	Wyoming	4
Idaho	6	Missouri	4.225	Pennsylvania	6	Average	5.07 %

Sales Tax by Province (Canada)

GST - a value-added tax, which the government imposes on most goods and services provided in or imported into Canada. PST - a retail sales tax, which three of the provinces impose on the price of most goods and some services. QST - a value-added tax, similar to the federal GST, which Quebec imposes. HST - Five provinces have combined their retail sales tax with the federal GST into one harmonized tax.

Province	PST (%)	QST (%)	GST(%)	HST(%)
Alberta	0	0	5	0
British Columbia	0	0	0	12
Manitoba	7	0	5	0
New Brunswick	0	0	0	13
Newfoundland	0	0	0	13
Northwest Territories	0	0	5	0
Nova Scotia	0	0	0	15
Ontario	0	0	0	13
Prince Edward Island	10	0	5	0
Quebec	0	8.5	5	0
Saskatchewan	5	0	5	0
Yukon	0	0	5	0

R012909-85 Unemployment Taxes and Social Security Taxes

State Unemployment Tax rates vary not only from state to state, but also with the experience rating of the contractor. The Federal Unemployment Tax rate is 6.2% of the first $7,000 of wages. This is reduced by a credit of up to 5.4% for timely payment to the state. The minimum Federal Unemployment Tax is 0.8% after all credits.

Social Security (FICA) for 2012 is estimated at time of publication to be 7.65% of wages up to $110,100.

R013113-40 Builder's Risk Insurance

Builder's Risk Insurance is insurance on a building during construction. Premiums are paid by the owner or the contractor. Blasting, collapse and underground insurance would raise total insurance costs above those listed. Floater policy for materials delivered to the job runs $.75 to $1.25 per $100 value. Contractor equipment insurance runs $.50 to $1.50 per $100 value. Insurance for miscellaneous tools to $1,500 value runs from $3.00 to $7.50 per $100 value.

Tabulated below are New England Builder's Risk insurance rates in dollars per $100 value for $1,000 deductible. For $25,000 deductible, rates can be reduced 13% to 34%. On contracts over $1,000,000, rates may be lower than those tabulated. Policies are written annually for the total completed value in place. For "all risk" insurance (excluding flood, earthquake and certain other perils) add $.025 to total rates below.

Coverage	Frame Construction (Class 1)		Brick Construction (Class 4)		Fire Resistive (Class 6)	
	Range	Average	Range	Average	Range	Average
Fire Insurance	$.350 to $ .850	$.600	$.158 to $.189	$.174	$.052 to $.080	$.070
Extended Coverage	.115 to .200	.158	.080 to .105	.101	.081 to .105	.100
Vandalism	.012 to .016	.014	.008 to .011	.011	.008 to .011	.010
Total Annual Rate	$.477 to $1.066	$.772	$.246 to $.305	$.286	$.141 to $.196	$.180

R013113-50 General Contractor's Overhead

There are two distinct types of overhead on a construction project: Project Overhead and Main Office Overhead. Project Overhead includes those costs at a construction site not directly associated with the installation of construction materials. Examples of Project Overhead costs include the following:

1. Superintendent
2. Construction office and storage trailers
3. Temporary sanitary facilities
4. Temporary utilities
5. Security fencing
6. Photographs
7. Clean up
8. Performance and payment bonds

The above Project Overhead items are also referred to as General Requirements and therefore are estimated in Division 1. Division 1 is the first division listed in the CSI MasterFormat but it is usually the last division estimated. The sum of the costs in Divisions 1 through 49 is referred to as the sum of the direct costs.

All construction projects also include indirect costs. The primary components of indirect costs are the contractor's Main Office Overhead and profit. The amount of the Main Office Overhead expense varies depending on the the following:

1. Owner's compensation
2. Project managers and estimator's wages
3. Clerical support wages
4. Office rent and utilities
5. Corporate legal and accounting costs
6. Advertising
7. Automobile expenses
8. Association dues
9. Travel and entertainment expenses

These costs are usually calculated as a percentage of annual sales volume. This percentage can range from 35% for a small contractor doing less than $500,000 to 5% for a large contractor with sales in excess of $100 million.

R013113-60 Workers' Compensation Insurance Rates by Trade

The table below tabulates the national averages for Workers' Compensation insurance rates by trade and type of building. The average "Insurance Rate" is multiplied by the "% of Building Cost" for each trade. This produces the "Workers' Compensation Cost" by % of total labor cost, to be added for each trade by building type to determine the weighted average Workers' Compensation rate for the building types analyzed.

Trade	Insurance Rate (% Labor Cost) Range			Insurance Rate Average	% of Building Cost Office Bldgs.	% of Building Cost Schools & Apts.	% of Building Cost Mfg.	Workers' Compensation Office Bldgs.	Workers' Compensation Schools & Apts.	Workers' Compensation Mfg.
Excavation, Grading, etc.	3.9 %	to	18.1%	9.0%	4.8%	4.9%	4.5%	0.43%	0.44%	0.41%
Piles & Foundations	5.1	to	59.5	16.8	7.1	5.2	8.7	1.19	0.87	1.46
Concrete	4.9	to	29.7	12.7	5.0	14.8	3.7	0.64	1.88	0.47
Masonry	4.4	to	28.7	12.1	6.9	7.5	1.9	0.83	0.91	0.23
Structural Steel	5.1	to	124.1	36.9	10.7	3.9	17.6	3.95	1.44	6.49
Miscellaneous & Ornamental Metals	4.2	to	22.6	10.1	2.8	4.0	3.6	0.28	0.40	0.36
Carpentry & Millwork	5.1	to	37	15.0	3.7	4.0	0.5	0.56	0.60	0.08
Metal or Composition Siding	4.9	to	43.3	14.9	2.3	0.3	4.3	0.34	0.04	0.64
Roofing	5.1	to	75.2	28.8	2.3	2.6	3.1	0.66	0.75	0.89
Doors & Hardware	2.8	to	37	10.5	0.9	1.4	0.4	0.09	0.15	0.04
Sash & Glazing	4.9	to	33.1	12.6	3.5	4.0	1.0	0.44	0.50	0.13
Lath & Plaster	2.5	to	30.5	10.8	3.3	6.9	0.8	0.36	0.75	0.09
Tile, Marble & Floors	2.3	to	18.9	8.0	2.6	3.0	0.5	0.21	0.24	0.04
Acoustical Ceilings	3.2	to	30.3	7.8	2.4	0.2	0.3	0.19	0.02	0.02
Painting	4.3	to	28.9	10.7	1.5	1.6	1.6	0.16	0.17	0.17
Interior Partitions	5.1	to	37	15.0	3.9	4.3	4.4	0.59	0.65	0.66
Miscellaneous Items	2.5	to	90.6	14.6	5.2	3.7	9.7	0.76	0.54	1.42
Elevators	1.8	to	12.8	5.9	2.1	1.1	2.2	0.12	0.06	0.13
Sprinklers	1.9	to	17.2	7.0	0.5	—	2.0	0.04	—	0.14
Plumbing	2.7	to	17.1	6.9	4.9	7.2	5.2	0.34	0.50	0.36
Heat., Vent., Air Conditioning	3.8	to	19.3	8.5	13.5	11.0	12.9	1.15	0.94	1.10
Electrical	2.6	to	13.2	5.6	10.1	8.4	11.1	0.57	0.47	0.62
Total	1.8 %	to	124.1%	—	100.0%	100.0%	100.0%	13.90%	12.32%	15.95%
				Overall Weighted Average	14.06%					

Workers' Compensation Insurance Rates by States

The table below lists the weighted average Workers' Compensation base rate for each state with a factor comparing this with the national average of 13.7%.

State	Weighted Average	Factor	State	Weighted Average	Factor	State	Weighted Average	Factor
Alabama	19.6%	143	Kentucky	17.0%	124	North Dakota	10.8%	79
Alaska	16.3	119	Louisiana	15.6	114	Ohio	8.7	64
Arizona	9.0	66	Maine	14.0	102	Oklahoma	14.7	107
Arkansas	10.2	74	Maryland	15.6	114	Oregon	12.0	88
California	27.9	204	Massachusetts	12.4	91	Pennsylvania	16.3	119
Colorado	7.1	52	Michigan	20.6	150	Rhode Island	11.9	87
Connecticut	23.1	169	Minnesota	24.1	176	South Carolina	20.1	147
Delaware	9.3	68	Mississippi	13.1	96	South Dakota	16.0	117
District of Columbia	11.2	82	Missouri	17.4	127	Tennessee	13.2	96
Florida	10.8	79	Montana	14.2	104	Texas	10.6	77
Georgia	27.8	203	Nebraska	18.1	132	Utah	7.8	57
Hawaii	9.0	66	Nevada	10.4	76	Vermont	14.6	107
Idaho	10.8	79	New Hampshire	21.1	154	Virginia	8.7	64
Illinois	22.6	165	New Jersey	13.4	98	Washington	11.1	81
Indiana	5.5	40	New Mexico	17.8	130	West Virginia	12.6	92
Iowa	10.8	79	New York	5.9	43	Wisconsin	14.9	109
Kansas	8.0	58	North Carolina	17.6	128	Wyoming	5.7	42
			Weighted Average for U.S. is	14.1% of payroll = 100%				

Rates in the following table are the base or manual costs per $100 of payroll for Workers' Compensation in each state. Rates are usually applied to straight time wages only and not to premium time wages and bonuses.

The weighted average skilled worker rate for 35 trades is 13.7%. For bidding purposes, apply the full value of Workers' Compensation directly to total labor costs, or if labor is 38%, materials 42% and overhead and profit 20% of total cost, carry 38/80 x 13.7% =6.5% of cost (before overhead and profit) into overhead. Rates vary not only from state to state but also with the experience rating of the contractor.

Rates are the most current available at the time of publication.

R013113-60 Workers' Compensation Insurance Rates by Trade and State (cont.)

State	Carpentry — 3 stories or less 5651	Carpentry — interior cab. work 5437	Carpentry — general 5403	Concrete Work — NOC 5213	Concrete Work — flat (flr., sdwk.) 5221	Electrical Wiring — inside 5190	Excavation — earth NOC 6217	Excavation — rock 6217	Glaziers 5462	Insulation Work 5479	Lathing 5443	Masonry 5022	Painting & Decorating 5474	Pile Driving 6003	Plastering 5480	Plumbing 5183	Roofing 5551	Sheet Metal Work (HVAC) 5538	Steel Erection — door & sash 5102	Steel Erection — inter., ornam. 5102	Steel Erection — structure 5040	Steel Erection — NOC 5057	Tile Work — (interior ceramic) 5348	Waterproofing 9014	Wrecking 5701
AL	29.28	13.11	25.00	12.94	9.22	8.58	11.91	11.91	20.78	14.99	8.43	22.04	16.81	18.81	14.91	8.97	48.49	9.00	10.13	10.13	46.10	24.11	14.15	8.68	46.10
AK	12.44	9.33	13.86	11.33	10.80	6.86	9.06	9.06	32.79	17.54	8.30	16.08	11.68	33.22	13.00	7.53	25.29	7.47	7.41	7.41	25.75	41.66	5.15	5.48	25.75
AZ	11.78	5.44	15.76	6.99	4.24	4.37	4.81	4.81	6.09	10.18	3.64	7.06	7.21	8.53	5.15	4.56	18.34	6.49	9.20	9.20	21.81	10.32	2.89	3.58	21.81
AR	11.58	5.85	10.93	6.39	5.50	4.31	6.73	6.73	7.95	8.55	4.39	7.75	8.88	10.93	12.95	4.70	17.06	5.13	6.99	6.99	28.58	21.04	6.21	2.97	28.58
CA	36.98	36.98	36.98	19.36	19.34	13.20	18.05	18.05	27.35	17.74	16.02	27.71	21.08	23.37	30.46	17.05	64.15	19.29	22.60	22.60	41.64	23.39	2.30	21.08	23.39
CO	7.62	4.14	5.79	6.03	4.35	2.67	4.88	4.88	4.93	7.66	3.52	7.28	5.10	7.13	5.24	3.62	13.42	3.77	5.01	5.01	24.23	9.35	3.77	2.76	21.38
CT	19.80	17.42	27.18	22.71	13.56	7.85	14.62	14.62	19.83	24.28	12.69	27.26	21.51	23.20	18.28	10.85	48.23	13.52	15.52	15.52	64.75	22.28	13.05	6.77	65.75
DE	9.15	9.15	7.05	7.35	6.12	3.18	5.54	5.54	7.90	7.05	7.90	8.28	9.53	11.07	7.90	4.76	17.93	6.13	8.83	8.83	17.30	8.83	5.60	8.28	17.30
DC	8.25	8.16	7.71	12.10	7.65	4.34	9.06	9.06	13.00	6.41	6.70	10.21	5.60	15.02	8.73	7.96	16.80	7.20	15.77	15.77	19.23	10.98	9.00	4.18	19.23
FL	9.33	7.33	10.95	11.45	5.16	5.13	5.96	5.96	8.48	8.55	4.12	8.77	7.91	38.24	11.10	4.82	17.10	6.49	8.06	8.06	19.04	10.13	4.43	4.41	19.04
GA	43.26	19.77	27.19	20.25	15.13	10.28	16.31	16.31	18.41	19.10	12.62	28.72	28.86	23.32	21.44	12.2	75.23	17.15	19.64	19.64	73.28	34.09	12.76	10.18	73.28
HI	7.84	6.87	14.67	7.64	4.33	4.13	5.18	5.18	8.07	5.61	6.55	7.82	6.96	9.14	8.22	4.26	13.77	4.11	6.51	6.51	28.23	9.49	5.22	5.37	20.79
ID	12.77	6.43	10.55	11.76	5.76	3.47	7.06	7.06	9.63	7.67	6.29	7.80	8.43	9.30	13.57	4.27	22.62	7.21	8.55	8.55	32.97	8.02	6.84	4.80	27.50
IL	23.60	15.01	21.72	29.69	12.06	8.24	11.22	11.22	18.09	20.32	12.88	20.85	13.07	21.77	22.24	10.77	36.73	12.98	19.24	19.24	79.82	21.35	16.91	5.78	29.82
IN	8.38	4.15	6.38	4.95	3.41	2.68	4.15	4.15	5.28	6.02	3.24	4.43	4.34	5.34	3.28	2.65	10.05	4.13	4.21	4.21	11.63	5.11	2.89	2.59	11.63
IA	10.34	8.42	12.90	12.72	8.24	4.47	6.88	6.88	9.20	5.82	4.64	8.40	6.76	8.73	10.98	6.40	19.92	4.80	5.70	5.70	36.26	14.65	6.70	4.60	24.10
KS	12.27	5.30	8.09	6.46	5.62	3.67	4.59	4.59	8.35	8.35	3.41	6.73	6.60	7.80	5.04	4.43	13.94	4.61	4.75	4.75	18.81	13.14	5.31	3.99	13.14
KY	12.63	12.30	24.88	12.96	7.51	5.36	11.65	11.65	16.68	13.04	8.83	7.70	11.00	21.98	12.55	5.87	34.41	11.42	10.07	10.07	60.00	17.23	18.88	4.22	60.00
LA	18.20	12.54	20.92	11.92	9.35	6.46	12.02	12.02	14.16	10.50	6.61	13.71	15.78	15.66	10.45	6.08	32.48	12.91	12.19	12.19	44.77	9.67	7.08	6.58	44.77
ME	14.80	10.05	21.30	15.69	9.73	5.57	10.34	10.34	14.40	14.60	5.58	10.56	14.17	12.31	10.83	7.41	24.74	8.22	7.40	7.40	38.54	15.25	5.97	5.13	38.54
MD	17.56	13.32	14.28	19.37	5.51	7.07	7.60	7.60	12.50	14.18	5.58	10.00	7.28	21.24	10.47	7.02	30.67	7.73	10.49	10.49	59.74	22.48	6.69	5.43	22.48
MA	8.68	5.23	9.61	18.85	6.24	2.84	4.35	4.35	9.58	7.78	5.27	10.55	5.09	12.92	4.68	3.50	30.99	5.72	6.89	6.89	54.08	33.00	5.81	2.48	23.75
MI	21.01	11.98	21.01	17.15	10.48	6.35	14.00	14.00	15.12	13.11	13.59	18.32	13.30	59.47	10.34	7.71	42.64	12.58	12.03	12.03	59.47	16.23	11.70	6.95	59.47
MN	19.68	20.00	29.85	10.35	13.68	5.50	10.18	10.18	33.05	22.80	10.63	15.00	14.53	30.90	10.63	8.08	62.73	17.35	9.73	9.73	124.08	8.93	11.70	7.20	8.93
MS	17.32	9.93	13.14	9.84	5.57	4.62	9.84	9.84	11.42	9.11	4.97	12.50	9.60	17.78	10.10	8.54	33.19	6.45	13.01	13.01	30.00	8.32	7.92	4.45	30.00
MO	18.26	12.25	13.57	17.90	11.00	7.12	11.34	11.34	10.74	13.27	9.16	15.43	11.27	18.34	14.38	10.70	36.70	9.28	11.58	11.58	38.70	38.11	13.39	6.50	38.70
MT	13.20	10.71	20.41	12.22	10.37	5.23	12.92	12.92	10.67	19.74	7.41	9.78	8.66	15.08	10.75	7.27	39.34	8.38	9.64	9.64	24.14	9.36	8.36	7.78	9.36
NE	20.53	15.68	20.38	19.08	12.10	7.83	15.30	15.30	12.18	19.13	8.00	18.20	11.23	15.73	13.68	9.48	34.05	11.63	13.15	13.15	44.88	22.70	8.23	6.70	63.83
NV	10.76	7.25	11.35	10.52	7.16	4.80	8.48	8.48	6.75	8.06	3.45	6.62	6.07	8.88	6.41	6.22	17.81	8.84	7.73	7.73	25.26	15.85	13.93	5.16	15.85
NH	20.77	11.63	18.79	25.37	13.01	5.70	13.22	13.22	12.59	13.98	7.47	19.78	24.64	15.55	10.58	10.41	35.43	8.15	11.93	11.93	99.47	28.64	12.59	6.79	99.47
NJ	15.50	9.86	15.50	14.21	9.79	4.45	8.75	8.75	7.99	10.68	12.42	14.48	10.11	14.56	12.42	6.24	38.07	6.71	10.85	10.85	16.29	15.14	7.82	5.82	15.93
NM	19.58	8.17	18.99	13.33	10.11	8.08	8.28	8.28	18.17	11.75	6.68	15.36	11.43	18.20	14.39	8.14	37.47	9.92	12.79	12.79	54.49	40.63	7.66	6.66	54.49
NY	4.93	2.81	6.01	7.46	5.60	2.64	3.94	3.94	6.56	3.28	4.63	6.16	4.85	6.74	2.51	3.31	12.36	4.57	5.46	5.46	10.44	6.29	3.12	3.09	4.57
NC	15.50	11.55	18.63	17.12	8.14	9.96	13.11	13.11	11.87	12.57	9.34	11.05	11.85	16.82	13.63	8.92	33.88	11.87	11.97	11.97	70.40	20.61	8.12	5.41	70.40
ND	10.78	10.78	10.78	4.92	4.92	3.79	4.18	4.18	10.78	10.78	8.88	6.00	6.57	12.68	8.88	6.21	19.53	6.21	12.68	12.68	12.68	12.68	10.78	19.53	8.53
OH	6.82	4.77	6.03	4.98	4.96	3.17	4.83	4.83	9.26	9.83	30.29	7.62	7.64	7.67	7.20	3.94	15.01	4.87	6.17	6.17	12.91	6	6.11	4.90	6.00
OK	13.42	9.17	11.76	11.76	6.34	5.73	10.83	10.83	16.07	19.47	8.45	10.02	8.47	20.25	12.19	7.09	21.13	9.05	13.76	13.76	39.72	23.97	6.50	5.71	39.72
OR	19.79	8.26	13.53	10.34	8.96	5.18	9.03	9.03	10.65	10.89	6.79	11.80	10.85	12.18	9.08	5.38	26.12	6.49	6.82	6.82	20.28	15.61	6.68	5.72	20.28
PA	16.83	16.83	13.22	16.61	11.93	6.69	9.79	9.79	12.75	13.22	12.75	14.24	14.78	17.14	12.75	8.62	33.19	8.74	16.63	16.63	25.25	16.63	9.61	14.24	25.25
RI	9.69	10.13	12.42	13.18	10.51	4.12	8.56	8.56	10.26	13.56	4.93	8.95	9.48	20.95	8.02	5.77	19.37	6.38	6.06	6.06	28.84	20.64	4.60	4.61	28.84
SC	21.43	18.13	22.49	16.89	9.59	11.45	11.78	11.78	16.99	18.58	11.06	13.93	15.50	20.61	15.75	12.36	50.08	13.06	13.91	13.91	41.22	32.53	10.50	6.33	41.22
SD	18.86	13.19	18.65	20.78	8.58	4.84	9.13	9.13	14.61	14.35	6.47	11.88	10.04	15.98	10.23	9.60	31.05	9.72	11.92	11.92	45.18	23.68	6.07	6.02	45.18
TN	16.91	10.01	10.79	11.98	5.04	5.70	11.30	11.30	11.59	8.82	7.16	12.26	9.19	21.55	10.85	5.70	30.21	7.60	8.74	8.74	22.74	23.16	7.39	4.11	22.74
TX	9.68	7.76	9.68	8.61	6.25	5.65	7.33	7.33	7.82	10.32	4.91	8.32	8.04	12.94	8.04	5.47	17.96	13.94	7.16	7.16	29.32	11.95	5.11	5.97	8.11
UT	9.81	5.76	7.93	7.13	5.38	3.09	6.77	6.77	7.13	5.43	4.01	6.86	5.93	7.69	4.67	3.28	21.47	5.41	4.91	4.91	17.18	8.18	3.74	4.02	18.30
VT	13.01	12.79	15.72	11.55	8.77	4.46	10.51	10.51	13.83	14.05	6.01	13.93	9.03	12.33	8.89	8.77	26.46	7.73	10.42	10.42	45.01	23.47	7.90	7.15	45.01
VA	8.91	6.79	7.37	8.34	4.22	3.84	6.23	6.23	6.35	5.77	6.19	6.15	7.43	7.98	6.26	4.34	21.14	5.67	6.19	6.19	25.35	11.51	4.06	2.65	25.35
WA	8.93	8.93	8.93	8.93	8.93	3.27	5.79	5.79	13.83	8.00	8.93	11.74	14.05	20.72	10.68	4.12	19.61	4.58	8.62	8.62	8.62	8.62	7.66	19.61	8.62
WV	13.52	9.63	10.91	11.26	6.25	6.07	9.18	9.18	11.22	9.76	7.21	10.37	10.25	15.47	9.74	6.22	25.10	7.93	8.31	8.31	34.41	16.71	7.86	3.96	32.85
WI	13.04	10.81	16.27	12.16	8.90	5.66	7.72	7.72	12.52	12.83	4.69	15.44	12.07	26.59	10.07	5.82	31.84	8.36	11.02	11.02	21.68	30.03	16.47	5.33	21.68
WY	5.07	5.07	5.07	5.07	5.07	5.07	5.07	5.07	5.07	5.07	5.07	5.07	5.07	5.07	5.07	5.07	5.07	5.07	5.07	5.07	5.07	5.07	5.07	5.07	5.07
AVG.	14.90	10.53	14.96	12.70	8.24	5.58	9.01	9.01	12.57	11.85	7.82	12.10	10.70	16.76	10.78	6.91	28.83	8.47	10.07	10.07	36.86	17.78	8.01	6.41	30.43

R013113-60 Workers' Compensation (cont.) (Canada in Canadian dollars)

Province		Alberta	British Columbia	Manitoba	Ontario	New Brunswick	Newfndld. & Labrador	Northwest Territories	Nova Scotia	Prince Edward Island	Quebec	Saskat-chewan	Yukon
Carpentry—3 stories or less	Rate	6.07	3.93	3.96	4.44	4.62	9.44	3.92	5.70	6.62	12.61	3.74	8.55
	Code	42143	721028	40102	723	238130	4226	4-41	4226	403	80110	B1226	202
Carpentry—interior cab. work	Rate	2.16	2.68	3.96	4.44	4.37	4.24	3.92	4.83	3.69	12.61	2.24	8.55
	Code	42133	721021	40102	723	238350	4279	4-41	4274	402	80110	B1127	202
CARPENTRY—general	Rate	6.07	3.93	3.93	4.44	4.62	4.24	3.92	5.70	6.62	12.61	3.74	8.55
	Code	42143	721028	40102	723	238130	4299	4-41	4226	403	80110	B1202	202
CONCRETE WORK—NOC	Rate	3.86	3.28	6.27	17.86	4.62	9.44	3.92	5.03	3.69	12.86	4.28	8.55
	Code	42104	721010	40110	748	238110	4224	4-41	4224	402	80100	B1314	203
CONCRETE WORK—flat (flr. sidewalk)	Rate	3.86	3.28	6.27	17.86	4.62	9.44	3.92	5.03	3.69	12.86	4.28	8.55
	Code	42104	721010	40110	748	238110	4224	4-41	4224	402	80100	B1314	203
ELECTRICAL Wiring—inside	Rate	1.72	1.71	1.74	3.60	2.09	2.57	2.67	2.68	3.69	4.55	2.24	4.96
	Code	42124	721019	40203	704	238210	4261	4-46	4261	402	80170	B1105	206
EXCAVATION—earth NOC	Rate	2.18	3.62	4.52	5.16	4.62	3.59	3.90	3.16	3.69	8.33	2.21	4.96
	Code	40604	721031	4	711	238190	4214	4-43	4214	402	80030	R1106	207
EXCAVATION—rock	Rate	2.18	3.62	4.19	5.16	4.62	3.59	3.90	3.16	3.69	8.33	2.21	4.96
	Code	40604	721031	40706	711	238190	4214	4-43	4214	402	80030	R1106	207
GLAZIERS	Rate	2.34	3.05	3.96	10.00	4.62	5.25	3.92	5.70	3.69	14.52	4.28	4.96
	Code	42121	715020	40109	751	238150	4233	4-41	4233	402	80150	B1304	212
INSULATION WORK	Rate	1.74	6.88	3.96	10.00	4.37	5.25	3.92	5.70	3.69	12.61	3.74	8.55
	Code	42184	721029	40102	751	238310	4234	4-41	4234	402	80110	B1207	202
LATHING	Rate	3.78	7.08	3.96	4.44	4.37	4.24	3.92	4.83	3.69	12.61	4.28	8.55
	Code	42135	721042	40102	723	238390	4279	4-41	4271	402	80110	B1316	202
MASONRY	Rate	3.84	3.02	3.96	12.39	4.62	5.25	3.92	5.70	3.69	12.86	4.28	8.55
	Code	42102	721037	40102	741	238140	4231	4-41	4231	402	80100	B1318	202
PAINTING & DECORATING	Rate	2.90	3.12	3.44	7.33	4.37	4.24	3.92	4.83	3.69	12.61	3.74	8.55
	Code	42111	721041	40105	719	238320	4275	4-41	4275	402	80110	B1201	202
PILE DRIVING	Rate	3.77	3.92	4.19	6.86	4.62	3.59	3.90	5.03	6.62	8.33	4.28	8.55
	Code	42159	722004	40706	732	238190	4129	4-43	4221	403	80030	B1310	202
PLASTERING	Rate	3.78	7.08	4.48	7.33	4.37	4.24	3.92	4.83	3.69	12.61	3.74	8.55
	Code	42135	721042	40108	719	238390	4271	4-41	4271	402	80110	B1221	202
PLUMBING	Rate	1.72	2.26	2.93	4.06	2.09	2.68	2.67	3.14	2.21	6.44	2.24	4.96
	Code	42122	721043	40204	707	238220	4241	4-46	4241	401	80160	B1101	214
ROOFING	Rate	5.85	6.52	6.74	14.44	4.62	9.44	3.92	7.76	3.69	19.78	4.28	8.55
	Code	42118	721036	40403	728	238160	4236	4-41	4236	402	80130	B1320	202
SHEET METAL WORK (HVAC)	Rate	1.72	2.26	6.74	4.06	2.09	2.68	2.67	3.14	2.21	6.44	2.24	4.96
	Code	42117	721043	40402	707	238220	4244	4-46	4244	401	80160	B1107	208
STEEL ERECTION—door & sash	Rate	1.74	12.03	8.24	17.86	4.62	9.44	3.92	5.70	3.69	20.90	4.28	8.55
	Code	42106	722005	40502	748	238120	4227	4-41	4227	402	80080	B1322	202
STEEL ERECTION—inter., ornam.	Rate	1.74	12.03	8.24	17.86	4.62	9.44	3.92	5.70	3.69	20.90	4.28	8.55
	Code	42106	722005	40502	748	238120	4227	4-41	4227	402	80080	B1322	202
STEEL ERECTION—structure	Rate	1.74	12.03	8.24	17.86	4.62	9.44	3.92	5.70	3.69	20.90	4.28	8.55
	Code	42106	722005	40502	748	238120	4227	4-41	4227	402	80080	B1322	202
STEEL ERECTION—NOC	Rate	1.74	12.03	8.24	17.86	4.62	9.44	3.92	5.70	3.69	20.90	4.28	8.55
	Code	42106	722005	40502	748	238120	4227	4-41	4227	402	80080	B1322	202
TILE WORK—inter. (ceramic)	Rate	2.80	3.34	1.90	7.33	4.37	4.24	3.92	4.83	3.69	12.61	4.28	8.55
	Code	42113	721054	40103	719	238340	4276	4-41	4276	402	80110	B1301	202
WATERPROOFING	Rate	2.90	3.10	3.96	4.44	4.62	4.24	3.92	5.70	3.69	17.81	3.74	8.55
	Code	42139	721016	40102	723	238190	4299	4-41	4239	402	80130	B1217	202
WRECKING	Rate	2.18	4.81	7.34	17.86	4.62	3.59	3.90	3.16	5.61	12.61	2.21	8.55
	Code	40604	721005	40106	748	238190	4211	4-43	4211	402	80110	R1125	202

General Requirements R0131 Project Management & Coordination

R013113-80 Performance Bond

This table shows the cost of a Performance Bond for a construction job scheduled to be completed in 12 months. Add 1% of the premium cost per month for jobs requiring more than 12 months to complete. The rates are "standard" rates offered to contractors that the bonding company considers financially sound and capable of doing the work. Preferred rates are offered by some bonding companies based upon financial strength of the contractor. Actual rates vary from contractor to contractor and from bonding company to bonding company. Contractors should prequalify through a bonding agency before submitting a bid on a contract that requires a bond.

Contract Amount	Building Construction Class B Projects	Highways & Bridges: Class A New Construction	Highways & Bridges: Class A-1 Highway Resurfacing
First $ 100,000 bid	$25.00 per M	$15.00 per M	$9.40 per M
Next 400,000 bid	$ 2,500 plus $15.00 per M	$ 1,500 plus $10.00 per M	$ 940 plus $7.20 per M
Next 2,000,000 bid	8,500 plus 10.00 per M	5,500 plus 7.00 per M	3,820 plus 5.00 per M
Next 2,500,000 bid	28,500 plus 7.50 per M	19,500 plus 5.50 per M	15,820 plus 4.50 per M
Next 2,500,000 bid	47,250 plus 7.00 per M	33,250 plus 5.00 per M	28,320 plus 4.50 per M
Over 7,500,000 bid	64,750 plus 6.00 per M	45,750 plus 4.50 per M	39,570 plus 4.00 per M

General Requirements R0154 Construction Aids

R015423-10 Steel Tubular Scaffolding

On new construction, tubular scaffolding is efficient up to 60′ high or five stories. Above this it is usually better to use a hung scaffolding if construction permits. Swing scaffolding operations may interfere with tenants. In this case, the tubular is more practical at all heights.

In repairing or cleaning the front of an existing building the cost of tubular scaffolding per S.F. of building front increases as the height increases above the first tier. The first tier cost is relatively high due to leveling and alignment.

The minimum efficient crew for erecting and dismantling is three workers. They can set up and remove 18 frame sections per day up to 5 stories high. For 6 to 12 stories high, a crew of four is most efficient. Use two or more on top and two on the bottom for handing up or hoisting. They can also set up and remove 18 frame sections per day. At 7' horizontal spacing, this will run about 800 S.F. per day of erecting and dismantling. Time for placing and removing planks must be added to the above. A crew of three can place and remove 72 planks per day up to 5 stories. For over 5 stories, a crew of four can place and remove 80 planks per day.

The table below shows the number of pieces required to erect tubular steel scaffolding for 1000 S.F. of building frontage. This area is made up of a scaffolding system that is 12 frames (11 bays) long by 2 frames high.

For jobs under twenty-five frames, add 50% to rental cost. Rental rates will be lower for jobs over three months duration. Large quantities for long periods can reduce rental rates by 20%.

Description of Component	Number of Pieces for 1000 S.F. of Building Front	Unit
5′ Wide Standard Frame, 6′-4″ High	24	Ea.
Leveling Jack & Plate	24	↓
Cross Brace	44	
Side Arm Bracket, 21″	12	
Guardrail Post	12	
Guardrail, 7′ section	22	
Stairway Section	2	
Stairway Starter Bar	1	
Stairway Inside Handrail	2	
Stairway Outside Handrail	2	
Walk-Thru Frame Guardrail	2	↓

Scaffolding is often used as falsework over 15′ high during construction of cast-in-place concrete beams and slabs. Two foot wide scaffolding is generally used for heavy beam construction. The span between frames depends upon the load to be carried with a maximum span of 5′.

Heavy duty shoring frames with a capacity of 10,000#/leg can be spaced up to 10′ O.C. depending upon form support design and loading.

Scaffolding used as horizontal shoring requires less than half the material required with conventional shoring.

On new construction, erection is done by carpenters.

Rolling towers supporting horizontal shores can reduce labor and speed the job. For maintenance work, catwalks with spans up to 70′ can be supported by the rolling towers.

R015423-20 Pump Staging

Pump staging is generally not available for rent. The table below shows the number of pieces required to erect pump staging for 2400 S.F. of building frontage. This area is made up of a pump jack system that is 3 poles (2 bays) wide by 2 poles high.

Item	Number of Pieces for 2400 S.F. of Building Front	Unit
Aluminum pole section, 24′ long	6	Ea.
Aluminum splice joint, 6′ long	3	↓
Aluminum foldable brace	3	↓
Aluminum pump jack	3	↓
Aluminum support for workbench/back safety rail	3	↓
Aluminum scaffold plank/workbench, 14″ wide x 24′ long	4	↓
Safety net, 22′ long	2	↓
Aluminum plank end safety rail	2	↓

The cost in place for this 2400 S.F. will depend on how many uses are realized during the life of the equipment.

R015433-10 Contractor Equipment

Rental Rates shown elsewhere in the book pertain to late model high quality machines in excellent working condition, rented from equipment dealers. Rental rates from contractors may be substantially lower than the rental rates from equipment dealers depending upon economic conditions; for older, less productive machines, reduce rates by a maximum of 15%. Any overtime must be added to the base rates. For shift work, rates are lower. Usual rule of thumb is 150% of one shift rate for two shifts; 200% for three shifts.

For periods of less than one week, operated equipment is usually more economical to rent than renting bare equipment and hiring an operator.

Costs to move equipment to a job site (mobilization) or from a job site (demobilization) are not included in rental rates, nor in any Equipment costs on any Unit Price line items or crew listings. These costs can be found elsewhere. If a piece of equipment is already at a job site, it is not appropriate to utilize mob/demob costs in an estimate again.

Rental rates vary throughout the country with larger cities generally having lower rates. Lease plans for new equipment are available for periods in excess of six months with a percentage of payments applying toward purchase.

Rental rates can also be treated as reimbursement costs for contractor-owned equipment. Owned equipment costs include depreciation, loan payments, interest, taxes, insurance, storage, and major repairs.

Monthly rental rates vary from 2% to 5% of the cost of the equipment depending on the anticipated life of the equipment and its wearing parts. Weekly rates are about 1/3 the monthly rates and daily rental rates about 1/3 the weekly rate.

The hourly operating costs for each piece of equipment include costs to the user such as fuel, oil, lubrication, normal expendables for the equipment, and a percentage of mechanic's wages chargeable to maintenance. The hourly operating costs listed do not include the operator's wages.

The daily cost for equipment used in the standard crews is figured by dividing the weekly rate by five, then adding eight times the hourly operating cost to give the total daily equipment cost, not including the operator. This figure is in the right hand column of the Equipment listings under Equipment Cost/Day.

Pile Driving rates shown for pile hammer and extractor do not include leads, crane, boiler or compressor. Vibratory pile driving requires an added field specialist during set-up and pile driving operation for the electric model. The hydraulic model requires a field specialist for set-up only. Up to 125 reuses of sheet piling are possible using vibratory drivers. For normal conditions, crane capacity for hammer type and size are as follows.

Crane Capacity	Hammer Type and Size		
	Air or Steam	Diesel	Vibratory
25 ton	to 8,750 ft.-lb.		70 H.P.
40 ton	15,000 ft.-lb.	to 32,000 ft.-lb.	170 H.P.
60 ton	25,000 ft.-lb.		300 H.P.
100 ton		112,000 ft.-lb.	

Cranes should be specified for the job by size, building and site characteristics, availability, performance characteristics, and duration of time required.

Backhoes & Shovels rent for about the same as equivalent size cranes but maintenance and operating expense is higher. Crane operators rate must be adjusted for high boom heights. Average adjustments: for 150′ boom add 2% per hour; over 185′, add 4% per hour; over 210′, add 6% per hour; over 250′, add 8% per hour and over 295′, add 12% per hour.

Tower Cranes of the climbing or static type have jibs from 50′ to 200′ and capacities at maximum reach range from 4,000 to 14,000 pounds. Lifting capacities increase up to maximum load as the hook radius decreases.

Typical rental rates, based on purchase price are about 2% to 3% per month.

Erection and dismantling runs between 500 and 2000 labor hours. Climbing operation takes 10 labor hours per 20′ climb. Crane dead time is about 5 hours per 40′ climb. If crane is bolted to side of the building add cost of ties and extra mast sections. Climbing cranes have from 80′ to 180′ of mast while static cranes have 80′ to 800′ of mast.

Truck Cranes can be converted to tower cranes by using tower attachments. Mast heights over 400′ have been used.

A single 100′ high material **Hoist and Tower** can be erected and dismantled in about 400 labor hours; a double 100′ high hoist and tower in about 600 labor hours. Erection times for additional heights are 3 and 4 labor hours per vertical foot respectively up to 150′, and 4 to 5 labor hours per vertical foot over 150′ high. A 40′ high portable Buck hoist takes about 160 labor hours to erect and dismantle. Additional heights take 2 labor hours per vertical foot to 80′ and 3 labor hours per vertical foot for the next 100′. Most material hoists do not meet local code requirements for carrying personnel.

A 150′ high **Personnel Hoist** requires about 500 to 800 labor hours to erect and dismantle. Budget erection time at 5 labor hours per vertical foot for all trades. Local code requirements or labor scarcity requiring overtime can add up to 50% to any of the above erection costs.

Earthmoving Equipment: The selection of earthmoving equipment depends upon the type and quantity of material, moisture content, haul distance, haul road, time available, and equipment available. Short haul cut and fill operations may require dozers only, while another operation may require excavators, a fleet of trucks, and spreading and compaction equipment. Stockpiled material and granular material are easily excavated with front end loaders. Scrapers are most economically used with hauls between 300′ and 1-1/2 miles if adequate haul roads can be maintained. Shovels are often used for blasted rock and any material where a vertical face of 8′ or more can be excavated. Special conditions may dictate the use of draglines, clamshells, or backhoes. Spreading and compaction equipment must be matched to the soil characteristics, the compaction required and the rate the fill is being supplied.

R024119-10 Demolition Defined

Whole Building Demolition - Demolition of the whole building with no concern for any particular building element, component, or material type being demolished. This type of demolition is accomplished with large pieces of construction equipment that break up the structure, load it into trucks and haul it to a disposal site, but disposal or dump fees are not included. Demolition of below-grade foundation elements, such as footings, foundation walls, grade beams, slabs on grade, etc., is not included. Certain mechanical equipment containing flammable liquids or ozone-depleting refrigerants, electric lighting elements, communication equipment components, and other building elements may contain hazardous waste, and must be removed, either selectively or carefully, as hazardous waste before the building can be demolished.

Foundation Demolition - Demolition of below-grade foundation footings, foundation walls, grade beams, and slabs on grade. This type of demolition is accomplished by hand or pneumatic hand tools, and does not include saw cutting, or handling, loading, hauling, or disposal of the debris.

Gutting - Removal of building interior finishes and electrical/mechanical systems down to the load-bearing and sub-floor elements of the rough building frame, with no concern for any particular building element, component, or material type being demolished. This type of demolition is accomplished by hand or pneumatic hand tools, and includes loading into trucks, but not hauling, disposal or dump fees, scaffolding, or shoring. Certain mechanical equipment containing flammable liquids or ozone-depleting refrigerants, electric lighting elements, communication equipment components, and other building elements may contain hazardous waste, and must be removed, either selectively or carefully, as hazardous waste, before the building is gutted.

Selective Demolition - Demolition of a selected building element, component, or finish, with some concern for surrounding or adjacent elements, components, or finishes (see the first Subdivision (s) at the beginning of appropriate Divisions). This type of demolition is accomplished by hand or pneumatic hand tools, and does not include handling, loading, storing, hauling, or disposal of the debris, scaffolding, or shoring. "Gutting" methods may be used in order to save time, but damage that is caused to surrounding or adjacent elements, components, or finishes may have to be repaired at a later time.

Careful Removal - Removal of a piece of service equipment, building element or component, or material type, with great concern for both the removed item and surrounding or adjacent elements, components or finishes. The purpose of careful removal may be to protect the removed item for later re-use, preserve a higher salvage value of the removed item, or replace an item while taking care to protect surrounding or adjacent elements, components, connections, or finishes from cosmetic and/or structural damage. An approximation of the time required to perform this type of removal is 1/3 to 1/2 the time it would take to install a new item of like kind (see Reference Number R220105-10). This type of removal is accomplished by hand or pneumatic hand tools, and does not include loading, hauling, or storing the removed item, scaffolding, shoring, or lifting equipment.

Cutout Demolition - Demolition of a small quantity of floor, wall, roof, or other assembly, with concern for the appearance and structural integrity of the surrounding materials. This type of demolition is accomplished by hand or pneumatic hand tools, and does not include saw cutting, handling, loading, hauling, or disposal of debris, scaffolding, or shoring.

Rubbish Handling - Work activities that involve handling, loading or hauling of debris. Generally, the cost of rubbish handling must be added to the cost of all types of demolition, with the exception of whole building demolition.

Minor Site Demolition - Demolition of site elements outside the footprint of a building. This type of demolition is accomplished by hand or pneumatic hand tools, or with larger pieces of construction equipment, and may include loading a removed item onto a truck (check the Crew for equipment used). It does not include saw cutting, hauling or disposal of debris, and, sometimes, handling or loading.

R024119-20 Dumpsters

Dumpster rental costs on construction sites are presented in two ways.

The cost per week rental includes the delivery of the dumpster; its pulling or emptying once per week, and its final removal. The assumption is made that the dumpster contractor could choose to empty a dumpster by simply bringing in an empty unit and removing the full one. These costs also include the disposal of the materials in the dumpster.

The Alternate Pricing can be used when actual planned conditions are not approximated by the weekly numbers. For example, these lines can be used when a dumpster is needed for 4 weeks and will need to be emptied 2 or 3 times per week. Conversely the Alternate Pricing lines can be used when a dumpster will be rented for several weeks or months but needs to be emptied only a few times over this period.

Existing Conditions R0265 Underground Storage Tank Removal

R026510-20 Underground Storage Tank Removal

Underground Storage Tank Removal can be divided into two categories: Non-Leaking and Leaking. Prior to removing an underground storage tank, tests should be made, with the proper authorities present, to determine whether a tank has been leaking or the surrounding soil has been contaminated.

To safely remove Liquid Underground Storage Tanks:

1. Excavate to the top of the tank.
2. Disconnect all piping.
3. Open all tank vents and access ports.
4. Remove all liquids and/or sludge.
5. Purge the tank with an inert gas.
6. Provide access to the inside of the tank and clean out the interior using proper personal protective equipment (PPE).
7. Excavate soil surrounding the tank using proper PPE for on-site personnel.
8. Pull and properly dispose of the tank.
9. Clean up the site of all contaminated material.
10. Install new tanks or close the excavation.

Existing Conditions — R0282 Asbestos Remediation

R028213-20 Asbestos Removal Process

Asbestos removal is accomplished by a specialty contractor who understands the federal and state regulations regarding the handling and disposal of the material. The process of asbestos removal is divided into many individual steps. An accurate estimate can be calculated only after all the steps have been priced.

The steps are generally as follows:

1. Obtain an asbestos abatement plan from an industrial hygienist.
2. Monitor the air quality in and around the removal area and along the path of travel between the removal area and transport area. This establishes the background contamination.
3. Construct a two part decontamination chamber at entrance to removal area.
4. Install a HEPA filter to create a negative pressure in the removal area.
5. Install wall, floor and ceiling protection as required by the plan, usually 2 layers of fireproof 6 mil polyethylene.
6. Industrial hygienist visually inspects work area to verify compliance with plan.
7. Provide temporary supports for conduit and piping affected by the removal process.
8. Proceed with asbestos removal and bagging process. Monitor air quality as described in Step #2. Discontinue operations when contaminate levels exceed applicable standards.
9. Document the legal disposal of materials in accordance with EPA standards.
10. Thoroughly clean removal area including all ledges, crevices and surfaces.
11. Post abatement inspection by industrial hygienist to verify plan compliance.
12. Provide a certificate from a licensed industrial hygienist attesting that contaminate levels are within acceptable standards before returning area to regular use.

Existing Conditions — R0283 Lead Remediation

R028319-60 Lead Paint Remediation Methods

Lead paint remediation can be accomplished by the following methods.

1. Abrasive blast
2. Chemical stripping
3. Power tool cleaning with vacuum collection system
4. Encapsulation
5. Remove and replace
6. Enclosure

Each of these methods has strengths and weakness depending on the specific circumstances of the project. The following is an overview of each method.

1. **Abrasive blasting** is usually accomplished with sand or recyclable metallic blast. Before work can begin, the area must be contained to ensure the blast material with lead does not escape to the atmosphere. The use of vacuum blast greatly reduces the containment requirements. Lead abatement equipment that may be associated with this work includes a negative air machine. In addition, it is necessary to have an industrial hygienist monitor the project on a continual basis. When the work is complete, the spent blast sand with lead must be disposed of as a hazardous material. If metallic shot was used, the lead is separated from the shot and disposed of as hazardous material. Worker protection includes disposable clothing and respiratory protection.
2. **Chemical stripping** requires strong chemicals be applied to the surface to remove the lead paint. Before the work can begin, the area under/adjacent to the work area must be covered to catch the chemical and removed lead. After the chemical is applied to the painted surface it is usually covered with paper. The chemical is left in place for the specified period, then the paper with lead paint is pulled or scraped off. The process may require several chemical applications. The paper with chemicals and lead paint adhered to it, plus the containment and loose scrapings collected by a HEPA (High Efficiency Particulate Air Filter) vac, must be disposed of as a hazardous material. The chemical stripping process usually requires a neutralizing agent and several wash downs after the paint is removed. Worker protection includes a neoprene or other compatible protective clothing and respiratory protection with face shield. An industrial hygienist is required intermittently during the process.
3. **Power tool cleaning** is accomplished using shrouded needle blasting guns. The shrouding with different end configurations is held up against the surface to be cleaned. The area is blasted with hardened needles and the shroud captures the lead with a HEPA vac and deposits it in a holding tank. An industrial hygienist monitors the project, protective clothing and a respirator is required until air samples prove otherwise. When the work is complete the lead must be disposed of as a hazardous material.
4. **Encapsulation** is a method that leaves the well bonded lead paint in place after the peeling paint has been removed. Before the work can begin, the area under/adjacent to the work must be covered to catch the scrapings. The scraped surface is then washed with a detergent and rinsed. The prepared surface is covered with approximately 10 mils of paint. A reinforcing fabric can also be embedded in the paint covering. The scraped paint and containment must be disposed of as a hazardous material. Workers must wear protective clothing and respirators.
5. **Remove and replace** is an effective way to remove lead paint from windows, gypsum walls and concrete masonry surfaces. The painted materials are removed and new materials are installed. Workers should wear a respirator and tyvek suit. The demolished materials must be disposed of as hazardous waste if it fails the TCLP (Toxicity Characteristic Leachate Process) test.
6. **Enclosure** is the process that permanently seals lead painted materials in place. This process has many applications such as covering lead painted drywall with new drywall, covering exterior construction with tyvek paper then residing, or covering lead painted structural members with aluminum or plastic. The seams on all enclosing materials must be securely sealed. An industrial hygienist monitors the project, and protective clothing and a respirator is required until air samples prove otherwise.

All the processes require clearance monitoring and wipe testing as required by the hygienist.

R031113-10 Wall Form Materials

Aluminum Forms

Approximate weight is 3 lbs. per S.F.C.A. Standard widths are available from 4″ to 36″ with 36″ most common. Standard lengths of 2′, 4′, 6′ to 8′ are available. Forms are lightweight and fewer ties are needed with the wider widths. The form face is either smooth or textured.

Metal Framed Plywood Forms

Manufacturers claim over 75 reuses of plywood and over 300 reuses of steel frames. Many specials such as corners, fillers, pilasters, etc. are available. Monthly rental is generally about 15% of purchase price for first month and 9% per month thereafter with 90% of rental applied to purchase for the first month and decreasing percentages thereafter. Aluminum framed forms cost 25% to 30% more than steel framed.

After the first month, extra days may be prorated from the monthly charge. Rental rates do not include ties, accessories, cleaning, loss of hardware or freight in and out. Approximate weight is 5 lbs. per S.F. for steel; 3 lbs. per S.F. for aluminum.

Forms can be rented with option to buy.

Plywood Forms, Job Fabricated

There are two types of plywood used for concrete forms.

1. Exterior plyform which is completely waterproof. This is face oiled to facilitate stripping. Ten reuses can be expected with this type with 25 reuses possible.
2. An overlaid type consists of a resin fiber fused to exterior plyform. No oiling is required except to facilitate cleaning. This is available in both high density (HDO) and medium density overlaid (MDO). Using HDO, 50 reuses can be expected with 200 possible.

Plyform is available in 5/8″ and 3/4″ thickness. High density overlaid is available in 3/8″, 1/2″, 5/8″ and 3/4″ thickness.

5/8″ thick is sufficient for most building forms, while 3/4″ is best on heavy construction.

Plywood Forms, Modular, Prefabricated

There are many plywood forming systems without frames. Most of these are manufactured from 1-1/8″ (HDO) plywood and have some hardware attached. These are used principally for foundation walls 8′ or less high. With care and maintenance, 100 reuses can be attained with decreasing quality of surface finish.

Steel Forms

Approximate weight is 6-1/2 lbs. per S.F.C.A. including accessories. Standard widths are available from 2″ to 24″, with 24″ most common. Standard lengths are from 2′ to 8′, with 4′ the most common. Forms are easily ganged into modular units.

Forms are usually leased for 15% of the purchase price per month prorated daily over 30 days.

Rental may be applied to sale price, and usually rental forms are bought. With careful handling and cleaning 200 to 400 reuses are possible.

Straight wall gang forms up to 12′ x 20′ or 8′ x 30′ can be fabricated. These crane handled forms usually lease for approx. 9% per month.

Individual job analysis is available from the manufacturer at no charge.

R031113-40 Forms for Reinforced Concrete

Design Economy

Avoid many sizes in proportioning beams and columns.

From story to story avoid changing column dimensions. Gain strength by adding steel or using a richer mix. If a change in size of column is necessary, vary one dimension only to minimize form alterations. Keep beams and columns the same width.

From floor to floor in a multi-story building vary beam depth, not width, as that will leave slab panel form unchanged. It is cheaper to vary the strength of a beam from floor to floor by means of steel area than by 2″ changes in either width or depth.

Cost Factors

Material includes the cost of lumber, cost of rent for metal pans or forms if used, nails, form ties, form oil, bolts and accessories.

Labor includes the cost of carpenters to make up, erect, remove and repair, plus common labor to clean and move. Having carpenters remove forms minimizes repairs.

Improper alignment and condition of forms will increase finishing cost. When forms are heavily oiled, concrete surfaces must be neutralized before finishing. Special curing compounds will cause spillages to spall off in first frost. Gang forming methods will reduce costs on large projects.

Materials Used

Boards are seldom used unless their architectural finish is required. Generally, steel, fiberglass and plywood are used for contact surfaces. Labor on plywood is 10% less than with boards. The plywood is backed up with 2 x 4′s at 12″ to 32″ O.C. Walers are generally 2 - 2 x 4′s. Column forms are held together with steel yokes or bands. Shoring is with adjustable shoring or scaffolding for high ceilings.

Reuse

Floor and column forms can be reused four or possibly five times without excessive repair. Remember to allow for 10% waste on each reuse.

When modular sized wall forms are made, up to twenty uses can be expected with exterior plyform.

When forms are reused, the cost to erect, strip, clean and move will not be affected. 10% replacement of lumber should be included and about one hour of carpenter time for repairs on each reuse per 100 S.F.

The reuse cost for certain accessory items normally rented on a monthly basis will be lower than the cost for the first use.

After fifth use, new material required plus time needed for repair prevent form cost from dropping further and it may go up. Much depends on care in stripping, the number of special bays, changes in beam or column sizes and other factors.

Costs for multiple use of formwork may be developed as follows:

2 Uses

$$\frac{\text{(1st Use + Reuse)}}{2} = \text{avg. cost/2 uses}$$

3 Uses

$$\frac{\text{(1st Use + 2 Reuse)}}{3} = \text{avg. cost/3 uses}$$

4 Uses

$$\frac{\text{(1st use + 3 Reuse)}}{4} = \text{avg. cost/4 uses}$$

R031113-60 Formwork Labor-Hours

Item	Hours Required				Total Hours	Multiple Use		
	Unit	Fabricate	Erect & Strip	Clean & Move	1 Use	2 Use	3 Use	4 Use
Beam and Girder, interior beams, 12″ wide	100 S.F.	6.4	8.3	1.3	16.0	13.3	12.4	12.0
Hung from steel beams		5.8	7.7	1.3	14.8	12.4	11.6	11.2
Beam sides only, 36″ high		5.8	7.2	1.3	14.3	11.9	11.1	10.7
Beam bottoms only, 24″ wide		6.6	13.0	1.3	20.9	18.1	17.2	16.7
Box out for openings		9.9	10.0	1.1	21.0	16.6	15.1	14.3
Buttress forms, to 8′ high		6.0	6.5	1.2	13.7	11.2	10.4	10.0
Centering, steel, 3/4″ rib lath			1.0		1.0			
3/8″ rib lath or slab form			0.9		0.9			
Chamfer strip or keyway	100 L.F.		1.5		1.5	1.5	1.5	1.5
Columns, fiber tube 8″ diameter			20.6		20.6			
12″			21.3		21.3			
16″			22.9		22.9			
20″			23.7		23.7			
24″			24.6		24.6			
30″			25.6		25.6			
Columns, round steel, 12″ diameter			22.0		22.0	22.0	22.0	22.0
16″			25.6		25.6	25.6	25.6	25.6
20″			30.5		30.5	30.5	30.5	30.5
24″			37.7		37.7	37.7	37.7	37.7
Columns, plywood 8″ x 8″	100 S.F.	7.0	11.0	1.2	19.2	16.2	15.2	14.7
12″ x 12″		6.0	10.5	1.2	17.7	15.2	14.4	14.0
16″ x 16″		5.9	10.0	1.2	17.1	14.7	13.8	13.4
24″ x 24″		5.8	9.8	1.2	16.8	14.4	13.6	13.2
Columns, steel framed plywood 8″ x 8″			10.0	1.0	11.0	11.0	11.0	11.0
12″ x 12″			9.3	1.0	10.3	10.3	10.3	10.3
16″ x 16″			8.5	1.0	9.5	9.5	9.5	9.5
24″ x 24″			7.8	1.0	8.8	8.8	8.8	8.8
Drop head forms, plywood		9.0	12.5	1.5	23.0	19.0	17.7	17.0
Coping forms		8.5	15.0	1.5	25.0	21.3	20.0	19.4
Culvert, box			14.5	4.3	18.8	18.8	18.8	18.8
Curb forms, 6″ to 12″ high, on grade		5.0	8.5	1.2	14.7	12.7	12.1	11.7
On elevated slabs		6.0	10.8	1.2	18.0	15.5	14.7	14.3
Edge forms to 6″ high, on grade	100 L.F.	2.0	3.5	0.6	6.1	5.6	5.4	5.3
7″ to 12″ high	100 S.F.	2.5	5.0	1.0	8.5	7.8	7.5	7.4
Equipment foundations		10.0	18.0	2.0	30.0	25.5	24.0	23.3
Flat slabs, including drops		3.5	6.0	1.2	10.7	9.5	9.0	8.8
Hung from steel		3.0	5.5	1.2	9.7	8.7	8.4	8.2
Closed deck for domes		3.0	5.8	1.2	10.0	9.0	8.7	8.5
Open deck for pans		2.2	5.3	1.0	8.5	7.9	7.7	7.6
Footings, continuous, 12″ high		3.5	3.5	1.5	8.5	7.3	6.8	6.6
Spread, 12″ high		4.7	4.2	1.6	10.5	8.7	8.0	7.7
Pile caps, square or rectangular		4.5	5.0	1.5	11.0	9.3	8.7	8.4
Grade beams, 24″ deep		2.5	5.3	1.2	9.0	8.3	8.0	7.9
Lintel or Sill forms		8.0	17.0	2.0	27.0	23.5	22.3	21.8
Spandrel beams, 12″ wide		9.0	11.2	1.3	21.5	17.5	16.2	15.5
Stairs			25.0	4.0	29.0	29.0	29.0	29.0
Trench forms in floor		4.5	14.0	1.5	20.0	18.3	17.7	17.4
Walls, Plywood, at grade, to 8′ high		5.0	6.5	1.5	13.0	11.0	9.7	9.5
8′ to 16′		7.5	8.0	1.5	17.0	13.8	12.7	12.1
16′ to 20′		9.0	10.0	1.5	20.5	16.5	15.2	14.5
Foundation walls, to 8′ high		4.5	6.5	1.0	12.0	10.3	9.7	9.4
8′ to 16′ high		5.5	7.5	1.0	14.0	11.8	11.0	10.6
Retaining wall to 12′ high, battered		6.0	8.5	1.5	16.0	13.5	12.7	12.3
Radial walls to 12′ high, smooth		8.0	9.5	2.0	19.5	16.0	14.8	14.3
2′ chords		7.0	8.0	1.5	16.5	13.5	12.5	12.0
Prefabricated modular, to 8′ high		—	4.3	1.0	5.3	5.3	5.3	5.3
Steel, to 8′ high		—	6.8	1.2	8.0	8.0	8.0	8.0
8′ to 16′ high		—	9.1	1.5	10.6	10.3	10.2	10.2
Steel framed plywood to 8′ high		—	6.8	1.2	8.0	7.5	7.3	7.2
8′ to 16′ high		—	9.3	1.2	10.5	9.5	9.2	9.0

R032110-10 Reinforcing Steel Weights and Measures

Bar Designation No.**	Nominal Weight Lb./Ft.	U.S. Customary Units			SI Units			
		Nominal Dimensions*			Nominal Dimensions*			
		Diameter in.	Cross Sectional Area, in.^{2}	Perimeter in.	Nominal Weight kg/m	Diameter mm	Cross Sectional Area, cm^2	Perimeter mm
3	.376	.375	.11	1.178	.560	9.52	.71	29.9
4	.668	.500	.20	1.571	.994	12.70	1.29	39.9
5	1.043	.625	.31	1.963	1.552	15.88	2.00	49.9
6	1.502	.750	.44	2.356	2.235	19.05	2.84	59.8
7	2.044	.875	.60	2.749	3.042	22.22	3.87	69.8
8	2.670	1.000	.79	3.142	3.973	25.40	5.10	79.8
9	3.400	1.128	1.00	3.544	5.059	28.65	6.45	90.0
10	4.303	1.270	1.27	3.990	6.403	32.26	8.19	101.4
11	5.313	1.410	1.56	4.430	7.906	35.81	10.06	112.5
14	7.650	1.693	2.25	5.320	11.384	43.00	14.52	135.1
18	13.600	2.257	4.00	7.090	20.238	57.33	25.81	180.1

* The nominal dimensions of a deformed bar are equivalent to those of a plain round bar having the same weight per foot as the deformed bar.

** Bar numbers are based on the number of eighths of an inch included in the nominal diameter of the bars.

R032110-80 Shop-Fabricated Reinforcing Steel

The material prices for reinforcing, shown in the unit cost sections of the book, are for 50 tons or more of shop-fabricated reinforcing steel and include:

1. Mill base price of reinforcing steel
2. Mill grade/size/length extras
3. Mill delivery to the fabrication shop
4. Shop storage and handling
5. Shop drafting/detailing
6. Shop shearing and bending
7. Shop listing
8. Shop delivery to the job site

Both material and installation costs can be considerably higher for small jobs consisting primarily of smaller bars, while material costs may be slightly lower for larger jobs.

Concrete R0322 Welded Wire Fabric Reinforcing

R032205-30 Common Stock Styles of Welded Wire Fabric

This table provides some of the basic specifications, sizes, and weights of welded wire fabric used for reinforcing concrete.

	New Designation Spacing — Cross Sectional Area (in.) — (Sq. in. 100)	Old Designation Spacing — Wire Gauge (in.) — (AS & W)		Steel Area per Foot Longitudinal in.	Longitudinal cm	Transverse in.	Transverse cm	Approximate Weight per 100 S.F. lbs	kg
Rolls	6 x 6 — W1.4 x W1.4	6 x 6 — 10 x 10		.028	.071	.028	.071	21	9.53
	6 x 6 — W2.0 x W2.0	6 x 6 — 8 x 8	1	.040	.102	.040	.102	29	13.15
	6 x 6 — W2.9 x W2.9	6 x 6 — 6 x 6		.058	.147	.058	.147	42	19.05
	6 x 6 — W4.0 x W4.0	6 x 6 — 4 x 4		.080	.203	.080	.203	58	26.91
	4 x 4 — W1.4 x W1.4	4 x 4 — 10 x 10		.042	.107	.042	.107	31	14.06
	4 x 4 — W2.0 x W2.0	4 x 4 — 8 x 8	1	.060	.152	.060	.152	43	19.50
	4 x 4 — W2.9 x W2.9	4 x 4 — 6 x 6		.087	.227	.087	.227	62	28.12
	4 x 4 — W4.0 x W4.0	4 x 4 — 4 x 4		.120	.305	.120	.305	85	38.56
Sheets	6 x 6 — W2.9 x W2.9	6 x 6 — 6 x 6		.058	.147	.058	.147	42	19.05
	6 x 6 — W4.0 x W4.0	6 x 6 — 4 x 4		.080	.203	.080	.203	58	26.31
	6 x 6 — W5.5 x W5.5	6 x 6 — 2 x 2	2	.110	.279	.110	.279	80	36.29
	4 x 4 — W1.4 x W1.4	4 x 4 — 4 x 4		.120	.305	.120	.305	85	38.56

NOTES: 1. Exact W—number size for 8 gauge is W2.1
2. Exact W—number size for 2 gauge is W5.4

R033105-10 Proportionate Quantities

The tables below show both quantities per S.F. of floor areas as well as form and reinforcing quantities per C.Y. Unusual structural requirements would increase the ratios below. High strength reinforcing would reduce the steel weights. Figures are for 3000 psi concrete and 60,000 psi reinforcing unless specified otherwise.

Type of Construction	Live Load	Span	Per S.F. of Floor Area				Per C.Y. of Concrete		
			Concrete	Forms	Reinf.	Pans	Forms	Reinf.	Pans
Flat Plate	50 psf	15 Ft.	.46 C.F.	1.06 S.F.	1.71lb.		62 S.F.	101 lb.	
		20	.63	1.02	2.40		44	104	
		25	.79	1.02	3.03		35	104	
	100	15	.46	1.04	2.14		61	126	
		20	.71	1.02	2.72		39	104	
		25	.83	1.01	3.47		33	113	
Flat Plate (waffle construction) 20″ domes	50	20	.43	1.00	2.10	.84 S.F.	63	135	53 S.F.
		25	.52	1.00	2.90	.89	52	150	46
		30	.64	1.00	3.70	.87	42	155	37
	100	20	.51	1.00	2.30	.84	53	125	45
		25	.64	1.00	3.20	.83	42	135	35
		30	.76	1.00	4.40	.81	36	160	29
Waffle Construction 30″ domes	50	25	.69	1.06	1.83	.68	42	72	40
		30	.74	1.06	2.39	.69	39	87	39
		35	.86	1.05	2.71	.69	33	85	39
		40	.78	1.00	4.80	.68	35	165	40
Flat Slab (two way with drop panels)	50	20	.62	1.03	2.34		45	102	
		25	.77	1.03	2.99		36	105	
		30	.95	1.03	4.09		29	116	
	100	20	.64	1.03	2.83		43	119	
		25	.79	1.03	3.88		35	133	
		30	.96	1.03	4.66		29	131	
	200	20	.73	1.03	3.03		38	112	
		25	.86	1.03	4.23		32	133	
		30	1.06	1.03	5.30		26	135	
One Way Joists 20″ Pans	50	15	.36	1.04	1.40	.93	78	105	70
		20	.42	1.05	1.80	.94	67	120	60
		25	.47	1.05	2.60	.94	60	150	54
	100	15	.38	1.07	1.90	.93	77	140	66
		20	.44	1.08	2.40	.94	67	150	58
		25	.52	1.07	3.50	.94	55	185	49
One Way Joists 8″ x 16″ filler blocks	50	15	.34	1.06	1.80	.81 Ea.	84	145	64 Ea.
		20	.40	1.08	2.20	.82	73	145	55
		25	.46	1.07	3.20	.83	63	190	49
	100	15	.39	1.07	1.90	.81	74	130	56
		20	.46	1.09	2.80	.82	64	160	48
		25	.53	1.10	3.60	.83	56	190	42
One Way Beam & Slab	50	15	.42	1.30	1.73		84	111	
		20	.51	1.28	2.61		68	138	
		25	.64	1.25	2.78		53	117	
	100	15	.42	1.30	1.90		84	122	
		20	.54	1.35	2.69		68	154	
		25	.69	1.37	3.93		54	145	
	200	15	.44	1.31	2.24		80	137	
		20	.58	1.40	3.30		65	163	
		25	.69	1.42	4.89		53	183	
Two Way Beam & Slab	100	15	.47	1.20	2.26		69	130	
		20	.63	1.29	3.06		55	131	
		25	.83	1.33	3.79		43	123	
	200	15	.49	1.25	2.70		41	149	
		20	.66	1.32	4.04		54	165	
		25	.88	1.32	6.08		41	187	

R033105-10 Proportionate Quantities (cont.)

4000 psi Concrete and 60,000 psi Reinforcing—Form and Reinforcing Quantities per C.Y.					
Item	Size	Forms	Reinforcing	Minimum	Maximum
Columns (square tied)	10″ x 10″	130 S.F.C.A.	#5 to #11	220 lbs.	875 lbs.
	12″ x 12″	108	#6 to #14	200	955
	14″ x 14″	92	#7 to #14	190	900
	16″ x 16″	81	#6 to #14	187	1082
	18″ x 18″	72	#6 to #14	170	906
	20″ x 20″	65	#7 to #18	150	1080
	22″ x 22″	59	#8 to #18	153	902
	24″ x 24″	54	#8 to #18	164	884
	26″ x 26″	50	#9 to #18	169	994
	28″ x 28″	46	#9 to #18	147	864
	30″ x 30″	43	#10 to #18	146	983
	32″ x 32″	40	#10 to #18	175	866
	34″ x 34″	38	#10 to #18	157	772
	36″ x 36″	36	#10 to #18	175	852
	38″ x 38″	34	#10 to #18	158	765
	40″ x 40″	32 ▼	#10 to #18	143	692

Item	Size	Form	Spiral	Reinforcing	Minimum	Maximum
Columns (spirally reinforced)	12″ diameter	34.5 L.F.	190 lbs.	#4 to #11	165 lbs.	1505 lb.
		34.5	190	#14 & #18	—	1100
	14″	25	170	#4 to #11	150	970
		25	170	#14 & #18	800	1000
	16″	19	160	#4 to #11	160	950
		19	160	#14 & #18	605	1080
	18″	15	150	#4 to #11	160	915
		15	150	#14 & #18	480	1075
	20″	12	130	#4 to #11	155	865
		12	130	#14 & #18	385	1020
	22″	10	125	#4 to #11	165	775
		10	125	#14 & #18	320	995
	24″	9	120	#4 to #11	195	800
		9	120	#14 & #18	290	1150
	26″	7.3	100	#4 to #11	200	729
		7.3	100	#14 & #18	235	1035
	28″	6.3	95	#4 to #11	175	700
		6.3	95	#14 & #18	200	1075
	30″	5.5	90	#4 to #11	180	670
		5.5	90	#14 & #18	175	1015
	32″	4.8	85	#4 to #11	185	615
		4.8	85	#14 & #18	155	955
	34″	4.3	80	#4 to #11	180	600
		4.3	80	#14 & #18	170	855
	36″	3.8	75	#4 to #11	165	570
		3.8	75	#14 & #18	155	865
	40″	3.0	70	#4 to #11	165	500
		3.0	70	#14 & #18	145	765

R033105-10 Proportionate Quantities (cont.)

3000 psi Concrete and 60,000 psi Reinforcing—Form and Reinforcing Quantities per C.Y.

Item	Type	Loading	Height	C.Y./L.F.	Forms/C.Y.	Reinf./C.Y.
Retaining Walls	Cantilever	Level Backfill	4 Ft.	0.2 C.Y.	49 S.F.	35 lbs.
			8	0.5	42	45
			12	0.8	35	70
			16	1.1	32	85
			20	1.6	28	105
		Highway Surcharge	4	0.3	41	35
			8	0.5	36	55
			12	0.8	33	90
			16	1.2	30	120
			20	1.7	27	155
		Railroad Surcharge	4	0.4	28	45
			8	0.8	25	65
			12	1.3	22	90
			16	1.9	20	100
			20	2.6	18	120
	Gravity, with Vertical Face	Level Backfill	4	0.4	37	None
			7	0.6	27	↓
			10	1.2	20	↓
		Sloping Backfill	4	0.3	31	↓
			7	0.8	21	↓
			10	1.6	15	↓

Item	Span	Live Load in Kips per Linear Foot							
		Under 1 Kip		2 to 3 Kips		4 to 5 Kips		6 to 7 Kips	
		Forms	Reinf.	Forms	Reinf.	Forms	Reinf.	Forms	Reinf.
Beams	10 Ft.	—	—	90 S.F.	170 #	85 S.F.	175 #	75 S.F.	185 #
	16	130 S.F.	165 #	85	180	75	180	65	225
	20	110	170	75	185	62	200	51	200
	26	90	170	65	215	62	215	—	—
	30	85	175	60	200	—	—	—	—

Item	Size	Type	Forms per C.Y.	Reinforcing per C.Y.
Spread Footings	Under 1 C.Y.	1,000 psf soil	24 S.F.	44 lbs.
		5,000	24	42
		10,000	24	52
	1 C.Y. to 5 C.Y.	1,000	14	49
		5,000	14	50
		10,000	14	50
	Over 5 C.Y.	1,000	9	54
		5,000	9	52
		10,000	9	56
Pile Caps (30 Ton Concrete Piles)	Under 5 C.Y.	shallow caps	20	65
		medium	20	50
		deep	20	40
	5 C.Y. to 10 C.Y.	shallow	14	55
		medium	15	45
		deep	15	40
	10 C.Y. to 20 C.Y.	shallow	11	60
		medium	11	45
		deep	12	35
	Over 20 C.Y.	shallow	9	60
		medium	9	45
		deep	10	40

R033105-10 Proportionate Quantities (cont.)

3000 psi Concrete and 60,000 psi Reinforcing — Form and Reinforcing Quantities per C.Y.

Item	Size	Pile Spacing	50 T Pile	100 T Pile	50 T Pile	100 T Pile
Pile Caps (Steel H Piles)	Under 5 C.Y.	24″ O.C.	24 S.F.	24 S.F.	75 lbs.	90 lbs.
		30″	25	25	80	100
		36″	24	24	80	110
	5 C.Y. to 10 C.Y.	24″	15	15	80	110
		30″	15	15	85	110
		36″	15	15	75	90
	Over 10 C.Y.	24″	13	13	85	90
		30″	11	11	85	95
		36″	10	10	85	90

Item		8″ Thick		10″ Thick		12″ Thick		15″ Thick	
	Height	Forms	Reinf.	Forms	Reinf.	Forms	Reinf.	Forms	Reinf.
Basement Walls	7 Ft.	81 S.F.	44 lbs.	65 S.F.	45 lbs.	54 S.F.	44 lbs.	41 S.F.	43 lbs.
	8		44		45		44		43
	9		46		45		44		43
	10		57		45		44		43
	12		83		50		52		43
	14		116		65		64		51
	16				86		90		65
	18	↓		↓		↓	106	↓	70

R033105-20 Materials for One C.Y. of Concrete

This is an approximate method of figuring quantities of cement, sand and coarse aggregate for a field mix with waste allowance included.

With crushed gravel as coarse aggregate, to determine barrels of cement required, divide 10 by total mix; that is, for 1:2:4 mix, 10 divided by 7 = 1-3/7 barrels.

If the coarse aggregate is crushed stone, use 10-1/2 instead of 10 as given for gravel.

To determine tons of sand required, multiply barrels of cement by parts of sand and then by 0.2; that is, for the 1:2:4 mix, as above, 1-3/7 x 2 x .2 = .57 tons.

Tons of crushed gravel are in the same ratio to tons of sand as parts in the mix, or 4/2 x .57 = 1.14 tons.

1 bag cement = 94#
4 bags = 1 barrel

1 C.Y. sand or crushed gravel = 2700#
1 ton sand or crushed gravel = 20 C.F.

1 C.Y. crushed stone = 2575#
1 ton crushed stone = 21 C.F.

Average carload of cement is 692 bags; of sand or gravel is 56 tons.

Do not stack stored cement over 10 bags high.

R033105-70 Placing Ready-Mixed Concrete

For ground pours allow for 5% waste when figuring quantities.

Prices in the front of the book assume normal deliveries. If deliveries are made before 8 A.M. or after 5 P.M. or on Saturday afternoons add 30%. Negotiated discounts for large volumes are not included in prices in front of book.

For the lower floors without truck access, concrete may be wheeled in rubber-tired buggies, conveyer handled, crane handled or pumped. Pumping is economical if there is top steel. Conveyers are more efficient for thick slabs.

At higher floors the rubber-tired buggies may be hoisted by a hoisting tower and wheeled to location. Placement by a conveyer is limited to three floors and is best for high-volume pours. Pumped concrete is best when building has no crane access. Concrete may be pumped directly as high as thirty-six stories using special pumping techniques. Normal maximum height is about fifteen stories.

Best pumping aggregate is screened and graded bank gravel rather than crushed stone.

Pumping downward is more difficult than pumping upward. Horizontal distance from pump to pour may increase preparation time prior to pour. Placing by cranes, either mobile, climbing or tower types, continues as the most efficient method for high-rise concrete buildings.

R033105-85 Lift Slabs

The cost advantage of the lift slab method is due to placing all concrete, reinforcing steel, inserts and electrical conduit at ground level and in reduction of formwork. Minimum economical project size is about 30,000 S.F. Slabs may be tilted for parking garage ramps.

It is now used in all types of buildings and has gone up to 22 stories high in apartment buildings. Current trend is to use post-tensioned flat plate slabs with spans from 22′ to 35′. Cylindrical void forms are used when deep slabs are required. One pound of prestressing steel is about equal to seven pounds of conventional reinforcing.

To be considered cured for stressing and lifting, a slab must have attained 75% of design strength. Seven days are usually sufficient with four to five days possible if high early strength cement is used. Slabs can be stacked using two coats of a non-bonding agent to insure that slabs do not stick to each other. Lifting is done by companies specializing in this work. Lift rate is 5′ to 15′ per hour with an average of 10′ per hour. Total areas up to 33,000 S.F. have been lifted at one time. 24 to 36 jacking columns are common. Most economical bay sizes are 24′ to 28′ with four to fourteen stories most efficient. Continuous design reduces reinforcing steel cost. Use of post-tensioned slabs allows larger bay sizes.

R034105-30 Prestressed Precast Concrete Structural Units

<table>
<tr><th>Type</th><th>Location</th><th>Depth</th><th colspan="2">Span in Ft.</th><th>Live Load Lb. per S.F.</th></tr>
<tr><td rowspan="3">Double Tee
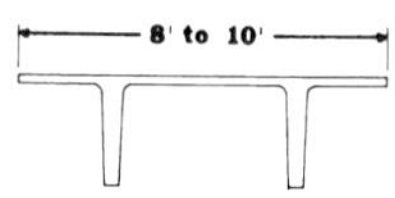</td><td>Floor</td><td>28″ to 34″</td><td colspan="2">60 to 80</td><td>50 to 80</td></tr>
<tr><td>Roof</td><td>12″ to 24″</td><td colspan="2">30 to 50</td><td>40</td></tr>
<tr><td>Wall</td><td>Width 8′</td><td colspan="2">Up to 55′ high</td><td>Wind</td></tr>
<tr><td rowspan="2">Multiple Tee
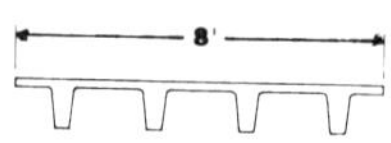</td><td>Roof</td><td>8″ to 12″</td><td colspan="2">15 to 40</td><td>40</td></tr>
<tr><td>Floor</td><td>8″ to 12″</td><td colspan="2">15 to 30</td><td>100</td></tr>
<tr><td rowspan="2">Plank

or
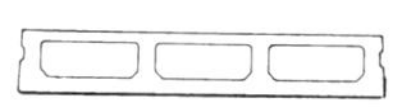</td><td rowspan="2">Roof
or
Floor</td><td></td><td>Roof</td><td>Floor</td><td rowspan="2">40 for Roof
100 for Floor</td></tr>
<tr><td>4″
6″
8″
10″
12″</td><td>13
22
26
33
42</td><td>12
18
25
29
32</td></tr>
<tr><td>Single Tee
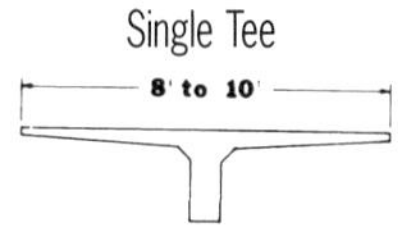</td><td>Roof</td><td>28″
32″
36″
48″</td><td colspan="2">40
80
100
120</td><td>40</td></tr>
<tr><td>AASHO Girder
</td><td>Bridges</td><td>Type 4
5
6</td><td colspan="2">100
110
125</td><td>Highway</td></tr>
<tr><td>Box Beam
</td><td>Bridges</td><td>15″
27″
33″</td><td colspan="2">40 to 100</td><td>Highway</td></tr>
</table>

The majority of precast projects today utilize double tees rather than single tees because of speed and ease of installation. As a result casting beds at manufacturing plants are normally formed for double tees. Single tee projects will therefore require an initial set up charge to be spread over the individual single tee costs.

For floors, a 2″ to 3″ topping is field cast over the shapes. For roofs, insulating concrete or rigid insulation is placed over the shapes.

Member lengths up to 40′ are standard haul, 40′ to 60′ require special permits and lengths over 60′ must be escorted. Over width and/or over length can add up to 100% on hauling costs.

Large heavy members may require two cranes for lifting which would increase erection costs by about 45%. An eight man crew can install 12 to 20 double tees, or 45 to 70 quad tees or planks per day.

Grouting of connections must also be included.

Several system buildings utilizing precast members are available. Heights can go up to 22 stories for apartment buildings. Optimum design ratio is 3 S.F. of surface to 1 S.F. of floor area.

Concrete — R0352 Lightweight Concrete Roof Insulation

R035216-10 Lightweight Concrete

Lightweight aggregate concrete is usually purchased ready mixed, but it can also be field mixed.

Vermiculite or Perlite comes in bags of 4 C.F. under various trade names. Weight is about 8 lbs. per C.F. For insulating roof fill use 1:6 mix. For structural deck use 1:4 mix over gypsum boards, steeltex, steel centering, etc., supported by closely spaced joists or bulb trees. For structural slabs use 1:3:2 vermiculite sand concrete over steeltex, metal lath, steel centering, etc., on joists spaced 2′-0″ O.C. for maximum L.L. of 80 P.S.F. Use same mix for slab base fill over steel flooring or regular reinforced concrete slab when tile, terrazzo or other finish is to be laid over.

For slabs on grade use 1:3:2 mix when tile, etc., finish is to be laid over. If radiant heating units are installed use a 1:6 mix for a base. After coils are in place, cover with a regular granolithic finish (mix 1:3:2) to a minimum depth of 1-1/2″ over top of units.

Reinforce all slabs with 6 x 6 or 10 x 10 welded wire mesh.

Masonry — R0421 Clay Unit Masonry

R042110-20 Common and Face Brick

Common building brick manufactured according to ASTM C62 and facing brick manufactured according to ASTM C216 are the two standard bricks available for general building use.

Building brick is made in three grades; SW, where high resistance to damage caused by cyclic freezing is required; MW, where moderate resistance to cyclic freezing is needed; and NW, where little resistance to cyclic freezing is needed. Facing brick is made in only the two grades SW and MW. Additionally, facing brick is available in three types; FBS, for general use; FBX, for general use where a higher degree of precision and lower permissible variation in size than FBS is needed; and FBA, for general use to produce characteristic architectural effects resulting from non-uniformity in size and texture of the units.

In figuring the material cost of brickwork, an allowance of 25% mortar waste and 3% brick breakage was included. If bricks are delivered palletized with 280 to 300 per pallet, or packaged, allow only 1-1/2% for breakage. Packaged or palletized delivery is practical when a job is big enough to have a crane or other equipment available to handle a package of brick. This is so on all industrial work but not always true on small commercial buildings.

The use of buff and gray face is increasing, and there is a continuing trend to the Norman, Roman, Jumbo and SCR brick.

Common red clay brick for backup is not used that often. Concrete block is the most usual backup material with occasional use of sand lime or cement brick. Building brick is commonly used in solid walls for strength and as a fire stop.

Brick panels built on the ground and then crane erected to the upper floors have proven to be economical. This allows the work to be done under cover and without scaffolding.

R042110-50 Brick, Block & Mortar Quantities

Running Bond					
Number of Brick per S.F. of Wall - Single Wythe with 3/8″ Joints				C.F. of Mortar per M Bricks, Waste Included	
Type Brick	Nominal Size (incl. mortar) L H W	Modular Coursing	Number of Brick per S.F.	3/8″ Joint	1/2″ Joint
Standard	8 x 2-2/3 x 4	3C=8″	6.75	10.3	12.9
Economy	8 x 4 x 4	1C=4″	4.50	11.4	14.6
Engineer	8 x 3-1/5 x 4	5C=16″	5.63	10.6	13.6
Fire	9 x 2-1/2 x 4-1/2	2C=5″	6.40	550 # Fireclay	—
Jumbo	12 x 4 x 6 or 8	1C=4″	3.00	23.8	30.8
Norman	12 x 2-2/3 x 4	3C=8″	4.50	14.0	17.9
Norwegian	12 x 3-1/5 x 4	5C=16″	3.75	14.6	18.6
Roman	12 x 2 x 4	2C=4″	6.00	13.4	17.0
SCR	12 x 2-2/3 x 6	3C=8″	4.50	21.8	28.0
Utility	12 x 4 x 4	1C=4″	3.00	15.4	19.6

For Other Bonds Standard Size Add to S.F. Quantities in Table to Left		
Bond Type	Description	Factor
Common	full header every fifth course	+20%
	full header every sixth course	+16.7%
English	full header every second course	+50%
Flemish	alternate headers every course	+33.3%
	every sixth course	+5.6%
Header = W x H exposed		+100%
Rowlock = H x W exposed		+100%
Rowlock stretcher = L x W exposed		+33.3%
Soldier = H x L exposed		—
Sailor = W x L exposed		-33.3%

Concrete Blocks Nominal Size	Approximate Weight per S.F. Standard	Approximate Weight per S.F. Lightweight	Blocks per 100 S.F.	Mortar per M block, waste included Partitions	Mortar per M block, waste included Back up
2″ x 8″ x 16″	20 PSF	15 PSF	113	27 C.F.	36 C.F.
4″	30	20		41	51
6″	42	30		56	66
8″	55	38		72	82
10″	70	47		87	97
12″	85	55		102	112

R050521-20 Welded Structural Steel

Usual weight reductions with welded design run 10% to 20% compared with bolted or riveted connections. This amounts to about the same total cost compared with bolted structures since field welding is more expensive than bolts. For normal spans of 18′ to 24′ figure 6 to 7 connections per ton.

Trusses — For welded trusses add 4% to weight of main members for connections. Up to 15% less steel can be expected in a welded truss compared to one that is shop bolted. Cost of erection is the same whether shop bolted or welded.

General — Typical electrodes for structural steel welding are E6010, E6011, E60T and E70T. Typical buildings vary between 2# to 8# of weld rod per ton of steel. Buildings utilizing continuous design require about three times as much welding as conventional welded structures. In estimating field erection by welding, it is best to use the average linear feet of weld per ton to arrive at the welding cost per ton. The type, size and position of the weld will have a direct bearing on the cost per linear foot. A typical field welder will deposit 1.8# to 2# of weld rod per hour manually. Using semiautomatic methods can increase production by as much as 50% to 75%.

R051223-10 Structural Steel

The bare material prices for structural steel, shown in the unit cost sections of the book, are for 100 tons of shop-fabricated structural steel and include:

1. Mill base price of structural steel
2. Mill scrap/grade/size/length extras
3. Mill delivery to a metals service center (warehouse)
4. Service center storage and handling
5. Service center delivery to a fabrication shop
6. Shop storage and handling
7. Shop drafting/detailing
8. Shop fabrication
9. Shop coat of primer paint
10. Shop listing
11. Shop delivery to the job site

In unit cost sections of the book that contain items for field fabrication of steel components, the bare material cost of steel includes:

1. Mill base price of structural steel
2. Mill scrap/grade/size/length extras
3. Mill delivery to a metals service center (warehouse)
4. Service center storage and handling
5. Service center delivery to the job site

R051223-15 Structural Steel Estimating for Repair and Remodeling Projects

The correct approach to estimating structural steel is dependent upon the amount of steel required for the particular project. If the project is a sizable addition to a building, the data can be used directly from the cost data book. This is not the case however if the project requires a small amount of steel, for instance to reinforce existing roof or floor structural systems. To better understand this, please refer to the unit price line for a W16x31 beam with bolted connections. Assume your project requires the reinforcement of the structural members of the roof system of a 3 story building required for the installation a new piece of HVAC equipment. The project will need 4 pieces of W16x31 that are 30′ long. After pricing this 120 L.F. job using the unit prices for a W16x31, an analysis will reveal that the price is wholly inadequate.

The first problem is apparent if you examine the amount of steel that can be installed per day. The unit price line indicates 900 linear feet per day can be installed by a 5 man crew with an 90 ton crane. This productivity is correct for new construction but certainly is not for repair and remodeling work. Installation of new structural steel considers that each member is installed from the foundation to the roof of the structure in a planned and systematic manner with a crane having unrestricted access to all parts of the project. Additionally each connection is planned and detailed with full field access for fit-up and final bolting. The erection is planned and progresses such that interferences and conflicts with other structural members are minimized, if not completely eliminated. All of these assumptions are clearly not the case with a repair and remodeling job, and a significant decrease in the stated productivity will be observed.

A crane will certainly be needed to lift the members into the general area of the project but in most cases will not be able to place the beams into their final position. An opening in the existing roof may not be large enough to permit the beams to pass through and it may be necessary to bring them into the building through existing windows or doors. Moving the beams to the actual area where they will be installed may involve hand labor and the use of dollies. Finally, hoists and/or jacks may be needed for final positioning.

The connection of new members to existing can often be accomplished by field bolting with accurate field measurements and good planning but in many cases access to both sides of existing members is not possible and field welding becomes the only alternative. In addition to the cost of the actual welding, protection of existing finishes, systems and structure and fire protection must be considered.

New beams can never be installed tight to the existing decks or floors which they must support and the use of shims and tack welding becomes necessary. Additionally, further planning and cost is involved in assuring that existing loads are minimized during the installation and shimming process.

It is apparent that installation of structural steel as part of a repair and remodeling project involves more than simply installing the members and estimating the cost in the same manner as new construction. The best procedure for estimating the total cost is adequate planning and coordination of each process and activity that will be needed. Unit costs for the materials, labor and equipment can then be attached to each needed activity and a final, complete price can be determined.

R053100-10 Decking Descriptions

General - All Deck Products

Steel deck is made by cold forming structural grade sheet steel into a repeating pattern of parallel ribs. The strength and stiffness of the panels are the result of the ribs and the material properties of the steel. Deck lengths can be varied to suit job conditions, but because of shipping considerations, are usually less than 40 feet. Standard deck width varies with the product used but full sheets are usually 12″, 18″, 24″, 30″, or 36″. Deck is typically furnished in a standard width with the ends cut square. Any cutting for width, such as at openings or for angular fit, is done at the job site.

Deck is typically attached to the building frame with arc puddle welds, self-drilling screws, or powder or pneumatically driven pins. Sheet to sheet fastening is done with screws, button punching (crimping), or welds.

Composite Floor Deck

After installation and adequate fastening, floor deck serves several purposes. It (a) acts as a working platform, (b) stabilizes the frame, (c) serves as a concrete form for the slab, and (d) reinforces the slab to carry the design loads applied during the life of the building. Composite decks are distinguished by the presence of shear connector devices as part of the deck. These devices are designed to mechanically lock the concrete and deck together so that the concrete and the deck work together to carry subsequent floor loads. These shear connector devices can be rolled-in embossments, lugs, holes, or wires welded to the panels. The deck profile can also be used to interlock concrete and steel.

Composite deck finishes are either galvanized (zinc coated) or phosphatized/painted. Galvanized deck has a zinc coating on both the top and bottom surfaces. The phosphatized/painted deck has a bare (phosphatized) top surface that will come into contact with the concrete. This bare top surface can be expected to develop rust before the concrete is placed. The bottom side of the deck has a primer coat of paint.

Composite floor deck is normally installed so the panel ends do not overlap on the supporting beams. Shear lugs or panel profile shape often prevent a tight metal to metal fit if the panel ends overlap; the air gap caused by overlapping will prevent proper fusion with the structural steel supports when the panel end laps are shear stud welded.

Adequate end bearing of the deck must be obtained as shown on the drawings. If bearing is actually less in the field than shown on the drawings, further investigation is required.

Roof Deck

Roof deck is not designed to act compositely with other materials. Roof deck acts alone in transferring horizontal and vertical loads into the building frame. Roof deck rib openings are usually narrower than floor deck rib openings. This provides adequate support of rigid thermal insulation board.

Roof deck is typically installed to endlap approximately 2″ over supports. However, it can be butted (or lapped more than 2″) to solve field fit problems. Since designers frequently use the installed deck system as part of the horizontal bracing system (the deck as a diaphragm), any fastening substitution or change should be approved by the designer. Continuous perimeter support of the deck is necessary to limit edge deflection in the finished roof and may be required for diaphragm shear transfer.

Standard roof deck finishes are galvanized or primer painted. The standard factory applied paint for roof deck is a primer paint and is not intended to weather for extended periods of time. Field painting or touching up of abrasions and deterioration of the primer coat or other protective finishes is the responsibility of the contractor.

Cellular Deck

Cellular deck is made by attaching a bottom steel sheet to a roof deck or composite floor deck panel. Cellular deck can be used in the same manner as floor deck. Electrical, telephone, and data wires are easily run through the chase created between the deck panel and the bottom sheet.

When used as part of the electrical distribution system, the cellular deck must be installed so that the ribs line up and create a smooth cell transition at abutting ends. The joint that occurs at butting cell ends must be taped or otherwise sealed to prevent wet concrete from seeping into the cell. Cell interiors must be free of welding burrs, or other sharp intrusions, to prevent damage to wires.

When used as a roof deck, the bottom flat plate is usually left exposed to view. Care must be maintained during erection to keep good alignment and prevent damage.

Cellular deck is sometimes used with the flat plate on the top side to provide a flat working surface. Installation of the deck for this purpose requires special methods for attachment to the frame because the flat plate, now on the top, can prevent direct access to the deck material that is bearing on the structural steel. It may be advisable to treat the flat top surface to prevent slipping.

Cellular deck is always furnished galvanized or painted over galvanized.

Form Deck

Form deck can be any floor or roof deck product used as a concrete form. Connections to the frame are by the same methods used to anchor floor and roof deck. Welding washers are recommended when welding deck that is less than 20 gauge thickness.

Form deck is furnished galvanized, prime painted, or uncoated. Galvanized deck must be used for those roof deck systems where form deck is used to carry a lightweight insulating concrete fill.

Wood, Plastics & Comp. R0611 Wood Framing

R061110-30 Lumber Product Material Prices

The price of forest products fluctuates widely from location to location and from season to season depending upon economic conditions. The bare material prices in the unit cost sections of the book show the National Average material prices in effect Jan. 1 of this book year. It must be noted that lumber prices in general may change significantly during the year.

Availability of certain items depends upon geographic location and must be checked prior to firm-price bidding.

Wood, Plastics & Comp. R0616 Sheathing

R061636-20 Plywood

There are two types of plywood used in construction: interior, which is moisture-resistant but not waterproofed, and exterior, which is waterproofed.

The grade of the exterior surface of the plywood sheets is designated by the first letter: A, for smooth surface with patches allowed; B, for solid surface with patches and plugs allowed; C, which may be surface plugged or may have knot holes up to 1″ wide; and D, which is used only for interior type plywood and may have knot holes up to 2-1/2″ wide. "Structural Grade" is specifically designed for engineered applications such as box beams. All CC & DD grades have roof and floor spans marked on them.

Underlayment-grade plywood runs from 1/4″ to 1-1/4″ thick. Thicknesses 5/8″ and over have optional tongue and groove joints which eliminate the need for blocking the edges. Underlayment 19/32″ and over may be referred to as Sturd-i-Floor.

The price of plywood can fluctuate widely due to geographic and economic conditions.

Typical uses for various plywood grades are as follows:

AA-AD Interior — cupboards, shelving, paneling, furniture

BB Plyform — concrete form plywood

CDX — wall and roof sheathing

Structural — box beams, girders, stressed skin panels

AA-AC Exterior — fences, signs, siding, soffits, etc.

Underlayment — base for resilient floor coverings

Overlaid HDO — high density for concrete forms & highway signs

Overlaid MDO — medium density for painting, siding, soffits & signs

303 Siding — exterior siding, textured, striated, embossed, etc.

Thermal & Moist. Protec. R0751 Built-Up Bituminous Roofing

R075113-20 Built-Up Roofing

Asphalt is available in kegs of 100 lbs. each; coal tar pitch in 560 lb. kegs. Prepared roofing felts are available in a wide range of sizes, weights and characteristics. However, the most commonly used are #15 (432 S.F. per roll, 13 lbs. per square) and #30 (216 S.F. per roll, 27 lbs. per square).

Inter-ply bitumen varies from 24 lbs. per sq. (asphalt) to 30 lbs. per sq. (coal tar) per ply, MF4@ 25%. Flood coat bitumen also varies from 60 lbs. per sq. (asphalt) to 75 lbs. per sq. (coal tar), MF4@ 25%. Expendable equipment (mops, brooms, screeds, etc.) runs about 16% of the bitumen cost. For new, inexperienced crews this factor may be much higher.

Rigid insulation board is typically applied in two layers. The first is mechanically attached to nailable decks or spot or solid mopped to non-nailable decks; the second layer is then spot or solid mopped to the first layer. Membrane application follows the insulation, except in protected membrane roofs, where the membrane goes down first and the insulation on top, followed with ballast (stone or concrete pavers). Insulation and related labor costs are NOT included in prices for built-up roofing.

R075213-30 Modified Bitumen Roofing

The cost of modified bitumen roofing is highly dependent on the type of installation that is planned. Installation is based on the type of modifier used in the bitumen. The two most popular modifiers are atactic polypropylene (APP) and styrene butadiene styrene (SBS). The modifiers are added to heated bitumen during the manufacturing process to change its characteristics. A polyethylene, polyester or fiberglass reinforcing sheet is then sandwiched between layers of this bitumen. When completed, the result is a pre-assembled, built-up roof that has increased elasticity and weatherablility. Some manufacturers include a surfacing material such as ceramic or mineral granules, metal particles or sand.

The preferred method of adhering SBS-modified bitumen roofing to the substrate is with hot-mopped asphalt (much the same as built-up roofing). This installation method requires a tar kettle/pot to heat the asphalt, as well as the labor, tools and equipment necessary to distribute and spread the hot asphalt.

The alternative method for applying APP and SBS modified bitumen is as follows. A skilled installer uses a torch to melt a small pool of bitumen off the membrane. This pool must form across the entire roll for proper adhesion. The installer must unroll the roofing at a pace slow enough to melt the bitumen, but fast enough to prevent damage to the rest of the membrane.

Modified bitumen roofing provides the advantages of both built-up and single-ply roofing. Labor costs are reduced over those of built-up roofing because only a single ply is necessary. The elasticity of single-ply roofing is attained with the reinforcing sheet and polymer modifiers. Modifieds have some self-healing characteristics and because of their multi-layer construction, they offer the reliability and safety of built-up roofing.

Thermal & Moist. Protec. R0784 Firestopping

R078413-30 Firestopping

Firestopping is the sealing of structural, mechanical, electrical and other penetrations through fire-rated assemblies. The basic components of firestop systems are safing insulation and firestop sealant on both sides of wall penetrations and the top side of floor penetrations.

Pipe penetrations are assumed to be through concrete, grout, or joint compound and can be sleeved or unsleeved. Costs for the penetrations and sleeves are not included. An annular space of 1″ is assumed. Escutcheons are not included.

Metallic pipe is assumed to be copper, aluminum, cast iron or similar metallic material. Insulated metallic pipe is assumed to be covered with a thermal insulating jacket of varying thickness and materials.

Non-metallic pipe is assumed to be PVC, CPVC, FR Polypropylene or similar plastic piping material. Intumescent firestop sealant or wrap strips are included. Collars on both sides of wall penetrations and a sheet metal plate on the underside of floor penetrations are included.

Ductwork is assumed to be sheet metal, stainless steel or similar metallic material. Duct penetrations are assumed to be through concrete, grout or joint compound. Costs for penetrations and sleeves are not included. An annular space of 1/2″ is assumed.

Multi-trade openings include costs for sheet metal forms, firestop mortar, wrap strips, collars and sealants as necessary.

Structural penetrations joints are assumed to be 1/2″ or less. CMU walls are assumed to be within 1-1/2″ of metal deck. Drywall walls are assumed to be tight to the underside of metal decking.

Metal panel, glass or curtain wall systems include a spandrel area of 5′ filled with mineral wool foil-faced insulation. Fasteners and stiffeners are included.

Openings — R0813 Metal Doors

R081313-20 Steel Door Selection Guide

Standard steel doors are classified into four levels, as recommended by the Steel Door Institute in the chart below. Each of the four levels offers a range of construction models and designs, to meet architectural requirements for preference and appearance, including full flush, seamless, and stile & rail. Recommended minimum gauge requirements are also included.

For complete standard steel door construction specifications and available sizes, refer to the Steel Door Institute Technical Data Series, ANSI A250.8-98 (SDI-100), and ANSI A250.4-94 Test Procedure and Acceptance Criteria for Physical Endurance of Steel Door and Hardware Reinforcements.

Level		Model	Construction	For Full Flush or Seamless		
				Min. Gauge	Thickness (in)	Thickness (mm)
I	Standard Duty	1 2	Full Flush Seamless	20	0.032	0.8
II	Heavy Duty	1 2	Full Flush Seamless	18	0.042	1.0
III	Extra Heavy Duty	1 2 3	Full Flush Seamless *Stile & Rail	16	0.053	1.3
IV	Maximum Duty	1 2	Full Flush Seamless	14	0.067	1.6

*Stiles & rails are 16 gauge; flush panels, when specified, are 18 gauge

Openings — R0851 Metal Windows

R085123-10 Steel Sash

Ironworker crew will erect 25 S.F. or 1.3 sash unit per hour, whichever is less.

Mechanic will point 30 L.F. per hour.

Painter will paint 90 S.F. per coat per hour.

Glazier production depends on light size.

Allow 1 lb. special steel sash putty per 16″ x 20″ light.

Openings — R0853 Plastic Windows

R085313-20 Replacement Windows

Replacement windows are typically measured per United Inch.

United Inches are calculated by rounding the width and height of the window opening up to the nearest inch, then adding the two figures.

The labor cost for replacement windows includes removal of sash, existing sash balance or weights, parting bead where necessary and installation of new window.

Debris hauling and dump fees are not included.

R087120-10 Hinges

All closer equipped doors should have ball bearing hinges. Lead lined or extremely heavy doors require special strength hinges. Usually 1-1/2 pair of hinges are used per door up to 7′-6″ high openings. Table below shows typical hinge requirements.

Use Frequency	Type Hinge Required	Type of Opening	Type of Structure
High	Heavy weight ball bearing	Entrances Toilet Rooms	Banks, Office buildings, Schools, Stores & Theaters Office buildings and Schools
Average	Standard weight ball bearing	Entrances Corridors Toilet Rooms	Dwellings Office buildings and Schools Stores
Low	Plain bearing	Interior	Dwellings

Door Thickness	Weight of Doors in Pounds per Square Foot				
	White Pine	Oak	Hollow Core	Solid Core	Hollow Metal
1-3/8″	3psf	6psf	1-1/2psf	3-1/2 — 4psf	6-1/2psf
1-3/4″	3-1/2	7	2-1/2	4-1/2 — 5-1/4	6-1/2
2-1/4″	4-1/2	9	—	5-1/2 — 6-3/4	6-1/2

R088110-10 Glazing Productivity

Some glass sizes are estimated by the "united inch" (height + width). The table below shows the number of lights glazed in an eight-hour period by the crew size indicated, for glass up to 1/4″ thick. Square or nearly square lights are more economical on a S.F. basis. Long slender lights will have a high S.F. installation cost. For insulated glass reduce production by 33%. For 1/2″ float glass reduce production by 50%. Production time for glazing with two glaziers per day averages: 1/4″ float glass 120 S.F.; 1/2″ float glass 55 S.F.; 1/2″ insulated glass 95 S.F.; 3/4″ insulated glass 75 S.F.

Glazing Method	United Inches per Light							
	40″	60″	80″	100″	135″	165″	200″	240″
Number of Men in Crew	1	1	1	1	2	3	3	4
Industrial sash, putty	60	45	24	15	18	—	—	—
With stops, putty bed	50	36	21	12	16	8	4	3
Wood stops, rubber	40	27	15	9	11	6	3	2
Metal stops, rubber	30	24	14	9	9	6	3	2
Structural glass	10	7	4	3	—	—	—	—
Corrugated glass	12	9	7	4	4	4	3	—
Storefronts	16	15	13	11	7	6	4	4
Skylights, putty glass	60	36	21	12	16	—	—	—
Thiokol set	15	15	11	9	9	6	3	2
Vinyl set, snap on	18	18	13	12	12	7	5	4
Maximum area per light	2.8 S.F.	6.3 S.F.	11.1 S.F.	17.4 S.F.	31.6 S.F.	47 S.F.	69 S.F.	100 S.F.

R092000-50 Lath, Plaster and Gypsum Board

Gypsum board lath is available in 3/8″ thick x 16″ wide x 4′ long sheets as a base material for multi-layer plaster applications. It is also available as a base for either multi-layer or veneer plaster applications in 1/2″ and 5/8″ thick–4′ wide x 8′, 10′ or 12′ long sheets. Fasteners are screws or blued ring shank nails for wood framing and screws for metal framing.

Metal lath is available in diamond mesh pattern with flat or self-furring profiles. Paper backing is available for applications where excessive plaster waste needs to be avoided. A slotted mesh ribbed lath should be used in areas where the span between structural supports is greater than normal. Most metal lath comes in 27″ x 96″ sheets. Diamond mesh weighs 1.75, 2.5 or 3.4 pounds per square yard, slotted mesh lath weighs 2.75 or 3.4 pounds per square yard. Metal lath can be nailed, screwed or tied in place.

Many **accessories** are available. Corner beads, flat reinforcing strips, casing beads, control and expansion joints, furring brackets and channels are some examples. Note that accessories are not included in plaster or stucco line items.

Plaster is defined as a material or combination of materials that when mixed with a suitable amount of water, forms a plastic mass or paste. When applied to a surface, the paste adheres to it and subsequently hardens, preserving in a rigid state the form or texture imposed during the period of elasticity.

Gypsum plaster is made from ground calcined gypsum. It is mixed with aggregates and water for use as a base coat plaster.

Vermiculite plaster is a fire-retardant plaster covering used on steel beams, concrete slabs and other heavy construction materials. Vermiculite is a group name for certain clay minerals, hydrous silicates or aluminum, magnesium and iron that have been expanded by heat.

Perlite plaster is a plaster using perlite as an aggregate instead of sand. Perlite is a volcanic glass that has been expanded by heat.

Gauging plaster is a mix of gypsum plaster and lime putty that when applied produces a quick drying finish coat.

Veneer plaster is a one or two component gypsum plaster used as a thin finish coat over special gypsum board.

Keenes cement is a white cementitious material manufactured from gypsum that has been burned at a high temperature and ground to a fine powder. Alum is added to accelerate the set. The resulting plaster is hard and strong and accepts and maintains a high polish, hence it is used as a finishing plaster.

Stucco is a Portland cement based plaster used primarily as an exterior finish.

Plaster is used on both interior and exterior surfaces. Generally it is applied in multiple-coat systems. A three-coat system uses the terms scratch, brown and finish to identify each coat. A two-coat system uses base and finish to describe each coat. Each type of plaster and application system has attributes that are chosen by the designer to best fit the intended use.

Gypsum Plaster Quantities for 100 S.Y.	2 Coat, 5/8″ Thick		3 Coat, 3/4″ Thick		
	Base	Finish	Scratch	Brown	Finish
	1:3 Mix	2:1 Mix	1:2 Mix	1:3 Mix	2:1 Mix
Gypsum plaster	1,300 lb.		1,350 lb.	650 lb.	
Sand	1.75 C.Y.		1.85 C.Y.	1.35 C.Y.	
Finish hydrated lime		340 lb.			340 lb.
Gauging plaster		170 lb.			170 lb.

Vermiculite or Perlite Plaster Quantities for 100 S.Y.	2 Coat, 5/8″ Thick		3 Coat, 3/4″ Thick		
	Base	Finish	Scratch	Brown	Finish
Gypsum plaster	1,250 lb.		1,450 lb.	800 lb.	
Vermiculite or perlite	7.8 bags		8.0 bags	3.3 bags	
Finish hydrated lime		340 lb.			340 lb.
Gauging plaster		170 lb.			170 lb.

Stucco-Three-Coat System Quantities for 100 S.Y.	On Wood Frame	On Masonry
Portland cement	29 bags	21 bags
Sand	2.6 C.Y.	2.0 C.Y.
Hydrated lime	180 lb.	120 lb.

R092910-10 Levels of Gypsum Drywall Finish

In the past, contract documents often used phrases such as "industry standard" and "workmanlike finish" to specify the expected quality of gypsum board wall and ceiling installations. The vagueness of these descriptions led to unacceptable work and disputes.

In order to resolve this problem, four major trade associations concerned with the manufacture, erection, finish and decoration of gypsum board wall and ceiling systems have developed an industry-wide *Recommended Levels of Gypsum Board Finish.*

The finish of gypsum board walls and ceilings for specific final decoration is dependent on a number of factors. A primary consideration is the location of the surface and the degree of decorative treatment desired. Painted and unpainted surfaces in warehouses and other areas where appearance is normally not critical may simply require the taping of wallboard joints and 'spotting' of fastener heads. Blemish-free, smooth, monolithic surfaces often intended for painted and decorated walls and ceilings in habitated structures, ranging from single-family dwellings through monumental buildings, require additional finishing prior to the application of the final decoration.

Other factors to be considered in determining the level of finish of the gypsum board surface are (1) the type of angle of surface illumination (both natural and artificial lighting), and (2) the paint and method of application or the type and finish of wallcovering specified as the final decoration. Critical lighting conditions, gloss paints, and thin wall coverings require a higher level of gypsum board finish than do heavily textured surfaces which are subsequently painted or surfaces which are to be decorated with heavy grade wall coverings.

The following descriptions were developed jointly by the Association of the Wall and Ceiling Industries-International (AWCI), Ceiling & Interior Systems Construction Association (CISCA), Gypsum Association (GA), and Painting and Decorating Contractors of America (PDCA) as a guide.

Level 0: Used in temporary construction or wherever the final decoration has not been determined. Unfinished. No taping, finishing or corner beads are required. Also could be used where non-predecorated panels will be used in demountable-type partitions that are to be painted as a final finish. .

Level 1 Frequently used in plenum areas above ceilings, in attics, in areas where the assembly would generally be concealed, or in building service corridors and other areas not normally open to public view. Some degree of sound and smoke control is provided; in some geographic areas, this level is referred to as "fire-taping," although this level of finish does not typically meet fire-resistant assembly requirements. Where a fire resistance rating is required for the gypsum board assembly, details of construction should be in accordance with reports of fire tests of assemblies that have met the requirements of the fire rating acceptable.

All joints and interior angles shall have tape embedded in joint compound. Accessories are optional at specifier discretion in corridors and other areas with pedestrian traffic. Tape and fastener heads need not be covered with joint compound. Surface shall be free of excess joint compound. Tool marks and ridges are acceptable.

Level 2 It may be specified for standard gypsum board surfaces in garages, warehouse storage, or other similar areas where surface appearance is not of primary importance.

All joints and interior angles shall have tape embedded in joint compound and shall be immediately wiped with a joint knife or trowel, leaving a thin coating of joint compound over all joints and interior angles. Fastener heads and accessories shall be covered with a coat of joint compound. Surface shall be free of excess joint compound. Tool marks and ridges are acceptable.

Level 3 Typically used in areas that are to receive heavy texture (spray or hand applied) finishes before final painting, or where commercial-grade (heavy duty) wall coverings are to be applied as the final decoration. This level of finish should not be used where smooth painted surfaces or where lighter weight wall coverings are specified. The prepared surface shall be coated with a drywall primer prior to the application of final finishes.

All joints and interior angles shall have tape embedded in joint compound and shall be immediately wiped with a joint knife or trowel, leaving a thin coating of joint compound over all joints and interior angles. One additional coat of joint compound shall be applied over all joints and interior angles. Fastener heads and accessories shall be covered with two separate coats of joint compound. All joint compounds shall be smooth and free of tool marks and ridges. The prepared surface shall be covered with a drywall primer prior to the application of the final decoration.

Level 4 This level should be used where residential grade (light duty) wall coverings, flat paints, or light textures are to be applied. The prepared surface shall be coated with a drywall primer prior to the application of final finishes. Release agents for wall coverings are specifically formulated to minimize damage if coverings are subsequently removed.

The weight, texture, and sheen level of the wall covering material selected should be taken into consideration when specifying wall coverings over this level of drywall treatment. Joints and fasteners must be sufficiently concealed if the wall covering material is lightweight, contains limited pattern, has a glossy finish, or has any combination of these features. In critical lighting areas, flat paints applied over light textures tend to reduce joint photographing. Gloss, semi-gloss, and enamel paints are not recommended over this level of finish.

All joints and interior angles shall have tape embedded in joint compound and shall be immediately wiped with a joint knife or trowel, leaving a thin coating of joint compound over all joints and interior angles. In addition, two separate coats of joint compound shall be applied over all flat joints and one separate coat of joint compound applied over interior angles. Fastener heads and accessories shall be covered with three separate coats of joint compound. All joint compounds shall be smooth and free of tool marks and ridges. The prepared surface shall be covered with a drywall primer like Sheetrock first coat prior to the application of the final decoration.

Level 5 The highest quality finish is the most effective method to provide a uniform surface and minimize the possibility of joint photographing and of fasteners showing through the final decoration. This level of finish is required where gloss, semi-gloss, or enamel is specified; when flat joints are specified over an untextured surface; or where critical lighting conditions occur. The prepared surface shall be coated with a drywall primer prior to the application of final decoration.

All joints and interior angles shall have tape embedded in joint compound and be immediately wiped with a joint knife or trowel, leaving a thin coating of joint compound over all joints and interior angles. Two separate coats of joint compound shall be applied over all flat joints and one separate coat of joint compound applied over interior angles. Fastener heads and accessories shall be covered with three separate coats of joint compound.

A thin skim coat of joint compound shall be trowel applied to the entire surface. Excess compound is immediately troweled off, leaving a film or skim coating of compound completely covering the paper. As an alternative to a skim coat, a material manufactured especially for this purpose may be applied such as Sheetrock Tuff-Hide primer surfacer. The surface must be smooth and free of tool marks and ridges. The prepared surface shall be covered with a drywall primer prior to the application of the final decoration.

R142000-10 Freight Elevators

Capacities run from 2,000 lbs. to over 100,000 lbs. with 3,000 lbs. to 10,000 lbs. most common. Travel speeds are generally lower and control less intricate than on passenger elevators. Costs in the unit price sections are for hydraulic and geared elevators.

R142000-20 Elevator Selective Costs **See R142000-40 for cost development.**

	Passenger		Freight		Hospital	
A. Base Unit	**Hydraulic**	**Electric**	**Hydraulic**	**Electric**	**Hydraulic**	**Electric**
Capacity	1,500 lb.	2,000 lb.	2,000 lb.	4,000 lb.	4,000 lb.	4,000 lb.
Speed	100 F.P.M.	200 F.P.M.	100 F.P.M.	200 F.P.M.	100 F.P.M.	200 F.P.M.
#Stops/Travel Ft.	2/12	4/40	2/20	4/40	2/20	4/40
Push Button Oper.	Yes	Yes	Yes	Yes	Yes	Yes
Telephone Box & Wire	"	"	"	"	"	"
Emergency Lighting	"	"	No	No	"	"
Cab	Plastic Lam. Walls	Plastic Lam. Walls	Painted Steel	Painted Steel	Plastic Lam. Walls	Plastic Lam. Walls
Cove Lighting	Yes	Yes	No	No	Yes	Yes
Floor	V.C.T.	V.C.T.	Wood w/Safety Treads	Wood w/Safety Treads	V.C.T.	V.C.T.
Doors, & Speedside Slide	Yes	Yes	Yes	Yes	Yes	Yes
Gates, Manual	No	No	No	No	No	No
Signals, Lighted Buttons	Car and Hall	Car and Hall	Car and Hall	Car and Hall	Car and Hall	Car and Hall
O.H. Geared Machine	N.A.	Yes	N.A.	Yes	N.A.	Yes
Variable Voltage Contr.	"	"	N.A.	"	"	"
Emergency Alarm	Yes	"	Yes	"	Yes	"
Class "A" Loading	N.A.	N.A.	"	"	N.A.	N.A.

R142000-30 Passenger Elevators

Electric elevators are used generally but hydraulic elevators can be used for lifts up to 70′ and where large capacities are required. Hydraulic speeds are limited to 200 F.P.M. but cars are self leveling at the stops. On low rises, hydraulic installation runs about 15% less than standard electric types but on higher rises this installation cost advantage is reduced. Maintenance of hydraulic elevators is about the same as electric type but underground portion is not included in the maintenance contract.

In electric elevators there are several control systems available, the choice of which will be based upon elevator use, size, speed and cost criteria. The two types of drives are geared for low speeds and gearless for 450 F.P.M. and over.

The tables on the preceding pages illustrate typical installed costs of the various types of elevators available.

R142000-40 Elevator Cost Development

To price a new car or truck from the factory, you must start with the manufacturer's basic model, then add or exchange optional equipment and features. The same is true for pricing elevators.

Requirement: One-passenger elevator, five-story hydraulic, 2,500 lb. capacity, 12′ floor to floor, speed 150 F.P.M., emergency power switching and maintenance contract.

Example:

Description	Adjustment
A. Base Elevator: Hydraulic Passenger, 1500 lb. Capacity, 100 fpm, 2 Stops, Standard Finish	1 Ea.
B. Capacity Adjustment (2,500 lb.)	1 Ea.
C. Excess Travel Adjustment: 48′ Total Travel (4 x 12′) minus 12′ Base Unit Travel =	36 V.L.F.
D. Stops Adjustment: 5 Total Stops minus 2 Stops (Base Unit) =	3 Stops
E. Speed Adjustment (150 F.P.M.)	1 Ea.
F. Options:	
1. Intercom Service	1 Ea.
2. Emergency Power Switching, Automatic	1 Ea.
3. Stainless Steel Entrance Doors	5 Ea.
4. Maintenance Contract (12 Months)	1 Ea.
5. Position Indicator for main floor level (none indicated in Base Unit)	1 Ea.

R220102-20 Labor Adjustment Factors

Labor Adjustment Factors are provided for Divisions 21, 22, and 23 to assist the mechanical estimator account for the various complexities and special conditions of any particular project. While a single percentage has been entered on each line of Division 22 01 02.20, it should be understood that these are just suggested midpoints of ranges of values commonly used by mechanical estimators. They may be increased or decreased depending on the severity of the special conditions.

The group for "existing occupied buildings", has been the subject of requests for explanation. Actually there are two stages to this group: buildings that are existing and "finished" but unoccupied, and those that also are occupied. Buildings that are "finished" may result in higher labor costs due to the workers having to be more careful not to damage finished walls, ceilings, floors, etc. and may necessitate special protective coverings and barriers. Also corridor bends and doorways may not accommodate long pieces of pipe or larger pieces of equipment. Work above an already hung ceiling can be very time consuming. The addition of occupants may force the work to be done on premium time (nights and/or weekends), eliminate the possible use of some preferred tools such as pneumatic drivers, powder charged drivers etc. The estimator should evaluate the access to the work area and just how the work is going to be accomplished to arrive at an increase in labor costs over "normal" new construction productivity.

R220105-10 Demolition (Selective vs. Removal for Replacement)

Demolition can be divided into two basic categories.

One type of demolition involves the removal of material with no concern for its replacement. The labor-hours to estimate this work are found under "Selective Demolition" in the Fire Protection, Plumbing and HVAC Divisions. It is selective in that individual items or all the material installed as a system or trade grouping such as plumbing or heating systems are removed. This may be accomplished by the easiest way possible, such as sawing, torch cutting, or sledge hammer as well as simple unbolting.

The second type of demolition is the removal of some item for repair or replacement. This removal may involve careful draining, opening of unions, disconnecting and tagging of electrical connections, capping of pipes/ducts to prevent entry of debris or leakage of the material contained as well as transport of the item away from its in-place location to a truck/dumpster. An approximation of the time required to accomplish this type of demolition is to use half of the time indicated as necessary to install a new unit. For example; installation of a new pump might be listed as requiring 6 labor-hours so if we had to estimate the removal of the old pump we would allow an additional 3 hours for a total of 9 hours. That is, the complete replacement of a defective pump with a new pump would be estimated to take 9 labor-hours.

R221113-50 Pipe Material Considerations

1. Malleable fittings should be used for gas service.
2. Malleable fittings are used where there are stresses/strains due to expansion and vibration.
3. Cast fittings may be broken as an aid to disassembling of heating lines frozen by long use, temperature and minerals.
4. Cast iron pipe is extensively used for underground and submerged service.
5. Type M (light wall) copper tubing is available in hard temper only and is used for nonpressure and less severe applications than K and L.
6. Type L (medium wall) copper tubing, available hard or soft for interior service.
7. Type K (heavy wall) copper tubing, available in hard or soft temper for use where conditions are severe. For underground and interior service.
8. Hard drawn tubing requires fewer hangers or supports but should not be bent. Silver brazed fittings are recommended, however soft solder is normally used.
9. Type DMV (very light wall) copper tubing designed for drainage, waste and vent plus other non-critical pressure services.

Domestic/Imported Pipe and Fittings Cost

The prices shown in this publication for steel/cast iron pipe and steel, cast iron, malleable iron fittings are based on domestic production sold at the normal trade discounts. The above listed items of foreign manufacture may be available at prices of 1/3 to 1/2 those shown. Some imported items after minor machining or finishing operations are being sold as domestic to further complicate the system.

Caution: Most pipe prices in this book also include a coupling and pipe hangers which for the larger sizes can add significantly to the per foot cost and should be taken into account when comparing "book cost" with quoted supplier's cost.

Plumbing — R2240 Plumbing Fixtures

R224000-40 Plumbing Fixture Installation Time

Item	Rough-In	Set	Total Hours	Item	Rough-In	Set	Total Hours
Bathtub	5	5	10	Shower head only	2	1	3
Bathtub and shower, cast iron	6	6	12	Shower drain	3	1	4
Fire hose reel and cabinet	4	2	6	Shower stall, slate		15	15
Floor drain to 4 inch diameter	3	1	4	Slop sink	5	3	8
Grease trap, single, cast iron	5	3	8	Test 6 fixtures			14
Kitchen gas range		4	4	Urinal, wall	6	2	8
Kitchen sink, single	4	4	8	Urinal, pedestal or floor	6	4	10
Kitchen sink, double	6	6	12	Water closet and tank	4	3	7
Laundry tubs	4	2	6	Water closet and tank, wall hung	5	3	8
Lavatory wall hung	5	3	8	Water heater, 45 gals. gas, automatic	5	2	7
Lavatory pedestal	5	3	8	Water heaters, 65 gals. gas, automatic	5	2	7
Shower and stall	6	4	10	Water heaters, electric, plumbing only	4	2	6

Fixture prices in front of book are based on the cost per fixture set in place. The rough-in cost, which must be added for each fixture, includes carrier, if required, some supply, waste and vent pipe connecting fittings and stops. The lengths of rough-in pipe are nominal runs which would connect to the larger runs and stacks. The supply runs and DWV runs and stacks must be accounted for in separate entries. In the eastern half of the United States it is common for the plumber to carry these to a point 5′ outside the building.

Heating, Ventilating & A.C. — R2331 HVAC Ducts & Casings

R233100-20 Ductwork

Duct Weight in Pounds per L.F., Straight Runs

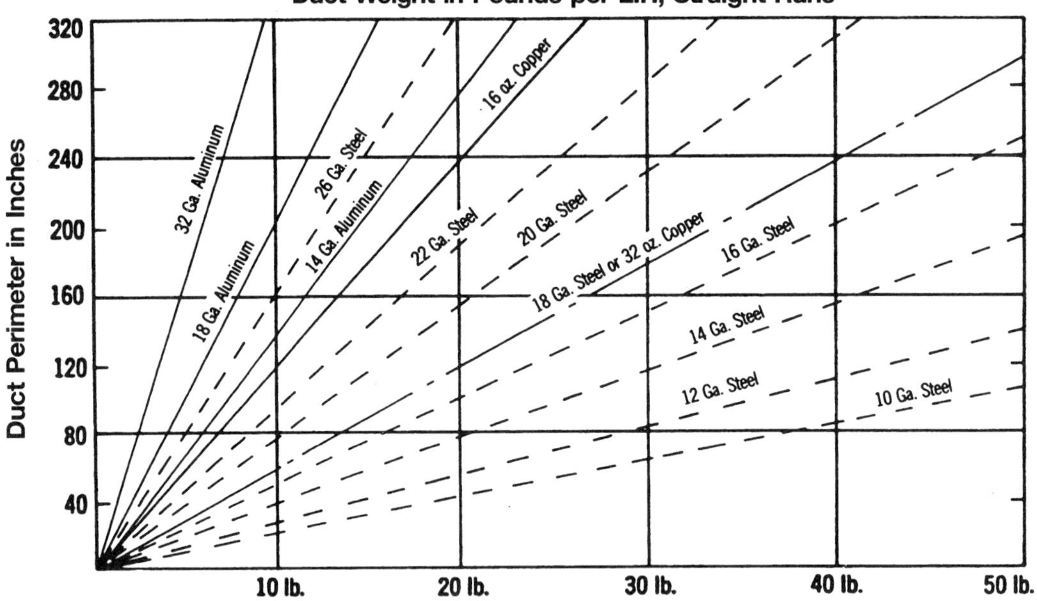

Add to the above for fittings; 90° elbow is 3 L.F.; 45° elbow is 2.5 L.F.; offset is 4 L.F.; transition offset is 6 L.F.; square-to-round transition is 4 L.F.; 90° reducing elbow is 5 L.F. For bracing and waste, add 20% to aluminum and copper, 15% to steel.

R260105-30 Electrical Demolition (Removal for Replacement)

The purpose of this reference number is to provide a guide to users for electrical "removal for replacement" by applying the rule of thumb: 1/3 of new installation time (typical range from 20% to 50%) for removal. Remember to use reasonable judgment when applying the suggested percentage factor. For example:

Contractors have been requested to remove an existing fluorescent lighting fixture and replace with a new fixture utilizing energy saver lamps and electronic ballast:

In order to fully understand the extent of the project, contractors should visit the job site and estimate the time to perform the renovation work in accordance with applicable national, state and local regulation and codes.

The contractor may need to add extra labor hours to his estimate if he discovers unknown concealed conditions such as: contaminated asbestos ceiling, broken acoustical ceiling tile and need to repair, patch and touch-up paint all the damaged or disturbed areas, tasks normally assigned to general contractors. In addition, the owner could request that the contractors salvage the materials removed and turn over the materials to the owner or dispose of the materials to a reclamation station. The normal removal item is 0.5 labor hour for a lighting fixture and 1.5 labor-hours for new installation time. Revise the estimate times from 2 labor-hours work up to a minimum 4 labor-hours work for just fluorescent lighting fixture.

For removal of large concentrations of lighting fixtures in the same area, apply an "economy of scale" to reduce estimating labor hours.

R260533-22 Conductors in Conduit

Table below lists maximum number of conductors for various sized conduit using THW, TW or THWN insulations.

Copper Wire Size	Conduit Size																										
	½″			¾″			1″			1-¼″			1-½″			2″			2-½″			3″		3-½″		4″	
	TW	THW	THWN	TW	THW	THWN	TW	THW	THWN	TW	THW	THWN	TW	THW	THWN	TW	THW	THWN	TW	THW	THWN	THW	THWN	THW	THWN	THW	THWN
#14	9	6	13	15	10	24	25	16	39	44	29	69	60	40	94	99	65	154	142	93		143		192			
#12	7	4	10	12	8	18	19	13	29	35	24	51	47	32	70	78	53	114	111	76	164	117		157			
#10	5	4	6	9	6	11	15	11	18	26	19	32	36	26	44	60	43	73	85	61	104	95	160	127		163	
#8	2	1	3	4	3	5	7	5	9	12	10	16	17	13	22	28	22	36	40	32	51	49	79	66	106	85	136
#6		1	1		2	4		4	6		7	11		10	15		16	26		23	37	36	57	48	76	62	98
#4		1	1		1	2		3	4		5	7		7	9		12	16		17	22	27	35	36	47	47	60
#3		1	1		1	1		2	3		4	6		6	8		10	13		15	19	23	29	31	39	40	51
#2		1	1		1	1		2	3		4	5		5	7		9	11		13	16	20	25	27	33	34	43
#1					1	1		1	1		3	3		4	5		6	8		9	12	14	18	19	25	25	32
1/0					1	1		1	1		2	3		3	4		5	7		8	10	12	15	16	21	21	27
2/0					1	1		1	1		1	2		3	3		5	6		7	8	10	13	14	17	18	22
3/0					1	1		1	1		1	1		2	3		4	5		6	7	9	11	12	14	15	18
4/0						1		1	1		1	1		1	2		3	4		5	6	7	9	10	12	13	15
250 kcmil								1	1		1	1		1	1		2	3		4	4	6	7	8	10	10	12
300								1	1		1	1		1	1		2	3		3	4	5	6	7	8	9	11
350									1		1	1		1	1		1	2		3	3	4	5	6	7	8	9
400											1	1		1	1		1	1		2	3	4	5	5	6	7	8
500											1	1		1	1		1	1		1	2	3	4	4	5	6	7
600												1		1	1		1	1		1	1	3	3	4	4	5	5
700														1	1		1	1		1	1	2	3	3	4	4	5
750 ▼														1	1		1	1		1	1	2	2	3	3	4	4

R312316-40 Excavating

The selection of equipment used for structural excavation and bulk excavation or for grading is determined by the following factors.

1. Quantity of material
2. Type of material
3. Depth or height of cut
4. Length of haul
5. Condition of haul road
6. Accessibility of site
7. Moisture content and dewatering requirements
8. Availability of excavating and hauling equipment

Some additional costs must be allowed for hand trimming the sides and bottom of concrete pours and other excavation below the general excavation.

When planning excavation and fill, the following should also be considered.

1. Swell factor
2. Compaction factor
3. Moisture content
4. Density requirements

A typical example for scheduling and estimating the cost of excavation of a 15′ deep basement on a dry site when the material must be hauled off the site is outlined below.

Assumptions:

1. Swell factor, 18%
2. No mobilization or demobilization
3. Allowance included for idle time and moving on job
4. No dewatering, sheeting, or bracing
5. No truck spotter or hand trimming

Number of B.C.Y. per truck = 1.5 C.Y. bucket x 8 passes = 12 loose C.Y.

$$= 12 \times \frac{100}{118} = 10.2 \text{ B.C.Y. per truck}$$

Truck Haul Cycle:

Load truck, 8 passes	=	4 minutes
Haul distance, 1 mile	=	9 minutes
Dump time	=	2 minutes
Return, 1 mile	=	7 minutes
Spot under machine	=	1 minute
		23 minute cycle

Fleet Haul Production per day in B.C.Y.

$$4 \text{ trucks} \times \frac{50 \text{ min. hour}}{23 \text{ min. haul cycle}} \times 8 \text{ hrs.} \times 10.2 \text{ B.C.Y.}$$

$$= 4 \times 2.2 \times 8 \times 10.2 = 718 \text{ B.C.Y./day}$$

Add the mobilization and demobilization costs to the total excavation costs. When equipment is rented for more than three days, there is often no mobilization charge by the equipment dealer. On larger jobs outside of urban areas, scrapers can move earth economically provided a dump site or fill area and adequate haul roads are available. Excavation within sheeting bracing or cofferdam bracing is usually done with a clamshell and production is low, since the clamshell may have to be guided by hand between the bracing. When excavating or filling an area enclosed with a wellpoint system, add 10% to 15% to the cost to allow for restricted access. When estimating earth excavation quantities for structures, allow work space outside the building footprint for construction of the foundation and a slope of 1:1 unless sheeting is used.

R312316-45 Excavating Equipment

The table below lists THEORETICAL hourly production in C.Y./hr. bank measure for some typical excavation equipment. Figures assume 50 minute hours, 83% job efficiency, 100% operator efficiency, 90° swing and properly sized hauling units, which must be modified for adverse digging and loading conditions. Actual production costs in the front of the book average about 50% of the theoretical values listed here.

Equipment	Soil Type	B.C.Y. Weight	% Swell	1 C.Y.	1-1/2 C.Y.	2 C.Y.	2-1/2 C.Y.	3 C.Y.	3-1/2 C.Y.	4 C.Y.
Hydraulic Excavator "Backhoe" 15′ Deep Cut	Moist loam, sandy clay	3400 lb.	40%	165	195	200	275	330	385	440
	Sand and gravel	3100	18	140	170	225	240	285	330	380
	Common earth	2800	30	150	180	230	250	300	350	400
	Clay, hard, dense	3000	33	120	140	190	200	240	260	320
Power Shovel Optimum Cut (Ft.)	Moist loam, sandy clay	3400	40	170	245	295	335	385	435	475
				(6.0)	(7.0)	(7.8)	(8.4)	(8.8)	(9.1)	(9.4)
	Sand and gravel	3100	18	165	225	275	325	375	420	460
				(6.0)	(7.0)	(7.8)	(8.4)	(8.8)	(9.1)	(9.4)
	Common earth	2800	30	145	200	250	295	335	375	425
				(7.8)	(9.2)	(10.2)	(11.2)	(12.1)	(13.0)	(13.8)
	Clay, hard, dense	3000	33	120	175	220	255	300	335	375
				(9.0)	(10.7)	(12.2)	(13.3)	(14.2)	(15.1)	(16.0)
Drag Line Optimum Cut (Ft.)	Moist loam, sandy clay	3400	40	130	180	220	250	290	325	385
				(6.6)	(7.4)	(8.0)	(8.5)	(9.0)	(9.5)	(10.0)
	Sand and gravel	3100	18	130	175	210	245	280	315	375
				(6.6)	(7.4)	(8.0)	(8.5)	(9.0)	(9.5)	(10.0)
	Common earth	2800	30	110	160	190	220	250	280	310
				(8.0)	(9.0)	(9.9)	(10.5)	(11.0)	(11.5)	(12.0)
	Clay, hard, dense	3000	33	90	130	160	190	225	250	280
				(9.3)	(10.7)	(11.8)	(12.3)	(12.8)	(13.3)	(12.0)
				Wheel Loaders				Track Loaders		
				3 C.Y.	4 C.Y.	6 C.Y.	8 C.Y.	2-1/4 C.Y.	3 C.Y.	4 C.Y.
Loading Tractors	Moist loam, sandy clay	3400	40	260	340	510	690	135	180	250
	Sand and gravel	3100	18	245	320	480	650	130	170	235
	Common earth	2800	30	230	300	460	620	120	155	220
	Clay, hard, dense	3000	33	200	270	415	560	110	145	200
	Rock, well-blasted	4000	50	180	245	380	520	100	130	180

R312323-30 Compacting Backfill

Compaction of fill in embankments, around structures, in trenches, and under slabs is important to control settlement. Factors affecting compaction are:

1. Soil gradation
2. Moisture content
3. Equipment used
4. Depth of fill per lift
5. Density required

Production Rate:

$$\frac{1.75' \text{ plate width x 50 F.P.M. x 50 min./hr. x .67' lift}}{27 \text{ C.F. per C.Y.}} = 108.5 \text{ C.Y./hr.}$$

Production Rate for 4 Passes:

$$\frac{108.5 \text{ C.Y.}}{4 \text{ passes}} = 27.125 \text{ C.Y./hr. x 8 hrs.} = 217 \text{ C.Y./day}$$

Example:

Compact granular fill around a building foundation using a 21″ wide x 24″ vibratory plate in 8″ lifts. Operator moves at 50 F.P.M. working a 50 minute hour to develop 95% Modified Proctor Density with 4 passes.

Earthwork — R3141 Shoring

R314116-40 Wood Sheet Piling

Wood sheet piling may be used for depths to 20′ where there is no ground water. If moderate ground water is encountered Tongue & Groove sheeting will help to keep it out. When considerable ground water is present, steel sheeting must be used.

For estimating purposes on trench excavation, sizes are as follows:

Depth	Sheeting	Wales	Braces	B.F. per S.F.
To 8′	3 x 12′s	6 x 8′s, 2 line	6 x 8′s, @ 10′	4.0 @ 8′
8′ x 12′	3 x 12′s	10 x 10′s, 2 line	10 x 10′s, @ 9′	5.0 average
12′ to 20′	3 x 12′s	12 x 12′s, 3 line	12 x 12′s, @ 8′	7.0 average

Sheeting to be toed in at least 2′ depending upon soil conditions. A five person crew with an air compressor and sheeting driver can drive and brace 440 SF/day at 8′ deep, 360 SF/day at 12′ deep, and 320 SF/day at 16′ deep. For normal soils, piling can be pulled in 1/3 the time to install. Pulling difficulty increases with the time in the ground. Production can be increased by high pressure jetting.

Earthwork — R3163 Drilled Caissons

R316326-60 Caissons

The three principal types of cassions are:

(1) **Belled Caissons,** which except for shallow depths and poor soil conditions, are generally recommended. They provide more bearing than shaft area. Because of its conical shape, no horizontal reinforcement of the bell is required.

(2) **Straight Shaft Caissons** are used where relatively light loads are to be supported by caissons that rest on high value bearing strata. While the shaft is larger in diameter than for belled types this is more than offset by the saving in time and labor.

(3) **Keyed Caissons** are used when extremely heavy loads are to be carried. A keyed or socketed caisson transfers its load into rock by a combination of end-bearing and shear reinforcing of the shaft. The most economical shaft often consists of a steel casing, a steel wide flange core and concrete. Allowable compressive stresses of .225 f′c for concrete, 16,000 psi for the wide flange core, and 9,000 psi for the steel casing are commonly used. The usual range of shaft diameter is 18″ to 84″. The number of sizes specified for any one project should be limited due to the problems of casing and auger storage. When hand work is to be performed, shaft diameters should not be less than 32″. When inspection of borings is required a minimum shaft diameter of 30″ is recommended. Concrete caissons are intended to be poured against earth excavation so permanent forms which add to cost should not be used if the excavation is clean and the earth sufficiently impervious to prevent excessive loss of concrete.

Soil Conditions for Belling		
Good	**Requires Handwork**	**Not Recommended**
Clay	Hard Shale	Silt
Sandy Clay	Limestone	Sand
Silty Clay	Sandstone	Gravel
Clayey Silt	Weathered Mica	Igneous Rock
Hard-pan		
Soft Shale		
Decomposed Rock		

Exterior Improvements — R3292 Turf & Grasses

R329219-50 Seeding

The type of grass is determined by light, shade and moisture content of soil plus intended use. Fertilizer should be disked 4″ before seeding. For steep slopes disk five tons of mulch and lay two tons of hay or straw on surface per acre after seeding. Surface mulch can be staked, lightly disked or tar emulsion sprayed. Material for mulch can be wood chips, peat moss, partially rotted hay or straw, wood fibers and sprayed emulsions. Hemp seed blankets with fertilizer are also available. For spring seeding, watering is necessary. Late fall seeding may have to be reseeded in the spring. Hydraulic seeding, power mulching, and aerial seeding can be used on large areas.

R331113-80 Piping Designations

There are several systems currently in use to describe pipe and fittings. The following paragraphs will help to identify and clarify classifications of piping systems used for water distribution.

Piping may be classified by schedule. Piping schedules include 5S, 10S, 10, 20, 30, Standard, 40, 60, Extra Strong, 80, 100, 120, 140, 160 and Double Extra Strong. These schedules are dependent upon the pipe wall thickness. The wall thickness of a particular schedule may vary with pipe size.

Ductile iron pipe for water distribution is classified by Pressure Classes such as Class 150, 200, 250, 300 and 350. These classes are actually the rated water working pressure of the pipe in pounds per square inch (psi). The pipe in these pressure classes is designed to withstand the rated water working pressure plus a surge allowance of 100 psi.

The American Water Works Association (AWWA) provides standards for various types of **plastic pipe.** C-900 is the specification for polyvinyl chloride (PVC) piping used for water distribution in sizes ranging from 4″ through 12″. C-901 is the specification for polyethylene (PE) pressure pipe, tubing and fittings used for water distribution in sizes ranging from 1/2″ through 3″. C-905 is the specification for PVC piping sizes 14″ and greater.

PVC pressure-rated pipe is identified using the standard dimensional ratio (SDR) method. This method is defined by the American Society for Testing and Materials (ASTM) Standard D 2241. This pipe is available in SDR numbers 64, 41, 32.5, 26, 21, 17, and 13.5. Pipe with an SDR of 64 will have the thinnest wall while pipe with an SDR of 13.5 will have the thickest wall. When the pressure rating (PR) of a pipe is given in psi, it is based on a line supplying water at 73 degrees F.

The National Sanitation Foundation (NSF) seal of approval is applied to products that can be used with potable water. These products have been tested to ANSI/NSF Standard 14.

Valves and strainers are classified by American National Standards Institute (ANSI) Classes. These Classes are 125, 150, 200, 250, 300, 400, 600, 900, 1500 and 2500. Within each class there is an operating pressure range dependent upon temperature. Design parameters should be compared to the appropriate material dependent, pressure-temperature rating chart for accurate valve selection.

Change Orders

Change Order Considerations

A change order is a written document, usually prepared by the design professional, and signed by the owner, the architect/engineer, and the contractor. A change order states the agreement of the parties to: an addition, deletion, or revision in the work; an adjustment in the contract sum, if any; or an adjustment in the contract time, if any. Change orders, or "extras" in the construction process occur after execution of the construction contract and impact architects/engineers, contractors, and owners.

Change orders that are properly recognized and managed can ensure orderly, professional, and profitable progress for all who are involved in the project. There are many causes for change orders and change order requests. In all cases, change orders or change order requests should be addressed promptly and in a precise and prescribed manner. The following paragraphs include information regarding change order pricing and procedures.

The Causes of Change Orders

Reasons for issuing change orders include:

- Unforeseen field conditions that require a change in the work
- Correction of design discrepancies, errors, or omissions in the contract documents
- Owner-requested changes, either by design criteria, scope of work, or project objectives
- Completion date changes for reasons unrelated to the construction process
- Changes in building code interpretations, or other public authority requirements that require a change in the work
- Changes in availability of existing or new materials and products

Procedures

Properly written contract documents must include the correct change order procedures for all parties—owners, design professionals and contractors—to follow in order to avoid costly delays and litigation.

Being "in the right" is not always a sufficient or acceptable defense. The contract provisions requiring notification and documentation must be adhered to within a defined or reasonable time frame.

The appropriate method of handling change orders is by a written proposal and acceptance by all parties involved. Prior to starting work on a project, all parties should identify their authorized agents who may sign and accept change orders, as well as any limits placed on their authority.

Time may be a critical factor when the need for a change arises. For such cases, the contractor might be directed to proceed on a "time and materials" basis, rather than wait for all paperwork to be processed—a delay that could impede progress. In this situation, the contractor must still follow the prescribed change order procedures including, but not limited to, notification and documentation.

All forms used for change orders should be dated and signed by the proper authority. Lack of documentation can be very costly, especially if legal judgments are to be made, and if certain field personnel are no longer available. For time and material change orders, the contractor should keep accurate daily records of all labor and material allocated to the change.

Owners or awarding authorities who do considerable and continual building construction (such as the federal government) realize the inevitability of change orders for numerous reasons, both predictable and unpredictable. As a result, the federal government, the American Institute of Architects (AIA), the Engineers Joint Contract Documents Committee (EJCDC) and other contractor, legal, and technical organizations have developed standards and procedures to be followed by all parties to achieve contract continuance and timely completion, while being financially fair to all concerned.

In addition to the change order standards put forth by industry associations, there are also many books available on the subject.

Pricing Change Orders

When pricing change orders, regardless of their cause, the most significant factor is when the change occurs. The need for a change may be perceived in the field or requested by the architect/engineer *before* any of the actual installation has begun, or may evolve or appear *during* construction when the item of work in question is partially installed. In the latter cases, the original sequence of construction is disrupted, along with all contiguous and supporting systems. Change orders cause the greatest impact when they occur *after* the installation has been completed and must be uncovered, or even replaced. Post-completion changes may be caused by necessary design changes, product failure, or changes in the owner's requirements that are not discovered until the building or the systems begin to function.

Specified procedures of notification and record keeping must be adhered to and enforced regardless of the stage of construction: *before, during,* or *after* installation. Some bidding documents anticipate change orders by requiring that unit prices including overhead and profit percentages—for additional as well as deductible changes—be listed. Generally these unit prices do not fully take into account the ripple effect, or impact on other trades, and should be used for general guidance only.

When pricing change orders, it is important to classify the time frame in which the change occurs. There are two basic time frames for change orders: *pre-installation change orders,* which occur before the start of construction, and *post-installation change orders,* which involve reworking after the original installation. Change orders that occur between these stages may be priced according to the extent of work completed using a combination of techniques developed for pricing *pre-* and *post-installation* changes.

The following factors are the basis for a check list to use when preparing a change order estimate.

Factors To Consider When Pricing Change Orders

As an estimator begins to prepare a change order, the following questions should be reviewed to determine their impact on the final price.

General

- *Is the change order work* pre-installation *or* post-installation?

  Change order work costs vary according to how much of the installation has been completed. Once workers have the project scoped in their minds, even though they have not started, it can be difficult to refocus. Consequently they may spend more than the normal amount of time understanding the change. Also, modifications to work in place, such as trimming or refitting, usually take more time than was initially estimated. The greater the amount of work in place, the more reluctant workers are to change it. Psychologically they may resent the change and as a result the rework takes longer than normal. Post-installation change order estimates must include demolition of existing work as required to accomplish the change. If the work is performed at a later time, additional obstacles, such as building finishes, may be present which must be protected. Regardless of whether the change occurs pre-installation or post-installation, attempt to isolate the identifiable factors and price them separately. For example, add shipping costs that may be required pre-installation or any demolition required post-installation. Then analyze the potential impact on productivity of psychological and/or learning curve factors and adjust the output rates accordingly. One approach is to break down the typical workday into segments and quantify the impact on each segment. The following chart may be useful as a guide:

Activities (Productivity) Expressed as Percentages of a Workday			
Task	**Means Mechanical Cost Data (for New Construction)**	**Pre-Installation Change Orders**	**Post-Installation Change Orders**
1. Study plans	3%	6%	6%
2. Material procurement	3%	3%	3%
3. Receiving and storing	3%	3%	3%
4. Mobilization	5%	5%	5%
5. Site movement	5%	5%	8%
6. Layout and marking	8%	10%	12%
7. Actual installation	64%	59%	54%
8. Clean-up	3%	3%	3%
9. Breaks—non-productive	6%	6%	6%
Total	100%	100%	100%

Change Order Installation Efficiency

The labor hours expressed (for new construction) are based on average installation time, using an efficiency level of approximately 60%–65%. For change order situations, adjustments to this efficiency level should reflect the daily labor-hour allocation for that particular occurrence.

If any of the specific percentages expressed in the above chart do not apply to a particular project situation, then those percentage points should be reallocated to the appropriate task(s). Example: Using data for new construction, assume there is no new material being utilized. The percentages for Tasks 2 and 3 would therefore be reallocated to other tasks. If the time required for Tasks 2 and 3 can now be applied to installation, we can add the time allocated for *Material Procurement* and *Receiving and Storing* to the *Actual Installation* time for new construction, thereby increasing the Actual Installation percentage.

This chart shows that, due to reduced productivity, labor costs will be higher than those for new construction by 5%–15% for pre-installation change orders and by 15%–25% for post-installation change orders. Each job and change order is unique and must be examined individually. Many factors, covered elsewhere in this section, can each have a significant impact on productivity and change order costs. All such factors should be considered in every case.

- *Will the change substantially delay the original completion date?*

  A significant change in the project may cause the original completion date to be extended. The extended schedule may subject the contractor to new wage rates dictated by relevant labor contracts. Project supervision and other project overhead must also be extended beyond the original completion date. The schedule extension may also put installation into a new weather season. For example, underground piping scheduled for October installation was delayed until January. As a result, frost penetrated the trench area, thereby changing the degree of difficulty of the task. Changes and delays may have a ripple effect throughout the project. This effect must be analyzed and negotiated with the owner.

- *What is the net effect of a deduct change order?*

  In most cases, change orders resulting in a deduction or credit reflect only bare costs. The contractor may retain the overhead and profit based on the original bid.

Materials

- *Will you have to pay more or less for the new material, required by the change order, than you paid for the original purchase?*

  The same material prices or discounts will usually apply to materials purchased for change orders as new construction. In some instances, however, the contractor may forfeit the advantages of competitive pricing for change orders. Consider the following example:

  A contractor purchased over $20,000 worth of fan coil units for an installation, and obtained the maximum discount. Some time later it was determined the project required an additional matching unit. The contractor has to purchase this unit from the original supplier to ensure a match. The supplier at this time may not discount the unit because of the small quantity, and the fact that he is no longer in a competitive situation. The impact of quantity on purchase can add between 0% and 25% to material prices and/or subcontractor quotes.

- *If materials have been ordered or delivered to the job site, will they be subject to a cancellation charge or restocking fee?*

  Check with the supplier to determine if ordered materials are subject to a cancellation charge. Delivered materials not used as result of a change order may be subject to a restocking fee if returned to the supplier. Common restocking charges run between 20% and 40%. Also, delivery charges to return the goods to the supplier must be added.

Labor

- *How efficient is the existing crew at the actual installation?*

  Is the same crew that performed the initial work going to do the change order? Possibly the change consists of the installation of a unit identical to one already installed; therefore, the change should take less time. Be sure to consider this potential productivity increase and modify the productivity rates accordingly.

- *If the crew size is increased, what impact will that have on supervision requirements?*

  Under most bargaining agreements or management practices, there is a point at which a working foreman is replaced by a nonworking foreman. This replacement increases project overhead by adding a nonproductive worker. If additional workers are added to accelerate the project or to perform changes while maintaining the schedule, be sure to add additional supervision time if warranted. Calculate the hours involved and the additional cost directly if possible.

- *What are the other impacts of increased crew size?*

  The larger the crew, the greater the potential for productivity to decrease. Some of the factors that cause this productivity loss are: overcrowding (producing restrictive conditions in the working space), and possibly a shortage of any special tools and equipment required. Such factors affect not only the crew working on the elements directly involved in the change order, but other crews whose movement may also be hampered.

  As the crew increases, check its basic composition for changes by the addition or deletion of apprentices or nonworking foreman, and quantify the potential effects of equipment shortages or other logistical factors.

- *As new crews, unfamiliar with the project, are brought onto the site, how long will it take them to become oriented to the project requirements?*

  The orientation time for a new crew to become 100% effective varies with the site and type of project. Orientation is easiest at a new construction site, and most difficult at existing, very restrictive renovation sites. The type of work also affects orientation time. When all elements of the work are exposed, such as concrete or masonry work, orientation is decreased. When the work is concealed or less visible, such as existing electrical systems, orientation takes longer. Usually orientation can be accomplished in one day or less. Costs for added orientation should be itemized and added to the total estimated cost.

- *How much actual production can be gained by working overtime?*

  Short term overtime can be used effectively to accomplish more work in a day. However, as overtime is scheduled to run beyond several weeks, studies have shown marked decreases in output. The following chart shows the effect of long term overtime on worker efficiency. If the anticipated change requires extended overtime to keep the job on schedule, these factors can be used as a guide to predict the impact on time and cost. Add project overhead, particularly supervision, that may also be incurred.

Days per Week	Hours per Day	Production Efficiency					Payroll Cost Factors	
		1 Week	2 Weeks	3 Weeks	4 Weeks	Average 4 Weeks	@ 1-1/2 Times	@ 2 Times
5	8	100%	100%	100%	100%	100%	100%	100%
	9	100	100	95	90	96.25	105.6	111.1
	10	100	95	90	85	91.25	110.0	120.0
	11	95	90	75	65	81.25	113.6	127.3
	12	90	85	70	60	76.25	116.7	133.3
6	8	100	100	95	90	96.25	108.3	116.7
	9	100	95	90	85	92.50	113.0	125.9
	10	95	90	85	80	87.50	116.7	133.3
	11	95	85	70	65	78.75	119.7	139.4
	12	90	80	65	60	73.75	122.2	144.4
7	8	100	95	85	75	88.75	114.3	128.6
	9	95	90	80	70	83.75	118.3	136.5
	10	90	85	75	65	78.75	121.4	142.9
	11	85	80	65	60	72.50	124.0	148.1
	12	85	75	60	55	68.75	126.2	152.4

Effects of Overtime

Caution: Under many labor agreements, Sundays and holidays are paid at a higher premium than the normal overtime rate.

The use of long-term overtime is counterproductive on almost any construction job; that is, the longer the period of overtime, the lower the actual production rate. Numerous studies have been conducted, and while they have resulted in slightly different numbers, all reach the same conclusion. The figure above tabulates the effects of overtime work on efficiency.

As illustrated, there can be a difference between the *actual* payroll cost per hour and the *effective* cost per hour for overtime work. This is due to the reduced production efficiency with the increase in weekly hours beyond 40. This difference between actual and effective cost results from overtime work over a prolonged period. Short-term overtime work does not result in as great a reduction in efficiency and, in such cases, effective cost may not vary significantly from the actual payroll cost. As the total hours per week are increased on a regular basis, more time is lost due to fatigue, lowered morale, and an increased accident rate.

As an example, assume a project where workers are working 6 days a week, 10 hours per day. From the figure above (based on productivity studies), the average effective productive hours over a 4-week period are:

$$0.875 \times 60 = 52.5$$

Depending upon the locale and day of week, overtime hours may be paid at time and a half or double time. For time and a half, the overall (average) *actual* payroll cost (including regular and overtime hours) is determined as follows:

$$\frac{40 \text{ reg. hrs.} + (20 \text{ overtime hrs.} \times 1.5)}{60 \text{ hrs.}} = 1.167$$

Based on 60 hours, the payroll cost per hour will be 116.7% of the normal rate at 40 hours per week. However, because the effective production (efficiency) for 60 hours is reduced to the equivalent of 52.5 hours, the effective cost of overtime is calculated as follows:

For time and a half:

$$\frac{40 \text{ reg. hrs.} + (20 \text{ overtime hrs.} \times 1.5)}{52.5 \text{ hrs.}} = 1.33$$

Installed cost will be 133% of the normal rate (for labor).

Thus, when figuring overtime, the actual cost per unit of work will be higher than the apparent overtime payroll dollar increase, due to the reduced productivity of the longer workweek. These efficiency calculations are true only for those cost factors determined by hours worked. Costs that are applied weekly or monthly, such as equipment rentals, will not be similarly affected.

Equipment

- *What equipment is required to complete the change order?*

  Change orders may require extending the rental period of equipment already on the job site, or the addition of special equipment brought in to accomplish the change work. In either case, the additional rental charges and operator labor charges must be added.

Summary

The preceding considerations and others you deem appropriate should be analyzed and applied to a change order estimate. The impact of each should be quantified and listed on the estimate to form an audit trail.

Change orders that are properly identified, documented, and managed help to ensure the orderly, professional and profitable progress of the work. They also minimize potential claims or disputes at the end of the project.

Estimating Tips

- The cost figures in this Square Foot Cost section were derived from approximately 11,200 projects contained in the RSMeans database of completed construction projects. They include the contractor's overhead and profit, but do not generally include architectural fees or land costs. The figures have been adjusted to January of the current year. New projects are added to our files each year, and outdated projects are discarded. For this reason, certain costs may not show a uniform annual progression. In no case are all subdivisions of a project listed.
- These projects were located throughout the U.S. and reflect a tremendous variation in square foot (S.F.) and cubic foot (C.F.) costs. This is due to differences, not only in labor and material costs, but also in individual owners' requirements. For instance, a bank in a large city would have different features than one in a rural area. This is true of all the different types of buildings analyzed. Therefore, caution should be exercised when using these Square Foot costs. For example, for court houses, costs in the database are local court house costs and will not apply to the larger, more elaborate federal court houses. As a general rule, the projects in the 1/4 column do not include any site work or equipment, while the projects in the 3/4 column may include both equipment and site work. The median figures do not generally include site work.
- None of the figures "go with" any others. All individual cost items were computed and tabulated separately. Thus, the sum of the median figures for Plumbing, HVAC and Electrical will not normally total up to the total Mechanical and Electrical costs arrived at by separate analysis and tabulation of the projects.
- Each building was analyzed as to total and component costs and percentages. The figures were arranged in ascending order with the results tabulated as shown. The 1/4 column shows that 25% of the projects had lower costs and 75% had higher. The 3/4 column shows that 75% of the projects had lower costs and 25% had higher. The median column shows that 50% of the projects had lower costs and 50% had higher.
- There are two times when square foot costs are useful. The first is in the conceptual stage when no details are available. Then square foot costs make a useful starting point. The second is after the bids are in and the costs can be worked back into their appropriate units for information purposes. As soon as details become available in the project design, the square foot approach should be discontinued and the project priced as to its particular components. When more precision is required, or for estimating the replacement cost of specific buildings, the current edition of *RSMeans Square Foot Costs* should be used.
- In using the figures in this section, it is recommended that the median column be used for preliminary figures if no additional information is available. The median figures, when multiplied by the total city construction cost index figures (see City Cost Indexes) and then multiplied by the project size modifier at the end of this section, should present a fairly accurate base figure, which would then have to be adjusted in view of the estimator's experience, local economic conditions, code requirements, and the owner's particular requirements. There is no need to factor the percentage figures, as these should remain constant from city to city. All tabulations mentioning air conditioning had at least partial air conditioning.
- The editors of this book would greatly appreciate receiving cost figures on one or more of your recent projects, which would then be included in the averages for next year. All cost figures received will be kept confidential, except that they will be averaged with other similar projects to arrive at Square Foot cost figures for next year's book. See the last page of the book for details and the discount available for submitting one or more of your projects.

		50 17 00 | S.F. Costs	UNIT	UNIT COSTS 1/4	UNIT COSTS MEDIAN	UNIT COSTS 3/4	% OF TOTAL 1/4	% OF TOTAL MEDIAN	% OF TOTAL 3/4	
01	0010	**APARTMENTS Low Rise (1 to 3 story)**	S.F.	69.50	87	116				01
	0020	Total project cost	C.F.	6.25	8.25	10.20				
	0100	Site work	S.F.	5.95	8.10	14.30	6.05%	10.55%	14.05%	
	0500	Masonry		1.37	3.18	5.50	1.54%	3.67%	6.35%	
	1500	Finishes		7.35	10	12.50	9.05%	10.75%	12.85%	
	1800	Equipment		2.27	3.45	5.10	2.73%	4.03%	5.95%	
	2720	Plumbing		5.40	6.95	8.85	6.65%	8.95%	10.05%	
	2770	Heating, ventilating, air conditioning		3.45	4.25	6.25	4.20%	5.60%	7.60%	
	2900	Electrical		4.02	5.35	7.20	5.20%	6.65%	8.40%	
	3100	Total: Mechanical & Electrical	↓	13.95	17.75	22	15.90%	18.05%	23%	
	9000	Per apartment unit, total cost	Apt.	65,000	99,000	146,000				
	9500	Total: Mechanical & Electrical	″	12,300	19,300	25,200				
02	0010	**APARTMENTS Mid Rise (4 to 7 story)**	S.F.	92.50	111	138				02
	0020	Total project costs	C.F.	7.20	9.95	13.60				
	0100	Site work	S.F.	3.69	7.30	13.15	5.25%	6.70%	9.15%	
	0500	Masonry		6.15	8.45	11.55	5.10%	7.25%	10.50%	
	1500	Finishes		11.60	15.15	19.10	10.55%	13.45%	17.70%	
	1800	Equipment		2.68	4.01	5.25	2.54%	3.48%	4.31%	
	2500	Conveying equipment		1.99	2.53	3.07	1.94%	2.27%	2.69%	
	2720	Plumbing		5.40	8.65	9.15	5.70%	7.20%	8.95%	
	2900	Electrical		6.10	8.25	10.05	6.65%	7.20%	8.95%	
	3100	Total: Mechanical & Electrical	↓	19.45	24.50	29.50	18.50%	21%	23%	
	9000	Per apartment unit, total cost	Apt.	104,500	123,000	204,000				
	9500	Total: Mechanical & Electrical	″	19,700	22,800	28,800				
03	0010	**APARTMENTS High Rise (8 to 24 story)**	S.F.	105	120	145				03
	0020	Total project costs	C.F.	10.15	11.80	15.10				
	0100	Site work	S.F.	3.79	6.15	8.60	2.58%	4.84%	6.15%	
	0500	Masonry		6.05	11	13.70	4.74%	9.65%	11.05%	
	1500	Finishes		11.60	14.50	17.15	9.75%	11.80%	13.70%	
	1800	Equipment		3.36	4.14	5.50	2.78%	3.49%	4.35%	
	2500	Conveying equipment		2.38	3.61	4.91	2.23%	2.78%	3.37%	
	2720	Plumbing		7.70	9.10	12.65	6.80%	7.20%	10.45%	
	2900	Electrical		7.20	9.10	12.25	6.45%	7.65%	8.80%	
	3100	Total: Mechanical & Electrical	↓	21.50	27.50	33	17.95%	22.50%	24.50%	
	9000	Per apartment unit, total cost	Apt.	108,500	119,500	166,000				
	9500	Total: Mechanical & Electrical	″	23,500	26,800	28,400				
04	0010	**AUDITORIUMS**	S.F.	109	148	212				04
	0020	Total project costs	C.F.	6.80	9.45	13.60				
	2720	Plumbing	S.F.	6.45	9.50	11.40	5.85%	7.20%	8.70%	
	2900	Electrical		8.40	12.25	20	6.80%	9.05%	11.35%	
	3100	Total: Mechanical & Electrical	↓	52.50	68.50	94	24.50%	29%	31.50%	
05	0010	**AUTOMOTIVE SALES**	S.F.	80.50	109	135				05
	0020	Total project costs	C.F.	5.30	6.35	8.25				
	2720	Plumbing	S.F.	3.67	6.35	6.95	2.89%	6.05%	6.50%	
	2770	Heating, ventilating, air conditioning		5.65	8.65	9.35	4.61%	10%	10.35%	
	2900	Electrical		6.35	9.95	13.80	7.25%	8.80%	12.15%	
	3100	Total: Mechanical & Electrical	↓	20.50	29	35	18.90%	20.50%	22%	
06	0010	**BANKS**	S.F.	158	197	250				06
	0020	Total project costs	C.F.	11.25	15.30	20				
	0100	Site work	S.F.	18.05	27.50	40	7.85%	12.95%	16.95%	
	0500	Masonry		8.20	15.70	28.50	3.36%	6.95%	10.05%	
	1500	Finishes		14.60	21.50	26.50	5.85%	8.60%	11.55%	
	1800	Equipment		5.95	13.05	26.50	1%	5.55%	10.50%	
	2720	Plumbing		4.94	7.05	10.30	2.82%	3.90%	4.93%	
	2770	Heating, ventilating, air conditioning		9.40	12.55	16.70	4.86%	7.15%	8.50%	
	2900	Electrical		15	20	26	8.20%	10.20%	12.20%	
	3100	Total: Mechanical & Electrical	↓	36	48.50	58	16.25%	19.45%	23%	
	3500	See also division 11020 & 11030 (MF2004 11 16 00 & 11 17 00)								

	50 17 00 | S.F. Costs		UNIT	UNIT COSTS 1/4	UNIT COSTS MEDIAN	UNIT COSTS 3/4	% OF TOTAL 1/4	% OF TOTAL MEDIAN	% OF TOTAL 3/4	
13	0010	**CHURCHES**	S.F.	106	135	175				13
	0020	Total project costs	C.F.	6.60	8.30	10.95				
	1800	Equipment	S.F.	1.27	3.04	6.45	.95%	2.11%	4.31%	
	2720	Plumbing		4.13	5.80	8.50	3.51%	4.96%	6.25%	
	2770	Heating, ventilating, air conditioning		9.65	12.60	17.85	7.50%	10%	12%	
	2900	Electrical		8.95	12.30	16.60	7.30%	8.80%	10.95%	
	3100	Total: Mechanical & Electrical		27.50	37	49	18.30%	22%	25%	
	3500	See also division 11040 (MF2004 11 91 00)								
15	0010	**CLUBS, COUNTRY**	S.F.	114	137	173				15
	0020	Total project costs	C.F.	9.20	11.20	15.45				
	2720	Plumbing	S.F.	6.90	10.20	23.50	5.60%	7.90%	10%	
	2900	Electrical		8.95	12.25	16	7%	8.95%	11%	
	3100	Total: Mechanical & Electrical		47.50	59.50	63	19%	26.50%	29.50%	
17	0010	**CLUBS, SOCIAL Fraternal**	S.F.	96	131	175				17
	0020	Total project costs	C.F.	5.70	8.60	10.30				
	2720	Plumbing	S.F.	5.70	7.10	10.75	5.60%	6.90%	8.55%	
	2770	Heating, ventilating, air conditioning		8.25	9.95	12.80	8.20%	9.25%	14.40%	
	2900	Electrical		6.85	11.25	12.90	5.95%	9.30%	10.55%	
	3100	Total: Mechanical & Electrical		35.50	38.50	48.50	21%	23%	23.50%	
18	0010	**CLUBS, Y.M.C.A.**	S.F.	119	154	194				18
	0020	Total project costs	C.F.	5.25	8.80	13.10				
	2720	Plumbing	S.F.	7.20	14.35	16.10	5.65%	7.60%	10.85%	
	2900	Electrical		9.35	11.40	17.05	6.25%	7.95%	9.20%	
	3100	Total: Mechanical & Electrical		35	42	71	20.50%	21.50%	28.50%	
19	0010	**COLLEGES Classrooms & Administration**	S.F.	115	156	207				19
	0020	Total project costs	C.F.	8.25	12.25	18.90				
	0500	Masonry	S.F.	8.45	16	19.55	5.65%	8.05%	10.50%	
	2720	Plumbing		5.85	12	22.50	5.10%	6.60%	8.95%	
	2900	Electrical		9.40	14.55	20	7.70%	9.75%	12%	
	3100	Total: Mechanical & Electrical		39.50	54	72	23.50%	28%	32%	
21	0010	**COLLEGES Science, Engineering, Laboratories**	S.F.	214	251	291				21
	0020	Total project costs	C.F.	12.30	17.95	20.50				
	1800	Equipment	S.F.	11.95	27	29.50	2%	6.45%	12.65%	
	2900	Electrical		17.70	25.50	38.50	7.10%	9.40%	12.10%	
	3100	Total: Mechanical & Electrical		65.50	78	121	28.50%	31.50%	41%	
	3500	See also division 11600 (MF2004 11 53 00)								
23	0010	**COLLEGES Student Unions**	S.F.	137	185	225				23
	0020	Total project costs	C.F.	7.65	10	12.35				
	3100	Total: Mechanical & Electrical	S.F.	51.50	55.50	66	23.50%	26%	29%	
25	0010	**COMMUNITY CENTERS**	S.F.	113	139	189				25
	0020	Total project costs	C.F.	7.35	10.55	13.65				
	1800	Equipment	S.F.	2.73	4.51	7.35	1.48%	3.01%	5.30%	
	2720	Plumbing		5.35	9.35	12.75	4.85%	7%	8.95%	
	2770	Heating, ventilating, air conditioning		8.55	12.45	17.85	6.80%	10.35%	12.90%	
	2900	Electrical		9.25	12.50	17.95	7.15%	8.95%	10.45%	
	3100	Total: Mechanical & Electrical		31.50	40	57	19.05%	24%	31%	
28	0010	**COURT HOUSES**	S.F.	163	192	265				28
	0020	Total project costs	C.F.	12.50	14.95	18.85				
	2720	Plumbing	S.F.	7.75	10.80	12.25	5.95%	7.45%	8.20%	
	2900	Electrical		17.30	19.25	28	8.90%	10.45%	11.85%	
	3100	Total: Mechanical & Electrical		48.50	63.50	69.50	22.50%	27.50%	30%	
30	0010	**DEPARTMENT STORES**	S.F.	60	81.50	103				30
	0020	Total project costs	C.F.	3.23	4.18	5.70				
	2720	Plumbing	S.F.	1.87	2.37	3.59	1.82%	4.21%	5.90%	
	2770	Heating, ventilating, air conditioning		5.45	8.45	12.70	8.20%	9.10%	14.80%	

50 17 00 | S.F. Costs

		50 17 00 | S.F. Costs	UNIT	UNIT COSTS 1/4	UNIT COSTS MEDIAN	UNIT COSTS 3/4	% OF TOTAL 1/4	% OF TOTAL MEDIAN	% OF TOTAL 3/4
30	2900	Electrical	S.F.	6.90	9.50	11.20	9.05%	12.15%	14.95%
	3100	Total: Mechanical & Electrical		12.15	15.55	27	13.20%	21.50%	50%
31	0010	**DORMITORIES Low Rise (1 to 3 story)**	S.F.	114	158	198			
	0020	Total project costs	C.F.	6.45	10.45	15.65			
	2720	Plumbing	S.F.	6.85	9.20	11.60	8.05%	9%	9.65%
	2770	Heating, ventilating, air conditioning		7.25	8.70	11.60	4.61%	8.05%	10%
	2900	Electrical		7.55	11.50	15.75	6.40%	8.65%	9.50%
	3100	Total: Mechanical & Electrical		39.50	42	65.50	22%	25%	27%
	9000	Per bed, total cost	Bed	48,400	53,500	115,500			
32	0010	**DORMITORIES Mid Rise (4 to 8 story)**	S.F.	140	183	226			
	0020	Total project costs	C.F.	15.45	16.95	20.50			
	2900	Electrical	S.F.	14.90	16.90	23	8.20%	10.20%	11.95%
	3100	Total: Mechanical & Electrical	″	41.50	83	84.50	25.50%	34.50%	37.50%
	9000	Per bed, total cost	Bed	20,000	45,600	263,500			
34	0010	**FACTORIES**	S.F.	53	79	122			
	0020	Total project costs	C.F.	3.40	5.10	8.40			
	0100	Site work	S.F.	6.05	11.05	17.50	6.95%	11.45%	17.95%
	2720	Plumbing		2.86	5.30	8.80	3.73%	6.05%	8.10%
	2770	Heating, ventilating, air conditioning		5.55	8	10.80	5.25%	8.45%	11.35%
	2900	Electrical		6.60	10.45	15.90	8.10%	10.50%	14.20%
	3100	Total: Mechanical & Electrical		18.85	29	38	21%	28.50%	35.50%
36	0010	**FIRE STATIONS**	S.F.	106	145	197			
	0020	Total project costs	C.F.	6.15	8.45	11.25			
	0500	Masonry	S.F.	15.45	28	36.50	8.35%	11.30%	15.70%
	1140	Roofing		3.43	9.30	10.55	1.90%	4.94%	5.05%
	1580	Painting		2.66	3.98	4.07	1.37%	1.57%	2.07%
	1800	Equipment		1.31	2.51	4.63	.62%	1.63%	3.42%
	2720	Plumbing		5.90	9.40	13.30	5.85%	7.35%	9.45%
	2770	Heating, ventilating, air conditioning		5.85	9.45	14.60	5.15%	7.40%	9.40%
	2900	Electrical		7.60	13.20	18	6.85%	8.60%	10.60%
	3100	Total: Mechanical & Electrical		38.50	49.50	56	18.40%	23%	27%
37	0010	**FRATERNITY HOUSES & Sorority Houses**	S.F.	105	136	186			
	0020	Total project costs	C.F.	10.50	10.95	13.15			
	2720	Plumbing	S.F.	7.95	9.10	16.65	6.80%	8%	10.85%
	2900	Electrical		6.95	15	18.40	6.60%	9.90%	10.65%
	3100	Total: Mechanical & Electrical		8.55	26.50	32		15.10%	15.90%
38	0010	**FUNERAL HOMES**	S.F.	111	151	275			
	0020	Total project costs	C.F.	11.35	12.60	24			
	2900	Electrical	S.F.	4.91	9	9.85	3.58%	4.44%	5.95%
	3100	Total: Mechanical & Electrical		17.45	25.50	35.50	12.90%	12.90%	12.90%
39	0010	**GARAGES, COMMERCIAL (Service)**	S.F.	63	97	134			
	0020	Total project costs	C.F.	4.14	6.10	8.90			
	1800	Equipment	S.F.	3.55	7.95	12.40	2.21%	4.62%	6.80%
	2720	Plumbing		4.36	6.70	12.25	5.45%	7.85%	10.65%
	2730	Heating & ventilating		5.70	7.45	9.45	5.25%	6.85%	8.20%
	2900	Electrical		5.95	9.10	13.15	7.15%	9.25%	10.85%
	3100	Total: Mechanical & Electrical		13.20	25.50	37.50	12.35%	17.40%	26%
40	0010	**GARAGES, MUNICIPAL (Repair)**	S.F.	97.50	123	173			
	0020	Total project costs	C.F.	5.75	7.30	12.60			
	0500	Masonry	S.F.	8.65	16.95	26.50	4.03%	9.15%	12.50%
	2720	Plumbing		4.14	7.95	15	3.59%	6.70%	7.95%
	2730	Heating & ventilating		7.10	10.25	19.80	6.15%	7.45%	13.50%
	2900	Electrical		6.80	10.75	15.60	6.65%	8.15%	11.15%
	3100	Total: Mechanical & Electrical		31.50	44	63.50	21.50%	23%	28.50%

	50 17 00 | S.F. Costs		UNIT	UNIT COSTS 1/4	UNIT COSTS MEDIAN	UNIT COSTS 3/4	% OF TOTAL 1/4	% OF TOTAL MEDIAN	% OF TOTAL 3/4	
41	0010	**GARAGES, PARKING**	S.F.	35.50	52.50	90				41
	0020	Total project costs	C.F.	3.38	4.59	6.70				
	2720	Plumbing	S.F.	1.02	1.58	2.44	1.72%	2.70%	3.85%	
	2900	Electrical		1.97	2.42	3.80	4.33%	5.20%	6.30%	
	3100	Total: Mechanical & Electrical		4.04	5.60	7	7%	8.90%	11.05%	
	3200									
	9000	Per car, total cost	Car	15,200	19,100	24,300				
43	0010	**GYMNASIUMS**	S.F.	100	133	176				43
	0020	Total project costs	C.F.	4.99	6.75	8.30				
	1800	Equipment	S.F.	2.38	4.46	8.55	1.98%	3.30%	6.70%	
	2720	Plumbing		6.35	7.40	9.70	4.65%	6.40%	7.75%	
	2770	Heating, ventilating, air conditioning		6.80	10.35	21	5.15%	9.05%	11.10%	
	2900	Electrical		7.70	10.25	13.80	6.60%	8.30%	10.30%	
	3100	Total: Mechanical & Electrical		28	37	44	19.60%	23.50%	28%	
	3500	See also division 11480 (MF2004 11 67 00)								
46	0010	**HOSPITALS**	S.F.	192	240	325				46
	0020	Total project costs	C.F.	14.55	18.10	26				
	1800	Equipment	S.F.	4.87	9.35	16.15	.80%	2.53%	4.80%	
	2720	Plumbing		16.55	23.50	30	7.60%	9.10%	10.85%	
	2770	Heating, ventilating, air conditioning		24.50	31	43.50	7.80%	12.95%	16.65%	
	2900	Electrical		21	28.50	40.50	9.90%	11.75%	14.10%	
	3100	Total: Mechanical & Electrical		62	86	128	28%	33.50%	37%	
	9000	Per bed or person, total cost	Bed	222,500	306,500	353,000				
	9900	See also division 11700 (MF2004 11 71 00)								
48	0010	**HOUSING For the Elderly**	S.F.	94.50	120	147				48
	0020	Total project costs	C.F.	6.75	9.35	11.95				
	0100	Site work	S.F.	6.60	10.20	14.95	5.05%	7.90%	12.10%	
	0500	Masonry		2.88	10.75	15.70	1.30%	6.05%	11%	
	1800	Equipment		2.28	3.14	5	1.88%	3.23%	4.43%	
	2510	Conveying systems		2.30	3.09	4.18	1.78%	2.20%	2.81%	
	2720	Plumbing		7	8.95	11.25	8.15%	9.55%	10.50%	
	2730	Heating, ventilating, air conditioning		3.60	5.10	7.60	3.30%	5.60%	7.25%	
	2900	Electrical		7.05	9.55	12.25	7.30%	8.50%	10.25%	
	3100	Total: Mechanical & Electrical		26	30	38.50	18.10%	22.50%	29%	
	9000	Per rental unit, total cost	Unit	88,000	102,500	114,500				
	9500	Total: Mechanical & Electrical	"	19,600	22,500	26,300				
50	0010	**HOUSING Public (Low Rise)**	S.F.	79.50	110	143				50
	0020	Total project costs	C.F.	7.05	8.80	10.95				
	0100	Site work	S.F.	10.10	14.55	23.50	8.35%	11.75%	16.50%	
	1800	Equipment		2.16	3.52	5.35	2.26%	3.03%	4.24%	
	2720	Plumbing		5.70	7.55	9.55	7.15%	9.05%	11.60%	
	2730	Heating, ventilating, air conditioning		2.88	5.60	6.10	4.26%	6.05%	6.45%	
	2900	Electrical		4.80	7.15	9.90	5.10%	6.55%	8.25%	
	3100	Total: Mechanical & Electrical		23	29.50	33	14.50%	17.55%	26.50%	
	9000	Per apartment, total cost	Apt.	87,000	99,000	124,500				
	9500	Total: Mechanical & Electrical	"	18,600	22,900	25,400				
51	0010	**ICE SKATING RINKS**	S.F.	68	159	174				51
	0020	Total project costs	C.F.	4.99	5.10	5.90				
	2720	Plumbing	S.F.	2.53	4.75	4.86	3.12%	3.23%	5.65%	
	2900	Electrical		7.25	11.15	11.80	6.30%	10.15%	15.05%	
	3100	Total: Mechanical & Electrical		12	17	21	18.95%	18.95%	18.95%	
52	0010	**JAILS**	S.F.	206	267	345				52
	0020	Total project costs	C.F.	18.60	26	30.50				
	1800	Equipment	S.F.	8.05	24	40.50	2.80%	6.05%	10.35%	
	2720	Plumbing		19.90	26.50	35	7%	8.90%	13.35%	
	2770	Heating, ventilating, air conditioning		18.65	25	48	7.50%	9.45%	17.75%	
	2900	Electrical		21.50	29.50	36.50	8.80%	11.65%	14.95%	

	50 17 00 | S.F. Costs		UNIT	UNIT COSTS 1/4	UNIT COSTS MEDIAN	UNIT COSTS 3/4	% OF TOTAL 1/4	% OF TOTAL MEDIAN	% OF TOTAL 3/4	
52	3100	Total: Mechanical & Electrical	S.F.	56	103	122	27.50%	30%	34%	52
53	0010	**LIBRARIES**	S.F.	132	168	218				53
	0020	Total project costs	C.F.	8.95	11.20	14.30				
	0500	Masonry	S.F.	10.45	18.05	30.50	5.80%	7.60%	11.60%	
	1800	Equipment		1.78	4.80	7.25	.34%	1.50%	4.07%	
	2720	Plumbing		4.82	6.85	9.50	3.38%	4.60%	5.70%	
	2770	Heating, ventilating, air conditioning		10.65	18.05	22	7.80%	10.95%	12.80%	
	2900	Electrical		13.40	17.35	22.50	8.35%	10.30%	11.95%	
	3100	Total: Mechanical & Electrical		39.50	51.50	62	20.50%	23%	26.50%	
54	0010	**LIVING, ASSISTED**	S.F.	121	142	168				54
	0020	Total project costs	C.F.	10.15	11.85	13.50				
	0500	Masonry	S.F.	3.50	4.23	5.20	2.36%	3.16%	3.90%	
	1800	Equipment		2.68	3.13	4.09	2.06%	2.45%	2.87%	
	2720	Plumbing		10.10	13.55	14.05	6.05%	8.15%	10.60%	
	2770	Heating, ventilating, air conditioning		12	12.55	13.75	7.95%	9.35%	9.70%	
	2900	Electrical		11.80	13.05	15	8.95%	9.95%	10.65%	
	3100	Total: Mechanical & Electrical		33	38.50	44.50	25%	28.50%	31.50%	
55	0010	**MEDICAL CLINICS**	S.F.	123	151	192				55
	0020	Total project costs	C.F.	9	11.65	15.50				
	1800	Equipment	S.F.	3.32	7	10.85	1.05%	2.94%	6.35%	
	2720	Plumbing		8.15	11.50	15.40	6.15%	8.40%	10.10%	
	2770	Heating, ventilating, air conditioning		9.75	12.75	18.75	6.65%	8.85%	11.35%	
	2900	Electrical		10.55	15	19.60	8.10%	10%	12.25%	
	3100	Total: Mechanical & Electrical		33.50	46	63	22.50%	27%	33.50%	
	3500	See also division 11700 (MF2004 11 71 00)								
57	0010	**MEDICAL OFFICES**	S.F.	115	143	175				57
	0020	Total project costs	C.F.	8.60	11.70	15.80				
	1800	Equipment	S.F.	3.81	7.55	10.70	.68%	4.97%	7.05%	
	2720	Plumbing		6.40	9.85	13.25	5.60%	6.80%	8.50%	
	2770	Heating, ventilating, air conditioning		7.70	11.15	14.70	6.10%	8%	9.70%	
	2900	Electrical		9.25	13.45	18.80	7.60%	9.80%	11.40%	
	3100	Total: Mechanical & Electrical		25	36	52	19.35%	23%	28.50%	
59	0010	**MOTELS**	S.F.	73	106	138				59
	0020	Total project costs	C.F.	6.50	8.70	14.20				
	2720	Plumbing	S.F.	7.40	9.40	11.25	9.45%	10.60%	12.55%	
	2770	Heating, ventilating, air conditioning		4.50	6.70	12.05	5.60%	5.60%	10%	
	2900	Electrical		6.90	8.70	10.85	7.45%	9.05%	10.45%	
	3100	Total: Mechanical & Electrical		22	29.50	50.50	18.50%	24%	25.50%	
	5000									
	9000	Per rental unit, total cost	Unit	37,100	70,500	76,500				
	9500	Total: Mechanical & Electrical	"	7,225	10,900	12,700				
60	0010	**NURSING HOMES**	S.F.	114	147	181				60
	0020	Total project costs	C.F.	9	11.25	15.35				
	1800	Equipment	S.F.	3.60	4.77	7.95	2.02%	3.62%	4.99%	
	2720	Plumbing		9.80	14.85	17.90	8.75%	10.10%	12.70%	
	2770	Heating, ventilating, air conditioning		10.30	15.65	21	9.70%	11.45%	11.80%	
	2900	Electrical		11.35	14.15	19.25	9.40%	10.55%	12.45%	
	3100	Total: Mechanical & Electrical		27	38	63.50	26%	29.50%	30.50%	
	9000	Per bed or person, total cost	Bed	50,500	63,500	82,000				
61	0010	**OFFICES Low Rise (1 to 4 story)**	S.F.	96.50	125	163				61
	0020	Total project costs	C.F.	6.85	9.50	12.50				
	0100	Site work	S.F.	7.75	13.70	20	6.20%	9.70%	13.55%	
	0500	Masonry		3.84	7.65	13.80	2.62%	5.45%	8.50%	
	1800	Equipment		1.02	2.01	5.45	.60%	1.50%	3.50%	
	2720	Plumbing		3.44	5.30	7.80	3.66%	4.50%	6.10%	
	2770	Heating, ventilating, air conditioning		7.60	10.60	15.55	7.20%	10.30%	11.70%	
	2900	Electrical		7.90	11.25	16.05	7.45%	9.65%	11.40%	

	50 17 00 | S.F. Costs		UNIT	UNIT COSTS 1/4	UNIT COSTS MEDIAN	UNIT COSTS 3/4	% OF TOTAL 1/4	% OF TOTAL MEDIAN	% OF TOTAL 3/4	
61	3100	Total: Mechanical & Electrical	S.F.	21.50	30	45	18.20%	22%	27%	61
62	0010	**OFFICES Mid Rise (5 to 10 story)**	S.F.	102	124	168				62
	0020	Total project costs	C.F.	7.20	9.20	13.10				
	2720	Plumbing	S.F.	3.08	4.78	6.85	2.83%	3.74%	4.50%	
	2770	Heating, ventilating, air conditioning		7.75	11.05	17.65	7.65%	9.40%	11%	
	2900	Electrical		7.55	9.70	14.65	6.35%	7.80%	10%	
	3100	Total: Mechanical & Electrical	↓	19.60	25.50	48	18.95%	21%	27.50%	
63	0010	**OFFICES High Rise (11 to 20 story)**	S.F.	125	158	194				63
	0020	Total project costs	C.F.	8.75	10.95	15.70				
	2900	Electrical	S.F.	7.60	9.30	13.80	5.80%	7.85%	10.50%	
	3100	Total: Mechanical & Electrical	↓	24.50	33	55.50	16.90%	23.50%	34%	
64	0010	**POLICE STATIONS**	S.F.	151	197	249				64
	0020	Total project costs	C.F.	11.95	14.65	20				
	0500	Masonry	S.F.	14.10	25	31.50	6.70%	9.10%	11.35%	
	1800	Equipment		2.39	10.50	16.65	.98%	3.35%	6.70%	
	2720	Plumbing		8.40	16.75	21	5.65%	6.90%	10.75%	
	2770	Heating, ventilating, air conditioning		13.15	17.45	24.50	5.85%	10.55%	11.70%	
	2900	Electrical		16.45	23.50	31	9.80%	11.85%	14.80%	
	3100	Total: Mechanical & Electrical	↓	54	64.50	87.50	25%	31.50%	32.50%	
65	0010	**POST OFFICES**	S.F.	119	146	187				65
	0020	Total project costs	C.F.	7.15	9	10.30				
	2720	Plumbing	S.F.	5.35	6.60	8.35	4.24%	5.30%	5.60%	
	2770	Heating, ventilating, air conditioning		8.35	10.30	11.45	6.65%	7.15%	9.35%	
	2900	Electrical		9.80	13.75	16.35	7.25%	9%	11%	
	3100	Total: Mechanical & Electrical	↓	28.50	37	42	16.25%	18.80%	22%	
66	0010	**POWER PLANTS**	S.F.	825	1,075	2,000				66
	0020	Total project costs	C.F.	22.50	49.50	106				
	2900	Electrical	S.F.	58	123	184	9.30%	12.75%	21.50%	
	8100	Total: Mechanical & Electrical	↓	145	470	1,050	32.50%	32.50%	52.50%	
67	0010	**RELIGIOUS EDUCATION**	S.F.	95.50	125	154				67
	0020	Total project costs	C.F.	5.30	7.60	9.50				
	2720	Plumbing	S.F.	3.99	5.65	7.95	4.40%	5.30%	7.10%	
	2770	Heating, ventilating, air conditioning		10.10	11.45	16.15	10.05%	11.45%	12.35%	
	2900	Electrical		7.60	10.70	14.15	7.60%	9.05%	10.35%	
	3100	Total: Mechanical & Electrical	↓	31	40.50	49.50	22%	23%	27%	
69	0010	**RESEARCH Laboratories & Facilities**	S.F.	146	207	300				69
	0020	Total project costs	C.F.	10.65	21	25				
	1800	Equipment	S.F.	6.35	12.50	30.50	.94%	4.58%	9.10%	
	2720	Plumbing		14.35	18.35	29.50	6.15%	8.30%	10.80%	
	2770	Heating, ventilating, air conditioning		12.85	43	51	7.25%	16.50%	17.50%	
	2900	Electrical		17.20	27.50	45	9.55%	11.50%	15.40%	
	3100	Total: Mechanical & Electrical	↓	55	95.50	137	29.50%	37%	41%	
70	0010	**RESTAURANTS**	S.F.	139	179	233				70
	0020	Total project costs	C.F.	11.65	15.30	20				
	1800	Equipment	S.F.	8.95	22	33.50	6.10%	13%	15.65%	
	2720	Plumbing		11	13.30	17.50	6.10%	8.15%	9%	
	2770	Heating, ventilating, air conditioning		13.95	19.30	23.50	9.20%	12%	12.40%	
	2900	Electrical		14.65	18.10	23.50	8.35%	10.55%	11.55%	
	3100	Total: Mechanical & Electrical	↓	45.50	48.50	63	21%	25%	29.50%	
	9000	Per seat unit, total cost	Seat	5,100	6,775	8,025				
	9500	Total: Mechanical & Electrical	"	1,275	1,700	2,000				
72	0010	**RETAIL STORES**	S.F.	65	87	115				72
	0020	Total project costs	C.F.	4.39	6.25	8.70				
	2720	Plumbing	S.F.	2.35	3.92	6.70	3.26%	4.60%	6.80%	
	2770	Heating, ventilating, air conditioning		5.05	6.95	10.45	6.75%	8.75%	10.15%	
	2900	Electrical		5.85	7.95	11.50	7.25%	9.90%	11.60%	
	3100	Total: Mechanical & Electrical	↓	15.50	19.85	27.50	17.05%	21%	23.50%	

50 17 00 | S.F. Costs

		50 17 00 | S.F. Costs	UNIT	UNIT COSTS 1/4	UNIT COSTS MEDIAN	UNIT COSTS 3/4	% OF TOTAL 1/4	% OF TOTAL MEDIAN	% OF TOTAL 3/4	
74	0010	**SCHOOLS Elementary**	S.F.	106	132	161				74
	0020	Total project costs	C.F.	6.90	8.85	11.45				
	0500	Masonry	S.F.	9.20	16.15	24	5.25%	10.65%	14.05%	
	1800	Equipment		2.89	4.88	9	1.83%	3.13%	4.61%	
	2720	Plumbing		6.10	8.60	11.50	5.70%	7.15%	9.35%	
	2730	Heating, ventilating, air conditioning		9.10	14.50	20.50	8.15%	10.80%	14.90%	
	2900	Electrical		10	13.25	16.90	8.45%	10.05%	11.80%	
	3100	Total: Mechanical & Electrical	↓	36.50	44.50	55	25%	27.50%	30%	
	9000	Per pupil, total cost	Ea.	12,100	18,000	40,400				
	9500	Total: Mechanical & Electrical	"	3,425	4,325	10,900				
76	0010	**SCHOOLS Junior High & Middle**	S.F.	109	134	163				76
	0020	Total project costs	C.F.	6.90	8.95	10.05				
	0500	Masonry	S.F.	13.60	17.80	21.50	8%	11.60%	14.35%	
	1800	Equipment		3.48	5.60	8.25	1.80%	3.09%	4.86%	
	2720	Plumbing		6.35	7.80	9.70	5.30%	6.80%	7.25%	
	2770	Heating, ventilating, air conditioning		12.70	15.45	27	8.90%	11.55%	14.20%	
	2900	Electrical		10.70	12.90	16.60	7.90%	9.35%	10.60%	
	3100	Total: Mechanical & Electrical	↓	35	44.50	55.50	23.50%	26.50%	29.50%	
	9000	Per pupil, total cost	Ea.	13,800	18,100	24,400				
78	0010	**SCHOOLS Senior High**	S.F.	113	139	174				78
	0020	Total project costs	C.F.	6.85	9.85	16.25				
	1800	Equipment	S.F.	2.98	7	9.75	1.88%	2.91%	4.56%	
	2720	Plumbing		6.35	9.55	17.45	5.60%	6.90%	8.30%	
	2770	Heating, ventilating, air conditioning		13	14.85	22.50	8.95%	11.60%	15%	
	2900	Electrical		11.45	14.85	22.50	8.65%	10.20%	12.25%	
	3100	Total: Mechanical & Electrical	↓	37.50	44	74	23.50%	26.50%	28.50%	
	9000	Per pupil, total cost	Ea.	10,700	21,700	27,200				
80	0010	**SCHOOLS Vocational**	S.F.	92	133	165				80
	0020	Total project costs	C.F.	5.70	8.20	11.35				
	0500	Masonry	S.F.	5.40	13.30	20.50	3.20%	4.61%	10.95%	
	1800	Equipment		2.87	7.15	9.85	1.24%	3.10%	4.26%	
	2720	Plumbing		5.85	8.75	12.85	5.40%	6.90%	8.55%	
	2770	Heating, ventilating, air conditioning		8.20	15.30	25.50	8.60%	11.90%	14.65%	
	2900	Electrical		9.55	13.05	17.85	8.45%	11%	13.20%	
	3100	Total: Mechanical & Electrical	↓	36	36.50	63	27.50%	29.50%	31%	
	9000	Per pupil, total cost	Ea.	12,800	34,200	51,000				
83	0010	**SPORTS ARENAS**	S.F.	80	107	165				83
	0020	Total project costs	C.F.	4.35	7.80	10.05				
	2720	Plumbing	S.F.	4.65	7.05	14.90	4.35%	6.35%	9.40%	
	2770	Heating, ventilating, air conditioning		10	11.85	16.45	8.80%	10.20%	13.55%	
	2900	Electrical		8.35	11.35	14.65	8.60%	9.90%	12.25%	
	3100	Total: Mechanical & Electrical	↓	21	37	49.50	21.50%	25%	27.50%	
85	0010	**SUPERMARKETS**	S.F.	74	85.50	101				85
	0020	Total project costs	C.F.	4.13	4.99	7.55				
	2720	Plumbing	S.F.	4.14	5.20	6.05	5.40%	6%	7.45%	
	2770	Heating, ventilating, air conditioning		6.10	8.10	9.85	8.60%	8.65%	9.60%	
	2900	Electrical		9.30	10.60	12.60	10.40%	12.45%	13.60%	
	3100	Total: Mechanical & Electrical	↓	23.50	25.50	34	23.50%	27.50%	28.50%	
86	0010	**SWIMMING POOLS**	S.F.	120	201	430				86
	0020	Total project costs	C.F.	9.60	12	13.10				
	2720	Plumbing	S.F.	11.10	12.70	17	4.80%	9.70%	20.50%	
	2900	Electrical		9	14.60	21.50	5.75%	6.95%	7.60%	
	3100	Total: Mechanical & Electrical	↓	22	56	76.50	11.15%	14.10%	23.50%	
87	0010	**TELEPHONE EXCHANGES**	S.F.	159	227	296				87
	0020	Total project costs	C.F.	9.90	15.90	22				
	2720	Plumbing	S.F.	6.70	10.40	15.20	4.52%	5.80%	6.90%	
	2770	Heating, ventilating, air conditioning		15.60	31.50	39	11.80%	16.05%	18.40%	

	50 17 00 | S.F. Costs		UNIT	UNIT COSTS 1/4	UNIT COSTS MEDIAN	UNIT COSTS 3/4	% OF TOTAL 1/4	% OF TOTAL MEDIAN	% OF TOTAL 3/4	
87	2900	Electrical	S.F.	16.20	25.50	45.50	10.90%	14%	17.85%	87
	3100	Total: Mechanical & Electrical	↓	48	91	128	29.50%	33.50%	44.50%	
91	0010	**THEATERS**	S.F.	100	129	189				91
	0020	Total project costs	C.F.	4.63	6.85	10.05				
	2720	Plumbing	S.F.	3.34	3.62	14.80	2.92%	4.70%	6.80%	
	2770	Heating, ventilating, air conditioning		9.75	11.80	14.60	8%	12.25%	13.40%	
	2900	Electrical		8.75	11.80	24	8.05%	9.35%	12.25%	
	3100	Total: Mechanical & Electrical	↓	22.50	30.50	36	21.50%	25.50%	27.50%	
94	0010	**TOWN HALLS City Halls & Municipal Buildings**	S.F.	112	142	185				94
	0020	Total project costs	C.F.	10.20	11.95	17.20				
	2720	Plumbing	S.F.	4.67	8.75	15.55	4.31%	5.95%	7.95%	
	2770	Heating, ventilating, air conditioning		8.45	16.75	24.50	7.05%	9.05%	13.45%	
	2900	Electrical		10.60	14.75	19.95	8.05%	9.45%	11.65%	
	3100	Total: Mechanical & Electrical	↓	37	46.50	71.50	22%	26.50%	31%	
97	0010	**WAREHOUSES & Storage Buildings**	S.F.	41.50	60.50	88.50				97
	0020	Total project costs	C.F.	2.19	3.42	5.65				
	0100	Site work	S.F.	4.31	8.55	12.95	6.05%	12.95%	19.85%	
	0500	Masonry		2.47	5.95	12.80	3.73%	7.40%	12.30%	
	1800	Equipment		.67	1.45	8.10	.72%	1.69%	5.55%	
	2720	Plumbing		1.39	2.50	4.67	2.90%	4.80%	6.55%	
	2730	Heating, ventilating, air conditioning		1.59	4.48	6	2.41%	5%	8.90%	
	2900	Electrical		2.47	4.65	7.70	5.15%	7.20%	10.10%	
	3100	Total: Mechanical & Electrical	↓	6.90	10.55	21	13.30%	19.10%	26%	
99	0010	**WAREHOUSE & OFFICES Combination**	S.F.	51.50	68.50	93.50				99
	0020	Total project costs	C.F.	2.63	3.81	5.65				
	1800	Equipment	S.F.	.89	1.72	2.56	.52%	1.20%	2.40%	
	2720	Plumbing		1.98	3.45	5.15	3.74%	4.76%	6.30%	
	2770	Heating, ventilating, air conditioning		3.13	4.89	6.85	5%	5.65%	10.05%	
	2900	Electrical		3.44	5.10	8.10	5.75%	8%	10%	
	3100	Total: Mechanical & Electrical	↓	9.55	14.80	23.50	14.40%	19.95%	24.50%	

Square Foot Project Size Modifier

One factor that affects the S.F. cost of a particular building is the size. In general, for buildings built to the same specifications in the same locality, the larger building will have the lower S.F. cost. This is due mainly to the decreasing contribution of the exterior walls plus the economy of scale usually achievable in larger buildings. The Area Conversion Scale shown below will give a factor to convert costs for the typical size building to an adjusted cost for the particular project.

The Square Foot Base Size lists the median costs, most typical project size in our accumulated data, and the range in size of the projects.

The Size Factor for your project is determined by dividing your project area in S.F. by the typical project size for the particular Building Type. With this factor, enter the Area Conversion Scale at the appropriate Size Factor and determine the appropriate cost multiplier for your building size.

Example: Determine the cost per S.F. for a 100,000 S.F. Mid-rise apartment building.

$$\frac{\text{Proposed building area} = 100{,}000 \text{ S.F.}}{\text{Typical size from below} = 50{,}000 \text{ S.F.}} = 2.00$$

Enter Area Conversion scale at 2.0, intersect curve, read horizontally the appropriate cost multiplier of .94. Size adjusted cost becomes .94 x \$111.00 = \$104.00 based on national average costs.

Note: For Size Factors less than .50, the Cost Multiplier is 1.1
For Size Factors greater than 3.5, the Cost Multiplier is .90

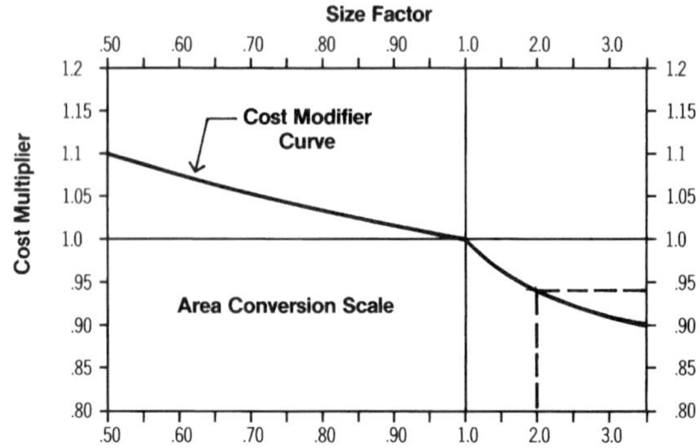

Square Foot Base Size							
Building Type	**Median Cost per S.F.**	**Typical Size Gross S.F.**	**Typical Range Gross S.F.**	**Building Type**	**Median Cost per S.F.**	**Typical Size Gross S.F.**	**Typical Range Gross S.F.**
Apartments, Low Rise	\$ 87.00	21,000	9,700 - 37,200	Jails	\$ 267.00	40,000	5,500 - 145,000
Apartments, Mid Rise	111.00	50,000	32,000 - 100,000	Libraries	168.00	12,000	7,000 - 31,000
Apartments, High Rise	120.00	145,000	95,000 - 600,000	Living, Assisted	142.00	32,300	23,500 - 50,300
Auditoriums	148.00	25,000	7,600 - 39,000	Medical Clinics	151.00	7,200	4,200 - 15,700
Auto Sales	109.00	20,000	10,800 - 28,600	Medical Offices	143.00	6,000	4,000 - 15,000
Banks	197.00	4,200	2,500 - 7,500	Motels	106.00	40,000	15,800 - 120,000
Churches	135.00	17,000	2,000 - 42,000	Nursing Homes	147.00	23,000	15,000 - 37,000
Clubs, Country	137.00	6,500	4,500 - 15,000	Offices, Low Rise	125.00	20,000	5,000 - 80,000
Clubs, Social	131.00	10,000	6,000 - 13,500	Offices, Mid Rise	124.00	120,000	20,000 - 300,000
Clubs, YMCA	154.00	28,300	12,800 - 39,400	Offices, High Rise	158.00	260,000	120,000 - 800,000
Colleges (Class)	156.00	50,000	15,000 - 150,000	Police Stations	197.00	10,500	4,000 - 19,000
Colleges (Science Lab)	251.00	45,600	16,600 - 80,000	Post Offices	146.00	12,400	6,800 - 30,000
College (Student Union)	185.00	33,400	16,000 - 85,000	Power Plants	1075.00	7,500	1,000 - 20,000
Community Center	139.00	9,400	5,300 - 16,700	Religious Education	125.00	9,000	6,000 - 12,000
Court Houses	192.00	32,400	17,800 - 106,000	Research	207.00	19,000	6,300 - 45,000
Dept. Stores	81.50	90,000	44,000 - 122,000	Restaurants	179.00	4,400	2,800 - 6,000
Dormitories, Low Rise	158.00	25,000	10,000 - 95,000	Retail Stores	87.00	7,200	4,000 - 17,600
Dormitories, Mid Rise	183.00	85,000	20,000 - 200,000	Schools, Elementary	132.00	41,000	24,500 - 55,000
Factories	79.00	26,400	12,900 - 50,000	Schools, Jr. High	134.00	92,000	52,000 - 119,000
Fire Stations	145.00	5,800	4,000 - 8,700	Schools, Sr. High	139.00	101,000	50,500 - 175,000
Fraternity Houses	136.00	12,500	8,200 - 14,800	Schools, Vocational	133.00	37,000	20,500 - 82,000
Funeral Homes	151.00	10,000	4,000 - 20,000	Sports Arenas	107.00	15,000	5,000 - 40,000
Garages, Commercial	97.00	9,300	5,000 - 13,600	Supermarkets	85.50	44,000	12,000 - 60,000
Garages, Municipal	123.00	8,300	4,500 - 12,600	Swimming Pools	201.00	20,000	10,000 - 32,000
Garages, Parking	52.50	163,000	76,400 - 225,300	Telephone Exchange	227.00	4,500	1,200 - 10,600
Gymnasiums	133.00	19,200	11,600 - 41,000	Theaters	129.00	10,500	8,800 - 17,500
Hospitals	240.00	55,000	27,200 - 125,000	Town Halls	142.00	10,800	4,800 - 23,400
House (Elderly)	120.00	37,000	21,000 - 66,000	Warehouses	60.50	25,000	8,000 - 72,000
Housing (Public)	110.00	36,000	14,400 - 74,400	Warehouse & Office	68.50	25,000	8,000 - 72,000
Ice Rinks	159.00	29,000	27,200 - 33,600				

Abbreviations

A	Area Square Feet; Ampere
ABS	Acrylonitrile Butadiene Stryrene; Asbestos Bonded Steel
A.C.	Alternating Current; Air-Conditioning; Asbestos Cement; Plywood Grade A & C
A.C.I.	American Concrete Institute
AD	Plywood, Grade A & D
Addit.	Additional
Adj.	Adjustable
af	Audio-frequency
A.G.A.	American Gas Association
Agg.	Aggregate
A.H.	Ampere Hours
A hr.	Ampere-hour
A.H.U.	Air Handling Unit
A.I.A.	American Institute of Architects
AIC	Ampere Interrupting Capacity
Allow.	Allowance
alt.	Altitude
Alum.	Aluminum
a.m.	Ante Meridiem
Amp.	Ampere
Anod.	Anodized
Approx.	Approximate
Apt.	Apartment
Asb.	Asbestos
A.S.B.C.	American Standard Building Code
Asbe.	Asbestos Worker
ASCE.	American Society of Civil Engineers
A.S.H.R.A.E.	American Society of Heating, Refrig. & AC Engineers
A.S.M.E.	American Society of Mechanical Engineers
A.S.T.M.	American Society for Testing and Materials
Attchmt.	Attachment
Avg., Ave.	Average
A.W.G.	American Wire Gauge
AWWA	American Water Works Assoc.
Bbl.	Barrel
B&B	Grade B and Better; Balled & Burlapped
B.&S.	Bell and Spigot
B.&W.	Black and White
b.c.c.	Body-centered Cubic
B.C.Y.	Bank Cubic Yards
BE	Bevel End
B.F.	Board Feet
Bg. cem.	Bag of Cement
BHP	Boiler Horsepower; Brake Horsepower
B.I.	Black Iron
Bit., Bitum.	Bituminous
Bit., Conc.	Bituminous Concrete
Bk.	Backed
Bkrs.	Breakers
Bldg.	Building
Blk.	Block
Bm.	Beam
Boil.	Boilermaker
B.P.M.	Blows per Minute
BR	Bedroom
Brg.	Bearing
Brhe.	Bricklayer Helper
Bric.	Bricklayer
Brk.	Brick
Brng.	Bearing
Brs.	Brass
Brz.	Bronze
Bsn.	Basin
Btr.	Better
Btu	British Thermal Unit
BTUH	BTU per Hour
B.U.R.	Built-up Roofing
BX	Interlocked Armored Cable
°C	degree centegrade
c	Conductivity, Copper Sweat
C	Hundred; Centigrade
C/C	Center to Center, Cedar on Cedar
C-C	Center to Center
Cab.	Cabinet
Cair.	Air Tool Laborer
Calc	Calculated
Cap.	Capacity
Carp.	Carpenter
C.B.	Circuit Breaker
C.C.A.	Chromate Copper Arsenate
C.C.F.	Hundred Cubic Feet
cd	Candela
cd/sf	Candela per Square Foot
CD	Grade of Plywood Face & Back
CDX	Plywood, Grade C & D, exterior glue
Cefi.	Cement Finisher
Cem.	Cement
CF	Hundred Feet
C.F.	Cubic Feet
CFM	Cubic Feet per Minute
c.g.	Center of Gravity
CHW	Chilled Water; Commercial Hot Water
C.I.	Cast Iron
C.I.P.	Cast in Place
Circ.	Circuit
C.L.	Carload Lot
Clab.	Common Laborer
Clam	Common maintenance laborer
C.L.F.	Hundred Linear Feet
CLF	Current Limiting Fuse
CLP	Cross Linked Polyethylene
cm	Centimeter
CMP	Corr. Metal Pipe
C.M.U.	Concrete Masonry Unit
CN	Change Notice
Col.	Column
CO_2	Carbon Dioxide
Comb.	Combination
Compr.	Compressor
Conc.	Concrete
Cont.	Continuous; Continued, Container
Corr.	Corrugated
Cos	Cosine
Cot	Cotangent
Cov.	Cover
C/P	Cedar on Paneling
CPA	Control Point Adjustment
Cplg.	Coupling
C.P.M.	Critical Path Method
CPVC	Chlorinated Polyvinyl Chloride
C.Pr.	Hundred Pair
CRC	Cold Rolled Channel
Creos.	Creosote
Crpt.	Carpet & Linoleum Layer
CRT	Cathode-ray Tube
CS	Carbon Steel, Constant Shear Bar Joist
Csc	Cosecant
C.S.F.	Hundred Square Feet
CSI	Construction Specifications Institute
C.T.	Current Transformer
CTS	Copper Tube Size
Cu	Copper, Cubic
Cu. Ft.	Cubic Foot
cw	Continuous Wave
C.W.	Cool White; Cold Water
Cwt.	100 Pounds
C.W.X.	Cool White Deluxe
C.Y.	Cubic Yard (27 cubic feet)
C.Y./Hr.	Cubic Yard per Hour
Cyl.	Cylinder
d	Penny (nail size)
D	Deep; Depth; Discharge
Dis., Disch.	Discharge
Db.	Decibel
Dbl.	Double
DC	Direct Current
DDC	Direct Digital Control
Demob.	Demobilization
d.f.u.	Drainage Fixture Units
D.H.	Double Hung
DHW	Domestic Hot Water
DI	Ductile Iron
Diag.	Diagonal
Diam., Dia	Diameter
Distrib.	Distribution
Div.	Division
Dk.	Deck
D.L.	Dead Load; Diesel
DLH	Deep Long Span Bar Joist
Do.	Ditto
Dp.	Depth
D.P.S.T.	Double Pole, Single Throw
Dr.	Drive
Drink.	Drinking
D.S.	Double Strength
D.S.A.	Double Strength A Grade
D.S.B.	Double Strength B Grade
Dty.	Duty
DWV	Drain Waste Vent
DX	Deluxe White, Direct Expansion
dyn	Dyne
e	Eccentricity
E	Equipment Only; East
Ea.	Each
E.B.	Encased Burial
Econ.	Economy
E.C.Y	Embankment Cubic Yards
EDP	Electronic Data Processing
EIFS	Exterior Insulation Finish System
E.D.R.	Equiv. Direct Radiation
Eq.	Equation
EL	elevation
Elec.	Electrician; Electrical
Elev.	Elevator; Elevating
EMT	Electrical Metallic Conduit; Thin Wall Conduit
Eng.	Engine, Engineered
EPDM	Ethylene Propylene Diene Monomer
EPS	Expanded Polystyrene
Eqhv.	Equip. Oper., Heavy
Eqlt.	Equip. Oper., Light
Eqmd.	Equip. Oper., Medium
Eqmm.	Equip. Oper., Master Mechanic
Eqol.	Equip. Oper., Oilers
Equip.	Equipment
ERW	Electric Resistance Welded
E.S.	Energy Saver
Est.	Estimated
esu	Electrostatic Units
E.W.	Each Way
EWT	Entering Water Temperature
Excav.	Excavation
Exp.	Expansion, Exposure
Ext.	Exterior
Extru.	Extrusion
f.	Fiber stress
F	Fahrenheit; Female; Fill
Fab.	Fabricated

Abbreviations

FBGS	Fiberglass
F.C.	Footcandles
f.c.c.	Face-centered Cubic
f'c.	Compressive Stress in Concrete; Extreme Compressive Stress
F.E.	Front End
FEP	Fluorinated Ethylene Propylene (Teflon)
F.G.	Flat Grain
F.H.A.	Federal Housing Administration
Fig.	Figure
Fin.	Finished
Fixt.	Fixture
Fl. Oz.	Fluid Ounces
Flr.	Floor
F.M.	Frequency Modulation; Factory Mutual
Fmg.	Framing
Fdn.	Foundation
Fori.	Foreman, Inside
Foro.	Foreman, Outside
Fount.	Fountain
fpm	Feet per Minute
FPT	Female Pipe Thread
Fr.	Frame
F.R.	Fire Rating
FRK	Foil Reinforced Kraft
FRP	Fiberglass Reinforced Plastic
FS	Forged Steel
FSC	Cast Body; Cast Switch Box
Ft.	Foot; Feet
Ftng.	Fitting
Ftg.	Footing
Ft lb.	Foot Pound
Furn.	Furniture
FVNR	Full Voltage Non-Reversing
FXM	Female by Male
Fy.	Minimum Yield Stress of Steel
g	Gram
G	Gauss
Ga.	Gauge
Gal., gal.	Gallon
gpm, GPM	Gallon per Minute
Galv.	Galvanized
Gen.	General
G.F.I.	Ground Fault Interrupter
Glaz.	Glazier
GPD	Gallons per Day
GPH	Gallons per Hour
GPM	Gallons per Minute
GR	Grade
Gran.	Granular
Grnd.	Ground
H	High Henry
H.C.	High Capacity
H.D.	Heavy Duty; High Density
H.D.O.	High Density Overlaid
H.D.P.E.	high density polyethelene
Hdr.	Header
Hdwe.	Hardware
Help.	Helper Average
HEPA	High Efficiency Particulate Air Filter
Hg	Mercury
HIC	High Interrupting Capacity
HM	Hollow Metal
HMWPE	high molecular weight polyethylene
H.O.	High Output
Horiz.	Horizontal
H.P.	Horsepower; High Pressure
H.P.F.	High Power Factor
Hr.	Hour
Hrs./Day	Hours per Day
HSC	High Short Circuit
Ht.	Height
Htg.	Heating
Htrs.	Heaters
HVAC	Heating, Ventilation & Air-Conditioning
Hvy.	Heavy
HW	Hot Water
Hyd.;Hydr.	Hydraulic
Hz.	Hertz (cycles)
I.	Moment of Inertia
IBC	International Building Code
I.C.	Interrupting Capacity
ID	Inside Diameter
I.D.	Inside Dimension; Identification
I.F.	Inside Frosted
I.M.C.	Intermediate Metal Conduit
In.	Inch
Incan.	Incandescent
Incl.	Included; Including
Int.	Interior
Inst.	Installation
Insul.	Insulation/Insulated
I.P.	Iron Pipe
I.P.S.	Iron Pipe Size
I.P.T.	Iron Pipe Threaded
I.W.	Indirect Waste
J	Joule
J.I.C.	Joint Industrial Council
K	Thousand; Thousand Pounds; Heavy Wall Copper Tubing, Kelvin
K.A.H.	Thousand Amp. Hours
KCMIL	Thousand Circular Mils
KD	Knock Down
K.D.A.T.	Kiln Dried After Treatment
kg	Kilogram
kG	Kilogauss
kgf	Kilogram Force
kHz	Kilohertz
Kip.	1000 Pounds
KJ	Kiljoule
K.L.	Effective Length Factor
K.L.F.	Kips per Linear Foot
Km	Kilometer
K.S.F.	Kips per Square Foot
K.S.I.	Kips per Square Inch
kV	Kilovolt
kVA	Kilovolt Ampere
K.V.A.R.	Kilovar (Reactance)
KW	Kilowatt
KWh	Kilowatt-hour
L	Labor Only; Length; Long; Medium Wall Copper Tubing
Lab.	Labor
lat	Latitude
Lath.	Lather
Lav.	Lavatory
lb.; #	Pound
L.B.	Load Bearing; L Conduit Body
L. & E.	Labor & Equipment
lb./hr.	Pounds per Hour
lb./L.F.	Pounds per Linear Foot
lbf/sq.in.	Pound-force per Square Inch
L.C.L.	Less than Carload Lot
L.C.Y.	Loose Cubic Yard
Ld.	Load
LE	Lead Equivalent
LED	Light Emitting Diode
L.F.	Linear Foot
L.F. Nose	Linear Foot of Stair Nosing
L.F. Rsr	Linear Foot of Stair Riser
Lg.	Long; Length; Large
L & H	Light and Heat
LH	Long Span Bar Joist
L.H.	Labor Hours
L.L.	Live Load
L.L.D.	Lamp Lumen Depreciation
lm	Lumen
lm/sf	Lumen per Square Foot
lm/W	Lumen per Watt
L.O.A.	Length Over All
log	Logarithm
L-O-L	Lateralolet
long.	longitude
L.P.	Liquefied Petroleum; Low Pressure
L.P.F.	Low Power Factor
LR	Long Radius
L.S.	Lump Sum
Lt.	Light
Lt. Ga.	Light Gauge
L.T.L.	Less than Truckload Lot
Lt. Wt.	Lightweight
L.V.	Low Voltage
M	Thousand; Material; Male; Light Wall Copper Tubing
M^2CA	Meters Squared Contact Area
m/hr.; M.H.	Man-hour
mA	Milliampere
Mach.	Machine
Mag. Str.	Magnetic Starter
Maint.	Maintenance
Marb.	Marble Setter
Mat; Mat'l.	Material
Max.	Maximum
MBF	Thousand Board Feet
MBH	Thousand BTU's per hr.
MC	Metal Clad Cable
M.C.F.	Thousand Cubic Feet
M.C.F.M.	Thousand Cubic Feet per Minute
M.C.M.	Thousand Circular Mils
M.C.P.	Motor Circuit Protector
MD	Medium Duty
M.D.O.	Medium Density Overlaid
Med.	Medium
MF	Thousand Feet
M.F.B.M.	Thousand Feet Board Measure
Mfg.	Manufacturing
Mfrs.	Manufacturers
mg	Milligram
MGD	Million Gallons per Day
MGPH	Thousand Gallons per Hour
MH, M.H.	Manhole; Metal Halide; Man-Hour
MHz	Megahertz
Mi.	Mile
MI	Malleable Iron; Mineral Insulated
m	Meter
mm	Millimeter
Mill.	Millwright
Min., min.	Minimum, minute
Misc.	Miscellaneous
ml	Milliliter, Mainline
M.L.F.	Thousand Linear Feet
Mo.	Month
Mobil.	Mobilization
Mog.	Mogul Base
MPH	Miles per Hour
MPT	Male Pipe Thread

Abbreviations

MRT	Mile Round Trip
ms	Millisecond
M.S.F.	Thousand Square Feet
Mstz.	Mosaic & Terrazzo Worker
M.S.Y.	Thousand Square Yards
Mtd., mtd.	Mounted
Mthe.	Mosaic & Terrazzo Helper
Mtng.	Mounting
Mult.	Multi; Multiply
M.V.A.	Million Volt Amperes
M.V.A.R.	Million Volt Amperes Reactance
MV	Megavolt
MW	Megawatt
MXM	Male by Male
MYD	Thousand Yards
N	Natural; North
nA	Nanoampere
NA	Not Available; Not Applicable
N.B.C.	National Building Code
NC	Normally Closed
N.E.M.A.	National Electrical Manufacturers Assoc.
NEHB	Bolted Circuit Breaker to 600V.
N.L.B.	Non-Load-Bearing
NM	Non-Metallic Cable
nm	Nanometer
No.	Number
NO	Normally Open
N.O.C.	Not Otherwise Classified
Nose.	Nosing
N.P.T.	National Pipe Thread
NQOD	Combination Plug-on/Bolt on Circuit Breaker to 240V.
N.R.C.	Noise Reduction Coefficient/ Nuclear Regulator Commission
N.R.S.	Non Rising Stem
ns	Nanosecond
nW	Nanowatt
OB	Opposing Blade
OC	On Center
OD	Outside Diameter
O.D.	Outside Dimension
ODS	Overhead Distribution System
O.G.	Ogee
O.H.	Overhead
O&P	Overhead and Profit
Oper.	Operator
Opng.	Opening
Orna.	Ornamental
OSB	Oriented Strand Board
O.S.&Y.	Outside Screw and Yoke
Ovhd.	Overhead
OWG	Oil, Water or Gas
Oz.	Ounce
P.	Pole; Applied Load; Projection
p.	Page
Pape.	Paperhanger
P.A.P.R.	Powered Air Purifying Respirator
PAR	Parabolic Reflector
Pc., Pcs.	Piece, Pieces
P.C.	Portland Cement; Power Connector
P.C.F.	Pounds per Cubic Foot
P.C.M.	Phase Contrast Microscopy
P.E.	Professional Engineer; Porcelain Enamel; Polyethylene; Plain End
Perf.	Perforated
PEX	Cross linked polyethylene
Ph.	Phase
P.I.	Pressure Injected
Pile.	Pile Driver
Pkg.	Package
Pl.	Plate
Plah.	Plasterer Helper
Plas.	Plasterer
Pluh.	Plumbers Helper
Plum.	Plumber
Ply.	Plywood
p.m.	Post Meridiem
Pntd.	Painted
Pord.	Painter, Ordinary
pp	Pages
PP, PPL	Polypropylene
P.P.M.	Parts per Million
Pr.	Pair
P.E.S.B.	Pre-engineered Steel Building
Prefab.	Prefabricated
Prefin.	Prefinished
Prop.	Propelled
PSF, psf	Pounds per Square Foot
PSI, psi	Pounds per Square Inch
PSIG	Pounds per Square Inch Gauge
PSP	Plastic Sewer Pipe
Pspr.	Painter, Spray
Psst.	Painter, Structural Steel
P.T.	Potential Transformer
P. & T.	Pressure & Temperature
Ptd.	Painted
Ptns.	Partitions
Pu	Ultimate Load
PVC	Polyvinyl Chloride
Pvmt.	Pavement
Pwr.	Power
Q	Quantity Heat Flow
Qt.	Quart
Quan., Qty.	Quantity
Q.C.	Quick Coupling
r	Radius of Gyration
R	Resistance
R.C.P.	Reinforced Concrete Pipe
Rect.	Rectangle
Reg.	Regular
Reinf.	Reinforced
Req'd.	Required
Res.	Resistant
Resi.	Residential
Rgh.	Rough
RGS	Rigid Galvanized Steel
R.H.W.	Rubber, Heat & Water Resistant; Residential Hot Water
rms	Root Mean Square
Rnd.	Round
Rodm.	Rodman
Rofc.	Roofer, Composition
Rofp.	Roofer, Precast
Rohe.	Roofer Helpers (Composition)
Rots.	Roofer, Tile & Slate
R.O.W.	Right of Way
RPM	Revolutions per Minute
R.S.	Rapid Start
Rsr	Riser
RT	Round Trip
S.	Suction; Single Entrance; South
SC	Screw Cover
SCFM	Standard Cubic Feet per Minute
Scaf.	Scaffold
Sch., Sched.	Schedule
S.C.R.	Modular Brick
S.D.	Sound Deadening
S.D.R.	Standard Dimension Ratio
S.E.	Surfaced Edge
Sel.	Select
S.E.R., S.E.U.	Service Entrance Cable
S.F.	Square Foot
S.F.C.A.	Square Foot Contact Area
S.F. Flr.	Square Foot of Floor
S.F.G.	Square Foot of Ground
S.F. Hor.	Square Foot Horizontal
S.F.R.	Square Feet of Radiation
S.F. Shlf.	Square Foot of Shelf
S4S	Surface 4 Sides
Shee.	Sheet Metal Worker
Sin.	Sine
Skwk.	Skilled Worker
SL	Saran Lined
S.L.	Slimline
Sldr.	Solder
SLH	Super Long Span Bar Joist
S.N.	Solid Neutral
S-O-L	Socketolet
sp	Standpipe
S.P.	Static Pressure; Single Pole; Self-Propelled
Spri.	Sprinkler Installer
spwg	Static Pressure Water Gauge
S.P.D.T.	Single Pole, Double Throw
SPF	Spruce Pine Fir
S.P.S.T.	Single Pole, Single Throw
SPT	Standard Pipe Thread
Sq.	Square; 100 Square Feet
Sq. Hd.	Square Head
Sq. In.	Square Inch
S.S.	Single Strength; Stainless Steel
S.S.B.	Single Strength B Grade
sst, ss	Stainless Steel
Sswk.	Structural Steel Worker
Sswl.	Structural Steel Welder
St.;Stl.	Steel
S.T.C.	Sound Transmission Coefficient
Std.	Standard
Stg.	Staging
STK	Select Tight Knot
STP	Standard Temperature & Pressure
Stpi.	Steamfitter, Pipefitter
Str.	Strength; Starter; Straight
Strd.	Stranded
Struct.	Structural
Sty.	Story
Subj.	Subject
Subs.	Subcontractors
Surf.	Surface
Sw.	Switch
Swbd.	Switchboard
S.Y.	Square Yard
Syn.	Synthetic
S.Y.P.	Southern Yellow Pine
Sys.	System
t.	Thickness
T	Temperature; Ton
Tan	Tangent
T.C.	Terra Cotta
T & C	Threaded and Coupled
T.D.	Temperature Difference
Tdd	Telecommunications Device for the Deaf
T.E.M.	Transmission Electron Microscopy
TFE	Tetrafluoroethylene (Teflon)
T. & G.	Tongue & Groove; Tar & Gravel
Th., Thk.	Thick
Thn.	Thin
Thrded	Threaded
Tilf.	Tile Layer, Floor
Tilh.	Tile Layer, Helper
THHN	Nylon Jacketed Wire
THW.	Insulated Strand Wire

Abbreviations

THWN	Nylon Jacketed Wire
T.L.	Truckload
T.M.	Track Mounted
Tot.	Total
T-O-L	Threadolet
T.S.	Trigger Start
Tr.	Trade
Transf.	Transformer
Trhv.	Truck Driver, Heavy
Trlr	Trailer
Trlt.	Truck Driver, Light
TTY	Teletypewriter
TV	Television
T.W.	Thermoplastic Water Resistant Wire
UCI	Uniform Construction Index
UF	Underground Feeder
UGND	Underground Feeder
U.H.F.	Ultra High Frequency
U.I.	United Inch
U.L.	Underwriters Laboratory
Uld.	unloading
Unfin.	Unfinished
URD	Underground Residential Distribution
US	United States
USP	United States Primed
UTP	Unshielded Twisted Pair
V	Volt
V.A.	Volt Amperes
V.C.T.	Vinyl Composition Tile
VAV	Variable Air Volume
VC	Veneer Core
Vent.	Ventilation
Vert.	Vertical
V.F.	Vinyl Faced
V.G.	Vertical Grain
V.H.F.	Very High Frequency
VHO	Very High Output
Vib.	Vibrating
V.L.F.	Vertical Linear Foot
Vol.	Volume
VRP	Vinyl Reinforced Polyester
W	Wire; Watt; Wide; West
w/	With
W.C.	Water Column; Water Closet
W.F.	Wide Flange
W.G.	Water Gauge
Wldg.	Welding
W. Mile	Wire Mile
W-O-L	Weldolet
W.R.	Water Resistant
Wrck.	Wrecker
W.S.P.	Water, Steam, Petroleum
WT., Wt.	Weight
WWF	Welded Wire Fabric
XFER	Transfer
XFMR	Transformer
XHD	Extra Heavy Duty
XHHW, XLPE	Cross-Linked Polyethylene Wire Insulation
XLP	Cross-linked Polyethylene
Y	Wye
yd	Yard
yr	Year
Δ	Delta
%	Percent
~	Approximately
Ø	Phase; diameter
@	At
#	Pound; Number
<	Less Than
>	Greater Than
Z	zone

A

B

C

D

E

F

G

H

I

J

K

N

O

Q

R

S

T

Notes

Notes

Division Notes

		CREW	DAILY OUTPUT	LABOR-HOURS	UNIT	2012 BARE COSTS MAT.	LABOR	EQUIP.	TOTAL	TOTAL INCL O&P

Division Notes

		CREW	DAILY OUTPUT	LABOR-HOURS	UNIT	2012 BARE COSTS				TOTAL INCL O&P
						MAT.	LABOR	EQUIP.	TOTAL	

Reed Construction Data/RSMeans— a tradition of excellence in construction cost information and services since 1942

Table of Contents

For more information visit the RSMeans website at **www.rsmeans.com**

Unit prices according to the latest MasterFormat!

Book Selection Guide

The following table provides definitive information on the content of each cost data publication. The number of lines of data provided in each unit price or assemblies division, as well as the number of reference tables and crews, is listed for each book. The presence of other elements such as equipment rental costs, historical cost indexes, city cost indexes, square foot models, or cross-referenced indexes is also indicated. You can use the table to help select the RSMeans book that has the quantity and type of information you most need in your work.

Unit Cost Divisions	Building Construction	Mechanical	Electrical	Commercial Renovation	Square Foot	Site Work Landsc.	Green Building	Interior	Concrete Masonry	Open Shop	Heavy Construction	Light Commercial	Facilities Construction	Plumbing	Residential
1	571	392	407	495		515	124	319	461	570	505	231	1001	408	174
2	745	264	83	684		965	180	363	219	742	704	439	1174	288	240
3	1539	222	212	906		1342	874	313	1847	1539	1481	386	1567	189	366
4	902	24	0	698		701	183	573	1137	878	593	490	1149	0	405
5	1912	263	168	1022		871	1799	1021	834	1903	1118	893	1905	339	724
6	1986	67	69	1573		401	534	1294	300	1965	380	1667	1602	22	2179
7	1456	208	125	1484		555	778	543	505	1453	26	1161	1516	217	915
8	2081	61	10	2277		296	1155	1765	658	2036	0	1661	2495	0	1412
9	1875	72	26	1670		306	444	1966	386	1817	15	1589	2111	54	1400
10	988	17	10	612		215	27	808	159	988	29	494	1074	231	221
11	1029	208	166	493		126	54	885	28	1014	0	235	1049	169	110
12	562	0	2	318		248	150	1707	12	551	19	364	1757	23	317
13	732	117	111	247		347	128	249	69	731	261	80	760	72	79
14	280	36	0	229		28	0	259	0	279	0	12	299	21	6
21	78	0	16	30		0	0	249	0	78	0	60	430	436	224
22	1144	7289	155	1108		1536	1100	609	20	1106	1660	834	7150	9008	662
23	1201	7048	605	784		157	900	691	38	1121	110	688	5126	1870	411
26	1249	496	9967	1010		763	572	1085	55	1237	539	1095	9903	441	559
27	67	0	261	34		9	0	59	0	67	35	48	263	0	4
28	96	58	124	71		0	21	78	0	99	0	40	139	44	25
31	1477	735	608	798		3186	290	7	1208	1421	3257	600	1540	660	606
32	767	54	0	838		4324	343	359	274	738	1797	380	1645	166	428
33	506	1008	465	205		1964	41	0	228	505	1887	117	1549	1197	115
34	109	0	20	4		144	0	0	31	64	122	0	129	0	0
35	18	0	0	0		166	0	0	0	18	281	0	83	0	0
41	52	0	0	24		7	0	22	0	52	30	0	59	14	0
44	75	83	0	0		0	0	0	0	0	0	0	75	75	0
46	23	16	0	0		162	149	0		23	152	0	33	33	0
48	0	0	0	0		0	19	0		0	0	0	0	0	0
Totals	23520	18738	13610	17614		19334	9865	15224	8469	22995	15001	13564	47583	15977	11582

Assem Div	Building Construction	Mechanical	Elec-trical	Commercial Renovation	Square Foot	Site Work Landscape	Assemblies	Green Building	Interior	Concrete Masonry	Heavy Construction	Light Commercial	Facilities Construction	Plumbing	Asm Div	Residential
A		15	0	188	150	577	598	0	0	536	571	154	24	0	1	368
B		0	0	825	2472	0	5572	56	330	1914	368	2032	143	0	2	211
C		0	0	647	864	0	1233	0	1558	146	0	763	245	0	3	588
D		1067	812	712	1798	72	2425	281	769	0	0	1272	1027	1083	4	859
E		0	0	86	259	0	297	0	5	0	0	256	5	0	5	392
F		0	0	0	122	0	122	0	0	0	0	122	3	0	6	357
G		527	294	317	202	3006	668	0	0	535	1372	200	293	685	7	300
															8	760
															9	80
															10	0
															11	0
															12	0
Totals		1609	1106	2775	5867	3655	10915	337	2662	3131	2311	4799	1740	1768		3915

Reference Section	Building Construction Costs	Mechanical	Electrical	Commercial Renovation	Square Foot	Site Work Landscape	Assem.	Green Building	Interior	Concrete Masonry	Open Shop	Heavy Construction	Light Commercial	Facilities Construction	Plumbing	Resi.
Reference Tables	yes	yes	yes	yes	no	yes	yes	yes	yes	yes	yes	yes	yes	yes	yes	yes
Models					107								46			28
Crews	527	527	527	505		527		527	527	527	503	527	503	505	527	503
Equipment Rental Costs	yes	yes	yes	yes		yes		yes	yes	yes	yes	yes	yes	yes	yes	yes
Historical Cost Indexes	yes	yes	yes	yes	yes	yes	yes	yes	yes	yes	yes	yes	yes	yes	yes	no
City Cost Indexes	yes	yes	yes	yes	yes	yes	yes	yes	yes	yes	yes	yes	yes	yes	yes	yes

Annual Cost Guides

Cost Guides also available online at meanscostworks.com

For more information visit the RSMeans website at **www.rsmeans.com**

Unit prices according to the latest MasterFormat!

RSMeans Building Construction Cost Data 2012

Offers you unchallenged unit price reliability in an easy-to-use format. Whether used for verifying complete, finished estimates or for periodic checks, it supplies more cost facts better and faster than any comparable source. Over 23,000 unit prices have been updated for 2012. The City Cost Indexes and Location Factors cover over 930 areas, for indexing to any project location in North America. Order and get *RSMeans Quarterly Update Service* FREE. You'll have year-long access to the RSMeans Estimating **HOTLINE** FREE with your subscription. Expert assistance when using RSMeans data is just a phone call away.

$177.95 per copy | Available Sept. 2011 | Catalog no. 60012

RSMeans Green Building Cost Data 2012

Estimate, plan, and budget the costs of green building for both new commercial construction and renovation work with this first edition of RSMeans Green Building Cost Data. More than 9,000 unit costs for a wide array of green building products plus assemblies costs. Easily identified cross references to LEED and Green Globes building rating systems criteria.

$147.95 per copy | Available Nov. 2011 | Catalog no. 60552

RSMeans Mechanical Cost Data 2012

Total unit and systems price guidance for mechanical construction. . . materials, parts, fittings, and complete labor cost information. Includes prices for piping, heating, air conditioning, ventilation, and all related construction.

Plus new 2012 unit costs for:

- Thousands of installed HVAC/ controls assemblies components
- "On-site" Location Factors for over 930 cities and towns in the U.S. and Canada
- Crews, labor, and equipment

$174.95 per copy | Available Oct. 2011 | Catalog no. 60022

RSMeans Facilities Construction Cost Data 2012

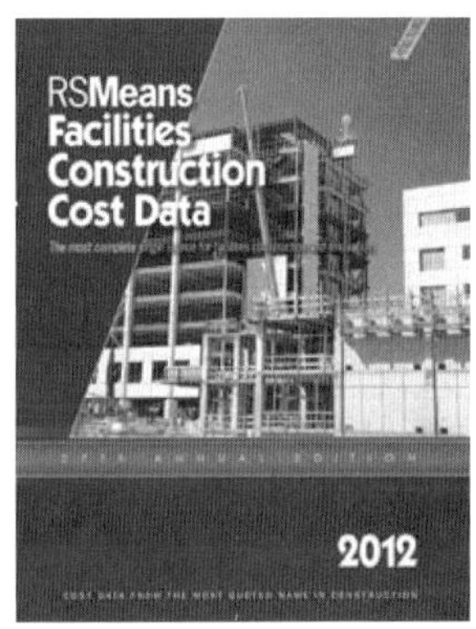

For the maintenance and construction of commercial, industrial, municipal, and institutional properties. Costs are shown for new and remodeling construction and are broken down into materials, labor, equipment, and overhead and profit. Special emphasis is given to sections on mechanical, electrical, furnishings, site work, building maintenance, finish work, and demolition.

More than 47,000 unit costs, plus assemblies costs and a comprehensive Reference Section are included.

$453.95 per copy | Available Nov. 2011 | Catalog no. 60202

RSMeans Square Foot Costs 2012

Accurate and Easy To Use

- **Updated price information** based on nationwide figures from suppliers, estimators, labor experts, and contractors
- "How-to-Use" sections, with **clear examples** of commercial, residential, industrial, and institutional structures
- Realistic graphics, offering true-to-life illustrations of building projects
- Extensive information on using square foot cost data, including sample estimates and alternate pricing methods

$187.95 per copy | Available Oct. 2011 | Catalog no. 60052

RSMeans Commercial Renovation Cost Data 2012

Commercial/Multifamily Residential

Use this valuable tool to estimate commercial and multifamily residential renovation and remodeling.

Includes: New costs for hundreds of unique methods, materials, and conditions that only come up in repair and remodeling, PLUS:

- Unit costs for more than 17,000 construction components
- Installed costs for more than 2,000 assemblies
- More than 930 "on-site" localization factors for the U.S. and Canada

$144.95 per copy | Available Oct. 2011 | Catalog no. 60042

Annual Cost Guides

Cost Guides also available online at meanscostworks.com

For more information visit the RSMeans website at **www.rsmeans.com**

Unit prices according to the latest MasterFormat!

RSMeans Facilities Maintenance & Repair Cost Data 2012

RSMeans Facilities Maintenance & Repair Cost Data gives you a complete system to manage and plan your facility repair and maintenance costs and budget efficiently. Guidelines for auditing a facility and developing an annual maintenance plan. Budgeting is included, along with reference tables on cost and management, and information on frequency and productivity of maintenance operations.

The only nationally recognized source of maintenance and repair costs. Developed in cooperation with the Civil Engineering Research Laboratory (CERL) of the Army Corps of Engineers.

$399.95 per copy | Available Nov. 2011 | Catalog no. 60302

RSMeans Light Commercial Cost Data 2012

Specifically addresses the light commercial market, which is a specialized niche in the construction industry. Aids you, the owner/designer/contractor, in preparing all types of estimates—from budgets to detailed bids. Includes new advances in methods and materials.

Assemblies Section allows you to evaluate alternatives in the early stages of design/planning.

Over 13,000 unit costs ensure that you have the prices you need. . . when you need them.

$132.95 per copy | Available Nov. 2011 | Catalog no. 60182

RSMeans Concrete & Masonry Cost Data 2012

Provides you with cost facts for virtually all concrete/masonry estimating needs, from complicated formwork to various sizes and face finishes of brick and block—all in great detail. The comprehensive Unit Price Section contains more than 8,000 selected entries. Also contains an Assemblies [Cost] Section, and a detailed Reference Section that supplements the cost data.

$162.95 per copy | Available Dec. 2011 | Catalog no. 60112

RSMeans Labor Rates for the Construction Industry 2012

Complete information for estimating labor costs, making comparisons, and negotiating wage rates by trade for more than 300 U.S. and Canadian cities. With 46 construction trades listed by local union number in each city, and historical wage rates included for comparison. Each city chart lists the county and is alphabetically arranged with handy visual flip tabs for quick reference.

$395.95 per copy | Available Dec. 2011 | Catalog no. 60122

RSMeans Construction Cost Indexes 2012

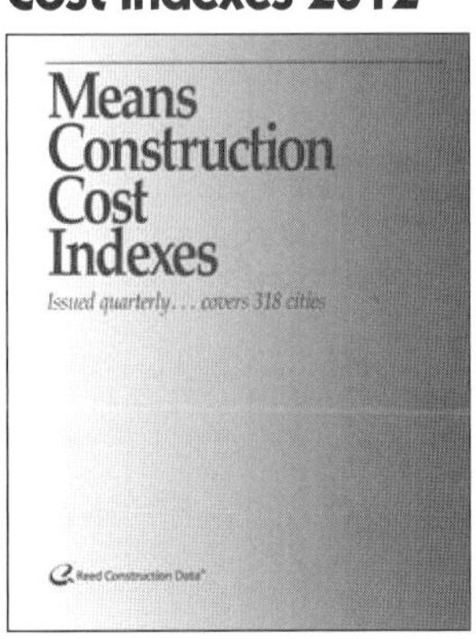

What materials and labor costs will change unexpectedly this year? By how much?

- Breakdowns for 318 major cities
- National averages for 30 key cities
- Expanded five major city indexes
- Historical construction cost indexes

$362.00 per year (subscription) | Catalog no. 50142
$90.50 individual quarters | Catalog no. 60142 A,B,C,D

RSMeans Interior Cost Data 2012

Provides you with prices and guidance needed to make accurate interior work estimates. Contains costs on materials, equipment, hardware, custom installations, furnishings, and labor costs. . . for new and remodel commercial and industrial interior construction, including updated information on office furnishings, and reference information.

$182.95 per copy | Available Nov. 2011 | Catalog no. 60092

Annual Cost Guides

Cost Guides also available online at meanscostworks.com

For more information visit the RSMeans website at **www.rsmeans.com**

Unit prices according to the latest MasterFormat!

RSMeans Heavy Construction Cost Data 2012

A comprehensive guide to heavy construction costs. Includes costs for highly specialized projects such as tunnels, dams, highways, airports, and waterways. Information on labor rates, equipment, and materials costs is included. Features unit price costs, systems costs, and numerous reference tables for costs and design.

$177.95 per copy | Available Dec. 2011 | Catalog no. 60162

RSMeans Plumbing Cost Data 2012

Comprehensive unit prices and assemblies for plumbing, irrigation systems, commercial and residential fire protection, point-of-use water heaters, and the latest approved materials. This publication and its companion, *RSMeans Mechanical Cost Data*, provide full-range cost estimating coverage for all the mechanical trades.

Now contains updated costs for potable water and radiant heat systems and high efficiency fixtures.

$179.95 per copy | Available Oct. 2011 | Catalog no. 60212

2012 RSMeans Seminar Schedule

Note: call for exact dates and details.

Location	Dates
Las Vegas, NV	March
Washington, DC	April
San Francisco, CA	June
Washington, DC	September
Dallas, TX	September
Las Vegas, NV	October
Orlando, FL	November
San Diego, CA	December

1-800-334-3509, ext. 925115

2012 Regional Seminar Schedule

Note: call for exact dates and details.

Location	Dates
Seattle, WA	January and August
Dallas/Ft. Worth, TX	January
Austin, TX	February
Oklahoma City, OK	February
Anchorage, AK	March and September
Phoenix, AZ	March
San Bernardino, CA	April
Aiea, HI	April
Kansas City, MO	April
Toronto	May
Denver, CO	May
Bethesda, MD	June
Columbus, GA	June
El Segundo, CA	August
Jacksonville FL	September
Birmingham, AL	October
St. Louis, MO	October
Houston, TX	October
Portland, OR	November
Baltimore, MD	November
San Antonio, TX	December
Raleigh, NC	December

1-800-334-3509, ext. 925115

For more information visit the RSMeans website at **www.rsmeans.com**

Registration Information

Register early... Save up to $100! Register 30 days before the start date of a seminar and save $100 off your total fee. *Note: This discount can be applied only once per order. It cannot be applied to team discount registrations or any other special offer.*

How to register Register by phone today! The RSMeans toll-free number for making reservations is **1-800-334-3509, ext. 925115.**

Individual seminar registration fee - $935. *Means CostWorks*® training registration fee - $375. To register by mail, complete the registration form and return, with your full fee, to: RSMeans Seminars, 700 Longwater Drive, Norwell, MA 02061.

Government pricing All federal government employees save off the regular seminar price. Other promotional discounts cannot be combined with the government discount.

Team discount program for two to four seminar registrations. Call for pricing: 1-800-334-3509, ext. 925115

Multiple course discounts When signing up for two or more courses, call for pricing.

Refund policy Cancellations will be accepted up to ten business days prior to the seminar start. There are no refunds for cancellations received later than ten working days prior to the first day of the seminar. A $150 processing fee will be applied for all cancellations. Written notice of cancellation is required. Substitutions can be made at any time before the session starts. **No-shows are subject to the full seminar fee.**

AACE approved courses Many seminars described and offered here have been approved for 14 hours (1.4 recertification credits) of credit by the AACE International Certification Board toward meeting the continuing education requirements for recertification as a Certified Cost Engineer/Certified Cost Consultant.

AIA Continuing Education We are registered with the AIA Continuing Education System (AIA/CES) and are committed to developing quality learning activities in accordance with the CES criteria. Many seminars meet the AIA/CES criteria for Quality Level 2. AIA members may receive (14) learning units (LUs) for each two-day RSMeans course.

Daily course schedule The first day of each seminar session begins at 8:30 a.m. and ends at 4:30 p.m. The second day begins at 8:00 a.m. and ends at 4:00 p.m. Participants are urged to bring a hand-held calculator, since many actual problems will be worked out in each session.

Continental breakfast Your registration includes the cost of a continental breakfast, and a morning and afternoon refreshment break. These informal segments allow you to discuss topics of mutual interest with other seminar attendees. (You are free to make your own lunch and dinner arrangements.)

Hotel/transportation arrangements RSMeans arranges to hold a block of rooms at most host hotels. To take advantage of special group rates when making your reservation, be sure to mention that you are attending the RSMeans seminar. You are, of course, free to stay at the lodging place of your choice. **(Hotel reservations and transportation arrangements should be made directly by seminar attendees.)**

Important Class sizes are limited, so please register as soon as possible.

Note: *Pricing subject to change.*

Registration Form

ADDS-1000

Call 1-800-334-3509, ext. 925115 to register or FAX this form 1-800-632-6732. Visit our website: www.rsmeans.com

Please register the following people for the RSMeans construction seminars as shown here. We understand that we must make our own hotel reservations if overnight stays are necessary.

☐ Full payment of $____________________enclosed.

☐ Bill me.

Please print name of registrant(s).

(To appear on certificate of completion)

____

____

____

____

P.O. #: ________________________

GOVERNMENT AGENCIES MUST SUPPLY PURCHASE ORDER NUMBER OR TRAINING FORM.

Firm name________________________

Address________________________

City/State/Zip________________________

Telephone no.______________ Fax no.______________

E-mail address________________________

Charge registration(s) to: ☐ MasterCard ☐ VISA ☐ American Express

Account no.______________Exp. date______________

Cardholder's signature________________________

Seminar name________________________

Seminar City ________________________

Please mail check to: RSMeans Seminars, 700 Longwater Drive, Norwell, MA 02061 USA

For more information visit the RSMeans website at **www.rsmeans.com**

Means CostWorks.com Training

Construction estimating is vital to the decision-making process at each state of every project. Means CostWorks.com works the way you do. It's systematic, flexible and intuitive. In this one day class you will see how you can estimate any phase of any project faster and better.

Some of what you'll learn:
- Customizing Means CostWorks.com
- Making the most of RSMeans "Circle Reference" numbers
- How to integrate your cost data
- Generate reports, exporting estimates to MS Excel, sharing, collaborating and more

Who should attend: architects & engineers, facility managers, contractors, construction estimators, and anyone using this powerful web-based tool.

Site Work & Heavy Construction Estimating

This *two-day* course teaches attendees how to estimate earthwork, site utilities, foundations, and site improvements, using the assemblies and the unit price methods.

Some of what you'll learn:
- Basic site work and heavy construction estimating skills
- Estimating foundations, utilities, earthwork, and site improvements
- Correct equipment usage, quality control, and site investigation for estimating purposes

Who should attend: project managers, design engineers, estimators, and contractors doing site work and heavy construction.

Facilities Construction Estimating

In this *two-day* course, professionals working in facilities management can get help with their daily challenges to establish budgets for all phases of a project.

Some of what you'll learn:
- Determining the full scope of a project
- Identifying the scope of risks and opportunities
- Creative solutions to estimating issues
- Organizing estimates for presentation and discussion
- Special techniques for repair/remodel and maintenance projects
- Negotiating project change orders

Who should attend: facility managers, engineers, contractors, facility tradespeople, planners, and project managers.

Practical Project Management for Construction Professionals

In this *two-day* course, acquire the essential knowledge and develop the skills to effectively and efficiently execute the day-to-day responsibilities of the construction project manager.

Covers:
- General conditions of the construction contract
- Contract modifications: change orders and construction change directives
- Negotiations with subcontractors and vendors
- Effective writing: notification and communications
- Dispute resolution: claims and liens

Who should attend: architects, engineers, owner's representatives, project managers.

Scheduling with Microsoft Project

Two days of hands-on training gives you a look at the basics to putting it all together with MS Project to create a complete project schedule. Learn to work better and faster, manage changes and updates, enhance tracking, generate actionable reports, and boost control of your time and budget.

Some of what you'll learn:
- Defining task/activity relationships: link tasks, establish predecessor/successor relationships and create logs
- Baseline scheduling: the essential what/when
- Using MS Project to determine the critical path
- Integrate RSMeans data with MS Project
- The dollar-loaded schedule: enhance the reliability of this "must-know" technique for scheduling public projects.

Who should attend: Contractors' project-management teams, architects and engineers, project owners and their representatives, and those interested in improving their project planning and scheduling and management skills.

Mechanical & Electrical Estimating

This *two-day* course teaches attendees how to prepare more accurate and complete mechanical/electrical estimates, avoiding the pitfalls of omission and double-counting, while understanding the composition and rationale within the RSMeans mechanical/electrical database.

Some of what you'll learn:
- The unique way mechanical and electrical systems are interrelated
- M&E estimates—conceptual, planning, budgeting, and bidding stages
- Order of magnitude, square foot, assemblies, and unit price estimating
- Comparative cost analysis of equipment and design alternatives

Who should attend: architects, engineers, facilities managers, mechanical and electrical contractors, and others who need a highly reliable method for developing, understanding, and evaluating mechanical and electrical contracts.

Professional Development

For more information visit the RSMeans website at **www.rsmeans.com**

Unit Price Estimating

This interactive *two-day* seminar teaches attendees how to interpret project information and process it into final, detailed estimates with the greatest accuracy level.

The most important credential an estimator can take to the job is the ability to visualize construction and estimate accurately.

Some of what you'll learn:

- Interpreting the design in terms of cost
- The most detailed, time-tested methodology for accurate pricing
- Key cost drivers—material, labor, equipment, staging, and subcontracts
- Understanding direct and indirect costs for accurate job cost accounting and change order management

Who should attend: Corporate and government estimators and purchasers, architects, engineers, and others who need to produce accurate project estimates.

RSMeans CostWorks® CD Training

This one-day course helps users become more familiar with the functionality of *RSMeans CostWorks* program. Each menu, icon, screen, and function found in the program is explained in depth. Time is devoted to hands-on estimating exercises.

Some of what you'll learn:

- Searching the database using all navigation methods
- Exporting RSMeans data to your preferred spreadsheet format
- Viewing crews, assembly components, and much more
- Automatically regionalizing the database

You are required to bring your own laptop computer for this course.

When you register for this course you will receive an outline for your laptop requirements.

Also offering web training for CostWorks CD!

Facilities Maintenance & Repair Estimating

This *two-day* course teaches attendees how to plan, budget, and estimate the cost of ongoing and preventive maintenance and repair for existing buildings and grounds.

Some of what you'll learn:

- The most financially favorable maintenance, repair, and replacement scheduling and estimating
- Auditing and value engineering facilities
- Preventive planning and facilities upgrading
- Determining both in-house and contract-out service costs
- Annual, asset-protecting M&R plan

Who should attend: Facility managers, maintenance supervisors, buildings and grounds superintendents, plant managers, planners, estimators, and others involved in facilities planning and budgeting.

Square Foot Estimating using Means CostWorks

This *two day* class uses the leading industry data and a powerful software package to develop highly accurate conceptual estimates for your construction projects. All attendees must bring a laptop computer loaded with the current year *Square Foot Model Costs* and the *Assemblies Cost Data* CostWorks titles.

Some of what you'll learn:

- Introduction to conceptual estimating
- Types of conceptual estimates
- Helpful hints
- Order of magnitude estimating
- Square foot estimating
- Assemblies estimating

Who should attend: architects, engineers, contractors, construction estimators, owner's representatives, and anyone looking for an electronic method for performing square foot estimating.

Assessing Scope of Work for Facility Construction Estimating

This *two-day* practical training program addresses the vital importance of understanding the SCOPE of projects in order to produce accurate cost estimates for facility repair and remodeling.

Some of what you'll learn:

- Discussions of site visits, plans/specs, record drawings of facilities, and site-specific lists
- Review of CSI divisions, including means, methods, materials, and the challenges of scoping each topic
- Exercises in SCOPE identification and SCOPE writing for accurate estimating of projects
- Hands-on exercises that require SCOPE, take-off, and pricing

Who should attend: corporate and government estimators, planners, facility managers, and others who need to produce accurate project estimates.

Reference Books

For more information visit the RSMeans website at **www.rsmeans.com**

Value Engineering: Practical Applications

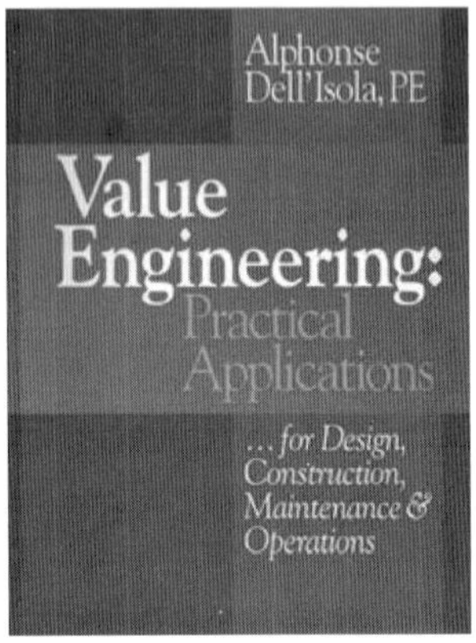

For Design, Construction, Maintenance & Operations

by Alphonse Dell'Isola, PE

A tool for immediate application—for engineers, architects, facility managers, owners, and contractors. Includes making the case for VE—the management briefing; integrating VE into planning, budgeting, and design; conducting life cycle costing; using VE methodology in design review and consultant selection; case studies; VE workbook; and a life cycle costing program on disk.

$79.95 per copy | Over 450 pages, illustrated, softcover | Catalog no. 67319A

Facilities Operations & Engineering Reference

by the Association for Facilities Engineering and RSMeans

An all-in-one technical reference for planning and managing facility projects and solving day-to-day operations problems. Selected as the official certified plant engineer reference, this handbook covers financial analysis, maintenance, HVAC and energy efficiency, and more.

$109.95 per copy | Over 700 pages, illustrated, hardcover | Catalog no. 67318

The Building Professional's Guide to Contract Documents

3rd Edition

by Waller S. Poage, AIA, CSI, CVS

A comprehensive reference for owners, design professionals, contractors, and students

- Structure your documents for maximum efficiency.
- Effectively communicate construction requirements.
- Understand the roles and responsibilities of construction professionals.
- Improve methods of project delivery.

$64.95 per copy | 400 pages Diagrams and construction forms, hardcover Catalog no. 67261A

Complete Book of Framing

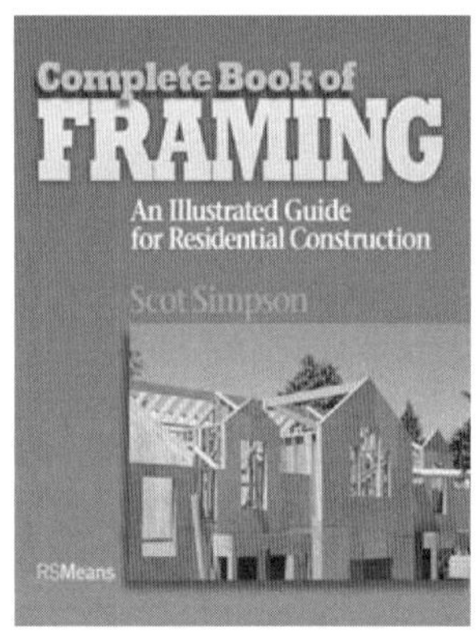

by Scot Simpson

This straightforward, easy-to-learn method will help framers, carpenters, and handy homeowners build their skills in rough carpentry and framing. Shows how to frame all the parts of a house: floors, walls, roofs, door & window openings, and stairs—with hundreds of color photographs and drawings that show every detail.

The book gives beginners all the basics they need to go from zero framing knowledge to a journeyman level.

$29.95 per copy | 336 pages, softcover | Catalog no. 67353

Cost Planning & Estimating for Facilities Maintenance

In this unique book, a team of facilities management authorities shares their expertise on:

- Evaluating and budgeting maintenance operations
- Maintaining and repairing key building components
- Applying *RSMeans Facilities Maintenance & Repair Cost Data* to your estimating

Covers special maintenance requirements of the ten major building types

$89.95 per copy | Over 475 pages, hardcover | Catalog no. 67314

Life Cycle Costing for Facilities

by Alphonse Dell'Isola and Dr. Steven Kirk

Guidance for achieving higher quality design and construction projects at lower costs! Cost-cutting efforts often sacrifice quality to yield the cheapest product. Life cycle costing enables building designers and owners to achieve both. The authors of this book show how LCC can work for a variety of projects — from roads to HVAC upgrades to different types of buildings.

$99.95 per copy | 396 pages, hardcover | Catalog no. 67341

Reference Books

For more information visit the RSMeans website at **www.rsmeans.com**

Interior Home Improvement Costs

9th Edition

Updated estimates for the most popular remodeling and repair projects—from small, do-it-yourself jobs to major renovations and new construction. Includes: kitchens & baths; new living space from your attic, basement, or garage; new floors, paint, and wallpaper; tearing out or building new walls; closets, stairs, and fireplaces; new energy-saving improvements, home theaters, and more!

$24.95 per copy | 250 pages, illustrated, softcover | Catalog no. 67308E

Exterior Home Improvement Costs

9th Edition

Updated estimates for the most popular remodeling and repair projects—from small, do-it-yourself jobs, to major renovations and new construction. Includes: curb appeal projects—landscaping, patios, porches, driveways, and walkways; new windows and doors; decks, greenhouses, and sunrooms; room additions and garages; roofing, siding, and painting; "green" improvements to save energy & water.

$24.95 per copy | Over 275 pages, illustrated, softcover | Catalog no. 67309E

Builder's Essentials: Plan Reading & Material Takeoff

For Residential and Light Commercial Construction

by Wayne J. DelPico

A valuable tool for understanding plans and specs, and accurately calculating material quantities. Step-by-step instructions and takeoff procedures based on a full set of working drawings.

$35.95 per copy | Over 420 pages, softcover | Catalog no. 67307

Means Unit Price Estimating Methods

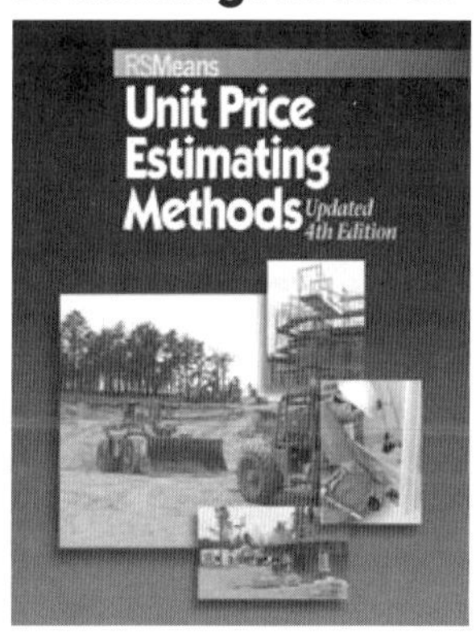

New 4th Edition

This new edition includes up-to-date cost data and estimating examples, updated to reflect changes to the CSI numbering system and new features of RSMeans cost data. It describes the most productive, universally accepted ways to estimate, and uses checklists and forms to illustrate shortcuts and timesavers. A model estimate demonstrates procedures. A new chapter explores computer estimating alternatives.

$59.95 per copy | Over 350 pages, illustrated, softcover | Catalog no. 67303B

Concrete Repair and Maintenance Illustrated

by Peter Emmons

Hundreds of illustrations show users how to analyze, repair, clean, and maintain concrete structures for optimal performance and cost effectiveness. From parking garages to roads and bridges to structural concrete, this comprehensive book describes the causes, effects, and remedies for concrete wear and failure. Invaluable for planning jobs, selecting materials, and training employees, this book is a must-have for concrete specialists, general contractors, facility managers, civil and structural engineers, and architects.

$69.95 per copy | 300 pages, illustrated, softcover | Catalog no. 67146

Total Productive Facilities Management

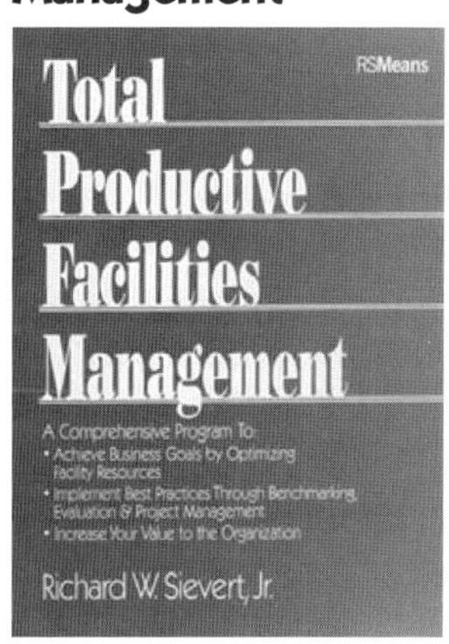

by Richard W. Sievert, Jr.

Today, facilities are viewed as strategic resources. . . elevating the facility manager to the role of asset manager supporting the organization's overall business goals. Now, Richard Sievert Jr., in this well-articulated guidebook, sets forth a new operational standard for the facility manager's emerging role. . . a comprehensive program for managing facilities as a true profit center.

$79.95 per copy | 275 pages, softcover | Catalog no. 67321

Means Illustrated Construction Dictionary

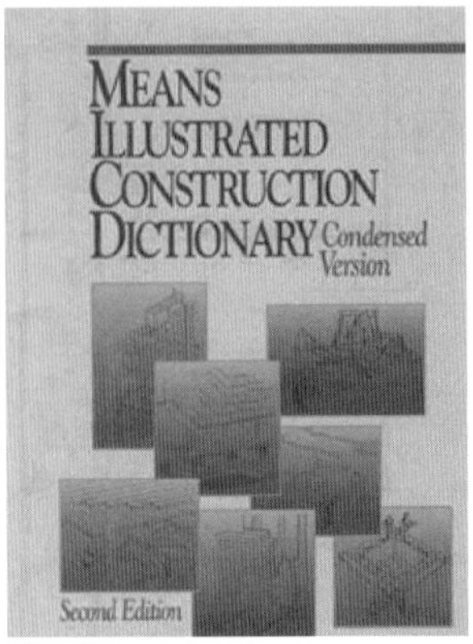

Condensed, 2nd Edition

The best portable dictionary for office or field use—an essential tool for contractors, architects, insurance and real estate personnel, facility managers, homeowners, and anyone who needs quick, clear definitions for construction terms. The second edition has been further enhanced with updates and hundreds of new terms and illustrations . . . in keeping with the most recent developments in the construction industry.

Now with a quick-reference Spanish section. Includes tools and equipment, materials, tasks, and more!

$59.95 per copy | Over 500 pages, softcover | Catalog no. 67282A

Square Foot & UNIFORMAT Assemblies Estimating Methods

3rd Edition

Develop realistic square foot and assemblies costs for budgeting and construction funding. The new edition features updated guidance on square foot and assemblies estimating using UNIFORMAT II. An essential reference for anyone who performs conceptual estimates.

$69.95 per copy | Over 300 pages, illustrated, softcover | Catalog no. 67145B

Facilities Planning & Relocation

by David D. Owen

A complete system for planning space needs and managing relocations. Includes step-by-step manual, over 50 forms, and extensive reference section on materials and furnishings.

New lower price and user-friendly format.

$89.95 per copy | Over 450 pages, softcover | Catalog no. 67301

Mechanical Estimating Methods

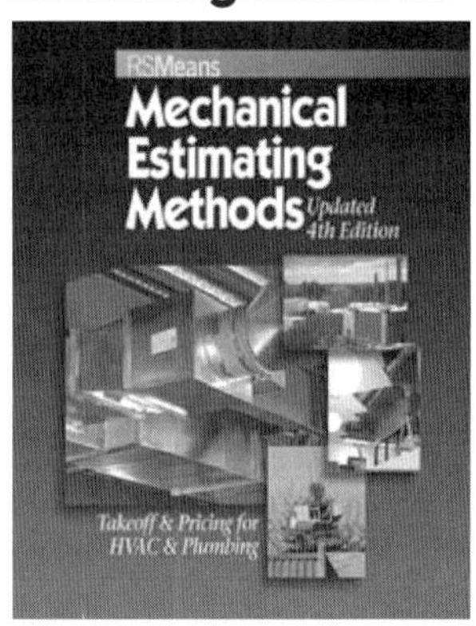

4th Edition

Completely updated, this guide assists you in making a review of plans, specs, and bid packages, with suggestions for takeoff procedures, listings, substitutions, and pre-bid scheduling for all components of HVAC. Includes suggestions for budgeting labor and equipment usage. Compares materials and construction methods to allow you to select the best option.

$64.95 per copy | Over 350 pages, illustrated, softcover | Catalog no. 67294B

Electrical Estimating Methods

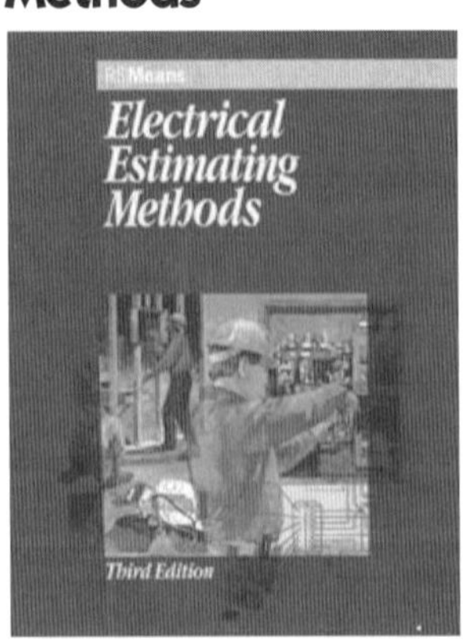

3rd Edition

Expanded edition includes sample estimates and cost information in keeping with the latest version of the CSI MasterFormat and UNIFORMAT II. Complete coverage of fiber optic and uninterruptible power supply electrical systems, broken down by components, and explained in detail. Includes a new chapter on computerized estimating methods. A practical companion to *RSMeans Electrical Cost Data.*

$64.95 per copy | Over 325 pages, hardcover | Catalog no. 67230B

Residential & Light Commercial Construction Standards

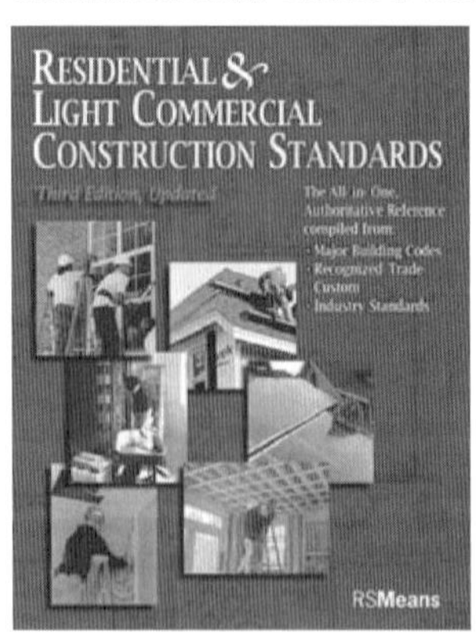

3rd Edition, Updated by RSMeans and contributing authors

This book provides authoritative requirements and recommendations compiled from leading professional associations, industry publications, and building code organizations. This all-in-one reference helps establish a standard for workmanship, quickly resolve disputes, and avoid defect claims. Updated third edition includes new coverage of green building, seismic, hurricane, and mold-resistant construction.

$59.95 | Over 550 pages, illustrated, softcover | Catalog no. 67322B

Reference Books

For more information visit the RSMeans website at **www.rsmeans.com**

Construction Business Management

by Nick Ganaway

Only 43% of construction firms stay in business after four years. Make sure your company thrives with valuable guidance from a pro with 25 years of success as a commercial contractor. Find out what it takes to build all aspects of a business that is profitable, enjoyable, and enduring. With a bonus chapter on retail construction.

$49.95 per copy | 200 pages, softcover | Catalog no. 67352

Project Scheduling & Management for Construction

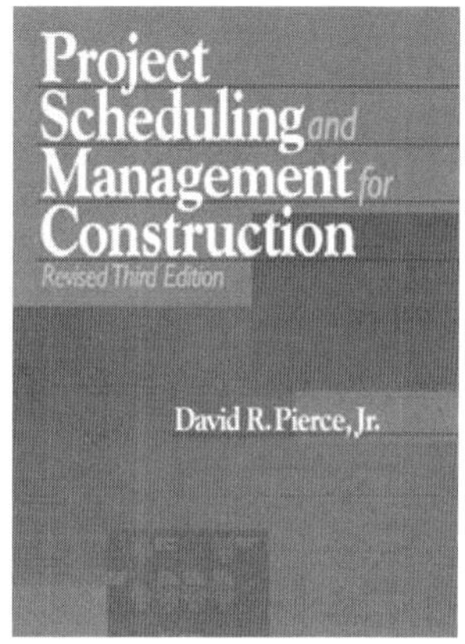

3rd Edition
by David R. Pierce, Jr.

A comprehensive yet easy-to-follow guide to construction project scheduling and control—from vital project management principles through the latest scheduling, tracking, and controlling techniques. The author is a leading authority on scheduling, with years of field and teaching experience at leading academic institutions. Spend a few hours with this book and come away with a solid understanding of this essential management topic.

$64.95 per copy | Over 300 pages, illustrated, hardcover | Catalog no. 67247B

The Practice of Cost Segregation Analysis

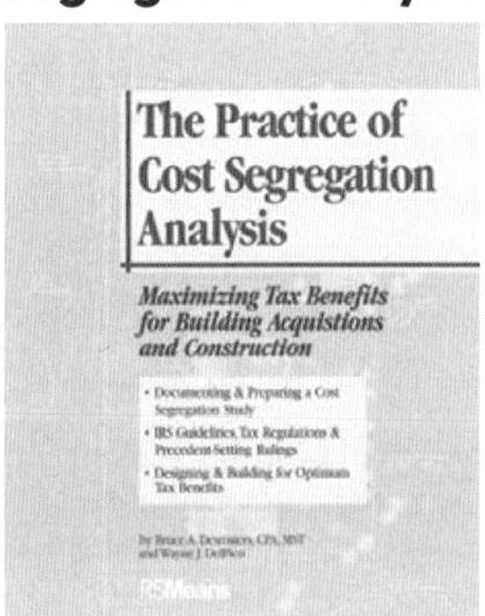

by Bruce A. Desrosiers and Wayne J. DelPico

This expert guide walks you through the practice of cost segregation analysis, which enables property owners to defer taxes and benefit from "accelerated cost recovery" through depreciation deductions on assets that are properly identified and classified.

With a glossary of terms, sample cost segregation estimates for various building types, key information resources, and updates via a dedicated website, this book is a critical resource for anyone involved in cost segregation analysis.

$99.95 per copy | Over 225 pages | Catalog no. 67345

Preventive Maintenance for Multi-Family Housing

by John C. Maciha

Prepared by one of the nation's leading experts on multi-family housing.

This complete PM system for apartment and condominium communities features expert guidance, checklists for buildings and grounds maintenance tasks and their frequencies, a reusable wall chart to track maintenance, and a dedicated website featuring customizable electronic forms. A must-have for anyone involved with multi-family housing maintenance and upkeep.

$89.95 per copy | 225 pages | Catalog no. 67346

How to Estimate with Means Data & CostWorks

New 3rd Edition
by RSMeans and Saleh A. Mubarak, Ph.D.

New 3rd Edition—fully updated with new chapters, plus new CD with updated *CostWorks* cost data and MasterFormat organization. Includes all major construction items—with more than 300 exercises and two sets of plans that show how to estimate for a broad range of construction items and systems—including general conditions and equipment costs.

$59.95 per copy | 272 pages, softcover | Includes CostWorks CD
Catalog no. 67324B

Job Order Contracting Expediting Construction Project Delivery

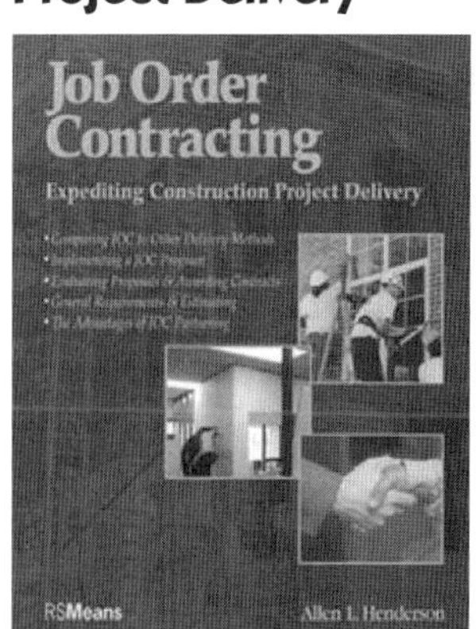

by Allen Henderson

Expert guidance to help you implement JOC—fast becoming the preferred project delivery method for repair and renovation, minor new construction, and maintenance projects in the public sector and in many states and municipalities. The author, a leading JOC expert and practitioner, shows how to:

- Establish a JOC program
- Evaluate proposals and award contracts
- Handle general requirements and estimating
- Partner for maximum benefits

$89.95 per copy | 192 pages, illustrated, hardcover | Catalog no. 67348

Reference Books

For more information visit the RSMeans website at **www.rsmeans.com**

Estimating Building Costs

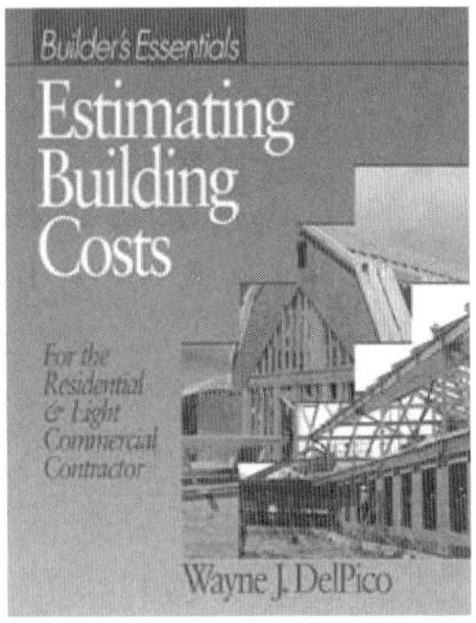

For the Residential & Light Commercial Contractor
by Wayne J. DelPico

Step-by-step estimating methods for residential and light commercial contractors. Includes a detailed look at every construction specialty—explaining all the components, takeoff units, and labor needed for well-organized, complete estimates. Covers correctly interpreting plans and specifications, and developing accurate and complete labor and material costs.

$29.95 per copy | Over 400 pages, illustrated, softcover | Catalog no. 67343

Building & Renovating Schools

This all-inclusive guide covers every step of the school construction process—from initial planning, needs assessment, and design, right through moving into the new facility. A must-have resource for anyone concerned with new school construction or renovation. With square foot cost models for elementary, middle, and high school facilities, and real-life case studies of recently completed school projects.

The contributors to this book—architects, construction project managers, contractors, and estimators who specialize in school construction—provide start-to-finish, expert guidance on the process.

$99.95 per copy | Over 425 pages, hardcover | Catalog no. 67342

The Homeowner's Guide to Mold

By Michael Pugliese

Expert guidance to protect your health and your home.

Mold, whether caused by leaks, humidity or flooding, is a real health and financial issue—for homeowners and contractors.

This full-color book explains:

- Construction and maintenance practices to prevent mold
- How to inspect for and remove mold
- Mold remediation procedures and costs
- What to do after a flood
- How to deal with insurance companies

$21.95 per copy | 144 pages, softcover | Catalog no. 67344

Universal Design Ideas for Style, Comfort & Safety

by RSMeans and Lexicon Consulting, Inc.

Incorporating universal design when building or remodeling helps people of any age and physical ability more fully and safely enjoy their living spaces. This book shows how universal design can be artfully blended into the most attractive homes. It discusses specialized products like adjustable countertops and chair lifts, as well as simple ways to enhance a home's safety and comfort. With color photos and expert guidance, every area of the home is covered. Includes budget estimates that give an idea how much projects will cost.

$21.95 | 160 pages, illustrated, softcover | Catalog no. 67354

Landscape Estimating Methods

5th Edition

Answers questions about preparing competitive landscape construction estimates, with up-to-date cost estimates and the new MasterFormat classification system. Expanded and revised to address the latest materials and methods, including new coverage on approaches to green building. Includes:

- Step-by-step explanation of the estimating process
- Sample forms and worksheets that save time and prevent errors

$64.95 per copy | Over 350 pages, softcover | Catalog no. 67295C

Plumbing Estimating Methods

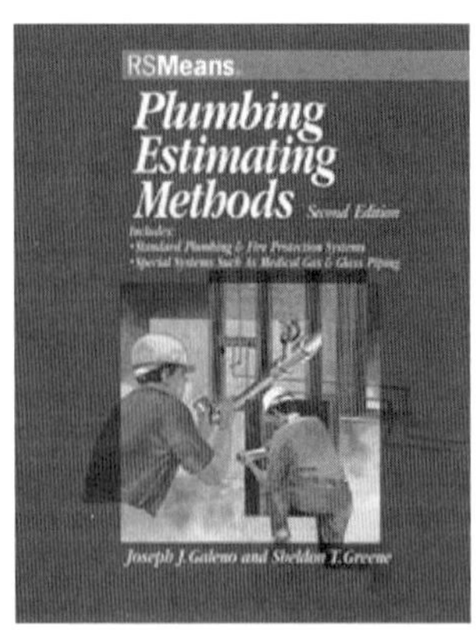

3rd Edition
by Joseph Galeno and Sheldon Greene

Updated and revised! This practical guide walks you through a plumbing estimate, from basic materials and installation methods through change order analysis. *Plumbing Estimating Methods* covers residential, commercial, industrial, and medical systems, and features sample takeoff and estimate forms and detailed illustrations of systems and components.

$59.95 per copy | 330+ pages, softcover | Catalog no. 67283B

Reference Books

For more information visit the RSMeans website at **www.rsmeans.com**

Understanding & Negotiating Construction Contracts

by Kit Werremeyer

Take advantage of the author's 30 years' experience in small-to-large (including international) construction projects. Learn how to identify, understand, and evaluate high risk terms and conditions typically found in all construction contracts—then negotiate to lower or eliminate the risk, improve terms of payment, and reduce exposure to claims and disputes. The author avoids "legalese" and gives real-life examples from actual projects.

$69.95 per copy | 300 pages, softcover | Catalog no. 67350

Spanish/English Construction Dictionary

2nd Edition
by RSMeans and the International Code Council

This expanded edition features thousands of the most common words and useful phrases in the construction industry with easy-to-follow pronunciations in both Spanish and English. Over 800 new terms, phrases, and illustrations have been added. It also features a new stand-alone "Safety & Emergencies" section, with colored pages for quick access.

Unique to this dictionary are the systems illustrations showing the relationship of components in the most common building systems for all major trades.

$23.95 per copy | Over 400 pages | Catalog no. 67327A

How Your House Works

by Charlie Wing

A must-have reference for every homeowner, handyman, and contractor—for repair, remodeling, and new construction. This book uncovers the mysteries behind just about every major appliance and building element in your house—from electrical, heating and AC, to plumbing, framing, foundations and appliances. Clear, "exploded" drawings show exactly how things should be put together and how they function—what to check if they don't work, and what you can do that might save you having to call in a professional.

$21.95 per copy | 160 pages, softcover | Catalog no. 67351

The Gypsum Construction Handbook

by USG

An invaluable reference of construction procedures for gypsum drywall, cement board, veneer plaster, and conventional plaster. This new edition includes the newest product developments, installation methods, fire- and sound-rated construction information, illustrated framing-to-finish application instructions, estimating and planning information, and more. Great for architects and engineers; contractors, builders, and dealers; apprentices and training programs; building inspectors and code officials; and anyone interested in gypsum construction. Features information on tools and safety practices, a glossary of construction terms, and a list of important agencies and associations. Also available in Spanish.

$29.95 | Over 575 pages, illustrated, softcover | Catalog no. 67357A

RSMeans Estimating Handbook

3rd Edition

Widely used in the industry for tasks ranging from routine estimates to special cost analysis projects, this handbook has been completely updated and reorganized with new and expanded technical information.

RSMeans Estimating Handbook will help construction professionals:

- Evaluate architectural plans and specifications
- Prepare accurate quantity takeoffs
- Compare design alternatives and costs
- Perform value engineering
- Double-check estimates and quotes
- Estimate change orders

$99.95 per copy | Over 900 pages, hardcover | Catalog No. 67276B

Means Illustrated Construction Dictionary

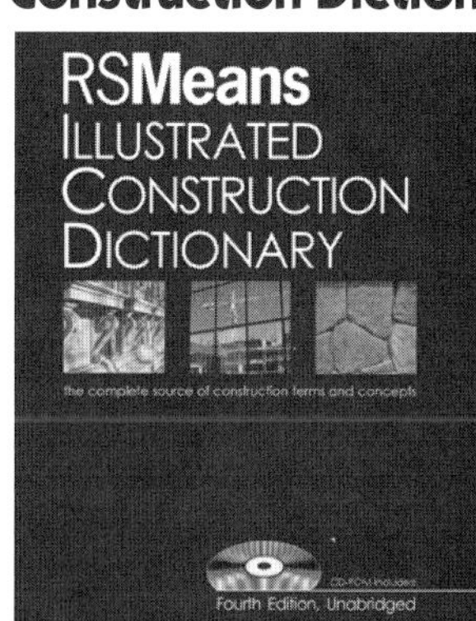

Unabridged 4th Edition, with CD-ROM

Long regarded as the industry's finest, *Means Illustrated Construction Dictionary* is now even better. With nearly 20,000 terms and more than 1,400 illustrations and photos, it is the clear choice for the most comprehensive and current information. The companion CD-ROM that comes with this new edition adds extra features such as: larger graphics and expanded definitions.

$99.95 per copy | Over 790 pages, illust., hardcover | Catalog no. 67292B

Reference Books

For more information visit the RSMeans website at **www.rsmeans.com**

Green Home Improvement

by Daniel D. Chiras, PhD

With energy costs rising and environmental awareness increasing, people are looking to make their homes greener. This book, with 65 projects and actual costs and projected savings help homeowners prioritize their green improvements.

Projects range from simple water savers that cost only a few dollars, to bigger-ticket items such as HVAC systems. With color photos and cost estimates, each project compares options and describes the work involved, the benefits, and the savings.

$24.95 | 320 pages, illustrated, softcover | Catalog no. 67355

New Titles

Construction Supervision

This brand new title inspires supervisory excellence with proven tactics and techniques applied by thousands of construction supervisors over the past decade. Recognizing the unique and critical role the supervisor plays in project success, the book's leadership guidelines carve out a practical blueprint for motivating work performance and increasing productivity through effective communication.

Features:

- A unique focus on field supervision and crew management
- Coverage of supervision from the foreman to the superintendent level
- An overview of technical skills whose mastery will build confidence and success for the supervisor
- A detailed view of "soft" management and communication skills

$90.00 per copy | Over 450 pages, softcover | Catalog no. 67358

Risk Management for Design and Construction

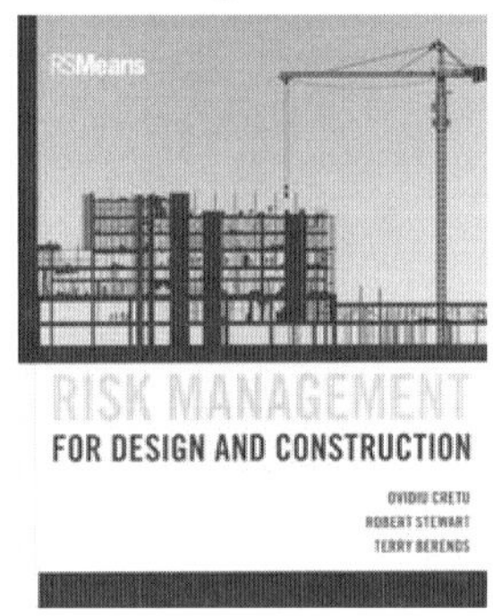

This book introduces risk as a central pillar of project management and shows how a project manager can be prepared for dealing with uncertainty. Written by experts in the field, *Risk Management for Design and Construction* uses clear, straightforward terminology to demystify the concepts of project uncertainty and risk.

Highlights include:

- Integrated cost and schedule risk analysis
- An introduction to a ready-to-use system of analyzing a project's risks and tools to proactively manage risks
- A methodology that was developed and used by the Washington State Department of Transportation
- Case studies and examples on the proper application of principles
- Information about combining value analysis with risk analysis

$125.00 per copy | Over 250 pages, softcover | Catalog no. 67359

Green Building: Project Planning & Cost Estimating

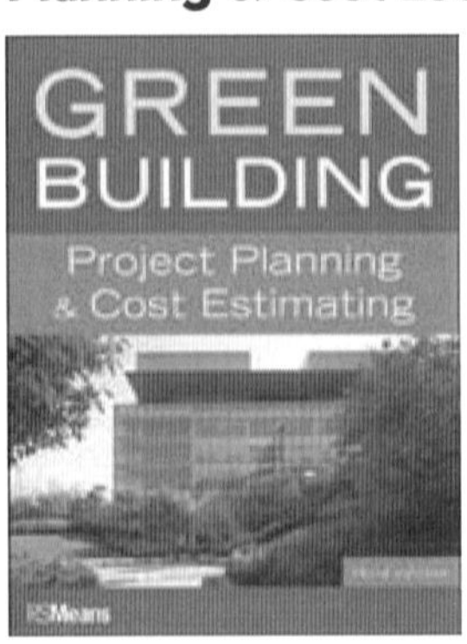

3rd Edition

Since the widely read first edition of this book, green building has gone from a growing trend to a major force in design and construction.

This new edition has been updated with the latest in green building technologies, design concepts, standards, and costs. Full-color with all new case studies—plus a new chapter on commercial real estate.

$99.95 per copy | Over 450 pages, softcover | Catalog no. 67338B

2012 Order Form

ORDER TOLL FREE 1-800-334-3509 OR FAX 1-800-632-6732

For more information visit the RSMeans website at **www.rsmeans.com**

Qty.	Book no.	COST ESTIMATING BOOKS	Unit Price	Total
	60062	Assemblies Cost Data 2012	$290.95	
	60012	Building Construction Cost Data 2012	177.95	
	60112	Concrete & Masonry Cost Data 2012	162.95	
	50142	Construction Cost Indexes 2012 (subscription)	362.00	
	60142A	Construction Cost Index–January 2012	90.50	
	60142B	Construction Cost Index–April 2012	90.50	
	60142C	Construction Cost Index–July 2012	90.50	
	60142D	Construction Cost Index–October 2012	90.50	
	60342	Contr. Pricing Guide: Resid. R & R Costs 2012	39.95	
	60232	Electrical Change Order Cost Data 2012	174.95	
	60032	Electrical Cost Data 2012	179.95	
	60202	Facilities Construction Cost Data 2012	453.95	
	60302	Facilities Maintenance & Repair Cost Data 2012	399.95	
	60552	Green Building Cost Data 2012	147.95	
	60162	Heavy Construction Cost Data 2012	177.95	
	60092	Interior Cost Data 2012	182.95	
	60122	Labor Rates for the Const. Industry 2012	395.95	
	60182	Light Commercial Cost Data 2012	132.95	
	60022	Mechanical Cost Data 2012	174.95	
	60152	Open Shop Building Const. Cost Data 2012	152.95	
	60212	Plumbing Cost Data 2012	179.95	
	60042	Commercial Renovation Cost Data 2012	144.95	
	60172	Residential Cost Data 2012	127.95	
	60282	Site Work & Landscape Cost Data 2012	172.95	
	60052	Square Foot Costs 2012	187.95	
	62011	Yardsticks for Costing (2011)	174.95	
	62012	Yardsticks for Costing (2012)	176.95	
		REFERENCE BOOKS		
	67329	Bldrs Essentials: Best Bus. Practices for Bldrs	29.95	
	67298A	Bldrs Essentials: Framing/Carpentry 2nd Ed.	24.95	
	67298AS	Bldrs Essentials: Framing/Carpentry Spanish	24.95	
	67307	Bldrs Essentials: Plan Reading & Takeoff	35.95	
	67342	Building & Renovating Schools	99.95	
	67339	Building Security: Strategies & Costs	89.95	
	67353	Complete Book of Framing	29.95	
	67146	Concrete Repair & Maintenance Illustrated	69.95	
	67352	Construction Business Management	49.95	
	67358	Construction Supervision	90.00	
	67314	Cost Planning & Est. for Facil. Maint.	89.95	
	67328A	Designing & Building with the IBC, 2nd Ed.	99.95	
	67230B	Electrical Estimating Methods, 3rd Ed.	64.95	
	67343	Estimating Bldg. Costs for Resi. & Lt. Comm.	29.95	
	67276B	Estimating Handbook, 3rd Ed.	99.95	
	67318	Facilities Operations & Engineering Reference	109.95	
	67301	Facilities Planning & Relocation	89.95	
	67338B	Green Building: Proj. Planning & Cost Est., 3rd Ed.	99.95	
	67355	Green Home Improvement	24.95	
	67357A	The Gypsum Construction Handbook	29.95	
	67357S	The Gypsum Construction Handbook (Spanish)	29.95	
	67148	Heavy Construction Handbook	99.95	
	67323	Historic Preservation: Proj. Planning & Est.	99.95	

Qty.	Book no.	REFERENCE BOOKS (Cont.)	Unit Price	Total
	67308E	Home Improvement Costs–Int. Projects, 9th Ed.	24.95	
	67309E	Home Improvement Costs–Ext. Projects, 9th Ed.	24.95	
	67344	Homeowner's Guide to Mold	21.95	
	67324B	How to Est.w/Means Data & CostWorks, 3rd Ed.	59.95	
	67351	How Your House Works	21.95	
	67282A	Illustrated Const. Dictionary, Condensed, 2nd Ed.	59.95	
	67292B	Illustrated Const. Dictionary, w/CD-ROM, 4th Ed.	99.95	
	67348	Job Order Contracting	89.95	
	67295C	Landscape Estimating Methods, 5th Ed.	64.95	
	67341	Life Cycle Costing for Facilities	99.95	
	67294B	Mechanical Estimating Methods, 4th Ed.	64.95	
	67283B	Plumbing Estimating Methods, 3rd Ed.	59.95	
	67345	Practice of Cost Segregation Analysis	99.95	
	67346	Preventive Maint. for Multi-Family Housing	89.95	
	67326	Preventive Maint. Guidelines for School Facil.	149.95	
	67247B	Project Scheduling & Manag. for Constr. 3rd Ed.	64.95	
	67322B	Resi. & Light Commercial Const. Stds., 3rd Ed.	59.95	
	67359	Risk Management for Design & Construction	125.00	
	67327A	Spanish/English Construction Dictionary, 2nd Ed.	23.95	
	67145B	Sq. Ft. & Assem. Estimating Methods, 3rd Ed.	69.95	
	67321	Total Productive Facilities Management	79.95	
	67350	Understanding and Negotiating Const. Contracts	69.95	
	67303B	Unit Price Estimating Methods, 4th Ed.	59.95	
	67354	Universal Design	21.95	
	67319A	Value Engineering: Practical Applications	79.95	

MA residents add 6.25% state sales tax	
Shipping & Handling**	
Total (U.S. Funds)*	

Prices are subject to change and are for U.S. delivery only. *Canadian customers may call for current prices. **Shipping & handling charges: Add 7% of total order for check and credit card payments. Add 9% of total order for invoiced orders.

Send order to: ADDV-1000

Name (please print) ______________________________

Company ______________________________

☐ Company

☐ Home Address ______________________________

____

City/State/Zip ______________________________

Phone # ____________________ P.O. # ____________________

(Must accompany all orders being billed)

Mail to: RSMeans, 700 Longwater Drive, Norwell, MA 02061

RSMeans Project Cost Report

By filling out this report, your project data will contribute to the database that supports the RSMeans Project Cost Square Foot Data. When you fill out this form, RSMeans will provide a $30 discount off one of the RSMeans products advertised in the preceding pages. Please complete the form including all items where you have cost data, and all the items marked (☑).

$30.00 Discount per product for each report you submit

Project Description (no remodeling projects, please)

☑ Building Use (Office School...) ______________________

☑ Address (City, State) ______________________

____

☑ Total Building Area (SF) ______________________

☑ Ground Floor (SF) ______________________

☑ Frame (Wood, Steel...) ______________________

☑ Exterior Wall (Brick, Tilt-up...) ______________________

☑ Basement: (checkone) ☐ Full ☐ Partial ☐ None

☑ Number of Stories ______________________

☑ Floor-to-Floor Height ______________________

☑ Volume (C.F.) ______________________

% Air Conditioned ____________ Tons ____________

Total Project Cost $ ____________

Owner ______________________

Architect ______________________

General Contractor ______________________

☑ Bid Date ______________________

☑ Typical Bay Size ______________________

☑ Capacity ______________________

☑ Labor Force: _____ % Union _____ % Non-Union

☑ Project Description (Circle one number in each line.)

1. Economy 2. Average 3. Custom 4. Luxury

1. Square 2. Rectangular 3. Irregular 4. Very Irregular

Comments ______________________

____

A	☑	General Conditions	$
B	☑	Site Work	$
C	☑	Concrete	$
D	☑	Masonry	$
E	☑	Metals	$
F	☑	Wood & Plastics	$
G	☑	Thermal &Moisture Protection	$
GR		Roofing & Flashing	$
H	☑	Doors and Windows	$
J	☑	Finishes	$
JP		Painting & Wall Covering	$

K	☑	Specialties	$
L	☑	Equipment	$
M	☑	Furnishings	$
N	☑	SpecialConstruction	$
P	☑	ConveyingSystems	$
Q	☑	Mechanical	$
QP		Plumbing	$
QB		HVAC	$
R	☑	Electrical	$
S	☑	Mech./Elec. Combined	$

Please specify the RSMeans product you wish to receive. Complete the address information.

Product Name ______________________

Product Number ______________________

Your Name ______________________

Title ______________________

Company ______________________

☐ Company

☐ Home Street Address ______________________

City, State, Zip ______________________

Return by mail or fax 888-492-6770.

Method of Payment:

Credit Card # ______________________

Expiration Date ______________________

Check ______________________

Purchase Order ______________________

Phone Number ______________________

(for those not comfortable mailing in credit card information)

RSMeans
a division of Reed Construction Data
Square Foot Costs Department
700 Longwater Drive
Norwell, MA 02061